COMMON ABBREVIATIONS USED IN ORGANIC CHEMISTRY

ORGANIC GROUPS

Abbreviation	*Meaning*	*Structure*
Ac	acetyl	$CH_3—\overset{O}{\overset{\parallel}{C}}—R$
	allyl	$H_2C{=}CH—CH_2—R$
Bz	benzoyl	$Ph—\overset{O}{\overset{\parallel}{C}}—R$
Boc	*t*-butyloxycarbonyl	$(CH_3)_3C—O—\overset{O}{\overset{\parallel}{C}}—R$
Bn	benzyl	$Ph—CH_2—R$
n-Bu	*n*-butyl	$CH_3—CH_2—CH_2—CH_2—R$
i-Bu	isobutyl	$(CH_3)_2CH—CH_2—R$
s-Bu	*sec*-butyl	$CH_3—CH_2—\underset{CH_3}{\underset{\mid}{CH}}—R$
t-Bu	*tert*-butyl	$(CH_3)_3C—R$
Cbz (or Z)	benzyloxycarbonyl	$Ph—CH_2—O—\overset{O}{\overset{\parallel}{C}}—R$
Me	methyl	$CH_3—R$
Et	ethyl	$CH_3—CH_2—R$
c-Hx	cyclohexyl	—R
Ph	phenyl	—R
Pr	propyl	$CH_3—CH_2—CH_2—R$
i-Pr	isopropyl	$(CH_3)_2CH—R$
Sia	*secondary* isoamyl	$(CH_3)_2CH—\underset{CH_3}{\underset{\mid}{CH}}—R$
THP	tetrahydropyranyl	O, R
Ts	*para*-toluenesulfonyl, "tosyl"	CH_3— —$\overset{O}{\overset{\parallel}{\underset{O}{\underset{\parallel}{S}}}}$—R

Not all of these abbreviations are used in this text, but they are provided for reference.

REAGENTS AND SOLVENTS

Abbreviation	*Meaning*	*Structure*
Ac_2O	acetic anhydride	$CH_3—\overset{O}{\overset{\parallel}{C}}—O—\overset{O}{\overset{\parallel}{C}}—CH_3$
DCC	dicyclohexylcarbodiimide	—N=C=N—
DIBAL or DIBAH	diisobutylaluminum hydride	$[(CH_3)_2CHCH_2]_2AlH$
DME, "glyme"	1,2-dimethoxyethane	$CH_3—O—CH_2CH_2—O—CH_3$
diglyme	bis(2-methoxyethyl) ether	$(CH_3—O—CH_2CH_2)_2O$
DMF	dimethylformamide	$H—\overset{O}{\overset{\parallel}{C}}—N(CH_3)_2$
DMSO	dimethyl sulfoxide	$CH_3—\overset{O}{\overset{\parallel}{S}}—CH_3$
EtOH	ethanol	CH_3CH_2OH
EtO^-	ethoxide ion	$CH_3CH_2—O^-$
Et_2O	diethyl ether	$CH_3CH_2—O—CH_2CH_3$
HMPA, HMPT	hexamethylphosphoric triamide	$[(CH_3)_2N]_3P{=}O$
LAH	lithium aluminum hydride	$LiAlH_4$
LDA	lithium diisopropylamide	$[(CH_3)_2CH]_2N—Li$
MCPBA	*meta*-chloroperoxybenzoic acid	Cl; $—\overset{O}{\overset{\parallel}{C}}—O—O—H$
MeOH	methanol	CH_3OH
MeO^-	methoxide ion	$CH_3—O^-$
MVK	methyl vinyl ketone	$CH_3—\overset{O}{\overset{\parallel}{C}}—CH{=}CH_2$
NBS	*N*-bromosuccinimide	O, N—Br, O
PCC	pyridinium chlorochromate	$pyr\cdot CrO_3\cdot HCl$
Pyr	pyridine	N:
t-BuOH	*tertiary* butyl alcohol	$(CH_3)_3C—OH$
t-BuOK	potassium *tertiary*-butoxide	$(CH_3)_3C—O^-\ {}^+K$
THF	tetrahydrofuran	O
TMS	tetramethylsilane	$(CH_3)_4Si$

L. G. WADE, JR.
WHITMAN COLLEGE

ORGANIC CHEMISTRY

SECOND EDITION

PRENTICE HALL, ENGLEWOOD CLIFFS, NEW JERSEY 07632

Library of Congress Cataloging-in-Publication Data

Wade, L. G., (date)
Organic chemistry/L. G. Wade, Jr.—2nd. ed.
p. cm.
Includes index.
ISBN 0-13-642588-7 :
1. Chemistry, Organic. I. Title.
QD251.2.W33 1991 90-44281
547—dc20 CIP

Acquisitions Editor: Dan Joraanstad
Editorial/production supervision: Debra Wechsler
Interior and cover design: Margaret Kenselaar
Prepress buyer: Paula Massenaro
Manufacturing buyer: Lori Bulwin
Page layout: Carol Ann Hyland

COVER PHOTOGRAPH
A computer-generated section of a DNA molecule; blue indicates nitrogen; white, carbon; red, oxygen; and yellow, phosphorus. The image is courtesy of Evans and Sutherland Computer Corporation.

Printed in the United States of America

10 9 8 7 6 5 4 3 2 1

ISBN 0-13-642588-7

Prentice-Hall International (UK) Limited, *London*
Prentice-Hall of Australia Pty. Limited, *Sydney*
Prentice-Hall Canada Inc., *Toronto*
Prentice-Hall Hispanoamericana, S.A., *Mexico*
Prentice-Hall of India Private Limited, *New Delhi*
Prentice-Hall of Japan, Inc., *Tokyo*
Simon & Schuster Asia Pte. Ltd., *Singapore*
Editora Prentice-Hall do Brasil, Ltda., *Rio de Janeiro*

To
Betsy,
Christine,
and Jennifer

BRIEF CONTENTS

CONTENTS

25 LIPIDS 1146

26 SYNTHETIC POLYMERS 1166

APPENDICES 1187

PREFACE

Organic chemistry students are often overwhelmed by the number of compounds, names, reactions, and mechanisms that confront them. They wonder whether they can learn all this material in a single year. Perhaps the most important function of a textbook is to organize the material to show students that most of organic chemistry consists of a few basic principles and a large number of extensions and applications of these principles. Relatively little memorization is required if the student grasps each major concept and develops flexibility in applying that concept.

In writing the first edition, my goal was to produce a modern, readable text that uses the most effective techniques of presentation and review. This second edition extends and refines that goal, differing from the first edition in several respects:

1. The chapters are reorganized to place stereochemistry and spectroscopy earlier, with these topics then integrated into all the functional group chapters.
2. Several topics have been added or have received increased emphasis in this edition. For example, sections on periodic acid cleavage of glycols and DIBAH reduction of esters have been added. Phenols, quinones, and thiols are a few of the topics that have received increased emphasis.
3. A number of new features have been added, including about a dozen problem-solving discussions, summary tables for comparing and contrasting important concepts, additional spectroscopy problems, and appendices on nomenclature, problem solving, and UV spectroscopy.

The entire book has been extensively microedited to tighten up the style and ensure clarity. In some cases, redundant examples have been eliminated. The chapters on stereochemistry have received particular attention, having been combined from two chapters into one. Similarly, the two chapters covering carboxylic acid derivatives have been combined into one, and most of the chemistry is now presented by reaction type (comparing the reactivity of the different acid derivatives) rather than one acid derivative at a time.

As in the first edition, each new topic is introduced carefully and explained thoroughly. Whenever possible, illustrations are used to help the student visualize each physical concept, and many in-chapter problems give immediate reinforcement as the student works through each chapter.

Throughout the book, the emphasis is on *chemical reactivity*. Chemical reactions are introduced as soon as possible, and each functional group is considered in view of its reactivity toward electrophiles, nucleophiles, oxidants, reductants, and other reagents. "Electron-pushing" mechanisms are stressed throughout as a means of explaining and predicting this reactivity. Structural concepts such as stereochemistry and spectroscopy are thoroughly covered as well, but they are presented as useful techniques that enhance the fundamental study of chemical reactivity.

ORGANIZATION

This book maintains the traditional organization that concentrates on one functional group at a time while comparing and contrasting the reactivity of different functional groups. Reactions are emphasized beginning in Chapter 4. Most of the important substitution, addition, and elimination reactions appear before the other major theory and structure chapters.

Some stereospecific substitutions and eliminations are first encountered before the stereochemistry chapter. These reactions are initially discussed using examples in which stereochemical changes are apparent from the cis or trans arrangement of substituents on a ring or double bond. When chiral compounds and nongeometric diastereomers are introduced in Chapter 6, stereospecific reactions are reviewed using these compounds as examples. This organization helps students learn these difficult topics by first emphasizing the reactions using simple examples and concentrating on reactivity. More esoteric stereochemical examples are used later as a review, once stereochemistry has been covered in detail.

Spectroscopy appears earlier in this edition than it did in the first. Spectroscopic techniques (IR, MS, and NMR) are covered in Chapters 11 and 12, where they will be encountered in the first semester. Still, a large amount of organic chemistry has been covered before this digression into structure determination. The principles of spectroscopy are reinforced in later chapters, where the characteristic spectral features of each functional group are summarized and representative spectroscopy problems appear.

KEY FEATURES

FLEXIBILITY OF COVERAGE

No two instructors teach organic chemistry in exactly the same way. This book covers all the fundamental topics in detail, building each new concept on those that come before. Many topics may be given more or less emphasis at the discretion of the instructor. Examples of these topics are ^{13}C NMR spectroscopy, ultraviolet spectroscopy, conservation of orbital symmetry, nucleic acids, and the special topics chapters: lipids and synthetic polymers.

Another area of flexibility is in the problems. The wide-ranging problem sets review the material from several viewpoints, and more study problems are provided than most students would be able to complete. This large variety allows the instructor to select the most appropriate problems for the individual course.

UP-TO-DATE TREATMENT

In addition to the classical reactions, this book covers many techniques and reactions that have more recently gained wide use among practicing chemists. Molecular-orbital theory is introduced early and used to explain electronic effects in conjugated and aromatic systems, pericyclic reactions, and ultraviolet spectroscopy. Carbon-13 NMR spectroscopy is treated as the routine tool it has become in most research laboratories. Many of the newer synthetic techniques are also included, such as the Birch reduction, DIBAH reduction of esters, alkylation of 1,3-dithianes, and oxidations using pyridinium chlorochromate.

REACTION MECHANISMS

Reaction mechanisms are important in all areas of organic chemistry, but they are difficult for many students. Students fall into the trap of memorizing a mechanism

while not understanding why it proceeds as it does. To avoid this problem, this book stresses the principles used to predict mechanisms. Problem-solving sections develop the basic techniques of approaching a mechanism problem and work to minimize rote memorization. These techniques emphasize deciding whether the reaction is acidic, basic, or free radical in nature and then breaking it down into the Lewis acid-base interactions between nucleophilic sites and electrophilic sites and the "arrow pushing" that helps to illustrate these individual steps.

INTRODUCTION TO MECHANISMS USING FREE-RADICAL HALOGENATION

The advantages and disadvantages of using free-radical halogenation to introduce reaction mechanisms have been debated for many years. The principal objection to free-radical halogenation is that it is not a useful synthetic reaction. But more useful reactions such as nucleophilic substitution and addition to alkenes are complicated by participation of the solvent and other effects. Gas-phase free-radical halogenation allows a clearer treatment of kinetics and thermodynamics, as long as its disadvantages as a synthetic reaction are carefully discussed and the student is aware of the limitations.

ORGANIC SYNTHESIS

Organic synthesis is stressed throughout this book, with progressive discussions of the process involved in developing a synthesis. The method of *retrosynthetic analysis* is emphasized, and the student is shown how to work backward from the target compound and forward from the starting materials and to find a common intermediate. Several new problem-solving discussions of organic synthesis have been added, emphasizing how one approaches a multistep synthesis.

Typical yields have been provided for many synthetic reactions, although I hope students will not misuse these numbers. Too often students consider the yield of a reaction to be a fixed characteristic just as the melting point of a compound is fixed. In practice, many factors affect product yields, and literature values for apparently similar reactions often differ by a factor of 2 or more. The yields given in this book are *typical* yields that a good student with excellent technique might expect to obtain.

SPECTROSCOPY

Spectroscopy is one of the most important tools of the organic chemist. This book develops the theory for each type of spectroscopy and then discusses the characteristic spectral features. Then the most useful and dependable characteristics are summarized into a small number of rules of thumb that allow the student to interpret most spectra without looking up or memorizing large tables of data. For reference use, extensive tables of NMR and IR data and a more complete version of the Woodward-Fieser rules for UV are provided as appendices.

This approach is particularly effective with ultraviolet spectroscopy, ^{13}CNMR, and mass spectrometry. Practical rules are given to help students see what information is available in the spectrum and what spectral characteristics usually correspond to what structural features. Sample problems show how the information from various spectra is combined to propose a structure. The emphasis is on helping students develop an intuitive feel for using spectroscopy to solve structural problems.

NOMENCLATURE

IUPAC nomenclature is stressed throughout the book, but common nomenclature is also discussed and used to develop students' familiarity. Teaching only the simpler IUPAC nomenclature might be justifiable, but such an approach would handicap students in their further study and use of the literature. Much of the literature of chemistry, biology, and medicine uses common names such as methyl ethyl ketone, isovaleric acid, methyl *t*-butyl ether, γ-aminobutyric acid, and ε-caprolactam. This book emphasizes why systematic nomenclature is often preferred, yet it encourages familiarity with common names as well.

STUDY AIDS

Several kinds of study aids are provided to emphasize and review the important points.

Tables. Extensive tables summarize the names and physical properties of related compounds.

Summary tables. Whenever a large amount of material lends itself to a concise summary, a summary table (highlighted by a blue background) is provided to compare and contrast this material. For example, a summary table on p. 195 compares the factors affecting S_N1 and S_N2 reactions, and another summary table (p. 206) contrasts E1 and E2 reactions.

Reaction summaries. Extensive reaction summaries review the syntheses and reactions of the functional groups. Each summary, highlighted by a blue background, appears at the end of the section discussing these reactions. The summaries include cross-references to reactions that are discussed elsewhere.

Glossaries. A complete glossary at the end of each chapter defines and explains technical terms introduced in the chapter. New terms defined in the glossary are printed in boldface the first time they appear in the chapter. The glossaries are particularly helpful in jogging the student's memory when reviewing the chapter.

Problems. The large number of in-chapter problems provides immediate reinforcement of material as it is learned. Later, end-of-chapter problems promote additional review and practice. The instructor should choose and assign problems that reflect the emphasis of the lectures.

Solved problems. Where appropriate, solved problems show students how a particular type of problem can be approached and what kind of answer constitutes a solution. The study problems can also serve as solved problems if students compare their answers with those given in the *Solutions Manual.*

Problem-solving strategies. New in the second edition, the problem-solving strategies suggest methods for approaching problems, especially those that require proposing mechanisms and developing multistep syntheses. Students often have difficulty with organic chemistry because they cannot see how to approach problem solving. Clearly one cannot convert organic chemistry problem solving to an algorithm where one simply follows a prescribed process to obtain a solution. Yet, after many discussions with colleagues about how a chemist approaches synthesis and mechanism problems, some general patterns began to emerge.

The procedures recommended in the problem-solving discussions approximate what an organic chemist habitually does in approaching these problems. They suggest ways of organizing one's thoughts to serve as a starting point, not a guaranteed route to the answers. They show some of the thought processes a chemist instinctively uses in approaching problems.

PEDAGOGY

One of the main goals in writing this text was to use the most effective techniques of presentation and review to facilitate the student's organization and understanding of organic chemistry. Many features of this book are derived from established pedagogical techniques, while others are newly applied to organic chemistry.

Style. The writing style has been kept clear and simple, avoiding stilted language and long, complex sentences.

Illustrations. Difficult concepts are presented visually, using many different types of illustrations. A good illustration often elicits a level of understanding that is impossible with any amount of verbal explanation.

Reinforcement. Important principles are reinforced throughout the book. Nearly every important rule, reaction, or synthesis is mentioned again in another chapter, accompanied by a cross-reference to the original discussion. The glossaries and summaries provide further reinforcement. Although these features add to the length of the book, they provide a much needed review of the important points and emphasize the interconnected, cumulative nature of organic chemistry.

Four-color printing. The use of four ink colors is intended to enhance the student's understanding. Major features of the text are highlighted for easy location:

1. Key definitions and rules are in blue type:

MARKOVNIKOV'S RULE The addition of a proton acid to the double bond of an alkene results in a product with the acid proton bonded to the carbon atom that already holds the greater number of hydrogen atoms.

2. Curved red arrows are used throughout for "electron pushing," to show the flow of electrons through the course of a reaction:

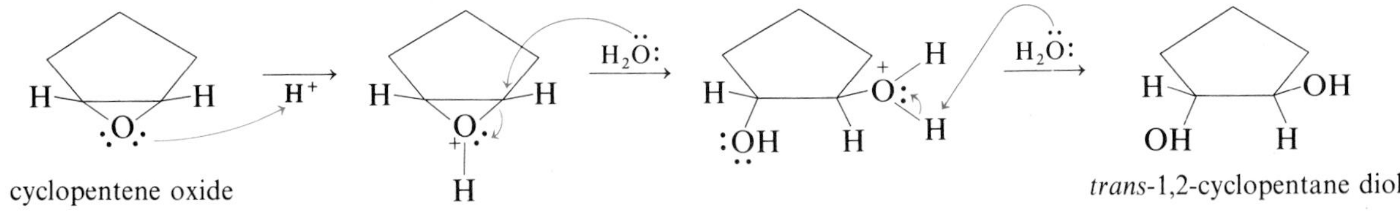

cyclopentene oxide   *trans*-1,2-cyclopentane diol

3. In-text problems are outlined in blue, and in-text solved problems are outlined in red:

PROBLEM 9-13

Show how you would synthesize each of the following alcohols by the addition of an appropriate Grignard reagent to an aldehyde.

(a) (b) (c)

SOLVED PROBLEM 7-4

A norbornene molecule that is labeled with deuterium is subjected to hydroboration-oxidation. Give the structures of the intermediates and products.

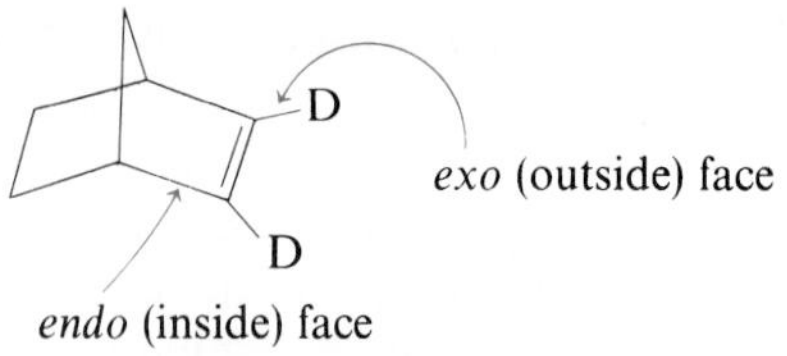

D

exo (outside) face

D

endo (inside) face

deuterium-labeled norbornene

SOLUTION The syn addition of BH_3 across the double bond of norbornene takes place mostly from the more accessible outer side of the double bond. This is called the *exo* face of the norbornene molecule. Oxidation gives a product with both the hydrogen atom and the hydroxyl group added to exo positions.

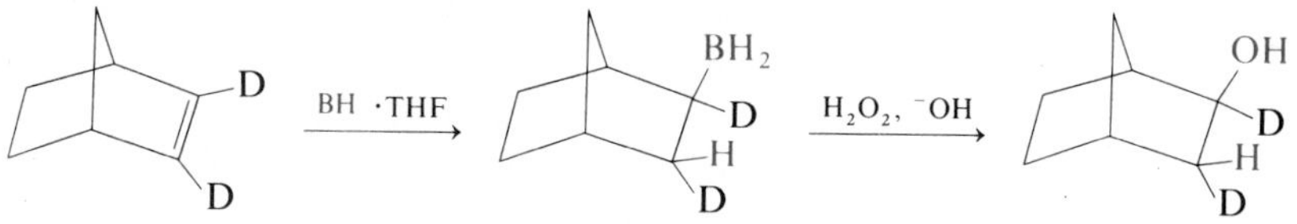

D, D — reactant $\xrightarrow{BH \cdot THF}$ BH_2, D, H, D — alkylborane $\xrightarrow{H_2O_2,\ ^-OH}$ OH, D, H, D — alcohol

4. Summaries and summary tables are printed on a blue background:

SUMMARY OF THE REACTIONS OF ALKENES

1. Electrophilic additions

a. *Addition of hydrogen halides* (Section 8-3)

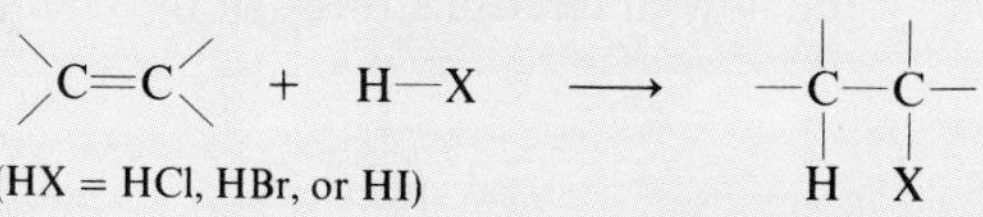

$$\mathrm{>C{=}C<} \; + \; \mathrm{H{-}X} \longrightarrow \mathrm{-\overset{|}{\underset{H}{C}}-\overset{|}{\underset{X}{C}}-}$$

(HX = HCl, HBr, or HI)

(Markovnikov orientation)
(anti-Markovnikov with HBr and peroxides)

Example

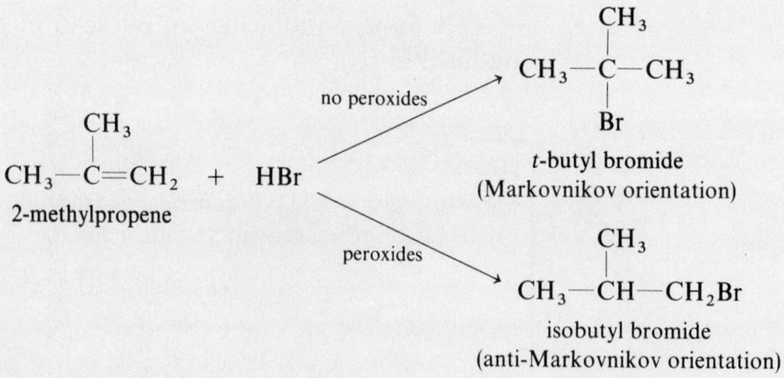

$$CH_3-\overset{CH_3}{\overset{|}{C}}{=}CH_2 \; + \; HBr \xrightarrow{\text{no peroxides}} CH_3-\overset{CH_3}{\overset{|}{\underset{Br}{\underset{|}{C}}}}-CH_3$$

2-methylpropene

t-butyl bromide
(Markovnikov orientation)

$$CH_3-\overset{CH_3}{\overset{|}{C}}{=}CH_2 \; + \; HBr \xrightarrow{\text{peroxides}} CH_3-\overset{CH_3}{\overset{|}{C}H}-CH_2Br$$

isobutyl bromide
(anti-Markovnikov orientation)

5. Problem-solving discussions are printed on a cream background:

PROBLEM SOLVING ORGANIC SYNTHESIS

Alkyl halides are readily made from other compounds, and the halogen atom is easily converted to other functional groups. This flexibility makes alkyl halides useful as reagents and intermediates for organic synthesis. **Organic synthesis** is the preparation of desired compounds from readily available materials. Synthesis is one of the major areas of organic chemistry, and nearly every chapter of this book involves organic synthesis in some way. A synthesis may be a simple one-step reaction, or it may involve many steps and incorporate a subtle strategy for assembling the correct carbon skeleton with all the functional groups in the right positions.

Many of the problems in this book are synthesis problems. In some synthesis problems, you are asked to show how to convert a given starting material to the desired product. There are obvious one-step answers to some of these prob-

6. Important general reactions are highlighted by a check mark in the margin:

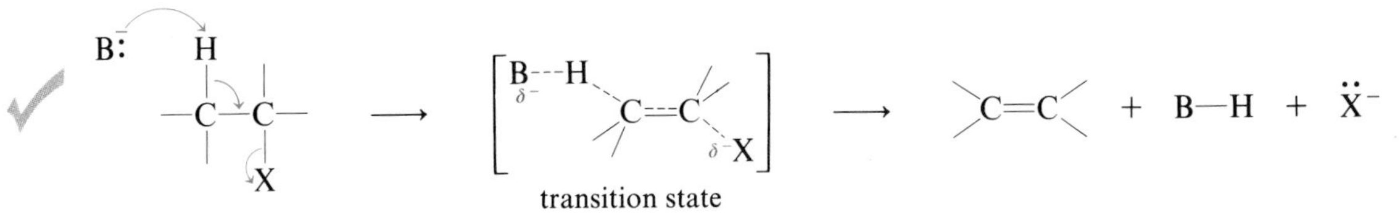

B: H —C—C— X → [B---H (δ−) C==C X (δ−)] → C=C + B—H + X⁻

transition state

E2 rate = k_r[R—X][B^-]

In addition, the variety of available colors makes it possible to highlight and distinguish key aspects of reactions, structures, and molecular drawings:

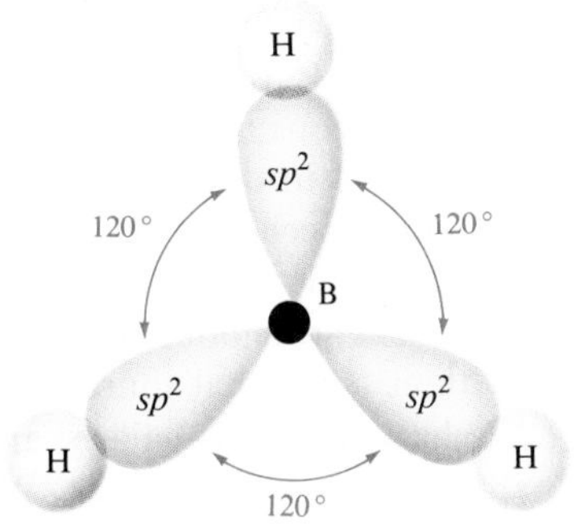

H
sp^2
120°
120°
B
sp^2
sp^2
H
H
120°

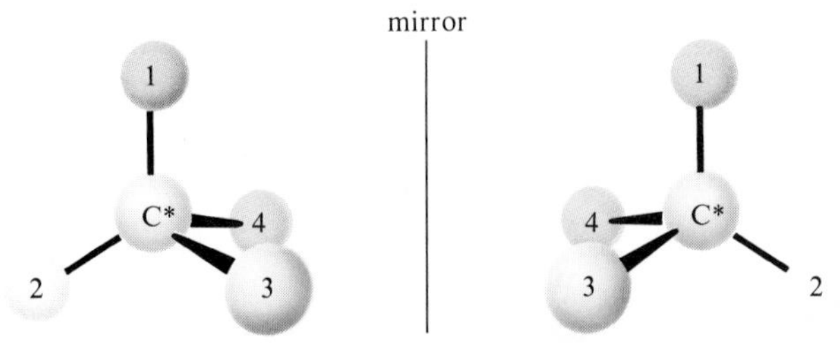

mirror

1 C* 4 2 3 | 1 C* 4 3 2

SUPPLEMENTS

SOLUTIONS MANUAL

Brief answers to many of the in-chapter problems are given at the back of this book. These answers are sufficient for a student who is on the right track, but they are of limited use to one who is having difficulty working the problems. The *Solutions Manual,* prepared by Jan W. Simek of California Polytechnic State University, contains worked-out solutions to all the problems. These solutions also give helpful hints on how to approach each kind of problem. This supplement is a useful aid for any student, and it is particularly valuable for students who feel that they understand the material but need more help with problem solving.

MOLECULAR MODELS

Every organic chemistry student needs a set of molecular models. These models are used to demonstrate a multitude of principles, including stereochemistry, ring strain, conformations of cyclic and acyclic systems, and many others. The Universal Molecular Model (UMM) Sets offered with this book are chosen because their flexible bonds allow the construction of strained systems, with the amount of bend in the bonds providing a qualitative idea of the amount of strain. In addition, the space-filling features of this new set help to show the origins of steric effects with bulky substituents.

INSTRUCTOR SUPPLEMENTS

An instructor's manual, transparencies, demonstration models, and a test bank are available for instructors using this textbook. Contact your Prentice Hall representative for details.

ACKNOWLEDGMENTS

I am pleased to thank the many talented people who helped with this revision. Particular thanks are due to Mary Castellion, who made thousands of useful suggestions throughout the process of revision and whose vision and persistence resulted in development of the problem-solving approach used throughout the book. Special thanks are also due to Jan W. Simek, author of the *Solutions Manual,* who made many useful and perceptive suggestions.

Throughout the production process, I have relied on technical advice from three additional sources: Albert W. Burgstahler (University of Kansas) helped by checking all the galleys, and William D. Closson (SUNY-Albany) checked all the page proofs. Brian Lian (Whitman College) checked the galleys from the viewpoint of a student. Although I did not adopt all their suggestions, most of them were helpful and contributed to the quality of the final product.

I would also like to thank the other reviewers of both editions for their valuable insight and commentary.

Ronald Blankespour
Calvin College

Warren Bosch
Elgin Community College

Guilford Jones, II
Boston University

James Keeffe
San Francisco State University

Robert Carlson
University of Kansas

Dr. Sheldon Clare
University of Pittsburgh, Johnstown

George B. Clemans
Bowling Green State University

Brian P. Coppola
University of Wisconsin, Whitewater

Melvin Druelinger
University of Southern Colorado

Charles D. Duncan
University of Alabama at Birmingham

Harry E. Ensley
Tulane University

Steven L. Fedder
University of Santa Clara

George W. Hay
Queen's University

John L. Hogg
Texas A&M University

Charles B. Rose
University of Nevada at Reno

Allen Schoffstall
University of Colorado at Colorado Springs

Jan William Simek
California Polytechnic State University

Walter S. Trahanovsky
Iowa State University

George H. Wahl, Jr.
North Carolina State University, Raleigh

James K. Whitesell
University of Texas at Austin

Joseph Kurz
Washington University in St. Louis

Linda Magid
University of Tennessee

Kenneth Marsi
California State University at Long Beach

Donald L. McLaugherty
University of Texas, Tyler

John L. Meisenheimer
Eastern Kentucky University

Nathan Miller
University of South Alabama

Roger K. Murry
University of Delaware

Michael A. Ogliaruso
Virginia Polytechnic Institute and State University

Michael Rathke
Michigan State University

Ulrich Hollstein
University of New Mexico

Stuart Rosenfeld
Smith College

Neil E. Schore
University of California at Davis

Malcolm Stevens
University of Hartford

C. A. VanderWerf
University of Florida

Danny White
American River College

Hans Zimmer
University of Cincinnati

Finally, I want to thank the people at Prentice Hall whose dedication and flexibility contributed to the completion of this project. As managing editor, Dan Joraanstad kept the project moving, ensured that the necessary resources were made available, and made many useful comments and suggestions on organization, format, and design. Ray Mullaney, editor-in-chief of College Book Editorial Development, made many useful suggestions in the revision, as he had also done for the first edition. Production editor Debra Wechsler kept the production process organized, on track, and on schedule. It has been a pleasure working with all of these thoroughly professional and competent people.

L. G. Wade, Jr.

ABOUT THE AUTHOR

L. G. "Skip" Wade decided to become a chemistry major during his sophomore year at Rice University, while taking organic chemistry from Professor Ronald M. Magid. After receiving his B.A. from Rice in 1969, Wade went on to Harvard University, where he did research with Professor James D. White. While at Harvard, he served as the Head Teaching Fellow for the organic laboratories and was strongly influenced by the teaching methods of two master educators, Professors Leonard K. Nash and Frank H. Westheimer.

After completing his Ph.D. at Harvard in 1974, Dr. Wade joined the chemistry faculty at Colorado State University. Over the course of fifteen years at Colorado State, Dr. Wade taught organic chemistry to thousands of students working toward careers in all areas of biology, chemistry, human medicine, veterinary medicine, and environmental studies. He also authored research papers in organic synthesis and in chemical education, as well as eleven books reviewing current research in organic synthesis. Since 1989, Dr. Wade has been a chemistry professor at Whitman College, where he teaches organic chemistry and pursues research interests in organic synthesis and forensic chemistry.

Dr. Wade's interest in forensic science has led him to testify as an expert witness in court cases involving drugs and firearms, and he has worked as a police firearms instructor and drug consultant. He also enjoys repairing and restoring old violins and bows, which he does professionally.

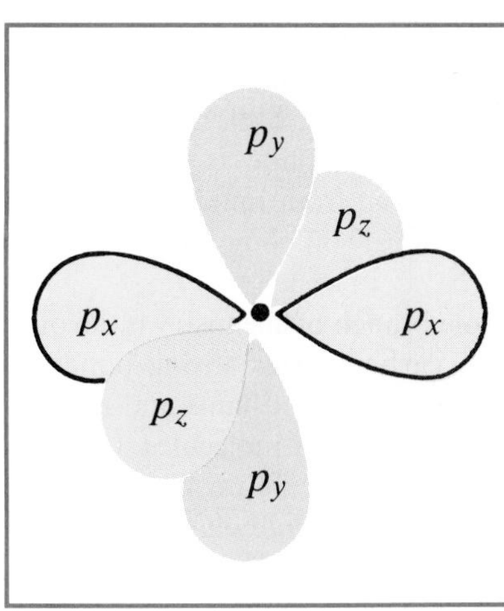

1

INTRODUCTION AND REVIEW

1-1 THE ORIGINS OF ORGANIC CHEMISTRY

The modern definition of **organic chemistry** is *the chemistry of carbon compounds.* The term **organic** literally means "derived from living organisms." Originally, the science of organic chemistry was the study of compounds extracted from living organisms and their natural products. Compounds such as sugar, urea, starch, waxes, and plant oils were considered "organic," and people believed that such natural products needed a "vital force" to create them. Organic chemistry, then, was the study of compounds having the vital force. Inorganic chemistry was the study of gases, rocks, and minerals and the compounds that could be made from them.

In the nineteenth century, the definition of organic chemistry had to be rewritten. Experiments had shown that organic compounds could be synthesized from inorganic compounds. One of these famous experiments was performed by the German chemist Friedrich Wöhler in 1828. He converted ammonium cyanate, made from ammonia and other inorganic chemicals, to urea simply by heating it in the absence of oxygen.

$$NH_4^+ \ {}^-OCN \xrightarrow{\text{heat}} H_2N{-}\overset{\overset{\displaystyle O}{\|}}{C}{-}NH_2$$

ammonium cyanate (inorganic) → urea (organic)

Urea had always come from living organisms and was presumed to contain the vital force, yet ammonium cyanate is inorganic and thus lacks the vital force. Some chemists claimed that a trace of vital force from Wöhler's hands must have contaminated the reaction, but most recognized the possibility of synthesizing organic compounds from inorganics. Many other syntheses were carried out, and the vital force theory was eventually discarded.

Even though organic compounds do not need a vital force, they are still distinguished from inorganic compounds. The distinctive feature of organic compounds is that they *all* contain one or more carbon atoms. Still, not all carbon compounds are organic; substances such as diamond, graphite, carbon dioxide, ammonium cyanate, and sodium carbonate are derived from minerals and have typical inorganic properties. Most of the millions of carbon compounds are considered to be organic, however.

What is so special about carbon that a whole branch of chemistry is devoted to its compounds? Unlike most other elements, carbon forms strong bonds to other carbon atoms and to a wide variety of other elements. Chains and rings of carbon atoms can be built up to form an endless variety of molecules. It is this diversity of carbon compounds that provides the basis for life on Earth. Living creatures are largely composed of complex organic compounds that serve structural, chemical, or genetic functions.

Chemists have learned to synthesize or simulate many of these complex molecules. The synthetic products serve as drugs, medicines, plastics, pesticides, paints, and fibers. Many of the most important advances in medicine are actually advances in organic chemistry. New synthetic drugs are developed to combat disease, and new polymers are molded to replace failing organs. Organic chemistry has gone through a full circle. It began as the study of compounds derived from "organs," and now it gives us the drugs and materials we need to save or replace those organs.

1-2 PRINCIPLES OF ATOMIC STRUCTURE

Before we begin our study of organic chemistry, we must review some basic principles. Although you have seen some of this material in general chemistry, we consider it from a slightly different viewpoint. Many of these concepts of atomic and molecular structure are crucial to your understanding of the structure and bonding of organic compounds.

1-2A STRUCTURE AND MASS OF THE ATOM

Atoms are made up of protons, neutrons, and electrons. Protons are positively charged and are found together with (uncharged) neutrons in the nucleus. Electrons, which have a negative charge that is equal in magnitude to the positive charge on the proton, move in the space surrounding the nucleus. Protons and neutrons have similar masses, about 1800 times the mass of an electron. Almost all the atom's mass is in the nucleus, but it is the electrons that take part in chemical bonding and reactions.

Each element is distinguished by the number of protons in the nucleus. The number of neutrons is usually similar to the number of protons, although the number of neutrons may vary. Atoms with the same number of protons but different numbers of neutrons are called **isotopes.** For example, the most common kind of carbon atom has six protons and six neutrons in its nucleus. Its mass number (the sum of the protons and neutrons) is 12, and we write its symbol as ^{12}C. About 1 percent of carbon atoms have seven neutrons; the mass number is 13, written ^{13}C. A very small fraction of carbon atoms are ^{14}C, a radioactive nucleus having eight neutrons. The ^{14}C isotope has a half-life of 5730 years, and the decay of this nucleus is used for determining the age of organic materials up to about 50,000 years old.

1-2B ELECTRONIC STRUCTURE OF THE ATOM

An element's chemical properties are determined by the number of protons in its nucleus and the corresponding number of electrons around the nucleus. It is the electrons that form bonds and determine the structure of the resulting molecules. Because they are small and light, electrons show properties of both particles and waves; in many ways, the electrons in atoms and molecules behave more like waves than like particles.

Electrons that are bound to nuclei are found in **orbitals.** The Heisenberg uncertainty principle states that we can never determine exactly where the electron is; but even though we do not know its exact location, we can speak of the **electron density,** the probability of finding the electron in a particular part of the orbital. An orbital, then, is an allowed energy state for an electron, with an associated probability function that defines the distribution of electron density in space.

For now, we need to review the atomic orbitals of an isolated atom. Most of the common elements in organic compounds are found in the first two rows of the periodic table, indicating that their electrons are contained in the first two electron shells, corresponding to values of the principal quantum number $n = 1$ and 2. We will concentrate on the orbitals of these two shells.

The first electron shell contains just the 1*s* orbital. All *s* orbitals are spherically symmetrical, meaning that they are nondirectional. The electron density is only a function of the distance from the nucleus. The electron density of the 1*s* orbital is graphed in Figure 1-1. Notice the exponential falloff in electron density with increasing distance from the nucleus. The electron density is highest (actually, it approaches infinity) *at* the nucleus, and it rapidly drops off at increasing distances from the nucleus. The 1*s* orbital might be imagined as a cotton boll, with the cottonseed at the middle representing the nucleus. The density of the cotton is highest nearest the seed, and it becomes less dense at greater distances from this "nucleus."

The second electron shell has two different kinds of orbitals, the 2*s* and 2*p* orbitals. Like the 1*s* orbital, the 2*s* is spherically symmetrical. Its electron density is not a simple exponential function, however. The 2*s* orbital has a smaller amount of electron density close to the nucleus. Most of the electron density is farther away, beyond a **node,** or region of zero electron density. Because most of the 2*s*

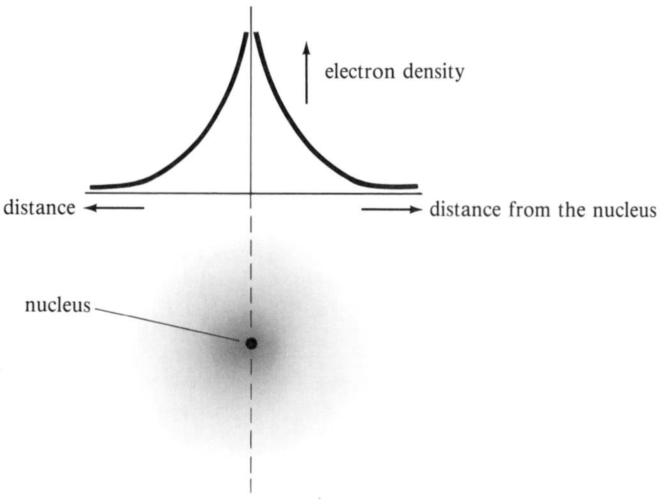

FIGURE 1-1 Graph and diagram of the 1*s* atomic orbital. The electron density is highest at the nucleus and drops off exponentially with increasing distance from the nucleus in any direction.

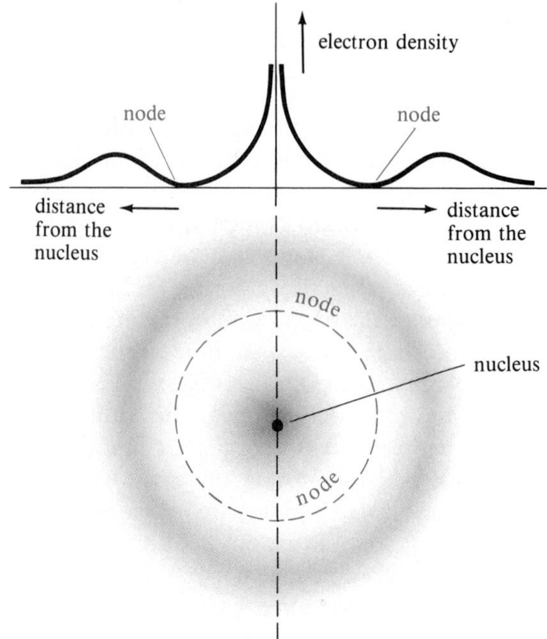

FIGURE 1-2 The 2*s* orbital has a small region of high electron density close to the nucleus, but most of the electron density is farther from the nucleus, beyond a node, or region of zero electron density.

electron density is farther from the nucleus than that of the 1*s*, the 2*s* orbital is higher in energy. Figure 1-2 shows a graph of the 2*s* orbital.

In addition to the 2*s* orbital, the second shell also contains three 2*p* atomic orbitals, one oriented in each of the three spatial directions. These orbitals are called the $2p_x$, the $2p_y$, and the $2p_z$, according to their direction along the *x*, *y*, or *z* axis. The 2*p* orbitals are slightly higher in energy than the 2*s*, because the average

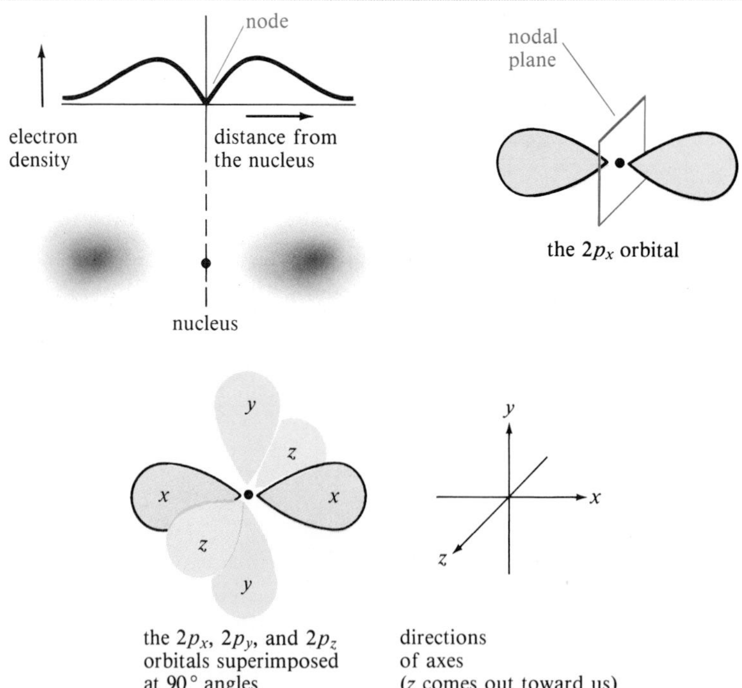

FIGURE 1-3 The 2*p* orbitals. There are three 2*p* orbitals, oriented at right angles to each other. Each is labeled according to its orientation along the *x*, *y*, or *z* axis.

location of the electron in a $2p$ orbital is farther from the nucleus. Each p orbital consists of two lobes, one on either side of the nucleus, with a **nodal plane** at the nucleus. The nodal plane is a flat (planar) region of space, including the nucleus, with zero electron density. The three $2p$ orbitals differ only in their spatial orientation, so they have identical energies. Orbitals with identical energies are called **degenerate orbitals.** Figure 1-3 shows the shapes of the three degenerate $2p$ atomic orbitals.

The *Pauli exclusion principle* tells us that each orbital can hold a maximum of two electrons, provided that their spins are paired. The first shell (one $1s$ orbital) can accommodate two electrons. The second shell (one $2s$ orbital and three $2p$ orbitals) can accommodate eight electrons, and the third shell (one $3s$ orbital, three $3p$ orbitals, and five $3d$ orbitals) can accommodate 18 electrons.

1-2C ELECTRONIC CONFIGURATIONS OF ATOMS

Aufbau means "building up" in German, and the *aufbau principle* tells us how to build up the electronic configuration of an atom's ground (most stable) state. Starting with the lowest-energy orbital, we fill the orbitals in order until we have added the proper number of electrons. Figure 1-4 and Table 1-1 show the application of the aufbau principle to the elements of the first three rows of the periodic table.

Two additional concepts are illustrated in Table 1-1. The **valence electrons** are those electrons that are in the outermost shell. Helium has two valence electrons and neon has eight, corresponding to a filled first shell and second shell, respectively. In general (for the representative elements), the column or group number of the periodic table corresponds to the number of valence electrons. Hydrogen and lithium have one valence electron, and they are both in the first column (group IA) of the periodic table. Carbon has four valence electrons, and it is in group IVA of the periodic table.

Notice that carbon's third and fourth valence electrons are not paired; they occupy separate orbitals. Although the Pauli exclusion principle says that two electrons can occupy the same orbital, the electrons repel each other and pairing requires additional energy. **Hund's rule** states that when there are two or more orbitals of the same energy, electrons will go into *different* orbitals rather than pair up in the same orbital. The first $2p$ electron (boron) goes into one $2p$ orbital,

TABLE 1-1
Electronic configurations of the elements of the first and second rows

Element	*Configuration*	*Valence electrons*
H	$1s^1$	1
He	$1s^2$	2
Li	$1s^2\ 2s^1$	1
Be	$1s^2\ 2s^2$	2
B	$1s^2\ 2s^2\ 2p_x^1$	3
C	$1s^2\ 2s^2\ 2p_x^1\ 2p_y^1$	4
N	$1s^2\ 2s^2\ 2p_x^1\ 2p_y^1\ 2p_z^1$	5
O	$1s^2\ 2s^2\ 2p_x^2\ 2p_y^1\ 2p_z^1$	6
F	$1s^2\ 2s^2\ 2p_x^2\ 2p_y^2\ 2p_z^1$	7
Ne	$1s^2\ 2s^2\ 2p_x^2\ 2p_y^2\ 2p_z^2$	8

FIGURE 1-4 First three rows of the periodic table. The organization of the periodic table results from the filling of atomic orbitals in order of increasing energy. For these representative elements, the number of the column corresponds to the number of valence electrons.

Partial periodic table

IA							noble gases (VIII)
H	IIA	IIIA	IVA	VA	VIA	VIIA	He
Li	Be	B	C	N	O	F	Ne
Na	Mg	Al	Si	P	S	Cl	Ar

energy ↑ — $1s$ — $2s$ — $2p_x$—$2p_y$—$2p_z$

relative orbital energies

the second (carbon) goes into a different orbital, and the third (nitrogen) occupies the last $2p$ orbital. The fourth, fifth, and sixth $2p$ electrons must pair up with the first three electrons.

PROBLEM 1-1

Write the electronic configurations of the third-row elements shown in the partial periodic table in Figure 1-4.

1-3 BOND FORMATION: THE OCTET RULE

In 1915, G. N. Lewis proposed several new theories describing how atoms bond to form molecules. One of these theories states that a filled shell of electrons is especially stable, and *atoms transfer or share electrons in such a way as to attain a filled shell of electrons:* the electron configuration of a noble gas such as He, Ne, or Ar. This principle has come to be called the **octet rule** because a filled shell implies eight valence electrons for the elements in the second row of the periodic table.

1-3A IONIC BONDING

There are two ways that atoms can interact to attain noble-gas configurations. Sometimes atoms attain noble-gas configurations by transferring electrons from one atom to another. For example, lithium has one electron more than the helium configuration, and fluorine has one electron less than the neon configuration. Lithium easily loses its valence electron, and fluorine easily gains one:

Li· ·F:	⟶	Li^+	:F:$^-$	⟶	Li^+ :F:$^-$
electron transfer		He configuration	Ne configuration		ionic bond

A transfer of one electron gives each element a noble-gas configuration. The resulting ions have opposite charges, and they attract each other to form an **ionic bond.** Ionic bonding usually results in the formation of a large crystal lattice rather than individual molecules. Ionic bonding is common in inorganic compounds, but relatively uncommon in organic compounds.

1-3B COVALENT BONDING

Covalent bonding, in which electrons are shared rather than transferred, is the most common form of bonding in organic compounds. Consider the hydrogen molecule, for example. The noble-gas configuration sought by hydrogen is that of helium, with two electrons in the first shell. If two hydrogen atoms come together and form a bond, they "share" their two electrons, and each atom has two electrons in its valence shell.

$$H\cdot + H\cdot \longrightarrow H{:}H$$ each H shares two electrons (He configuration)

Covalent bonding is particularly important in organic chemistry, and we will study covalent bonding in detail in Chapter 2.

1-4 LEWIS STRUCTURES

The simplest way to symbolize the bonding in a covalent molecule is to use **Lewis structures** as we did above, using H:H for the hydrogen molecule. In a Lewis structure, each valence electron is symbolized by a dot, or a bonding pair of electrons is symbolized by a dash (—). We try to arrange all the atoms so they have their appropriate noble-gas configurations: two electrons for hydrogen and octets for the second-row elements.

A more interesting structure is that of methane, CH_4.

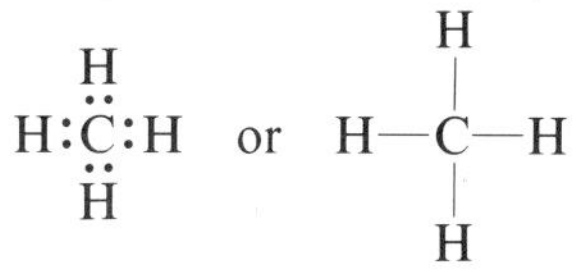

```
    H                H
   ..                |
 H:C:H    or    H—C—H
   ..                |
    H                H
```

methane

Carbon contributes four valence electrons, and each hydrogen contributes one, to give a total of eight electrons. All eight electrons surround carbon to give it an octet, and each hydrogen atom shares two of the electrons.

The Lewis structure for ethane, C_2H_6, is more complex.

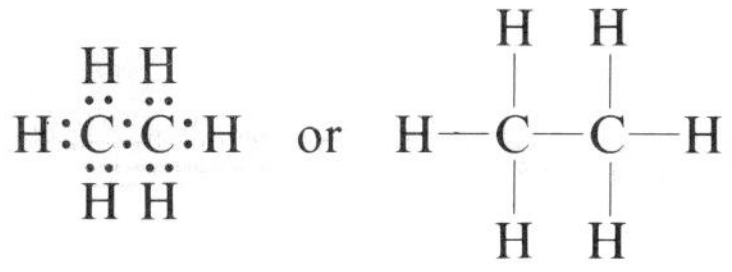

```
   H H                H  H
   .. ..              |  |
 H:C:C:H    or   H—C—C—H
   .. ..              |  |
   H H                H  H
```

ethane

Once again, we have computed the total number of electrons (14) and distributed them so that each carbon atom is surrounded by eight and each hydrogen by two. The only possible structure for ethane is the one shown, with the two carbon atoms sharing a pair of electrons and each hydrogen atom sharing a pair with one of the carbons. The ethane structure shows the most important characteristic of carbon—its ability to form strong carbon-carbon bonds.

We will encounter many structures with **nonbonding electrons** in the valence shell. A pair of nonbonding electrons is often called a **lone pair.** These are electrons that are not shared. Oxygen atoms, nitrogen atoms, and the halogens (F, Cl, Br, I) usually have nonbonding electrons in their stable compounds. These lone pairs of nonbonding electrons help to determine the reactivity of their parent compounds. As the following structures show, there is a lone pair of electrons on the nitrogen atom of methylamine, and there are two lone pairs on the oxygen atom of ethanol. Halogen atoms usually have three lone pairs, as shown in the structure of chloromethane.

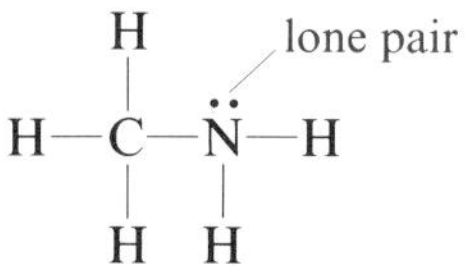

```
   H      lone pair        H  H       lone pairs          H
   |       ..              |  |        ..                  |    ..
H—C—N—H               H—C—C—O:                      H—C—Cl:   lone pairs
   |   |                   |  |  |                         |    ..
   H   H                   H  H  H                         H
```

methylamine 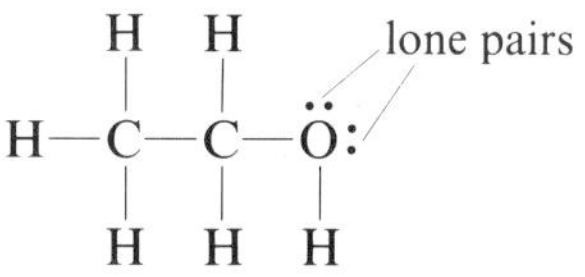 — ethanol 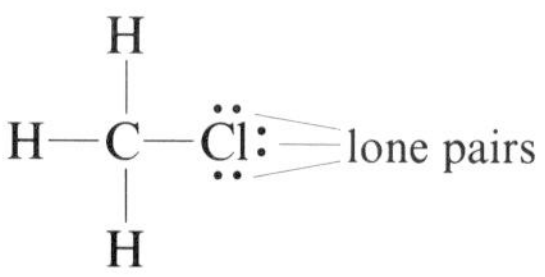 — chloromethane

PROBLEM 1-2

Draw Lewis structures for the following compounds.

(a) ammonia, NH_3
(b) water, H_2O
(c) hydronium ion, H_3O^+
(d) propane, C_3H_8
(e) ethylamine, $CH_3CH_2NH_2$
(f) dimethyl ether, CH_3OCH_3
(g) fluoroethane, CH_3CH_2F
(h) borane, BH_3
(i) boron trifluoride, BF_3

Explain what is unusual about the bonding in compounds in parts (h) and (i).

1-5 MULTIPLE BONDING

In drawing Lewis structures in Section 1-4, we always placed just one pair of electrons between any two atoms. The sharing of one pair between two atoms is called a **single bond.** Many molecules have adjacent atoms sharing two or even three electron pairs. The sharing of two pairs is called a **double bond,** and the sharing of three pairs is called a **triple bond.** Ethylene, C_2H_4, is an example of an organic compound with a double bond. When we draw a Lewis structure for ethylene, the only way to provide both of the carbon atoms with octets is to allow them to share two pairs of electrons. The following are examples of organic compounds with double bonds. In each case, four electrons (two pairs) are shared between two atoms to give them octets. A double dash (=) is used to symbolize a double bond.

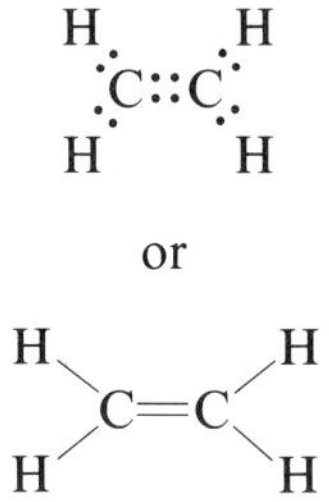

```
H     H         H                H
 C::C            C::O:            C::N:
H     H         H                H     H

   or              or               or

H     H         H                H
 C=C             C=O:             C=N:
H     H         H                H     H

 ethylene      formaldehyde     formaldimine
```

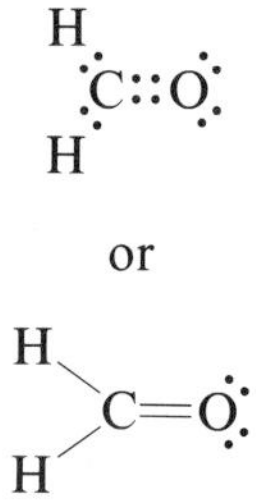

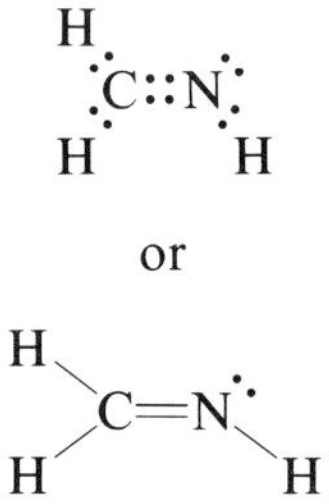

Acetylene, C_2H_2, is an example of an organic compound with a triple bond. When we draw the structure of acetylene, three pairs of electrons must be placed between the carbon atoms to give them octets. The following are examples of organic compounds with triple bonds. A triple dash (≡) is used to symbolize a triple bond.

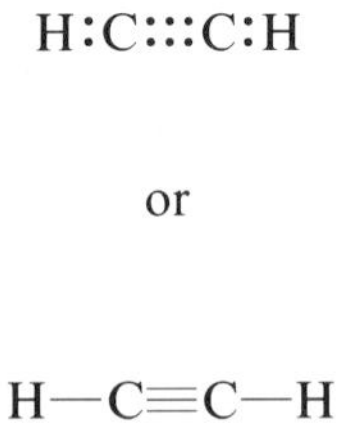

```
                    H       H             H
H:C:::C:H         H:C:C:::C:C:H         H:C:C:::N:
                    H       H             H

   or                   or                 or

                    H       H             H
                    |       |             |
H—C≡C—H           H—C—C≡C—C—H           H—C—C≡N:
                    |       |             |
                    H       H             H

 acetylene        dimethylacetylene     acetonitrile
```

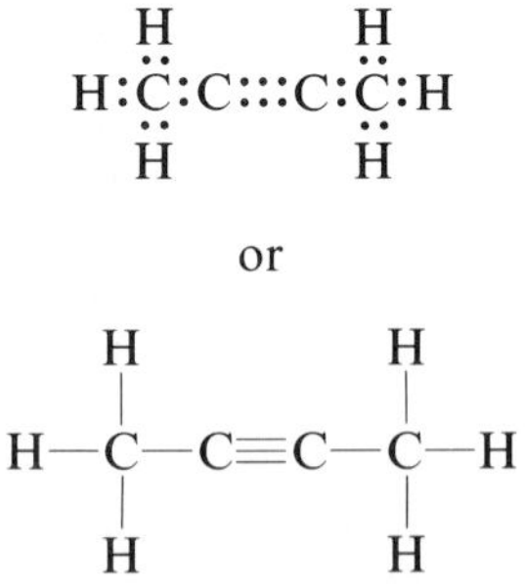

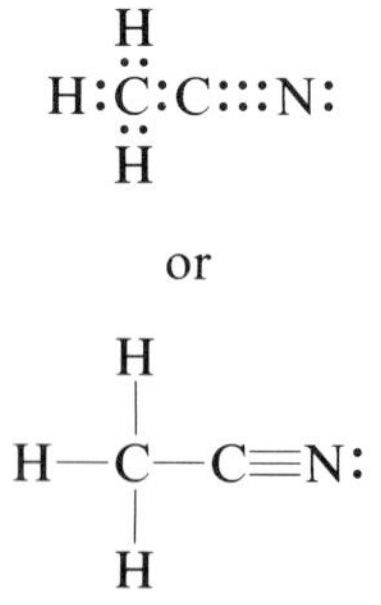

From the Lewis structures we have presented, it can be seen that carbon normally forms four bonds in neutral organic compounds. Nitrogen generally forms three bonds, and oxygen usually forms two. Hydrogen and the halogens usually form only one bond. The number of bonds an atom usually forms is called its **valence.** Carbon is tetravalent, nitrogen is trivalent, oxygen is divalent, and hydrogen and the halogens are monovalent. By remembering the usual number of bonds for these common elements, we can write organic structures more easily. If we draw a structure with each atom having its usual number of bonds, the correct Lewis structure usually results.

SUMMARY OF COMMON BONDING PATTERNS

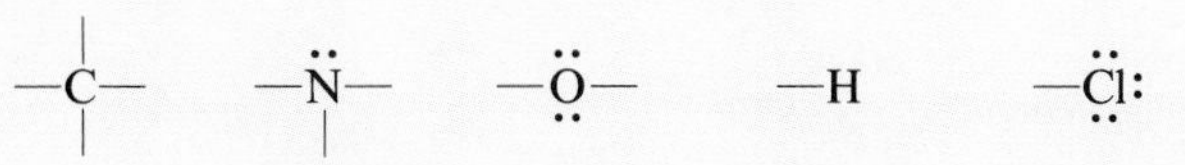

	—C—	—N̈—	—Ö—	—H	—C̈l:
	carbon	nitrogen	oxygen	hydrogen	halogens
valence:	4	3	2	1	1
lone pairs:	0	1	2	0	3

PROBLEM 1-3
Write a Lewis structure for each of the following molecular formulas.

(a) N_2 (b) HCN (c) HONO
(d) CO_2 (e) H_2CNH (f) HCO_2H
(g) C_2H_3Cl (h) HNNH (i) C_3H_6
(j) C_3H_4 (two double bonds) (k) C_3H_4 (one triple bond)

PROBLEM 1-4
Circle any lone pairs (pairs of nonbonding electrons) in the structures you drew for Problem 1-3.

1-6 ELECTRONEGATIVITY AND BOND POLARITY

A bond with the bonding electrons shared equally between the two bonded atoms is called a **nonpolar bond.** Examples of nonpolar bonds are the bond in H_2 and the C—C bond in ethane. In most bonds between two different elements, the bonding electrons are attracted more strongly by one of the two nuclei. An unequally shared pair of bonding electrons is called a **polar bond.**

When carbon is bonded to chlorine, for example, the bonding pair of electrons is attracted more strongly to the chlorine atom. The carbon atom bears a small partial positive charge, and the chlorine atom bears an equal amount of negative charge. The structure below shows the polar carbon-chlorine bond in chloromethane. We symbolize the bond polarity by an arrow with its head at the negative end of the polar bond and a plus sign at the positive end. The bond polarity is measured by its dipole moment (μ), defined to be the amount of charge separation (δ^+ and δ^-) multiplied by the bond length. The symbol δ^+ means "a small amount of positive charge"; δ^- means "a small amount of negative charge."

μ (arrow ⟶ from C to Cl), δ^+ on C, δ^- on Cl: H_3C—Cl

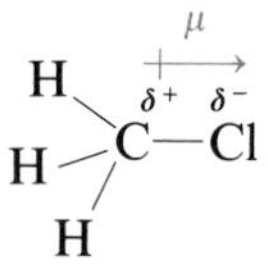

We often use **electronegativities** as a guide in predicting whether a given bond will be polar and the direction of its dipole moment. The Pauling electronegativity scale, most commonly used by organic chemists, is based on bonding

H						
2.2						
Li 1.0	Be 1.6	B 1.8	C 2.5	N 3.0	O 3.4	F 4.0
Na 0.9	Mg 1.3	Al 1.6	Si 1.9	P 2.2	S 2.6	Cl 3.2
						Br 3.0
						I 2.7

FIGURE 1-5 The electronegativities of some of the elements found in organic compounds.

properties, and it is useful for predicting the polarity of covalent bonds. Elements with higher electronegativities generally have more attraction for the bonding electrons. Therefore, in a bond between two different atoms, the atom with the higher electronegativity is the negative end of the dipole. Figure 1-5 shows the Pauling electronegativities for some of the important elements in organic compounds. Notice that the electronegativities increase from left to right across the periodic table. Nitrogen, oxygen, and the halogens are all more electronegative than carbon; sodium and lithium are less electronegative. Hydrogen's electronegativity is similar to that of carbon, and we usually consider C—H bonds to be nonpolar.

PROBLEM 1-5

Use electronegativities to predict the direction of the dipole moments of the following bonds.

(a) C—Cl (b) C—O (c) C—N (d) C—S (e) C—B
(f) N—Cl (g) N—O (h) N—S (i) N—B (j) B—Cl

1-7 FORMAL CHARGES

In polar bonds, the partial charges (δ^+ or δ^-) on the bonded atoms are *real.* **Formal charges** provide a method for keeping track of electrons, but they may or may not correspond to real charges. In most cases, if the Lewis structure shows that an atom has a formal charge, it actually bears at least a part of that charge. The concept of formal charge helps us to see which atoms bear most of the charge in a charged molecule, and it also helps us to see charged atoms in molecules that are neutral overall.

To calculate formal charges, we simply count how many electrons contribute to the charge of each atom and compare that number with the number of valence electrons in the free, neutral atom (given by the group number in the periodic table). The electrons that contribute to an atom's charge are:

1. *All* its unshared (nonbonding) electrons; plus
2. *Half* the (bonding) electrons it shares with other atoms, or one electron of each bonding pair.

The formal charge of a given atom can be calculated by the formula:

$$\text{formal charge} = (\text{group number}) - (\text{nonbonding electrons}) - \tfrac{1}{2}(\text{shared electrons})$$

SOLVED PROBLEM 1-1

Compute the formal charge on each atom in the following structures.

(a) Methane, CH_4

$$\begin{array}{c} H \\ H\!:\!\overset{\cdot\cdot}{\underset{\cdot\cdot}{C}}\!:\!H \\ H \end{array}$$

SOLUTION Each of the hydrogen atoms in methane has one bonding pair of electrons (two shared electrons). Half of two shared electrons is one electron, and one valence electron is what hydrogen needs to be neutral. Hydrogen atoms with one bond are formally neutral: $FC = 1 - 0 - 1 = 0$.

The carbon atom has four bonding pairs of electrons (eight electrons). Half of eight shared electrons is four electrons, and four electrons are what carbon (group IVA) needs to be neutral. Carbon is formally neutral whenever it has four bonds: $FC = 4 - 0 - \frac{1}{2}(8) = 0$.

(b) The hydronium ion, H_3O^+

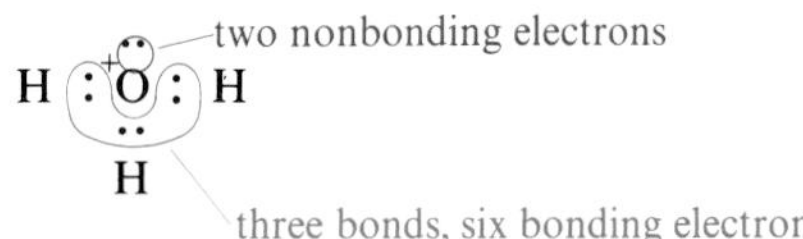

SOLUTION In drawing the Lewis structure for this ion, we use eight electrons: six from oxygen plus three from the hydrogens, minus one because the ion has a positive charge. Each hydrogen has one bond and is neutral. Oxygen is surrounded by an octet, with six bonding electrons and two nonbonding electrons. Half the bonding electrons plus all the nonbonding electrons contribute to its charge: $6/2 + 2 = 5$; but oxygen (group VIA) needs six valence electrons to be neutral. Consequently, oxygen atom has a formal charge of $+1$: $FC = 6 - 2 - \frac{1}{2}(6) = +1$.

(c) $H_3N—BH_3$

$$\begin{array}{c} H\ H \\ H\!:\!\overset{+}{N}\!:\!\overset{-}{B}\!:\!H \\ H\ H \end{array}$$

Nitrogen has four bonds, eight bonding electrons

Boron has four bonds, eight bonding electrons

SOLUTION This is an example of a neutral compound where the individual atoms are formally charged. The Lewis structure shows that both nitrogen and boron have four shared bonding pairs of electrons. Both boron and nitrogen have $8/2 = 4$ electrons contributing to their charges. Nitrogen (group V) needs five valence electrons to be neutral, so it bears a formal charge of $+1$. Boron (group III) needs only three valence electrons to be neutral, so it bears a formal charge of -1.

Nitrogen: $FC = 5 - 0 - \frac{1}{2}(8) = +1$

Boron: $FC = 3 - 0 - \frac{1}{2}(8) = -1$

(d) $[H_2CNH_2]^+$

$$H_2C{=}\overset{+}{N}H_2$$

SOLUTION This Lewis structure shows that both carbon and nitrogen have four shared pairs of bonding electrons. With four bonds, carbon is formally neutral; however, nitrogen is in group V, and it bears a formal positive charge: $FC = 5 - 0 - 4 = +1$.

Notice that this compound might also be drawn with the following Lewis structure:

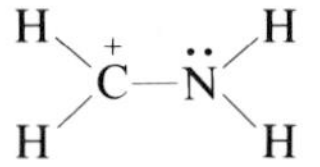

$$\underset{H}{\overset{H}{\diagdown}}\overset{+}{C}-\ddot{N}\underset{H}{\overset{H}{\diagup}}$$

In this structure, the carbon atom has three bonds with six bonding electrons. We calculate that $6/2 = 3$ electrons, one short of the four needed for a neutral carbon atom: $FC = 4 - 0 - \frac{1}{2}(6) = +1$.

Nitrogen has six bonding electrons and two nonbonding electrons. We calculate that $6/2 + 2 = 5$, and the nitrogen is uncharged in this second structure: $FC = 5 - 2 - \frac{1}{2}(6) = 0$.

The significance of these two Lewis structures is discussed in Section 1-9.

Most organic compounds contain only a few common elements, and most of these atoms have complete octets of electrons. The following summary table shows some of the most commonly occurring bonding structures, using dashes to represent bonding pairs of electrons. Use the rules for calculating formal charges discussed above to verify the charges shown on these structures. A good understanding of the structures shown here will help you to draw organic compounds and their ions quickly and correctly.

SUMMARY OF COMMON BONDING PATTERNS FOUND IN ORGANIC COMPOUNDS AND IONS

Atom	*Valence electrons*	*Positively charged*	*Neutral*	*Negatively charged*
B	3		—B— (no octet)	—B⁻—
C	4	—C⁺— (no octet)	—C—	—C̈⁻—
N	5	—N⁺—	—N̈—	—N̈⁻— (two lone pairs)
O	6	—Ö⁺—	—Ö— (two lone pairs)	—Ö:⁻ (three lone pairs)
halogen	7	—Cl⁺— (two lone pairs)	—Cl: (three lone pairs)	:Cl:⁻ (four lone pairs)

1-8 IONIC STRUCTURES

Some organic compounds contain ionic bonds. For example, the structure of methylammonium chloride (CH_3NH_3Cl) cannot be drawn using just covalent bonds because that would require nitrogen to have five bonds, implying ten electrons in its valence shell. The correct structure shows chloride ion to be ionically bonded to the rest of the structure.

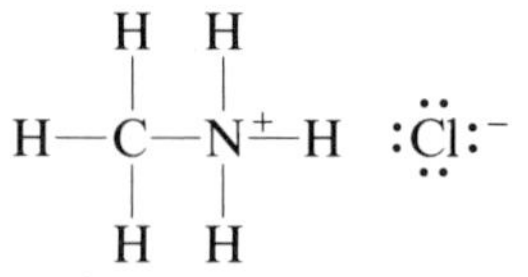

$H_3C-\overset{+}{N}H_3\ :\ddot{Cl}:^-$ — methylammonium chloride

H_3C-NH_3Cl (N with five bonds; too many electrons around nitrogen) — cannot be drawn covalently

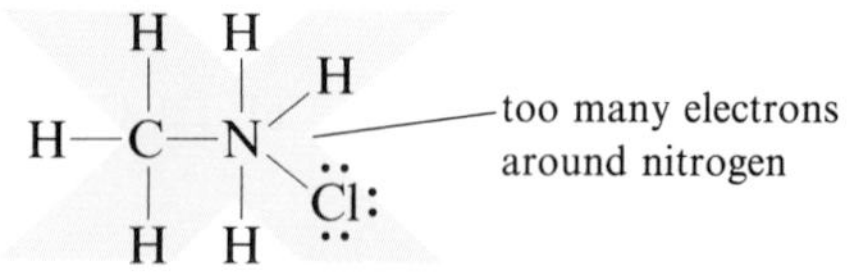

Some molecules can be drawn either covalently or ionically. For example, sodium methoxide ($NaOCH_3$) may be drawn with either a covalent bond or an ionic bond between sodium and oxygen. Because sodium generally forms ionic bonds with oxygen (as in NaOH), the ionically bonded structure is usually preferred. In general, bonds between atoms with very large electronegativity differences (about 2 or more) are often drawn as ionic.

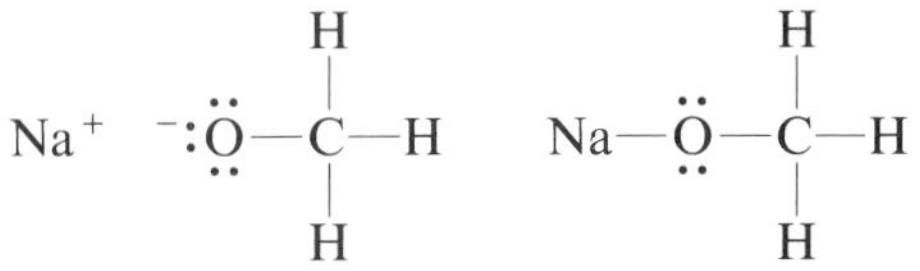

$$\mathrm{Na^{+}\ {}^{-}\!:\!\ddot{\underset{\cdot\cdot}{O}}\!-\!CH_2\!-\!H}\qquad\qquad \mathrm{Na\!-\!\ddot{\underset{\cdot\cdot}{O}}\!-\!CH_2\!-\!H}$$

more common — less common

PROBLEM 1-6

Determine the formal charge of each atom (except hydrogen) in the following compounds and ions.

(a) $[CH_3OH_2]^+$ (b) NH_4Cl (c) $(CH_3)_2NH_2Cl$ (d) NaOH
(e) $^{+}CH_3$ (f) $^{-}CH_3$ (g) $NaBH_4$ (h) $NaBH_3CN$
(i) $(CH_3)_2O{-}BF_3$ (j) $[HONH_3]^+$ (k) $KOC(CH_3)_3$ (l) $[H_2C{=}OH]^+$

1-9 RESONANCE STRUCTURES

1-9A RESONANCE HYBRIDS

In Solved Problem 1-1(d) we encountered a compound whose structure could be represented by either of two valid Lewis structures. In general, when two or more valence bond structures are possible and they differ only in the placement of electrons, the molecule will have characteristics of both structures. The different structures are called **resonance structures,** and the actual molecule is a **resonance hybrid** of these structures. In the example above, we saw that the ion $[H_2CNH_2]^+$ might be represented by either of the following structures:

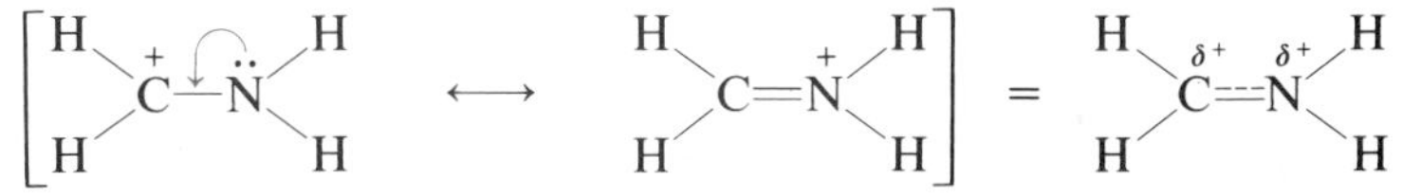

$$\left[\mathrm{H_2\overset{+}{C}\!-\!\ddot{N}H_2} \longleftrightarrow \mathrm{H_2C\!=\!\overset{+}{N}H_2}\right] = \mathrm{H_2\overset{\delta+}{C}\!\overset{\cdots}{=}\!\overset{\delta+}{N}H_2}$$

resonance structures — combined representation

The actual structure of this ion is a resonance hybrid of two resonance structures. Part of the positive charge is on carbon and part is on nitrogen. Notice that nitrogen's nonbonding electrons (in the structure at the far left) can be moved into the bond to give a double bond and an octet on carbon. In the combined representation, the positive charge is shown *delocalized* (spread out) onto both the carbon atom and the nitrogen atom. In the left resonance structure, the positive charge is on carbon, but carbon does not have an octet. In the right resonance structure, both atoms have octets, but the positive charge is on the electronegative nitrogen atom. This spreading of the positive charge over two atoms makes the ion more stable than it would be if the entire charge were localized only on the carbon or only on the nitrogen. We call this a **resonance-stabilized** cation. Resonance is most important when it allows a charge to be delocalized over two or more atoms, as in this example.

There are many examples of resonance hybrids in organic chemistry. When acetic acid loses a proton, the resulting acetate ion has a negative charge delocalized over both of the oxygen atoms. Each oxygen atom bears half of the negative charge, and this delocalization stabilizes the ion. Each of the carbon-oxygen bonds is halfway between a single bond and a double bond, and they are said to have a *bond order* of $1\frac{1}{2}$.

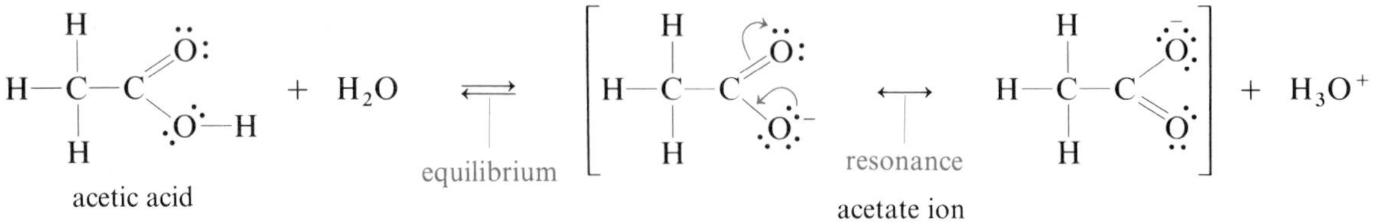

H—C(H)(H)—C(=O:)—Ö—H + H_2O ⇌ [H—C(H)(H)—C(=O:)—Ö:⁻ ⟷ H—C(H)(H)—C(—Ö:⁻)=O:] + H_3O^+

acetic acid — equilibrium — resonance — acetate ion

We use a single double-headed arrow between resonance structures (and often enclose the structures in brackets) to indicate that the actual structure is a hybrid of the Lewis structures we have drawn. By contrast, an equilibrium is represented by two arrows in different directions.

Some uncharged molecules actually have resonance-stabilized, equal positive and negative formal charges. We can draw two Lewis structures for nitromethane (CH_3NO_2), but both of them have a formal positive charge on nitrogen. Nitromethane must have a positive charge on the nitrogen atom and a negative charge spread equally over the two oxygen atoms. The N—O bonds are midway between single and double bonds, as indicated in the combined representation:

[H—C(H)(H)—$\overset{+}{N}$(=O:)—Ö:⁻ ⟷ H—C(H)(H)—$\overset{+}{N}$(—Ö:⁻)=O:] = H—C(H)(H)—$\overset{+}{N}$(⋯$O^{\delta-}$)⋯$O^{\delta-}$

resonance structures — combined representation

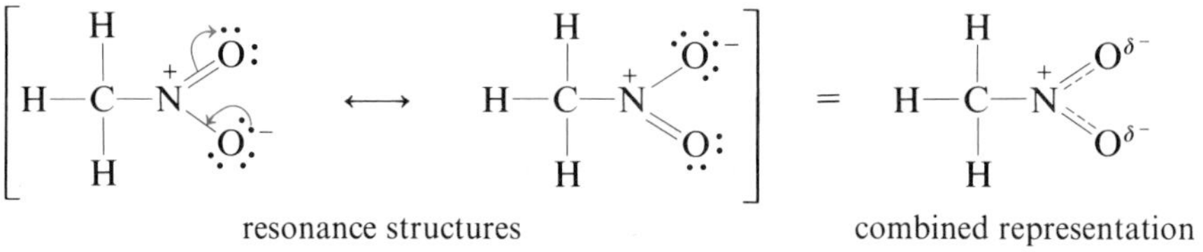

Remember that individual resonance structures do not exist. The molecule does not "resonate" between these structures. It is a hybrid with some characteristics of both. An analogy is a mule, which is a hybrid of a horse and a donkey. The mule does not "resonate" between looking like a horse and looking like a donkey; it looks like a mule all the time, with the broad back of the horse and the long ears of the donkey.

1-9B MAJOR AND MINOR RESONANCE CONTRIBUTORS

Two or more correct Lewis structures for the same compound may or may not represent electron distributions of equal energy. The two structures given above for the acetate ion have similar bonding, and they are of identical energy. The following two structures are bonded differently, however.

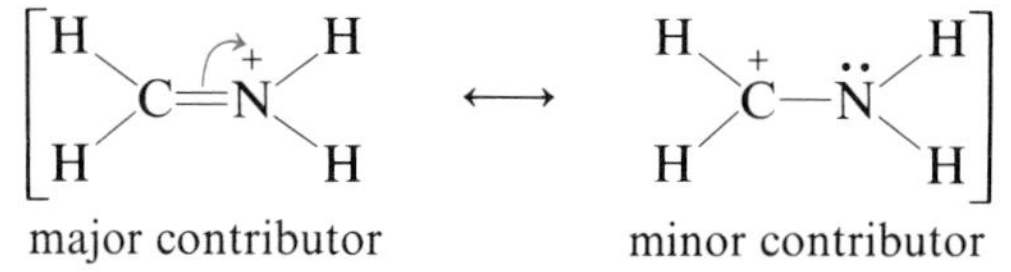

[$H_2C=\overset{+}{N}H_2$ ⟷ $H_2\overset{+}{C}—\ddot{N}H_2$]

major contributor — minor contributor

These structures are not of equal energy. The first structure has the positive charge on nitrogen. The second has the positive charge on carbon, and the carbon atom does not have an octet. The first structure is more stable, because it has an additional bond and all the atoms have octets. Many stable ions have a positive charge on a nitrogen atom with four bonds (see Summary Table, page 12). We call the more stable resonance structure the **major contributor,** and the less stable structure the **minor contributor.** The structure of the actual compound resembles the major contributor more than it does the minor contributor.

Many organic molecules have major and minor resonance contributors. Formaldehyde ($H_2C=O$) can be written with a negative charge on oxygen balanced by a positive charge on carbon. This polar resonance structure is higher in energy than the double-bonded structure, because it has charge separation, fewer bonds, and the positively charged carbon atom does not have an octet. The charge-separated structure is a minor contributor; but it helps to explain why the formaldehyde C=O bond is very polar, with a partial positive charge on carbon and a partial negative charge on oxygen.

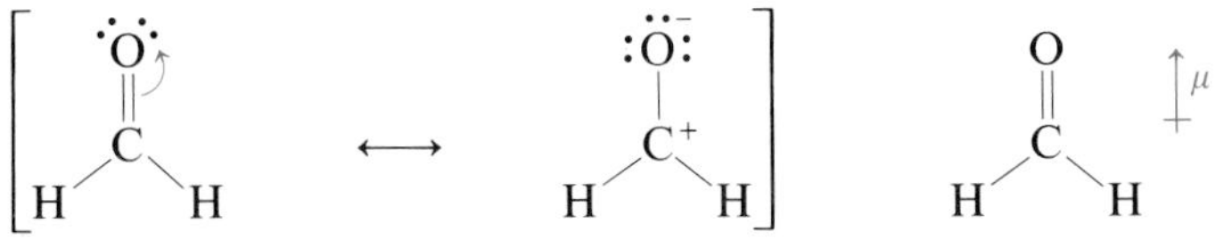

all octets
no charge separation
(major contributor)

no octet on C
charge separation
(minor contributor)

dipole moment

In drawing resonance structures, we will try to draw structures that are as low in energy as possible. The best candidates are those that have the maximum number of bonds and the maximum number of octets. Also, we should look for structures with the minimum amount of charge separation.

Only electrons can be delocalized. Unlike electrons, nuclei cannot be delocalized. They must remain in the same places, with the same bond distances and angles, in all the resonance contributors. Some general rules will help us to draw realistic resonance structures.

1. All the resonance structures must be valid Lewis structures for the compound.
2. Only the placement of the electrons may be shifted from one structure to another. (Electrons in double bonds and lone pairs are the ones that are most commonly shifted.) Nuclei cannot be moved, and the bond angles must remain the same.
3. The number of unpaired electrons (if any) must remain the same. Most stable compounds have no unpaired electrons, and all the electrons must remain paired in all the resonance structures.
4. The major resonance contributor is the one with the lowest energy; good contributors generally have all octets satisfied, as many bonds as possible, and as little charge separation as possible. Negative charges are more stable on the more electronegative atoms.
5. Resonance stabilization is most important when it serves to delocalize a charge over two or more atoms.

SOLVED PROBLEM 1-2

For each of the following compounds, draw the important resonance structures. Indicate which structures are major and minor contributors or whether they have the same energy.

(a)

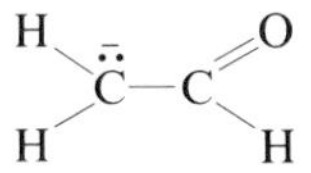

SOLUTION

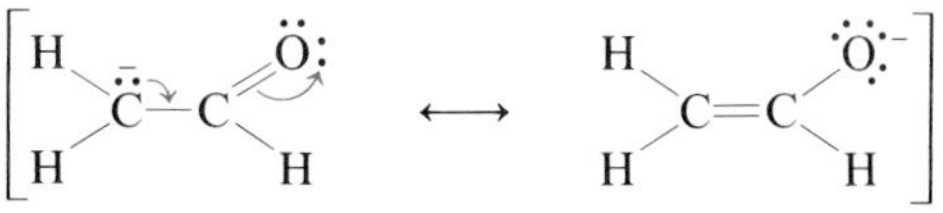

minor contributor

major contributor

Both of these structures have octets on oxygen and both carbon atoms, and they have the same number of bonds. The first structure has the negative charge on carbon; the second has it on oxygen. Oxygen is the more electronegative element, so the second structure is the major contributor.

(b) $H_2C{=}CH{-}NO_2$

SOLUTION

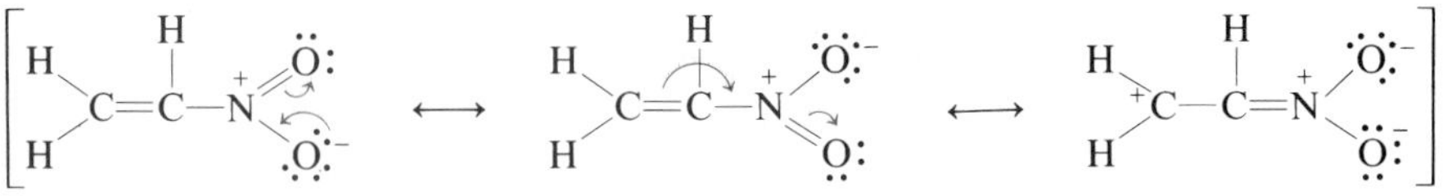

major contributor ⟷ major contributor ⟷ minor contributor

The first two structures simply show the charge separation of the nitro group, with a positive charge on the nitrogen atom and a negative charge spread over the two oxygen atoms. These two structures have identical bonding, and they are equal in energy. The third structure is higher in energy because it has fewer bonds, more charge separation, and the carbon with the positive charge lacks an octet. This third structure is useful because it shows how a partial positive charge arises on one of the carbon atoms.

PROBLEM 1-7

Draw the important resonance structures for the following molecules and ions.

(a) CO_3^{2-} (b) NO_3^- (c) NO_2^-
(d) $H_2C{=}CH{-}CH_2^+$ (e) $H_2C{=}CH{-}CH_2^-$ (f) SO_4^{2-}

PROBLEM 1-8

For each of the following compounds, draw the important resonance structures. Indicate which structures are major and minor contributors or whether they have the same energy.

(a) $[H_2CNO_2]^-$ (b) $H{-}C({=}O){-}NH_2$ (c) $[H_2COH]^+$ (d) H_2CNN

(e) $[H_2CCN]^-$ (f) $H_2\overset{-}{C}{-}C({=}O){-}O{-}CH_3$ (g) $H{-}C({=}O){-}\overset{-}{C}H{-}C({=}O){-}H$

1-10 STRUCTURAL FORMULAS

Several kinds of formulas are used by organic chemists to represent organic compounds. Some of these formulas involve a shorthand notation that requires some explanation. The most obvious formulas are the **structural formulas,** which actually show which atoms are bonded to which. There are two types of structural formulas, complete Lewis structures and condensed structural formulas. As we have seen, a Lewis structure symbolizes a bonding pair of electrons as a pair of dots or as a dash (—). Lone pairs of electrons are shown as pairs of dots in Lewis structures.

1-10A CONDENSED STRUCTURAL FORMULAS

Condensed structural formulas (Table 1-2) are written without showing all the individual bonds. In a condensed structure, each of the central atoms is shown together with the atoms that are bonded to it. The atoms bonded to a central

TABLE 1-2
Examples of condensed structural formulas

Compound	*Lewis structure*	*Condensed structural formula*
ethane	H H │ │ H—C—C—H │ │ H H	CH_3CH_3
isobutane	H H H │ │ │ H—C—C—C—H │ │ │ H │ H H—C—H │ H	$(CH_3)_3CH$
n-hexane	H H H H H H │ │ │ │ │ │ H—C—C—C—C—C—C—H │ │ │ │ │ │ H H H H H H	$CH_3(CH_2)_4CH_3$
diethyl ether	H H H H │ │ ·· │ │ H—C—C—O—C—C—H │ │ ·· │ │ H H H H	$CH_3CH_2OCH_2CH_3$ or $CH_3CH_2—O—CH_2CH_3$ or $(CH_3CH_2)_2O$
ethanol	H H │ │ ·· H—C—C—O—H │ │ ·· H H	CH_3CH_2OH
isopropyl alcohol	·· H :O—H │ │ H—C—C—H │ │ H │ H—C—H │ H	$(CH_3)_2CHOH$
dimethylamine	H H │ ·· │ H—C—N—C—H │ │ │ H H H	$(CH_3)_2NH$

atom are often listed after the central atom (as in CH_3CH_3 rather than $H_3C—CH_3$) even if that is not their actual bonding order. In many cases, if there are two or more identical groups, parentheses and a subscript may be used to represent all the identical groups. Nonbonding electrons are rarely shown in condensed structural formulas.

When a condensed structural formula is written for a compound containing double or triple bonds, the multiple bonds are often drawn as they would be in a Lewis structure. Table 1-3 shows examples of condensed structural formulas containing multiple bonds.

As you can see from Tables 1-2 and 1-3, the distinction between a complete Lewis structural formula and a condensed structural formula is a hazy one. Chemists often draw formulas with some parts condensed and other parts completely drawn out. You should work with these different types of formulas so that you understand what all of them mean.

TABLE 1-3
Condensed structural formulas for double and triple bonds

Compound	*Lewis structure*	*Condensed structural formula*
2-butene	H H H H—C—C=C—C—H H H H	$CH_3CHCHCH_3$ or $CH_3CH{=}CHCH_3$
acetonitrile	H H—C—C≡N: H	CH_3CN or $H_3C{-}C{\equiv}N$
acetaldehyde	H :O: H—C—C—H H	CH_3CHO or $CH_3\overset{O}{\overset{\|\|}{C}}H$
acetone	H :O: H H—C—C—C—H H H	CH_3COCH_3 or $CH_3\overset{O}{\overset{\|\|}{C}}CH_3$
acetic acid	H :O: H—C—C—Ö—H H	CH_3COOH or $CH_3\overset{O}{\overset{\|\|}{C}}{-}OH$ or CH_3CO_2H

TABLE 1-4
Examples of line-angle drawings

Compound	*Condensed structure*	*Line-angle formula*
hexane	$CH_3(CH_2)_4CH_3$	
2-hexene	$CH_3CH{=}CHCH_2CH_2CH_3$	
3-hexanol	$CH_3CH_2CH(OH)CH_2CH_2CH_3$	OH
cyclohexene	CH_2 CH_2 CH CH_2 CH CH_2	
2-methylcyclohexanol	CH_2 CH_2 CHOH CH_2 $CHCH_3$ CH_2	OH, CH_3 or OH
pyridine	H H C H C C C C H N H	N

PROBLEM 1-9

Draw a complete Lewis structure for each of the following condensed structural formulas.

(a) $CH_3(CH_2)_3CH(CH_3)_2$ (b) $(CH_3)_2CHCH_2Cl$ (c) $CH_3CH_2COCH_2CH_3$
(d) CH_3CH_2CHO (e) CH_3COCN (f) $(CH_3)_3CCOOH$
(g) $(CH_3CH_2)_2CO$

1-10B LINE-ANGLE FORMULAS

Another kind of shorthand used for organic structures is the **line-angle formula,** sometimes called a "stick figure." Line-angle formulas are often used for cyclic compounds and occasionally for noncyclic ones. In a stick figure, bonds are represented by lines, and carbon atoms are assumed to be present wherever two lines meet or a line begins or ends. Nitrogen, oxygen, and halogen atoms are shown, but hydrogen atoms are not usually drawn unless they are bonded to a drawn atom. Each carbon atom is assumed to have enough hydrogen atoms to give it a total of four bonds. Table 1-4 shows some examples of line-angle drawings.

PROBLEM 1-10

Give a Lewis structure for each of the following line-angle structures.

(a) 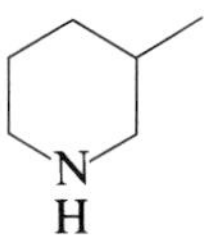 N H (b) 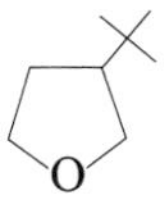 O (c) 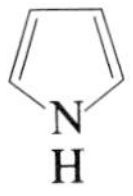 N H (d) 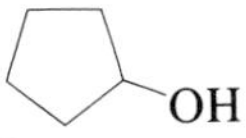 OH

(e) 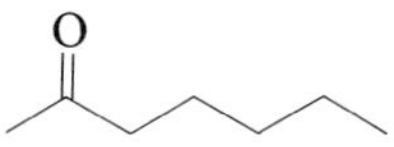 O (f) 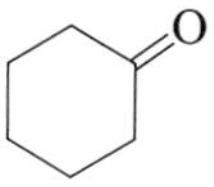 O (g)

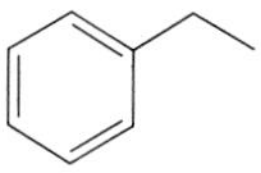

1-11 MOLECULAR FORMULAS AND EMPIRICAL FORMULAS

Before we can write possible structural formulas for a compound, we need to know its molecular formula. The **molecular formula** simply gives the number of atoms of each element in one molecule of the compound. For example, the molecular formula for 1-butanol is $C_4H_{10}O$.

$$CH_3CH_2CH_2CH_2OH$$

1-butanol, molecular formula $C_4H_{10}O$

Calculation of the empirical formula Molecular formulas are determined by a two-step process. The first step is the determination of an **empirical formula,** simply the relative ratios of the elements present. Suppose, for example, that an unknown compound was found by quantitative elemental analysis to contain 40.0 percent carbon and 6.67 percent hydrogen. The remainder of the weight is assumed to be oxygen, giving 53.3 percent oxygen. To convert these numbers to an empirical formula, we can follow a simple procedure:

1. Assume the sample contains 100 g, so the percent value gives the number of grams of each element. Divide that number of grams by the atomic weight to get the number of moles of that atom in the 100-g sample.

2. Divide each of these numbers of moles by the smallest one. This step should give recognizable ratios.

For this example, we do the following computations:

$$\frac{40.0 \text{ g C}}{12.0 \text{ g/mol}} = 3.33 \text{ mol C}; \quad \frac{3.33 \text{ mol}}{3.33 \text{ mol}} = 1$$

$$\frac{6.67 \text{ g H}}{1.01 \text{ g/mol}} = 6.60 \text{ mol H}; \quad \frac{6.60 \text{ mol}}{3.33 \text{ mol}} = 1.98 \cong 2$$

$$\frac{53.3 \text{ g O}}{16.0 \text{ g/mol}} = 3.33 \text{ mol O}; \quad \frac{3.33 \text{ mol}}{3.33 \text{ mol}} = 1$$

The first computation divides the number of grams of carbon by 12, the number of grams of hydrogen by 1, and the number of grams of oxygen by 16. We then compare these numbers by dividing them by the smallest number, 3.33. The final result is a ratio of one carbon to two hydrogens to one oxygen. This result gives the empirical formula $C_1H_2O_1$ or CH_2O, which simply shows the ratios of the elements. The molecular formula can be any multiple of this empirical formula, because any multiple also has the same ratio of elements. Possible molecular formulas are CH_2O, $C_2H_4O_2$, $C_3H_6O_3$, $C_4H_8O_4$, and so on.

Calculation of the molecular formula How do we know which of these possible molecular formulas is correct? We can choose the right one only if we know the molecular weight. Molecular weights can be determined by several physical methods. Freezing point depressions depend on the concentration of a solute; we can dissolve a known amount of the compound in a solvent and see how much the solvent's freezing point is depressed. Boiling points also depend on the concentration of a solute, so we can measure the solvent's boiling point elevation. If our sample is volatile, we can convert it to a gas and use its volume to determine the number of moles according to the gas law.

These physical methods for determination of molecular weights have largely been replaced by the use of *mass spectrometry.* As we will see in Chapter 11, mass spectrometry gives an accurate molecular weight using only a minute amount of the sample.

For our example (empirical formula CH_2O), let's assume that the molecular weight is determined to be about 60. The weight of one CH_2O unit is 30, so our unknown compound must contain twice this many atoms. The molecular formula must be $C_2H_4O_2$. The compound might have the following structure, that of acetic acid.

$$CH_3-\overset{\overset{\displaystyle O}{\|}}{C}-OH$$

acetic acid, $C_2H_4O_2$

In Chapters 11, 12, and 15 we will see how other techniques are used to determine the complete structure for a compound once we have its molecular formula.

PROBLEM 1-11

Compute the empirical and molecular formulas for each of the following elemental analyses. In each case, propose a structure that fits the molecular formula.

	C	H	N	Cl	MW
(a)	40.0%	6.67%	0	0	90
(b)	32.0%	6.67%	18.7%	0	75
(c)	37.2%	7.75%	0	55.0%	64
(d)	38.4%	4.80%	0	56.8%	125

1-12 ARRHENIUS ACIDS AND BASES

We will often use the concept of acids and bases in our study of organic chemistry. We need to consider exactly what is meant by the terms *acid* and *base*. Most people would agree that H_2SO_4 is an acid and NaOH is a base. Is BF_3 an acid or a base? Is ethylene ($H_2C{=}CH_2$) an acid or a base? To answer these questions, we need to understand the three different definitions of acids and bases: The Arrhenius definition, the Brønsted-Lowry definition, and the Lewis definition. Acidic compounds were first classified on the basis of their sour taste. The Latin terms *acidus* (sour) and *acetum* (vinegar) gave rise to our modern terms *acid* and *acetic acid*. Alkaline compounds (bases) were considered to be substances that neutralize acids, such as limestone and plant ashes (*al kalai* in Arabic).

The Arrhenius theory, developed at the end of the nineteenth century, helped to provide a better understanding of acids and bases. Acids were defined as substances that dissociate in water to give H_3O^+ ions. The stronger acids, such as sulfuric acid (H_2SO_4), were assumed to dissociate to a greater degree than weaker acids, such as acetic acid (CH_3COOH).

$$\underset{\text{sulfuric acid}}{H_2SO_4} + H_2O \rightleftharpoons H_3O^+ + HSO_4^-$$

$$\underset{\text{acetic acid}}{CH_3{-}\overset{\overset{\displaystyle O}{\|}}{C}{-}OH} + H_2O \rightleftharpoons H_3O^+ + CH_3{-}\overset{\overset{\displaystyle O}{\|}}{C}{-}O^-$$

Using the Arrhenius definition, bases are substances that dissociate in water to give hydroxide ions. Strong bases, such as NaOH, were assumed to dissociate more completely than weaker, sparingly-soluble bases such as $Mg(OH)_2$.

$$NaOH \rightleftharpoons Na^+ + {}^-OH$$

$$Mg(OH)_2 \rightleftharpoons Mg^{2+} + 2\,{}^-OH$$

The acidity or basicity of an aqueous (water) solution is measured by the concentration of H_3O^+. This value also implies the concentration of ^-OH, because these two concentrations are related by the water autoprotolysis constant

$$K_w = [H_3O^+][^-OH] = 1.00 \times 10^{-14} \quad \text{(at 24°C)}$$

In a neutral solution, the concentrations of H_3O^+ and ^-OH are equal.

$$[H_3O^+] = [^-OH] = 1.0 \times 10^{-7}\,M \quad \text{in a neutral solution}$$

Acidic and basic solutions are defined by an excess of H_3O^+ or ^-OH.

$$\text{acidic:} \quad [H_3O^+] > 10^{-7}\,M \quad \text{and} \quad [^-OH] < 10^{-7}\,M$$

$$\text{basic:} \quad [H_3O^+] < 10^{-7}\,M \quad \text{and} \quad [^-OH] > 10^{-7}\,M$$

Because these concentrations can span a wide range of values, the acidity or basicity of a solution is usually measured on a logarithmic scale. The **pH** is defined as the negative logarithm (base 10) of the H_3O^+ concentration.

$$pH = -\log_{10}[H_3O^+]$$

A neutral solution has a pH of 7, an acidic solution has a pH less than 7, and a basic solution has a pH greater than 7.

PROBLEM 1-12

Calculate the pH of the following solutions.

(a) 5.00 g of HBr in 100 mL of aqueous solution
(b) 2.00 g of NaOH in 50 mL of aqueous solution

The Arrhenius definition was an important contribution to understanding many acids and bases, but it does not explain the reactivity of compounds such as ammonia (NH_3). Ammonia is known to neutralize acids, yet it has no hydroxide ion in its molecular formula. A more versatile theory of acids and bases is necessary to include ammonia and many organic acids and bases.

1-13 BRØNSTED-LOWRY ACIDS AND BASES

In 1923 the Brønsted-Lowry definition of acids and bases was developed, based on the transfer of protons. *A* **Brønsted-Lowry acid** *is any species that can donate a proton, and a* **Brønsted-Lowry base** *is any species that can accept a proton.* All the Arrhenius acids and bases fit under this definition, because compounds that dissociate to give H_3O^+ are proton donors, and compounds that dissociate to give ^-OH are proton acceptors. The hydroxide ion accepts a proton to form H_2O. In addition to the Arrhenius acids and bases, the Brønsted-Lowry definition includes many bases that have no hydroxide ions, yet can accept protons. Consider the following examples of acids donating protons to bases. NaOH is a base under either the Arrhenius or Brønsted-Lowry definition. The other three bases are included under the Brønsted-Lowry definition but not under the Arrhenius definition (they have no hydroxide ions).

$$\underset{\text{proton donor}}{HCl} + \underset{\text{proton acceptor}}{NaOH} \rightleftharpoons NaCl + H_2O$$

$$\underset{\text{proton donor}}{H_2SO_4} + \underset{\text{proton acceptor}}{:NH_3} \rightleftharpoons HSO_4^- + {}^+NH_4$$

$$\underset{\text{proton donor}}{HCl} + \underset{\text{proton acceptor}}{H_2C{=}CH(CH_3)} \rightleftharpoons H_3C{-}C^+H(CH_3) + Cl^-$$

$$\underset{\text{proton donor}}{HNO_3} + \underset{\text{proton acceptor}}{H_2C{=}\ddot{N}H} \rightleftharpoons H_2C{=}N^+H_2 + {}^-NO_3$$

When a base accepts a proton, it becomes capable of returning that proton: It becomes an acid. When an acid donates its proton, it becomes capable of accepting that proton back: It becomes a base. One of the most important principles of the Brønsted-Lowry definition is this concept of **conjugate acids and bases.** For example, NH_4^+ and NH_3 are a conjugate acid-base pair. NH_3 is the base; when it accepts a proton it is transformed into its conjugate acid, NH_4^+. Many compounds (water, for instance) can react either as an acid or as a base. Here are some additional examples of conjugate acid-base pairs.

$$\underset{\text{acid}}{H_2SO_4} + \underset{\text{base}}{H_2O} \rightleftharpoons \underset{\text{conjugate base}}{HSO_4^-} + \underset{\text{conjugate acid}}{H_3O^+}$$

$$\underset{\text{acid}}{H_2O} + \underset{\text{base}}{:NH_3} \rightleftharpoons \underset{\text{conjugate base}}{^-OH} + \underset{\text{conjugate acid}}{NH_4^+}$$

$$\underset{\text{acid}}{H{-}\overset{O}{\overset{\|}{C}}{-}OH} + \underset{\text{base}}{CH_3{-}\ddot{\underset{\cdot\cdot}{O}}:^-} \rightleftharpoons \underset{\text{conjugate base}}{H{-}\overset{O}{\overset{\|}{C}}{-}\ddot{\underset{\cdot\cdot}{O}}:^-} + \underset{\text{conjugate acid}}{CH_3{-}\ddot{\underset{\cdot\cdot}{O}}{-}H}$$

1-13A ACID STRENGTH

The strength of a Brønsted-Lowry acid is expressed as it is in the Arrhenius definition, by the extent of its ionization in water. The general reaction of an acid (HA) with water is the following:

$$\underset{\text{acid}}{HA} + H_2O \overset{K_a}{\rightleftharpoons} H_3O^+ + \underset{\text{base}}{A^-} \qquad K_a = \frac{[H_3O^+][A^-]}{[HA]}$$

(HA and A^-: conjugate acid-base pair)

K_a is called the *acid dissociation constant,* and its size indicates the relative strength of the acid. The stronger the acid is, the more it dissociates, giving a larger value of K_a. Acid dissociation constants vary over a wide range. Strong acids are almost completely ionized in water, and their dissociation constants are greater than 1. Most organic acids are weak acids, with values of K_a that are less than 10^{-4}. Many organic compounds are extremely weak acids; for example, methane and ethane are essentially nonacidic, with K_a values less than 10^{-40}.

Because they span such a wide range, acid dissociation constants are often expressed on a logarithmic scale. The pK_a of an acid is defined just like the pH of a solution: as the negative logarithm (base 10) of K_a.

$$pK_a = -\log_{10} K_a$$

SOLVED PROBLEM 1-3

Calculate K_a and pK_a for water.

SOLUTION The equilibrium that defines K_a for water is

$$\underset{\text{acid (HA)}}{H_2O} + \underset{\text{solvent}}{H_2O} \overset{K_a}{\rightleftharpoons} H_3O^+ + \underset{\text{conjugate base }(A^-)}{^-OH}$$

Water serves as both the acid and the solvent in this acid dissociation. The equilibrium expression is

$$K_a = \frac{[H_3O^+][A^-]}{[HA]} = \frac{[H_3O^+][^-OH]}{[H_2O]}$$

We already know that $[H_3O^+][^-OH] = 1.00 \times 10^{-14}$, the ion-product constant for water.

The concentration of H_2O in water is simply the number of moles of water in 1 L (about 1 kg).

$$\frac{1000 \text{ g/L}}{18 \text{ g/mol}} = 55.6 \text{ mol/L}$$

Substitution gives

$$K_a = \frac{[H_3O^+][^-OH]}{[H_2O]} = \frac{1.00 \times 10^{-14}}{55.6} = 1.8 \times 10^{-16}\ M$$

The logarithm of 1.8×10^{-16} is -15.7, and the pK_a of water is 15.7.

Strong acids generally have values of pK_a around 0, and weak acids, such as most organic acids, have values of pK_a that are greater than 4. *Weaker acids have larger values of* pK_a. Table 1-5 gives values of K_a and pK_a for some common inorganic and organic compounds. Notice that the values of pK_a increase as the values of K_a decrease.

TABLE 1-5

Values of K_a and pK_a for some common inorganic and organic acids

Acid	K_a	pK_a
strong acids (HCl, H_2SO_4, etc.)	>1	<0
$HF + H_2O \rightleftharpoons H_3O^+ + F^-$ hydrofluoric acid	6.8×10^{-4}	3.17
$H-C(=O)-OH + H_2O \rightleftharpoons H_3O^+ + H-C(=O)-O^-$ formic acid	1.7×10^{-4}	3.76
$CH_3-C(=O)-OH + H_2O \rightleftharpoons H_3O^+ + CH_3-C(=O)-O^-$ acetic acid	1.8×10^{-5}	4.74
$H-C\equiv N\!: + H_2O \rightleftharpoons H_3O^+ + {}^-\!:C\equiv N\!:$ hydrocyanic acid	6.0×10^{-10}	9.22
$CH_3-OH + H_2O \rightleftharpoons H_3O^+ + CH_3O^-$ methyl alcohol	3.2×10^{-16}	15.5
$H_2O + H_2O \rightleftharpoons H_3O^+ + HO^-$ water	1.8×10^{-16}	15.7
$NH_3 + H_2O \rightleftharpoons H_3O^+ + {}^-\!:\ddot{N}H_2$ ammonia	10^{-33}	33
$CH_4 + H_2O \rightleftharpoons H_3O^+ + {}^-\!:CH_3$ methane	$<10^{-40}$	>40

1-13B BASE STRENGTH

There is a relationship between the strength of an acid and that of its conjugate base. For an acid (HA) to be strong, its conjugate base (A^-) must be stable in its anionic form; otherwise, HA would be reluctant to lose its proton. Therefore, the conjugate base of a strong acid must be a weak base. On the other hand, if an acid is weak, its conjugate is a strong base.

$$\underset{\text{strong acid}}{HCl} + H_2O \rightleftarrows H_3O^+ + \underset{\text{weak base}}{Cl^-}$$

$$\underset{\text{weak acid}}{CH_3\text{—}\ddot{\underset{\cdot\cdot}{O}}H} + H_2O \rightleftarrows H_3O^+ + \underset{\text{strong base}}{CH_3\ddot{\underset{\cdot\cdot}{O}}{:}^-}$$

In the reaction of an acid with a base, the equilibrium generally favors the *weaker* acid and base. For example, in the reactions above, H_3O^+ is a weaker acid than HCl but a stronger acid than CH_3OH. It also follows that H_2O is a stronger base than Cl^- but a weaker base than CH_3O^-.

PROBLEM 1-13

Predict the products of the following acid-base reactions. Use the information in Table 1-5 to predict whether the equilibrium will favor the reactants or the products.

(a) $HCOOH + {}^-CN$ (b) $CH_3COO^- + CH_3OH$ (c) $CH_3OH + NaNH_2$
(d) $NaOCH_3 + HCN$ (e) $HCl + H_2O$ (f) $H_3O^+ + CH_3O^-$

The strength of a base is measured much like the strength of an acid, by using the equilibrium constant of the hydrolysis reaction.

$$\underset{\substack{\text{conjugate}\\ \text{base}}}{A^-} + H_2O \overset{K_b}{\rightleftarrows} \underset{\substack{\text{conjugate}\\ \text{acid}}}{HA} + {}^-OH$$

The equilibrium constant (K_b) for this reaction is called the *base-dissociation constant* for the base A^-. Because this constant spans a wide range of values, it is often given in logarithmic form. The negative logarithm (base 10) of K_b is defined as pK_b.

$$K_b = \frac{[HA][{}^-OH]}{[A^-]} \qquad pK_b = -\log_{10} K_b$$

When we multiply K_a by K_b, we can see how the acidity of an acid is related to the basicity of its conjugate base.

$$(K_a)(K_b) = \frac{[H_3O^+][A^-]}{[HA]}\frac{[HA][{}^-OH]}{[A^-]} = \underset{\text{water ion-product constant}}{[H_3O^+][{}^-OH]} = 1.0 \times 10^{-14}$$

$$(K_a)(K_b) = 10^{-14}$$

Logarithmically,

$$pK_a + pK_b = -\log 10^{-14} = 14$$

The product of K_a and K_b must always equal the ion-product constant of water, 10^{-14}. If the value of K_a is large, the value of K_b must be small; that is, the stronger an acid, the weaker its conjugate base. Similarly, if the value of K_a is small, K_b must be correspondingly large.

SOLVED PROBLEM 1-4

Each of the following compounds can act as an acid. Show the reaction of each compound with a general base (A^-) and show the structure of the conjugate base that results.

(a) CH_3CH_2OH (b) CH_3NH_2 (c) CH_3COOH

SOLUTION (a) Ethanol (CH_3CH_2OH) can lose the O—H proton to give a conjugate base that is an organic analog of hydroxide ion:

$$CH_3CH_2—\ddot{\underset{\cdot\cdot}{O}}—H + A^- \rightleftharpoons CH_3CH_2—\ddot{\underset{\cdot\cdot}{O}}:^- + HA$$

ethanol (acid) — base — conjugate base

(C—H protons are much less acidic than O—H protons, because carbon is less electronegative than oxygen, and the negative charge is therefore less stable on carbon.)

(b) Methylamine (CH_3NH_2) is a very weak acid. A very strong base can abstract a proton to give a powerful conjugate base.

$$CH_3—\underset{\cdot\cdot}{\overset{H}{N}}—H + A^- \rightleftharpoons CH_3—\ddot{\underset{\cdot\cdot}{N}}^-—H + HA$$

methylamine — very strong base — powerful conjugate base

(c) Acetic acid (CH_3COOH) is a moderately strong acid, giving the resonance-stabilized acetate ion as its conjugate base.

$$CH_3—\overset{O}{\overset{\|}{C}}—\ddot{\underset{\cdot\cdot}{O}}—H + A^- \rightleftharpoons \left[CH_3—\overset{O}{\overset{\|}{C}}—\ddot{\underset{\cdot\cdot}{O}}:^- \longleftrightarrow CH_3—\overset{:\ddot{O}:^-}{\overset{|}{C}}=\ddot{O}\right] + HA$$

acetic acid — acetate ion (conjugate base)

SOLVED PROBLEM 1-5

Each of the compounds in Solved Problem 1-4 can also react as a base. Show the reaction of each compound with a general acid (HA), and show the structure of the conjugate acid that results.

SOLUTION (a) Ethanol can undergo protonation on its oxygen atom. Notice that it is one of the lone pairs of the oxygen that forms the new O—H bond.

$$CH_3CH_2—\ddot{\underset{\cdot\cdot}{O}}—H + HA \rightleftharpoons CH_3CH_2—\underset{\cdot\cdot}{\overset{H}{O^+}}—H + A:^-$$

ethanol — acid — conjugate acid

(b) The nitrogen atom of methylamine has a pair of electrons that can bond to a proton.

$$CH_3—\ddot{N}H_2 + HA \rightleftharpoons CH_3—\overset{H}{\underset{+}{N}}H_2 + A:^-$$

methylamine — acid — conjugate acid

(c) Acetic acid has nonbonding electrons on both its oxygen atoms. Either of these oxygen atoms might become protonated, but protonation of the double-bonded oxygen is favored because protonation of this oxygen gives a symmetrical resonance-stabilized conjugate acid.

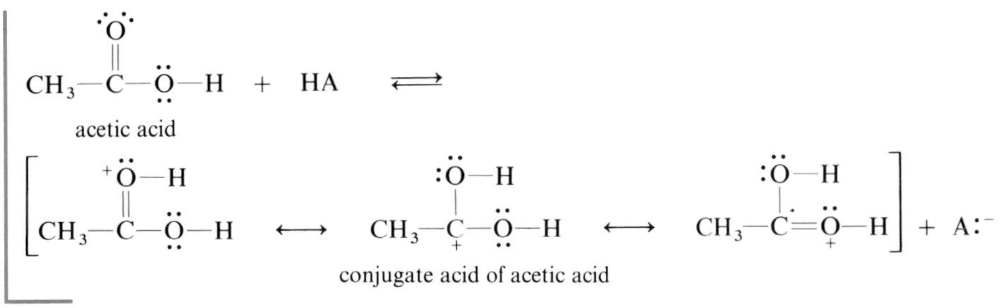

$$CH_3-\overset{\overset{\displaystyle :\ddot{O}:}{\|}}{C}-\ddot{O}-H + HA \rightleftharpoons$$

acetic acid

$$\left[CH_3-\overset{\overset{\displaystyle ^{+}\ddot{O}-H}{\|}}{C}-\ddot{O}-H \longleftrightarrow CH_3-\overset{\overset{\displaystyle :\ddot{O}-H}{|}}{\underset{+}{C}}-\ddot{O}-H \longleftrightarrow CH_3-\overset{\overset{\displaystyle :\ddot{O}-H}{|}}{C}=\underset{+}{\ddot{O}}-H \right] + A:^-$$

conjugate acid of acetic acid

PROBLEM 1-14

Show the product of protonation on the other (—OH) oxygen of acetic acid. Explain why protonation of the double-bonded oxygen is favored.

PROBLEM 1-15

(a) Rank ethanol, methylamine, and acetic acid in decreasing order of acidity. (b) Rank ethanol, methylamine (pK_b 3.36), and ethoxide ion ($CH_3CH_2O^-$) in decreasing order of basicity. In each case, explain your ranking.

PROBLEM 1-16

Predict the products for each of the following acid-base reactions. Label the conjugate acids and bases and show any resonance stabilization.

(a) $CH_3CH_2OH + CH_3NH^-$ (b) $CH_3CH_2COOH + CH_3NHCH_3$
(c) $CH_3OH + H_2SO_4$ (d) $HCOOH + H_2SO_4$
(e) $CH_3NH_3^+ + CH_3O^-$

1-14 LEWIS ACIDS AND BASES

In Solved Problem 1-5, each of the proton-transfer reactions involved a proton acceptor (a base) using a pair of nonbonding electrons to bond to the proton. G. N. Lewis reasoned that this kind of reaction does not need a proton; a base could use its lone pair of electrons to bond to some other kind of electron-deficient species. In effect, we can look at an acid-base reaction from the viewpoint of the *bonds* that are being formed and broken rather than the proton that is transferred. The following reaction shows the proton transfer with emphasis on the bonds being broken and formed. Notice how curved arrows are used to show the movement of the participating electrons.

$$B:^- \quad H:A \rightleftharpoons B:H + {}^-:A$$

Lewis defined bases as species with nonbonding electrons that can be donated to form new bonds. Lewis acids are species that can accept these electron pairs to form new bonds. Since a Lewis acid *accepts* a pair of electrons, it is called an **electrophile,** from the Greek words meaning "lover of electrons." A Lewis base is called a **nucleophile,** or "lover of nuclei," because it donates electrons to a nucleus with an empty orbital.

The Lewis acid-base definitions allow reactions having nothing to do with protons to be considered as acid-base reactions. Below are some examples of Lewis acid-base reactions. Notice that the common Brønsted-Lowry acids and bases also fall under the Lewis definition, with a proton serving as the electrophile. Arrows are used to show the movement of electrons, generally from the nucleophile to the electrophile.

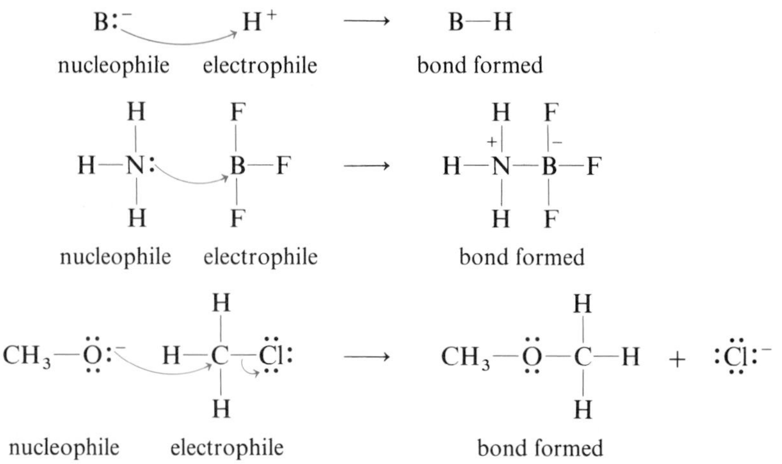

$$B{:}^- + H^+ \longrightarrow B{-}H$$

nucleophile electrophile bond formed

$$H_3N{:} + BF_3 \longrightarrow H_3\overset{+}{N}{-}\overset{-}{B}F_3$$

nucleophile electrophile bond formed

$$CH_3{-}\ddot{\underset{..}{O}}{:}^- + H{-}CH_2{-}\ddot{\underset{..}{Cl}}{:} \longrightarrow CH_3{-}\ddot{\underset{..}{O}}{-}CH_2{-}H + {:}\ddot{\underset{..}{Cl}}{:}^-$$

nucleophile electrophile bond formed

Some of the terms associated with acids and bases have evolved specific meanings in organic chemistry. When organic chemists use the term *base,* they usually mean a proton acceptor (a Brønsted-Lowry base). Similarly, the term *acid* usually means a proton acid (a Brønsted-Lowry acid). When the acid-base reaction involves formation of a bond to some other element (especially carbon), organic chemists refer to the electron donor as a *nucleophile* (Lewis base) and the electron acceptor as an *electrophile* (Lewis acid).

The **curved arrow formalism** is used to show the flow of an electron pair *from the electron donor to the electron acceptor.* The movement of each pair of electrons involved in making or breaking bonds is indicated by its own separate arrow, as shown above. In this book, these curved arrows are always printed in red. In the reaction of CH_3O^- with CH_3Cl above, one curved arrow shows the lone pair on oxygen forming a bond to carbon. Another curved arrow shows that the C—Cl bonding pair detaches from carbon and becomes a lone pair on the Cl^- product.

The curved arrow formalism is universally used as a symbolic device for keeping track of the flow of electrons in reactions. We have also used this device (in Section 1-9, for example) to keep track of electrons in resonance structures as we imagined their "flow" in going from one resonance structure to another. Of course we know that electrons do not "flow" in resonance structures: They are simply delocalized. Still, the curved arrow formalism helps our *minds* flow from one resonance structure to another. We will find ourselves constantly using these (red) curved arrows to keep track of electrons both as reactants change to products and as we imagine additional resonance structures of a hybrid.

PROBLEM 1-17 (Partially Solved)

In the following acid-base reactions

(1) Determine which species are acting as acids and which are acting as bases.

(2) Use the curved-arrow formalism to show the movement of electron pairs in these reactions, as well as the imaginary movement in the resonance hybrids of the products.

(3) Indicate which reactions are best termed Brønsted-Lowry acid-base reactions.

(a)

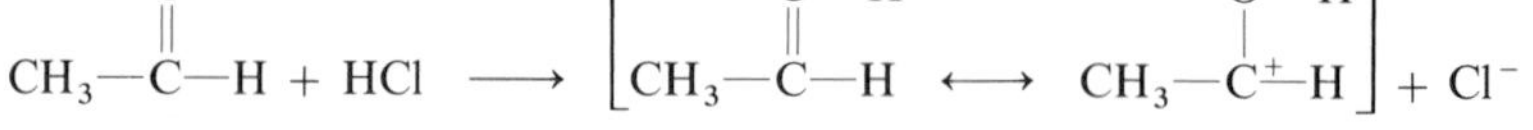

$$\underset{\text{acetaldehyde}}{CH_3{-}\overset{\overset{\displaystyle O}{\|}}{C}{-}H} + HCl \longrightarrow \left[CH_3{-}\overset{\overset{\displaystyle O^+{-}H}{\|}}{C}{-}H \longleftrightarrow CH_3{-}\overset{\overset{\displaystyle O{-}H}{|}}{C^+}{-}H\right] + Cl^-$$

This reaction is clearly a proton transfer from HCl to the C=O group; therefore, it is a Brønsted-Lowry acid-base reaction, with HCl acting as the acid (proton donor) and acetaldehyde acting as the base (proton acceptor). Before drawing any curved arrows, remember that arrows must show the movement of electrons: *from* the electron pair donor (the base) *to* the electron pair acceptor (the acid). An arrow must go *from* the electrons on acetaldehyde that form the bond *to* the hydrogen atom, and the bond to chlorine must break with the chloride ion taking these electrons. Drawing these arrows is easier once we draw good Lewis structures for all the reactants and products.

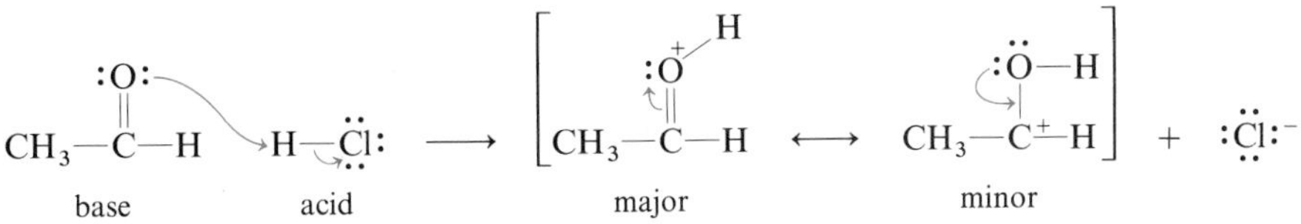

$$\underset{\text{base}}{CH_3-\overset{\overset{:O:}{\|}}{C}-H} \quad \underset{\text{acid}}{H-\ddot{\underset{..}{Cl}}:} \longrightarrow \left[\underset{\text{major}}{CH_3-\overset{\overset{:\overset{+}{O}-H}{\|}}{C}-H} \longleftrightarrow \underset{\text{minor}}{CH_3-\overset{\overset{:\ddot{O}-H}{|}}{\overset{+}{C}}-H}\right] + :\ddot{\underset{..}{Cl}}:^-$$

The resonance structures of the product show that a pair of electrons can be moved between the oxygen atom and the C=O pi bond. The positive charge is delocalized over the carbon and oxygen atoms, with most of the positive charge on oxygen because all octets are satisfied in that resonance structure.

(b)
$$\underset{\text{acetaldehyde}}{CH_3-\overset{\overset{O}{\|}}{C}-H} + CH_3-O^- \longrightarrow CH_3-\overset{\overset{O^-}{|}}{\underset{\underset{O-CH_3}{|}}{C}}-H$$

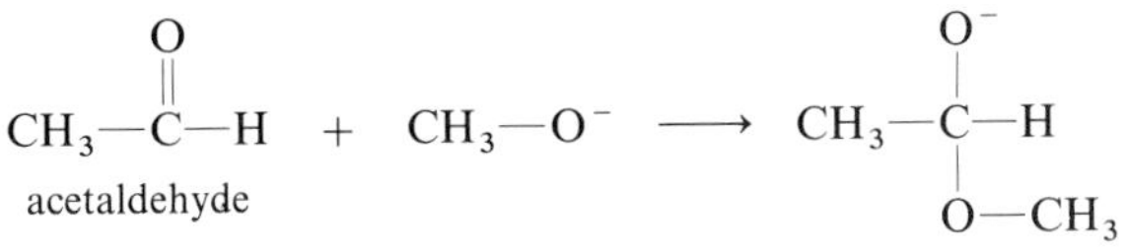

In this case no proton has been transferred, so this is not a Brønsted-Lowry acid-base reaction; but a bond has been formed between the C=O carbon atom and the oxygen of the CH_3-O^- group. Drawing the Lewis structures makes it clear that the CH_3-O^- group (the nucleophile in this reaction) donates the electrons to form the new bond to acetaldehyde (the electrophile). This result agrees with our intuition that a negatively charged ion is likely to be electron-rich and therefore an electron donor.

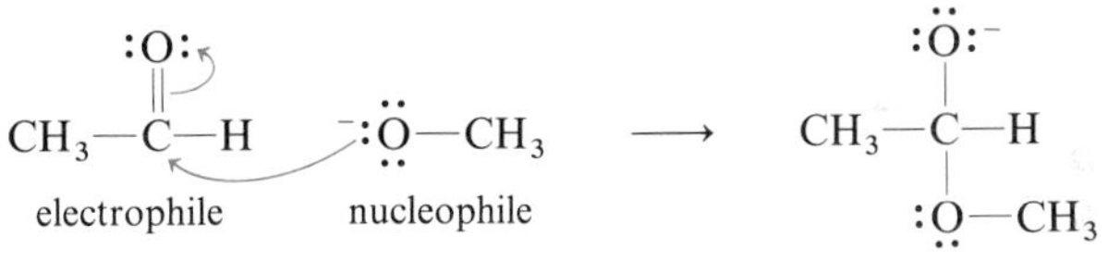

$$\underset{\text{electrophile}}{CH_3-\overset{\overset{:O:}{\|}}{C}-H} \quad \underset{\text{nucleophile}}{^-:\ddot{\underset{..}{O}}-CH_3} \longrightarrow CH_3-\overset{\overset{:\ddot{O}:^-}{|}}{\underset{\underset{:\underset{..}{O}-CH_3}{|}}{C}}-H$$

Notice that acetaldehyde acts as nucleophile (base) in part (a) and electrophile (acid) in part (b). Like most organic compounds, acetaldehyde is both acidic and basic. It acts as a base if you add a strong enough acid to make it donate electrons or accept a proton. It acts as an acid if the base you add is strong enough to donate an electron pair or abstract a proton.

(c) 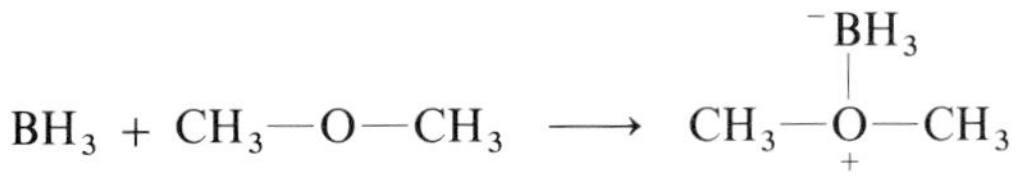 $BH_3 + CH_3-O-CH_3 \longrightarrow CH_3-\overset{\overset{^-BH_3}{|}}{\underset{+}{O}}-CH_3$

(d) 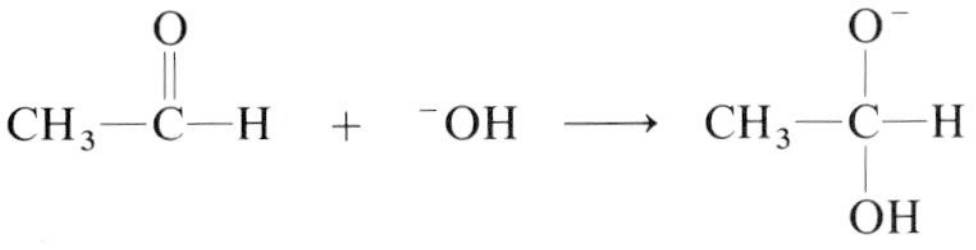 $CH_3-\overset{\overset{O}{\|}}{C}-H + {}^-OH \longrightarrow CH_3-\overset{\overset{O^-}{|}}{\underset{\underset{OH}{|}}{C}}-H$

(e) 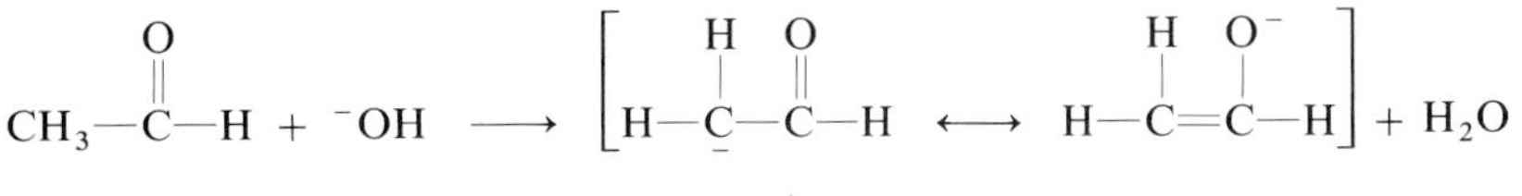 $CH_3-\overset{\overset{O}{\|}}{C}-H + {}^-OH \longrightarrow \left[H-\overset{\overset{H}{|}}{\underset{-}{C}}-\overset{\overset{O}{\|}}{C}-H \longleftrightarrow H-\overset{\overset{H}{|}}{C}=\overset{\overset{O^-}{|}}{C}-H\right] + H_2O$

(f) $CH_3-NH_2 + CH_3-Cl \longrightarrow CH_3-\overset{+}{N}H_2-CH_3 + Cl^-$

Each chapter ends with a glossary that summarizes the most important new terms in the chapter. You should use these glossaries as more than simply a dictionary to look up unfamiliar terms as you encounter them; the index serves that purpose. The most important use of the glossary is as one of the tools for reviewing the chapter. You can read carefully through the glossary to see if you understand and remember all the terms and associated chemistry mentioned there. Anything that seems unfamiliar can be reviewed by turning to the page number given at the end of each glossary listing.

GLOSSARY

acids and bases (pp. 21–29)

(Arrhenius definitions)	**acid:** dissociates in water to give H_3O^+
	base: dissociates in water to give ^-OH
(Brønsted-Lowry definitions)	**acid:** proton donor
	base: proton acceptor
(Lewis definitions)	**acid:** electron-pair acceptor (electrophile)
	base: electron-pair donor (nucleophile)

conjugate acid The acid that results from protonation of a base. (p. 23)

conjugate base The base that results from loss of a proton from an acid. (p. 23)

covalent bonding Bonding that occurs by the sharing of electrons in the region between two nuclei. (p. 6)

- **single bond** A covalent bond that involves the sharing of one pair of electrons. (p. 8)
- **double bond** A covalent bond that involves the sharing of two pairs of electrons. (p. 8)
- **triple bond** A covalent bond that involves the sharing of three pairs of electrons. (p. 8)

curved arrow formalism A method of drawing curved arrows to keep track of electron movement from nucleophile to electrophile (or within a molecule) during the course of a reaction. (p. 28)

degenerate orbitals Orbitals with identical energies. (p. 5)

electron density The relative probability of finding an electron in a certain region of space. (p. 3)

electronegativity A measure of an element's affinity for electrons. Elements with higher electronegativities have more attraction for their bonding electrons. (p. 9)

electrophile An electron-pair acceptor (Lewis acid). (p. 27)

formal charges A method for keeping track of charges, showing what charge would be on an atom in a particular Lewis structure. (p. 10)

Hund's rule When there are two or more unfilled orbitals of the same energy (degenerate orbitals), the lowest-energy configuration places the electrons in different orbitals (with parallel spins) rather than paired in the same orbital. (p. 5)

ionic bonding Bonding that occurs by the attraction of oppositely charged ions. Ionic bonding usually results in the formation of a large three-dimensional crystal lattice. (p. 6)

isotopes Atoms with the same number of protons but different numbers of neutrons; atoms of the same element but with different atomic masses. (p. 2)

Lewis structure A structural formula that shows all valence electrons, with the bonds symbolized by dashes (—) or by pairs of dots, and nonbonding electrons, symbolized by dots. (p. 7)

line-angle formula ("stick figure") A shorthand structural formula with bonds represented by lines and carbon atoms wherever two lines meet or a line begins or bends. Nitrogen, oxygen, and halogen atoms are shown, but hydrogen atoms are not; each carbon atom is assumed to have enough hydrogens to give it four bonds. (pp. 18, 19)

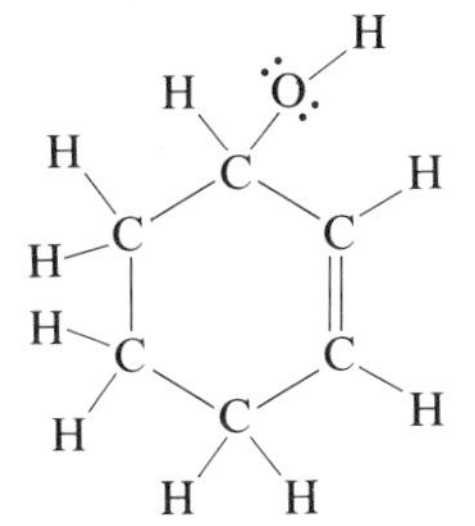

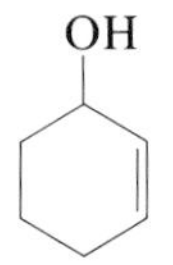

Lewis structure of cyclohex-2-enol equivalent line-angle formula

molecular formula The number of atoms of each element in one molecule of a compound. The **empirical formula** simply gives the ratios of atoms of the different elements. For example, the molecular formula of glucose is $C_6H_{12}O_6$. Its empirical formula is CH_2O. Neither the molecular formula nor the empirical formula gives any structural information. (p. 19)

node In an orbital, a region with zero electron density. (p. 3)

nodal plane A flat (planar) region of space with zero electron density. (p. 5)

nonbonding electrons Valence electrons that are not used for bonding. A pair of nonbonding electrons is often called a **lone pair.** (p. 7)

nucleophile An electron-pair donor (Lewis base). (p. 27)

octet rule Atoms generally form bonding arrangements that give them filled shells of electrons (noble-gas configurations). For the second-row elements, this configuration has eight valence electrons. (p. 6)

orbital An allowed energy state for an electron bound to a nucleus; the probability function that defines the distribution of electron density in space. The *Pauli exclusion principle* states that up to two electrons can occupy each orbital if their spins are paired. (p. 3)

organic chemistry (new definition) The chemistry of carbon compounds. (old definition) The study of compounds derived from living organisms and their natural products. (p. 1)

polar bond A covalent bond in which electrons are shared unequally. A bond with equal sharing of electrons is called a **nonpolar bond.** (p. 9)

resonance hybrid A molecule or ion for which two or more valid Lewis structures can be drawn, differing only in the placement of the valence electrons. These Lewis structures are called **resonance structures.** The more important (lower-energy) structures are called **major contributors,** and the less important (higher-energy) structures are called **minor contributors.** When a charge is spread over two or more atoms by resonance, it is said to be **delocalized,** and the molecule is said to be **resonance stabilized.** (pp. 13–16)

structural formulas A **complete structural formula** (such as a Lewis structure) shows all the atoms and bonds in the molecule. A **condensed structural formula** shows each central atom along with the atoms bonded to it. A **line-angle formula** ("stick figure") assumes that there is a carbon atom wherever a line begins or ends. See Section 1-10 for examples. (p. 16)

valence The number of bonds an atom usually forms. (p. 9)

valence electrons Those electrons that are in the outermost shell. (p. 5)

ESSENTIAL PROBLEM-SOLVING SKILLS IN CHAPTER 1

1. Draw and interpret Lewis, condensed, and line-angle structural formulas.
2. Draw resonance structures and use them to predict stabilities.
3. Calculate empirical and molecular formulas from elemental compositions.
4. Predict relative acidities and basicities based on structure, bonding, and resonance of conjugate acid-base pairs.
5. Calculate, use, and interpret values of K_a and pK_a.
6. Identify nucleophiles and electrophiles, and predict Lewis acid-base reactions.

STUDY PROBLEMS

It is easy to fool yourself into thinking that you understand organic chemistry when you actually do not. As you read through this book, all the facts and ideas may make sense, yet you have not learned to combine and *use* those facts and ideas. An examination is a painful time to learn that you do not really understand the material.

The best way to learn organic chemistry is to use it. You will certainly need to read and reread all the material in the chapter, but this level of understanding is just the beginning. You need to work with the ideas, applying them to new compounds and new reactions that you have never seen before. This is why problems are provided with each chapter. By working problems, you force yourself to use the material and fill in the gaps in your understanding. You also increase your level of self-confidence and your ability to do well on exams both in this course and in your other courses.

Several kinds of problems are included in each chapter. There are problems within the chapters, giving examples and drill over the material as it is covered. Working these problems as you read through the chapter will ensure your understanding as you go along. Answers to many of these in-chapter problems are found at the back of this book. At the end of each chapter, there are additional problems. These "study problems" give you additional experience using the material, and they force you to think in more depth about the ideas. Some of the study problems have short answers in the back of this book, and all of them have detailed answers in the accompanying study guide.

Taking organic chemistry without working the problems is something like skydiving without a parachute. Initially, there is a carefree sense of freedom and daring. But then, there is the inevitable jolt that comes at the end for those who went unprepared.

1-18. Define and give an example for each of the following terms.

(a) isotopes **(b)** orbital **(c)** node
(d) degenerate orbitals **(e)** valence electrons **(f)** ionic bonding
(g) covalent bonding **(h)** Lewis structure **(i)** nonbonding electrons
(j) single bond **(k)** double bond **(l)** triple bond
(m) polar bond **(n)** formal charges **(o)** resonance structures
(p) molecular formula **(q)** empirical formula **(r)** Arrhenius acid and base
(s) Brønsted-Lowry acid and base **(t)** Lewis acid and base **(u)** electrophile
(v) nucleophile

1-19. Name the element that corresponds to each of the following electronic configurations.

(a) $1s^2\ 2s^2\ 2p^2$ **(b)** $1s^2\ 2s^2\ 2p^4$ **(c)** $1s^2\ 2s^2\ 2p^6\ 3s^2\ 3p^3$ **(d)** $1s^2\ 2s^2\ 2p^6\ 3s^2\ 3p^5$

1-20. There is a small portion of the periodic table that you must know to do organic chemistry. Construct this part from memory, using the following steps.

(a) From memory, make a list of the elements in the first two rows of the periodic table, together with their numbers of valence electrons.
(b) Use this list to construct the first two rows of the periodic table.
(c) Organic compounds often contain sulfur, phosphorus, chlorine, bromine, and iodine. Add these elements to your periodic table.

1-21. For each of the following compounds, state whether its bonding is covalent, ionic, or a mixture of covalent and ionic.

(a) NaCl **(b)** NaOH **(c)** CH_3Li **(d)** CH_2Cl_2 **(e)** $NaOCH_3$ **(f)** HCO_2Na **(g)** CF_4

1-22. **(a)** Both PCl_3 and PCl_5 are stable compounds. Draw Lewis structures for these two compounds.
(b) NCl_3 is a known compound, but all attempts to synthesize NCl_5 have failed. Draw Lewis structures for NCl_3 and a hypothetical NCl_5, and explain why NCl_5 is an unlikely structure.

1-23. Draw a Lewis structure for each of the following species.

(a) N_2H_4 **(b)** N_2H_2 **(c)** $(CH_3)_4NCl$ **(d)** CH_3CN **(e)** CH_3CHO
(f) $CH_3S(O)CH_3$ **(g)** H_2SO_4 **(h)** CH_3NCO **(i)** $CH_3OSO_2OCH_3$ **(j)** $CH_3C(NH)CH_3$
(k) $(CH_3)_3CNO$

1-24. Draw a Lewis structure for each of the following compounds. Include all nonbonding pairs of electrons.

(a) $CH_3CHCHCH_2CHCHCOOH$ **(b)** $NCCH_2COCH_2CHO$
(c) $CH_2CHCH(OH)CH_2CO_2H$ **(d)** $CH_3CH(CH_3)CH_2C(CH_2CH_3)_2CHO$

1-25. Draw a line-angle formula for each compound in Problem 1-24.

1-26. Draw Lewis structures for
(a) two compounds of formula C_4H_{10}.
(b) two compounds of formula C_2H_7N.
(c) three compounds of formula $C_3H_8O_2$.
(d) two compounds of formula C_2H_4O.

1-27. Draw a complete structural formula and a condensed structural formula for
(a) three compounds of formula C_3H_8O.
(b) five compounds of formula C_3H_6O.

1-28. Some of the following molecular formulas correspond to stable compounds. When possible, draw a stable structure for each formula.

CH_2 CH_3 CH_4 CH_5
C_2H_2 C_2H_3 C_2H_4 C_2H_5 C_2H_6 C_2H_7
C_3H_3 C_3H_4 C_3H_5 C_3H_6 C_3H_7 C_3H_8 C_3H_9

Can you propose a general rule for the numbers of hydrogen atoms in stable hydrocarbons?

1-29. Draw a complete Lewis structure, including lone pairs, for each of the following compounds.

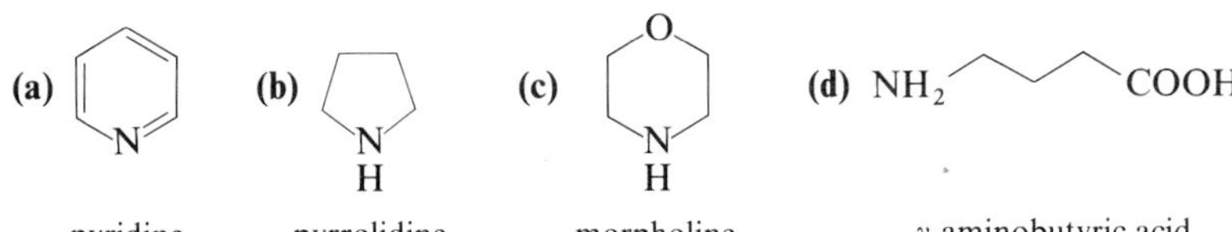

pyridine pyrrolidine morpholine γ-aminobutyric acid

1-30. Give the molecular formula of each compound shown in Problem 1-29.

1-31. A careful analysis has shown that compound X contains 62.0 percent carbon and 10.4 percent hydrogen. No nitrogen or halogen was found.
(a) Compute an empirical formula for compound X.
(b) A molecular weight determination showed that compound X has a molecular weight of approximately 117. Find the molecular formula of compound X.
(c) There are many possible structures that have this molecular formula. Draw complete structural formulas for four of them.

1-32. For each of the following structures
(1) Draw a Lewis structure; fill in any nonbonding electrons.
(2) Calculate the formal charge on each atom other than hydrogen.
All are electrically neutral except as noted.

(a) $\left[H_2C{=}N{=}N \longleftrightarrow H_2C{-}N{\equiv}N \right]$ **(b)** $(CH_3)_3NO$ (trimethylamine oxide)

(c) $[CH_2{=}CH{-}CH_2]^+$ **(d)** CH_3NO_2 **(e)** $[(CH_3)_3O]^+$

1-33. (1) From what you remember of electronegativities, show the direction of the dipole moments of the following bonds.
(2) In each case, predict whether the dipole moment is large or small.
(a) C—Cl **(b)** C—H **(c)** C—Li **(d)** C—N **(e)** C—O
(f) C—B **(g)** C—Mg **(h)** N—H **(i)** O—H **(j)** C—Br

1-34. Determine whether the following pairs of structures are actually different compounds or simply resonance structures of the same compounds.

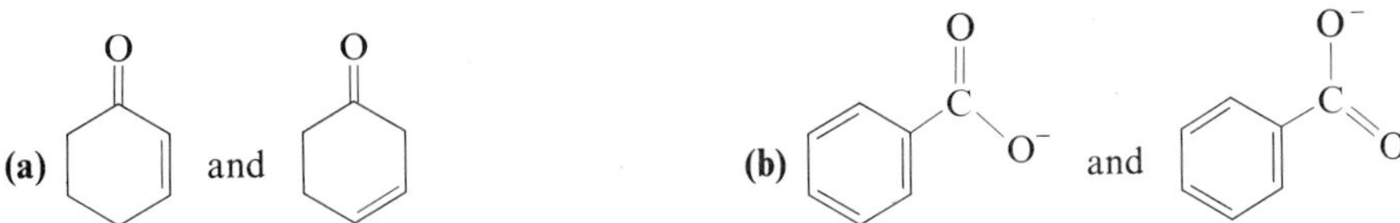

(c) [benzene Kekulé structure] and [benzene Kekulé structure]

(d) $CH_2{=}C(O^-){-}H$ and $\bar{C}H_2{-}C({=}O){-}H$

(e) $H{-}C({=}O){-}CH_3$ and $H{-}C(O{-}H){=}CH_2$

(f) $H{-}\overset{+}{C}(O{-}H){-}H$ and $H{-}C({=}\overset{+}{O}{-}H){-}H$

(g) $H{-}C({=}O){-}NH_2$ and $H{-}C(O^-){=}\overset{+}{N}H_2$

(h) $CH_2{=}C{=}O$ and $H{-}C{\equiv}C{-}OH$

(i) $CH_2{=}CH{-}\overset{+}{C}H_2$ and $\overset{+}{C}H_2{-}CH{=}CH_2$

(j) $CH_3{-}C({=}O){-}CH{=}CH_2$ and $CH_3{-}C(O^-){=}CH{-}\overset{+}{C}H_2$

1-35. Draw the important resonance structures for the following molecules and ions.

(a) $CH_3{-}C({=}O){-}\bar{C}H_2$

(b) $H{-}C({=}O){-}CH{=}CH{-}\bar{C}H_2$

(c) [benzene ring]$-\overset{+}{C}H_2$

(d) [cyclopentadienyl ring with H and + on one carbon]

(e) [benzene ring]$-O^-$

(f) [pyridine ring with lone pair on N]

(g) $CH_3{-}CH{=}CH{-}CH{=}CH{-}\overset{+}{C}H{-}CH_3$

(h) $CH_3{-}CH{=}CH{-}CH{=}CH{-}CH_2{-}\overset{+}{C}H_2$

1-36. (a) Draw the resonance structures for SO_2 (bonded O—S—O).
(b) Draw the resonance structures for ozone (bonded O—O—O).
(c) Sulfur dioxide has one more resonance structure than ozone. Explain why this structure is not possible for ozone.

*** 1-37.** The following compound can become protonated on any of the three nitrogen atoms. One of these nitrogens is much more basic than the others, however.

(a) Draw the important resonance structures of the products of protonation on each of the three nitrogen atoms.
(b) Determine which nitrogen atom is the most basic.

$$CH_3{-}NH{-}C({=}NH){-}NH_2$$

1-38. In the following sets of resonance structures, label the major and minor contributors and state which structures are of equal energy. Add any missing structures.

(a) $\left[CH_3{-}\ddot{\bar{C}}H{-}C{\equiv}N{:} \longleftrightarrow CH_3{-}CH{=}C{=}\ddot{N}{:}^-\right]$

(b) $\left[CH_3{-}C(O^-){=}CH{-}\overset{+}{C}H{-}CH_3 \longleftrightarrow CH_3{-}\overset{+}{C}(O^-){-}CH{=}CH{-}CH_3\right]$

(c) $\left[CH_3{-}C({=}O){-}\bar{C}H{-}C({=}O){-}CH_3 \longleftrightarrow CH_3{-}C(O^-){=}CH{-}C({=}O){-}CH_3\right]$

(d) $\left[CH_3{-}\bar{C}H{-}CH{=}CH{-}NO_2 \longleftrightarrow CH_3{-}CH{=}CH{-}\bar{C}H{-}NO_2\right]$

(e) $\left[CH_3{-}CH_2{-}\overset{+}{C}(NH_2){-}NH_2 \longleftrightarrow CH_3{-}CH_2{-}C(NH_2){=}\overset{+}{N}H_2\right]$

1-39. In each of the following pairs of ions, determine which ion is more stable. Use resonance structures to explain your answers.

(a) $CH_3{-}\overset{+}{C}H{-}CH_3$ or $CH_3{-}\overset{+}{C}H{-}OCH_3$

(b) $CH_2{=}CH{-}\overset{+}{C}H{-}CH_3$ or $CH_2{=}CH{-}CH_2{-}\overset{+}{C}H_2$

(c) $\overset{-}{C}H_2{-}CH_3$ or $\overset{-}{C}H_2{-}C{\equiv}N\!:$

(d) 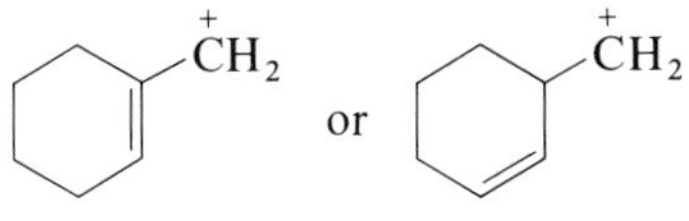 (cyclohexenyl $\overset{+}{C}H_2$, double bond at the ring carbon bearing CH_2) or (cyclohexenyl $\overset{+}{C}H_2$, double bond away from the ring carbon bearing CH_2)

(e) $CH_3{-}N(CH_3){-}\underset{+}{C}(CH_3){-}CH_3$ or $CH_3{-}CH(CH_3){-}\underset{+}{C}(CH_3){-}CH_3$

1-40. Rank the following species in order of increasing acidity. Explain your reasons for ordering them as you do.

$$NH_3 \qquad H_2SO_4 \qquad CH_3OH \qquad CH_3COOH$$

1-41. Rank the following species in order of increasing basicity. Explain your reasons for ordering them as you do.

$$CH_3O^- \qquad CH_3COOH \qquad CH_3COO^- \qquad NaOH \qquad NH_2^- \qquad HSO_4^-$$

1-42. The K_a of phenylacetic acid is 5.2×10^{-5}, and the pK_a of propionic acid is 4.87.

$$C_6H_5{-}CH_2{-}\overset{O}{\overset{\|}{C}}{-}OH \qquad\qquad CH_3{-}CH_2{-}\overset{O}{\overset{\|}{C}}{-}OH$$

phenylacetic acid, $K_a = 5.2 \times 10^{-5}$ propionic acid, $pK_a = 4.87$

(a) Calculate the pK_a of phenylacetic acid and the K_a of propionic acid.
(b) Which of these is the stronger acid? Calculate how much stronger an acid it is.

1-43. Label the reactants in these acid-base reactions as Lewis acids (electrophiles) or Lewis bases (nucleophiles). Use curved arrows to show the movement of electron pairs in the reactions.

(a) $CH_3\ddot{\underset{..}{O}}\!:^- + CH_3{-}\ddot{\underset{..}{Cl}}\!: \longrightarrow CH_3{-}\ddot{\underset{..}{O}}{-}CH_3 + :\ddot{\underset{..}{Cl}}\!:^-$

(b) $CH_3{-}\ddot{O}^+(CH_3){-}CH_3 + :\ddot{O}(H){-}H \longrightarrow CH_3{-}\ddot{O}(CH_3)\!: + CH_3{-}\ddot{O}^+(H){-}H$

(c) $H{-}\overset{O}{\overset{\|}{C}}{-}H + :NH_3 \longrightarrow H{-}C(:\ddot{O}:^-)(^+NH_3){-}H$

(d) $CH_3{-}\ddot{N}H_2 + CH_3{-}CH_2{-}\ddot{\underset{..}{Cl}}\!: \longrightarrow CH_3{-}\overset{+}{N}H_2{-}CH_2CH_3 + :\ddot{\underset{..}{Cl}}\!:^-$

(e) $CH_3{-}\overset{\ddot{O}}{\overset{\|}{C}}{-}CH_3 + H_2SO_4 \longrightarrow CH_3{-}\overset{\dot{O}^+{-}H}{\overset{\|}{C}}{-}CH_3 + HSO_4^-$

(f) $(CH_3)_3CCl + AlCl_3 \longrightarrow (CH_3)_3C^+ + AlCl_4^-$

(g) $CH_3{-}\overset{\overset{\ddot{O}}{\|}}{C}{-}CH_3 + {}^{-}\!:\!\ddot{\underset{..}{O}}{-}H \longrightarrow CH_3{-}\overset{\overset{:\ddot{O}:^-}{|}}{C}{=}CH_2 + H{-}\ddot{\underset{..}{O}}{-}H$

(h) $CH_2{=}CH_2 + BF_3 \longrightarrow \overset{-}{B}F_3{-}CH_2{-}\overset{+}{C}H_2$

(i) $\overset{-}{B}F_3{-}CH_2{-}\overset{+}{C}H_2 + CH_2{=}CH_2 \longrightarrow \overset{-}{B}F_3{-}CH_2{-}CH_2{-}CH_2{-}\overset{+}{C}H_2$

1-44. Predict the products of the following acid-base reactions.

(a) $H_2SO_4 + CH_3COO^- \rightleftharpoons$

(b) $CH_3COOH + (CH_3)_3N: \rightleftharpoons$

(c) $C_6H_5{-}\overset{\overset{O}{\|}}{C}{-}OH + {}^{-}OH \rightleftharpoons$

(d) $(CH_3)_3\overset{+}{N}H + {}^{-}OH \rightleftharpoons$

(e) $HO{-}\overset{\overset{O}{\|}}{C}{-}OH + 2\,{}^{-}OH \rightleftharpoons$

(f) $H_2O + NH_3 \rightleftharpoons$

(g) $HCOOH + CH_3O^- \rightleftharpoons$

***1-45.** Methyllithium (CH_3Li) is often used as a base in organic reactions.
(a) Predict the products of the following acid-base reaction.

$$CH_3CH_2{-}OH + CH_3{-}Li \longrightarrow$$

(b) What is the conjugate acid of CH_3Li? Would you except CH_3Li to be a strong base or a weak base?

***1-46.** In 1934, Edward A. Doisy of Washington University extracted 3000 pounds of hog ovaries to isolate a few milligrams of pure estradiol, a potent female hormone. Doisy burned 5.00 mg of this precious sample in oxygen and found that 14.54 mg of CO_2 and 3.97 mg of H_2O were generated.
(a) Determine the empirical formula of estradiol.
(b) The molecular weight of estradiol was later determined to be 272. Determine the molecular formula of estradiol.

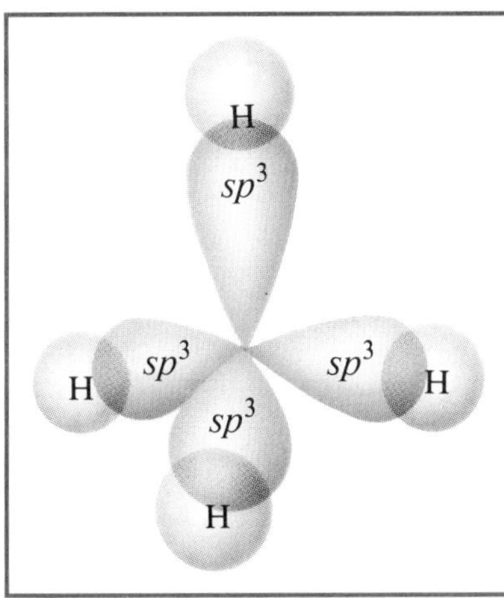

2

STRUCTURE AND PROPERTIES OF ORGANIC MOLECULES

In Chapter 1 we considered how atoms bond together to gain noble-gas configurations, forming molecules in the process. Using the octet rule, we drew Lewis structures for organic molecules and used these diagrams to determine which bonds are single bonds, double bonds, and triple bonds. We discussed various ways of drawing organic structures, and we saw how resonance structures represent molecules whose actual bonding cannot be shown by a single Lewis structure.

The material covered in Chapter 1 explains little about the actual shapes and properties of organic molecules. These aspects of molecular structure can be explained by considering how the atomic orbitals on different atoms mix to form **molecular orbitals.** We use these molecular orbitals to explain the shapes and properties of organic molecules.

2-1 WAVE PROPERTIES OF ELECTRONS IN ORBITALS

Most of us like to envision the atom as a miniature solar system, with the electrons orbiting about the nucleus. Although this solar system picture satisfies our intuition, it does not correspond with what we know about the atom. About 1923, Louis de Broglie suggested that the properties of electrons in atoms are better explained by treating the electrons as waves rather than as particles.

There are two general kinds of waves, *traveling waves* and *standing waves.* Examples of traveling waves are the sound waves that carry a thunderclap and the water waves that form the wake of a boat. Standing waves vibrate in a fixed location. Standing waves are found inside an organ pipe, where the rush of air creates a vibrating air column, and in the wave pattern of a guitar string when it is plucked. An electron in an atomic orbital is like a stationary, bound vibration: a standing wave.

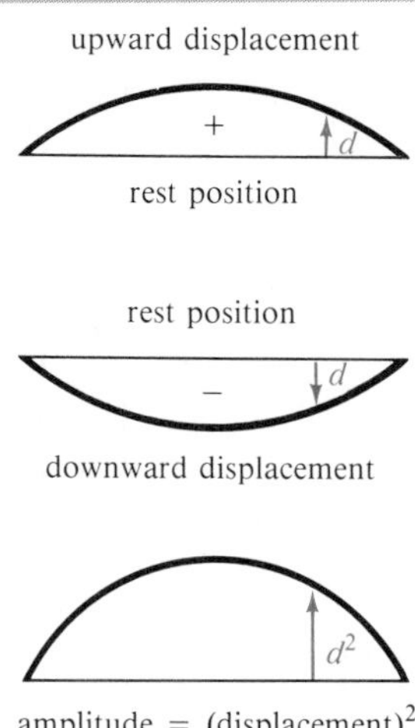

FIGURE 2-1 A standing wave. The fundamental frequency of a guitar string is a standing wave with the string alternately displaced upward and downward. The amplitude at any point is the square of the displacement at that point.

We can understand the features of an orbital (a three-dimensional standing wave) more easily by using a guitar string as a one-dimensional analogy (see Fig. 2-1). If you pluck a guitar string at its middle, a standing wave results. This vibration has all of the string displaced upward for a fraction of a second, then downward for an equal time. If we draw an instantaneous picture of the wave-form, it shows the string displaced in a smooth curve either upward or downward, depending on the exact instant of the picture. The *amplitude* of the wave is the square of its displacement. Whether the instantaneous displacement of the wave is upward or downward, its amplitude is positive.

A 1*s* orbital is like this guitar string, except that it is three-dimensional. The orbital can be described by its **wave function,** which is the mathematical description of the shape of the wave as it vibrates. All of the wave is positive in sign for a brief instant; then it is negative in sign. The electron density at any point is simply the amplitude of the wave at that point: that is, the square of the wave function. *Notice that the plus sign and the minus sign of these wave functions are not charges. The plus or minus sign is the instantaneous sign of the constantly changing wave function.* The 1*s* orbital is spherically symmetrical, and it is often represented by a circle (representing a sphere) with a nucleus in the center and with a plus or minus sign to indicate the instantaneous sign of the wave function (Fig. 2-2).

If we gently place a finger at the center of the guitar string while plucking the string, the finger keeps the midpoint of the string from moving. The amplitude at the midpoint is zero, and this point is a **node.** The string vibrates in two parts, with the two halves vibrating in opposite directions. We say that the two halves of the string are *out of phase:* When one is displaced upward, the other is displaced downward. Figure 2-3 shows this first harmonic of the guitar string.

The first harmonic of the guitar string resembles the 2*p* orbital (Fig. 2-4). We have drawn the 2*p* orbital as two "lobes," separated by a node (a nodal plane). The two lobes of the *p* orbital are out of phase with each other. Whenever the wave function has a plus sign in one lobe, it has a minus sign in the other lobe.

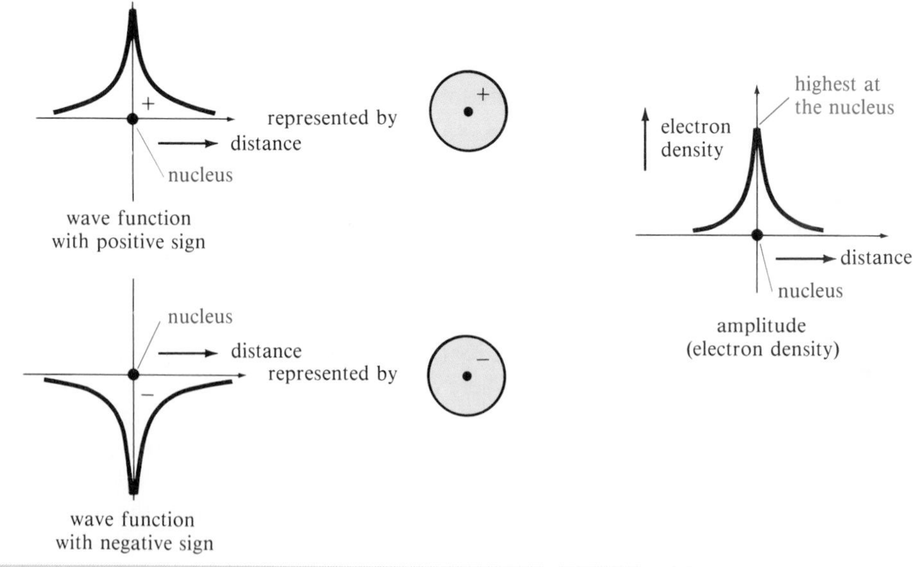

FIGURE 2-2 The 1*s* orbital is similar to the fundamental vibration of a guitar string. The wave function is instantaneously all positive or all negative. The amplitude (the square of the wave function) is the electron density. A circle with a nucleus is used to represent the spherically symmetrical *s* orbital.

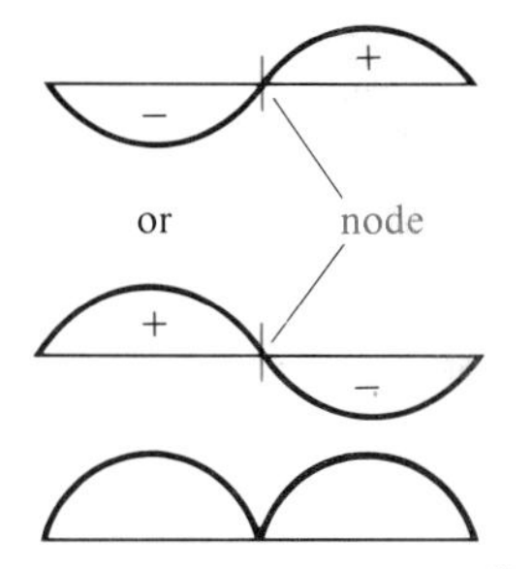

FIGURE 2-3 First harmonic of a guitar string. The two halves of the string are separated by a node, a point with zero amplitude. The two halves vibrate out of phase with each other.

2-1A LINEAR COMBINATION OF ATOMIC ORBITALS

The most important wave property of atomic orbitals is their ability to combine and overlap to give more complex standing waves. This process is called the **linear combination of atomic orbitals** (LCAO): Wave functions are added and subtracted to give the wave functions of new orbitals. The number of new orbitals generated always equals the number of orbitals we started with.

1. When orbitals on *different* atoms interact, they produce **molecular orbitals** that lead to bonding (or antibonding).
2. When orbitals on the *same* atom interact, they give **hybrid atomic orbitals** that define the geometry of the bonds formed.

We begin by looking at how atomic orbitals on different atoms interact to give molecular orbitals. Then we consider how atomic orbitals on the same atom can interact to give hybrid atomic orbitals.

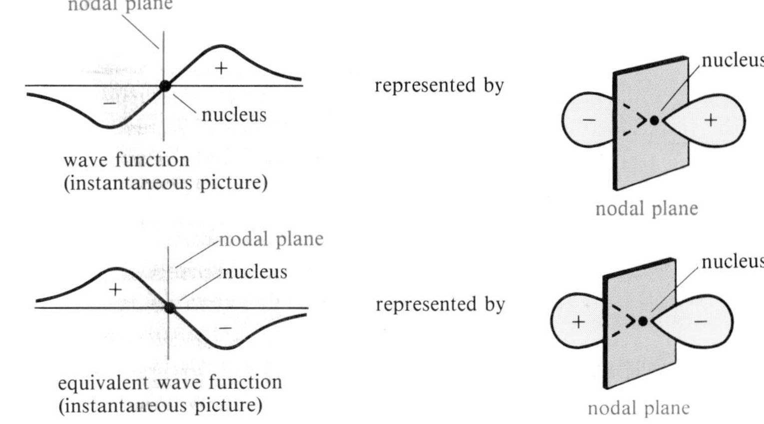

FIGURE 2-4 The 2*p* orbital has two lobes, separated by a nodal plane. The two lobes are out of phase with each other. When one has a plus sign, the other has a minus sign.

2-2 MOLECULAR ORBITALS

The stability of a covalent bond results from a large amount of electron density in the bonding region, the space between the two nuclei (Fig. 2-5). In the bonding region, the electrons are close to both nuclei, resulting in a lowering of the overall energy. The bonding electrons also mask the positive charges of the nuclei, so that they do not repel each other as much as they would otherwise.

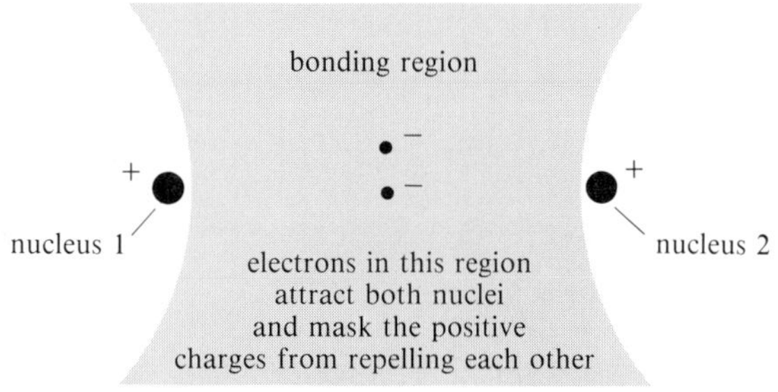

bonding region

nucleus 1 · nucleus 2

electrons in this region attract both nuclei and mask the positive charges from repelling each other

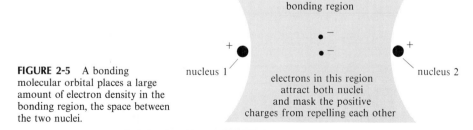

FIGURE 2-5 A bonding molecular orbital places a large amount of electron density in the bonding region, the space between the two nuclei.

There is always an optimum distance for the two bonded nuclei. If they are too far apart, their attraction for the bonding electrons is diminished. If they are too close together, their electrostatic repulsion pushes them apart. The internuclear distance that gives the minimum energy (the strongest bond) is called the *bond length.*

2-2A THE HYDROGEN MOLECULE; SIGMA BONDING

The hydrogen molecule is the simplest example of molecular bonding. As two hydrogen atoms approach each other, their 1*s* wave functions can add *constructively* so that they reinforce each other, or *destructively* so that they cancel out where they overlap. Figure 2-6 shows how the wave functions interact constructively when they are in phase and have the same sign in the region between the nuclei. The wave functions reinforce each other and increase the electron density in this bonding region. The result is a **bonding molecular orbital** (bonding MO).

The bonding MO depicted in Figure 2-6 has most of its electron density centered *along* the line connecting the nuclei. This type of bond is called a **cylindrically symmetrical bond** or a **sigma bond** (**σ bond**). Sigma bonds are the most common bonds in organic compounds. All single bonds in organic compounds are sigma bonds, and every double or triple bond contains one sigma bond.

When two hydrogen 1*s* orbitals overlap out of phase with each other, an **antibonding molecular orbital** results (Figure 2-7). The two 1*s* wave functions have opposite signs, and they tend to cancel out where they overlap. The result is a node (actually a nodal plane) separating the two atoms. The presence of a node separating the two nuclei usually indicates that the orbital is antibonding.

Figure 2-8 shows the relative energies of the atomic orbitals and the molecular orbitals of the H_2 system. When the 1*s* orbitals are in phase, the resulting molecular orbital is a sigma bonding MO, with lower energy than that of a 1*s* atomic orbital. Overlap of two 1*s* orbitals out of phase gives an antibonding (σ^*) orbital with higher energy than that of a 1*s* atomic orbital. The two electrons in the H_2 system are found with paired spins in the sigma bonding MO, giving a stable H_2 molecule. In stable molecules, the antibonding orbitals (such as σ^*) are usually vacant.

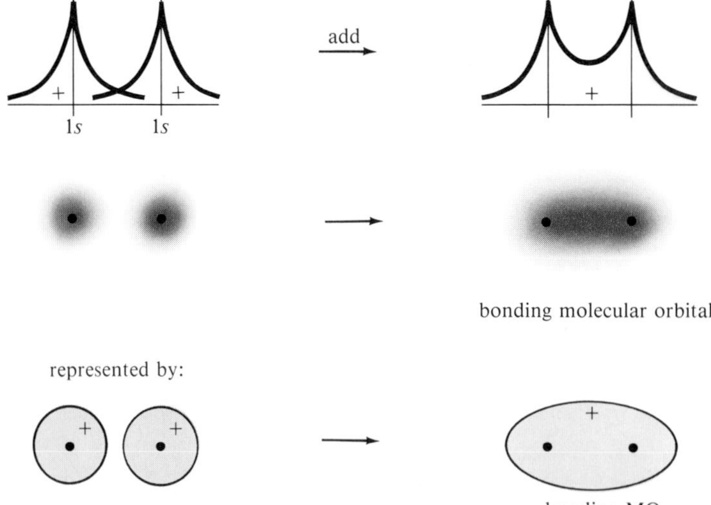

FIGURE 2-6 Formation of a σ-bonding MO. When the 1*s* orbitals of two hydrogen atoms overlap in phase with each other, they interact constructively to form a bonding MO. The electron density in the bonding region (between the nuclei) is increased. The result is a cylindrically symmetrical bond, or sigma bond.

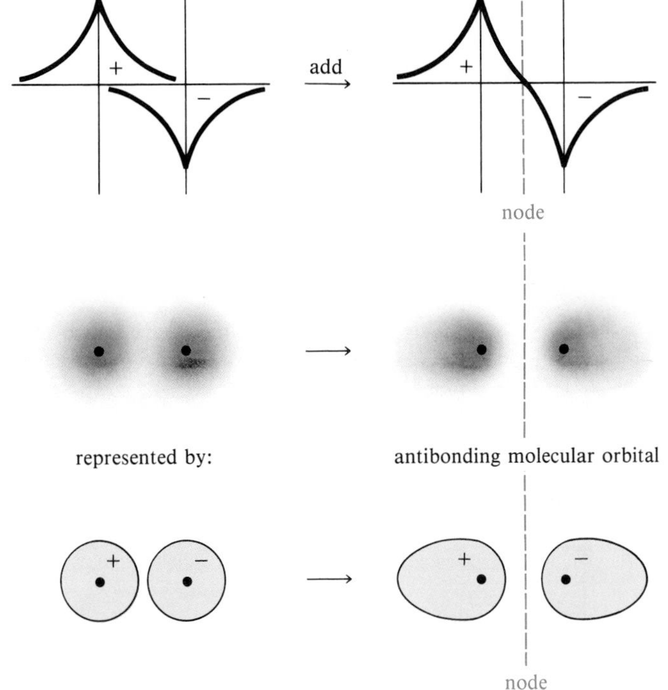

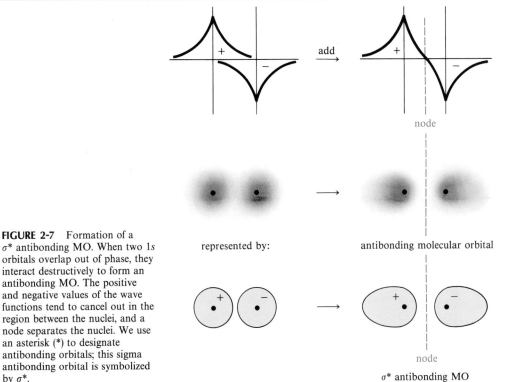

FIGURE 2-7 Formation of a σ^* antibonding MO. When two 1*s* orbitals overlap out of phase, they interact destructively to form an antibonding MO. The positive and negative values of the wave functions tend to cancel out in the region between the nuclei, and a node separates the nuclei. We use an asterisk (*) to designate antibonding orbitals; this sigma antibonding orbital is symbolized by σ^*.

antibonding

σ^*

energy

1*s*

1*s*

atomic orbital

atomic orbital

σ

bonding

molecular orbitals

FIGURE 2-8 When the two hydrogen 1*s* orbitals overlap, a sigma bonding MO and a sigma antibonding MO result. Two electrons go into the bonding MO, forming a stable H_2 molecule.

2-2B SIGMA OVERLAP INVOLVING *p* ORBITALS

When two *p* orbitals overlap along the line between the nuclei, a bonding orbital and an antibonding orbital result. Once again, most of the electron density is centered along the line between the nuclei; this is the linear overlap of a sigma bonding MO. The constructive overlap of two *p* orbitals along the line joining the nuclei forms a σ bond represented as follows:

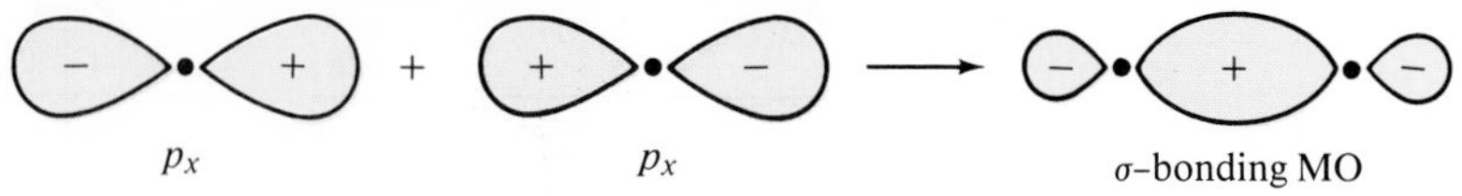

p_x + p_x → σ–bonding MO

SOLVED PROBLEM 2-1

Draw the σ^* antibonding orbital that results from destructive overlap of the two p_x orbitals shown above.

SOLUTION This orbital involves the destructive overlap of lobes of the two *p* orbitals with opposite phases. If the signs are reversed on one of the orbitals, adding the two orbitals gives an antibonding orbital with a node separating the two nuclei:

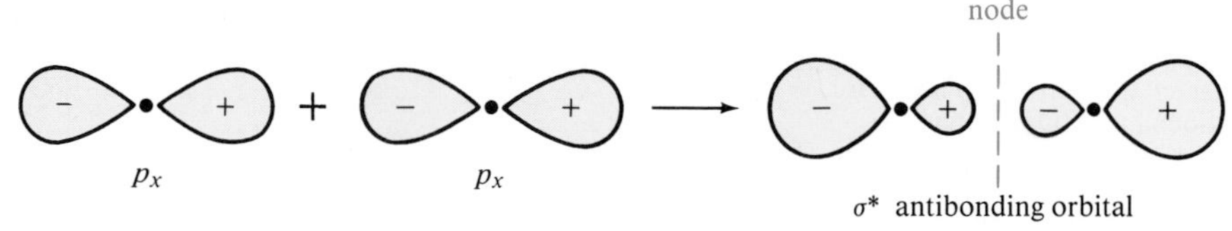

Overlap of an *s* orbital with a *p* orbital also gives a bonding MO and an antibonding MO, as shown below. Constructive overlap of the *s* orbital with the p_x orbital gives a σ-bonding MO with its electron density centered along the line between the nuclei. Destructive overlap gives a σ^* antibonding orbital with a node separating the nuclei.

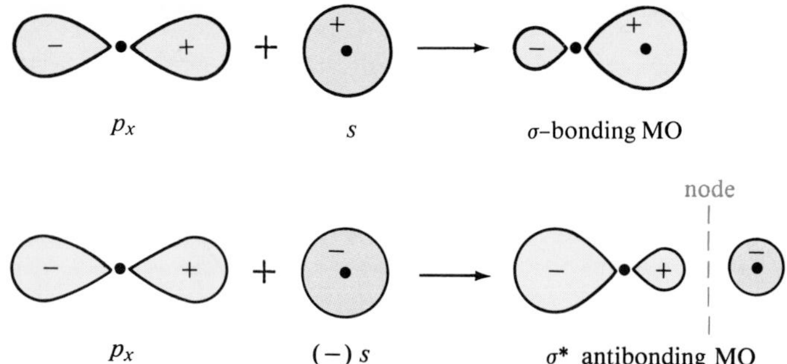

2-3 PI BONDING

A **pi bond** (**π bond**) results from overlap between two *p* orbitals oriented perpendicular to the line connecting the nuclei (Fig. 2-9). These parallel orbitals overlap sideways, with most of the electron density centered *above and below* the line connecting the nuclei. A pi molecular orbital is *not* cylindrically symmetrical; it involves parallel overlap rather than the linear overlap of a sigma bond. Figure 2-9 shows the pi bonding MO and the corresponding π^* antibonding MO.

2-3A SINGLE AND DOUBLE BONDS

A double bond requires the presence of four electrons in the bonding region between the nuclei. The first pair of electrons goes into the sigma bonding MO, forming the stronger sigma bond. The second pair of electrons cannot go into the

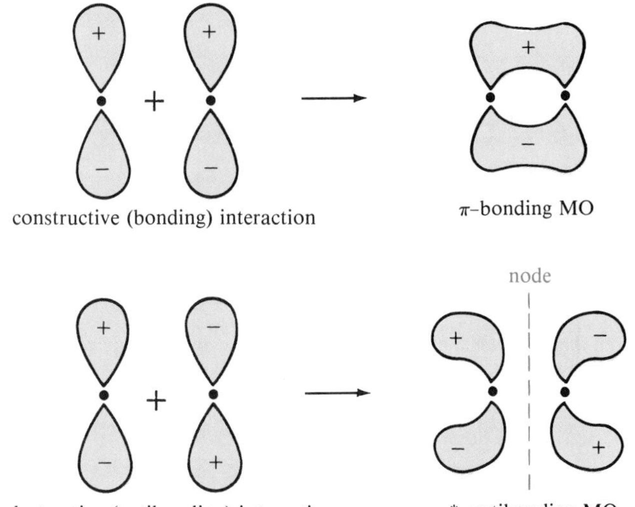

FIGURE 2-9 The sideways overlap of two *p* orbitals leads to a pi bonding MO and a pi antibonding MO. A pi bond is not as strong as most sigma bonds.

same orbital or the same space. It goes into a pi bonding MO, with its electron density centered above and below that of the sigma bond.

The combination of a sigma bond and a pi bond is the normal structure of a double bond. Figure 2-10 shows the structure of ethylene, an organic molecule containing a carbon-carbon double bond.

Thus far, we have discussed only bonds involving overlap of simple *s* and *p* atomic orbitals. Although these simple bonds are sometimes seen in organic compounds, they are not as common as bonds formed using **hybrid atomic orbitals.** Hybrid atomic orbitals result from the mixing of orbitals on the *same* atom. The geometry of these hybridized orbitals helps us to predict the actual structures and bond angles observed in organic compounds.

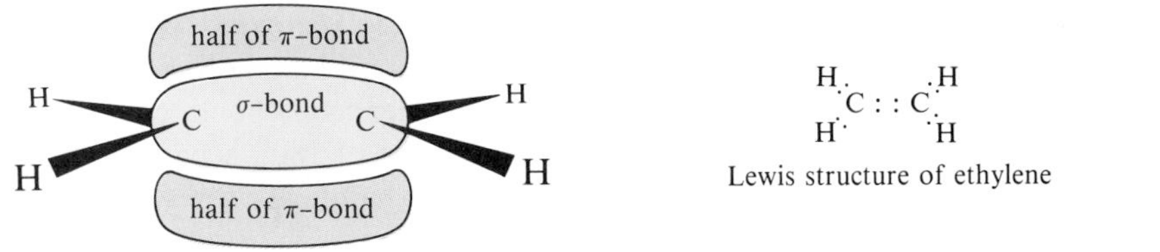

FIGURE 2-10 The second bond of a double bond is a pi bond. The pi bond has its electron density centered in two lobes, above and below the sigma bond. Together, the two lobes of the pi bonding molecular orbital constitute one bond.

2-4 HYBRIDIZATION AND MOLECULAR SHAPES

If we predict the bond angles of organic molecules using just the simple *s* and *p* orbitals, we must predict bond angles of about 90°. The *s* orbitals are nondirectional, and the *p* orbitals are oriented at 90° to one another (see Figure 1-3). Figure 2-11 shows that this prediction is entirely wrong. Bond angles in organic compounds are usually close to 109°, 120°, or 180°. These bond angles are explained by the *valence-shell electron pair repulsion theory* (VSEPR theory): Electron pairs repel each other, and the bonds and lone pairs around a central atom generally are separated by the largest possible angles. An angle of 109.5° is the largest possible separation for four pairs of electrons; 120° is the largest separation for three pairs; and 180° is the largest separation for two pairs. All the structures in Figure 2-11 have bond angles that separate their bonds about as far apart as possible.

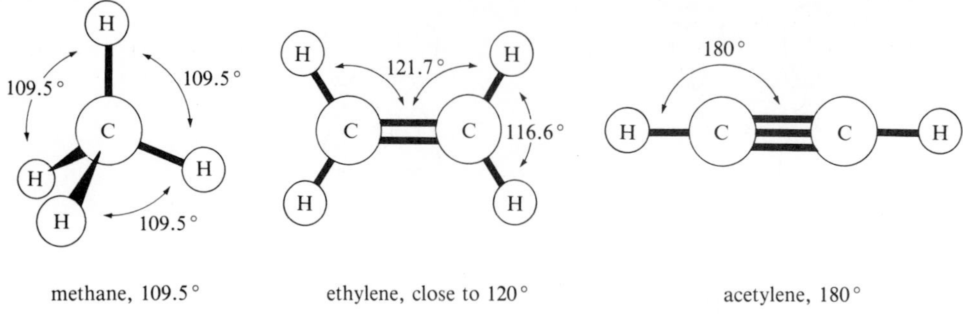

FIGURE 2-11 The angles between the p orbitals are all 90°, but few organic compounds have bond angles of 90°. Their bond angles are usually close to 109°, 120°, or 180°.

The shapes of these molecules cannot result from bonding between simple s and p atomic orbitals. Although these orbitals have the lowest energies for isolated atoms in space, they are not the best for forming bonds. To explain the shapes of common organic molecules, we assume that the s and p orbitals combine to form **hybrid atomic orbitals** that separate the electron pairs more widely in space and place more electron density in the bonding region between the nuclei.

2-4A *sp* HYBRID ORBITALS

Orbitals can interact to form new orbitals. We have used this principle to form molecular orbitals by adding and subtracting atomic orbitals on *different* atoms, but we can also add and subtract orbitals on the *same* atom. Consider the result when we add a p orbital to an s orbital on the same atom (Fig. 2-12).

The resulting orbital is called an ***sp* hybrid orbital.** Its electron density is concentrated toward one side of the atom. We started with two orbitals (s and p), so we must finish with two sp hybrid orbitals. The second sp hybrid orbital results if we add the p orbital with the opposite phase (Fig. 2-12).

The result of this hybridization process is a pair of sp hybrid orbitals, one directed toward the left and one toward the right. These hybridized orbitals provide enhanced electron density in the bonding region for a sigma bond toward the left of the atom and for another sigma bond toward the right. In addition, these hybrid orbitals give a bond angle of 180°, separating the bonding electrons as much as possible. In general, sp hybridization results in this **linear** bonding arrangement.

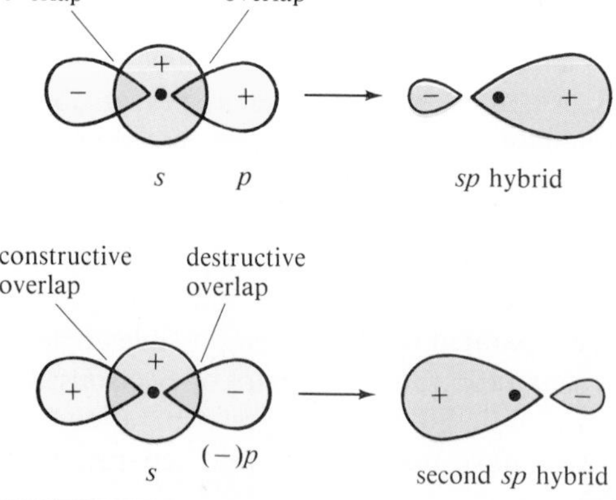

FIGURE 2-12 Addition of an s orbital to a p orbital gives an sp hybrid atomic orbital, with most of its electron density on one side of the nucleus. Adding the p orbital with opposite phase gives the other sp hybrid atomic orbital, with most of its electron density on the opposite side of the nucleus from the first hybrid.

SOLVED PROBLEM 2-2

Draw the Lewis structure for beryllium hydride, BeH_2. Draw the orbitals that overlap in the bonding of BeH_2, and label the hybridization of each orbital. Predict the H—Be—H bond angle.

SOLUTION First, we draw a Lewis structure for BeH_2.

$$H:Be:H$$

There are only four valence electrons in BeH_2 (two from Be and one from each H), so it is impossible to provide the Be atom with an octet of electrons. The bonding must involve orbitals on Be that give the strongest bonds (the most electron density in the bonding region) and also allow the two pairs of electrons to be separated as far as possible.

Hybrid orbitals concentrate the electron density in the bonding region, and *sp* hybrids give 180° separation for two pairs of electrons. Hydrogen cannot use hybridized orbitals, since the closest available *p* orbitals are the 2*p*'s, much higher in energy than the 1*s*. The bonding in BeH_2 results from overlap of *sp* hybrid orbitals on Be with the 1*s* orbitals on hydrogen. Figure 2-13 shows how this occurs.

First bond

Be (*sp*) H (1*s*)

Second bond

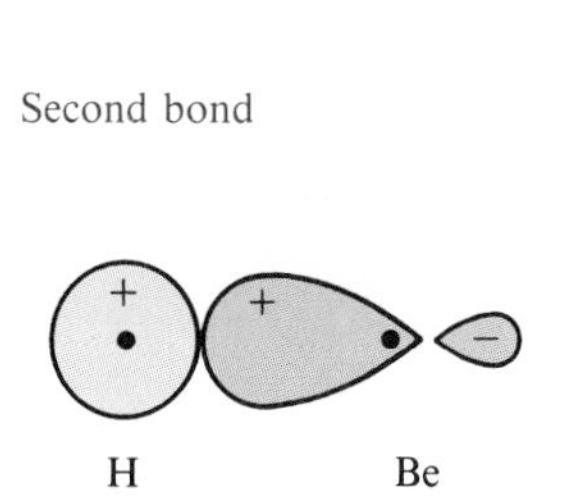

Superimposed picture

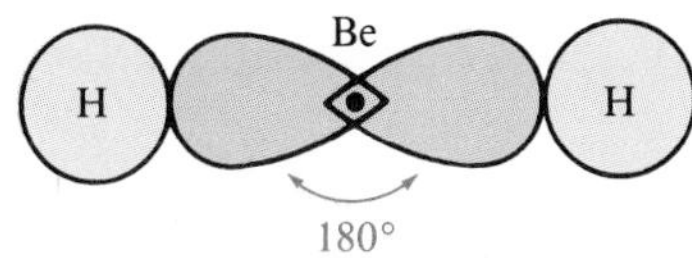

180° (linear) bond angle for *sp* hybrid

FIGURE 2-13 The bonding of BeH_2. To form two sigma bonds, the two *sp* hybridized atomic orbitals on Be overlap with the 1*s* orbitals of hydrogen. The bond angle is 180° (linear).

2-4B sp^2 HYBRID ORBITALS

To orient three bonds as far apart as possible, bond angles of 120° are required. When an *s* orbital is combined with two *p* orbitals, the resulting three hybrid orbitals are oriented at 120° angles to each other (Fig. 2-14). These orbitals are called **sp^2 hybrid orbitals,** because they are composed of one *s* and two *p* orbitals. The 120° arrangement is called **trigonal** geometry, in contrast to the linear geometry associated with *sp* hybrid orbitals.

SOLVED PROBLEM 2-3

Borane (BH_3) is not stable under normal conditions, but it has been detected at low pressure.

(a) Draw the Lewis structure for borane.

(b) Draw a diagram of the bonding in this molecule, and label the hybridization of each orbital.

(c) Predict the H—B—H bond angle.

SOLUTION There are only six valence electrons in borane. Boron has a single bond to each of the three hydrogen atoms.

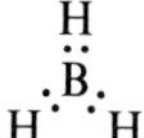

The best bonding orbitals are those that provide the greatest electron density in the bonding region while keeping the three pairs of bonding electrons as far apart as possible. Hybridization of an *s* orbital with two *p* orbitals gives three sp^2 hybrid orbitals directed 120° apart. Overlap of these orbitals with the hydrogen 1*s* orbitals gives a planar, trigonal molecule.

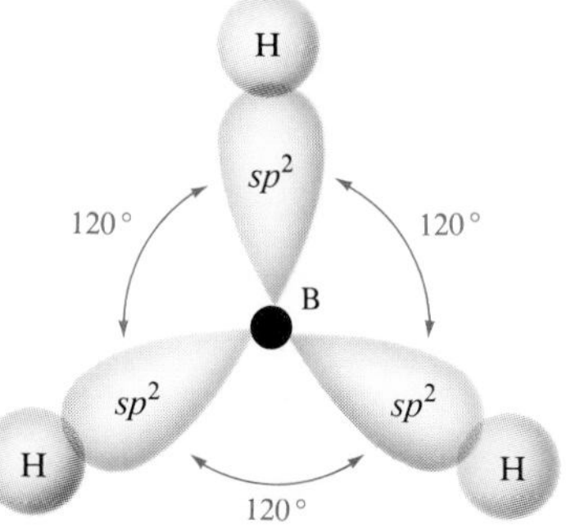

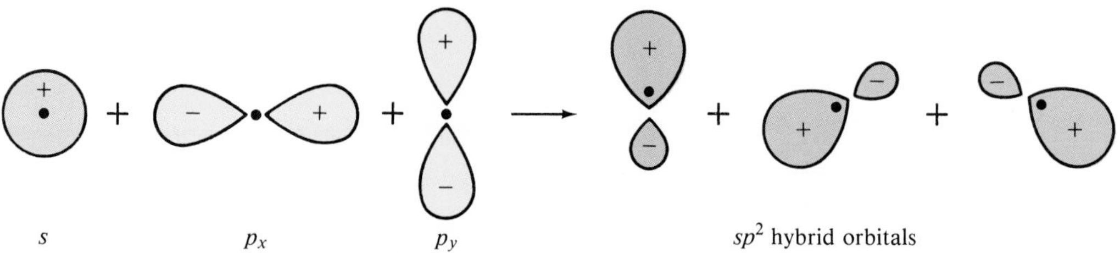

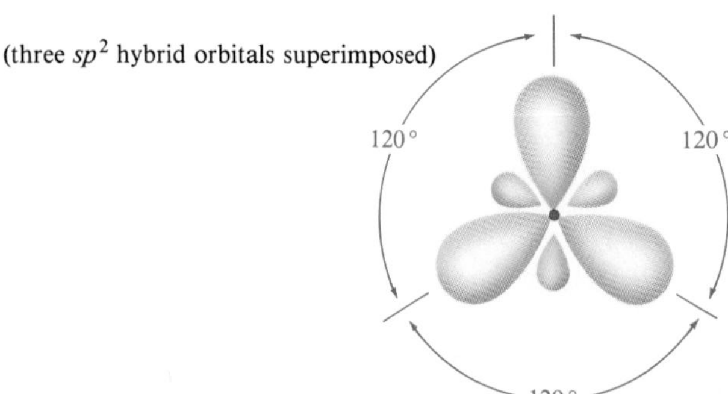

FIGURE 2-14 Hybridization of an *s* orbital with two *p* orbitals gives a set of three sp^2 hybrid orbitals. The bond angles associated with this trigonal structure are about 120°.

2-4C sp^3 HYBRID ORBITALS

Many organic compounds contain carbon atoms that are bonded to four other atoms. When four bonds are oriented as far apart as possible, they form a regular tetrahedron (109.5° bond angles), as pictured in Figure 2-15. This **tetrahedral** arrangement can be explained by hybridization of the *s* orbital with all three *p* orbitals. The resulting four orbitals are called ***sp^3* hybrid orbitals,** because they are composed of one *s* and three *p* orbitals.

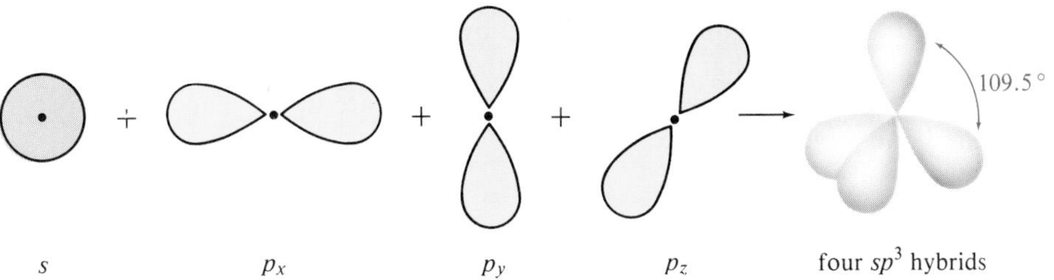

FIGURE 2-15 Hybridization of an *s* orbital with all three *p* orbitals gives four sp^3 hybrid orbitals with tetrahedral geometry and 109.5° bond angles.

Methane (CH_4) is the simplest example of sp^3 hybridization (Fig. 2-16). The Lewis structure for methane has eight valence electrons (four from carbon and one from each hydrogen), corresponding to four C—H single bonds. Tetrahedral geometry separates these bonds by the largest possible angle, 109.5°.

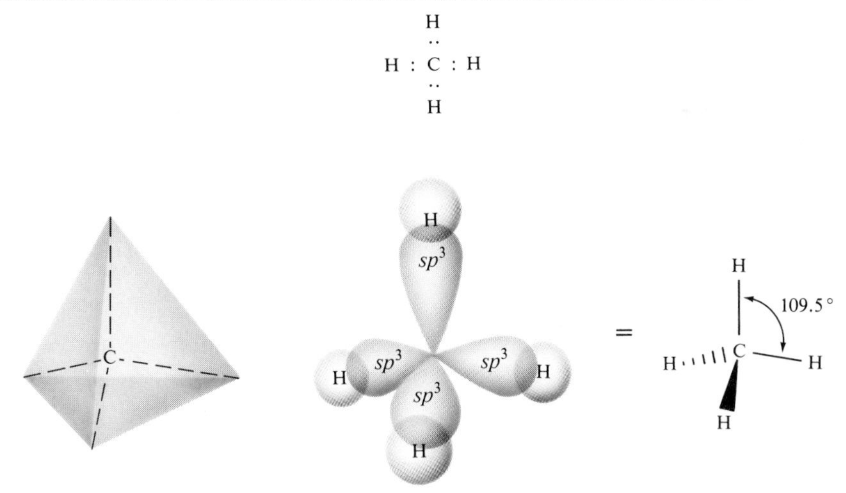

FIGURE 2-16 Methane has tetrahedral geometry, using four sp^3 hybrid orbitals to form sigma bonds to the four hydrogen atoms.

2-5 DRAWING THREE-DIMENSIONAL MOLECULES

Figures 2-15 and 2-16 were more difficult to draw than the earlier figures, because they depict three-dimensional objects on a two-dimensional piece of paper. The p_z orbital should look like it points in and out of the page, and the tetrahedron should look three-dimensional. These drawings use perspective to add the third dimension.

Most organic molecules are three-dimensional, but the use of perspective is difficult when the molecule is large and complicated. Organic chemists have developed a shorthand notation to simplify three-dimensional drawings. Dashed lines are used to indicate bonds that go backward, away from the reader. Wedge-shaped bonds are used to depict bonds that come forward, toward the reader. Straight lines are used for bonds in the plane of the paper. The last drawing of methane in Figure 2-16 uses dashed lines and wedges in this manner to show perspective in a three-dimensional molecule.

The three-dimensional structure of ethane, C_2H_6, shows that the molecule has the shape of two tetrahedra joined together. Each of the carbon atoms is sp^3 hybridized, with four sigma bonds formed by the four sp^3 hybrid orbitals. Dashed lines represent bonds that go away from the viewer, wedges represent bonds that come out toward the viewer, and other bond lines are in the plane of the paper. All the bond angles are close to 109.5°.

PROBLEM 2-1

(a) Use your molecular models to make ethane, and compare the model with the structure given above.

(b) Make a model of propane (C_3H_8), and draw this model using dashed lines and wedges to represent bonds going back and coming forward.

2-6 GENERAL RULES OF HYBRIDIZATION AND GEOMETRY

At this point we can consider some general rules for determining the hybridization of orbitals and the bond angles of atoms in organic molecules. After stating these rules, we solve some problems to show how the rules are used.

RULE 1 Both sigma-bonding electrons and lone pairs occupy hybrid orbitals. The number of hybrid orbitals on an atom is computed by adding the number of sigma bonds and the number of lone pairs of electrons on that atom.

Since the first bond to another atom is always a sigma bond, the number of hybrid orbitals may be computed by adding the number of lone pairs to the number of atoms bonded.

RULE 2 Once the number of hybrid orbitals has been calculated, use the hybridization and geometry that give the widest possible separation of the bond and lone pairs:

SUMMARY OF HYBRIDIZATION AND GEOMETRY

Hybrid orbitals	*Hybridization*	*Geometry*	*Approximate bond angles*
2	$s + p = sp$	linear	180°
3	$s + p + p = sp^2$	trigonal	120°
4	$s + p + p + p = sp^3$	tetrahedral	109.5°

Lone pairs of electrons take up more space than bonding pairs of electrons, and they compress the bond angles.

RULE 3 If two or three pairs of electrons form a multiple bond between two atoms, the first bond is a sigma bond formed by a hybrid orbital. The second bond is a pi bond, consisting of two lobes above and below the sigma bond, formed by two *p* orbitals. The third bond of a triple bond is another pi bond (see Fig. 2-18).

We illustrate these rules by working through some sample problems.

SOLVED PROBLEM 2-4

Determine the hybridization of the nitrogen atom in ammonia, NH_3. Draw a picture of the three-dimensional structure of ammonia, and predict the bond angles.

SOLUTION The hybridization depends on the number of sigma bonds plus lone pairs. A Lewis structure provides this information.

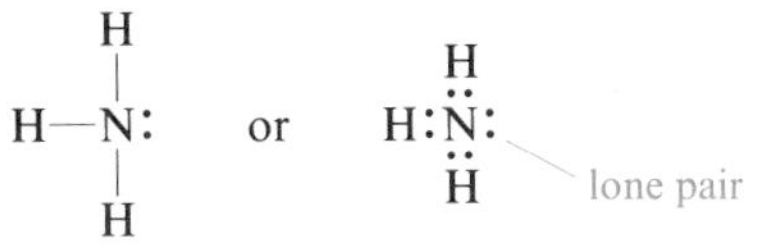

H—N(—H)(—H): or H:N̈:H (lone pair)

In this structure there are three sigma bonds and one pair of nonbonding electrons. Four hybrid orbitals are required, implying sp^3 hybridization and tetrahedral geometry around the nitrogen atom, with bond angles of about 109.5°. The resulting structure is much like that of methane, except that one of the sp^3 hybrid orbitals is occupied by a lone pair of electrons.

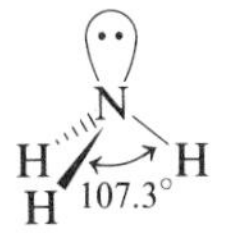

107.3°

Notice that the bond angles in ammonia (107.3°) are slightly smaller than the ideal tetrahedral angle, 109.5°. The nonbonding electrons are more diffuse than a bonding pair of electrons, and they take up more space. The lone pair repels the electrons in the N—H bonds, compressing the bond angle.

PROBLEM 2-2

Determine the hybridization of water, H_2O. Draw a picture of its three-dimensional structure, and explain why its bond angle is 104.5°.

SOLVED PROBLEM 2-5

Predict the hybridization, geometry, and bond angles for ethylene, C_2H_4.

SOLUTION The Lewis structure of ethylene is

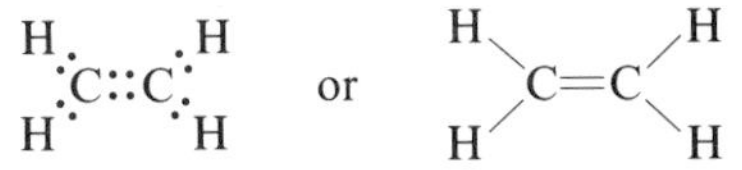

$H_2C::CH_2$ or $H_2C{=}CH_2$

Each carbon atom has an octet, and there is a double bond between the carbon atoms. Each carbon is bonded to three other atoms (three sigma bonds), and there are no lone pairs. The carbon atoms are sp^2 hybridized, and the bond angles are trigonal: about 120°. The double bond is composed of a sigma bond formed by overlap of two sp^2 hybridized orbitals, plus a pi bond formed by overlap of the unhybridized *p* orbitals remaining on the carbon atoms. Because the pi bond requires parallel alignment of its two *p* orbitals, the ethylene molecule must be planar (Fig. 2-17).

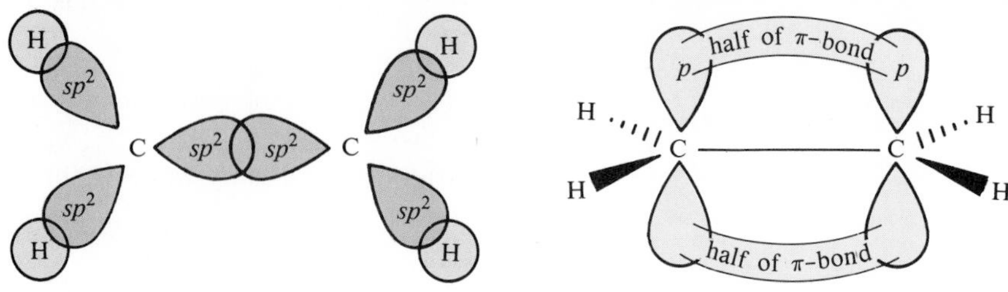

FIGURE 2-17 The carbon atoms in ethylene are sp^2 hybridized, with trigonal bond angles of about 120°. All the carbon and hydrogen atoms lie in the same plane.

PROBLEM 2-3

Predict the hybridization, geometry, and bond angles for 2-butene, $CH_3CH{=}CHCH_3$.

SOLVED PROBLEM 2-6

Predict the hybridization, geometry, and bond angles for acetylene, C_2H_2.

SOLUTION The Lewis structure of acetylene is

$$H{:}C{:::}C{:}H \qquad \text{or} \qquad H{-}C{\equiv}C{-}H$$

In this structure, both carbon atoms have octets, but each carbon is bonded to just two other atoms, requiring two sigma bonds. There are no lone pairs. Each carbon atom is *sp* hybridized, and linear (180° bond angles). Notice that the *sp* hybrid orbitals are generated by the *s* orbital and the p_x orbital (the *p* orbital directed along the line joining the nuclei). The p_y orbitals and the p_z orbitals are unhybridized.

The triple bond is composed of one sigma bond, formed by overlap of *sp* hybrid orbitals, plus two pi bonds. One pi bond results from overlap of the two p_y orbitals and another from overlap of the two p_z orbitals (Fig. 2-18).

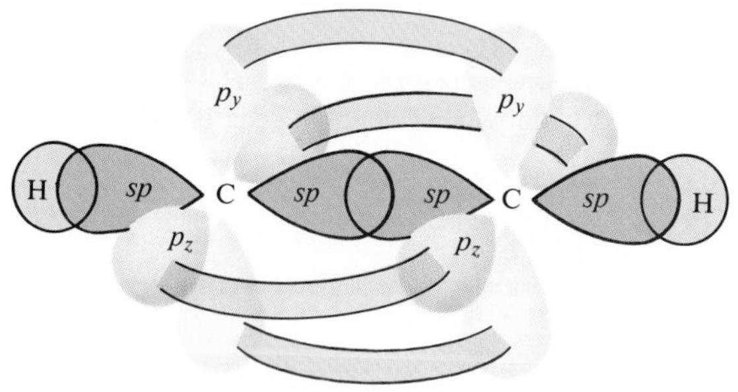

FIGURE 2-18 The carbon atoms in acetylene are *sp* hybridized, with linear (180°) bond angles. The triple bond contains one sigma bond and two perpendicular pi bonds.

PROBLEM 2-4

Predict the hybridization, geometry, and bond angles for the atoms in acetonitrile, $CH_3{-}C{\equiv}N{:}$.

SOLVED PROBLEM 2-7

Predict the hybridization, geometry, and bond angles for the atoms in acetaldehyde, CH_3CHO.

SOLUTION The Lewis structure for acetaldehyde is

$$\mathrm{H{:}\overset{\displaystyle H}{\underset{\displaystyle H}{\ddot{C}}}{:}C{\overset{\cdot\cdot\ddot{O}:}{\underset{\cdot H}{}}}} \quad \text{or} \quad \mathrm{H{-}\overset{\displaystyle H}{\underset{\displaystyle H}{\overset{|}{\underset{|}{C}}}}{-}C\overset{\mathbin{/\!/}\ddot{O}:}{\underset{\backslash H}{}}}$$

The oxygen atom and both carbon atoms have octets of electrons. The CH_3 carbon atom is sigma bonded to four atoms, so it is sp^3 hybridized (and tetrahedral). The C=O carbon is bonded to three atoms (no lone pairs), so it is sp^2 hybridized and its bond angles are about 120°.

The oxygen atom is probably sp^2 hybridized, because it is bonded to one atom (carbon) and has two lone pairs, requiring a total of three hybridized orbitals. We cannot experimentally measure the angles of the lone pairs on oxygen, however, so it is impossible to confirm whether the oxygen atom is really sp^2 hybridized.

The double bond between carbon and oxygen looks just like the double bond in ethylene. There is a sigma bond formed by overlap of sp^2 hybrid orbitals, and a pi bond formed by overlap of the unhybridized p orbitals on carbon and oxygen (see Fig. 2-19).

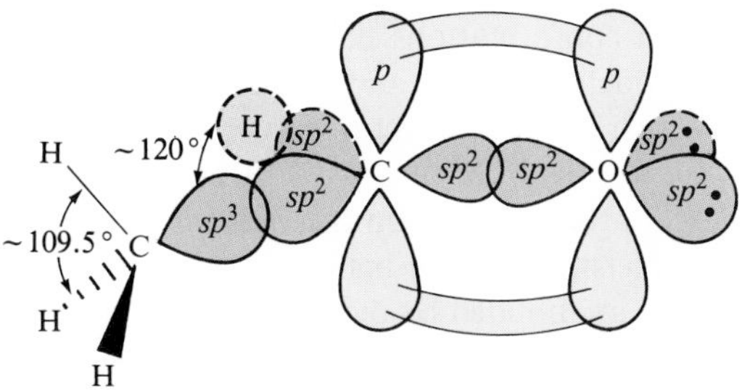

FIGURE 2-19 The CH_3 carbon in acetaldehyde is sp^3 hybridized, with tetrahedral bond angles of about 109.5°. The carbonyl (C=O) carbon is sp^2 hybridized, with bond angles of about 120°. The oxygen atom is probably sp^2 hybridized, but we cannot measure any bond angles to verify this prediction.

PROBLEM 2-5

(1) Draw a Lewis structure for each of the following compounds.

(2) Label the hybridization, geometry, and bond angles around each atom other than hydrogen.

(3) Draw a three-dimensional representation (using wedges and dashed lines) of the structure.

(a) CO_2 (b) $CH_3{-}O{-}CH_3$ (c) $(CH_3)_3N$
(d) HCOOH (e) HCN (f) $CH_3CH{=}CH_2$
(g) ozone (O_3), bonded OOO

PROBLEM 2-6

Allene, $CH_2{=}C{=}CH_2$, has the structure shown below. Explain how the bonding in allene requires the two $={CH_2}$ groups at its ends to be at right angles to each other.

$$\mathrm{\overset{H}{\underset{H}{}}\!\!>C{=}C{=}C\!<\!\!\overset{H}{\underset{H}{}}}$$

allene

2-7 ROTATION OF SINGLE BONDS

The Lewis structure of ethane ($CH_3{-}CH_3$) shows that both carbon atoms are sp^3 hybridized and tetrahedral. Ethane looks like two methane molecules that have had a hydrogen plucked off (to form a methyl group) and are joined by overlap of their sp^3 orbitals (Fig. 2-20).

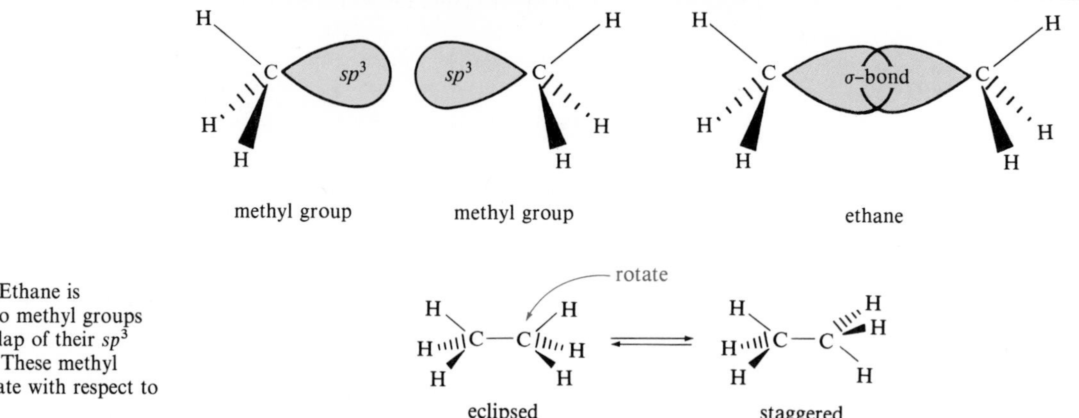

FIGURE 2-20 Ethane is composed of two methyl groups bonded by overlap of their sp^3 hybrid orbitals. These methyl groups may rotate with respect to each other.

We can draw many structures for ethane, differing only in how one methyl group is twisted in relation to the other methyl group. Such structures, differing only in rotations about a single bond, are called *conformations*. Two of the infinite number of conformations of ethane are shown in Figure 2-20. Construct a molecular model of ethane, and twist the model into these two conformations.

Which of these structures for ethane is the "right" one? Are the two methyl groups lined up so that their C—H bonds are parallel (*eclipsed*), or are they *staggered,* as in the drawing on the right? The answer is that both structures, and all the possible structures in between, are correct structures for ethane. The two carbon atoms are bonded by overlap of their sp^3 orbitals to form a sigma bond along the line between the carbons. This sigma bond can be twisted without destroying the linear overlap of the two sp^3 orbitals. No matter how you turn one of the methyl groups, its sp^3 orbital still overlaps with the sp^3 orbital of the other carbon atom.

2-8 RIGIDITY OF DOUBLE BONDS

Not all bonds allow free rotation. In ethylene, the double bond between the two CH_2 groups has a sigma bond and a pi bond. The sigma bond is unaffected when we twist one of the two CH_2 groups, but the pi bond loses its overlap when it is twisted. The two *p* orbitals cannot overlap when the two ends of the molecule are at right angles, and the pi bond is effectively broken.

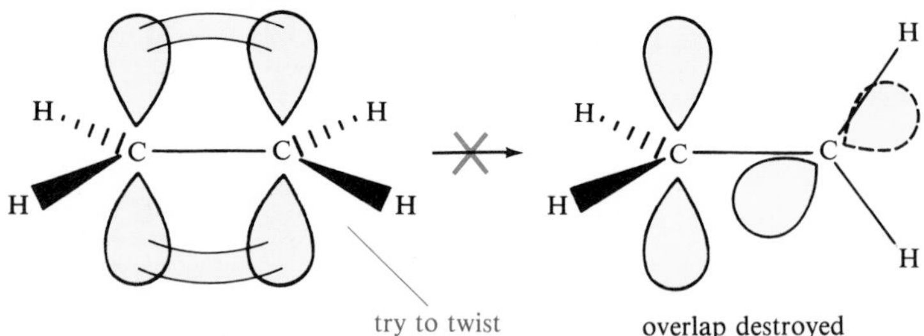

We can make a generalization:

Rotation is allowed by single bonds, but double bonds are rigid and cannot be twisted.

Because double bonds are rigid, we can separate and isolate compounds that differ only in how their substituents are arranged on a double bond. For example, the double bond in 2-butene ($CH_3—CH═CH—CH_3$) prevents the two ends of the molecule from rotating. Two different compounds are possible, and they have different physical properties. The molecule with the methyl groups on the same side of the double bond is called *cis*-2-butene, and the one with the methyl groups on opposite sides is called *trans*-2-butene. Such molecules are discussed further in the next section.

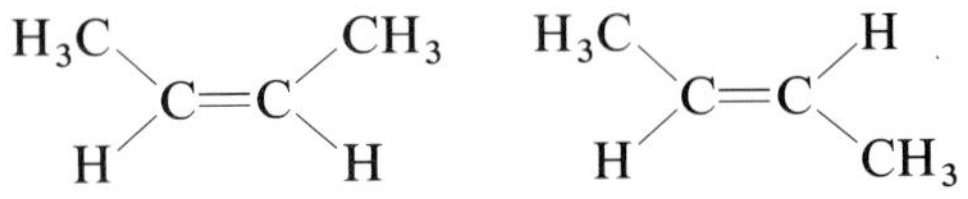

$H_3C(H)C{=}C(H)CH_3$ (methyl groups on the same side) — *cis*-2-butene

$H_3C(H)C{=}C(CH_3)H$ (methyl groups on opposite sides) — *trans*-2-butene

PROBLEM 2-7

Two compounds are known with the formula $CH_3—CH═N—CH_3$.

(a) Draw a Lewis structure for this molecule, and label the hybridization of each carbon and nitrogen atom.
(b) What two compounds have this formula?
(c) Explain why there is only one compound known with the formula $(CH_3)_2CNCH_3$.

2-9 STRUCTURAL ISOMERISM

If you were asked to draw a structural formula for C_4H_{10}, either of the following structures would be a correct answer.

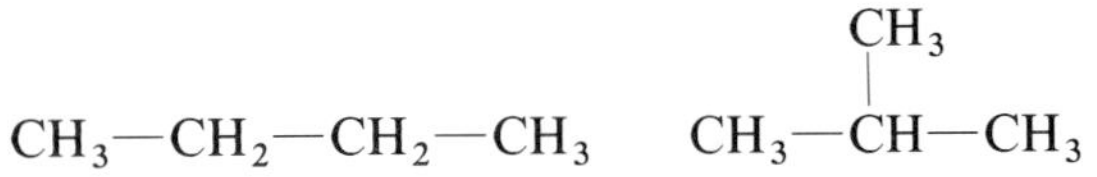

$CH_3—CH_2—CH_2—CH_3$ — *n*-butane

$CH_3—CH(CH_3)—CH_3$ — isobutane

These two compounds are isomers. **Isomers** are different compounds with the same molecular formula. *n*-Butane and isobutane are called **structural isomers,** because they differ in their order of bonding; that is, different atoms are bonded together. The first compound (*n*-butane for "normal" butane) has its carbon atoms in a straight chain four carbons long. The second compound ("isobutane" for "an isomer of butane") has a branched structure with a longest chain of three carbon atoms and a methyl side chain.

2-10 STEREOISOMERISM

We have also seen another kind of isomerism. Above we showed two isomers of 2-butene with exactly the same order of bonding. These two compounds differ only in the spatial orientations of the groups attached to the double bond. They are stereoisomers. **Stereoisomers** are isomers that differ only in how their atoms are oriented in space. *cis*- and *trans*-2-Butene are stereoisomers, because they differ only in how their atoms are oriented in space. 1-Butene is a structural isomer of *cis*- and *trans*-2-butene.

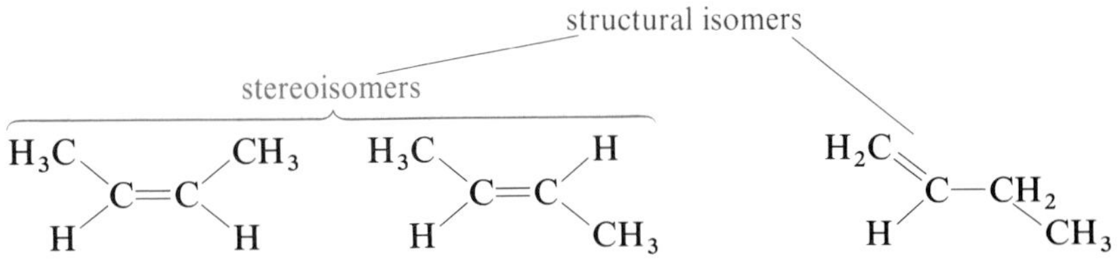

structural isomers

stereoisomers

cis-2-butene ($H_3C(H)C{=}C(H)CH_3$) — *trans*-2-butene ($H_3C(H)C{=}C(CH_3)H$) — 1-butene ($H_2C{=}C(H)—CH_2—CH_3$)

Cis and trans isomers are only one example of stereoisomerism. The study of the structure and chemistry of stereoisomers is called **stereochemistry.** We will encounter stereochemistry throughout our study of organic chemistry, and Chapter 6 is devoted entirely to this field.

Cis-trans isomers are also called **geometric isomers** because they differ in the geometry of the groups on a double bond. The cis isomer is always the one with similar groups on the same side of the double bond, and the trans isomer has similar groups on opposite sides of the double bond. To have geometric isomerism, there must be two different groups on each end of the double bond; otherwise, there can be no cis or trans isomers (see Problem 2-8). We study geometric isomers in more detail in Chapters 6 and 7.

PROBLEM 2-8

2-Methyl-2-butene has the structure $(CH_3)_2C{=}CHCH_3$. Are there geometric (cis and trans) isomers of 2-methyl-2-butene? Explain your answer.

PROBLEM 2-9

Which of the following compounds show geometric isomerism? Draw the cis and trans isomers of those that do.

(a) $CHF{=}CHF$ (b) $F_2C{=}CH_2$ (c) $CH_2{=}CH{-}CH_2{-}CH_3$

(d) $=CHCH_3$ (e) $-CHCHCH_3$ (f) $=CHCH_3$

PROBLEM 2-10

Give the relationship between the following pairs of structures. The possible relationships are the following:

same compound	structural isomers
geometric isomers	not isomers (different molecular formula)

(a) $\begin{array}{c}CH_3CH_2CHCH_2CH_3\\ |\\ CH_2CH_3\end{array}$ and $\begin{array}{c}CH_3CH_2CHCH_2CH_2CH_3\\ |\\ CH_3\end{array}$

(b) $\begin{array}{ccc}Br & & H\\ & C{=}C & \\ H & & Br\end{array}$ and $\begin{array}{ccc}Br & & Br\\ & C{=}C & \\ H & & H\end{array}$

(c) $\begin{array}{ccc}Br & & H\\ & C{=}C & \\ H & & Br\end{array}$ and $\begin{array}{ccc}Br & & H\\ & C{=}C & \\ Br & & H\end{array}$

(d) $\begin{array}{ccc}Br & & H\\ & C{=}C & \\ H & & Br\end{array}$ and $\begin{array}{ccc}H & & Br\\ & C{=}C & \\ Br & & H\end{array}$

(e) $\begin{array}{cccc} & H & Cl & \\ & | & | & \\ H{-} & C{-} & C{-} & H\\ & | & | & \\ & Cl & H & \end{array}$ and $\begin{array}{cccc} & Cl & Cl & \\ & | & | & \\ H{-} & C{-} & C{-} & H\\ & | & | & \\ & H & H & \end{array}$

(f) $\begin{array}{cccc} & H & CH_3 & \\ & | & | & \\ H{-} & C{-} & C{-} & H\\ & | & | & \\ & CH_3 & H & \end{array}$ and $\begin{array}{cccc} & CH_3 & H & \\ & | & | & \\ H{-} & C{-} & C{-} & CH_3\\ & | & | & \\ & H & H & \end{array}$

(g) $CH_3—CH_2—CH_2—CH_3$ and $CH_3—CH=CH—CH_3$
(h) $CH_2=CH—CH_2CH_2CH_3$ and $CH_3—CH=CH—CH_2CH_3$
(i) $CH_2=CHCH_2CH_2CH_3$ and $CH_3CH_2CH_2CH=CH_2$

(j) 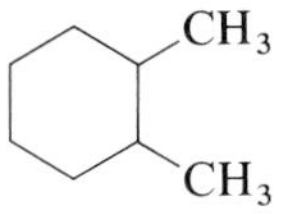 CH_3 CH_3 and CH_3 CH_3 (k) 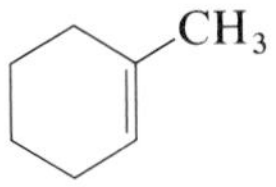 CH_3 and CH_3

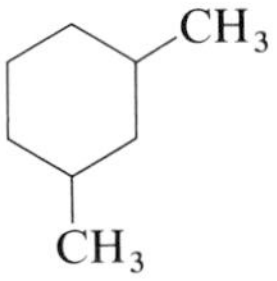

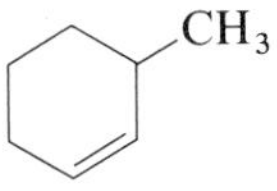

2-11 POLARITY OF BONDS AND MOLECULES

In Chapter 1 we reviewed the concept of polar covalent bonds between atoms with different electronegativities. Now we are ready to combine this concept with molecular geometry to study the polarity of entire molecules.

2-11A BOND DIPOLE MOMENTS

The polarity of an individual bond is measured as its dipole moment in units of debye, abbreviated D. The **bond dipole moment,** μ, is defined as

$$\mu = 4.8 \times \delta \times d$$

where 4.8 represents the charge on an electron, δ is the amount of charge separation on the two atoms (in units of the charge on an electron) and d is the bond length (in angstrom units, Å). Dipole moments are measured experimentally, and they can be used to calculate other information such as bond lengths and charge separations.

For example, the dipole moment can be used to calculate the charge separation of a typical C—O single bond. The bond length is 1.43 Å, and the dipole moment is measured as 0.86 D. We calculate that the amount δ of charge separation is about 0.125 electronic charge, so the carbon atom has about an eighth of a full positive charge and the oxygen atom has about an eighth of a full negative charge.

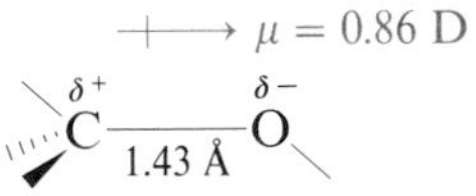

$\mu = 0.86$ D

$\delta+$ C—O $\delta-$ (1.43 Å)

$$0.86\text{ D} = 4.8 \times \delta \times 1.43\text{ Å}$$
$$\delta = 0.125\text{ e}$$

PROBLEM 2-11

The C=O double bond has a dipole moment of about 2.4 D and a bond length of about 1.21 Å.

(a) Calculate the amount of charge separation in this bond.
(b) Use this information to explain the relative importance of the following two resonance contributors.

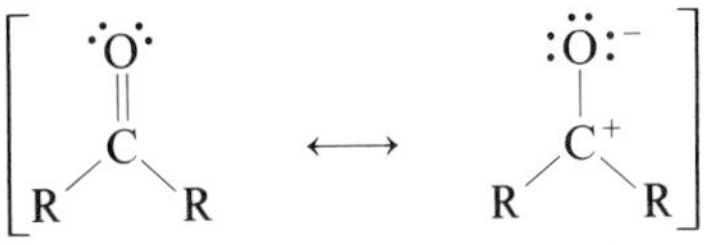

[$R_2C=\ddot{O}:$ ⟷ $R_2C^+—\ddot{O}:^-$]

Bond dipole moments in organic compounds range from zero in symmetrical bonds to about 3.6 D for the strongly polar C≡N: triple bond. Table 2-1 shows

TABLE 2-1
Bond moments (debye) for some common covalent bonds

Bond	*Dipole moment, μ*	*Bond*	*Dipole moment, μ*
C—N	0.22	H—C	0.3
C—O	0.86	H—N	1.31
C—F	1.51	H—O	1.53
C—Cl	1.56	C=O	2.4
C—Br	1.48	C≡N	3.6
C—I	1.29		

typical dipole moments for some of the bonds common in organic molecules. Recall that the positive end of the crossed arrow corresponds to the less electronegative (partial positive charge) end of the dipole.

2-11B MOLECULAR DIPOLE MOMENTS

A **molecular dipole moment** is the dipole moment of the molecule taken as a whole. The molecular dipole moment is a good indicator of the overall polarity of a molecule. Its value is equal to the *vector* sum of the individual bond dipole moments. This vector sum reflects both the magnitude and the direction of each individual bond dipole moment. Figure 2-21 gives some examples of molecular dipole moments. The figure shows two compounds, carbon tetrachloride and carbon dioxide, that have very polar bonds but zero molecular dipole moments. This canceling of dipole moments occurs in symmetrical molecules, where the individual bond dipole moments are oriented in opposing directions.

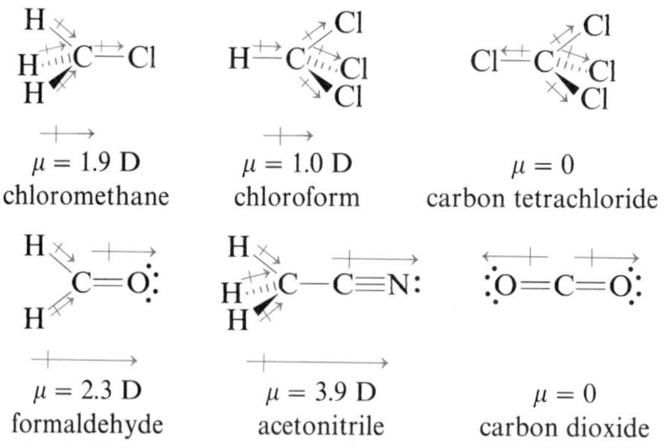

$\mu = 1.9$ D
chloromethane

$\mu = 1.0$ D
chloroform

$\mu = 0$
carbon tetrachloride

$\mu = 2.3$ D
formaldehyde

$\mu = 3.9$ D
acetonitrile

$\mu = 0$
carbon dioxide

FIGURE 2-21 A molecular dipole moment is the vector sum of the individual bond dipole moments. If the bond dipole moments are symmetrically arranged so that they cancel out, the molecule will have an overall dipole moment of zero.

Some molecules, such as water and ammonia (Fig. 2-22), have larger dipole moments than we would expect from the vector sum of their bond dipole moments. This result is explained by the polarity of the lone pairs of electrons. Each lone pair corresponds to a charge separation, with the nucleus having a partial positive charge balanced by the negative charge of the lone pair. The contribution of lone pairs also helps to explain the large dipole moments of C=O and C≡N bonds.

Although we can often make a close prediction of a molecular dipole moment, it is usually necessary to consult a table or make the measurement to obtain an accurate value.

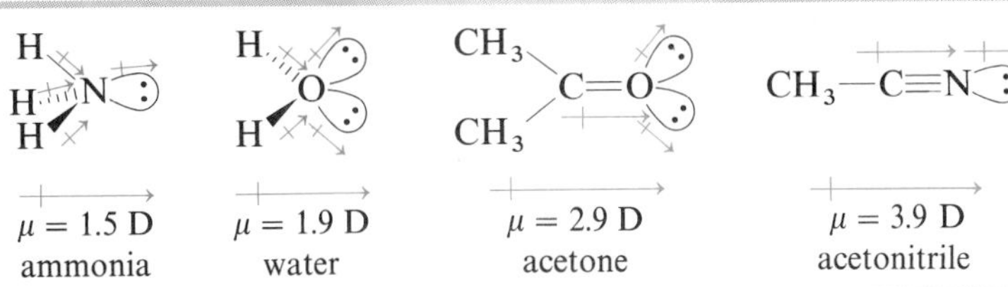

FIGURE 2-22 The presence of lone pairs may have a significant effect on the molecular dipole moment.

PROBLEM 2-12

The N—F bond is more polar than the N—H bond, but NF_3 has a *smaller* dipole moment than NH_3. Explain this curious result.

NH_3	NF_3
$\mu = 1.5$ D	$\mu = 0.2$ D

PROBLEM 2-13

For each of the following compounds

(1) Draw the structure.

(2) Show how the bond dipole moments (and those of any nonbonding pairs of electrons) contribute to the molecular dipole moment.

(3) Predict whether the compound has a large (>1 D), small, or zero dipole moment.

(a) CH_2Cl_2	(b) CH_3F	(c) CF_4	(d) CH_3OH
(e) O_3	(f) HCN	(g) CH_3CHO	(h) $H_2C{=}NH$
(i) $(CH_3)_3N$	(j) $CH_2{=}CHCl$	(k) BF_3	(l) $BeCl_2$
(m) NH_4^+			

PROBLEM 2-14

Two isomers of 1,2-dichloroethene are known. One has a dipole moment of 2.95 D; the other has zero dipole moment. Draw the two isomers and explain why one has zero dipole moment.

CHCl=CHCl

1,2-dichloroethene

2-12 INTERMOLECULAR ATTRACTIONS AND REPULSIONS

Whenever two molecules approach, they attract or repel each other. This interaction can be described fairly simply in the case of atoms (like the noble gases) or simple molecules such as H_2 or Cl_2. In general, the forces are attractive until the molecules come so close that they infringe on each other's atomic radius. When this happens, the small attractive force quickly becomes a large repulsive force, and the molecules "bounce" off each other. With complicated organic molecules, these attractive and repulsive forces are difficult to predict. We can still describe the nature of the forces, however, and we can show how they affect the physical properties of organic compounds.

The attractions between molecules are particularly important in solids and liquids. In these "condensed" phases, the molecules are continuously in contact with each other. The melting points, boiling points, and solubilities of organic compounds show the effects of these forces. There are three major kinds of attractive forces that cause molecules to associate into solids and liquids: the dipole-dipole forces of polar molecules, the London forces that affect all molecules, and the "hydrogen bonds" that link molecules having —OH or —NH groups.

2-12A DIPOLE-DIPOLE FORCES

Most molecules have permanent dipole moments as a result of their polar bonds. Each molecular dipole moment has a positive end and a negative end. The most stable arrangement has the positive end of one dipole close to the negative end of another. When two negative ends or two positive ends approach each other, they experience a mild repulsion. The molecules may turn and orient themselves in the more stable positive-to-negative arrangement. **Dipole-dipole forces,** therefore, are generally attractive intermolecular forces resulting from the attraction of the positive and negative ends of the molecular dipole moments of polar molecules. Figure 2-23 shows the attractive and repulsive orientations of polar molecules, using chloromethane as the example.

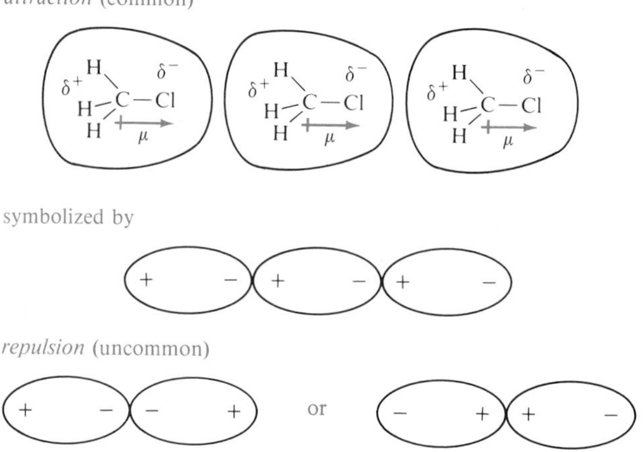

FIGURE 2-23 Dipole-dipole interactions result from the approach of two polar molecules. If their positive and negative ends approach, the interaction is an attractive one. If two negative ends or two positive ends approach, the interaction is repulsive. In a liquid or a solid, the molecules are mostly oriented with the positive and negative ends together, and the net force is attractive.

Polar molecules are mostly oriented in the lower-energy positive-to-negative arrangement, and the net force is attractive. This attraction must be overcome when the liquid vaporizes, resulting in larger heats of vaporization and higher boiling points for compounds with strongly polar molecules.

2-12B THE LONDON DISPERSION FORCE

Carbon tetrachloride (CCl_4) has zero dipole moment, yet its boiling point is higher than that of chloroform ($\mu = 1.0$ D). Clearly, there must be some kind of force other than dipole-dipole forces holding together the molecules of carbon tetrachloride.

$\mu = 0$	$\mu = 1.0$ D
carbon tetrachloride, b.p. = 77°C	chloroform, b.p. = 62°C

In nonpolar molecules such as carbon tetrachloride, the principal attractive force is the **London dispersion force,** one of the **van der Waals forces** (Fig. 2-24). The London force arises from temporary dipole moments that are induced in a molecule by other nearby molecules. Even though carbon tetrachloride has no permanent dipole moment, the electrons are not always evenly distributed. A small temporary dipole moment is induced when one molecule approaches another mole-

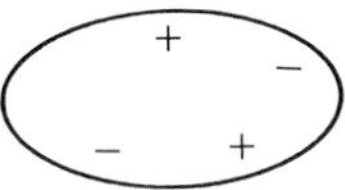

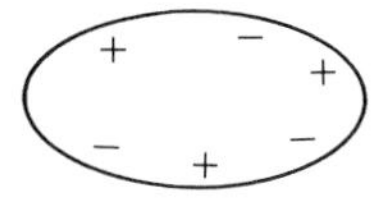

random temporary dipoles when separated

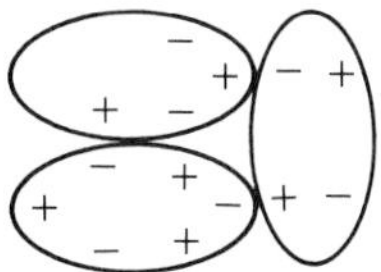

coordinated temporary dipoles when in contact

FIGURE 2-24 London dispersion forces result from the attraction of coordinated temporary dipole moments.

cule in which the electrons are slightly displaced from a symmetrical arrangement. The electrons in the approaching molecule will be displaced slightly so that an attractive dipole-dipole interaction results.

These temporary dipoles last only a fraction of a second, and they change continuously, yet they are correlated so that their net force is attractive. This attractive force depends on close surface contact of two molecules, so it is roughly proportional to the molecular surface area. Carbon tetrachloride has a larger surface area than chloroform (a chlorine atom is much larger than a hydrogen atom), and the intermolecular van der Waals attractions between carbon tetrachloride molecules are stronger than they are between chloroform molecules.

We can see the effects of London forces in the boiling points of simple hydrocarbons, as well. If we compare the boiling points of several different isomers, the isomers with larger surface areas (and greater potential for London force attraction) have higher boiling points. The boiling points of three isomers of molecular formula C_5H_{12} are given below. The long-chain isomer (*n*-pentane) has the greatest surface area and the highest boiling point. As the amount of chain branching increases, the molecule becomes more spherical and its surface area decreases. The most highly branched isomer (neopentane) has the smallest surface area and the lowest boiling point.

$$CH_3-CH_2-CH_2-CH_2-CH_3$$

n-pentane, b.p. = 36°C

$$CH_3-\underset{}{\overset{CH_3}{\overset{|}{C}}}H-CH_2-CH_3$$

isopentane, b.p. = 28°C

$$CH_3-\underset{\underset{CH_3}{|}}{\overset{\overset{CH_3}{|}}{C}}-CH_3$$

neopentane, b.p. = 10°C

2-12C HYDROGEN BONDING

A **hydrogen bond** is not a true bond but a particularly strong form of dipole-dipole attraction. A hydrogen atom can participate in hydrogen bonding if it is bonded to oxygen, nitrogen, or fluorine. Organic compounds do not contain H—F bonds, so we consider only N—H and O—H hydrogens to be hydrogen bonded (Fig. 2-25).

The O—H and N—H bonds are strongly polarized, leaving the hydrogen atom with a partial positive charge. This electrophilic hydrogen atom has a strong affinity for nonbonding electrons, and it forms intermolecular attachments with the nonbonding electrons on oxygen or nitrogen atoms.

Although hydrogen bonding is a strong form of intermolecular attraction, it is much weaker than a normal C—H, N—H, or O—H covalent bond. Breaking

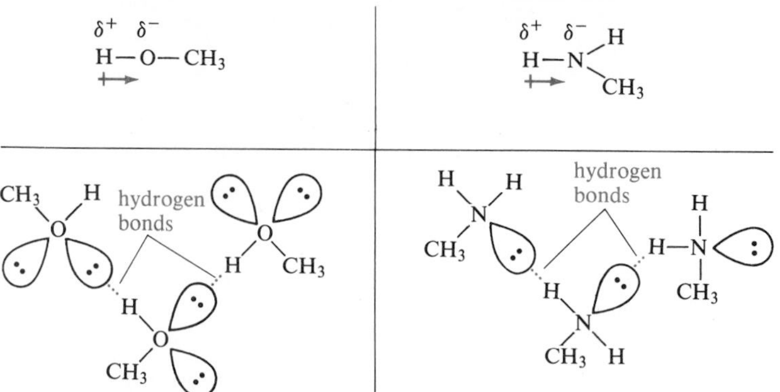

FIGURE 2-25 Hydrogen bonding is a strong intermolecular attraction between an electrophilic O—H or N—H hydrogen atom and a pair of nonbonding electrons.

a hydrogen bond requires about 5 kcal/mol (20 kJ/mol), compared with about 100 kcal/mol (about 400 kJ/mol) required to break a C—H, N—H, or O—H bond.

Hydrogen bonding has a large effect on the physical properties of organic compounds. The structures and boiling points of ethanol and dimethyl ether, two isomers of molecular formula C_2H_6O, are as follows:

$CH_3—CH_2—OH$	$CH_3—O—CH_3$
ethanol, b.p. 78°C	dimethyl ether, b.p. −25°C

These two isomers have the same size and the same molecular weight; however, ethanol has an O—H hydrogen, and it is extensively hydrogen bonded. Dimethyl ether has no O—H hydrogen, and it cannot form hydrogen bonds. As a result of its hydrogen bonding, ethanol has a boiling point over 100°C higher than that of dimethyl ether.

The effect of N—H hydrogen bonding on boiling points can be seen in the isomers of formula C_3H_9N shown below. Trimethylamine has no N—H hydrogens, and it is not hydrogen bonded. Ethylmethylamine has one N—H hydrogen atom, and the resulting hydrogen bonding raises its boiling point about 34°C above that of trimethylamine. Propylamine, with two N—H hydrogens, is more extensively hydrogen bonded and has the highest boiling point of these three isomers.

$CH_3—\ddot{N}(CH_3)—CH_3$	$CH_3CH_2—\ddot{N}(H)—CH_3$	$CH_3CH_2CH_2—\ddot{N}(H)—H$
trimethylamine, b.p. 3.5°C	ethylmethylamine, b.p. 37°C	propylamine, b.p. 49°C

PROBLEM 2-15

Draw the hydrogen bonding that takes place between

(a) two molecules of ethanol.
(b) two molecules of propylamine.

SOLVED PROBLEM 2-8

Rank the following compounds in order of increasing boiling points. Explain the reasons for your chosen order.

$$\begin{array}{c} CH_3 \\ | \\ CH_3-C-CH_3 \\ | \\ CH_3 \end{array}$$

neopentane   *n*-hexane   2,3-dimethylbutane

OH

1-pentanol   2-methyl-2-butanol

SOLUTION Except for neopentane, these compounds have similar molecular weights. Neopentane is the lightest, and it is a compact spherical structure that minimizes van der Waals attractions. Neopentane is the lowest-boiling compound.

Neither *n*-hexane nor 2,3-dimethylbutane is hydrogen bonded, so they will be next higher in boiling points. Because 2,3-dimethylbutane is more highly branched (and has a lower surface area) than *n*-hexane, 2,3-dimethylbutane will be lower boiling than *n*-hexane. So far, we have

neopentane < 2,3-dimethylbutane < *n*-hexane < the others

The two remaining compounds are both hydrogen bonded, and 1-pentanol has more area for van der Waals forces. Therefore, 1-pentanol should be the highest-boiling compound. We predict the following order:

neopentane <	2,3-dimethylbutane <	*n*-hexane <	2-methyl-2-butanol <	1-pentanol
10°C	58°C	69°C	102°C	138°C

The actual boiling points are given here to show that our prediction is correct.

PROBLEM 2-16

For each pair of compounds, circle the compound you expect to have the higher boiling point. Explain your reasoning.

(a) $(CH_3)_3C—C(CH_3)_3$ and $(CH_3)_2CH—CH_2CH_2—CH(CH_3)_2$
(b) $CH_3(CH_2)_6CH_3$ and $CH_3(CH_2)_5CH_2OH$
(c) $HOCH_2—(CH_2)_4—CH_2OH$ and $(CH_3)_3CCH(OH)CH_3$
(d) $(CH_3CH_2CH_2)_2NH$ and $(CH_3CH_2)_3N$
(e) NH and $—NH_2$

2-13 POLARITY EFFECTS ON SOLUBILITIES

In addition to affecting boiling points and melting points, intermolecular forces determine the solubility properties of organic compounds. The general rule is that polar substances dissolve in polar solvents, and nonpolar substances dissolve in nonpolar solvents. We discuss the reasons for this "*like dissolves like*" rule now, and then apply the rule in later chapters as we discuss the solvent properties of organic compounds.

There are four different cases to consider in discussing the effects of polarity on solubility: (1) a polar solute with a polar solvent, (2) a polar solute with a nonpolar solvent, (3) a nonpolar solute with a nonpolar solvent, and (4) a nonpolar solute with a polar solvent. We will use sodium chloride and water as our polar solute and solvent, remembering that very polar organic compounds have similar solubility properties.

Polar solute in a polar solvent (dissolves) When you think about sodium chloride dissolving in water, it seems remarkable that the oppositely charged ions could be separated from each other. A great deal of energy is required to separate these ions. A polar solvent (such as water) can separate the ions because it *solvates* them (Fig. 2-26). If water is the solvent, the solvation process is called *hydration.* As the salt dissolves, water molecules surround each ion, with the appropriate end of the water dipole moment next to the ion. In the case of the positive (sodium) ion, the oxygen atom of the water molecule approaches. The negative ions (chloride) are approached by the hydrogen atoms of the water molecules.

Because water molecules are very polar, a large amount of energy is released in the hydration of the sodium and chloride ions. This energy is nearly sufficient to overcome the lattice energy of the crystal. The salt dissolves, partly as a result of the strong solvation by water molecules and partly as a result of the increase in entropy (randomness or freedom of movement) when it dissolves.

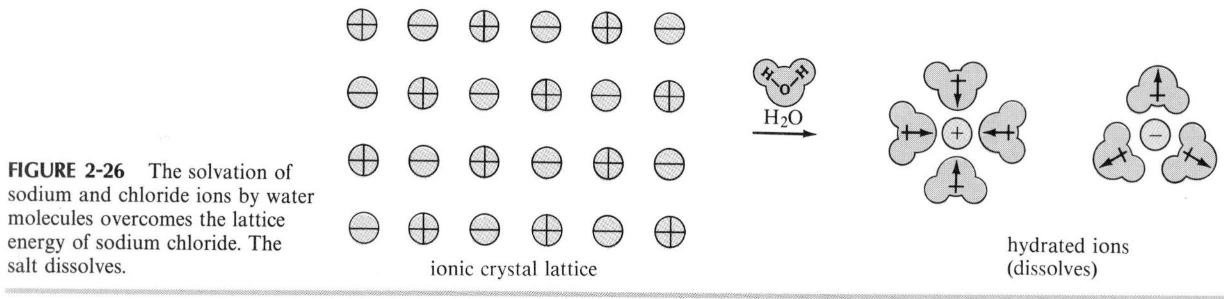

FIGURE 2-26 The solvation of sodium and chloride ions by water molecules overcomes the lattice energy of sodium chloride. The salt dissolves.

Polar solute in a nonpolar solvent (does not dissolve) If you stir sodium chloride with a nonpolar solvent such as turpentine or gasoline, you will find that the salt does not dissolve (Fig. 2-27). The nonpolar molecules of these solvents do not solvate ions very strongly, and they cannot overcome the large lattice energy of the salt crystal. This is a case where the attractions of the ions in the solid for each other are much greater than their attractions for the solvent.

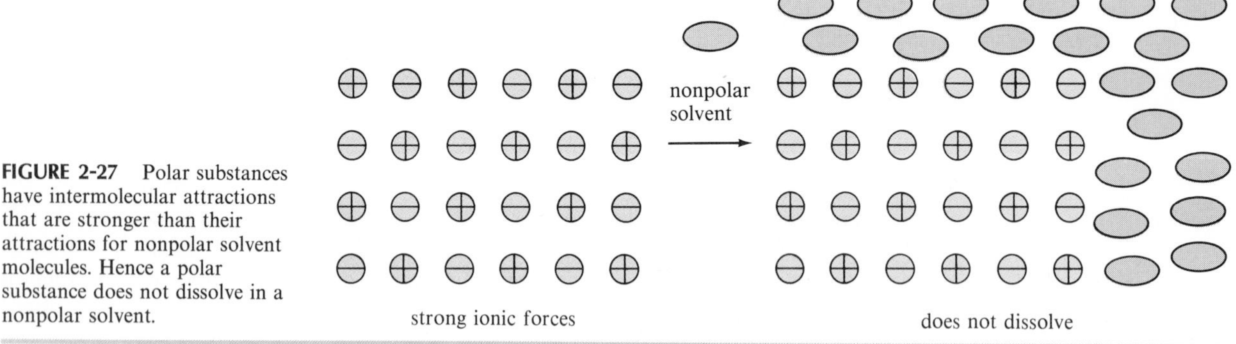

FIGURE 2-27 Polar substances have intermolecular attractions that are stronger than their attractions for nonpolar solvent molecules. Hence a polar substance does not dissolve in a nonpolar solvent.

Nonpolar solute in a nonpolar solvent (dissolves) Paraffin "wax" dissolves easily in gasoline. Both paraffin and gasoline consist of nonpolar hydrocarbons (Fig. 2-28). The molecules of a nonpolar substance (paraffin) are weakly attracted to each other, and these van der Waals attractions are easily overcome by van der

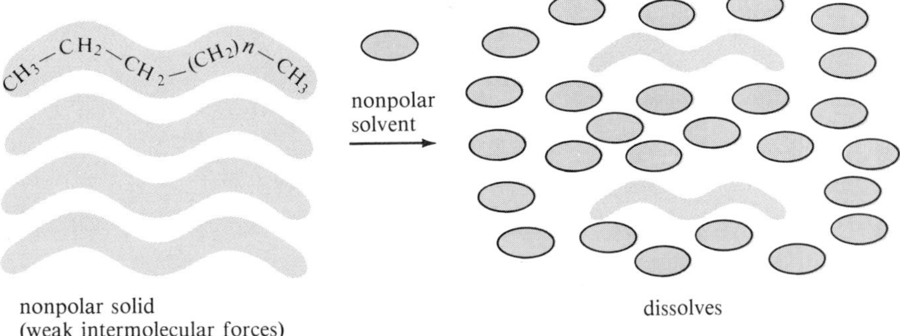

FIGURE 2-28 The weak intermolecular attractions of a nonpolar substance are overcome by the weak attractions for a nonpolar solvent. The nonpolar substance dissolves.

Waals attractions with the solvent. Although there is little change in energy when the nonpolar substance dissolves in a nonpolar solvent, there is a large increase in entropy.

Nonpolar solute in a polar solvent (does not dissolve) Everyone who does home canning knows that a nonpolar solid such as paraffin does not dissolve in a polar solvent such as water. Why not? The nonpolar molecules are only weakly attracted to each other, and little energy is required to separate them. The problem is that the water molecules are strongly attracted to each other by their hydrogen bonding. If a nonpolar paraffin molecule were to dissolve, it would have to displace some of these hydrogen bonds, yet there is almost no energy released from solvation of the nonpolar molecule. In effect, the hydrogen-bonded network of water molecules *excludes* the paraffin molecules (Fig. 2-29).

Figures 2-26 through 2-29 show why the saying "like dissolves like" is generally true. Polar substances dissolve in polar solvents, and nonpolar substances dissolve in nonpolar solvents. This general rule also applies to the mixing of liquids. Everyone knows that water and gasoline (or oil) do not mix. Gasoline and oil are both nonpolar hydrocarbons, however, and they mix freely with each other. They do not dissolve in water because they would have to break up the hydrogen bonds of the water molecules.

Ethanol is a polar molecule, and it is miscible with water: It mixes freely with water in all proportions. Ethanol has an O—H group that forms hydrogen bonds

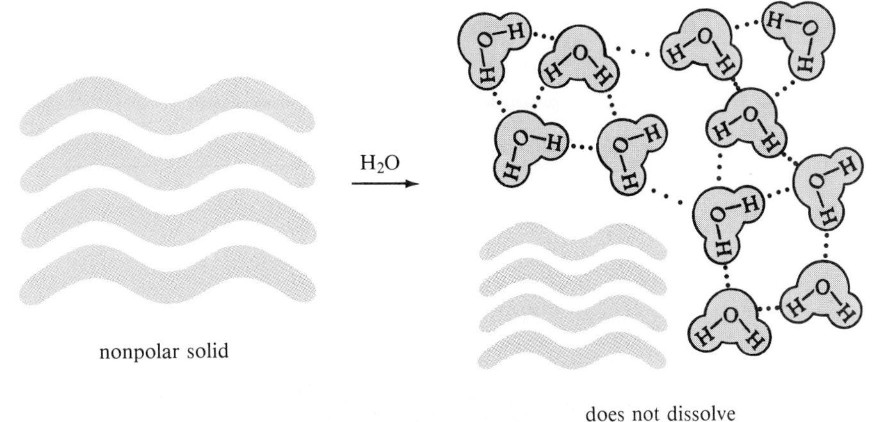

FIGURE 2-29 If a nonpolar molecule were to dissolve in water, it would break up the hydrogen bonds between the water molecules. Therefore, nonpolar substances do not dissolve in water.

with the water molecules. When ethanol dissolves in water, it forms new ethanol-water hydrogen bonds to replace the water-water hydrogen bonds that are broken:

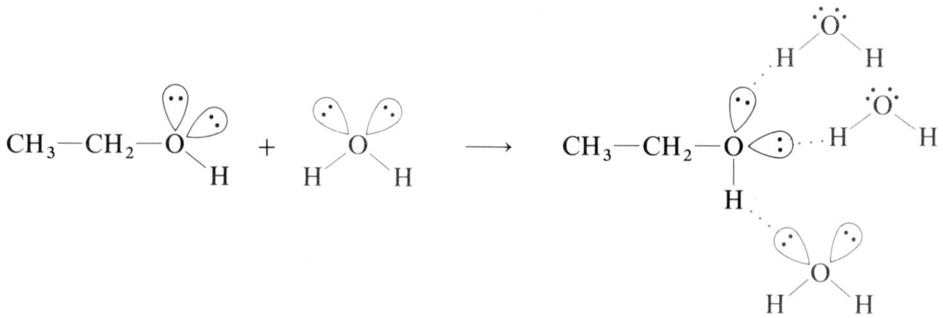

$$CH_3-CH_2-\ddot{O}-H + H-\ddot{O}-H \longrightarrow CH_3-CH_2-\ddot{O}(-H\cdots OH_2)\cdots H-OH$$

In the next section we will see many different kinds of organic compounds with a wide variety of "functional groups." As you encounter these new compounds, you should look to see whether the molecules are polar or nonpolar and whether they can engage in hydrogen bonding.

PROBLEM 2-17

Circle the member of each pair that is more soluble in water.

(a) $CH_3CH_2OCH_2CH_3$ or $CH_3CH_2CH_2CH_2CH_3$
(b) $CH_3CH_2NHCH_3$ or $CH_3CH_2CH_2CH_3$
(c) CH_3CH_2OH or $CH_3CH_2CH_2CH_2OH$

(d) 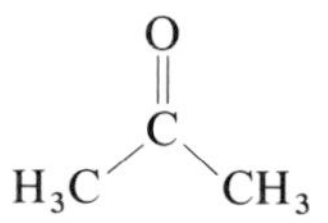 $H_3C-C(=O)-CH_3$ or cyclohexanone

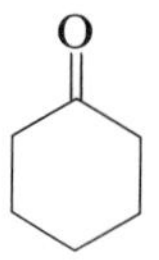

2-14 HYDROCARBONS

There are many different kinds of organic compounds. In future chapters we will go through each major class carefully and in detail. The various kinds of organic compounds are briefly described here, so that you will recognize them as you encounter them. For the purposes of this brief survey, we divide the simple organic compounds into three classes: (1) hydrocarbons, (2) compounds containing oxygen, and (3) compounds containing nitrogen. As their name implies, the **hydrocarbons**

TABLE 2-2
Correspondence of prefixes and numbers of carbon atoms

Alkane name	*Number of carbons*	*Alkane name*	*Number of carbons*
*meth*ane	1	*hex*ane	6
*eth*ane	2	*hept*ane	7
*prop*ane	3	*oct*ane	8
*but*ane	4	*non*ane	9
*pent*ane	5	*dec*ane	10

CH_4 methane; CH_3-CH_3 ethane; $CH_3-CH_2-CH_3$ propane; $CH_3-CH_2-CH_2-CH_3$ butane; $CH_3-CH(CH_3)-CH_3$ isobutane are compounds composed exclusively of carbon and hydrogen. The major classes of hydrocarbons are the alkanes, the alkenes, the alkynes, and the aromatic hydrocarbons.

2-14A ALKANES

The **alkanes** are hydrocarbons that contain only single bonds. Alkane names generally have the *-ane* suffix, while the first part of the name gives the number of carbon atoms. Table 2-2 shows how the prefixes in the names correspond with the number of carbon atoms.

The **cycloalkanes** are a special class of alkanes, those that are in the form of a ring. Figure 2-30 shows some examples of cycloalkanes.

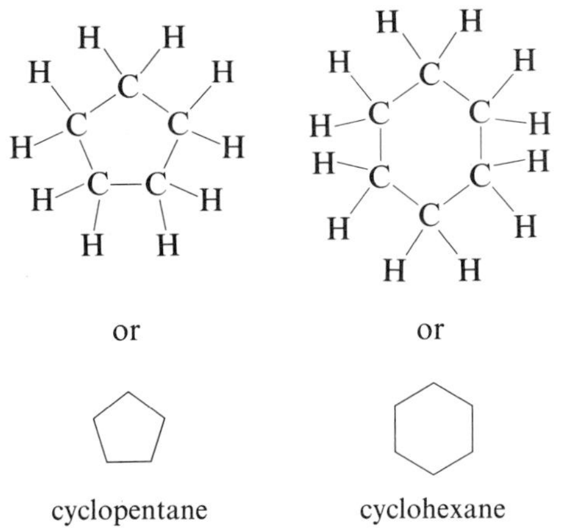

or　　　　or

cyclopentane　　　　cyclohexane

FIGURE 2-30 Cycloalkanes are alkanes in the form of a ring.

Alkanes are the major components of heating gases (natural gas and liquefied petroleum gas), gasoline, motor oil, fuel oil, and paraffin "wax." Other than combustion, alkanes undergo few reactions. In fact, when a molecule contains an alkane portion and a nonalkane portion, we often ignore the presence of the alkane portion because it is relatively unreactive.

The reactive nonalkane part of the molecule is called the **functional group,** since that is where reactions usually occur. Most nonalkane compounds are characterized and classified by the functional group or groups they contain.

In molecules with functional groups, the molecule may be represented as the functional group with alkyl groups attached. An **alkyl group** is an alkane portion of a molecule, with one hydrogen atom removed to allow bonding to the functional group. Figure 2-31 shows how an alkyl group is named, using a substituted cycloalkane as the example.

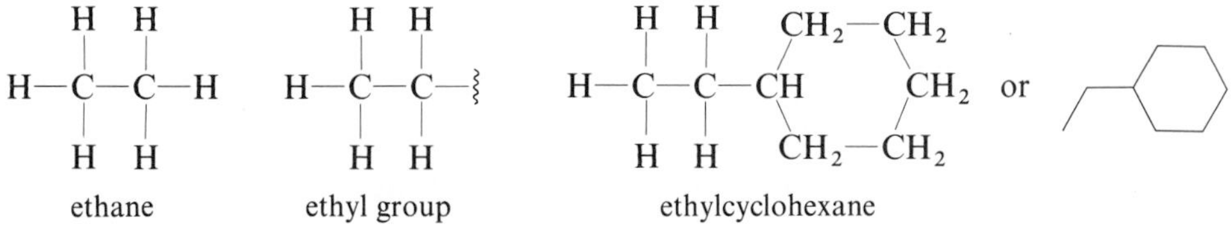

ethane　　　ethyl group　　　ethylcyclohexane

FIGURE 2-31 Alkyl groups are named like the alkanes they are derived from, with a *-yl* suffix.

In many cases we are concerned only with the structure of one part of a molecule; the rest of the structure is unimportant. In these cases we often use the symbol R as a substituent. The R group is simply any alkyl group. We presume that the exact nature of the R group is not terribly important.

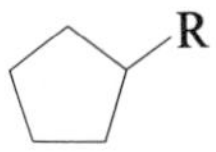

[cyclopentane]—R might be [cyclopentane]—CH_3 or [cyclopentane]—$CH(CH_3)$—CH_3 or other compounds.

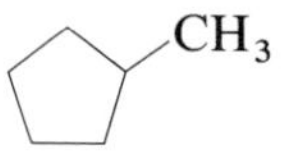

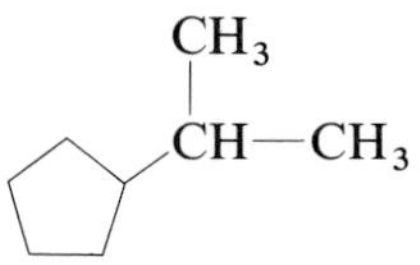

an alkylcyclopentane — methylcyclopentane — isopropylcyclopentane

2-14B ALKENES

Alkenes are hydrocarbons that contain carbon-carbon double bonds. A carbon-carbon double bond is the most reactive part of an alkene, so we say that the double bond is the *functional group* of the alkene. Alkene names end in the *-ene* suffix, as shown by the following examples:

$CH_2{=}CH_2$ ethene (ethylene)

$CH_2{=}CH{-}CH_3$ propene (propylene)

$CH_2{=}CH{-}CH_2{-}CH_3$ 1-butene

$CH_3{-}CH{=}CH{-}CH_3$ 2-butene

Carbon-carbon double bonds cannot rotate, and many alkenes show geometric (cis-trans) isomerism (Section 2-10). The following are the cis-trans isomers of some simple alkenes:

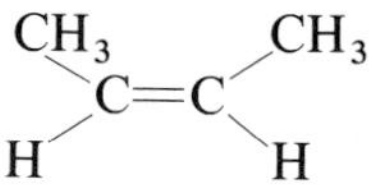

$(CH_3)(H)C{=}C(CH_3)(H)$	$(CH_3)(H)C{=}C(H)(CH_3)$	$(Cl)(H)C{=}C(Cl)(H)$	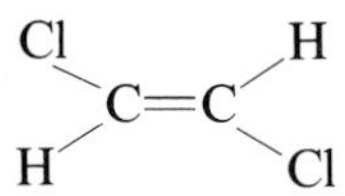 $(Cl)(H)C{=}C(H)(Cl)$
cis-2-butene	*trans*-2-butene	*cis*-1,2-dichloroethene	*trans*-1,2-dichloroethene

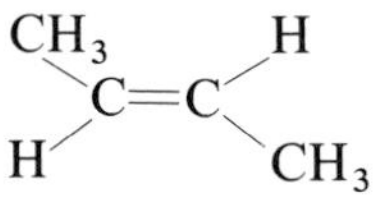

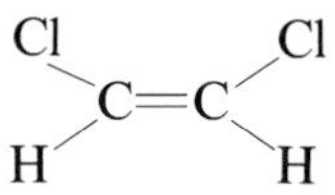

Cycloalkenes are also common. Unless the rings are very large, cycloalkenes are always the cis isomers, and the term *cis* is omitted from the names. In a large ring, a trans double bond may occur, giving a trans-cycloalkene.

cyclopentene 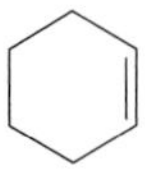 — cyclohexene 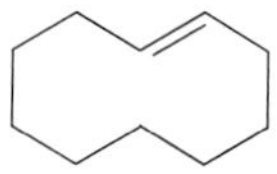 — *trans*-cyclodecene

2-14C ALKYNES

Alkynes are hydrocarbons with carbon-carbon triple bonds. The $C{\equiv}C$ triple bond is the characteristic functional group of the alkynes. Alkyne names generally have the *-yne* suffix, although some of their common names (*acetylene,* for example) do not conform to this rule. The triple bond is linear, so there is no possibility of geometric isomerism in alkynes.

$H{-}C{\equiv}C{-}H$ ethyne (acetylene)

$H{-}C{\equiv}C{-}CH_3$ propyne (methylacetylene)

$H{-}C{\equiv}C{-}CH_2{-}CH_3$ 1-butyne

$CH_3{-}C{\equiv}C{-}CH_3$ 2-butyne

In an alkyne, four atoms must be in a straight line. These four collinear atoms are not easily bent into a ring, and cycloalkynes are rare. Cycloalkynes are stable only if the ring is fairly large.

2-14D AROMATIC HYDROCARBONS

The compounds below may look like cycloalkenes, but their properties are quite different from those of the simple alkenes. These **aromatic hydrocarbons** are all derivatives of *benzene,* a six-membered ring with three double bonds. This bonding arrangement is unusually stable. The reasons for this unusual stability are presented in Chapter 16.

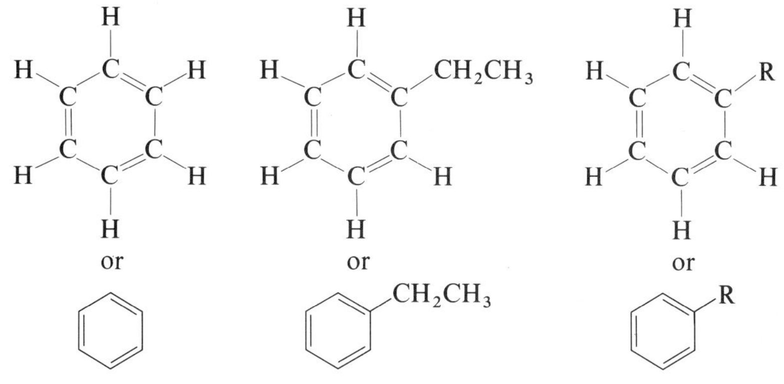

benzene

ethylbenzene (CH_2CH_3)

an alkylbenzene (R)

PROBLEM 2-18

Classify the following hydrocarbons and draw a complete structural formula for each one. A compound may fit into more than one of the following possible classifications:

alkane	cycloalkane	aromatic hydrocarbon
alkene	cycloalkene	
alkyne	cycloalkyne	

(a) $CH_3(CH_2)_3CH_3$ (b) $CH_3CHCHCH_2CH_3$ (c) $CH_3CCCH_2CH_2CH_3$

(d) 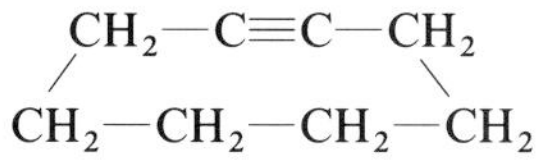 $CH_2—C≡C—CH_2$ / $CH_2—CH_2—CH_2—CH_2$ (e) 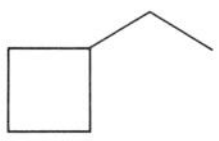 (f) 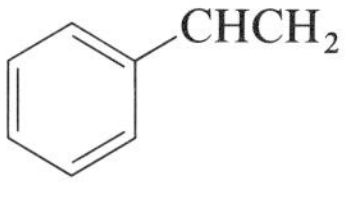 $CHCH_2$

(g) 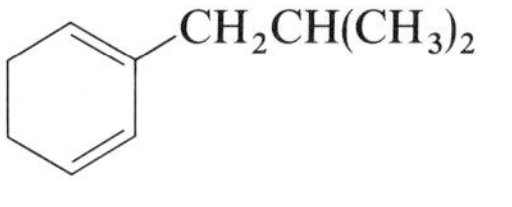 $CH_2CH(CH_3)_2$ (h) 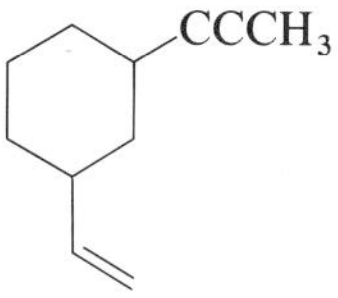 $CCCH_3$ (i)

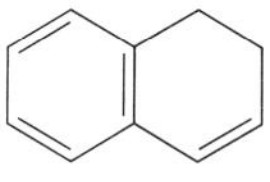

2-15 ORGANIC COMPOUNDS CONTAINING OXYGEN

Many organic compounds contain oxygen atoms bonded to alkyl groups. The major classes of oxygen-containing compounds are the alcohols, the ethers, the ketones and aldehydes, and the carboxylic acids and their derivatives.

2-15A ALCOHOLS

Alcohols are organic compounds that contain the **hydroxyl group** (—OH) as their functional group. The general formula for an alcohol is R—OH. Because the hydroxyl group is strongly polar and can participate in hydrogen bonding, alcohols are among the most polar organic compounds. Some of the simple alcohols like ethanol and methanol are miscible (soluble in all proportions) with water. Four of the most common alcohols are given below. Notice that the suffix of each name is the *-ol* suffix from the word "alcoh*ol*."

$CH_3—OH$	$CH_3—CH_2—OH$	$CH_3—CH_2—CH_2—OH$	$CH_3—CH(OH)—CH_3$
methanol (methyl alcohol)	ethanol (ethyl alcohol)	1-propanol (*n*-propyl alcohol)	2-propanol (isopropyl alcohol)

Alcohols are common organic compounds. Methyl alcohol (methanol), also known as "wood alcohol," is used as an industrial solvent and as an automobile racing fuel. Ethyl alcohol (ethanol) is sometimes called "grain alcohol" because it is produced by the fermentation of grain or almost any other organic material. "Isopropyl alcohol" is the common name for 2-propanol, used as "rubbing alcohol."

2-15B ETHERS

Ethers are composed of two alkyl groups bonded to an oxygen atom. The general formula for an ether is R—O—R′. (The symbol R′ represents another alkyl group, except that the second alkyl group might be different from the first.) Like alcohols, ethers are much more polar than hydrocarbons. Ethers have no O—H hydrogens, however, and they cannot hydrogen bond with themselves. Ether names are often formed from the names of the alkyl groups and the word "ether." Diethyl ether is the common "ether" used for surgical anesthesia and for starting engines in cold weather. The following are three simple ethers.

$CH_3—O—CH_3$	$CH_3—CH_2—O—CH_2—CH_3$	$C_6H_{11}—O—CH_3$
dimethyl ether	diethyl ether	cyclohexyl methyl ether

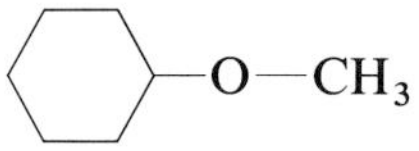

2-15C ALDEHYDES AND KETONES

The functional group for both aldehydes and ketones is the **carbonyl group,** C=O. A **ketone** has two alkyl groups bonded to the carbonyl group; an **aldehyde** carbonyl group has one alkyl group and a hydrogen atom. Ketone names generally have the *-one* suffix; aldehyde names use either the *-al* suffix or the *-aldehyde* suffix.

The carbonyl group is strongly polar, and most ketones and aldehydes are more soluble in water than are comparable hydrocarbons. In fact, both acetone and acetaldehyde are miscible with water. Acetone, often used as nail polish remover, is a common solvent with low toxicity.

Examples of ketones

$R—C(=O)—R'$	$CH_3—C(=O)—CH_3$	$CH_3—C(=O)—CH_2—CH_3$	$C_6H_{10}O$
RCOR′, a ketone	2-propanone (acetone)	2-butanone (methyl ethyl ketone)	cyclohexanone

Examples of aldehydes

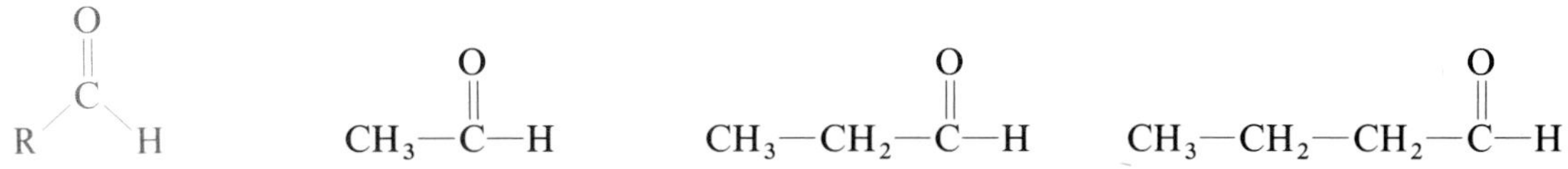

$$\underset{\text{RCHO, an aldehyde}}{\mathrm{R{-}C({=}O){-}H}} \qquad \underset{\text{ethanal (acetaldehyde)}}{\mathrm{CH_3{-}\overset{\overset{\displaystyle O}{\|}}{C}{-}H}} \qquad \underset{\text{propanal (propionaldehyde)}}{\mathrm{CH_3{-}CH_2{-}\overset{\overset{\displaystyle O}{\|}}{C}{-}H}} \qquad \underset{\text{butanal (butyraldehyde)}}{\mathrm{CH_3{-}CH_2{-}CH_2{-}\overset{\overset{\displaystyle O}{\|}}{C}{-}H}}$$

2-15D CARBOXYLIC ACIDS

Carboxylic acids contain the **carboxyl group,** —COOH, as their functional group. The general formula for a carboxylic acid is R—COOH (or RCO_2H). The carboxyl group is a combination of a carbonyl group and a hydroxyl group, but this combination has different properties from those of typical ketones and alcohols. Carboxylic acids owe their acidity (pK_a of about 5) to the resonance-stabilized *carboxylate anions* formed by deprotonation. The following reaction shows the dissociation of a general carboxylic acid.

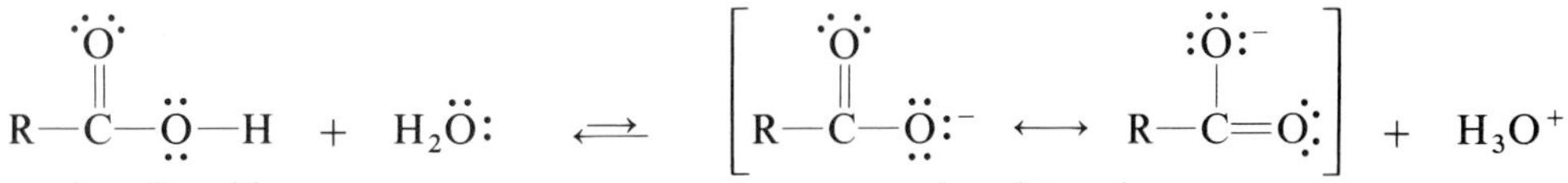

$$\underset{\text{carboxylic acid}}{\mathrm{R{-}\overset{\overset{\displaystyle \ddot{O}}{\|}}{C}{-}\ddot{O}{-}H}} + \mathrm{H_2\ddot{O}:} \rightleftharpoons \underset{\text{carboxylate anion}}{\left[\mathrm{R{-}\overset{\overset{\displaystyle \ddot{O}}{\|}}{C}{-}\ddot{O}:^-} \longleftrightarrow \mathrm{R{-}\overset{\overset{\displaystyle :\ddot{O}:^-}{|}}{C}{=}\ddot{O}:}\right]} + \mathrm{H_3O^+}$$

Historical names are commonly used for carboxylic acids. Formic acid was first isolated from ants, genus *Formica.* Acetic acid, found in vinegar, gets its name from the Latin word for "sour" (*acetum*). Propionic acid gives the tangy flavor to "sharp" cheeses, and butyric acid provides the pungent aroma of rancid butter. The following are some common carboxylic acids.

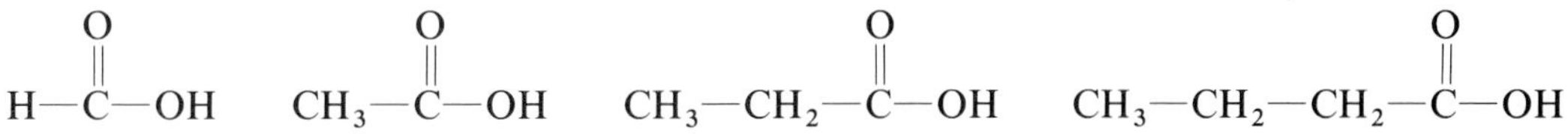

$$\underset{\substack{\text{methanoic acid}\\\text{(formic acid)}}}{\mathrm{H{-}\overset{\overset{\displaystyle O}{\|}}{C}{-}OH}} \qquad \underset{\substack{\text{ethanoic acid}\\\text{(acetic acid)}}}{\mathrm{CH_3{-}\overset{\overset{\displaystyle O}{\|}}{C}{-}OH}} \qquad \underset{\substack{\text{propanoic acid}\\\text{(propionic acid)}}}{\mathrm{CH_3{-}CH_2{-}\overset{\overset{\displaystyle O}{\|}}{C}{-}OH}} \qquad \underset{\substack{\text{butanoic acid}\\\text{(butyric acid)}}}{\mathrm{CH_3{-}CH_2{-}CH_2{-}\overset{\overset{\displaystyle O}{\|}}{C}{-}OH}}$$

Like the ketones, aldehydes, and alcohols, carboxylic acids are strongly polar, and they are much more soluble in water than comparable hydrocarbons. All four of the carboxylic acids shown above are miscible with water.

PROBLEM 2-19

Draw a complete structural formula and classify each of the following compounds. The possible classifications are

alcohol	ketone	carboxylic acid
ether	aldehyde	

(a) CH_3CH_2CHO (b) $CH_3CH_2CH(OH)CH_3$ (c) $CH_3COCH_2CH_3$

(d) $CH_3CH_2OCH_2CH_3$ (e) 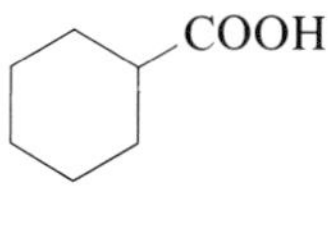 cyclohexane–COOH (f) 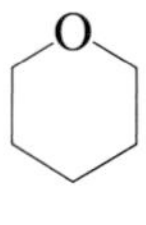 tetrahydropyran (six-membered ring containing O)

(g) 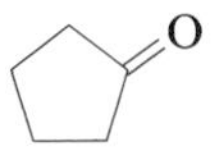 cyclopentanone (=O) (h) 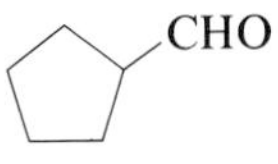 cyclopentane–CHO (i) 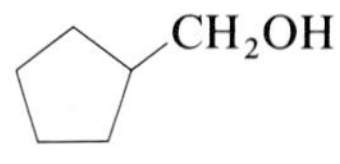 cyclopentane–CH_2OH

2-15E CARBOXYLIC ACID DERIVATIVES

Several related functional groups can be formed from carboxylic acids. Each contains the carbonyl group bonded to an oxygen or other electron-withdrawing element. Among these functional groups are the **acid chlorides,** the **esters,** and the **amides.** All these groups can be converted back to carboxylic acids by acidic or basic hydrolysis.

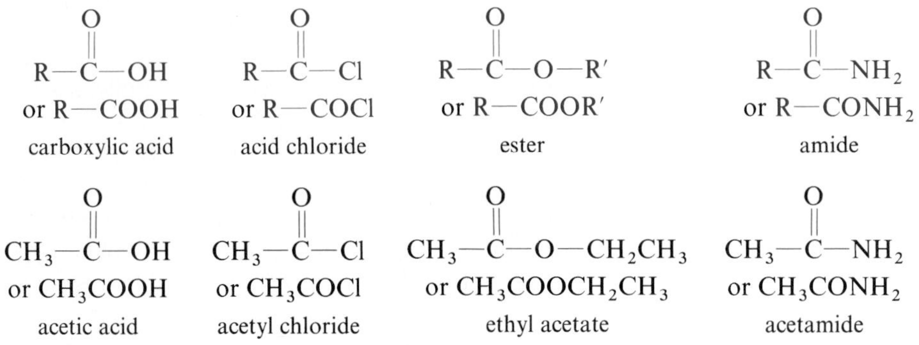

$$\underset{\text{carboxylic acid}}{\mathrm{R{-}C({=}O){-}OH} \text{ or } \mathrm{R{-}COOH}} \qquad \underset{\text{acid chloride}}{\mathrm{R{-}C({=}O){-}Cl} \text{ or } \mathrm{R{-}COCl}} \qquad \underset{\text{ester}}{\mathrm{R{-}C({=}O){-}O{-}R'} \text{ or } \mathrm{R{-}COOR'}} \qquad \underset{\text{amide}}{\mathrm{R{-}C({=}O){-}NH_2} \text{ or } \mathrm{R{-}CONH_2}}$$

$$\underset{\text{acetic acid}}{\mathrm{CH_3{-}C({=}O){-}OH} \text{ or } \mathrm{CH_3COOH}} \qquad \underset{\text{acetyl chloride}}{\mathrm{CH_3{-}C({=}O){-}Cl} \text{ or } \mathrm{CH_3COCl}} \qquad \underset{\text{ethyl acetate}}{\mathrm{CH_3{-}C({=}O){-}O{-}CH_2CH_3} \text{ or } \mathrm{CH_3COOCH_2CH_3}} \qquad \underset{\text{acetamide}}{\mathrm{CH_3{-}C({=}O){-}NH_2} \text{ or } \mathrm{CH_3CONH_2}}$$

2-16 ORGANIC COMPOUNDS CONTAINING NITROGEN

Nitrogen is another element that is often found in the functional groups of organic compounds. The most common "nitrogenous" organic compounds are the amines, the amides, and the nitriles.

2-16A AMINES

Amines are alkylated derivatives of ammonia. Like ammonia, amines are basic. Because of their basicity, naturally occurring amines are often called *alkaloids.* Simple amines are named by naming the alkyl groups bonded to nitrogen and adding the word "amine." The structures of some simple amines and also the structure of nicotine, a toxic alkaloid found in tobacco leaves, are shown below:

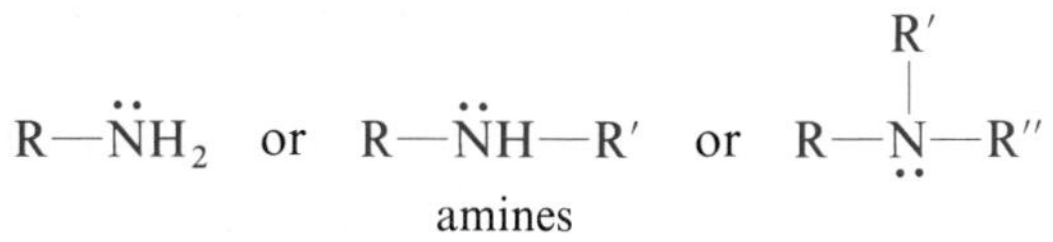

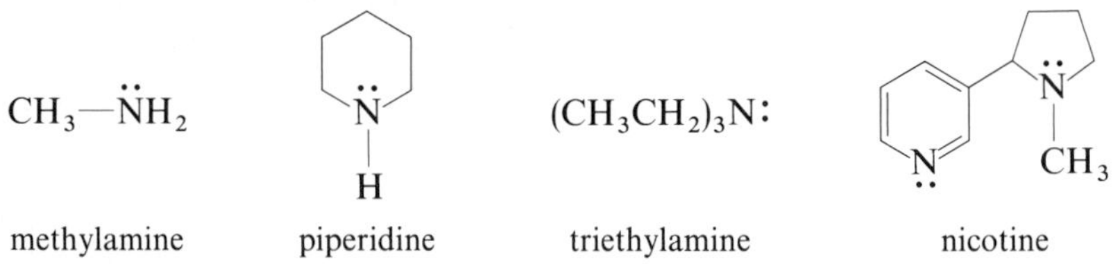

$$\underset{\text{amines}}{\mathrm{R{-}\ddot{N}H_2} \text{ or } \mathrm{R{-}\ddot{N}H{-}R'} \text{ or } \mathrm{R{-}\underset{..}{N}(R'){-}R''}}$$

$$\underset{\text{methylamine}}{\mathrm{CH_3{-}\ddot{N}H_2}} \qquad \underset{\text{piperidine}}{\text{(ring)}\ddot{N}{-}H} \qquad \underset{\text{triethylamine}}{\mathrm{(CH_3CH_2)_3N}\!:} \qquad \underset{\text{nicotine}}{\text{pyridine ring } \ddot{N} \text{ bonded to pyrrolidine ring } \ddot{N}{-}\mathrm{CH_3}}$$

2-16B AMIDES

Amides are acid derivatives that result from a combination of an acid with ammonia or an amine. Proteins have the structure of long-chain, complicated amides.

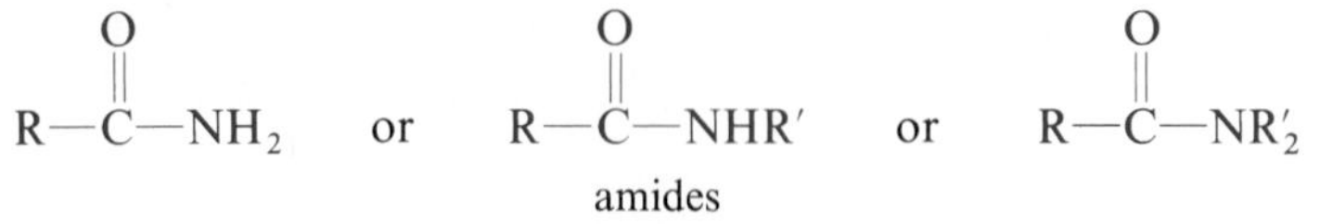

$$\underset{\text{amides}}{\mathrm{R{-}C({=}O){-}NH_2} \text{ or } \mathrm{R{-}C({=}O){-}NHR'} \text{ or } \mathrm{R{-}C({=}O){-}NR'_2}}$$

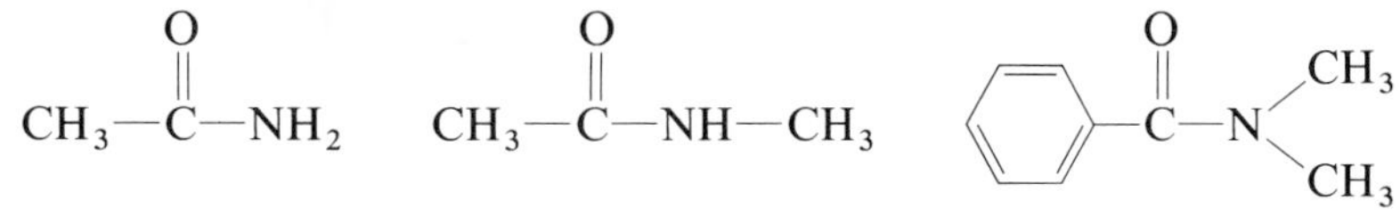

$$CH_3-\overset{O}{\overset{\|}{C}}-NH_2 \qquad CH_3-\overset{O}{\overset{\|}{C}}-NH-CH_3 \qquad C_6H_5-\overset{O}{\overset{\|}{C}}-N(CH_3)_2$$

acetamide — *N*-methylacetamide — *N*,*N*-dimethylbenzamide

2-16C NITRILES

A **nitrile** is a compound containing the **cyano** group, $-C{\equiv}N$. We have used the cyano group as an example of *sp* hybridized bonding in Section 2.6.

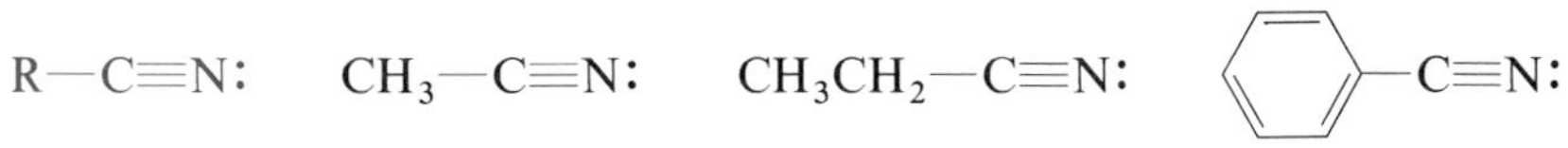

$$R-C{\equiv}N: \qquad CH_3-C{\equiv}N: \qquad CH_3CH_2-C{\equiv}N: \qquad C_6H_5-C{\equiv}N:$$

a nitrile — acetonitrile — propionitrile — benzonitrile

All these classes of compounds are summarized in the table of **Common Organic Compounds and Functional Groups** given on the front inside cover for convenient reference.

PROBLEM 2-20

Draw a complete structural formula and classify each of the following compounds.

(a) $CH_3CH_2CONHCH_3$ (b) $(CH_3CH_2)_2NH$ (c) $(CH_3)_2CHCOOCH_3$
(d) $(CH_3CH_2)_3CCH_2COCl$ (e) $(CH_3CH_2)_2O$ (f) $CH_3CH_2CH_2CN$

(g) $(CH_3)_3CCH_2CH_2COOH$ (h) 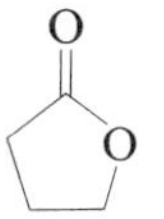 [five-membered ring with ring O and C=O adjacent] (i) [five-membered ring with O and C=O]

(j) [six-membered ring with N—CH₃] (k) [six-membered ring with N—CH₃ and adjacent C=O] (l) [six-membered ring N—C(=O)—CH₃]

(m) [six-membered ring with C=O, N—CH₃] (n) [six-membered ring with O and C=O] (o) [cyclohexane—CN] (p) 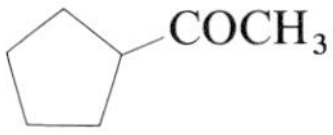 [cyclopentane—COCH₃]

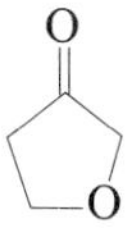

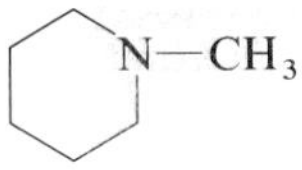

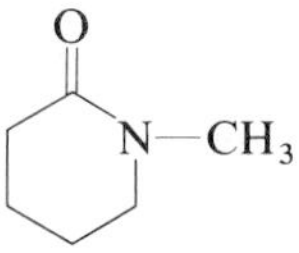

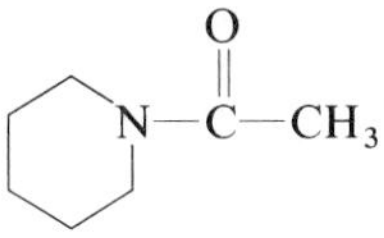

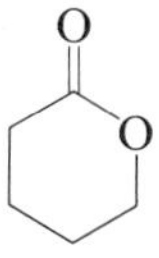

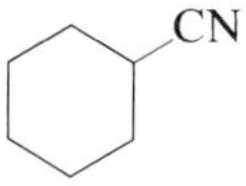

PROBLEM 2-21

Circle the functional groups in the following structures. State to which class (or classes) of compounds the structure belongs.

(a) $CH_2{=}CHCH_2CH_3$ (b) CH_3-O-CH_3 (c) CH_3CHO
(d) $HCONH_2$ (e) CH_3NHCH_3 (f) $R-COOH$

(g) 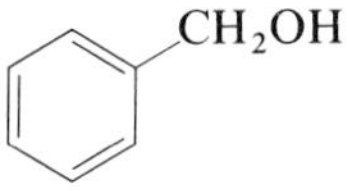 [benzene—CH₂OH] (h) 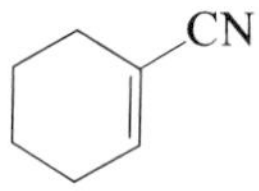 [cyclohexene—CN] (i) 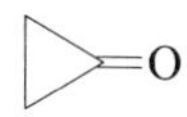 [cyclopropanone, C=O]

(j) 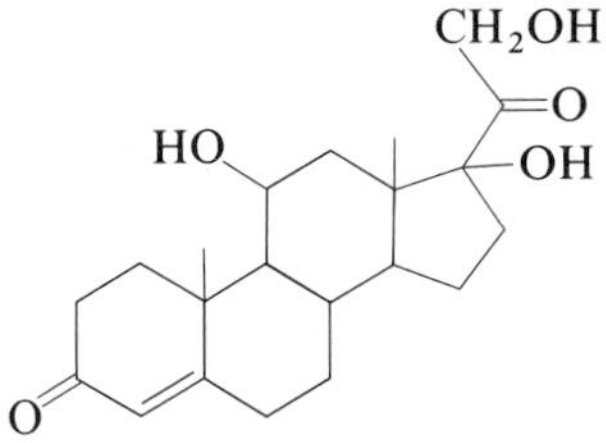 [structure: hydrocortisone, with CH_2OH, =O, OH, HO and O labels]

hydrocortisone

(k) 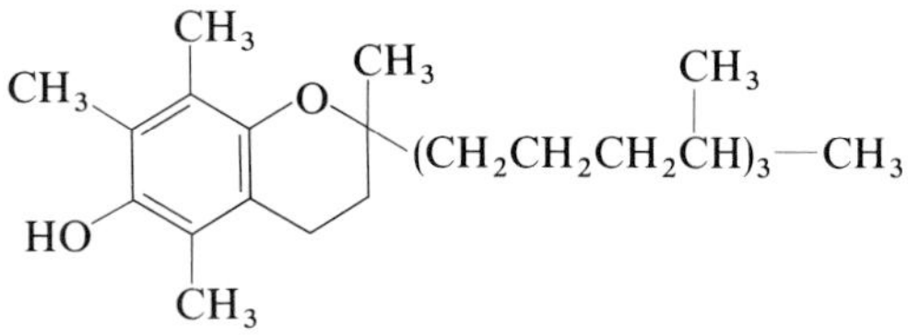 [structure: vitamin E, with CH_3, CH_3, CH_3, HO, O, CH_3, $(CH_2CH_2CH_2CH)_3$—CH_3 labels]

vitamin E

GLOSSARY

acid chloride An acid derivative with a chlorine atom in place of the hydroxyl group. (p. 70)

$$R-\overset{O}{\overset{\|}{C}}-Cl$$

alcohol A compound that contains a hydroxyl group; R—OH. (p. 67)

aldehyde A carbonyl group with one alkyl group and one hydrogen; $R-\overset{O}{\overset{\|}{C}}-H$. (p. 68)

alkyl group A hydrocarbon group with only single bonds; an alkane with one hydrogen removed, to allow bonding to another group; symbolized R. (p. 65)

amide An acid derivative that contains an amine instead of the hydroxyl group of the acid. (p. 70)

$$R-\overset{O}{\overset{\|}{C}}-NH_2 \quad R-\overset{O}{\overset{\|}{C}}-NHR' \quad R-\overset{O}{\overset{\|}{C}}-NR'_2$$

amine An alkylated analog of ammonia; $R-NH_2$, R_2NH, or R_3N. (p. 70)

bond dipole moment A measure of the polarity of an individual bond in a molecule, defined as $\mu = (4.8 \times d \times \delta)$. μ is the dipole moment in debye (10^{-10} esu-Å), d is the bond length in angstrom units, and δ is the effective amount of charge separated, in units of the electronic charge. (p. 55)

carbonyl group The >C=O functional group, as in a ketone or aldehyde. (p. 68)

carboxyl group The —COOH functional group, as in a carboxylic acid. (p. 69)

carboxylic acid A compound that contains the carboxyl group; $R-\overset{O}{\overset{\|}{C}}-OH$. (p. 69)

cyano group The —C≡N functional group, as in a nitrile. (p. 70)

dipole-dipole forces Attractive intermolecular forces resulting from the attraction of the positive and negative ends of the molecular dipole moments of polar molecules. (p. 58)

double bond A bond consisting of four electrons between two nuclei. One pair of electrons forms a sigma bond, and the other pair forms a pi bond. (pp. 42 and 49)

ester An acid derivative with an alkyl group replacing the acid proton; $R-\overset{O}{\overset{\|}{C}}-OR'$. (p. 70)

ether A compound with an oxygen bonded between two alkyl groups; R—O—R′. (p. 68)

functional group The reactive, nonalkane part of an organic molecule. (p. 65)

hybrid atomic orbital A directional orbital formed from a combination of *s* and *p* orbitals on the same atom. Two orbitals are formed by *sp* hybridization, three orbitals by sp^2 hybridization, and four orbitals by sp^3 hybridization. (p. 44)

sp hybrid orbitals give a bond angle of 180° (**linear** geometry).

sp^2 hybrid orbitals give bond angles of 120° (**trigonal** geometry).

sp^3 hybrid orbitals give bond angles of 109.5° (**tetrahedral** geometry).

hydrocarbons Compounds composed exclusively of carbon and hydrogen. (p. 64)

alkanes: Hydrocarbons containing only single bonds.

alkenes: Hydrocarbons containing C═C double bonds.

alkynes: Hydrocarbons containing C≡C triple bonds.

cycloalkanes, cycloalkenes, cycloalkynes: Alkanes, alkenes, and alkynes in the form of a ring.

aromatic hydrocarbons: Hydrocarbons containing a *benzene ring*, a six-membered ring with three double bonds.

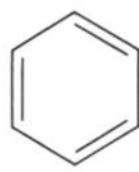

benzene

hydrogen bond A particularly strong intermolecular attraction between a nonbonding pair of electrons and an electrophilic O—H or N—H hydrogen. Hydrogen bonds have bond energies of about 5 kcal/mol (21 kJ/mol), compared with about 100 kcal/mol (about 400 kJ/mol) for typical C—H bonds. (p. 59)

hydroxyl group The —OH functional group, as in an alcohol. (p. 67)

isomers Different compounds with the same molecular formula. (p. 53)

Structural isomers differ in their order of bonding.

Stereoisomers differ only in how their atoms are oriented in space.

Geometric isomers are stereoisomers that differ in their cis-trans arrangement on a ring or a double bond. The cis isomer has similar groups on the same side, while the trans isomer has similar groups on opposite sides.

Stereochemistry is the study of the structure and chemistry of stereoisomers.

ketone A carbonyl group with two alkyl groups attached; R—C(═O)—R′. (p. 68)

linear combination of atomic orbitals (LCAO) Wave functions can add to each other to produce the wave functions of new orbitals. The number of new orbitals generated must always equal the original number of orbitals. (p. 39)

London dispersion forces Intermolecular forces resulting from the attraction of coordinated temporary dipole moments induced in adjacent molecules. (p. 58)

molecular dipole moment The vector sum of the bond dipole moments (and any nonbonding pairs of electrons) in a molecule; a measure of the polarity of a molecule. (p. 56)

molecular orbital (MO) An orbital formed by overlap of atomic orbitals on different atoms. MOs can be either bonding or antibonding, but only the bonding MOs are filled in most stable molecules. (p. 39)

A **bonding molecular orbital** places a large amount of electron density in the bonding region between the nuclei. The energy of an electron in a bonding MO is lower than it is in an atomic orbital.

An **antibonding molecular orbital** places most of the electron density outside the bonding region. The energy of an electron in an antibonding MO is higher than it is in an atomic orbital.

nitrile A compound containing a cyano group, —C≡N. (p. 70)

node In an orbital, a region of space with zero electron density. (p. 38)

pi bond (π bond) A bond formed by sideways overlap of two *p* orbitals. A pi bond has its electron density in two lobes, one above and one below the line joining the nuclei. (p. 42)

sigma bond (σ bond) A bond with most of its electron density centered along the line joining the nuclei; a cylindrically symmetrical bond. Single bonds are normally sigma bonds. (p. 40)

stereochemistry The study of the structure and chemistry of stereoisomers. (p. 53)

triple bond A bond consisting of six electrons between two nuclei. One pair of electrons forms a sigma bond, and the other two pairs form two pi bonds at right angles to each other. (p. 50)

van der Waals forces The attractive forces between neutral molecules, including dipole-dipole forces and London forces. (p. 58)

> **dipole-dipole forces:** The forces between polar molecules resulting from attraction of their permanent dipole moments.
>
> **London forces:** Intermolecular forces resulting from the attraction of coordinated temporary dipole moments induced in adjacent molecules.

wave function (ψ) The mathematical description of an orbital. The *amplitude* of the wave, given by the square of the wave function (ψ^2), is proportional to the electron density. (p. 38)

ESSENTIAL PROBLEM-SOLVING SKILLS IN CHAPTER 2

1. Draw the structure of a single bond, double bond, and triple bond.
2. Predict the hybridization and geometry of an atom in a molecule.
3. Draw a good three-dimensional representation of a given molecule.
4. Identify structural isomers and stereoisomers.
5. Identify polar and nonpolar molecules and predict which ones can engage in hydrogen bonding.
6. Predict general trends in the boiling points and solubilities of compounds, based on their polarity and hydrogen-bonding ability.
7. Identify and draw examples of the general classes of hydrocarbons.
8. Identify and draw examples of the classes of compounds containing oxygen or nitrogen.

STUDY PROBLEMS

2-22. Define and give examples of the following terms.

(a) bonding MO **(b)** antibonding MO **(c)** hybrid atomic orbital **(d)** sigma bond
(e) pi bond **(f)** double bond **(g)** triple bond **(h)** structural isomers
(i) geometric isomers **(j)** stereoisomers **(k)** bond dipole moment **(l)** molecular dipole moment
(m) dipole-dipole forces **(n)** London forces **(o)** hydrogen bonding **(p)** miscible liquids
(q) hydrocarbons **(r)** alkyl group **(s)** functional group **(t)** alkane
(u) alkene **(v)** alkyne **(w)** alcohol **(x)** ether
(y) ketone **(z)** aldehyde **(A)** aromatic hydrocarbon **(B)** carboxylic acid
(C) ester **(D)** amine **(E)** amide **(F)** nitrile

2-23. If the carbon atom in CH_2Cl_2 were flat, there would be two stereoisomers: The carbon atom in CH_2Cl_2 is actually tetrahedral. Make a model of this compound and determine whether there are any stereoisomers of CH_2Cl_2.

```
     H             H
     |             |
  H—C—Cl      Cl—C—Cl
     |             |
     Cl            H
```

2-24. Cyclopropane (C_3H_6, a three-membered ring) is more reactive than most other cycloalkanes.
(a) Draw a Lewis structure for cyclopropane.
(b) Compare the bond angles of the carbon atoms in cyclopropane with those in an acyclic (noncyclic) alkane.
(c) Suggest why cyclopropane is so reactive.

2-25. For each of the following compounds
(1) Give the hybridization and approximate bond angles around each atom except hydrogen.
(2) Draw a three-dimensional diagram, including any lone pairs of electrons.
(a) H_3O^+ **(b)** ^-OH **(c)** NH_2NH_2 **(d)** $(CH_3)_3N$ **(e)** $(CH_3)_4N^+$
(f) CH_3COOH **(g)** $CH_3CH{=}NH$ **(h)** CH_3OH **(i)** CH_2O

2-26. For each of the following compounds
(1) Draw a Lewis structure.
(2) Show the kinds of orbitals that overlap to form each bond.
(3) Give approximate bond angles around each atom except hydrogen.
(a) $(CH_3)_2NH$ **(b)** CH_3CH_2OH **(c)** $CH_3{-}CH{=}N{-}CH_3$
(d) $CH_3{-}CH{=}C(CH_3)_2$ **(e)** $CH_3{-}C{\equiv}C{-}CHO$ **(f)** $H_2N{-}CH_2{-}CN$
(g) 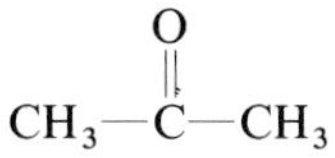 $CH_3{-}C({=}O){-}CH_3$ **(h)** 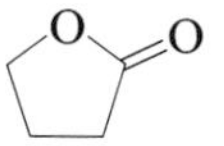 [five-membered ring lactone structure] **(i)** 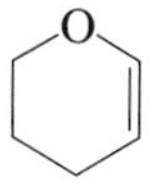 [six-membered ring with O and double bond structure]

2-27. In most amines, the nitrogen atom is sp^3 hybridized, with a pyramidal structure and bond angles close to 109°. In formamide, the nitrogen atom is found to be planar, with bond angles close to 120°. Explain this surprising finding. (*Hint:* Consider resonance structures and the overlap needed in them.)

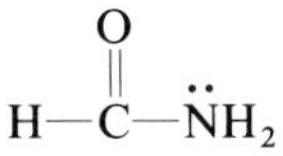

$$H{-}C({=}O){-}\ddot{N}H_2$$

formamide

2-28. Predict the hybridization and geometry of the carbon atoms in the following anion. (Hint: Resonance)

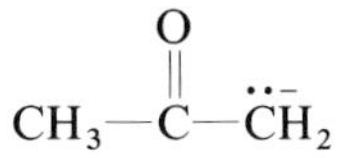

$$CH_3{-}C({=}O){-}\ddot{C}H_2^{-}$$

2-29. Draw orbital pictures of the pi bonding in the following compounds.
(a) CH_3COCH_3 **(b)** HCN **(c)** $CH_2{=}CH{-}CH{=}CH{-}CN$ **(d)** $CH_3{-}C{\equiv}C{-}CHO$

2-30. **(a)** Draw the structure of *cis*-$CH_3{-}CH{=}CH{-}CH_2CH_3$ showing the pi bond with its proper geometry.
(b) Circle the six coplanar atoms in this compound.
(c) Draw the trans isomer and circle the coplanar atoms. Are there still six?
(d) Circle the coplanar atoms in the following structure.

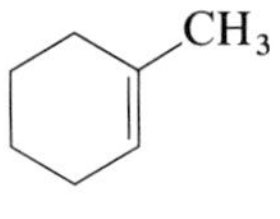

CH_3

2-31. In 2-pentyne, $CH_3CCCH_2CH_3$, there are four atoms in a straight line. Use dashed lines and wedges to draw a three-dimensional representation of this molecule, and circle the four atoms that are in a straight line.

2-32. Which of the following compounds show geometric isomerism? Draw the cis and trans isomers of the ones that do.
(a) $CH_2{=}C(CH_3)_2$ **(b)** $CH_3CH{=}CHCH_3$ **(c)** $CH_3{-}C{\equiv}C{-}CH_3$
(d) cyclopentene, [cyclopentene structure] **(e)** 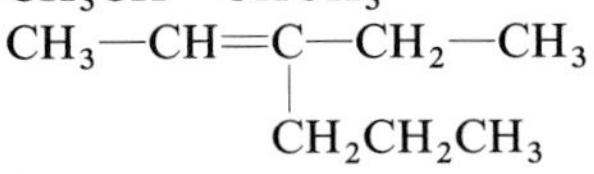 $CH_3{-}CH{=}C(CH_2CH_2CH_3){-}CH_2{-}CH_3$

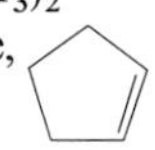

2-33. Give the relationships between the following pairs of structures. The possible relationships are the following: same compound; geometric isomers; structural isomers; not isomers (different molecular formula).

(a) $CH_3CH_2CH_2CH_3$ and $(CH_3)_3CH$ **(b)** $CH_2{=}CH{-}CH_2Cl$ and $CHCl{=}CH{-}CH_3$

(c) CH_3 CH_3 (cis) and CH_3 … CH_3 (trans) **(d)** CH_3 CH_3 (cis) and $(CH_3)_2C{=}CH_2$

(e) CH_3 … and … CH_3 **(f)** $CH_3{-}CH_2{-}CH_2{-}CH_2CH_3$ and $CH_3{-}CH_2{-}CH_2{-}CH_2CH_3$

(g) and

2-34. Sulfur dioxide has a dipole moment of 1.60 D. Carbon dioxide has a dipole moment of zero, even though C—O bonds are more polar than S—O bonds. Explain this apparent contradiction.

2-35. For each of the following compounds
(1) Draw the Lewis structure.
(2) Show how the bond dipole moments (and those of any nonbonding pairs of electrons) contribute to the molecular dipole moment.
(3) Predict whether the compound will have a large (>1 D), small, or zero dipole moment.

(a) $CH_3{-}CH{=}N{-}CH_3$ **(b)** $CH_3{-}CN$ **(c)** CBr_4

(d) $CH_3{-}C({=}O){-}CH_3$ **(e)** $(NC)_2C{=}C(CN)_2$ **(f)** **(g)**

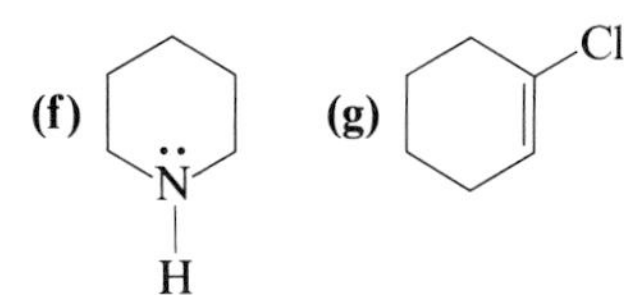

2-36. Diethyl ether and 1-butanol are isomers, and they have similar solubilities in water. Their boiling points are very different, however. Explain why these two compounds have similar solubility properties but dramatically different boiling points.

$CH_3CH_2{-}O{-}CH_2CH_3$	$CH_3CH_2CH_2CH_2{-}OH$
diethyl ether, b.p. 35°C	1-butanol, b.p. 118°C
8.4 mL dissolves in 100 mL H_2O	9.1 mL dissolves in 100 mL H_2O

2-37. *N*-methylpyrrolidine has a boiling point of 81°C, and piperidine has a boiling point of 106°C.
(a) Explain this large difference (25°C) in boiling point for these two isomers.
(b) Tetrahydropyran has a boiling point of 88°C, and cyclopentanol has a boiling point of 141°C. These two isomers have a boiling point difference of 53°C. Explain why the two oxygen-containing isomers have a much larger boiling point difference than the two amine isomers.

N—CH_3 N—H

N-methylpyrrolidine, b.p. 81°C piperidine, b.p. 106°C

tetrahydropyran, b.p. 88°C cyclopentanol, b.p. 141°C

2-38. Which of the following pure compounds can form hydrogen bonds? Which can form hydrogen bonds with water?

(a) $(CH_3CH_2)_2NH$ **(b)** $(CH_3CH_2)_3N$ **(c)** $CH_3CH_2CH_2OH$
(d) $(CH_3CH_2CH_2)_2O$ **(e)** $CH_3(CH_2)_3CH_3$ **(f)** $CH_2{=}CH{-}CH_2CH_3$
(g) CH_3COCH_3 **(h)** CH_3CH_2COOH **(i)** CH_3CH_2CHO

(j) 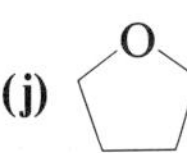

(k) 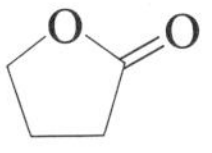 (five-membered ring lactone: O, =O)

(l) $CH_3{-}C({=}O){-}NH_2$

2-39. Predict which compound in each pair has a higher boiling point.

(a) $CH_3CH_2OCH_3$ or $CH_3CH(OH)CH_3$
(b) $CH_3CH_2CH_2CH_3$ or $CH_3CH_2CH_2CH_2CH_3$
(c) $CH_3CH_2CH_2CH_2CH_3$ or $(CH_3)_2CHCH_2CH_3$
(d) $CH_3CH_2CH_2CH_2CH_3$ or $CH_3CH_2CH_2CH_2CH_2Cl$

2-40. Circle the functional groups in the following structures. State to which class (or classes) of compounds the structure belongs.

(a) 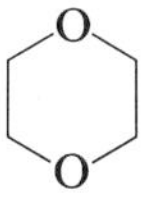 (six-membered ring with two O) **(b)** 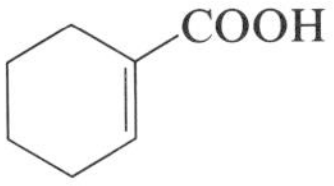 (cyclohexene with COOH) **(c)** 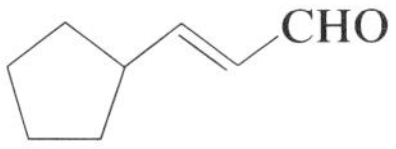 (cyclopentyl–CH=CH–CHO)

(d) 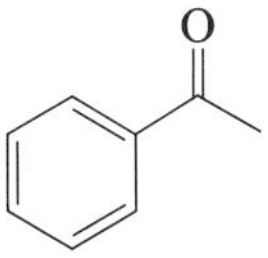 (phenyl–C(=O)–CH₃) **(e)** 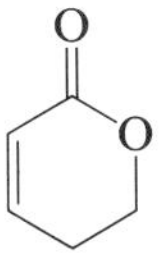 (six-membered lactone, O, =O) **(f)** 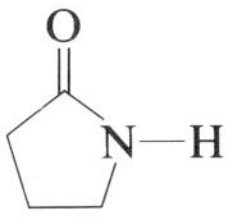 (five-membered lactam, =O, N—H)

(g) 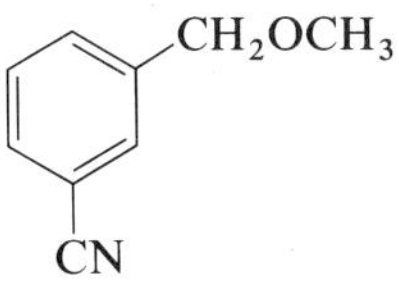 (benzene ring with CH_2OCH_3 and CN) **(h)** $CH_3{-}CH(NH_2){-}COOCH_3$

2-41. Dimethyl sulfoxide (DMSO) has been used as an antiinflammatory rub for racehorses. DMSO and acetone seem to have similar structures, but the C=O carbon atom in acetone is planar while the S=O sulfur atom in DMSO is pyramidal. Draw correct Lewis structures for DMSO and acetone, predict the hybridizations, and explain these observations.

$CH_3{-}S({=}O){-}CH_3$ DMSO $CH_3{-}C({=}O){-}CH_3$ acetone

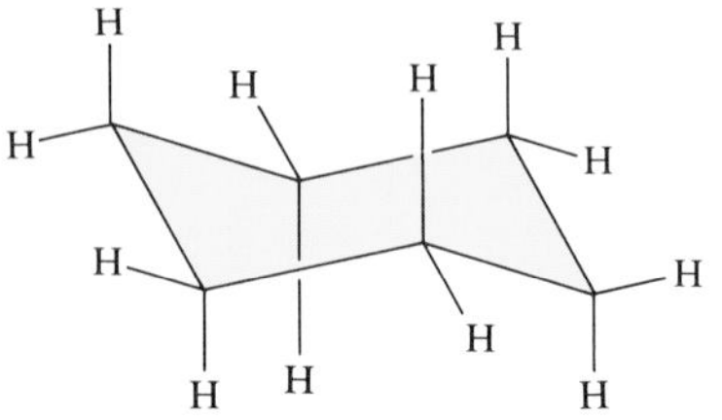

3

STRUCTURE AND STEREOCHEMISTRY OF ALKANES

Whenever possible, we will study organic chemistry using families of compounds to organize the material. The properties and reactions of the compounds in a family are similar, as their structures are similar. By considering how the structural features of a class of compounds determine their properties, we can predict the properties and reactions of similar new compounds. This organization elevates organic chemistry from a catalog of individual compounds to a systematic study of a few types of compounds. Organic molecules are classified according to their reactive parts, called *functional groups.* We considered some of the common functional groups in Sections 2-14 through 2-16.

An **alkane** is a hydrocarbon that contains only single bonds. The alkanes are the simplest and least reactive class of organic compounds, because they contain only carbon and hydrogen and they have no reactive functional groups. Although alkanes undergo reactions such as cracking and combustion at high temperatures, they are much less reactive under most conditions than other classes of compounds having functional groups.

3-1 CLASSIFICATION OF HYDROCARBONS (REVIEW)

In Section 2-14 we learned that hydrocarbons are classified according to their bonding. A hydrocarbon with a carbon-carbon double bond (such as ethylene) is called an *alkene.* If a hydrocarbon has a carbon-carbon triple bond (like acetylene), it is called an *alkyne.* Hydrocarbons with aromatic (benzenelike) rings are called *aromatic hydrocarbons.* If a hydrocarbon has no double or triple bonds, it is said to be **saturated.** Another way to describe alkanes, then, is as the class of **saturated hydrocarbons.** The following table reviews the classification of hydrocarbons.

SUMMARY OF HYDROCARBON CLASSIFICATION ACCORDING TO BONDING

Compound type	*Functional group*	*Example*
alkanes	none (no double or triple bonds)	$CH_3—CH_2—CH_3$, propane
alkenes	$>C=C<$ double bond	$CH_2=CH—CH_3$, propene
alkynes	$—C≡C—$ triple bond	$H—C≡C—CH_3$, propyne
aromatics	benzene ring	$C_6H_5CH_2CH_3$, ethylbenzene

3-2 MOLECULAR FORMULAS OF ALKANES

The structures and formulas of the first 16 straight-chain alkanes are shown in Table 3-1. Any isomers of these compounds have the same molecular formulas even though their structures are different. Notice how the molecular formulas increase by two hydrogen atoms each time a carbon atom is added.

The structures of the alkanes in Table 3-1 are purposely written in an unusual manner. The *n*-alkanes are a homologous series (Section 3-5), and they can be represented as a chain of CH_2 groups terminated by a hydrogen atom at each end. Each carbon atom is bonded to two hydrogen atoms, and there are two more hydrogens at the ends of the chain. If the molecule contains n carbon atoms, it must contain $(2n + 2)$ hydrogen atoms. Figure 3-1 shows this representation of alkane structure and how it leads to formulas of the form C_nH_{2n+2}.

Although we have derived the C_nH_{2n+2} formula using the *n*-alkanes, its use is not limited to straight-chain molecules. Any isomer of one of these *n*-alkanes has the same molecular formula. Butane and pentane follow the C_nH_{2n+2} rule; therefore, branched alkanes such as isobutane, isopentane, and neopentane also follow the rule.

TABLE 3-1
Formulas and physical properties of the straight-chain alkanes, called the *n*-alkanes

Alkane	*Number of carbons*	*Structure*	*Formula*	*Boiling point (°C)*	*Melting point (°C)*	*Density*[a]
methane	1	$H—CH_2—H$	CH_4	−162	−183	0.47
ethane	2	$H—(CH_2)_2—H$	C_2H_6	−89	−183	0.57
propane	3	$H—(CH_2)_3—H$	C_3H_8	−42	−188	0.50
butane	4	$H—(CH_2)_4—H$	C_4H_{10}	0	−138	0.58
pentane	5	$H—(CH_2)_5—H$	C_5H_{12}	36	−130	0.56
hexane	6	$H—(CH_2)_6—H$	C_6H_{14}	69	−95	0.66
heptane	7	$H—(CH_2)_7—H$	C_7H_{16}	98	−91	0.68
octane	8	$H—(CH_2)_8—H$	C_8H_{18}	126	−57	0.70
nonane	9	$H—(CH_2)_9—H$	C_9H_{20}	151	−54	0.72
decane	10	$H—(CH_2)_{10}—H$	$C_{10}H_{22}$	174	−30	0.74
undecane	11	$H—(CH_2)_{11}—H$	$C_{11}H_{24}$	196	−26	0.75
dodecane	12	$H—(CH_2)_{12}—H$	$C_{12}H_{26}$	216	−10	0.76
tridecane	13	$H—(CH_2)_{13}—H$	$C_{13}H_{28}$	235	−5	0.76
tetradecane	14	$H—(CH_2)_{14}—H$	$C_{14}H_{30}$	254	6	0.77
pentadecane	15	$H—(CH_2)_{15}—H$	$C_{15}H_{32}$	271	10	0.79
hexadecane	16	$H—(CH_2)_{16}—H$	$C_{16}H_{34}$	287	18	0.77

[a] Densities are given in g/mL at 20°C, except for methane and ethane, whose densities are given at their boiling points.

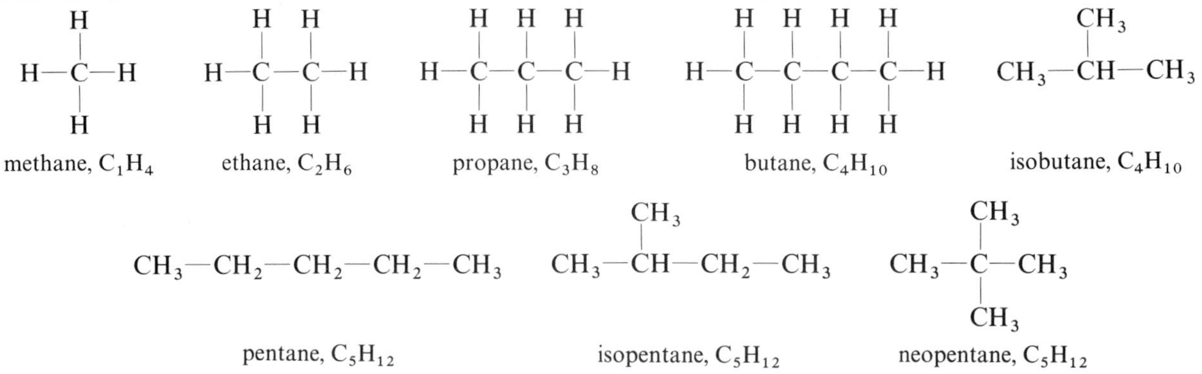

methane, C_1H_4 ethane, C_2H_6 propane, C_3H_8 butane, C_4H_{10} isobutane, C_4H_{10}

$CH_3-CH_2-CH_2-CH_2-CH_3$ pentane, C_5H_{12}

$CH_3-CH(CH_3)-CH_2-CH_3$ isopentane, C_5H_{12}

$CH_3-C(CH_3)_2-CH_3$ neopentane, C_5H_{12}

FIGURE 3-1 Examples of the general alkane molecular formula, C_nH_{2n+2}.

PROBLEM 3-1

Using the general molecular formula for alkanes

(a) Predict the molecular formula of triacontane, the C_{30} straight-chain alkane.
(b) Predict the molecular formula of 4,6-diethyl-12-(3,5-dimethyloctyl)triacontane, an alkane containing 44 carbon atoms.

3-3 PHYSICAL PROPERTIES OF ALKANES

Because alkanes are relatively unreactive, they are used primarily as solvents, fuels, and lubricants. Natural gas, gasoline, kerosene, heating oil, lubricating oil, and paraffin "wax" are all composed primarily of alkanes, with different physical properties resulting from their different structures.

3-3A SOLUBILITIES OF ALKANES

Alkanes are nonpolar, so they dissolve in nonpolar or weakly polar organic solvents. Alkanes are said to be **hydrophobic** ("water hating") because they do not dissolve in water. Their hydrophobic nature makes alkanes good lubricants and preservatives for metal, because they keep water from reaching the metal surface and causing corrosion.

3-3B DENSITIES OF ALKANES

The densities of some straight-chain alkanes are listed in Table 3-1. These straight-chain alkanes are called the **normal alkanes** or ***n*-alkanes** because their carbon atoms are bonded in a single chain, with no branching. Alkanes have densities of approximately 0.7 g/mL, compared with a density of 1.0 g/mL for water. Because alkanes are less dense than water and insoluble in water, a mixture of an alkane (such as gasoline or oil) and water quickly separates into two phases, with the alkane on top.

3-3C BOILING POINTS OF ALKANES

Table 3-1 also shows the boiling points and melting points of the straight-chain alkanes. The boiling points increase smoothly with the increasing number of carbon atoms and the increasing molecular weight of the alkane. Larger molecules have

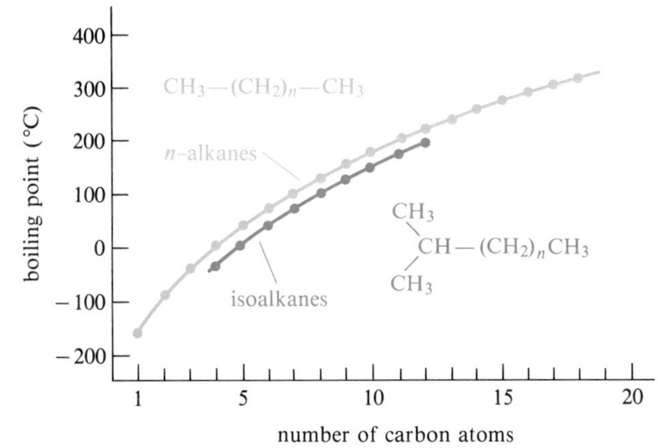

FIGURE 3-2 Alkane boiling points. Comparison of the boiling points of the straight-chain alkanes (blue) with those of some branched alkanes (red). Because of their smaller surface areas, branched alkanes have lower boiling points than those of straight-chain alkanes.

larger surface areas, resulting in increased intermolecular van der Waals attractions. These increased attractions must be overcome for vaporization and boiling to occur. A larger molecule, with greater surface area and greater van der Waals attractions, therefore boils at a higher temperature.

A graph of *n*-alkane boiling points versus the number of carbon atoms (the blue line in Fig. 3-2) shows the increase in boiling points with increasing molecular weight. The addition of a CH_2 group increases the boiling point by about 30°C up to about ten carbons, and by about 20°C in the higher alkanes.

The red line in Figure 3-2 represents the boiling points of some branched alkanes. In general, a branched alkane boils at a lower temperature than the *n*-alkane with the same number of carbon atoms. This difference in boiling points is also explained by the intermolecular van der Waals forces. Branched alkanes are more compact, with less surface area for London force interactions.

3-3D MELTING POINTS OF ALKANES

The blue line in Figure 3-3 is a graph of the melting points of the *n*-alkanes. Like the boiling points, the melting points generally increase with increasing molecular weight. The melting point graph is not smooth, however. Alkanes with even numbers of carbon atoms pack better into a solid structure, and higher temperatures

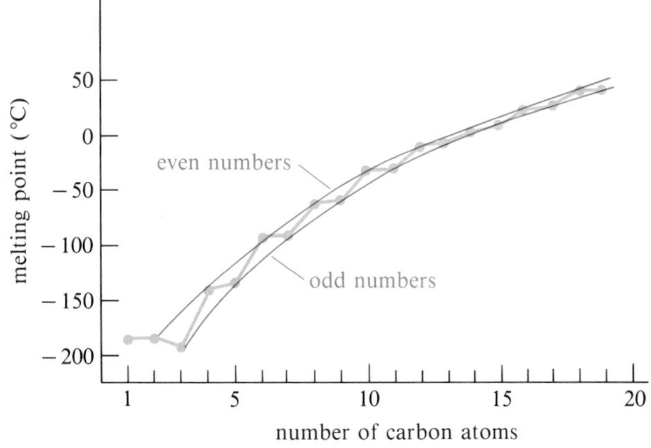

FIGURE 3-3 Alkane melting points. The melting point curve for *n*-alkanes with even numbers of carbon atoms is slightly higher than that for alkanes with odd numbers of carbons.

TABLE 3-2
Boiling and melting points of three isomers of C_6H_{14}[a]

Isomer	*Boiling point (°C)*	*Melting point (°C)*
$(CH_3)_2CH-CH_2-CH_2-CH_3$	60	−154
$(CH_3)_2CH-CH(CH_3)_2$	58	−135
$CH_3-C(CH_3)_2-CH_2-CH_3$	50	−98

[a] As molecules become more spherical in shape, their surface areas decrease, giving weaker London forces and lower boiling points. The more compact shape packs better into a solid, however, giving a higher melting point.

are needed to melt them. Alkanes with similar but odd numbers of carbon atoms do not pack as well and melt at lower temperatures. The sawtooth-shaped graph of melting points is smoothed by drawing separate lines (red) for the alkanes with even and odd numbers of carbon atoms.

Branching of the alkane chain also affects its melting point. A branched alkane generally melts *higher* than the *n*-alkane with the same number of carbon atoms. Branching of an alkane gives it a more compact three-dimensional structure, which packs more easily into a solid structure and increases its melting point. Table 3-2 shows the boiling points and melting points of three isomers of formula C_6H_{14}. The boiling points decrease and the melting points increase as the shape of the molecule becomes more highly branched and compact.

PROBLEM 3-2
List each of the following sets of compounds in order of increasing boiling point.

(a) octane, nonane, and decane
(b) octane, $(CH_3)_3C-C(CH_3)_3$, and $CH_3CH_2C(CH_3)_2CH_2CH_2CH_3$

PROBLEM 3-3
Repeat Problem 3-2, listing the compounds in order of increasing melting point.

3-4 USES AND SOURCES OF ALKANES

The differences in their physical properties are the features that distinguish the most important uses of each group of alkanes.

3-4A MAJOR USES OF ALKANES

The first four alkanes (methane, ethane, propane, and butane) are gases at room temperature and atmospheric pressure. Methane and ethane are difficult to liquefy, so they are usually handled as compressed gases. Upon cooling to cryogenic (very low) temperatures, however, methane and ethane become liquids. *Liquefied natural gas,* mostly methane, is transported in special refrigerated tankers much more easily than it can be transported as a compressed gas.

Propane and butane are easily liquefied at room temperature, under a modest amount of pressure. These gases, often obtained along with liquid petroleum, are stored in low-pressure cylinders of *liquefied petroleum gas (LPG)*. Propane and butane are used as fuels, both for heating and for internal combustion engines. They burn very cleanly, and pollution-control equipment is rarely necessary. In many agricultural areas, propane and butane are more cost-effective tractor fuels than are gasoline and diesel fuel. Propane and butane have also largely replaced the Freons (see p. 168) as propellants in aerosol cans. Unlike alkanes, the chlorofluorocarbon Freon propellants are suspected of damaging the earth's protective ozone layer.

The next four alkanes are free-flowing, volatile liquids. Isomers of pentane, hexane, heptane, and octane are the primary constituents of gasoline. Their volatility is crucial for this use, because the carburetor simply squirts a stream of gasoline into the intake air as it rushes through. If gasoline did not evaporate easily, it would reach the cylinder in the form of droplets. Droplets cannot burn as efficiently as a vapor, and the engine would smoke and give low mileage.

The nonanes (C_9) through about the hexadecanes (C_{16}) are higher-boiling liquids that begin to show appreciable viscosity, especially in the higher-molecular-weight compounds. These alkanes are found in kerosene and jet fuel, and the higher alkanes of this group are found in diesel fuel. Diesel fuel is not very volatile, so it would not function well in a carburetor. In a diesel engine, the fuel is sprayed directly into the cylinder right at the top of the compression stroke. The hot, highly compressed air in the cylinder causes the diesel fuel to burn quickly, swirling and vaporizing as it burns.

Some of the alkanes in diesel fuel have fairly high freezing points, and they may solidify in cold weather. This partial solidification causes the diesel fuel to turn into a waxy, semisolid mass. Owners of diesel engines in cold climates often mix gasoline or kerosene with their diesel fuel in the winter. The added gasoline dissolves the frozen alkanes, diluting the slush and allowing it to be pumped to the cylinders.

Alkanes with more than 16 or 18 carbon atoms are most often used as lubricating and heating oils. These are sometimes called "mineral" oils, because they come from petroleum, which was once considered a mineral.

Paraffin "wax" is not a true wax, but a purified mixture of high-molecular-weight alkanes with melting points well above room temperature. The true waxes are long-chain esters, discussed in Chapter 25.

3-4B ALKANE SOURCES; PETROLEUM REFINING

Alkanes are mostly derived from petroleum and petroleum by-products. *Petroleum,* often called *crude oil,* is pumped from wells that reach into pockets containing the remains of prehistoric plants. The principal constituents of crude oil are the alkanes, some aromatics, and some undesirable compounds containing sulfur and nitrogen. The composition of petroleum and the amounts of contaminants vary from one source to another, and a refinery must be carefully adjusted to process a particular type of crude oil. Because of their different qualities, different prices are paid for light Arabian crude, West Texas crude, and other classes of crude petroleum.

The first step in refining petroleum is a careful fractional distillation. The products of that distillation are not pure alkanes but mixtures of alkanes with useful ranges of boiling points. Table 3-3 shows the major fractions obtained from distillation of crude petroleum.

TABLE 3-3
Major fractions obtained from distillation of crude petroleum

Boiling range (°C)	*Number of carbons*	*Fraction*	*Use*
under 30	2–4	petroleum gas	LP gas for heating
30–180	4–9	gasoline	motor fuel
160–230	8–16	kerosene	heat, jet fuel
200–320	10–18	diesel	motor fuel
300–450	16–30	heavy oil	heating

Distills under a vacuum: motor oils, petroleum "jelly," paraffin "wax"
Does not distill: asphalt, petroleum "coke"

After the petroleum is distilled, **catalytic cracking** converts some of the less valuable fractions to more valuable products. Catalytic cracking involves heating alkanes in the presence of materials that catalyze the cleavage of large molecules into smaller ones. Cracking is often used to convert higher-boiling fractions into mixtures that can be blended with gasoline. When cracking is done in the presence of hydrogen **(hydrocracking),** the product is a mixture of alkanes, free of sulfur and nitrogen impurities. The following reaction shows the catalytic hydrocracking of a molecule of tetradecane into two molecules of heptane.

$$CH_3{-}(CH_2)_{12}{-}CH_3 + H_2 \xrightarrow[\text{SiO}_2 \text{ or Al}_2\text{O}_3 \text{ catalyst}]{\text{heat}} 2\ CH_3{-}(CH_2)_5{-}CH_3$$

3-4C NATURAL GAS

Natural gas was once treated as a waste product of petroleum production and was destroyed by flaring it off. Now, however, natural gas is an equally valuable natural resource, pumped and stored throughout the developed world. Natural gas is about 70 percent methane, 10 percent ethane, and 15 percent propane, depending on the source of the gas. Small amounts of other hydrocarbons and contaminants are also present. Natural gas is sometimes found above pockets of petroleum, although it is usually found in places where there is little, if any, recoverable petroleum. Natural gas is used primarily as a fuel to heat buildings and to generate electricity. It is also important as a starting material for the production of fertilizers.

3-5 NOMENCLATURE OF ALKANES

From pentane on, the names of the alkanes (Tables 3-1 and 3-4) are composed of a prefix derived from the Greek word for the number of carbon atoms present, plus the suffix **-ane.** The general formula for the *n*-alkanes is a chain of $-CH_2-$ groups **(methylene groups),** terminated at each end by a hydrogen atom. The *n*-alkanes differ from each other only by the number of methylene groups in the chain. Such a series of compounds, differing only by the number of $-CH_2-$ groups, is called a **homologous series,** and the individual members of the series are called **homologs.** For example, butane is a homolog of propane, and both of these are homologs of hexane.

3-5A COMMON NAMES

If all alkanes had straight-chain structures, their nomenclature would be very simple. Most alkanes have structural isomers, however, and we need a way of naming all the different isomers. For example, there are two isomers of formula

TABLE 3-4
Names of the straight-chain alkanes (the *n*-alkanes)[a]

Name	*Carbon atoms*	*Formula*	*Name*	*Carbon atoms*	*Formula*
methane	1	CH_4	dodecane	12	$CH_3(CH_2)_{10}CH_3$
ethane	2	CH_3CH_3	tridecane	13	$CH_3(CH_2)_{11}CH_3$
propane	3	$CH_3CH_2CH_3$	tetradecane	14	$CH_3(CH_2)_{12}CH_3$
butane	4	$CH_3(CH_2)_2CH_3$	pentadecane	15	$CH_3(CH_2)_{13}CH_3$
pentane	5	$CH_3(CH_2)_3CH_3$	hexadecane	16	$CH_3(CH_2)_{14}CH_3$
hexane	6	$CH_3(CH_2)_4CH_3$	heptadecane	17	$CH_3(CH_2)_{15}CH_3$
heptane	7	$CH_3(CH_2)_5CH_3$	octadecane	18	$CH_3(CH_2)_{16}CH_3$
octane	8	$CH_3(CH_2)_6CH_3$	nonadecane	19	$CH_3(CH_2)_{17}CH_3$
nonane	9	$CH_3(CH_2)_7CH_3$	eicosane	20	$CH_3(CH_2)_{18}CH_3$
decane	10	$CH_3(CH_2)_8CH_3$	triacontane	30	$CH_3(CH_2)_{28}CH_3$
undecane	11	$CH_3(CH_2)_9CH_3$	tetracontane	40	$CH_3(CH_2)_{38}CH_3$

[a] The names are composed of the Greek prefix for the number of carbon atoms plus the *-ane* suffix.

C_4H_{10}. The straight-chain isomer is simply called *butane* (or *n-butane*), and the branched isomer is called *isobutane,* meaning an "isomer of butane."

$$CH_3{-}CH_2{-}CH_2{-}CH_3 \qquad CH_3{-}\underset{\displaystyle}{\overset{\displaystyle CH_3 \atop |}{CH}}{-}CH_3$$

butane (*n*-butane) isobutane

The three isomers of C_5H_{12} are called *pentane* (or *n-pentane*), *isopentane,* and *neopentane.*

$$CH_3{-}CH_2{-}CH_2{-}CH_2{-}CH_3 \qquad CH_3{-}\overset{\displaystyle CH_3 \atop |}{CH}{-}CH_2{-}CH_3 \qquad CH_3{-}\underset{\displaystyle | \atop CH_3}{\overset{\displaystyle CH_3 \atop |}{C}}{-}CH_3$$

pentane (*n*-pentane) isopentane neopentane

Isobutane, isopentane, and *neopentane* are examples of **common names** or **trivial names,** historical names arising from common usage. Common names cannot easily describe the larger, more complicated molecules having many isomers, however. The number of isomers for any molecular formula grows rapidly as the number of carbon atoms increases. For example, there are five structural isomers of hexane, 18 isomers of octane, and 75 isomers of decane! We need a system of nomenclature that enables us to name complicated molecules without having to memorize hundreds of these historical common names.

3-5B IUPAC OR SYSTEMATIC NAMES

A group of chemists representing the countries of the world met in 1892 to devise a system for naming compounds that would be simple to use, require a minimum of memorization, and yet was flexible enough to name even the most complicated organic compounds. This was the first meeting of the group that came to be known as the International Union of Pure and Applied Chemistry, abbreviated **IUPAC.** This international group has developed a detailed system of nomenclature that we call the **IUPAC rules.** The IUPAC rules are accepted throughout the world as the standard method for naming organic compounds. The names that are generated using this system are called **IUPAC names** or **systematic names.**

The IUPAC system works in a similar manner to name many different families of compounds. We will consider the naming of alkanes in detail, and later extend these rules to other kinds of compounds as we encounter them. The IUPAC system for naming alkanes uses the longest chain of carbon atoms as the main chain, which is numbered to give the locations of side chains.

The main chain The first rule of nomenclature gives the base name of the compound:

> RULE 1 Find the longest continuous chain of carbon atoms, and use the name of this chain as the base name of the compound.

For example, the longest chain of carbon atoms in the following compound contains six carbons. The compound is therefore named as a *hexane* derivative. The longest chain is rarely drawn in a straight line, so you should expect to have to look carefully to find it.

$$\overset{2}{C}H_2\overset{1}{C}H_3$$
$$CH_3-\underset{3}{C}H-\underset{4}{C}H_2-\underset{5}{C}H_2-\underset{6}{C}H_3$$

3-methyl*hexane*

The groups attached to the main chain are called **substituents** because they are substituted (in place of a hydrogen atom) on the main chain. When there are two longest chains of equal length, use the chain with the greater number of substituents. The following compound contains two different seven-carbon chains and is named as a *heptane*. We choose the chain on the right as the main chain because it has more substituents (in red) attached to the chain.

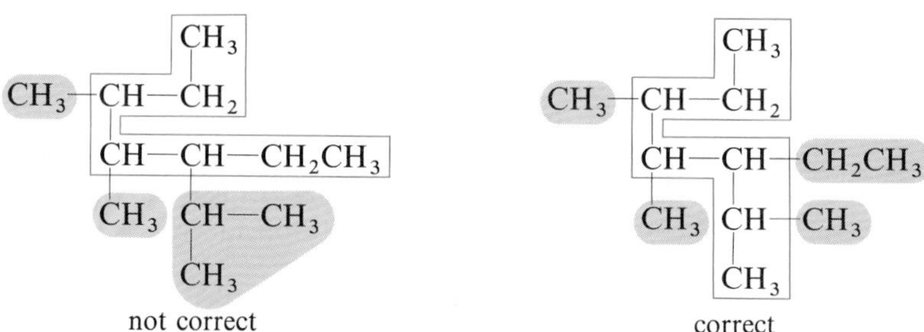

not correct — longest chain, but only three substituents

correct — longest chain, four substituents

Numbering the main chain In order to give the locations of all the substituents, each carbon atom on the main chain must be given a number.

> RULE 2 Number the longest chain beginning with the end of the chain nearest a substituent.

We start the numbering from the end nearest a branch so that the numbers of the substituted carbons will be as low as possible. In the structure of the heptane given above at the right, numbering from top to bottom gives the first branch at C3 (carbon atom 3), while numbering from bottom to top gives the first branch at C2. Numbering from bottom to top is correct.

$$\begin{array}{c}\text{incorrect: } CH_3-{}^3CH({}^2CH_2-{}^1CH_3)-{}^4CH(CH_3)-{}^5CH(CH_2CH_3)-{}^6CH(CH_3)-{}^7CH_3\end{array}$$

incorrect

$$\begin{array}{c}\text{correct: } CH_3-{}^5CH({}^6CH_2-{}^7CH_3)-{}^4CH(CH_3)-{}^3CH(CH_2CH_3)-{}^2CH(CH_3)-{}^1CH_3\end{array}$$

correct

3-ethyl-2,4,5-trimethyl heptane

Naming alkyl groups Next, the substituent groups must be given names.

RULE 3 Name the substituent groups attached to the longest chain as **alkyl groups.** Give the location of each alkyl group by the number of the main-chain carbon atom to which it is attached.

Alkyl groups are named by replacing the *-ane* suffix of the alkane name with *-yl*. *Methane* becomes *methyl; ethane* becomes *ethyl.*

CH_4, methane	CH_3-, methyl group
CH_3-CH_3, ethane	CH_3-CH_2-, ethyl group
$CH_3-CH_2-CH_3$, propane	$CH_3-CH_2-CH_2-$, propyl group

The following alkanes show the use of the alkyl group nomenclature.

$${}^1CH_3-{}^2CH_2-{}^3CH(CH_3)-{}^4CH_2-{}^5CH_2-{}^6CH_3$$

3-methylhexane

$${}^1CH_3-{}^2CH_2-{}^3CH(CH_2CH_3)-{}^4CH_2-{}^5CH_2-{}^6CH(CH_3)-{}^7CH_2-{}^8CH_2-{}^9CH_3$$

3-ethyl-6-methylnonane

Figure 3-4 gives the names of the most common alkyl groups, those having up to four carbon atoms. The propyl and butyl groups are simply straight-chain three- and four-carbon alkyl groups. These groups are often named as the "*n*-propyl" and "*n*-butyl" groups, however, to eliminate any question about which propyl group or butyl group is meant.

One carbon

CH_3- methyl group

Two carbons

CH_3-CH_2- ethyl group

Three carbons

$CH_3-CH_2-CH_2-$ *n*-propyl group (or "propyl group")

$CH_3-CH(CH_3)-$ isopropyl group

Four carbons

$CH_3-CH_2-CH_2-CH_2-$ *n*-butyl group (or "butyl group")

$CH_3-CH(CH_3)-CH_2-$ isobutyl group

$CH_3-CH_2-CH(CH_3)-$ *sec*-butyl group

$CH_3-C(CH_3)_2-$ *tert*-butyl group (or "*t*-butyl group")

FIGURE 3-4 Some common alkyl groups.

The simple branched alkyl groups are usually known by common names. The isopropyl and isobutyl groups have a characteristic "iso" $(CH_3)_2CH$ grouping, just as in isobutane.

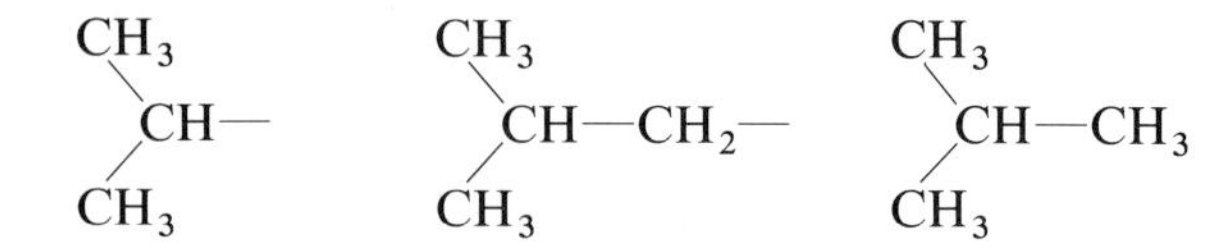

$$\begin{array}{ccc} (CH_3)_2CH- & (CH_3)_2CH-CH_2- & (CH_3)_2CH-CH_3 \\ \text{isopropyl group} & \text{isobutyl group} & \text{isobutane} \end{array}$$

The names of the *secondary*-butyl (*sec*-butyl) and *tertiary*-butyl (*tert*-butyl or *t*-butyl) groups are based on the **degree of alkyl substitution** of the carbon atom attached to the main chain. In the *sec*-butyl group, the carbon atom bonded to the main chain is **secondary** (2°), or bonded to two other carbon atoms. In the *t*-butyl group, it is **tertiary** (3°), or bonded to three other carbon atoms. In both the *n*-butyl group and the isobutyl group, the carbon atoms bonded to the main chain are **primary** (1°), bonded to only one other carbon atom.

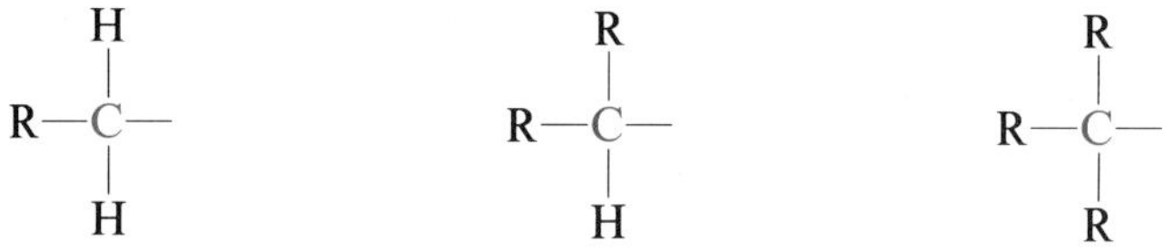

$$\begin{array}{ccc} R-CH_2- & R-CHR- & R-CR_2- \\ \text{a } \textit{primary } (1^\circ) \text{ carbon} & \text{a } \textit{secondary } (2^\circ) \text{ carbon} & \text{a } \textit{tertiary } (3^\circ) \text{ carbon} \end{array}$$

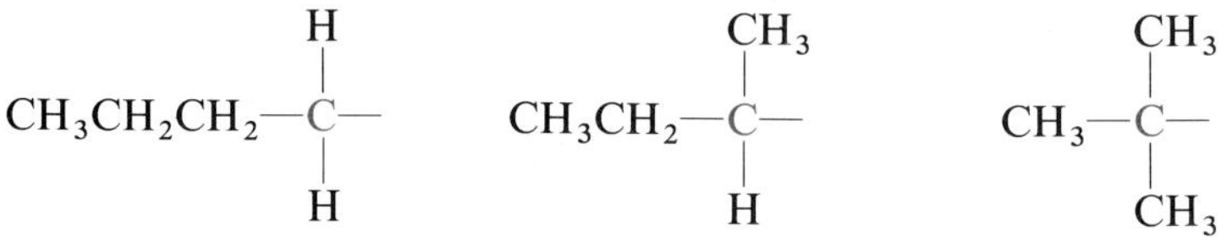

$$\begin{array}{ccc} CH_3CH_2CH_2-CH_2- & CH_3CH_2-CH(CH_3)- & CH_3-C(CH_3)_2- \\ n\text{-butyl group } (1^\circ) & \textit{sec}\text{-butyl group } (2^\circ) & t\text{-butyl group } (3^\circ) \end{array}$$

SOLVED PROBLEM 3-1

Give the structures of 4-isopropyloctane and 5-*t*-butyldecane.

SOLUTION 4-Isopropyloctane has a chain of eight carbons, with an isopropyl group on the fourth carbon. 5-*t*-Butyldecane has a chain of ten carbons, with a *t*-butyl group on the fifth.

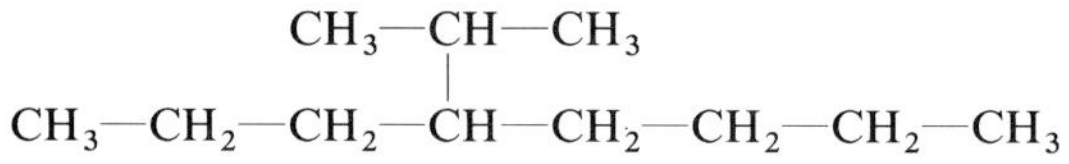

$$CH_3-CH_2-CH_2-CH(CH(CH_3)_2)-CH_2-CH_2-CH_2-CH_3$$

4-isopropyloctane

$$CH_3-CH_2-CH_2-CH_2-CH(C(CH_3)_3)-CH_2-CH_2-CH_2-CH_2-CH_3$$

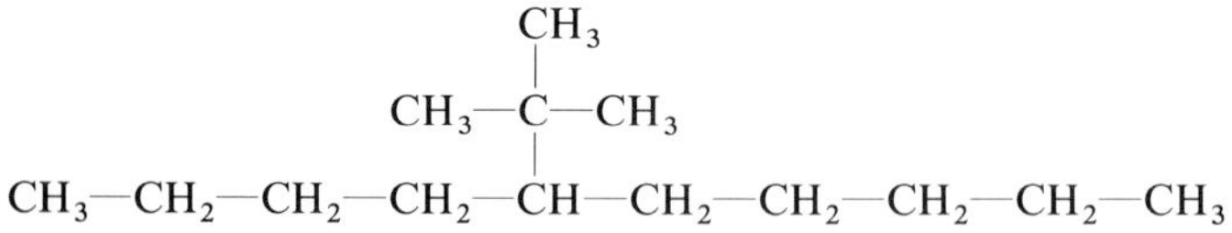

5-*t*-butyldecane

PROBLEM 3-4

Name the following alkanes.

(a) 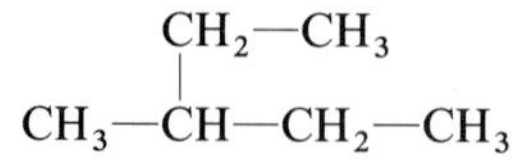 $CH_3-CH(CH_2-CH_3)-CH_2-CH_3$

```
                                            CH3—CH—CH3
                                                |
                                                CH2  CH3
                                                |    |
(b) CH3—CH2—CH2—CH2—CH2—CH2—CH—CH—CH3
```

Multiple groups The final rule tells how to organize the names of compounds with more than one substituent.

RULE 4 When two or more substituents are present, list them in alphabetical order. When two or more of the *same* alkyl substituent are present, use the prefixes *di-*, *tri-*, *tetra-*, and so on, to avoid having to name the alkyl group twice.

di- means 2	*penta-* means 5
tri- means 3	*hexa-* means 6
tetra- means 4	

Using this rule, we are ready to construct names for some fairly complicated structures. Let us finish naming the heptane on p. 86. This compound has an ethyl group on C3 and three methyl groups on C2, C4, and C5. The ethyl group is listed alphabetically before the methyl groups.

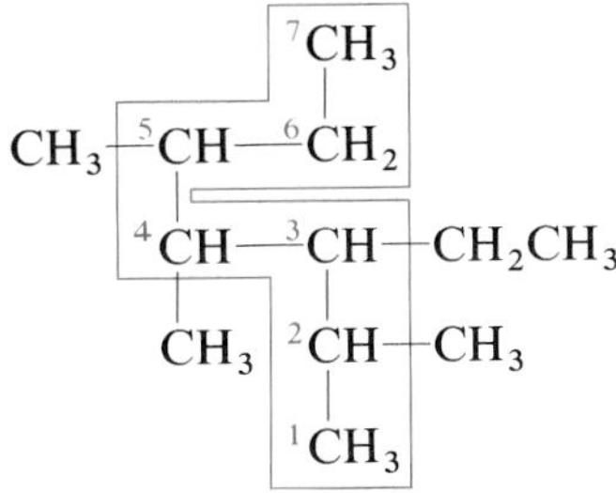

3-ethyl-2,4,5-trimethylheptane

SOLVED PROBLEM 3-2

Give a systematic (IUPAC) name for the following compound.

```
            CH3
            |
            CH—CH3   CH2CH3
            |        |
   CH3—CH—CH—CH2—CH—CH3
        |
   CH3—C—CH3
        |
        CH3
```

SOLUTION The longest carbon chain contains eight carbon atoms, so this compound is named as an octane. Numbering from left to right gives a branch on C2; numbering from right to left gives a branch on C3, so we number from left to right.

```
            CH3
            |                8
            CH—CH3   7CH2CH3
            |         |
   CH3—3CH—4CH—5CH2—6CH—CH3
        |
  1CH3—2C—CH3
        |
        CH3
```

There are four methyl groups: Two are on C2, one is on C3, and one is on C6. These four groups will be listed as "2,2,3,6-tetramethyl . . . ." There is an isopropyl group on C4. Listing the isopropyl group and the methyl groups alphabetically, we have

4-isopropyl-2,2,3,6-tetramethyloctane

SUMMARY OF RULES FOR NAMING ALKANES

To name an alkane, we follow four rules:

1. Find the longest continuous chain of carbon atoms, and use this chain as the base name of the compound.
2. Number the longest chain beginning with the end of the chain nearest a branch.
3. Name the substituent groups attached to the longest chain as alkyl groups. Give the location of each alkyl group by the number of the main-chain carbon atom to which it is attached.
4. When two or more substituents are present, list them in alphabetical order. When two or more of the *same* alkyl substituent are present, use the prefixes *di-*, *tri*, *tetra-*, and so on (ignored in alphabetizing), to avoid having to name the alkyl group twice.

PROBLEM 3-5

Write structures for the following compounds.

(a) 3-ethyl-3-methylpentane
(b) 3-methyl-5-propylnonane
(c) 4-*t*-butyl-2-methylheptane
(d) 5-isopropyl-3,3,4-trimethyloctane

PROBLEM 3-6

Provide IUPAC names for the following compounds.

(a) $(CH_3)_2CHCH_2CH_3$

(b) $CH_3{-}C(CH_3)_2{-}CH_3$

(c) 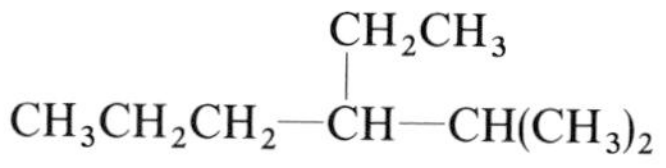
$$CH_3CH_2CH_2{-}\underset{}{\overset{CH_2CH_3}{\overset{|}{CH}}}{-}CH(CH_3)_2$$

(d) 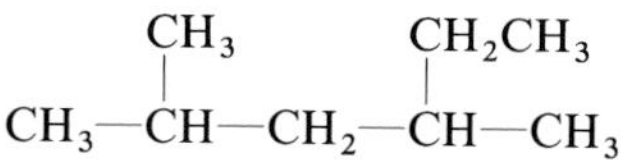
$$CH_3{-}\overset{CH_3}{\overset{|}{CH}}{-}CH_2{-}\overset{CH_2CH_3}{\overset{|}{CH}}{-}CH_3$$

(e) 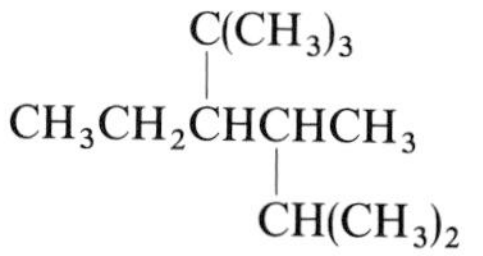
$$CH_3CH_2\overset{C(CH_3)_3}{\overset{|}{CH}}\underset{\underset{CH(CH_3)_2}{|}}{CH}CH_3$$

(f) 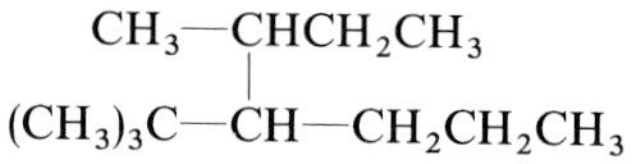
$$(CH_3)_3C{-}\overset{CH_3{-}CHCH_2CH_3}{\overset{|}{CH}}{-}CH_2CH_2CH_3$$

PROBLEM 3-7

Give structures and names for

(a) the five isomers of C_6H_{14}.
(b) the nine isomers of C_7H_{16}.

Complex substituents More complex alkyl groups are occasionally encountered. They are named by a systematic method using the longest alkyl chain as the base alkyl group. The base alkyl group is numbered beginning with the carbon atom bonded to the main chain (the "head carbon"). The substituents on the base alkyl group are listed with the appropriate numbers, and parentheses are used to set off the name of the complex alkyl group. The following examples illustrate the systematic method for naming alkyl groups.

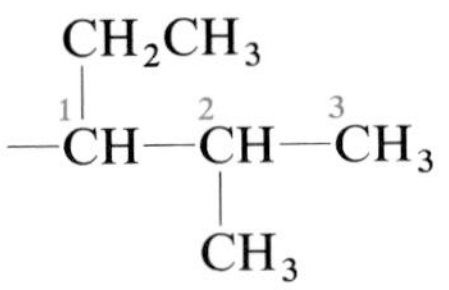

$$\begin{array}{c} CH_2CH_3 \\ | \\ \overset{1}{-CH}-\overset{2}{CH}-\overset{3}{CH_3} \\ | \\ CH_3 \end{array}$$

a (1-ethyl-2-methylpropyl) group

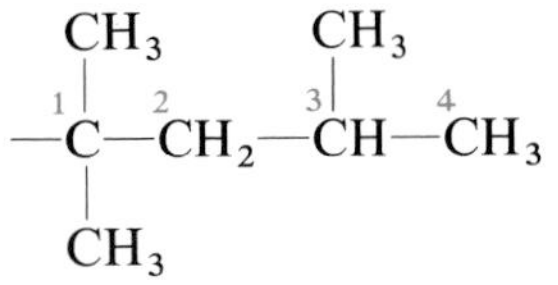

$$\begin{array}{c} CH_3 \quad\quad CH_3 \\ | \quad\quad\quad | \\ \overset{1}{-C}-\overset{2}{CH_2}-\overset{3}{CH}-\overset{4}{CH_3} \\ | \\ CH_3 \end{array}$$

a (1,1,3-trimethylbutyl) group

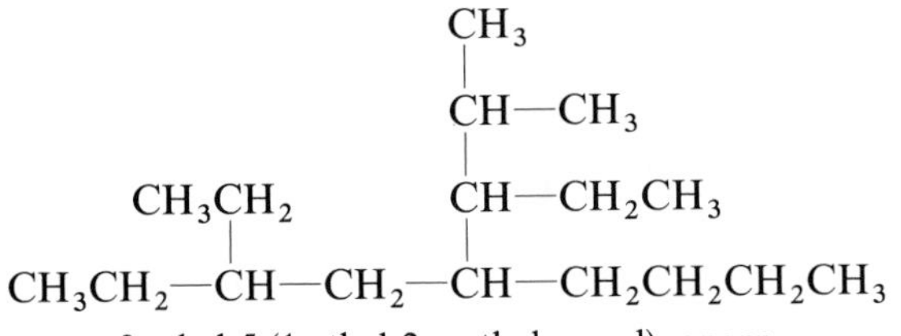

$$\begin{array}{c} CH_3 \\ | \\ CH-CH_3 \\ | \\ CH_3CH_2 \quad\quad CH-CH_2CH_3 \\ | \quad\quad\quad\quad | \\ CH_3CH_2-CH-CH_2-CH-CH_2CH_2CH_2CH_3 \end{array}$$

3-ethyl-5-(1-ethyl-2-methylpropyl)nonane

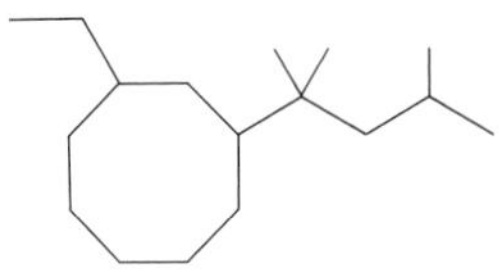

1-ethyl-3-(1,1,3-trimethylbutyl)cyclooctane

PROBLEM 3-8

Draw the structures of the following groups and give their more common names.

(a) the (1-methylethyl) group (b) the (2-methylpropyl) group
(c) the (1-methylpropyl) group (d) the (1,1-dimethylethyl) group

PROBLEM 3-9

Draw the structures of the following compounds.

(a) 3-(1-methylethyl)hexane (b) 5-(1,2,2-trimethylpropyl)nonane

PROBLEM 3-10

Without looking at the structures, give a molecular formula for each compound in Problem 3-9. Use the names of the groups to determine the number of carbon atoms, then use the $(2n + 2)$ rule.

3-6 STRUCTURE AND CONFORMATIONS OF ALKANES

Although alkanes are not as reactive as other classes of organic compounds, they have many of the same structural characteristics. We use the simple alkanes as examples to study some of the fundamental properties of organic compounds, including the structure of the sp^3 hybridized carbon atom and properties of C—C and C—H single bonds.

3-6A STRUCTURE OF METHANE

The simplest alkane is *methane,* CH_4. Methane is perfectly tetrahedral, with the 109.5° bond angles predicted for an sp^3 hybridized carbon atom. The four hydrogen atoms are covalently bonded to the central carbon atom, with bond lengths of 1.09 Å.

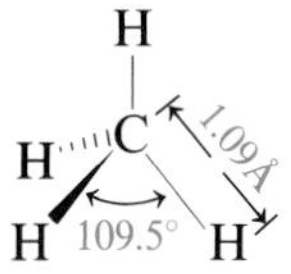

H
H C 1.09Å
H 109.5° H

3-6B CONFORMATIONS OF ETHANE

Ethane, the two-carbon alkane, is composed of two *methyl* groups (methane molecules with one hydrogen removed) with overlapping sp^3 hybrid orbitals forming a sigma bond between them. The two methyl groups are not fixed in a single position, but are relatively free to rotate about the sigma bond connecting the two carbon atoms. The bond maintains its linear bonding overlap as the carbon atoms turn.

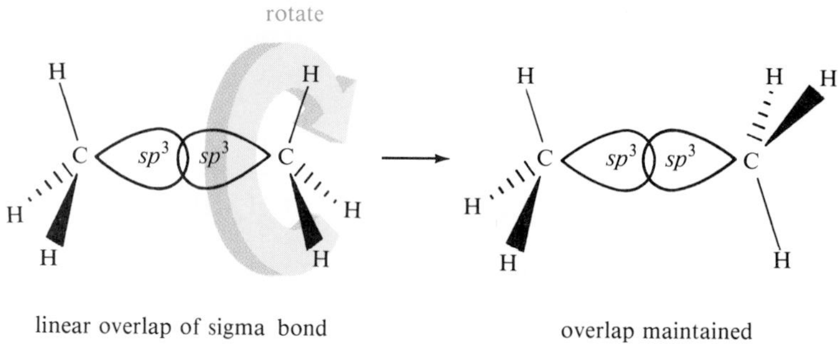

rotate

linear overlap of sigma bond

overlap maintained

The different arrangements formed by rotations about a single bond are called **conformations** or **conformers.** Pure, single conformations cannot be isolated in most cases, because the molecules are constantly rotating through all the possible conformations.

In drawing conformations, we will often use **Newman projections,** a way of drawing a molecule looking straight down the bond connecting two carbon atoms (Fig. 3-5). The front carbon atom is represented by three lines (three bonds) coming together in a Y shape. The back carbon is represented by a circle with three bonds pointing out from it. Until you become familiar with the Newman projection, you should make a model of each example and compare your model with the drawings.

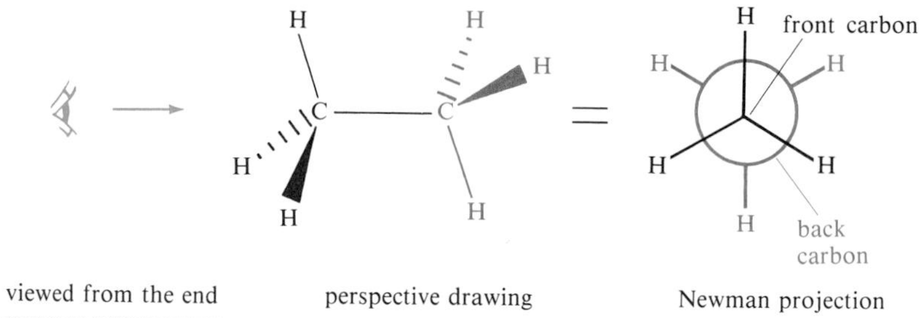

viewed from the end — perspective drawing — Newman projection (front carbon, back carbon)

FIGURE 3-5 The Newman projection looks straight down the carbon-carbon bond.

An infinite number of conformations are possible for ethane, because the angle between the hydrogen atoms on the front and back carbon atoms can take on an infinite number of values. Figure 3-6 uses Newman projections and sawhorse structures to illustrate some of these ethane conformations. **Sawhorse structures** picture the molecule looking down at an angle toward the carbon-carbon bond. These structures can be misleading, depending on how the eye sees them. We will generally use perspective or Newman projections to draw molecular conformations.

Any of these conformations can be specified by its **dihedral angle** (θ), the angle between the C—H bonds on the front carbon atom and the C—H bonds on the back carbon in the Newman projection. Two of the conformations have special names. The conformation with $\theta = 0°$ is called the **eclipsed conformation** because the Newman projection shows the hydrogen atoms on the back carbon to be hidden (eclipsed) by those on the front carbon. The **staggered conformation,** with $\theta = 60°$, has the hydrogen atoms on the back carbon staggered halfway

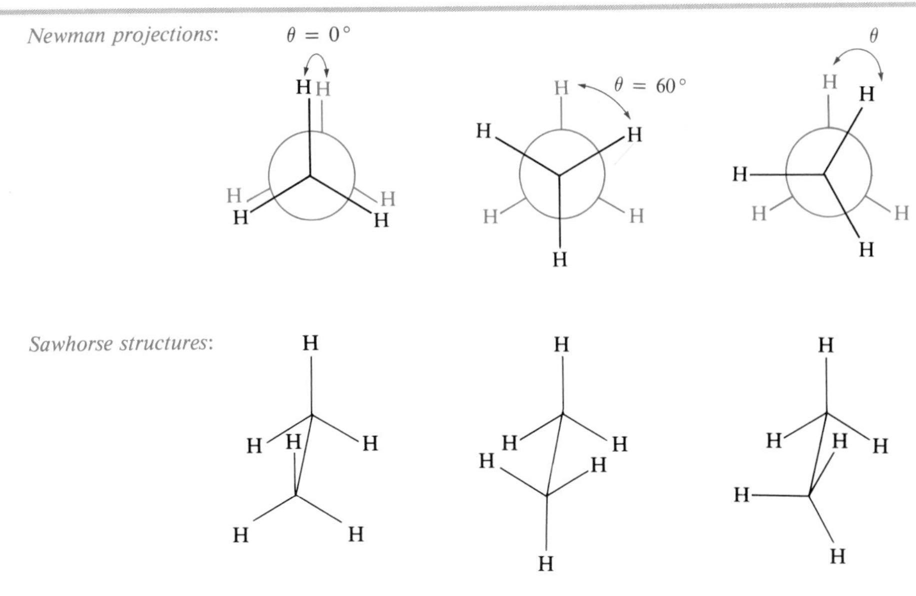

FIGURE 3-6 Ethane conformations. The eclipsed conformation has a dihedral angle $\theta = 0°$, and the staggered conformation has $\theta = 60°$. Any other conformation is called a skew conformation.

between the hydrogens on the front carbon. Any other, intermediate conformation is called a **skew conformation.**

All the possible conformations exist in a sample of ethane gas. At room temperature, the ethane molecules are rotating and their conformations are constantly changing. These conformations are not all equally favored, however. The lowest-energy conformation is the staggered conformation, with the electron clouds in the C—H bonds separated as much as possible. The eclipsed conformation places the C—H electron clouds closer together; it is about 3 kcal/mol (12 kJ/mol) higher in energy than the staggered conformation. Three kilocalories is not a large amount of energy, and at room temperature most molecules have enough kinetic energy to overcome this small rotational barrier.

Figure 3-7 is a graph showing how the potential energy of ethane changes as the carbon-carbon bond rotates. The y axis shows the potential energy relative to the most stable (staggered) conformation. The x axis shows the dihedral angle

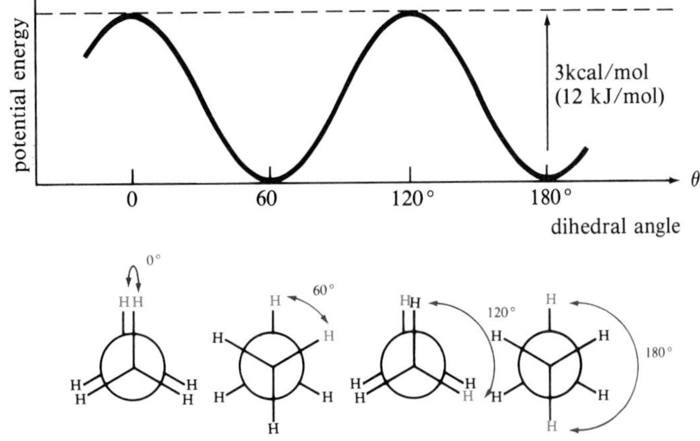

FIGURE 3-7 The torsional energy of ethane is lowest in the staggered conformation. The eclipsed conformation is about 3 kcal/mol (12 kJ/mol) higher in energy. At room temperature, this 3-kcal (12 kJ) barrier is easily overcome, and the molecules rotate constantly.

as it increases from 0° (eclipsed) through 60° (staggered) and on through additional eclipsed and staggered conformations as θ continues to increase. The study of the energetics of different conformations is called **conformational analysis.** The products of many reactions depend on the conformation of the reactant molecule; therefore, conformational analysis is useful to determine which conformations are energetically favored.

As ethane rotates toward the eclipsed conformation, its potential energy increases, and there is resistance to the rotation. This resistance to twisting (torsion) is called **torsional strain,** and the 3 kcal/mol (12 kJ/mol) of energy required is called **torsional energy.**

3-6C CONFORMATIONS OF PROPANE

Propane is the three-carbon alkane, of formula C_3H_8. Figure 3-8 shows a three-dimensional representation of propane and a Newman projection looking down one of the carbon-carbon bonds.

viewed from the end

perspective drawing

Newman projection

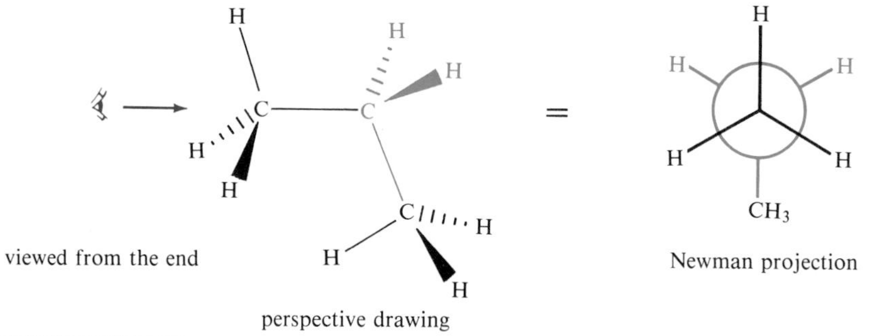

FIGURE 3-8 Propane is the three-carbon alkane. It is shown here as a perspective drawing and as a Newman projection looking down one of the carbon-carbon bonds.

Figure 3-9 shows a graph of the torsional energy of propane as one of the carbon-carbon bonds rotates. The torsional energy of the eclipsed conformation is about 3.3 kcal/mol (14 kJ/mol), only 0.3 kcal (1.3 kJ) more than that required for ethane. Apparently, the torsional strain resulting from eclipsing a carbon-hydrogen bond with a carbon-methyl bond is only 0.3 kcal (1.3 kJ) more than the strain of eclipsing two carbon-hydrogen bonds.

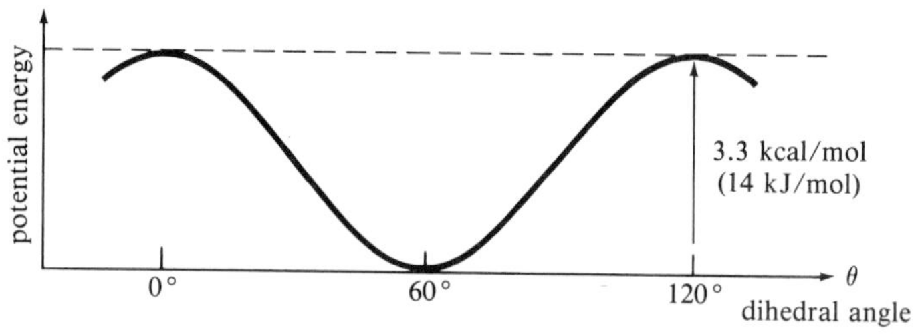

FIGURE 3-9 Torsional energy of propane. When a C—C bond of propane rotates, the torsional energy varies much like it does in ethane, but with 0.3 kcal/mol (1.3 kJ/mol) of additional torsional energy in the eclipsed conformation.

PROBLEM 3-11

Draw a graph, similar to Figure 3-9, of the torsional strain of 2-methylpropane as it rotates about the bond between C1 and C2. Show the dihedral angle and draw a Newman projection for each staggered and eclipsed conformation.

3-7 CONFORMATIONS OF BUTANE

Butane is the four-carbon alkane, of molecular formula C_4H_{10}. We refer to *n*-butane as a straight-chain alkane, but the chain of carbon atoms is not really straight. The angles between the carbon atoms are close to the tetrahedral angle, about 109.5°. Rotations about any of the carbon-carbon bonds are possible; Figure 3-10 shows Newman projections, looking along the central C2—C3 bond, for four conformations of butane. Notice that we have defined the dihedral angle θ as the angle between the two end methyl groups.

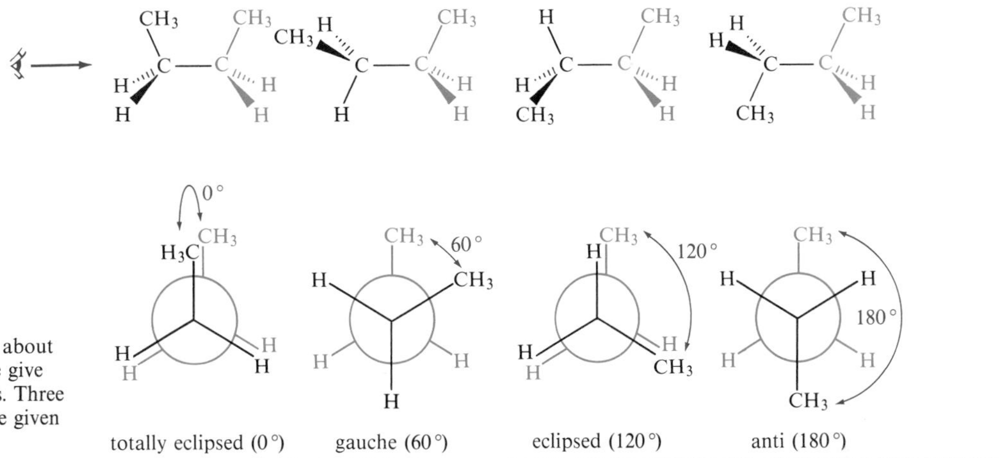

FIGURE 3-10 Butane conformations. Rotations about the center bond in butane give different molecular shapes. Three of these conformations are given specific names.

Three of the conformations shown in Figure 3-10 are given special names. When the methyl groups are pointed in the same direction ($\theta = 0°$), they eclipse each other. This conformation is called **totally eclipsed** to distinguish it from the other eclipsed conformations, like the one at $\theta = 120°$. At a dihedral angle of 60°, the butane molecule is staggered and the methyl groups are toward the left and right of each other. This 60° conformation is called **gauche,** the French word for "left." Another staggered conformation occurs at $\theta = 180°$, with the methyl groups pointing in opposite directions. This conformation is called **anti** because the methyl groups are "opposed."

3-7A TORSIONAL ENERGY OF BUTANE

A graph of the relative torsional energies of the butane conformations is shown in Figure 3-11. All the staggered conformations (anti and gauche) are lower in energy than any of the eclipsed conformations. The anti conformation is lowest in energy because it places the bulky methyl groups as far apart as possible. The gauche conformations, with the methyl groups separated by just 60°, are 0.9 kcal (3.8 kJ) higher in energy than the anti conformation because the methyl groups are close enough that their electron clouds begin to repel each other.

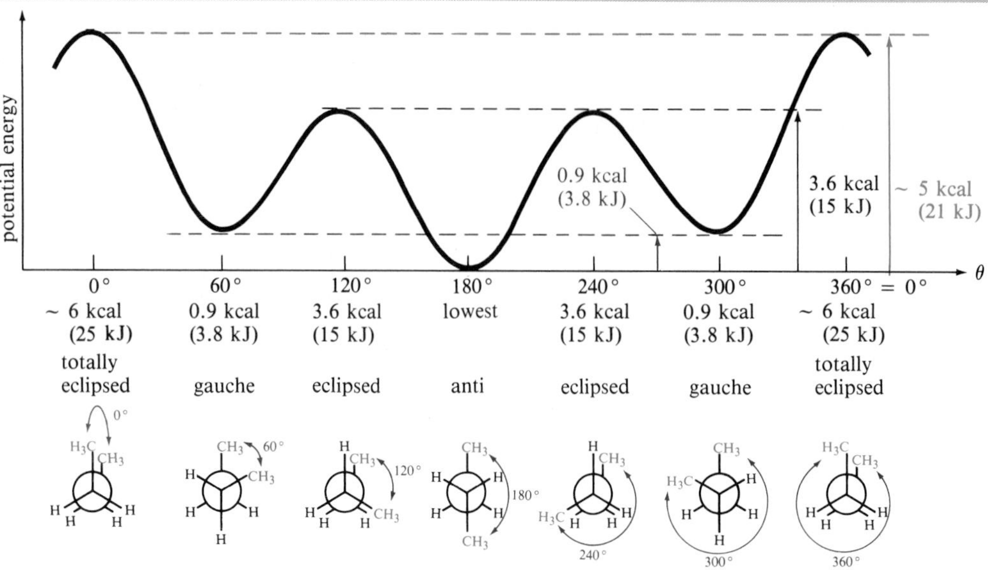

FIGURE 3-11 Torsional energy of butane. The anti conformation is lowest in energy, and the totally eclipsed conformation is highest in energy.

3-7B STERIC HINDRANCE

The totally eclipsed conformation is about 1.4 kcal (6 kJ) higher in energy than the other eclipsed conformations, because it forces the two end methyl groups so close together that their electron clouds experience a strong repulsion. This kind of interference between two bulky groups is called **steric strain** or **steric hindrance.** The following structure shows the interference between the methyl groups in the totally eclipsed conformation.

steric
hindrance

Rotating the totally eclipsed conformation 30° to a gauche conformation releases most, but not all, of this steric strain. The gauche conformation is still 0.9 kcal/mol (3.8 kJ/mol) higher in energy than the most stable anti conformation.

What we have learned about the conformations of butane can be applied to other alkanes. We can predict that carbon-carbon single bonds will assume staggered conformations whenever possible to avoid eclipsing of the groups attached to them. Among the staggered conformations, the anti conformation is preferred. since it has the lowest torsional energy. We must remember, however, that there is enough thermal energy present at room temperature for the molecules to rotate rapidly among all the different conformations. The relative stabilities are important, because more molecules will be found in the more stable conformations than in the less stable ones.

3-8 CONFORMATIONS OF HIGHER ALKANES

The higher alkanes resemble butane in their preference for anti and gauche conformations about the carbon-carbon bonds. The lowest-energy conformation for any straight-chain alkane is the one with all the internal carbon-carbon bonds in their anti conformations. These anti conformations give the alkane chain a zigzag shape. At room temperature, the internal carbon-carbon bonds undergo rotation, and many molecules contain gauche conformations. These gauche conformations make kinks in the zigzag structure. Nevertheless, we frequently draw alkane chains in a zigzag structure to represent the most stable arrangement.

anti conformation　　gauche conformation　　octane, all anti conformation

PROBLEM 3-12

Draw a perspective representation of the most stable conformation of 3-methylhexane.

3-9 CYCLOALKANES

Many organic compounds are **cyclic:** They contain rings of atoms. The carbohydrates we eat are cyclic, the nucleotides that make up our DNA and RNA are cyclic, and the antibiotics we use to treat diseases are cyclic. In this chapter we use the cycloalkanes as examples to discuss the properties and stability of cyclic compounds.

Cycloalkanes are alkanes that contain rings of carbon atoms. Simple cycloalkanes are named like acyclic (noncyclic) alkanes, with the prefix *cyclo-* indicating the presence of a ring. For example, the cycloalkane with four carbon atoms in a ring is called *cyclobutane.* The cycloalkane with seven carbon atoms in a ring is *cycloheptane.* Line-angle formulas are often used for drawing the structures of cycloalkanes (Fig. 3-12).

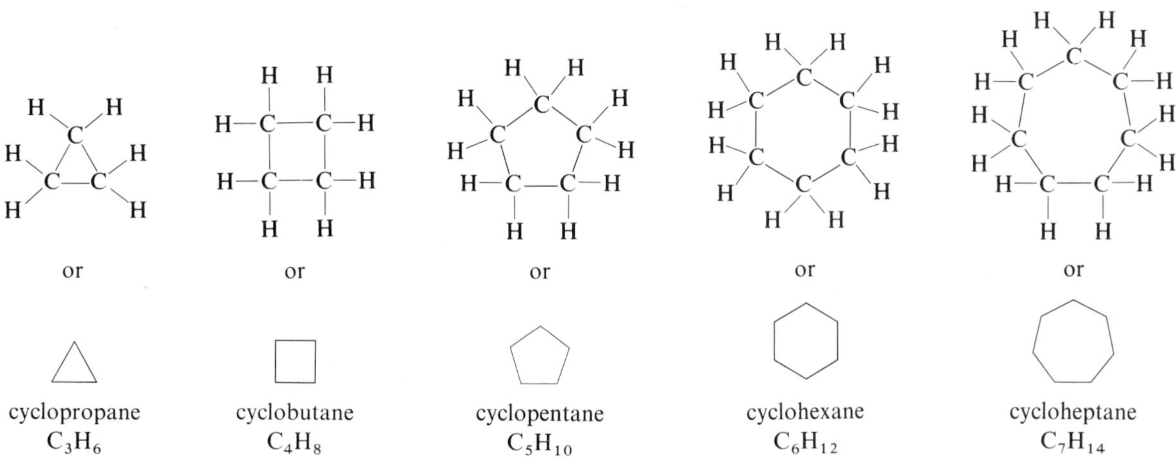

FIGURE 3-12 Structures of some simple cycloalkanes.

3-9A GENERAL MOLECULAR FORMULAS OF CYCLOALKANES

These simple cycloalkanes are rings of CH_2 groups (methylene groups). Each of these cycloalkanes has exactly twice as many hydrogen atoms as carbon atoms, giving the general molecular formula C_nH_{2n}. This general formula has two fewer hydrogen atoms than the $(2n + 2)$ formula for an acyclic alkane. A cyclic chain has no ends, and no hydrogens are needed to cap off the ends of the chain.

3-9B PHYSICAL PROPERTIES OF CYCLOALKANES

Most cycloalkanes resemble the acyclic, open-chain alkanes in their physical properties and in their chemistry. They are nonpolar, relatively inert compounds with boiling points and melting points that depend on their molecular weights. The cycloalkanes are held in a more compact cyclic shape, so their physical properties are similar to those of the compact, branched alkanes. The physical properties of the common cycloalkanes are listed in Table 3-5.

TABLE 3-5
Physical properties of some simple cycloalkanes

Cycloalkane	*Formula*	*Boiling point (°C)*	*Melting point (°C)*	*Density*
cyclopropane	C_3H_6	−33	−128	0.72
cyclobutane	C_4H_8	−12	−50	0.75
cyclopentane	C_5H_{10}	49	−94	0.74
cyclohexane	C_6H_{12}	81	7	0.78
cycloheptane	C_7H_{14}	118	−12	0.81
cyclooctane	C_8H_{16}	148	14	0.83

3-9C NOMENCLATURE OF CYCLOALKANES

Cycloalkanes are named much like acyclic alkanes. Substituted cycloalkanes use the cycloalkane for the base name, with the alkyl groups named as substituents. If there is just one substituent, no numbering is needed.

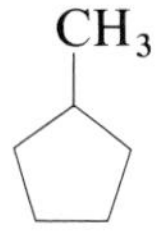

methylcyclopentane 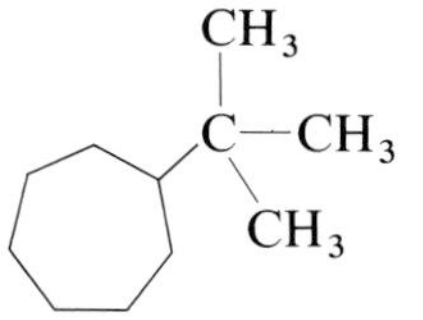   *t*-butylcycloheptane 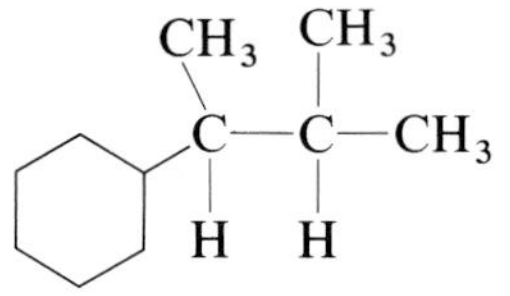   (1,2-dimethylpropyl)cyclohexane

If there are two or more substituents on the ring, the ring carbon atoms are numbered. The numbering begins with one of the substituted ring carbons and continues in the direction that gives the lowest possible numbers to the other substituents. In the name, the substituents are listed in alphabetical order.

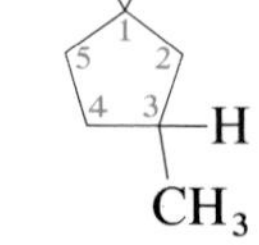

1,1,3-trimethylcyclopentane 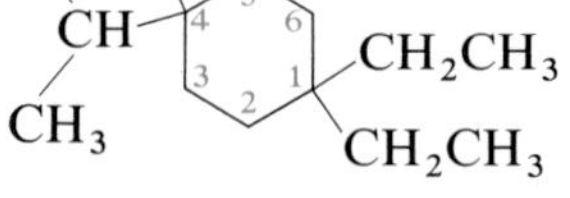   1,1-diethyl-4-isopropylcyclohexane

When the acyclic portion of the molecule contains more carbon atoms than the cyclic portion (or when it contains an important functional group), the cyclic portion is sometimes named as a cycloalkyl substituent on the acyclic chain.

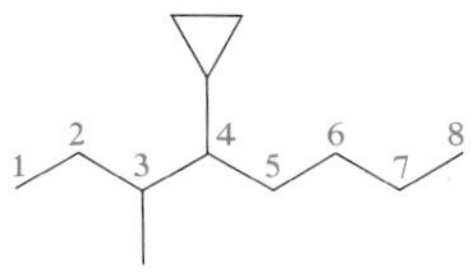

$H-\overset{1}{C}\equiv\overset{2}{C}-\overset{3}{C}H_2-\overset{4}{C}H_2-\underset{5}{C}H_2-$(cyclobutyl)

4-cyclopropyl-3-methyloctane 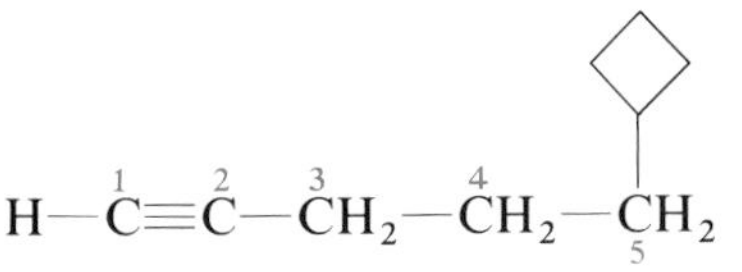   5-cyclobutyl-1-pentyne 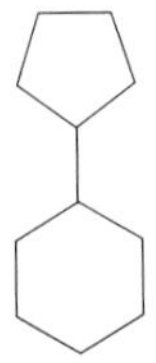   cyclopentylcyclohexane

PROBLEM 3-13

Give an IUPAC name for each of the following compounds.

(a) $CH_3-CH-CH_2CH_3$ on a cyclopentane ring bearing CH_3 and CH_3

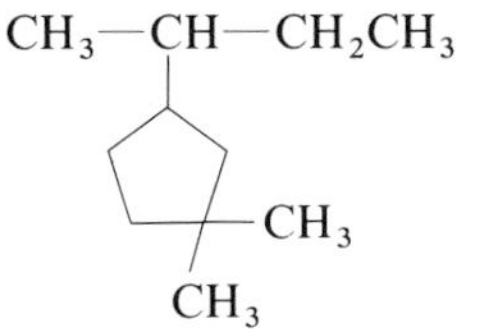

(b) cyclohexane with cyclopropyl and CH_3, CH_3

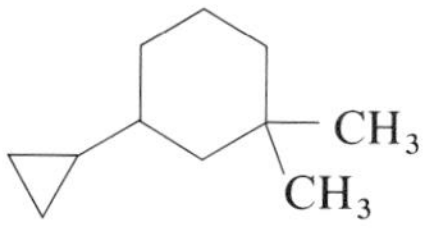

(c)

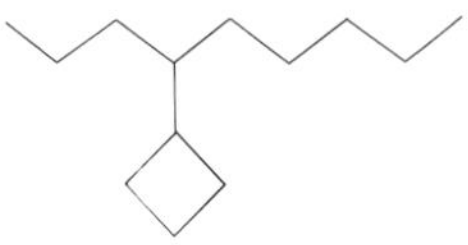

PROBLEM 3-14

Draw the structure and give the molecular formula for each of the following compounds.

(a) cyclododecane (b) propylcyclohexane
(c) cyclopropylcyclopentane (d) 3-ethyl-1,1-dimethylcyclohexane

3-10 GEOMETRIC ISOMERISM IN CYCLOALKANES

Open-chain alkanes undergo rotations about their carbon-carbon single bonds, and they are free to assume any of an infinite number of conformations. Alkenes have rigid double bonds that prevent rotation, giving rise to cis and trans isomers with different orientations of the groups on the double bond (Section 2-10). Cycloalkanes are similar to alkenes in this respect. A cycloalkane has two distinct faces. If two substituents point toward the same face, they are **cis.** If they point toward opposite faces, they are **trans.** These **geometric isomers** cannot interconvert without breaking and re-forming bonds.

Figure 3-13 compares the geometric isomers of 2-butene with those of 1,2-dimethylcyclopentane. You should make models of these compounds to convince yourself that *cis*- and *trans*-1,2-dimethylcyclopentane cannot interconvert by simple rotations about the bonds.

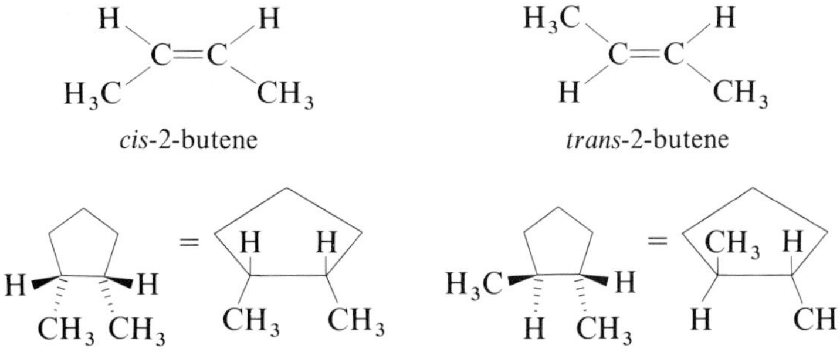

cis-2-butene   *trans*-2-butene

cis-1,2-dimethylcyclopentane   *trans*-1,2-dimethylcyclopentane

FIGURE 3-13 Geometric isomerism in cycloalkanes. Like alkenes, cycloalkane rings are restricted from free rotation. Two substituents on a cycloalkane must be either on the same side (cis) or on opposite sides (trans) of the ring.

PROBLEM 3-15

Which of the following cycloalkanes are capable of geometric (cis-trans) isomerism? Draw the cis and trans isomers.

(a) 3-ethyl-1,1-dimethylcyclohexane (b) 1,4-dimethylcyclohexane
(c) 1-ethyl-3-methylcyclopentane (d) 1-cyclopropyl-2-methylcyclohexane

PROBLEM 3-16

Give a complete IUPAC name for each of the following cycloalkanes.

(a) 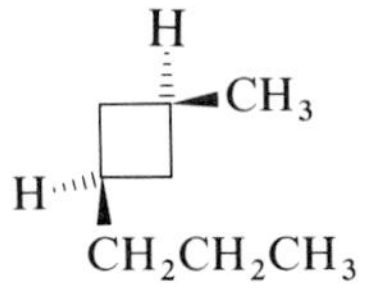 H, CH_3, H, $CH_2CH_2CH_3$ (b) 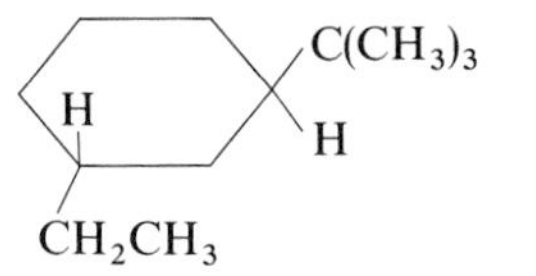 H, $C(CH_3)_3$, H, CH_2CH_3 (c) H, CH_3, H_3C, H

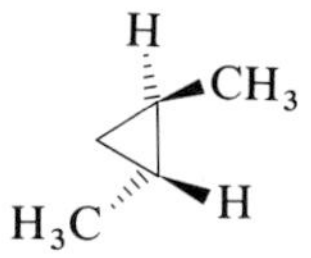

3-11 STABILITIES OF CYCLOALKANES; RING STRAIN

Although all the simple cycloalkanes (up to about C_{20}) have been synthesized, the most common ring sizes contain five and six carbon atoms. We will study the stabilities and conformations of these rings in detail because they help to determine the properties of many important organic compounds.

Why are five-membered and six-membered rings more common than the other ring sizes? Adolf von Baeyer first attempted to explain the relative stabilities of cyclic molecules in the late nineteenth century, and he was awarded a Nobel Prize for this work in 1905. Baeyer reasoned that the carbon atoms in acyclic alkanes have bond angles of 109.5°. We now explain this bond angle by the tetrahedral geometry of the sp^3 hybridized carbon atoms.

If a cycloalkane requires bond angles other than 109.5°, the orbitals of its carbon-carbon bonds cannot achieve optimum overlap, and the cycloalkane must have some **angle strain** (sometimes called **Baeyer strain**) associated with it. Figure 3-14 shows that a planar cyclobutane, with its 90° bond angles, is expected to have significant angle strain.

In addition to this angle strain, the Newman projection in Figure 3-14 shows that the bonds are eclipsed, resembling the *totally eclipsed* conformation of butane (Section 3-7). This eclipsing of bonds gives rise to torsional strain. Together, the angle strain and the torsional strain add to give what we call the **ring strain** of the cyclic compound. The amount of ring strain associated with a particular compound depends primarily on the size of the ring.

Before we discuss the ring strain of different cycloalkanes, we need to consider how ring strain is measured. In theory, we should measure the total amount of energy in the cyclic compound and then subtract the amount of energy in a similar,

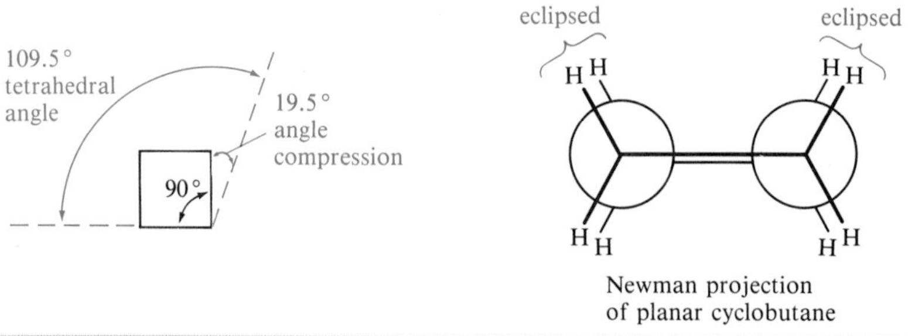

FIGURE 3-14 The ring strain of a planar cyclobutane molecule results from two factors. It has angle strain from the compressing of the bond angles to 90° rather than the tetrahedral angle of 109.5°. It also has torsional strain from the eclipsing of the bonds.

strain-free reference compound. The difference should be the amount of extra energy due to ring strain in the cyclic compound. These measurements are commonly made using heats of combustion.

3-11A HEATS OF COMBUSTION

The **heat of combustion** is the amount of heat released when a compound is burned with an excess of oxygen in a sealed container called a *bomb calorimeter*. If the compound has extra energy as a result of ring strain, that extra energy is released in the combustion. The heat of combustion is usually measured by a temperature rise in the water bath surrounding the "bomb."

A cycloalkane can be represented by the molecular formula $(CH_2)_n$, so the general reaction in the bomb calorimeter is:

$$(CH_2)_n + \tfrac{3}{2}n\, O_2 \longrightarrow n\, CO_2 + n\, H_2O + n\,(\text{energy per } CH_2)$$

cycloalkane, $(CH_2)_n$ — heat of combustion

The molar heat of combustion of cyclohexane is found to be nearly twice that of cyclopropane, simply because cyclohexane contains twice as many methylene (CH_2) groups per mole. To compare the relative stabilities of cycloalkanes, we divide the heat of combustion by the number of methylene (CH_2) groups. The result is the energy per CH_2 group. A comparison of these energies provides values for the amounts of ring strain (per methylene group) in the cycloalkanes.

Table 3-6 shows the heats of combustion, in kcal/mol and kJ/mol, for some simple cycloalkanes. The reference value of 157.4 kcal (659 kJ) per mole of CH_2 groups comes from a (presumably unstrained) long-chain alkane. The values show large amounts of ring strain in cyclopropane and cyclobutane. Cyclopentane, cycloheptane, and cyclooctane have much smaller amounts of ring strain, and cyclohexane has no ring strain at all. We will discuss several of these molecules in detail to explain how this pattern of ring strain arises.

3-11B CYCLOPROPANE

The information in Table 3-6 shows that cyclopropane bears more ring strain per methylene group than do any of the other cycloalkanes. There are two factors that contribute to this large ring strain. First is the angle strain required to force the

TABLE 3-6
Heats of combustion (per mole) for some simple alkanes[a]

Ring size	*Cycloalkane*	*Molar heat of combustion*	*Heat of combustion per CH_2 group*	*Ring strain per CH_2 group*	*Total ring strain*
3	cyclopropane	499.8 kcal	166.6 kcal	9.2 kcal	27.6 kcal (115 kJ)
4	cyclobutane	655.9 kcal	164.0 kcal	6.6 kcal	26.4 kcal (110 kJ)
5	cyclopentane	793.5 kcal	158.7 kcal	1.3 kcal	6.5 kcal (27 kJ)
6	cyclohexane	944.5 kcal	157.4 kcal	0.0 kcal	0.0 kcal (0.0 kJ)
7	cycloheptane	1108.3 kcal	158.3 kcal	0.9 kcal	6.3 kcal (26 kJ)
8	cyclooctane	1268.9 kcal	158.6 kcal	1.2 kcal	9.6 kcal (40 kJ)
reference: long-chain alkane			157.4 kcal	0.0 kcal	0.0 kcal (0.0 kJ)

[a] By using heats of combustion (per mole of CH_2 groups), we can compute the ring strain of cycloalkanes. The reference value is the heat of combustion per CH_2 group for a long-chain alkane.

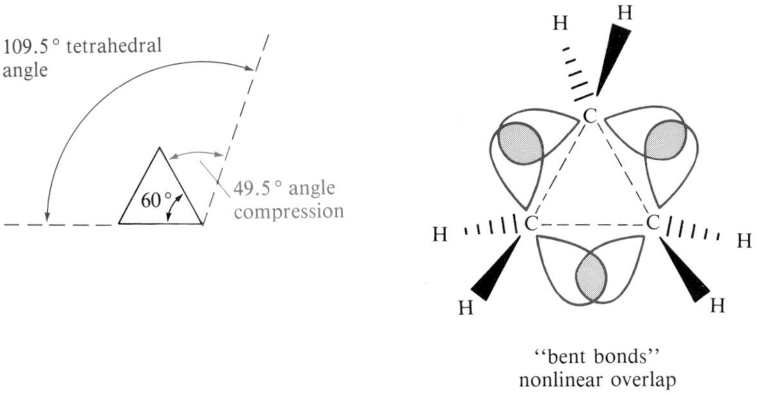

109.5° tetrahedral angle

60°

49.5° angle compression

"bent bonds" nonlinear overlap

FIGURE 3-15 Angle strain. The bond angles in cyclopropane are compressed to 60° from the usual 109.5° bond angle of sp^3 hybridized carbon atoms. This severe angle strain leads to nonlinear overlap of the sp^3 orbitals and "bent bonds."

bond angles from the tetrahedral angle of 109.5° to the 60° angles of cyclopropane. The bonding overlap of the carbon-carbon sp^3 orbitals is weakened when the bond angles differ so much from the tetrahedral angle. The sp^3 orbitals cannot point directly toward each other, and they overlap at an angle to form a weaker "bent bond" (Fig. 3-15).

Torsional strain is the second factor in cyclopropane's large ring strain. The three-membered ring is planar and all the bonds are eclipsed. A Newman projection of one of the carbon-carbon bonds (Fig. 3-16) shows that the conformation of the cyclopropane bonds resembles the totally eclipsed conformation of butane. The torsional strain in cyclopropane is not as great as its angle strain, but it helps to account for the large total ring strain.

Cyclopropane is generally more reactive than other alkanes. Reactions that open the cyclopropane ring release 27.6 kcal (115 kJ) of ring strain, which provides an additional driving force for these reactions.

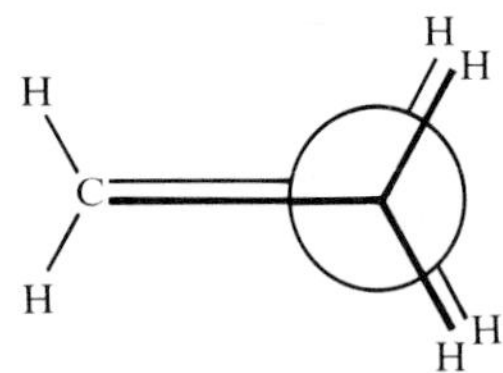

Newman projection of cyclopropane

FIGURE 3-16 Torsional strain. All the carbon-carbon bonds in cyclopropane are eclipsed, generating torsional strain that contributes to the total ring strain.

PROBLEM 3-17

The heat of combustion of *cis*-1,2-dimethylcyclopropane is larger than that of the trans isomer. Which isomer is more stable? Use drawings to explain this difference in stability.

3-11C CYCLOBUTANE

The total ring strain in cyclobutane is almost as great as that in cyclopropane, but distributed over four carbon atoms. If cyclobutane were perfectly planar and square, it would have 90° bond angles. A planar geometry requires eclipsing of all the bonds, however, as in cyclopropane. To reduce this torsional strain, cyclobutane actually assumes a slightly folded form with bond angles of 88°. These smaller bond angles require slightly more angle strain than 90° angles, but the relief of some of the torsional strain appears to compensate for a small increase in angle strain (Fig. 3-17).

PROBLEM 3-18

trans-1,2-Dimethylcyclobutane is more stable than *cis*-1,2-dimethylcyclobutane, but *cis*-1,3-dimethylcyclobutane is more stable than *trans*-1,3-dimethylcyclobutane. Use drawings to explain these observations.

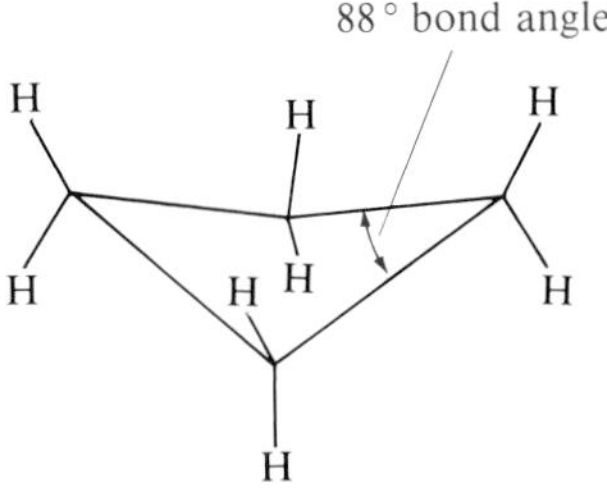

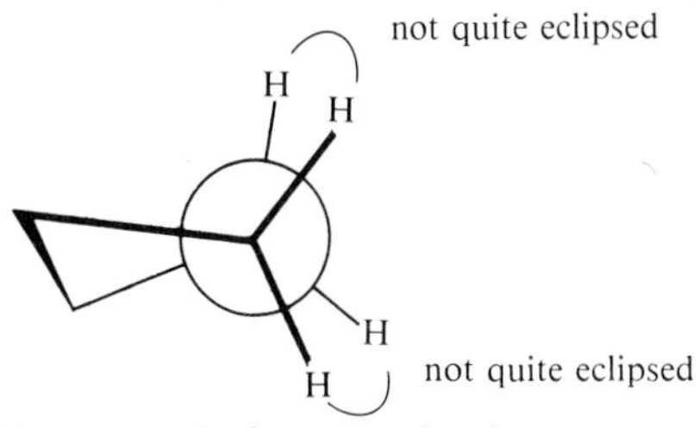

FIGURE 3-17 The conformation of cyclobutane is slightly folded. Folding gives partial relief from the eclipsing of bonds, as shown in the Newman projection. Compare this actual structure with the hypothetical planar structure in Figure 3-14.

3-11D CYCLOPENTANE

If cyclopentane had the shape of a planar, regular pentagon, its bond angles would be 108°, a value very close to the tetrahedral angle of 109.5°. A planar structure would require all the bonds to be eclipsed, however. Cyclopentane actually assumes a slightly puckered "envelope" conformation that reduces the eclipsing and lowers the torsional strain (Fig. 3-18). This puckered shape is not fixed in one conformation but undulates around the ring by the thermal up-and-down motion of the five methylene groups. The "flap" of the envelope seems to move around the ring as the molecule undulates.

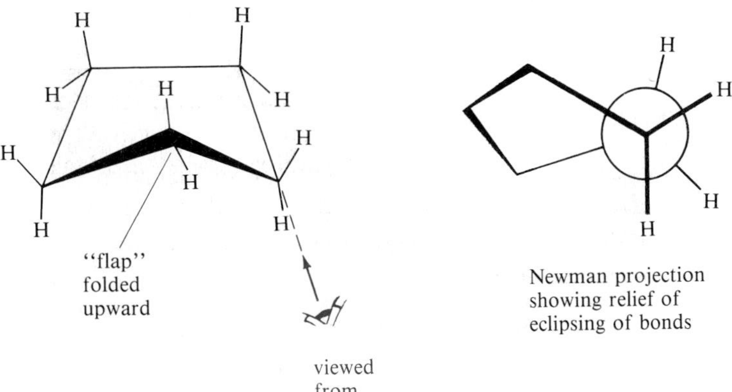

FIGURE 3-18 The conformation of cyclopentane is slightly folded, like the shape of an envelope. This puckered conformation reduces the eclipsing of adjacent CH_2 groups.

3-12 CYCLOHEXANE CONFORMATIONS

The combustion data (Table 3-6) show that cyclohexane has *no* ring strain. Cyclohexane must have bond angles that are very near the tetrahedral angle (no angle strain) and also have no eclipsing of bonds (no torsional strain). A planar, regular hexagon would have bond angles of 120° rather than 109.5°, implying some angle strain. A planar ring would also have torsional strain from eclipsing of the bonds on adjacent CH_2 groups. The cyclohexane ring clearly cannot be planar.

3-12A CHAIR AND BOAT CONFORMATIONS

The cyclohexane molecule achieves tetrahedral bond angles and staggered conformations by assuming a puckered conformation. The most stable conformation is the **chair conformation** shown in Figure 3-19. Build a molecular model of cyclohexane, and compare its shape with the drawings in Figure 3-19. In the chair conformation, the angles between the carbon-carbon bonds are all 109.5°. The

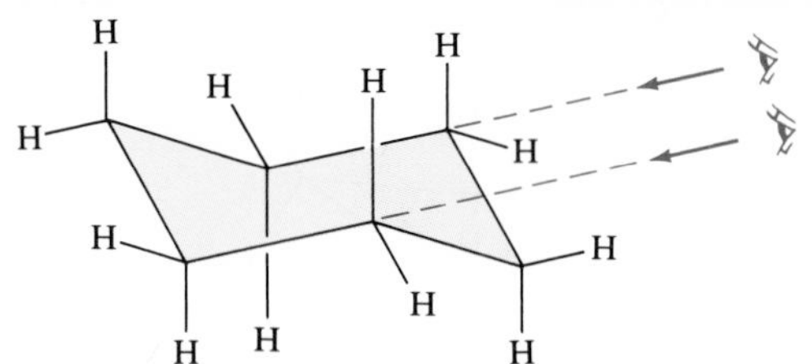

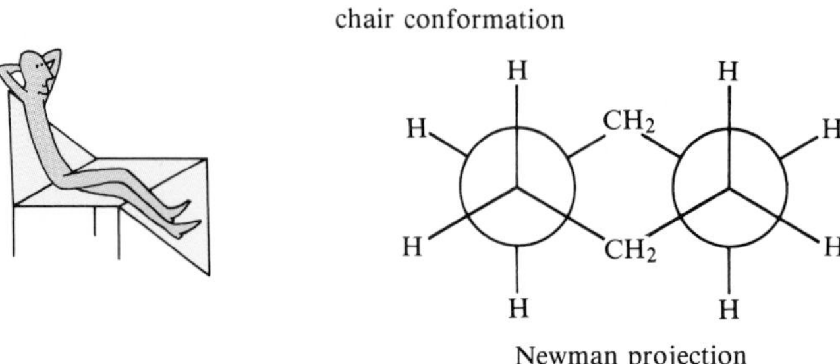

FIGURE 3-19 The chair conformation of cyclohexane has one methylene group puckered upward and another puckered downward. Viewed from the Newman projection, the chair conformation has no eclipsing of the carbon-carbon bonds. The bond angles are 109.5°.

Newman projection looking down the "seat" bonds show that the bonds are all in staggered conformations.

The **boat conformation** of cyclohexane (Fig. 3-20) also has bond angles of 109.5° and avoids angle strain. The boat conformation resembles the chair conformation except that the "footrest" methylene group is folded upward. The boat conformation suffers from torsional strain, however, because there is eclipsing of bonds. This eclipsing forces two of the hydrogens on the ends of the "boat" to interfere with each other. These hydrogens are called **flagpole hydrogens,** because they point upward from the ends of the boat like two flagpoles. The second drawing in Figure 3-20 uses a Newman projection of the carbon-carbon bonds along the sides of the boat to show this eclipsing.

A cyclohexane molecule in the boat conformation actually exists as a slightly twisted **twist boat conformation,** also shown in Figure 3-20. If you assemble your molecular model in the boat conformation and twist it slightly, the flagpole hydrogens move away from each other and the eclipsing of the bonds is reduced. Even though the twist boat is lower in energy than the symmetrical boat, it is still about 5.5 kcal/mol (23 kJ/mol) higher in energy than the chair conformation. When someone refers to the "boat conformation," the twist boat is often intended.

At any instant, most of the molecules in a cyclohexane sample are in chair conformations. The energy barrier between the boat and chair is sufficiently low, however, that the conformations interconvert many times each second. The inter-

"flagpole" hydrogens — eclipsed — symmetrical boat — Newman projection — "twist" boat

FIGURE 3-20 In the symmetrical boat conformation of cyclohexane, eclipsing of bonds results in torsional strain. In the actual molecule, the boat conformation is twisted to give the twist boat, a conformation with less eclipsing of the bonds and less interference of the two flagpole hydrogens.

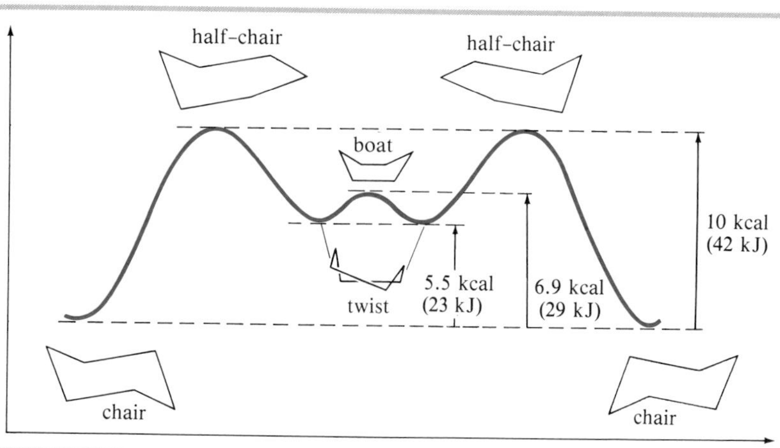

FIGURE 3-21 Conformational energy of cyclohexane. The chair conformation is most stable, followed by the twist boat. To convert between these two conformations, the molecule must pass through the unstable half-chair conformation.

conversion from the chair to the boat takes place by the "footrest" of the chair flipping upward and forming the boat. The highest-energy point in this process is the conformation where the footrest is planar with the sides of the molecule. This unstable arrangement is called the **half-chair conformation.** Figure 3-21 shows how the energy of cyclohexane varies as it interconverts between the boat and chair forms.

3-12B AXIAL AND EQUATORIAL POSITIONS

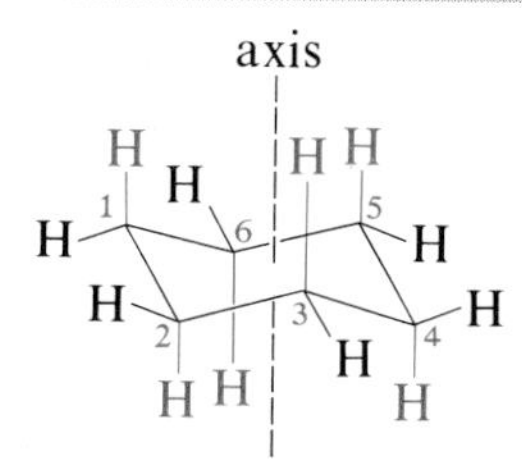

FIGURE 3-22 The axial bonds are directed vertically, parallel to the axis of the ring. The equatorial bonds are directed outward, toward the equator of the ring. As they are numbered here, the odd-numbered carbons have their *upward* bonds axial and their *downward* bonds equatorial. The even-numbered carbons have their *downward* bonds axial and their *upward* bonds equatorial.

If we could freeze cyclohexane in the chair conformation, we would see that there are two distinctly different kinds of carbon-hydrogen bonds. Six of the bonds (one on each carbon atom) are directed up and down, parallel to the axis of the ring. These are called **axial bonds.** The other six bonds point out from the ring, along the "equator" of the ring. These are called **equatorial bonds.** The axial bonds are shown in red in Figure 3-22, and the equatorial bonds are shown in black.

Each carbon atom in cyclohexane is bonded to two hydrogen atoms, one directed upward and one downward. As the carbon atoms are numbered in Figure 3-22, C1 has an axial bond upward and an equatorial bond downward. C2 has an equatorial bond upward and an axial bond downward. The pattern alternates. The odd-numbered carbon atoms have axial bonds up and equatorial bonds down, like C1. The even-numbered carbons have equatorial bonds up and axial bonds down, like C2. This pattern of alternating axial and equatorial bonds is helpful for predicting the conformations of substituted cyclohexanes, as described in Sections 3-13 and 3-14.

3-12C DRAWING CHAIR CONFORMATIONS

Drawing realistic pictures of cyclohexane conformations is not difficult, but certain rules must be followed to show the actual positions of the substituents on the ring. Make a cyclohexane ring with your models, put it in the chair conformation, and use it to follow along with this discussion.

We draw the carbon-carbon bond framework using lines with only three different slopes. (Look at your model!) These slopes are labeled *a*, *b*, and *c* in Figure 3-23. We can draw the chair with the headrest to the left and the footrest to the right, or vice versa.

Now we fill in the axial and equatorial bonds. The axial bonds are drawn vertically (Fig. 3-24). When a vertex of the chair points upward, its axial bond also points upward. A downward-pointing vertex has a downward-pointing axial

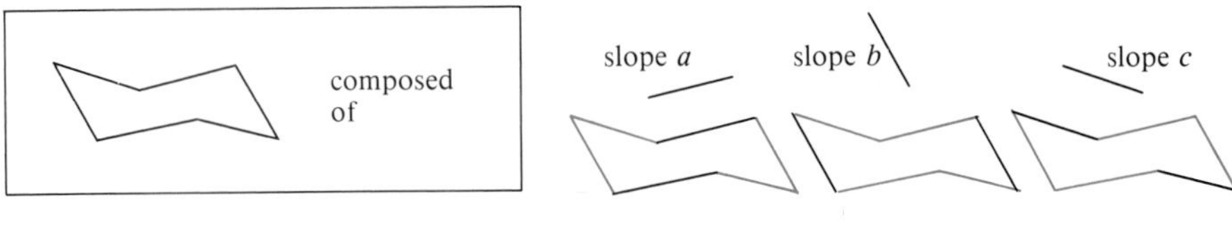

FIGURE 3-23 The carbon-carbon bond framework of the chair conformation is drawn using three slopes of lines.

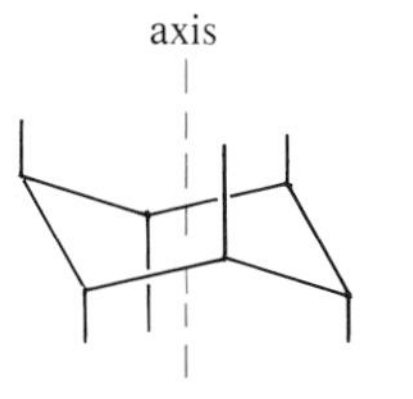

FIGURE 3-24 Axial bonds. Cyclohexane chair with the axial bonds drawn in. The axial bonds are directed vertically.

bond. For example, C1 is an upward-pointing vertex, so its axial bond points upward. C2 points downward, and its axial bond also points downward.

The equatorial bonds are more difficult. Each carbon atom is represented by a vertex formed by the intersection of two lines, having two of the possible slopes *a*, *b*, and *c*. Each equatorial bond should have the third slope—that is the slope that is *not* represented by the two lines forming the vertex.

Look at your model and also at Figure 3-25. The vertex representing C1 is formed by lines of slopes *b* and *c*. Its equatorial bond should have slope *a*. C2 is represented by a vertex formed by lines of slopes *a* and *b*. The equatorial bond on C2 should have slope *c*. Figure 3-25 shows cyclohexane with the equatorial bonds filled in. Notice the W- and M-shaped patterns that result when these bonds are drawn correctly.

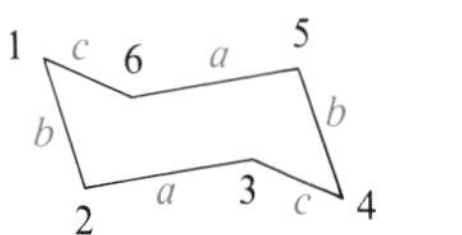

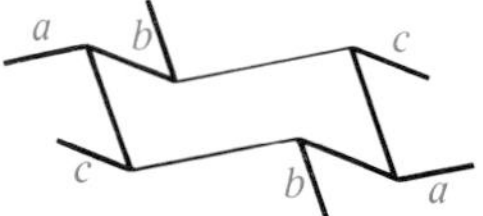

FIGURE 3-25 Equatorial bonds. The equatorial bond from each vertex follows the third slope *not* used to form that vertex. When the equatorial bonds are correctly drawn, there is a W-shaped pattern at one end and an M-shaped pattern at the other.

PROBLEM 3-19

Draw 1,2,3,4,5,6-hexamethylcyclohexane with all the methyl groups

(a) in axial positions.
(b) in equatorial positions.

3-13 CONFORMATIONS OF MONOSUBSTITUTED CYCLOHEXANES

A substituent on a cyclohexane ring (in the chair conformation) can occupy either an axial or an equatorial position. In many cases, the reactivity of the substituent depends on whether its position is axial or equatorial. The two possible chair conformations for methylcyclohexane are shown below. These conformations are in equilibrium, because they interconvert at room temperature. The twist boat conformation serves as an intermediate in this **chair-chair interconversion.** Notice that the chair-chair interconversion inverts the axial and equatorial relationships of the substituents.

Chair-chair interconversion of methylcyclohexane

H equatorial
CH_3 axial
⇌
H
CH_3
⇌
axial H
CH_3 equatorial

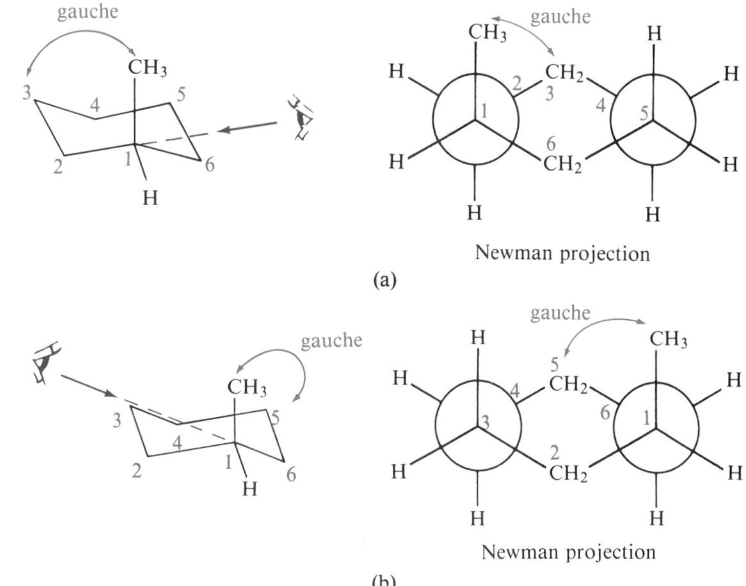

FIGURE 3-26 (a) When the methyl substituent is in an axial position on C1, it is gauche to C3. (b) The axial methyl group on C1 is also gauche to C5 of the ring.

Because the two chair conformations interconvert at room temperature, the one that is lower in energy predominates. Careful measurements have shown that the chair conformation with the methyl group in an equatorial position is the most stable conformation. It is about 1.7 kcal/mol (7.1 kJ/mol) lower in energy than the conformation with the methyl group in an axial position. Both of these chair conformations are lower in energy than any boat conformation. We can show how the 1.7-kcal energy difference between the axial and equatorial positions arises by examining molecular models and Newman projections of the two conformations. First, make a model of methylcyclohexane and use it to follow this discussion.

Consider a Newman projection looking along the armrest bonds of the conformation with the methyl group axial (Fig. 3-26a): The methyl group is on C1, and we are looking from C1 toward C2. There is a 60° angle between the bond to the methyl group and the bond from C2 to C3, and the methyl substituent and C3 are in a gauche relationship. In our analysis of torsional strain in butane, we saw that a gauche interaction raises the energy of a conformation by 0.9 kcal/mol (3.8 kJ/mol) relative to the anti conformation.

This axial methyl group is also gauche to C5, as you will see if you look along the C1—C6 bond in your model. Figure 3-26b shows this second gauche relationship.

The Newman projection for the cyclohexane conformation with the methyl group equatorial shows that the methyl group has an anti relationship to both C3 and C5. Figure 3-27 shows the Newman projection along the C1 to C2 bond, with the anti relationship of the methyl group to C3.

PROBLEM 3-20

Draw a Newman projection, similar to Figure 3-27, down the C1 to C6 bond in the equatorial conformation of methylcyclohexane. Show that the equatorial methyl group is also anti to C5.

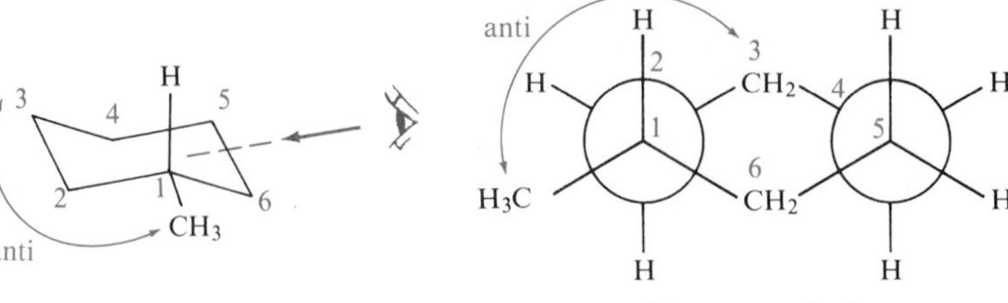

FIGURE 3-27 Looking down the C1—C2 bond of the equatorial conformation, we find that the methyl group is anti to C3.

The axial methylcyclohexane conformation has two gauche interactions, each representing about 0.9 kcal (3.8 kJ) of additional energy. The equatorial methyl group has no gauche interactions. We predict that the axial conformation is higher in energy by 1.8 kcal (7.6 kJ) per mole, in good agreement with the experimental value of 1.7 kcal (7.1 kJ) per mole. Figure 3-28 shows that the gauche relationship of the axial methyl group with C3 and C5 places the methyl hydrogens close to the axial hydrogens on these carbons, and their electron clouds begin to interfere. This form of steric hindrance is called a **1,3-diaxial interaction** because it involves substituents on the carbon atom that would be numbered C3 if the carbon bearing the methyl group is numbered C1. These 1,3-diaxial interactions are not present in the equatorial conformation.

1,3-diaxial interactions

more stable by 1.7 kcal/mol (7.1 kJ/mol)

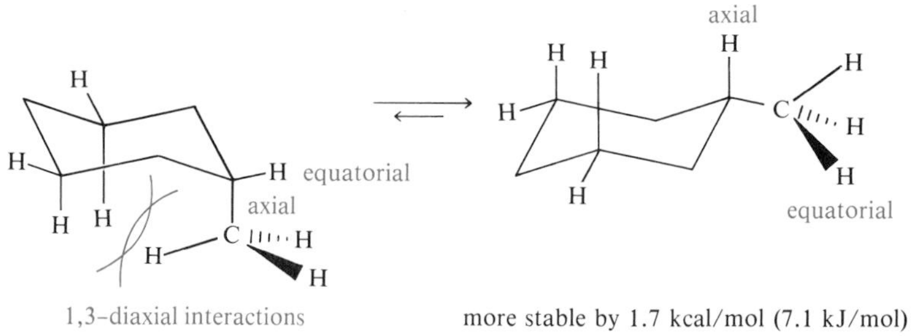

FIGURE 3-28 The axial substituent interferes with the axial hydrogens on C3 and C5. This interference is called a 1,3-diaxial interaction.

In most cases a larger group has a larger energy difference between the axial and equatorial positions. This is because the 1,3-diaxial interaction shown in Figure 3-28 is stronger for larger groups. Table 3-7 shows the energy differences between the axial and equatorial positions for several different alkyl groups and functional groups. The axial position is higher in energy in each case.

PROBLEM 3-21

Table 3-7 shows that the axial-equatorial energy difference for methyl, ethyl, and isopropyl groups increases gradually: 1.7, 1.8, and 2.1 kcal/mol (7.1, 7.5, and 8.8 kJ/mol). The *t*-butyl group jumps to an energy difference of 5.4 kcal/mol (23 kJ/mol), over twice the value for the isopropyl group. Draw pictures of the axial conformations of isopropylcyclohexane and *t*-butylcyclohexane, and show why the *t*-butyl substituent experiences such a large increase in axial energy over the isopropyl group.

PROBLEM 3-22

Draw the most stable conformation of

(a) ethylcyclohexane. (b) isopropylcyclohexane.
(c) *t*-butylcyclohexane.

TABLE 3-7
Energy differences between the axial and equatorial conformations of monosubstituted cyclohexanes

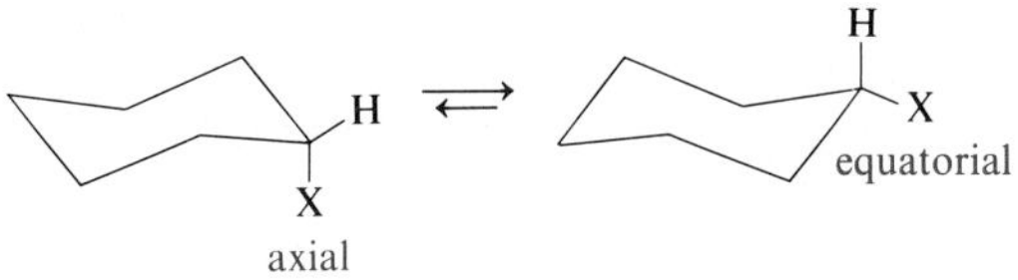

X	$E(axial) - E(equatorial)$ (kcal/mol)	(kJ/mol)
—F	0.2	0.8
—CN	0.2	0.8
—Cl	0.5	2.1
—Br	0.6	2.5
—OH	1.0	4.1
—COOH	1.4	5.9
$—CH_3$	1.7	7.1
$—CH_2CH_3$	1.8	7.5
$—CH(CH_3)_2$	2.1	8.8
$—C(CH_3)_3$	5.4	23

3-14 CONFORMATIONS OF DISUBSTITUTED CYCLOHEXANES

The steric interference between substituents in axial positions is particularly severe when there are large substituents on two carbon atoms that bear a 1,3-diaxial relationship (C1 and C3, or C1 and C5). Figure 3-29 shows the large 1,3-diaxial interaction between the two methyl groups in the unfavorable diaxial conformation of *cis*-1,3-dimethylcyclohexanne. The 1,3-diaxial interaction is relieved when the molecule converts to the diequatorial conformation.

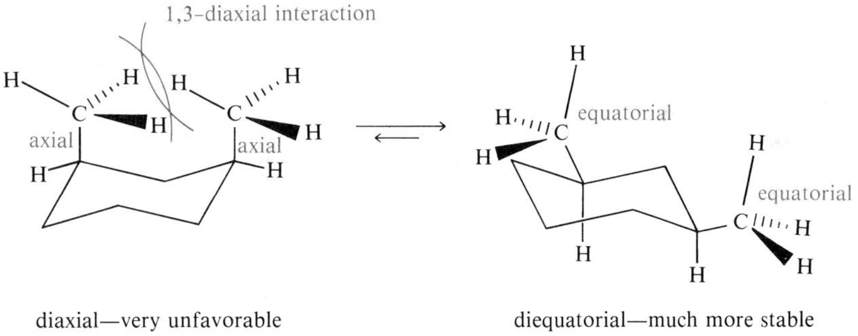

FIGURE 3-29 Two chair conformations are possible for *cis*-1,3-dimethylcyclohexane. The unfavorable conformation has both methyl groups in axial positions, with a 1,3-diaxial interaction between them. The more stable conformation has both methyl groups in equatorial positions.

Either of the two chair conformations of *trans*-1,3-dimethylcyclohexane has one methyl group in an axial position and one in an equatorial position. These conformations have equal energies, and they are present in equal amounts.

Chair conformations of trans-1,3-dimethylcyclohexane

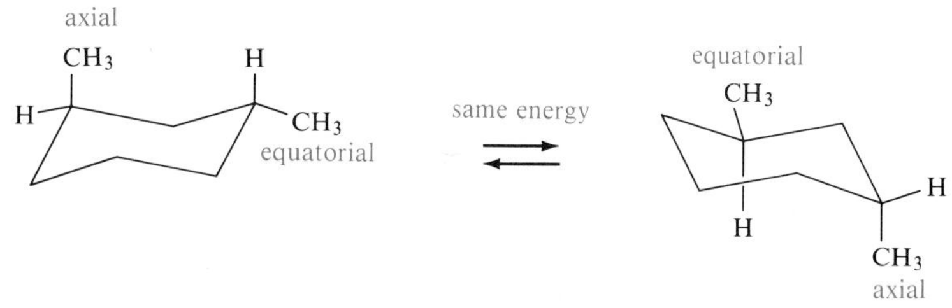

Now we can compare the relative stabilities of the cis and trans isomers of 1,3-dimethylcyclohexane. The most stable conformation of the cis isomer has both methyl groups in equatorial positions, while either conformation of the trans isomer places one methyl group in an axial position. The trans isomer is therefore higher in energy than the cis isomer by about 1.7 kcal/mol, the energy difference between axial and equatorial methyl groups. It is important to remember that the cis and trans isomers cannot interconvert, and there is no equilibrium between these isomers.

SOLVED PROBLEM 3-3

(a) Draw both chair conformations of *cis*-1,2-dimethylcyclohexane, and determine which conformer is more stable.
(b) Repeat for the trans isomer.
(c) Predict which isomer (cis or trans) is more stable.

SOLUTION (a) There are two chair conformations possible for the cis isomer, and these two conformations interconvert at room temperature. Each of these conformations places one methyl group axial and one equatorial, giving them the same energy.

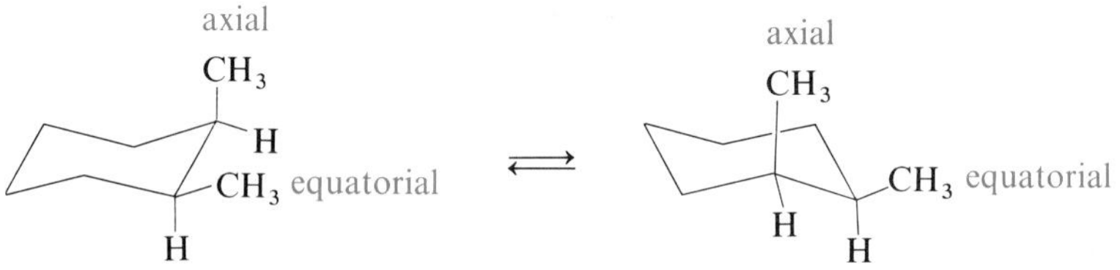

axial CH_3 / H / CH_3 equatorial / H ⇌ axial CH_3 / CH_3 equatorial / H / H

same energy

(b) There are two chair conformations of the trans isomer that interconvert at room temperature. One of these has both methyl groups axial, and the other has both equatorial. The diequatorial conformation is more stable, because neither methyl group occupies the more hindered axial position.

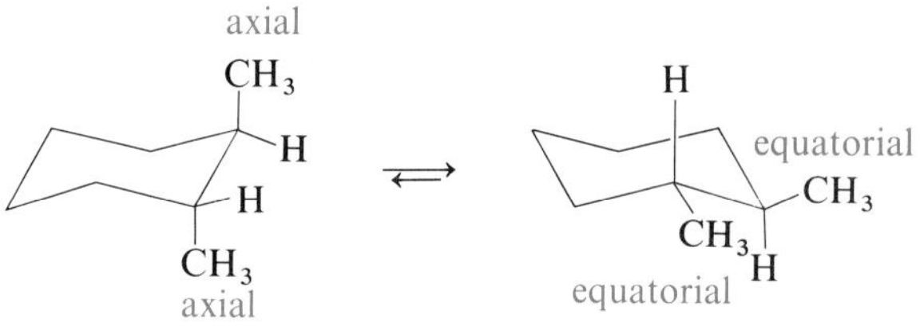

axial CH_3 / H / H / CH_3 axial ⇌ H / equatorial CH_3 / CH_3 equatorial / H

higher energy (diaxial) lower energy (diequatorial)

(c) The trans isomer is more stable. The most stable conformation of the trans isomer is diequatorial, and therefore about 1.7 kcal/mol (7.1 kJ/mol) lower in energy than either conformation of the cis isomer, each having one methyl axial and one equatorial. Remember, however, that cis and trans are distinct isomers and cannot interconvert.

PROBLEM 3-23

(a) Draw both chair conformations of *cis*-1,4-dimethylcyclohexane, and determine which conformer is more stable.
(b) Repeat for the trans isomer.
(c) Predict which isomer (cis or trans) is more stable.

PROBLEM 3-24

Use your results from Problem 3-23 to complete the following table. Each entry shows the positions of two groups arranged as shown. For example, two groups that are trans on adjacent carbons (*trans*-1,2) must be both equatorial (e,e) or both axial (a,a).

Positions	*cis*	*trans*
1,2	(e,a) or (a,e)	(e,e) or (a,a)
1,3		
1,4		

3-14A SUBSTITUENTS OF DIFFERENT SIZES

In many substituted cyclohexanes, the substituents are of different sizes. As shown in Table 3-7 (page 109), the energy difference between the axial and equatorial positions for a larger group is greater than that for a smaller group. In general, if both groups cannot be equatorial, the most stable conformation has the larger group equatorial and the smaller group axial.

SOLVED PROBLEM 3-4

Draw the most stable conformation of *trans*-1-ethyl-3-methylcyclohexane.

SOLUTION First, we draw the two conformations.

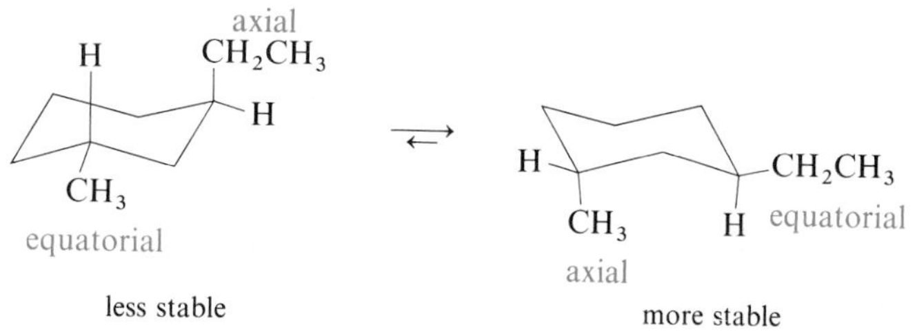

axial CH_2CH_3, H; CH_3 equatorial, H — less stable ⇄ H, CH_3 axial; CH_2CH_3 equatorial, H — more stable

Both of these conformations require one group to be axial while the other is equatorial. The ethyl group is bulkier than the methyl group, so the conformation with the ethyl group equatorial is more stable. These chair conformations are in equilibrium at room temperature, and the one with the equatorial ethyl group predominates.

PROBLEM 3-25

Draw the two chair conformations of each of the following substituted cyclohexanes. In each case, label the more stable conformation.

(a) *cis*-1-ethyl-2-methylcyclohexane
(b) *trans*-1-ethyl-2-methylcyclohexane
(c) *cis*-1-ethyl-4-isopropylcyclohexane
(d) *trans*-1-ethyl-4-methylcyclohexane

3-14B RECOGNIZING CIS AND TRANS ISOMERS

Some students find it difficult to look at a chair conformation and tell whether a disubstituted cyclohexane is the cis isomer or the trans isomer. In the following drawing the two methyl groups appear to be oriented in similar directions. They are actually trans but are sometimes mistakenly identified as cis.

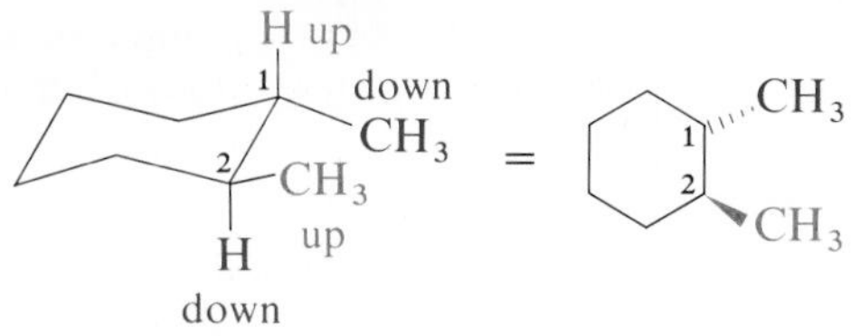

H up
1 down
CH_3
2 CH_3
H up
down

CH_3
1
2
CH_3

trans-1,2-dimethylcyclohexane

This ambiguity is eliminated by recognizing that each of the ring carbon atoms has two available bonds, one directed upward and one directed downward. In this drawing the methyl group on C1 is on the downward bond, and the methyl on C2 is on the upward bond. Because one is on a downward bond and one on an upward bond, their relationship is trans. A cis relationship would require both groups to be upward or both to be downward.

PROBLEM 3-26

Name each of the following compounds, remembering that two up bonds are cis; two down bonds are cis; one up bond and one down bond are trans.

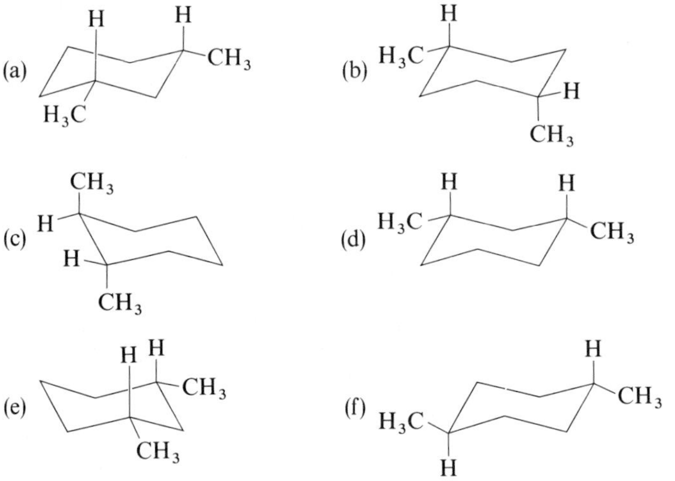

(a) H H CH_3 H_3C

(b) H H_3C H CH_3

(c) CH_3 H H CH_3

(d) H H H_3C CH_3

(e) H H CH_3 CH_3

(f) H H_3C CH_3 H

3-14C EXTREMELY BULKY GROUPS

Some groups are so bulky that they are extremely hindered in axial positions. Cyclohexanes with *tertiary*-butyl substituents show the strong hindrance of an axial *t*-butyl group. Regardless of the other groups present, the most stable conformation has a *t*-butyl group in an equatorial position. The following figure shows the severe steric interactions in a chair conformation with a *t*-butyl group axial.

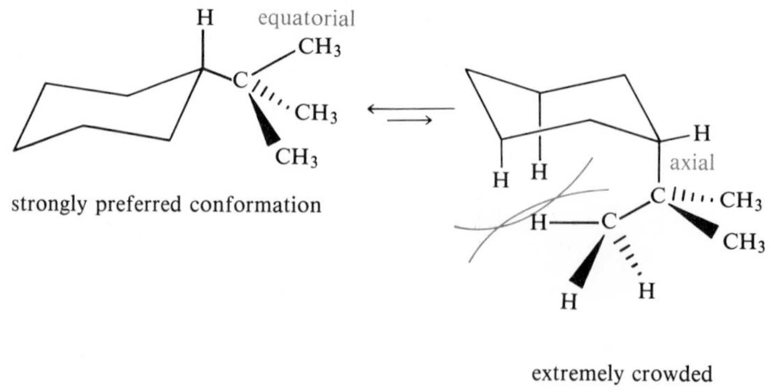

H equatorial
CH_3
C
CH_3
CH_3

strongly preferred conformation

H
axial
H H
C CH_3
CH_3
H—C
H H

extremely crowded

If two *t*-butyl groups are attached to the ring, both of them are much less hindered in equatorial positions. When neither chair conformation allows both bulky groups to be equatorial, they may force the ring into a twist boat conformation. For example, *cis*-1,4-di-*t*-butylcyclohexane (Fig. 3-30) is most stable in the twist boat conformation.

axial

t-butyl group moves out of the axial position

twist boat

FIGURE 3-30 The most stable conformation of *cis*-1,4-di-*t*-butylcyclohexane is a twist boat conformation. Either of the two chair conformations requires one of the bulky *t*-butyl groups to occupy an axial position.

PROBLEM 3-27
Draw the most stable conformation of

(a) *cis*-1-*t*-butyl-3-ethylcyclohexane.
(b) *trans*-1-*t*-butyl-2-methylcyclohexane.
(c) *trans*-3-*t*-butyl-1-(1,1-dimethylpropyl)cyclohexane.

3-15 BICYCLIC MOLECULES

Two or more rings can be joined into *bicyclic* or *polycyclic* systems. In principle, there are three ways that two rings may be joined. **Fused rings** are the most common, sharing two adjacent carbon atoms and the bond between them. **Bridged rings** are also common, sharing two nonadjacent carbon atoms (the **bridgehead carbons**) and one or more carbon atoms (the bridge) between them. **Spirocyclic compounds,** in which the two rings share only one carbon atom, are relatively rare.

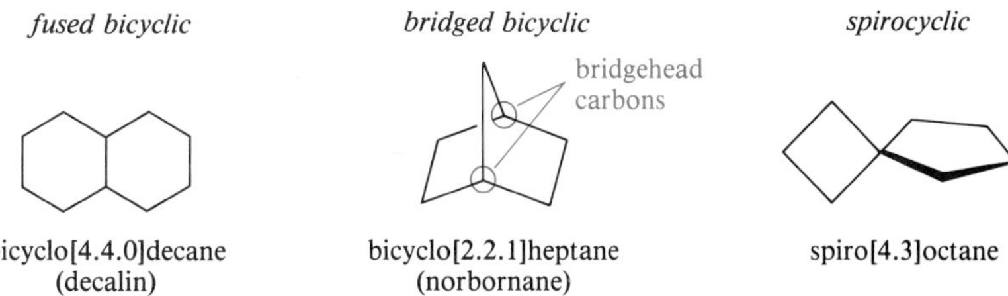

3-15A NOMENCLATURE OF BICYCLIC ALKANES

The name of a bicyclic compound is based on the name of the alkane having the same number of carbons as there are in the ring system. This name follows the prefix *bicyclo* and a set of brackets. The following examples contain eight carbon atoms and are named bicyclo[4.2.0]octane and bicyclo[3.2.1]octane, respectively.

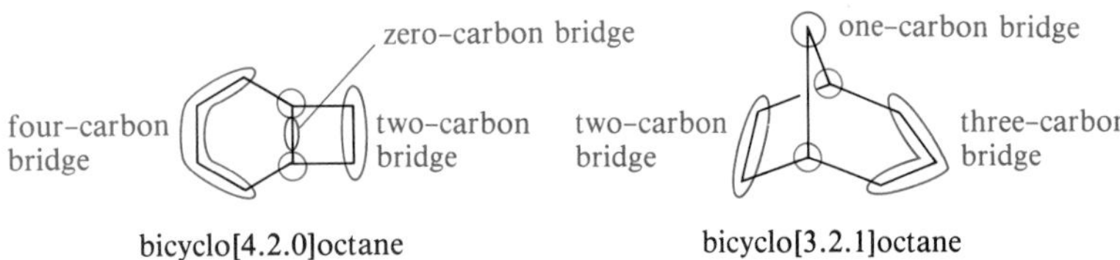

All fused and bridged bicyclic systems have three "bridges" connecting the two bridgehead atoms (red circles) where the rings connect. The numbers in the brackets give the number of carbon atoms in each of the three bridges connecting the bridgehead carbons, in order of decreasing size.

PROBLEM 3-28
Name the following compounds.

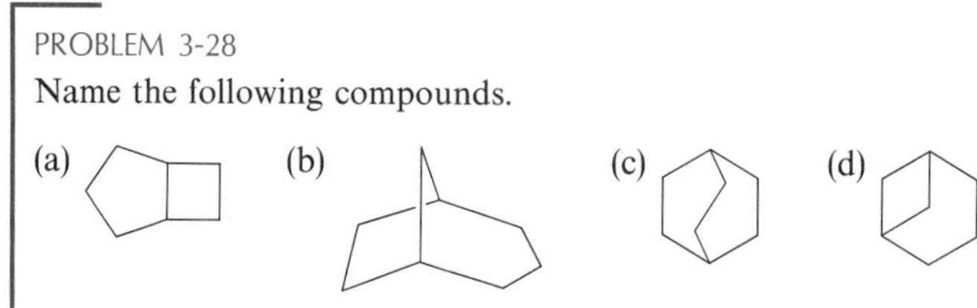

3-15B CIS- AND TRANS-DECALIN

Decalin (bicyclo[4.4.0]decane) is the most common example of a fused ring system. Two geometric isomers of decalin exist, as shown in Figure 3-31. One has the rings fused using two cis bonds, while the other is fused using two trans bonds. You should make a model of decalin to follow this discussion.

If we consider the left ring in the drawing of *cis*-decalin, the bonds to the right ring are both directed downward (and the attached hydrogens are directed upward). These bonds are therefore cis, and this is a cis ring fusion. In *trans*-decalin, one of the bonds to the right ring is directed upward and the other downward. These bonds are trans, and this is a trans ring fusion. The six-membered rings in both isomers assume chair conformations, as shown in Figure 3.31.

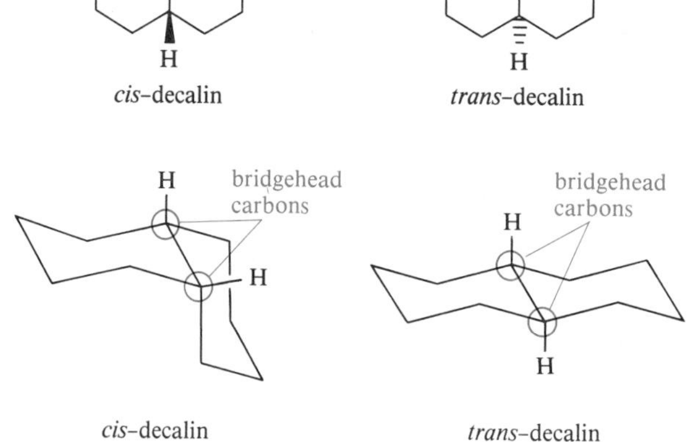

FIGURE 3-31 *cis*-Decalin has a ring fusion where the second ring is attached by two cis bonds. *trans*-Decalin is fused using two trans bonds. The six-membered rings in *cis*- and *trans*-decalin assume chair conformations.

The conformation of *cis*-decalin is somewhat flexible, while the trans isomer is quite rigid. If one of the rings in the trans isomer did a chair-chair interconversion, the bonds to the second ring would both become axial and would be directed 180° apart. This is an impossible conformation, and it prevents any chair-chair interconversion in *trans*-decalin.

PROBLEM 3-29
Use your models to do a chair-chair interconversion on each ring of the conformation of *cis*-decalin shown in Figure 3-31. Draw the conformation that results.

3-16 REACTIONS OF ALKANES

Alkanes are the least reactive class of organic compounds. This low reactivity is reflected in another term for alkanes: **paraffins.** The name *paraffin* comes from two Latin terms, *para,* meaning "opposed to," and *affinitas,* meaning "affinity." Chemists found that alkanes did not react with strong acids or bases, nor with most other reagents. They attributed this low reactivity to lack of an affinity for other reagents, and they coined the name "paraffins."

Most of the useful reactions of alkanes take place under energetic or high-temperature conditions. These conditions are inconvenient to use in a laboratory, because they require specialized equipment and they make it difficult to control the rate of the reaction. Alkane reactions usually form mixtures of products that are difficult to separate. These mixtures may be of commercial importance for an industry, however, where the products may be separated and sold separately. Newer methods of selective functionalization may eventually change this picture. For now, however, the following alkane reactions are rarely seen in laboratory applications, but they are widely used in the chemical industry and even in your home and car.

3-16A COMBUSTION

Combustion is a rapid oxidation that takes place at high temperatures, converting alkanes to carbon dioxide and water. Little control over the reaction is possible, except for moderating the temperature and controlling the fuel/air ratio to achieve efficient burning.

$$C_nH_{(2n+2)} \quad + \quad \text{excess } O_2 \quad \xrightarrow{\text{heat}} \quad n\,CO_2 \quad + \quad (n+1)\,H_2O$$

Example

$$CH_3CH_2CH_3 \quad + \quad 5\,O_2 \quad \xrightarrow{\text{heat}} \quad 3\,CO_2 \quad + \quad 4\,H_2O$$

Unfortunately, the burning of gasoline and fuel oil pollutes the air and depletes the petroleum resources needed for lubricants and chemical feedstocks. Solar and nuclear heat sources cause less pollution, and they do not deplete these important natural resources. Facilities that use these more modern heat sources are more expensive than those that rely on the combustion of alkanes, however.

3-16B CRACKING AND HYDROCRACKING

As discussed in Section 3-4B, the catalytic "cracking" of large hydrocarbons at high temperatures gives smaller hydrocarbons. The cracking process is usually operated under conditions that give the maximum yields of products needed for use in gasoline. Hydrocracking involves the addition of hydrogen to give saturated hydrocarbons; cracking without hydrogen given mixtures of alkanes and alkenes.

Catalytic hydrocracking

$C_{12}H_{26}$ long-chain alkane $\xrightarrow[\text{catalyst}]{H_2,\ \text{heat}}$ C_5H_{12} + C_7H_{16} shorter-chain alkanes

Catalytic cracking

$$\underset{\text{long-chain alkane}}{C_{12}H_{26}} \xrightarrow[\text{catalyst}]{\text{heat}} \underset{\text{shorter-chain alkanes and alkenes}}{C_5H_{10} + C_7H_{16}}$$

3-16C HALOGENATION

Under the proper conditions, alkanes react with the halogens (F_2, Cl_2, Br_2, I_2) to form alkyl halides. For example, methane reacts with chlorine (Cl_2) to form a mixture of chloromethane (methyl chloride), dichloromethane (methylene chloride), trichloromethane (chloroform), and tetrachloromethane (carbon tetrachloride).

$$CH_4 + Cl_2 \xrightarrow{\text{heat or light}} CH_3Cl + CH_2Cl_2 + CHCl_3 + CCl_4 + HCl$$

Heat or light is usually needed to initiate this **halogenation.** The reaction of alkanes with chlorine or bromine proceeds at a moderate rate and is easily controlled. The reaction with fluorine is often too fast to control, however, while iodine reacts very slowly or not at all. We discuss the halogenation of alkanes in detail in Chapter 4.

GLOSSARY

acyclic Not cyclic. (p. 97)

alkane A hydrocarbon having only single bonds; a **saturated hydrocarbon;** general formula: C_nH_{2n+2}. (p. 78)

alkyl group The group of atoms remaining after a hydrogen atom is removed from an alkane; an alkanelike substituent. Symbolized by R. (p. 87)

angle strain or **Baeyer strain** The strain associated with compressing bond angles to smaller (or larger) angles. (p. 100)

anti conformation A conformation with a 180° dihedral angle between the largest groups. Usually the lowest-energy conformation. (p. 95)

aromatic hydrocarbon A hydrocarbon having a benzenelike aromatic ring. (p. 78)

axial position One of six positions (three up and three down) on the cyclohexane ring that are parallel to the "axis" of the ring. (p. 105)

bridged bicyclic compound A compound containing two rings joined at nonadjacent carbon atoms. (p. 113)

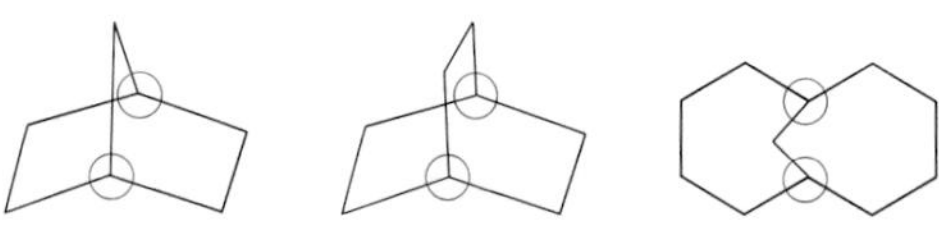

bridged bicyclic systems (bridgeheads circled)

bridgehead carbons The carbon atoms shared by two or more rings. There are three chains of carbon atoms (bridges) connecting the bridgeheads. (p. 113)

catalytic cracking The heating of large alkane molecules over a catalyst to cleave them into smaller molecules. (p. 115)

chair-chair interconversion The process of one chair conformation of a cyclohexane flipping into another one, with all of the axial and equatorial positions reversed. The boat conformation is an intermediate for the chair-chair interconversion. (p. 106)

chair (methyls axial) boat chair (methyls equatorial)

combustion A rapid oxidation at high temperatures in the presence of air or oxygen. (p. 115)

common names The names that have developed historically, generally with a specific name for each compound; also called **trivial names**. (p. 85)

conformational analysis The study of the energetics of different conformations. (p. 94)

conformations or **conformers** Structures that are related by rotations about single bonds. In most cases, conformations interconvert at room temperature, and they are not true isomers. (p. 92)

totally eclipsed conformation gauche conformation anti conformation

conformations of cyclohexane. (p. 103)

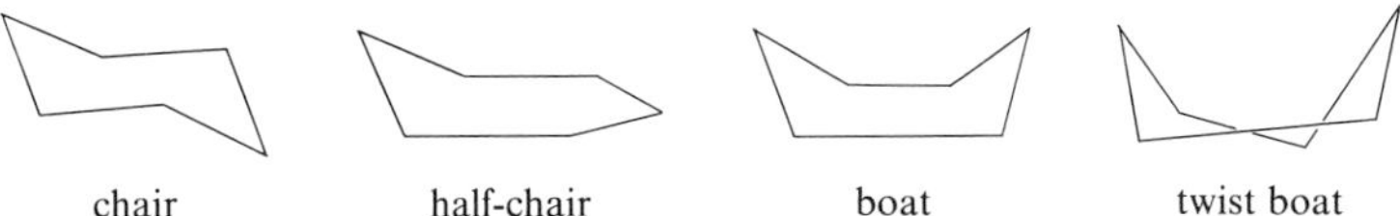

> **chair conformation:** The most stable conformation of cyclohexane, with one part (the "headrest") puckered upward and another part (the "footrest") puckered downward.
>
> **boat conformation:** The less stable puckered conformation of cyclohexane, with both parts puckered upward. The most stable boat conformation is actually the **twist boat** (or simply **twist**) conformation. Twisting minimizes the torsional strain and the steric strain.
>
> **flagpole hydrogens:** Two hydrogens in the boat conformation point upward like flagpoles. The twist boat reduces the steric repulsion of the flagpole hydrogens.
>
> **half-chair conformation:** The unstable conformation halfway between the chair conformation and the boat conformation. Part of the ring is flat in the half-chair conformation.

cyclic Containing a ring of atoms. (p. 97)

cycloalkane An alkane that contains a ring of carbon atoms; general formula: C_nH_{2n}. (p. 97)

degree of alkyl substitution The number of alkyl groups bonded to a carbon atom in a compound or in an alkyl group. (p. 88)

primary (1°) carbon atom **secondary (2°) carbon atom** **tertiary (3°) carbon atom** **quaternary (4°) carbon atom**

1,3-diaxial interaction The strong steric hindrance between two axial groups on cyclohexane carbons with one carbon between them. (p. 108)

dihedral angle (θ) The angle between two specified groups in a Newman projection. (p. 92)

eclipsed conformation Any conformation with bonds directly lined up with each other, one behind the other in the Newman projection. The conformation with $\theta = 0°$ is an eclipsed conformation. (p. 92)

equatorial position One of the six bonds (three down and three up) on the cyclohexane ring that are directed out toward the "equator" of the ring. (p. 105)

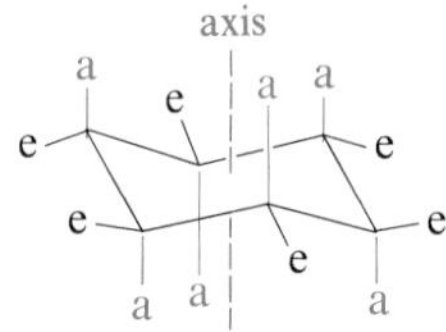

axial bonds in red, equatorial bonds in black

fused ring system A molecule in which two or more rings share two adjacent carbon atoms. (p. 113)

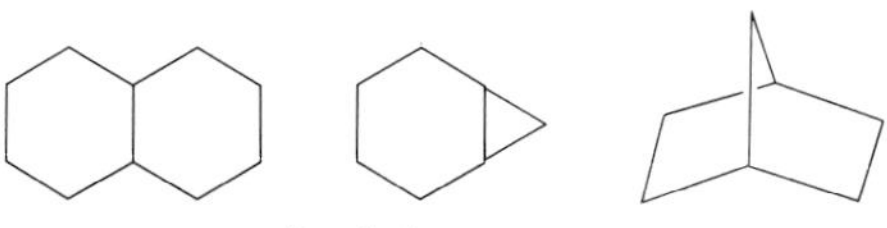

fused ring systems

gauche conformation A conformation with a 60° dihedral angle between the largest groups. (p. 95)

geometric isomers Stereoisomers that differ only with respect to their cis or trans arrangement on a ring or double bond. (p. 99)

> **cis:** Having two similar groups directed toward the same face of a ring or double bond.
>
> **trans:** Having two similar groups directed toward opposite faces of a ring or double bond.

H_3C, CH_3 / C=C / H, H — *cis*-2-butene

H_3C, H / C=C / H, CH_3 — *trans*-2-butene

H, H / CH_3, CH_3 — *cis*-1,2-dimethylcyclopentane

CH_3, H / H, CH_3 — *trans*-1,2-dimethylcyclopentane

halogenation The reaction of alkanes with halogens, in the presence of heat or light, to give products with halogen atoms substituted for hydrogen atoms. (p. 116)

$$R{-}H + X_2 \xrightarrow{\text{heat or light}} R{-}X + HX \qquad X = F, Cl, Br, I$$

heat of combustion The heat given off when a mole of a compound is burned with excess oxygen to give CO_2 and H_2O in a *bomb calorimeter*. A measure of the energy content of a molecule. (p. 101)

homologs Two compounds that differ only by one or more $-CH_2-$ groups. (p. 84)

hydrocracking Catalytic cracking in the presence of hydrogen to give mixtures of alkanes. (p. 115)

hydrophilic Attracted to water; soluble in water.

hydrophobic Repelled by water; insoluble in water. (p. 80)

IUPAC names The systematic names that follow the rules adopted by the International Union of Pure and Applied Chemistry. (p. 85)

methine group The $-\overset{|}{C}H-$ group.

methylene group The $-CH_2-$ group. (p. 84)

methyl group The $—CH_3$ group. (p. 87)

***n*-alkane, normal alkane,** or **straight-chain alkane** An alkane with all its carbon atoms in a single chain, with no branching or alkyl substituents. (p. 80)

Newman projections A way of drawing the conformations of a molecule by looking straight down the bond connecting two carbon atoms. (p. 92)

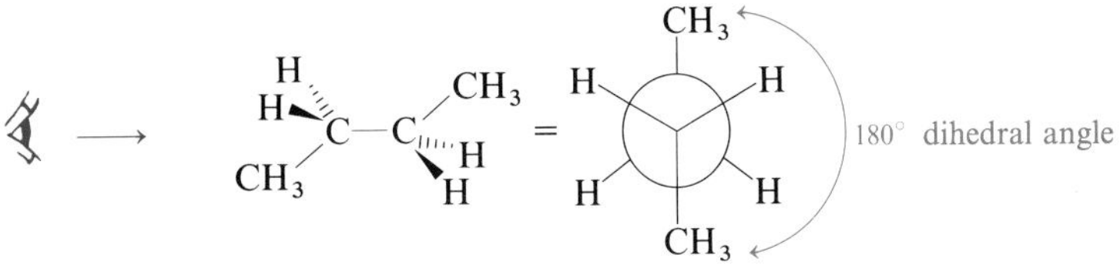

180° dihedral angle

a Newman projection of butane in the anti conformation

paraffins Another term for alkanes. (p. 113)

ring strain The extra strain associated with the cyclic structure of a compound, as compared with a similar acyclic compound. Composed of angle strain and torsional strain. (p. 100)

> **angle strain** or **Baeyer strain:** The strain associated with compressing bond angles to smaller (or larger) angles.
>
> **torsional strain:** The strain associated with eclipsing of bonds in the ring.

saturated Having no double or triple bonds. (p. 78)

sawhorse structures A way of picturing conformations by looking down at an angle toward the carbon-carbon bond. (p. 92)

skew conformation Any conformation that is not precisely staggered or eclipsed. (p. 93)

spirocyclic compounds Bicyclic compounds in which the two rings share only one carbon atom. (p. 113)

staggered conformation Any conformation with the bonds equally spaced in the Newman projection. The conformation with $\theta = 60°$ is a staggered conformation. (p. 92)

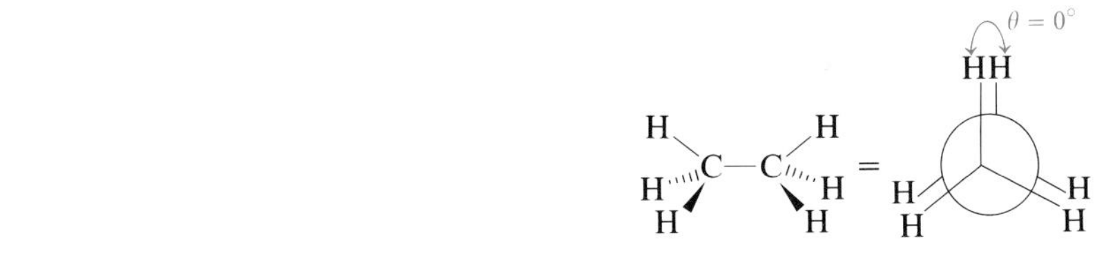

$\theta = 0°$ $\theta = 60°$

eclipsed conformation of ethane

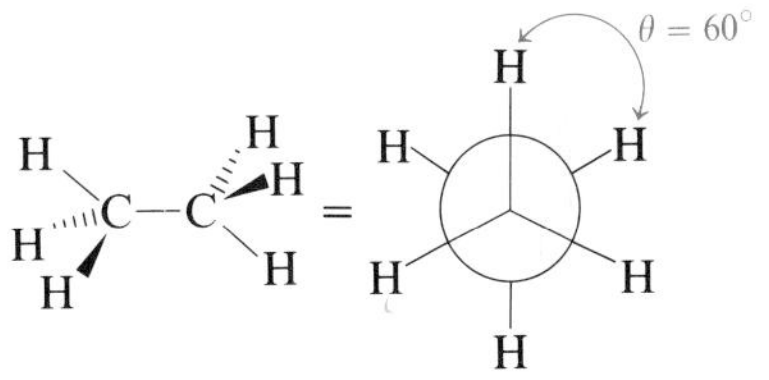

staggered conformation of ethane

steric hindrance or **steric strain** The interference between two bulky groups that are so close together that their electron clouds experience a strong repulsion. (p. 96)

substituent A side chain or appendage on the main chain. (p. 86)

systematic names Same as IUPAC names, the systematic names that follow the rules adopted by the International Union of Pure and Applied Chemistry. (p. 85)

torsional energy or **conformational energy** The energy required to twist a bond into a specific conformation. (p. 94)

torsional strain The resistance to twisting about a bond. (p. 94)

totally eclipsed conformation A conformation with a 0° dihedral angle between the largest groups. Usually, the highest-energy conformation. (p. 95)

ESSENTIAL PROBLEM-SOLVING SKILLS IN CHAPTER 3

1. Explain and predict trends in physical properties of alkanes.
2. Correctly name alkanes, cycloalkanes, and bicyclic alkanes.

3. Given the name of an alkane, draw the structure and give the molecular formula.
4. Compare the energies of alkane conformations and predict the most stable conformation.
5. Compare the energies of cycloalkanes and explain ring strain.
6. Identify and draw cis and trans stereoisomers of cycloalkanes.
7. Draw accurate cyclohexane conformations, and predict the most stable conformations of substituted cyclohexanes.

STUDY PROBLEMS

3-30. Define and give an example of each of the following terms.

(a) alkane
(b) alkene
(c) alkyne
(d) saturated
(e) hydrophobic
(f) aromatic
(g) hydrophilic
(h) *n*-alkane
(i) methylene group
(j) methyl group
(k) common name
(l) systematic name
(m) conformers
(n) eclipsed
(o) Newman projection
(p) staggered
(q) gauche
(r) anti conformation
(s) an acyclic alkane
(t) geometric isomers on a ring
(u) chair conformation
(v) boat conformation
(w) twist boat
(x) half-chair conformation
(y) axial position
(z) equatorial position
(A) catalytic cracking
(B) chair-chair interconversion
(C) fused ring system
(D) bridged bicyclic compound
(E) bridgehead carbon atoms
(F) combustion

3-31. Which of the following Lewis structures represent the same compound? Which ones represent different compounds?

```
        H  H  H             H  H  H  H            H  CH3
        |  |  |             |  |  |  |            |  |
(a) H—C—C—C—H        H—C—C—C—C—H         H—C—C—H
        |  |  |             |  |  |  |            |  |
        H  H  CH3           H  H  H  H            H  CH3

       CH3 CH3          H   CH3
        |   |           |   |
     H—C—C—H        H—C—C—H
        |   |           |   |
        H   H          CH3  H
```

(b)

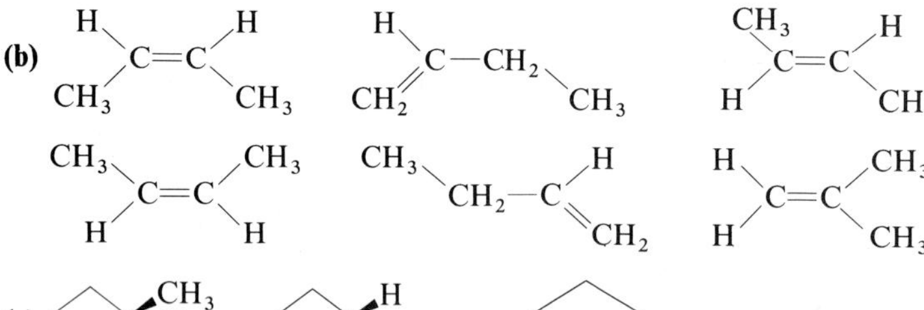

(c)

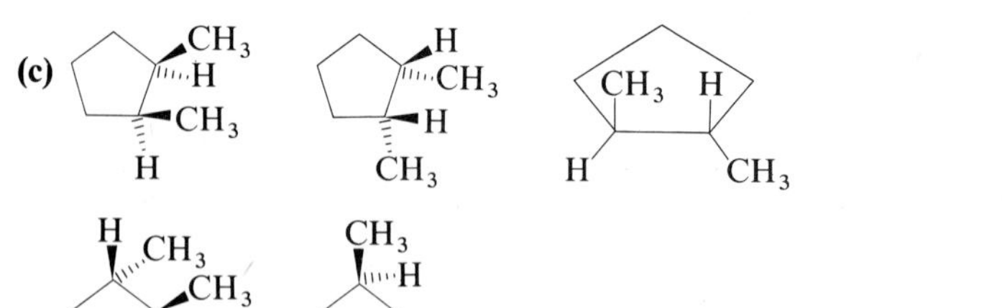

3-32. Draw the structure that corresponds with each of the following names.

(a) 3-ethyloctane
(b) 4-isopropyldecane
(c) *sec*-butylcycloheptane
(d) 2,3-dimethyl-4-propylnonane
(e) 2,2,4,4,-tetramethylhexane
(f) *trans*-1,3-diethylcyclopentane
(g) *cis*-1-ethyl-4-methylcyclohexane
(h) isobutylcyclopentane
(i) *t*-butylcyclohexane
(j) pentylcyclohexane
(k) cyclobutylcyclohexane

3-33. Each of the following descriptions applies to more than one alkane. In each case, draw and name two structures that match the description.
(a) a methylheptane **(b)** a diethyldecane **(c)** a *cis*-diethylcycloheptane
(d) a *trans*-dimethylcyclopentane **(e)** a (2,3-dimethylpentyl)cycloalkane

3-34. Write structures for a homologous series of alcohols (R—OH) having from one to six carbons.

3-35. Give the IUPAC names of the following alkanes.
(a) $CH_3C(CH_3)_2CH(CH_2CH_3)CH_2CH_2CH(CH_3)_2$

(b) 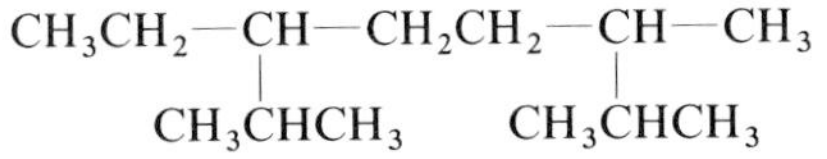 $CH_3CH_2—CH(CH_3CHCH_3)—CH_2CH_2—CH(CH_3CHCH_3)—CH_3$

(c)

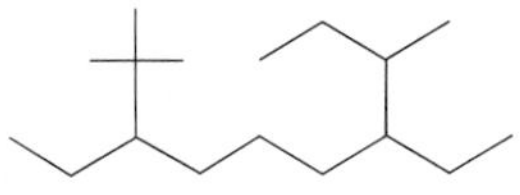

(d) 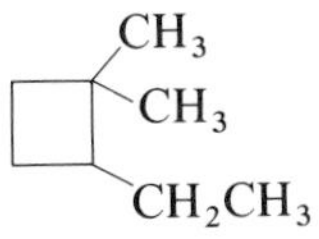 CH_3, CH_3, CH_2CH_3

(e)

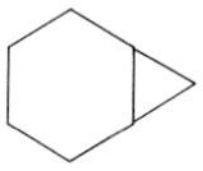

(f) 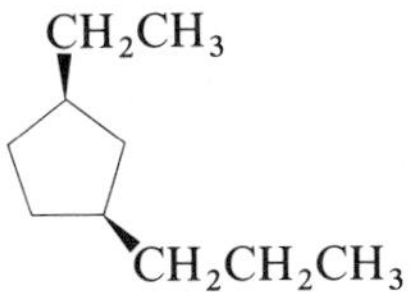 CH_2CH_3, $CH_2CH_2CH_3$

(g) 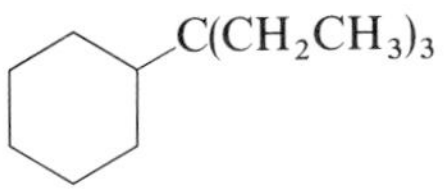 $C(CH_2CH_3)_3$

(h) 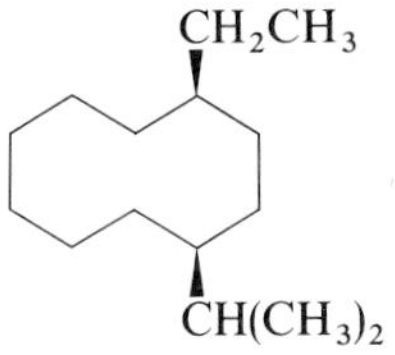 CH_2CH_3, $CH(CH_3)_2$

3-36. Draw and name eight isomers of molecular formula C_8H_{18}

3-37. The following names are all incorrect or incomplete, but they represent real structures. Draw each structure and name it correctly.
(a) 2-ethylpentane **(b)** 3-isopropylhexane **(c)** 4-methylhexane
(d) 2-dimethylbutane **(e)** 2-cyclohexylbutane **(f)** 2,3-diethylcyclopentane

3-38. In each of the following pairs of compounds, which compound has the higher boiling point? Explain your reasoning.
(a) octane or 2,2,3-trimethylpentane **(b)** heptane or 2-methylnonane **(c)** 2,2,5-trimethylhexane or nonane

3-39. There are eight different five-carbon alkyl groups.
(a) Draw them.
(b) Give them systematic names.
(c) In each case, label the degree of substitution (primary, secondary, or tertiary) of the head carbon atom, bonded to the main chain.

3-40. **(a)** Draw the two chair conformations of *cis*-1,3-dimethylcyclohexane, and label all the positions as axial or equatorial.
(b) Label the higher-energy conformation and the lower-energy conformation.
(c) The energy difference in these two conformations has been measured as about 5.4 kcal (23 kJ) per mole. How much of this energy difference is due to the torsional energy of gauche relationships?
(d) How much energy is due to the additional steric strain of the 1,3-diaxial interaction?

3-41. Draw the two chair conformations of each of the following compounds and label the substituents as axial and equatorial. In each case, determine which conformation is more stable.
(a) *cis*-1-ethyl-2-isopropylcyclohexane **(b)** *trans*-1-ethyl-2-isopropylcyclohexane
(c) *cis*-1-ethyl-3-methylcyclohexane **(d)** *trans*-1-ethyl-3-methylcyclohexane
(e) *cis*-1-ethyl-4-methylcyclohexane

3-42. Using what you know about the conformational energetics of substituted cyclohexanes, predict which of the two decalin isomers is more stable. Explain your reasoning.

3-43. The most stable form of the common sugar glucose contains a six-membered ring in the chair conformation with all the substituents equatorial. Draw this most stable conformation of glucose.

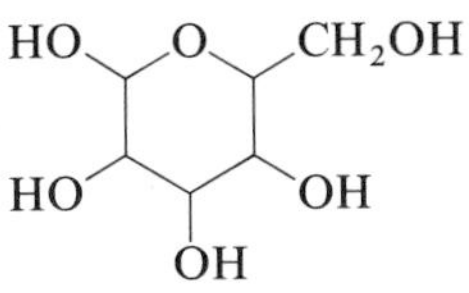

HO O CH_2OH
HO OH
OH
glucose

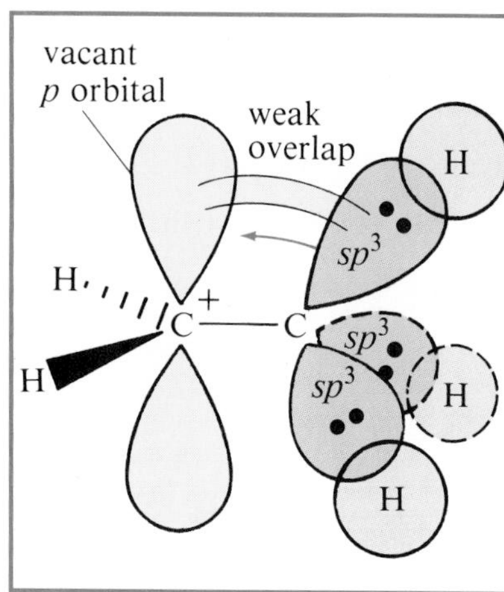

4

THE STUDY OF CHEMICAL REACTIONS

4-1 INTRODUCTION

The most interesting and useful aspect of organic chemistry is the study of reactions. To understand the reactions of organic compounds, you must first understand the ways that reactions occur and how one studies them.

Any chemical reaction is governed by both the **equilibrium** and the **rate** of the reaction. Once the reactant concentrations have stopped decreasing and the product concentrations have stopped increasing, the reaction has come to equilibrium. The amounts of reactants and products present at equilibrium depend on their relative stabilities. Even though the equilibrium may favor the formation of a product, however, it does not imply that that reaction will take place at a useful rate. The rate of a reaction depends on several other factors, including the **mechanism,** or step-by-step pathway from reactants to products.

Much of this chapter will consider the reactions of alkanes, because we have already studied the structure and properties of alkanes. In practice, alkanes are so unreactive that they are rarely used as starting materials for most organic syntheses. Once we have developed the tools for the study of alkane reactions, we will use them to study more complicated and more useful reactions.

4-2 THE CHLORINATION OF METHANE

The reaction of methane with chlorine produces a mixture of chlorinated products, whose composition depends on the amount of chlorine added and also on the reaction conditions. Either light or heat is needed for the reaction to take place at a useful rate. The first reaction with chlorine is as follows:

$$\underset{\text{methane}}{\mathrm{H{-}\overset{\displaystyle H}{\underset{\displaystyle H}{C}}{-}H}} \quad + \quad \underset{\text{chlorine}}{\mathrm{Cl{-}Cl}} \quad \xrightarrow{\text{heat or light}} \quad \underset{\substack{\text{chloromethane}\\ \text{(methyl chloride)}}}{\mathrm{H{-}\overset{\displaystyle H}{\underset{\displaystyle H}{C}}{-}Cl}} \quad + \quad \underset{\text{hydrogen chloride}}{\mathrm{H{-}Cl}}$$

This reaction may continue (heat or light is needed for each step):

$$\mathrm{H{-}\overset{\displaystyle H}{\underset{\displaystyle H}{C}}{-}Cl} \xrightarrow{Cl_2} \underset{+\,\mathrm{HCl}}{\mathrm{H{-}\overset{\displaystyle Cl}{\underset{\displaystyle H}{C}}{-}Cl}} \xrightarrow{Cl_2} \underset{+\,\mathrm{HCl}}{\mathrm{Cl{-}\overset{\displaystyle Cl}{\underset{\displaystyle H}{C}}{-}Cl}} \xrightarrow{Cl_2} \underset{+\,\mathrm{HCl}}{\mathrm{Cl{-}\overset{\displaystyle Cl}{\underset{\displaystyle Cl}{C}}{-}Cl}}$$

This sequence raises several questions about the chlorination of methane. Why is heat or light needed for the reaction to go? Why do we get a mixture of products? Is there any way to modify the reaction to get just one pure product? Are the observed products formed because they are the most stable products possible? Or are they favored because they are formed faster than any other products?

To answer such questions, we must understand three aspects of the reaction: the mechanism, the thermodynamics, and the kinetics.

1. The **mechanism** is the complete, step-by-step description of exactly which bonds break and which bonds form in what order to give the observed products.
2. **Thermodynamics** is the study of the energy changes that accompany chemical and physical transformations. It allows us to compare the stability of reactants and products and predict which compounds are favored by the equilibrium.
3. **Kinetics** is the study of reaction rates, determining which products are formed fastest. Kinetics also helps to predict how the rate will change if we change the reaction conditions.

We use the chlorination of methane to show how we study a reaction. Before we can propose a detailed mechanism for the chlorination, we must learn everything we can about how the reaction works and what factors affect the reaction rate and the product distribution.

A careful study of the chlorination reaction has established three important characteristics:

1. The chlorination does not occur at room temperature in the absence of light. The reaction begins when light falls on the mixture or when it is heated. Therefore, we know that this reaction requires some form of energy to *initiate* it.
2. The most effective wavelength of light is a blue color that is strongly absorbed by chlorine gas. This finding implies that light is absorbed by the chlorine molecule, activating the chlorine so that it initiates the reaction with methane.
3. The light-initiated reaction has a high *quantum yield*. This means that many molecules of the product are formed for every photon of light absorbed. Our

mechanism must explain how hundreds of individual reactions of methane with chlorine result from the absorption of a single photon by a single molecule of chlorine.

4-3 THE FREE-RADICAL CHAIN REACTION

A **chain reaction** mechanism has been proposed to explain the chlorination of methane. A chain reaction consists of three kinds of steps:

1. The **initiation step,** which generates a reactive intermediate.
2. **Propagation steps,** in which the reactive intermediate reacts with a stable molecule to form another reactive intermediate, allowing the chain to continue until the supply of reactants is exhausted or the reactive intermediate is destroyed.
3. **Termination steps,** side reactions that destroy reactive intermediates and tend to slow or stop the reaction.

In studying the chlorination of methane, we will consider just the first reaction to form chloromethane (common name *methyl chloride*). This reaction is a **substitution:** Chlorine does not add to methane, but a chlorine atom *substitutes* for one of the hydrogen atoms, and the missing hydrogen atom appears in the HC1 by-product.

$$\mathrm{CH_3{-}H} + \mathrm{Cl{-}Cl} \xrightarrow{\text{heat or light } (h\nu)} \mathrm{CH_3{-}Cl} + \mathrm{H{-}Cl}$$

methane chlorine chloromethane (methyl chloride)

4-3A THE INITIATION STEP: GENERATION OF RADICALS

Chlorination requires the formation of a **reactive intermediate:** a short-lived species that is never present in high concentration because it reacts as quickly as it is formed. The promotion of the reaction by blue light suggests that initiation results from the absorption of a photon by a chlorine molecule. The energy of a photon of light is related to its frequency ν by the relationship $E = h\nu$, where h is Planck's constant. Blue light has an energy of about 60 kcal (250 kJ) per einstein (an einstein is a mole of photons), and the energy required to split a chlorine molecule (Cl_2) into two chlorine atoms is 58 kcal (242 kJ) per mole.

The splitting of a chlorine molecule by the absorption of a photon of light is shown below. Notice the use of fishhook-shaped half-arrows to show the movement of single unpaired electrons in this reaction. Just as we use curved arrows to represent the movement of electron *pairs,* we use these half-arrows to represent the movement of single unpaired electrons. These half-arrows show that the two electrons in the Cl—Cl bond are separated, and one leaves with each chlorine atom.

$$:\ddot{\underset{..}{\mathrm{Cl}}}:\ddot{\underset{..}{\mathrm{Cl}}}: \;+\; \text{photon } (h\nu) \longrightarrow \;:\ddot{\underset{..}{\mathrm{Cl}}}\cdot \;+\; \cdot\ddot{\underset{..}{\mathrm{Cl}}}:$$

This initiation step produces two highly reactive chlorine atoms. These Cl· atoms have an odd number of valence electrons (seven), one of which is unpaired. The unpaired electron is often called the *odd electron* or the *radical electron.* Species with unpaired electrons are called **radicals** or **free radicals.** Radicals are extremely reactive because they lack on octet of electrons. Therefore, they are electron deficient and strongly electrophilic. The odd electron readily combines with an electron in another atom to complete an octet and give a stable bond. Figure 4-1 shows the Lewis structures of some free radicals. Radicals are often represented by a structure with a single dot representing the unpaired odd electron.

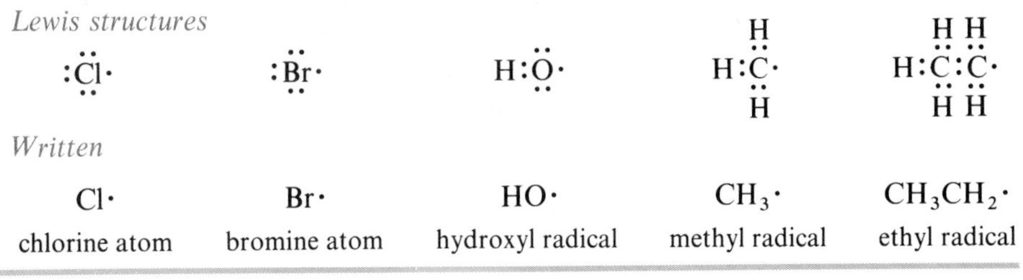

Lewis structures

$:\ddot{\underset{\cdot\cdot}{Cl}}\cdot$	$:\ddot{\underset{\cdot\cdot}{Br}}\cdot$	$H:\ddot{\underset{\cdot\cdot}{O}}\cdot$	$H:\overset{H}{\underset{H}{\ddot{\underset{\cdot\cdot}{C}}}}\cdot$	$H:\overset{H\ H}{\underset{H\ H}{\ddot{\underset{\cdot\cdot}{C}}:\ddot{\underset{\cdot\cdot}{C}}}}\cdot$
Written				
$Cl\cdot$	$Br\cdot$	$HO\cdot$	$CH_3\cdot$	$CH_3CH_2\cdot$
chlorine atom	bromine atom	hydroxyl radical	methyl radical	ethyl radical

FIGURE 4-1 Free radicals are reactive species with odd numbers of electrons. The unpaired electron is represented by a dot in the formula.

PROBLEM 4-1

Draw Lewis structures for the following free radicals.

(a) the *n*-propyl radical, $CH_3—CH_2—\dot{C}H_2$
(b) the *t*-butyl radical, $(CH_3)_3C\cdot$
(c) the isopropyl radical
(d) the iodine atom

4-3B THE PROPAGATION STEPS

When a chlorine radical collides with a methane molecule, it abstracts (removes) a hydrogen atom from methane. One of the electrons in the C—H bond remains on carbon, while the other combines with the odd electron on the chlorine atom to form the H—C1 bond.

First propagation step

$$\underset{\text{methane}}{H—\overset{H}{\underset{H}{\overset{|}{\underset{|}{C}}}}—H} \quad + \quad \underset{\text{chlorine atom}}{Cl\cdot} \longrightarrow \underset{\text{methyl radical}}{H—\overset{H}{\underset{H}{\overset{|}{\underset{|}{C}}}}\cdot} \quad + \quad \underset{\text{hydrogen chloride}}{H—Cl}$$

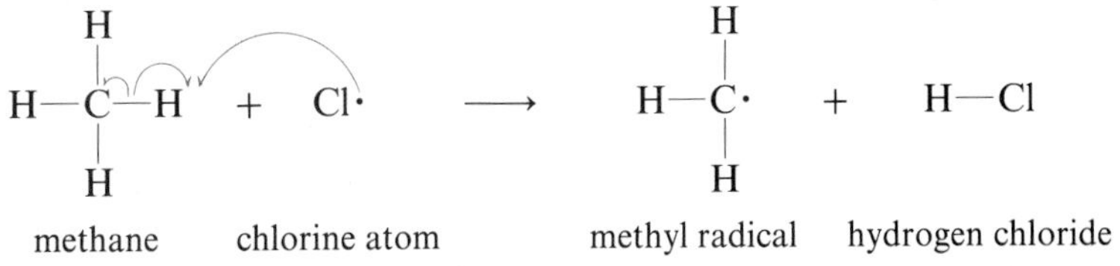

This step forms only one of the final products: the molecule of HC1. A later step must form chloromethane. Notice that the first propagation step begins with one free radical (the chlorine atom) and produces another free radical (the methyl radical). The regeneration of a free radical is characteristic of a propagation step of a chain reaction. Because another reactive intermediate is produced, the reaction can continue.

In the second propagation step, the methyl radical reacts with a molecule of chlorine to form chloromethane. The odd electron of the methyl radical combines with one of the two electrons in the C1—C1 bond to give the C1—CH_3 bond, and the chlorine atom is left with the other electron.

Second propagation step

$$\mathrm{H_3C\cdot} + \mathrm{Cl{-}Cl} \longrightarrow \mathrm{H_3C{-}Cl} + \mathrm{Cl\cdot}$$

methyl radical chlorine molecule chloromethane chlorine atom

In addition to forming chloromethane, the second propagation step produces another chlorine atom. The chlorine atom can react with another molecule of methane, giving HC1 and a methyl radical, which reacts with $C1_2$ to give chloromethane and regenerate yet another chlorine atom. In this way the chain reaction continues until the supply of the reactants is exhausted or some other reaction consumes the radical intermediates. The chain reaction explains the formation of many molecules of methyl chloride and HC1 by each photon of light that is absorbed. We summarize the reaction mechanism as follows:

Initiation

$$\mathrm{Cl{-}Cl} + h\nu\ \text{(light)} \longrightarrow 2\ \mathrm{Cl\cdot}$$

Propagation

$$\mathrm{H{-}CH_2{-}H} + \mathrm{Cl\cdot} \longrightarrow \mathrm{H_3C\cdot} + \mathrm{H{-}Cl}$$

$$\mathrm{H_3C\cdot} + \mathrm{Cl{-}Cl} \longrightarrow \mathrm{H_3C{-}Cl} + \mathrm{Cl\cdot}$$

continues the chain

Overall reaction

$$\mathrm{H{-}CH_2{-}H} + \mathrm{Cl{-}Cl} \longrightarrow \mathrm{H_3C{-}Cl} + \mathrm{H{-}Cl}$$

Notice that the overall reaction is simply the sum of the propagation steps.

PROBLEM 4-2

(a) Write the propagation steps leading to the formation of dichloromethane (CH_2Cl_2) from chloromethane.

(b) Explain why free-radical halogenation usually gives mixtures of products.

4-3C TERMINATION REACTIONS

If anything happens to consume some of the free-radical intermediates without generating new ones, the chain reaction will slow or stop. Therefore, the most important side reactions in a chain reaction are the ones that consume free radicals. Such a side reaction is called a **termination reaction:** a step that produces fewer reactive intermediates (free radicals) than it consumes. The following are some of the possible termination reactions in the chlorination of methane:

$$H_3C\cdot + Cl\cdot \longrightarrow H_3C{-}Cl$$

$$Cl\cdot + Cl\cdot \longrightarrow Cl{-}Cl$$

$$H_3C\cdot + \cdot CH_3 \longrightarrow H_3C{-}CH_3$$

$$H_3C\cdot \text{ collides with wall} \longrightarrow H_3C{-}\text{wall}$$

$$Cl\cdot \text{ collides with wall} \longrightarrow Cl{-}\text{wall}$$

Combination of any two free radicals is a termination step because it decreases the number of free radicals. Other termination steps involve reactions of the free radicals with the walls of the vessel or other contaminants. One might object that the first of these termination steps actually gives chloromethane, one of the products. Yet this step consumes the free radicals that are necessary for the reaction to continue, thus breaking the chain. Its contribution to the amount of product obtained from the reaction is extremely small in comparison to the contribution of the propagation steps.

While a chain reaction is in progress, the concentration of radicals is very low. The probability of two radicals combining in a termination step is much lower than the probability that they will encounter a molecule of reactant and give a propagation step. The termination steps become important toward the end of the reaction, when there are relatively few molecules of reactants available. At this point the free radicals are less likely to encounter a molecule of reactant than they are to encounter each other (or the wall of the container). The chain reaction quickly stops.

PROBLEM 4-3

Each of the following proposed mechanisms for the free-radical chlorination of methane is wrong. Explain how the experimental evidence disproves each mechanism.

(a) $Cl_2 + h\nu \longrightarrow Cl_2^*$ (an "activated" form of Cl_2)
$Cl_2^* + CH_4 \longrightarrow HCl + CH_3Cl$
(b) $CH_4 + h\nu \longrightarrow CH_3\cdot + H\cdot$
$CH_3\cdot + Cl_2 \longrightarrow CH_3Cl + Cl\cdot$
$Cl\cdot + H\cdot \longrightarrow HCl$

PROBLEM 4-4

Free-radical chlorination of hexane gives very poor yields of 1-chlorohexane, while cyclohexane can be converted to chlorocyclohexane in good yield.

(a) How do you account for this difference?
(b) What ratio of reactants (cyclohexane and chlorine) would you use for the synthesis of chlorocyclohexane?

4-4 EQUILIBRIUM CONSTANTS AND FREE ENERGY

Having determined a mechanism for the chlorination of methane, we can consider the energetics of the individual steps in more detail. First, however, we must review some of the principles needed for the discussion.

Thermodynamics is the branch of chemistry that deals with the energy changes accompanying chemical and physical transformations. These energy changes are most useful for describing the properties of systems at equilibrium. Let's review how energy and entropy variables describe an equilibrium.

The equilibrium concentrations of reactants and products are governed by the **equilibrium constant** of the reaction. For example, if A and B react to give C and D, then the equilibrium constant K_{eq} is defined by the following equation:

$$A + B \rightleftharpoons C + D$$

$$K_{eq} = \frac{[\text{products}]}{[\text{reactants}]} = \frac{[C][D]}{[A][B]}$$

The value of K_{eq} tells us the position of the equilibrium: whether the products or the reactants are more stable, and therefore energetically favored. If K_{eq} is larger than 1, the reaction is favored as written from left to right. If K_{eq} is less than 1, the reverse reaction is favored (from right to left as written); when the products C and D are mixed, the reactants A and B will form.

The chlorination of methane has a very large equilibrium constant of about 1.1×10^{19}.

$$CH_4 \quad + \quad Cl_2 \quad \rightleftharpoons \quad CH_3Cl \quad + \quad HCl$$

$$K_{eq} = \frac{[CH_3Cl][HCl]}{[CH_4][Cl_2]} = 1.1 \times 10^{19}$$

The equilibrium constant for chlorination is so large that the remaining amounts of the reactants are close to zero at equilibrium. Such a reaction is said to *go to completion,* and the value of K_{eq} is a measure of the reaction's tendency to go to completion.

From the value of K_{eq} we can calculate the change in **free energy** (sometimes called **Gibbs free energy**) that accompanies the reaction. Free energy is represented by G, and the change (Δ) in free energy associated with a reaction is represented by ΔG, the difference between the free energy of the products and the free energy of the reactants.

$$\Delta G = (\text{free energy of products}) - (\text{free energy of reactants})$$

The **standard Gibbs free energy change** ΔG° is most commonly used. The symbol $^\circ$ designates a reaction involving reactants and products in their standard states (pure substances in their most stable states at 25°C and 1 atm pressure). The relationship between ΔG° and K_{eq} is given by the expression

$$K_{eq} = e^{-\Delta G^\circ/RT}$$

or conversely, by

$$\Delta G^\circ = -RT(\ln K_{eq}) = -2.303RT(\log_{10} K_{eq})$$

R = 1.987 cal/kelvin-mol (8.314 J/kelvin-mol), the gas constant

T = absolute temperature, in kelvins

e = 2.718, the base of natural logarithms

The value of RT at 25°C is about 0.592 kcal/mol (2.48 kJ/mol).

The formula shows that a reaction is favored (large K_{eq}) if it has a *negative* value of ΔG° (energy is released). A reaction that has a positive value of ΔG° (energy must be added) is an unfavorable reaction. These predictions agree with our intuition that reactions should go from higher-energy states to lower-energy states, with a net decrease in free energy.

SOLVED PROBLEM 4-1

Calculate the value of ΔG° for the chlorination of methane.

SOLUTION

$$\Delta G^\circ = -2.303RT(\log K_{eq})$$

K_{eq} for the chlorination is 1.1×10^{19}, and $\log K_{eq} = 19.04$

At 25°C (about 298 K), the value of RT is

$$RT = (1.987 \text{ cal/kelvin-mol})(298 \text{ kelvins}) = 592 \text{ cal/mol, or } 0.592 \text{ kcal/mol}$$

Substituting, we have:

$$\Delta G^\circ = (-2.303)(0.592 \text{ kcal/mol})(19.04) = -25.9 \text{ kcal/mol } (-108 \text{ kJ/mol})$$

This is a large negative value for ΔG°, showing that this chlorination has a large driving force that pushes it toward completion.

Table 4-1 shows what percentages of the starting materials are converted to products at equilibrium for reactions with various values of ΔG°. In general, a reaction goes nearly to completion (>99 percent) for values of ΔG° that are more negative than about -3.0 kcal.

TABLE 4-1
Product composition as a function of ΔG° at 25°C

ΔG°	$K = e^{-\Delta G^\circ/RT}$	*Conversion to products*
+1.0 kcal/mol (+4.2 kJ/mol)	0.18	15%
+0.5 kcal/mol (+2.1 kJ/mol)	0.43	30%
0.0 kcal/mol (0.0 kJ/mol)	1.0	50%
−0.5 kcal/mol (−2.1 kJ/mol)	2.3	70%
−1.0 kcal/mol (−4.2 kJ/mol)	5.4	84%
−2.0 kcal/mol (−8.4 kJ/mol)	29	97%
−3.0 kcal/mol (−13. kJ/mol)	159	99.4%
−4.0 kcal/mol (−17. kJ/mol)	860	99.88%
−5.0 kcal/mol (−21. kJ/mol)	4660	99.98%

PROBLEM 4-5

The following reaction has a value of $\Delta G^\circ = -0.50$ kcal/mol (-2.1 kJ/mol).

$$CH_3Br + H_2S \rightleftharpoons CH_3SH + HBr$$

(a) Calculate K_{eq} at room temperature for this reaction as written.
(b) Starting with a 1 M solution of CH_3Br and H_2S, calculate the final concentrations of all four species at equilibrium.

PROBLEM 4-6

At room temperature, the reaction of two molecules of acetone to form diacetone alcohol proceeds to an extent of about 5 percent. Determine the value of ΔG° for this reaction.

$$2\,CH_3{-}\overset{\overset{\displaystyle O}{\|}}{C}{-}CH_3 \underset{}{\overset{^{-}OH}{\rightleftharpoons}} CH_3{-}\overset{\overset{\displaystyle O}{\|}}{C}{-}CH_2{-}\overset{\overset{\displaystyle OH}{|}}{C}(CH_3)_2$$

acetone   diacetone alcohol

4-5 ENTHALPY AND ENTROPY

The change in free energy is composed of two terms: the change in **enthalpy** and the change in **entropy** multiplied by the temperature.

$$\Delta G^\circ = \Delta H^\circ - T\,\Delta S^\circ$$

ΔG° = change in free energy = (free energy of products) − (free energy of reactants)

ΔH° = change in enthalpy = (enthalpy of products) − (enthalpy of reactants)

ΔS° = change in entropy = (entropy of products) − (entropy of reactants)

At low temperatures, the enthalpy term (ΔH°) is usually much larger than the entropy term ($-T\,\Delta S^\circ$), and the entropy term is sometimes ignored.

4-5A ENTHALPY

The **change in enthalpy** (ΔH°) is the heat of reaction—the amount of heat evolved or consumed in the course of a reaction, usually given in kilocalories per mole. The enthalpy change is a measure of the relative strength of the bonding in the products and the reactants. Reactions tend to favor products with the lowest enthalpy (those with the strongest bonds).

If weaker bonds are broken and stronger bonds are formed, heat is evolved and the reaction is **exothermic** (negative value of ΔH°). In an exothermic reaction, the enthalpy term contributes to a favorable negative value of ΔG°. If stronger bonds are broken and weaker bonds are formed, then energy is consumed in the reaction, and the reaction is **endothermic** (positive value of ΔH°). In an endothermic reaction, the enthalpy term contributes to an unfavorable positive value of ΔG°.

The value of ΔH° for the chlorination of methane is about -25 kcal/mol (-104.5 kJ/mol). This is a highly exothermic reaction, with the decrease in enthalpy serving as the primary driving force.

4-5B ENTROPY

Entropy is often described as randomness, or freedom of motion. Reactions tend to favor products with the greatest entropy. Notice the negative sign in the entropy term of the free-energy expression. A positive value of the **entropy change** (ΔS°), indicating that the products have more freedom of motion than the reactants, contributes to a favorable (negative) value of ΔG°.

In many cases, the enthalpy change is much larger than the entropy change, and the enthalpy term dominates the equation for ΔG°. Therefore, a negative value of ΔS° does not necessarily mean that the reaction has an unfavorable value of ΔG°. The formation of strong bonds (the change in enthalpy) is usually the most important component in the driving force for a reaction.

In the chlorination of methane, the value of ΔS° is $+2.9$ eu (entropy units or cal/kelvin-mole). The $-T\,\Delta S^\circ$ term in the free energy is

$$\begin{aligned} -T\,\Delta S^\circ &= -(298^\circ\ \text{K})(2.9\ \text{cal/K-mol}) = -860\ \text{cal/mol} \\ &= -0.86\ \text{kcal/mol}\ (3.6\ \text{kJ/mol}) \end{aligned}$$

The value of $\Delta G^\circ = -26$ kcal/mol is divided into enthalpy and entropy terms:

$$\begin{aligned} \Delta G^\circ = \Delta H^\circ - T\,\Delta S^\circ &= -25\ \text{kcal/mol} - 0.86\ \text{kcal/mol} \\ &= -26\ \text{kcal/mol}\ (-108\ \text{kJ/mol}) \end{aligned}$$

The enthalpy change is the largest factor in the driving force for chlorination. This is the case in most organic reactions: The entropy term is often small in relation to the enthalpy term. When we discuss chemical reactions involving the breaking and forming of bonds, we can often use the values of the enthalpy changes (ΔH°), under the assumption that $\Delta G^\circ \cong \Delta H^\circ$. We must be cautious in making this approximation, however, since some reactions have relatively small changes in enthalpy and larger changes in entropy.

SOLVED PROBLEM 4-2

Predict whether the value of ΔS° for the dissociation of Cl_2 is positive (favorable) or negative (unfavorable). What effect does the entropy term have on the sign of the value of ΔG° for this reaction?

$$Cl_2 \xrightarrow{h\nu} 2\,Cl\cdot$$

SOLUTION Two isolated chlorine atoms have much more freedom of motion than a single chlorine molecule. Therefore, the change in entropy is positive. This positive (favorable) value of $T\,\Delta S^\circ$ is small, however, compared with the much larger, positive (unfavorable) value of ΔH°. The enthalpy term predominates, showing that the chlorine molecule is much more stable than two chlorine atoms.

PROBLEM 4-7

When ethene is treated with hydrogen and a platinum catalyst is added, hydrogen adds across the double bond to form ethane. At room temperature, the reaction goes to completion. Predict the signs of ΔH° and ΔS° for this reaction. Explain these signs in terms of bonding and freedom of motion.

$$\underset{\text{ethene}}{H_2C{=}CH_2} + H_2 \xrightleftharpoons{\text{Pt catalyst}} \underset{\text{ethane}}{H{-}CH_2{-}CH_2{-}H}$$

PROBLEM 4-8

For each of the following reactions, estimate whether ΔS° for the reaction is positive, negative, or impossible to predict. In general, two smaller molecules have more freedom of motion (greater entropy) than one larger molecule.

(a) $\underset{n\text{-decane}}{C_{10}H_{22}} \xrightarrow[\text{catalyst}]{\text{heat}} \underset{\text{propene}}{C_3H_6} + \underset{\text{heptane}}{C_7H_{16}}$

(b) The formation of diacetone alcohol:

$$2\,CH_3{-}\overset{O}{\overset{\|}{C}}{-}CH_3 \xrightleftharpoons{^-OH} CH_3{-}\overset{O}{\overset{\|}{C}}{-}CH_2{-}\overset{OH}{\overset{|}{C}}(CH_3)_2$$

(c) $CH_3{-}\overset{O}{\overset{\|}{C}}{-}OH + CH_3OH \xrightleftharpoons{H^+} CH_3{-}\overset{O}{\overset{\|}{C}}{-}OCH_3 + H_2O$

4-6 BOND DISSOCIATION ENERGIES

We can put known amounts of methane and chlorine into a bomb calorimeter and use a hot wire to initiate the reaction. The temperature rise in the calorimeter is used to calculate the precise value of the heat of reaction, ΔH°. This calculation shows that 25 kcal (105 kJ) of heat is evolved (exothermic) for each mole of methane converted to chloromethane. Thus ΔH° for the reaction is negative, and the heat of reaction is given as:

$$\Delta H^\circ = -25 \text{ kcal/mol } (105 \text{ kJ/mol})$$

In many cases, we would like to be able to predict whether a particular reaction will be endothermic or exothermic without actually measuring the heat of reaction. We can calculate an approximate heat of reaction by adding and subtracting the energies involved in the breaking and forming of bonds. To do this calculation, we need to know the energies of the affected bonds.

The **bond dissociation energy (BDE)** is the amount of energy required to break a particular bond *homolytically,* that is, in such a way that each bonded atom retains one of the bond's two electrons. In contrast, when a bond is broken and one of the two atoms retains both electrons, we say that **heterolytic cleavage** has occurred.

Homolytic cleavage (free radicals result):

$$\text{A:B} \longrightarrow \text{A}\cdot + \cdot\text{B} \qquad \Delta H^\circ = \text{bond dissociation energy}$$

$$:\ddot{\underset{..}{\text{Cl}}}:\ddot{\underset{..}{\text{Cl}}}: \longrightarrow 2\,:\ddot{\underset{..}{\text{Cl}}}\cdot \qquad \Delta H^\circ = 58 \text{ kcal/mol } (242 \text{ kJ/mol})$$

Heterolytic cleavage (ions result):

$$\text{A:B} \longrightarrow \text{A}^+ + {}^-\text{:B}$$

$$(CH_3)_3C{-}\ddot{\underset{..}{\text{Cl}}}: \longrightarrow (CH_3)_3C^+ + :\ddot{\underset{..}{\text{Cl}}}:^-$$

Homolytic cleavage (radical cleavage) forms free radicals, while heterolytic cleavage forms ions. A heterolytic cleavage is sometimes called an **ionic cleavage.** Note that a curved arrow is used to show the movement of the electron pair in an ionic cleavage, and half-arrows to show the separation of the individual electrons in a homolytic cleavage.

Energy is released when bonds are formed, and energy is consumed to break bonds. Therefore, bond dissociation energies are always positive (endothermic). By studying the heats of reaction for many different reactions, chemists have developed reliable tables of bond dissociation energies. Table 4-2 gives the bond dissociation energies for the homolysis of bonds in a variety of common molecules.

4-7 CALCULATION OF ENTHALPY CHANGES IN CHLORINATION

Using values from Table 4-2, we can predict the heat of reaction for the chlorination of methane. The overall enthalpy change is the sum of the dissociation energies of the bonds broken minus the sum of the dissociation energies of the bonds formed. This reaction involves the breaking of a $CH_3{-}H$ bond and a $Cl{-}Cl$ bond, and the formation of a $CH_3{-}Cl$ bond and a $H{-}Cl$ bond.

TABLE 4-2
Bond dissociation energies for homolytic cleavages

$$A{:}B \longrightarrow A\cdot + \cdot B$$

Bond	Bond dissociation energy kcal/mol	kJ/mol
H—X bonds		
H—H	104	435
D—D	106	444
F—F	38	159
Cl—Cl	58	242
Br—Br	46	192
I—I	36	151
H—F	136	569
H—Cl	103	431
H—Br	88	368
H—I	71	297
HO—H	119	498
HO—OH	51	213
Methyl bonds		
CH_3—H	104	435
CH_3—F	109	456
CH_3—Cl	84	351
CH_3—Br	70	293
CH_3—I	56	234
CH_3—OH	91	381
Bonds to primary carbons		
CH_3CH_2—H	98	410
CH_3CH_2—F	107	448
CH_3CH_2—Cl	81	339
CH_3CH_2—Br	68	285
CH_3CH_2—I	53	222
CH_3CH_2—OH	91	381
$CH_3CH_2CH_2$—H	98	410
$CH_3CH_2CH_2$—F	107	448
$CH_3CH_2CH_2$—Cl	81	339
$CH_3CH_2CH_2$—Br	68	285
$CH_3CH_2CH_2$—I	53	222
$CH_3CH_2CH_2$—OH	91	381
Bonds to secondary carbons		
$(CH_3)_2CH$—H	95	397
$(CH_3)_2CH$—F	106	444
$(CH_3)_2CH$—Cl	80	335
$(CH_3)_2CH$—Br	68	285
$(CH_3)_2CH$—I	53	222
$(CH_3)_2CH$—OH	91	381
Bonds to tertiary carbons		
$(CH_3)_3C$—H	91	381
$(CH_3)_3C$—F	106	444
$(CH_3)_3C$—Cl	79	331
$(CH_3)_3C$—Br	65	272
$(CH_3)_3C$—I	50	209
$(CH_3)_3C$—OH	91	381
Other C—H bonds		
Ph—CH_2—H	85	356
CH_2=$CHCH_2$—H	87	364
CH_2=CH—H	108	452
Ph—H	110	460
C—C bonds		
CH_3—CH_3	88	368
CH_3CH_2—CH_3	85	356
CH_3CH_2—CH_2CH_3	82	343
$(CH_3)_2CH$—CH_3	84	351
$(CH_3)_3C$—CH_3	81	339

$$CH_3\text{—}H \;+\; Cl\text{—}Cl \longrightarrow CH_3\text{—}Cl \;+\; H\text{—}Cl$$

Bonds broken	ΔH° (per mole)	*Bonds formed*	ΔH° (per mole)
Cl—Cl	+58 kcal (242 kJ)	H—Cl	−103 kcal (431 kJ)
CH_3—H	+104 kcal (435 kJ)	CH_3—Cl	−84 kcal (351 kJ)
total	+162 kcal (677 kJ)	total	−187 kcal (782 kJ)

$$\Delta H^\circ = +162 \text{ kcal} + (-187) \text{ kcal} = -25 \text{ kcal/mol } (-105 \text{ kJ/mol})$$

The bond dissociation energies also provide the information we need to compute the heat of reaction for each individual step:

First propagation step

$$Cl\cdot \;+\; CH_4 \longrightarrow CH_3\cdot \;+\; HCl$$

Breaking a CH_3—H bond	+104 kcal/mol (+435 kJ/mol)
Forming the HCl bond	−103 kcal/mol (−431 kJ/mol)
Step total	+1 kcal/mol (+4 kJ/mol)

Second propagation step

$$CH_3\cdot \quad + \quad Cl_2 \quad \longrightarrow \quad CH_3Cl \quad + \quad Cl\cdot$$

Breaking a Cl—Cl bond	+58 kcal/mol (+243 kJ/mol)
Forming a CH_3—Cl bond	−84 kcal/mol (−352 kJ/mol)
Step total	−26 kcal/mol (−109 kJ/mol)

Grand total = +1 kcal/mol + (−26 kcal/mol) = −25 kcal/mol (−105 kJ/mol)

This summary shows that the sum of the values of $\Delta H°$ for the individual propagation steps gives the overall enthalpy change for the reaction. The initiation step, $Cl_2 \rightarrow 2\ Cl\cdot$ is not added to give the overall enthalpy change because it is not necessary for each molecule of product formed. The first splitting of a chlorine molecule simply begins the chain reaction, which generates hundreds or thousands of molecules of chloromethane.

PROBLEM 4-9

(a) Propose a mechanism for the free-radical chlorination of ethane,

$$CH_3{-}CH_3 \quad + \quad Cl_2 \quad \xrightarrow{h\nu} \quad CH_3{-}CH_2Cl \quad + \quad HCl$$

(b) Calculate $\Delta H°$ for each step in this reaction.

(c) Calculate the overall value of $\Delta H°$ for this reaction.

Alternate mechanism The mechanism we have used is not the only one that might be proposed to explain the reaction of methane with chlorine. We know that the initiating step must be the splitting of a molecule of Cl_2, but there are other propagation steps that would form the correct products:

(a) $Cl\cdot + CH_3{-}H \longrightarrow CH_3{-}Cl + H\cdot$ $\quad \Delta H° = +104\text{ kcal} - 84\text{ kcal} = +20\text{ kcal}\ (+84\text{ kJ})$

(b) $H\cdot + Cl{-}Cl \longrightarrow H{-}Cl + Cl\cdot$ $\quad \Delta H° = +58\text{ kcal} - 103\text{ kcal} = -45\text{ kcal}\ (-189\text{ kJ})$

Total −25 kcal (−105 kJ)

This alternate mechanism seems plausible, but step (a) is endothermic by 20 kcal/mol (84 kJ/mol). The previous mechanism provides a lower-energy alternative. When a chlorine atom collides with a methane molecule, it will not react to give methyl chloride and a hydrogen atom ($\Delta H° = +20$ kcal = +84 kJ); it will react to give HCl and a methyl radical ($\Delta H° = +1$ kcal = +4 kJ), the first propagation step of the correct mechanism.

PROBLEM 4-10

(a) Using bond dissociation energies from Table 4-2 (page 133), calculate the heat of reaction for each of the steps in the free-radical bromination of methane,

$$Br_2 \quad + \quad CH_4 \quad \xrightarrow{\text{heat or light}} \quad CH_3Br + HBr$$

(b) Calculate the overall heat of reaction.

4-8 KINETICS AND THE RATE EQUATION

Kinetics is the study of reaction rates. How fast a reaction goes is just as important as the position of its equilibrium. The fact that thermodynamics favors a reaction (negative $\Delta G°$) does not necessarily mean the reaction will actually occur. For example, a mixture of gasoline and oxygen does not react without a spark or a catalyst. Similarly, a mixture of methane and chlorine does not react if it is kept cold and dark.

The **rate** of a reaction is how fast the products appear and the reactants disappear. We can determine the rate by measuring the increase in the concentrations of the products with time, or the decrease in the concentrations of the reactants with time.

Reaction rates depend on the concentrations of the reactants. The greater the concentrations, the more often the reactants collide and the greater the chance of reaction. A **rate equation** (sometimes called a **rate law**) is the relationship between the concentrations of the reactants and the observed reaction rate. Each reaction has its own rate equation, *determined experimentally* by changing the concentrations of the reactants and measuring the change in the rate. For example, consider the general reaction

$$A + B \longrightarrow C + D$$

The reaction rate is usually proportional to the concentrations of the reactants ([A] and [B]) raised to some powers, a and b. We can use a general rate expression to represent this relationship:

$$\text{rate} = k_r[A]^a[B]^b$$

where k_r is the **rate constant,** and the values of the powers (a and b) must be determined experimentally. From just the stoichiometry of the reaction, we *cannot* guess or calculate the rate equation. The rate equation depends on the mechanism of the reaction and on the individual rates of the steps in the mechanism.

In the general rate equation, the power a is called the **order** of the reaction with respect to reactant A, and b is the order of the reaction with respect to B. The sum of these powers ($a + b$) is called the **overall order** of the reaction.

The following reaction has a simple rate equation:

$$CH_3{-}Br \quad + \quad {}^-OH \quad \xrightarrow{H_2O/\text{acetone}} \quad CH_3{-}OH \quad + \quad Br^-$$

Experiments show that doubling the concentration of methyl bromide, $[CH_3Br]$, doubles the rate of reaction. Doubling the concentration of hydroxide ion, $[{}^-OH]$, also doubles the rate. Thus, the rate is proportional to both $[CH_3Br]$ and $[{}^-OH]$, and the rate equation has the following form:

$$\text{rate} = k_r[CH_3Br][{}^-OH]$$

This rate equation is *first order* in each of the two reagents, because it is proportional to the first power of their concentrations. The rate equation is *second order overall,* because the sum of the powers of the concentrations in the rate equation is 2; that is, (first order) + (first order) = second order overall.

Reactions of the same overall type do not necessarily have the same form of rate equation. For example, the following reaction has a different kinetic order:

$$(CH_3)_3C{-}Br \quad + \quad {}^-OH \quad \xrightarrow{H_2O/\text{acetone}} \quad (CH_3)_3{-}OH \quad + \quad Br^-$$

Doubling the concentration of *t*-butyl bromide ($[(CH_3)_3C{-}Br]$) causes the rate to double, but doubling the concentration of hydroxide ion ($[{}^-OH]$) has no effect on the rate of this particular reaction. The rate equation is

$$\text{rate} = k_r[(CH_3)_3C{-}Br]$$

This reaction is first order in *t*-butyl bromide, and zeroth order in hydroxide ion (proportional to $[{}^-OH]$ to the zeroth power). It is first order overall.

The most important fact to remember is that *the rate equation must be determined experimentally*. We cannot predict the form of the rate equation from the stoichiometry of the reaction. First we determine the rate equation experimentally, then use that information to propose consistent mechanisms.

PROBLEM 4-11

The reaction of *t*-butyl chloride with methanol is found to follow the rate equation given below:

$$\underset{t\text{-butyl chloride}}{(CH_3)_3C\text{—}Cl} + \underset{\text{methanol}}{CH_3\text{—}OH} \longrightarrow \underset{\text{methyl } t\text{-butyl ether}}{(CH_3)_3C\text{—}OCH_3} + HCl$$

$$\text{rate} = k_r[(CH_3)_3C\text{—}Cl]$$

(a) What is the kinetic order with respect to *t*-butyl chloride?
(b) What is the kinetic order with respect to methanol?
(c) What is the kinetic order overall?

PROBLEM 4-12

Chloromethane reacts with dilute sodium cyanide ($Na^{+}\ {}^{-}C{\equiv}N$) according to the following equation:

$$\underset{\text{chloromethane}}{CH_3\text{—}Cl} + \underset{\text{cyanide}}{{}^{-}C{\equiv}N} \longrightarrow \underset{\text{acetonitrile}}{CH_3\text{—}C{\equiv}N} + \underset{\text{chloride}}{Cl^{-}}$$

When the concentration of chloromethane is doubled, the rate is observed to double. When the concentration of cyanide ion is tripled, the rate is observed to triple.

(a) What is the kinetic order with respect to chloromethane?
(b) What is the kinetic order with respect to cyanide ion?
(c) Write the rate equation for this reaction.
(d) What is the kinetic order overall?

PROBLEM 4-13

When a small piece of platinum is added to a mixture of ethylene and hydrogen, a reaction takes place:

$$\underset{\text{ethylene}}{H_2C{=}CH_2} + \underset{\text{hydrogen}}{H_2} \xrightarrow{\text{Pt catalyst}} \underset{\text{ethane}}{H\text{—}CH_2\text{—}CH_2\text{—}H}$$

Doubling the concentration of hydrogen has no effect on the reaction rate. Doubling the concentration of ethylene also has no effect.

(a) What is the kinetic order of this reaction with respect to ethylene? Hydrogen? What is the overall order?
(b) Write the unusual rate equation for this reaction.
(c) Explain this strange rate equation, and suggest what one might do to accelerate the reaction.

4-9 ACTIVATION ENERGY AND THE TEMPERATURE DEPENDENCE OF RATES

The rate constant k_r is a characteristic of each particular reaction. Its value depends on the conditions of the reaction, especially the reaction temperature. This temperature dependence is expressed by the Arrhenius equation,

$$k_r = Ae^{-E_a/RT}$$

A = a constant (the "preexponential" term)

E_a = activation energy

R = the gas constant, 1.987 cal/kelvin-mole

T = the absolute temperature

The **activation energy, E_a,** is the minimum kinetic energy the molecules must possess to overcome the repulsions between their electron clouds when they collide. The exponential term $e^{-E_a/RT}$ corresponds to the fraction of collisions in which the particles have the minimum energy E_a needed to react. The "preexponential" factor A accounts for the frequency of collisions and the fraction of collisions with the proper orientation for the reaction to occur.

The Arrhenius equation implies that the rate of a reaction depends on the fraction of molecules with kinetic energy of at least E_a. Figure 4-2 shows how the distribution of kinetic energies in a sample of a gas depends on the temperature. The black curved line shows the molecular energy distribution at room temperature, and the dashed lines show the energy needed to overcome barriers of 1 kcal (4 kJ), 10 kcal (42 kJ), and 19 kcal (79 kJ). The area under the curve to the right of each barrier corresponds to the number of molecules with enough energy to overcome that barrier. The red curve shows how the energy distribution is shifted at 100°C. At 100°C many more molecules have the energy needed to overcome the energy barriers, especially the 19-kcal/mol (79-kJ/mol) barrier.

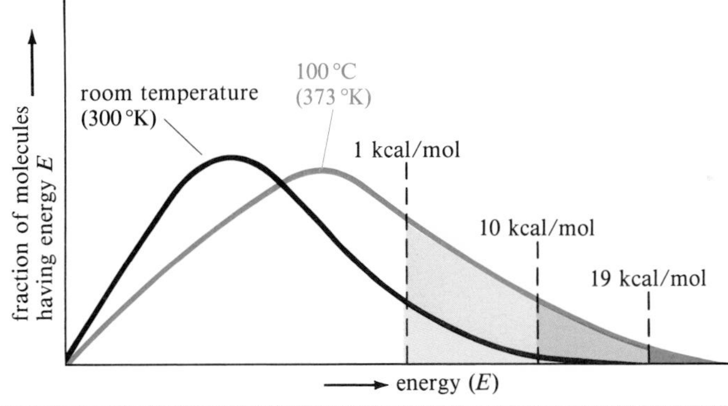

FIGURE 4-2 Graph showing how the number of molecules having a given activation energy decreases as the activation energy increases. At a higher temperature (red curve), more collisions have the needed energy.

Table 4-3 shows the dependence of reaction rates on the temperature by listing values of the relative rate constant k_{rel}, which is just the exponential term $e^{-E_a/RT}$, for some typical values of E_a and some convenient temperatures. With a typical activation energy of about 10 to 15 kcal/mol (40 to 60 kJ/mol), the re-

TABLE 4-3
Variation of the relative rate constant k_{rel} with temperature

	Values of $k_{rel} = e^{-E_a/RT}$ (units of 10^{-9})		
E_a (per mole)	*27°C (300°K)*	*37°C (310°K)*	*100°C (373°K)*
5 kcal (21 kJ)	240,000	320,000	1,200,000
10 kcal (42 kJ)	58	99	1,500
15 kcal (63 kJ)	0.014	0.031	1.9
20 kcal (84 kJ)	0.0000033	0.0000098	0.0023

action rate approximately doubles when the temperature is raised by 10°C, as from 27°C (room temperature) to 37°C (body temperature).

Table 4-3 shows that the relative rate constant k_{rel} increases quickly when the temperature is raised. It might seem that raising the temperature would always be a good way to save time by making the reaction go faster. The problem with raising the temperature is that *all* reactions are accelerated, including all the unwanted side reactions. We try to find a temperature that allows the desired reaction to go at a reasonable rate, while not producing unacceptable rates of side reactions.

4-10 TRANSITION STATES

The activation energy E_a represents the energy difference between the reactants and the **transition state,** the highest-energy state in the course of the reaction. In effect, the activation energy is the barrier that must be overcome for the reaction to take place. The value of E_a is always positive, and its magnitude depends on the relative energy of the transition state. The term *transition state* implies that this configuration is the transition between the reactants and products, and the molecules can either go on to products or return to reactants.

Unlike the reactants or products, a transition state is unstable and cannot be isolated. It is not an intermediate, because an **intermediate** is a species that exists for some finite length of time, even if it is very short. An intermediate has at least some stability, but the transition state is a transient on the path from one intermediate to another. The transition state is often symbolized by a double dagger superscript (‡), and the changes in variables such as free energy, enthalpy, and entropy involved in achieving the transition state are symbolized $\Delta G^‡$, $\Delta H^‡$, and $\Delta S^‡$. $\Delta G^‡$ is similar to E_a, and the symbol $\Delta G^‡$ is often used in speaking of the activation energy.

Transition states have high energies because bonds must begin to break before other bonds can form. The following equation shows the reaction of a chlorine radical with methane, followed by the transition state with the C—H bond partially broken and the H—Cl bond partially formed. Transition states are often enclosed by brackets to emphasize their transient nature.

$$CH_4 + \cdot Cl \rightleftharpoons \left[H_3C\text{---}H\text{---}Cl \right]^‡ \longrightarrow H-\dot{C}H_2 + H-Cl$$

transition state

Reaction-energy profiles The concepts of transition state and activation energy are easier to understand graphically. Figure 4-3 shows a **reaction-energy profile** for a one-step exothermic reaction. The vertical axis of the energy profile represents the total potential energy of all the species involved in the reaction. The horizontal axis is called the **reaction coordinate.** The reaction coordinate symbolizes the progress of the reaction, going from the reactants on the left to the products on the right. The transition state is the highest point on the graph, and the activation energy is the energy difference between the reactants and the transition state. The heat of reaction ($\Delta H°$) is the difference in energy between the reactants and the products. If a **catalyst** were added to the reaction in Figure 4-3, it would create a transition state of lower energy, thereby lowering the activation energy. Addition of a catalyst would not change the energies of the reactants and products, however; therefore, the heat of reaction would be unaffected.

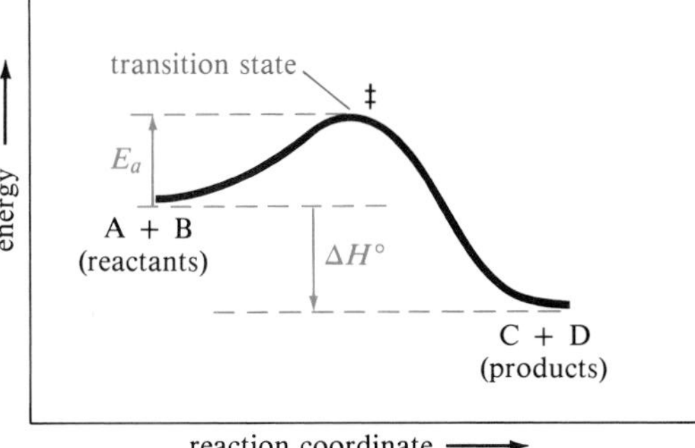

FIGURE 4-3 Reaction-energy profile for a one-step exothermic reaction. The reactants are toward the left and the products are toward the right. The vertical axis represents the total energy. The transition state is the highest point on the graph, and the activation energy is the energy difference between the reactants and the transition state.

SOLVED PROBLEM 4-3

Consider the following reaction:

$$CH_4 + Cl\cdot \longrightarrow \cdot CH_3 + HCl$$

This reaction has an activation energy (E_a) of +4 kcal/mol (+17 kJ/mol) and a $\Delta H°$ of +1 kcal/mol (+4 kJ/mol). Draw a reaction-energy profile for this reaction.

SOLUTION We draw a diagram that shows the products to be 1 kcal *higher* in energy than the reactants. The barrier is made to be 4 kcal higher in energy than the reactants.

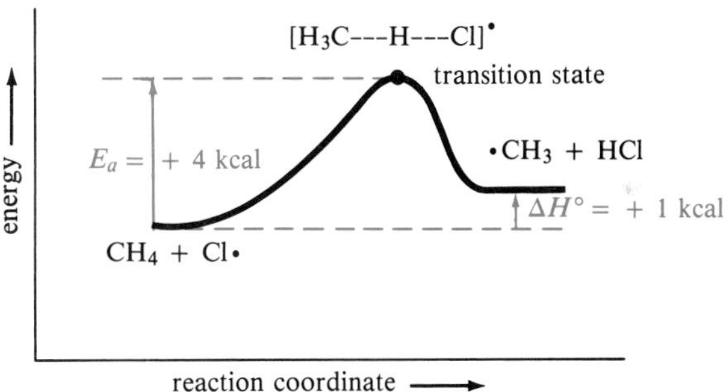

PROBLEM 4-14

(a) Draw the reaction-energy profile for the reverse reaction:

$$CH_3\cdot + HCl \longrightarrow CH_4 + Cl\cdot$$

(b) What is the activation energy for this reverse reaction?
(c) What is the heat of reaction ($\Delta H°$) for this reverse reaction?

PROBLEM 4-15

(a) Draw a reaction-energy profile for the following reaction:

$$CH_3\cdot + Cl_2 \longrightarrow CH_3Cl + Cl\cdot$$

The activation energy is 1 kcal/mol (4 kJ/mol), and the overall $\Delta H°$ for the reaction is −26 kcal/mol (−109 kJ/mol).

(b) Give the equation for the reverse reaction.
(c) What is the activation energy for the reverse reaction?

4-11 RATES OF MULTISTEP REACTIONS

Many reactions proceed by mechanisms involving several steps and several intermediates. The reaction of methane with chlorine, for example, goes through two propagation steps. The propagation steps are shown below, along with their heats of reaction and their activation energies. Just the propagation steps are shown, because the rate of the initiation step is controlled by the amount of light or heat available to split chlorine molecules.

Step	$\Delta H°$ (per mole)	E_a (per mole)
$CH_4 + Cl\cdot \longrightarrow CH_3\cdot + HCl$	+1 kcal (+4 kJ)	4 kcal (17 kJ)
$CH_3\cdot + Cl_2 \longrightarrow CH_3Cl + Cl\cdot$	−26 kcal (−109 kJ)	1 kcal (4 kJ)

In this reaction, $Cl\cdot$ and $CH_3\cdot$ are *reactive intermediates.* Unlike transition states, these intermediates are stable so long as they do not collide with other atoms or molecules. They are free radicals, however, and they are quite reactive toward other molecules. Figure 4-4 shows a single reaction-energy profile that includes both propagation steps of the chlorination. It is easy to tell the difference between transition states and intermediates in Figure 4-4. The energy maxima (high points) are the unstable transition states, and the energy minima (low points) are the intermediates. This complete energy profile provides most of the important information about the energetics of the reaction.

The rate-determining step In a multistep reaction, each step has its own characteristic rate. There can be only one overall reaction rate, however, and it is controlled by the **rate-determining step.** In general, the *highest-energy* step of a multistep reaction is the "bottleneck," and it determines the overall rate. How can we tell which step is rate determining? If we have the reaction-energy profile, it is simple: The highest point in the energy profile is the transition state with the highest energy—the transition state for the rate-determining step.

The highest point in the energy profile of the chlorination of methane (Figure 4-4) is the transition state for the reaction of methane with a chlorine radical. This step must be rate determining. If we calculate a rate for this slow step, it will be the rate for the overall reaction. The second, faster step will consume the products of the slow step at the same rate as they are formed.

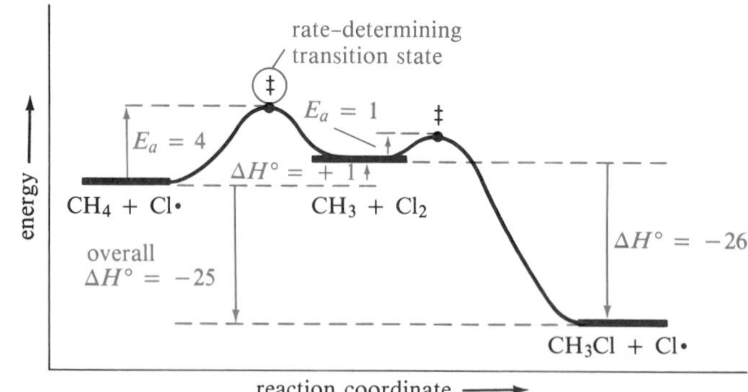

FIGURE 4-4 Combined reaction-energy profile for the chlorination of methane. The transition states are the energy maxima, and the intermediates are the energy minima.

4-12 ISOTOPE EFFECTS

One experimental method for finding the rate-determining step is using **isotope effects,** which are changes in reaction rates resulting from using different isotopes. The isotope most commonly used is deuterium (D), the isotope of hydrogen with mass number 2, which has a proton and a neutron in its nucleus. Although the chemistry of deuterium is nearly identical to that of hydrogen, bonds to deuterium are slightly stronger than bonds to hydrogen, and the activation energy for abstraction of a deuterium atom is slightly higher than that for a hydrogen atom.

If a hydrogen atom is being abstracted in the rate-determining step, a compound with deuterium in place of that hydrogen atom will react more slowly. For example, methane undergoes free-radical chlorination 12 times as fast as tetradeuteriomethane (CD_4). This evidence implies that a C—H (or C—D) bond is being broken in the rate-determining step of the reaction.

Faster

$$CH_4 + Cl\cdot \longrightarrow CH_3Cl + HCl \qquad \text{relative rate} = 12$$

Slower

$$CD_4 + Cl\cdot \longrightarrow CD_3Cl + DCl \qquad \text{relative rate} = 1$$

Rate-determining step

$$CH_4 + Cl\cdot \rightleftharpoons [H_3C\text{---}H\text{---}Cl]^{\ddagger} \longrightarrow H_3C\cdot + H\text{—}Cl$$

bond to H is breaking in transition state

Slower with deuterium

$$CD_4 + Cl\cdot \rightleftharpoons [D_3C\text{---}D\text{---}Cl]^{\ddagger} \longrightarrow D_3C\cdot + D\text{—}Cl$$

bond to D is breaking in transition state

4-13 TEMPERATURE DEPENDENCE OF HALOGENATION

We now apply what we know about rates to the reaction of methane with the halogens. The rate-determining step for chlorination is the endothermic reaction of the chlorine atom with methane to form a methyl radical and a molecule of HCl.

$$\text{Rate-determining step:} \quad CH_4 + Cl\cdot \longrightarrow CH_3\cdot + HCl$$

The activation energy for this step is 4 kcal/mol (17 kJ/mol). At room temperature, the value of $e^{-E_a/RT}$ is 1300 (units of 10^{-6}). This value represents a rate that is fast but controllable.

In a free-radical chain reaction, it is important that each propagation step occur quickly, or the free radicals will undergo many collisions and become involved in termination steps. We can predict how quickly the various halogen atoms react with methane by using the measured activation energies of the slow steps:

Reaction	E_a *(per mole)*	*Relative rate* ($e^{-E_a/RT} \times 10^6$) *300°K (27°C)*	*500°K (227°C)*
$F\cdot + CH_4 \longrightarrow HF + CH_3\cdot$	1.2 kcal (5 kJ)	140,000	300,000
$Cl\cdot + CH_4 \longrightarrow HCl + CH_3\cdot$	4 kcal (16 kJ)	1300	18,000
$Br\cdot + CH_4 \longrightarrow HBr + CH_3\cdot$	18 kcal (75 kJ)	9×10^{-8}	0.015
$I\cdot + CH_4 \longrightarrow HI + CH_3\cdot$	34 kcal (140 kJ)	2×10^{-19}	2×10^{-9}

Based on these relative rates, we can make predictions about the reactions of methane with halogen radicals. The reaction with fluorine should be difficult to control, because its relative rate is very high. The reaction with chlorine should have a moderate rate at room temperature, but it may become difficult to control if the temperature rises much (the rate at 500°K is rather high). The reaction with bromine is very slow, but heating might give an observable rate. Iodination is probably out of the question, since its rate is exceedingly slow even at 500°K.

Laboratory halogenations show that our predictions are right. In fact, fluorine reacts explosively with methane, and chlorine reacts at a moderate rate. A mixture of bromine and methane must be heated to react, and iodine does not react at all.

PROBLEM 4-16

The bromination of methane proceeds through the following steps:

	$\Delta H°$ (per mole)	E_a (per mole)
$Br_2 \longrightarrow 2\,Br\cdot$	+46 kcal (192 kJ)	46 kcal (192 kJ)
$CH_4 + Br\cdot \longrightarrow CH_3\cdot + HBr$	+16 kcal (67 kJ)	18 kcal (75 kJ)
$CH_3\cdot + Br_2 \longrightarrow CH_3Br + Br\cdot$	−24 kcal (−101 kJ)	1 kcal (4 kJ)

(a) Draw a complete reaction energy profile for this reaction.
(b) Label the rate-determining step.
(c) Draw the structure of each transition state.
(d) Compute the overall value of $\Delta H°$ for the bromination.

PROBLEM 4-17

(a) Using the BDEs in Table 4-2 (p. 133), compute the value of $\Delta H°$ for each step in the iodination of methane.
(b) Compute the overall value of $\Delta H°$ for iodination.
(c) Suggest *two* reasons why iodine is not observed to react with methane.

4-14 HALOGENATION OF HIGHER ALKANES

Up to now, our discussions of halogenation have used methane as the starting material. Using such a simple compound, we could concentrate on the thermodynamic and kinetic aspects of the reaction. Now we consider the halogenation of the "higher" alkanes, meaning those of "higher molecular weight."

4-14A CHLORINATION OF PROPANE: PRODUCT RATIOS

Halogenation is a substitution reaction, where a halogen atom replaces a hydrogen.

$$R{-}H + X_2 \longrightarrow R{-}X + H{-}X$$

In methane, all four hydrogen atoms are identical, and it does not matter which hydrogen is replaced. In the higher alkanes, the replacement of different hydrogen atoms leads to different products. As an example, consider the chlorination of propane. Two monochlorinated (just one chlorine atom) products are possible. One has the chlorine atom on a primary carbon atom, and the other has the chlorine atom on the secondary carbon atom.

$$\underset{\text{propane}}{CH_3{-}CH_2{-}CH_3} + Cl_2 \xrightarrow{h\nu,\ 25°C} \underset{\substack{\text{1-chloropropane, 40\%}\\(n\text{-propyl chloride})}}{\overset{Cl}{\overset{|}{CH_2}}{-}CH_2{-}CH_3} + \underset{\substack{\text{2-chloropropane, 60\%}\\\text{(isopropyl chloride)}}}{CH_3{-}\overset{Cl}{\overset{|}{CH}}{-}CH_3}$$

The product ratio in this reaction shows that the replacement of hydrogen atoms by chlorine is not random. Propane has six primary hydrogens (hydrogens bonded to a primary carbon) and only two secondary hydrogens (bonded to the secondary carbon), yet the major product results from substitution of a secondary hydrogen. We can calculate how reactive each kind of hydrogen is by dividing the amount of product observed by the number of hydrogens that can be replaced to give that product. Figure 4-5 shows the definition of primary, secondary, and tertiary hydrogens and the calculation of their relative reactivity. The secondary hydrogens are 4.5 times as reactive as the primary hydrogens. To explain this preference for reaction at the secondary position, we must look carefully at the reaction mechanism (Fig. 4-6).

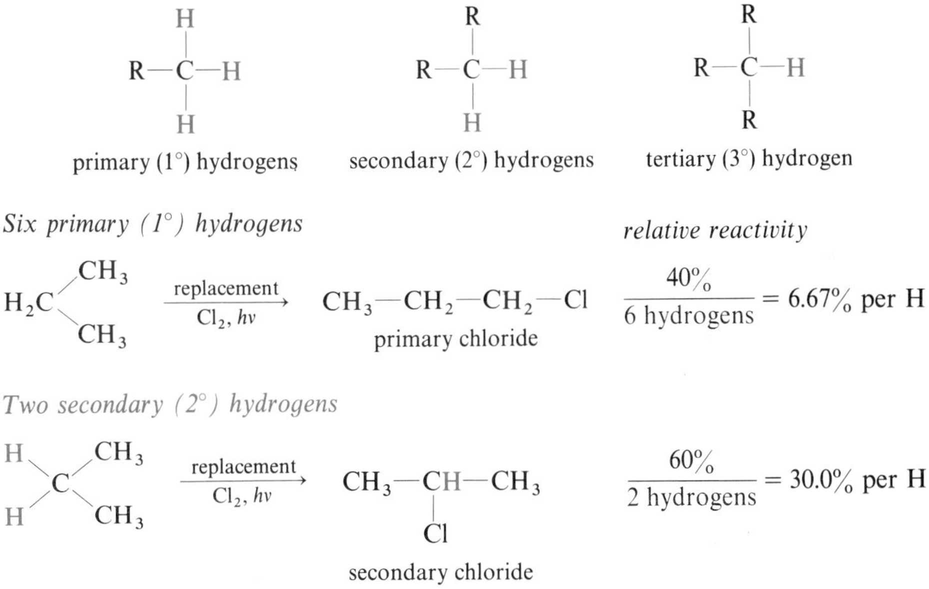

R—CH(H)—H: primary (1°) hydrogens

R—C(R)(H)—H: secondary (2°) hydrogens

R—C(R)(R)—H: tertiary (3°) hydrogen

Six primary (1°) hydrogens — *relative reactivity*

$$H_2C(CH_3)_2 \xrightarrow[Cl_2,\ h\nu]{\text{replacement}} CH_3—CH_2—CH_2—Cl \qquad \frac{40\%}{6\ \text{hydrogens}} = 6.67\%\ \text{per H}$$

primary chloride

Two secondary (2°) hydrogens

$$H_2C(CH_3)_2 \xrightarrow[Cl_2,\ h\nu]{\text{replacement}} CH_3—CH(Cl)—CH_3 \qquad \frac{60\%}{2\ \text{hydrogens}} = 30.0\%\ \text{per H}$$

secondary chloride

The 2° hydrogens are $\frac{30.0}{6.67} = 4.5$ times as reactive as the 1° hydrogens.

FIGURE 4-5 There are six primary hydrogens in propane and only two secondary hydrogens, yet the major product results from replacement of a secondary hydrogen.

When a chlorine atom reacts with propane, abstraction of a hydrogen atom can give either a primary radical or a secondary radical. The structure of the radical formed in this step determines the structure of the observed product, either 1-chloropropane or 2-chloropropane. The product ratio shows that the secondary radical is formed preferentially. This preference for reaction at the secondary position results from the greater stability of the secondary free radical and the transition state leading to it.

PROBLEM 4-18

What would be the product ratio in the chlorination of propane if all the hydrogens were abstracted at equal rates?

PROBLEM 4-19

Classify each hydrogen atom in the following compounds as primary (1°), secondary (2°), or tertiary (3°).

(a) butane (b) isobutane (c) 2-methylbutane
(d) cyclohexane (e) norbornane (see page 113)

Initiation: Splitting of the chlorine molecule

$$Cl_2 + h\nu \longrightarrow 2\,Cl\cdot$$

First propagation step: Abstraction (removal) of a primary or secondary hydrogen

$$CH_3{-}CH_2{-}CH_3 + Cl\cdot \longrightarrow \underset{\text{primary radical}}{\cdot CH_2{-}CH_2{-}CH_3} \text{ or } \underset{\text{secondary radical}}{CH_3{-}\dot{C}H{-}CH_3} + HCl$$

Second propagation step: Reaction with chlorine to form the alkyl chloride

$$\underset{\text{primary radical}}{\cdot CH_2{-}CH_2{-}CH_3} + Cl_2 \longrightarrow \underset{\substack{\text{primary chloride}\\ \text{(1-chloropropane)}}}{Cl{-}CH_2{-}CH_2{-}CH_3} + Cl\cdot$$

or

$$\underset{\text{secondary radical}}{CH_3{-}\dot{C}H{-}CH_3} + Cl_2 \longrightarrow \underset{\substack{\text{secondary chloride}\\ \text{(2-chloropropane)}}}{CH_3{-}\overset{\overset{\displaystyle Cl}{|}}{C}H{-}CH_3} + Cl\cdot$$

FIGURE 4-6 The mechanism for free-radical chlorination of propane. The first propagation step forms either a primary radical or a secondary radical. This radical determines whether the final product will be the primary chloride or the secondary chloride.

4-14B FREE-RADICAL STABILITIES

Figure 4-7 shows the energy required (the bond dissociation energy) to form a free radical by breaking a bond between a hydrogen atom and a carbon atom. This energy is greatest for a methyl carbon, and it decreases for a primary carbon,

Formation of a methyl radical — *Bond dissociation energy*

$$CH_4 \longrightarrow H\cdot + \cdot CH_3 \qquad \Delta H^\circ = 104 \text{ kcal (435 kJ)}$$

Formation of a primary (1°) radical

$$CH_3{-}CH_2{-}CH_3 \longrightarrow H\cdot + CH_3{-}CH_2{-}\dot{C}H_2 \qquad \Delta H^\circ = 98 \text{ kcal (410 kJ)}$$

Formation of a secondary (2°) radical

$$CH_3{-}CH_2{-}CH_3 \longrightarrow H\cdot + CH_3{-}\dot{C}H{-}CH_3 \qquad \Delta H^\circ = 95 \text{ kcal (397 kJ)}$$

Formation of a tertiary (3°) radical

$$CH_3{-}\overset{\overset{\displaystyle CH_3}{|}}{\underset{\underset{\displaystyle CH_3}{|}}{C}}{-}H \longrightarrow CH_3{-}\overset{\overset{\displaystyle CH_3}{|}}{\underset{\underset{\displaystyle CH_3}{|}}{C}}\cdot + H\cdot \qquad \Delta H^\circ = 91 \text{ kcal (381 kJ)}$$

FIGURE 4-7 The bond dissociation energies show that the more highly substituted free radicals are more stable than the less highly substituted ones.

a secondary carbon, and a tertiary carbon. The more highly substituted the carbon atom is, the less energy is required to form the free radical.

From the information in Figure 4-7 we conclude that free radicals are more stable if they are more highly substituted. The following free radicals are listed in decreasing order of stability.

$$\underset{\text{R}}{\overset{\text{R}}{\text{R—}\overset{|}{\underset{|}{\text{C}}}\cdot}} > \underset{\text{H}}{\overset{\text{R}}{\text{R—}\overset{|}{\underset{|}{\text{C}}}\cdot}} > \underset{\text{H}}{\overset{\text{H}}{\text{R—}\overset{|}{\underset{|}{\text{C}}}\cdot}} > \underset{\text{H}}{\overset{\text{H}}{\text{H—}\overset{|}{\underset{|}{\text{C}}}\cdot}}$$

$$3^\circ > 2^\circ > 1^\circ > \text{Me}\cdot$$

tertiary > secondary > primary > methyl

In the chlorination of propane, the secondary hydrogen atom is abstracted more often because the secondary radical, and the transition state leading to it, are lower in energy than the primary radical and its transition state. Using the bond dissociation energies in Table 4-2 (page 133), we can calculate ΔH° for each of the possible reaction steps. Abstraction of the secondary hydrogen is 3 kcal/mol (13 kJ/mol) more exothermic than abstraction of the primary hydrogen.

$$CH_3—CH_2—CH_3 \;+\; Cl\cdot \;\longrightarrow\; CH_3—CH_2—CH_2\cdot \;+\; H—Cl$$

Energy required to break the $CH_3—CH_2—CH_2 \wr H$ bond	+98 kcal/mol (+410 kJ/mol)
Energy released in forming the $H \wr Cl$ bond	−103 kcal/mol (−431 kJ/mol)
Total energy for reaction at the primary position	−5 kcal/mol (−21 kJ/mol)

$$CH_3—\overset{\overset{CH_3}{|}}{CH_2} \;+\; Cl\cdot \;\longrightarrow\; CH_3—\overset{\overset{CH_3}{|}}{CH}\cdot \;+\; H—Cl$$

Energy required to break the $CH_3—\overset{\overset{CH_3}{|}}{CH} \wr H$ bond	+95 kcal/mol (+397 kJ/mol)
Energy released in forming the $H \wr Cl$ bond	−103 kcal/mol (−431 kJ/mol)
Total energy for reaction at the secondary position	−8 kcal/mol (−34 kJ/mol)

A reaction energy profile for this first propagation step appears in Figure 4-8. The activation energy to form the secondary radical is slightly lower, and the secondary radical is therefore formed faster than the primary radical.

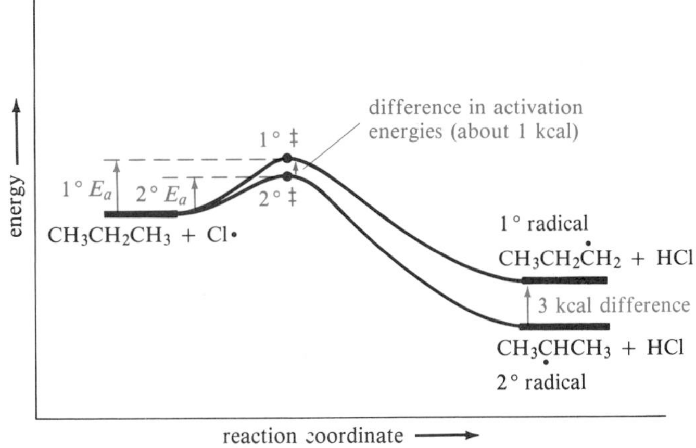

FIGURE 4-8 Energy profile for the first propagation step in the chlorination of propane. Formation of the secondary radical has a lower activation energy than formation of the primary radical.

SOLVED PROBLEM 4-4

Tertiary hydrogen atoms react with Cl· about 5.5 times as fast as primary ones. Predict the product ratios for chlorination of isobutane.

SOLUTION There are nine primary hydrogens and one tertiary hydrogen in isobutane.

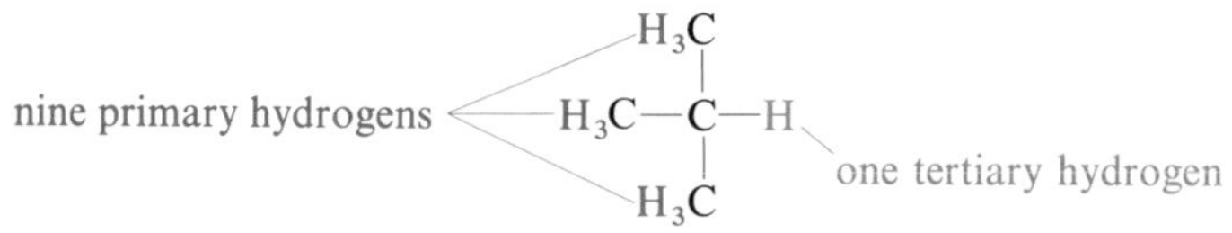

$$\text{nine primary hydrogens} \rightarrow (H_3C)_3C\text{—H} \leftarrow \text{one tertiary hydrogen}$$

$$(9 \text{ primary hydrogens}) \times (\text{reactivity } 1.0) = 9.0 \text{ relative amount of reaction}$$

$$(1 \text{ tertiary hydrogen}) \times (\text{reactivity } 5.5) = 5.5 \text{ relative amount of reaction}$$

Even though the primary hydrogens are less reactive, there are so many of them that the primary product is the major product. The product ratio will be 9.0:5.5 or about 1.6:1.

$$\text{Fraction of primary} = \frac{9.0}{9.0 + 5.5} = 62\%$$

$$\text{Fraction of tertiary} = \frac{5.5}{9.0 + 5.5} = 38\%$$

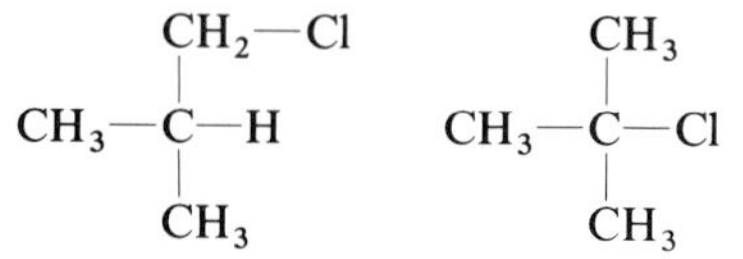

$$CH_3\text{—}CH(CH_3)\text{—}CH_2\text{—}Cl \qquad CH_3\text{—}C(CH_3)_2\text{—}Cl$$

major product 62% — minor product 38%

PROBLEM 4-20

Use the bond dissociation energies in Table 4-2 (page 133) to calculate the heats of reaction for the two possible first propagation steps in the chlorination of isobutane. Use this information to draw a reaction-energy profile, like Figure 4-8, comparing the activation energies for formation of the two radicals.

PROBLEM 4-21

Predict the ratios of products that result when isopentane (2-methylbutane) is chlorinated.

PROBLEM 4-22

(a) When *n*-heptane burns in a gasoline engine, the combustion process takes place too quickly. The explosive detonation makes a noise called "knocking." When 2,2,4-trimethylpentane (isooctane) is burned, combustion takes place in a slower, more controlled manner. Combustion is a free-radical chain reaction, and its rate depends on the reactivity of the free-radical intermediates. Explain why isooctane has less tendency to knock than does *n*-heptane.

(b) Alkoxy radicals (R—O·) are generally more stable than alkyl (R·) radicals. Explain why *t*-butyl alcohol, $(CH_3)_3COH$, works as an antiknock additive for gasoline.

4-14C BROMINATION OF PROPANE

Figure 4-9 shows the free-radical reaction of propane with bromine. Notice that this reaction is both heated to 125°C and irradiated with light to achieve a mod-

$$CH_3{-}CH_2{-}CH_3 + Br_2 \xrightarrow{h\nu,\ 125^\circ C} \underset{\text{primary bromide, 3\%}}{CH_3{-}CH_2{-}\overset{\overset{Br}{|}}{C}H_2} + \underset{\text{secondary bromide, 97\%}}{CH_3{-}\overset{\overset{Br}{|}}{C}H{-}CH_3}$$

Relative reactivity

six primary hydrogens $\dfrac{3\%}{6} = 0.5\%$ per H

two secondary hydrogens $\dfrac{97\%}{2} = 48.5\%$ per H

The 2° hydrogens are $\dfrac{48.5}{0.5} = 97$ times as reactive as the 1° hydrogens.

FIGURE 4-9 This 97:3 ratio of products shows that a secondary hydrogen is abstracted 97 times as rapidly as a primary hydrogen. Bromination (reactivity ratio 97:1) is much more selective than chlorination (reactivity ratio 4.5:1).

erate rate. The secondary product (2-bromopropane) is favored by a 97:3 product ratio. From this product ratio, we calculate that the secondary hydrogens are 97 times as reactive as the primary hydrogens.

The 97:1 reactivity ratio for bromination is much larger than the 4.5:1 ratio for chlorination. We say that bromination is more *selective* than chlorination because the major reaction is favored by a larger amount. To explain this enhanced selectivity, we must consider the transition states and activation energies for the rate-determining step.

As with chlorination, the rate-determining step in bromination is the first propagation step: abstraction of a hydrogen atom by a bromine radical. The energetics of the two possible hydrogen abstractions are shown below. Compare these numbers with the energetics of the first propagation step of chlorination shown on page 145. The bond energies are taken from Table 4-2 (page 133.)

$$CH_3{-}CH_2{-}CH_3 + Br\cdot \longrightarrow CH_3{-}CH_2{-}CH_2\cdot + H{-}Br$$

Energy required to break the $CH_3{-}CH_2{-}CH_2{\wr}H$ bond	+98 kcal/mol (+410 kJ/mol)
Energy released in forming the $H{\wr}Br$ bond	−88 kcal/mol (−368 kJ/mol)
Total energy for reaction at the primary position	+10 kcal/mol (+42 kJ/mol)

$$CH_3{-}\overset{\overset{CH_3}{|}}{C}H_2 + Br\cdot \longrightarrow CH_3{-}\overset{\overset{CH_3}{|}}{C}H\cdot + H{-}Br$$

Energy required to break the $CH_3{-}\overset{\overset{CH_3}{|}}{C}H{\wr}H$ bond	+95 kcal/mol (+397 kJ/mol)
Energy released in forming the $H{\wr}Br$ bond	−88 kcal/mol (−368 kJ/mol)
Total energy for reaction at the secondary position	+7 kcal/mol (+29 kJ/mol)

The differences in the energies for chlorination and those for bromination (above) result from the difference in the bond dissociation energies of H—Cl (103 kcal) and H—Br (88 kcal). The HBr bond is weaker, and abstraction of a hydrogen atom by Br· is endothermic. This endothermic step explains why bromination is much slower than chlorination, but it still does not explain the enhanced selectivity observed with bromination.

Consider the reaction energy profile for the first propagation step in the bromination of propane (Fig. 4-10). Although the difference in values of ΔH° between abstraction of a primary hydrogen and a secondary hydrogen is still 3 kcal/mol (13 kJ/mol), this energy profile for bromination shows a much larger difference in activation energies for abstraction of the primary and secondary hydrogens than we saw in the chlorination reaction (Fig. 4-8).

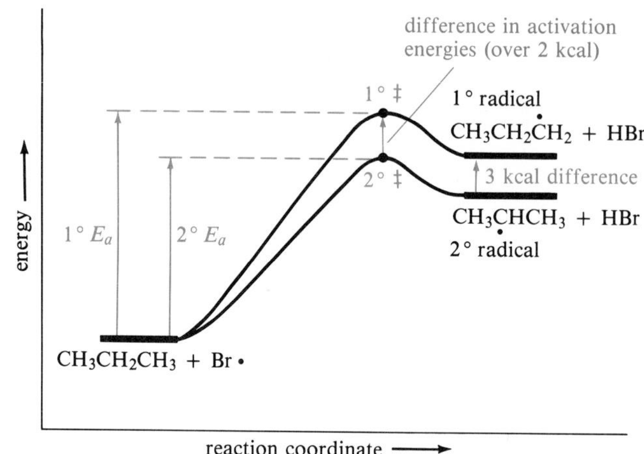

FIGURE 4-10 Reaction energy profile for the first propagation step in the bromination of propane. The energy difference in the transition states is nearly as large as the energy difference in the products.

4-15 THE HAMMOND POSTULATE

Figure 4-11 summarizes the energy profiles for bromination and chlorination of propane. Together, these energy profiles provide the explanation for the enhanced selectivity observed in bromination.

When we compare the reaction energy profiles for the first propagation steps of chlorination and bromination, there are two important differences.

1. The first propagation step is endothermic for bromination but exothermic for chlorination.
2. The transition states forming the 1° and 2° radicals for the endothermic bromination have a larger energy difference than those for the exothermic chlorination, even though the energy difference of the products is the same (3 kcal or 13 kJ) in both reactions.

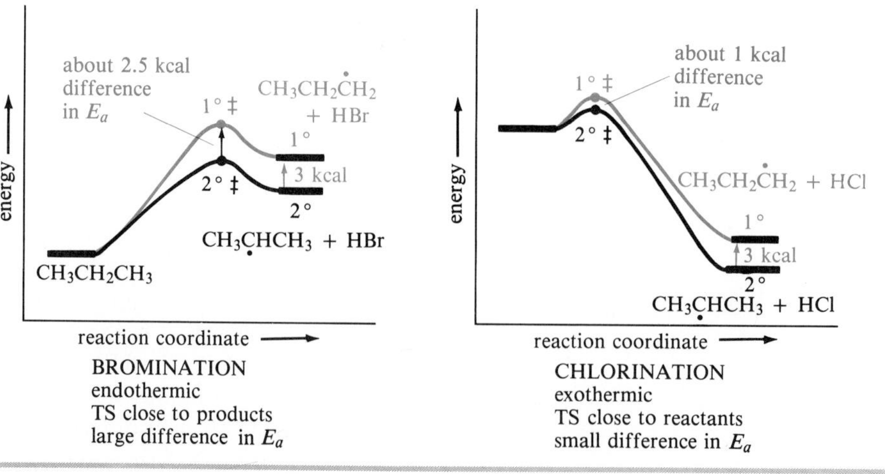

FIGURE 4-11 In the endothermic bromination, the transition state is closer to the products in energy and in structure. The difference in activation energies is about 2.5 kcal (10 kJ), nearly the entire energy difference in the radicals. In the exothermic chlorination, the transition state is closer to the reactants in energy and in structure. The difference in activation energies for chlorination is about 1 kcal (4 kJ), only a small part of the energy difference of the radicals.

In general, we will find that these differences are related:

> In an endothermic reaction, the transition state is closer to the products in energy and in structure. In an exothermic reaction, the transition state is closer to the reactants in energy and in structure.

Figure 4-12 contrasts the transition states for bromination and chlorination. The productlike transition state for bromination has the C—H bond nearly broken and a great deal of radical character on the carbon atom. In the reactantlike transition state for chlorination, the C—H bond has just begun to break. The transition state of the exothermic chlorination reflects only a small part of the energy difference of the products. Therefore, chlorination is less selective.

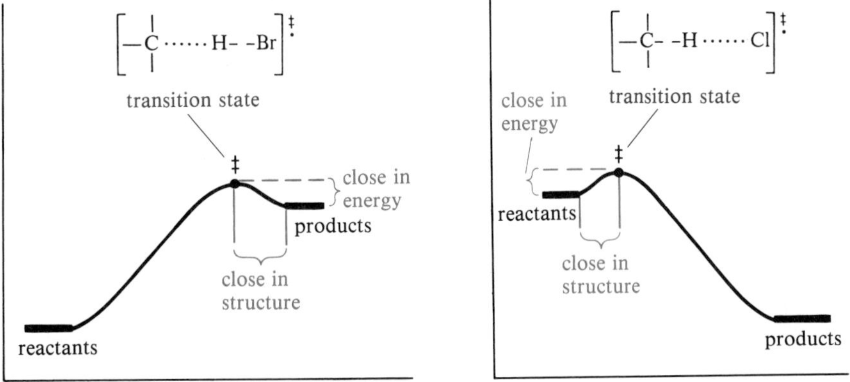

FIGURE 4-12 In the endothermic bromination, the transition state resembles the free radical. In the chlorination, the free radical has just begun to form.

These reactions are examples of a more general principle, called the **Hammond postulate.**

> HAMMOND POSTULATE Related species that are similar in energy are also similar in structure.

This general rule tells us something about the transition states in endothermic and exothermic reactions. The transition state is always the point of highest energy on the energy profile. Its structure resembles either the reactants or the products, whichever ones are higher in energy. In an endothermic reaction, the products are higher in energy, and the transition state is productlike. In an exothermic reaction, the reactants are higher in energy and the transition state is reactantlike.

PROBLEM 4-23

(a) Compute the heats of reaction for abstraction of a primary hydrogen and a secondary hydrogen from propane by a fluorine radical.

$$CH_3—CH_2—CH_3 \; + \; F\cdot \; \longrightarrow \; CH_3—CH_2—CH_2\cdot \; + \; HF$$

$$CH_3—CH_2—CH_3 \; + \; F\cdot \; \longrightarrow \; CH_3—\dot{C}H—CH_3 \; + \; HF$$

(b) What product distribution would you expect to obtain from the free-radical fluorination of propane?

PROBLEM 4-24

We showed earlier (page 141) that there is a large deuterium isotope effect in the chlorination of methane; methane undergoes free-radical chlorination about 12 times as fast as tetradeuteriomethane (CD_4). Monochlorination of deuterioethane (C_2H_5D) leads to a mixture containing 93 percent C_2H_4DCl and 7 percent C_2H_5Cl.

(a) Calculate the relative rates of abstraction of hydrogen and deuterium in the chlorination of ethane.
(b) Consider the thermodynamics of the chlorination of methane and the chlorination of ethane, and use the Hammond postulate to explain why one of these reactions has a much larger isotope effect than the other.

PROBLEM SOLVING: PROPOSING REACTION MECHANISMS

Throughout this course, you will need to propose mechanisms to explain reactions. Methods for dealing with different types of mechanisms will be discussed as we encounter them. These techniques for dealing with a wide variety of mechanisms are collected in Appendix 4, but at this point we consider only free-radical mechanisms like those in this chapter.

FREE-RADICAL REACTIONS General principles: Free-radical reactions generally proceed by chain-reaction mechanisms, using an initiator with an easily broken bond (such as chlorine, bromine, or a peroxide) to start the chain reaction. In drawing the mechanism, expect free-radical intermediates (especially highly substituted or resonance-stabilized intermediates). Watch for the most stable free radicals and avoid any high-energy radicals such as hydrogen atoms.

1. **Draw a step that breaks the weak bond in the initiator.**
   A free-radical reaction usually begins with an initiation step in which the initiator undergoes homolytic (free-radical) cleavage to give two radicals.
2. **Draw a reaction of the initiator with one of the starting materials.**
   One of the initiator radicals reacts with one of the starting materials to give a free-radical version of the starting material. The initiator might abstract a hydrogen atom or add to a double bond, depending on what reaction leads to formation of the observed product. You might want to consider bond dissociation energies to see which reaction is energetically favored.
3. **Draw a reaction of the free-radical version of the starting material with another starting material molecule to form a bond needed in the product and generate a new radical intermediate.**
   Check your intermediates to be sure that you have used the most stable radical intermediates. For a realistic chain reaction, no new initiation steps should be required; a radical should be regenerated in each propagation step.
4. **Draw termination step(s).**
   The reaction ends with termination steps, which are side reactions rather than part of the product-forming mechanism. The reaction of any two free radicals to give a stable molecule is a termination step, as is a collision of a free radical with the container.

Before we illustrate this procedure, we want to call your attention to a few common mistakes. Avoiding these mistakes will be a great help in drawing correct mechanisms. Making these mistakes will cause difficulty in solving the problem and could make it impossible.

COMMON MISTAKES TO AVOID IN DRAWING MECHANISMS:

1. Do not use condensed or line-angle formulas for reaction sites. Draw all the bonds and all the substituents of each carbon atom affected throughout the mechanism. Three-bonded carbon atoms in intermediates are most likely to be radicals in the free-radical reactions we have studied. If you draw condensed formulas or line-angle formulas, you will likely misplace a hydrogen atom and show a reactive species on the wrong carbon.

2. Do not show more than one step occurring at once, unless they really do occur at once.

Our sample problem is to draw the mechanism of the reaction of methylcyclopentane with bromine under irradiation with light and predict the major product formed.

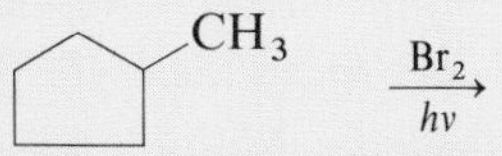

$$\text{(cyclopentane)}\text{—}CH_3 \xrightarrow[h\nu]{Br_2}$$

First, as in every mechanism problem, we draw what we know, showing all the bonds and all the substituents of each carbon atom that may be affected throughout the mechanism.

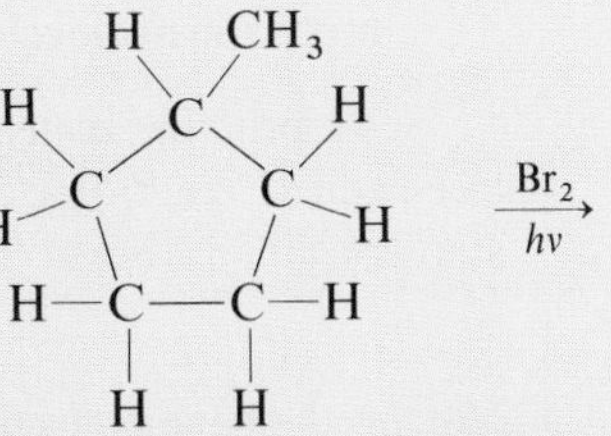

(Structural formula of methylcyclopentane showing all H atoms) $\xrightarrow[h\nu]{Br_2}$

1. **Draw a step involving cleavage of the weak bond in the initiator.**
   The use of light with bromine suggests a free-radical reaction, with the light providing the energy for dissociation of Br_2. This homolytic cleavage initiates the chain reaction by generating two Br· radicals.

   Initiation step

   $$Br\text{—}Br \xrightarrow{h\nu} Br\cdot \; + \; \cdot Br$$

2. **Draw a reaction of the initiator with one of the starting materials.**
   One of these initiator radicals should react with methylcyclopentane to give a free-radical version of methylcyclopentane. As we have seen, a bromine or chlorine radical can abstract a hydrogen atom from an alkane to generate an alkyl radical. The bromine radical is relatively selective, and the most stable alkyl radical should result. Abstraction of the tertiary hydrogen atom gives a tertiary radical.

   First propagation step

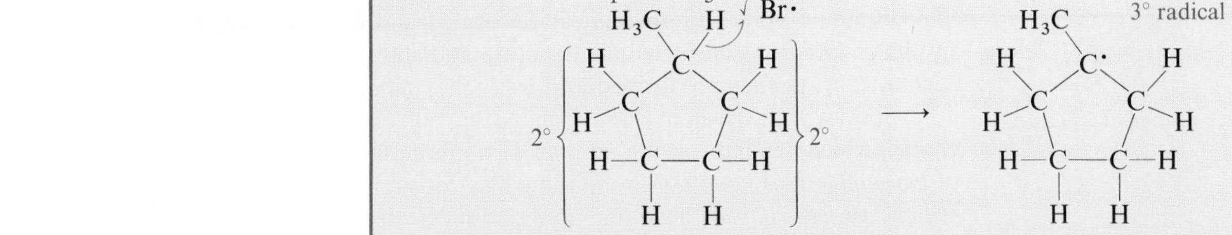

   (Methylcyclopentane with hydrogens labeled 1° (H_3C), 3° (H), and 2° (ring H's), attacked by Br·) $\longrightarrow$ (3° radical: H_3C—C· in ring)

3. **Draw a reaction of the free-radical version of the starting material with another starting material molecule to form a bond needed in the product and generate a new radical intermediate.**

The alkyl radical should react with another starting-material molecule in another propagation step to generate a product and another radical. Reaction of the alkyl radical with Br_2 gives 1-bromo-1-methylcyclopentane (the major product) and another bromine radical to continue the chain.

Second propagation step

$$\text{1-methylcyclopentyl radical} + Br\text{—}Br \longrightarrow \text{1-bromo-1-methylcyclopentane} + Br\cdot$$

4. **Draw termination step(s).**
It is left to you to add some possible termination steps and summarize the mechanism developed above.

As practice in using a systematic approach to proposing mechanisms for free-radical reactions, work Problem 4-25 by going through each of the four steps outlined above

PROBLEM 4-25

2,3-Dimethylbutane reacts with bromine in the presence of light to give a good yield of a monobrominated product. Further reaction gives a dibrominated product. Predict the structures of these products, and propose a mechanism for formation of the monobrominated product.

4-16 REACTIVE INTERMEDIATES

Reactive intermediates are short-lived species that are never present in high concentrations because they react as quickly as they are formed. In most cases, reactive intermediates are fragments of molecules (like free radicals), often having atoms with unusual numbers of bonds. For example, some of the common reactive intermediates contain carbon atoms with only two or three bonds, compared with carbon's four bonds in its stable compounds. Such species react quickly with a wide variety of compounds to give more stable products with tetravalent carbon atoms.

Although reactive intermediates are not stable compounds, they are extremely important to our study of organic chemistry. Most of the reaction mechanisms we will encounter involve reactive intermediates. If you are to understand these mechanisms and propose mechanisms of your own, you need to know how the common reactive intermediates are formed and how they are likely to react. In this chapter we consider their structure and stability, and in later chapters we encounter ways they are formed and ways they react to give stable compounds.

Species with trivalent (three-bonded) carbon are classified according to their charge, which depends on the number of nonbonding electrons. The *carbocations* or *carbonium ions* have no nonbonding electrons and are positively charged. The *radicals* or *free radicals* have one nonbonding electron and are neutral. The *carbanions* have a pair of nonbonding electrons and are negatively charged. The most common intermediates with a divalent (two-bonded) carbon atom are the *carbenes*. Carbenes have two nonbonding electrons on the divalent carbon atom, making it uncharged.

$$\begin{array}{cccc} \text{H}-\overset{\text{H}}{\underset{\text{H}}{\text{C}}}{}^{+} & \text{H}-\overset{\text{H}}{\underset{\text{H}}{\text{C}}}\cdot & \text{H}-\overset{\text{H}}{\underset{\text{H}}{\text{C}}}:^{-} & \text{H}_2\text{C}: \\ \text{carbocation} & \text{radical} & \text{carbanion} & \text{carbene} \end{array}$$

4-16A CARBOCATIONS

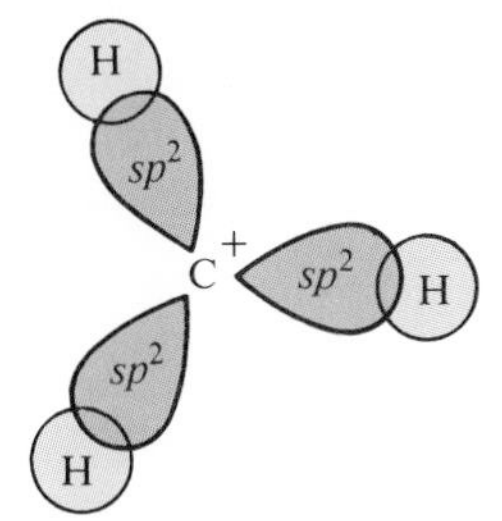

top view

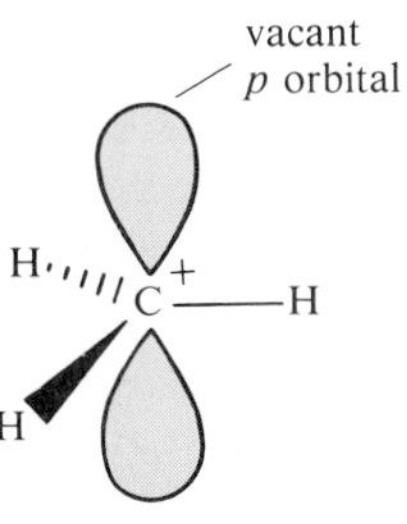

side view

FIGURE 4-13 The structure of the methyl cation is similar to the structure of BH_3. The carbon atom is sigma bonded to three hydrogen atoms by overlap of its sp^2 hybrid orbitals with the s orbitals of hydrogen. There is a vacant p orbital perpendicular to the plane of the three C—H bonds.

A **carbocation** (often called a **carbonium ion**) is a species that contains a carbon atom bearing a positive charge. The positively charged carbon atom is bonded to three other atoms and has no nonbonding electrons. It is sp^2 hybridized, with a planar structure and bond angles of about 120°. For example, the methyl cation (CH_3^+) is planar, with bond angles of exactly 120°. The unhybridized p orbital is vacant, and lies perpendicular to the plane of the C—H bonds (Fig. 4-13). The structure of CH_3^+ is similar to the structure of BH_3, discussed in Chapter 2.

The carbocation carbon atom has only six electrons in its valence shell, and the positive charge enhances its reactivity toward nucleophiles. Therefore, a carbocation is a particularly strong electrophile. Most carbocations react with the first nucleophile they encounter, which is often a molecule of the solvent. We study the reactions of carbocations with nucleophiles in Chapter 5.

Like free radicals, carbocations are *electron-deficient* species: They have fewer than eight electrons in the valence shell of the carbon atom. Also like free radicals, carbocations are stabilized by alkyl substituents. An alkyl group stabilizes an electron-deficient carbocation in two ways: (1) through an inductive effect, and (2) through the partial overlap of filled orbitals with empty ones. The **inductive effect** is a donation of electron density through the sigma bonds of the molecule. The positively charged carbon atom withdraws some electron density from the alkyl groups bonded to it. This inductive donation of electron density by alkyl substituents can be represented as follows:

$$^{\delta+}CH_3 \rightarrow \overset{+}{C}(\leftarrow CH_3^{\delta+})(\uparrow CH_3^{\delta+})$$

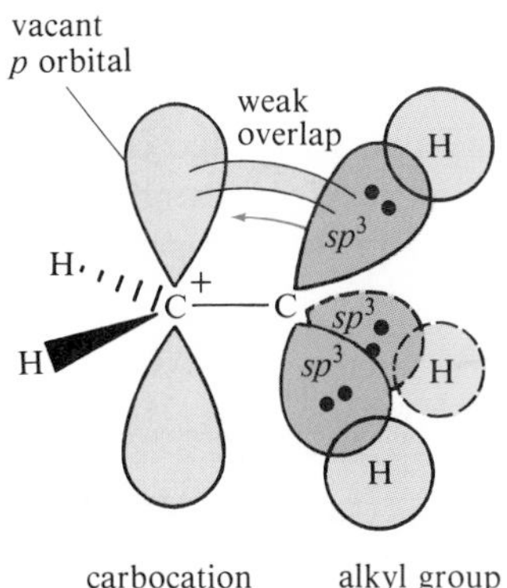

FIGURE 4-14 A carbocation is stabilized by overlap of filled orbitals on an adjacent alkyl group with the vacant p orbital of the carbocation. Overlap between a sigma bond and a p orbital is sometimes called hyperconjugation.

Alkyl substituents also have filled sp^3 orbitals that can overlap with the empty p-orbital on the positively charged carbon atom, further stabilizing the carbocation. Even though the attached alkyl group rotates, one of its sigma bonds is always aligned with the empty p orbital on the carbocation. The pair of electrons in this sigma bond spreads out slightly into the empty p orbital, stabilizing the electron-deficient carbon atom. This type of overlap between a p orbital and a sigma bond is sometimes called *hyperconjugation.* Figure 4-14 shows how hyperconjugation with an alkyl group stabilizes a carbocation.

In general, more highly substituted carbocations are more stable.

$$\begin{array}{ccccccc} \text{R}-\overset{\text{R}}{\underset{\text{R}}{\text{C}}}{}^{+} & > & \text{R}-\overset{\text{R}}{\underset{\text{H}}{\text{C}}}{}^{+} & > & \text{R}-\overset{\text{H}}{\underset{\text{H}}{\text{C}}}{}^{+} & > & \text{H}-\overset{\text{H}}{\underset{\text{H}}{\text{C}}}{}^{+} \\ \text{most stable} & & & & & & \text{least stable} \end{array}$$

Stability of carbocations

3° > 2° > 1° > methyl

Unsaturated carbocations may also be stabilized by *resonance stabilization.* If a pi bond is adjacent to a carbocation, the filled *p* orbitals of the pi bond will overlap with the empty *p* orbital of the carbocation. The result is a delocalized ion, with the positive charge shared by two atoms. This resonance delocalization is particularly effective in stabilizing a carbocation.

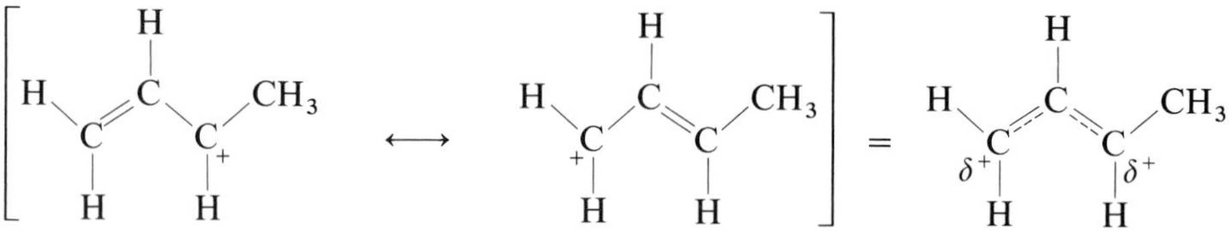

PROBLEM 4-26

The triphenylmethyl cation is so stable that some of its salts can be stored for months. Explain why this cation is so stable.

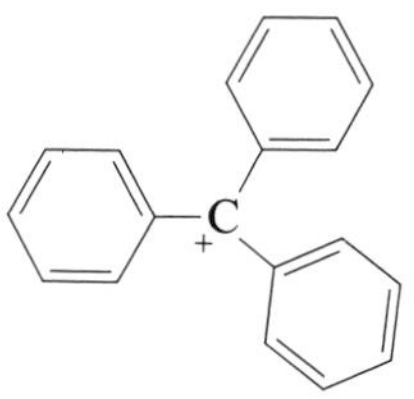

triphenylmethyl cation

PROBLEM 4-27

Rank the following carbocations in decreasing order of stability. Classify each as primary, secondary, or tertiary.

(a) The isopentyl cation, $(CH_3)_2CHCH_2—\overset{+}{C}H_2$

(b) The 3-methyl-2-butyl cation, $CH_3—\overset{+}{C}H—CH(CH_3)_2$

(c) The 2-methyl-2-butyl cation, $CH_3—\overset{+}{C}(CH_3)CH_2CH_3$

4-16B FREE RADICALS

Like carbocations, **free radicals** are sp^2 hybridized and planar. Unlike carbocations, however, the *p* orbital perpendicular to the plane of the C—H bonds of the radical is not empty; it contains the odd electron. Figure 4-15 shows the structure of the methyl radical.

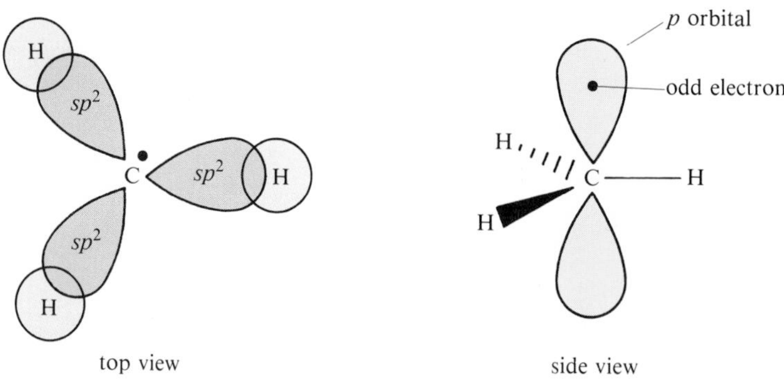

FIGURE 4-15 The structure of the methyl radical is like that of the methyl cation (Fig. 4-13) except that there is an additional electron. The odd electron is in the *p* orbital perpendicular to the plane of the three C—H bonds.

Radicals and carbocations are both electron-deficient because both lack an octet around the carbon atom. Like carbocations, radicals are stabilized by the electron-donating effect of alkyl groups, making the more highly substituted radicals more stable. This effect is confirmed by the bond dissociation energies shown in Figure 4-7: Less energy is required to break a C—H bond to form a more highly substituted radical.

$$R_3C\cdot \;>\; R_2HC\cdot \;>\; RH_2C\cdot \;>\; H_3C\cdot$$

most stable ... least stable

Stability of radicals

3° > 2° > 1° > methyl

Like carbocations, radicals can be stabilized by resonance. Overlap with the *p* orbitals of a pi bond allows the odd electron to be delocalized over two carbon atoms. Resonance delocalization is particularly effective in stabilizing a radical.

$$\left[H_2C{=}CH{-}\dot{C}H{-}CH_3 \;\longleftrightarrow\; H_2\dot{C}{-}CH{=}CH{-}CH_3 \right] \quad \text{or} \quad H_2C\overset{\delta\cdot}{}{\cdots}CH{\cdots}\overset{\delta\cdot}{C}H{-}CH_3$$

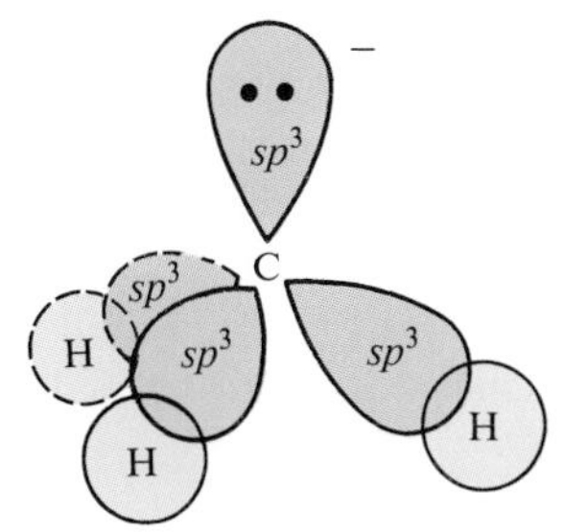

methyl anion

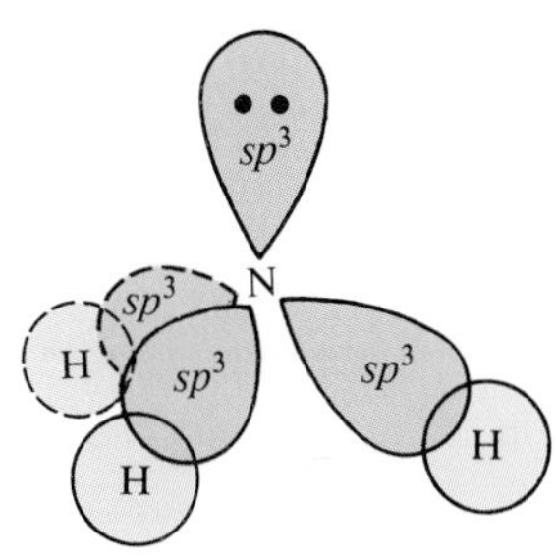

ammonia

FIGURE 4-16 Both the methyl anion and ammonia have an sp^3 hybridized central atom, with a nonbonding pair of electrons occupying one of the tetrahedral positions.

PROBLEM 4-28

Rank the following radicals in decreasing order of stability. Classify each as primary, secondary, or tertiary.

(a) The isopentyl radical, $(CH_3)_2CHCH_2{-}\dot{C}H_2$

(b) The 3-methyl-2-butyl radical, $CH_3{-}\dot{C}H{-}CH(CH_3)_2$

(c) The 2-methyl-2-butyl radical, $CH_3{-}\dot{C}(CH_3)CH_2CH_3$

4-16C CARBANIONS

A **carbanion** has a trivalent carbon atom that bears a negative charge. There are eight electrons around the carbon atom (three bonds and one lone pair), so it is not electron deficient; rather, it is electron rich and nucleophilic. A carbanion has the same electronic structure as an amine. Compare the structures of a methyl carbanion and ammonia:

$$H_3C{:}^- \qquad H_3N{:}$$

methyl anion ammonia

The structure and hybridization of a carbanion also resemble the structure and hybridization of an amine. The carbon atom is sp^3 hybridized and tetrahedral. One of the tetrahedral positions is occupied by an unshared lone pair of electrons. Figure 4-16 compares the orbital structures and geometry of ammonia and the methyl anion.

Like the amines, carbanions are nucleophilic and basic. A carbanion has a negative charge on its carbon atom, however, making it a more powerful base and a stronger nucleophile than an amine. For example, a carbanion is sufficiently basic to remove a proton from ammonia.

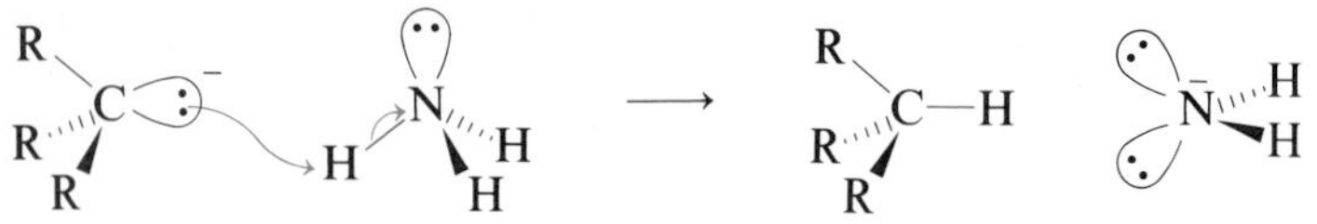

$$R_3C{:}^- \quad + \quad {:}NH_3 \quad \longrightarrow \quad R_3CH \quad + \quad {}^-{:}\ddot{N}H_2$$

The stability order of carbanions reflects their high electron density. Alkyl groups and other electron-donating groups actually destabilize a carbanion. The order of stability is the opposite of that for carbocations and free radicals, which are electron deficient and are stabilized by alkyl groups.

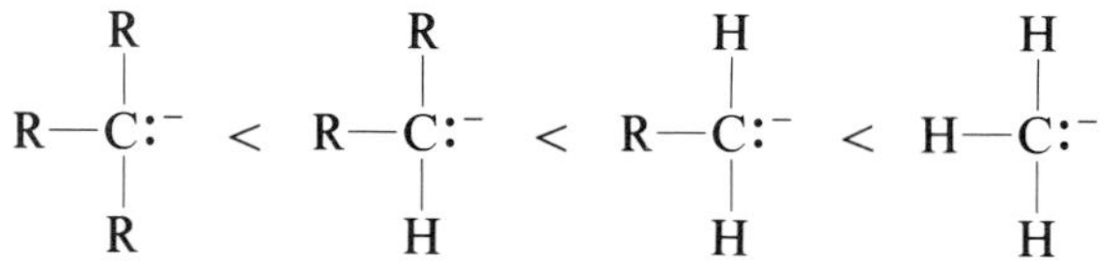

$$R_3C{:}^- < R_2HC{:}^- < RH_2C{:}^- < H_3C{:}^-$$

least stable ... most stable

Stability of carbanions

3° < 2° < 1° < methyl

Carbanions that occur as intermediates in organic reactions are almost always bonded to stabilizing groups. They can be stabilized either by inductive effects or by resonance. For example, halogen atoms are electron withdrawing, and they stabilize carbanions through the inductive withdrawal of electron density. Resonance can also play an important role in stabilizing carbanions. A carbonyl group (C=O) stabilizes an adjacent carbanion by overlap of its pi bond with the nonbonding electrons of the carbanion. The negative charge is delocalized onto the electronegative oxygen atom of the carbonyl group.

$$\left[HC(=\ddot{O}{:})-\ddot{C}H_2^{-} \longleftrightarrow HC({:}\ddot{O}{:}^-)=CH_2 \right]$$

resonance-stabilized carbanion

PROBLEM 4-29

Acetylacetone (2,4-pentanedione) reacts with sodium hydroxide to give water and the sodium salt of a carbanion. Write a complete structural formula for the carbanion, and use resonance structures to show the stabilization of the carbanion.

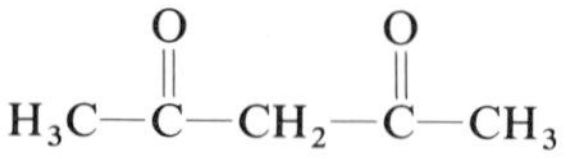

$$H_3C-C(=O)-CH_2-C(=O)-CH_3$$

acetylacetone (2,4-pentanedione)

PROBLEM 4-30

Acetonitrile ($CH_3C{\equiv}N$) is deprotonated by very strong bases. Write resonance structures to show the stabilization of the carbanion that results.

4-16D CARBENES

Carbenes are uncharged reactive intermediates containing a divalent carbon atom. The simplest carbene has the formula $:CH_2$ and is called *methylene,* just as a $—CH_2—$ group in a molecule is called a methylene group. One way of generating carbenes is to form a carbanion that can expel a halide ion. For example, a strong base can abstract a proton from bromoform ($CHBr_3$) to give an inductively stabilized carbanion. This carbanion expels bromide ion to give dibromocarbene.

$$CHBr_3 \text{ (bromoform)} + {}^-OH \rightleftharpoons H_2O + Br_3C:^- \text{ (a carbanion)} \longrightarrow Br_2C: \text{ (dibromocarbene)} + Br^-$$

bromoform — a carbanion — dibromocarbene

The electronic structure of dibromocarbene formed in this reaction is shown below. The carbon atom is sp^2 hybridized, with trigonal geometry. An unshared pair of electrons occupies one of the sp^2 hybrid orbitals, and there is an empty p orbital extending above and below the plane of the three atoms. Because a carbene has both a lone pair of electrons and an empty p orbital, it can react as both a nucleophile and an electrophile.

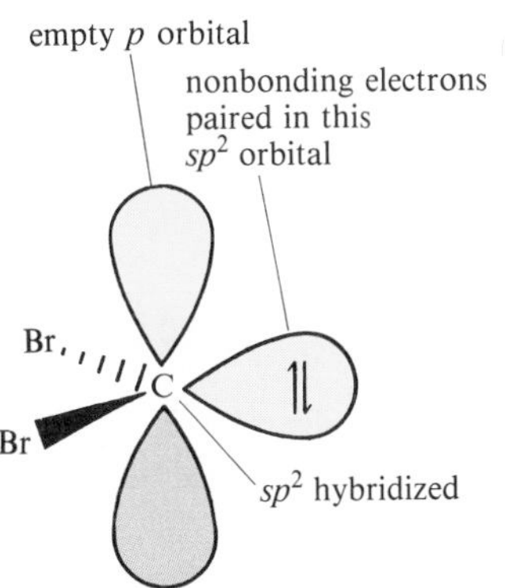

Methylene itself is formed when diazomethane, CH_2N_2, is heated or irradiated with light. The diazomethane molecule splits to form a stable nitrogen molecule and the very reactive carbene.

$$H_2\bar{C}—\overset{+}{N}{\equiv}N: \xrightarrow{\text{heat or light}} H_2C: + :N{\equiv}N:$$

diazomethane — methylene — nitrogen

Carbenes have never been purified or even made in a high concentration, because when two carbenes collide, they immediately dimerize to give an alkene.

$$R_2C: + :CR_2 \xrightarrow{\text{very fast}} R_2C{=}CR_2$$

Carbenes and carbenoids (carbenelike reagents) are useful both for the synthesis of other compounds and for the investigation of reaction mechanisms. The carbene intermediate is generated in the presence of its target compound, so that it can react immediately and the concentration of the carbene is always low. We study reactions using carbenes in Chapter 8.

PROBLEM 4-31

When it is strongly heated, ethyl diazoacetate decomposes to give nitrogen gas and a carbene. Draw a Lewis structure of the carbene.

$$:N\equiv \overset{+}{N}-\overset{-}{C}H-\overset{\overset{O}{\|}}{C}-O-CH_2CH_3$$

ethyl diazoacetate

SUMMARY OF REACTIVE INTERMEDIATES

	structure	*stability*	*acid/base*
carbocations	$-\overset{\vert}{\underset{\vert}{C}}{}^{+}$	$3° > 2° > 1° > Me^+$	electrophilic strong acids
radicals	$-\overset{\vert}{\underset{\vert}{C}}\cdot$	$3° > 2° > 1° > Me\cdot$	electrophilic
carbanions	$-\overset{\vert}{\underset{\vert}{C}}{:}^{-}$	$Me^- > 1° > 2° > 3°$	nucleophilic strong bases
carbenes	$>C:$		both nucleophilic and electrophilic

GLOSSARY

activation energy (E_a) The difference in energy between the reactants and the transition state; the minimum energy the reactants must have for the reaction to occur. (p. 137)

bond dissociation energy (BDE) The amount of energy required to break a particular bond homolytically, to give radicals. (p. 132)

$$A{:}B \longrightarrow A\cdot + B\cdot \qquad \Delta H° = \text{BDE}$$

carbanion A strongly nucleophilic species with a negatively charged carbon atom having only three bonds. The carbon atom has a nonbonding pair of electrons. (p. 155)

carbene A highly reactive species with an uncharged carbon atom with two bonds and a nonbonding pair of electrons. The simplest carbene is methylene, $:CH_2$. (p. 157)

carbocation (carbonium ion) A strongly electrophilic species with a positively charged carbon atom having only three bonds. (p. 153)

catalyst A substance that increases the rate of a reaction (by lowering E_a) without being consumed by the reaction. (p. 138)

chain reaction A multistep reaction where a reactive intermediate formed in one step brings about a second step that generates the intermediate needed for the first step. (p. 124)

> **initiation step:** The preliminary step in a chain reaction, where the reactive intermediate is first formed.
>
> **propagation steps:** The steps in a chain reaction that are repeated over and over to form the product. The sum of the propagation steps should give the net reaction.
>
> **termination steps:** Any steps where a reactive intermediate is consumed without another one being generated.

enthalpy (heat content; H) A measure of the heat energy in a system. In a reaction, the heat absorbed or evolved is called the *heat of reaction*, $\Delta H°$. A decrease in enthalpy (negative $\Delta H°$) is favorable for a reaction. (p. 130)

> **endothermic:** Consuming heat (having a positive $\Delta H°$).

exothermic: Giving off heat (having a negative $\Delta H°$).

entropy (S) A measure of disorder or freedom of motion. An increase in entropy (positive $\Delta S°$) is favorable for a reaction. (p. 130)

equilibrium A state of a system such that no more change is taking place; the rate of the forward reaction equals the rate of the reverse reaction. (p. 128)

equilibrium constant A quantity calculated from the relative amounts of the products and reactants present at equilibrium. (p. 128) For the reaction

$$aA + bB \rightleftharpoons cC + dD$$

the equilibrium constant is

$$K_{eq} = \frac{[C]^c[D]^d}{[A]^a[B]^b}$$

free energy (Gibbs free energy; G) A measure of a reaction's tendency to go in the direction written. A decrease in free energy (negative ΔG) is favorable for a reaction. (p. 128)

Free-energy change is defined: $\Delta G = \Delta H - T\Delta S$

standard Gibbs free energy change: ($\Delta G°$) The free-energy change corresponding to reactants and products in their standard states (pure substances in their most stable states) at 25°C and 1 atm pressure. $\Delta G°$ is related to K_{eq} by

$$K_{eq} = e^{-\Delta G°/RT}$$

Hammond postulate Related species (on a reaction energy profile) that are similar in energy are also similar in structure. In an exothermic reaction, the transition state is closer to the reactants in energy and in structure. In an endothermic reaction, the transition state is closer to the products in energy and in structure. (p. 149)

heterolytic cleavage (ionic cleavage) The breaking of a bond in such a way that one of the atoms retains both of the bond's electrons. A heterolytic cleavage forms two ions. (p. 132)

$$A{:}B \longrightarrow A{:}^- + B^+$$

homolytic cleavage (radical cleavage) The breaking of a bond in such a way that each atom retains one of the bond's two electrons. A homolytic cleavage produces two radicals. (p. 132)

$$A{:}B \longrightarrow A\cdot + \cdot B$$

inductive effect A donation (or withdrawal) of electron density through sigma bonds. (p. 153)

intermediate A molecule or a fragment of a molecule that is formed in a reaction and exists for a finite length of time before it reacts in the next step. An intermediate corresponds to a relative minimum (a low point) in the reaction energy profile. (p. 138)

reactive intermediate: A short-lived species that is never present in high concentration because it reacts as quickly as it is formed. (p. 152)

isotope effect A change in a reaction rate resulting from using a different isotope. For example, breaking a bond to deuterium is generally slower than breaking the same bond to hydrogen. (p. 140)

kinetics The study of reaction rates. (p. 134)

mechanism The step-by-step pathway from reactants to products showing which bonds break and which bonds form in what order. The mechanism should include the structures of all intermediates and arrows to show the movement of electrons. (p. 123)

radical (free radical) A highly reactive species in which one of the atoms has an odd number of electrons. Most commonly, a radical contains a carbon atom with three bonds and an "odd" (unshared) electron. (pp. 125 and 154)

rate-determining step The slowest step in a multistep sequence of reactions. In general, the rate-determining step is the step with the highest-energy transition state. (p. 140)

rate equation (rate law) The relationship between the concentrations of the reagents and the observed reaction rate. (p. 135) A general rate law for the reaction $A + B \rightarrow C + D$ is

$$\text{rate} = k_r[\text{A}]^a[\text{B}]^b$$

kinetic order: The power of a concentration term in the rate equation. The rate equation above is ath order in [A], bth order in [B], and $(a + b)$th order overall.

rate constant: The constant k_r in the rate equation.

rate of a reaction The amount of product formed or reactant consumed per unit of time. (p. 135)

reaction-energy profile (potential-energy profile) A plot of potential-energy changes as the reactants are converted to products. The vertical axis is potential energy (usually free energy, but occasionally enthalpy). The horizontal axis is the **reaction coordinate,** a measure of the progress of the reaction. (p. 138)

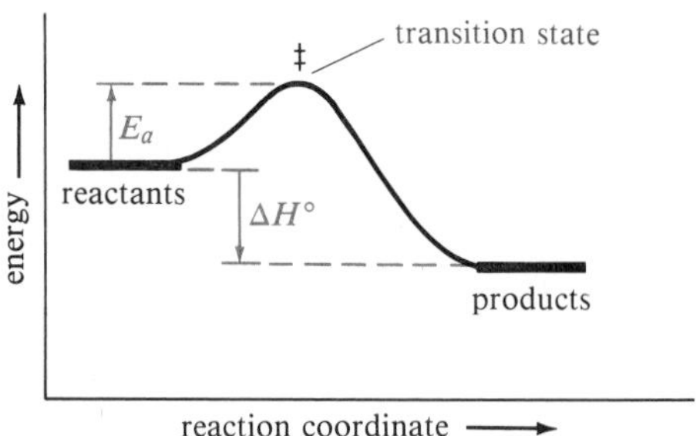

substitution reaction A reaction in which one atom replaces another, usually as a substituent on a carbon atom. (p. 124)

thermodynamics The study of the energy changes accompanying chemical transformations. Thermodynamics is generally concerned with systems at equilibrium. (p. 128)

transition state (activated complex) The state of highest energy between reactants and products. A relative maximum (high point) on the reaction energy profile. (p. 138)

ESSENTIAL PROBLEM-SOLVING SKILLS IN CHAPTER 4

1. Explain the mechanism and energetics of the free-radical halogenation of alkanes.
2. Based on the selectivity of halogenation, predict the products of halogenation of an alkane.
3. Calculate free-energy changes from equilibrium constants.
4. Calculate enthalpy changes from bond dissociation energies.
5. Determine the order of a reaction and suggest a possible mechanism based on its rate equation.
6. Use energy profiles to discuss transition states, activation energies, intermediates, and the rate-determining step of a multistep reaction.
7. Explain how to use isotope effects to determine whether a C—H bond is being broken in the rate-determining step of a reaction.
8. Use the Hammond postulate to predict whether a transition state will be reactantlike or productlike.
9. Describe the structures of carbocations, carbanions, free radicals, and carbenes, and the structural features that stabilize them. Explain which are electrophilic and which are nucleophilic.

4-32. Define and give an example of each of the following terms.

(a) homolytic cleavage	**(b)** heterolytic cleavage	**(c)** free radical
(d) carbocation	**(e)** carbanion	**(f)** carbene
(g) carbonium ion	**(h)** intermediate	**(i)** catalyst
(j) transition state	**(k)** rate equation	**(l)** equilibrium constant
(m) rate constant	**(n)** reaction mechanism	**(o)** chain reaction
(p) substitution reaction	**(q)** activation energy	**(r)** bond dissociation energy

4.33. Consider the following reaction-energy profile.

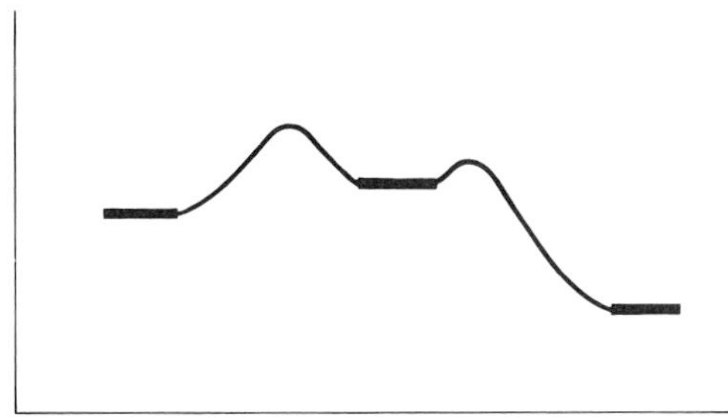

(a) Label the activation energy for the first step and the second step.
(b) Is the overall reaction endothermic or exothermic? What is the sign of $\Delta H°$?
(c) Which points in the curve correspond to intermediates? Which correspond to transition states?
(d) Label the transition state of the rate-determining step.

4.34. Draw a reaction-energy profile for a one-step exothermic reaction. Label the parts that represent the reactants, products, transition state, activation energy, and heat of reaction.

4-35. Draw a reaction-energy profile for a two-step endothermic reaction with a rate-determining second step.

4-36. Treatment of *t*-butyl alcohol with concentrated HCl gives *t*-butyl chloride.

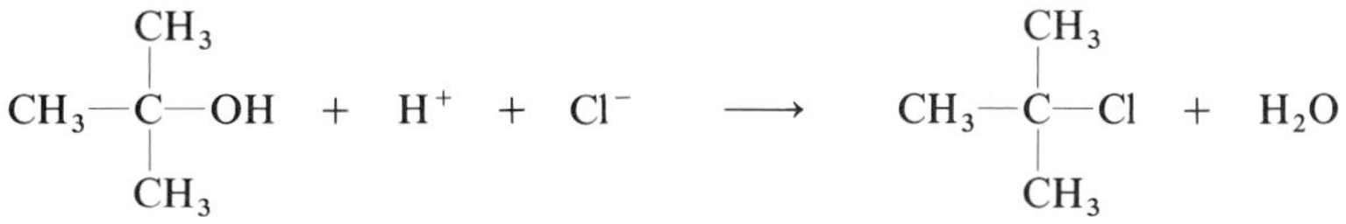

$$\underset{t\text{-butyl alcohol}}{(CH_3)_3C{-}OH} + H^+ + Cl^- \longrightarrow \underset{t\text{-butyl chloride}}{(CH_3)_3C{-}Cl} + H_2O$$

When the concentration of H^+ is doubled, the reaction rate doubles. When the concentration of *t*-butyl alcohol is tripled, the reaction rate triples. When the chloride ion concentration is quadrupled, however, the reaction rate is unchanged. Write the rate equation for this reaction.

4-37. Label each hydrogen atom in the following compounds as primary (1°), secondary (2°), or tertiary (3°).

(a) $CH_3CH_2CH(CH_3)_2$ **(b)** $(CH_3)_3CCH_2C(CH_3)_3$ **(c)** (methylcyclohexane, CH_3)

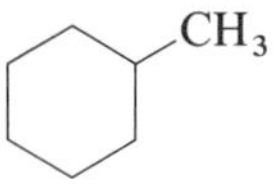

(d) (bicyclic structure) **(e)** (bicyclic structure, H CH_3)

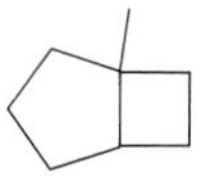

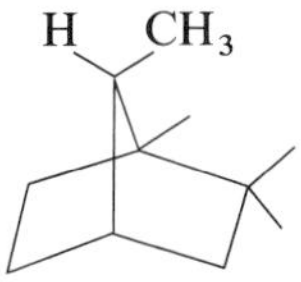

4-38. Use bond dissociation energies to calculate values of $\Delta H°$ for the following reactions.

(a) $CH_3{-}CH_3 + I_2 \longrightarrow CH_3CH_2I + HI$
(b) $CH_3CH_2Cl + HI \longrightarrow CH_3CH_2I + HCl$
(c) $(CH_3)_3C{-}OH + HCl \longrightarrow (CH_3)_3C{-}Cl + H_2O$
(d) $CH_3CH_2CH_3 + H_2 \longrightarrow CH_3CH_3 + CH_4$
(e) $CH_3CH_2OH + HBr \longrightarrow CH_3CH_2{-}Br + H_2O$

4-39. Use the information in Table 4-2 (p. 133) to rank the following radicals in decreasing order of stability.

$\cdot CH_3$ $\quad CH_3\dot{C}H_2$ $\quad C_6H_5\text{—}\dot{C}H_2$ 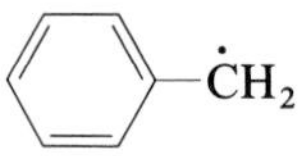 $\quad (CH_3)_3C\cdot$

$(CH_3)_2\dot{C}H$ $\quad CH_2{=}CH\text{—}\dot{C}H_2$

4-40. For each of the following alkanes
(1) Draw all the possible monochlorinated derivatives.
(2) Determine whether free-radical chlorination would be a good way to make any of these monochlorinated derivatives.
(a) cyclopentane **(b)** methylcyclopentane
(c) 2,3-dimethylbutane **(d)** 2,2,3,3-tetramethylbutane

4-41. Write a mechanism for the light-initiated reaction of cyclohexane with chlorine to give chlorocyclohexane. Label the initiation and propagation steps.

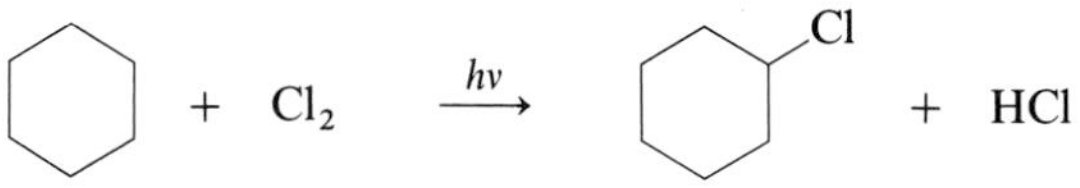

$$\text{cyclohexane} + Cl_2 \xrightarrow{h\nu} \text{chlorocyclohexane} + HCl$$

cyclohexane chlorocyclohexane

4-42. Draw the important resonance structures of the following free radicals.

(a) $CH_2{=}CH\text{—}\dot{C}H_2$ **(b)** 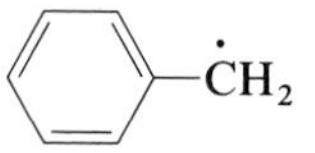 $C_6H_5\text{—}\dot{C}H_2$ **(c)** $CH_3\text{—}\overset{O}{\overset{\|}{C}}\text{—}O\cdot$ **(d)** 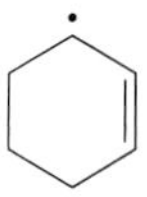 (cyclohexenyl radical) **(e)** 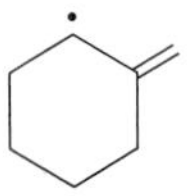 (methylenecyclohexyl radical)

4-43. The light-catalyzed reaction of cyclohexene with a low concentration of bromine gives exclusively 3-bromocyclohexene. Propose a mechanism to explain this preference for substitution at the methylene group next to the double bond.

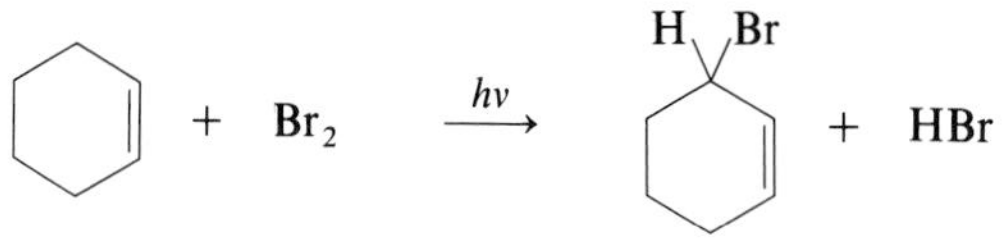

$$\text{cyclohexene} + Br_2 \xrightarrow{h\nu} \text{3-bromocyclohexene} + HBr$$

cyclohexene 3-bromocyclohexene

4-44. For each of the following compounds, predict the major product of free-radical bromination. Remember that bromination is highly selective and only the most stable radical will be formed.
(a) cyclohexane **(b)** methylcyclopentane **(c)** 2,2,3-trimethylbutane
(d) decalin **(e)** 3-methyloctane **(f)** hexane

(g) 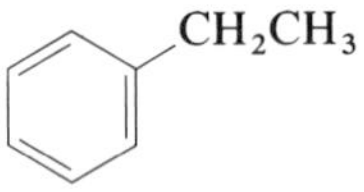 $C_6H_5\text{—}CH_2CH_3$ ethylbenzene **(h)** 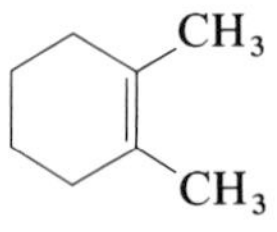 1,2-dimethylcyclohexene

4-45. When exactly 1 mole of methane is mixed with exactly 1 mole of chlorine and light is shone on the mixture, a chlorination reaction occurs. The products are found to contain substantial amounts of di-, tri-, and tetrachloromethane.
(a) Explain how a mixture is formed from this stoichiometric mixture of reactants, and propose mechanisms for the formation of these compounds from methyl chloride.
(b) How would you run this reaction to get a good conversion of methane to CH_3Cl? Methane to CCl_4?

4-46. The chlorination of pentane gives a mixture of three monochlorinated products.
(a) Draw their structures.
(b) Predict the ratios in which these monochlorination products will be formed, remembering that a chlorine atom abstracts a secondary hydrogen about 4.5 times as fast as it abstracts a primary hydrogen.

4-47. **(a)** Draw the structure of the transition state for the second propagation step in the chlorination of methane.

$$\cdot CH_3 \quad + \quad Cl_2 \quad \longrightarrow \quad CH_3Cl \quad + \quad Cl\cdot$$

Be careful to show whether the transition state is productlike or reactantlike and which of the two partial bonds is stronger.
(b) Repeat for the second propagation step in the bromination of methane.

4-48. Peroxides are often added to free-radical reactions as initiators because the oxygen-oxygen bond is homolytically cleaved rather easily. For example, the bond dissociation energy of the O—O bond in hydrogen peroxide (H—O—O—H) is only 51 kcal/mol (213 kJ/mol). Give a mechanism for the hydrogen peroxide-initiated reaction of cyclopentane with chlorine.

4-49. When dichloromethane is treated with strong NaOH, an intermediate is generated that reacts like a carbene. Draw the structure of this reactive intermediate and give a mechanism for its formation.

* **4-50.** When ethene is treated in a calorimeter with H_2 and a Pt catalyst, the heat of reaction is found to be -32.7 kcal/mol (-137 kJ/mol), and the reaction goes to completion. When the reaction takes place at 1400°K, the equilibrium is found to be evenly balanced, with $K_{eq} = 1$. Compute the value of ΔS for this reaction.

$$\underset{\text{ethene}}{CH_2{=}CH_2} \quad + \quad H_2 \quad \underset{}{\overset{\text{Pt catalyst}}{\rightleftarrows}} \quad CH_3{-}CH_3 \qquad \Delta H = -32.7 \text{ kcal/mol}$$

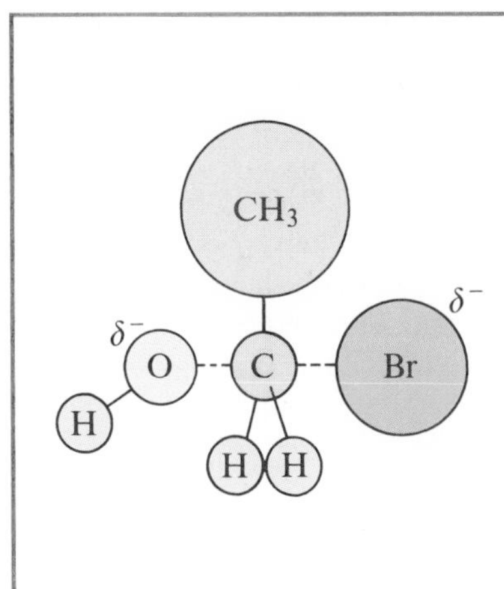

5

ALKYL HALIDES: NUCLEOPHILIC SUBSTITUTION AND ELIMINATION

5-1 INTRODUCTION

Our study of organic chemistry is organized into families of compounds according to their functional groups. The alkyl halides are compounds that contain halogen atoms as their functional groups. In Chapter 4 we saw that alkyl halides may be formed by the free-radical halogenation of alkanes. Free-radical halogenation is not a particularly good method for synthesizing most alkyl halides, but it is useful for studying free-radical reaction mechanisms. Better syntheses of alkyl halides will be covered in later chapters, together with the chemistry of the functional groups involved.

In this chapter we consider the physical properties and reactions of alkyl halides. We use their reactions to introduce substitution and elimination, two of the most important types of reactions in organic chemistry. Many other reactions show similarities to the substitution and elimination reactions of alkyl halides, and the techniques used to study these mechanisms will be used throughout our study of organic reactions.

There are three major classes of organohalogen compounds: the alkyl halides, the vinyl halides, and the aryl halides. An **alkyl halide** simply has a halogen atom bonded to an alkyl group. A **vinyl halide** has a halogen atom bonded to one of the sp^2 hybrid carbon atoms of an alkene. An **aryl halide** has a halogen atom bonded to one of the sp^2 hybrid carbon atoms of an aromatic ring. The chemistry of vinyl halides and aryl halides is quite different from that of alkyl halides because their bonding and hybridization are different. We will consider the reactions of vinyl halides and aryl halides in later chapters. The structures of some representative alkyl halides, vinyl halides, and aryl halides are shown below, with their most common names and uses.

Alkyl halides

$CHCl_3$	CF_2Cl_2	$CCl_3—CH_3$	$CF_3—CHClBr$
chloroform	Freon-12	1,1,1-trichloroethane	Halothane
solvent	refrigerant	cleaning fluid	nonflammable anesthetic

Vinyl halides

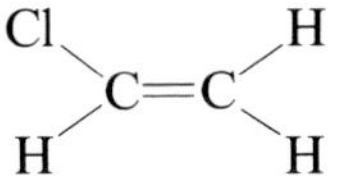

vinyl chloride
monomer for poly(vinyl chloride)

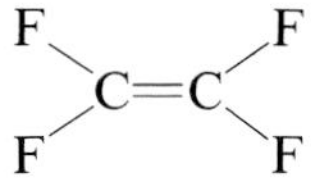

tetrafluoroethylene (TFE)
monomer for Teflon

Aryl halides

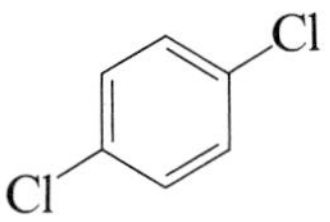

para-dichlorobenzene
mothballs

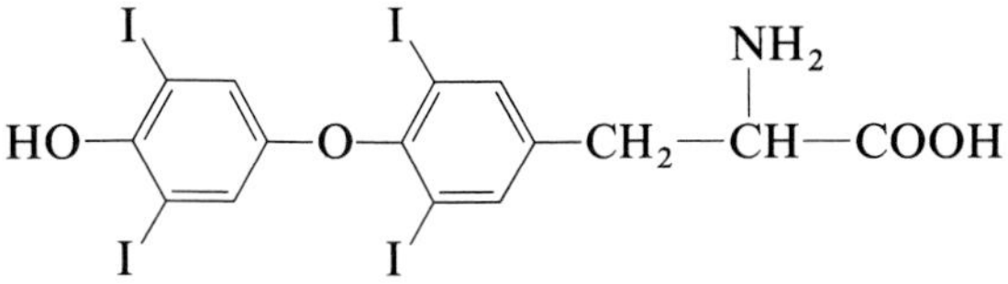

thyroxine
thyroid hormone

The carbon-halogen bond in an alkyl halide is polar because halogen atoms are more electronegative than carbon atoms. Most reactions of alkyl halides result from breaking this polarized bond. Under the right conditions, a halogen atom can leave as a halide ion, taking the bonding pair of electrons with it. By serving as a leaving group, the halogen can be eliminated from the alkyl halide, or it can be replaced (substituted for) by a wide variety of functional groups. This versatility allows alkyl halides to serve as intermediates in the synthesis of many other functional groups.

PROBLEM 5-1

Classify each of the following compounds as an alkyl halide, a vinyl halide, or an aryl halide.

(a) $CH_3CHCFCH_3$ (b) $(CH_3)_3C—Br$ (c) $C_6H_5—Br$

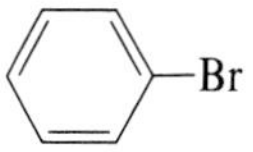

(d) 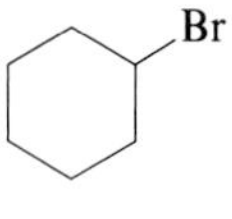 bromocyclohexane (e) 1-bromocyclohexene (f)

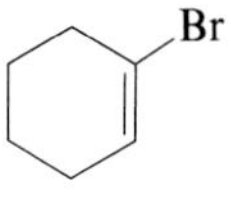

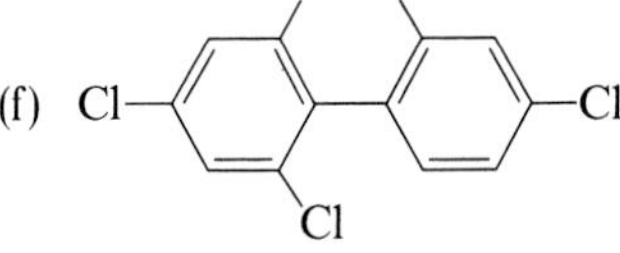

a PCB (polychlorinated biphenyl)

5-2 NOMENCLATURE OF ALKYL HALIDES

There are two ways of naming alkyl halides. The systematic (IUPAC) nomenclature treats an alkyl halide as an alkane with a *halo-* substituent: Fluorine is *fluoro-*, chlorine is *chloro-*, bromine is *bromo-*, and iodine is *iodo-*. The result is a systematic **haloalkane** name, as in 1-chlorobutane or 2-bromopropane. Common or "trivial" names of alkyl halides are constructed by naming the alkyl group and then the halide, as in "isopropyl bromide." This is the origin of the term *alkyl halide.*

Common names are useful only for simple alkyl halides, such as those named below.

	$CH_3—CH_2F$	$\underset{\text{}}{CH_2(Cl)}—CH_2CH_2CH_3$	$CH_3—CH(Br)—CH_3$
IUPAC name:	fluoroethane	1-chlorobutane	2-bromopropane
common name:	ethyl fluoride	*n*-butyl chloride	isopropyl bromide

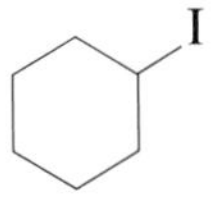

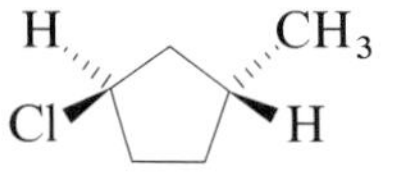

	(cyclohexane ring with I)	(cyclopentane ring: Cl and H at C1; CH_3 and H at C3)
IUPAC name:	iodocyclohexane	*trans*-1-chloro-3-methylcyclopentane
common name:	cyclohexyl iodide	(none)

	$CH_3CH_2—CH(CH_2I)—CH_2CH_3$	$CH_3CH_2CH_2—CH(CH_2CH_2F)—CH_2CH_2CH_3$
IUPAC name:	3-(iodomethyl)pentane	4-(2-fluoroethyl)heptane

Some of the halomethanes have acquired common names that are not clearly related to their structures. Some historical common names are shown below. A compound of formula CH_2X_2 (a methylene group with two halogens) is called a *methylene halide,* a compound of formula CHX_3 is called a *haloform,* and a compound of formula CX_4 is called a *carbon tetrahalide.*

	CH_2Cl_2	$CHCl_3$	CCl_4
IUPAC name:	dichloromethane	trichloromethane	tetrachloromethane
common name:	methylene chloride	chloroform	carbon tetrachloride

PROBLEM 5-2

Give the structures of the following compounds.

(a) methylene iodide (b) carbon tetrabromide
(c) iodoform (d) 3-bromo-2-methylpentane
(e) *cis*-1-fluoro-3-(fluoromethyl)cyclohexane

Alkyl halides are classified according to the nature of the carbon atom bonded to the halogen. If the halogen-bearing carbon is bonded to one carbon atom, it is primary (1°) and the alkyl halide is a **primary halide.** If two carbon atoms are bonded to the halogen-bearing carbon, it is secondary (2°) and the compound is a **secondary halide.** A **tertiary halide** (3°) has three other carbon atoms bonded to the halogen-bearing carbon atom. If the halogen-bearing carbon atom is in a methyl group (bonded to no other carbon atoms), the compound is a *methyl halide.*

$CH_3—X$	$R—CH_2—X$	$R—CH(R)—X$	$R—C(R)(R)—X$
methyl halide	primary (1°) halide	secondary (2°) halide	tertiary (3°) halide

Example

	$CH_3—Br$	$CH_3CH_2CH_2—F$	$CH_3—CH(I)—CH_2CH_3$	$(CH_3)_3C—Cl$
		1°	2°	3°
IUPAC name:	bromomethane	1-fluoropropane	2-iodobutane	2-chloro-2-methylpropane
common name:	methyl bromide	*n*-propyl fluoride	*sec*-butyl iodide	*t*-butyl chloride

A **geminal dihalide** (Latin, *geminus,* "twin") has the two halogen atoms bonded to the same carbon atom. A **vicinal dihalide** (Latin, *vicinus,* "neighboring") has the two halogens bonded to adjacent carbon atoms. Later we discuss several reactions that are specific to geminal and vicinal dihalides.

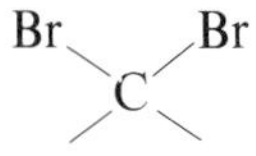

a geminal dibromide 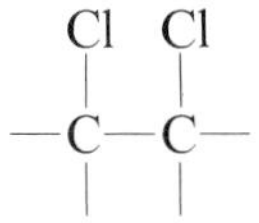 a vicinal dichloride

PROBLEM 5-3
For each of the following compounds
(1) Give the IUPAC name.
(2) Give the common name (if possible).
(3) Classify the compound as a methyl, primary, secondary, or tertiary halide.

(a) $(CH_3)_2CH—CH_2Cl$ (b) $CH_3—CH(—CH(Br)—CH_2—CH_3)—CH_2CH_3$ (c) $CH_3—CHCl_2$

(d) $BrCH_2—CCl_3$ (e) $CHCl_3$ (f) $(CH_3)_3C—Br$

(g) $CH_3—CH_2—CHBr—CH_3$ (h) $CH_3—CH(CH_2CH_3)—CH_2Cl$ (i) CH_2I_2

(j) 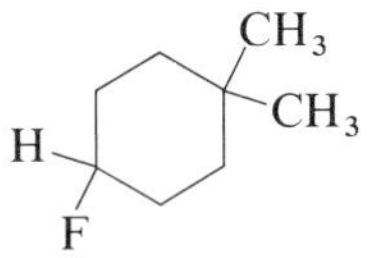 (k) 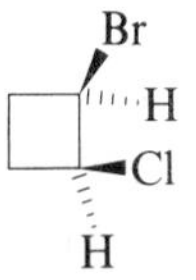 (l)

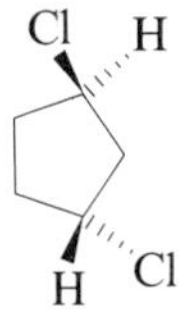

5-3 COMMON USES OF ALKYL HALIDES

5-3A SOLVENTS

Alkyl halides are used primarily as industrial and household solvents. Carbon tetrachloride (CCl_4) was once used for dry cleaning, spot removing, and other domestic cleaning. Carbon tetrachloride is toxic and carcinogenic (causes cancer), however, and dry cleaners now use 1,1,1-trichloroethane and other solvents instead.

Methylene chloride (CH_2Cl_2) and chloroform ($CHCl_3$) are also good solvents for cleaning and degreasing work. Methylene chloride was once used to dissolve the caffeine from coffee beans to produce "decaffeinated" coffee. Concerns about the safety of coffee with residual traces of methylene chloride prompted a change to the use of liquid carbon dioxide instead of methylene chloride. Chloroform is more toxic and carcinogenic than methylene chloride (but less so than carbon tetrachloride); it has been replaced by methylene chloride and other solvents in most industrial degreasers and paint removers.

Even the safest halogenated solvents such as methylene chloride and 1,1,1-trichloroethane should be used carefully, however. They are all potentially toxic and carcinogenic, and they can dissolve the fatty oils that protect skin, causing a form of dermatitis.

5-3B REAGENTS

Alkyl halides also serve as reagents for the synthesis of other compounds. We will often encounter syntheses using alkyl halides as starting materials for making more complex molecules. The conversion of alkyl halides to organometallic reagents, compounds containing carbon-metal bonds, is a particularly important tool for organic synthesis. We discuss the formation of organometallic compounds in Section 9-8.

5-3C ANESTHETICS

Chloroform ($CHCl_3$) was the first substance found to produce general anesthesia, opening new possibilities for careful surgery with a patient who is unconscious and relaxed. Chloroform is toxic and carcinogenic, however, and it was soon abandoned in favor of safer anesthetics. A less toxic anesthetic is a mixed alkyl halide, CF_3—CHClBr, with the trade name Halothane. Ethyl chloride is often used as a topical anesthetic for minor procedures. When sprayed on the skin, its evaporation (b.p. 12°C) cools the area and enhances the numbing effect.

5-3D REFRIGERANTS

The **Freons** are fluorinated haloalkanes that were developed to replace ammonia as a refrigerant gas. Ammonia is toxic, and leaking refrigerators often killed people who were working or sleeping nearby. Freon-12, CF_2Cl_2, is now the most widely used refrigerant. The release of Freons into the atmosphere has raised concerns about their reactions with the earth's protective ozone layer. It appears that Freon-12 gradually diffuses up into the stratosphere, where its chlorine atoms catalyze the decomposition of ozone (O_3) into oxygen (O_2). The Freon-catalyzed depletion of ozone has been blamed for the "hole" in the ozone layer that was recently found to occur over the South Pole.

Recent international treaties have limited the future production and use of the ozone-destroying Freons. Freon-12 can be replaced by low-boiling hydrocarbons or carbon dioxide in aerosol cans, where it was once used as a propellant. In refrigerators and automotive air conditioners, Freon-12 can be replaced by Freon-22, $CHClF_2$. Freons with C—H bonds (such as Freon-22) are generally destroyed at lower altitudes before they reach the stratosphere.

5-3E PESTICIDES

Alkyl halides have contributed to human welfare through their use as insecticides. Since antiquity, people have died from famine and disease caused or carried by insects, fleas, and other vermin. The "black death" of the Middle Ages wiped out nearly a third of the population of Europe through infection by the flea-borne bubonic plague. Whole regions of Africa and tropical America were uninhabited and unexplored because people could not survive insect-borne diseases such as malaria, yellow fever, and sleeping sickness.

Arsenic compounds, nicotine, and other crude insecticides were developed

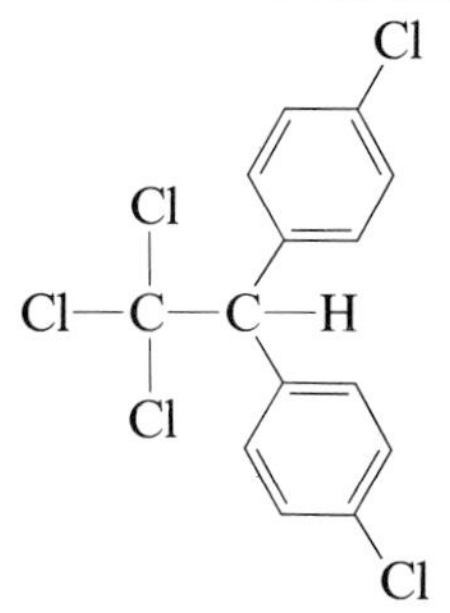

Cl

Cl

Cl—C—C—H

Cl

Cl

DDT

FIGURE 5-1 DDT is Dichlorodiphenyltrichloroethane, or 1,1,1-trichloro-2,2-bis-(p-chlorophenyl)ethane. DDT was the first chlorinated insecticide. Its use rendered large parts of the world safe from insect-borne disease and starvation.

in the nineteenth century, but these compounds are just as toxic to birds, animals, and people as they are to insects. Their use is extremely hazardous; but a hazardous insecticide was still preferable to certain death by starvation or disease.

The war against insects changed dramatically in 1939 with the discovery of DDT (Fig. 5-1). DDT is extremely toxic to insects, but its toxicity in mammals is quite low. About an ounce of DDT is required to kill a person, but that same amount of insecticide protects an acre of land against locusts or mosquitos. DDT has saved over 100 million human lives by controlling mosquito-borne and flea-borne diseases and by protecting food crops. As with most inventions, DDT showed some undesired side effects. It is a long-lasting insecticide, and its residues accumulate in the environment. The widespread use of DDT as an agricultural insecticide led to the development of substantial concentrations in wildlife, causing declines in several species. In 1972, DDT was banned by the U.S. Environmental Protection Agency for use as an agricultural insecticide. It is still used throughout the world, however, whenever insect-borne diseases threaten human life.

Many other chlorinated insecticides have been developed. Some of them also accumulate in the environment, gradually producing toxic effects in wildlife. Others can be used with little adverse impact if they are applied properly. Because of their persistent toxic effects, chlorinated insecticides are rarely used in agriculture. They are generally used as a last resort when a potent insecticide is needed to protect life or property. The structures of some chlorinated insecticides are shown below.

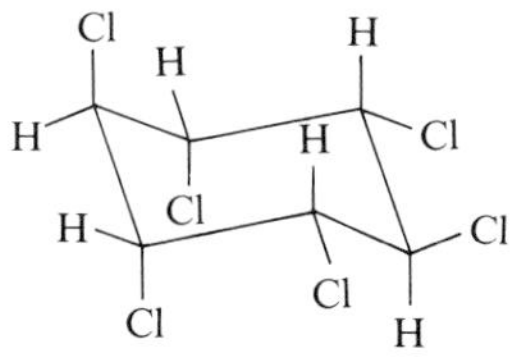

Lindane

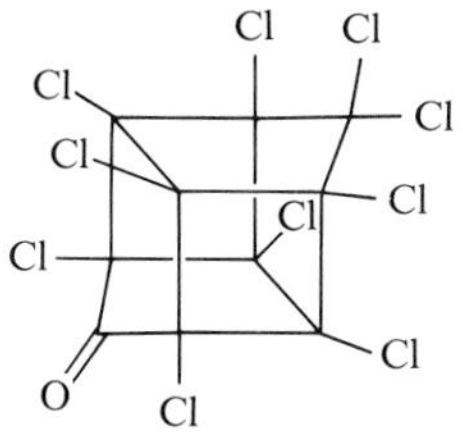

Kepone

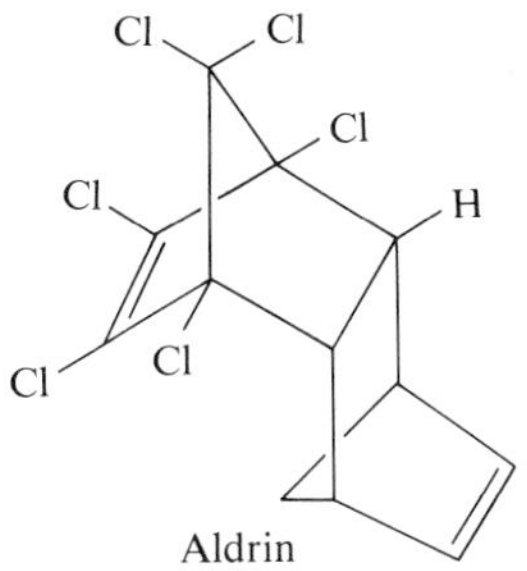

Aldrin

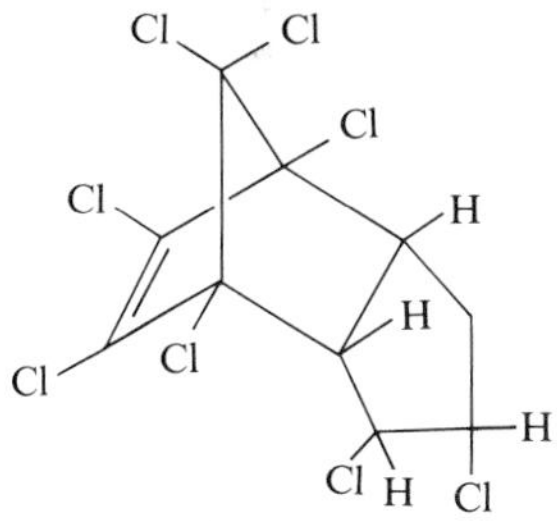

Chlordane

PROBLEM 5-4

Kepone and chlordane are synthesized from hexachlorocyclopentadiene and other five-membered-ring compounds. Show how these two pesticides are composed of two five-membered rings.

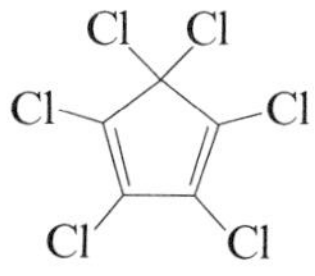

hexachlorocyclopentadiene

5-4 STRUCTURE OF ALKYL HALIDES

In an alkyl halide, the halogen atom is bonded to an sp^3 hybrid carbon atom. The halogen atom is more electronegative than carbon, and the C—X bond is polarized with a partial positive charge on carbon and a partial negative charge on the halogen.

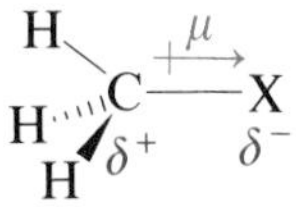

δ^+ δ^-

The electronegativities of the halogens *decrease* in the order

$$F > Cl > Br > I$$

The carbon-halogen bond lengths *increase* as the halogen atoms become bigger (larger atomic radii) in the order

$$C—F < C—Cl < C—Br < C—I$$

The overall result is that the bond dipole moments decrease in the order

	C—Cl >	C—F >	C—Br >	C—I
dipole moment, μ:	1.56D	1.51D	1.48D	1.29D

A *molecular* dipole moment is the vector sum of the individual bond dipole moments. Molecular dipole moments are not easy to predict, because they depend on the bond angles and other factors that vary with the specific molecule. Table 5-1 shows the experimentally measured dipole moments of the halogenated methanes. Notice how the four symmetrically oriented polar bonds of the carbon tetrahalides cancel to give a molecular dipole moment of zero.

TABLE 5-1
Molecular dipole moments of methyl halides

X	CH_3X	CH_2X_2	CHX_3	CX_4
F	1.82	1.97	1.65	0
Cl	1.94	1.60	1.03	0
Br	1.79	1.45	1.02	0
I	1.64	1.11	1.00	0

Cl
Cl C Cl
Cl

carbon tetrachloride

PROBLEM 5-5
Although the C—I bond is longer than a C—Cl bond, the C—Cl bond has a larger dipole moment. Explain.

5-5 PHYSICAL PROPERTIES OF ALKYL HALIDES

5-5A BOILING POINTS

There are two types of intermolecular forces that strongly influence the boiling points of alkyl halides. The London force attraction is the strongest intermolecular force in an alkyl halide, and the dipole-dipole attraction is an additional force. Because the London force is a *surface* attraction, molecules with larger surface areas have larger London attractions, resulting in higher boiling points.

Molecules with higher molecular weights are generally higher boiling because they have greater surface area and are heavier (and therefore slower moving). The surface areas of the alkyl halides also vary with the surface areas of the halogens. We can get an idea of the relative surface areas of the halogen atoms by considering their van der Waals radii. Table 5-2 shows that an alkyl fluoride has about the same surface area as the corresponding alkane; thus its London attractive forces are similar. The alkyl fluoride has a larger dipole moment, however, so the total attractive forces are slightly greater in the alkyl fluoride, giving it a higher boiling point. For example, the boiling point of *n*-butane is 0°C, while that of *n*-butyl fluoride is 33°C.

TABLE 5-2
Van der Waals radii of halogen atoms in alkyl halides

Halogen	*Van der Waals radius (10^{-8} cm)*
F	1.35
Cl	1.8
Br	1.95
I	2.15
H (for comparison)	1.2

The other halogens are considerably larger than fluorine, giving them more surface area and raising the boiling points of their alkyl halides. With a boiling point of 78°C, *n*-butyl chloride shows the influence of chlorine's much larger surface area. This trend continues with *n*-butyl bromide (b.p. 102°C) and *n*-butyl iodide (b.p. 131°C). Table 5-3 shows the boiling points and densities of some simple

TABLE 5-3
Molecular weights, boiling points, and densities of some simple alkyl halides

Compound	*Molecular weight*	*Boiling point (°C)*	*Density (g/mL)*
CH_3—F	34	−78	
CH_3—Cl	50.5	−24	0.92
CH_3—Br	95	4	1.68
CH_3—I	142	42	2.28
CH_2Cl_2	85	40	1.34
$CHCl_3$	119	61	1.50
CCl_4	154	77	1.60
CH_3CH_2—F	48	−38	0.72
CH_3CH_2—Cl	64.5	12	0.90
CH_3CH_2—Br	109	38	1.46
CH_3CH_2—I	156	72	1.94
$CH_3CH_2CH_2$—F	62	3	0.80
$CH_3CH_2CH_2$—Cl	78.5	47	0.89
$CH_3CH_2CH_2$—Br	123	71	1.35
$CH_3CH_2CH_2$—I	170	102	1.75
$(CH_3)_2CH$—Cl	78.5	36	0.86
$(CH_3)_2CH$—Br	123	59	1.31
$(CH_3)_2CH$—I	170	89	1.70
$CH_3CH_2CH_2CH_2$—F	76	33	0.78
$CH_3CH_2CH_2CH_2$—Cl	92.5	78	0.89
$CH_3CH_2CH_2CH_2$—Br	137	102	1.28
$CH_3CH_2CH_2CH_2$—I	184	131	1.62
$(CH_3)_3C$—Cl	92.5	52	0.84
$(CH_3)_3C$—Br	137	73	1.23
$(CH_3)_3C$—I	184	100	1.54

alkyl halides. Notice that compounds with branched, more spherical shapes have lower boiling points as a result of their smaller surface areas. For example, *n*-butyl bromide has a boiling point of 102°C, while the more spherical *t*-butyl bromide has a boiling point of only 73°C. This effect is similar to the one we saw with alkanes.

PROBLEM 5-6
For each of the following pairs of compounds, predict which compound will have the higher boiling point. Check Table 5-3 to see if your prediction was right, then explain why that compound has the higher boiling point.

(a) isopropyl bromide and *n*-butyl bromide
(b) isopropyl chloride and *t*-butyl bromide
(c) *n*-butyl bromide and *n*-butyl chloride

5-5B DENSITIES

Table 5-3 also shows the densities of these alkyl halides. Like their boiling points, the densities follow a predictable trend. The alkyl fluorides and the alkyl chlorides (those with just one chlorine atom) are less dense than water (1.00 g/mL). Alkyl chlorides with two or more chlorine atoms are denser than water, and all the alkyl bromides and alkyl iodides are denser than water.

PROBLEM 5-7
When water is shaken with hexane, the two liquids separate into two phases. Show which compound is present in the top and which in the bottom phase. When water is shaken with chloroform, a similar two-phase system results. Again, show which compound is present in each phase. Explain the difference in the two experiments.

5-6 PREPARATION OF ALKYL HALIDES

Many syntheses of alkyl halides use the chemistry of functional groups that we have not yet covered. For now, we review the use of free-radical halogenation and briefly summarize the other, often more useful syntheses of alkyl halides. These other syntheses will be discussed in subsequent chapters.

5-6A FREE-RADICAL HALOGENATION

Although we discussed its mechanism at length in Section 4-3, free-radical halogenation is not an effective general method for the synthesis of alkyl halides. It usually produces mixtures of products because there are different kinds of hydrogen atoms that can be abstracted. Also, there is often no way to ensure that each molecule acquires just one halogen atom. For example, the chlorination of propane gives a messy mixture of products.

$$CH_3{-}CH_2{-}CH_3 + Cl_2 \xrightarrow{h\nu} \begin{cases} CH_3{-}CH_2{-}CH_2Cl + CH_3{-}CHCl{-}CH_3 \\ + CH_3{-}CHCl{-}CH_2Cl + CH_3{-}CCl_2{-}CH_3 \\ + CH_3{-}CH_2{-}CHCl_2 + \text{others} \end{cases}$$

In industry, free-radical halogenation is sometimes useful because the reagents are cheap, the mixture of products can be carefully distilled, and each of the individual products is sold separately. In a laboratory, however, we need a good yield of one particular product. Free-radical halogenation rarely provides good selectivity and yield, so it is seldom used in the laboratory. Laboratory syntheses using free-radical halogenation are generally limited to specialized compounds that give a single major product, such as the following examples.

$$\text{cyclohexane} + Cl_2 \xrightarrow{h\nu} \text{chlorocyclohexane } (50\%)$$

$$\underset{\text{isobutane}}{(CH_3)_3C{-}H} + Br_2 \xrightarrow{h\nu} \underset{t\text{-butyl bromide } (90\%)}{(CH_3)_3C{-}Br}$$

All the hydrogen atoms in cyclohexane are equivalent, and free-radical chlorination gives a good yield of chlorocyclohexane. Formation of dichlorides and trichlorides is possible, but these side reactions are controlled by using only a small amount of chlorine and an excess of cyclohexane. Free-radical bromination is highly selective (Section 4-14), and it gives good yields of products that have one type of hydrogen atom that is more reactive than the others. Isobutane has only one tertiary hydrogen atom, and this atom is preferentially abstracted to give a tertiary free radical.

5-6B ALLYLIC HALOGENATION

Bromination of cyclohexene gives a good yield of 3-bromocyclohexene, where bromine has substituted for an **allylic** hydrogen on the carbon atom next to the double bond.

$$\text{cyclohexene} + Br_2 \xrightarrow{h\nu} \text{3-bromocyclohexene } (80\%) + HBr$$

This selective allylic bromination occurs because the intermediate leading to this product is resonance stabilized. Abstraction of an allylic hydrogen atom gives a resonance-stabilized allylic radical. This radical reacts with Br_2, regenerating a bromine radical.

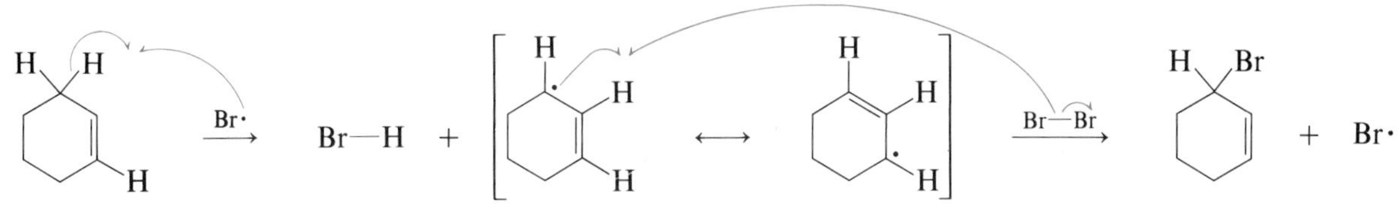

cyclohexene $\xrightarrow{Br\cdot}$ Br—H + [allylic radical ⟷ allylic radical] $\xrightarrow{Br—Br}$ allylic bromide + Br·

cyclohexene   abstraction of allylic H   allylic radical   allylic bromide

Overall reaction

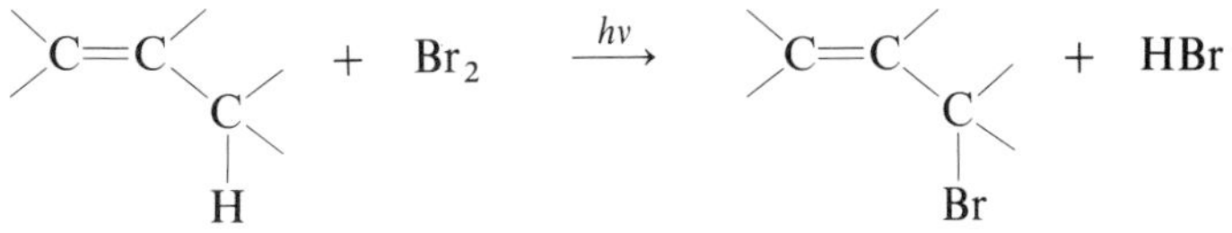

$$\text{an allylic hydrogen} + Br_2 \xrightarrow{h\nu} \text{an allylic bromide} + HBr$$

an allylic hydrogen   an allylic bromide

A large excess of bromine must be avoided, because bromine can add to the double bond (Chapter 7). *N*-Bromosuccinimide is often used as the bromine source in free-radical brominations because it combines with the HBr side product formed in the reaction to provide a nearly constant low concentration of bromine.

$$\text{N—Br} + HBr \longrightarrow \text{N—H} + Br_2$$

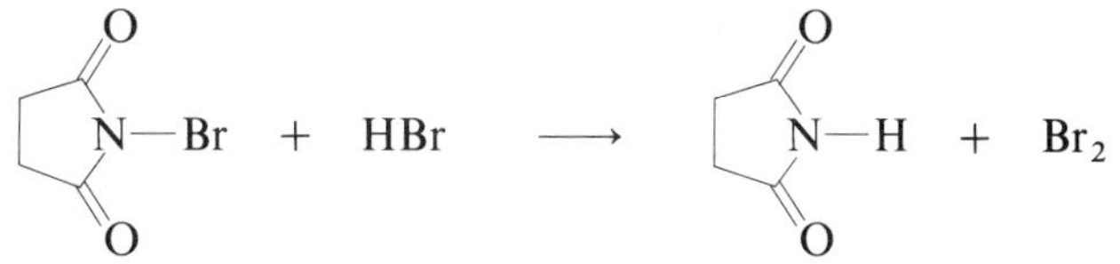

N-bromosuccinimide (NBS)   regenerates a low concentration of Br_2

Allylic halogenation is discussed in more detail in Chapter 14.

PROBLEM 5-8

Show how free-radical halogenation might be used to synthesize each of the following compounds. In each case, explain why we expect to get a single major product.

(a) 1-chloro-2,2-dimethylpropane (neopentyl chloride)
(b) 2-bromo-2-methylbutane
(c) C_6H_5—CH(Br)—$CH_2CH_2CH_3$

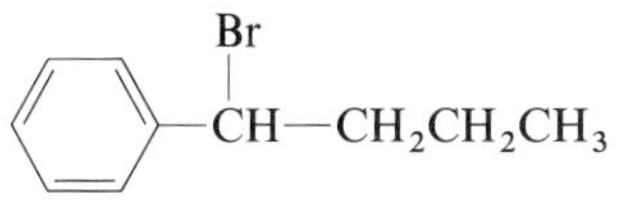

1-bromo-1-phenylbutane

PROBLEM 5-9

The light-catalyzed reaction of 2,3-dimethyl-2-butene with a small concentration of bromine gives two products:

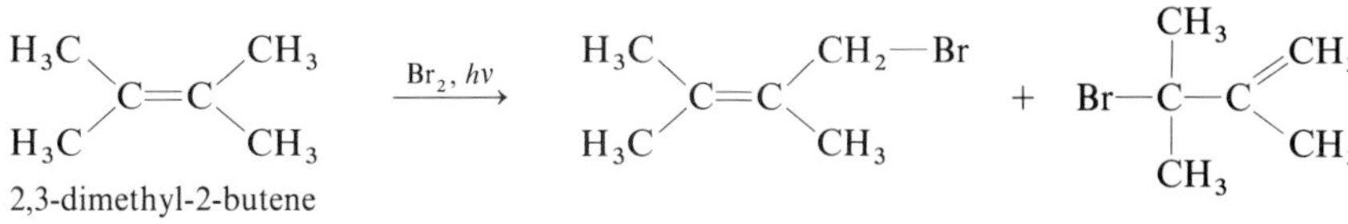

$$(H_3C)_2C{=}C(CH_3)_2 \xrightarrow{Br_2,\ h\nu} (H_3C)_2C{=}C(CH_3)CH_2{-}Br + Br{-}C(CH_3)_2{-}C(CH_3){=}CH_2$$

2,3-dimethyl-2-butene

Give a mechanism for this reaction, showing how the two products arise as a consequence of the resonance-stabilized intermediate.

Following is a brief summary of the most important methods of making alkyl halides. Several of these methods are not discussed until later, but they are listed here so that you can use this summary for reference throughout the course.

SUMMARY OF METHODS FOR PREPARING ALKYL HALIDES

1. From alkanes: free-radical halogenation (synthetically useful only in certain cases) (Section 4-14)

$$\text{R—H} \xrightarrow{X_2,\ \text{heat or light}} \text{R—X} + \text{H—X}$$

Example

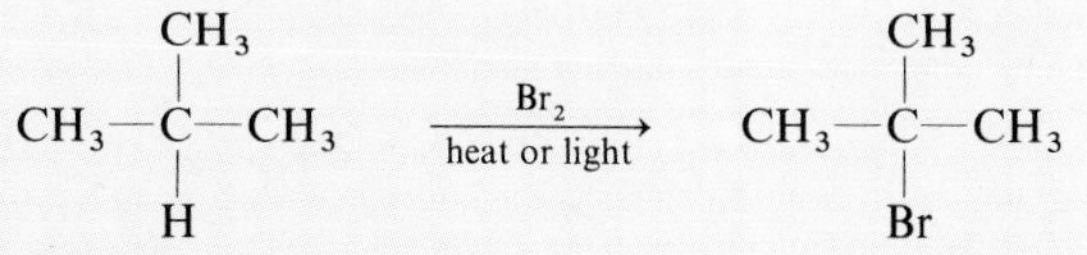

$$\underset{\text{isobutane}}{(CH_3)_3C\text{—H}} \xrightarrow[\text{heat or light}]{Br_2} \underset{t\text{-butyl bromide}}{(CH_3)_3C\text{—Br}}$$

2. From alkenes and alkynes

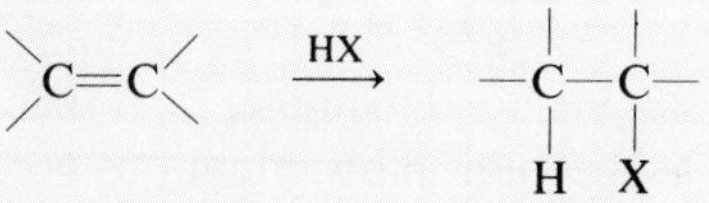

$$\text{>C=C<} \xrightarrow{HX} \text{—C(H)—C(X)—} \quad \text{(Section 8-3)}$$

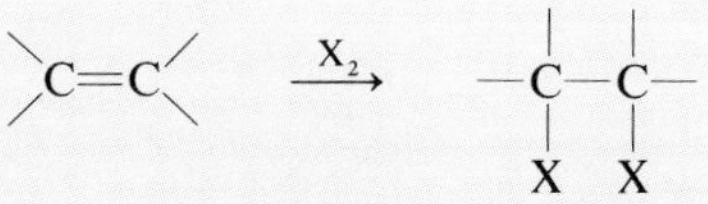

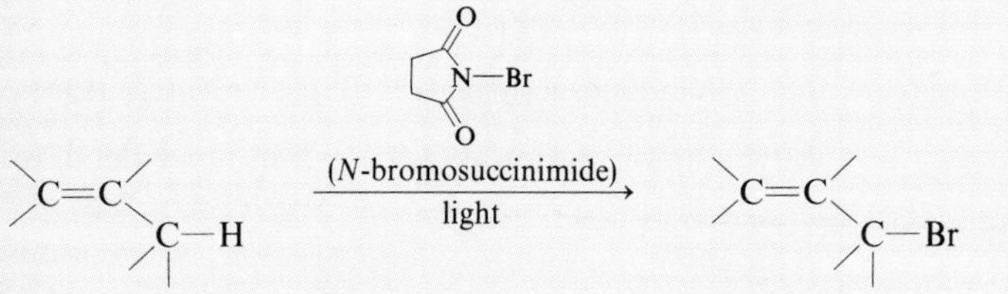

$$\text{>C=C<} \xrightarrow{X_2} \text{—C(X)—C(X)—}$$

$$\text{>C=C—C—H} \xrightarrow[\text{light}]{\text{(}N\text{-bromosuccinimide)}} \text{>C=C—C—Br} \quad \text{(Section 5-6B and Section 15-7)}$$

(N—Br succinimide reagent)

(allylic bromination)

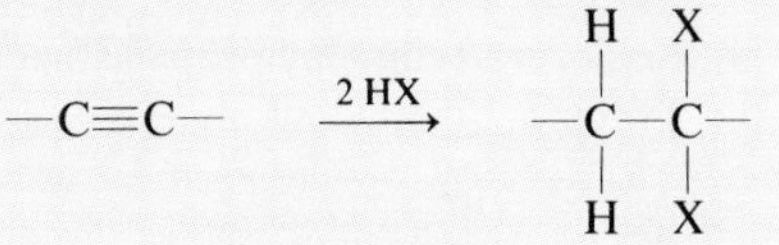

$$\text{—C≡C—} \xrightarrow{2\,HX} \text{—C(H)(H)—C(X)(X)—} \quad \text{(Section 14-10E)}$$

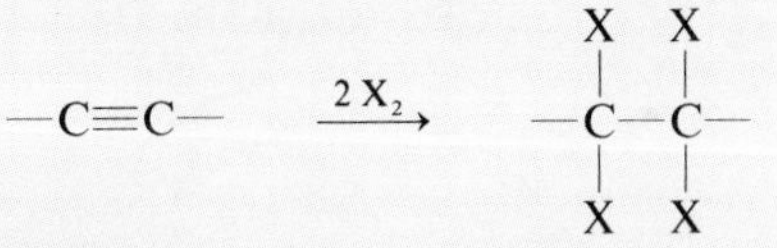

$$\text{—C≡C—} \xrightarrow{2\,X_2} \text{—C(X)(X)—C(X)(X)—} \quad \text{(Section 14-10D)}$$

Examples

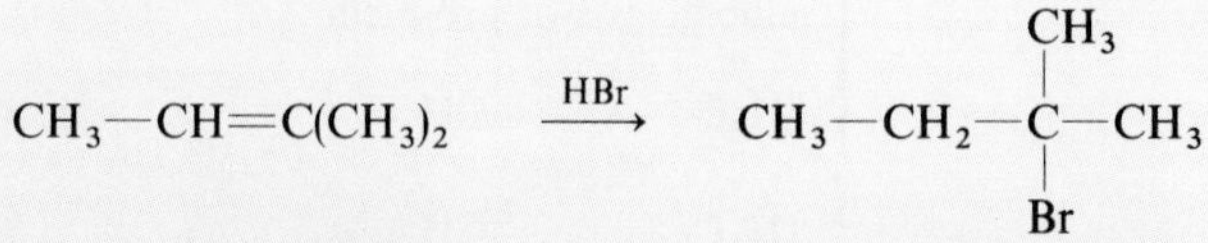

$$\underset{\text{2-methyl-2-butene}}{CH_3\text{—}CH\text{=}C(CH_3)_2} \xrightarrow{HBr} \underset{\text{2-bromo-2-methylbutane}}{CH_3\text{—}CH_2\text{—}C(CH_3)_2\text{—}Br}$$

$$\underset{\text{2-butene}}{CH_3\text{—}CH\text{=}CH\text{—}CH_3} \xrightarrow{Cl_2} \underset{\text{2,3-dichlorobutane}}{CH_3\text{—}CHCl\text{—}CHCl\text{—}CH_3}$$

$$\mathrm{H{-}C{\equiv}C{-}CH_2CH_2CH_3} \xrightarrow{\text{2 HBr}} \mathrm{CH_3{-}CBr_2{-}CH_2CH_2CH_3}$$

1-pentyne → 2,2-dibromopentane

3. *From alcohols* (Sections 10-8, 10-9, 10-10)

$$\mathrm{R{-}OH} \xrightarrow{\text{HX, PX}_3\text{, or others}} \mathrm{R{-}X}$$

Example

$$\mathrm{CH_3CH_2CH_2CH_2OH} \xrightarrow{\text{HBr, H}_2\text{SO}_4} \mathrm{CH_3CH_2CH_2CH_2Br}$$

1-butanol → 1-bromobutane

4. *From other halides* (Section 5-10)

$$\mathrm{R{-}X} + \mathrm{I^-} \xrightarrow{\text{acetone}} \mathrm{R{-}I} + \mathrm{X^-}$$

$$2\,\mathrm{R{-}Cl} + \mathrm{Hg_2F_2} \longrightarrow 2\,\mathrm{R{-}F} + \mathrm{Hg_2Cl_2}$$

Example

$$\mathrm{H_2C{=}CH{-}CH_2Cl} + \mathrm{NaI} \xrightarrow{\text{acetone}} \mathrm{H_2C{=}CH{-}CH_2I}$$

allyl chloride → allyl iodide

5-7 REACTIONS OF ALKYL HALIDES: SUBSTITUTION AND ELIMINATION

Alkyl halides are versatile organic compounds, easily converted to most of the other functional groups. Alkyl halides can also be coupled, forming new carbon-carbon bonds and enlarging the size of the alkyl group. The most important property of alkyl halides is the ability of the halogen atom to leave with its bonding pair of electrons to form a stable halide ion. When another atom or ion comes in to replace the halide ion, the reaction is called a **substitution.** When the halide ion leaves together with another atom or ion (often H^+), the reaction is called an **elimination.** In many eliminations, a molecule of H—X is lost from the alkyl halide to give an alkene. Such eliminations are called **dehydrohalogenations** to indicate that a hydrogen halide has been removed from the alkyl halide. Substitution and elimination reactions often occur in competition with each other.

In a nucleophilic substitution, the halide ion ($X\!:^-$) is replaced by some other **nucleophile** ($Nuc\!:^-$).

Nucleophilic substitution

✓ $$\underset{\mathrm{H}\quad\mathrm{X}}{\mathrm{-C-C-}} + \mathrm{Nuc\!:^-} \longrightarrow \underset{\mathrm{H}\quad\mathrm{Nuc}}{\mathrm{-C-C-}} + \mathrm{X\!:^-}$$

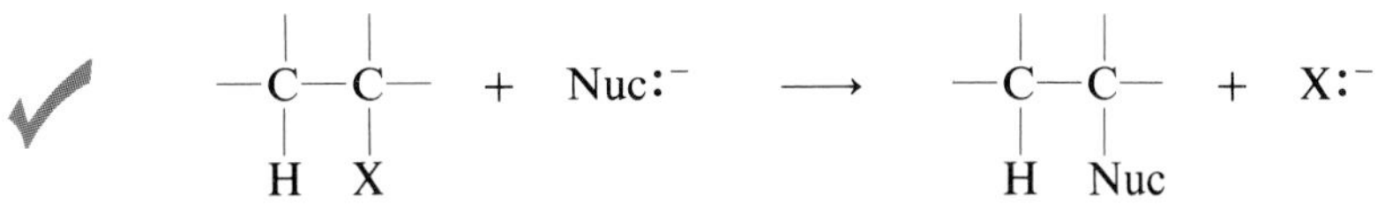

In an elimination, both the halide ion and another substituent are lost.

Elimination

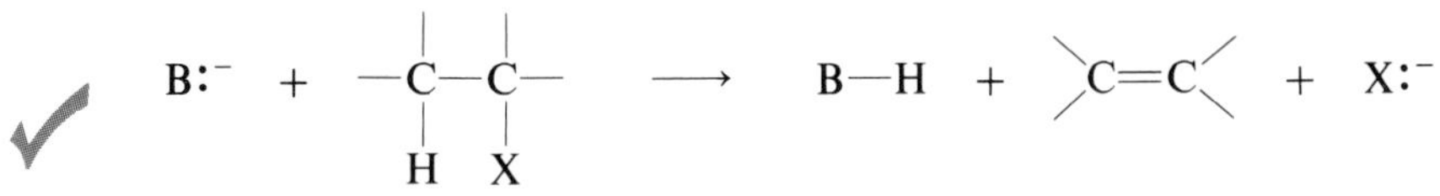

✓ $$\mathrm{B\!:^-} + \underset{\mathrm{H}\quad\mathrm{X}}{\mathrm{-C-C-}} \longrightarrow \mathrm{B{-}H} + \mathrm{>C{=}C<} + \mathrm{X\!:^-}$$

The substitution shown above is called a **nucleophilic substitution** because the reagent ($Nuc\!:^-$) reacts as a nucleophile, using its lone pair of electrons to form a new bond to a carbon atom. In the elimination (a dehydrohalogenation), the reagent ($B\!:^-$) reacts as a base, abstracting a proton from the alkyl halide. Many nucleophiles are also basic and can engage in either substitution or elimination, depending on the alkyl halide and the reaction conditions.

In addition to alkyl halides, many other types of compounds undergo substitution and elimination reactions. Substitutions and eliminations are introduced in this chapter using the alkyl halides as examples. In later chapters, substitutions and eliminations of other types of compounds will frequently be encountered.

PROBLEM 5-10

Classify each of the following reactions as a substitution, elimination, or neither.

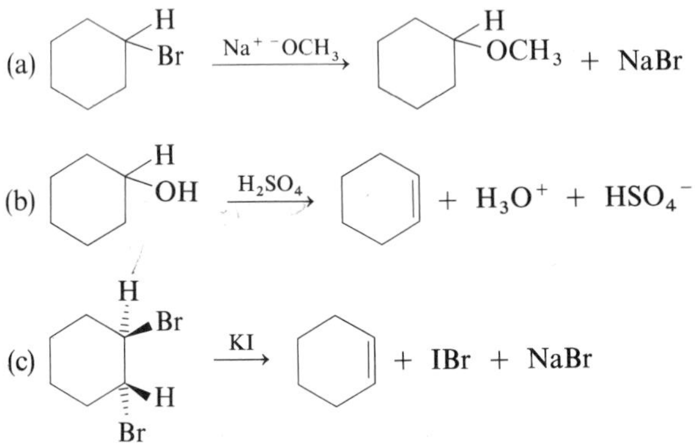

(a) cyclohexyl bromide $\xrightarrow{Na^{+}\ ^{-}OCH_3}$ cyclohexyl methyl ether (OCH_3) + NaBr

(b) cyclohexanol $\xrightarrow{H_2SO_4}$ cyclohexene + H_3O^+ + HSO_4^-

(c) trans-1,2-dibromocyclohexane $\xrightarrow{KI}$ cyclohexene + IBr + NaBr

PROBLEM 5-11

Give the structures of the substitution products expected when 1-bromohexane reacts with

(a) $Na^{+}\ ^{-}OCH_2CH_3$ (b) NaCN (c) NaOH

5-8 SECOND-ORDER NUCLEOPHILIC SUBSTITUTION: THE S_N2 REACTION

A nucleophilic substitution has the general form

$$\text{Nuc:}^- + -\overset{|}{\underset{|}{C}}-X \longrightarrow \text{Nuc}-\overset{|}{\underset{|}{C}}- + X:^-$$

where Nuc:^- is the nucleophile and $X:^-$ is the leaving halide ion. An example is the reaction of methyl iodide (CH_3I) with potassium hydroxide. The product is methyl alcohol.

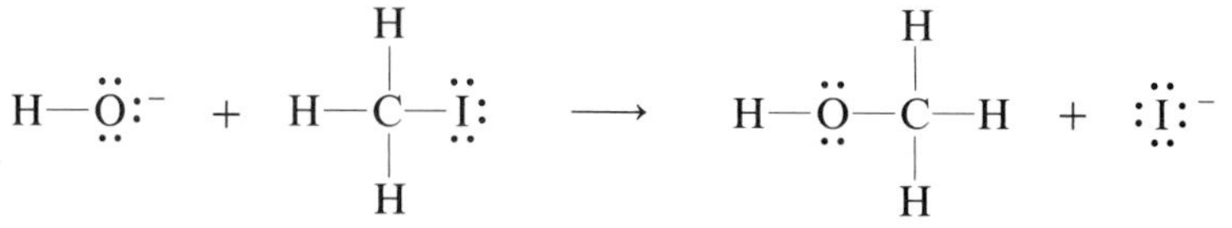

$$H-\ddot{\underset{..}{O}}:^- + H-\overset{H}{\underset{H}{C}}-\ddot{\underset{..}{I}}: \longrightarrow H-\ddot{\underset{..}{O}}-\overset{H}{\underset{H}{C}}-H + :\ddot{\underset{..}{I}}:^-$$

Hydroxide ion is the nucleophile and methyl iodide is the electrophile in this substitution. Hydroxide is a good nucleophile (donor of an electron pair) because the oxygen atom has unshared pairs of electrons and a negative charge. The carbon atom of methyl iodide is electrophilic because it is bonded to an electronegative iodine atom. Electron density is drawn away from carbon by the halogen atom, giving the carbon atom a partial positive charge. The negative charge of the hydroxide ion is attracted to this partial positive charge.

$$H-\ddot{\underset{..}{O}}:^- \qquad H-\overset{H}{\underset{H}{\overset{\delta+}{C}}}-\overset{\delta-}{I}$$

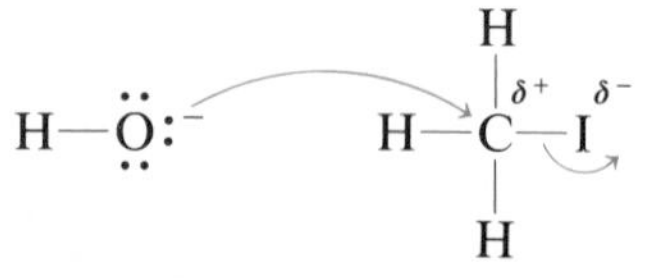

nucleophile electrophile

The hydroxide ion attacks the electrophilic carbon atom by donating a pair of electrons to form a new bond; in general, nucleophiles are said to attack electrophiles, not the other way around. Notice that arrows are used to show the movement of pairs of electrons, from the electron-rich nucleophile to the electron-deficient carbon atom of the electrophile. Carbon can accommodate only eight electrons in its valence shell, so the carbon-iodine bond must begin to break as the oxygen-carbon bond begins to form. Iodide ion is the **leaving group;** it leaves with the pair of electrons that once bonded it to the carbon atom.

The following mechanism shows attack by the nucleophile (hydroxide), the transition state, and the departure of the leaving group (iodide).

$$\mathrm{H\ddot{O}:^-} \;+\; \mathrm{H_3C{-}\ddot{I}:} \;\longrightarrow\; \left[\mathrm{H\ddot{O}\text{---}CH_3\text{---}\ddot{I}:}\right]^- \;\longrightarrow\; \mathrm{H\ddot{O}{-}CH_3} \;+\; \mathrm{:\ddot{I}:^-}$$

nucleophile electrophile transition state product leaving group

The reaction of methyl iodide with hydroxide ion is a **concerted reaction,** taking place in a single step with bonds breaking and forming at the same time. The middle structure is a transition state, a point of maximum energy rather than an intermediate. In this transition state, the bond to the nucleophile (hydroxide) is partially formed, and the bond to the leaving group (iodide) is partially broken. Remember that a transition state is not a discrete molecule that can be isolated; it exists only for an instant.

The potential-energy profile for this substitution is shown in Figure 5-2. This profile shows only one transition state and no intermediates between the reactants and the products. The reactants are seen to be slightly higher in energy than the products because this reaction is known to be exothermic. The transition state is much higher in energy because it involves a five-coordinate carbon atom with two partial bonds.

The one-step mechanism shown in Figure 5-2 was determined largely from kinetic information. One can vary the concentrations of the reactants and observe the effects on the reaction rate (how much methanol is formed per second). The rate is found to double when the concentration of *either* reactant is doubled. The reaction is therefore first order in each of the reactants, and second order overall. The rate equation has the following form:

$$\text{rate} = k_r[\mathrm{CH_3I}][^{-}\mathrm{:\ddot{O}H}]$$

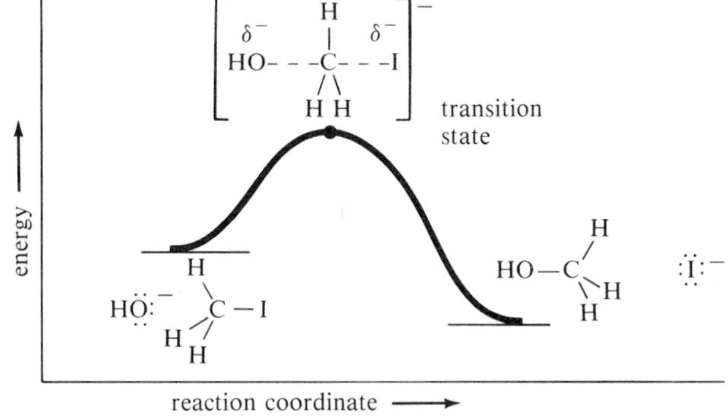

FIGURE 5-2 The potential-energy profile for the reaction of methyl iodide with hydroxide shows only one energy maximum, the transition state. There are no intermediates.

This rate equation is consistent with the one-step mechanism shown. This mechanism requires a collision between a molecule of methyl iodide and a hydroxide ion. Both of these species are present in the transition state, and the collision frequency is proportional to both concentrations. The value of the rate constant k_r depends on several factors, including the energy of the transition state and the temperature. (Section 4-4).

This one-step nucleophilic substitution is an example of the **S_N2 mechanism.** The abbreviation S_N2 stands for *Substitution, Nucleophilic, bimolecular*. The term *bimolecular* means that the transition state of the rate-determining step (the only step in this reaction) involves the collision of *two* molecules. Most bimolecular reactions have rate equations that are second order overall.

PROBLEM 5-12

Under certain conditions, the reaction of 0.5 *M* methyl iodide with 1.0 *M* sodium hydroxide forms methanol at a rate of 0.05 mol/L per second. What would be the rate if 0.1 *M* methyl iodide and 2.0 *M* NaOH were used?

5-9 STEREOCHEMISTRY OF THE S_N2 REACTION

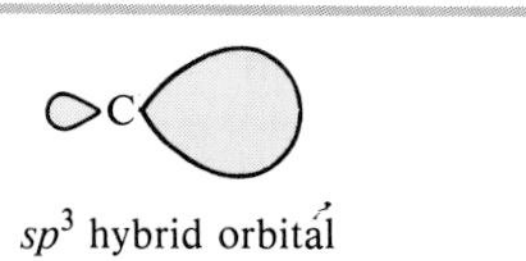

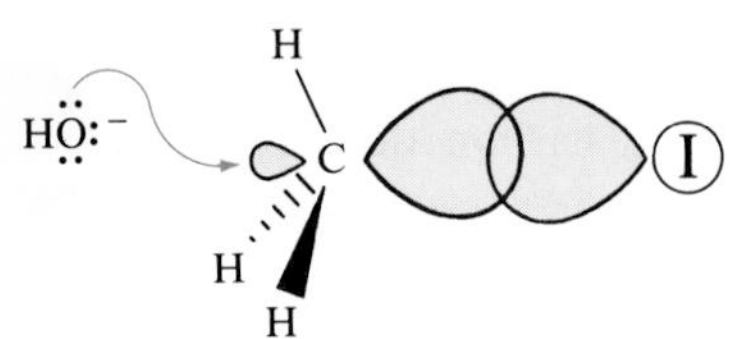

FIGURE 5-3 In the S_N2 reaction, back-side attack takes place on the small back lobe of the sp^3 hybrid orbital.

Figure 5-3 shows the orbital picture of the S_N2 reaction of hydroxide ion with methyl iodide. The attacking nucleophile *must* attack from the side of the carbon atom opposite the position of the leaving group. A carbon atom can have only four filled bonding orbitals (an octet), and the three bonds to the hydrogen atoms leave only one orbital for partial bonding to the attacking and leaving groups. The attacking group must begin to form a bond using the same orbital that bonds to the leaving group. Since the iodine atom blocks its side of the molecule, the only available part of carbon's sp^3 orbital is the small back lobe, the part of the orbital that we usually neglect to draw.

Attack by a nucleophile on the small back lobe is called **back-side attack** (Fig. 5-4). Back-side attack literally turns the tetrahedron of the carbon atom inside out, like an umbrella that is caught by the wind. This process is called **inversion of configuration.**

In the iodomethane reaction, it is impossible to tell from the structure of the product that an inversion of configuration has occurred. In other cases, however, the stereochemistry of the product shows inversion. For example, when *cis*-1-bromo-3-methylcyclopentane reacts with hydroxide ion, the product of S_N2 substitution is found to be *trans*-3-methylcyclopentanol. This product shows inversion of configuration. By looking at the transition state, we can see how the change in stereochemistry takes place.

Inversion of configuration in the S_N2 reaction

$^-$:ÖH; H, H, Br, CH_3 → [δ^-OH, H, H, Br, CH_3]$^-$ → OH, H, H, CH_3, Br^-

cis-1-bromo-3-methylcyclopentane — transition state — *trans*-3-methylcyclopentanol

To study the mechanism of a nucleophilic substitution, we often look at the product to see if inversion of configuration has taken place. If it has, the S_N2

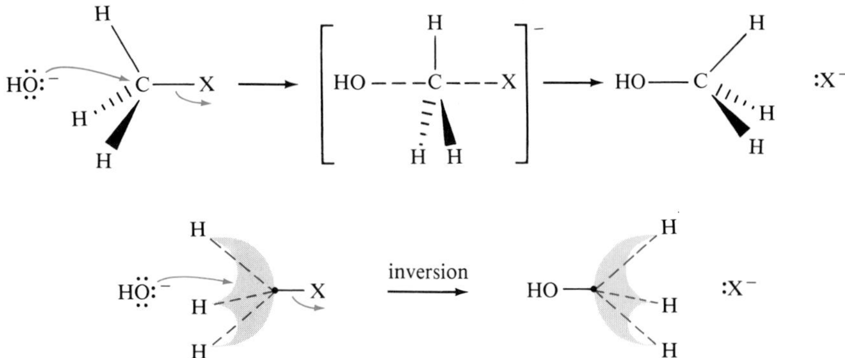

FIGURE 5-4 Back-side attack results in an inversion of the carbon atom's tetrahedron, like the inversion of an umbrella when caught by the wind.

mechanism is a good possibility, especially if the reaction kinetics are second order. In many cases it is difficult or impossible to determine whether inversion has occurred. The reaction of methyl iodide with hydroxide ion gives the same product whether the configuration is inverted or retained. In these cases we use kinetics and other evidence to help determine the reaction mechanism.

PROBLEM 5-13

Draw the structure of the product expected from the S_N2 substitution of *cis*-1-bromo-4-methylcyclohexane with potassium hydroxide.

5-10 GENERALITY OF THE S_N2 REACTION

Many useful reactions take place by the S_N2 mechanism. The reaction of an alkyl halide, such as methyl iodide, with hydroxide ion gives an alcohol. Other nucleophiles convert alkyl halides to a wide variety of functional groups. The following table summarizes some of the types of compounds that can be formed by nucleophilic displacement of alkyl halides.

SUMMARY OF S_N2 REACTIONS OF ALKYL HALIDES

$$\text{Nuc:}^- + \text{R—X} \longrightarrow \text{Nuc—R} + \text{X}^-$$

Nucleophile		*Product*	*Class of product*
R—X + $^-$:I	⟶	R—I	alkyl halide
R—X + $^-$:OH	⟶	R—OH	alcohol
R—X + $^-$:OR′	⟶	R—OR′	ether
R—X + $^-$:SH	⟶	R—SH	thiol (mercaptan)
R—X + $^-$:SR′	⟶	R—SR′	thioether (sulfide)
R—X + :NH_3	⟶	R—NH_3^+ X^-	amine
R—X + $^-$:C≡C—R′	⟶	R—C≡C—R′	alkyne
R—X + $^-$:C≡N	⟶	R—C≡N	nitrile
R—X + R′—COO:$^-$	⟶	R′—COO—R	ester
R—X + :P(Ph)$_3$	⟶	[R—PPh_3]$^+$ $^-$X	phosphonium salt

Halogen exchange reactions The S_N2 reaction provides a useful method for synthesizing alkyl iodides and fluorides, which are more difficult to make than alkyl chlorides and bromides. Halides can be converted to other halides by **halogen exchange reactions,** in which one halide displaces another.

Iodide is a strong nucleophile, and many alkyl chlorides react with sodium iodide to give alkyl iodides. Alkyl fluorides are difficult to synthesize directly, and they are often made by treating alkyl chlorides with fluoride salts such as Hg_2F_2.

$$R{-}\ddot{\underset{..}{X}}{:} \; + \; {:}\ddot{\underset{..}{I}}{:}^- \longrightarrow R{-}\ddot{\underset{..}{I}}{:} \; + \; {:}\ddot{\underset{..}{X}}{:}^-$$

$$R{-}Cl \; + \; Hg_2F_2 \longrightarrow R{-}F \; + \; Hg_2Cl_2$$

Examples

$$\underset{\text{allyl chloride}}{H_2C{=}CH{-}CH_2Cl} \; + \; NaI \longrightarrow \underset{\text{allyl iodide}}{H_2C{=}CH{-}CH_2I} \; + \; NaCl$$

$$\underset{\text{ethyl chloride}}{2\,CH_3CH_2Cl} \; + \; Hg_2F_2 \longrightarrow \underset{\text{ethyl fluoride}}{2\,CH_3CH_2F} \; + \; Hg_2Cl_2$$

PROBLEM 5-14

Predict the major product for each of the following substitution reactions.

(a) $\underset{\text{ethyl bromide}}{CH_3CH_2Br} \; + \; \underset{\text{potassium } t\text{-butoxide}}{(CH_3)_3C{-}O^-\,{}^+K}$

(b) $\underset{\text{sodium acetylide}}{H{-}C{\equiv}C{:}^-\,{}^+Na} \; + \; \underset{\text{1-chlorobutane}}{CH_3CH_2CH_2CH_2{-}Cl}$

(c) $(CH_3)_2CH{-}CH_2{-}Br$ + excess NH_3

(d) $CH_3CH_2I + NaCN$

(e) 1-chloropentane + NaI

(f) 1-chloropentane + Hg_2F_2

PROBLEM 5-15

Show how you would convert 1-chlorobutane into each of the following compounds.

(a) 1-butanol

(b) 1-fluorobutane

(c) 1-iodobutane

(d) $CH_3{-}(CH_2)_3{-}CN$

(e) $CH_3{-}(CH_2)_3{-}C{\equiv}CH$

(f) $CH_3CH_2{-}O{-}(CH_2)_3{-}CH_3$

(g) $CH_3{-}(CH_2)_3{-}NH_2$

5-11 STRUCTURAL EFFECTS ON THE S_N2 REACTION

Many alkyl halides undergo substitution by the S_N2 mechanism. Some of these reactions are fast, but others are extremely slow. The reaction rate depends on several factors. One important factor in S_N2 reactivity is the structure of the alkyl halide undergoing substitution. The S_N2 reaction goes rapidly with methyl halides and most primary alkyl halides. The reaction is more sluggish with secondary alkyl halides, and tertiary halides fail to react by the S_N2 mechanism. Table 5-4 shows the effect of alkyl substitution on the rate of the S_N2 displacement.

For simple alkyl halides, the relative rates in the S_N2 reaction are as follows:

$$\text{Relative rates for } S_N2\text{:} \quad CH_3X > 1° > 2° > 3°$$

The physical explanation for this order of reactivity is suggested by the information in Table 5-4. All the slow-reacting compounds have one property in common: The back side of the electrophilic carbon atom is crowded by the presence of bulky groups. Tertiary halides are more hindered than secondary halides, which are more hindered than primary halides. Even a bulky primary halide (like neopentyl bromide) undergoes the S_N2 reaction at a rate similar to that of a tertiary halide. The relative rates show that it is the bulk of the alkyl groups, rather than an electronic effect, that is responsible for the variation in reactivity of alkyl halides in the S_N2 displacement.

This effect on the rate is another example of **steric hindrance.** When the nucleophile approaches the back side of the electrophilic carbon atom, it must come within bonding distance of the small back lobe of the C—X sp^3 orbital. If there are two alkyl groups bonded to the carbon atom, this is difficult. Three alkyl groups make it impossible. Just one alkyl group can produce a large amount of

TABLE 5-4
Effect of substituents on the rate of S_N2 reactions[a]

Class of halide	*Example*	*Relative rate*
methyl	CH_3—Br	>1000
primary (1°)	CH_3CH_2—Br	50
secondary (2°)	$(CH_3)_2CH$—Br	1
tertiary (3°)	$(CH_3)_3C$—Br	<0.001
n-butyl (1°)	$CH_3CH_2CH_2CH_2$—Br	20
isobutyl (1°)	$(CH_3)_2CHCH_2$—Br	2
neopentyl (1°)	$(CH_3)_3CCH_2$—Br	0.0005

[a] Two or three alkyl groups, or even a single bulky alkyl group, slow the reaction rate. The rates listed are compared to the secondary case (isopropyl bromide), assigned a relative rate of 1.

steric hindrance if it is an unusually bulky group such as the *t*-butyl group of neopentyl bromide.

In terms of the activation energy, it is the steric hindrance in the transition state that governs the effect on the reaction rate. Figure 5-5 depicts the transition states for the S_N2 reaction of a primary halide and a tertiary halide. The crowding in the tertiary transition state raises its energy, increasing the activation energy and slowing the reaction.

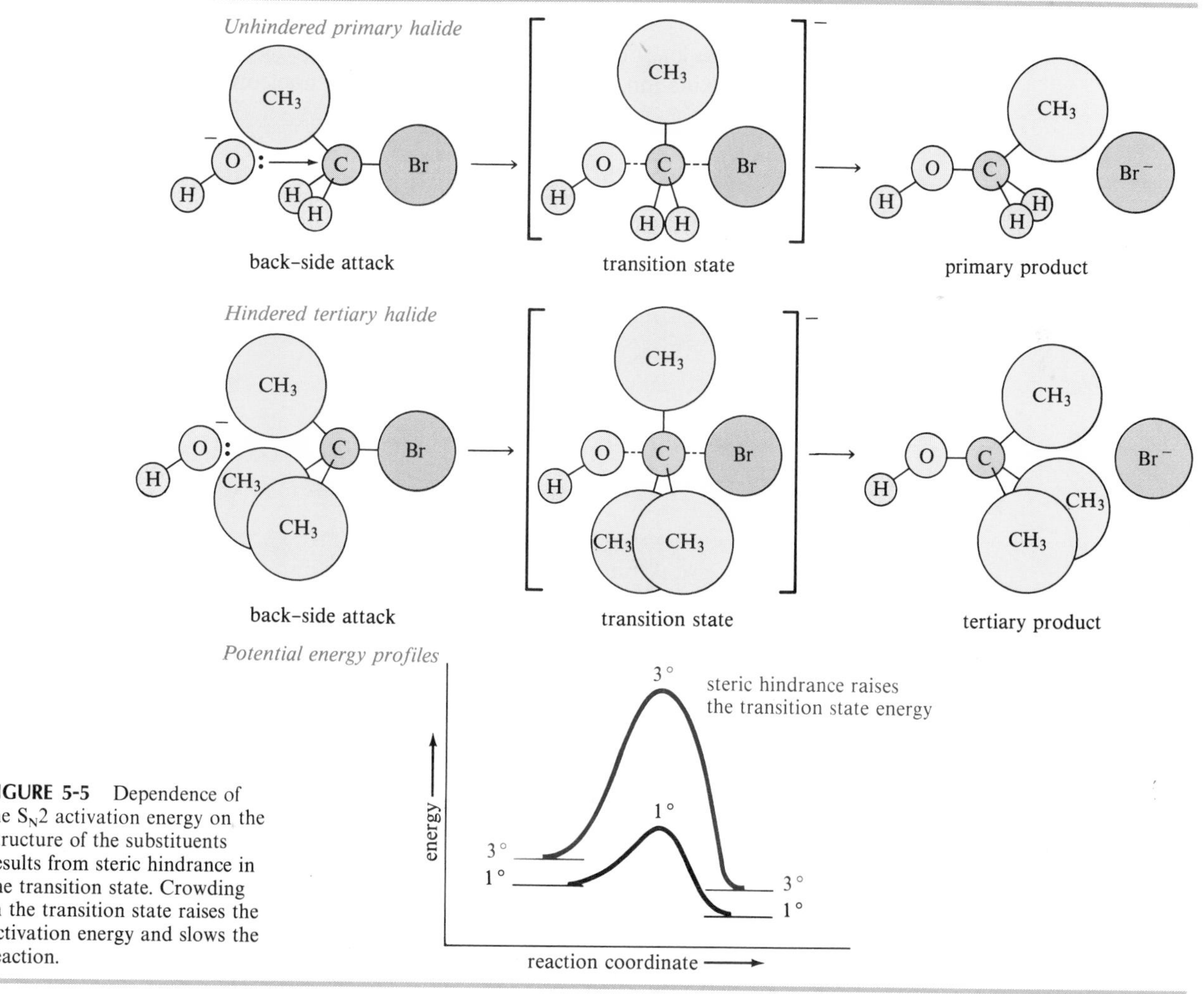

FIGURE 5-5 Dependence of the S_N2 activation energy on the structure of the substituents results from steric hindrance in the transition state. Crowding in the transition state raises the activation energy and slows the reaction.

PROBLEM 5-16

For each of the following pairs of compounds, state which compound would be the better S_N2 substrate.

(a) 2-methyl-1-iodopropane or *t*-butyl iodide
(b) cyclohexyl bromide or 1-bromo-1-methylcyclohexane
(c) 2-bromobutane or isopropyl bromide
(d) 2,2-dimethyl-1-chlorobutane or 2-chlorobutane
(e) 1-iodo-2,2-dimethylpropane or isopropyl iodide

5-12 STRENGTH OF THE NUCLEOPHILE IN S_N2 REACTIONS

The nature of the nucleophile also affects the rate of the S_N2 reaction. A strong nucleophile is much more effective than a weak one in attacking an electrophilic carbon atom. For example, both methanol (CH_3OH) and methoxide ion (CH_3O^-) are nucleophilic; but methoxide ion reacts with electrophiles in the S_N2 reaction about 1 million times faster than methanol.

Methoxide ion has nonbonding electrons that are readily available for bonding. In the transition state, the negative charge is shared by the oxygen of the methoxide ion and by the halide leaving group. Methanol, however, has no negative charge; the transition state has a partial negative charge on the halide but a partial positive charge on the methanol oxygen atom.

It is generally true that a species with a negative charge is a stronger nucleophile than a similar species without a negative charge. As in the case of methanol and the methoxide ion, a base is always a stronger nucleophile than its conjugate acid.

$$CH_3{-}\ddot{O}{:}^- + CH_3{-}I \longrightarrow \left[CH_3{-}\overset{\delta^-}{O}\cdots CH_3 \cdots \overset{\delta^-}{I}\right]^- \longrightarrow CH_3{-}O{-}CH_3 + {:}I{:}^-$$

base, stronger nucleophile, lower E_a

$$CH_3{-}\ddot{O}H + CH_3{-}I \longrightarrow \left[CH_3{-}\overset{\delta^+}{O}(H)\cdots CH_3 \cdots \overset{\delta^-}{I}\right] \longrightarrow CH_3{-}\overset{+}{O}(H){-}CH_3 + {:}I{:}^-$$

conjugate acid, weaker nucleophile, higher E_a

We might be tempted to say that methoxide is a much better nucleophile because it is so much more basic. This would be a mistake, because basicity and nucleophilicity are two fundamentally different properties. **Basicity** is defined by the *equilibrium constant* for abstracting a proton. **Nucleophilicity** is defined by the *rate* of attack on an electrophilic carbon atom. In both cases, the nucleophile (or base) forms a new bond. If the new bond is to a proton, it has reacted as a base; if the new bond is to carbon, it has reacted as a nucleophile.

Basicity

$$B{:}^- + H{-}A \overset{K_{eq}}{\rightleftharpoons} B{-}H + A{:}^-$$

$$B\!:^- + \; -\overset{|}{\underset{|}{C}}-X \xrightarrow{k_r} B-\overset{|}{\underset{|}{C}}- + X\!:^-$$

Table 5-5 lists some common nucleophiles in decreasing order of their nucleophilicity in common solvents such as water and the alcohols. The order of nucleophiles shows three major trends:

TABLE 5-5
Some common nucleophiles, listed in decreasing order of nucleophilicity in common solvents such as water and the alcohols

strong nucleophiles	$(CH_3CH_2)_3P\!:$	moderate nucleophiles	$:\ddot{Br}:^-$
	$^-:\ddot{S}-H$		$:NH_3$
	$:\ddot{I}:^-$		$CH_3-\ddot{S}-CH_3$
	$(CH_3CH_2)_2\ddot{N}H$		$:\ddot{Cl}:^-$
	$^-:C{\equiv}N$		$CH_3C(=O)-\ddot{O}:^-$
	$(CH_3CH_2)_3N\!:$		
	$H-\ddot{O}:^-$	weak nucleophiles	$:\ddot{F}:^-$
	$CH_3-\ddot{O}:^-$		$H-\ddot{O}-H$
			$CH_3-\ddot{O}-H$

SUMMARY OF TRENDS IN NUCLEOPHILICITY

1. A species with a negative charge is a stronger nucleophile than a similar species without a negative charge. In particular, a base is a stronger nucleophile than its conjugate acid.

$$^-:\ddot{O}H > H_2\ddot{O}: \qquad ^-:\ddot{S}H > (CH_3)_2\ddot{S}: \qquad ^-:\ddot{N}H_2 > :NH_3$$

2. Nucleophilicity decreases going from left to right in the periodic table, following the increase in electronegativity from left to right. The more electronegative elements have more tightly held nonbonding electrons that are less reactive toward forming new bonds.

$$^-:\ddot{O}H > :\ddot{F}:^- \qquad :NH_3 > H_2\ddot{O}: \qquad (CH_3CH_2)_3P: > (CH_3CH_2)_2\ddot{S}:$$

3. Nucleophilicity increases going down the periodic table, following the increase in size and polarizability.

$$:\ddot{I}:^- > :\ddot{Br}:^- > :\ddot{Cl}:^- > :\ddot{F}:^- \qquad ^-:\ddot{Se}H > ^-:\ddot{S}H > ^-:\ddot{O}H$$

$$(CH_3CH_2)_3P: > (CH_3CH_2)_3N:$$

Going down a column in the periodic table, the atoms become larger, with more electrons at a greater distance from the nucleus. The electrons are more loosely held, and the atom is more **polarizable:** Its electrons can move more freely toward a positive charge, resulting in stronger bonding in the transition state. The increased mobility of its electrons enhances the atom's ability to begin to form a bond at a relatively long distance.

Figure 5-6 illustrates this polarizability effect by comparing the attack of iodide ion and the attack of fluoride ion on methyl chloride. The outer shell of

Figure 5-6 Comparison of fluoride ion and iodide ion as nucleophiles in the S_N2 reaction. Fluoride has tightly bound electrons that cannot begin to form a C—F bond until the atoms are very close together. Iodide has more loosely bound outer electrons that begin bonding much earlier in the reaction.

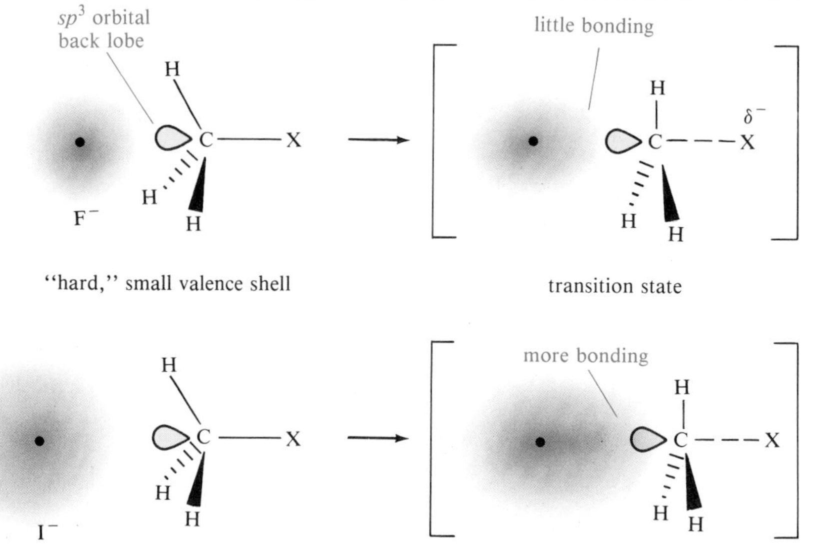

the fluoride ion is the second shell. These electrons are tightly held close to the nucleus. Fluoride is a "hard" (low-polarizability) nucleophile, and its nucleus must approach the carbon nucleus quite closely before the electrons can begin to overlap and form a bond. In the transition state, there is little bonding between fluorine and carbon. The outer shell of the iodide ion is the fifth shell. These electrons are held more loosely, making the iodide ion a "soft" (high-polarizability) nucleophile. The outer electrons begin to shift and overlap with the carbon atom from farther away. There is a great deal of bonding between iodine and carbon in the transition state.

Another factor in the nucleophilicity of these ions is their solvation in hydroxylic (containing —OH groups) solvents. Small anions are solvated much more strongly than large ones, especially in the hydrogen-bonded hydroxylic solvents. When the anion reacts as a nucleophile, energy is required to "strip off" some of the solvent molecules that are hydrogen bonded to its lone pairs of electrons. This energy is much larger for a small, strongly solvated ion such as fluoride than it is for a large, diffuse, less strongly solvated ion such as iodide.

solvent partially stripped off in the transition state

In general, the polarizability of atoms and ions increases with increasing atomic number, and the solvation of ions (in hydroxylic solvents) decreases with increasing atomic number. Therefore, we make the general statement that nucleophilicity increases going down a column of the periodic table, as long as we compare similar species with similar charges. Remember, however, that these trends in nucleophilicity are generalizations that apply under most conditions, but there are exceptions, especially in nonhydroxylic solvents. Nucleophilicity is a complex property, depending on the specific reagents, solvents, and conditions.

5-13 STERIC EFFECTS ON NUCLEOPHILICITY

In Section 5-11 we saw that the S_N2 reaction is sensitive to bulky groups on the electrophile. Bulky groups on the nucleophile also slow the reaction rate. For example, the *t*-butoxide ion is a stronger base than ethoxide ion, but the bulky *t*-butoxide is a weaker nucleophile. Steric hindrance has little effect on *basicity*, because basicity involves attack on an unhindered proton. When a *nucleophilic* attack at a carbon atom is involved, however, a bulky base is a less effective nucleophile.

$$(CH_3)_3C\text{—}\ddot{\underset{..}{O}}\text{:}^- \qquad\qquad CH_3\text{—}CH_2\text{—}\ddot{\underset{..}{O}}\text{:}^-$$

t-butoxide (hindered)
stronger base, yet weaker nucleophile

ethoxide (unhindered)
weaker base, yet stronger nucleophile

PROBLEM 5-17
From each of the following pairs, predict the stronger nucleophile in the S_N2 reaction (using an alcohol as the solvent). Explain your prediction.

(a) $(CH_3CH_2)_3N$ or $(CH_3CH_2)_2NH$
(b) $(CH_3)_2O$ or $(CH_3)_2S$
(c) NH_3 or PH_3
(d) CH_3S^- or H_2S
(e) $(CH_3)_3N$ or $(CH_3)_2O$
(f) $(CH_3)_2CH\text{—}O^-$ or $CH_3CH_2CH_2\text{—}O^-$
(g) I^- or Cl^-

5-14 LEAVING-GROUP EFFECTS

The bond between carbon and the leaving group is breaking in the transition state of the S_N2 reaction:

$$\text{Nuc:}^- + \text{C—X} \longrightarrow [\text{Nuc}\cdots\text{C}\cdots\text{X}]^- \longrightarrow \text{Nuc—C} + \text{:X}^-$$

a generic nucleophile; bond forming; bond breaking; transition state; leaving group

The S_N2 reaction is quite sensitive to the amount of energy needed to begin to break the bond between carbon and the leaving group. For a nucleophile to attack at a carbon atom and replace a leaving group, the leaving group must have three important characteristics.

First, a good leaving group is *electron withdrawing,* so that it creates a partial positive charge on the carbon atom. Another carbon atom or a hydrogen atom is not a good leaving group, because neither withdraws enough electron density to make the carbon electrophilic. The halogens are all more electronegative than carbon, however, and the polarized carbon-halogen bond puts a partial positive charge on carbon. This electron-withdrawing effect helps to stabilize the transition state for the S_N2 reaction. Other electronegative elements, such as oxygen, nitrogen, and sulfur, are also sufficiently electronegative to create an electrophilic carbon atom. Elements that polarize the C—X bond generally appear toward the right side of the periodic table.

C—H C—B C—C
not very polar bonds

C—N C—O C—F
C—S C—Cl
C—Br
C—I
sufficiently polar bonds

Second, the leaving group should be *stable* after leaving with the bonding pair of electrons (also implying stability in the transition state). In most of our examples, stable halide ions have been the leaving groups. In general, good leaving groups should be *weak bases,* and therefore they are the conjugate bases of strong acids (see Section 1-13). The hydrohalic acids (HF, HCl, HBr, and HI) are relatively strong acids, and their conjugate bases (F^-, Cl^-, Br^-, and I^-) are all weak.

Other weak bases can also act as good leaving groups. Examples of weak bases that are seen as leaving groups are sulfate ions, sulfonate ions, and phosphate ions. Hydroxide ion is a strong base, and the alkoxide ions (such as methoxide ion) are also strong bases. Hydroxide and alkoxide ions are poor leaving groups in the S_N2 reaction.

Ions that are weak bases and good leaving groups

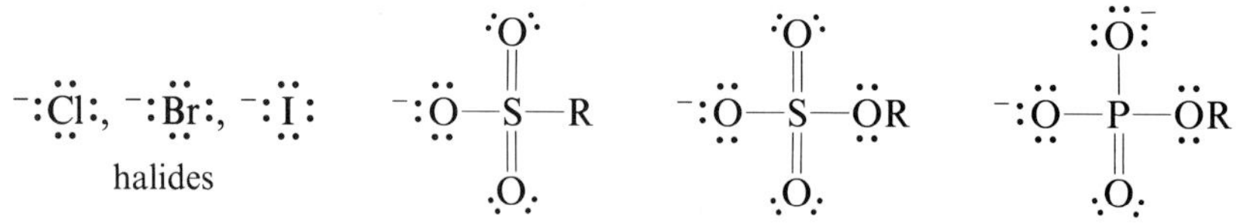

$^-$:C̈l:, $^-$:B̈r:, $^-$:Ï: — halides

$^-$:O—S(=O)(=O)—R — sulfonate

$^-$:O—S(=O)(=O)—OR — sulfate

$^-$:O—P(=O)(O$^-$)—OR — phosphate

Ions that are strong bases and poor leaving groups

$^-$:ÖH — hydroxide $^-$:ÖR — alkoxide $^-$:N̈H_2 — amide

Neutral molecules that are good leaving groups

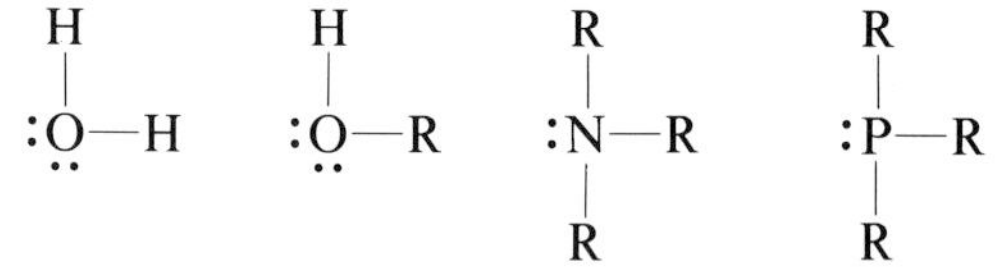

:O(H)—H — water :O(H)—R — alcohols :N(R)(R)—R — amines :P(R)(R)—R — phosphines

Neutral molecules can also be good leaving groups. A neutral molecule often serves as the leaving group from a *positively charged* electrophile. For example, the —OH group of an alcohol is not a good leaving group because it would have to leave as hydroxide ion.

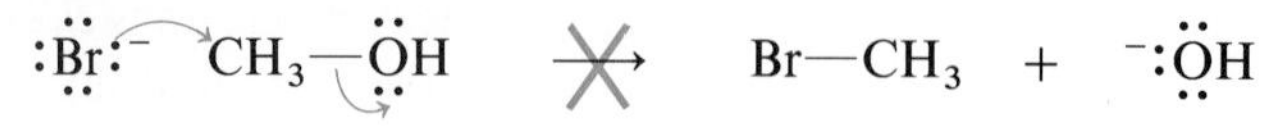

$$:\ddot{Br}:^- \quad CH_3—\ddot{O}H \quad \not\longrightarrow \quad Br—CH_3 \; + \; ^-:\ddot{O}H$$

This reaction fails because $^-$:ÖH is a poor leaving group.

If the alcohol is placed in an acidic solution, however, the hydroxyl group is protonated. Water can then serve as a leaving group.

$$CH_3{-}\ddot{O}H + H^+ \rightleftharpoons :\ddot{Br}:^- \rightarrow CH_3{-}\overset{H}{\overset{|}{\ddot{O}^+}}{-}H \longrightarrow :\ddot{Br}{-}CH_3 + :\overset{H}{\overset{|}{\ddot{O}}}{-}H$$

protonated alcohol — water

Finally, the ability of a species to act as a good leaving group depends on its *polarizability:* that is, its ability to continue bonding with a carbon atom while it is leaving. This bonding stabilizes the transition state, minimizing the activation energy. In many ways, the departure of a leaving group is like the attack of a nucleophile, except that the bond is breaking rather than forming. Highly polarizable nucleophiles and highly polarizable leaving groups both stabilize the transition state because they engage in more bonding at a longer distance. Iodide ion, one of the most polarizable of ions, is both a strong nucleophile and an excellent leaving group.

PROBLEM 5-18

When dimethyl ether ($CH_3{-}O{-}CH_3$) is treated with concentrated HBr, the initial products are CH_3Br and CH_3OH. Propose a mechanism to account for this reaction.

PROBLEM 5-19

Rank the following compounds in decreasing order of their reactivity toward the S_N2 reaction with sodium ethoxide ($Na^+\ {}^-OCH_2CH_3$) in ethanol.

methyl chloride	*t*-butyl iodide	neopentyl bromide
isopropyl bromide	methyl iodide	ethyl chloride

5-15 FIRST-ORDER NUCLEOPHILIC SUBSTITUTION: THE S_N1 REACTION

When *t*-butyl bromide is placed in boiling methanol, methyl *t*-butyl ether can be isolated from the reaction mixture. Because this reaction takes place with the solvent acting as the nucleophile, it is called a **solvolysis** (*solvo* for "solvent" plus *lysis*, meaning "cleavage").

$$\underset{t\text{-butyl bromide}}{(CH_3)_3C{-}Br} + \underset{\text{methanol}}{CH_3{-}OH} \xrightarrow{\text{boil}} \underset{\text{methyl }t\text{-butyl ether}}{(CH_3)_3C{-}O{-}CH_3} + HBr$$

This reaction is certainly a substitution, since methoxide has replaced bromide on the *t*-butyl group. It does not go through the S_N2 mechanism, however. The S_N2 reaction requires a strong nucleophile and a substrate that is not too hindered. Methanol is a weak nucleophile, and *t*-butyl bromide is a hindered tertiary halide and a poor S_N2 substrate.

An interesting characteristic of this reaction is that its rate does not depend on the concentration of methanol, the nucleophile. The rate depends only on the concentration of the starting material, *t*-butyl bromide.

$$\text{rate} = k_r[(CH_3)_3C{-}Br]$$

This rate equation is first order overall: first order in the concentration of the starting material and zeroth order in the concentration of the nucleophile. Because the rate does not depend on the concentration of the nucleophile, we infer that the nucleophile is not present in the transition state of the rate-determining step. The nucleophile must react *after* the slow step.

This type of substitution is called the **S_N1 reaction,** for *Substitution, Nucleophilic, unimolecular*. The term *unimolecular* means there is only one molecule involved in the transition state of the rate-determining step. The mechanism of the

S_N1 reaction of *t*-butyl bromide with methanol is shown below. The ionization of the alkyl halide (first step) is the rate-determining step.

Step 1: Formation of carbocation (rate-determining)

$$(CH_3)_3C{-}\ddot{\underset{\cdot\cdot}{Br}}{:} \rightleftharpoons (CH_3)_3C^+ + {:}\ddot{\underset{\cdot\cdot}{Br}}{:}^- \quad \text{(slow)}$$

Step 2: Nucleophilic attack

$$(CH_3)_3C^+ + {:}\ddot{O}(H){-}CH_3 \rightleftharpoons (CH_3)_3C{-}\ddot{O}^+(H){-}CH_3 \quad \text{(fast)}$$

Final step: Loss of proton to solvent

$$(CH_3)_3C{-}\ddot{O}^+(H){-}CH_3 + CH_3{-}\ddot{\underset{\cdot\cdot}{O}}H \rightleftharpoons (CH_3)_3C{-}\ddot{\underset{\cdot\cdot}{O}}{-}CH_3 + CH_3{-}\ddot{O}^+(H){-}H \quad \text{(fast)}$$

The S_N1 mechanism is a two-step process. The first step is a slow ionization to form a carbocation. The second step is a fast attack on the carbocation by the nucleophile. The carbocation is a very strong electrophile; it reacts very fast with both strong and weak nucleophiles. In the case of attack by water or an alcohol, loss of a proton gives the final uncharged product. Following is the general mechanism for the S_N1 reaction.

$$R{-}X \rightleftharpoons R^+ + X{:}^- \quad \text{rate-determining step}$$

$$R^+ + Nuc{:}^- \longrightarrow R{-}Nuc \quad \text{(very fast)}$$

The potential-energy profile of the S_N1 reaction (Fig. 5-7) shows why the rate does not depend on the strength or concentration of the nucleophile. The ionization (first step) is highly endothermic, and its large activation energy determines the overall reaction rate. The nucleophilic attack (second step) is strongly exothermic, with a lower-energy transition state. In effect, a nucleophile reacts with the carbocation almost as soon as it forms. The potential-energy profiles of the S_N1 mechanism and the S_N2 mechanism are compared in Figure 5-7. The S_N1 reaction has a true intermediate, the carbocation. The intermediate appears as a relative minimum (a low point) in the reaction-energy profile. Reagents and conditions that favor formation of the carbocation (the slow step) will accelerate the S_N1 reaction; reagents and conditions that hinder its formation will retard the reaction.

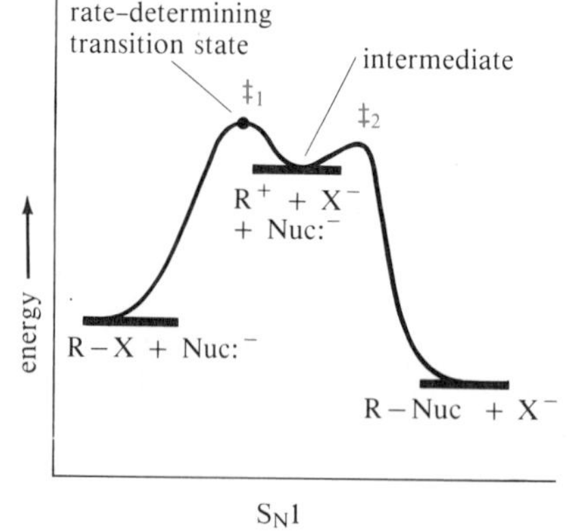

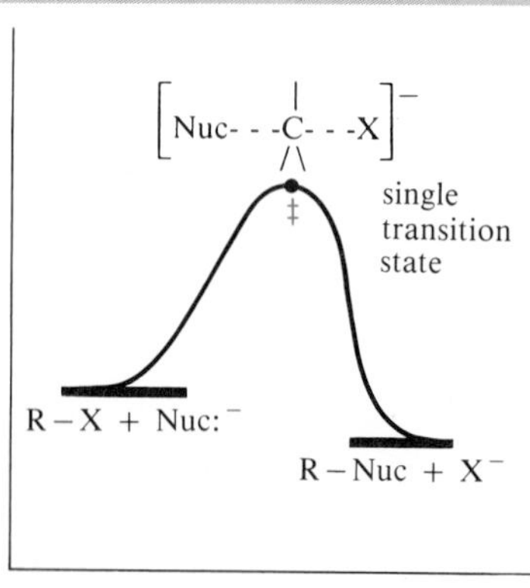

FIGURE 5-7 Potential-energy profiles of the S_N1 and S_N2 reactions. The S_N1 is a two-step mechanism with two transition states (‡1 and ‡2) and an intermediate (the carbocation). The S_N2 has only one transition state and no intermediate.

5-15A SUBSTITUENT EFFECTS

More highly substituted alkyl halides are more reactive toward the S_N1 reaction than less substituted ones. The reactivity order is

$$3^\circ > 2^\circ > 1^\circ > CH_3X$$

the same as the order of stability of carbocations. This order is *opposite* that for the S_N2 reaction. The rate-determining step of the S_N1 is an ionization to form a carbocation, and a more highly substituted alkyl halide gives a more stable carbocation. In this endothermic reaction the transition state resembles the carbocation, and the more stable carbocation is formed faster. The activation energies for ionization of primary and tertiary alkyl halides to form carbocations are compared in Figure 5-8.

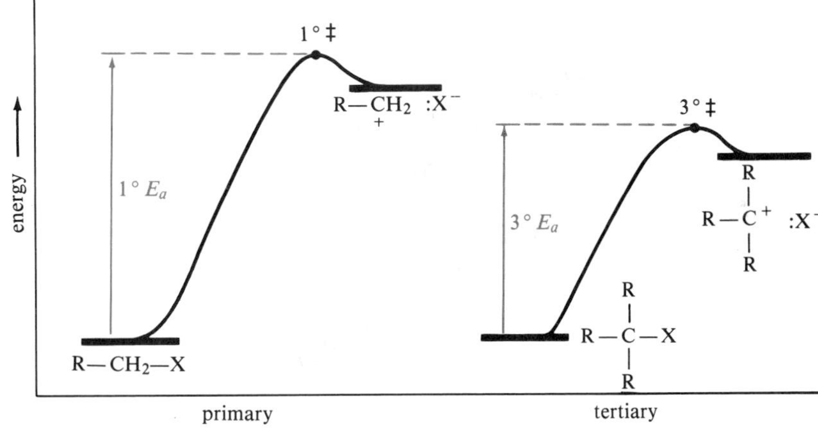

FIGURE 5-8 A tertiary carbocation is more stable than a primary carbocation. This difference in stabilities is also reflected in the activation energies for formation of the carbocations.

Resonance stabilization of the carbocation can also promote the S_N1 reaction. For example, allyl bromide is a primary halide, but it undergoes the S_N1 reaction about as fast as a secondary halide. The carbocation formed by ionization is resonance stabilized, with the positive charge spread equally over two carbon atoms.

allyl bromide ⇄ [resonance-stabilized carbocation] Br^- $\xrightarrow{\text{Nuc:}^-}$ product with Nuc

resonance-stabilized carbocation

PROBLEM 5-20

3-Bromocyclohexene is a secondary halide, and benzyl bromide is a primary halide. Both of these halides undergo S_N1 substitution about as fast as most tertiary halides. Use resonance structures to explain this enhanced reactivity.

Br

CH_2Br

3-bromocyclohexene   benzyl bromide

5-15B LEAVING-GROUP EFFECTS

Although the strength of the nucleophile has no effect on the rate of the S_N1 reaction, a good leaving group is necessary. The nucleophile is not involved in the rate-determining ionization step, but the leaving group *is leaving* in this step (Fig. 5-9). Therefore, the S_N1 reaction requires a good leaving group, for the same reasons that the S_N2 reaction requires a good leaving group.

When a good leaving group leaves, the energy of the transition state is lower (and the activation energy is lower) than when a poor leaving group leaves. A good leaving group helps to stabilize the developing negative charge, and its polarizability allows more bonding overlap in the transition state.

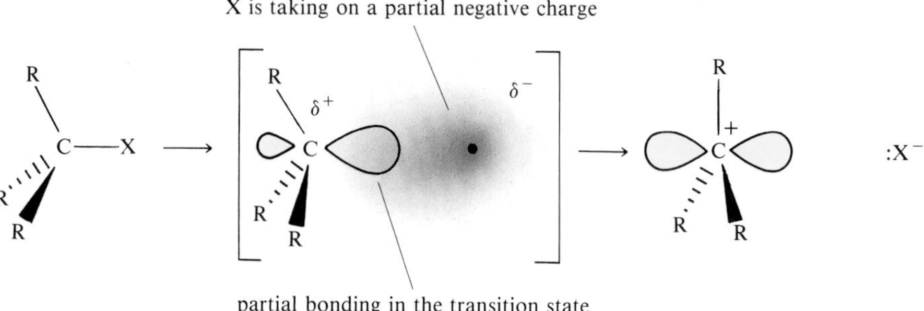

FIGURE 5-9 In the transition state of the S_N1 ionization, the leaving group is taking on a negative charge. The C—X bond is breaking, and a polarizable leaving group can still maintain substantial overlap.

PROBLEM 5-21

Choose the member of each pair that will react faster in the S_N1 reaction.

(a) 2-iodo-2-methylbutane and ethyl iodide
(b) 2-iodo-2-methylbutane and *t*-butyl chloride
(c) *n*-propyl bromide and 3-bromocyclohexene
(d) methyl iodide and cyclohexyl bromide

5-15C SOLVENT EFFECTS

The S_N1 reaction goes much more readily in polar solvents that stabilize ions. The rate-determining step involves the formation of two ions, and these ions take on their charges in the transition state. Polar solvents such as water and the alcohols strongly solvate these ions by an interaction of the solvent's dipole moment with the charge of the ion. Solvation lowers the energy of the ions and also the energy of the transition state leading to them.

$R—\ddot{X}: \rightleftharpoons R^+ \; :\ddot{X}:^-$

ionization

solvated ions

TABLE 5-6

Dielectric constants of common solvents and ionization rates of *t*-butyl chloride in those solvents

Solvent	ε[a]	*Relative rate*
water	78	8000
methanol	33	1000
ethanol	24	200
acetone	21	1
diethyl ether	4.3	0.001
hexane	2.0	<0.0001

[a] The dielectric constant ε is a measure of a solvent's ability to solvate ions. This property has a large effect on the rate of ionization of an alkyl halide.

The ability of a solvent to solvate ions is measured by its *dielectric constant* (ε), a measure of the solvent's polarity. Table 5-6 lists the dielectric constants of some common solvents and the relative rates of ionization of *t*-butyl chloride in these solvents.

5-16 STEREOCHEMISTRY OF THE S_N1 REACTION

The positively charged carbon atom of a carbocation is sp^2 hybridized. The groups bonded to this carbon atom are arranged in a plane, with an empty p orbital perpendicular to the plane. A nucleophile can attack this carbocation at either lobe of the empty p orbital, as shown below.

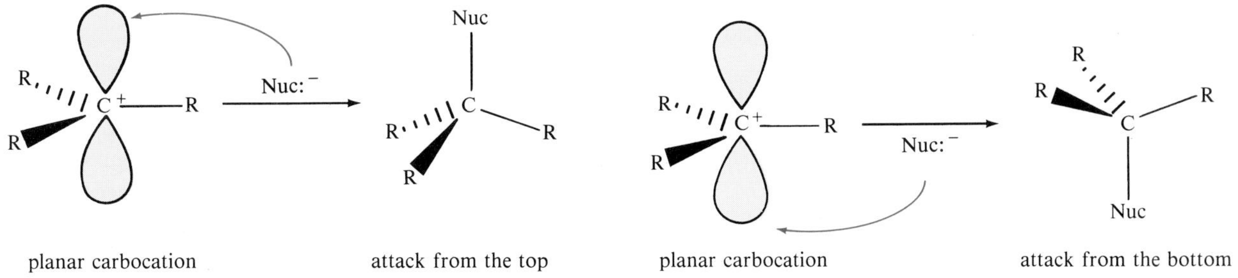

planar carbocation — attack from the top — planar carbocation — attack from the bottom

Unlike the S_N2 reaction, where back-side attack always gives inversion of configuration, the carbocation in the S_N1 reaction may be attacked from either face. Attack on the front side (the side the leaving group left) gives a product with **retention of configuration;** that is, the nucleophile occupies the same stereochemical position in the product as the leaving group did in the reactant. Attack from the back side is also possible, resulting in **inversion of configuration.**

Products obtained from S_N1 reactions are usually mixtures showing both retention and inversion of configuration. In many cases, the two products are indistinguishable. In some cases, however, two different products result. Figure 5-10 shows a cyclic case where one of the faces of the cyclopentane ring has been

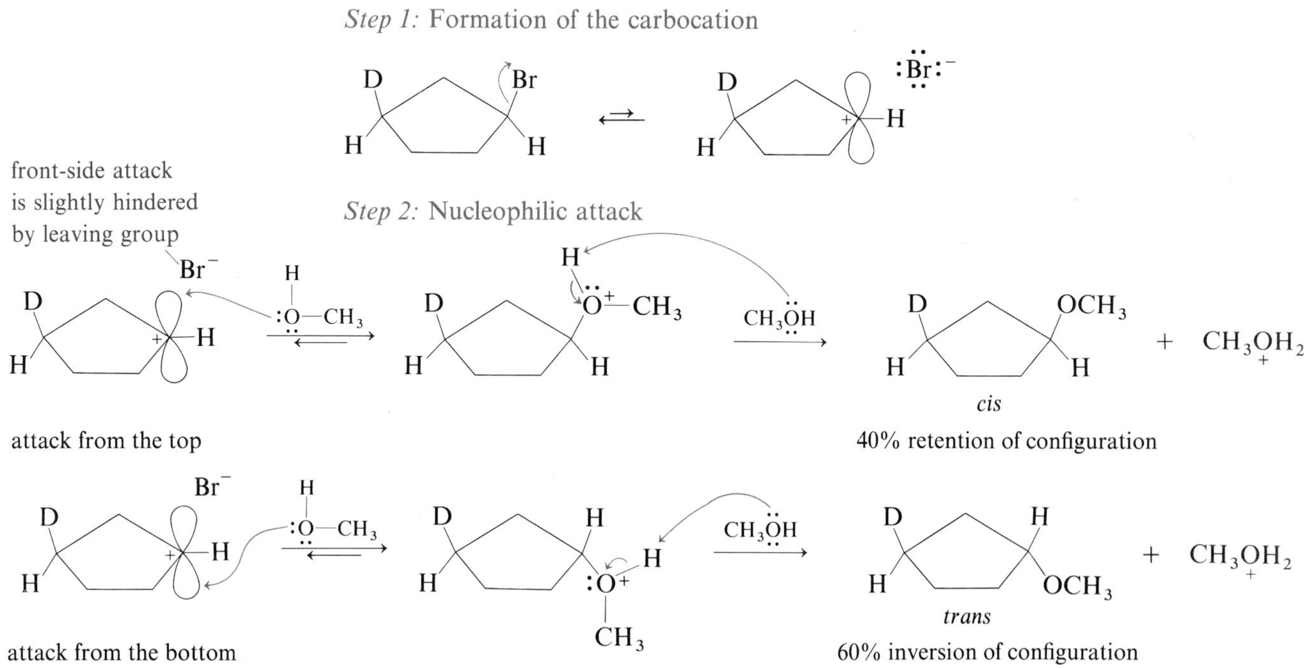

Step 1: Formation of the carbocation

Step 2: Nucleophilic attack

front-side attack is slightly hindered by leaving group

attack from the top

CH_3OH → cis — 40% retention of configuration + $CH_3OH_2^+$

attack from the bottom

CH_3OH → *trans* — 60% inversion of configuration + $CH_3OH_2^+$

FIGURE 5-10 In the S_N1 reaction of *cis*-1-bromo-3-deuteriocylopentane with methanol, the carbocation can be attacked from either face. Attack from the same face that bromide left gives retention of configuration and the cis product. Attack from the opposite face gives inversion of configuration and the trans product.

"labeled" using a deuterium atom, the isotope of hydrogen with mass number 2. Deuterium has the same size and shape as hydrogen, and it undergoes the same reactions. In this molecule it distinguishes between the two faces: The reactant has the bromine atom cis to the deuterium, and the product of retention has the nucleophile cis to the deuterium. The product of inversion has the nucleophile trans to the deuterium. The product mixture obtained from the reaction contains both the cis and trans isomers.

We might expect a symmetrical carbocation to give equal amounts of products from retention and inversion, and this is sometimes the case. In most S_N1 reactions, however, there is slightly more *inversion* of configuration. A carbocation is strongly electrophilic, and it usually reacts with the first nucleophile it encounters. Since the leaving group is departing from the front side, the front of the carbocation is briefly blocked by the leaving group. The cation is more likely to encounter a nucleophile attacking from the back side, giving inversion of configuration. The reaction in Figure 5-10 shows a slight preference for inversion of configuration.

5-17 REARRANGEMENTS IN THE S_N1 REACTION

Carbocations frequently undergo structural changes to form more stable ions. These structural changes are called **rearrangements.** A rearrangement may occur after a carbocation is formed, or it may occur as the leaving group is leaving. Rearrangements are not seen in the S_N2 reaction, where no carbocation is formed and the one-step mechanism allows no opportunity for rearrangement.

An example of a reaction with rearrangement is the S_N1 reaction of 2-bromo-3-methylbutane in boiling ethanol. The product is a mixture of 2-ethoxy-3-methylbutane (not rearranged) and 2-ethoxy-2-methylbutane (rearranged).

$$CH_3-\underset{}{\overset{Br}{\overset{|}{C}H}}-\underset{\underset{CH_3}{|}}{CH}-CH_3 \xrightarrow[\text{heat}]{CH_3CH_2OH} CH_3-\overset{OCH_2CH_3}{\overset{|}{C}H}-\underset{\underset{CH_3}{|}}{CH}-CH_3 + CH_3-CH_2-\overset{OCH_2CH_3}{\overset{|}{\underset{\underset{CH_3}{|}}{C}}}-CH_3 + HBr$$

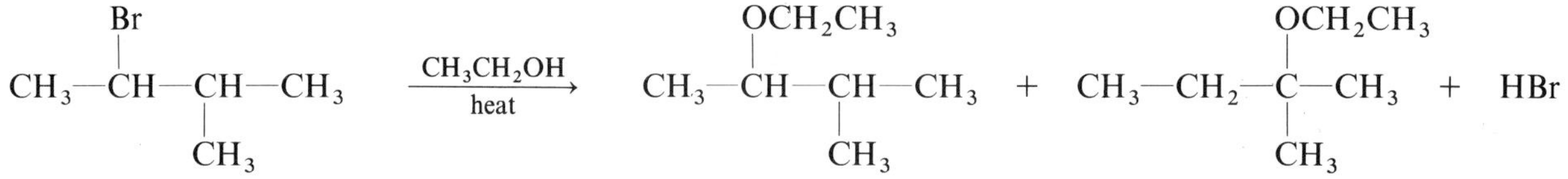

2-bromo-3-methylbutane | (not rearranged) 2-ethoxy-3-methylbutane | (rearranged) 2-ethoxy-2-methylbutane

PROBLEM 5-22

Give the S_N1 mechanism for the formation of 2-ethoxy-3-methylbutane, the unrearranged product in this reaction.

The rearranged product, 2-ethoxy-2-methylbutane, results from a **hydride shift:** the movement of a hydrogen atom with its bonding pair of electrons. A hydride shift is represented by the symbol $\sim H$. In this case, the hydride shift converts the initially formed secondary carbocation to a more stable tertiary carbocation. Attack by the solvent gives the rearranged product.

Step 1: Formation of the carbocation

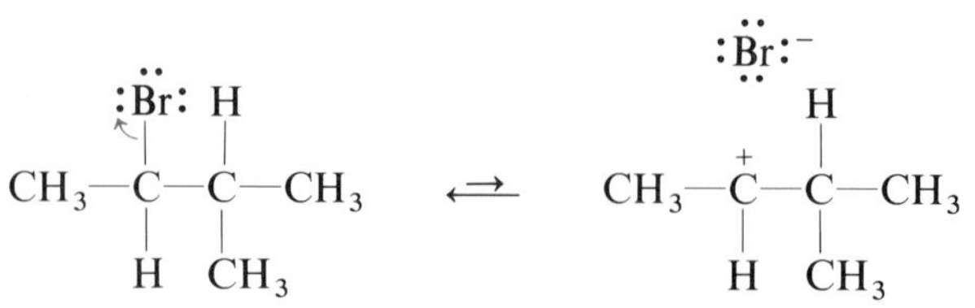

$$CH_3-\overset{:\ddot{Br}:}{\overset{|}{\underset{\underset{H}{|}}{C}}}-\overset{H}{\overset{|}{\underset{\underset{CH_3}{|}}{C}}}-CH_3 \rightleftarrows CH_3-\overset{+}{\underset{\underset{H}{|}}{C}}-\overset{H}{\overset{|}{\underset{\underset{CH_3}{|}}{C}}}-CH_3 \quad :\ddot{Br}:^-$$

Step 2: Rearrangement, followed by solvent attack

hydrogen moves with a pair of electrons

$$\underset{\text{secondary carbocation}}{CH_3-\overset{+}{C}H-C(H)(CH_3)-CH_3} \xrightarrow{\sim H} \underset{\text{tertiary carbocation}}{CH_3-CH_2-\overset{+}{C}(CH_3)-CH_3} \xrightarrow{CH_3CH_2OH} \xrightarrow{-H^+} \underset{\text{rearranged product}}{CH_3-CH_2-C(OCH_2CH_3)(CH_3)-CH_3}$$

When neopentyl bromide is boiled with ethanol, it gives *only* a rearranged substitution product. This product results from a **methyl shift** (represented by the symbol $\sim CH_3$), the migration of a methyl group together with its pair of electrons. Without rearrangement, ionization of neopentyl bromide would give a very unstable primary carbocation.

$$\underset{\text{neopentyl bromide}}{CH_3-C(CH_3)_2-CH_2-\ddot{Br}:} \not\leftrightarrow \underset{\text{primary carbocation}}{CH_3-C(CH_3)_2-CH_2^+ \quad :\ddot{Br}:^-}$$

The methyl shift occurs *while* the bromide ion is leaving, so that only the more stable tertiary carbocation is formed. Since the rearrangement is required for ionization, only rearranged products are observed.

$$CH_3-C(CH_3)_2-CH_2-\ddot{Br}: \xrightarrow{\sim CH_3} CH_3-\overset{+}{C}(CH_3)-CH_2-CH_3 \quad + \quad Br^-$$

Attack by ethanol on the tertiary carbocation gives the observed product.

$$CH_3-\overset{+}{C}(CH_3)-CH_2-CH_3 \xrightarrow{CH_3CH_2-\ddot{O}-H} \xrightarrow{-H^+} \underset{\text{rearranged product}}{CH_3-C(CH_3)(O-CH_2CH_3)-CH_2-CH_3}$$

PROBLEM 5-23

Propose a mechanism involving a hydride shift or an alkyl shift for each of the following solvolysis reactions. Explain how each rearrangement forms a more stable intermediate.

(a) $CH_3-C(CH_3)_2-CH(I)-CH_3 \xrightarrow[\text{heat}]{CH_3OH} CH_3-C(CH_3)_2-CH(OCH_3)-CH_3 + CH_3-C(CH_3)(O-CH_3)-CH(CH_3)-CH_3$

(b) 1-chloro-2-methylcyclohexane $\xrightarrow[\text{heat}]{CH_3CH_2OH}$ 1-(OCH_2CH_3)-2-(CH_3)cyclohexane + 1-(OCH_2CH_3)-1-(CH_3)cyclohexane

(c) [cyclohexyl iodide] + $CH_3-\overset{O}{\overset{\|}{C}}-OH$ $\xrightarrow{\text{heat}}$ [cyclohexyl]$-O-\overset{O}{\overset{\|}{C}}-CH_3$ + [cyclohexyl]$-O-\overset{O}{\overset{\|}{C}}-CH_3$

(d) [cyclohexyl]$-CH_2I$ $\xrightarrow[\text{heat}]{CH_3CH_2OH}$ [1-methylcyclohexyl]$-OCH_2CH_3$ + [cycloheptyl]$-OCH_2CH_3$

5-18 COMPARISON OF S_N1 AND S_N2 REACTIONS

Let's compare what we know about the S_N1 and S_N2 reactions, and then organize this material into a brief table.

Kinetics The rate of the S_N1 reaction is proportional to the concentration of the alkyl halide but not the concentration of the nucleophile. It follows a first-order rate equation.

The rate of the S_N2 reaction is proportional to the concentrations of both the alkyl halide [R—X] and the nucleophile [Nuc:⁻]. It follows a second-order rate equation.

$$S_N1 \text{ rate} = k_r[\text{R—X}]$$

$$S_N2 \text{ rate} = k_r[\text{R—X}][\text{Nuc:}^-]$$

Rearrangements The S_N1 reaction involves a carbocation intermediate. This intermediate can rearrange, usually by a hydride shift or an alkyl shift, to give a more stable carbocation.

The S_N2 reaction takes place in one step with no intermediates. No rearrangement is possible in the S_N2 reaction.

S_N1: Rearrangements are possible.

S_N2: Rearrangements are not possible.

Stereochemistry The S_N1 reaction involves a flat carbocation intermediate that can be attacked from either face. Therefore, the S_N1 usually gives a mixture of inversion and retention of configuration.

The S_N2 reaction takes place through a back-side attack, which inverts the stereochemistry of the carbon atom. Complete inversion of configuration is the result.

S_N1 stereochemistry: Mixture of retention and inversion.

S_N2 stereochemistry: Complete inversion.

Effect of the substrate The structure of the substrate (the alkyl halide) is the most important factor in determining which of these two substitution mechanisms will operate. Methyl halides and primary halides cannot easily ionize and undergo S_N1 substitution, because methyl and primary carbocations are quite high in energy. They are relatively unhindered, however, and they make good S_N2 substrates.

Tertiary halides are too hindered to undergo the S_N2 reaction, but they can ionize to form relatively stable tertiary carbocations. Tertiary halides undergo substitution exclusively through the S_N1 mechanism. Secondary halides can undergo substitution by either mechanism, depending on the conditions.

S_N1 substrates: $3° > 2°$ (1° and CH_3X are not suitable)

S_N2 substrates: $CH_3X > 1° > 2°$ (3° is not suitable)

Effect of the nucleophile The nucleophile takes part in the slow step (the only step) of the S_N2 reaction but not in the slow step of the S_N1. Therefore, a strong nucleophile promotes the S_N2 but not the S_N1. Weak nucleophiles fail to promote the S_N2 reaction; therefore, reactions with weak nucleophiles often go by the S_N1 mechanism if the substrate is secondary or tertiary.

S_N1: Nucleophile strength is unimportant.

S_N2: Strong nucleophiles are required.

Effect of the solvent The slow step of the S_N1 reaction involves the formation of two ions. Solvation of these ions is crucial to stabilizing them and lowering the activation energy of their formation. Very polar ionizing solvents such as water and the alcohols are needed for the S_N1.

Less charge separation is generated in the transition state of the S_N2 reaction. Strong solvation may weaken the strength of the nucleophile due to the energy needed to strip off the solvent molecules. Therefore, the S_N2 reaction often goes faster in less polar solvents.

S_N1: Good ionizing solvent required.

S_N2: May go faster in a less polar solvent.

SUMMARY OF NUCLEOPHILIC SUBSTITUTIONS

	S_N1	*S_N2*
characteristics		
kinetics	first-order, k_r[RX]	second-order, k_r[RX][Nuc:$^-$]
rearrangements	common	not possible
stereochemistry	mixture of inversion and retention	complete inversion
promoting factors		
substrate (RX)	$3° > 2°$	$CH_3X > 1° > 2°$
leaving group	good one required	good one required
nucleophile	weak nucleophiles are okay	strong nucleophile needed
solvent	good ionizing solvent needed	wide variety of solvents

PROBLEM 5-24

For each of the following reactions, give the expected substitution product and predict whether the mechanism will be predominantly first order or second order.

(a) 2-chloro-2-methylbutane + CH_3COOH
(b) isobutyl bromide + sodium methoxide
(c) 1-iodo-1-methylcyclohexane + ethanol
(d) cyclohexyl bromide + methanol
(e) cyclohexyl bromide + sodium ethoxide

PROBLEM 5-25

A reluctant first-order substrate can be forced to ionize by adding some silver nitrate (one of the few soluble silver salts) to the reaction. Silver ion reacts with the halogen to form a silver halide (a highly exothermic reaction), generating the cation of the alkyl group.

$$R{-}X + Ag^+ \longrightarrow R^+ + AgX\downarrow$$

Give a mechanism for each of the following silver-promoted rearrangements.

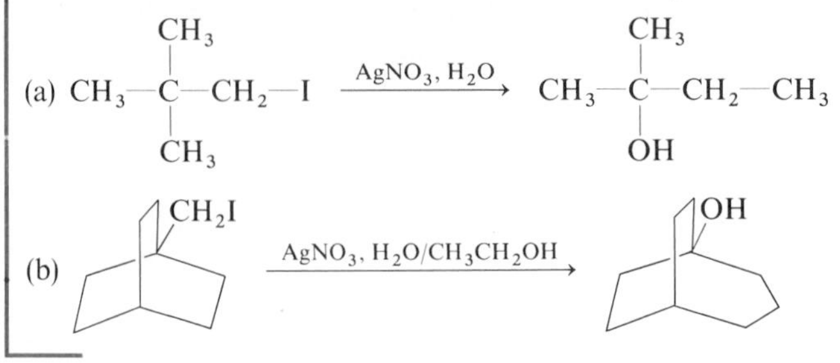

(a) $CH_3-C(CH_3)_2-CH_2-I \xrightarrow{AgNO_3,\ H_2O} CH_3-C(CH_3)(OH)-CH_2-CH_3$

(b) CH_2I $\xrightarrow{AgNO_3,\ H_2O/CH_3CH_2OH}$ OH

5-19

FIRST-ORDER ELIMINATION: THE E1 REACTION

An **elimination** involves the loss of two atoms or groups from the substrate, usually with formation of a pi bond. In many cases, substitution and elimination reactions can take place at the same time and under the same conditions, and the two reactions compete with each other. Depending on the reagents and conditions involved, an elimination might be either a first-order (E1) or second-order (E2) process.

5-19A MECHANISM AND KINETICS OF THE E1 REACTION

The abbreviation **E1** stands for *Elimination, unimolecular*. The term *unimolecular* means that the rate-determining transition state involves a single molecule rather than a collision between two molecules. The rate-determining step of an E1 reaction is the same as the first step in the S_N1 reaction: unimolecular ionization of the starting material to form a carbocation. In a fast second step, a base abstracts a proton on the carbon atom adjacent to the C^+. The electrons that once formed the carbon-hydrogen bond now form a pi bond between the two carbon atoms. The general mechanism for the E1 reaction is as follows:

Step 1: Formation of the carbocation (rate-determining)

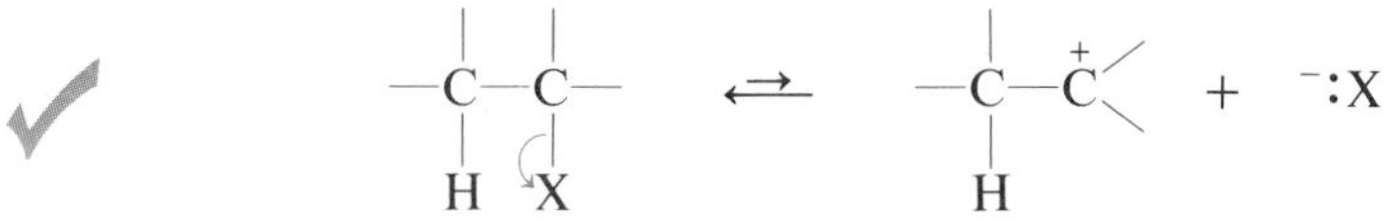

$$H-\overset{|}{\underset{|}{C}}-\overset{|}{\underset{|}{C}}-X \rightleftharpoons H-\overset{|}{\underset{|}{C}}-\overset{+}{C}\langle \; + \; {}^-:X$$

Step 2: A base abstracts a proton (fast)

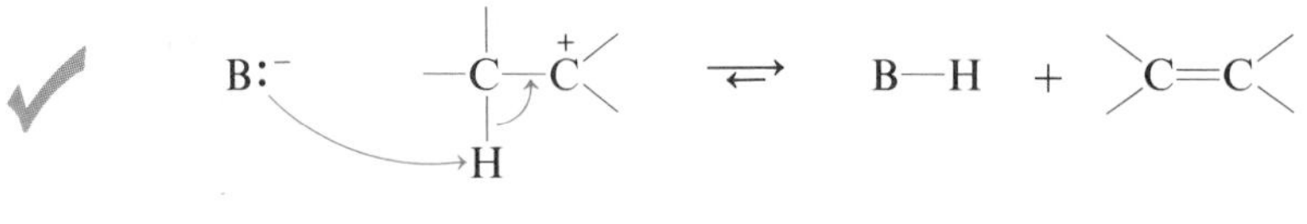

$$B:^- \; + \; H-\overset{|}{\underset{|}{C}}-\overset{+}{C}\langle \; \rightleftharpoons \; B-H \; + \; \rangle C=C\langle$$

5-19B COMPETITION WITH THE S_N1 REACTION

The E1 reaction almost always occurs in conjunction with the S_N1. Whenever a carbocation is formed, it can undergo either substitution or elimination, and mixtures of products result. The following reaction shows the competition in the reaction of *t*-butyl bromide with boiling ethanol.

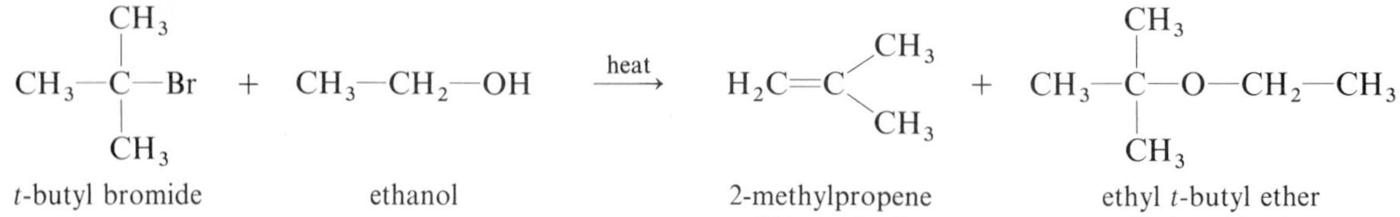

$$\underset{t\text{-butyl bromide}}{(CH_3)_3C-Br} + \underset{\text{ethanol}}{CH_3-CH_2-OH} \xrightarrow{\text{heat}} \underset{\substack{\text{2-methylpropene} \\ \text{(E1 product)}}}{H_2C=C(CH_3)_2} + \underset{\substack{\text{ethyl } t\text{-butyl ether} \\ (S_N1 \text{ product})}}{(CH_3)_3C-O-CH_2-CH_3}$$

The first product (2-methylpropene) has undergone **dehydrohalogenation.** Under these first-order conditions (the absence of a strong base) the dehydrohalogenation takes place by the E1 mechanism: Ionization of the alkyl halide gives a carbocation intermediate, which loses a proton to give the alkene. The substitution product results from a nucleophilic attack on the carbocation. Ethanol serves as a base in the elimination and as a nucleophile in the substitution.

Step 1: Ionization to form a carbocation

$$(CH_3)_3C{-}\ddot{\underset{..}{Br}}: \rightleftharpoons (CH_3)_3C^+ + :\ddot{\underset{..}{Br}}:^-$$

Step 2: Basic attack by the solvent (E1 reaction)

$$CH_3CH_2{-}\ddot{O}(H): + H{-}CH_2{-}\overset{+}{C}(CH_3){-}CH_3 \rightleftharpoons H_2C{=}C(CH_3)_2 + CH_3CH_2{-}\ddot{O}^+H_2$$

or: Nucleophilic attack by the solvent (S_N1 reaction)

$$CH_3{-}\overset{+}{C}(CH_3){-}CH_3 \underset{}{\overset{CH_3CH_2-\ddot{O}-H}{\rightleftharpoons}} CH_3{-}C(CH_3)(\overset{+}{\ddot{O}}(H)CH_2CH_3){-}CH_3 \overset{CH_3CH_2-\ddot{O}-H}{\rightleftharpoons} CH_3{-}C(CH_3)(:\ddot{O}CH_2CH_3){-}CH_3 \quad CH_3CH_2O^+H_2$$

Under ideal conditions, one of these first-order reactions provides a good yield of a single desired product. Often, however, the carbocation intermediate reacts in two or more ways to give a mixture of products. For this reason, the S_N1 and E1 reactions of alkyl halides are not routinely used for organic synthesis. They have been studied in great detail, however, to learn about the properties and stabilities of carbocations.

5-19C ORBITALS AND ENERGETICS

In the second step of the E1 mechanism, the adjacent carbon atom must rehybridize to sp^2 as the base attacks the proton and the electrons flow into the new pi bond.

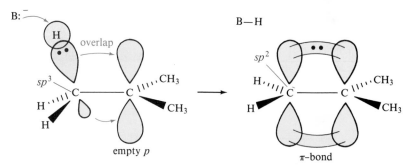

The E1 reaction has a potential-energy profile (Fig. 5-11) similar to that for the S_N1 reaction. The ionization step is strongly endothermic, with a rate-determining transition state. The second step is a fast exothermic deprotonation

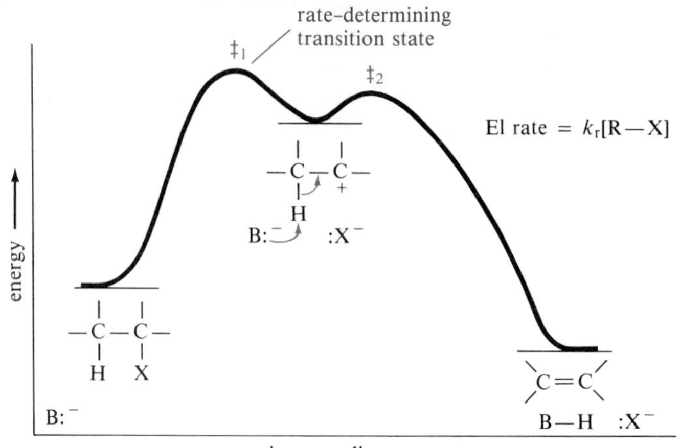

FIGURE 5-11 Potential-energy profile of the E1 reaction. The first step is a rate-determining ionization. Compare this energy profile with that of the S_N1 reaction, Figure 5-7.

by a base. The base is not involved in the reaction until *after* the rate-determining step, so the rate depends only on the concentration of the alkyl halide.

We have now seen four ways that a carbocation can react to become more stable:

SUMMARY OF CARBOCATION REACTIONS

1. React with its own leaving group to return to the reactant.
2. React with a nucleophile to form a substitution product.
3. Lose a proton to form an elimination product (an alkene).
4. Rearrange to a more stable carbocation and then react further.

PROBLEM 5-26

Give the substitution and elimination products you expect from each of the following reactions.

(a) 3-bromo-3-ethylpentane heated in methanol
(b) 1-iodo-1-methylcyclopentane heated in ethanol
(c) 3-bromo-2,2-dimethylbutane heated in ethanol
(d) iodocyclohexane + silver nitrate in water (see Problem 5-25)

5-20 SECOND-ORDER ELIMINATION: THE E2 REACTION

Elimination can also take place under second-order conditions with a strong base present. As an example, consider the reaction of *t*-butyl bromide with methoxide ion in methanol.

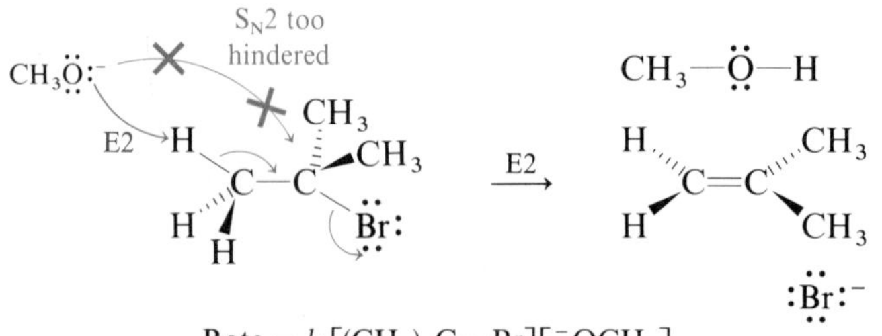

$$\text{Rate} = k_r[(CH_3)_3C\text{—}Br][^-OCH_3]$$

This is a second-order reaction because methoxide ion is a strong base and a strong nucleophile. It attacks the alkyl halide faster than the halide can ionize to give a first-order reaction. No substitution product (methyl *t*-butyl ether) is observed, however. The S_N2 mechanism is blocked because the tertiary alkyl halide is too hindered. The observed product is 2-methylpropene, resulting from elimination of HBr and formation of a double bond. In this reaction, methoxide has reacted as a *base* rather than as a *nucleophile.* Instead of attacking the back side of the hindered electrophilic carbon, methoxide has abstracted a proton from one of the methyl groups. This reaction takes place in one step, with bromide leaving as the base is abstracting a proton.

The rate of this elimination is proportional to the concentrations of both the alkyl halide and the base, giving a second-order rate equation. This is a *bimolecular* process, with both the base and the alkyl halide participating in the transition state. Therefore, this mechanism is abbreviated **E2** for *Elimination, bimolecular.*

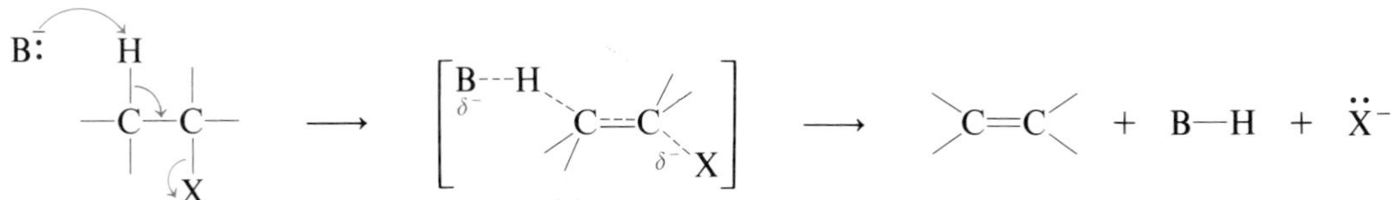

B: H —C—C— X → [B---H C==C X, δ⁻ δ⁻] → >C=C< + B—H + $\ddot{X}^-$

transition state

$$\text{E2 rate} = k_r[\text{R—X}][\text{B}^-]$$

PROBLEM 5-27

Predict the elimination products in the following reactions.

(a) *sec*-butyl bromide + $NaOCH_2CH_3$
(b) 3-bromo-3-ethylpentane + methanol
(c) 2-bromo-3-ethylpentane + $NaOCH_3$
(d) bromocyclohexane + $NaOCH_2CH_3$

5-21 STEREOCHEMISTRY OF THE E2 REACTION

Like the S_N2 reaction, the E2 follows a **concerted mechanism:** Bond breaking and bond formation take place at the same time, and the partial formation of new bonds lowers the energy of the transition state. Concerted mechanisms require specific geometric arrangements so that the orbitals of the bonds being broken can overlap with those being formed and the electrons can flow smoothly from one bond to another. The geometric arrangement required by the S_N2 reaction is a back-side attack; with the E2 reaction, a coplanar arrangement of orbitals is needed.

The E2 elimination requires the partial formation of a new pi bond, with its parallel *p* orbitals, in the transition state. The electrons that once formed a C—H bond must begin to overlap with the orbital that the leaving group is vacating. The formation of this new pi bond implies that these two sp^3 orbitals must be parallel so that pi overlap is possible as the hydrogen and halogen leave and the orbitals rehybridize to the *p* orbitals of the new pi bond.

Figure 5-12 shows two conformations that provide the necessary coplanar alignment of the leaving group, the departing hydrogen, and the two carbon atoms. When the hydrogen and the halogen eclipse each other ($\theta = 0°$), their orbitals are aligned. This is called the **syn-coplanar** conformation. When the hydrogen and the halogen are anti to each other ($\theta = 180°$), their orbitals are once again aligned. This is called the **anti-coplanar** conformation.

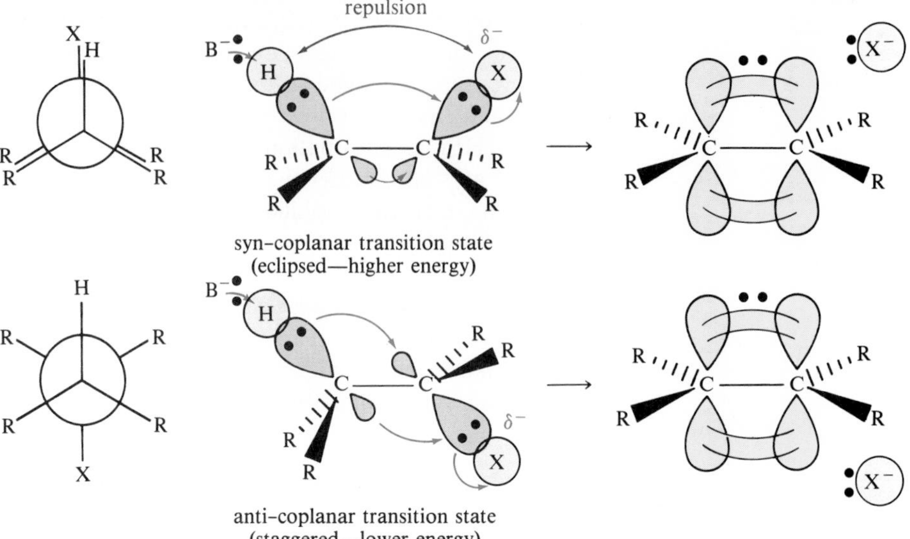

FIGURE 5-12 Concerted transition states of the E2 reaction. The orbitals of the hydrogen atom and the halide must be aligned so that they can begin to form a pi bond in the transition state. Either of these coplanar conformations provides the necessary overlap.

Of these possible conformations, the anti-coplanar arrangement is most commonly seen in the E2 reaction. The transition state for the anti-coplanar arrangement is a staggered conformation along the carbon-carbon bond, with the nucleophile far away from the leaving group. In most cases, this transition state is lower in energy than that for the syn-coplanar elimination.

The transition state for syn-coplanar elimination is an eclipsed conformation. In addition to the higher energy resulting from eclipsing interactions, the transition state suffers from interference between the attacking base and the leaving group. To abstract the proton, the base must approach quite close to the leaving group. In most cases, the leaving group is bulky and negatively charged, and the repulsion between the base and the leaving group raises the energy of the syn-coplanar transition state above that of the anti-coplanar arrangement.

Some molecules are rigidly held in eclipsed (or nearly eclipsed) conformations, with a hydrogen atom and a leaving group in a syn-coplanar arrangement. Such compounds are likely to undergo E2 elimination by a concerted syn-coplanar mechanism. Deuterium labeling (using D, the hydrogen isotope with mass number 2) is used in the following reaction to show which atom is abstracted by the base. Only the hydrogen atom is abstracted, because it is held in a syn-coplanar position with the bromine atom. The deuterium atom cannot easily become anti-coplanar with the bromine atom.

syn-coplanar

Br, H, H, D $\xrightarrow{Na^{+}\ {}^{-}OCH_3}$ H, D + $HOCH_3$

PROBLEM 5-28

When the compound shown below is treated with sodium methoxide, the only elimination product is the trans isomer. None of the cis isomer is observed. Use a careful drawing of the transition state to explain this result.

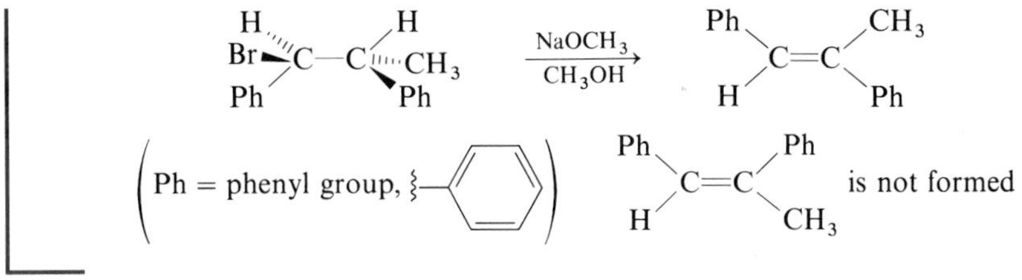

H, Br C—C CH$_3$ Ph, H Ph $\xrightarrow[\text{CH}_3\text{OH}]{\text{NaOCH}_3}$ Ph(H)C=C(CH$_3$)(Ph)

(Ph = phenyl group) Ph(H)C=C(Ph)(CH$_3$) is not formed

5-21A E2 REACTIONS IN CYCLOHEXANE SYSTEMS

Nearly all cyclohexanes are most stable in chair conformations. In the chair, all the carbon-carbon bonds are staggered, and E2 eliminations usually take place through staggered anti-coplanar transition states. Any two adjacent axial bonds are in an anti-coplanar conformation, ideally oriented for the E2 reaction. As drawn, the axial bonds are vertical. On any two adjacent carbon atoms, one has its axial bond pointing up and the other has its axial bond pointing down. These two bonds are trans to each other, and we refer to their geometry as **trans-diaxial.**

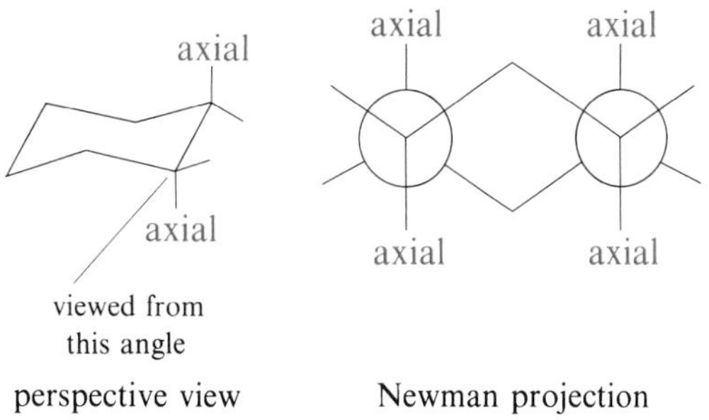

An E2 elimination can take place on this chair conformation only if the proton and the leaving group can get into a trans-diaxial arrangement. Figure 5-13 shows the E2 dehydrohalogenation of bromocyclohexane. The molecule must flip into the chair conformation with the bromine atom axial before elimination can occur.

(You should make models of the structures in the following examples and problems.)

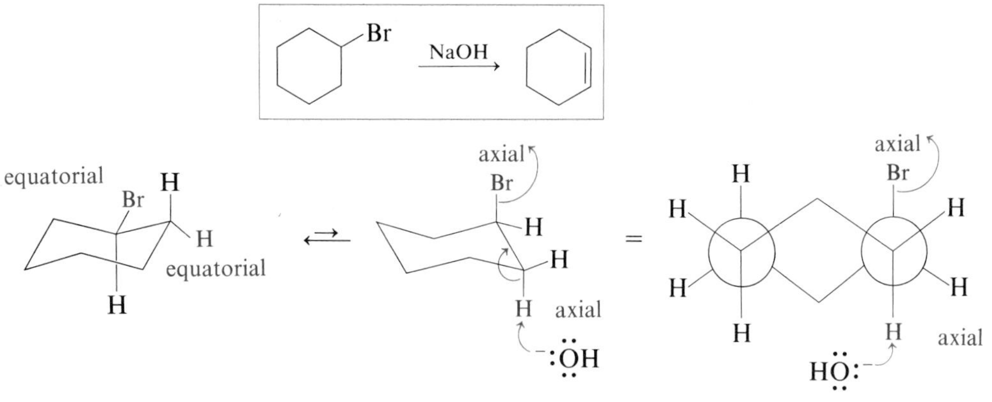

FIGURE 5-13 E2 elimination of bromocyclohexane requires that the proton and the leaving group both be axial.

SOLVED PROBLEM 5-1

Explain why the following deuterated 1-bromo-2-methylcyclohexane undergoes dehydrohalogenation by the E2 mechanism to give only the indicated product. Two other alkenes are not observed.

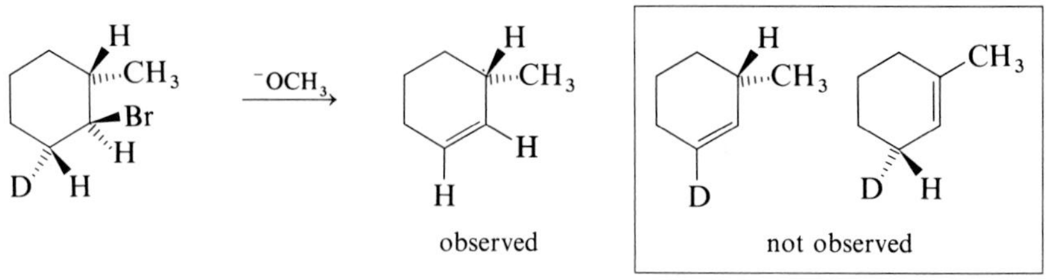

$\xrightarrow{^{-}OCH_3}$

observed

not observed

SOLUTION In an E2 elimination, the hydrogen atom and the leaving group must be in a trans-diaxial relationship. In this compound, only one hydrogen atom—the deuterium—is trans to the bromine atom. When the bromine atom is axial, the adjacent deuterium is also axial, providing a trans-diaxial arrangement.

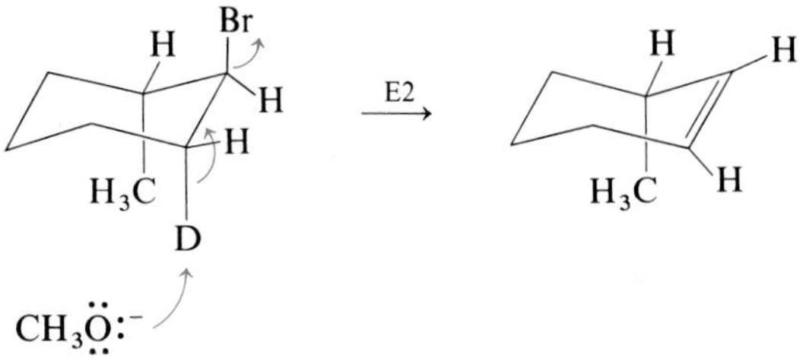

$CH_3\ddot{O}:^-$ $\xrightarrow{E2}$

PROBLEM 5-29

When the following stereoisomer of 2-bromo-1,3-dimethylcyclohexane is treated with sodium methoxide, no E2 reaction is observed. Explain why this compound cannot undergo the E2 reaction in the chair conformation.

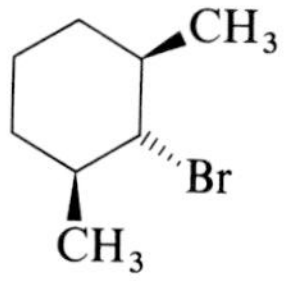

PROBLEM 5-30

Give the expected product(s) of E2 elimination for each of the following reactions. (*Hint:* Use models!)

(a) 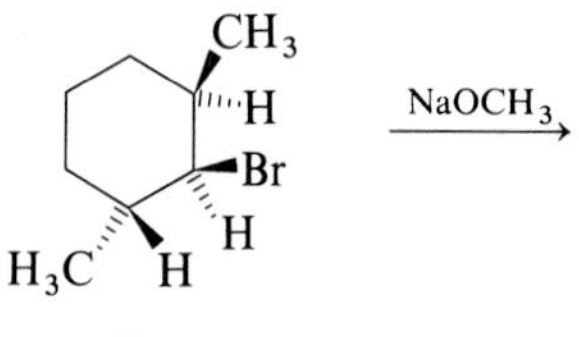 $\xrightarrow{NaOCH_3}$ one product

(b) 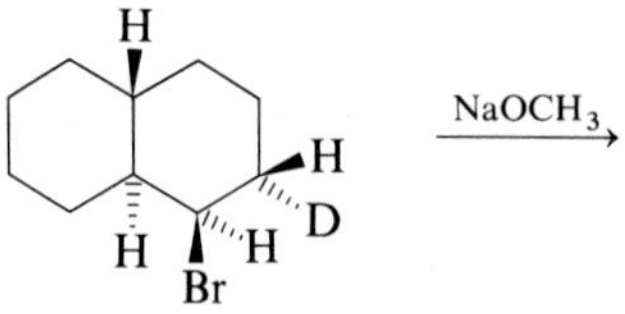 $\xrightarrow{NaOCH_3}$ two products

5-22 POSITIONAL ORIENTATION OF ELIMINATION: THE SAYTZEFF RULE

Many compounds can eliminate in more than one way. For example, the reaction of 2-bromobutane with potassium hydroxide gives two elimination products.

$$CH_3-CH_2-\underset{\displaystyle Br}{\underset{|}{CH}}-CH_3 \xrightarrow{^-:\ddot{O}H}$$

$$\underset{\text{19\%, monosubstituted}}{CH_3-CH_2-CH{=}CH_2} \quad + \quad \underset{\text{81\%, disubstituted}}{CH_3-CH{=}CH-CH_3}$$

The first product has a *monosubstituted* double bond, with one substituent on the doubly bonded carbons. It has the general formula $R-CH{=}CH_2$. The second product has a *disubstituted* double bond, with general formula $R-CH{=}CH-R$ (or $R_2C{=}CH_2$). In most E1 and E2 eliminations where there are two or more possible elimination products, *the product with the most highly substituted double bond will predominate*. This general principle is called the **Saytzeff rule,** and reactions that give the most highly substituted alkene are said to follow **Saytzeff orientation.**

SAYTZEFF RULE: In elimination reactions, the most highly substituted alkene usually predominates.

$$\underset{\text{tetrasubstituted}}{R_2C{=}CR_2} > \underset{\text{trisubstituted}}{R_2C{=}CHR} > \underset{\text{disubstituted}}{RHC{=}CHR \text{ and } R_2C{=}CH_2} > \underset{\text{monosubstituted}}{RHC{=}CH_2}$$

This order of preference is the same as the order of stability of alkenes. We consider the stability of alkenes in more detail in Section 7-7, but for now it is enough just to know that the more highly substituted alkenes are more stable. Later, we will study some unusual reactions where the Saytzeff rule does not apply.

SOLVED PROBLEM 5-2

When 3-iodo-2,2-dimethylbutane is treated with silver nitrate in ethanol, three elimination products are formed. Give their structures and predict which ones will be formed in larger amounts.

SOLUTION Silver nitrate reacts with the alkyl iodide to give silver iodide and a cation.

$$CH_3-\underset{CH_3}{\overset{CH_3}{C}}-CHI-CH_3 + Ag^+ \longrightarrow CH_3-\underset{CH_3}{\overset{CH_3}{C}}-\overset{+}{C}H-CH_3 + AgI\downarrow$$

This secondary cation can lose a proton to give an unrearranged alkene **(A),** or it can rearrange to give a more stable tertiary cation.

Loss of a proton

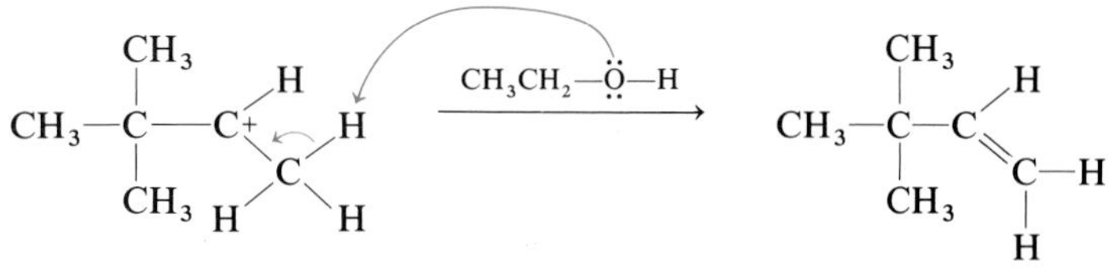

$$CH_3-\underset{CH_3}{\overset{CH_3}{C}}-\overset{+}{C}H-CH_3 \xrightarrow{CH_3CH_2-\ddot{O}-H} CH_3-\underset{CH_3}{\overset{CH_3}{C}}-CH{=}CH_2$$

Product **(A)**

Rearrangement

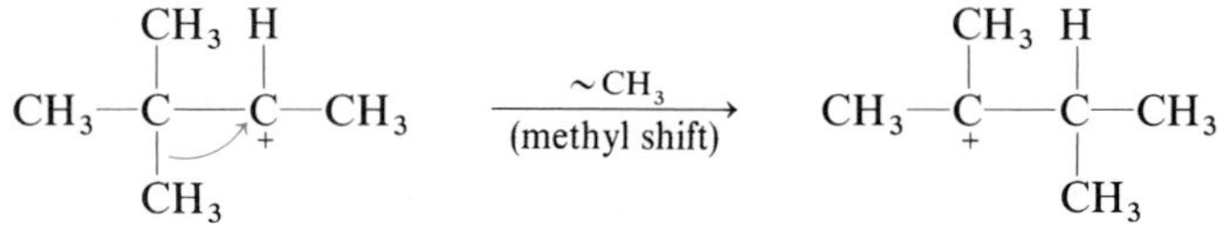

$$\underset{\text{2° carbocation}}{CH_3-\overset{\overset{CH_3}{|}}{\underset{\underset{CH_3}{|}}{C}}-\overset{\overset{H}{|}}{\overset{+}{C}}-CH_3} \xrightarrow[\text{(methyl shift)}]{\sim CH_3} \underset{\text{3° carbocation}}{CH_3-\overset{\overset{CH_3}{|}}{\overset{+}{C}}-\overset{\overset{H}{|}}{\underset{\underset{CH_3}{|}}{C}}-CH_3}$$

The tertiary cation can lose a proton in either of two positions. One of the products **(B)** is a tetrasubstituted alkene, and the other **(C)** is disubstituted.

Formation of a tetrasubstituted alkene

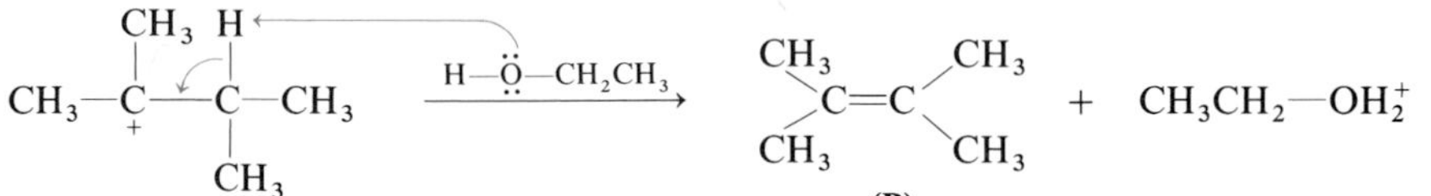

$$CH_3-\overset{\overset{CH_3}{|}}{\overset{+}{C}}-\overset{\overset{H}{|}}{\underset{\underset{CH_3}{|}}{C}}-CH_3 \xrightarrow{H-\ddot{O}-CH_2CH_3} \underset{\textbf{(B)}}{(CH_3)_2C{=}C(CH_3)_2} + CH_3CH_2-OH_2^+$$

Formation of a disubstituted alkene

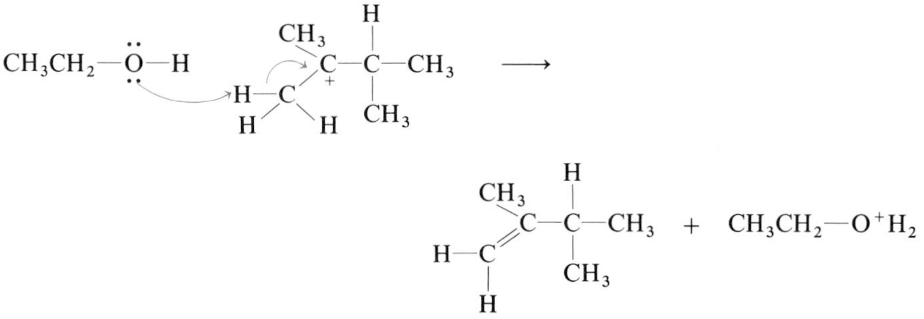

$$CH_3CH_2-\ddot{O}-H \quad H_2C(H)-\overset{+}{C}(CH_3)-\overset{\overset{H}{|}}{\underset{\underset{CH_3}{|}}{C}}-CH_3 \longrightarrow$$

$$\underset{\textbf{(C)}}{H_2C{=}C(CH_3)-\overset{\overset{H}{|}}{\underset{\underset{CH_3}{|}}{C}}-CH_3} + CH_3CH_2-O^+H_2$$

Product **B** will predominate over product **C** because the double bond in **B** is more highly substituted. Whether product **A** is a major product will depend on the specific reaction conditions and whether proton loss or rearrangement occurs faster.

PROBLEM 5-31

Each of the two carbocations in Solved Problem 5-2 can also react with ethanol to give a substitution product. Give the structures of the two substitution products formed in this reaction.

PROBLEM 5-32

The reaction of 2-bromobutane with potassium hydroxide on page 203 can also give substitution. Show the substitution product and give the mechanism of its formation.

PROBLEM 5-33

Give the structure of the elimination products for each of the following reactions, and label the major products. When a rearrangement occurs, show how a more stable intermediate is formed.

(a) 2-bromopentane + NaOH
(b) 1-(bromomethyl)-1-methylcyclopentane heated in methanol
(c) *cis*-1-bromo-2-methylcyclohexane + $AgNO_3$ in ethanol
(d) *cis*-1-bromo-2-methylcyclohexane + NaOEt
(e) neopentyl bromide + $AgNO_3$ in methanol
(f) neopentyl bromide + $NaOCH_3$

Reactivity of the substrate The order of reactivity of alkyl halides toward E2 dehydrohalogenation is found to be:

$$3^\circ > 2^\circ > 1^\circ$$

This order of reactivity reflects the greater stability of highly substituted double bonds. Elimination of a tertiary halide generally gives a more highly substituted alkene than elimination of a secondary halide, which gives a more highly substituted alkene than a primary halide. The stabilities of the alkene products are reflected in the transition states leading to them, giving lower activation energies and higher rates for elimination of alkyl halides that lead to more highly substituted alkenes.

5-23 COMPARISON OF E1 AND E2 ELIMINATION MECHANISMS

Let's summarize the major points we need to remember about the E1 and E2 reactions, concentrating on the factors that help us predict which of these mechanisms will operate under a given set of experimental conditions.

Kinetics The rate of the E1 reaction is proportional to the concentration of the alkyl halide [R—X] but not to the concentration of the nucleophile. It follows a first-order rate equation.

The rate of the E2 reaction is proportional to the concentrations of both the alkyl halide [R—X] and the base [B:⁻]. It follows a second-order rate equation.

$$\text{E1 rate} = k_r[\text{R—X}]$$

$$\text{E2 rate} = k_r[\text{R—X}][\text{B:}^-]$$

Effect of the base The nature of the base is the single most important factor in determining whether an elimination will go by the E1 or E2 mechanism. If a strong base is present, the rate of the bimolecular reaction will be greater than the rate of ionization, and the E2 reaction will predominate (perhaps accompanied by some S_N2).

If there is no strong base present, it is likely that a unimolecular ionization will occur, followed by proton abstraction by a weak base such as the solvent. Under these conditions, the E1 reaction usually predominates (always accompanied by some S_N1).

E1: Base strength is unimportant.

E2: Strong bases are required.

Effect of the substrate For both the E1 and the E2 reactions, the order of reactivity is

$$3^\circ > 2^\circ > 1^\circ$$

In the E1 reaction, the rate-determining step is the formation of a carbocation, and the reactivity order reflects the stability of carbocations. In the E2 reaction, the more highly substituted halides generally form more highly substituted, and therefore more stable alkenes.

E1: $3^\circ > 2^\circ$ (1° usually will not go E1)

E2: $3^\circ > 2^\circ > 1^\circ$

Effect of the solvent The slow step of the E1 reaction involves the formation of two ions. Like the S_N1, the E1 reaction is critically dependent on very polar ionizing solvents such as water and the alcohols.

In the E2 reaction, the transition state spreads out the negative charge of the base over the entire molecule. There is no more need for solvation in the E2 transition state than there is in the reactants. Therefore, the E2 is less sensitive to the solvent, and some reagents are more strongly basic in less polar solvents.

E1: Good ionizing solvent required.

E2: Solvent polarity is not so important.

Stereochemistry The E1 reaction begins with an ionization to give a flat carbocation. No particular geometry is required.

The E2 reaction takes place through a concerted mechanism that requires a coplanar arrangement of the bonds to the atoms being eliminated. The transition state is usually anti-coplanar, although it may be syn-coplanar in rigid systems. In the cyclohexanes, the anti-coplanar requirement means that the proton and the leaving group must have a trans-diaxial relationship on adjacent carbon atoms.

E1: No particular geometry required for the slow step.

E2: Coplanar arrangement (usually anti) required for the transition state.

Rearrangements The E1 reaction involves a carbocation intermediate. This intermediate can rearrange, usually by the shift of a hydride or an alkyl group, to give a more stable carbocation.

The E2 reaction takes place in one step with no intermediates. No rearrangement is possible in the E2 reaction.

E1: Rearrangements are common.

E2: Rearrangements are not possible.

SUMMARY OF ELIMINATION REACTIONS

	E1	*E2*
characteristics		
kinetics	first order, k_r[RX]	second order, k_r[RX][B:$^-$]
rearrangements	common	not possible
stereochemistry	no special geometry	coplanar TS required
orientation	most highly substituted alkene	most highly substituted alkene
promoting factors		
substrate	$3° > 2°$	$3° > 2° > 1°$
leaving group	good one required	good one required
base	weak bases work	strong base required
solvent	good ionizing solvent	wide variety of solvents

5-24 ELIMINATION VERSUS SUBSTITUTION

Now that we have studied a number of ways that alkyl halides react with nucleophiles and bases, the logical question is: How one can predict which of these possible reactions will take place in any given example?

If we mix an alkyl halide with a certain nucleophile in an appropriate solvent, will the S_N2 or the E2 reaction take place? Or will S_N1 or E1 be the predominant reaction? We cannot always answer this question. *Often, we cannot predict the course of the reaction.* In most cases, though, we can eliminate some of the possibilities and make some good predictions. Here are some general guidelines:

1. *The strength of the base or nucleophile determines the order of the reaction.* If a strong nucleophile (or base) is present, it will probably force second-order kinetics: either S_N2 or E2. A strong nucleophile is sufficiently reactive to attack the electrophilic carbon atom or abstract a proton faster than the molecule can undergo the ionization required for the first-order reactions.

If no strong base or nucleophile is present, the fastest reaction will probably be a first-order reaction, either S_N1 or E1. If we add silver salts to the reaction, we can *force* some kind of ionization to occur.

2. *Primary halides usually undergo the S_N2 reaction, occasionally the E2 reaction.* Most primary halides cannot undergo first-order reactions, since primary carbocations are not formed under normal conditions. With good nucleophiles, S_N2 substitution is usually observed. With a very strong base present, E2 elimination may also be observed, although this reaction is not as common as the S_N2 for primary halides.

   If the alkyl group on a primary halide is very bulky, the S_N2 reaction may be slowed enough that we may see more elimination. Also, if the primary alkyl halide can easily rearrange to give a more stable carbocation, the rearranged S_N1 and E1 products may be observed.

3. *Tertiary halides usually undergo the E2 reaction (strong base) or a mixture of E1 and S_N1 (weak base).* Tertiary halides cannot undergo the S_N2 reaction. A strong base forces second-order kinetics, resulting in elimination by the E2 mechanism. In the absence of a strong base, tertiary halides react by first-order processes, usually a mixture of S_N1 and E1. The specific reaction conditions determine the ratio of substitution to elimination, with high temperatures favoring elimination.

The reactions of secondary halides are the most difficult to predict. With a strong base, either the S_N2 or the E2 reaction is possible. With a weak base and a good ionizing solvent, either the S_N1 or the E1 reaction is possible. Mixtures of products are common, with high temperatures favoring elimination.

SOLVED PROBLEM 5-3

Predict the mechanisms and products of the following reactions.

(a) 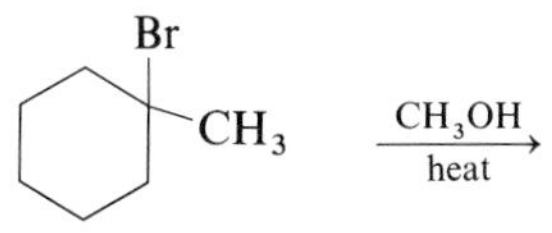 1-bromo-1-methylcyclohexane $\xrightarrow[\text{heat}]{CH_3OH}$

Br, CH_3

1-bromo-1-methylcyclohexane

(b) 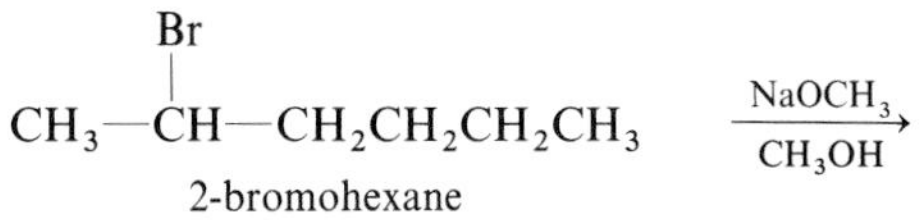 $CH_3-\underset{\text{Br}}{CH}-CH_2CH_2CH_2CH_3 \xrightarrow[CH_3OH]{NaOCH_3}$

2-bromohexane

SOLUTION (a) There is no strong base or nucleophile present, so this reaction must be first order, with an ionization of the alkyl halide as the slow step. Deprotonation of the carbocation gives either of two elimination products, and nucleophilic attack gives a substitution product.

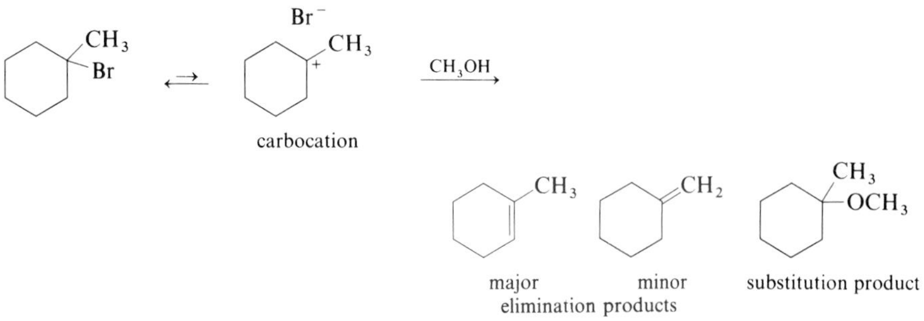

CH_3, Br $\rightleftarrows$ Br^- CH_3 (+) $\xrightarrow{CH_3OH}$

carbocation

CH_3 CH_2 CH_3, OCH_3

major   minor   substitution product

elimination products

SOLUTION (b) This reaction takes place with a strong base, so it is second order. This secondary halide can undergo both S_N2 substitution and E2 elimination. Both products will be formed, with the relative proportions of substitution and elimination depending on the reaction conditions.

$$CH_3-CH=CH-CH_2CH_2CH_3 \qquad CH_2=CH-CH_2CH_2CH_2CH_3$$

major — minor: E2 products

$$CH_3-\underset{}{CH}(OCH_3)-CH_2CH_2CH_2CH_3$$

S_N2 product

PROBLEM 5-34

Predict the products and mechanisms of the following reactions. When more than one product or mechanism is possible, explain which are most likely.

(a) ethyl bromide + sodium ethoxide
(b) *t*-butyl bromide + sodium ethoxide
(c) isopropyl bromide + sodium ethoxide
(d) isobutyl bromide + sodium hydroxide
(e) isobutyl bromide + silver nitrate in ethanol/water
(f) 1-bromo-1-methylcyclopentane heated in methanol
(g) (bromomethyl)cyclopentane + silver nitrate in methanol

PROBLEM 5-35

Potassium iodide reacts with vicinal dibromides to give alkenes by the E2 elimination of two bromine atoms.

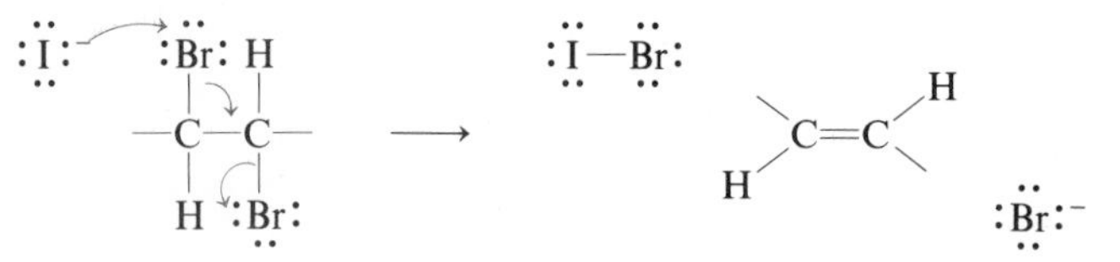

$$:\ddot{I}:^- + \text{Br}-\underset{H}{C}-\underset{Br}{C}-H \longrightarrow :\ddot{I}-\ddot{Br}: + HC=CH + :\ddot{Br}:^-$$

The following compounds show different rates of debromination. One reacts quite fast, and the other seems not to react at all. Explain this surprising difference in rates.

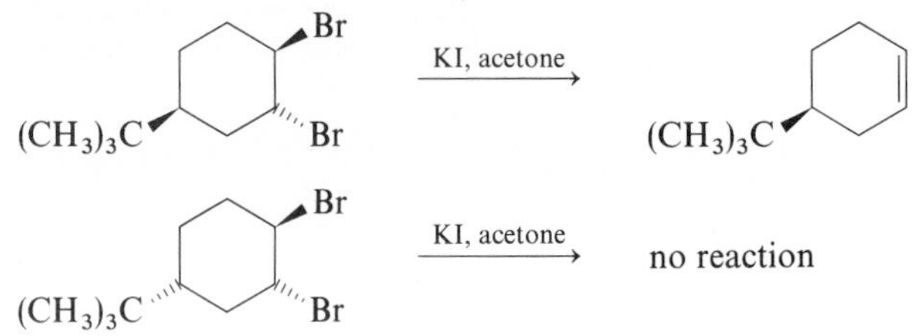

$(CH_3)_3C$-substituted 1,2-dibromocyclohexane (trans) $\xrightarrow{\text{KI, acetone}}$ $(CH_3)_3C$-cyclohexene

$(CH_3)_3C$-substituted 1,2-dibromocyclohexane (isomer) $\xrightarrow{\text{KI, acetone}}$ no reaction

5-25

REDUCTION OF ALKYL HALIDES

Most alkyl halides react with lithium aluminum hydride, $LiAlH_4$, to give alkanes. This reaction is a *reduction,* because it replaces the electronegative halogen atom with a hydrogen atom.

$$R-X \xrightarrow{LiAlH_4} R-H$$

Example

$$\underset{n\text{-decyl bromide}}{C_9H_{19}-CH_2-Br} \xrightarrow{LiAlH_4} \underset{n\text{-decane (70\%)}}{C_9H_{19}-CH_3}$$

This reduction may be viewed as a displacement of the halide ion by a hydride ion ($H:^-$) donated by the tetrahydroaluminate (AlH_4^-) ion. It is not clear that the reaction actually goes by this S_N2 mechanism, however.

$$H_3Al^- H + H{-}C(H)(C_9H_{19}){-}Br: \longrightarrow H{-}AlH_2 + H{-}C(H)(C_9H_{19}){-}H + :Br:^-$$

Under the proper conditions, electropositive elements such as zinc and magnesium also reduce alkyl halides. The solvents used in such reductions are usually alcohols or acetic acid. These reactions are called *dissolving metal* reductions because the metal gradually "dissolves" in the solvent (actually, it reacts with it). The combination of zinc and acetic acid is the most common method for reducing alkyl halides to alkanes. The reaction with magnesium is more commonly used to generate the useful organomagnesium reagent (called a Grignard reagent), which we discuss in Chapter 9.

$$R{-}X \xrightarrow[\text{or Mg, alcohol}]{\text{Zn, } CH_3COOH} R{-}H$$

Examples

$$\underset{n\text{-hexadecyl iodide}}{C_{16}H_{33}I} \xrightarrow{\text{Zn, } CH_3COOH} \underset{\substack{n\text{-hexadecane} \\ (85\%)}}{C_{16}H_{34}}$$

$$\underset{\text{chlorocyclohexane}}{C_6H_{11}Cl} \xrightarrow{\text{Mg, } (CH_3)_2CH{-}OH} \underset{\substack{\text{cyclohexane} \\ (83\%)}}{C_6H_{12}}$$

SUMMARY OF THE REACTIONS OF ALKYL HALIDES

1. *Nucleophilic substitutions* (Section 5-10)
   a. *Alcohol formation*

$$R{-}X + {}^-:\ddot{O}H \longrightarrow R{-}OH + :\ddot{X}:^-$$

Example

$$\underset{\text{ethyl bromide}}{CH_3{-}CH_2{-}Br} + NaOH \longrightarrow \underset{\text{ethyl alcohol}}{CH_3{-}CH_2{-}OH} + NaBr$$

   b. *Halide exchange*

$$R{-}X + :\ddot{I}:^- \longrightarrow R{-}I + :\ddot{X}:^-$$

$$R{-}Cl + Hg_2F_2 \longrightarrow R{-}F + Hg_2Cl_2$$

Example

$$\underset{\text{allyl chloride}}{H_2C{=}CH{-}CH_2Cl} + NaI \longrightarrow \underset{\text{allyl iodide}}{H_2C{=}CH{-}CH_2I} + NaCl$$

   c. *Williamson ether synthesis*

$$R{-}X + R'\ddot{O}:^- \longrightarrow R{-}\ddot{O}{-}R' + X:^- \quad \text{ether synthesis}$$

$$R{-}X + R'\ddot{S}:^- \longrightarrow R{-}\ddot{S}{-}R' + X:^- \quad \text{thioether synthesis}$$

Example

$$CH_3{-}I + CH_3{-}CH_2{-}O^-Na^+ \longrightarrow CH_3{-}O{-}CH_2{-}CH_3 + Na^+I^-$$

methyl iodide — sodium ethoxide — methyl ethyl ether

d. *Amine synthesis*

$$R{-}X + \text{excess } {:}NH_3 \longrightarrow R{-}NH_3^+X^- \xrightarrow{:NH_3} R{-}\ddot{N}H_2 + NH_4^+\,{}^-X$$

amine

Example

$$CH_3CH_2CH_2{-}Br + NH_3 \longrightarrow CH_3CH_2CH_2{-}NH_3^+\,{}^-Br \xrightarrow{NH_3}$$

n-propyl bromide

$$CH_3CH_2CH_2{-}NH_2 + NH_4^+\,{}^-Br$$

n-propylamine

e. *Nitrile synthesis*

$$R{-}X + {}^-{:}C{\equiv}N{:} \longrightarrow R{-}C{\equiv}N{:} + X^-$$

cyanide — nitrile

Example

$$(CH_3)_2CH{-}CH_2{-}CH_2{-}Cl + NaCN \longrightarrow (CH_3)_2CH{-}CH_2{-}CH_2{-}CN + NaCl$$

1-chloro-3-methylbutane — 4-methylpentanonitrile

f. *Alkyne synthesis*

$$R{-}C{\equiv}C{:}^- + R'{-}X \longrightarrow R{-}C{\equiv}C{-}R' + X^-$$

acetylide ion — alkyne

Examples

$$CH_3{-}C{\equiv}C{-}H + NaNH_2 \longrightarrow CH_3{-}C{\equiv}C{:}^-Na^+ + NH_3$$

propyne — sodium amide — sodium propynide

$$CH_3{-}C{\equiv}C{:}^-\,{}^+Na + CH_3{-}CH_2{-}I \longrightarrow CH_3{-}C{\equiv}C{-}CH_2{-}CH_3 + NaI$$

propynide ion — ethyl iodide — 2-pentyne

2. *Eliminations*

a. *Dehydrohalogenation* (Sections 5-20 and 7-9A)

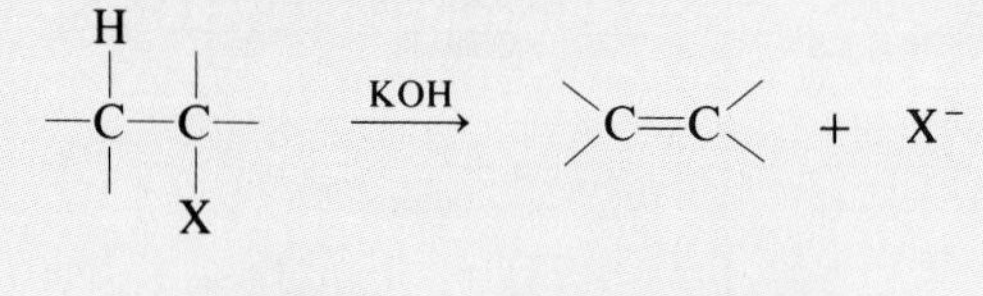

$$-\overset{H}{\underset{|}{\overset{|}{C}}}-\underset{X}{\underset{|}{\overset{|}{C}}}- \xrightarrow{KOH} \;\rangle C{=}C\langle\; + X^-$$

Examples

chlorocyclohexane $\xrightarrow{KOH}$ cyclohexene

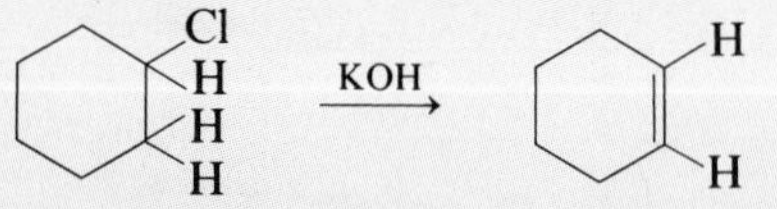

$$CH_3{-}CH_2{-}CH_2{-}CH_2{-}CHBr_2 \xrightarrow{KOH,\ heat} CH_3{-}CH_2{-}CH_2{-}C{\equiv}C{-}H$$

1,1-dibromopentane — 1-pentyne

b. *Dehalogenation* (Sections 5-24 and 7-9B)

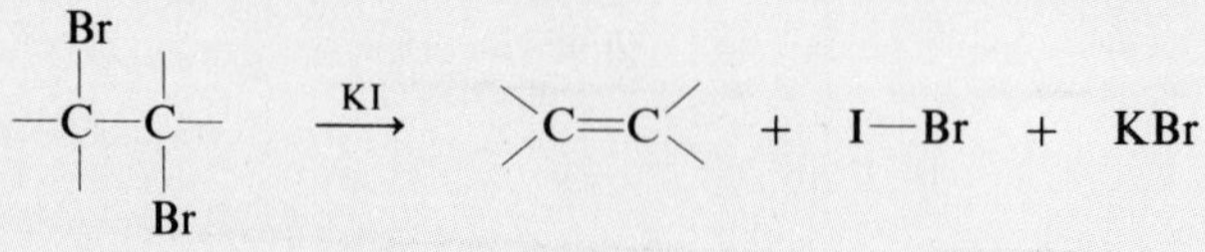

$$-\overset{Br}{\underset{|}{\overset{|}{C}}}-\underset{Br}{\underset{|}{\overset{|}{C}}}- \xrightarrow{KI} \;\rangle C{=}C\langle\; + I{-}Br + KBr$$

Example

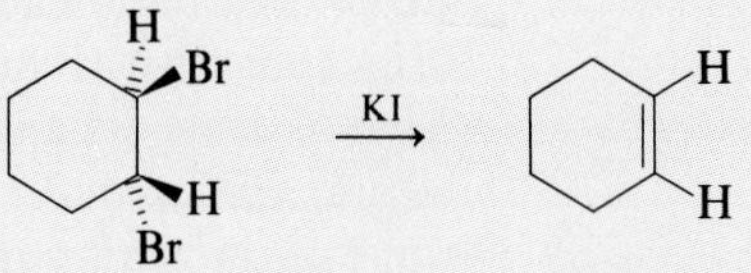

$$\text{trans-1,2-dibromocyclohexane} \xrightarrow{\text{KI}} \text{cyclohexene}$$

trans-1,2-dibromocyclohexane　　　　cyclohexene

3. *Reduction* (Section 5-25)

$$R{-}X \xrightarrow{\text{reducing agent}} R{-}H$$

Example

$$C_9H_{19}{-}CH_2{-}Br \xrightarrow{LiAlH_4} C_9H_{19}{-}CH_3$$

n-decyl bromide　　　　*n*-decane

$$C_{16}H_{33}I \xrightarrow{Zn,\ CH_3COOH} C_{16}H_{34}$$

n-hexadecyl iodide　　　　*n*-hexadecane

4. *Formation of organometallic reagents* (Section 9-8)
   a. *Grignard reagents*

$$R{-}X + Mg \xrightarrow{CH_3CH_2{-}O{-}CH_2CH_3} R{-}Mg{-}X$$

(X = Cl, Br, or I)　　　　organomagnesium halide (Grignard reagent)

Example

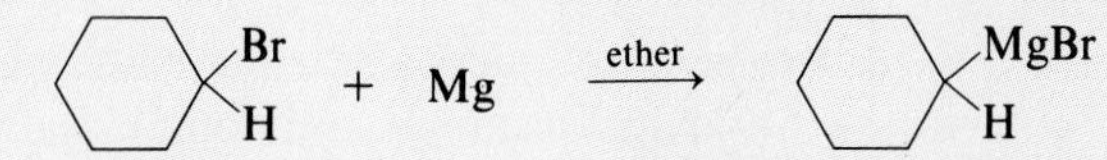

$$\text{bromocyclohexane} + Mg \xrightarrow{\text{ether}} \text{cyclohexylmagnesium bromide}$$

bromocyclohexane　　　　cyclohexylmagnesium bromide

   b. *Organolithium reagents*

$$R{-}X + 2\,Li \longrightarrow R{-}Li + Li^+X^-$$

(X = Cl, Br, or I)　　　　organolithium

Example

$$CH_3CH_2CH_2CH_2{-}Br + 2\,Li \xrightarrow{\text{hexane}} CH_3CH_2CH_2CH_2{-}Li + LiBr$$

n-butyl bromide　　　　*n*-butyllithium

5. *Coupling: The Corey-House reaction* (Section 9-10)

$$2\,R{-}Li + CuI \longrightarrow R_2CuLi + LiI$$

$$R_2CuLi + R'{-}X \longrightarrow R{-}R' + R{-}Cu + LiX$$

Example

$$2\,CH_3I \xrightarrow{4\,Li} 2\,CH_3Li + 2\,LiI \xrightarrow{CuI} (CH_3)_2CuLi$$

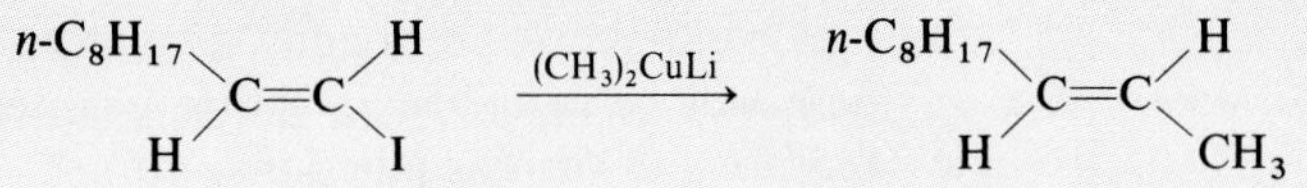

$$(n\text{-}C_8H_{17})HC{=}CHI \xrightarrow{(CH_3)_2CuLi} (n\text{-}C_8H_{17})HC{=}CH(CH_3)$$

PROBLEM SOLVING: ORGANIC SYNTHESIS

Alkyl halides are readily made from other compounds, and the halogen atom is easily converted to other functional groups. This flexibility makes alkyl halides useful as reagents and intermediates for organic synthesis. **Organic synthesis** is the preparation of desired compounds from readily available materials. Synthesis is one of the major areas of organic chemistry, and nearly every chapter of this book involves organic synthesis in some way. A synthesis may be a simple one-step reaction, or it may involve many steps and incorporate a subtle strategy for assembling the correct carbon skeleton with all the functional groups in the right positions.

Many of the problems in this book are synthesis problems. In some synthesis problems, you are asked to show how to convert a given starting material to the desired product. There are obvious one-step answers to some of these problems, while others may require several steps and there may be many correct answers. In solving multistep synthetic problems, it is often helpful to analyze the problem backwards: that is, to begin with the desired product (called the *target compound*) and see how it might be mentally changed or broken down to give the starting materials. This backwards approach to synthesis is called a **retrosynthetic analysis.**

Some problems allow you to begin with any compounds that meet a certain restriction; for example, you might be allowed to use any alcohols containing no more than four carbon atoms. A retrosynthetic analysis can be used to break the target compound down into fragments no larger than four carbon atoms; then those fragments could be formed from the appropriate alcohols by functional group chemistry.

The following suggestions may be helpful for solving synthesis problems.

1. Do not guess a starting material and try every possible reaction to convert it to the target compound. Rather, begin with the target compound and use a retrosynthetic analysis to simplify it.
2. You can use simple equations, with reagents written above and below the arrows, to show the reactions. The equations do not have to be balanced, but they should include all the reagents and conditions that are important to the success of the reaction.

$$\text{A} \xrightarrow{\text{Br}_2,\ \text{light}} \text{B} \xrightarrow[\text{heat}]{\text{NaOH, alcohol}} \text{C} \xrightarrow{\text{H}^+,\ \text{H}_2\text{O}} \text{D}$$

3. Focus your attention on the functional groups, since that is generally where reactions occur. Be careful not to use any reagents that are known to react with a functional group that you do not intend to modify.

In solving multistep synthesis problems, you will rarely be able to "see" the solution immediately. These problems are best approached systematically, working backwards and considering alternative routes for the synthesis. To illustrate a systematic approach that can guide you in solving synthesis problems, we will work out the synthesis of an ether from alkanes. By using alkanes as starting materials, we can illustrate reactions that were recently covered. Alkanes are rarely used as starting materials in actual practice, however, since functionalized compounds are readily available. The problem-solving method described here will be extended in future chapters to multistep syntheses based on the reactions of various functional groups.

To be systematic about retrosynthetic analysis requires beginning with an examination of the structure of the product. We will consider the synthesis of ethoxycyclohexane from alkanes containing up to six carbon atoms.

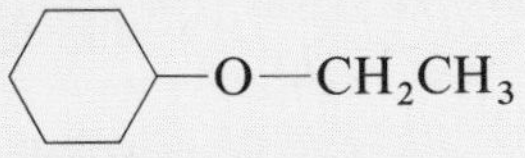

$$\text{C}_6\text{H}_{11}\text{—O—CH}_2\text{CH}_3$$

ethoxycyclohexane

1. **Review the functional groups and carbon skeleton of the target compound.** The target compound is an ether. The two alkyl groups in the ether are a six-carbon cyclohexane ring and a two-carbon ethyl group.
2. **Review the functional groups and carbon skeletons of the starting materials (if specified), and see how their skeletons might fit together in the target compound.**
The synthesis is to begin with alkanes containing six carbon atoms or fewer, so every functional group needed in the synthesis must be introduced into an alkane. Most likely, we will start with cyclohexane to give the six-carbon cyclohexyl group and ethane to give the two-carbon ethyl group in the product.
3. **Compare methods for synthesizing the functional groups in the target compound and select the reactions that are most likely to give the correct product. This step may require writing several possible reactions and evaluating them.**
Ethers can be synthesized by nucleophilic reactions between alkyl halides and alkoxides (Table 5-4). The target compound might be formed by S_N2 attack of an alkoxide ion on an alkyl halide in either of two ways:

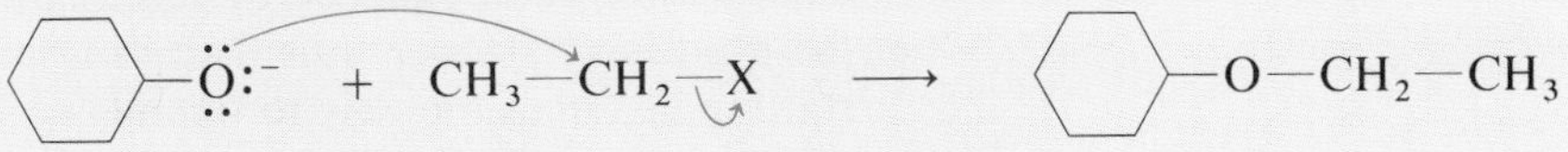

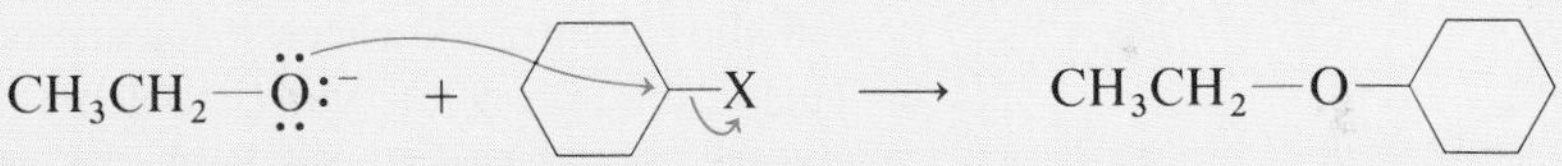

$$\text{C}_6\text{H}_{11}\text{—}\ddot{\underset{..}{\text{O}}}\text{:}^- + \text{CH}_3\text{—CH}_2\text{—X} \longrightarrow \text{C}_6\text{H}_{11}\text{—O—CH}_2\text{—CH}_3$$

$$\text{CH}_3\text{CH}_2\text{—}\ddot{\underset{..}{\text{O}}}\text{:}^- + \text{C}_6\text{H}_{11}\text{—X} \longrightarrow \text{CH}_3\text{CH}_2\text{—O—C}_6\text{H}_{11}$$

The first reaction is better because the S_N2 attack is on a primary alkyl halide. The second reaction requires attack on a secondary halide, which is more prone to side reactions such as E2 elimination. Assuming we can make the necessary reactants, we will choose the first reaction.
4. **Working backwards through as many steps as necessary, compare methods for synthesizing the reactants needed for the final step. This process may require writing several possible reaction sequences and evaluating them, keeping in mind the specified starting materials.**
Two reactants are needed for the selected final reaction: An ethyl halide and an alkoxide ion. The ethyl halide could be ethyl chloride, which can be made by direct chlorination of ethane because all the hydrogen atoms in ethane are equivalent. Abstraction of any hydrogen atom leads to ethyl chloride.

$$\text{CH}_3\text{—CH}_3 + \text{Cl}_2 \xrightarrow{\text{heat or light}} \text{CH}_3\text{—CH}_2\text{—Cl}$$

Alkoxide ions are commonly formed by the reaction of an alcohol with sodium metal.

$$\text{R—O—H} + \text{Na} \longrightarrow \text{Na}^{+\,-}\text{O—R} + \tfrac{1}{2}\text{H}_2\uparrow$$

The alkoxide needed in the final step is formed by adding sodium to cyclohexanol.

$$\text{cyclohexanol } (C_6H_{11}{-}O{-}H) + Na \longrightarrow C_6H_{11}{-}O^- \; Na^+ \text{ (alkoxide ion)} + \tfrac{1}{2}H_2\uparrow$$

cyclohexanol — alkoxide ion

To choose a route to cyclohexanol, methods for alcohol synthesis should be reviewed. So far, you know that alcohols can be made by S_N2 displacement of alkyl halides by hydroxide ion. For the synthesis of cyclohexanol, the reaction is

$$C_6H_{11}{-}X \xrightarrow{^-OH} C_6H_{11}{-}OH$$

a cyclohexyl halide — cyclohexanol

Chlorocyclohexane can be made efficiently by free-radical halogenation of cyclohexane because all the hydrogen atoms in cyclohexane are equivalent. Abstraction of any hydrogen gives chlorocyclohexane.

$$C_6H_{12} + Cl_2 \xrightarrow{\text{heat or light}} C_6H_{11}{-}Cl$$

cyclohexane — chlorocyclohexane

The S_N2 attack by hydroxide ion on chlorocyclohexane (a secondary alkyl halide) will be accompanied by some elimination. This step comes early in the synthesis, however, when we can accept a smaller yield because the starting materials are easy to obtain.

5. **Summarize the complete synthesis in the forward direction, including all steps and all reagents, and check it for errors and omissions.**
This summary is left to you (Problem 5-36) as a review of both the chemistry involved in the synthesis and the method used to develop multistep syntheses.

PROBLEM 5-36

Summarize the synthesis of ethoxycyclohexane from cyclohexane and ethane. This summary should be in the synthetic (forward) direction, showing each step and all reagents.

Problem 5-37 requires devising several multistep syntheses. As practice in working such problems, we suggest that you proceed in order through each of the five steps outlined above.

PROBLEM 5-37

Show how you would synthesize each of the following compounds, starting with any alkanes or cycloalkanes that contain no more than six carbon atoms. In using free-radical halogenation, be careful to use reactions that will give good yields of the correct products.

(a) $CH_2{=}CH{-}CH_2{-}F$ (b) $(CH_3)_2C{=}C(CH_3){-}CH_2{-}C{\equiv}N$ (c) 3-bromocyclohexene (H, Br on C3) (d) dicyclopentyl ether ($C_5H_9{-}O{-}C_5H_9$)

acid A species that can donate a proton.

acidity (acid strength): The thermodynamic reactivity of an acid.

alkyl halide (haloalkane) A derivative of an alkane in which one (or more) of the hydrogen atoms has been replaced by a halogen. (p. 164)

$$CH_3{-}CH_2{-}I$$

ethyl iodide, an alkyl halide

allylic The saturated position adjacent to a carbon-carbon double bond. (p. 172)

allylic halogenation: Substitution of a halogen for a hydrogen at the allylic position.

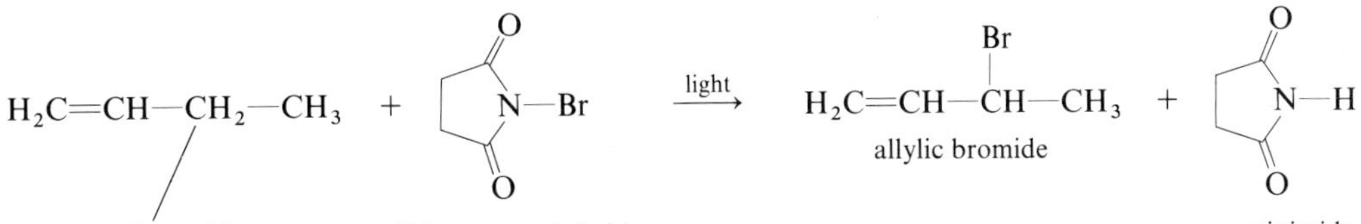

$$H_2C{=}CH{-}CH_2{-}CH_3 \;+\; \text{NBS} \xrightarrow{\text{light}} H_2C{=}CH{-}CH(Br){-}CH_3 \;+\; \text{succinimide}$$

allylic position   N-bromosuccinimide   allylic bromide   succinimide

anti-coplanar Having a dihedral angle of 180°. (p. 199)

syn-coplanar Having a dihedral angle of 0°. (p. 199)

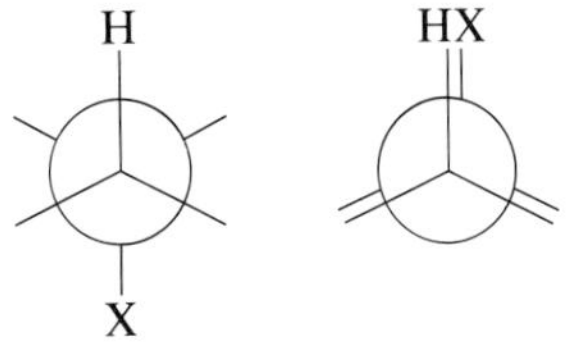

H, X   HX

anti-coplanar   syn-coplanar

aryl halide An aromatic compound (benzene derivative) in which one or more of the aryl hydrogen atoms has been replaced by a halogen. (p. 164)

Cl

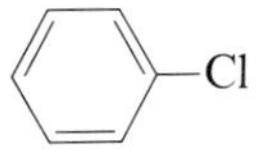

chlorobenzene, an aryl halide

base An electron-rich species that can abstract a proton. (p. 182)

basicity (base strength): The thermodynamic reactivity of a base.

concerted reaction A reaction in which the breaking of bonds and the formation of new bonds occur at the same time (in one step). (pp. 177 and 199)

dehydrohalogenation An elimination in which the two atoms lost are a hydrogen atom and a halogen atom. (p. 197)

electrophile (Lewis acid) A species that can accept an electron pair from a nucleophile, forming a bond. (p. 177)

electrophilicity (electrophile strength): The kinetic reactivity of an electrophile.

elimination A reaction that involves the loss of two atoms or groups from the substrate, usually resulting in the formation of a pi bond. (pp. 175 and 196)

E2 reaction (elimination, bimolecular): A concerted elimination involving a transition state where the base is abstracting a proton at the same time that the leaving group is leaving. The anti-coplanar transition state is generally preferred.

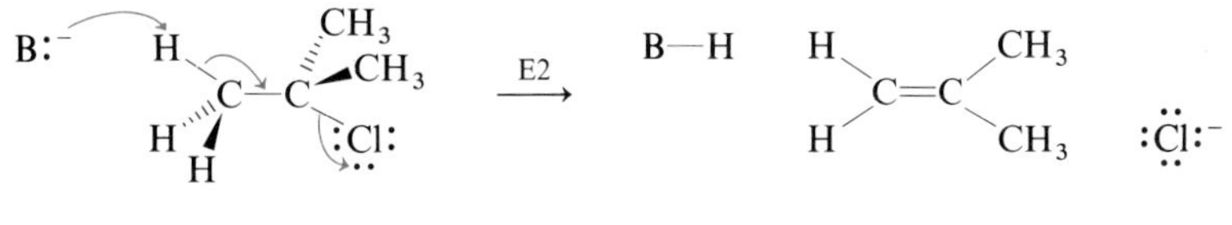

$$B:^- \; + \; H{-}CH_2{-}C(CH_3)_2{-}Cl \xrightarrow{\text{E2}} B{-}H \; + \; H_2C{=}C(CH_3)_2 \; + \; :\!Cl\!:^-$$

E1 reaction (elimination, unimolecular): A multistep elimination where the leaving group is lost in a slow ionization step, then a proton is lost in a second step. Saytzeff orientation is generally preferred.

$$\mathrm{H_2(H)C{-}C(CH_3)_2{-}Cl} \rightleftarrows \mathrm{B{:}^- \; H_2(H)C{-}C^+(CH_3)_2 \;\; {:}Cl{:}^-} \rightleftarrows \mathrm{B{-}H \;\; H_2C{=}C(CH_3)_2 \;\; {:}Cl{:}^-}$$

Freons A generic name for a group of chlorofluorocarbons used as refrigerants, propellants, and solvents. Freon-12 is CF_2Cl_2, and Freon-22 is $CHClF_2$. (p. 168)

geminal dihalide A dihalide with both halogens on the same carbon atom. (p. 167)

$$CH_3{-}CH_2{-}CBr_2{-}CH_3$$

a geminal dibromide

halogen exchange reaction A substitution where one halogen atom replaces another; commonly used to form fluorides and iodides. (p. 179)

hydride shift Movement of a hydrogen atom with a pair of electrons from one atom (usually carbon) to another. Hydride shifts are examples of rearrangements that convert carbocations into more stable carbocations. (p. 192)

inversion of configuration Formation of a product with the opposite configuration as the reactant. In a nucleophilic substitution, inversion of configuration occurs when the nucleophile assumes a stereochemical position in the product opposite the position of the leaving group in the reactant, usually as a result of **back-side attack.** (p. 178)

leaving group The atom or group of atoms that departs during a substitution or elimination reaction. The leaving group can be charged or uncharged, but it departs with the pair of electrons that originally bonded the group to the remainder of the molecule. (p. 177)

methyl shift Rearrangement of a methyl group with a pair of electrons from one atom (usually carbon) to another. A methyl shift (or any alkyl shift) in a carbocation generally results in a more stable carbocation. (p. 193)

nucleophile (Lewis base) An electron-rich species that can donate a pair of electrons to form a bond. (p. 182)

nucleophilicity (nucleophile strength): The kinetic reactivity of a nucleophile.

nucleophilic substitution A reaction where a nucleophile replaces another group or atom (the leaving group) group in a molecule. (p. 175)

polarizable Having electrons that are easily displaced toward a positive charge. Polarizable atoms can begin to form a bond at a relatively long distance. (p. 183)

primary halide, secondary halide, tertiary halide These terms specify the substitution of the halogen-bearing carbon atom (sometimes called the **head** *carbon*). If the head carbon is bonded to one other carbon, it is **primary.** If it is bonded to two carbons, it is **secondary,** and if bonded to three carbons, it is **tertiary.** (p. 166)

$CH_3{-}CH_2{-}Br$ — a primary halide (1°)

$CH_3{-}CH(CH_3){-}Br$ — a secondary halide (2°)

$CH_3{-}C(CH_3)_2{-}Br$ — a tertiary halide (3°)

reagent The compound that serves as the attacking species in a reaction.

rearrangement A reaction involving a change in the bonding sequence within a molecule. Rearrangements are common in reactions such as the S_N1 and E1 which involve carbocation intermediates. (p. 192)

retention of configuration Formation of a product with the same configuration as the reactant. In a nucleophilic substitution, retention of configuration occurs when the nucleophile assumes the same stereochemical position in the product as the leaving group occupied in the reactant. (p. 191)

retrosynthetic analysis A method of working backwards to solve multistep synthetic problems. (p. 212)

Saytzeff rule An elimination usually gives the most highly substituted alkene product. The Saytzeff rule does not always apply, but when it does, the reaction is said to give Saytzeff orientation. (p. 203)

solvolysis A nucleophilic substitution where the solvent serves as the attacking reagent. "Solvolysis" literally means "cleavage by the solvent." (p. 187)

$$(CH_3)_3C\text{—}Br \xrightarrow{CH_3OH,\ heat} (CH_3)_3C\text{—}OCH_3 \ + \ (CH_3)_2C\text{=}CH_2 \ + \ HBr$$

steric hindrance or **steric strain** Hindrance because of interference by bulky groups when they approach a position where their electron clouds begin to repel each other. (p. 180)

substitution (displacement) A reaction in which an attacking species (nucleophile, electrophile, or free radical) replaces another group of the same type. (p. 175)

> **S_N2 reaction** (substitution, nucleophilic, bimolecular): The concerted displacement of one nucleophile by another on an sp^3 hybrid carbon atom. (p. 176)
>
> **S_N1 reaction** (substitution, nucleophilic, unimolecular): A two-step interchange of nucleophiles, with bond breaking preceding bond formation. The first step is an ionization to form a carbocation. The second step is the reaction of the carbocation with a nucleophile. (p. 187)

substrate The compound that is attacked by the reagent. (p. 194)

trans-diaxial An anti and coplanar arrangement allowing E2 elimination of two adjacent substituents on a cyclohexane ring. The substituents must be trans to each other, and both must be in axial positions on the ring. (p. 201)

vicinal dihalide A dihalide with the halogens on adjacent carbon atoms. (p. 167)

$$CH_3\text{—}CHBr\text{—}CHBr\text{—}CH_3$$

a vicinal dibromide

vinyl halide A derivative of an alkene in which one (or more) of the hydrogen atoms on the doubly bonded carbon atoms has been replaced by a halogen. (p. 164)

$$CH_3\text{—}CH\text{=}\underset{}{\overset{\overset{Br}{|}}{C}}\text{—}CH_3$$

a vinyl bromide

ESSENTIAL PROBLEM-SOLVING SKILLS IN CHAPTER 5

1. Correctly name alkyl halides and identify them as 1°, 2°, or 3°.
2. Predict the products of S_N1, S_N2, E1, and E2 reactions, including stereochemistry.
3. Draw the mechanisms and energy profiles of S_N1, S_N2, E1, and E2 reactions.
4. Predict and explain the rearrangement of cations in first-order reactions.
5. Predict which substitutions or eliminations will be faster, based on differences in substrate, base/nucleophile, leaving group, or solvent.
6. Predict whether a reaction will be first order or second order.
7. When possible, predict predominance of substitution or elimination.
8. Use the Saytzeff rule to predict major and minor elimination products.
9. Use retrosynthetic analysis to solve multistep synthesis problems with alkyl halides as reagents, intermediates, or products.

STUDY PROBLEMS

5-38. Define and give an example of each of the following terms.

(a) nucleophile **(b)** electrophile **(c)** leaving group **(d)** substitution
(e) S_N2 reaction **(f)** S_N1 reaction **(g)** solvolysis **(h)** elimination
(i) E2 reaction **(j)** E1 reaction **(k)** rearrangement **(l)** base
(m) steric hindrance **(n)** alkyl halide **(o)** aryl halide **(p)** vinyl halide
(q) allylic halide **(r)** primary halide **(s)** secondary halide **(t)** tertiary halide

5-39. Give the structures of the following compounds.

(a) isopropyl chloride **(b)** isobutyl bromide **(c)** 1,2-dibromo-3-methylpentane
(d) 2,2,2-trichloroethanol **(e)** *trans*-1,4-diiodocyclohexane

5-40. Give a systematic (IUPAC) name for each of the following compounds.

(a) 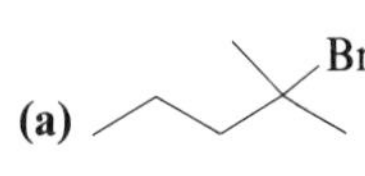

(b) 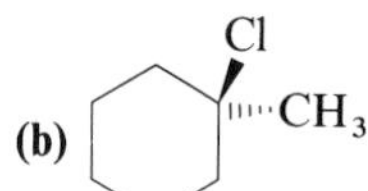

(c) 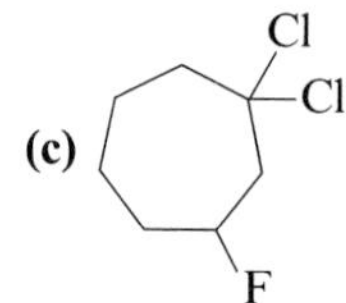

(d) 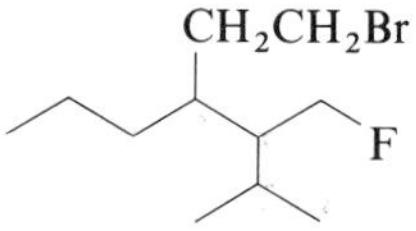 CH_2CH_2Br F

(e) 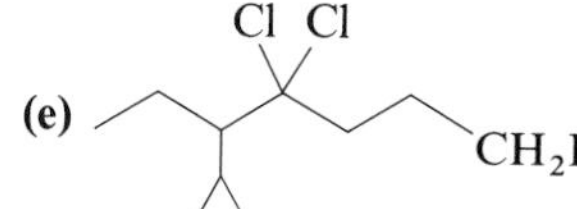

(f) 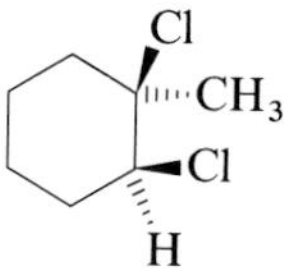 Cl CH_3 Cl H

5-41. Predict the compound in each pair that will undergo the S_N2 reaction faster.

(a)  Cl and Cl **(b)**  Cl and I

(c) 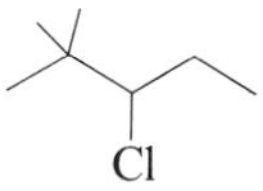 Cl and Cl **(d)** 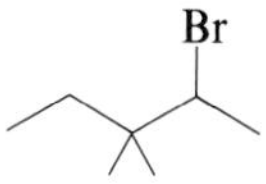 Br and Br

(e) 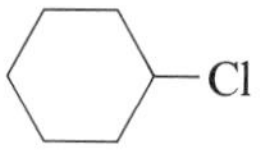 Cl and CH_2Cl

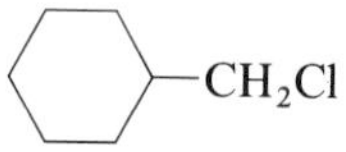

5-42. Predict the compound in each pair that will undergo solvolysis (in aqueous ethanol) more rapidly.

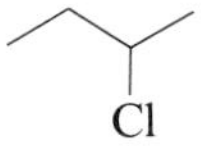

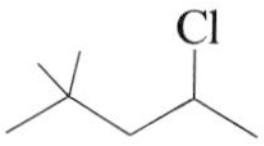

(a) $(CH_3CH_2)_2CH—Cl$ and $(CH_3)_3C—Cl$ **(b)** 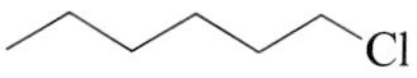 Cl and Cl

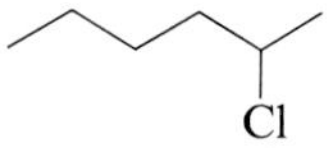

(c) 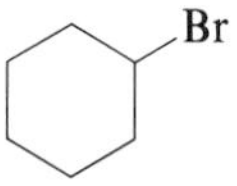 Br and CH_2Br **(d)** 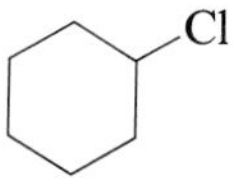 Cl and I

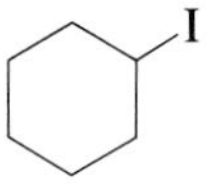

(e) 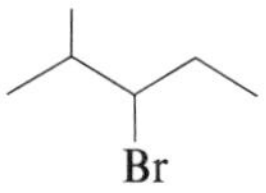 Br and Br **(f)** 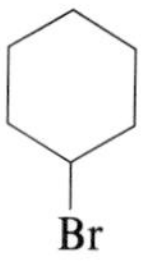 Br and Br

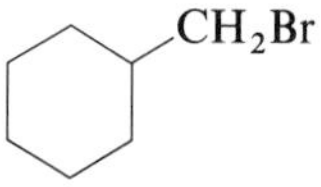

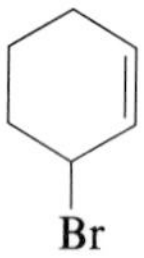

5-43. Show how each of the following compounds might be synthesized by the S_N2 displacement of an alkyl halide.

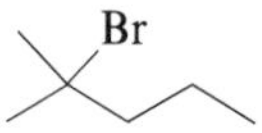

(a) 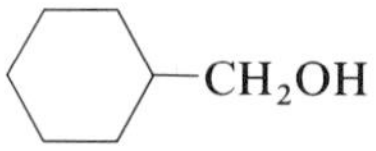 CH_2OH **(b)** 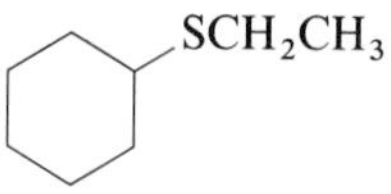 SCH_2CH_3 **(c)** $(CH_3)_3C—OCH_3$ **(d)** 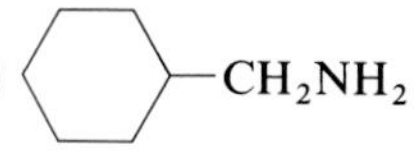 CH_2NH_2

(e)  $H_2C{=}CH{-}CH_2CN$ (f) 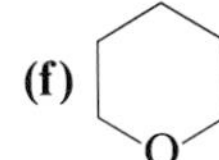 (g) $H{-}C{\equiv}C{-}CH_2CH_2CH_3$

* **5-44.** **(a)** Give two syntheses for $(CH_3)_2CH{-}O{-}CH_2CH_3$ and explain which synthesis is better.
(b) A student wanted to synthesize methyl *t*-butyl ether, $CH_3{-}O{-}C(CH_3)_3$. He attempted the synthesis by adding sodium methoxide (CH_3ONa) to *t*-butyl chloride, but he obtained none of the desired product. Show what product is formed in this reaction, and give a better synthesis of methyl *t*-butyl ether.

5-45. When ethyl bromide is added to potassium *t*-butoxide, the product is ethyl *t*-butyl ether.

$$\underset{\text{ethyl bromide}}{CH_3CH_2{-}Br} + \underset{\text{potassium } t\text{-butoxide}}{(CH_3)_3C{-}O^{-}\,{}^{+}K} \longrightarrow \underset{\text{ethyl } t\text{-butyl ether}}{(CH_3)_3C{-}O{-}CH_2CH_3}$$

(a) What happens to the reaction rate if the concentration of ethyl bromide is doubled?
(b) What happens to the rate if the concentration of potassium *t*-butoxide is tripled and the concentration of ethyl bromide is doubled?
(c) What happens to the rate if the temperature is raised?

5-46. Chlorocyclohexane reacts with sodium cyanide (NaCN) in ethanol to give cyanocyclohexane. The rate of formation of cyanocyclohexane increases when a small amount of sodium iodide is added to the solution. Explain this acceleration in the rate.

5-47. When *t*-butyl bromide is heated with ethanol, one of the products is ethyl *t*-butyl ether.
(a) What happens to the reaction rate if the concentration of ethanol is doubled?
(b) What happens to the rate if the concentration of *t*-butyl bromide is tripled and the concentration of ethanol is doubled?
(c) What happens to the rate if the temperature is raised?

5-48. Give the solvolysis products expected when each of the following compounds is heated in ethanol.

(a) Br (b) Br (c) Cl (d) Br

5-49. Allylic halides have the structure

X
C
C=C

(a) Show how the first-order ionization of an allylic halide leads to a resonance-stabilized cation.
(b) Draw the resonance structures of the allylic cations formed by ionization of the following allylic halides.

(i) CH_2Br (ii) Br (iii) Br

5-50. List the following carbocations in decreasing order of their stability.

CH_3 + ${}^{+}CH_2$ CH_3 + CH_3 +

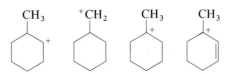

5-51. Two of the carbocations in Problem 5-50 are prone to rearrangement. Show how they might rearrange to more stable carbocations.

5-52. Predict the products of the following S_N2 reactions.

(a) $CH_3CH_2ONa + CH_3CH_2Cl \xrightarrow{CH_3CH_2OH}$

(b) 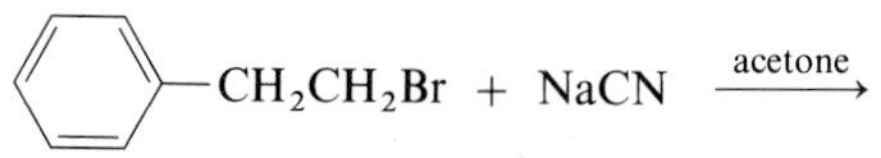 C_6H_5—$CH_2CH_2Br + NaCN \xrightarrow{\text{acetone}}$

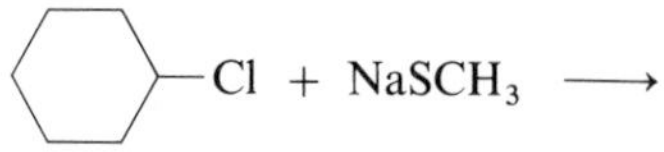

(c) cyclohexyl—Cl + $NaSCH_3 \longrightarrow$

(d) $CH_3(CH_2)_8CH_2Cl + NaI \xrightarrow{\text{acetone}}$

(e) 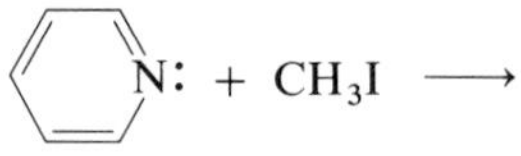 pyridine (N:) + $CH_3I \longrightarrow$

(f) $(CH_3)_3C$—CH_2CH_2Br + excess $NH_3 \longrightarrow$

(g) 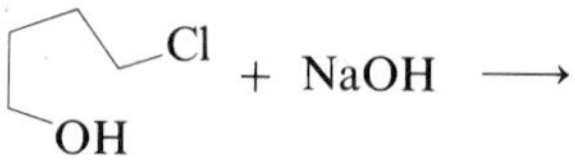 Cl—$CH_2CH_2CH_2CH_2$—OH + NaOH $\longrightarrow$

(h) 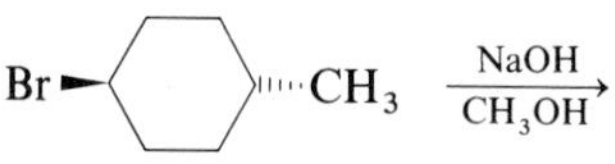 Br—cyclohexane—CH_3 $\xrightarrow[CH_3OH]{NaOH}$

5-53. Predict the dehydrohalogenation product(s) that result when the following alkyl halides are heated in alcoholic KOH. When more than one product is formed, predict the major and minor products.

(a) 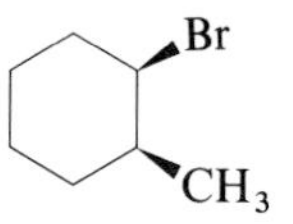 cyclohexane with Br and CH_3 (b) 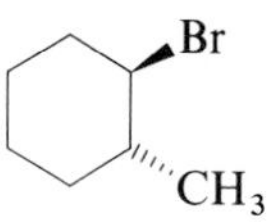 cyclohexane with Br and CH_3 (c) 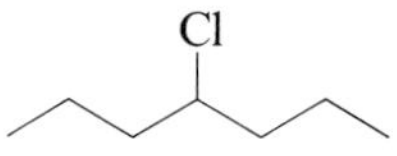 3-chloropentane (Cl)

5-54. Predict the major and minor products of the E2 dehydrohalogenation of each of the following.

(a) $(CH_3)_2CH$—$C(CH_3)_2$ with Br (b) $(CH_3)_2CH$—CH—CH_3 with Br (c) $(CH_3)_2C$—CH_2—CH_3 with Br (d) 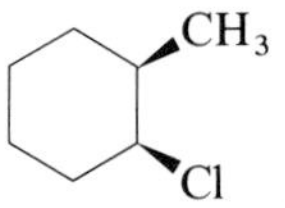 cyclohexane with CH_3 and Cl

5-55. Predict the products of E1 elimination of the following compounds. Label the major products.

(a) $(CH_3)_3C$—CH—CH_3 with Br (b) 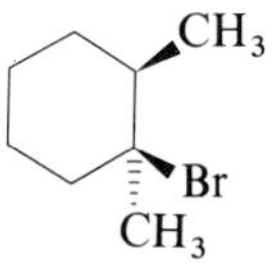 cyclohexane with CH_3, Br, CH_3 (c) 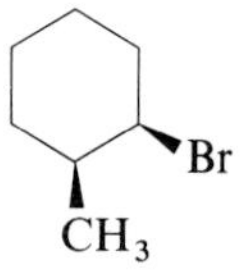 cyclohexane with Br and CH_3

5-56. When 1-bromomethylcyclohexene undergoes solvolysis in ethanol, three major products are formed. Give mechanisms to account for these three products.

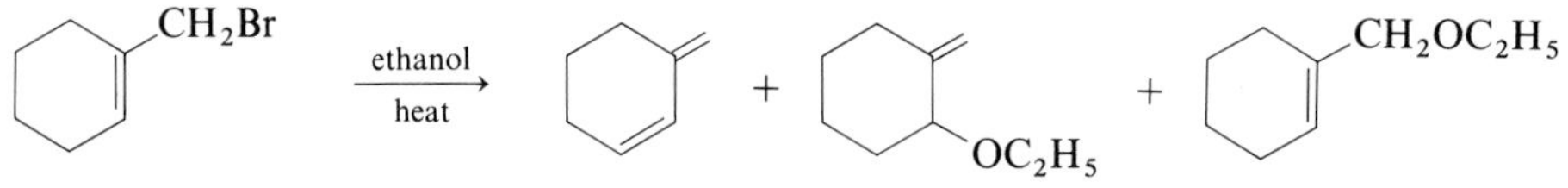

1-bromomethylcyclohexene (CH_2Br) $\xrightarrow[\text{heat}]{\text{ethanol}}$ methylenecyclohexene + methylenecyclohexane with OC_2H_5 + cyclohexene with $CH_2OC_2H_5$

* **5-57.** Protonation converts the hydroxyl group of an alcohol to a good leaving group. Suggest a mechanism for each of the following reactions.

(a) 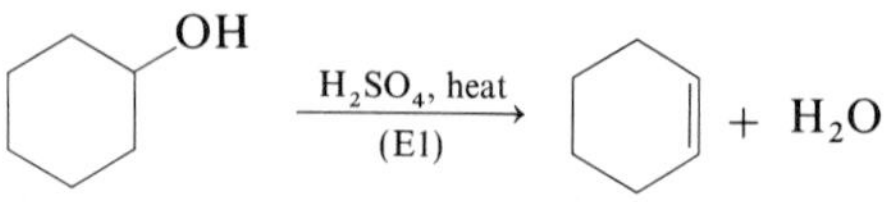 cyclohexanol (OH) $\xrightarrow[\text{(E1)}]{H_2SO_4\text{, heat}}$ cyclohexene + H_2O

(b) 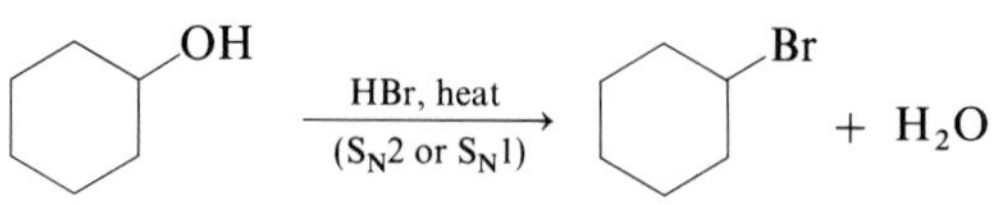 cyclohexanol (OH) $\xrightarrow[(S_N2 \text{ or } S_N1)]{\text{HBr, heat}}$ bromocyclohexane (Br) + H_2O

* **5-58.** Predict the products of the following eliminations of vicinal dibromides with potassium iodide. Remember to consider the geometric constraints of the E2 reaction (See Problem 5-35, page 208).

(a) CH_3—CH_2—CH(Br)—CH_2—Br (b) 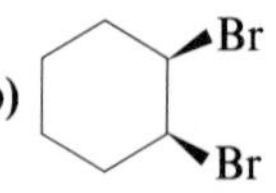 cyclohexane with Br and Br (c) 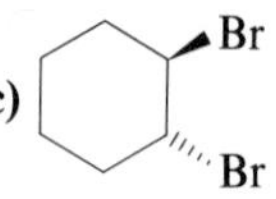 cyclohexane with Br and Br

(d) 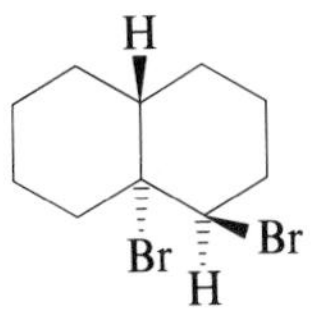 (e)

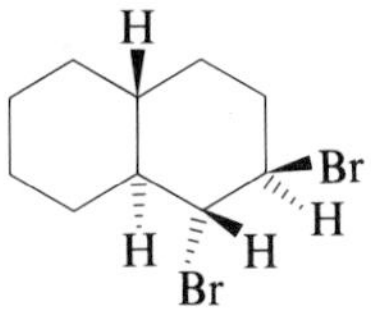

5-59. Give a mechanism to explain the two products formed in the following reaction.

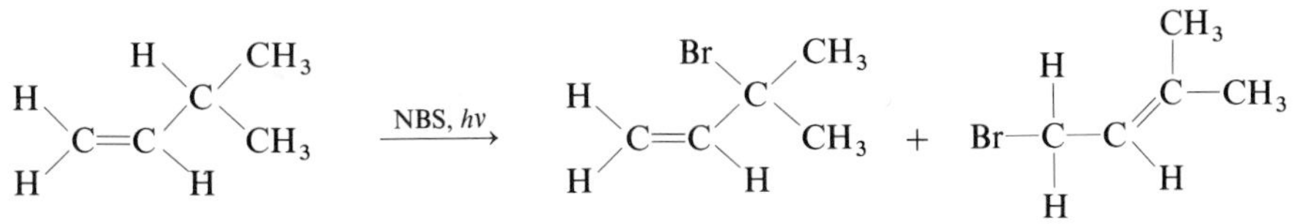

3-methyl-1-butene $\xrightarrow{\text{NBS, } h\nu}$ not rearranged + rearranged

5-60. Predict the major product of the following reaction and give a mechanism to support your prediction.

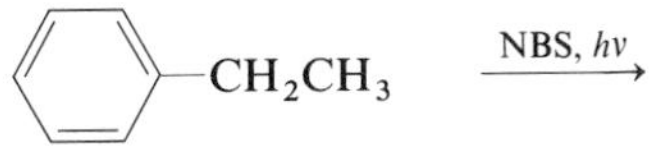

$C_6H_5-CH_2CH_3 \xrightarrow{\text{NBS, } h\nu}$

ethylbenzene

*** 5-61.** Upon solvolysis of bromomethylcyclopentane in methanol, a complex product mixture is formed. The five compounds shown below are found in the product mixture. Give a mechanism to account for the formation of each of these products.

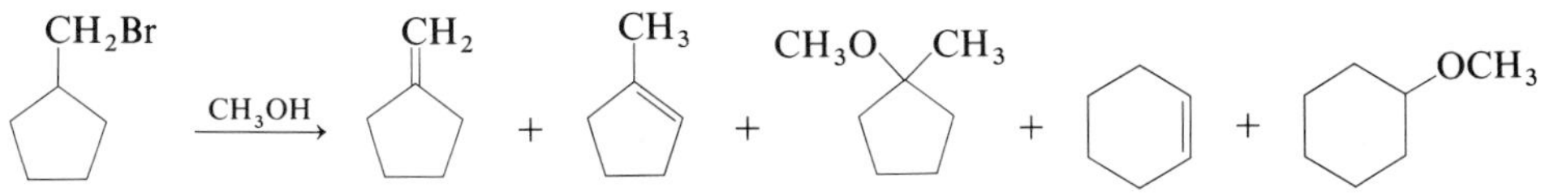

CH_2Br (cyclopentane) $\xrightarrow{CH_3OH}$ CH_2 + CH_3 + CH_3O CH_3 + (cyclohexene) + OCH_3

5-62. One of the following dichloronorbornanes undergoes elimination much faster than the other. Determine which one reacts faster, and explain the large difference in rates.

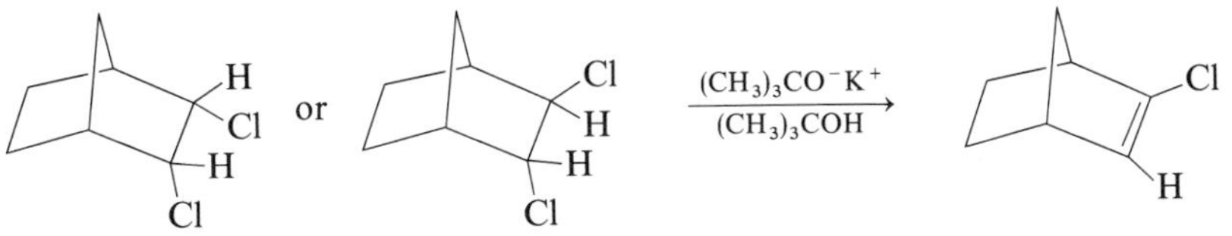

cis or trans $\xrightarrow[(CH_3)_3COH]{(CH_3)_3CO^-K^+}$

*** 5-63.** Deuterium (D) is the isotope of hydrogen of mass number 2, with a proton and a neutron in its nucleus. We have seen that the chemistry of deuterium is nearly identical to the chemistry of hydrogen, except that the C—D bond is slightly (1.2 kcal/mol or 5.0 kJ/mol) stronger than the C—H bond. Reaction rates tend to be slower if a C—D bond (as opposed to a C—H bond) is broken in a rate-determining step. This effect on the rate is called a *kinetic isotope effect*.

(a) Propose a mechanism to explain each product in the following reaction.

$$CH_3-CH(Br)-CH_3 \xrightarrow{\text{KOH, alcohol}} CH_2{=}CH-CH_3 + CH_3-CH(OH)-CH_3$$

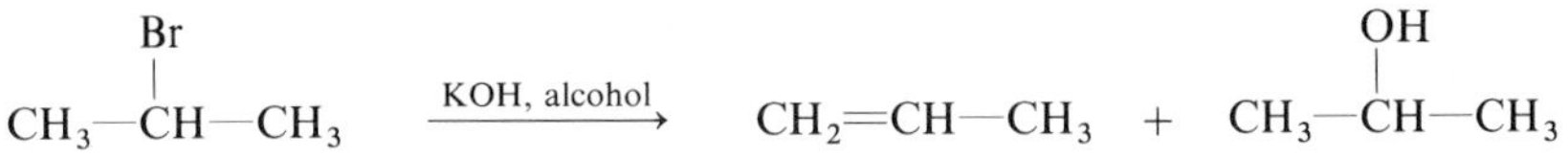

elimination product; substitution product

(b) When the following deuterated compound reacts under the same conditions, the rate of formation of the substitution product is unchanged, while the rate of formation of the elimination product is slowed by a factor of 7.

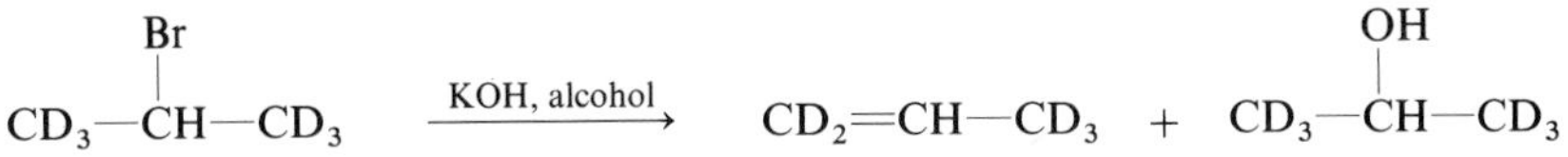

$$CD_3-CH(Br)-CD_3 \xrightarrow{\text{KOH, alcohol}} CD_2{=}CH-CD_3 + CD_3-CH(OH)-CD_3$$

seven times slower; rate unchanged

Explain why the elimination rate is slowed, but the substitution rate is unchanged.

(c) A similar reaction takes place upon heating the alkyl halide in an acetone/water mixture.

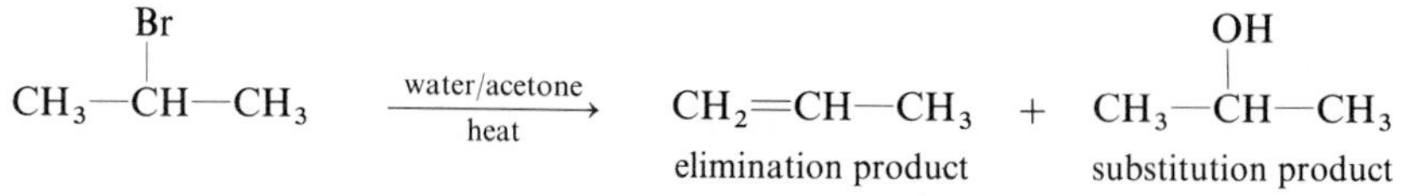

$$CH_3-\underset{\displaystyle Br}{\overset{}{CH}}-CH_3 \xrightarrow[\text{heat}]{\text{water/acetone}} \underset{\text{elimination product}}{CH_2{=}CH-CH_3} + \underset{\text{substitution product}}{CH_3-\overset{OH}{CH}-CH_3}$$

(Br on the central carbon of the reactant.)

Give a mechanism for the formation of each product under these conditions, and predict how the rate of formation of each product will change when the deuterated halide reacts. Explain your prediction.

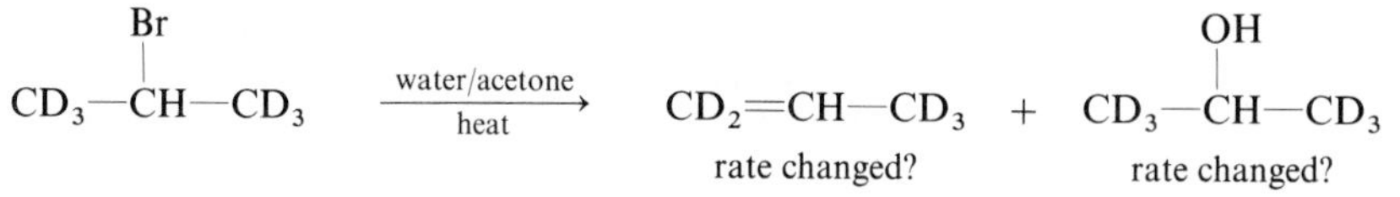

$$CD_3-\overset{Br}{CH}-CD_3 \xrightarrow[\text{heat}]{\text{water/acetone}} \underset{\text{rate changed?}}{CD_2{=}CH-CD_3} + \underset{\text{rate changed?}}{CD_3-\overset{OH}{CH}-CD_3}$$

★ 5-64. When the following compound is treated with sodium methoxide in methanol, two elimination products are possible. Explain why the deuterated product predominates by about a 7:1 ratio (refer to Problem 5-63).

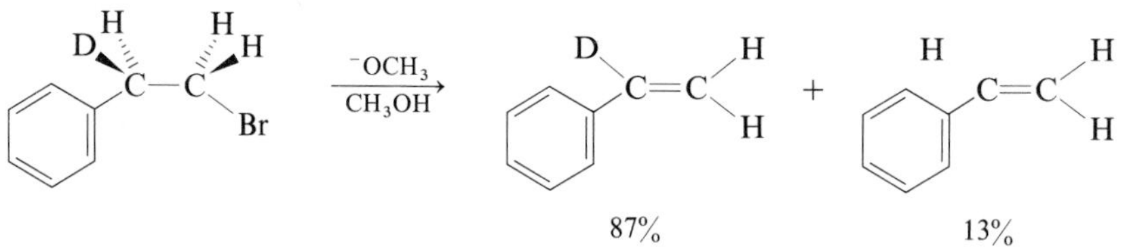

$$\text{Ph–CHD–CH}_2\text{Br} \xrightarrow[CH_3OH]{^-OCH_3} \underset{87\%}{\text{Ph(D)C=CH}_2} + \underset{13\%}{\text{Ph(H)C=CH}_2}$$

★ 5-65. The reaction of an amine with an alkyl halide gives an ammonium salt.

$$\underset{\text{amine}}{R_3N:} + \underset{\text{alkyl halide}}{R'{-}X} \longrightarrow \underset{\text{ammonium salt}}{R_3\overset{+}{N}{-}R'\ X^-}$$

The rate of this S_N2 reaction is very sensitive to the polarity of the solvent. Draw an energy profile for this reaction in a nonpolar solvent and another in a polar solvent. Consider the nature of the transition state, and explain why this reaction should be sensitive to the polarity of the solvent. Predict whether it will be faster or slower in a more polar solvent.

★ 5-66. The following reaction takes place under second-order conditions (strong nucleophile), yet the structure of the product shows rearrangement. Also, the rate of this reaction is several thousand times faster than the rate of substitution of hydroxide ion on 2-chlorobutane under similar conditions. Propose a mechanism to explain the enhanced rate and rearrangement observed in this unusual reaction. ("Et" is the abbreviation for ethyl.)

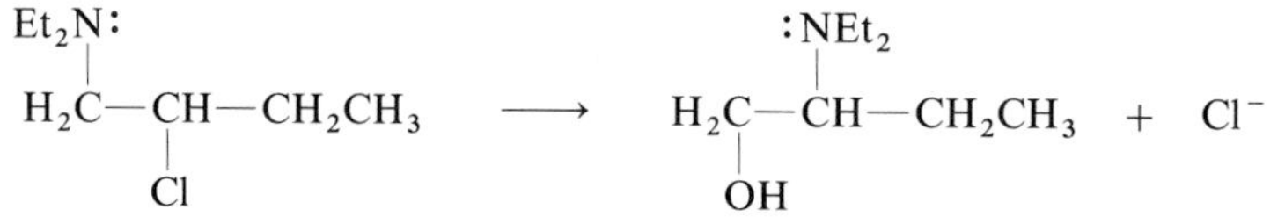

$$\underset{\displaystyle\quad}{\overset{Et_2N:}{H_2C}}-\underset{Cl}{CH}-CH_2CH_3 \longrightarrow \underset{OH}{H_2C}-\overset{:NEt_2}{CH}-CH_2CH_3 + Cl^-$$

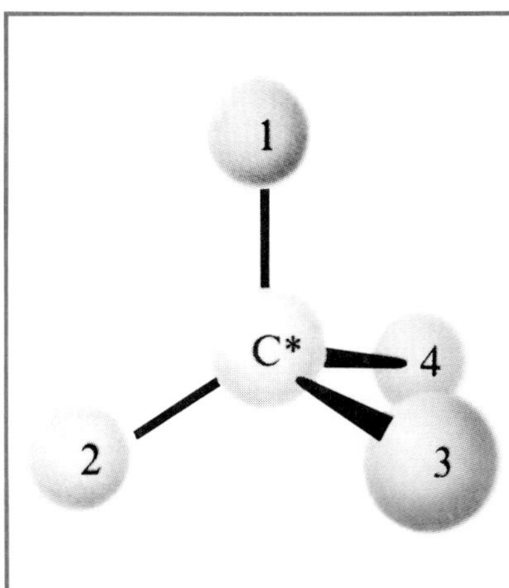

6

STEREOCHEMISTRY

6-1 INTRODUCTION

Stereochemistry is the study of the three-dimensional structure of molecules. It is impossible to study organic chemistry without using stereochemistry. For example, the cis and trans isomers of 2-butene are a special type of isomers called *geometric* (or *cis-trans*) *isomers*. Although both of these 2-butenes have the formula $CH_3—CH{=}CH—CH_3$, they differ in how their atoms are arranged in space. This kind of difference is characteristic of a class of isomers called **stereoisomers,** which differ only in their three-dimensional arrangement of atoms.

H_3C / H / C=C / H / CH_3 — *trans*-2-butene

H_3C / CH_3 / C=C / H / H — *cis*-2-butene

The discovery of stereoisomerism was one of the most important breakthroughs in the structural theory of organic chemistry. Stereochemistry explained why several different types of isomers can exist, and it also forced scientists to propose a tetrahedral arrangement of the bonds around a saturated carbon atom. Without understanding stereochemistry, it is impossible to comprehend the structures and reactions of organic compounds. In this chapter we study the three-dimensional structures of molecules to understand their stereochemical relationships. We compare the various types of stereoisomers and study reactions that can differentiate among stereoisomers. The substitutions and eliminations covered in Chapter 5 are particularly helpful for demonstrating these stereochemical relationships.

6-2 CHIRALITY

What is the difference between your left hand and your right hand? They look similar, yet a left-handed glove does not fit the right hand. The same principle applies to your feet. They look almost identical, yet the left shoe fits painfully on the right foot. The relationship between your two hands or your two feet is that they are nonsuperimposable (nonidentical) mirror images of each other. Objects that have left-handed and right-handed forms are called **chiral** (*kī′ rəl,* rhymes with "spiral"), the Greek word for "handed."

We can tell whether an object is chiral by looking at its mirror image (Fig. 6-1). Every physical object (with the possible exception of a vampire) has a mirror image, but *a chiral object has a mirror image that is different from the original object.* For example, a chair and a spoon and a glass of water all look the same in a mirror. Such objects are called **achiral,** meaning "not chiral." A hand looks different in the mirror. If the original hand were the right hand, it would look like a left hand in the mirror.

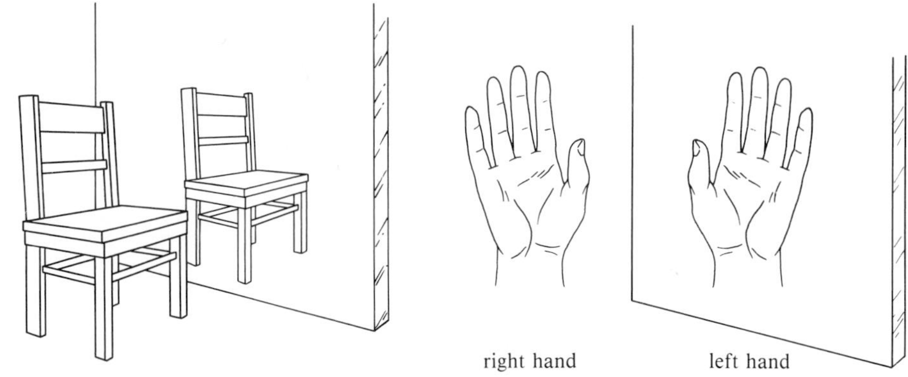

FIGURE 6-1 Use of a mirror to test for chirality. An object is chiral if its mirror image is different from the original object.

Besides shoes and gloves, which are obviously "handed," we see many other chiral objects every day (Fig. 6-2). What is the difference between an English car and an American car? The English car has the steering wheel on the right-hand

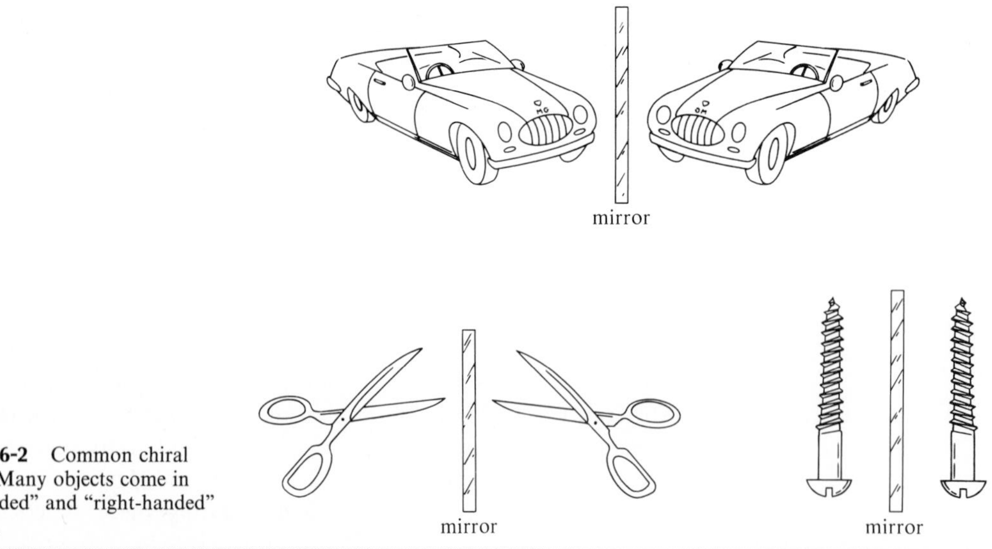

FIGURE 6-2 Common chiral objects. Many objects come in "left-handed" and "right-handed" versions.

side, while the American car has it on the left. To a first approximation, the English and American cars are nonsuperimposable mirror images. Most screws have right-hand threads and are turned clockwise to tighten. The mirror image of a right-handed screw is a left-handed screw, turned counterclockwise to tighten. Those of us who are left-handed realize that scissors are chiral. Most scissors are right-handed. If you use them in your left hand, they cut very poorly, if at all. A left-handed person must go to a well-stocked store to find a pair of left-handed scissors, the mirror image of the "standard" right-handed scissors.

PROBLEM 6-1

Determine whether the following objects are chiral or achiral.

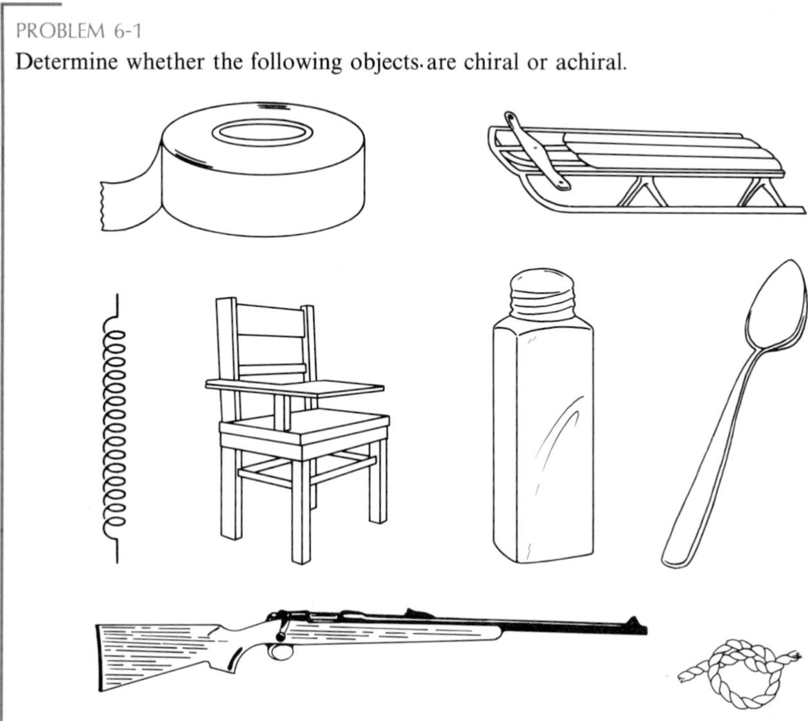

6-2A CHIRALITY AND ENANTIOMERISM IN ORGANIC MOLECULES

Like other objects, molecules are either chiral or achiral. For example, consider the two geometric isomers of 1,2-dichlorocyclopentane. The cis isomer is achiral ("not chiral"), since its mirror image is superimposable on the original molecule (Fig. 6-3). Two molecules are said to be **superimposable** if they can be placed on top of each other and the three-dimensional position of each atom of one molecule coincides with the equivalent atom of the other molecule. To draw the mirror

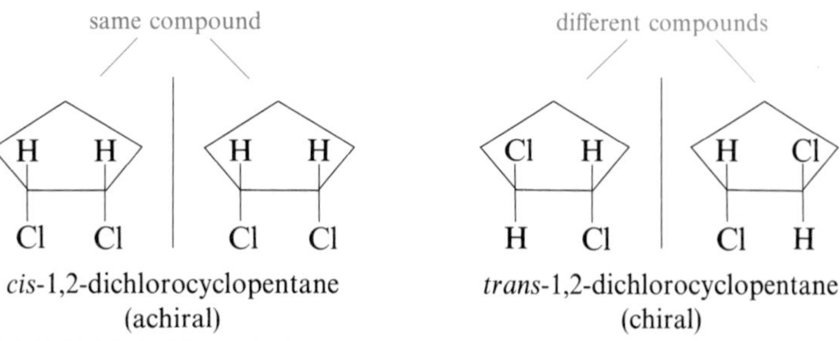

cis-1,2-dichlorocyclopentane (achiral)

trans-1,2-dichlorocyclopentane (chiral)

FIGURE 6-3 Stereoisomers of 1,2-dichlorocyclopentane. The cis isomer has no enantiomers; it is achiral. The trans isomer is chiral; it can exist in either of two nonsuperimposable enantiomeric forms.

image of a molecule, simply draw the same structure with left and right reversed. The up-and-down and front-and-back directions are unchanged. These two mirror-image structures are clearly identical (superimposable), and *cis*-1,2-dichlorocyclopentane is achiral.

The mirror image of *trans*-1,2-dichlorocyclopentane is different from (nonsuperimposable with) the original molecule. These are two different compounds, and we should expect to discover two mirror-image isomers of *trans*-1,2-dichlorocyclopentane. You should make models of these isomers to convince yourself that they are different no matter how you twist and turn them. Such nonsuperimposable mirror-image molecules are called **enantiomers.** A chiral compound always has an enantiomer (a nonsuperimposable mirror image). An achiral compound always has a mirror image that is the same as the original molecule. Let's review the definitions of these words.

enantiomers: Pairs of compounds that are nonsuperimposable mirror images.
chiral: ("handed") Different from its mirror image; having an enantiomer.
achiral: ("not handed") Identical with its mirror image; not chiral.

Any compound that is chiral must have an enantiomer. Any compound that is achiral cannot have an enantiomer.

PROBLEM 6-2

Make a model and draw a three-dimensional structure for each of the following compounds. Then draw the mirror image of your original structure and determine whether the mirror image is the same compound. Label each structure as being chiral or achiral, and label pairs of enantiomers.

(a) *cis*-1,2-dimethylcyclobutane
(b) *trans*-1,2-dimethylcyclobutane
(c) *cis*- and then *trans*-1,3-dimethylcyclobutane
(d) 2-bromobutane

(e) 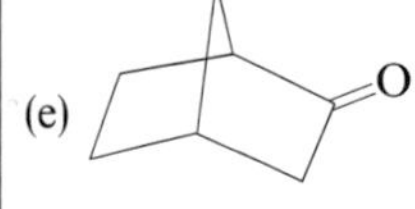

(f) 

6-2B CHIRAL CARBON ATOMS

Problem 6-2(d) shows that a ring is not necessary for a molecule to be chiral. What is it about a molecule that makes it chiral? The most common feature (but not the only one) that lends chirality is a carbon atom that is bonded to four different groups. Such a carbon atom is called a **chiral carbon atom,** an **asymmetric carbon atom,** or a **stereocenter.** The preferred usage varies, so we will use the simplest term: **chiral carbon atom.** Notice in Figure 6-4 how the tetrahedral arrangement around a chiral carbon atom gives it a nonsuperimposable mirror image.

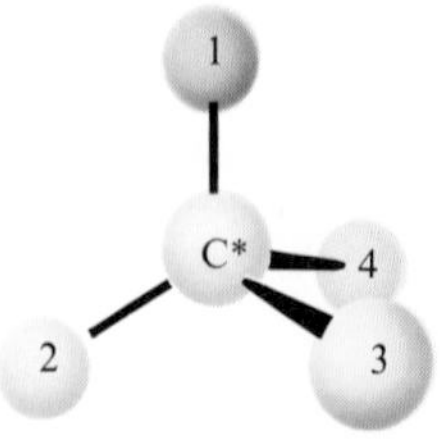

mirror

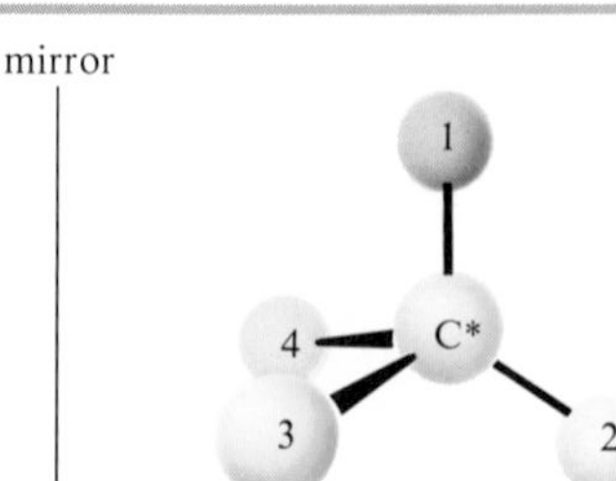

FIGURE 6-4 Enantiomers of a chiral carbon atom. These two mirror images are nonsuperimposable.

No matter how you twist and turn the structures in Figure 6-4, they can never look the same. You should make models of these structures (using different-colored balls for the four different groups) to get an intuitive feeling for how they are different. A chiral carbon atom is often labeled with an asterisk (*) in drawing the structure of a compound.

If two groups on a carbon atom are the same, however, the arrangement usually is not chiral. Figure 6-5 shows the mirror image of a tetrahedral structure with only three different groups attached; two of the four groups are the same. If the structure on the right is rotated 180°, it superimposes on the left structure.

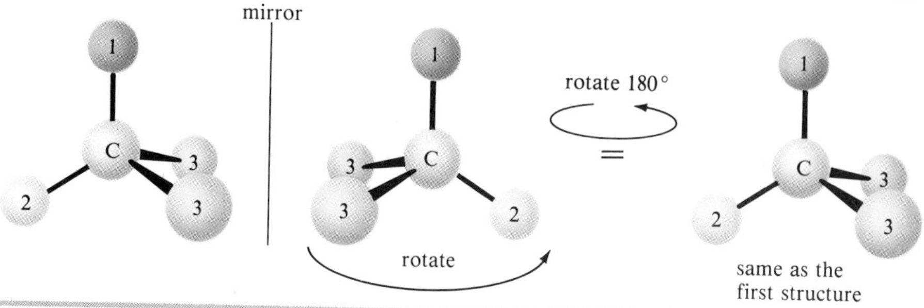

FIGURE 6-5 A carbon atom bonded to just three different types of groups is not chiral.

2-Bromobutane [Problem 6-2(d)] is chiral because it has a chiral carbon atom. Carbon atom 2 of 2-bromobutane is bonded to a hydrogen atom, a bromine atom, a methyl group, and an ethyl group.

$$CH_3-\overset{\overset{\displaystyle Br}{|}}{\underset{*}{C}H}-CH_2CH_3$$

A three-dimensional drawing shows that 2-bromobutane cannot be superimposed on its mirror image.

CH_3 / *C ··· H / CH_3CH_2 / Br | CH_3 / H ··· C* / Br / CH_2CH_3

We can make some generalizations at this point, as long as you remember that the ultimate test for chirality is always whether the molecule's mirror image is the same or different.

1. If a compound has no chiral carbon atom, it is usually achiral.
2. If a compound has just one chiral carbon atom, it is chiral.
3. If a compound has more than one chiral carbon, it may or may not be chiral. (We will see examples in Section 6-11.)

SOLVED PROBLEM 6-1

Star each chiral carbon atom in the following structure:

H OH / ring carbons 1*, 2, 3*, 4, 5, 6 — $\underset{*}{CH}-CH_3$ / H Cl

SOLUTION There are three chiral carbons, starred in red.

1. The (CHOH) carbon of the side chain is chiral. Its four substituents are the ring, a hydrogen atom, a hydroxyl group, and a methyl group.
2. Carbon atom C1 of the ring is chiral. Its four substituents are the side chain, a hydrogen atom, the part of the ring closer to the chlorine atom (—CH_2—CHCl—), and the part of the ring farther from the chlorine atom (—CH_2—CH_2—CH_2—CHCl).
3. The ring carbon bearing the chlorine atom is chiral. Its four substituents are the chlorine atom, a hydrogen atom, the part of the ring closer to the side chain, and the part of the ring farther from the side chain.

Notice that different groups may be different in any manner. For example, the ring carbon bearing the chlorine atom is chiral even though two of its ring substituents initially appear to be —CH_2— groups. These two parts of the ring are different because one is closer to the side chain and one is farther. The *entire* structure of the group must be considered.

PROBLEM 6-3

Draw a three-dimensional structure for each of the following compounds, and star all chiral carbon atoms. Draw the mirror image for each structure, and tell whether you have drawn a pair of enantiomers or just the same molecule twice. Build molecular models of any of these examples that seem difficult to you.

(a) 1-bromobutane (b) 1-pentanol 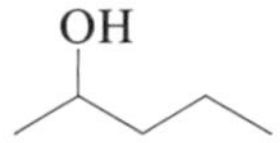 (c) 2-pentanol

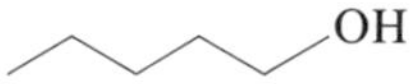

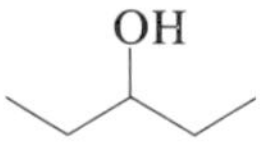

(d) 3-pentanol (e) chlorocyclohexane

(f) CH_3—CH(NH_2)—COOH (g) *cis*-1,2-dichlorocyclobutane

(h) *trans*-1,2-dichlorocyclobutane (i) *trans*-1,3-dichlorocyclobutane

(j) (k)

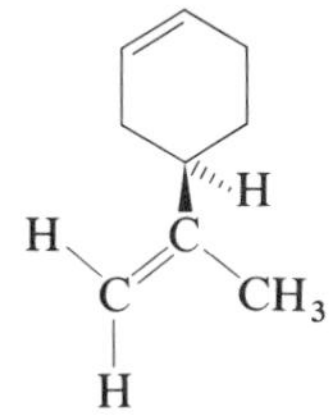

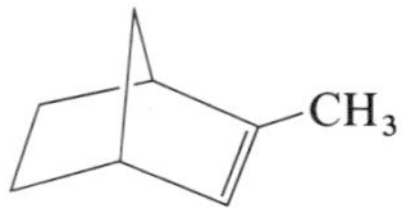

6-2C MIRROR PLANES OF SYMMETRY

In Figure 6-3 we saw that *cis*-1,2-dichlorocyclopentane is achiral. Its mirror image was found to be identical with the original molecule. Figure 6-6 shows a shortcut that sometimes shows whether a molecule is chiral.

If we draw a line down the middle of *cis*-1,2-dichlorocyclopentane, bisecting a carbon atom and two hydrogen atoms, the part of the molecule that appears to the right of the line is the mirror image of the part on the left. This kind of symmetry is called an **internal mirror plane,** sometimes symbolized by the Greek lowercase letter sigma (σ). Since the right-hand side of the molecule is the reflec-

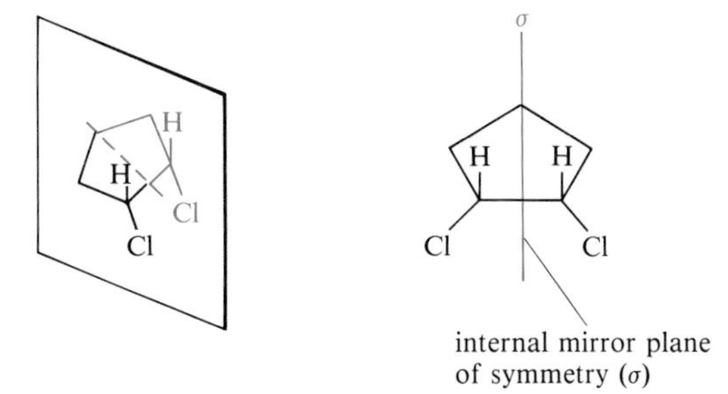

FIGURE 6-6 *cis*-1,2-Dichlorocyclopentane has a mirror plane of symmetry. Any compound with an internal mirror plane of symmetry *cannot* be chiral.

tion of the left-hand side, the molecule's mirror image is the same as the original molecule.

Notice below that the chiral trans isomer of 1,2-dichlorocyclopentane does not have a mirror plane of symmetry. The chlorine atoms do not reflect into each other across our hypothetical mirror plane. One of them is directed up, the other down.

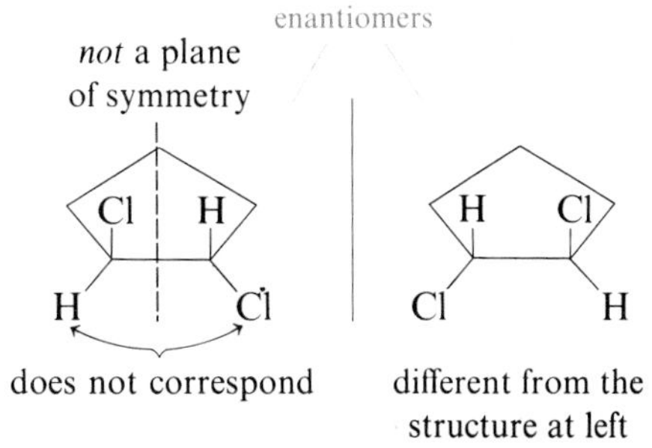

In a nonrigorous way, these examples show the following principle:

> *Any molecule that has an internal mirror plane of symmetry cannot be chiral,* even though it may contain chiral carbon atoms.

On the other hand, the converse is not true. When we cannot find a mirror plane of symmetry, that does not necessarily mean that the molecule must be chiral. The following example has no internal mirror plane of symmetry, yet the mirror image is superimposable on the original molecule. You may need to make models to show that the mirror images are just two drawings of the same compound.

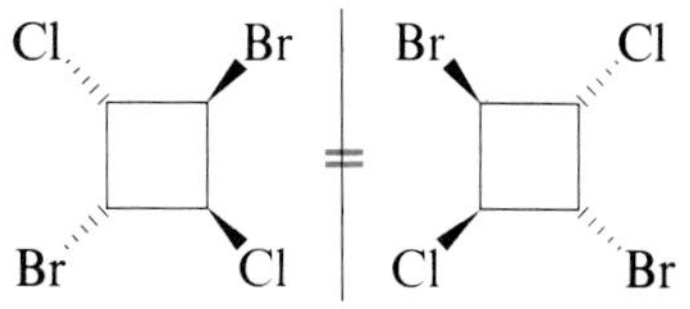

(hydrogens are omitted for clarity)

Using what we know about mirror planes of symmetry, we can see why a chiral (asymmetric) carbon atom is special. Figure 6-4 showed that a chiral carbon has a mirror image that is nonsuperimposable on the original structure. If a carbon atom has only three different kinds of substituents, however, it has an internal mirror plane of symmetry (Fig. 6-7). Therefore, it cannot contribute to chirality in a molecule.

FIGURE 6-7 A carbon atom with two identical substituents (only three different kinds of substituents) usually has an internal mirror plane of symmetry.

viewed from this angle

PROBLEM 6-4

For each of the following compounds, determine whether the molecule has an internal mirror plane of symmetry. If it does, draw the mirror plane on a three-dimensional drawing of the molecule. If the molecule does not have an internal mirror plane, determine whether or not the structure is chiral.

(a) methane
(b) *cis*-1,2-dibromocyclobutane
(c) *trans*-1,2-dibromocyclobutane
(d) 1,2-dichloropropane
(e) glyceraldehyde, $HOCH_2-CH(OH)-CHO$
(f) 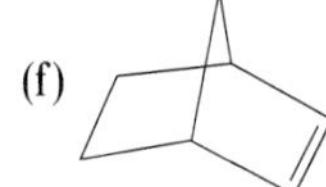
(g) 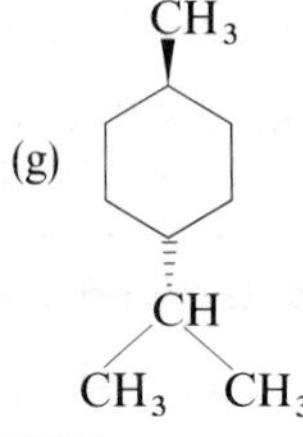

6-3 (R) AND (S) NOMENCLATURE OF CHIRAL CARBON ATOMS

Alanine is one of the amino acids that are linked to form proteins. Alanine has a chiral carbon atom and exists in two enantiomeric forms.

natural alanine | unnatural alanine

These mirror images are different, and this difference is reflected in their biochemistry. Only the enantiomer on the left can be metabolized by the usual enzyme; the one on the right is not recognized as a useful amino acid. We need a simple way to distingush between enantiomers and to give each of them a unique name.

The most widely accepted system for naming the configuration of a chiral carbon atom is the **Cahn-Ingold-Prelog convention,** which assigns to each chiral carbon atom a letter (*R*) or (*S*). To name the configuration of a chiral carbon atom, follow this two-step procedure:

1. Assign a "priority" to each group bonded to the chiral carbon. We speak of group 1 as having the highest priority, group 2 second, group 3 third, and group 4 as having the lowest priority.

   (a) Atoms with higher atomic numbers receive higher priorities. For example, if the four groups bonded to a chiral carbon atom were H, CH_3, NH_2, and F, the fluorine atom (atomic number 9) would have the highest priority, followed by the nitrogen atom of the NH_2 group (atomic number 7), then by the carbon atom of the methyl group (atomic number 6). Note that we look only at the atomic number of the atom directly attached to the chiral carbon, not the entire group. Needless to say, hydrogen comes last.

   In the case of different isotopes of the same element, the heavier isotopes have higher priorities. For example, tritium (^3H) receives a higher priority than deuterium (^2H), followed by hydrogen (^1H).

   Examples of priority for atoms bonded to a chiral carbon:

$$I > S > O > N > {}^{13}C > {}^{12}C > Li > {}^{3}H > {}^{2}H > {}^{1}H$$

   (b) In case of ties, use the next atoms along the chain as tiebreakers. For example, we assign a higher priority to isopropyl $-CH(CH_3)_2$ than to ethyl $-CH_2CH_3$. The first carbon atom in the ethyl group is bonded to two hydrogens and one carbon, while the first carbon in the isopropyl group is bonded to two carbons and one hydrogen. An ethyl group and a $-CH_2CH_2Br$ have identical first atoms and second atoms, but the bromine atom in the third position gives $-CH_2CH_2Br$ a higher priority than $-CH_2CH_3$.

   Examples

$$-CHBr_2 > -CH_2Cl > -C(CH_3)_3 > -CH(CH_3)CH_2F > -CH(CH_3)_2 > -CH_2CH_3$$

   (c) Treat double and triple bonds as if each bond were a bond to a separate atom. For this method imagine that each pi bond were broken and the atoms at both ends duplicated. Note that when you break a bond you always add *two* imaginary atoms.

$-CH{=}CH_2$ becomes $R-CH(Ⓒ)-CH_2(Ⓒ)$ — break and duplicate

$R-CH{=}NH$ becomes $R-CH(Ⓝ)-NH(Ⓒ)$ — break and duplicate

$R-C{\equiv}C-H$ becomes $R-C(Ⓒ)(Ⓒ)-C(Ⓒ)(Ⓒ)-H$ — break and duplicate

$R-C(OH){=}O$ becomes $R-C(OH)(Ⓞ)-O(Ⓒ)$ — break and duplicate

2. Using a three-dimensional drawing or a model, put the fourth priority group in back and view the molecule along the bond from the chiral carbon to the fourth priority group. Draw an arrow from the first priority group, through the second, to the third. If the arrow points clockwise, the chiral carbon atom is called (*R*) (Latin, *rectus,* "upright"). If the arrow points counterclockwise, the chiral carbon atom is called (*S*) (Latin, *sinister,* "left").

   Alternatively, you can draw the arrow and imagine turning a car's steering wheel in that direction. If the car would go to the left, the chiral carbon atom is designated (*S*). If the car would go to the right, the chiral carbon atom is designated (*R*).

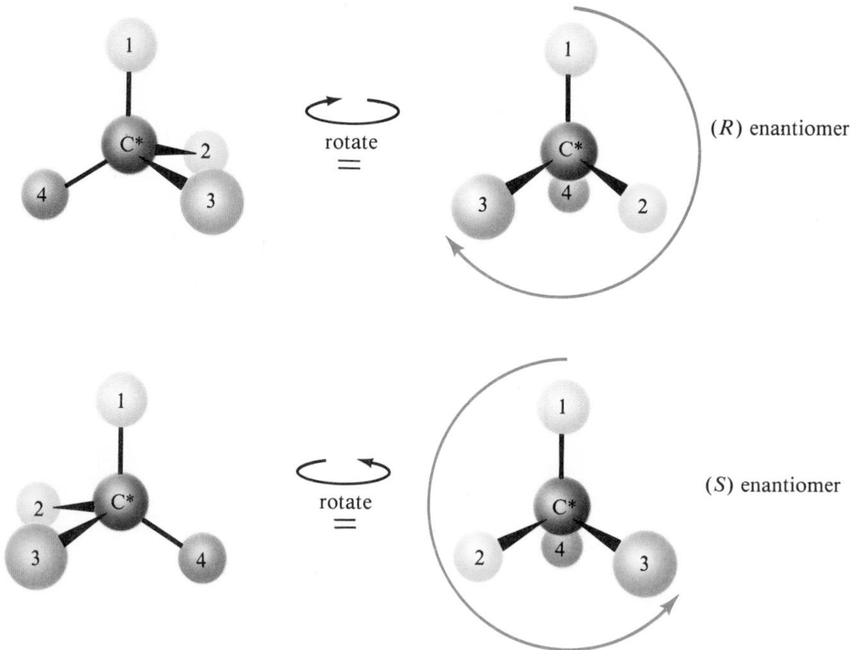

As an example, let's use the enantiomers of alanine. The naturally occurring enantiomer is the one on the left, determined to have the (*S*) configuration.

Of the four atoms attached to the chiral carbon in alanine, nitrogen has the largest atomic number, giving it the highest priority. Next is the —COOH carbon atom, since it is bonded to oxygen atoms. Third is the methyl group, followed by the hydrogen atom. When we position the natural enantiomer with its hydrogen atom pointing away from us, the arrow from $-NH_2$ to —COOH to $-CH_3$ points counterclockwise. Thus the naturally occurring enantiomer of alanine has the (*S*) configuration. Make models of these enantiomers to illustrate how they are named (*R*) and (*S*).

SOLVED PROBLEM 6-2

Draw the enantiomers of 1,3-dibromobutane and label them as (*R*) and (*S*). (Making a model is particularly helpful for this type of problem.)

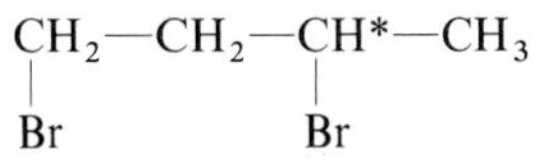

$$\underset{\text{Br}}{\text{CH}_2}-\text{CH}_2-\underset{\text{Br}}{\text{CH}^*}-\text{CH}_3$$

SOLUTION The third carbon atom in 1,3-dibromobutane is chiral. The bromine atom receives first priority, the ($-CH_2CH_2Br$) group second priority, the methyl group third, and the hydrogen fourth. The following mirror images are drawn with the hydrogen atom back, ready to assign (*R*) or (*S*) as shown.

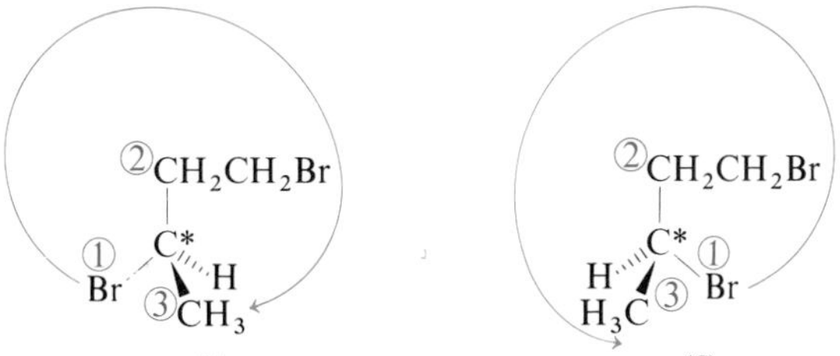

(1) Br, (2) CH_2CH_2Br, C*, H, (3) CH_3
(*R*)

(2) CH_2CH_2Br, C*, H, (1) Br, (3) H_3C
(*S*)

SOLVED PROBLEM 6-3

The structure of one of the enantiomers of carvone is shown below. Find the chiral carbon atom and determine whether it has the (*R*) or the (*S*) configuration.

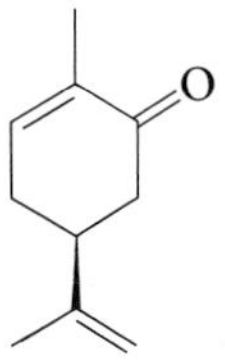

SOLUTION The chiral carbon atom is one of the ring carbons, as indicated by the asterisk in the structure below. Although there are two $-CH_2-$ groups bonded to the carbon, they are different $-CH_2-$ groups. One is a $-CH_2-CO-$ group, and the other is a $-CH_2-CH{=}C$ group. The groups are assigned priorities, and this is found to be the (*S*) enantiomer.

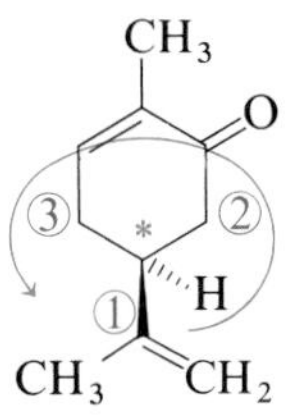

CH_3, O, (3), (2), *, H, (1), CH_3, CH_2

(*S*)-carvone

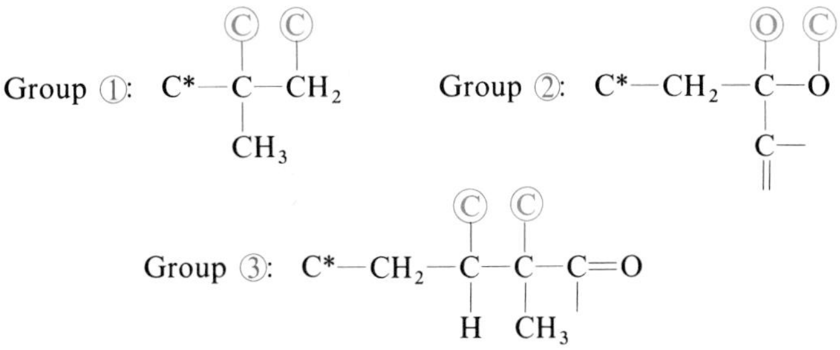

Group (1): C*—C—CH_2 (with (C) (C); CH_3)

Group (2): C*—CH_2—C—O (with (O) (C); C—)

Group (3): C*—CH_2—C—C—C=O (with (C) (C); H, CH_3)

PROBLEM 6-5

Star each chiral carbon atom in the following examples. For each chiral carbon, determine whether it has the (*R*) or the (*S*) configuration.

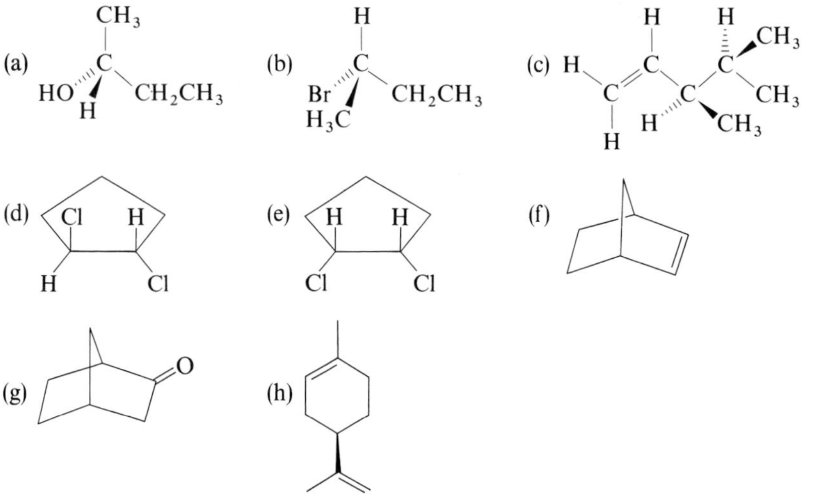

(a) CH_3, HO, H, C, CH_2CH_3 (b) H, Br, H_3C, C, CH_2CH_3 (c) H, H, H, CH_3, C, C, C, C, H, H, CH_3, CH_3

(d) Cl, H, H, Cl (e) H, H, Cl, Cl (f)

(g) O (h)

PROBLEM 6-6

In Problem 6-3 you drew the enantiomers for a number of chiral compounds. Now go back and designate each chiral carbon atom as either (*R*) or (*S*).

6-4 OPTICAL ACTIVITY

Mirror-image molecules have very similar physical properties. Compare the following properties of (*R*)-2-bromobutane and (*S*)-2-bromobutane.

	(*R*)-2-bromobutane	(*S*)-2-bromobutane
boiling point (°C)	91.2	91.2
melting point (°C)	−112	−112
refractive index	1.436	1.436
density	1.253	1.253

How, then, can enantiomerism be observed and measured in the laboratory? There is one physical property of enantiomers for which there is an easily measured difference: the direction in which they rotate the plane of polarized light.

6-4A PLANE-POLARIZED LIGHT

Most of what we see is unpolarized light, vibrating randomly in all directions. **Plane-polarized light** is composed of waves that vibrate in only one plane. Although there are other types of "polarized light," the term usually refers to plane-polarized light.

When unpolarized light passes through a polarizing filter, the randomly vibrating light waves are filtered so that most of the light passing through is vibrating in one direction (Fig. 6-8). The direction of vibration of light passing through a polarizing filter is called the *axis* of the filter. Polarizing filters may be made from carefully cut calcite crystals or from plastic sheets that have been treated in a special way. Plastic polarizing filters are often used as lenses in sun-

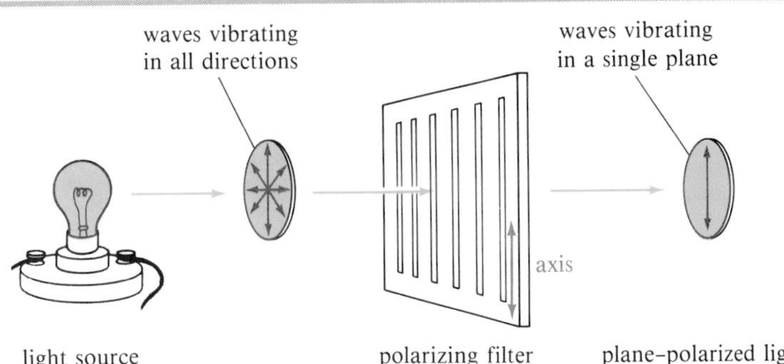

FIGURE 6-8 The waves of plane-polarized light vibrate primarily in a single plane.

glasses, since the axis of the filters can be positioned to filter out most reflected glare.

When light passes first through one polarizing filter and then through another, the amount of light emerging depends on the relationship between the axes of the two filters (Fig. 6-9). If the axes of the two filters are lined up (parallel), then nearly all the light that passes through the first filter also passes through the second. If the axes of the two filters are perpendicular (*crossed poles*), however, all the polarized light that emerges from the first filter is stopped by the second. At intermediate angles of rotation, intermediate amounts of light pass through.

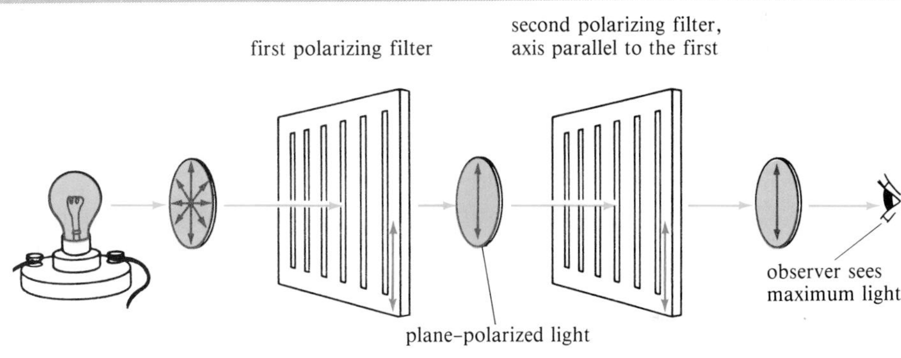

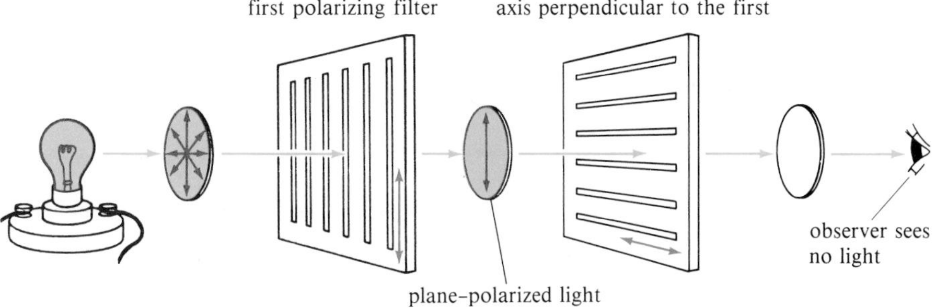

FIGURE 6-9 When the axis of a second polarizing filter is parallel to that of the first, a maximum amount of light passes through. When the axes of the filters are perpendicular, no light passes through.

You can demonstrate this effect for yourself by wearing a pair of polarized sunglasses while looking at a light source through another pair (Fig. 6-10). The second pair seems to be quite transparent, as long as its axis is lined up with the pair you are wearing. When the second pair is turned at 90°, however, the lenses become opaque, as if they were covered with black ink.

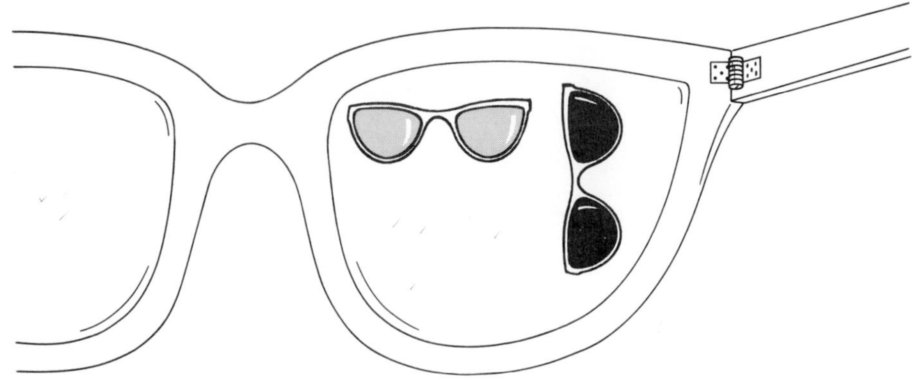

FIGURE 6-10 Using sunglasses to demonstrate parallel axes of polarization and crossed poles. When the two pairs of sunglasses are parallel, a maximum amount of light passes through. When they are perpendicular, very little light passes through.

6-4B ROTATION OF PLANE-POLARIZED LIGHT

When polarized light passes through a solution containing a chiral compound, the chiral compound causes the plane of vibration to rotate (Fig. 6-11). For example, if polarized light vibrating in the vertical plane passes through a solution of one of the enantiomers of 2-butanol, it might emerge with its plane of vibration rotated 30° from the vertical. This rotation can be detected by using a second polarizing filter and rotating it until the maximum amount of light is observed.

When this rotation of polarized light was first discovered, the field of stereochemistry was quite primitive, and the relationship between chirality and the ability to rotate the plane of polarized light was unknown. The rotation of the plane of polarized light was called **optical activity,** and substances that could rotate the plane of polarized light were said to be **optically active.** These terms are still in use today.

Before the relationship between chirality and optical activity was known, enantiomers were called **optical isomers** because they were distinguished by their optical activity. The term was loosely applied to more than one type of isomerism among optically active compounds, however, and this ambiguous term has been replaced by the well-defined term *enantiomers.*

Two enantiomers have identical physical properties, except for the direction in which they rotate the plane of polarized light.

> Enantiomeric compounds rotate the plane of polarized light by exactly the same amount but in opposite directions.

If the (*R*) isomer rotates the plane 30° clockwise, the (*S*) isomer will rotate it 30°

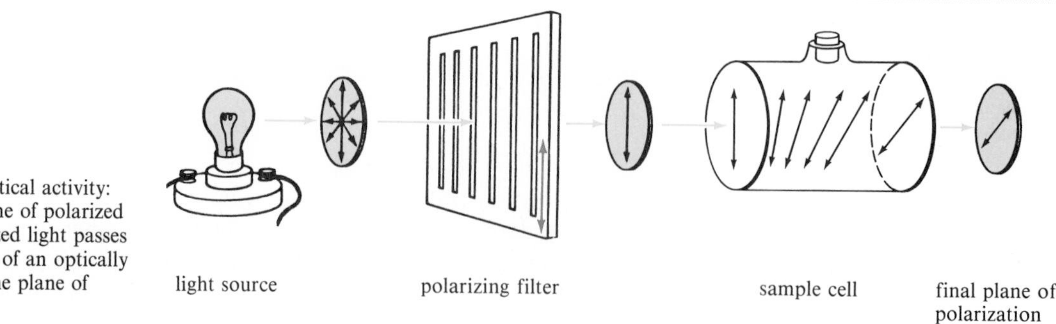

FIGURE 6-11 Optical activity: rotation of the plane of polarized light. When polarized light passes through a solution of an optically active substance, the plane of vibration rotates.

counterclockwise. If the (R) enantiomer rotates the plane 5° counterclockwise, the (S) enantiomer will rotate it 5° clockwise.

We cannot predict which direction a particular enantiomer [either (R) or (S)] will rotate the plane of polarized light.

(R) and (S) are simply terms of nomenclature, while the direction and magnitude of rotation are physical properties that must be measured.

6-4C POLARIMETRY

An instrument that measures the rotation of polarized light is called a **polarimeter.** It has a cell to contain a solution of the optically active material and a system for passing polarized light through the solution and measuring the rotation as the light emerges (Fig. 6-12). In the polarimeter, the light from a sodium lamp is first filtered so that it consists of just one wavelength (one color). This is necessary because most compounds rotate different wavelengths of light by different amounts. The wavelength of light most commonly used for polarimetry is one of the yellow emission lines in the spectrum of sodium, called the *sodium D line.* A sodium lamp with a yellow filter provides monochromatic (one color) light for polarimetry.

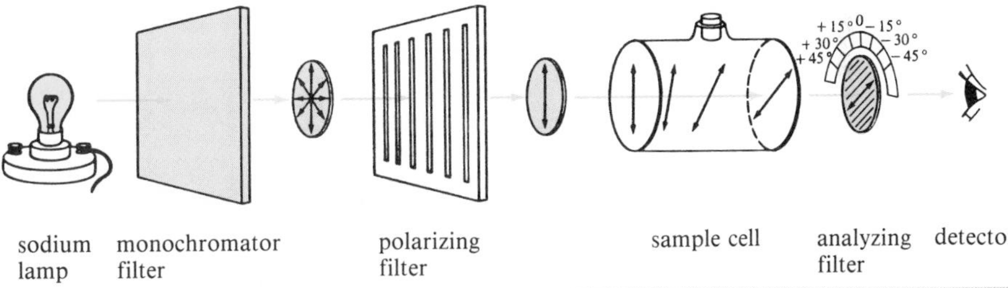

FIGURE 6-12 Schematic diagram of a polarimeter. The light originates at a source (usually a sodium lamp) and passes through a polarizing filter and the sample cell. The analyzing filter is another polarizing filter equipped with a protractor. It is turned until a maximum amount of light is observed, and the rotation is read from the protractor.

Monochromatic light from the source passes through a polarizing filter, polarizing the light in a known plane. The polarized light then passes through the sample cell containing a solution of the optically active compound. On leaving the sample cell, the polarized light encounters another polarizing filter. This filter is movable, with a scale allowing the operator to read the angle between the axis of the second (analyzing) filter and the axis of the first (polarizing) filter. The operator rotates the analyzing filter until the maximum amount of light is transmitted, and then the observed rotation is read from the protractor. The observed rotation is symbolized by α, the Greek letter alpha.

Compounds that rotate the plane of polarized light toward the right (clockwise) are called **dextrorotatory,** from the Greek word *dexios,* meaning "toward the right." Compounds that rotate the plane toward the left (counterclockwise) are called **levorotatory,** from the Latin word *laevus,* meaning "toward the left." These terms are sometimes abbreviated by a lowercase *d* or *l*. They are more commonly given using the sign of the rotation:

dextrorotatory (clockwise) rotations are (+);
levorotatory (counterclockwise) rotations are (−).

For example, the isomer of 2-butanol that rotates the plane of polarized light clockwise is named *d*-2-butanol or (+)-2-butanol. Its enantiomer, *l*-2-butanol or (−)-2-butanol, rotates the plane counter-clockwise by exactly the same amount.

You can see the principle of polarimetry by using two pairs of polarized sunglasses, a beaker, and some corn syrup. Wear one pair of sunglasses, look down at a light, and hold another pair of sunglasses above the light. Notice that the most light is transmitted through the two pairs of sunglasses when their axes are parallel. Very little light is transmitted when their axes are perpendicular.

Put syrup into the beaker, and hold the beaker above the bottom pair of sunglasses so that the light passes through one pair of sunglasses (the polarizing filter), then the beaker (the optically active sample), and then the other pair of sunglasses (the analyzing filter); see Figure 6-13.

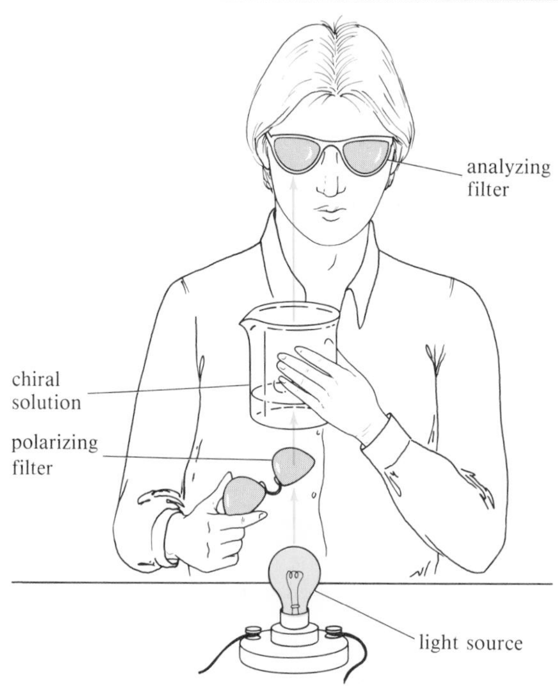

FIGURE 6-13 Apparatus for using a light bulb and two pairs of polarized sunglasses as a simple polarimeter.

Again, check the angles giving maximum and minimum light transmission. Is the syrup solution dextrorotatory or levorotatory? Did you notice the color variation as you rotated the filter? You can see why just one color of light should be used for accurate work.

6-4D SPECIFIC ROTATION

The rotation of polarized light by an optically active compound is a characteristic physical property of that compound, just like the boiling point or the density. The rotation (α) observed in a polarimeter depends on the concentration of the sample solution, the path length of the cell, and how strongly optically active the compound is. For example, twice as concentrated a solution would give twice the original rotation. Similarly, a 20-cm cell gives twice the rotation observed using a similar concentration in a 10-cm cell.

We define a **specific rotation** $[\alpha]$ whose value is independent of the conditions of measurement. Because the observed rotation α is proportional to how far the light travels through the solution, the observed rotation must be divided by the path length of the cell (l). The observed rotation is also proportional to the concentration (c) of the optically active compound in the sample solution, so the observed rotation is divided by c as well:

$$[\alpha] = \frac{\alpha(\text{observed})}{c \cdot l}$$

where

α(observed) = rotation observed in the polarimeter
c = concentration, g/ml
l = length of sample cell (path length), decimeters (dm)

Since many polarimeters have 10-cm (1-dm) cells, the units used for l are decimeters. Because the molecular weight of a new compound is often unknown, the concentration c is not measured using molarity. Instead, it is measured in grams of solute per milliliter of solution (g/mL), a concentration unit that can be used with any known or unknown compound.

SOLVED PROBLEM 6-4

When one of the enantiomers of 2-butanol is placed in a polarimeter, the observed rotation is 4.05° counterclockwise. The solution was made by diluting 6 g of (−)-2-butanol to a total of 40 mL, and the solution was placed into a 200-mm polarimeter tube for the measurement. Determine the specific rotation for this enantiomer of 2-butanol.

SOLUTION Since it is levorotatory, this must be a sample of (−)-2-butanol. The concentration is 6 g per 40 mL = 0.15 g/mL, and the path length is 200 mm = 2 dm. The specific rotation is

$$[\alpha]_{\text{D}}^{25} = \frac{-4.05^\circ}{(0.15)(2)} = -13.5^\circ$$

A rotation depends on the wavelength of light used and also on the temperature, so these data are given together with the rotation. In the expression above, the "25" means that the measurement was made at 25°C, and the "D" means that the light used was the D line of the sodium spectrum.

Without even measuring it, we can predict that the specific rotation for the other enantiomer of 2-butanol will be

$$[\alpha]_{\text{D}}^{25} = +13.5^\circ$$

where the (+) refers to the clockwise direction of the rotation. This enantiomer would be called (+)-2-butanol. We could refer to this pair of enantiomers as (+)-2-butanol and (−)-2-butanol or as (*R*)-2-butanol and (*S*)-2-butanol.

Does this mean that (*R*)-2-butanol is the dextrorotatory isomer and (*S*)-2-butanol is levorotatory? Not at all! The rotation of a compound, (+) or (−), is something that we measure in the polarimeter, depending on how the molecule interacts with light. The (*R*) and (*S*) nomenclature is our own artificial way of describing how the atoms are arranged in space.

In the laboratory it is simple to measure a rotation and see whether a particular substance is (+) or (−). On paper it is simple to determine whether a particular drawing is named (*R*) or (*S*). But it is difficult to predict whether a structure we call (*R*) will rotate polarized light clockwise or counterclockwise. Similarly, it is difficult to predict whether a dextrorotatory substance in a flask has the (*R*) or (*S*) configuration.

PROBLEM 6-7

A solution of 2.0 g of (+)-glyceraldehyde, $HOCH_2—CHOH—CHO$, in 10 mL of water was placed in a 100-mm cell. Using the sodium D line, a rotation of $+1.74°$ was found at 25°C. Determine the specific rotation of (+)-glyceraldehyde.

PROBLEM 6-8

A solution of 0.5 g of (−)-epinephrine (see Fig. 6-14) dissolved in 10 mL of dilute HCl was placed in a 20-cm polarimeter tube. Using the sodium D line, the rotation was found to be $-5.0°$ at 25°C. Determine the specific rotation of epinephrine.

PROBLEM 6-9

A chiral sample gives a rotation that is close to 180°. How can one tell whether this rotation is $+180°$ or $-180°$?

6-5 BIOLOGICAL DISCRIMINATION OF ENANTIOMERS

If only a polarimeter could differentiate between enantiomers, one might ask whether the difference is an important one. Any **chiral probe** can distinguish between enantiomers, however. For example, if you had some left-handed gloves and some right-handed ones, you could distinguish between them by checking to see which ones fit onto your right hand.

Enzymes in living systems are chiral, and they are capable of distinguishing between enantiomers. Usually, only one enantiomer of a pair can fit properly into the chiral active site of an enzyme. For instance, the levorotatory form of epinephrine is one of the principal hormones secreted by the adrenal medulla. When synthetic epinephrine is given to a patient, the (−) form has the same stimulating effect as the natural hormone. The (+) form lacks this effect and is mildly toxic. Figure 6-14 shows a simplified picture of how only the (−) enantiomer fits into the enzyme's active site.

Biological systems are capable of distinguishing between the enantiomers of many different chiral compounds. In general, just one of the enantiomers produces the characteristic effect; the other either produces no effect or has a totally different effect. Even your nose is capable of distinguishing between some enantiomers. For example, (−)-carvone is the fragrance associated with spearmint oil, while (+)-carvone has the tangy odor of caraway seed. We can conclude that the receptor sites for the sense of smell are chiral, just as the active sites in most enzymes are chiral. In general, enantiomers do not interact identically with other *chiral* molecules, whether or not they are of biological origin.

PROBLEM 6-10

Find the chiral carbon atoms in the enantiomers of carvone, and label each as (*R*) or (*S*).

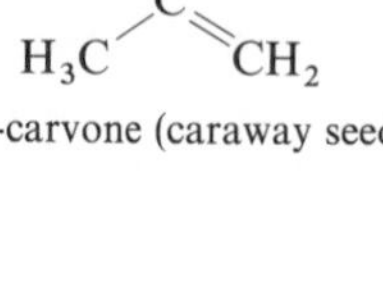

(+)-carvone (caraway seed) 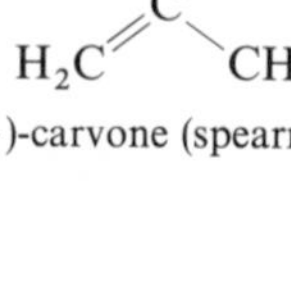 (−)-carvone (spearmint)

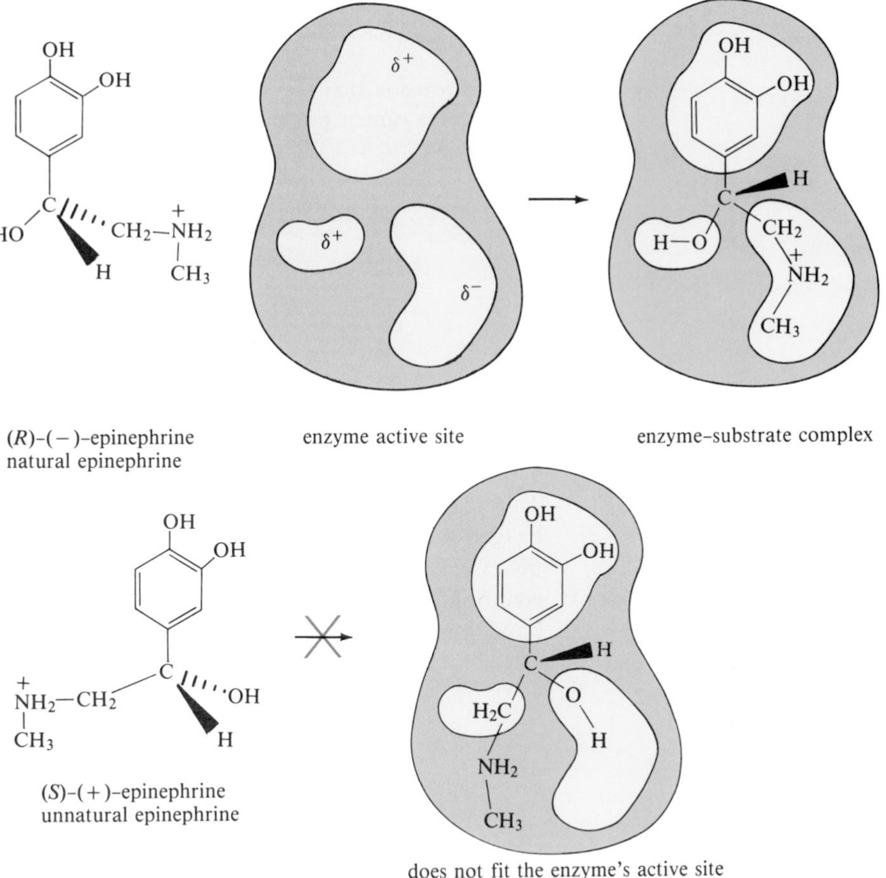

FIGURE 6-14 Chiral recognition of epinephrine by an enzyme. Only the levorotatory enantiomer fits into the active site of the enzyme.

6-6 RACEMIC MIXTURES

Suppose we had a mixture of equal amounts of (+)-2-butanol and (−)-2-butanol. The (+) isomer would rotate polarized light clockwise with a specific rotation of +13.5° and the (−) isomer would rotate the polarized light counterclockwise by exactly the same amount. We would observe a rotation of zero, just as though 2-butanol were achiral. A solution of equal amounts of two enantiomers, so that the mixture is optically inactive, is called a **racemic mixture.**

Sometimes a racemic mixture is called a **racemic modification,** a **racemate,** a ***(d,l)* pair,** or a **(±) pair.** A racemic mixture is symbolized by placing (±) or (*d,l*) in front of the name of the compound. For example, racemic 2-butanol would be symbolized as "(±)-2-butanol" or "(*d,l*)-2-butanol."

CH_3 — C (H, HO, CH_2CH_3) and CH_3 — C (CH_3CH_2, H, OH)

(*S*)-(+)-2-butanol
+13.5° rotation

(*R*)-(−)-2-butanol
−13.5° rotation

A racemic mixture contains equal amounts of the two enantiomers.

You might guess that a racemic mixture would be unusual, since it requires exactly equal quantities of the two enantiomers. This is not the case, however;

many reactions lead to racemic products, especially when an achiral molecule is converted to a chiral molecule.

> A reaction that uses optically inactive starting materials and optically inactive reagents cannot produce a product that is optically active. Any chiral product must be formed as a racemic mixture.

For example, hydrogen adds across the C=O double bond of a ketone to produce an alcohol.

$$RR'C{=}O \xrightarrow{H_2,\ Ni} R'{-}\underset{H}{\overset{R}{C}}{-}O{-}H$$

Because the carbonyl group is flat, a simple ketone such as 2-butanone is achiral. Hydrogenation of 2-butanone gives 2-butanol, a chiral molecule (Fig. 6-15). This reaction involves adding hydrogen atoms to the C=O carbon atom and oxygen atom. If the hydrogen atoms are added to one face of the double bond, the (*S*) enantiomer results. Addition of hydrogen to the other face forms the (*R*) enantiomer. It is equally probable for hydrogen to add to either face of the double bond, and equal amounts of the (*S*) and the (*R*) enantiomers are formed.

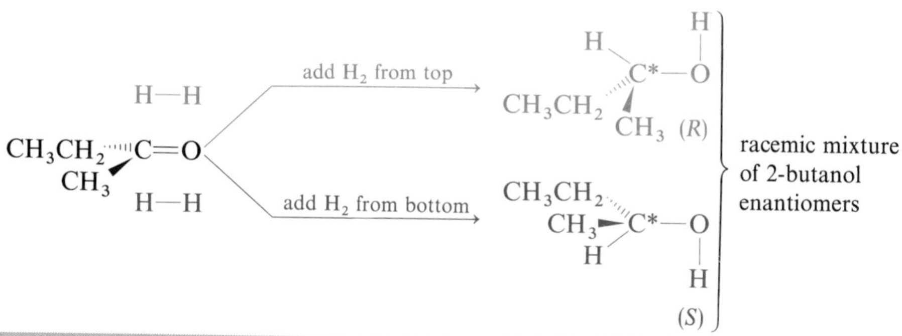

FIGURE 6-15 Hydrogenation of 2-butanone forms racemic 2-butanol. Hydrogen adds to either face of the double bond. Addition of H_2 to one face gives the (*R*) product, while addition to the other face gives the (*S*) product.

Logically, it makes sense that optically inactive reagents and catalysts cannot form optically active products. If the starting materials and reagents are optically inactive, how could we obtain a product that is dextrorotatory? There is no reason for a dextrorotatory product to be favored over a levorotatory one. The (+) product and the (−) product are favored equally, and they are formed in equal amounts: a racemic mixture.

6-7 ENANTIOMERIC EXCESS AND OPTICAL PURITY

Sometimes we deal with mixtures that are neither optically pure (all one enantiomer) nor racemic (equal amounts of the two enantiomers). In these cases, we specify its **optical purity** (o.p.). The optical purity of a mixture is defined as the ratio of its rotation to the rotation of a pure enantiomer. For example, if we have some [mostly (+)] 2-butanol with a specific rotation of +9.54°, we compare this rotation with the +13.5° rotation of the pure (+) enantiomer.

$$\text{o.p.} = \frac{\text{observed rotation}}{\text{rotation of pure enantiomer}} \times 100\% = \frac{9.54^\circ}{13.5^\circ} \times 100\% = 70.7\%$$

The **enantiomeric excess** (e.e.) is a similar method for expressing the relative amounts of enantiomers in a mixture. To compute the enantiomeric excess of a mixture, we calculate the *excess* of the predominant enantiomer as a percentage of the entire mixture. The calculation of enantiomeric excess generally gives the same result as the calculation of optical purity, and we often use the two terms interchangeably. Algebraically, we use the following formula:

$$\text{o.p.} = \text{e.e.} = \frac{d - l}{d + l} \times 100\% = \frac{\text{(excess of one over the other)}}{\text{(entire mixture)}} \times 100\%$$

The units cancel out in the calculation of either e.e. or o.p., so these formulas can be used whether the amounts of the enantiomers are expressed in concentrations or in grams.

SOLVED PROBLEM 6-5

Calculate the e.e. and the specific rotation of a mixture containing 6 g of (+)-2-butanol and 4 g of (−)-2-butanol.

SOLUTION In this mixture there is a 2-g excess of the (+) isomer and a total of 10 g, for an e.e. of 20%.

$$\text{o.p.} = \text{e.e.} = \frac{6 - 4}{6 + 4} = \frac{2}{10} = 20\%$$

The specific rotation of enantiomerically pure (+)−2-butanol is +13.5°. The rotation of this mixture is

$$\begin{aligned}\text{observed rotation} &= \text{(rotation of pure enantiomer)} \times \text{(o.p.)}\\ &= (+13.5^\circ) \times (20\%) = +2.7^\circ\end{aligned}$$

PROBLEM 6-11

When (*R*)-2-bromobutane is heated with water, the S_N1 substitution proceeds twice as fast as the S_N2. Calculate the e.e. and the specific rotation expected for the product. The rotations of the butanol enantiomers are shown on page 241.

PROBLEM 6-12

A chemist finds that the addition of (+)-epinephrine to the catalytic reduction of 2-butanone (Fig. 6-15) gives a product that is slightly optically active, with a specific rotation of +0.45°. Calculate the percentages of (+)-2-butanol and (−)-2-butanol formed in this reaction.

6-8 CHIRALITY OF CONFORMATIONALLY MOBILE SYSTEMS

Let's consider whether *cis*-1,2-dibromocyclohexane is chiral. If we did not know about chair conformations, we might draw a flat cyclohexane ring. With a flat ring, the molecule has an internal mirror plane of symmetry, and it is achiral.

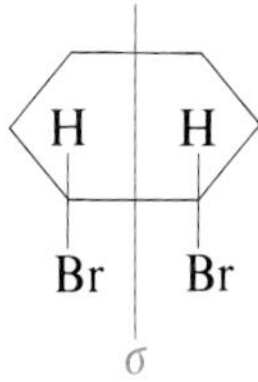

But we know that cyclohexane rings are not flat. In reality, the ring is puckered into a chair conformation with one bromine atom axial and one equatorial. A chair conformation of *cis*-1,2-dibromocyclohexane and its mirror image are shown below. These two mirror-image structures are nonsuperimposable. You may be able to see the difference more easily if you make models of these two conformations.

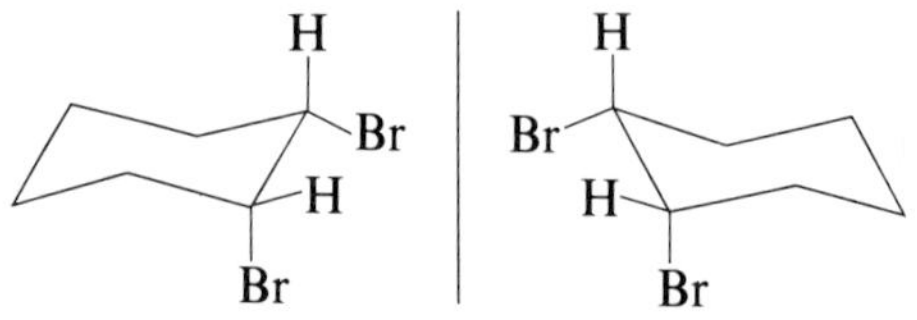

H Br H Br | H Br H Br

Does this mean that *cis*-1,2-dibromocyclohexane is chiral? No, it does not, because the chair-chair interconversion is rapid at room temperature. If we had a bottle of just the conformation on the left above, the molecules would quickly undergo chair-chair interconversions. Since the two mirror-image conformations are constantly interconverting and they have identical energies, any sample of *cis*-1,2-dibromocyclohexane must contain equal amounts of the two mirror images.

A molecule cannot be optically active if it is in equilibrium with its mirror image.

cis-1,2-Dibromocyclohexane exists as a racemic mixture, but with a major difference: It is impossible to have an optically active sample of *cis*-1,2-dibromocyclohexane. The molecule is simply incapable of showing optical activity. We could have predicted the correct result by imagining that the cyclohexane ring is flat.

This finding leads to a general principle that we can use with conformationally mobile systems:

To determine whether a conformationally mobile molecule can be optically active, consider its most symmetric conformation.

An alternative statement of this rule is that a molecule cannot be optically active if it is in equilibrium with a structure (or a conformation) that is achiral. Because **conformers** differ only by rotations about single bonds, they are generally in equilibrium at room temperature. We can consider cyclohexane rings as though they were flat (the most symmetric conformation), and we should consider straight-chain compounds in their most symmetric conformations, usually an eclipsed conformation.

SOLVED PROBLEM 6-6

Draw each of the following compounds in its most symmetric conformation, and determine whether it is capable of showing optical activity.

(a) 1,2-dichloroethane

SOLUTION The most symmetric conformation is the one with the two chlorine atoms eclipsed. This conformation has two internal mirror planes of symmetry, and it is definitely achiral.

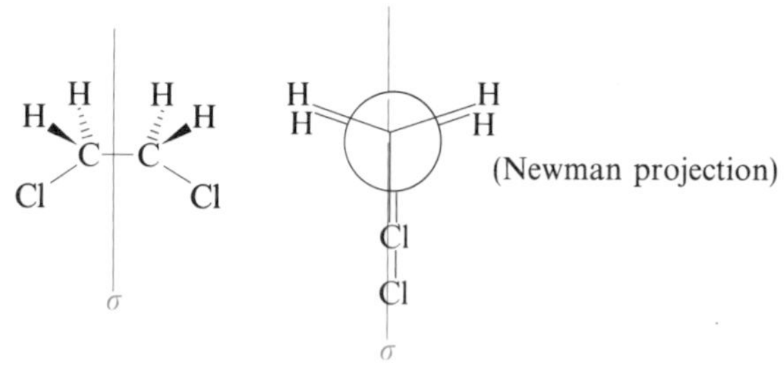

(Newman projection)

(b) *trans*-1,2-dibromocyclohexane

SOLUTION Drawn in its most symmetric conformation, this molecule is still nonsuperimposable on its mirror image. It is a chiral compound, and no chair-chair interconversion can interconvert the two enantiomers.

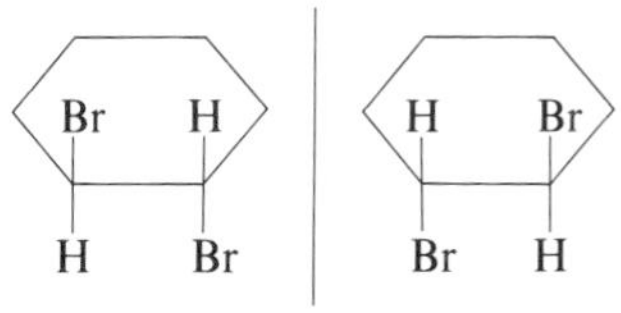

PROBLEM 6-13

Draw each of the following compounds in its most symmetric conformation and determine whether it is capable of showing optical activity.

(a) 1-bromo-1-chloroethane
(b) 1-bromo-2-chloroethane
(c) 1-bromo-1,2-dichloroethane
(d) *cis*-1,3-dibromocyclohexane
(e) *trans*-1,3-dibromocyclohexane
(f) *trans*-1,4-dibromocyclohexane

6-9 CHIRAL COMPOUNDS WITHOUT CHIRAL CARBON ATOMS

Most chiral organic compounds have at least one chiral carbon atom. There are some compounds that are chiral because they have another chiral atom, such as phosphorus, sulfur, or nitrogen, serving as a stereocenter. Some compounds are chiral even though they have no chiral atoms at all. In these types of compounds there are characteristics of the molecules' shapes that lend chirality to the molecular structure.

6-9A CONFORMATIONAL ENANTIOMERISM

Some molecules are so bulky or so highly strained that they cannot easily convert from one chiral conformation to the mirror-image conformation. They cannot achieve the most symmetric conformation because the most symmetric conformation has too much steric hindrance or ring strain. Since these molecules are "locked" into one conformation, we must evaluate the individual locked-in conformation to determine whether the molecule is chiral.

Figure 6-16 shows three conformations of a sterically crowded derivative of biphenyl. The center drawing shows the molecule in its most symmetric conformation. This conformation is planar, and it has a mirror plane of symmetry. If the molecule could achieve this conformation, or even pass through it for an instant, it could not be optically active. This planar conformation is very high in energy, however, because the iodine and bromine atoms are too large to be forced so close together. The molecule can exist only in one of the two staggered conformations

impossible, too crowded

staggered conformation (chiral)

eclipsed conformation (symmetric, achiral)

staggered conformation (chiral)

FIGURE 6-16 This crowded biphenyl cannot pass through its symmetric conformation because there is too much crowding of the iodine and bromine atoms. The molecule is "locked" into one of the two chiral, enantiomeric staggered conformations.

FIGURE 6-17 *trans*-Cyclooctene is strained, unable to achieve a symmetric, planar conformation. It is locked into one of these two enantiomeric conformations. *Trans*-cyclooctene has been isolated in optically active form.

shown on the left and right in Figure 6-16. These conformations are nonsuperimposable mirror images, and they cannot interconvert. They are enantiomers, and they can be separated and isolated. Each of them is optically active, and they have equal and opposite specific rotations.

Even a simple but strained molecule can show conformational enantiomerism. *trans*-Cyclooctene is the smallest stable *trans*-cycloalkene, and it is strained. If *trans*-cyclooctene existed as a planar ring, even for an instant, it could not be chiral. Make a molecular model of *trans*-cyclooctene, however, and you will see that it cannot exist as a planar ring. In fact, its ring is folded into the three-dimensional structure pictured in Figure 6-17. The mirror image of this structure is different, and *trans*-cyclooctene is a chiral molecule. In fact, the enantiomers of *trans*-cyclooctene have been separated and characterized, and they are optically active.

6-9B THE ALLENES

The *sp* hybrid carbon atom requires a linear arrangement of atoms (Section 2-4). There are two kinds of *sp* hybrid carbon atoms. The most common are those that participate in triple bonds, such as in alkynes or nitriles:

$$H—C≡C—H \qquad CH_3—C≡N:$$

sp hybridized

A carbon atom that participates simultaneously in two double bonds is alsó *sp* hybridized. Compounds containing a C=C=C unit are called **allenes.** The parent compound, propadiene, has the common name *allene*.

sp hybridized

$$H_2C=C=CH_2$$

allene

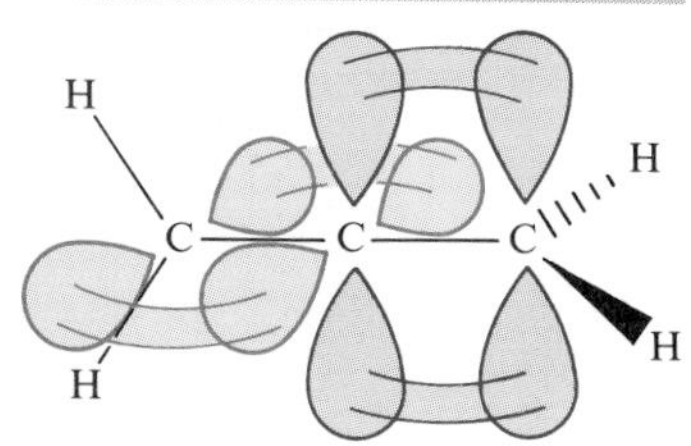

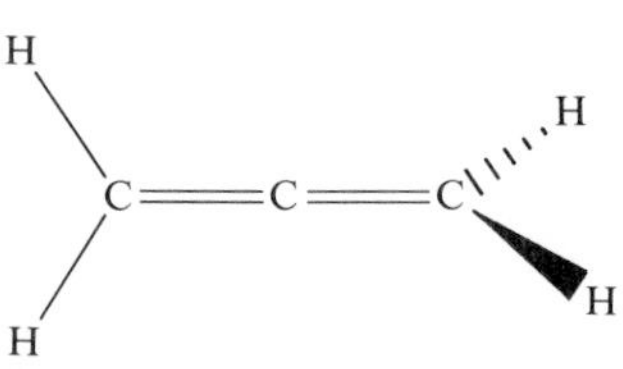

FIGURE 6-18 The two ends of the allene molecule are perpendicular.

In allene, the central carbon atom is *sp* hybridized and linear, while the two outer carbon atoms are sp^2 hybridized and trigonal. We might imagine that the whole molecule lies in a plane, but this is not correct. The central *sp* hybrid carbon atom must use different *p* orbitals to form the pi bonds with the two outer carbon atoms. The two unhybridized *p* orbitals on a *sp* hybrid carbon atom are perpendicular, so the two pi bonds must also be perpendicular. Figure 6-18 shows the three-dimensional structure of allene.

Allene itself is achiral. If you draw its mirror image, you will find it identical with the original molecule. If we add some substituents to allene, however, the situation is different.

Make a model of the following compound:

$$\overset{1}{CH_3}-\overset{2}{CH}=\overset{3}{C}=\overset{4}{CH}-\overset{5}{CH_3}$$

2,3-pentadiene

Carbon atom 3 is the *sp* hybrid allene carbon atom. Carbons 2 and 4 are both sp^2 and planar, but their planes are perpendicular to each other. None of the carbon atoms are attached to four different atoms, so there is no chiral carbon atom in this molecule. Nevertheless, 2,3-pentadiene is chiral, as you should see from your models and from the following drawings of the enantiomers.

H, H_3C C=C=C CH_3, H | H_3C, H C=C=C H, CH_3

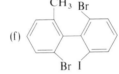

enantiomers of 2,3-pentadiene

PROBLEM 6-14

Draw three-dimensional representations of the following compounds. Which have chiral carbon atoms? Which have no chiral carbons but are chiral anyway? Use your models for any of these examples that seem unclear.

(a) 1,3-dichloropropadiene
(b) 1-chloro-1,2-butadiene
(c) 1-chloro-3-methyl-1,2-butadiene
(d) 1-chloro-1,3-butadiene
(e) bromocyclohexane
(f) CH_3 Br, Br I

(g) 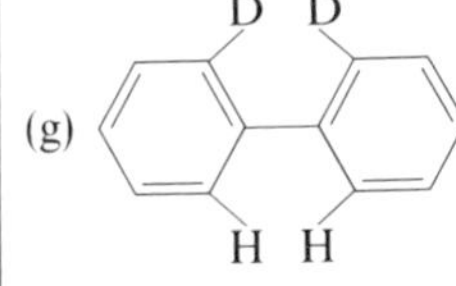

(h) 

6-10 FISCHER PROJECTIONS

We have been using dashed lines and wedges to indicate perspective in drawing the stereochemistry of chiral carbon atoms. When we draw molecules with several chiral carbons, the perspective drawings become time-consuming and cumbersome. In addition, the complicated perspective drawings make it difficult to see the similarities and differences in groups of stereoisomers.

At the turn of the century, Emil Fischer was studying the stereochemistry of sugars (Chapter 23), which contain as many as seven chiral carbon atoms. To draw these structures in perspective would have been difficult, and to pick out minor stereochemical differences in the drawings would have been nearly impossible. Fischer developed a symbolic way of drawing molecules with chiral carbon atoms, allowing them to be drawn rapidly. The **Fischer projection** also facilitates comparison of stereoisomers, holding them in their most symmetrical conformation and emphasizing any differences in their stereochemistry.

6-10A DRAWING FISCHER PROJECTIONS

The Fischer projection looks like a cross, with the chiral carbon (usually not drawn in) at the point where the lines cross. The horizontal lines are assumed to be wedges, that is, bonds that project out toward the viewer. The vertical lines are assumed to project away from the viewer, as dashed lines. Figure 6-19 shows the perspective that is implied by the Fischer projection. The center drawing illustrates why this projection is sometimes called the "bow tie convention." Problem 6-15 should help you to visualize what is meant by the Fischer projection formula.

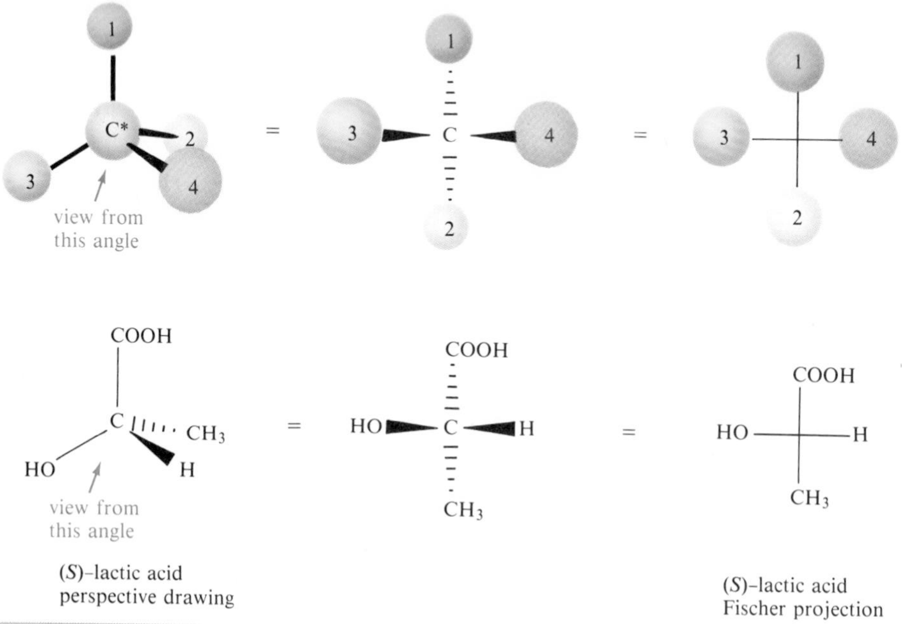

view from this angle

COOH, C, CH_3, HO, H — view from this angle — (*S*)-lactic acid perspective drawing

=

COOH, HO—C—H, CH_3

=

COOH, HO—+—H, CH_3 — (*S*)-lactic acid Fischer projection

FIGURE 6-19 The Fischer projection uses a cross to represent a chiral carbon atom. The horizontal lines project toward the viewer, and the vertical lines project away from the viewer.

PROBLEM 6-15

For each of the following sets of examples, make a model of the first structure and tell what the relationship of each of the other structures is to the first structure. Examples of relationships: same compound, enantiomer, structural isomer.

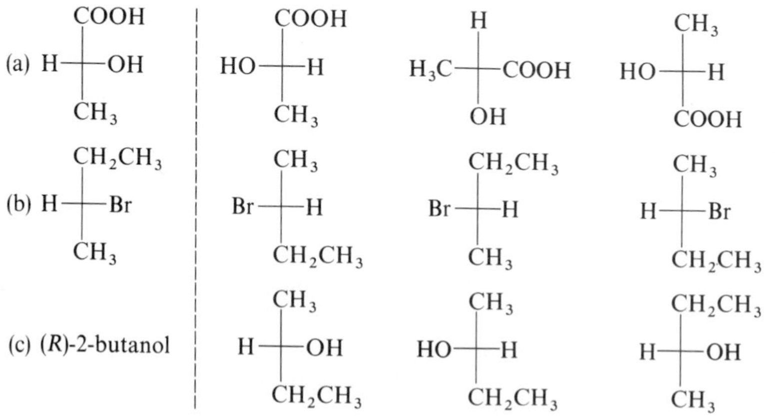

(a) COOH H—+—OH CH_3	COOH HO—+—H CH_3	H H_3C—+—COOH OH	CH_3 HO—+—H COOH
(b) CH_2CH_3 H—+—Br CH_3	CH_3 Br—+—H CH_2CH_3	CH_2CH_3 Br—+—H CH_3	CH_3 H—+—Br CH_2CH_3
(c) (*R*)-2-butanol	CH_3 H—+—OH CH_2CH_3	CH_3 HO—+—H CH_2CH_3	CH_2CH_3 H—+—OH CH_3

In working Problem 6-15, you may have noticed that Fischer projections that differ only by a 180° rotation are the same. When we rotate a Fischer projection by 180°, the vertical (dashed line) bonds still end up vertical, and the horizontal (wedged) lines still end up horizontal. The "horizontal lines forward, vertical lines back" convention is maintained. Rotation by 180° is allowed.

$$\begin{array}{c} COOH \\ H-\!\!\!+\!\!\!-OH \\ CH_3 \end{array} = \begin{array}{c} COOH \\ H\blacktriangleright C\blacktriangleleft OH \\ CH_3 \end{array} \overset{\text{rotate } 180^\circ}{=} \begin{array}{c} CH_3 \\ HO\blacktriangleright C\blacktriangleleft H \\ COOH \end{array} = \begin{array}{c} CH_3 \\ HO-\!\!\!+\!\!\!-H \\ COOH \end{array}$$

On the other hand, if we were to rotate a Fischer projection by 90°, we would change the configuration and confuse the viewer. The original projection has the vertical groups back (dashed lines) and the horizontal groups forward. When we rotate the projection by 90°, the vertical bonds become horizontal and the horizontal bonds become vertical. The viewer assumes that the horizontal bonds come forward and that the vertical bonds go back. The viewer sees a different molecule (actually, the enantiomer of the original molecule). A 90° rotation is NOT allowed.

$$\begin{array}{c} COOH \\ H-\!\!\!+\!\!\!-OH \\ CH_3 \end{array} = \begin{array}{c} COOH \\ H\blacktriangleright C\blacktriangleleft OH \\ CH_3 \end{array} \overset{90^\circ}{=} \begin{array}{c} H \\ H_3C\cdots C\cdots COOH \\ OH \end{array} \neq \begin{array}{c} H \\ H_3C-\!\!\!+\!\!\!-COOH \\ OH \end{array}$$

incorrect orientation — enantiomer

In comparing Fischer projections, we cannot rotate them by 90° and we cannot turn them over. Either of these operations gives an incorrect representation of the molecule. The Fischer projection must be kept in the plane of the paper, and it may be rotated only by 180°.

The final rule for drawing Fischer projections helps to ensure that we do not rotate the drawing by 90°. This rule is that the carbon chain is drawn along the vertical line of the Fischer projection, usually with the most highly oxidized end carbon atom at the top. For example, to represent (*R*)-1,2-propanediol with a Fischer projection, we should arrange the three carbon atoms along the vertical. The two ends of the molecule are a methyl group and a $-CH_2OH$ group. The $-CH_2OH$ group is more highly oxidized and is placed at the top.

$$\begin{array}{c} CH_2OH \\ | \\ H\diagup C \cdots CH_3 \\ \quad OH \end{array} = \begin{array}{c} CH_2OH \\ H\blacktriangleright C\blacktriangleleft OH \\ CH_3 \end{array} = \begin{array}{c} CH_2OH \\ H-\!\!\!+\!\!\!-OH \\ CH_3 \end{array}$$

viewed from this angle

PROBLEM 6-16

Draw a Fischer projection diagram for each of the following compounds. Remember that the cross represents the chiral carbon atom, and the carbon chain should be along the vertical, with the more highly oxidized end at the top.

(a) (*S*)-1,2-propanediol (b) (*R*)-2-bromo-1-butanol (c) (*S*)-1,2-dibromobutane

(d) (*R*)-2-butanol (e) (*R*)-glyceraldehyde, $HO-CH_2-\underset{}{\overset{OH}{\overset{|}{C}H}}-CHO$

6-10B DRAWING MIRROR IMAGES OF FISCHER PROJECTIONS

How does one draw the mirror image of a molecule drawn in Fischer projection? With our perspective drawings, the rule was to reverse left and right while keeping the other directions (up and down, front and back) in their same positions. This rule still applies to Fischer projections. Interchanging the groups on the horizontal part of the cross reverses left and right while leaving the other directions unchanged.

COOH / H—+—OH / CH_3 — (*R*)-lactic acid

COOH / HO—+—H / CH_3 — (*S*)-lactic acid

Testing for enantiomerism is particularly simple using Fischer projections. If the Fischer projections are properly drawn (carbon chain along the vertical), and if the mirror image cannot be made to look the same as the original structure with a 180° rotation in the plane of the paper, the two mirror images are enantiomers. In the following examples, any groups that fail to superimpose after a 180° rotation are circled in red.

Original: CH_3 / σ---H—+—OH--- / CH_3 | *Mirror image*: CH_3 / ---HO—+—H---σ / CH_3 — 180° → *180° rotation*: CH_3 / H—+—OH / CH_3

These mirror images are the same. 2-Propanol is achiral.

CH_2OH / H—+—OH / CH_3 | CH_2OH / HO—+—H / CH_3 — 180° → (CH_3) / H—+—OH / (CH_2OH)

These mirror images are different. 1,2-Propanediol is chiral.

CH_3 / H—+—Br / Br—+—H / CH_3 | CH_3 / Br—+—H / H—+—Br / CH_3 — 180° → CH_3 / (Br)—+—(H) / (H)—+—(Br) / CH_3

These mirror images are different. This structure is chiral.

Mirror planes of symmetry are particularly easy to identify from the Fischer projection, because this projection is normally the most symmetric conformation. In the first example above and in the following example, the symmetry planes are indicated in red; these examples clearly cannot be chiral.

CH$_3$ / H—Br / H—Br / CH$_3$ (σ) | CH$_3$ / Br—H / Br—H / CH$_3$ (σ) ⟶ 180° ⟶ CH$_3$ / H—Br / H—Br / CH$_3$

These mirror images are the same. This structure is achiral.

PROBLEM 6-17

For each of the following Fischer projections

(1) Draw the mirror image.

(2) Determine whether the mirror image is the same as, or different from, the original structure.

(3) Draw in any mirror planes of symmetry that are apparent from your Fischer projections.

(a) CHO / H—OH / CH$_2$OH

(b) CH$_2$OH / H—Br / CH$_2$OH

(c) CH$_2$Br / Br—Br / CH$_3$

(d) CHO / H—OH / H—OH / CH$_2$OH

(e) CH$_2$OH / H—OH / H—OH / CH$_2$OH

(f) CH$_2$OH / HO—H / H—OH / CH$_2$OH

6-10C ASSIGNING (*R*) AND (*S*) CONFIGURATIONS FROM FISCHER PROJECTIONS

The Cahn-Ingold-Prelog convention (Section 6-3) can be applied to structures drawn using Fischer projections. Let's review the two rules for assigning (*R*) and (*S*): (1) Assign priorities to the groups bonded to the chiral carbon atom; and (2) put the lowest-priority group (usually H) in back and draw an arrow from group 1 to group 2 to group 3. Clockwise is (*R*), and counterclockwise is (*S*).

The (*R*) or (*S*) configuration can also be determined directly from the Fischer projection, without having to convert it to a perspective drawing. The lowest-priority atom is usually hydrogen. In the Fischer projection, the carbon chain is along the vertical line, so the hydrogen atom is on the horizontal line and projects out in front. Once we have assigned priorities, we can draw an arrow from group 1 to group 2 to group 3 and see which way it goes. If the molecule were turned around so that the hydrogen is in back, the arrow would rotate in the other direction. By mentally turning the arrow around, we can assign the configuration.

As an example, consider the Fischer projection formula of one of the enantiomers of glyceraldehyde. First priority goes to the —OH group, followed by the —CHO group and the —CH$_2$OH group. The hydrogen atom receives the lowest priority. The arrow from group 1 to group 2 to group 3 appears counterclockwise in the Fischer projection. If the molecule is turned over so that the hydrogen is in back, the arrow is clockwise, and this is the (*R*) enantiomer of glyceraldehyde.

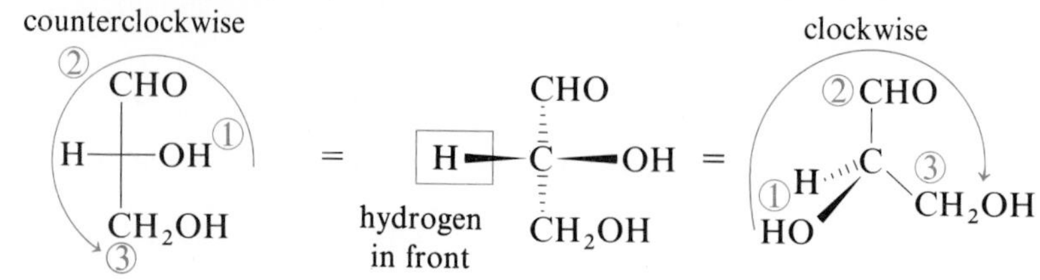

counterclockwise

Fischer projection
(*R*)-(+)-glyceraldehyde

hydrogen in front

clockwise

perspective drawing
(*R*)-(+)-glyceraldehyde

PROBLEM 6-18

For each of the following Fischer projections, determine whether each chiral carbon atom is of the (*R*) or the (*S*) configuration.

(a) the structures in Problem 6-17

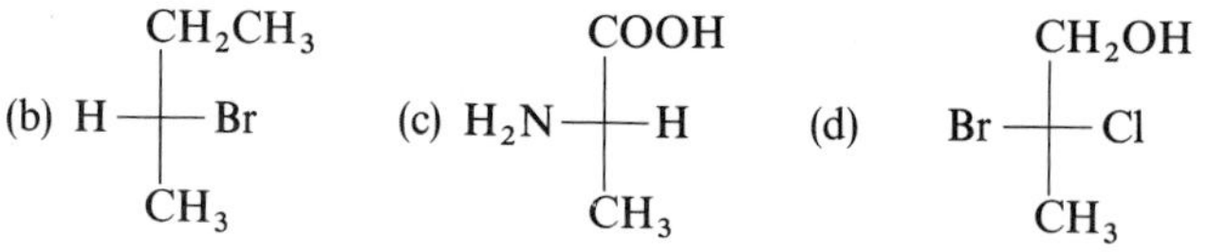

(b) H—C(—Br) with CH_2CH_3 on top, CH_3 on bottom
(c) H_2N—C(—H) with COOH on top, CH_3 on bottom
(d) Br—C(—Cl) with CH_2OH on top, CH_3 on bottom
(careful—no hydrogen)

SUMMARY OF FISCHER PROJECTIONS AND THEIR USE

1. Most useful for compounds with two or more chiral carbon atoms.
2. Chiral carbons are at the centers of crosses.
3. The vertical lines project away from the viewer, the horizontal lines toward the viewer.
4. The carbon chain is usually placed along the vertical, most oxidized end (the carbon with the most bonds to O or halogen) at the top.
5. The entire projection can be rotated 180° (but not 90°) in the plane of the paper without changing its stereochemistry.
6. Interchanging any two groups on a chiral carbon (for example, those on the horizontal line) inverts its stereochemistry.

6-11 DIASTEREOMERS

We have defined *stereoisomers* as isomers whose atoms are bonded together in the same order but differ in how the atoms are directed in space. We have also considered enantiomers, the mirror-image isomers, in detail. All other stereoisomers are classified as **diastereomers,** which are defined as *stereoisomers that are not mirror images.* Most diastereomers are either geometric isomers or compounds containing two or more chiral carbon atoms.

6-11A GEOMETRIC ISOMERISM ON DOUBLE BONDS

We have already seen one class of diastereomers, the **geometric,** or **cis-trans, isomers.** For example, there are two isomers of 2-butene:

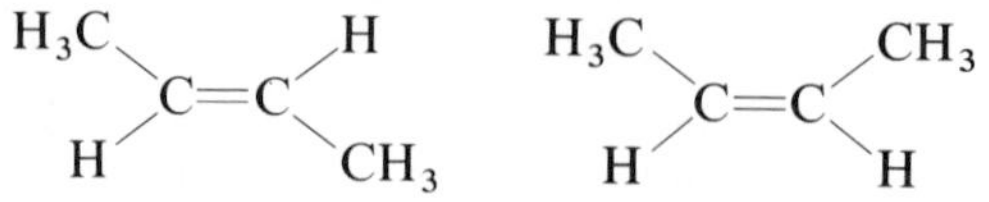

trans-2-butene

cis-2-butene

These stereoisomers are not mirror images of each other, so they are not enantiomers. They are diastereomers.

6-11B GEOMETRIC ISOMERISM ON RINGS

Geometric isomerism is also possible when there is a ring present. *cis*- and *trans*-1,2-Dimethylcyclopentane are geometric isomers, and they are also diastereomers. The trans diastereomer has an enantiomer, but the cis diastereomer has an internal mirror plane of symmetry, and it is achiral.

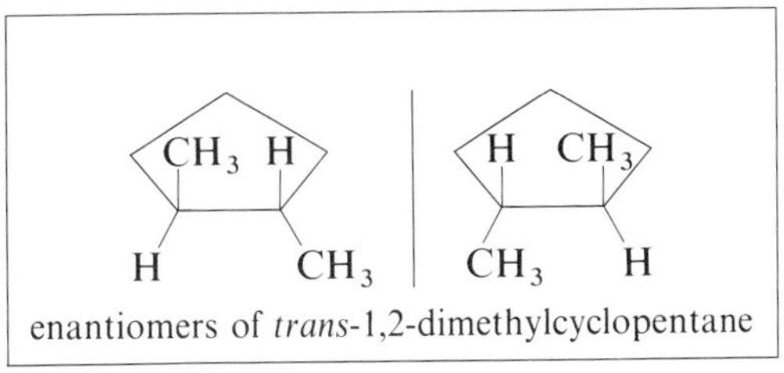

enantiomers of *trans*-1,2-dimethylcyclopentane

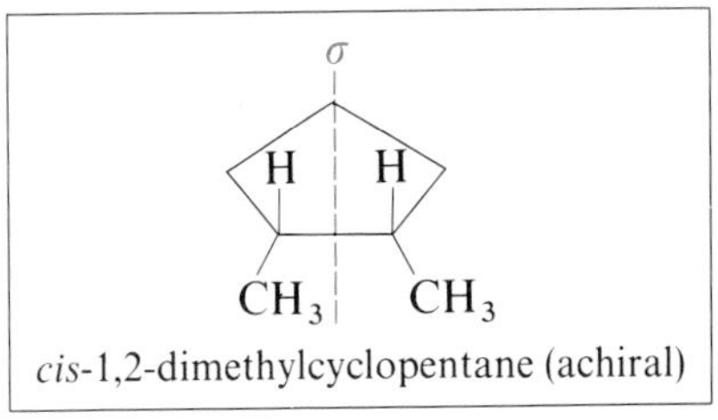

cis-1,2-dimethylcyclopentane (achiral)

6-11C DIASTEREOMERS OF MOLECULES WITH TWO OR MORE CHIRAL CARBON ATOMS

Apart from geometric isomers, most other compounds that show diastereomerism have two or more chiral carbon atoms. For example, 2-bromo-3-chlorobutane has two chiral carbon atoms, and it exists in two diastereomeric forms. The two diastereomers of 2-bromo-3-chlorobutane are shown below. Make molecular models of these two stereoisomers.

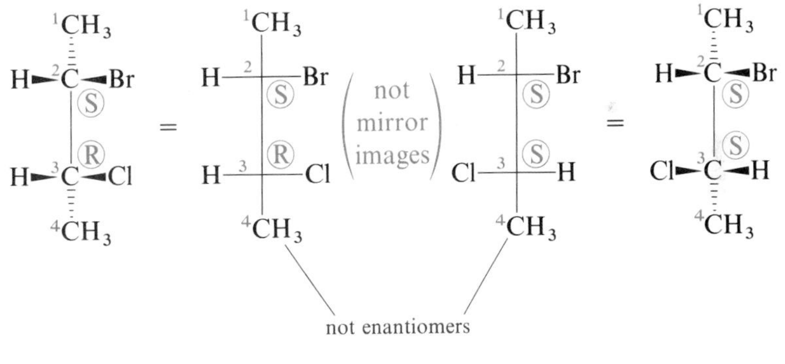

not mirror images

not enantiomers

These two structures are not the same; they differ in the orientation of their atoms in space and therefore are stereoisomers. They are not enantiomers, however, because they are not mirror images of each other: C2 has the (*S*) configuration in both structures, while C3 is (*R*) in the structure on the left and (*S*) in the structure on the right. The number 3 carbon atoms are mirror images of each other, but the number 2 carbon atoms are not. If these two compounds were mirror images of each other, both chiral carbons would have to be mirror images of each other.

Since these two compounds are stereoisomers but not enantiomers, they must be diastereomers. In fact, both of these diastereomers are chiral and each has an enantiomer. Thus, there are a total of four stereoisomeric 2-bromo-3-chlorobutanes: two pairs of enantiomers. Either member of one pair of enantiomers is a diastereomer of either member of the other pair.

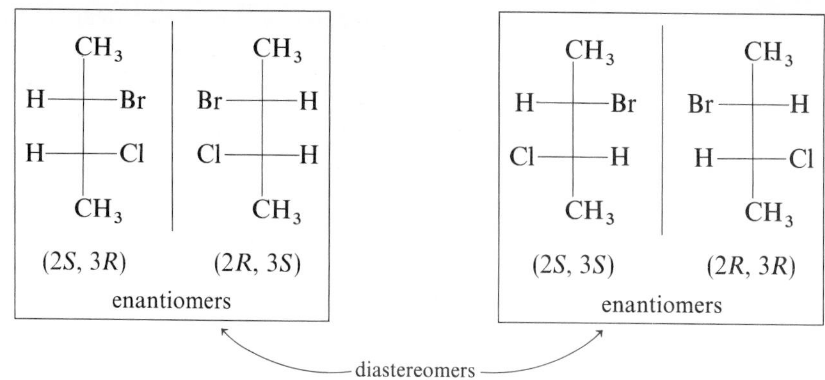

We have now seen all the types of isomers that we need to study, and we can diagram their relationships and summarize their definitions.

SUMMARY OF TYPES OF ISOMERS

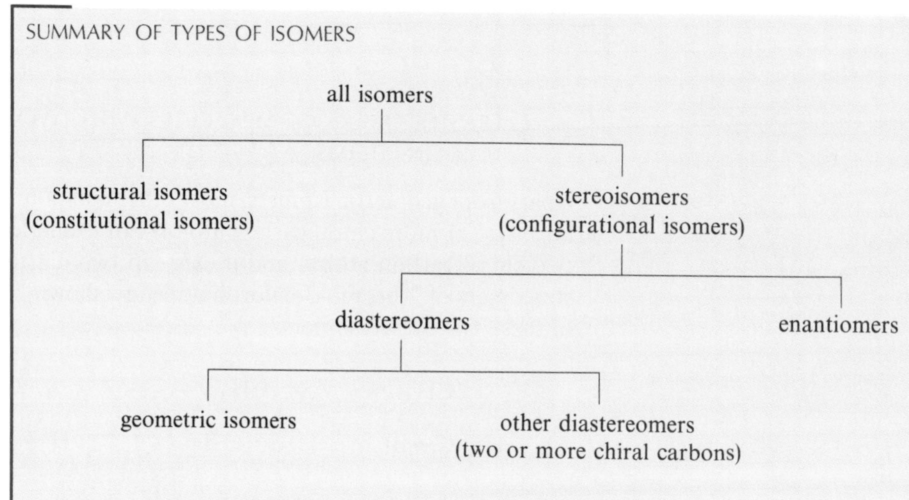

Isomers are different compounds with the same molecular formula.

Structural isomers are isomers that differ in the order in which atoms are bonded together. Structural isomers are sometimes called **constitutional isomers** because they differ in their pattern of bonding.

Stereoisomers are isomers that differ only in the orientation of the atoms in space. Stereoisomers are sometimes called **configurational isomers** as a counterpart to the term "constitutional isomers."

Enantiomers are mirror-image isomers.

Diastereomers are stereoisomers that are not mirror images of each other.

Geometric isomers are diastereomers that differ in their cis-trans arrangement on a ring or a double bond.

PROBLEM 6-19

For each of the following pairs, give the relationship between the two compounds. Making models will be helpful.

(a) (2*R*,3*S*)-2,3-dibromohexane and (2*S*,3*R*)-2,3-dibromohexane
(b) (2*R*,3*S*)-2,3-dibromohexane and (2*R*,3*R*)-2,3-dibromohexane

(c) 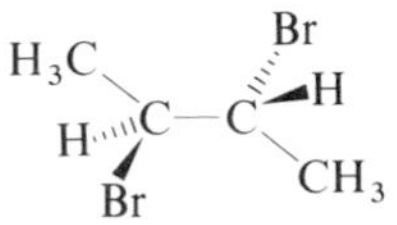 and

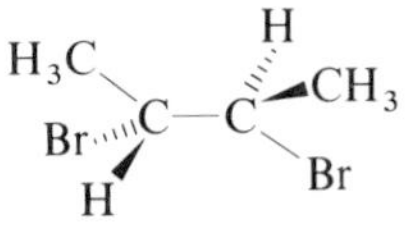

(d) 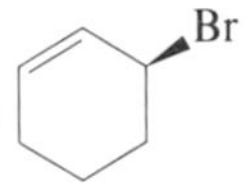 and 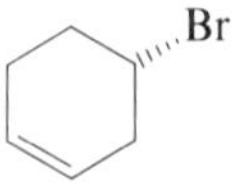 (e) 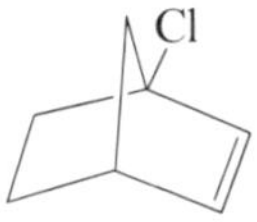 and

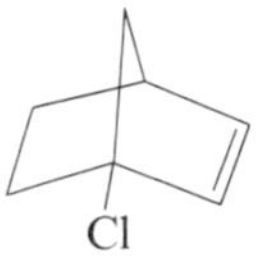

(f) 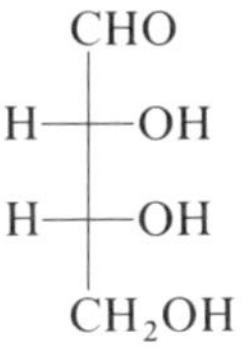 and 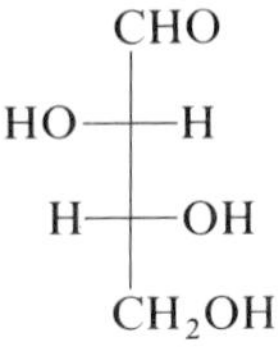 (g) 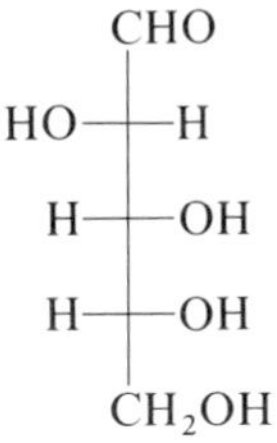 and

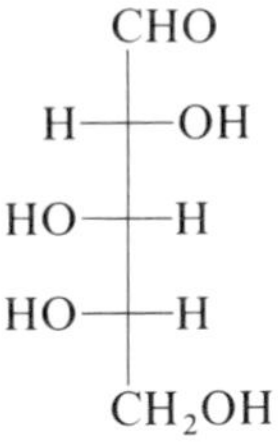

(h) 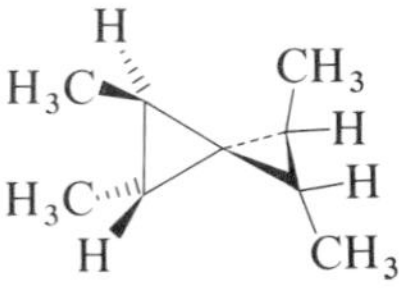 and its mirror image

6-12 STEREOCHEMISTRY OF MOLECULES WITH TWO OR MORE CHIRAL CARBONS

In the preceding section we saw there are four stereoisomers (two pairs of enantiomers) of 2-bromo-3-chlorobutane. These four isomers are simply all the permutations of (*R*) and (*S*) configurations at the two chiral carbon atoms, C2 and C3:

diastereomers

(2*R*,3*R*) (2*S*,3*S*)	(2*R*,3*S*) (2*S*,3*R*)
enantiomers	enantiomers

We might suspect that a compound with *n* chiral carbon atoms would have 2^n stereoisomers. This formula often works, and it is called the **2^n rule,** where *n* is the number of chiral carbon atoms. The 2^n rule suggests we should look for a *maximum* of 2^n stereoisomers, although we may not always find 2^n isomers.

2,3-Dibromobutane is an example of a compound having fewer than 2^n stereoisomers. It has two chiral carbons, C2 and C3, and the 2^n rule predicts a maximum of four stereoisomers. The four permutations of (*R*) and (*S*) configurations at C2 and C3 are shown below. Make molecular models of these structures to compare them.

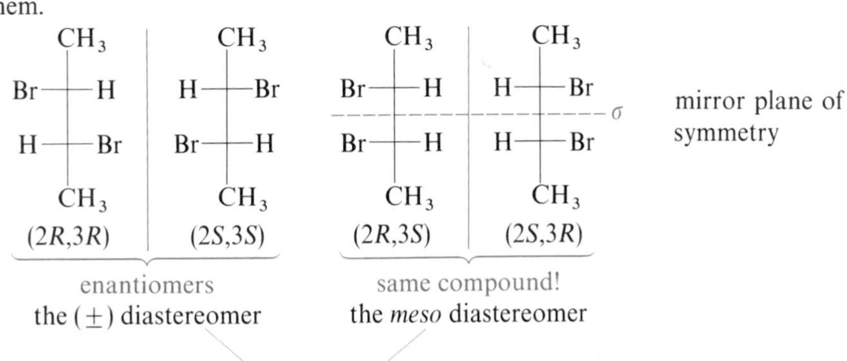

There are only three stereoisomers of 2,3-dibromobutane, because two of the four structures are identical. The diastereomer on the right is achiral, having a mirror plane of symmetry. The chiral carbon atom with (*R*) configuration reflects into the other chiral carbon having (*S*) configuration, almost as though the molecule were a racemic mixture within itself.

6-13 MESO COMPOUNDS

Compounds that are achiral even though they have chiral carbon atoms are called **meso compounds.** The (2*R*,3*S*) isomer of 2,3-dibromobutane is a meso compound; most meso compounds have this kind of symmetric structure, with two similar halves of the molecule having opposite configurations. In speaking of the two diastereomers of 2,3-dibromobutane, the symmetric one is called the *meso diastereomer,* and the chiral one is called the (±) *diastereomer,* since one enantiomer is (+) and the other is (−).

MESO COMPOUND An achiral compound that has chiral carbon atoms.

We have already seen other meso compounds, although we have not yet called them that. For example, the cis isomer of 1,2-dichlorocyclopentane has two chiral carbon atoms, yet it is achiral. Therefore, it is a meso compound. The same is true of *cis*-1,2,-dibromocyclohexane. Even though it is not symmetric in its chair conformation, *cis*-1,2-dibromocyclohexane consists of equal amounts of the two enantiomeric chair conformations in a rapid equilibrium. We are justified in looking at the molecule in its flat conformation to determine whether it is chiral.

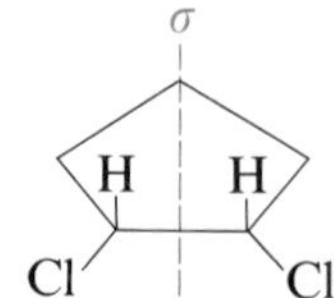

cis-1,2-dichlorocyclopentane

σ
H H
Br Br
cis-1,2-dibromocyclohexane

SOLVED PROBLEM 6-7

Determine which of the following compounds are optically active. Star any chiral carbon atoms, and draw in any mirror planes. Label any meso compounds. (Use your molecular models to follow along.)

(a) CH_3 / H—OH / HO—H / CH_3

(b) CH_2OH / Br—Cl / Br—Cl / CH_2OH

(c) CH_3 / H—OH / HO—H / HO—H / H—OH / CH_3

(d) H / Br / H / Cl

SOLUTION

(a) CH_3 / H—*—OH / HO—*—H / CH_3 | CH_3 / HO—*—H / H—*—OH / CH_3

This compound does *not* have a plane of symmetry, and we might suspect that it is chiral. Drawing the mirror image shows that it is nonsuperimposable on the original structure. These are the enantiomers of a chiral compound.

(b) Fischer projection: CH_2OH top; Br—C*—Cl; mirror plane σ; Br—C*—Cl; CH_2OH bottom — meso

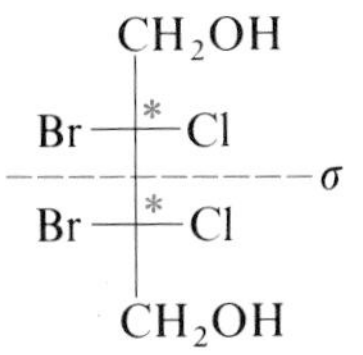

(c) Fischer projection: CH_3 top; H—C*—OH; HO—C*—H; mirror plane σ; HO—C*—H; H—C*—OH; CH_3 bottom — meso

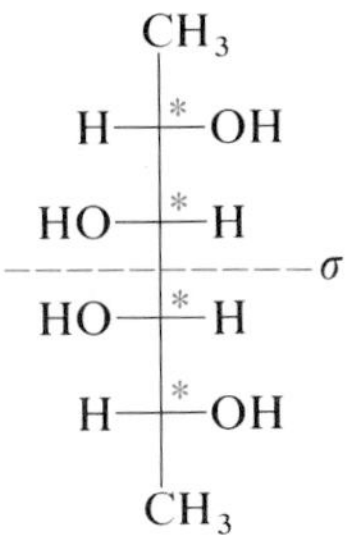

Both (b) and (c) have mirror planes of symmetry and are achiral. Because they have chiral carbon atoms yet are achiral, they are meso.

(d) [ring with H, H on starred carbons; Cl and Br substituents] | [mirror image: ring with H, H; Br and Cl substituents]

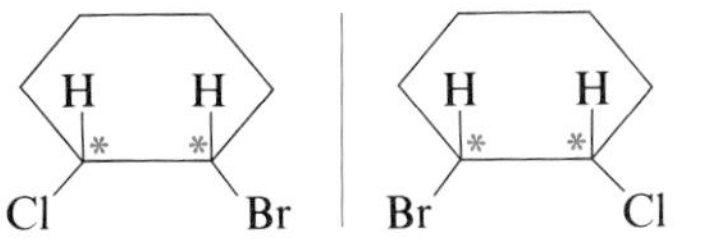

Drawing this compound in its most symmetric conformation (flat) shows that it does not have a mirror plane of symmetry. When we draw the mirror image, it is found to be an enantiomer.

PROBLEM 6-20

Which of the following compounds are optically active? Draw each compound in its most symmetric conformation, star any chiral carbon atoms, and draw in any mirror planes. Label any meso compounds. You may use Fischer projections if you prefer.

(a) *meso*-2,3-dibromo-2,3-dichlorobutane (b) (±)-2,3-dibromo-2,3-dichlorobutane

(c) butane (d) (2*R*,3*S*) $\overset{4}{H}OCH_2$—$\overset{3}{C}HBr$—$\overset{2}{C}HOH$—$\overset{1}{C}H_2OH$

(e) Fischer projection: CHO top; H—OH; HO—H; H—OH; H—OH; CH_2OH bottom

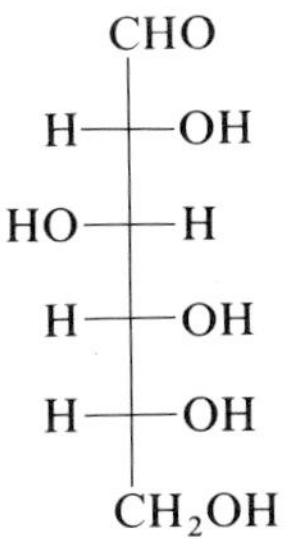

(f) [chair cyclohexane with substituents: H, Br, CH_3, H, H, Br, CH_3, H]

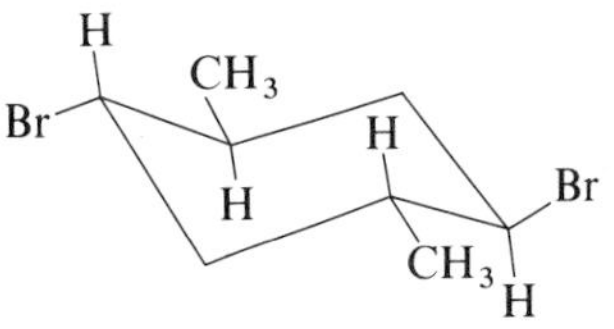

(g) [chair cyclohexane with substituents: Br, H, H, Br, H, CH_3, CH_3, H]

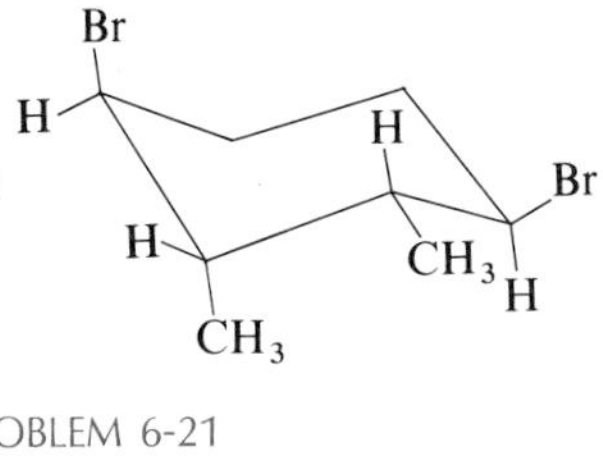

PROBLEM 6-21

Draw all the distinct stereoisomers for each of the following structures. Show the relationships (enantiomers, diastereomers, etc.) between the isomers. Label any meso isomers and draw in any mirror planes of symmetry.

(a) CH_3—CHCl—CHOH—COOH
(b) tartaric acid, HOOC—CHOH—CHOH—COOH
(c) HOOC—CHBr—CHOH—CHOH—COOH

(d) HO— [cyclopropane ring bearing two CH_3 groups]

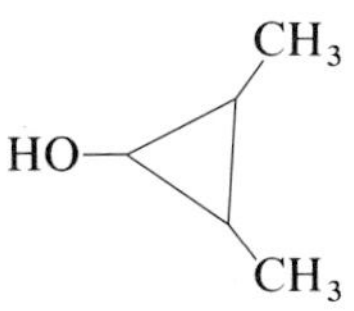

6-14 PHYSICAL PROPERTIES OF DIASTEREOMERS

We have seen that enantiomers have identical physical properties except for the direction in which they rotate polarized light. Diastereomers, on the other hand, generally have different physical properties. For example, consider the diastereomers of 2-butene (shown below). The symmetry of *trans*-2-butene causes the dipole moments of the bonds to cancel. The dipole moments in *cis*-2-butene do not cancel, but add together to create a molecular dipole moment. The dipole-dipole attractions of *cis*-2-butene give it a higher boiling point than that of *trans*-2-butene.

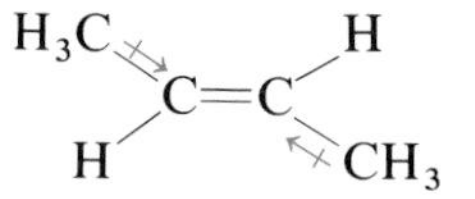

$H_3C(H)C{=}C(H)CH_3$	$H_3C(H)C{=}C(H)CH_3$
$\mu = 0$	$\mu = 0.33$ D
trans-2-butene	*cis*-2-butene
bond dipoles cancel	vector sum dipole ↕
b.p. = 0.9°C	b.p. = 3.7°C

Diastereomers that are not geometric isomers also have different physical properties. The two diastereomers of 2,3-dibromosuccinic acid have melting points that differ by over 100°C!

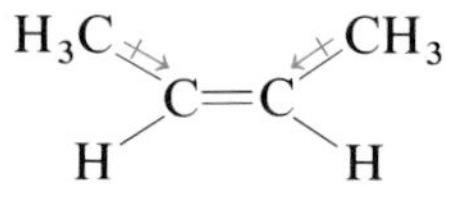

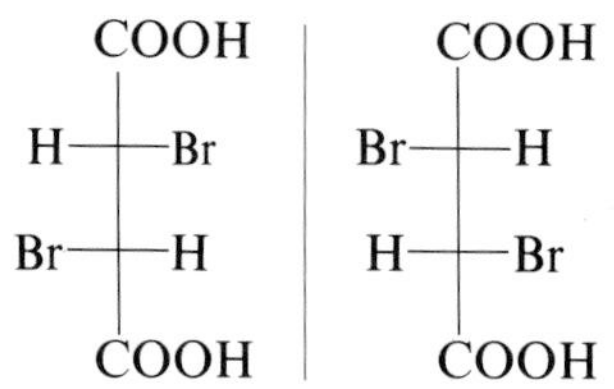

COOH; H–C–Br; Br–C–H; COOH	COOH; Br–C–H; H–C–Br; COOH	COOH; H–C–Br; H–C–Br; COOH
(+) and (−)-2,3-dibromosuccinic acid		*meso*-2,3-dibromosuccinic acid
m.p. for either is 158°C		m.p. = 273°C

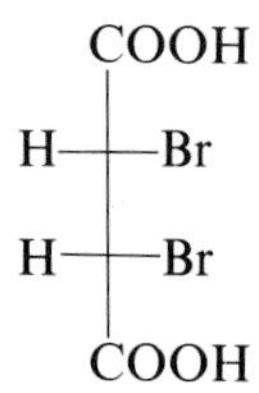

Most of the common sugars are diastereomers of glucose. All these diastereomers have different physical properties. For example, glucose and galactose are diastereomeric sugars that differ only in the stereochemistry of one chiral carbon atom, C4.

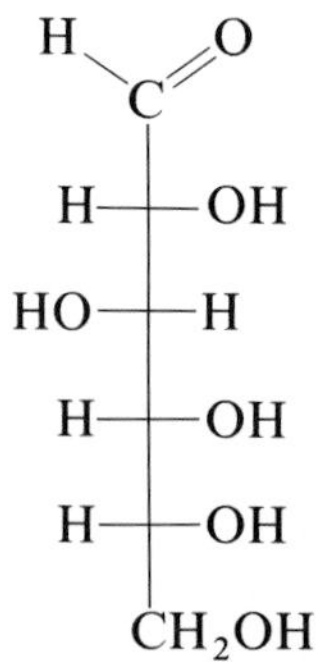

(+)-glucose	(+)-galactose
CHO	CHO
H–C–OH	H–C–OH
HO–C–H	HO–C–H
H–C–OH	HO–C–H
H–C–OH	H–C–OH
CH_2OH	CH_2OH

(+)-glucose, m.p. 148°C 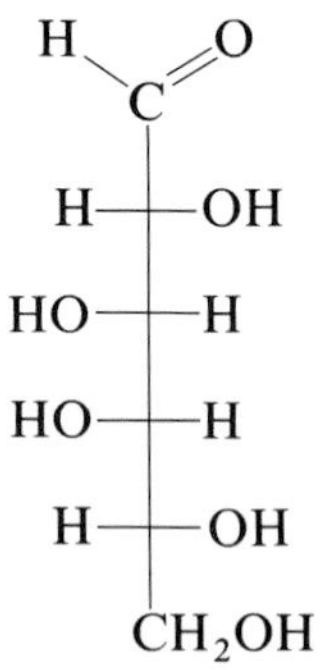 (+)-galactose, m.p. 167°C

The fact that diastereomers have different physical properties means that they may be separated by ordinary means such as distillation, recrystallization, and chromatography. As we will see in the next section, the separation of enantiomers is a more difficult process.

PROBLEM 6-22

Which of the following pairs of compounds could be separated by recrystallization or distillation?

(a) *meso*-tartaric acid and (*d,l*)-tartaric acid (HOOC—CHOH—CHOH—COOH)

(b) 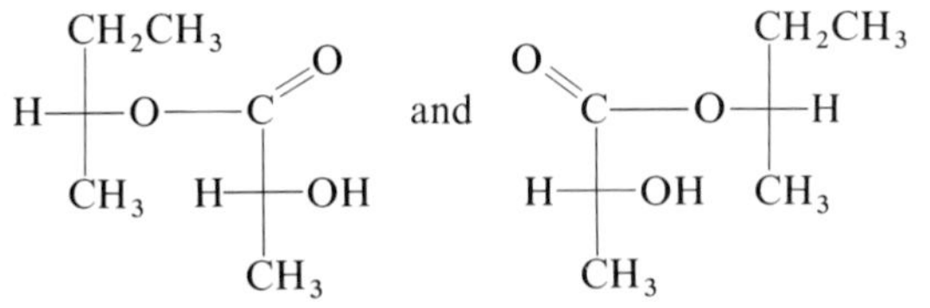 [Fischer-projection ester structures] and [Fischer-projection ester structures]

(c) 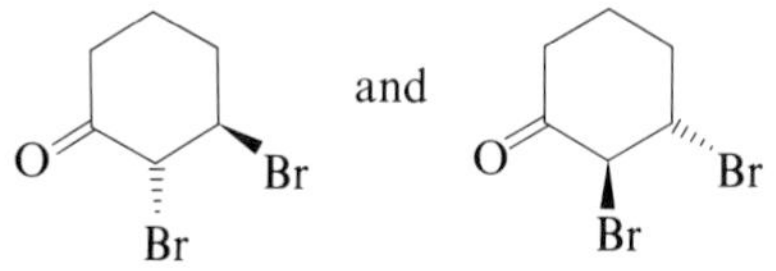 [2,3-dibromocyclohexanone stereoisomers] and [2,3-dibromocyclohexanone stereoisomers]

(d) 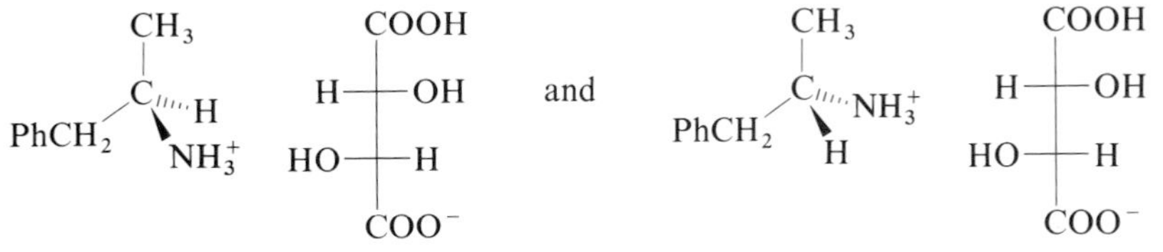 $PhCH_2$–C(H)(CH_3)–NH_3^+ with tartrate (COO^-) (an acid-base salt) and $PhCH_2$–C(H)(CH_3)–NH_3^+ with tartrate (COO^-)

6-15 RESOLUTION OF ENANTIOMERS

Pure enantiomers of optically active compounds are often obtained by isolation from biological sources. Most optically active molecules are found as only one enantiomer in living organisms. For example, pure (+)-tartaric acid can be isolated from the precipitate formed by yeast during the fermentation of wine. Pure (+)-glucose is obtained from many different sugar sources, such as grapes, sugar beets, sugar cane, and honey.

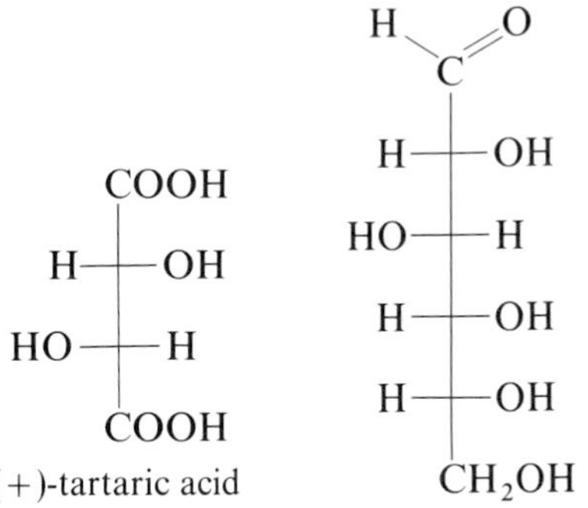

(+)-tartaric acid
(2*R*,3*R*)-tartaric acid

(+)-glucose

When a chiral compound is synthesized from achiral reagents, however, a racemic mixture of enantiomers is obtained. For example, we saw that the reduction of 2-butanone (achiral) to 2-butanol (chiral) gives a racemic mixture:

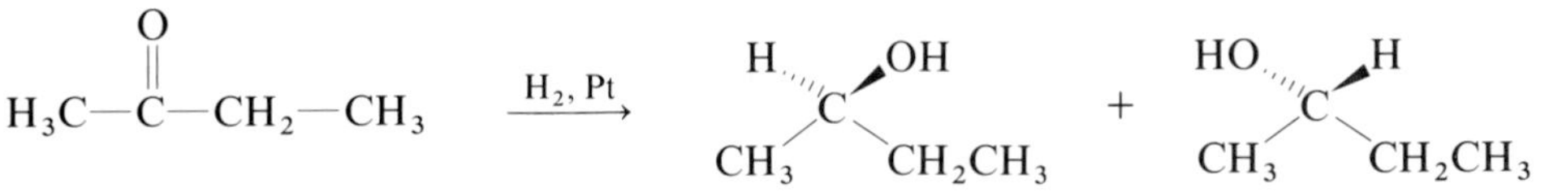

$$H_3C-\overset{O}{\overset{\|}{C}}-CH_2-CH_3 \xrightarrow{H_2,\ Pt} (R)\text{-2-butanol} + (S)\text{-2-butanol}$$

2-butanone — (*R*)-2-butanol — (*S*)-2-butanol

If we need just one pure enantiomer of 2-butanol, we must find a way of separating it from the other enantiomer. The separation of enantiomers is called **resolution,** and it is a very different process from the usual physical separations. A chiral probe is necessary for the resolution of enantiomers; such a chiral compound or apparatus is called a **resolving agent.**

In 1848, Louis Pasteur noticed that racemic (±)-tartaric acid crystallizes into mirror-image crystals. Using a microscope and a pair of tweezers, he physically separated the enantiomeric crystals. He found that solutions made from the "left-handed" crystals rotate polarized light in one direction, and solutions made from the "right-handed" crystals rotate polarized light in the opposite direction. Pasteur had accomplished the first artificial resolution of enantiomers. Unfortunately, few racemic compounds crystallize as separate enantiomers, and other methods are required.

6-15A CHEMICAL RESOLUTION OF ENANTIOMERS

The most common method for resolving a racemic mixture into its enantiomers is to use an enantiomerically pure natural product that bonds with the compound to be resolved. When the enantiomers of the racemic compound bond to the pure resolving agent, a pair of diastereomers results. The diastereomers are separated, then the resolving agent is cleaved from the separated enantiomers.

Let's consider how a racemic mixture of (*R*)- and (*S*)-2-butanol might be resolved. We need a resolving agent that reacts with an alcohol and that is readily available in an enantiomerically pure state. A carboxylic acid combines with an alcohol to form an ester. Although we have not yet studied the chemistry of esters (Chapter 21), you can see how an acid and an alcohol can combine with the loss of water:

$$\underset{\text{acid}}{\mathrm{R{-}\overset{\overset{\displaystyle O}{\|}}{C}{-}OH}} + \underset{\text{alcohol}}{\mathrm{R'{-}OH}} \xrightleftharpoons{(\mathrm{H^+}\ \text{catalyst})} \underset{\text{ester}}{\mathrm{R{-}\overset{\overset{\displaystyle O}{\|}}{C}{-}O{-}R'}} + \underset{\text{water}}{\mathrm{H{-}O{-}H}}$$

For our resolving agent, we need an optically active chiral acid to react with 2-butanol. Any winery can provide large amounts of pure (+)-tartaric acid. Figure 6-20 shows that diastereomeric esters are formed when (*R*)- and (*S*)-2-butanol react with (+)-tartaric acid. We can represent the reaction schematically as follows:

(*R*)- and (*S*)-2-butanol plus (*R*,*R*)-tartaric acid	$\xrightarrow{H^+}$	(*R*)-2-butyl (*R*,*R*)-tartrate	+	(*S*)-2-butyl (*R*,*R*)-tartrate
		diastereomers, *not* mirror images		

The diastereomers of 2-butyl tartrate have different physical properties, and they can be separated by conventional distillation, recrystallization, or chromatography. Separation of the diastereomers leaves us with two flasks, each containing one of the diastereomeric esters. The resolving agent is then cleaved from the separated enantiomers of 2-butanol by the reverse of the reaction used to make the ester. Adding an acid catalyst and an excess of water to an ester drives the equilibrium toward the acid and the alcohol:

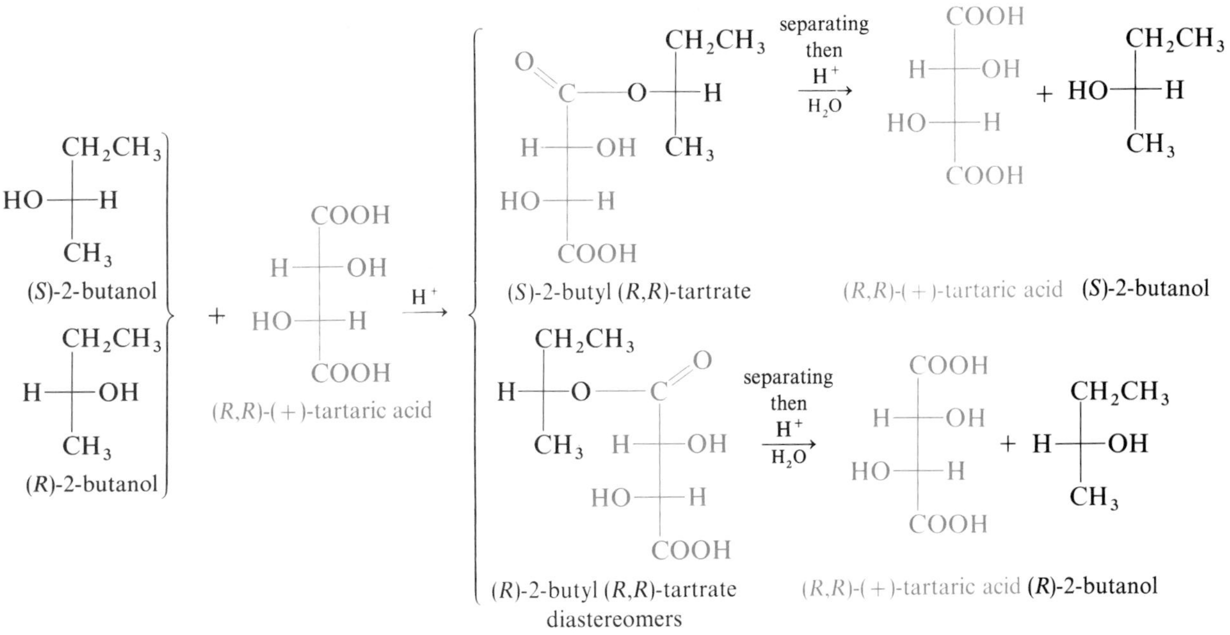

$$\left.\begin{matrix}\text{(S)-2-butanol: } CH_2CH_3, \; HO\text{—}C\text{—}H, \; CH_3 \\ \text{(R)-2-butanol: } CH_2CH_3, \; H\text{—}C\text{—}OH, \; CH_3\end{matrix}\right\} + \text{(R,R)-(+)-tartaric acid} \xrightarrow{H^+} \left\{\begin{matrix}\text{(S)-2-butyl (R,R)-tartrate} \\ \text{(R)-2-butyl (R,R)-tartrate}\end{matrix}\right.$$

(*S*)-2-butanol

(*R*)-2-butanol

(*R*,*R*)-(+)-tartaric acid

(*S*)-2-butyl (*R*,*R*)-tartrate $\xrightarrow[H_2O]{\text{separating then } H^+}$ (*R*,*R*)-(+)-tartaric acid + (*S*)-2-butanol

(*R*)-2-butyl (*R*,*R*)-tartrate $\xrightarrow[H_2O]{\text{separating then } H^+}$ (*R*,*R*)-(+)-tartaric acid + (*R*)-2-butanol

diastereomers

FIGURE 6-20 Formation of (*R*)- and (*S*)-2-butyl tartrate. The reaction of a pure enantiomer of one compound with a racemic mixture of another compound produces a mixture of diastereomers. Separation of the diastereomers, followed by hydrolysis, gives the resolved enantiomers.

$$\underset{\text{ester}}{R\text{—}\overset{O}{\overset{\|}{C}}\text{—}O\text{—}R} + \underset{\text{water}}{H\text{—}O\text{—}H} \xrightleftharpoons{(H^+ \text{ catalyst})} \underset{\text{acid}}{R\text{—}\overset{O}{\overset{\|}{C}}\text{—}OH} + \underset{\text{alcohol}}{R'\text{—}OH}$$

Hydrolysis of (*R*)-2-butyl tartrate gives (*R*)-2-butanol and (+)-tartaric acid, and hydrolysis of (*S*)-2-butyl tartrate gives (*S*)-2-butanol and (+)-tartaric acid. The recovered tartaric acid would probably be thrown away since it is cheap and nontoxic. Many other chiral resolving agents are expensive, requiring that they be carefully recovered and recycled.

PROBLEM 6-23

To show that (*R*)-2-butyl-(*R*,*R*)-tartrate and (*S*)-2-butyl-(*R*,*R*)-tartrate are not enantiomers, draw and name the mirror images of these compounds.

6-15B CHROMATOGRAPHIC RESOLUTION OF ENANTIOMERS

A more convenient method for resolving enantiomers than by using chemical reactions involves passing the racemic mixture through a column containing particles whose surface is coated with chiral molecules (Fig. 6-21). As the solution passes through the column, the enantiomers form weak complexes, usually through hydrogen bonding, with the chiral column packing. The solvent flows continually through the column, and the dissolved enantiomers gradually move along, retarded by the time they spend complexed with the column packing.

The special feature of this chromatography is the fact that the enantiomers form diastereomeric complexes with the chiral column packing. These diastereo-

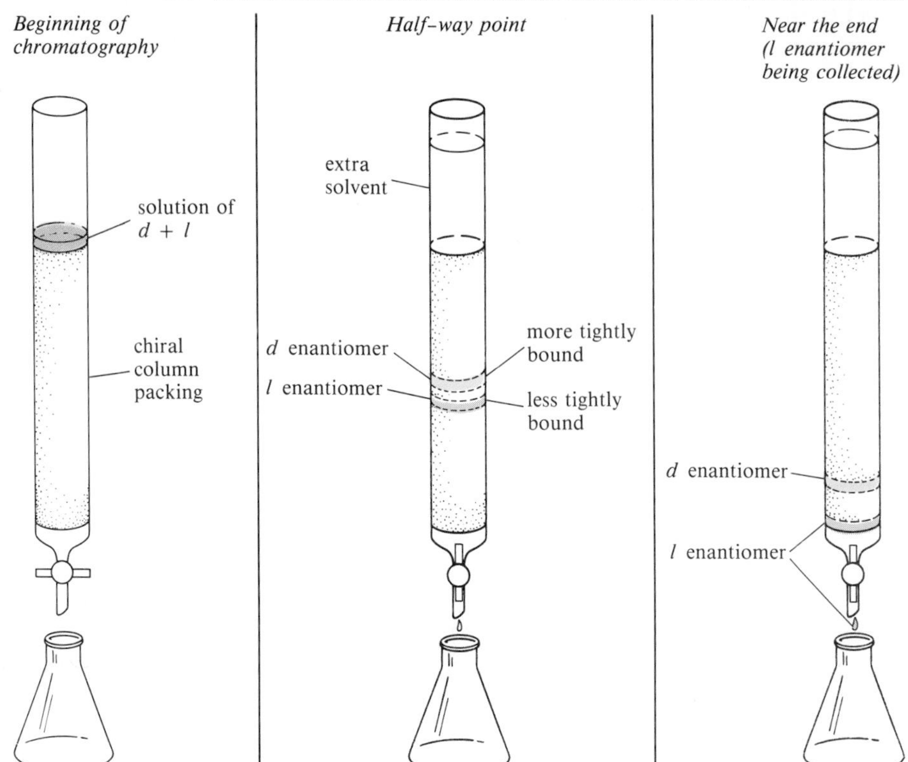

FIGURE 6-21 Chromatographic resolution of enantiomers. The enantiomers of the racemic compound form diastereomeric complexes with the chiral material on the column packing. One of the enantiomers binds more tightly than the other, and its movement through the column is slower. The more tightly bound enantiomer leaves the column after the one that is less tightly bound.

meric complexes have different physical properties. They also have different binding energies and different equilibrium constants for complexation. One of the two enantiomers spends more time complexed with the chiral column packing. The more strongly complexed enantiomer passes through the column more slowly and emerges from the column after the faster-moving (more weakly complexed) enantiomer.

6-16 ABSOLUTE AND RELATIVE CONFIGURATION

Throughout our study of stereochemistry, we have drawn three-dimensional representations and Fischer projections, and we have spoken of chiral carbons having the (*R*) or (*S*) configuration. All these ways of describing the configuration about a chiral carbon atom are absolute; that is, they give the actual orientation of the atoms in space. We say that these methods specify the **absolute configuration** of the molecule. For example, given the name "(*R*)-2-butanol" any chemist can construct an accurate molecular model or draw a three-dimensional representation.

> ABSOLUTE CONFIGURATION The detailed stereochemical picture of a molecule, including how the atoms are arranged in space. Alternatively, the (*R*) or (*S*) configuration at each chiral carbon atom.

Chemists have determined the absolute configurations of many chiral compounds since 1951, when X-ray crystallography was first used to find the orientation of atoms in space. Before 1951, there was no way to link the stereochemical drawings with the actual enantiomers and their observed rotations. No absolute configurations were known. It was possible, however, to correlate the configuration of one compound with that of another and to show that two compounds had the same or opposite configurations. For example, when we convert one compound into another using a reaction that does not affect the stereochemistry at the chiral carbon atom, we know that the product must have the same **relative configuration** as the reactant, even if we cannot determine the absolute configuration of either compound.

RELATIVE CONFIGURATION The experimentally determined relationship between the configurations of two molecules, even though we may not know the absolute configuration of either.

Use of relative configurations to study the S_N2 reaction Using relative configurations, it was shown that the S_N2 reaction proceeds with inversion of a chiral carbon atom. When two British chemists (J. Kenyon and H. Phillips) attacked this problem in 1930, no absolute configuration was known for any optically active compound. These chemists first showed that the conversion of a chiral alcohol to a tosylate ester (Section 10-7) goes without breaking the bond between the chiral carbon atom and oxygen. They did this by using 2-butanol labeled with ^{18}O. The 2-butyl tosylate formed in the reaction still contained the labeled oxygen atom, implying that the C—^{18}O bond remains intact. This result implies retention of configuration.

$$CH_3-\underset{\displaystyle H}{\overset{\displaystyle {}^{18}O-H}{C}}-CH_2CH_3 \xrightarrow[\text{pyridine}]{H_3C-C_6H_4-SO_2-Cl\ (TsCl)} CH_3-\underset{\displaystyle H}{\overset{\displaystyle {}^{18}O-SO_2-C_6H_4-CH_3}{C}}-CH_2CH_3 + HCl$$

For their next reaction, Kenyon and Phillips prepared some optically active (−)-2-butanol with a specific rotation of −13°. Tosylation gave an optically active tosylate ester with the same relative configuration as (−)-2-butanol. Reaction of this tosylate with potassium hydroxide gave 2-butanol as a product, but this product was (+)-2-butanol, with a specific rotation of +13°. The final product has undergone an inversion of configuration (a **Walden inversion**).

$$\underset{(-)\text{-2-butanol}}{CH_3-\underset{\displaystyle H}{\overset{\displaystyle O-H}{C}}-CH_2CH_3} \xrightarrow[\text{retention}]{Cl-SO_2-C_6H_4-CH_3,\ \text{pyridine}} \underset{\text{same relative configuration}}{CH_3-\underset{\displaystyle H}{\overset{\displaystyle O-SO_2-C_6H_4-CH_3}{C}}-CH_2CH_3} \xrightarrow[\text{inversion}]{KOH} \underset{(+)\text{-2-butanol}}{CH_3-\underset{\displaystyle O-H}{\overset{\displaystyle H}{C}}-CH_2CH_3}$$

Since the product has the opposite configuration from that of the starting alcohol and the tosylate, the reaction with potassium hydroxide must have inverted the configuration of the chiral carbon. This was one of the most important pieces of evidence leading to the formulation of the mechanism of the S_N2 reaction. Now we know that (−)-2-butanol has the (*R*) absolute configuration, and we can write out the Kenyon-Phillips experiment using three-dimensional detail.

Tosylation step (retention of configuration)

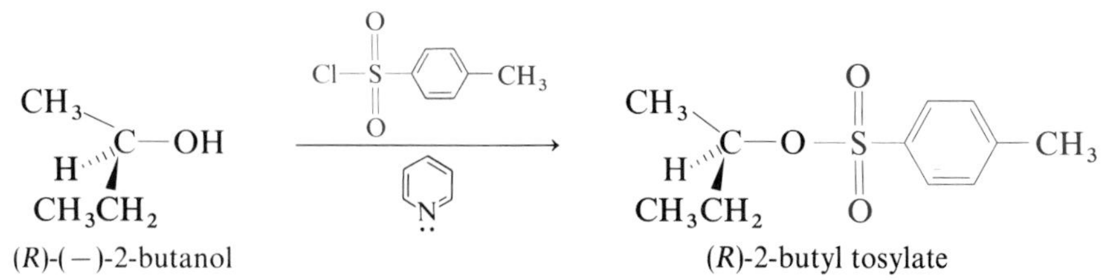

$$\underset{(R)\text{-}(-)\text{-2-butanol}}{CH_3(H)(CH_3CH_2)C{-}OH} \xrightarrow[\text{pyridine}]{Cl{-}SO_2{-}C_6H_4{-}CH_3} \underset{(R)\text{-2-butyl tosylate}}{CH_3(H)(CH_3CH_2)C{-}O{-}SO_2{-}C_6H_4{-}CH_3}$$

Nucleophilic substitution (inversion of configuration)

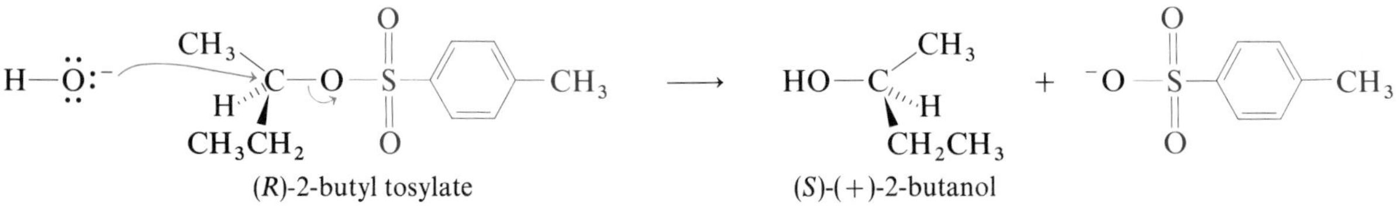

$$H{-}\ddot{O}{:}^{-} + \underset{(R)\text{-2-butyl tosylate}}{CH_3(H)(CH_3CH_2)C{-}O{-}SO_2{-}C_6H_4{-}CH_3} \longrightarrow \underset{(S)\text{-}(+)\text{-2-butanol}}{HO{-}C(CH_3)(H)(CH_2CH_3)} + {}^{-}O{-}SO_2{-}C_6H_4{-}CH_3$$

PROBLEM 6-24

Draw a perspective structure or a Fischer projection for the expected substitution products of the following reactions.

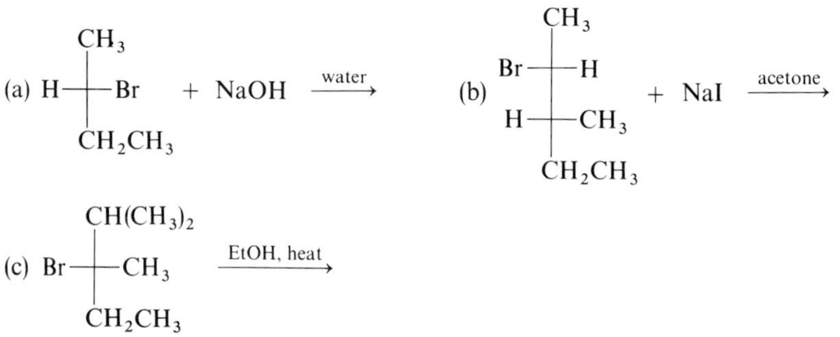

(a) Fischer projection: top CH_3; left H, right Br; bottom CH_2CH_3 $+ \; NaOH \xrightarrow{\text{water}}$

(b) Fischer projection: top CH_3; first carbon: left Br, right H; second carbon: left H, right CH_3; bottom CH_2CH_3 $+ \; NaI \xrightarrow{\text{acetone}}$

(c) Fischer projection: top $CH(CH_3)_2$; left Br, right CH_3; bottom CH_2CH_3 $\xrightarrow{\text{EtOH, heat}}$

6-17 REACTIONS AT THE CHIRAL CARBON ATOM: S_N1 AND S_N2 MECHANISMS

One of the most interesting and useful aspects of stereochemistry is the study of what happens to optically active molecules when they react. The products isolated from the reaction of a chiral starting material can tell us a great deal about the reaction mechanism. Also, we can use our knowledge of the stereochemical consequences of a reaction to develop useful synthetic methods for chiral compounds. We divide our discussion of the reactions of chiral compounds into three parts:

1. Reactions that take place at the chiral carbon atom
2. Reactions that do not involve the chiral carbon atom
3. Reactions that generate a new chiral carbon atom

Our study of the reactions of optically active compounds will highlight the stereochemical implications of the reactions encountered in Chapter 5. If the mechanisms and stereochemistry of the S_N1, S_N2, E1, and E2 reactions are not clear in your mind, you might like to review them at this time.

6-17A INVERSION OF CONFIGURATION IN THE S_N2 REACTION

When a reaction takes place at a chiral carbon atom, it may change the configuration of the chiral carbon. For example, the S_N2 mechanism involves an *inversion of configuration* of the carbon atom under attack by the nucleophile. An inversion of configuration gives a product whose stereochemistry is opposite that of the reactant. The reaction of hydroxide ion with optically active 2-butyl tosylate proceeds with inversion of configuration:

(*R*)-2-butyl tosylate → (*S*)-(+)-2-butanol + tosylate ion

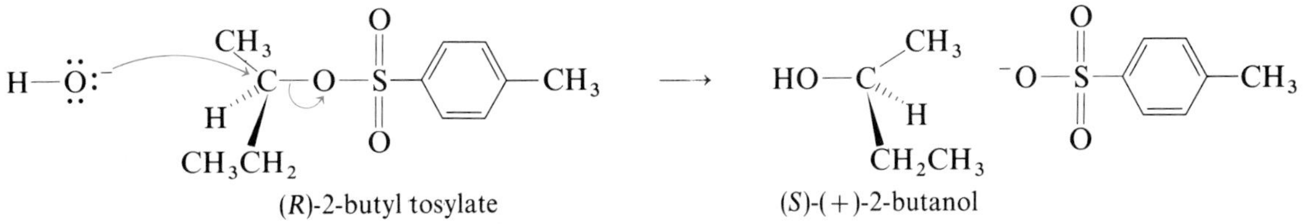

P. Walden first proved that a substitution reaction had inverted the configuration of a chiral carbon in 1893. In his honor, a substitution that inverts the configuration at a chiral carbon atom is called a **Walden inversion.** The S_N2 family of reactions are common examples of Walden inversions since the back-side attack of the nucleophile inverts the configuration of the electrophilic carbon atom.

6-17B RACEMIZATION IN THE S_N1 REACTION

Reactions of optically active compounds sometimes show neither clean inversion of configuration nor clean retention of configuration. This result is called **racemization.** If the product is optically inactive (a racemate), we say that the reaction proceeds with *complete racemization.* If some optical activity remains, the compound has undergone *partial racemization.* In the case of partial racemization, we say that the reaction has produced *predominant retention* or *predominant inversion,* depending on whether the major product is the retention product or the inversion product.

An example of a mechanism that usually proceeds with racemization is the S_N1 reaction (Section 5-15). Since a flat, achiral carbocation is formed as an intermediate, the product is found to be racemic. The following reaction shows the first-order hydrolysis of a tertiary alkyl halide, (*R*)-3-bromo-2,3-dimethylpentane.

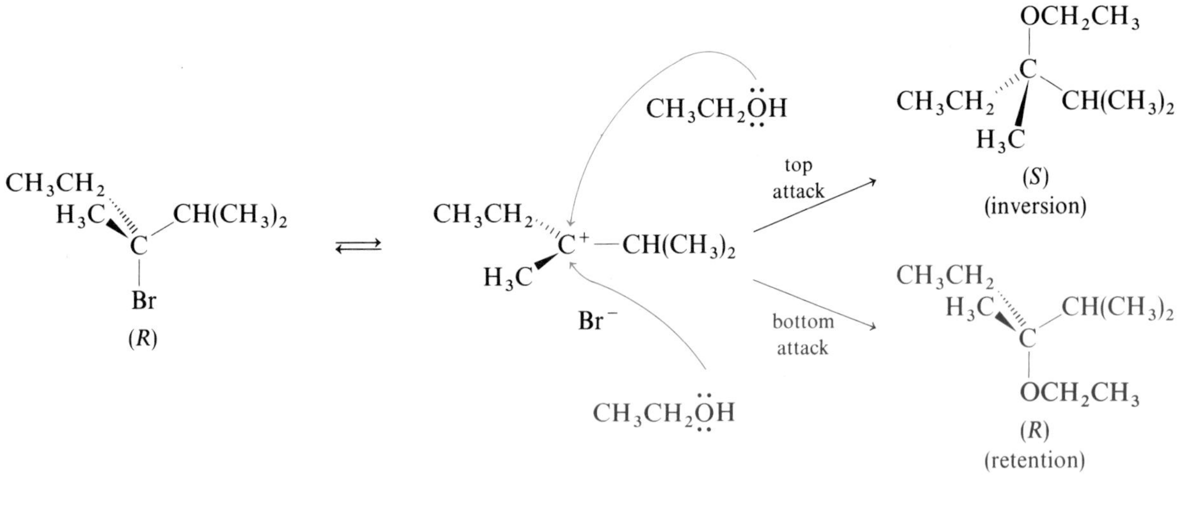

(*R*) ⇌ carbocation + Br^-; CH_3CH_2OH top attack → (*S*) (inversion); CH_3CH_2OH bottom attack → (*R*) (retention)

Although the starting material is optically active, the intermediate carbocation is planar and achiral. Ethanol can attack the carbocation on either face, leading to racemization. Attack on the top face leads to a product with the (*S*) configuration (inversion of configuration); attack on the bottom face gives the (*R*) configuration (retention of configuration).

The racemization is not complete, however, because the leaving bromide ion partially blocks the bottom side of the carbocation. Ethanol can attack more easily from the top, giving predominant inversion of configuration. This result shows that the leaving group in the S_N1 reaction does not always have time to diffuse away from the carbocation before attack by the nucleophile occurs.

6-17C RETENTION OF CONFIGURATION WITH AN INTERNAL NUCLEOPHILE

Inversion of configuration and racemization are the most common stereochemical results of reactions that take place at a chiral carbon atom. A few reactions at chiral carbon atoms give products that have the same configuration as the starting material; this result is called **retention of configuration.** An example is the reaction of an alcohol with thionyl chloride, $SOCl_2$. This reaction provides a method for converting alcohols to alkyl chlorides with retention of configuration.

CH_3CH_2, H, CH_3—C—OH + Cl—S(=O)—Cl ⟶ CH_3CH_2, H, CH_3—C—O—S(=O)—Cl + HCl

CH_3CH_2, H, CH_3—C—Ö:—S(=O)—Cl $\xrightarrow{\text{(slow)}}$ CH_3CH_2, H, CH_3—C^+ Cl—S=O :O: $\xrightarrow{\text{(fast)}}$ CH_3CH_2, H, CH_3—C—Cl + SO_2

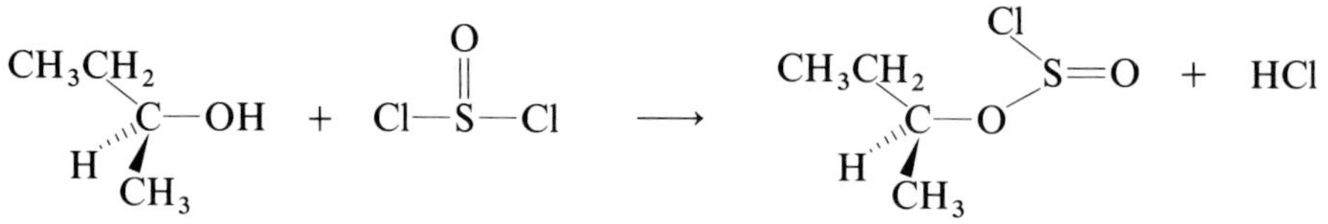

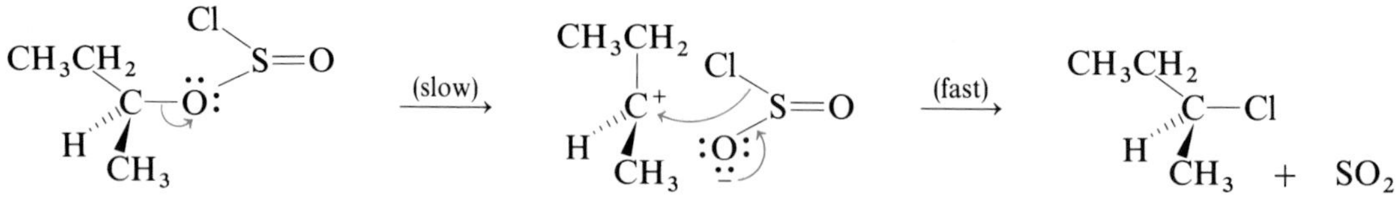

This mechanism is a modified version of the S_N1. The hydroxyl group is converted to an excellent leaving group, which leaves to form a carbocation. A chloride donor converts the carbocation to the alkyl chloride. Why isn't the product racemized as in the standard S_N1 reaction? The difference lies in the source of the nucleophile. In the S_N1 reaction the nucleophile is randomly distributed throughout the solution. In the thionyl chloride reaction the nucleophile is a part of the leaving group. The carbocation and the leaving group form a closely associated ion pair, and the chloride ion immediately attacks the nearby face of the carbocation.

PROBLEM 6-25

Draw the stereochemistry of the products of the following reactions.

(a) (*R*)-2-bromobutane + aqueous KOH
(b) (*S*)-3-bromo-3-methylhexane + ethanol, heat
(c)  cyclopentane with OH and CH_3 (wedged) + $SOCl_2$

6-18 REACTIONS THAT DO NOT INVOLVE A CHIRAL CARBON ATOM

When a reaction takes place in such a way that the chiral carbon atom maintains its four bonds intact, retention of configuration is observed. In most cases, the term "retention of configuration" refers to reactions that actually take place at the chiral carbon atom, yet give products with retention of configuration. When a reaction does not involve the chiral carbon atom, it is generally assumed that the stereochemistry of the chiral carbon is not affected.

In Section 6-16 we saw that tosylation of an alcohol goes with retention of configuration. We showed that 2-butanol labeled with ^{18}O retains the labeled oxygen atom throughout the tosylation reaction. This result implies that the $^{*}C{-}^{18}O$ bond is not broken during the tosylation and the original configuration of the chiral carbon atom is retained.

Even the S_N2 reaction can take place with retention of configuration, provided that the chiral carbon atom is not the carbon being attacked by the nucleophile. For example, in the following S_N2 reaction, the chiral carbon atom is a part of the nucleophile, and its four bonds are never broken. It is the iodomethane that is attacked and undergoes inversion of configuration. Since iodomethane is not chiral, however, its inversion of configuration is unnoticed in the product.

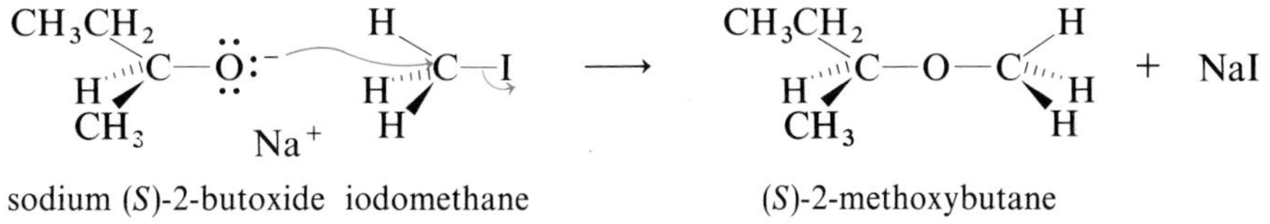

$$CH_3CH_2(H)(CH_3)C{-}\ddot{O}{:}^{-}\ Na^+ \ + \ H_3C{-}I \longrightarrow CH_3CH_2(H)(CH_3)C{-}O{-}CH_3 \ + \ NaI$$

sodium (*S*)-2-butoxide iodomethane (*S*)-2-methoxybutane

6-19 REACTIONS THAT GENERATE A NEW CHIRAL CARBON ATOM

Many reactions form new chiral carbon atoms; yet if a reaction begins with achiral (or racemic) reagents, then equal amounts of the two mirror images of the new chiral carbon will result, and the product will be racemic. For example, we saw (Section 6-6) that hydrogenation of 2-butanone (achiral) gives a racemic mixture of the chiral product, 2-butanol.

Just as living things come only from other living things, optically active products come only from optically active reagents or catalysts. In every reaction that gives a chiral product from an achiral reagent, there will be two equally probable mirror-image mechanisms giving the two enantiomers of the product.

6-19A ASYMMETRIC INDUCTION

Asymmetric induction is the use of an optically active reagent or catalyst to convert an optically inactive starting material to an optically active product. For example, the reduction of 2-butanone can be accomplished in a stereospecific manner by an enzyme. In this case, an achiral starting material is converted to an optically active product by a chiral catalyst. The enzyme selectively catalyzes the addition of hydrogen to just one of the faces of the C=O double bond.

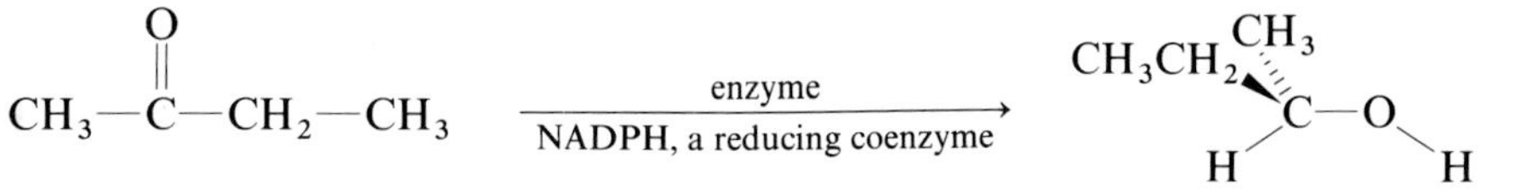

$$CH_3{-}\overset{\overset{\displaystyle O}{\|}}{C}{-}CH_2{-}CH_3 \xrightarrow[\text{NADPH, a reducing coenzyme}]{\text{enzyme}} CH_3CH_2(CH_3)(H)C{-}O{-}H$$

(*R*)-2-butanol, $[\alpha]_D^{25} = -13°$

Chemists have begun to imitate nature's use of chiral catalysts for asymmetric induction. For example, a complex of rhodium with a chiral ligand called DIOP catalyzes the hydrogenation of an achiral starting material to the biologically active (−) form of *dopa*. Because the catalyst is chiral, the transition state leading to the two enantiomers of dopa are diastereomeric. They have different energies, and the transition state leading to the (−) enantiomer of the product is favored.

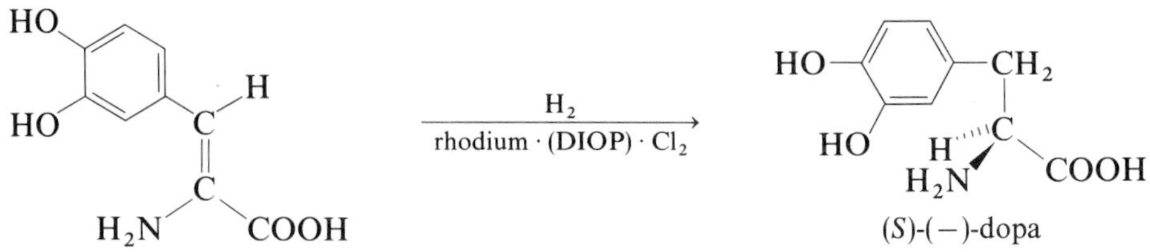

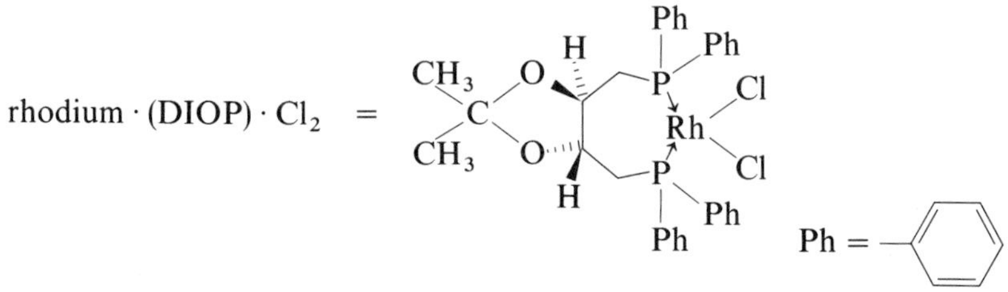

$\xrightarrow[\text{rhodium} \cdot (\text{DIOP}) \cdot \text{Cl}_2]{\text{H}_2}$

(*S*)-(−)-dopa

rhodium · (DIOP) · Cl_2 =

Ph =

Levodopa [*l*-dopa or (−)-dopa] is used in patients with Parkinson's disease to counteract a deficiency of dopamine, one of the neurotransmitters in the brain. Dopamine itself is useless as a drug, because it cannot cross the "blood-brain barrier" that is, it cannot get into the cerebrospinal fluid from the bloodstream. (−)-Dopa, on the other hand, is an amino acid related to tyrosine. It crosses the blood-brain barrier into the cerebrospinal fluid, where it undergoes enzymatic conversion to dopamine. Only the (−) enantiomer of dopa can be transformed into dopamine; the other enantiomer, (+)-dopa, is toxic to the patient.

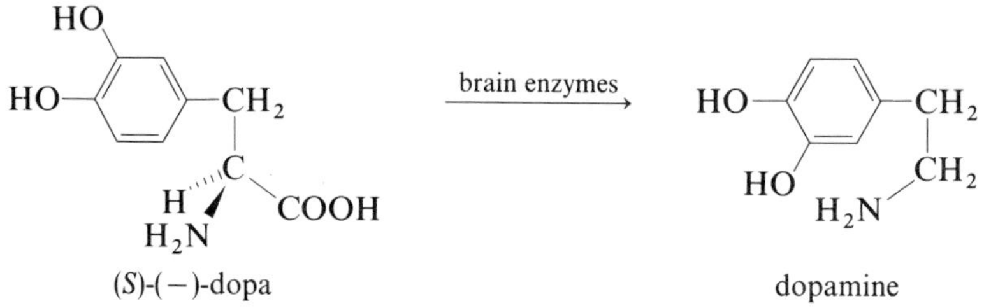

(*S*)-(−)-dopa $\xrightarrow{\text{brain enzymes}}$ dopamine

6-19B DIRECTING INFLUENCE OF A CHIRAL CARBON

If a molecule already has one chiral center, that chiral center can "direct" a reagent to form a second chiral carbon atom in the desired manner. The reagent does not need to be chiral because the chirality is contained within the starting material itself. Figure 6-22 shows an example of the directing influence of one chiral carbon atom in the formation of another. The catalytic reduction of 2-methylcyclopentanone gives almost entirely *cis*-2-methylcyclopentanol, because hydrogen is added preferentially to the less hindered face of the molecule.

The reagent in Figure 6-22 is not distinguishing between two equivalent faces of the molecule but between one relatively unhindered face and a face with a methyl group in the way. The products are not enantiomers but diastereomers. *cis*- and *trans*-2-Methylcyclopentanol are different compounds with different properties, and the transition states leading to their formation are diastereomeric with different energies.

FIGURE 6-22 The catalytic reduction of 2-methylcyclopentanone gives primarily *cis*-2-methylcyclopentanol. The reduction occurs more readily on the face of the molecule that is not hindered by the bulky methyl group.

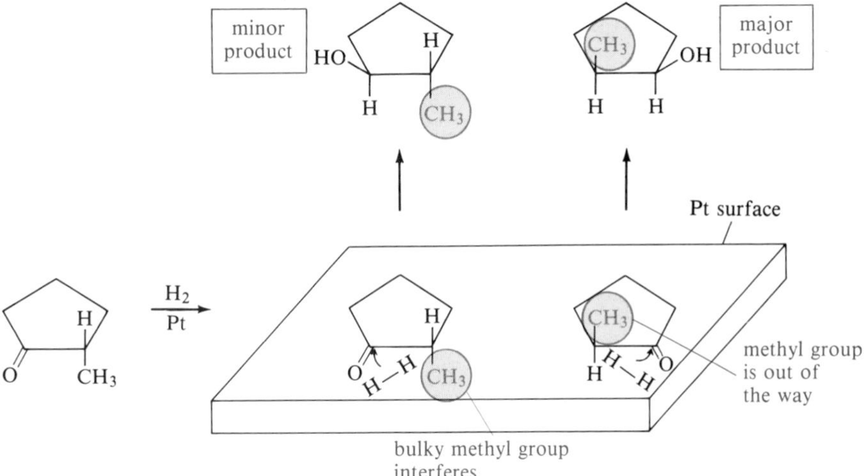

6-20 STEREOSPECIFICITY OF ANTI ELIMINATIONS

A **stereospecific reaction** is one in which a particular stereoisomer reacts to give one specific stereoisomer [or (*d,l*) pair] of the product. We have already seen several stereospecific reactions, particularly the S_N2 reaction and the E2 reaction. The S_N2 reaction is stereospecific in that it always gives inversion of the configuration of the carbon atom that is attacked. Attack by hydroxide ion on (*R*)-2-bromobutane gives (*S*)-2-butanol, while attack on (*S*)-2-bromobutane gives (*R*)-2-butanol.

The E2 reaction is stereospecific because it usually goes through an anti and coplanar transition state. The products are alkenes, and different diastereomers of starting materials commonly give different diastereomers of alkenes. Figure 6-23 shows the reaction of 2,3-dibromobutane with iodide ion, an E2 elimination of the two bromine atoms. This E2 elimination goes through an anti and coplanar transition state. Notice that *meso*-2,3-dibromobutane eliminates to give *trans*-2-butene, while (±)-2,3-dibromobutane (either enantiomer) eliminates to give *cis*-2-butene. Use your models to follow the stereochemistry of this elimination.

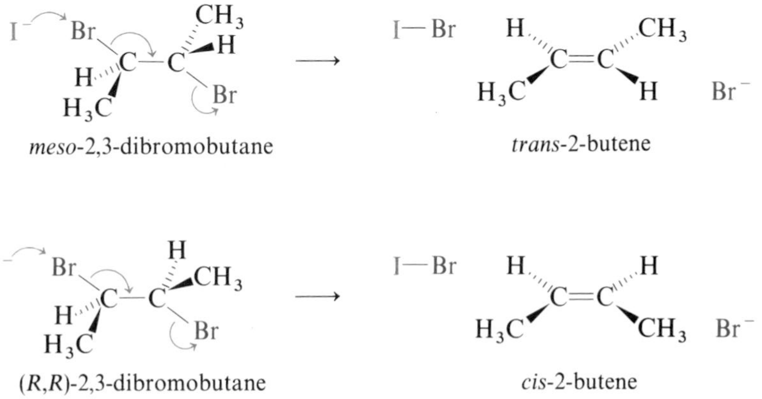

meso-2,3-dibromobutane → I—Br + *trans*-2-butene + Br^-

(*R,R*)-2,3-dibromobutane → I—Br + *cis*-2-butene + Br^-

FIGURE 6-23 In the presence of iodide, the *meso* diastereomer of 2,3-dibromobutane eliminates to give *trans*-2-butene. Under the same conditions, the (±) diastereomer eliminates to give *cis*-2-butene.

PROBLEM 6-26

In the debromination of (*d,l*)-2,3-dibromobutane, we have seen that the (*R,R*) enantiomer gives the cis product. Draw the same reaction using the (*S,S*) enantiomer and show that it gives the same product.

PROBLEM 6-27

Predict the products of the following reactions.

(a) 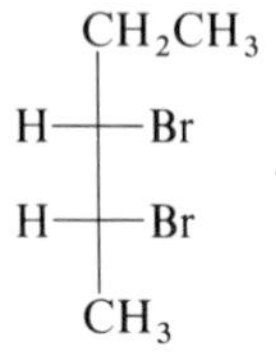 Fischer projection: top CH_2CH_3; H—C—Br; H—C—Br; bottom CH_3 $+ \text{KI} \longrightarrow$ (elimination)

(b) 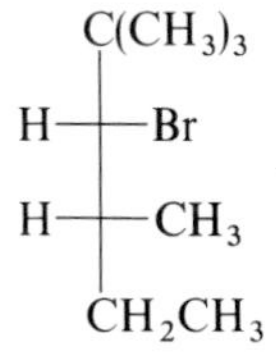 Fischer projection: top $C(CH_3)_3$; H—C—Br; H—C—CH_3; bottom CH_2CH_3 $+ \text{KOH} \longrightarrow$ (substitution and elimination)

(c) (3*R*,4*R*)-3,4-dibromoheptane + KI $\longrightarrow$ (elimination)

GLOSSARY

absolute configuration The detailed stereochemical picture of a molecule, including how the atoms are arranged in space. Alternatively, the actual (*R*) or the (*S*) configuration at each chiral carbon atom. (p. 262)

achiral Not chiral. (p. 224)

anti Adding to (or eliminating from) opposite faces of a molecule. (pp. 199, 269)

asymmetric carbon atom (chiral carbon atom) A carbon atom that is bonded to four different groups. (p. 226)

asymmetric induction (asymmetric synthesis) The formation of an optically active product from an optically inactive starting material. Such a process requires the use of an optically active reagent or catalyst. (p. 267)

Cahn-Ingold-Prelog convention The accepted method for designating the absolute configuration of a chiral carbon atom as either (*R*) or (*S*). (p. 230)

chiral Different from its mirror image. (p. 224)

chiral carbon atom (asymmetric carbon atom) A carbon atom that is bonded to four different groups. (p. 226)

chiral probe A molecule or an object that is chiral and can use its own chirality to differentiate between mirror images. (p. 240)

cis On the same side of a ring or double bond. (p. 252)

configurational isomers (stereoisomers) Isomers whose atoms are bonded together in the same order but which differ in how the atoms are oriented in space. (p. 254)

conformers (conformational isomers) Structures that differ only by rotations about sigma bonds. In most cases, conformers interconvert at room temperature, thus are not different compounds and not true isomers. (p. 244)

constitutional isomers (structural isomers) Isomers that differ in the order in which their atoms are bonded together. (p. 254)

dextrorotatory Rotating the plane of polarized light clockwise. (p. 237)

diastereomers Stereoisomers that are not mirror images. (p. 252)

enantiomeric excess (e.e.) The excess of one enantiomer in a mixture of enantiomers expressed as a percentage of the mixture. Algebraically,

$$\text{e.e.} = \frac{R - S}{R + S} \times 100\%$$

Similar to optical purity. (p. 243)

enantiomers A pair of nonsuperimposable mirror image molecules; mirror-image isomers. (p. 226)

Fischer projection A method for drawing a chiral carbon atom as a cross. The carbon chain is kept along the vertical, with the most oxidized carbon at the top. Vertical bonds project away from the viewer, and horizontal bonds project toward the viewer. (p. 247)

geometric isomers (cis-trans isomers) Isomers that differ in their cis-trans arrangement on a ring or double bond. Geometric isomers are a subclass of diastereomers. (p. 252)

internal mirror plane A plane of symmetry through the middle of a molecule, dividing the molecule into two mirror-image halves. A molecule with an internal mirror plane of symmetry cannot be chiral. (p. 228)

inversion of configuration (*see also* Walden inversion) A process in which the groups around a chiral carbon atom are changed to the opposite spatial configuration. (p. 265)

$$HO^- + \underset{(S)}{(R^1)(R^2)(H)C{-}Br} \longrightarrow \underset{(R)}{HO{-}C(R^1)(R^2)(H)} + Br^-$$

The S_N2 reaction usually goes with *inversion of configuration.*

isomers Different compounds with the same molecular formula. (p. 254)

levorotatory Rotating the plane of polarized light counterclockwise. (p. 237)

meso compound An achiral compound that contains chiral carbon atoms. (p. 256)

optical isomers (archaic; see *enantiomers*) Compounds with identical properties except for the direction in which they rotate polarized light. (p. 236)

optically active Capable of rotating the plane of polarized light. (p. 236)

optical purity (o.p.) The specific rotation of a mixture of two enantiomers, expressed as a percentage of the specific rotation of one of the pure enantiomers. Algebraically,

$$\text{o.p.} = \frac{\text{observed rotation}}{\text{rotation of pure enantiomer}} \times 100\%$$

Similar to enantiomeric excess. (p. 242)

plane-polarized light Light composed of waves that vibrate in only one plane. (p. 234)

polarimeter An instrument that quantitatively measures the rotation of plane-polarized light by an optically active compound. (p. 237)

racemic mixture [racemate, racemic modification, (*d,l*) pair, (±) pair] A mixture of equal quantities of enantiomers, such that the mixture is optically inactive. (p. 241)

racemization The process when an optically active reactant gives a product that shows neither clean retention of configuration nor clean inversion of configuration. (p. 265)

relative configuration The experimentally determined relationship between the configurations of two molecules even though we may not know the absolute configuration of either. (p. 262)

resolution The process of separating a racemic mixture into the pure enantiomers. Resolution requires some kind of a chiral resolving agent. (p. 259)

resolving agent A chiral compound (or chiral material on a chromatographic column) used for separating enantiomers. (p. 260)

retention of configuration A process in which the groups around a chiral carbon atom are unchanged in their spatial orientation. (p. 266)

$$\underset{(S)}{R^1R^2HC{-}OH} + Cl{-}\overset{\overset{\displaystyle O}{\|}}{S}{-}Cl \longrightarrow \underset{(S)}{R^1R^2HC{-}Cl}$$

2^n rule A molecule with n chiral carbon atoms *might* have as many as 2^n stereoisomers. (p. 255)

specific rotation A measure of a compound's ability to rotate the plane of polarized light.

$$[\alpha]_D^{25} = \frac{\alpha(\text{observed})}{c \cdot l}$$

where c is concentration in g/mL and l is length of sample cell (path length) in decimeters. (p. 238)

stereochemistry The study of the three-dimensional structure of molecules. (p. 223)

stereoisomers (configurational isomers) Isomers whose atoms are bonded together in the same order but which differ in how the atoms are oriented in space. (p. 223)

stereospecific reaction A reaction in which a particular stereoisomer reacts to give one specific stereoisomer [or (*d,l*) pair] of the product. (p. 269)

structural isomers (constitutional isomers) Isomers that differ in the order in which their atoms are bonded together. (p. 254)

superimposable Identical in all respects. The three-dimensional positions of all atoms coincide when the molecules are placed on top of each other. (p. 225)

syn Adding to (or eliminating from) the same face of a molecule. (p. 199)

trans On opposite sides of a ring or double bond. (p. 252)

Walden inversion (*see also* inversion of configuration) A step in a reaction sequence in which an inversion in the configuration of a chiral carbon atom takes place. (p. 265)

ESSENTIAL PROBLEM-SOLVING SKILLS IN CHAPTER 6

1. Classify molecules as chiral or achiral, and identify mirror planes of symmetry.
2. Identify chiral carbon atoms and name them using the (*R*) and (*S*) nomenclature.
3. Calculate specific rotations from polarimetry data.
4. Draw all stereoisomers of a given structure.
5. Identify enantiomers, diastereomers, and meso compounds.
6. Draw correct Fischer projections of chiral carbon atoms.
7. Predict the stereochemistry of products of reactions such as substitutions and eliminations on optically active compounds.
8. Predict the differences in products of stereospecific reactions of diastereomers.

STUDY PROBLEMS

6-28. Briefly define each of the following terms and give an example.

(a) (*R*) and (*S*) **(b)** chiral and achiral **(c)** chiral carbon atom
(d) cis and trans **(e)** anti elimination **(f)** isomers
(g) structural isomers **(h)** stereoisomers **(i)** enantiomers
(j) Fischer projection **(k)** diastereomers **(l)** geometric isomers

(m) optical isomers
(n) meso
(o) racemization
(p) racemic mixture
(q) specific rotation
(r) dextrorotatory
(s) (*d,l*) and (±)
(t) absolute configuration
(u) relative configuration
(v) inversion of configuration
(w) retention of configuration

6-29. For each of the following structures
(1) Star any chiral carbon atoms.
(2) Label each chiral carbon as (*R*) or (*S*).
(3) Draw in any internal mirror planes of symmetry.
(4) Label the structure as chiral or achiral.
(5) Label any meso structures.

(a) 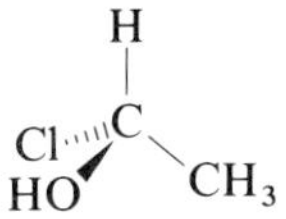 (b) 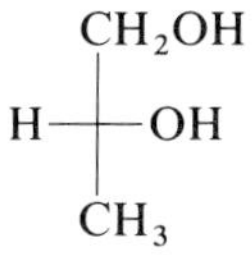 (c) 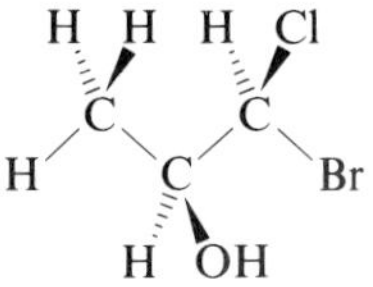 (d)

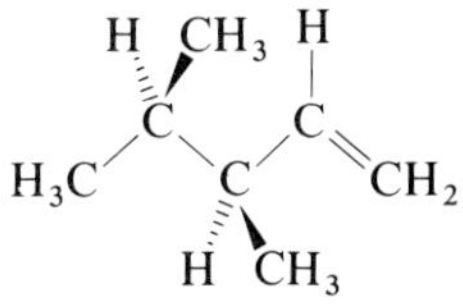

(e) 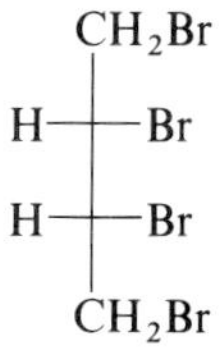 (f) 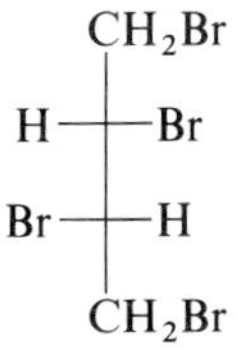 (g) 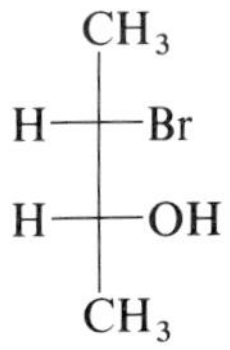 (h)

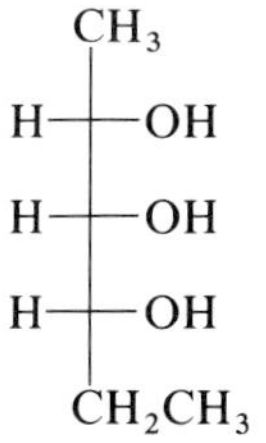

(i) 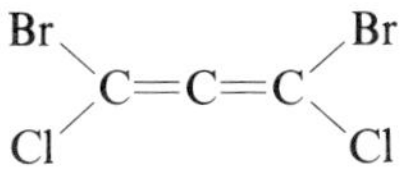 (j) 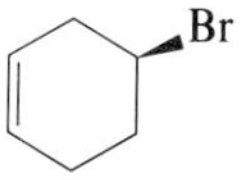 (k) 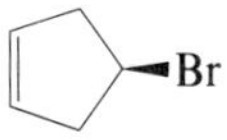 (l)

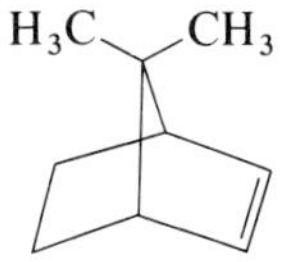

(m) 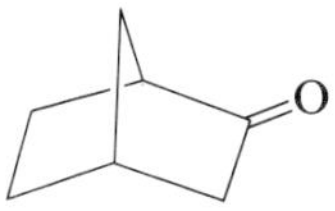 (n)

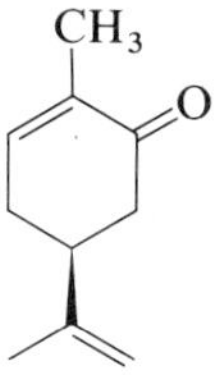

6-30. Draw a three-dimensional representation that corresponds to each of the following descriptions.
(a) (*S*)-2-chlorobutane
(b) (*R*)-1,1,2-trimethylcyclohexane
(c) (2*R*,3*S*)-2,3-dibromohexane
(d) (1*R*,2*R*)-1,2-dibromocyclohexane
(e) *meso*-3,4-hexanediol, $CH_3CH_2CH(OH)CH(OH)CH_2CH_3$
(f) (±)-3,4-hexanediol

6-31. Convert the following perspective formulas to Fischer projections.

(a) 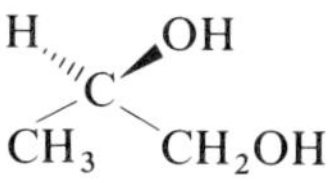 (b) 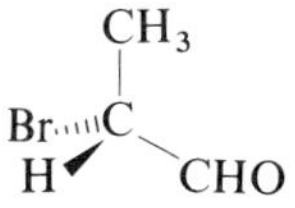 (c) 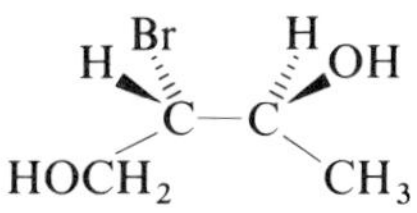 (d)

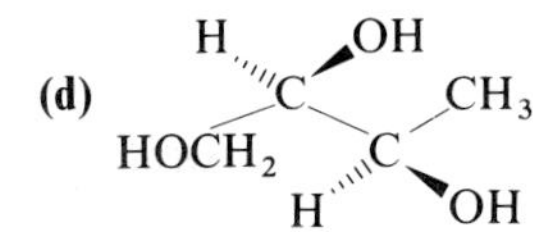

6-32. Convert the following Fischer projections to perspective formulas.

(a) 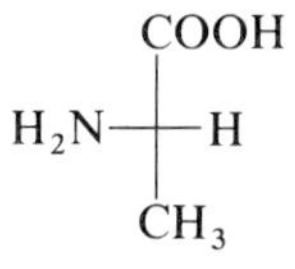 (b) 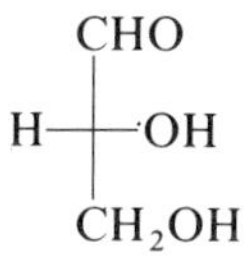 (c) 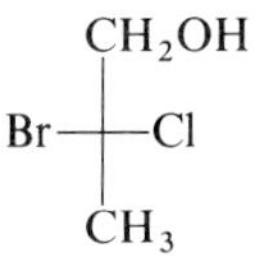 (d)

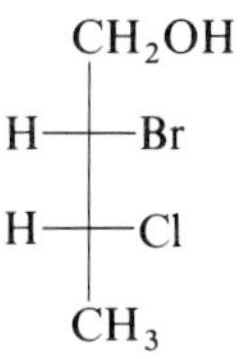

6-33. Give the stereochemical relationships between the following pairs of isomers. Examples are: same compound, structural isomers, enantiomers, diastereomers.

(a) 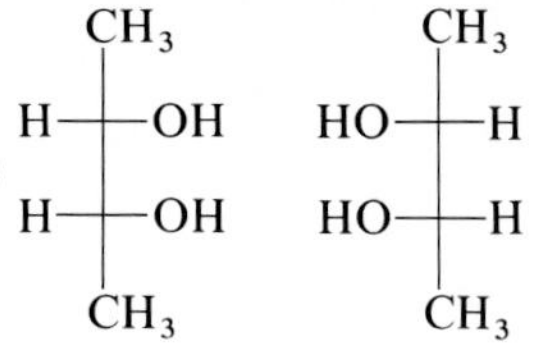 (b) 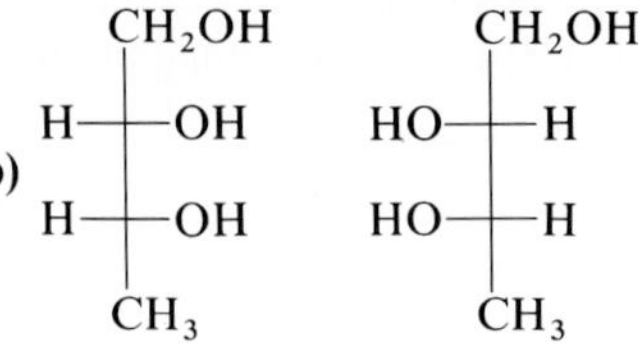 (c) 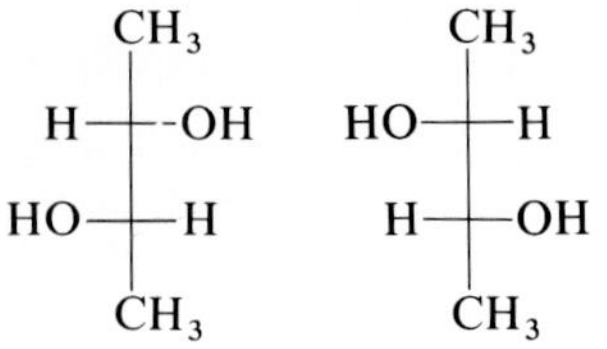 (d) 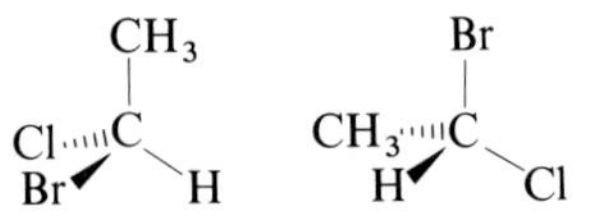 (e) 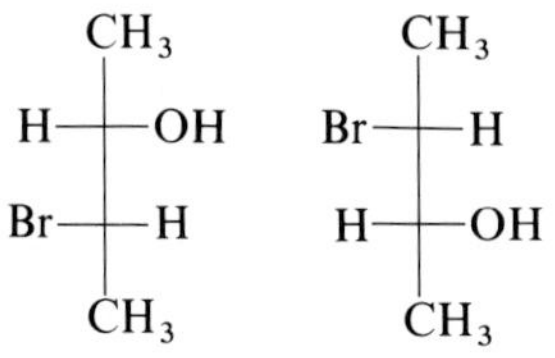 (f) 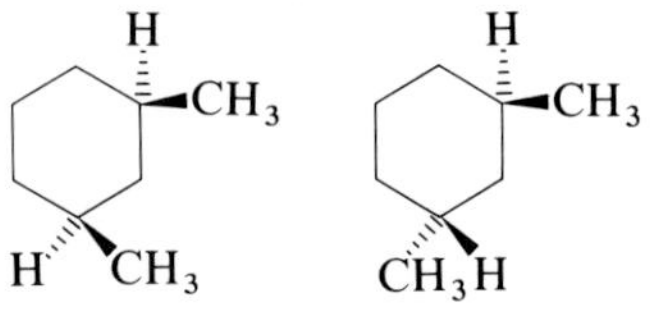 (g) 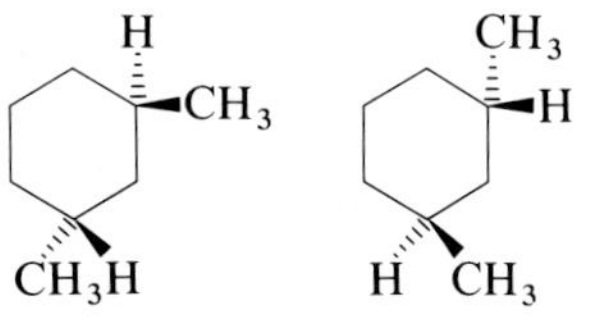 (h)

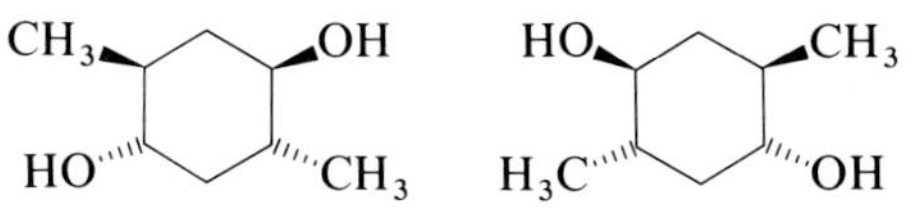

6-34. Draw the enantiomer, if any, for each of the following structures.

(a) 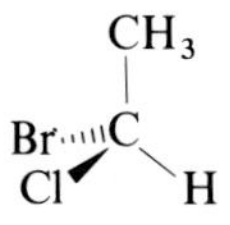 (b) 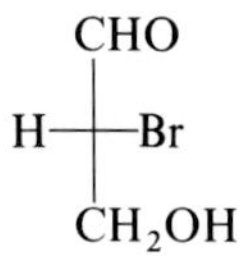 (c) 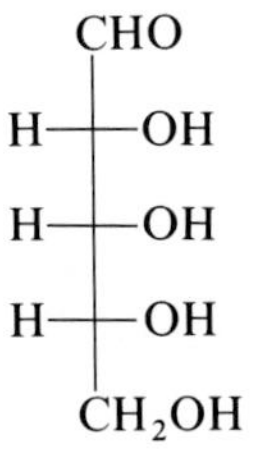 (d) 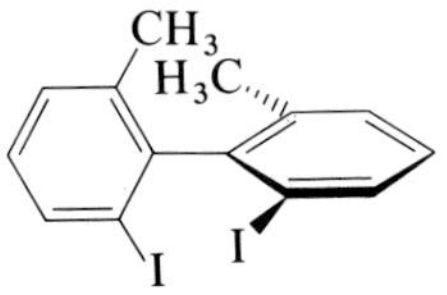 (e) 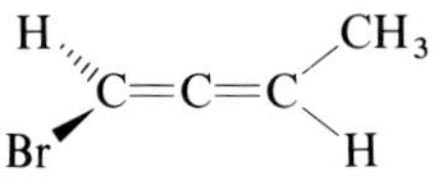 (f) 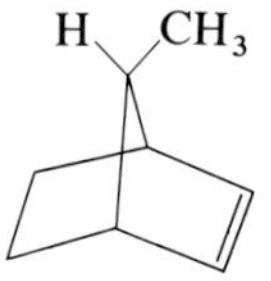 (g) 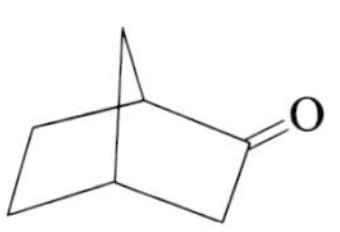 (h) 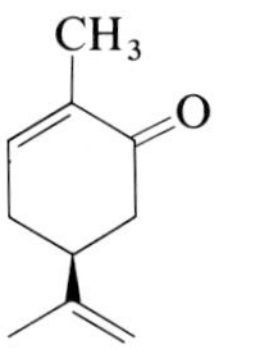 (i)

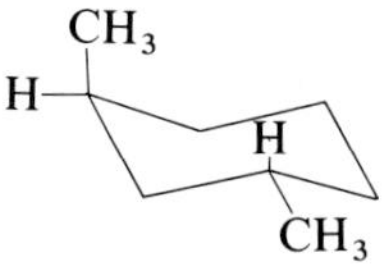

6-35. When (*d,l*)-2,3-dibromobutane reacts with one equivalent of potassium hydroxide, some of the products are (2*S*,3*R*)-3-bromo-2-butanol and its enantiomer and *trans*-2-bromo-2-butene. Give mechanisms to account for these products.

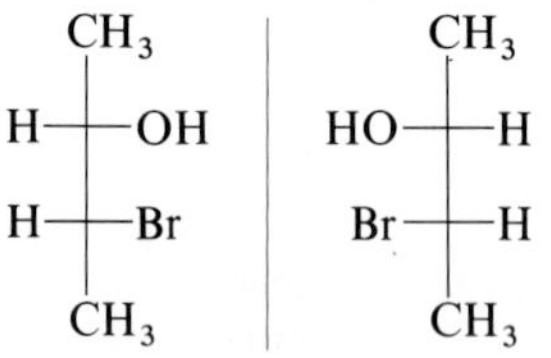

(2*S*, 3*R*) (2*R*, 3*S*)

3-bromo-2-butanol

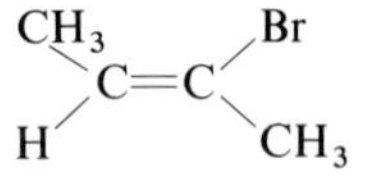

trans-2-bromo-2-butene

6-36. Calculate the specific rotations of the following samples taken at 25°C using the sodium D line.

(a) 1.00 g of sample is dissolved in 20.0 mL of ethanol. 5.00 mL of this solution is placed in a 20.0-cm polarimeter tube. The observed rotation is 1.25° counterclockwise.

(b) 0.050 g of sample is dissolved in 2.0 mL of ethanol, and this solution is placed in a 2.0-cm polarimeter tube. The observed rotation is clockwise 0.043°.

6-37. (+)-Tartaric acid has a specific rotation of +12.0°. Calculate the specific rotation of a mixture of 60 percent (+)-tartaric acid and 40 percent (−)-tartaric acid.

6-38. For each of the following structures

(1) Draw all of the stereoisomers.

(2) Label each structure chiral or achiral.

(3) Give the relationships between the stereoisomers (enantiomers, diastereomers).

(a) CHO; H—C—OH; H—C—OH; CH_2OH (Fischer projection)

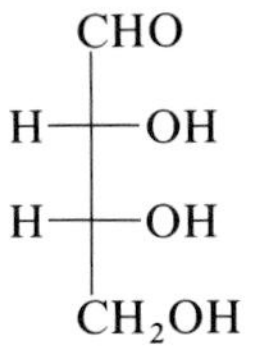

(b) CH_3, CH_3-substituted cyclopentane =C(H)(CH_3)

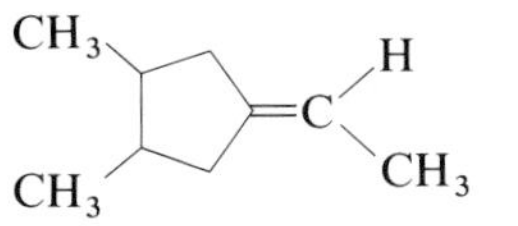

6-39. The specific rotation of (*S*)-2-iodobutane is +15.90°.

(a) Draw the structure of (*S*)-2-iodobutane.

(b) Predict the specific rotation of (*R*)-2-iodobutane.

(c) Determine the percentage composition of a mixture of (*R*)- and (*S*)-2-iodobutane with a specific rotation of −7.95°.

6-40. A solution of pure (*S*)-2-iodobutane ($[\alpha] = +15.90°$) in acetone is allowed to react with radioactive iodide, $^{137}I^-$, until 1.0 percent of the iodobutane contains radioactive iodine. The specific rotation of this recovered iodobutane is found to be +15.58°.

(a) Determine the percentages of (*R*)- and (*S*)-2-iodobutane in the product mixture.

(b) What does this result suggest about the mechanism of the reaction of 2-iodobutane with iodide ion?

6-41. **(a)** Optically active 2-bromobutane undergoes racemization on treatment with a solution of KBr. Give a mechanism for this racemization.

(b) In contrast, optically active 2-butanol does not racemize on treatment with a solution of KOH. Explain why a reaction like that in (a) does not occur.

(c) Optically active 2-butanol does racemize in dilute acid. Propose a mechanism for this racemization.

* **6-42.** When the tosylate of 3-phenyl-2-butanol is treated with sodium methoxide, two alkenes result (by E2 elimination). The Saytzeff product predominates.

(a) Draw the reaction, showing the major and minor products.

(b) When one pure stereoisomer of 3-phenyl-2-butyl tosylate reacts, one pure stereoisomer of the major product results. For example, when the tosylate of (2*R*,3*R*)-3-phenyl-2-butanol reacts, the product is the stereoisomer with the methyl groups cis. Use your models to draw a Newman projection of the transition state to show why this stereospecificity is observed.

(c) Use a Newman projection of the transition state to predict the major product of elimination of (2*S*,3*R*)-3-phenyl-2-butyl tosylate.

(d) Predict the major product from elimination of (2*S*,3*S*)-3-phenyl-2-butyl tosylate. This prediction can be made without drawing any structures, by considering the results in part (b).

6-43. Because the S_N1 reaction goes through a flat carbocation, we expect an optically active starting material to give a completely racemized product. In many cases, however, S_N1 reactions actually give more of the product of *inversion* of configuration. In general, as the stability of the carbocation increases, the excess inversion product decreases. Extremely stable carbocations give completely racemic products. Explain these observations.

✱ 6-44. Kenyon and Phillips used the following cycle to study the stereochemistry of the solvolysis of 1-phenyl-2-propyl tosylate in ethanol. Consider the differences in the rotations of the two samples of 2-ethoxy-1-phenylpropane obtained, and explain the stereochemistry observed in each step. Also explain why the products have rotations that differ both in direction and in magnitude.

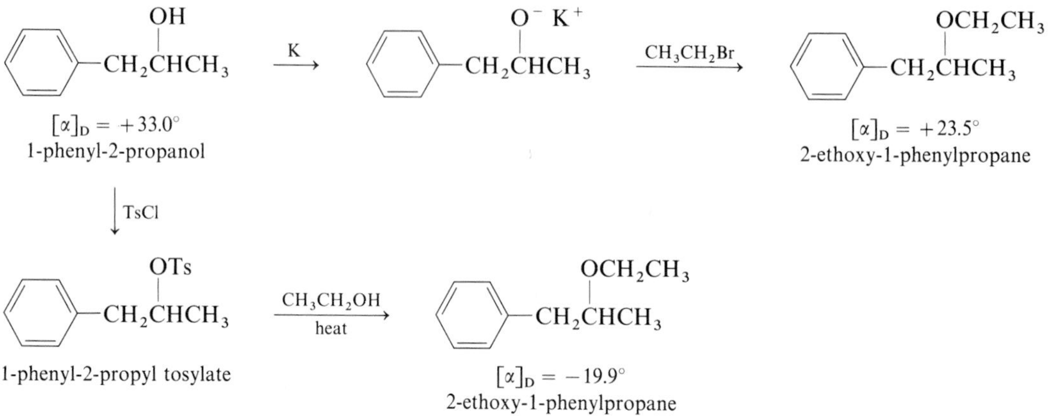

$$C_6H_5-CH_2CH(OH)CH_3 \xrightarrow{K} C_6H_5-CH_2CH(O^-K^+)CH_3 \xrightarrow{CH_3CH_2Br} C_6H_5-CH_2CH(OCH_2CH_3)CH_3$$

1-phenyl-2-propanol, $[\alpha]_D = +33.0°$ → 2-ethoxy-1-phenylpropane, $[\alpha]_D = +23.5°$

1-phenyl-2-propanol $\xrightarrow{TsCl}$ 1-phenyl-2-propyl tosylate

$$C_6H_5-CH_2CH(OTs)CH_3 \xrightarrow[\text{heat}]{CH_3CH_2OH} C_6H_5-CH_2CH(OCH_2CH_3)CH_3$$

1-phenyl-2-propyl tosylate → 2-ethoxy-1-phenylpropane, $[\alpha]_D = -19.9°$

✱ 6-45. **(a)** Design an alkyl halide that will give *only* 2,4-diphenyl-2-pentene upon treatment with potassium *t*-butoxide (a bulky base that promotes E2 elimination).

(b) What stereochemistry is required in your alkyl halide so that *only* the following stereoisomer of the product is formed?

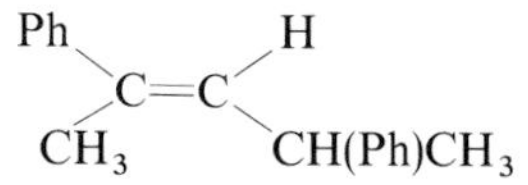

$$(Ph)(CH_3)C=C(H)(CH(Ph)CH_3)$$

(Ph and H on the upper side; CH_3 and $CH(Ph)CH_3$ on the lower side)

✱ 6-46. **(a)** Draw all the stereoisomers of 2,3,4-tribromopentane. (Use of Fischer projections will be helpful.) Label each structure as chiral or meso.

(b) In each structure, label C2 and C4 as (*R*) or as (*S*).

(c) Under what circumstances is C3 chiral?

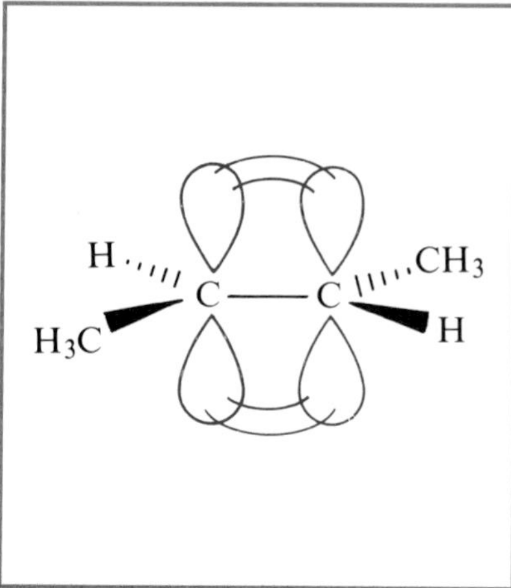

7 STRUCTURE AND SYNTHESIS OF ALKENES

7-1 INTRODUCTION

Alkenes are hydrocarbons with carbon-carbon double bonds. Alkenes are sometimes called **olefins,** a term derived from *olefiant gas,* meaning "oil-forming gas." This term originated with early experimentalists who noticed that the alkene derivatives they made had an oily appearance. The simplest alkene is *ethylene,* with molecular formula C_2H_4:

$$\mathrm{H_2C{=}CH_2}$$

ethylene (ethene)

The bond energy of a carbon-carbon double bond is about 146 kcal/mol (611 kJ/mol), compared with the single-bond energy of about 83 kcal/mol (347 kJ/mol). We can calculate the approximate energy of a pi bond:

double bond dissociation energy	146 kcal/mol	(611 kJ/mol)
subtract sigma bond dissociation energy	(−) 83 kcal/mol	(−)(347 kJ/mol)
pi bond dissociation energy	63 kcal/mol	(264 kJ/mol)

This value of 63 kcal/mol is much less than the sigma-bond energy of 83 kcal/mol, making pi bonds more reactive than sigma bonds.

Because a carbon-carbon double bond is relatively reactive, it is considered to be a functional group. The reactions of alkenes are characterized by the reactions of their carbon-carbon double bonds. We have seen several reactions of alkenes in previous chapters, and we have seen alkenes formed in elimination reactions. In this chapter we study alkenes in more detail, concentrating on their properties and on the ways they are synthesized.

7-2 THE ORBITAL DESCRIPTION OF THE ALKENE DOUBLE BOND

In a Lewis structure, the double bond of an alkene is represented by two pairs of electrons between the carbon atoms. The Pauli exclusion principle tells us that two pairs of electrons can go into one region of space between the carbon nuclei only if each pair has its own orbital. Using ethylene as an example, let's consider how the electrons are distributed in the double bond.

7-2A THE SIGMA-BOND FRAMEWORK

The sigma-bond framework of ethylene is constructed using hybridized orbitals. Each carbon atom is bonded to three other atoms (one carbon and two hydrogens) and has no nonbonding pairs of electrons. Three hybrid orbitals are needed, implying sp^2 hydridization. We have seen (Section 2-4) that an sp^2 hybrid carbon atom has bond angles of about 120°, giving optimum separation of three atoms bonded to a carbon atom.

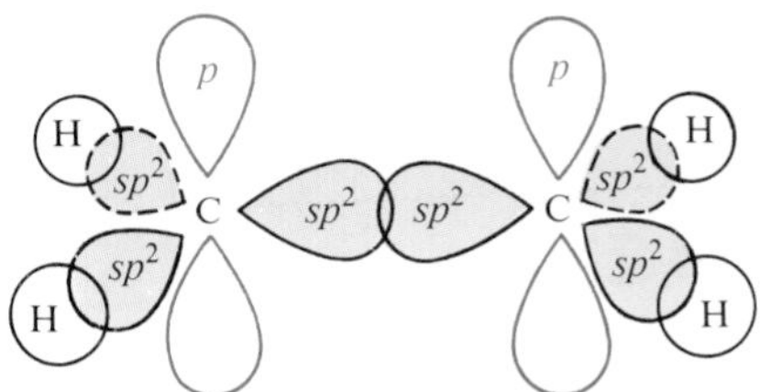

sigma bonding orbitals of ethylene

Each of the carbon-hydrogen bonds is formed by overlap of an sp^2 hybrid orbital on carbon with the 1*s* orbital of a hydrogen atom. The C—H bond length in ethylene (1.08 Å) is slightly shorter than the C—H bond in ethane (1.09 Å), because the sp^2 orbital in ethylene has more *s* character (one-third *s*) than an sp^3 orbital (one-fourth *s*). The *s* orbital is closer to the nucleus than the *p* orbital, contributing to shorter bonds.

H 1.33 Å H; C=C; 116.6°; 121.7°; 1.08 Å; H H

ethylene

H 1.54 Å H; H; C—C; H; 1.09 Å; H H

ethane

The remaining sp^2 orbitals overlap in the region between the carbon nuclei, providing a bonding orbital for one pair of electrons. The electrons in these sp^2 orbitals form one bond between the double-bonded carbon atoms. This bond is a sigma bond because its electron density is centered along the line joining the nuclei. The C=C bond in ethylene (1.33 Å) is much shorter than the C—C bond (1.54 Å) in ethane, partly because the sigma bond of ethylene is formed from sp^2 orbitals (with more *s* character), and partly because there is an additional bond drawing the atoms together.

7-2B THE PI BOND

Two more electrons must go into the carbon-carbon bonding region, and each carbon atom still has an additional unhybridized *p* orbital. These *p* orbitals overlap to form a pi-bonding molecular orbital, with half of the orbital above the line connecting the carbon atoms and the other half below the line. The two electrons in this pi-bonding molecular orbital form the second bond between the double-bonded carbon atoms. For pi overlap to occur, these *p* orbitals must be parallel, which requires that the two carbon atoms be oriented with all their C—H bonds in a single plane (Fig. 7-1).

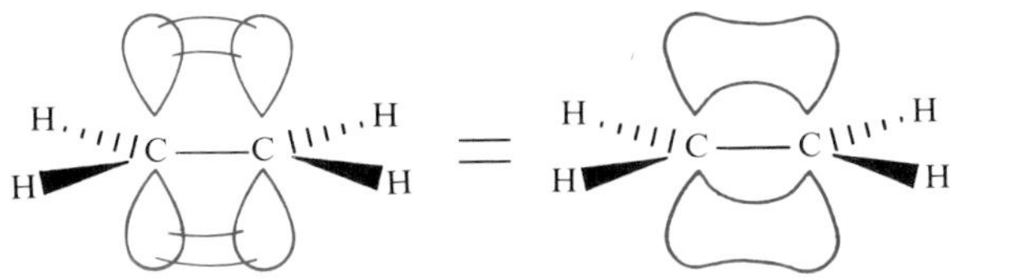

FIGURE 7-1 The pi bond in ethylene is formed by overlap of the unhybridized *p* orbitals on the sp^2 hybrid carbon atoms. This overlap requires the two ends of the molecule to be coplanar, preventing rotation about the double bond.

Figure 7-1 shows that the two ends of the ethylene molecule cannot be twisted with respect to each other without disrupting the pi bond. Unlike single bonds, there is no rotation about a carbon-carbon double bond. This is the origin of cis-trans isomerism. If two groups are on the same side of a double bond (cis), they cannot rotate to opposite sides (trans). Figure 7-2 shows that there are two distinct isomers of 2-butene: *cis*-2-butene and *trans*-2-butene.

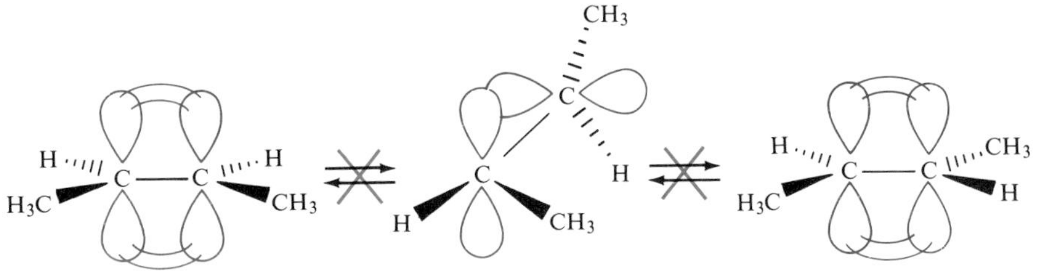

cis — no overlap with the ends perpendicular — trans

FIGURE 7-2 The two isomers of 2-butene cannot interconvert by rotation about the carbon-carbon double bond.

7-3 ELEMENTS OF UNSATURATION

7-3A ELEMENTS OF UNSATURATION IN HYDROCARBONS

Alkenes are called **unsaturated** because they react with hydrogen in the presence of a catalyst. The product is an alkane, which is called **saturated** because it cannot react with any more hydrogen. Either the pi bond of an alkene (or an alkyne) or the ring of a cyclic compound decreases the number of hydrogen atoms in a molecular formula. These structural features are called **elements of unsaturation.** Each element of unsaturation corresponds to two fewer hydrogen atoms than in the "saturated" formula.

$CH_3—CH_2—CH_3$	$CH_3—CH{=}CH_2$	CH_2 (ring) $CH_2—CH_2$
propane, C_3H_8	propene, C_3H_6	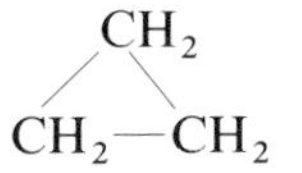 cyclopropane, C_3H_6
saturated	one element of unsaturation	one element of unsaturation

Consider, for example, the formula C_4H_8. A saturated alkane would have a formula $C_nH_{(2n+2)}$, C_4H_{10}. The formula C_4H_8 is missing two hydrogen atoms, so it has one **element of unsaturation,** either a pi bond or a ring. There are five structural isomers of formula C_4H_8:

$CH_2{=}CH{-}CH_2CH_3$ (1-butene) $\quad CH_3{-}CH{=}CH{-}CH_3$ (2-butene)

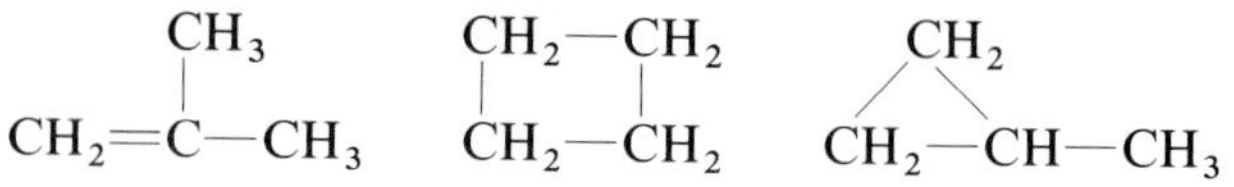

$CH_2{=}C(CH_3){-}CH_3$ (isobutylene) $\quad$ cyclobutane ($CH_2{-}CH_2$ / $CH_2{-}CH_2$ ring) $\quad$ methylcyclopropane (CH_2 bridging $CH_2{-}CH{-}CH_3$)

Whenever you need to determine a structure for a particular molecular formula, it is valuable to find the number of elements of unsaturation. Calculate the number of hydrogen atoms that would be in a saturated formula $C_nH_{(2n+2)}$ and see how many are missing. The number of elements of unsaturation is simply half the number of missing hydrogens. If you prefer to use a formula, then

$$\text{elements of unsaturation} = \frac{2C + 2 - H}{2} \qquad \begin{array}{l} C = \text{number of carbons} \\ H = \text{number of hydrogens} \end{array}$$

This simple calculation allows you to consider possible structures quickly, without always having to check for the correct molecular formula.

PROBLEM 7-1

(a) Calculate the number of elements of unsaturation implied by the molecular formula C_6H_{12}.
(b) Give five examples of structures with this formula (C_6H_{12}). At least one should contain a ring, and at least one a double bond.

PROBLEM 7-2

Determine the number of elements of unsaturation in the molecular formula C_4H_6. Give all nine possible structures having this formula. Remember that

a double bond = one element of unsaturation

a ring = one element of unsaturation

a triple bond = two elements of unsaturation

7-3B ELEMENTS OF UNSATURATION WITH HETEROATOMS

Heteroatoms (*hetero,* "different") are any atoms other than carbon and hydrogen. The simple rule above for calculating elements of unsaturation in hydrocarbons can be extended to include heteroatoms. Let's consider how the addition of a heteroatom affects the number of hydrogen atoms in the formula.

Halogens Halogens simply substitute for hydrogen atoms in the molecular formula. The formula C_2H_6 is saturated, so the formula $C_2H_4F_2$ is also saturated. C_4H_8 has one element of unsaturation, and $C_4H_5Br_3$ also has one element of unsaturation. In calculating the number of elements of unsaturation, simply *count halogens as hydrogen atoms.*

$CH_3{-}CHF_2$	$CH_3{-}CH{=}CH{-}CBr_3$	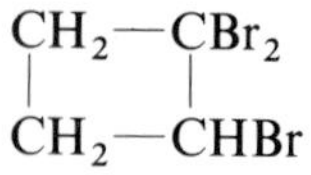 $CH_2{-}CBr_2$ / $CH_2{-}CHBr$ (ring)
$C_2H_4F_2$	$C_4H_5Br_3$	$C_4H_5Br_3$
saturated	one element of unsaturation	one element of unsaturation

Oxygen An oxygen atom can be added to the chain without changing the number of hydrogen atoms or carbon atoms. In calculating the number of elements of unsaturation, *ignore the oxygen atoms.*

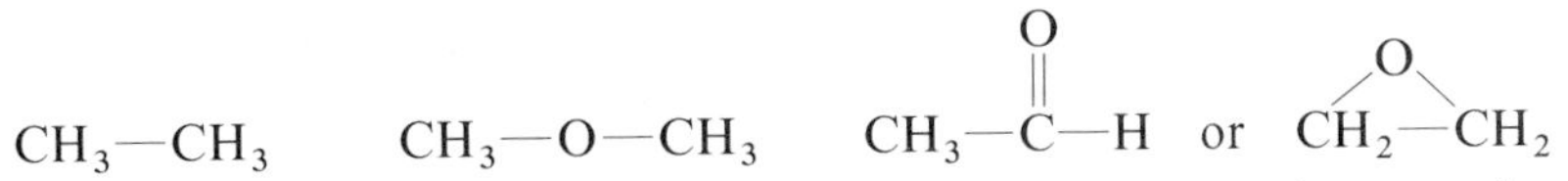

CH_3-CH_3 — C_2H_6, saturated

CH_3-O-CH_3 — C_2H_6O, saturated

$CH_3-C(=O)-H$ or CH_2-CH_2 (ring with O) — C_2H_4O, one element of unsaturation

Nitrogen A nitrogen atom can take the place of a carbon atom in the chain, but nitrogen is trivalent, having only one additional hydrogen atom, compared with two hydrogens for each additional carbon atom. In computing the elements of unsaturation, we *count nitrogen as half a carbon atom.*

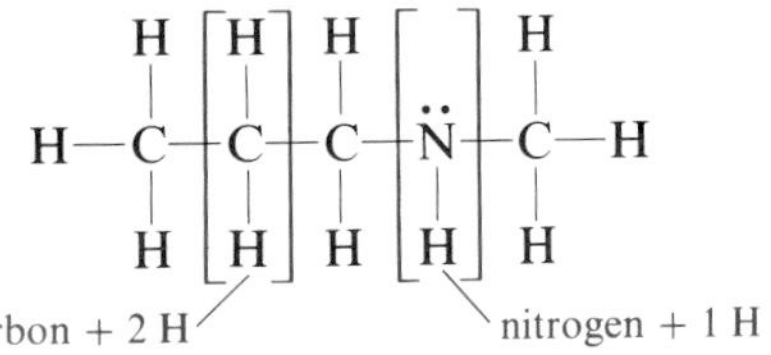

H–C(H)(H)–[C(H)(H)]–C(H)(H)–[N̈(H)]–C(H)(H)–H

carbon + 2 H   nitrogen + 1 H

The formula C_4H_9N is like a formula with $4\frac{1}{2}$ carbon atoms, with saturated formula $C_{4.5}H_{9+2}$. The formula C_4H_9N has one element of unsaturation, because it is two hydrogen atoms short of the saturated formula.

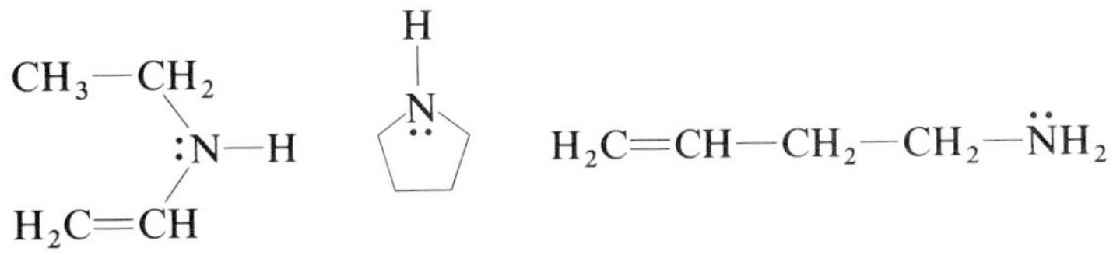

$CH_3-CH_2-\ddot{N}H-CH=CH_2$ pyrrolidine (N–H) $H_2C=CH-CH_2-CH_2-\ddot{N}H_2$

examples of formula C_4H_9N, one element of unsaturation

SOLVED PROBLEM 7-1

Draw at least four compounds of formula C_4H_6NOCl.

SOLUTION Counting the nitrogen as $\frac{1}{2}$ carbon, ignoring the oxygen, and counting chlorine as a hydrogen shows the formula is equivalent to $C_{4.5}H_7$. The saturated formula for 4.5 carbon atoms is $C_{4.5}H_{11}$, showing that C_4H_6NOCl has two elements of unsaturation. These could be two double bonds, two rings, one triple bond, or a ring and a double bond. There are many possibilities, four of which are listed below.

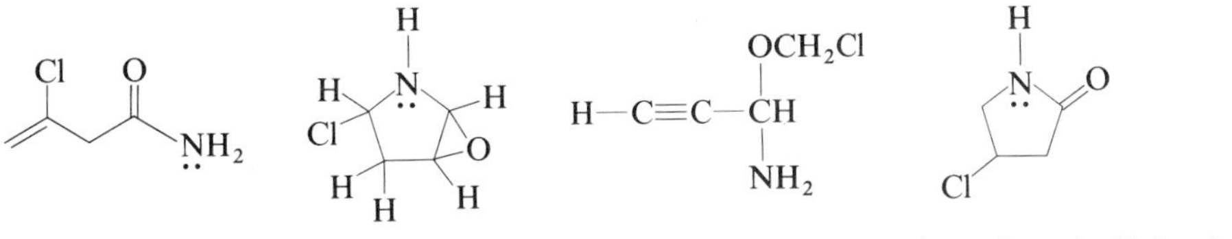

Cl, O, $\ddot{N}H_2$ — two double bonds

H, N, H, Cl, O, H, H, H — two rings

$H-C\equiv C-CH(OCH_2Cl)(NH_2)$ — one triple bond

H, N, O, Cl — one ring and one double bond

PROBLEM 7-3

Draw five more compounds of formula C_4H_6NOCl.

PROBLEM 7-4

For each of the following molecular formulas, determine the number of elements of unsaturation and give three examples.

(a) $C_3H_4Cl_2$ (b) C_4H_8O (c) $C_4H_4O_2$ (d) $C_5H_5NO_2$ (e) C_6H_3NClBr

7-4 NOMENCLATURE OF ALKENES

Simple alkenes are named much like alkanes, using the root name of the longest chain containing the double bond. The ending is changed from *-ane* to *-ene*. For example, "ethane" becomes "ethene," "propane" becomes "propene," and "cyclohexane" becomes "cyclohexene."

$CH_3—CH_3$ $CH_2=CH_2$ $CH_3—CH_2—CH_3$ $CH_2=CH—CH_3$ 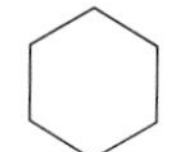 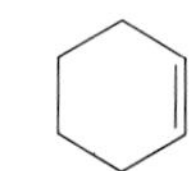

IUPAC names: ethane, ethene, propane, propene, cyclohexane, cyclohexene

When the chain contains more than three carbon atoms, a number is used to give the location of the double bond. The chain is numbered starting from the end closest to the double bond, and the double bond is given the *lower* number of its two double-bonded carbon atoms. Cycloalkenes are assumed to have the double bond in the number 1 position.

$\overset{1}{CH_2}=\overset{2}{CH}—\overset{3}{CH_2}—\overset{4}{CH_3}$ $\overset{1}{CH_3}—\overset{2}{CH}=\overset{3}{CH}—\overset{4}{CH_3}$ (cyclohexene ring numbered 1–6)

IUPAC names: 1-butene, 2-butene, cyclohexene

$\overset{1}{CH_2}=\overset{2}{CH}—\overset{3}{CH_2}—\overset{4}{CH_2}—\overset{5}{CH_3}$ $\overset{1}{CH_3}—\overset{2}{CH}=\overset{3}{CH}—\overset{4}{CH_2}—\overset{5}{CH_3}$ $\overset{6}{CH_3}—\overset{5}{CH_2}—\overset{4}{CH_2}—\overset{3}{CH}=\overset{2}{CH}—\overset{1}{CH_3}$

IUPAC names: 1-pentene, 2-pentene, 2-hexene

A compound with two double bonds is called a **diene.** A **triene** has three double bonds, and a **tetraene** has four. Numbers are used to specify the locations of the double bonds.

$\overset{1}{CH_2}=\overset{2}{CH}—\overset{3}{CH}=\overset{4}{CH_2}$ $\overset{7}{CH_3}—\overset{6}{CH}=\overset{5}{CH}—\overset{4}{CH}=\overset{3}{CH}—\overset{2}{CH}=\overset{1}{CH_2}$ (cyclooctatetraene ring numbered 1–8)

IUPAC names: 1,3-butadiene, 1,3,5-heptatriene, 1,3,5,7-cyclooctatetraene

If there are alkyl groups attached to the main chain, each is listed with a number to give its location. Note that the double bond is still given preference in numbering, however.

$\overset{4}{CH_3}—\overset{3}{CH}=\overset{2}{C}(CH_3)—\overset{1}{CH_3}$ $\overset{4}{CH_3}—\overset{3}{CH}(CH_3)—\overset{2}{CH}=\overset{1}{CH_2}$ $\overset{1}{CH_3}—\overset{2}{CH}=\overset{3}{C}(CH_3)—\overset{4}{CH_2}—\overset{5}{CH_2}—\overset{6}{CH}(CH_3)—\overset{7}{CH_3}$

IUPAC names: 2-methyl-2-butene, 3-methyl-1-butene, 3,6-dimethyl-2-heptene

CH_3 CH_2CH_3 Br

IUPAC names: 1-methylcyclopentene, 2-ethyl-1,3-cyclohexadiene, 7-bromo-1,3,5-cycloheptatriene

Alkenes as substituents Alkenes named as substituents are called *alkenyl groups.* Alkenyl groups are named either systematically (ethenyl, propenyl, etc.), or using common names. The vinyl group, the allyl group, the methylene group, and the phenyl group are frequently named as substituents. The phenyl group (Ph) is different from the others, because it is aromatic (see Chap. 16) and does not undergo the typical reactions of alkenes.

—CH=CH_2	$\text{—CH}_2\text{—CH=CH}_2$	=CH_2	—C_6H_5
vinyl group (ethenyl group)	allyl group (2-propenyl group)	methylene group (methylidene group)	phenyl group (Ph)
$\text{CH}_2\text{=CHCH(CH=CH}_2\text{)CH}_2\text{CH=CH}_2$	$\text{CH}_2\text{=CH—CH}_2\text{—Cl}$		
3-vinyl-1,5-hexadiene	allyl chloride (3-chloropropene)	3-methylenecyclohexene	2-phenyl-1,3-cyclopentadiene

Common names Most alkenes are conveniently named by the IUPAC system, but common names are sometimes used for the simplest compounds.

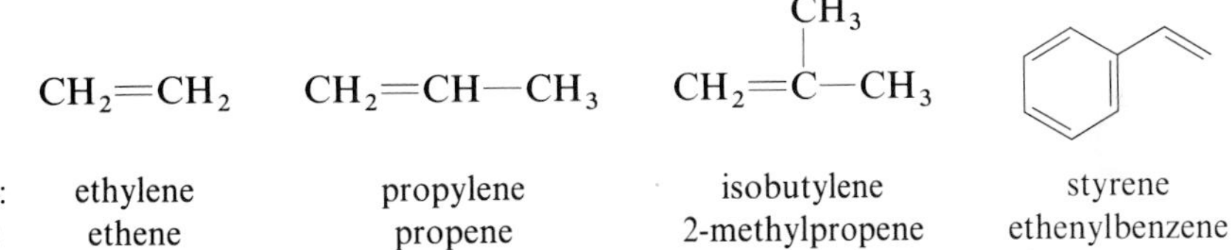

7-5 NOMENCLATURE OF GEOMETRIC ISOMERS

7-5A CIS-TRANS NOMENCLATURE

In Chapter 2 we saw that the rigidity and lack of rotation of carbon-carbon double bonds gives rise to cis-trans isomerism, a variety of stereoisomerism also called **geometric isomerism.** If two similar groups bonded to the carbons of the double bond are on the same side of the bond, the alkene is the *cis* isomer. If the similar groups are on opposite sides of the bond, the alkene is *trans.* Not all alkenes are capable of showing geometric isomerism. If either carbon of the double bond holds two identical groups, the molecule cannot have cis and trans forms. Following are some cis and trans alkenes and some alkenes that cannot show cis-trans isomerism.

$(CH_3)(H)C=C(CH_2CH_3)(H)$	$(CH_3)(H)C=C(H)(CH_2CH_3)$	$(CH_3)(CH_3)C=C(H)(CH_2CH_3)$	$(H)(H)C=C(CH_2CH_2CH_3)(H)$
cis-2-pentene	*trans*-2-pentene	2-methyl-2-pentene	1-pentene

(neither *cis* nor *trans*)

trans-Cycloalkenes are unstable unless the ring is large enough (at least eight carbon atoms) to accommodate the trans double bond. Therefore, all cycloalkenes are assumed to be cis unless they are specifically named trans. The cis name is rarely used with cycloalkenes, except to distinguish a large cycloalkene from its trans isomer.

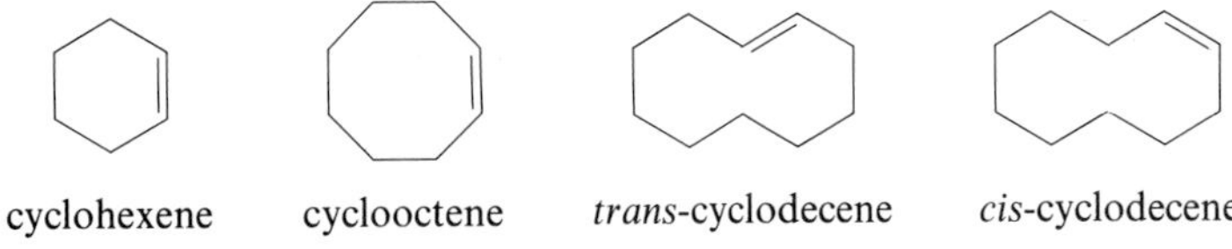

cyclohexene cyclooctene *trans*-cyclodecene *cis*-cyclodecene

7-5B *E-Z* NOMENCLATURE

The cis-trans nomenclature for geometric isomers sometimes fails to give an unambiguous name. The isomers of 1-bromo-1-chloropropene are not clearly cis or trans because it is not obvious which substituents are referred to as being cis or trans.

Br(Cl)C=C(CH$_3$)(H)   Cl(Br)C=C(CH$_3$)(H)

geometric isomers of 1-bromo-1-chloropropene

In response to this problem, another system of nomenclature is used. The **E-Z system** of nomenclature for geometric isomers is patterned after the Cahn-Ingold-Prelog convention for chiral carbon atoms (Section 6-3). It assigns a unique configuration of either *E* or *Z* to any double bond capable of geometric isomerism.

To use the *E-Z* system, we mentally separate the alkene into its two ends. Assign priorities [just as you would for determining the (*R*) or (*S*) configuration of a chiral carbon atom] to the two substituents on each end of the double bond. The isomer with the two atoms numbered 1 *together* (cis) on the same side of the double bond is called the *Z* isomer, from the German word *zusammen* ("together"). The isomer with the two atoms numbered 1 *opposite* (trans) is called the *E* isomer, from the German word *entgegen* ("opposite").

①(2)C= =C(①)(2)   ①(2)C= =C(2)(①)

Zusammen   *Entgegen*

For example,

Br(Cl)C=C(CH$_3$)(H) becomes ① Br, ② Cl C= =C CH$_3$ ①, H ② (①'s together) = *Z*

(*Z*)-1-bromo-1-chloropropene

The other isomer is named similarly:

Cl(Br)C=C(CH$_3$)(H) becomes ② Cl, ① Br C= =C CH$_3$ ①, H ② (①'s opposite) = *E*

(*E*)-1-bromo-1-chloropropene

If the alkene has more than one double bond, the stereochemistry about each double bond should be specified. The following compound is properly named 3-bromo-(3*Z*,5*E*)-octadiene.

3-bromo-(3Z,5*E*)-octadiene

SUMMARY OF RULES FOR NAMING ALKENES

The following rules summarize the IUPAC system for naming alkenes.

1. Select the longest chain or largest ring that contains *the largest possible number of double bonds,* and name it with the *-ene* suffix. If there are two double bonds, the suffix is -diene; for three, -triene; for four, -tetraene; and so on.

2. Number the chain from the end closest to the double bonds. Number a ring so that the double bond is between carbons 1 and 2. The numbers giving the locations of the double bonds are placed in front of the root name.
3. Substituent groups are named as they are in alkanes, with their locations indicated by the number of the main-chain carbon to which they are attached. The ethenyl group and the propenyl group are usually called the *vinyl* group and the *allyl* group, respectively.
4. For compounds that show geometric isomerism, add the appropriate prefix: *cis* or *trans*, or *E* or *Z*. Cycloalkenes are assumed to be cis unless named otherwise.

PROBLEM 7-5

Give the systematic (IUPAC) names of the following alkenes.

(a) $CH_2{=}CH{-}CH_2{-}CH(CH_3)_2$ (b) $CH_3(CH_2)_3{-}C({=}CH_2){-}CH_2CH_3$

(c) $CH_2{=}CH{-}CH_2{-}CH{=}CH_2$ (d) $CH_2{=}C{=}CH{-}CH{=}CH_2$

(e) CH_3 CH_3 (f) (g) (h) CH_3 CH_3 (i) CH_2

PROBLEM 7-6

(1) Determine which of the following compounds show cis-trans isomerism.
(2) Draw and name the cis and trans isomers of those that do.

(a) 3-hexene (b) 1,3-butadiene (c) 2,4-hexadiene
(d) 3-methyl-2-pentene (e) 2,3-dimethyl-2-pentene (f) cyclopentene

PROBLEM 7-7

Each of the following names is incorrect. Draw the structure represented by the incorrect name (or a consistent structure if the name is ambiguous) and give your drawing the correct name.

(a) *cis*-2,3-dimethyl-2-pentene (b) 3-vinyl-4-hexene
(c) 2-methylcyclopentene (d) 6-chlorocyclohexadiene
(e) 3,4-dimethylcyclohexene (f) *cis*-2,5-dibromo-3-ethyl-2-pentene

PROBLEM 7-8

Some of the following examples can show geometric isomerism, and some cannot. For the ones that can, draw all the geometric isomers and assign complete names using the E-Z system.

(a) 3-bromo-2-chloro-2-pentene (b) 3-ethyl-2,4-hexadiene
(c) 3-bromo-2-methyl-2-butene (d) 1,3-pentadiene
(e) 3-(*t*-butyl)-4-methyl-3-hexene (f) 3,7-dichloro-2,5-octadiene

(g) 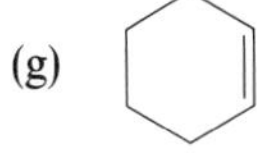
cyclohexene

(h) 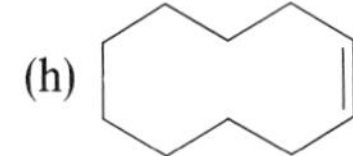
cyclodecene

(i) 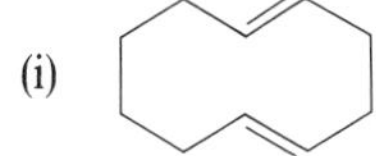
1,5-cyclodecadiene

7-6 COMMERCIAL IMPORTANCE OF ALKENES

Because the carbon-carbon double bond is easily converted to other functional groups, alkenes are important intermediates in the synthesis of drugs, pesticides, and other valuable chemicals. These uses account for only a small fraction of the billions of pounds of alkenes used annually to make polymers, however.

A **polymer** (Greek, *poly,* "many," and *meros,* "parts") is a large molecule made up of many **monomer** (Greek, *mono,* "one") molecules. Polymers are important industrial products; their synthesis and properties are covered in detail in Chapter 26. Alkenes can be **polymerized** by a chain reaction that causes additional alkene (monomer) molecules to add to the end of the growing polymer chain. Many of the common polymers are made by polymerizing alkenes. Because these polymers result from the addition of many individual alkene units, they are called **addition polymers. Polyolefins** are polymers made from monofunctional alkenes such as ethylene and propylene. Figure 7-3 shows the structures of some addition polymers made from simple alkenes.

$$\mathrm{H_2C{=}CH_2} \xrightarrow{\text{polymerize}} \mathrm{-CH_2-CH_2-[CH_2-CH_2]_n-CH_2-CH_2-}$$

ethylene (monomer) → polyethylene (polymer)

$$\mathrm{H_2C{=}CHCl} \xrightarrow{\text{polymerize}} \mathrm{-CH_2-CHCl-[CH_2-CHCl]_n-CH_2-CHCl-}$$

vinyl chloride → poly(vinyl chloride) PVC, "vinyl"

$$\mathrm{F_2C{=}CF_2} \xrightarrow{\text{polymerize}} \mathrm{-CF_2-CF_2-[CF_2-CF_2]_n-CF_2-CF_2-}$$

tetrafluoroethylene → poly(tetrafluoroethylene) PTFE, Teflon

FIGURE 7-3 Under the proper conditions, alkenes polymerize to form addition polymers. Many of the common polymers are produced in this way.

7-7 STABILITY OF ALKENES

In many alkene syntheses the major product is the most stable alkene. Before considering alkene syntheses, we should consider which of several possible alkene products is the most stable. Stabilities of compounds can be compared by converting each to a common product and comparing the amounts of heat given off. We might convert several alkenes to CO_2 and H_2O and measure heats of combustion; however, heats of combustion are large numbers (several hundred kcal/mol), and it is difficult to measure accurately the small differences in these large numbers. Heats of hydrogenation can be measured similarly, yet they are smaller numbers and provide more accurate energy differences.

7-7A HEATS OF HYDROGENATION

Alkene energies are often compared by measuring the **heat of hydrogenation:** the heat given off ($\Delta H°$) during catalytic hydrogenation. The alkene is treated with hydrogen in the presence of a platinum catalyst. Hydrogen adds to the double bond, reducing the alkene to an alkane. Hydrogenation is mildly exothermic, with about 20 to 30 kcal (80 to 120 kJ) of heat evolved per mole of hydrogen consumed. For example, for 1-butene and *trans*-2-butene,

$$\underset{\substack{\text{1-butene}\\ \text{(monosubstituted)}}}{H_2C{=}CH{-}CH_2{-}CH_3} + H_2 \xrightarrow{Pt} \underset{\text{butane}}{CH_3{-}CH_2{-}CH_2{-}CH_3} \qquad \Delta H^\circ = -30.3 \text{ kcal/mol} \; (-127 \text{ kJ/mol})$$

$$\underset{\substack{\textit{trans}\text{-2-butene}\\ \text{(disubstituted)}}}{(H_3C)(H)C{=}C(H)(CH_3)} + H_2 \xrightarrow{Pt} \underset{\text{butane}}{CH_3{-}CH_2{-}CH_2{-}CH_3} \qquad \Delta H^\circ = -27.6 \text{ kcal/mol} \; (-115 \text{ kJ/mol})$$

Figure 7-4 shows these heats of hydrogenation on a potential-energy profile. The difference in the stabilities of 1-butene and *trans*-2-butene is shown to be the difference in their heats of hydrogenation. Figure 7-5 shows that *trans*-2-butene is more stable by

$$30.3 \text{ kcal/mol} - 27.6 \text{ kcal/mol} = 2.7 \text{ kcal/mol} \quad (12 \text{ kJ/mol})$$

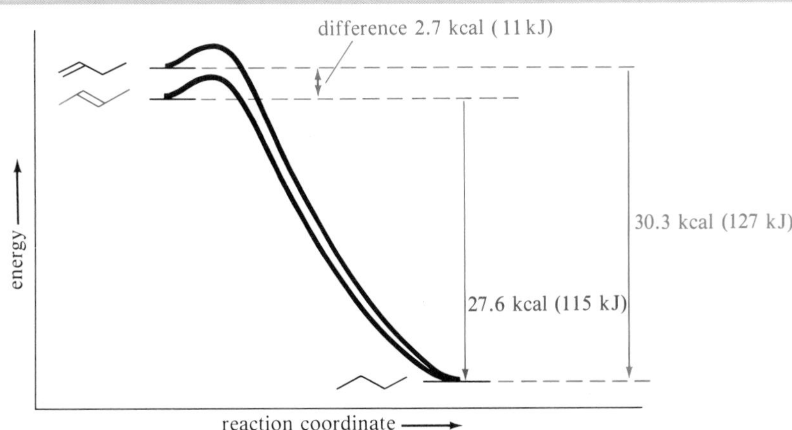

FIGURE 7-4 *trans*-2-Butene is more stable than 1-butene by 2.7 kcal/mol (12 kJ/mol).

7-7B SUBSTITUTION EFFECTS

A 2.7-kcal/mol (12 kJ/mol) stability difference is typical for a monosubstituted alkene (1-butene) and a trans-disubstituted alkene (*trans*-2-butene). In the equations below we compare the monosubstituted double bond of 3-methyl-1-butene with trisubstituted double bond of 2-methyl-2-butene. The trisubstituted alkene is more stable by 3.4 kcal/mol (14 kJ/mol).

$$\underset{\substack{\text{3-methyl-1-butene}\\ \text{(monosubstituted)}}}{CH_2{=}CH{-}CH(CH_3){-}CH_3} \xrightarrow{H_2,\ Pt} \underset{\text{2-methylbutane}}{CH_3{-}CH_2{-}CH(CH_3){-}CH_3} \qquad \Delta H^\circ = -30.3 \text{ kcal} \; (-127 \text{ kJ})$$

$$\underset{\substack{\text{2-methyl-2-butene}\\ \text{(trisubstituted)}}}{CH_3{-}CH{=}C(CH_3){-}CH_3} \xrightarrow{H_2,\ Pt} \underset{\text{2-methylbutane}}{CH_3{-}CH_2{-}CH(CH_3){-}CH_3} \qquad \Delta H^\circ = -26.9 \text{ kcal} \; (-113 \text{ kJ})$$

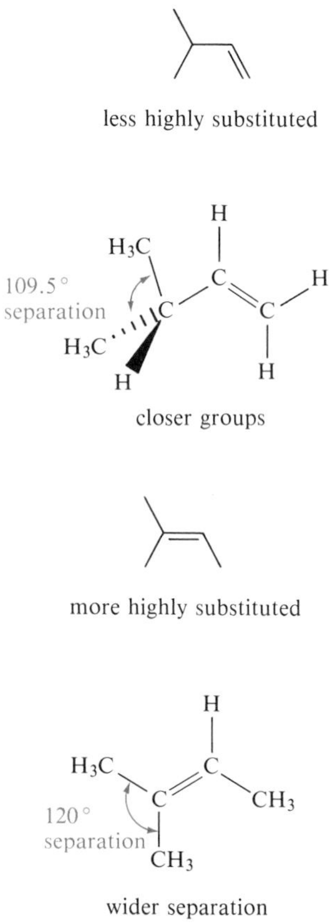

less highly substituted

H

H_3C

109.5° separation

H

C

C

C

H_3C

H

H

closer groups

more highly substituted

H

H_3C

C

C

CH_3

120° separation

CH_3

wider separation

FIGURE 7-5 The isomer with the more highly substituted double bond has a larger angular separation between the bulky alkyl groups.

To be completely correct, we should compare heats of hydrogenation only for compounds that give the same alkane, as 3-methyl-1-butene and 2-methyl-2-butene do. Yet, most alkenes with similar substitution patterns give similar heats of hydrogenation. For example, 3,3-dimethyl-1-butene (below) hydrogenates to give a different alkane than 3-methyl-1-butene or 1-butene (above); yet all three of these monosubstituted alkenes have similar heats of hydrogenation, because the different alkanes formed have similar energies.

$$CH_2{=}CH{-}C(CH_3)_2{-}CH_3 \xrightarrow{H_2,\ Pt} CH_3{-}CH_2{-}C(CH_3)_2{-}CH_3 \qquad \Delta H^\circ = -30.3\text{ kcal}\ (-127\text{ kJ})$$

3,3-dimethyl-1-butene (monosubstituted) → 2,2-dimethylbutane

In practice, heats of hydrogenation can be used to compare the stabilities of quite different alkenes as long as they hydrogenate to give alkanes of similar energies. Most open-chain alkanes and unstrained cycloalkanes are of similar energies, and we can use this approximation. Table 7-1 shows the heats of hydrogenation of a variety of alkenes with different substitution. The compounds are ranked in decreasing order of their heats of hydrogenation; that is, from the least stable double bonds to the most stable. Note that the values are quite similar for alkenes with similar substitution patterns.

The most stable double bonds are those with the most alkyl groups attached. For example, hydrogenation of ethylene (no alkyl groups attached) evolves 32.8 kcal/mol, while propene and 1-pentene (one alkyl group for each) give off 30.1 kcal/mol. Double bonds with two alkyl groups hydrogenate to produce about 28 kcal/mol. Three of four alkyl substituents further stabilize the double bond, as with 2-methyl-2-butene (trisubstituted, 26.9 kcal) and 2,3-dimethyl-2-butene (tetrasubstituted, 26.6 kcal).

The values in Table 7-1 confirm the Saytzeff rule, which we encountered in our study of elimination reactions: *More highly substituted double bonds are usually more stable.* In other words, the alkyl groups attached to the double-bonded carbon help to stabilize the alkene.

Two factors are probably responsible for the stabilizing effect that alkyl groups have on a double bond. Alkyl groups are electron donating, and they contribute electron density to the pi bond. In addition, bulky substituents like alkyl groups are best situated as far apart as possible. In an alkane, they are separated by the tetrahedral bond angle, about 109.5°. A double bond increases this separation to about 120°. In general, the alkyl groups are separated best by the most highly substituted double bond. This steric effect is illustrated in Figure 7-5 for two **double-bond isomers** (isomers that differ only in the position of the double bond). The isomer with the monosubstituted double bond separates the alkyl groups by only 109.5°, while the disubstituted double bond separates them by about 120°.

PROBLEM 7-9

Use the data in Table 7-1 to determine the energy difference between 2,3-dimethyl-1-butene and 2,3-dimethyl-2-butene. Which of these double-bond isomers is more stable?

TABLE 7-1
Molar heats of hydrogenation of alkenes[a]

Name	Structure	Molar heat of hydrogenation ($-\Delta H°$) kcal	kJ	General structure
ethene (ethylene)	$H_2C{=}CH_2$	32.8	137	unsubstituted
propene (propylene)	$CH_3{-}CH{=}CH_2$	30.1	126	monosubstituted $R{-}CH{=}CH_2$
1-butene	$CH_3{-}CH_2{-}CH{=}CH_2$	30.3	127	
1-pentene	$CH_3{-}CH_2{-}CH_2{-}CH{=}CH_2$	30.1	126	
1-hexene	$CH_3{-}(CH_2)_3{-}CH{=}CH_2$	30.1	126	
3-methyl-1-butene	$(CH_3)_2CH{-}CH{=}CH_2$	30.3	127	
3,3-dimethyl-1-butene	$(CH_3)_3C{-}CH{=}CH_2$	30.3	127	
cis-2-butene	$(CH_3)HC{=}CH(CH_3)$ (cis)	28.6	120	disubstituted (cis) $RHC{=}CHR$ (R, R cis)
cis-2-pentene	$(CH_3{-}CH_2)HC{=}CH(CH_3)$ (cis)	28.6	120	
2-methylpropene (isobutylene)	$(CH_3)_2C{=}CH_2$	28.0	117	disubstituted (geminal) $R_2C{=}CH_2$
2-methyl-1-butene	$CH_3{-}CH_2{-}C(CH_3){=}CH_2$	28.5	119	
2,3-dimethyl-1-butene	$(CH_3)_2CH{-}C(CH_3){=}CH_2$	28.0	117	
trans-2-butene	$(CH_3)HC{=}CH(CH_3)$ (trans)	27.6	116	disubstituted (trans) $RHC{=}CHR$ (R, R trans)
trans-2-pentene	$(CH_3{-}CH_2)HC{=}CH(CH_3)$ (trans)	27.6	116	
2-methyl-2-butene	$CH_3{-}C(CH_3){=}CH{-}CH_3$	26.9	113	trisubstituted $R_2C{=}CHR$
2,3-dimethyl-2-butene	$(CH_3)_2C{=}C(CH_3)_2$	26.6	111	tetrasubstituted $R_2C{=}CR_2$

[a] A lower value of the heat of hydrogenation corresponds to a lower energy and greater stability in the alkene.

7-7C ENERGY DIFFERENCES IN CIS-TRANS ISOMERS

The heats of hydrogenation in Table 7-1 also show that trans isomers are generally more stable than the corresponding cis isomers. This trend seems reasonable because the alkyl substituents are separated farther in the trans isomers than they are in the cis isomers. The greater stability of the trans isomer is evident in the 2-pentenes, which show a 1.0 kcal (4 kJ) difference between the cis and trans isomers.

$$\text{cis-}(H_3C)HC{=}CH(CH_2CH_3) + H_2 \xrightarrow{Pt} CH_3{-}CH_2{-}CH_2{-}CH_2CH_3 \quad \Delta H° = -28.6\ \text{kcal/mol}\ (-120\ \text{kJ/mol})$$

$$\text{trans-}(H_3C)HC{=}CH(CH_2CH_3) + H_2 \xrightarrow{Pt} CH_3{-}CH_2{-}CH_2{-}CH_2CH_3 \quad \Delta H° = -27.6\ \text{kcal/mol}\ (-116\ \text{kJ/mol})$$

A 1-kcal/mol (4-kJ-mol) difference between cis and trans isomers is typical for disubstituted alkenes. Figure 7-6 summarizes the relative stabilities of alkenes, comparing them with ethylene, the least stable of the simple alkenes.

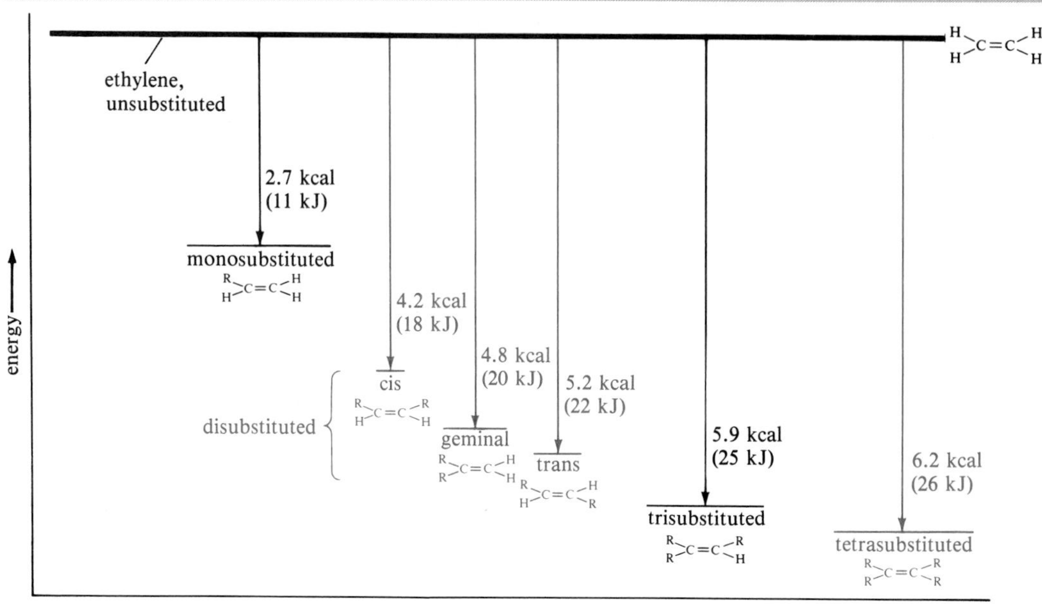

FIGURE 7-6 Relative stabilities of alkenes compared with ethylene.

PROBLEM 7-10

For each of the following pairs of isomers, tell which one would be more stable and by about how many kcal/mol or kJ/mol.

(a) *cis,cis*-2,4-hexadiene and *trans,trans*-2,4-hexadiene
(b) 2-methyl-1-butene and 3-methyl-1-butene
(c) 2-methyl-1-butene and 2-methyl-2-butene
(d) 2,3-dimethyl-1-butene and 2,3-dimethyl-2-butene

7-7D STABILITY OF CYCLOALKENES

Most cycloalkenes react like acyclic (noncyclic) alkenes. The presence of a ring makes a major difference only if there is ring strain involved, either by the small size of the ring or by the presence of a trans double bond. Rings that are five-membered or larger can easily accommodate double bonds, and these cycloalkenes react much like the straight-chain alkenes. The three- and four-membered rings show evidence of ring strain, however.

Cyclobutene Cyclobutene has a heat of hydrogenation of -30.7 kcal/mol (-128 kJ/mol), compared with cyclopentene's value of -26.6 kcal/mol (-111 kJ/mol).

$$\text{cyclobutene} + H_2 \xrightarrow{Pt} \text{cyclobutane} \qquad \Delta H^\circ = -30.7 \text{ kcal/mol } (-128 \text{ kJ/mol})$$

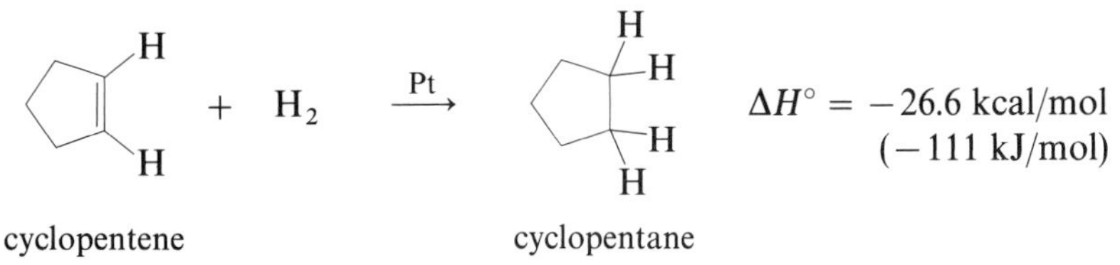

$$\text{cyclopentene} + H_2 \xrightarrow{Pt} \text{cyclopentane} \qquad \Delta H^\circ = -26.6\ \text{kcal/mol}\ (-111\ \text{kJ/mol})$$

The double bond in cyclobutene thus has about 4 kcal/mol of *extra* ring strain (in addition to the ring strain in cyclobutane) by virtue of the small ring. The 90° bond angles in cyclobutene compress the angles of the sp^2 hybrid carbons (normally 120°) more than they compress the sp^3 hybrid angles (normally 109.5°) in cyclobutane. The extra ring strain in cyclobutene makes its double bond more reactive than that of a typical double bond.

Cyclopropene Cyclopropene has bond angles of about 60°, compressing the bond angles of the carbon-carbon double bond to half their usual value of 120°. The double bond in cyclopropene is highly strained.

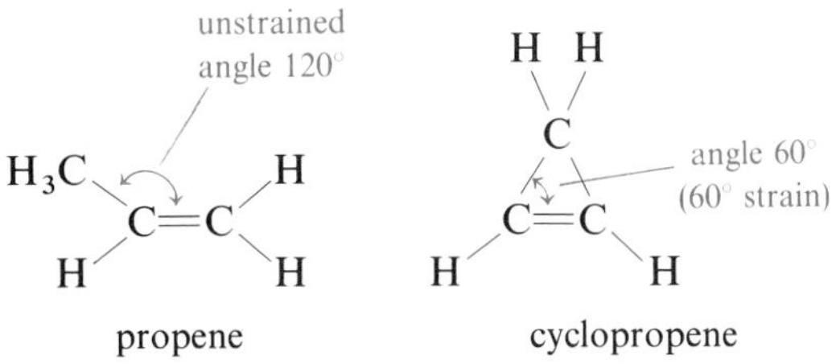

unstrained angle 120°

angle 60° (60° strain)

propene

cyclopropene

Many chemists once believed that a cyclopropene could never be made because it would snap open immediately from the large ring strain. Cyclopropene itself was eventually synthesized, however, and it could be stored in the cold. Cyclopropenes were still considered to be strange, highly unusual compounds. Natural product chemists were surprised when they examined the kernel oil of *Sterculis foetida,* a tropical tree. One of the constituents of this oil is *sterculic acid,* a carboxylic acid that contains a cyclopropene ring.

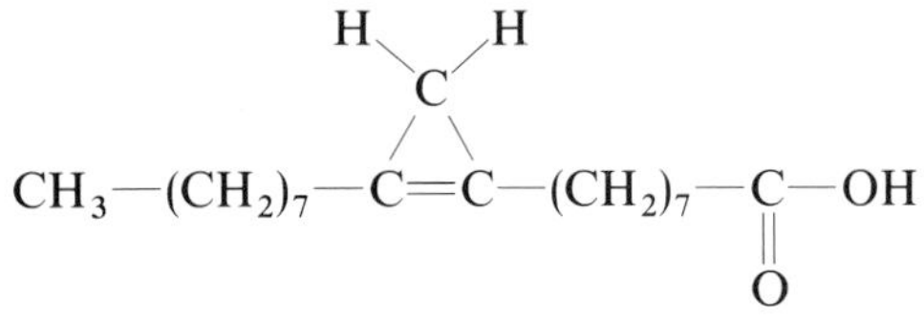

$$CH_3-(CH_2)_7-C{=}C-(CH_2)_7-C(=O)-OH \quad (\text{ring } CH_2 \text{ bridging the } C{=}C)$$

sterculic acid

trans-Cycloalkenes Another difference between cyclic and acyclic alkenes involves the relationship between cis and trans isomers. In the acyclic alkenes, the trans isomers are usually more stable; but the trans isomers of lower cycloalkenes are rare, and those with fewer than eight carbon atoms are unstable at room temperature. The problem with making a *trans*-cycloalkene lies in the geometry of the trans double bond. The two alkyl groups on a trans double bond are so far apart that several carbon atoms are needed to complete the ring.

Try to make a model of *trans*-cyclohexene, being careful that the large amount of ring strain does not break your models. *trans*-Cyclohexene is too strained to be isolated, while *trans*-cycloheptene can be isolated at low temperatures. *trans*-Cyclooctene is stable at room temperature, although its cis isomer is still more stable.

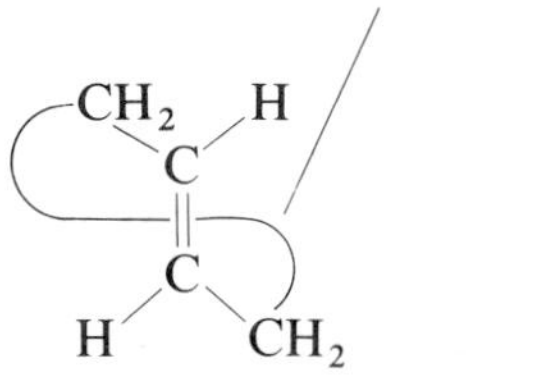

ring connects behind the double bond

trans cyclic system

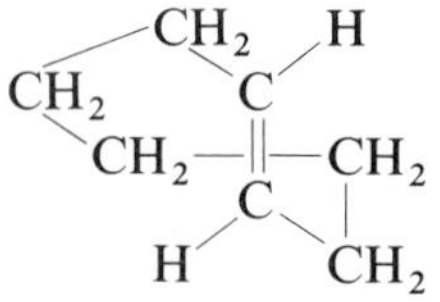

trans-cycloheptene
marginally stable

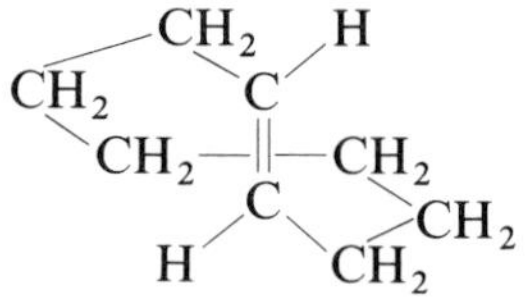

trans-cyclooctene
stable

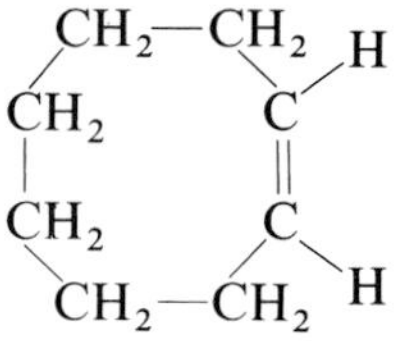

cis-cyclooctene
more stable

Once a cycloalkene contains at least ten or more carbon atoms, it can easily accommodate a trans double bond. For cyclodecene and larger cycloalkenes, the trans isomer is nearly as stable as the cis isomer.

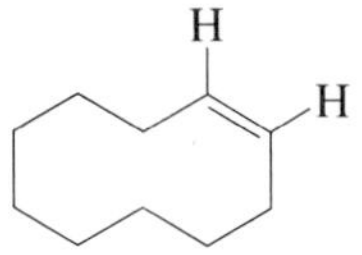

cis-cyclodecene

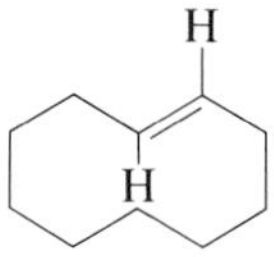

trans-cyclodecene

Bredt's rule We have seen that a *trans*-cycloalkene is not stable unless there are at least eight carbon atoms in the ring. An interesting extension of this principle is called Bredt's rule.

> BREDT'S RULE A bridged bicyclic compound cannot have a double bond at a bridgehead position unless one of the rings contains at least eight carbon atoms.

Let's review exactly what Bredt's rule means. A **bicyclic** compound is one that contains two rings. The **bridgehead carbon atoms** are the carbon atoms that are part of both rings, with three links connecting them. A **bridged bicyclic** compound has at least one carbon atom in each of the three links between the bridgehead carbons. In the following examples, the bridgehead carbon atoms are circled in red.

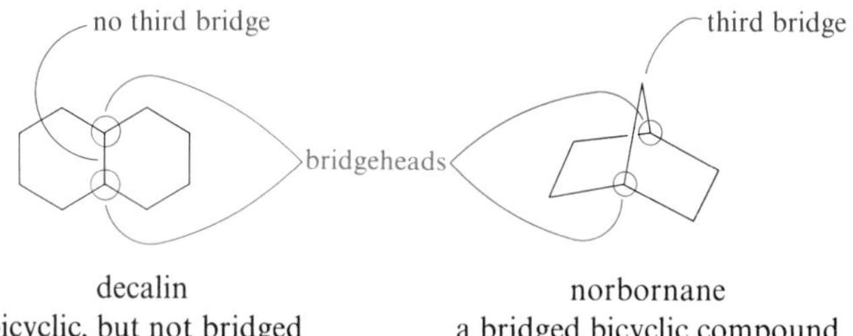

decalin
bicyclic, but not bridged

norbornane
a bridged bicyclic compound

If there is a double bond at the bridgehead carbon of a bridged bicyclic system, then one of the two rings contains a cis double bond and the other must contain a trans double bond. For example, the following structures show that norbornane contains a five-membered ring and a six-membered ring. If there is a double bond at the bridgehead carbon atom, the five-membered ring contains a cis double bond, and the six-membered ring contains a trans double bond. This is an unstable arrangement, called a "Bredt's rule violation." If the larger ring contains at least eight carbon atoms, then it can contain the trans double bond and the bridgehead double bond is stable.

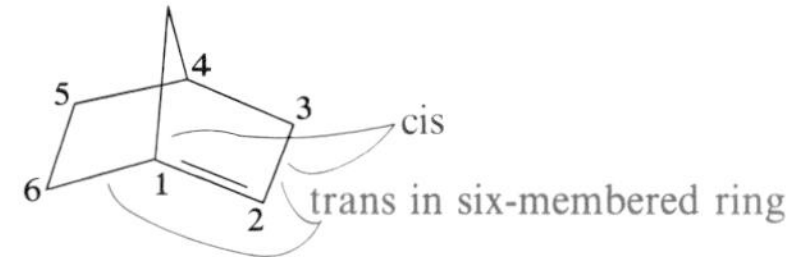

Bredt's rule violation

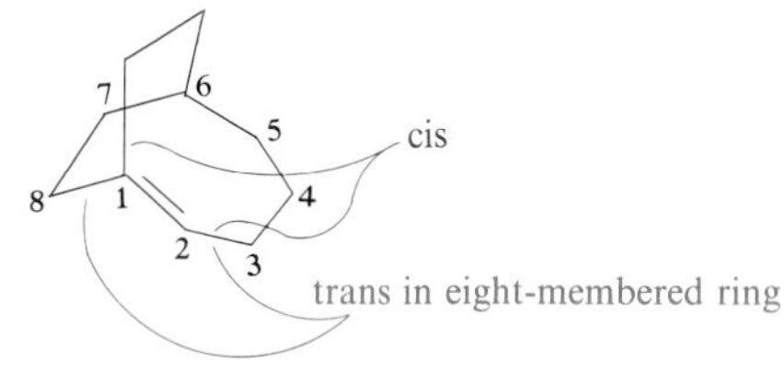

Stable: trans in an eight-membered ring

SOLVED PROBLEM 7-2

Which of the following alkenes are stable?

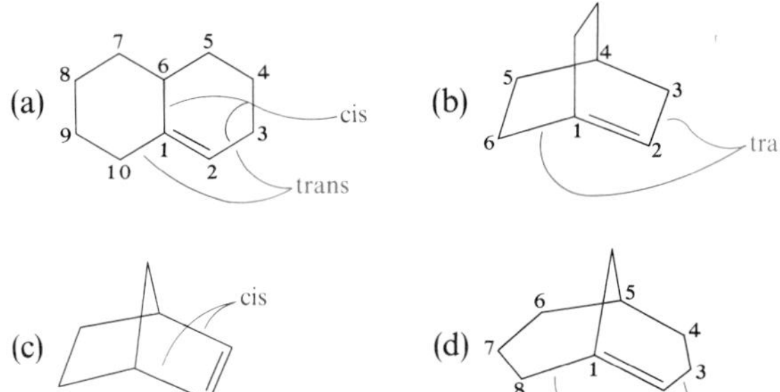

SOLUTION Compound (a) is stable. Although the double bond is at a bridgehead, it is not a bridged bicyclic system. As you can see, the trans double bond is in a ten-membered ring.

Compound (b) is a Bredt's rule violation and is not stable. The largest ring contains six carbon atoms, and the trans double bond cannot be stable in this bridgehead position.

Compound (c) (norbornene) is stable. The (cis) double bond is not at a bridgehead carbon.

Compound (d) is stable. Although the double bond is at the bridgehead of a bridged bicyclic system, there is an eight-membered ring to accommodate the trans double bond.

PROBLEM 7-11

Explain why each of the following alkenes is or is not stable.

(a) 1,2-dimethylcyclobutene
(b) *trans*-1,2-dimethylcyclobutene
(c) *trans*-3,4-dimethylcyclobutene
(d) *trans*-1,2-dimethylcyclodecene

(e) 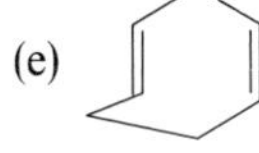 (f) 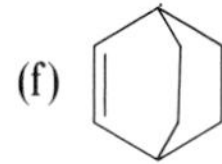 (g) 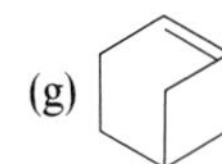 (h) 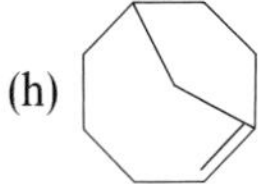

7-8 PHYSICAL PROPERTIES OF ALKENES

7-8A BOILING POINTS AND DENSITIES

Most physical properties of alkenes are similar to those of the corresponding alkanes. For example, the boiling points of 1-butene, *cis*-2-butene, *trans*-2-butene, and *n*-butane are all close to 0°C. Also like the alkanes, alkenes have densities that range around 0.6 or 0.7 g/cm^3. The boiling points and densities of some representative alkenes are listed in Table 7-2. The table shows that the boiling

TABLE 7-2
Physical properties of some representative alkenes

Name	*Structure*	*Carbons*	*Boiling point, °C*	*Density (g/cm³)*
ethene (ethylene)	$CH_2{=}CH_2$	2	−104	
propene (propylene)		3	−47	0.52
isobutylene	$(CH_3)_2C{=}CH_2$	4	−7	0.59
1-butene	$CH_3CH_2CH{=}CH_2$	4	−6	0.59
trans-2-butene	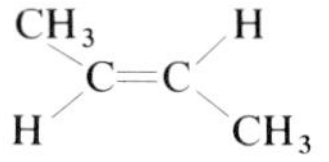 CH_3 H C=C H CH_3	4	1	0.60
cis-2-butene	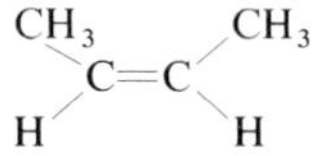 CH_3 CH_3 C=C H H	4	4	0.62
3-methyl-1-butene	$(CH_3)_2CH{-}CH{=}CH_2$	5	25	0.65
1-pentene	$CH_3CH_2CH_2{-}CH{=}CH_2$	5	30	0.64
trans-2-pentene	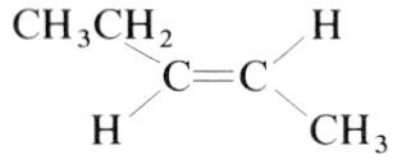 CH_3CH_2 H C=C H CH_3	5	36	0.65
cis-2-pentene	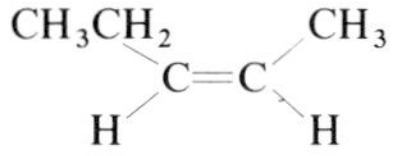 CH_3CH_2 CH_3 C=C H H	5	37	0.66
2-methyl-2-butene	$(CH_3)_2C{=}CH{-}CH_3$	5	39	0.66
1-hexene	$CH_3(CH_2)_3{-}CH{=}CH_2$	6	64	0.68
2,3-dimethyl-2-butene	$(CH_3)_2C{=}C(CH_3)_2$	6	73	0.71
1-heptene	$CH_3(CH_2)_4{-}CH{=}CH_2$	7	93	0.70
1-octene	$CH_3(CH_2)_5{-}CH{=}CH_2$	8	122	0.72
1-nonene	$CH_3(CH_2)_6{-}CH{=}CH_2$	9	146	0.73
1-decene	$CH_3(CH_2)_7{-}CH{=}CH_2$	10	171	0.74

points of alkenes increase smoothly with molecular weight. As with the alkanes, the more highly branched alkenes are more volatile and have lower boiling points. For example, isobutylene has a boiling point of −7°C, which is lower than the boiling points of any of the unbranched butenes.

7-8B POLARITY

Like the alkanes, alkenes are relatively nonpolar. They are insoluble in water but soluble in nonpolar solvents such as hexane, gasoline, halogenated solvents, and the ethers. Alkenes tend to be slightly more polar than alkanes, however, because the more weakly held electrons in the pi bond are more polarizable (contributing to instantaneous dipole moments) and because the vinylic bonds tend to be slightly polar (contributing to a permanent dipole moment).

Alkyl groups are slightly electron donating toward a double bond, helping to stabilize it. This donation of electron density slightly polarizes the vinylic bond, with a small partial positive charge on the alkyl group and a small negative charge on the double-bond carbon atom. For example, propene has a small dipole moment of 0.35 D.

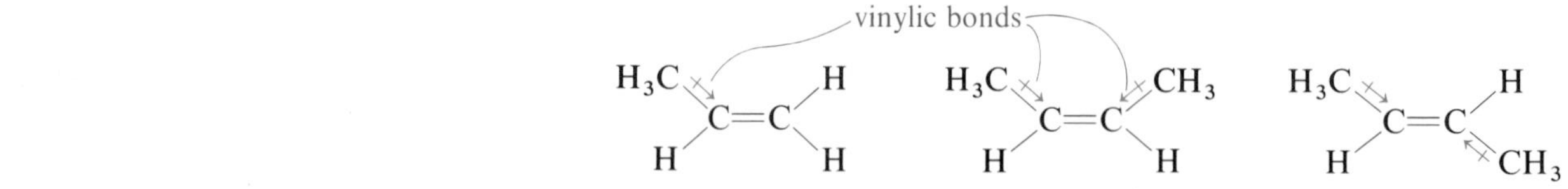

vinylic bonds

H_3C H / C=C / H H
propene, $\mu = 0.35$ D

H_3C CH_3 / C=C / H H
vector sum = ↕
$\mu = 0.33$ D
b.p. 4°C

H_3C H / C=C / H CH_3
vector sum = 0
$\mu = 0$
b.p. 1°C

In a cis-disubstituted alkene, the vector sum of the two dipole moments is directed perpendicular to the double bond. In a trans-disubstituted alkene, however, the two dipole moments tend to cancel out. If the alkene is symmetrically trans-disubstituted, the dipole moment will be exactly zero. For example, *cis*-2-butene has a nonzero dipole moment, while the trans isomer has no measurable dipole moment.

The boiling points of the 2-butenes reflect this difference in their dipole moments. Compounds with permanent dipole moments engage in dipole-dipole attractions, while those without permanent dipole moments engage only in van der Waals attractions. *cis*- and *trans*-2-Butene have similar van der Waals attractions, but only the cis isomer has dipole-dipole attractions. Because of its increased intermolecular attractions, *cis*-2-butene must be heated to a slightly higher temperature (4°C versus 1°C) before it begins to boil.

The effect of bond polarity is even more apparent in the 1,2-dichloroethenes, with their strongly polar carbon-chlorine bonds. The cis isomer has a large dipole moment (2.95 D), giving it a boiling point 12°C higher than that of the trans isomer, with zero dipole moment.

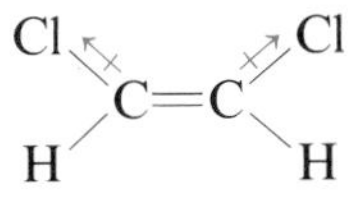

cis	trans
Cl(H)C=C(Cl)H	Cl(H)C=C(H)Cl
vector sum = ↑	vector sum = 0
μ = 2.95 D	μ = 0
b.p. = 60°C	b.p. = 48°C

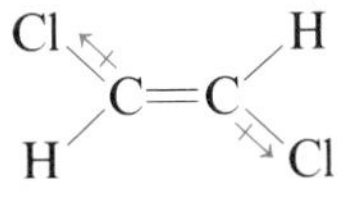

PROBLEM 7-12

For each pair of compounds, predict the one with a higher boiling point. Which compounds have zero dipole moments?

(a) *cis*-1,2-dichloroethene and *cis*-1,2-dibromoethene
(b) *cis*- and *trans*-2,3-dichloro-2-butene
(c) cyclohexene and 1,2-dichlorocyclohexene

7-9 SYNTHESIS OF ALKENES

7-9A ALKENE SYNTHESIS BY DEHYDROHALOGENATION OF ALKYL HALIDES

Dehydrohalogenation is the formation of an alkene by the elimination of a hydrogen atom and a halogen from an alkyl halide. We discussed the mechanism of this elimination in Sections 5-19 through 5-23. Dehydrohalogenation can take place by either of two mechanisms: the E2 mechanism or the E1 mechanism.

Dehydrohalogenation by the E2 mechanism Second-order elimination is a more reliable synthetic tool. The E2 dehydrohalogenation takes place through a one-step mechanism, where a strong base abstracts a proton from one carbon atom as the leaving group leaves the adjacent carbon.

Elimination by the E2 mechanism

$$B:^- \;+\; H{-}\overset{|}{\underset{|}{C}}{-}\overset{|}{\underset{|}{C}}{-}X \longrightarrow \;>C{=}C<\; + \; B{-}H \; + \; :X^-$$

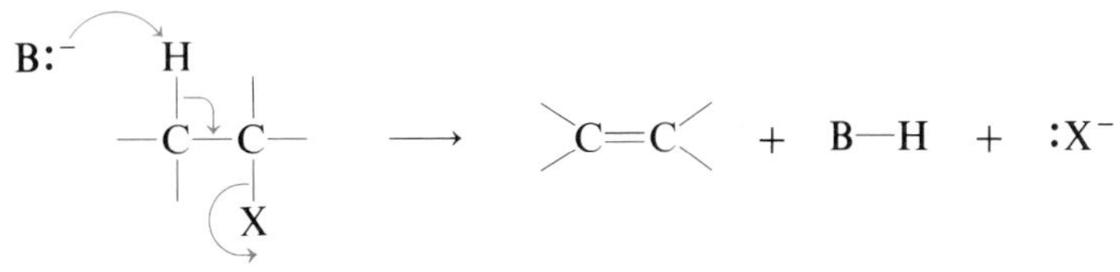

In using the E2 dehydrohalogenation to synthesize alkenes, remember that the elimination must take place in a one-step, coplanar arrangement (review Section 5-21). This requirement makes elimination difficult in some compounds, but it leads to pure products with other compounds.

The E2 dehydrohalogenation gives excellent yields with bulky secondary and tertiary alkyl halides that are poor S_N2 substrates. A strong base forces second-order elimination by abstracting a proton. The molecule's bulkiness hinders second-order substitution and a relatively pure elimination product results. Tertiary halides are the best E2 substrates because they are prone to elimination and cannot undergo S_N2 substitution.

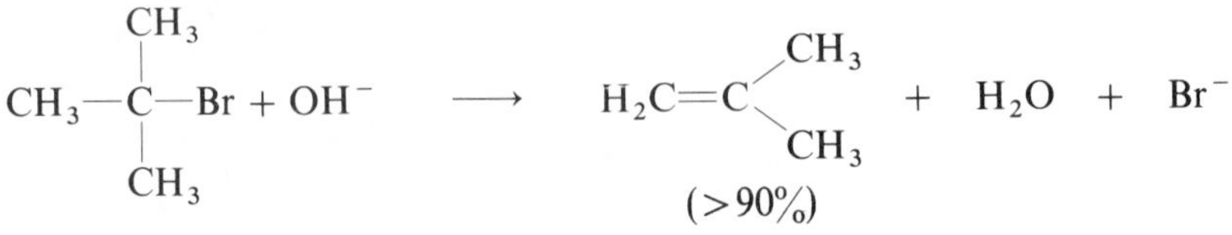

$$(CH_3)_3C{-}Br + OH^- \longrightarrow H_2C{=}C(CH_3)_2 + H_2O + Br^-$$

(>90%)

Use of a bulky base If the substrate is prone to substitution, we can minimize the amount of substitution by using a bulky base. A bulky base has large alkyl groups that hinder its approach to attack a carbon atom (substitution), yet it can easily abstract a proton from the substrate (elimination). Some of the bulky strong bases that are commonly used for elimination are *t*-butoxide ion, diisopropylamine, triethylamine, and 2,6-dimethylpyridine.

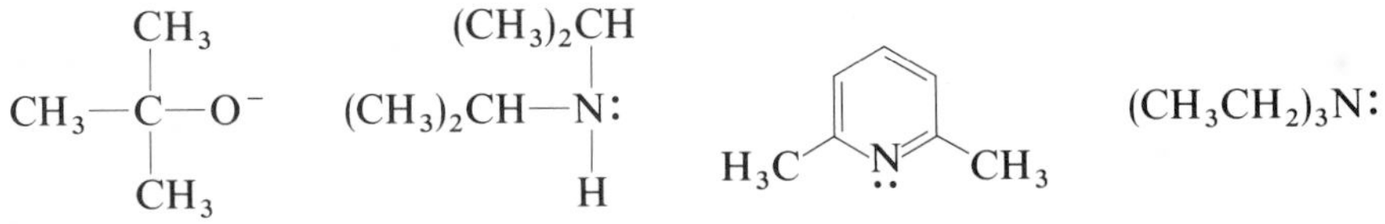

$(CH_3)_3C{-}O^-$ — *t*-butoxide

$[(CH_3)_2CH]_2N{:}{-}H$ — diisopropylamine

2,6-dimethylpyridine

$(CH_3CH_2)_3N{:}$ — triethylamine

The dehydrohalogenation of bromocyclohexane by diisopropylamine illustrates the use of a bulky base for elimination. Bromocyclohexane, a secondary alkyl halide, can undergo both substitution and elimination reactions. Elimination (E2) is favored over substitution (S_N2) by using diisopropylamine as the base. Diisopropylamine is too bulky to be an effective nucleophile, but it acts as a strong base to abstract a proton.

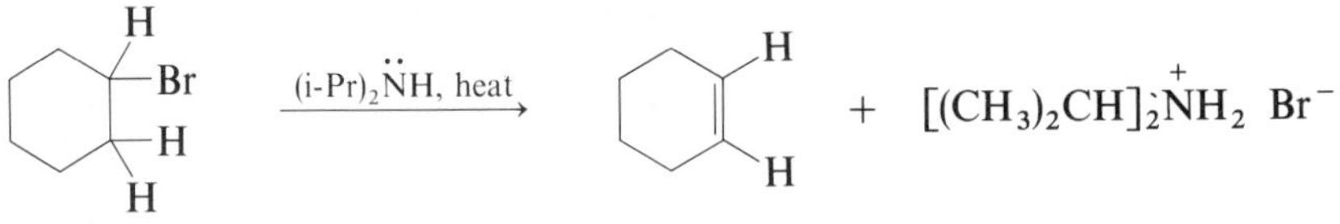

$$\text{cyclohexyl bromide} \xrightarrow{(i\text{-}Pr)_2\ddot{N}H,\ \text{heat}} \text{cyclohexene} + [(CH_3)_2CH]_2\overset{+}{N}H_2\ Br^-$$

cyclohexyl bromide

cyclohexene (93%)

Formation of the Hofmann product Bulky bases are also used to accomplish dehydrohalogenations that do not follow the **Saytzeff rule.** A bulky base is often sterically hindered from abstracting the proton that leads to the most highly substituted alkene. In these cases it abstracts a less hindered proton, often the one that leads to formation of the least highly substituted product, called the **Hofmann product.** The following reaction gives mostly the **Saytzeff product** with the relatively unhindered ethoxide ion, but mostly the Hofmann product with the bulky *t*-butoxide ion.

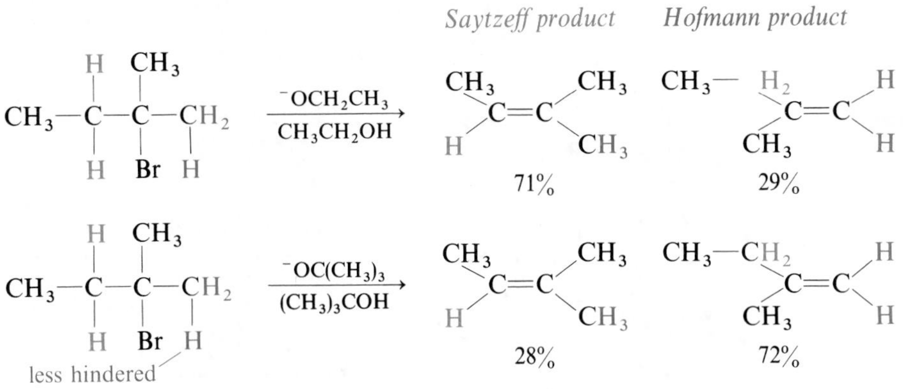

Saytzeff product — Hofmann product

$CH_3-CH(H)-C(CH_3)(Br)-CH_2(H)$ $\xrightarrow[CH_3CH_2OH]{^-OCH_2CH_3}$ $(CH_3)(H)C{=}C(CH_3)_2$ (71%) + $CH_3-CH_2(CH_3)C{=}CH_2$ (29%)

$CH_3-CH(H)-C(CH_3)(Br)-CH_2(H)$ $\xrightarrow[(CH_3)_3COH]{^-OC(CH_3)_3}$ $(CH_3)(H)C{=}C(CH_3)_2$ (28%) + $CH_3-CH_2(CH_3)C{=}CH_2$ (72%)

less hindered

SOLVED PROBLEM 7-3

Predict the elimination product(s) of the following reaction.

[Structure: H, Ph, CH_3 on one carbon; H, Ph, Br on the other; H and Br labeled anti, coplanar] $\xrightarrow{(i\text{-}Pr)_2NH}$

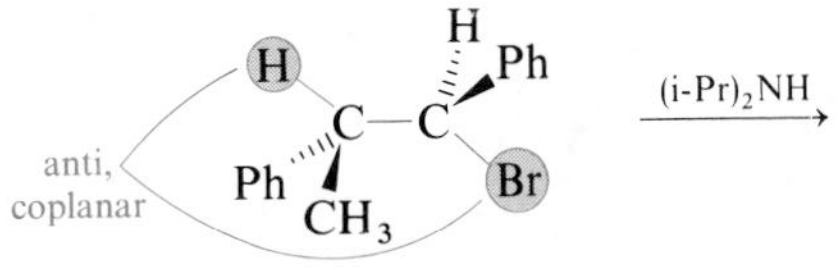

SOLUTION The conformation shown in the reactant places the proton and the halide in an anti-coplanar arrangement. Concerted elimination gives only the *trans*-alkene. Use your models to show that the reaction must take place from this conformation and that only the following trans product results:

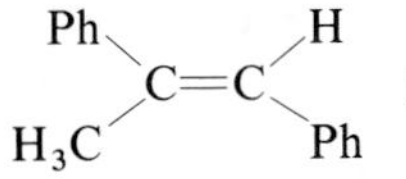

$(Ph)(H_3C)C{=}C(H)(Ph)$ pure trans

SOLVED PROBLEM 7-4

Predict the elimination product(s) of the following reaction.

[Structure: cyclohexane with substituents H, CH_3, Br, H, H, CH_3] $\xrightarrow{(CH_3)_3C{-}O^-}$

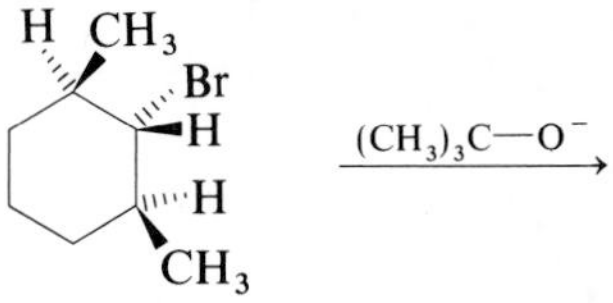

SOLUTION This compound cannot eliminate, because no hydrogen atoms can achieve a coplanar arrangement with the (bromide) leaving group. Cyclohexyl halides normally eliminate the proton and leaving group from trans-diaxial positions of the chair conformation (review Section 5-21). If there are no adjacent protons trans to the leaving group, the E2 reaction may be impossible.

PROBLEM 7-13

Predict the products of the following reactions.

(a) 1-bromo-1-methylcyclohexane + NaOH in acetone
(b) 1-bromo-1-methylcyclohexane + ethanol, heat
(c) chlorocyclohexane + NaOH in acetone
(d) chlorocyclohexane + triethylamine, $(CH_3CH_2)_3N$:
(e) 1-bromo-1-methylcyclohexane + $(CH_3CH_2)_3N$:
(f) *meso*-1,2-dibromo-1,2-diphenylethane + NaOH in acetone
(g) (*d,l*)-1,2-dibromo-1,2-diphenylethane + NaOH in acetone

(h) 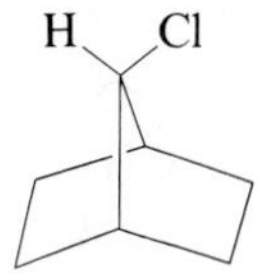 [bicyclic structure with H and Cl on bridge carbon] + NaOH in acetone

(i) 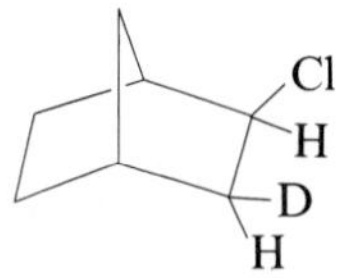 [bicyclic structure with Cl, H, D, H substituents] + $(CH_3)_3C{-}O^-$ in $(CH_3)_3C{-}OH$

Dehydrohalogenation by the E1 mechanism First-order dehydrohalogenation usually takes place in a good ionizing solvent (such as an alcohol or water), without a strong nucleophile or base present to force second-order kinetics. The substrate is usually a secondary or tertiary alkyl halide. First-order elimination involves ionization to form a carbocation which loses a proton to a weak base (usually the solvent). The E1 dehydrohalogenation is generally accompanied by the S_N1 substitution because the nucleophilic solvent can also attack the carbocation directly, forming the substitution product.

Elimination by the E1 mechanism

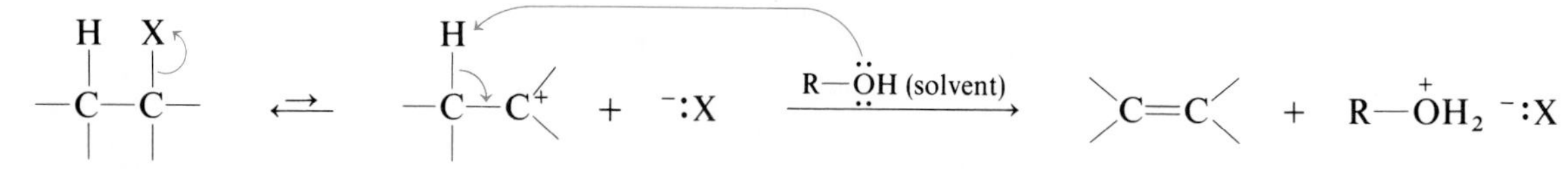

$$-\overset{H}{C}-\overset{X}{C}- \rightleftarrows -\overset{H}{C}-C^+ + {}^-{:}X \xrightarrow{R-\ddot{O}H\ (solvent)} >C{=}C< + R-\overset{+}{\ddot{O}}H_2\ {}^-{:}X$$

formation of the carbocation — proton abstraction — elimination products

Accompanied by S_N1 substitution

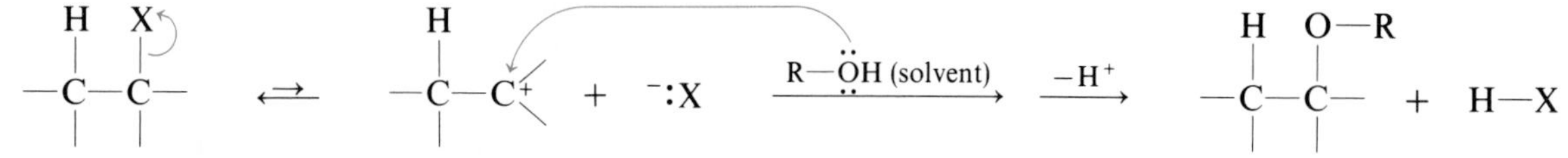

$$-\overset{H}{C}-\overset{X}{C}- \rightleftarrows -\overset{H}{C}-C^+ + {}^-{:}X \xrightarrow{R-\ddot{O}H\ (solvent)} \xrightarrow{-H^+} -\overset{H}{C}-\overset{O-R}{C}- + H-X$$

formation of the carbocation — nucleophilic attack — substitution products

7-9B ALKENE SYNTHESIS BY DEHALOGENATION OF VICINAL DIBROMIDES

A compound with two halogens on the same carbon is called a **geminal** dihalide; one with two halogens on adjacent carbons is called a **vicinal** dihalide.

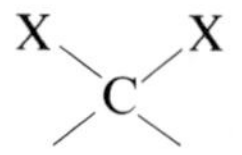

$$\overset{X\ \ X}{>C<} \qquad -\overset{X}{C}-\overset{X}{C}-$$

a geminal dihalide 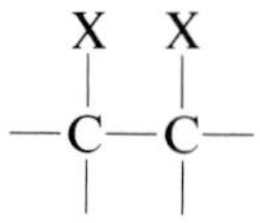 — a vicinal dihalide

Vicinal dibromides are converted to alkenes by reduction with either iodide ion or zinc in acetic acid. A molecule of Br_2 (an oxidizing agent) is removed from the dihalide to form an alkene. Zinc serves as a reducing agent in zinc/acetic acid **dehalogenation.** The reaction is heterogeneous (part solid and part liquid), with the actual reduction taking place at the surface of the metallic zinc. Zinc is oxidized from the 0 oxidation state to the +2 oxidation state, forming $ZnBr_2$.

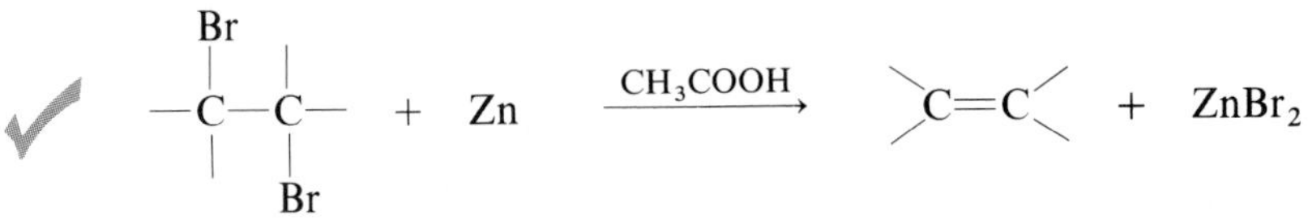

$$-\overset{\text{Br}}{\overset{|}{C}}-\underset{\text{Br}}{\underset{|}{C}}- + Zn \xrightarrow{CH_3COOH} \;\rangle C=C\langle\; + ZnBr_2$$

Example

$$CH_3-\underset{\text{Br}}{\underset{|}{CH}}-\underset{\text{Br}}{\underset{|}{CH}}-CH_3 + Zn \xrightarrow{CH_3COOH} CH_3-CH=CH-CH_3 + ZnBr_2$$

The reaction with iodide takes place through the E2 mechanism, with the same geometric constraints as the E2 dehydrohalogenation. Elimination usually takes place through an anti-coplanar arrangement, as shown in the example below.

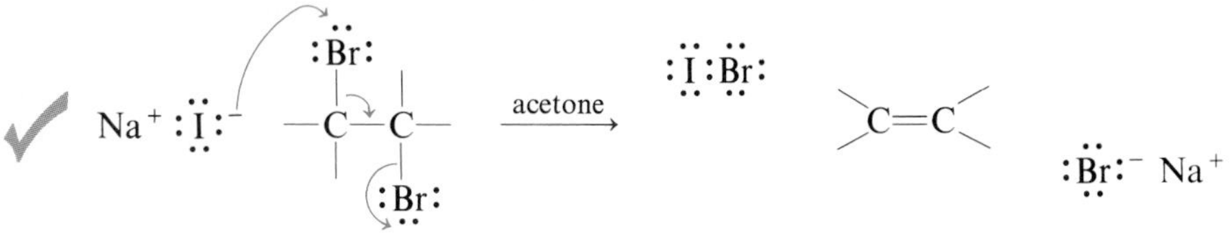

$$Na^+ \;:\ddot{I}:^- \quad -\overset{:\ddot{Br}:}{\overset{|}{C}}-\underset{:\ddot{Br}:}{\underset{|}{C}}- \xrightarrow{\text{acetone}} \;:\ddot{I}:\ddot{Br}: \quad \rangle C=C\langle \quad :\ddot{Br}:^- \;Na^+$$

Example

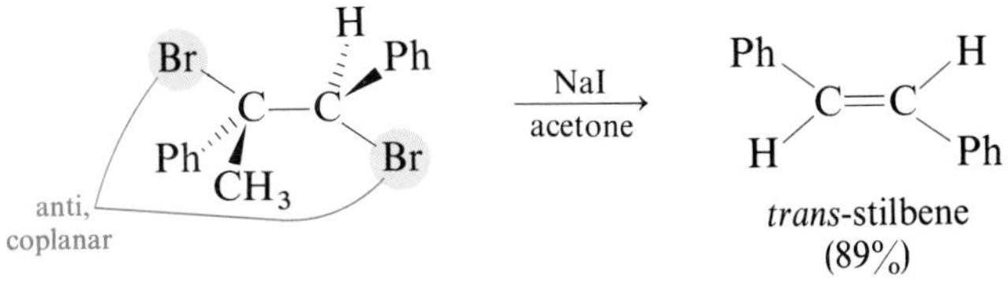

Br, Ph, CH$_3$ on one carbon; H, Ph, Br on the other (anti, coplanar Br atoms) $\xrightarrow[\text{acetone}]{\text{NaI}}$ Ph(H)C=C(H)Ph

trans-stilbene (89%)

You should use your models to show that only the trans isomer of stilbene is formed in this example by elimination through the anti-coplanar transition state.

PROBLEM 7-14

Predict the products of the following reactions.

(a) 1,2-dibromodecane + Zn in CH_3COOH
(b) *trans*-1,2-dibromocyclohexane + NaI in acetone
(c) *trans*-1,2-dibromocyclodecane + NaI in acetone
(d) [cyclohexane bearing Br, Br and $(CH_3)_3C$] + NaI in acetone
(e) [cyclodecane bearing two Br] + NaI in acetone

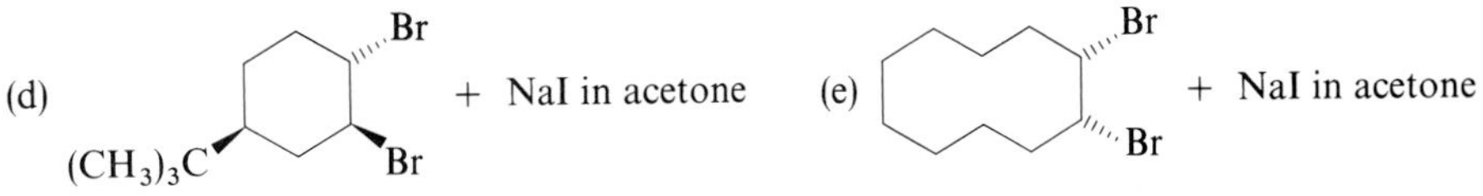

7-9C ALKENE SYNTHESIS BY DEHYDRATION OF ALCOHOLS

One of the best methods for the synthesis of alkenes is the **dehydration** of alcohols. The word *dehydration* literally means the removal of water.

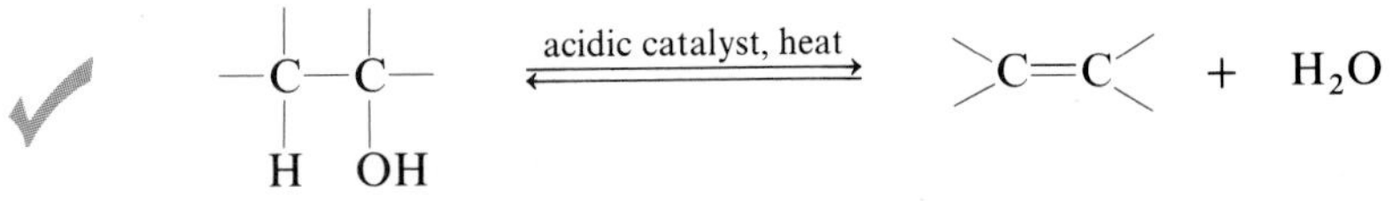

$$-\underset{\text{H}}{\underset{|}{C}}-\underset{\text{OH}}{\underset{|}{C}}- \;\underset{}{\overset{\text{acidic catalyst, heat}}{\rightleftharpoons}}\; \rangle C=C\langle + H_2O$$

The dehydration reaction is reversible, and in most cases the equilibrium constant is not large. In fact, the reverse reaction is a method for converting alkenes to alcohols (see Section 8-4). The dehydration is forced to completion by removing the products from the reaction mixture as they form. The alkene boils at a lower temperature than the alcohol because the alcohol molecules participate in hydrogen bonding. A carefully controlled distillation removes the alkene while leaving the alcohol in the reaction mixture.

Concentrated sulfuric acid and concentrated phosphoric acid are often used as reagents for dehydration, because these acids act both as acidic catalysts and as dehydrating agents. The hydration of these acids is highly exothermic. The overall reaction (using sulfuric acid) is

$$\underset{\text{H}\quad\text{OH}}{-\text{C}-\text{C}-} + H_2SO_4 \rightleftharpoons \text{C}{=}\text{C} + H_3O^+ + HSO_4^-$$

The mechanism of dehydration resembles the E1 mechanism covered in Chapter 5. The hydroxyl group of the alcohol is a poor leaving group (^-OH), but protonation by the acidic catalyst converts it to a good leaving group (H_2O). Ionization of the protonated alcohol gives a carbocation that loses a proton to give the alkene. Any weak base such as H_2O or HSO_4^- can abstract the proton in the final step.

Step 1: Protonation of the hydroxyl group (fast equilibrium)

$$\underset{|\quad|}{\overset{\text{H}\ :\ddot{\text{O}}-\text{H}}{-\text{C}-\text{C}-}} + \text{H}-\text{O}-\underset{\text{O}}{\overset{\text{O}}{\text{S}}}-\text{O}-\text{H} \rightleftharpoons \overset{\text{H}\ :\overset{\text{H}}{\text{O}^+}-\text{H}}{-\text{C}-\text{C}-} \quad HSO_4^-$$

Step 2: Ionization (slow; rate-determining) *Step 3:* Proton abstraction (fast)

$$\overset{\text{H}\ :\overset{\text{H}}{\text{O}^+}-\text{H}}{-\text{C}-\text{C}-} \rightleftharpoons \overset{\text{H}}{-\text{C}-\text{C}^+} + H_2\ddot{O}: \longrightarrow \text{C}{=}\text{C} + H_3O^+$$

Because the dehydration of an alcohol goes through the E1 mechanism, the usual order of reactivity for the E1 is observed: 3° alcohols react faster than 2° alcohols, and 1° alcohols are the least reactive. Rearrangements of the carbocation intermediate are also common.

SOLVED PROBLEM 7-5

Give a mechanism for the sulfuric acid-catalyzed dehydration of *t*-butyl alcohol.

SOLUTION The first step is the protonation of the hydroxyl group, converting it to a good leaving group.

$$CH_3-\underset{CH_3}{\overset{CH_3}{C}}-\ddot{O}-H + H_2SO_4 \rightleftharpoons CH_3-\underset{CH_3}{\overset{CH_3}{C}}-\overset{+}{O}H_2 + HSO_4^-$$

The second step is the ionization of the protonated alcohol to give a carbocation.

$$CH_3-\underset{CH_3}{\overset{CH_3}{C}}-\overset{+}{O}H_2 \rightleftharpoons CH_3-\overset{+}{C}(CH_3)_2 + H_2\ddot{O}:$$

Abstraction of a proton completes the mechanism.

$$H_2\ddot{O}: \quad H-\overset{\overset{H}{|}}{\underset{\underset{H}{|}}{C}}-C^{+}\begin{matrix}CH_3\\ CH_3\end{matrix} \quad \rightleftharpoons \quad \begin{matrix}H\\ H\end{matrix}\!\!>C=C<\!\!\begin{matrix}CH_3\\ CH_3\end{matrix} \quad + \quad H_3O^+$$

PROBLEM SOLVING: PROPOSING REACTION MECHANISMS

In Chapter 4, we considered general principles for drawing free-radical reaction mechanisms. Reactions involving strong bases and strong nucleophiles, or strong acids and strong electrophiles, are much more common than free-radical reactions. In general, when organic chemists speak of an *acid* or a *base,* they mean a *proton acid* (proton donor) or a *proton base* (proton acceptor). When they speak of an *electrophile* or a *nucleophile,* they mean an *electron pair acceptor* or an *electron pair donor.* (These usages overlap somewhat, because the transfer of a proton involves donating and accepting a pair of electrons.)

The first step in proposing a reaction mechanism is to classify the reaction by examining what is known about the reactants and the reaction conditions. A free-radical initiator such as chlorine, bromine, or a peroxide suggests that a free-radical chain reaction like those discussed in Chapter 4 is most likely. In the presence of a strong acid or a reactant that can dissociate to give a strong electrophile, a mechanism involving strong electrophiles (such as the S_N1, the E1, alcohol dehydration, and so on) is most likely. In the presence of a strong base or a strong nucleophile, a mechanism involving strong nucleophiles (such as the S_N2 or the E2) is most likely. (Appendix 4 contains more complete methods for approaching mechanism problems.)

Once you have decided which type of mechanism is most likely, there are general principles that can guide you in proposing the mechanism. Some of these principles for free-radical reactions were discussed in Chapter 4. Now we consider reactions that involve either strong nucleophiles or strong electrophiles as intermediates. In later chapters, we will apply these principles to more complex mechanisms.

Whenever you start to work out a mechanism, draw all the bonds and all the substituents of each carbon atom affected throughout the mechanism. The three-bonded carbon atoms are likely to be the reactive intermediates. If you attempt to draw condensed formulas or line-angle formulas, you will likely misplace a hydrogen atom and show the wrong carbon atom as a radical, cation, or anion.

Show only one step occurring at once; never try to combine steps, unless two or more bonds really do change position in one step (as in the E2 reaction, for example). Protonation of an alcohol and loss of water to give a carbocation, for example, must be shown as two steps. You must not simply circle the hydroxyl and the proton to show water falling off.

Use curved arrows to show the *movement of electrons* in each step of the reaction. This movement is always from the nucleophile (electron donor) to the electrophile (electron acceptor). For example, protonation of an alcohol must show the arrow going from the electrons of the hydroxyl oxygen to the proton—never from the proton to the hydroxyl group. *Don't* use curved arrows to try to "point out" where the proton (or other reagent) goes.

REACTIONS INVOLVING STRONG NUCLEOPHILES When a strong base or nucleophile is present, expect to see intermediates that are also strong bases and strong nucleophiles; anionic intermediates are common. Acids and electrophiles in such a reaction are generally weak, however. Avoid drawing carbocations, H_3O^+, and other strong acids. They are unlikely to coexist with strong bases and strong nucleophiles.

Functional groups are often converted to alkoxides, carbanions, or other strong nucleophiles by deprotonation or reaction with a strong nucleophile. Then, the carbanion or other strong nucleophile reacts with a weak electrophile such as a carbonyl group or an alkyl halide.

Consider, for example, the mechanism for the dehydrohalogenation of 3-bromopentane:

$$CH_3{-}CH_2{-}\underset{\displaystyle Br}{\underset{|}{CH}}{-}CH_2{-}CH_3 \xrightarrow{CH_3CH_2O^-} CH_3{-}CH{=}CH{-}CH_2{-}CH_3$$

Someone who has not read Chapter 5 or the guidelines above for classifying mechanisms might propose an ionization, followed by loss of a proton:

Incorrect mechanism

$$CH_3{-}\overset{\displaystyle H}{\underset{\displaystyle H}{C}}{-}\overset{\displaystyle H}{\underset{\displaystyle Br}{C}}{-}CH_2{-}CH_3 \rightleftarrows$$

$$CH_3{-}\overset{\displaystyle H}{\underset{\displaystyle H}{C}}{-}\overset{\displaystyle H}{\underset{+}{C}}{-}CH_2{-}CH_3 \quad Br^- \xrightarrow{\text{unlikely}} CH_3{-}\overset{\displaystyle H}{C}{=}\overset{\displaystyle H}{C}{-}CH_2{-}CH_3 \quad H^+ \quad Br^-$$

This would be a very bad mechanism, violating several general principles of proposing mechanisms. First, in the presence of ethoxide ion (a strong base), both the carbocation and the H^+ ion are unlikely. Second, the mechanism fails to explain why the strong base is required; the rate of ionization would be unaffected by the presence of ethoxide ion. Also, H^+ doesn't just fall off (even in an acidic reaction); it must be removed by a base.

The presence of ethoxide ion (a strong base and a strong nucleophile) in the reaction suggests that the mechanism involves only strong bases and nucleophiles and not any strongly acidic intermediates. As shown in Section 7-9A, the reaction occurs by the E2 mechanism, a simple example of a reaction involving a strong nucleophile. In this concerted reaction, ethoxide ion removes a proton as the electron pair left behind forms a pi bond and expels bromide ion.

Correct mechanism

$$CH_3CH_2O^- \quad CH_3{-}\overset{\displaystyle H}{\underset{\displaystyle H}{C}}{-}\overset{\displaystyle H}{\underset{\displaystyle Br}{C}}{-}CH_2{-}CH_3 \longrightarrow CH_3CH_2O{-}H \quad CH_3{-}\underset{\displaystyle H}{C}{=}\overset{\displaystyle H}{C}{-}CH_2{-}CH_3 \quad Br^-$$

REACTIONS INVOLVING STRONG ELECTROPHILES When a strong acid or electrophile is present, expect to see intermediates that are also strong acids and strong electrophiles; cationic intermediates are common. Bases and nucleophiles in such a reaction are generally weak, however. Avoid drawing carbanions, alkoxide ions, and other strong bases. They are unlikely to coexist with strong acids and strong electrophiles.

Functional groups are often converted to carbocations or other strong electrophiles by protonation or by reaction with a strong electrophile. Then the carbocation or other strong electrophile reacts with a weak nucleophile such as an alkene or the solvent.

For example, consider the dehydration of 2,2-dimethyl-1-propanol:

$$CH_3-\underset{\underset{CH_3}{|}}{\overset{\overset{CH_3}{|}}{C}}-CH_2-OH \xrightarrow{H_2SO_4,\ 150°C} (CH_3)_2C=CH-CH_3$$

The presence of sulfuric acid indicates that the reaction is acidic and should involve strong electrophiles. The hydroxyl group is a poor leaving group; it certainly cannot ionize to give a carbocation and ^-OH (and we do not expect to see a strong base like ^-OH in this acidic reaction). Yet, the hydroxyl group is weakly basic, and in the presence of a strong acid it can become protonated.

Step 1: Protonation of the hydroxyl group

$$\underset{\text{starting alcohol}}{CH_3-\underset{\underset{CH_3}{|}}{\overset{\overset{CH_3}{|}}{C}}-CH_2-\ddot{\underset{..}{O}}-H} + H_2SO_4 \longrightarrow \underset{\text{protonated alcohol}}{CH_3-\underset{\underset{CH_3}{|}}{\overset{\overset{CH_3}{|}}{C}}-CH_2-\overset{\overset{H}{|}}{\underset{..}{O^+}}-H} + HSO_4^-$$

The protonated hydroxyl group ($-\overset{+}{O}H_2$) is a good leaving group. A simple ionization to a carbocation would form a primary carbocation. Primary carbocations, however, are very unstable; therefore, a methyl shift occurs as water leaves, so that a primary carbocation is never formed. A tertiary carbocation results.

Step 2: Ionization with rearrangement

$$\underset{\text{protonated alcohol}}{CH_3-\underset{\underset{CH_3}{|}}{\overset{\overset{CH_3}{|}}{C}}-CH_2-\overset{\overset{H}{|}}{\underset{..}{O^+}}-H} \xrightarrow[\text{with } CH_3 \text{ shift}]{H_2O \text{ leaving}} \underset{\text{tertiary carbocation}}{CH_3-\overset{\overset{CH_3}{|}}{\underset{+}{C}}-\underset{\underset{CH_3}{|}}{CH_2}} + H_2\ddot{O}:$$

$(\sim CH_3)$

The final step is loss of a proton to a weak base, such as HSO_4^- or H_2O (but *not* ^-OH, which is incompatible with the acidic solution). Either of two types of protons, labeled 1 and 2 below, could be lost to give alkenes. Loss of proton 2 gives the required product.

Step 3: Abstraction of a proton to form the required product

$$H-\underset{\underset{H\,(1)}{|}}{\overset{\overset{H}{|}}{C}}-\underset{+}{\overset{\overset{CH_3}{|}}{C}}-\underset{\underset{H\,(2)}{|}}{\overset{\overset{H}{|}}{C}}-CH_3 \quad (H_2\ddot{O}:) \longrightarrow \underset{\text{abstract proton 1}}{H_2C=C(CH_3)-CH_2-CH_3} \quad \text{or} \quad \underset{\substack{\text{abstract proton 2}\\ \text{observed product}}}{(CH_3)_2C=C(H)CH_3}$$

Because abstraction of proton 2 gives the most highly substituted (therefore most stable) product, the Saytzeff rule predicts it will be the major product. Note that in other problems, however, you may be asked to propose mechanisms to explain unusual compounds that are minor products.

PROBLEM 7-15

For practice in recognizing mechanisms, classify each reaction according to the type of mechanism you expect:

(i) Free-radical chain reaction
(ii) Reaction involving strong bases and strong nucleophiles
(iii) Reaction involving strong acids and strong electrophiles

(a) 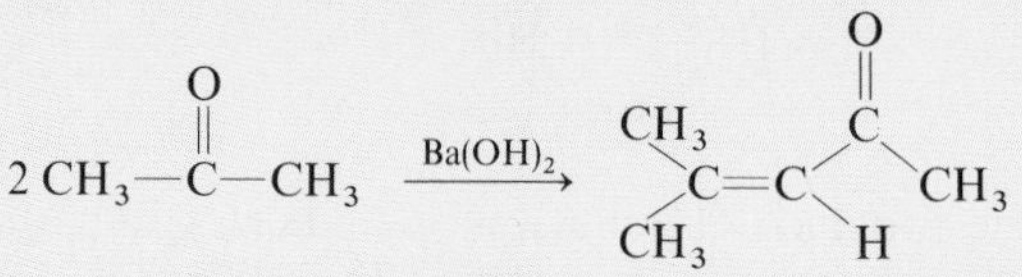 $2\ CH_3-\overset{O}{\overset{\|}{C}}-CH_3 \xrightarrow{Ba(OH)_2} (CH_3)_2C=CH-\overset{O}{\overset{\|}{C}}-CH_3$

(b) 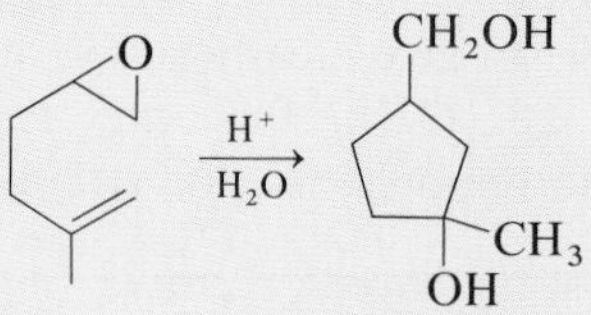 [epoxide with pendant trisubstituted alkene] $\xrightarrow[H_2O]{H^+}$ [cyclopentane bearing CH_2OH and, on another carbon, OH and CH_3]

(c) styrene $\xrightarrow[\text{heat}]{C_6H_5-\overset{O}{\overset{\|}{C}}-O-O-\overset{O}{\overset{\|}{C}}-C_6H_5}$ polystyrene

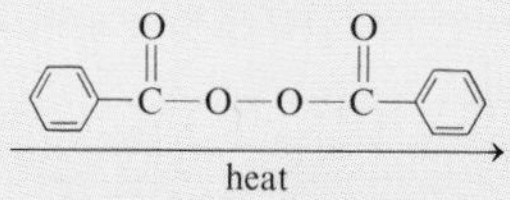

(d) ethylene $\xrightarrow{BF_3}$ polyethylene

Problems 7-16 and 7-17 provide practice in proposing mechanisms by applying the general principles described above.

PROBLEM 7-16

Propose mechanisms for the following reactions:

(a) $CH_3-CH_2-CH_2-CH_2-OH \xrightarrow{H_2SO_4,\ 140\ C} CH_3-CH=CH-CH_3$

(*Hint:* Hydride shift)

(b) 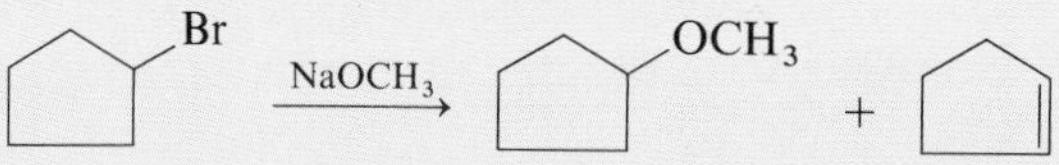 [bromocyclopentane] $\xrightarrow{NaOCH_3}$ [methoxycyclopentane, OCH_3] + [cyclopentene]

PROBLEM 7-17

Propose a mechanism for each of the following reactions.

(a) 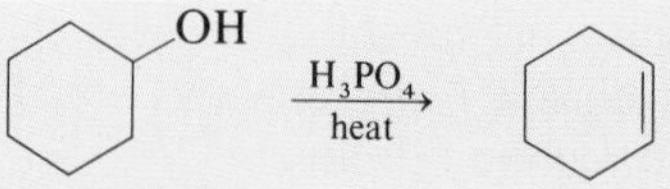 [cyclohexanol, OH] $\xrightarrow[\text{heat}]{H_3PO_4}$ [cyclohexene]

(b) 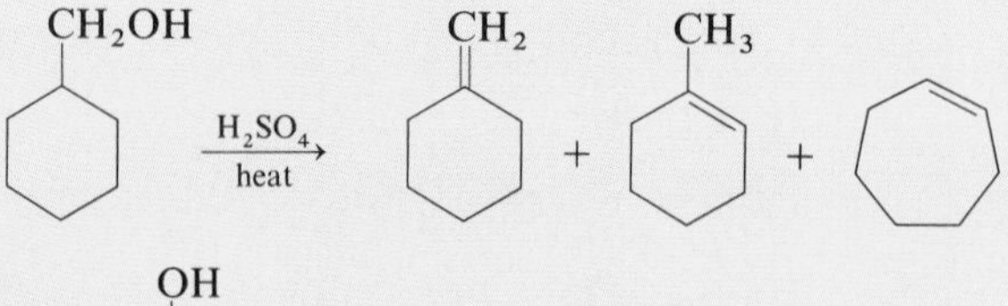 [cyclohexylmethanol, CH_2OH] $\xrightarrow[\text{heat}]{H_2SO_4}$ [methylenecyclohexane, CH_2] + [1-methylcyclohexene, CH_3] + [cycloheptene]

(c) 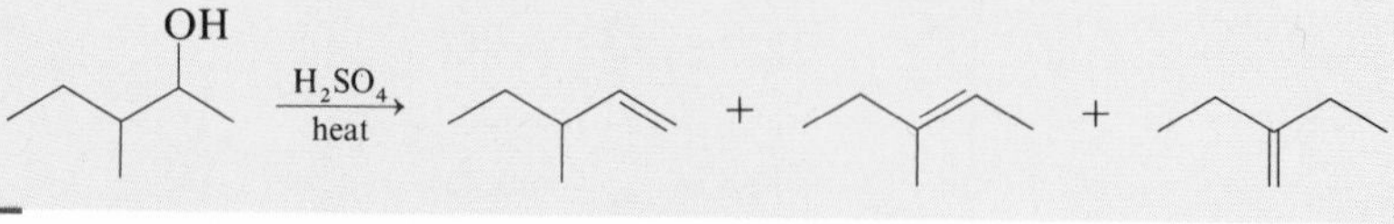 [3-methyl-2-butanol, OH] $\xrightarrow[\text{heat}]{H_2SO_4}$ [3-methyl-1-butene] + [2-methyl-2-butene] + [2-methyl-1-butene]

7-9D CATALYTIC CRACKING

The least expensive way to make alkenes on a large scale is by the **catalytic cracking** of petroleum: heating a mixture of alkanes in the presence of a catalyst. One of the ways alkenes are formed is through a bond cleavage to give an alkene and a shortened alkane.

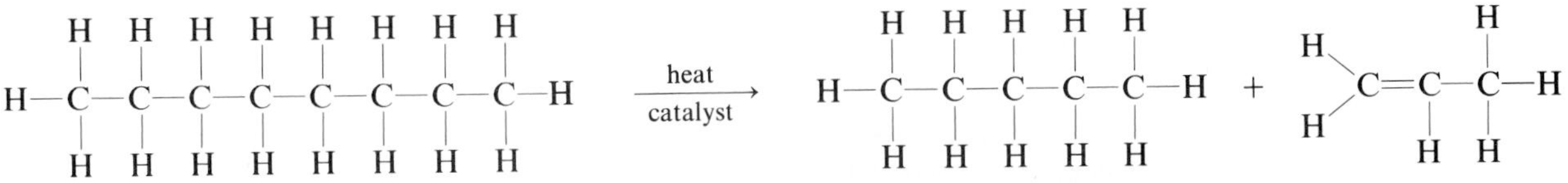

$$\underset{\text{long-chain alkane}}{H-\underset{H}{\overset{H}{C}}-\underset{H}{\overset{H}{C}}-\underset{H}{\overset{H}{C}}-\underset{H}{\overset{H}{C}}-\underset{H}{\overset{H}{C}}-\underset{H}{\overset{H}{C}}-\underset{H}{\overset{H}{C}}-\underset{H}{\overset{H}{C}}-H} \xrightarrow[\text{catalyst}]{\text{heat}} \underset{\text{shorter alkane}}{H-\underset{H}{\overset{H}{C}}-\underset{H}{\overset{H}{C}}-\underset{H}{\overset{H}{C}}-\underset{H}{\overset{H}{C}}-\underset{H}{\overset{H}{C}}-H} + \underset{\text{alkene}}{H_2C{=}\underset{H}{C}-\underset{H}{\overset{H}{C}}-H}$$

Cracking is used primarily to make small alkenes, up to about six carbon atoms. Its value depends on having a market for all the different alkenes and alkanes produced. The average molecular weight and the relative amounts of alkanes and alkenes can be controlled by varying the temperature, catalyst, and concentration of hydrogen in the cracking process. A careful distillation on a huge column separates the mixture into its pure components, ready to be packaged and sold.

Because mixtures are always formed, catalytic cracking is not suitable for the laboratory synthesis of alkenes. Better methods are available for synthesizing relatively pure alkenes from a variety of other functional groups. Several of these methods were discussed previously.

7-9E DEHYDROGENATION OF ALKANES

Dehydrogenation is the removal of H_2 from a molecule, just the reverse of hydrogenation. Dehydrogenation of an alkane gives an alkene. This reaction has an unfavorable enthalpy change, but a favorable entropy change. A general dehydrogenation and a specific example follow:

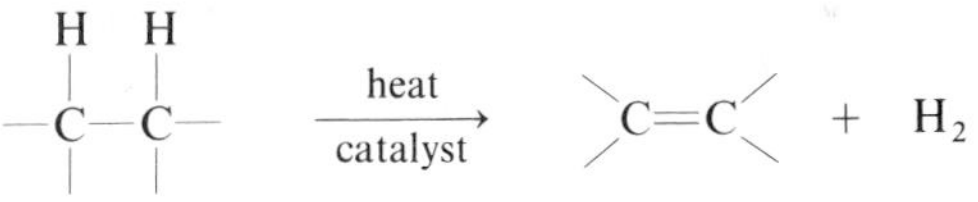

$$-\underset{|}{\overset{H}{C}}-\underset{|}{\overset{H}{C}}- \xrightarrow[\text{catalyst}]{\text{heat}} \;{>}C{=}C{<} \;+\; H_2$$

$$\Delta H^\circ = +20 \text{ to } +30 \text{ kcal } (+80 \text{ to } +120 \text{ kJ}) \qquad \Delta S^\circ = +30 \text{ eu}$$

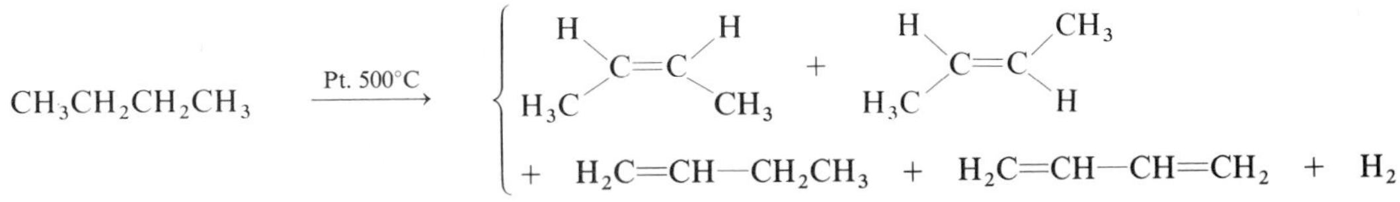

$$CH_3CH_2CH_2CH_3 \xrightarrow{\text{Pt, 500°C}} \left\{ \begin{array}{l} \textit{cis-}\;(H)(H_3C)C{=}C(H)(CH_3) \;+\; \textit{trans-}\;(H)(H_3C)C{=}C(CH_3)(H) \\ +\; H_2C{=}CH{-}CH_2CH_3 \;+\; H_2C{=}CH{-}CH{=}CH_2 \;+\; H_2 \end{array} \right.$$

The hydrogenation of alkenes (Section 7-7) is exothermic with values of ΔH° around -20 to -30 kcal (-80 to -120 kJ). Therefore, dehydrogenation is endothermic and has an unfavorable (positive) value of ΔH°. The entropy change for dehydrogenation is strongly favorable ($\Delta S^\circ = +30$ eu), however, because one alkane molecule is converted into two molecules (the alkene and hydrogen), and two molecules are more disordered than one.

The equilibrium constant for the hydrogenation-dehydrogenation equilibrium depends on the change in free energy, $\Delta G = \Delta H - T\Delta S$. At room temperature, the enthalpy term predominates and hydrogenation is favored. When the temperature is raised, however, the $(-T\Delta S)$ entropy term becomes larger and eventually dominates the expression. At a sufficiently high temperature, dehydrogenation is favored.

PROBLEM 7-18

The dehydrogenation of butane to *trans*-2-butene has $\Delta H^\circ = +27.6$ kcal/mol (+116 kJ/mol) and $\Delta S^\circ = +28.0$ eu. (1 eu = 1 cal/kelvin)

(a) Compute the value of ΔG° for dehydrogenation at room temperature (25°C or 298 K). Is dehydrogenation favored or disfavored?
(b) Compute the value of ΔG for dehydrogenation at 1000°C, assuming ΔS and ΔH are constant. Is dehydrogenation favored or disfavored?

In many ways, dehydrogenation is similar to catalytic cracking. In both cases a catalyst is used to lower the activation energy, and both reactions use high temperatures to increase a favorable entropy term ($-T\Delta S$) and overcome an unfavorable enthalpy term (ΔH). Unfortunately, dehydrogenation and catalytic cracking also share a tendency to produce mixtures of products, and neither reaction is well suited for the laboratory synthesis of alkenes.

7-9F SUMMARY OF ALKENE SYNTHESES

The following is a summary of methods for synthesizing alkenes. References are given to the sections in which the reactions are discussed in detail. Some of the reactions listed in this summary are not discussed in this chapter but are covered later when their mechanisms will be more easily understood.

SUMMARY OF METHODS FOR SYNTHESIS OF ALKENES

1. *Dehydrohalogenation of alkyl halides* (Section 7-9A)

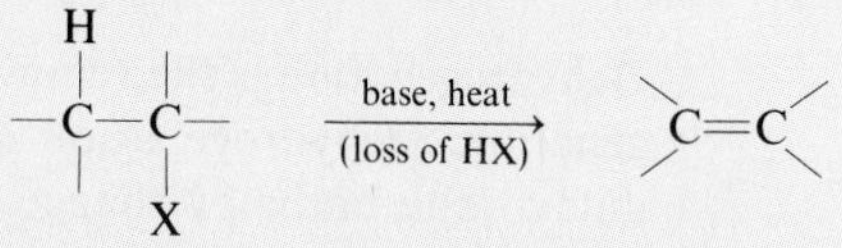

$$\mathrm{-\overset{H}{\underset{|}{\overset{|}{C}}}-\overset{|}{\underset{X}{\underset{|}{C}}}-} \xrightarrow[\text{(loss of HX)}]{\text{base, heat}} \mathrm{>C=C<}$$

Example

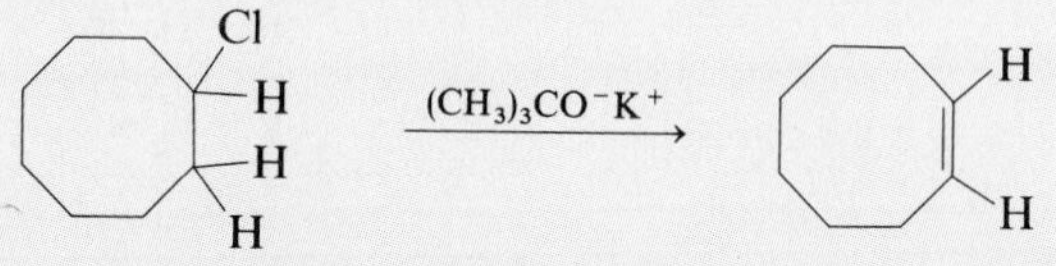

$$\text{chlorocyclooctane} \xrightarrow{(CH_3)_3CO^-K^+} \text{cyclooctene}$$

2. *Dehalogenation of vicinal dibromides* (Section 7-9B)

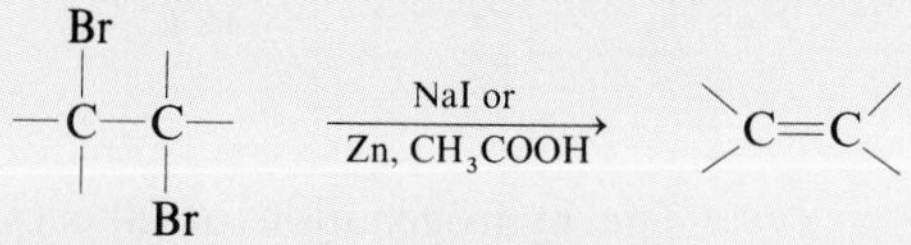

$$\mathrm{-\overset{Br}{\underset{|}{\overset{|}{C}}}-\overset{|}{\underset{Br}{\underset{|}{C}}}-} \xrightarrow[\text{Zn, }CH_3COOH]{\text{NaI or}} \mathrm{>C=C<}$$

Example

$$\mathrm{Ph-\overset{Br}{\underset{H}{C}}-\overset{H}{\underset{Br}{C}}-Ph} \xrightarrow[\text{acetone}]{\text{NaI}} \mathrm{(Ph)(H)C=C(H)(Ph)}$$

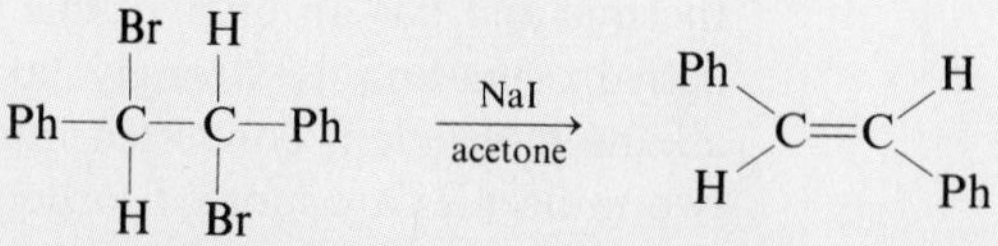

3. *Dehydration of alcohols* (Section 7-9C)

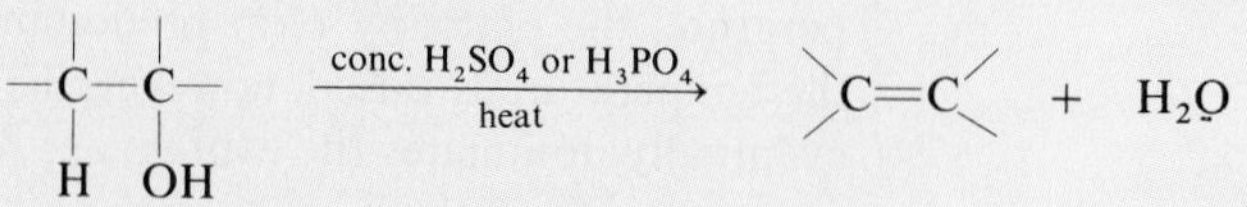

$$\mathrm{-\underset{H}{C}-\underset{OH}{C}-} \xrightarrow[\text{heat}]{\text{conc. }H_2SO_4\text{ or }H_3PO_4} \mathrm{>C=C<} + H_2O$$

Example

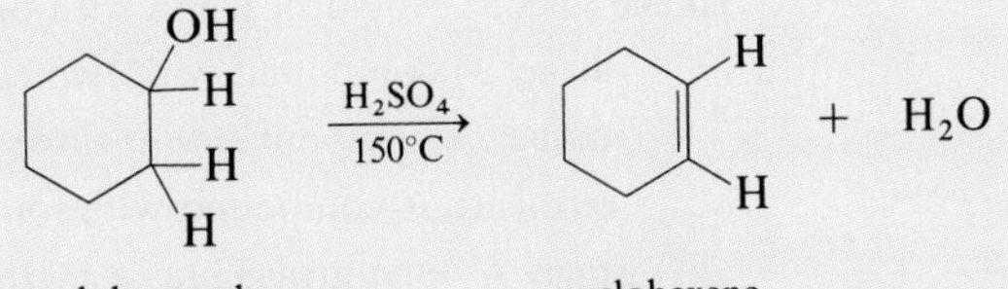

$$\text{cyclohexanol (C}_6\text{H}_{10}\text{(OH)H)} \xrightarrow[150°C]{H_2SO_4} \text{cyclohexene} + H_2O$$

cyclohexanol cyclohexene

4. Dehydrogenation of alkanes (Section 7-9E)

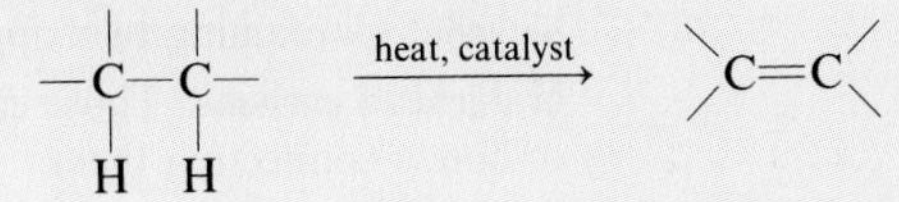

$$-\underset{H}{\overset{|}{C}}-\underset{H}{\overset{|}{C}}- \xrightarrow{\text{heat, catalyst}} >C=C<$$

Example

$$CH_3CH_2CH_2CH_3 \xrightarrow{\text{Pt, 500°C}} \begin{cases} \text{1-butene} + \textit{cis-} \text{ and } \textit{trans-}\text{2-butene} + \\ \text{1,3-butadiene} + H_2 \end{cases}$$

(Useful only for small alkenes; commonly gives mixtures.)

5. Hofmann elimination (Section 19-15)

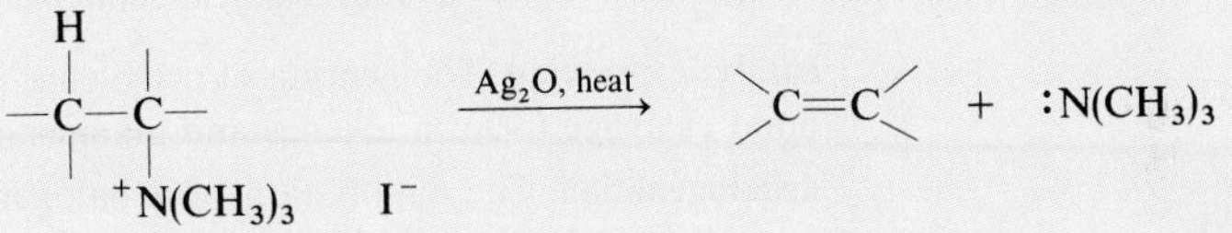

$$-\overset{H}{\underset{|}{\overset{|}{C}}}-\underset{^+N(CH_3)_3}{\overset{|}{C}}- \quad I^- \xrightarrow{Ag_2O,\text{ heat}} >C=C< + :N(CH_3)_3$$

(Usually gives the least highly substituted alkene.)

Example

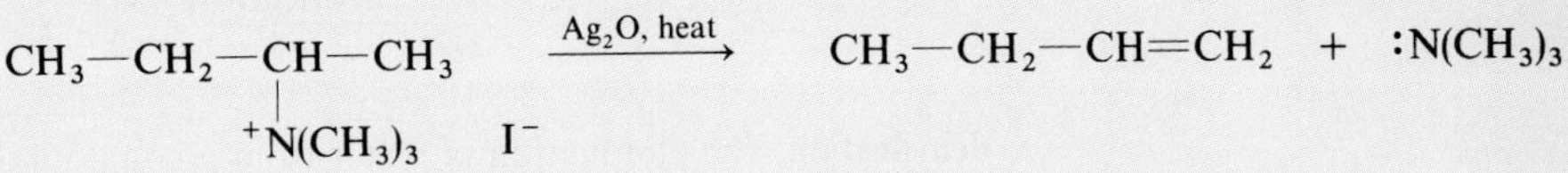

$$CH_3-CH_2-\underset{^+N(CH_3)_3}{\underset{|}{CH}}-CH_3 \quad I^- \xrightarrow{Ag_2O,\text{ heat}} CH_3-CH_2-CH=CH_2 + :N(CH_3)_3$$

6. Reduction of alkynes (Section 14-10)

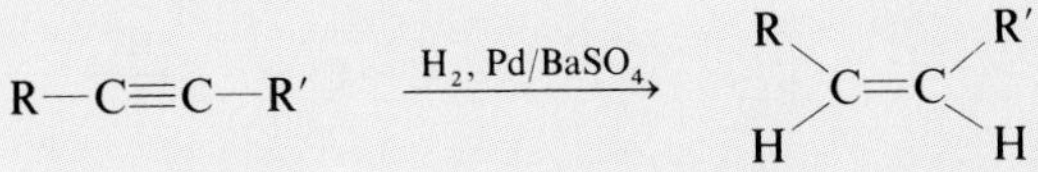

$$R-C\equiv C-R' \xrightarrow{H_2,\ Pd/BaSO_4} \underset{H}{\overset{R}{>}}C=C\underset{H}{\overset{R'}{<}} \quad \textit{cis}\text{-alkene}$$

$$R-C\equiv C-R' \xrightarrow{Na,\ NH_3} \underset{H}{\overset{R}{>}}C=C\underset{R'}{\overset{H}{<}} \quad \textit{trans}\text{-alkene}$$

Example

$$CH_3CH_2-C\equiv C-CH_2CH_3 \xrightarrow{H_2,\ Pd/BaSO_4} \underset{H}{\overset{CH_3CH_2}{>}}C=C\underset{H}{\overset{CH_2CH_3}{<}}$$

$$CH_3CH_2-C\equiv C-CH_2CH_3 \xrightarrow{Na,\ NH_3} \underset{H}{\overset{CH_3CH_2}{>}}C=C\underset{CH_2CH_3}{\overset{H}{<}}$$

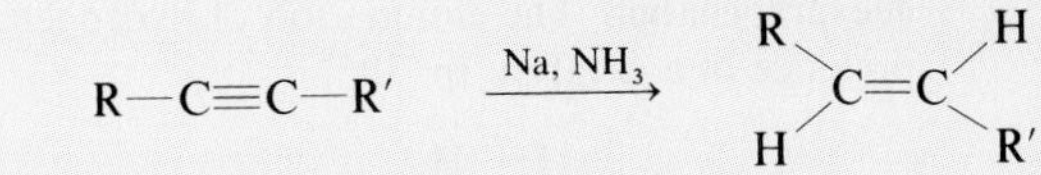

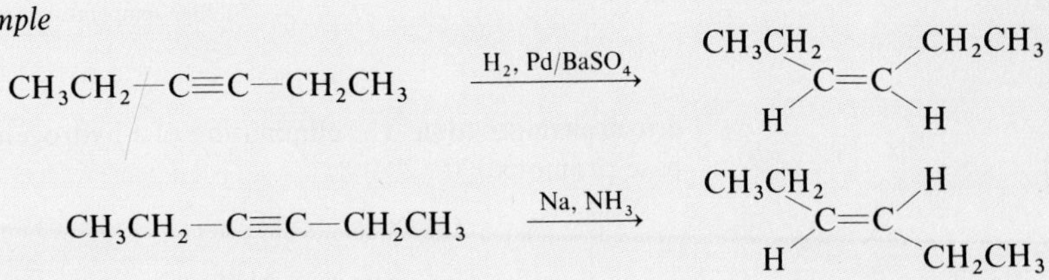

7. Wittig reaction (Section 22-12)

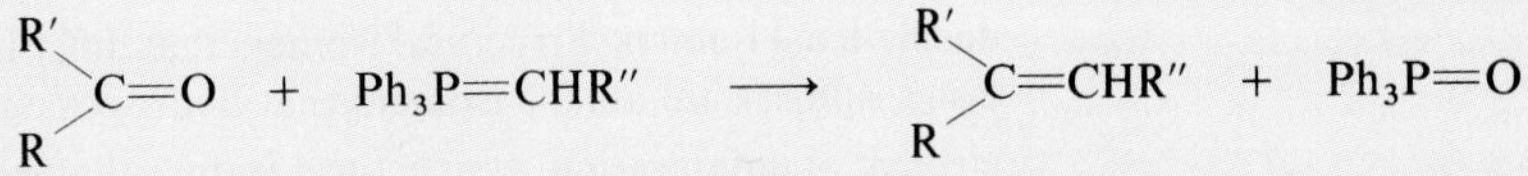

$$\underset{R}{\overset{R'}{>}}C=O + Ph_3P=CHR'' \longrightarrow \underset{R}{\overset{R'}{>}}C=CHR'' + Ph_3P=O$$

Example

$$\text{(cyclopentanone) } C=O + Ph_3P=CHCH_3 \longrightarrow \text{(cyclopentylidene) } C=C\underset{CH_3}{\overset{H}{<}}$$

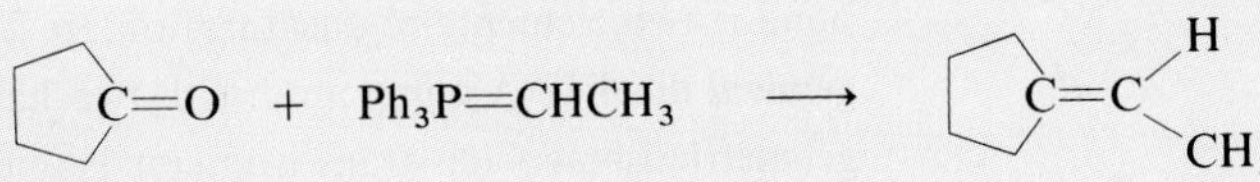

cyclopentanone

alkene (olefin) A hydrocarbon with one or more carbon-carbon double bonds. (p. 277)

diene: A compound with two carbon-carbon double bonds. (p. 282)

triene: A compound with three carbon-carbon double bonds.

tetraene: A compound with four carbon-carbon double bonds.

allyl group A vinyl group plus a methylene group: $CH_2{=}CH{-}CH_2{-}$ (p. 283)

Bredt's rule A stable bridged bicyclic compound cannot have a double bond at a bridgehead position unless one of the rings contains at least eight carbon atoms. (p. 292)

bicyclic: Containing two rings.

bridgehead carbons: Those carbon atoms that are part of both rings, with three arms of bonds connecting them.

bridged bicyclic: Having at least one carbon atom in each of the three links connecting the bridgehead carbons.

a bridged bicyclic compound　　a Bredt's rule violation

catalytic cracking The heating of petroleum products in the presence of a catalyst, causing bond cleavage to form alkenes and alkanes of lower molecular weight. (p. 305)

dehalogenation The elimination of a halogen (X_2) from a compound. Dehalogenation is formally a reduction. (p. 298)

$$-\overset{X}{\underset{|}{\overset{|}{C}}}-\overset{X}{\underset{|}{\overset{|}{C}}}- \xrightarrow{\text{Zn, } CH_3COOH} \;\; >C{=}C< \;+\; ZnX_2$$

dehydration The elimination of water from a compound, usually acid catalyzed. (p. 299)

$$-\overset{H}{\underset{|}{\overset{|}{C}}}-\overset{OH}{\underset{|}{\overset{|}{C}}}- \xrightarrow{H^+} \;\; >C{=}C< \;+\; H_2O$$

dehydrogenation The elimination of hydrogen (H_2) from a compound, usually done in the presence of a catalyst. (p. 305)

$$-\overset{H}{\underset{|}{\overset{|}{C}}}-\overset{H}{\underset{|}{\overset{|}{C}}}- \xrightarrow{\text{Pt, high temperature}} \;\; >C{=}C< \;+\; H_2$$

dehydrohalogenation The elimination of a hydrogen halide (HX) from a compound (usually base promoted). (p. 295)

$$-\overset{H}{\underset{|}{\overset{|}{C}}}-\overset{X}{\underset{|}{\overset{|}{C}}}- \xrightarrow{KOH} \;\; >C{=}C< \;+\; H_2O \;+\; K^+X^-$$

double-bond isomers Structural isomers that differ in the position of a double bond. Double-bond isomers are alkene isomers that hydrogenate to give the same alkane. (p. 288)

element of unsaturation A structural feature that causes a reduction of two hydrogen atoms in the molecular formula. A double bond or a ring is one element of unsaturation; a triple bond is two elements of unsaturation. (p. 279)

geminal dihalide A compound with two halogens on the same carbon atom. (p. 298)

geometric isomers (cis-trans isomers) Isomers that differ in their cis-trans arrangement on a ring or double bond. Geometric isomers are a subclass of diastereomers. (p. 283)

cis: Having similar groups on the same side of a double bond or a ring.

trans: Having similar groups on opposite sides of a double bond or a ring.

***Z*:** Having the higher-priority groups on the same side of a double bond.

***E*:** Having the higher-priority groups on opposite sides of a double bond.

heteroatom Any atom other than carbon or hydrogen. (p. 280)

Hofmann product The least highly substituted alkene product. (p. 296)

hydrogenation The addition of hydrogen to a molecule. The most common hydrogenation is the addition of H_2 across a double bond in the presence of a catalyst (*catalytic hydrogenation*). The value of $(-\Delta H^\circ)$ for this reaction is called the *heat of hydrogenation.* (p. 286)

$$\text{>C=C<} + H_2 \xrightarrow{\text{Pt}} \text{-CH-CH-} \qquad -\Delta H^\circ = \text{heat of hydrogenation}$$

olefin An alkene. (p. 277)

polymer A substance of high molecular weight made by linking together many small molecules, called the *monomer*. (p. 286)

addition polymer: A polymer formed by simple addition of monomer units.

polyolefin: A type of addition polymer with an olefin serving as the monomer.

saturated Having only single bonds (alkanes, for example); incapable of undergoing addition reactions. (p. 279)

Saytzeff rule An elimination usually gives the most stable alkene product, commonly the most highly substituted alkene product. The Saytzeff rule does not always apply, especially with a bulky base or a bulky leaving group. (pp. 203, 288)

Saytzeff elimination: An elimination that gives the Saytzeff product.

Saytzeff product: The most highly substituted alkene product. (p. 296)

thermal cracking The heating of petroleum products, causing bond cleavage to form products of lower molecular weight. Cracking is often used to form mixtures that are rich in ethylene and propylene. *Catalytic cracking* involves the use of a catalyst, usually an acidic aluminosilicate mineral. (p. 305)

unsaturated Having multiple bonds that can undergo addition reactions (alkenes, for example). (p. 279)

vicinal dihalide A compound with two halogens on adjacent carbon atoms. (p. 298)

vinyl group An ethenyl group, $CH_2{=}CH{-}$ (p. 283)

ESSENTIAL PROBLEM-SOLVING SKILLS IN CHAPTER 7

1. Draw and name all alkenes with a given molecular formula.
2. Use the *E-Z* and cis-trans systems to name geometric isomers.
3. Use heats of hydrogenation to compare stabilities of alkenes.
4. Predict relative stabilities of alkenes and cycloalkenes based on structure and stereochemistry.
5. Predict the products of dehydrohalogenation of alkyl halides, dehalogenation of dibromides, and dehydration of alcohols, including major and minor products.
6. Propose logical mechanisms for dehydrohalogenation, dehalogenation, and dehydration reactions.
7. Predict and explain the stereochemistry of E2 eliminations to form alkenes.
8. Propose effective single-step and multistep syntheses of alkenes.

STUDY PROBLEMS

7-19. Define each of the following terms and give an example.

(a) double-bond isomers **(b)** Saytzeff elimination **(c)** element of unsaturation
(d) Hofmann product **(e)** Bredt's rule violation **(f)** hydrogenation
(g) dehydrogenation **(h)** dehydrohalogenation **(i)** dehydration
(j) dehalogenation **(k)** geminal dihalide **(l)** vicinal dihalide
(m) heteroatom **(n)** polymer

7-20. Draw a structure for each of the following compounds.

(a) 3-methyl-1-pentene **(b)** 3,4-dibromo-1-butene **(c)** 1,3-cyclohexadiene
(d) (*Z*)-3-methyl-2-octene **(e)** vinylcyclopropane **(f)** (*Z*)-2-bromo-2-pentene
(g) (3*Z*,6*E*)-1,3,6-octatriene

7-21. Give a correct name for each of the following compounds.

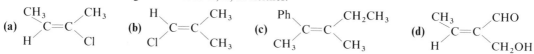

(a) $CH_3-CH_2-C(=CH_2)-CH_2-CH_2-CH_3$ **(b)** $(CH_2CH_3)_2C{=}CHCH_3$ **(c)**

(d) 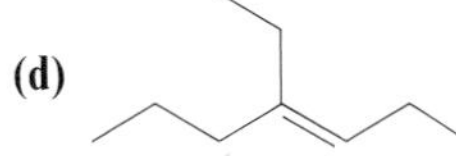

(e) 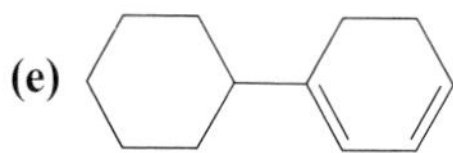

7-22. Label each of the following structures as *Z*, *E*, or neither.

(a) $(CH_3)(H)C{=}C(CH_3)(Cl)$ **(b)** $(H)(Cl)C{=}C(CH_3)(CH_3)$ **(c)** $(Ph)(CH_3)C{=}C(CH_2CH_3)(CH_3)$ **(d)** $(CH_3)(H)C{=}C(CHO)(CH_2OH)$

7-23. **(a)** Draw and name all five isomers of formula C_3H_5F.
(b) Cholesterol, $C_{27}H_{46}O$, has only one pi bond. What else can you say about its structure?

7-24. Draw and name all stereoisomers of 3-methyl-2,4-hexadiene
(a) using the cis-trans nomenclature. **(b)** using the *E-Z* nomenclature.

7-25. Determine which of the following compounds show cis-trans isomerism. Draw and label the isomers, using both the cis-trans and *E-Z* nomenclatures where applicable.

(a) 1-pentene **(b)** 2-pentene **(c)** 3-hexene
(d) 1,1-dibromopropane **(e)** 1,2-dibromopropane **(f)** 2,4-hexadiene

7-26. For each of the following alkenes, indicate the direction of the dipole moment. For each pair, determine which compound has the larger dipole moment.

(a) *cis*-1,2-difluoroethene and *trans*-1,2-difluoroethene
(b) *cis*-1,2-dibromoethene and *trans*-2,3-dibromo-2-butene
(c) *cis*- and *trans*-1,2-dibromo-1,2-dichloroethene
(d) *cis*-1,2-dibromo-1,2-dichloroethene and *cis*-1,2-dichloroethene

7-27. Predict the products of the following reactions. When more than one product is expected, predict which product will be the major product.

(a) OH $\xrightarrow[\text{heat}]{H_2SO_4}$ **(b)** 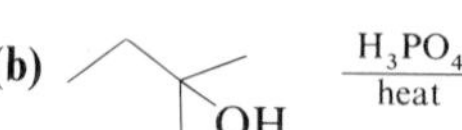 $\xrightarrow[\text{heat}]{H_3PO_4}$

(c) 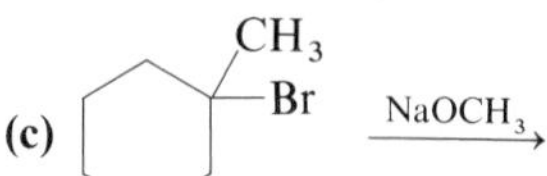 $\xrightarrow{NaOCH_3}$ **(d)** CH_3, Br $\xrightarrow{NaOC(CH_3)_3}$

7-28. Write a balanced equation for each of the following reactions.

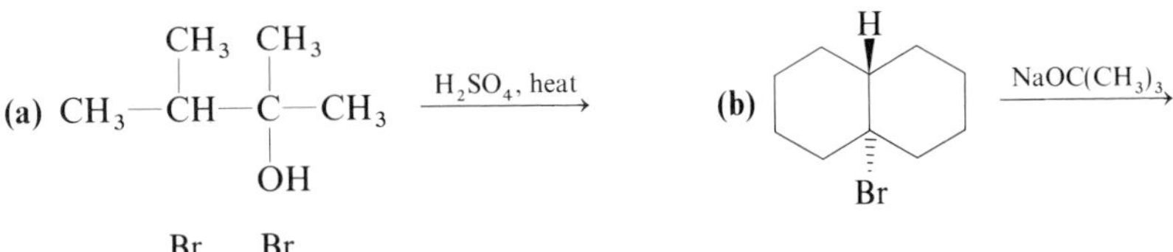

(c) $CH_3-CHBr-CHBr-CH_3 \xrightarrow{\text{Zn, } CH_3COOH}$

7-29. Show how you would prepare cyclopentene from each of the following compounds.
(a) *trans*-1,2-dibromocyclopentane **(b)** cyclopentanol
(c) cyclopentyl bromide **(d)** cyclopentane (not by dehydrogenation)

7-30. Predict the products formed by sodium hydroxide-promoted dehydrohalogenation of the following compounds. In each case, predict which product will be the major product.
(a) 1-bromobutane **(b)** 2-chlorobutane **(c)** 3-bromopentane
(d) 1-bromo-1-methylcyclohexane **(e)** 1-bromo-2-methylcyclohexane

7-31. What halides would undergo dehydrohalogenation to give the following pure alkenes?
(a) 1-butene **(b)** isobutylene **(c)** 2-pentene **(d)** methylenecyclohexane **(e)** 4-methylcyclohexene

7-32. Predict the major products of dehydration of the following alcohols.
(a) 2-pentanol **(b)** 1-methylcyclopentanol **(c)** 2-methylcyclohexanol **(d)** 2,2-dimethyl-1-propanol

★ 7-33. Give a mechanism to explain the formation of the following product. In your mechanism, explain the cause of the rearrangement and explain the failure to form the Saytzeff product.

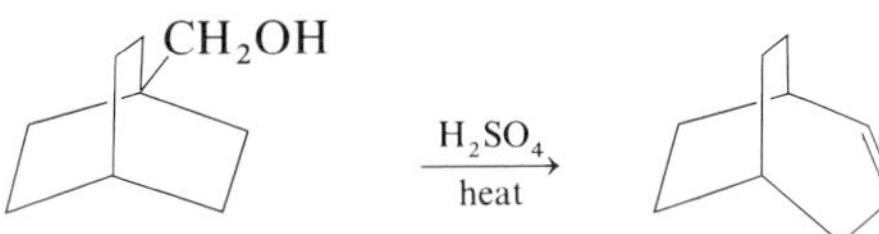

7-34. In the dehydrohalogenation of alkyl halides, a strong base such as *t*-butoxide usually gives the best results via the E2 mechanism.
(a) Explain why a strong base such as *t*-butoxide cannot dehydrate an alcohol through the E2 mechanism.
(b) Explain why strong acid, used in the dehydration of an alcohol, is not effective in the dehydrohalogenation of an alkyl halide.

7-35. E1 eliminations of alkyl halides are not very useful for synthetic purposes because they give mixtures of substitution and elimination products. Explain why the sulfuric acid-catalyzed dehydration of cyclohexanol gives a good yield of cyclohexene even though the reaction goes by an E1 mechanism. (*Hint:* What are the nucleophiles in the reaction mixture? What products are formed if these nucleophiles attack the carbocation? What further reactions can these substitution products undergo?)

7-36. The following reaction is called the **pinacol rearrangement.** The reaction begins with an acid-promoted ionization to give a carbocation. This carbocation undergoes a methyl shift to give a more stable, resonance-stabilized cation. Loss of a proton gives the observed product. Propose a mechanism for the pinacol rearrangement.

$$CH_3-C(CH_3)(OH)-C(CH_3)(OH)-CH_3 \xrightarrow{H_2SO_4,\ \text{heat}} CH_3-C(=O)-C(CH_3)_2-CH_3$$

pinacol → pinacolone

7-37. A chemist allows some pure (2*S*,3*R*)-3-bromo-2,3-diphenylpentane to react with a solution of sodium ethoxide ($NaOCH_2CH_3$) in ethanol. The products are two alkenes: **A** (cis/trans mixture) and **B,** a single pure isomer. Under the same conditions, the reaction of (2*S*,3*S*)-3-bromo-2,3-diphenylpentane gives two alkenes, **A** (cis/trans mixture) and **C.** Upon catalytic hydrogenation, all three of these alkenes (**A, B,** and **C**) give 2,3-diphenylpentane. Determine the structures of **A, B,** and **C,** give equations for their reactions with sodium ethoxide in ethanol, and explain the stereospecificity of these reactions.

7-38. Treatment of 2-bromopropane with a solution of sodium ethoxide ($NaOCH_2CH_3$) in ethanol gives a product mixture containing about 75% propene and 25% 2-ethoxypropane. The same reaction with hexadeuterio-2-bromopropane, $CD_3CHBrCD_3$, gives 31% $CD_2{=}CH{-}CD_3$ and 69% $(CD_3)_2CHOCH_2CH_3$. Explain why using the deuterated starting material reverses the product ratio. (You might want to review Section 4-12.)

7-39. A double bond in a six-membered ring is usually more stable in an endocyclic position than in an exocyclic position. Hydrogenation data on two pairs of compounds are given below. One pair suggests that the energy difference between endocyclic and exocyclic double bonds is about 2.1 kcal. The other pair suggests an energy difference of about 1.2 kcal. Which number do you trust as being more representative of the actual energy difference? Explain your answer.

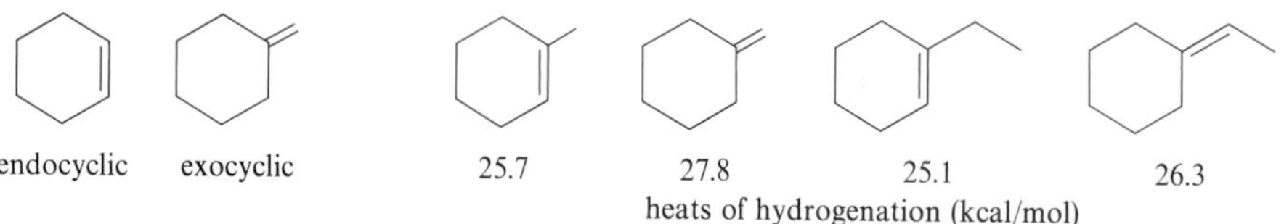

heats of hydrogenation (kcal/mol)

7-40. The energy difference between *cis*- and *trans*-2-butene is about 1 kcal/mol; however, the trans isomer of 4,4-dimethyl-2-pentene is 3.8 kcal/mol more stable than the cis isomer. Explain this large difference.

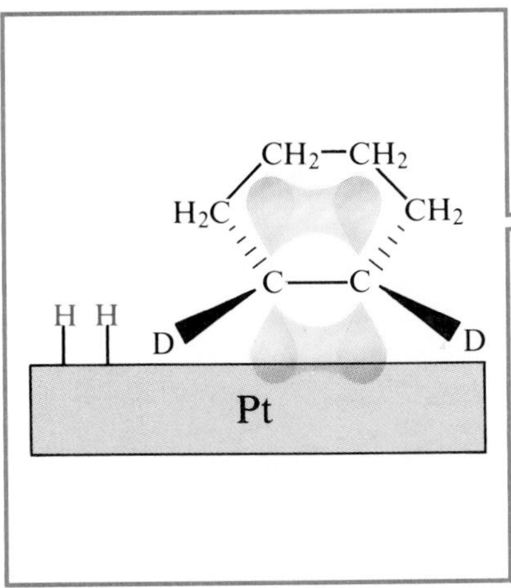

8

REACTIONS OF ALKENES

All alkenes have one common feature: a carbon-carbon double bond. Study of the reactions of alkenes is really the study of the reactivity of the carbon-carbon double bond. Once again, the concept of the functional group helps us to organize and simplify the study of chemical reactions. By studying the characteristic reactions of the double bond, we can predict the reactions of many alkenes that we have never seen before.

8-1 REACTIVITY OF THE CARBON-CARBON DOUBLE BOND

Since single bonds (sigma bonds) are more stable than pi bonds, we might expect the double bond to react in such a way that the pi bond is transformed into a sigma bond. In fact, this is the most common reaction of double bonds. We have already seen an example of such a reaction. Catalytic hydrogenation converts the C=C pi bond and the H—H sigma bond into two C—H sigma bonds. The reaction is exothermic (ΔH° = about -20 to -30 kcal/mol or about -80 to -120 kJ/mol), showing that the product is more stable than the reactants.

$$\gt C{=}C\lt \quad + \quad H{-}H \quad \xrightarrow{\text{catalyst}} \quad -\underset{H}{\overset{|}{\underset{|}{C}}}-\underset{H}{\overset{|}{\underset{|}{C}}}- \quad + \quad \text{energy}$$

Hydrogenation of an alkene is an example of an **addition** reaction, one of the three major reaction types we have studied: addition, elimination, and substitution. In an addition, two molecules combine to form one product molecule. When an alkene undergoes addition, two groups add to the carbon atoms of the double bond, and the carbons become saturated. In many ways, addition is the reverse of **elimination,** in which one molecule is split into two fragment molecules. In a **substitution,** one fragment replaces another fragment in a molecule.

Addition

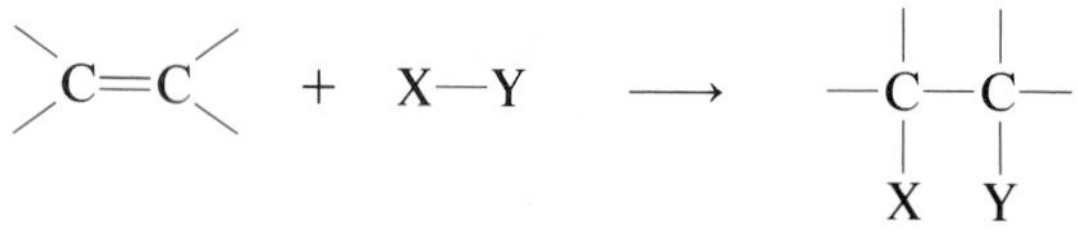

$$\gt C{=}C\lt \; + \; X{-}Y \;\longrightarrow\; -\underset{X}{\overset{|}{\underset{|}{C}}}-\underset{Y}{\overset{|}{\underset{|}{C}}}-$$

Elimination

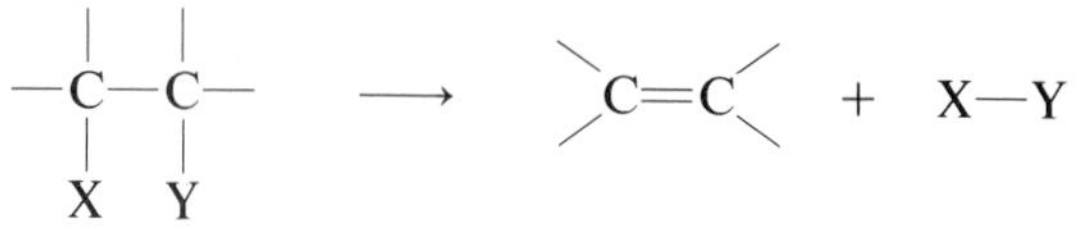

$$-\underset{X}{\overset{|}{\underset{|}{C}}}-\underset{Y}{\overset{|}{\underset{|}{C}}}- \;\longrightarrow\; \gt C{=}C\lt \; + \; X{-}Y$$

Substitution

$$-\overset{|}{\underset{|}{C}}-X \; + \; Y^- \;\longrightarrow\; -\overset{|}{\underset{|}{C}}-Y \; + \; X^-$$

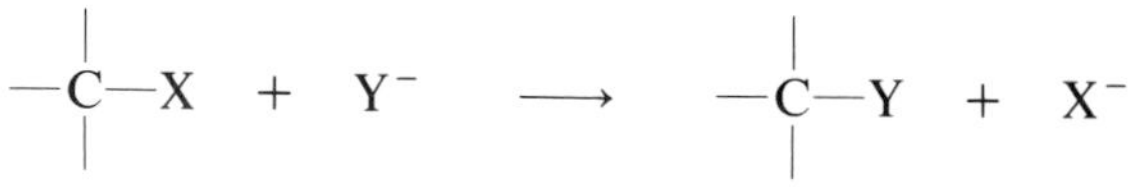

Addition is the most common reaction of alkenes, and we will consider additions in detail. A wide variety of functional groups can be formed by the addition of suitable reagents to the double bonds of alkenes.

8-2 ELECTROPHILIC ADDITION TO ALKENES

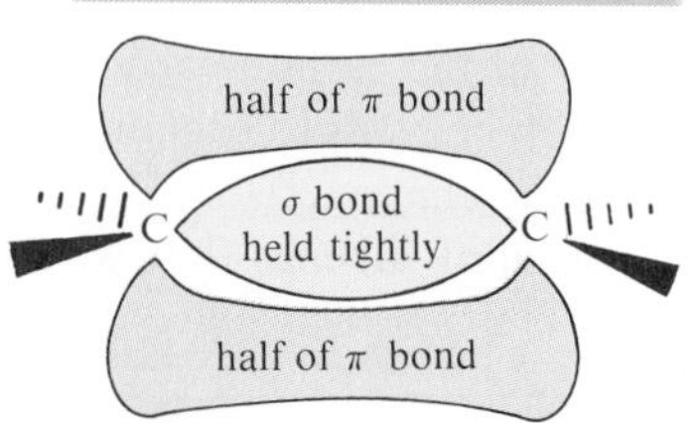

FIGURE 8-1 The electrons in the pi bond are spread farther from the carbon nuclei than the sigma electrons, and they are more loosely held.

In principle, many different reagents could add to a double bond to form more stable products; that is, the reactions are energetically favorable. Not all of these reactions have convenient rates, however. For example, the reaction of ethylene with hydrogen (to give ethane) is strongly exothermic, but the rate is very slow. A mixture of ethylene and hydrogen can remain for years without any appreciable reaction taking place. The addition of a catalyst such as platinum, palladium, or nickel allows the reaction to take place at a rapid rate.

Some reagents react with carbon-carbon double bonds without the aid of a catalyst. To understand what types of reagents react with double bonds, consider the structure of the pi bond. Although the electrons in the sigma-bond framework are very tightly held, the pi bond is delocalized above and below the sigma bond (Fig. 8-1). The pi-bonding electrons are spread farther from the carbon nuclei and they are more loosely held. A strong electrophile has an affinity for these loosely held electrons, and it can pull them away to form a new bond (Fig. 8-2). Reaction with an electrophile leaves one of the carbon atoms of the double bond with only three bonds and a positive charge: a carbocation.

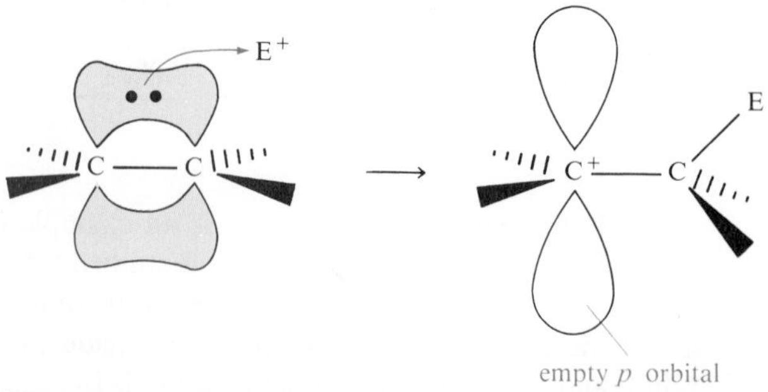

FIGURE 8-2 When a strong electrophile approaches the weakly held pi electrons, it pulls the electrons out of the pi bond to form a new sigma bond. A carbocation results. Notice that the arrow shows the movement of electrons, from the electron-rich pi bond to the electron-poor electrophile.

In most additions, the carbocation is attacked by a nucleophile as in the second step of the S_N1 reaction. A stable addition product is formed, in which the electrophile and the nucleophile are bonded to the carbon atoms that were in the double bond. The following schematic reaction uses E^+ as the electrophile and $Nuc{:}^-$ as the nucleophile.

Step 1: Attack of the pi bond on the electrophile

$$>C=C< \; + \; E^+ \longrightarrow -\overset{|}{\underset{E}{C}}-C^+<$$

Step 2: Attack by the nucleophile

$$-\overset{|}{\underset{E}{C}}-C^+< \; + \; Nuc{:}^- \longrightarrow -\overset{|}{\underset{E}{C}}-\overset{|}{\underset{Nuc}{C}}-$$

This type of reaction depends on the presence of a strong electrophile to attract the electrons of the pi bond and generate a carbocation in the rate-determining step. Most of the alkene reactions we will study fall into this large class of **electrophilic additions** to alkenes.

To illustrate the electrophilic addition, consider what happens when gaseous HBr adds to 2-butene. The proton in HBr is electrophilic; it reacts with the alkene to form a carbocation.

$$CH_3-\overset{H}{\overset{|}{C}}=\overset{H}{\overset{|}{C}}-CH_3 \quad \underset{H-Br}{} \rightleftarrows CH_3-\overset{H}{\overset{|}{\underset{H}{\underset{|}{C}}}}-\overset{H}{\overset{|}{\underset{+}{C}}}-CH_3 + Br^-$$

Bromide ion reacts rapidly with the carbocation to give a stable product in which the elements of HBr have added to the ends of the double bond.

$$CH_3-\overset{H}{\overset{|}{\underset{H}{\underset{|}{C}}}}-\overset{H}{\overset{|}{\underset{+}{C}}}-CH_3 + Br^- \rightleftarrows CH_3-\overset{H}{\overset{|}{\underset{H}{\underset{|}{C}}}}-\overset{H}{\overset{|}{\underset{Br}{\underset{|}{C}}}}-CH_3$$

To learn more about these electrophilic additions, we will consider additions to alkenes using a wide variety of reagents: hydrogen halides, water, borane, hydrogen, carbenes, halogens, oxidizing agents, and even other alkenes. Two aspects you should watch for in each reaction are the *regiochemistry* or *orientation* of addition (how to predict which part of the reagent will add to which end of the double bond) and the *stereochemistry* of the addition, if it is stereospecific.

8-3 ADDITION OF HYDROGEN HALIDES TO ALKENES

8-3A ORIENTATION OF ADDITION: MARKOVNIKOV'S RULE

The simple electrophilic addition mechanism shown above applies to a large number of reactions. We can use this mechanism to predict the outcome of some fairly complicated reactions. For example, the addition of HBr to 2-methyl-2-butene could lead to either of two products, yet only one is observed.

$$CH_3-\overset{CH_3}{\overset{|}{C}}=CH-CH_3 + H-Br \longrightarrow \underset{\text{observed}}{CH_3-\overset{CH_3}{\overset{|}{\underset{Br}{\underset{|}{C}}}}-\underset{H}{\underset{|}{CH}}-CH_3} \quad \text{or} \quad \underset{\text{not observed}}{CH_3-\overset{CH_3}{\overset{|}{\underset{H}{\underset{|}{C}}}}-\underset{Br}{\underset{|}{CH}}-CH_3}$$

The first step in the mechanism is protonation of the double bond. If the proton added to the secondary carbon, the product would be different from the one formed if the proton added to the tertiary carbon.

$$CH_3-\overset{\overset{\displaystyle CH_3}{|}}{C}=CH-CH_3 + H-Br \xrightarrow{\text{add } H^+ \text{ to secondary carbon}} CH_3-\underset{+}{\overset{\overset{\displaystyle CH_3}{|}}{C}}-\underset{\underset{\displaystyle H}{|}}{CH}-CH_3 \quad Br^-$$

tertiary carbocation

$$CH_3-\overset{\overset{\displaystyle CH_3}{|}}{C}=CH-CH_3 + H-Br \xrightarrow{\text{add } H^+ \text{ to tertiary carbon}} \!\!\!\!\!\!\!\times\;\; CH_3-\overset{\overset{\displaystyle CH_3}{|}}{\underset{\underset{\displaystyle H}{|}}{C}}-\underset{+}{CH}-CH_3 \quad Br^-$$

secondary carbocation

In the first equation, the proton adds to the secondary carbon of the double bond, forming a tertiary carbocation. In the second equation, the proton adds to the tertiary carbon atom, forming a secondary carbocation. Clearly, the tertiary carbocation is more stable (see Section 4-16A), so the first reaction is favored.

The second half of the mechanism shows the final product of the reaction of 2-methyl-2-butene with HBr:

$$CH_3-\underset{+}{\overset{\overset{\displaystyle CH_3}{|}}{C}}-\underset{\underset{\displaystyle H}{|}}{CH}-CH_3 \quad :\!Br^- \longrightarrow CH_3-\underset{\underset{\displaystyle Br}{|}}{\overset{\overset{\displaystyle CH_3}{|}}{C}}-\underset{\underset{\displaystyle H}{|}}{CH}-CH_3$$

Note that protonation of one carbon atom of a double bond gives a carbocation with its positive charge on the carbon atom that was *not* protonated. Therefore, the proton adds to the end of the double bond that is *less* highly substituted to give the *more highly substituted carbocation* (the more stable carbocation).

Figure 8-3 shows some examples of additions where the proton has added to the less highly substituted carbon atom of the double bond. We say that the addition of HBr is **regiospecific,** because in each case only one of the two possible orientations of addition is observed.

Markovnikov's rule A Russian chemist, Vladimir Markovnikov, first showed the orientation of addition of HBr to alkenes in 1869. Markovnikov stated:

> MARKOVNIKOV'S RULE The addition of a proton acid to the double bond of an alkene results in a product with the acid proton bonded to the carbon atom that already holds the greater number of hydrogen atoms.

This is the original statement of **Markovnikov's rule.** Additions that follow this rule are said to follow **Markovnikov orientation** and give the **Markovnikov product.** We are often interested in adding electrophiles other than proton acids to the double bonds of alkenes. Markovnikov's rule can be extended to include a wide variety of other additions, based on the addition of the electrophile in such a way as to produce the most stable carbocation intermediate.

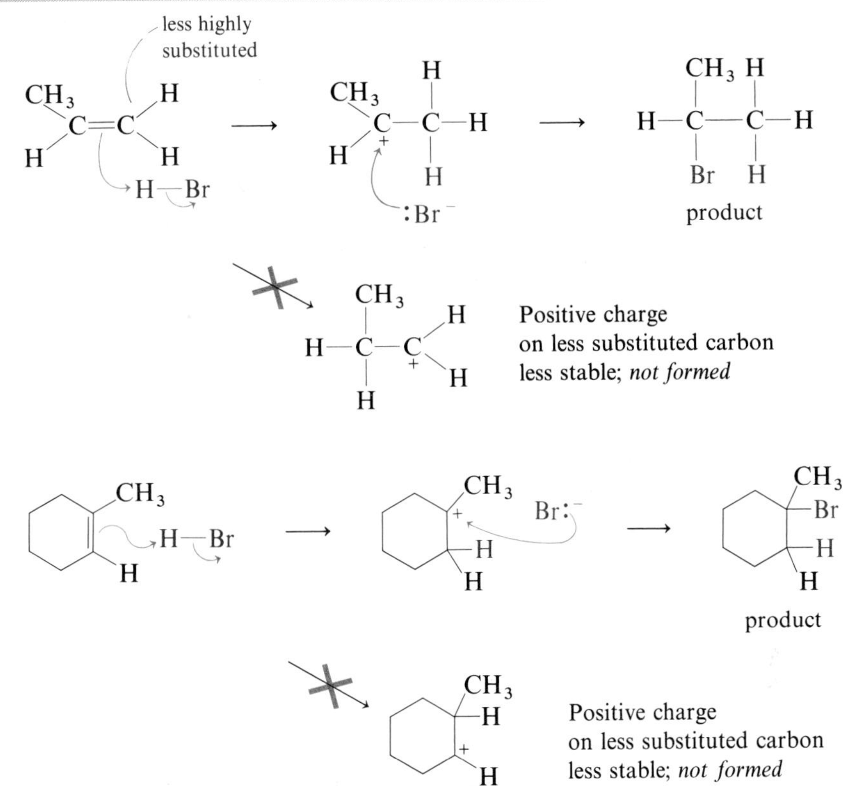

FIGURE 8-3 An electrophile adds to the less substituted end of the double bond to give the more highly substituted (and therefore more stable) carbocation.

MARKOVNIKOV'S RULE (extended) In an electrophilic addition to an alkene, the electrophile adds in such a way as to generate the most stable intermediate.

Like HBr, both HCl and HI add to the double bonds of alkenes, and they also follow Markovnikov's rule; for example,

$$CH_3-\overset{CH_3}{\overset{|}{C}}=CH-CH_2CH_3 + HCl \longrightarrow CH_3-\overset{CH_3}{\overset{|}{\underset{Cl}{\underset{|}{C}}}}-\underset{H}{\underset{|}{CH}}-CH_2CH_3$$

(methylenecyclohexane, CH_2) + HI ⟶ (1-iodo-1-methylcyclohexane, I)

(octahydronaphthalene, H) + HCl ⟶ (decahydronaphthalene, Cl, H, H)

PROBLEM 8-1

Predict the major products for the following reactions.

(a) $CH_3-CH=CH_2$ + HBr
(b) 2-methylpropene + HCl
(c) 1-methylcyclohexene + HI
(d) 4-methylcyclohexene + HBr

PROBLEM 8-2

When 1,3-butadiene reacts with 1 mole of HBr, both 3-bromo-1-butene and 1-bromo-2-butene are formed. Give a detailed mechanism to account for this mixture of products.

8-3B FREE-RADICAL ADDITION OF HBr: ANTI-MARKOVNIKOV ADDITION

In 1933, M. S. Kharasch and F. W. Mayo showed that **anti-Markovnikov** products result from addition of HBr (but not HCl or HI) in the presence of peroxides. Peroxides give rise to free radicals that act as catalysts to accelerate the addition, causing it to occur by a different mechanism. The oxygen-oxygen bond in peroxides is rather weak. It can break to give two radicals.

$$\mathrm{R{-}\ddot{\underset{..}{O}}{-}\ddot{\underset{..}{O}}{-}R \xrightarrow{\text{heat}} R{-}\ddot{\underset{..}{O}}\cdot + \cdot\ddot{\underset{..}{O}}{-}R} \qquad \Delta H^\circ = +36 \text{ kcal } (+150 \text{ kJ})$$

Alkoxy radicals (R—O·) catalyze the anti-Markovnikov addition of HBr. The mechanism of this free-radical chain reaction is shown below.

Initiation

$$\mathrm{R{-}O{-}O{-}R \xrightarrow{\text{heat}} R{-}O\cdot + \cdot O{-}R}$$

$$\mathrm{R{-}O\cdot + H{-}Br \longrightarrow R{-}O{-}H + Br\cdot}$$

Propagation

$$\mathrm{{>}C{=}C{<} + Br\cdot \longrightarrow {-}\underset{Br}{\overset{|}{C}}{-}\dot{C}{<}}$$

$$\mathrm{{-}\underset{Br}{\overset{|}{C}}{-}\dot{C}{<} + H{-}Br \longrightarrow {-}\underset{Br}{\overset{|}{C}}{-}\underset{H}{\overset{|}{C}}{-} + Br\cdot}$$

Let's consider each step of this mechanism. In the initiation step, free radicals generated from the peroxide react with HBr to form bromine radicals.

$$\mathrm{R{-}\ddot{\underset{..}{O}}\cdot + H{-}\ddot{\underset{..}{Br}}: \longrightarrow R{-}\ddot{\underset{..}{O}}{-}H + :\ddot{\underset{..}{Br}}\cdot} \qquad \Delta H^\circ = -15 \text{ kcal } (-63 \text{ kJ})$$

The bromine radical lacks an octet of electrons in its valence shell, making it electron deficient and electrophilic. It adds to a double bond, forming a new free radical with the odd electron on a carbon atom.

$$\mathrm{:\ddot{\underset{..}{Br}}\cdot + {>}C{=}C{<} \longrightarrow {-}\underset{:\underset{..}{Br}:}{\overset{|}{C}}{-}\dot{C}{<}} \qquad \Delta H^\circ = -3 \text{ kcal } (-12 \text{ kJ})$$

This free radical reacts with an HBr molecule to generate another bromine radical.

$$\mathrm{{-}\underset{Br}{\overset{|}{C}}{-}\dot{C}{<} + H{-}\ddot{\underset{..}{Br}}: \longrightarrow {-}\underset{Br}{\overset{|}{C}}{-}\underset{H}{\overset{|}{C}}{-} + :\ddot{\underset{..}{Br}}\cdot} \qquad \Delta H^\circ = -6 \text{ kcal } (-25 \text{ kJ})$$

The regenerated bromine radical reacts with another molecule of the alkene, continuing the chain reaction. Notice that each of these propagation steps starts with one free radical and ends with another free radical. The number of free radicals is constant, until two free radicals come together and terminate the chain reaction.

Radical addition of HBr to unsymmetrical alkenes Now we must explain the anti-Markovnikov orientation found in the products of the peroxide-catalyzed reaction. When the alkene is unsymmetrical, if the bromine radical adds to the secondary end of the double bond, a tertiary radical results. Addition to the tertiary end forms a less stable secondary radical.

$$CH_3-\overset{\overset{CH_3}{|}}{C}=CH-CH_3 \;+\; Br\cdot \;\longrightarrow\; CH_3-\overset{\overset{CH_3}{|}}{\underset{\bullet}{C}}-\underset{\underset{Br}{|}}{CH}-CH_3$$

tertiary radical (more stable)

$$CH_3-\overset{\overset{CH_3}{|}}{C}=CH-CH_3 \;+\; Br\cdot \;\not\longrightarrow\; CH_3-\overset{\overset{CH_3}{|}}{\underset{\underset{Br}{|}}{C}}-\underset{\bullet}{CH}-CH_3$$

secondary radical (less stable)

This reaction is similar to the addition of a proton to an alkene. If the electrophile (in this case, Br·) adds to the more highly substituted end of the double bond, the radical electron appears on the less highly substituted carbon. Therefore Br· adds to the *less* highly substituted end of the double bond to give the more stable free radical. This intermediate reacts with HBr to give the anti-Markovnikov product, in which H has added to the end of the double bond that started with *fewer* hydrogens.

$$CH_3-\overset{\overset{CH_3}{|}}{\underset{\bullet}{C}}-\underset{\underset{Br}{|}}{CH}-CH_3 \;+\; H-Br \;\longrightarrow\; CH_3-\overset{\overset{CH_3}{|}}{\underset{\underset{H}{|}}{C}}-\underset{\underset{Br}{|}}{CH}-CH_3 \;+\; Br\cdot$$

anti-Markovnikov product

Note that *both* mechanisms for the addition of HBr to an alkene (with and without peroxides) follow our extended statement of Markovnikov's rule: In both cases, the *electrophile* adds to the less substituted end of the double bond to give the more stable carbocation or free radical. In the ionic reaction, the electrophile is H^+. In the peroxide-catalyzed free-radical reaction, the electrophile is Br·.

Many students wonder why the reaction with Markovnikov orientation does not take place in the presence of peroxides, together with the free-radical chain reaction. It actually does take place, but the peroxide-catalyzed reaction is much faster. If just a tiny bit of peroxide is present, a mixture of Markovnikov and anti-Markovnikov products results. If an appreciable amount of peroxide is present, the catalyzed chain reaction is so much faster than the uncatalyzed ionic reaction that only the anti-Markovnikov product is observed.

The reversal of orientation in the presence of peroxides is called the **peroxide effect.** It occurs only in the addition of HBr to alkenes. The reaction of an alkyl radical with HCl is strongly endothermic, so the free-radical chain reaction is not effective for the addition of HCl.

$$Cl-\underset{|}{\overset{|}{C}}-\dot{C}\!\!< \;+\; H-Cl \;\longrightarrow\; Cl-\underset{|}{\overset{|}{C}}-\underset{|}{\overset{|}{C}}-H \;+\; Cl\cdot \qquad \Delta H^\circ = +10\text{ kcal }(+42\text{ kJ})$$

Similarly, the reaction of an iodine atom with an alkene is strongly endothermic, and the free-radical addition of HI is not observed. Only HBr has just the right reactivity for this interesting free-radical chain reaction to take place.

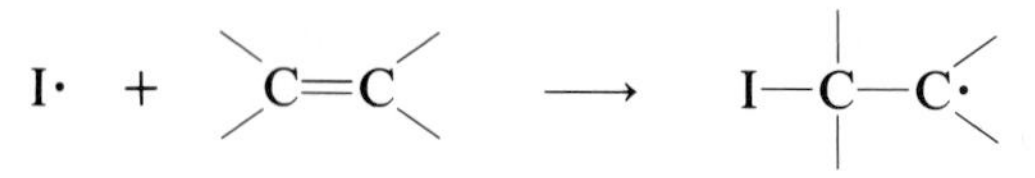

$$\text{I}\cdot + \text{>C=C<} \longrightarrow \text{I–}\overset{|}{\underset{|}{\text{C}}}\text{–C}\cdot\text{<} \qquad \Delta H^\circ = +13 \text{ kcal } (+54 \text{ kJ})$$

SOLVED PROBLEM 8-1

Show how you would accomplish the following synthetic conversions.

(a) Convert 1-methylcyclohexene to 1-bromo-1-methylcyclohexane.

SOLUTION This synthesis requires the addition of HBr to an alkene with Markovnikov orientation. Ionic addition of HBr gives the correct product.

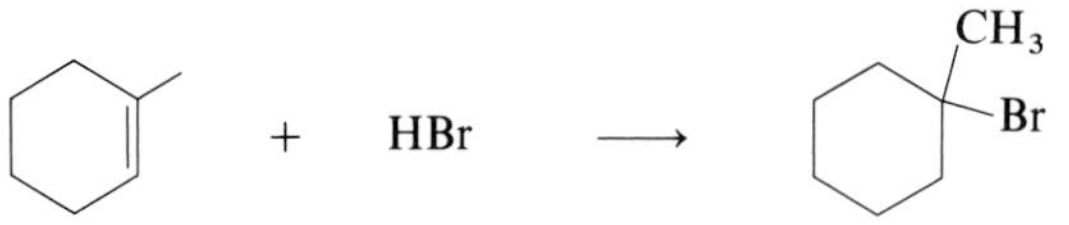

1-methylcyclohexene + HBr ⟶ 1-bromo-1-methylcyclohexane

(b) Convert 1-methylcyclohexanol to 1-bromo-2-methylcyclohexane.

SOLUTION This synthesis requires the conversion of an alcohol to an alkyl bromide with the bromine atom at the neighboring carbon atom. This is the anti-Markovnikov product, which could be formed by the radical-catalyzed addition of HBr to 1-methylcyclohexene.

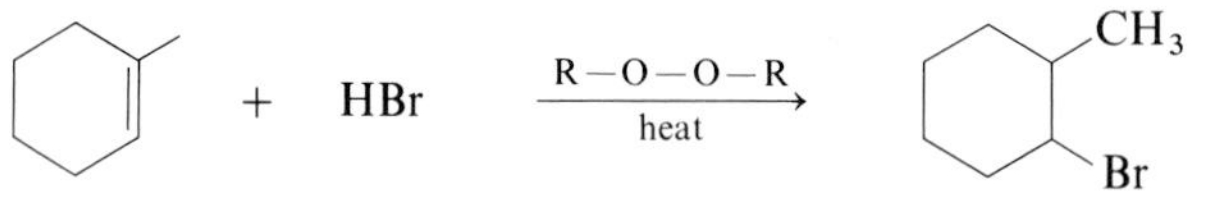

1-methylcyclohexene + HBr $\xrightarrow[\text{heat}]{\text{R–O–O–R}}$ 1-bromo-2-methylcyclohexane

1-Methylcyclohexene is easily synthesized by the dehydration of 1-methylcyclohexanol. The most highly substituted alkene is the desired product.

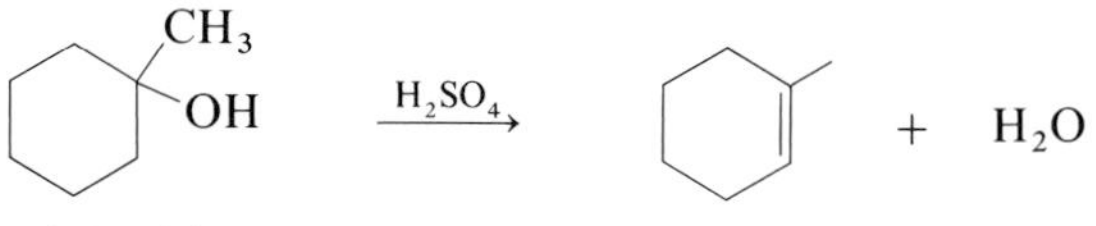

1-methylcyclohexanol $\xrightarrow{H_2SO_4}$ 1-methylcyclohexene + H_2O

The two-step synthesis is summarized as follows.

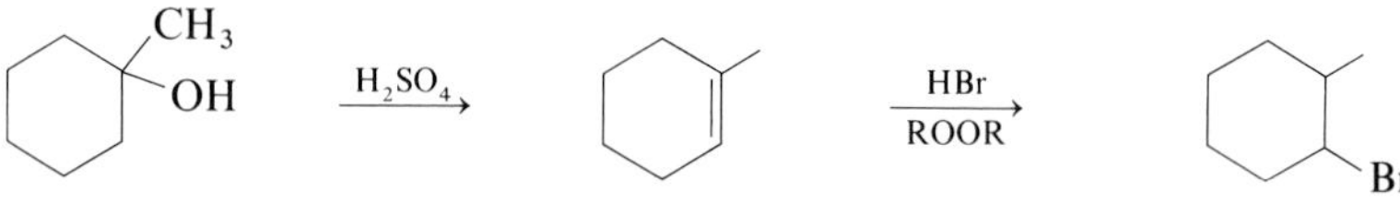

1-methylcyclohexanol $\xrightarrow{H_2SO_4}$ 1-methylcyclohexene $\xrightarrow[\text{ROOR}]{\text{HBr}}$ 1-bromo-2-methylcyclohexane

PROBLEM 8-3

Predict the major products of the following reactions.

(a) 2-methylpropene + HBr + $CH_3\text{–}\overset{O}{\overset{\|}{C}}\text{–O–O–}\overset{O}{\overset{\|}{C}}\text{–}CH_3$

(b) 1-methylcyclohexene + HBr + $CH_3CH_2—O—O—CH_2CH_3$

(c) 1-phenylpropene + HBr + di-*t*-butyl peroxide (phenyl = Ph = C_6H_5—)

PROBLEM 8-4

Show how you would accomplish the following synthetic conversions.

(a) 1-butene → 1-bromobutane
(b) 1-butene → 2-bromobutane
(c) 2-methylcyclohexanol → 1-bromo-1-methylcyclohexane
(d) 2-methyl-2-butanol → 2-bromo-3-methylbutane

8-4 ADDITION OF WATER: HYDRATION OF ALKENES

Treatment of an alkene with water in the presence of a strongly acidic catalyst can convert an alkene to an alcohol. Formally, this reaction is a **hydration** (the addition of water), with a hydrogen atom adding to one carbon and a hydroxyl group adding to the other. Hydration of an alkene is the reverse of the dehydration of alcohols that we studied in Section 7-9C.

Hydration of an alkene

$$\underset{\text{alkene}}{>C{=}C<} + H_2O \xrightleftharpoons{H^+} \underset{\substack{\text{alcohol} \\ \text{(Markovnikov orientation)}}}{-\overset{H}{\underset{|}{\overset{|}{C}}}-\overset{OH}{\underset{|}{\overset{|}{C}}}-}$$

Dehydration of an alcohol

$$\underset{\text{alcohol}}{-\overset{H}{\underset{|}{\overset{|}{C}}}-\overset{OH}{\underset{|}{\overset{|}{C}}}-} \xrightleftharpoons{H^+} \underset{\text{alkene}}{>C{=}C<} + H_2O$$

For dehydrating alcohols, a concentrated dehydrating acid (such as H_2SO_4 or H_3PO_4) is used to drive the equilibrium to favor the alkene. Hydration of an alkene is accomplished by adding excess water to drive the equilibrium to favor the alcohol.

8-4A MECHANISM OF HYDRATION

The *principle of microscopic reversibility* states that a forward reaction and a reverse reaction taking place under the same conditions (as in an equilibrium) must follow the same reaction pathway in microscopic detail. The hydration and dehydration reactions are the two complementary reactions in an equilibrium; therefore, they must follow the same reaction pathway. It makes sense that the lowest-energy transition states and intermediates for the reverse reaction are the same as those for the forward reaction except in reverse order.

According to the principle of microscopic reversibility, we can write the hydration mechanism by arranging the steps of the dehydration (Section 7-9C) in reverse order. Protonation of the double bond forms a carbocation. Nucleophilic attack by water, followed by loss of a proton, gives the alcohol.

Mechanism of acid-catalyzed hydration

Step 1

$$\mathrm{>C{=}C<} + \mathrm{H{-}\ddot{O}^{+}(H){-}H} \rightleftarrows \mathrm{{-}C(H){-}C^{+}<} + \mathrm{H_2\ddot{O}:}$$

Step 2

$$\mathrm{{-}C(H){-}C^{+}<} + \mathrm{H_2\ddot{O}:} \rightleftarrows \mathrm{{-}C(H){-}C(:O^{+}H_2){-}}$$

Step 3

$$\mathrm{{-}C(H){-}C(:O^{+}H_2){-}} + \mathrm{H_2\ddot{O}:} \rightleftarrows \mathrm{{-}C(H){-}C(:\ddot{O}H){-}} + \mathrm{H_3\ddot{O}^{+}}$$

Step 1 shows the protonation of the double bond to form a carbocation, the same reaction as the first step in the addition of other electrophilic reagents, such as HBr.

Step 2 is the attack of water on the carbocation formed in step 1. Water is the solvent for the hydration (dilute acid), and it is the nucleophile that the carbocation is mostly likely to encounter. The product of step 2 is the protonated alcohol.

Step 3 is a proton transfer from the protonated alcohol to water, regenerating the proton catalyst consumed in step 1.

8-4B ORIENTATION OF HYDRATION

Step 1 of the hydration mechanism is identical to the first step in the addition of HBr, HCl, or HI. Since Markovnikov's rule governs this step in the addition of the hydrogen halides, it should also determine the orientation of hydration. Consider the hydration of 2-methyl-2-butene.

$$\mathrm{CH_3{-}C(CH_3){=}CH{-}CH_3} + \mathrm{H{-}\ddot{O}^{+}(H){-}H} \rightleftarrows \underset{3^\circ,\ \text{more stable}}{\mathrm{CH_3{-}\overset{+}{C}(CH_3){-}CH(H){-}CH_3}} \quad \text{but not} \quad \underset{2^\circ,\ \text{less stable}}{\mathrm{CH_3{-}C(CH_3)(H){-}\overset{+}{C}H{-}CH_3}}$$

The proton adds to the less highly substituted end of the double bond, so that the positive charge appears at the more highly substituted end. Water attacks the carbocation to give the protonated alcohol.

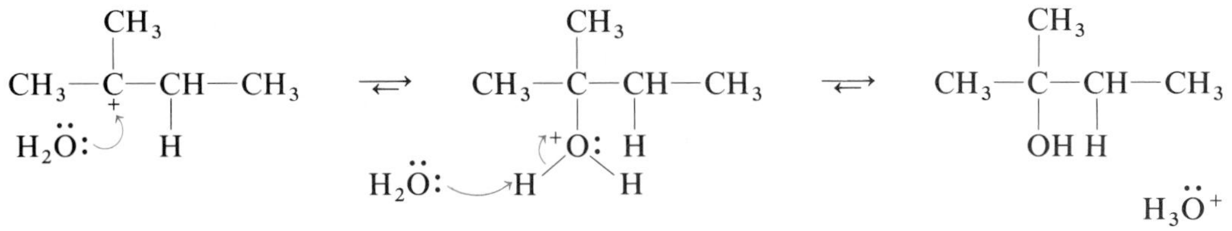

$$CH_3-\overset{CH_3}{\underset{+}{C}}-\underset{H}{CH}-CH_3 \quad H_2\ddot{O}: \rightleftharpoons CH_3-\overset{CH_3}{\underset{H-\overset{+}{O}:}{C}}-\underset{H}{CH}-CH_3 \;(H_2\ddot{O}: \to H) \rightleftharpoons CH_3-\overset{CH_3}{\underset{OH}{C}}-\underset{H}{CH}-CH_3 \quad H_3\ddot{O}^+$$

The final product shows that the reaction has obeyed Markovnikov's rule. The proton has added to the end of the double bond that already had more hydrogens (that is, the less highly substituted end), and the —OH group has added to the more highly substituted end.

PROBLEM 8-5
Predict the products of the following hydration reactions.

(a) 1-methylcyclopentene + dilute acid
(b) 2-phenylpropene + dilute acid
(c) 1-phenylcyclohexene + dilute acid

PROBLEM 8-6
An inexperienced graduate student wanted to make 3,3-dimethyl-2-butanol. She treated 3,3-dimethyl-1-butene with dilute acid and recovered a mixture of 2,3-dimethyl-2-butanol and 2,3-dimethyl-2-butene. Using a detailed mechanism, show why these products are formed rather than the desired alcohol.

8-5 INDIRECT HYDRATION OF ALKENES

Many alkenes do not undergo hydration easily in dilute aqueous acid, because they are only slightly soluble and very small concentrations of the alkene are obtained. In many cases, the overall equilibrium favors the alkene rather than the alcohol. No amount of catalysis can cause a reaction to occur if the energetics are unfavorable. Two more powerful methods can be used to form alcohols with Markovnikov orientation from alkenes: addition of sulfuric acid followed by hydrolysis, and the oxymercuration-demercuration process. In effect, these reactions simply provide the means for hydration in difficult cases.

8-5A ADDITION OF SULFURIC ACID, THEN HYDROLYSIS

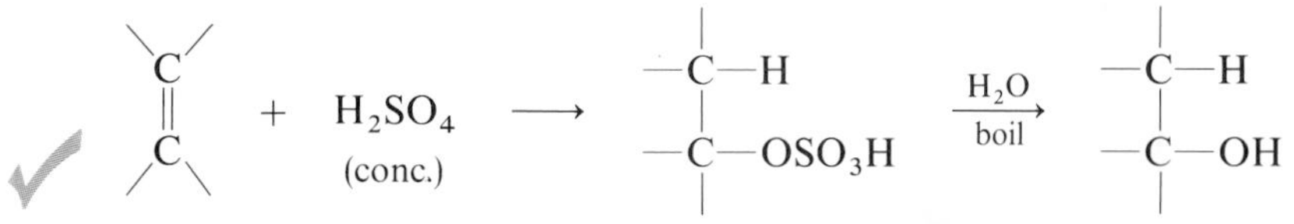

$$\text{>C=C<} + \underset{\text{(conc.)}}{H_2SO_4} \longrightarrow -\overset{|}{C}-H,\; -\underset{|}{C}-OSO_3H \xrightarrow[\text{boil}]{H_2O} -\overset{|}{C}-H,\; -\underset{|}{C}-OH$$

alkene — alkyl hydrogen sulfate — alcohol (Markovnikov orientation)

Alkenes react and dissolve in concentrated sulfuric acid. Sulfuric acid protonates the alkene, and the resulting carbocation dissolves.

$$\text{>C=C<} + H_2SO_4 \rightleftharpoons -\overset{H}{\underset{|}{C}}-\overset{+}{C}< + HSO_4^-$$

alkene — carbocation (soluble) — bisulfate

This protonation takes place in *concentrated* sulfuric acid, where the only nucleophile available is the bisulfate ion, HSO_4^-. Bisulfate is a weak nucleophile, but the carbocation is a strong electrophile. Attack by bisulfate gives an alkyl hydrogen sulfate.

$$-\overset{H}{\underset{|}{\overset{|}{C}}}-C^+ \;+\; {}^-O-\overset{O}{\underset{O}{\overset{\|}{\underset{\|}{S}}}}-O-H \;\longrightarrow\; -\overset{H}{\underset{|}{\overset{|}{C}}}-\underset{|}{\overset{|}{C}}-O-\overset{O}{\underset{O}{\overset{\|}{\underset{\|}{S}}}}-O-H$$

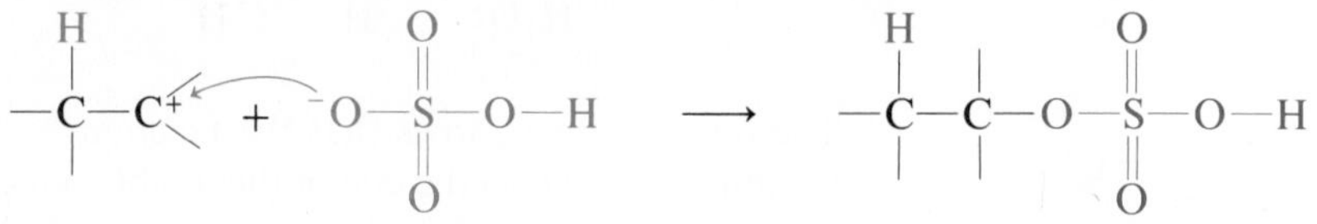

carbocation bisulfate an alkyl hydrogen sulfate

As an example, consider the addition of sulfuric acid to 2-methyl-2-butene. The proton adds (with Markovnikov orientation) in the first step, followed by attack of the bisulfate ion. The overall reaction is the electrophilic addition of sulfuric acid across the double bond, with Markovnikov orientation.

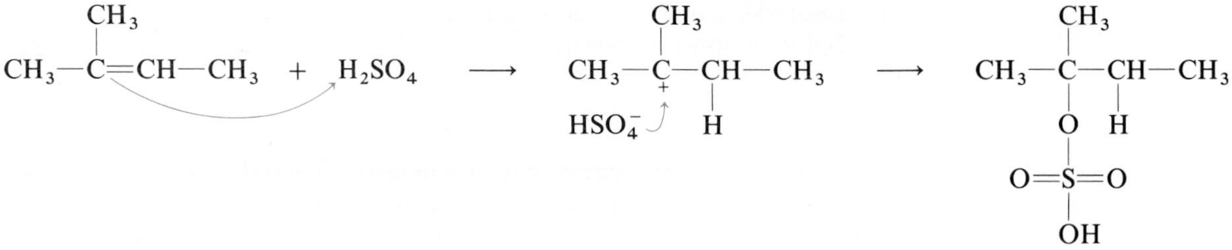

$$CH_3-\overset{CH_3}{\overset{|}{C}}=CH-CH_3 + H_2SO_4 \longrightarrow CH_3-\overset{CH_3}{\overset{|}{\underset{+}{C}}}-\underset{H}{\underset{|}{CH}}-CH_3 \;\; HSO_4^- \longrightarrow CH_3-\overset{CH_3}{\overset{|}{C}}-\underset{H}{\underset{|}{CH}}-CH_3 \text{ (C bonded to } O-SO_2-OH)$$

An alkyl hydrogen sulfate can be converted to an alcohol by boiling in water. This substitution is usually an S_N1 reaction, with the bisulfate ion serving as the leaving group. Ionization gives a carbocation that is quickly attacked by the solvent (water) to give the Markovnikov alcohol, the same product that would be formed by direct acid-catalyzed hydration.

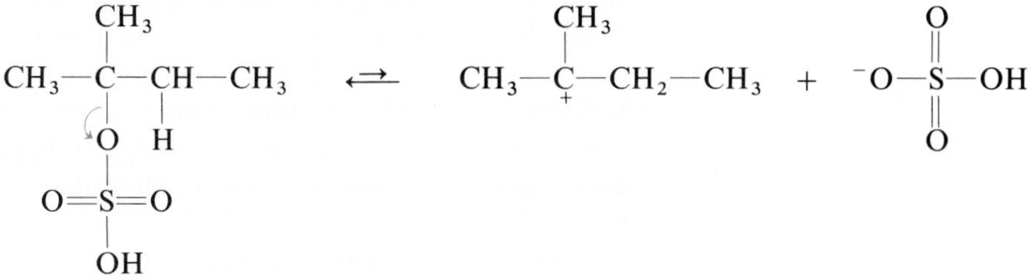

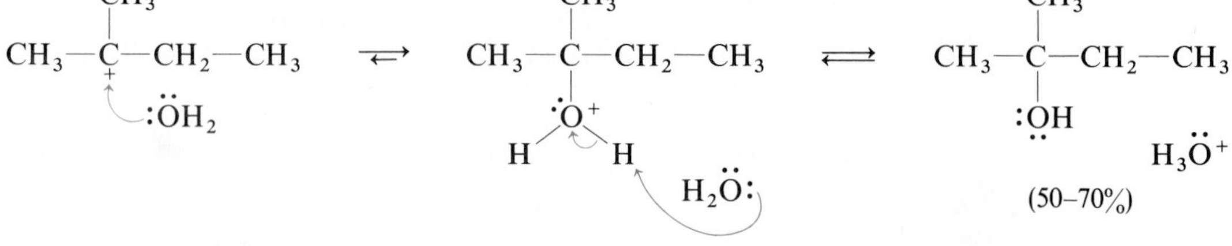

$$CH_3-\overset{CH_3}{\overset{|}{C}}(OSO_2OH)-\underset{H}{\underset{|}{CH}}-CH_3 \rightleftarrows CH_3-\overset{CH_3}{\overset{|}{\underset{+}{C}}}-CH_2-CH_3 + {}^-O-\overset{O}{\underset{O}{\overset{\|}{\underset{\|}{S}}}}-OH$$

$$CH_3-\overset{CH_3}{\overset{|}{\underset{+}{C}}}-CH_2-CH_3 \;\; :\ddot{O}H_2 \rightleftarrows CH_3-\overset{CH_3}{\overset{|}{C}}(\dot{O}^+H_2)-CH_2-CH_3 \;\; H_2\ddot{O}: \rightleftarrows CH_3-\overset{CH_3}{\overset{|}{C}}(:\ddot{O}H)-CH_2-CH_3 \;\; H_3\ddot{O}^+$$

(50–70%)

8-5B OXYMERCURATION-DEMERCURATION

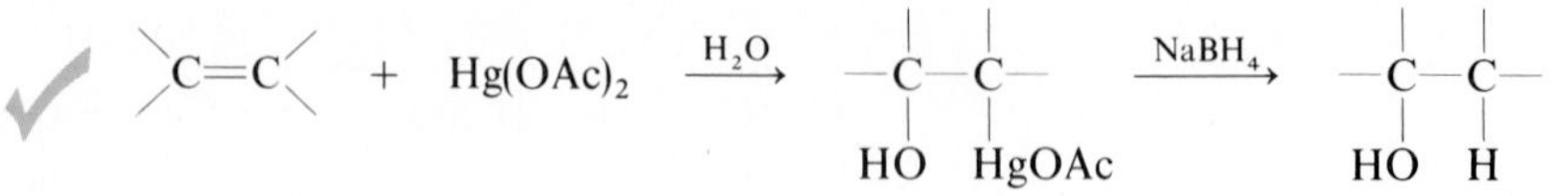

$$>C=C< + Hg(OAc)_2 \xrightarrow{H_2O} -\underset{HO}{\underset{|}{\overset{|}{C}}}-\underset{HgOAc}{\underset{|}{\overset{|}{C}}}- \xrightarrow{NaBH_4} -\underset{HO}{\underset{|}{\overset{|}{C}}}-\underset{H}{\underset{|}{\overset{|}{C}}}-$$

alkene alcohol (Markovnikov orientation)

Oxymercuration-demercuration is another method for converting alkenes to alcohols with Markovnikov orientation, but with the advantage that it does not involve a free carbocation and there is no opportunity for rearrangements. Carbocation rearrangements are common in both acid-catalyzed hydration and the formation of alkyl hydrogen sulfates.

The reagent for mercuration is mercuric acetate, $Hg(OCOCH_3)_2$, abbreviated $Hg(OAc)_2$. There are several theories as to how this reagent acts as an electrophile, but the simplest one is that mercuric acetate dissociates slightly to form a positively charged mercury species, $^+Hg(OAc)$.

$$CH_3-\overset{O}{\overset{\|}{C}}-O-Hg-O-\overset{O}{\overset{\|}{C}}-CH_3 \rightleftharpoons CH_3-\overset{O}{\overset{\|}{C}}-O-Hg^+ + CH_3-\overset{O}{\overset{\|}{C}}-O^-$$

$Hg(OAc)_2$ $\quad$ $^+Hg(OAc)$ $\quad$ ^-OAc

Oxymercuration involves an electrophilic attack on the double bond by the positively charged mercury species. The product is a *mercurinium ion*, an organometallic cation containing a three-membered ring.

$$>C=C< \; ^+Hg(OAc) \longrightarrow -\underset{|}{C}\overset{Hg^+(OAc)}{\frown}\underset{|}{C}-$$

mercurinium ion

Mercuration commonly takes place in a solution containing water and an organic solvent to dissolve the alkene. Attack on the mercurinium ion by water gives (after deprotonation) an organomercurial alcohol.

$$H_2\ddot{O}: + \text{mercurinium ion } (^+Hg(OAc)) \longrightarrow -\overset{|}{\underset{H-\overset{+}{O}:-H}{C}}-\overset{Hg(OAc)}{\overset{|}{\underset{|}{C}}}- \;(H_2\ddot{O}:) \xrightarrow{-H^+} -\overset{|}{\underset{:\ddot{O}H}{C}}-\overset{Hg(OAc)}{\overset{|}{\underset{|}{C}}}-$$

organomercurial alcohol

The second step is **demercuration,** in which sodium borohydride ($NaBH_4$, a reducing agent) replaces the mercuric acetate fragment with hydrogen. This step converts the organomercurial alcohol to the alcohol.

$$4\; -\overset{|}{\underset{HO}{C}}-\overset{Hg(OAc)}{\overset{|}{\underset{|}{C}}}- + NaBH_4 + 4\,^-OH \longrightarrow 4\; -\overset{|}{\underset{HO}{C}}-\overset{H}{\overset{|}{\underset{|}{C}}}- + NaB(OH)_4 + 4\,Hg\downarrow + 4\,^-OAc$$

organomercurial alcohol $\quad$ alcohol

Oxymercuration-demercuration of an unsymmetrical alkene generally gives Markovnikov orientation of addition, as shown by the oxymercuration of 2-methyl-2-butene. In this unsymmetrical case, the mercurinium ion has a considerable amount of positive charge on the more highly substituted carbon atom. Attack by water occurs on this more electrophilic carbon, giving Markovnikov orientation of addition. The electrophile, $^+Hg(OAc)$, remains bonded to the less highly substituted end of the double bond. Reduction of the organomercurial alcohol then gives the Markovnikov alcohol, 2-methyl-2-butanol.

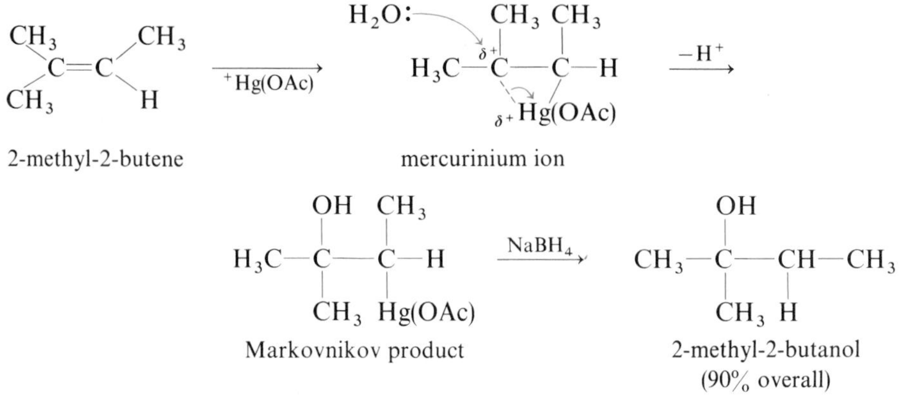

$$(CH_3)_2C{=}CH(CH_3) \xrightarrow{^{+}Hg(OAc)} \text{mercurinium ion} \xrightarrow{H_2\ddot{O}:,\ -H^+}$$

2-methyl-2-butene → mercurinium ion

$$H_3C{-}C(OH)(CH_3){-}CH(CH_3){-}Hg(OAc) \xrightarrow{NaBH_4} CH_3{-}C(OH)(CH_3){-}CH_2{-}CH_3$$

Markovnikov product → 2-methyl-2-butanol (90% overall)

The following reaction shows the use of oxymercuration-demercuration to convert cyclopentene to cyclopentanol. Notice that the attack by water on the mercurinium ion must come from the opposite side of the ring, resulting in addition of the hydroxyl group and the mercury atom to opposite sides of the ring to give the trans product. Such an addition to opposite faces of a double bond is called an **anti addition.** Similarly, addition of two groups to the same face of a double bond is called a **syn addition.**

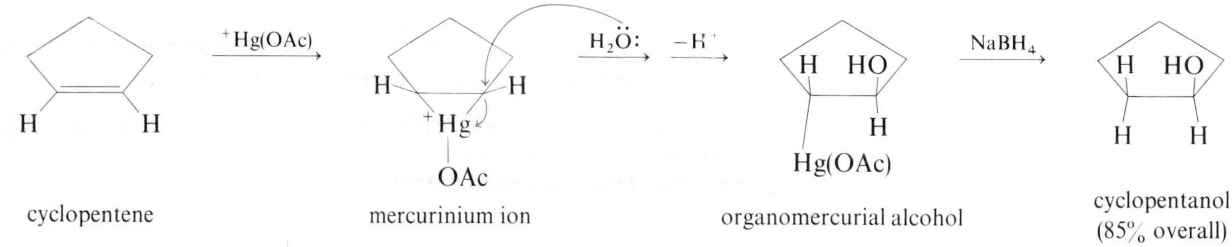

cyclopentene $\xrightarrow{^{+}Hg(OAc)}$ mercurinium ion $\xrightarrow{H_2\ddot{O}:}$ $\xrightarrow{-H^+}$ organomercurial alcohol $\xrightarrow{NaBH_4}$ cyclopentanol (85% overall)

Of the three methods we have seen for Markovnikov hydration of alkenes, oxymercuration-demercuration is most commonly used in the laboratory. It gives better yields than direct acid-catalyzed hydration, it avoids the possibility of rearrangements, and it does not involve such harsh conditions as the concentrated sulfuric acid required to make the alkyl hydrogen sulfate.

8-6 ALKOXYMERCURATION-DEMERCURATION

Alkoxymercuration-demercuration converts alkenes to ethers by adding an alcohol across the double bond of the alkene.

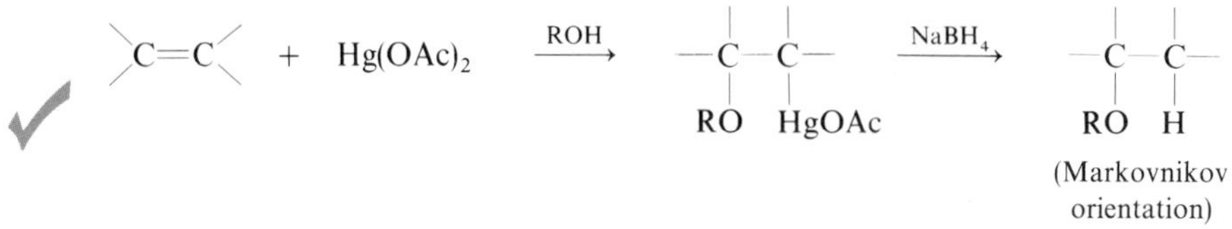

$$\rangle C{=}C\langle \;+\; Hg(OAc)_2 \xrightarrow{ROH} {-}C(OR){-}C(HgOAc){-} \xrightarrow{NaBH_4} {-}C(OR){-}C(H){-}$$

(Markovnikov orientation)

Mercuration involves formation of a mercurinium cation that is then attacked by the nucleophilic solvent. If the solvent is water, an —OH group is added to the cation. If the solvent is an alcohol (R—O—H), however, the product contains the alkoxyl (—O—R) group. Addition of an alkoxy group and a mercury species is called **alkoxymercuration.**

$$\text{R—}\ddot{\text{O}}\text{H} + \text{mercurinium ion } (\text{C—C bridged by }{}^{+}\text{Hg(OAc)}) \longrightarrow \text{R—O}^{+}\text{(H)—C—C—Hg(OAc)} \xrightarrow{\text{R—}\ddot{\text{O}}\text{—H}} \text{R—}\ddot{\text{O}}\text{—C—C—Hg(OAc)} + \text{R—O}^{+}\text{H}_2$$

When the organomercurial intermediate is reduced by sodium borohydride ($NaBH_4$), the Hg(OAc) group is replaced by hydrogen. The product is an ether:

$$\text{R—O—C—C—Hg(OAc)} \xrightarrow{NaBH_4} \text{R—O—C—C—H}$$

an ether

Because the mercurinium ion is attacked at the more highly substituted end of the double bond, this reaction gives Markovnikov orientation of addition. The Hg(OAc) group appears at the *less* highly substituted end of the double bond, and reduction gives the Markovnikov product with hydrogen at the less highly substituted end of the double bond.

SOLVED PROBLEM 8-2

Show the intermediates and products that result from alkoxymercuration-demercuration of 1-methylcyclopentene, using methanol as the alcohol in the alkoxymercuration step.

SOLUTION The mercuric acetate cation adds to cyclopentene to give the cyclic mercurinium ion. This ion has a considerable amount of positive charge on the more highly substituted tertiary carbon atom, and this carbon is attacked by methanol. Notice the anti orientation of the addition.

$$\text{1-methylcyclopentene} \xrightarrow{{}^{+}\text{Hg(OAc)}} \text{mercurinium ion } (\delta^+ \text{ at C–CH}_3;\ \text{attacked by } CH_3\text{—}\ddot{\text{O}}\text{—H}) \xrightarrow{-H^+} \text{trans intermediate}$$

1-methylcyclopentene — mercurinium ion — trans intermediate (product of anti addition)

Reduction of this intermediate gives the Markovnikov product, 1-methoxy-1-methylcyclopentane.

$$\text{intermediate (H, Hg(OAc), CH}_3\text{, OCH}_3) \xrightarrow{NaBH_4} \text{1-methoxy-1-methylcyclopentane (H, H, CH}_3\text{, OCH}_3)$$

intermediate — 1-methoxy-1-methylcyclopentane

PROBLEM 8-7

(a) Give a detailed mechanism for the following reaction.

$$CH_3-\underset{\displaystyle}{\overset{\displaystyle CH_3}{\overset{|}{C}}}=CH-CH_3 \xrightarrow{Hg(OAc)_2,\ CH_3CH_2OH} CH_3-\underset{\displaystyle CH_3CH_2O}{\underset{|}{\overset{\displaystyle CH_3}{\overset{|}{C}}}}-\underset{\displaystyle Hg(OAc)}{\underset{|}{CH}}-CH_3$$

(90%)

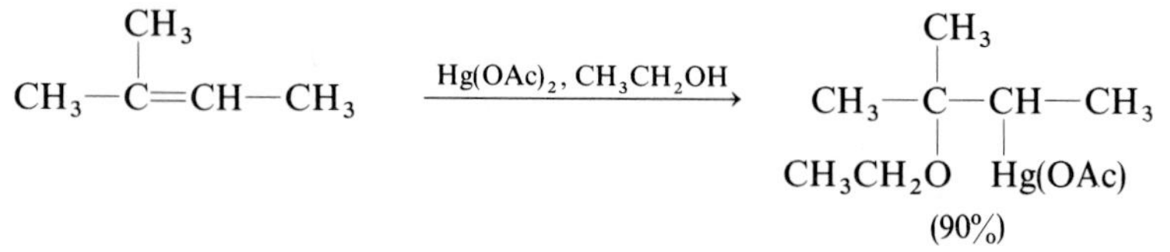

(b) Give the structure of the product that results when this intermediate is reduced by sodium borohydride.

PROBLEM 8-8

Predict the major products of the following reactions.

(a) 2-methylpropene + cold, concentrated H_2SO_4
(b) the product from part (a), boiled with water
(c) 1-methylcyclohexene + aqueous $Hg(OAc)_2$
(d) the product from part (c), treated with $NaBH_4$
(e) 4-chlorocycloheptene + $Hg(OAc)_2$ in CH_3OH
(f) the product from part (e), treated with $NaBH_4$

PROBLEM 8-9

Show how you would accomplish the following synthetic conversions.

(a) 1-butene → 2-butyl hydrogen sulfate
(b) 1-butene → 2-methyoxybutane
(c) 2-iodo-1-methylcyclopentane → 1-methylcyclopentanol

8-7 HYDROBORATION OF ALKENES

Hydroboration-oxidation converts alkenes to ethers by adding water across the double bond with anti-Markovnikov orientation.

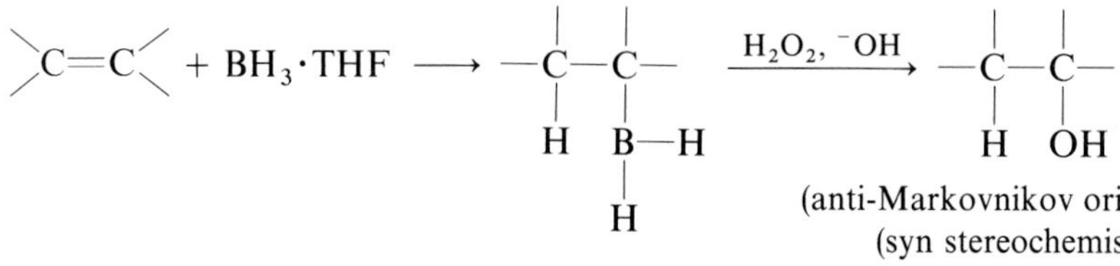

$$\gt C=C \lt + BH_3\cdot THF \longrightarrow -\underset{\displaystyle H}{\underset{|}{\overset{|}{C}}}-\underset{\displaystyle B-H \atop {| \atop H}}{\underset{|}{\overset{|}{C}}}- \xrightarrow{H_2O_2,\ ^{-}OH} -\underset{\displaystyle H}{\underset{|}{\overset{|}{C}}}-\underset{\displaystyle OH}{\underset{|}{\overset{|}{C}}}-$$

(anti-Markovnikov orientation)
(syn stereochemistry)

We have seen three methods for the hydration of an alkene with Markovnikov orientation. Suppose, however, that we want to convert an alkene to the anti-Markovnikov alcohol. For example, the following transformations cannot be accomplished using any of the hydration procedures covered thus far.

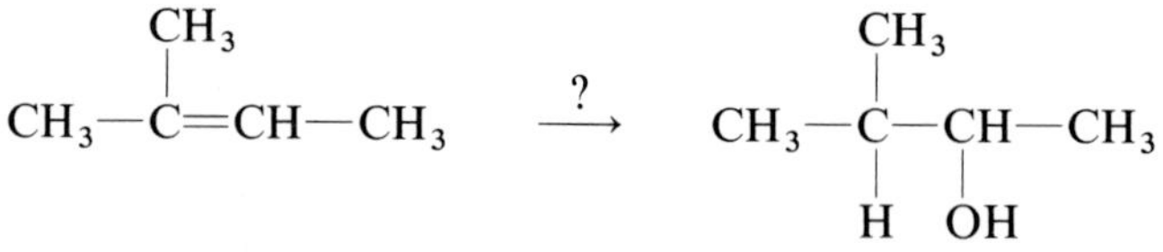

$$CH_3-\overset{\displaystyle CH_3}{\overset{|}{C}}=CH-CH_3 \xrightarrow{?} CH_3-\underset{\displaystyle H}{\underset{|}{\overset{\displaystyle CH_3}{\overset{|}{C}}}}-\underset{\displaystyle OH}{\underset{|}{CH}}-CH_3$$

2-methylbutene → 3-methyl-2-butanol

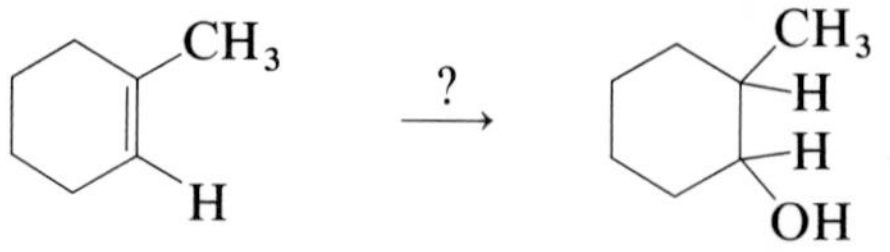

1-methylcyclohexene $\xrightarrow{?}$ 2-methylcyclohexanol

Such anti-Markovnikov hydrations were impossible until H. C. Brown, of Purdue University, discovered that diborane (B_2H_6) adds to alkenes with anti-Markovnikov orientation to form alkylboranes, which can be oxidized to give anti-Markovnikov alcohols. This discovery led to the development of a large field of borane chemistry, for which Brown received the Nobel Prize in 1979.

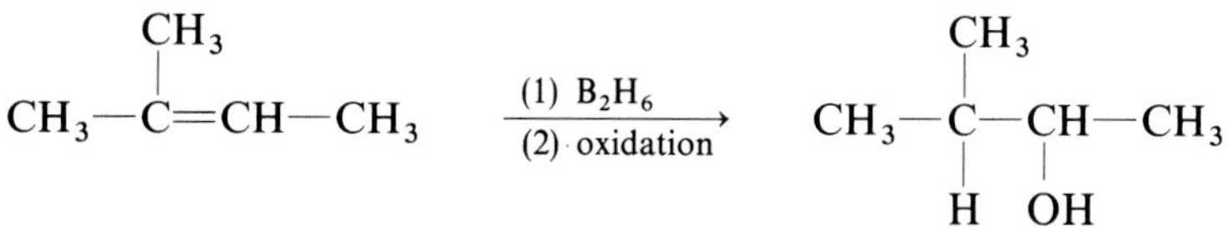

$$CH_3-\overset{\overset{\Large CH_3}{|}}{C}=CH-CH_3 \xrightarrow[\text{(2) oxidation}]{\text{(1) } B_2H_6} CH_3-\underset{\underset{\Large H}{|}}{\overset{\overset{\Large CH_3}{|}}{C}}-\underset{\underset{\Large OH}{|}}{CH}-CH_3$$

2-methyl-2-butene — 3-methyl-2-butanol (>90%)

Diborane (B_2H_6) is a dimer composed of two molecules of borane (BH_3). The bonding in diborane is unconventional, using three-centered (banana-shaped) bonds with protons in the middle of them. Diborane is in equilibrium with a small amount of borane, BH_3.

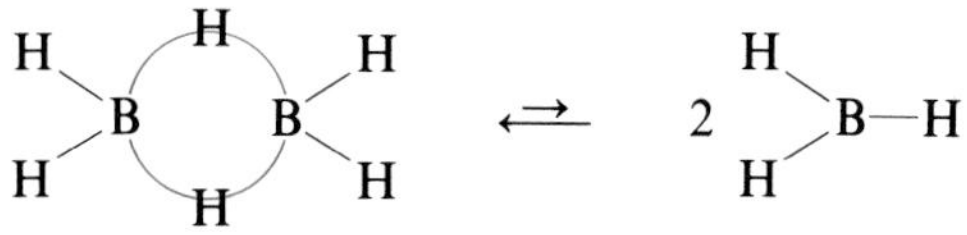

$$\text{diborane } (H_2B(\mu\text{-}H)_2BH_2) \rightleftharpoons 2\ H_2B-H \text{ (borane)}$$

diborane — borane

Diborane is an inconvenient reagent: a toxic, flammable, and explosive gas. It is more easily used as a complex with tetrahydrofuran (THF), a cyclic ether. This complex reacts like diborane, yet the solution is easily measured and transferred. The $BH_3 \cdot THF$ reagent is the form of borane that is commonly used in organic reactions.

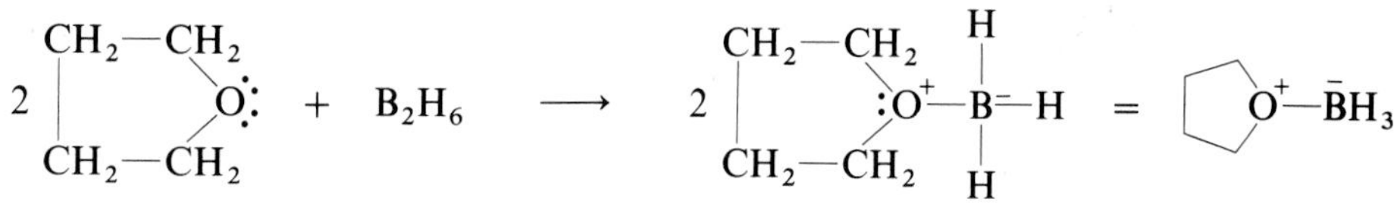

$$2\ (CH_2)_4\ddot{O}: + B_2H_6 \longrightarrow 2\ (CH_2)_4\overset{+}{O}-\overset{-}{B}H_3 = BH_3\cdot THF$$

tetrahydrofuran (THF) — diborane — borane-THF complex = $BH_3 \cdot THF$

8-7A MECHANISM OF HYDROBORATION

Borane is an electron-deficient compound. It has only six valence electrons, and the boron atom lacks an octet. This lack of an octet is the driving force for the unusual bonding structures ("banana" bonds, for example) found in boron compounds. As an electron-deficient compound, BH_3 is a strong electrophile, capable of adding to a double bond (Fig. 8-4). This **hydroboration** of the double bond is thought to occur in one step, with the boron atom adding to the less highly substituted end of the double bond.

In the transition state, the carbon atom at the end of the double bond where boron is *not* adding has a partial positive charge. This partial charge is more stable on the more highly substituted carbon atom of the double bond. The product shows boron bonded to the less highly substituted end of the double bond and hydrogen bonded to the more highly substituted end.

The second step of the hydroboration-oxidation process is the oxidation of the boron atom, removing it from carbon and replacing it with a hydroxyl (—OH)

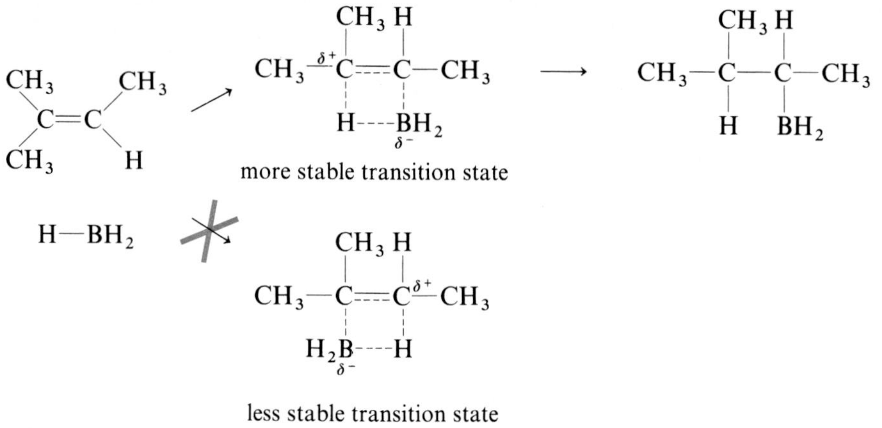

$$CH_3(CH_3)C{=}CH(CH_3) + H{-}BH_2 \rightarrow \text{more stable transition state: } CH_3{-}\overset{\delta+}{C}(CH_3){=}C(H){-}CH_3 \cdots H\cdots\overset{\delta-}{BH_2} \longrightarrow CH_3{-}C(CH_3)(H){-}C(H)(BH_2){-}CH_3$$

more stable transition state

$$\text{(not formed)}\quad CH_3{-}C(CH_3){=}\overset{\delta+}{C}(H){-}CH_3 \cdots \overset{\delta-}{H_2B}\cdots H$$

less stable transition state

FIGURE 8-4 Borane adds to the double bond in a single step, with boron adding to the less highly substituted carbon and hydrogen adding to the more highly substituted carbon. This orientation of addition places the partial positive charge in the transition state on the more highly substituted carbon atom.

group. Aqueous sodium hydroxide and hydrogen peroxide (HOOH or H_2O_2) are used for the oxidation.

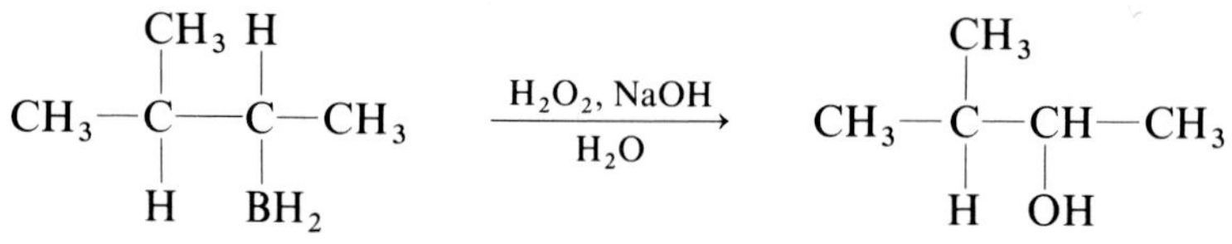

$$CH_3{-}C(CH_3)(H){-}C(H)(BH_2){-}CH_3 \xrightarrow[H_2O]{H_2O_2,\ NaOH} CH_3{-}C(CH_3)(H){-}CH(OH){-}CH_3$$

This hydration of an alkene by hydroboration-oxidation is another example of a reaction that does not follow the original statement of Markovnikov's rule (so the product is called "anti-Markovnikov") but still follows our extended understanding of the reasoning behind Markovnikov's rule. The electrophilic boron atom adds to the *less* highly substituted end of the double bond, placing the positive charge (and the hydrogen atom) at the more highly substituted end.

SOLVED PROBLEM 8-3

Show how you would convert 1-methylcyclopentanol to 2-methylcyclopentanol.

SOLUTION Working backward, use hydroboration-oxidation to form 2-methyl-cyclopentanol from 1-methylcyclopentene.

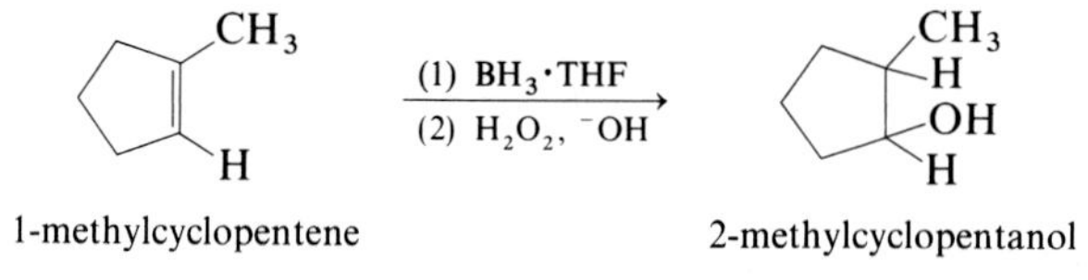

$$\text{1-methylcyclopentene} \xrightarrow[(2)\ H_2O_2,\ ^-OH]{(1)\ BH_3\cdot THF} \text{2-methylcyclopentanol}$$

1-methylcyclopentene 2-methylcyclopentanol

1-Methylcyclopentene is the most highly substituted alkene that results from dehydration of 1-methylcyclopentanol.

$$\text{1-methylcyclopentanol} \xrightarrow{H_2SO_4} \text{1-methylcyclopentene} + H_2O$$

1-methylcyclopentanol 1-methylcyclopentene

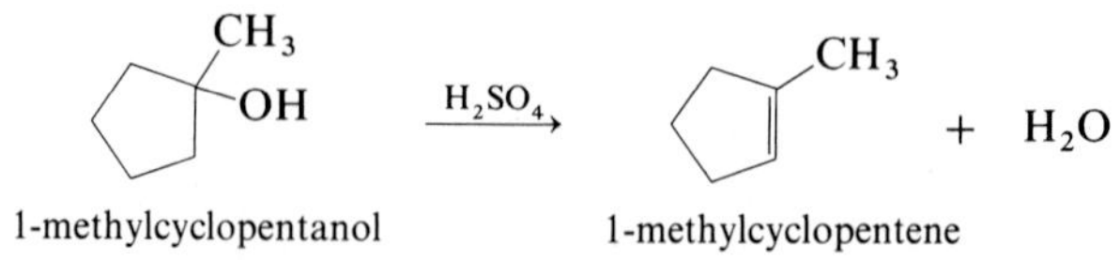

The 2-methylcyclopentanol that results from this synthesis is the pure trans isomer. This stereochemical result is discussed in Section 8-7C.

PROBLEM 8-10

Predict the major products of the following reactions.

(a) propene + $BH_3 \cdot THF$
(b) the product from part (a) + H_2O_2/OH^-
(c) 2-methyl-2-pentene + $BH_3 \cdot THF$
(d) the product from part (c) + H_2O_2/OH^-
(e) 1-methylcyclohexene + $BH_3 \cdot THF$
(f) the product from part (e) + H_2O_2/OH^-

PROBLEM 8-11

Show how you would accomplish the following synthetic conversions.

(a) 1-butene → 1-butanol
(b) 1-butene → 2-butanol
(c) 2-bromo-2,4-dimethylpentane → 2,4-dimethyl-3-pentanol

8-7B STOICHIOMETRY OF HYDROBORATION

For simplicity, we have neglected the fact that three moles of an alkene react with each mole of BH_3. Each of the bonds in BH_3 can add across the double bond on an alkene molecule. The first addition forms an alkylborane, the second, a dialkylborane, and the third a trialkylborane.

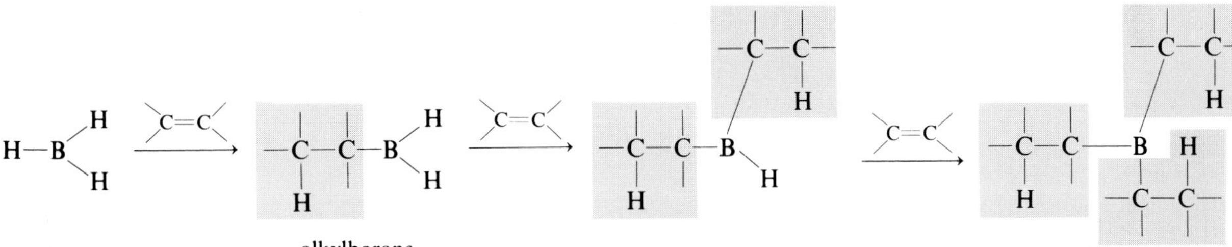

alkylborane

dialkylborane

trialkylborane

Summary

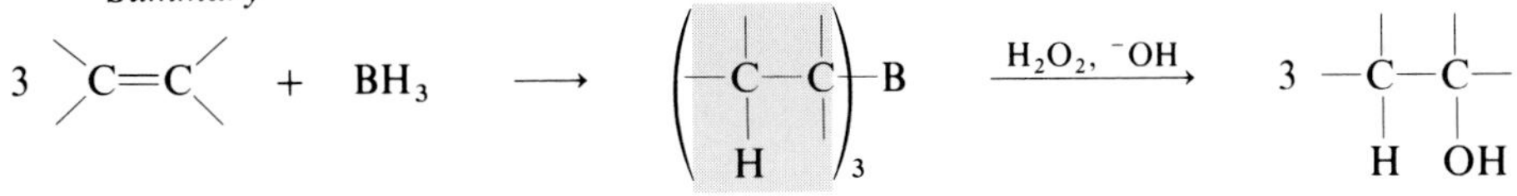

$$3\ \mathrm{C{=}C} + BH_3 \longrightarrow \left(\mathrm{H{-}C{-}C{-}}\right)_3\mathrm{B} \xrightarrow{H_2O_2,\ ^-OH} 3\ \mathrm{H{-}C{-}C{-}OH}$$

Trialkylboranes react exactly as we have discussed, and they oxidize to give anti-Markovnikov alcohols. Boranes are often drawn as the 1:1 monoalkylboranes to simplify their structure and emphasize the structure of the organic part of the molecule.

8-7C STEREOCHEMISTRY OF HYDROBORATION

The simultaneous addition of boron and hydrogen to the double bond, as shown in Figure 8-4, leads to a **syn** addition—the two atoms (boron and hydrogen) add across the double bond from the *same* side of the molecule. (If they had added from opposite sides of the molecule, the addition would be called **anti.**)

The stereochemistry of the hydroboration-oxidation of 1-methylcyclopentene is shown below. Boron and hydrogen add from the same face of the double bond (syn) to form a trialkylborane. Oxidation of the trialkylborane replaces the boron atom with a hydroxyl group in exactly the same stereochemical position. The product is *trans*-2-methylcyclopentanol. Because the product is chiral and the reactants are achiral, a racemic mixture is expected.

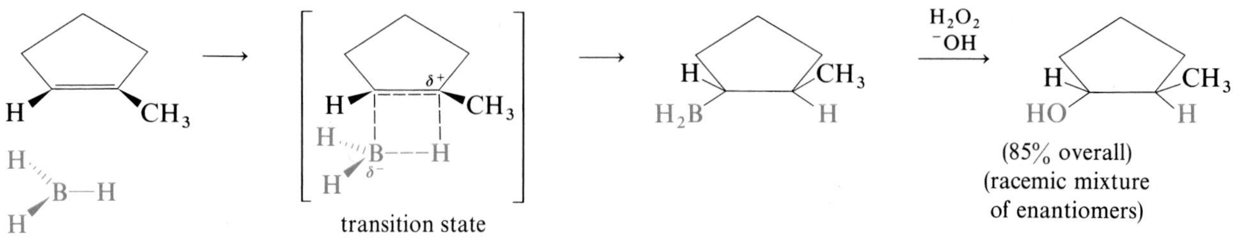

H CH$_3$ H B—H H → [H $\delta+$ CH$_3$ H B---H $\delta-$ H] transition state → H CH$_3$ H$_2$B H $\xrightarrow{H_2O_2,\ ^-OH}$ H CH$_3$ HO H

(85% overall)
(racemic mixture of enantiomers)

The hydroboration of alkenes is another example of a **stereospecific reaction,** where a particular stereoisomer of the starting compound reacts to give just one stereoisomer [or (±) pair] of the product. Problem 8-14 considers the different products formed by the hydroboration-oxidation of two acyclic diastereomers.

SOLVED PROBLEM 8-4

A norbornene molecule labeled with deuterium is subjected to hydroboration-oxidation. Give the structures of the intermediates and products.

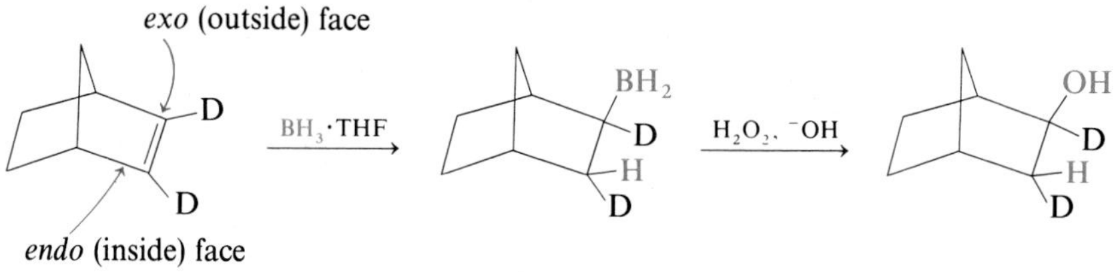

exo (outside) face

D D $\xrightarrow{BH_3\cdot THF}$ BH$_2$ D H D $\xrightarrow{H_2O_2,\ ^-OH}$ OH D H D

endo (inside) face

deuterium-labeled norbornene

alkylborane

alcohol (racemic mixture)

SOLUTION The syn addition of BH_3 across the double bond of norbornene takes place mostly from the more accessible outer side of the double bond. This is called the *exo* face of the norbornene molecule. Oxidation gives a product with both the hydrogen atom and the hydroxyl group added to exo positions. (The less accessible inner side of the double bond is called the *endo* face.)

PROBLEM 8-12

In the hydroboration of 1-methylcyclopentene shown above, the reagents are achiral and the products are chiral. The product is a racemic mixture of *trans*-2-methylcyclopentanol, but only one enantiomer is shown. Show how the other enantiomer is formed.

PROBLEM 8-13

Predict the major products of the following reactions. Include stereochemistry where applicable.

(a) 1-methylcycloheptene + $BH_3\cdot THF$, then H_2O_2, OH^-
(b) *trans*-4,4-dimethyl-2-pentene + $BH_3\cdot THF$, then H_2O_2, OH^-

(c) 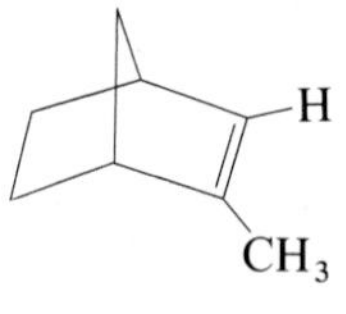 [norbornene with H and CH$_3$ on the double bond] + $BH_3\cdot THF$, then H_2O_2, OH^-

PROBLEM 8-14

(a) When (Z)-3-methyl-3-hexene undergoes hydroboration-oxidation, two isomeric products are formed. Give their structures, and label each chiral carbon atom as (R) or (S). What is the relationship between these isomers?

(b) Repeat part (a) for (E)-3-methyl-3-hexene. What is the relationship between the products formed from (Z)-3-methyl-3-hexene and those formed from (E)-3-methyl-3-hexene?

PROBLEM 8-15

Show how you would accomplish the following transformations.

(a) 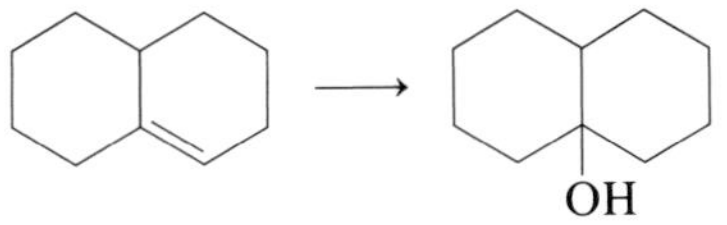 [bicyclic alkene] $\longrightarrow$ [bicyclic alcohol, OH] (b) 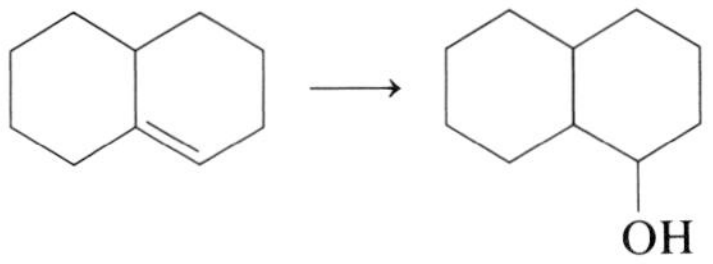 [bicyclic alkene] $\longrightarrow$ [bicyclic alcohol, OH]

(c) 1-methylcycloheptanol $\longrightarrow$ 2-methylcycloheptanol

PROBLEM 8-16

When HBr adds across the double bond of 1,2-dimethylcyclopentene, the product is a mixture of the cis and trans isomers. Show why this addition is not stereospecific.

8-8 CATALYTIC HYDROGENATION OF ALKENES

Although we have mentioned **catalytic hydrogenation** before, we now consider the mechanism and stereochemistry in more detail. The hydrogenation of an alkene is formally a reduction, involving the addition of H_2 across the double bond to give an alkane. A catalyst containing Pt, Pd, or Ni is usually required.

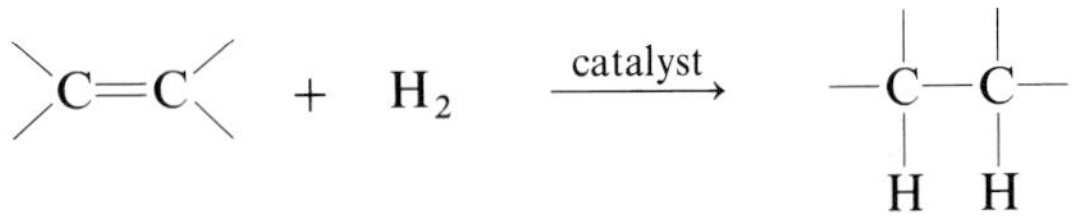

$$>C=C< \; + \; H_2 \xrightarrow{\text{catalyst}} -\underset{H}{\overset{|}{C}}-\underset{H}{\overset{|}{C}}-$$

Example

$$CH_3-CH=CH-CH_3 + H_2 \xrightarrow{Pt} CH_3-CH_2-CH_2-CH_3$$

Most alkenes are hydrogenated at room temperature using hydrogen gas at atmospheric pressure. The alkene is usually dissolved in an alcohol, an alkane, or acetic acid. A small amount of platinum, palladium, or nickel catalyst is added, and the container is shaken or stirred while the reaction proceeds. Hydrogenation actually takes place at the surface of the metal, where the liquid solution of the alkene comes into contact with hydrogen and the catalyst.

Hydrogen gas is adsorbed onto the surface of these metal catalysts, and the catalyst weakens the H—H bond. In fact, if H_2 and D_2 are mixed in the presence of a platinum catalyst, the two isotopes quickly scramble to produce a random mixture of HD, H_2, and D_2. No scrambling occurs in the absence of the catalyst. Hydrogenation is an example of **heterogeneous catalysis,** with the (solid) catalyst in a different phase from the reactant solution. In contrast, **homogeneous catalysis** involves the reactants and the catalyst in the same phase, as in the acid-catalyzed dehydration of an alcohol.

Because the two hydrogen atoms are added from a solid surface, they add with syn stereochemistry. For example, when 1,2-dideuteriocyclopentene is treated with hydrogen gas over a catalyst, the product initially isolated is the cis isomer resulting from syn addition.

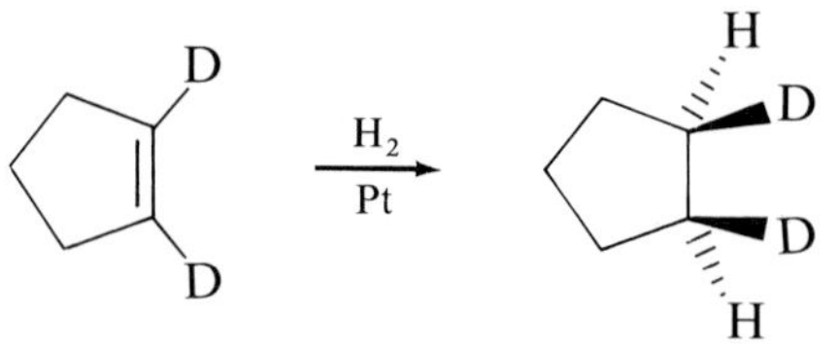

$$\text{1,2-dideuteriocyclopentene} \xrightarrow[\text{Pt}]{H_2} \text{cis-1,2-dideuteriocyclopentane}$$

One face of the alkene pi bond binds to the catalyst, which has hydrogen adsorbed on its surface. Hydrogen is inserted into the pi bond, and the product is freed from the catalyst. Both hydrogen atoms are added to the face of the double bond that is complexed with the catalyst.

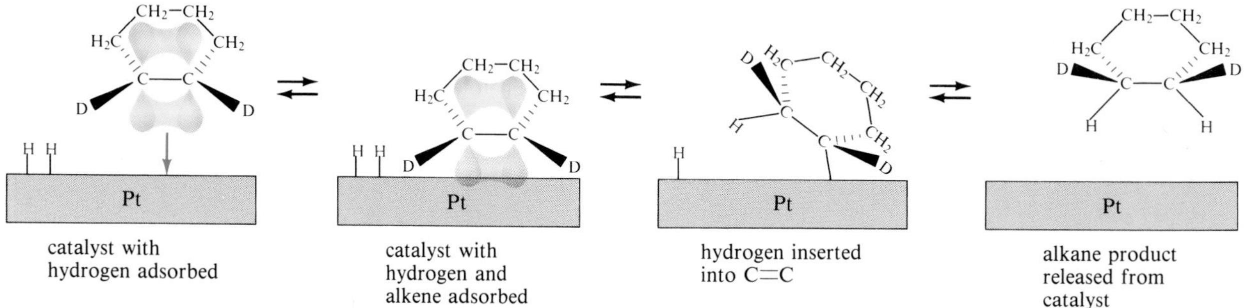

PROBLEM 8-17

Give the expected major product for each of the following reactions, including stereochemistry where applicable.

(a) 1-butene + H_2/Pt (b) *cis*-2-butene + H_2/Ni

(c) 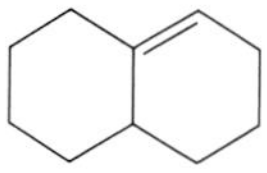 [bicyclic alkene] + H_2/Pt (d)  [diene] + excess H_2/Pt

PROBLEM 8-18

One of the principal components of lemon oil is *limonene*, $C_{10}H_{16}$. When limonene is treated with excess hydrogen and a platinum catalyst, the product is an alkane of formula $C_{10}H_{20}$. What can you conclude about the structure of limonene?

8-9 ADDITION OF CARBENES TO ALKENES

Methylene ($:CH_2$) is the simplest of the **carbenes:** uncharged, very reactive intermediates that have a carbon atom with two bonds and two nonbonding electrons. Like borane (BH_3), methylene is a potent electrophile because it has an unfilled octet. It adds to the electron-rich pi bond of an olefin to form a cyclopropane.

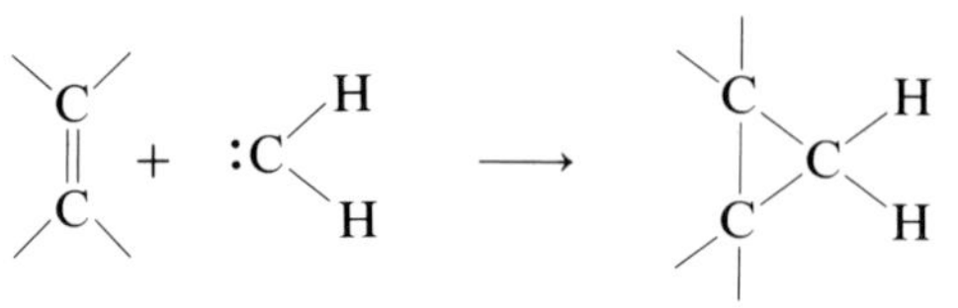

$$\text{C=C} + :CH_2 \longrightarrow \text{cyclopropane}$$

Heating or photolysis of diazomethane (CH_2N_2) gives nitrogen gas and methylene:

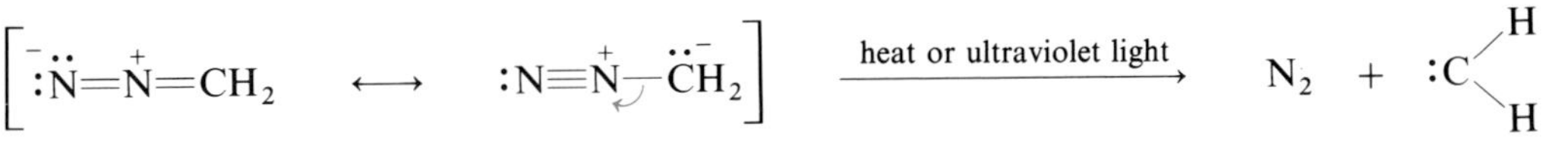

$$\left[:\overset{-}{\ddot{N}}=\overset{+}{N}=CH_2 \longleftrightarrow :N\equiv\overset{+}{N}-\overset{..-}{C}H_2 \right] \xrightarrow{\text{heat or ultraviolet light}} N_2 + :CH_2$$

diazomethane — methylene

There are two difficulties with using diazomethane to cyclopropanate double bonds. First, diazomethane is extremely toxic and explosive. A safer reagent would be more convenient for routine use. Second, methylene generated from diazomethane is so reactive that it inserts into C—H bonds as well as C=C bonds. In the reaction of propene with diazomethane-generated methylene, for example, several side products are formed.

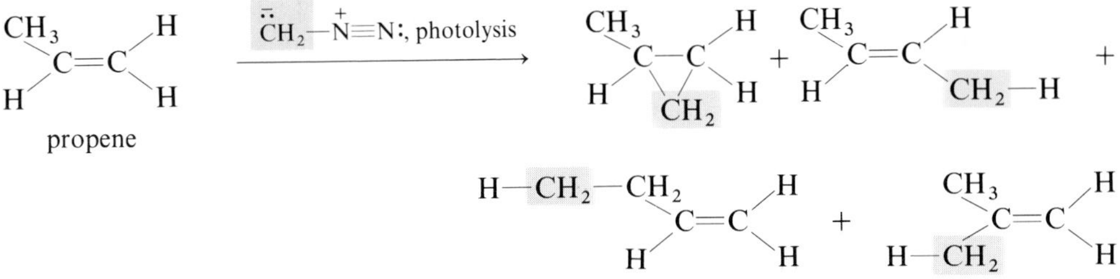

$$CH_3CH=CH_2 \xrightarrow{\ddot{C}H_2-\overset{+}{N}\equiv N:,\ \text{photolysis}} \text{methylcyclopropane} + CH_3CH=CH-CH_2-H + H-CH_2-CH_2-CH=CH_2 + H-CH_2-C(CH_3)=CH_2$$

propene

PROBLEM 8-19

Show how the insertion of methylene into a bond of cyclohexene can produce each of the following.

(a) 1-methylcyclohexene (b) 3-methylcyclohexene (c) norcarane,

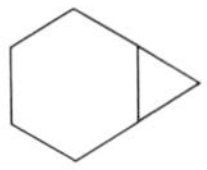

8-9A THE SIMMONS-SMITH REACTION

$$>C=C< + ICH_2ZnI \longrightarrow -\overset{|}{C}\underset{CH_2}{-}\overset{|}{C}- + ZnI_2$$

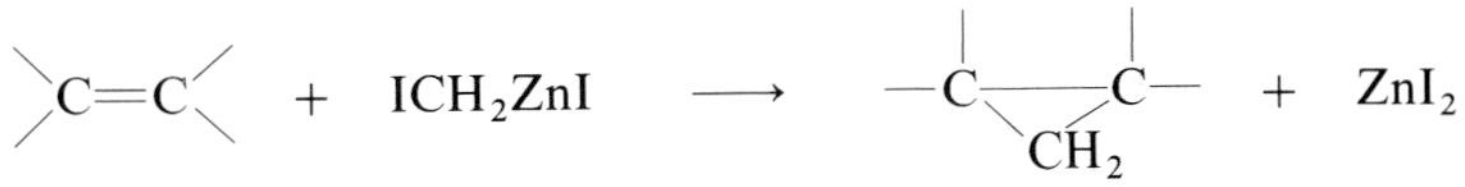

Two DuPont chemists discovered a reagent that converts alkenes to cyclopropanes in better yields than diazomethane, yet with fewer side reactions. The **Simmons-Smith reaction,** named in their honor, is one of the best ways of making cyclopropanes.

The Simmons-Smith reagent is made by adding methylene iodide to the "zinc-copper couple," zinc dust that has been activated with an impurity of copper. The reagent probably resembles iodomethyl zinc iodide, ICH_2ZnI. This kind of reagent is often called a **carbenoid** because it reacts very much like a carbene, although it does not contain an actual divalent carbon atom.

$$CH_2I_2 + Zn(Cu) \longrightarrow ICH_2ZnI$$

Simmons-Smith reagent (a carbenoid)

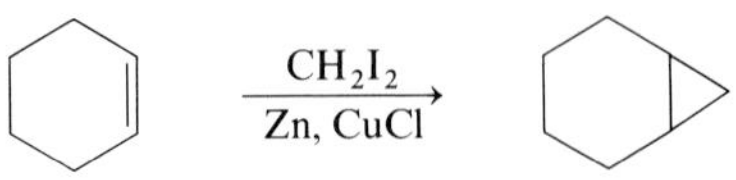

$$\text{cyclohexene} \xrightarrow[\text{Zn, CuCl}]{CH_2I_2} \text{norcarane}$$

(59%)

8-9B FORMATION OF CARBENES BY ALPHA-ELIMINATION

Carbenes are also formed by the reactions of halogenated compounds with bases. If a carbon atom has bonds to at least one hydrogen and to enough halogen atoms to make the hydrogen slightly acidic, it may be possible to form a carbene. For example, bromoform ($CHBr_3$) reacts with a 50 percent aqueous solution of potassium hydroxide to form dibromocarbene.

$$\underset{\text{bromoform}}{CHBr_3} + K^{+}\,{}^{-}OH \rightleftarrows {}^{-}\!:\!CBr_3\ K^{+} + H_2O$$

$${}^{-}\!:\!CBr_3 \rightleftarrows \underset{\text{dibromocarbene}}{:\!CBr_2} + Br^{-}$$

This dehydrohalogenation is called **alpha elimination** because the hydrogen atom and the halogen are lost from the same carbon atom. The more common dehydrohalogenations to form alkenes are called **beta eliminations** because the hydrogen and the halogen are lost from adjacent carbon atoms. Dibromocarbene formed from $CHBr_3$ can add to a double bond to form a dibromocyclopropane.

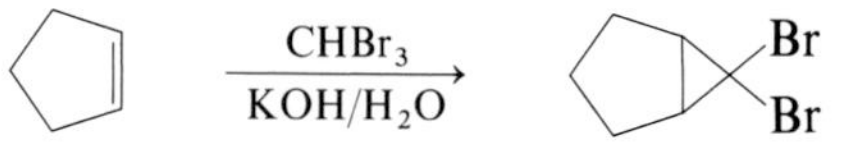

[cyclopentene] $\xrightarrow[KOH/H_2O]{CHBr_3}$ [6,6-dibromobicyclo[3.1.0]hexane: Br, Br]

The products of these cyclopropanations retain any cis or trans stereochemistry of the reactants.

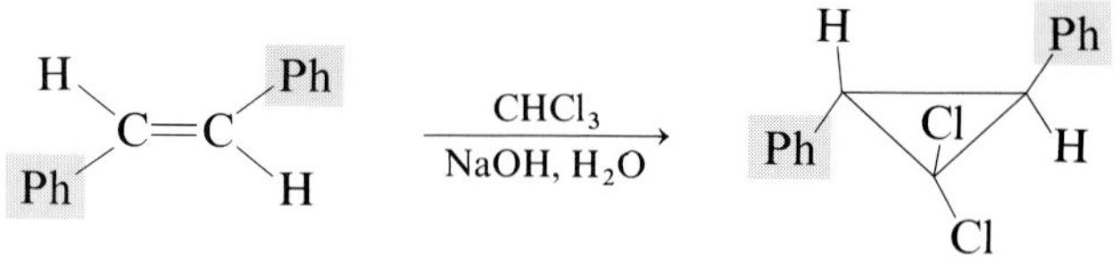

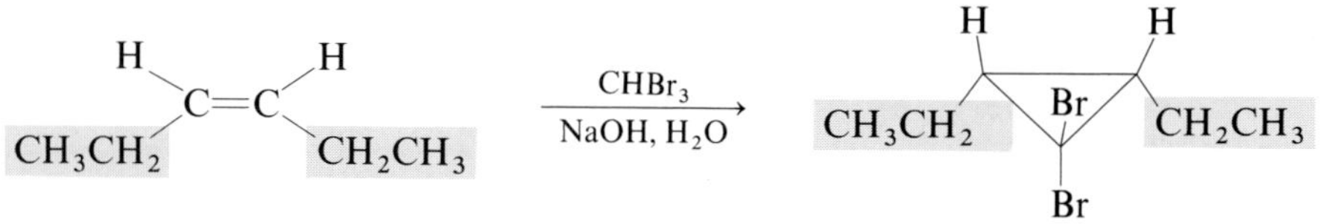

[trans-stilbene: H, Ph / C=C / Ph, H] $\xrightarrow[NaOH,\ H_2O]{CHCl_3}$ [cyclopropane: H, Ph, Ph, H, Cl, Cl]

[cis-3-hexene: H, H / C=C / CH_3CH_2, CH_2CH_3] $\xrightarrow[NaOH,\ H_2O]{CHBr_3}$ [cyclopropane: H, H, CH_3CH_2, CH_2CH_3, Br, Br]

PROBLEM 8-20

Predict the major products of the following reactions.

(a) cyclohexene + $CHCl_3$, 50% $NaOH/H_2O$

(b) 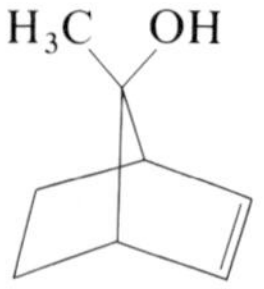 [bicyclic alkene with H_3C, OH] + CH_2I_2, Zn(Cu)

(c) 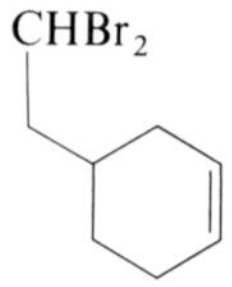 [$CHBr_2$-CH$_2$-cyclohexene] + 50% $NaOH/H_2O$

PROBLEM 8-21

Show how you would accomplish each of the following synthetic conversions.

(a) *trans*-2-butene ⟶ *trans*-1,2-dimethylcyclopropane

(b) cyclopentene ⟶ [bicyclo[3.1.0]hexane with H, Br]

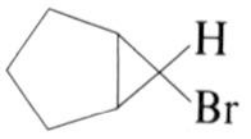

(c) cyclohexanol ⟶ [bicyclo[4.1.0]heptane with Cl, Cl]

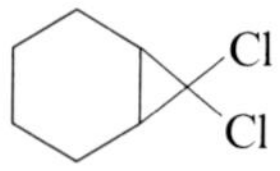

8-10 ADDITION OF HALOGENS TO ALKENES

Halogens add to alkenes to form vicinal dihalides.

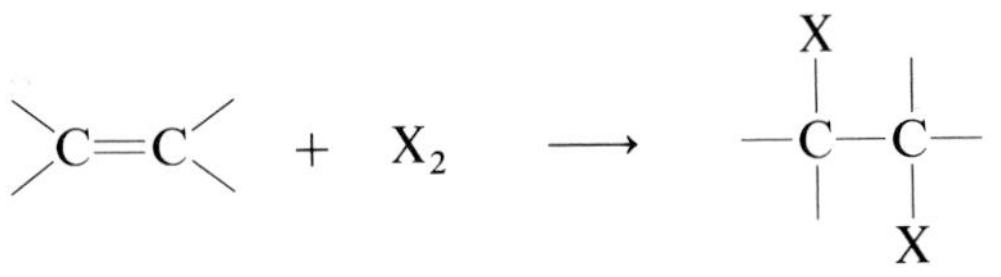

$$>C=C< \; + \; X_2 \longrightarrow -\overset{X}{\underset{|}{\overset{|}{C}}}-\underset{X}{\underset{|}{\overset{|}{C}}}-$$

($X_2 = Cl_2$, Br_2, sometimes I_2) usually anti addition

8-10A MECHANISM OF HALOGEN ADDITION

A halogen molecule (Br_2, Cl_2, or I_2) is electrophilic; a nucleophile can react, displacing a halide ion:

$$Nuc:^- + :\ddot{Br}-\ddot{Br}: \longrightarrow Nuc-\ddot{Br}: + :\ddot{Br}:^-$$

In this reaction the nucleophile attacks the electrophilic nucleus of one bromine atom. The other bromine atom serves as the leaving group, departing as bromide ion. Many reactions fit this general pattern, for example:

$$H\ddot{O}:^- + :\ddot{Br}-\ddot{Br}: \longrightarrow HO-\ddot{Br}: + :\ddot{Br}:^-$$

$$H_3N: + :\ddot{Cl}-\ddot{Cl}: \longrightarrow H_3\overset{+}{N}-\ddot{Cl}: + :\ddot{Cl}:^-$$

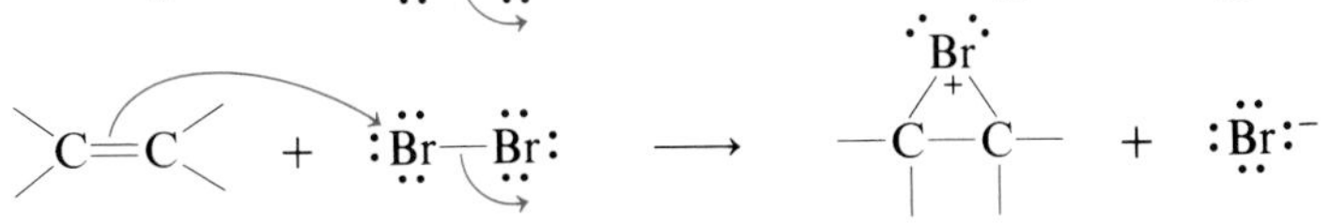

$$>C=C< + :\ddot{Br}-\ddot{Br}: \longrightarrow -C\overset{Br^+}{-}C- + :\ddot{Br}:^-$$

bromonium ion

In the last reaction the pi electrons of an alkene attack the bromine molecule, expelling a bromide ion. This reaction generates a cation called a **bromonium ion** which contains a three-membered ring with a positive charge on the bromine atom, similar in structure to the mercurinium ion discussed in Section 8-5B. The following reaction shows the formation and opening of a general halonium ion and the structures of a **chloronium ion,** a **bromonium ion,** and an **iodonium ion.**

Formation of halonium ion *Opening of halonium ion*

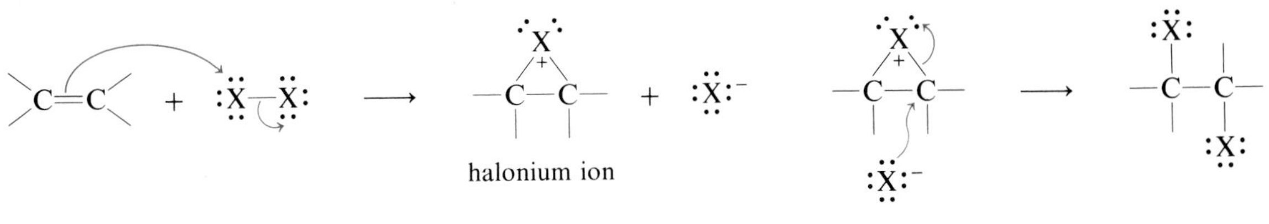

$$>C=C< + :\ddot{X}-\ddot{X}: \longrightarrow -C\overset{X^+}{-}C- + :\ddot{X}:^- \qquad -C\overset{X^+}{-}C- + :\ddot{X}:^- \longrightarrow -\overset{:\ddot{X}:}{\overset{|}{C}}-\underset{:\ddot{X}:}{\underset{|}{C}}-$$

halonium ion

Example

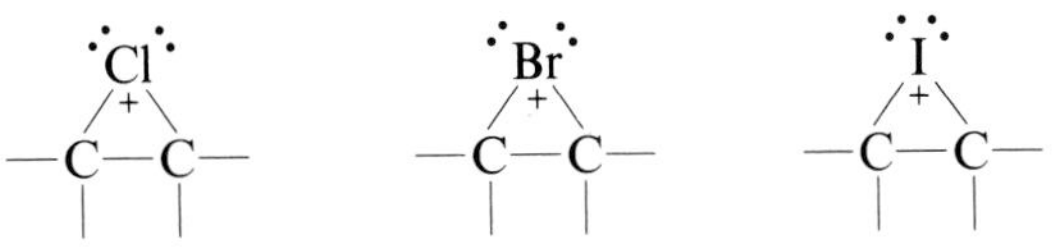

chloronium ion bromonium ion iodonium ion

In a halonium ion, unlike a normal carbocation, all atoms have filled octets. There is considerable ring strain, however, combined with a positive charge on the electronegative halogen atom, making the halonium ion strongly electrophilic. Attack by a nucleophile such as halide ion, as shown above, opens a halonium ion to give a stable product.

Chlorine and bromine commonly add to alkenes by the halonium ion mechanism. Iodination is used less frequently because the diiodide products decompose easily. The solvents used must be inert to the halogens; methylene chloride (CH_2Cl_2), chloroform ($CHCl_3$), and carbon tetrachloride (CCl_4) are the most frequent choices.

8-10B STEREOCHEMISTRY OF HALOGEN ADDITION

The addition of bromine to cyclopentene is a stereospecific anti addition.

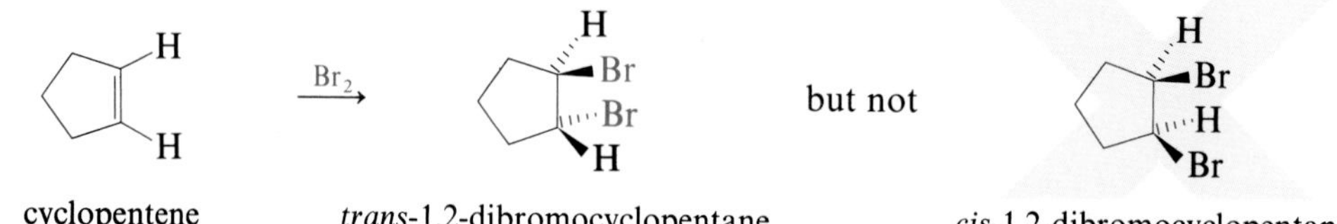

cyclopentene — *trans*-1,2-dibromocyclopentane (92%) — *cis*-1,2-dibromocyclopentane (not formed)

This anti stereochemistry is explained by the bromonium ion mechanism. When a nucleophile attacks a halonium ion, it must do so from the back side, in a manner similar to the S_N2 displacement. This back-side attack assures anti orientation of addition.

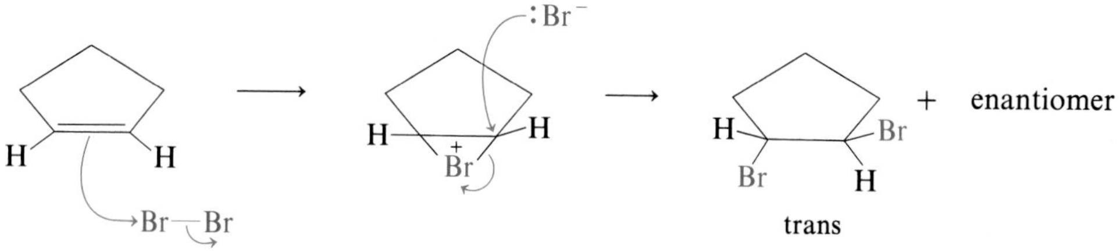

Halogen addition is another example of a stereospecific reaction, where a particular stereoisomer of the starting material gives only one stereoisomer of the product. Figure 8-5 shows additional examples of this anti addition of halogens to alkenes.

The addition of bromine is often used as a simple chemical test for the presence of olefinic double bonds. A solution of bromine in carbon tetrachloride is a clear, deep red color. When an alkene is added to this solution, the red bromine color disappears (we say it is "decolorized") and the solution becomes clear and colorless.

PROBLEM 8-22

Give mechanisms to account for the stereochemistry of the products observed from the addition of bromine to *cis*- and *trans*-2-butene (Figure 8-5). Why are two products formed from the cis isomer but only one from the trans? (Making models will be helpful.)

PROBLEM 8-23

Give mechanisms and predict the major products of the following reactions. Include stereochemistry where appropriate.

(a) cycloheptene + Br_2 in CH_2Cl_2

(b) (*E*)-3-decene + Br_2 in CCl_4

(c) 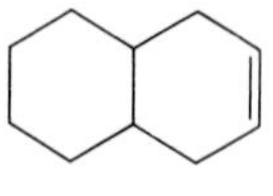 + Cl_2 in $CHCl_3$

(d) 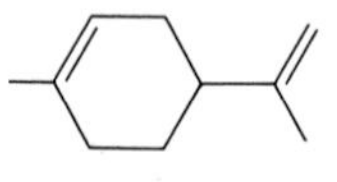 + 2 Cl_2 in CCl_4

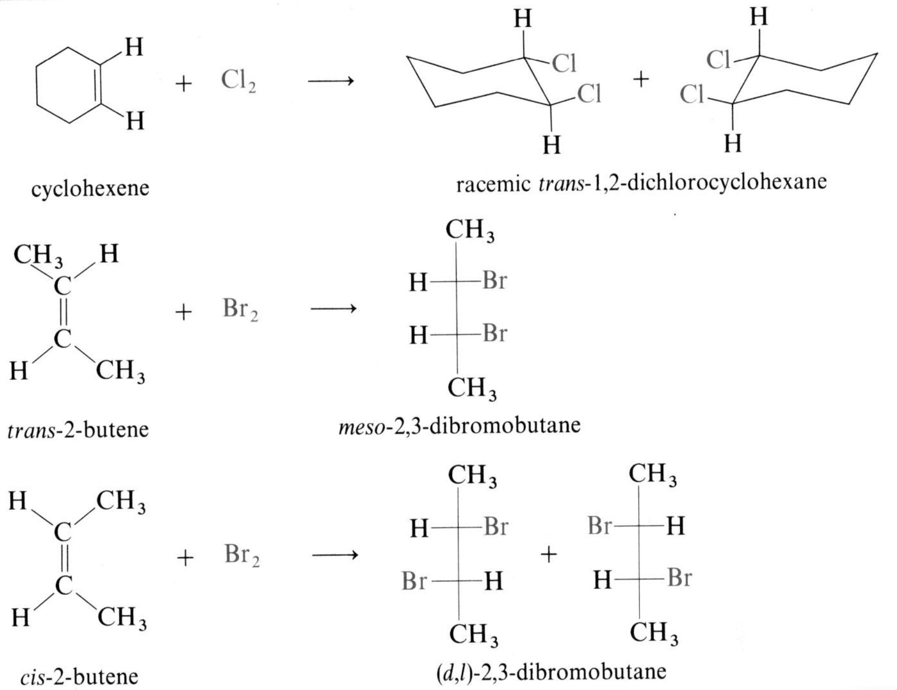

cyclohexene + Cl_2 ⟶ racemic *trans*-1,2-dichlorocyclohexane

trans-2-butene + Br_2 ⟶ *meso*-2,3-dibromobutane

cis-2-butene + Br_2 ⟶ (*d,l*)-2,3-dibromobutane

FIGURE 8-5 The stereospecific addition of halogens to alkenes gives products of anti addition to the double bond.

8-11 FORMATION OF HALOHYDRINS

In the presence of water, halogens add to alkenes to form halohydrins.

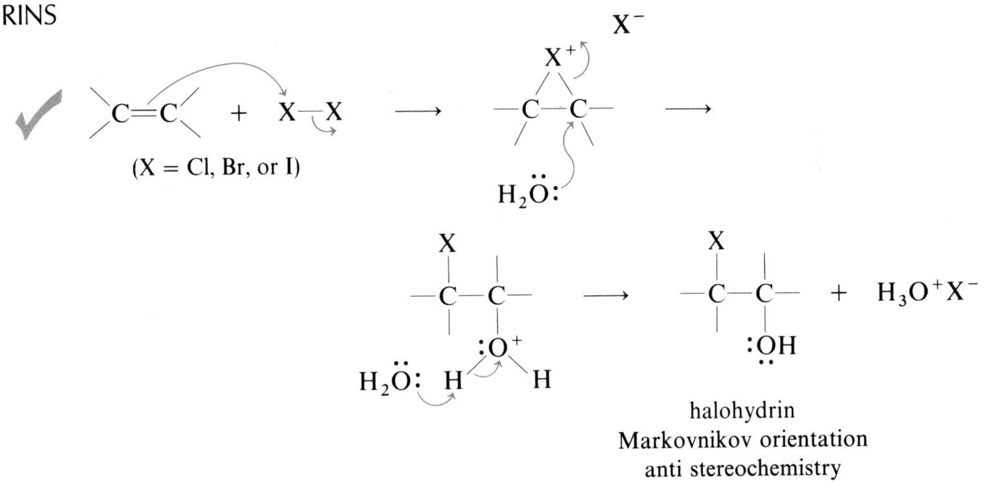

$$\text{>C=C<} + X\text{—}X \longrightarrow \text{halonium ion } (X^+) + X^- \xrightarrow{H_2\ddot{O}:} \text{—C(X)—C(}^+OH_2\text{)—} \longrightarrow \text{—C(X)—C(:}\ddot{O}H\text{)—} + H_3O^+X^-$$

(X = Cl, Br, or I)

halohydrin
Markovnikov orientation
anti stereochemistry

We have assumed that halogenation takes place with no solvent or with an inert solvent such as carbon tetrachloride (CCl_4) or chloroform ($CHCl_3$). Under these conditions, only the halide ion is available as a nucleophile to attack the halonium ion.

When an alkene reacts with a halogen in the presence of a nucleophilic solvent such as water, a solvent molecule is the most likely nucleophile to attack the halonium ion. If a water molecule attacks the halonium ion, the final product has a halogen on one carbon atom and a hydroxyl group on the adjacent carbon. Such a compound is called a **halohydrin:** a *chlorohydrin,* a *bromohydrin,* or an *iodohydrin,* depending on the halogen atom.

Stereochemistry of halohydrin formation Because the mechanism involves a halonium ion, the orientation of addition is anti as in halogenation. For example, the addition of bromine water to cyclopentene gives *trans*-2-bromocyclopentanol, the product of anti addition across the double bond.

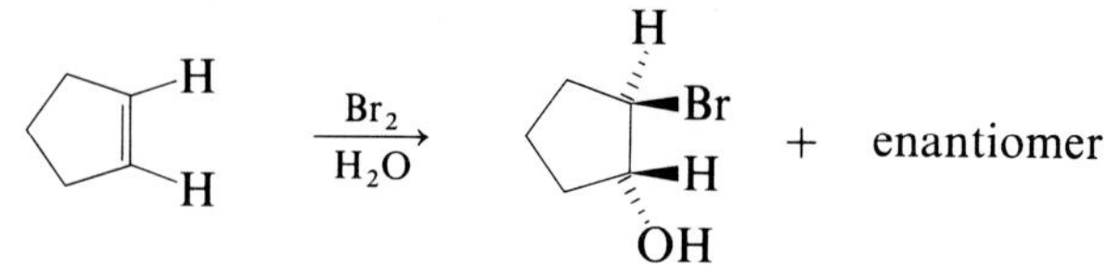

$$\text{cyclopentene} \xrightarrow[H_2O]{Br_2} \textit{trans}\text{-2-bromocyclopentanol (cyclopentene bromohydrin)} + \text{enantiomer}$$

PROBLEM 8-24

Give a detailed mechanism for the addition of bromine water to cyclopentene, being careful to show why the trans product results and how equal amounts of two enantiomers are formed.

Orientation of halohydrin formation Even though a halonium ion is involved, the extended version of Markovnikov's rule applies to halohydrin formation. When propene reacts with chlorine water, the major product has the electrophile (the chlorine atom) bonded to the less highly substituted carbon of the double bond. The nucleophile (the hydroxyl group) is bonded to the more highly substituted carbon.

$$H_2C{=}CH{-}CH_3 + Cl_2 + H_2O \longrightarrow H_2C(Cl){-}CH(OH){-}CH_3 + HCl$$

The Markovnikov orientation observed in halohydrin formation is explained by the structure of the halonium ion intermediate. The two carbon atoms bonded to the halogen have partial positive charges, with a larger charge (and a weaker bond to the halogen) on the more highly substituted carbon atom (Fig. 8-6). The nucleophile (water) attacks this more highly substituted, more electrophilic carbon atom. The result is both anti stereochemistry and Markovnikov orientation. This halonium ion mechanism can be used to explain and predict a wide variety of reactions in both nucleophilic and nonnucleophilic solvents. The halonium ion mechanism is similar to the mercurinium ion mechanism for the oxymercuration of an alkene with Markovnikov orientation (Section 8-5B).

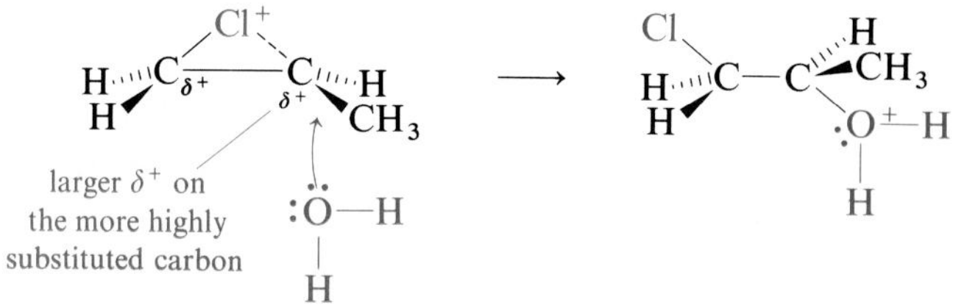

FIGURE 8-6 The more highly substituted carbon of the chloronium ion bears more positive charge than the less substituted carbon. Attack by water occurs on the more highly substituted carbon to give the Markovnikov product.

SOLVED PROBLEM 8-5

When cyclohexene is treated with bromine in saturated aqueous sodium chloride, a mixture of *trans*-2-bromocyclohexanol and *trans*-1-bromo-2-chlorocyclohexane results. Give a mechanism to account for these two products.

SOLUTION Cyclohexene reacts with bromine to give a bromonium ion. The most abundant nucleophiles in saturated aqueous sodium chloride solution are water and chloride ions. Attack by water gives the bromohydrin, and attack by chloride gives the dihalide. Either of these attacks gives anti stereochemistry.

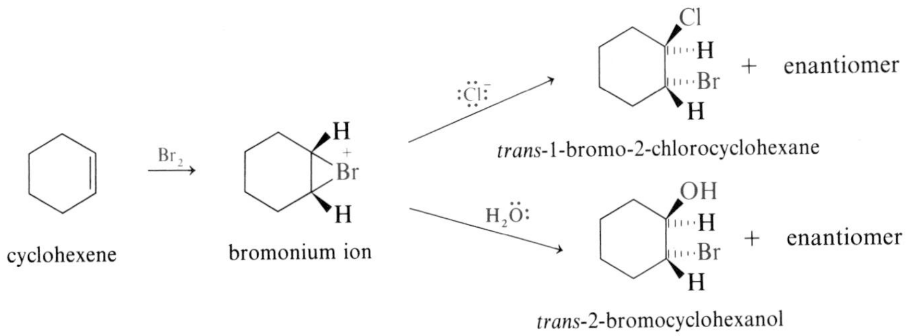

cyclohexene $\xrightarrow{Br_2}$ bromonium ion $\xrightarrow{:\ddot{Cl}:^-}$ *trans*-1-bromo-2-chlorocyclohexane + enantiomer

bromonium ion $\xrightarrow{H_2\ddot{O}:}$ *trans*-2-bromocyclohexanol + enantiomer

PROBLEM 8-25

Predict the major product(s) for each of the following reactions. Include stereochemistry where appropriate.

(a) 1-methylcyclopentene + Cl_2/H_2O
(b) 2-methyl-2-butene + Br_2/H_2O
(c) *cis*-2-butene + Cl_2/H_2O
(d) *trans*-2-butene + Cl_2/H_2O
(e) 1-methylcyclopentene + Br_2 in saturated aqueous NaCl

PROBLEM 8-26

Show how you would accomplish the following synthetic conversions.

(a) 3-methyl-2-pentene → 2-chloro-3-methyl-3-pentanol
(b) chlorocyclohexane → *trans*-2-chlorocyclohexanol
(c) 1-methylcyclopentanol → 2-chloro-1-methylcyclopentanol

8-12 EPOXIDATION OF ALKENES

Halogens are oxidizing agents, and the addition of a halogen molecule across a double bond is an oxidation. When we speak of the oxidation of alkenes, however, we usually mean the reactions of alkenes that form carbon-oxygen bonds. These reactions are particularly important because many of the common functional groups contain oxygen, and alkene oxidations are some of the best methods for introducing oxygen into organic molecules. We will consider methods for epoxidation, hydroxylation, and oxidative cleavage of the double bonds of alkenes.

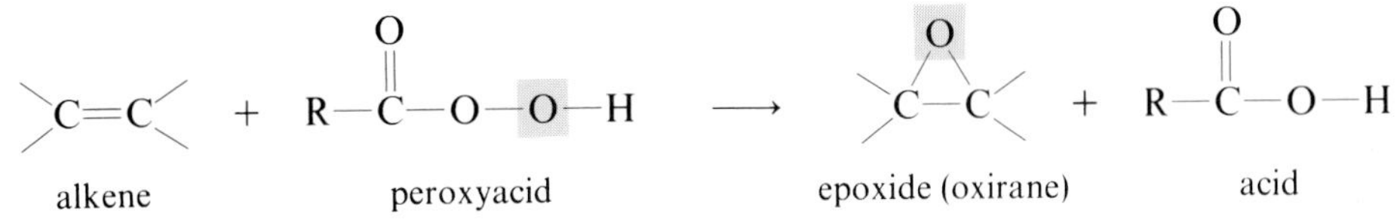

$$\text{C=C (alkene)} + \text{R–C(=O)–O–O–H (peroxyacid)} \longrightarrow \text{C–C with bridging O (epoxide (oxirane))} + \text{R–C(=O)–O–H (acid)}$$

An **epoxide** is a three-membered cyclic ether, sometimes called an **oxirane.** Epoxides are valuable synthetic intermediates, used to convert alkenes to a variety of other functional groups. An alkene is converted to an epoxide by a **peroxyacid,** a carboxylic acid that has an extra oxygen atom in a —O—O—(peroxy) linkage. The epoxidation of an alkene is clearly an oxidation, since an oxygen atom is added. Some common peroxyacids (sometimes called **peracids**) and their corresponding carboxylic acids are shown below.

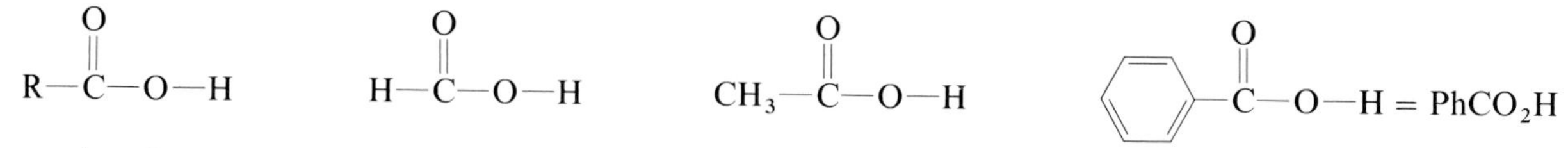

$$\mathrm{R{-}\overset{\overset{\large O}{\|}}{C}{-}O{-}H} \qquad \mathrm{H{-}\overset{\overset{\large O}{\|}}{C}{-}O{-}H} \qquad \mathrm{CH_3{-}\overset{\overset{\large O}{\|}}{C}{-}O{-}H} \qquad \mathrm{C_6H_5{-}\overset{\overset{\large O}{\|}}{C}{-}O{-}H = PhCO_2H}$$

a carboxylic acid — formic acid — acetic acid — benzoic acid

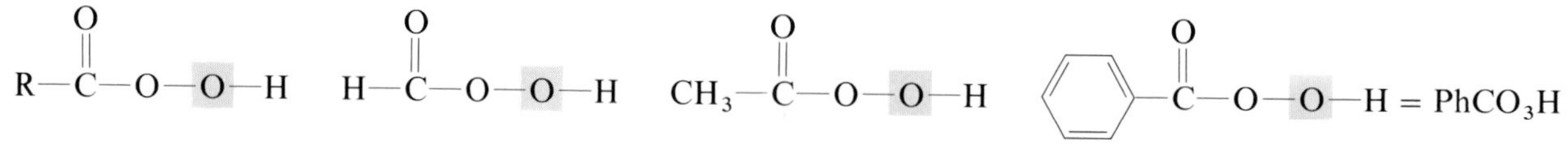

$$\mathrm{R{-}\overset{\overset{\large O}{\|}}{C}{-}O{-}O{-}H} \qquad \mathrm{H{-}\overset{\overset{\large O}{\|}}{C}{-}O{-}O{-}H} \qquad \mathrm{CH_3{-}\overset{\overset{\large O}{\|}}{C}{-}O{-}O{-}H} \qquad \mathrm{C_6H_5{-}\overset{\overset{\large O}{\|}}{C}{-}O{-}O{-}H = PhCO_3H}$$

a peroxyacid — peroxyformic acid — peroxyacetic acid — peroxybenzoic acid

A peroxyacid epoxidizes an alkene by a concerted electrophilic reaction where several bonds are broken and several are formed at the same time. Starting with the alkene and the peroxyacid, a one-step reaction gives the epoxide and the acid directly, without any intermediates.

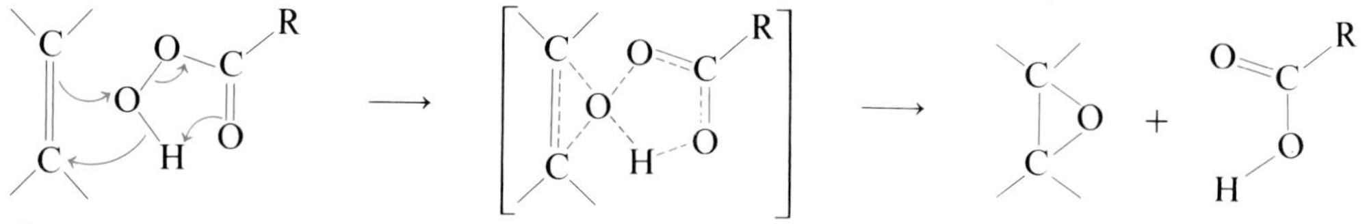

alkene + peroxyacid → [transition state] → epoxide + acid

Because the epoxidation takes place in one step, there is no opportunity for the alkene molecule to rotate and change its cis or trans geometry. Whatever stereochemistry is present in the alkene is retained in the epoxide. Peroxybenzoic acid (Ph—CO_3H) is a common epoxidizing reagent in peroxyacid epoxidations.

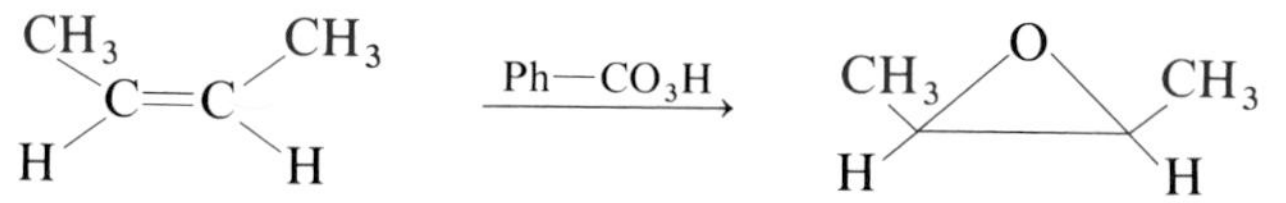

cis $CH_3CH{=}CHCH_3$ $\xrightarrow{\text{Ph—CO}_3\text{H}}$ cis epoxide

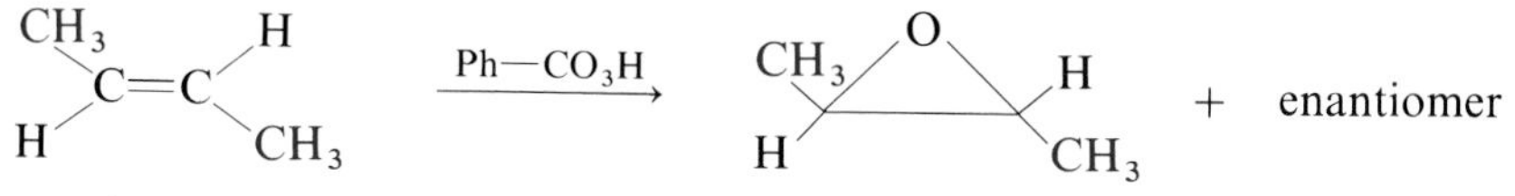

trans $CH_3CH{=}CHCH_3$ $\xrightarrow{\text{Ph—CO}_3\text{H}}$ trans epoxide + enantiomer

PROBLEM 8-27

Predict the products, including stereochemistry where appropriate, for the peroxybenzoic acid epoxidations of the following alkenes.

(a) *cis*-2-hexene (b) *trans*-2-hexene
(c) *cis*-cyclodecene (d) *trans*-cyclodecene

8-13 ACID-CATALYZED OPENING OF EPOXIDES

Most epoxides are easily isolated as stable products if the solution is not too acidic. Any moderately strong acid protonates the epoxide, however. Water can attack the protonated epoxide, opening the ring and forming a 1,2-diol, commonly called a **glycol.** Following is the mechanism of the acid-catalyzed opening of an epoxide to give a glycol.

epoxide

protonated epoxide

a glycol
(anti orientation)

Since glycol formation involves a back-side attack on a protonated epoxide, this reaction leads to anti orientation of the hydroxyl groups on the double bond. For example, when 1,2-epoxycyclopentane ("cyclopentene oxide") is treated with dilute mineral acid, the product is pure *trans*-1,2-cyclopentanediol.

cyclopentene oxide

trans-1,2-cyclopentanediol
(racemic)

PROBLEM 8-28

(a) Give a detailed mechanism for the conversion of *cis*-3-hexene to the epoxide (3,4-epoxyhexane) and the ring-opening reaction to give the glycol, 3,4-hexanediol. In your mechanism, pay particular attention to the stereochemistry of the intermediates and products.
(b) Repeat part (a) for *trans*-3-hexene. Compare the products obtained from *cis*- and *trans*-3-hexene.

Reagents can be chosen to give either the epoxide or the glycol. Peroxyacetic acid and peroxyformic acid are strongly acidic, and they dissolve in water. When an aqueous solution of peroxyacetic acid or peroxyformic acid is used as the epoxidizing agent, the epoxide is protonated and opens to the glycol. Peroxybenzoic acid is a weaker acid, commonly used in nonnucleophilic solvents such as carbon tetrachloride. Peroxybenzoic acid generally gives good yields of the epoxides. Figure 8-7 compares the use of strong and weak peroxyacids for the oxidation of alkenes.

PROBLEM 8-29

Because of its desirable solubility properties, *meta*-chloroperoxybenzoic acid (MCPBA) is often used in peroxyacid epoxidations. Give a mechanism for the reaction of *trans*-2-methyl-3-heptene with MCPBA, and predict the structure of the product.

MCPBA

PROBLEM 8-30

Predict the major products of the following reactions.

(a) *cis*-2-butene + peroxybenzoic acid ($PhCO_3H$) in chloroform
(b) 1-methylcyclooctene + peroxyformic acid (HCO_3H) in water
(c) *trans*-cyclodecene + MCPBA in methylene chloride
(d) *trans*-2-butene + peroxyacetic acid (CH_3CO_3H) in water

PROBLEM 8-31

When 1,2-epoxycyclohexane (cyclohexene oxide) is treated with anhydrous HCl in methanol, the principal product is *trans*-2-methoxycyclohexanol. Give a detailed mechanism to account for the formation of this product.

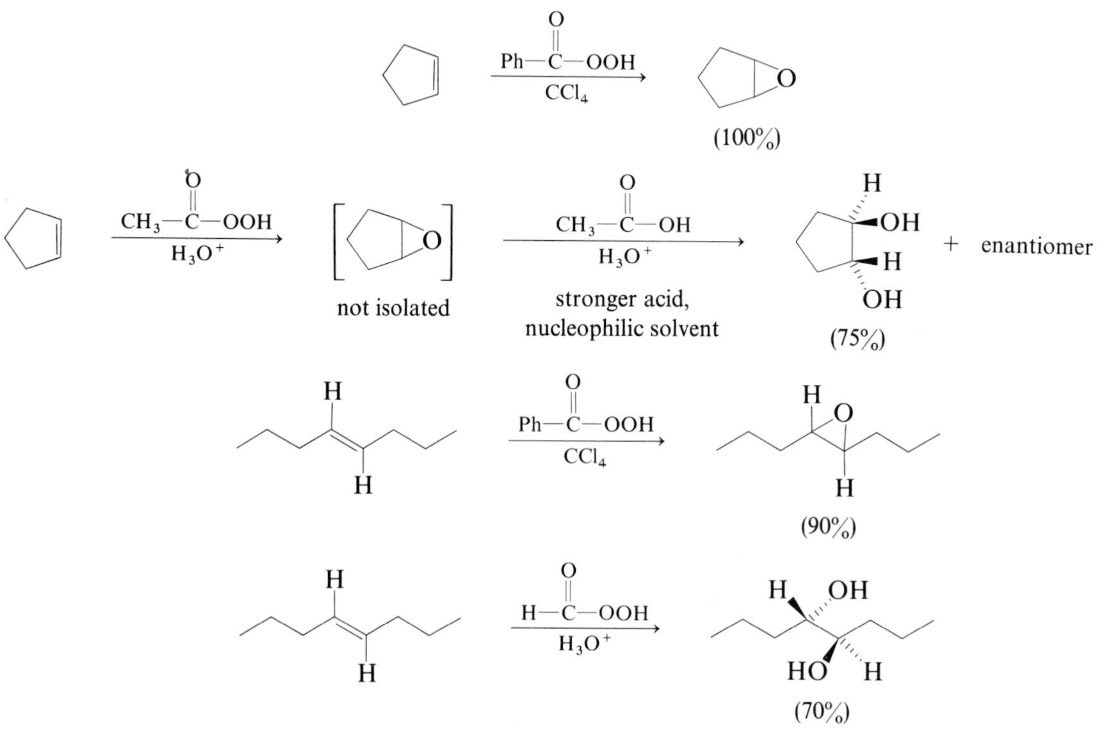

Ph—C(=O)—OOH, CCl_4 → (100%)

CH_3—C(=O)—OOH, H_3O^+ → [epoxide] not isolated → CH_3—C(=O)—OH, H_3O^+ (stronger acid, nucleophilic solvent) → H, OH, H, OH + enantiomer (75%)

Ph—C(=O)—OOH, CCl_4 → H, O, H (90%)

H—C(=O)—OOH, H_3O^+ → H, OH, HO, H (70%)

FIGURE 8-7 A weakly acidic peroxyacid, such as peroxybenzoic acid, is used to obtain a good yield of an epoxide. Stronger acids such as peroxyacetic acid and peroxyformic acid are used to form the glycol in one step.

8-14 SYN HYDROXYLATION OF ALKENES

Either osmium tetroxide or potassium permanganate hydroxylates an alkene with syn stereochemistry to give a glycol.

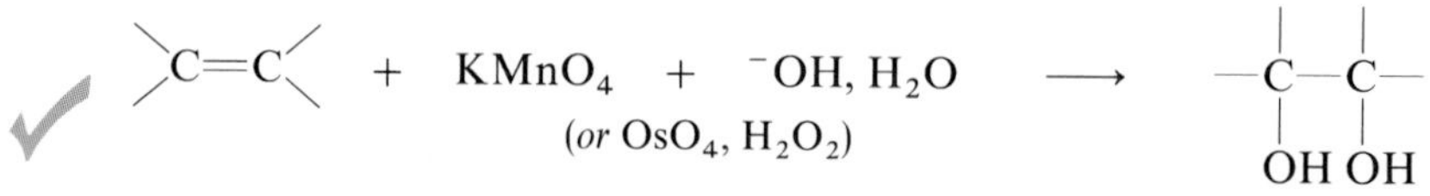

$$\text{C=C} + KMnO_4 + {}^-OH, H_2O \;(or\; OsO_4, H_2O_2) \longrightarrow \underset{OH}{-C}-\underset{OH}{C-}$$

syn addition

The conversion of an alkene to a glycol involves the addition of a hydroxyl group to each end of the double bond: the **hydroxylation** of the double bond. We have seen that epoxidation of an alkene, followed by acidic hydrolysis, gives anti hydroxylation of the double bond. There are also reagents available for the hydroxylation of alkenes with syn stereochemistry. The two most common reagents for this purpose are osmium tetroxide and potassium permanganate.

8-14A OSMIUM TETROXIDE HYDROXYLATION

Osmium tetroxide (OsO_4, sometimes called *osmic acid*) reacts with alkenes in a concerted step to form a cyclic osmate ester. Hydrogen peroxide hydrolyzes the osmate ester and reoxidizes the osmium to osmium tetroxide. The regenerated osmium tetroxide catalyst continues to hydroxylate more molecules of the alkene.

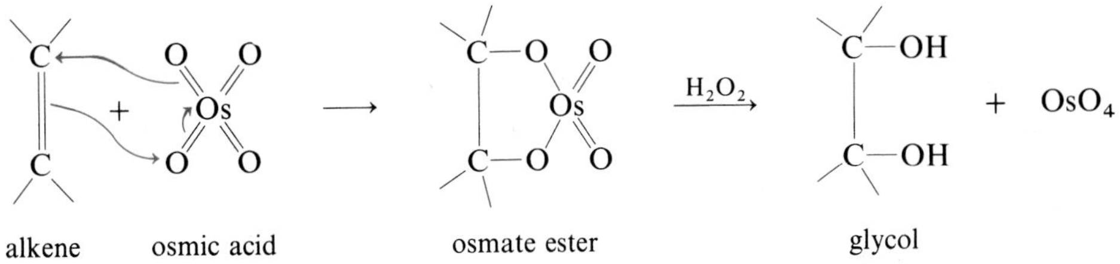

alkene + osmic acid → osmate ester $\xrightarrow{H_2O_2}$ glycol + OsO_4

Because the two carbon-oxygen bonds are formed simultaneously with the cyclic osmate ester, the oxygen atoms add to the same face of the double bond: that is, with syn stereochemistry. The following reactions show the use of OsO_4 and H_2O_2 for the syn hydroxylation of alkenes.

concerted formation of osmate ester $\xrightarrow{H_2O_2}$ *cis*-glycol (65%) + OsO_4

$\xrightarrow{OsO_4,\ H_2O_2}$

cis-3-hexene $\xrightarrow{OsO_4,\ H_2O_2}$ *meso*-3,4-hexanediol

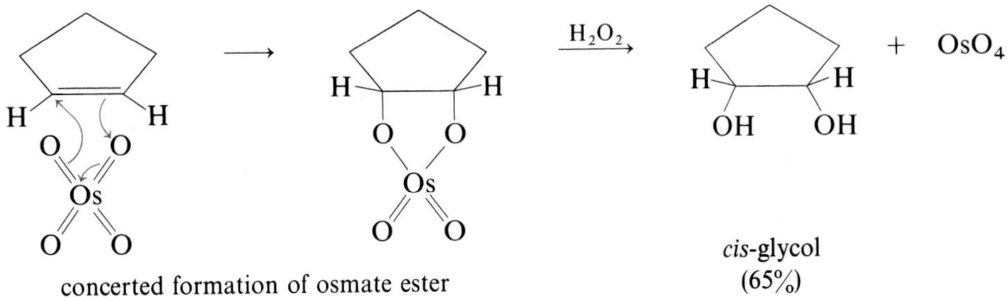

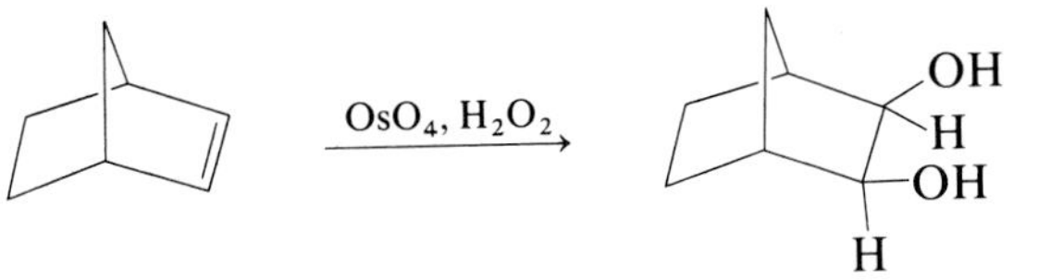

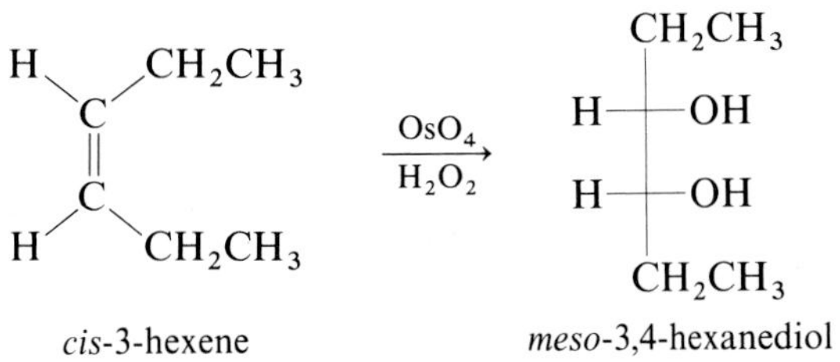

8-14B PERMANGANATE HYDROXYLATION

Osmium tetroxide is an expensive and highly toxic reagent. Hydroxylation of alkenes with syn stereochemistry is also accomplished by a cold, dilute basic solution of potassium permanganate, with slightly reduced yields in most cases. Like osmium tetroxide, permanganate adds to the olefinic double bond to form a cyclic ester: a manganate ester in this case. The basic solution hydrolyzes the manganate ester, liberating the glycol and producing a brown precipitate of manganese dioxide, MnO_2.

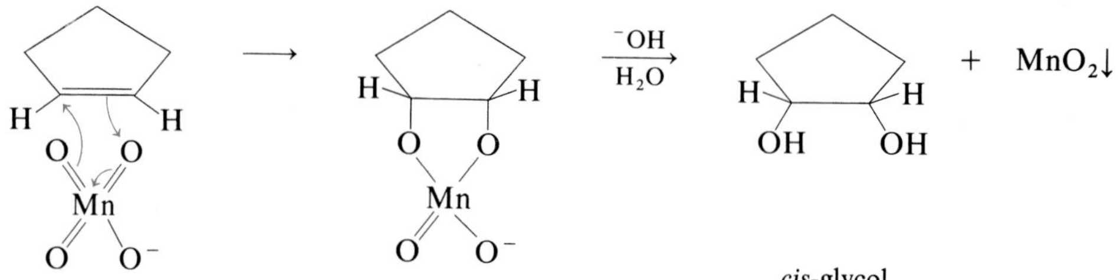

concerted formation of manganate ester

cis-glycol (49%)

In addition to its synthetic value, the permanganate oxidation of alkenes provides a simple chemical test for the presence of an alkene. When an alkene is added to a clear, deep-purple aqueous solution of potassium permanganate, the solution loses its purple color and becomes the murky, opaque brown color of MnO_2.

8-14C CHOOSING A REAGENT

To hydroxylate an alkene with syn stereochemistry, which is the better reagent: osmium tetroxide or potassium permanganate? Osmium tetroxide gives better yields, but permanganate is cheaper and safer to use. The answer depends on the circumstances.

If the starting material is only 2 mg of a compound 15 steps along in a difficult synthesis, we use osmium tetroxide. The better yield is crucial because the starting material is precious, and little osmic acid is needed. If the hydroxylation is the first step in a synthesis and involves 5 kg of the starting material, we use potassium permanganate. The cost of buying enough osmium tetroxide would be prohibitive, and dealing with such a large amount of a volatile, toxic reagent would be inconvenient. On such a large scale, we can accept the lower yield of the permanganate oxidation.

PROBLEM 8-32

Predict the major products of the following reactions, including stereochemistry.

(a) cyclohexene + OsO_4/H_2O_2
(b) cyclohexene + peroxyacetic acid in water
(c) *cis*-2-pentene + OsO_4/H_2O_2
(d) *cis*-2-pentene + peroxyacetic acid in water
(e) *trans*-2-pentene + OsO_4/H_2O_2
(f) *trans*-2-pentene + peroxyacetic acid in water

PROBLEM 8-33

Show how you would accomplish each of the following conversions.

(a) *cis*-3-hexene to *meso*-3,4-hexanediol
(b) *cis*-3-hexene to (*d,l*)-3,4-hexanediol
(c) *trans*-3-hexene to *meso*-3,4-hexanediol
(d) *trans*-3-hexene to (*d,l*)-3,4-hexanediol

8-15 OXIDATIVE CLEAVAGE OF ALKENES

8-15A CLEAVAGE BY PERMANGANATE

In the potassium permanganate hydroxylation, if the solution is warm or acidic or too concentrated, **oxidative cleavage** of the glycol may occur. Mixtures of ketones and carboxylic acids are formed, depending on whether there are any oxidizable aldehyde C—H bonds in the initial fragments. The following reaction shows the oxidative cleavage of a double bond by hot or concentrated permanganate.

$$\underset{\text{R}}{\overset{\text{R}}{}}\text{C=C}\underset{\text{H}}{\overset{\text{R}'}{}} \xrightarrow{\text{KMnO}_4} \left[\text{R}-\underset{\text{OH}}{\overset{\text{R}}{\text{C}}}-\underset{\text{OH}}{\overset{\text{R}'}{\text{C}}}-\text{H}\right] \longrightarrow R_2C{=}O + [O{=}C(R')H]$$

glycol — ketone (stable) — aldehyde (oxidizable)

$$\longrightarrow R_2C{=}O + O{=}C(R')OH$$

ketone — acid

Example

$$\text{butylidenecyclohexane} \xrightarrow[\text{(warm, conc.)}]{\text{KMnO}_4} \text{cyclohexanone} + \text{butanoic acid}$$

8-15B OZONOLYSIS

$$R_2C{=}CR'H + O_3 \longrightarrow \text{ozonide} \xrightarrow{(CH_3)_2S} R_2C{=}O + O{=}CR'H$$

ozonide

Ozonolysis is used for oxidative cleavage of alkenes more often than permanganate because the yields are better and ozonolysis is a gentler and more versatile reaction. Ozone (O_3) is a high-energy form of oxygen that is produced when ultraviolet light or an electrical discharge passes through oxygen gas. Ultraviolet light from the sun converts oxygen to ozone in the upper atmosphere, where the ozone helps to shield the earth from some of the high-energy ultraviolet irradiation it would otherwise receive.

$$\tfrac{3}{2}\,O_2 + 34 \text{ kcal (142 kJ)} \longrightarrow O_3$$

Ozone has 34 kcal/mol (142 kJ/mol) of excess energy over oxygen, and it is much more reactive. A Lewis structure of ozone shows that the central oxygen atom bears a positive charge, and each of the outer oxygen atoms bears half a negative charge.

$$O_3 = \left[\,{}^{-}\!:\ddot{\text{O}}-\overset{+}{\text{O}}=\ddot{\text{O}}: \longleftrightarrow :\ddot{\text{O}}=\overset{+}{\text{O}}-\ddot{\text{O}}:^{-}\right]$$

Ozone reacts with an alkene to form a cyclic compound called a *primary ozonide* or *molozonide* (because 1 mol of ozone has been added). The molozonide has two peroxy (—O—O—) linkages, and it is quite unstable. It rearranges rapidly, even at very low temperatures, to form an ozonide.

molozonide (primary ozonide) — ozonide

Ozonides are not very stable, and they are rarely isolated. They are immediately reduced in most cases by a reducing agent such a dimethyl sulfide. The products of this reduction are ketones and aldehydes.

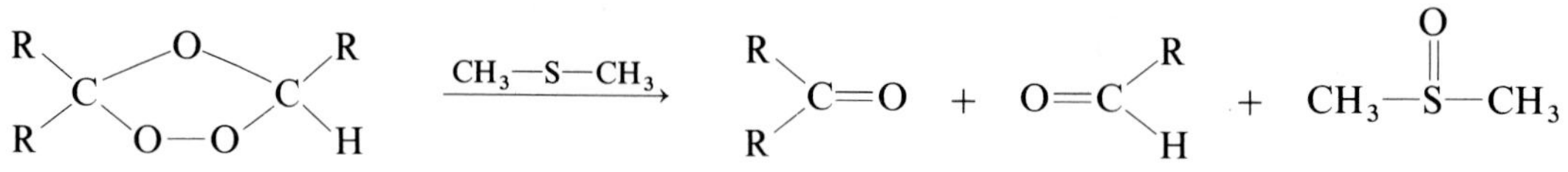

$$\text{ozonide} \xrightarrow{CH_3-S-CH_3} R_2C=O + O=CHR + CH_3-\overset{O}{\overset{\|}{S}}-CH_3$$

ozonide — dimethyl sulfide — ketones, aldehydes — dimethyl sulfoxide (DMSO)

To predict the products from the ozonolysis-reduction of any given alkene, simply erase the double bond and add two oxygen atoms as carbonyl (C=O) groups where the double bond used to be. (The third oxygen atom of ozone reacts with dimethyl sulfide to form dimethyl sulfoxide.)

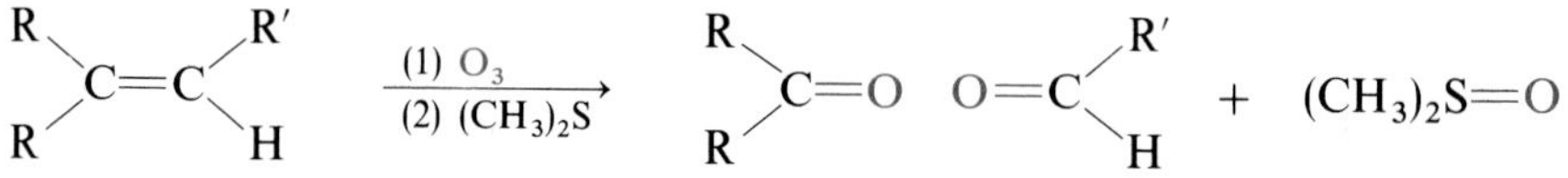

$$R_2C=CHR' \xrightarrow[(2)\ (CH_3)_2S]{(1)\ O_3} R_2C=O \quad O=CHR' + (CH_3)_2S=O$$

The following reactions show the products that are obtained from the ozonolysis of some representative alkenes.

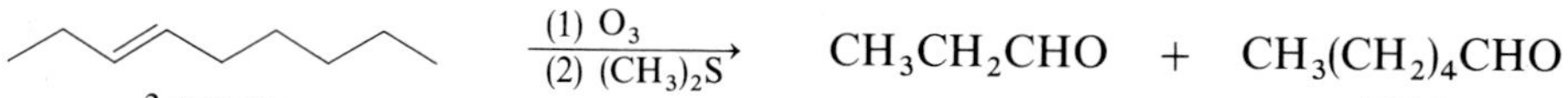

$$\text{3-nonene} \xrightarrow[(2)\ (CH_3)_2S]{(1)\ O_3} CH_3CH_2CHO + CH_3(CH_2)_4CHO \quad (65\%)$$

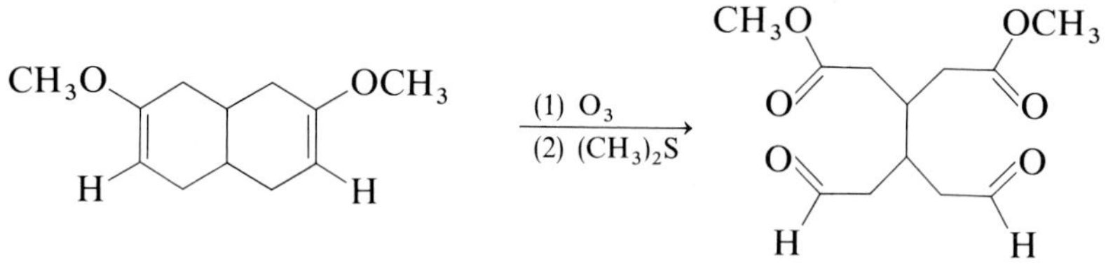

$$\xrightarrow[(2)\ (CH_3)_2S]{(1)\ O_3} \quad (55\%)$$

$$\xrightarrow[(2)\ (CH_3)_2S]{(1)\ O_3}$$

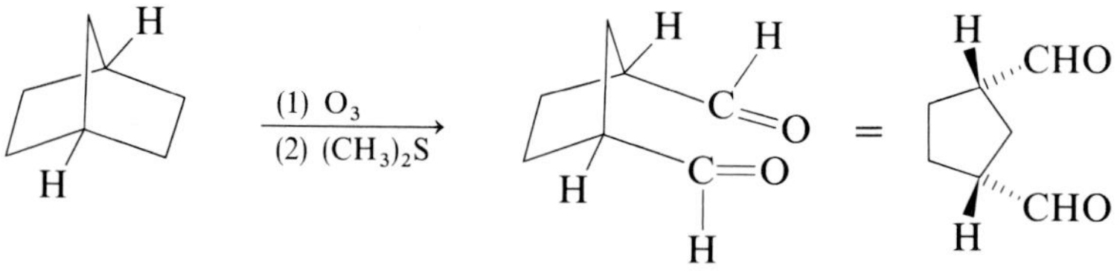

One of the most common uses of ozonolysis is for determining the positions of double bonds in alkenes. For example, if we were uncertain of the position of the methyl group in a methylcyclopentene, the products of ozonolysis-reduction would confirm the structure of the original alkene.

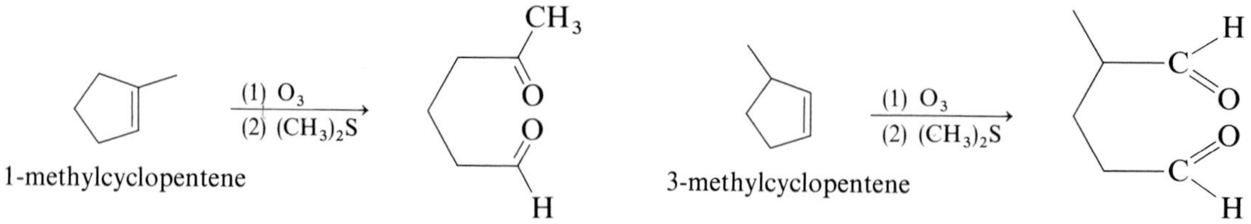

$$\text{1-methylcyclopentene} \xrightarrow[(2)\ (CH_3)_2S]{(1)\ O_3} \qquad \text{3-methylcyclopentene} \xrightarrow[(2)\ (CH_3)_2S]{(1)\ O_3}$$

SOLVED PROBLEM 8-6

Ozonolysis-reduction of an unknown alkene gives an equimolar mixture of cyclohexanecarbaldehyde and 2-butanone. Determine the structure of the original alkene.

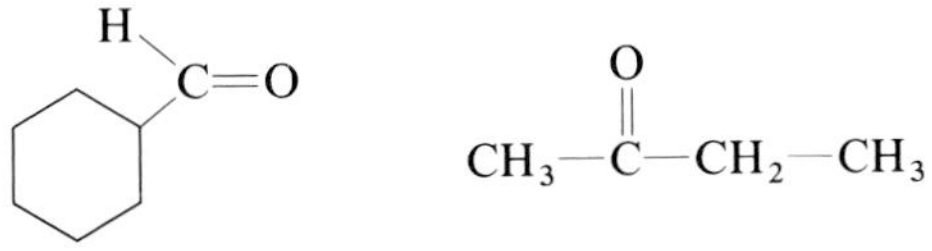

cyclohexanecarbaldehyde

$CH_3-\overset{O}{\overset{\|}{C}}-CH_2-CH_3$

2-butanone

SOLUTION We can reconstruct the alkene by removing the two oxygen atoms of the carbonyl groups (C=O) and connecting the remaining carbon atoms with a double bond. One uncertainty remains, however: The original alkene might be either of two possible geometric isomers.

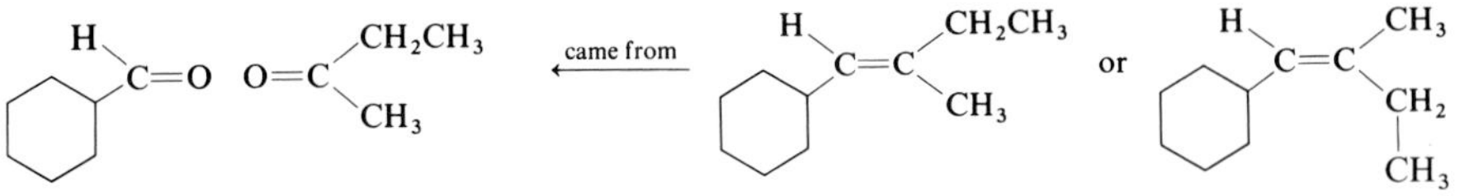

came from

or

(remove oxygen atoms and reconnect the double bond)

PROBLEM 8-34

Give structures of the alkenes that would give the following products upon ozonolysis-reduction.

(a) $CH_3-\overset{O}{\overset{\|}{C}}-CH_2-CH_2-CH_2-\overset{O}{\overset{\|}{C}}-CH_2-CH_3$

(b) 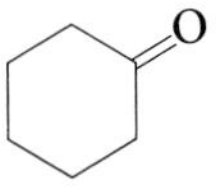 cyclohexanone and $CH_3-CH_2-CH_2-\overset{O}{\overset{\|}{C}}-H$

(c) $CH_3-CH_2-\overset{O}{\overset{\|}{C}}-CH_2-CH_2-CH_2-CH_3$ and $CH_3-CH_2-\overset{O}{\overset{\|}{C}}-H$

8-15C COMPARISON OF PERMANGANATE CLEAVAGE AND OZONOLYSIS

Both permanganate cleavage and ozonolysis break the carbon-carbon double bond and replace it with carbonyl (C=O) groups. In the permanganate cleavage, any aldehyde products are further oxidized to carboxylic acids. In the ozonolysis-reduction procedure, the aldehyde products are generated in the dimethyl sulfide reduction step, and they are not oxidized.

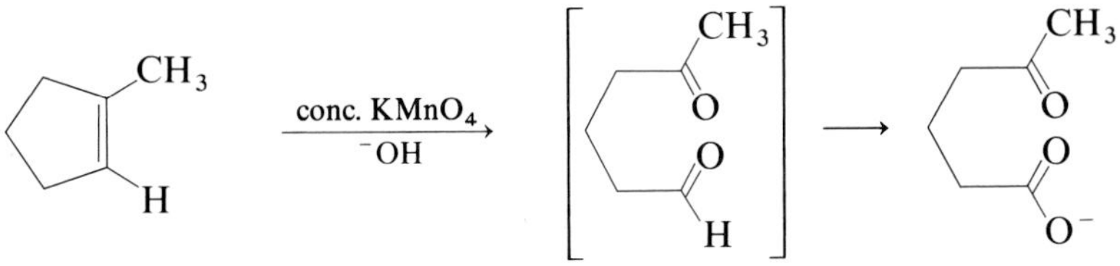

conc. $KMnO_4$, ^-OH

(not isolated)

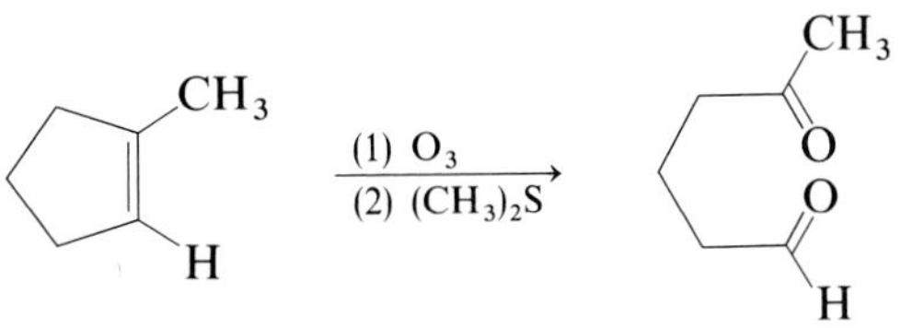

(1) O_3 (2) $(CH_3)_2S$

PROBLEM 8-35

Predict the major products of the following reactions.

(a) (E)-3-methyl-3-octene + ozone, then $(CH_3)_2S$
(b) (Z)-3-methyl-3-octene + warm, concentrated $KMnO_4$

(c) 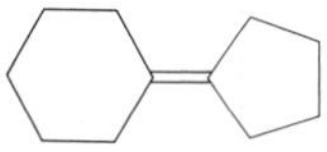 [cyclohexylidenecyclopentane] + O_3, then $(CH_3)_2S$

(d) 1-ethylcycloheptene + ozone, then $(CH_3)_2S$
(e) 1-ethylcycloheptene + warm, concentrated $KMnO_4$
(f) 1-ethylcycloheptene + cold, dilute $KMnO_4$

8-16 DIMERIZATION AND POLYMERIZATION OF ALKENES

8-16A DIMERIZATION AND TRIMERIZATION

Most electrophilic additions to alkenes occur by a two-step mechanism. In the first step, the alkene acts as a nucleophile, attacking the electrophile (E^+) and forming a carbocation. In the second step, a nucleophile ($Nuc:^-$) attacks the carbocation. Under the right conditions (no other good nucleophile present), another molecule of the alkene can serve as the nucleophile to attack the carbocation. This step forms a bond between two molecules of the alkene, with the positive charge on a carbon atom of the second molecule.

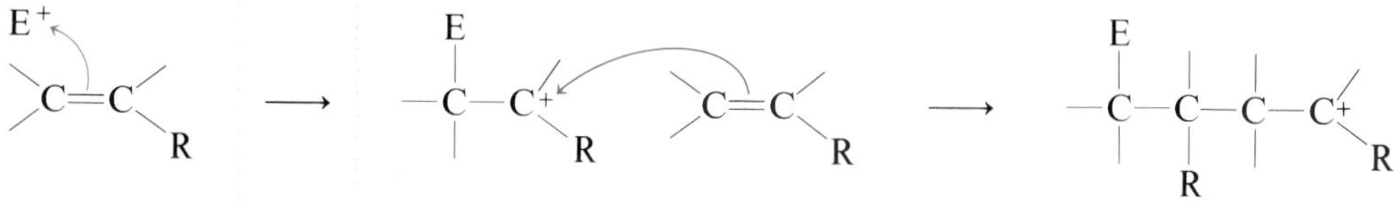

$$E^+ + >C=C(R)- \longrightarrow E-C-C^+(R) + >C=C(R)- \longrightarrow E-C-C(R)-C-C^+(R)$$

Consider what happens when pure isobutylene is treated with a trace of concentrated sulfuric acid. Protonation of the alkene forms a carbocation. If a large concentration of isobutylene is available, another molecule of the alkene may act as the nucleophile, attacking the carbocation. If the resulting carbocation loses a proton, the product is an alkene that contains eight carbon atoms, formed from two molecules of isobutylene. If the carbocation reacts with a third molecule of isobutylene, a trimer may result.

Attack by the second molecule of isobutylene

$$H_2SO_4 + H_2C=C(CH_3)_2 \longrightarrow CH_3-C^+(CH_3)_2 \quad H_2C=C(CH_3)_2 \longrightarrow CH_3-C(CH_3)_2-CH_2-C^+(CH_3)_2$$

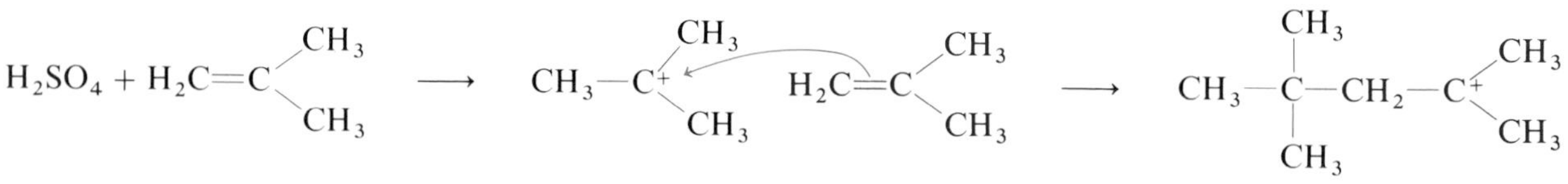

Deprotonation to give the dimer

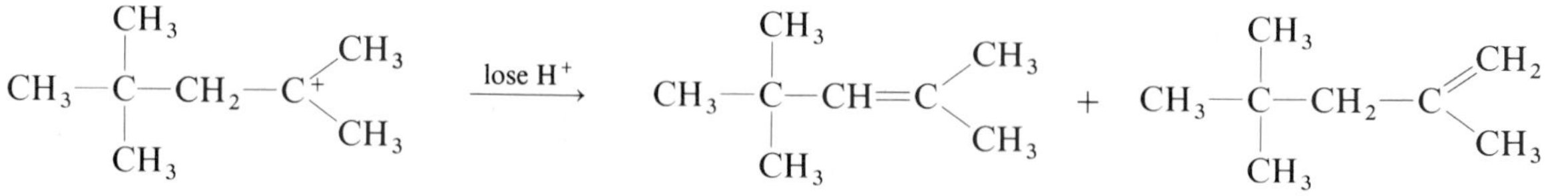

$$CH_3-C(CH_3)_2-CH_2-C^+(CH_3)_2 \xrightarrow{\text{lose } H^+} CH_3-C(CH_3)_2-CH=C(CH_3)_2 + CH_3-C(CH_3)_2-CH_2-C(CH_3)=CH_2$$

"diisobutylene" (80%) (Saytzeff product) "diisobutylene" (20%)

Attack by a third molecule to give a trimer

$$\underset{\text{dimer}}{CH_3-\overset{CH_3}{\underset{CH_3}{|}}{C}-CH_2-C^+(CH_3)_2} + \underset{\text{third monomer}}{H_2C{=}C(CH_3)_2} \longrightarrow CH_3-C(CH_3)_2-CH_2-C(CH_3)_2-CH_2-C^+(CH_3)_2 \xrightarrow{-H^+}$$

$$\underset{\text{trimer}}{CH_3-C(CH_3)_2-CH_2-C(CH_3)_2-CH{=}C(CH_3)_2}$$

The starting material, isobutylene, is called the **monomer,** from the Greek *mono* ("one") and *meros* ("parts"). The product derived from two molecules of the starting material is called a **dimer,** from the Greek *di* ("two") and *meros* ("parts"). Similarly, a **trimer** is composed of three molecules of the monomer.

8-16B ALKENE POLYMERS

When isobutylene is treated with a small amount of acid catalyst in a reaction mixture with no other bases or nucleophiles present, many molecules of isobutylene may react to form a **polymer** before the loss of a proton ends the process. Polymerization gives products with higher molecular weights (more monomer units) if the electrophilic catalyst is completely nonnucleophilic and nonbasic.

Boron trifluoride (BF_3) is an excellent nonnucleophilic and nonbasic polymerization catalyst. Like borane (BH_3), BF_3 is electron deficient and a strong Lewis acid. These characteristics are enhanced by the three electron-withdrawing bonds to fluorine. Boron trifluoride adds to the less highly substituted end of an alkene double bond to give the more stable carbocation.

$$F-BF_2 + H_2C{=}C(CH_3)_2 \longrightarrow F-\overset{F}{\underset{F}{|}}{B^-}-CH_2-C^+(CH_3)_2$$

When BF_3 is used as a polymerization catalyst, the alkene is the most likely nucleophile available to attack the carbocation. The chain-lengthening reaction takes place many times before something happens to terminate the process.

First chain-lengthening step

$$F_3B^--CH_2-C^+(CH_3)_2 + H_2C{=}C(CH_3)_2 \longrightarrow F_3B^--CH_2-C(CH_3)_2-CH_2-C^+(CH_3)_2$$

The polymerization continues

$$(P)-CH_2-C(CH_3)_2-CH_2-C^+(CH_3)_2 + H_2C{=}C(CH_3)_2 \longrightarrow (P)-CH_2-C(CH_3)_2-CH_2-C(CH_3)_2-CH_2-C^+(CH_3)_2$$

(P)— = growing polymer chain

The most likely ending of this BF_3-catalyzed polymerization is the loss of a proton from the carbocation at the end of the growing chain. This side reaction protonates another molecule of isobutylene, however, initiating a new polymer chain.

Termination of a polymer chain

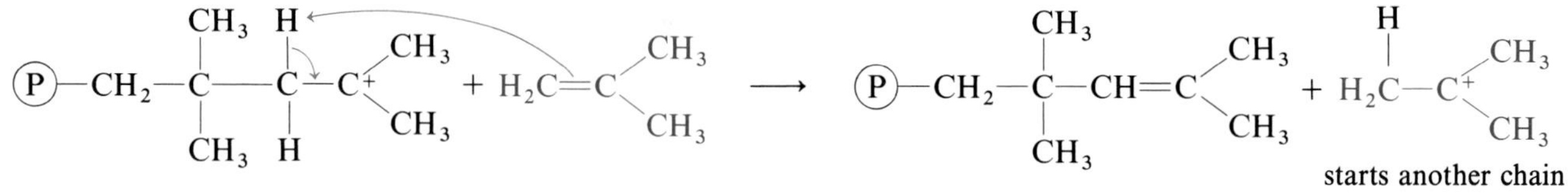

$$\text{(P)}-CH_2-\underset{CH_3}{\overset{CH_3}{|}}C-\underset{H}{\overset{H}{|}}C-C^{+}(CH_3)_2 + H_2C{=}C(CH_3)_2 \longrightarrow \text{(P)}-CH_2-\underset{CH_3}{\overset{CH_3}{|}}C-CH{=}C(CH_3)_2 + \underset{\text{starts another chain}}{H_3C-C^{+}(CH_3)_2}$$

The polymer of isobutylene is *polyisobutylene,* one of the constituents of the *butyl rubber* used in inner tubes and other synthetic rubber products. This type of polymer, in which one molecule after another adds to the chain, is called an **addition polymer.** We will study other types of polymers and other techniques for polymerization in Chapter 26.

PROBLEM 8-36

Give a detailed mechanism for the following reaction.

$$2\,(CH_3)_2C{=}CH-CH_3 \quad + \quad \text{cat. } H^{+} \quad \longrightarrow \quad \text{2,3,4,4-tetramethyl-2-hexene}$$

PROBLEM 8-37

Show the first three steps (as far as the tetramer) in the BF_3-catalyzed polymerization of propylene to form polypropylene.

PROBLEM 8-38

When cyclohexanol is dehydrated to cyclohexene, a gummy green substance forms on the bottom of the flask. Suggest what this residue might be and give a mechanism for its formation (as far as the dimer).

SUMMARY OF THE REACTIONS OF ALKENES

1. Electrophilic additions

a. *Addition of hydrogen halides* (Section 8-3)

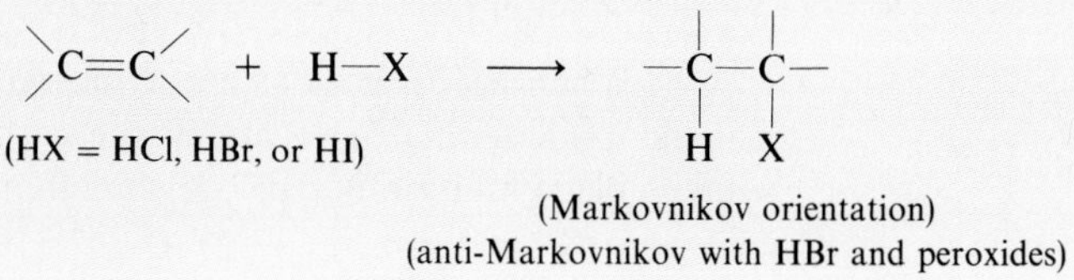

$$\rangle C{=}C\langle \quad + \quad H-X \quad \longrightarrow \quad -\underset{H}{\overset{|}{C}}-\underset{X}{\overset{|}{C}}-$$

(HX = HCl, HBr, or HI)

(Markovnikov orientation)
(anti-Markovnikov with HBr and peroxides)

Example

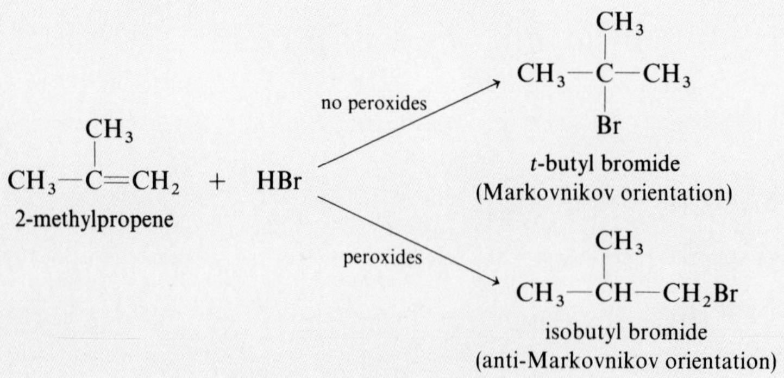

$$CH_3-\overset{CH_3}{\overset{|}{C}}{=}CH_2 + HBr \xrightarrow{\text{no peroxides}} CH_3-\underset{Br}{\overset{CH_3}{\overset{|}{\underset{|}{C}}}}-CH_3$$

2-methylpropene → *t*-butyl bromide (Markovnikov orientation)

$$CH_3-\overset{CH_3}{\overset{|}{C}}{=}CH_2 + HBr \xrightarrow{\text{peroxides}} CH_3-\overset{CH_3}{\overset{|}{C}}H-CH_2Br$$

isobutyl bromide (anti-Markovnikov orientation)

b. *Acid-catalyzed hydration* (Section 8-4)

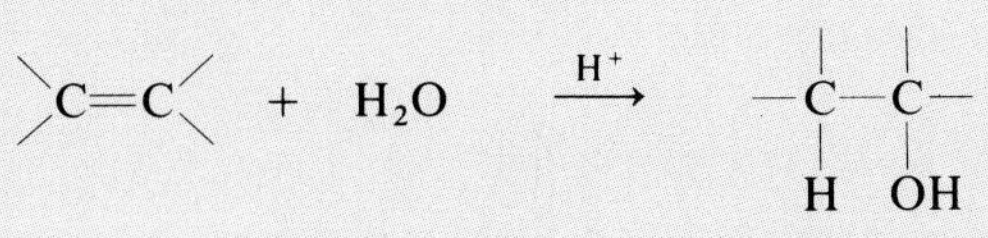

$$\mathrm{>C{=}C<} + H_2O \xrightarrow{H^+} \mathrm{-\underset{H}{C}-\underset{OH}{C}-}$$

(Markovnikov orientation)

Example

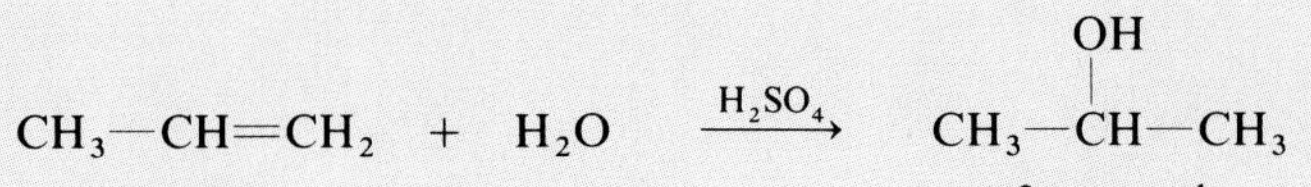

$$\underset{\text{propene}}{CH_3-CH{=}CH_2} + H_2O \xrightarrow{H_2SO_4} \underset{\text{2-propanol}}{CH_3-\overset{OH}{CH}-CH_3}$$

c. *Addition of sulfuric acid* (Section 8-5A)

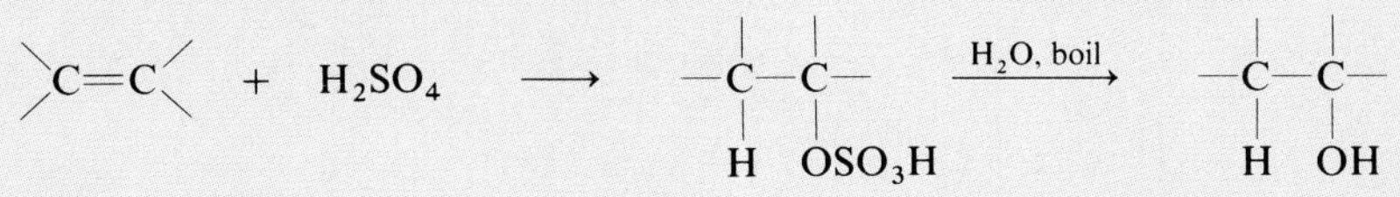

$$\mathrm{>C{=}C<} + H_2SO_4 \longrightarrow \mathrm{-\underset{H}{C}-\underset{OSO_3H}{C}-} \xrightarrow{H_2O,\ \text{boil}} \mathrm{-\underset{H}{C}-\underset{OH}{C}-}$$

(Markovnikov orientation)

Example

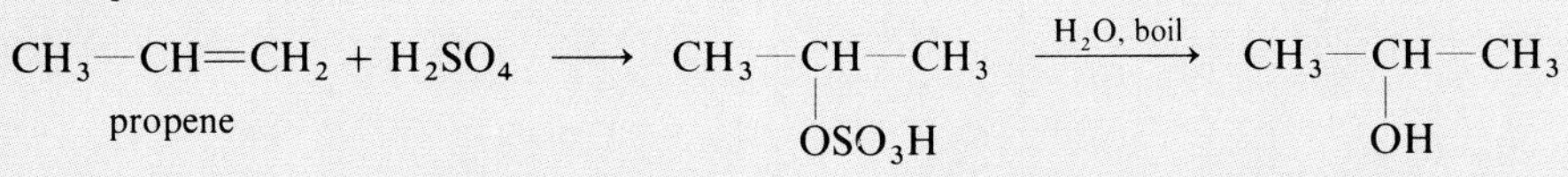

$$\underset{\text{propene}}{CH_3-CH{=}CH_2} + H_2SO_4 \longrightarrow CH_3-\underset{OSO_3H}{CH}-CH_3 \xrightarrow{H_2O,\ \text{boil}} \underset{\text{2-propanol}}{CH_3-\underset{OH}{CH}-CH_3}$$

d. *Oxymercuration-demercuration* (Section 8-5B)

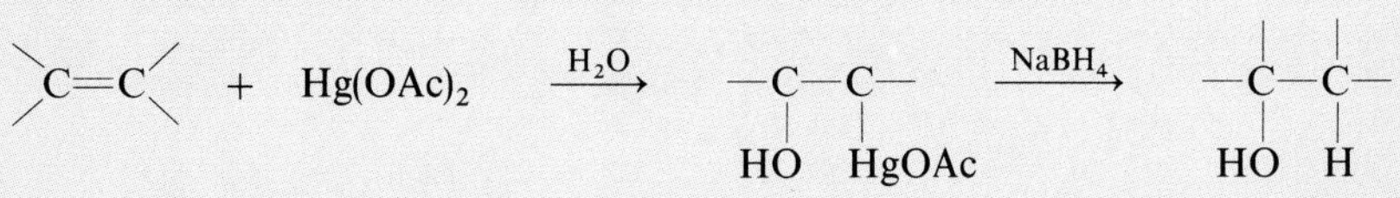

$$\mathrm{>C{=}C<} + Hg(OAc)_2 \xrightarrow{H_2O} \mathrm{-\underset{HO}{C}-\underset{HgOAc}{C}-} \xrightarrow{NaBH_4} \mathrm{-\underset{HO}{C}-\underset{H}{C}-}$$

(Markovnikov orientation)

Example

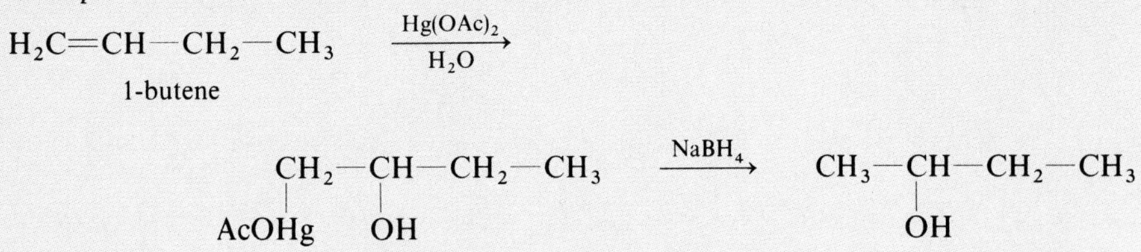

$$\underset{\text{1-butene}}{H_2C{=}CH-CH_2-CH_3} \xrightarrow[H_2O]{Hg(OAc)_2}$$

$$\underset{AcOHg}{CH_2}-\underset{OH}{CH}-CH_2-CH_3 \xrightarrow{NaBH_4} \underset{\text{2-butanol}}{CH_3-\underset{OH}{CH}-CH_2-CH_3}$$

e. *Alkoxymercuration-demercuration* (Section 8-6)

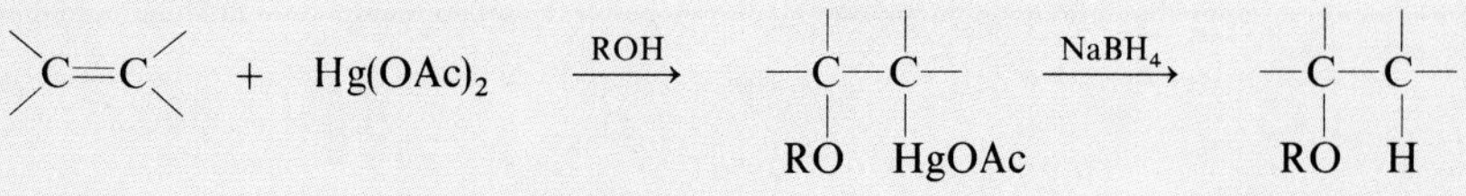

$$\mathrm{>C{=}C<} + Hg(OAc)_2 \xrightarrow{ROH} \mathrm{-\underset{RO}{C}-\underset{HgOAc}{C}-} \xrightarrow{NaBH_4} \mathrm{-\underset{RO}{C}-\underset{H}{C}-}$$

(Markovnikov orientation)

Example

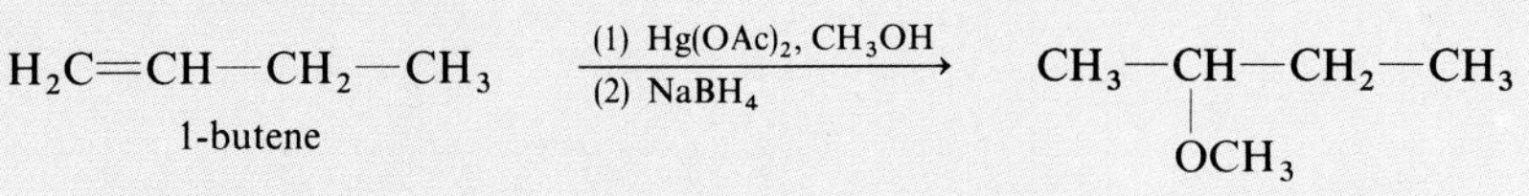

$$\underset{\text{1-butene}}{H_2C{=}CH-CH_2-CH_3} \xrightarrow[\text{(2) } NaBH_4]{\text{(1) } Hg(OAc)_2,\ CH_3OH} \underset{\text{2-methoxybutane}}{CH_3-\underset{OCH_3}{CH}-CH_2-CH_3}$$

f. *Hydroboration-oxidation* (Section 8-7)

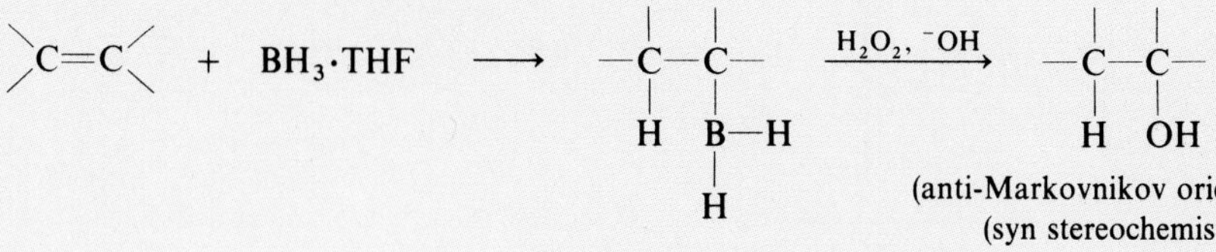

$$>C=C< \; + \; BH_3\cdot THF \longrightarrow -\overset{|}{\underset{H}{C}}-\overset{|}{\underset{B-H}{C}}- \xrightarrow{H_2O_2,\; ^-OH} -\overset{|}{\underset{H}{C}}-\overset{|}{\underset{OH}{C}}-$$

(anti-Markovnikov orientation)
(syn stereochemistry)

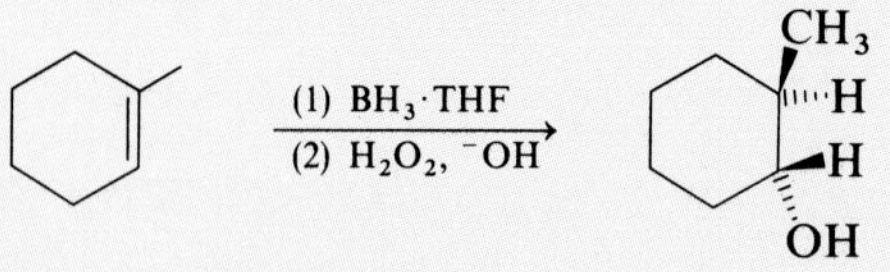

$$\text{1-methylcyclohexene} \xrightarrow[(2)\; H_2O_2,\; ^-OH]{(1)\; BH_3\cdot THF} \text{trans-2-methylcyclohexanol } (CH_3, H, H, OH)$$

g. *Dimerization (and polymerization)* (Section 8-16)

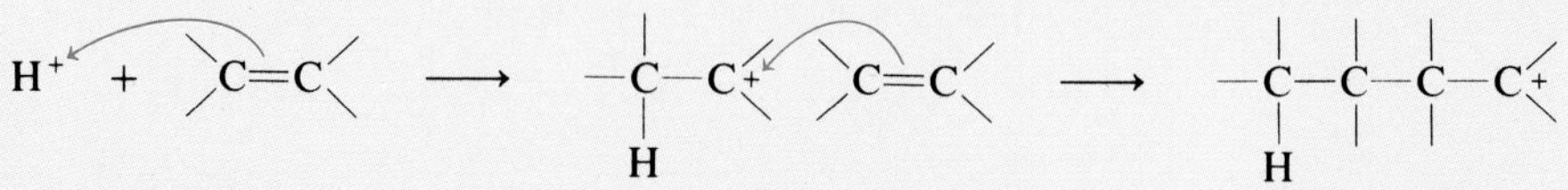

$$H^+ + >C=C< \longrightarrow -\overset{|}{\underset{H}{C}}-C^+< \;\; >C=C< \longrightarrow -\overset{|}{\underset{H}{C}}-\overset{|}{C}-\overset{|}{C}-C^+<$$

(may lose a proton or continue)

Example

$$2\; CH_3-CH=CH_2 \xrightarrow{H_2SO_4} CH_3-\overset{CH_3}{\overset{|}{CH}}-CH=CH-CH_3$$

propene → 4-methyl-2-pentene

2. *Reduction: catalytic hydrogenation* (Section 8-8)

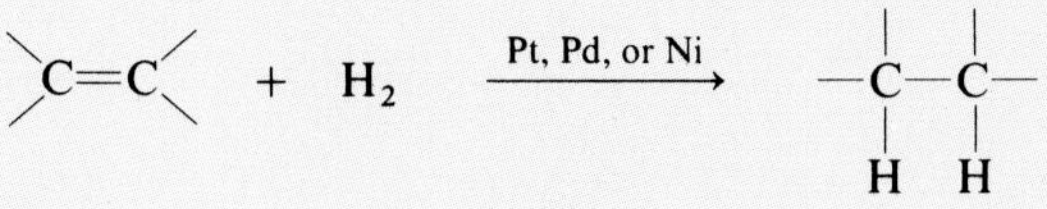

$$>C=C< \; + \; H_2 \xrightarrow{\text{Pt, Pd, or Ni}} -\overset{|}{\underset{H}{C}}-\overset{|}{\underset{H}{C}}-$$

(syn addition)

3. *Addition of carbenes: cyclopropanation* (Section 8-9)

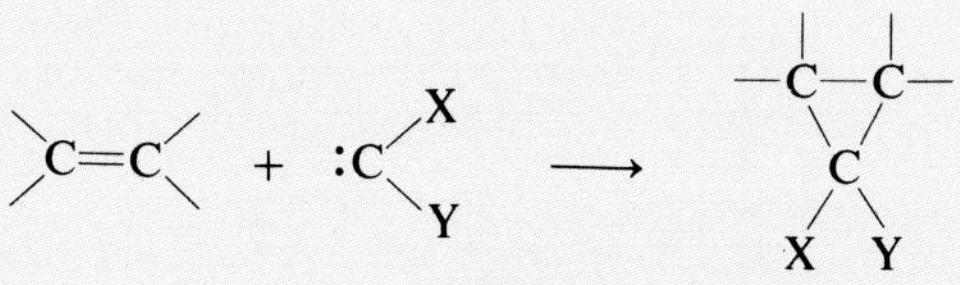

$$>C=C< \; + \; :C\begin{smallmatrix} X \\ Y \end{smallmatrix} \longrightarrow \text{cyclopropane with } C(X)(Y)$$

(X, Y = H, Cl, Br, I, or –COOEt)

Example

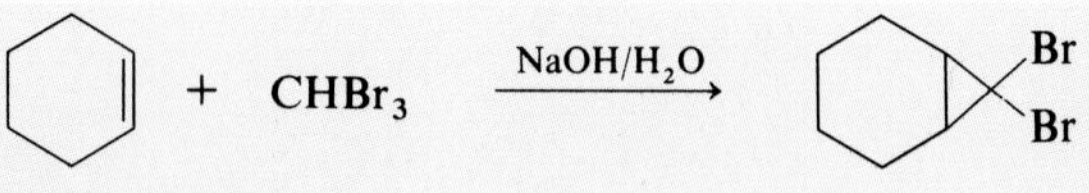

$$\text{cyclohexene} + CHBr_3 \xrightarrow{NaOH/H_2O} \text{7,7-dibromobicyclo[4.1.0]heptane (Br, Br)}$$

cyclohexene

4. *Oxidative additions*

a. *Addition of halogens* (Section 8-10)

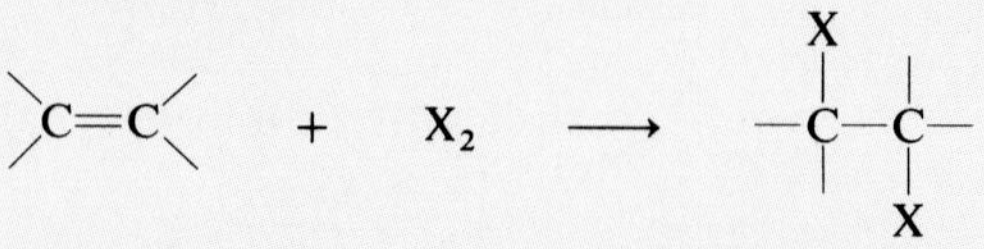

$$>C=C< \; + \; X_2 \longrightarrow -\overset{X}{\overset{|}{C}}-\underset{X}{\underset{|}{C}}-$$

($X_2 = Cl_2, Br_2$, sometimes I_2) (anti addition)

Example

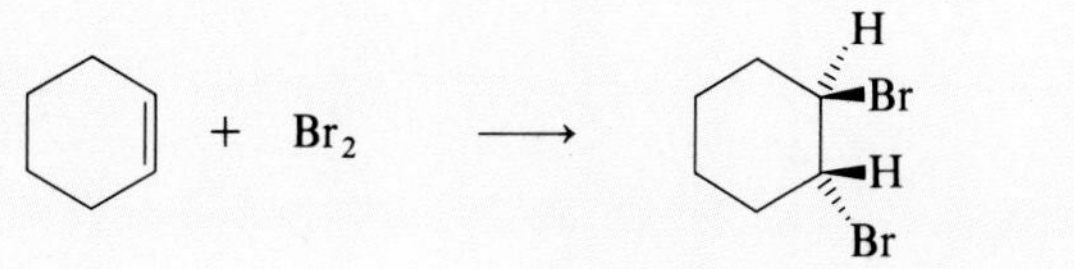

$$\text{cyclohexene} + Br_2 \longrightarrow \textit{trans}\text{-1,2-dibromocyclohexane}$$

b. *Halohydrin formation:* (Section 8-11)

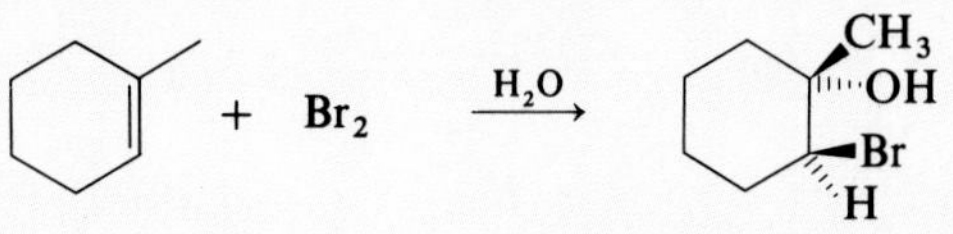

$$\text{1-methylcyclohexene} + Br_2 \xrightarrow{H_2O} \text{2-bromo-1-methylcyclohexanol}$$

(anti addition)
(Markovnikov orientation)

c. *Epoxidation* (Section 8-12)

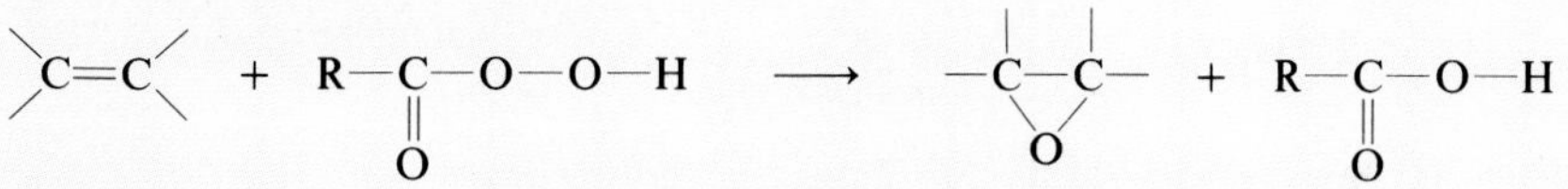

$$\underset{\text{alkene}}{>C{=}C<} + \underset{\text{peroxyacid}}{R{-}\overset{\overset{\displaystyle O}{\|}}{C}{-}O{-}O{-}H} \longrightarrow \underset{\text{syn addition}}{{-}\overset{|}{C}{-}\overset{|}{C}{-}\ (\text{bridged by } O)} + R{-}\overset{\overset{\displaystyle O}{\|}}{C}{-}O{-}H$$

Example

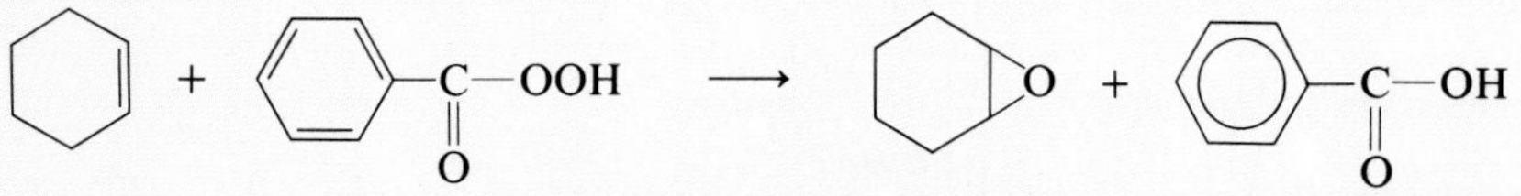

$$\underset{\text{cyclohexene}}{C_6H_{10}} + \underset{\text{peroxybenzoic acid}}{C_6H_5{-}\overset{\overset{\displaystyle O}{\|}}{C}{-}OOH} \longrightarrow \underset{\substack{\text{epoxycyclohexane}\\ \text{(cyclohexene oxide)}}}{C_6H_{10}O} + \underset{\text{benzoic acid}}{C_6H_5{-}\overset{\overset{\displaystyle O}{\|}}{C}{-}OH}$$

d. *Anti hydroxylation* (Section 8-13)

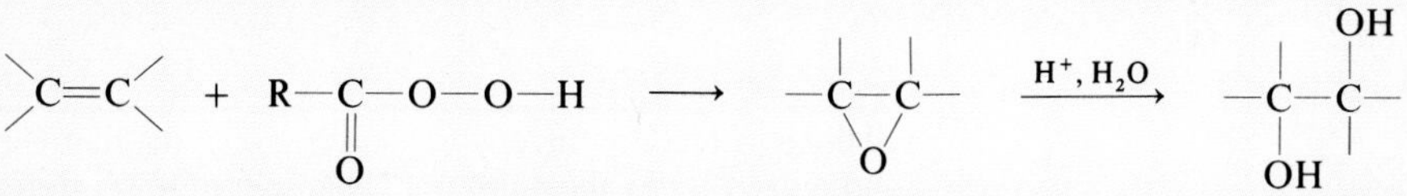

$$>C{=}C< + R{-}\overset{\overset{\displaystyle O}{\|}}{C}{-}O{-}O{-}H \longrightarrow {-}\overset{|}{C}{-}\overset{|}{C}{-}\ (\text{bridged by } O) \xrightarrow{H^+,\ H_2O} {-}\underset{OH}{\overset{|}{C}}{-}\underset{|}{\overset{OH}{C}}{-}$$

(anti addition)

Example

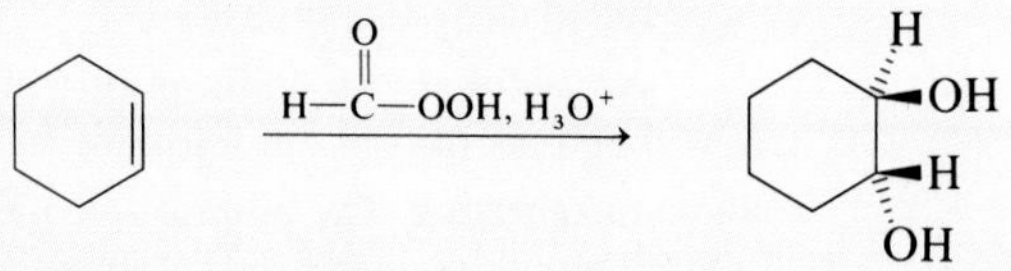

$$\text{cyclohexene} \xrightarrow{H{-}\overset{\overset{\displaystyle O}{\|}}{C}{-}OOH,\ H_3O^+} \textit{trans}\text{-cyclohexane-1,2-diol}$$

e. *Syn hydroxylation* (Section 8-14)

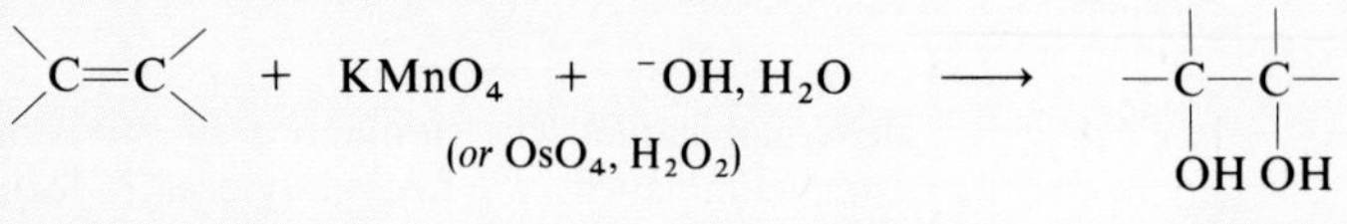

$$>C{=}C< + KMnO_4 + {}^-OH, H_2O \longrightarrow {-}\underset{OH}{\overset{|}{C}}{-}\underset{OH}{\overset{|}{C}}{-}$$

(*or* OsO_4, H_2O_2)

(syn addition)

Example

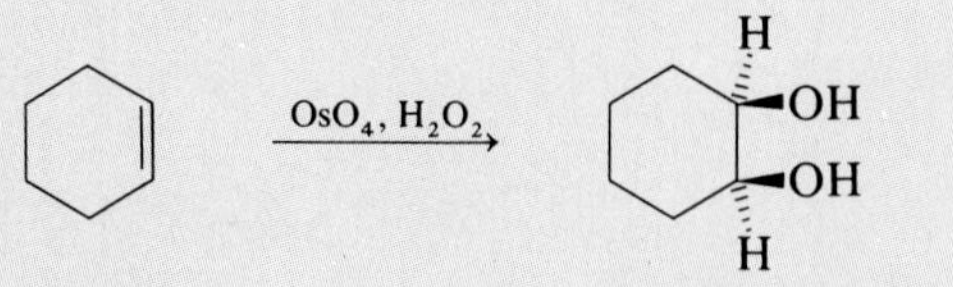

$$\text{cyclohexene} \xrightarrow{OsO_4,\ H_2O_2} \textit{cis}\text{-cyclohexane-1,2-diol}$$

cyclohexene   *cis*-cyclohexane-1,2-diol

5. *Oxidative cleavage of alkenes* (Section 8-15)

a. *Ozonolysis*

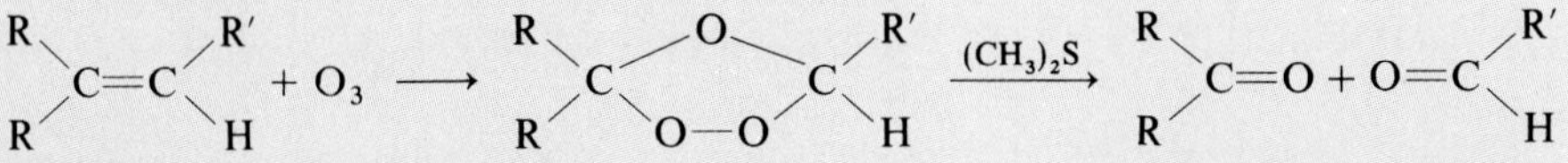

$$R_2C{=}CHR' + O_3 \longrightarrow \text{ozonide} \xrightarrow{(CH_3)_2S} R_2C{=}O + O{=}CHR'$$

ozonide   ketones and aldehydes

Example

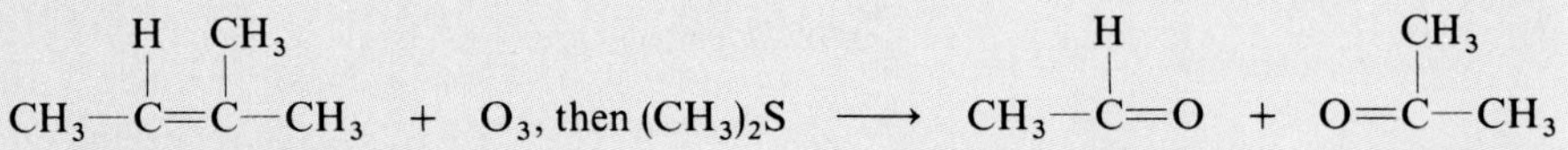

$$CH_3{-}CH{=}C(CH_3){-}CH_3 + O_3\text{, then } (CH_3)_2S \longrightarrow CH_3{-}CH{=}O + O{=}C(CH_3){-}CH_3$$

2-methyl-2-butene   acetaldehyde   acetone

b. *Potassium permanganate*

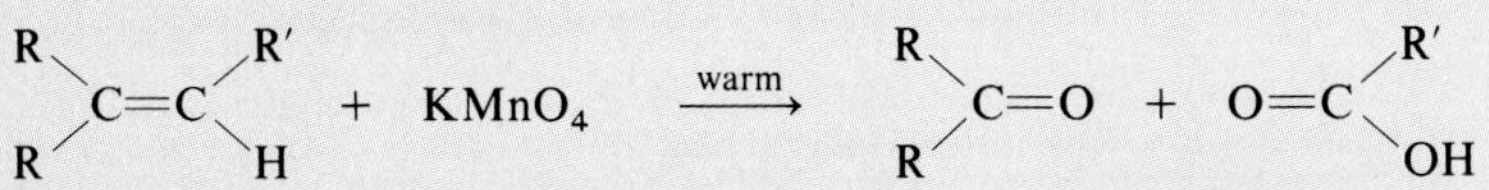

$$R_2C{=}CHR' + KMnO_4 \xrightarrow{\text{warm}} R_2C{=}O + O{=}C(OH)R'$$

ketones and acids
(aldehydes are oxidized)

Example

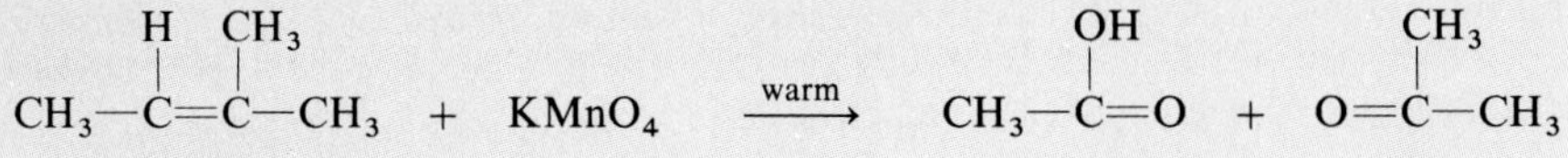

$$CH_3{-}CH{=}C(CH_3){-}CH_3 + KMnO_4 \xrightarrow{\text{warm}} CH_3{-}C(OH){=}O + O{=}C(CH_3){-}CH_3$$

2-methyl-2-butene   acetic acid   acetone

GLOSSARY

addition A reaction involving an increase in the number of groups attached to the alkene, and a decrease in the number of elements of unsaturation. (p. 313)

anti addition: An addition in which two groups add to opposite faces of the double bond (as in addition of Br_2). (p. 326)

electrophilic addition: An addition in which the electrophile bonds first, followed by the nucleophile. (p. 314)

syn addition: An addition in which two groups add to the same face of the double bond (as in osmium tetroxide hydroxylation). (p. 326)

alkoxymercuration The addition of mercuric acetate to an alkene in an alcohol solution, forming an alkoxymercurial intermediate. (p. 326)

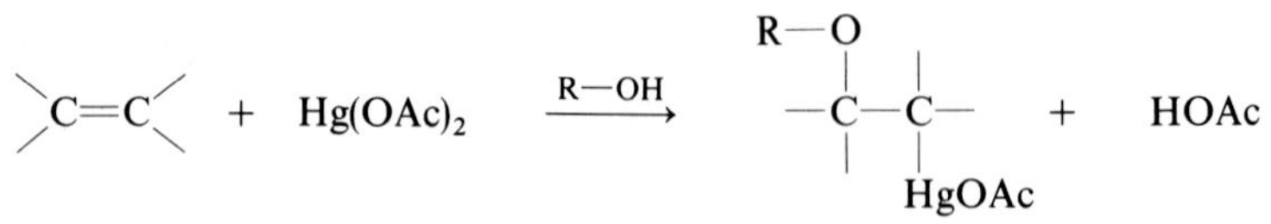

$$>C{=}C< + Hg(OAc)_2 \xrightarrow{R-OH} -C(OR){-}C(HgOAc)- + HOAc$$

alpha elimination The elimination of two atoms or groups from the same carbon atom. Alpha eliminations are frequently used to form carbenes. (p. 336)

$$CHBr_3 + KOH \longrightarrow :CBr_2 + H_2O + KBr$$

beta elimination The elimination of two atoms or groups from adjacent carbon atoms. This is the most common type of elimination. (p. 336)

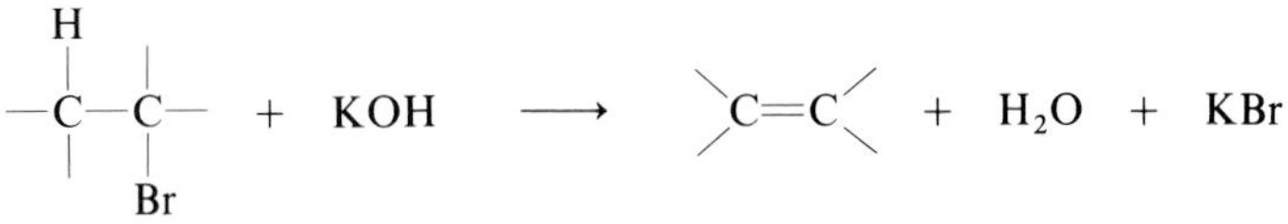

$$\mathrm{H{-}\overset{|}{\underset{|}{C}}{-}\overset{|}{\underset{|}{C}}{-}Br} + KOH \longrightarrow \;\rangle C{=}C\langle\; + H_2O + KBr$$

carbene A reactive intermediate with a neutral carbon atom having only two bonds and two nonbonding electrons. Methylene ($:CH_2$) is the simplest carbene. (p. 334)

demercuration The removal of a mercury species from a molecule. Demercuration of the products of oxymercuration or alkoxymercuration is usually accomplished using sodium borohydride. (p. 325)

dimer A compound composed of two molecules of a smaller, simpler compound called the **monomer.** A **trimer** is composed of three molecules of the monomer. (p. 350)

dimerization: The formation of a dimer.

epoxide (oxirane) A three-membered cyclic ether. (p. 341)

epoxidation: The formation of an epoxide, usually from an alkene. A peroxyacid is generally used for alkene epoxidations.

glycol A 1,2-diol. (p. 343)

halogenation The addition of a halogen (X_2) to a molecule. (p. 337)

halohydrin A beta-haloalcohol, with a halogen and a hydroxyl group on adjacent carbon atoms. (p. 339)

halonium ion A reactive, cationic intermediate with a three-membered ring containing a halogen atom. (p. 337)

heterogeneous catalysis Use of a catalyst that is a separate phase from the reactants. For example, the platinum hydrogenation catalyst is a solid, in a separate phase from the liquid alkene. (p. 333)

homogeneous catalysis Use of a catalyst that is in the same phase as the reactants. For example, the acid catalyst in hydration is in the liquid phase with the alkene. (p. 333)

hydration The addition of water to a molecule. The hydration of an alkene forms an alcohol. (p. 321)

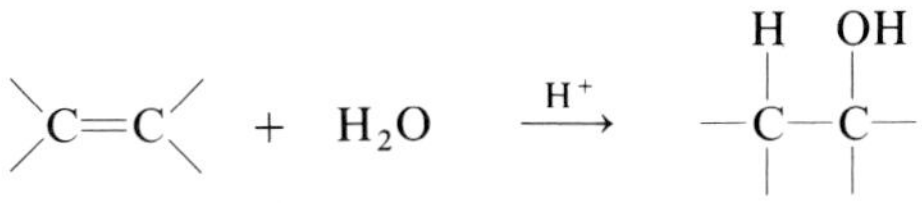

$$\rangle C{=}C\langle \; + H_2O \xrightarrow{H^+} -\overset{\overset{\displaystyle H}{|}}{\underset{|}{C}}-\overset{\overset{\displaystyle OH}{|}}{\underset{|}{C}}-$$

hydroboration The addition of borane (BH_3) or one of its derivatives ($BH_3 \cdot THF$, for example) to a molecule. (p. 328)

hydrogenation The addition of hydrogen to a molecule. The most common hydrogenation is the addition of H_2 across a double bond in the presence of a catalyst (**catalytic hydrogenation** or **catalytic reduction**). (p. 333)

hydroxylation The addition of two hydroxyl groups, one at each carbon of the double bond (p. 344)

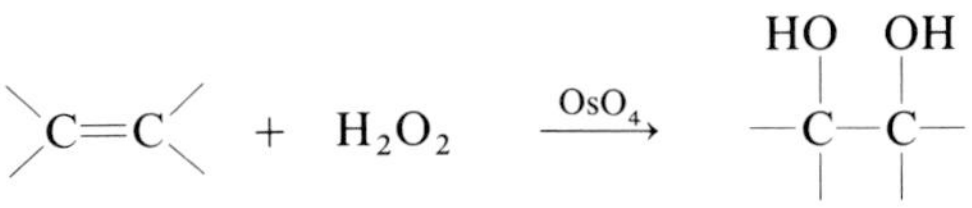

$$\rangle C{=}C\langle \; + H_2O_2 \xrightarrow{OsO_4} -\overset{\overset{\displaystyle HO}{|}}{\underset{|}{C}}-\overset{\overset{\displaystyle OH}{|}}{\underset{|}{C}}-$$

Markovnikov's rule (p. 316)

(*original statement*) When a proton acid adds to the double bond of an alkene, the acid proton bonds to the carbon atom that is already bonded to more hydrogen atoms.

(*extended statement*) In an electrophilic addition to an alkene, the electrophile adds in such a way as to generate the most stable intermediate.

Markovnikov orientation: An orientation of addition that obeys the original statement of Markovnikov's rule; gives the **Markovnikov product.**

anti-Markovnikov orientation: An orientation of addition that is opposite to that predicted by the original statement of Markovnikov's rule; gives the **anti-Markovnikov product.**

oxidative cleavage The cleavage of a carbon-carbon bond through oxidation. Carbon-carbon double bonds are commonly cleaved by ozonolysis/reduction or by warm, concentrated permanganate. (p. 346)

oxymercuration The addition of aqueous mercuric acetate to an alkene. (p. 325)

$$\rangle C{=}C\langle \; + \; Hg(OAc)_2 \xrightarrow{H_2O} -\overset{HO}{\underset{|}{\overset{|}{C}}}-\underset{HgOAc}{\underset{|}{\overset{|}{C}}}- \; + \; HOAc$$

ozonolysis The use of ozone, usually followed by reduction, to cleave a double bond. (p. 347)

peroxide effect The reversal of orientation of HBr addition to alkenes in the presence of peroxides. A free-radical mechanism is responsible for the peroxide effect. (p. 319)

peroxyacid (peracid) A carboxylic acid with an extra oxygen atom and a peroxy (—O—O—) linkage, general formula RCO_3H. (p. 341)

polymer A high-molecular-weight compound composed of many molecules of a smaller, simpler compound called the **monomer.** (p. 350)

polymerization: The reaction of the monomer molecules to form a polymer.

regiospecific reaction A reaction that always gives the same orientation on an unsymmetrical starting material. For example, the addition of HCl is regiospecific, predicted by Markovnikov's rule. Hydroboration/oxidation is regiospecific because it consistently gives anti-Markovnikov orientation. (p. 316)

Simmons-Smith reaction A cyclopropanation of an alkene using the carbenoid generated from diiodomethane and the zinc-copper couple. (p. 335)

stereospecific reaction A reaction that converts a particular stereoisomer of the starting material to only one stereoisomer (or (±)pair) of the product. (p. 332)

anti addition: A reaction where two groups add to opposite faces of the double bond. Examples are halogenation and hydroxylation using a peroxyacid.

syn addition: A reaction where two groups add to the same face of the double bond. Examples are catalytic hydrogenation and hydroboration.

ESSENTIAL PROBLEM-SOLVING SKILLS IN CHAPTER 8

1. Predict the products of additions, oxidations, reductions, and cleavages of alkenes, including
   (a) Orientation of reaction (regiochemistry)
   (b) Stereochemistry.
2. Propose logical mechanisms to explain the observed products of alkene reactions, including regiochemistry and stereochemistry.
3. Use alkenes as starting materials and intermediates in devising one-step and multistep syntheses.
4. When more than one method is usable for a chemical transformation, choose the better method and explain its advantages.
5. Use clues provided by products of reactions such as ozonolysis to determine the structure of an unknown alkene.

In studying these reaction-intensive chapters, students ask whether they should "memorize" all the reactions. Doing organic chemistry is like speaking a foreign language, and the reactions are our vocabulary. Without knowing the words, how can you construct sentences? Making flash cards often helps.

In organic chemistry, the mechanisms, regiochemistry, and stereochemistry are our conjugations, declensions, and adjective endings. You must develop *facility* with the reactions, as you develop facility with words you use in speaking. Problems and multistep syntheses are the syntax and sentences of organic chemistry. You must *practice* combining all aspects of your vocabulary in solving these problems. Students who fail exams often do so because they have memorized the vocabulary, but they have not practiced doing problems. Others fail because they think they can do problems, but they lack the vocabulary. If you understand the reactions and can do the end-of-chapter problems without looking back, you should do well on your exams.

STUDY PROBLEMS

8-39. Define each of the following terms and give an example.

(a) dimerization **(b)** polymerization **(c)** electrophilic addition
(d) stereospecific addition **(e)** syn addition **(f)** anti addition
(g) Markovnikov addition **(h)** anti-Markovnikov addition **(i)** peroxide effect
(j) hydrogenation **(k)** hydration **(l)** homogeneous catalysis
(m) heterogeneous catalysis **(n)** halogenation **(o)** halohydrin
(p) hydroxylation **(q)** epoxidation **(r)** oxidative cleavage
(s) hydroboration **(t)** oxymercuration-demercuration **(u)** alkoxymercuration-demercuration
(v) carbene addition **(w)** alpha elimination **(x)** beta elimination

8-40. Predict the major products of the following reactions and give the structures of any intermediates. Include stereochemistry where appropriate.

(a) 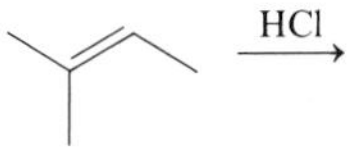 $\xrightarrow{HCl}$ **(b)** 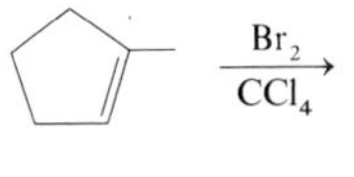 $\xrightarrow[CCl_4]{Br_2}$ **(c)** 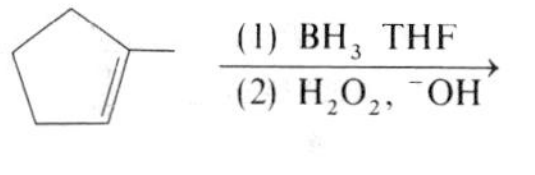 $\xrightarrow[(2)\ H_2O_2,\ ^-OH]{(1)\ BH_3\ THF}$

(d) $\xrightarrow[(2)\ (CH_3)_2S]{(1)\ O_3}$ **(e)** $\xrightarrow[ROOR]{HBr}$ **(f)** $\xrightarrow[ROOR]{HCl}$

(g) $\xrightarrow{PhCO_3H}$ **(h)** $\xrightarrow[H_2O_2]{OsO_4}$ **(i)** $\xrightarrow[(cold,\ dil.)]{KMnO_4,\ ^-OH}$

(j) $\xrightarrow[H^+,\ H_2O]{CH_3CO_3H}$ **(k)** $\xrightarrow[(warm,\ conc.)]{KMnO_4,\ ^-OH}$ **(l)** 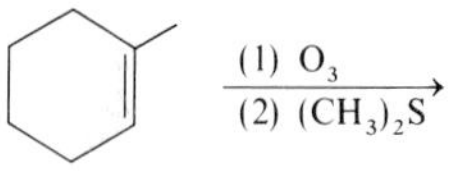 $\xrightarrow[(2)\ (CH_3)_2S]{(1)\ O_3}$

(m) 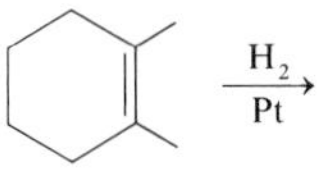 $\xrightarrow[Pt]{H_2}$ **(n)** $\xrightarrow{H^+,\ H_2O}$ **(o)** $\xrightarrow[(2)\ NaBH_4]{(1)\ Hg(OAc)_2,\ H_2O}$

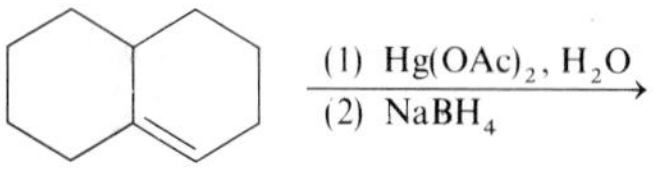

(p) 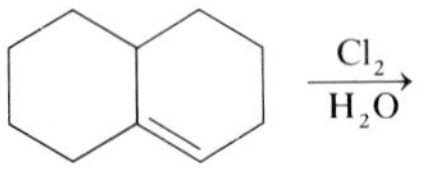 $\xrightarrow[H_2O]{Cl_2}$

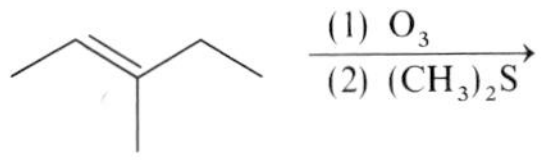

8-41. Propose mechanisms consistent with the following reactions.

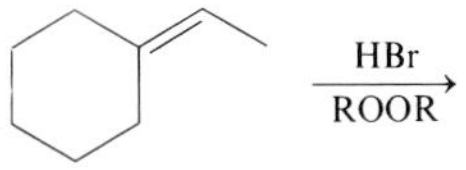

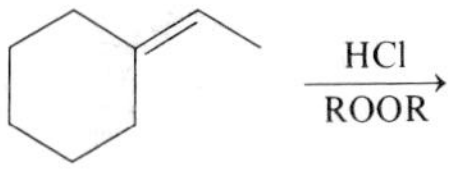

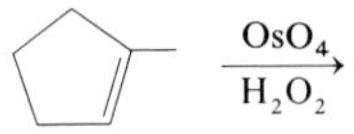

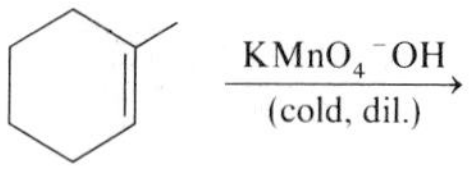

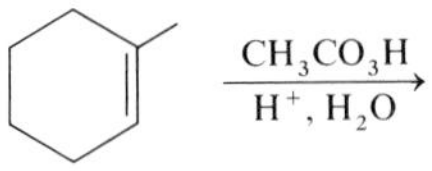

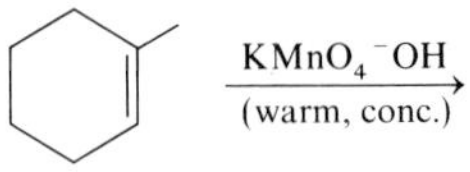

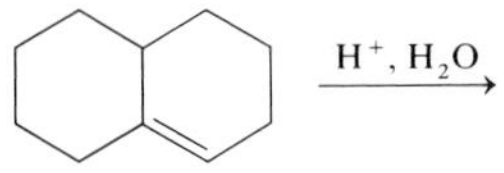

(a)  $\xrightarrow[ROOR]{HBr}$ Br **(b)** 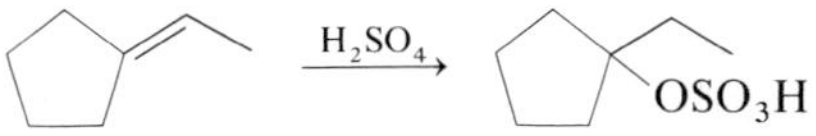 $\xrightarrow{H_2SO_4}$ OSO_3H

(c) 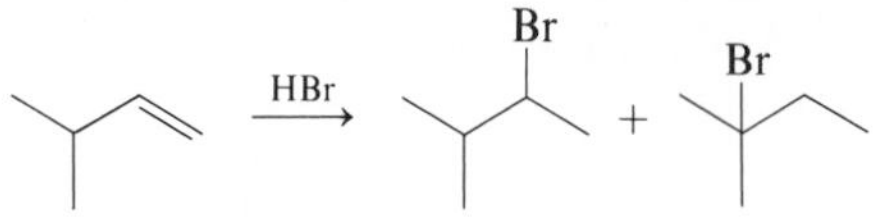 $\xrightarrow{\text{HBr}}$ + (d) 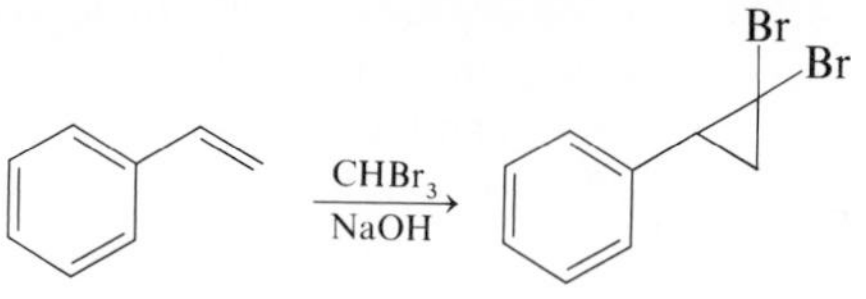 $\xrightarrow[\text{NaOH}]{\text{CHBr}_3}$

(e) 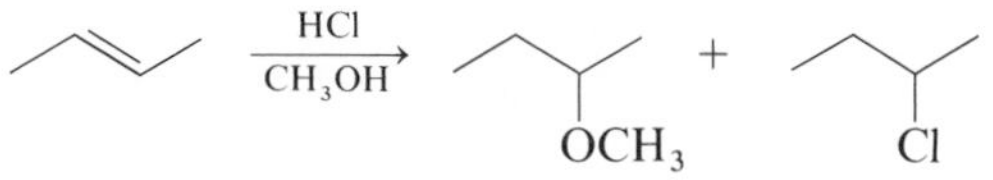 $\xrightarrow[\text{CH}_3\text{OH}]{\text{HCl}}$ +

(f) 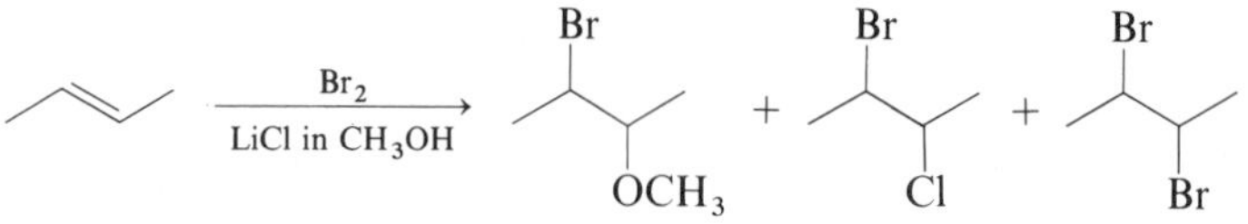 $\xrightarrow[\text{LiCl in CH}_3\text{OH}]{\text{Br}_2}$ + +

(g) 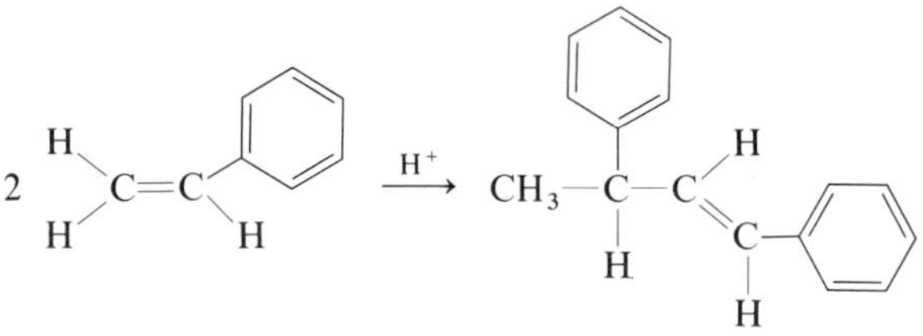 2 $\xrightarrow{\text{H}^+}$

8-42. Show how you would synthesize each of the following compounds using methylenecyclohexane as your starting material.

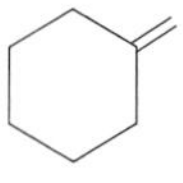

(a) 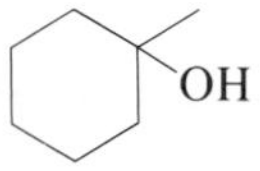 (b) 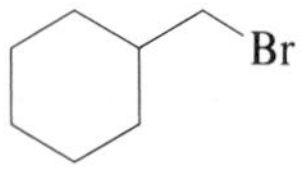 (c)

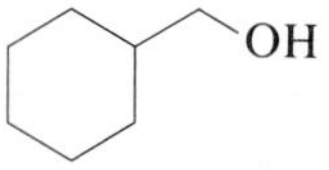

(d) 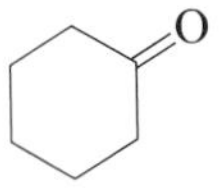 (e) 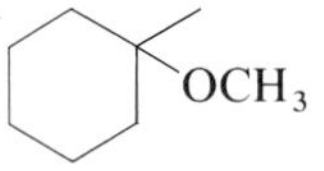 (f)

methylenecyclohexane

(g) 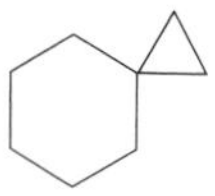 (h) 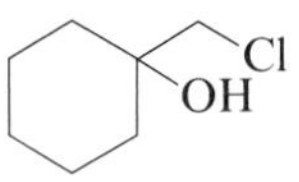 (i)

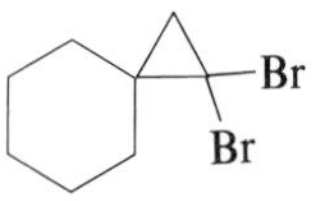

8-43. Limonene is one of the compounds that give lemons their tangy odor. Show the structures of the products expected when limonene reacts with an excess of each of these reagents.

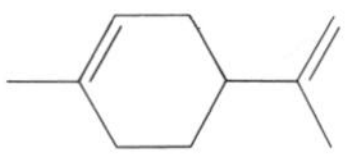

(a) borane in tetrahydrofuran, followed by basic hydrogen peroxide
(b) peroxybenzoic acid
(c) ozone, then dimethyl sulfide
(d) a mixture of osmic acid and hydrogen peroxide
(e) hot, concentrated potassium permanganate
(f) peroxyacetic acid in water
(g) hydrogen and a platinum catalyst
(h) hydrogen bromide gas
(i) hydrogen bromide gas in a solution containing dimethyl peroxide
(j) bromine water
(k) chlorine gas
(l) mercuric acetate in methanol, followed by sodium borohydride
(m) methylene iodide pretreated with the zinc-copper couple

limonene

8-44. Boron trifluoride catalyzes the polymerization of styrene (vinylbenzene). Show the structure of polystyrene and give a mechanism for its formation.

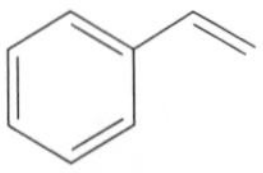

styrene

8-45. Poly(ethyl acrylate) has the formula

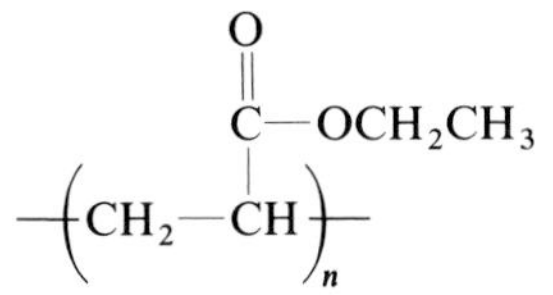

$$\left(\mathrm{CH_2{-}CH}\right)_n \text{ with } \mathrm{C(=O){-}OCH_2CH_3} \text{ on CH}$$

Give the structure of the ethyl acrylate monomer.

8-46. Draw the structures of the following compounds and determine which member of each pair is more reactive toward the addition of HBr.

(a) propene or 2-methylpropene **(b)** cyclohexene or 1-methylcyclohexene
(c) 1-butene or 1,3-butadiene

8-47. Cyclohexene is dissolved in a solution of lithium chloride in chloroform. To this solution is added one equivalent of bromine. The material isolated from this reaction contains primarily a mixture of *trans*-1,2-dibromocyclohexane and *trans*-1-bromo-2-chlorocyclohexane. Use a detailed mechanism to show how these compounds are formed.

8-48. Draw a potential energy profile for the propagation steps of the free-radical addition of HBr to isobutylene. Draw curves representing the reactions leading to both the Markovnikov and the anti-Markovnikov products. Compare the values of ΔG° and E_a for the rate-determining steps, and explain why only one of these products is observed.

8-49. Give the products expected when the following compounds are ozonized and reduced.

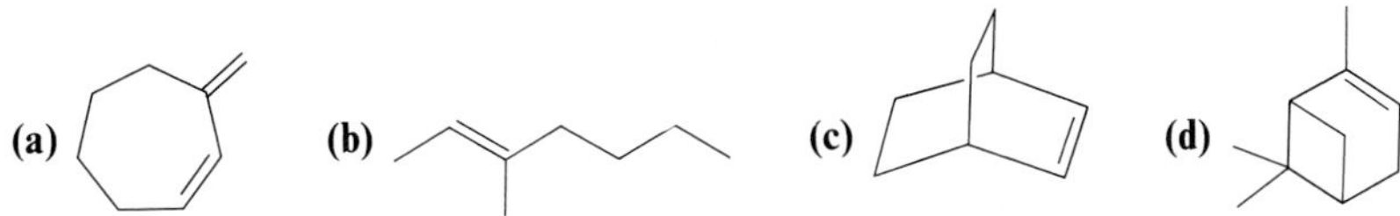

8-50. Show how you would make each of the following compounds from a suitable cyclic alkene.

(a) 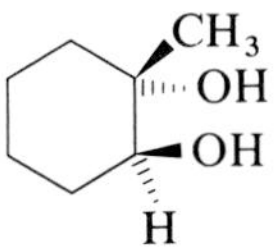 CH_3, OH, OH, H
(b) 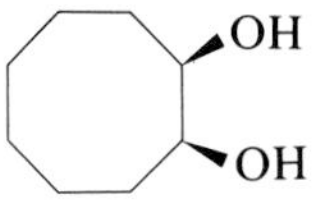 OH, OH
(c) 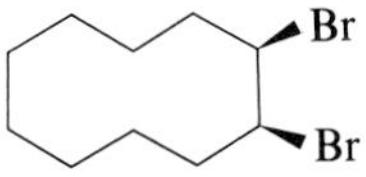 Br, Br

(d) 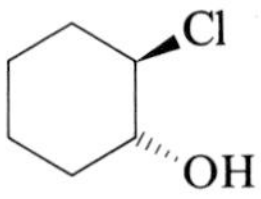 Cl, OH
(e) 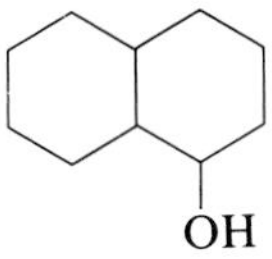 OH
(f) 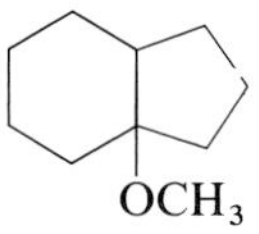 OCH_3

8-51. One of the constituents of turpentine is α-pinene, formula $C_{10}H_{16}$. The following scheme (called a "road map") gives some reactions of α-pinene. Determine the structure of α-pinene and of the reaction products **A** through **E**.

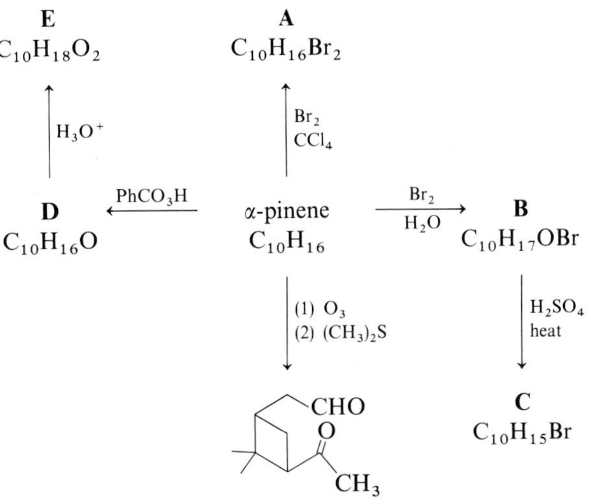

8-52. The sex attractant of the housefly has the formula $C_{23}H_{46}$. When treated with warm potassium permanganate, this pheromone gives two products: $CH_3(CH_2)_{12}COOH$ and $CH_3(CH_2)_7COOH$. Suggest a structure for this sex attractant. Is there any part of the structure that is uncertain?

8-53. In contact with a platinum catalyst, an unknown alkene reacts with 3 moles of hydrogen gas to give 1-isopropyl-4-methylcyclohexane. When the unknown alkene is ozonized and reduced, the products are the following:

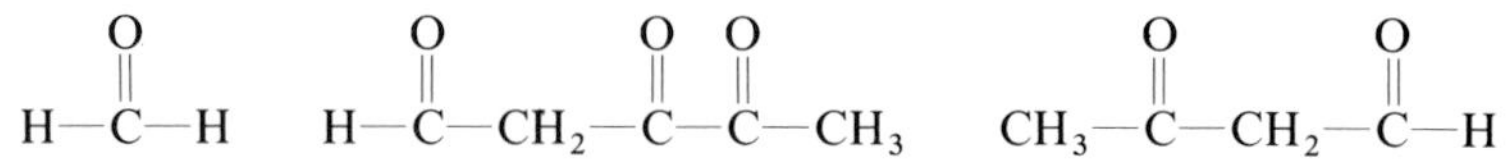

$$H-\overset{\overset{O}{\|}}{C}-H \qquad H-\overset{\overset{O}{\|}}{C}-CH_2-\overset{\overset{O}{\|}}{C}-\overset{\overset{O}{\|}}{C}-CH_3 \qquad CH_3-\overset{\overset{O}{\|}}{C}-CH_2-\overset{\overset{O}{\|}}{C}-H$$

Deduce the structure of the unknown alkene.

***8-54.** Propose a mechanism for the following reaction.

$\xrightarrow{H^+}$

O HO

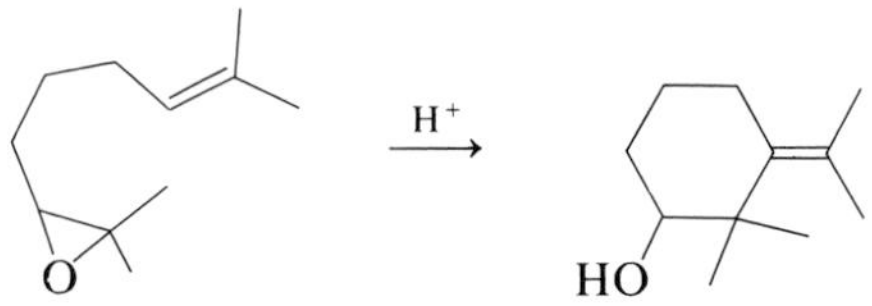

8-55. The two butenedioic acids are called *fumaric acid* (trans) and *maleic acid* (cis). 2,3-Dihydroxybutanedioic acid is called *tartaric acid.*

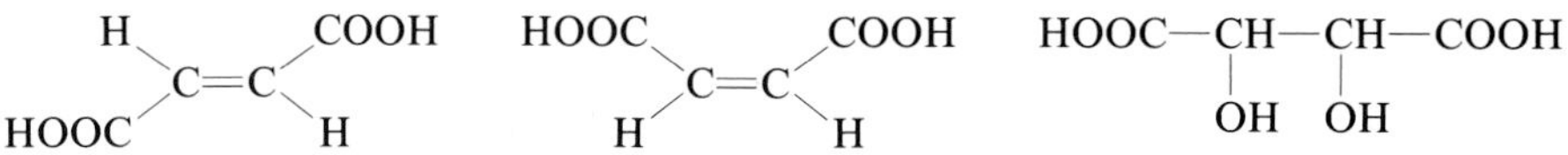

H, COOH, C=C, HOOC, H — fumaric acid

HOOC, COOH, C=C, H, H — maleic acid

HOOC—CH—CH—COOH (OH OH) — tartaric acid

Show how you would convert:

(a) fumaric acid to (*d,l*)-tartaric acid. **(b)** fumaric acid to *meso*-tartaric acid.

(c) maleic acid to (*d,l*)-tartaric acid. **(d)** maleic acid to *meso*-tartaric acid.

8-56. The compound BD_3 is a deuterated form of borane. Predict the product formed when 1-methylcyclohexene reacts with $BD_3 \cdot THF$ followed by basic hydrogen peroxide.

***8-57.** Many enzymes catalyze reactions that are similar to reactions we might use for organic synthesis. Enzymes tend to be stereospecific in their reactions, and asymmetric induction is common. The following reaction, part of the tricarboxylic acid cycle of cell respiration, resembles a reaction we might use in the laboratory; however, the enzyme-catalyzed reaction gives only the (*S*) enantiomer of the product, malic acid.

H, COOH, C=C, HOOC, H — fumaric acid $\xrightarrow[H_2O,\ pH\ 7.4]{\text{fumarase}}$ HO, H, C, COOH, CH_2COOH — (*S*)-malic acid

Product in D_2O: COOD; DO—H; D—H; COOD

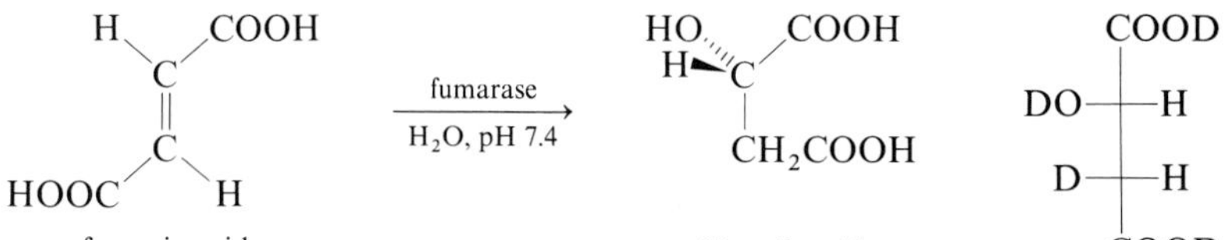

(a) What type of reaction does fumarase catalyze?

(b) Is fumaric acid chiral? Is malic acid chiral? In the enzyme-catalyzed reaction, is the product (malic acid) optically active?

(c) If we were to run the above reaction in the laboratory using sulfuric acid as the catalyst, would the product (malic acid) be optically active?

(d) Do you expect the fumarase enzyme to be a chiral molecule?

(e) When the enzyme-catalyzed reaction takes place in D_2O, the *only* product is the stereoisomer pictured above. No enantiomer or diastereomer of this compound is formed. Is the enzyme-catalyzed reaction a syn or anti addition?

(f) Assume we found conditions to convert fumaric acid to deuterated malic acid using hydroboration with $BD_3 \cdot THF$ followed by oxidation with D_2O_2 and NaOD. Use Fischer projections to show the stereoisomer(s) of deuterated malic acid you would expect to be formed.

8-58. An unknown compound decolorizes bromine in carbon tetrachloride, and it undergoes catalytic reduction to give decalin. When treated with warm, concentrated potassium permanganate, this compound gives *cis*-cyclohexane-1,2-dicarboxylic acid and oxalic acid. Propose a structure for the unknown compound.

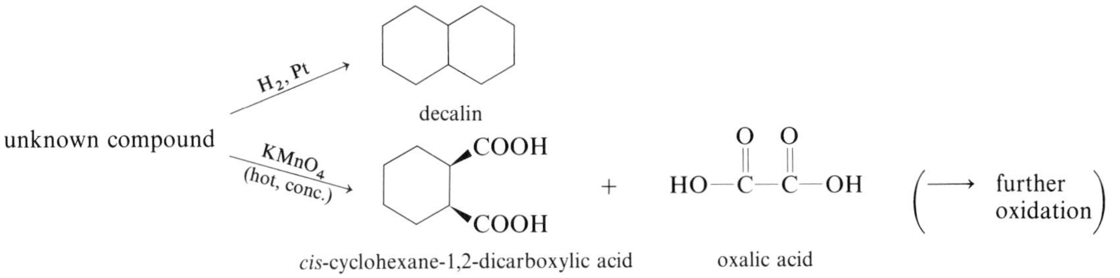

unknown compound $\xrightarrow{H_2, Pt}$ decalin

unknown compound $\xrightarrow{KMnO_4 \text{ (hot, conc.)}}$ *cis*-cyclohexane-1,2-dicarboxylic acid + oxalic acid (HO—C(=O)—C(=O)—OH) ($\longrightarrow$ further oxidation)

8-59. **(a)** The following cyclization has been observed in the oxymercuration-demercuration of this unsaturated alcohol. Propose a mechanism for this reaction.

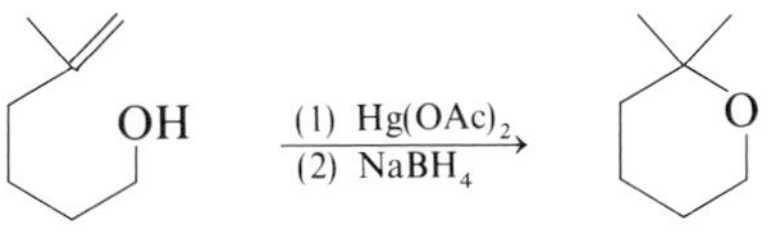

$\xrightarrow[(2)\ NaBH_4]{(1)\ Hg(OAc)_2}$

(b) Predict the product of formula $C_7H_{13}BrO$ from the reaction of this same unsaturated alcohol with bromine. Give a mechanism to support your prediction.

8-60. An inexperienced graduate student treated 5-decene with borane in THF, placed the flask in a refrigerator, and left for a party. When he returned from the party, he discovered that the refrigerator was broken and it had gotten quite warm inside. Although all the THF had evaporated from the flask, he treated the residue with basic hydrogen peroxide. To his surprise, he recovered a fair yield of 1-decanol. Use a mechanism to show how this reaction might have occurred. (*Hint:* The addition of BH_3 is reversible.)

8-61. We have seen many examples where halogens add to alkenes with anti stereochemistry via the halonium ion mechanism. However, when 1-phenylcyclohexene reacts with chlorine in carbon tetrachloride, a mixture of the cis and trans isomers of the product is recovered. Propose a mechanism and explain this lack of stereospecificity.

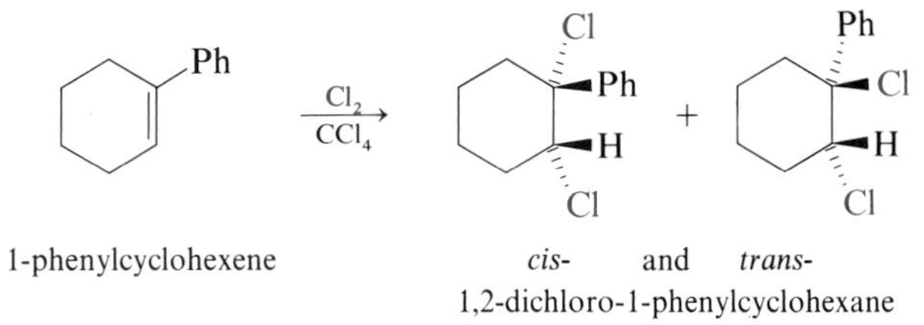

1-phenylcyclohexene $\xrightarrow[CCl_4]{Cl_2}$ *cis*- and *trans*- 1,2-dichloro-1-phenylcyclohexane

8-62. Unknown **X**, C_5H_9Br, does not react with bromine nor with dilute $KMnO_4$. Upon treatment with potassium *t*-butoxide, **X** gives only one product, **Y**, C_5H_8. Unlike **X**, **Y** decolorizes bromine and changes $KMnO_4$ from purple to brown. Ozonolysis/reduction of **Y** gives **Z**, $C_5H_8O_2$. Propose consistent structures for **X**, **Y**, and **Z**.

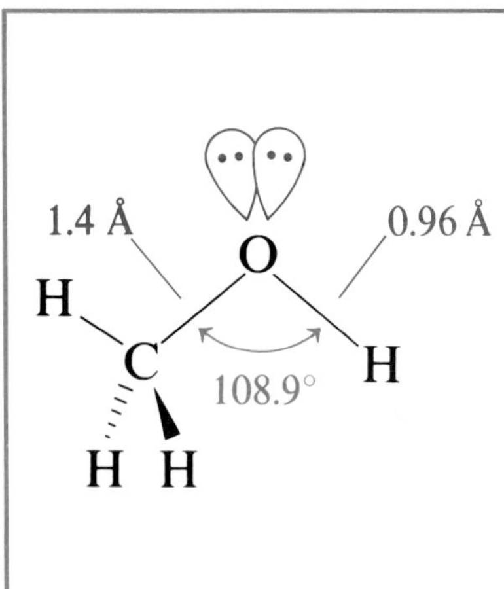

9 STRUCTURE AND SYNTHESIS OF ALCOHOLS

9-1 INTRODUCTION

Alcohols are common organic substances that have many practical, everyday applications. Ethyl alcohol (grain alcohol) is found in alcoholic beverages, cosmetics, and drug preparations. Methyl alcohol (wood alcohol) is used as a fuel and a solvent. Isopropyl alcohol (rubbing alcohol) is used as a skin cleanser for injections and minor cuts.

$$CH_3-CH_2-OH \qquad CH_3-OH \qquad CH_3-\underset{}{\overset{\overset{\displaystyle OH}{|}}{C}H}-CH_3$$

ethyl alcohol — methyl alcohol — isopropyl alcohol

Alcohols are synthesized by a wide variety of methods, and the —OH group may be converted to most other functional groups. For these reasons, alcohols are one of the most useful classes of organic compounds. In this chapter we discuss the physical properties of alcohols and summarize the methods used to synthesize them. In Chapter 10 (Reactions of Alcohols) we continue our study of the central role that alcohols play in organic chemistry as reagents, solvents, and synthetic intermediates.

9-2 STRUCTURE AND CLASSIFICATION OF ALCOHOLS

The structure of an alcohol resembles the structure of water, with an alkyl group replacing one of the hydrogen atoms of water. Figure 9-1 compares the structures of water and methanol. Both have sp^3 hybridized oxygen atoms, but the C—O—H bond angle in methanol (108.9°) is considerably larger than the H—O—H bond angle in water (104.5°) because the methyl group is much larger than a hydrogen atom. The bulky methyl group counteracts the bond angle compression caused by

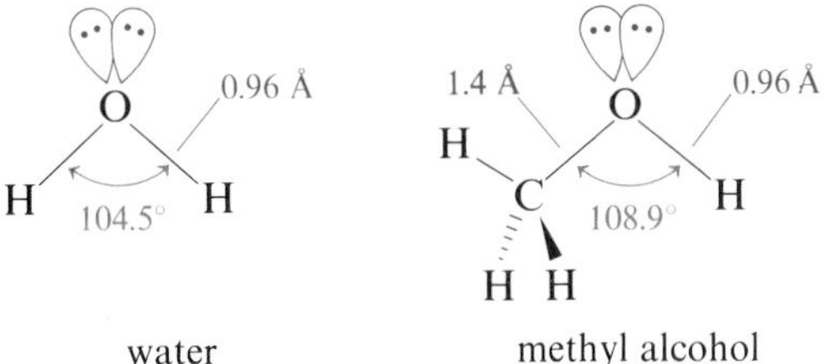

FIGURE 9-1 Comparison of the structures of water and methyl alcohol.

the nonbonding pairs of electrons. The O—H bond lengths are about the same in water and methanol (0.96 Å), but the C—O bond is considerably longer (1.4 Å), reflecting the larger covalent radius of carbon compared to hydrogen.

One way of organizing the alcohol family is to classify each alcohol according to the type of carbon atom bonded to the —OH group. If this carbon atom is primary (bonded to one other carbon atom), the compound is a **primary alcohol.** A **secondary alcohol** has the —OH group attached to a secondary carbon atom, and a **tertiary alcohol** has it bonded to a tertiary carbon. When we studied alkyl halides, we saw that primary, secondary, and tertiary halides react differently. The same principle holds for alcohols. We need to learn how these classes of alcohols are similar and under what conditions they react differently. Figure 9-2 shows examples of primary, secondary, and tertiary alcohols.

Alcohols with the hydroxyl group bonded directly to an aromatic (benzene) ring are called **phenols.** Phenols have many properties similar to those of other alcohols, while other properties derive from their aromatic character. In this chapter we consider the properties of phenols that are similar to those of other alcohols, while noting some of the differences. In Chapter 16 we consider the aromatic nature of phenols and the reactions that result from that aromaticity.

9-3 NOMENCLATURE OF ALCOHOLS AND PHENOLS

9-3A IUPAC NAMES ("ALKANOL" NAMES)

The IUPAC system provides unique names for alcohols based on rules that are similar to those for other classes of compounds. In general, the name carries the *-ol* suffix, together with a number to give the location of the hydroxyl group. The formal rules are summarized in a three-step procedure:

1. Name the longest carbon chain that contains the carbon atom bearing the —OH group. Drop the final *-e* from the alkane name and add the suffix *-ol* to give the root name.
2. Number the longest carbon chain starting at the end nearest the hydroxyl group, and use the appropriate number to indicate the position of the —OH group. (The hydroxyl group takes precedence over double and triple bonds.)
3. Name all the substituents and give their numbers, as you would for an alkane or an alkene.

In the following example, the longest carbon chain has four carbons, and the root name is *butanol.* The —OH group is on the second carbon atom, so this is a 2-butanol. The complete IUPAC name is 1-chloro-3,3-dimethyl-2-butanol.

$$\begin{array}{ccccccc} & & CH_3 & OH & & & \\ & & | & | & & & \\ {}^4CH_3 & — & {}^3C & — {}^2CH & — & {}^1CH_2 & — Cl \\ & & | & & & & \\ & & CH_3 & & & & \end{array}$$

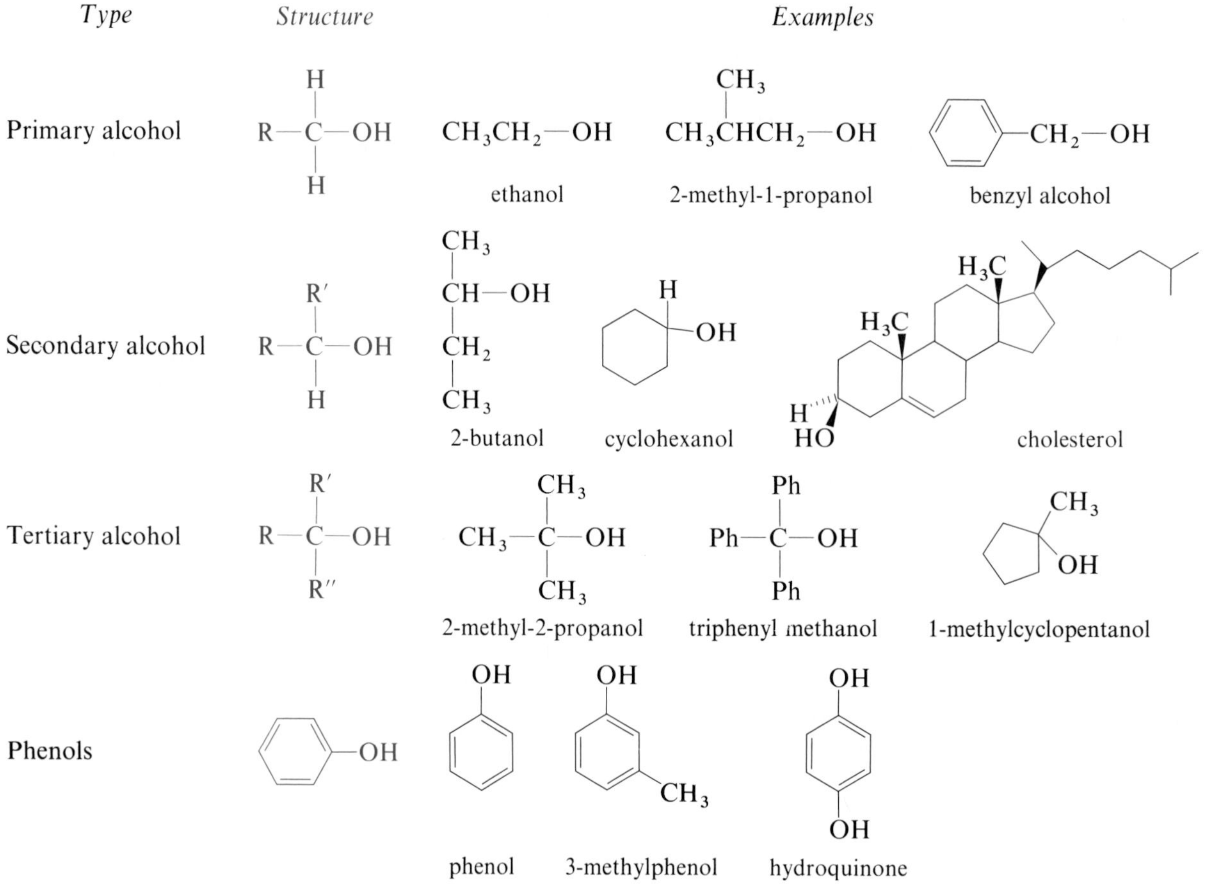

Type	*Structure*	*Examples*		
Primary alcohol	R—CH₂—OH	CH_3CH_2—OH ethanol	$CH_3CH(CH_3)CH_2$—OH 2-methyl-1-propanol	C_6H_5—CH_2—OH benzyl alcohol
Secondary alcohol	R—CHR′—OH	CH_3—CH(OH)—CH_2—CH_3 2-butanol	cyclohexanol	cholesterol
Tertiary alcohol	R—CR′R″—OH	CH_3—C(CH_3)₂—OH 2-methyl-2-propanol	Ph—C(Ph)(Ph)—OH triphenyl methanol	1-methylcyclopentanol
Phenols	C_6H_5—OH	phenol	3-methylphenol	hydroquinone

FIGURE 9-2 Alcohols are classified according to the type of carbon atom (primary, secondary, or tertiary) bonded to the hydroxyl group. Phenols are alcohols with the hydroxyl group bonded to a carbon atom in a benzene ring.

Cyclic alcohols are named using the prefix *cyclo-*; the hydroxyl group is assumed to be on C1.

IUPAC name: *trans*-2-chlorocyclohexanol 1-ethylcyclopropanol

SOLVED PROBLEM 9-1

Give the systematic (IUPAC) name for the following alcohol.

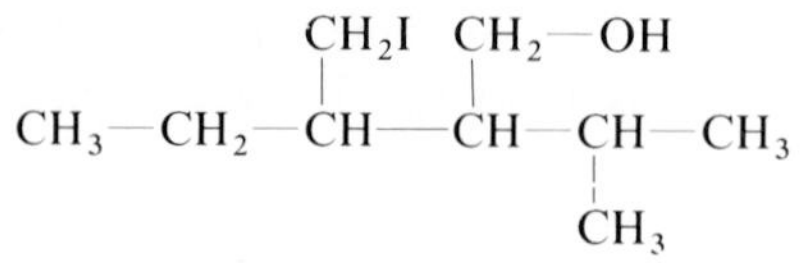

$$CH_3-CH_2-CH(CH_2I)-CH(CH_2-OH)-CH(CH_3)-CH_3$$

SOLUTION The longest chain contains six carbon atoms, but it does not contain the carbon bonded to the hydroxyl group. The longest chain containing the carbon bonded to the

—OH group is the one indicated in red, containing five carbon atoms. This chain is numbered from right to left so as to give the hydroxyl-bearing carbon atom the lowest possible number.

$$\begin{array}{c} \quad\quad\quad\quad CH_2I \quad {}^1CH_2—OH \\ {}^5CH_3—{}^4CH_2—{}^3CH—{}^2CH—CH—CH_3 \\ \quad\quad\quad\quad\quad\quad\quad\quad CH_3 \end{array}$$

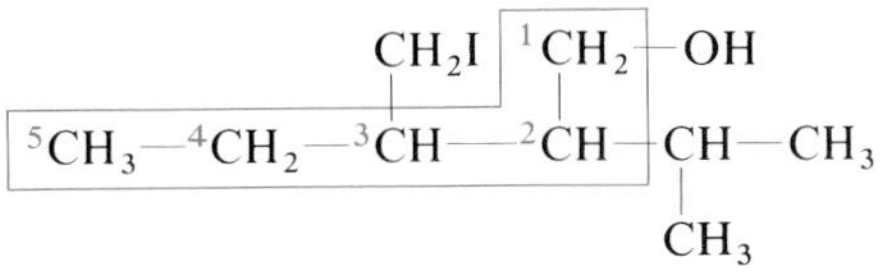

The correct name for this compound is 3-(iodomethyl)-2-isopropyl-1-pentanol.

Alcohols containing double and triple bonds are named using the *-ol* suffix on the alkene or alkyne name. The carbon chain is numbered so as to give the lowest possible number to the carbon atom bonded to the hydroxyl group. If numbers are needed to give the location of the multiple bonds, the position of the —OH group can be given by putting its number right next to the *-ol* suffix.

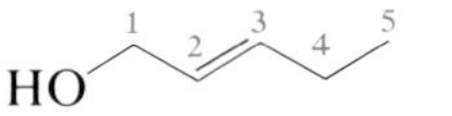

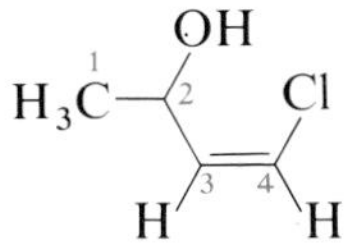

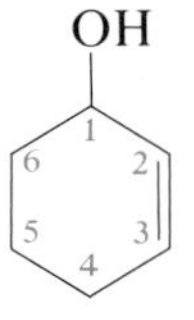

IUPAC name: *trans*-2-penten-1-ol (Z)-4-chloro-3-buten-2-ol 2-cyclohexen-1-ol

If a structure is too difficult to name as an alcohol, or if the hydroxyl group is a minor part of a more significant structure, the —OH functional group may be named as a *hydroxy* substituent.

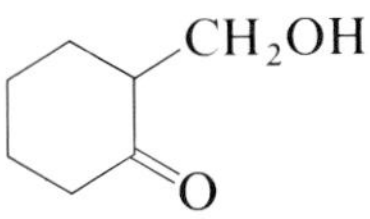

2-hydroxymethylcyclohexanone 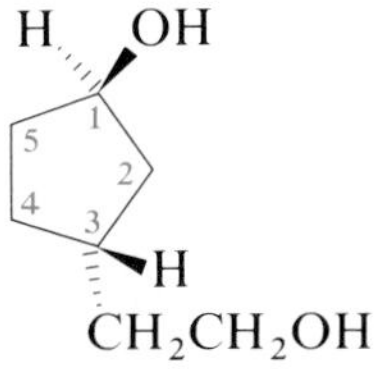 *trans*-3-(2-hydroxyethyl)cyclopentanol 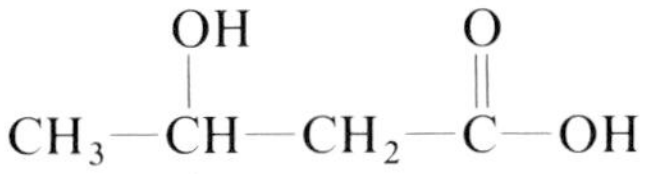 3-hydroxybutanoic acid

PROBLEM 9-1

Give the IUPAC names of the following alcohols.

(a) 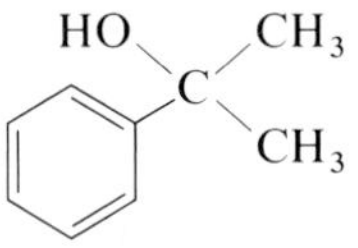 (b) 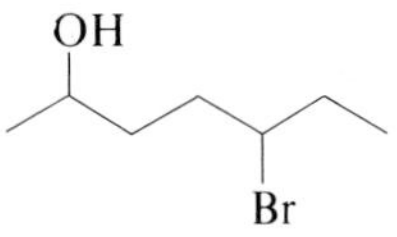 (c)

(d) 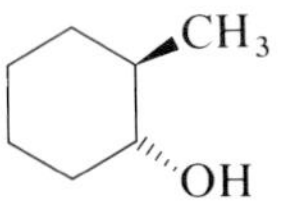 (e) 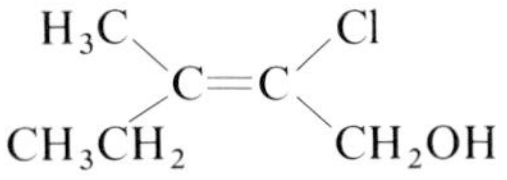 (f)

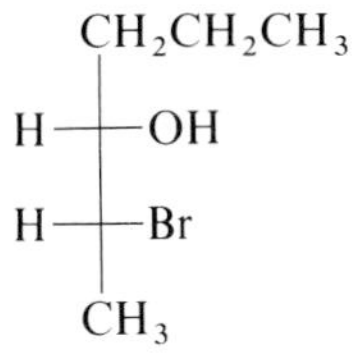

9-3B COMMON NAMES OF ALCOHOLS

The common name of an alcohol is derived from the common name of the alkyl group and the word "alcohol." This system pictures an alcohol as a molecule of water with an alkyl group replacing one of the hydrogen atoms. If the structure is complex, the common nomenclature becomes awkward and the IUPAC nomenclature should be used.

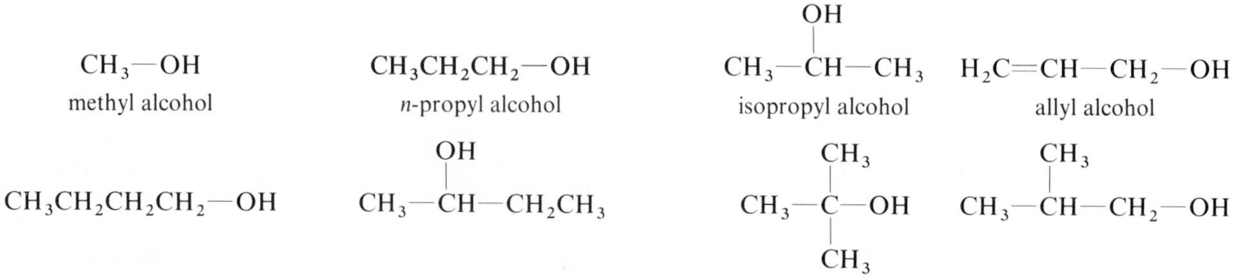

$CH_3—OH$ methyl alcohol

$CH_3CH_2CH_2—OH$ *n*-propyl alcohol

$CH_3—CH(OH)—CH_3$ isopropyl alcohol

$H_2C{=}CH—CH_2—OH$ allyl alcohol

$CH_3CH_2CH_2CH_2—OH$ *n*-butyl alcohol

$CH_3—CH(OH)—CH_2CH_3$ *sec*-butyl alcohol

$CH_3—C(CH_3)_2—OH$ *t*-butyl alcohol

$CH_3—CH(CH_3)—CH_2—OH$ isobutyl alcohol

PROBLEM 9-2

Give both the IUPAC name and the common name for each of the following alcohols.

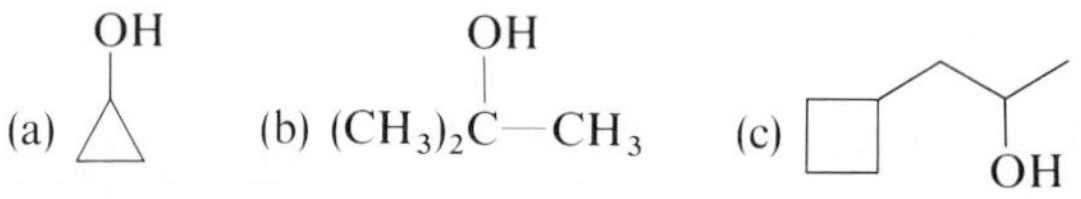

(a) cyclopropanol (OH on cyclopropane ring) (b) $(CH_3)_2C(OH)—CH_3$ (c) cyclobutyl—CH_2—$CH(OH)$—CH_3 (d) $(CH_3)_2CHCH_2CH_2OH$

PROBLEM 9-3

For each of the following molecular formulas, draw all the possible alcohols with that formula. Give the IUPAC name for each alcohol.

(a) C_3H_8O (b) $C_4H_{10}O$ (c) C_4H_8O

9-3C NOMENCLATURE OF DIOLS

Alcohols with two —OH groups are called **diols** or **glycols.** They are named like other alcohols except that the suffix *diol* is used and two numbers are needed to tell where the two hydroxyl groups are located. This is the preferred, systematic (IUPAC) method for naming diols.

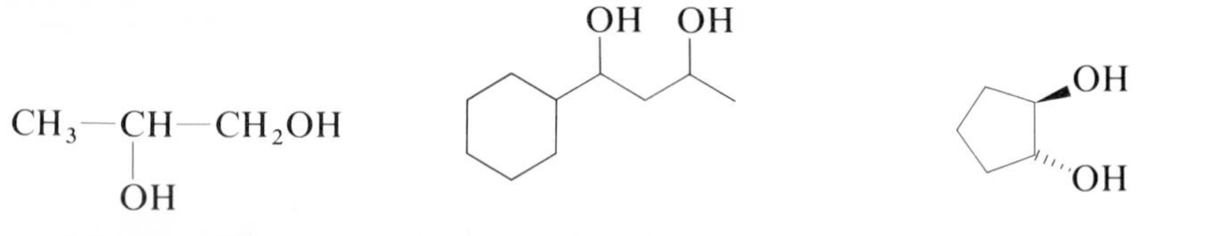

$CH_3—CH(OH)—CH_2OH$ (cyclohexyl)$—CH(OH)—CH_2—CH(OH)—CH_3$ (cyclopentane with OH wedge and OH hash)

IUPAC name: 1,2-propanediol 1-cyclohexyl-1,3-butanediol *trans*-1,2-cyclopentanediol

The term *glycol* generally means a 1,2-diol, or **vicinal diol,** with its two hydroxyl groups on adjacent carbon atoms. Glycols are usually synthesized by the hydroxylation of alkenes, using peroxyacids, osmium tetroxide, or potassium permanganate (Section 8-14).

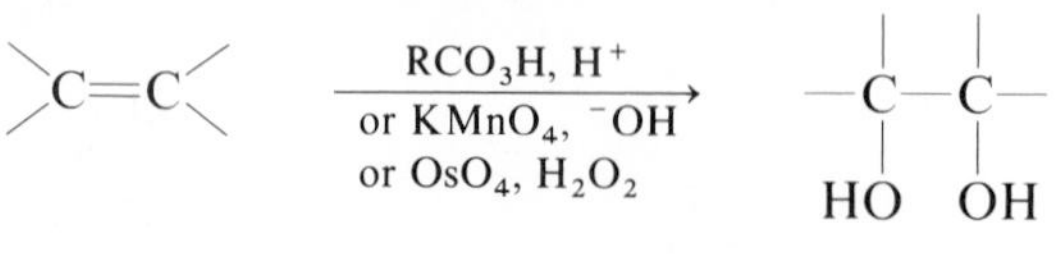

$$>C{=}C< \xrightarrow[\text{or } KMnO_4,\ ^-OH \text{ or } OsO_4,\ H_2O_2]{RCO_3H,\ H^+} —\underset{HO}{\overset{|}{C}}—\underset{OH}{\overset{|}{C}}—$$

This synthesis of glycols is reflected in their common names. The glycol is named for the alkene from which it is synthesized:

	$CH_2(OH)—CH_2(OH)$	$CH_2(OH)—CH(OH)—CH_3$	cyclohexane ring with two cis OH groups
common name:	ethylene glycol	propylene glycol	*cis*-cyclohexene glycol
IUPAC name:	1,2-ethanediol	1,2-propanediol	*cis*-1,2-cyclohexanediol

The common names of glycols can be awkward and confusing, because the *ene* portion of the name should imply the presence of an alkene double bond, but the glycol does not contain a double bond. We will generally use the IUPAC "diol" nomenclature for diols; however, you should be aware that the names "ethylene glycol" (automotive antifreeze) and "propylene glycol" (used in medicines and foods) are universally accepted for these common diols.

PROBLEM 9-4

Give a systematic (IUPAC) name for each of the following diols.

(a) $CH_3CH(OH)(CH_2)_4CH(OH)C(CH_3)_3$ (b) $HO—(CH_2)_8—OH$

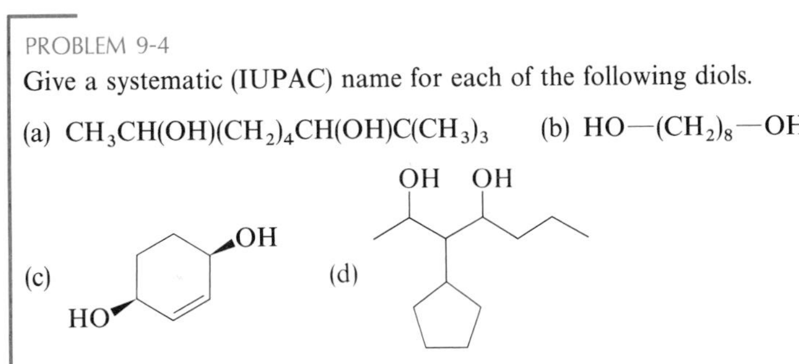

9-3D NOMENCLATURE OF PHENOLS

Because the phenol structure involves a benzene ring, the terms *ortho* (1,2-disubstituted), *meta* (1,3-disubstituted), and *para* (1,4-disubstituted) are often used in the common names. The following examples illustrate the systematic names and the common names of some simple phenols.

	(phenol ring with OH and Br)	(phenol ring with OH and NO_2)	(phenol ring with OH and CH_3CH_2)
IUPAC name:	2-bromophenol	3-nitrophenol	4-ethylphenol
common name:	*ortho*-bromophenol	*meta*-nitrophenol	*para*-ethylphenol

The methylphenols are called *cresols,* while the benzenediols have names based on their historical uses and sources rather than their structures. We will generally use the systematic names of phenolic compounds.

(OH, CH_3)	(OH, OH)	(OH, OH)	(OH, HO)
2-methylphenol (*ortho*-cresol)	1,2-benzenediol (catechol)	1,3-benzenediol (resorcinol)	1,4-benzenediol (hydroquinone)

9-4 PHYSICAL PROPERTIES OF ALCOHOLS

Most of the common alcohols, up to about 11 or 12 carbon atoms, are liquids at room temperature. Methanol and ethanol are free-flowing volatile liquids with characteristic fruity odors. The higher alcohols (the butanols through the decanols) are somewhat viscous, and some of the highly branched isomers are solids at room temperature. These higher alcohols have heavier but still fruity odors. 1-Propanol and 2-propanol fall in the middle, with a barely noticeable viscosity and a characteristic odor often associated with a physician's office. Table 9-1 lists the physical properties of some common alcohols.

9-4A BOILING POINTS OF ALCOHOLS

Since we often deal with liquid alcohols, we forget how surprising it *should* be that the lower-molecular-weight alcohols are liquids. For example, ethyl alcohol and propane have similar molecular weights, yet their boiling points differ by about 120°C.

$H-O-CH_2CH_3$ $\mu = 1.69$ D
ethanol, MW 46
b.p. 78°C

$H_3C-O-CH_3$ $\mu = 1.30$ D
dimethyl ether, MW 46
b.p. −25°C

$H_3C-CH_2-CH_3$ $\mu = 0.08$ D
propane, MW 44
b.p. −42°C

Such a large difference in boiling points indicates that ethanol molecules are attracted to each other much more strongly than propane molecules. Two important intermolecular forces are responsible: dipole-dipole attractions and hydrogen bonding.

The polarized C—O and H—O bonds and the nonbonding electrons add to produce a dipole moment of 1.69 D in ethanol, compared with a dipole moment of only 0.08 D in propane. The positive and negative ends of these dipoles align to produce attractive interactions in ethanol. Dimethyl ether has nearly as large

TABLE 9-1
Physical properties of selected alcohols

IUPAC name	*Common name*	*Formula*	*m.p. (°C)*	*b.p. (°C)*	*density*
methanol	methyl alcohol	CH_3OH	−97	65	0.79
ethanol	ethyl alcohol	CH_3CH_2OH	−114	78	0.79
1-propanol	*n*-propyl alcohol	$CH_3CH_2CH_2OH$	−126	97	0.80
2-propanol	isopropyl alcohol	$(CH_3)_2CHOH$	−89	82	0.79
1-butanol	*n*-butyl alcohol	$CH_3(CH_2)_3OH$	−90	118	0.81
2-butanol	*sec*-butyl alcohol	$CH_3CH(OH)CH_2CH_3$	−114	100	0.81
2-methyl-1-propanol	isobutyl alcohol	$(CH_3)_2CHCH_2OH$	−108	108	0.80
2-methyl-2-propanol	*t*-butyl alcohol	$(CH_3)_3COH$	25	83	0.79
1-pentanol	*n*-pentyl alcohol	$CH_3(CH_2)_4OH$	−79	138	0.82
3-methyl-1-butanol	isopentyl alcohol	$(CH_3)_2CHCH_2CH_2OH$	−117	132	0.81
2,2-dimethyl-1-propanol	neopentyl alcohol	$(CH_3)_3CCH_2OH$	52	113	0.81
cyclopentanol	cyclopentyl alcohol	*cyclo*-C_5H_9OH	−19	141	0.95
1-hexanol	*n*-hexanol	$CH_3(CH_2)_5OH$	−52	156	0.82
cyclohexanol	cyclohexyl alcohol	*cyclo*-$C_6H_{11}OH$	25	162	0.96
1-heptanol	*n*-heptyl alcohol	$CH_3(CH_2)_6OH$	−34	176	0.82
1-octanol	*n*-octyl alcohol	$CH_3(CH_2)_7OH$	−16	194	0.83
1-nonanol	*n*-nonyl alcohol	$CH_3(CH_2)_8OH$	−6	214	0.83
1-decanol	*n*-decyl alcohol	$CH_3(CH_2)_9OH$	6	233	0.83
2-propen-1-ol	allyl alcohol	$H_2C{=}CH{-}CH_2OH$	−129	97	0.86
phenylmethanol	benzyl alcohol	Ph—CH_2OH	−15	205	1.05
diphenylmethanol	diphenylcarbinol	Ph_2CHOH	69	298	
triphenylmethanol	triphenylcarbinol	Ph_3COH	162	380	1.20

a dipole moment as ethanol (1.30 D), yet its boiling point is 103°C below that of ethanol. Clearly there must be another form of interaction between ethanol molecules.

Hydrogen bonding is the major intermolecular attraction responsible for ethanol's high boiling point. Ethanol has a hydrogen atom, strongly polarized by its bond to oxygen, that can form a hydrogen bond with a pair of nonbonding electrons (Section 2-12C). Ethers have two alkyl groups bonded to their oxygen atoms, so they have no O—H hydrogen atoms to form hydrogen bonds. Hydrogen bonds have a strength of about 5 kcal (21 kJ) per mole: much weaker than typical bonds of 70 to 110 kcal, but much stronger than dipole-dipole attractions.

9-4B SOLUBILITY PROPERTIES OF ALCOHOLS

Water and alcohols have similar properties because they all contain hydroxyl groups that can form hydrogen bonds. Alcohols form hydrogen bonds with water, and several of the lower-molecular-weight alcohols are *miscible* (soluble in any proportions) with water. Similarly, alcohols are much better solvents than hydrocarbons for polar substances. Significant amounts of ionic compounds such as sodium chloride can dissolve in some of the lower alcohols. We call the hydroxyl group **hydrophilic,** meaning "water loving," because of its affinity for water and other polar substances.

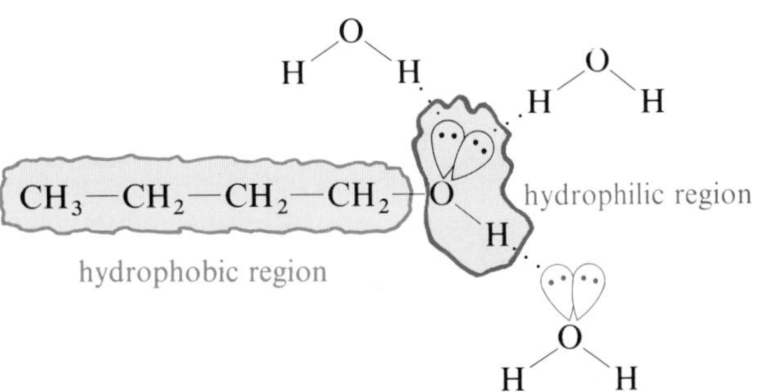

TABLE 9-2
Water solubility of alcohols (at 25°C)

Alcohol	*Solubility in water*
methyl	miscible
ethyl	miscible
n-propyl	miscible
t-butyl	miscible
isobutyl	10.0%
n-butyl	9.1%
n-pentyl	2.7%
cyclohexyl	3.6%
n-hexyl	0.6%
phenol	9.3%
hexane-1,6-diol	miscible

The alcohol's alkyl group is called **hydrophobic** ("water hating") because it acts like an alkane: It disrupts the network of dipole-dipole attractions and hydrogen bonds of a polar solvent such as water. The alkyl group makes the alcohol less hydrophilic, yet it lends solubility in nonpolar organic solvents. Many alcohols are miscible with a wide range of nonpolar organic solvents.

Table 9-2 lists the solubility of some simple alcohols in water. The water solubility decreases as the alkyl group becomes larger. With one-, two-, or three-carbon alkyl groups, alcohols are miscible with water. A four-carbon alkyl group is large enough that some isomers are not miscible, yet *t*-butyl alcohol, with a compact spherical shape, is miscible. Phenol is unusually soluble for a six-carbon alcohol because of its compact shape and the particularly strong hydrogen bonds formed between phenolic —OH groups and water molecules.

PROBLEM 9-5

Predict which member of each pair will be more soluble in water. Explain the reasons for your answers.

(a) 1-hexanol and cyclohexanol
(b) 1-heptanol and 4-methylphenol
(c) 3-ethyl-3-hexanol and 2-octanol
(d) 2-hexanol and cyclooctane-1,4-diol

(e) H OH OH H and H OH OH H

PROBLEM 9-6

Dimethylamine, $(CH_3)_2NH$, has a molecular weight of 45 and a boiling point of 7.4°C. Trimethylamine, $(CH_3)_3N$, has a higher molecular weight (59) but a *lower* boiling point (3.5°C). Explain this apparent discrepancy.

9-5 COMMERCIALLY IMPORTANT ALCOHOLS

9-5A METHANOL

Methanol (methyl alcohol) was originally produced by the destructive distillation of wood chips in the absence of air. This source led to the name **wood alcohol.** During Prohibition (1919–1933), when the manufacture of alcoholic beverages was prohibited, anything called "alcohol" was often used for mixing drinks. Since methanol is quite toxic, this practice resulted in many cases of blindness and death.

Methanol is a superb fuel for internal combustion engines. For many years, all the winning cars at Indianapolis used methanol-burning Offenhauser racing engines. Only in the past few years have gasoline engines surpassed the old Offenhausers. Methanol is becoming attractive as a motor fuel once again because of the increased price and uncertain availability of crude oil.

Today, most methanol is synthesized by a catalytic process in which carbon monoxide and hydrogen react to form methanol. This reaction uses high temperatures and pressures and requires large, complicated industrial reactors.

$$\underset{\text{synthesis gas}}{CO + 2\,H_2} \xrightarrow[\text{ZnO—Cr}_2\text{O}_3\text{ catalyst}]{300\text{–}400°\text{C, }200\text{–}300\text{ atm H}_2} CH_3OH$$

Synthesis gas containing the hydrogen and carbon monoxide needed to make methanol can be generated by the partial burning of coal in the presence of water. By carefully regulating the amount of water added, the correct ratio of carbon monoxide to hydrogen can be obtained in the synthesis gas produced.

$$3\,C + 4\,H_2O \xrightarrow{\text{high temperature}} CO_2 + \underset{\text{synthesis gas}}{2\,CO + 4\,H_2}$$

Synthesis gas can also be formed underground, without mining the coal. Two wells are drilled into the coal seam, and the coal between the wells is shattered by explosives. The coal is ignited, and compressed air and water are forced into one well (Fig. 9-3). The burning coal provides the heat and the carbon needed, producing synthesis gas which leaves through the second well. This is called the *in situ* ("in place") *process,* since it is done without having to move the coal.

9-5B ETHANOL

The prehistoric discovery of ethanol probably occurred when rotten fruit was consumed and found to have an intoxicating effect. This discovery presumably led to the intentional fermentation of fruit juices. The primitive wine that resulted could be stored (in a sealed container) without danger of decomposition, and it also served as a safe, unpolluted source of drinking water.

Ethanol can be produced by the fermentation of sugars and starches from

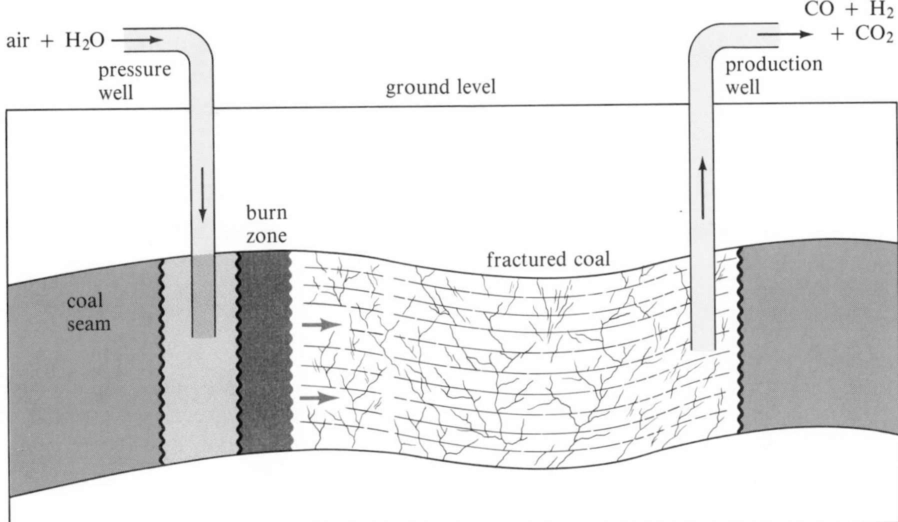

FIGURE 9-3 *In situ* formation of synthesis gas. A mixture of air and water is forced into a burning coal seam, producing CO_2, CO, and H_2.

many different sources. Grains such as corn, wheat, rye, and barley are common sources, giving ethanol the name **grain alcohol.** Cooking the grain, followed by addition of sprouted barley, called *malt,* converts some of the starches to simpler sugars. Brewer's yeast is then added, and the solution is incubated while the yeast cells convert simple sugars such as glucose to ethanol and carbon dioxide.

$$\underset{\text{glucose}}{C_6H_{12}O_6} \xrightarrow{\text{yeast enzymes}} \underset{\text{ethanol}}{2\,C_2H_5OH} + 2\,CO_2$$

The alcoholic solution that results from fermentation contains only 12 to 15 percent alcohol, because yeast cells cannot survive higher concentrations. Distillation increases the alcohol concentration to about 40 to 50 percent (80 to 100 "proof") for "hard" liquors. Distillation of ethanol-water solutions cannot increase the ethanol concentration above 95 percent because the solution of 95 percent ethanol and 5 percent water boils at a lower temperature (78.15°C) than either pure water (100°C) or pure ethanol (78.3°C). Such a mixture of liquids that boils at a lower temperature than either of its components is called a minimum-boiling **azeotrope.**

The 95 percent alcohol produced by distillation is well suited for use as a solvent and a reagent when traces of water do not affect the reaction. When **absolute alcohol** (100 percent ethanol) is required, the 95 percent azeotrope is passed through a dehydrating agent such as anhydrous calcium oxide (CaO), which removes the final 5 percent water.

Since World War II, most industrial ethanol has been synthesized directly by the catalyzed high-temperature, high-pressure, gas-phase reaction of water with ethylene. Catalysts such as P_2O_5, tungsten oxide, or various specially treated clays are used.

$$H_2C{=}CH_2 + H_2O \xrightarrow[\text{catalyst}]{\text{100–300 atm, 300°C}} CH_3{-}CH_2{-}OH$$

Like methanol, ethanol is an excellent motor fuel, with an octane rating above 100. A car's carburetor must be adjusted if it is to run on pure ethanol, but solutions of about 10 percent ethanol in gasoline ("gasohol") work well without any adjustments.

9-5C 2-PROPANOL

2-Propanol (isopropyl alcohol) is made by the catalytic hydration of propylene. 2-Propanol is commonly used as rubbing alcohol rather than ethanol, because 2-propanol has less of a drying effect on the skin and is not regulated and taxed by the government.

$$\underset{\text{propylene}}{CH_3{-}CH{=}CH_2} + H_2O \xrightarrow[\text{catalyst}]{100\text{–}300\text{ atm, }300^\circ C} \underset{\text{2-propanol}}{CH_3{-}\underset{|}{\overset{}{CH}}(OH){-}CH_3}$$

9-6 ACIDITY OF ALCOHOLS AND PHENOLS

Like the hydroxyl proton of water, the hydroxyl proton of an alcohol is weakly acidic. A strong base can remove the hydroxyl proton of an alcohol to give an **alkoxide ion.**

$$\underset{\text{alcohol}}{R{-}\ddot{\underset{..}{O}}{-}H} + B{:}^- \rightleftharpoons \underset{\text{alkoxide ion}}{R{-}\ddot{\underset{..}{O}}{:}^-} + B{-}H$$

Example

$$\underset{\text{ethanol}}{CH_3CH_2{-}\ddot{\underset{..}{O}}{-}H} + B{:}^- \rightleftharpoons \underset{\text{ethoxide ion}}{CH_3CH_2{-}\ddot{\underset{..}{O}}{:}^-} + B{-}H$$

The acidities of alcohols vary widely, from alcohols that are about as acidic as water to some that are much less acidic. The acid dissociation constant, K_a, of an alcohol is defined by the equilibrium:

$$R{-}O{-}H + H_2O \overset{K_a}{\rightleftharpoons} R{-}O^- + H_3O^+$$

$$K_a = \frac{[H_3O^+][RO^-]}{[ROH]} \qquad pK_a = -\log(K_a)$$

9-6A EFFECTS ON ACIDITY

The acid dissociation constants for alcohols vary according to their structure, from about 10^{-16} for methanol down to about 10^{-19} for most tertiary alcohols. The acidity decreases as the substitution on the alkyl group increases because the more highly substituted alkyl group inhibits solvation of the alkoxide ion and drives the dissociation equilibrium to the left. Table 9-3 compares the acid dissociation constants for some representative alcohols with those of water and other acids.

Table 9-3 also shows that substitution by electron-withdrawing halogen atoms enhances the acidity of alcohols. For example, 2-chloroethanol is more acidic than ethanol because the electron-withdrawing chlorine atom helps to stabilize the 2-chloroethoxide ion.

$$\underset{\text{ethanol}}{CH_3{-}CH_2{-}OH} + H_2O \rightleftharpoons \underset{\substack{\text{ethoxide ion}\\\text{(less stable)}}}{CH_3{-}CH_2{-}O^-} + H_3O^+ \qquad K_a = 1.3 \times 10^{-16}$$

$$\underset{\text{2-chloroethanol}}{Cl{-}CH_2{-}CH_2{-}OH} + H_2O \rightleftharpoons \underset{\substack{\text{2-chloroethoxide ion}\\\text{(stabilized by Cl)}}}{Cl{-}CH_2{-}CH_2{-}O^-} + H_3O^+ \qquad K_a = 5.0 \times 10^{-15}$$

TABLE 9-3
Acid dissociation constants of representative alcohols

Alcohol	*Structure*	K_a	pK_a
methanol	CH_3—OH	3.2×10^{-16}	15.5
ethanol	CH_3CH_2—OH	1.3×10^{-16}	15.9
2-chloroethanol	Cl—CH_2CH_2—OH	5.0×10^{-15}	14.3
2,2,2-trifluoroethanol	F_3C—CH_2—OH	4.0×10^{-13}	12.4
isopropyl alcohol	$(CH_3)_2CH$—OH	1.0×10^{-18}	18.0
t-butyl alcohol	$(CH_3)_3C$—OH	1.0×10^{-19}	19.0
cyclohexanol	C_6H_{11}—OH	1.0×10^{-18}	18.0
phenol	C_6H_5—OH	1.0×10^{-10}	10.0
	Comparison with other acids		
water	H_2O	1.8×10^{-16}	15.7
acetic acid	CH_3COOH	1.6×10^{-5}	4.8
hydrochloric acid	HCl	$1.6 \times 10^{+2}$	−2.2

PROBLEM 9-7
Predict which member of each pair will be more acidic. Explain your answers.

(a) methanol and *t*-butyl alcohol
(b) 1-chloroethanol and 2-chloroethanol
(c) 2-chloroethanol and 2,2-dichloroethanol

9-6B FORMATION OF SODIUM AND POTASSIUM ALKOXIDES

In Chapter 10 we will see many useful reactions of alkoxide ions. When an alkoxide ion is needed in a synthesis, it is usually formed by the reaction of sodium or potassium metal with the alcohol. This is a reduction, with the hydrogen ion reduced to form hydrogen gas, which bubbles out of the solution, leaving the sodium or potassium salt of the alkoxide ion.

$$R—O—H \quad + \quad Na \quad \longrightarrow \quad R—O^{-}\,{}^{+}Na \quad + \quad \tfrac{1}{2}H_2\uparrow$$

Example

$$\underset{\text{ethanol}}{CH_3CH_2OH} \quad + \quad \underset{\text{sodium metal}}{Na} \quad \longrightarrow \quad \underset{\text{sodium ethoxide}}{CH_3CH_2O^{-}\,{}^{+}Na} \quad + \quad \underset{\text{hydrogen gas}}{\tfrac{1}{2}H_2\uparrow}$$

The more acidic alcohols, like methanol and ethanol, react rapidly with sodium to form sodium methoxide and sodium ethoxide. Secondary alcohols, such as 2-propanol, react at a more moderate pace. Tertiary alcohols, such as *t*-butyl alcohol, react slowly with sodium. Potassium is often used with tertiary alcohols because it is more reactive than sodium and the reaction can be completed in a convenient amount of time.

$$\underset{t\text{-butyl alcohol}}{(CH_3)_3C—OH} \quad + \quad \underset{\text{potassium}}{K} \quad \longrightarrow \quad \underset{\text{potassium } t\text{-butoxide}}{(CH_3)_3C—O^{-}\,{}^{+}K} \quad + \quad \tfrac{1}{2}H_2\uparrow$$

9-6C ACIDITY OF PHENOLS

We might expect that phenol would have about the same acidity as cyclohexanol, since their structures are rather similar. Our prediction is wrong; phenol is nearly 100 million (10^8) times as acidic as cyclohexanol.

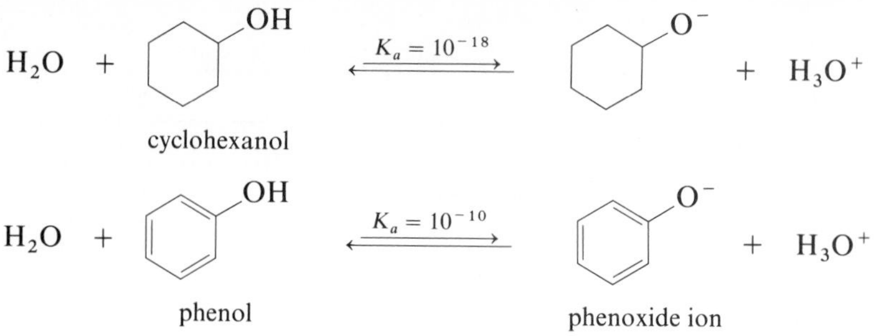

$$H_2O + \text{cyclohexanol} \underset{}{\overset{K_a = 10^{-18}}{\rightleftharpoons}} \text{cyclohexoxide} + H_3O^+$$

cyclohexanol

$$H_2O + \text{phenol} \overset{K_a = 10^{-10}}{\rightleftharpoons} \text{phenoxide ion} + H_3O^+$$

phenol — phenoxide ion

Cyclohexanol is a typical secondary alcohol, with a typical acid dissociation constant. There must be something special about phenol, making it an unusually strong acid. If we consider the phenoxide ion in more detail, it becomes apparent that the negative charge is not confined to the oxygen atom but is delocalized over the oxygen and three carbon atoms of the benzene ring.

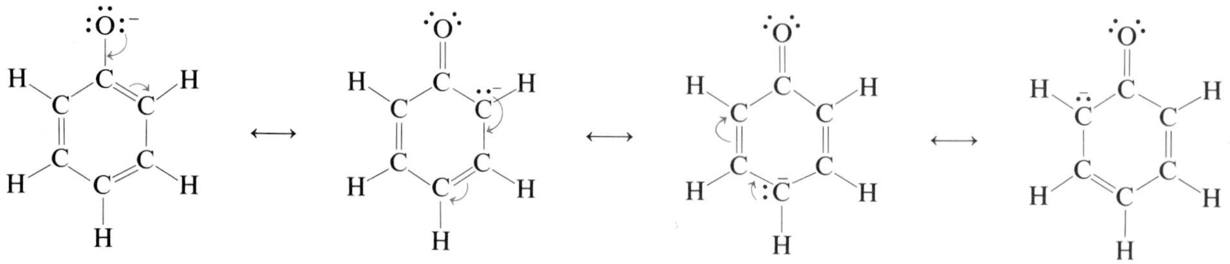

A large part of the negative charge in the resonance hybrid still resides on the oxygen atom, since it is the most electronegative of the four atoms sharing the charge. But the ability to spread the negative charge over four atoms rather than concentrating it on just one atom produces a more stable ion. The reaction of phenol with sodium hydroxide is exothermic, and the following equilibrium lies far to the right.

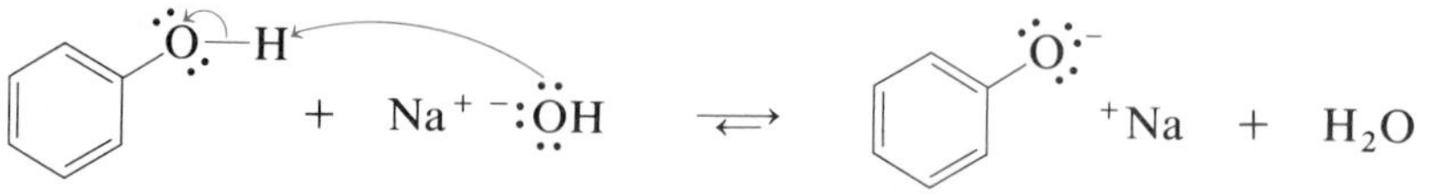

$$\text{phenol} + Na^+ \ {}^-{:}\ddot{O}H \rightleftharpoons \text{sodium phenoxide} \ ({}^+Na) + H_2O$$

phenol — sodium phenoxide

Phenoxide anions are prepared simply by adding the phenol to an aqueous solution of sodium hydroxide or potassium hydroxide. There is no need to use sodium or potassium metal.

PROBLEM 9-8

A nitro group ($-NO_2$) effectively stabilizes a negative charge on an adjacent carbon atom through resonance:

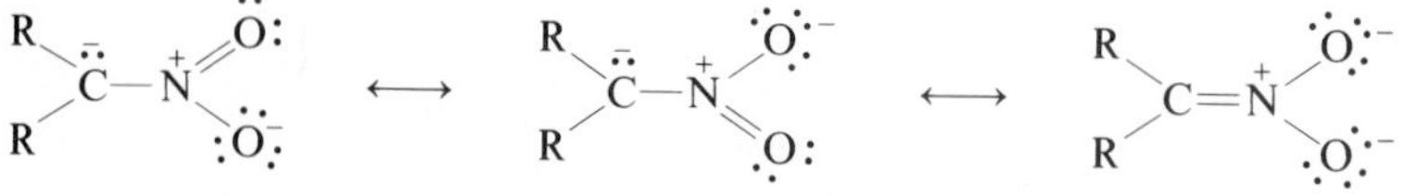

$$R_2\bar{C}-\overset{+}{N}(=O)O^- \longleftrightarrow R_2\bar{C}-\overset{+}{N}(O^-)=O \longleftrightarrow R_2C=\overset{+}{N}(O^-)_2$$

Two of the following nitrophenols are much more acidic than phenol itself. The third compound is only slightly more acidic than phenol. Use resonance structures of the appropriate phenoxide ions to show why two of these anions should be unusually stable.

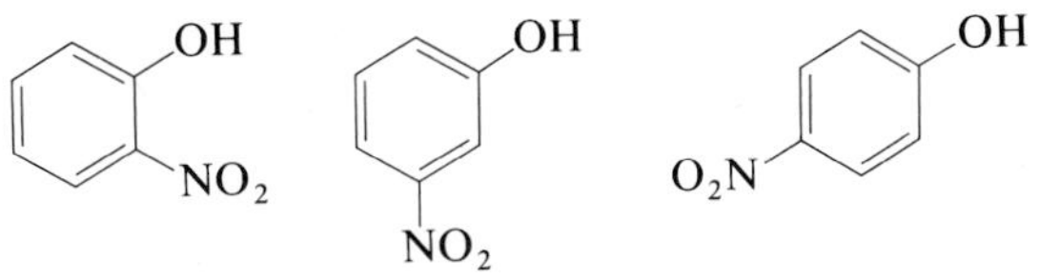

2-nitrophenol 3-nitrophenol 4-nitrophenol

PROBLEM 9-9

The following compounds are slightly soluble in water. One of these compounds is very soluble in a dilute aqueous solution of sodium hydroxide, however. The other is still only slightly soluble.

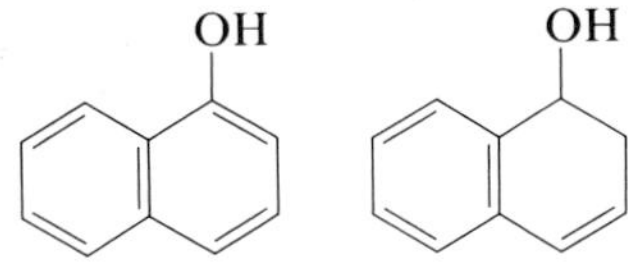

(a) Explain the difference in solubility of these compounds in dilute sodium hydroxide.
(b) Show how this difference might be exploited to separate a mixture of these two compounds using a separatory funnel.

9-7 SYNTHESIS OF ALCOHOLS: INTRODUCTION AND REVIEW

One of the reasons alcohols are important synthetic intermediates is that they can be synthesized directly from a wide variety of other functional groups. In previous chapters we have seen the conversion of alkyl halides to alcohols by substitution reactions and the conversion of alkenes to alcohols by hydration, hydroboration, and hydroxylation. These reactions are summarized below, with references for review if needed.

Nucleophilic substitution on an alkyl halide or tosylate. See Chapter 5. (Usually via the S_N2 mechanism; competes with elimination.)

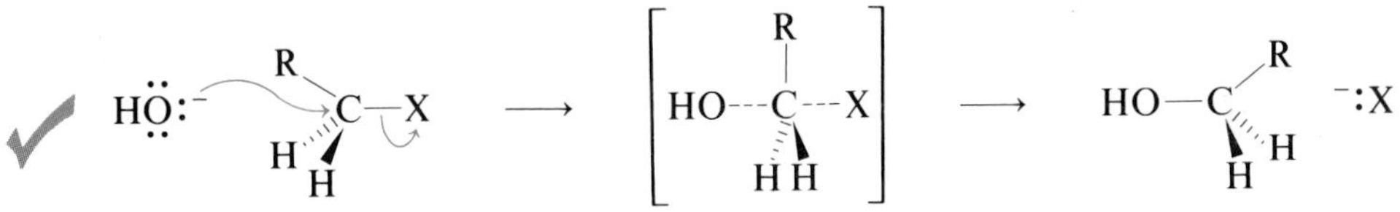

transition state

Example

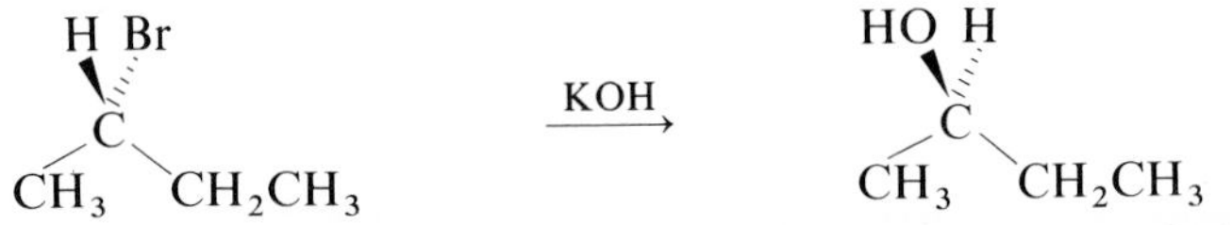

KOH

(*S*)-2-bromobutane

(*R*)-2-butanol, 100% inverted configuration (plus elimination products)

9-7A SYNTHESIS OF ALCOHOLS FROM ALKENES. (See Chapter 8.)

Acid-catalyzed hydration (Section 8-4)

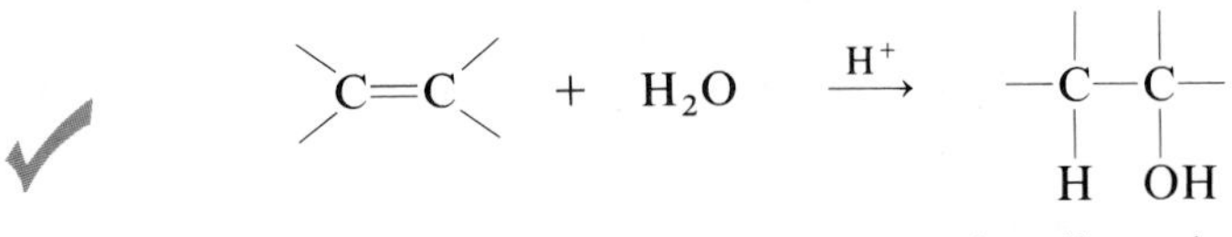

$$\text{C=C} + H_2O \xrightarrow{H^+} -\underset{H}{C}-\underset{OH}{C}-$$

Markovnikov orientation

Oxymercuration-demercuration (Section 8-5B)

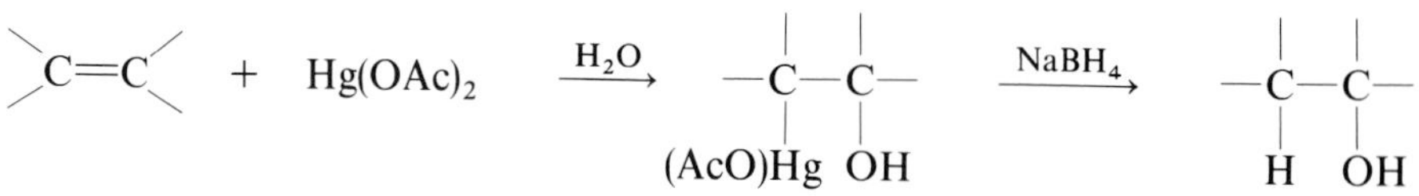

$$\mathrm{>C{=}C<} + \mathrm{Hg(OAc)_2} \xrightarrow{H_2O} \mathrm{-\overset{|}{\underset{(AcO)Hg}{C}}-\overset{|}{\underset{OH}{C}}-} \xrightarrow{NaBH_4} \mathrm{-\overset{|}{\underset{H}{C}}-\overset{|}{\underset{OH}{C}}-}$$

Markovnikov orientation

Example

$$\mathrm{(H_3C)(H)C{=}C(CH_3)_2} \xrightarrow[H_2O]{Hg(OAc)_2} \mathrm{H-\overset{H_3C}{\underset{(AcO)Hg}{C}}-\overset{OH}{\underset{CH_3}{C}}-CH_3} \xrightarrow{NaBH_4} \mathrm{H-\overset{H_3C}{\underset{H}{C}}-\overset{OH}{\underset{CH_3}{C}}-CH_3}$$

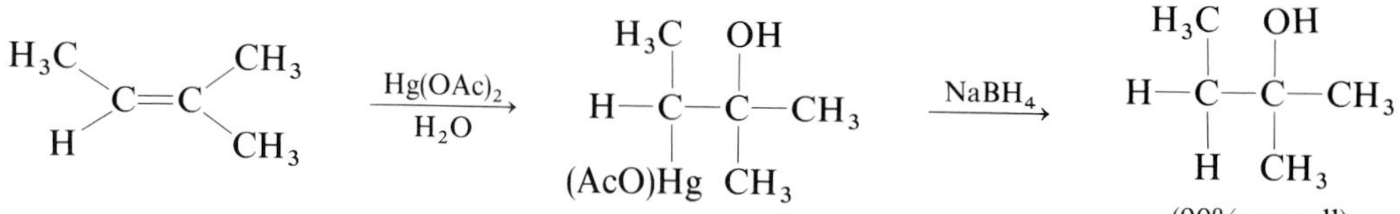

(90% overall)

Hydroboration-oxidation (Section 8-7)

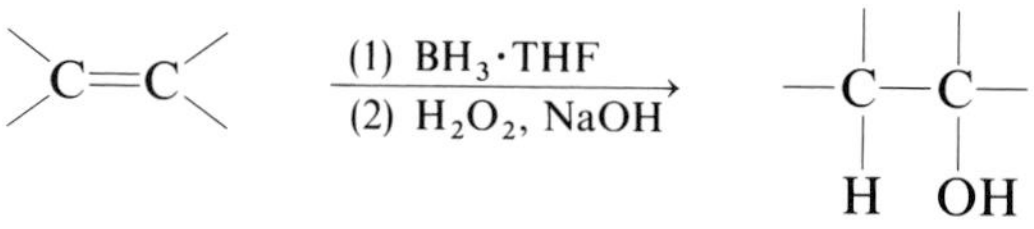

$$\mathrm{>C{=}C<} \xrightarrow[(2)\ H_2O_2,\ NaOH]{(1)\ BH_3\cdot THF} \mathrm{-\overset{|}{\underset{H}{C}}-\overset{|}{\underset{OH}{C}}-}$$

syn addition, anti-Markovnikov orientation

Example

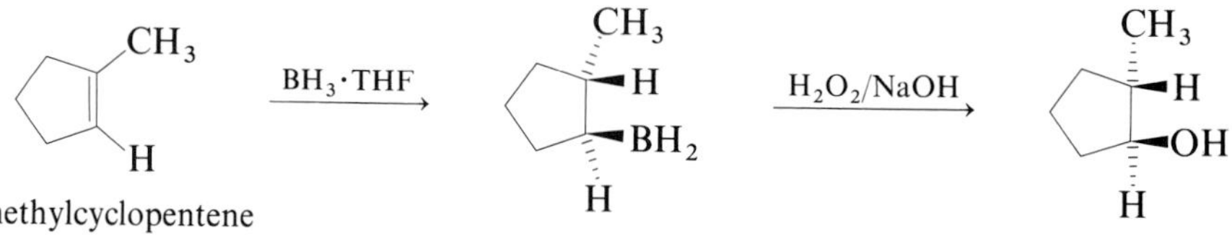

1-methylcyclopentene $\xrightarrow{BH_3\cdot THF}$ (CH$_3$, H, BH$_2$, H adduct) $\xrightarrow{H_2O_2/NaOH}$ *trans*-2-methylcyclopentanol (85%)

Hydroxylation: synthesis of 1,2-diols from alkenes (Sections 8-13 and 8-14)

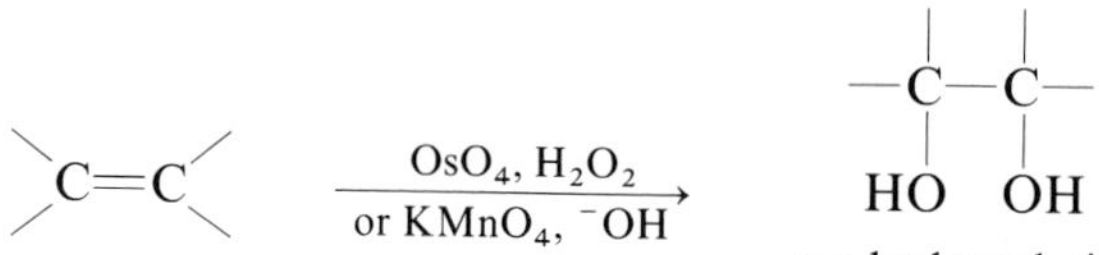

$$\mathrm{>C{=}C<} \xrightarrow[\text{or } KMnO_4,\ ^-OH]{OsO_4,\ H_2O_2} \mathrm{-\overset{|}{\underset{HO}{C}}-\overset{|}{\underset{OH}{C}}-}$$

syn hydroxylation

Example

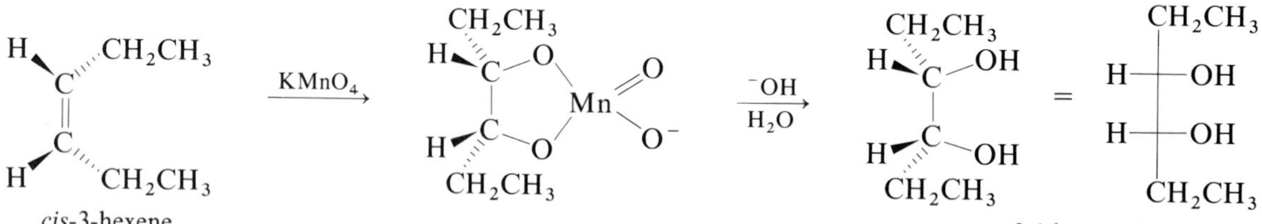

cis-3-hexene $\xrightarrow{KMnO_4}$ (cyclic manganate ester, $Mn(=O)O^-$) $\xrightarrow[H_2O]{^-OH}$ *meso*-3,4-hexanediol (60%)

$$\mathrm{CH_2CH_3},\ \mathrm{H-C-OH},\ \mathrm{H-C-OH},\ \mathrm{CH_2CH_3}$$

$$\mathrm{>C{=}C<} \xrightarrow{\mathrm{H-\overset{O}{\overset{\|}{C}}-OOH\ or\ CH_3-\overset{O}{\overset{\|}{C}}-OOH,\ H_2O}} \mathrm{-\overset{|}{\underset{OH}{C}}-\overset{OH}{\underset{|}{C}}-}$$

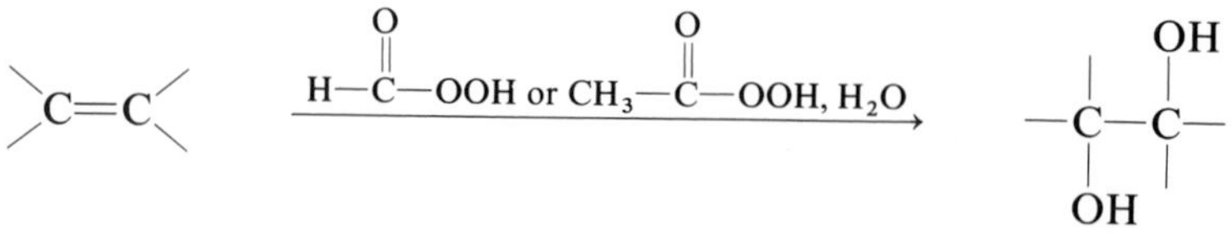

anti hydroxylation

Example

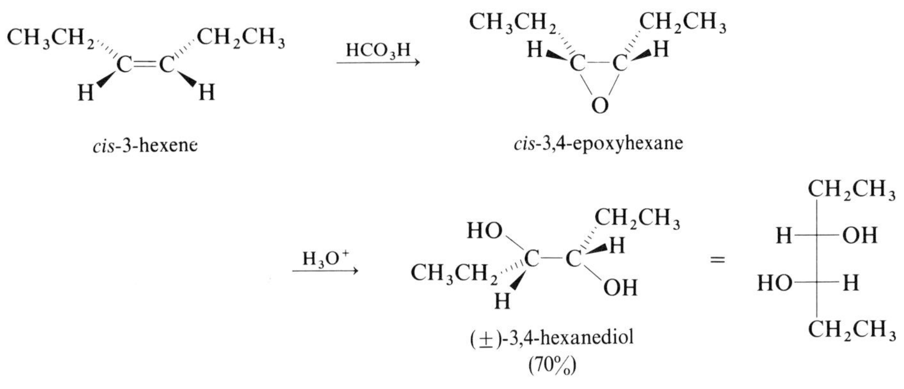

CH_3CH_2, H / C=C \ CH_2CH_3, H (cis) $\xrightarrow{HCO_3H}$ CH_3CH_2, H–C—C–H, CH_2CH_3 (epoxide O bridging)

cis-3-hexene — *cis*-3,4-epoxyhexane

$\xrightarrow{H_3O^+}$ HO, CH_3CH_2, H –C—C– CH_2CH_3, H, OH = Fischer projection: CH_2CH_3 / H—OH / HO—H / CH_2CH_3

(±)-3,4-hexanediol
(70%)

9-8 ORGANOMETALLIC REAGENTS FOR ALCOHOL SYNTHESIS

Organometallic compounds contain covalent bonds between carbon atoms and metal atoms. Organometallic reagents are useful because they have nucleophilic carbon atoms, in contrast to the electrophilic carbon atoms of alkyl halides. Most metals are more electropositive than carbon, and the C—M bond is polarized with a partial positive charge on the metal and a partial negative charge on carbon. The following table shows the electronegativities of some metals used in making organometallic compounds.

Electronegativities

Li 1.0			C 2.5
Na 0.9	Mg 1.3	Al 1.6	
K 0.8			

C—M bond

$\overset{\longleftarrow\!+}{C—Li}$ (δ^- δ^+)

$\overset{\longleftarrow\!+}{C—Mg}$ (δ^- δ^+)

9-8A GRIGNARD REAGENTS

Organometallic compounds of lithium and magnesium are used most frequently for the synthesis of alcohols. The organomagnesium halides, of empirical formula R—Mg—X, are called **Grignard reagents** in honor of the French chemist Victor Grignard, who discovered their utility around 1905 and received the Nobel Prize in 1912. Grignard reagents result from the reaction of an alkyl halide with magnesium metal. This reaction is always carried out in an ether solvent, because ether must be present to solvate and stabilize the Grignard reagent as it forms. Although we write the Grignard reagent as R—Mg—X, the actual species in solution usually contains two, three, or four of these units associated together with several molecules of the ether solvent. Diethyl ether, $CH_3CH_2—O—CH_2CH_3$, is the most common solvent for these reactions, although other ethers can also be used.

$$\text{R—X} + \text{Mg} \xrightarrow{CH_3CH_2—O—CH_2CH_3} \text{R—Mg—X}$$

(X = Cl, Br, or I) — organomagnesium halide (Grignard reagent)

Grignard reagents may be made from primary, secondary, and tertiary alkyl halides, as well as from vinyl and aryl halides. Alkyl iodides are the most reactive halides, followed by bromides and chlorides. Alkyl fluorides generally do not react. The following reactions show the formation of some typical Grignard reagents.

$$\underset{\text{iodomethane}}{CH_3—I} + Mg \xrightarrow{\text{ether}} \underset{\text{methylmagnesium iodide}}{CH_3—Mg—I}$$

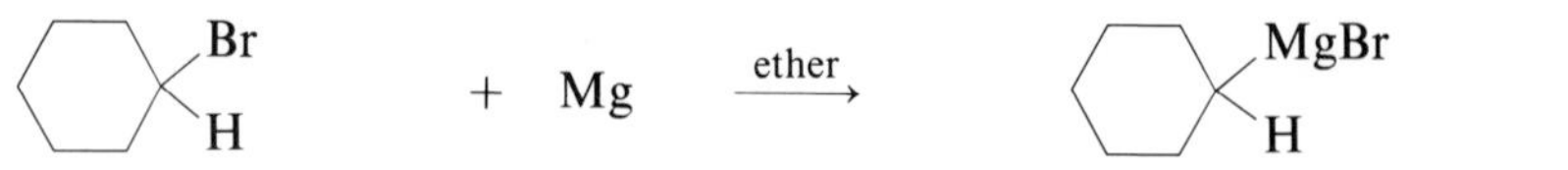

$$\underset{\text{bromocyclohexane}}{C_6H_{11}Br} + Mg \xrightarrow{\text{ether}} \underset{\text{cyclohexylmagnesium bromide}}{C_6H_{11}MgBr}$$

$$\underset{\text{allyl bromide}}{H_2C{=}CH—CH_2—Br} + Mg \xrightarrow{\text{ether}} \underset{\text{allylmagnesium bromide}}{H_2C{=}CH—CH_2—MgBr}$$

9-8B ORGANOLITHIUM REAGENTS

Like magnesium, lithium reacts with alkyl halides, vinyl halides, and aryl halides to form organometallic compounds. Ether is not necessary for this reaction; organolithium reagents are made and used in a wide variety of solvents.

$$\underset{(X = Cl, Br, or I)}{R—X + 2\,Li} \longrightarrow \underset{\text{organolithium}}{R—Li} + Li^+X^-$$

Examples

$$\underset{n\text{-butyl bromide}}{CH_3CH_2CH_2CH_2—Br} + 2\,Li \xrightarrow{\text{hexane}} \underset{n\text{-butyllithium}}{CH_3CH_2CH_2CH_2—Li} + LiBr$$

$$\underset{\text{vinyl chloride}}{H_2C{=}CH—Cl} + 2\,Li \xrightarrow{\text{pentane}} \underset{\text{vinyllithium}}{H_2C{=}CH—Li} + LiCl$$

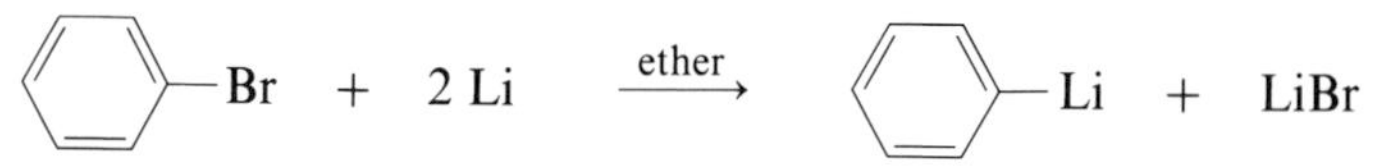

$$\underset{\text{bromobenzene}}{C_6H_5—Br} + 2\,Li \xrightarrow{\text{ether}} \underset{\text{phenyllithium}}{C_6H_5—Li} + LiBr$$

PROBLEM 9-10

Which of the following compounds would be suitable solvents for Grignard reactions?

(a) n-hexane (b) $CH_3—O—CH_3$ (c) $CHCl_3$
(d) cyclohexane (e) benzene (f) $CH_3OCH_2CH_2OCH_3$
(g)  THF (tetrahydrofuran) (h) 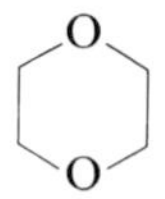 1,4-dioxane

PROBLEM 9-11

Predict the products of the following reactions.

(a) $CH_3CH_2Br + Mg \xrightarrow{\text{ether}}$

(b) isobutyl iodide + Li $\xrightarrow{\text{hexane}}$

(c) 1-bromo-4-fluorocyclohexane + Mg $\xrightarrow{\text{THF}}$

(d) $CH_2{=}CCl—CH_2—CH_3 + Li \xrightarrow{\text{ether}}$

9-9 ADDITION OF ORGANOMETALLIC REAGENTS TO CARBONYL COMPOUNDS

Because they are somewhat like carbanions, Grignard and organolithium reagents are strong nucleophiles and strong bases. Their most useful nucleophilic reactions are those in which they add to carbonyl groups. Figure 9-4 shows the structure of the carbonyl group. Since oxygen is more electronegative than carbon, the double bond between carbon and oxygen is strongly polarized. The oxygen atom has a partial negative charge balanced by an equal amount of positive charge on the carbon atom.

FIGURE 9-4 The C=O double bond of a carbonyl group resembles the C=C double bond of an alkene; however, the carbonyl double bond is strongly polarized. The oxygen atom bears a partial negative charge, and the carbon atom bears a partial positive charge.

The positively charged carbon is electrophilic; attack by a nucleophile places a negative charge on the electronegative oxygen atom.

$$\text{Nuc:}^- \;\; \rangle C{=}\ddot{O}: \;\longrightarrow\; \text{Nuc}{-}\overset{|}{\underset{|}{C}}{-}\ddot{O}:^-$$

The product of this nucleophilic attack is an alkoxide ion, a strong base. The addition of water or a dilute acid protonates the alkoxide to the alcohol.

$$\text{Nuc}{-}\overset{|}{\underset{|}{C}}{-}\ddot{O}:^- \;\; \underset{(\text{or } H_3O^+)}{H{-}\ddot{O}{-}H} \;\longrightarrow\; \text{Nuc}{-}\overset{|}{\underset{|}{C}}{-}\ddot{O}H \;+\; ^-{:}\ddot{O}H$$

Either a Grignard reagent or an organolithium reagent can serve as the nucleophile in this addition to a carbonyl group. The following discussions refer to Grignard reagents, but they also apply to organolithium reagents. The Grignard reagent adds to the carbonyl group to form an alkoxide ion. Addition of an acid (in a separate step) protonates the alkoxide to give the alcohol.

$$\overset{\delta-}{\boxed{R'}}{-}\overset{\delta+}{MgX} + \begin{matrix}R\\ \\R\end{matrix}\rangle C{=}\ddot{O}: \xrightarrow{(1)\text{ ether}} \underset{\text{magnesium alkoxide salt}}{\boxed{R'}{-}\overset{R}{\underset{R}{C}}{-}\ddot{O}:^-\;^+MgX} \xrightarrow{(2)\ H_3O^+} \underset{\text{alcohol}}{\boxed{R'}{-}\overset{R}{\underset{R}{C}}{-}\ddot{O}{-}H} + XMgOH$$

We are interested primarily in the reactions of Grignard reagents with ketones and aldehydes. **Ketones** are compounds with two alkyl groups bonded to a carbonyl group. **Aldehydes** have one alkyl group and one hydrogen atom bonded to the carbonyl group. **Formaldehyde** has two hydrogen atoms bonded to the carbonyl group.

$$\underset{\text{a ketone}}{\begin{matrix}R\\ \\R\end{matrix}\rangle C{=}O} \qquad \underset{\text{an aldehyde}}{\begin{matrix}H\\ \\R\end{matrix}\rangle C{=}O} \qquad \underset{\text{formaldehyde}}{\begin{matrix}H\\ \\H\end{matrix}\rangle C{=}O}$$

9-9A ADDITION OF GRIGNARD REAGENTS TO FORMALDEHYDE: FORMATION OF PRIMARY ALCOHOLS

$$\boxed{R}\!-\!MgX + \underset{\text{formaldehyde}}{H_2C{=}O} \xrightarrow{\text{ether}} \boxed{R}\!-\!CH_2\!-\!O^-\,{}^+MgX \xrightarrow{H_3O^+} \underset{\text{primary alcohol}}{\boxed{R}\!-\!CH_2\!-\!OH}$$

Grignard reagent — formaldehyde — primary alcohol

Addition of a Grignard reagent to formaldehyde, followed by protonation, gives a primary alcohol with one more carbon atom than in the Grignard reagent.

$$\underset{\text{butylmagnesium bromide}}{CH_3CH_2CH_2CH_2\!-\!MgBr} + \underset{\text{formaldehyde}}{H_2C{=}O} \xrightarrow[\text{(2) then } H_3O^+]{\text{(1) ether}} \underset{\text{1-pentanol (92\%)}}{CH_3CH_2CH_2CH_2\!-\!CH_2\!-\!OH}$$

PROBLEM 9-12

Show how you would synthesize each of the following alcohols by the addition of an appropriate Grignard reagent to formaldehyde.

(a) cyclohexyl–CH_2OH (b) [branched chain]–OH (c) cyclopentenyl–CH_2OH

9-9B ADDITION OF GRIGNARD REAGENTS TO ALDEHYDES: FORMATION OF SECONDARY ALCOHOLS

$$\boxed{R}\!-\!MgX + \underset{\text{aldehyde}}{R'(H)C{=}O} \xrightarrow{\text{ether}} \boxed{R}\!-\!CR'H\!-\!O^-\,{}^+MgX \xrightarrow{H_3O^+} \underset{\text{secondary alcohol}}{\boxed{R}\!-\!CR'H\!-\!OH}$$

Grignard reagent — aldehyde — secondary alcohol

Grignard reagents add to aldehydes to give, after protonation, secondary alcohols. The two alkyl groups of the secondary alcohol are the alkyl group from the Grignard reagent and the alkyl group that was bonded to the carbonyl group of the aldehyde.

$$CH_3CH_2\!-\!MgBr + \underset{\text{acetaldehyde}}{H_3C(H)C{=}O} \xrightarrow{\text{ether}} CH_3\!-\!CH_2\!-\!C(CH_3)(H)\!-\!O^-\,{}^+MgX \xrightarrow{H_3O^+} \underset{\text{2-butanol (85\%)}}{CH_3CH_2\!-\!C(CH_3)(H)\!-\!OH}$$

PROBLEM 9-13

Show how you would synthesize each of the following alcohols by the addition of an appropriate Grignard reagent to an aldehyde.

(a) [branched chain with OH] (b) phenyl–CH(OH)–CH_2CH_3 (c) dicyclohexyl–C(H)(OH)

9-9C ADDITION OF GRIGNARD REAGENTS TO KETONES: FORMATION OF TERTIARY ALCOHOLS

$$\boxed{R}\!-\!MgX \;+\; \underset{\text{ketone}}{R'R''C{=}O} \;\xrightarrow{\text{ether}}\; \boxed{R}\!-\!\underset{R''}{\overset{R'}{C}}\!-\!O^-\;{}^+MgX \;\xrightarrow{H_3O^+}\; \boxed{R}\!-\!\underset{R''}{\overset{R'}{C}}\!-\!OH$$

Grignard reagent — ketone — tertiary alcohol

A ketone has two alkyl groups bonded to its carbonyl carbon atom. Addition of a Grignard reagent, followed by protonation, gives a tertiary alcohol, with three alkyl groups bonded to the carbinol carbon atom. Two of the alkyl groups are the two originally bonded to the ketone carbonyl group. The third alkyl group comes from the Grignard reagent.

$$CH_3CH_2\!-\!MgBr \;+\; \underset{\text{2-pentanone}}{(CH_3CH_2CH_2)(H_3C)C{=}O} \;\xrightarrow[(2)\ H_3O^+]{(1)\ \text{ether}}\; \underset{\text{3-methyl-3-hexanol (90\%)}}{CH_3CH_2\!-\!\underset{CH_3}{\overset{CH_3CH_2CH_2}{C}}\!-\!OH}$$

PROBLEM 9-14

Show how you would synthesize each of the following alcohols by the addition of an appropriate Grignard reagent to a ketone.

(a) Ph_3C—OH (b) 1-methylcyclohexanol (c) 1,1-dicyclohexyl-1-butanol

9-9D ADDITION OF GRIGNARD REAGENTS TO ACID CHLORIDES AND ESTERS

Acid chlorides and **esters** are derivatives of carboxylic acids. In such **acid derivatives** the —OH group of a carboxylic acid is replaced by other electron-withdrawing groups. In acid chlorides, the hydroxyl group of the acid is replaced by a chlorine atom. In esters, the hydroxyl group is replaced by an alkoxyl (—O—R) group.

$$\underset{\text{carboxylic acid}}{R\!-\!\overset{O}{\overset{\|}{C}}\!-\!OH} \qquad \underset{\text{acid chloride}}{R\!-\!\overset{O}{\overset{\|}{C}}\!-\!Cl} \qquad \underset{\text{ester}}{R\!-\!\overset{O}{\overset{\|}{C}}\!-\!O\!-\!R'}$$

Acid chlorides and esters react with two equivalents of Grignard reagents to give (after protonation) tertiary alcohols.

$$2\,\boxed{R}\!-\!MgX \;+\; \underset{\text{acid chloride}}{R'\!-\!\overset{O}{\overset{\|}{C}}\!-\!Cl} \;\xrightarrow[(2)\ H_3O^+]{(1)\ \text{ether}}\; \underset{\text{tertiary alcohol}}{R'\!-\!\underset{\boxed{R}}{\overset{\boxed{R}}{C}}\!-\!OH}$$

$$2\,\boxed{R}\!-\!MgX \;+\; \underset{\text{ester}}{R'\!-\!\overset{O}{\overset{\|}{C}}\!-\!OR''} \;\xrightarrow[(2)\ H_3O^+]{(1)\ \text{ether}}\; \underset{\text{tertiary alcohol}}{R'\!-\!\underset{\boxed{R}}{\overset{\boxed{R}}{C}}\!-\!OH}$$

Addition of the first equivalent of the Grignard reagent produces an unstable intermediate. This intermediate expels a chloride ion (in the acid chloride) or an alkoxide ion (in the ester), giving a ketone. The alkoxide ion is a suitable leaving group in this reaction, because its leaving stabilizes a negatively charged intermediate in a strongly exothermic step.

Attack on an acid chloride

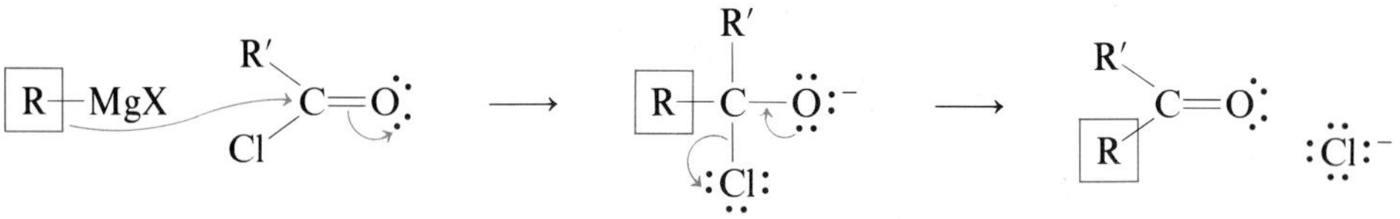

$$\boxed{R}\text{—MgX} + R'\text{—C(=O)—Cl} \longrightarrow \boxed{R}\text{—}\overset{R'}{\underset{:\ddot{Cl}:}{C}}\text{—}\ddot{O}:^- \longrightarrow R'\text{—C(=}\ddot{O}\text{:)—}\boxed{R} \quad :\ddot{Cl}:^-$$

Attack on an ester

$$\boxed{R}\text{—MgX} + R'\text{—C(=O)—}\ddot{O}\text{—}R'' \longrightarrow \boxed{R}\text{—}\overset{R'}{\underset{:\ddot{O}-R''}{C}}\text{—}\ddot{O}:^- \longrightarrow R'\text{—C(=O)—}\boxed{R} \quad ^-:\ddot{O}\text{—}R''$$

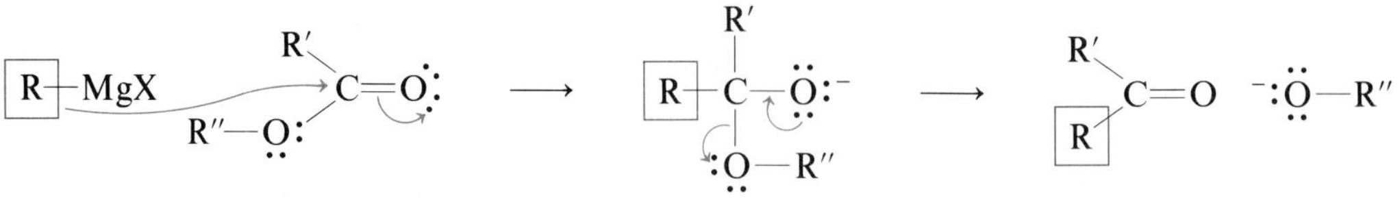

The ketone reacts with a second equivalent of the Grignard reagent, forming the magnesium salt of a tertiary alkoxide. Protonation gives a tertiary alcohol with one of its alkyl groups derived from the acid chloride or ester, and the other two derived from the Grignard reagent.

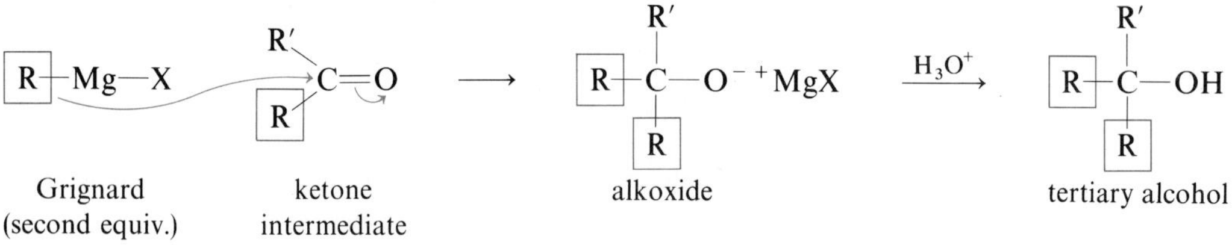

$$\boxed{R}\text{—Mg—X} + R'\text{—C(=O)—}\boxed{R} \longrightarrow \boxed{R}\text{—}\overset{R'}{\underset{\boxed{R}}{C}}\text{—O}^-\ {}^+\text{MgX} \xrightarrow{H_3O^+} \boxed{R}\text{—}\overset{R'}{\underset{\boxed{R}}{C}}\text{—OH}$$

Grignard (second equiv.) — ketone intermediate — alkoxide — tertiary alcohol

An example using an ester is shown below. When an excess of ethylmagnesium bromide is added to methyl benzoate, the first equivalent adds and methoxide is expelled to give propiophenone. Addition of a second equivalent, followed by protonation, gives a tertiary alcohol: 3-phenyl-3-pentanol.

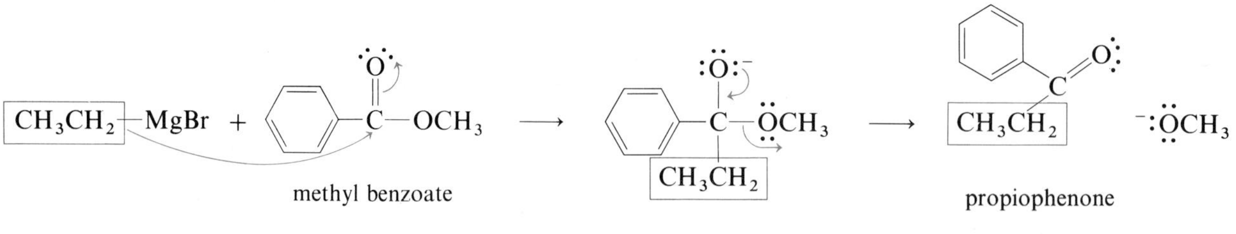

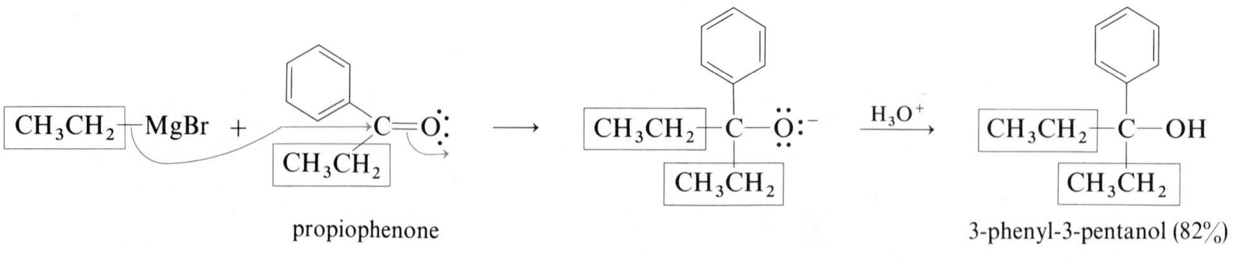

$$\boxed{CH_3CH_2}\text{—MgBr} + Ph\text{—C(=}\ddot{O}\text{)—OCH}_3 \longrightarrow Ph\text{—}\overset{:\ddot{O}:^-}{\underset{\boxed{CH_3CH_2}}{C}}\text{—}\ddot{O}CH_3 \longrightarrow Ph\text{—C(=}\ddot{O}\text{)—}\boxed{CH_3CH_2} \quad ^-:\ddot{O}CH_3$$

methyl benzoate — propiophenone

$$\boxed{CH_3CH_2}\text{—MgBr} + Ph\text{—C(=}\ddot{O}\text{)—}\boxed{CH_3CH_2} \longrightarrow \boxed{CH_3CH_2}\text{—}\overset{Ph}{\underset{\boxed{CH_3CH_2}}{C}}\text{—}\ddot{O}:^- \xrightarrow{H_3O^+} \boxed{CH_3CH_2}\text{—}\overset{Ph}{\underset{\boxed{CH_3CH_2}}{C}}\text{—OH}$$

propiophenone — 3-phenyl-3-pentanol (82%)

PROBLEM 9-15

Propose a mechanism for the reaction of acetyl chloride with phenylmagnesium bromide to give 1,1-diphenylethanol.

$$\underset{\text{acetyl chloride}}{CH_3-\overset{\overset{O}{\|}}{C}-Cl} + 2\ \underset{\text{phenylmagnesium bromide}}{C_6H_5-MgBr} \xrightarrow[(2)\ H_3O^+]{(1)\ \text{ether}} \underset{\text{1,1-diphenylethanol}}{CH_3-\overset{\overset{OH}{|}}{\underset{\underset{C_6H_5}{|}}{C}}-C_6H_5}$$

PROBLEM 9-16

Show how you would use the addition of Grignard reagents to acid chlorides and esters to synthesize the following alcohols.

(a) Ph_3C-OH (b) 3-ethyl-2-methyl-3-pentanol
(c) dicyclohexylphenylmethanol

PROBLEM 9-17

A formate ester, such as ethyl formate, reacts with an excess of a Grignard reagent to give (after protonation) secondary alcohols with two identical alkyl groups.

$$2\,R-MgX + \underset{\text{ethyl formate}}{H-\overset{\overset{O}{\|}}{C}-O-CH_2CH_3} \xrightarrow[(2)\ H_3O^+]{(1)\ \text{ether}} \underset{\text{secondary alcohol}}{R-\overset{\overset{OH}{|}}{CH}-R}$$

(a) Give a mechanism to show how the reaction of ethyl formate with an excess of allylmagnesium bromide gives, after protonation, 1,6-heptadien-4-ol.

$$2\,\underset{\text{allylmagnesium bromide}}{H_2C{=}CH-CH_2MgBr} + \underset{\text{ethyl formate}}{H-\overset{\overset{O}{\|}}{C}-OCH_2CH_3} \xrightarrow[(2)\ H_3O^+]{(1)\ \text{ether}} \underset{\text{1,6-heptadien-4-ol (80\%)}}{(H_2C{=}CH-CH_2)_2CH-OH}$$

(b) Show how you would use the reaction of a Grignard reagent with ethyl formate to synthesize each of the following secondary alcohols.
   (i) 3-pentanol
   (ii) diphenylmethanol
   (iii) *trans,trans*-2,7-nonadien-5-ol

9-9E ADDITION OF GRIGNARD REAGENTS TO ETHYLENE OXIDE

$$\boxed{R}-MgX + \underset{\text{ethylene oxide}}{\overset{\ddot{O}}{CH_2-CH_2}} \xrightarrow{\text{ether}} \underset{\text{alkoxide}}{\boxed{R}-CH_2-\overset{\overset{\ddot{O}:^-\ ^+MgX}{|}}{CH_2}} \xrightarrow{H_3O^+} \underset{\text{primary alcohol}}{\boxed{R}-CH_2-\overset{\overset{OH}{|}}{CH_2}}$$

Although Grignard reagents usually do not react with ethers, epoxides are unusually reactive ethers because of their ring strain. Ethylene oxide reacts with Grignard reagents to give, after protonation, primary alcohols with *two* additional carbon atoms. Notice that the nucleophilic attack by the Grignard reagent opens the ring and relieves the ring strain.

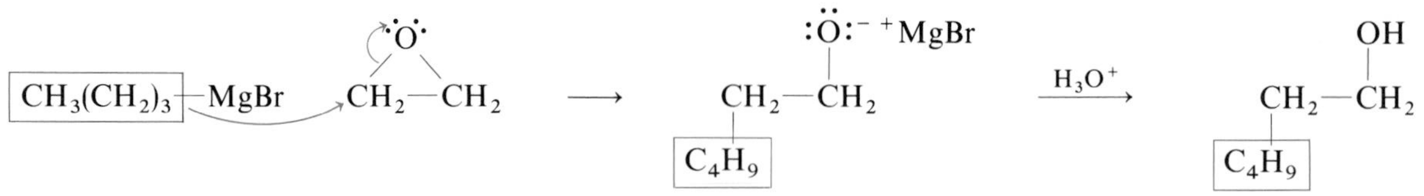

$$CH_3(CH_2)_3{-}MgBr \quad + \quad \underset{\text{ethylene oxide}}{CH_2{-}CH_2\ (\text{with }\ddot{O})} \longrightarrow \underset{C_4H_9}{CH_2}{-}CH_2{-}\ddot{O}{:}^{-}\ {}^{+}MgBr \xrightarrow{H_3O^+} \underset{C_4H_9}{CH_2}{-}CH_2{-}OH$$

butylmagnesium bromide — ethylene oxide — 1-hexanol (61%)

PROBLEM 9-18

Show how you would synthesize the following alcohols by the addition of a Grignard reagent to ethylene oxide.

(a) 2-phenylethanol (b) 4-methyl-1-pentanol (c) cyclohexane bearing CH_2CH_2OH and CH_3 on adjacent carbons

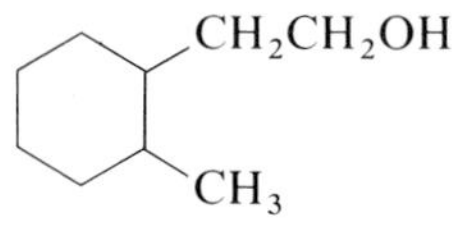

9-10 OTHER REACTIONS OF ORGANOMETALLIC REAGENTS: THE COREY-HOUSE REACTION

Organometallic reagents are especially useful for making alcohols, so they are covered here with alcohol syntheses. We also need to consider other important reactions of organometallic reagents that do not synthesize alcohols, but are useful for making other compounds. One of the best ways of joining alkyl groups is the **Corey-House reaction,** named in honor of its inventors E. J. Corey and Herbert House. The Corey-House reaction uses an organocopper reagent, a **lithium dialkylcuprate,** to couple with an alkyl halide.

$$\underset{\text{a lithium dialkylcuprate}}{R_2CuLi} + R'{-}X \longrightarrow R{-}R' + R{-}Cu + LiX$$

The lithium dialkylcuprate is formed by the reaction of two equivalents of the corresponding organolithium reagent with cuprous iodide:

$$2\,R{-}Li + CuI \longrightarrow R_2CuLi + LiI$$

The Corey-House reaction takes place as if a carbanion ($R{:}^-$) were present and the carbanion attacked the alkyl halide to displace the halide ion. This is not necessarily the actual mechanism, however.

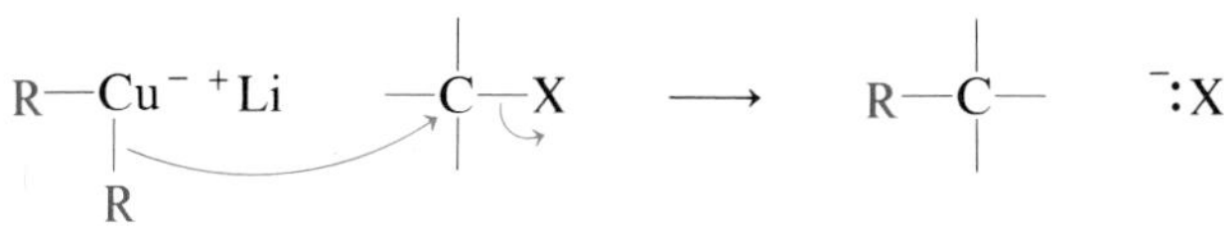

$$R{-}\underset{R}{Cu^-}\ {}^{+}Li \quad -\overset{|}{\underset{|}{C}}{-}X \longrightarrow R{-}\overset{|}{\underset{|}{C}}{-} \quad {}^{-}{:}X$$

The Corey-House reaction is a general one, coupling a wide variety of alkyl and vinyl halides. Some examples follow; note that some of these examples are compounds that cannot react by the S_N2 mechanism. Because of examples such as these, we know that the Corey-House reaction cannot always occur by the S_N2 mechanism.

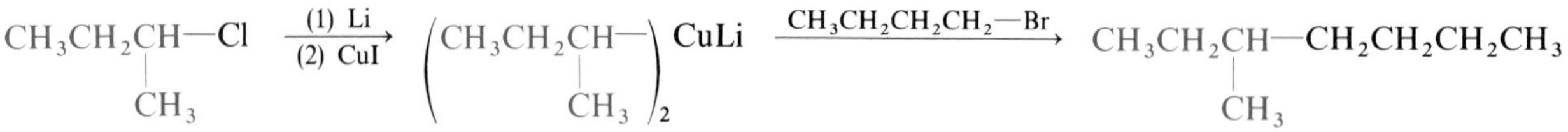

$$\underset{\text{2-chlorobutane}}{CH_3CH_2\underset{CH_3}{CH}{-}Cl} \xrightarrow[(2)\ CuI]{(1)\ Li} \underset{\text{lithium dialkylcuprate}}{\left(CH_3CH_2\underset{CH_3}{CH}{-}\right)_2CuLi} \xrightarrow{CH_3CH_2CH_2CH_2{-}Br} \underset{\text{3-methylheptane (70\%)}}{CH_3CH_2\underset{CH_3}{CH}{-}CH_2CH_2CH_2CH_3}$$

The following reactions cannot go by the S_N2 mechanism, yet they succeed.

$$\text{(cyclohexane-fused cyclopropane)}CBr_2 \xrightarrow{(CH_3)_2CuLi,\ (CH_3CH_2)_2O} \text{(cyclohexane-fused cyclopropane)}C(CH_3)_2$$

(65%)

$$n\text{-}C_8H_{17}(H)C{=}C(H)(I) \xrightarrow{(C_4H_9)_2CuLi} n\text{-}C_8H_{17}(H)C{=}C(H)(C_4H_9)$$

(74%)

PROBLEM 9-19

Show how you would synthesize the following compounds from alkyl halides, vinyl halides, and aryl halides containing no more than six carbon atoms.

(a) *n*-octane (b) 3-methylheptane
(c) *n*-butylcyclohexane (d) *trans*-3-octene

9-11 SIDE REACTIONS OF ORGANOMETALLIC REAGENTS

Grignard and organolithium reagents are strong nucleophiles and strong bases. In addition to their useful additions to carbonyl compounds, they react with other acidic or electrophilic compounds. In some cases these are useful reactions, but they are commonly seen as annoying side reactions where a small impurity of water or an alcohol destroys the reagent.

9-11A REACTIONS WITH ACIDIC COMPOUNDS

Grignard and organolithium reagents react vigorously and irreversibly with water. For example, consider the reaction of ethyllithium with water.

$$\underset{\text{ethyllithium}}{CH_3{-}\overset{\delta^-}{CH_2}{-}\overset{\delta^+}{Li}} + H{-}O{-}H \longrightarrow \underset{\text{ethane}}{CH_3{-}CH_2{-}H} + Li^+\ {}^-OH$$

The products are strongly favored in this reaction. Ethane is a *very* weak acid (K_a of about 10^{-50}), so the reverse reaction (abstraction of a proton from ethane by lithium hydroxide) is unlikely. When ethyllithium is added to water, ethane instantly bubbles to the surface.

Why would we ever *want* to add an organometallic reagent to water? This is simply another way of reducing an alkyl halide:

$$R{-}X + 2\,Li \longrightarrow R{-}Li + LiX \xrightarrow{H_2O} R{-}H + LiOH$$

In particular, this reaction provides a way to "label" a compound with deuterium at any position where a halogen is present.

$$CH_3{-}\underset{CH_3}{CH}{-}\underset{Br}{CH}{-}CH_3 \xrightarrow{Li} CH_3{-}\underset{CH_3}{CH}{-}\underset{Li}{CH}{-}CH_3 \xrightarrow{D-\ddot{O}-D} CH_3{-}\underset{CH_3}{CH}{-}\underset{D}{CH}{-}CH_3 \quad :\ddot{O}{-}D$$

In addition to O—H groups, the protons of N—H and S—H groups and the hydrogen atom of a terminal alkyne, —C≡C—H, are sufficiently acidic to protonate Grignard and organolithium reagents. Unless we want to protonate the reagent, compounds with these groups are considered incompatible with Grignard and organolithium reagents.

PROBLEM 9-20

Predict the products of the following reactions.

(a) methylmagnesium iodide + D_2O $\longrightarrow$
(b) *n*-butyllithium + CH_3CH_2OH $\longrightarrow$

(c) 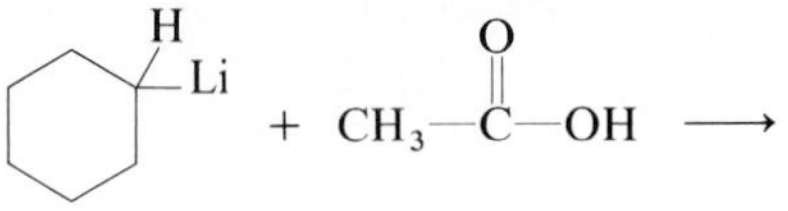 cyclohexyllithium + acetic acid (CH_3—C(=O)—OH) $\longrightarrow$

(d) 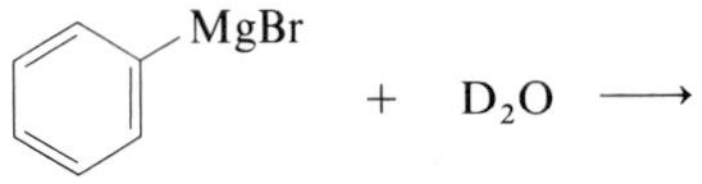 phenylmagnesium bromide + D_2O $\longrightarrow$

9-11B ELECTROPHILIC MULTIPLE BONDS

Grignard reagents are useful because they add to the electrophilic double bonds of carbonyl groups. However, we must make sure that the *only* electrophilic double bond in the solution is the one we want the reagent to attack. There must not be any electrophilic double (or triple) bonds in the solvent or in the Grignard reagent itself, or these will be attacked as well. Any multiple bond involving a strongly electronegative element is likely to be attacked, including C=O, S=O, C=N, N=O, and C≡N bonds.

In later chapters we encounter methods for *protecting* susceptible groups to prevent the reagent from attacking them. For now, simply remember that the following groups react with a Grignard or organolithium reagent, and avoid compounds containing these groups except for the one carbonyl group that gives the desired reaction.

Protonate the Grignard or organolithium:	O—H, N—H, S—H, —C≡C—H
Attacked by the Grignard or organolithium:	C=O, C=N, C≡N, S=O, N=O

PROBLEM 9-21

Point out the flaw in each of the following incorrect Grignard syntheses.

(a) 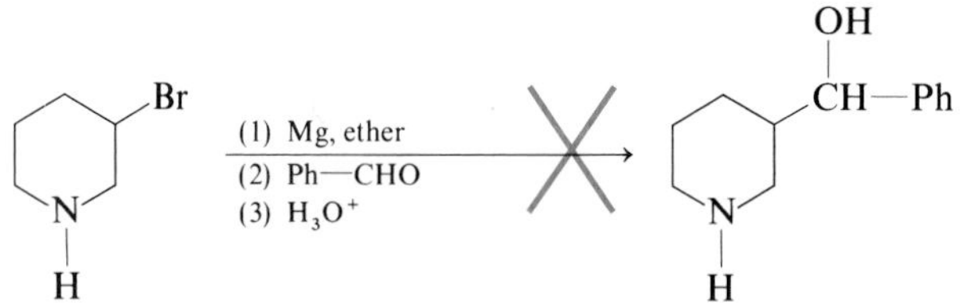 3-bromopiperidine (N—H) $\xrightarrow{\text{(1) Mg, ether; (2) Ph—CHO; (3) } H_3O^+}$ ✗ piperidine with CH(OH)—Ph (N—H)

(b) 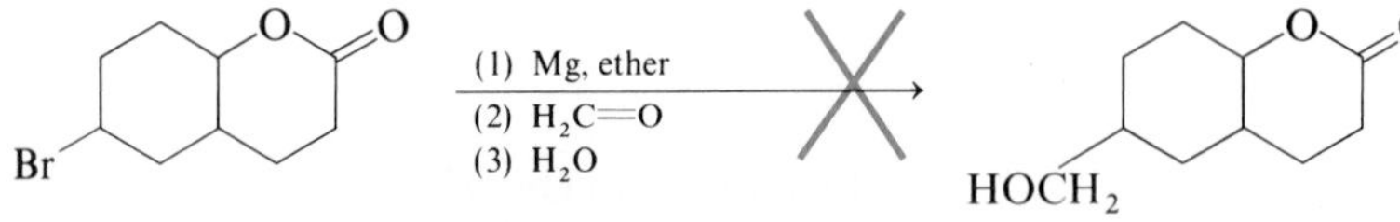 Br-substituted bicyclic lactone (O, =O) $\xrightarrow{\text{(1) Mg, ether; (2) } H_2C{=}O\text{; (3) } H_2O}$ ✗ $HOCH_2$-substituted bicyclic lactone

(c) 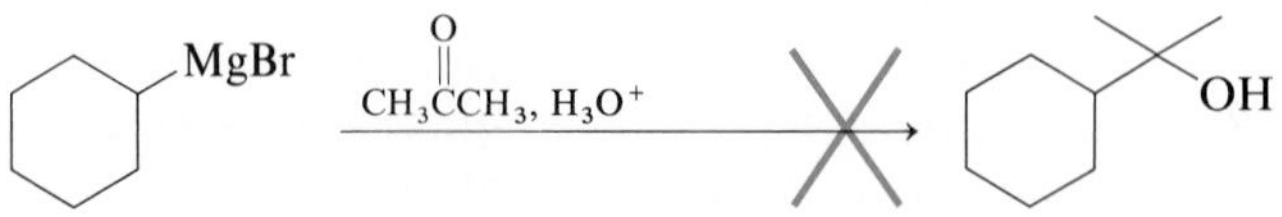 cyclohexyl-MgBr $\xrightarrow{CH_3CCH_3 (=O),\ H_3O^+}$ ✗ cyclohexyl-C(CH_3)$_2$OH

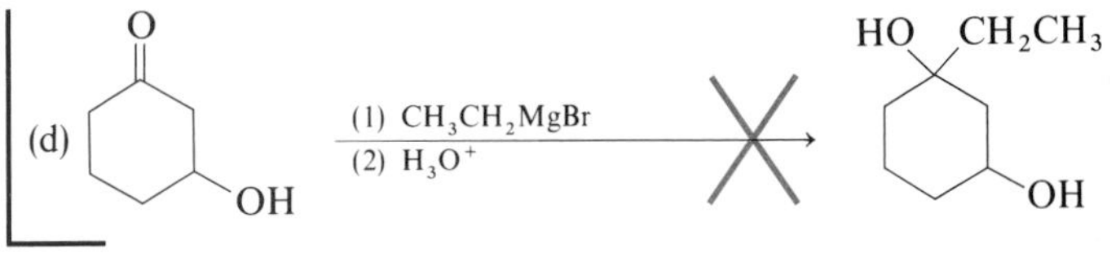

(d) 3-hydroxycyclohexanone $\xrightarrow[(2)\ H_3O^+]{(1)\ CH_3CH_2MgBr}$ ✗ 1-ethylcyclohexane-1,3-diol (HO, CH_2CH_3; OH)

9-12

REDUCTION OF THE CARBONYL GROUP: SYNTHESIS OF PRIMARY AND SECONDARY ALCOHOLS

Grignard reagents add to carbonyl compounds to form alcohols with additional carbon atoms. **Hydride reagents** add a hydride ion (H:$^-$), reducing the carbonyl group to an alkoxide ion with no additional carbon atoms. Subsequent protonation gives the alcohol.

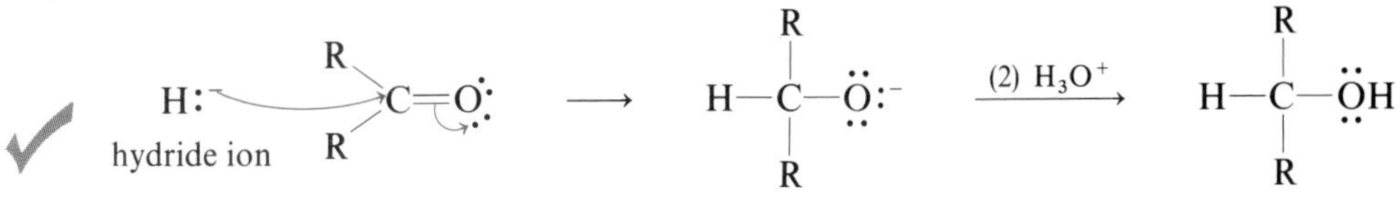

$$H:^- \text{ (hydride ion)} + R_2C{=}\ddot{O}: \longrightarrow H{-}CR_2{-}\ddot{O}:^- \text{ (alkoxide ion)} \xrightarrow{(2)\ H_3O^+} H{-}CR_2{-}\ddot{O}H$$

alkoxide ion

The two most commonly used hydride reagents, sodium borohydride ($NaBH_4$) and lithium aluminum hydride ($LiAlH_4$), reduce carbonyl groups in excellent yields. These reagents are called *complex hydrides* because they do not have a simple hydride structure such as $Na^+ \ {}^-H$ or $Li^+ \ {}^-H$. Instead, their hydrogen atoms, bearing partial negative charges, are covalently bonded to boron and aluminum atoms. This arrangement makes the hydride a better nucleophile, while reducing its basicity.

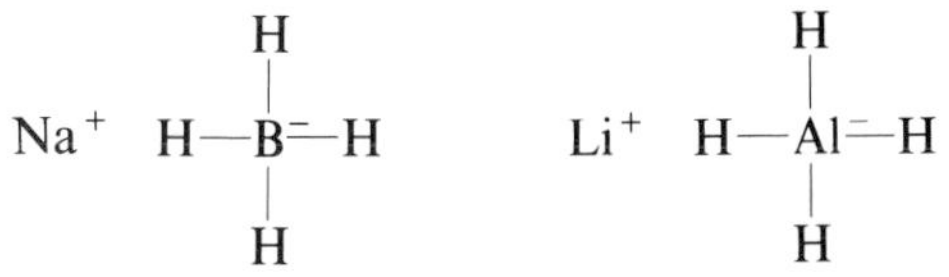

$Na^+ \ H{-}\overset{H}{\underset{H}{B^-}}{-}H$ sodium borohydride

$Li^+ \ H{-}\overset{H}{\underset{H}{Al^-}}{-}H$ lithium aluminium hydride

Of these common hydride reducing agents, lithium aluminum hydride (LAH) is much stronger and more difficult to work with. LAH reacts explosively with water and alcohols, liberating hydrogen gas and sometimes starting fires. Sodium borohydride reacts slowly with alcohols and with water as long as the pH is high (basic). Sodium borohydride is a convenient and highly selective reducing agent.

9-12A USES OF SODIUM BOROHYDRIDE

Sodium borohydride reduces aldehydes to primary alcohols and ketones to secondary alcohols. The reactions take place in a wide variety of solvents, including alcohols, ethers, and water. The yields are generally excellent.

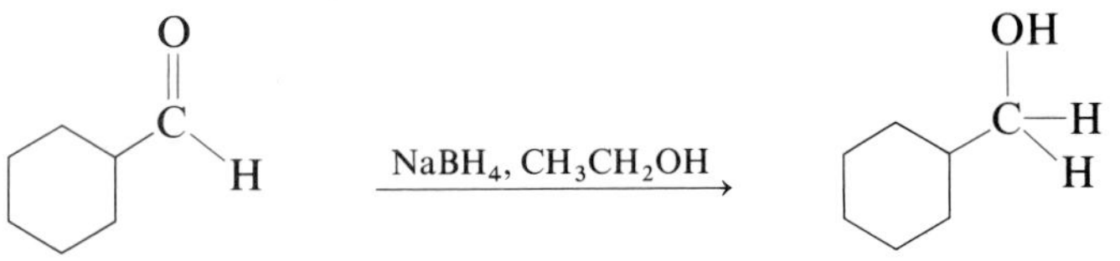

cyclohexane carbaldehyde ($C_6H_{11}{-}CHO$) $\xrightarrow{NaBH_4,\ CH_3CH_2OH}$ cyclohexyl carbinol ($C_6H_{11}{-}CH_2OH$) (95%)

$$\underset{\text{2-butanone}}{CH_3-\overset{\overset{\large O}{\|}}{C}-CH_2CH_3} \xrightarrow{NaBH_4,\ CH_3OH} \underset{\text{2-butanol (100\%)}}{CH_3-\overset{\overset{\large OH}{|}}{CH}-CH_2CH_3}$$

Sodium borohydride is very selective, and usually does not react with carbonyl groups that are less reactive than ketones and aldehydes. For example, carboxylic acids and esters are unreactive toward borohydride reduction. Sodium borohydride can reduce a ketone or an aldehyde in the presence of an acid or ester.

$$O{=}C_6H_{9}(H)-CH_2-\overset{\overset{\large O}{\|}}{C}-OCH_3 \xrightarrow{NaBH_4} HO(H)C_6H_{9}(H)-CH_2-\overset{\overset{\large O}{\|}}{C}-OCH_3$$

9-12B USES OF LITHIUM ALUMINUM HYDRIDE

Lithium aluminum hydride ($LiAlH_4$, abbreviated LAH) is a much stronger reagent than sodium borohydride. It easily reduces ketones and aldehydes and also the less reactive carbonyl groups: those in acids, esters, and other acid derivatives (see Chapter 20). LAH reduces ketones to secondary alcohols and aldehydes, acids, and esters to primary alcohols. The lithium salt of the alkoxide ion is initially formed, then the (cautious!) addition of dilute acid protonates the alkoxide. For example, LAH reduces both of the functional groups of the keto ester in the previous example.

$$O{=}C_6H_{10}-CH_2-\overset{\overset{\large O}{\|}}{C}-OCH_3 \xrightarrow[(2)\ H_3O^+]{(1)\ LiAlH_4} HO(H)C_6H_{9}(H)-CH_2-CH_2OH$$

In summary, sodium borohydride is the best reagent for reduction of just a ketone or an aldehyde. Using $NaBH_4$, we can reduce a ketone or an aldehyde in the presence of an acid or an ester, but we do not have a method (so far) for reducing an acid or an ester in the presence of a ketone or an aldehyde. The sluggish acid or ester requires the use of $LiAlH_4$, and this reagent also reduces the ketone or aldehyde.

SUMMARY OF THE REACTIONS OF LITHIUM ALUMINUM HYDRIDE AND SODIUM BOROHYDRIDE[a]

		$NaBH_4$	*$LiAlH_4$*
aldehyde	$R-\overset{\overset{O}{\|}}{C}-H$	$R-CH_2OH$	$R-CH_2-OH$
ketone	$R-\overset{\overset{O}{\|}}{C}-R'$	$R-\overset{\overset{OH}{\mid}}{CH}-R$	$R-\overset{\overset{OH}{\mid}}{CH}-R'$
alkene	$>C{=}C<$	no reaction	no reaction
acid anion	$R-\overset{\overset{O}{\|}}{C}-O^-$ (anion in base)	no reaction	$R-CH_2-OH$
ester	$R-\overset{\overset{O}{\|}}{C}-OR'$	no reaction	$R-CH_2-OH$

[a] The products shown are the final products, after hydrolysis of the alkoxide.

PROBLEM 9-22

Predict the products you expect from the reaction of $NaBH_4$ with the following compounds.

(a) $CH_3—(CH_2)_8—CHO$ (b) $CH_3CH_2—\overset{O}{\overset{\|}{C}}—OCH_3$ (c) Ph—COOH

(d) 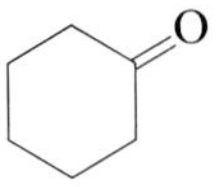   (e)

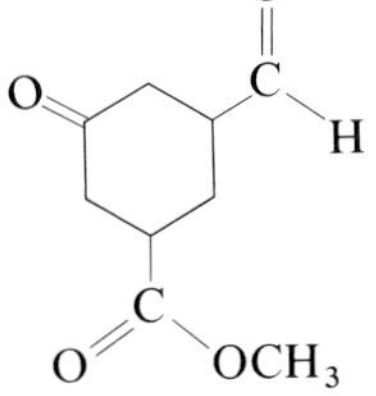

PROBLEM 9-23

Repeat Problem 9-22 using $LiAlH_4$ (followed by hydrolysis) as the reagent.

PROBLEM 9-24

Show how you would synthesize each of the following alcohols by reducing an appropriate carbonyl compound.

(a) 1-heptanol   (b) 2-heptanol   (c) 2-methyl-3-hexanol   (d)

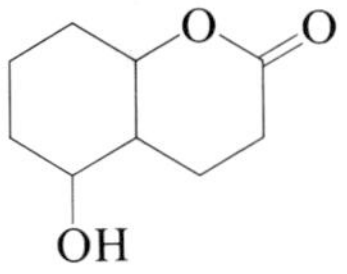

9-12C CATALYTIC HYDROGENATION OF KETONES AND ALDEHYDES

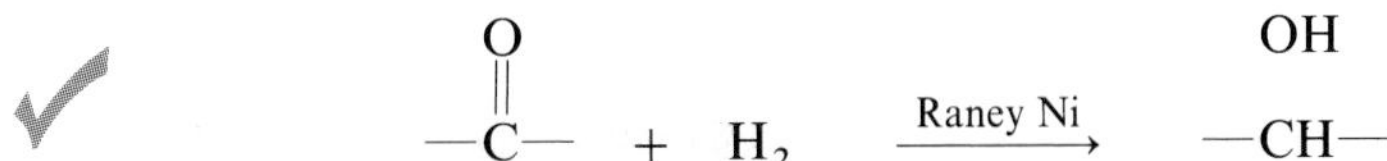

The conversion of a ketone or an aldehyde to an alcohol involves adding two hydrogen atoms across the C=O bond. This addition can also be accomplished by catalytic hydrogenation, commonly using Raney nickel as the catalyst. Carbon-carbon double bonds are reduced faster than the carbonyl group, however, so any alkene double bonds present will be reduced. In most cases, sodium borohydride is more convenient for the reduction of simple ketones and aldehydes.

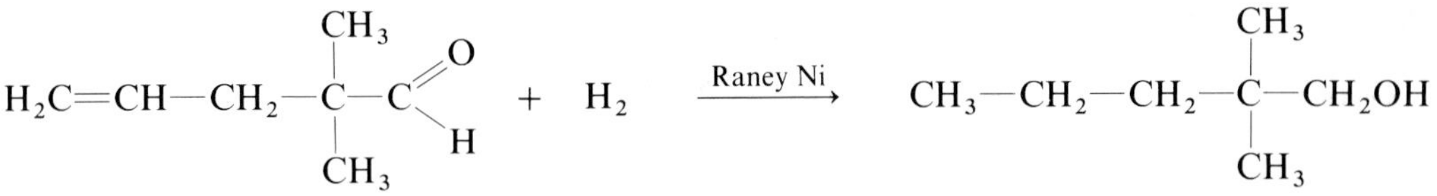

$$H_2C{=}CH—CH_2—C(CH_3)_2—CHO + H_2 \xrightarrow{\text{Raney Ni}} CH_3—CH_2—CH_2—C(CH_3)_2—CH_2OH$$

2,2-dimethyl-4-pentenal → 2,2-dimethyl-1-pentanol (94%)

9-13 THIOLS (MERCAPTANS)

Thiols are the sulfur equivalent of alcohols, with a —SH group in place of the hydroxyl group. Thiols are also called **mercaptans** ("captures mercury") because they form stable heavy-metal derivatives. IUPAC names are derived from the alkane names, using the suffix *—thiol.* Common names are formed like those of alcohols, using the name of the alkyl group with the word *mercaptan.* The —SH group itself is called a *mercapto* group.

	$CH_3—SH$	$CH_3CH_2CH_2CH_2—SH$	$CH_3—CH=CH—CH_2—SH$	$HS—CH_2—CH_2—OH$
IUPAC name:	methanethiol	1-butanethiol	2-butene-1-thiol	2-mercaptoethanol
common name:	methyl mercaptan	*n*-butyl mercaptan		

The odor of thiols is their strongest characteristic. Skunk scent is composed mainly of 1-butanethiol and 2-butene-1-thiol, with small amounts of other thiols. Methanethiol is added to natural gas (odorless methane) to give it the characteristic "gassy" odor useful for detecting leaks.

Thiols can be prepared by the S_N2 reaction of sodium hydrosulfide with an unhindered alkyl halide. The thiol product is still nucleophilic, so a large excess of hydrosulfide is used to avoid the product undergoing a second alkylation to give a sulfide (R—S—R).

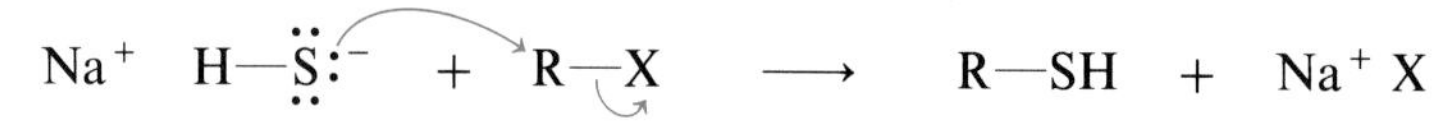

$$Na^+ \; H—\ddot{\underset{..}{S}}:^- \; + \; R—X \longrightarrow R—SH \; + \; Na^+ \, X^-$$

sodium hydrosulfide; alkyl halide or tosylate; thiol

Unlike alcohols, thiols are easily oxidized to give a dimer called a **disulfide.** The reverse reaction, reduction of the disulfide to the thiol, takes place under reducing conditions. Formation and cleavage of disulfide linkages is an important aspect of protein chemistry (Chapter 24), where disulfide "bridges" between cysteine amino acid residues hold the protein chain in its active conformation.

$$R—SH \; + \; HS—R \underset{Zn,\, HCl}{\overset{Br_2}{\rightleftharpoons}} R—S—S—R \; + \; 2\,HBr$$

two molecules of thiol; disulfide

Example

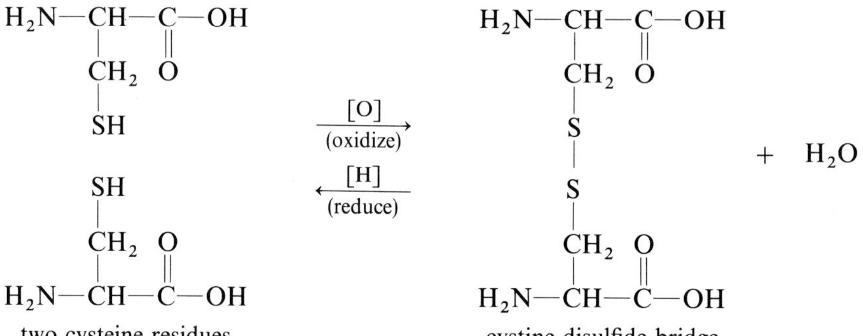

$$2\; H_2N—CH(CH_2SH)—C(=O)—OH \underset{[H]\ (reduce)}{\overset{[O]\ (oxidize)}{\rightleftharpoons}} H_2N—CH(COOH)—CH_2—S—S—CH_2—CH(COOH)—NH_2 \; + \; H_2O$$

two cysteine residues; cystine disulfide bridge

PROBLEM 9-25

Give IUPAC names for the following compounds.

(a) 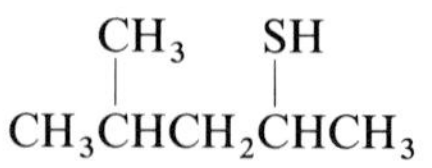 $CH_3CH(CH_3)CH_2CH(SH)CH_3$

(b) 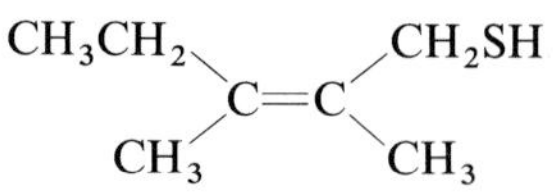 $(CH_3CH_2)(CH_3)C=C(CH_2SH)(CH_3)$

(c) cyclohexene ring bearing SH

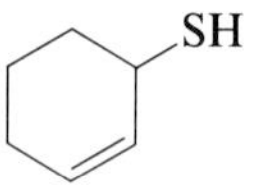

PROBLEM 9-26

Authentic skunk spray has become valuable for use in scent-masking products. Show how you would synthesize the two major components of skunk spray from any of the readily available butenes or from 1,3-butadiene.

SUMMARY OF ALCOHOL SYNTHESES

I. *From alkenes*

1. *Hydration* (Sections 8-4 through 8-7)
   a. *Acid-catalyzed: forms Markovnikov alcohols; or*
   b. *Oxymercuration-demercuration: forms Markovnikov alcohols*

$$H_2C{=}C(CH_3)_2 \xrightarrow[\text{(1) Hg(OAc)}_2,\ H_2O \quad \text{(2) NaBH}_4]{H_2SO_4,\ H_2O \text{ or}} H_3C{-}C(CH_3)(OH){-}CH_3$$

   c. *Hydroboration-oxidation: forms anti-Markovnikov alcohols*

$$H_2C{=}C(CH_3)_2 \xrightarrow[\text{(2) } H_2O_2,\ NaOH]{\text{(1) } BH_3\cdot THF} HO{-}CH_2{-}CH(CH_3){-}CH_3$$

2. *Hydroxylation: forms vicinal diols (glycols)* (Section 8-13 and 8-14)
   a. *Syn hydroxylation, using $KMnO_4/NaOH$ or using OsO_4/H_2O_2*

$$\text{cyclopentene} \xrightarrow[or\ KMnO_4/NaOH]{OsO_4/H_2O_2} \textit{cis}\text{-cyclopentane-1,2-diol}$$

cyclopentene — *cis*-cyclopentane-1,2-diol

   b. *Anti hydroxylation, using peracids*

$$\text{cyclopentene} \xrightarrow[H_2O]{R{-}CO_3H,\ H^+} \textit{trans}\text{-cyclopentane-1,2-diol} \quad (+\ \text{enantiomer})$$

cyclopentene — *trans*-cyclopentane-1,2-diol

II. *From alkyl halides: nucleophilic substitution* (Section 5-10 and 5-15)

1. *Second-order substitution: primary (and some secondary) halides*

$$(CH_3)_2CHCH_2CH_2{-}Br \xrightarrow[H_2O]{KOH} (CH_3)_2CHCH_2CH_2{-}OH$$

2. *First-order substitution: tertiary (and some secondary) halides*

$$CH_3{-}C(CH_3)(Cl){-}CH_3 \xrightarrow[\text{heat}]{\text{acetone/water}} CH_3{-}C(CH_3)(OH){-}CH_3 + H_2C{=}C(CH_3){-}CH_3$$

t-butyl chloride — *t*-butyl alcohol — isobutylene

III. *From carbonyl compounds: nucleophilic addition to the carbonyl group* (Section 9-10)

1. *Addition of a Grignard or organolithium reagent*

$$-\overset{O}{\overset{\|}{C}}- + R{-}MgX \xrightarrow{\text{ether}} -\overset{O^-\ {}^+MgX}{\overset{|}{\underset{R}{\underset{|}{C}}}}- \xrightarrow{H_3O^+} -\overset{OH}{\overset{|}{\underset{R}{\underset{|}{C}}}}-$$

a. *Addition to formaldehyde gives a primary alcohol*

$$CH_3CH_2MgBr + H_2C{=}O \xrightarrow[(2)\ H_2O]{(1)\ \text{ether}} CH_3CH_2{-}CH_2{-}OH$$

ethylmagnesium bromide → 1-propanol

b. *Addition to an aldehyde gives a secondary alcohol*

$$C_6H_5{-}MgBr + CH_3{-}\overset{O}{\overset{\|}{C}}{-}H \xrightarrow[(2)\ H_2O]{(1)\ \text{ether}} C_6H_5{-}\overset{H}{\underset{CH_3}{C}}{-}OH$$

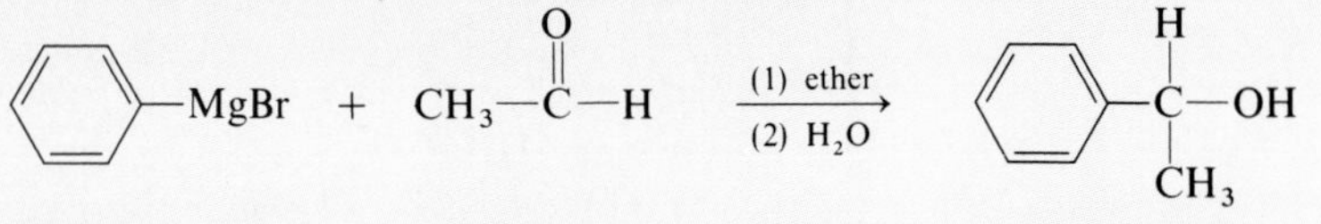

phenylmagnesium bromide, acetaldehyde → 1-phenylethanol

c. *Addition to a ketone gives a tertiary alcohol*

$$CH_3CH_2MgCl + \text{cyclohexanone} \xrightarrow[(2)\ H_2O]{(1)\ \text{ether}} \text{1-ethylcyclohexanol (OH, } CH_2CH_3\text{)}$$

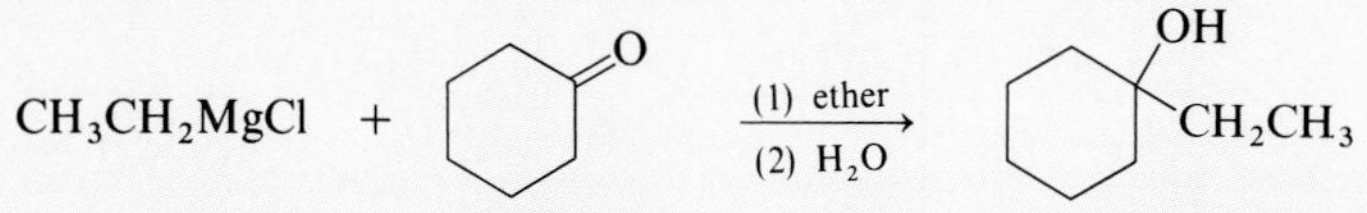

cyclohexanone → 1-ethylcyclohexanol

d. *Addition to an acid halide or an ester gives a tertiary alcohol*

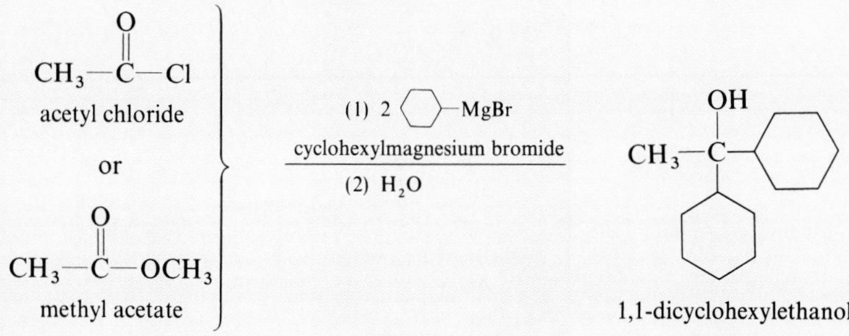

$$CH_3{-}\overset{O}{\overset{\|}{C}}{-}Cl$$

acetyl chloride

or

$$CH_3{-}\overset{O}{\overset{\|}{C}}{-}OCH_3$$

methyl acetate

$$\xrightarrow[(2)\ H_2O]{(1)\ 2\ C_6H_{11}{-}MgBr\ \text{(cyclohexylmagnesium bromide)}} CH_3{-}\overset{OH}{\underset{C_6H_{11}}{C}}{-}C_6H_{11}$$

1,1-dicyclohexylethanol

e. *Addition to ethylene oxide gives a primary alcohol (with two carbon atoms added)*

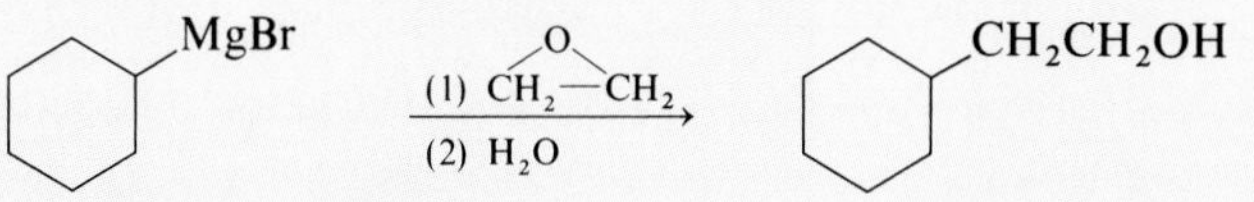

$$C_6H_{11}{-}MgBr \xrightarrow[(2)\ H_2O]{(1)\ CH_2{-}CH_2\ (\text{O bridge})} C_6H_{11}{-}CH_2CH_2OH$$

cyclohexylmagnesium bromide → 2-cyclohexylethanol

2. *Reduction of carbonyl compounds* (Section 9-13)

a. *Catalytic hydrogenation of aldehydes and ketones*

$$-\overset{O}{\overset{\|}{C}}- + H_2 \xrightarrow{\text{Raney Ni}} -\overset{OH}{\overset{|}{CH}}-$$

This method is usually not as selective or as effective as the use of hydride reagents.

b. *Use of hydride reagents*

(1) Reduction of an aldehyde gives a primary alcohol

$$C_6H_5{-}\overset{O}{\overset{\|}{C}}{-}H \xrightarrow{NaBH_4} C_6H_5{-}CH_2OH$$

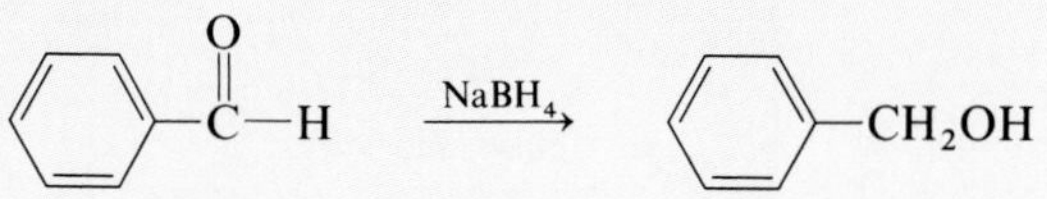

benzaldehyde → benzyl alcohol

(2) Reduction of a ketone gives a secondary alcohol

$$\text{cyclohexanone} \xrightarrow{NaBH_4} \text{cyclohexanol (H, OH)}$$

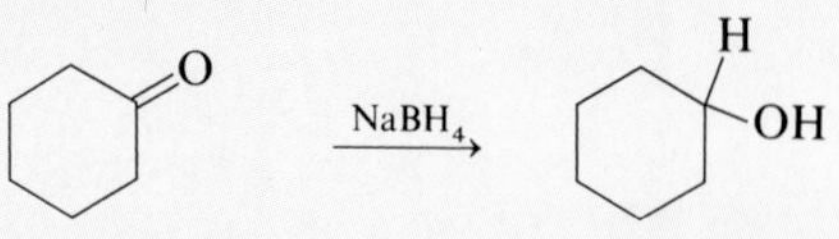

cyclohexanone → cyclohexanol

(3) Reduction of an acid, ester, or acid chloride gives a primary alcohol

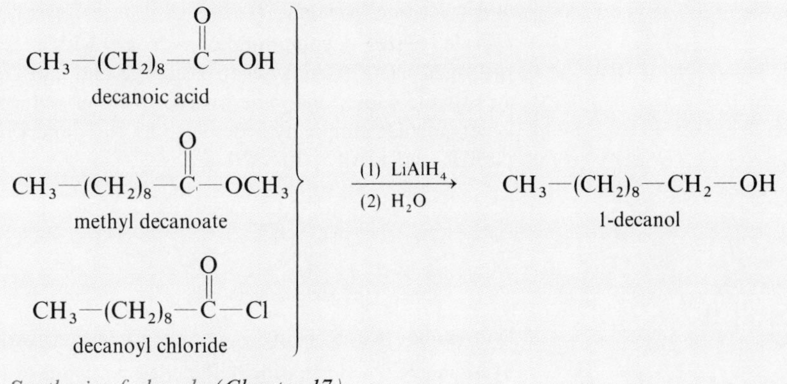

$$\left.\begin{array}{c} CH_3-(CH_2)_8-\overset{\overset{\displaystyle O}{\|}}{C}-OH \\ \text{decanoic acid} \\ \\ CH_3-(CH_2)_8-\overset{\overset{\displaystyle O}{\|}}{C}-OCH_3 \\ \text{methyl decanoate} \\ \\ CH_3-(CH_2)_8-\overset{\overset{\displaystyle O}{\|}}{C}-Cl \\ \text{decanoyl chloride} \end{array}\right\} \xrightarrow[(2)\ H_2O]{(1)\ LiAlH_4} \begin{array}{c} CH_3-(CH_2)_8-CH_2-OH \\ \text{1-decanol} \end{array}$$

IV. Synthesis of phenols (Chapter 17)

GLOSSARY

acid derivatives Compounds that are related to carboxylic acids, but have other electron-withdrawing groups in place of the —OH group of the acid. Three examples are acid chlorides, esters, and amides. (p. 383)

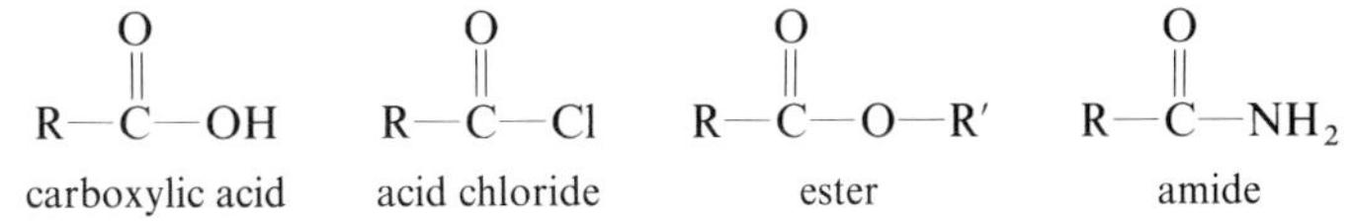

$$\underset{\text{carboxylic acid}}{R-\overset{\overset{\displaystyle O}{\|}}{C}-OH} \qquad \underset{\text{acid chloride}}{R-\overset{\overset{\displaystyle O}{\|}}{C}-Cl} \qquad \underset{\text{ester}}{R-\overset{\overset{\displaystyle O}{\|}}{C}-O-R'} \qquad \underset{\text{amide}}{R-\overset{\overset{\displaystyle O}{\|}}{C}-NH_2}$$

alcohol A compound in which a hydrogen atom of a hydrocarbon has been replaced by a hydroxyl group, —OH. (p. 364)

Alcohols are classified as **primary, secondary,** or **tertiary** depending on whether the hydroxyl group is bonded to a primary, secondary, or tertiary carbon atom. (p. 365)

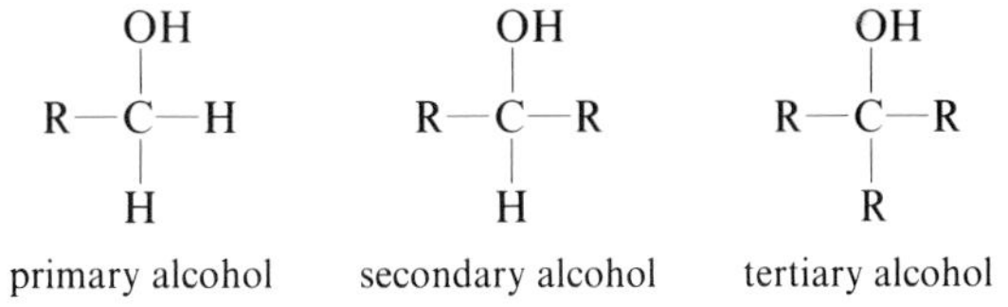

$$\underset{\text{primary alcohol}}{R-\overset{\overset{\displaystyle OH}{|}}{\underset{\underset{\displaystyle H}{|}}{C}}-H} \qquad \underset{\text{secondary alcohol}}{R-\overset{\overset{\displaystyle OH}{|}}{\underset{\underset{\displaystyle H}{|}}{C}}-R} \qquad \underset{\text{tertiary alcohol}}{R-\overset{\overset{\displaystyle OH}{|}}{\underset{\underset{\displaystyle R}{|}}{C}}-R}$$

aldehyde A carbonyl compound with one alkyl group and one hydrogen on the carbonyl group. (p. 381). **Formaldehyde** has two hydrogens on the carbonyl group.

alkoxide ion The anion ($R-\ddot{\underset{\cdot\cdot}{O}}\!:^-$) formed by deprotonation of an alcohol. (p. 374)

azeotrope A mixture of two or more liquids that distills at a constant temperature and gives a distillate of definite composition. For example, a mixture of 95 percent ethanol and 5 percent water boils at a lower temperature than does either pure ethanol or pure water. (p. 373)

Corey-House reaction A coupling reaction using a **lithium dialkylcuprate** and an alkyl halide. (p. 386)

$$\underset{\text{a lithium dialkylcuprate}}{R_2CuLi} + R'-X \longrightarrow R-R' + R-Cu + LiX$$

diol A compound with two alcohol —OH groups. (p. 368)

glycol: Synonymous with **diol**. The term "glycol" is most commonly applied to the 1,2-diols, also called vicinal diols.

epoxides (*oxiranes*) Compounds containing oxygen in a three-membered ring. (p. 385)

grain alcohol Ethanol, ethyl alcohol. **Absolute alcohol** is 100 percent ethanol. (p. 373)

Grignard reagent An organomagnesium halide, written in the form R—Mg—X. The actual reagent is more complicated in structure, usually a dimer or trimer complexed with several molecules of ether. (p. 379)

hydride reagent A compound of hydrogen with a less electronegative element, such that the hydrogen has a partial negative charge and can be donated with its pair of electrons to an organic compound. The hydride transfer reduces the organic compound. Hydride reagents include simple hydrides such as NaH and LiH, as well as complex hydrides such as $NaBH_4$ and $LiAlH_4$. (p. 389)

$$M^+ \ H:^- + \ \rangle C{=}\ddot{O}: \longrightarrow H{-}\overset{|}{\underset{|}{C}}{-}\ddot{\underset{..}{O}}:^- \ {}^+M$$

hydride reagent — reduced

hydrophilic ("water loving") Attracted to water; water-soluble (p. 371)

hydrophobic ("water hating") Repelled by water; water-insoluble. (p. 371)

ketone A carbonyl compound with two alkyl groups on the carbonyl group. (p. 381)

miscible Mutually soluble in any proportions. (p. 371)

organolithium reagent An organometallic reagent of the form R—Li. (p. 380)

organometallic compounds (organometallic reagents) Compounds containing metal atoms directly bonded to carbon. (p. 379)

phenol A compound with a hydroxyl group bonded directly to an aromatic ring. (p. 365)

rubbing alcohol 2-propanol, isopropyl alcohol. (p. 374)

skunk (*noun*) A digitigrade omnivorous quadruped that effectively synthesizes thiols; (*verb*) to prevent from scoring in a game or contest. (p. 392)

thiol (mercaptan) The sulfur analog of an alcohol, R—S—H (p. 391)

disulfide The oxidized dimer of a thiol, R—S—S—R (p. 392)

wood alcohol Methanol, methyl alcohol. (p. 372)

ESSENTIAL PROBLEM-SOLVING SKILLS IN CHAPTER 9

1. Draw and name alcohols, phenols, diols, and thiols.
2. Predict relative boiling points, acidities, and solubilities of alcohols.
3. Show how to convert alkenes, alkyl halides, and carbonyl compounds to alcohols.
4. Predict the alcohol products of hydration, hydroboration, and hydroxylation of alkenes.
5. Use Grignard and organolithium reagents effectively for the synthesis of primary, secondary, and tertiary alcohols with the required carbon skeletons.
6. Predict the products from reactions of lithium dialkylcuprates with alkyl and alkenyl halides.
7. Propose syntheses of simple thiols.

STUDY PROBLEMS

9-27. Briefly define each of the following terms and give an example.

(a) primary alcohol **(b)** secondary alcohol **(c)** tertiary alcohol
(d) phenol **(e)** diol **(f)** glycol
(g) alkoxide ion **(h)** epoxide **(i)** Grignard reagent
(j) organolithium reagent **(k)** ketone **(l)** aldehyde
(m) carboxylic acid **(n)** acid chloride **(o)** ester
(p) hydride reagents **(q)** thiol **(r)** disulfide

9-28. Give a systematic (IUPAC) name for each of the following alcohols. Classify each alcohol as primary, secondary, or tertiary.

(a) 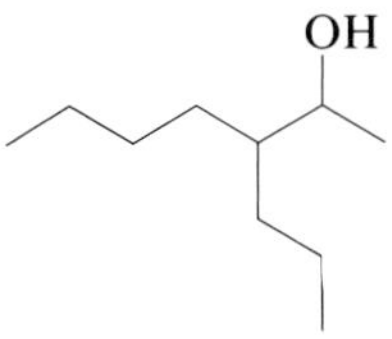 OH

(b) 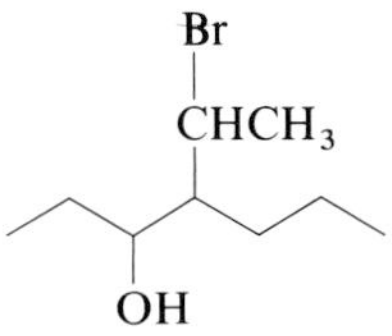 Br, $CHCH_3$, OH

(c) 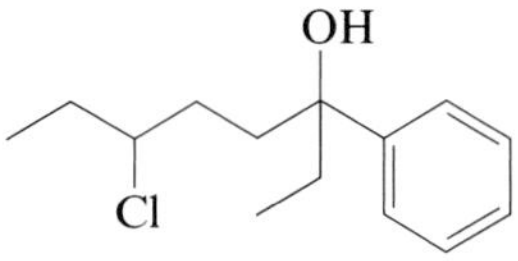 OH, Cl

(d) 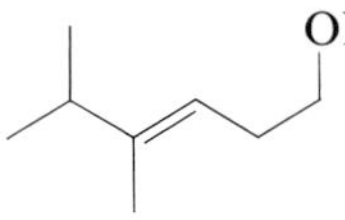 OH

(e) 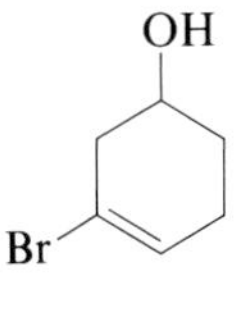 OH, Br

(f) 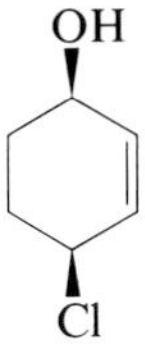 OH, Cl

9-29. Give a systematic (IUPAC) name for each of the following diols and phenols.

(a) 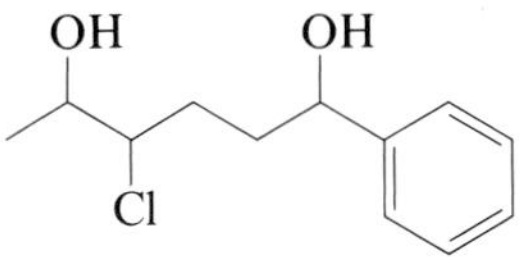 OH, OH, Cl

(b) 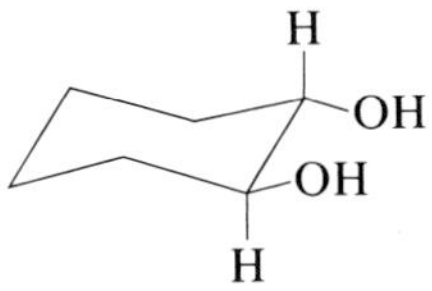 H, OH, OH, H

(c) 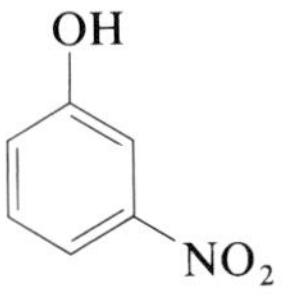 OH, NO_2

(d) 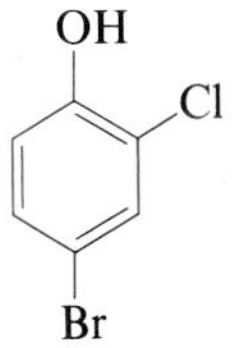 OH, Cl, Br

9-30. Draw the structures of the following compounds.

(a) triphenylmethanol
(b) 3-(bromomethyl)-4-octanol
(c) 3-cyclopenten-1-ol
(d) 3-cyclohexyl-3-pentanol
(e) *meso*-2,4-pentanediol
(f) cyclopentene glycol
(g) 4-iodophenol
(h) (2R,3R)-2,3-hexanediol
(i) 3-cyclopentenethiol
(j) dimethyl disulfide

9-31. Predict which member of each pair has the higher boiling point and explain the reasons for your predictions.

(a) 1-hexanol and 3,3-dimethyl-1-butanol
(b) 2-hexanone and 2-hexanol
(c) 2-hexanol and 1,5-hexanediol

9-32. Predict which member of each pair is more acidic and explain the reasons for your predictions.

(a) cyclopentanol and 3-chlorophenol
(b) cyclohexanol and 2-chlorocyclohexanol
(c) cyclohexanol and cyclohexanecarboxylic acid
(d) 2,2-dimethyl-1-butanol and 1-butanol

9-33. Predict which member of each group is most soluble in water and explain the reasons for your predictions.

(a) 1-butanol, 2-methyl-1-propanol, and 2-methyl-2-propanol
(b) chlorocyclohexane, cyclohexanol, and 1,2-cyclohexanediol
(c) chlorocyclohexane, cyclohexanol, and 4-methylcyclohexanol

9-34. Show how you would synthesize the following alcohols from appropriate alkenes.

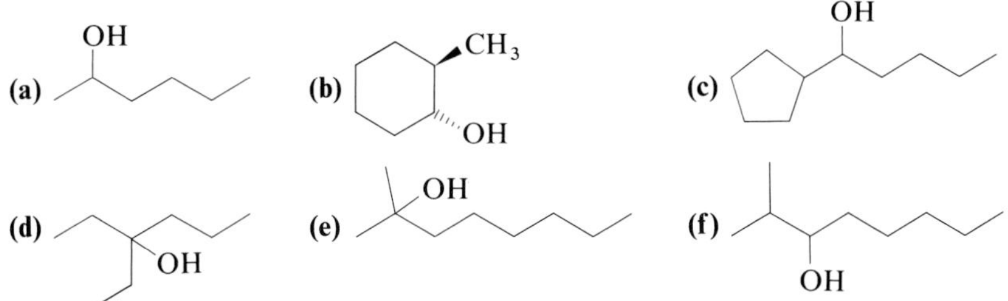

9-35. State the organic products you would expect to isolate from the following reactions (after hydrolysis).

(a) 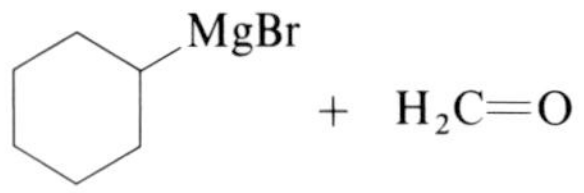 cyclohexyl–MgBr + $H_2C{=}O$ **(b)** 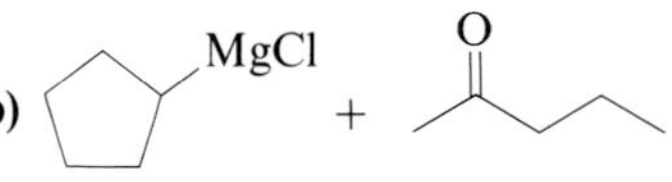 cyclopentyl–MgCl + 2-butanone **(c)** $CH_3-CH(MgI)-CH_3$ + Ph—CHO

(d) CH_3MgI + 3-hydroxycyclohexanone **(e)** 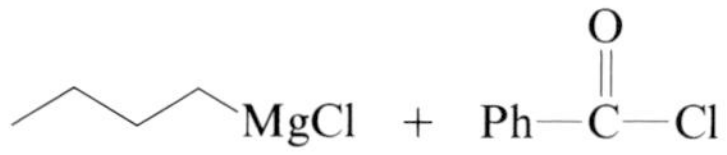 2 $CH_3CH_2CH_2CH_2MgCl$ + $Ph-C(=O)-Cl$ **(f)** Ph—MgBr + $Ph-C(=O)-Ph$

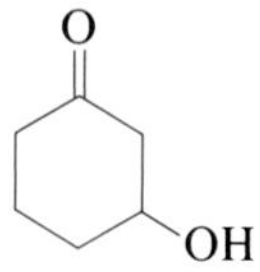

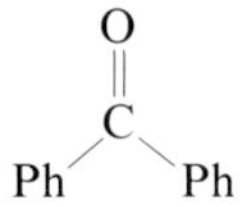

(g) 2 Ph—MgBr + $CH_3CH_2-C(=O)-OCH_3$ **(h)** 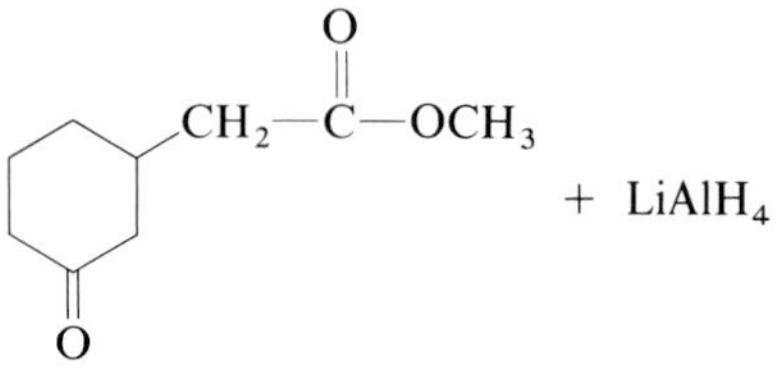 3-oxocyclohexyl–$CH_2-C(=O)-OCH_3$ + $LiAlH_4$

(i) 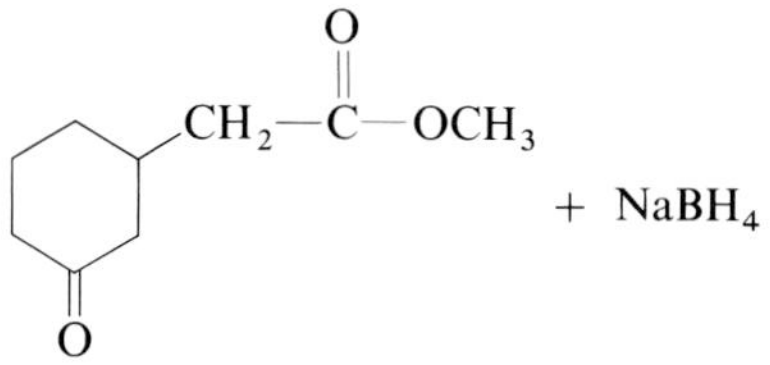 3-oxocyclohexyl–$CH_2-C(=O)-OCH_3$ + $NaBH_4$ **(j)** $CH_3-CH(CH_3)-CHO$ + $NaBH_4$

(k) 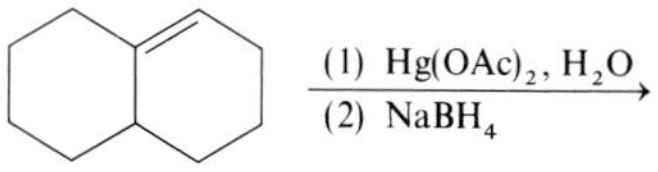 octahydronaphthalene $\xrightarrow[(2)\ NaBH_4]{(1)\ Hg(OAc)_2,\ H_2O}$ **(l)** 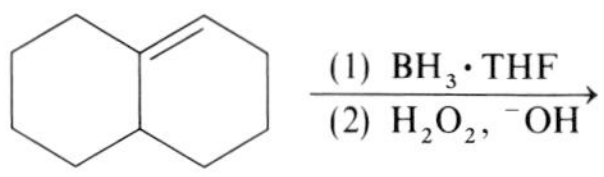 octahydronaphthalene $\xrightarrow[(2)\ H_2O_2,\ ^-OH]{(1)\ BH_3\cdot THF}$

(m) 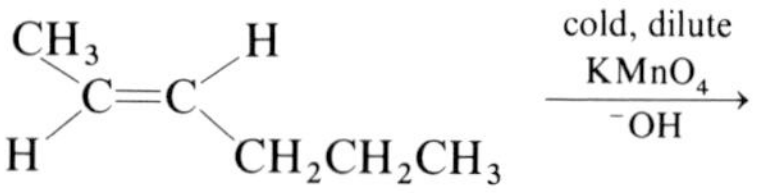 $(CH_3)(H)C{=}C(H)(CH_2CH_2CH_3)$ $\xrightarrow[^-OH]{\text{cold, dilute } KMnO_4}$ **(n)** 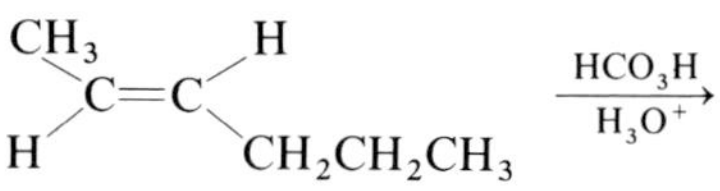 $(CH_3)(H)C{=}C(H)(CH_2CH_2CH_3)$ $\xrightarrow[H_3O^+]{HCO_3H}$

(o) $(CH_2{=}CH)_2\,Cu\,Li$ + $CH_3CH_2CH{=}CHCH_2Br$

9-36. Show how you would use Grignard syntheses to prepare the following alcohols from the indicated starting material and any other necessary reagents.

(a) 3-octanol from hexanal, $CH_3(CH_2)_4CHO$
(b) 1-octanol from 1-bromoheptane
(c) 1-cyclohexylethanol from acetaldehyde, CH_3CHO
(d) 2-cyclohexylethanol from bromocyclohexane
(e) benzyl alcohol (Ph—CH_2—OH) from bromobenzene (Ph—Br)
(f) 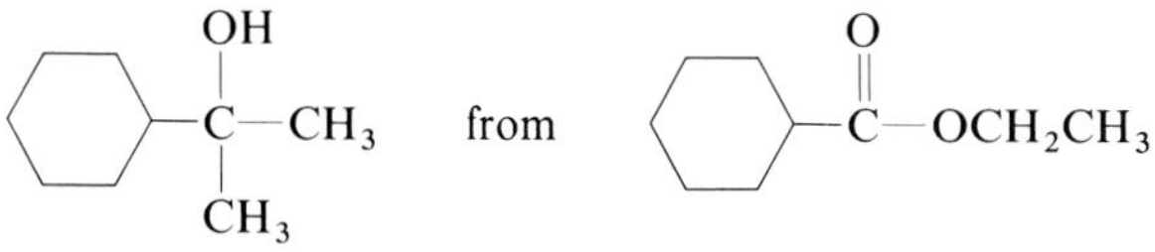 cyclohexyl–$C(OH)(CH_3)-CH_3$ from cyclohexyl–$C(=O)-OCH_2CH_3$
(g) cyclopentylphenylmethanol from benzaldehyde (Ph—CHO)

9-37. Show how you would accomplish each of the following transformations in good yield. You may use any additional reagents that you need.

(a) 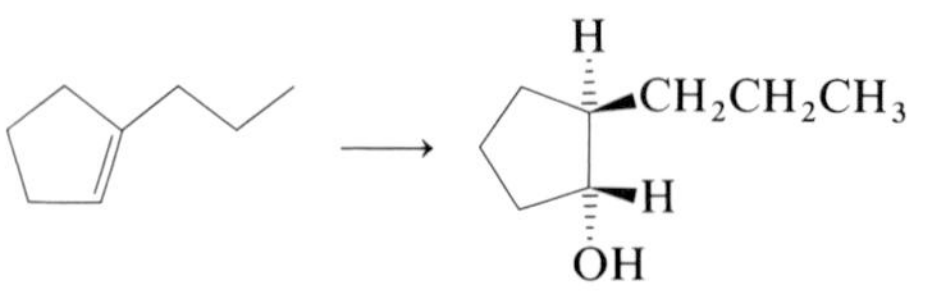 1-propylcyclopentene ⟶ trans-2-propylcyclopentanol (H, $CH_2CH_2CH_3$; H, OH)

(b) Ph—CH_2CH_2Cl ⟶ Ph—$CH_2CH_2CH_2OH$

(c) 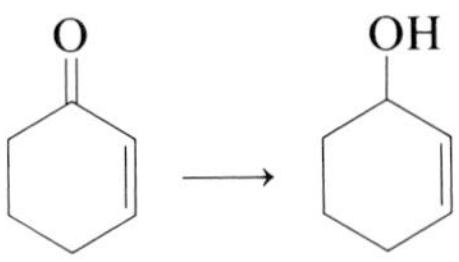 2-cyclohexen-1-one $\longrightarrow$ 2-cyclohexen-1-ol   (d) 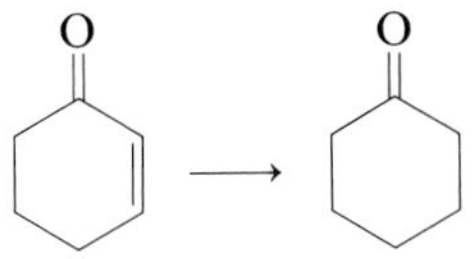 2-cyclohexen-1-one $\longrightarrow$ cyclohexanone

(e) $CH_3-\overset{O}{\overset{\|}{C}}-CH_2CH_2-\overset{O}{\overset{\|}{C}}-OCH_2CH_3 \longrightarrow CH_3-\overset{OH}{\overset{|}{CH}}-CH_2CH_2-\overset{O}{\overset{\|}{C}}-OCH_2CH_3$

(f) $CH_3-\overset{O}{\overset{\|}{C}}-CH_2CH_2-\overset{O}{\overset{\|}{C}}-OCH_2CH_3 \longrightarrow CH_3-\overset{OH}{\overset{|}{CH}}-CH_2CH_2-\overset{OH}{\overset{|}{CH_2}}$

9-38. Show how you would synthesize
(a) 2-phenylethanol by the addition of formaldehyde to a suitable Grignard reagent
(b) 2-phenylethanol from a suitable alkene
(c) cyclohexylmethanol from an alkyl halide using the S_N2 reaction
(d) 3-cyclohexyl-1-propanol by the addition of ethylene oxide to a suitable Grignard reagent
(e) *cis*-2-penten-1-thiol from a suitable alkenyl halide.
(f) 2,5-dimethylhexane from a four-carbon alkyl halide.

9-39. Complete the following acid-base reactions. In each case indicate whether the equilibrium favors the reactants or the products.

(a) $CH_3CH_2-O^- +$ C_6H_5–OH $\rightleftarrows$

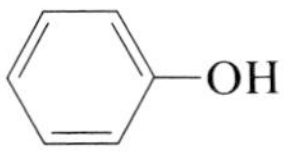

(b) KOH + CH_3CH_2OH $\rightleftarrows$

(c) $(CH_3)_3C-O^- + CH_3CH_2OH$ $\rightleftarrows$

(d) KOH + Cl–C_6H_3(Cl)–OH $\rightleftarrows$ (3,4-dichlorophenol)

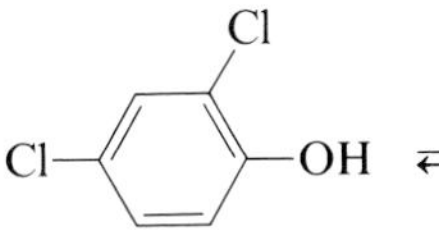

(e) 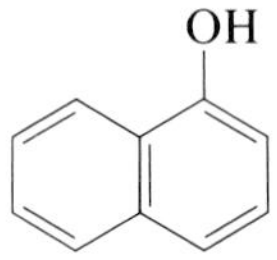 1-naphthol (OH) $+ CH_3O^-$ $\rightleftarrows$

(f) $(CH_3)_3C-O^- + H_2O$ $\rightleftarrows$

9-40. Suggest carbonyl compounds and reducing agents that might be used to form the following alcohols.
(a) *n*-octanol **(b)** 1-cyclohexyl-1-propanol **(c)** 1-phenyl-1-butanol

(d) 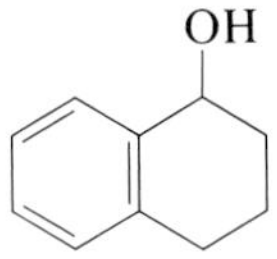 1,2,3,4-tetrahydronaphthalen-1-ol (OH)   **(e)** 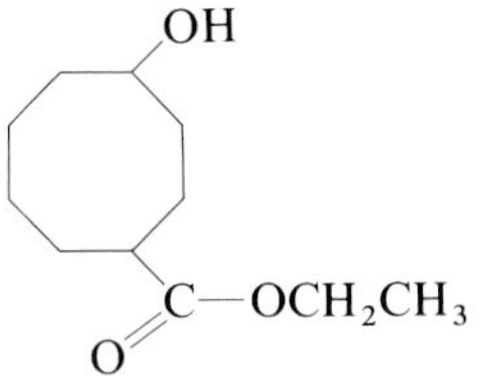 cyclooctane with OH and $C(=O)-OCH_2CH_3$ **(f)** 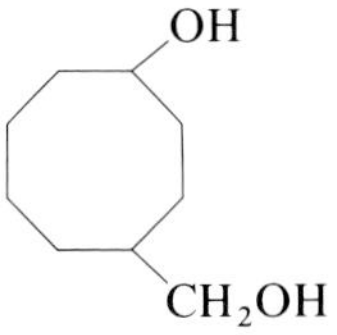 cyclooctane with OH and CH_2OH

✱ 9-41. Geminal diols, or 1,1-diols, are usually unstable, spontaneously losing water to give carbonyl compounds. Therefore, geminal diols are regarded as hydrated forms of ketones and aldehydes. Propose a mechanism for the acid-catalyzed loss of water from propane-2,2-diol to give acetone.

$$CH_3-\overset{HO\ \ OH}{C}-CH_3 \xrightleftharpoons{H^+} CH_3-\overset{O}{\overset{\|}{C}}-CH_3 + H_2O$$

propane-2,2-diol   acetone

✱ 9-42. Vinyl alcohols are generally unstable, quickly isomerizing to carbonyl compounds. Propose mechanisms for the following isomerizations.

(a) 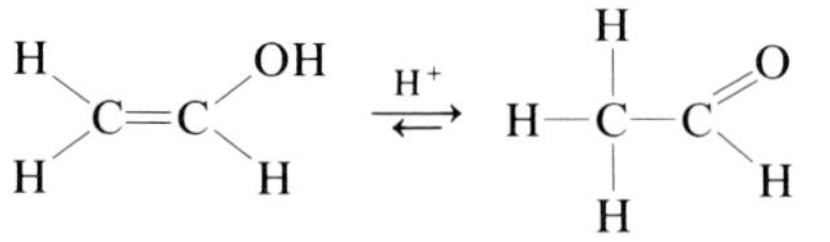 $H_2C{=}CH{-}OH \underset{}{\overset{H^+}{\rightleftharpoons}} H{-}CH_2{-}CH{=}O$

vinyl alcohol — acetaldehyde

(b) 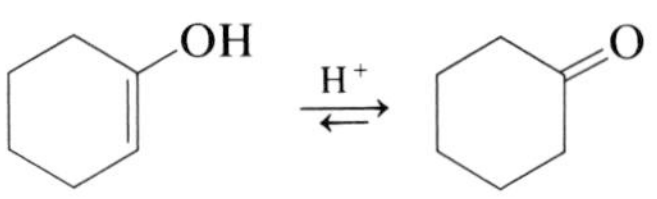 cyclohexenol (1-cyclohexen-1-ol) $\overset{H^+}{\rightleftharpoons}$ cyclohexanone

✱ 9-43. Compound **A** ($C_7H_{11}Br$) was treated with magnesium in ether to give **B** ($C_7H_{11}MgBr$), which reacts violently with D_2O to give 1-methylcyclohexene with a deuterium atom on the methyl group **(C)**. Reaction of **B** with acetone (CH_3COCH_3) followed by hydrolysis gives **D** ($C_{10}H_{18}O$). Heating **D** with concentrated H_2SO_4 gives **E** ($C_{10}H_{16}$), which decolorizes 2 equivalents of Br_2 to give **F** ($C_{10}H_{16}Br_4$). **E** undergoes hydrogenation with excess H_2 and a Pt catalyst to give isobutylcyclohexane. Determine the structures of compounds **A** through **F**, and show your reasoning throughout.

✱ 9-44. Grignard reagents react slowly with oxetane to produce primary alcohols. Give a mechanism for this reaction and suggest why oxetane reacts with Grignard reagents even though most ethers do not.

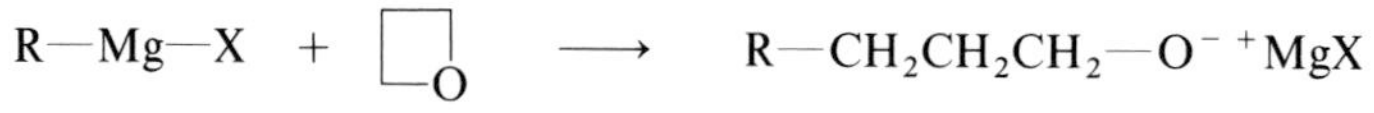

$$R{-}Mg{-}X \;+\; \text{oxetane} \longrightarrow R{-}CH_2CH_2CH_2{-}O^- \;^+MgX$$

Grignard reagent — oxetane — salt of 1° alcohol

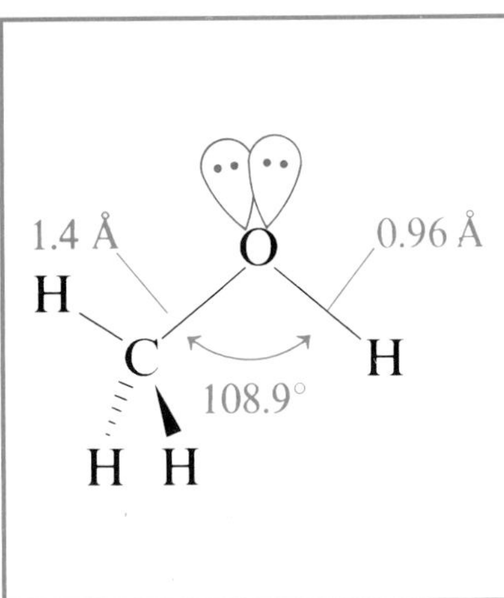

10

REACTIONS OF ALCOHOLS

Alcohols are a particularly important class of organic compounds because the hydroxyl group is easily converted to almost any other functional group. In Chapter 9 we studied reactions that form alcohols. In this chapter our aim is to understand how alcohols react and which reagents are most useful for converting them to other kinds of compounds.

10-1 OXIDATION STATES OF ALCOHOLS AND RELATED FUNCTIONAL GROUPS

Oxidation of alcohols leads to ketones, aldehydes, and carboxylic acids. These functional groups, in turn, undergo a wide variety of additional reactions. For these reasons, alcohol oxidations are some of the most common organic reactions. Many reagents are available for oxidizing alcohols, but we study only the ones that have the widest range of uses and the best selectivity. An understanding of the more common oxidants can later be extended to include additional reagents.

We can tell that an oxidation or a reduction of an alcohol has taken place by considering the number of bonds between the carbinol (C—OH) carbon atom and oxygen atoms. For example, in a primary alcohol the carbinol carbon atom has one bond to an oxygen atom; in an alkane it has none; in an aldehyde it has two; and in an acid it has three. Figure 10-1 compares the oxidation states of primary, secondary, and tertiary alcohols with those of the functional groups obtained by oxidation or reduction. The symbol [O] indicates an unspecified oxidizing agent.

Figure 10-1 shows that the oxidation of an alcohol results in the removal of two hydrogen atoms: one from the carbinol carbon and one from the hydroxyl

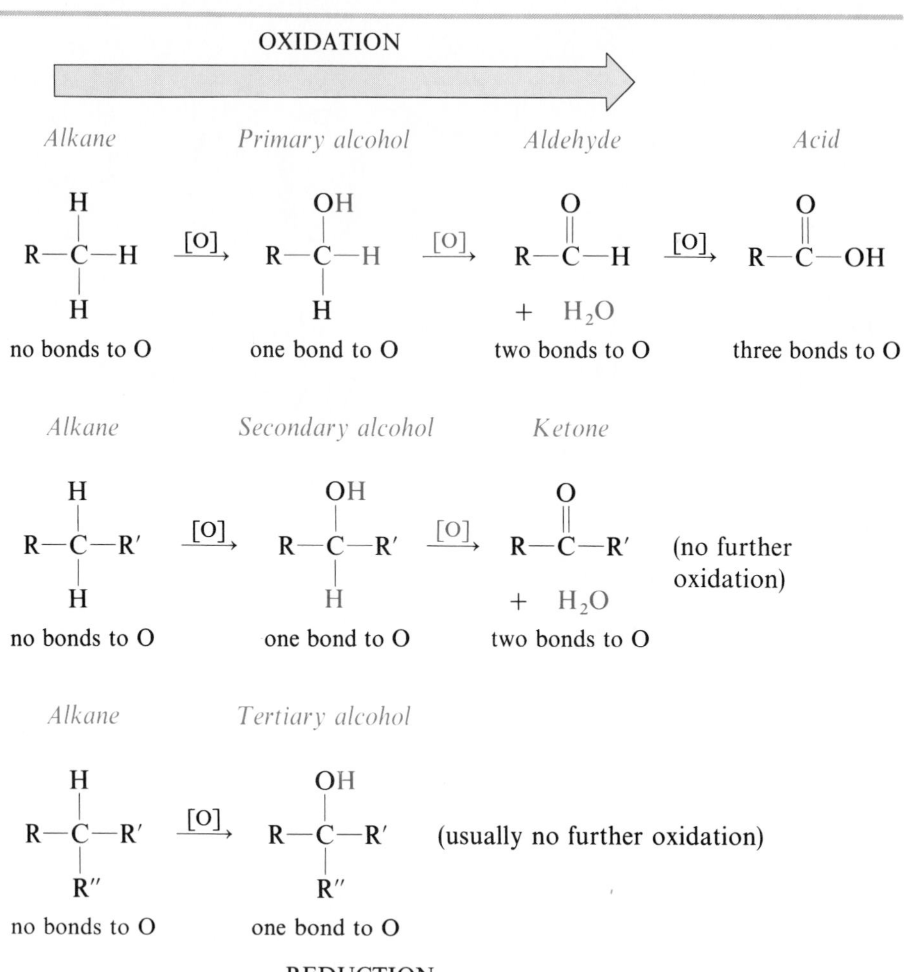

FIGURE 10-1 An alcohol is more oxidized than an alkane, yet less oxidized than carbonyl compounds such as ketones, aldehydes, and acids. Oxidation of a primary alcohol leads to an aldehyde, and further oxidation leads to an acid. Secondary alcohols are oxidized to ketones. Tertiary alcohols cannot be oxidized without breaking carbon-carbon bonds.

oxygen. A tertiary alcohol cannot be oxidized easily because there is no hydrogen atom available on the carbinol carbon. Alcohols can be reduced to alkanes by replacement of the hydroxyl group by a hydrogen atom.

PROBLEM 10-1

Classify each of the following reactions as an oxidation, a reduction, or neither.

(a) $CH_3{-}CH_2OH \xrightarrow{CrO_3 \cdot \text{pyridine}} CH_3{-}C({=}O){-}H \xrightarrow{H_2CrO_4} CH_3{-}C({=}O){-}OH$

(b) $CH_4 \longrightarrow CH_3OH \longrightarrow H{-}C({=}O){-}OH \longrightarrow H{-}C({=}O){-}H \longrightarrow HO{-}C({=}O){-}OH$

(c) $CH_3{-}C(CH_3)(OH){-}C(CH_3)(OH){-}CH_3 \xrightarrow{H^+} CH_3{-}C({=}O){-}C(CH_3)_2{-}CH_3 + H_2O$

(d) $CH_3{-}CH_2{-}OH \xrightarrow{LiAlH_4/TiCl_4} CH_3{-}CH_3$

(e) cyclopentyl–CHO $\xrightarrow{H^+, CH_3OH}$ cyclopentyl–CH(OCH_3)$_2$ + H_2O

10-2 OXIDATION OF ALCOHOLS

10-2A OXIDATION OF SECONDARY ALCOHOLS

Secondary alcohols are easily oxidized to give excellent yields of ketones. The **chromic acid** reagent usually provides the most efficient procedure for laboratory oxidations of secondary alcohols.

$$\underset{\text{secondary alcohol}}{R-\overset{OH}{\overset{|}{C}H}-R'} \xrightarrow{Na_2Cr_2O_7/H_2SO_4} \underset{\text{ketone}}{R-\overset{O}{\overset{\|}{C}}-R'}$$

Example

$$\underset{\text{cyclohexanol}}{C_6H_{11}OH} \xrightarrow[H_2SO_4]{Na_2Cr_2O_7} \underset{\substack{\text{cyclohexanone} \\ (90\%)}}{C_6H_{10}{=}O}$$

The chromic acid reagent is prepared by dissolving sodium dichromate ($Na_2Cr_2O_7$) in a mixture of sulfuric acid and water. The active species in the mixture is probably chromic acid, H_2CrO_4, or the acid chromate ion, $HCrO_4^-$. The same result is achieved by adding the more expensive chromium trioxide (CrO_3) to dilute sulfuric acid.

$$\underset{\text{sodium dichromate}}{Na_2Cr_2O_7} + H_2O + 2\,H_2SO_4 \longrightarrow \underset{\text{chromic acid } (H_2CrO_4)}{2\,H-O-\overset{O}{\overset{\|}{\underset{O}{\underset{\|}{Cr}}}}-O-H} + 2\,Na^+ + 2\,HSO_4^-$$

$$\underset{\text{chromium trioxide}}{CrO_3} + H_2O \longrightarrow \underset{\text{chromic acid}}{H-O-\overset{O}{\overset{\|}{\underset{O}{\underset{\|}{Cr}}}}-O-H} \rightleftharpoons H^+ + \underset{\text{acid chromate ion}}{{}^-O-\overset{O}{\overset{\|}{\underset{O}{\underset{\|}{Cr}}}}-O-H}$$

The mechanism of chromic acid oxidation probably involves the formation of a chromate ester. Elimination of the chromate ester gives the ketone. In the elimination, the carbinol carbon retains its oxygen atom but loses its hydrogen and gains the second bond to oxygen.

Formation of the chromate ester

$$\underset{\text{alcohol}}{R-\overset{R'}{\overset{|}{\underset{H}{\underset{|}{C}}}}-O-H} + \underset{\text{chromic acid}}{H-O-\overset{O}{\overset{\|}{\underset{O}{\underset{\|}{Cr}}}}-O-H} \longrightarrow \underset{\text{chromate ester}}{R-\overset{R'}{\overset{|}{\underset{H}{\underset{|}{C}}}}-O-\overset{O}{\overset{\|}{\underset{O}{\underset{\|}{Cr}}}}-O-H} + H_2O$$

Elimination of the chromate ester and oxidation of the carbinol carbon

$$\mathrm{R\!-\!\underset{H_2\ddot{O}:\;H}{\overset{R'}{C}}\!-\!\ddot{O}\!-\!\underset{O}{\overset{O}{\underset{\|}{\overset{\|}{Cr}}}}\!-\!O\!-\!H \longrightarrow R\!-\!\overset{R'}{C}\!=\!\ddot{O}: \;+\; H_3O^+ \;+\; {}^{-}\!:\!O\!-\!\overset{O}{\overset{\|}{Cr}}\!-\!O\!-\!H}$$

The chromium(IV) species formed above reacts further to give the stable reduced form, chromium(III). Both sodium dichromate and chromic acid are orange in color, while chromic ion (Cr^{3+}) is a deep blue. One can follow the progress of a chromic acid oxidation by observing the color change from orange through various shades of green to a greenish blue. In fact, the color change observed with chromic acid can be used as a test for the presence of an oxidizable alcohol.

10-2B OXIDATION OF PRIMARY ALCOHOLS

Oxidation of a secondary alcohol gives a ketone, and a similar oxidation of a primary alcohol forms an aldehyde. Unlike a ketone, however, an aldehyde is easily oxidized further to give a carboxylic acid.

$$\underset{\text{primary alcohol}}{\mathrm{R\!-\!\overset{OH}{\overset{|}{C}H}\!-\!H}} \xrightarrow{[O]} \underset{\text{aldehyde}}{\mathrm{R\!-\!\overset{O}{\overset{\|}{C}}\!-\!H}} \xrightarrow{[O]} \underset{\text{acid}}{\mathrm{R\!-\!\overset{O}{\overset{\|}{C}}\!-\!OH}}$$

Obtaining the aldehyde is usually difficult, since most oxidizing agents that are strong enough to oxidize the primary alcohol also oxidize the aldehyde. Chromic acid generally oxidizes a primary alcohol all the way to the carboxylic acid.

$$\underset{\text{cyclohexyl methanol}}{C_6H_{11}\!-\!CH_2OH} \xrightarrow[H_2SO_4]{Na_2Cr_2O_7} \underset{\substack{\text{cyclohexanecarboxylic acid}\\(92\%)}}{C_6H_{11}\!-\!\overset{O}{\overset{\|}{C}}\!-\!OH}$$

The **Jones reagent,** a solution of diluted chromic acid in acetone, is milder than the usual chromic acid reagent. It oxidizes some primary alcohols to aldehydes in acceptable yields. The **Jones oxidation** requires a careful procedure, however, since an excess of the reagent will oxidize the aldehyde.

$$\underset{\text{primary alcohol}}{\mathrm{R\!-\!\underset{H}{\underset{|}{\overset{OH}{\overset{|}{C}}}}\!-\!H}} \xrightarrow[\text{(acetone)}]{K_2Cr_2O_7,\ H_2SO_4,\ CH_3-\overset{O}{\overset{\|}{C}}-CH_3} \underset{\text{aldehyde}}{\mathrm{R\!-\!\overset{O}{\overset{\|}{C}}\!-\!H}} \xrightarrow{\text{excess reagent}} \underset{\text{acid}}{\mathrm{R\!-\!\overset{O}{\overset{\|}{C}}\!-\!OH}}$$

Example

$$\underset{\text{1-heptanol}}{CH_3(CH_2)_5\!-\!CH_2OH} \xrightarrow[\text{acetone}]{K_2Cr_2O_7,\ H_2SO_4} \underset{\substack{\text{heptanal}\\(50\%)}}{CH_3(CH_2)_5\!-\!\overset{O}{\overset{\|}{C}}\!-\!H}$$

A better reagent for the limited oxidation of primary alcohols to aldehydes is the **Collins reagent,** a complex of chromium trioxide with two molecules of pyridine. **Pyridinium chlorochromate (PCC)** is a more soluble version of the Collins reagent, in which chromium trioxide is complexed with one molecule of pyridine and one molecule of HCl. Both of these reagents oxidize most primary alcohols to aldehydes in excellent yields.

Collins reagent:	*Pyridinium chlorochromate (PCC):*
$CrO_3 \cdot 2$:N(pyridine)	(pyridine)N: $\cdot CrO_3 \cdot HCl$
chromium trioxide·2 pyridine	pyridine·chromium trioxide·HCl

$$\underset{\text{primary alcohol}}{R-\overset{\displaystyle OH}{\underset{\displaystyle H}{C}}-H} \xrightarrow[\textit{or } CrO_3\cdot\text{pyridine}\cdot HCl \text{ (PCC)}]{CrO_3\cdot 2 \text{ pyridine (Collins reagent)}, CH_2Cl_2} \underset{\text{aldehyde}}{R-\overset{\displaystyle O}{\overset{\|}{C}}-H}$$

Examples

$$\underset{\text{1-heptanol}}{CH_3(CH_2)_5-CH_2OH} \xrightarrow[\text{(Collins reagent)}]{CrO_3\cdot 2 \text{ pyridine}} \underset{\substack{\text{heptanal} \\ (93\%)}}{CH_3(CH_2)_5-\overset{\displaystyle O}{\overset{\|}{C}}-H}$$

$$\underset{\text{2-ethyl-1-hexanol}}{\text{(skeletal structure)}-OH} \xrightarrow[\text{(pyridinium chlorochromate, PCC)}]{\ddot{N}(\text{pyridine})\cdot CrO_3\cdot HCl} \underset{\substack{\text{2-ethylhexanal} \\ (87\%)}}{\text{(skeletal structure)}-\overset{\displaystyle O}{\overset{\|}{C}}-H}$$

10-2C RESISTANCE OF TERTIARY ALCOHOLS TO OXIDATION

The oxidation of tertiary alcohols is not an important reaction in organic chemistry. Tertiary alcohols have no hydrogen atoms on the carbinol carbon atom, and any oxidation must take place by breaking carbon-carbon bonds. Such oxidations require severe conditions and result in mixtures of products.

The **chromic acid test** for primary and secondary alcohols makes use of tertiary alcohols' resistance to oxidation. When a primary or secondary alcohol is added to the chromic acid reagent, the orange color changes to green or blue. When a nonoxidizable substance (such as a tertiary alcohol, a ketone, or an alkane) is added to the reagent, no immediate color change occurs.

In summary, the chromium reagents are the most commonly used oxidants for alcohols. Either chromic acid or the Jones reagent is appropriate for the oxidation of a secondary alcohol to a ketone. The chromic acid reagent is also suitable for oxidizing a primary alcohol all the way to the carboxylic acid. If the aldehyde is required, the Collins reagent (or PCC) is capable of oxidizing a primary alcohol selectively to the aldehyde.

SUMMARY OF ALCOHOL OXIDATIONS

To oxidize	*to*	*reagent to use*
2° alcohol	ketone	chromic acid or Jones reagent
1° alcohol	aldehyde	Collins reagent or PCC
1° alcohol	acid	chromic acid

PROBLEM 10-2

Predict the products of the reactions of the following compounds with chromic acid.

(a) cyclohexanol
(b) 1-methylcyclohexanol
(c) 2-methylcyclohexanol
(d) cyclohexanone
(e) cyclohexane
(f) acetic acid, CH_3COOH
(g) ethanol
(h) acetaldehyde, CH_3CHO

PROBLEM 10-3

Predict the products of the reaction of cyclohexylmethanol with each of the following reagents.

(a) $Na_2Cr_2O_7/H_2SO_4$
(b) dilute CrO_3/H_2SO_4/acetone (Jones reagent)
(c) CrO_3/pyridine (Collins reagent)
(d) pyridinium chlorochromate

10-3 ADDITIONAL METHODS FOR ALCOHOL OXIDATIONS

Many other reagents and procedures have been developed for oxidizing alcohols. For example, potassium permanganate is often used as a less expensive alternative to the chromium oxidizing agents. Permanganate oxidizes secondary alcohols to ketones and primary alcohols to carboxylic acids. Permanganate oxidations must be carefully controlled, or the strong oxidizing agent will cleave carbon-carbon bonds.

$$C_6H_5-CH(OH)-CH_3 \xrightarrow[H_2O,\ \text{buffer}]{KMnO_4} C_6H_5-C(=O)-CH_3 + MnO_2$$

1-phenylethanol → acetophenone (72%)

$$CH_3(CH_2)_4-CH_2OH \xrightarrow{KMnO_4,\ ^-OH} CH_3(CH_2)_4-C(=O)-O^- + MnO_2$$

1-hexanol → hexanoate ion (92%) (the basic form of hexanoic acid)

Perhaps the least expensive method for oxidation of alcohols is dehydrogenation: literally the removal of two hydrogen atoms. This industrial reaction is done at high temperature using a copper or copper oxide catalyst. The hydrogen by-product may be sold or used in the plant. The primary limitation of the dehydrogenation process is the inability of many organic compounds to survive a reaction at 300°C. Dehydrogenation is not well suited for laboratory syntheses.

$$\text{R—}\overset{\overset{\boxed{\text{H}}}{|}\text{O}}{\underset{\underset{\boxed{\text{H}}}{|}}{\text{C}}}\text{—R}' \xrightarrow{\text{heat, CuO}} \text{R—}\overset{\overset{\text{O}}{\|}}{\text{C}}\text{—R}' + \boxed{\text{H}_2}$$

Example

$$\text{cyclohexanol} \xrightarrow[300^\circ\text{C}]{\text{CuO}} \text{cyclohexanone (90\%)} + H_2$$

cyclohexanol

cyclohexanone
(90%)

PROBLEM 10-4

What is it about dehydrogenation that enables it to take place at 300°C but not at 25°C?

(a) Would you expect the kinetics, thermodynamics, or both to be unfavorable at 25°C? (*Hint:* Is the reverse reaction favorable at 25°C?)
(b) Which of these factors (kinetics or thermodynamics) improves as the temperature is raised?
(c) Explain the changes in the kinetics and thermodynamics of this reaction as the temperature increases.

PROBLEM 10-5

Give the structure of the principal product(s) when each of the following alcohols reacts with (1) $Na_2Cr_2O_7/H_2SO_4$; (2) Jones reagent; (3) $CrO_3 \cdot$pyridine; (4) $KMnO_4$, ^-OH.

(a) 1-octanol (b) 3-octanol
(c) 2-cyclohexen-1-ol (d) 1-methylcyclohexanol

PROBLEM 10-6

Suggest the method that would work best for each of the following *laboratory* syntheses.

(a) 1-butanol → butanal, $CH_3CH_2CH_2CHO$
(b) 1-butanol → butanoic acid, $CH_3CH_2CH_2COOH$
(c) 2-butanol → 2-butanone, $CH_3COCH_2CH_3$
(d) 2-buten-1-ol → 2-butenal, $CH_3CH{=}CH{-}CHO$
(e) 2-buten-1-ol → 2-butenoic acid $CH_3CH{=}CH{-}COOH$
(f) 1-methylcyclohexanol → 2-methylcyclohexanone (several steps)

10-4 BIOLOGICAL OXIDATION OF ALCOHOLS

Although it is the least toxic of the alcohols, ethanol is still a poisonous substance. When someone is suffering from a mild case of ethanol poisoning, we say that he or she is in*toxic*ated. Animals often consume food that has fermented and contains alcohol. Their bodies must be equipped to detoxify any alcohol in the food, to keep it from building up in the blood and poisoning the brain. To detoxify ethanol, the liver produces an enzyme called **alcohol dehydrogenase (ADH).**

Alcohol dehydrogenase catalyzes an oxidation: the removal of two hydrogen atoms from the alcohol molecule. The oxidizing agent for this reaction is called **nicotinamide adenine dinucleotide (NAD).** NAD exists in two forms: the oxidized form, called NAD^+, and the reduced form, called NADH. The following equation shows that ethanol is oxidized to acetaldehyde and NAD^+ is reduced to NADH. A subsequent oxidation, catalyzed by another enzyme, converts acetaldehyde to acetate ion. Acetate is the nontoxic anion of acetic acid.

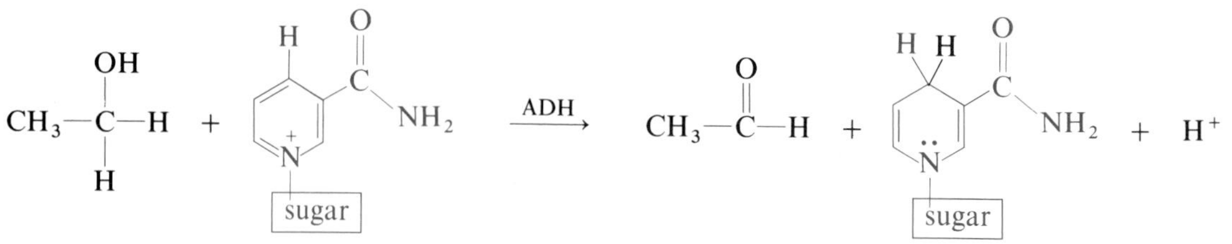

$$CH_3-CH(OH)-H + NAD^+ \xrightarrow{ADH} CH_3-C(=O)-H + NADH + H^+$$

ethanol reduced form; NAD^+ oxidized form; acetaldehyde oxidized form; NADH reduced form

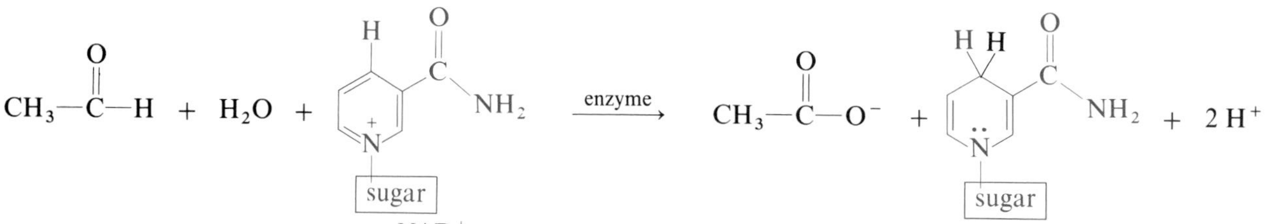

$$CH_3-C(=O)-H + H_2O + NAD^+ \xrightarrow{enzyme} CH_3-C(=O)-O^- + NADH + 2\,H^+$$

NAD^+; acetate ion; NADH

The oxidations shown above take place with most small primary alcohols. Unfortunately, the oxidation products of other alcohols are not always as nontoxic as the acetate ion. Methanol is oxidized first to formaldehyde and then to formic acid. Both of these compounds are more toxic than methanol itself.

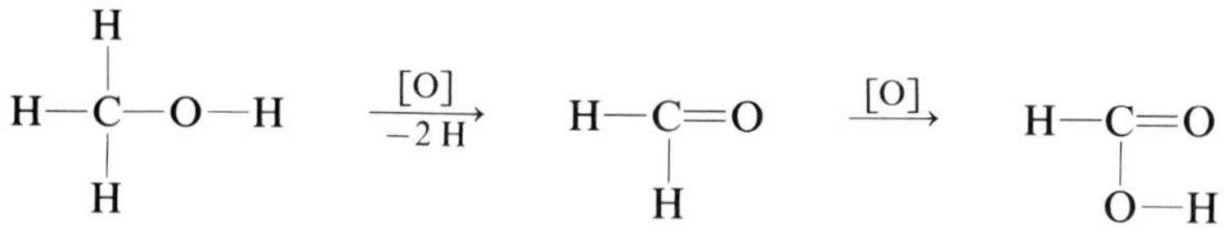

$$H-CH_2-O-H \xrightarrow[-2\,H]{[O]} H-C(H)=O \xrightarrow{[O]} H-C(O-H)=O$$

methanol; formaldehyde; formic acid

Ethylene glycol is a toxic diol. Its oxidation product is oxalic acid, the toxic compound found in rhubarb leaves and in many other plant leaves.

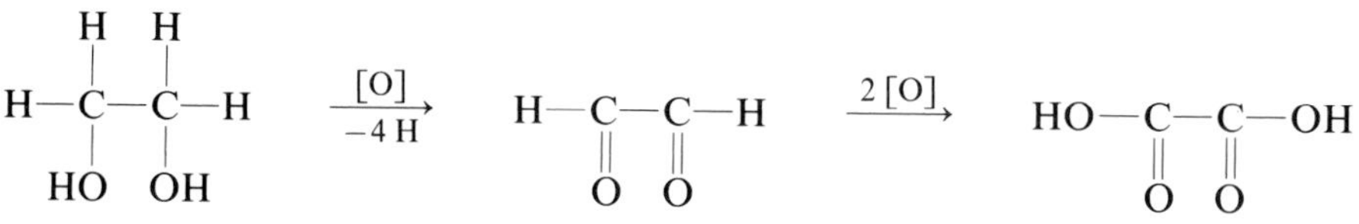

$$H-CH(OH)-CH(OH)-H \xrightarrow[-4\,H]{[O]} H-C(=O)-C(=O)-H \xrightarrow{2\,[O]} HO-C(=O)-C(=O)-OH$$

ethylene glycol; oxalic acid

Many poisonings by methanol and ethylene glycol occur each year. Alcoholics often drink ethanol that has been "denatured" by the addition of methanol. The methanol is oxidized to formic acid, which may cause blindness and death. Dogs are often poisoned by sweet-tasting ethylene glycol when antifreeze is left in an open container. Once the glycol is metabolized to oxalic acid, the dog's kidneys fail, causing death.

The treatment for methanol or ethylene glycol poisoning is the same. The patient is given intravenous infusions of diluted ethanol. The ADH enzyme is swamped by all the ethanol, and most of the methanol (or ethylene glycol) is excreted by the kidneys before it can be oxidized to formic acid (or oxalic acid). This is an example of the *competitive inhibition* of an enzyme. The enzyme catalyzes the oxidation of both ethanol and methanol, but the large quantity of added ethanol ties up the enzyme, allowing time for excretion of the methanol before it is oxidized.

PROBLEM 10-7

As an antidote to methanol poisoning, a chronic alcoholic requires a much larger dose of ethanol than does a nonalcoholic patient. Suggest a reason for the larger dose of the antidote in an alcoholic.

PROBLEM 10-8

Unlike ethylene glycol, propylene glycol (propane-1,2-diol) is nontoxic because it oxidizes to a common metabolic intermediate. Give the structure of the biological oxidation product of propylene glycol.

10-5 REDUCTION OF ALCOHOLS

The reduction of alcohols to alkanes is not a common reaction, because it removes a functional group, leaving fewer options for further reactions.

$$R{-}OH \xrightarrow{\text{reduction}} R{-}H$$

In some cases, alcohols are reduced directly by using $LiAlH_4$ in conjunction with titanium chloride or aluminum chloride.

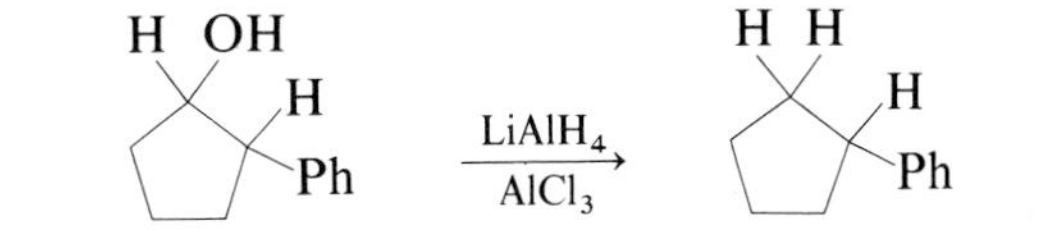

$$\text{2-phenylcyclopentanol} \xrightarrow[AlCl_3]{LiAlH_4} \text{phenylcyclopentane (78\%)}$$

Unfortunately, this reduction does not work well with some alcohols. A more general method for reducing an alcohol involves converting the alcohol to the tosylate ester, then using a hydride reducing agent to displace the tosylate leaving group. This reaction is more general in its scope, working with most primary and secondary alcohols.

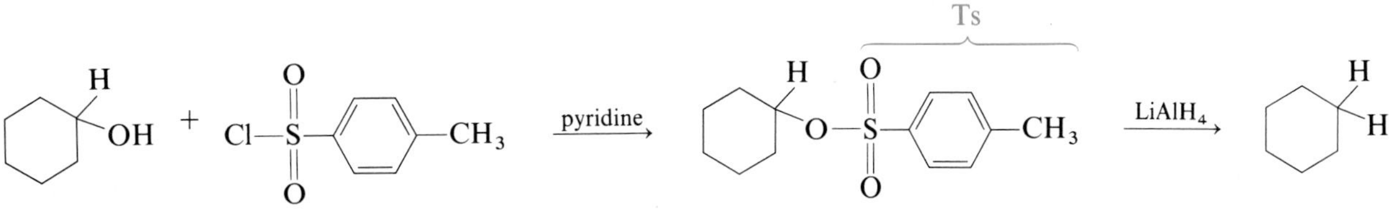

$$\text{cyclohexanol} + \underset{\text{tosyl chloride, TsCl}}{Cl{-}SO_2{-}C_6H_4{-}CH_3} \xrightarrow{\text{pyridine}} \underset{\text{cyclohexyl tosylate}}{C_6H_{11}{-}O{-}\underbrace{SO_2{-}C_6H_4{-}CH_3}_{Ts}} \xrightarrow{LiAlH_4} \underset{\text{cyclohexane (75\%)}}{C_6H_{12}}$$

PROBLEM 10-9

Predict the products of the following reactions.

(a) cyclopentanol + $LiAlH_4/TiCl_4$
(b) cyclopentanol + TsCl/pyridine
(c) product of (b) + $LiAlH_4$
(d) cyclopentanol + H_2SO_4, heat
(e) product of (d) + H_2, Pt

10-6 ALCOHOLS AS NUCLEOPHILES AND ELECTROPHILES

One of the reasons alcohols are versatile chemical intermediates is that the hydroxyl group may be replaced by a variety of other functional groups. In general, this can happen in one of two ways.

This bond is broken when alcohols react as nucleophiles.

—C—O—H

This bond is broken when alcohols react as electrophiles.

—C—O—H

An alcohol is a weak nucleophile; the nonbonding electrons on the oxygen atom can attack a strong electrophile. For example, a carbocation is attacked by an alcohol (as in the S_N1 reaction). Or, the alcohol can be converted to the alkoxide ion, a strong nucleophile that attacks weaker electrophiles.

$$R—\ddot{O}(H): + C^+ \longrightarrow R—\ddot{O}^+(H)—C—$$

alcohol

$$R—\ddot{O}:^- + —C—X \longrightarrow R—\ddot{O}—C— \quad X^-$$

alkoxide

An alcohol is a poor electrophile, because the hydroxyl group is a poor leaving group. The hydroxyl group can be converted to a good leaving group either by tosylation or by protonation. Tosylate esters and protonated alcohols undergo substitution and elimination reactions by all the first-order and second-order mechanisms we have studied with alkyl halides.

$$R—OH \;+\; TsCl \longrightarrow HCl \;+\; R—OTs \rightleftarrows R^+ \;\; ^-OTs$$

(or reacts with a nucleophile or base)

$$R—\ddot{O}H \;+\; H^+ \rightleftarrows R—\overset{H}{O^+}—H \rightleftarrows R^+ \;+\; H_2O$$

(or reacts with a nucleophile or base)

10-7 FORMATION AND USE OF TOSYLATE ESTERS

Tosylate esters are easily made from alcohols in very high yields, often using tosyl chloride as the reagent and pyridine as the solvent. The tosylate group is an excellent leaving group, and alkyl tosylates undergo substitution and elimination reactions muct like alkyl halides. In many cases, a tosylate is more reactive than the equivalent alkyl halide.

$$—C—C(OH)— \xrightarrow[\text{pyridine}]{\text{TsCl}} —C—C(OTs)— \xrightarrow[\text{(substitution)}]{\text{Nuc:}^-} —C—C(Nuc)— \;+\; ^-OTs$$

or elimination:

$$—C(H)—C(OTs)— \xrightarrow{\text{base (B:}^-)} \rangle C{=}C\langle \;+\; B—H \;+\; ^-OTs$$

B:

The two-step reduction of cyclohexanol to cyclohexane shown in the preceding section is an example of the reactivity of tosylate esters. Tosylation of the alcohol converts the hydroxyl group to a good leaving group. Lithium aluminum hydride displaces the tosylate group, forming an alkane. The Kenyon-Phillips experiment (Section 6-16) also used the tosylate ester as a leaving group. Kenyon and Phillips used 2-butanol labeled with ^{18}O, and showed that the labeled oxygen atom is present in the product. They concluded that the C—O bond is not broken in the tosylation step, and the 2-butyl tosylate formed must have the same configuration as the starting material.

The mechanism of tosylate formation shows how the C—O bond of the alcohol remains intact throughout the reaction, and the alcohol retains its stereochemical configuration:

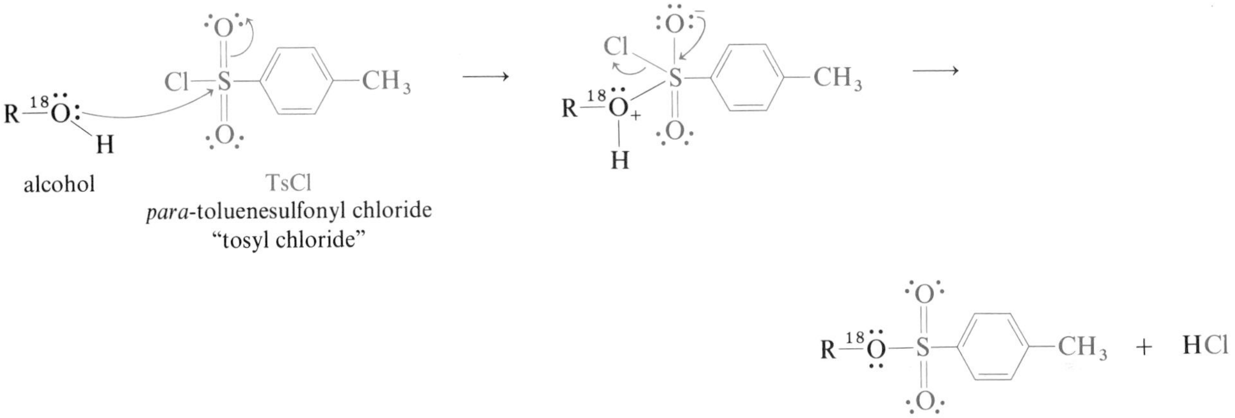

$$R{-}^{18}\ddot{O}{-}H + Cl{-}SO_2{-}C_6H_4{-}CH_3 \longrightarrow R{-}^{18}\ddot{O}^{+}(H){-}S(Cl)(O^-)(=O){-}C_6H_4{-}CH_3 \longrightarrow R{-}^{18}\ddot{O}{-}SO_2{-}C_6H_4{-}CH_3 + HCl$$

alcohol

TsCl
para-toluenesulfonyl chloride
"tosyl chloride"

R-OTs
a *para*-toluenesulfonate ester
a "tosylate" ester

The following reaction shows the S_N2 displacement of tosylate ion ($^-$OTs) from (*S*)-2-butyl tosylate with inversion of configuration. The tosylate ion is a particularly stable anion, with its negative charge delocalized over three oxygen atoms.

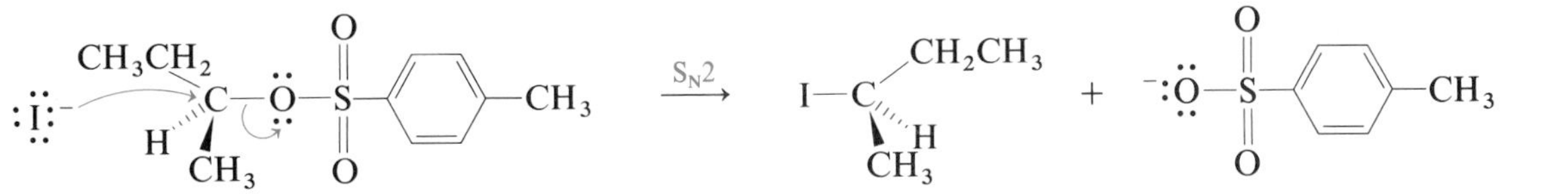

$$:\ddot{I}:^- + CH_3CH_2(H)(CH_3)C{-}O{-}SO_2{-}C_6H_4{-}CH_3 \xrightarrow{S_N2} I{-}C(CH_2CH_3)(H)(CH_3) + {}^-:\ddot{O}{-}SO_2{-}C_6H_4{-}CH_3$$

iodide — (*S*)-2-butyl tosylate — (*R*)-2-butyl iodide — tosylate ion

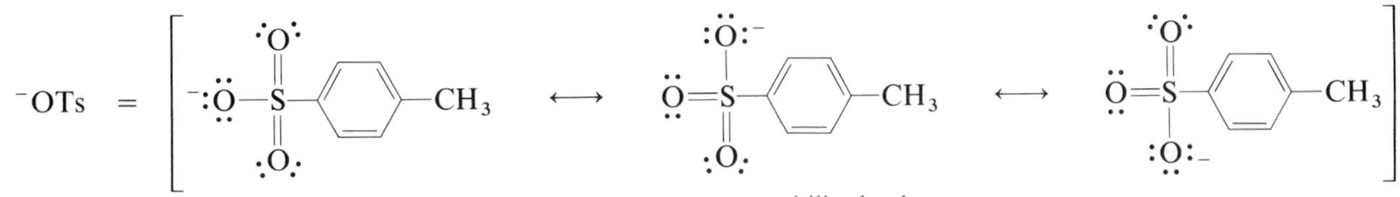

$$^-OTs = \left[{}^-:\ddot{O}{-}S(=O)_2{-}C_6H_4{-}CH_3 \longleftrightarrow O{=}S(O^-)(=O){-}C_6H_4{-}CH_3 \longleftrightarrow O{=}S(=O)(O^-){-}C_6H_4{-}CH_3 \right]$$

resonance-stabilized anion

Like the halide group of an alkyl halide, the tosylate leaving group can be displaced by a wide variety of nucleophiles. These substitutions take place by either the S_N1 mechanism or the S_N2 mechanism, although the S_N2 mechanism (strong nucleophile) is more commonly used in synthetic preparations. The following reactions show the generality of the S_N2 displacements of tosylates. In each case, the alkyl group R must be primary or unhindered secondary if substitution is to predominate over elimination.

S_N2 reactions of tosylate esters

R—OTs	+	$^-$OH hydroxide	⟶	R—OH alcohol	+	$^-$OTs
R—OTs	+	$^-$C≡N cyanide	⟶	R—C≡N nitrile	+	$^-$OTs
R—OTs	+	Br^- halide	⟶	R—Br alkyl halide	+	$^-$OTs
R—OTs	+	R′—O^- alkoxide	⟶	R—O—R′ ether	+	$^-$OTs
R—OTs	+	:NH_3 ammonia	⟶	R—NH_3^+ $^-$OTs amine salt		
R—OTs	+	$LiAlH_4$ LAH	⟶	R—H alkane	+	$^-$OTs

PROBLEM 10-10

Predict the major products of the following reactions.

(a) ethyl tosylate + potassium *t*-butoxide
(b) isobutyl tosylate + NaI
(c) (*R*)-2-hexyl tosylate + NaCN
(d) the tosylate of cyclohexylmethanol + excess NH_3
(e) *n*-butyl tosylate + sodium acetylide, H—C≡C:$^-$ $^+$Na

PROBLEM 10-11

Show how you would convert 1-propanol (and whatever reagents are needed) to the following compounds using a tosylate intermediate.

(a) 1-bromopropane
(b) *n*-propylamine, $CH_3CH_2CH_2NH_2$
(c) $CH_3CH_2CH_2$—O—CH_2CH_3 ethyl propyl ether
(d) $CH_3CH_2CH_2$—CN butyronitrile

10-8 REACTIONS OF ALCOHOLS WITH HYDROHALIC ACIDS

In an acidic solution, an alcohol is in equilibrium with its protonated form. Protonation converts the hydroxyl group from a poor leaving group to a good leaving group.

$$R-\ddot{O}-H + H^+ \rightleftharpoons R-\overset{H}{\overset{|}{O^+}}-H$$

poor leaving group — good leaving group

In the protonated form, the hydroxyl group can leave as a molecule of water—a particularly good leaving group. Once the alcohol is protonated, all the usual substitution and elimination reactions are feasible, depending on the structure (primary, secondary, or tertiary) of the alcohol.

Most of the really strong nucleophiles are strong bases, and they cannot be used in an acidic solution, as protonation makes them less nucleophilic. The halide ions are exceptions, however. They are good nucleophiles, yet they are anions of strong acids and therefore very weak bases. Hydrobromic acid (HBr) and hydrochloric acid (HCl) are commonly used to convert alcohols to the corresponding alkyl halides.

Reactions with hydrobromic acid

$$R{-}OH \quad + \quad HBr/H_2O \quad \longrightarrow \quad R{-}Br$$

Concentrated hydrobromic acid rapidly converts *t*-butyl alcohol to *t*-butyl bromide. The strong acid protonates the hydroxyl group, converting it to a good leaving group. The hindered tertiary carbon atom cannot undergo an S_N2 displacement, but it can ionize to form a tertiary carbocation. Attack by bromide ion gives the alkyl bromide. The mechanism is similar to the other S_N1 mechanisms we have studied, except that water serves as the leaving group from the protonated alcohol.

Protonation and loss of the hydroxyl group

$$(CH_3)_3C{-}\ddot{\underset{..}{O}}{-}H \quad H^+ \rightleftharpoons (CH_3)_3C{-}\overset{+}{\underset{..}{O}}H_2 \rightleftharpoons (CH_3)_3C^+ + :\ddot{O}H_2$$

t-butyl alcohol — protonated — tertiary carbocation

Attack by bromide

$$(CH_3)_3C^+ \quad {}^-:\ddot{\underset{..}{Br}}: \longrightarrow (CH_3)_3C{-}Br$$

tertiary carbocation — *t*-butyl bromide

Many other alcohols react with HBr, with the reaction mechanism depending on the structure of the alcohol. For example, 1-butanol reacts with sodium bromide in concentrated sulfuric acid to give 1-bromobutane by an S_N2 displacement. The sodium bromide/sulfuric acid reagent generates HBr in the solution.

$$CH_3(CH_2)_2{-}CH_2OH \xrightarrow{NaBr,\ H_2SO_4} CH_3(CH_2)_2{-}CH_2Br$$

1-butanol — 1-bromobutane (90%)

Protonation converts the hydroxyl group to a good leaving group, but ionization to a primary carbocation is very unfavorable. The protonated primary alcohol is well suited for the S_N2 displacement, however. A back-side attack by bromide ion gives 1-bromobutane.

Protonation of the alcohol

$$CH_3CH_2CH_2{-}CH_2{-}\ddot{\underset{..}{O}}{-}H + H^+ \rightleftharpoons CH_3CH_2CH_2{-}CH_2{-}\overset{+}{\ddot{O}}H_2$$

Displacement by bromide

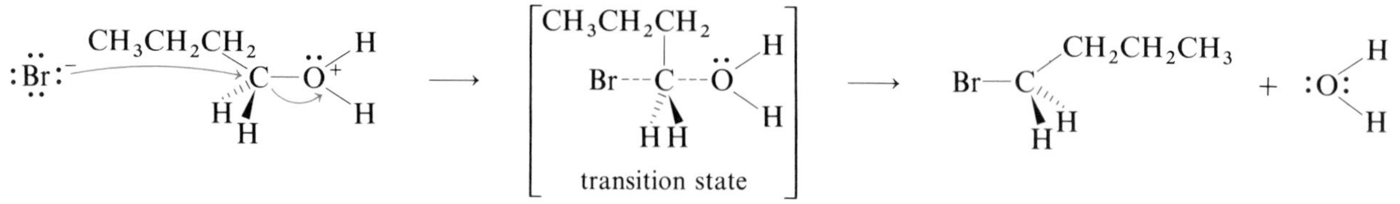

$$:\ddot{Br}:^- + CH_3CH_2CH_2-CH_2-\overset{+}{O}H_2 \longrightarrow \left[Br\cdots C(H)(H)(CH_2CH_2CH_3)\cdots OH_2 \right]^{\ddagger} \longrightarrow Br-CH_2CH_2CH_3 + :\ddot{O}H_2$$

transition state

Secondary alcohols also react with HBr to form alkyl bromides, usually by the S_N1 mechanism. Under some conditions the S_N2 mechanism may also be involved. For example, cyclohexanol is converted to bromocyclohexane using HBr as the reagent.

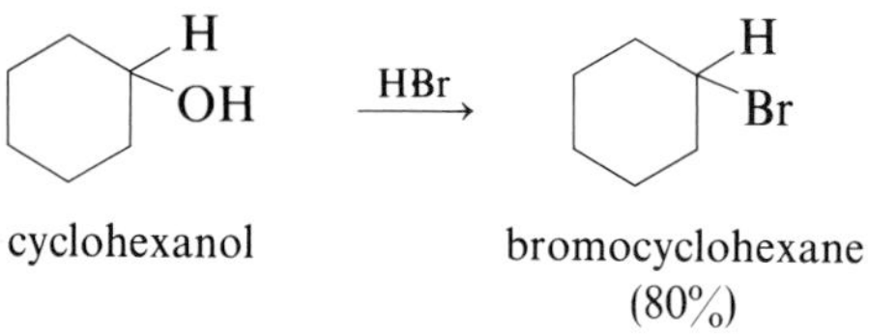

$$\text{cyclohexanol} \xrightarrow{HBr} \text{bromocyclohexane (80\%)}$$

cyclohexanol — bromocyclohexane (80%)

PROBLEM 10-12
Propose a mechanism for the reaction of

(a) cyclohexanol with HBr to form bromocyclohexane.
(b) 2-cyclohexylethanol with HBr to form 1-bromo-2-cyclohexylethane.

Reactions with hydrochloric acid

$$R-OH \quad + \quad HCl/H_2O \xrightarrow{ZnCl_2} R-Cl$$

Hydrochloric acid (HCl) reacts with alcohols in much the same way that hydrobromic acid does. For example, concentrated aqueous HCl reacts with *t*-butyl alcohol to give *t*-butyl chloride.

$$(CH_3)_3C-OH \quad + \quad HCl/H_2O \longrightarrow (CH_3)_3C-Cl \quad + \quad H_2O$$

t-butyl alcohol — *t*-butyl chloride (98%)

PROBLEM 10-13
The reaction of *t*-butyl alcohol with concentrated HCl goes by the S_N1 mechanism. Write a mechanism for this reaction.

The chloride ion is smaller and less polarizable than bromide ion, making chloride a weaker nucleophile. An additional Lewis acid, such as zinc chloride ($ZnCl_2$), is sometimes necessary to promote the reaction of HCl with primary and secondary alcohols. Zinc chloride coordinates with the oxygen atom of the alcohol in the same way that a proton does—except that zinc chloride coordinates more strongly.

The reagent composed of HCl and $ZnCl_2$ is called the **Lucas reagent.** Secondary and tertiary alcohols generally react with the Lucas reagent by the S_N1 mechanism.

$$(CH_3)_2CH{-}\ddot{O}{-}H + ZnCl_2 \rightleftharpoons (CH_3)_2CH{-}\overset{+}{O}(H){-}\overset{-}{Z}nCl_2 \rightleftharpoons$$

$$:\ddot{O}(H){-}\overset{-}{Z}nCl_2 + H{-}\overset{+}{C}(CH_3)_2 \xrightarrow{Cl^-} H{-}C(CH_3)_2{-}Cl$$

When a primary alcohol reacts with the Lucas reagent, ionization is not possible—the primary carbocation is too unstable. Primary substrates react by the S_N2 mechanism, more slowly than the S_N1 reaction of secondary and tertiary substrates. For example, when 1-butanol reacts with the Lucas reagent, chloride ion attacks the complex from the back, displacing the leaving group.

$$:\ddot{Cl}:^- + CH_3CH_2CH_2CH_2{-}\overset{+}{O}(H){-}\overset{-}{Z}nCl_2 \longrightarrow \left[\overset{\delta-}{Cl}\cdots C(H)(H)(CH_2CH_2CH_3)\cdots \overset{\delta+}{O}(H){-}\overset{-}{Z}nCl_2 \right]^{\ddagger}_{\text{transition state}} \longrightarrow Cl{-}CH_2CH_2CH_2CH_3 + :\ddot{O}(H){-}\overset{-}{Z}nCl_2$$

The Lucas test The Lucas reagent reacts with primary, secondary, and tertiary alcohols at fairly predictable rates, and these rates can be used to distinguish among the three types of alcohols. When the reagent is first added to the alcohol, the mixture forms a single homogeneous phase. The concentrated HCl solution is very polar, and the polar alcohol-zinc chloride complex dissolves. Once the alcohol has reacted to form the alkyl halide, the relatively nonpolar halide separates into a second phase.

The **Lucas test** involves adding the Lucas reagent to an unknown alcohol and watching for the second phase to separate (see Table 10-1). Tertiary alcohols react almost instantaneously, because they form relatively stable tertiary carbocations. Secondary alcohols react in about 1 to 5 minutes, because their secondary carbocations are less stable than the tertiary ones. Primary alcohols react very slowly. Since they cannot form carbocations, the activated primary alcohol simply remains in solution until it is attacked by the chloride ion. With a primary alcohol, the reaction may take from 10 minutes to several days.

TABLE 10-1
Reactions of alcohols with the Lucas reagent

Alcohol type	*Time to react (min)*
primary	>6
secondary	1–5
tertiary	<1

PROBLEM 10-14
Show how you would use a simple chemical test to distinguish between the following pairs of compounds. Tell what you would observe with each compound.

(a) isopropyl alcohol and *t*-butyl alcohol
(b) isopropyl alcohol and 2-butanone, $CH_3COCH_2CH_3$
(c) 1-hexanol and cyclohexanol
(d) allyl alcohol and 1-propanol
(e) 2-butanone and *t*-butyl alcohol

Limitations on the use of hydrohalic acids with alcohols The reactions of alcohols with hydrohalic acids do not always give good yields of the expected alkyl halides. Four principal limitations restrict the generality of this technique.

1. *Limited ability to make alkyl iodides.* Most alcohols do not react with HI to give acceptable yields of alkyl iodides. Alkyl iodides are valuable intermedi-

ates, however, because the iodides are the most reactive of the alkyl halides. We discuss a better technique for making alkyl iodides in the next section.

2. *Poor yields of alkyl chlorides from primary and secondary alcohols.* Primary and secondary alcohols react with HCl much slower than tertiary alcohols, even with zinc chloride added. Under these conditions, side reactions give rise to poor yields of the alkyl halides.
3. *Eliminations.* Heating an alcohol in a concentrated acid such as HCl or HBr often leads to elimination. Once the hydroxyl group of the alcohol has been protonated and converted to a good leaving group, it becomes a candidate for both substitution and elimination.
4. *Rearrangements.* Carbocation intermediates are always prone to rearrangements. We have seen (Section 5-17) that hydrogen atoms and alkyl groups can migrate from one carbon atom to another to form a more stable carbocation. This rearrangement may occur as the leaving group leaves, or it may occur once the cation has formed.

SOLVED PROBLEM 10-1

When 3-methyl-2-butanol is treated with concentrated HBr, the major product is 2-bromo-2-methylbutane. Propose a mechanism for the formation of this product.

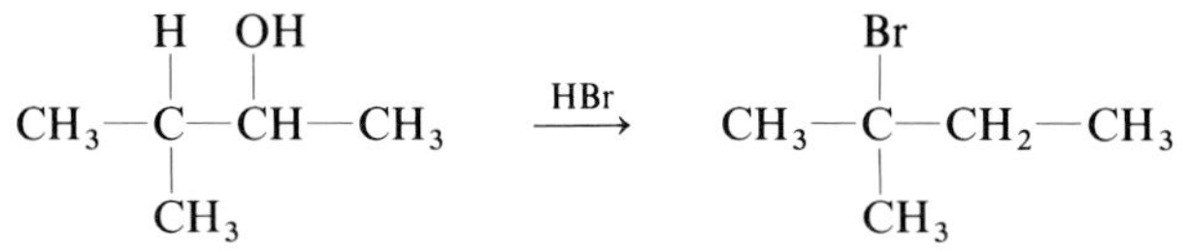

$$CH_3-C(H)(CH_3)-CH(OH)-CH_3 \xrightarrow{HBr} CH_3-C(Br)(CH_3)-CH_2-CH_3$$

3-methyl-2-butanol → 2-bromo-2-methylbutane

SOLUTION The alcohol is protonated by the strong acid. This protonated secondary alcohol loses water to form a secondary carbocation.

$$CH_3-C(H)(CH_3)-CH(\ddot{O}H)-CH_3 + H^+ \rightleftharpoons CH_3-C(H)(CH_3)-CH(\overset{+}{O}H_2)-CH_3 \rightleftharpoons CH_3-C(H)(CH_3)-\overset{+}{C}H-CH_3 + H_2\ddot{O}:$$

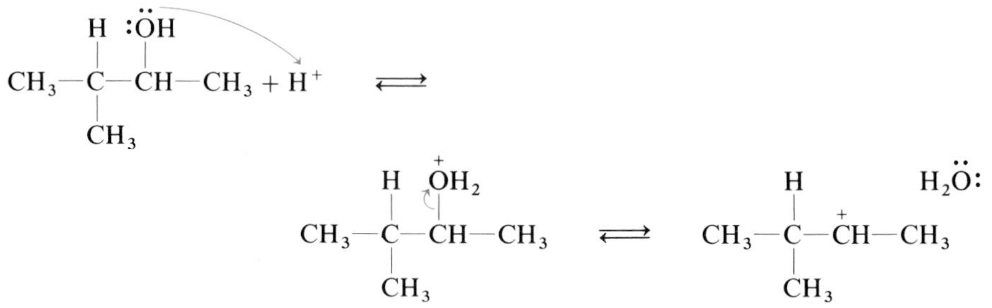

protonated alcohol; secondary carbocation

A hydride shift transforms the secondary carbocation into a more stable tertiary cation. Attack by bromide leads to the observed product.

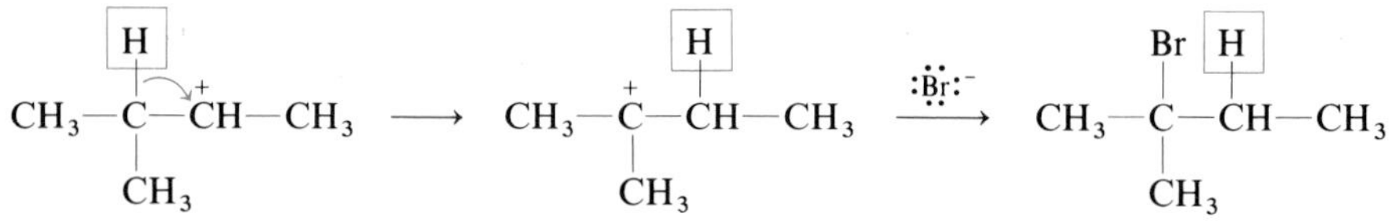

$$CH_3-C(H)(CH_3)-\overset{+}{C}H-CH_3 \longrightarrow CH_3-\overset{+}{C}(CH_3)-CH(H)-CH_3 \xrightarrow{:\ddot{Br}:^-} CH_3-C(Br)(CH_3)-CH(H)-CH_3$$

secondary carbocation → tertiary carbocation → observed product

Although rearrangements are usually seen as annoying side reactions, a clever chemist can use a rearrangement to accomplish a synthetic goal. Problem 10-15 shows how an alcohol substitution with rearrangement might be used in a synthesis.

PROBLEM 10-15

Neopentyl alcohol, $(CH_3)_3CCH_2OH$, reacts with concentrated HBr to give 2-bromo-2-methylbutane, a rearranged product. Give a mechanism for the formation of this product.

PROBLEM 10-16

When cyclohexylmethanol reacts with the Lucas reagent, one of the products is chlorocycloheptane. Propose a mechanism to explain the formation of this product.

PROBLEM 10-17

When *cis*-2-methylcyclohexanol reacts with the Lucas reagent, the major product is 1-chloro-1-methylcyclohexane. Propose a mechanism to explain the formation of this product.

10-9 REACTIONS OF ALCOHOLS WITH PHOSPHORUS HALIDES

Several phosphorus halides are useful for converting alcohols to alkyl halides. Phosphorus tribromide, phosphorus trichloride, and phosphorus pentachloride work well and are commercially available.

$$3\,R{-}OH \;+\; PCl_3 \;\longrightarrow\; 3\,R{-}Cl \;+\; P(OH)_3$$

$$3\,R{-}OH \;+\; PBr_3 \;\longrightarrow\; 3\,R{-}Br \;+\; P(OH)_3$$

$$R{-}OH \;+\; PCl_5 \;\longrightarrow\; R{-}Cl \;+\; POCl_3 \;+\; HCl$$

Phosphorus triiodide is not sufficiently stable to be prepared in advance or stored, but it can be generated *in situ* (in the reaction mixture) by the reaction of phosphorus with iodine.

$$6\,R{-}OH \;+\; 2\,P \;+\; 3\,I_2 \;\longrightarrow\; 6\,R{-}I \;+\; 2\,P(OH)_3$$

Phosphorus halides produce good yields of most primary and secondary alkyl halides, but none works well with tertiary alcohols. The two phosphorus halides used most often are PBr_3 and the phosphorus/iodine combination. Phosphorus tribromide is often the best reagent for converting a primary or secondary alcohol to the alkyl bromide, especially if the alcohol might rearrange in strong acid. Phosphorus and iodine is one of the best reagents for converting a primary or secondary alcohol to the alkyl iodide. For the synthesis of alkyl chlorides, the thionyl chloride/pyridine reagent discussed below generally gives better yields than PCl_3 or PCl_5, especially with tertiary alcohols.

The following examples show the conversion of primary and secondary alcohols to bromides and iodides by treatment with PBr_3 and P/I_2, respectively. Rearrangements are not a severe problem with these reagents, as demonstrated by the following synthesis.

$$CH_3{-}\underset{CH_3}{\overset{CH_3}{C}}{-}CH_2OH \;+\; PBr_3 \;\longrightarrow\; CH_3{-}\underset{CH_3}{\overset{CH_3}{C}}{-}CH_2Br$$

neopentyl alcohol → neopentyl bromide (60%)

$$CH_3(CH_2)_{14}{-}CH_2OH \;+\; P/I_2 \;\longrightarrow\; CH_3(CH_2)_{14}{-}CH_2I \quad (85\%)$$

$$CH_3OH \;+\; P/I_2 \;\longrightarrow\; CH_3I \quad (93\%)$$

PROBLEM 10-18

Write a balanced equation for each reaction shown above.

Mechanism of the reaction with phosphorus trihalides The mechanism of the reaction of alcohols with phosphorus trihalides explains why rearrangements are uncommon and why the phosphorus halides work so poorly with tertiary alcohols. The mechanism is shown below using PBr_3 as the reagent; PCl_3 and PI_3 (generated from phosphorus and iodine) react in a similar manner.

Displacement of bromide ion

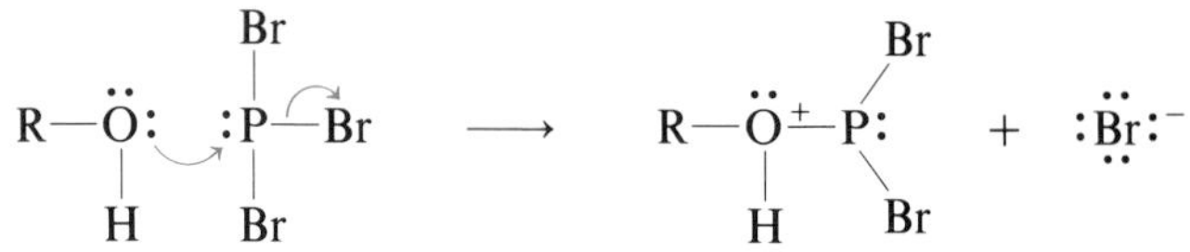

$$R-\ddot{O}(H): + :PBr_3 \longrightarrow R-\ddot{O}^{+}(H)-PBr_2 + :\ddot{Br}:^-$$

S_N2 attack on the alkyl group

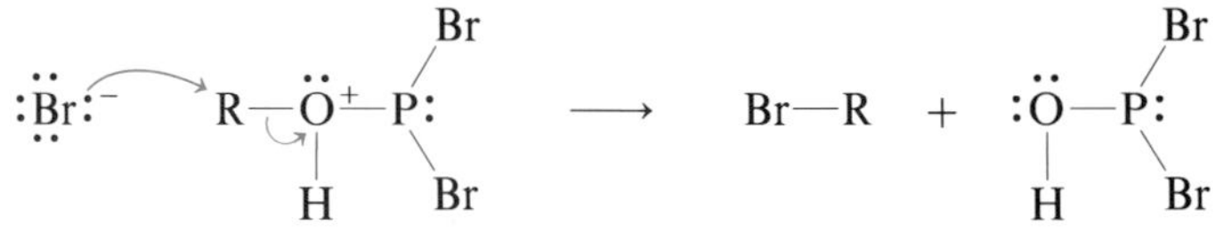

$$:\ddot{Br}:^- + R-\ddot{O}^{+}(H)-PBr_2 \longrightarrow Br-R + :\ddot{O}(H)-PBr_2$$

leaving group

The mechanism shown above explains why rearrangements are uncommon in the reactions of alcohols with PBr_3 and the other phosphorus halides. No carbocation is involved, so there is no opportunity for rearrangement. It also explains the poor yields with tertiary alcohols. The final step is an S_N2 displacement where bromide ion attacks the back side of the alkyl group. This attack is hindered if the alkyl group is tertiary. In the case of a tertiary alcohol, an ionization to a carbocation is needed. Such an ionization is slow, and it invites side reactions.

10-10 REACTIONS OF ALCOHOLS WITH THIONYL CHLORIDE

Thionyl chloride, $SOCl_2$, is often the best reagent for converting an alcohol to an alkyl chloride. The gaseous SO_2 and HCl by-products leave the reaction mixture and ensure that there can be no reverse reaction.

$$R-OH + Cl-S(=O)-Cl \xrightarrow{\text{heat}} R-Cl + SO_2 + HCl$$

Under the proper conditions, thionyl chloride reacts by an interesting mechanism summarized below. In the first step, the nonbonding electrons of the hydroxyl oxygen atom attack the electrophilic sulfur atom of thionyl chloride. Chloride ion is expelled and a proton is lost to give a chlorosulfite ester. In the second step, the chlorosulfite ester ionizes and the sulfur atom quickly delivers chloride to the carbocation.

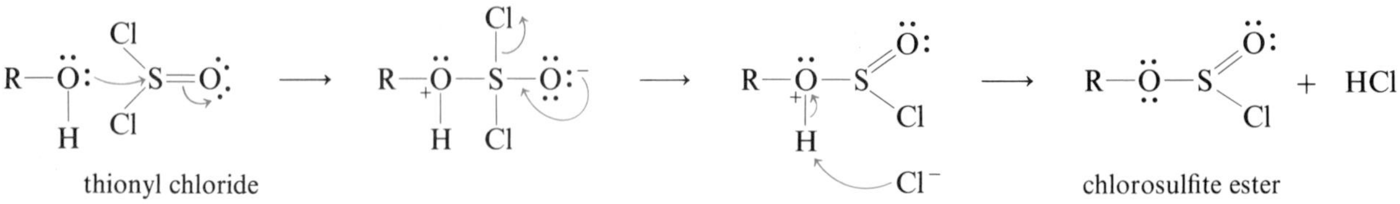

$$R-\ddot{O}(H): + Cl_2S=\ddot{O}: \longrightarrow R-\overset{+}{O}(H)-SCl(Cl)-\ddot{O}:^- \longrightarrow R-\overset{+}{O}(H)-S(=O)Cl + Cl^- \longrightarrow R-\ddot{O}-S(=O)Cl + HCl$$

thionyl chloride — chlorosulfite ester

$$R-O-S(=O)-Cl \longrightarrow R^{+}\ {}^{-}O-S(=O)-Cl \ \text{(ion pair)} \xrightarrow{\text{(fast)}} R-Cl + O=S=O$$

The preceding mechanism resembles the S_N1 mechanism, except that the nucleophile is delivered to the carbocation by the leaving group, giving retention of configuration as shown in the following example.

$$\underset{(R)\text{-2-octanol}}{CH_3(CH_2)_4CH_2-CH(OH)-CH_3} \xrightarrow[\text{dioxane (solvent)}]{SOCl_2} \underset{(R)\text{-2-chlorooctane } (84\%)}{CH_3(CH_2)_4CH_2-CHCl-CH_3}$$

SUMMARY OF THE MOST PROMISING REAGENTS FOR CONVERSION OF ALCOHOLS TO THE CORRESPONDING ALKYL HALIDES

Class of alcohol	*Chloride*	*Bromide*	*Iodide*
primary	$SOCl_2$	PBr_3 or HBr[a]	P/I_2
secondary	$SOCl_2$	PBr_3	P/I_2[a]
tertiary	HCl	HBr	HI[a]

[a] Works only in selected cases.

PROBLEM 10-19

Suggest how you would convert *trans*-4-methylcyclohexanol to

(a) *trans*-1-chloro-4-methylcyclohexane.
(b) *cis*-1-chloro-4-methylcyclohexane.

PROBLEM 10-20

Two products are observed in the reaction

$$\text{1-D-cyclohex-2-en-1-ol} \xrightarrow{SOCl_2} \text{1-chloro-1-D-cyclohex-2-ene} + \text{3-chloro-1-D-cyclohexene}$$

(a) Suggest a mechanism to explain how these two products are formed.
(b) Your mechanism for part (a) should be different from the usual mechanism of the reaction of $SOCl_2$ with alcohols. Explain why the reaction follows a different mechanism in this case.

PROBLEM 10-21

Give the structures of the products you would expect when each of the following alcohols reacts with (1) HCl, $ZnCl_2$; (2) HBr; (3) PBr_3; (4) P/I_2; (5) $SOCl_2$.

(a) 1-butanol (b) 2-butanol
(c) 2-methyl-2-butanol (d) 2,2-dimethyl-1-butanol

10-11 DEHYDRATION REACTIONS OF ALCOHOLS

10-11A FORMATION OF ALKENES

We have seen the mechanism for dehydration of alcohols to alkenes in Chapter 7 together with other syntheses of alkenes. Dehydration requires an acidic catalyst to protonate the hydroxyl group of the alcohol and convert it to a good leaving group. An equilibrium is established between reactants and products.

$$\text{H–C–C–OH} \overset{H^+}{\rightleftharpoons} \text{C=C} + H_2O$$

To drive this equilibrium toward the products, it is necessary to remove one or more of the products as they form. This can be done either by distilling the products out of the reaction mixture or by adding a dehydrating agent to remove water.

In practice, a combination of distillation and a dehydrating agent is often used. The alcohol is mixed with a dehydrating acid, and the reaction mixture is heated to boiling. The alkene boils at a lower temperature than the alcohol (because the alcohol is hydrogen-bonded), and the alkene distills out of the mixture. For example,

$$\text{cyclohexanol} \xrightleftharpoons{\text{conc. } H_2SO_4} \text{cyclohexene} + H_2O$$

cyclohexanol, b.p. 161°C

cyclohexene, b.p. 83°C (80%) (distilled from the mixture)

Alcohol dehydrations generally take place through the E1 mechanism; the E2 mechanism is rarely observed. Protonation of the hydroxyl group converts it to a good leaving group. Water leaves, forming a carbocation. Loss of a proton gives the alkene.

$$\text{cyclohexanol} \overset{H_2SO_4}{\rightleftharpoons} \text{protonated alcohol} + HSO_4^- \rightleftharpoons \text{carbocation} \overset{H_2\ddot{O}:}{\rightleftharpoons} \text{cyclohexene} + H_3O^+$$

Figure 10-2 shows the reaction energy profile for the E1 dehydration of an alcohol. The first step is a mildly exothermic protonation, followed by an endothermic, rate-determining ionization. A fast, strongly exothermic deprotonation gives the alkene.

Because the rate-determining step is the formation of a carbocation, the ease of dehydration of alcohols follows the same order as the ease of formation of carbocations: $3° > 2° > 1°$. As in other carbocation reactions, rearrangements are common.

In the dehydration of a primary alcohol, a rearrangement *must* occur as a carbocation is formed to avoid producing an energetically unfavorable primary carbocation. The following mechanism shows how 1-butanol undergoes dehydration with rearrangement of give a mixture of 1-butene and 2-butene. In accordance with the Saytzeff rule, 2-butene (the more highly substituted alkene) is the major product.

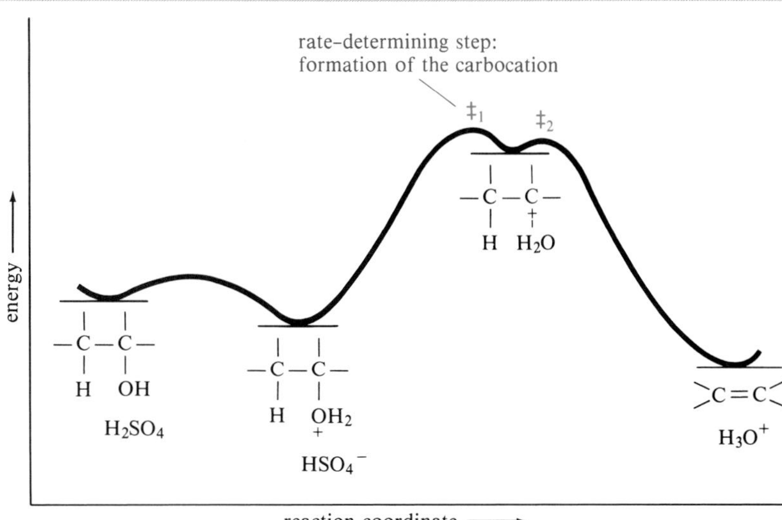

FIGURE 10-2 Reaction energy profile for dehydration of an alcohol.

Ionization of the protonated alcohol, with rearrangement

H_2O leaves
H:⁻ migrates

secondary carbocation

secondary carbocation

loss of H_a^+
2-butene (major, 70%)
a disubstituted alkene

loss of H_b^+
1-butene (minor, 30%)
a monosubstituted alkene

We can summarize the utility of dehydration as a synthetic reaction and how to predict the products:

1. Dehydration works best with tertiary alcohols and almost as well with secondary alcohols. Rearrangements and poor yields are common with primary alcohols.
2. (Saytzeff rule) If two or more alkenes might be formed by deprotonation of the carbocation, the most highly substituted alkene usually predominates.

The following problems show how these rules are used to predict the products of dehydrations. The carbocations are drawn to show how rearrangements occur and more than one product may result.

SOLVED PROBLEM 10-2

Predict the products of dehydration of the following alcohols catalyzed by sulfuric acid.

(a) 1-methylcyclohexanol (b) neopentyl alcohol

SOLUTION (a) 1-Methylcyclohexanol reacts to form a tertiary carbocation. Abstraction of a proton may occur on any one of three carbon atoms. The two secondary atoms are equivalent, and abstraction of a proton from one of these carbons leads to the trisubstituted double bond of the major product. Abstraction of a methyl proton leads to the disubstituted double bond of the minor product.

$$\text{1-methylcyclohexanol} \xrightarrow{H^+} \text{cation} \longrightarrow \text{major product (trisubstituted)} + \text{minor product (disubstituted)}$$

1-methylcyclohexanol cation major product (trisubstituted) minor product (disubstituted)

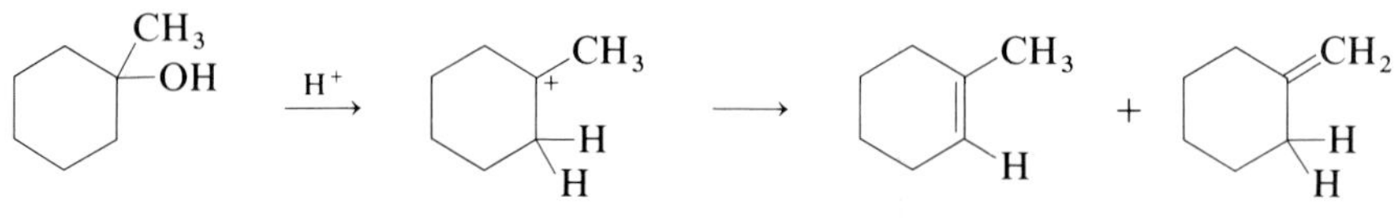

(b) Neopentyl alcohol cannot simply ionize to form a carbocation, because it would form a primary cation. Rearrangement occurs as the leaving group leaves, giving a tertiary carbocation. Loss of a proton from the adjacent secondary carbon gives the trisubstituted double bond of the major product. Loss of a proton from the methyl group gives the monosubstituted double bond of the minor product.

$$CH_3-C(CH_3)_2-CH_2\ddot{O}H \xrightarrow{H^+} CH_3-C(CH_3)_2-CH_2-\overset{+}{O}H_2 \longrightarrow CH_3-\overset{+}{C}(CH_3)-CH_2CH_3 \quad :\ddot{O}H_2$$

neopentyl alcohol (2,2-dimethyl-1-propanol) ionization with rearrangement 3° cation

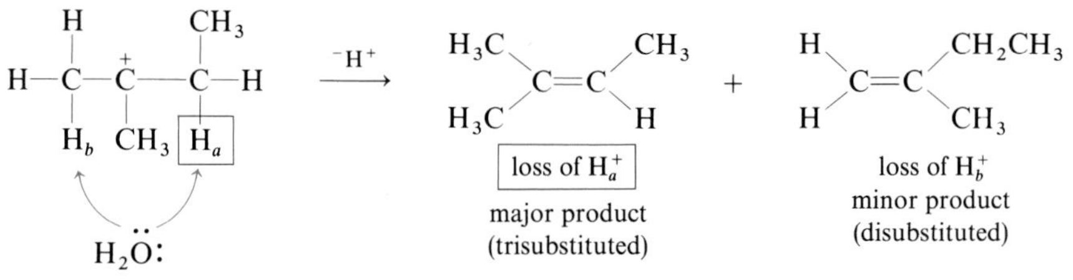

$$H_b-CH_2-\overset{+}{C}(CH_3)-CH(CH_3)H_a \quad (H_2\ddot{O}:) \xrightarrow{-H^+} (H_3C)_2C{=}C(CH_3)H + H_2C{=}C(CH_2CH_3)(CH_3)$$

loss of H_a^+ major product (trisubstituted)

loss of H_b^+ minor product (disubstituted)

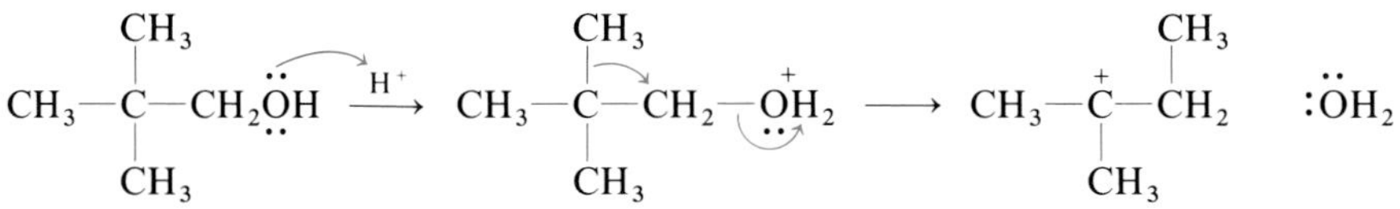

PROBLEM 10-22

Predict the principal products of the sulfuric acid-catalyzed dehydration of the following alcohols. When more than one product is expected, label the major and minor products.

(a) 2-methyl-2-butanol (b) 1-pentanol
(c) 2-pentanol (d) 1-isopropylcyclohexanol
(e) 2-methylcyclohexanol

10-11B BIMOLECULAR DEHYDRATION TO FORM ETHERS

In some cases a protonated primary alcohol may be attacked by another molecule of the alcohol and undergo an S_N2 displacement. The net reaction is a dehydration taking place between two molecules to form an ether. The attack by ethanol on a protonated molecule of ethanol gives diethyl ether.

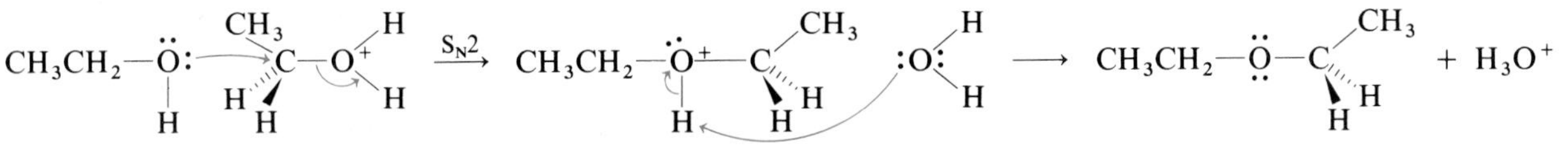

$$CH_3CH_2-\ddot{O}: \longrightarrow C(CH_3)(H)(H)-O^+H_2 \xrightarrow{S_N2} CH_3CH_2-\ddot{O}^+(H)-C(CH_3)(H)(H) \quad :\ddot{O}H_2 \longrightarrow CH_3CH_2-\ddot{O}-C(CH_3)(H)(H) + H_3O^+$$

nucleophilic electrophilic protonated ether water diethyl ether

Bimolecular dehydration can be used to synthesize symmetrical dialkyl ethers from simple, unhindered primary alcohols. This method is used for the industrial synthesis of diethyl ether ($CH_3CH_2-O-CH_2CH_3$) and dimethyl ether (CH_3-O-CH_3). Under the acidic dehydration conditions, two reactions compete: elimination to give an alkene and substitution to give an ether.

Substitution to give the ether, a bimolecular dehydration

$$\underset{\text{ethanol}}{2\ CH_3CH_2OH} \xrightarrow{H_2SO_4,\ 140°C} \underset{\text{diethyl ether}}{CH_3CH_2-O-CH_2CH_3} + H_2O$$

Elimination to give the alkene, a unimolecular dehydration

$$\underset{\text{ethanol}}{CH_3CH_2OH} \xrightarrow{H_2SO_4,\ 180°C} \underset{\text{ethylene}}{CH_2{=}CH_2} + H_2O$$

PROBLEM 10-23

Contrast the mechanisms of the two dehydrations of ethanol shown above.

How can we control these two competing dehydrations? The equation for substitution (the ether synthesis) shows two molecules of alcohol going to two molecules: one molecule of diethyl ether and one of water. The equation for elimination shows one molecule of alcohol going to two molecules: one of ethylene and one of water. The elimination results in an *increase* in the number of molecules, and therefore an increase in the randomness (entropy) of the system. Clearly, the elimination has a more positive change in entropy (ΔS) than the substitution, and the $-T\Delta S$ term in the Gibbs free energy becomes more favorable for the elimination as the temperature increases. Substitution to give the ether is favored at about 140°C and below, and elimination is favored at about 180°C and above. Therefore, diethyl ether is produced industrially by heating ethanol with an acidic catalyst at about 140°C.

PROBLEM 10-24

Explain why the acid-catalyzed dehydration is not a good method for the synthesis of an unsymmetrical ether such as ethyl methyl ether, $CH_3CH_2-O-CH_3$.

PROBLEM 10-25

Give a detailed mechanism for the following reaction.

$$2\ CH_3OH \xrightarrow{H_2SO_4,\ heat} CH_3-O-CH_3 + H_2O$$

10-12 UNIQUE REACTIONS OF DIOLS

10-12A THE PINACOL REARRANGEMENT

Using our knowledge of alcohol reactions, we can explain some reactions that seem strange at first glance. The following reaction is an example of the **pinacol rearrangement.**

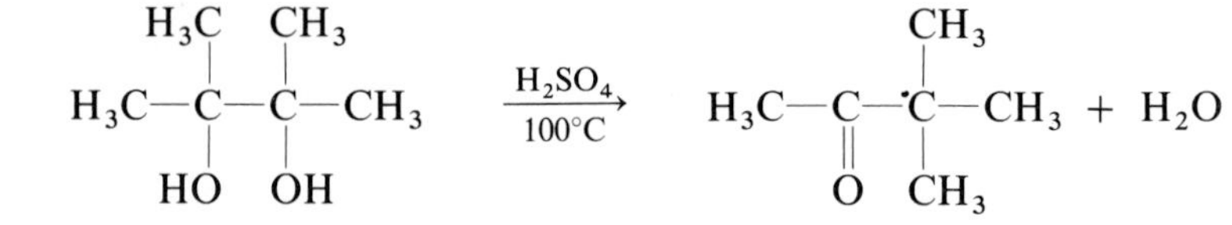

$$\underset{\text{pinacol (2,3-dimethyl-2,3-butanediol)}}{(CH_3)_2C(OH)\text{—}C(OH)(CH_3)_2} \xrightarrow[100^\circ C]{H_2SO_4} \underset{\text{pinacolone (3,3-dimethyl-2-butanone)}}{H_3C\text{—}C(=O)\text{—}C(CH_3)_3} + H_2O$$

The pinacol rearrangement is formally a dehydration. The reaction is acid catalyzed, and the first step is protonation of one of the hydroxyl oxygens. Loss of water gives a tertiary carbocation, as we would expect for any tertiary alcohol.

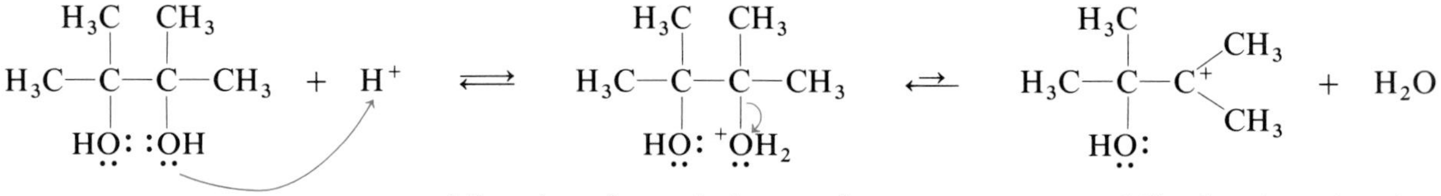

$$(CH_3)_2C(OH)\text{—}C(OH)(CH_3)_2 + H^+ \rightleftharpoons (CH_3)_2C(OH)\text{—}C(\overset{+}{O}H_2)(CH_3)_2 \rightleftharpoons (CH_3)_2C(OH)\text{—}\overset{+}{C}(CH_3)_2 + H_2O$$

Migration of a methyl group forms a resonance-stabilized carbocation that is even more stable than a tertiary carbocation.

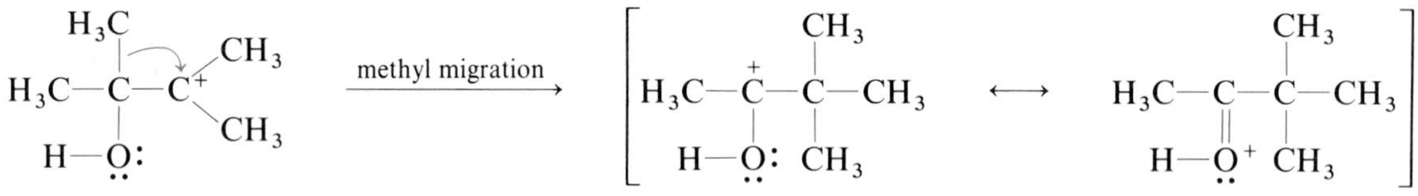

$$(CH_3)_2C(OH)\text{—}\overset{+}{C}(CH_3)_2 \xrightarrow{\text{methyl migration}} \left[H_3C\text{—}\overset{+}{C}(\text{—}\ddot{O}\text{—}H)\text{—}C(CH_3)_3 \longleftrightarrow H_3C\text{—}C(=\overset{+}{O}\text{—}H)\text{—}C(CH_3)_3 \right]$$

resonance-stabilized carbocation

The second resonance structure is particularly stable because all of the atoms are surrounded by octets of electrons. This extra stability is the driving force for the rearrangement. Deprotonation of the resonance-stabilized cation gives the observed product, pinacolone.

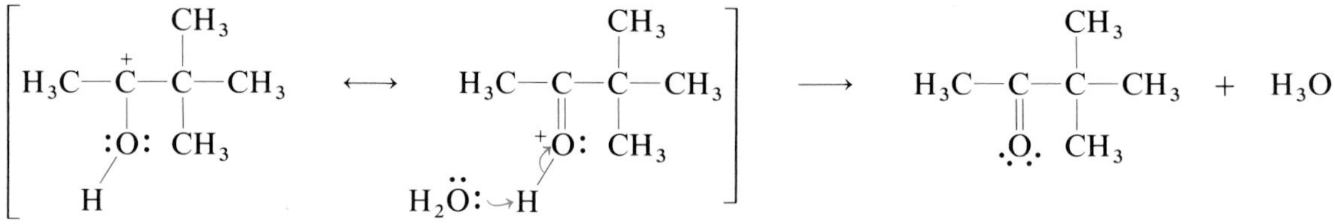

$$\underset{\text{resonance-stabilized carbocation}}{\left[H_3C\text{—}\overset{+}{C}(\text{—}\ddot{O}\text{—}H)\text{—}C(CH_3)_3 \longleftrightarrow H_3C\text{—}C(=\overset{+}{O}\text{—}H)\text{—}C(CH_3)_3 \right]} \xrightarrow{H_2\ddot{O}:} \underset{\text{pinacolone}}{H_3C\text{—}C(=O)\text{—}C(CH_3)_3} + H_3O^+$$

> **PROBLEM SOLVING: PROPOSING REACTION MECHANISMS**
>
> In view of the large number of reactions we've covered, proposing mechanisms for reactions you have never seen before may seem nearly impossible. As you gain experience in working mechanism problems, seeing similarities to known reactions will become much easier. Let's consider how an organic chemist systematically approaches a mechanism problem. A more complete version of this method appears in Appendix 4. Although this stepwise approach cannot solve all such problems, it should provide a starting point to begin building your experience and confidence.

DETERMINING THE TYPE OF MECHANISM First, determine what kinds of conditions and catalysts are involved. In general, reactions may be classified as involving (a) strong electrophiles (including acid-catalyzed reactions), (b) strong nucleophiles (including base-catalyzed reactions), or (c) free radicals. These three types of mechanisms are quite distinct, and you should first try to determine which type is involved. If uncertain, you can develop more than one type of mechanism and see which fits the facts better.

(a) In the presence of a strong acid or a reactant that can dissociate to give a strong electrophile, the mechanism probably involves strong electrophiles as intermediates. Acid-catalyzed reactions and reactions involving carbocations (such as the S_N1, the E1, and most alcohol dehydrations) generally fall in this category.

(b) In the presence of a strong base or a strong nucleophile, the mechanism probably involves strong nucleophiles as intermediates. Base-catalyzed reactions and those depending on base strength (such as the S_N2 and the E2) generally fall in this category.

(c) Free-radical reactions usually require a free-radical initiator such as chlorine, bromine, NBS, or a peroxide. In most free-radical reactions there is no need for a strong acid or base.

Once you have determined which type of mechanism you will write, there are general methods for approaching the problem. At this point, we consider mostly the electrophilic reactions covered in recent chapters. Suggestions for drawing the mechanisms of reactions involving strong nucleophiles and free-radical reactions are collected in Appendix 4.

REACTIONS INVOLVING STRONG ELECTROPHILES General principles: When a strong acid or electrophile is present, expect to see intermediates that are strong acids and strong electrophiles; cationic intermediates are common. Bases and nucleophiles in such a reaction are generally weak, however. Avoid drawing carbanions, alkoxide ions, and other strong bases. They are unlikely to coexist with strong acids and strong electrophiles.

Functional groups are often converted to carbocations or other strong electrophiles by protonation or reaction with a strong electrophile. Then the carbocation or other strong electrophile reacts with a weak nucleophile such as an alkene or the solvent.

1. **Consider the carbon skeletons of the reactants and products, and decide which carbon atoms in the products are most likely derived from which carbon atoms in the reactants.**
2. **Consider whether any of the reactants is a strong enough electrophile to react without being activated. If not, consider how one of the reactants might be converted to a strong electrophile by protonation of a Lewis basic site (or complexation with a Lewis acid).**
   Protonation of an alcohol, for example, converts it to a strong electrophile, which can undergo attack or lose water to give a carbocation, an even stronger electrophile. Protonation of an alkene converts it to a carbocation.
3. **Consider how a nucleophilic site on another reactant (or, in a cyclization, in another part of the same molecule) can attack the strong electrophile to form a bond needed in the product. Draw the product of this bond formation.**
   If the intermediate is a carbocation, consider whether it is likely to rearrange in a manner that would form a bond present in the product. If there isn't any possible nucleophilic attack that leads in the direction of the product, consider other ways of converting one of the reactants to a strong electrophile.

4. **Consider how the product of nucleophilic attack might be converted to the final product (if it has the right carbon skeleton) or reactivated to form another bond needed in the product.**
   To move a proton from one atom to another (as in an isomerization), try adding a proton to the new position, then removing it from the old position.
5. **Draw out all steps of the mechanism using curved arrows to show the movement of electrons.**
   Be careful to show only one step at a time.

COMMON MISTAKES TO AVOID IN DRAWING MECHANISMS

1. Do not use condensed or line-angle formulas for reaction sites. Draw all the bonds and all the substituents of each carbon atom affected throughout the mechanism. Three-bonded carbon atoms in intermediates are most likely to be carbocations in reactions involving strong electrophiles and acidic conditions. If you draw condensed formulas or line-angle formulas, you will likely misplace a hydrogen atom and show a reactive species on the wrong carbon.
2. Do not show more than one step occurring at once. Do not show two or three bonds changing position in one step unless the changes really are concerted (take place simultaneously). For example, protonation of an alcohol and loss of water to give a carbocation are two steps. You must not show the hydroxyl group "jumping" off the alcohol to join up with an anxiously waiting proton.
3. Remember that curved arrows show *movement of electrons,* always from the nucleophile (electron donor) to the electrophile (electron acceptor). For example, protonation of a double bond must show the arrow going from the electrons of the double bond to the proton—never from the proton to the double bond. Resist the urge to use an arrow to "point out" where the proton (or other reagent) goes.

To illustrate the stepwise method for reactions involving strong electrophiles, we will develop a mechanism to account for the cyclization shown below. The cyclized product is a minor product in this reaction. Note that a mechanism problem is different from a synthesis problem, because a mechanism problem may deal with how an unusual or unexpected minor product is formed.

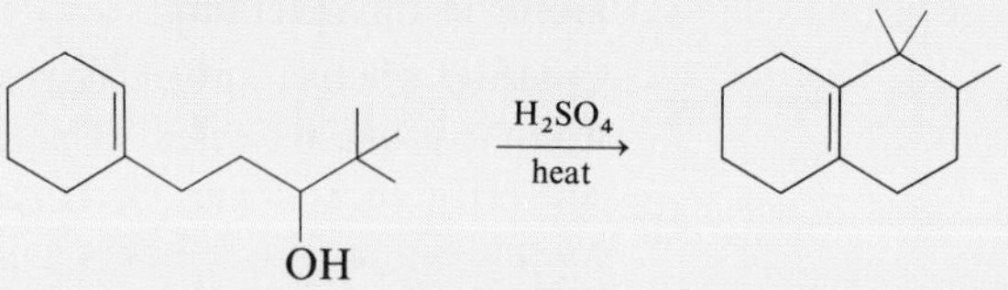

OH

$\xrightarrow[\text{heat}]{H_2SO_4}$

In the presence of sulfuric acid, this is clearly an acid-catalyzed mechanism. We expect strong electrophiles, cationic intermediates (possibly carbocations), and strong acids. Carbanions, alkoxide ions, and other strong bases and strong nucleophiles are unlikely.

1. **Consider the carbon skeletons of the reactants and products, and decide which carbon atoms in the products are most likely derived from which carbon atoms in the reactants.**
   Drawing the starting material and the product with all the substituents of the affected carbon atoms, we see the major changes circled below. A vinyl

hydrogen must be lost, a vinyl =C—C bond must be formed, a methyl group must move over one carbon atom, and the hydroxyl group must be lost.

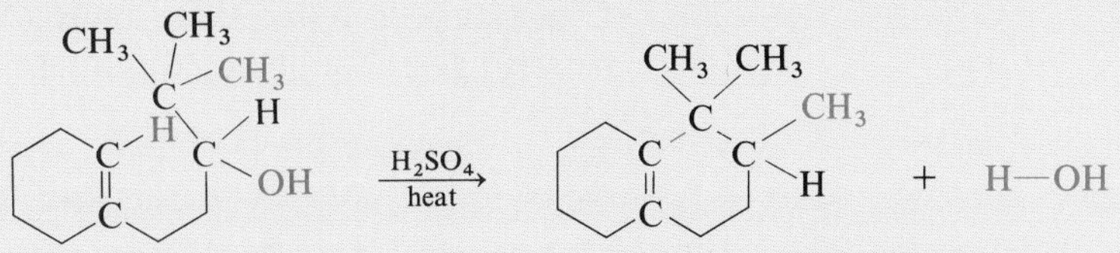

$$\xrightarrow[\text{heat}]{H_2SO_4} \quad + \quad H{-}OH$$

2. **Consider whether any of the reactants is a strong enough electrophile to react without being activated. If not, consider how one of the reactants might be converted to a strong electrophile by protonation of a Lewis basic site (or complexation with a Lewis acid).**
The starting material is not a strong electrophile, so it must be activated. Sulfuric acid could generate a strong electrophile either by protonating the double bond or by protonating the hydroxyl group. Protonating the double bond would form the tertiary carbocation, activating the wrong end of the double bond. Also, there is no good nucleophilic site on the side chain to attack this carbocation to form the correct ring. Protonating the double bond is a dead end.

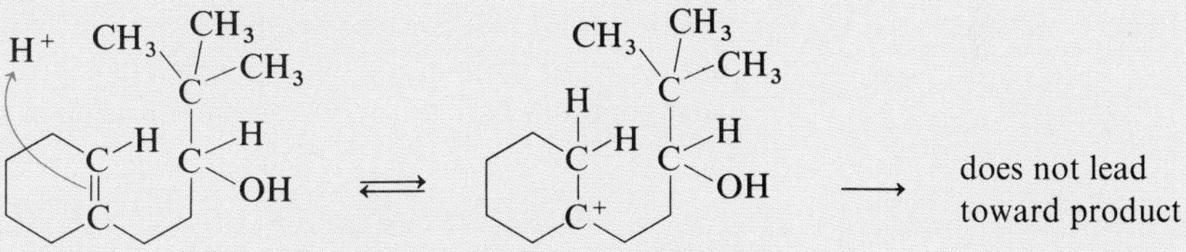

does not lead toward product

The other basic site is the hydroxyl group. An alcohol can protonate on the hydroxyl group and lose water to form a carbocation.

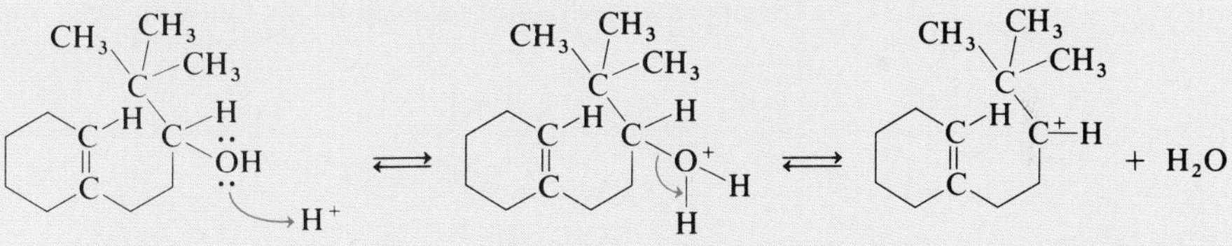

$+ \; H_2O$

3. **Consider how a nucleophilic site on another reactant (or, in a cyclization, in another part of the same molecule) can attack the strong electrophile to form a bond needed in the product. Draw the product of this bond formation.**
The carbocation can be attacked by the electrons in the double bond to form a ring; but the positive charge is on the wrong carbon atom to give a six-membered ring. A favorable rearrangement of the secondary carbocation to a tertiary one shifts the positive charge to the correct carbon atom and accomplishes the methyl shift we identified in step 1. Attack by the (weakly) nucleophilic electrons in the double bond gives the correct six-membered ring.

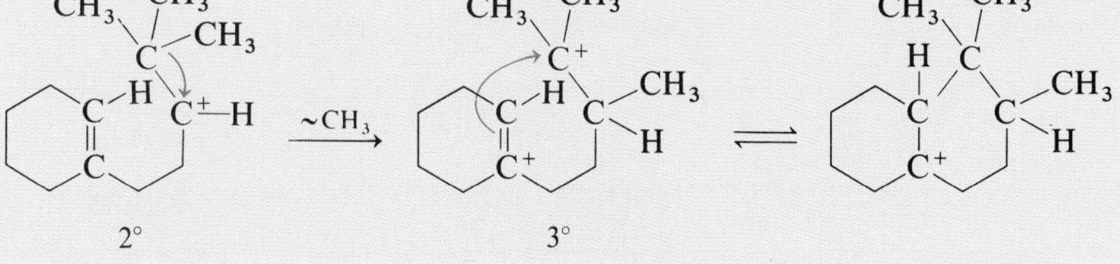

$\xrightarrow{\sim CH_3}$

2°　　3°

4. **Consider how the product of nucleophilic attack might be converted to the final product (if it has the right carbon skeleton) or reactivated to form another bond needed in the product.**
Loss of a proton (to HSO_4^- or H_2O, but *not* to ^-OH, which is not compatible!) gives the observed product.

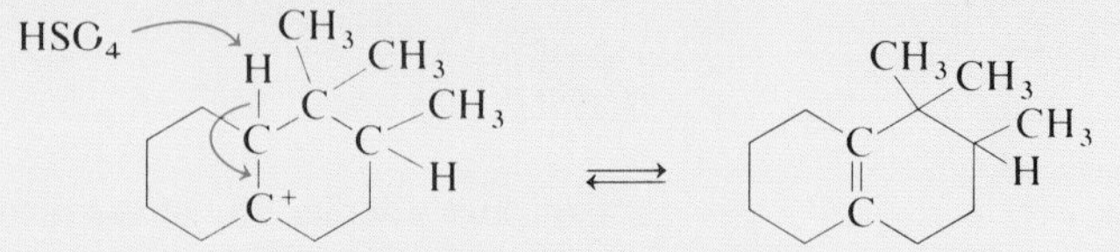

5. **Draw out all steps of the mechanism using curved arrows to show the movement of electrons.**
Combining the equations written immediately above gives the complete mechanism for this reaction.

Following are two problems that require proposing mechanisms for reactions involving strong electrophiles. Work each one by completing the five steps described above.

PROBLEM 10-26

The following reaction also involves a starting material with a double bond and a hydroxyl group, yet its mechanism follows a different course. Propose a mechanism for this reaction, and point out how this rearrangement resembles that seen in the pinacol rearrangement.

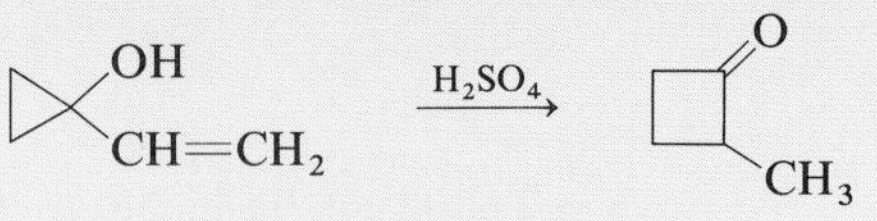

PROBLEM 10-27

Propose a mechanism for each of the following reactions.

(a) 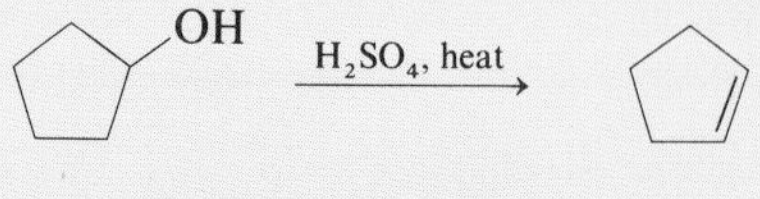 $\xrightarrow{H_2SO_4,\text{ heat}}$

(b) 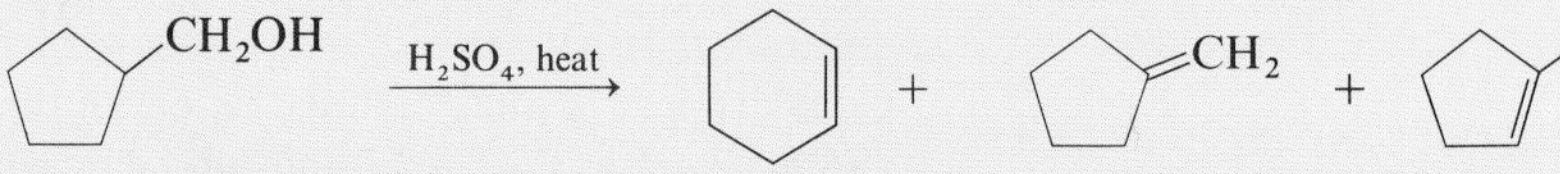 $\xrightarrow{H_2SO_4,\text{ heat}}$ + +

(c) 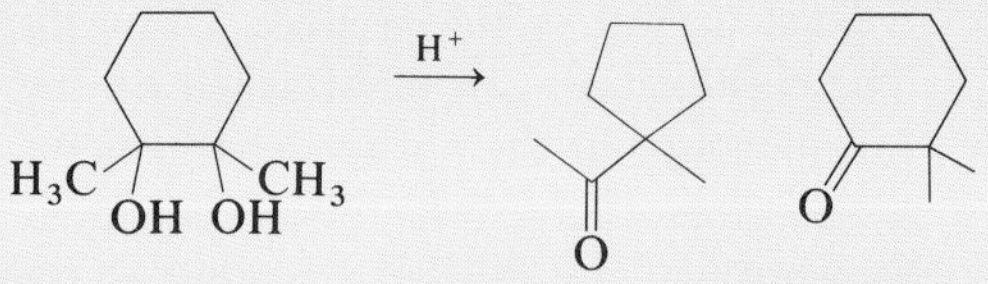 $\xrightarrow{H^+}$

10-12B PERIODIC ACID CLEAVAGE OF GLYCOLS

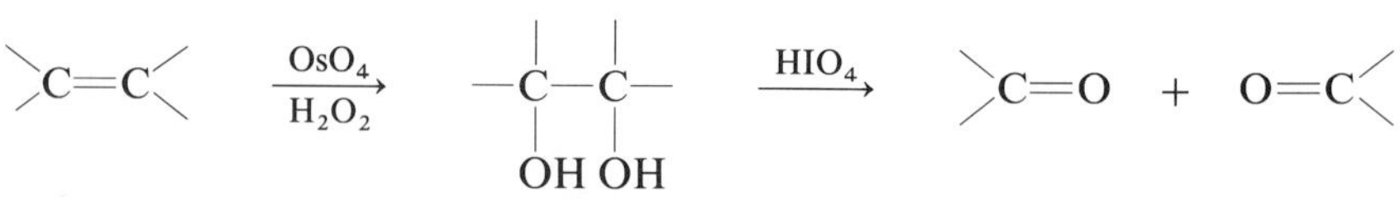

$$\underset{\text{alkene}}{>C=C<} \xrightarrow[H_2O_2]{OsO_4} \underset{\text{glycol}}{-\underset{OH}{\overset{|}{C}}-\underset{OH}{\overset{|}{C}}-} \xrightarrow{HIO_4} \underset{\text{ketones and aldehydes}}{>C=O \ + \ O=C<}$$

1,2-Diols (glycols), such as those formed by hydroxylation of alkenes, are cleaved by periodic acid (HIO_4). The products are the same ketones and aldehydes that would be formed by ozonolysis-reduction of the alkene. Hydroxylation followed by periodic acid cleavage serves as a useful alternative to ozonolysis, and the periodate cleavage by itself is useful for determining the structures of sugars (Chapter 23).

Periodic acid cleavage of a glycol is believed to involve a cyclic periodate intermediate like that shown below.

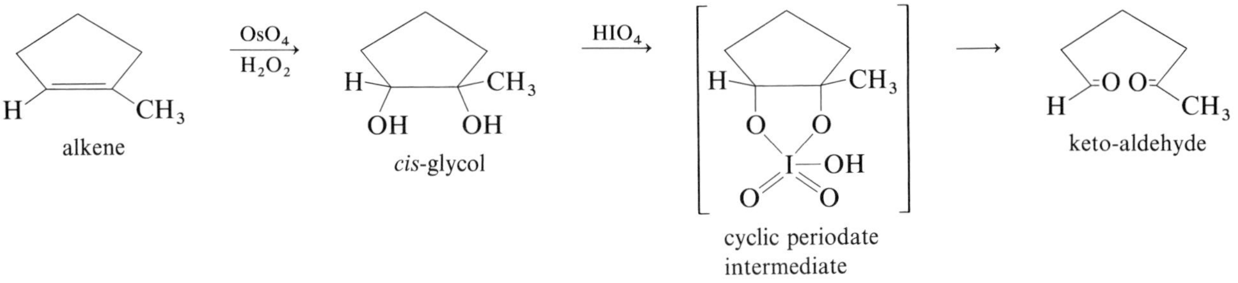

alkene $\xrightarrow[H_2O_2]{OsO_4}$ *cis*-glycol $\xrightarrow{HIO_4}$ [cyclic periodate intermediate] $\longrightarrow$ keto-aldehyde

PROBLEM 10-28

Predict the products formed by periodic acid cleavage of the following diols.

(a) $CH_3CH(OH)CH(OH)CH_3$

(b) 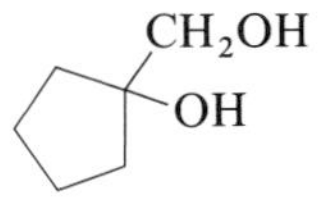 CH_2OH OH

(c) Ph—C(OH)(CH_3)—$CH(OH)CH_2CH_3$

(d) 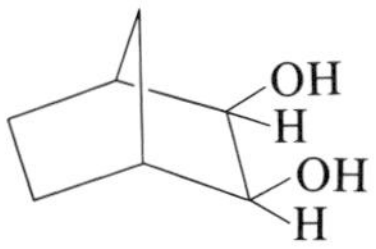 bicyclic diol (OH, H; OH, H)

10-13 ESTERIFICATION OF ALCOHOLS

Although we have used *tosylate esters* on several occasions, to an organic chemist the term **ester** normally means an ester of a carboxylic acid. Replacing the —OH group of a carboxylic acid with the —OR group of an alcohol gives a carboxylic ester. The following reaction, called the **Fischer esterification,** shows the relationship between the alcohol and the acid on the left, and the ester and water on the right. The mechanism of the Fischer esterification is closely related to other mechanisms involving carboxylic acids, and is covered in Chapter 20.

The Fischer esterification

$$R{-}O{-}H + H{-}O{-}\overset{\overset{\displaystyle O}{\|}}{C}{-}R' \overset{H^+}{\rightleftharpoons} R{-}O{-}\overset{\overset{\displaystyle O}{\|}}{C}{-}R' + H{-}O{-}H$$

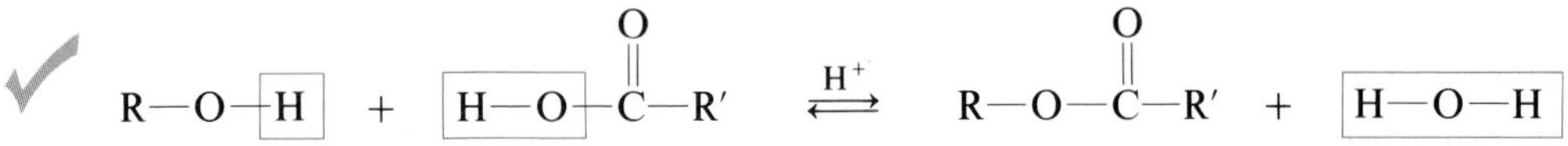

For example, if we mix isopropyl alcohol with acetic acid and add a drop of sulfuric acid as a catalyst, the following equilibrium results.

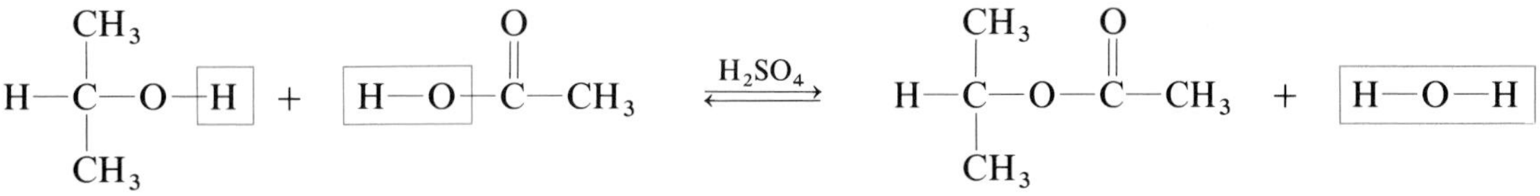

$$(CH_3)_2CH{-}O{-}H + H{-}O{-}\overset{\overset{\displaystyle O}{\|}}{C}{-}CH_3 \overset{H_2SO_4}{\rightleftharpoons} (CH_3)_2CH{-}O{-}\overset{\overset{\displaystyle O}{\|}}{C}{-}CH_3 + H{-}O{-}H$$

isopropyl alcohol + acetic acid ⇌ isopropyl acetate + water

Because the Fischer esterification is an equilibrium (often with an unfavorable equilibrium constant), some clever techniques are often required to achieve a good yield of the ester. We cover some of those techniques in Chapter 20. There is a more powerful way to form an ester, however, without having to deal with an unfavorable equilibrium. An alcohol reacts with an acid chloride in an exothermic reaction to give an ester.

$$\underset{\text{alcohol}}{R-O-\boxed{H}} + \underset{\text{acid chloride}}{\boxed{Cl}-\overset{O}{\overset{\|}{C}}-R'} \longrightarrow \underset{\text{ester}}{R-O-\overset{O}{\overset{\|}{C}}-R'} + \boxed{HCl}$$

The mechanism of this reaction is covered with similar mechanisms in Chapter 21.

PROBLEM 10-29

Show the alcohol and the acid chloride you would use to form each of the following esters.

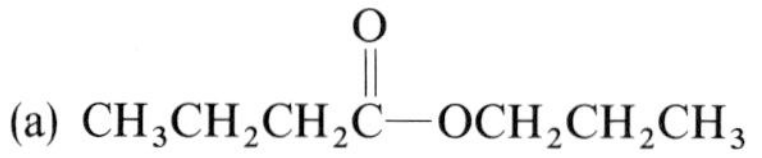

(a) $CH_3CH_2CH_2\overset{O}{\overset{\|}{C}}-OCH_2CH_2CH_3$ *n*-propyl butyrate

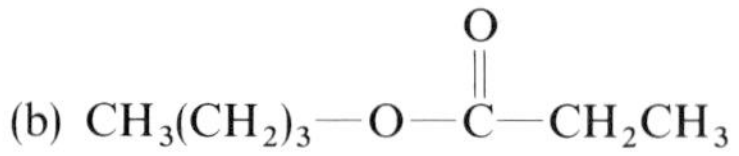

(b) $CH_3(CH_2)_3-O-\overset{O}{\overset{\|}{C}}-CH_2CH_3$ *n*-butyl propionate

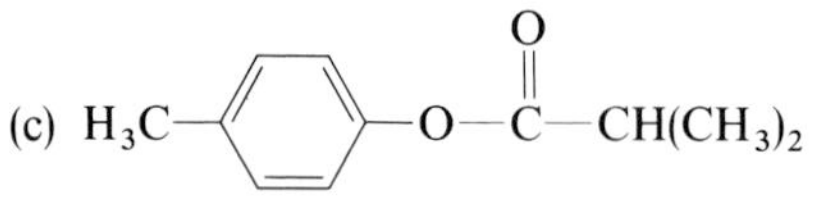

(c) $H_3C-C_6H_4-O-\overset{O}{\overset{\|}{C}}-CH(CH_3)_2$ *p*-tolyl isobutyrate

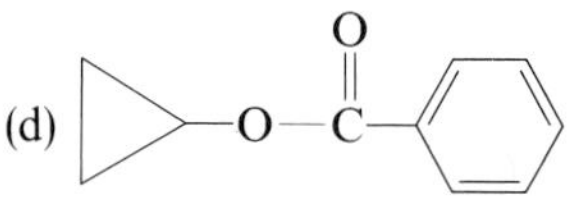

(d) $C_3H_5-O-\overset{O}{\overset{\|}{C}}-C_6H_5$ cyclopropyl benzoate

10-14 ESTERS OF INORGANIC ACIDS

10-14A SULFONATE ESTERS

In addition to esters of carboxylic acids, alcohols can form **inorganic esters** with inorganic acids. For example, the tosylate of an alcohol is derived from *para*-toluenesulfonic acid plus the alcohol, with loss of a molecule of water. In practice, tosylate esters are synthesized by the reaction of alcohols with tosyl chloride in pyridine (Section 10-7).

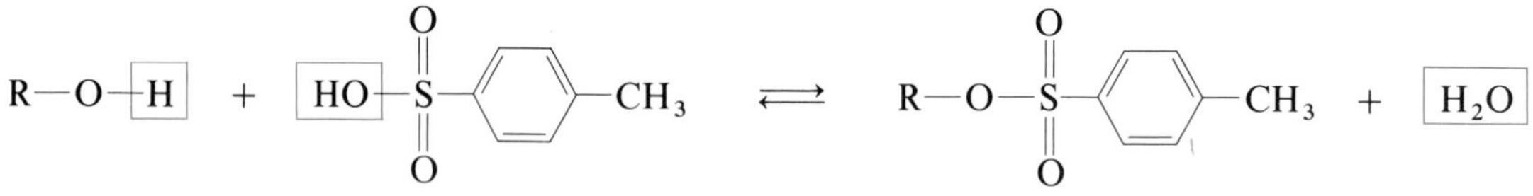

$$\underset{\text{alcohol}}{R-O-\boxed{H}} + \underset{\substack{\textit{para}\text{-toluenesulfonic acid}\\ \text{(TsOH)}}}{\boxed{HO}-\overset{O}{\overset{\|}{\underset{O}{\underset{\|}{S}}}}-C_6H_4-CH_3} \rightleftharpoons \underset{\substack{\textit{para}\text{-toluenesulfonate ester}\\ \text{(ROTs)}}}{R-O-\overset{O}{\overset{\|}{\underset{O}{\underset{\|}{S}}}}-C_6H_4-CH_3} + \boxed{H_2O}$$

10-14B SULFATE ESTERS

A **sulfate ester** is like a sulfonate ester, except there is no alkyl group directly bonded to the sulfur atom. In an alkyl sulfate ester, there are alkoxy groups bonded to sulfur through oxygen atoms.

$$HO-SO_2-OH \xrightleftharpoons{CH_3O-H} CH_3-O-SO_2-OH + H_2O \xrightleftharpoons{CH_3OH} CH_3-O-SO_2-O-CH_3 + H_2O$$

sulfuric acid — methyl sulfate — dimethyl sulfate

The sulfate ion is an excellent leaving group. Like sulfonate esters, sulfate esters are good electrophiles. Nucleophiles react with sulfate esters to give alkylated products. For example, the reaction of dimethyl sulfate with ammonia gives a sulfate salt of methylamine, $CH_3NH_3^+\ CH_3OSO_3^-$.

$$H_3N\colon + CH_3-O-SO_2-O-CH_3 \longrightarrow H_3N^+-CH_3 \quad ^-O-SO_2-O-CH_3$$

ammonia — dimethyl sulfate — methylammonium — methylsulfate

PROBLEM 10-30
Use resonance structures to show that the negative charge in the methylsulfate anion is shared equally by three oxygen atoms.

10-14C NITRATE ESTERS

Esters formed from alcohols and nitric acid are called **nitrate esters.**

$$R-O-H + H-O-N^+(=O)O^- \longrightarrow R-O-N^+(=O)O^- + H-O-H$$

alcohol — nitric acid — alkyl nitrate ester

The best known nitrate ester is "nitroglycerine," whose systematic name is *glyceryl trinitrate.* Glyceryl trinitrate results from the reaction of glycerol (1,2,3-propanetriol) with three molecules of nitric acid.

$$\begin{matrix} CH_2-O-H \\ | \\ CH-O-H \\ | \\ CH_2-O-H \end{matrix} + 3\ HO-NO_2 \longrightarrow \begin{matrix} CH_2-O-NO_2 \\ | \\ CH-O-NO_2 \\ | \\ CH_2-O-NO_2 \end{matrix} + 3\ H_2O$$

glycerol (glycerine) — nitric acid — glyceryl trinitrate (nitroglycerine)

10-14D PHOSPHATE ESTERS

Alkyl phosphates are esters that combine 1 mole of phosphoric acid with 1, 2, or 3 moles of an alcohol. For example, methanol forms three **phosphate esters.**

$$HO-\overset{\overset{\displaystyle O}{\|}}{\underset{\underset{\displaystyle OH}{|}}{P}}-OH \xrightarrow{CH_3OH,\ -H_2O} CH_3-O-\overset{\overset{\displaystyle O}{\|}}{\underset{\underset{\displaystyle OH}{|}}{P}}-OH \xrightarrow{CH_3OH,\ -H_2O}$$

phosphoric acid — monomethyl phosphate

$$CH_3-O-\overset{\overset{\displaystyle O}{\|}}{\underset{\underset{\displaystyle O-CH_3}{|}}{P}}-OH \xrightarrow{CH_3OH,\ -H_2O} CH_3-O-\overset{\overset{\displaystyle O}{\|}}{\underset{\underset{\displaystyle O-CH_3}{|}}{P}}-O-CH_3$$

dimethyl phosphate — trimethyl phosphate

Phosphate esters play a central role in biochemistry. For example, Figure 10-3 shows how phosphate ester linkages compose the "backbone" of the nucleic acids RNA (ribonucleic acid) and DNA (deoxyribonucleic acid), which carry the genetic information in the nucleus of a cell (see Chapter 23).

10-15 FORMATION AND REACTIONS OF ALKOXIDES

In Section 9-6B we saw that the hydroxyl proton may be removed from an alcohol by reduction with an "active" metal such as sodium or potassium. This reaction generates a sodium or potassium salt of an **alkoxide ion** and hydrogen gas.

$$R-\ddot{\underset{..}{O}}-H \ + \ Na \longrightarrow R-\ddot{\underset{..}{O}}:^- \ {}^+Na \ + \ \tfrac{1}{2}H_2\uparrow$$

$$R-\ddot{\underset{..}{O}}-H \ + \ K \longrightarrow R-\ddot{\underset{..}{O}}:^- \ {}^+K \ + \ \tfrac{1}{2}H_2\uparrow$$

The reactivity of alcohols toward sodium and potassium decreases in the following order: methyl $> 1° > 2° > 3°$. Sodium reacts readily with primary alcohols and some secondary alcohols. Potassium is more reactive than sodium and is commonly used with tertiary alcohols and some secondary alcohols.

The alkoxide ion is a strong nucleophile as well as a powerful base. Unlike the alcohol itself, the alkoxide ion reacts with primary alkyl halides and tosylates to form ethers. This general reaction is called the **Williamson ether synthesis.** The Williamson ether synthesis is an S_N2 displacement, and the alkyl halide (or tosylate) must be primary so that a back-side attack is not hindered. If the alkyl halide is not primary, elimination usually results.

Williamson ether synthesis

$$R-\ddot{\underset{..}{O}}:^- \ {}^+Na \quad R'-CH_2-X \longrightarrow R-O-CH_2-R' + NaX$$

alkoxide ion — primary alkyl halide — ether

Example

$$CH_3-CH_2-O^- \ {}^+Na \ + \ CH_3I \longrightarrow CH_3-CH_2-O-CH_3 \ + \ NaI$$

sodium ethoxide — methyl iodide — ethyl methyl ether

PROBLEM 10-31

What is wrong with the following proposed synthesis of ethyl methyl ether? First, ethanol is treated with acid to protonate the hydroxyl group, and then methoxide is added to displace water.

$$CH_3CH_2-OH + H^+ \longrightarrow CH_3CH_2-\overset{+}{O}H_2 \xrightarrow{CH_3O^-} CH_3CH_2-O-CH_3$$

(incorrect synthesis of ethyl methyl ether)

PROBLEM 10-32

Show how to use the Williamson ether synthesis to make butyl isopropyl ether.

PROBLEM 10-33

(a) Show how ethanol and cyclohexanol may be used to synthesize cyclohexyl ethyl ether (tosylation followed by the Williamson ether synthesis).
(b) Why can't we synthesize this product simply by mixing the two alcohols, adding some sulfuric acid, and heating?

PROBLEM 10-34

The anions of phenols (phenoxide ions) may be used in the Williamson ether synthesis, especially with very reactive alkylating reagents such as dimethyl sulfate. Using phenol, dimethyl sulfate, and other necessary reagents, show how you would synthesize methyl phenyl ether.

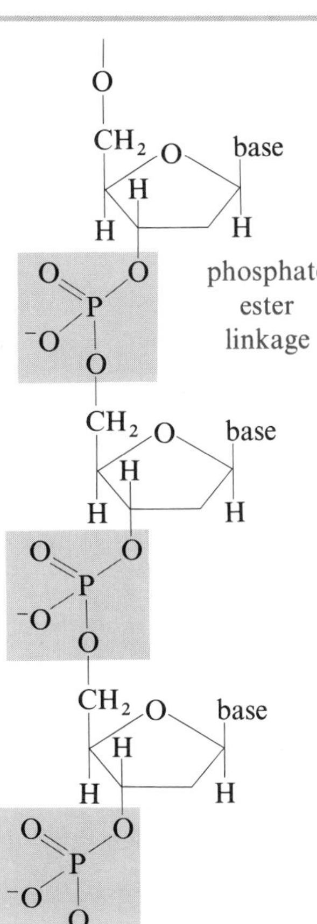

FIGURE 10-3 This schematic representation of a small segment of DNA shows the phosphate ester groups that bond the individual nucleotides together. The "base" on each of the nucleotides corresponds to one of the four heterocyclic bases present in DNA (see Section 23-20).

PROBLEM SOLVING: MULTISTEP SYNTHESIS

Organic synthesis is used both to make larger amounts of useful natural compounds and to invent totally new compounds in search of improved properties and biological effects. Synthesis also serves as one of the best methods for developing a firm command of organic chemistry. Planning a practical multistep synthesis requires a working knowledge of both the applications and the limitations of a wide variety of organic reactions. We will often use synthesis problems for reviewing and reinforcing the reactions covered in future chapters.

A systematic approach to solving multistep synthesis problems by working backward, in the "retrosynthetic" direction, was introduced in Chapter 5. This method requires studying the structure of the target molecule and considering what final reactions might be used to create the molecule from simpler or more easily obtainable intermediate compounds. Comparing two or more of the most likely intermediates and the steps for their synthesis is usually necessary. Eventually, this retrosynthetic analysis should lead back to starting materials that are readily available or meet the requirements defined in the problem.

We can now extend our systematic retrosynthetic analysis to problems involving alcohols and the Grignard reaction. As examples, we consider the syntheses of an acyclic diol and a disubstituted cyclohexane, concentrating on the crucial steps that assemble the carbon skeletons and generate the final functional groups.

Our first problem is to synthesize 3-ethyl-2,3-pentanediol from compounds containing no more than three carbon atoms.

$$CH_3-CH(OH)-C(OH)(CH_2CH_3)-CH_2-CH_3$$

3-ethyl-2,3-pentanediol

1. **Review the functional groups and carbon skeleton of the target compound.** The compound is a vicinal diol (glycol) containing seven carbon atoms. Glycols are commonly made by hydroxylation of alkenes, and this glycol would be made by hydroxylation of 3-ethyl-2-pentene, which effectively becomes the target compound.

$$\underset{\text{3-ethyl-2-pentene}}{CH_3-CH{=}C(CH_2-CH_3)-CH_2-CH_3} \xrightarrow[\text{(or other methods)}]{KMnO_4\text{, cold, dilute}} \underset{\text{3-ethyl-2,3-pentanediol}}{CH_3-CH(OH)-C(OH)(CH_2-CH_3)-CH_2-CH_3}$$

2. **Review the functional groups and carbon skeletons of the starting materials (if specified), and see how their skeletons might fit together into the target compound.**
The limitation is that the starting materials contain no more than three carbon atoms. To form a seven-carbon product will require at least three fragments, probably a three-carbon fragment and two two-carbon fragments. A functional group that can be converted to an alkene will be needed on either C2 or C3 of the chain, since 3-ethyl-2-pentene has a double bond between C2 and C3.

3. **Compare methods for assembling the carbon skeleton of the target compound to determine which methods provide a key intermediate with the correct carbon skeleton and functional groups at the correct positions where they can be converted to the functionality in the target molecule.**
At this point, the Grignard reaction is our most powerful method for assembling a carbon skeleton, and Grignards can be used to make primary, secondary, and tertiary alcohols. The secondary alcohol 3-ethyl-2-pentanol has its functional group on C2, while the tertiary alcohol 3-ethyl-3-pentanol has it on C3. Either of these alcohols can be synthesized by an appropriate Grignard reaction, but 3-ethyl-2-pentanol may dehydrate to give a mixture of products. Because of its symmetry, 3-ethyl-3-pentanol dehydrates to give only the desired alkene, 3-ethyl-2-pentene. It also dehydrates more easily because it is a tertiary alcohol.

$$\underset{\text{3-ethyl-2-pentanol}}{CH_3-\underset{OH}{CH}-\overset{CH_2-CH_3}{CH}-CH_2-CH_3} \xrightarrow{H_2SO_4} \underset{\substack{\text{(major)}\\ \text{3-ethyl-2-pentene}}}{CH_3-CH=\overset{CH_2-CH_3}{C}-CH_2-CH_3} + \underset{\substack{\text{(minor)}\\ \text{3-ethyl-1-pentene}}}{CH_2=CH-\overset{CH_2-CH_3}{CH}-CH_2-CH_3}$$

$$\underset{\text{3-ethyl-3-pentanol}}{CH_3-CH_2-\overset{CH_2-CH_3}{\underset{OH}{C}}-CH_2-CH_3} \xrightarrow{H_2SO_4} \underset{\substack{\text{(only product)}\\ \text{3-ethyl-2-pentene}}}{CH_3-CH=\overset{CH_2-CH_3}{C}-CH_2-CH_3} \quad \text{(preferred synthesis)}$$

4. **Working backward through as many steps as necessary, compare methods for synthesizing the reactants needed for assembly of the key intermediate with the correct carbon skeleton and functionality in the proper position. (This process may require writing several possible reaction sequences and evaluating them, keeping in mind the specified starting materials.)**
The key intermediate, 3-ethyl-3-pentanol, is simply methanol substituted by three ethyl groups. The last step in its synthesis must add an ethyl group. Addition of an ethyl Grignard reagent to 3-pentanone gives 3-ethyl-3-pentanol.

$$\underset{\text{3-pentanone}}{CH_3-CH_2-\underset{\underset{O}{\|}}{C}-CH_2-CH_3} \xrightarrow[\text{(2) } H_3O^+]{\text{(1) } CH_3-CH_2-MgBr} \underset{\text{3-ethyl-3-pentanol}}{CH_3-CH_2-\overset{CH_2-CH_3}{\underset{OH}{C}}-CH_2-CH_3}$$

The synthesis of 3-pentanone from a 3-carbon fragment and a 2-carbon fragment requires several steps (see problem 10-35). Perhaps there is a better alternative, considering that the key intermediate has three ethyl groups on a carbinol carbon atom. Two similar alkyl groups can be added in one Grignard reaction with an acid chloride or ester (Section 9-9D). Addition of two moles of the ethyl Grignard reagent to a three-carbon acid chloride gives 3-ethyl-3-pentanol.

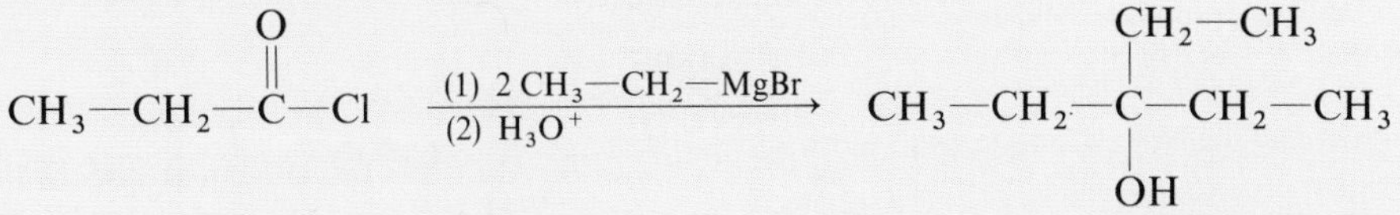

$$CH_3{-}CH_2{-}\overset{\overset{\large O}{\|}}{C}{-}Cl \xrightarrow[(2)\ H_3O^+]{(1)\ 2\,CH_3{-}CH_2{-}MgBr} CH_3{-}CH_2{-}\underset{\underset{\large OH}{|}}{\overset{\overset{\large CH_2{-}CH_3}{|}}{C}}{-}CH_2{-}CH_3$$

propionyl chloride — 3-ethyl-3-pentanol

5. **Summarize the complete synthesis in the forward direction, including all steps and all reagents, and check it for errors and omissions.**

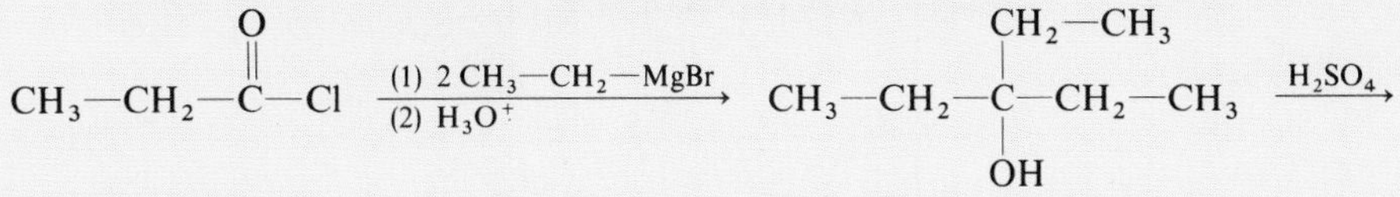

$$CH_3{-}CH_2{-}\overset{\overset{\large O}{\|}}{C}{-}Cl \xrightarrow[(2)\ H_3O^+]{(1)\ 2\,CH_3{-}CH_2{-}MgBr} CH_3{-}CH_2{-}\underset{\underset{\large OH}{|}}{\overset{\overset{\large CH_2{-}CH_3}{|}}{C}}{-}CH_2{-}CH_3 \xrightarrow{H_2SO_4}$$

propionyl chloride — 3-ethyl-3-pentanol

$$CH_3{-}CH{=}\overset{\overset{\large CH_2{-}CH_3}{|}}{C}{-}CH_2{-}CH_3 \xrightarrow[\text{(cold, dilute)}]{KMnO_4} CH_3{-}\underset{\underset{\large OH}{|}}{CH}{-}\underset{\underset{\large OH}{|}}{\overset{\overset{\large CH_2{-}CH_3}{|}}{C}}{-}CH_2{-}CH_3$$

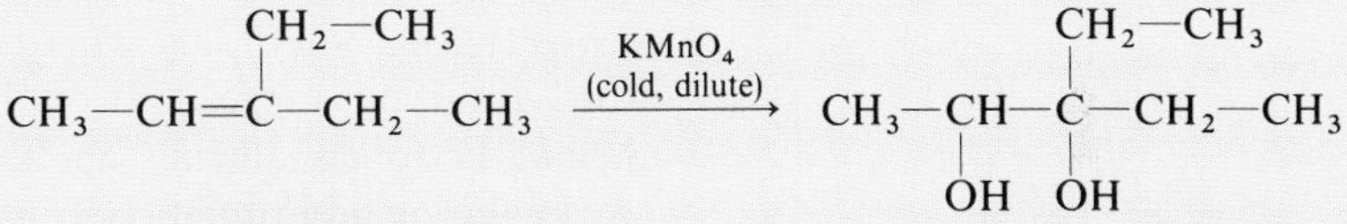

3-ethyl-2-pentene — 3-ethyl-2,3-pentanediol

PROBLEM 10-35

To practice working through the early parts of a multistep synthesis, devise syntheses of

(a) 3-ethyl-2-pentanol from compounds containing no more than three carbon atoms.
(b) 3-pentanone from alcohols containing no more than three carbon atoms.

As another example of the systematic approach to multistep synthesis, we next consider the synthesis of 1-bromo-2-methylcyclohexane from cyclohexanol.

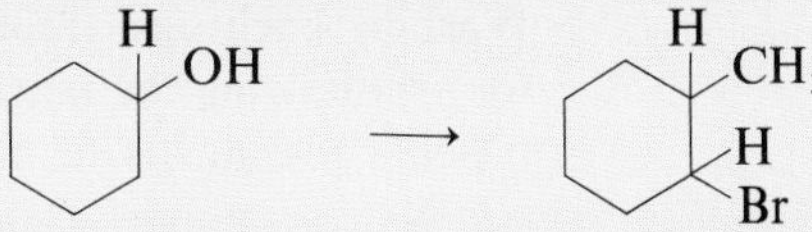

[cyclohexanol (H, OH) → 1-bromo-2-methylcyclohexane (H, CH_3; H, Br)]

1. **Review the functional groups and carbon skeleton of the target compound.**
The skeleton has seven carbon atoms: a cyclohexyl ring with a methyl group. It is an alkyl bromide, with the bromine atom on the ring carbon next to the methyl group.

2. **Review the functional groups and carbon skeletons of the starting materials (if specified), and see how their skeletons might fit together into the target compound.**
The starting compound has only six carbon atoms; clearly, the methyl group must be added, presumably at the functional group. There are no restrictions on the methylating reagent, but it must provide a product with a functional group that can be converted to an adjacent halide.

3. **Compare methods for assembling the carbon skeleton of the target compound to determine which methods provide a key intermediate with the correct carbon skeleton and functional groups at the correct positions where they can be converted to the functionality in the target molecule.**

Once again, the clear choice is a Grignard reaction, but there are two possible reactions that give the methylcyclohexane skeleton. A cyclohexyl Grignard reagent can add to formaldehyde, or a methyl Grignard reagent can add to cyclohexanone. (There are other possibilities, but none that are more direct.)

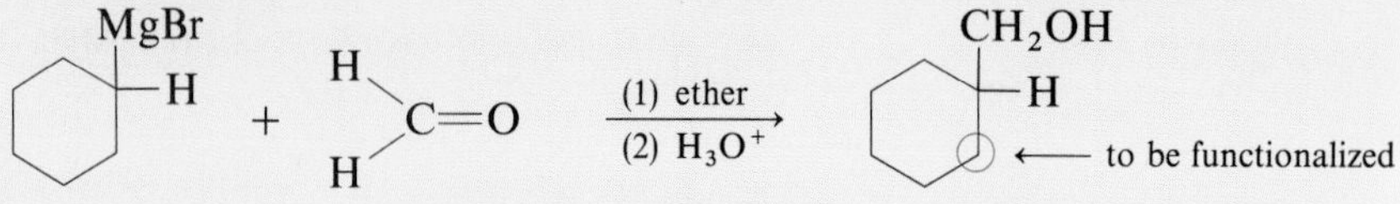

MgBr H + H H C=O (1) ether (2) H_3O^+ → CH_2OH H ← to be functionalized

alcohol C

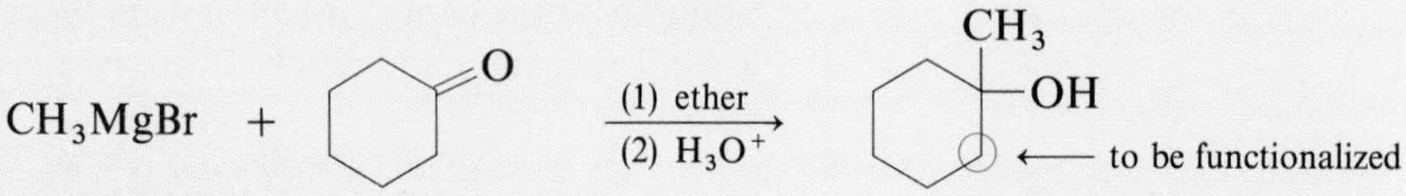

CH_3MgBr + O (1) ether (2) H_3O^+ → CH_3 OH ← to be functionalized

alcohol D

Neither product has its functional group on the carbon atom that is functionalized in the target compound. Alcohol C needs its functional group moved two carbon atoms, while alcohol D needs it moved only one carbon atom. Intuition suggests alcohol D is a better intermediate. Converting alcohol D to an alkene functionalizes the correct carbon atom. Anti-Markovnikov addition of HBr converts the alkene to an alkyl halide with the bromine atom on the correct carbon atom.

CH_3 OH H H → H_2SO_4, heat → CH_3 H → HBr, ROOR → H CH_3 Br H

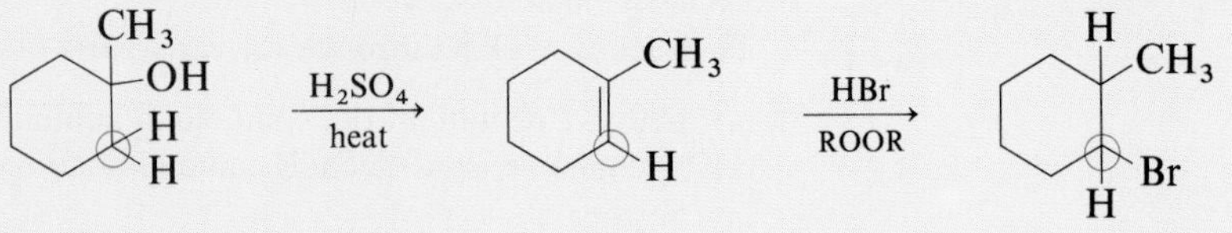

alcohol D — target compound

4. **Working backward through as many steps as necessary, compare methods for synthesizing the reactants needed for assembly of the key intermediate.** All that remains is to make cyclohexanone by oxidation of cyclohexanol.

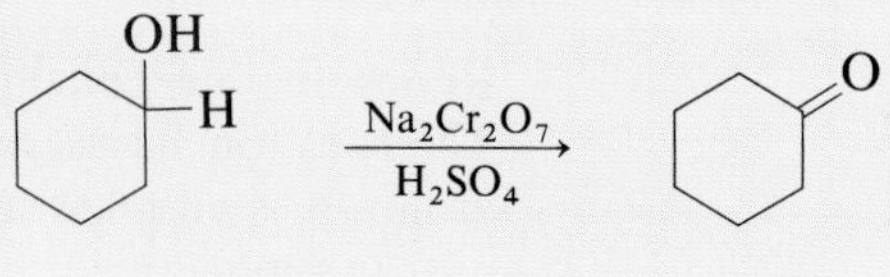

OH H → $Na_2Cr_2O_7$, H_2SO_4 → O

5. **Summarize the complete synthesis in the forward direction, including all steps and all reagents, and check it for errors and omissions.**

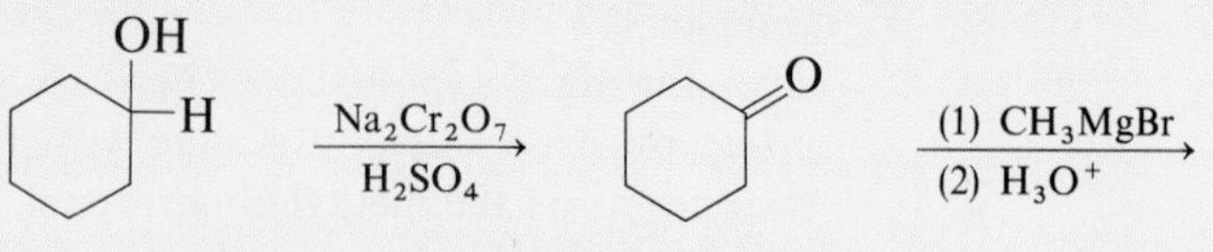

OH H → $Na_2Cr_2O_7$, H_2SO_4 → O → (1) CH_3MgBr (2) H_3O^+ →

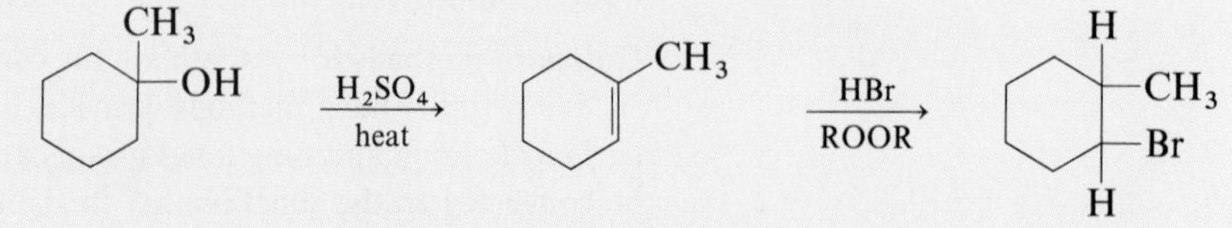

CH_3 OH → H_2SO_4, heat → CH_3 → HBr, ROOR → H CH_3 Br H

Problem 10-36 provides practice in multistep syntheses and using alcohols as intermediates.

PROBLEM 10-36

Show how the following compounds might be formed from cyclohexanol.

(a) 1-chloro-1-phenylcyclohexane (b) 2-methylcyclohexanol
(c) *trans*-cyclohexane-1,2-diol (work backward!)

SUMMARY OF THE REACTIONS OF ALCOHOLS

1. Oxidation-reduction reactions

a. ***Oxidation of secondary alcohols to ketones*** (Section 10-2A)

$$R{-}CH(OH){-}R' \xrightarrow{Na_2Cr_2O_7,\ H_2SO_4} R{-}C(=O){-}R'$$

Example: Jones oxidation of 2-butanol

$$\underset{\text{2-butanol}}{CH_3{-}CH(OH){-}CH_2CH_3} \xrightarrow[\text{acetone}]{Na_2Cr_2O_7,\ H_2SO_4} \underset{\text{2-butanone}}{CH_3{-}C(=O){-}CH_2CH_3}$$

b. ***Oxidation of primary alcohols to carboxylic acids*** (Section 10-2B)

$$R{-}CH_2{-}OH \xrightarrow{Na_2Cr_2O_7,\ H_2SO_4} R{-}C(=O){-}OH$$

Example: Oxidation of 1-hexanol

$$\underset{\text{1-hexanol}}{CH_3(CH_2)_4{-}CH_2{-}OH} \xrightarrow{Na_2Cr_2O_7,\ H_2SO_4} \underset{\text{hexanoic acid}}{CH_3(CH_2)_4{-}C(=O){-}OH}$$

c. ***Oxidation of primary alcohols to aldehydes*** (Section 10-2B)

$$R{-}CH_2{-}OH \xrightarrow[\text{or PCC}]{CrO_3\cdot\text{pyridine}} R{-}C(=O){-}H$$

Example: Oxidation of 1-hexanol

$$\underset{\text{1-hexanol}}{CH_3(CH_2)_4{-}CH_2{-}OH} \xrightarrow{CrO_3\cdot\text{pyridine}} \underset{\text{hexanal}}{CH_3(CH_2)_4{-}C(=O){-}H}$$

d. ***Reduction of alcohols to alkanes*** (Section 10-5)

$$R{-}OH \xrightarrow{LiAlH_4,\ TiCl_4} R{-}H$$

(or TsCl/pyridine, then $LiAlH_4$)

Example: Reduction of cyclohexanol

$$\underset{\text{cyclohexanol}}{C_6H_{10}(H)(OH)} \xrightarrow{LiAlH_4,\ TiCl_4} \underset{\text{cyclohexane}}{C_6H_{10}(H)(H)}$$

2. *Cleavage of the alcohol hydroxyl group* —C⌇O—H

a. *Conversion of alcohols to alkyl halides* (Sections 10-8 through 10-10)

$$R{-}OH \xrightarrow{\text{HCl or SOCl}_2\text{/pyridine}} R{-}Cl$$

$$R{-}OH \xrightarrow{\text{HBr or PBr}_3} R{-}Br$$

$$R{-}OH \xrightarrow{\text{HI or P/I}_2} R{-}I$$

Examples

$$\underset{t\text{-butyl alcohol}}{(CH_3)_3C{-}OH} \xrightarrow{\text{HCl}} \underset{t\text{-butyl chloride}}{(CH_3)_3C{-}Cl}$$

$$\underset{\text{isobutyl alcohol}}{(CH_3)_2CH{-}CH_2OH} \xrightarrow{\text{PBr}_3} \underset{\text{isobutyl bromide}}{(CH_3)_2CH{-}CH_2Br}$$

$$\underset{\text{1-hexanol}}{CH_3(CH_2)_4{-}CH_2OH} \xrightarrow{\text{P/I}_2} \underset{\text{1-iodohexane}}{CH_3(CH_2)_4{-}CH_2I}$$

b. *Dehydration of alcohols to form alkenes* (Section 10-11A)

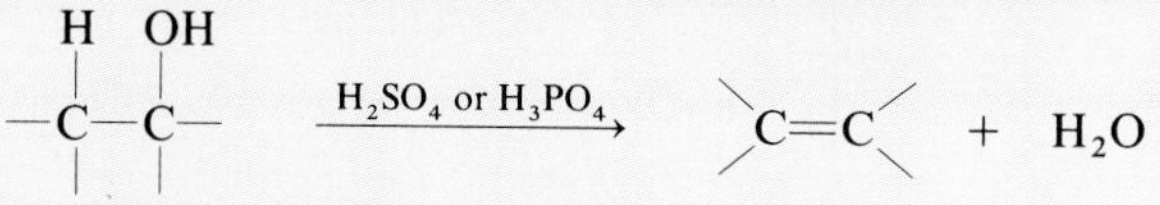

$$-\overset{H}{\underset{|}{\overset{|}{C}}}-\overset{OH}{\underset{|}{\overset{|}{C}}}- \xrightarrow{H_2SO_4 \text{ or } H_3PO_4} >C{=}C< + H_2O$$

Example: Dehydration of cyclohexanol

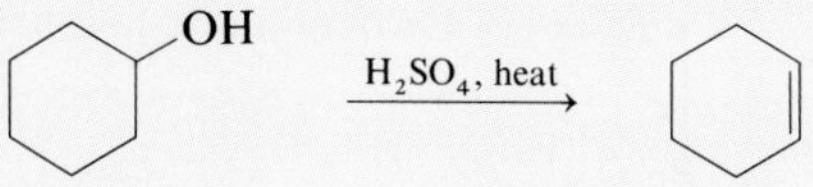

$$\underset{\text{cyclohexanol}}{C_6H_{11}OH} \xrightarrow{H_2SO_4\text{, heat}} \underset{\text{cyclohexene}}{C_6H_{10}}$$

c. *Dehydration of alcohols to form ethers* (Section 10-11B)

$$2\,R{-}OH \xrightarrow{H^+} R{-}O{-}R + H_2O$$

Example

$$\underset{\text{ethanol}}{2\,CH_3CH_2OH} \xrightarrow[140^\circ]{H_2SO_4} \underset{\text{diethyl ether}}{CH_3CH_2{-}O{-}CH_2CH_3} + H_2O$$

3. *Cleavage of the hydroxyl proton* —C—O⌇H

a. *Tosylation* (Section 10-7)

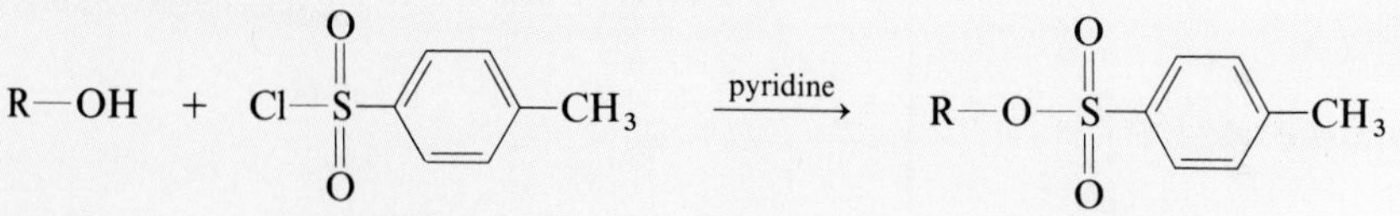

$$\underset{\text{alcohol}}{R{-}OH} + \underset{\text{tosyl chloride (TsCl)}}{Cl{-}SO_2{-}C_6H_4{-}CH_3} \xrightarrow{\text{pyridine}} \underset{\text{alkyl tosylate}}{R{-}O{-}SO_2{-}C_6H_4{-}CH_3}$$

Example

$$\underset{\text{isopropyl alcohol}}{(CH_3)_2CH{-}OH} \xrightarrow{\text{TsCl, pyridine}} \underset{\text{isopropyl tosylate}}{(CH_3)_2CH{-}OTs}$$

b. *Acylation to form esters* (Section 10-13)

$$R{-}OH \xrightarrow[\text{(acyl chloride)}]{R'{-}\overset{O}{\overset{\|}{C}}{-}Cl} R{-}O{-}\overset{O}{\overset{\|}{C}}{-}R'$$

ester

Example

$$\text{cyclohexanol (C}_6\text{H}_{10}\text{(H)OH)} \xrightarrow[\text{(acetyl chloride)}]{CH_3{-}COCl} \text{C}_6\text{H}_{10}\text{(H)}{-}O{-}\overset{O}{\overset{\|}{C}}{-}CH_3 + HCl$$

cyclohexanol — cyclohexyl acetate

c. *Deprotonation to form an alkoxide* (Section 10-15)

$$R{-}OH + Na \longrightarrow R{-}O^{-}\,{}^{+}Na + \tfrac{1}{2}H_2$$

$$R{-}OH + K \longrightarrow R{-}O^{-}\,{}^{+}K + \tfrac{1}{2}H_2$$

Example

$$CH_3{-}CH_2{-}OH + Na \longrightarrow Na^{+}\,{}^{-}O{-}CH_2{-}CH_3$$

ethanol — sodium ethoxide

d. *Williamson ether synthesis* (Sections 10-15 and 13-5)

$$R{-}O^{-} + R'{-}X \longrightarrow R{-}O{-}R' + X^{-}$$

(R′ must be primary)

Example

$$Na^{+}\,{}^{-}O{-}CH_2{-}CH_3 + CH_3I \longrightarrow CH_3{-}CH_2{-}O{-}CH_3 + NaI$$

sodium ethoxide — methyl iodide — ethyl methyl ether

GLOSSARY

alcohol dehydrogenase (ADH) An enzyme used by living cells to oxidize ethyl alcohol to acetic acid. (p. 407)

alkoxide ion The anion formed by deprotonating an alcohol. (p. 432)

$$R{-}\ddot{\underset{..}{O}}{-}H + Na \longrightarrow R{-}\ddot{\underset{..}{O}}{:}^{-}\,{}^{+}Na + \tfrac{1}{2}H_2\uparrow$$

chromic acid reagent The solution formed by adding sodium or potassium dichromate (and a small amount of water) to concentrated sulfuric acid. (p. 403)

> **chromic acid test:** When a primary or secondary alcohol is added to the chromic acid reagent, the orange color changes to green or blue. A nonoxidizable compound (such as a tertiary alcohol, a ketone, or an alkane) produces no color change. (p. 405)

Collins reagent ($CrO_3 \cdot 2$ pyridine) A complex of chromium trioxide with pyridine, used to oxidize primary alcohols selectively to aldehydes. (p. 405)

ester An acid derivative formed by the reaction of an acid with an alcohol with loss of water. The most common esters are carboxylic esters, composed of carboxylic acids and alcohols. (p. 429)

> **Fischer esterification:** The acid-catalyzed reaction of an alcohol with a carboxylic acid to give an ester. (p. 429)

$$R{-}\overset{O}{\overset{\|}{C}}{-}\boxed{OH} + \boxed{H}{-}O{-}R' \overset{H^+}{\rightleftharpoons} R{-}\overset{O}{\overset{\|}{C}}{-}O{-}R' + \boxed{H_2O}$$

acid — alcohol — ester

ether A compound containing on oxygen atom bonded to two alkyl or aryl groups. (p. 432)

inorganic esters Compounds derived from alcohols and inorganic acids with loss of water. (p. 430) Examples are:

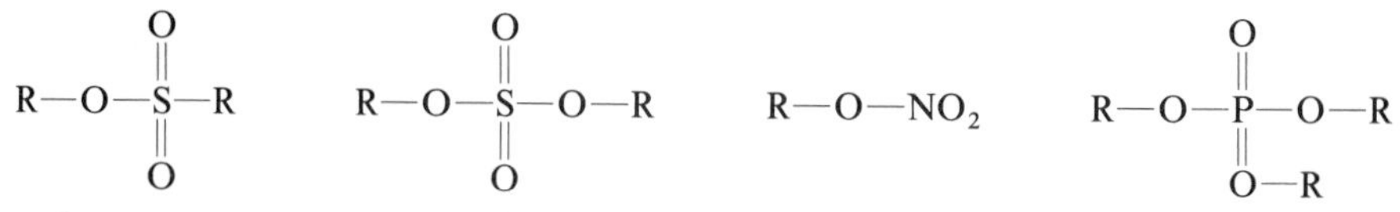

$R{-}O{-}S(=O)_2{-}R$	$R{-}O{-}S(=O)_2{-}O{-}R$	$R{-}O{-}NO_2$	$R{-}O{-}P(=O)(O{-}R){-}O{-}R$
sulfonate esters	sulfate esters	nitrate esters	phosphate esters

Jones oxidation The oxidation of 1° and 2° alcohols using dilute chromic acid dissolved in acetone. (p. 404)

Lucas test A test used to determine whether an alcohol is primary, secondary, or tertiary. The test measures the rate of reaction of the alcohol with the *Lucas reagent,* $ZnCl_2$ in concentrated HCl. Tertiary alcohols react very fast, secondary alcohols react more slowly, and primary alcohols react very slowly. (p. 415)

nicotinamide adenine dinucleotide (NAD) A biological oxidizing/reducing reagent that operates in conjunction with enzymes such as alcohol dehydrogenase. (p. 407)

pinacol rearrangement A dehydration of a glycol in which one of the groups migrates and a proton is lost to give a ketone. (p. 424)

pyridinium chlorochromate (PCC) A complex of chromium trioxide with pyridine and HCl. PCC is a more soluble version of the Collins reagent. (p. 405)

tosylate ester An ester of an alcohol with *para*-toluenesulfonic acid. Like halide ions, the tosylate anion is an excellent leaving group. (p. 410)

Williamson ether synthesis The S_N2 reaction between an alkoxide ion and a primary alkyl halide or tosylate. The product is an ether. (p. 432)

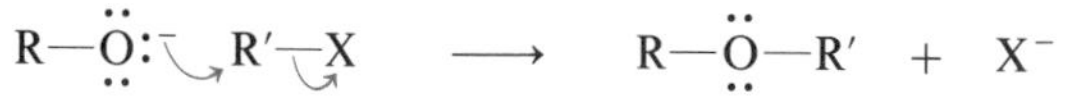

$$R{-}\ddot{\underset{..}{O}}{:}^{-}\ \ R'{-}X \longrightarrow R{-}\ddot{\underset{..}{O}}{-}R' + X^-$$

ESSENTIAL PROBLEM-SOLVING SKILLS IN CHAPTER 10

1. Identify whether oxidation or reduction is needed to interconvert alkanes, alcohols, aldehydes, ketones, and acids, and identify reagents that will accomplish the conversion.

2. Predict the products of the reactions of alcohols with
   (a) Oxidizing and reducing agents
   (b) Carboxylic acids and acid chlorides
   (c) Dehydrating reagents, especially H_2SO_4 and H_3PO_4
   (d) Inorganic acids and acid chlorides
   (e) Sodium and potassium.

3. Predict the products of reactions of alkoxide ions.

4. Propose chemical tests to distinguish alcohols from the other types of compounds we have studied.

5. Use your knowledge of alcohol and diol reactions to propose mechanisms and products of similar reactions you have never seen before.

6. Show how to convert an alcohol to a related compound with a different functional groups.

7. Predict the products of pinacol rearrangement and periodate cleavage of glycols.

8. Use retrosynthetic analysis to propose effective single-step and multistep syntheses of compounds using alcohols as intermediates (especially those using Grignard and organolithium reagents to assemble the carbon skeletons).

STUDY PROBLEMS

10-37. Briefly define each of the following terms and give an example.
(a) Collins oxidation **(b)** carboxylic ester **(c)** tosylate ester
(d) ether **(e)** Williamson ether synthesis **(f)** Jones oxidation
(g) Lucas test **(h)** pinacol rearrangement **(i)** chromic acid oxidation
(j) alkoxide ion

10-38. In each case, show how you would synthesize the chloride, bromide, and iodide from the corresponding alcohol.
(a) 1-halobutane (halo = chloro, bromo, iodo) **(b)** halocyclopentane
(c) 1-halo-1-methylcyclohexane **(d)** 1-halo-2-methylcyclohexane

10-39. Predict the major products of the following reactions.
(a) (*R*)-2-butanol + TsCl in pyridine **(b)** (*S*)-2-butyl tosylate + NaBr
(c) cyclooctanol + CrO_3/H_2SO_4 in acetone **(d)** cyclopentylmethanol + $CrO_3 \cdot$ pyridine
(e) cyclopentylmethanol + $Na_2Cr_2O_7/H_2SO_4$ **(f)** cyclopentanol + $HCl/ZnCl_2$
(g) cycloheptanol + $LiAlH_4/TiCl_4$ **(h)** cyclooctylmethanol + CH_3CH_2MgBr
(i) potassium *t*-butoxide + methyl iodide **(j)** sodium methoxide + *t*-butyl iodide
(k) *n*-butanol + HBr **(l)** cyclopentanol + H_2SO_4, heat
(m) product from **(l)** + OsO_4/H_2O_2, then HIO_4 **(n)** sodium ethoxide + 1-bromobutane
(o) sodium ethoxide + 2-methyl-2-bromobutane

10-40. Show how you would accomplish the following synthetic conversions.

(a) cyclohexane ring with OH and H → cyclohexane ring with OCH_2CH_3 and H

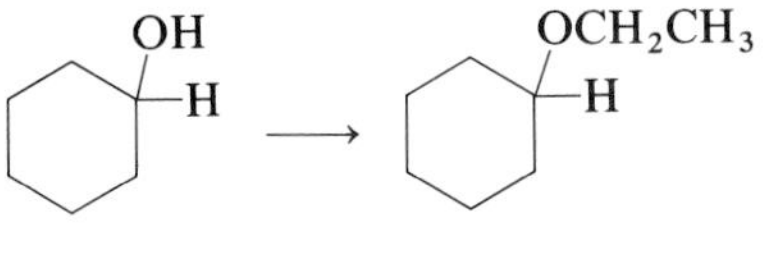

(b) 2 benzene ring–CH_2Br → benzene–CH=CH–benzene

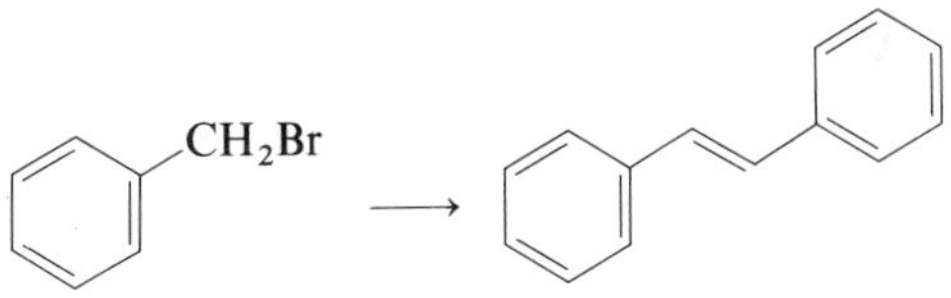

(c) cyclohexane ring with Br and H → cyclohexane ring with $\overset{O}{\overset{\|}{C}}-CH_2CH_2CH_3$ and H

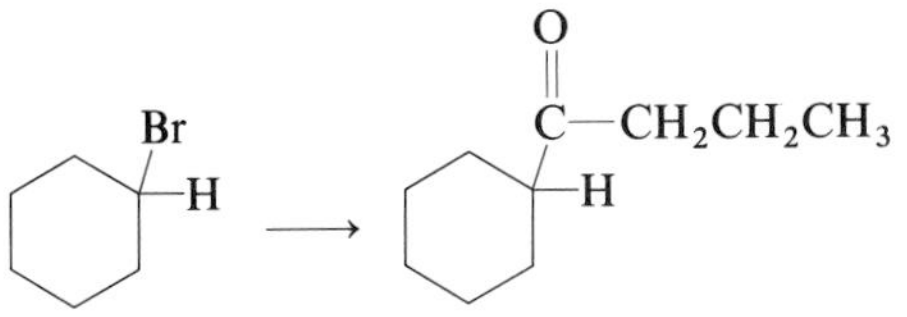

(d) cyclohexane ring with CH_2OH and H → cyclohexane ring–$\overset{OH}{\overset{|}{C}}HCH_2CH_3$

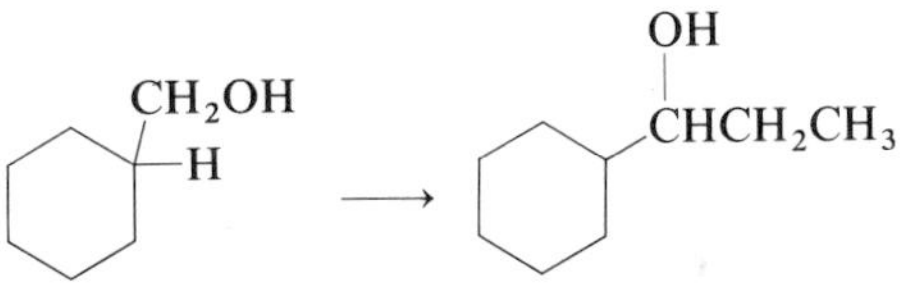

10-41 Predict the major products of dehydration catalyzed by sulfuric acid.
(a) 1-hexanol **(b)** 2-hexanol **(c)** 3-pentanol
(d) 1-methylcyclopentanol **(e)** cyclopentylmethanol **(f)** 2-methylcyclopentanol

10-42. Predict the esterification products of the following acid/alcohol pairs.

(a) $CH_3CH_2CH_2COOH + CH_3OH$ **(b)** benzene ring–OH + $CH_3-\overset{O}{\overset{\|}{C}}-OH$

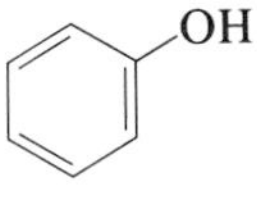

(c) benzene ring–COOH + CH_3CH_2OH **(d)** $2\ CH_3CH_2OH + H_3PO_4$ **(e)** $CH_3OH + HNO_3$

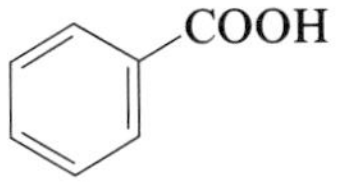

10-43. Show how you would make the methanesulfonate ester of cyclohexanol, beginning with cyclohexanol and an appropriate acid chloride.

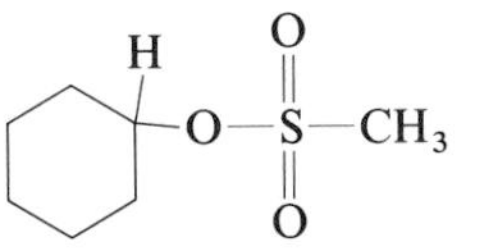

cyclohexane ring with H and $-O-\overset{O}{\overset{\|}{S}}(=O)-CH_3$

cyclohexyl methanesulfonate

10-44. Show how you would convert (*S*)-2-hexanol to
(a) (*S*)-2-chlorohexane. **(b)** (*R*)-2-bromohexane.

10-45. When 1-cyclohexylethanol is treated with concentrated aqueous HBr, the major product is 1-bromo-1-ethylcyclohexane.

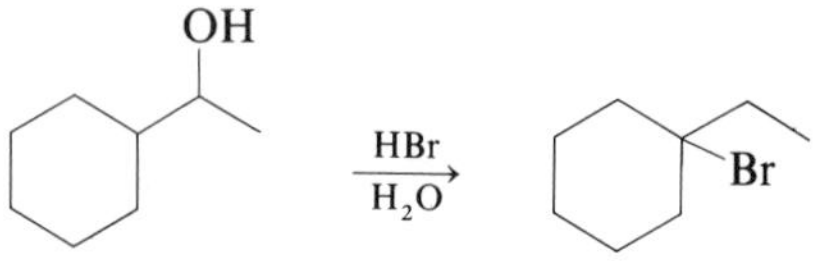

OH — $\xrightarrow[H_2O]{HBr}$ — Br

(a) Give a mechanism for this reaction.
(b) How would you convert 1-cyclohexylethanol to (1-bromoethyl)cyclohexane in good yield?

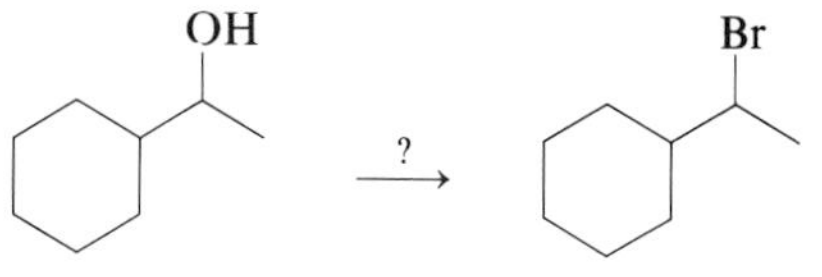

OH $\xrightarrow{?}$ Br

10-46. Show how you would make each of the following compounds, beginning with an alcohol of your choice.

(a) 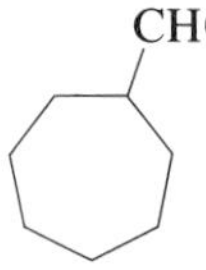 CHO (b) 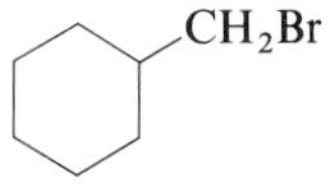 CH_2Br (c) 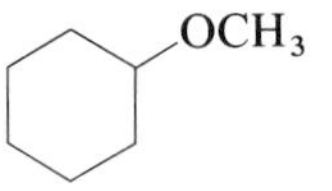 OCH_3 (d) 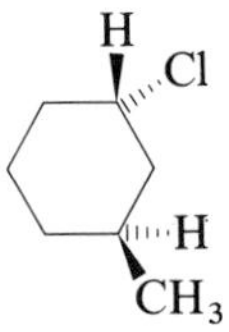 H, Cl, H, CH_3 (e) 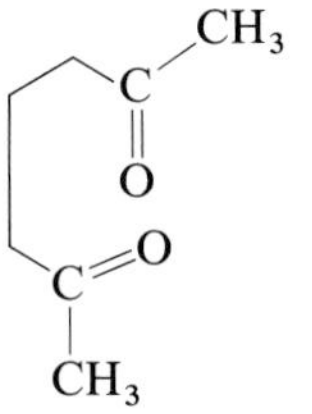 CH_3, C, O, C=O, CH_3

(f) 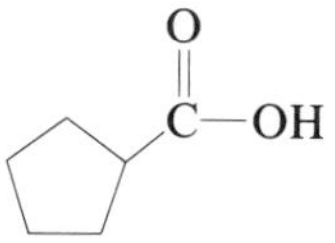 O, C—OH (g) 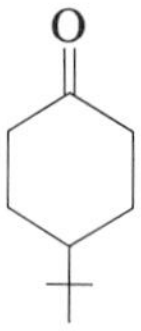 O (h) 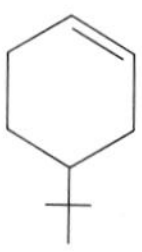 (i) 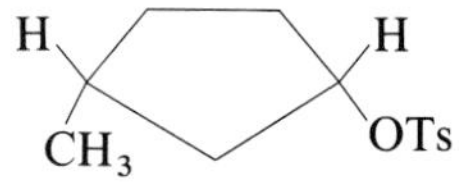 H, CH_3, H, OTs (j) 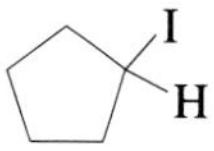 I, H

10-47. Predict the major products (including stereochemistry) when *cis*-3-methylcyclohexanol reacts with each of the following reagents.
(a) PBr_3 **(b)** $SOCl_2$ **(c)** Lucas reagent **(d)** concentrated HBr **(e)** TsCl/pyridine, then NaBr

10-48. Show how you would use simple chemical tests to distinguish between the following pairs of compounds. In each case, describe what you would do and what you would observe.
(a) 1-butanol and 2-butanol **(b)** 2-butanol and 2-methyl-2-butanol
(c) cyclohexanol and cyclohexene **(d)** cyclohexanol and cyclohexanone
(e) cyclohexanone and 1-methylcyclohexanol

10-49. Write the important resonance structures of the following anions.

(a) 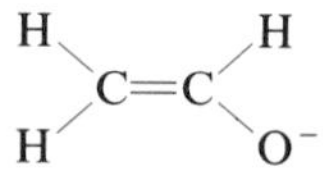 H, H, C=C, H, O^- (b) 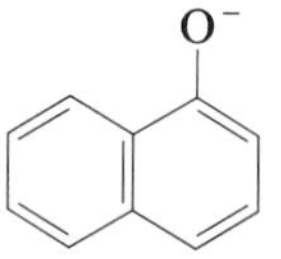 O^- (c) $^-O—\overset{O}{\underset{O}{S}}—CH_3$

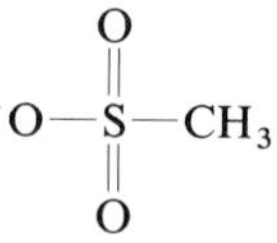

10-50. Compound **A** is an optically active alcohol. Treatment with chromic acid converts **A** into a ketone, **B**. In a separate reaction, **A** is treated with PBr_3, converting **A** into compound **C**. Compound **C** is purified, and then it is allowed to react with magnesium in ether. Compound **B** is added to the resulting solution of a Grignard reagent. After hydrolysis, this solution is found to contain 3,4-dimethyl-3-hexanol. Propose structure for compounds **A**, **B**, and **C**.

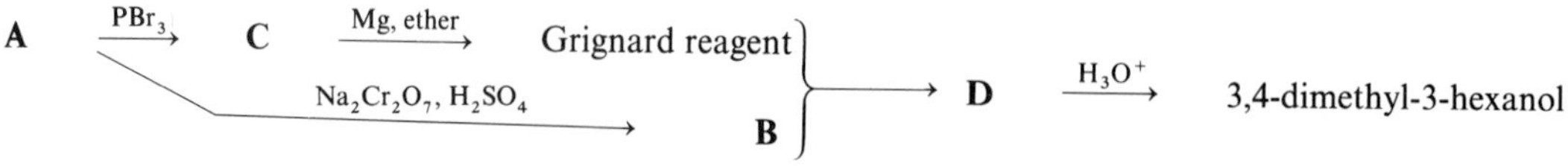

A $\xrightarrow{PBr_3}$ **C** $\xrightarrow{\text{Mg, ether}}$ Grignard reagent
A $\xrightarrow{Na_2Cr_2O_7,\ H_2SO_4}$ **B**
Grignard reagent + **B** $\longrightarrow$ **D** $\xrightarrow{H_3O^+}$ 3,4-dimethyl-3-hexanol

10-51. Give the structures of the intermediates and products **A** through **E**.

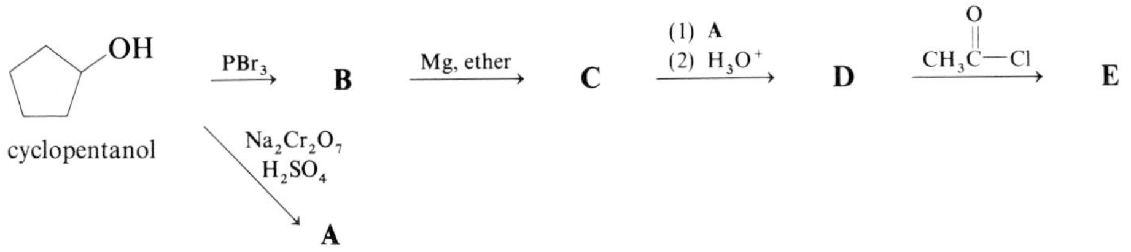

cyclopentanol $\xrightarrow{PBr_3}$ **B** $\xrightarrow{\text{Mg, ether}}$ **C** $\xrightarrow{(1)\ \mathbf{A}\quad (2)\ H_3O^+}$ **D** $\xrightarrow{CH_3C(=O)-Cl}$ **E**

cyclopentanol $\xrightarrow{Na_2Cr_2O_7,\ H_2SO_4}$ **A**

10-52. Under acid catalysis, tetrahydrofurfuryl alcohol reacts to give surprisingly good yields of dihydropyran. Propose a mechanism to explain this useful synthesis.

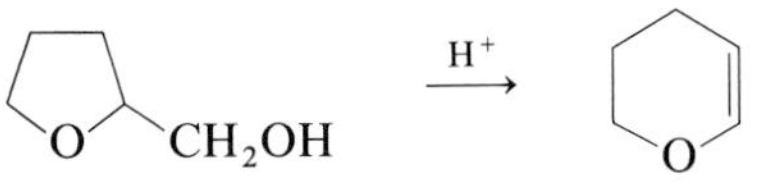

tetrahydrofurfuryl alcohol (CH_2OH) $\xrightarrow{H^+}$ dihydropyran

* **10-53.** The following compound is first treated with OsO_4/H_2O_2 and then with concentrated H_2SO_4. Propose structures for both the intermediate product and the final product, which has undergone pinacol rearrangement.

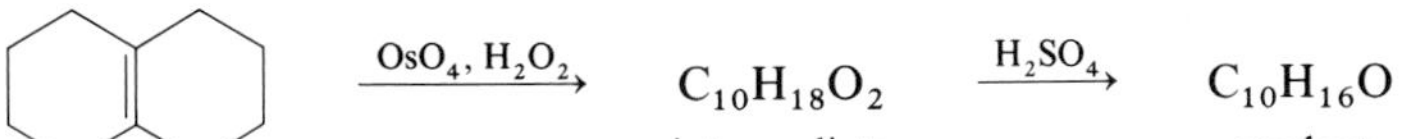

$\xrightarrow{OsO_4,\ H_2O_2}$ $C_{10}H_{18}O_2$ (intermediate) $\xrightarrow{H_2SO_4}$ $C_{10}H_{16}O$ (product)

10-54. Show how you would synthesize each of the following compounds. As starting materials, you may use any alcohols containing four or fewer carbon atoms, cyclopentanol, and any necessary solvents and inorganic reagents.

(a) 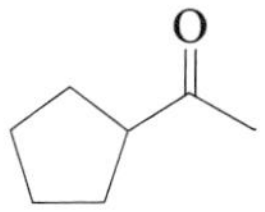 (O) (b) 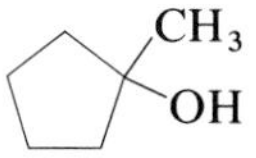 CH_3, OH (c) 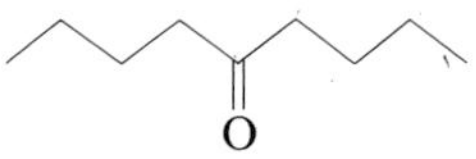 O (d) 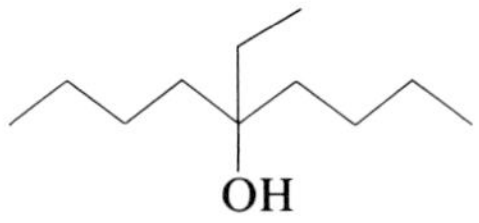 OH

(e) 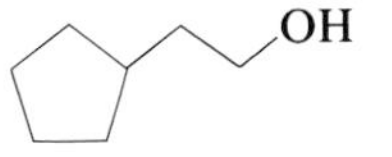 OH (f) 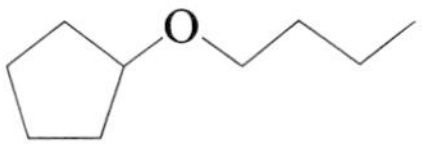 O (g) 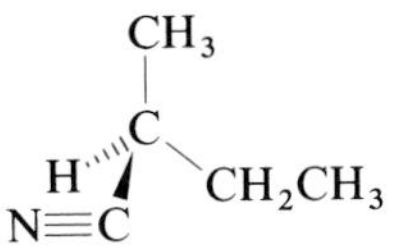 CH_3, C, H, CH_2CH_3, N≡C (h) Br

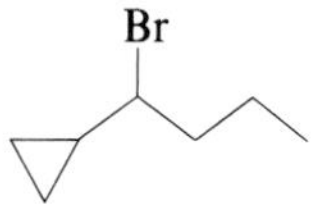

* **10-55.** Many alcohols undergo dehydration at 0°C when treated with phosphorus oxychloride ($POCl_3$) in the basic solvent pyridine. (Phosphorus oxychloride is the acid chloride of phosphoric acid, with chlorine atoms in place of the hydroxyl groups of phosphoric acid.)

(a) Propose a mechanism for the dehydration of cyclopentanol using $POCl_3$ and pyridine. The first half of the mechanism, formation of a dichlorophosphate ester, is similar to the first half of the mechanism of reaction of an alcohol with thionyl chloride. Like a tosylate, the dichlorophosphate group is a good leaving group. The second half of the mechanism might be either first order or second order; draw both alternatives for now.

(b) When *trans*-2-methylcyclopentanol undergoes dehydration using $POCl_3$ in pyridine, the major product is 3-methylcyclopentene, and not the Saytzeff product. What is the stereochemistry of the dehydration? What does this stereochemistry imply about the correct mechanism in part (a)? Explain your reasoning.

* **10-56.** Two unknowns, **X** and **Y**, both having molecular formula C_4H_8O, give the following results with four chemical tests. Propose structures for **X** and **Y** consistent with this information.

	bromine	*Na metal*	*chromic acid*	*Lucas reagent*
Compound **X**	decolorizes	bubbles	orange to green	no reaction
Compound **Y**	no reaction	no reaction	no reaction	no reaction

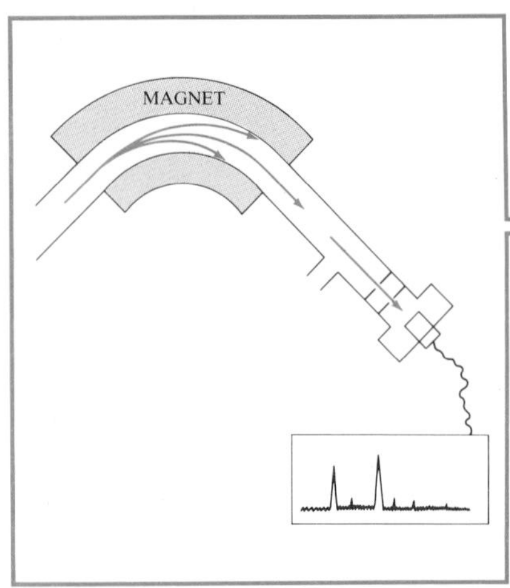

11

INFRARED SPECTROSCOPY AND MASS SPECTROMETRY

11-1 INTRODUCTION

One of the most important tasks of organic chemistry is the determination of organic structures. When an interesting compound is isolated from a natural source, its structure must be completely determined before a synthesis is begun. Whenever we run a reaction, we must determine whether the product has the desired structure. The structure of an unwanted product must be known so that the reaction conditions can be altered to favor the desired product.

In many cases, a compound can be identified by chemical means. The molecular formula is found by analyzing the elemental composition and determining the molecular weight. If the compound has been characterized before, we can compare its physical properties (melting point, boiling point, etc.) with the published values. Chemical tests can be performed to suggest the functional groups present and narrow the range of possible structures before the physical properties are used to make an identification.

These procedures are not sufficient, however, for complex compounds that have never been synthesized and characterized. They are also impractical with compounds that are difficult to obtain, because a relatively large sample is required to complete the elemental analysis and all the functional group tests. We need analytical techniques that work with tiny samples and that do not damage the sample in any way.

Absorption spectroscopy is the measurement of the amount of light absorbed by a compound as a function of the wavelength of light. In general, a sample is irradiated by a light source, and the amount of light transmitted at various wavelengths is measured by a detector. Unlike chemical tests, spectroscopic techniques are *nondestructive;* that is, none of the sample is destroyed. Many different kinds of spectra can be measured with no loss of the sample.

We will cover four spectroscopic or related techniques that serve as powerful tools for structure determination in organic chemistry:

Infrared spectroscopy, covered in this chapter, observes the vibrations of bonds and provides evidence of the functional groups present.

Mass spectrometry, also covered in this chapter, bombards molecules with electrons and breaks them apart. Analysis of the masses of the fragments gives the molecular weight, possibly the molecular formula, and clues to the structure and functional groups.

Nuclear magnetic resonance spectroscopy, covered in Chapter 12, observes the chemical environments of the hydrogen atoms (or the carbon atoms) and provides evidence of the structure of the alkyl groups and clues to the functional groups.

Ultraviolet spectroscopy, covered in Chapter 15, observes electronic transitions and provides information on the electronic bonding in the sample.

These spectroscopic techniques are complementary, and they are most powerful when used together. In many cases, an unknown compound cannot be completely identified from one spectrum without additional information, yet the structure can be determined with confidence using two or more different types of spectra. In Chapter 12 we consider how information from different types of spectroscopy is combined to provide a reliable structure.

11-2 THE ELECTROMAGNETIC SPECTRUM

Visible light, infrared light, ultraviolet light, microwaves, and radio waves are examples of electromagnetic radiation. They all travel at the speed of light, about 3×10^{10} cm/sec, but they differ in frequency and wavelength. The **frequency** of a wave is the number of complete wave cycles that pass a fixed point in a second. Frequency, represented by the Greek letter ν (nu), is usually given in hertz (Hz), meaning cycles per second. The **wavelength,** represented by the Greek letter λ (lambda), is the distance between any two peaks (or any two troughs) of the wave.

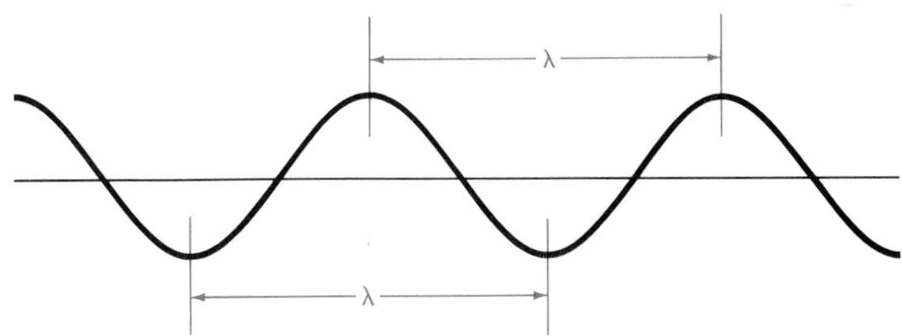

The wavelength and frequency, which are inversely proportional, are related by the equation

$$\nu\lambda = c \qquad \text{or} \qquad \lambda = \frac{c}{\nu}$$

where

c = speed of light (3×10^{10} cm/sec)

ν = frequency in hertz

λ = wavelength in centimeters

Electromagnetic waves travel as **photons,** which are massless packets of energy. The energy of a photon is proportional to its frequency and inversely proportional to its wavelength. A photon of frequency ν (or wavelength λ) has an energy given by

$$\varepsilon = h\nu = \frac{hc}{\lambda}$$

where h is Planck's constant, 1.58×10^{-37} kcal-sec or 6.62×10^{-37} kJ-sec.

Multiplying by Avogadro's number (N) gives the energy (E) of a mole of photons of frequency ν or wavelength λ given in centimeters.

$$E = Nh\nu = \frac{Nhc}{\lambda} = \frac{2.86 \times 10^{-3} \text{ kcal/mol}}{\lambda \text{ (cm)}}$$

Under certain conditions, a molecule struck by a photon may absorb the photon's energy. In this case, the molecule's energy is increased by an amount equal to the photon's energy $h\nu$. For this reason, we often represent the irradiation of a reaction mixture by the symbol $h\nu$.

The **electromagnetic spectrum** is the range of all possible frequencies, ranging from zero to infinity. In practice, the spectrum ranges from the very low radio frequencies used to communicate with submarines to the very high frequencies of gamma rays. Figure 11-1 shows the frequency, wavelength, and energy relationships of the various parts of the electromagnetic spectrum.

higher frequency
shorter wavelength

ENERGY

Wavelength (λ)		*Energy per mole*	*Molecular effects*
10^{-10} meter	gamma rays	10^6 kcal	
10^{-8} meter	X rays	10^4 kcal	ionization
	vacuum UV	10^2 kcal	
	near UV		electronic transitions
10^{-6} meter	visible	10 kcal	
10^{-4} meter	infrared (IR)	1 kcal	molecular vibrations
		10^{-2} kcal	
10^{-2} meter	microwave	10^{-4} kcal	rotational motion
10^0 (1 meter) 10^2 meters	radio	10^{-6} kcal	nuclear spin transitions

FIGURE 11-1 The electromagnetic spectrum

The electromagnetic spectrum is continuous, and the exact positions of the dividing lines between the different regions are somewhat arbitrary. Toward the top of the spectrum in Figure 11-1 are the higher frequencies, shorter wavelengths, and higher energies. Toward the bottom are the lower frequencies, longer wavelengths, and lower energies. X rays (very high energy) are so energetic they excite electrons past all the energy levels, causing ionization. Energies in the ultraviolet-visible range excite electrons to higher energy levels within molecules. Infrared energies excite molecular vibrations, and microwave energies excite rotations.

Radio-wave frequencies (very low energy) excite the nuclear spin transitions observed in nuclear magnetic resonance (NMR) spectroscopy.

11-3 THE INFRARED REGION

The **infrared** (from the Latin, *infra*, meaning "below" red) region of the spectrum corresponds to frequencies from just below the visible frequencies to just above the highest microwave and radar frequencies: wavelengths of about 8×10^{-5} cm to 1×10^{-2} cm. Common infrared spectrometers operate in the middle of this region, at wavelengths between 2.5×10^{-4} cm and 25×10^{-4} cm, corresponding to energies of about 1.1 to 11 kcal (4.6 to 46 kJ) per mole. Although infrared photons do not have enough energy to cause electronic transitions, they can cause groups of atoms to vibrate with respect to the bonds that connect them. Like electronic transitions, these vibrational transitions correspond to certain distinct energies, and molecules absorb infrared radiation only at certain wavelengths and frequencies.

The position of an infrared absorption band is specified by its wavelength (λ), measured in *microns* (μm). A micron (or *micrometer*) corresponds to one millionth (10^{-6}) of a meter, or 10^{-4} cm. A more common unit, however, is the **wavenumber** ($\bar{\nu}$), which corresponds to the number of cycles (wavelengths) of the wave in a centimeter. The wavenumber is simply the reciprocal of the wavelength (in centimeters). Since 1 cm = 10,000 μm, the wavenumber can also be calculated by dividing 10,000 by the wavelength in microns. The units of the wavenumber are cm^{-1} (*reciprocal centimeters*).

$$\bar{\nu}\,(\text{cm}^{-1}) = \frac{1}{\lambda\,(\text{cm})} = \frac{10{,}000\ \mu\text{m/cm}}{\lambda\,(\mu\text{m})} \qquad \text{or} \qquad \lambda\,(\mu\text{m}) = \frac{10{,}000\ \mu\text{m/cm}}{\bar{\nu}\,(\text{cm}^{-1})}$$

For example, an absorption at a wavelength of 4 μm corresponds to a wavenumber of 2500 cm^{-1}.

$$\bar{\nu} = \frac{10{,}000\ \mu\text{m/cm}}{4\ \mu\text{m}} = 2500\ \text{cm}^{-1} \qquad \text{or} \qquad \lambda = \frac{10{,}000\ \mu\text{m/cm}}{2500\ \text{cm}^{-1}} = 4\ \mu\text{m}$$

Wavenumbers (in cm^{-1}) have become the most common method for specifying IR absorptions, so we will use wavenumbers throughout this book. The wavenumber is proportional to the frequency (ν) of the wave, so it is also proportional to the energy of a photon of this frequency ($E = h\nu$). Many reference works still use micron units, however, so you should be prepared to interconvert these units.

PROBLEM 11-1
Complete the following conversion table.

$\bar{\nu}$ (cm^{-1})	1600	1640	1700		2500		
λ (μm)				4.55	4.00	3.33	3.03

11-4 MOLECULAR VIBRATIONS

Before discussing characteristic infrared absorptions, we should understand some theory about the vibrational energies of molecules. The figure below shows that a covalent bond between two atoms acts like a spring. If the bond is stretched, a restoring force pulls the two atoms together toward their equilibrium bond length.

If the bond is compressed, the restoring force pushes the two atoms apart. If the bond is stretched or compressed and then released, the atoms vibrate.

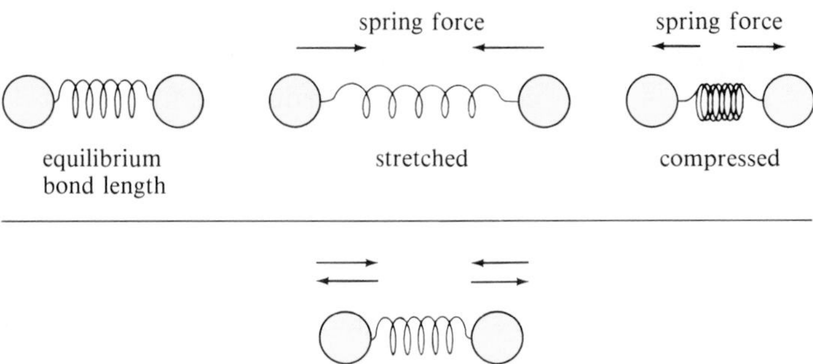

The frequency of the stretching vibration depends on two quantities: the masses of the atoms and the stiffness of the bond. Heavier atoms vibrate more slowly than lighter ones; for example, a C—D bond has a lower characteristic frequency than a C—H bond. In a group of bonds with similar bond energies, the *frequency decreases with increasing atomic weight.*

Stronger bonds are generally stiffer, requiring more force to stretch or compress them. Thus, stronger bonds usually vibrate faster than weaker bonds (assuming the atoms have similar masses). For example, O—H bonds are stronger than C—H bonds, and O—H bonds vibrate at higher frequencies. Triple bonds are stronger than double bonds, so triple bonds vibrate at higher frequencies than double bonds. Similarly, double bonds vibrate at higher frequencies than single bonds. In a group of bonds having atoms of similar masses, the *frequency increases with bond energy.*

Table 11-1 lists some common types of bonds together with their stretching frequencies as examples of the variation in frequency with the masses of the atoms and the strength of the bonds.

Even with simple compounds, infrared spectra contain many different absorptions, not just one absorption for each bond. Many of these absorptions result

TABLE 11-1

Bond stretching frequencies[a]

Bond		*Approximate bond energy [kcal (kJ)]*		*Approximate stretching frequency (cm⁻¹)*	
		Frequency dependence on atomic masses			
C—H	heavier atoms ↓	100 (420)		3000	$\bar{\nu}$ decreases ↓
C—D		100 (420)		2100	
C—C		83 (350)		1200	
		Frequency dependence on bond energies			
C—C		83 (350)	stronger bond ↓	1200	$\bar{\nu}$ increases ↓
C=C		146 (611)		1660	
C≡C		200 (840)		2200	
C—N		73 (305)	↓	1200	↓
C=N		147 (615)		1650	
C≡N		213 (891)		2200	
C—O		86 (360)	↓	1100	↓
C=O		178 (745)		1700	

[a] Bond vibration frequencies depend primarily on the masses of the atoms and on the stiffness of the bond (generally, the higher the bond energy, the stiffer the bond). In a group of bonds with similar bond energies, the *frequency decreases with increasing atomic weight.* In a group of bonds between similar atoms, the *frequency increases with bond energy.*

from stretching vibrations of the molecule as a whole, or from bending vibrations. In a bending vibration, the bond lengths stay constant, but the bond angles vibrate about their equilibrium values.

Consider the fundamental vibrational modes of a water molecule (Fig. 11-2). The two O—H bonds can stretch in phase with each other (symmetric stretching), or they can stretch out of phase (antisymmetric stretching). The H—O—H bond angle can also change in a bending vibration, making a scissoring motion.

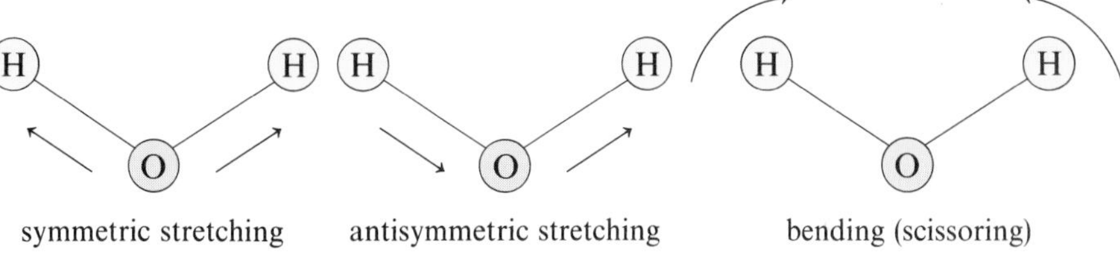

FIGURE 11-2 A nonlinear molecule with n atoms has $3n - 6$ fundamental vibrational modes. Water has $3(3) - 6 = 3$ modes. Two of these are stretching modes, and one is a bending mode.

A nonlinear molecule with n atoms generally has $3n - 6$ fundamental vibrational modes. Methane, then, has $3(5) - 6 = 9$ fundamental modes, and ethane has $3(8) - 6 = 18$ fundamental modes. Combinations and multiples of these simple, "fundamental" modes of vibration are also observed. As you can see, the number of absorptions in an infrared spectrum can be quite large, even for simple molecules.

It is very unlikely that two different compounds (except enantiomers) will have the same frequencies for all their various complex vibrations. For this reason, the infrared spectrum is considered to provide a "fingerprint" of a molecule. In fact, the region of the IR spectrum containing most of these complex vibrations (600 to 1400 cm^{-1}) is commonly called the **fingerprint region** of the spectrum.

Because the simple stretching vibrations are the most characteristic and predictable, our study of infrared spectroscopy will concentrate on them. Although we will largely ignore bending vibrations at this stage, you should remember that their absorptions generally appear in the 600 to 1400 cm^{-1} region of the spectrum. Experienced spectroscopists can tell a great deal about the structure of a molecule from these "wagging," "scissoring," "rocking," and "twisting" vibrations that absorb in the fingerprint region. The reference table of IR frequencies (Appendix 2) lists both stretching and bending characteristic frequencies.

11-5 IR-ACTIVE AND IR-INACTIVE VIBRATIONS

Not all molecular vibrations absorb infrared radiation. To understand which ones do and which do not, we need to consider how an electromagnetic field interacts with a molecular bond. The key to this interaction lies with the dipole moment of the bond. A bond with a dipole moment can be visualized as a positive charge and a negative charge separated by a spring. If this bond is placed in an electric field (Fig. 11-3), it is either stretched or compressed, depending on the direction of the field.

One of the components of an electromagnetic wave is a rapidly reversing electric field (**E**). This field alternately stretches and compresses a polar bond, as shown in Figure 11-3. When the electric field is in the same direction as the dipole moment, the bond is compressed and its dipole moment decreases. When the field is opposite to the dipole moment, the bond stretches and its dipole moment increases. If this alternate stretching and compressing of the bond by the electromagnetic wave occurs at the frequency of the molecule's natural rate of vibration,

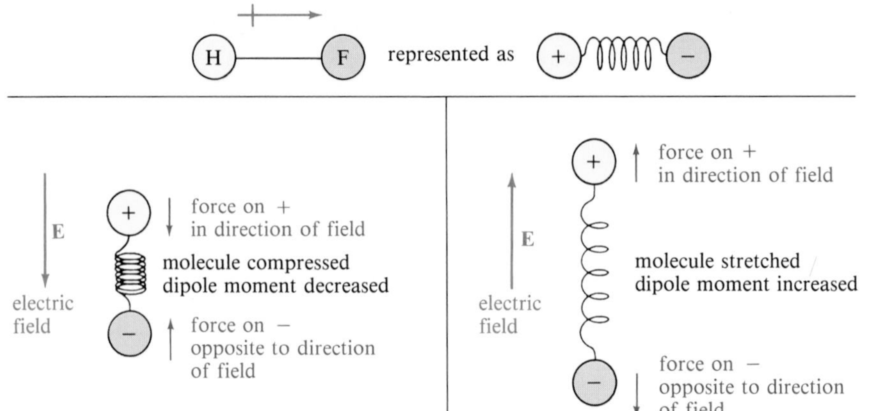

FIGURE 11-3 A bond with a dipole moment (as in HF, for instance) is either stretched or compressed by an electric field, depending on the direction of the field. Notice that the force on the positive charge is in the direction of the electric field (**E**), and the force on the negative charge is in the opposite direction.

energy may be absorbed. Vibrations of bonds with dipole moments generally result in IR absorptions and are said to be **IR active.**

If a bond is symmetrical and has zero dipole moment, the electric field does not interact with the bond. For example, the triple bond of acetylene (H—C≡C—H) has zero dipole moment, and the dipole moment remains zero if the bond is stretched or compressed. Since the vibration produces no change in the dipole moment, there can be no absorption of energy. This vibration is said to be **IR inactive,** and its characteristic frequency is not seen in the IR spectrum. The key to an IR-active vibration is that *the vibration must change the dipole moment of the molecule.*

In general, if a bond has a dipole moment, its stretching frequency causes an absorption in the IR spectrum. If a bond is symmetrically substituted and has zero dipole moment, its stretching vibration is weak or absent in the spectrum. Bonds with zero dipole moments sometimes produce absorptions (usually weak) because molecular collisions, rotations, and other vibrations make them unsymmetrical part of the time.

PROBLEM 11-2

Which of the bonds shown in red are expected to have IR-active stretching frequencies?

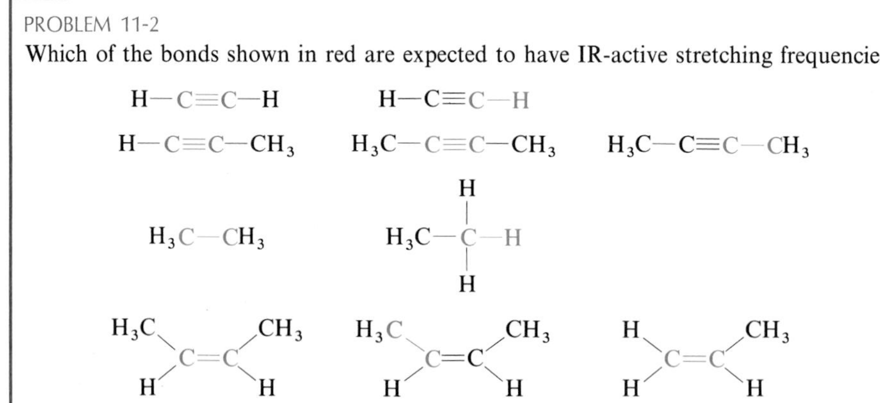

11-6 MEASUREMENT OF THE IR SPECTRUM

An **infrared spectrometer** measures the frequencies of infrared light absorbed by a compound. In a typical infrared spectrometer (Fig. 11-4), two beams of light are used. The *sample beam* passes through the sample cell, which holds the sample as a thin film or dissolved in a solvent. The *reference beam* passes through a reference

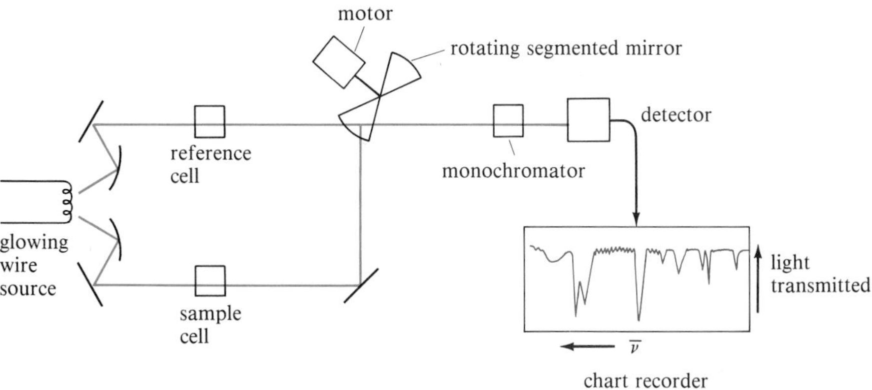

FIGURE 11-4 Block diagram of an infrared spectrometer.

cell that contains only the solvent. A rotating mirror alternately allows light from each of the two beams to enter the monochromator.

The monochromator uses prisms or diffraction gratings to allow only one frequency of light to enter the detector at a time. It scans the range of infrared frequencies, moving a pen along the corresponding frequencies on the x axis of the chart paper. Higher frequencies (shorter wavelengths) appear toward the left of the chart paper. The detector signal is proportional to the *difference* in the intensity of light in the sample and reference beams, with the reference beam compensating for any absorption by air or by the solvent. The detector signal controls movement of the pen along the y axis, with 100 percent transmittance (no absorption) at the top of the paper, and 0 percent transmittance (absorption of all the light) at the bottom.

In the **infrared spectrum** of n-octane (Fig. 11-5), there are four major absorption bands. The broad band between 2800 and 3000 cm^{-1} results from C—H stretching vibrations, and the band at 1460 cm^{-1} results from a scissoring vibration of the CH_2 groups. The absorptions at 1385 and 695 cm^{-1} result from the rocking of CH_3 and CH_2 groups, respectively. Since most organic compounds contain at least some saturated C—H bonds and some CH_2 and CH_3 groups, all these bands are quite common. In fact, without an authentic spectrum for comparison, *we could not look at this spectrum and conclude that the compound is octane.* We could be fairly certain that the compound is an alkane, however.

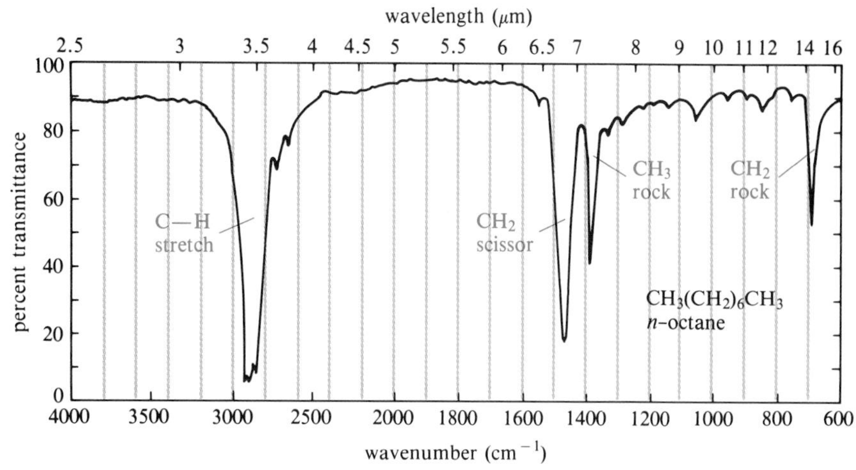

FIGURE 11-5 Infrared spectrum of n-octane. Notice that the frequencies scanned by a typical IR spectrometer range from about 600 cm^{-1} to about 4000 cm^{-1}.

One further characteristic to notice in the octane spectrum is the absence of any identifiable C—C stretching absorptions. (Table 11-1 shows that C—C stretching absorptions occur around 1200 cm^{-1}.) Although there are seven C—C bonds in octane, their dipole moments are small, and their absorptions are weak and indistinguishable. This result is common for alkanes with no functional groups to polarize the C—C bonds.

11-7 INFRARED SPECTROSCOPY OF HYDROCARBONS

Hydrocarbons contain only carbon-carbon bonds and carbon-hydrogen bonds. Although an infrared spectrum does not provide enough information to identify a structure conclusively (unless an authentic spectrum is available to compare "fingerprints"), the absorption frequencies of the carbon-carbon and carbon-hydrogen bonds can indicate the presence of carbon-carbon double and triple bonds.

11-7A CARBON-CARBON BOND STRETCHING

Stronger bonds generally absorb at higher frequencies than do weaker bonds, because of the greater stiffness associated with a stronger bond. Carbon-carbon single bonds absorb around 1200 cm^{-1}, C=C double bonds absorb around 1660 cm^{-1}, and C≡C triple bonds absorb around 2200 cm^{-1}.

Carbon-carbon bond stretching frequencies

C—C	1200 cm^{-1}
C=C	1660 cm^{-1}
C≡C	2200 cm^{-1}

As discussed for the octane spectrum above, C—C single bond absorptions (and most other absorptions in the fingerprint region) are not very reliable. We use the fingerprint region primarily to confirm the identity of an unknown compound by comparison with an authentic spectrum.

The absorptions of C=C double bonds, however, are very useful for structure determination. Most unsymmetrically substituted double bonds produce observable stretching absorptions in the region 1600 to 1680 cm^{-1}. The specific frequency of the double-bond stretching vibration depends on whether there is another double bond nearby. When two double bonds are one bond apart (as in 1,3-cyclohexadiene, below), they are said to be **conjugated.** As we will see in Chapter 15, conjugated double bonds are slightly more stable than isolated double bonds, because there is a small amount of pi-bonding between them. This overlap between the pi bonds leaves a little less electron density in the double bonds themselves; as a result, they are just a little less stiff and vibrate a little more slowly than an isolated double bond. Isolated double bonds absorb around 1640 to 1680 cm^{-1}, and conjugated double bonds absorb around 1620 to 1640 cm^{-1}.

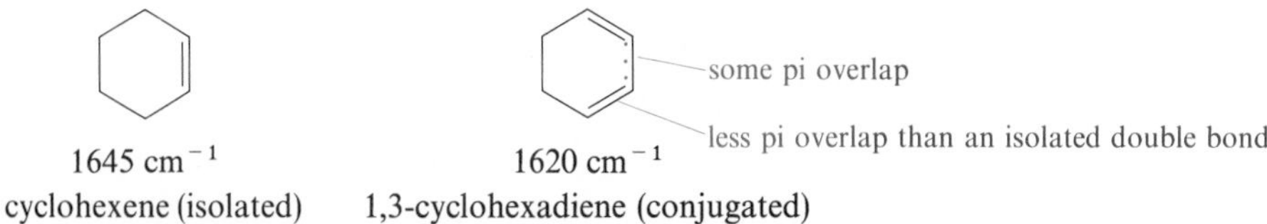

The effect of conjugation is even more pronounced in aromatic compounds. Aromatic C=C bonds are more like $1\frac{1}{2}$ bonds than true double bonds, and their reduced pi-bonding overlap results in less stiff bonds with lower stretching frequencies, around 1600 cm^{-1}.

$$\left[\text{benzene} \longleftrightarrow \text{benzene} \right] \qquad \bar{\nu} = 1600\ \text{cm}^{-1}$$

bond order = $1\frac{1}{2}$

Characteristic C=C stretching frequencies

isolated C=C	1640–1680 cm^{-1}
conjugated C=C	1620–1640 cm^{-1}
aromatic C=C	approx. 1600 cm^{-1}

Carbon-carbon triple bonds in alkynes are stronger (and stiffer) than carbon-carbon single or double bonds, and they absorb infrared light at higher frequencies. Most alkyne C≡C triple bonds have stretching frequencies between 2100 and 2200 cm^{-1}. Terminal alkynes usually give sharp C≡C stretching signals of moderate intensity. The C≡C stretching absorption of an internal alkyne may be weak or absent, however, due to the symmetry of the disubstituted triple bond with a very small or zero dipole moment.

μ

R—C≡C—H C≡C stretch observed around 2100 to 2200 cm^{-1}

terminal alkyne

μ is small or zero

R—C≡C—R′ C≡C stretch may be weak or absent

internal alkyne

11-7B CARBON-HYDROGEN BOND STRETCHING

Alkanes, alkenes, and alkynes are also characterized on the basis of their C—H stretching frequencies. Carbon-hydrogen bonds involving sp^3 hybrid carbon atoms generally absorb at frequencies just *below* 3000 cm^{-1}, while those involving sp^2 hybrid carbons absorb just *above* 3000 cm^{-1}. We explain this difference by the amount of *s* character in the carbon orbital used to form the bond. The *s* orbital is closer to the nucleus than the *p* orbitals, and stronger, stiffer bonds result from orbitals with more *s* character.

An sp^3 orbital is one-fourth *s* character, while an sp^2 orbital is one-third *s* character. We expect the bond using the sp^2 orbital to be slightly stronger, with a larger spring constant and a higher vibration frequency. The C—H bond of a terminal alkyne is formed using an *sp* hybrid orbital, with about one-half *s* character. This bond is stiffer than a C—H bond using an sp^3 or sp^2 hybrid carbon, and it absorbs at a higher frequency: about 3300 cm^{-1}.

C—H bond stretching frequencies

—C—C—H	sp^3 hybridized, one-fourth *s* character	2800–3000 cm^{-1}
>C=C< H	sp^2 hybridized, one-third *s* character	3000–3100 cm^{-1}
—C≡C—H	*sp* hybridized, one-half *s* character	3300 cm^{-1}

Figure 11-6 compares the IR spectra of *n*-hexane, 1-hexene, and *cis*-4-octene. The hexane spectrum is similar to that of *n*-octane (Fig. 11-5). The C—H stretching frequencies form a band between 2800 and 3000 cm^{-1}, and the bands in the fingerprint region are due to the bending vibrations discussed for Figure 11-5. This spectrum simply indicates the *absence* of any IR-active functional groups.

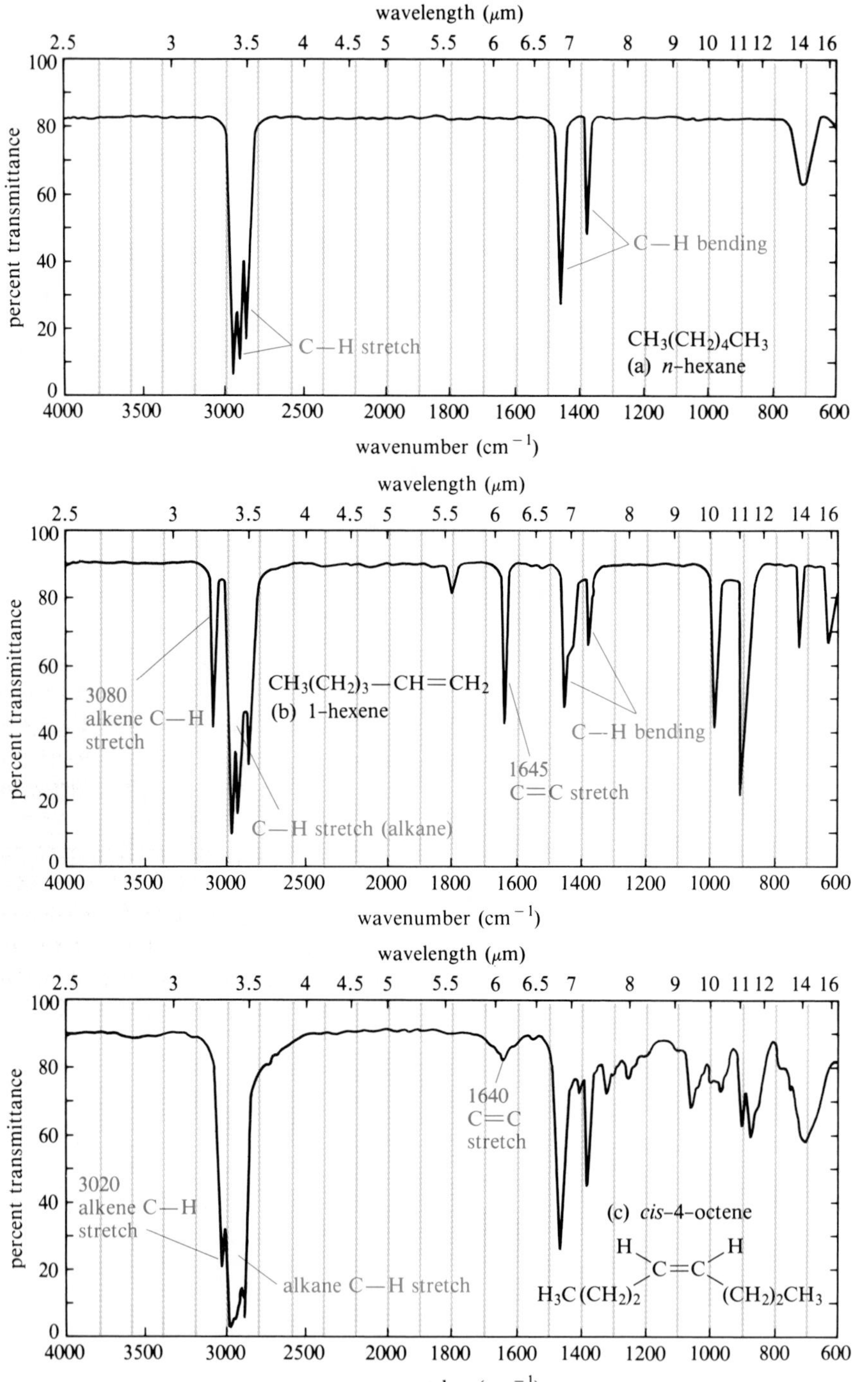

FIGURE 11-6 Comparison of the IR spectra of (a) *n*-hexane, (b) 1-hexene, and (c) *cis*-4-octene. The most characteristic absorptions in the 1-hexene spectrum are the C=C double bond stretch at 1645 cm^{-1} and the unsaturated =C—H stretch at 3080 cm^{-1}. The symmetrically substituted double bond in *cis*-4-octene gives a very weak C=C stretching absorption. The unsaturated =C—H stretch at 3020 cm^{-1} is still apparent, however.

The spectrum of 1-hexene shows additional absorptions characteristic of a double bond. The C—H stretch at 3080 cm^{-1} corresponds to stretching of the alkene =C—H bonds involving sp^2 hybrid carbons. The absorption at 1645 cm^{-1} results from stretching of the (isolated) C=C double bond.

The spectrum of *cis*-4-octene (Figure 11-6c) resembles the spectrum of 1-hexene, except the C=C stretching absorption at 1640 cm^{-1} is very weak in *cis*-4-octene because the symmetrically substituted double bond has nearly zero dipole moment. The weak absorption that does occur is the result of small, transient dipole moments due to molecular distortions caused by collisions. Even without the weak C=C stretching absorption, the unsaturated =C—H stretching absorption just above 3000 cm^{-1} suggests the presence of an alkene double bond.

Figure 11-7 compares the IR spectra of 1-octyne and 2-octyne. In addition to the alkane absorptions, the 1-octyne spectrum shows sharp peaks at 3320 and 2120 cm^{-1}. The absorption at 3320 cm^{-1} results from stretching of the stiff ≡C—H bond formed by the *sp* hybrid alkyne carbon. The 2120 cm^{-1} absorption results from the stretching vibration of the C≡C triple bond.

The spectrum of 2-octyne might confuse an inexperienced person. Since there is no acetylenic hydrogen, there is no ≡C—H stretching absorption around 3300 cm^{-1}. There is no visible C≡C stretching absorption around 2100 to 2200 cm^{-1}, either, because the disubstituted triple bond has a very small dipole moment. This spectrum would fail to alert us to the presence of a triple bond.

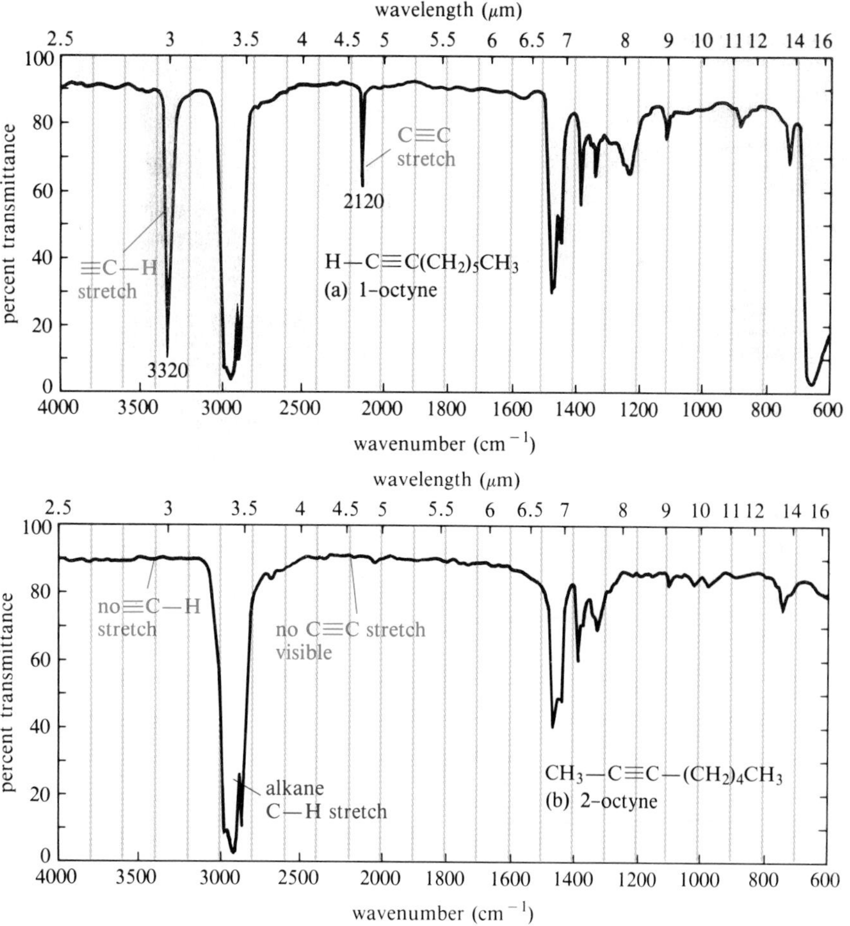

FIGURE 11-7 (a) The IR spectrum of 1-octyne shows characteristic absorptions at 3320 cm^{-1} (alkynyl ≡C—H stretch) and at 2120 cm^{-1} (alkynyl C≡C stretch). (b) Neither of these absorptions is seen in the spectrum of 2-octyne.

PROBLEM 11-3

For each of the following hydrocarbon spectra, determine whether the compound is an alkane, an alkene, an alkyne, or an aromatic hydrocarbon. In some cases, more than one unsaturated group may be present.

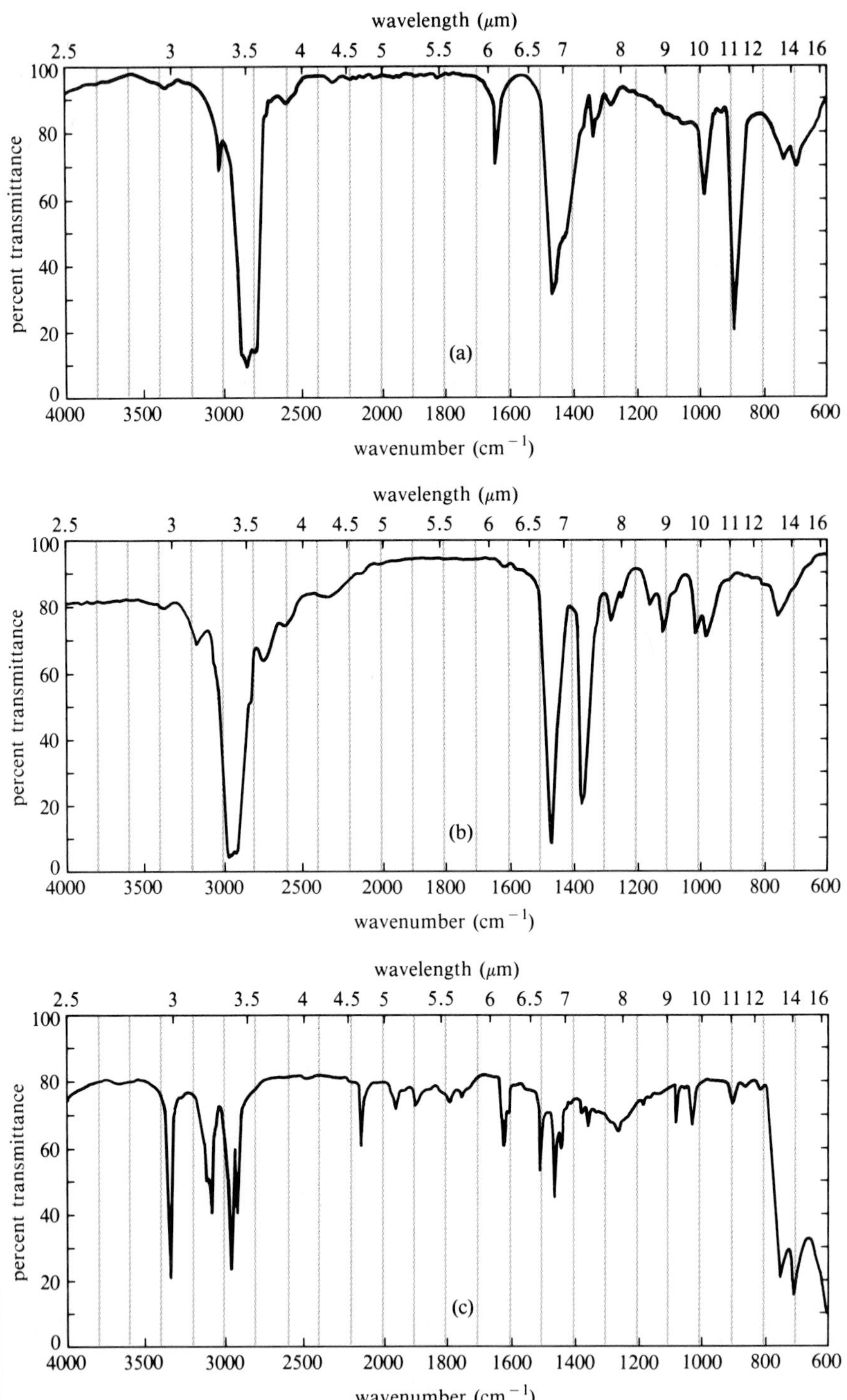

11-8 CHARACTERISTIC ABSORPTIONS OF ALCOHOLS AND AMINES

The O—H bonds of alcohols and the N—H bonds of amines are strong and stiff. The vibration frequencies of O—H and N—H bonds therefore occur at higher frequencies than those of most C—H bonds (except for alkynyl ≡C—H bonds).

R—O—H	R—NH—H	R—NH—R′	R—NR″—R′
alcohol	primary	secondary	tertiary
	amines		

O—H and N—H stretching frequencies.

alcohol O—H	3300 cm^{-1},	broad
acid O—H	3000 cm^{-1},	broad
amine N—H	3300 cm^{-1},	broad with spikes

Alcohol O—H bonds absorb over a wide range of frequencies, centered around 3300 cm^{-1}. Alcohol molecules are always involved in hydrogen bonding, with different molecules having different instantaneous arrangements. The O—H stretching frequencies reflect this diversity of hydrogen-bonding arrangements, resulting in very broad absorptions. Notice the broad O—H absorption centered around 3300 cm^{-1} in the infrared spectrum of 1-butanol (Fig. 11-8).

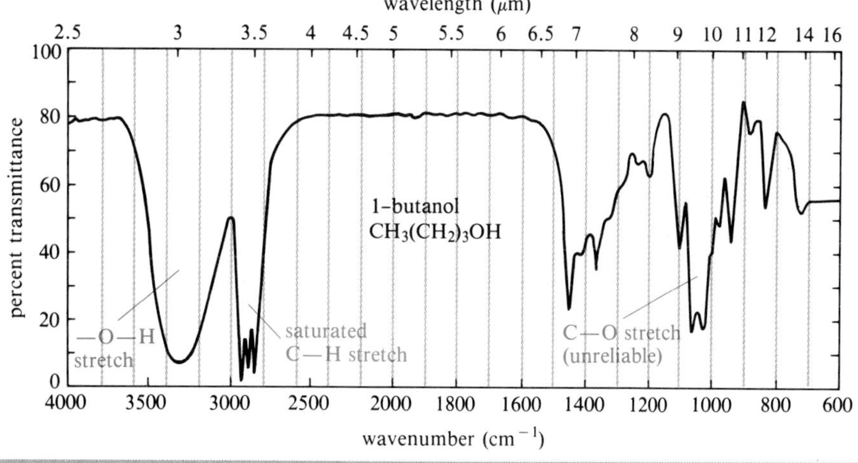

FIGURE 11-8 The IR spectrum of 1-butanol shows a broad, intense O—H stretching absorption centered around 3300 cm^{-1}. The broad shape of this peak is due to the diverse nature of the hydrogen-bonding interactions of alcohol molecules.

Figure 11-8 also shows a strong C—O stretching absorption centered near 1050 cm^{-1}. Compounds with C—O bonds (alcohols and ethers, for example) generally show strong absorptions in the range 1000 to 1200 cm^{-1}; however, there are other functional groups that also absorb in this region. Therefore, a strong peak between 1000 and 1200 cm^{-1} does not necessarily imply a C—O bond, but the *absence* of an absorption in this region suggests that there is no C—O bond in a molecule. For simple ethers, this unreliable C—O absorption is usually the only clue that the compound might be an ether.

Amine N—H bonds also have stretching frequencies in the 3300 cm^{-1} region, or even slightly higher. Like alcohols, amines participate in hydrogen bonding that can broaden the N—H absorptions. With amines, however, there may be one or more sharp spikes superimposed on the broad N—H stretching absorption: often one N—H spike for a secondary amine (R_2NH) and two N—H spikes for a primary amine (RNH_2). These sharp spikes, combined with the presence of nitrogen in the molecular formula, help us to distinguish amines from alcohols. Tertiary amines (R_3N) have no N—H bonds, and they do not give rise to N—H stretching absorptions in the IR spectrum. Figure 11-9 shows the spectrum of dipropylamine, a secondary amine.

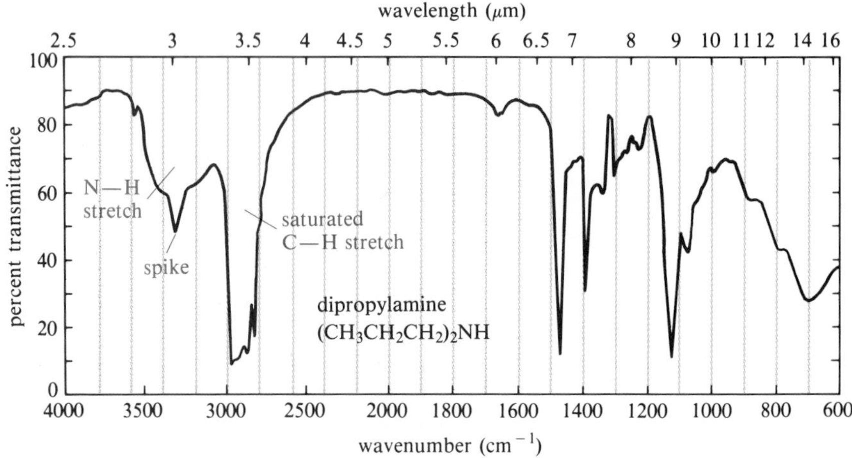

FIGURE 11-9 The IR spectrum of dipropylamine shows a broad N—H stretching absorption centered around 3300 cm^{-1}. Notice the spike present in this broad absorption.

11-9 CHARACTERISTIC ABSORPTIONS OF CARBONYL COMPOUNDS

Because it has a large dipole moment, the C=O double bond produces intense stretching absorptions in the infrared spectrum. Carbonyl groups absorb at frequencies around 1700 cm^{-1}, but the exact frequency depends on the specific functional group and the rest of the molecule containing the carbonyl. For these reasons, infrared spectroscopy is often the best method for detecting and identifying the type of carbonyl group in an unknown compound. To simplify our discussion of carbonyl absorptions, we will consider a "normal" stretching frequency for simple ketones, aldehydes, and carboxylic acids. Then we examine the types of carbonyl groups that deviate from this frequency.

11-9A SIMPLE KETONES, ALDEHYDES, AND ACIDS

The C=O stretching vibrations of simple ketones, aldehydes, and carboxylic acids occur at frequencies around 1710 cm^{-1}. These frequencies are higher than those for C=C double bonds because the C=O double bond is stronger and stiffer.

ketone	aldehyde	acid
R—C(=O)—R′ (C=O: 1710 cm^{-1})	R—C(=O)—H (C=O: 1710 cm^{-1}; C—H: 2700, 2800 cm^{-1})	R—C(=O)—O—H (C=O: 1710 cm^{-1}; O—H: broad, 2500–3500 cm^{-1})

In addition to the strong C=O stretching absorption, an aldehyde shows a characteristic set of two low-frequency C—H stretching frequencies around 2700 and 2800 cm^{-1}. Neither a ketone nor an acid produces these characteristic absorptions. Figure 11-10 compares the IR spectra of a simple ketone and a simple aldehyde. Notice the characteristic carbonyl stretching absorptions in both spectra, as well as the aldehyde C—H absorptions at 2715 and 2820 cm^{-1} in the aldehyde spectrum.

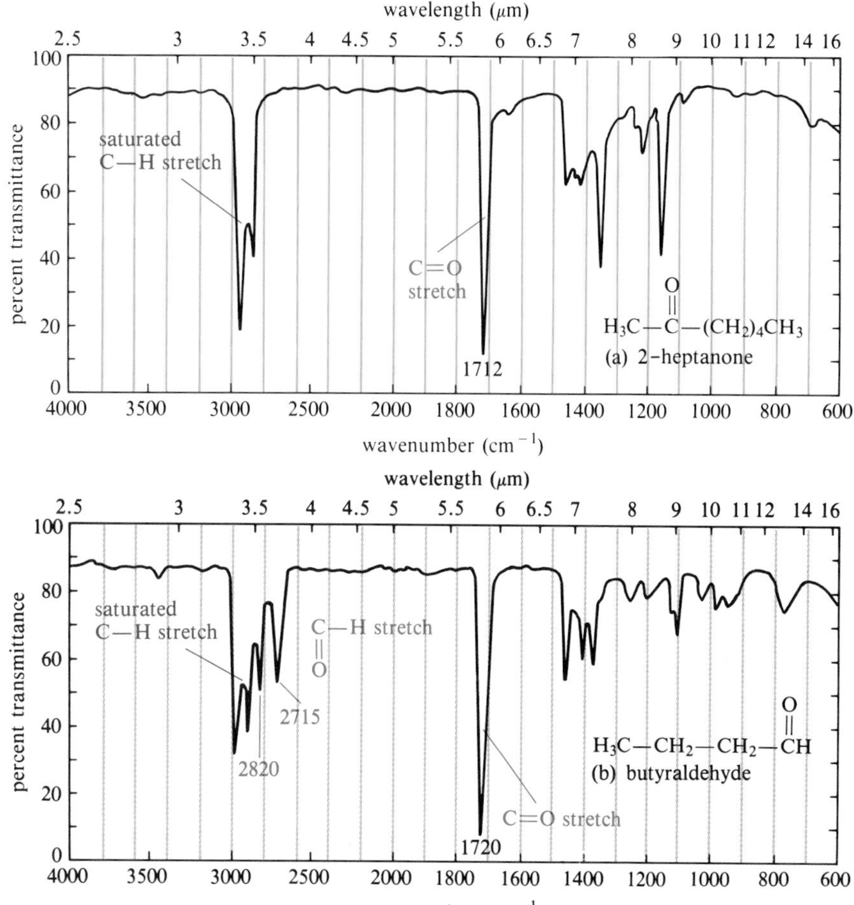

FIGURE 11-10 Infrared spectra of (a) 2-heptanone and (b) butyraldehyde. Both the ketone and the aldehyde show intense carbonyl absorptions around 1710 cm^{-1}. In the aldehyde spectrum there are two peaks (2715 and 2820 cm^{-1}) that are characteristic of the stretching vibration of the aldehyde C—H bond.

A carboxylic acid produces a characteristic O—H absorption which accompanies the intense carbonyl stretching absorption (Fig. 11-11). Because of the un-

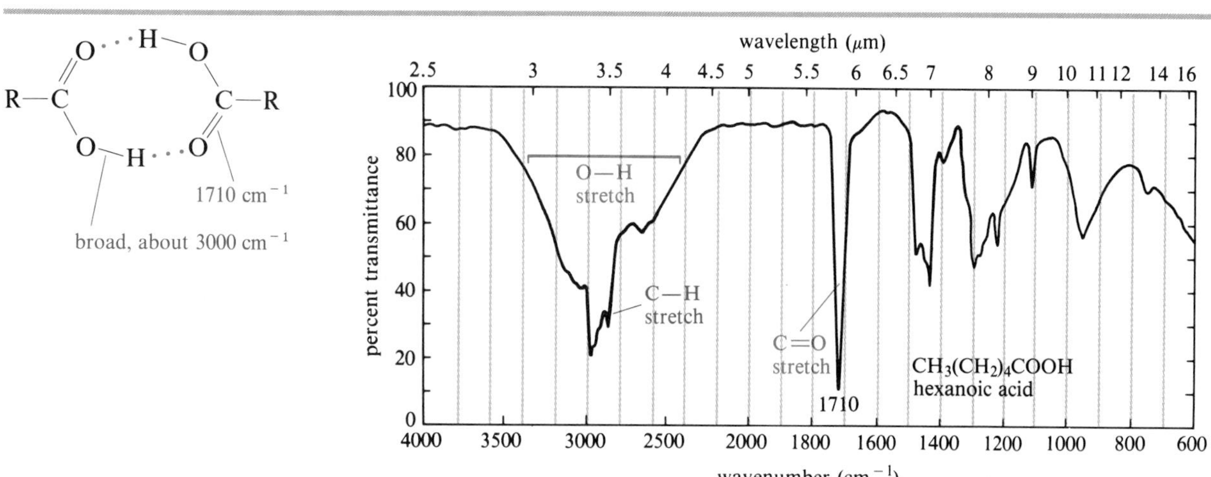

FIGURE 11-11 Infrared spectrum of hexanoic acid. Carboxylic acids show a characteristic broad O—H absorption from about 2500 to 3500 cm^{-1}. This broad absorption gives the entire C—H stretching region a broad appearance, punctuated by sharper C—H stretching absorptions.

usually strong hydrogen bonding in carboxylic acids, the broad O—H stretching frequency is shifted to about 3000 cm^{-1}, centered on top of the usual C—H absorption. This broad O—H absorption gives a characteristic overinflated shape to the peaks in the C—H stretching region. Participation of the acid carbonyl group in hydrogen bonding frequently results in broadening of its strong carbonyl absorption as well.

SOLVED PROBLEM 11-1

Determine the functional group(s) in the compound whose IR spectrum appears below.

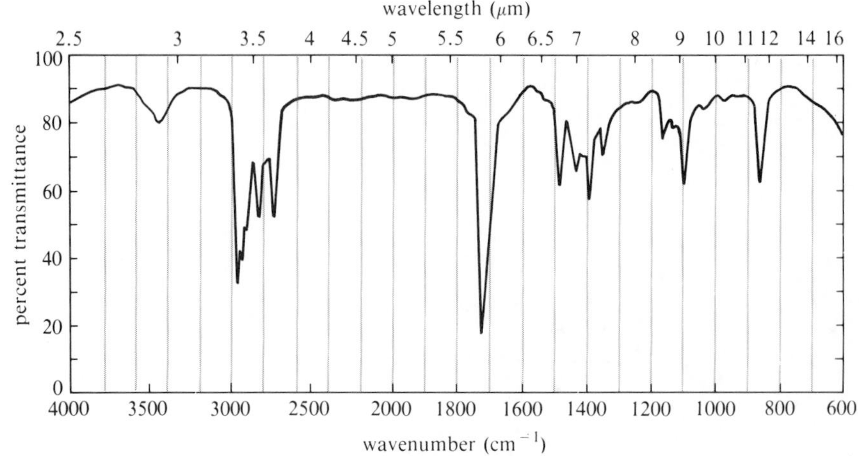

SOLUTION First, look at the spectrum and see what peaks (outside the fingerprint region) don't look like alkane peaks: a weak peak around 3400 cm^{-1}, a strong peak about 1720 cm^{-1}, and an unusual C—H stretching region. The C—H region has two additional peaks around 2720 and 2820 cm^{-1}. The strong peak at 1720 cm^{-1} must be a C=O, and the peaks at 2720 and 2820 cm^{-1} suggest an aldehyde. The broad peak around 3400 cm^{-1} looks like an alcohol O—H, but it is misleading. From experience, we know alcohols give much stronger O—H absorptions. This small peak might be from an impurity of water or from a small amount of the hydrate of the aldehyde (see Chap. 18). Many IR spectra show small, unexplained absorptions in the O—H region.

PROBLEM 11-4

Spectra are given for three compounds. Each of these compounds has one or more of the following functional groups: alcohol, amine, ketone, aldehyde, acid. Determine the functional group(s) in each compound.

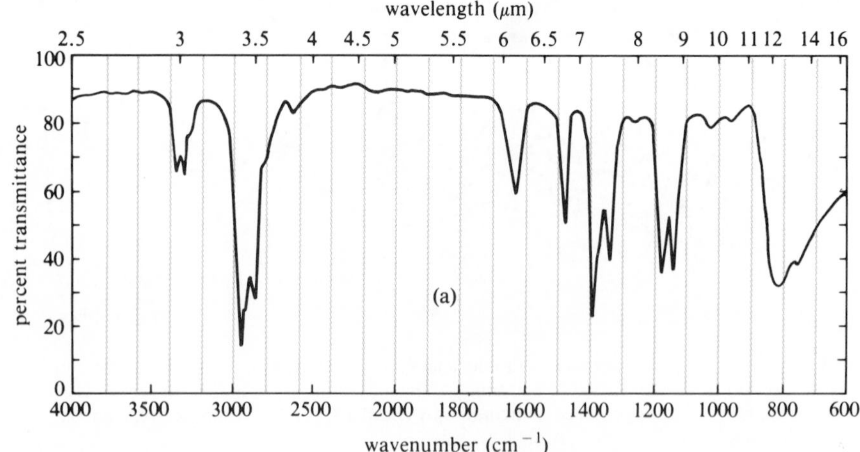

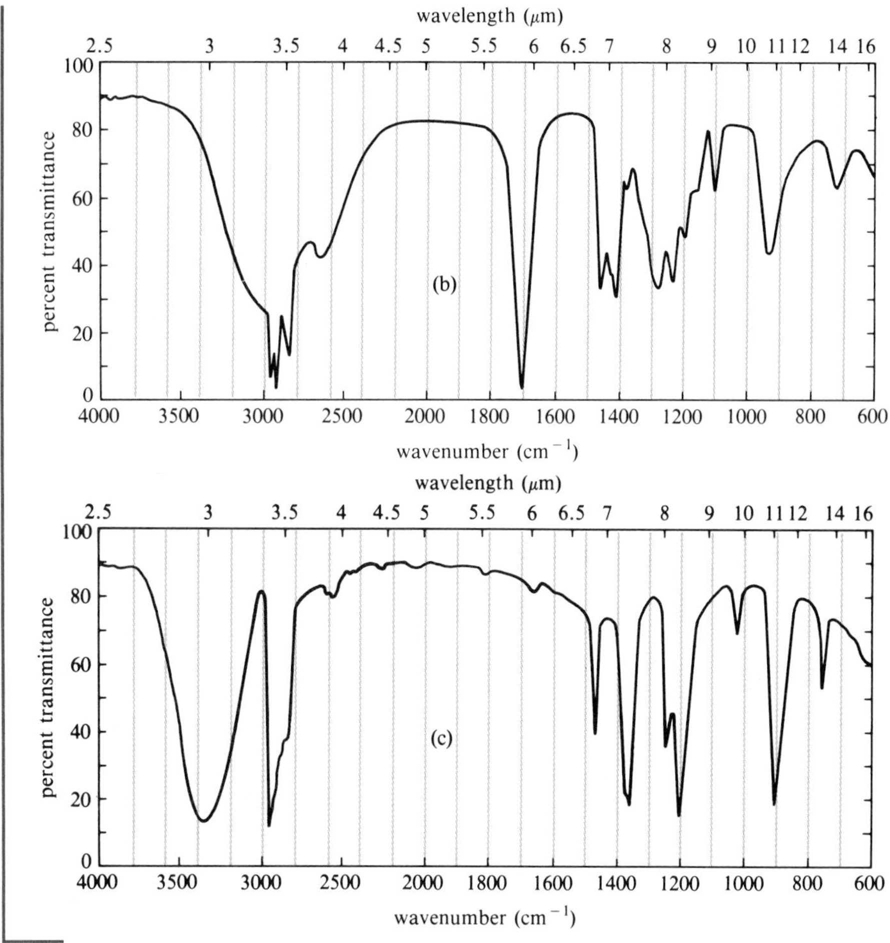

11-9B RESONANCE LOWERING OF CARBONYL FREQUENCIES

In Section 11-7A we saw that conjugation of a C=C double bond lowers its stretching frequency. This is also true of conjugated carbonyl groups, as shown below. The partial pi-bonding character of the bond between the conjugated double bonds reduces the electron density of the carbonyl pi bond, weakening it and lowering the stretching frequency to about 1680 cm^{-1} for conjugated ketones, aldehydes, and acids.

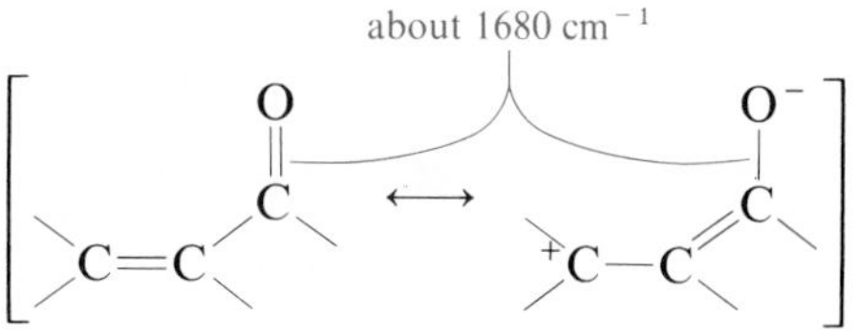

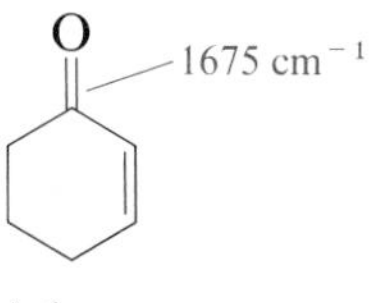

about 1680 cm^{-1}

[C=C–C=O ⟷ $^{+}$C–C=C–O^{-}]

1675 cm^{-1} — 2-cyclohexenone

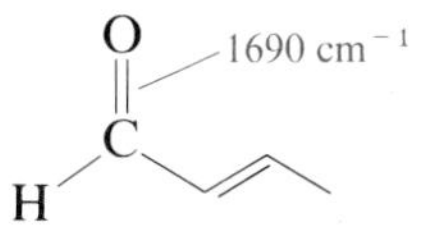

1690 cm^{-1} — 2-butenal

1685 cm^{-1} — benzoic acid

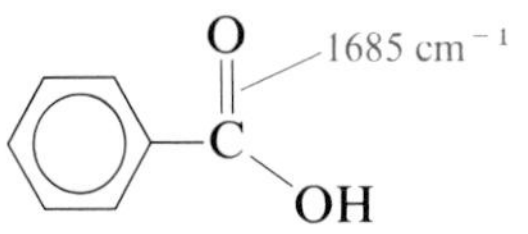

The C=C absorption of a conjugated carbonyl compound may not be apparent in the IR spectrum, because it is so much weaker than the C=O absorption. The presence of the C=C double bond is often inferred from its effect on the C=O frequency and the presence of any unsaturated =C—H absorptions above 3000 cm^{-1}.

The carbonyl groups of amides absorb at particularly low IR frequencies: about 1640 to 1680 cm^{-1}. The dipolar resonance structure (shown below) places part of the pi bond between carbon and nitrogen, leaving less than a full C=O double bond.

about 1640–1680 cm^{-1}

$$\left[\overset{O}{\underset{}{\overset{\|}{C}}} - \ddot{N} \longleftrightarrow \overset{O^-}{\overset{|}{C}} = N^+ \right]$$

1640 cm^{-1}

$$CH_3CH_2CH_2-\overset{O}{\overset{\|}{C}}-NH_2$$

The very low frequency of the amide carbonyl stretch might be mistaken for an alkene C=C stretch. For example, consider the spectra of butyramide (C=O at 1640 cm^{-1}) and 1-methylcyclopentene (C=C at 1650 cm^{-1}) in Figure 11-12. There are three striking differences between these spectra: (1) the amide carbonyl absorption is much stronger than the absorption of the alkene double bond; (2) there are prominent N—H stretching absorptions in the amide spectrum; and (3) there is an unsaturated C—H stretching absorption (just to the left of 3000 cm^{-1}) in the alkene spectrum. These examples show that we can distinguish between

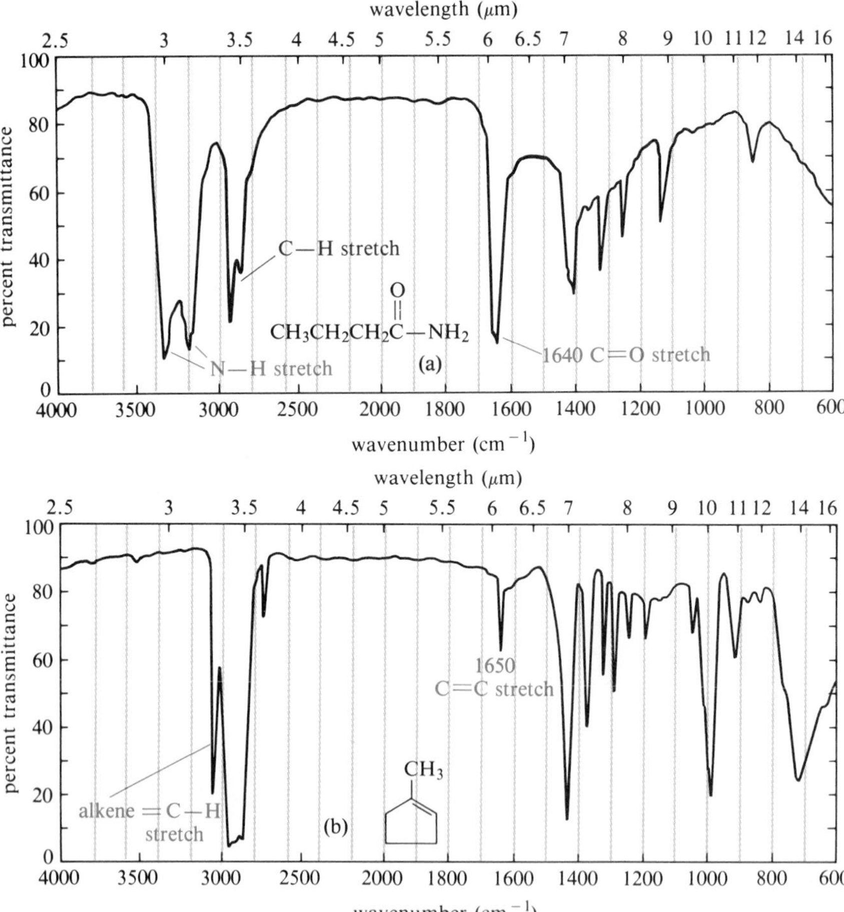

FIGURE 11-12 Although the carbonyl group of butyramide (a) and the olefinic double bond of 1-methylcyclopentene (b) absorb in the same region of the spectrum, there are three clues to distinguish the alkene from the amide: (1) the C=O absorption is much stronger than the C=C; (2) there are N—H absorptions (3300 cm^{-1}) in the amide; and (3) there is an unsaturated =C—H absorption in the alkene.

C=O and C=C absorptions, even when they appear in the same region of the spectrum.

PROBLEM 11-5

Most primary amides ($R-\overset{O}{\overset{\|}{C}}-NH_2$) show *two* absorptions in the N—H stretching region (about 3300 cm^{-1}), as in the butyramide spectrum (Fig. 11-12). Consider the resonance structures of the amide and suggest a reason for this double peak.

11-9C CARBONYL ABSORPTIONS ABOVE 1710 cm^{-1}

There are also compounds whose carbonyl groups absorb at frequencies *higher* than 1710 cm^{-1}. For example, simple carboxylic esters absorb around 1735 cm^{-1}. These higher-frequency absorptions are also seen in strained cyclic ketones (in a five-membered ring or smaller). In a small ring, the angle strain on the carbonyl group forces more pi-electron density into the C=O double bond, resulting in a stronger, stiffer bond.

$R-\overset{O}{\overset{\|}{C}}-O-R'$ (C=O about 1735 cm^{-1})
a carboxylic ester

$CH_3(CH_2)_6\overset{O}{\overset{\|}{C}}-OCH_2CH_3$ (C=O 1738 cm^{-1})
ethyl octanoate

cyclobutanone: C=O (1785 cm^{-1}) with ring C, CH_2, CH_2, CH_2
cyclobutanone

11-10 CHARACTERISTIC ABSORPTIONS OF CARBON-NITROGEN BONDS

The absorptions of carbon-nitrogen bonds are similar to those of carbon-carbon bonds, except that carbon-nitrogen bonds are more polar and give stronger absorptions. Carbon-nitrogen single bonds absorb around 1200 cm^{-1}, in a region close to many C—C and C—O absorptions. Therefore, the C—N single bond stretch is rarely useful for structure determination.

Carbon-nitrogen double bonds resemble C=C double bonds, absorbing around 1660 cm^{-1}; however, the C=N bond gives rise to a much stronger absorption due to its greater dipole moment. The C=N absorptions often resemble carbonyl absorptions in their intensity, except at a lower frequency than most carbonyl absorptions.

The most readily recognized carbon-nitrogen bond is the triple bond of a nitrile (Fig. 11-13). The stretching frequency of the nitrile C≡N bond is close to that of an acetylenic C≡C triple bond, about 2200 cm^{-1}. Nitriles and alkynes can usually be distinguished by two differences:

1. Nitrile triple bonds are more polar than C≡C triple bonds. Therefore, nitriles usually produce stronger absorptions than alkynes in the 2200 cm^{-1} region.
2. Nitriles usually absorb at frequencies that are slightly *higher* than 2200 cm^{-1} (2200 to 2300 cm^{-1}), while alkynes usually absorb slightly *lower* than 2200 cm^{-1} (2100 to 2200 cm^{-1}).

C—N bond stretching frequencies

	C—N	1200 cm^{-1}	usually strong
	C=N	1660 cm^{-1}	usually strong
	C≡N	>2200 cm^{-1}	usually strong
for comparison:	C≡C	<2200 cm^{-1}	(usually moderate or weak)

wavelength (μm)

percent transmittance

C≡N stretch 2225

$CH_3CH_2CH_2{-}C{\equiv}N$

wavenumber (cm^{-1})

FIGURE 11-13 Nitrile triple bonds have stretching frequencies that are slightly higher (and usually more intense) than those of alkyne triple bonds. Compare this spectrum of butyronitrile with that of 1-octyne in Figure 11-7.

PROBLEM 11-6

The infrared spectra for three compounds are provided. Each of these compounds has one or more of the following functional groups: conjugated ketone, ester, amide, nitrile, and C≡C triple bond. Determine the functional group(s) in each compound.

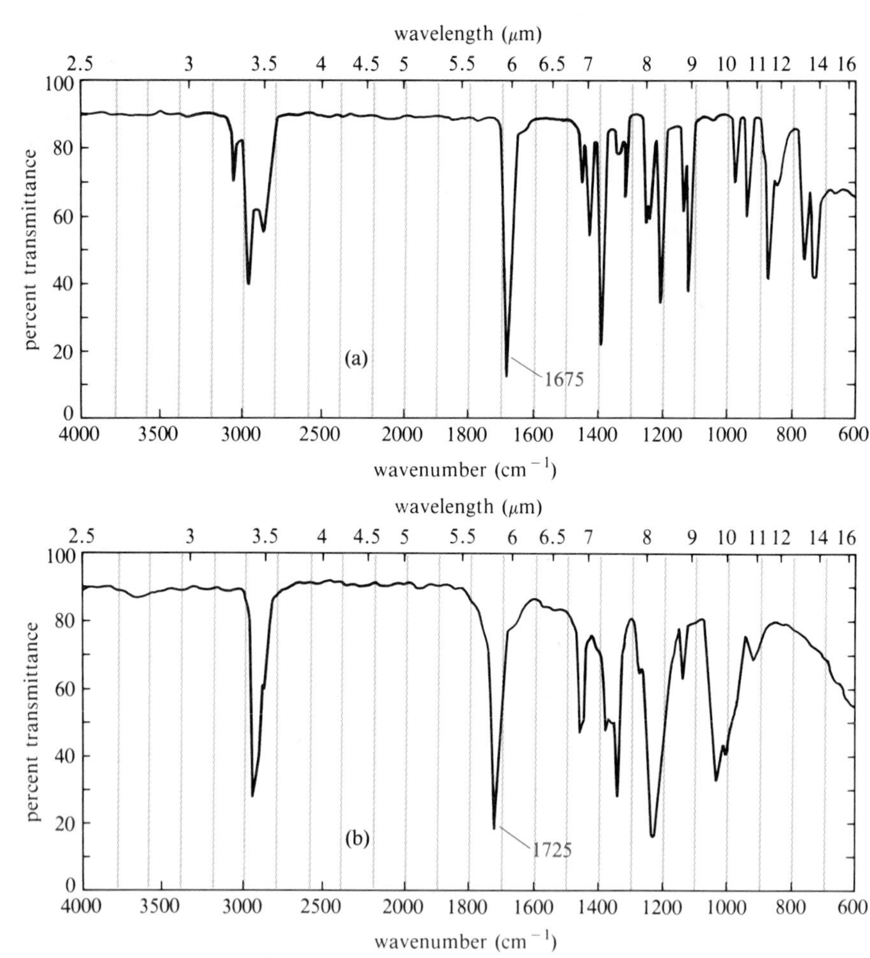

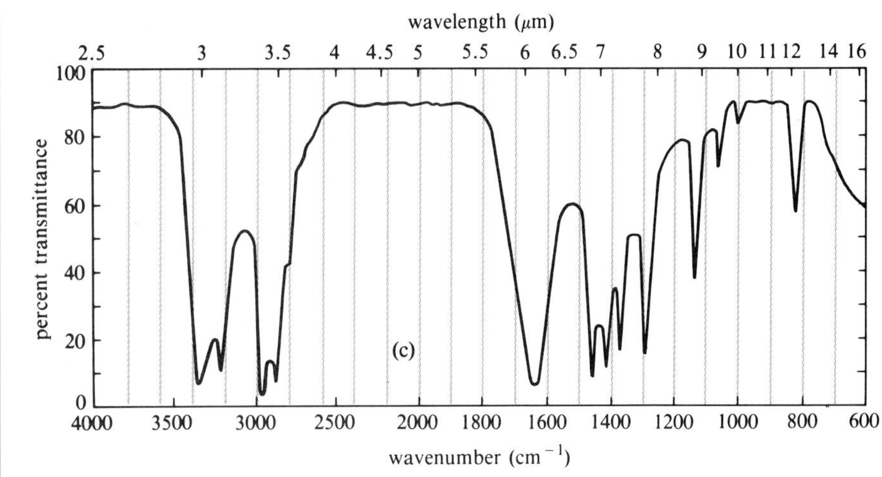

11-11 SIMPLIFIED SUMMARY OF IR STRETCHING FREQUENCIES

To a student learning about infrared spectroscopy it may seem there are too many numbers to memorize. There are hundreds of characteristic absorptions for different kinds of compounds, and a detailed reference table of characteristic frequencies is given in Appendix 2. For everyday use, however, we can memorize only a few stretching frequencies, those shown in Table 11-2. In using Table 11-2, keep in mind that the numbers are approximate and that they do not give ranges to cover all the unusual cases. Also, remember how the frequencies change as a result of conjugation, ring strain, and other factors.

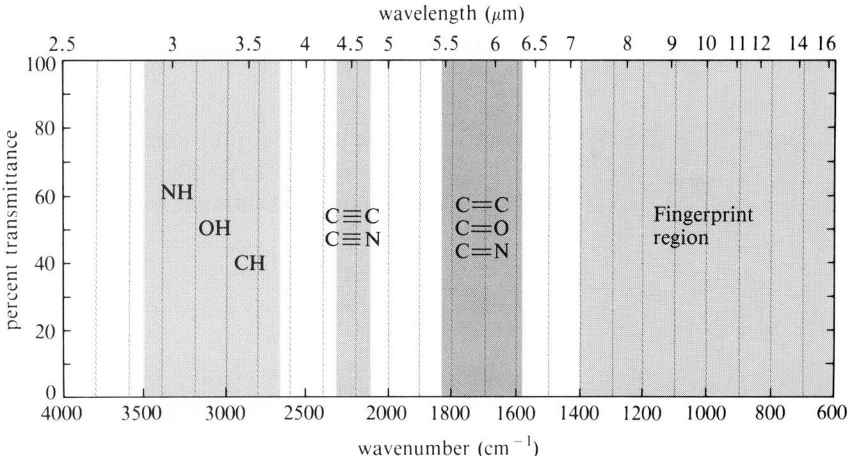

Strengths and limitations of infrared spectroscopy The most useful aspect of infrared spectroscopy is its ability to identify the functional groups in a compound. Yet, IR does not provide much information about the carbon skeleton or the alkyl groups in the compound. These aspects of the structure are more easily determined by NMR, as we will see in Chapter 12. Even an expert spectroscopist can rarely determine a structure based only on the IR spectrum.

Ambiguities arise often in the interpretation of IR spectra. For example, a strong absorption at 1680 cm^{-1} might arise from an amide, an isolated double bond, a conjugated ketone, a conjugated aldehyde, or a conjugated carboxylic acid. Familiarity with other regions of the spectrum usually enables us to determine which of these functional groups is present. In some cases, we cannot be entirely

TABLE 11-2
Simplified summary of IR stretching frequencies[a]

Frequency (cm^{-1})	*Functional group*	*Comments*
3300	alcohol O—H amine, amide N—H alkyne≡C—H	always broad may be broad, sharp, or broad with spikes always sharp
3000	alkane—C—H (with single bonds above and below the C)	just below 3000 cm^{-1}
	alkene=C—H	just above 3000 cm^{-1}
	acid O—H	very broad
2200	alkyne—C≡C— nitrile—C≡N	just below 2200 cm^{-1} just above 2200 cm^{-1}
1710	>C=O (very strong)	ketones, aldehydes, acids esters higher, about 1735 cm^{-1} conjugation lowers frequency amides lower, about 1650 cm^{-1}
1660	>C=C<	conjugation lowers frequency aromatic C=C about 1600 cm^{-1}
	>C=N—	stronger than C=C
	amide >C=O	stronger than C=C (see above)

[a] Ethers, esters, and alcohols show C—O stretching between 1000 and 1200 cm^{-1}.

certain of the functional group without additional information, usually provided by other types of spectroscopy.

Infrared spectroscopy *can* provide conclusive proof that two compounds are either the same or different. The peaks in the fingerprint region depend on complex vibrations involving the entire molecule, and it is impossible for any two compounds (except enantiomers) to have precisely the same infrared spectrum.

To summarize, an infrared spectrum is valuable in three ways:

1. It provides an indication of the functional groups in the compound.
2. It shows the *absence* of other functional groups that would give strong absorptions if they were present.
3. It can confirm the identity of a compound by comparison of its spectrum with that of a known sample.

SOLVED PROBLEM 11-2

You have an unknown with an absorption at 1680 cm^{-1}; it might be an amide, an isolated double bond, a conjugated ketone, a conjugated aldehyde, or a conjugated carboxylic acid. Describe what spectral characteristics you would look for to help you determine which of these possible functional groups might be causing the 1680 peak.

SOLUTION

Amide: (1680 peak is strong.) Look for N—H absorptions (with spikes) around 3300 cm^{-1}.

Isolated double bond: (1680 peak is weak or moderate.) Look for =C—H absorptions just above 3000 cm^{-1}.

Conjugated ketone: (1680 peak is strong.) There must be a double bond nearby, conjugated with the C=O, to lower the C=O frequency to 1680 cm^{-1}. Look for the C=C of the nearby double bond (moderate, 1620 to 1640 cm^{-1}) and its =C—H above 3000 cm^{-1}.

Conjugated aldehyde: (1680 peak is strong.) Look for the aldehyde C—H stretch about 2700 and 2800 cm^{-1}. Also look for the C=C and =C—H of the nearby double bond (1620 to 1640 cm^{-1} and just above 3000 cm^{-1}).

Conjugated carboxylic acid: (1680 peak is strong.) Look for the characteristic acid O—H stretch centered *on top of* the C—H stretch around 3000 cm^{-1}. Also look for the C=C and =C—H of the nearby double bond (1620 to 1640 cm^{-1} and just above 3000 cm^{-1}).

11-12 READING AND INTERPRETING INFRARED SPECTRA (SOLVED PROBLEMS)

Many students are unsure how much information they should be able to obtain from an infrared spectrum. In Chapter 12, we will use IR together with NMR and other information to determine the entire structure. For the present, concentrate on getting as much information as you can from the IR spectrum by itself. Several solved spectra are included below to show what information can be inferred. An experienced spectroscopist could obtain more information from these spectra, but we will concentrate on the major, most reliable, features.

Study this section by looking at each spectrum and writing down the important frequencies and your proposed functional groups. Then look at the solution and compare it with your solution. The actual structures of these compounds are shown at the end of this section. They are not given with the solutions because *you cannot determine these structures using only the infrared spectrum*, so the structure is not a part of a realistic solution.

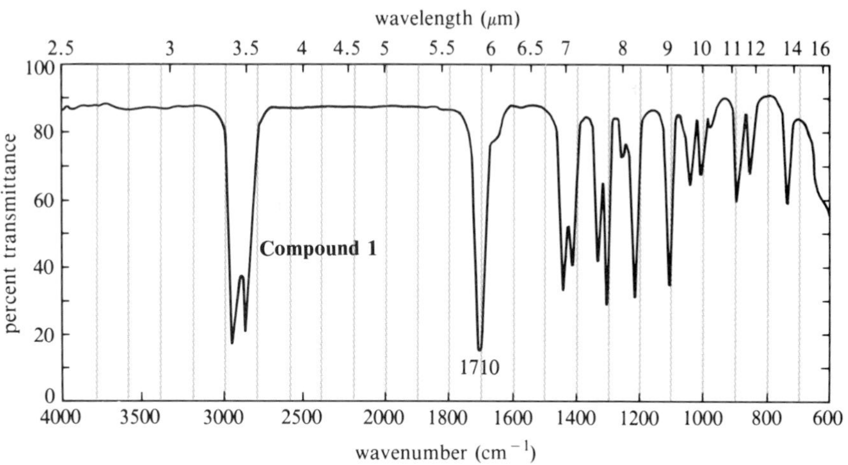

Compound 1. This spectrum is most useful for what it does *not* show. There is a carbonyl absorption at 1710 cm^{-1} and very little else. There is no aldehyde proton, no hydroxyl proton, and no N—H proton. The carbonyl absorption could indicate an aldehyde, ketone, or acid, except that the lack of aldehyde C—H stretch eliminates an aldehyde, and the lack of O—H stretch eliminates an acid. There is no visible C=C stretch and no unsaturated C—H absorption above 3000 cm^{-1}, so the compound appears to be otherwise saturated. The compound is probably a simple aliphatic ketone.

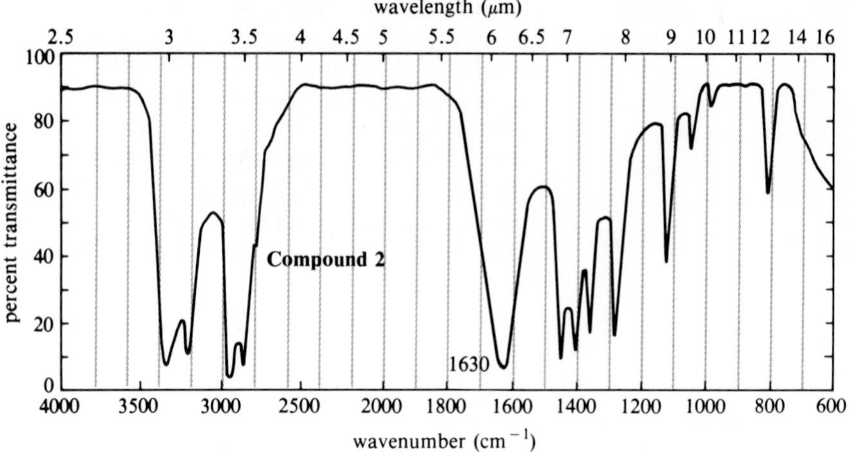

Compound 2. The carbonyl absorption at 1630 cm^{-1} suggests either an amide or a conjugated C═C double bond. This peak is so intense that it probably indicates a carbonyl group. The doublet (a pair of peaks) of N—H absorption around 3300 cm^{-1} also suggests a primary amide, R—$CONH_2$. Since there is no C—H absorption above 3000 cm^{-1}, this is probably a saturated amide.

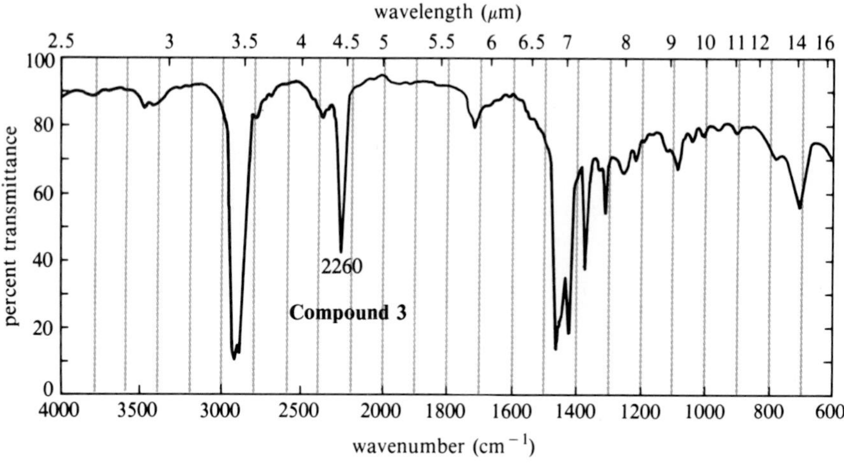

Compound 3. The sharp peak at 2260 cm^{-1} results from a nitrile C≡N stretch. (An alkyne C≡C absorption would be weaker, and below 2200 cm^{-1}.) The absence of C═C stretch or C—H stretch above 3000 cm^{-1} suggests that the nitrile is otherwise saturated.

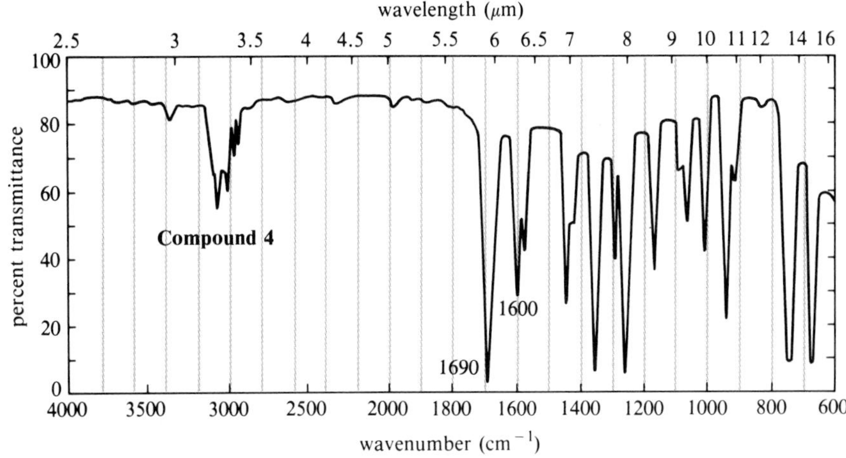

Compound 4. The carbonyl absorption at 1690 cm^{-1} is about right for a conjugated ketone, aldehyde, or acid. (An amide would be lower in frequency, and a C=C double bond would not be so strong.) The absence of any N—H stretch, O—H stretch, or aldehyde C—H stretch leaves a conjugated ketone as the best possibility. The C=C stretch at 1600 cm^{-1} indicates an aromatic ring. We presume that the aromatic ring is conjugated with the carbonyl group of the ketone.

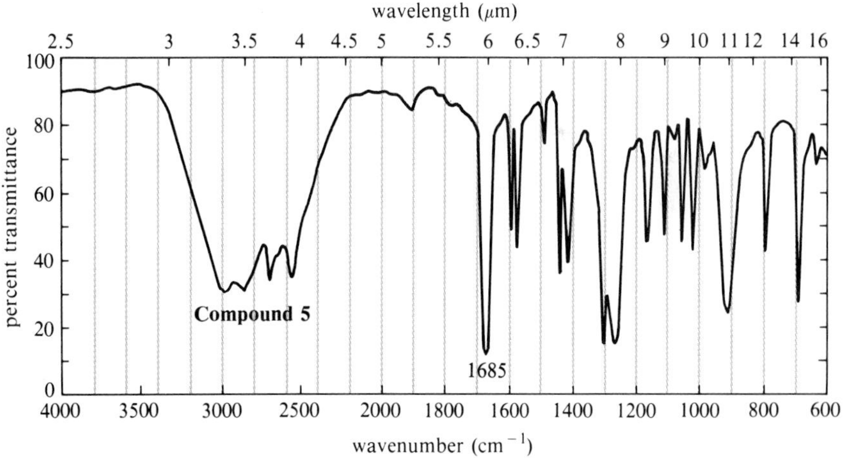

Compound 5. The broad O—H stretch that obliterates most of the C—H stretching region suggests a carboxylic acid. The C=O stretch is low for an acid (1685 cm^{-1}), implying a conjugated acid. The aromatic C=C absorption at 1600 cm^{-1} suggests that the acid may be conjugated with an aromatic ring.

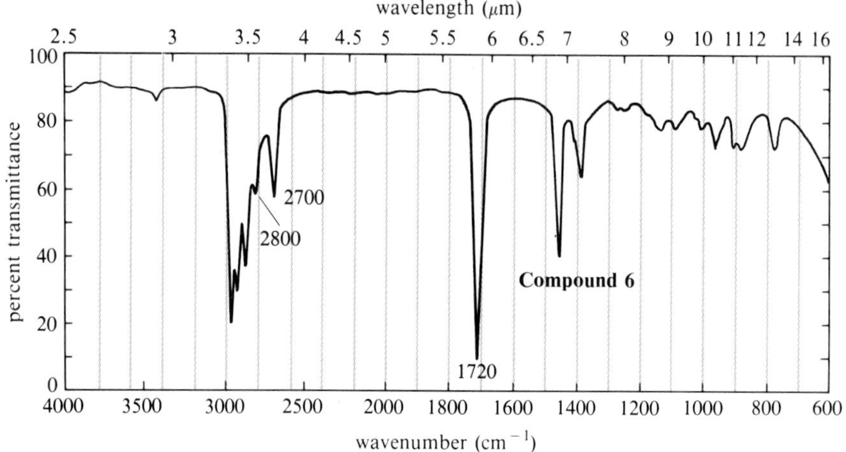

Compound 6. The carbonyl absorption at 1720 cm^{-1} suggests a ketone, aldehyde, or acid. The evidence of aldehyde C—H stretching at about 2700 and 2800 cm^{-1} confirms an aldehyde. Since all of the C—H stretch is below 3000 cm^{-1}, and since there is no visible C=C stretch, the aldehyde is probably saturated.

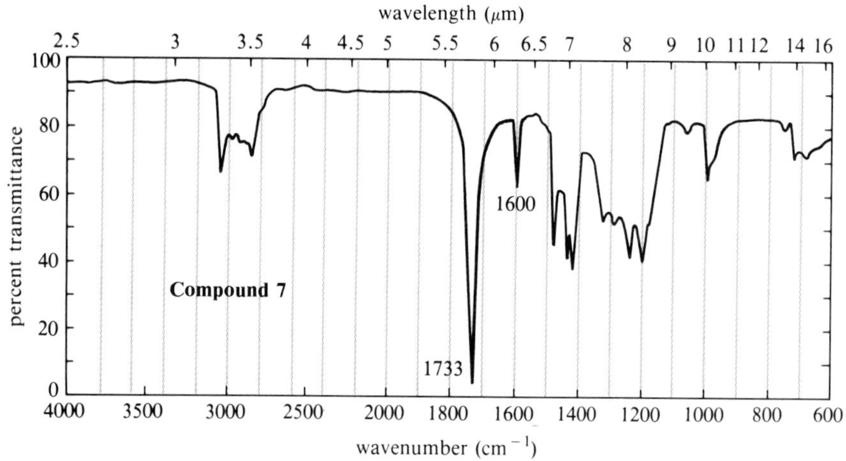

Compound 7. The carbonyl absorption at 1733 cm^{-1} suggests an ester. The weak peak at 1600 cm^{-1} indicates an aromatic ring, but it cannot be conjugated with the ester because the ester absorption is close to its usual (unconjugated) position. The presence of both saturated (below 3000 cm^{-1}) and unsaturated (above 3000 cm^{-1}) C—H stretching in the 3000 cm^{-1} region confirms the presence of both alkyl and unsaturated portions of the molecule.

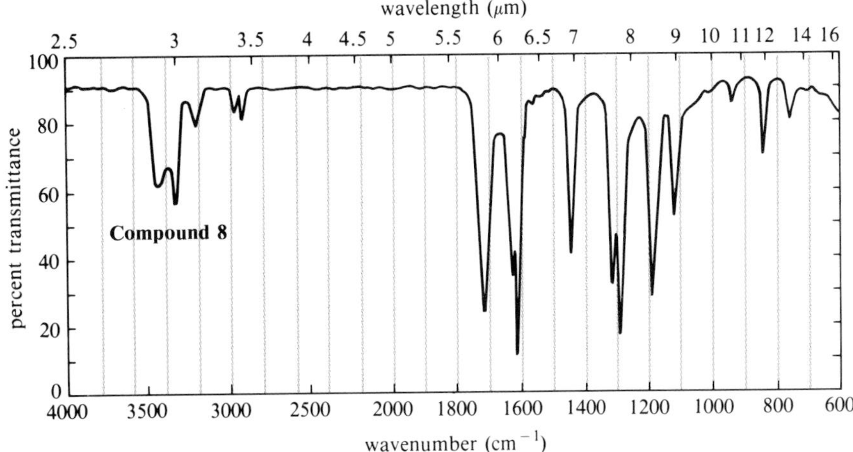

Compound 8 (a difficult problem). The absorption around 3400 to 3500 cm^{-1} has two relatively sharp peaks: probably N—H rather than O—H. Combined with a carbonyl group (strong at 1705 cm^{-1}), we would suspect an amide, except amide frequencies are not usually that high. The carbonyl absorption must be from a separate functional group: perhaps an aldehyde, ketone, or acid? We can rule out an aldehyde (no aldehyde C—H at 2700 and 2800 cm^{-1}) and an acid (no acid O—H centered at 3000 cm^{-1}), but we cannot rule out a ketone.

There is a C═C absorption (probably an aromatic ring) at 1605 cm^{-1}. If this ring were conjugated with the carbonyl group, it would have *lowered* the C═O stretch to 1705 cm^{-1}, probably from about 1725 or 1735 cm^{-1}, the correct range for an ester. This compound contains an amine, an aromatic ring, and some kind of carbonyl group. If the carbonyl is conjugated with the aromatic ring, it is probably an ester. If it is not conjugated, it is probably a ketone.

(An experienced spectroscopist might guess that the aromatic C═C is unusually strong because it is polarized by conjugation with a carbonyl group.)

Structures of the compounds

(These structures cannot be determined from their IR spectra alone.)

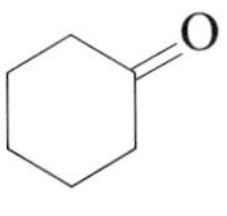

(cyclohexanone)
compound 1

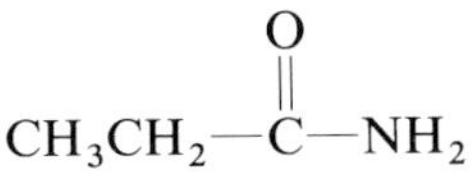

$CH_3CH_2—C(═O)—NH_2$
compound 2

$CH_3(CH_2)_4—C≡N$
compound 3

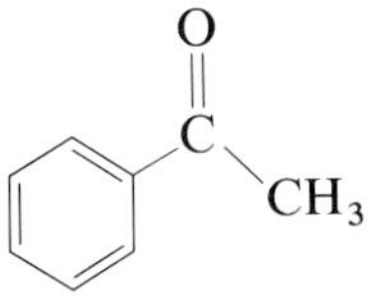

$C_6H_5—C(═O)—CH_3$
compound 4

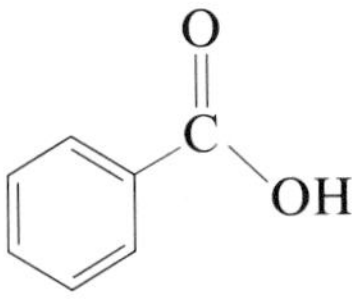

$C_6H_5—C(═O)—OH$
compound 5

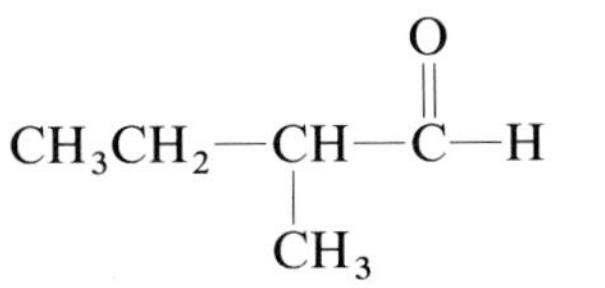

$CH_3CH_2—CH(CH_3)—C(═O)—H$
compound 6

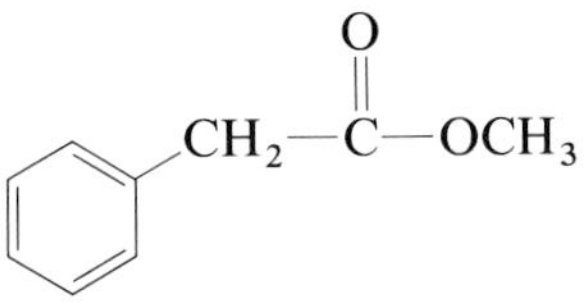

$C_6H_5—CH_2—C(═O)—OCH_3$
compound 7

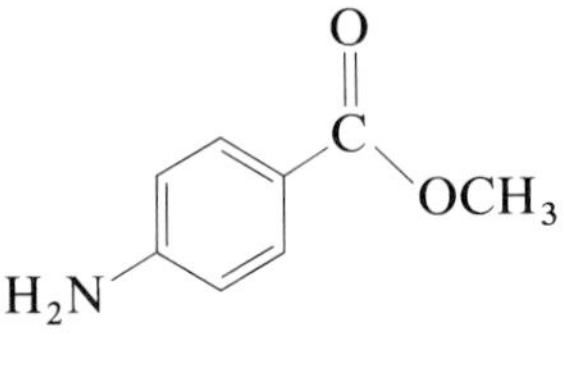

$H_2N—C_6H_4—C(═O)—OCH_3$
compound 8

PROBLEM 11-7

For each of the following infrared spectra, interpret all of the significant stretching frequencies above 1580 cm^{-1}.

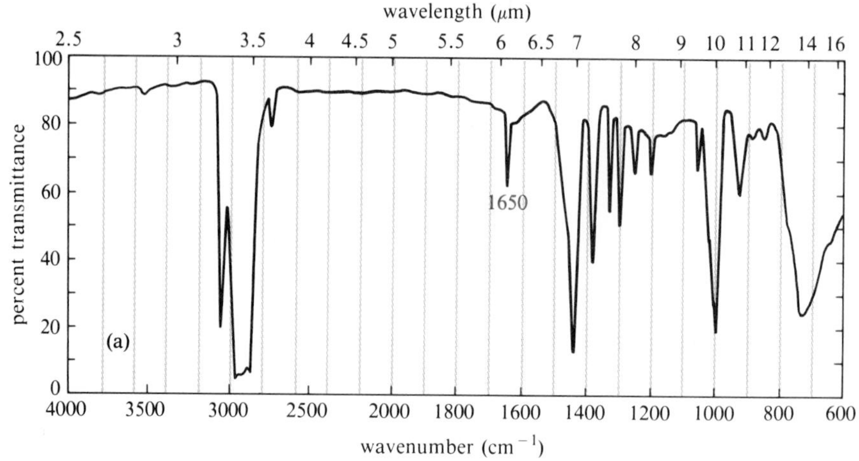

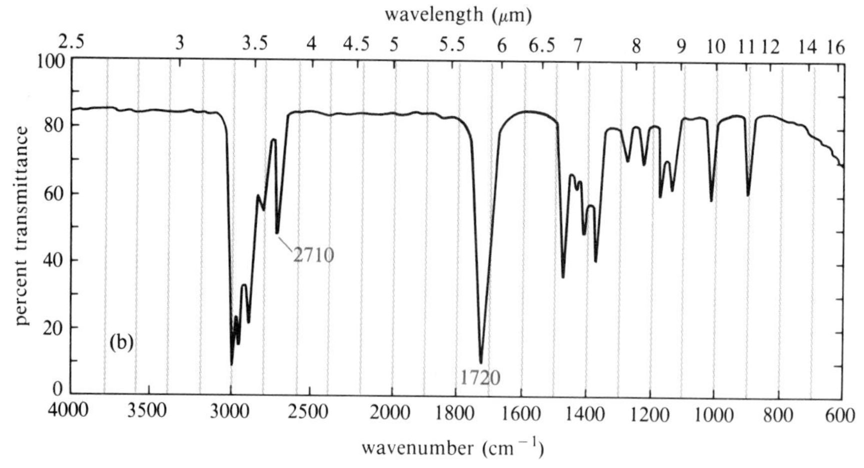

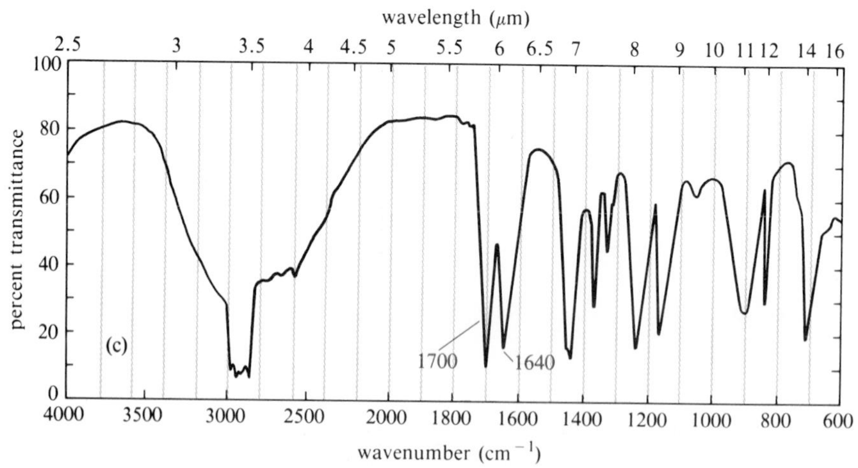

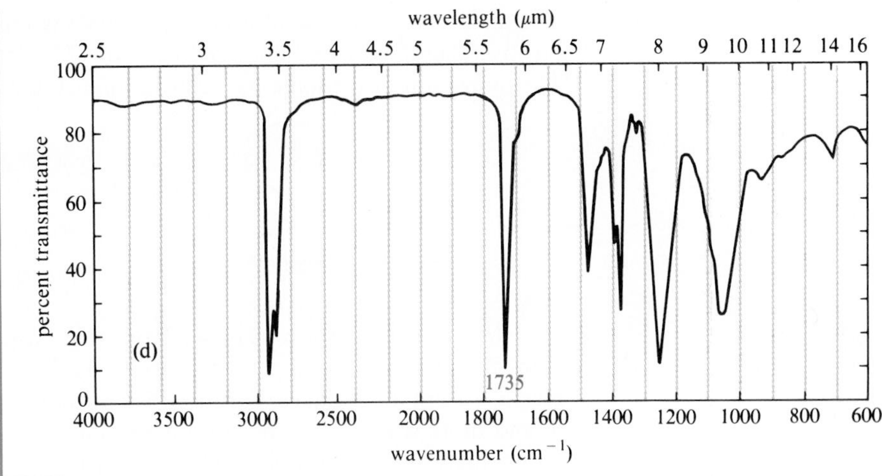

11-13 INTRODUCTION TO MASS SPECTROMETRY

Infrared spectroscopy gives information about the functional groups in a molecule, but it tells us little about the size of the molecule or what heteroatoms are present. To determine a structure, we need a molecular weight and, if possible, a molecular formula. Molecular formulas were once obtained by careful analysis of the elemental composition, together with a molecular weight determined by freezing point depression or some other technique. This is a long and tedious process, and it requires a large amount of pure material. Many important compounds are available only in tiny quantities, and they may be impure.

Mass spectrometry (MS) provides the molecular weight and valuable information about the molecular formula, using a very small amount of sample. High-resolution mass spectrometry (HRMS) can provide an accurate molecular formula. The mass spectrum also provides structural information that can be used to confirm a structure that was arrived at from NMR and IR spectroscopy.

Mass *spectrometry* is fundamentally different from *spectroscopy*. Spectroscopy involves the absorption (or emission) of light over a range of wavelengths. Mass spectrometry does not use light at all. In the mass spectrometer, a sample is struck by high-energy electrons, breaking the molecules apart. The masses of the fragments are measured, and this information is used to reconstruct the molecule. The process is similar to analyzing a vase by shooting it with a rifle, then weighing all the pieces.

11-13A THE MASS SPECTROMETER

A **mass spectrometer** ionizes molecules in a high vacuum, sorts the ions according to their masses, and records the abundance of ions of each mass. A **mass spectrum** is the graph plotted by the mass spectrometer, with the masses plotted as the x axis and the relative number of ions of each mass on the y axis. Several different methods are used for ionizing samples, and there are also several methods for separating the ions according to their masses. We will discuss only the most common techniques, *electron impact ionization* for forming the ions, and *magnetic deflection* for separating the ions.

Electron impact ionization When an electron strikes a neutral molecule, it may ionize that molecule by knocking out an additional electron.

$$e^- + M \longrightarrow [M]^{+\cdot} + 2e^-$$

The most common type of ionized molecule is missing one electron, so it has a positive charge and one unpaired electron. The ion is therefore a **radical cation.** The ionization of methane by electron impact, for example, is shown below.

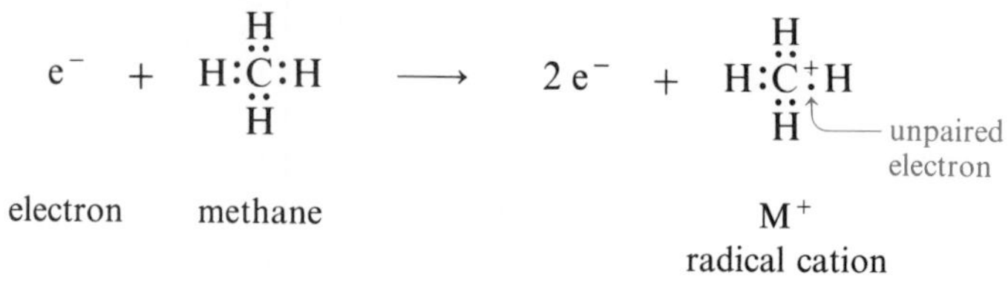

$$\underset{\text{electron}}{e^-} + \underset{\text{methane}}{\mathrm{H{:}\overset{\displaystyle H}{\underset{\displaystyle H}{\ddot{\underset{\cdot\cdot}{C}}}}{:}H}} \longrightarrow 2\,e^- + \underset{\text{radical cation}}{\underset{M^{+}}{\mathrm{H{:}\overset{\displaystyle H}{\underset{\displaystyle H}{\ddot{\underset{\cdot\cdot}{C}}}}{\overset{+}{\cdot}}H}}}$$

unpaired electron

Most of the carbocations we have seen have a three-bonded carbon atom with six paired electrons in its valence shell. The radical cation in the preceding reaction is not a normal carbocation. The carbon atom has seven electrons around it, and those seven electrons bond it to four other atoms. This unusual cation is represented by the formula $[CH_4]^{+\cdot}$, with the + indicating the positive charge and the · indicating the unpaired electron.

In addition to ionizing a molecule, the impact of an energetic electron may break it apart. This **fragmentation** process gives a characteristic mixture of different ions. The radical cation corresponding to the mass of the original molecule is called the **molecular ion,** abbreviated M^+. The ions of smaller molecular weights are called *fragments.* Bombardment of ethane molecules by energetic electrons, for example, produces the molecular ion and several fragments. Both charged and uncharged fragments are formed, but *only the positively charged fragments are detected by the mass spectrometer.*

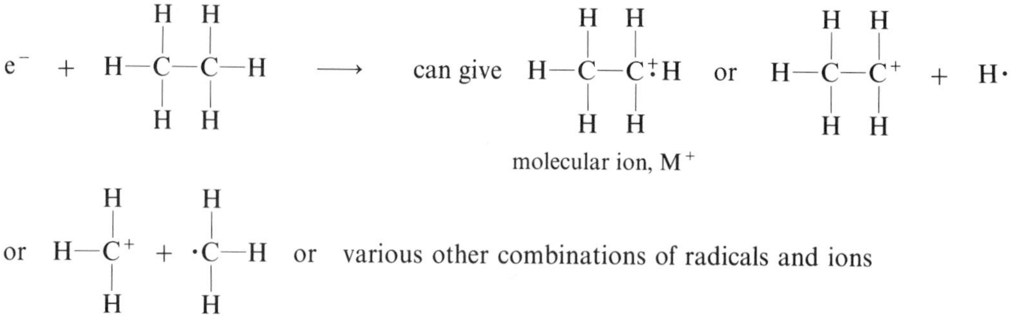

$$e^- + \mathrm{H{-}\overset{H}{\underset{H}{C}}{-}\overset{H}{\underset{H}{C}}{-}H} \longrightarrow \text{can give } \underset{\text{molecular ion, } M^+}{\mathrm{H{-}\overset{H}{\underset{H}{C}}{-}\overset{H}{\underset{H}{C}}\overset{+}{\cdot}H}} \text{ or } \mathrm{H{-}\overset{H}{\underset{H}{C}}{-}\overset{H}{\underset{H}{C}}{}^{+}} + \mathrm{H}\cdot$$

$$\text{or } \mathrm{H{-}\overset{H}{\underset{H}{C}}{}^{+}} + \cdot\mathrm{\overset{H}{\underset{H}{C}}{-}H} \text{ or various other combinations of radicals and ions}$$

We discuss the common modes of fragmentation in Section 11-15.

Separation of ions of different masses Once ionization and fragmentation have formed a mixture of ions, these ions must be separated and detected. The most common type of mass spectrometer, shown in Figure 11-14, separates ions by a technique called *magnetic deflection.*

The sample molecules are ionized by an electron beam passing through a vacuum chamber. The positively charged ions are attracted to a negatively charged plate, which has a narrow slit to allow some of the ions to pass through. The beam of ions enters an evacuated flight tube, with a curved portion positioned between the poles of a large magnet. When a charged particle passes through a magnetic field, there is a transverse force on the particle, and its path is bent. The path of a heavier ion is bent less than the path of a lighter ion.

The exact radius of curvature of an ion's path depends on its mass-to-charge ratio, symbolized by $\boldsymbol{m/z}$ (or by m/e in the older literature). In this expression, m is the mass of the ion (in amu) and z is its charge in units of the electronic charge. The vast majority of ions have a charge of $+1$, however, so we can consider their paths to be curved by an amount that depends only on their masses.

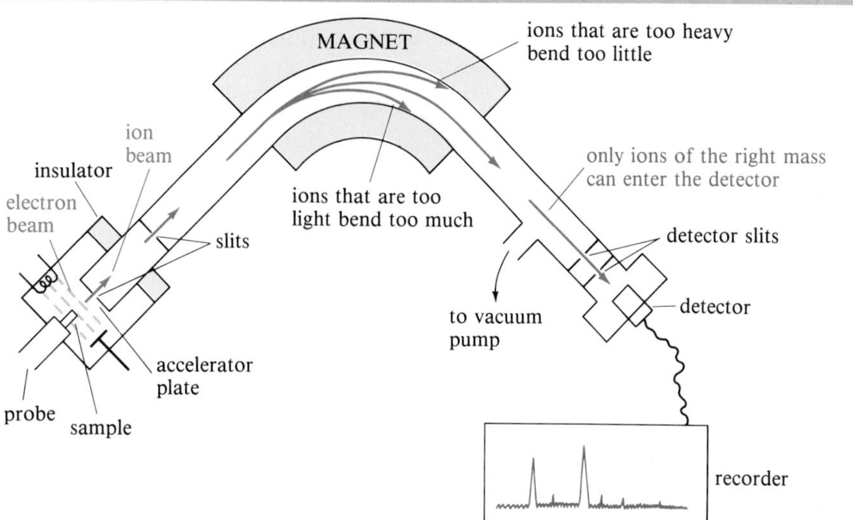

FIGURE 11-14 Diagram of a mass spectrometer. The mass spectrometer bombards the sample with a beam of electrons, causing the molecules to ionize and fragment. The mixture of ions is accelerated and passes through a magnetic field, where the paths of the lighter ions are bent more than those of the heavier ions. By varying the magnetic field, the spectrometer plots the abundance of ions of each mass.

At the end of the flight tube is another slit, followed by an ion detector connected to an amplifier. At any given magnetic field, only ions of one particular mass will be bent exactly the right amount to pass through the slit and enter the detector. The detector produces a signal that is proportional to the number of ions striking it. By varying the magnetic field, the spectrometer scans all the possible ion masses and produces a graph of the number of ions of each mass.

11-13B THE MASS SPECTRUM

The mass spectrometer usually produces the spectrum as a trace on light-sensitive paper or a graph on a computer screen. This information is tabulated, and the spectrum is printed as a bar graph or as a table of relative intensities (Fig. 11-15). In the printed mass spectrum all the masses are rounded off to the nearest whole-number mass unit. The peaks are assigned intensities as percentages of the strongest peak, called the **base peak.** Notice that *the base peak does not necessarily correspond to the mass of the molecular ion.* It is simply the strongest peak, making it easy for other peaks to be expressed as percentages.

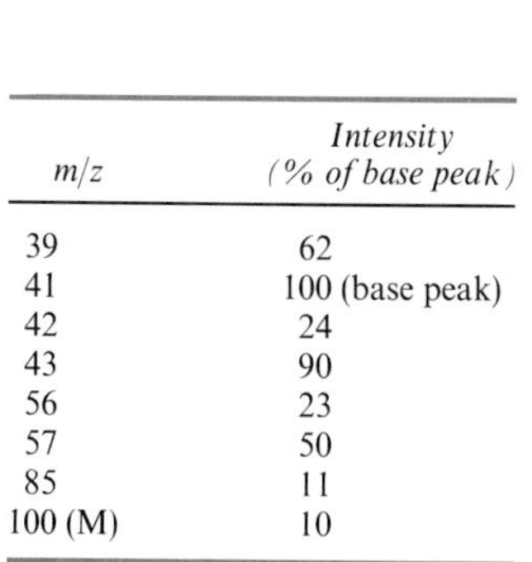

m/z	*Intensity (% of base peak)*
39	62
41	100 (base peak)
42	24
43	90
56	23
57	50
85	11
100 (M)	10

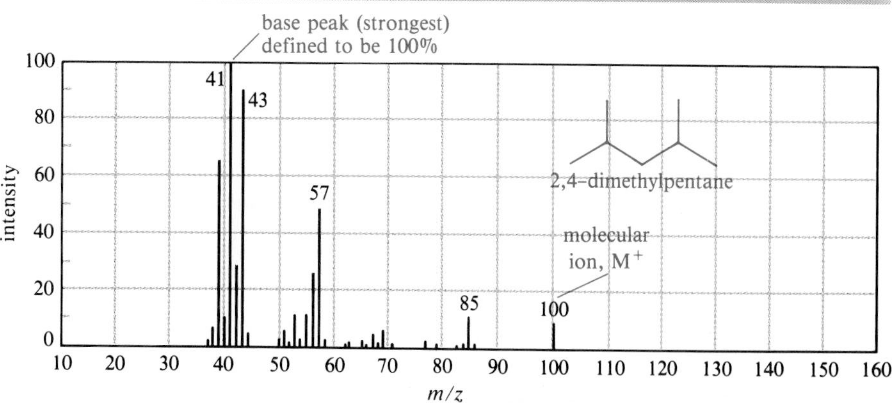

FIGURE 11-15 Mass spectrum of 2,4-dimethylpentane, given both as a bar graph and in tabular form. Notice that all the intensities are given as percentages of the strongest peak (the base peak). In this example, the base peak is at m/z 41 and the molecular ion ("parent peak") is at m/z 100.

The molecular ion peak is observed in most mass spectra, meaning that a detectable number of molecular ions (M^+) reach the detector without fragmenting. These molecular ions are usually the particles of highest mass in the mass spectrum. The value of m/z for the molecular ion immediately gives the molecular weight of the compound.

11-14 DETERMINATION OF THE MOLECULAR FORMULA BY MASS SPECTROMETRY

11-14A HIGH-RESOLUTION MASS SPECTROMETRY

Although mass spectra usually show the particle masses rounded to the nearest whole number, the masses are not really integral. The ^{12}C nucleus is *defined* to have a mass of exactly 12 atomic mass units (amu), and all other nuclei have masses based on this standard. For example, a proton has a mass of about 1, but not exactly: Its mass is 1.007825 amu. Table 11-3 shows the atomic masses for the most common isotopes found in organic compounds.

TABLE 11-3
Exact masses of common isotopes[a]

Isotope	*Atomic mass (amu)*
^{12}C	12.000000
^{1}H	1.007825
^{16}O	15.994914
^{14}N	14.003050

[a] These masses are used with a high-resolution mass spectrometer to determine molecular formulas.

Determination of a molecular formula is possible using a **high-resolution mass spectrometer** (HRMS), one that uses extra stages of electrostatic or magnetic focusing to form a very precise beam and to detect the particle masses to an accuracy of about 1 part in 20,000. A mass determined to several significant figures using a HRMS is referred to as an *exact mass*. Although it is not really exact, it is much more accurate than the usual integral mass numbers. Comparison of the exact mass with calculated masses listed in tables by molecular formulas allows identification of the correct formula.

Consider a molecular ion with a mass of 44. This approximate molecular weight might correspond to C_3H_8 (propane), C_2H_4O (acetaldehyde), CO_2, or CN_2H_4. Each of these molecular formulas corresponds to a different "exact" mass:

	C_3H_8		C_2H_4O		CO_2		CN_2H_4
3 C	36.00000	2 C	24.00000	1 C	12.00000	1 C	12.00000
8 H	8.06260	4 H	4.03130			4 H	4.03130
		1 O	15.99490	2 O	31.98983	2 N	28.00610
	44.06260		44.02620		43.98983		44.03740

If the HRMS measured the "exact" mass of this ion as 44.029 mass units, we would conclude that the compound has a molecular formula of C_2H_4O because the mass corresponding to this formula is closest to the observed value. It is rarely necessary to calculate the masses for the possible molecular formulas as we have done above. Published tables of "exact" masses are available for comparison with the value obtained from the HRMS. Depending on the completeness of the tables, they may include sulfur, halogens, or other elements.

11-14B USE OF HEAVIER ISOTOPE PEAKS

Whether or not a high-resolution mass spectrometer is available, there are characteristics of molecular ion peaks that provide information about the molecular formula. Most elements do not consist of a single isotope, but contain heavier isotopes in varying amounts. These heavier isotopes give rise to small peaks at higher mass numbers than the major M^+ molecular ion peak. A peak that is one mass unit heavier than the M^+ peak is called the **M + 1 peak;** two units heavier, the **M + 2 peak;** and so on. Table 11-4 gives the isotopic composition of some common elements, showing how they contribute to the M + 1 and M + 2 peaks.

Ideally, the isotopic compositions in Table 11-4 could be used to determine the entire molecular formula of a compound, by carefully measuring the intensity

TABLE 11-4
Isotopic composition of some common elements

Element	M^+		*M + 1*		*M + 2*	
hydrogen	^1H	100.0%				
carbon	^{12}C	98.9%	^{13}C	1.1%		
nitrogen	^{14}N	99.6%	^{15}N	0.4%		
oxygen	^{16}O	99.8%			^{18}O	0.2%
sulfur	^{32}S	95.0%	^{33}S	0.8%	^{34}S	4.2%
chlorine	^{35}Cl	75.5%			^{37}Cl	24.5%
bromine	^{79}Br	50.5%			^{81}Br	49.5%
iodine	^{127}I	100.0%				

of the M^+, M + 1, and M + 2 peaks. In practice, however, there are background peaks at every mass number. Since these background peaks are often similar in intensity to the M + 1 peak, an accurate measurement of the M + 1 peak is impossible. High-resolution mass spectrometry is much more reliable.

Some elements (particularly S, Cl, Br, I, and N) are recognizable using the MS molecular-ion peaks, however. A compound with no sulfur, chlorine, or bromine has a small M + 1 peak and an even smaller M + 2 peak. If a compound contains sulfur, the M + 2 peak is larger than the M + 1 peak: about 4 percent of the M^+ peak. If chlorine is present, the M + 2 peak is about a third as large as the M^+ peak. If bromine is present, the M^+ and M + 2 ions have about equal intensities; the molecular ion appears as a doublet separated by two mass units, with one mass corresponding to ^{79}Br and one to ^{81}Br.

Iodine is recognized by the presence of the iodonium ion, I^+, at m/z 127 in the spectrum. This clue is combined with a characteristic 127-unit gap in the spectrum corresponding to loss of the iodine radical. Nitrogen (or an odd number of nitrogen atoms) is suggested by an odd molecular weight. Stable compounds containing only carbon, hydrogen, and oxygen have even molecular weights.

Recognizable elements in the MS

Br	M + 2 as large as M^+
Cl	M + 2 a third as large as M^+
I	I^+ at 127; large gap
N	odd M^+
S	M + 2 larger than usual (4% of M^+)

The following spectra show compounds containing sulfur, chlorine, and bromine.

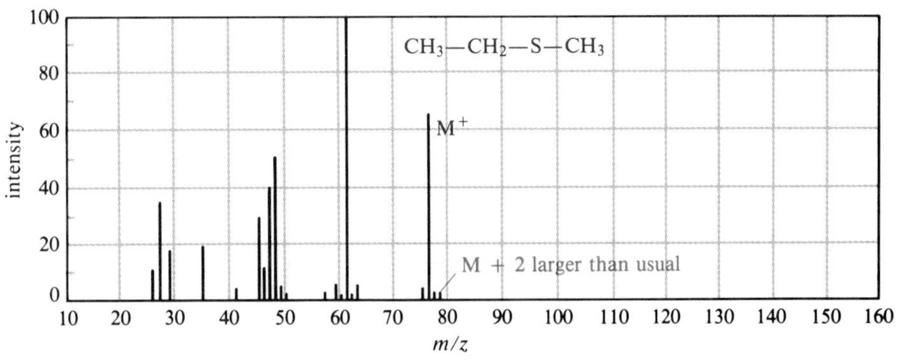

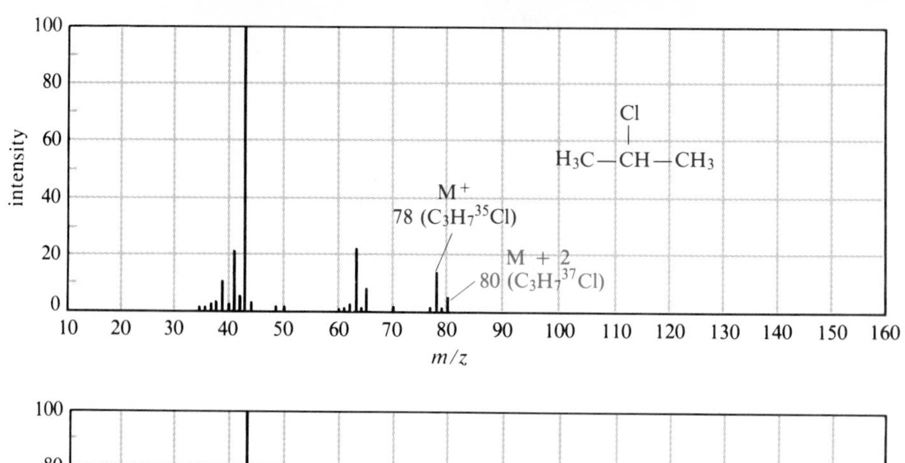

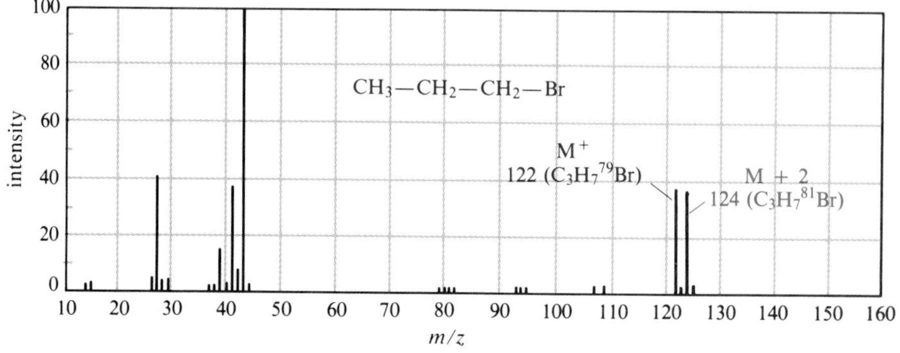

PROBLEM 11-8

Point out which of these four mass spectra indicate the presence of sulfur, chlorine, bromine, iodine, or nitrogen. Suggest a molecular formula for each.

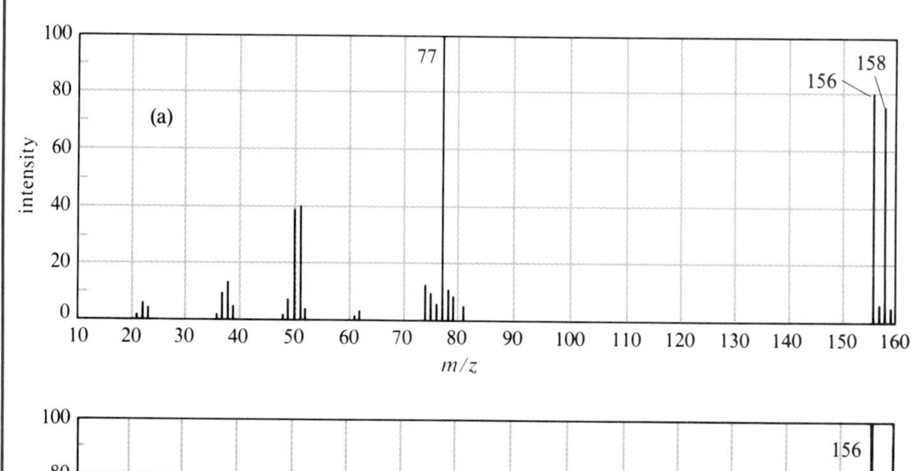

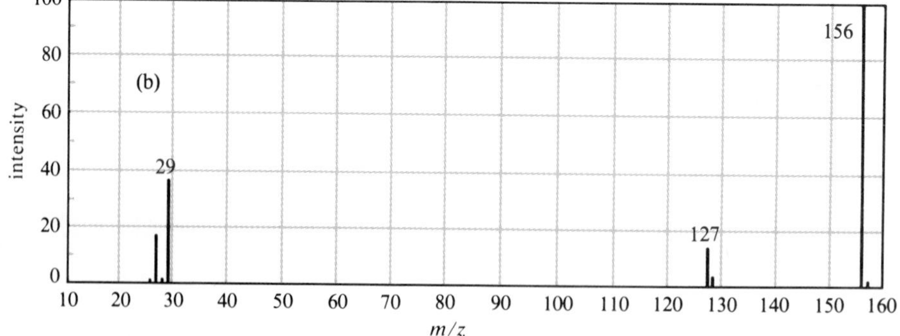

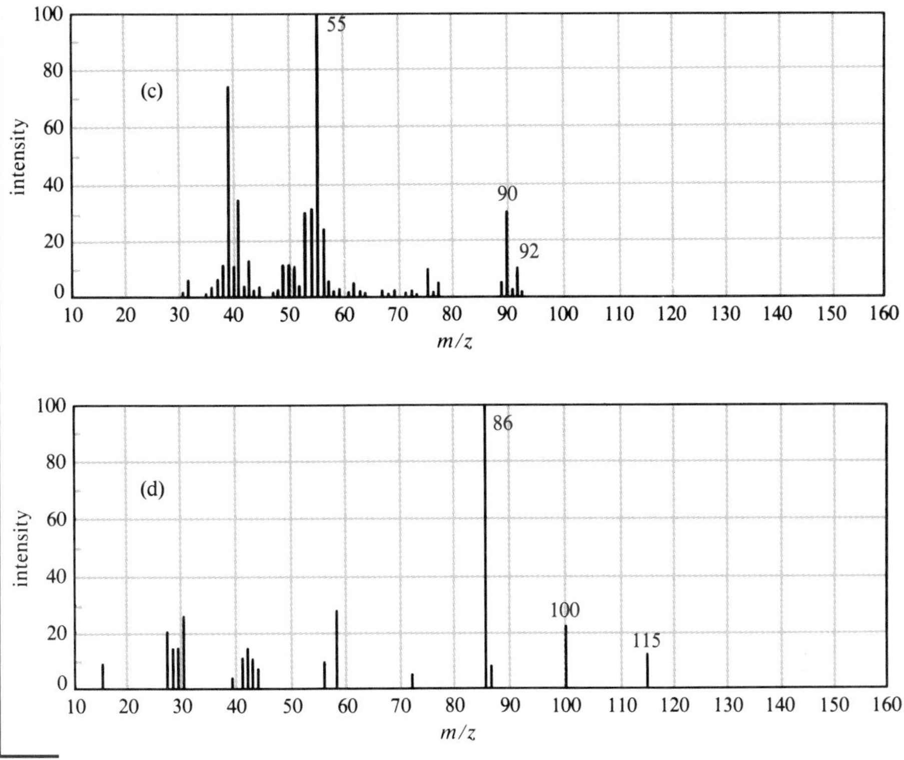

11-15 FRAGMENTATION PATTERNS IN MASS SPECTROMETRY

In addition to the molecular formula, the mass spectrum provides structural information. An electron with a typical energy of 70 eV (1610 kcal/mol or 6740 kJ/mol) has far more energy than that required to ionize a molecule. The impact of the 70 eV electron forms the radical cation, and it often breaks a bond to give a cation and a radical. The resulting cation is observed by the mass spectrometer, but the uncharged radical is not accelerated or detected. We can infer the mass of the uncharged radical from the amount of mass lost from the molecular ion to give the observed cation fragment.

Ionization

$$\mathrm{R{:}R'} + e^- \longrightarrow \underset{\substack{\text{radical cation} \\ \text{(molecular ion)}}}{[\mathrm{R\cdot R'}]^{+\cdot}} + 2\,e^-$$

Fragmentation

$$[\mathrm{R\cdot R'}]^{+\cdot} \longrightarrow \underset{\substack{\text{cation fragment} \\ \text{(observed)}}}{\mathrm{R}^+} + \underset{\substack{\text{radical fragment} \\ \text{(not observed)}}}{\cdot\mathrm{R'}}$$

This bond breaking does not occur randomly; it tends to form the most stable fragments. By knowing the stable fragments that result from different kinds of compounds, we can recognize structural features and use the mass spectrum to confirm a proposed structure.

11-15A MASS SPECTRA OF ALKANES

The mass spectrum of *n*-hexane (Figure 11-16) shows several characteristics typical of straight-chain alkanes. The base peak (m/z 57) corresponds to loss of an ethyl group, giving an ethyl radical and a butyl cation. The neutral ethyl radical is not detected, since it is not charged and is not deflected by the magnetic field.

m/z of the charged fragment on this side of the broken bond — 57

$$[CH_3CH_2CH_2CH_2 | CH_2CH_3]^{+\cdot} \longrightarrow CH_3CH_2CH_2CH_2^+ + \cdot CH_2CH_3$$

hexane radical cation M^+ 86 → 1-butyl cation, detected at m/z 57 + ethyl radical (29), not detected

A similar fragmentation gives an ethyl cation and a butyl radical. In this case, the ethyl fragment (m/z 29) is detected.

29

$$[CH_3CH_2CH_2CH_2 | CH_2CH_3]^{+\cdot} \longrightarrow CH_3CH_2CH_2CH_2\cdot + {}^+CH_2CH_3$$

hexane radical cation M^+ 86 → 1-butyl radical (57), not detected + ethyl cation, detected at m/z 29

Symmetric cleavage of hexane gives a propyl cation and a propyl radical.

43

$$[CH_3CH_2CH_2 | CH_2CH_2CH_3]^{+\cdot} \longrightarrow CH_3CH_2CH_2^+ + \cdot CH_2CH_2CH_3$$

hexane radical cation M^+ 86 → propyl cation, detected at m/z 43 + propyl radical (43), not detected

The cleavage to give a pentyl cation (m/z 71) and a methyl radical is very weak because the methyl radical is less stable than a substituted radical. The cleavage to give a methyl cation (m/z 15) and a pentyl radical is not visible because the methyl cation is less stable than a substituted cation. It appears that the stability of the cation is more important than the stability of the radical, since a weak peak appears corresponding to the loss of a methyl radical, but cleavage to give a methyl cation is not observed.

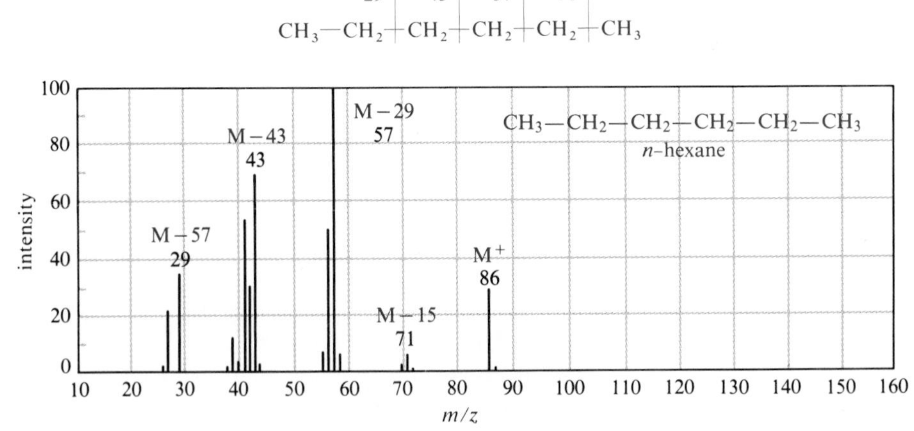

FIGURE 11-16 Mass spectrum of *n*-hexane. There are groups of ions corresponding to loss of one-, two-, three-, and four-carbon fragments.

$$[CH_3CH_2CH_2CH_2CH_2 \overset{71}{|} CH_3]^{+\cdot} \longrightarrow CH_3CH_2CH_2CH_2CH_2{}^+ + \cdot CH_3$$

hexane radical cation M^+ 86 → pentyl cation, weak at m/z 71 + methyl radical (15), not detected

$$[CH_3CH_2CH_2CH_2CH_2 \overset{15\ \text{(not seen)}}{|} CH_3]^{+\cdot} \not\longrightarrow CH_3CH_2CH_2CH_2CH_2\cdot + {}^+CH_3$$

hexane radical cation M^+ 86 ✗→ pentyl radical (71), not detected + methyl cation (too unstable)

Cation and radical stabilities can be used to explain the mass spectra of branched alkanes as well. Figure 11-17 shows the mass spectrum of 2-methylpentane. Fragmentation of a branched alkane commonly occurs at the branch carbon atom to give the most highly substituted cation and radical. Fragmentation of 2-methylpentane at the branched carbon atom can give a secondary carbocation in either of two ways:

$$\left[CH_3CH_2CH_2 \wr \underset{}{\overset{CH_3}{\overset{|}{C}H}} \wr CH_3\right]^{+\cdot}$$

2-methylpentane radical cation, m/z 86

$$\nearrow \; CH_3CH_2CH_2-\overset{CH_3}{\overset{|}{C}H^+} + \cdot CH_3$$

2-pentyl cation, m/z 71 + methyl radical

$$\searrow \; {}^+\overset{CH_3}{\overset{|}{C}H}-CH_3 + CH_3CH_2CH_2\cdot$$

isopropyl cation, m/z 43 (base peak) + propyl radical

Both fragmentations give secondary cations, but the second gives a primary radical instead of a methyl radical. Therefore, the second fragmentation accounts for the base (largest) peak, while the first accounts for the other large peak at m/z 71. Other fragmentations to give primary cations account for the weaker peaks.

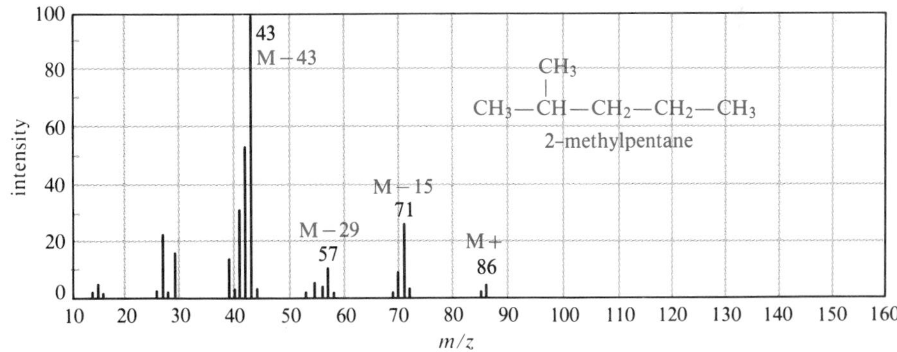

FIGURE 11-17 Mass spectrum of 2-methylpentane. The base peak corresponds to the loss of a propyl radical to give an isopropyl cation.

PROBLEM 11-9

Show the fragmentation that accounts for the cation at m/z 57 in the mass spectrum of 2-methylpentane.

PROBLEM 11-10

Show the fragmentations that give rise to the peaks at m/z 43, 57, and 85 in the mass spectrum of 2,4-dimethylpentane (Fig. 11-15).

11-15B FRAGMENTATION SPLITTING OUT A SMALL MOLECULE; MASS SPECTRA OF ALCOHOLS

Peaks are commonly seen in the mass spectrum corresponding to loss of small, stable molecules. Loss of a small molecule is usually indicated by a peak for an ion that has lost a fragment with an even mass number. A radical cation may lose water (18), CO (28), CO_2 (44), and even ethene (28) or other alkenes. The most common example of the loss of a small molecule is the loss of water from alcohols, which occurs so readily that the molecular ion is usually weak or absent. The peak corresponding to loss of water (the M − 18 peak) is usually strong, however.

The mass spectrum of 3-methyl-1-butanol (Fig. 11-18) is typical for alcohols. The peak at *m/z* 70 that *appears* to be the molecular ion is actually the intense M − 18 peak. The molecular ion (*m/z* 88) is not observed because it loses water so readily. The base peak at *m/z* 55 corresponds to loss of water and a methyl group.

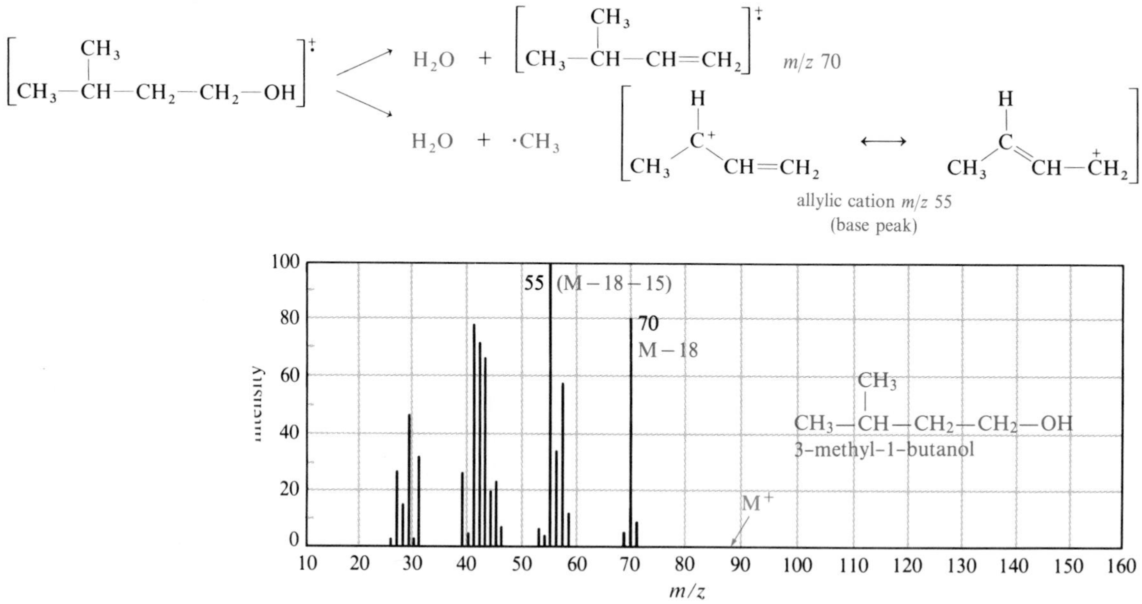

$$[CH_3-CH(CH_3)-CH_2-CH_2-OH]^{+\cdot} \longrightarrow H_2O + [CH_3-C(CH_3)H-CH=CH_2]^{+\cdot} \quad m/z\ 70$$

$$[CH_3-CH(CH_3)-CH_2-CH_2-OH]^{+\cdot} \longrightarrow H_2O + \cdot CH_3 + [CH_3-\overset{+}{C}H-CH=CH_2 \longleftrightarrow CH_3-CH=CH-\overset{+}{C}H_2]$$

allylic cation *m/z* 55 (base peak)

FIGURE 11-18 The strong peak at *m/z* 70 in the mass spectrum of 3-methyl-1-butanol is actually the M − 18 peak, corresponding to loss of water. The molecular ion is not visible because it loses water very easily.

In addition to losing water, alcohols tend to fragment next to the carbinol carbon atom to give a reasonance-stabilized carbocation.

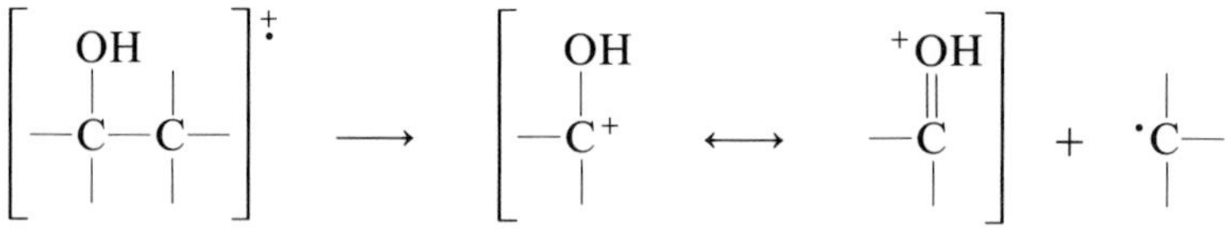

$$\left[-\overset{OH}{\underset{|}{C}}-\overset{|}{\underset{|}{C}}-\right]^{+\cdot} \longrightarrow \left[-\overset{OH}{\underset{|}{C^+}} \longleftrightarrow -\overset{^+OH}{\underset{|}{\overset{\|}{C}}}\right] + \cdot\overset{|}{\underset{|}{C}}-$$

This type of *alpha-cleavage* is prominent in the spectrum of 2,6-dimethyl-4-heptanol shown in Problem 11-11.

PROBLEM 11-11

Account for the peaks at *m/z* 87, 111, and 126 in the mass spectrum of 2,6-dimethyl-4-heptanol.

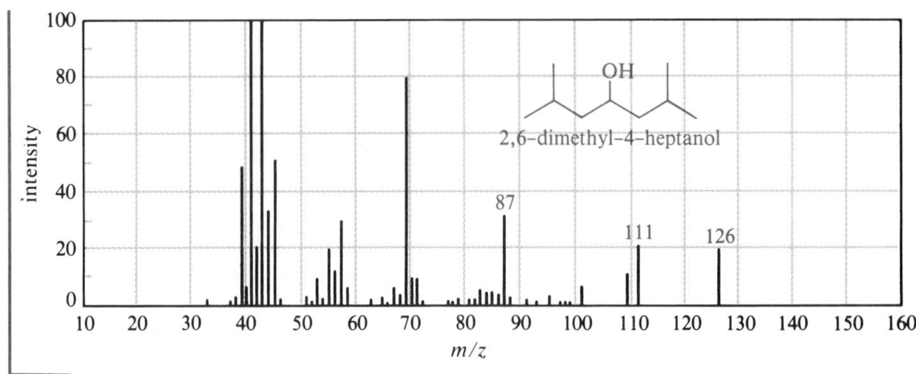

11-15C FRAGMENTATION GIVING RESONANCE-STABILIZED CATIONS; MASS SPECTRA OF ALKENES

Fragmentation in the mass spectrometer occurs to give resonance-stabilized cations whenever possible. The most common fragmentation of alkenes is the cleavage of an allylic bond to give a resonance-stabilized allylic cation. For example, Figure 11-19 shows how the radical cation of 2-hexene undergoes allylic cleavage to give the resonance-stabilized cation responsible for the base peak at m/z 55. We will encounter other types of resonance-stabilized cations in the mass spectra of ethers, amines, and carbonyl compounds in later chapters covering the chemistry of these functional groups.

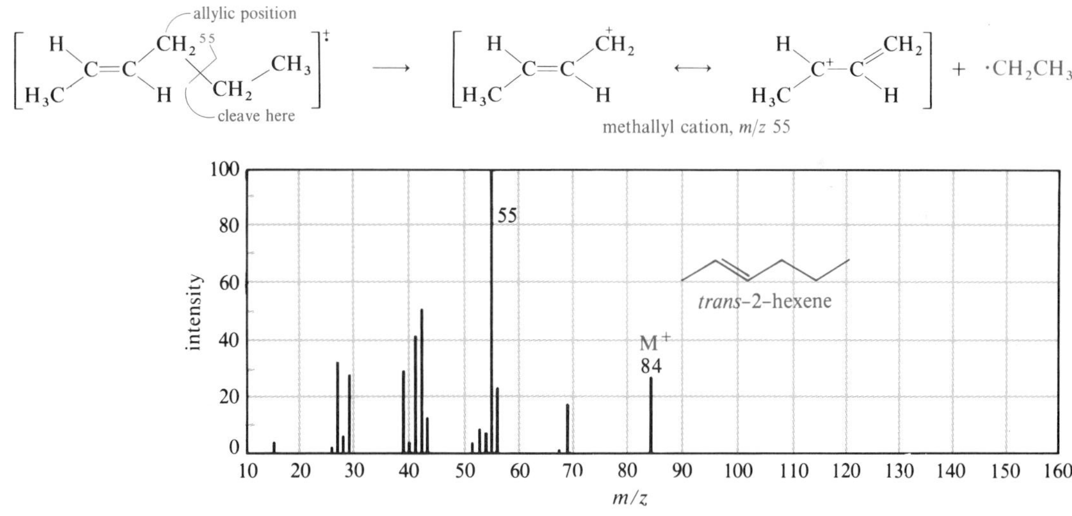

FIGURE 11-19 The radical cation of 2-hexene undergoes cleavage of an allylic bond to give a methallyl cation of m/z 55. The other fragment from this cleavage is an ethyl radical, undetected because it is uncharged.

PROBLEM 11-12

Catalytic hydrogenation of compound X gives 2,6-dimethyloctane as the only product. The mass spectrum of compound X shows a molecular ion at m/z 140 and prominent peaks at m/z 57 and m/z 83. Suggest a structure for compound X and justify your answer.

SUMMARY OF COMMON FRAGMENTATION PATTERNS

This summary is provided for rapid reference to the common fragmentation patterns of simple functional groups. Some of these functional groups are discussed in later chapters, but they are included here so this reference table can be used throughout the course.

1. *Alkanes:* cleavage to give the most stable carbocations

$$\left[\begin{array}{c} R' \\ | \\ R-C-R'' \\ | \\ H \end{array}\right]^{+\cdot} \longrightarrow \begin{array}{c} R' \\ | \\ R-C^{+} \\ | \\ H \end{array} + \cdot R''$$

2. *Alcohols:* loss of water

$$\left[\begin{array}{c} H\quad OH \\ |\quad\ | \\ -C-C- \\ |\quad\ | \end{array}\right]^{+\cdot} \longrightarrow \left[\ \rangle C{=}C\langle\ \right]^{+\cdot} + H_2O$$

α cleavage

$$\left[\begin{array}{c} OH \\ | \\ -C-C- \\ |\quad\ | \end{array}\right]^{+\cdot} \longrightarrow \left[\begin{array}{c} OH \\ | \\ -C^{+} \\ | \end{array} \longleftrightarrow \begin{array}{c} ^{+}OH \\ \| \\ -C \\ | \end{array}\right] + \cdot\overset{|}{\underset{|}{C}}-$$

3. *Alkenes and aromatics:* cleavage to give allylic and benzylic carbocations

$$[R-CH{=}CH-CH_2 \mid R']^{+\cdot} \longrightarrow R-CH{=}CH-\overset{+}{C}H_2 + \cdot R'$$

allylic cation

(The following fragmentations are covered in later chapters.)

$$\left[C_6H_5-CH_2 \mid R\right]^{+\cdot} \longrightarrow C_6H_5-\overset{+}{C}H_2 + \cdot R \longrightarrow C_7H_7^{+} + \cdot R$$

benzylic cation m/z 91 — tropylium ion m/z 91

4. *Amines:* α cleavage to give stabilized cations

$$[R_2N-CH_2-R']^{+\cdot} \longrightarrow R_2\overset{+}{N}{=}CH_2 + \cdot R'$$

iminium ion

5. *Ethers:*

Loss of an alkyl group

$$[R-CH_2-O \mid R']^{+\cdot} \longrightarrow R-CH{=}\overset{+}{O}H + \cdot R'$$

stabilized cation

or

$$[R-CH_2-O \mid R']^{+\cdot} \longrightarrow R-CH_2-\ddot{\underset{..}{O}}\cdot + {}^{+}R$$

alkyl cation

α cleavage

$$[R \mid CH_2-O-R']^{+\cdot} \xrightarrow{\alpha\ \text{cleavage}} H_2C{=}\overset{+}{O}-R' + \cdot R$$

stabilized cation

6. *Ketones and aldehydes:* loss of alkyl groups to give acylium ions

$$\left[R-\overset{\overset{\displaystyle O}{\|}}{C}-R' \right]^{+\cdot} \longrightarrow \underset{\text{acylium ion}}{R-C\equiv O^+} + \cdot R'$$

The McLafferty rearrangement splits out olefins.

γ-hydrogen

$$\left[R-\overset{\overset{\displaystyle O}{\|}}{C}-\underset{\alpha}{C}-\underset{\beta}{C}-\underset{\gamma}{\overset{\overset{\displaystyle H}{|}}{C}}-R' \right]^{+\cdot} \longrightarrow \left[\overset{\overset{\displaystyle O-H}{|}}{C}(R)=\underset{\alpha}{C} \right]^{+\cdot} + \underset{\beta}{C}=\underset{\gamma}{C}-R'$$

GLOSSARY

absorption spectroscopy The measurement of the amount of light absorbed by a compound as a function of the wavelength of light. (p. 444)

base peak The strongest peak in a mass spectrum. (p. 475)

conjugated double bonds Two double bonds that are one bond apart, so that their pi-bonding orbitals can overlap with each other. (pp. 452 and 461)

electromagnetic spectrum The range of all possible electromagnetic frequencies from zero to infinity. In practice, it ranges from radio waves up to gamma rays. (p. 445)

fingerprint region The portion of the infrared spectrum between 600 and 1400 cm^{-1}, where many complex vibrations occur; so named because no two different compounds (except enantiomers) have exactly the same absorptions in this region. (p. 449)

fragmentation The breaking apart of a molecular ion upon ionization in a mass spectrometer. (p. 479)

frequency (ν) The number of complete wave cycles that pass a fixed point in a second, or the number of reversals of the electromagnetic field per second. (p. 445)

high-resolution mass spectrometer A mass spectrometer that can measure masses very accurately, usually to 1 part in 20,000. This high precision allows calculation of molecular formulas using the known atomic masses of the elements. (p. 476)

infrared spectrum A graph of the infrared energy absorbed by a sample as a function of the wavelength (λ, expressed in μm) or the frequency ($\bar{\nu}$, expressed as a wavenumber, cm^{-1}). (p. 447)

IR active A vibration that changes the dipole moment of the molecule and thus can absorb infrared light. (p. 450)

IR inactive A vibration that does not change the dipole moment of the molecule and thus cannot absorb infrared light. (p. 450)

mass spectrometer An instrument that ionizes molecules, sorts the ions according to their masses, and records the abundance of ions of each mass. (p. 473)

mass spectrum The graph produced by a mass spectrometer, showing the masses along the x axis and their abundance along the y axis. (p. 473)

> ***m/z*** (formerly ***m/e***): The mass-to-charge ratio of an ion. Most ions have a charge of +1, and m/z simply represents their masses.

molecular ion In mass spectrometry, the ion with the same mass as the molecular weight of the original compound; no fragmentation has occurred. (p. 474)

> **M + 1 peak:** An isotopic peak that is one mass unit heavier than the major molecular ion peak. (p. 476)

M + 2 peak: An isotopic peak that is two mass units heavier than the major molecular ion peak. (p. 476)

photon A massless packet of electromagnetic energy. (p. 446)

radical cation A positively charged ion with an unpaired electron; commonly formed by electron-impact ionization, when the impinging electron knocks out an additional electron. (p. 474)

$$\text{R:R} \;+\; e^- \;\longrightarrow\; [\text{R}\cdot\text{R}]^{\ddagger} \;+\; 2\,e^-$$

radical cation

wavelength (λ) The distance between any two peaks (or any two troughs) of a wave. (p. 445)

wavenumber ($\bar{\nu}$) The number of wavelengths that fit into one centimeter (cm^{-1} or reciprocal centimeters); proportional to the frequency. The product of the wavenumber (in cm^{-1}) and the wavelength (in μm) is 10,000. (p. 447)

ESSENTIAL PROBLEM-SOLVING SKILLS IN CHAPTER 11

1. Given an IR spectrum, identify the reliable characteristic peaks.
2. Explain why some characteristic peaks are usually strong or weak, and why some may be absent.
3. Predict the stretching frequencies of common functional groups.
4. Identify functional groups from IR spectra.
5. Identify conjugated and strained C=O bonds and conjugated and aromatic C=C bonds from their absorptions in the IR spectrum.
6. Determine molecular weights from mass spectra.
7. When possible, recognize from mass spectra the presence of Br, Cl, I, N, and S atoms.
8. Predict the major ions from fragmentation of alkanes, alkenes, and alcohols.
9. Use the fragmentation pattern to determine whether a proposed structure is consistent with the mass spectrum.

STUDY PROBLEMS

11-13. Define and give an example of each of the following terms.
(a) fingerprint region **(b)** an IR-active vibration **(c)** an IR-inactive vibration
(d) wavelength **(e)** conjugated double bonds **(f)** a radical cation
(g) a molecular ion **(h)** wavenumber

11-14. Convert the following infrared wavelengths to cm^{-1}.
(a) 6.24 μm, typical for an aromatic C=C **(b)** 3.38 μm, typical for a saturated C—H bond
(c) 5.85 μm, typical for a ketone carbonyl **(d)** 5.75 μm, typical for an ester carbonyl
(e) 4.52 μm, typical for a nitrile **(f)** 3.03 μm, typical for an alcohol O—H

11-15. All the following compounds absorb infrared radiation between 1600 and 1800 cm^{-1}. In each case
(1) Show which bonds absorb in this region.
(2) Predict the approximate absorption frequencies.
(3) Predict which compound of each pair absorbs more strongly in this region.

(a) $H_2C=CH-CH_2CH_3$ and $O=CH-CH_2CH_3$

(b) $H_2C=CH-CH_2CH_3$ and $H_2C=CH-OCH_2CH_3$

(c) 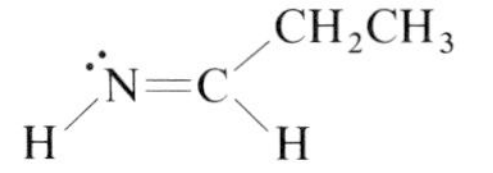 $\ddot{N}H{=}CH(CH_2CH_3)$ [H—N=C(H)—CH₂CH₃] and $H_2C{=}CH{-}CH_2CH_3$

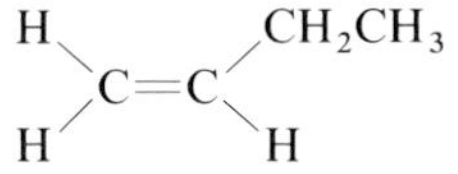

(d) 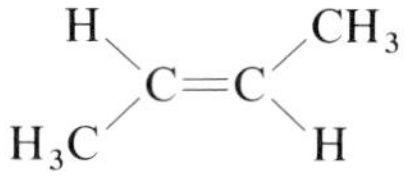 $H_3C{-}CH{=}CH{-}CH_3$ (trans) and $H_2C{=}CH{-}CH_2CH_3$

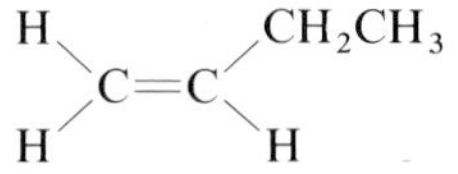

11-16. Describe the characteristic infrared absorption frequencies that would allow you to distinguish between the following pairs of compounds.

(a) 2,3-dimethyl-2-butene and 2,3-dimethyl-l-butene **(b)** 1,3-cyclohexadiene and 1,4-cyclohexadiene

(c) $CH_3(CH_2)_5{-}C{\equiv}C{-}H$ (1-octyne) and $CH_3(CH_2)_3{-}C{\equiv}C{-}CH_2CH_3$ (3-octyne) **(d)** 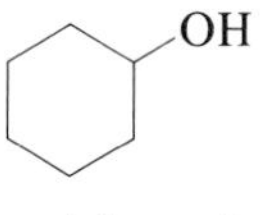 cyclohexanol and cyclohexanone

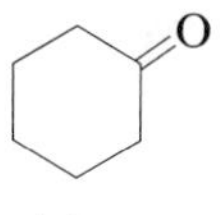

(e) $CH_3(CH_2)_3{-}C(=O){-}H$ (pentanal) and $CH_3(CH_2)_2{-}C(=O){-}CH_3$ (2-pentanone) **(f)** cyclohexanol and cyclohexene

(g) $CH_3CH_2CH_2{-}C(=O){-}OH$ (butanoic acid) and $CH_3{-}CH(OH){-}CH_2{-}C(=O){-}H$ (3-hydroxybutanal)

(h) $CH_3CH_2CH_2{-}C(=O){-}NH_2$ (butanamide) and $CH_3CH_2{-}C(=O){-}CH_2CH_3$ (3-pentanone)

11-17. Four infrared spectra are shown below, corresponding to four of the following compounds. For each spectrum, determine the correct structure and explain how the peaks in the spectrum correspond to the structure you have chosen.

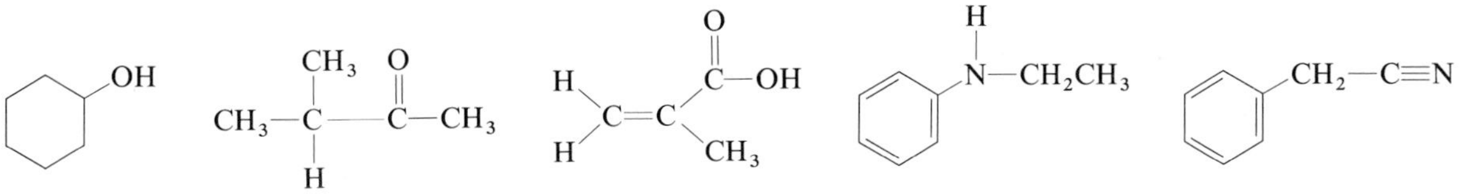

cyclohexanol; $CH_3{-}CH(CH_3){-}C(=O){-}CH_3$; $H_2C{=}C(CH_3){-}C(=O){-}OH$; $C_6H_5{-}NH{-}CH_2CH_3$; $C_6H_5{-}CH_2{-}C{\equiv}N$

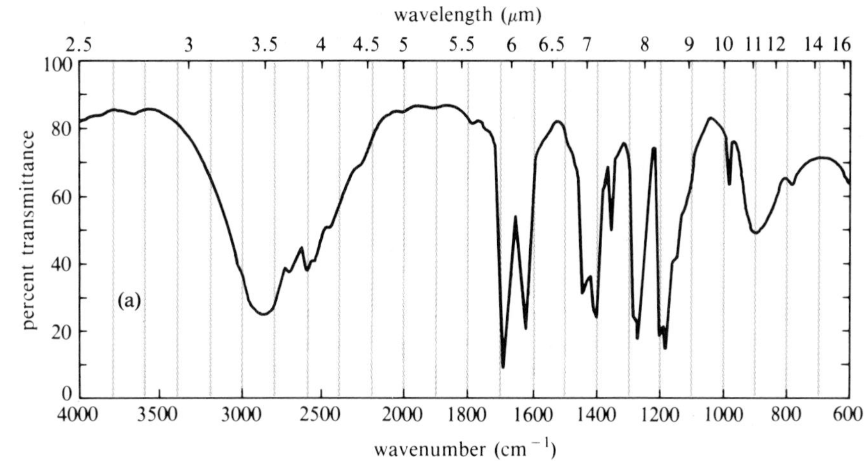

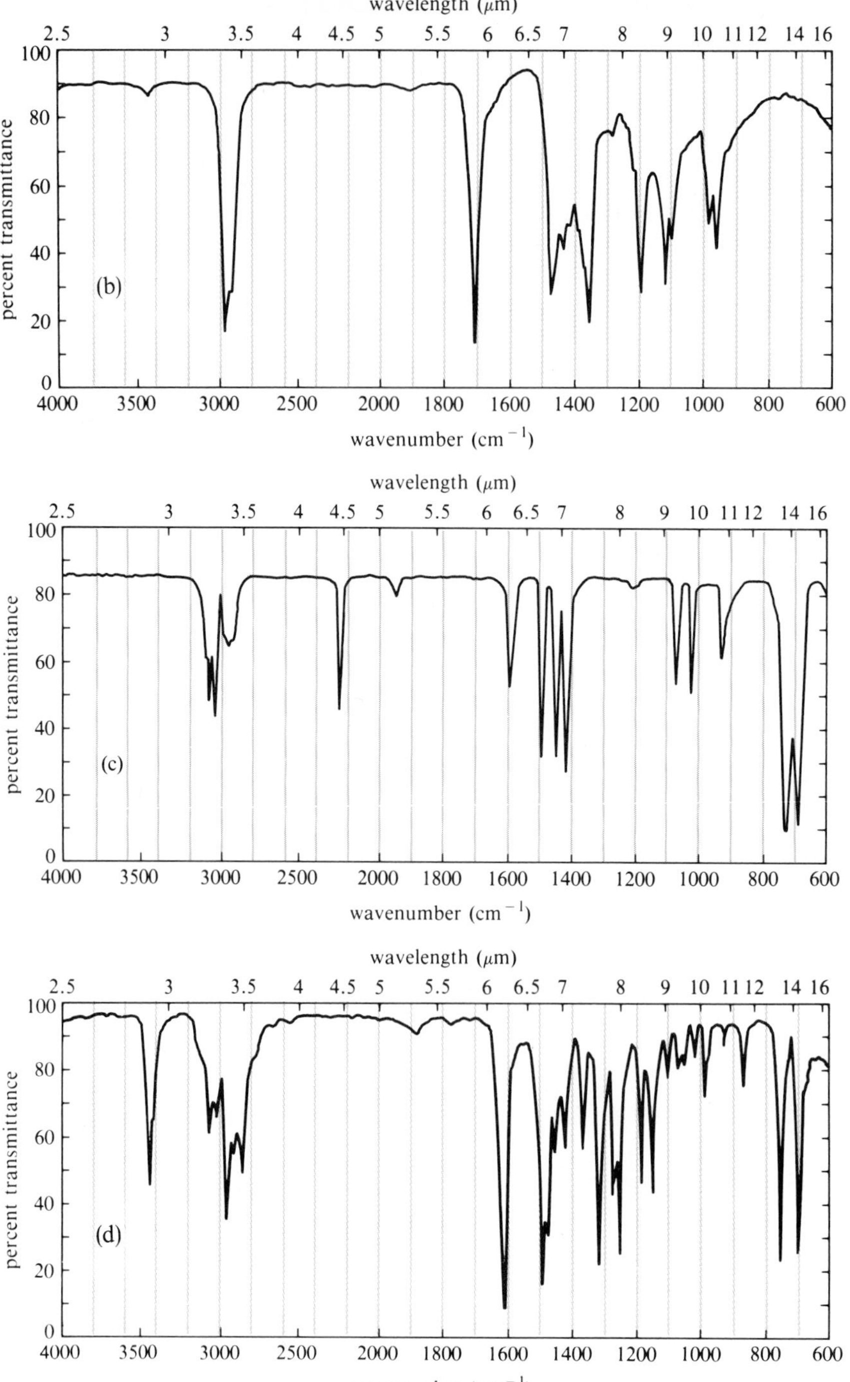

11-18. Predict the values of m/z and the structures of the most abundant fragments observed in the mass spectra of the following compounds.
(a) 2-methylpentane **(b)** 3-methyl-2-hexene **(c)** 4-methyl-2-pentanol

11-19. Give logical fragmentation reactions to account for the following ions observed in these mass spectra.
(a) *n*-octane: 114, 85, 71, 57 **(b)** methylcyclohexane: 98, 83
(c) 2-methyl-2-pentene: 84, 69 **(d)** 1-pentanol: 70, 55, 41, 31

✻ 11-20. A C—D (carbon-deuterium) bond is electronically much like a C—H bond, and it has a similar stiffness, measured by the *spring constant*, k. The deuterium atom has twice the mass of a hydrogen atom, however.

(a) Use the fact that the infrared absorption frequency is proportional to $\sqrt{k/m}$ to calculate the IR absorption frequency of a typical C—D bond.

(b) A chemist has dissolved a sample in deuterochloroform ($CDCl_3$), then decides to take the IR spectrum and simply evaporates most of the $CDCl_3$. What functional group will *appear* to be present in this IR spectrum as a result of the $CDCl_3$ impurity?

✻ 11-21. The mass spectrum of *n*-octane shows a prominent molecular ion peak (m/z 114). There is also a large peak at m/z 57, but it is not the base peak. The mass spectrum of 3,4-dimethylhexane shows a smaller molecular ion, and the peak at mass 57 is the base peak. Explain these trends in abundance of the molecular ions and the ions at mass 57 and predict the intensities of the peaks at masses 57 and 114 in the spectrum of 2,2,3,3-tetramethylbutane.

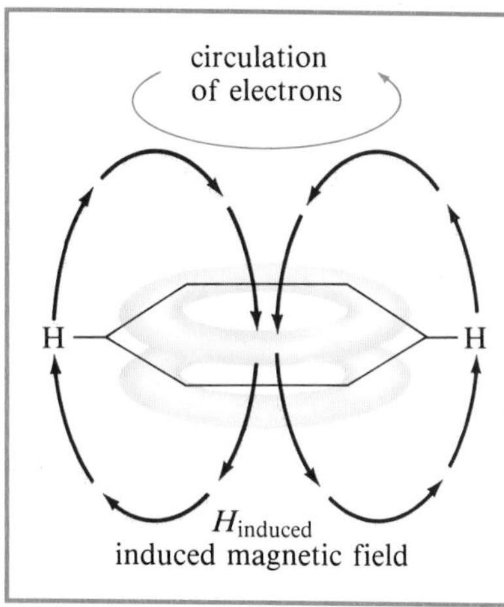

12

NUCLEAR MAGNETIC RESONANCE SPECTROSCOPY

12-1 INTRODUCTION

Nuclear magnetic resonance spectroscopy (NMR) is the most powerful tool available for organic structure determination. Like IR spectroscopy, NMR can be used with a very small sample, and it does not harm the sample. The NMR spectrum provides a great deal of information about the structure of the compound, and some structures can be determined using only the NMR spectrum. More commonly, however, the NMR spectrum is used in conjunction with other forms of spectroscopy and chemical analysis to determine the structures of complicated organic molecules.

NMR is used to study a wide variety of nuclei, including ^{1}H, ^{13}C, ^{15}N, ^{19}F, and ^{31}P. Since hydrogen and carbon are major components of organic compounds, organic chemists find proton (^{1}H) and carbon-13 (^{13}C) NMR to be most useful. Historically, NMR was first used to study protons (the nuclei of hydrogen atoms), and proton magnetic resonance (PMR) spectrometers are the most common. "Nuclear magnetic resonance" is assumed to mean "proton magnetic resonance" unless a different nucleus is specified. We begin our study of NMR with proton magnetic resonance and conclude with a discussion of ^{13}C NMR.

12-2 THEORY OF NUCLEAR MAGNETIC RESONANCE

A nucleus with an odd atomic number or an odd mass number has a *nuclear spin* that can be observed by the NMR spectrometer. A proton is the simplest nucleus, and its odd atomic number of 1 indicates it has a spin. We can visualize a spinning proton as a rotating sphere of positive charge (Fig. 12-1). This movement of charge is like an electric current in a loop of wire. It generates a magnetic

FIGURE 12-1 A spinning proton generates a magnetic field, called its magnetic moment. This magnetic field (H) resembles that of a small bar magnet.

spinning proton

loop of current

bar magnet

field (symbolized by H) called the **magnetic moment,** that looks like the field of a small bar magnet.

When a small bar magnet is placed in the field of a larger magnet (Fig. 12-2), it twists to align itself with the field of the larger magnet—a lower-energy arrangement than an orientation against the field. The same effect is seen when an external magnetic field (H_0) is applied to a proton, as shown below. Quantum mechanics requires the proton's magnetic moment to be aligned either *with* the external field or *against* the field. The lower-energy state with the proton aligned with the field is called the *alpha-spin* (α-spin) *state.* The higher-energy state with the proton aligned against the external magnetic field is called the *beta-spin* (β-spin) *state.*

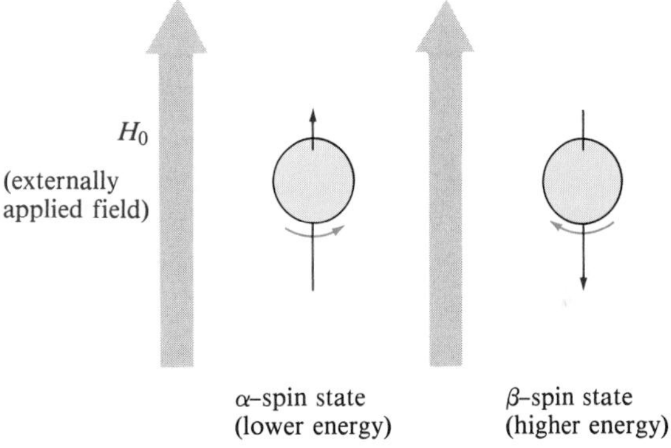

In the absence of a magnetic field, proton magnetic moments have random orientations. When an external magnetic field is applied, each proton in a sample assumes the α state or the β state. Because the α-spin state is lower in energy, there are more α spins than β spins.

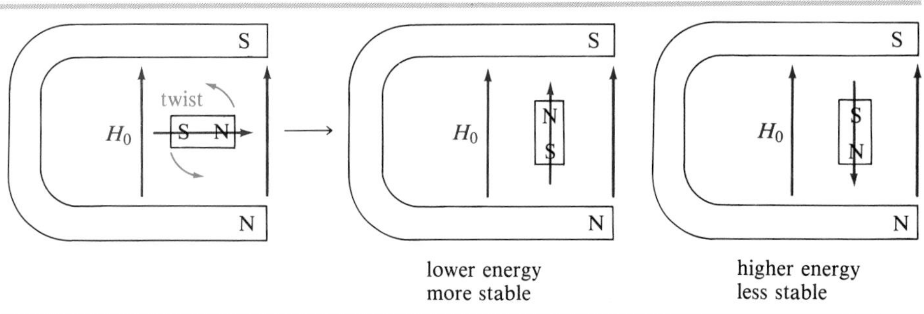

FIGURE 12-2 An external magnetic field (H_0) applies a force to a small bar magnet, twisting the bar magnet to align it with the external field. The arrangement of the bar magnet aligned *with* the field is lower in energy than the arrangement aligned *against* the field.

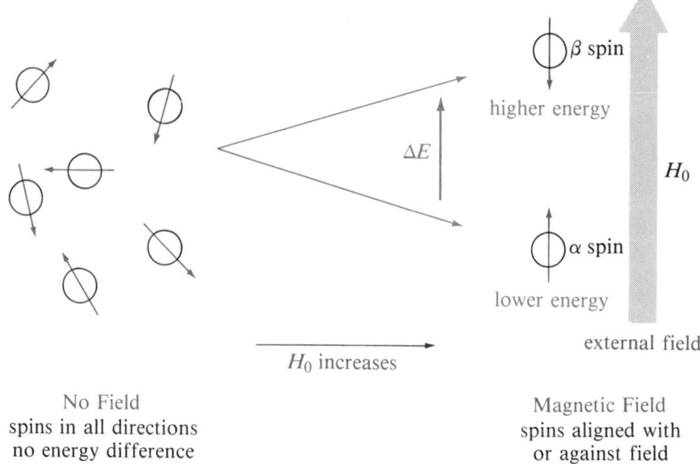

In a strong magnetic field, the energy difference between the two spin states is larger than it is in a weaker field. In fact, the energy difference is proportional to the strength of the magnetic field, as expressed in the equation

$$\Delta E = \gamma \frac{h}{2\pi} H_0$$

ΔE = energy difference between α and β states

h = Planck's constant

H_0 = strength of the external magnetic field

γ = gyromagnetic ratio, 26,753 sec^{-1} gauss^{-1} for a proton

The **gyromagnetic ratio** (γ) is a constant that depends on the magnetic moment of the nucleus under study. Magnetic fields are measured in *gauss;* for example, the strength of the earth's magnetic field is about 0.57 gauss.

The energy difference between the two spin states is not very large. For a strong external magnetic field of 25,000 gauss, it is only about 10^{-5} kcal (4×10^{-5} kJ) per mole. Even this small energy difference can be detected by the NMR technique, however. When a proton interacts with a photon with just the right amount of electromagnetic energy the proton's spin can flip from α to β or from β to α. A nucleus aligned with the field can absorb the energy needed to "flip" and become aligned against the field.

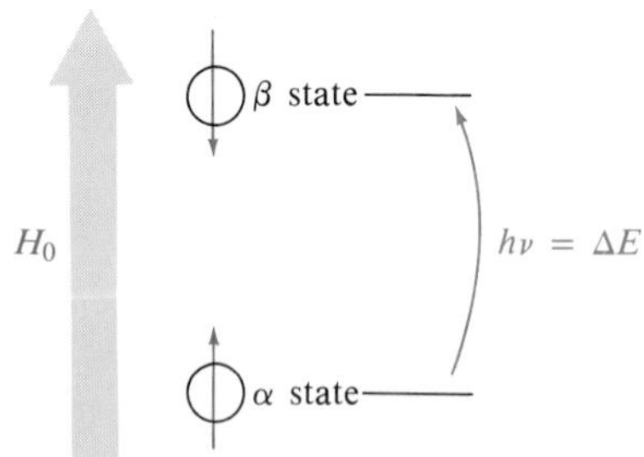

FIGURE 12-3 A nucleus is "in resonance" when it is irradiated with radio-frequency photons having energy equal to the energy difference between the spin states. Under these conditions, a proton in the α-spin state can absorb a photon and change to the β-spin state.

When a nucleus is subjected to the right combination of magnetic field and electromagnetic radiation to flip its spin, it is said to be "in resonance" (Fig. 12-3), and its absorption of energy is detected by the NMR spectrometer. This is the origin of the term "nuclear magnetic resonance."

A photon's energy is given by $E = h\nu$, showing that the energy E is proportional to ν, the frequency of the electromagnetic wave. This equation can be combined with the equation for the energy difference between the spin states:

$$\Delta E = h\nu = \gamma \frac{h}{2\pi} H_0$$

Rearranging to solve for ν shows that the resonance frequency ν is proportional to the applied magnetic field (H_0) and the gyromagnetic ratio (γ).

$$\nu = \frac{1}{2\pi} \gamma H_0$$

For a proton, $\gamma = 26{,}753\ \text{sec}^{-1}\ \text{gauss}^{-1}$, and

$$\nu = \frac{(26{,}753\ \text{sec}^{-1}\ \text{gauss}^{-1})}{2\pi} \times H_0 = (4257.8\ \text{sec}^{-1}\ \text{gauss}^{-1}) \times H_0$$

Using the fields of currently available magnets, proton resonance frequencies occur in the radio-frequency region of the spectrum. NMR spectrometers are usually designed for the most powerful magnet that is practical for the price range of the spectrometer (to make ΔE as large and easily detected as possible), and the radio frequency needed for resonance is calculated based on the field. The most common operating frequency for student spectrometers is 60 MHz (megahertz; 1 million cycles per second), corresponding to a magnetic field of 14,092 gauss. Higher-resolution research instruments commonly operate at frequencies of 100 to 300 MHz (and even higher), corresponding to fields of 23,486 to 70,458 gauss. Most of the NMR spectra we will consider were obtained using a 60-MHz instrument.

SOLVED PROBLEM 12-1

Calculate the magnetic fields that correspond to proton resonance frequencies of 60 MHz and 100 MHz.

SOLUTION We substitute into the equation $\nu = (1/2\pi)\gamma H_0$.

$$60\ \text{MHz} = 60 \times 10^6\ \text{sec}^{-1} = (4257.8\ \text{sec}^{-1}\ \text{gauss}^{-1}) \times H_0$$
$$H_0 = 14{,}092\ \text{gauss}$$
$$100\ \text{MHz} = 100 \times 10^6\ \text{sec}^{-1} = (4257.8\ \text{sec}^{-1}\ \text{gauss}^{-1}) \times H_0$$
$$H_0 = 23{,}486\ \text{gauss}$$

12-3 MAGNETIC SHIELDING BY ELECTRONS

Up to now we have considered the resonance of a naked proton in a magnetic field; but real protons in organic compounds are not naked. They are surrounded by electrons that partially shield them from the external magnetic field. The electrons circulate and generate a small "induced" magnetic field that opposes the externally applied field.

A similar effect occurs when a loop of wire is moved into a magnetic field. The electrons in the wire are induced to flow around the loop in the direction shown in Figure 12-4; this is the principle of the electric generator. The induced electric current creates a magnetic field that opposes the external field.

In a molecule, the electron cloud around each nucleus acts like a loop of wire, rotating in response to the external field. This induced rotation is a circular current whose magnetic field opposes the external field. The result is that the magnetic field at the nucleus is weaker than the external field, and we say the nucleus is **shielded.** The effective magnetic field *at the shielded proton* is always weaker than the external field, so the applied field must be made stronger for resonance to occur at a given frequency (Fig. 12-5).

At 60 MHz, an unshielded naked proton absorbs at 14,092 gauss; but a shielded proton requires a stronger field. For example, if a proton is shielded by 0.3 gauss when the external field is 14,092.0 gauss, the effective magnetic field *at the proton* is 14,091.7 gauss. If the external field is increased to 14,092.3 gauss, the effective magnetic field at the proton is increased to 14,092.0 gauss, which brings this proton into resonance.

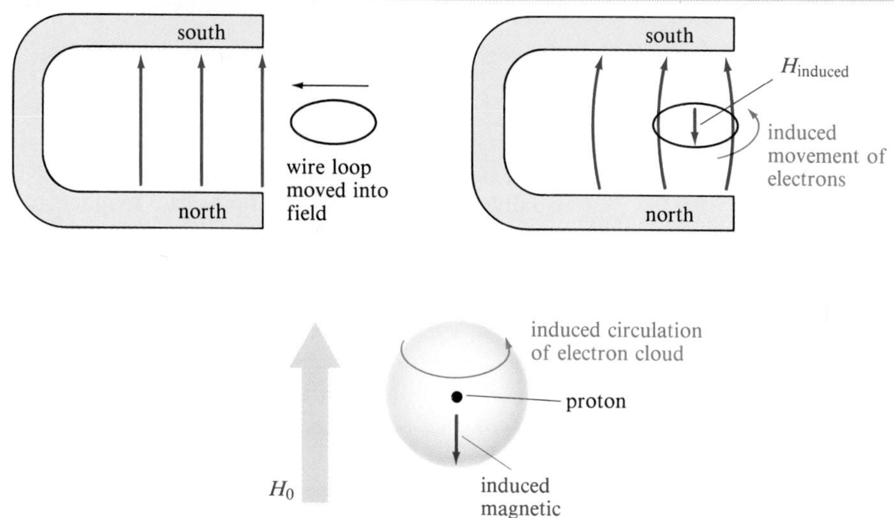

FIGURE 12-4 When a loop of wire is moved into a magnetic field, a current is induced in the wire. This current produces its own smaller magnetic field, in the opposite direction from the applied field. In a molecule, electrons can circulate around a nucleus. The resulting "current" sets up a magnetic field that opposes the external field, so that the nucleus feels a slightly weaker field.

If all protons were shielded by the same amount, they would all be in resonance at the same combination of frequency and magnetic field. Fortunately, protons in different chemical environments are shielded by different amounts. In methanol, for example, the electronegative oxygen atom withdraws some electron density from around the hydroxyl proton. The hydroxyl proton is not shielded as much as the methyl protons, so the hydroxyl proton absorbs at a lower field than the methyl protons (but still at a higher field than a naked proton). We say that the hydroxyl protons are **deshielded** somewhat by the presence of the electronegative oxygen atom.

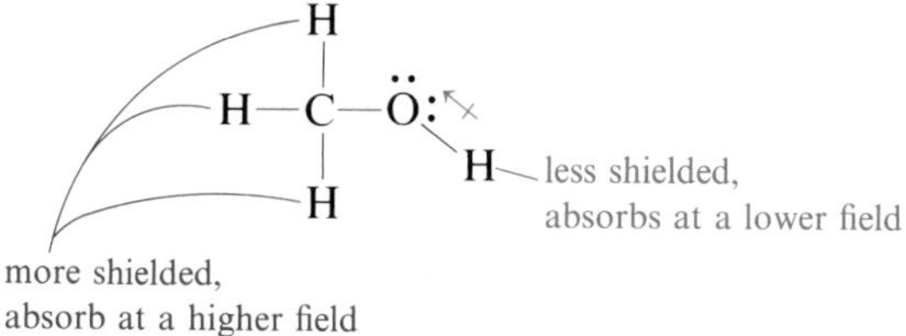

Because of the diverse and complex structures of organic molecules, the

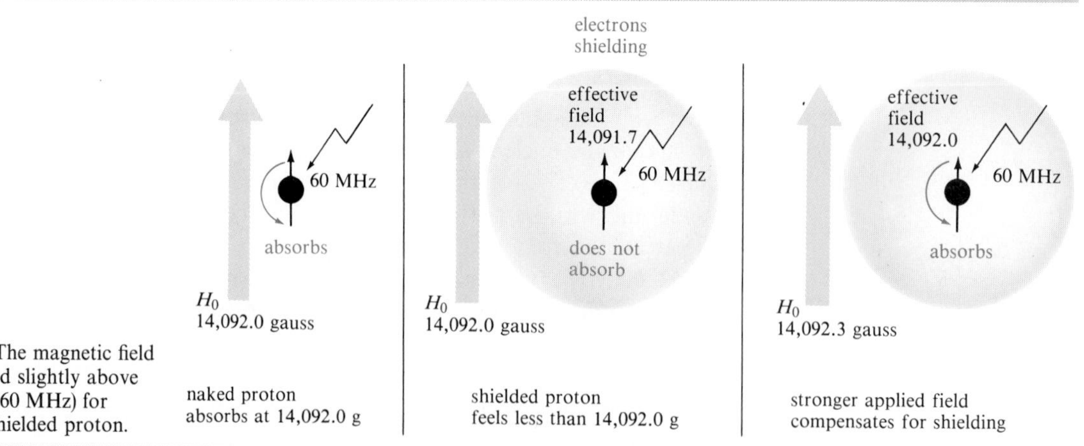

FIGURE 12-5 The magnetic field must be increased slightly above 14,092 gauss (at 60 MHz) for resonance of a shielded proton.

shielding effects of electrons at various positions are generally different. A careful measurement of the field strengths required for resonance of the various protons in a molecule provides us with two important types of information:

1. The *number of different absorptions* implies how many different types of protons there are.
2. The *amount of shielding* shown by these absorptions often implies the electronic structure of the molecule close to each type of proton.

Two other aspects of the NMR spectrum we will consider are the intensities of the signals and their splitting patterns:

3. The *intensities of the signals* imply how many protons of each type are present.
4. The *splitting of the signals* gives information about other nearby protons.

Before discussing the design of spectrometers, let's review what happens in an NMR spectrometer. Protons (in the sample compound) are placed in a magnetic field, where they align either with the field or against it. While still in the magnetic field, the protons are subjected to radiation of a frequency they can absorb by changing their orientation relative to the field. All isolated protons absorb at the same frequency, proportional to the magnetic field.

But protons in a molecule are partially shielded from the magnetic field, and this shielding depends on the proton's environment within a molecule. Thus protons in different environments within a molecule exposed to a constant frequency absorb the radiation at different magnetic field strengths. The NMR spectrometer must be equipped to vary the magnetic field and plot a graph of energy absorption as a function of the magnetic field strength. Such a graph is called a **nuclear magnetic resonance spectrum.**

12-4 THE NMR SPECTROMETER

The simplest type of NMR spectrometer (Fig. 12-6) consists of four parts:

1. A stable magnet, with a sensitive controller to produce a precise magnetic field
2. A radio-frequency (RF) transmitter, capable of emitting a precise frequency

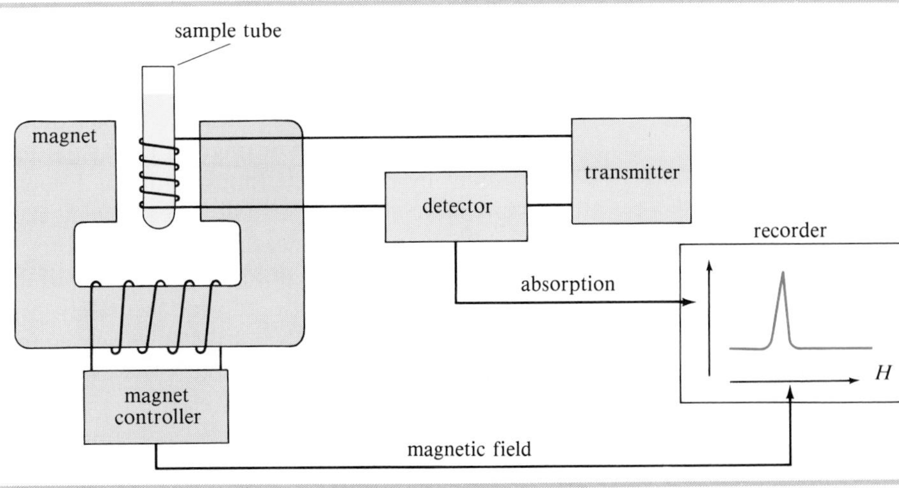

FIGURE 12-6 Block diagram of a nuclear magnetic resonance spectrometer.

3. A detector to measure the absorption of RF energy by the sample
4. A recorder to plot the output from the detector against the applied magnetic field

The recorder prints a graph of absorption (on the y axis) as a function of the applied magnetic field (on the x axis). Higher values of the magnetic field are toward the right **(upfield),** and lower values are toward the left **(downfield).** The absorptions of more shielded protons appear upfield, toward the right of the spectrum; and less shielded protons appear downfield, toward the left. The actual spectrum of methanol is shown in Figure 12-7.

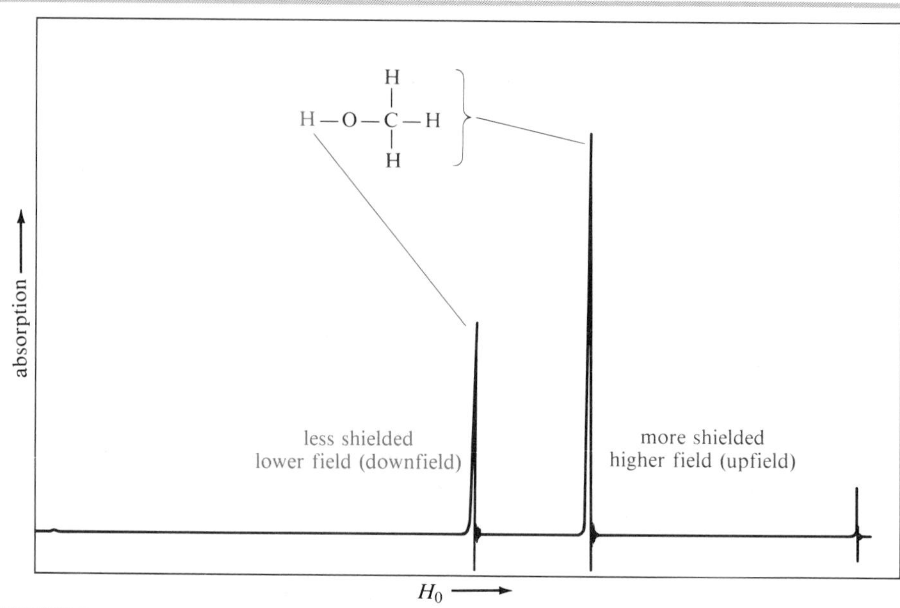

FIGURE 12-7 Proton NMR spectrum of methanol. The more shielded methyl protons appear toward the right of the spectrum (higher field); the less shielded hydroxyl proton appears toward the left (lower field).

12-5 THE CHEMICAL SHIFT

12-5A MEASUREMENT OF CHEMICAL SHIFTS

The variations in the positions of NMR absorptions, arising from electron shielding and deshielding, are called **chemical shifts.**

> **chemical shift** The difference (in parts per million) between the resonance frequency of the proton being observed and that of tetramethylsilane (TMS).

In practice, it is difficult to measure the absolute field where a proton absorbs with enough accuracy to distinguish individual protons, because the absorptions often differ by only a few thousandths of a gauss at an applied field of 14,092 gauss. A more accurate method for expressing chemical shifts is to determine the value relative to a reference compound added to the sample. The *difference* in the magnetic field strength required for resonance of the sample protons and the reference protons can be measured very accurately.

$$CH_3-\underset{CH_3}{\overset{CH_3}{Si}}-CH_3$$

tetramethylsilane (TMS)

The most common NMR reference compound is *tetramethylsilane,* $(CH_3)_4Si$, abbreviated TMS. Because silicon is less electronegative than carbon, the methyl groups of TMS are relatively electron rich, and their protons are very well shielded. They absorb at a higher field strength than most hydrogens bonded to carbon or other elements, so most NMR signals appear *downfield* (to the left) of the TMS signal. All the protons in TMS absorb at exactly the same chemical shift, giving one strong absorption.

A small amount of TMS is added to the sample, and the instrument measures the difference in magnetic field between where the protons in the sample absorb and where the TMS absorbs. For each type of proton in the sample, the distance downfield of TMS is the chemical shift of those protons.

Chemical shifts are measured in *parts per million* (ppm), a dimensionless fraction of the total applied field. By custom, the difference in field (the chemical shift) between the NMR signal of a proton and that of TMS is not measured in gauss, but in frequency units (hertz or Hz). Remember, in NMR, frequency units and magnetic field units are always proportional, with $\nu = \gamma H_0/2\pi$. The horizontal axis of the NMR spectrum is calibrated in hertz. A chemical shift in ppm can be calculated by dividing the shift measured in hertz by the spectrometer frequency measured in millions of hertz (megahertz or MHz).

$$\text{chemical shift (ppm)} = \frac{\text{shift downfield from TMS (Hz)}}{\text{total spectrometer frequency (MHz)}}$$

The chemical shift (in ppm) of a given proton is the same regardless of the operating field and frequency of the spectrometer.

The most common scale of chemical shifts is the δ (delta) scale, which we will use (Fig. 12-8). The absorption of tetramethylsilane (TMS) is *defined* as 0.00 ppm on the δ scale. Most protons absorb at lower fields than TMS, so the δ scale increases toward the lower field (toward the left of the spectrum). The spectrum is calibrated in both frequency and ppm δ.

$$\text{chemical shift, ppm } \delta = \frac{\text{shift downfield from TMS (in Hz)}}{\text{spectrometer frequency (in MHz)}}$$

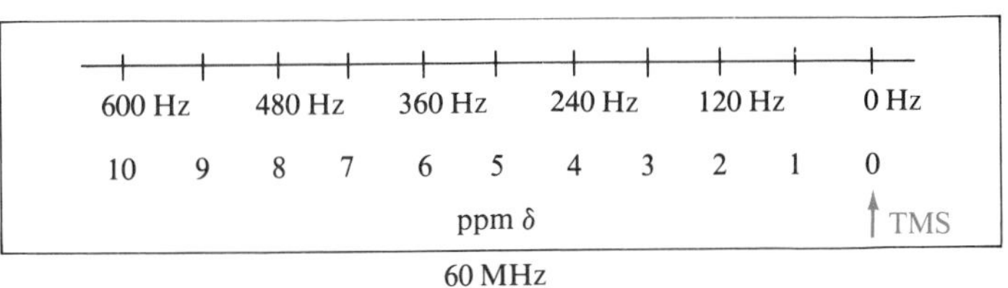

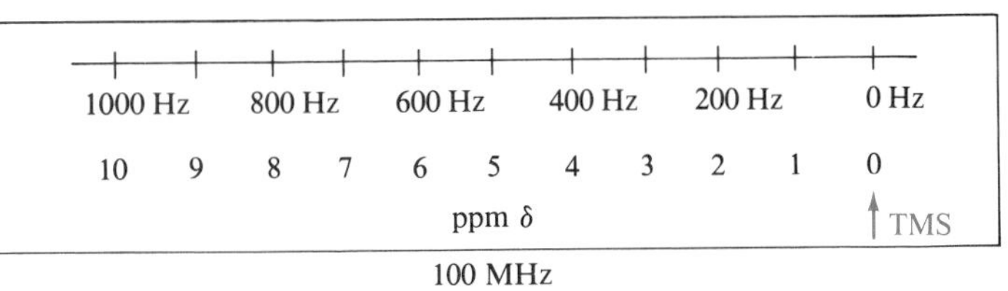

FIGURE 12-8 Use of the δ scale with 60- and 100-MHz spectrometers. The absorption of TMS is defined as 0, with the scale increasing from right to left (toward the lower field). Each δ unit is 1 ppm difference from TMS: 60 Hz at 60 MHz, and 100 Hz at 100 MHz.

SOLVED PROBLEM 12-2

A 60-MHz spectrometer records a proton that absorbs at a frequency 426 Hz downfield from TMS.

(a) Determine its chemical shift, and express this shift as a magnetic field difference.
(b) Predict this proton's chemical shift at 100 MHz. Using a 100-MHz spectrometer, how far downfield (in gauss and in hertz) from TMS would this proton absorb?

SOLUTION (a) The chemical shift is the fraction

$$\frac{\text{shift downfield (Hz)}}{\text{spectrometer frequency (MHz)}} = \frac{426\ \text{Hz}}{60.0\ \text{MHz}} = 7.10\ \text{ppm}$$

The chemical shift of this proton is $\delta 7.10$. The field shift is 14,092 gauss $\times$ $(7.10 \times 10^{-6}) =$ 0.100 gauss.

(b) The chemical shift is unchanged at 100 MHz: $\delta 7.10$.
The field shift is 23,486 gauss $\times$ $(7.10 \times 10^{-6}) = 0.167$ gauss.
The frequency shift is 100 MHz $\times$ $(7.10 \times 10^{-6}) = 710$ Hz.

PROBLEM 12-1

Using a 60-MHz spectrometer, the protons in iodomethane absorb at a position 130 Hz downfield from TMS.

(a) What is the chemical shift of these protons?
(b) Determine the difference in the magnetic field required for resonance of the iodomethane protons compared with the TMS protons.
(c) What is the chemical shift of the iodomethane protons using a 100-MHz spectrometer?
(d) How many hertz downfield from TMS would they absorb at 100 MHz?

Most student NMR spectrometers operate at 60 MHz, and the proton spectra used in this text were recorded at 60 MHz unless labeled otherwise. The 60-MHz NMR spectrum of methanol (Fig. 12-9) shows the two absorptions of methanol together with the TMS reference peak at $\delta 0.0$. The methyl protons absorb 205 Hz (0.048 gauss) downfield from TMS. Their chemical shift is 3.4 ppm, so we say that the methyl protons absorb at $\delta 3.4$. The hydroxyl proton absorbs farther downfield, at a position around 290 Hz (0.068 gauss) from TMS. Its chemical shift is $\delta 4.8$.

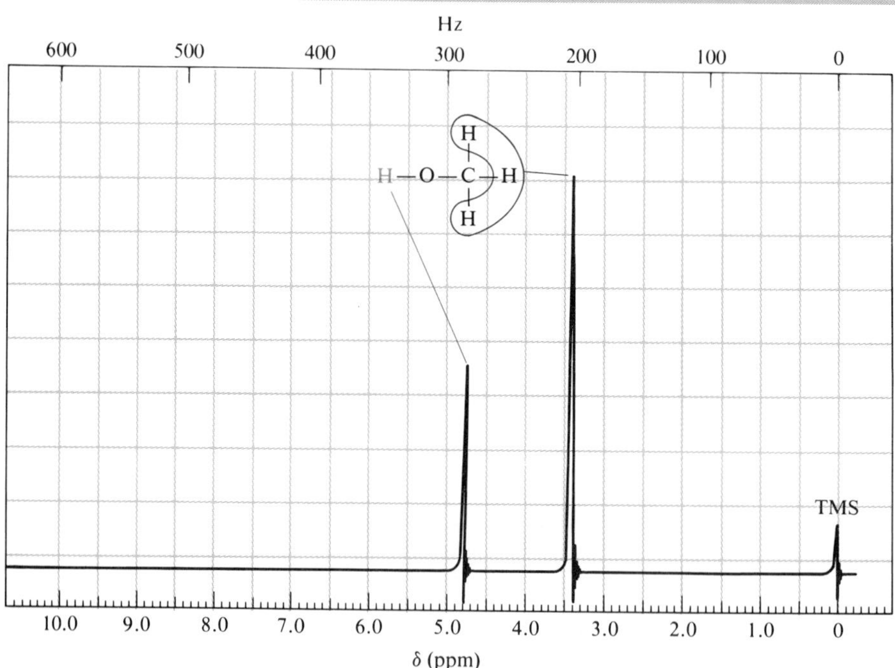

FIGURE 12-9 60-MHz NMR spectrum of methanol. The methyl protons absorb at $\delta 3.4$, and the hydroxyl proton absorbs at $\delta 4.8$.

Both the hydroxyl proton and the methyl protons of methanol show the deshielding effects of the electronegative oxygen atom. The chemical shift of a methyl group in an alkane is about $\delta 0.9$. Therefore, the methanol oxygen deshields the methyl protons by an additional 2.6 ppm. Other electronegative atoms produce similar deshielding effects. Table 12-1 compares the chemical shifts of methanol with those of the methyl halides. Notice that the chemical shift of the methyl protons depends on the electronegativity of the substituent, with more electronegative substituents resulting in more deshielding and larger chemical shifts.

TABLE 12-1
Variation of chemical shift with electronegativity

	X in CH_3—X				
	F	OH	Cl	Br	I
electronegativity of X	4.1	3.5	2.8	2.7	2.2
chemical shift of CH_3—X	$\delta 4.3$	$\delta 3.4$	$\delta 3.1$	$\delta 2.7$	$\delta 2.2$

The effect of an electronegative group on the chemical shift also depends on its distance from the protons. In methanol, the hydroxyl proton is separated from oxygen by one bond, and its chemical shift is $\delta 4.8$. The methyl protons are separated from oxygen by two bonds, and their chemical shift is $\delta 3.4$. In general, the effect of an electron-withdrawing substituent decreases with increasing distance, and the effects are usually negligible on protons that are separated from the electronegative group by four or more bonds.

This decreasing effect can be seen by comparing the chemical shifts of the various protons in 1-bromobutane with those in butane. The deshielding effect of an electronegative substituent drops off rapidly with distance. In 1-bromobutane, protons on the α-carbon are deshielded by about 2.5 ppm, and the β protons are deshielded by about 0.4 ppm. Protons that are more distant than the β protons are deshielded by a negligible amount.

TABLE 12-2
Chemical shifts of the chloromethanes[a]

Compound	Chemical shift	Difference
CH_4 (H—C—H with H above and below)	$\delta 0.2$	
		2.8 ppm
CH_3Cl (H—C—Cl with H above and below)	$\delta 3.0$	
		2.3 ppm
CH_2Cl_2 (H—C—Cl with Cl above, H below)	$\delta 5.3$	
		1.9 ppm
$CHCl_3$ (H—C—Cl with Cl above and below)	$\delta 7.2$	

[a] Each chlorine atom added changes the chemical shift of the remaining methyl protons by about 2 to 3 ppm. These changes are nearly additive.

	butane				1-bromobutane			
	H—CH_2—CH_2—CH_2—CH_2—H (H—C—C—C—C—H)				H—C—C—C—C—Br			
chemical shift:	0.9	1.3	1.3	0.9	0.9	1.3	1.7	3.4
deshielding resulting from Br, ppm:					0.0	0.0	0.4	2.5

If more than one electron-withdrawing group is present, the deshielding effects are nearly (but not quite) additive. In the chloromethanes (Table 12-2), the addition of the first chlorine atom causes a shift to $\delta 3.0$, the second chlorine shifts the absorption further to $\delta 5.3$ and the third chlorine moves the chemical shift to $\delta 7.2$ for chloroform. The chemical shift *difference* is about 2 to 3 ppm each time another chlorine atom is added, but each additional chlorine moves the peak slightly less than the previous one did.

12-5B CHARACTERISTIC VALUES OF CHEMICAL SHIFTS

Since the chemical shift of a proton is determined by its environment, we can construct a table of approximate chemical shifts for many types of compounds. Let's begin with a short table of representative chemical shifts (Table 12-3) and then consider the reasons for some of the more interesting and unusual values. A comprehensive table of chemical shifts appears in Appendix 1.

TABLE 12-3
Typical values of chemical shifts[a]

Type of proton	*Approximate δ*
alkane ($-CH_3$)	0.9
alkane ($-CH_2-$)	1.3
alkane ($-\overset{|}{\underset{|}{C}}H-$)	1.4
$-\overset{O}{\overset{\|}{C}}-CH_3$	2.1
$-C{\equiv}C-H$	2.5
$R-CH_2-X$ (X = halogen, O)	3–4
$>C{=}C<^{H}$ (vinyl H)	5–6
$>C{=}C<^{CH_3}$ (allylic CH_3)	1.7
$Ph-H$	7.2
$Ph-CH_3$	2.3
$R-CHO$	9–10
$R-COOH$	10–12
$R-OH$	variable, about 2–5
$Ar-OH$	variable, about 4–7
$R-NH_2$	variable, about 1.5–4

[a] These values are approximate, as all chemical shifts are affected by neighboring substituents. The numbers given here assume that alkyl groups are the only other substituents present. A more complete table of chemical shifts appears in Appendix 1.

SOLVED PROBLEM 12-3

Using Table 12-3, predict the chemical shifts of the protons in the following compounds.

(a) $CH_3-\overset{O}{\overset{\|}{C}}-OH$ (b) $Cl-CH_2^a-CH_2^b-CH_3^c$ (c) $(CH_3^a)_3CCH^b{=}CH_2^c$

SOLUTION (a) The methyl group in acetic acid is adjacent to a carbonyl group; Table 12-3 predicts a chemical shift of about δ2.1. (The experimental value is δ2.10.) The acid proton ($-COOH$) should absorb between δ10 and δ12. (The experimental value is δ11.4, variable.)

(b) Protons *a* are on the carbon atom bearing the chlorine, and they absorb between δ3 and δ4 (experimental: δ3.7). Protons *b* are one carbon removed, and they are predicted to absorb about δ1.7, like the β protons in 1-bromobutane (experimental: δ1.8). The methyl protons *c* will be nearly unaffected, absorbing around δ0.9 ppm (experimental: δ1.0).

(c) Methyl protons *a* are expected to absorb around δ0.9 (experimental: δ1.0). The vinyl protons *b* and *c* are expected to absorb between δ5 and δ6 (experimental δ5.8 for *b* and δ4.9 for *c*).

Vinyl and aromatic protons Table 12-3 shows that double bonds and aromatic rings produce large deshielding effects on their vinyl and aromatic protons. These deshielding effects result from the same type of circulation of electrons that normally shields nuclei from the magnetic field. In benzene and its derivatives, the aromatic ring of electrons acts as a conductor, and the external magnetic field induces a ring current (Fig. 12-10). At the center of the ring, the induced field acts

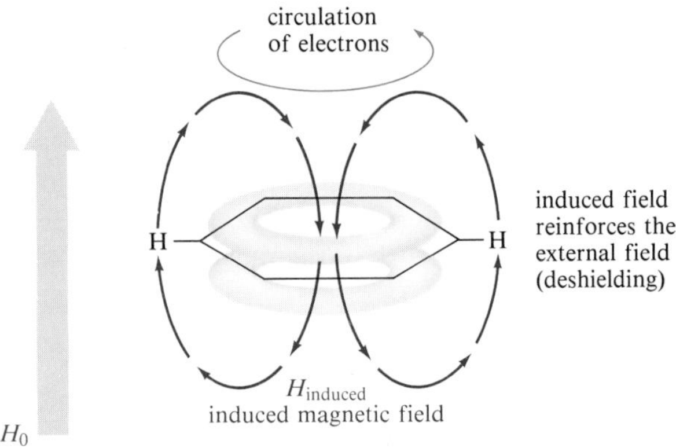

FIGURE 12-10 The induced magnetic field of the circulating aromatic electrons opposes the applied magnetic field along the axis of the ring. The aromatic hydrogens are on the equator of the ring, where the induced field lines curve around and reinforce the applied field.

to oppose the external field. These induced field lines curve around, however, and on the edge of the ring the induced field *adds to* the external field. As a result, the aromatic protons are actually *deshielded,* resulting in absorption at low values of the applied magnetic field. Benzene absorbs at δ7.2, and most aromatic protons absorb in the range δ7 to δ8.

We should remember that the benzene molecule is not always lined up in the position shown in Figure 12-10. Because it is constantly tumbling in the solution, the chemical shift observed for its protons is an average of all the possible orientations. If we could hold a benzene molecule in the position shown in Figure 12-10, its protons would absorb at a field even lower than δ7.2. Other orientations, such as the one with the benzene ring edge-on to the magnetic field, would be less deshielded and would absorb at a higher field. It is the *average* of all these orientations that is observed by the resonance at δ7.2.

Figure 12-11 shows the NMR spectrum of toluene (methylbenzene). The aromatic protons absorb at a chemical shift of δ7.2. The methyl protons are deshielded by a smaller amount, absorbing at δ2.3.

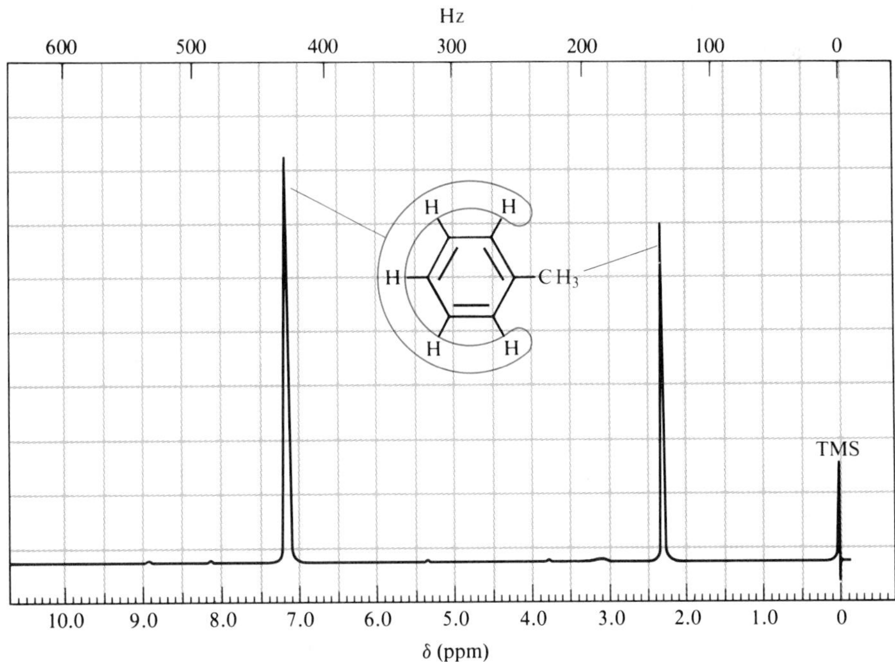

FIGURE 12-11 Proton NMR spectrum of toluene. The aromatic protons absorb at a chemical shift of δ7.2, and the methyl protons absorb at δ2.3.

The vinyl protons of an alkene are deshielded by the pi electrons in the same way that aromatic protons are deshielded. The effect is not so large in the alkene, however, because there is not such a large, effective ring of electrons as there is in benzene. Once again, the motion of the pi electrons generates an induced magnetic field that opposes the applied field at the middle of the double bond. The vinyl protons are on the periphery of this field, however, where the induced field reinforces the external field (Fig. 12-12). As a result of this deshielding effect, most vinyl protons absorb in the range $\delta 5$ to $\delta 6$.

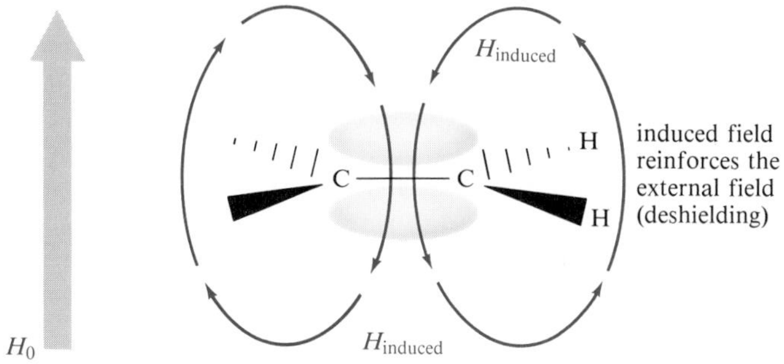

$H_{induced}$
H
C
C
H
induced field reinforces the external field (deshielding)
H_0
$H_{induced}$

FIGURE 12-12 Vinyl protons are positioned on the periphery of the induced magnetic field of the pi electrons. In this position they are deshielded by the induced magnetic field.

Acetylenic hydrogens Since the pi bond of an olefin deshields the vinyl protons, we might expect an acetylenic hydrogen (—C≡C—H) to be even more deshielded by the two pi bonds of the triple bond. The opposite is true: Acetylenic hydrogens absorb around $\delta 2.5$, compared with $\delta 5$ to $\delta 6$ for vinyl protons. Figure 12-13 shows that the triple bond has a cylinder of electron density surrounding the sigma bond. As the molecules tumble in solution, in some orientations this cylinder of electrons can circulate to produce an induced magnetic field. The acetylenic proton lies along the *axis* of this induced field, and the induced field is shielding. When this shielded orientation is averaged with all other possible orientations of the molecule, the result is a resonance around $\delta 2.5$.

$H_{induced}$ shields the proton

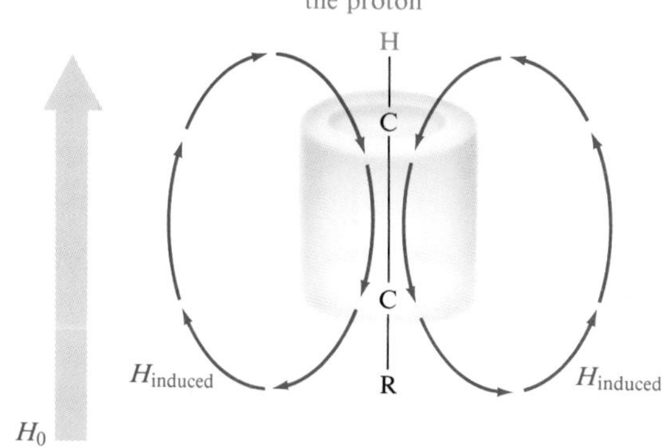

FIGURE 12-13 When the acetylenic triple bond is aligned with the magnetic field, the cylinder of electrons circulates to create an induced magnetic field. The acetylenic proton lies along the axis of this field, where the induced magnetic field opposes the external field.

Aldehyde protons Aldehyde protons (—CHO) absorb at even lower fields than vinyl protons and aromatic protons: between $\delta 9$ and $\delta 10$. Figure 12-14 shows that the aldehyde proton is deshielded both by the circulation of the electrons in the double bond and by the inductive electron-withdrawing effect of the carbonyl oxygen atom.

Hydrogen-bonded protons The chemical shifts of O—H protons in alcohols and

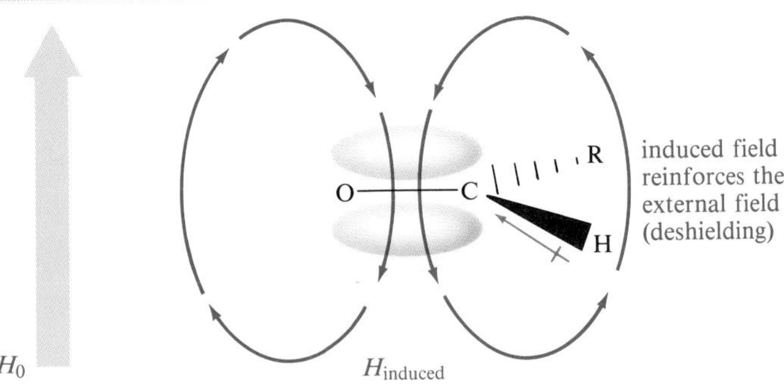

FIGURE 12-14 Like a vinyl proton, the aldehyde proton is deshielded by the circulation of electrons in the pi bond. It is also deshielded by the electron-withdrawing effect of the carbonyl (C=O) group, giving a resonance between $\delta 9$ and $\delta 10$.

N—H protons in amines depend on the concentration. In concentrated solutions, these protons are deshielded by hydrogen bonding, and they absorb at a relatively low field: about $\delta 3.5$ for an amine N—H and about $\delta 4.5$ for an alcohol O—H. When the alcohol or amine is diluted with a non-hydrogen-bonding solvent such as CCl_4, hydrogen bonding becomes less important. In dilute solutions, these resonances are commonly observed around $\delta 2$.

Hydrogen bonding and the proton exchange that accompanies it may contribute to a broadening of the peak corresponding to the resonance of an O—H or N—H proton. A broad peak is observed because protons are exchanging from one molecule to another during the NMR resonance (see Section 12-10). The protons pass through a variety of environments during this exchange, giving absorptions over a wider range of frequencies and field strengths.

Carboxylic acid protons Since carboxylic acid protons are adjacent to a carbonyl group, they have considerable positive character. They are strongly deshielded, and absorb at chemical shifts greater than $\delta 10$. Carboxylic acids frequently exist as hydrogen-bonded dimers, with moderate rates of proton exchange that broaden the absorption of the acid proton.

The proton NMR spectrum of acetic acid is shown in Figure 12-15. As we expect, the methyl group adjacent to the carbonyl absorbs at a chemical shift of $\delta 2.1$. The acid proton absorption appears at a chemical shift that is not scanned in the usual range of the NMR spectrum. It is seen in a second trace with a 150-Hz offset, meaning that this trace actually corresponds to frequencies with chemical shifts 150 Hz *larger* than that shown on the trace. The acid proton absorption appears around 710 Hz: 560 Hz read from the trace, plus the 150 Hz offset. 710 Hz corresponds to a chemical shift of $\delta 11.8$.

PROBLEM 12-2

Predict the chemical shifts of the protons in the following compounds.

(a) 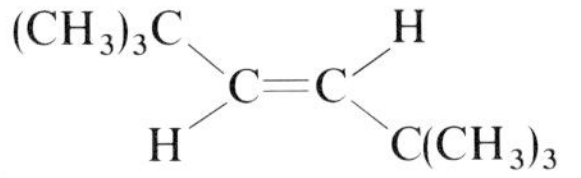 $(CH_3)_3C(H)C{=}C(H)C(CH_3)_3$ (trans)

(b) 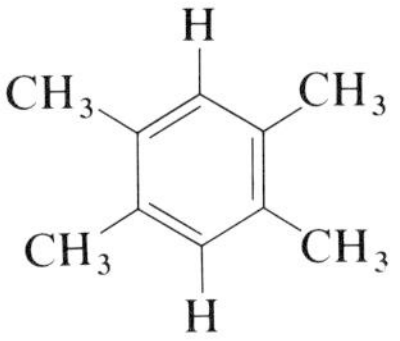 1,2,4,5-tetramethylbenzene (benzene ring with four CH_3 groups and two H)

(c) CH_3O—C_6H_4—OCH_3 (para-dimethoxybenzene, ring H's shown)

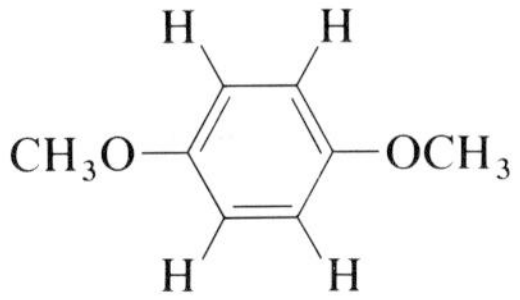

(d) CH_3—$C(CH_3)(OH)$—C≡C—H

(e) C_6H_5—CH_2—C(=O)—OH

(f) CH_3—$C(CH_3)(Br)$—CH_2Br

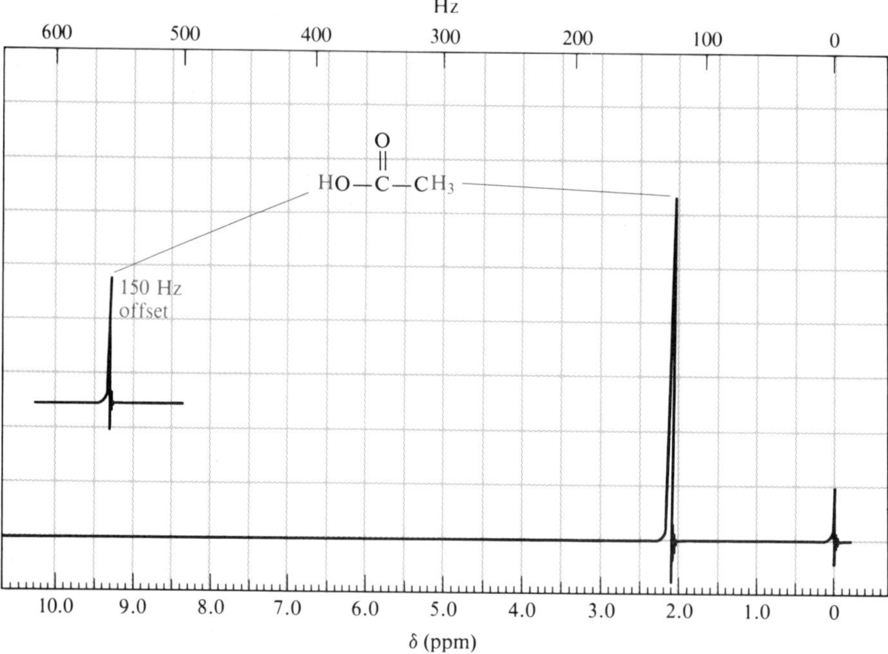

FIGURE 12-15 In the NMR spectrum of acetic acid, the methyl protons are deshielded to about $\delta 2.1$ by the adjacent carbonyl group. The acid proton appears at $\delta 11.8$, shown on an offset trace.

12-6 THE NUMBER OF SIGNALS

In general, the number of NMR signals corresponds with the number of different kinds of protons present in the molecule. For example, methyl *t*-butyl ether has two types of protons (Fig. 12-16). The three methyl protons are chemically identical, and they give rise to a single absorption at $\delta 3.4$. The *t*-butyl protons are chemically different from the methyl protons, absorbing at $\delta 1.2$.

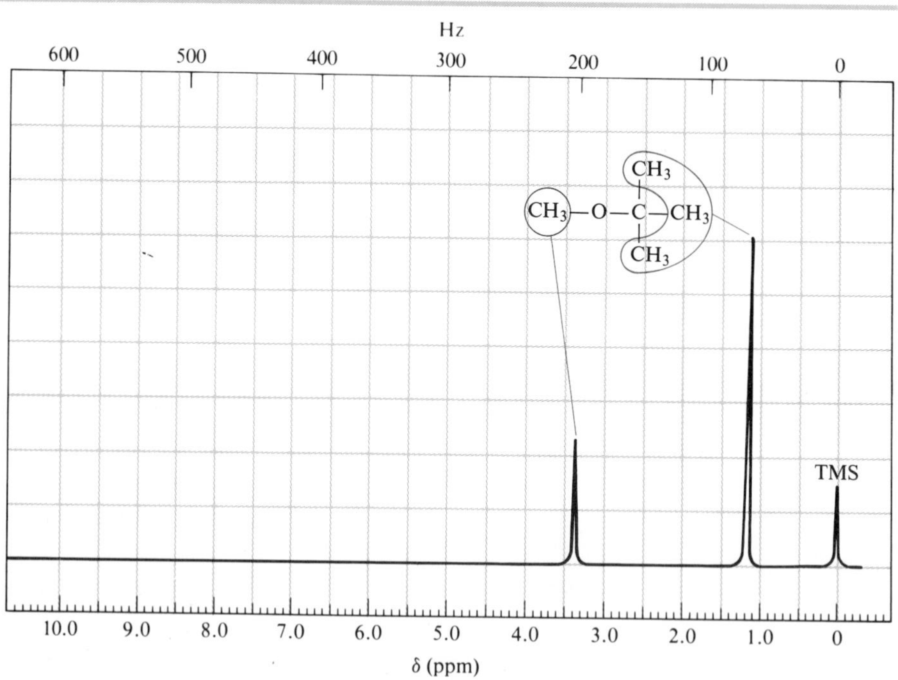

FIGURE 12-16 There are two types of protons in methyl *t*-butyl ether, giving two NMR signals.

Protons in identical chemical environments with the same shielding have the same chemical shift. Such protons are said to be **chemically equivalent.** This is what is meant whenever we use the term *equivalent* in discussing NMR spectroscopy. In methyl *t*-butyl ether, the three methyl protons are chemically equivalent and the nine *t*-butyl protons are chemically equivalent.

Another example is methyl acetoacetate, whose spectrum is shown in Figure 12-17. This ester has three types of protons: the methoxyl protons (*a*), with a chemical shift of δ3.8; the methylene protons (*b*), deshielded by two adjacent carbonyl groups, with a chemical shift of δ3.5; and the methyl protons (*c*), at δ2.3.

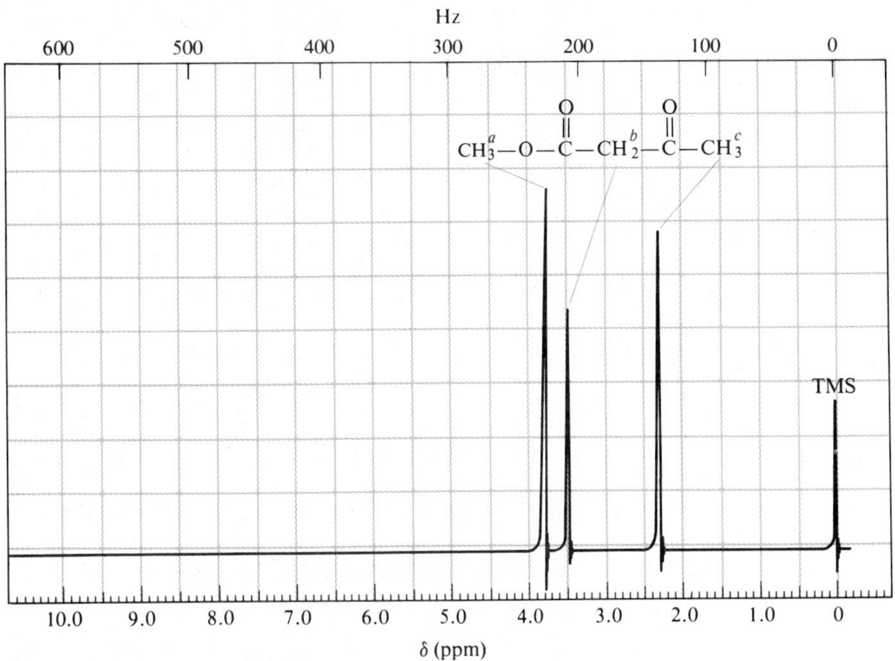

FIGURE 12-17 There are three types of protons in methyl acetoacetate, giving three signals in the NMR spectrum.

In some cases, fewer signals may appear in the NMR spectrum than there are different types of protons in the molecule. For example, Figure 12-18 shows the structure and spectrum of *o*-xylene (1,2-dimethylbenzene). There are three different types of protons, labeled *a* for the two equivalent methyl groups, *b* for the protons adjacent to the methyl groups, and *c* for the protons two carbons removed. The spectrum shows only two absorptions, however.

The upfield signal at δ2.3 corresponds to the six methyl protons, H^a. The absorption at δ7.2 corresponds to all four of the aromatic protons, H^b and H^c. Although the two types of aromatic protons are different, the methyl groups do not strongly influence the electron density of the ring or the amount of shielding felt by any of the substituents on the ring. The aromatic protons produce two signals, but these signals happen to occur at the same chemical shift. These protons are said to be **accidentally equivalent.**

PROBLEM 12-3

Determine the number of different kinds of protons in each of the following compounds.

(a) 1-chloropropane (b) 2-chloropropane
(c) 2,2-dimethylbutane (d) 2-bromo-1-methylbenzene,

CH_3

Br

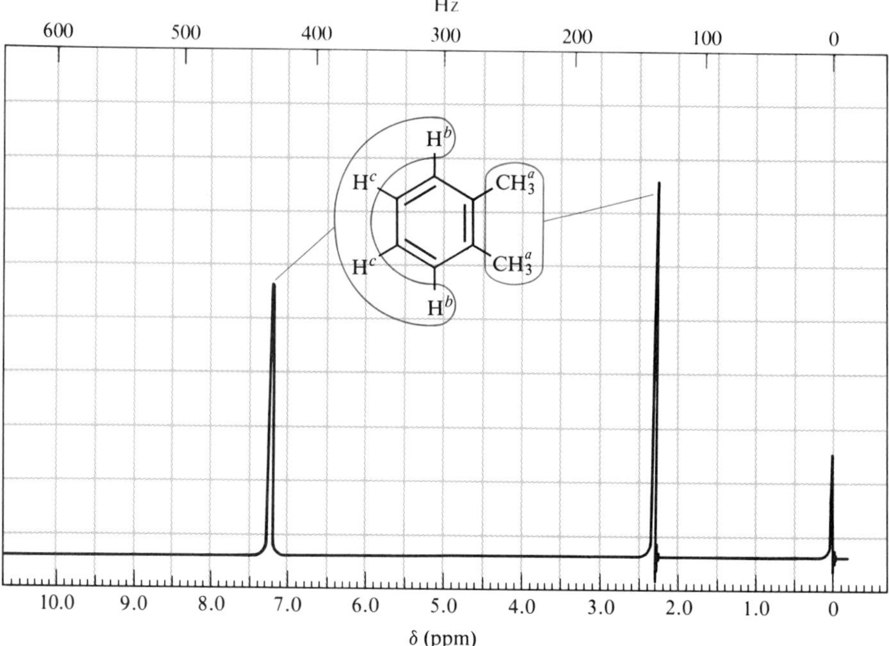

FIGURE 12-18 There are three types of protons in *o*-xylene, but only two absorptions are seen in the spectrum. The aromatic protons H^b and H^c are accidentally equivalent, producing a single peak at $\delta 7.2$.

12-7 AREAS OF THE PEAKS

The area under a peak is proportional to the number of hydrogens contributing to that peak. For example, in the methyl *t*-butyl ether spectrum (Fig. 12-19) the absorption of the *t*-butyl protons is larger and stronger than that of the methyl protons. This difference reflects the fact that there are three times as many *t*-butyl protons as methyl protons. We cannot simply compare peak heights, however; it is the *area* under the peak that is important.

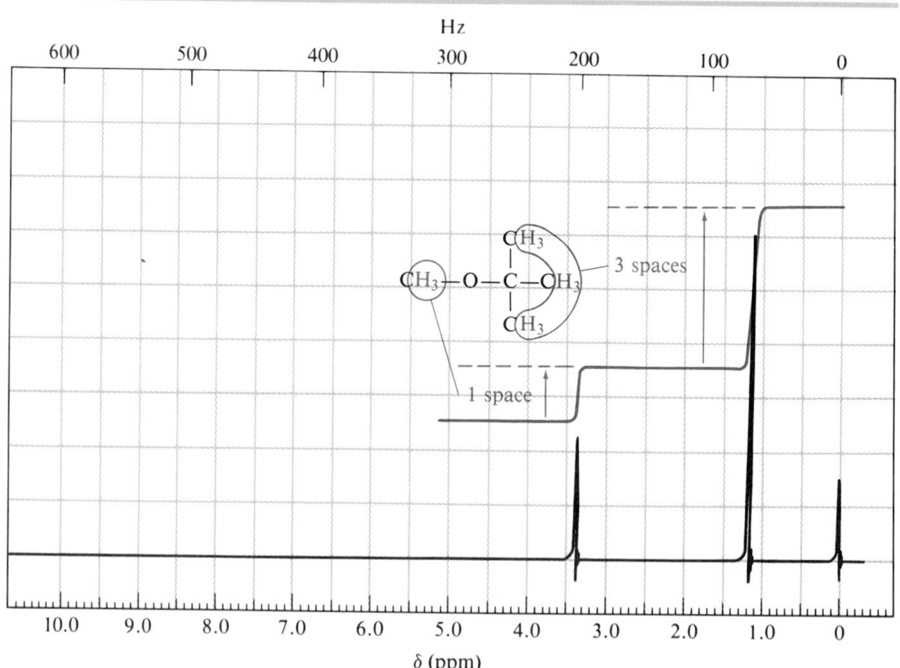

FIGURE 12-19 Integrated NMR spectrum of methyl *t*-butyl ether. In going over a peak, the integrator trace (blue) rises by an amount that is proportional to the area under the peak.

NMR spectrometers have **integrators** that compute the relative areas of peaks. In the integration mode the instrument produces a second trace that rises when it goes over a peak. The amount that the integrator trace rises is proportional to the area of that peak. The integrator trace, shown in blue in Figure 12-19, does not specifically indicate that methyl *t*-butyl ether has three methyl hydrogens and nine *t*-butyl hydrogens. It simply shows that about three times as many hydrogens are represented by the absorption at $\delta 1.2$ as are represented by the absorption at $\delta 3.4$.

Consider another example: Figure 12-20 shows the integrated spectrum of a compound with molecular formula $C_6H_{12}O_2$.

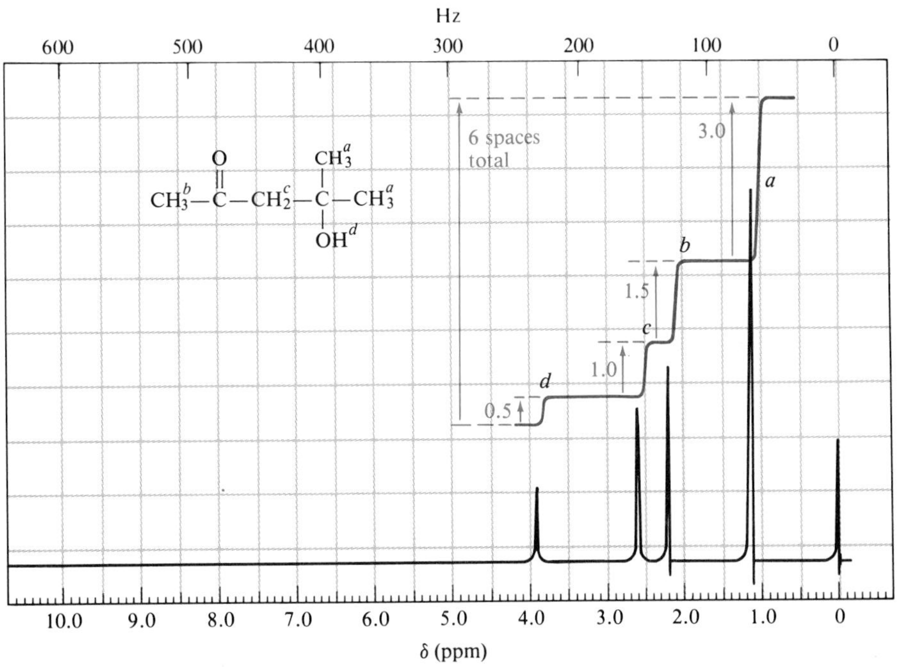

FIGURE 12-20 Proton NMR spectrum for a compound of molecular formula $C_6H_{12}O_2$.

Because we know the molecular formula, we can use the integral trace to determine exactly how many protons are responsible for each peak. The integrator has moved a total of 6 spaces vertically in integrating the 12 protons in the molecule. Each proton is represented by

$$\frac{6 \text{ spaces}}{12 \text{ hydrogens}} = \text{about 0.5 space per hydrogen}$$

The signal at $\delta 3.9$ has an integral of 0.5 space, so it must represent one proton. At $\delta 2.6$, the integrator moves 1 space, corresponding to two protons. The signal at $\delta 2.2$ has an integral of 1.5 spaces, for three protons; and the signal at $\delta 1.2$ (3 spaces) corresponds to six protons. Considering the expected chemical shifts together with the information provided by the integrator leaves little doubt which protons are responsible for which signals in the spectrum. We can make the following peak assignments:

$H_3^bC-C(=O)-CH_2^c-C(CH_3^a)(OH^d)-CH_3^a$

area = 3, $\delta 2.2$ (CH_3^b)

area = 2, $\delta 2.6$ (CH_2^c)

area = 6, $\delta 1.2$ (CH_3^a)

area = 1, $\delta 3.9$ (OH^d)

PROBLEM 12-4

Draw the integral trace expected for the NMR spectrum of methyl acetoacetate, shown in Figure 12-17.

PROBLEM 12-5

Determine the ratios of the peak areas in the following spectra. Then use this information, together with the chemical shifts, to pair up the compounds with their spectra. Assign the peaks in each spectrum to the protons they represent in the molecular structure. Possible structures:

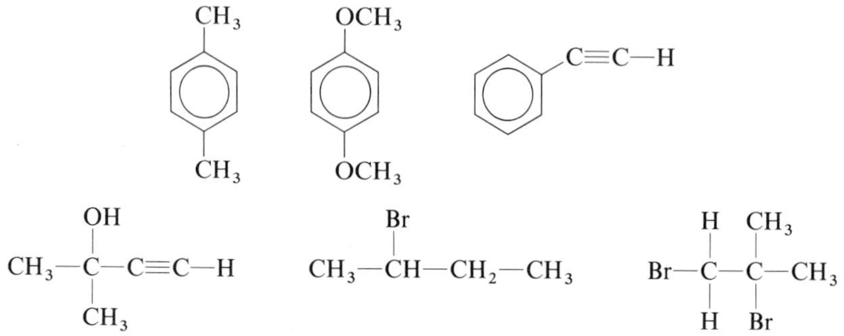

CH_3–C_6H_4–CH_3 (para) OCH_3–C_6H_4–OCH_3 (para) C_6H_5–$C{\equiv}C$–H

CH_3–C(OH)(CH_3)–$C{\equiv}C$–H CH_3–CH(Br)–CH_2–CH_3 Br–CH_2–C(Br)(CH_3)–CH_3

Suggestions: (1) If you are having trouble counting the fractional spaces, use a millimeter ruler to measure the integrals. (2) You don't know the total number of hydrogens, so try setting the smallest integral equal to one hydrogen and the others proportionally. If some of the other integrals are not whole numbers of hydrogens, then set the smallest equal to 2 or 3 as required. For example, 1:1.3:2 would become 3:4:6 and we would look for a compound with this ratio or 6:8:12 or 9:12:18, and so on.

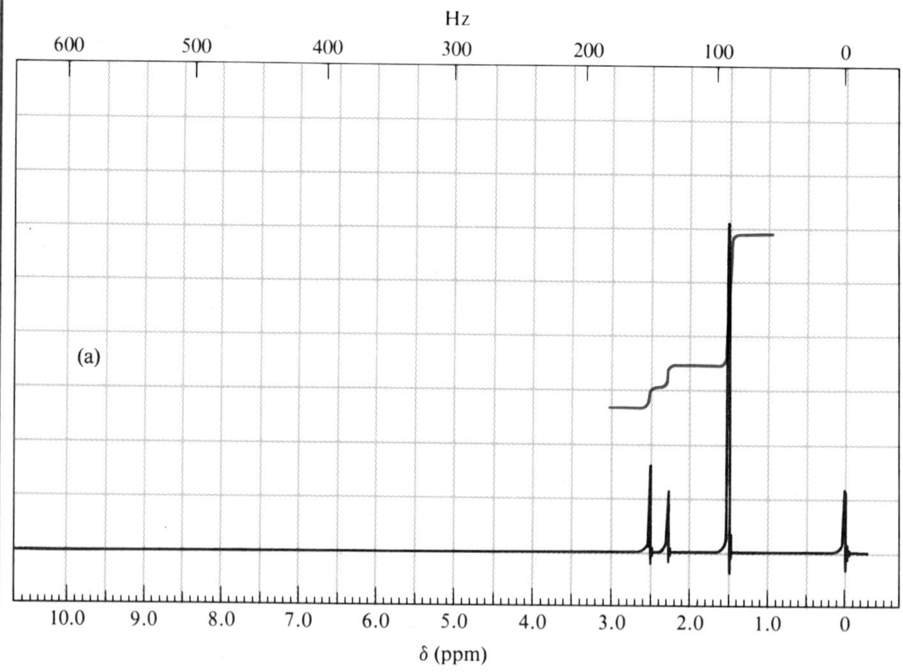

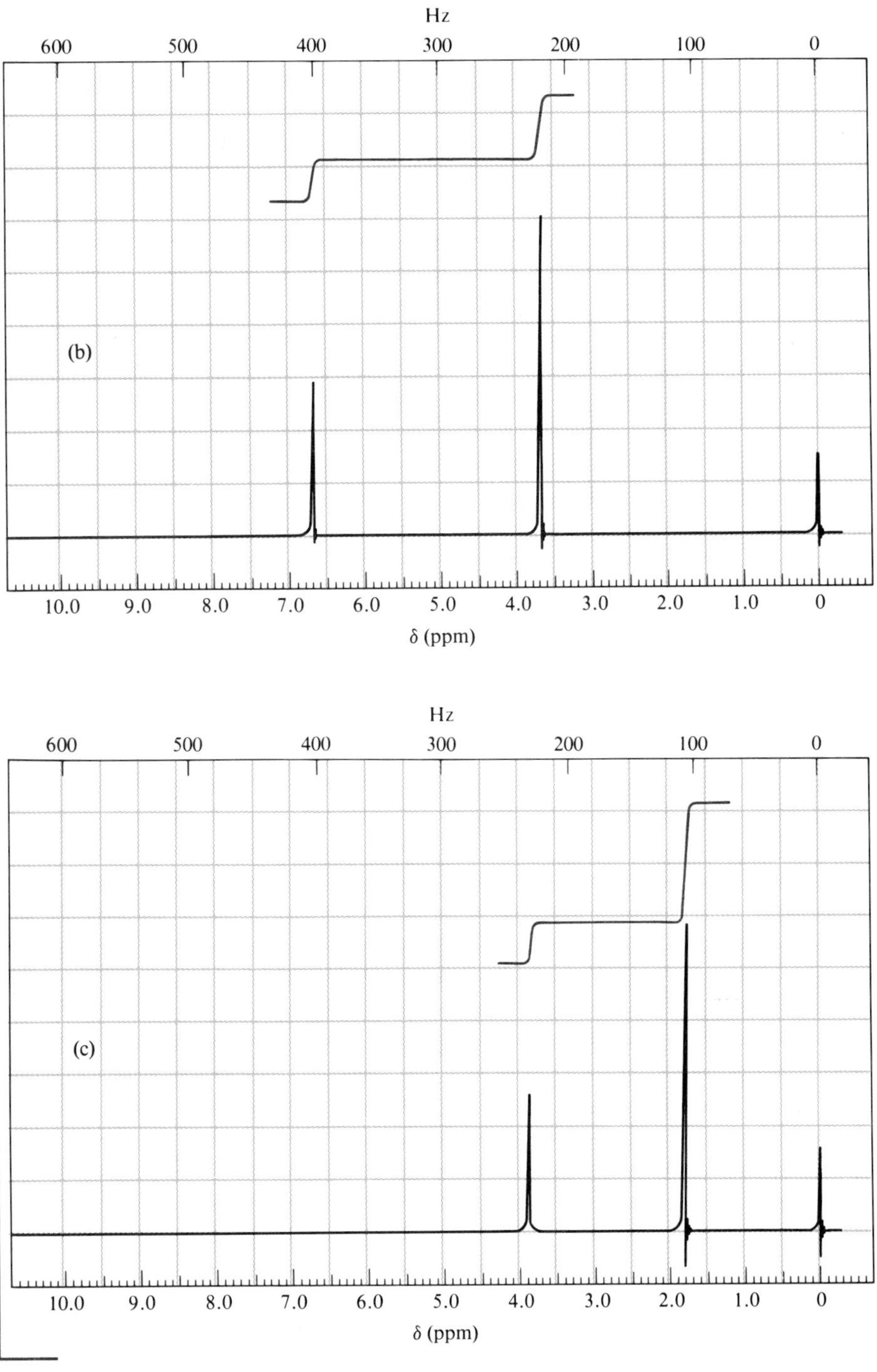

12-8 SPIN-SPIN SPLITTING

12-8A THEORY OF SPIN-SPIN SPLITTING

A proton in the NMR spectrometer is subjected to both the external magnetic field and the induced field of the shielding electrons. If there are other protons nearby, their small magnetic fields also affect the absorption frequency of the proton we are observing. Consider the spectrum of 1,1,2-tribromoethane (Fig. 12-21). As expected, there are two signals with areas in the ratio of 1:2. The smaller signal (H^a) appears at δ5.7, shifted downfield by the two adjacent bromine atoms. The larger signal (H^b) appears at δ4.1. These signals do not appear as single peaks, however, but rather as a triplet (three peaks) and a doublet (two peaks), respectively. This splitting of signals into multiplets, called spin-spin splitting, results when two different types of protons are close enough that their magnetic fields influence each other. Such protons are said to be **magnetically coupled.**

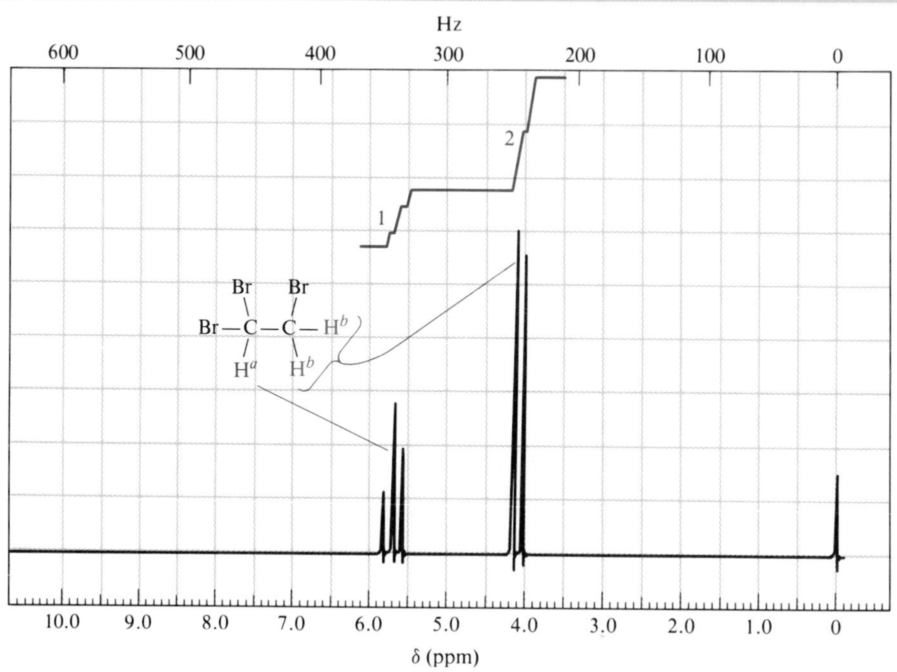

FIGURE 12-21 The proton NMR spectrum of 1,1,2-tribromoethane consists of a triplet of area 1 at δ5.7 and a doublet of area 2 at δ4.1.

Spin-spin splitting is explained by considering the possible individual spins of the magnetically coupled protons. Assume that our spectrometer is scanning the signal for the H^b protons of 1,1,2-tribromoethane at δ4.1. These protons are under the influence of the small magnetic field of the adjacent proton, H^a. The field produced by H^a is not the same for every molecule in the sample. In some molecules H^a is aligned with the external magnetic field, and in others it is aligned against the field (Fig. 12-22).

When the H^a proton is aligned with the field, the H^b protons feel a slightly stronger total field: They are effectively deshielded, and they absorb at a lower field. When the H^a proton is aligned against the field, the H^b protons are shielded, and they absorb at a higher field. These are the two absorptions of the doublet seen for the H^b protons. About half of the molecules have H^a aligned with the field and about half against the field, so the two absorptions of the doublet are nearly equal in area.

Spin-spin splitting is a reciprocal property: If one proton splits another, the second proton must split the first. Proton *a* appears as a triplet (at δ5.7) because there are four permutations of the two H^b proton spins, with two of them giving the same magnetic field. When both H^b spins are aligned with the applied field, proton *a* is deshielded; when both H^b spins are aligned against the field, proton *a*

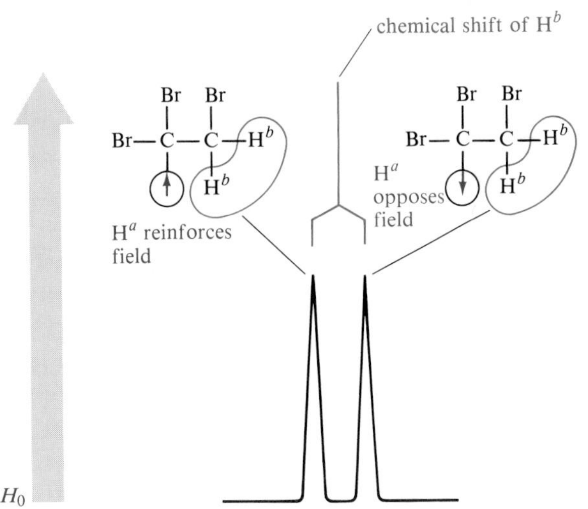

FIGURE 12-22 When the nearby H^a proton is aligned with the external magnetic field, it deshields H^b; when H^a is aligned against the field, it shields H^b.

is shielded; and when the two H^b spins are opposite each other (two possible permutations), they cancel each other out. Three signals result, with the middle signal twice as large as the others because it corresponds to two possible spin permutations (Fig. 12-23).

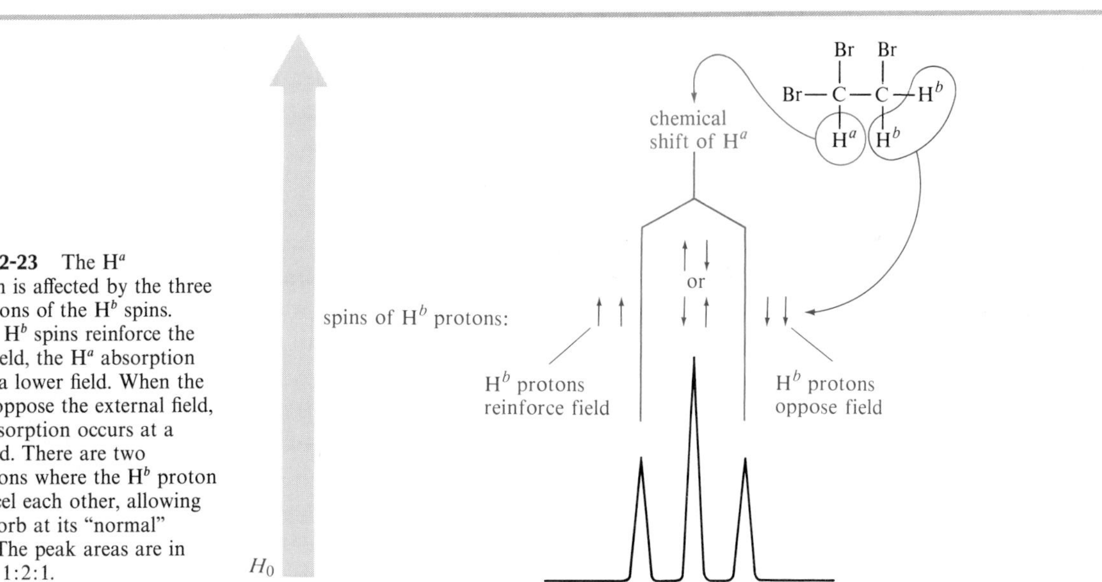

FIGURE 12-23 The H^a absorption is affected by the three combinations of the H^b spins. When the H^b spins reinforce the external field, the H^a absorption occurs at a lower field. When the H^b spins oppose the external field, the H^a absorption occurs at a higher field. There are two permutations where the H^b proton spins cancel each other, allowing H^a to absorb at its "normal" position. The peak areas are in the ratios 1:2:1.

12-8B THE $N + 1$ RULE

The type of analysis used above for the spin-spin splitting of 1,1,2-tribromoethane can be extended to more complicated systems. In general, the multiplicity (number of peaks) of a signal is given by the ***N* + 1 rule:**

> ***N* + 1 rule:** If a signal is split by N equivalent protons, it is split into $N + 1$ peaks.

The relative areas of the $N + 1$ **multiplet** that results are approximately given by the appropriate line of Pascal's triangle:

Relative peak intensities of symmetrical multiplets

Number of equivalent protons causing splitting	*Number of peaks (multiplicity)*	*Area ratios (Pascal's triangle)*
0	1 (singlet)	1
1	2 (doublet)	1 1
2	3 (triplet)	1 2 1
3	4 (quartet)	1 3 3 1
4	5 (quintet)	1 4 6 4 1
5	6 (sextet)	1 5 10 10 5 1
6	7 (septet)	1 6 15 20 15 6 1

In ethylbenzene (CH_3—CH_2—Ph), for example (Fig. 12-24), the methyl protons are split by two adjacent protons, and they appear upfield as a triplet of areas 1:2:1. The methylene (CH_2) protons are split by three protons, appearing at a lower field as a quartet of areas 1:3:3:1. This splitting pattern is typical for an ethyl group. Because ethyl groups are common, you should learn to recognize this familiar pattern. All five of the aromatic protons absorb close to 7.2 ppm because the alkyl substituent has little effect on the chemical shifts of the aromatic protons.

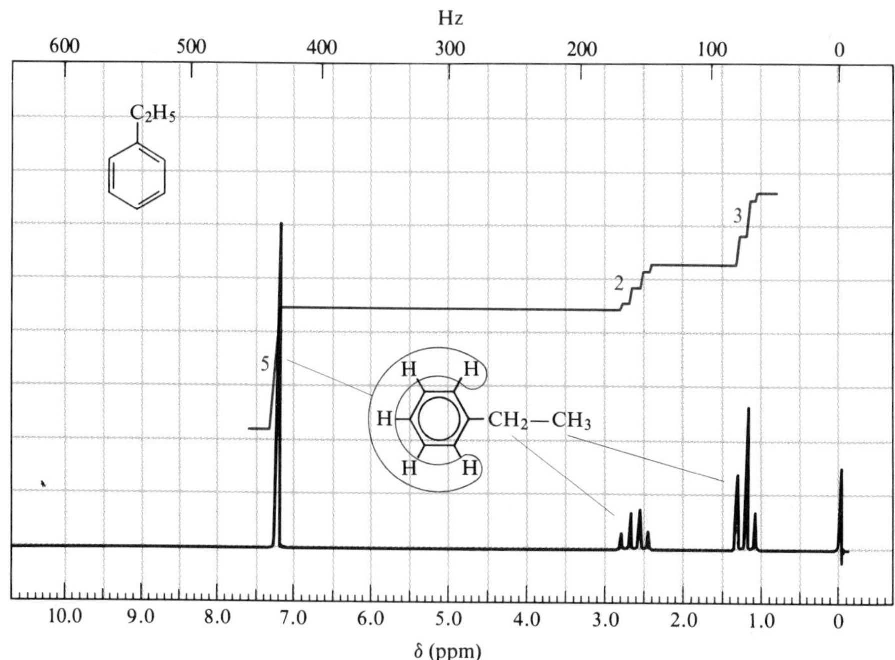

FIGURE 12-24 Proton NMR spectrum of ethylbenzene. The ethyl group appears as a triplet at $\delta 1.2$ (—CH_3) and a quartet at $\delta 2.6$ (—CH_2—). The aromatic protons, although theoretically not all equivalent, appear as a singlet at $\delta 7.2$.

Figure 12-24 also shows that the adjacent aromatic protons do not split each other. In general, *protons that are in resonance at the same field strength cannot produce observable spin-spin splitting.* Therefore, *splitting is never observed between equivalent protons.* The aromatic protons in ethylbenzene are not chemically equivalent, but they are "accidentally equivalent" and absorb at nearly the same field strength. The accidental equivalence prevents any observable spin-spin splitting between these aromatic protons.

12-8C THE RANGE OF MAGNETIC COUPLING

In ethylbenzene there is no spin-spin splitting between the aromatic protons and the protons of the ethyl group. These protons are not on adjacent carbon atoms, so they are too far away from each other to be magnetically coupled.

not on adjacent carbons; no splitting observed

protons on adjacent carbon atoms split each other

The magnetic coupling that causes spin-spin splitting takes place primarily through the bonds of the molecule. Most examples of spin-spin splitting involve coupling between protons that are separated by no more than three bonds: They are therefore bonded to adjacent carbon atoms (vicinal protons).

Most spin-spin splitting is between protons on adjacent carbon atoms.

Protons bonded to the same carbon atom (geminal protons) can split each other *only if they are nonequivalent.* In most cases, protons on the same carbon atom are equivalent, and they do not split each other.

Bonded to the same carbon: two bonds between protons

spin-spin splitting is normally observed (if nonequivalent)

Bonded to adjacent carbons: three bonds between protons

spin-spin splitting is normally observed (this is the most common case)

Bonded to nonadjacent carbons: four or more bonds between protons

spin-spin splitting is *not* normally observed

Protons separated by more than three bonds usually do not produce observable spin-spin splitting. Occasionally, such "long-range coupling" does occur, but these cases are unusual. For now, we consider only nonequivalent protons on adjacent carbon atoms (or closer) to be magnetically coupled.

You may have noticed that the two multiplets in the upfield part of the ethylbenzene spectrum are not quite symmetrical. In general, a multiplet "points" upward toward the signal of the protons responsible for the splitting. In the ethyl signal (Fig. 12-25), the quartet at lower field "points" toward the triplet at higher field, and vice versa.

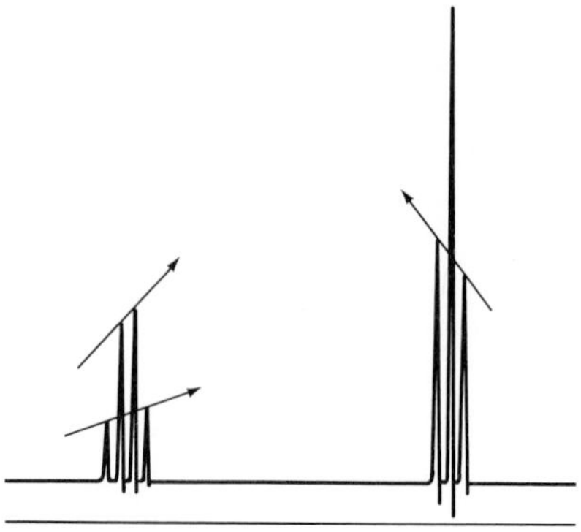

characteristic ethyl group

FIGURE 12-25 A multiplet often "points" upward toward the absorption of the protons that are causing the splitting. The ethyl multiplets in the ethylbenzene spectrum "point" toward each other.

As another example of a splitting pattern, the NMR spectrum of isopropyl methyl ketone (3-methyl-2-butanone) is shown in Figure 12-26.

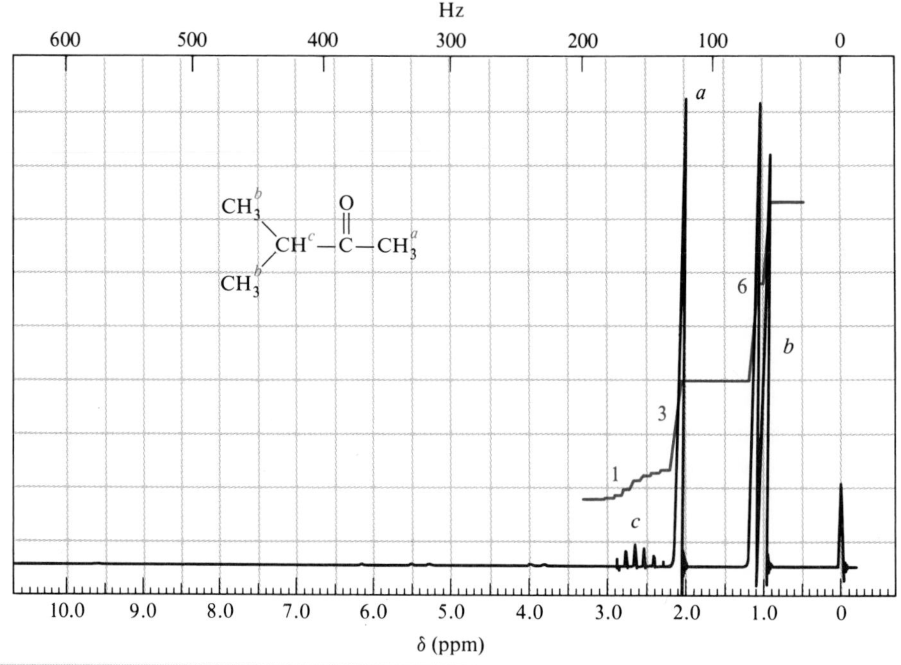

FIGURE 12-26 Proton NMR spectrum of isopropyl methyl ketone. The isopropyl group appears as a characteristic pattern of a strong doublet at a higher field and a weak multiplet (a septet) at a lower field. The methyl group appears as a singlet at $\delta 2.1$.

The three protons (a) of the methyl group bonded to the carbonyl appear as a singlet of relative area 3, at about $\delta 2.1$. Methyl ketones and acetate esters characteristically give such singlets around $\delta 2.1$, since there are no protons on the adjacent carbon atom.

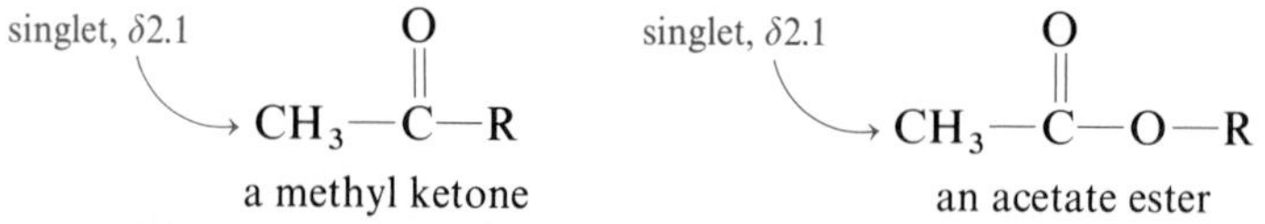

singlet, $\delta 2.1$ → CH_3—C(=O)—R
a methyl ketone

singlet, $\delta 2.1$ → CH_3—C(=O)—O—R
an acetate ester

The six methyl protons (b) of the isopropyl group are equivalent. They appear as an absorption of relative area 6 at about $\delta 1.1$, slightly deshielded by the carbonyl group two bonds away. This absorption is a downfield-pointing doublet because these protons are magnetically coupled to the methine proton c.

The methine proton H^c appears as an absorption of relative area 1, at $\delta 2.5$. This absorption is a septet (seven peaks), because it is coupled to the six adjacent methyl protons (*b*). Some very small peaks are part of this septet, so we may not see all seven unless the spectrum is amplified, as shown in Figure 12-27, where all seven peaks are visible. The pattern seen in this spectrum is typical for an isopropyl group: The methyl protons give a strong doublet at higher field, and the methine proton gives a weak multiplet (usually difficult to count the peaks) at a lower field. It is useful to be able to recognize an isopropyl group from this characteristic pattern.

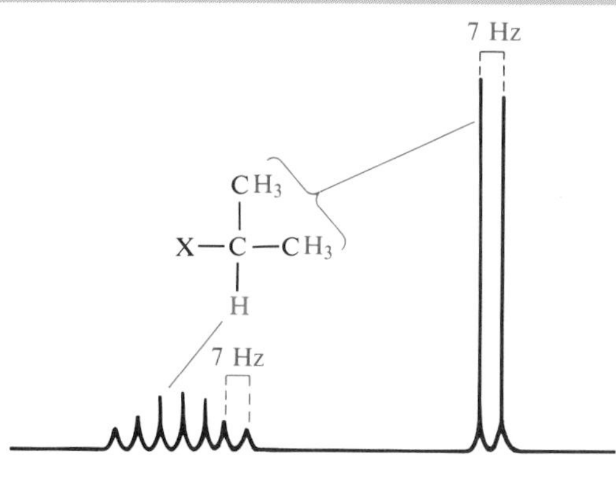

FIGURE 12-27 Characteristic isopropyl group pattern.

PROBLEM SOLVING: DRAWING AN NMR SPECTRUM

In learning about NMR spectra, you have seen that the chemical shift of a proton is determined by its environment, that typical chemical shift values can be assigned to specific types of protons, that the areas under peaks are proportional to the numbers of protons of each type, and that nearby protons of different types cause spin-spin splitting. By analyzing the structure of a molecule with these principles in mind, you can predict the general features of an NMR spectrum. The process is not difficult if a systematic approach is used. The stepwise method is illustrated here by drawing the NMR spectrum of the compound shown below.

$$(CH_3^a)_2CH^b{-}O{-}\overset{\overset{\displaystyle O}{\|}}{C}{-}CH_2^c{-}CH_3^d$$

1. **Determine how many types of protons are present, together with their proportions.** In the example above, there are four types of protons, labeled *a*, *b*, *c*, and *d*. The area ratios should be 6:1:2:3.
2. **Estimate the chemical shifts of the protons. (Table 12.3 and Appendix 1 serve as guides.)**
   Proton *b* is on a carbon atom bonded to oxygen; it should absorb around $\delta 3$ to $\delta 4$. Protons *a* are less deshielded by the oxygen, probably around $\delta 1$ to $\delta 2$. Protons *c* are on a carbon bonded to a carbonyl group; they should absorb around $\delta 2.1$ to $\delta 2.5$. Protons *d*, one carbon removed from a carbonyl, will be deshielded less than protons *c*, and also less than

protons *a*, which are next to a more strongly deshielded carbon atom. Protons *d* should absorb around $\delta 1.0$.

3. **Determine the splitting patterns.**
Protons *a* and *b* split each other into a doublet and a septet, respectively. (The result is a typical isopropyl group pattern.) Protons *c* and *d* split each other into a quartet and a triplet, respectively (a typical ethyl group pattern).

4. **Summarize each absorption in order, from the lowest field to the highest.**

	proton b	*protons c*	*protons a*	*protons d*
area	1	2	6	3
chemical shift	3–4	2.1–2.5	1–2	1
splitting	septet	quartet	doublet	triplet

5. **Draw the spectrum, using the information from your summary.**

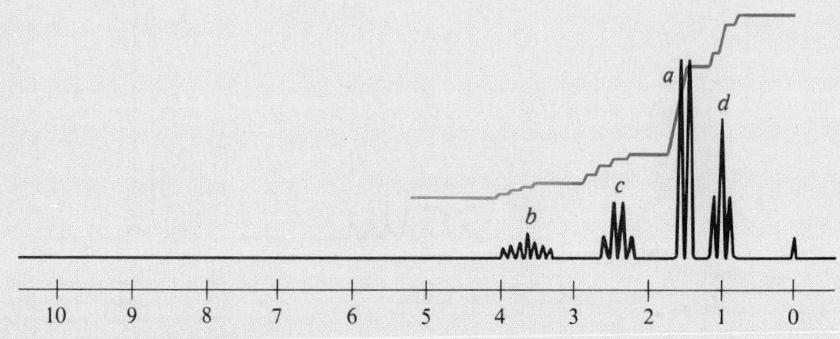

Working through the following problem should help you become comfortable with predicting NMR spectra.

PROBLEM 12-6

Draw the NMR spectra you expect for the following compounds.

(a) $(CH_3)_2CH{-}O{-}CH(CH_3)_2$

(b) $Cl{-}CH_2{-}CH_2{-}\overset{O}{\overset{\|}{C}}{-}O{-}CH_3$

(c) $Ph{-}CH(CH_3)_2$

(d) $CH_3CH_2O{-}C_6H_4{-}OCH_2CH_3$

(e) $\begin{array}{l} CH_2{-}COOCH_2CH_3 \\ | \\ CH_2{-}COOCH_2CH_3 \end{array}$

12-8D COUPLING CONSTANTS

The distances between the peaks of multiplets can provide additional structural information. These distances are all about 7 Hz in the methyl isopropyl ketone spectrum (Figs. 12-26 and 12-27). These splittings are equal, because *any two magnetically coupled protons must have equal effects on each other*. The distance between the peaks of the H^c multiplet (split by H^b) must equal the distance between the peaks of the H^b doublet (split by H^c).

The distance between the peaks of a multiplet (measured in hertz) is called the **coupling constant** between the magnetically coupled protons. Coupling constants are often represented by J, and the coupling constant between H^a and H^b is represented by J_{ab}. In complicated spectra with many types of protons, groups of neighboring protons can sometimes be identified by measuring their coupling constants. Multiplets that have the same coupling constant may arise from adjacent groups of protons that split each other.

The magnetic effect that one proton has on another depends on the nature of the bonds connecting the protons, but it does not depend on the strength of the external magnetic field. For this reason, the coupling constant (measured in hertz) does not vary with the field strength of the spectrometer. A spectrometer operating at 100 MHz records the same coupling constants as a 60-MHz instrument.

Figure 12-28 shows some typical values of coupling constants. The paper used in NMR spectrometers usually has a fine grid calibrated in hertz to allow reading of coupling constants.

Structure		Approx. J	Structure	Approx. J
H–C–C–H	(free rotation)	7 Hz[a]	(ortho)	8 Hz
H–C=C–H	(cis)	10 Hz	(meta)	2 Hz
H–C=C–H	(trans)	15 Hz	(allylic)	6 Hz
C=CH₂	(geminal)	2 Hz		

FIGURE 12-28 Typical values of proton coupling constants.

[a] The value of 7 Hz in an alkyl group is averaged for rapid rotation about the carbon-carbon bond. If rotation is hindered by a ring or bulky groups, other splitting constants may be observed.

Coupling constants help to distinguish among the possible isomers of a compound, as in the spectrum of *p*-nitrotoluene (Fig. 12-29). The methyl protons (*c*) absorb as a singlet at $\delta 2.5$, and the aromatic protons appear as a pair of doublets. The doublet centered around $\delta 7.3$ corresponds to the two aromatic protons ortho to the methyl group (*a*). The doublet centered around $\delta 8.0$ corresponds to the two protons ortho to the electron-withdrawing nitro group (*b*).

Each proton *a* is magnetically coupled to one *b* proton, splitting the H^a absorption into a doublet. Similarly, each proton *b* is magnetically coupled to one proton *a*, splitting the H^b absorption into a doublet. The coupling constant is 8 Hz, suggesting that the magnetically coupled protons H^a and H^b are ortho to each other.

Both the ortho and meta isomers of nitrotoluene have four distinct types of aromatic protons, and the spectra for these isomers are more complex. It is clear that Figure 12-29 corresponds to the para isomer of nitrotoluene.

Coupling constants also help to distinguish stereoisomers. In Figure 12-30, the 9-Hz coupling constant between the two vinyl protons of methyl (*Z*)-3-chloroacrylate shows that they are cis to one another.

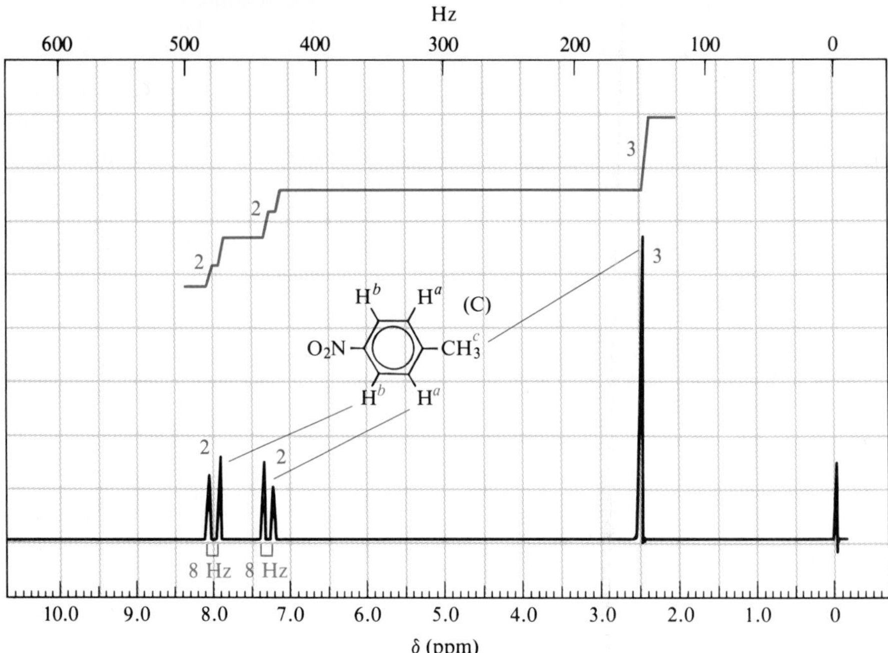

FIGURE 12-29 Proton NMR spectrum of *p*-nitrotoluene.

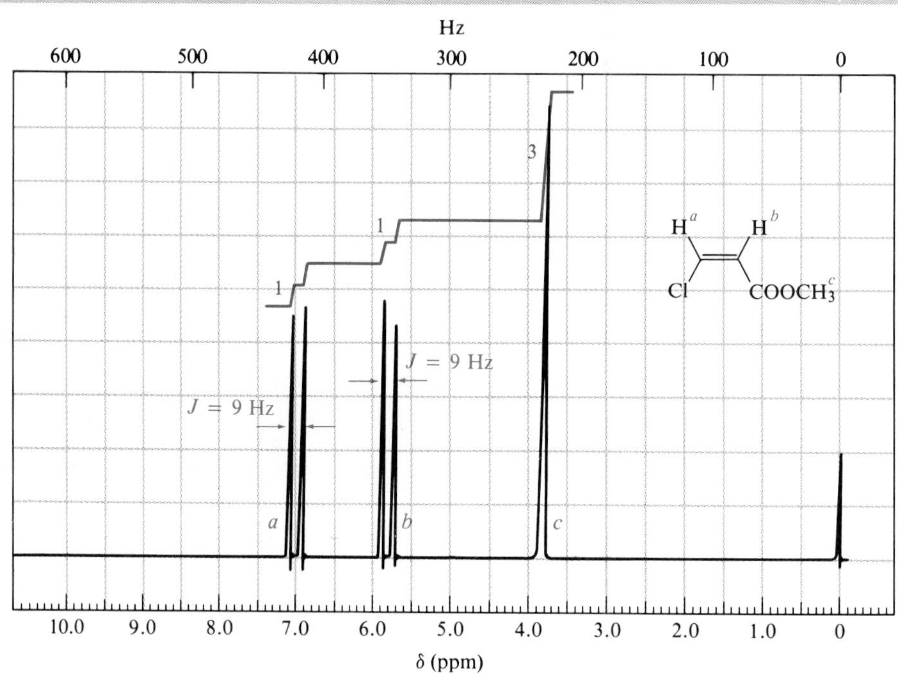

FIGURE 12-30 Proton NMR *spectrum of methyl* (*Z*)-3-chloroacrylate.

PROBLEM 12-7

Draw the expected NMR spectrum for methyl (*E*)-3-chloroacrylate, the geometric isomer of the example in Figure 12-30.

PROBLEM 12-8

Draw the NMR spectra you expect for the following compounds.

(a) 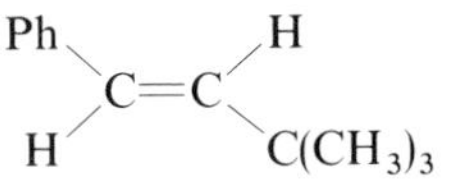

(b) 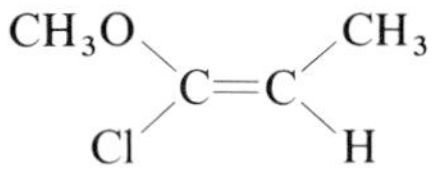

(c) 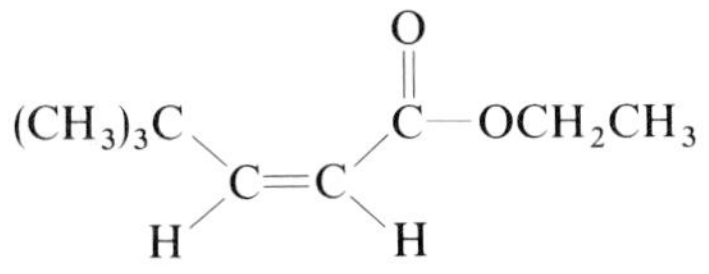

(d) 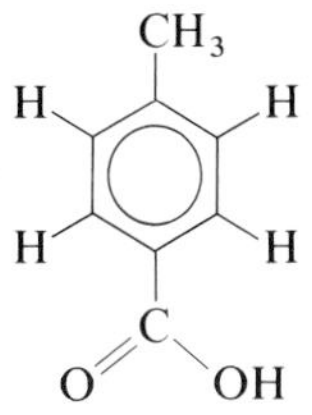

PROBLEM 12-9

An unknown compound (C_3H_2NCl) shows moderately strong IR absorptions around 1650 cm^{-1} and 2200 cm^{-1}. Its NMR spectrum consists of two doublets (J = 14 Hz) at δ5.9 and δ7.1. Propose a structure consistent with these data.

PROBLEM 12-10

Two spectra are given below. Propose a structure that corresponds to each spectrum.

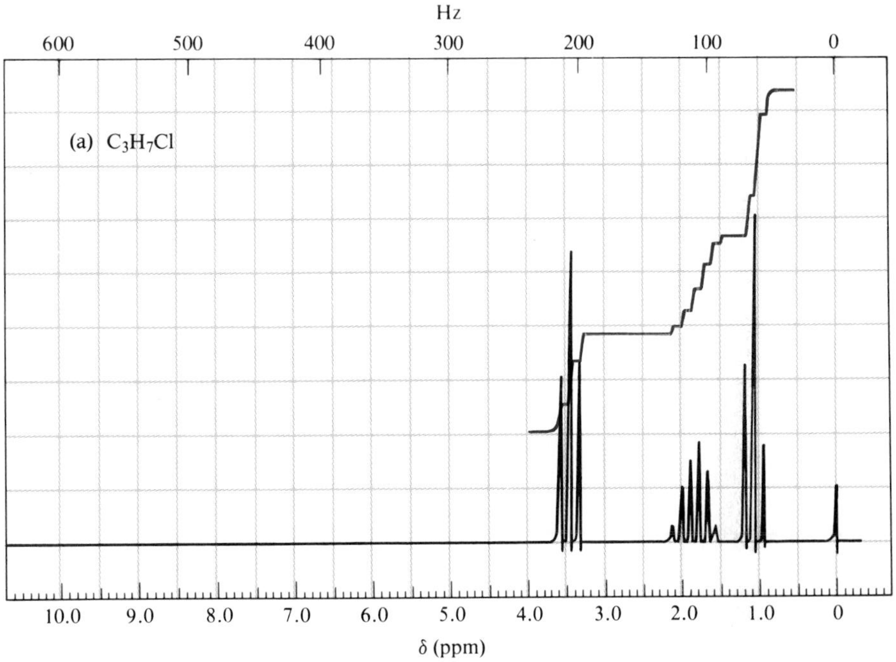

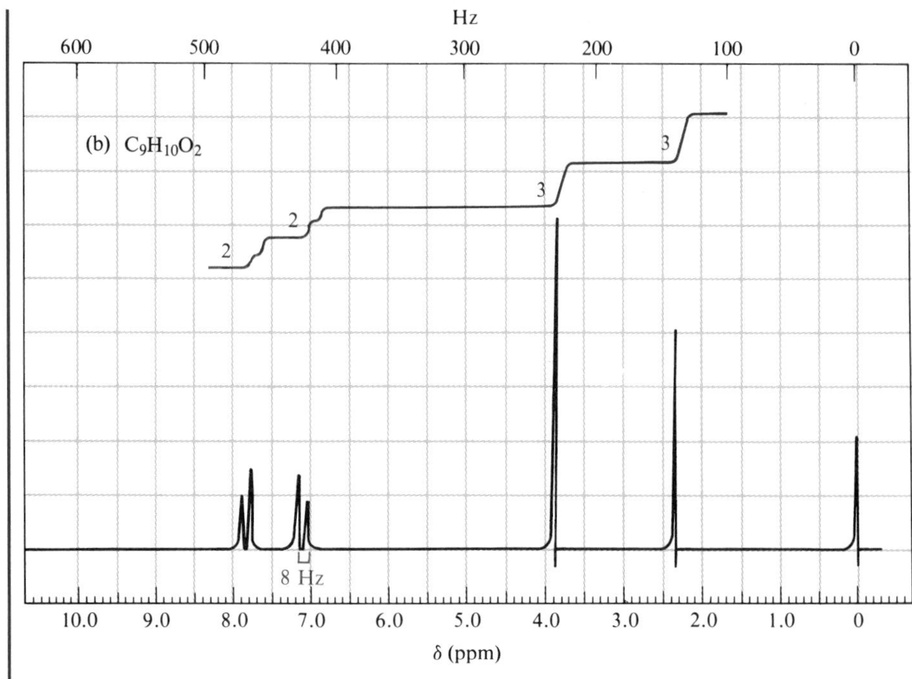

12-8E COMPLEX SPLITTING

There are many cases of **complex splitting,** in which signals are split by adjacent protons of more than one type, with different coupling constants. Consider the vinyl proton H^a, adjacent to the phenyl ring of styrene. The chemical shift of H_a is $\delta 6.6$, deshielded by both the vinyl group and the aromatic ring.

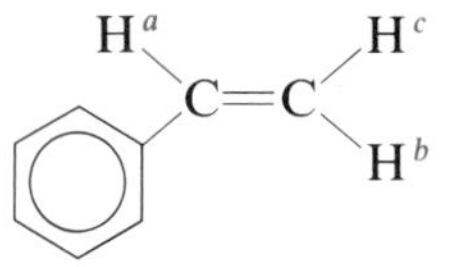

H^a H^c

C=C

H^b

styrene

H^a is coupled to H^b with a typical trans coupling constant $J_{ab} = 17$ Hz. It is also coupled to proton H^c with a constant $J_{ac} = 11$ Hz. The observed H^a signal is split into a doublet of spacing 17 Hz, and each of those peaks is further split into a doublet of spacing 11 Hz, for a total of four peaks. This complex splitting, called a *doublet of doublets,* can be analyzed by a diagram called a *tree,* as shown in Figure 12-31.

The proton NMR spectrum of styrene is shown in Figure 12-32. The absorption of H^a, with splitting as in Figure 12-31, is centered at $\delta 6.6$. H^b is also split by two nonequivalent protons: It is split by H^a with a trans coupling constant $J_{ab} =$ 17 Hz, and further split by H^c with a geminal coupling constant $J_{bc} = 1.4$ Hz. The H^b doublet of doublets, centered at $\delta 5.65$, is shown in Figure 12-33.

PROBLEM 12-11

Draw a splitting tree, similar to Figures 12-31 and 12-33, for proton H^c in styrene. What is the chemical shift of proton H^c?

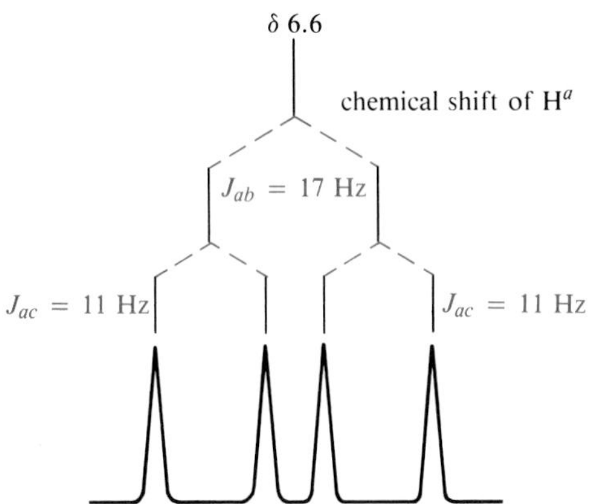

FIGURE 12-31 The H^a signal in styrene is split ($J_{ab} = 17$ Hz) by coupling with H^b, and further split ($J_{ac} = 11$ Hz) by coupling with H^c.

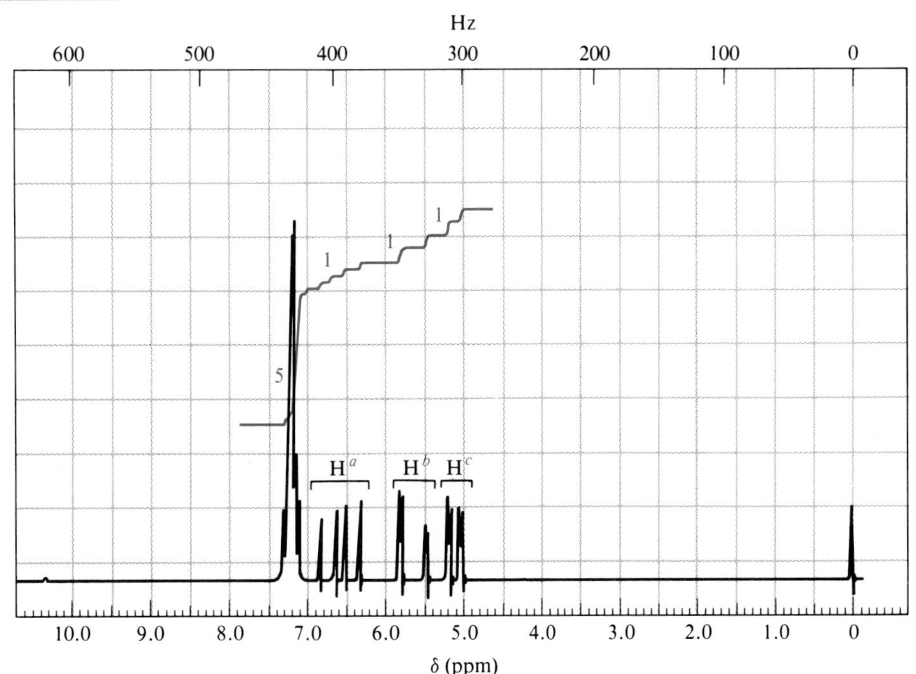

FIGURE 12-32 Proton NMR spectrum of styrene.

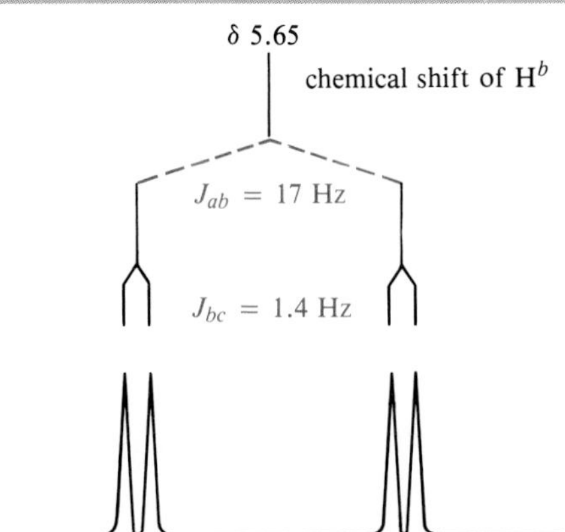

FIGURE 12-33 Tree showing the splitting of the H^b proton in styrene. The signal is split by coupling with H^a ($J_{ab} = 17$ Hz, and further split by coupling with H^c($J_{bc} = 1.4$ Hz).

Sometimes a signal is split by two or more different kinds of protons with similar coupling constants. Consider *n*-propyl iodide, where the *b* protons on the middle carbon atom are split by two types of protons: the methyl protons (H^c) and the CH_2I protons (H^a) (Fig. 12-34).

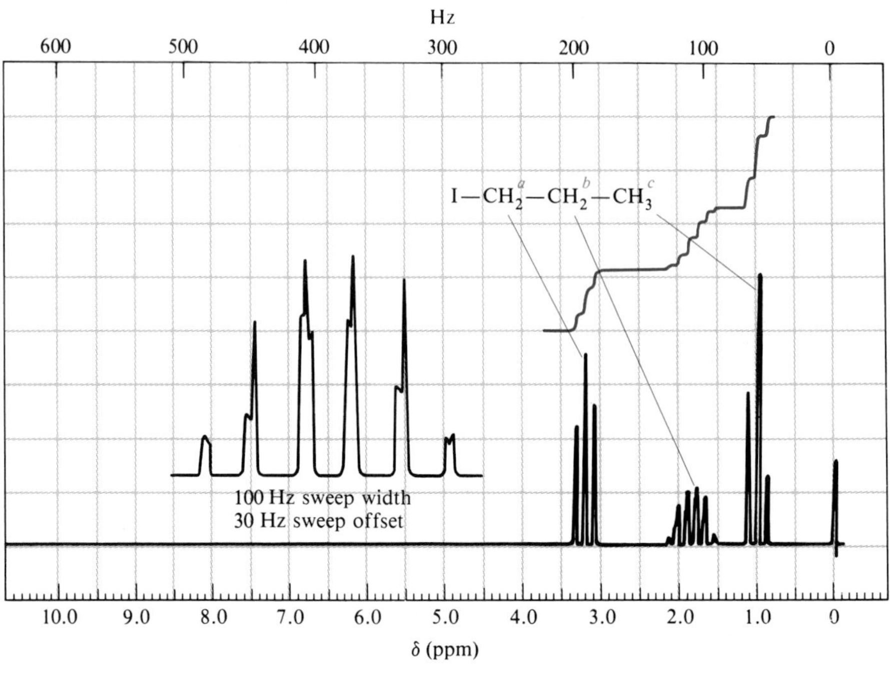

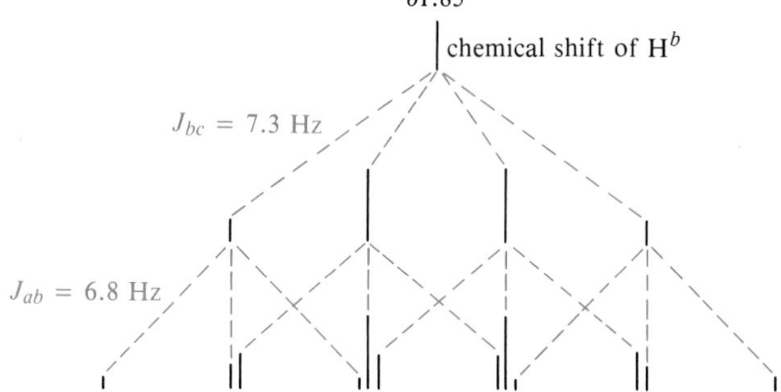

FIGURE 12-34 The NMR spectrum of *n*-propyl iodide seems to show the H^b signal split into a sextet by the five hydrogens on the adjacent carbon atoms. On closer inspection, however, the multiplet is seen to be an imperfect sextet, the result of complex splitting by two sets of protons (*a* and *c*) with similar splitting constants.

The coupling constants for these two interactions are similar: $J_{ab} = 7.3$ Hz, and $J_{bc} = 6.8$ Hz. The spectrum shows the H^b signal as a sextet, almost as though there were five equivalent protons coupled with H^b. The second trace, enlarged and offset, shows that the pattern is not a perfect sextet. The analysis of the splitting pattern appears in Figure 12-34, serving as a reminder that the $(N + 1)$ rule for predicting the peaks in a multiplet works only when the signal is split by *equivalent* protons.

PROBLEM 12-12

The spectrum of *trans*-2-hexenoic acid is shown below.

(a) Give peak assignments to show which protons give rise to which peaks in the spectrum.

(b) Draw a tree to show the complex splitting of the vinyl proton centered around 7 ppm. Estimate the values of the coupling constants.

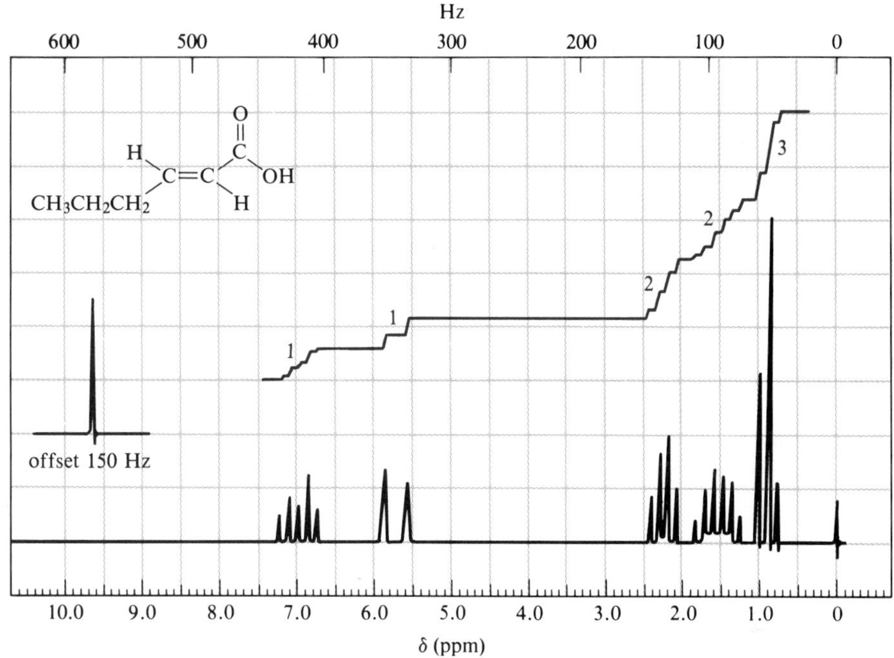

PROBLEM 12-13

The NMR spectrum of cinnamaldehyde is shown below.

(a) Determine the chemical shifts of H^a, H^b, and H^c. The absorption of one of these protons is difficult to see; look carefully at the integrals.
(b) Estimate the coupling constants J_{ab} and J_{bc}.
(c) Draw a tree to analyze the complex splitting of the proton centered at δ6.6.

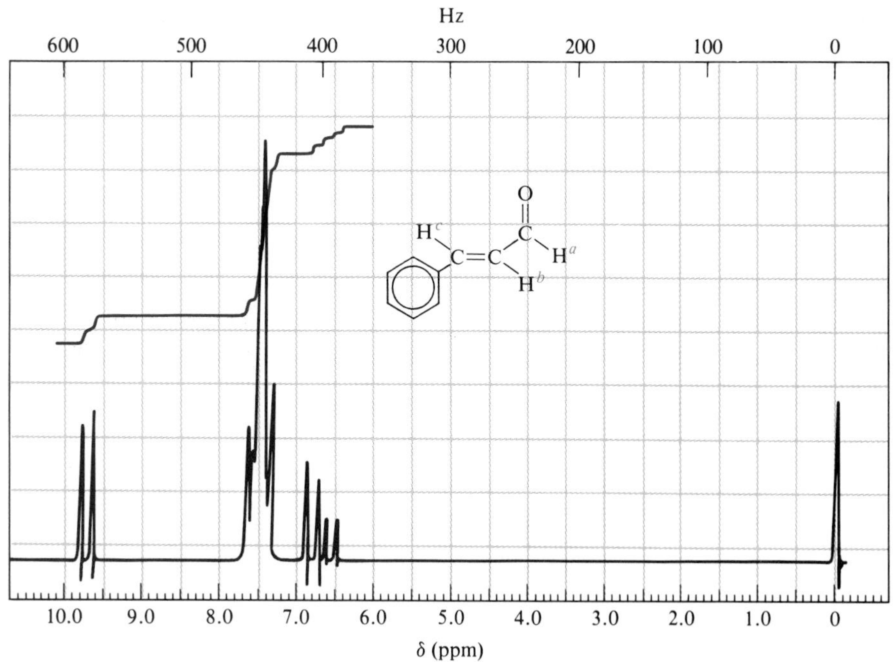

PROBLEM 12-14

Consider the proton NMR spectrum of the following ketone.

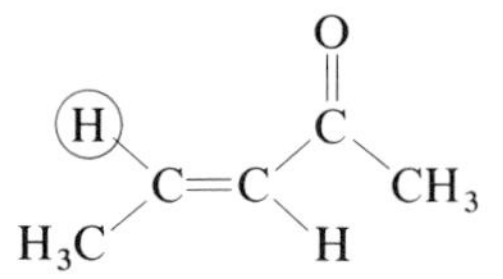

$$\underset{H_3C}{\overset{\textcircled{H}}{}}\!\!>C{=}C<\!\!\underset{H}{\overset{C(=O)CH_3}{}}$$

(a) Predict the approximate chemical shift of each type of proton.
(b) Predict the number of NMR peaks for each type of proton.
(c) Draw a tree to show the splitting predicted for the absorption of the proton circled.

12-9 STEREOCHEMICAL NONEQUIVALENCE OF PROTONS

Stereochemical differences often result in different chemical shifts for protons on the same carbon atom. For example, the two protons on Cl of allyl bromide (3-bromopropene) are not equivalent. Proton *a* is cis to the $-CH_2Br$ group, and H^b is trans. H^a absorbs at $\delta 5.3$; H^b absorbs at $\delta 5.1$. There are four different (by NMR) types of protons in allyl bromide, as shown in the structure

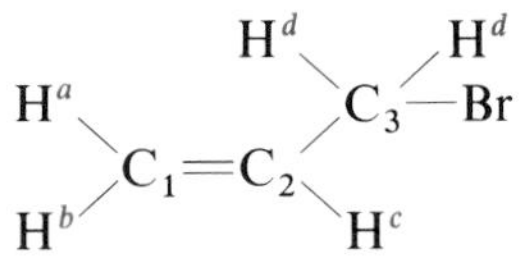

$$H^a(H^b)C_1{=}C_2(H^c)-C_3H^dH^d-Br$$

To determine whether similar-appearing protons are equivalent, mentally substitute another atom for each of the protons in question. *If the same product is formed by imaginary replacement of either of two protons, those protons are chemically equivalent.*

For example, the replacement of any of the three methyl protons in ethanol by an imaginary Z atom gives the same compound; these hydrogens are chemically equivalent.

$$H^1H^2H^3C-CH_2OH$$

replace H^1 → ZH_2C-CH_2OH; replace H^2 → $HZHC-CH_2OH$; replace H^3 → $HHZC-CH_2OH$

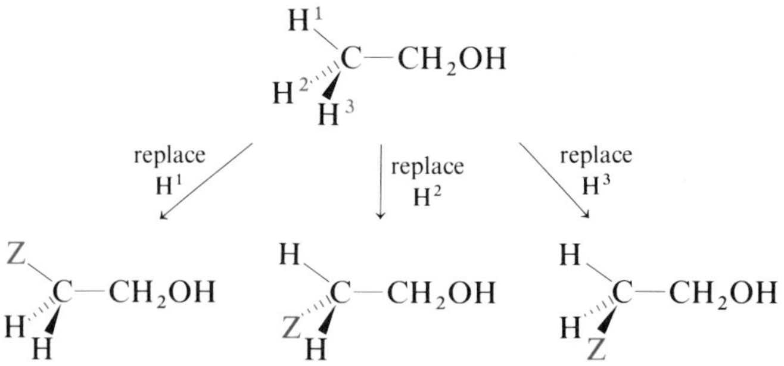

different conformations of the same compound

When this imaginary replacement test is applied to the protons on Cl of allyl bromide, the imaginary products are different. Replacement of the cis hydrogen gives the cis diastereomer, and replacement of the trans hydrogen gives the trans diastereomer. Because the two imaginary products are diastereomers, these protons on Cl are called **diastereotopic** protons.

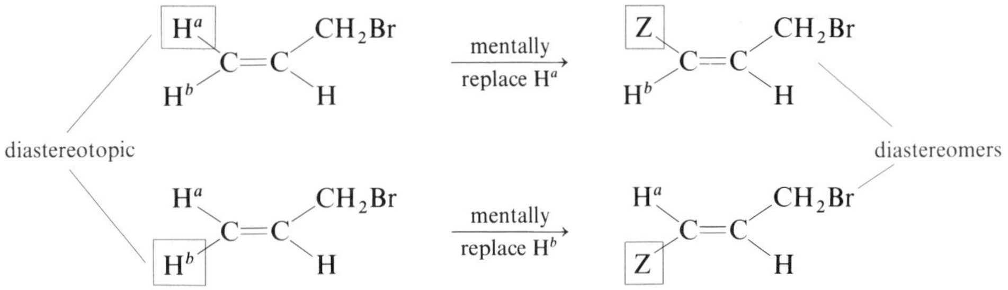

diastereotopic

H^a (boxed), CH_2Br, H^b, H on C=C —mentally replace H^a→ Z (boxed), CH_2Br, H^b, H on C=C

H^a, CH_2Br, H^b (boxed), H on C=C —mentally replace H^b→ H^a, CH_2Br, Z (boxed), H on C=C

diastereomers

Cyclobutanol shows these stereochemical relationships in a cyclic system. The hydroxyl proton H^a is clearly unique; it absorbs between $\delta 3$ and $\delta 5$, depending on the solvent and concentration. H^b is also unique, absorbing between $\delta 3$ and $\delta 4$. Protons H^e and H^f are diastereotopic (and absorb at different fields) because H^e is cis to the hydroxyl group; H^f is trans.

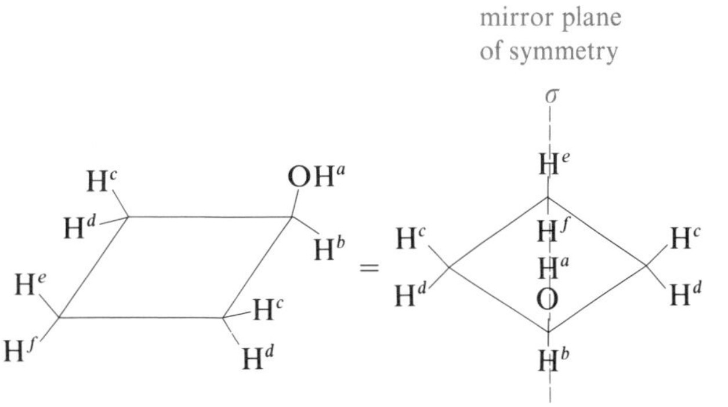

mirror plane of symmetry

σ

H^c, H^d, OH^a, H^b, H^e, H^f, H^c, H^d = H^c, H^d, H^e, H^f, H^a, O, H^b, H^c, H^d

To distinguish among the other four protons, notice that cyclobutanol has an internal mirror plane of symmetry. Protons H^c are cis to the hydroxyl group, while protons H^d are trans. Therefore, protons H^c are diastereotopic to protons H^d, and the two sets of protons absorb at different magnetic fields and are capable of splitting each other.

PROBLEM 12-15

Use the imaginary replacement technique to show that protons H^c and H^d in cyclobutanol are diastereotopic.

Diastereomerism also occurs in saturated, acyclic compounds; for example, 1,2-dichloropropane is a simple compound that contains diastereotopic protons. The two protons on the $-CH_2Cl$ group are diastereotopic; their imaginary replacement gives diastereomers.

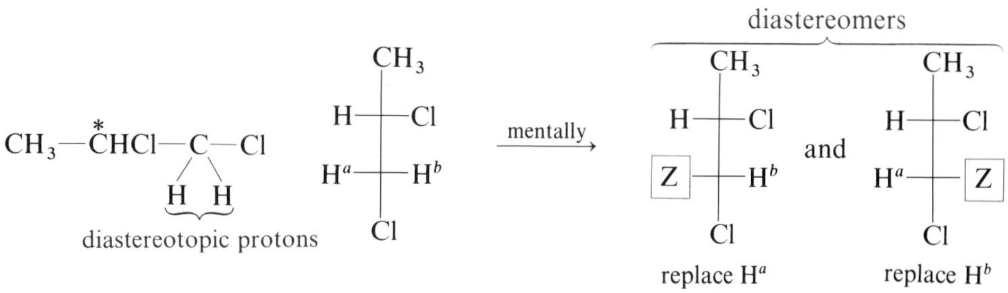

$CH_3-\overset{*}{C}HCl-C(H)(H)-Cl$

diastereotopic protons

CH_3 / H—Cl / H^a—H^b / Cl —mentally→

diastereomers

CH_3 / H—Cl / Z—H^b / Cl (replace H^a) and CH_3 / H—Cl / H^a—Z / Cl (replace H^b)

diastereotopic,
different environments

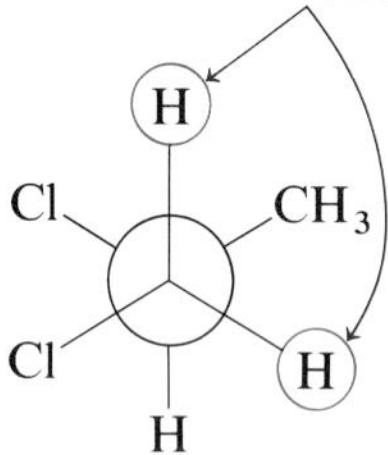

Since the two protons on C1 are diastereotopic, they are nonequivalent by NMR. They absorb at different chemical shifts ($\delta 3.6$ and $\delta 3.7$) and they split each other. The most stable conformation of 1,2-dichloropropane has these two protons in different chemical environments. The presence of a chiral carbon atom adjacent to the $—CH_2Cl$ group gives rise to the different chemical environments and the nonequivalence of these two protons. When a molecule contains a chiral carbon atom, the protons on any methylene ($—CH_2—$) groups are often diastereotopic.

PROBLEM 12-16 *

Predict the theoretical number of different NMR signals produced by each of the following compounds and give approximate chemical shifts. Point out any diastereotopic relationships.

(a) 2-bromobutane (b) cyclopentanol (c) Ph—CHBr—CH_2Br (d) vinyl chloride

12-10 TIME DEPENDENCE OF NMR SPECTROSCOPY

We have already seen evidence that NMR does not provide an instantaneous picture of a molecule. For example, a terminal alkyne does not give a spectrum where the molecules oriented along the field absorb at high field and those oriented perpendicular to the field absorb at lower field (see Fig. 12-13). What we see is one signal whose position is averaged over the chemical shifts of all the orientations of a rapidly tumbling molecule. In general, any type of movement or change that takes place faster than about a tenth of a second will produce an averaged NMR spectrum.

12-10A CONFORMATIONAL CHANGES

This principle is illustrated by the cyclohexane spectrum. In the chair conformation, there are two kinds of protons: the axial hydrogens and the equatorial hydrogens. The axial hydrogens become equatorial and the equatorial hydrogens become axial by chair-chair interconversions, and these interconversions are fast on an NMR time scale at room temperature. The NMR spectrum of cyclohexane shows only one sharp, averaged peak (at $\delta 1.4$) at room temperature.

H_1 (ax)
H_2 (eq) $\rightleftarrows$ H_1 (eq)
H_2 (ax)

Low temperatures retard the chair-chair interconversion of cyclohexane. The NMR spectrum at $-89°C$ shows two nonequivalent types of protons that split each other, giving two broad bands corresponding to the absorptions of the axial and equatorial protons. The broadening of the bands results from spin-spin splitting between axial and equatorial protons on the same carbon atoms and on adjacent carbons. This technique of using low temperatures to stop conformational interconversions is called *freezing out* the conformations.

12-10B FAST PROTON TRANSFERS

Hydroxyl protons Like conformational interconversions, chemical processes often occur faster than the NMR technique can observe them. Figure 12-35 shows two NMR spectra for ethanol. Part (a) shows the splitting we would predict for ethanol, and part (b) shows the spectrum that is more commonly observed.

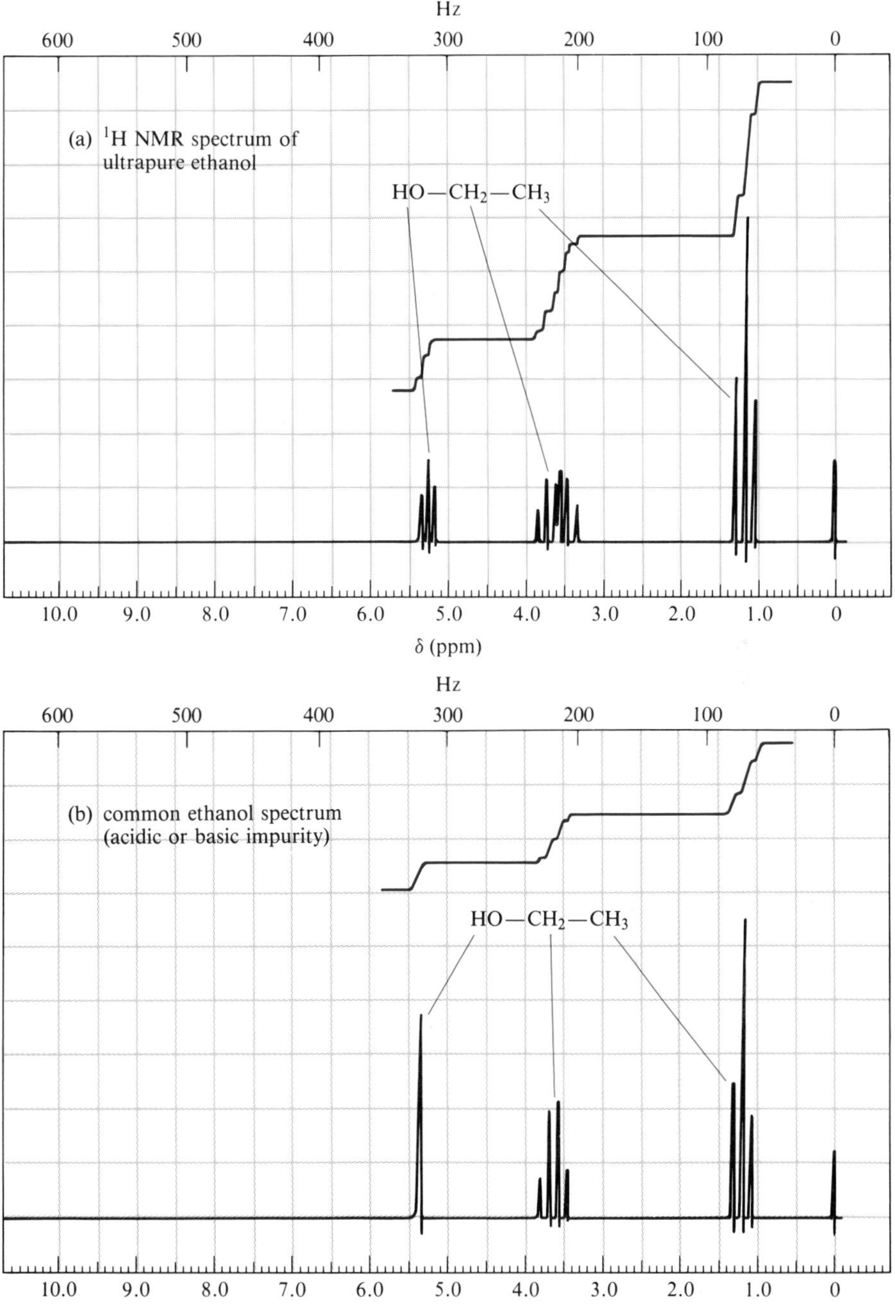

FIGURE 12-35 Comparison of the NMR spectrum of unusually pure ethanol and the spectrum of ethanol with a trace of an acidic (or basic) impurity. An acidic or basic impurity catalyzes a fast exchange of the —OH proton from one ethanol molecule to another. This rapidly exchanging proton produces a single, unsplit absorption at an averaged field.

Figure 12-35(a) shows the expected coupling between the hydroxyl (—OH) proton and the adjacent methylene (—CH$_2$—) protons, with a coupling constant of about 5 Hz, in an ultrapure sample of ethanol with no contamination of acid, base, or water. Part (b) shows a typical sample of ethanol, with some acid or base present to catalyze the interchange of the hydroxyl protons. No splitting between the hydroxyl proton and the methylene protons occurs. During the NMR measurement, each hydroxyl proton becomes attached to a large number of different ethanol molecules and experiences all possible spin arrangements of the methylene

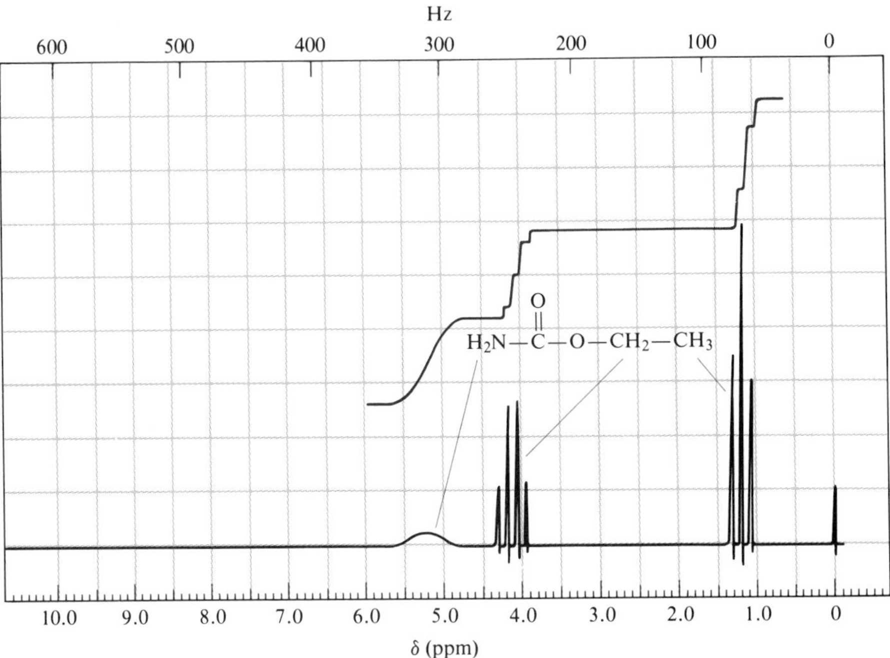

FIGURE 12-36 Example of a proton NMR spectrum with a very broad N—H absorption.

group. What we see is a single, unsplit hydroxyl absorption corresponding to the averaged field the proton experiences from bonding to many different ethanol molecules.

Proton exchange occurs in almost all alcohols and carboxylic acids, and in many amines and amides. If the exchange is fast (as it usually is for —OH protons), we see one sharp, averaged signal. If the exchange is very slow, we see splitting. If the exchange is moderately slow, we may see a broadened peak that is neither cleanly split nor cleanly averaged.

PROBLEM 12-17

Give mechanisms to show the interchange of protons between ethanol molecules under
(a) acid catalysis. (b) base catalysis.

N—H protons Protons on nitrogen often show broadened signals in the NMR, both because of moderate rates of exchange and because of the magnetic properties of the nitrogen nucleus. Depending on the rate of exchange and other factors, N—H protons may give absorptions that are sharp and cleanly split, sharp and unsplit (averaged), or broad and shapeless. Figure 12-36 illustrates an NMR spectrum where the $—NH_2$ protons produce a very broad absorption, the shapeless peak centered at $\delta 5.3$.

Because the chemical shifts of O—H and N—H protons depend on the concentration and the solvent, it is often difficult to tell whether or not a given peak corresponds to one of these types of protons. We can use proton exchange to identify their NMR signals. Shake the sample with deuterium oxide, D_2O. Any exchangeable hydrogens are quickly replaced by deuterium atoms, which are invisible in the proton NMR spectrum.

$$R—O—H \;+\; D—O—D \;\rightleftharpoons\; R—O—D \;+\; D—O—H$$

$$R—NH_2 \;+\; 2\,D—O—D \;\rightleftharpoons\; R—ND_2 \;+\; 2\,D—O—H$$

When a second NMR spectrum is recorded (after shaking with D_2O), the signals from any exchangeable protons are either absent or much less intense.

PROBLEM 12-18

Draw the NMR spectrum expected from ethanol that has been shaken with a drop of D_2O.

PROBLEM 12-19

Propose chemical structures consistent with the following NMR spectra and molecular formulas.

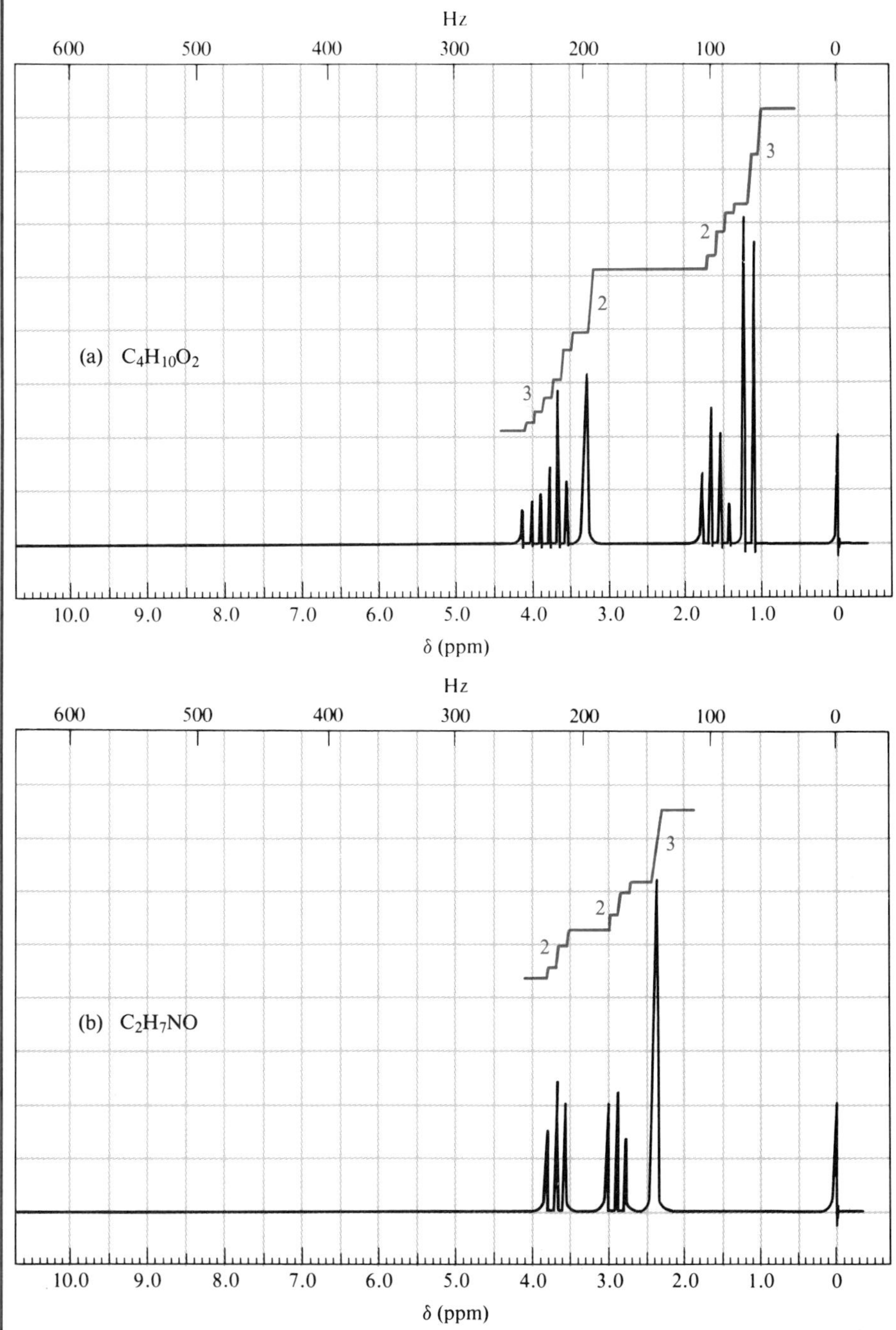

12-11 INTERPRETATION OF PROTON NMR SPECTRA

Learning to interpret NMR spectra requires practice with a large number of examples and problems. The problems at the end of this chapter should help you gain confidence in your ability to assemble a structure from the NMR spectrum combined with other information. This section provides some hints that can help to make spectral analysis a little easier.

When you first look at a spectrum, consider the major features before getting bogged down in the minor details. A few major characteristics you might notice are the following:

1. If the molecular formula is known, use it to determine the number of elements of unsaturation (see Section 6-3). The number of elements of unsaturation can suggest the presence of rings, double bonds, or triple bonds. Matching the integrated peak areas with the number of protons in the formula gives the numbers of protons represented by the individual peaks.

2. Any broadened singlets in the spectrum might be due to —OH or —NH protons. If the broad singlet is deshielded past 10 ppm, an acid —OH group is likely.

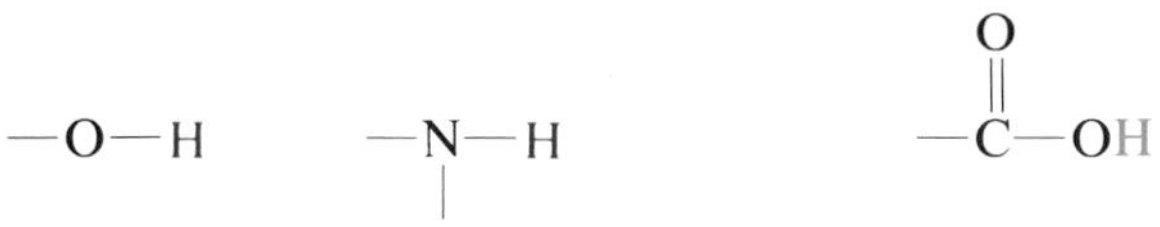

—O—H —N—H —C(=O)—OH

frequently broad singlets broad (or sharp) singlet, $\delta > 10$

3. An absorption around $\delta 3$ to $\delta 4$ suggests protons on a carbon bearing an electronegative element such as oxygen or a halogen. Protons that are more distant from the electronegative atom will be less strongly deshielded.

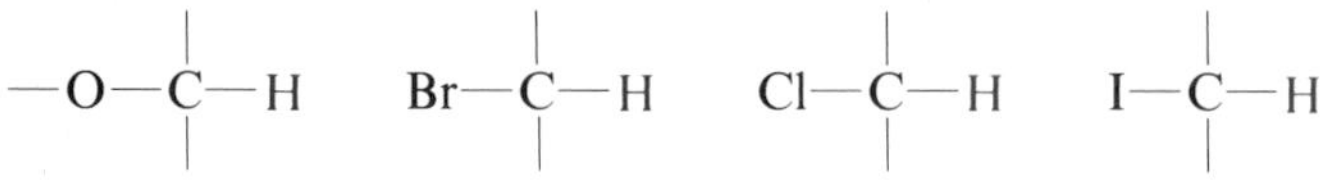

—O—C—H Br—C—H Cl—C—H I—C—H

approximately $\delta 3$ to $\delta 4$ for hydrogens on carbons bearing oxygen or halogen

4. Absorptions around $\delta 7$ to $\delta 8$ suggest the presence of an aromatic ring. If some of the aromatic absorptions are farther downfield than $\delta 7.2$, an electron-withdrawing substituent may be attached.

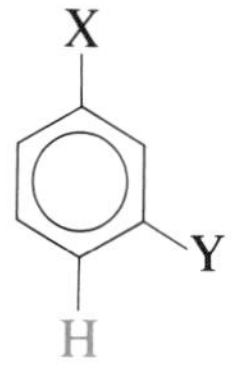

X, Y, H (benzene ring)

around $\delta 7$–$\delta 8$

5. Absorptions around $\delta 5$ to $\delta 6$ suggest vinyl protons. Splitting constants can be used to differentiate cis and trans.

H—C=C—H (cis) H—C=C—H (trans)

around $\delta 5$–$\delta 6$; $J = 10$ Hz around $\delta 5$–$\delta 6$; $J = 15$ Hz

6. You should learn to recognize ethyl groups and isopropyl groups (and structures that resemble these groups) by their characteristic splitting patterns.

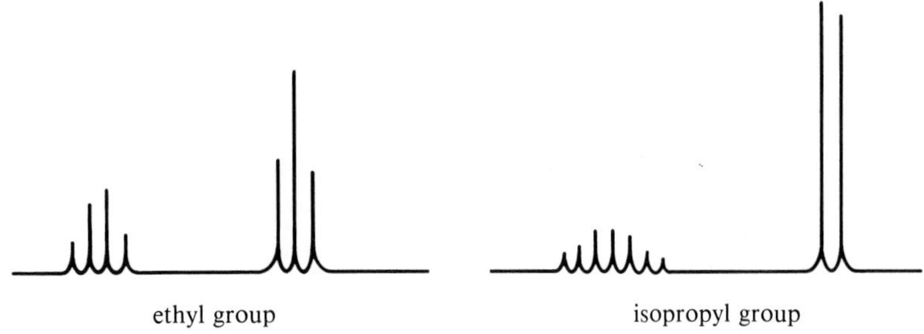

7. Absorptions around $\delta 2.1$ to $\delta 2.5$ may suggest protons adjacent to a carbonyl group. A singlet at $\delta 2.1$ often results from a methyl group bonded to a carbonyl group.

—C(=O)—C—H approx. $\delta 2.1$–$\delta 2.5$

—C(=O)—CH_3 singlet, $\delta 2.1$

8. Absorptions in the range $\delta 9$ to $\delta 10$ suggest the presence of an aldehyde.

—C(=O)—H

aldehyde, $\delta 9$–$\delta 10$

9. A sharp singlet around $\delta 2.5$ suggests a terminal alkyne.

—C≡C—H

approx. $\delta 2.5$

These hints are neither exact nor complete. They are simple methods for making educated guesses about the major features of a compound from its NMR spectrum. The hints can be used to draw partial structures to examine all the possible ways they might be combined to give a molecule that corresponds with the spectrum. Figure 12-37 gives a graphic presentation of some of the most common chemical shifts in the NMR spectrum. A more complete table of chemical shifts appears in Appendix 1.

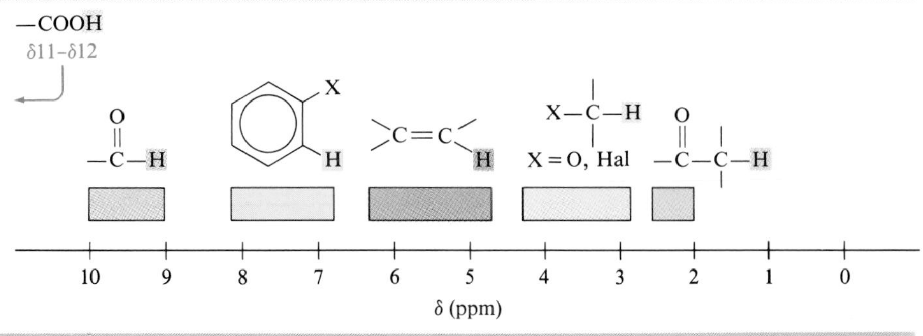

FIGURE 12-37 Common chemical shifts in the ¹H NMR spectrum.

Consider how you might approach the NMR spectrum shown in Figure 12-38. The molecular formula is known to be $C_4H_8O_2$, implying one element of unsaturation (the saturated formula would be $C_4H_{10}O_2$). There are three types of protons in this spectrum. The absorptions at $\delta 4.1$ and $\delta 1.2$ resemble an ethyl group—confirmed by the 2:3 ratio of the integrals of these absorptions.

partial structure: $-CH_2-CH_3$

The ethyl group is probably bonded to an electronegative element, since its methylene ($-CH_2-$) protons absorb close to $\delta 4$. Because the molecular formula contains oxygen, an ethoxy group is suggested.

partial structure: $-O-CH_2-CH_3$

The singlet at $\delta 2.15$ (area = 3) might be a methyl group bonded to a carbonyl group. A carbonyl group would also account for the element of unsaturation.

partial structure: $-\overset{\overset{\displaystyle O}{\|}}{C}-CH_3$

We have accounted for all eight hydrogen atoms in the spectrum. Putting together all the clues, we arrive at a proposed structure.

$$\overset{a}{CH_3}-\overset{b}{CH_2}-O-\overset{\overset{\displaystyle O}{\|}}{C}-\overset{c}{CH_3}$$

ethyl acetate

At this point, the structure should be rechecked to make sure it is consistent with the proton ratios given by the integrals, the chemical shifts of the signals, and the spin-spin splitting. In ethyl acetate, the H^a protons give a triplet (split by the adjacent CH_2 group, $J = 7$ Hz) of area 3 at $\delta 1.2$; the H^b protons give a quartet (split

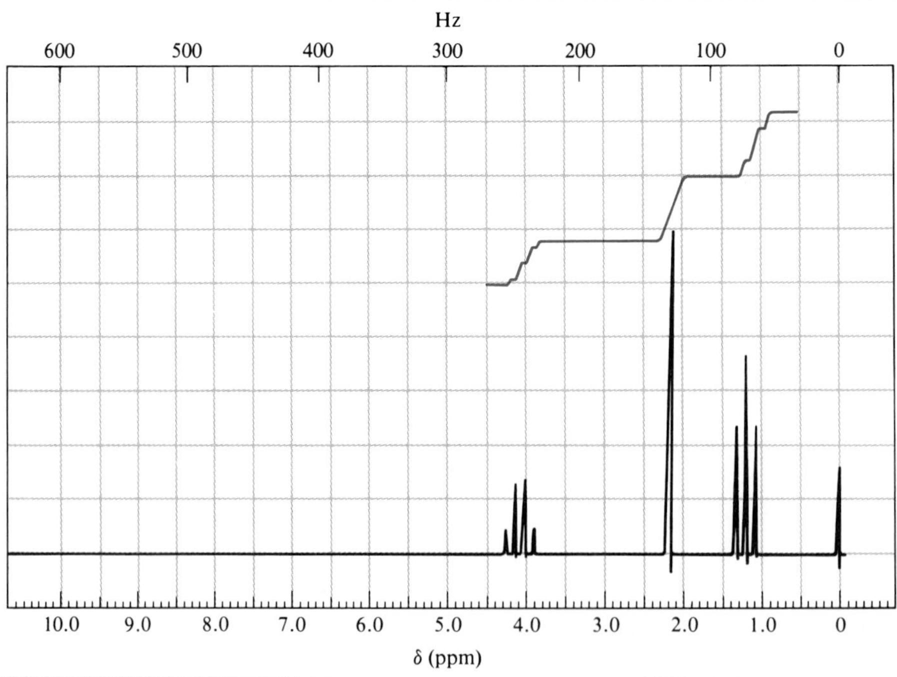

FIGURE 12-38 Proton NMR spectrum for a compound of formula $C_4H_8O_2$.

by the adjacent CH_3 group, $J = 7$ Hz) of area 2 at δ4.1; and the H^c protons give a singlet of area 3 at δ2.15.

PROBLEM 12-20

Draw the expected NMR spectrum of methyl propionate, and point out how it differs from the spectrum of ethyl acetate.

$$CH_3-O-\overset{\overset{\displaystyle O}{\|}}{C}-CH_2-CH_3$$

methyl propionate

SOLVED PROBLEM 12-4

Propose a structure for the compound of molecular formula $C_4H_{10}O$ whose proton NMR spectrum appears below.

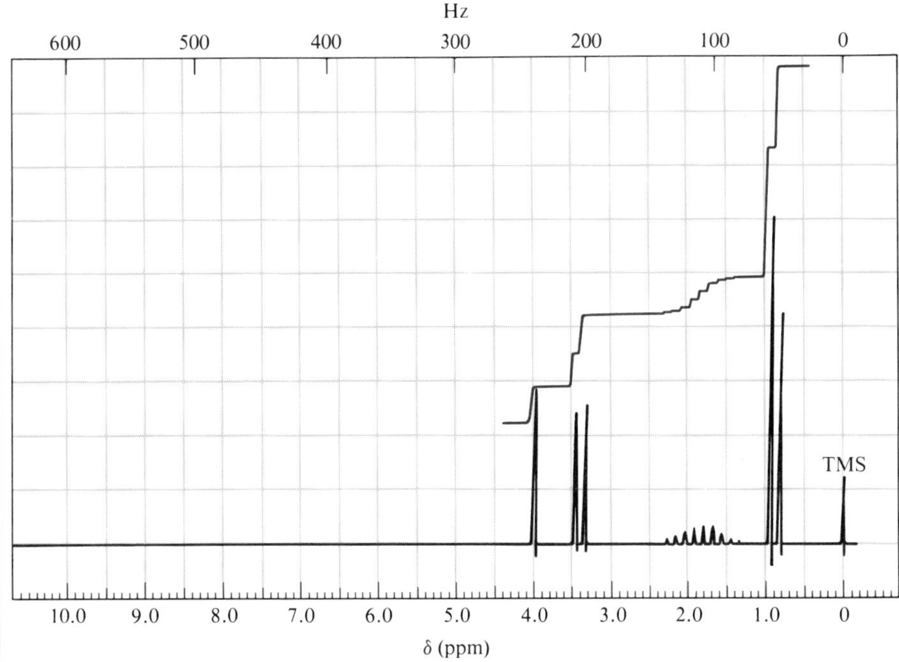

SOLUTION The molecular formula $C_4H_{10}O$ indicates there are no elements of unsaturation. There are four types of hydrogens in this spectrum, in the ratio 1:2:1:6. The singlet (one proton) at δ4.0 might be a hydroxyl group, and the absorption (two protons) at δ3.4 corresponds to protons on a carbon atom bonded to an oxygen. The δ3.4-ppm signal is a doublet, implying that the adjacent carbon atom bears one hydrogen.

$$\text{partial structure:}\quad H-O-CH_2-\overset{\overset{\displaystyle H}{|}}{\underset{|}{C}}-$$

(Since we cannot be certain that the δ4.0 absorption is actually a hydroxyl group, we might consider shaking the sample with D_2O. If the 4.0-ppm absorption represents a hydroxyl group, it will shrink or vanish after shaking with D_2O.)

The absorptions at δ1.8 and δ0.9 resemble the pattern for an isopropyl group. The integral ratio of 1:6 supports this assumption. Since the methine ($-\underset{|}{C}H-$) proton of the isopropyl group absorbs at a fairly high field, the isopropyl group must be bonded to a carbon atom rather than an oxygen.

partial structure: $-\overset{|}{\underset{|}{C}}-CH(CH_3)_2$

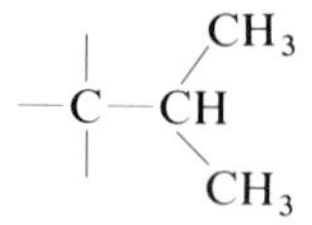

Our two partial structures add to a total of six carbon atoms (compared with the four in the molecular formula), because two of the carbon atoms appear in both partial structures. Drawing the composite of the partial structures, we have isobutyl alcohol:

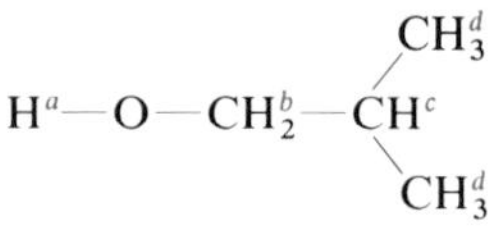

$$H^a-O-CH_2^b-CH^c(CH_3^d)_2$$

isobutyl alcohol

This structure must be rechecked to make sure that it has the correct molecular formula and that it accounts for all the structural evidence provided by the spectrum (Problem 12-21).

PROBLEM 12-21

(a) Give the spectral assignments for the protons in isobutyl alcohol (Solved Problem 12-4). For example,

H^a is a singlet, area = 1, at $\delta 4.0$

(b) Explain the ragged appearance of the methine proton multiplet around $\delta 1.8$.

PROBLEM 12-22

Five proton NMR spectra are given below, together with molecular formulas. In each case, propose a structure that is consistent with the spectrum.

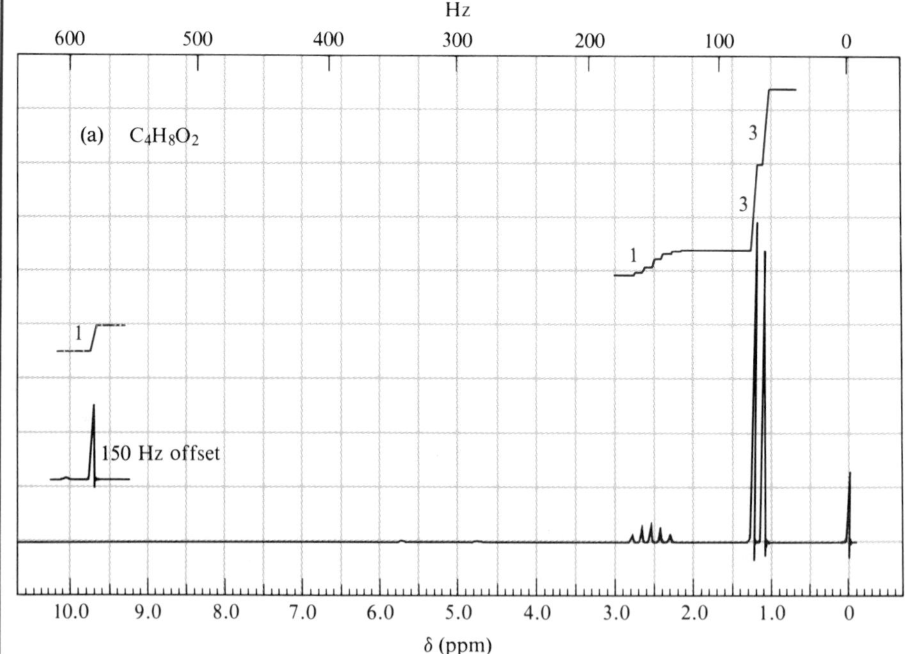

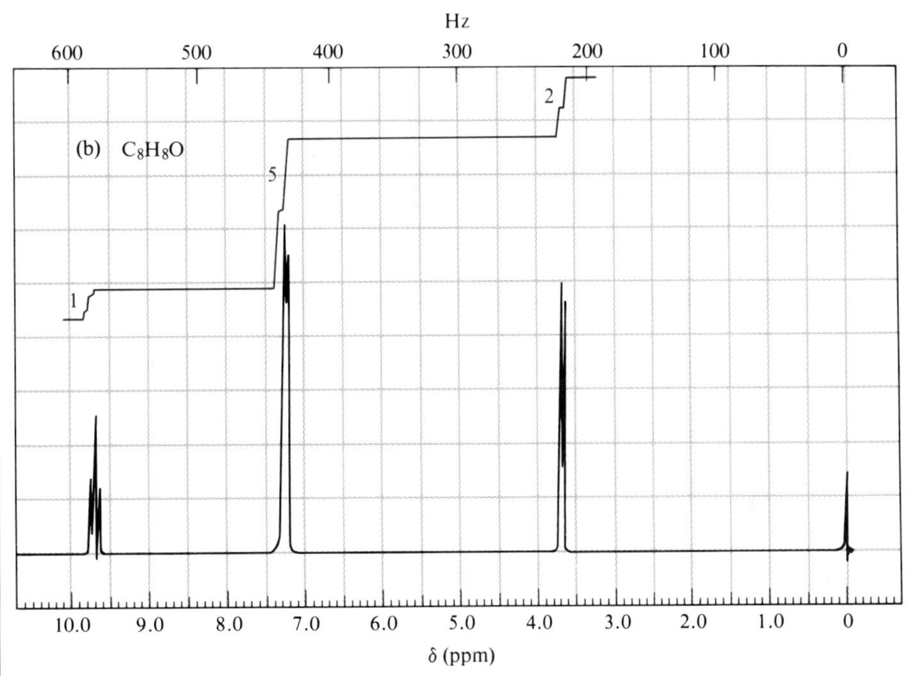

Hz
600
500
400
300
200
100
0
(b) C_8H_8O
1
5
2
10.0
9.0
8.0
7.0
6.0
5.0
4.0
3.0
2.0
1.0
0
δ (ppm)

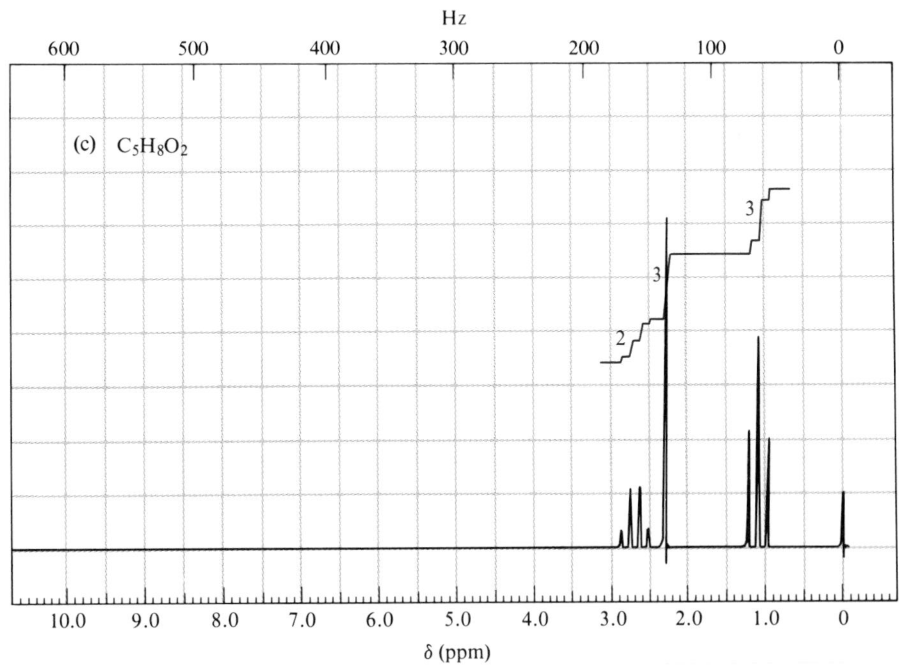

Hz
600
500
400
300
200
100
0
(c) $C_5H_8O_2$
2
3
3
10.0
9.0
8.0
7.0
6.0
5.0
4.0
3.0
2.0
1.0
0
δ (ppm)

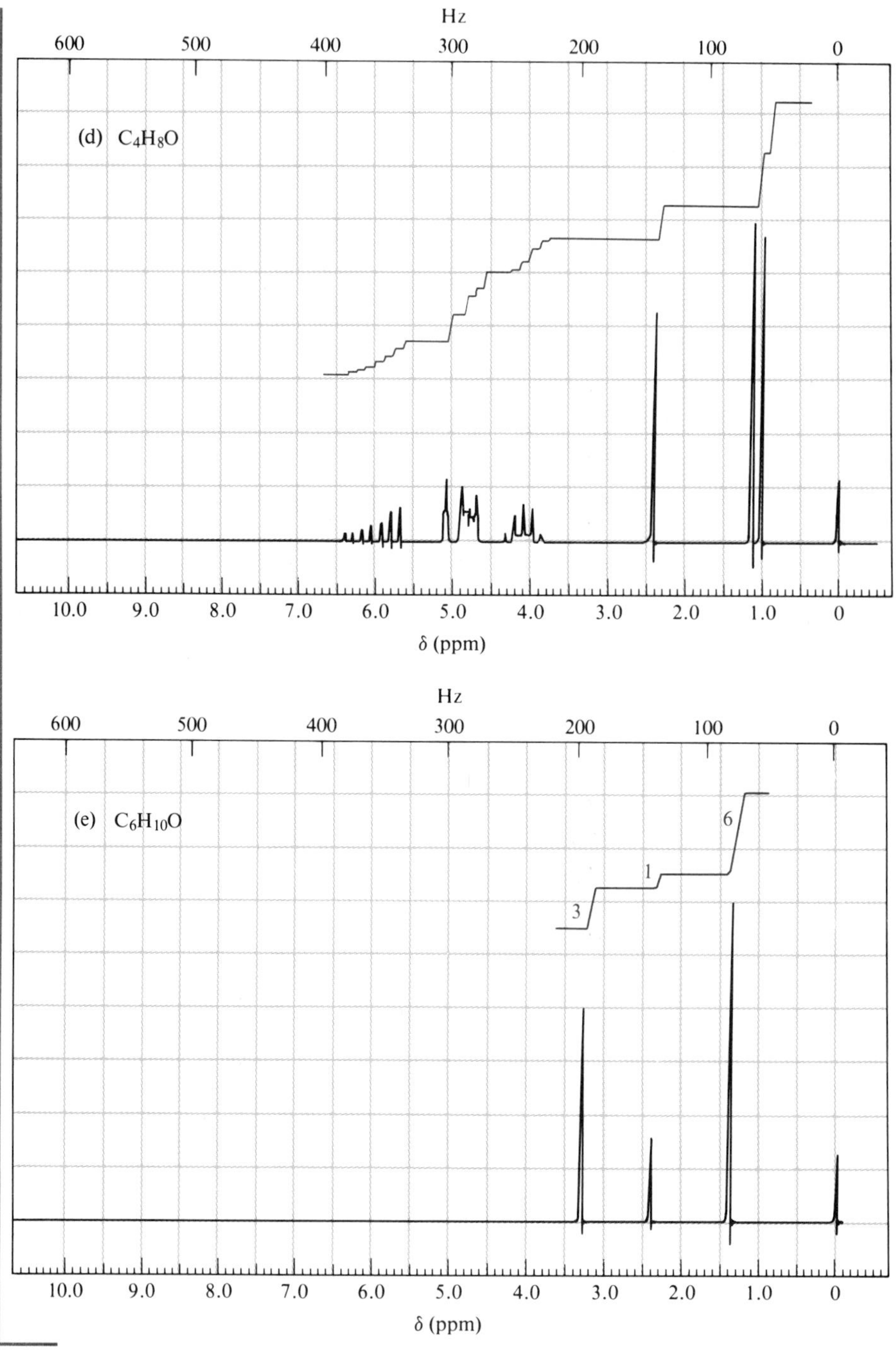

12-12 CARBON-13 NMR SPECTROSCOPY

Proton NMR (^{1}H NMR) was the first routine NMR technique, but the introduction of new instrumentation has allowed carbon NMR (^{13}C NMR) to become nearly as common. In many ways, ^{13}C NMR is complementary to ^{1}H NMR. Using ^{1}H NMR, we infer the structure of the carbon skeleton by observing the magnetic environments of the hydrogen atoms; but ^{13}C NMR determines the magnetic environments of the carbon atoms themselves. The delay in the establishment of carbon NMR as a convenient, routine technique stems from the weakness of carbon NMR signals.

12-12A SENSITIVITY OF CARBON NMR

About 99 percent of the carbon atoms in a natural sample are the isotope ^{12}C. This isotope has an even number of protons and an even number of neutrons, and therefore has no magnetic spin and cannot give rise to NMR signals. The less abundant isotope, ^{13}C, has an odd number of neutrons, giving it a magnetic spin of $\frac{1}{2}$, just like the proton. Because only 1 percent of the carbon atoms in a sample are the magnetic ^{13}C isotope, the sensitivity of ^{13}C NMR is decreased by a factor of 100. The gyromagnetic ratio of ^{13}C is only one-fourth that of the proton, and the ^{13}C resonance frequency (at a given magnetic field) is only one-fourth of that for ^{1}H NMR. The smaller gyromagnetic ratio leads to a further decrease in sensitivity.

Because ^{13}C NMR is less sensitive than ^{1}H NMR, special techniques are needed to obtain a spectrum. If we simply operate the spectrometer in a normal [called *continuous wave* or CW] manner, the desired signals are very weak and are lost in the noise. If several spectra are taken and stored in a computer, however, they can be averaged and the accumulated spectrum plotted by the computer. When many spectra are averaged, the random noise tends to cancel while the desired signals are reinforced. Since the ^{13}C NMR technique is much less sensitive than the ^{1}H NMR technique, hundreds of spectra are commonly averaged to produce a usable result. Several minutes are required to scan each CW spectrum, and this averaging procedure is long and tedious. Fortunately, there is a better technique.

12-12B FOURIER TRANSFORM NMR SPECTROSCOPY

When magnetic nuclei are placed in a magnetic field and irradiated with a pulse of radio frequency close to their resonant frequency, the nuclei absorb some of the energy and precess like little tops at their resonant frequencies (Fig. 12-39). This precession of many nuclei at slightly different frequencies produces a complex signal that decays as the nuclei lose the energy they gained from the pulse. This signal is called a *free induction decay* (or *transient*) and it contains all the information needed to calculate a spectrum. The free induction decay (FID) can be recorded by a radio receiver and a computer in 1 to 2 sec, and many FIDs can be averaged in a few minutes. A computer converts the averaged transients into a spectrum.

The mathematical technique used to compute the spectrum from the free

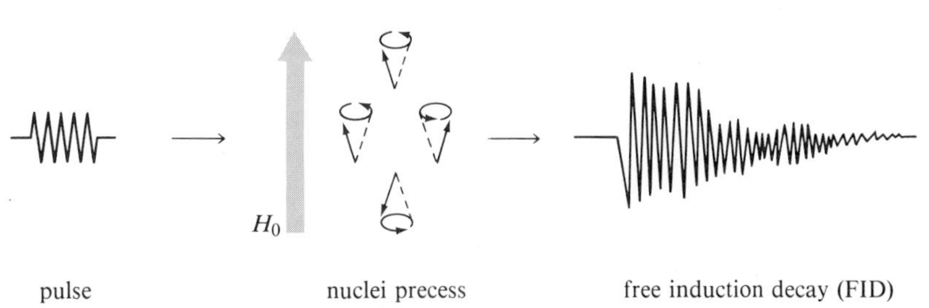

FIGURE 12-39 The Fourier transform NMR spectrometer delivers a radio-frequency pulse close to the resonance frequency of the nuclei. Each nucleus precesses at its own resonance frequency, generating a free induction decay. Many of these transient FIDs are accumulated and averaged in a short period of time. A computer does the Fourier transform operation on the averaged FID, producing the spectrum printed on the recorder.

induction decay is called a *Fourier transform,* and this technique of using pulses and collecting transients is called **Fourier transform spectroscopy.** A spectrometer equipped to do Fourier transform spectroscopy is usually more expensive than a normal continuous wave spectrometer, since it must have a fairly sophisticated computer with the capability of storing thousands of complicated transients. A good ^{13}C NMR instrument usually has the capability to do ^{1}H NMR spectra as well. Use of the Fourier transform technique with proton spectroscopy allows good spectra to be obtained with a very small amount (less than a milligram) of the sample.

12-12C CARBON CHEMICAL SHIFTS

Carbon chemical shifts are usually about 15 to 20 times larger than comparable proton chemical shifts, which is not surprising because the carbon atom is one atom closer than its attached hydrogen to a shielding or deshielding group. For example, an aldehyde proton absorbs around $\delta 9.4$ in the ^{1}H NMR spectrum; the carbonyl carbon atom absorbs around 180 ppm downfield from TMS in the ^{13}C spectrum. Figure 12-40 shows the proton and carbon spectra of an aldehyde for comparison.

The proton and carbon spectra in Figure 12-40 are calibrated so the full width of the proton spectrum is 10 ppm, while the width of the ^{13}C spectrum is 200 ppm, 20 times as large. Notice how the corresponding peaks in the two spectra almost line up vertically. This proportionality of ^{13}C NMR and ^{1}H NMR chemical shifts is an approximation that allows us to make a first estimate of a carbon atom's chemical shift. For example, since the peak for the aldehyde proton in Figure 12-40 is at $\delta 9.6$ in the proton spectrum, we can expect the peak for the aldehyde carbon to appear at a chemical shift between 15 and 20 times as large (between $\delta 144$ and $\delta 192$) in the carbon spectrum. The actual position is at $\delta 180$.

Figure 12-41 gives some typical ranges of chemical shifts for carbon atoms in organic molecules. Because chemical shift effects are larger in CMR, an electron-withdrawing group has a substantial effect on the chemical shift of a carbon atom beta (one carbon removed) to the group. For example, Figure 12-42 shows the PMR and CMR spectra of 1,2,2-trichloropropane. The methyl (CH_3) carbon absorbs at 33 ppm downfield from TMS, because the two chlorine atoms on the adjacent $—CCl_2—$ carbon have a substantial effect on the methyl carbon. The chemical shift of this methyl carbon is about 15 times that of its attached protons ($\delta 2.1$), in accordance with our prediction. Similarly, the chemical shift of the $—CH_2Cl$ carbon (56 ppm) is about 15 times that of its protons ($\delta 4.0$). Although the CCl_2 carbon has no protons, the proton in a $—CHCl_2$ group generally absorbs around $\delta 5.8$. The carbon absorption at 87 ppm is about 15 times this proton shift.

12-12D IMPORTANT DIFFERENCES BETWEEN PROTON AND CARBON TECHNIQUES

Most of the characteristics of ^{13}C NMR spectroscopy are similar to those of the ^{1}H NMR technique. There are some important differences, however.

Operating frequency The gyromagnetic ratio for ^{13}C is about one-fourth that of the proton, so the resonance frequency is also about one-fourth. A spectrometer with a 14,092-gauss magnet needs a 60-MHz transmitter for protons and a 15.1-MHz transmitter for ^{13}C. A spectrometer with a 23,486-gauss magnet needs a 100-MHz transmitter for protons and a 25.2-MHz transmitter for ^{13}C.

Peak areas The areas of peaks obtained using Fourier transform techniques are

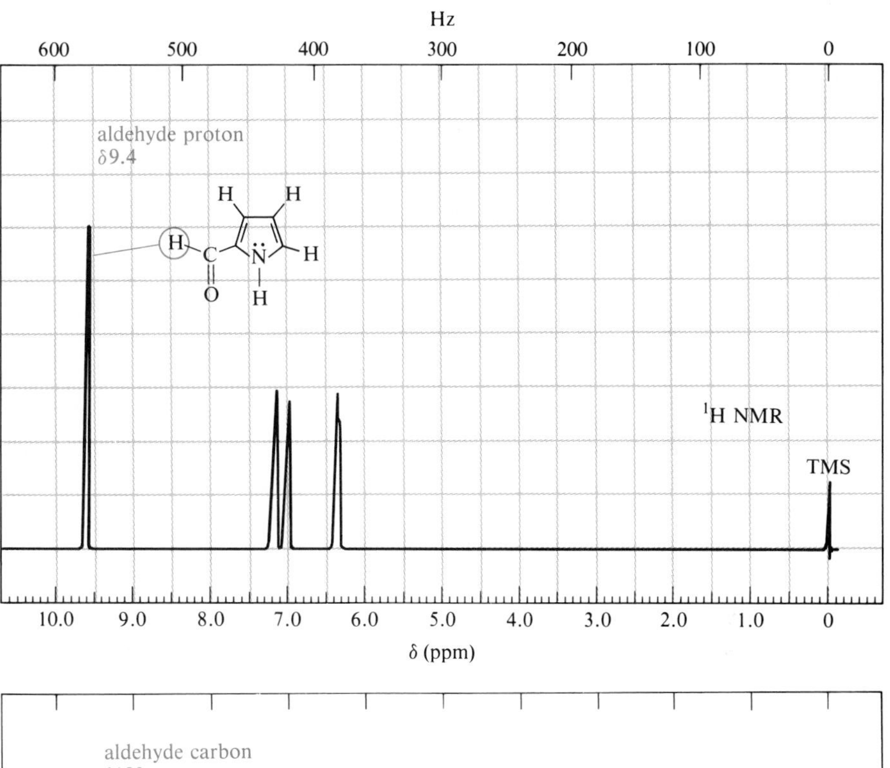

FIGURE 12-40 Proton and ^{13}C NMR spectra of a heterocyclic aldehyde. Notice the correlation of the chemical shifts in the two spectra. The proton spectrum has a sweep width of 10 ppm, and the carbon spectrum has a width of 200 ppm.

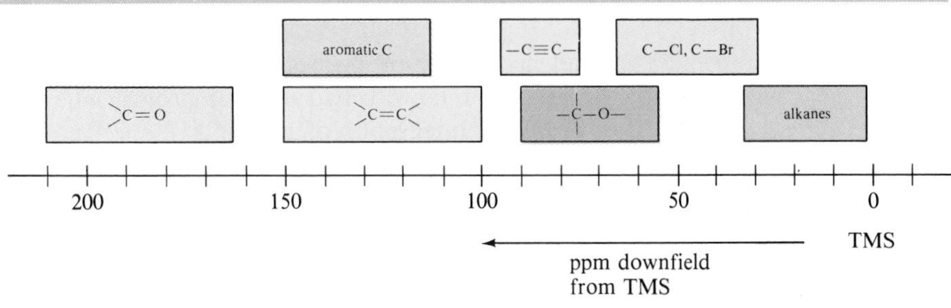

FIGURE 12-41 Table of approximate chemical shift values for ^{13}C NMR. Most of these chemical shift values for a carbon nucleus are about 15 to 20 times the chemical shift of a proton if it were bonded to the carbon atom.

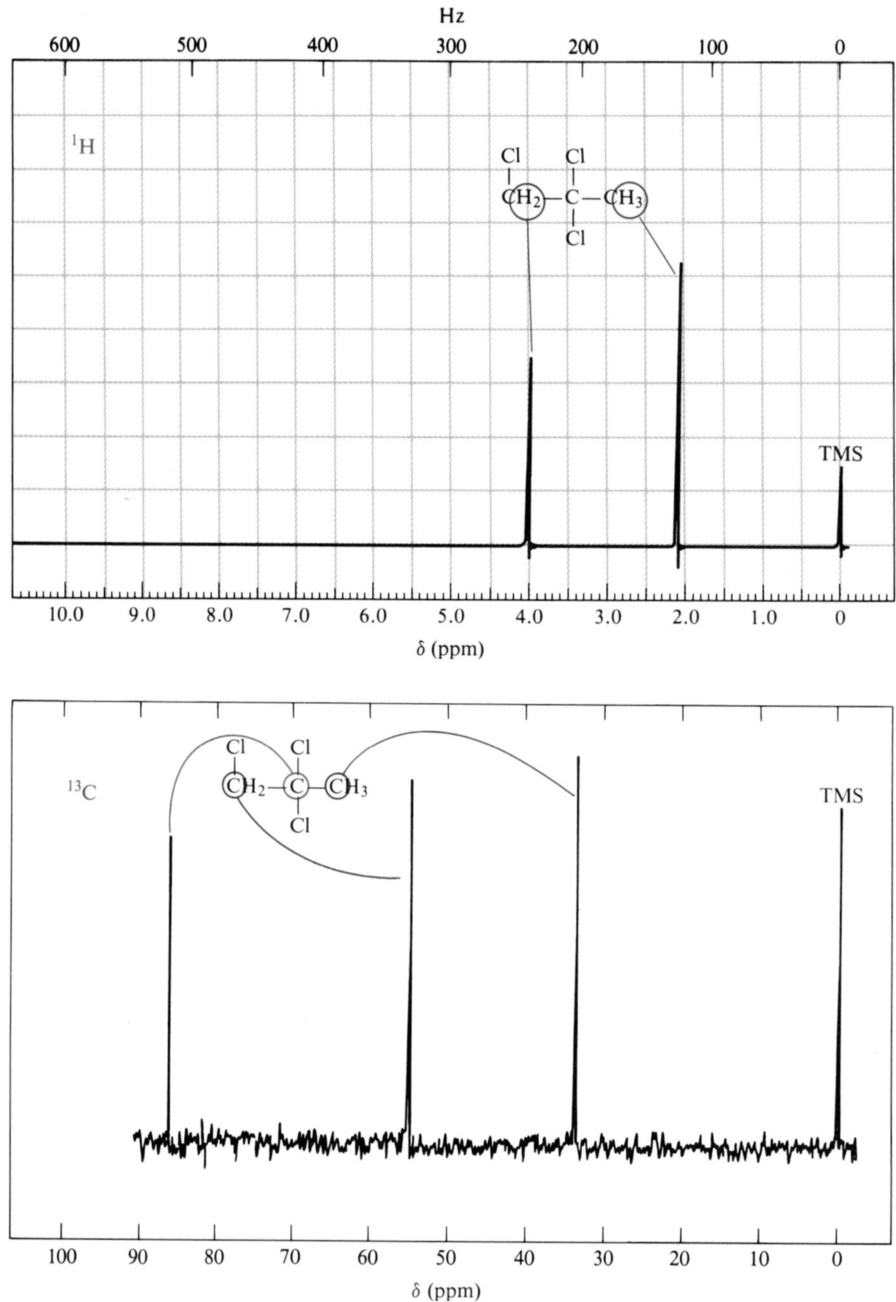

FIGURE 12-42 Proton and ^{13}C NMR spectra of 1,2,2-trichloropropane.

not necessarily proportional to the number of carbons giving rise to the peaks. Carbon atoms with two or three protons attached usually give the strongest absorptions. Integration of a ^{13}C NMR spectrum (or any Fourier-transformed spectrum) therefore does not give reliable ratios of carbon atoms.

Spin-spin splitting ^{13}C NMR splitting patterns are quite different from those observed in ^1H NMR. Only 1 percent of the carbon atoms in the ^{13}C NMR sample are magnetic, so there is a small probability that an observed ^{13}C nucleus is adjacent to another ^{13}C nucleus. Therefore, carbon-carbon splitting can be ignored. Most carbon atoms are bonded directly to hydrogen atoms or are sufficiently close to hydrogen atoms for carbon-hydrogen spin-spin coupling to be observed, however. This carbon-hydrogen coupling produces splitting patterns that can be complicated and difficult to interpret.

Proton spin decoupling To simplify ^{13}C NMR spectra, they are commonly recorded using **proton spin decoupling,** where the protons are continuously irradiated with a broad-band ("noise") proton transmitter. As a result, all the protons are continuously in resonance, and they rapidly flip their spins. The carbon nuclei see an *average* of the possible combinations of proton spin states. Each carbon signal appears as a single, unsplit peak because any carbon-hydrogen splitting has been eliminated. The spectra in Figures 12-40 and 12-42 were generated in this manner.

PROBLEM 12-23

Draw the expected noise-decoupled ^{13}C NMR spectra of the following compounds.

(a) [2,3-dihydrofuran ring with O] (b) [cyclohexene ring] (c) $CH_3-C(=O)-CH(CH_3)-CH_3$ (d) $H_2C=CH-C(=O)H$

Off-resonance decoupling Proton noise decoupling produces spectra that are very simple, but some valuable information is lost in the process. **Off-resonance decoupling** simplifies the spectrum but allows some of the splitting information to be retained (Fig. 12-43). Using off-resonance decoupling, the ^{13}C nuclei are split

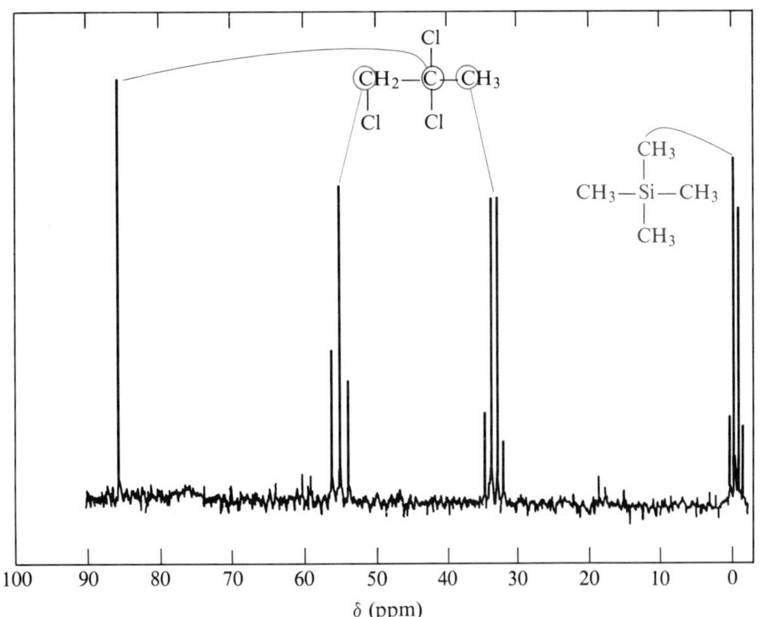

FIGURE 12-43 Off-resonance decoupled CMR spectrum of 1,2,2-trichloropropane. The CCl_2 group appears as a singlet, the CH_2Cl group as a triplet, and the CH_3 group as a quartet. Compare this spectrum with Figure 12-42.

only by the protons directly bonded to them. A carbon atom with one attached proton (a methine) appears as a doublet, while a carbon with two attached protons (a methylene) gives a triplet. A methyl carbon is split into a quartet. Off-resonance decoupled spectra are easily recognized by the appearance of TMS as a quartet at 0 ppm, split by the three protons of each methyl group.

PROBLEM 12-24
Repeat Problem 12-23, sketching the off-resonance decoupled ^{13}C spectra of these compounds.

The best procedure for obtaining a ^{13}C NMR is to run the spectrum twice: First, the singlets in the noise-decoupled spectrum indicate the number of non-equivalent carbon atoms and their chemical shifts. Second, the multiplicities of the absorptions in the off-resonance decoupled spectrum indicate the number of hydrogen atoms bonded to each carbon atom. ^{13}C spectra are often given with two traces, one noise decoupled and the other off-resonance decoupled. If just one trace is given, it is usually noise decoupled.

12-13 INTERPRETATION OF CARBON NMR SPECTRA

Interpretation of ^{13}C NMR spectra involves the same principles as those used with ^{1}H NMR spectra. In fact, ^{13}C NMR spectra are often easier to interpret. For example, consider the ^{13}C NMR spectrum of δ-valerolactone in Figure 12-44. Notice that the CH_2 groups in the upper (off-resonance decoupled) spectrum are split into triplets, but they appear as singlets in the lower (noise-decoupled) spectrum.

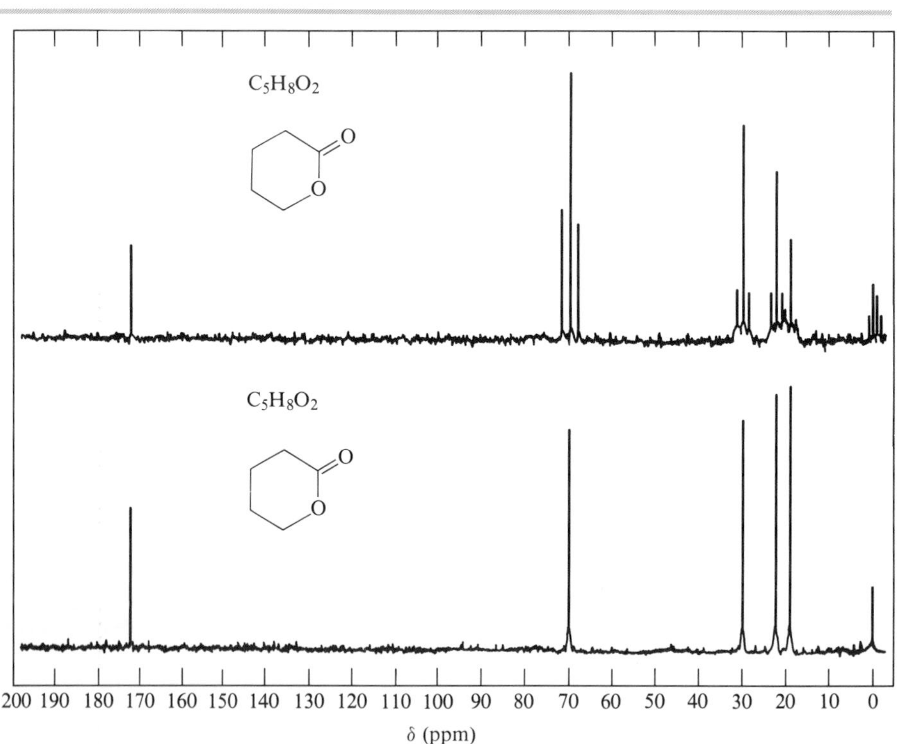

FIGURE 12-44 Off-resonance-decoupled and noise-decoupled spectra of δ-valerolactone, molecular formula $C_5H_8O_2$.

As we have seen in Figures 12-40 and 12-41, the absorption at 173 ppm is appropriate for a carbonyl carbon. The off-resonance decoupled spectrum shows a singlet at 173 ppm, implying that no hydrogens are bonded to the carbonyl carbon.

>C=O

173 ppm

H δ3–δ4

—C—O—

70 ppm

The chemical shift of the next absorption is about 70 ppm. This is about 20 times the chemical shift of a proton on a carbon bonded to an electronegative element. The molecular formula implies that the electronegative element must be oxygen. Since the absorption at 70 ppm is a triplet in the off-resonance spectrum, this carbon must be a methylene (—CH_2—) group.

partial structures: —CH_2—O— —C(=O)—

δ2.1 H O

30 ppm —C—C—

H

The absorption at 30 ppm corresponds to a carbon atom bonded to a carbonyl group. Remember that a proton on a carbon adjacent to a carbonyl group absorbs around 2.1 ppm, so we expect the carbon to have a chemical shift about 15 to 20 times as large. This carbon atom is in the form of a methylene group, as shown by the triplet in the off-resonance decoupled spectrum.

partial structures: —CH_2—C(=O)— —CH_2—O—

The two signals at 19 and 22 ppm are from carbon atoms that are not directly bonded to any deshielding group, although the carbon at 22 ppm is probably closer to one of the oxygen atoms. These are also triplets in the off-resonance spectrum, and therefore correspond to methylene groups. We can propose

partial structures: —C—CH_2—CH_2—CH_2—O— —CH_2—C(=O)—

The molecular formula $C_5H_8O_2$ implies the presence of two elements of unsaturation. The carbonyl (C=O) group accounts for one, but there are no more carbonyl groups and no double-bonded olefinic carbon atoms. The other element of unsaturation must be a ring. Combining the partial structures into a ring gives the complete structure.

30 173

19 O

22 O

70

In the following problems, only the noise-decoupled spectra are provided. In cases where the off-resonance spectra are available, the off-resonance multiplicity of each peak is indicated: (s) = singlet, (d) = doublet, (t) = triplet, and (q) = quartet.

PROBLEM 12-25

A bottle of allyl bromide was found to contain a large amount of an impurity. A careful distillation separated the impurity, which was found to have the molecular formula C_3H_6O. The following ^{13}C NMR spectrum of the impurity was obtained.

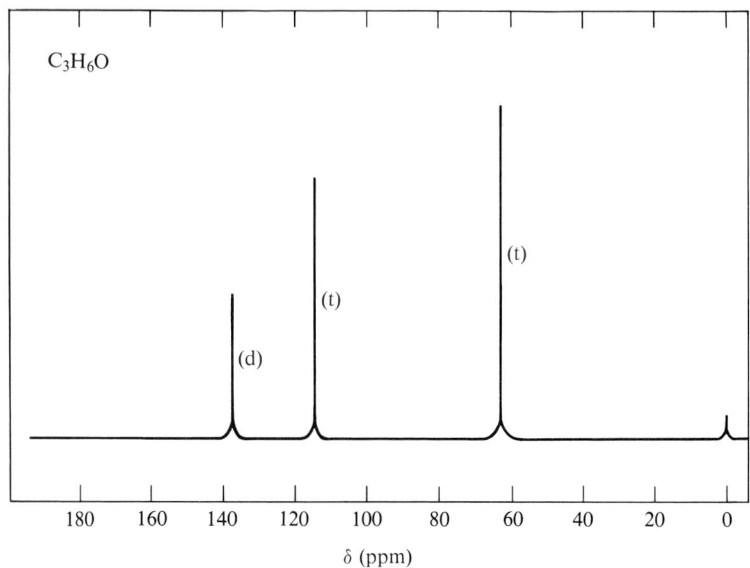

(a) Propose a structure for this impurity.
(b) Assign the peaks in the ^{13}C NMR spectrum to the carbon atoms in the structure.
(c) Suggest how this impurity arose in the allyl bromide sample.

PROBLEM 12-26

An inexperienced graduate student was making some 4-hydroxybutanoic acid. He obtained an excellent yield of a different compound, whose ^{13}C NMR spectrum appears below.

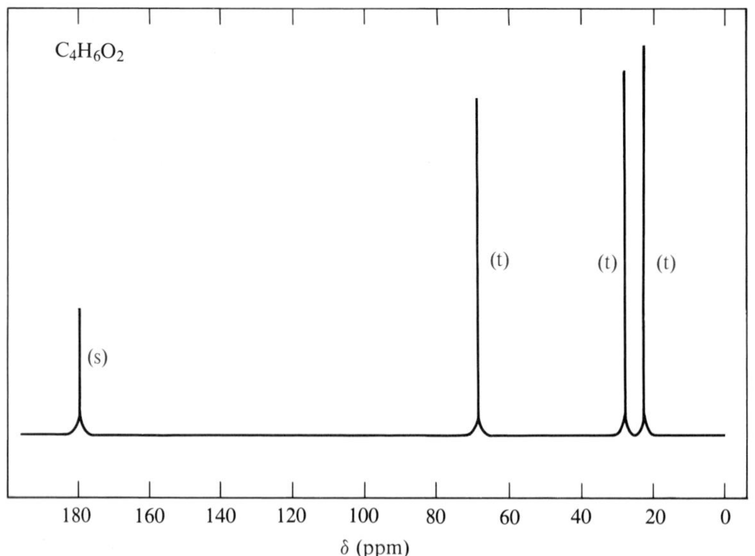

(a) Propose a structure for this product.
(b) Assign the peaks in the ^{13}C NMR spectrum to the carbon atoms in the structure.

PROBLEM 12-27

A laboratory student was converting cyclohexanol to cyclohexyl bromide using 1 equivalent of sodium bromide in a large excess of concentrated sulfuric acid. The major product recovered was not cyclohexyl bromide but the compound of formula C_6H_{10} whose ^{13}C NMR spectrum appears below.

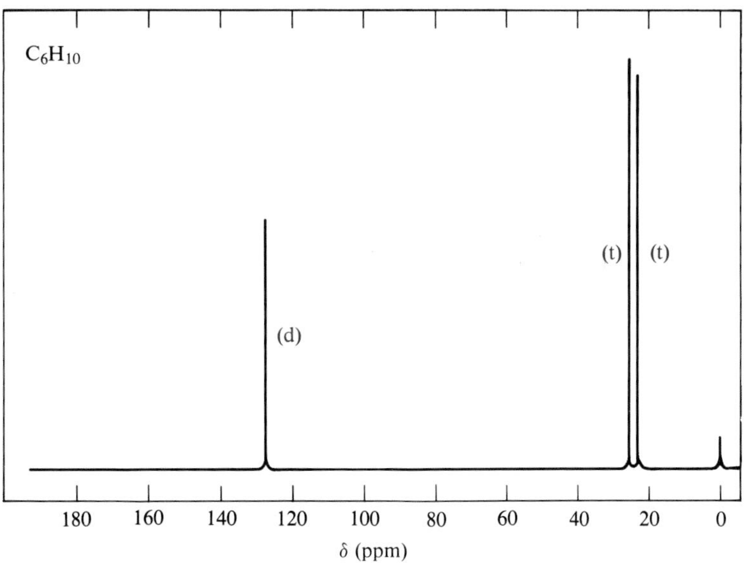

(a) Propose a structure for this product.
(b) Assign the peaks in the ^{13}C NMR spectrum to the carbon atoms in the structure.
(c) Suggest modifications in the reaction to obtain a better yield of cyclohexyl bromide.

12-14 STRATEGY FOR SOLVING SPECTRAL PROBLEMS

We have now learned to use both IR and NMR spectroscopy, as well as mass spectrometry, to determine the structures of unknown organic compounds. These techniques usually provide a unique structure with little chance of error. A major part of successful spectral interpretation is using an effective strategy rather than simply looking at the spectra, hoping that something obvious will jump out. A systematic approach should take into account the strengths and weaknesses of each technique. The following table summarizes the information provided by each spectroscopic technique.

SUMMARY OF THE INFORMATION PROVIDED BY EACH TYPE OF SPECTROSCOPY

	MS	*IR*	*NMR*
molecular weight	✓		
molecular formula	✓ (HRMS)		
heteroatoms	✓	S	S
functional groups	S	✓	H
alkyl substituents	S		✓

✓, usually provides this information.
H, usually provides helpful information.
S, sometimes provides helpful information.

We can set out a guide to how you might go about identifying an unknown compound, but the actual process depends on what you already know about the chemistry of the compound, and on what you learn from each spectrum. Always go through the process with scratch paper and a pencil, since you need to keep track of mass numbers, possible functional groups, and carbon skeletons.

1. *Mass spectrum.* Look for a molecular ion, and determine a tentative molecular weight. Remember that some compounds (alcohols, for example) may fail to give a visible molecular ion peak. If the molecular weight is odd, a nitrogen atom is suggested. If a HRMS is available, compare the "exact" mass with the tables to find a molecular formula with a mass close to the experimental value.

   Look for anything unusual or characteristic about the mass spectrum: Does the M + 2 peak of the parent ion look larger than the M + 1 peak? It might contain S, Cl, or Br. Is there a large gap and a peak at 127 characteristic of I?

   Although we might look at the MS fragmentation patterns to help determine the structure, this is more time-consuming than going on to some of the other spectra. We can verify the fragmentation patterns more easily once we have a proposed structure.

2. *Infrared spectrum.* Look for N—H or O—H peaks in the 3300 cm^{-1} region, and also for unsaturated =C—H peaks to the left of 3000 cm^{-1}. Also look for C≡C or C≡N stretch around 2200 cm^{-1}, and for C=O, C=C, or C=N stretch between 1600 and 1800 cm^{-1}. The exact position of the peak, plus other spectral characteristics, should help to determine the functional groups in the molecule. For example, sharp peaks in the 3300 cm^{-1} region might imply N—H, or a broad O—H band centered over the C—H stretch at 3000 cm^{-1} might imply that a carbonyl peak is from —COOH.

   The combination of the IR with an odd molecular ion in the mass spectrum should confirm amines, amides, and nitriles. The presence of a strong —OH absorption might suggest that the apparent molecular ion in the MS could be low by 18 units, from loss of water.

3. *Nuclear magnetic resonance spectrum.* First look for strongly deshielded protons, such as in acids (δ10 to δ12), aldehydes (δ9 to δ10), and aromatic protons (δ6 to δ8). Moderately deshielded peaks might be vinyl protons (δ5 to δ6) or protons on a carbon bonded to an electronegative atom such as oxygen or halogen (δ3 to δ4). A peak around δ2.1 to δ2.5 might be an acetylenic proton or a proton adjacent to a carbonyl group, a benzene ring, or a vinyl group.

   These possibilities should be checked to see which are consistent with the IR spectrum. Finally, the spin-spin splitting patterns should be analyzed to suggest the structures of the alkyl groups present.

Once we have considered all the spectra, there should be one or two tentative structures. Each tentative structure should be checked to see whether it accounts for the major characteristics of each spectrum.

Are the molecular weight and formula of the tentative structure consistent with the appearance (or the absence) of the molecular ion in the mass spectrum? Are there peaks in the mass spectrum corresponding to the expected fragmentation products?

Does the tentative structure explain each of the characteristic stretching frequencies in the infrared spectrum? Does it account for any shifting of frequencies from their usual positions?

Does the tentative structure account for each proton (or carbon) in the NMR spectrum? Does it also account for the observed chemical shifts and spin-spin splitting patterns?

If the tentative structure successfully accounts for all these features of the spectra, we can be fairly confident that it is correct.

PROBLEM 12-28 (Partially Solved)

Sets of spectra are given for two different compounds. For each set of spectra

(1) Look at each individual spectrum individually and list the structural characteristics you can determine from that spectrum.

(2) Look at the set of spectra as a group and propose a tentative structure.

(3) Verify that your proposed structure accounts for the major features of each spectrum. The solution for compound 1 is given after the problem; but go as far as you can before looking at this solution.

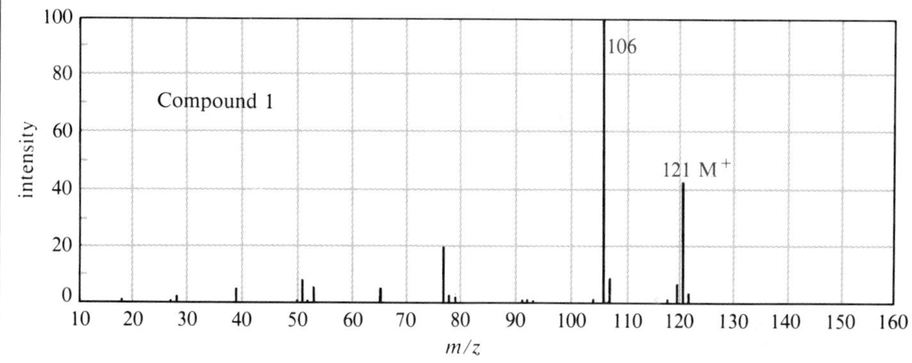

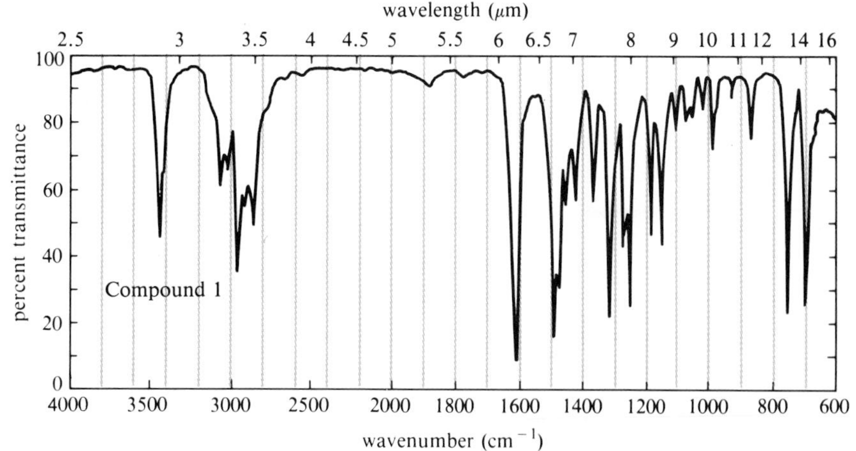

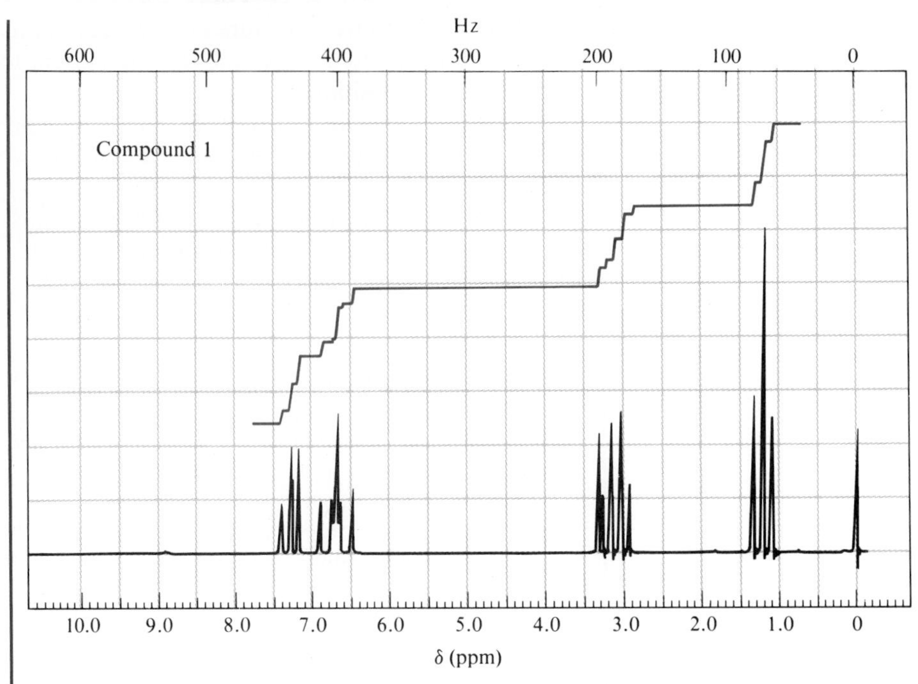
Hz
600
500
400
300
200
100
0
Compound 1
10.0
9.0
8.0
7.0
6.0
5.0
4.0
3.0
2.0
1.0
0
δ (ppm)

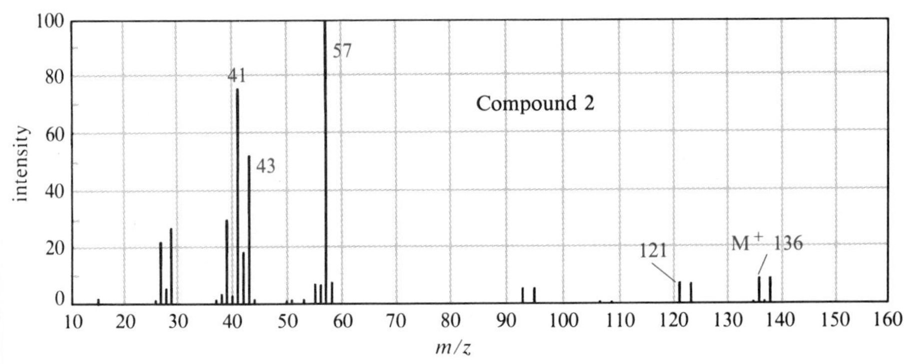
100
80
60
40
20
0
intensity
57
41
43
Compound 2
121
M+ 136
10
20
30
40
50
60
70
80
90
100
110
120
130
140
150
160
m/z

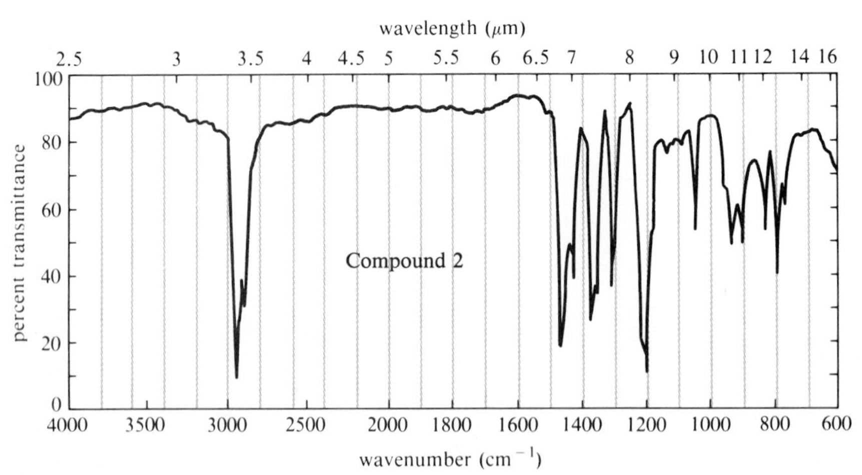
wavelength (μm)
2.5
3
3.5
4
4.5
5
5.5
6
6.5
7
8
9
10
11
12
14
16
100
80
60
40
20
0
percent transmittance
Compound 2
4000
3500
3000
2500
2000
1800
1600
1400
1200
1000
800
600
wavenumber (cm⁻¹)

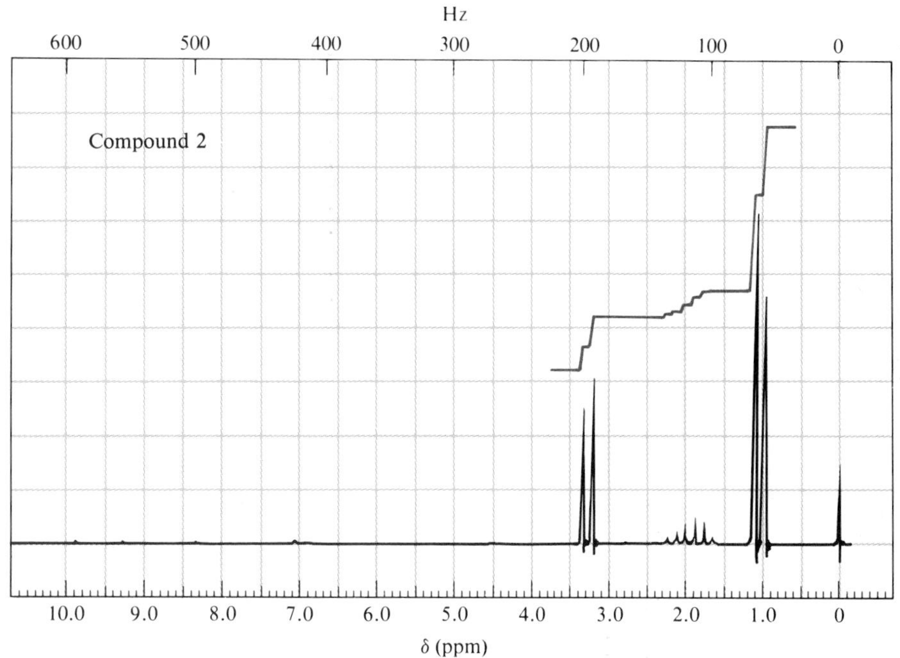

SOLUTION TO COMPOUND 1

Mass spectrum: The MS shows an odd molecular weight (121), possibly indicating the presence of a nitrogen atom.

Infrared spectrum: The IR shows a sharp peak around 3400 cm^{-1}, probably indicating the N—H of an amine or the ≡C—H of a terminal alkyne. Because there is probably a nitrogen atom present and there is no other evidence for an alkyne (no C≡C stretch around 2200 cm^{-1}), the 3400 cm^{-1} absorption probably indicates an N—H bond. The unsaturated =C—H absorptions above 3000 cm^{-1}, combined with an aromatic C=C stretch around 1600 cm^{-1}, indicate an aromatic ring.

NMR spectrum: The NMR shows complex splitting in the aromatic region, probably from a benzene ring; the integral of 5 suggests that the ring is monosubstituted. The fact that part of the aromatic absorption is shifted upfield of $\delta 7.2$ suggests that the substituent on the benzene ring is a pi-electron-donating group like an amine or an ether. An ethyl group (total area 5) is seen at $\delta 1.2$ and $\delta 3.1$, appropriate for protons on a carbon atom bonded to nitrogen. A sharp singlet of area 1 appears at $\delta 3.3$, probably resulting from the N—H seen in the IR spectrum. Combining this information, we propose a nitrogen atom bonded to a hydrogen atom, a benzene ring, and an ethyl group: total molecular weight 121, in agreement with the molecular ion in the mass spectrum.

Proposed structure for compound 1:

C₆H₅—N̈(H)—CH_2—CH_3

The proposed structure shows an aromatic ring with 5 protons, which explains the aromatic absorptions in the NMR and the C=C at 1600 cm^{-1} and the =C—H above 3000 cm^{-1} in the IR. The aromatic ring is bonded to an electron-donating —NHR group, which explains the odd molecular weight, the N—H absorption in the IR, and the aromatic absorptions shifted above $\delta 7.2$ in the NMR. The ethyl group bonded to nitrogen explains the ethyl absorptions in the NMR, deshielded to $\delta 3.1$ by the nitrogen atom. The base peak in the MS (M − 15 = 106) is explained by the loss of a methyl group to give a resonance-stabilized cation:

$$\left[\text{Ph}-\ddot{\text{N}}(\text{H})-\text{CH}_2 \overset{106}{+} \text{CH}_3\right]^{+\cdot} \longrightarrow \left[\text{Ph}-\ddot{\text{N}}(\text{H})-\overset{+}{\text{C}}\text{H}_2 \longleftrightarrow \text{Ph}-\overset{+}{\text{N}}(\text{H})=\text{CH}_2\right] + \cdot\text{CH}_3$$

m/z 106 — loss of 15

GLOSSARY

accidentally equivalent nuclei Nuclei that are nonequivalent by NMR, yet absorb at the same chemical shift. Nuclei that absorb at the same chemical shift cannot split each other, whether they are theoretically equivalent or accidentally equivalent. (p. 505)

chemically equivalent atoms Atoms that theoretically cannot be distinguished chemically. Using the replacement test, the replacement of chemically equivalent atoms gives identical compounds. (pp. 505, 524)

chemical shift The difference (in ppm) between the resonance frequency of the proton (or carbon nucleus) being observed and that of tetramethylsilane (TMS). Chemical shifts are usually given on the δ (delta) scale, the number of parts per million downfield from TMS. (p. 496)

complex splitting Splitting by two or more different kinds of protons with different coupling constants (p. 520)

coupling constant (J) The distance (in hertz) between any two peaks of a multiplet. (p. 516)

deshielded Bonded to a group that withdraws part of the shielding electron density from around the nucleus. (p. 494)

diastereotopic atoms Nuclei that occupy diastereomeric positions. Using the replacement test, the replacement of diastereotopic atoms gives diastereomers. Diastereotopic nuclei can be distinguished by NMR, and they can split each other unless they are *accidentally equivalent*. (p. 524)

downfield At a lower value of the applied magnetic field, toward the left on the conventional NMR spectrum. The more deshielded a nucleus is, the farther downfield it absorbs. (p. 496)

Fourier transform spectroscopy Spectroscopy that involves collecting transients (containing all the different resonance frequencies) and converting the averaged transients into a spectrum using the mathematical Fourier transform. (p. 537)

> **transient** (free-induction decay or FID): The signal that results when many nuclei are irradiated by a pulse of energy and precess at their resonance frequencies.

gyromagnetic ratio (γ) A measure of the magnetic properties of a nucleus. The resonance frequency (ν) is given by the equation $\nu = \gamma H_{eff}/2\pi$, where H_{eff} is the effective magnetic field at the nucleus. The gyromagnetic ratio of a proton is 26,753 sec^{-1} $gauss^{-1}$. The gyromagnetic ratio of a ^{13}C nucleus is 6728 sec^{-1} $gauss^{-1}$. (p. 492)

induced magnetic field The magnetic field set up by the motion of electrons in a molecule (or in a wire) in response to the application of an external magnetic field. (p. 493)

integration The measurement of the area under a peak, proportional to the number of protons giving rise to that peak. Integrations of Fourier-transformed spectra are unreliable. (p. 507)

magnetic moment The magnitude of a nuclear magnetic field, related to the gyromagnetic ratio γ. (p. 491)

magnetically coupled Protons that are close enough that their magnetic fields influence each other, resulting in spin-spin splitting. (p. 510)

nuclear magnetic resonance spectroscopy (NMR) A form of spectroscopy that measures the absorption of radio-frequency energy by nuclei in a magnetic field. The energy absorbed causes nuclear spin transitions. (p. 490)

carbon magnetic resonance (^{13}C NMR, CMR): NMR of the ^{13}C isotope of carbon.

proton magnetic resonance (^{1}H NMR, PMR): NMR of protons.

off-resonance decoupling A technique used with ^{13}C NMR whereby only the protons directly bonded to a carbon atom cause spin-spin splitting. (p. 541)

shielded Surrounded by electrons whose induced magnetic field opposes the externally applied magnetic field. The effective magnetic field at the shielded nucleus is less than the applied magnetic field. (p. 493)

spin decoupling Elimination of spin-spin splitting by constantly irradiating one type of nuclei at its resonance frequency. For example, in ^{13}C spectroscopy, the protons are often irradiated to eliminate their splitting of the carbon signals. The irradiated protons have rapidly changing spin states, and the attached carbon atom absorbs at an averaged value of the magnetic field. (p. 541)

spin-spin splitting (magnetic coupling) The interaction of the magnetic fields of two or more nuclei, usually through the bonds connecting them. Spin-spin splitting converts an absorption to a **multiplet,** a set of smaller peaks. (p. 509)

multiplet: A group of peaks resulting from the spin-spin splitting of the absorption of a single type of nucleus. A **doublet** has two peaks, a **triplet** has three peaks, a **quartet** has four peaks, and so on. (p. 512)

$N + 1$ **rule:** An absorption that is being split by N equivalent protons is split into a multiplet with $N + 1$ individual peaks. (p. 511)

upfield At a higher value of the applied magnetic field, toward the right on the conventional NMR spectrum. The more shielded a nucleus is, the farther upfield it absorbs. (p. 496)

ESSENTIAL PROBLEM-SOLVING SKILLS IN CHAPTER 12

1. Given a structure, determine which protons are equivalent and which are nonequivalent; predict the number of signals and their approximate chemical shifts.
2. Given the chemical shifts of absorptions, suggest likely types of protons.
3. Use the integral trace to determine the relative numbers of different types of protons.
4. Predict which protons in a structure will be magnetically coupled, and predict the number of peaks and approximate coupling constants of their multiplets.
5. Use the proton spin-spin splitting patterns to determine the structure of alkyl groups.
6. Draw the general features of the NMR spectrum of a given compound.
7. Predict the approximate chemical shifts of carbon atoms in a given compound; given the chemical shifts of ^{13}C absorptions, suggest likely types of carbons.
8. Use the off-resonance-decoupled ^{13}C spectrum to determine the number of hydrogens bonded to a given carbon.
9. Combine the chemical shifts, integrals, and spin-spin splitting patterns in the NMR spectrum with information from infrared and mass spectra to determine the structures of organic compounds.

STUDY PROBLEMS

12-29. An unknown compound has the molecular formula $C_9H_{11}Br$. Its proton NMR spectrum shows the following absorptions.

singlet, δ7.1, integral 4.4 cm
singlet, δ2.3, integral 13.0 cm
singlet, δ2.2, integral 6.7 cm

Propose a structure for this unknown compound.

12-30. Predict the multiplicity (the number of peaks as a result of splitting) for each shaded proton in the following compounds.

(a) $CH_3—CH_2—CCl_2—CH_3$ (b) $CH_3—CH(CH_3)—OH$ (c) $CH_3—CH(CH_3)—CH_3$

(d) H, H, H, H (ring positions) —CH_3 (e) —CH_3 (ring with O)

12-31. Predict the approximate chemical shifts of the protons in the following compounds.

(a) benzene (b) cyclohexane (c) $CH_3—O—CH_2—CH_2—CHCl_2$

(d) $CH_3—CH_2—C{\equiv}C—H$ (e) $CH_3—CH_2—C(=O)—CH_3$ (f) $(CH_3)_2CH—CH_2—CH_2—OH$

(g) $CH_3—CH_2—C(=O)—H$ (h) $CH_3—CH{=}CH—CHO$ (i) $HOOC—CH_2—CH_2—C(=O)—O—CH(CH_3)_2$

(j) methylenecyclohexane (k) indane (l) γ-butyrolactone

12-32. The following proton NMR spectrum is of a compound of molecular formula C_3H_8O.
(a) Propose a structure for this compound.
(b) Make peak assignments, showing which protons give rise to which absorptions in the spectrum.

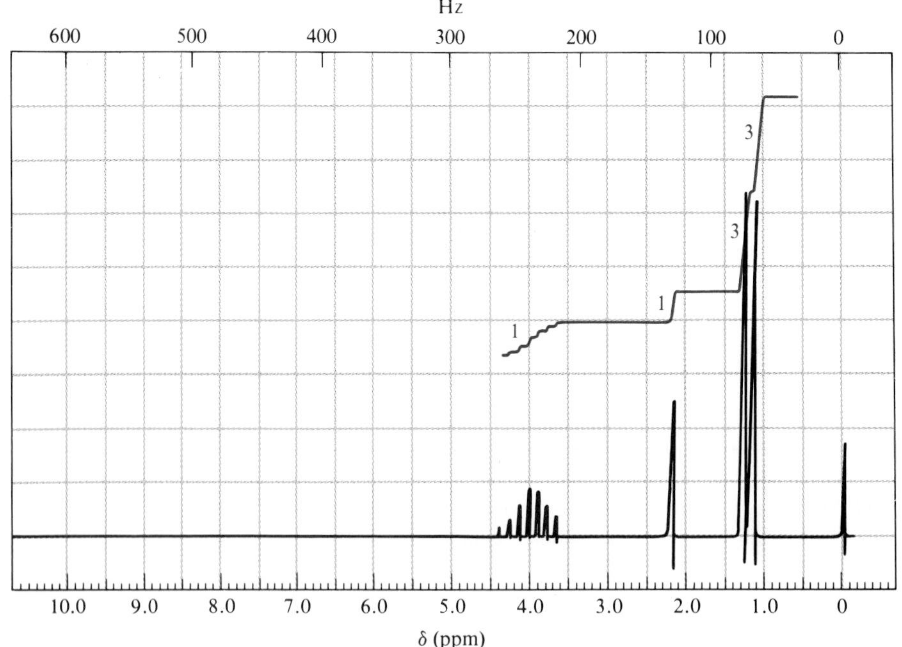

12-33. Using a 60-MHz spectrometer, the following absorption is observed.

doublet, $J = 7$ Hz, at $\delta 4.00$

(a) How many hertz from the TMS peak is this absorption?
(b) Where would this peak be located (in ppm and in hertz) in the 100-MHz spectrum of this sample?
(c) What would J be in the 100-MHz spectrum?

12-34. A compound ($C_{10}H_{12}O_2$) whose spectrum appears below was isolated from a reaction mixture containing 2-phenylethanol and acetic acid.
(a) Propose a structure for this compound.
(b) Make peak assignments, showing which protons give rise to which absorptions in the spectrum.

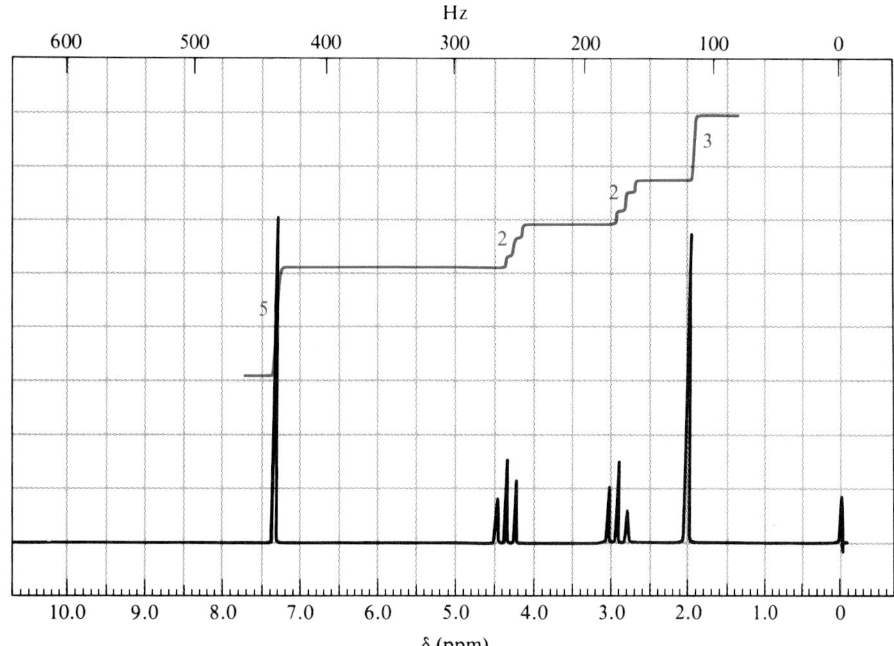

12-35. Sketch your predictions of the proton NMR spectra of the following compounds.

(a) $CH_3—O—CH_2CH_3$ **(b)** $(CH_3)_2CH—\overset{O}{\overset{\|}{C}}—CH_3$ **(c)** 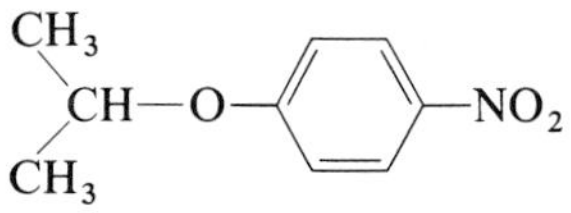 $(CH_3)_2CH—O—C_6H_4—NO_2$ (para)

(d) 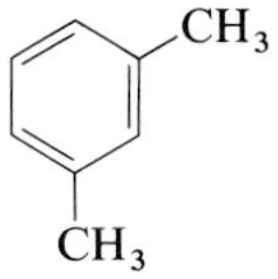 benzene ring with two CH_3 groups (meta) **(e)** $Cl—CH_2—CH_2—CH_2—Cl$

12-36. Tell precisely how you would use the proton NMR spectra to distinguish between the following pairs of compounds.

(a) 1-bromopropane and 2-bromopropane **(b)** $CH_3—CH_2—\overset{O}{\overset{\|}{C}}—H$ and $CH_3—\overset{O}{\overset{\|}{C}}—CH_3$

(c) $CH_3—CH_2—O—\overset{O}{\overset{\|}{C}}—CH_3$ and $CH_3—CH_2—\overset{O}{\overset{\|}{C}}—O—CH_3$

(d) $CH_3—CH_2—C\equiv C—H$ and $CH_3—C\equiv C—CH_3$

12-37. Give the approximate chemical shift and multiplicity of each absorption band in the off-resonance decoupled ^{13}C NMR spectra of the following compounds.

(a) $CH_3—\overset{\overset{O}{\|}}{C}—O—CH_2—CH_3$ (ethyl acetate) **(b)** $H_2C=CH—CH_2Cl$ (3-chloropropene)

12-38. The following off-resonance decoupled carbon NMR was obtained from a compound of formula $C_3H_5Cl_3$. Propose a structure for this compound and show which carbon atoms give rise to which peaks in the spectrum.

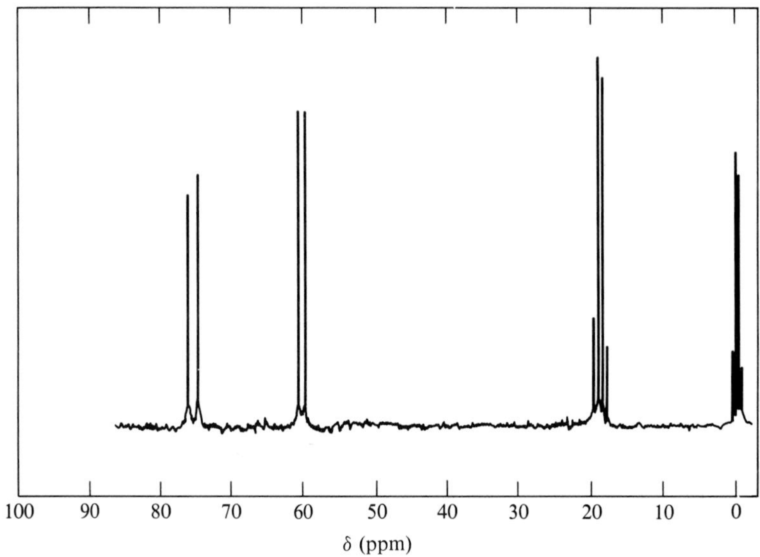

12-39. The following proton NMR spectra are given together with the molecular formulas of the compounds they represent. In each case, propose a structure for the compound and give peak assignments.

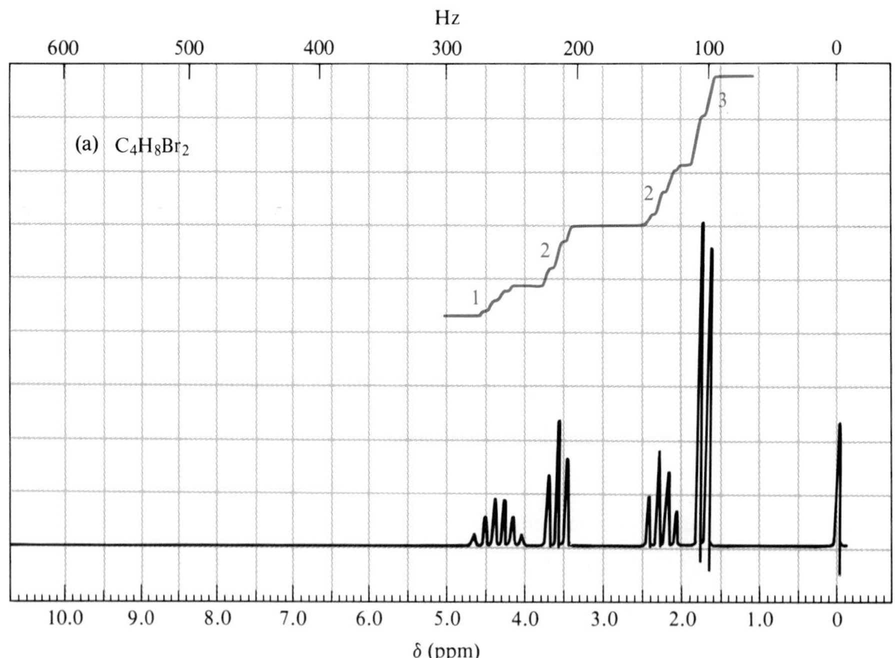

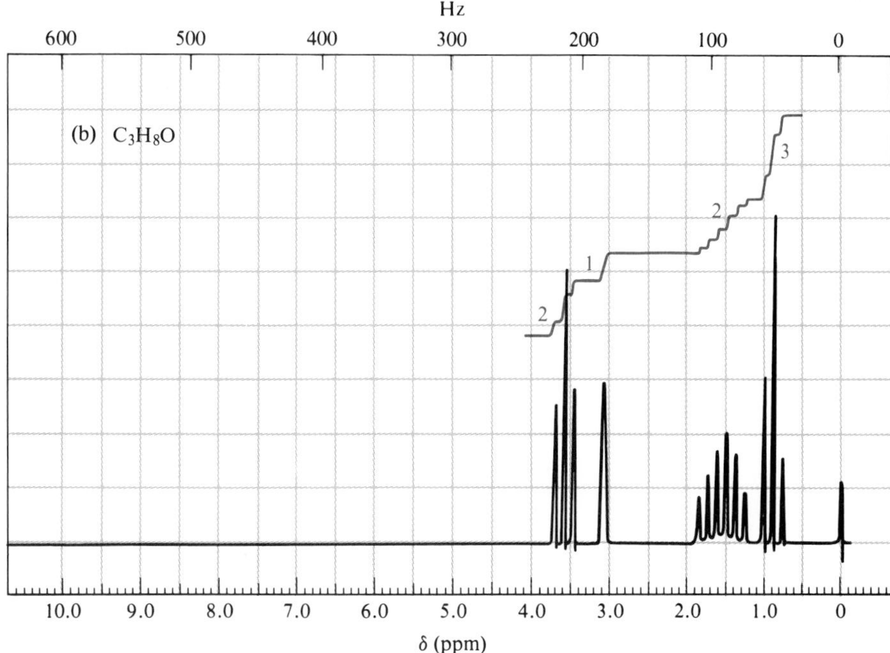

12-40. When 2-chloro-2-methylbutane is treated with a variety of strong bases, the products always seem to contain two isomers (A and B) of formula C_5H_{10}. When sodium hydroxide is used as the base, isomer A predominates. When potassium *t*-butoxide is used as the base, isomer B predominates. The proton NMR spectra of A and B are given below.

(a) Determine the structures of isomers A and B.

(b) Explain why A is the major product using sodium hydroxide as the base, and why B is the major product using potassium *t*-butoxide as the base.

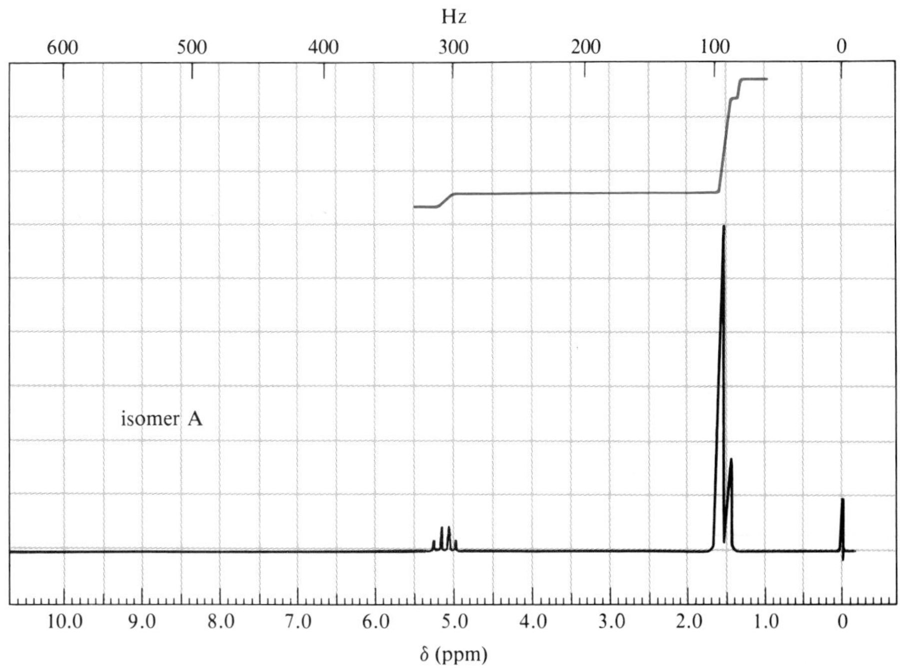

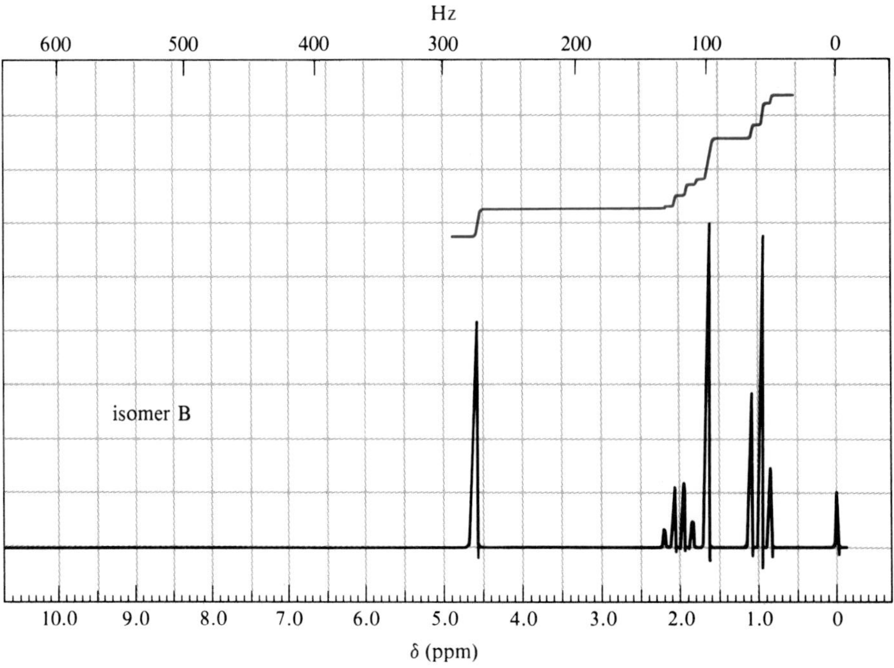

12-41. Sets of spectra are given for two different compounds. For each set of spectra

(1) Look at each individual spectrum individually and list the structural characteristics which you can determine from that spectrum.

(2) Look at the set of spectra as a group and propose a tentative structure.

(3) Verify that your proposed structure accounts for the major features of each spectrum.

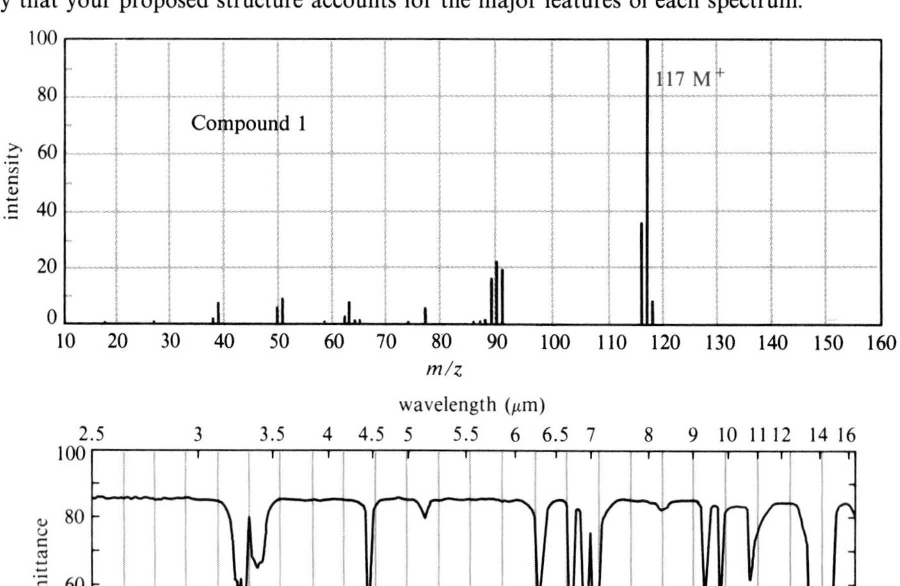

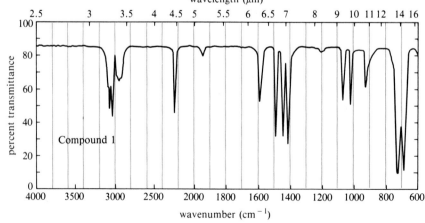

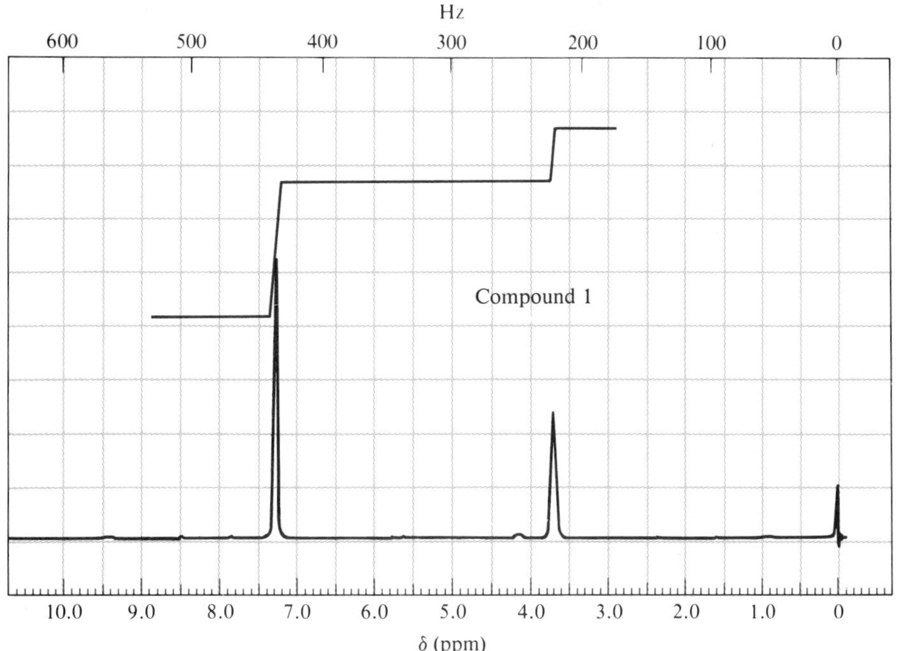
Hz
600
500
400
300
200
100
0
Compound 1
10.0
9.0
8.0
7.0
6.0
5.0
4.0
3.0
2.0
1.0
0
δ (ppm)

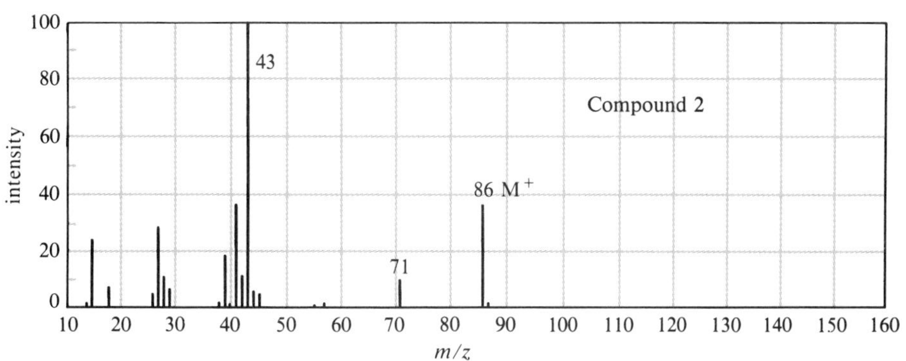
100
80
60
40
20
0
intensity
43
Compound 2
86 M+
71
10
20
30
40
50
60
70
80
90
100
110
120
130
140
150
160
m/z

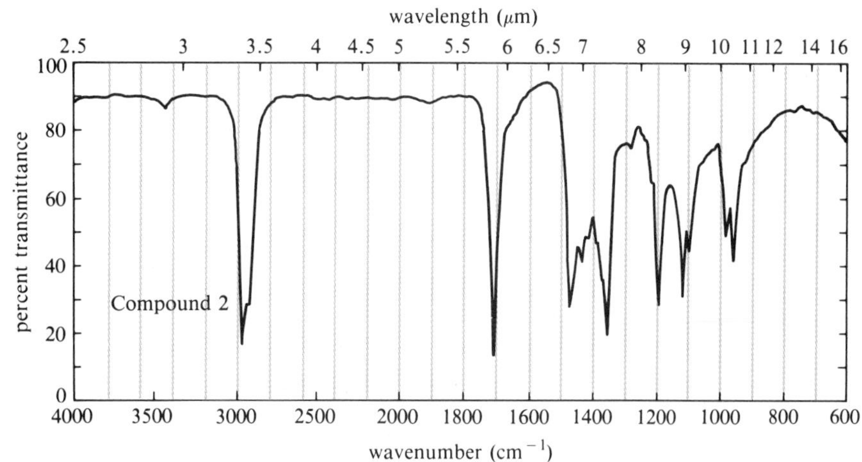
wavelength (μm)
2.5
3
3.5
4
4.5
5
5.5
6
6.5
7
8
9
10
11
12
14
16
100
80
60
40
20
0
percent transmittance
Compound 2
4000
3500
3000
2500
2000
1800
1600
1400
1200
1000
800
600
wavenumber (cm−1)

Problem 12-41 ***continued***

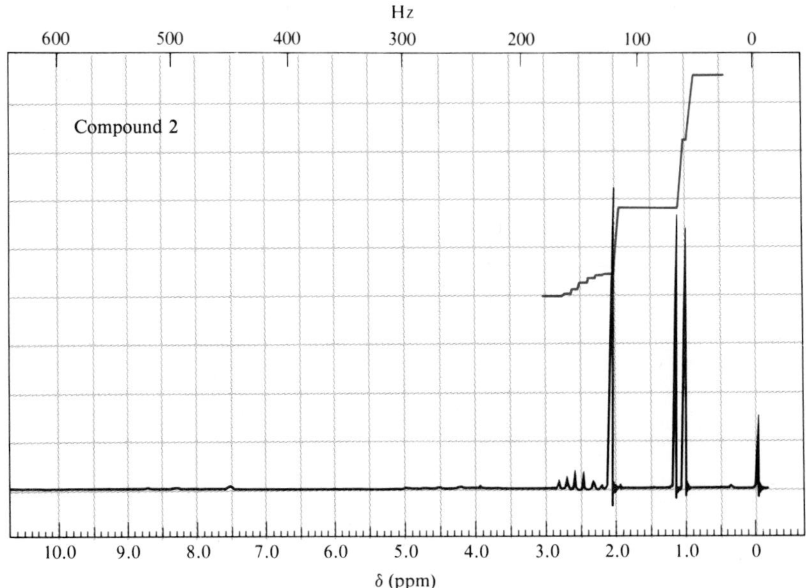

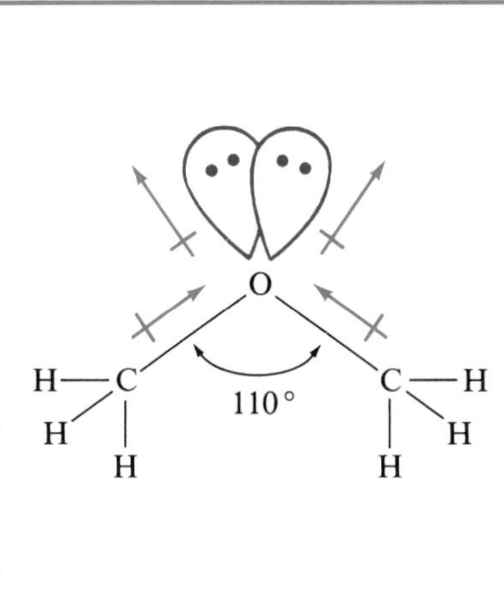

13

ETHERS AND EPOXIDES

13-1 INTRODUCTION

Ethers are compounds of formula R—O—R′, where R and R′ may be alkyl groups or aryl (benzene ring) groups. Like alcohols, ethers are relatives of water with alkyl groups replacing the hydrogen atoms. In the alcohols, one hydrogen atom of water is replaced by an alkyl group. In an ether, both hydrogens are replaced by alkyl groups. The two alkyl groups are the same in a **symmetrical ether** and different in an **unsymmetrical ether.**

H—O—H	R—O—H	R—O—R′
water	alcohol	ether

Examples of ethers

CH_3CH_2—O—CH_2CH_3	C_6H_5—O—CH_3	$(CH_3)_2CH$—O—$CH(CH_3)_2$
diethyl ether (a symmetrical ether)	methyl phenyl ether (an unsymmetrical ether)	diisopropyl ether (a symmetrical ether)

As with other functional groups, we discuss how ethers are formed and how they react. Ethers (other than epoxides) are relatively unreactive, however, and they are not frequently used as synthetic intermediates. The most common use of ethers is as solvents for organic reactions. We consider in detail the properties of ethers and how these properties make ethers such valuable solvents for organic reactions.

The most important commercial ether is diethyl ether, often called "ethyl ether" or simply "ether." Ether is a good solvent for reactions and extractions, and it is often used as a volatile starting fluid for diesel and gasoline engines. Ether was used as a surgical anesthetic for over a hundred years (since 1842), but it is highly flammable, and patients often vomit as they regain consciousness. Several compounds that are less flammable and more easily tolerated are now in use, including nitrous oxide (N_2O) and halothane ($BrClCH—CF_3$).

13-2 PHYSICAL PROPERTIES OF ETHERS

13-2A POLARITY OF ETHERS

Although ethers lack the polar hydroxyl group of the alcohols, they are still strongly polar compounds. Table 13-1 compares the dipole moments of diethyl ether and tetrahydrofuran with those of an alkene and an alcohol of similar molecular weights. An ether such as tetrahydrofuran provides a highly polar solvent without the reactivity of the alcohol hydroxyl group.

TABLE 13-1 Comparison of the dipole moments of two ethers, an alkane and an alcohol.

	Dipole moment (debyes)
pentane	0.1 D
diethyl ether	1.2 D
tetrahydrofuran	1.6 D
1-butanol	1.7 D

The dipole moment of an ether is the sum of four individual dipole moments (Fig. 13-1). Each of the two C—O bonds is polarized, with the carbon atoms bearing a partial positive charge. In addition, the two nonbonding electron pairs contribute to the dipole moment. The vector sum of these four individual moments is the overall molecular dipole moment.

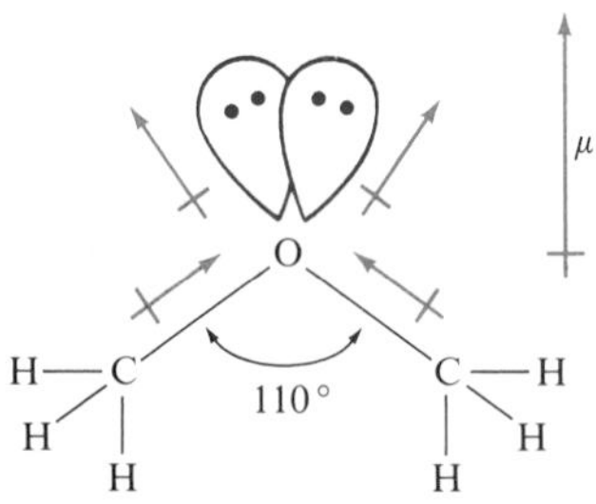

FIGURE 13-1 Ethers have large molecular dipole moments because the individual moments of the C—O bonds and the lone pairs reinforce to give large vector sums.

13-2B BOILING POINTS OF ETHERS; HYDROGEN BONDING

Figure 13-2 shows that a hydrogen bond requires both a hydrogen bond *donor* and a hydrogen bond *acceptor*. The donor is the molecule with an O—H or N—H group. The acceptor is the molecule whose lone pair of electrons forms a weak partial bond to the hydrogen atom provided by the donor. An ether molecule has the lone pair to form a hydrogen bond with an alcohol (or other hydrogen bond donor), but it cannot form a hydrogen bond with another ether molecule. Because ether molecules are not held together by hydrogen bonds, they are more volatile than alcohols with similar molecular weights.

Ethers do have relatively large dipole moments that result in dipole-dipole attractions (Section 2-12A), but these attractions are much weaker than hydrogen bonding. Table 13-2 compares the boiling points of several ethers, alcohols, and alkanes. Notice that the boiling points of dimethyl ether and diethyl ether are nearly 100°C lower than those of alcohols with similar molecular weights. The ethers' dipole moments have relatively little effect on their boiling points, which are similar to those of alkanes of similar molecular weights. Table 13-3 lists the physical properties of a representative group of common ethers.

13-2C ETHERS AS POLAR SOLVENTS

Ethers are ideally suited as solvents for many organic reactions. They dissolve a wide range of polar and nonpolar substances, and their relatively low boiling

Alcohol

O—H ••• O
R
H
R
hydrogen bond
donor acceptor

Ether

O—R
O
R
R
R
no hydrogen bond
no donor acceptor

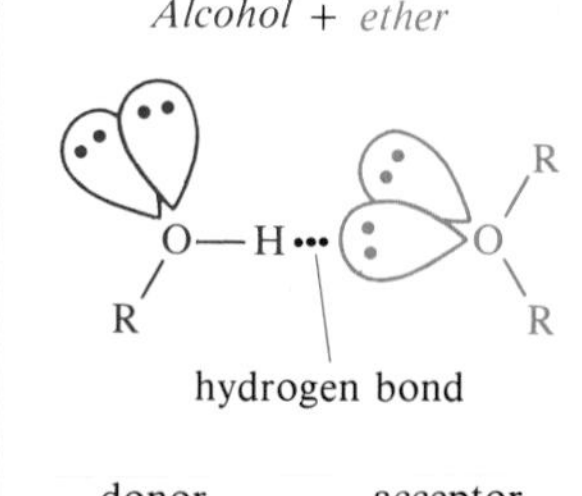

FIGURE 13-2 A molecule of water or an alcohol can serve as both a hydrogen bond donor and acceptor. Ether molecules have no hydroxyl groups, so they are not hydrogen bond donors. If a hydrogen bond donor is present, ethers can serve as hydrogen bond acceptors.

TABLE 13-2
Comparison of the boiling points of ethers, alkanes, and alcohols of similar molecular weights[a]

Compound	*Formula*	*MW*	*b.p. (°C)*	*Dipole moment (D)*
water	H_2O	18	100	1.9
ethanol	CH_3CH_2-OH	46	78	1.7
dimethyl ether	CH_3-O-CH_3	46	−25	1.3
propane	$CH_3CH_2CH_3$	44	−42	0.1
n-butanol	$CH_3CH_2CH_2CH_2-OH$	74	118	1.7
diethyl ether	$CH_3CH_2-O-CH_2CH_3$	74	35	1.2
pentane	$CH_3CH_2CH_2CH_2CH_3$	72	36	0.1

[a] The alcohols are hydrogen bonded, giving them much higher boiling points. The ethers have boiling points that are closer to those of alkanes with similar molecular weights.

TABLE 13-3
Physical properties of some representative ethers

Name	*Structure*	*m.p. (°C)*	*b.p. (°C)*	*Density*
dimethyl ether	CH_3-O-CH_3	−140	−25	0.66
ethyl methyl ether	$CH_3CH_2-O-CH_3$		8	0.72
diethyl ether	$CH_3CH_2-O-CH_2CH_3$	−116	35	0.71
di-*n*-propyl ether	$CH_3CH_2CH_2-O-CH_2CH_2CH_3$	−122	91	0.74
diisopropyl ether	$(CH_3)_2CH-O-CH(CH_3)_2$	−60	68	0.74
1,2-dimethoxyethane (DME)	$CH_3-O-CH_2CH_2-O-CH_3$	−58	83	0.86
methyl phenyl ether (anisole)	CH_3-O-	−37	154	0.99
diphenyl ether	O	27	259	1.07
furan	O	−86	32	0.94
tetrahydrofuran (THF)	O	−108	65	0.89
pyran	O		80	
1,4-dioxane	O O	11	101	1.03

points simplify their evaporation from the reaction products. Nonpolar substances tend to be more soluble in ethers than in alcohols because ethers have no hydrogen-bonding network to be broken up by the nonpolar solute.

Polar substances tend to be nearly as soluble in ethers as in alcohols because of ethers' large dipole moments and their ability to serve as hydrogen bond acceptors. The nonbonding electron pairs of an ether effectively solvate cations, as shown in Figure 13-3. Ethers do not solvate anions as well as alcohols do, however. Ionic substances with small, "hard" anions requiring strong solvation to overcome their strong ionic bonding forces are generally insoluble in ether solvents. Substances with large, diffuse anions, such as iodides, acetates, and other organic

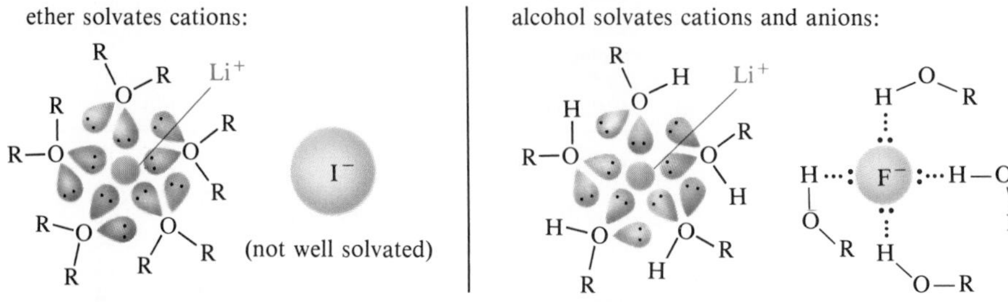

FIGURE 13-3 An ionic substance such as lithium iodide (LiI) is moderately soluble in ether solvents because the small lithium cation is strongly solvated by the ether's lone pairs of electrons. Unlike alcohols, ethers cannot serve as hydrogen bond donors, and they do not solvate small anions well.

anions tend to be more soluble in ether solvents than substances with smaller, harder anions.

Alcohols cannot be used as solvents for reagents that are more strongly basic than the alkoxide ion. The hydroxyl group quickly protonates the base, destroying the basic reagent.

$$\underset{\text{strong base}}{B{:}^-} + \underset{\text{alcohol}}{R{-}\ddot{\underset{\cdot\cdot}{O}}H} \rightleftharpoons \underset{\text{protonated base}}{B{-}H} + \underset{\text{alkoxide ion}}{R{-}\ddot{\underset{\cdot\cdot}{O}}{:}^-}$$

Ethers are nonhydroxylic (no hydroxyl group), and they are normally unreactive toward strong bases. For this reason, ethers are frequently used as solvents for very strong, polar bases (like the Grignard reagent) that require polar solvents. The following four ethers are commonly used as solvents for organic reactions. DME, THF, and dioxane are miscible with water, while diethyl ether is sparingly soluble in water.

$CH_3CH_2{-}O{-}CH_2CH_3$	$CH_3{-}O{-}CH_2CH_2{-}O{-}CH_3$	(five-membered ring with O)	(six-membered ring with two O)
diethyl ether	1,2-dimethoxyethane	tetrahydrofuran	1,4-dioxane
"ether"	DME, "glyme"	THF, oxolane	dioxane
b.p. 35°C	b.p. 82°C	b.p. 65°C	b.p. 101°C

PROBLEM 13-1

For each of the following compounds, rank the given solvents in decreasing order of their ability to dissolve the given compound.

Solutes

(a) NaOAc

(b) naphthalene

(c) 2-naphthol

Solvents

ethyl ether
water
ethanol
dichloromethane

FIGURE 13-4 Complexation of an ether with a Grignard reagent stabilizes the reagent and helps to keep it in solution.

13-2D STABLE COMPLEXES OF ETHERS WITH REAGENTS

The special properties of ethers (polarity, lone pairs, but no hydroxyl group) are necessary for the formation and use of many reagents. For example, Grignard reagents cannot form unless there is an ether present. Although the exact structure of the Grignard reagent is still unknown, it appears that one or more molecules of ether must share its lone pairs of electrons with the magnesium atom. This sharing of electrons stabilizes the reagent, and it helps to keep the reagent in solution (Fig. 13-4).

Ether's nonbonding electrons also stabilize borane, BH_3. Pure borane exists as a dimer called diborane, B_2H_6. Diborane is a toxic, flammable, and explosive gas, whose use is both dangerous and inconvenient. Borane forms a stable complex with tetrahydrofuran. This $BH_3 \cdot THF$ complex is commercially available as a 1 *M* solution, easily measured and transferred like any other liquid reagent. The availability of the $BH_3 \cdot THF$ complex has contributed greatly to the convenience of the hydroboration procedure (Section 8-7).

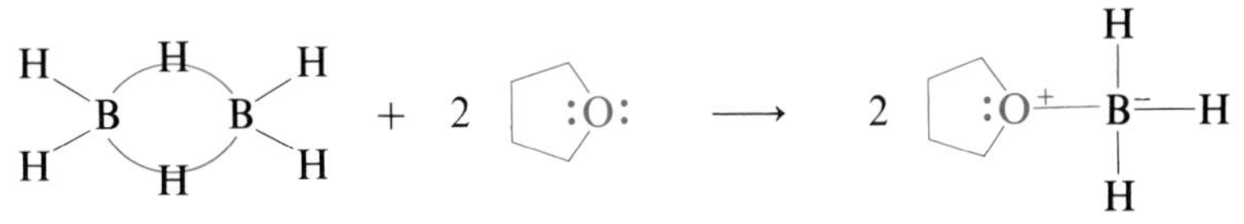

diborane + 2 tetrahydrofuran ⟶ 2 $BH_3 \cdot THF$

Boron trifluoride is used as a Lewis acid catalyst in a wide variety of reactions. Like diborane, BF_3 is a toxic gas, but BF_3 forms a stable complex with ethers, allowing it to be conveniently stored and measured. The complex of BF_3 with diethyl ether is called "boron trifluoride ethyl etherate."

$$BF_3 + :O(CH_2CH_3)_2 \longrightarrow F_3B^{-}\text{—}O^{+}(CH_2CH_3)_2$$

boron trifluoride + diethyl ether ⟶ $BF_3 \cdot OEt_2$ "boron trifluoride ethyl etherate"

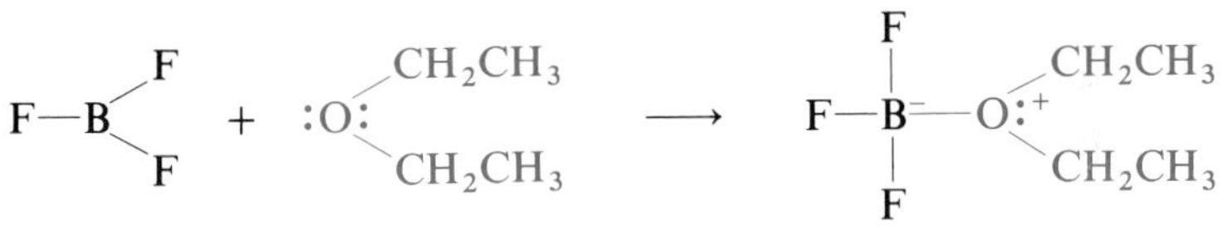

PROBLEM 13-2

Aluminum trichloride ($AlCl_3$) dissolves in ether with the evolution of a large amount of heat. Show the structure of the resulting aluminum chloride etherate complex.

13-3 NOMENCLATURE OF ETHERS

We have been using the common nomenclature of ethers, which is sometimes called the *alkyl alkyl ether* system. The IUPAC system, generally used with more complicated ethers, is sometimes called the *alkoxy alkane* method. Common names are almost always used for simple ethers.

13-3A COMMON NAMES (ALKYL ALKYL ETHER NAMES)

The common names of ethers are constructed by naming the two alkyl groups bonded to the oxygen atom and adding the word "ether." The two alkyl groups are named in alphabetical order. For example, if one of the alkyl groups is methyl and the other is ethyl, the common name is "ethyl methyl ether." If both groups are methyl, the name is "dimethyl ether." If just one alkyl group is described in the name of an ether, it implies that the ether is symmetrical, with two of the groups as in "ethyl ether."

13-3B IUPAC NAMES (ALKOXY ALKANE NAMES)

The IUPAC names of ethers use the more complex alkyl group as the base compound and name the rest of the ether as an alkoxy group. For example, cyclohexyl methyl ether is named methoxycyclohexane. This systematic nomenclature is often the only good way of naming more complex ethers.

	$CH_3—O—CH_2CH_3$	$CH_3—O—CH_3$	$Cl—CH_2—O—CH_3$
IUPAC name:	methoxyethane	methoxymethane	chloromethoxymethane
common name:	ethyl methyl ether	dimethyl ether (methyl ether)	chloromethyl methyl ether

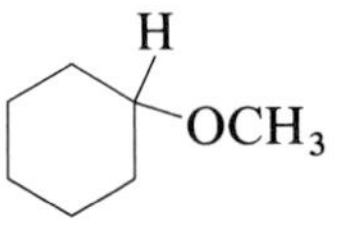

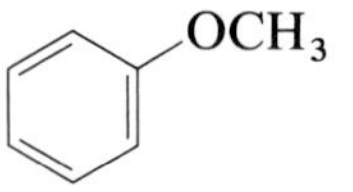

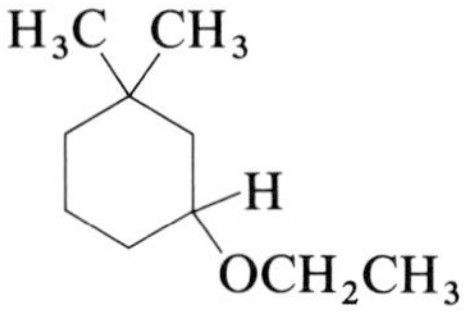

	[cyclohexane with H and OCH_3]	[benzene with OCH_3]	[cyclohexane with H_3C, CH_3, H, OCH_2CH_3]
IUPAC name:	methoxycyclohexane	methoxybenzene	3-ethoxy-1,1-dimethylcyclohexane
common name:	cyclohexyl methyl ether	methyl phenyl ether, anisole	(none)

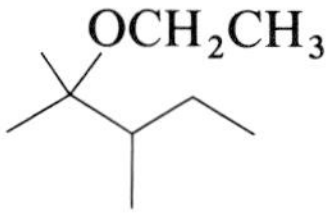

	[OCH_2CH_3 skeletal structure]	[cyclobutane with Cl, H, OCH_3, H]	$CH_2—OH$ / $CH_2—O—CH_2CH_3$
IUPAC name:	2-ethoxy-2,3-dimethylpentane	*trans*-2-chloro-1-methoxycyclobutane	2-ethoxyethanol

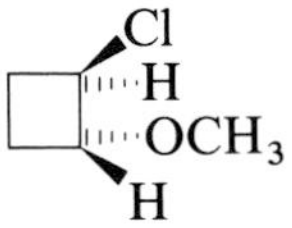

PROBLEM 13-3

Give a common name and a systematic name for each of the following compounds.

(a) 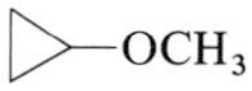 [cyclopropyl]$—OCH_3$ (b) $CH_3CH_2—O—CH(CH_3)_2$ (c) $ClCH_2CH_2OCH_3$

(d) $(CH_3)_3C—O—CH(CH_3)(CH_2CH_3)$ (e) 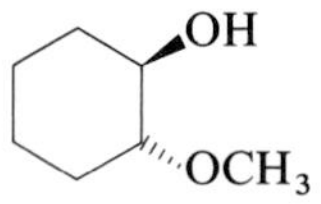 [cyclohexane with OH and OCH_3]

13-3C NOMENCLATURE OF CYCLIC ETHERS

The cyclic ethers are our first examples of **heterocyclic compounds,** containing a ring in which one or more of the ring atoms is an element other than carbon. This atom, called the **heteroatom,** is numbered 1 in numbering the ring atoms. Heterocyclic ethers are an especially important and useful class of ethers.

The epoxides (oxiranes) We have already encountered some of the chemistry of the epoxides (Section 8-12). **Epoxides** are three-membered cyclic ethers, usually formed by peroxyacid oxidation of the corresponding alkenes. The common name of an epoxide is formed by adding "oxide" to the name of the alkene that is oxidized. The following reactions show the synthesis and common names of some simple epoxides.

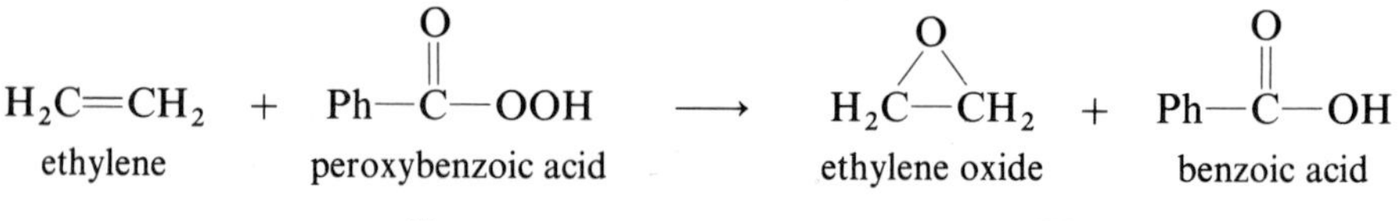

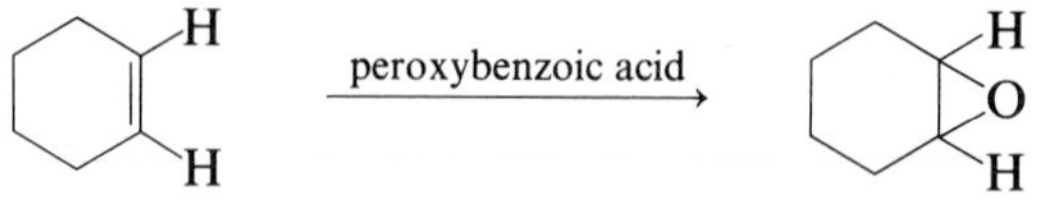

$$H_2C{=}CH_2 + Ph—\overset{O}{\overset{\|}{C}}—OOH \longrightarrow H_2C\overset{O}{—}CH_2 + Ph—\overset{O}{\overset{\|}{C}}—OH$$

ethylene + peroxybenzoic acid → ethylene oxide + benzoic acid

cyclohexene —(peroxybenzoic acid)→ cyclohexene oxide

One systematic method for naming epoxides is to name the rest of the molecule and use the term "epoxy" as a substituent. The numbers of the two carbon atoms bonded to the epoxide oxygen are given.

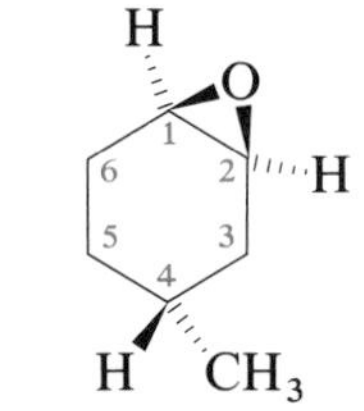

trans-4-methyl-1,2-epoxycyclohexane

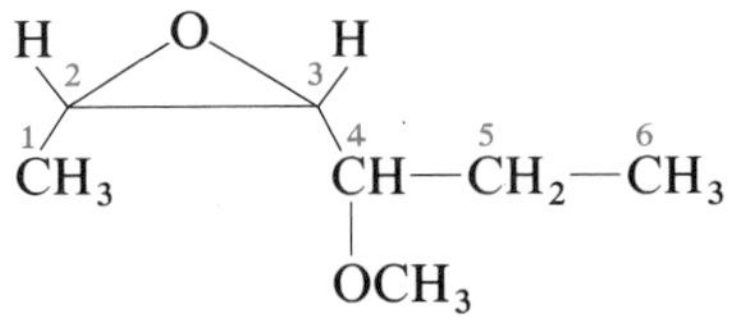

cis-2,3-epoxy-4-methoxyhexane

Another systematic method names epoxides as derivatives of the parent compound, ethylene oxide. The systematic name for ethylene oxide is "oxirane." Notice that the ring atoms of a heterocyclic compound are numbered starting with the heteroatom and numbering in the direction to give the lowest substituent numbers.

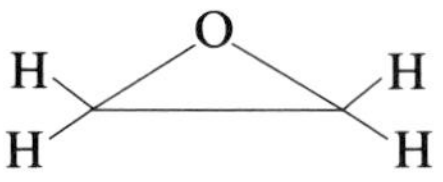

oxirane

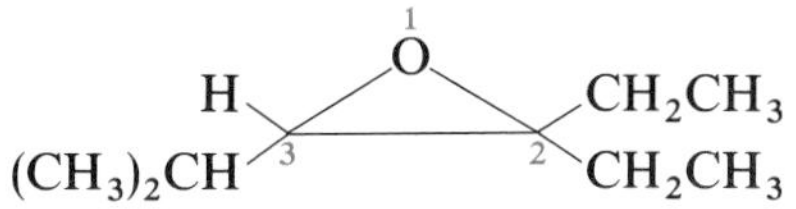

2,2-diethyl-3-isopropyloxirane

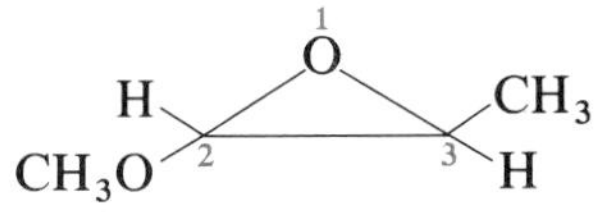

trans-2-methoxy-3-methyloxirane

The oxetanes The least common simple cyclic ethers are the four-membered **oxetanes.** Because these four-membered rings are strained, they are more reactive than are larger cyclic ethers and open-chain ethers. They are not as reactive as the highly strained oxiranes (epoxides), however.

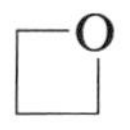

oxetane

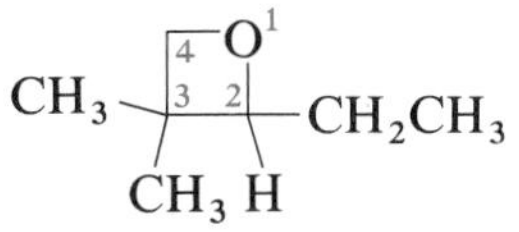

2-ethyl-3,3-dimethyloxetane

The furans The five-membered cyclic ethers are commonly named after an aromatic member of this group, **furan.** We consider the aromaticity of furan and other heterocycles in Chapter 16.

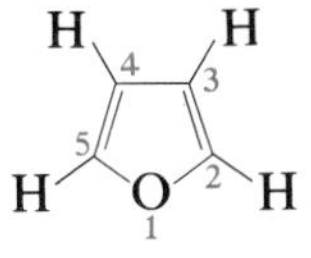

furan

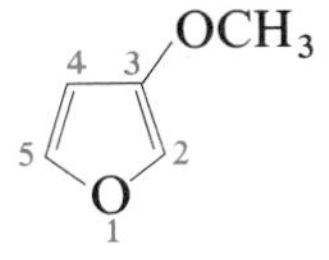

3-methoxyfuran

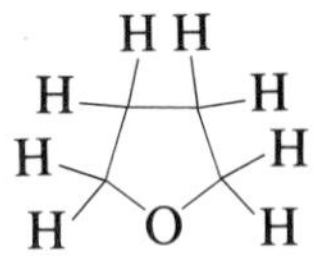

tetrahydrofuran (THF)

The saturated five-membered cyclic ether resembles furan but has four additional hydrogen atoms. Therefore, it is called *tetrahydrofuran* (THF). One of the most polar ethers, tetrahydrofuran is an excellent nonhydroxylic organic solvent for polar reagents. Grignard reactions sometimes succeed in THF even when they fail in diethyl ether.

The pyrans The six-membered cyclic ethers are commonly named as derivatives of **pyran,** an unsaturated ether. The saturated compound has four more hydrogen atoms, so it is called *tetrahydropyran* (THP).

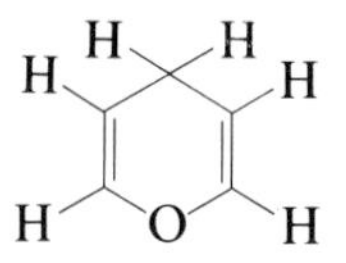

pyran 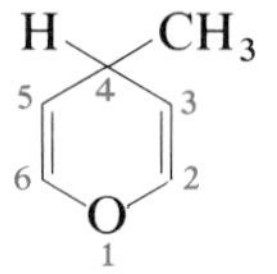 4-methylpyran 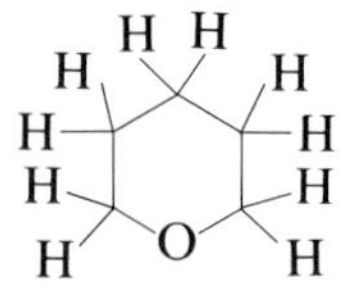 tetrahydropyran (THP)

The dioxanes Heterocyclic ethers with two oxygen atoms in a six-membered ring are called **dioxanes.** The most common form of dioxane is the one with the two oxygen atoms in a 1,4-relationship. 1,4-Dioxane is miscible with water, and it is widely used as a polar solvent for organic reactions.

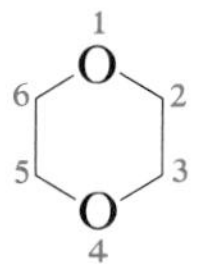

1,4-dioxane 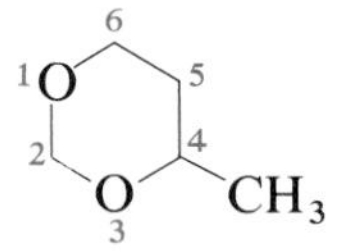 4-methyl-1,3-dioxane 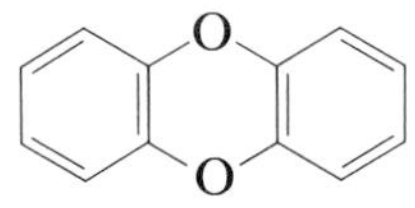 dibenzodioxin (dioxin)

Dioxin is a common name for dibenzodioxin, the 1,4-dioxane that is fused with two benzene rings. The name "dioxin" is often used incorrectly in the news media for 2,3,7,8-tetrachlorodibenzodioxin (TCDD), a toxic contaminant in the synthesis of the herbicide 2,4,5-trichlorophenoxyacetic acid, called 2,4,5-T or Agent Orange. Many dioxins are toxic, because they associate with DNA and cause a misreading of the genetic code.

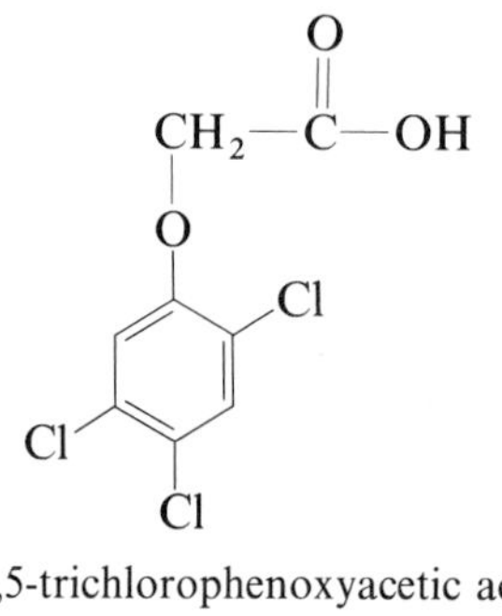

2,4,5-trichlorophenoxyacetic acid
(2,4,5-T or Agent Orange)

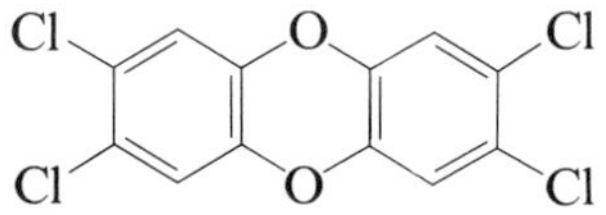

2,3,7,8-tetrachlorodibenzodioxin
(TCDD, incorrectly "dioxin")

PROBLEM 13-4

1,4-Dioxane is made commercially by the acid-catalyzed dehydration of an alcohol.

(a) Show what alcohol would dehydrate to give 1,4-dioxane.
(b) Give a mechanism for this reaction.

PROBLEM 13-5

Name the following heterocyclic ethers.

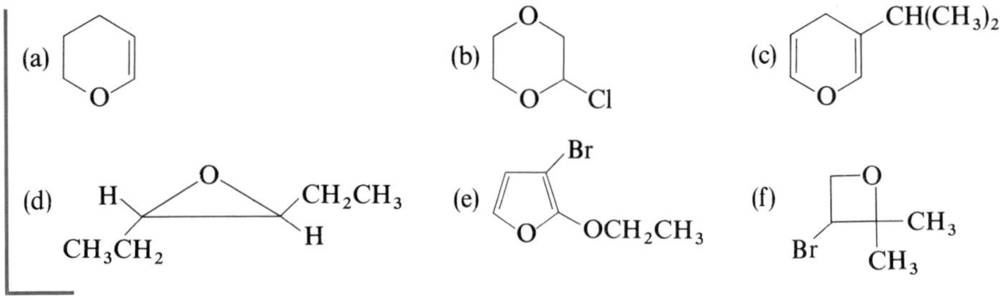

(a) [3,4-dihydro-2H-pyran structure]

(b) [2-chloro-1,4-dioxane structure]

(c) [ring ether with $CH(CH_3)_2$ substituent]

(d) H, CH_3CH_2 / O / CH_2CH_3, H [epoxide structure]

(e) [furan with Br and OCH_2CH_3]

(f) [oxetane with Br, CH_3, CH_3]

13-4 SPECTROSCOPY OF ETHERS

Infrared spectroscopy of ethers Infrared spectra do not show obvious or reliable absorptions for ethers. Most ethers give a moderate to strong C—O stretch around 1000 to 1200 cm^{-1} (in the fingerprint region), but many compounds other than ethers give similar absorptions. Nevertheless, the IR spectrum can be useful, because it shows the *absence* of carbonyl (C=O) groups and hydroxyl (O—H) groups. If the molecular formula is known to contain an oxygen atom, the lack of carbonyl or hydroxyl absorptions in the IR suggests the presence of an ether.

Mass spectrometry of ethers The most common fragmentation of ethers involves cleavage next to one of the carbon atoms bonded to oxygen. Since this carbon is *alpha* to the oxygen atom, this mode of fragmentation is called **α cleavage.** The resulting *oxonium ion* is resonance stabilized by the nonbonding electrons on oxygen.

α Cleavage

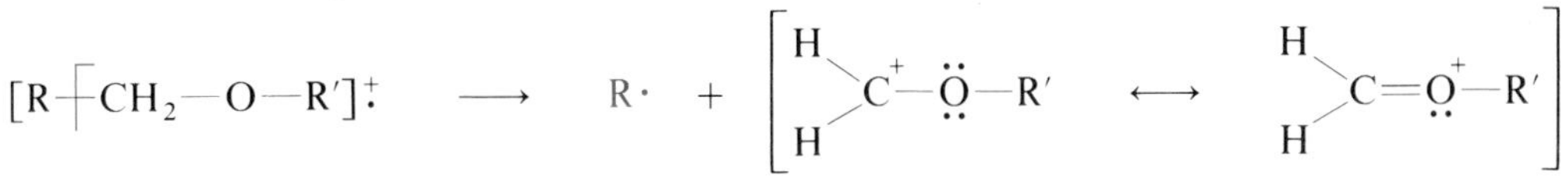

$$[R \nmid CH_2-O-R']^{+\cdot} \longrightarrow \underset{\text{not observed}}{R\cdot} + \underset{\text{oxonium ion}}{\left[H_2\overset{+}{C}-\ddot{O}-R' \longleftrightarrow H_2C=\overset{+}{\ddot{O}}-R' \right]}$$

Another common cleavage of ethers is the loss of either of the two alkyl groups to give either another oxonium ion or an alkyl cation.

Loss of an alkyl group

$$[R-CH_2-O \nmid R']^{+\cdot} \xrightarrow{\sim H} \underset{\text{oxonium ion}}{[R-\overset{+}{C}H-\ddot{O}-H \longleftrightarrow R-CH=\overset{+}{\ddot{O}}-H]} + \underset{\text{not observed}}{\cdot R'}$$

or

$$[R-CH_2-O \nmid R']^{+\cdot} \longrightarrow \underset{\text{not observed}}{R-CH_2-O\cdot} + \underset{\text{alkyl cation}}{{}^{+}R'}$$

The mass spectrum of methyl *n*-propyl ether appears in Figure 13-5. The four most abundant ions correspond to the molecular ion, α cleavage, and loss of each of the two alkyl groups. All these modes of cleavage form resonance-stabilized oxonium ions.

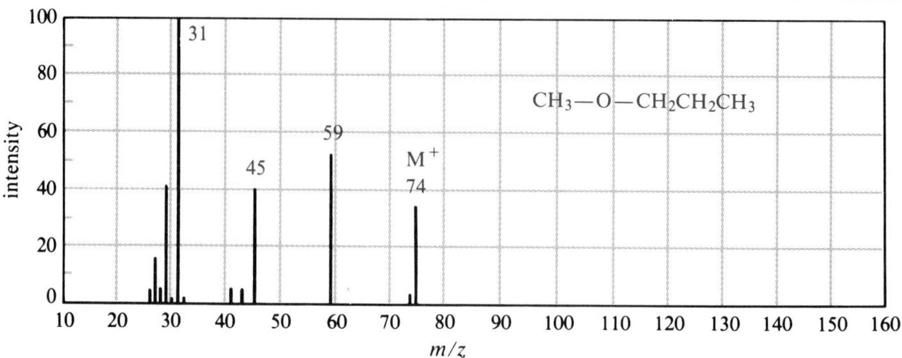

α Cleavage

$$[CH_3{-}O{-}CH_2 \mid CH_2{-}CH_3]^{+\cdot} \xrightarrow{\alpha\text{ cleavage}} CH_3{-}\overset{+}{O}{=}CH_2 + \cdot CH_2CH_3$$

(cleavage at 45; m/z 74 → m/z 45, loss of 29)

Loss of methyl group

$$[CH_3 \mid O{-}CH_2{-}CH_2{-}CH_3]^{+\cdot} \longrightarrow H{-}\overset{+}{O}{=}CH{-}CH_2{-}CH_3 + \cdot CH_3$$

(cleavage at 59; m/z 74 → m/z 59, loss of 15)

Loss of propyl group

$$[CH_3{-}O \mid CH_2{-}CH_2{-}CH_3]^{+\cdot} \longrightarrow H_2C{=}\overset{+}{O}{-}H + \cdot CH_2CH_2CH_3$$

(cleavage at 31; m/z 74 → m/z 31, loss of 43)

FIGURE 13-5 The mass spectrum of methyl *n*-propyl ether shows major peaks for the molecular ion, α cleavage, loss of the methyl group, and loss of the propyl group.

PROBLEM 13-6

Give a fragmentation reaction to account for each of the numbered peaks in the mass spectrum of *n*-butyl isopropyl ether.

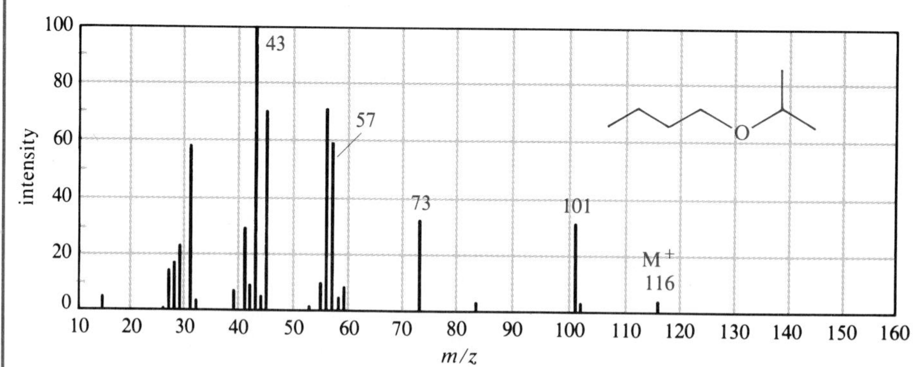

NMR spectroscopy of ethers In the ^{13}C spectrum, a carbon atom bonded to oxygen generally absorbs between $\delta 65$ and $\delta 90$. Protons on carbon atoms bonded to oxygen usually absorb at chemical shifts between $\delta 3$ and $\delta 4$ in the ^1H NMR spectrum. Both alcohols and ethers have resonances in this range. See, for example, the NMR spectrum of methyl *t*-butyl ether (page 506) and that of ethanol (page 527). If a compound containing C, H, and O has resonances in the correct range, and if there is no O—H stretch or C=O stretch in the IR spectrum, an ether is the most likely functional group.

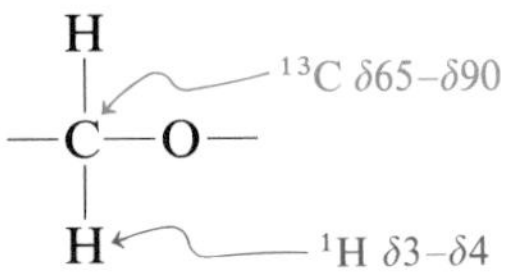

$$\mathrm{H{-}C(H){-}O{-}} \quad \text{(}^{13}\text{C } \delta 65\text{–}\delta 90;\ ^{1}\text{H } \delta 3\text{–}\delta 4\text{)}$$

13-5 THE WILLIAMSON ETHER SYNTHESIS

We have already seen most of the common methods for synthesizing ethers. We review them at this time, looking more closely at the mechanisms to see which methods are most suitable for preparing various kinds of ethers. The **Williamson ether synthesis** is the most reliable and versatile ether synthesis. This method involves the S_N2 attack of an alkoxide ion on an unhindered primary alkyl halide or tosylate. Secondary alkyl halides and tosylates are occasionally used in the Williamson synthesis, but such examples are rare and the yields are generally poor.

$$R{-}\ddot{O}{:}^{-} \quad R'{-}X \longrightarrow R{-}O{-}R' + {:}\ddot{X}{:}^{-}$$

Examples

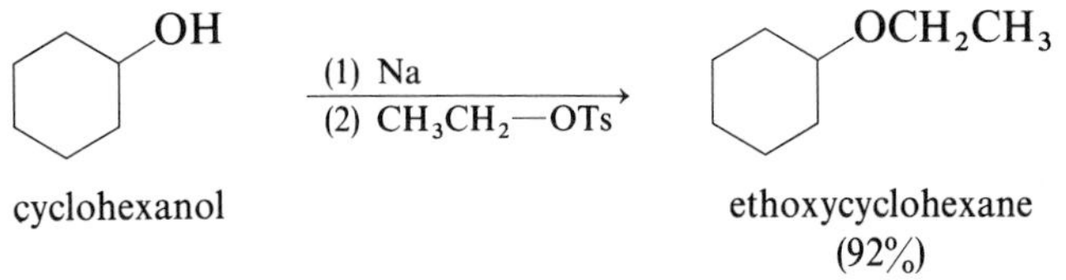

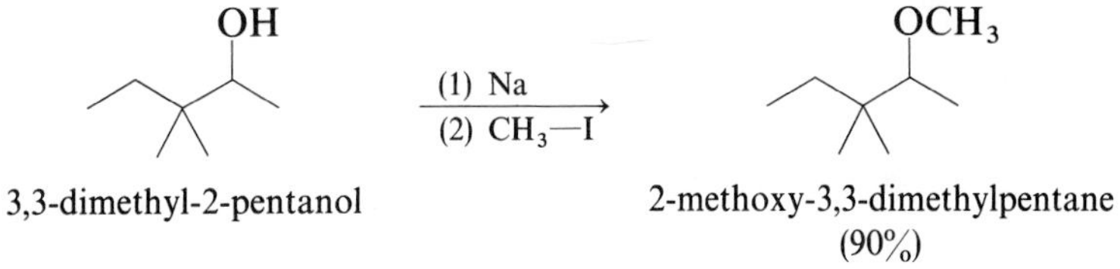

$$\text{cyclohexanol} \xrightarrow[\text{(2) } CH_3CH_2{-}OTs]{\text{(1) Na}} \text{ethoxycyclohexane (92\%)}$$

$$\text{3,3-dimethyl-2-pentanol} \xrightarrow[\text{(2) } CH_3{-}I]{\text{(1) Na}} \text{2-methoxy-3,3-dimethylpentane (90\%)}$$

SOLVED PROBLEM 13-1

(a) Why would the following reaction be a poor method for the synthesis of *t*-butyl propyl ether?

(b) What would be the major product from this reaction?

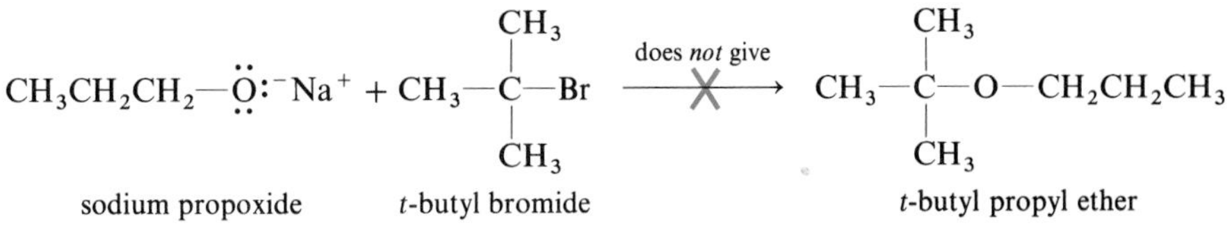

$$\underset{\text{sodium propoxide}}{CH_3CH_2CH_2{-}\ddot{O}{:}^{-}\,Na^+} + \underset{t\text{-butyl bromide}}{(CH_3)_3C{-}Br} \xrightarrow{\text{does } not \text{ give}} \underset{t\text{-butyl propyl ether}}{(CH_3)_3C{-}O{-}CH_2CH_2CH_3}$$

SOLUTION The desired S_N2 reaction cannot occur on the tertiary alkyl halide. The alkoxide ion is a strong base as well as a nucleophile, and elimination prevails.

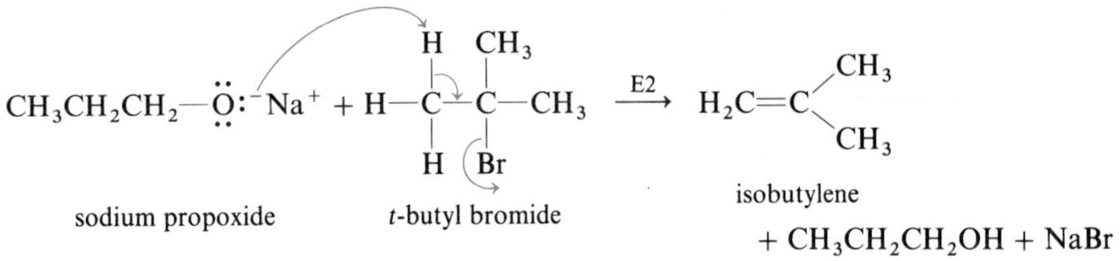

$$\underset{\text{sodium propoxide}}{CH_3CH_2CH_2{-}\ddot{O}{:}^{-}\,Na^+} + \underset{t\text{-butyl bromide}}{H{-}CH_2{-}C(CH_3)_2{-}Br} \xrightarrow{\text{E2}} \underset{\text{isobutylene}}{H_2C{=}C(CH_3)_2} + CH_3CH_2CH_2OH + NaBr$$

PROBLEM 13-7

Propose a better Williamson synthesis of *t*-butyl propyl ether than the poor method given in Solved Problem 13-1.

Synthesis of phenyl ethers A phenol (aromatic alcohol) can be used as the alcohol fragment (but not the halide fragment) for the Williamson ether synthesis. Phenols are more acidic than aliphatic alcohols (Section 9-6), and sodium hydroxide is sufficiently basic to form the phenoxide ion. As with other alkoxides, the electrophile should have an unhindered primary alkyl group and a good leaving group.

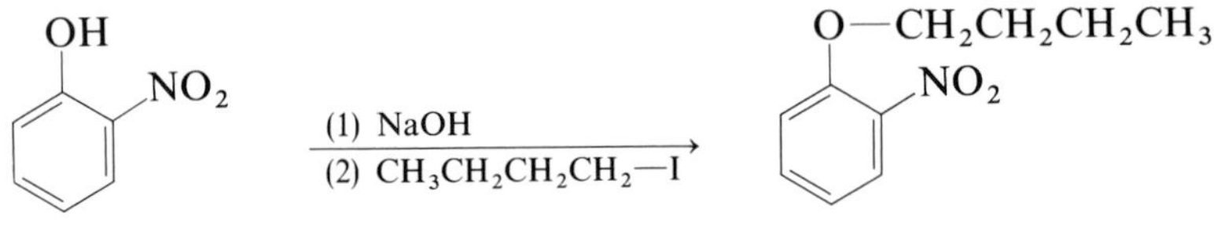

$$\text{2-nitrophenol} \xrightarrow[\text{(2) } CH_3CH_2CH_2CH_2\text{—I}]{\text{(1) NaOH}} \text{2-butoxynitrobenzene (80\%)}$$

OH, NO_2 (2-nitrophenol) → $O—CH_2CH_2CH_2CH_3$, NO_2 (2-butoxynitrobenzene (80%))

PROBLEM 13-8

Show how you would use the Williamson ether synthesis to prepare each of the following ethers. You may use any alcohols or phenols as your organic starting materials.

(a) cyclohexyl propyl ether (b) isopropyl methyl ether
(c) 1-methoxy-4-nitrobenzene (d) ethyl *n*-propyl ether (two ways)
(e) benzyl *t*-butyl ether (benzyl = Ph—CH_2—)

13-6 SYNTHESIS OF ETHERS BY ALKOXYMERCURATION-DEMERCURATION

The **alkoxymercuration-demercuration** process adds a molecule of an alcohol across the double bond of an alkene (Section 8-6). The product is an ether, as shown below.

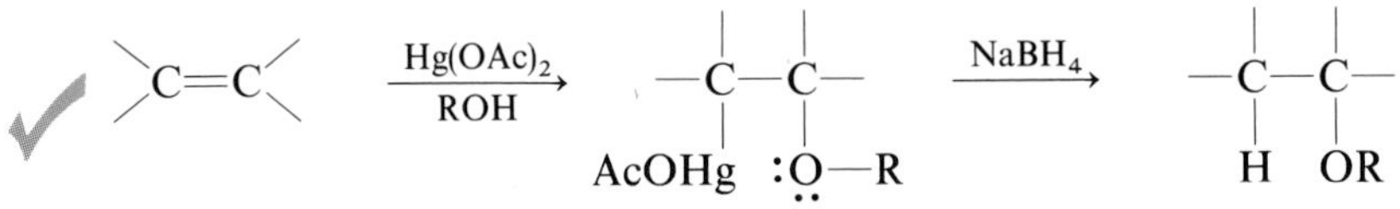

$$>C=C< \xrightarrow[\text{ROH}]{Hg(OAc)_2} \text{—C(HgOAc)—C(:\ddot{O}—R)—} \xrightarrow{NaBH_4} \text{—C(H)—C(OR)—}$$

AcOHg :Ö—R — mercurial ether

Example

$$\underset{\text{1-hexene}}{CH_3(CH_2)_3—CH=CH_2} \xrightarrow[\text{(2) } NaBH_4]{\text{(1) } Hg(OAc)_2,\ CH_3OH} CH_3(CH_2)_3—CH(OCH_3)—CH_3$$

2-methoxyhexane
(Markovnikov product)
(80%)

PROBLEM 13-9

Show how the following ethers might be synthesized using (1) alkoxymercuration-demercuration, and (2) the Williamson synthesis. (When one of these methods cannot be used for the given ether, point out why it would not work.)

(a) 2-methoxybutane (b) ethyl cyclohexyl ether
(c) 2-methyl-1-methoxycyclopentane (d) 1-methyl-1-methoxycyclopentane
(e) 1-methyl-1-isopropoxycyclopentane (f) *t*-butyl phenyl ether

13-7 SYNTHESIS OF ETHERS BY BIMOLECULAR DEHYDRATION OF ALCOHOLS (Industrial Method)

The least expensive method for synthesizing simple symmetrical ethers is the acid-catalyzed bimolecular dehydration, discussed in Section 10-11B. Unimolecular dehydration (to give an alkene) competes with the bimolecular dehydration; the alcohol must have an unhindered primary alkyl group, and the temperature must be kept low for the reaction to give mostly the ether. If the alcohol is hindered, or the temperature too high, the delicate balance between substitution and elimination shifts in favor of elimination, and very little ether is formed. Because the dehydration procedure is so limited in its scope, it finds little use in the laboratory synthesis of ethers.

Bimolecular dehydration

✓

$$2\,R{-}OH \xrightleftharpoons{H^+} R{-}O{-}R + H_2O$$

Examples

$$2\,CH_3{-}OH \xrightarrow{H_2SO_4,\ 140^\circ C} CH_3{-}O{-}CH_3 + H_2O$$

methyl alcohol → dimethyl ether (100%)

$$2\,CH_3CH_2OH \xrightarrow{H_2SO_4,\ 140^\circ C} CH_3CH_2{-}O{-}CH_2CH_3 + H_2O$$

ethyl alcohol → diethyl ether (88%)

$$2\,CH_3CH_2CH_2OH \xrightarrow{H_2SO_4,\ 140^\circ C} CH_3CH_2CH_2{-}O{-}CH_2CH_2CH_3 + H_2O$$

n-propyl alcohol → *n*-propyl ether (75%)

$$CH_3{-}\underset{\displaystyle OH}{\underset{|}{CH}}{-}CH_3 \xrightarrow{H_2SO_4,\ heat} H_2C{=}CH{-}CH_3 + H_2O$$

isopropyl alcohol → unimolecular dehydration (no ether is formed)

If the reaction conditions are carefully controlled, the intermolecular dehydration is a cheap synthesis of diethyl ether. In fact, this is the industrial method used to produce millions of gallons of diethyl ether each year.

PROBLEM 13-10

Explain why intermolecular dehydration is not a good method for the synthesis of unsymmetrical ethers such as ethyl methyl ether.

PROBLEM 13-11

Propose a mechanism for the acid-catalyzed dehydration of 1-propanol to *n*-propyl ether, as shown above. When the temperature of this reaction is allowed to rise too high, propene is formed. Give a mechanism for the formation of propene and explain why its formation is favored at higher temperatures.

PROBLEM 13-12

Which of the following ethers could be formed in good yield by dehydration of the corresponding alcohols? For those that could not be formed by dehydration, suggest an alternative method that would work.

(a) dibutyl ether (b) ethyl *n*-propyl ether (c) di-*sec*-butyl ether

SUMMARY OF THE SYNTHESIS OF ETHERS

1. The Williamson ether synthesis (Sections 10-15 and 13-5)

$$\mathrm{R{-}\ddot{\underset{\cdot\cdot}{O}}{:}^{-} \quad + \quad R'{-}X \quad \longrightarrow \quad R{-}\ddot{\underset{\cdot\cdot}{O}}{-}R' \quad + \quad X^{-}}$$

X = Cl, Br, I, OTs, etc. R′ must be primary

2. Addition of an alcohol across a double bond: Alkoxymercuration-demercuration (Sections 8-6 and 13-6)

$$\mathrm{{>}C{=}C{<} \quad + \quad Hg(OAc)_2 \xrightarrow{R-OH} -\underset{AcOHg}{\overset{|}{\underset{|}{C}}}-\underset{OR}{\overset{|}{\underset{|}{C}}}- \xrightarrow{NaBH_4} -\underset{H}{\overset{|}{\underset{|}{C}}}-\underset{OR}{\overset{|}{\underset{|}{C}}}-}$$

Markovnikov orientation

3. Bimolecular dehydration of alcohols (Sections 10-11B and 13-7)

$$\mathrm{2\,R{-}OH \overset{H^+}{\rightleftarrows} R{-}O{-}R \quad + \quad H_2O}$$

R must be primary

13-8 CLEAVAGE OF ETHERS BY HBr AND HI

Unlike alcohols, ethers are not commonly used as synthetic intermediates because they do not undergo a wide range of reactions. It is this unreactivity that makes ethers so attractive as solvents. Even so, ethers do undergo a limited number of characteristic reactions.

Ethers are cleaved by heating with HBr or HI to give alkyl bromides or alkyl iodides.

$$\mathrm{R{-}O{-}R' \xrightarrow[(X\,=\,Br\ or\ I)]{excess\ HX} R{-}X \quad + \quad R'{-}X}$$

Ethers are unreactive toward bases, but they can react under acidic conditions. A protonated ether can undergo substitution or elimination with the expulsion of a stable alcohol molecule. Ethers react with concentrated HBr and HI because these reagents are sufficiently acidic to protonate the ether, while bromide and iodide are good nucleophiles for the substitution. In effect, this reaction converts a dialkyl ether into two alkyl halides. The conditions are very strong, however, and the molecule must not contain any functional groups that would be sensitive to the strong acid.

$$\underset{\text{ether}}{\mathrm{R{-}\ddot{\underset{\cdot\cdot}{O}}{-}R'}} + \mathrm{H^+X^-} \rightleftarrows \underset{\text{protonated ether}}{\mathrm{X^-\ \ R{-}\overset{H}{\overset{|}{\underset{\cdot\cdot}{O^+}}}{-}R'}} \longrightarrow \underset{\text{alkyl halide}}{\mathrm{X{-}R}} + \underset{\text{alcohol}}{\mathrm{{:}\overset{H}{\overset{|}{\underset{\cdot\cdot}{O}}}{-}R'}} \xrightarrow{HX} \mathrm{X{-}R + X{-}R'}$$

(X = Br or I)

Iodide and bromide ions are good nucleophiles but weak bases, so they are more likely to substitute by the S_N2 mechanism than to eliminate by the E2 mechanism. The reaction of diethyl ether with HBr is an example of this displacement. The ethanol produced by the cleavage reacts with HBr (see Section 10-8), and the final products are two molecules of ethyl bromide.

Protonation and cleavage of the ether

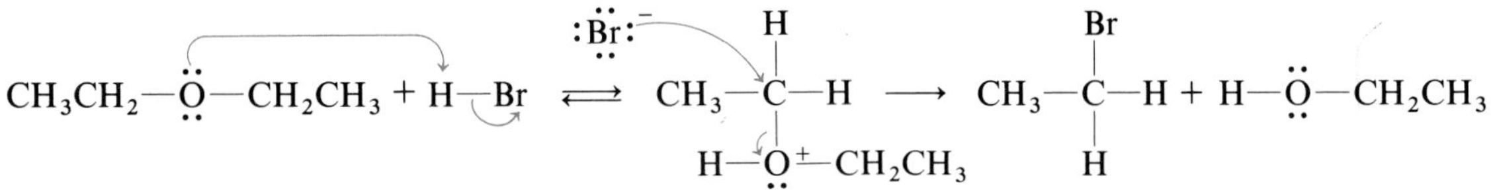

$$CH_3CH_2-\ddot{O}-CH_2CH_3 + H-Br \rightleftharpoons CH_3-\underset{H-\ddot{O}^+-CH_2CH_3}{\overset{H}{C}}-H \xrightarrow{:\ddot{Br}:^-} CH_3-\underset{H}{\overset{Br}{C}}-H + H-\ddot{O}-CH_2CH_3$$

diethyl ether | protonated ether | ethyl bromide | ethanol

Conversion of ethanol to ethyl bromide

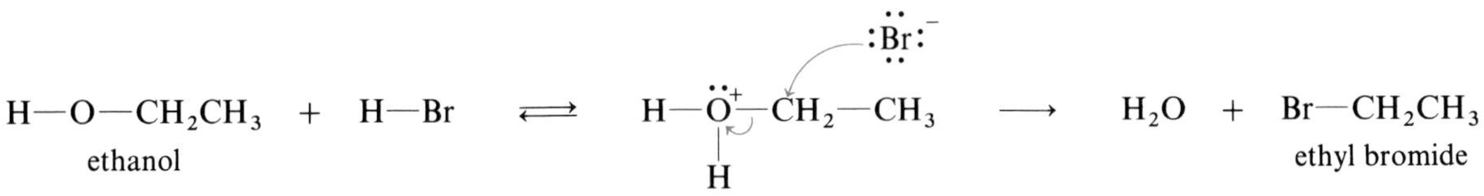

$$H-O-CH_2CH_3 + H-Br \rightleftharpoons H-\underset{H}{\ddot{O}^+}-CH_2-CH_3 \xrightarrow{:\ddot{Br}:^-} H_2O + Br-CH_2CH_3$$

ethanol | ethyl bromide

Overall reaction

$$CH_3CH_2-O-CH_2CH_3 \xrightarrow{\text{excess } HBr/H_2O} 2\ CH_3CH_2-Br$$

diethyl ether | ethyl bromide

Hydroiodic acid (HI) reacts with ethers the same way that HBr does. Aqueous iodide is a stronger nucleophile than aqueous bromide, and iodide reacts at a faster rate. We can rank the hydrohalic acids in order of their reactivity toward the cleavage of ethers:

$$HI > HBr \gg HCl$$

PROBLEM 13-13

Propose a mechanism for the following reaction.

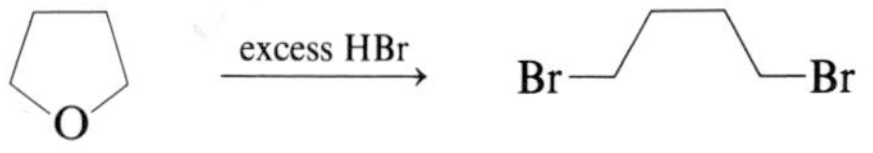

tetrahydrofuran $\xrightarrow{\text{excess HBr}}$ Br–$CH_2CH_2CH_2CH_2$–Br (1,4-dibromobutane)

Phenyl ethers Phenyl ethers (one of the groups bonded to oxygen is a benzene ring) react with HBr or HI to give alkyl halides and phenols. Phenols do not react further to give halides, because the sp^2 hybridized carbon atom of the phenol cannot undergo the S_N2 (or S_N1) reaction needed for conversion to the halide.

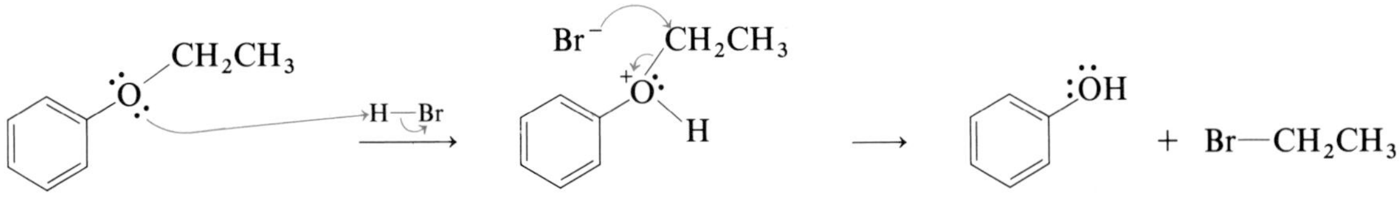

$$Ph-\ddot{O}-CH_2CH_3 \xrightarrow{H-Br} Ph-\overset{+}{\ddot{O}}(H)-CH_2CH_3 \xrightarrow{Br^-} Ph-\ddot{O}H + Br-CH_2CH_3$$

ethyl phenyl ether | protonated ether | phenol | ethyl bromide

(no further reaction)

PROBLEM 13-14

Predict the products of the following reactions. An excess of acid is available in each case.

(a) ethoxycyclohexane + HBr (b) tetrahydropyran + HI

(c) anisole (methoxybenzene) + HBr (d) [chromane structure: benzene ring fused to ring containing O] + HI

(e) $Ph—O—CH_2CH_2—CH(CH_3)—CH_2—O—CH_2CH_3$ + HBr

13-9

AUTOXIDATION OF ETHERS

When ethers are stored in the presence of atmospheric oxygen, they slowly oxidize to produce hydroperoxides and dialkyl peroxides, both of which are explosive. Such a spontaneous oxidation by atmospheric oxygen is called an **autoxidation.**

$$R—O—\overset{|}{\underset{|}{C}}—H \xrightarrow[\text{(slow)}]{\text{excess } O_2} R—O—\overset{|}{\underset{|}{C}}—O—O—H + R—O—O—\overset{|}{\underset{|}{C}}—H$$

hydroperoxide dialkyl peroxide

Example

$$(CH_3)_2CH—O—CH(CH_3)_2 \xrightarrow[\text{(weeks or months)}]{\text{excess } O_2} (CH_3)_2CH—O—C(CH_3)_2—OOH + (CH_3)_2CH—O—O—CH(CH_3)_2$$

diisopropyl ether hydroperoxide diisopropyl peroxide

Organic chemists often buy large containers of ethers and use small quantities over several months. Once a container has been opened, it contains atmospheric oxygen, and the autoxidation process begins. After several months, a large amount of peroxide is present. If the ether is distilled or evaporated, the peroxide becomes concentrated and an explosion may occur.

Such an explosion may be avoided by taking a few simple precautions. Ethers should be bought in small quantities, kept in tightly sealed containers, and used promptly. Any procedure requiring evaporation or distillation of an ether should use only peroxide-free ether. Any ether that might be contaminated with peroxides should be discarded or treated to destroy the peroxides.

SUMMARY OF REACTIONS OF ETHERS

1. *Cleavage by HBr and HI* (Section 13-8)

$$R—O—R' \xrightarrow[\text{(X = Br, I)}]{\text{excess HX}} R—X + R'—X$$

$$Ar—O—R \xrightarrow[\text{(X = Br, I)}]{\text{excess HX}} Ar—OH + R—X$$

Ar = aromatic ring

Example

$$CH_3CH_2—O—CH_3 \xrightarrow{\text{excess HI}} CH_3CH_2I + CH_3I$$

ethyl methyl ether ethyl iodide methyl iodide

2. *Autoxidation* (Section 13-9)

$$R—O—\overset{|}{\underset{|}{C}}—H \xrightarrow[\text{(slow)}]{\text{excess } O_2} R—O—\overset{|}{\underset{|}{C}}—O—O—H + R—O—O—\overset{|}{\underset{|}{C}}—H$$

hydroperoxide dialkyl peroxide

13-10 SYNTHESIS OF EPOXIDES

The synthesis and reactions of epoxides are unlike those of other ethers. Here we review the epoxidation techniques already covered (Section 8-12) and see how the Williamson ether synthesis is applied to the synthesis of epoxides.

13-10A PEROXYACID EPOXIDATION

Peroxyacids (sometimes called *peracids*) are used to convert alkenes to epoxides. If the peroxyacid is strongly acidic, the epoxide opens to a glycol. Therefore, to make an epoxide we use a weak peroxyacid such as peroxybenzoic acid. Because of its desirable solubility properties, *meta*-chloroperoxybenzoic acid (MCPBA) is often used for these epoxidations.

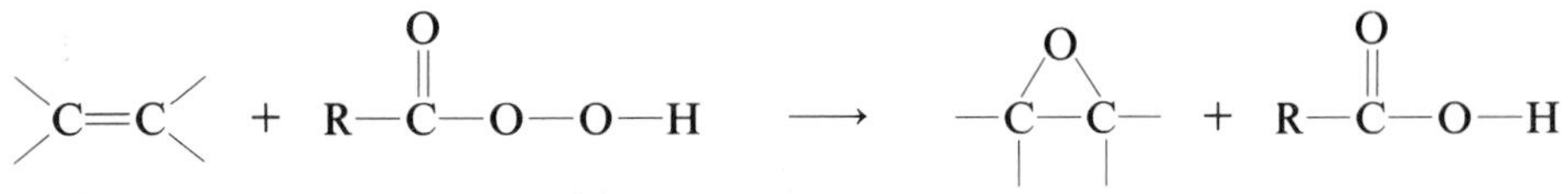

>C=C< + R—C(=O)—O—O—H ⟶ epoxide + R—C(=O)—O—H

alkene peroxyacid epoxide acid

Example

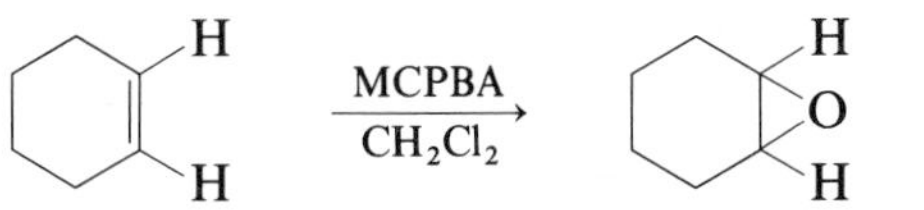

cyclohexene —MCPBA, CH_2Cl_2→ epoxycyclohexane (100%)

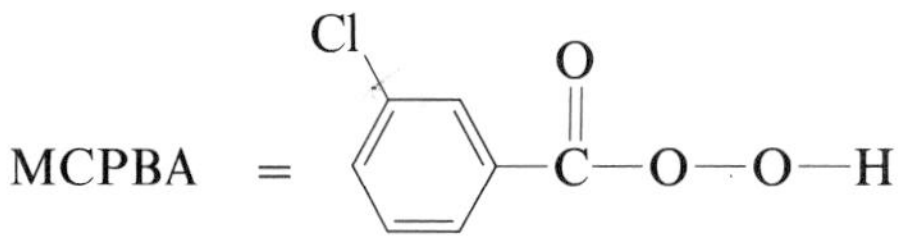

MCPBA = *meta*-chloroperoxybenzoic acid

The epoxidation takes place in a one-step, **concerted** reaction that maintains the stereochemistry of any substituents on the double bond.

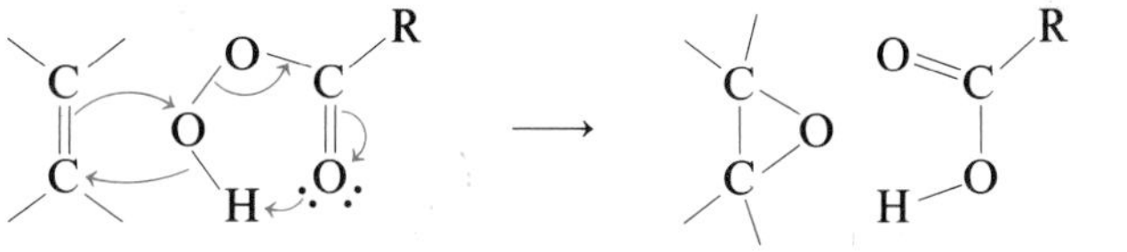

alkene peroxyacid ⟶ epoxide acid

The peroxyacid epoxidation is quite general, with electron-rich double bonds reacting fastest. The following reactions are difficult transformations made possible by this selective, stereospecific epoxidation procedure.

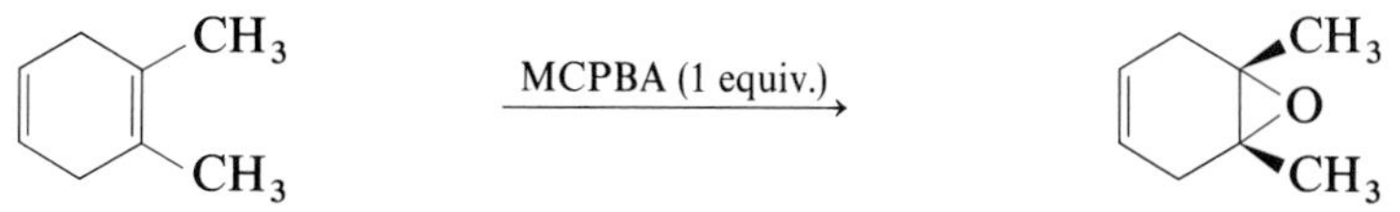

1,2-dimethyl-1,4-cyclohexadiene —MCPBA (1 equiv.)→ *cis*-4,5-epoxy-4,5-dimethylcyclohexene

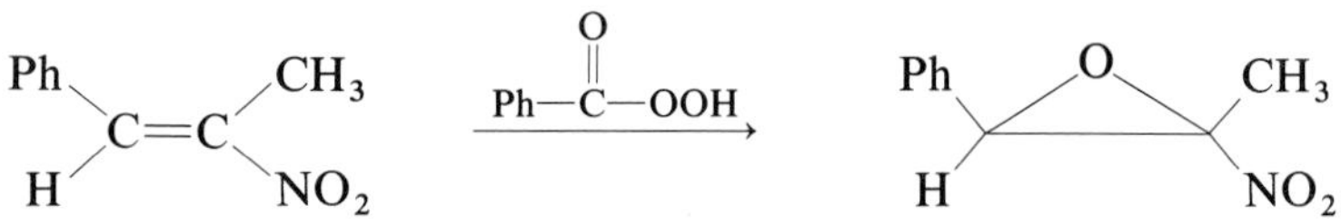

E-2-nitro-1-phenylpropene —Ph—C(=O)—OOH→ *E*-1-methyl-1-nitro-2-phenyloxirane

13-10B DISPLACEMENT OF HALOHYDRINS

A second synthesis of epoxides and other cyclic ethers involves a variant of the Williamson ether synthesis. If an alkoxide ion and a halogen atom are located in the same molecule, the alkoxide may displace a halide ion and form a ring. The treatment of a **halohydrin** with a base leads to an epoxide through this internal S_N2 attack.

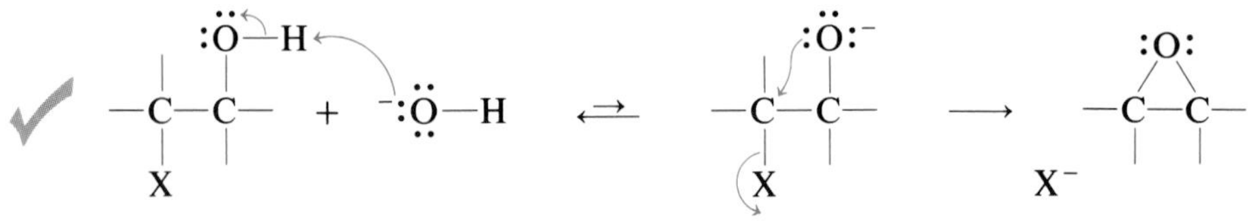

(X = Cl, Br, I)

Halohydrins are easily generated by treating alkenes with aqueous solutions of halogens. Bromine water and chlorine water add across double bonds with Markovnikov orientation (Section 8-11). The following reaction shows the conversion of cyclopentene to the chlorohydrin by the reaction with chlorine water. Treatment of the chlorohydrin with aqueous sodium hydroxide gives the epoxide.

Formation of the chlorohydrin

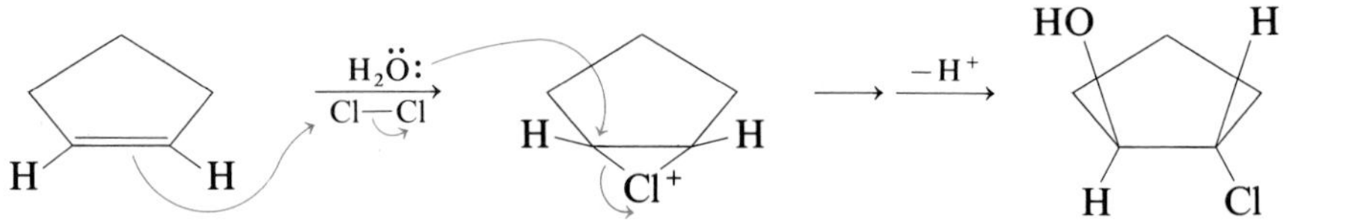

cyclopentene — chlorine water (H_2O:, Cl—Cl) — chloronium ion — ($-H^+$) — *trans*-chlorohydrin (mixture of enantiomers)

Displacement of the chlorohydrin

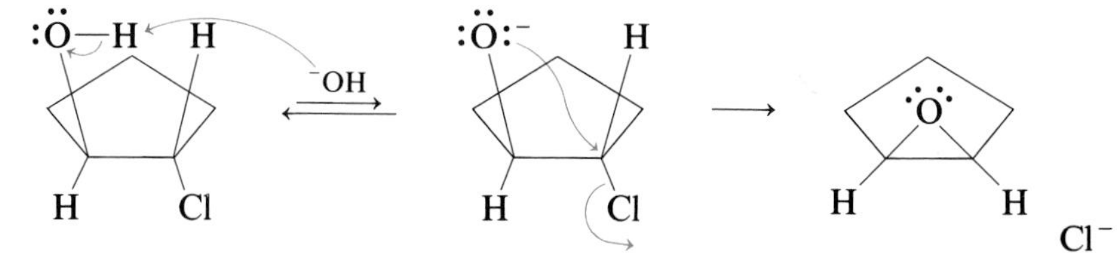

trans-chlorohydrin — (^-OH) ⇌ alkoxide → epoxide (50% overall) + Cl^-

This reaction can be used to synthesize cyclic ethers with larger rings. The difficulty lies in preventing the base (added to deprotonate the alcohol) from attacking and displacing the halide. 2,6-Lutidine, a bulky base that cannot easily attack a carbon atom, can be used to deprotonate the hydroxyl group to give a five-membered cyclic ether.

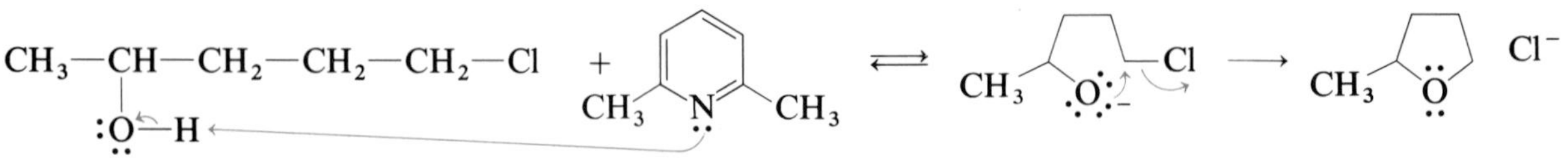

$CH_3-CH(OH)-CH_2-CH_2-CH_2-Cl$ + 2,6-lutidine (2,6-dimethylpyridine) ⇌ alkoxide → 2-methyl tetrahydrofuran (85%) + Cl^-

PROBLEM 13-15

Show how you would accomplish each of the following transformations. Some of these examples require more than one step.

(a) 2-methylpropene ⟶ 2,2-dimethyloxirane
(b) 1-phenylethanol ⟶ 2-phenyloxirane
(c) 5-chloro-1-pentene ⟶ tetrahydropyran
(d) 5-chloro-1-pentene ⟶ 2-methyl tetrahydrofuran
(e) 2-chloro-1-hexanol ⟶ 1,2-epoxyhexane

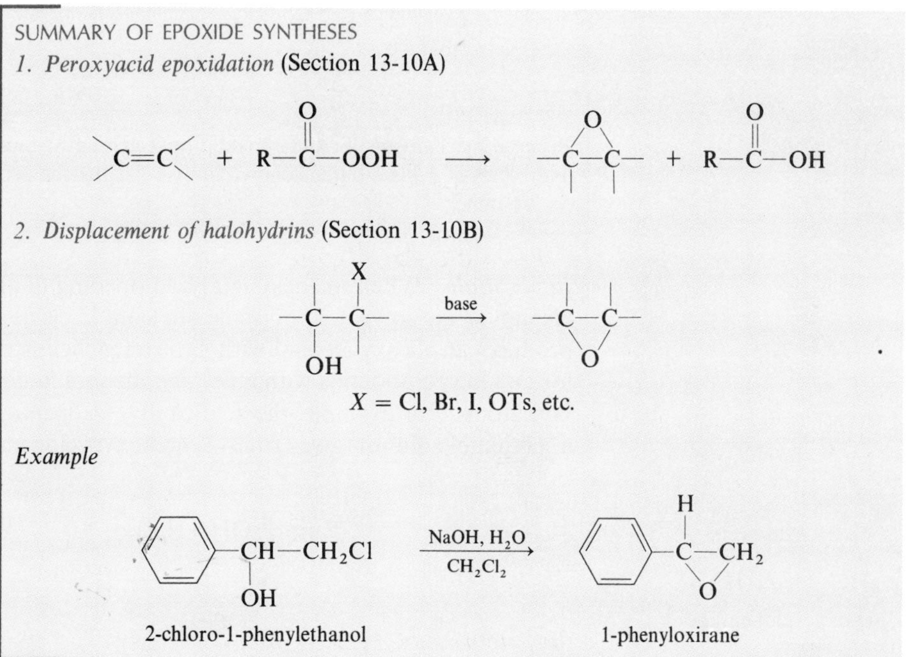

SUMMARY OF EPOXIDE SYNTHESES
1. Peroxyacid epoxidation (Section 13-10A)
2. Displacement of halohydrins (Section 13-10B)
base
X = Cl, Br, I, OTs, etc.
Example
NaOH, H2O
CH2Cl2
2-chloro-1-phenylethanol
1-phenyloxirane

13-11 ACID-CATALYZED RING OPENING OF EPOXIDES

Epoxides are much more reactive than common dialkyl ethers, because of the large strain energy (about 25 kcal/mol or 105 kJ/mol) associated with the three-membered ring. Unlike other ethers, epoxides react under both acidic and basic conditions. The products of acid-catalyzed opening depend on the solvent used.

In water In Section 8-13 we saw that acid-catalyzed hydrolysis of epoxides gives glycols with anti stereochemistry. The mechanism of this hydrolysis involves protonation of the oxygen (forming a good leaving group), then a nucleophilic attack by water. The anti stereochemistry results from the back-side attack of water on the protonated epoxide.

H_3O^+ ⇌ ; $H_2\ddot{O}:$ ⟶ ; $H_2\ddot{O}:$ ⟶

1,2-epoxycyclopentane

trans-cyclopentane-1,2-diol (mixture of enantiomers)

Direct anti hydroxylation of an alkene (without isolation of the epoxide intermediate) is possible by using an acidic aqueous solution of a peroxyacid. As soon as the epoxide is formed, it hydrolyzes to the glycol. Peroxyacetic acid (CH_3CO_3H) and peroxyformic acid (HCO_3H) are often used for the anti hydroxylation of alkenes.

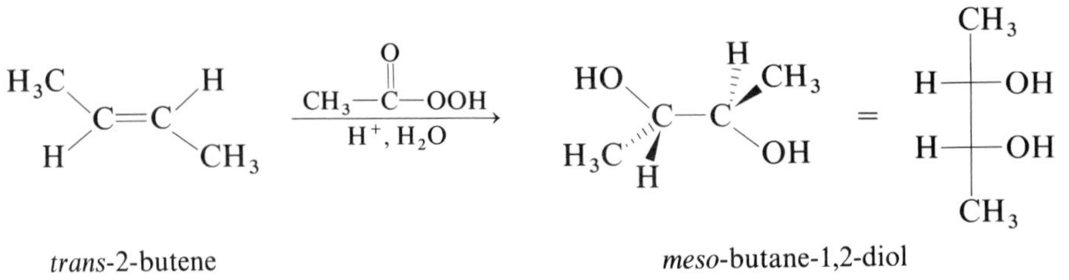

$$\underset{\textit{trans}\text{-2-butene}}{H_3C(H)C{=}C(H)CH_3} \xrightarrow[H^+,\ H_2O]{CH_3-\overset{O}{\overset{\|}{C}}-OOH} \underset{\textit{meso}\text{-butane-1,2-diol}}{HO(H_3C)(H)C-C(H)(CH_3)OH}$$

trans-2-butene — meso-butane-1,2-diol

PROBLEM 13-16

Propose mechanisms for the epoxidation and ring-opening steps of the epoxidation and hydrolysis of *trans*-2-butene shown above. Predict the product of the same reaction with *cis*-2-butene.

In alcohols When the acid-catalyzed opening of an epoxide takes place with an alcohol as the solvent, a molecule of alcohol acts as the nucleophile. This reaction produces an alkoxy alcohol with anti stereochemistry. This is an excellent method for making compounds with ether and alcohol functional groups on adjacent carbon atoms. For example, the acid-catalyzed opening of 1,2-epoxycyclopentane in a methanol solution gives *trans*-2-methoxycyclopentanol.

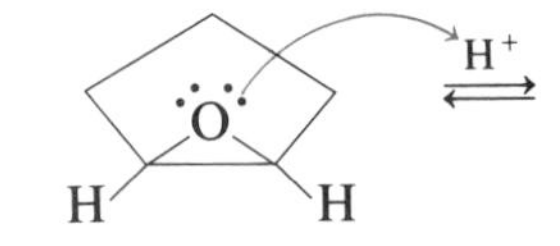

1,2-epoxycyclopentane $\overset{H^+}{\rightleftharpoons}$ (protonated epoxide) $\xrightarrow{CH_3\ddot{O}H}$ (protonated ether) $\xrightarrow{CH_3\ddot{O}H}$ *trans*-2-methoxycyclopentanol (82%) (mixture of enantiomers)

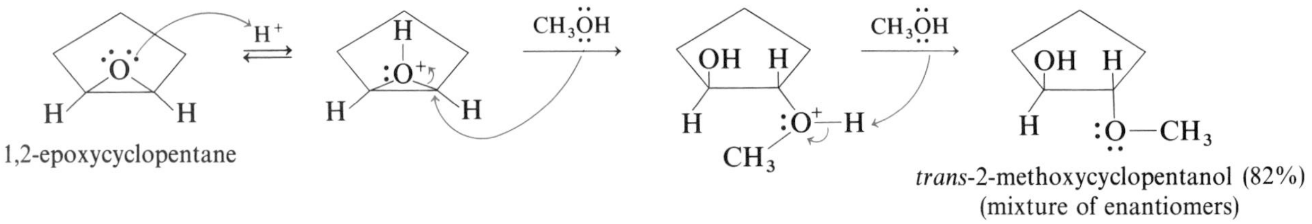

PROBLEM 13-17

Cellosolve is the trade name for 2-ethoxyethanol, a common industrial solvent. This compound is produced in chemical plants that use ethylene as their only organic feedstock. Show how you would accomplish this industrial process.

Using hydrohalic acids If an epoxide is allowed to react with a hydrohalic acid (HCl, HBr, or HI), the protonated epoxide is attacked by a halide ion. This reaction is analogous to the cleavage of ethers by HBr or HI. The halohydrin initially formed reacts further with HX to give a 1,2-dihalide. This is rarely a useful synthetic reaction, since the 1,2-dihalide can be made directly from the alkene by electrophilic addition of X_2.

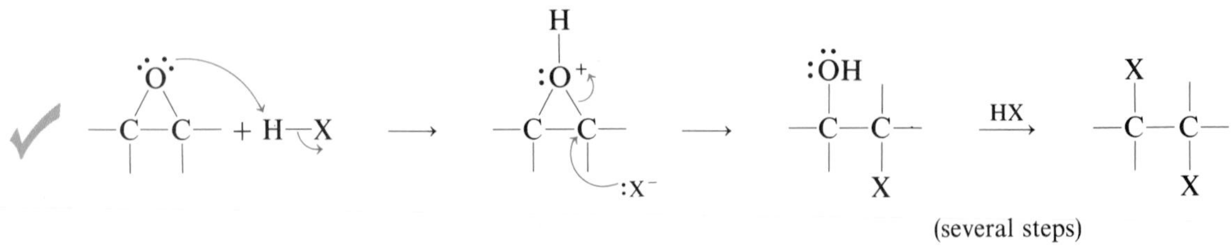

$$\text{epoxide} + H{-}X \longrightarrow \text{protonated epoxide} + :X^- \longrightarrow -\overset{:\ddot{O}H}{\overset{|}{C}}-\underset{X}{\underset{|}{C}}- \xrightarrow[\text{(several steps)}]{HX} -\overset{X}{\overset{|}{C}}-\underset{X}{\underset{|}{C}}-$$

PROBLEM 13-18

When ethylene oxide is treated with anhydrous HBr gas, the major product is 1,2-dibromoethane. When ethylene oxide is treated with concentrated aqueous HBr, the major product is ethylene glycol. Use mechanisms to explain these results.

The acid-catalyzed opening of squalene-2,3-epoxide The biosynthesis of steroid hormones is believed to involve an acid-catalyzed opening of squalene-2,3-epoxide (Fig. 13-6). Squalene is a member of the class of olefinic natural products called *terpenes* (see Section 25-8). An enzyme, squalene epoxidase, oxidizes squalene to the epoxide, which opens and forms a carbocation that cyclizes under the control of another enzyme. The cyclized intermediate rearranges to lanosterol, which is later converted to cholesterol and other steroids.

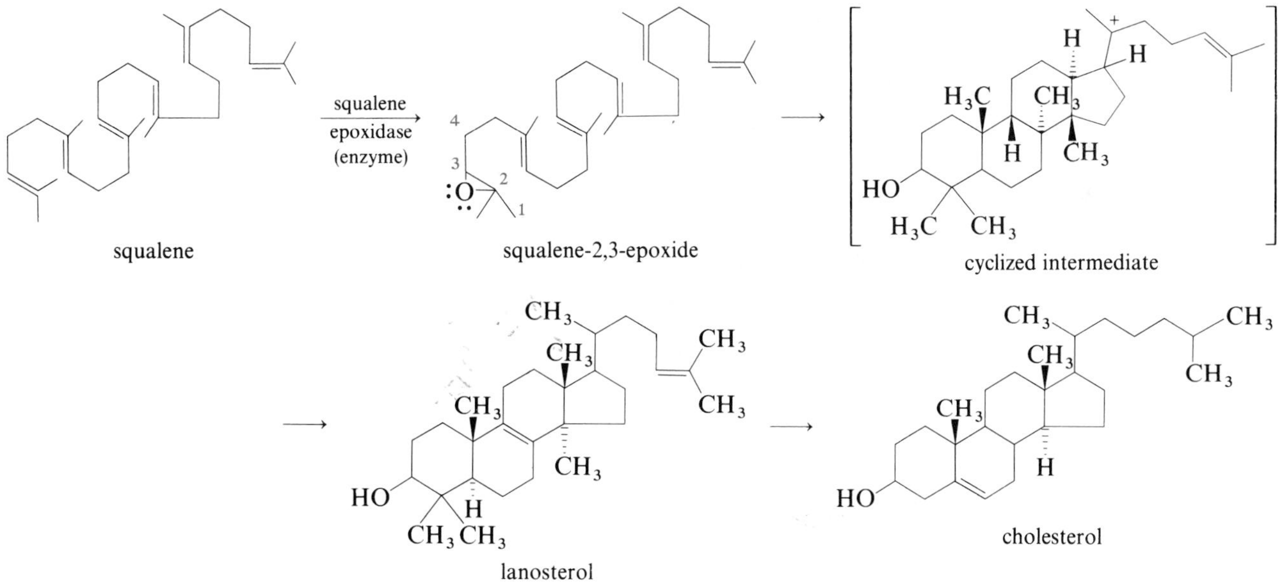

squalene

squalene epoxidase (enzyme)

squalene-2,3-epoxide

cyclized intermediate

lanosterol

cholesterol

FIGURE 13-6 The biosynthesis of steroids starts with the epoxidation of squalene to squalene-2, 3-epoxide. The opening of this epoxide promotes the cyclization of the carbon skeleton under the control of an enzyme. The cyclized intermediate is converted to lanosterol, then to other steroids.

Although the cyclization of squalene-2,3-epoxide takes place under the control of an enzyme, its mechanism is similar to the acid-catalyzed opening of other epoxides. The epoxide oxygen becomes protonated, and then it is attacked by a nucleophile. In this case, the nucleophile is a pi bond. The initial result is a tertiary carbocation (Fig. 13-7).

This initial carbocation is attacked by another double bond, leading to the formation of another ring and another tertiary carbocation. A repetition of this process leads to the cyclized intermediate shown in Figure 13-6.

PROBLEM 13-19

Show the rest of the mechanism for formation of the cyclized intermediate in Figure 13-6.

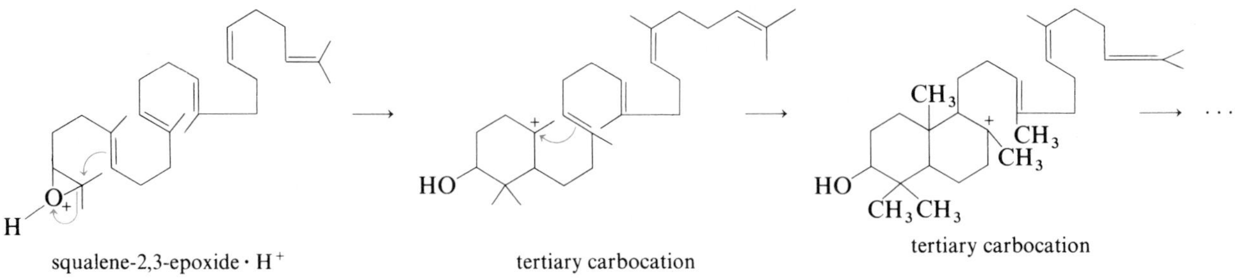

H

squalene-2,3-epoxide · H^+

HO

tertiary carbocation

HO

CH_3

CH_3

CH_3

CH_3CH_3

tertiary carbocation

FIGURE 13-7 The cyclization of squalene epoxide begins with the acid-catalyzed opening of the epoxide. Each additional cyclization step forms another carbocation.

13-12 BASE-CATALYZED RING OPENING OF EPOXIDES

Most ethers do not undergo nucleophilic substitutions or eliminations under basic conditions because the alkoxide ion is not a good leaving group. Epoxides have about 25 kcal (105 kJ) per mole of ring strain that is released upon ring opening, however, and this strain is enough to compensate for the poor alkoxide leaving group. Figure 13-8 compares the energy profiles for nucleophilic attack on an ether and on an epoxide. The starting epoxide is about 25 kcal/mol (105 kJ/mol) higher in energy than the ether, and its displacement has a lower activation energy.

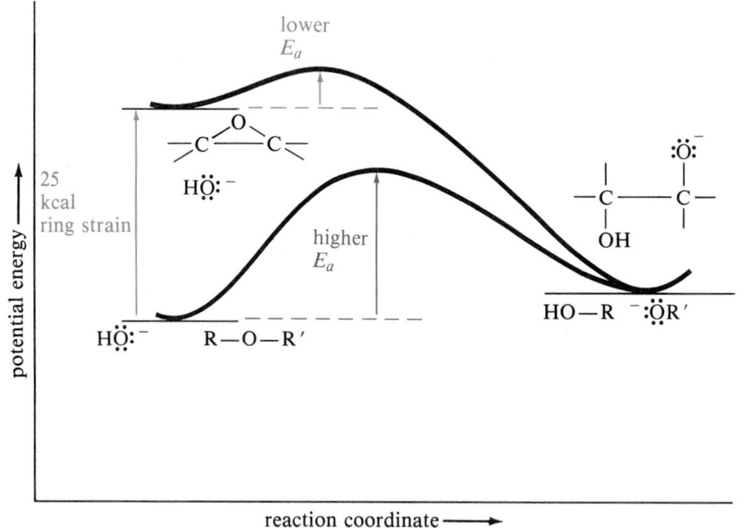

FIGURE 13-8 An epoxide is higher in energy than an acyclic ether by about 25 kcal (105 kJ) per mole of ring strain. The ring strain is released in the product, giving it an energy similar to the products from the acyclic ether. The release of the ring strain makes the displacement of an epoxide a thermodynamically favorable reaction.

The reaction of the epoxide with hydroxide ion leads to the same product as the acid-catalyzed opening of the epoxide: the 1,2-diol (glycol), with anti stereochemistry. In fact, either the acid-catalyzed or base-catalyzed reaction may be used to open an epoxide, but the acid-catalyzed reaction takes place under milder conditions. Unless there is an acid-sensitive functional group present, the acid-catalyzed hydrolysis is preferred.

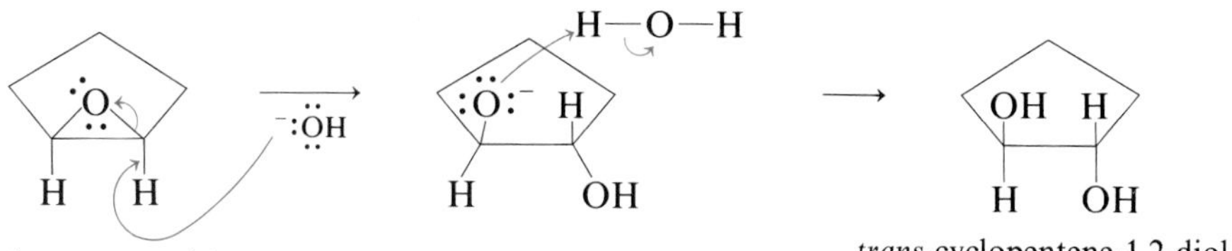

cyclopentene oxide

trans-cyclopentene-1,2-diol
(mixture of enantiomers)

Like hydroxide, alkoxide ions react with epoxides to form ring-opened products. For example, cyclopentene oxide reacts with sodium methoxide in methanol to give the same *trans*-2-methoxycyclopentanol as the acid-catalyzed opening in methanol.

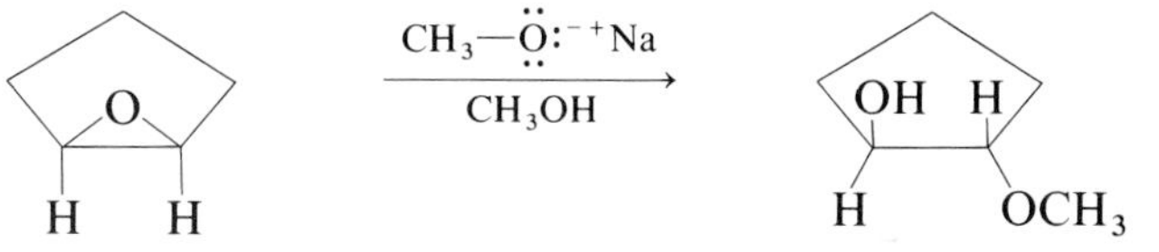

$CH_3-\ddot{O}:^- \, ^+Na$, CH_3OH

cyclopentene oxide

trans-2-methoxycyclopentanol
(mixture of enantiomers)

PROBLEM 13-20
Give a complete mechanism for the reaction of cyclopentene oxide with sodium methoxide.

PROBLEM 13-21
Predict the major product when each of the following reagents reacts with ethylene oxide.

(a) sodium ethoxide (b) sodium amide, $NaNH_2$ (c) 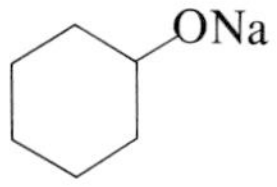 [cyclohexyl–ONa]

13-13 ORIENTATION OF EPOXIDE RING OPENING

Symmetrically substituted epoxides (such as cyclopentene oxide, above) give the same product in both the acid-catalyzed and base-catalyzed ring opening. An unsymmetrical epoxide gives different products under acid-catalyzed and base-catalyzed conditions, however.

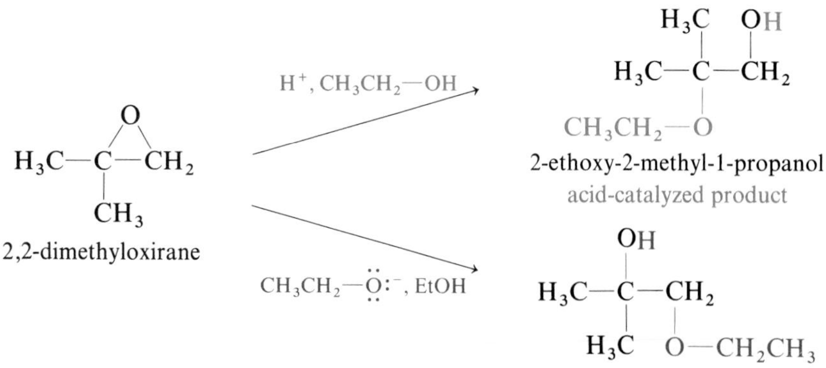

2,2-dimethyloxirane

H^+, CH_3CH_2-OH

2-ethoxy-2-methyl-1-propanol
acid-catalyzed product

$CH_3CH_2-\ddot{O}:^-$, EtOH

1-ethoxy-2-methyl-2-propanol
base-catalyzed product

We can apply what we know about reaction mechanisms to explain these different products. Under basic conditions, an S_N2 displacement is involved. The alkoxide ion simply attacks the less hindered carbon atom.

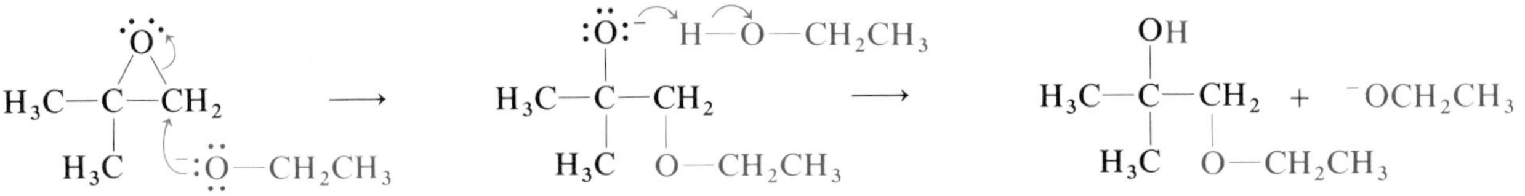

Under acidic conditions, the attack of the alcohol on the protonated epoxide determines the orientation of the product. It might seem that the alcohol would attack at the less hindered oxirane carbon, but this is not the case. In the protonated epoxide, there is a balancing act between ring strain and the energy it costs to put some of the positive charge on the carbon atoms. We can represent this sharing of positive charge by drawing three resonance structures.

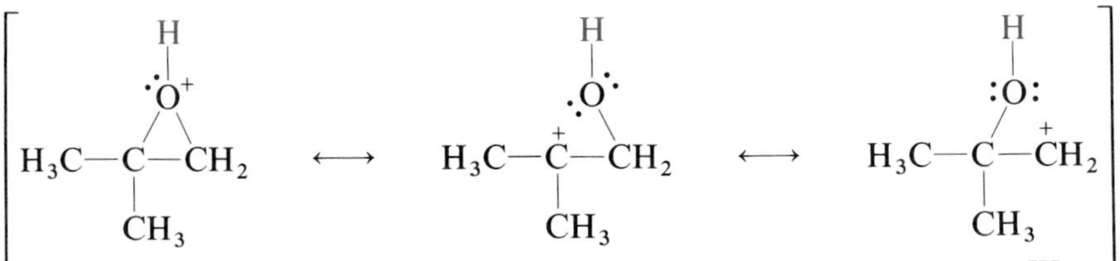

structure I ⟷ structure II ⟷ structure III

Structure I is the conventional structure for the protonated epoxide, while structures II and III show that the oxirane carbon atoms share part of the positive charge. The tertiary carbon atom bears a larger part of the positive charge, and it is more strongly electrophilic; that is, structure II is more important than structure III. The bond between this tertiary carbon and oxygen is weaker, implying a lower transition state energy for attack at the tertiary carbon atom. Attack by the weak nucleophile (ethanol in this case) is sensitive to the strength of the electrophile, and it occurs at the more electrophilic tertiary carbon.

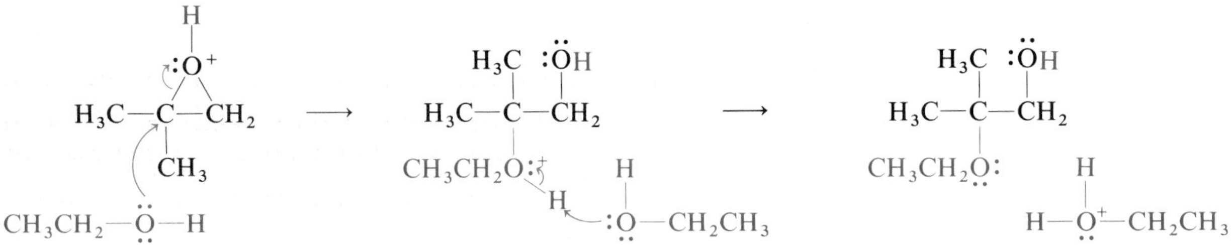

This ring opening at the more highly substituted carbon is similar to the opening of a bromonium ion in the formation of a bromohydrin (Section 8-11) and the opening of the mercurinium ion during oxymercuration (Section 8-5B). All three of these reactions involve the opening of an electrophilic three-membered ring by a weak nucleophile, and attack takes place at the carbon atom that is more stable with a positive charge: usually the more highly substituted carbon. Most base-catalyzed epoxide openings, on the other hand, involve attack by a strong nucleophile at the less hindered carbon atom of the epoxide.

SOLVED PROBLEM 13-2

Predict the major products for the reaction of 1-methyl-1,2-epoxycyclopentane with

(a) sodium ethoxide in ethanol. (b) H_2SO_4 in methanol.

SOLUTION (a) Sodium ethoxide attacks the less hindered secondary carbon to give (*E*)-2-ethoxy-1-methylcyclopentanol.

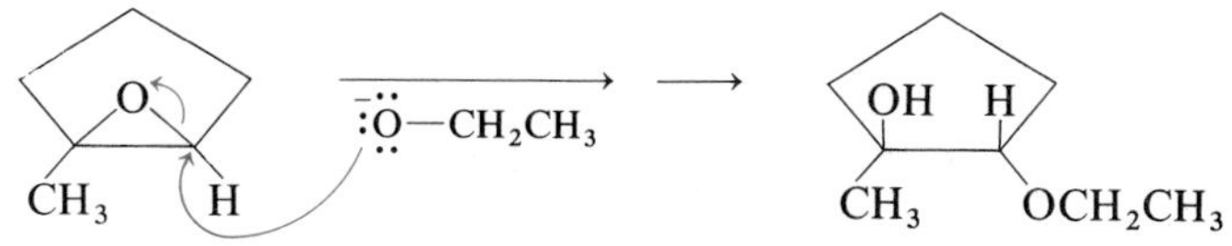

$$\text{epoxide (CH}_3\text{, H)} + :\ddot{O}{-}CH_2CH_3 \longrightarrow \longrightarrow \text{product (OH, H; CH}_3\text{, OCH}_2CH_3)$$

(b) Under acidic conditions, the alcohol attacks the more electrophilic tertiary carbon atom of the protonated epoxide. The product is (*E*)-2-ethoxy-2-methylcyclopentanol.

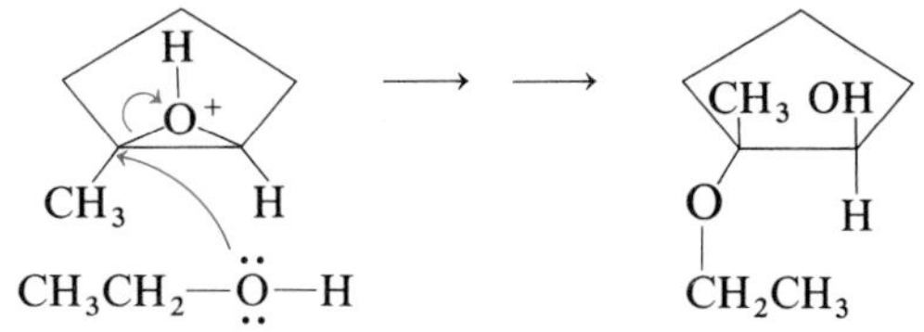

$$\text{protonated epoxide (CH}_3\text{, H)} + CH_3CH_2{-}\ddot{O}{-}H \longrightarrow \longrightarrow \text{product (CH}_3\text{, OH; O–CH}_2CH_3\text{, H)}$$

PROBLEM 13-22

Predict the major products of the following reactions, including stereochemistry where appropriate.

(a) 2,2-dimethyloxirane + H^+/H_2O^{18} (oxygen-labeled water)
(b) 2,2-dimethyloxirane + $^-O^{18}H/H_2O^{18}$
(c) (*Z*)-2-ethyl-2,3-dimethyloxirane + CH_3O^-/CH_3OH
(d) (*Z*)-2-ethyl-2,3-dimethyloxirane + H^+/CH_3OH

13-14 REACTIONS OF GRIGNARD AND ORGANOLITHIUM REAGENTS WITH EPOXIDES

Like other strong nucleophiles, Grignard and organolithium reagents attack epoxides to give (after hydrolysis) ring-opened alcohols.

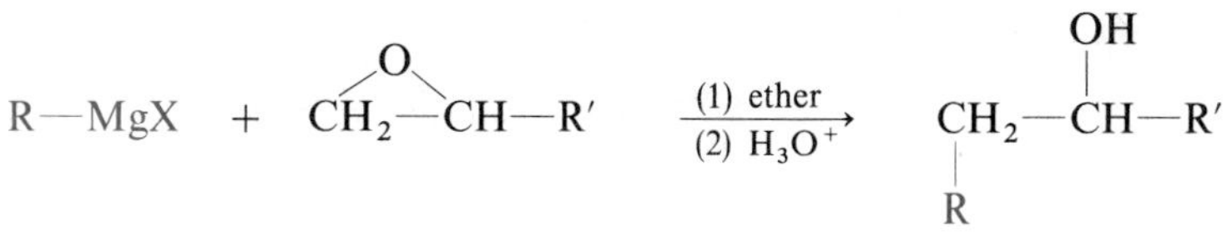

$$R{-}MgX + \overset{O}{CH_2{-}CH}{-}R' \xrightarrow[(2)\ H_3O^+]{(1)\ \text{ether}} \underset{R}{CH_2}{-}\overset{OH}{CH}{-}R'$$

One of the most useful examples of nucleophilic attack on epoxides is the attack by an organometallic reagent, either a Grignard or organolithium reagent. Like other strong nucleophiles, the Grignard reagent attacks an epoxide and opens the ring. The product, after hydrolysis, is an alcohol. Ethylmagnesium bromide reacts with oxirane (ethylene oxide) to form the magnesium salt of 1-butanol.

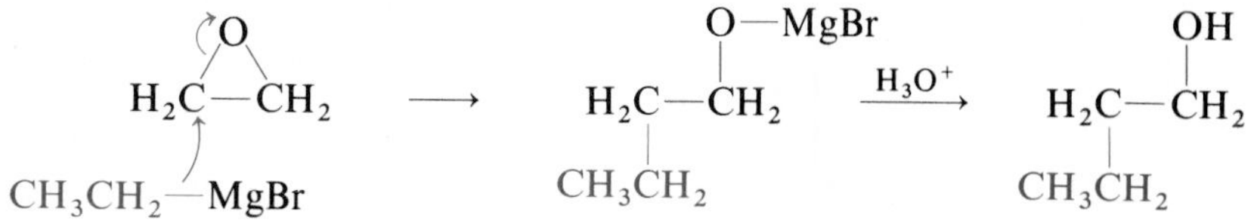

$$\overset{O}{H_2C{-}CH_2} + CH_3CH_2{-}MgBr \longrightarrow \underset{CH_3CH_2}{H_2C}{-}\overset{O-MgBr}{CH_2} \xrightarrow{H_3O^+} \underset{CH_3CH_2}{H_2C}{-}\overset{OH}{CH_2}$$

1-butanol

More complicated epoxides can be used in this reaction, with the carbanion attacking the less hindered epoxide carbon atom. This reaction works best if one of the oxirane carbons is unsubstituted, to allow an unhindered nucleophilic attack.

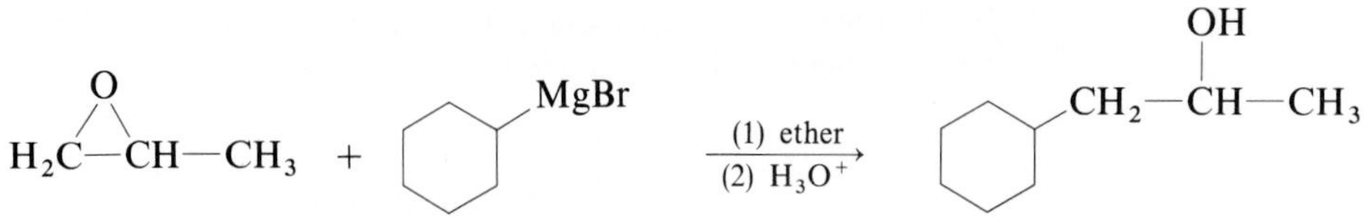

$$\underset{\text{methyloxirane}}{H_2C\overset{O}{\triangle}CH-CH_3} + \underset{\text{cyclohexylmagnesium bromide}}{C_6H_{11}-MgBr} \xrightarrow[(2)\ H_3O^+]{(1)\ \text{ether}} \underset{\text{1-cyclohexyl-2-propanol}}{C_6H_{11}-CH_2-\underset{|}{\overset{OH}{CH}}-CH_3}$$

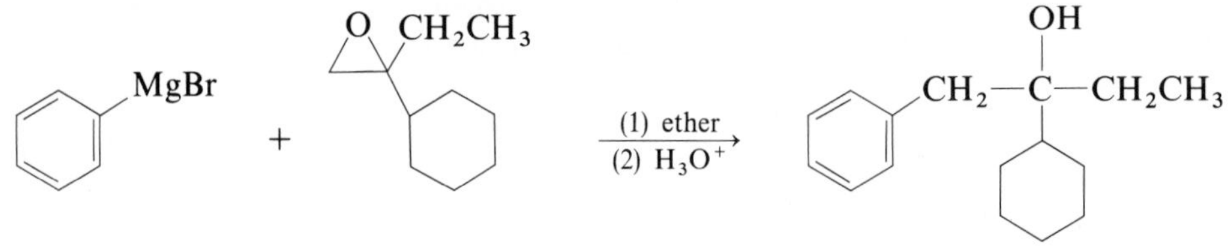

$$\underset{\text{phenylmagnesium bromide}}{C_6H_5-MgBr} + \underset{\text{2-cyclohexyl-2-ethyloxirane}}{\text{(oxirane with } CH_2CH_3 \text{ and cyclohexyl on C-2)}} \xrightarrow[(2)\ H_3O^+]{(1)\ \text{ether}} \underset{\text{2-cyclohexyl-1-phenyl-2-butanol}}{C_6H_5-CH_2-\overset{OH}{\underset{C_6H_{11}}{C}}-CH_2CH_3}$$

PROBLEM 13-23

Give the expected products of the following reactions. Include a hydrolysis step where necessary.

(a) ethylene oxide + isopropylmagnesium bromide
(b) 2,2-dimethyloxirane + methyllithium
(c) cyclopentyloxirane + ethylmagnesium bromide

SUMMARY OF THE REACTIONS OF EPOXIDES

1. Acid-Catalyzed Opening (Section 13-11)

a. *In water*

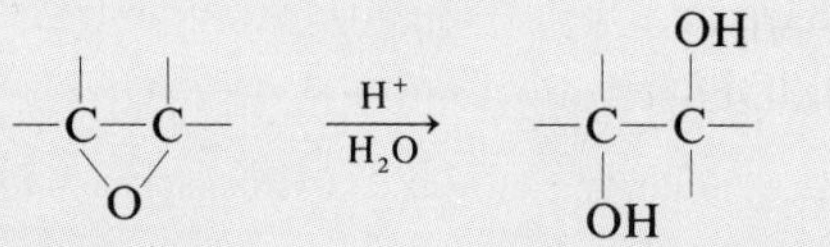

$$-\overset{|}{C}-\overset{|}{C}- \text{(epoxide, O bridging)} \xrightarrow[H_2O]{H^+} -\underset{OH}{\overset{|}{C}}-\overset{OH}{\underset{|}{C}}-$$

anti stereochemistry

b. *In alcohols*

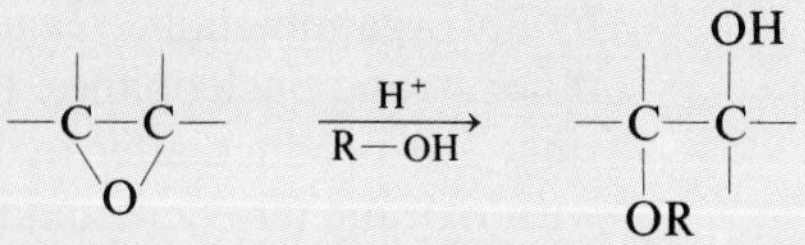

$$-\overset{|}{C}-\overset{|}{C}- \text{(epoxide, O bridging)} \xrightarrow[R-OH]{H^+} -\underset{OR}{\overset{|}{C}}-\overset{OH}{\underset{|}{C}}-$$

The alkoxy group bonds to the more highly substituted carbon.

Example

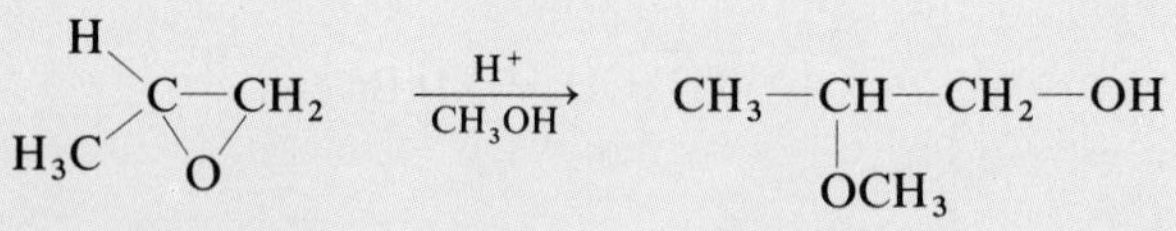

$$\underset{\substack{\text{methyl oxirane}\\\text{(propylene oxide)}}}{(H)(H_3C)C\overset{O}{\triangle}CH_2} \xrightarrow[CH_3OH]{H^+} \underset{\text{2-methoxy-1-propanol}}{CH_3-\underset{OCH_3}{\underset{|}{CH}}-CH_2-OH}$$

c. *Using hydrohalic acids* (X = Cl, Br, I)

$$\underset{\text{(epoxide)}}{-\overset{|}{C}-\overset{|}{C}-} \xrightarrow{H-X} -\underset{X}{\overset{|}{C}}-\overset{OH}{\overset{|}{C}}- \xrightarrow{H-X} -\underset{X}{\overset{|}{C}}-\overset{X}{\overset{|}{C}}-$$

2. *Base-catalyzed opening*

a. *With alkoxides* (Section 13-12)

$$-\overset{|}{C}-CH_2 \text{ (epoxide, O bridge)} \xrightarrow[R-OH]{R-\ddot{\underset{..}{O}}:^-} -\underset{OH}{\overset{|}{C}}-CH_2-OR$$

The alkoxy group bonds to the less highly substituted carbon.

Example

$$\underset{\text{propylene oxide}}{(H)(H_3C)C-CH_2 \text{ (O bridge)}} \xrightarrow[CH_3OH]{CH_3\ddot{\underset{..}{O}}:^-\ {}^+Na} \underset{\text{1-methoxy-2-propanol}}{CH_3-\underset{OH}{CH}-CH_2-OCH_3}$$

b. *With organometallics* (Section 13-14)

$$-\overset{|}{C}-CH_2 \text{ (epoxide, O bridge)} \xrightarrow[(2)\ H_2O]{(1)\ R-M} -\underset{OH}{\overset{|}{C}}-CH_2-R$$

M = Li or MgX R bonds to the less substituted carbon

Example

$$\underset{\text{propylene oxide}}{(H)(H_3C)C-CH_2 \text{ (O bridge)}} \xrightarrow[(2)\ H_2O]{(1)\ C_6H_{11}-MgBr} \underset{\text{1-cyclohexyl-2-propanol}}{CH_3-\underset{OH}{CH}-CH_2-C_6H_{11}}$$

GLOSSARY

alkoxymercuration Addition of mercury and an alkoxy group to a double bond, usually by a solution of mercuric acetate in an alcohol. Usually followed by sodium borohydride reduction to give an ether. (p. 570)

$$>C=C< \xrightarrow[R-O-H]{Hg(OAc)_2} \underset{\text{(alkoxymercuration)}}{-\overset{R-O}{\overset{|}{\underset{|}{C}}}-\overset{Hg(OAc)}{\overset{|}{\underset{|}{C}}}-} \xrightarrow{NaBH_4} \underset{\text{(reduction)}}{-\overset{R-O}{\overset{|}{\underset{|}{C}}}-\overset{H}{\overset{|}{\underset{|}{C}}}-}$$

alpha cleavage The breaking of a bond between the first and second carbon atoms adjacent to the ether oxygen atom (or other functional group). (p. 567)

autoxidation Any oxidation that proceeds spontaneously using the oxygen in the air. Autoxidation of ethers gives hydroperoxides and dialkyl peroxides. (p. 574)

concerted reaction A reaction that takes place in one step, with simultaneous bond breaking and bond forming. (p. 575)

dioxane A heterocyclic ether with two oxygen atoms in a six-membered ring (p. 566).

epoxidation Oxidation of an alkene to an epoxide. Usually accomplished by treating the alkene with a peroxyacid. (p. 575)

epoxide (oxirane) A compound containing a three-membered heterocyclic ether. (p. 564)

ether A compound with two alkyl groups bonded to an oxygen atom, R—O—R′. (p. 559)

> **symmetrical ether:** An ether with two identical alkyl groups bonded to an oxygen atom.
>
> **unsymmetrical ether:** An ether with two different alkyl groups bonded to an oxygen atom.

furan The five-membered heterocyclic ether with two carbon-carbon double bonds, or a derivative of furan. (p. 565)

halohydrin A compound containing a halogen atom and a hydroxyl group on adjacent carbon atoms. Chlorohydrins, bromohydrins, and iodohydrins are most common. (p. 576)

heterocyclic compound (heterocycle) A compound containing a ring in which one or more of the ring atoms are elements other than carbon. The noncarbon ring atoms are called **heteroatoms.** (p. 564)

> **heterocyclic ethers:**

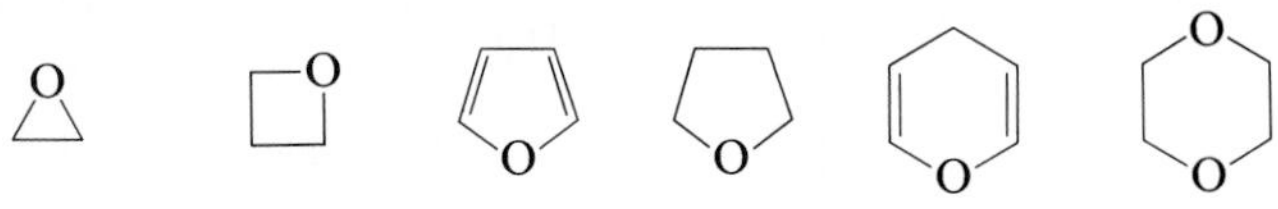

epoxide or oxirane	oxetane	furan	THF	pyran	1,4-dioxane

MCPBA An abbreviation for *meta*-chloroperoxybenzoic acid, a common epoxidizing agent. (p. 575)

oxetane A compound containing a four-membered heterocyclic ether. (p. 565)

peroxide Any compound containing the —O—O— linkage. The oxygen-oxygen bond is easily cleaved, and organic peroxides are prone to explosions. (p. 575)

H—O—O—H	R—O—OH	R—O—O—R
hydrogen peroxide	and alkyl hydroperoxide	a dialkyl peroxide

peroxyacid (peracid) A carboxylic acid with an extra oxygen in the hydroxyl group. (p. 575)

$$\mathrm{R{-}C({=}O){-}O{-}O{-}H}$$

pyran The six-membered heterocyclic ether with two carbon-carbon double bonds, or a derivative of pyran. (p. 566)

Williamson ether synthesis The formation of an ether by the S_N2 reaction of an alkoxide ion with an alkyl halide or tosylate. In general, the electrophile must be primary, or occasionally secondary. (p. 569)

$$\mathrm{R{-}\ddot{\underset{..}{O}}{:}^{-}\quad R'{-}X \longrightarrow R{-}O{-}R' + X^{-}}$$

ESSENTIAL PROBLEM-SOLVING SKILLS IN CHAPTER 13

1. Draw and name ethers and heterocyclic ethers, including epoxides.
2. Predict relative boiling points and solubilities of ethers.
3. Explain the stabilizing effects of ether solvents on electrophilic reagents and their compatibility with organometallic reagents.

4. Determine the structures of ethers from their spectra, and explain the characteristic absorptions and fragmentations in their spectra.

5. Devise efficient laboratory syntheses of ethers and epoxides, including those using:
   (a) The Williamson ether synthesis
   (b) Alkoxymercuration-demercuration
   (c) Peroxyacid epoxidation
   (d) Displacement of halohydrins.

6. Predict the products of the reactions of ethers and epoxides, including:
   (a) Cleavage and autoxidation of ethers
   (b) Acid- and base-promoted opening of epoxides
   (c) Reactions of epoxides with organometallic reagents.

7. Use your knowledge of the mechanisms of ether and epoxide reactions to propose mechanisms and products of similar reactions you have never seen before.

STUDY PROBLEMS

13-24. Briefly define each of the following terms and give an example.
(a) autoxidation **(b)** Williamson ether synthesis **(c)** alkoxymercuration-demercuration
(d) heterocyclic compound **(e)** epoxidation **(f)** concerted reaction
(g) unsymmetrical ether

13-25. Write structural formulas for the following compounds.
(a) ethyl isopropyl ether **(b)** di-*n*-butyl ether **(c)** 2-ethoxyoctane
(d) divinyl ether **(e)** allyl methyl ether **(f)** cyclohexene oxide
(g) *cis*-2,3-epoxyhexane **(h)** (2*R*, 3*S*)-2-methoxy-3-pentanol

13-26. Give common names for the following compounds.
(a) $(CH_3)_2CH-O-CH(CH_3)CH_2CH_3$ **(b)** $(CH_3)_3C-O-CH_2CH(CH_3)_2$
(c) $Ph-O-CH_2CH_3$ **(d)** $Cl-CH_2-O-CH_2CH_2CH_3$

(e) H, OH, OH, H **(f)** OCH_3 **(g)** O, CH_3 **(h)** O

13-27. Give IUPAC names for the following compounds.
(a) $CH_3-O-CH(CH_3)CH_2OH$ **(b)** $Ph-O-CH_2CH_3$ **(c)** OCH_3 **(d)** OCH_3, OCH_3, OH

(e)
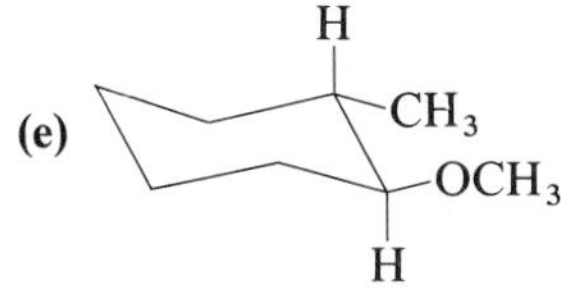

(f) H, O, H, Cl **(g)** H, O, CH_2CH_3, CH_3O, H **(h)** O, Br **(i)** O, O

13-28. Predict the products of the following reactions.
(a) *sec*-butyl isopropyl ether + conc. HBr, heat **(b)** *t*-butyl ethyl ether + conc. HBr, heat
(c) di-*n*-butyl ether + hot conc. NaOH **(d)** di-*n*-butyl ether + Na metal
(e) ethoxybenzene + conc. HI, heat **(f)** 1,2-epoxyhexane + H^+, CH_3OH
(g) *trans*-2,3-epoxyoctane + H^+, H_2O **(h)** propylene oxide + methylamine (CH_3NH_2)

(i) potassium *t*-butoxide + *n*-butyl bromide

(j) 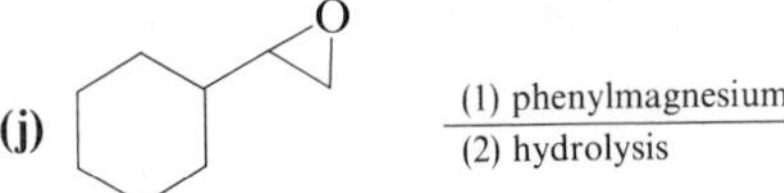 $\xrightarrow[\text{(2) hydrolysis}]{\text{(1) phenylmagnesium bromide}}$

(k) $\xrightarrow{\text{MCPBA, } CH_2Cl_2}$ **(l)** $\xrightarrow{\text{HBr}}$

(m) $\xrightarrow{CH_3O^-,\ CH_3OH}$ **(n)** $\xrightarrow{H^+/CH_3OH}$

13-29. (A true story.) An inexperienced graduate student moved into a laboratory and began work. He needed some diethyl ether for a reaction, so he opened an old, rusty 1-gallon can marked "ethyl ether" and found that there was half a gallon left. To purify the ether, the student set up a distillation apparatus, started a careful distillation, and went to the stockroom for the other reagents he needed. While he was at the stockroom, the student heard a muffled "boom." He quickly returned to his lab to find a worker from another laboratory putting out a fire. Most of the distillation apparatus was embedded in the ceiling.
(a) Explain what probably happened. **(b)** Explain how this near-disaster might have been prevented.

13-30. **(a)** Predict the values of *m/z* and the structures of the most abundant fragments observed in the mass spectrum of di-*n*-propyl ether.
(b) Give logical fragmentation reactions that account for the following ions observed in the mass spectrum of 2-methoxypentane: 102, 87, 71, 59, 31.

13-31. The following reaction resembles the acid-catalyzed cyclization of squalene oxide. Give a mechanism for this reaction.

$\xrightarrow[H_2O]{H^+}$ CH_2OH, CH_3, OH

13-32. Show how you would accomplish the following synthetic transformations in good yield.
(a) 1-hexene ⟶ 1-phenyl-2-hexanol **(b)** 1-hexene ⟶ 1-methoxy-2-hexanol
(c) 1-hexene ⟶ 2-methoxy-1-hexanol

13-33. Give the structures of intermediates *A* through *H* in the following synthesis of *trans*-1-cyclohexyl-2-methoxycyclohexane.

OH $\xrightarrow[\text{heat}]{H_2SO_4}$ *A* $\xrightarrow{\text{MCPBA}}$ *F* $\xrightarrow[\text{(2) } H_3O^+]{\text{(1) } E}$ *G* $\xrightarrow{\text{Na}}$ *H* $\xrightarrow{C}$ product

A $\xrightarrow{\text{(1) Hg(OAc)}_2\text{, } CH_3OH\text{; (2) } NaBH_4}$ *B* $\xrightarrow[\text{heat}]{\text{conc. HBr}}$ *C* (gas) + *D*

D $\xrightarrow{\text{Mg, ether}}$ *E*

✱13-34. (Another true story.) An organic lab student carried out the reaction of methylmagnesium iodide with acetone (CH_3COCH_3), followed by hydrolysis. During the distillation to isolate the product, he forgot to mark the vials he used to collect the fractions. He turned in a product of formula $C_4H_{10}O$ that gave the NMR spectrum below. The IR spectrum showed only a weak O—H stretch at about 3300 cm^{-1}, and the mass spectrum showed a base peak at *m/z* 59. Propose a structure for this product, explain how it corresponds to the observed spectra, and suggest how the student isolated this compound.

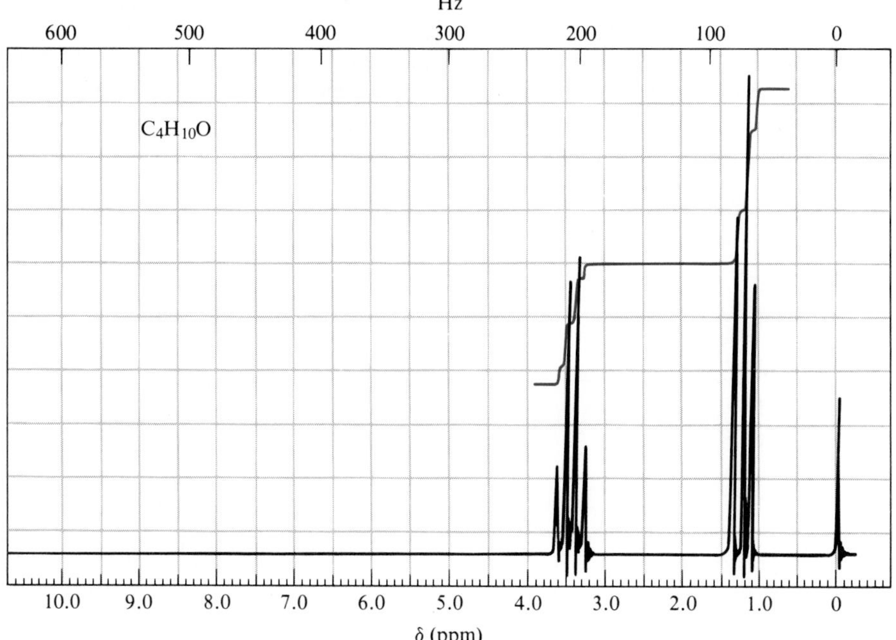

13-35. Show how you would synthesize the following ethers in good yield from the indicated starting materials and any additional reagents that are needed.

(a) cyclopentyl *n*-propyl ether from cyclopentanol and 1-propanol
(b) *n*-butyl phenyl ether from phenol and 1-butanol
(c) 2-methoxydecane from a decene
(d) 1-methoxydecane from a decene
(e) 1-ethoxy-1-methylcyclohexane from 1-methylcyclohexene
(f) *trans*-2,3-epoxyoctane from *trans*-2-octene

13-36. There are two different ways of making 2-ethoxyoctane from 2-octanol using the Williamson ether synthesis. When pure (−)-2-octanol of specific rotation −8.24° is treated with sodium metal and then ethyl iodide, the product is 2-ethoxyoctane with a specific rotation of −15.6°. When pure (−)-2-octanol is treated with thionyl chloride and then with sodium ethoxide, the product is also 2-ethoxyoctane. Predict the rotation of the 2-ethoxyoctane made using the thionyl chloride/sodium ethoxide procedure, and give a detailed mechanism to support your prediction.

13-37. Under base-catalyzed conditions, several molecules of propylene oxide can polymerize to give short polymers. Give a mechanism for the base-catalyzed formation of the following trimer.

$$3\ H_2C\overset{O}{—}CH—CH_3 \xrightarrow{^-OH} HO—CH_2—\underset{}{\overset{CH_3}{\overset{|}{CH}}}—O—CH_2—\overset{CH_3}{\overset{|}{CH}}—O—CH_2—\overset{CH_3}{\overset{|}{CH}}—OH$$

13-38. Under the right conditions, the following acid-catalyzed double cyclization proceeds in remarkably good yields. Propose a mechanism; does this reaction resemble a biological process you have seen?

$\xrightarrow{H^+}$ OH

13-39. Propylene oxide is a chiral molecule. Hydrolysis of propylene oxide gives propylene glycol, another chiral molecule.

(a) Draw the enantiomers of propylene oxide.
(b) Give a mechanism for the acid-catalyzed hydrolysis of pure (*R*)-propylene oxide.
(c) Give a mechanism for the base-catalyzed hydrolysis of pure (*R*)-propylene oxide.
(d) Explain why the acid-catalyzed hydrolysis of optically active propylene oxide gives product with a rotation in the opposite direction from the product of the base-catalyzed hydrolysis.

13-40. An industrial chemist carried out an acid-catalyzed reaction using methyl cellosolve (2-methoxyethanol) as the solvent. When the 2-methoxyethanol was redistilled, a higher-boiling fraction (b.p. 162°) was also recovered. The mass spectrum of this fraction showed the molecular weight to be 134, and the IR and NMR spectra appear below. Determine the structure of this compound, and propose a mechanism for its formation.

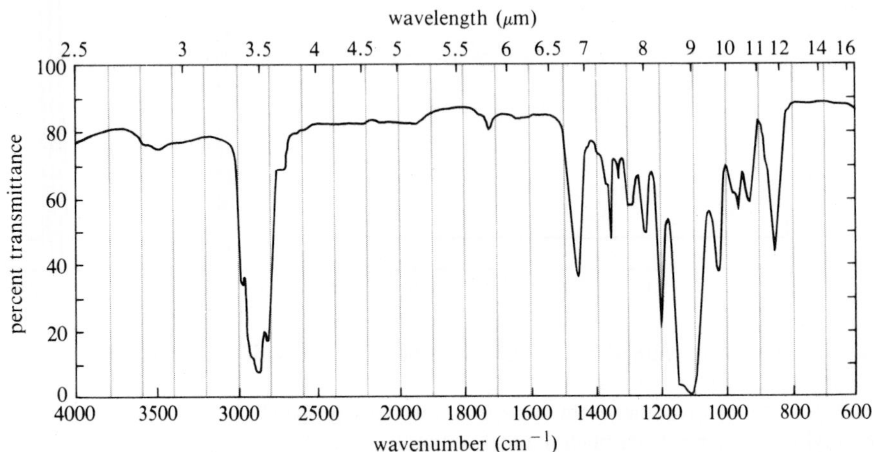

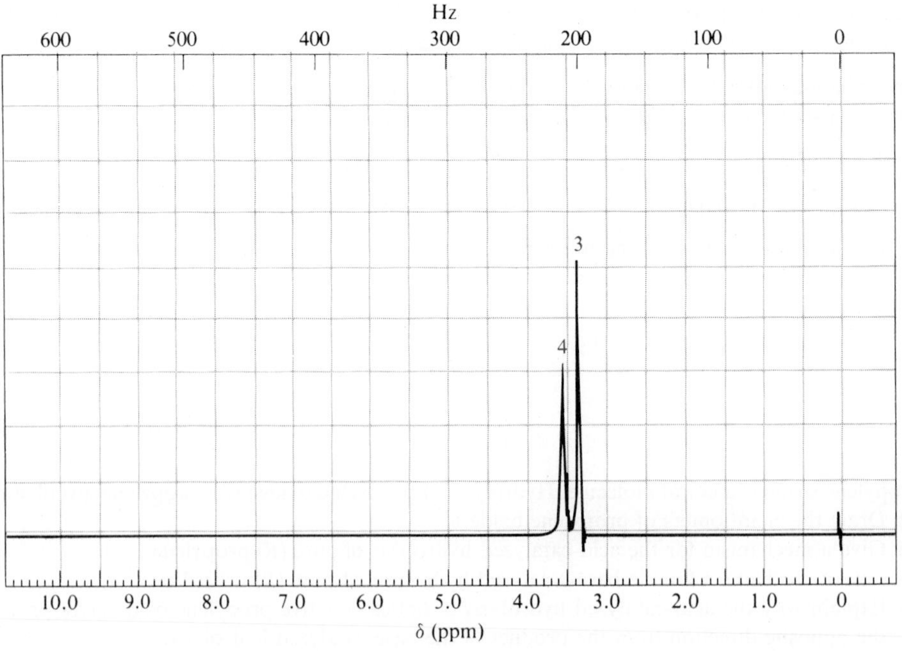

13-41. A compound of molecular formula C_8H_8O gives the NMR and IR spectra shown below. Propose a structure and show how it is consistent with the observed absorptions.

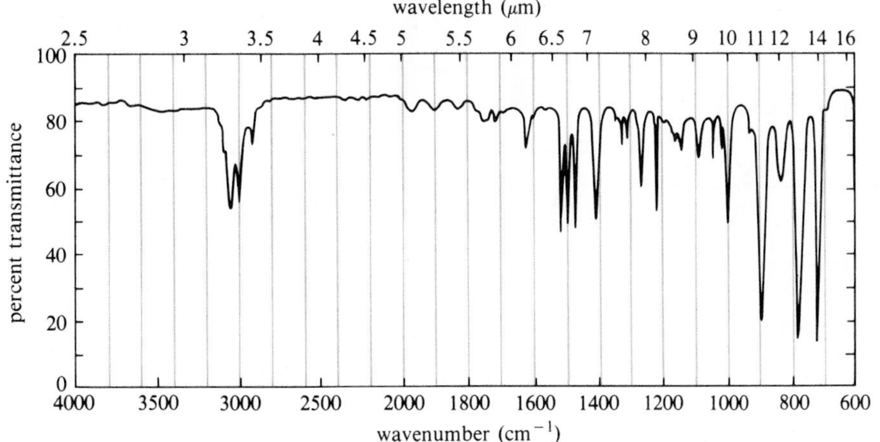

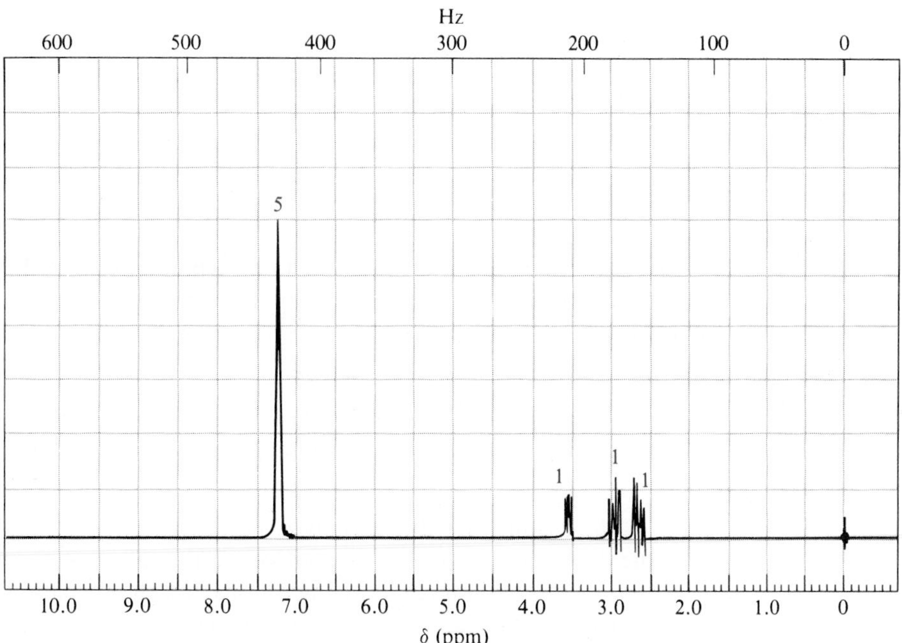

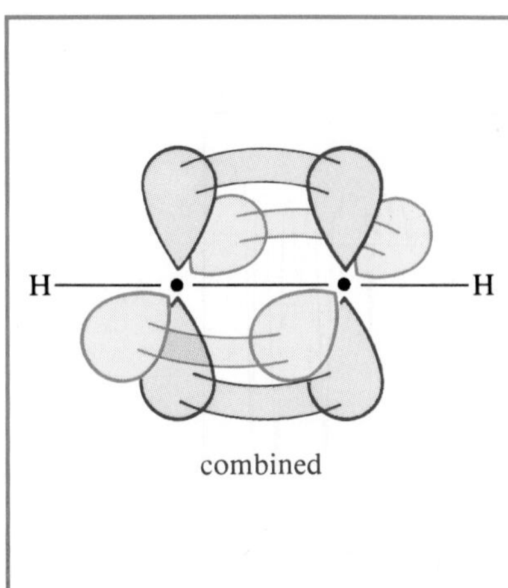

14

ALKYNES

14-1 INTRODUCTION

Alkynes are hydrocarbons that contain carbon-carbon triple bonds. Alkynes are also called **acetylenes,** because they are derivatives of acetylene, the simplest alkyne.

$$\underset{\text{acetylene}}{H-C\equiv C-H} \qquad \underset{\text{ethylacetylene}}{CH_3CH_2-C\equiv C-H} \qquad \underset{\text{dimethylacetylene}}{CH_3-C\equiv C-CH_3}$$

The chemistry of the carbon-carbon triple bond is similar to the chemistry of the double bond. In this chapter we see that alkynes undergo most of the reactions of alkenes, especially the additions and the oxidations. We also consider several reactions that are specific to alkynes: some that depend on the unique characteristics of the C≡C triple bond, and others that depend on the unusual acidity of the acetylenic C—H bond.

The presence of a triple bond gives an alkyne four fewer hydrogens than the corresponding alkane. Its molecular formula is like that of a molecule with two double bonds: C_nH_{2n-2}. Therefore, the triple bond contributes two elements of unsaturation (e.u.) (Section 7-3).

$$\underset{\substack{\text{ethane, } C_2H_6 \\ 0 \text{ e.u., } C_nH_{2n+2}}}{H_3C-CH_3} \qquad \underset{\substack{\text{ethene, } C_2H_4 \\ 1 \text{ e.u., } C_nH_{2n}}}{H_2C=CH_2} \qquad \underset{\substack{\text{ethyne, } C_2H_2 \\ 2 \text{ e.u., } C_nH_{2n-2}}}{H-C\equiv C-H}$$

PROBLEM 14-1

Draw structural formulas for at least two alkynes with each of the following molecular formulas.

(a) C_6H_{10} (b) C_8H_{12} (c) C_7H_{10}

14-2 NOMENCLATURE OF ALKYNES

IUPAC names The IUPAC nomenclature for alkynes is similar to that for alkenes. We find the longest continuous chain of carbon atoms that includes the triple bond and change the *-ane* ending of the parent alkane to *-yne.* The chain is numbered from the end closest to the triple bond, and the position of the triple bond is designated by its lower-numbered carbon atom. Substituents are given numbers to indicate their locations.

	$H-C\equiv C-H$	$CH_3-C\equiv C-H$	$CH_3-C\equiv C-CH_3$	$CH_3-CH(CH_3)-C\equiv C-CH_2-CH(Br)-CH_3$
IUPAC name:	ethyne (acetylene)	propyne	2-butyne	6-bromo-2-methyl-3-heptyne

When additional functional groups are present, the suffixes are combined to produce the compound names of the *alkenynes, alkynols,* and so on.

	$H_2C=C(CH_3)-C\equiv C-CH_3$	$CH_3-CH(OH)-C\equiv C-H$	$CH_3-C\equiv C-CH(OCH_3)-CH_2CH_3$
IUPAC name:	2-methyl-1-penten-3-yne	3-butyn-2-ol	4-methoxy-2-hexyne

Common names The common names of alkynes describe them as derivatives of acetylene. Most alkynes can be named as a molecule of acetylene with one or two alkyl substituents. This nomenclature is like the common nomenclature for ethers, where we name the two alkyl groups bonded to the oxygen atom.

$H-C\equiv C-H$	$R-C\equiv C-H$	$R-C\equiv C-R'$
acetylene	an alkylacetylene	a dialkylacetylene

	$CH_3-C\equiv C-H$	$Ph-C\equiv C-H$	$CH_3-C\equiv C-CH_2CH_3$
common name:	methylacetylene	phenylacetylene	ethylmethylacetylene
	$(CH_3)_2CH-C\equiv C-CH(CH_3)_2$	$Ph-C\equiv C-Ph$	$H-C\equiv C-CH_2OH$
common name:	diisopropylacetylene	diphenylacetylene	hydroxymethylacetylene (propargyl alcohol)

Many of an alkyne's chemical properties depend on whether there is an acetylenic hydrogen ($H-C\equiv C$), that is, if the triple bond comes at the end of a carbon chain. Such an alkyne is called a **terminal alkyne** or a **terminal acetylene.** If the triple bond is located somewhere other than the end of the carbon chain, the alkyne is called an **internal alkyne** or an **internal acetylene.**

acetylenic hydrogen	(no acetylenic hydrogen)
$\boxed{H}-C\equiv C-CH_2CH_3$	$CH_3-C\equiv C-CH_3$
1-butyne, a *terminal* alkyne	2-butyne, an *internal* alkyne

PROBLEM 14-2

For each of the following molecular formulas, draw all the isomeric alkynes and give their IUPAC names. Circle the acetylenic hydrogen of each terminal alkyne.

(a) C_4H_6 (two isomers) (b) C_5H_8 (three isomers)

14-3 ELECTRONIC STRUCTURE OF ALKYNES

In Section 2-4 we saw the electronic structure of a triple bond. Let's review this structure, using acetylene as the example. The Lewis structure of acetylene shows three pairs of electrons in the region between the carbon nuclei:

$$H:C:::C:H$$

Each of the carbon atoms is bonded to two other atoms, and there are no nonbonding pairs of valence electrons. Each carbon atom needs two hybrid orbitals to form the sigma-bond framework. Hybridization of the *s* orbital with one *p* orbital gives two hybrid orbitals, directed 180° apart, for each carbon atom. Overlap of these *sp* hybrid orbitals with each other and with the hydrogen *s* orbitals gives the sigma-bond framework of acetylene. Experimental results have confirmed this linear (180°) structure.

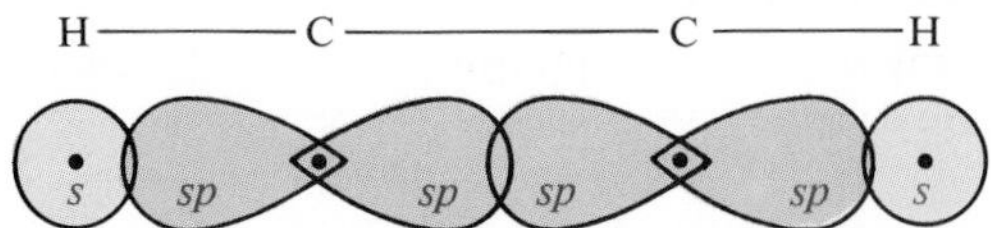

Two pi bonds result from overlap of the two remaining unhybridized *p* orbitals on each of the carbon atoms. These orbitals overlap at right angles to each other, forming one pi bond with electron density above and below the C—C sigma bond, and the other with electron density in front and in back of the sigma bond. The shape of these pi bonds is such that they blend to form a cylinder of electron density encircling the sigma bond between the two carbon atoms.

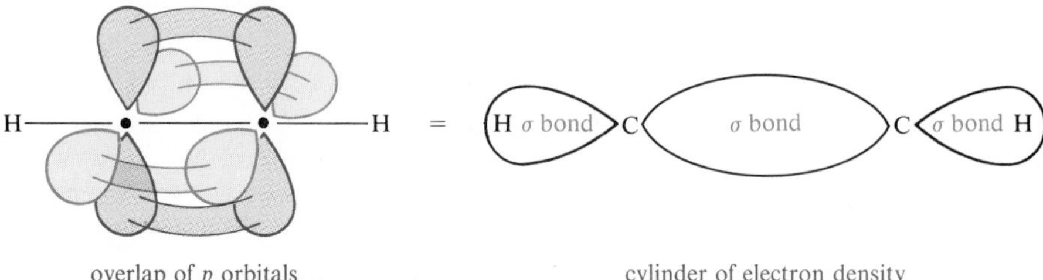

overlap of *p* orbitals

cylinder of electron density

The carbon-carbon bond length in acetylene is 1.20 Å, and the carbon-hydrogen bonds are each 1.06 Å. Both of these bonds are shorter than the corresponding bonds in ethane and in ethene.

1.54 Å, 1.09 Å — ethane

1.33 Å, 1.08 Å — ethene

H—C≡C—H: 1.20 Å, 1.06 Å — ethyne

The triple bond is relatively short because of the attractive overlap of three bonding pairs of electrons and the high **s character** of the *sp* hybrid orbitals. The *sp* hybrid orbitals are about one-half *s* character (as opposed to $\frac{1}{3}$ *s* character of sp^2 hybrids and $\frac{1}{4}$ of sp^3 hybrids), using more of the closer, tightly-held *s* orbitals. The use of *sp* hybrid orbitals also explains the slightly shorter C—H bonds in acetylene compared with ethylene.

14-4 PHYSICAL PROPERTIES OF ALKYNES

The physical properties of alkynes (Table 14-1) are similar to those of the corresponding alkanes and alkenes. Alkynes are relatively nonpolar hydrocarbons, nearly insoluble in water. They are quite soluble in most organic solvents, including acetone, ether, methylene chloride, chloroform, and some alcohols.

Acetylene, propyne, and the butynes are gases at room temperature, just like the corresponding alkanes and alkenes. In fact, the boiling points of the alkynes are nearly the same as those of alkanes and alkenes with similar carbon skeletons.

TABLE 14-1
Physical properties of selected alkynes

Name	*Structure*	*m.p. (°C)*	*b.p. (°C)*	*Density (g/cm³)*
ethyne (acetylene)	$H{-}C{\equiv}C{-}H$	−82	−75	0.62
propyne	$H{-}C{\equiv}C{-}CH_3$	−101	−23	0.67
1-butyne	$H{-}C{\equiv}C{-}CH_2CH_3$	−122	8	0.67
2-butyne	$CH_3{-}C{\equiv}C{-}CH_3$	−28	27	0.69
1-pentyne	$H{-}C{\equiv}C{-}CH_2CH_2CH_3$	−98	40	0.70
2-pentyne	$CH_3{-}C{\equiv}C{-}CH_2CH_3$	−101	55	0.71
3-methyl-1-butyne	$CH_3{-}CH(CH_3){-}C{\equiv}C{-}H$		28	0.67
1-hexyne	$H{-}C{\equiv}C{-}(CH_2)_3{-}CH_3$	−124	71	0.72
2-hexyne	$CH_3{-}C{\equiv}C{-}CH_2CH_2CH_3$	−92	84	0.73
3-hexyne	$CH_3CH_2{-}C{\equiv}C{-}CH_2CH_3$	−51	82	0.73
3,3-dimethyl-1-butyne	$(CH_3)_3C{-}C{\equiv}C{-}H$	−81	38	0.67
1-heptyne	$H{-}C{\equiv}C{-}(CH_2)_4CH_3$	−80	100	0.73
1-octyne	$H{-}C{\equiv}C{-}(CH_2)_5CH_3$	−70	126	0.75
1-nonyne	$H{-}C{\equiv}C{-}(CH_2)_6CH_3$	−65	151	0.76
1-decyne	$H{-}C{\equiv}C{-}(CH_2)_7CH_3$	−36	182	0.77

14-5 COMMERCIAL IMPORTANCE OF ALKYNES

14-5A USES OF ACETYLENE AND METHYLACETYLENE

Acetylene is by far the most important commercial alkyne. Its largest use is as the fuel for the oxyacetylene welding torch. Acetylene is a colorless, foul-smelling gas that burns in air with a yellow, sooty flame. When the flame is supplied with pure oxygen, however, the color turns to light blue and its temperature increases dramatically. A comparison of the heat of combustion for acetylene with those of ethene and ethane shows why this gas makes an excellent fuel for a high-temperature flame.

$$CH_3CH_3 + \tfrac{7}{2}\,O_2 \longrightarrow 2\,CO_2 + 3\,H_2O \qquad \Delta H^\circ = -373 \text{ kcal } (-1561 \text{ kJ})$$

$$-373 \text{ kcal divided by 5 moles of products} = -75 \text{ kcal/mol of products } (-312 \text{ kJ/mol})$$

$$H_2C{=}CH_2 + 3\,O_2 \longrightarrow 2\,CO_2 + 2\,H_2O \qquad \Delta H^\circ = -337 \text{ kcal } (-1410 \text{ kJ})$$

$$-337 \text{ kcal divided by 4 moles of products} = -84 \text{ kcal/mol of products } (-352 \text{ kJ/mol})$$

$$HC{\equiv}CH + \tfrac{5}{2}\,O_2 \longrightarrow 2\,CO_2 + 1\,H_2O \qquad \Delta H^\circ = -317 \text{ kcal } (-1326 \text{ kJ})$$

$$-317 \text{ kcal divided by 3 moles of products} = -106 \text{ kcal/mol of products } (-442 \text{ kJ/mol})$$

If we were simply heating a house by burning one of these fuels, we might choose ethane as our fuel, because it produces the most heat per mole of ethane consumed. In the welding torch we want to achieve the highest possible *temperature* of the gaseous products. The heat of reaction must raise the temperature of the reaction products to the flame temperature. Roughly speaking, the increase in temperature of the products is proportional to the heat given off *per mole of products* formed. This rise in temperature is largest with acetylene, which gives off the most heat per mole of products. The oxyacetylene flame reaches temperatures as high as 2800°C.

When acetylene was first used for welding, it was considered a dangerous, explosive gas. Acetylene is thermodynamically unstable, and when the compressed gas is subjected to thermal or mechanical shock it decomposes to its elements, releasing 56 kcal (234 kJ) of energy per mole. This initial decomposition often splits the container, allowing the products (hydrogen and finely divided carbon) to burn in the air.

$$H{-}C{\equiv}C{-}H \longrightarrow 2\,C + H_2 \qquad \Delta H^\circ = -56 \text{ kcal/mol } (-234 \text{ kJ/mol})$$

$$2\,C + H_2 \xrightarrow{\frac{5}{2}O_2} 2\,CO_2 + H_2O \qquad \Delta H^\circ = -261 \text{ kcal/mol } (-1090 \text{ kJ/mol})$$

Acetylene is safely stored and handled in cylinders that are filled with crushed firebrick wet with acetone. Acetylene dissolves freely in acetone, and the dissolved gas is not so prone to decomposition. The firebrick helps to control the decomposition by minimizing the free volume of the cylinder, cooling and controlling any decomposition before it gets out of control.

Methylacetylene is another common alkyne used in welding torches. Methylacetylene does not decompose as easily as acetylene, and it burns better in air (as opposed to pure oxygen). Methylacetylene is well suited for household soldering and brazing that requires higher temperatures than those reached by propane torches. The industrial synthesis produces methylacetylene as a mixture with its isomer, propadiene (allene). This mixture is sold commercially under the trade name MAPP gas (*M*ethyl*A*cetylene-*P*ro*P*adiene).

$$\underset{\text{methylacetylene}}{CH_3{-}C{\equiv}C{-}H} \qquad \underset{\text{propadiene (allene)}}{H_2C{=}C{=}CH_2}$$

14-5B MANUFACTURE OF ACETYLENE

Acetylene, one of the cheapest organic chemicals, is made from coal or from natural gas. The synthesis from coal involves heating lime and coke in an electric furnace to produce calcium carbide. Addition of water to calcium carbide produces acetylene and hydrated lime.

$$\underset{\text{coke}}{3\,C} + \underset{\text{lime}}{CaO} \xrightarrow{\text{electric furnace, 2500°C}} \underset{\text{calcium carbide}}{CaC_2} + CO$$

$$CaC_2 + 2\,H_2O \longrightarrow \underset{\text{acetylene}}{H{-}C{\equiv}C{-}H} + \underset{\text{hydrated lime}}{Ca(OH)_2}$$

This second reaction was used as a light source in coal mines until battery-powered lights became available. A miner's lamp allows water to drip very slowly onto some calcium carbide. Acetylene is generated, feeding a small flame where the gas burns in air with a yellow flickering light. Unfortunately, this flame ignites the methane gas commonly found in coal seams, resulting in explosions. Battery-powered miners' lamps provide better light and reduce the danger of methane explosions.

The synthesis of acetylene from natural gas is a simple process. Natural gas consists mostly of methane, which forms acetylene when it is heated for a very short period of time.

$$2\ CH_4 \xrightarrow[0.01\ \text{sec}]{1500°\text{C}} H{-}C{\equiv}C{-}H \;+\; 3\ H_2$$

Although this reaction is endothermic, there are twice as many moles of products as there are of reactants. The increase in the number of moles results in an increase in entropy, and the $(-T\Delta S)$ term in the free energy $(\Delta G = \Delta H - T\Delta S)$ predominates at this high temperature.

PROBLEM 14-3
What reaction would acetylene likely undergo if it were kept at 1500°C for too long?

14-6 SPECTROSCOPY OF ALKYNES

14-6A NMR SPECTROSCOPY OF ALKYNES

The acetylenic proton of a terminal alkyne gives a characteristic peak of area 1 (usually a singlet) in the ^{1}H NMR between δ2.0 and δ2.5. Internal alkynes have no acetylenic proton; protons on a carbon atom next to the triple bond absorb between δ1.8 and δ2.2.

$$-CH_2{-}C{\equiv}C{-}H$$

proton chemical shifts: δ1.8–δ2.2 (CH$_2$) δ2.0–δ2.5 (H)

In the ^{13}C NMR, acetylenic carbons absorb around δ65 to δ85, with substituted carbons absorbing at slightly lower field than unsubstituted carbons.

$$R{-}C{\equiv}C{-}H$$

carbon chemical shifts: δ75–δ85 (R–C) δ65–δ70 (C–H)

14-6B INFRARED SPECTROSCOPY OF ALKYNES

The C≡C stretch of an alkyne appears slightly below 2200 cm^{-1} in the infrared spectrum. This absorption is of moderate strength in terminal alkynes, but it may be weak in internal alkynes because the vibration produces little change in the dipole moment. Terminal alkynes also show a strong ≡C—H stretch appearing as a sharp peak around 3300 cm^{-1}.

$$R{-}C{\equiv}C{-}H \qquad\qquad R{-}C{\equiv}C{-}R'$$

IR absorptions: 2100–2200 cm^{-1} (C≡C), 3300 cm^{-1} (C–H); 2100–2200 cm^{-1} (weak)

PROBLEM 14-4

Determine the structure of the compound (C_6H_{10}) that gives the following IR and NMR spectra:

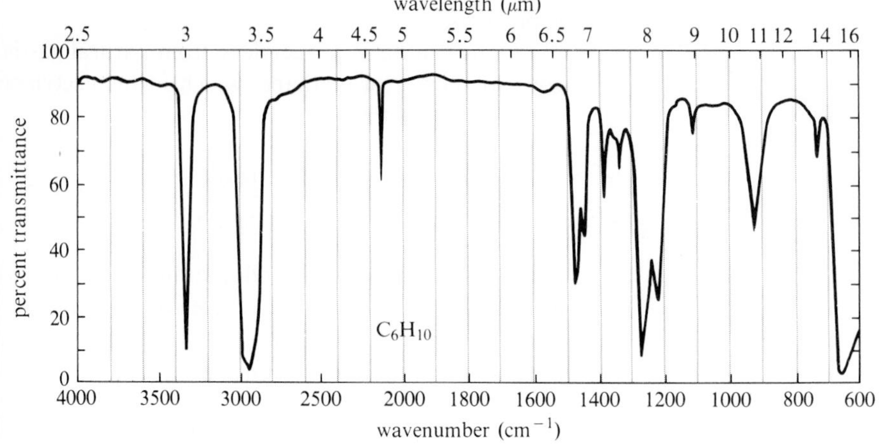

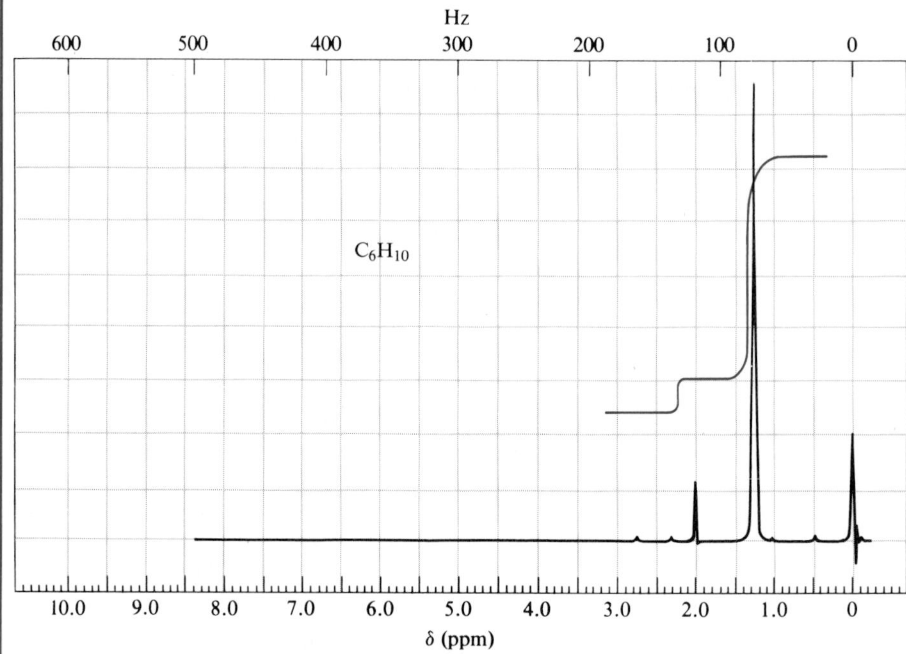

14-7 ACIDITY OF ALKYNES

The most distinctive property of alkynes is their unusual acidity. Terminal alkynes are much more acidic than other hydrocarbons. This acidity, resulting from the nature of the *sp* hybrid ≡C—H bond, facilitates the formation of acetylide ions whose use is a central feature of alkyne chemistry.

Table 14-2 shows that the acidity of a C—H bond varies with its hybridization, increasing with the increasing *s* character of the orbitals: $sp^3 < sp^2 < sp$. (Remember that a *smaller* value of pK_a corresponds to a stronger acid.) The acet-

TABLE 14-2
Acidities of alkanes, alkenes, and alkynes

Compound	*Conjugate base*	*Hybridization*	*s character*	pK_a
$H_3C{-}CH_3$	$H_3C{-}CH_2^-$	sp^3	25%	50
$H_2C{=}CH_2$	$H_2C{=}CH^-$	sp^2	33%	44
$H{-}C{\equiv}C{-}H$	$H{-}C{\equiv}C^-$	sp	50%	25
		For comparison:		
R—OH	$R{-}\ddot{O}{:}^-$			16–18
$:NH_3$	$:\ddot{N}H_2^-$			35

ylenic proton is about 10^{19} times as acidic as a vinyl proton. When an acetylenic proton is abstracted, the resulting carbanion has the lone pair of electrons in the *sp* hybrid orbital. Electrons in this orbital are close to the nucleus, and there is less charge separation than in carbanions with the lone pair in sp^2 or sp^3 hybrid orbitals. Note that acetylene can be deprotonated by the amide ($^-NH_2$) ion, but not by an alkoxide ion.

14-7A FORMATION OF ACETYLIDE IONS

Unlike alkanes and alkenes, terminal acetylenes are easily deprotonated to form carbanions called **acetylide ions.** The acetylenic proton is removed by a very strong base, such as a Grignard or organolithium reagent. Hydroxide ion and alkoxide ions are not strong enough bases to deprotonate alkynes. Internal alkynes do not have acetylenic protons, so they do not react.

$$CH_3CH_2{-}C{\equiv}C{-}\underset{\text{acidic proton}}{H} + CH_3{-}Li \longrightarrow CH_3CH_2{-}C{\equiv}C{-}Li + CH_4\uparrow$$

1-butyne, a terminal alkyne → a lithium acetylide

$$C_6H_{11}{-}C{\equiv}C{-}H + CH_3CH_2MgBr \longrightarrow C_6H_{11}{-}C{\equiv}C{-}MgBr + CH_3CH_3\uparrow$$

cyclohexylacetylene; ethylmagnesium bromide; a magnesium acetylide; ethane

$$CH_3{-}C{\equiv}C{-}CH_3 \xrightarrow{\text{base:}^-} \text{no reaction}$$

(no acetylenic proton)
2-butyne, an internal alkyne

Sodium amide ($Na^+\ ^-{:}NH_2$) is frequently used as the base to form sodium acetylide salts. The amide ion ($^-NH_2$) is the conjugate base of ammonia, a compound that is itself a base. Ammonia is also a very weak acid, however, with an acidity constant $K_a = 10^{-35}$ ($pK_a = 35$). One of its hydrogens can be reduced by sodium to give the sodium salt of the amide ion, a very strong conjugate base.

$$\underset{\text{ammonia}}{H-\overset{H}{\overset{|}{\underset{\cdot\cdot}{N}}}-H} + Na \xrightarrow{Fe^{3+}\text{ catalyst}} \underset{\text{sodium amide ("sodamide")}}{Na^{+}\ {}^{-}{:}\overset{H}{\overset{|}{\underset{\cdot\cdot}{N}}}-H} + \tfrac{1}{2}H_2\uparrow$$

Sodium amide reacts with terminal alkynes to form sodium acetylides. Unlike lithium and magnesium acetylides, a sodium acetylide has little covalent character and may be considered a true carbanion.

$$R-C\equiv C-H + Na^{+}\ {}^{-}{:}\ddot{N}H_2 \longrightarrow \underset{\text{a sodium acetylide}}{R-C\equiv C{:}^{-}\ {}^{+}Na} + {:}NH_3$$

Example

$$CH_3CH_2-C\equiv C-H + Na^{+}\ {}^{-}{:}\ddot{N}H_2 \longrightarrow CH_3CH_2-C\equiv C{:}^{-}\ {}^{+}Na + {:}NH_3$$

Acetylide ions are strong nucleophiles. In fact, one of the best methods for synthesizing substituted alkynes is a nucleophilic attack of an acetylide ion on an unhindered alkyl halide or tosylate. We consider this displacement reaction in detail in Section 14-8A.

$$CH_3CH_2-C\equiv C{:}^{-}\ {}^{+}Na + H_3C-I \longrightarrow CH_3CH_2-C\equiv C-CH_3 + NaI$$

PROBLEM 14-5

The boiling points of 1-hexene (64°C) and 1-hexyne (71°C) are sufficiently close that it is difficult to achieve a clean separation by distillation. Show how you might use its acidity to remove the last traces of 1-hexyne from a sample of 1-hexene.

PROBLEM 14-6

Predict the products of the following acid-base reactions, or indicate if no significant reaction would take place.

(a) $H-C\equiv C-H + NaNH_2$
(b) $H-C\equiv C-H + CH_3Li$
(c) $H-C\equiv C-H + NaOCH_3$
(d) $H-C\equiv C-H + NaOH$
(e) $H-C\equiv C{:}^{-}\ {}^{+}Na + CH_3OH$
(f) $H-C\equiv C{:}^{-}\ {}^{+}Na + H_2O$
(g) $H-C\equiv C{:}^{-}\ {}^{+}Na + H_2C=CH_2$
(h) $H_2C=CH_2 + NaNH_2$
(i) $CH_3OH + NaNH_2$

14-7B HEAVY-METAL ACETYLIDES

Silver(I) and copper(I) salts react with terminal alkynes to form silver and copper acetylides in much the same way as sodium amide, Grignard reagents, and organolithium reagents react to give acetylides. Silver and copper acetylides are bonded more covalently than other acetylides, however, and they are much less basic and less nucleophilic. Unlike the sodium acetylides, silver and copper acetylides are not very soluble; they form characteristic precipitates:

$$R-C\equiv C-H + Ag^{+} \longrightarrow \underset{\text{(light-colored precipitate)}}{R-C\equiv C-Ag\downarrow} + H^{+}$$

$$R-C\equiv C-H + Cu^{+} \longrightarrow \underset{\text{(brick-red precipitate)}}{R-C\equiv C-Cu\downarrow} + H^{+}$$

Internal acetylenes are unreactive toward Ag^+ and Cu^+ because they have no acetylenic protons. Adding a silver reagent or a copper(I) reagent to a solution of an alkyne shows whether the alkyne is terminal or internal. The terminal alkyne forms a precipitate, but the internal alkyne does not. This reaction provides a simple chemical test for terminal alkynes.

$$CH_3-CH_2-C{\equiv}C-H \quad + \quad Ag^+ \text{ (or } Cu^+) \quad \longrightarrow \quad \underset{\text{precipitate}}{CH_3-CH_2-C{\equiv}C-Ag\downarrow \text{ (or Cu)}} \quad + \quad H^+$$

$$CH_3-C{\equiv}C-CH_3 \quad + \quad Ag^+ \text{ (or } Cu^+) \quad \longrightarrow \quad \text{no reaction}$$

This qualitative test commonly uses $AgNO_3$ or $CuNO_3$, often in an alcoholic solution, or ammonia complexes of Ag(I) and Cu(I) ions made by adding a small amount of aqueous ammonia to the solution of silver or cuprous nitrate.

$$Ag^+ \; ^-NO_3 \quad + \quad 2\,NH_3 \quad \longrightarrow \quad Ag(NH_3)_2^+ \; ^-NO_3$$

$$Cu^+ \; ^-NO_3 \quad + \quad 2\,NH_3 \quad \longrightarrow \quad Cu(NH_3)_2^+ \; ^-NO_3$$

In addition to its use as a qualitative test, the formation of heavy metal acetylides can be used for the purification of alkynes—for example, when a mixture of terminal and internal alkyne isomers is formed in a double dehydrohalogenation (see Problem 14-7). Addition of Ag(I) or Cu(I) causes the terminal alkyne to precipitate as its acetylide, which can be filtered off from the internal alkyne.

$$R-CH_2-C{\equiv}C-H \quad + \quad Cu^+ \quad \longrightarrow \quad R-CH_2-C{\equiv}C-Cu\downarrow \text{ (precipitate, filtered out)}$$

$$R-C{\equiv}C-CH_3 \quad + \quad Cu^+ \quad \longrightarrow \quad \text{no reaction, remains in solution}$$

Addition of dilute acid regenerates the terminal alkyne from its acetylide.

$$R-CH_2-C{\equiv}C-Cu \quad + \quad HCl \quad \longrightarrow \quad R-CH_2-C{\equiv}C-H \quad + \quad CuCl$$

Because they are not strong nucleophiles, silver and copper acetylides are not commonly used in the alkylation and carbonyl addition reactions discussed in the next section. These heavy-metal acetylides tend to explode when dry, so they are always acidified while still wet.

14-8 SYNTHESIS OF ALKYNES FROM ACETYLIDES

Two different approaches are commonly used for the synthesis of alkynes. In the first, an appropriate electrophile undergoes nucleophilic attack by an acetylide ion. The electrophile may be an unhindered primary alkyl halide (undergoes S_N2), or it may be a carbonyl compound (undergoes attack similar to a Grignard reaction). This is a general approach used in most laboratory syntheses of alkynes.

The second approach forms the triple bond by a double dehydrohalogenation of a dihalide. Isomerization of the triple bond may occur (see Section 14-9), so dehydrohalogenation is useful only when the desired product has the triple bond in a thermodynamically favored position.

14-8A ALKYLATION OF ACETYLIDE IONS

An acetylide ion is a strong base and a powerful nucleophile. It can displace a halide or tosylate ion from a suitable substrate, giving a substituted acetylene.

$$R-C{\equiv}C:^- \quad + \quad R'-X \quad \xrightarrow{S_N2} \quad R-C{\equiv}C-R' \quad + \quad X^-$$

(R′—X must be a primary alkyl halide or tosylate)

If this S_N2 reaction is to produce a good yield of the substitution product, the alkyl halide must be an excellent S_N2 substrate: It must be primary, and there should be no bulky substituents or branches close to the reaction center. Following are examples of alkyne syntheses using displacement of primary halides by acetylide ions.

$$\underset{\text{sodium acetylide}}{H{-}C{\equiv}C{:}^{-}\ {}^{+}Na} + \underset{\text{1-bromobutane}}{CH_3CH_2CH_2CH_2{-}Br} \longrightarrow \underset{\substack{\text{1-hexyne}\\ \text{(butylacetylene)}\\ (75\%)}}{H{-}C{\equiv}C{-}CH_2CH_2CH_2CH_3} + NaBr$$

$$\underset{\substack{\text{cyclohexylacetylene}\\ \text{(ethynylcyclohexane)}}}{C_6H_{11}{-}C{\equiv}C{-}H} \xrightarrow[\text{(2) ethyl bromide}]{\text{(1) } NaNH_2} \underset{\substack{\text{ethylcyclohexylacetylene}\\ \text{(1-cyclohexyl-1-butyne)}\\ (70\%)}}{C_6H_{11}{-}C{\equiv}C{-}CH_2CH_3}$$

If the back-side approach is hindered, the acetylide ion may abstract a proton, giving elimination by the E2 mechanism.

$$\underset{\text{butynide ion}}{CH_3CH_2{-}C{\equiv}C{:}^{-}} + \underset{\text{isopropyl bromide}}{H_3C{-}\overset{\overset{\displaystyle Br}{|}}{CH}{-}CH_3} \longrightarrow \underset{\text{butyne}}{CH_3CH_2{-}C{\equiv}C{-}H} + \underset{\text{propene}}{H_2C{=}CH{-}CH_3} + Br^{-}$$

SOLVED PROBLEM 14-1

Show how to synthesize 3-decyne from acetylene and any necessary alkyl halides.

SOLUTION Another name for 3-decyne is ethyl *n*-hexylacetylene. It can be made by adding an ethyl group and a hexyl group to acetylene. This can be done in either order; we will begin by adding the hexyl group.

$$\underset{\text{acetylene}}{H{-}C{\equiv}C{-}H} \xrightarrow[\text{(2) } CH_3(CH_2)_5Br]{\text{(1) } NaNH_2} \underset{\text{1-octyne}}{CH_3(CH_2)_5{-}C{\equiv}C{-}H}$$

$$\underset{\text{1-octyne}}{CH_3(CH_2)_5{-}C{\equiv}C{-}H} \xrightarrow[\text{(2) } CH_3CH_2Br]{\text{(1) } NaNH_2} \underset{\text{3-decyne}}{CH_3(CH_2)_5{-}C{\equiv}C{-}CH_2CH_3}$$

PROBLEM 14-7

Show the reagents and intermediates involved in the synthesis of 3-decyne by adding the ethyl group first and the hexyl group last.

PROBLEM 14-8

Show how you might synthesize each of the following compounds, using acetylene and any suitable alkyl halides as your starting materials. If the compound given cannot be synthesized by this method, explain why.

(a) 1-hexyne (b) 2-hexyne
(c) 3-hexyne (d) 4-methyl-2-hexyne
(e) 5-methyl-2-hexyne (f) cyclodecyne

14-8B ADDITION OF ACETYLIDE IONS TO CARBONYL GROUPS AND EPOXIDES

Acetylide ions react like Grignard and organolithium reagents in most of the standard Grignard syntheses, providing convenient syntheses of primary, secondary, and tertiary acetylenic alcohols. Following is a brief summary of the use of acetylide ions in Grignard reactions.

Reaction with ketones and aldehydes

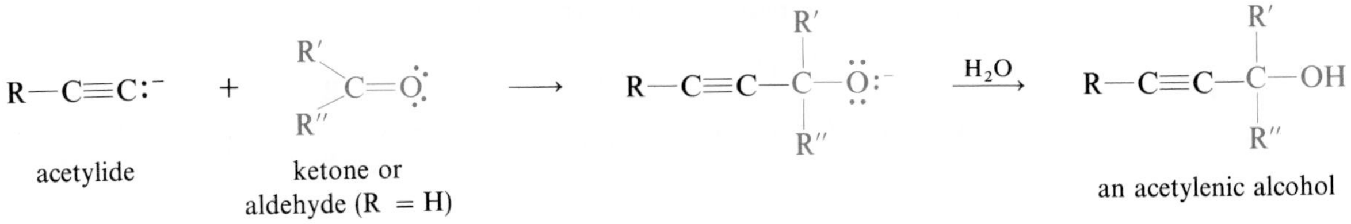

$$R{-}C{\equiv}C{:}^- \; + \; \underset{R''}{\overset{R'}{}}C{=}\ddot{O}{:} \longrightarrow R{-}C{\equiv}C{-}\underset{R''}{\overset{R'}{C}}{-}\ddot{O}{:}^- \xrightarrow{H_2O} R{-}C{\equiv}C{-}\underset{R''}{\overset{R'}{C}}{-}OH$$

acetylide; ketone or aldehyde (R = H); an acetylenic alcohol

Examples

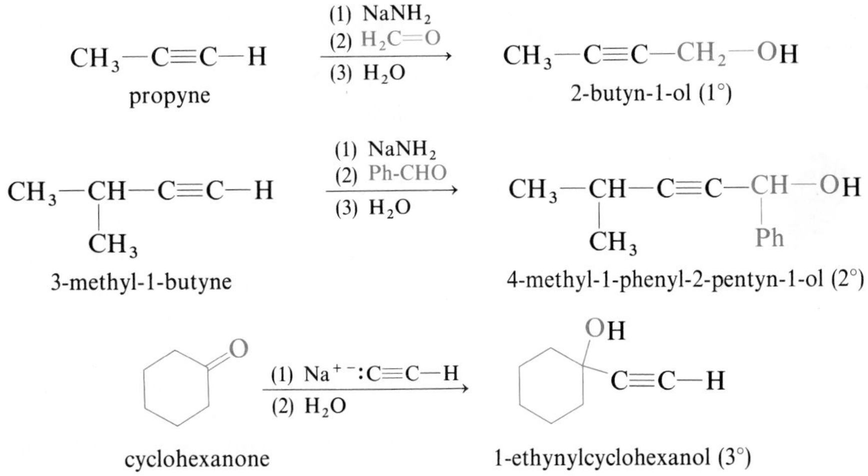

$$CH_3{-}C{\equiv}C{-}H \xrightarrow[(3)\ H_2O]{(1)\ NaNH_2 \quad (2)\ H_2C{=}O} CH_3{-}C{\equiv}C{-}CH_2{-}OH$$

propyne → 2-butyn-1-ol (1°)

$$CH_3{-}\underset{CH_3}{CH}{-}C{\equiv}C{-}H \xrightarrow[(3)\ H_2O]{(1)\ NaNH_2 \quad (2)\ Ph\text{-}CHO} CH_3{-}\underset{CH_3}{CH}{-}C{\equiv}C{-}\underset{Ph}{CH}{-}OH$$

3-methyl-1-butyne → 4-methyl-1-phenyl-2-pentyn-1-ol (2°)

$$\text{cyclohexanone} \xrightarrow[(2)\ H_2O]{(1)\ Na^{+}\ {}^{-}{:}C{\equiv}C{-}H} \text{1-ethynylcyclohexanol (OH and } C{\equiv}C{-}H \text{ on the same ring carbon)}$$

cyclohexanone → 1-ethynylcyclohexanol (3°)

Reaction with epoxides

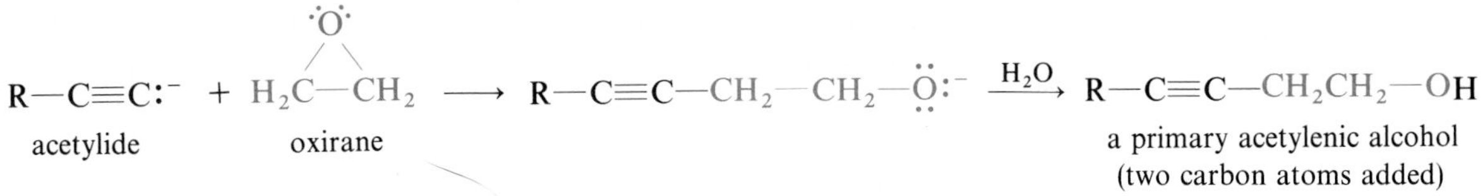

$$R{-}C{\equiv}C{:}^- + H_2C{-}CH_2 \text{ (with } \ddot{O} \text{ bridging)} \longrightarrow R{-}C{\equiv}C{-}CH_2{-}CH_2{-}\ddot{O}{:}^- \xrightarrow{H_2O} R{-}C{\equiv}C{-}CH_2CH_2{-}OH$$

acetylide; oxirane; a primary acetylenic alcohol (two carbon atoms added)

Example

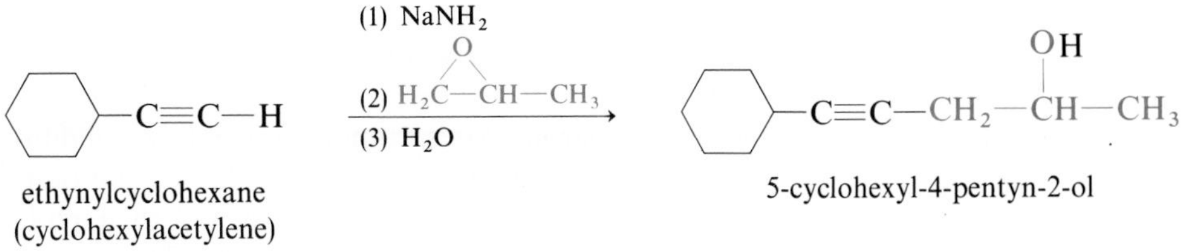

$$\text{C}_6\text{H}_{11}{-}C{\equiv}C{-}H \xrightarrow[(3)\ H_2O]{(1)\ NaNH_2 \quad (2)\ H_2C{-}CH{-}CH_3 \text{ (epoxide, O)}} \text{C}_6\text{H}_{11}{-}C{\equiv}C{-}CH_2{-}\underset{}{\overset{OH}{CH}}{-}CH_3$$

ethynylcyclohexane (cyclohexylacetylene) → 5-cyclohexyl-4-pentyn-2-ol

PROBLEM 14-9

For each of the following acetylenic compounds, propose a synthesis that begins with acetylene and uses whatever additional reagents are necessary.

(a) 2-heptyn-4-ol
(b) 3-butyn-1-ol
(c) 2-propyn-1-ol (propargyl alcohol)
(d) 3-methyl-4-hexyn-3-ol

14-9 SYNTHESIS OF ALKYNES BY ELIMINATION REACTIONS

It is sometimes possible to generate a carbon-carbon triple bond by eliminating two molecules of HX from a dihalide. Dehydrohalogenation of a *geminal* or *vicinal* dihalide gives a vinyl halide. Under strongly basic conditions, a second dehydrohalogenation may occur to form an alkyne.

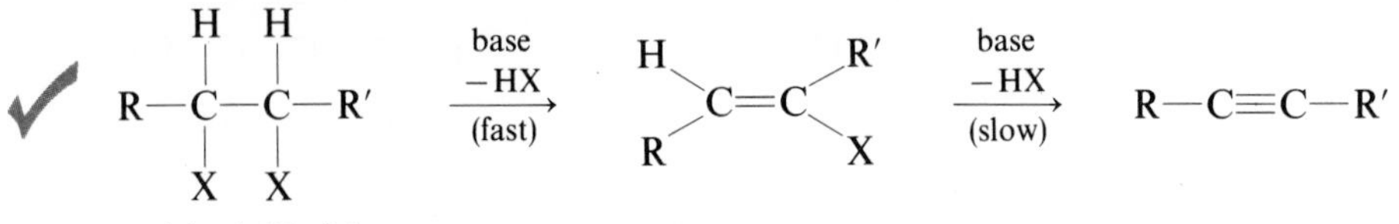

$$\underset{\text{a vicinal dihalide}}{R-\underset{X}{\overset{H}{C}}-\underset{X}{\overset{H}{C}}-R'} \xrightarrow[\text{(fast)}]{\text{base}\atop -HX} \underset{R}{\overset{H}{}}\!\!>C=C<\!\!\underset{X}{\overset{R'}{}} \xrightarrow[\text{(slow)}]{\text{base}\atop -HX} R-C\equiv C-R'$$

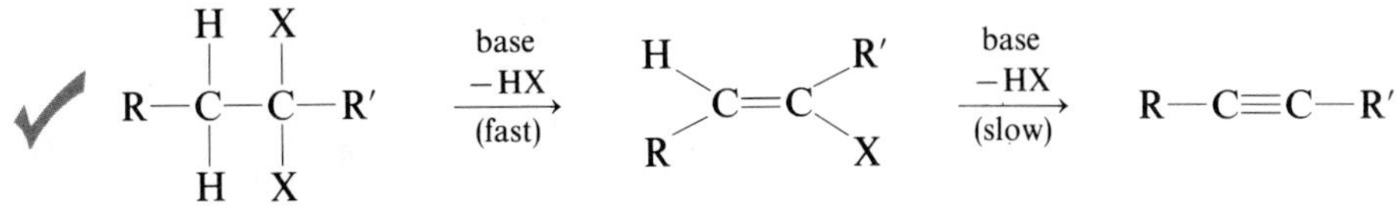

$$\underset{\text{a geminal dihalide}}{R-\underset{H}{\overset{H}{C}}-\underset{X}{\overset{X}{C}}-R'} \xrightarrow[\text{(fast)}]{\text{base}\atop -HX} \underset{R}{\overset{H}{}}\!\!>C=C<\!\!\underset{X}{\overset{R'}{}} \xrightarrow[\text{(slow)}]{\text{base}\atop -HX} R-C\equiv C-R'$$

Conditions for elimination We have already seen (Section 7-9A) many examples of the dehydrohalogenation of alkyl halides. The second step in the preceding reactions is new, however, because it involves the dehydrohalogenation of a vinyl halide to give an alkyne. This second dehydrohalogenation occurs only under extremely basic conditions. It usually involves molten KOH or alcoholic KOH in a sealed tube, usually heated to temperatures close to 200°C. Sodium amide is also used for the double dehydrohalogenation. Since the amide ion ($^{-}:NH_2$) is a much stronger base than hydroxide, the amide reaction takes place at a lower temperature than the reaction using KOH. The following reactions are carefully chosen to form products that do not rearrange (see below).

$$\underset{\text{2,3-dibromopentane}}{CH_3-CH_2-\overset{Br}{\overset{|}{C}H}-\overset{Br}{\overset{|}{C}H}-CH_3} \xrightarrow[200°C]{\text{KOH (fused)}} \underset{\text{2-pentyne (45\%)}}{CH_3-CH_2-C\equiv C-CH_3}$$

$$\underset{\text{1,1-dichloropentane}}{CH_3-CH_2-CH_2-CH_2-CHCl_2} \xrightarrow[(2)\ H_2O]{(1)\ NaNH_2,\ 150°} \underset{\text{1-pentyne (55\%)}}{CH_3-CH_2-CH_2-C\equiv C-H}$$

Base-catalyzed rearrangements Unfortunately, the double dehydrohalogenation is limited by the severe conditions required. Any functional groups that are sensitive to strong bases cannot survive these conditions; also, the alkyne products may rearrange under these extremely basic conditions. Figure 14-1 shows how the loss of protons at one carbon atom and their replacement in another position leads to isomerization of the triple bond. This ability to isomerize implies that all the possible triple-bond isomers will equilibrate, and the most stable isomer will predominate. The most stable alkyne isomer is generally the one with the most highly substituted triple bond.

Any of several isomers of dibromopentane give 2-pentyne upon dehydrohalogenation with fused KOH at 200°C. In each case the alkyne formed initially rearranges to the most stable isomer, 2-pentyne.

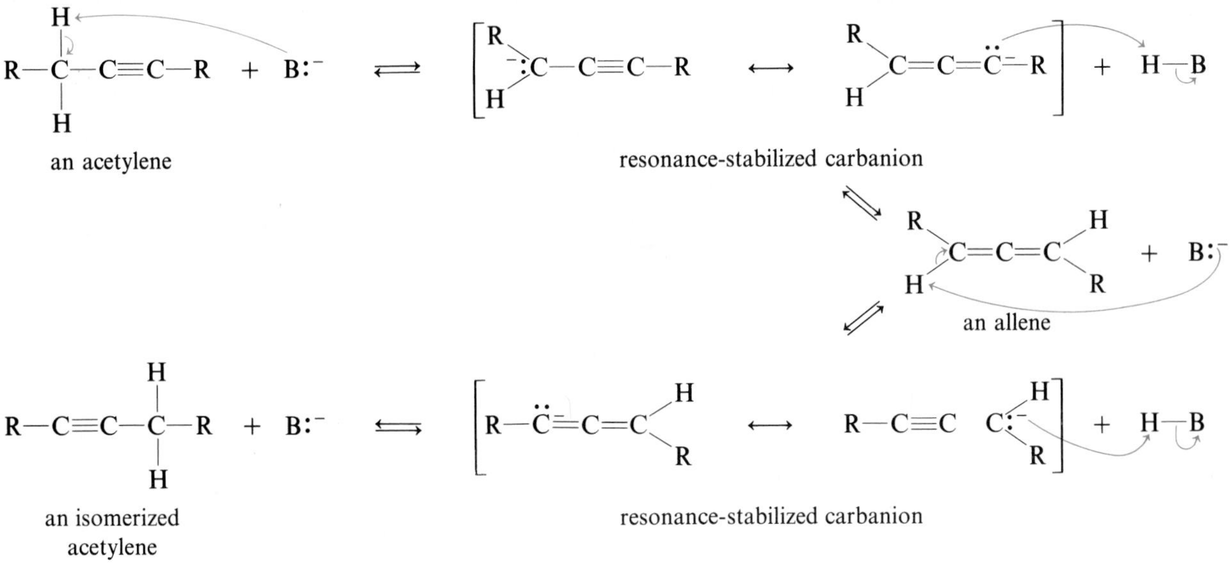

$$R-\underset{H}{\overset{H}{C}}-C\equiv C-R + B:^- \rightleftharpoons \left[\begin{matrix} R \\ {}^-:C-C\equiv C-R \\ H \end{matrix} \longleftrightarrow \begin{matrix} R \\ C=C=\ddot{C}^--R \\ H \end{matrix} \right] + H-B$$

an acetylene

resonance-stabilized carbanion

$$\rightleftharpoons \quad \begin{matrix} R & & H \\ & C=C=C & \\ H & & R \end{matrix} + B:^-$$

an allene

$$R-C\equiv C-\underset{H}{\overset{H}{C}}-R + B:^- \rightleftharpoons \left[R-\ddot{C}^-=C=C\begin{matrix} H \\ R \end{matrix} \longleftrightarrow R-C\equiv C\ C\begin{matrix} H \\ :^- \\ R \end{matrix} \right] + H-B$$

an isomerized acetylene

resonance-stabilized carbanion

FIGURE 14-1 Under extremely basic conditions, an acetylenic triple bond can migrate along the carbon chain by repeated deprotonation and reprotonation.

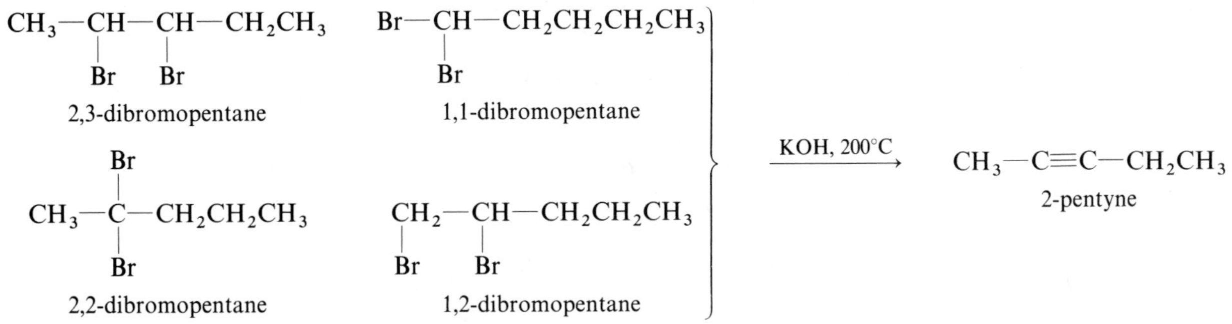

$$\left.\begin{matrix} CH_3-\underset{Br}{CH}-\underset{Br}{CH}-CH_2CH_3 & Br-\underset{Br}{CH}-CH_2CH_2CH_2CH_3 \\ \text{2,3-dibromopentane} & \text{1,1-dibromopentane} \\ CH_3-\underset{Br}{\overset{Br}{C}}-CH_2CH_2CH_3 & \underset{Br}{CH_2}-\underset{Br}{CH}-CH_2CH_2CH_3 \\ \text{2,2-dibromopentane} & \text{1,2-dibromopentane} \end{matrix}\right\} \xrightarrow{\text{KOH, 200°C}} \underset{\text{2-pentyne}}{CH_3-C\equiv C-CH_2CH_3}$$

PROBLEM 14-10

Give a mechanism to show how 1,1-dibromopentane reacts with fused KOH at 200°C to give 2-pentyne.

$$Br-\overset{Br}{CH}-CH_2CH_2CH_2CH_3 \xrightarrow{\text{KOH, 200°C}} CH_3-C\equiv C-CH_2CH_3$$

PROBLEM 14-11*

Using heats of hydrogenation, we can show that most internal alkynes are about 4 kcal (17 kJ) per mole more stable than their corresponding terminal alkynes. Calculate the ratio of terminal alkyne to internal alkyne present at equilibrium at 200°C.

Isomerization also results when sodium amide is used as the base in the double dehydrohalogenation. All possible triple-bond isomers are formed, but sodium amide is such a strong base that it deprotonates the terminal acetylene, removing it from the equilibrium and making the acetylide ion the favored product. When water is added to quench the reaction, the acetylide ion is protonated to give the terminal alkyne.

$$\underset{\text{one component of the mixture}}{R-C\equiv C-H} + {}^{-}:\ddot{N}H_2 \rightleftharpoons \underset{\substack{\text{acetylide ion}\\\text{(major component)}}}{R-C\equiv C:^{-}} + :NH_3 \xrightarrow{H_2O} \underset{\text{major product}}{R-C\equiv C-H}$$

$$\underset{\text{2-octyne}}{CH_3(CH_2)_4-C\equiv C-CH_3} \xrightarrow[\text{(2) } H_2O]{\text{(1) } NaNH_2\text{, 150°C}} \underset{\substack{\text{1-octyne}\\(80\%)}}{CH_3(CH_2)_4-CH_2-C\equiv C-H}$$

SUMMARY OF DEHYDROHALOGENATIONS TO FORM ALKYNES

$$\text{R(H)C=C(X)R'} \xrightarrow[\text{(2) } H_2O]{\text{(1) } NaNH_2\text{, 150°}} \text{terminal alkyne}$$

$$\text{R(H)C=C(X)R'} \xrightarrow{NaOH\text{, 200°}} \text{most stable internal alkyne}$$

PROBLEM 14-12

(a) Give a mechanism to show how 2-pentyne reacts with sodium amide to give 1-pentyne.
(b) Explain how this reaction converts a more stable isomer (2-pentyne) to a less stable isomer (1-pentyne).
(c) How would you accomplish the opposite reaction, converting 1-pentyne into 2-pentyne?

PROBLEM 14-13

Show which of the following compounds could be synthesized in good yield by a double dehydrohalogenation from a dihalide. In each case

(1) Show which base you would use (KOH or $NaNH_2$).
(2) Show how your starting material might be synthesized from an alkene.

(a) 2-butyne (b) 1-octyne (c) 2-octyne (d) cyclodecyne

SUMMARY OF THE SYNTHESIS OF ALKYNES

1. *Alkylation of acetylide ions* (Section 14-8A)

$$R-C\equiv C:^{-} + R'-X \xrightarrow{S_N2} R-C\equiv C-R' + X^{-}$$

(R′—X must be an unhindered primary halide or tosylate)

Example

$$\underset{\text{sodium propynide}}{H_3C-C\equiv C:^{-}\,{}^{+}Na} + \underset{\text{1-bromopropane}}{CH_3CH_2CH_2-Br} \longrightarrow \underset{\text{2-hexyne}}{H_3C-C\equiv C-CH_2CH_2CH_3}$$

2. *Additions to carbonyl groups and epoxides* (Section 14-8B)

$$R-C\equiv C:^{-} + R'_2C=\ddot{O}: \longrightarrow R-C\equiv C-CR'_2-\ddot{O}:^{-} \xrightarrow{H_2O} R-C\equiv C-CR'_2-OH$$

$$\underset{\text{acetylide}}{R-C\equiv C:^{-}} + \underset{\text{oxirane}}{H_2C\overset{\ddot{O}}{—}CH_2} \longrightarrow$$

$$R-C\equiv C-CH_2-CH_2-\ddot{O}:^{-} \xrightarrow{H_2O} \underset{\text{primary acetylenic alcohol}}{R-C\equiv C-CH_2CH_2-OH}$$

Examples

$$\underset{\text{sodium acetylide}}{H-C\equiv C:^-} + \underset{\text{propanal}}{CH_3CH_2-\overset{\overset{\displaystyle O}{\|}}{C}-H} \xrightarrow[(2)\ H_2O]{} \underset{\text{1-pentyn-2-ol}}{H-C\equiv C-\overset{\overset{\displaystyle OH}{|}}{CH}-CH_2CH_3}$$

$$\underset{\text{sodium propynide}}{CH_3-C\equiv C:^-\ ^+Na} + \underset{\text{oxirane}}{H_2\overset{O}{C-}CH_2} \xrightarrow[(2)\ H_2O]{} \underset{\text{3-pentyn-1-ol}}{CH_3-C\equiv C-CH_2-CH_2-OH}$$

3. *Double dehydrohalogenation of alkyl dihalides* (Section 14-9)

$$R-\underset{H}{\overset{X}{C}}-\underset{H}{\overset{X}{C}}-R' \text{ or } R-\underset{H}{\overset{H}{C}}-\underset{X}{\overset{X}{C}}-R' \xrightarrow[\text{or } NaNH_2]{\text{fused KOH}} R-C\equiv C-R'$$

(KOH forms internal alkynes; $NaNH_2$ forms terminal alkynes.)

Examples

$$\underset{\text{2,2-dichloropentane}}{CH_3CH_2-CH_2-CCl_2-CH_3} + \text{KOH (fused)} \longrightarrow \underset{\text{2-pentyne}}{CH_3CH_2-C\equiv C-CH_3}$$

$$\underset{\text{2,2-dichloropentane}}{CH_3CH_2-CH_2-CCl_2-CH_3} + NaNH_2 \longrightarrow \underset{\text{1-pentyne}}{CH_3CH_2-CH_2-C\equiv C-H}$$

14-10 ADDITION REACTIONS OF ALKYNES

We have already discussed some of the most important reactions of alkynes. The nucleophilic attack of acetylide ions on electrophiles, for example, is one of the best methods for making more complicated alkynes. (Section 14-8). Now we consider reactions that involve transformations of the carbon-carbon triple bond itself.

Many of the reactions of alkynes are similar to the corresponding reactions of alkenes, because both involve pi bonds between two carbon atoms. Like the pi bond of an alkene, the pi bond of an alkyne readily undergoes addition reactions. Table 14-3 shows how the energy difference between the kinds of carbon-carbon bonds can be used to estimate how much energy it takes to break a particular bond. The bond energy of the alkyne triple bond is only about 54 kcal (226 kJ) more than the bond energy of an alkene double bond. This is the energy needed to break one of the pi bonds of an alkyne.

TABLE 14-3
Approximate bond energies of carbon-carbon bonds

Bond	*Total energy*	*Class of bond*	*Approximate energy*
C—C	83 kcal (347 kJ)	alkane sigma bond	83 kcal (347 kJ)
C=C	146 kcal (611 kJ)	alkene pi bond	63 kcal (264 kJ)
C≡C	200 kcal (837 kJ)	second alkyne pi bond	54 kcal (226 kJ)

Reagents add across the triple bonds of alkynes just as they add across the double bonds of alkenes. In effect, this reaction converts a pi bond into a sigma bond. Since sigma bonds are generally stronger than pi bonds, the reaction is usually exothermic. Alkynes have two pi bonds, so one or two molecules can add across the triple bond, depending on the reagents and the conditions.

$$\underset{\pi \text{ bond}}{R-C\equiv C-R} + \underset{\sigma \text{ bond}}{A-B} \longrightarrow \underset{2\sigma \text{ bonds}}{\overset{A\quad\quad B}{\underset{R\quad\quad R}{C=C}}} \xrightarrow{A-B} R-\overset{A}{\underset{A}{C}}-\overset{B}{\underset{B}{C}}-R'$$

We must consider the possibility of a double addition whenever we add a reagent across the triple bond of an alkyne. Some conditions might allow the reaction to stop after a single addition, while other conditions give double addition.

14-10A ADDITION OF HYDROGEN TO ALKYNES (REDUCTION TO ALKANES)

In the presence of a suitable catalyst, hydrogen adds to an alkyne, reducing it to an alkane. For example, when either of the butyne isomers reacts with hydrogen and a platinum catalyst, the product is *n*-butane. Platinum, palladium, and nickel catalysts are commonly used in this reduction.

$$R-C\equiv C-R + 2\,H_2 \xrightarrow{\text{Pt, Pd, or Ni}} R-\overset{H}{\underset{H}{C}}-\overset{H}{\underset{H}{C}}-R'$$

Examples

$$\underset{\text{1-butyne}}{H-C\equiv C-CH_2CH_3} + 2\,H_2 \xrightarrow{\text{Pt}} \underset{\substack{\text{butane}\\(100\%)}}{H-CH_2-CH_2-CH_2CH_3}$$

$$\underset{\text{2-butyne}}{CH_3-C\equiv C-CH_3} + 2\,H_2 \xrightarrow{\text{Pt}} \underset{\substack{\text{butane}\\(100\%)}}{CH_3-CH_2-CH_2-CH_3}$$

Catalytic hydrogenation takes place in two steps, with an alkene intermediate. With efficient catalysts such as platinum, palladium, or nickel, it is usually impossible to stop the reaction at the alkene stage.

$$R-C\equiv C-R' \xrightarrow{H_2,\ \text{Pt}} \left[\overset{R\quad\quad R'}{\underset{H\quad\quad H}{C=C}}\right] \xrightarrow{H_2,\ \text{Pt}} R-\overset{H}{\underset{H}{C}}-\overset{H}{\underset{H}{C}}-R$$

14-10B HYDROGENATION TO CIS ALKENES

The hydrogenation of an alkyne can be stopped at the alkene stage by using a "poisoned" catalyst, made by treating a good catalyst with a compound that makes the catalyst less effective. **Lindlar's catalyst** is a poisoned palladium catalyst, composed of powdered barium sulfate coated with palladium, poisoned with quinoline.

$$\underset{\text{alkyne}}{R-C\equiv C-R'} \xrightarrow[\substack{\text{quinoline, } CH_3OH\\ \text{(Lindlar's catalyst)}}]{H_2,\ Pd/BaSO_4} \underset{\text{cis alkene}}{\overset{R\quad\quad R'}{\underset{H\quad\quad H}{C=C}}}$$

The catalytic hydrogenation of alkynes is similar to the hydrogenation of alkenes, and both proceed with syn stereochemistry. In catalytic hydrogenation, the face of a pi bond contacts the solid catalyst, and the catalyst weakens the pi bond, allowing two hydrogen atoms to add (Fig. 14-2). This simultaneous (or nearly simultaneous) addition of two hydrogen atoms on the same face of the alkyne ensures syn stereochemistry.

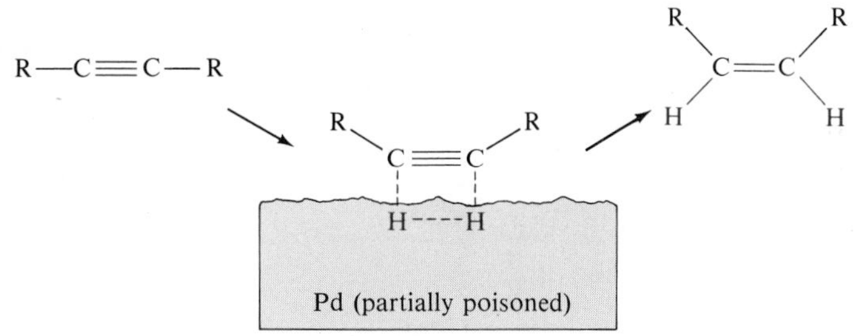

R—C≡C—R → (R)C≡C(R) on surface with H----H, Pd (partially poisoned) → (R)(H)C=C(R)(H)

FIGURE 14-2 Catalytic hydrogenation of alkynes using the Lindlar catalyst.

In an internal alkyne, the syn addition always gives a cis product. For example, when 2-hexyne is hydrogenated using the Lindlar catalyst, the product is *cis*-2-hexene.

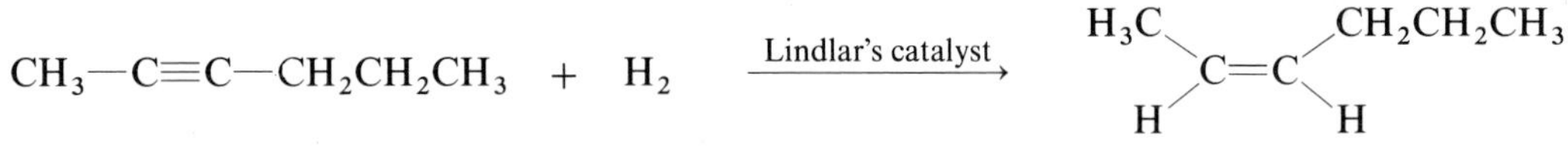

$$CH_3{-}C{\equiv}C{-}CH_2CH_2CH_3 + H_2 \xrightarrow{\text{Lindlar's catalyst}} (H_3C)(H)C{=}C(CH_2CH_2CH_3)(H)$$

2-hexyne → *cis*-2-hexene

$$H{-}C{\equiv}C{-}CH_2CH_2CH_2CH_3 + H_2 \xrightarrow{\text{Lindlar's catalyst}} (H)(H)C{=}C((CH_2)_3CH_3)(H)$$

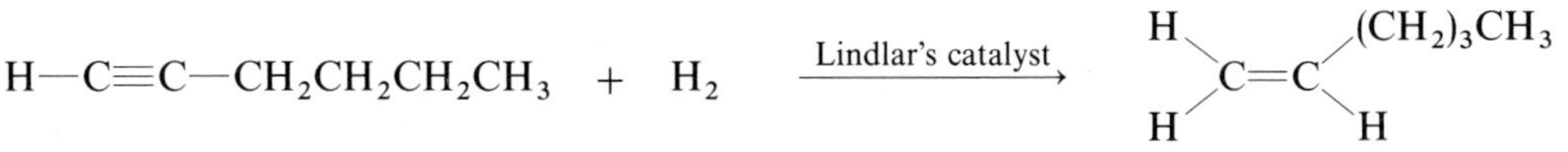

1-hexyne → 1-hexene

14-10C REDUCTION TO TRANS ALKENES

To form a trans alkene, the addition to the alkyne must take place with anti stereochemistry. Sodium metal in liquid ammonia reduces alkynes with anti stereochemistry, and this reduction is used to convert alkynes to trans alkenes.

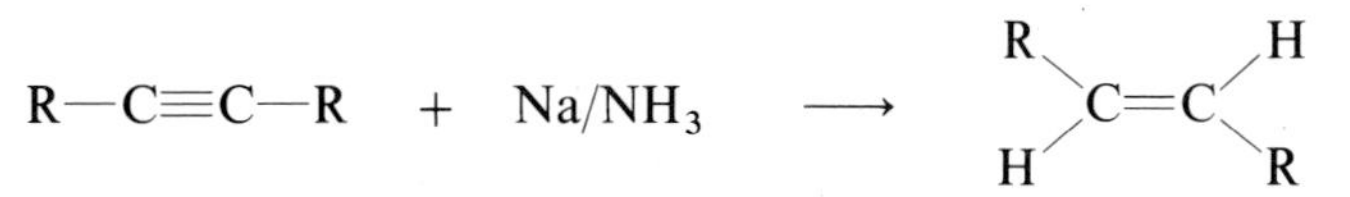

✓

$$R{-}C{\equiv}C{-}R + Na/NH_3 \longrightarrow (R)(H)C{=}C(H)(R)$$

alkyne → trans alkene

Example

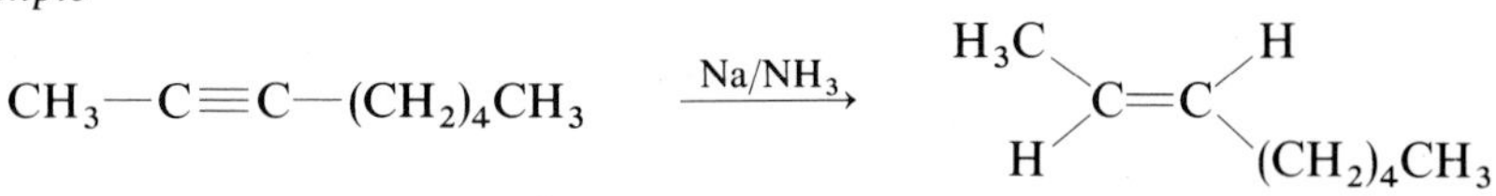

$$CH_3{-}C{\equiv}C{-}(CH_2)_4CH_3 \xrightarrow{Na/NH_3} (H_3C)(H)C{=}C(H)((CH_2)_4CH_3)$$

2-octyne → *trans*-2-octene (80%)

Ammonia (b.p. −33°C) is a gas at room temperature, but it is kept liquid by using dry ice to cool the reaction vessel. Sodium dissolves in liquid ammonia to produce the deep blue color of solvated electrons. Reduction is the gain of electrons, and it is these solvated electrons that actually reduce the alkyne.

$$NH_3 + Na \longrightarrow NH_3{\cdot}e^- \text{ (deep blue solution)} + Na^+$$

solvated electron

The reduction proceeds by addition of an electron to the alkyne to form a radical anion, followed by protonation to give a neutral radical. Protons are provided by the ammonia solvent or by an alcohol added as a cosolvent. Addition of another electron, followed by another proton, gives the product.

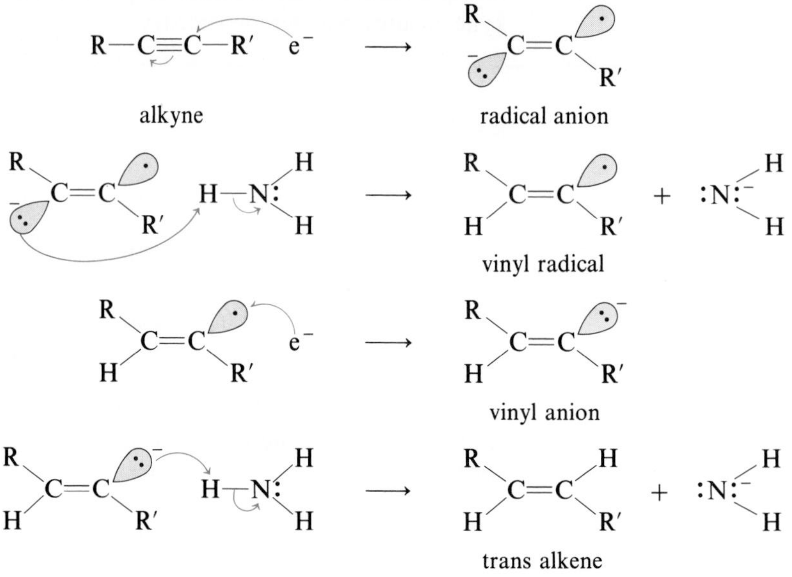

$$\text{R—C}\equiv\text{C—R}' \quad e^- \longrightarrow \text{radical anion (R(}^-\text{)C=C(}\cdot\text{)R}')$$

alkyne — radical anion

$$\text{R(}^-\text{)C=C(}\cdot\text{)R}' + \text{H—NH}_2 \longrightarrow \text{R(H)C=C(}\cdot\text{)R}' + :\ddot{N}H_2^-$$

vinyl radical

$$\text{R(H)C=C(}\cdot\text{)R}' \quad e^- \longrightarrow \text{R(H)C=C(}^-\text{)R}'$$

vinyl anion

$$\text{R(H)C=C(}^-\text{)R}' + \text{H—NH}_2 \longrightarrow \text{R(H)C=C(H)R}' + :\ddot{N}H_2^-$$

trans alkene

The anti stereochemistry of the sodium-ammonia reduction appears to result from the greater stability of the vinyl radical in the trans configuration. An electron is added to the trans radical to give a trans vinyl anion, which is quickly protonated to the trans alkene.

PROBLEM 14-14
Show how you would convert

(a) 2-pentyne to *cis*-2-pentene
(b) 2-pentyne to *trans*-2-pentene
(c) *cis*-cyclodecene to *trans*-cyclodecene.
(d) *trans*-3-octene to *cis*-3-octene.

14-10D ADDITION OF HALOGENS

Bromine and chlorine add to alkynes just as they add to alkenes. If 1 mole of halogen adds to an alkyne, the product is a dihaloalkene. The stereochemistry of addition may be either syn or anti, and the products are often mixtures of cis and trans isomers.

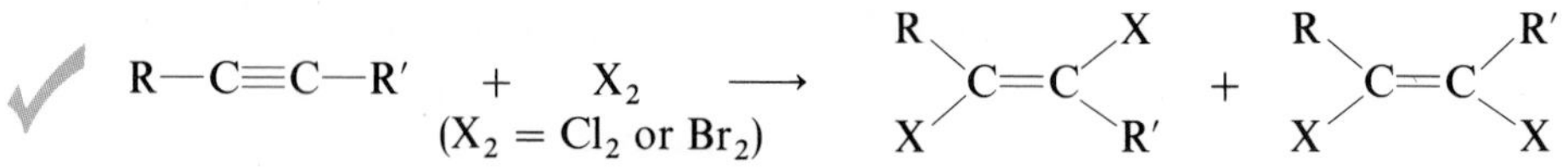

$$\text{R—C}\equiv\text{C—R}' + X_2 \longrightarrow \text{R(X)C=C(X)R}' + \text{R(X)C=C(R}'\text{)X}$$
$$(X_2 = Cl_2 \text{ or } Br_2)$$

Example

$$CH_3(CH_2)_3\text{—C}\equiv\text{C—H} + Br_2 \longrightarrow CH_3(CH_2)_3(Br)C{=}C(Br)H \; (72\%) + CH_3(CH_2)_3(Br)C{=}C(H)Br \; (28\%)$$

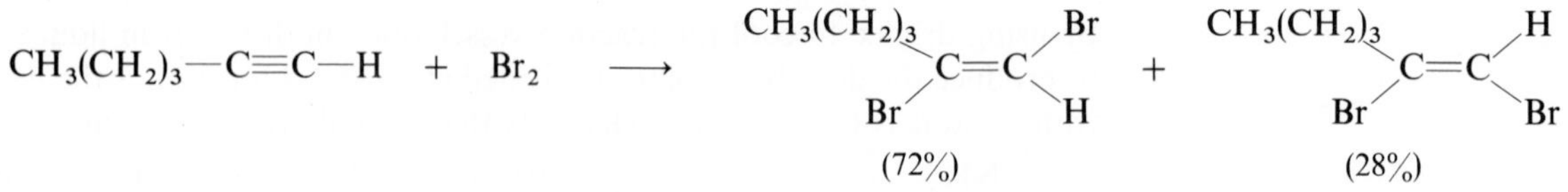

If 2 moles of halogen add to an alkyne, a tetrahalide results. Sometimes it is difficult to keep the reaction from proceeding all the way to the tetrahalide even when we want it to stop at the dihalide.

$$\text{R—C}{\equiv}\text{C—R}' \quad + \quad 2\,X_2 \quad \longrightarrow \quad \text{R—CX}_2\text{—CX}_2\text{—R}'$$
$$(X = \text{Cl or Br})$$

Example

$$CH_3(CH_2)_3\text{—C}{\equiv}\text{C—H} \quad + \quad 2\,Cl_2 \quad \longrightarrow \quad CH_3(CH_2)_3\text{—CCl}_2\text{—CCl}_2\text{—H}$$
$$(100\%)$$

PROBLEM 14-15
In the addition of just 1 mole of bromine to 1-hexyne, should the 1-hexyne be added to a bromine solution or should the bromine be added to the 1-hexyne? Explain your answer.

14-10E ADDITION OF HYDROGEN HALIDES

Hydrogen halides add across the alkyne triple bond in much the same way that they add across the alkene double bond. The initial product is a vinyl halide. When a hydrogen halide adds to a terminal alkyne, the product has the orientation predicted by Markovnikov's rule. A second molecule of HX can add, usually with the same orientation as the first.

$$\text{R—C}{\equiv}\text{C—H} + \boxed{\text{H}}\text{—X} \longrightarrow \text{(R)(X)C}{=}\text{C(H)}\boxed{\text{H}} \xrightarrow{\boxed{\text{H}}\text{—X}} \text{R—CX}_2\text{—C}\boxed{\text{H}}\boxed{\text{H}}\text{—H}$$
$$(HX = \text{HCl, HBr, or HI})$$

For example, when 1-pentyne reacts with HBr, the Markovnikov product is obtained. In an internal alkyne such as 2-pentyne, however, the acetylenic carbon atoms are equally substituted, and a mixture of products results.

$$\underset{\text{1-pentyne}}{\text{H—C}{\equiv}\text{C—}CH_2CH_2CH_3} \quad + \quad \text{HBr} \quad \longrightarrow \quad \underset{\substack{\text{2-bromo-1-pentene} \\ \text{(Markovnikov product)}}}{H_2C{=}C(Br)CH_2CH_2CH_3}$$

$$\underset{\text{2-pentyne}}{CH_3\text{—C}{\equiv}\text{C—}CH_2CH_3} \quad + \quad \text{HBr} \quad \longrightarrow \quad \underset{\substack{\text{2-bromo-2-pentene} \\ [(E)\text{ and }(Z)\text{ isomers}]}}{CH_3\text{—C(Br)}{=}\text{C(H)—}CH_2CH_3} \quad + \quad \underset{\substack{\text{3-bromo-2-pentene} \\ [(E)\text{ and }(Z)\text{ isomers}]}}{CH_3\text{—C(H)}{=}\text{C(Br)—}CH_2CH_3}$$

The mechanism of this reaction is similar to the mechanism of hydrogen halide addition to alkenes. The **vinyl cation** formed in the first step is more stable with the positive charge on the more highly substituted carbon atom. Attack by halide ion completes the reaction.

$$\text{R—C}\equiv\text{C—H} + \boxed{\text{H—X}} \longrightarrow \text{R—}\overset{+}{\text{C}}\text{=C(H)}\boxed{\text{H}} + :\ddot{\text{X}}:^- \longrightarrow \text{(R)(X)C=C(H)}\boxed{\text{H}}$$

alkyne — vinyl cation — Markovnikov orientation

When 2 moles of a hydrogen halide add to an alkyne, the second mole usually adds with the same orientation as the first. This consistent orientation leads to a geminal dihalide. For example, a double Markovnikov addition of HBr to 1-pentyne gives 2,2-dibromopentane.

$$\text{H—C}\equiv\text{C—CH}_2\text{CH}_2\text{CH}_3 \xrightarrow{\text{HBr}} \text{(H)}_2\text{C=C(Br)CH}_2\text{CH}_2\text{CH}_3 \xrightarrow{\text{HBr}} \text{H—CH}_2\text{—CBr}_2\text{—CH}_2\text{CH}_2\text{CH}_3$$

1-pentyne — 2-bromo-l-pentene — 2,2-dibromopentane

PROBLEM 14-16

Propose a mechanism for the entire reaction of 1-pentyne with 2 moles of HBr. Show why Markovnikov's rule should be observed in both the first and second additions of HBr.

PROBLEM 14-17

The reaction of 2-octyne with 2 equivalents of HCl gives a mixture of two products.

(a) Give the structures of the two products.
(b) Show why the second equivalent of HCl adds with the same orientation as the first in each case.

The effect of peroxides on the addition of HBr to alkenes (Section 8-3B) is also seen with alkynes: Peroxides catalyze the addition of HBr to alkynes in the anti-Markovnikov sense.

$$\text{H—C}\equiv\text{C—CH}_2\text{CH}_2\text{CH}_3 + \boxed{\text{H—Br}} \xrightarrow{\text{ROOR}} \text{(H)(Br)C=C(}\boxed{\text{H}}\text{)CH}_2\text{CH}_2\text{CH}_3$$

1-pentyne — 1-bromo-1-pentene (mixture of cis and trans isomers)

PROBLEM 14-18

Propose a mechanism for the reaction of 1-pentyne with HBr in the presence of peroxides. Show why anti-Markovnikov orientation results.

PROBLEM 14-19

Show how 1-hexyne might be converted to

(a) 1,2-dichlorohexene.
(b) 1-bromohexene.
(c) 2-bromohexene.
(d) 1,1,2,2-tetrabromohexane.
(e) 2-bromohexane.
(f) 2,2-dibromohexane.

Mercuric ion-catalyzed hydration Alkynes undergo acid-catalyzed addition of water across the triple bond in the presence of mercuric ion as a catalyst. A mixture of mercuric sulfate in aqueous sulfuric acid is commonly used as the reagent. The hydration of alkynes is similar to the hydration of alkenes, and it also goes with Markovnikov orientation. The products are not the alcohols we might expect, however.

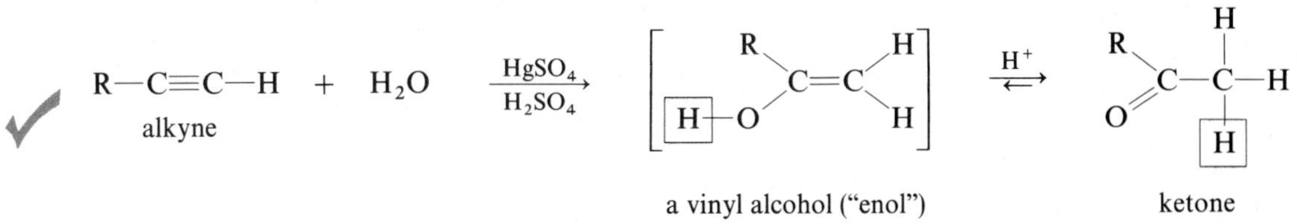

$$\underset{\text{alkyne}}{R-C\equiv C-H} + H_2O \xrightarrow[H_2SO_4]{HgSO_4} \underset{\text{a vinyl alcohol ("enol")}}{\left[(HO)(R)C=CH_2 \right]} \overset{H^+}{\rightleftharpoons} \underset{\text{ketone}}{R-C(=O)-CH_3}$$

The initial product of hydration is a vinyl alcohol, called an **enol.** Enols are usually unstable, and isomerize to the corresponding ketones. As shown above, this isomerization simply involves losing the hydroxyl proton of the enol and regaining a proton at the methyl position, with a corresponding shift of the pi bond. This type of rapid equilibrium between two isomeric functional groups is called **tautomerism,** and the one shown above is called **keto-enol tautomerism** (covered in more detail in Chapter 18). The more stable keto form predominates.

In acidic solution, the keto-enol tautomerism takes place by addition of a proton at the methylene position, followed by loss of the hydroxyl proton.

Addition of a proton at the methylene group

Loss of the hydroxyl proton

$$\underset{\text{enol form}}{(HO)(R)C=CH_2} \underset{}{\overset{H^+}{\rightleftharpoons}} \underset{\text{resonance-stabilized intermediate}}{\left[(HO)(R)\overset{+}{C}-CH_3 \longleftrightarrow (H\overset{+}{O}=)(R)C-CH_3 \right]} \overset{H_2\ddot{O}:}{\rightleftharpoons} \underset{\text{keto form}}{R-C(=O)-CH_3} + H_3O^+$$

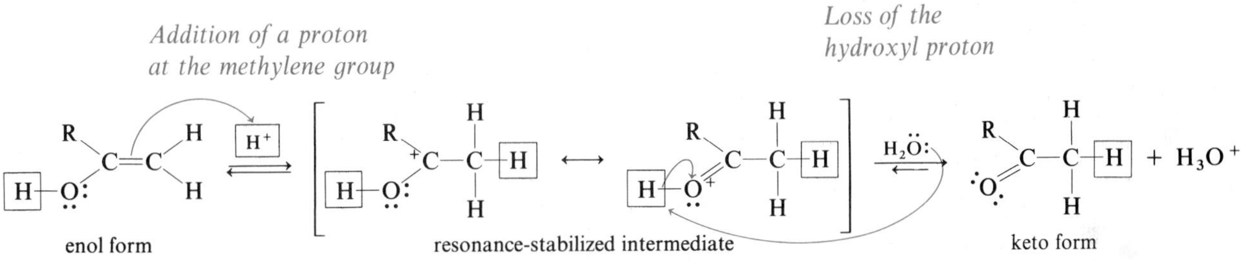

The mercuric-catalyzed hydration of 1-butyne, for example, gives 1-buten-2-ol as an intermediate. In the acidic solution, the intermediate quickly equilibrates to its more stable keto tautomer, 2-butanone.

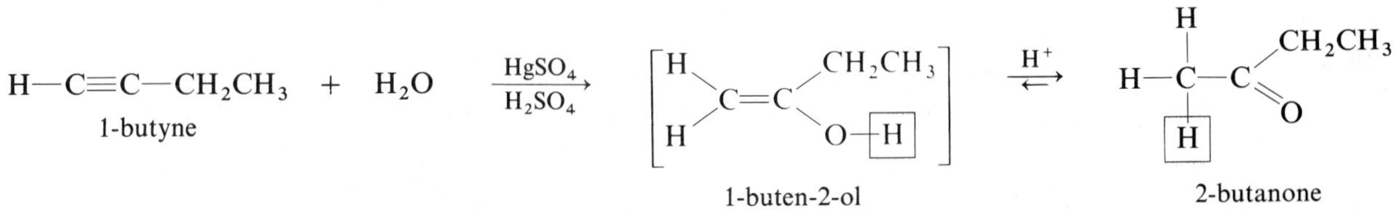

$$\underset{\text{1-butyne}}{H-C\equiv C-CH_2CH_3} + H_2O \xrightarrow[H_2SO_4]{HgSO_4} \underset{\text{1-buten-2-ol}}{\left[H_2C=C(CH_2CH_3)(O-H) \right]} \overset{H^+}{\rightleftharpoons} \underset{\text{2-butanone}}{H_3C-C(=O)-CH_2CH_3}$$

When this same procedure is applied to 2-butyne, the product is once again 2-butanone.

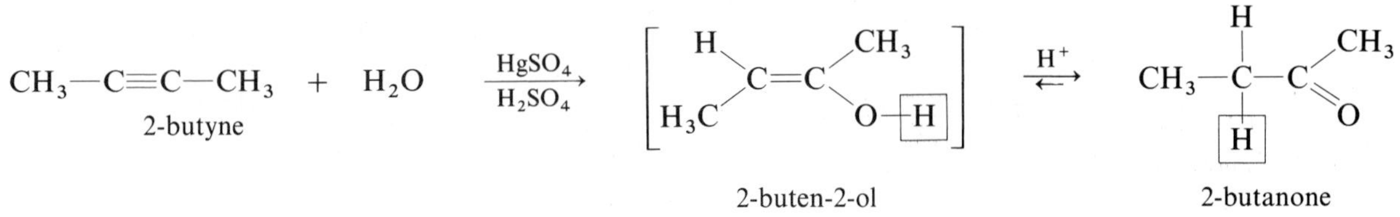

$$\underset{\text{2-butyne}}{CH_3-C\equiv C-CH_3} + H_2O \xrightarrow[H_2SO_4]{HgSO_4} \underset{\text{2-buten-2-ol}}{\left[(H)(H_3C)C=C(CH_3)(O-H) \right]} \overset{H^+}{\rightleftharpoons} \underset{\text{2-butanone}}{CH_3-CH_2-C(=O)-CH_3}$$

PROBLEM 14-20

When 2-pentyne is treated with mercuric sulfate in dilute sulfuric acid, the product is a mixture of two ketones. Give the structures of these products and use mechanisms to show how they are formed.

Hydroboration-oxidation In Section 8-7 we saw that hydroboration-oxidation adds water across the double bonds of alkenes with anti-Markovnikov orientation. A similar reaction takes place with alkynes, except that a hindered dialkylborane is used to prevent addition of two molecules of borane across the triple bond. Di(secondary isoamyl)borane, called "disiamylborane," adds to the triple bond only once to give a vinylborane. In a terminal alkyne, the boron atom bonds to the terminal carbon atom.

$$R'-C\equiv C-H \;+\; Sia_2BH \longrightarrow R'(H)C=C(H)BSia_2$$

alkyne; disiamylborane; a vinylborane

Sia = $CH(CH_3)_2-CH(CH_3)-$ "*sec*-isoamyl" or **"siamyl"**

Oxidation of the vinylborane using basic hydrogen peroxide gives a vinyl alcohol (enol), resulting from anti-Markovnikov addition of water across the triple bond. This enol quickly tautomerizes to its more stable carbonyl (keto) form. In the case of a terminal alkyne, the keto product is an aldehyde. This sequence is an excellent method for converting terminal alkynes to aldehydes.

$$R'(H)C=C(H)BSia_2 \xrightarrow[NaOH]{H_2O_2} \left[R'(H)C=C(H)O-\boxed{H}\right] \overset{^-OH}{\rightleftharpoons} H-C(R')(\boxed{H})-C(H)=O$$

vinylborane; unstable enol form; aldehyde

SOLVED PROBLEM 14-2

Under basic reaction conditions, the keto-enol tautomerism operates by a different mechanism. Propose a mechanism for the tautomerism of the enol formed in the hydroboration-oxidation to its keto form, the aldehyde.

SOLUTION In acid, the enol was first protonated and then it lost a proton. Under basic conditions, the enol first loses its hydroxyl proton, then regains a proton on the adjacent carbon atom.

$$R(H)C=C(H)\ddot{O}-\boxed{H} \xrightarrow{^-:\ddot{O}H} \left[R(H)C=C(H)\ddot{O}:^- \longleftrightarrow R(H)\,^-\!:C-C(H)=\ddot{O}:\right] \xrightarrow{H-\ddot{O}:-H} H-C(R)(\boxed{H})-C(H)=\ddot{O}: \;+\; ^-OH$$

enol form; stabilized "enolate" ion; keto form

Hydroboration of 1-hexyne, for example, gives the vinylborane with boron on the less highly substituted carbon. Oxidation of this intermediate gives an enol that quickly tautomerizes to hexanal.

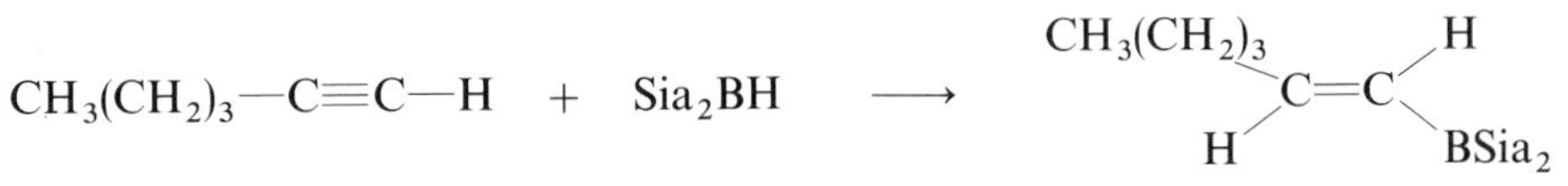

$$\underset{\text{1-hexyne}}{CH_3(CH_2)_3{-}C{\equiv}C{-}H} + Sia_2BH \longrightarrow \underset{\text{a vinylborane}}{(CH_3(CH_2)_3)(H)C{=}C(H)(BSia_2)}$$

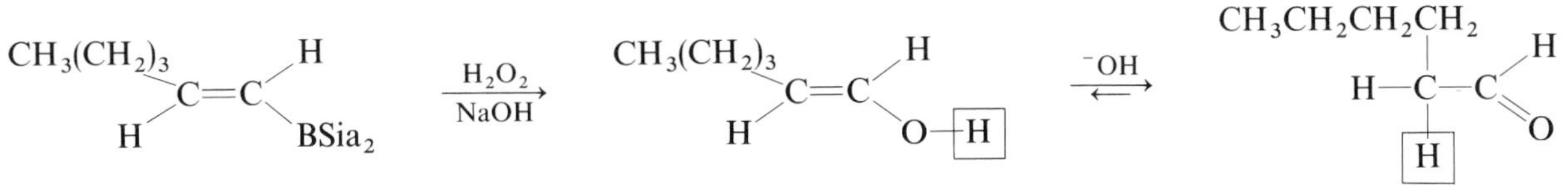

$$\underset{\text{vinylborane}}{(CH_3(CH_2)_3)(H)C{=}C(H)(BSia_2)} \xrightarrow[NaOH]{H_2O_2} \underset{\text{enol}}{(CH_3(CH_2)_3)(H)C{=}C(H)(O{-}\boxed{H})} \overset{^{-}OH}{\rightleftharpoons} \underset{\text{hexanal (65\%)}}{CH_3CH_2CH_2CH_2{-}CH(\boxed{H}){-}CH{=}O}$$

PROBLEM 14-21

The hydroboration-oxidation of internal alkynes produces ketones.

(a) When hydroboration-oxidation is applied to 2-butyne, a single pure product is obtained. Determine the structure of this product and show the intermediates in its formation.

(b) When the hydroboration-oxidation process is applied to 2-pentyne, two products are obtained. Show why such a mixture of products should be expected with an unsymmetrical internal alkyne.

PROBLEM 14-22

For each of the following compounds, give the product(s) expected from (1) $HgSO_4/H_2SO_4$ catalyzed hydrolysis and (2) hydroboration-oxidation.

(a) 1-hexyne (b) 2-hexyne (c) 3-hexyne (d) cyclodecyne

PROBLEM 14-23

Disiamylborane adds only once to alkynes by virtue of its two bulky secondary isoamyl groups. Disiamylborane is prepared by the reaction of $BH_3 \cdot THF$ with an alkene.

(a) Draw the structural formulas of the reagents and the products in the preparation of disiamylborane.

(b) Explain why the reaction in part (a) goes only as far as the dialkylborane. Why is Sia_3B not formed?

14-11 OXIDATION OF ALKYNES

14-11A PERMANGANATE OXIDATIONS

Under mild conditions, potassium permanganate oxidizes alkenes to diols (Section 8-14B). A similar reaction occurs with alkynes. If an alkyne is treated with aqueous potassium permanganate under nearly neutral conditions, an α-diketone results. This is conceptually the same as hydroxylating each of the two pi bonds of the alkyne, then losing two molecules of water to give the diketone.

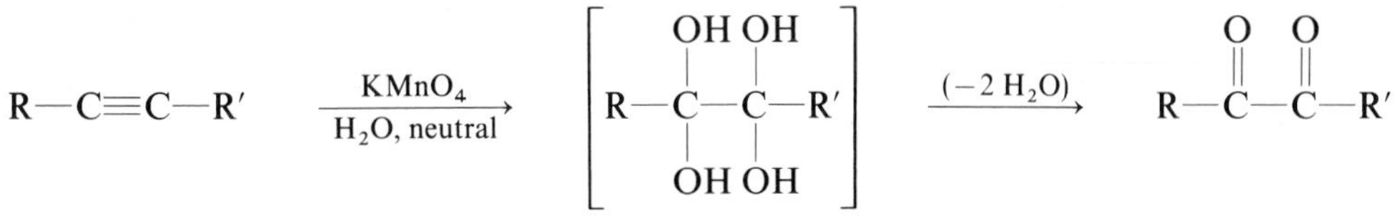

$$R{-}C{\equiv}C{-}R' \xrightarrow[H_2O,\ \text{neutral}]{KMnO_4} \left[R{-}C(OH)_2{-}C(OH)_2{-}R' \right] \xrightarrow{(-2\,H_2O)} R{-}\overset{O}{\overset{\|}{C}}{-}\overset{O}{\overset{\|}{C}}{-}R'$$

If the reaction mixture is allowed to become too warm or too basic, the diketone undergoes oxidative cleavage. The products are the salts of carboxylic acids, which can be converted to the free acids by the addition of dilute acid.

$$\text{R—C}\equiv\text{C—R}' \xrightarrow[\text{H}_2\text{O, heat}]{\text{KMnO}_4\text{, KOH}} \text{R—}\overset{\overset{\displaystyle\text{O}}{\|}}{\text{C}}\text{—O}^{-}\,{}^{+}\text{K} + \text{K}^{+}\,{}^{-}\text{O—}\overset{\overset{\displaystyle\text{O}}{\|}}{\text{C}}\text{—R}' \xrightarrow[\text{H}_2\text{O}]{\text{HCl}} \text{R—}\overset{\overset{\displaystyle\text{O}}{\|}}{\text{C}}\text{—OH} + \text{HO—}\overset{\overset{\displaystyle\text{O}}{\|}}{\text{C}}\text{—R}'$$

For example, when 2-pentyne is treated with a dilute solution of neutral permanganate, the product is 2,3-pentanedione.

$$\underset{\text{2-pentyne}}{\text{CH}_3\text{—C}\equiv\text{C—CH}_2\text{CH}_3} \xrightarrow[\text{H}_2\text{O, neutral}]{\text{KMnO}_4} \underset{\substack{\text{2,3-pentanedione}\\(90\%)}}{\text{CH}_3\text{—}\overset{\overset{\displaystyle\text{O}}{\|}}{\text{C}}\text{—}\overset{\overset{\displaystyle\text{O}}{\|}}{\text{C}}\text{—CH}_2\text{CH}_3}$$

Under harsher conditions, permanganate cleaves the triple bond, giving acetate and propionate ions. Upon acidification, these anions are reprotonated to acetic acid and propionic acid.

$$\underset{\text{2-pentyne}}{\text{CH}_3\text{—C}\equiv\text{C—CH}_2\text{CH}_3} \xrightarrow[\text{H}_2\text{O, heat}]{\text{KMnO}_4\text{, KOH}} \underset{\text{acetate}}{\text{CH}_3\text{—}\overset{\overset{\displaystyle\text{O}}{\|}}{\text{C}}\text{—O}^-} + \underset{\text{propionate}}{{}^-\text{O—}\overset{\overset{\displaystyle\text{O}}{\|}}{\text{C}}\text{—CH}_2\text{CH}_3}$$

$$\xrightarrow{\text{H}^+} \underset{\text{acetic acid}}{\text{CH}_3\text{—}\overset{\overset{\displaystyle\text{O}}{\|}}{\text{C}}\text{—OH}} + \underset{\text{propionic acid}}{\text{HO—}\overset{\overset{\displaystyle\text{O}}{\|}}{\text{C}}\text{—CH}_2\text{CH}_3}$$

Terminal alkynes are cleaved similarly to give a carboxylic acid and CO_2.

$$\underset{\text{1-hexyne}}{\text{CH}_3(\text{CH}_2)_3\text{—C}\equiv\text{C—H}} \xrightarrow[\text{(2) H}^+]{\text{(1) KMnO}_4\text{, KOH, H}_2\text{O}} \underset{\text{pentanoic acid}}{\text{CH}_3(\text{CH}_2)_3\text{—}\overset{\overset{\displaystyle\text{O}}{\|}}{\text{C}}\text{—OH}} + \text{CO}_2\uparrow$$

14-11B OZONOLYSIS

Ozonolysis of an alkyne, followed by hydrolysis, gives products that are similar to those obtained from the oxidative cleavage by permanganate. Either of these oxidative cleavages can be used to determine the position of the triple bond in an unknown alkyne (see Problem 14-25).

$$\text{R—C}\equiv\text{C—R}' \xrightarrow[\text{(2) H}_2\text{O}]{\text{(1) O}_3} \text{R—COOH} + \text{R}'\text{—COOH}$$

Example

$$\text{CH}_3\text{—C}\equiv\text{C—CH}_2\text{CH}_3 \xrightarrow[\text{(2) H}_2\text{O}]{\text{(1) O}_3} \text{CH}_3\text{—COOH} + \text{CH}_3\text{CH}_2\text{—COOH}$$

PROBLEM 14-24

Predict the product(s) you would expect from treatment of each of the following compounds with (1) dilute, neutral $KMnO_4$ and (2) hot, basic $KMnO_4$, then dilute acid.

(a) 1-hexyne (b) 2-hexyne (c) 3-hexyne
(d) 2-methyl-3-hexyne (e) cyclodecyne

PROBLEM 14-25

Oxidative cleavages can be used to determine the positions of the triple bonds in alkynes.

(a) An unknown alkyne undergoes oxidative cleavage to give adipic acid and 2 equivalents of acetic acid. Propose a structure for the alkyne.

$$\text{unknown alkyne} \xrightarrow[(2)\ H_2O]{(1)\ O_3} \underset{\text{adipic acid}}{HOOC{-}(CH_2)_4{-}COOH} \quad + \quad 2\,CH_3COOH$$

(b) An unknown alkyne undergoes oxidative cleavage to give the following triacid plus just 1 equivalent of propanoic acid. Propose a structure for the alkyne.

$$\text{unknown alkyne} \xrightarrow[(2)\ H_2O]{(1)\ O_3} \underset{\text{a triacid}}{HOOC{-}(CH_2)_7{-}\overset{\overset{COOH}{|}}{CH}{-}COOH} + \underset{\text{propionic acid}}{CH_3CH_2COOH}$$

PROBLEM SOLVING: MULTISTEP SYNTHESIS

We have pointed out the value of multistep synthesis problems in exercising your knowledge of organic reactions, and in Chapters 5 and 10 have illustrated a systematic approach to synthesis. Now we can apply this approach to a fairly difficult problem emphasizing alkyne chemistry. The compound to be synthesized is 3-octyn-2-one

$$\underset{\text{3-octyn-2-one}}{CH_3CH_2CH_2CH_2{-}C{\equiv}C{-}\overset{\overset{O}{\|}}{C}{-}CH_3}$$

and the starting materials are acetylene and alcohols containing no more than four carbon atoms. In this problem it is necessary to consider not only how to assemble the carbon skeleton and how to introduce the functional groups, but also when it is best to put in the functional groups. We begin, as usual, with an examination of the target compound and then examine possible intermediates and synthetic routes.

1. **Review the functional groups and carbon skeleton of the target compound.** The target compound contains eight carbon atoms and two functional groups: a carbon-carbon triple bond and a ketone.
2. **Review the functional groups and carbon skeletons of the starting materials, and see how their skeletons might fit together in the target compound.** It might be possible to put two four-carbon alcohols together into this eight-carbon skeleton, but getting the right functional groups in the right places would be a nightmare. We have good methods (Section 14-8) for forming carbon-carbon bonds next to triple bonds, however, by using acetylide ions as nucleophiles. Using a carbonyl group in the electrophile retains functionality at the carbon next to the triple bond.

$$\underset{\text{4 carbons}}{CH_3CH_2CH_2CH_2{-}} \qquad \underset{\text{acetylene}}{{-}C{\equiv}C{-}} \qquad \underset{\text{2 carbons (functionalized)}}{{-}\overset{\overset{O}{\|}}{C}{-}CH_3}$$

3. **Compare methods for assembling the carbon skeleton of the target compound to determine which methods provide a key intermediate with the**

correct carbon skeleton and functional groups at the correct positions where they can be converted to the functionality in the target molecule. Using two alkylations of acetylene gives the correct carbon skeleton, but there is no functional group at the position where the carbonyl group must appear in the product.

$$CH_3(CH_2)_3{-}X \xrightarrow{HC{\equiv}C:^-} CH_3(CH_2)_3{-}C{\equiv}CH$$

$$\xrightarrow[(2)\ CH_3CH_2X]{(1)\ NaNH_2} CH_3(CH_2)_3{-}C{\equiv}C{-}CH_2CH_3$$

no functional group

Functionality at the carbon next to the triple bond can be maintained by using a ketone or aldehyde in a Grignard-like reaction. The alcohol that results is easily oxidized to the target compound. We add the functionalized group *after* adding the butyl group, because the butyl group is less likely to be affected by subsequent reactions. (In general, we try to add less reactive groups earlier in a synthesis, and more reactive groups later.)

$$CH_3(CH_2)_3{-}X \xrightarrow{HC{\equiv}C:^-} CH_3(CH_2)_3{-}C{\equiv}CH$$

$$\xrightarrow[(2)\ CH_3{-}\overset{O}{\overset{\|}{C}}{-}H \quad (3)\ H_3O^+]{(1)\ NaNH_2} CH_3(CH_2)_3{-}C{\equiv}C{-}\overset{OH}{\overset{|}{C}H}{-}CH_3$$

key intermediate

4. **Working backward through as many steps as necessary, compare methods for synthesizing the reactants needed for assembly of the key intermediate with the correct carbon skeleton and functionality in the proper position.** 1-Bromobutane can be made from 1-butanol using HBr or PBr_3, and acetaldehyde can be made from ethanol by Collins oxidation.

$$CH_3(CH_2)_3{-}OH \xrightarrow{HBr} CH_3(CH_2)_3{-}Br$$

$$CH_3{-}CH_2{-}OH \xrightarrow{CrO_3\cdot pyridine} CH_3{-}\overset{O}{\overset{\|}{C}}{-}H$$

5. **Summarize the complete synthesis in the forward direction, including all steps and all reagents, and check it for errors and omissions.** This final step is left to you as an exercise. Try to do it without looking at this solution, reviewing each thought process as you summarize the synthesis.

Now try your hand at the syntheses in Problem 14-26, using them for practice in the systematic approach to multistep syntheses.

PROBLEM 14-26

Develop syntheses for the following compounds, using acetylene and alcohols containing no more than 4 carbon atoms as your organic starting materials.

(a) 3-methyl-4-nonyn-3-ol (b) *cis*-1-ethyl-2-methylcyclopropane

(c) $\overset{H}{\underset{CH_3}{}}\!\!>C{=}C<\!\!\overset{H}{\underset{}{}}\ \ CH(OH){-}CH(CH_3){-}CH_3$ (cis: CH_3 and $CH(OH)CH(CH_3)CH_3$ on the same side, H and H on the same side)

SUMMARY OF REACTIONS OF ALKYNES

1. Acetylide chemistry

1. Formation of acetylide anions (alkynides,

a. *Sodium, lithium, and magnesium acetylides* (Section 14-7A)

$$R{-}C{\equiv}C{-}H \;+\; NaNH_2 \longrightarrow R{-}C{\equiv}C{:}^{-}\,{}^{+}Na \;+\; NH_3$$

$$R{-}C{\equiv}C{-}H \;+\; R'{-}Li \longrightarrow R{-}C{\equiv}C{-}Li \;+\; R'{-}H$$

$$R{-}C{\equiv}C{-}H \;+\; R'{-}MgX \longrightarrow R{-}C{\equiv}C{-}MgX \;+\; R'{-}H$$

Example

$$\underset{\text{propyne}}{CH_3{-}C{\equiv}C{-}H} \;+\; \underset{\text{sodium amide}}{NaNH_2} \longrightarrow \underset{\substack{\text{sodium propynide}\\ \text{(propynyl sodium)}}}{CH_3{-}C{\equiv}C{:}^{-}\,{}^{+}Na} \;+\; NH_3$$

b. *Heavy-metal acetylides* (Section 14-7B)

$$R{-}C{\equiv}C{-}H \;+\; Ag^{+} \longrightarrow R{-}C{\equiv}C{-}Ag\downarrow$$

$$R{-}C{\equiv}C{-}H \;+\; Cu^{+} \longrightarrow R{-}C{\equiv}C{-}Cu\downarrow$$

(These reactions are used to test for the presence of a terminal alkyne.)

2. Alkylation of acetylide ions (Section 14-8A)

$$R{-}C{\equiv}C{:}^{-} \;+\; R'{-}X \longrightarrow R{-}C{\equiv}C{-}R'$$

(R′—X must be an unhindered primary halide or tosylate.)

Example

$$\underset{\text{sodium butynide}}{CH_3CH_2{-}C{\equiv}C{:}^{-}\,{}^{+}Na} + \underset{\text{1-bromopropane}}{CH_3CH_2CH_2{-}Br} \longrightarrow \underset{\text{3-heptyne}}{CH_3CH_2{-}C{\equiv}C{-}CH_2CH_2CH_3}$$

3. Reactions with carbonyl groups and epoxides (Section 14-8B)

$$R{-}C{\equiv}C{:}^{-} + R'_2C{=}\ddot{O}{:} \longrightarrow R{-}C{\equiv}C{-}CR'_2{-}\ddot{O}{:}^{-} \xrightarrow{H_2O} R{-}C{\equiv}C{-}CR'_2{-}OH$$

$$\underset{\text{acetylide}}{R{-}C{\equiv}C{:}^{-}} \;+\; \underset{\text{oxirane}}{H_2C{-}CH_2\ (\text{epoxide, } \ddot{O})} \longrightarrow$$

$$R{-}C{\equiv}C{-}CH_2{-}CH_2{-}\ddot{O}{:}^{-} \xrightarrow{H_2O} \underset{\substack{\text{a primary}\\ \text{acetylenic alcohol}}}{R{-}C{\equiv}C{-}CH_2CH_2{-}OH}$$

Example

$$\underset{\text{sodium propynide}}{CH_3{-}C{\equiv}C{:}^{-}\ Na^{+}} \xrightarrow[\text{(2) } H_2O]{\text{(1) } CH_3CH_2{-}C(=O){-}CH_3} \underset{\text{3-methyl-4-hexyn-3-ol}}{CH_3{-}C{\equiv}C{-}C(OH)(CH_3){-}CH_2CH_3}$$

II. Additions to the triple bond (Section 14-10)

1. Reduction to alkanes (Section 14-10A)

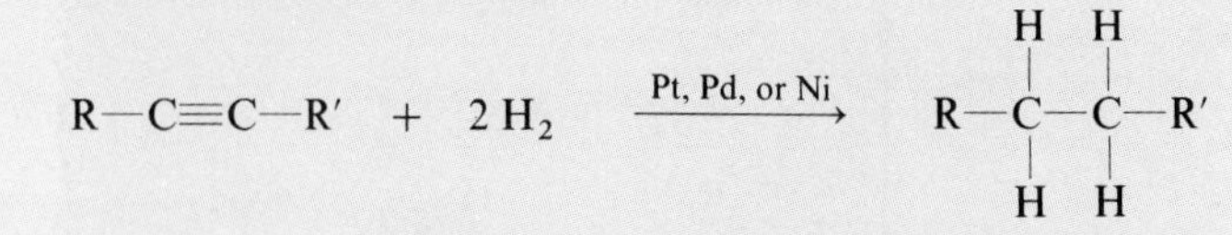

$$R{-}C{\equiv}C{-}R' + 2\,H_2 \xrightarrow{\text{Pt, Pd, or Ni}} R{-}CH_2{-}CH_2{-}R'$$

Example

$$\underset{\text{3-pentyn-1-ol}}{CH_3CH_2{-}C{\equiv}C{-}CH_2{-}OH} + 2H_2 \xrightarrow{\text{Pt}} \underset{\text{1-pentanol}}{CH_3CH_2{-}CH_2{-}CH_2{-}CH_2{-}OH}$$

2. Reduction to alkenes (Sections 14-10B and 14-10C)

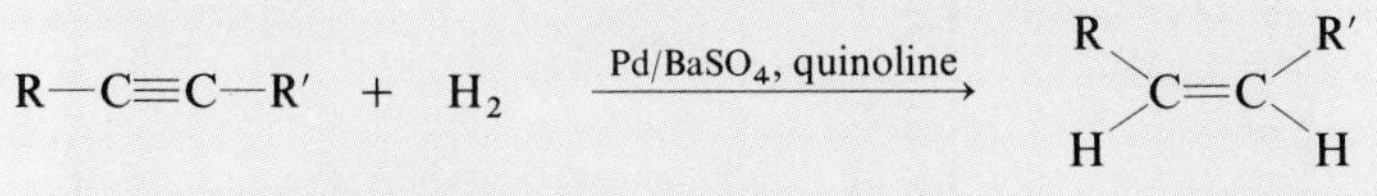

$$R{-}C{\equiv}C{-}R' + H_2 \xrightarrow{\text{Pd/BaSO}_4\text{, quinoline}} \underset{cis}{(R)(H)C{=}C(R')(H)}$$

$$R{-}C{\equiv}C{-}R' \xrightarrow{\text{Na, NH}_3} \underset{trans}{(R)(H)C{=}C(H)(R')}$$

Examples

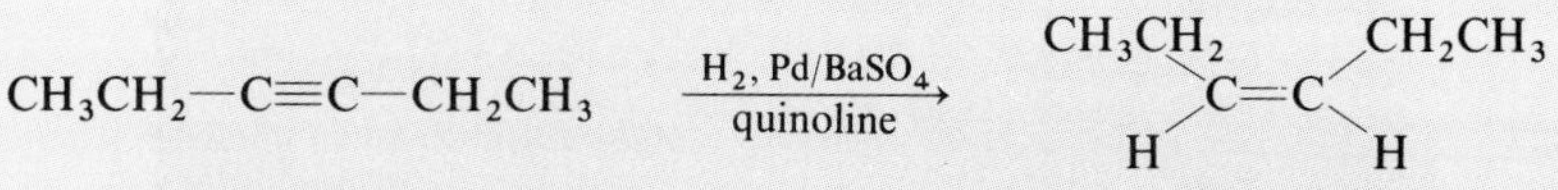

$$\underset{\text{3-hexyne}}{CH_3CH_2{-}C{\equiv}C{-}CH_2CH_3} \xrightarrow[\text{quinoline}]{H_2,\ \text{Pd/BaSO}_4} \underset{cis\text{-3-hexene}}{(CH_3CH_2)(H)C{=}C(CH_2CH_3)(H)}$$

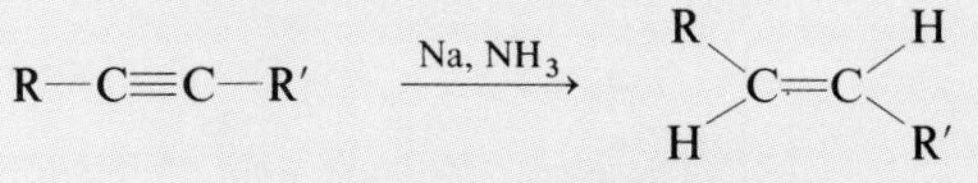

$$\underset{\text{3-hexyne}}{CH_3CH_2{-}C{\equiv}C{-}CH_2CH_3} \xrightarrow{\text{Na, NH}_3} \underset{trans\text{-3-hexene}}{(CH_3CH_2)(H)C{=}C(H)(CH_2CH_3)}$$

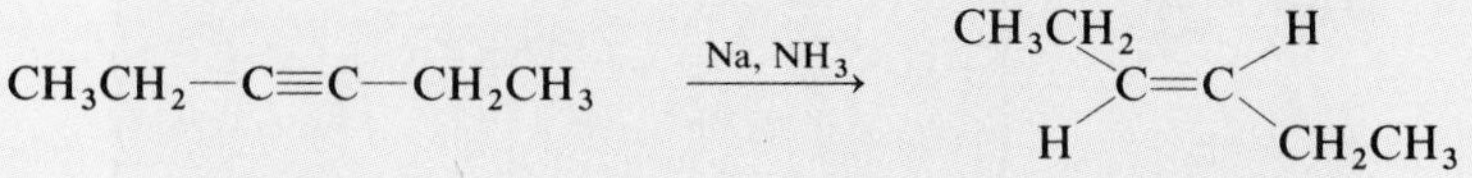

3. Addition of halogens ($X_2 = Cl_2, Br_2$) (Section 14-10D)

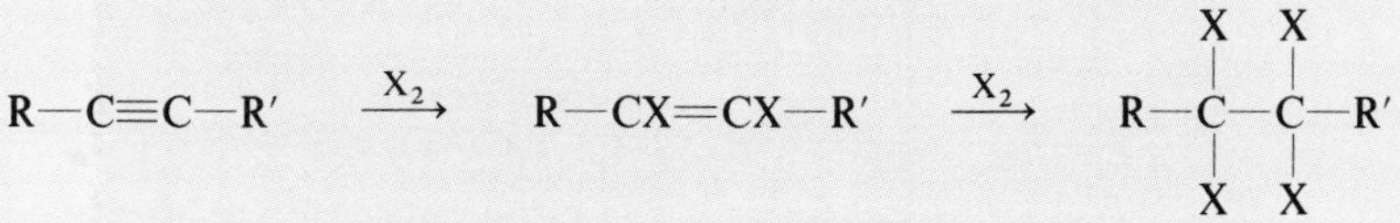

$$R{-}C{\equiv}C{-}R' \xrightarrow{X_2} R{-}CX{=}CX{-}R' \xrightarrow{X_2} R{-}CX_2{-}CX_2{-}R'$$

Example

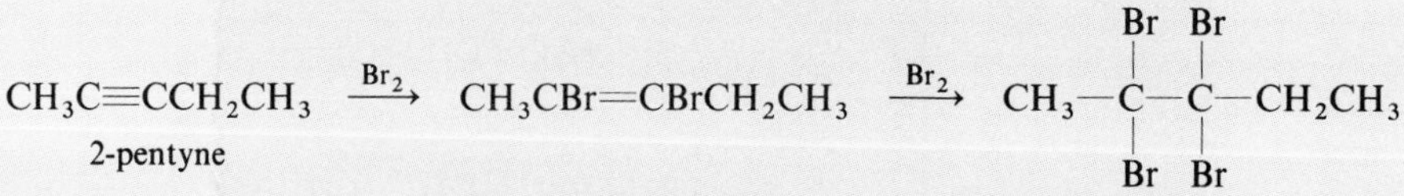

$$\underset{\text{2-pentyne}}{CH_3C{\equiv}CCH_2CH_3} \xrightarrow{Br_2} \underset{cis\text{- and } trans\text{-2,3-dibromo-2-pentene}}{CH_3CBr{=}CBrCH_2CH_3} \xrightarrow{Br_2} \underset{\text{2,2,3,3-tetrabromopentane}}{CH_3{-}CBr_2{-}CBr_2{-}CH_2CH_3}$$

4. Addition of hydrogen halides (where $HX = HCl, HBr$, or HI) (Section 14-10E)

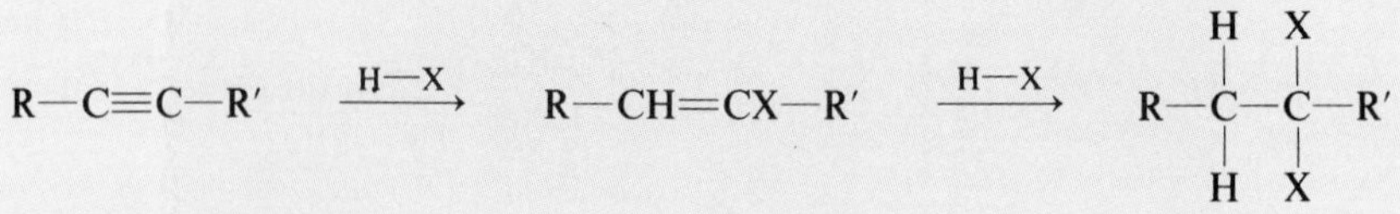

$$R{-}C{\equiv}C{-}R' \xrightarrow{H{-}X} R{-}CH{=}CX{-}R' \xrightarrow{H{-}X} R{-}CH_2{-}CX_2{-}R'$$

(This addition follows Markovnikov's rule.)

Example

$$CH_3CH_2{-}C{\equiv}C{-}H \xrightarrow{HCl} \underset{\text{2-chloro-1-butene}}{(CH_3CH_2)(Cl)C{=}CH_2} \xrightarrow{HCl} \underset{\text{2,2-dichlorobutane}}{CH_3CH_2{-}CCl_2{-}CH_3}$$

1-butyne

5. *Addition of water* (Section 14-10F)

   a. *Catalyzed by* $HgSO_4/H_2SO_4$

$$R{-}C{\equiv}C{-}R' + H_2O \xrightarrow{HgSO_4,\ H_2SO_4} \left[\underset{\substack{\text{vinyl alcohol}\\ \text{(unstable)}}}{(R)(HO)C{=}C(H)(R')}\right] \longrightarrow \underset{\substack{\text{ketone}\\ \text{(stable)}}}{R{-}\overset{O}{\overset{\|}{C}}{-}CH_2{-}R'}$$

(Markovnikov orientation)

Example

$$\underset{\text{propyne}}{CH_3{-}C{\equiv}C{-}H} + H_2O \xrightarrow{HgSO_4,\ H_2SO_4} \underset{\text{2-propanone (acetone)}}{CH_3{-}\overset{O}{\overset{\|}{C}}{-}CH_3}$$

   b. *Hydroboration-oxidation*

$$R{-}C{\equiv}C{-}R' \xrightarrow[\text{(2) } H_2O_2,\ NaOH]{\text{(1) } Sia_2BH{\cdot}THF} \left[\underset{\substack{\text{vinyl alcohol}\\ \text{(unstable)}}}{(R)(H)C{=}C(R')(OH)}\right] \longrightarrow \underset{\substack{\text{ketone or aldehyde}\\ \text{(stable)}}}{R{-}CH_2{-}\overset{O}{\overset{\|}{C}}{-}R'}$$

(anti-Markovnikov orientation.)

Example

$$\underset{\text{propyne}}{CH_3{-}C{\equiv}C{-}H} \xrightarrow[\text{(2) } H_2O_2,\ NaOH]{\text{(1) } Sia_2BH{\cdot}THF} \underset{\text{propanal}}{CH_3{-}CH_2{-}\overset{O}{\overset{\|}{C}}{-}H}$$

III. Oxidation of alkynes (Section 14-11)

1. *Oxidation to alpha-diketones* (Section 14-11A)

$$R{-}C{\equiv}C{-}R' \xrightarrow[H_2O,\ \text{neutral}]{KMnO_4} R{-}\overset{O}{\overset{\|}{C}}{-}\overset{O}{\overset{\|}{C}}{-}R'$$

Example

$$\underset{\text{2-pentyne}}{CH_3{-}C{\equiv}C{-}CH_2CH_3} \xrightarrow[H_2O,\ \text{neutral}]{KMnO_4} \underset{\text{pentane-2,3-dione}}{CH_3{-}\overset{O}{\overset{\|}{C}}{-}\overset{O}{\overset{\|}{C}}{-}CH_2CH_3}$$

2. *Oxidative cleavage* (Section 14-11B)

$$R{-}C{\equiv}C{-}R' \xrightarrow[\text{(2) } H^+]{\text{(1) } KMnO_4,\ ^-OH} R{-}\overset{O}{\overset{\|}{C}}{-}OH + HO{-}\overset{O}{\overset{\|}{C}}{-}R'$$

(or O_3, then H_2O)

Example

$$CH_3{-}C{\equiv}C{-}CH_2CH_3 \xrightarrow[\text{(2) } H^+]{\text{(1) } KMnO_4,\ NaOH} CH_3{-}\overset{O}{\overset{\|}{C}}{-}OH + HO{-}\overset{O}{\overset{\|}{C}}{-}CH_2CH_3$$

GLOSSARY

acetylene The simplest alkyne, H—C≡C—H. Also used as a synonym for *alkyne,* a generic term for a compound containing a C≡C triple bond. (p. 592)

acetylide The anionic salt of a terminal alkyne. Metal acetylides are organometallic compounds with a metal atom in place of the acetylenic hydrogen of a terminal alkyne. The metal-carbon bond may be covalent or ionic or partially covalent and partially ionic. (p. 599)

$R{-}C{\equiv}C{:}^{-}\ {}^{+}Na$ — a sodium acetylide

$R{-}C{\equiv}C{-}Ag$ — a silver acetylide

alkyne Any compound containing a carbon-carbon triple bond. (p. 592)

A **terminal alkyne** has a triple bond at the end of a chain, with an **acetylenic hydrogen.**

An **internal alkyne** has the triple bond somewhere other than at the end of the chain.

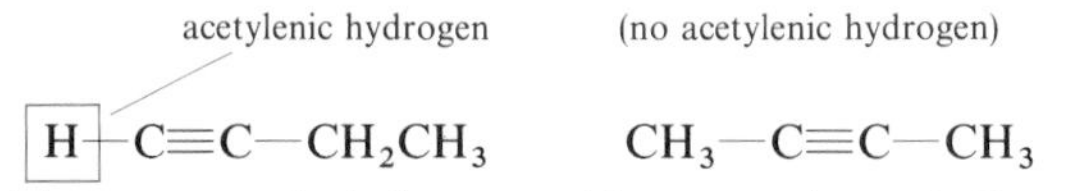

acetylic hydrogen: $H{-}C{\equiv}C{-}CH_2CH_3$ — 1-butyne, a terminal alkyne

(no acetylenic hydrogen): $CH_3{-}C{\equiv}C{-}CH_3$ — 2-butyne, an internal alkyne

enol An alcohol with the hydroxyl group bonded to a carbon atom of a carbon-carbon double bond. Most enols are unstable, spontaneously isomerizing to their carbonyl tautomers, called the **keto** form of the compound. (p. 613)

Lindlar's catalyst A heterogeneous catalyst for the hydrogenation of alkynes to cis alkenes. In its most common form, it consists of a thin coating of palladium on barium sulfate, with quinoline added to decrease the catalytic activity. (p. 608)

s character The fraction of a hybrid orbital that corresponds to an *s* orbital; about one-half for *sp* hybrids, one-third for sp^2 hybrids, and one-fourth for sp^3 hybrids. (p. 594)

siamyl group A contraction for "secondary isoamyl," abbreviated "Sia." This is the 1,2-dimethylpropyl group. Disiamylborane is used for hydroboration of terminal alkynes, because this bulky borane adds only once to the triple bond. (p. 614)

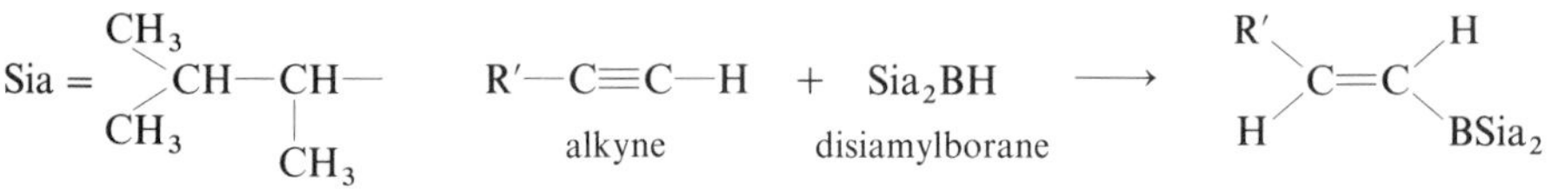

Sia = $(CH_3)_2CH{-}CH(CH_3){-}$ "*sec*-isoamyl" or "siamyl"

$$R'{-}C{\equiv}C{-}H + Sia_2BH \longrightarrow R'(H)C{=}C(H)(BSia_2)$$

alkyne + disiamylborane ⟶ a vinylborane

tautomers Isomers that can quickly interconvert by the movement of a proton (and another bond) from one site to another. An equilibrium between tautomers is called a **tautomerism.** (p. 613)

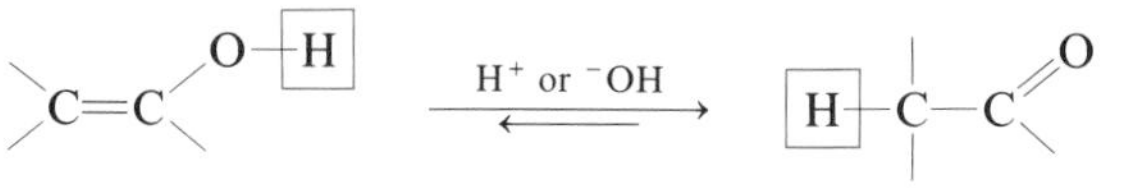

$$>C{=}C<^{O-H} \xrightleftharpoons{H^+ \text{ or } ^-OH} H{-}C{-}C(=O)$$

enol form ⇌ keto form

The **keto-enol tautomerism** is the equilibrium between these two tautomers.

vinyl cation A cation with a positive charge on one of the carbon atoms of a C═C double bond. The cationic carbon atom is usually *sp* hybridized. Vinyl cations are often generated by the addition of an electrophile to a carbon-carbon triple bond. (p. 611)

$$R{-}C{\equiv}C{-}R' + E^+ \longrightarrow R(E)C{=}\overset{+}{C}{-}R'$$

(sp^2, sp) a vinyl cation

ESSENTIAL PROBLEM-SOLVING SKILLS IN CHAPTER 14

1. Name alkynes and draw the structures from their names.
2. Explain why alkynes are acidic and show how to generate nucleophilic acetylide ions and heavy-metal acetylides.
3. Propose effective single-step and multistep syntheses of alkynes.
4. Predict the products of additions, oxidations, reductions, and cleavages of alkynes, including orientation of reaction (regiochemistry) and stereochemistry.
5. Use alkynes as starting materials and intermediates in one-step and multistep syntheses.
6. Show how the reduction of an alkyne leads to an alkene or alkene derivative with the desired stereochemistry.
7. Use clues provided by reactions and spectra to determine the structures of unknown alkynes.

STUDY PROBLEMS

14-27. Briefly define each of the following terms and give an example.
(a) alkyne **(b)** acetylide ion **(c)** enol
(d) tautomerism **(e)** Lindlar's catalyst **(f)** disiamylborane
(g) vinyl cation **(h)** oxidative cleavage of an alkyne **(i)** hydration of an alkyne
(j) hydroboration of an alkyne

14-28. Write structural formulas for the following compounds.
(a) 3-nonyne **(b)** methyl-*n*-pentylacetylene **(c)** ethynylbenzene
(d) cyclohexylacetylene **(e)** 5-methyl-3-octyne **(f)** *trans*-3,5-dibromocyclodecyne
(g) 3-octyn-2-ol **(h)** *cis*-6-ethyl-2-octen-4-yne **(i)** 1,4-heptadiyne
(j) vinylacetylene **(k)** (*S*)-3-methyl-1-penten-4-yne

14-29. Give common names for the following compounds.
(a) $CH_3—C{\equiv}C—CH_2CH_3$ **(b)** $Ph—C{\equiv}C—H$
(c) 3-methyl-4-octyne **(d)** $(CH_3)_3C—C{\equiv}C—CH(CH_3)CH_2CH_3$

14-30. Give IUPAC names for the following compounds.

(a) 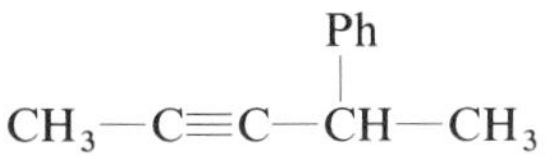 $CH_3—C{\equiv}C—CH(Ph)—CH_3$ (Ph on the CH carbon)

(b) $CH_3—CBr_2—C{\equiv}C—CH_3$

(c) $(CH_3)_2CH—C{\equiv}C—CH_2C(CH_3)_3$

(d) 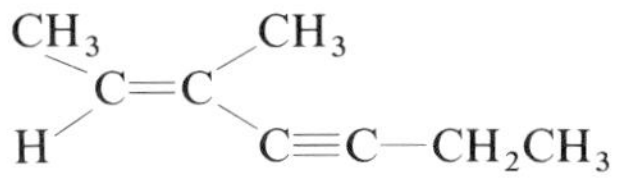 $(CH_3)(H)C{=}C(CH_3)(C{\equiv}C—CH_2CH_3)$ (CH₃ and CH₃ on the same side; H and C≡C—CH₂CH₃ below)

(e) 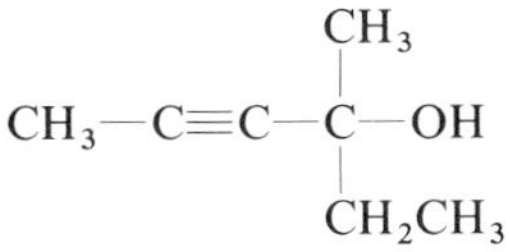 $CH_3—C{\equiv}C—C(CH_3)(CH_2CH_3)—OH$

(f) 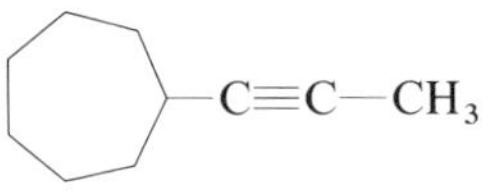 cycloheptyl—$C{\equiv}C—CH_3$

14-31. **(a)** Draw and name the seven alkynes of formula C_6H_{10}.
(b) Which compounds in part (a) will form precipitates when treated with a solution of cuprous ions?

14-32. *Muscalure,* the sex attractant of the common housefly, is *cis*-9-tricosene. Most syntheses of alkenes give mostly the more stable trans isomer. Devise a synthesis of muscalure from acetylene and alcohols of your choice. Your synthesis must give entirely the cis isomer of muscalure.

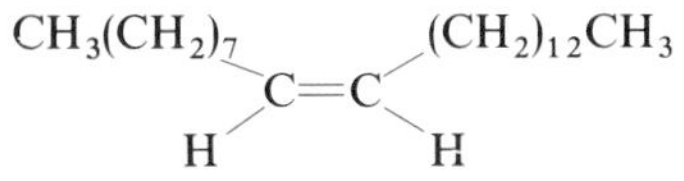

$$\underset{H}{CH_3(CH_2)_7}\!\!>\!C{=}C\!<\!\underset{H}{(CH_2)_{12}CH_3}$$

cis-9-tricosene, "muscalure"

14-33. When we synthesize an internal alkyne, it is often contaminated with small amounts of a terminal isomer. The boiling points are usually too close for a clean separation by distillation. Give equations to show how you might remove small amounts of 1-decyne from a sample of 2-decyne.

14-34. Predict the products of the reaction of 1-pentyne with each of the following reagents.
(a) 1 equivalent of HCl
(b) 2 equivalents of HCl
(c) excess H_2, Ni
(d) H_2, $Pd/BaSO_4$, quinoline
(e) 1 equivalent of Br_2
(f) 2 equivalents of Br_2
(g) cold, dilute $KMnO_4$
(h) warm, conc. $KMnO_4$, NaOH
(i) Na, liquid ammonia
(j) $NaNH_2$
(k) $Ag(NH_3)_2^+$
(l) $H_2SO_4/HgSO_4$, H_2O
(m) Sia_2BH, then H_2O_2, ^-OH

14-35. Show how you would accomplish the following synthetic transformations. Show all intermediates.
(a) 2,2-dibromobutane ⟶ 1-butyne
(b) 2,2-dibromobutane ⟶ 2-butyne
(c) 1-butyne ⟶ 3-octyne
(d) *trans*-2-hexene ⟶ 2-hexyne
(e) *cis*-2-hexene ⟶ 1-hexyne
(f) cyclodecyne ⟶ *cis*-cyclodecene
(g) cyclodecyne ⟶ *trans*-cyclodecene
(h) 1-hexyne ⟶ 2-hexanone, $CH_3COCH_2CH_2CH_2CH_3$
(i) 1-hexyne ⟶ hexanal, $CH_3(CH_2)_4CHO$
(j) *trans*-2-hexene   *cis*-2-hexene

14-36. Potassium hydroxide is mixed with 2,3-dibromohexane, and the mixture is heated at 200°C in a sealed tube for 1 hour. The product mixture **(A)** is mixed with a copper(I)-ammonia complex and a precipitate forms. The precipitate **(B)** and the liquid phase **(C)** are separated. The precipitate is acidified and the product **(D)** is distilled (b.p. 71°C). Product **D** is treated with sodium amide, followed by acetone, and then by dilute acid to give alcohol **(F).**

The liquid phase **(C)** is distilled and the products are collected over a boiling range of 80 to 85°C. This distillate is treated with sodium amide at 150°C for 1 hour, and the product mixture is distilled to give a pure alkyne **(E)** of boiling point 71°C. Give the structures of the alkynes present in products **A** through **E,** and give the structure of alcohol **F.**

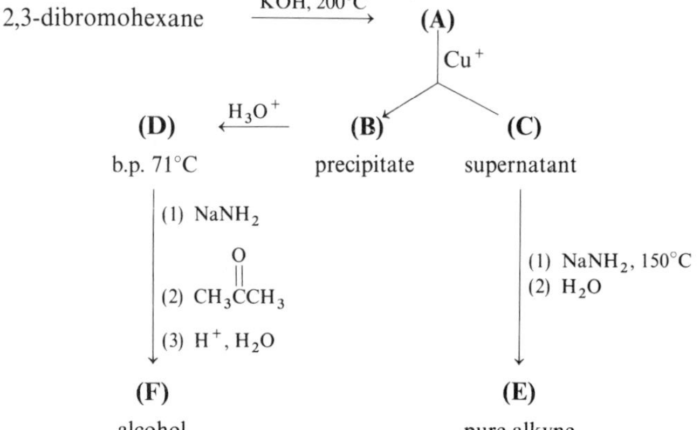

14-37. Predict the products formed when $CH_3CH_2—C≡C:^{-}\ ^{+}Na$ reacts with the following compounds.
(a) ethyl bromide
(b) *t*-butyl bromide
(c) formaldehyde
(d) cyclohexanone
(e) $CH_3CH_2CH_2CHO$
(f) ethylene oxide
(g) cyclohexanol
(h) 2-butanone, $CH_3CH_2—C(=O)—CH_3$

14-38. Show how you would synthesize the following compounds using acetylene and any alcohols containing no more than four carbon atoms as starting materials.
(a) 1-hexyne
(b) 2-hexyne
(c) *cis*-2-hexene
(d) *trans*-2-hexene

(e) hexane
(f) 2,2-dibromohexane
(g) pentanal, $CH_3CH_2CH_2CH_2CHO$
(h) 2-pentanone, $CH_3—CO—CH_2CH_2CH_3$
(i) (±)-3,4-dibromohexane
(j) *meso*-2,3-butanediol

14-39. When treated with hydrogen and a platinum catalyst, an unknown compound **(X)** absorbs 5 equivalents of hydrogen to give *n*-butylcyclohexane. When **X** is treated with silver nitrate in ethanol, a white precipitate forms. This precipitate is found to be soluble in dilute acid. Treatment of **X** with an excess of ozone, followed by dimethyl sulfide and water, gives the following products:

$$H—C(=O)—CH_2—CH_2—C(=O)—C(=O)—H \qquad H—C(=O)—C(=O)—H \qquad H—C(=O)—C(=O)—OH \qquad H—C(=O)—OH$$

Propose a structure for the unknown compound **(X)**. Is there any uncertainty in your structure?

14-40. An unknown compound **(Y)** forms a precipitate when treated with an aqueous silver(I) complex. Its IR spectrum shows a sharp absorption around 3300 cm^{-1}, several absorptions above and below 3000 cm^{-1}, a sharp absorption around 2130 cm^{-1}, and a moderate absorption around 1635 cm^{-1}. The proton NMR spectrum of **Y** appears below.

Ozonolysis of **Y**, followed by treatment with dimethyl sulfide and water, gives 1 equivalent of pyruvic acid, 1 equivalent of formaldehyde, and 1 equivalent of formic acid. Determine the structure of this unknown compound.

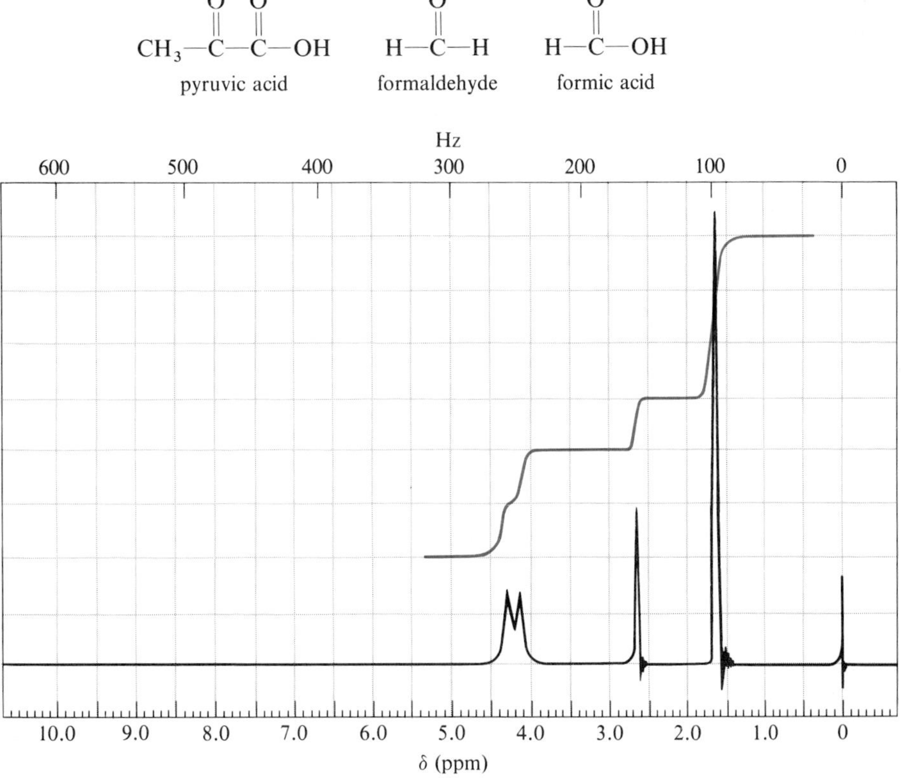

14-41. Using any necessary inorganic reagents, show how you would convert acetylene and isobutyl bromide to
(a) *meso*-2,7-dimethyl-4,5-octanediol.
(b) *d,l*-2,7-dimethyl-4,5-octanediol.

14-42. When compound **Z** is treated with a silver-ammonia complex, a white precipitate forms. Treatment of compound **Z** with ozone, followed by treatment with dimethyl sulfide and washing with water, gives formic acid, 3-oxobutanic acid, and hexanal.

$$\mathbf{Z} \xrightarrow[(2)\ (CH_3)_2S,\ H_2O]{(1)\ O_3} \underset{\text{formic acid}}{H-\overset{\overset{\displaystyle O}{\|}}{C}-OH} + \underset{\text{3-oxobutanoic acid}}{CH_3-\overset{\overset{\displaystyle O}{\|}}{C}-CH_2-\overset{\overset{\displaystyle O}{\|}}{C}-OH} + \underset{\text{hexanal}}{CH_3(CH_2)_4-\overset{\overset{\displaystyle O}{\|}}{C}-H}$$

Propose a structure for compound **Z.** What uncertainty is there in the structure you have proposed?

14-43. Propose structures for intermediates and products **(A)** through **(L)**.

$$\text{1-bromobutane} \xrightarrow{\text{Mg, ether}} \mathbf{(D)} \xrightarrow[(2)\ H_3O^+]{(1)\ CH_3(CH_2)_3CHO} \mathbf{(E)}$$

$$\text{1-bromobutane} \xrightarrow{KOH,\ H_2O} \mathbf{(A)} \xrightarrow{Na} \mathbf{(B)} \xrightarrow{CH_3CH_2Br} \mathbf{(C)}$$

$$\mathbf{(E)} \xrightarrow{H_2SO_4,\ \text{heat}} \mathbf{(F)}$$

$$\mathbf{(F)} \xrightarrow[(2)\ (CH_3)_2S]{(1)\ O_3} CH_3(CH_2)_3CHO + CH_3CH_2CH_2CHO$$

$$\mathbf{(F)} \xrightarrow{Br_2} \mathbf{(G)} \xrightarrow{\text{fused KOH}} \mathbf{(H)}\ \text{mixture} \xrightarrow[(2)\ H_2O]{(1)\ NaNH_2,\ 150°C} \text{1-nonyne}$$

$$\text{1-nonyne} \xrightarrow[(2)\ H_2O_2,\ NaOH]{(1)\ Sia_2BH} \mathbf{(I)}$$

$$\text{1-nonyne} \xrightarrow{NaNH_2} \mathbf{(J)} \xrightarrow[(2)\ H_3O^+]{(1)\ CH_3-\overset{\overset{\displaystyle O}{\|}}{C}-CH_2CH_3} \mathbf{(K)}$$

$$\text{1-nonyne} \xrightarrow{2\ HBr} \mathbf{(L)}$$

* **14-44.** The following functional group interchange is a useful synthesis of aldehydes.

$$\underset{\text{terminal alkyne}}{R-C\equiv C-H} \longrightarrow \underset{\text{aldehyde}}{R-CH_2-\overset{\overset{\displaystyle O}{\|}}{C}-H}$$

(a) What reagents were used in this chapter for this transformation? Give an example to illustrate this method.
(b) This functional group interchange can also be accomplished using the following sequence.

$$R-C\equiv C-H \xrightarrow[CH_3CH_2OH]{NaOCH_2CH_3} \overset{H}{\underset{R}{>}}C=C\overset{OCH_2CH_3}{\underset{H}{<}}$$

$$\overset{H}{\underset{R}{>}}C=C\overset{OCH_2CH_3}{\underset{H}{<}} \xrightarrow{H_3O^+} R-CH_2-\overset{\overset{\displaystyle O}{\|}}{C}-H$$

Propose mechanisms for these steps.
(c) Explain why a nucleophilic reagent such as ethoxide adds to an alkyne more easily than it adds to an alkene.

14-45. A foul-smelling compound of molecular weight 102 gives the NMR and IR spectra shown on the facing page. Propose a structure and show how it is consistent with the observed absorptions.

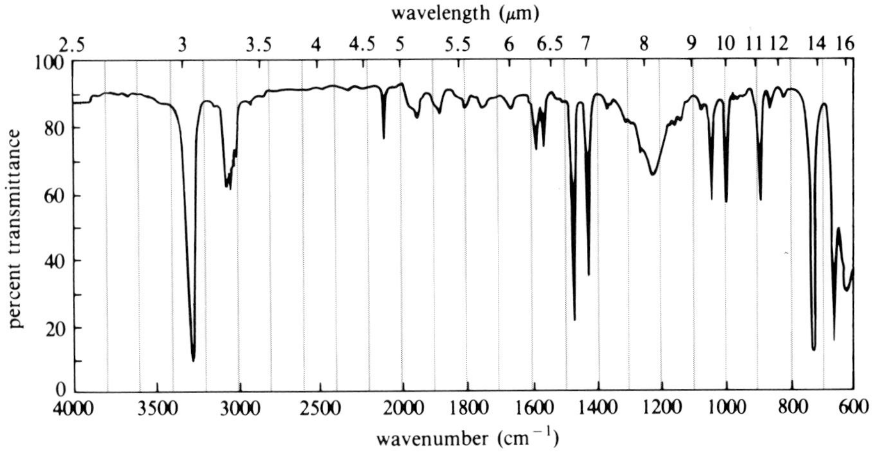

wavelength (μm)
2.5 3 3.5 4 4.5 5 5.5 6 6.5 7 8 9 10 11 12 14 16
percent transmittance
100 80 60 40 20 0
4000 3500 3000 2500 2000 1800 1600 1400 1200 1000 800 600
wavenumber (cm⁻¹)

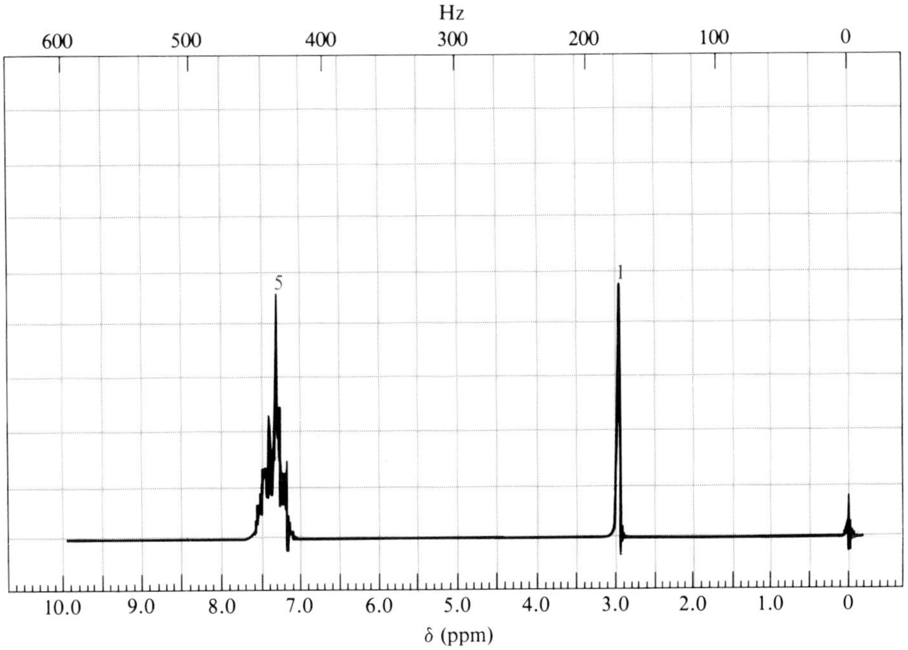

Hz
600 500 400 300 200 100 0
5
1
10.0 9.0 8.0 7.0 6.0 5.0 4.0 3.0 2.0 1.0 0
δ (ppm)

14-46. An unknown compound is found to have molecular formula $C_8H_{14}O$. Its proton NMR, ^{13}C NMR, and IR spectra appear below. The singlet at $\delta 3.1$ in the proton NMR disappears when the sample is shaken with D_2O. Determine the structure of this compound, make NMR peak assignments for the protons and carbon atoms, and show how this compound might be synthesized from acetylene and a six-carbon ketone.

wavelength (μm)
2.5 3 3.5 4 4.5 5 5.5 6 6.5 7 8 9 10 11 12 14 16
percent transmittance
100 80 60 40 20 0
4000 3500 3000 2500 2000 1800 1600 1400 1200 1000 800 600
wavenumber (cm⁻¹)

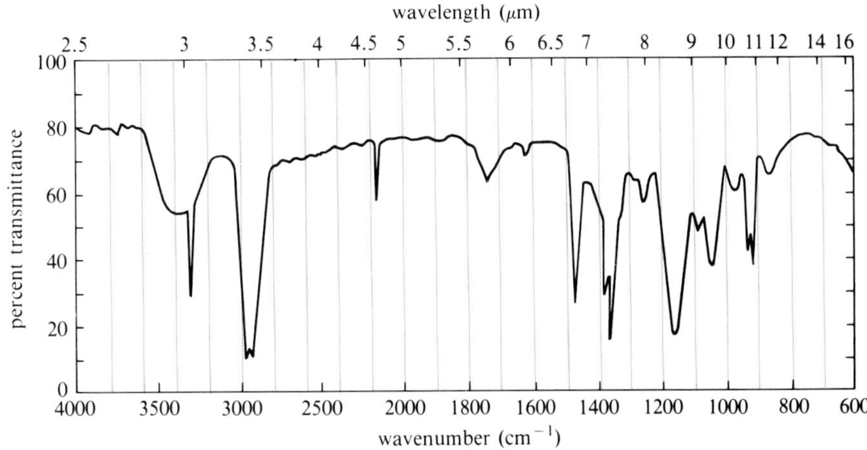

Problem 14-46 *continued*

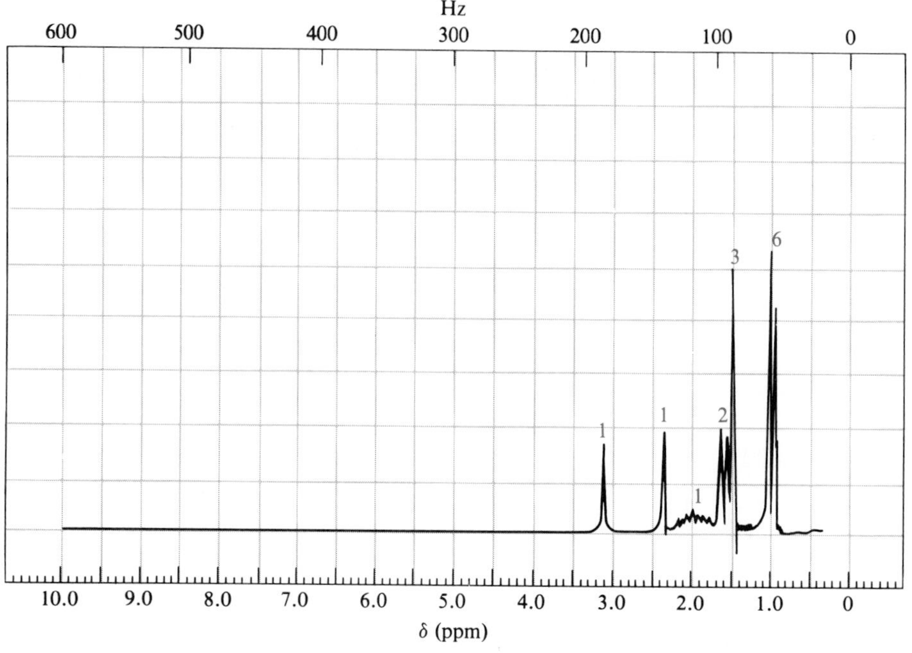

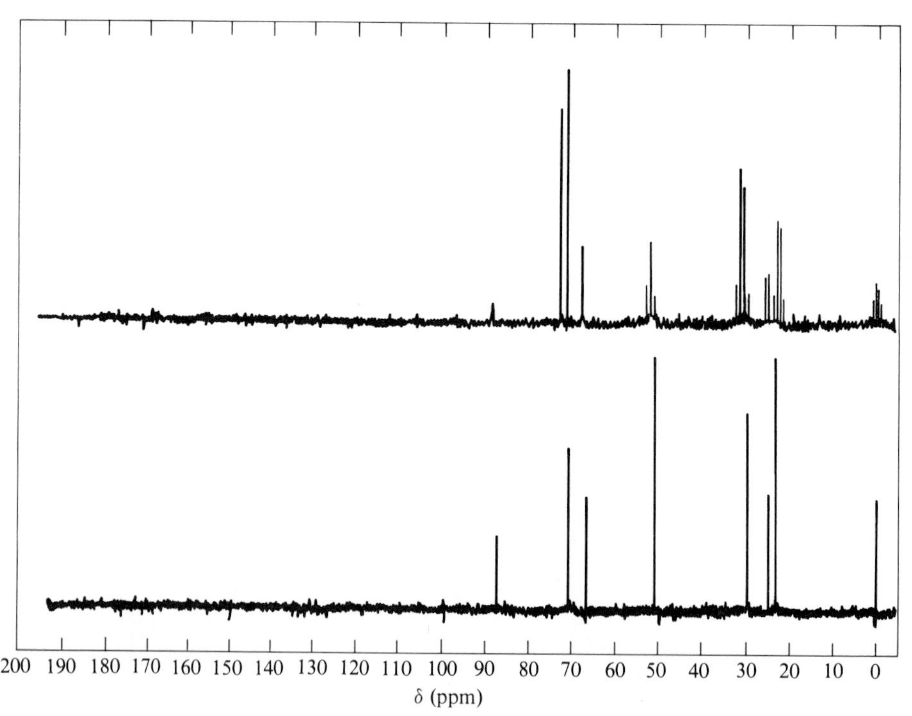

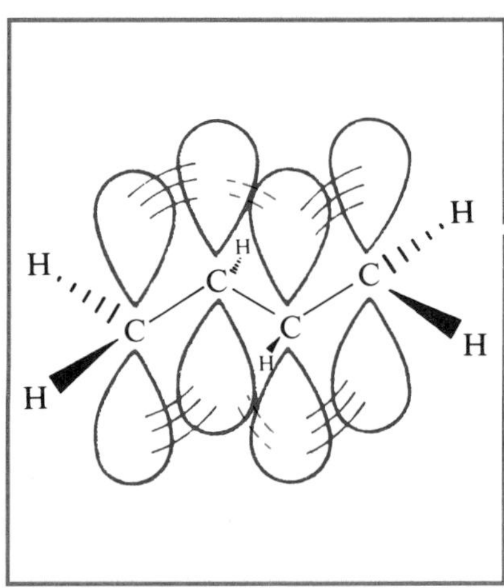

15

CONJUGATED SYSTEMS, ORBITAL SYMMETRY, AND ULTRAVIOLET SPECTROSCOPY

15-1 INTRODUCTION

Double bonds that are separated by just one single bond interact with each other and are called **conjugated double bonds.** Double bonds with two or more single bonds separating them have little interaction and are called **isolated double bonds.** For example, 1,3-pentadiene has conjugated double bonds, while 1,4-pentadiene has isolated double bonds.

$$>C{=}C<\ \text{—}\ >C{=}C<$$

conjugated double bonds
(more stable than isolated double bonds)

$$H_2C{=}CH{-}CH{=}CH{-}CH_3$$

1,3-pentadiene

$$>C{=}C<\ CH_2\ >C{=}C<$$

isolated double bonds

$$H_2C{=}CH{-}CH_2{-}CH{=}CH_2$$

1,4-pentadiene

Because of the interaction between the double bonds, systems containing conjugated double bonds tend to be more stable than similar systems with isolated double bonds. In this chapter we consider the unique properties of conjugated systems, the theoretical reasons for this extra stability, and some of the characteristic reactions of molecules containing conjugated double bonds. We also study ultraviolet spectroscopy, a tool for determining the structures of conjugated systems.

15-2 STABILITIES OF DIENES

In Chapter 7 we used **heats of hydrogenation** to compare the relative stabilities of alkenes. For example, the heats of hydrogenation of 1-pentene and *trans*-2-pentene show that the disubstituted double bond in *trans*-2-pentene is 2.6 kcal (10 kJ) per mole more stable than the monosubstituted double bond in 1-pentene.

1-pentene $\xrightarrow[\text{Pt}]{H_2}$ pentane $\Delta H° = -30.0$ kcal (-125 kJ)

trans-2-pentene $\xrightarrow[\text{Pt}]{H_2}$ pentane $\Delta H° = -27.4$ kcal (-115 kJ)

When a molecule has two isolated double bonds, the heat of hydrogenation is close to the sum of the heats of hydrogenation for the individual double bonds. For example, the heat of hydrogenation of 1,4-pentadiene is -60.2 kcal (-252 kJ), about twice that of 1-pentene.

1,4-pentadiene $\xrightarrow[\text{Pt}]{2\,H_2}$ pentane $\Delta H° = -60.2$ kcal (-252 kJ)

For conjugated dienes, the heat of hydrogenation is less than the sum for the individual double bonds. For example, *trans*-1,3-pentadiene has a monosubstituted double bond like the one in 1-pentene and a disubstituted double bond like the one in 2-pentene. The sum of the heats of hydrogenation of 1-pentene and 2-pentene is -57.4 kcal (-240 kJ), but the heat of hydrogenation of *trans*-1,3-pentadiene is -53.7 kcal (-225 kJ), showing that the conjugated diene has about 3.7 kcal (15 kJ) extra stability.

$$(-30.0 \text{ kcal}) + (-27.4 \text{ kcal}) = -57.4 \text{ kcal } (-240 \text{ kJ})$$

trans-1,3-pentadiene $\xrightarrow[\text{Pt}]{2\,H_2}$ pentane

actual value: -53.7 kcal (-225 kJ)
more stable by 3.7 kcal (15 kJ)

What happens if two double bonds are even closer together than in the conjugated case? Successive double bonds with no intervening single bonds are called **cumulated double bonds.** Consider 1,2-pentadiene, which contains cumulated double bonds. Such 1,2-diene systems are also called **allenes,** after the simplest member of the class, 1,2-propadiene or "allene," $H_2C{=}C{=}CH_2$. The heat of hydrogenation of 1,2-pentadiene is -69.8 kcal/mol (-292 kcal/mol).

$H_2C{=}C{=}CH(CH_2CH_3)$ $\xrightarrow[\text{Pt}]{H_2}$ $CH_3CH_2CH_2CH_2CH_3$ $\Delta H° = -69.8$ kcal (-292 kJ)

1,2-pentadiene (ethylallene)   pentane

sum of 1-pentene + 2-pentene $\Delta H° = -57.4$ kcal (-240 kJ)
1,2-pentadiene is *less* stable by 12.4 kcal (52 kJ)

Because 1,2-pentadiene has a larger heat of hydrogenation than 1,4-pentadiene, we conclude that the cumulated double bonds of allenes are less stable than isolated double bonds and much less stable than conjugated double bonds. Figure 15-1 summarizes the relative stability of isolated, conjugated, and cumulated dienes and compares them with alkynes.

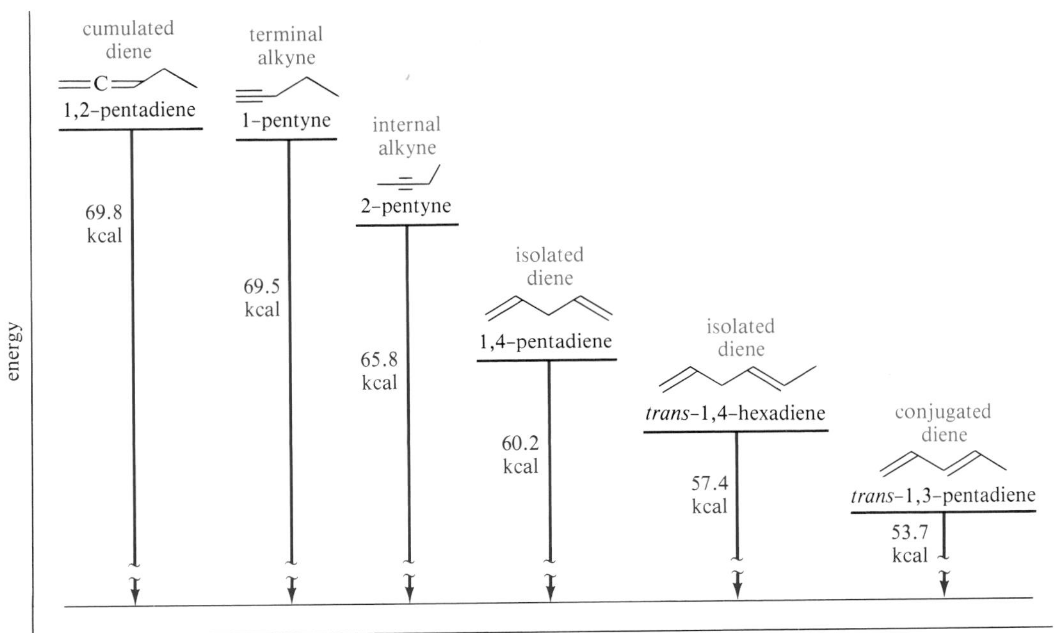

FIGURE 15-1 Relative energies of alkynes and conjugated, isolated, and cumulated dienes, based on heats of hydrogenation (kcal/mol).

PROBLEM 15-1

Rank each of the following groups of compounds in order of increasing heat of hydrogenation.

(a) 1,2-hexadiene; 1,3,5-hexatriene; 1,3-hexadiene; 1,4-hexadiene; 1,5-hexadiene; 2,4-hexadiene

(b) 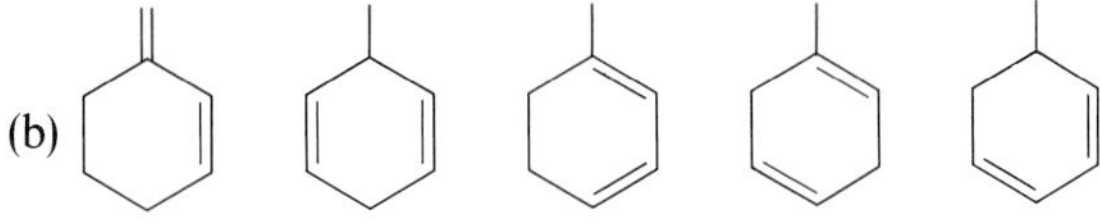

PROBLEM 15-2

In a strongly acidic solution, 1,4-cyclohexadiene tautomerizes to 1,3-cyclohexadiene. Give a mechanism for this rearrangement and explain why it is energetically favorable.

PROBLEM 15-3

(Review) The central carbon atom of an allene is a member of two double bonds, and it has an interesting orbital arrangement that holds the two ends of the molecule at right angles to each other.

(a) Draw an orbital diagram of allene, showing why the two ends are perpendicular.
(b) Draw the two enantiomers of 1,3-dichloroallene.

15-3 MOLECULAR ORBITAL PICTURE OF A CONJUGATED SYSTEM

In Figure 15-1 the compound with conjugated double bonds is 3.7 kcal/mol (15 kJ/mol) more stable than a similar compound with isolated double bonds. This 3.7 kcal (15 kJ) of extra stability in the conjugated molecule is called the **resonance energy** of the system. (Other terms favored by some chemists are *conjugation energy, delocalization energy,* and *stabilization energy.*) The origin of this extra stability of conjugated systems is best explained by examining their **molecular orbitals.** Let's start by considering the molecular orbitals of the simplest conjugated diene, 1,3-butadiene.

15-3A THE STRUCTURE AND BONDING OF 1,3-BUTADIENE

The heat of hydrogenation of 1,3-butadiene is about 3.6 kcal (15 kJ) less than twice that of 1-butene, showing that 1,3-butadiene has a resonance energy of 3.6 kcal.

$$\underset{\text{1,3-butadiene}}{H_2C{=}CH{-}CH{=}CH_2} \xrightarrow{H_2,\,Pt} CH_3{-}CH_2{-}CH_2{-}CH_3 \qquad \Delta H^\circ = -56.6 \text{ kcal } (-237 \text{ kJ})$$

$$\underset{\text{1-butene}}{H_2C{=}CH{-}CH_2{-}CH_3} \xrightarrow{H_2,\,Pt} CH_3{-}CH_2{-}CH_2{-}CH_3 \qquad \Delta H^\circ = -30.1 \text{ kcal } (-126 \text{ kJ})$$

$$\times 2 = -60.2 \text{ kcal } (-252 \text{ kJ})$$

resonance energy of 1,3-butadiene = 60.2 kcal − 56.6 kcal = 3.6 kcal (15 kJ)

The most stable conformation of 1,3-butadiene is shown in Figure 15-2. Note that this conformation is planar, with the p orbitals on the two pi bonds aligned.

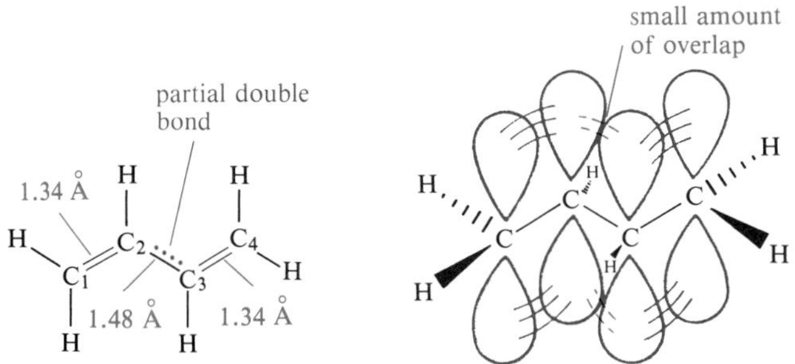

FIGURE 15-2 Structure of 1,3-butadiene in its most stable conformation. The 1.48-Å central carbon-carbon single bond is shorter than the 1.54-Å bonds typical of alkanes because of its partial double-bond character.

The C2—C3 bond in 1,3-butadiene is considerably shorter than a carbon-carbon single bond in an alkane: 1.48 versus 1.54 Å. Although this bond is shortened slightly by the increased s character of the sp^2 hybrid orbitals, the most important cause is the pi-bonding overlap and partial double-bond character of the bond. The planar conformation, with the p orbitals of the two double bonds aligned, allows overlap between the pi bonds. In effect, the electrons in the two double bonds are **delocalized** over the entire molecule, creating some pi overlap and pi bonding in the C2—C3 bond. The length of this bond is intermediate between the normal length of a single bond and that of a double bond.

Lewis structures are not adequate to represent delocalized molecules such as 1,3-butadiene. To represent the bonding in conjugated systems such as 1,3-butadiene accurately, we must consider molecular orbitals that represent the entire conjugated pi system, not just one bond at a time.

15-3B CONSTRUCTING THE MOLECULAR ORBITALS OF 1,3-BUTADIENE

All four carbon atoms of 1,3-butadiene are sp^2 hybridized, and (in the planar conformation) they all have overlapping p orbitals. Let's review how we constructed the pi molecular orbitals (MO's) of ethylene from the p atomic orbitals of the two carbon atoms (Figure 15-3). Each p orbital consists of two lobes, with opposite phases of the wave function in the two lobes. The plus and minus signs used in drawing these orbitals indicate the *phase of the wave function,* **not** electrical charges. To minimize confusion, we will use color in the lobes of the p orbitals to emphasize the phase difference.

In the π-bonding orbital of ethylene, there is overlap of lobes with the same

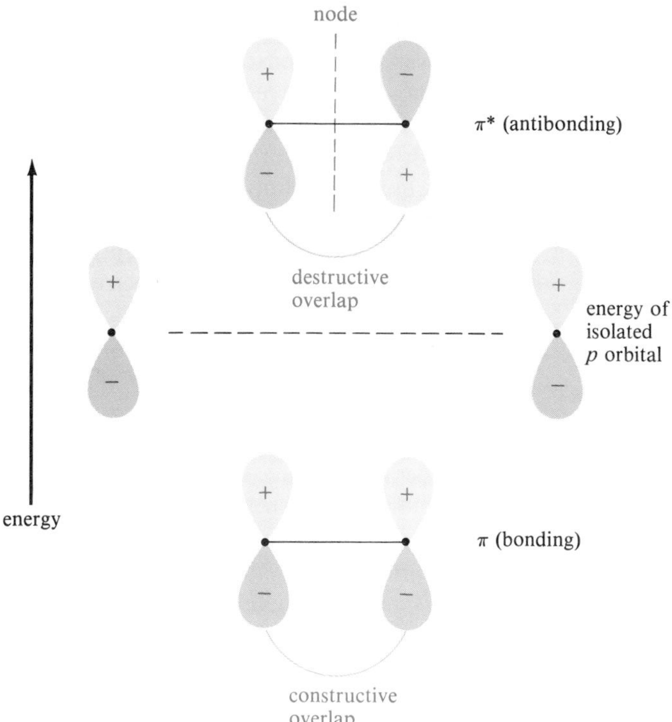

FIGURE 15-3 The pi bonding orbital of ethylene is formed by constructive overlap of the unhybridized p orbitals on the sp^2 hybrid carbon atoms. The antibonding pi orbital is formed by the destructive overlap of these two orbitals. The combination of two p orbitals must result in the formation of exactly two molecular orbitals.

sign (+ with + and − with −) in the bonding region between the nuclei. We call this reinforcement of the wave function **constructive overlap.** In the antibonding orbital, there is canceling of opposite signs (+ with −) in the bonding region. This canceling of the wave function is called **destructive overlap.** Electrons have lower potential energy in the **bonding MO** than in the original p orbitals, and higher potential energy in the **antibonding MO.** In the ground state of ethylene, two electrons are in the bonding MO, but the antibonding MO is vacant.

When viewing Figure 15-3, there are several important principles to keep in mind. Constructive overlap results in a bonding interaction; destructive overlap results in an antibonding interaction. Also, the number of molecular orbitals is always the same as the number of p orbitals used to form the MO's. These molecular orbitals have energies that are symmetrically distributed above and below the energy of the starting p orbitals. Half are bonding MO's and half are antibonding MO's.

Now we are ready to construct the molecular orbitals of 1,3-butadiene. The p orbitals on C1 through C4 overlap, giving an extended system of four p orbitals that form four pi molecular orbitals. Two MO's are bonding, and two are antibonding. To represent the four p orbitals on the carbon atoms of 1,3-butadiene, we draw four p orbitals in a line. Although 1,3-butadiene is not linear, this simple straight-line representation makes it easier to draw and visualize the molecular orbitals.

represented by

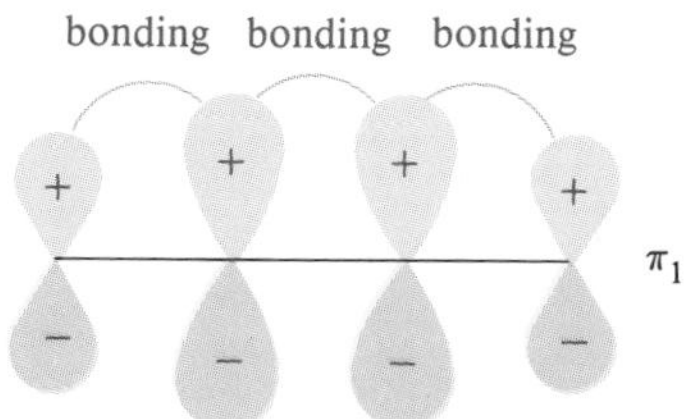

FIGURE 15-4 The lowest-energy orbital of 1,3-butadiene has a bonding interaction between each pair of adjacent carbon atoms. This orbital is termed π_1, because it is a pi bonding orbital and it has the lowest energy.

The lowest-energy molecular orbital always consists entirely of bonding interactions. We indicate such an orbital by drawing all the positive phases of the *p* orbitals overlapping constructively on one face of the molecule, and the negative phases overlapping constructively on the other face. Figure 15-4 shows the lowest-energy MO for 1,3-butadiene. This MO places electron density on all four *p* orbitals, with slightly more on C2 and C3. (In these figures, larger and smaller *p* orbitals are used to show which atoms bear more of the electron density in a particular MO.)

This lowest-energy orbital is exceptionally stable for two reasons: There are three bonding interactions, and the electrons are delocalized over four nuclei. This orbital helps to illustrate why the conjugated system is more stable than two isolated double bonds. It also shows there is some pi-bond character between C2 and C3, which lowers the energy of the planar conformation and helps to explain the short C2—C3 bond length.

As with ethylene, the second molecular orbital, π_2, of butadiene (Fig. 15-5) has one node in the center of the molecule. This molecular orbital represents the classical picture of a diene. There are bonding interactions at the C1—C2 and C3—C4 bonds, and there is a (weaker) antibonding interaction between C2 and C3. The pi bonding between the central carbon atoms produced by π_1 is partially offset by this antibonding interaction of π_2. It is not quite canceled, however, and there is still some double-bond character in this central bond.

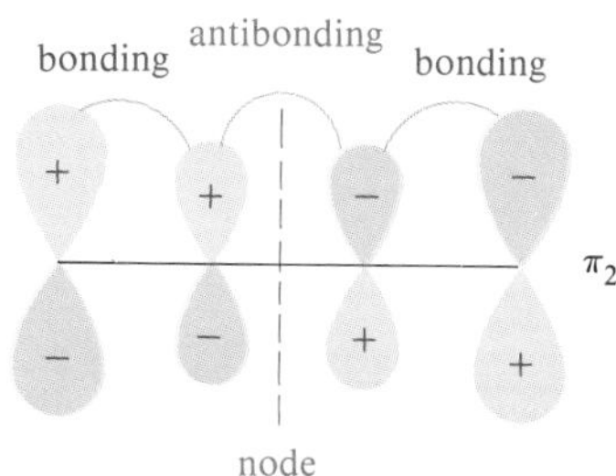

FIGURE 15-5 The second MO of 1,3-butadiene has one node in the center of the molecule. There are bonding interactions at the C1—C2 and C3—C4 bonds, and there is a (weaker) antibonding interaction between C2 and C3. This π_2 orbital is a bonding orbital but is not as strongly bonding as π_1.

The π_2 orbital has two bonding interactions and one antibonding interaction, so we expect it to be a bonding orbital (two bonding − one antibonding = one bonding). It is not as strongly bonding, nor as low in energy, as the all-bonding π_1 orbital. Adding and subtracting bonding and antibonding interactions is not a reliable method for calculating energies of molecular orbitals, but it is useful for predicting whether a given orbital is bonding or antibonding and for ranking orbitals in order of their energy.

The third butadiene MO, π_3, has two nodes (Fig. 15-6). There is a bonding interaction at the C2—C3 bond, and there are two antibonding interactions, one at the C1—C2 bond, and the other at the C3—C4 bond. This is an antibonding orbital, and it is vacant in the ground state.

The fourth, and last, molecular orbital (π_4) of 1,3-butadiene has three nodes and is totally antibonding (Fig. 15-7). This MO has the highest energy and is unoccupied in the molecule's ground state. This highest-energy MO (π_4) is typical: For most systems the highest-energy molecular orbital has an antibonding interaction at each bond.

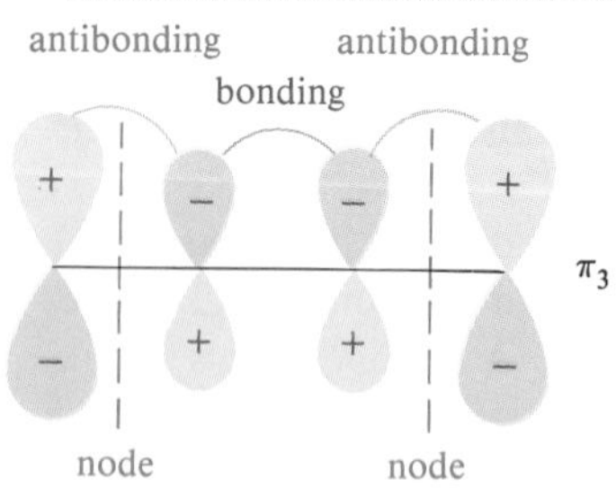

FIGURE 15-6 The third butadiene MO is an antibonding orbital, and it is vacant in the ground state.

Butadiene has four pi electrons (two electrons in each of the two double bonds of the Lewis structure) to be placed in the four MO's described above. Each MO can accommodate two electrons, and the lowest-energy MO's are filled first. Therefore, the four pi electrons go into the two lowest-energy MO's, π_1 and π_2. Figure 15-8 shows the electronic configuration of 1,3-butadiene. Both bonding MO's are filled, and both antibonding MO's are empty. Most stable molecules have this arrangement of filled bonding orbitals and vacant antibonding orbitals. Figure 15-8 also gives the relative energies of the ethylene MO's to show that the conjugated butadiene system is slightly more stable than two ethylene double bonds.

The partial double-bond character between C2 and C3 in 1,3-butadiene explains why the molecule is most stable in a planar conformation. There are actually two planar conformations that allow overlap between C2 and C3. The

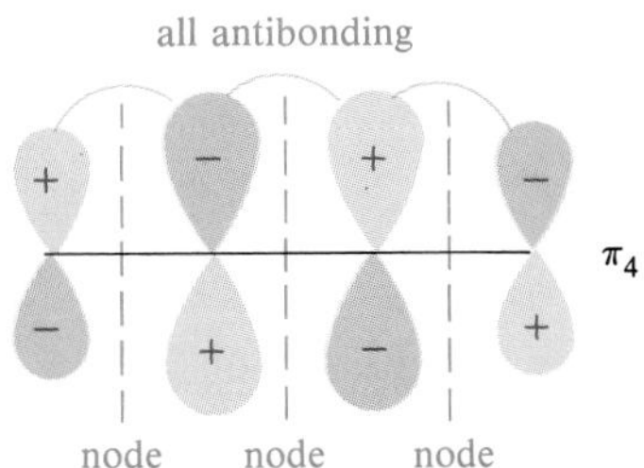

FIGURE 15-7 The highest-energy MO of 1,3-butadiene has three nodes and three antibonding interactions. It is strongly antibonding and its energy is very high.

***s*-trans conformation** is slightly more stable than the ***s*-cis conformation** (by 2.3 kcal or 9.6 kJ), in which there is interference between the two nearby hydrogen atoms.

mild interference

s-trans ⇌ *s*-cis

These conformations are named *s*-trans ("single"-trans) and *s*-cis ("single"-cis) because they are single-bond analogs of trans and cis isomers about a double bond. There is a much smaller rotational barrier, however: about 4.9 kcal/mol (20.5 kJ/mol) for rotation about the C2—C3 bond of butadiene, compared with a barrier of about 60 kcal/mol for rotation of a double bond in an alkene. The *s*-cis and *s*-trans conformers of butadiene (and all the skew conformations in between) easily interconvert at room temperature.

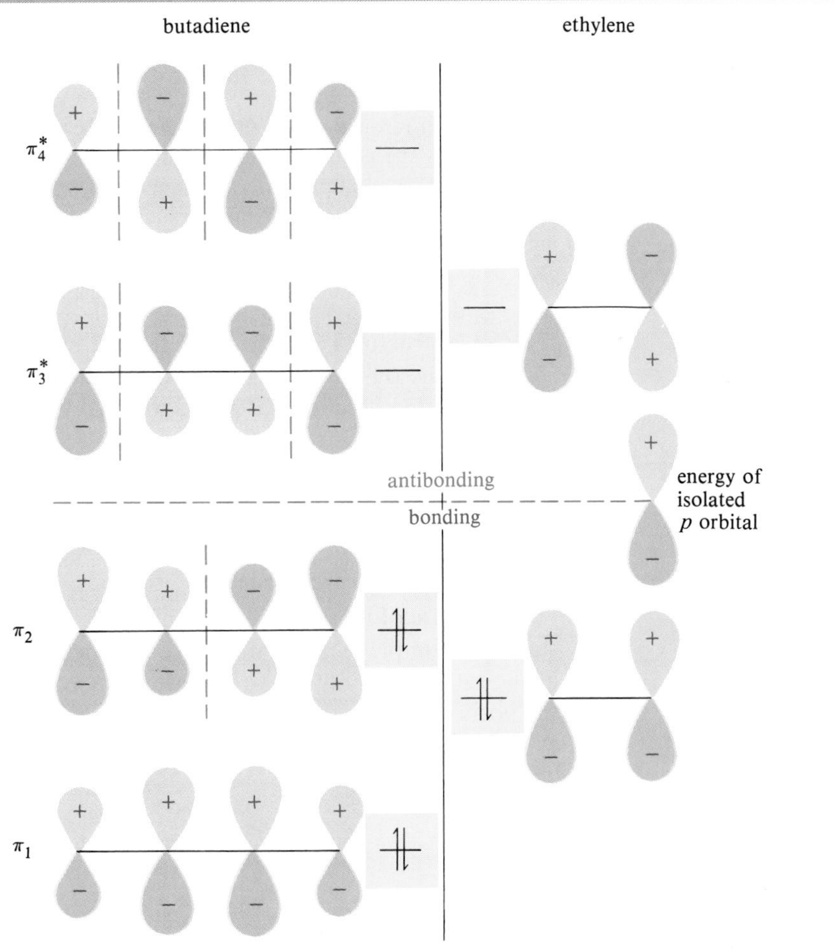

FIGURE 15-8 In both 1,3-butadiene and ethylene the bonding MO's are filled and the antibonding MO's are vacant. The average energy of the butadiene electrons is slightly lower than the average energy of the electrons in ethylene. This lower energy is the resonance stabilization of the conjugated diene.

15-4 ALLYLIC CATIONS

In Chapter 7 we saw that the $—CH_2—CH{=}CH_2$ group is called the **allyl group.** Many compounds have names that use this terminology.

$H_2C{=}CH—CH_2Br$ (allyl bromide) $\quad H_2C{=}CH—CH_2—O—CH_2—CH{=}CH_2$ (allyl ether) $\quad H_2C{=}CH—CH_2OH$ (allyl alcohol)

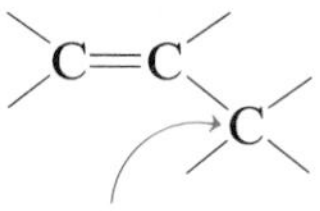

allylic position

When allyl bromide is heated with a good ionizing solvent, it ionizes to a carbocation that is an allyl group with a positive charge. This cation is called the **allyl cation.** More highly substituted analogs are called **allylic cations.** All allylic cations are stabilized by resonance with the adjacent double bond, which delocalizes the positive charge over two carbon atoms.

$$H_2C{=}CH—CH_2—\ddot{Br}: \rightleftharpoons [H_2C{=}CH—\overset{+}{C}H_2 \longleftrightarrow H_2\overset{+}{C}—CH{=}CH_2] + :\ddot{Br}:^-$$

allyl bromide — allyl cation

Substituted allylic cations

$$H_2C{=}CH—\overset{+}{C}H—CH_3 \qquad (CH_3)_2C{=}CH—\overset{+}{C}H_2$$

(cyclohexenyl cation)

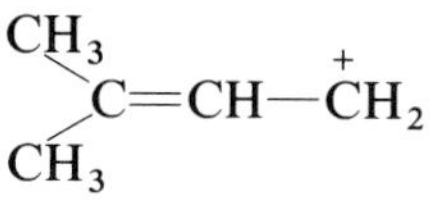

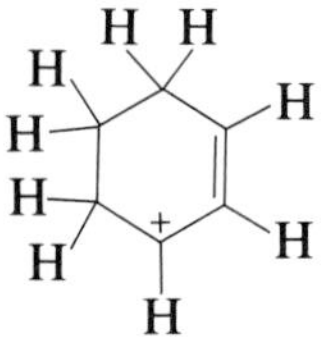

PROBLEM 15-4

Draw another resonance structure for each of the substituted allylic cations shown above, showing how the positive charge is shared by another carbon atom. In each case, state whether your second resonance structure is a more or a less important resonance contributor than the first structure. (Which structure places the positive charge on the more highly substituted carbon atom?)

A delocalized ion such as the allyl cation may be represented either by resonance structures, as shown on the left below, or by a combined structure as shown on the right. Although the combined structure is more concise, it is sometimes confusing because it attempts to convey all the information provided by two or more resonance structures.

$$\left[H_2\underset{1}{C}{=}\underset{2}{C}H—\underset{3}{\overset{+}{C}}H_2 \longleftrightarrow H_2\underset{1}{\overset{+}{C}}—\underset{2}{C}H{=}\underset{3}{C}H_2\right] \quad \text{or} \quad H_2\underset{1}{\overset{\frac{1}{2}+}{C}}{\cdots}\underset{2}{C}H{\cdots}\underset{3}{\overset{\frac{1}{2}+}{C}}H_2$$

resonance structures — combined representation

Because of its resonance stabilization, the (primary) allyl cation is about as stable as a simple secondary carbocation such as the isopropyl cation. Substituted allylic cations generally have at least one secondary carbon atom bearing part of the positive charge; they are about as stable as simple tertiary carbocations such as the *t*-butyl cation.

Stability of carbocations

$$H_3C^+ < 1° < 2°, \text{allyl} < 3°, \text{substituted allylic}$$

$$H_2\overset{\frac{1}{2}+}{C}{\cdots}CH{\cdots}\overset{\frac{1}{2}+}{C}H_2 \text{ is about as stable as } CH_3—\overset{+}{C}H—CH_3$$

$$CH_3—\overset{\delta+}{C}H{\cdots}CH{\cdots}\overset{\delta+}{C}H_2 \text{ is about as stable as } CH_3—\overset{+}{C}(CH_3)_2$$

15-5 1,2- AND 1,4-ADDITION TO CONJUGATED DIENES

An allylic cation can react with a nucleophile at either of the two positive centers. For example, when HBr adds to 1,3-butadiene, a mixture of two products results. One of the products, 3-bromo-1-butene, results from addition across one of the double bonds. The second product, 1-bromo-2-butene, has a double bond between the two central carbon atoms, resulting from a shift of the double bond.

$$H_2C{=}CH{-}CH{=}CH_2 + HBr \longrightarrow H_2C(H){-}CH(Br){-}CH{=}CH_2 + H_2C(H){-}CH{=}CH{-}CH_2(Br)$$

3-bromo-1-butene (1,2-addition) 1-bromo-2-butene (1,4-addition)

The first product results from a simple electrophilic addition of HBr across a double bond to a pair of adjacent carbon atoms. This process is called a **1,2-addition,** whether or not these two carbon atoms are numbered 1 and 2 in naming the compound. In the second product, the proton and bromide ion add at the ends of the conjugated system, to carbon atoms with a 1,4-relationship. Such an addition is called a **1,4-addition,** whether or not these carbon atoms are numbered 1 and 4 in naming the compound.

$$>C{=}C{-}C{=}C< \xrightarrow{A-B} {}^1C(A){-}{}^2C(B){-}C{=}C< + {}^1C(A){-}{}^2C{=}{}^3C{-}{}^4C(B)$$

1,2-addition 1,4-addition

HBr adds to 1,3-butadiene by a mechanism similar to other electrophilic additions to alkenes. The proton is the electrophile, adding to the alkene to give the most stable carbocation. Protonation of 1,3-butadiene gives an allylic cation, which is stabilized by resonance delocalization of the positive charge over two carbon atoms.

$$Br{-}H + H_2C{=}CH{-}CH{=}CH_2 \longrightarrow \left[H_3C{-}C^+H{-}CH{=}CH_2 \longleftrightarrow H_3C{-}CH{=}CH{-}C^+H_2 \right]$$

Bromide can attack this resonance-stabilized intermediate at either of the two carbon atoms sharing the positive charge. Attack at the secondary carbon gives 1,2-addition, while attack at the primary carbon gives 1,4-addition.

$$\left[H_3C{-}C^+H{-}CH{=}CH_2 \longleftrightarrow H_3C{-}CH{=}CH{-}C^+H_2 \right] \xrightarrow{:\ddot{Br}:^-} H_3C{-}CH(Br){-}CH{=}CH_2 \text{ and } H_3C{-}CH{=}CH{-}CH_2{-}Br$$

(1,2 and 1,4 attack shown by arrows) 1,2-addition 1,4-addition

The key to the formation of these two products is the presence of a double bond in the proper position for formation of a stabilized allylic cation. Molecules having such double bonds are likely to undergo reactions whose intermediates take advantage of the double bond's ability to delocalize charges and radical (unpaired) electrons.

PROBLEM 15-5

Treatment of an alkyl halide with alcoholic $AgNO_3$ often promotes ionization.

$$Ag^+ + R{-}Cl \longrightarrow AgCl + R^+$$

When 3-chloro-1-methylcyclopentene reacts with $AgNO_3$ in ethanol, two isomeric ethers are formed. Suggest structures for these ethers and give a mechanism for their formation.

PROBLEM 15-6

Give a detailed mechanism for each of the following reactions, showing explicitly how the observed mixtures of products are formed.

(a) 3-methyl-2-buten-1-ol + HBr →
1-bromo-3-methyl-2-butene + 3-bromo-3-methyl-1-butene

(b) 2-methyl-3-buten-2-ol + HBr →
1-bromo-3-methyl-2-butene + 3-bromo-3-methyl-1-butene

(c) 1,3-butadiene + Br_2 → 3,4-dibromo-1-butene + 1,4-dibromo-2-butene

(d) 1-chloro-2-butene + $AgNO_3$, H_2O → 2-buten-1-ol + 3-buten-2-ol

(e) 3-chloro-1-butene + $AgNO_3$, H_2O → 2-buten-1-ol + 3-buten-2-ol

15-6 KINETIC VERSUS THERMODYNAMIC CONTROL IN THE ADDITION OF HBr TO 1,3-BUTADIENE

One of the interesting peculiarities of the reaction of 1,3-butadiene with HBr is the effect of temperature on the product mixture. If the reagents are allowed to react briefly at −80°C, the 1,2-addition product predominates. If this reaction mixture is later allowed to warm to 40°C, however, or if the reaction itself is carried out at 40°C, the composition favors the product of 1,4-addition.

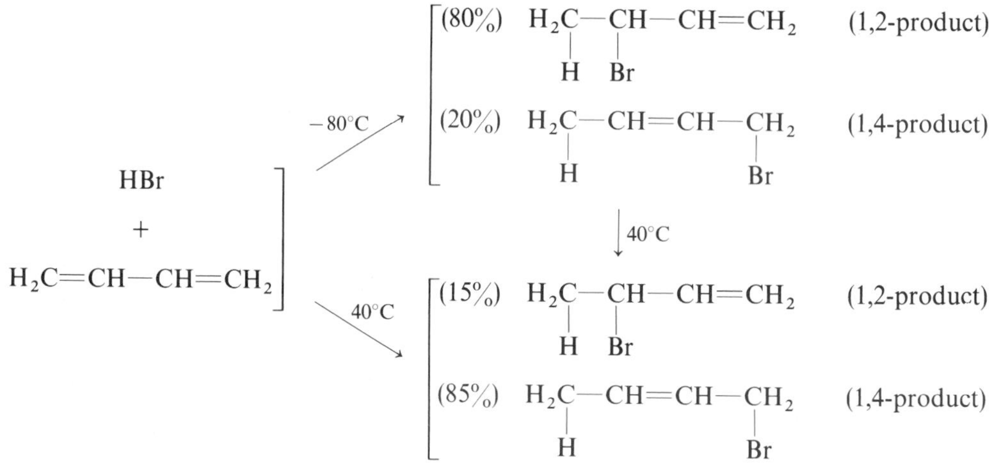

$$HBr + H_2C{=}CH{-}CH{=}CH_2 \xrightarrow{-80^\circ C} \begin{bmatrix} (80\%)\ H_2C(H){-}CH(Br){-}CH{=}CH_2 & \text{(1,2-product)} \\ (20\%)\ H_2C(H){-}CH{=}CH{-}CH_2Br & \text{(1,4-product)} \end{bmatrix}$$

$$\downarrow 40^\circ C$$

$$HBr + H_2C{=}CH{-}CH{=}CH_2 \xrightarrow{40^\circ C} \begin{bmatrix} (15\%)\ H_2C(H){-}CH(Br){-}CH{=}CH_2 & \text{(1,2-product)} \\ (85\%)\ H_2C(H){-}CH{=}CH{-}CH_2Br & \text{(1,4-product)} \end{bmatrix}$$

This variation in product composition with temperature reminds us that the most stable product is not always the major product. Of the two products, we expect 1-bromo-2-butene (the 1,4-product) to be more stable, since it has the more highly substituted double bond. This prediction is supported by the fact that this isomer predominates when the reaction mixture is warmed to 40°C and allowed to equilibrate.

A potential-energy diagram for the second step of this reaction (Fig. 15-9) helps to show why one product is favored at low temperatures and another at higher temperatures. The allylic cation is in the center of the diagram; it can react toward the left to give the 1,2-product or toward the right to give the 1,4-product. The initial product depends on where bromide attacks the resonance-stabilized allylic cation. Bromide can attack at either of the two carbon atoms that share

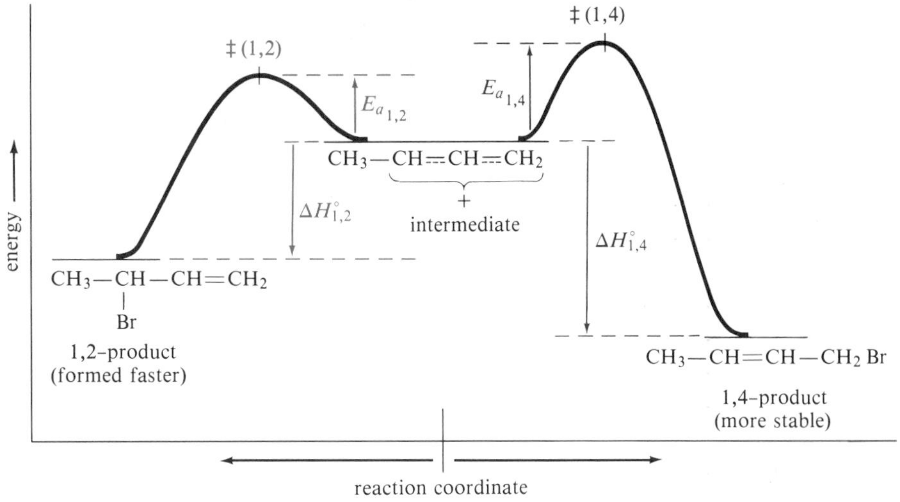

FIGURE 15-9 The potential-energy diagram for the reaction of 1,3-butadiene with HBr shows that the transition state leading to 1,2-addition has a lower energy than that leading to the 1,4-product, so the 1,2-product is formed faster (kinetic product). However, the 1,2-product itself is not as stable as the 1,4-product. If equilibrium is reached, the 1,4-product predominates (thermodynamic product).

the positive charge. Attack at the secondary carbon gives 1,2-addition, and attack at the primary carbon gives 1,4-addition.

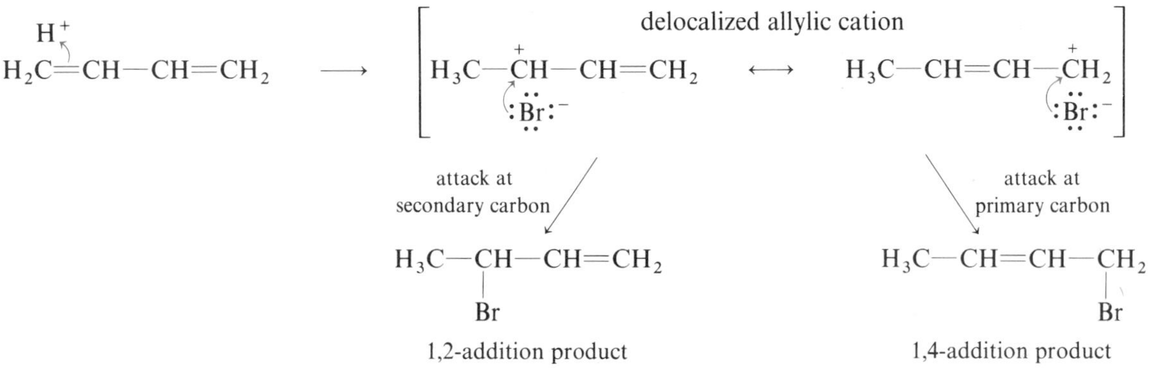

$$H^+ + H_2C{=}CH{-}CH{=}CH_2 \longrightarrow \left[H_3C{-}\overset{+}{C}H{-}CH{=}CH_2 \;\; :\ddot{Br}:^- \longleftrightarrow H_3C{-}CH{=}CH{-}\overset{+}{C}H_2 \;\; :\ddot{Br}:^- \right]$$

delocalized allylic cation

attack at secondary carbon → $H_3C{-}CH(Br){-}CH{=}CH_2$ (1,2-addition product)

attack at primary carbon → $H_3C{-}CH{=}CH{-}CH_2Br$ (1,4-addition product)

Kinetic control at −80°C The transition state for 1,2-addition has a lower energy than the transition state for 1,4-addition, giving the 1,2-addition a lower activation energy (E_a). This is not surprising, because the 1,2-addition results from bromide attack at the more highly substituted secondary carbocation, which bears more of the positive charge because it is better stabilized than the primary carbocation. Because the 1,2-addition has a lower activation energy than the 1,4-addition, the 1,2-addition takes place faster (at *all* temperatures).

The attack by bromide on the allylic cation is a strongly exothermic process, so the reverse reaction has a large activation energy. At −80°C few collisions take place with this much energy, and the rate of the reverse reaction is practically zero. Under these conditions, the product that is formed faster predominates. Because the kinetics of the reaction determine the results, this situation is called **kinetic control** of the reaction. The 1,2-product, favored under these conditions, is called the **kinetic product.**

Thermodynamic control at 40°C At 40°C a significant fraction of molecular collisions have enough energy for reverse reactions to occur. Notice that the activation energy for the reverse of the 1,2-addition is less than that for the reverse of the 1,4-addition. Although the 1,2-product is still formed faster, it also reverts to the allylic cation faster than the 1,4-product. At 40°C an equilibrium is set up, and the concentration of each species is determined by its relative energy. The 1,4-product is the most stable species, and it predominates. Since thermodynamics determine the results, this situation is called **thermodynamic control** (or **equilibrium control**) of the reaction. The 1,4-product, favored under these conditions, is called the **thermodynamic product.**

We will see many additional examples of reactions whose products may be determined by kinetic control or by thermodynamic control, depending on the conditions. In general, reactions that do not reverse easily are kinetically controlled, because an equilibrium is rarely established. In kinetically controlled reactions, the product with the lowest-energy transition state predominates. Reactions that are easily reversible are thermodynamically controlled, unless something happens to prevent equilibrium from being attained. In thermodynamically controlled reactions, the lowest-energy product predominates.

PROBLEM 15-7

When Br_2 is added to 1,3-butadiene at −15°C, the product mixture contains 60 percent of product A and 40 percent of product B. When the same reaction takes place at 60°C, the product ratio is 10 percent A and 90 percent B.

(a) Propose structures for products A and B. (*Hint:* In many cases, an allylic carbocation is more stable than a bromonium ion.)
(b) Give a mechanism to account for formation of both A and B.
(c) Show why A predominates at −15°C, yet B predominates at 60°C.
(d) If you had a solution of pure A, and its temperature were raised to 60°C, what would you expect to happen? Give a mechanism to support your prediction.

15-7 ALLYLIC RADICALS

We have seen that bromine can add to cyclohexene by an ionic, bromonium ion mechanism (Section 8-10).

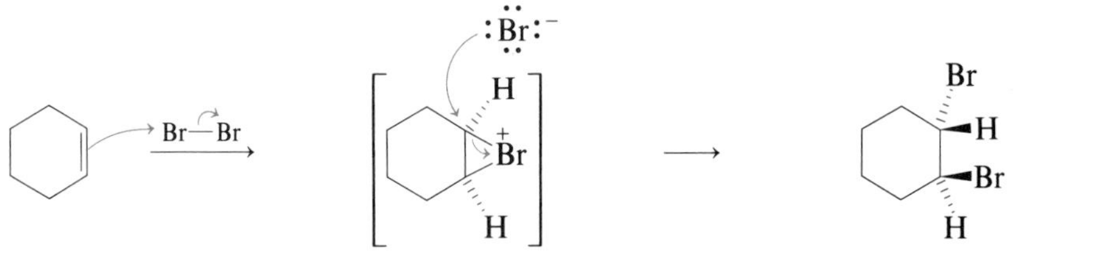

:Br:⁻ H Br H Br H Br H

cyclohexene — bromonium ion intermediate — *trans*-1,2-dibromocyclohexane

When only a small amount of bromine is available, the ionic addition proceeds slowly. If we shine light on the reaction mixture, however, a *substitution* occurs.

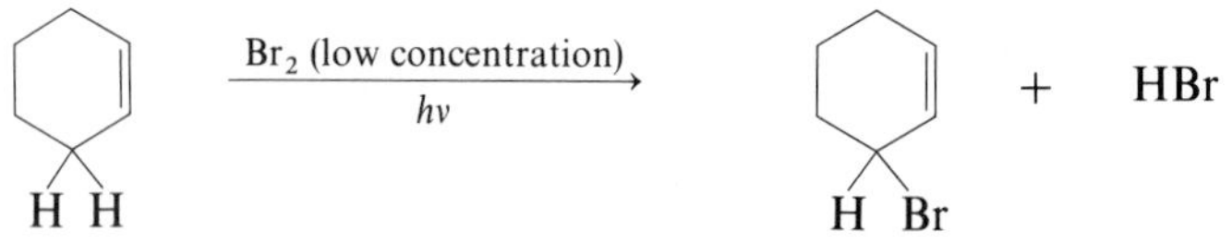

Br_2 (low concentration), $h\nu$; H H; H Br; + HBr

cyclohexene — 3-bromocyclohexene

One bromine atom has substituted for one of the allylic hydrogen atoms in cyclohexene. The displaced allylic hydrogen combines with the remaining bromine atom to form HBr gas. This substitution involves a free-radical chain reaction.

Because free radicals are highly reactive, even a small concentration of free radicals can produce a fast chain reaction. The reaction is initiated by the light-induced dissociation of Br_2 to two bromine radicals. A bromine radical abstracts a hydrogen atom from the allylic position, giving a resonance-stabilized allylic radical. This allylic radical reacts with another molecule of Br_2 to give the allylic bromide and another bromine radical.

Initiation step

$$Br_2 \xrightarrow{h\nu} 2\ Br\cdot$$

Propagation steps

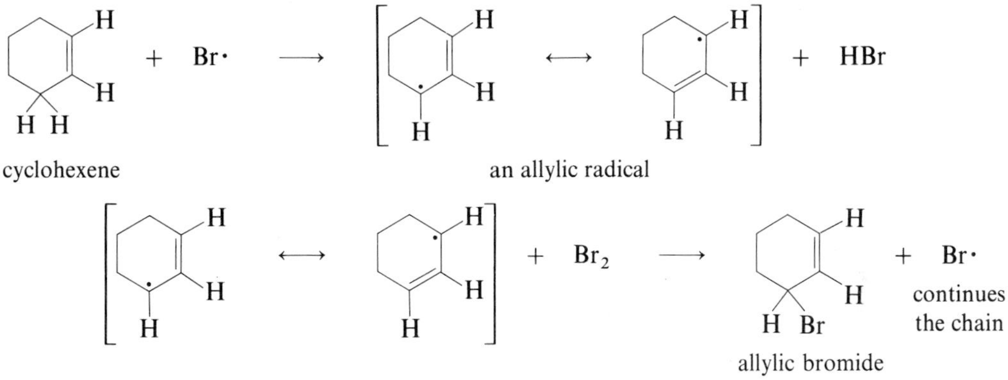

cyclohexene + Br· ⟶ [an allylic radical] + HBr

[allylic radical] + Br_2 ⟶ allylic bromide + Br· (continues the chain)

Stability of allylic radicals Why is it that in the first propagation step, a bromine radical abstracts only an allylic hydrogen atom, not one from another secondary site? This preference for abstraction of allylic hydrogens derives from the resonance stabilization of the allylic free radical that is formed. The bond dissociation energies required to generate several different kinds of free radicals are compared below. Notice that the allyl radical (a primary free radical) is actually 3 kcal (13 kJ) per mole *more* stable than the *tertiary*-butyl radical.

$$CH_3CH_2{-}H \longrightarrow CH_3CH_2\cdot + H\cdot \qquad \Delta H = +98\text{ kcal }(+410\text{ kL}) \quad \text{(primary)}$$
$$(CH_3)_2CH{-}H \longrightarrow (CH_3)_2CH\cdot + H\cdot \qquad \Delta H = +94\text{ kcal }(+393\text{ kJ}) \quad \text{(secondary)}$$
$$(CH_3)_3C{-}H \longrightarrow (CH_3)_3C\cdot + H\cdot \qquad \Delta H = +91\text{ kcal }(+381\text{ kJ}) \quad \text{(tertiary)}$$
$$H_2C{=}CH{-}CH_2{-}H \longrightarrow H_2C{=}CH{-}CH_2\cdot + H\cdot \qquad \Delta H = +88\text{ kcal }(+368\text{ kJ}) \quad \text{(allyl)}$$

The allylic 2-cyclohexenyl radical in the free-radical bromination of cyclohexene has its odd electron delocalized over two secondary carbon atoms, so it is even more stable than the unsubstituted allyl radical. The second propagation step may occur at either of the two radical carbons, but in this symmetrical case reaction at either position gives 3-bromocyclohexene as the product. Less symmetrical compounds often give mixtures of products resulting from an **allylic shift:** In the product, the double bond can appear at either of the two positions it occupies in the resonance structures of the allylic radical. An allylic shift in a radical reaction is similar to the 1,4-addition of an electrophilic reagent such as HBr to a diene (Section 15-5).

The following propagation steps show how a mixture of products results from the free-radical allylic bromination of 1-butene.

$$CH_3{-}CH_2{-}CH{=}CH_2 + Br\cdot \longrightarrow [CH_3{-}\dot{C}H{-}CH{=}CH_2 \longleftrightarrow CH_3{-}CH{=}CH{-}\dot{C}H_2] + HBr$$

resonance-stabilized allylic radical

$$\downarrow Br_2$$

$$\underbrace{CH_3{-}CH(Br){-}CH{=}CH_2 + CH_3{-}CH{=}CH{-}CH_2Br}_{\text{(mixture)}} + Br\cdot$$

(mixture)

PROBLEM 15-8

When methylenecyclohexane is treated with a low concentration of bromine under irradiation by a sunlamp, two products are formed.

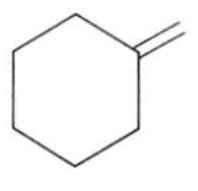

methylenecyclohexane + Br_2 $\longrightarrow$ two substitution products + HBr

(a) Propose structures for these two products.
(b) Give a mechanism to account for their formation.

Bromination using NBS Effective use of free-radical allylic bromination requires a very low concentration of bromine in the reaction mixture to enhance allylic substitution over ionic addition. Simply adding bromine would make the concentration too high, and ionic addition of bromine to the double bond would result. A convenient bromine source for allylic bromination is *N*-bromosuccinimide (NBS), a brominated derivative of succinimide, an amide of the four-carbon diacid, succinic acid.

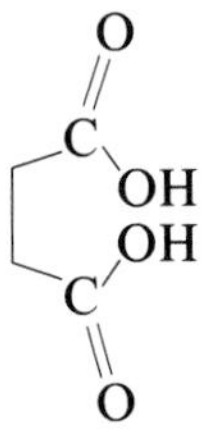

succinic acid 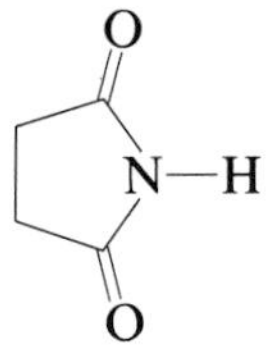 | succinimide 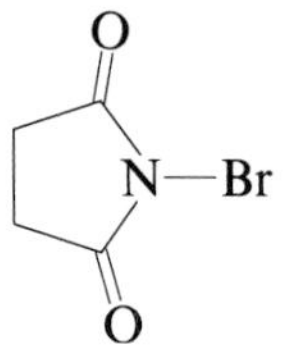 | *N*-bromosuccinimide (NBS)

NBS provides a fairly constant, low concentration of Br_2 because it reacts with the HBr liberated in the substitution, converting it back into Br_2. This reaction also removes the HBr by-product, preventing it from adding across the double bond (with anti-Markovnikov orientation) by its own free-radical chain reaction.

Step 1: Free-radical allylic substitution (mechanism on pg. 641)

$$R{-}H + Br_2 \xrightarrow{h\nu} R{-}Br + HBr$$

Step 2: NBS converts the HBr by-product back into Br_2

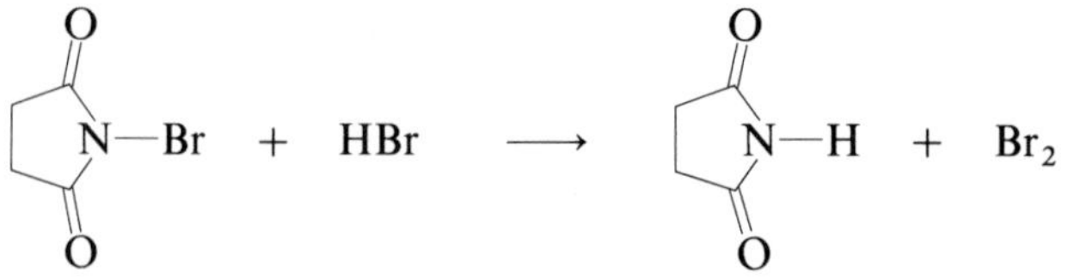

$$\text{NBS} + HBr \longrightarrow \text{succinimide} + Br_2$$

NBS | succinimide

This NBS reaction is carried out in a clever way. The allylic compound is dissolved in carbon tetrachloride, and 1 equivalent of NBS is added. NBS is denser than CCl_4 and not very soluble in it, so it sinks to the bottom of the CCl_4 solution. The reaction is initiated using a sunlamp for illumination or a radical initiator such as a peroxide. The NBS gradually *appears* to rise to the top of the CCl_4 layer. It is actually converted to succinimide, which is less dense than CCl_4. Once all the solid succinimide has risen to the top, the sunlamp is turned off, the solution is filtered to remove the succinimide, and the CCl_4 is evaporated to recover the product.

PROBLEM 15-9

Devise a complete mechanism for the light-initiated reaction of 1-hexene with *N*-bromosuccinimide in carbon tetrachloride solution.

PROBLEM 15-10

Predict the product(s) of light-initiated reaction with NBS in CCl_4 for each of the following starting materials.

(a) cyclopentene (b) *trans*-2-pentene (c) C_6H_5—CH_3 toluene

15-8 MOLECULAR ORBITALS OF THE ALLYLIC SYSTEM

Let's take a closer look at the electronic structure of allylic systems, using the allyl radical as our example. One resonance structure shows a pi bond between C1 and C2 with the radical electron on C3, and the other shows a pi bond between C2 and C3 with the radical electron on C1. These two resonance structures imply there is half a pi bond between C1 and C2 and half a pi bond between C2 and C3, with the radical electron half on C1 and half on C3.

$$\left[H_2\dot{C}_1{-}C_2H{=}C_3H_2 \longleftrightarrow H_2C_1{=}C_2H{-}\dot{C}_3H_2 \right] = H_2C_1^{\delta\cdot}{\cdots}C_2H{\cdots}C_3^{\delta\cdot}H_2$$

resonance structures — combined representation

Remember that a compound does not "resonate" among its resonance structures: It has characteristics of all at the same time, but no resonance structure has an independent existence. To have pi bonding overlap simultaneously between C1 and C2 and between C2 and C3, the *p* orbitals of all three carbon atoms must be parallel. The geometric structure of the allyl system is shown in Figure 15-10. The

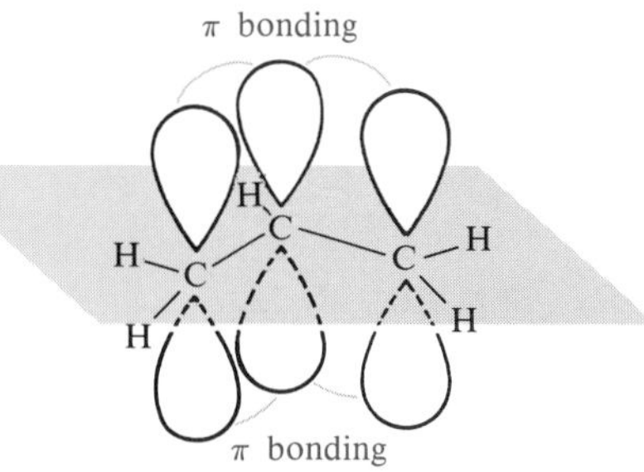

FIGURE 15-10 Geometric structure of the allyl cation, allyl radical, and allyl anion.

allyl cation, the allyl radical, and the allyl anion all have this same geometric structure, differing only in the number of pi electrons.

Just as the four p orbitals of 1,3-butadiene overlap to form four molecular orbitals, the three atomic p orbitals of the allyl system overlap to form three molecular orbitals, shown in Figure 15-11. These three MO's share several important features with the MO's of the butadiene system. The first MO is entirely bonding, the second has one node, and the third has two nodes and (because it is the highest-energy MO) is entirely antibonding. (An asterisk is often used to show that an orbital is antibonding, as in π_3^*.)

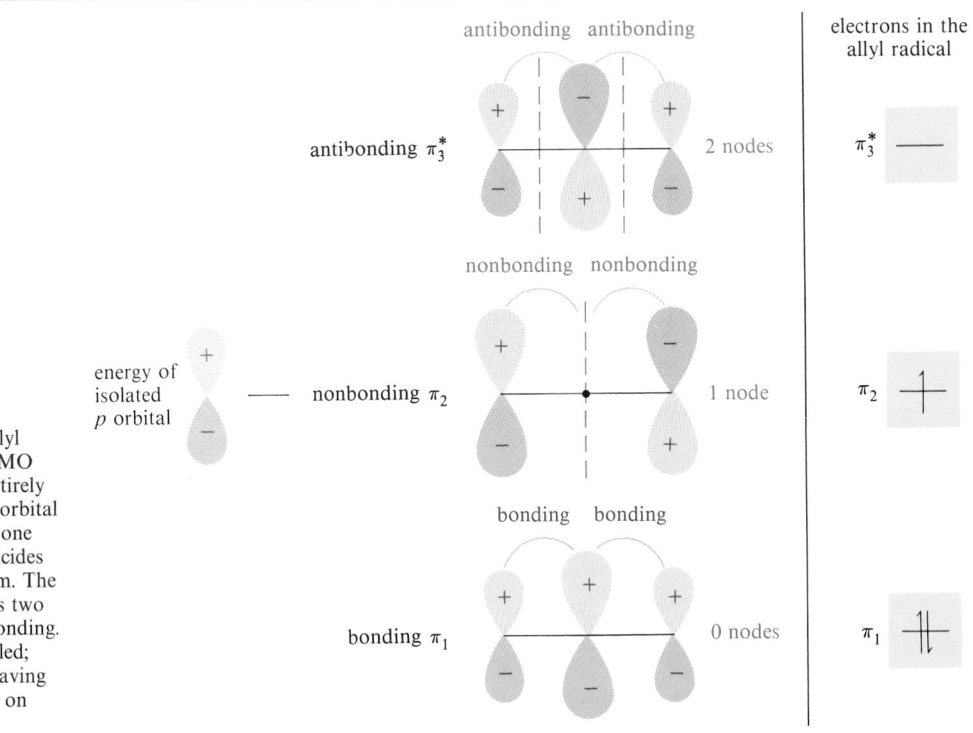

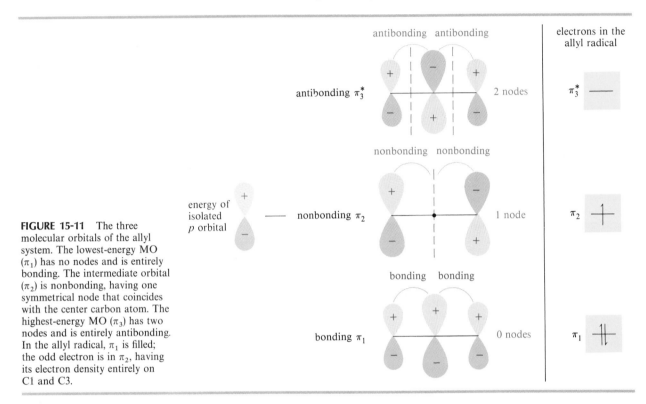

FIGURE 15-11 The three molecular orbitals of the allyl system. The lowest-energy MO (π_1) has no nodes and is entirely bonding. The intermediate orbital (π_2) is nonbonding, having one symmetrical node that coincides with the center carbon atom. The highest-energy MO (π_3) has two nodes and is entirely antibonding. In the allyl radical, π_1 is filled; the odd electron is in π_2, having its electron density entirely on C1 and C3.

As with butadiene, we expect that half of the MO's will be bonding, and half antibonding; but with an odd number of MO's, they cannot be symmetrically divided. One of the MO's must appear at the middle of the energy levels, neither bonding nor antibonding: It is a **nonbonding molecular orbital.** Electrons in a nonbonding orbital have the same energy as they do in an isolated p orbital.

The structure of the nonbonding orbital (π_2) may seem strange because there is zero electron density on the center p orbital (C2). This is the case because π_2 must have one node, and the only symmetrical position for one node is in the center of the molecule, crossing C2. We can tell from its structure that π_2 must be nonbonding, because C2's p orbital has zero overlap with C1 and zero overlap with C3. The total is zero bonding, or a nonbonding orbital.

15-9 ELECTRONIC CONFIGURATIONS OF THE ALLYL RADICAL, CATION, AND ANION

The right-hand column of Figure 15-11 shows the electronic structure for the allyl radical, with three pi electrons in the lowest available molecular orbitals. Two electrons are in the all-bonding MO, representing the pi bond shared between the C1—C2 bond and the C2—C3 bond. The odd electron goes into π_2, with zero electron density on the central carbon atom (C2). This MO representation agrees with the resonance picture showing the radical electron shared equally by C1 and C3, but not C2. Both the resonance and MO pictures successfully predict that the radical will react at either of the two end carbon atoms, C1 and C3.

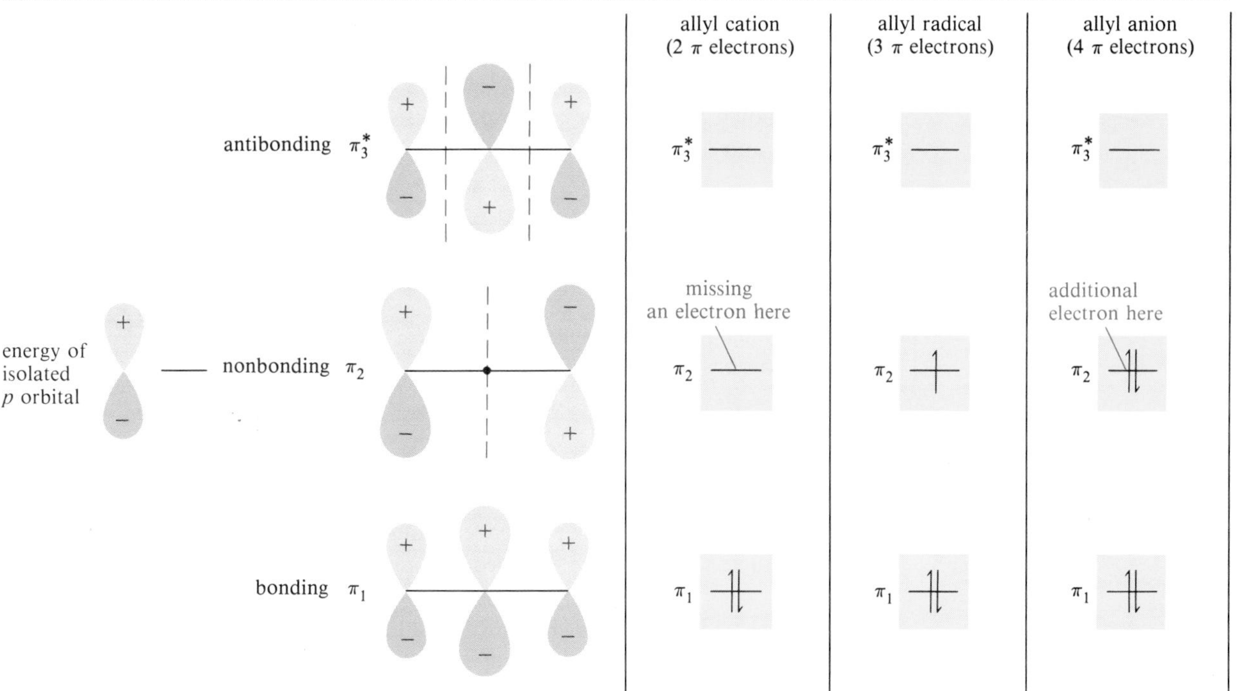

FIGURE 15-12 Comparison of the electronic structure of the allyl cation and allyl anion with the allyl radical. The allyl cation has no electron in π_2, leaving half a positive charge on each of C1 and C3. The allyl anion has another electron in π_2, giving half a negative charge to each of C1 and C3.

The electronic configuration of the allyl cation (Fig. 15-12) differs from that of the allyl radical by removal of the odd electron from π_2, which has half of its electron density on C1 and half on C3. In effect, we have removed half an electron from each of C1 and C3, and C2 remains unchanged. This MO picture is consistent with the resonance picture showing the positive charge shared by C1 and C3.

resonance structures — combined representation

Figure 15-12 also shows the electronic configuration of the allyl anion, which differs from the allyl radical by the addition of another electron to π_2, the non-bonding orbital with its electron density divided between C1 and C3. In agreement with the resonance picture, this electron's negative charge should be divided equally between C1 and C3.

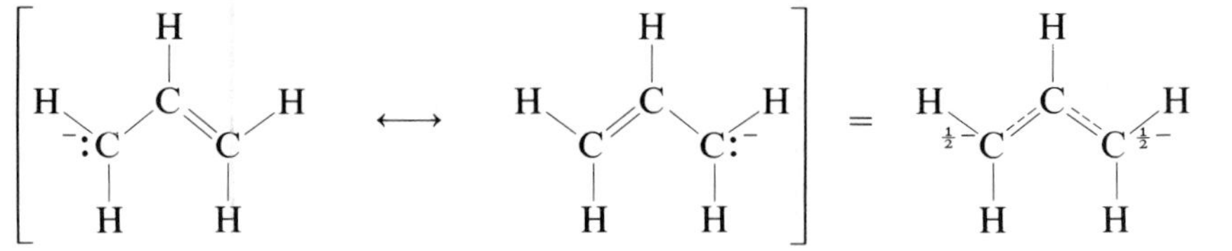

resonance structures combined representation

The molecular orbital representation shows the allyl anion has a pair of electrons in π_2, a nonbonding orbital. This picture is also consistent with the resonance structures shown above, with a lone pair of nonbonding electrons evenly divided between C1 and C3.

PROBLEM 15-11
When 1-bromo-2-butene is added to magnesium metal in dry ether, a Grignard reagent is formed. Addition of water to this Grignard reagent gives a mixture of 1-butene and 2-butene (cis and trans). When the Grignard reagent is made using 3-bromo-1-butene, addition of water produces exactly the same mixture of products in the same ratios. Explain this curious result.

15-10 S_N2 DISPLACEMENT REACTIONS OF ALLYLIC HALIDES AND TOSYLATES

Allylic halides and tosylates show enhanced reactivity toward nucleophilic displacement reactions by the S_N2 mechanism, usually undergoing second-order substitution without allylic shifts or other rearrangements. For example, allyl bromide reacts with nucleophiles by the S_N2 mechanism about 40 times faster than *n*-propyl bromide.

Figure 15-13 shows how this rate enhancement can be explained by allylic delocalization of electrons in the transition state. The transition state for the S_N2 reaction looks like a trigonal carbon atom with a *p* orbital perpendicular to the three substituents. The electrons of the attacking nucleophile are forming a bond

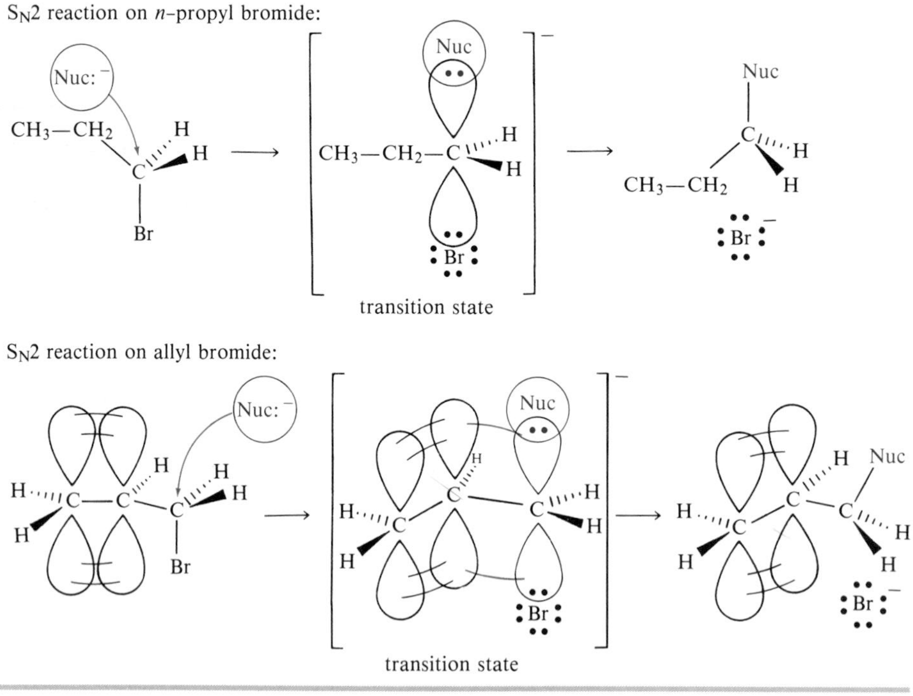

S_N2 reaction on *n*-propyl bromide:

transition state

S_N2 reaction on allyl bromide:

transition state

FIGURE 15-13 In the transition state for the S_N2 reaction of allyl bromide with a nucleophile, the double bond is conjugated with the *p* orbital that is momentarily present on the reacting carbon atom. The resulting overlap lowers the energy of the transition state, increasing the reaction rate.

using one lobe of the p orbital, while the leaving group's electrons are leaving from the other lobe.

When the substrate is allylic, the transition state receives resonance stabilization through conjugation with the p orbitals of the pi bond. This stabilization lowers the energy of the transition state, resulting in a lower activation energy and an enhanced rate.

The enhanced reactivity of allylic halides and tosylates makes them particularly attractive as electrophiles for S_N2 reactions. Allylic halides are so reactive that they couple with Grignard and organolithium reagents, a reaction that does not work well with unactivated halides.

$$\underset{\text{allyl bromide}}{H_2C{=}CH{-}CH_2Br} + \underset{n\text{-butyllithium}}{CH_3{-}(CH_2)_3{-}Li} \longrightarrow \underset{\substack{\text{1-heptene}\\(85\%)}}{H_2C{=}CH{-}CH_2{-}(CH_2)_3{-}CH_3} + LiBr$$

PROBLEM 15-12

Show how you might synthesize the following compounds starting with alkyl or alkenyl halides containing four carbon atoms or fewer.

(a) 1-heptene (b) 5-methyl-2-hexene

15-11 THE DIELS-ALDER REACTION

In 1928, the German chemists Otto Diels and Kurt Alder discovered that alkynes and alkenes with electron-withdrawing groups add to conjugated dienes to form six-membered rings. The **Diels-Alder reaction** proved to be an extremely useful synthetic tool, providing one of the best ways to make six-membered rings with diverse functionality and controlled stereochemistry. In 1950, Diels and Alder were awarded the Nobel Prize for their work.

The Diels-Alder reaction is also called a **[4 + 2] cycloaddition,** because a ring is formed by the interaction of four pi electrons in the diene with two pi electrons on the alkene or alkyne. Since the electron-poor alkene or alkyne is prone to react with a diene, it is called a **dienophile** ("lover of dienes"). In effect, the Diels-Alder reaction converts two pi bonds into two sigma bonds. We can symbolize the Diels-Alder reaction by using three arrows to show the movement of three pairs of electrons. The electron-withdrawing groups (—W) are usually cyano (—C≡N) groups or carbonyl-containing (C=O) groups.

H W C C H H Δ (heat) H C—W C—H H

diene dienophile

W C C H Δ W C C H

diene dienophile

The Diels-Alder reaction is analogous to a nucleophile-electrophile reaction. The diene is electron-rich (like a nucleophile), while the dienophile is electron-poor.

Simple dienes such as 1,3-butadiene are sufficiently electron rich to be effective dienes for the Diels-Alder reaction. The presence of electron-releasing groups such as alkyl groups or alkoxy (—OR) groups may further enhance the reactivity of the diene.

Simple alkenes and alkynes such as ethene and ethyne are not good dienophiles, however. A good dienophile generally has one or more electron-withdrawing groups pulling electron density away from the pi bond. Dienophiles commonly have carbonyl-containing (C=O) groups or cyano (—C≡N) groups to enhance their Diels-Alder reactivity. Figure 15-14 shows some representative Diels-Alder reactions involving a variety of different dienes and dienophiles.

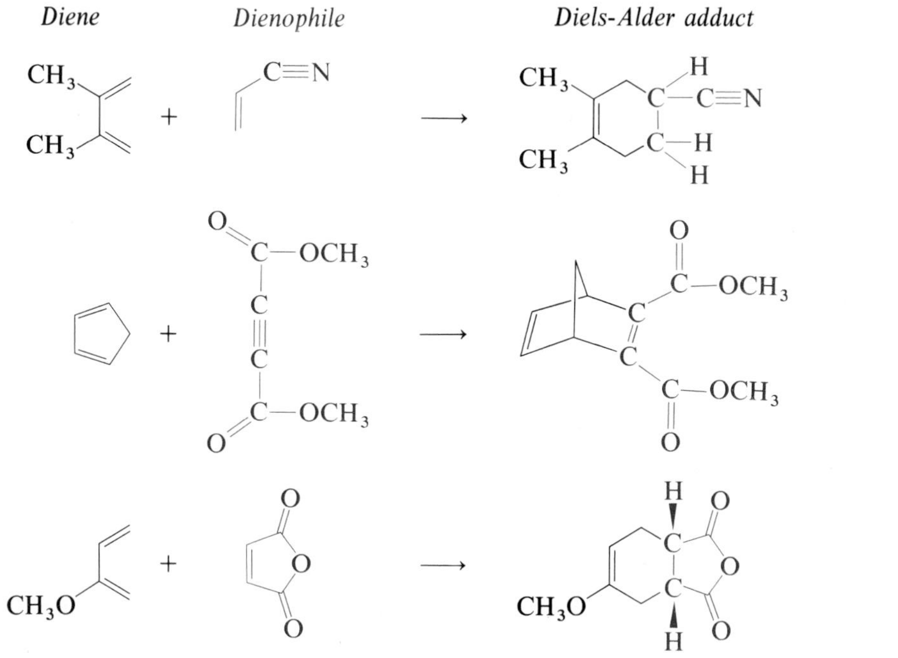

Diene | *Dienophile* | *Diels-Alder adduct*

FIGURE 15-14 Examples of the Diels-Alder reaction. Electron-releasing substituents activate the diene; electron-withdrawing substituents activate the dienophile.

PROBLEM 15-13

Predict the products of the following proposed Diels-Alder reactions.

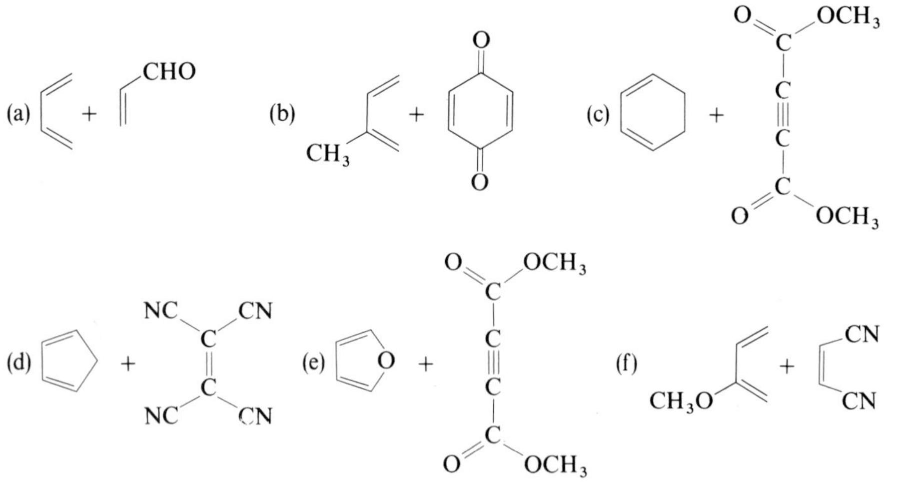

(a) (b) (c)

(d) (e) (f)

PROBLEM 15-14

What dienes and dienophiles would react to give the following Diels-Alder products?

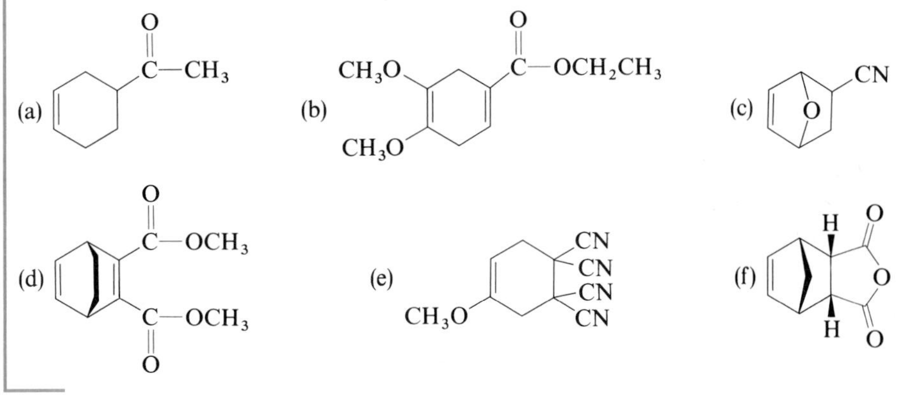

(a) (b) (c) (d) (e) (f)

The mechanism of the Diels-Alder reaction is a simultaneous cyclic movement of six electrons: four in the diene and two in the dienophile. The simple three-arrow representation of the mechanism shown on page 647 is fairly accurate. This is called a **concerted reaction** because all the bond making and bond breaking occurs simultaneously. For the three pairs of electrons to move simultaneously, however, the transition state must have a geometry that allows overlap of the two end *p* orbitals of the diene with those of the dienophile. Figure 15-15 shows the necessary geometry of the transition state.

overlap begins as these orbitals come together

reactants transition state product

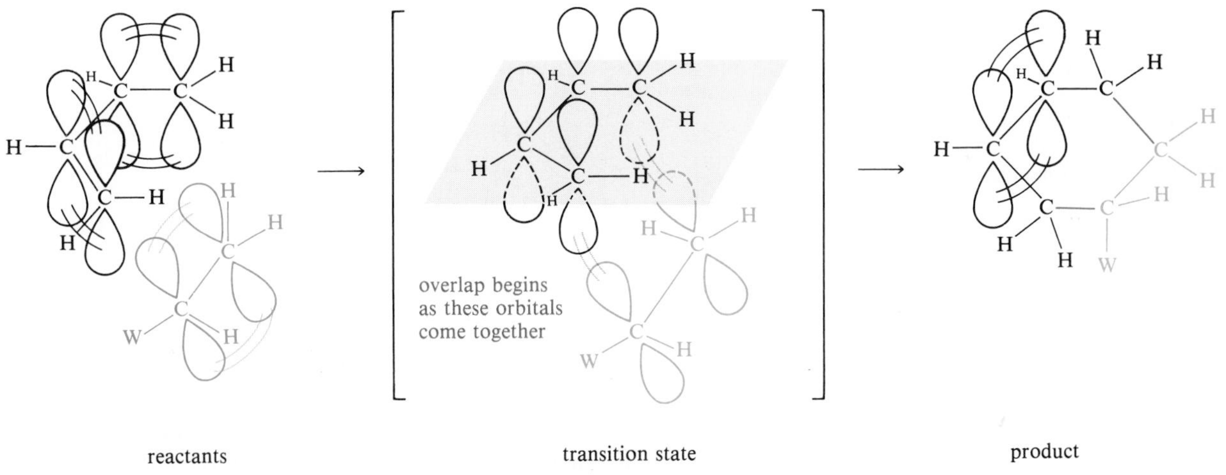

FIGURE 15-15 The Diels-Alder reaction has a concerted mechanism, with all the bond making and bond breaking occurring in a single step. Three pairs of electrons move simultaneously, requiring a transition state with overlap between the end *p* orbitals of the diene and those of the dienophile.

15-11A STEREOCHEMICAL REQUIREMENTS OF THE DIELS-ALDER TRANSITION STATE

The structure of the Diels-Alder transition state (Fig. 15-15) explains several characteristics of this useful reaction. In particular, it explains why some isomers react differently from others, and it enables us to predict the stereochemistry of the products. There are three major stereochemical features of the Diels-Alder reaction that are controlled by the requirements of the transition state.

***s-cis* conformation of the diene** The diene must be in the *s*-cis conformation to react. When the diene is in the *s*-trans conformation, the end *p* orbitals are too far apart to overlap with the *p* orbitals of the dienophile. The *s*-trans conformation usually has a lower energy than the *s*-cis, but this energy difference is not enough to prevent most dienes from undergoing the Diels-Alder reaction. For example, the *s*-trans conformation of butadiene is only 2.3 kcal (9.6 kJ) per mole lower in energy than the *s*-cis conformation.

s-cis

s-trans
2.3 kcal (9.6 kJ) more stable

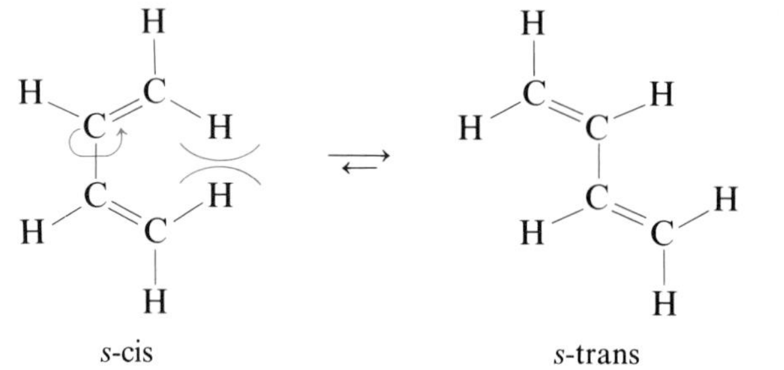

Structural features that aid or hinder the diene in achieving the *s*-cis conformation affect its ability to participate in the Diels-Alder reaction. Figure 15-16 shows that dienes with functional groups that hinder the *s*-cis conformation react slower than butadiene. Dienes with functional groups that hinder the *s*-trans conformation react faster than butadiene.

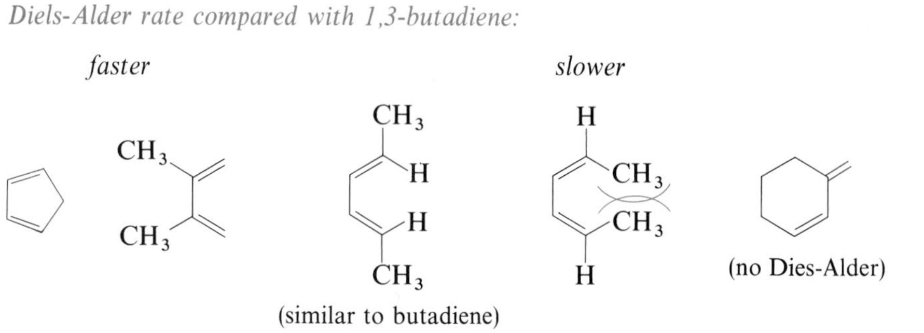

Diels-Alder rate compared with 1,3-butadiene:

faster *slower*

(similar to butadiene) (no Dies-Alder)

FIGURE 15-16 Dienes that easily adopt the *s*-cis conformation undergo the Diels-Alder reaction more readily.

Because cyclopentadiene is fixed in the *s*-cis conformation, it is highly reactive in the Diels-Alder reaction. It is so reactive, in fact, that at room temperature cyclopentadiene slowly reacts with itself to form dicyclopentadiene. Cyclopentadiene is regenerated by heating the dimer to about 200°C. Above 200°C the Diels-Alder reaction reverses, and the more volatile cyclopentadiene monomer distills over into a cold flask. The monomer can be stored indefinitely at dry ice temperatures.

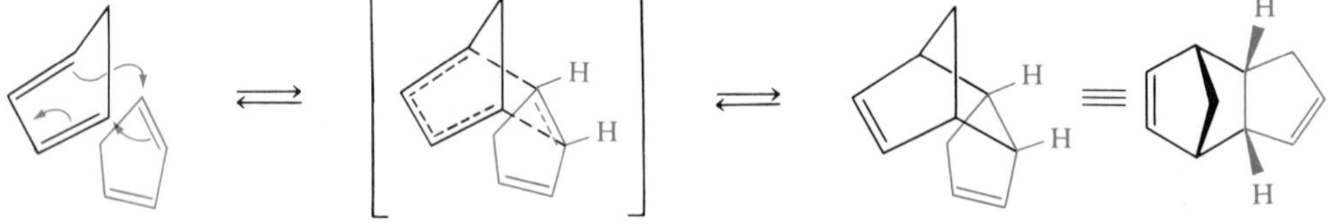

2-cyclopentadiene (monomer) dicyclopentadiene (dimer)

Syn stereochemistry The Diels-Alder reaction is a syn addition with respect to both the diene and the dienophile. The dienophile adds to one face of the diene, and the diene adds to one face of the dienophile. As you can see from the transition state in Figure 15-15, there is no opportunity for any of the substituents to change their stereochemical positions during the course of the reaction. Substituents that are on the same side of the diene or dienophile will be cis on the newly formed ring. The following examples show the results of this syn addition.

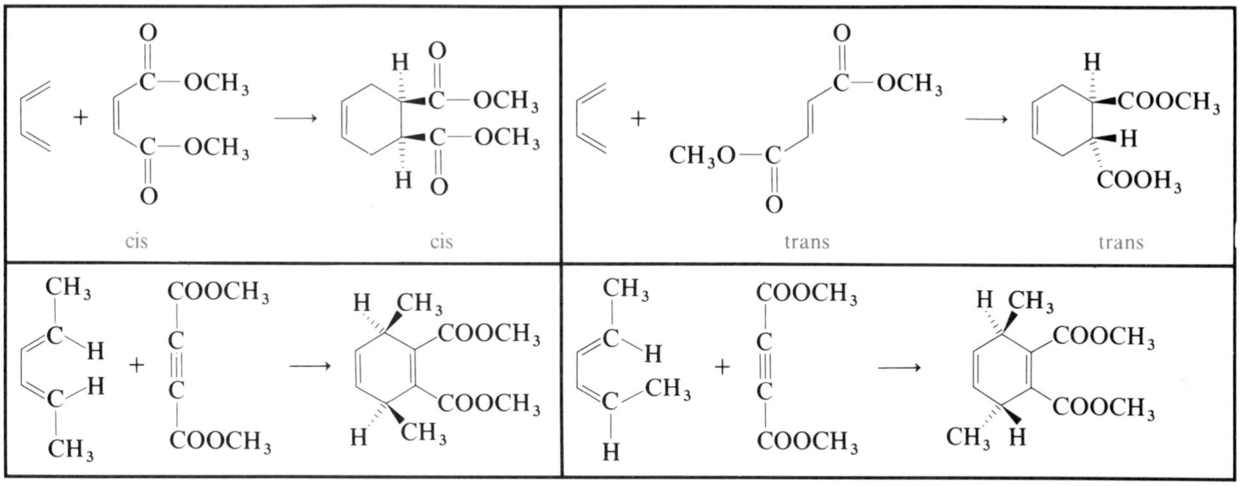

The endo rule When the dienophile has a pi bond in its electron-withdrawing group (as in a carbonyl group or a cyano group), the p orbitals in that electron-withdrawing group approach the central carbon atoms (C2 and C3) of the diene. This proximity results in **secondary overlap:** an overlap of the p orbitals of the electron-withdrawing group with the p orbitals of C2 and C3 of the diene (Fig. 15-17). Secondary overlap helps to stabilize the transition state.

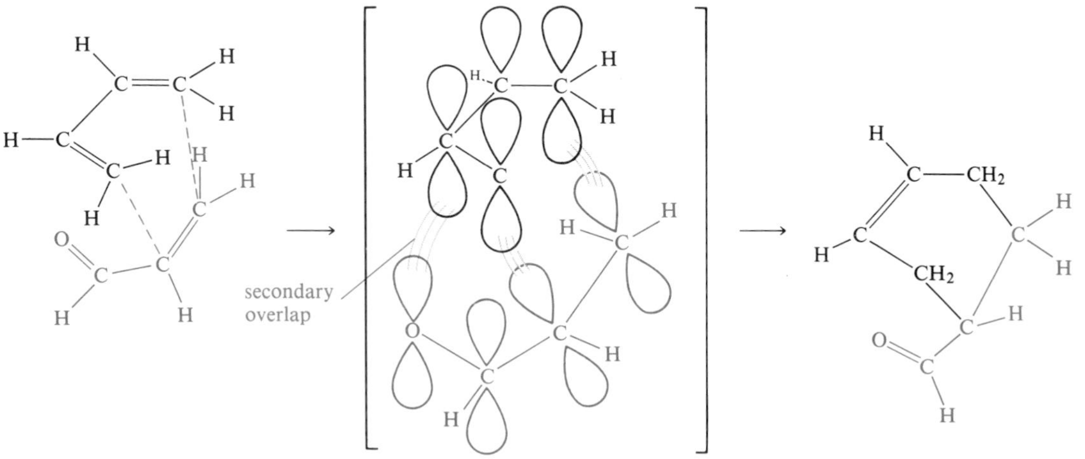

FIGURE 15-17 In most Diels-Alder reactions, there is *secondary overlap* between the p orbitals of the electron-withdrawing group and those of the central carbon atoms of the diene. Secondary overlap provides additional stabilization of the transition state.

The influence of secondary overlap was first observed in reactions using cyclopentadiene to form bicyclic ring systems. In the bicyclic product (called *norbornene*), the stereochemical position closest to the central atoms of the diene is called the **endo position,** because the substituent seems to be inside the pocket formed by the six-membered ring of norbornene. The stereochemical preference for the endo position is called the **endo rule.**

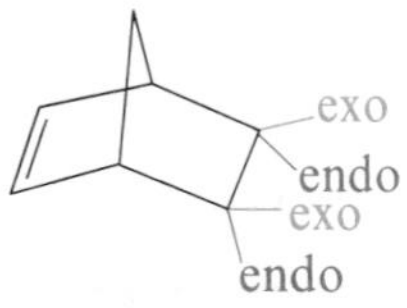

exo
endo
exo
endo

stereochemical positions of norbornene

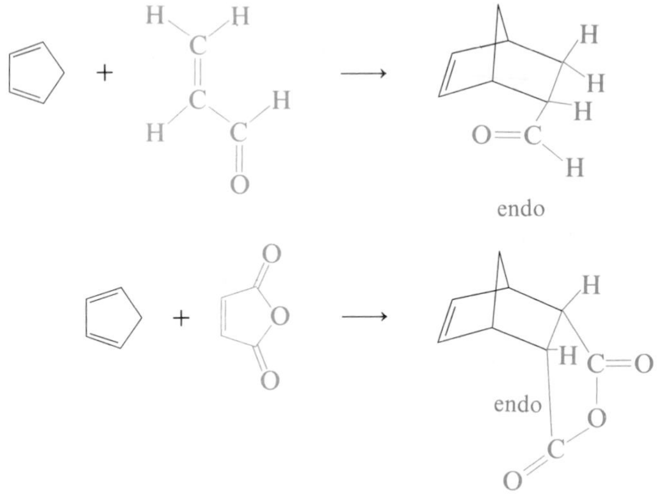

endo

endo

The endo rule is useful for predicting the products of many types of Diels-Alder reactions, regardless of whether they use cyclopentadiene to form norbornene systems. The following examples show the use of the endo rule with other types of Diels-Alder reactions.

but not

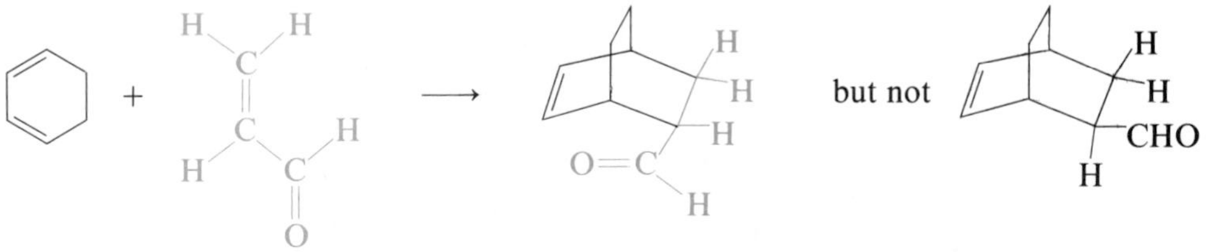

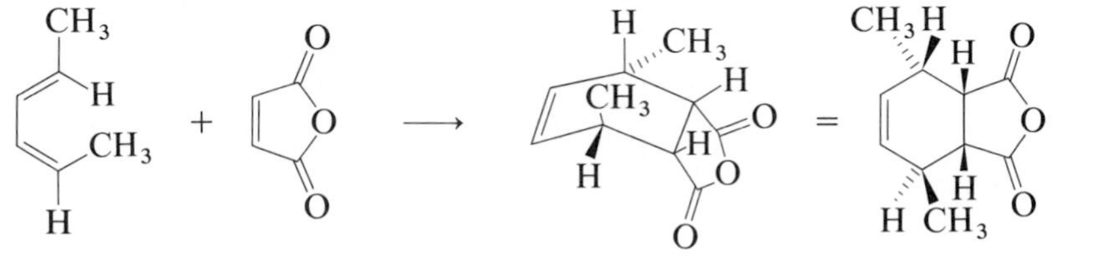

SOLVED PROBLEM 15-1

Use the endo rule to predict the product of the following cycloaddition.

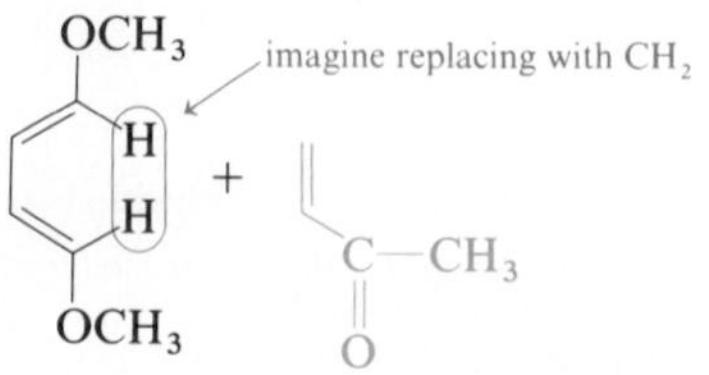

imagine replacing with CH_2

SOLUTION Imagine that this diene were a substituted cyclopentadiene; the endo product would be formed.

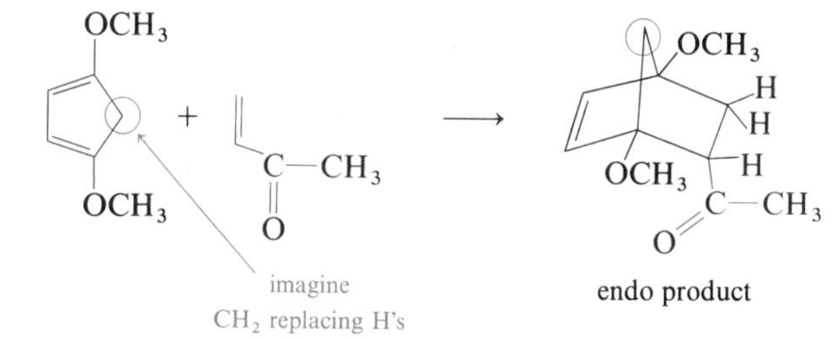

imagine
CH_2 replacing H's

endo product

In the imaginary reaction, we replaced the two inside hydrogens by the rest of the cyclopentadiene ring. Now we put them back and have the actual product.

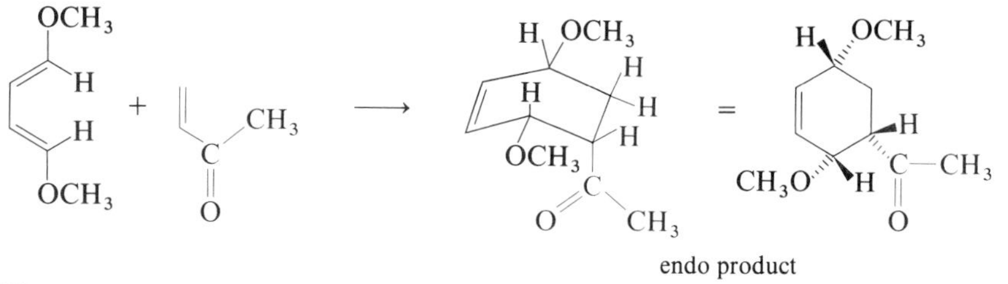

endo product

PROBLEM 15-15
For each of the following proposed Diels-Alder reactions, predict the major product. Include stereochemistry where appropriate.

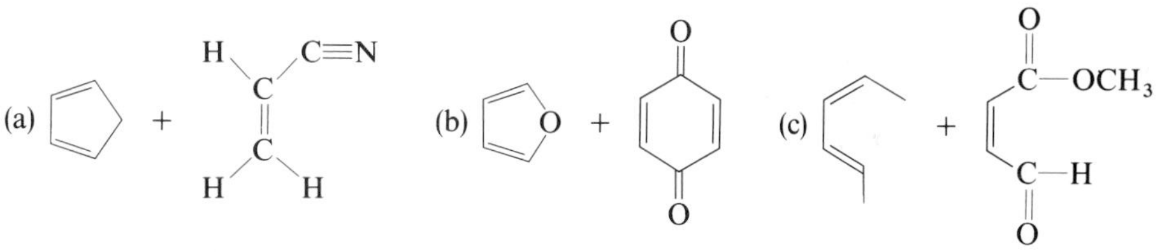

(a) (b) (c)

15-11B DIELS-ALDER REACTIONS USING UNSYMMETRICAL REAGENTS

Even when the diene and dienophile are both unsymmetrically substituted, the Diels-Alder reaction usually gives a single product rather than a mixture. The product is the isomer that results from orienting the diene and dienophile so that we can *imagine* a hypothetical reaction intermediate with a "push-pull" flow of electrons from the electron-donating group to the electron-withdrawing group. In the following schematic representation, D is an electron-donating substituent on the diene, and W is an electron-withdrawing substituent on the dienophile.

Formation of 1,4 product

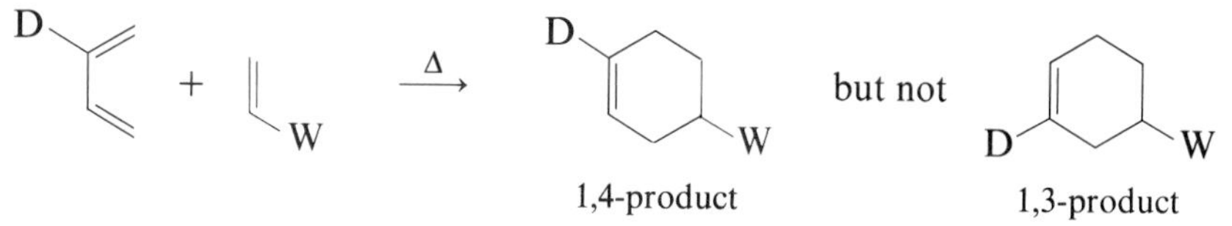

but not

1,4-product

1,3-product

imaginary flow of electrons

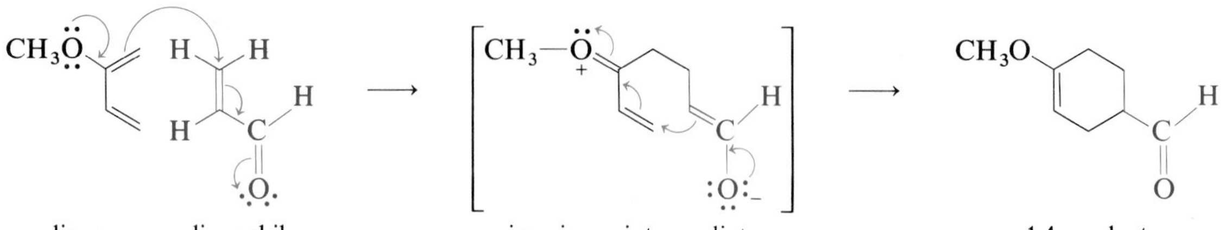

diene dienophile imaginary intermediate 1,4-product

Formation of 1,2 product

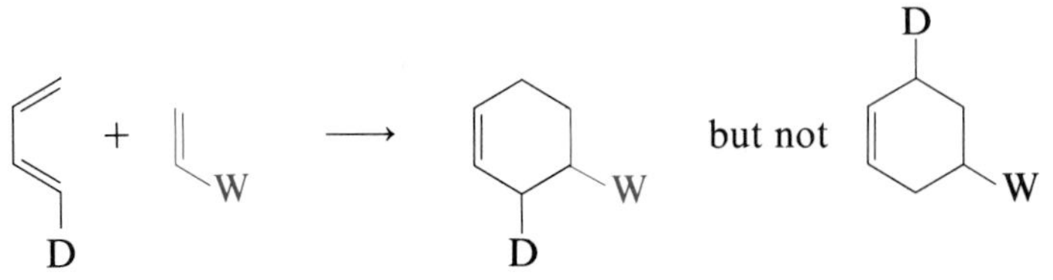

but not

1,2-product 1,3-product

imaginary flow of electrons

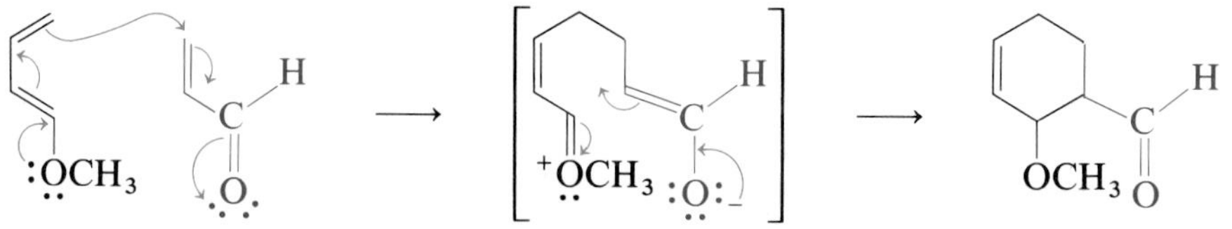

diene dienophile imaginary intermediate 1,2-product

The hypothetical intermediates are not completely imaginary, because their push-pull resonance helps to stabilize the concerted Diels-Alder transition state. The correct products of these unsymmetrical Diels-Alder reactions can be predicted simply by remembering that the electron-donating groups of the diene and the electron-withdrawing groups of the dienophile usually bear either a 1,2-relationship or a 1,4-relationship in the products, but not a 1,3-relationship.

SOLVED PROBLEM 15-2

Predict the products of the following proposed Diels-Alder reactions.

(a) 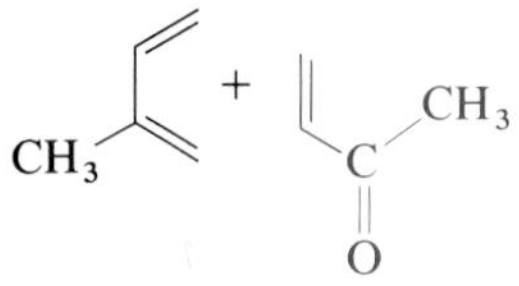 (b)

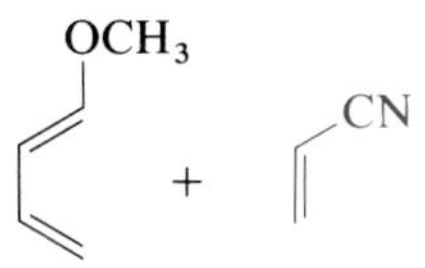

SOLUTION (a) The methyl group is weakly electron donating to the diene, and the carbonyl group is electron withdrawing from the dienophile. The two possible orientations place these groups in a 1,4-relationship or a 1,3-relationship. We select the 1,4-relationship for our predicted product. (The experimental results show a 70:30 preference for the 1,4-product.)

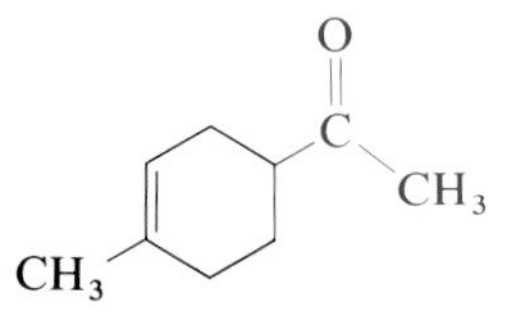

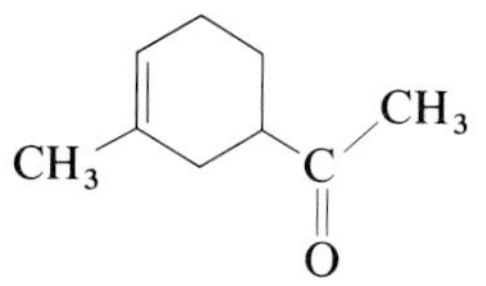

CH_3 — C(=O)CH_3 (structures)

1,4-relationship (major) (70%)

1,3-relationship (minor) (30%)

(b) The methoxyl group ($—O—CH_3$) is strongly electron donating to the diene, and the cyano group ($—C{\equiv}N$) is electron withdrawing from the dienophile. Depending on the orientation of addition, the product has either a 1,2- or a 1,3-relationship of these two groups. We select the 1,2-relationship, and the endo rule predicts cis stereochemistry of the two substituents.

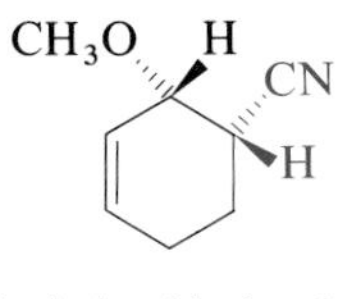

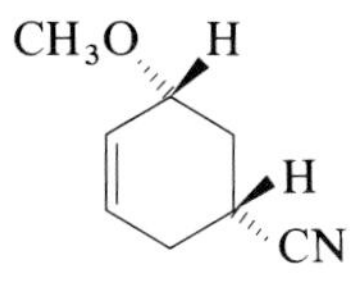

1,2-relationship (product)

1,3-relationship (not formed)

PROBLEM 15-16

In Solved Problem 15-2, we simply predicted the products having a 1,2- or 1,4-relationship of the proper substituents. Draw the "imaginary intermediates" for these reactions.

PROBLEM 15-17

Predict the products of the following Diels-Alder reactions.

(a) 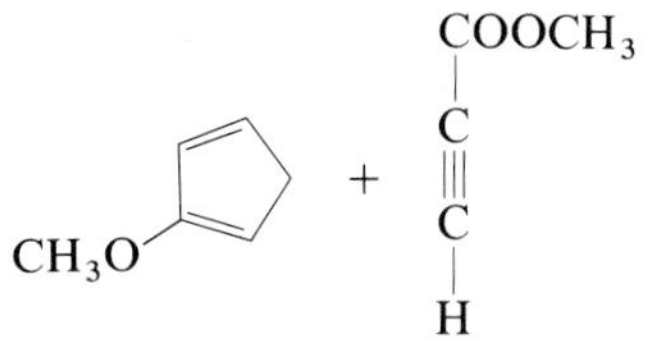 CH_3O-cyclopentadiene + $H—C{\equiv}C—COOCH_3$

(b) CH_3-substituted diene (CH_3) + $CH_2{=}CH—CN$

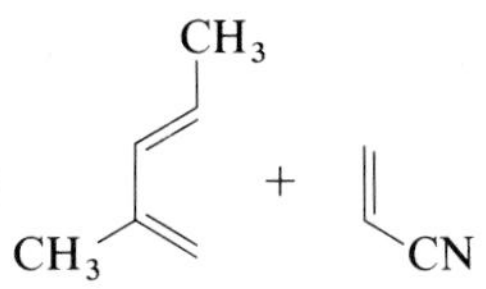

(c) CH_3O-cyclohexadiene + cyclohexenone

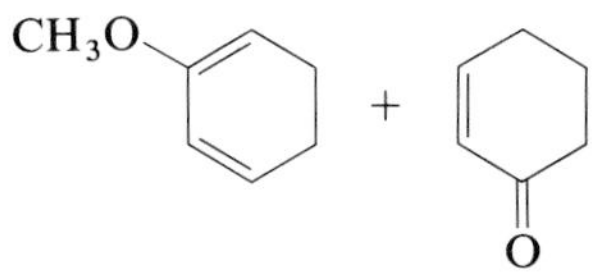

15-12 THE DIELS-ALDER AS AN EXAMPLE OF A PERICYCLIC REACTION

To understand why the Diels-Alder reaction takes place, we must consider the molecular orbitals involved. The Diels-Alder reaction is a **cycloaddition:** Two molecules combine in a one-step, **concerted** (simultaneous) **reaction** to form a new ring. Cycloadditions such as the Diels-Alder are one class of **pericyclic reactions,** which involve concerted breaking and formation of bonds within a closed ring of interacting orbitals. Figure 15-15 (page 649) shows the closed loop of interacting orbitals in the Diels-Alder transition state. Each carbon atom of the new ring has one orbital involved in this closed loop.

A concerted pericyclic reaction has a single transition state, whose activation energy may be supplied by heat (thermal induction) or by ultraviolet light (photochemical induction). Some pericyclic reactions proceed only under thermal induction, while others proceed only under photochemical induction. Some pericyclic reactions take place under both thermal and photochemical conditions, but the two sets of conditions give different products.

For many years, pericyclic reactions were poorly understood and unpredictable. Around 1965, Robert B. Woodward and Roald Hoffmann developed a theory for predicting the results of pericyclic reactions by considering the symmetry of the molecular orbitals of the reactants and products. Their theory, called **conservation of orbital symmetry,** says that the MO's of the reactants must flow smoothly into the MO's of the products without any drastic changes on symmetry: That is, there must be bonding interactions to help stabilize the transition state. If this cannot happen, the concerted cyclic reaction cannot occur. Conservation of symmetry has been used to develop "rules" to predict which pericyclic reactions are feasible and what products will result. These rules are often called the **Woodward-Hoffman rules.**

15-12A CONSERVATION OF ORBITAL SYMMETRY IN THE DIELS-ALDER REACTION

We will not develop all the Woodward-Hoffmann rules, but we will show how the molecular orbitals can indicate whether a cycloaddition will take place. The simple Diels-Alder reaction of butadiene with ethylene serves as our first example. The molecular orbitals of butadiene and ethylene are represented in Figure 15-18. Butadiene, with four atomic *p* orbitals, has four molecular orbitals: two bonding MO's (filled) and two antibonding MO's (vacant). Ethylene, with two atomic *p* orbitals, has two MO's: a bonding MO (filled) and an antibonding MO (vacant).

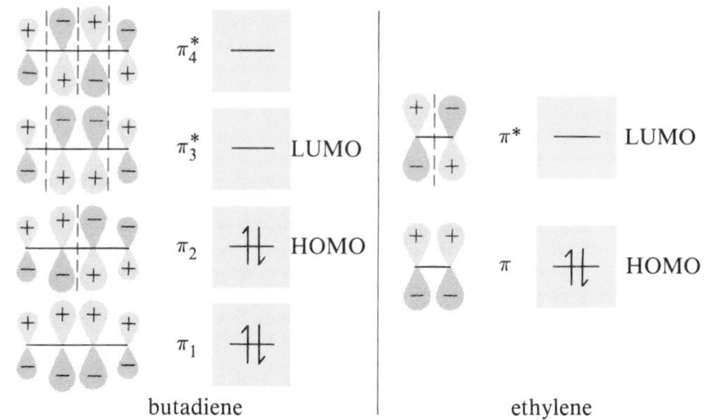

FIGURE 15-18 Molecular orbitals of butadiene and ethylene.

In the Diels-Alder reaction, the diene acts as the electron-rich nucleophile, and the dienophile acts as the electron-poor electrophile. If we imagine the diene contributing a pair of electrons to the dienophile, the highest-energy electrons of the diene require the least activation energy for such a donation. The electrons in the highest-energy occupied orbital, called the **Highest Occupied Molecular Orbital** (HOMO), are the important ones. The HOMO of butadiene is π_2, and its symmetry determines the course of the reaction.

The orbital in ethylene that receives these electrons is the lowest available unoccupied orbital, the **Lowest Unoccupied Molecular Orbital** (LUMO). In ethylene, the LUMO is the π^* antibonding orbital. If the electrons in the HOMO of butadiene can flow smoothly into the LUMO of ethylene, a concerted reaction can take place. Figure 15-19 shows the interaction of the HOMO of butadiene with the LUMO of ethylene. The figure shows that the HOMO of butadiene has the correct symmetry to overlap with the LUMO of ethylene. Having the correct symmetry means the orbitals that must overlap to form the new bonds can overlap constructively: plus with plus and minus with minus. These bonding interactions stabilize the transition state and promote the concerted reaction.

Figure 15-19 shows constructive overlap (bonding interactions) between the end orbitals of the HOMO of butadiene and the LUMO of ethylene, where the new sigma bonds will form in the Diels-Alder reaction. This favorable result shows the reaction is **symmetry-allowed.** The Diels-Alder reaction is common, and this theory correctly predicts a favorable transition state.

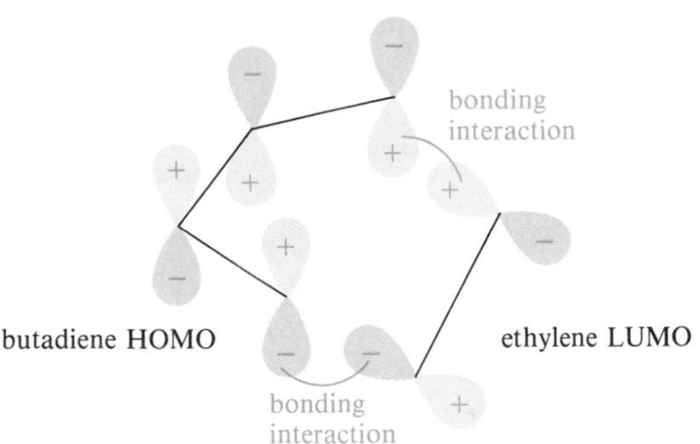

FIGURE 15-19 The HOMO of butadiene forms a bonding overlap with the LUMO of ethylene, because the orbitals have similar symmetry. This reaction is therefore symmetry-allowed.

15-12B THE "FORBIDDEN" [2 + 2] CYCLOADDITION

If the interaction shown in Figure 15-19 had produced an overlap of positive-phase orbitals with negative-phase orbitals (destructive overlap), antibonds would be generated, and the reaction would be called **symmetry-forbidden.** The [2 + 2] cycloaddition of two ethylenes to give cyclobutane is such a symmetry-forbidden reaction.

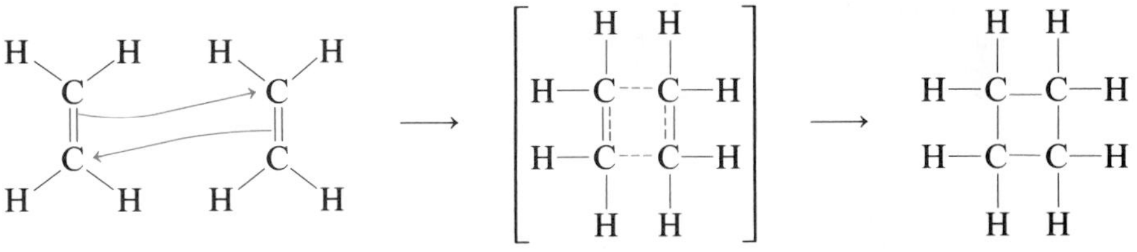

two ethylenes → (transition state) → cyclobutane

This [2 + 2] cycloaddition requires that the HOMO of one of the ethylenes overlap with the LUMO of the other. Figure 15-20 shows that an antibonding interaction results from this overlap. For a cyclobutane molecule to result, one of the MO's would have to change its symmetry: Orbital symmetry would not be conserved, and the reaction is therefore symmetry-forbidden. Such a symmetry-forbidden reaction can occasionally be made to occur, but it cannot occur in the concerted pericyclic manner shown above.

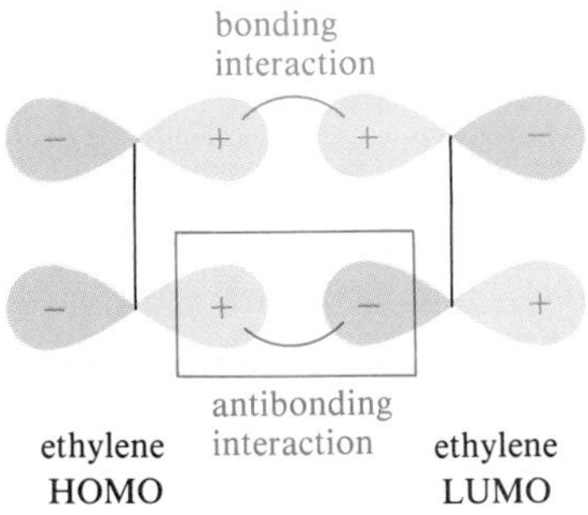

bonding interaction

antibonding interaction

ethylene HOMO

ethylene LUMO

FIGURE 15-20 The HOMO and LUMO of two ethylene molecules have different symmetries, and they overlap to form an antibonding interaction. The concerted [2 + 2] cycloaddition is therefore symmetry-forbidden.

15-12C PHOTOCHEMICAL INDUCTION OF CYCLOADDITIONS

When ultraviolet light rather than heat is used to induce pericyclic reactions, our predictions are generally reversed. For example, the [2 + 2] cycloaddition of two ethylenes is photochemically "allowed." When a photon with the correct energy of ultraviolet light strikes ethylene, one of the pi electrons is excited to the next higher molecular orbital (Fig. 15-21). This higher orbital, formerly the LUMO, is now occupied: It is the new HOMO*, the HOMO of the excited molecule.

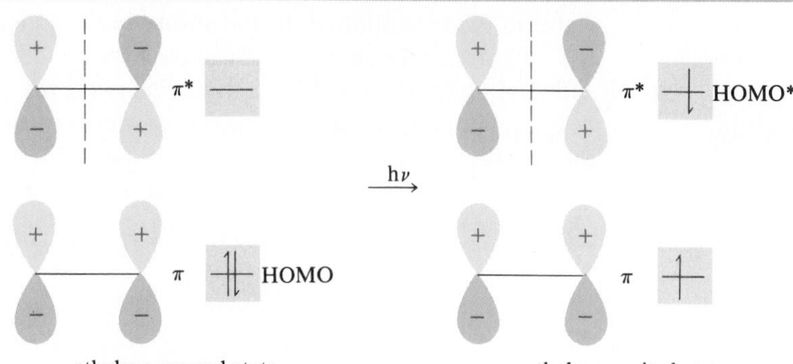

FIGURE 15-21 Ultraviolet light excites one of the ethylene pi electrons into the antibonding orbital. The antibonding orbital is now occupied, and it is the new HOMO*.

The HOMO* of the excited ethylene molecule has the same symmetry as the LUMO of a ground-state molecule. An excited molecule can react with a ground-state molecule to give cyclobutane (Fig. 15-22). The [2 + 2] cycloaddition is therefore *photochemically allowed* but *thermally forbidden*. In most cases, photochemically allowed reactions are thermally forbidden and thermally allowed reactions are photochemically forbidden.

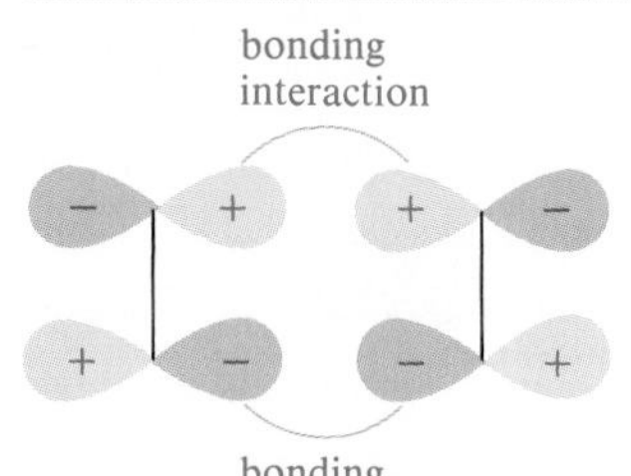

FIGURE 15-22 Photochemical [2 + 2] cycloaddition. The HOMO* of the excited ethylene overlaps favorably with the LUMO of an unexcited (ground-state) molecule.

PROBLEM 15-18

Show that the [4 + 2] Diels-Alder reaction is photochemically forbidden.

PROBLEM 15-19

(a) Show that the [4 + 4] cycloaddition of two butadiene molecules to give cycloocta-1,5-diene is thermally forbidden but photochemically allowed.

(b) There is a different, thermally allowed cycloaddition of two butadiene molecules. Show this reaction and explain why it is thermally allowed. (*Hint:* Consider the dimerization of cyclopentadiene.)

15-13 ULTRAVIOLET ABSORPTION SPECTROSCOPY

We have already encountered three powerful spectroscopic techniques used by organic chemists. Infrared spectroscopy (IR, Chapter 11) observes the vibrations of molecular bonds, providing information about the nature of the bonding and the functional groups present in a molecule. Nuclear magnetic resonance spectroscopy (NMR, Chapter 12) detects nuclear transitions, providing information about the electronic and molecular environment of the nuclei. From NMR information we can determine the structure of the alkyl groups present and often infer the functional groups. A mass spectrometer (MS, Chapter 11) bombards molecules with electrons, causing them to break apart. Analysis of the masses of the fragments provides a molecular weight (and perhaps a molecular formula), as well as structural information about the original compound.

We now study **ultraviolet (UV) spectroscopy,** which detects the electronic transitions of conjugated systems and provides information about the length and structure of the conjugated part of a molecule. UV spectroscopy gives more specialized information than IR or NMR, and it is less commonly used than the other techniques.

15-13A SPECTRAL REGION

In our study of infrared spectroscopy, we saw that an organic molecule can absorb electromagnetic radiation if the frequency of the waves corresponds to the frequency of the molecular motions of bonds in the molecule. Common IR spectrometers operate at wavelengths between 2.5×10^{-4} and 25×10^{-4} cm, corresponding to energies of about 1.1 to 11 kcal (4.6 to 46 kJ) per mole.

TABLE 15-1
Comparison of infrared and ultraviolet wavelengths

Spectral region	*Wavelength, λ*	*Energy range*
ultraviolet	200 – 400 nm (200 – 400 × 10^{-9} m)	70 – 140 kcal (300 – 600 kJ) per mole
infrared	2.5 – 25 μm (2.5 – 25 × 10^{-6} m)	1.1 – 11 kcal (4.6 – 46 kJ) per mole

Ultraviolet frequencies correspond to shorter wavelengths and much larger energies than infrared (Table 15-1). The UV region is a range of frequencies just beyond the visible: *ultra*, meaning "beyond," and *violet*, the highest-frequency visible light. Its wavelengths are given in units of nanometers (nm; 10^{-9} m). Common UV spectrometers operate in the range 200 to 400 nm (2×10^{-5} to 4×10^{-5} cm), corresponding to photon energies of about 70 to 140 kcal (300 to 600 kJ) per mole. These spectrometers often extend into the visible region (longer wavelength, lower energy) and are called **UV-visible spectrometers.** UV-visible energies correspond to electronic transitions: the energy needed to excite an electron from one molecular orbital to another.

15-13B ULTRAVIOLET LIGHT AND ELECTRONIC TRANSITIONS

The wavelengths of UV light absorbed by a molecule are determined by the electronic energy differences between orbitals in the molecule. Sigma bonds are very stable, and the electrons in sigma bonds are usually unaffected by wavelengths of light above 200 nm. Pi bonds have electrons that are more easily excited into higher-energy orbitals. Conjugated systems are particularly likely to have low-lying vacant orbitals, and electronic transitions into these orbitals produce characteristic absorptions in the ultraviolet region.

Ethylene, for example, has two pi orbitals: the bonding orbital (π) and the antibonding orbital (π^*). The ground state has two electrons in the bonding orbital and none in the antibonding orbital. An electron can be excited from the bonding orbital (π) to the antibonding orbital (π^*) by the absorption of a photon of light with the right amount of energy. This transition from a π bonding orbital to a π^* antibonding orbital is called a $\pi \rightarrow \pi^*$ *transition* (Fig. 15-23).

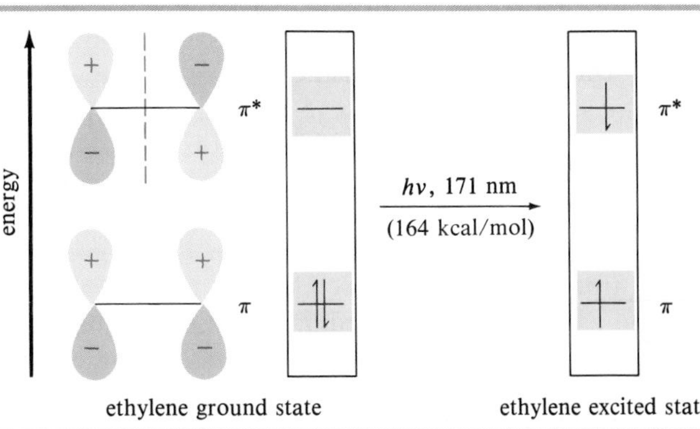

FIGURE 15-23 The absorption of a 171-nm photon excites an electron from the π-bonding MO of ethylene to the π^*-antibonding MO. This absorption requires light of greater energy (shorter wavelength) than the range covered by a typical UV spectrometer.

The $\pi \rightarrow \pi^*$ transition of ethylene requires absorption of light at 171 nm (164 kcal/mol or 686 kJ/mol). Most UV spectrometers cannot detect this absorption, because it is obliterated by the absorption caused by oxygen in the air. In conjugated systems, however, there are electronic transitions with lower energies that correspond to wavelengths longer than 200 nm. Figure 15-24 compares the MO energies of ethylene with those of butadiene to show that the HOMO and LUMO of butadiene are closer in energy than those of ethylene.

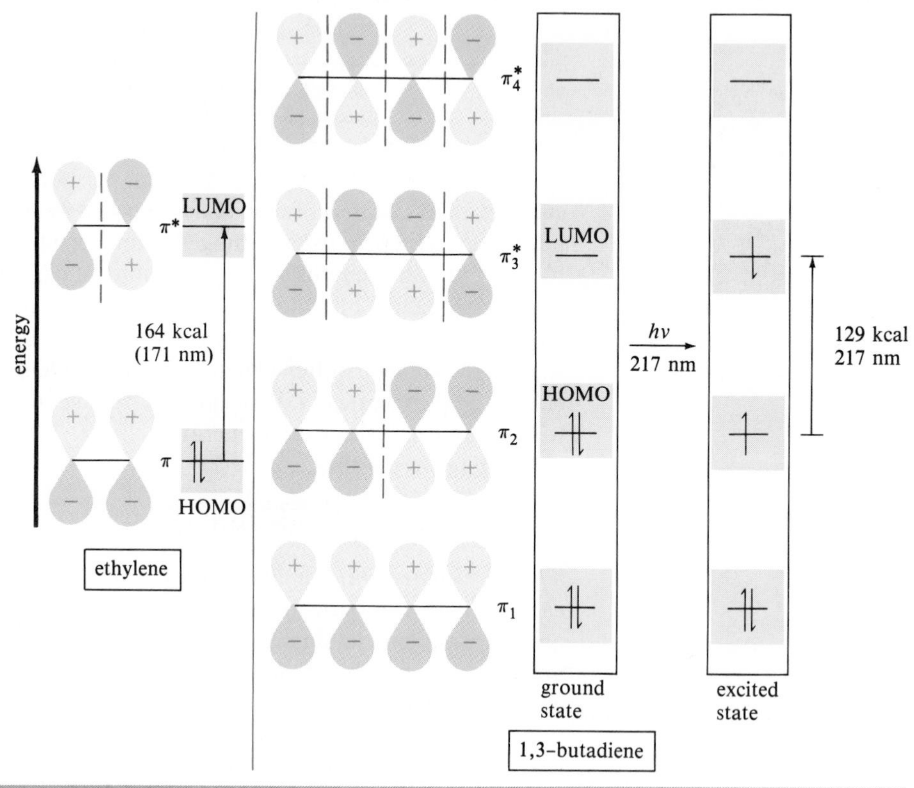

FIGURE 15-24 In 1,3-butadiene, the $\pi \rightarrow \pi^*$ transition absorbs at a wavelength of 217 nm (129 kcal/mol or 540 kJ/mol), compared with 171 nm (164 kcal/mol or 686 kJ/mol) for ethylene. This longer-wavelength (lower-energy) absorption results from a smaller energy difference between the HOMO and LUMO in butadiene than in ethylene.

In ethylene there is only one occupied π-MO and only one unoccupied MO. The only possible transition is the excitation of an electron from the occupied MO to the unoccupied MO. In butadiene there are four possible transitions, involving excitation of an electron from either of the filled orbitals into either of the empty ones. The lowest-energy transition, corresponding to absorption of light of the longest wavelength, is the excitation of an electron from the highest occupied molecular orbital (HOMO) to the lowest unoccupied molecular orbital (LUMO). This is a $\pi_2 \rightarrow \pi_3^*$ transition.

Notice in Figure 15-24 that the HOMO of butadiene is higher in energy than the HOMO of ethylene. Also, the LUMO of butadiene is lower in energy than the LUMO of ethylene. Both of these differences reduce the relative energy of the $\pi \rightarrow \pi^*$ transition in butadiene. The resulting absorption is at 217 nm (129 kcal/mol or 540 kJ/mol), which can be measured using a standard UV spectrometer.

Just as conjugated dienes absorb at longer wavelengths than simple alkenes, conjugated trienes absorb at even longer wavelengths. In 1,3,5-hexatriene, for example (Fig. 15-25), the HOMO is π_3, and the LUMO is π_4^*. The lowest-energy transition is the excitation of an electron from π_3 into π_4^*. The HOMO in 1,3,5-

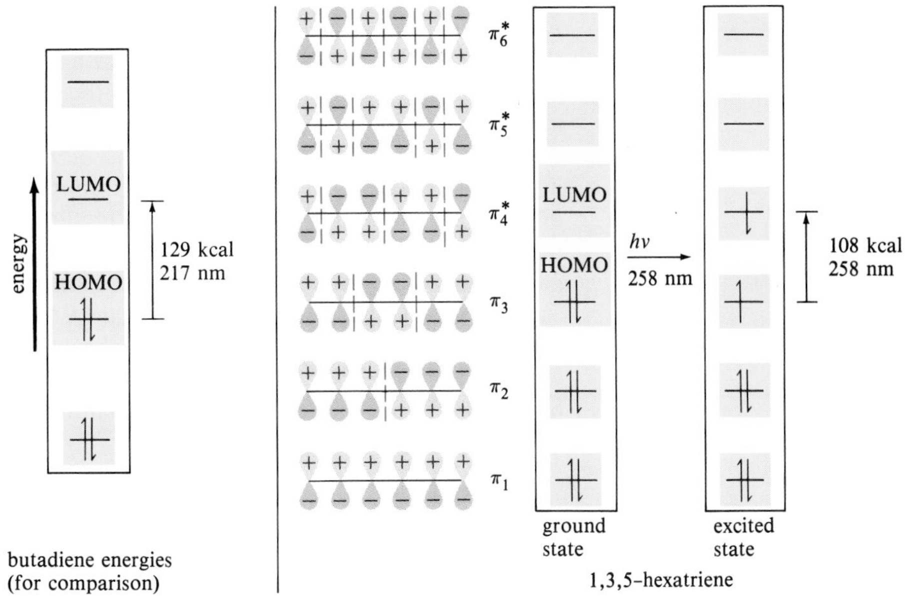

FIGURE 15-25 1,3,5-hexatriene has a smaller energy difference (108 kcal/mol or 452 kJ/mol) between its HOMO and LUMO than does 1,3-butadiene (129 kcal/mol or 540 kJ/mol). The $\pi \rightarrow \pi^*$ transition corresponding to this energy difference absorbs at a longer wavelength: 258 nm, compared with 217 nm for 1,3-butadiene.

hexatriene is slightly higher in energy than that for 1,3-butadiene, and the hexatriene LUMO is slightly lower in energy. Once again, the narrowing of the energy between the HOMO and the LUMO gives a lower-energy, longer-wavelength absorption. The principal $\pi \rightarrow \pi^*$ transition in 1,3,5-hexatriene occurs at 258 nm (108 kcal/mol or 452 kJ/mol).

We can summarize the effects of conjugation on the wavelength of UV absorption by stating a general rule: *A compound that contains a longer chain of conjugated double bonds absorbs light at a longer wavelength.* β-Carotene, which has 11 conjugated double bonds in its pi system, absorbs at 454 nm, well into the visible region of the spectrum, corresponding to absorption of blue light. White light from which blue has been removed appears orange. β-Carotene is the principal compound responsible for giving carrots their orange color.

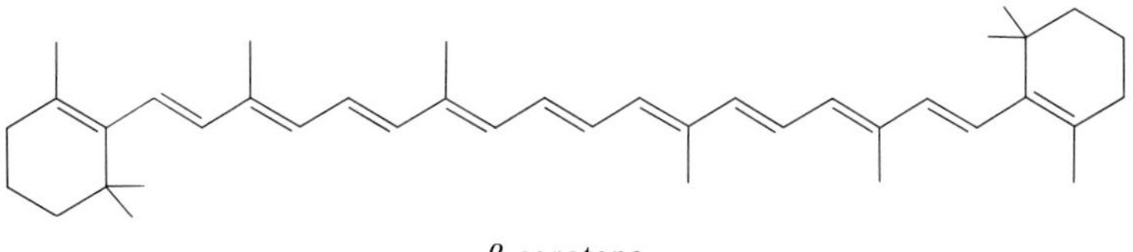

β-carotene

Because they have no interaction with each other, isolated double bonds do not contribute to shifting the UV absorption to longer wavelengths. Both their reactions and their UV absorptions are like those of simple alkenes. For example, 1,4-pentadiene absorbs at 178 nm, a value that is typical of simple alkenes rather than conjugated dienes.

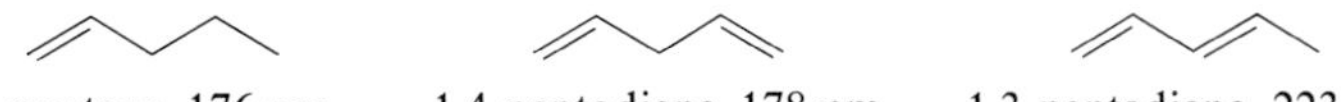

1-pentene, 176 nm 1,4-pentadiene, 178 nm 1,3-pentadiene, 223 nm

15-13C OBTAINING AN ULTRAVIOLET SPECTRUM

To measure the ultraviolet (or UV-visible) spectrum of a compound, the sample is dissolved in a solvent that does not absorb above 200 nm. The sample solution is placed in one quartz cell, and some of the solvent is placed in a **reference cell.** An ultraviolet spectrometer operates by comparing the amount of light transmitted through the sample (the **sample beam**) with the amount of light in a **reference beam.** The reference beam passes through the reference cell to compensate for any absorption of light by the cell and the solvent.

The spectrometer (Fig. 15-26) has a *source* that emits all frequencies of UV light (above 200 nm). This light passes through a *monochromator,* which uses a diffraction grating or a prism to spread the light into a spectrum and select one wavelength. This single wavelength of light is split into two beams, with one beam passing through the sample cell and the other passing through the reference (solvent) cell. The detector continuously measures the intensity ratio of the reference beam (I_r) compared with the sample beam (I_s). As the spectrometer scans the wavelengths in the UV region, a chart recorder draws a graph (called a **spectrum**) of the absorbance of the sample as a function of the wavelength.

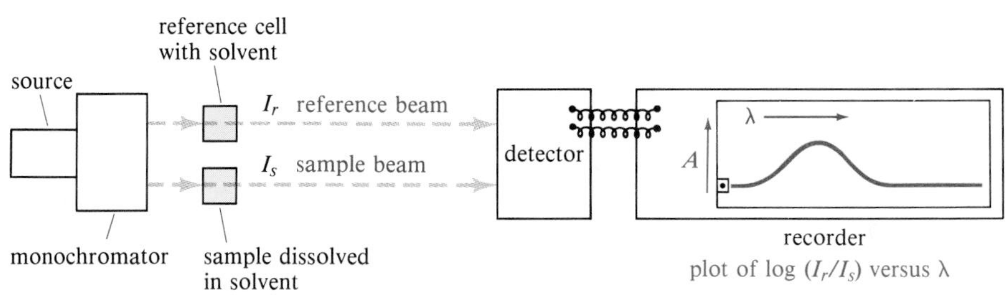

FIGURE 15-26 In the ultraviolet spectrometer, the source produces all frequencies of light in the UV region. The monochromator selects one wavelength of light, which is split into two beams. One beam passes through the sample cell, while the other passes through the reference cell. The detector measures the ratio of the two beams, and the chart recorder plots this ratio as a function of wavelength.

The *absorbance, A,* of the sample at a particular wavelength is governed by *Beer's law,*

$$A = \log\left(\frac{I_r}{I_s}\right) = \varepsilon c l$$

where c is the sample concentration in moles per liter, l is the path length of light through the cell in centimeters, and ε is called the **molar extinction coefficient** (or **molar absorptivity**) of the sample. ε is a measure of how strongly the sample absorbs light at that wavelength.

If the sample absorbs light at a particular wavelength, the sample beam (I_s) is less intense than the reference beam (I_r), and the ratio I_r/I_s is greater than 1. The ratio is equal to 1 when there is no absorption. The absorbance (the logarithm of the ratio) is therefore greater than zero when the sample absorbs and equal to zero when it does not. A UV spectrum is a plot of A, the absorbance of the sample, as a function of the wavelength.

UV-visible spectra tend to show rather broad peaks and valleys. The spectral data that are most characteristic of a sample are

1. The wavelength(s) of maximum absorbance, called λ_{max}
2. The value of the extinction coefficient ε at each maximum.

Since UV-visible spectra are broad and lacking in detail, they are rarely printed as actual spectra. The spectral information is simply given as a list of the value or values of λ_{max} together with the extinction coefficient for each value of λ_{max}.

The actual UV spectrum of isoprene (2-methyl-1,3-butadiene) is shown in Figure 15-27. This spectrum could be summarized as follows:

$$\lambda_{max} = 222 \text{ nm} \qquad \varepsilon = 20{,}000$$

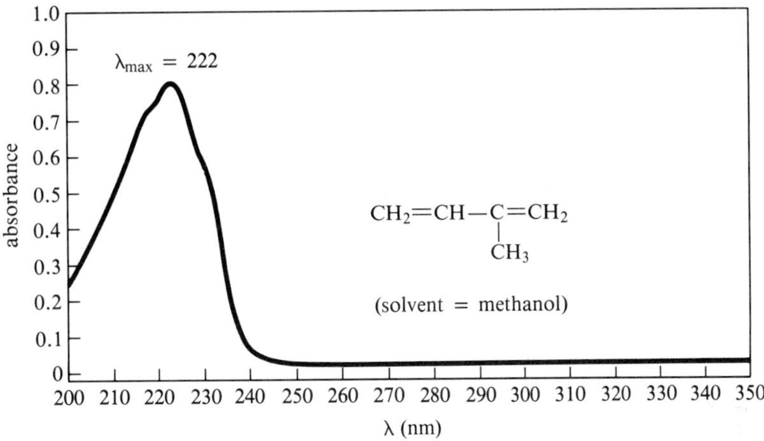

FIGURE 15-27 The UV spectrum of isoprene dissolved in methanol shows $\lambda_{max} = 222$ nm, $\varepsilon = 20{,}000$.

The value of λ_{max} is read directly from the spectrum, but the extinction coefficient ε must be calculated from the concentration of the solution and the path length of the cell. If the isoprene concentration was 4×10^{-5} M and a 1-cm cell was used, the extinction coefficient is found by rearranging Beer's law ($A = \varepsilon cl$).

$$\varepsilon = \frac{A}{cl} = \frac{0.8}{4 \times 10^{-5}} = 20{,}000$$

Molar extinction coefficients in the range 5000 to 30,000 are typical for the $\pi \rightarrow \pi^*$ transitions of conjugated polyene systems. Such a large extinction coefficient is helpful since a spectrum may be obtained with a very small amount of sample. On the other hand, samples and solvents for UV spectroscopy must be extremely pure. A minute impurity with a large extinction coefficient can easily obliterate the spectrum of the desired compound.

PROBLEM 15-20

One milligram of a compound of molecular weight 160 is dissolved in 10 mL of ethanol, and the solution is poured into a 1-cm UV cell. The UV spectrum is taken, and there is an absorption at $\lambda_{max} = 247$ nm. The maximum absorbance at 247 nm is 0.50. Calculate the value of ε for this absorption.

15-13D INTERPRETATION OF UV-VISIBLE SPECTRA

The values of λ_{max} and ε for a conjugated molecule depend on the exact nature of the conjugated system and its substituents. R. B. Woodward and L. F. Fieser developed an extensive set of correlations between molecular structures and absorption maxima, called the **Woodward-Fieser rules.** These rules are summarized in Appendix 3. For most purposes, however, we can use some simple generalizations for estimating approximate values of λ_{max} for common types of systems. Table 15-2 gives the values of λ_{max} for several types of isolated alkenes, conjugated dienes, conjugated trienes, and a conjugated tetraene.

TABLE 15-2
Ultraviolet absorption maxima of some representative molecules

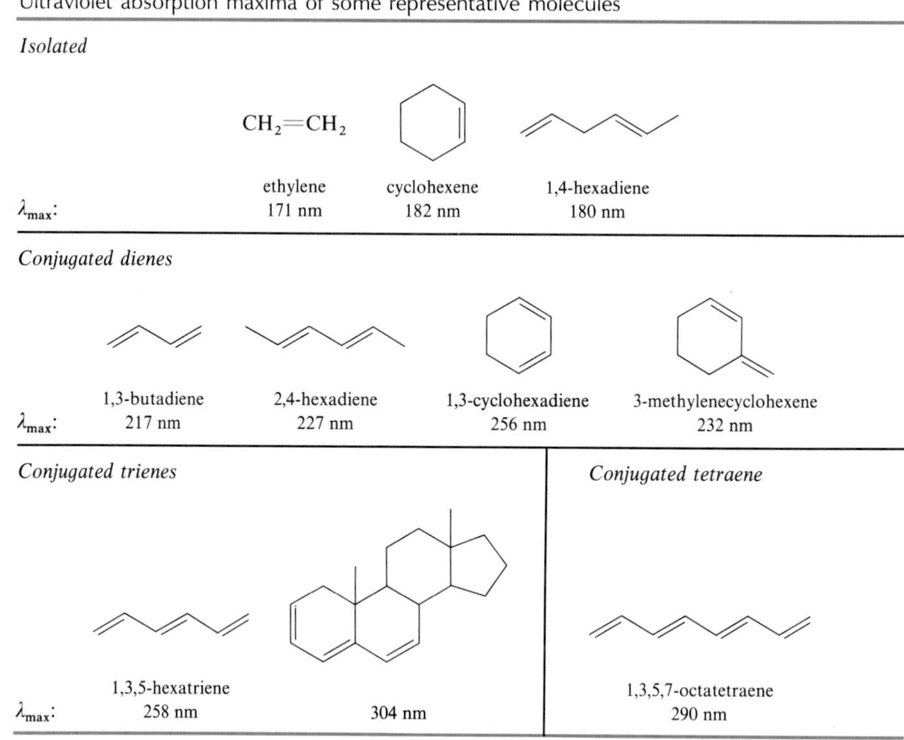

Isolated
$CH_2{=}CH_2$
ethylene
171 nm
cyclohexene
182 nm
1,4-hexadiene
180 nm
λ_{max}:
Conjugated dienes
1,3-butadiene
217 nm
2,4-hexadiene
227 nm
1,3-cyclohexadiene
256 nm
3-methylenecyclohexene
232 nm
λ_{max}:
Conjugated trienes
1,3,5-hexatriene
258 nm
304 nm
λ_{max}:
Conjugated tetraene
1,3,5,7-octatetraene
290 nm

The examples in Table 15-2 show that the addition of another conjugated double bond to a conjugated system has a large effect on λ_{max}. In going from ethylene (171 nm) to 1,3-butadiene (217 nm) to 1,3,5-hexatriene (258 nm) to 1,3,5,7-octatetraene (290 nm), the values of λ_{max} increase by about 30 to 40 nm for each double bond extending the conjugated system. Alkyl groups also increase the value of λ_{max}, by about 5 nm per alkyl group. For example, 2,4-dimethyl-1,3-pentadiene has the same conjugated system as 1,3-butadiene, but with three additional alkyl groups (circled below). Its absorption maximum is at 232 nm, which is a wavelength 15 nm longer than λ_{max} for 1,3-butadiene at 217 nm.

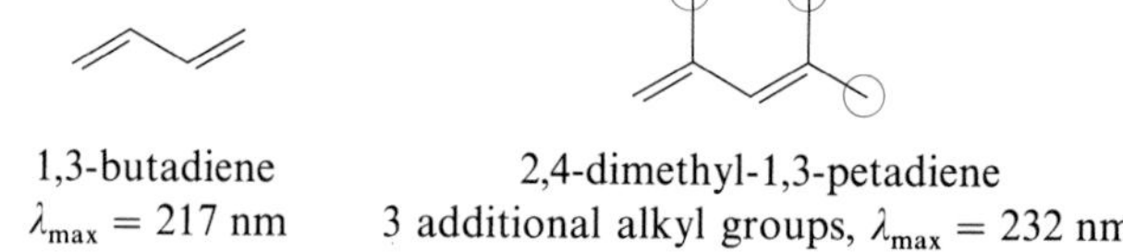

1,3-butadiene
$\lambda_{max} = 217$ nm

2,4-dimethyl-1,3-petadiene
3 additional alkyl groups, $\lambda_{max} = 232$ nm

Structural difference	*Approximate effect on λ_{max}*
additional conjugated C=C	30–40 nm longer
additional alkyl substituent	about 5 nm longer

SOLVED PROBLEM 15-3

Rank the following dienes in order of increasing values of λ_{max}. (Their actual absorption maxima are 185 nm, 235 nm, 273 nm, and 300 nm).

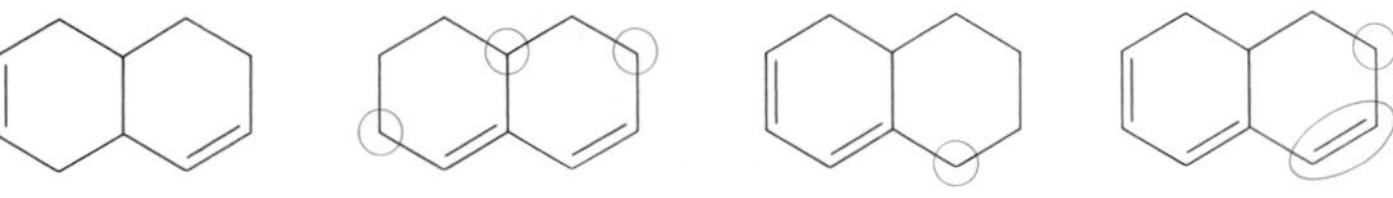

SOLUTION

λ_{max}: 185 nm 235 nm 273 nm 300 nm

These four compounds are an isolated diene, two conjugated dienes, and a conjugated triene. The isolated diene will have the shortest value of λ_{max} (185 nm), close to that of cyclohexene (182 nm).

The second compound looks like 1,3-butadiene (217 nm), but with three additional alkyl substituents (circled). Its absorption maximum should be around 230 to 240 nm, and 235 nm must be the correct value.

The third compound looks like 1,3-cyclohexadiene (256 nm), but with an additional alkyl substituent (circled) raising the value of λ_{max}. So 273 nm must be the correct value.

The fourth compound looks like 1,3-cyclohexadiene (256 nm), but with an additional conjugated double bond (circled) and another alkyl group (circled). We predict a value of λ_{max} about 35 nm longer than for 1,3-cyclohexadiene, and 300 nm must be the correct value.

PROBLEM 15-21

Using the examples in Table 15-2 to guide you, match the following UV absorption maxima (λ_{max}) with the corresponding compounds: (1) 232 nm; (2) 237 nm; (3) 273 nm; (4) 283 nm; (5) 313 nm; (6) 353 nm.

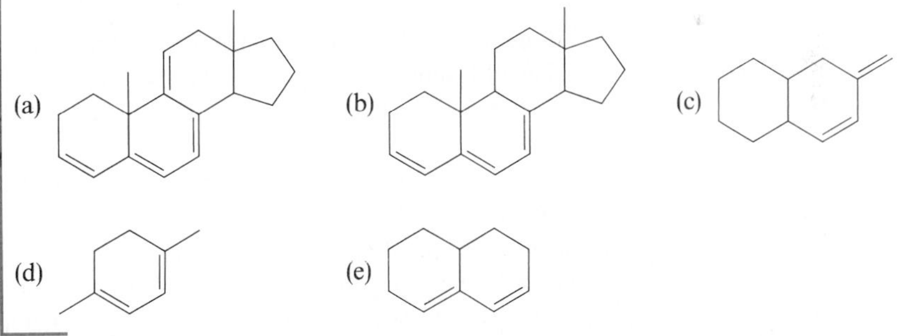

GLOSSARY

1,2-addition An addition in which two atoms or groups add to adjacent carbon atoms. (p. 637)

$$\overset{1}{C}=\overset{2}{C} \; + \; \boxed{A-B} \; \longrightarrow \; -\underset{\boxed{A}}{C}-\underset{\boxed{B}}{C}-$$

a 1,2-addition

1,4-addition An addition in which two atoms or groups add to carbon atoms that bear a 1,4-relationship. (p. 637)

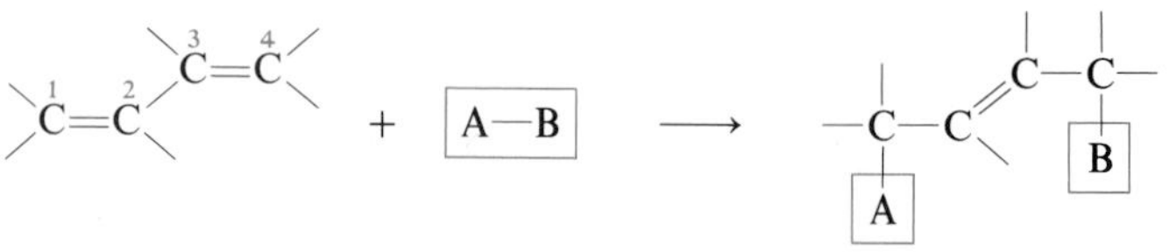

a 1,4-addition

allylic position The carbon atom adjacent to a carbon-carbon double bond. The term is used in naming compounds, such as an **allylic halide,** or in referring to reactive intermediates, such as an **allylic cation,** an **allylic radical,** or an **allylic anion.** (p. 636)

allylic position

$H_2C{=}CH{-}CHBr{-}CH_3$ — an allylic halide

$(CH_3)_2C{=}CH{-}\overset{+}{C}(CH_3)_2$ — an allylic cation

allylic shift The isomerization of a double bond that occurs through the delocalization of an allylic intermediate. (p. 641)

$$H_2C{=}CH{-}CH_2{-}CH_3 \xrightarrow[h\nu]{NBS} H_2C{=}CH{-}CHBr{-}CH_3 + BrCH_2{-}CH{=}CH{-}CH_3$$

allylic shift product

concerted reaction A reaction in which all bond making and bond breaking occurs in the same step. (p. 649)

conjugated double bonds Double bonds separated by a single bond, with interaction by overlap of the *p* orbitals in the two pi bonds. (p. 629)

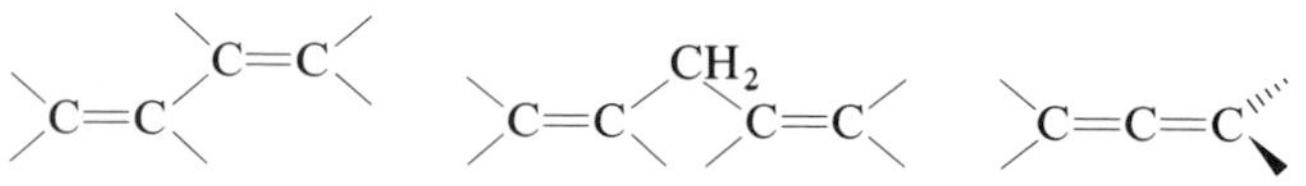

conjugated isolated cumulated

isolated double bonds: Double bonds separated by two or more single bonds. Isolated double bonds react independently, as they do in a simple alkene.

cumulated double bonds: Successive double bonds with no intervening single bonds.

an allene (a cumulene): A compound containing cumulated carbon-carbon double bonds.

conservation of orbital symmetry A theory of pericyclic reactions stating that the MO's of the reactants must flow smoothly into the MO's of the products without any drastic changes in symmetry: That is, there must be bonding interactions to help stabilize the transition state. (p. 656)

constructive overlap An overlap of orbitals that contributes to bonding. Overlap of lobes with similar phases (+ phase with + or − with −) is generally constructive overlap. (p. 633)

cycloaddition A reaction of two alkenes or polyenes to form a cyclic product. Cycloadditions often take place through concerted interaction of the pi electrons in two unsaturated fragments. (p. 655)

delocalized orbital A molecular orbital that results from the combination of three or more atomic orbitals. When filled, these orbitals spread electron density over all the atoms involved. (p. 632)

destructive overlap An overlap of orbitals that contributes to antibonding. Overlap of lobes with opposite phases (+ phase with − phase) is generally destructive overlap. (p. 633)

Diels-Alder reaction A synthesis of six-membered rings by a [4 + 2] cycloaddition. This means that four pi electrons in one fragment interact with two pi electrons in the other fragment to form a new ring. (p. 647)

dienophile ("diene lover"): The two-pi-electron component that reacts with a diene in the Diels-Alder reaction.

endo rule: The stereochemistry of a Diels-Alder product is often determined by **secondary overlap** of the electron-withdrawing group of the dienophile, keeping it in close proximity to the central atoms of the diene. With cyclic dienes, this geometry favors endo products. (p. 651)

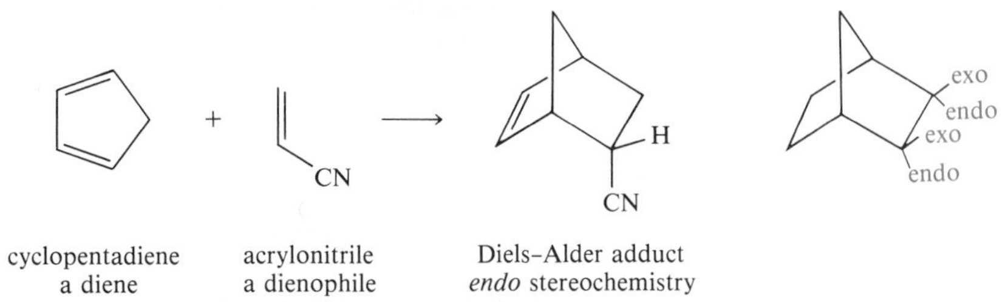

+ CN → H CN

exo endo exo endo

cyclopentadiene a diene | acrylonitrile a dienophile | Diels–Alder adduct *endo* stereochemistry

heat of hydrogenation The enthalpy of reaction that accompanies the addition of hydrogen to a mole of an unsaturated compound. (p. 630)

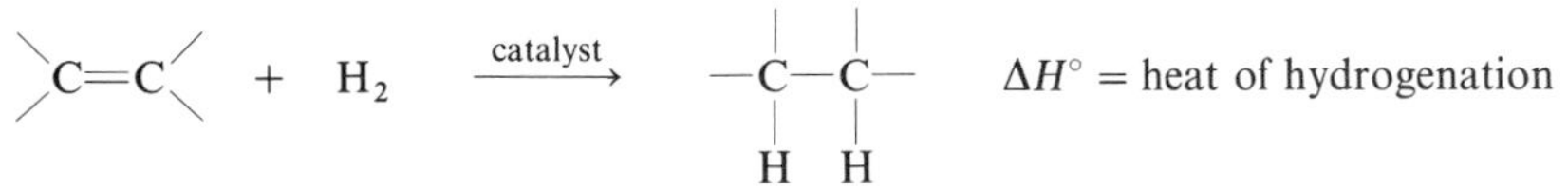

$$\mathrm{>C{=}C<} + \mathrm{H_2} \xrightarrow{\text{catalyst}} \mathrm{-\overset{|}{\underset{H}{C}}-\overset{|}{\underset{H}{C}}-} \qquad \Delta H^\circ = \text{heat of hydrogenation}$$

HOMO An abbreviation for "highest occupied molecular orbital." In a photochemically excited state, this is represented as HOMO*. (p. 656)

kinetic control The product distribution is governed by the rates at which the various products are formed (p. 639)

kinetic product: The product that is formed fastest; the major product under kinetic control.

LUMO An abbreviation for "lowest unoccupied molecular orbital." (p. 656)

molar extinction coefficient ε (molar absorptivity) A measure of how strongly a compound absorbs light at a particular wavelength. It is defined by *Beer's law,*

$$A = \log\left(\frac{I_r}{I_s}\right) = \varepsilon c l$$

where A is the absorbance, I_r and I_s are the amounts of light passing through the reference and sample beams, c is the sample concentration in moles per liter, and l is the path length of light through the cell. (p. 662)

molecular orbitals (MO's) Orbitals that include more than one atom in a molecule. Molecular orbitals can be bonding, antibonding, or nonbonding. (p. 632)

bonding molecular orbitals: MO's that are lower in energy than the isolated atomic orbitals from which they are made. Electrons in these orbitals serve to hold the atoms together.

antibonding molecular orbitals: MO's that are higher in energy than the isolated atomic orbitals from which they are made. Electrons in these orbitals tend to push the atoms apart.

nonbonding molecular orbitals: MO's that are similar in energy to the isolated atomic orbitals from which they are made. Electrons in these orbitals have no effect on the bonding of the atoms. (p. 644)

pericyclic reaction A reaction involving concerted reorganization of electrons within a closed loop of interacting orbitals. Cycloadditions are one class of pericyclic reactions. (p. 655)

reference beam A second beam in the spectrometer, passing through a **reference cell** containing only the solvent. The **sample beam** is compared with this beam to compensate for any absorption by the cell or the solvent. (p. 662)

resonance energy The extra stabilization provided by delocalization, compared with a localized structure. For dienes and polyenes, the resonance energy is the extra stability of the conjugated system compared with the energy of a compound with an equivalent number of isolated double bonds. (p. 631)

***s*-cis conformation** A cis-like conformation of a single bond in a conjugated diene or polyene. (p. 635)

***s*-trans conformation** A trans-like conformation of a single bond in a conjugated diene or polyene. (p. 635)

s-cis conformation *s*-trans conformation

symmetry-allowed The MO's of the reactants can flow into the MO's of the products in one concerted step according to the rules of the conservation of orbital symmetry. In a symmetry-allowed cycloaddition there is constructive overlap (+ phase with +, − phase with −) between the HOMO of one molecule and the LUMO of the other. (p. 657)

symmetry-forbidden The MO's of the reactants are of incorrect symmetries to flow into those of the products in one concerted step. (p. 657)

thermodynamic control (equilibrium control) The product distribution is governed by the stabilities of the products. Thermodynamic control operates when the reaction mixture is allowed to come to equilibrium. (p. 639)

> **thermodynamic product:** The most stable product; the major product under thermodynamic control.

UV-visible spectroscopy The measurement of the absorption of ultraviolet and visible light as a function of the wavelength of the light. Ultraviolet light consists of wavelengths from about 100 to 350 nm. Visible light is from about 350 nm (violet) to 700 nm (red). (p. 658)

Woodward-Fieser rules A set of rules that correlate values of λ_{max} in the UV-visible spectrum with structures of conjugated systems. (p. 663 and Appendix 3)

Woodward-Hoffman rules A set of symmetry rules that predict whether a particular pericyclic reaction is symmetry-allowed or symmetry-forbidden. (p. 656)

ESSENTIAL PROBLEM-SOLVING SKILLS IN CHAPTER 15

1. Show how to construct the molecular orbitals of ethylene, butadiene, and the allylic system. Show the electronic configurations of ethylene, butadiene, and the allyl cation, radical, and anion.

2. Recognize reactions, such as free-radical reactions and cationic reactions, that are enhanced by resonance stabilization of the intermediates. Develop mechanisms to explain the enhanced rates and observed products, and draw resonance structures of the stabilized intermediates.

3. Predict the products of Diels-Alder reactions, including the orientation of cycloaddition with unsymmetrical reagents and the stereochemistry of the products.

4. By comparing the molecular orbitals of the reactants, predict which cycloadditions will be thermally allowed and which will be photochemically allowed.

5. Use values of λ_{max} from UV-visible spectra to estimate the length of conjugated systems and compare compounds with similar structures.

15-22. Briefly define each of the following terms and give an example.

(a) λ_{max}
(b) an extinction coefficient
(c) an allylic alcohol
(d) an endo product
(e) conjugated double bonds
(f) cumulated double bonds
(g) isolated double bonds
(h) a substituted allene
(i) a molecular orbital
(j) an antibonding MO
(k) an allylic radical
(l) an *s*-trans conformation
(m) a 1,2-addition
(n) a 1,4-addition
(o) a cycloaddition
(p) kinetic control of a reaction
(q) a Diels-Alder reaction
(r) thermodynamic control
(s) a dienophile
(t) a concerted reaction
(u) a HOMO, a HOMO*, and a LUMO
(v) a symmetry-forbidden reaction

15-23. Classify the following dienes and polyenes as isolated, conjugated, cumulated, or some combination of these classifications.

(a) 1,5-cyclooctadiene
(b) 1,3-cyclooctadiene
(c) 1,2-cyclodecadiene
(d) 1,3,6-cyclooctatriene
(e) 1,3,5-cyclohexatriene (benzene)

15-24. Predict the products of the following reactions.

(a) allyl bromide + cyclohexyl magnesium bromide
(b) cyclopentadiene + anhydrous HCl
(c) 2-methylpropene + NBS, light
(d) 1-pentene + NBS, light
(e) 1,3-butadiene + bromine water
(f) 1,3,5-hexatriene + bromine in CCl_4
(g) 1-(bromomethyl)-2-methylcyclopentene, heated in methanol
(h) cyclopentadiene + methyl acrylate, $CH_2{=}CH{-}COOCH_3$
(i) 1,3-cyclohexadiene + $CH_3OOC{-}C{\equiv}C{-}COOCH_3$

15-25. Show how the reaction of an allylic halide with a Grignard reagent might be used to synthesize each of the following hydrocarbons.

(a) 5-methyl-1-hexene
(b) 2,5,5-trimethyl-2-heptene

15-26. Draw the important resonance contributors for the following cations, anions, and radicals.

(a) 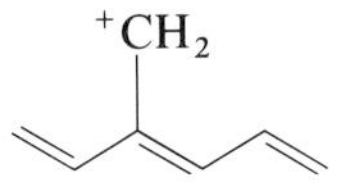 $^+CH_2$
(b) 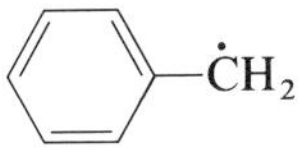 $\dot{C}H_2$
(c) 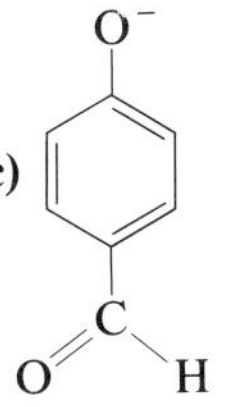 O^- C O H
(d) 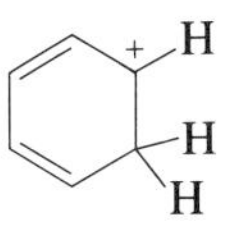 + H H H

15-27. A solution was prepared using 0.0010 g of an unknown compound (of molecular weight around 390) in 100 mL of ethanol. Some of this solution was placed in a 1-cm cell and the UV spectrum was measured. This solution was found to have $\lambda_{max} = 235$ nm, with $A = 0.74$.

(a) Compute the value of the extinction coefficient at 235 nm.
(b) Which of the following compounds might give this spectrum?

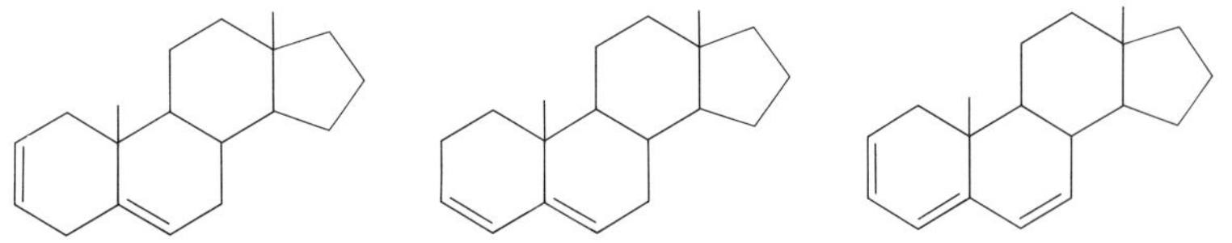

15-28. When *N*-bromosuccinimide is added to 1-hexene and a sunlamp is shone on the mixture, three products result.

(a) Give the structures of these three products.
(b) Give a mechanism that accounts for the formation of these three products.

15-29. Predict the products of the following Diels-Alder reactions. Include stereochemistry where appropriate.

(a) 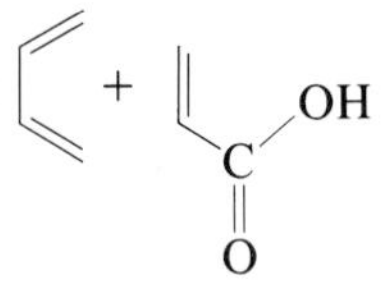 (b) 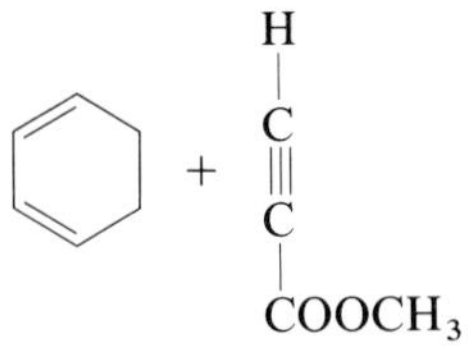 (c)

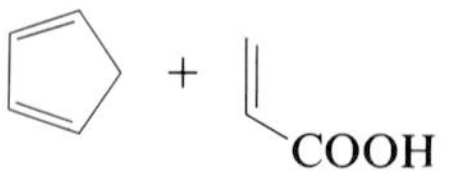

(d) 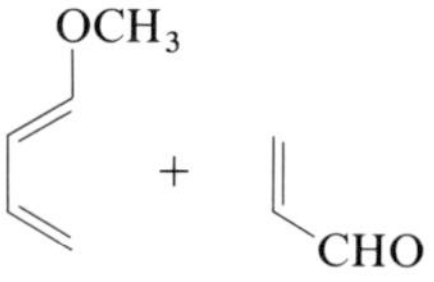 (e) 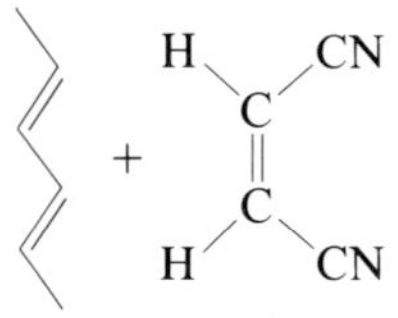 (f)

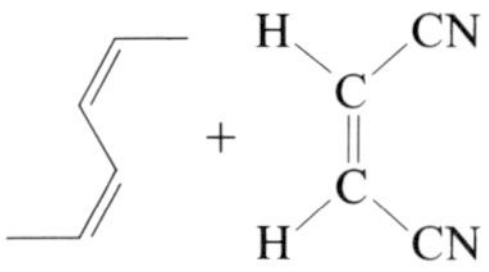

15-30. For each of the following structures
(1) Draw all the important resonance contributors.
(2) Evaluate the significance of each resonance contributor.

(a) 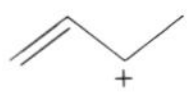 (b) 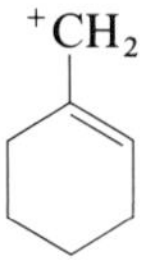 (c) 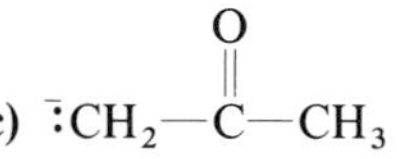 $\ddot{\text{:}}CH_2—\overset{O}{\overset{\|}{C}}—CH_3$ (d) 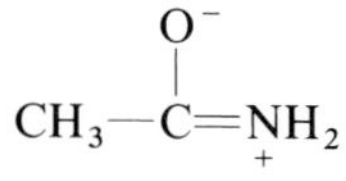 $CH_3—\overset{O^-}{\overset{|}{C}}=\overset{+}{N}H_2$

(e) 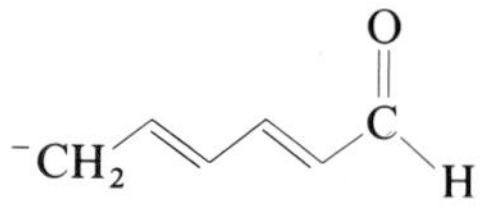 (f) 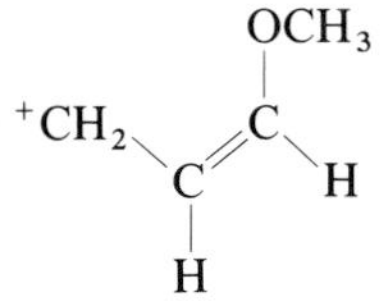 (g)

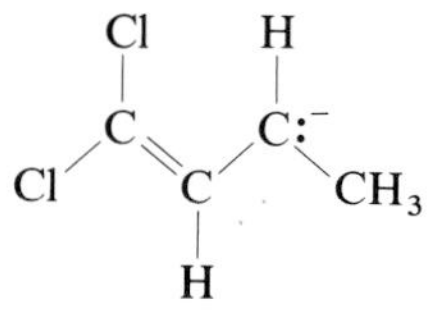

15-31. A graduate student was following a procedure to make 3-propyl-1,4-cyclohexadiene. During the workup procedure, his research adviser called him into her office. By the time he returned to his bench, the product had warmed to a higher temperature than recommended. He isolated the product, which gave the appropriate =C—H stretch in the IR, but the C=C stretch appeared around 1630 cm^{-1} as opposed to the literature value of 1650 cm^{-1} for the desired product. The mass spectrum showed the correct molecular weight, but the base peak was at M-29 rather than at M-43 as expected.

(a) Should he have his IR recalibrated, or should he repeat the experiment, watching the temperature more carefully? What does the 1630 cm^{-1} absorption suggest?

(b) Draw the structure of the desired product and propose a structure for the actual product.

(c) Show why he expected the MS base peak to be at M-43, and show how your proposed structure would give an intense peak at M-29.

15-32. Show how Diels-Alder reactions might be used to synthesize the following compounds.

(a) 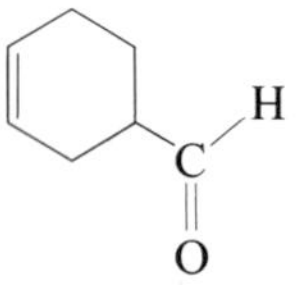 (b) 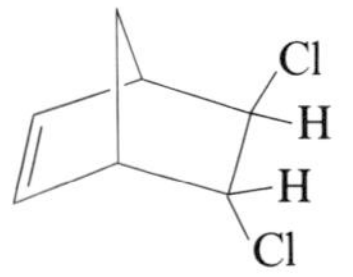 (c)

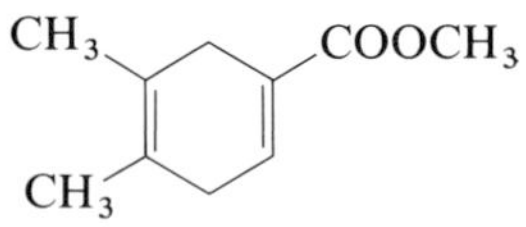

(d) 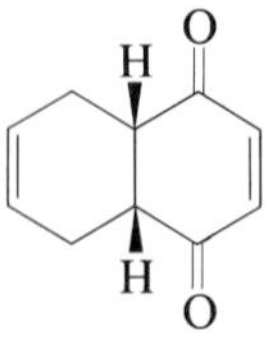 (e) 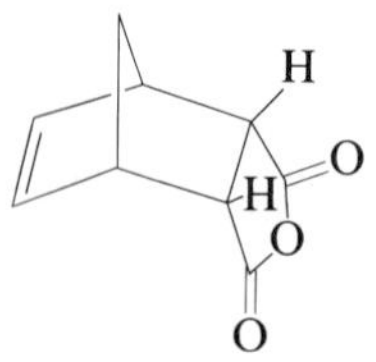 (f)

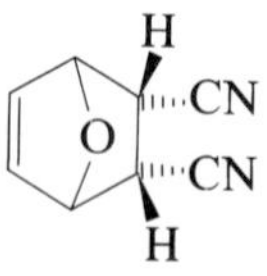

(g) chlordane (h) aldrin (i)

✱ **15-33.** **(a)** Sketch the pi molecular orbitals of 1,3,5-hexatriene (Figure 15-25).
(b) Show the electronic configuration of the ground state of 1,3,5-hexatriene.
(c) Show what product would result from the [6 + 2] cycloaddition of 1,3,5-hexatriene with maleic anhydride.

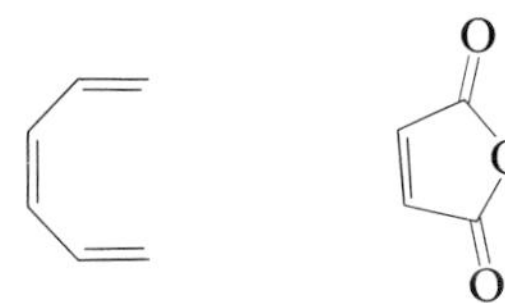

1,3,5-hexatriene maleic anhydride

(d) Show that the [6 + 2] cyclization of 1,3,5-hexatriene with maleic anhydride is thermally forbidden but photochemically allowed.
(e) Show the Diels-Alder product that would actually result from heating 1,3,5-hexatriene with maleic anhydride.

✱ **15-34.** The pentadienyl radical, $H_2C{=}CH{-}CH{=}CH{-}CH_2\cdot$, has its odd electron delocalized over three carbon atoms.
(a) Use resonance structures to show which three carbon atoms bear the odd electron.
(b) How many molecular orbitals are there in the molecular orbital picture of the pentadienyl radical?
(c) How many nodes are there in the lowest-energy MO of the pentadienyl system? How many in the highest-energy MO?
(d) Draw the MO's of the pentadienyl system in order of increasing energy.
(e) Show how many electrons are in each MO for the pentadienyl radical.
(f) Show how your molecular orbital picture agrees with the resonance picture showing delocalization of the odd electron onto three carbon atoms.
(g) Remove the highest-energy electron from the pentadienyl radical to give the pentadienyl cation. Which carbon atoms share the positive charge? Does this picture agree with the resonance picture?
(h) Add an electron to the pentadienyl radical to give the pentadienyl anion. Which carbon atoms share the negative charge? Does this picture agree with the resonance picture?

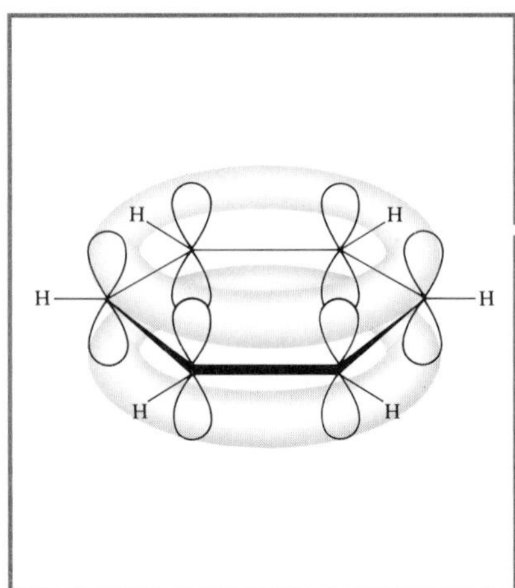

16

AROMATIC COMPOUNDS

16-1 INTRODUCTION; THE DISCOVERY OF BENZENE

In 1825, Michael Faraday isolated a pure compound of boiling point 80°C from the oily mixture that condensed from illuminating gas, the fuel burned in gas-lights. Elemental analysis showed an unusually small hydrogen-to-carbon ratio of 1:1, corresponding to an empirical formula of CH. Faraday named the new compound "bicarburet of hydrogen." Eilhard Mitscherlich synthesized the same compound in 1834 by heating benzoic acid, isolated from gum benzoin, in the presence of lime. Like Faraday, Mitscherlich found that the empirical formula was CH. He also used a vapor-density measurement to determine a molecular weight of about 78, for a molecular formula of C_6H_6. Since the new compound was derived from gum benzoin, he named it benzin, now called *benzene.*

Many other compounds discovered in the late nineteenth century seemed to be related to benzene. These compounds had low hydrogen-to-carbon ratios and pleasant aromas, and they could be converted to benzene or related compounds. This group of compounds was called **aromatic** because of their pleasant odors. Other organic compounds, without these properties, were called **aliphatic,** meaning "fatlike." As the unusual stability of aromatic compounds was investigated, the term "aromatic" came to be applied to compounds with this stability, regardless of their odors.

16-2 THE STRUCTURE AND PROPERTIES OF BENZENE

The Kekulé structure In 1866, Friedrich Kekulé proposed a cyclic structure for benzene with three double bonds. Considering that multiple bonds had been proposed only recently (1859), the cyclic structure with alternating single and double bonds was considered somewhat bizarre.

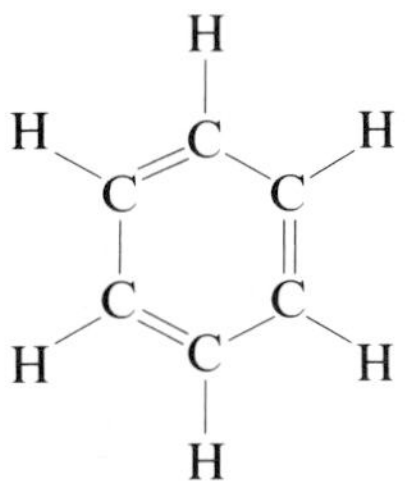

The **Kekulé structure** has its shortcomings, however. For example, it predicts two different 1,2-dichlorobenzenes, but only one is known to exist. Kekulé suggested (incorrectly) that a fast equilibrium interconverts the two isomers of 1,2-dichlorobenzene.

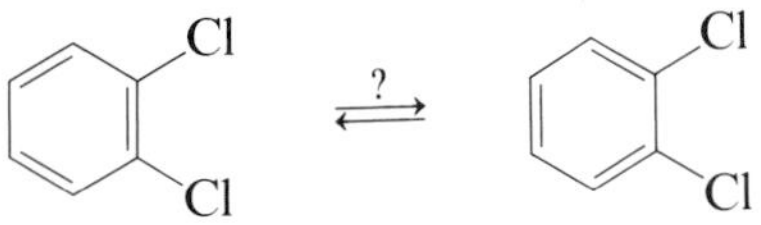

1,2-dichlorobenzene

The resonance representation The resonance picture of benzene is a natural extension of Kekulé's hypothesis. In a Kekulé structure, the single bonds would be longer than the double bonds. The bonds in benzene are known to be all the same length (1.397 Å), however, and the ring is known to be planar. Because the ring is planar and the carbon nuclei are positioned at equal distances, the two Kekulé structures differ only in the positioning of the pi electrons.

Benzene is actually a resonance hybrid of the two Kekulé structures. This representation suggests that the pi electrons are delocalized, with a bond order of $1\frac{1}{2}$ between adjacent carbon atoms. The carbon-carbon bond lengths in benzene are shorter than typical single-bond lengths, yet longer than typical double-bond lengths.

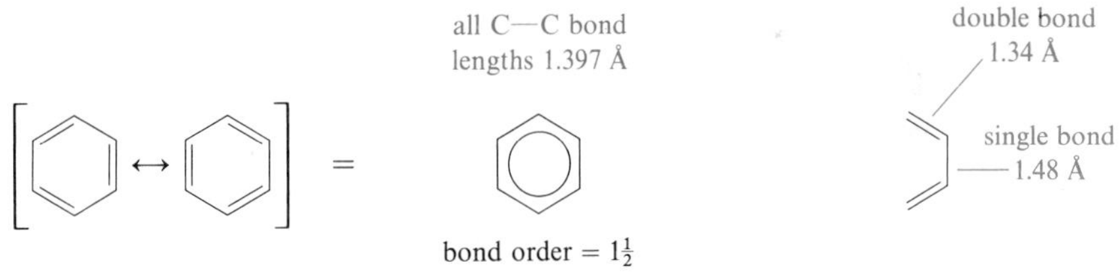

all C—C bond lengths 1.397 Å

double bond 1.34 Å

single bond 1.48 Å

bond order = $1\frac{1}{2}$

resonance representation combined representation butadiene

The resonance-delocalized picture explains most of the structural properties of benzene and its derivatives—the *benzenoid* aromatic compounds. Because the pi bonds are delocalized over the ring, we often inscribe a circle in the hexagon rather than drawing three localized double bonds. This representation helps us to remember that there are no localized single or double bonds, and it prevents us from trying to draw supposedly different isomers that differ only in the placement of double bonds in the ring. We often use Kekulé structures in drawing reaction mechanisms, however, to show the movement of individual pairs of electrons.

PROBLEM 16-1

Write Lewis structures for the Kekulé representations of benzene. Show all the valence electrons.

According to this resonance picture, we can draw a more realistic representation of benzene (Fig. 16-1). Benzene consists of a ring of sp^2 hybrid carbon atoms, each bonded to one hydrogen atom. All the carbon-carbon bonds are the same length, and all the bond angles are exactly 120°. Each sp^2 carbon atom has an unhybridized p orbital perpendicular to the plane of the ring, and six electrons occupy this circle of p orbitals.

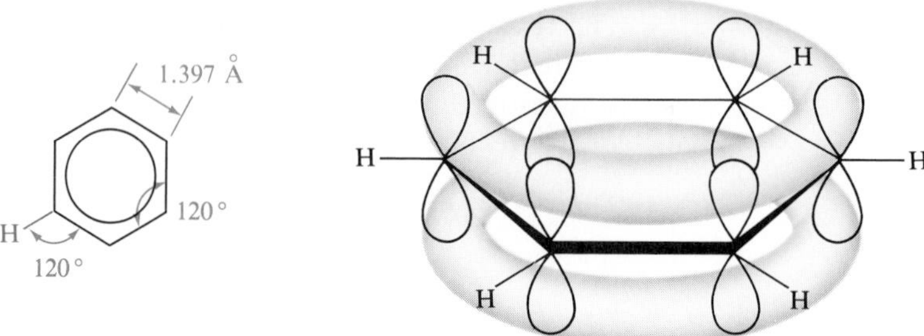

FIGURE 16-1 Benzene is a flat ring of sp^2 hybrid carbon atoms with their unhybridized p orbitals all aligned and overlapping. The ring of p orbitals contains six electrons. The carbon-carbon bond lengths are all 1.397 Å, and all the bond angles are exactly 120°.

For now, we will consider an **aromatic compound** to be a cyclic compound containing some number of conjugated double bonds and having an unusually large resonance energy. Using benzene as an example, we consider how aromatic compounds differ from aliphatic compounds. Then we discuss why the aromatic structure confers extra stability and how we can predict aromaticity in some interesting and unusual compounds.

The unusual reactions of benzene Benzene is actually much more stable than we would expect from the simple resonance-delocalized picture. First we consider the evidence for this unusual stability, and then we construct the molecular orbital picture to explain it.

Both the Kekulé structure and the resonance-delocalized picture show that benzene is a cyclic conjugated triene. We might expect benzene to undergo the typical reactions of polyenes. In fact, however, its reactions are quite unusual. For example, an alkene decolorizes potassium permanganate by reacting to form a glycol. The purple permanganate color disappears, and a precipitate of manganese dioxide forms. When permanganate is added to benzene, however, no reaction occurs.

$$\text{cyclohexene} \xrightarrow{KMnO_4,\ H_2O} \text{cis-cyclohexane-1,2-diol} + MnO_2$$

$$\text{benzene} \xrightarrow{KMnO_4,\ H_2O} \text{no reaction}$$

Most alkenes decolorize solutions of bromine in carbon tetrachloride. The red bromine color disappears as bromine adds across the double bond. When bromine is added to benzene, no reaction occurs, and the red bromine color remains.

$\xrightarrow[CCl_4]{Br_2}$ (trans-1,2-dibromocyclohexane)

$\xrightarrow[CCl_4]{Br_2}$ no reaction

The addition of a catalyst such as ferric bromide to the mixture of bromine and benzene causes the bromine color to disappear slowly. HBr gas is evolved as a by-product, and the expected addition of Br_2 has *not* taken place. Instead, the organic product results from *substitution* of a bromine atom for a hydrogen, and all three double bonds are retained.

$\xrightarrow[CCl_4]{Br_2, FeBr_3}$ (bromobenzene) + HBr↑ ; (dibromocyclohexadiene) is not formed

The unusual stability of benzene Benzene's reluctance to undergo typical alkene reactions suggests that it must be unusually stable. By comparing molar heats of hydrogenation, we can get a quantitative idea of its stability. Benzene, cyclohexene, and the cyclohexadienes all hydrogenate to cyclohexane. Figure 16-2 shows how heats of hydrogenation are used to compute the resonance energies of 1,3-cyclohexadiene and benzene, based on the following reasoning.

1. Hydrogenation of cyclohexene is exothermic by 28.6 kcal (120 kJ) per mole.
2. Hydrogenation of 1,4-cyclohexadiene is exothermic by 57.4 kcal (240 kJ) per mole, about twice the heat of hydrogenation of cyclohexene. The resonance

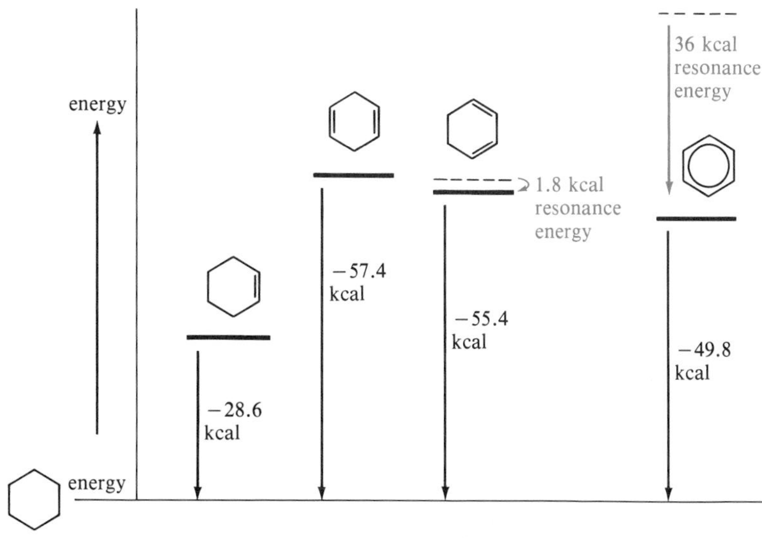

FIGURE 16-2 The molar heats of hydrogenation and the relative energies of cyclohexene, 1,4-cyclohexadiene, 1,3-cyclohexadiene, and benzene. The dashed lines represent the predicted energy if every double bond had the same energy as the double bond in cyclohexene.

energy of the isolated double bonds in 1,4-cyclohexadiene is about zero.

3. Hydrogenation of 1,3-cyclohexadiene is exothermic by 55.4 kcal (232 kJ) per mole, about 1.8 kcal (7.5 kJ) less than twice the value for cyclohexene. A resonance energy of 1.8 kcal (7.5 kJ) is typical for a conjugated diene.
4. Hydrogenation of benzene requires higher pressures of hydrogen and a more active catalyst. This hydrogenation is exothermic by 49.8 kcal (208 kJ) per mole, about 36.0 kcal (151 kJ) less than three times the value for cyclohexene.

The huge 36-kcal (151-kJ) resonance energy of benzene cannot be explained by simple conjugation effects alone. The heat of hydrogenation for benzene is actually smaller than that for 1,3-cyclohexadiene. The hydrogenation of the *first* double bond of benzene must be endothermic, the first endothermic hydrogenation we have encountered. In practice, this reaction is very difficult to stop after the addition of 1 mole of H_2, since the product (1,3-cyclohexadiene) hydrogenates more easily than benzene itself. Clearly, the benzene ring is exceptionally unreactive.

benzene + H_2 $\xrightarrow{\text{catalyst}}$ 1,3-cyclohexadiene

	$\Delta H^\circ_{\text{hydrogenation}}$
benzene:	-49.8 kcal (-208 kJ)
1,3-cyclohexadiene:	-55.4 kcal (-232 kJ)
$\Delta H^\circ =$	$+5.6$ kcal ($+23$ kJ)

PROBLEM 16-2

Using the information in Figure 16-2, calculate the values of ΔH° for the following reactions.

(a) benzene + H_2 $\xrightarrow{\text{catalyst}}$ 1,3-cyclohexadiene (b) benzene + 2 H_2 $\xrightarrow{\text{catalyst}}$ cyclohexene

(c) 1,3-cyclohexadiene + H_2 $\xrightarrow{\text{catalyst}}$ cyclohexene

Failures of the resonance picture For many years, chemists assumed that benzene's large resonance energy results from having two identical, stable resonance structures. It was thought that other hydrocarbons with analogous conjugated systems of alternating single and double bonds would show similar stability. These cyclic hydrocarbons with alternating single and double bonds are called **annulenes.** For example, benzene is the six-membered annulene, so it can be named [6]annulene. Cyclobutadiene is [4]annulene, cyclooctatetraene is [8]annulene, and larger annulenes are named similarly.

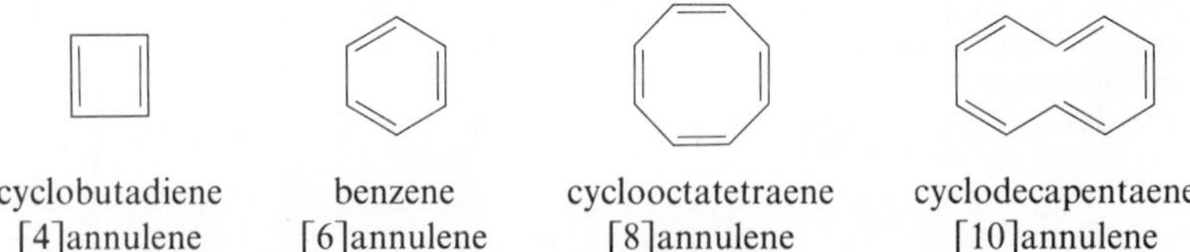

cyclobutadiene [4]annulene; benzene [6]annulene; cyclooctatetraene [8]annulene; cyclodecapentaene [10]annulene

For the double bonds to be completely conjugated, the annulene must be planar so that the p orbitals of the pi bonds overlap. As long as an annulene is

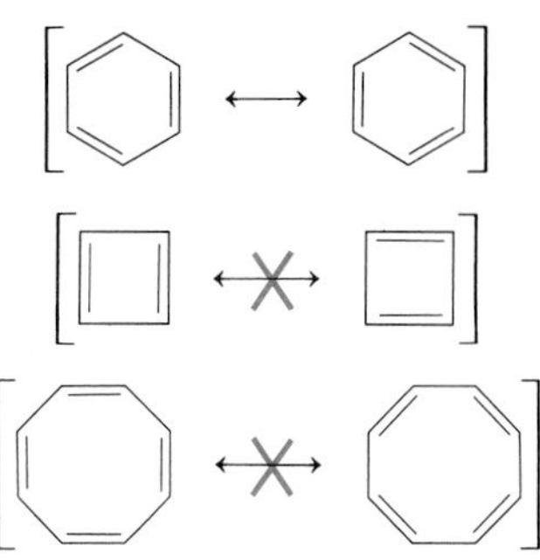

FIGURE 16-3 Cyclobutadiene and cyclooctatetraene have alternating single and double bonds similar to those of benzene. These compounds were mistakenly expected to be aromatic.

assumed to be planar, we can draw two Kekulé-like structures that seem to show a benzene-like resonance. Figure 16-3 shows proposed benzene-like resonance structures for cyclobutadiene and cyclooctatetraene. Do these resonance structures imply that these [4] and [8]annulenes are unusually stable, just like benzene? Experiments have shown that cyclobutadiene and cyclooctatetraene are not unusually stable, and that this simple resonance picture is incorrect.

Cyclobutadiene has never been isolated and purified. It undergoes an extremely fast Diels-Alder dimerization. To avoid the Diels-Alder reaction, cyclobutadiene has been prepared at low concentrations in the gas phase and as individual molecules trapped in frozen argon at very low temperatures. This is not the behavior that we expect from a molecule with exceptional stability!

In 1911, Richard Willstätter synthesized cyclooctatetraene and found that it reacts like a normal polyene. Bromine adds readily to cyclooctatetraene, and permanganate oxidizes its double bonds. This evidence shows that cyclooctatetraene is much less stable than benzene. In fact, recent structural studies have shown that cyclooctatetraene is not planar. It is most stable in a "tub" conformation, with poor overlap between adjacent pi bonds.

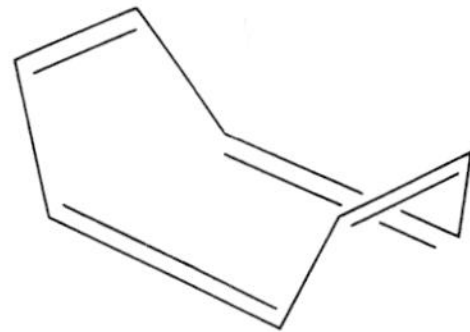

"tub" conformation of cyclooctatetraene

Visualizing benzene as a resonance hybrid of two Kekulé structures cannot explain the unusual stability of the aromatic ring. As we have seen with other conjugated systems, molecular orbital theory provides the key to understanding aromaticity and predicting which compounds will have the stability of an aromatic system.

PROBLEM 16-3

(a) Draw the resonance structures of benzene, cyclobutadiene, and cyclooctatetraene, showing all the carbon and hydrogen atoms.
(b) Assuming that these molecules are all planar, show how the *p* orbitals on the sp^2 hybrid carbon atoms form continuous rings of overlapping orbitals above and below the plane of the carbon atoms.

PROBLEM 16-4

Show the product of the Diels-Alder dimerization of cyclobutadiene. (This reaction is similar to the dimerization of cyclopentadiene, discussed in Section 15-11.)

16-3 THE MOLECULAR ORBITALS OF BENZENE

Benzene has a planar ring of six sp^2 hybrid carbon atoms, each with an unhybridized *p* orbital that overlaps with the *p* orbitals of its neighbors to form a continuous ring of orbitals above and below the plane of the carbon atoms. Six pi electrons are contained in this ring of overlapping *p* orbitals.

The six overlapping *p* orbitals create a cyclic system of molecular orbitals. Cyclic systems have molecular orbitals that differ from linear systems such as

1,3-butadiene and the allyl system. We can still follow the same principles in developing a molecular orbital representation for benzene, however.

1. There are six atomic *p* orbitals that overlap to form the benzene pi system. Therefore, there must be six molecular orbitals.
2. The lowest-energy molecular orbital is entirely bonding, with constructive overlap between all pairs of adjacent *p* orbitals. There are no nodes in this lowest-lying MO.
3. The number of nodes increases as the MO's increase in energy.
4. The MO's should be evenly divided between bonding and antibonding MO's, with the possibility of nonbonding MO's in some cases.

Figure 16-4 shows the six π molecular orbitals of benzene as viewed from above, showing the sign of the top lobe of each *p* orbital. The first MO (π_1) is entirely bonding with no nodes. It is a very low-energy orbital, because it has six bonding interactions and the electrons are delocalized over all six carbon atoms. The top lobes of the *p* orbitals all have the same sign, as do the bottom lobes. The six *p* orbitals overlap to form a continuously bonding ring of electron density.

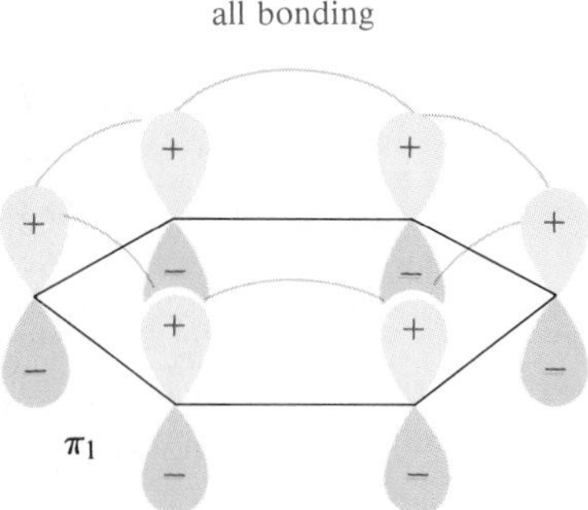

In a cyclic system of overlapping *p* orbitals, the intermediate energy levels are **degenerate,** with two orbitals at each energy level. Both π_2 and π_3 have one nodal plane, as we expect at the second energy level.

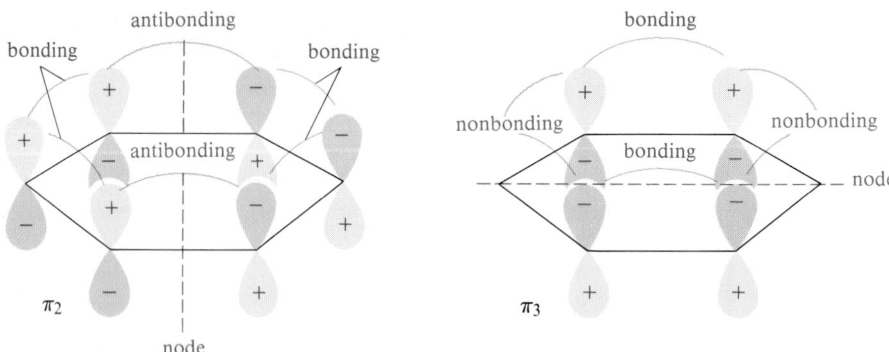

Notice that π_2 has four bonding interactions and two antibonding interactions, for a total of two net bonding interactions. Similarly, π_3 has two bonding interactions and four nonbonding interactions, also totaling two net bonding interactions. Although we cannot use the number of bonding and antibonding interactions as

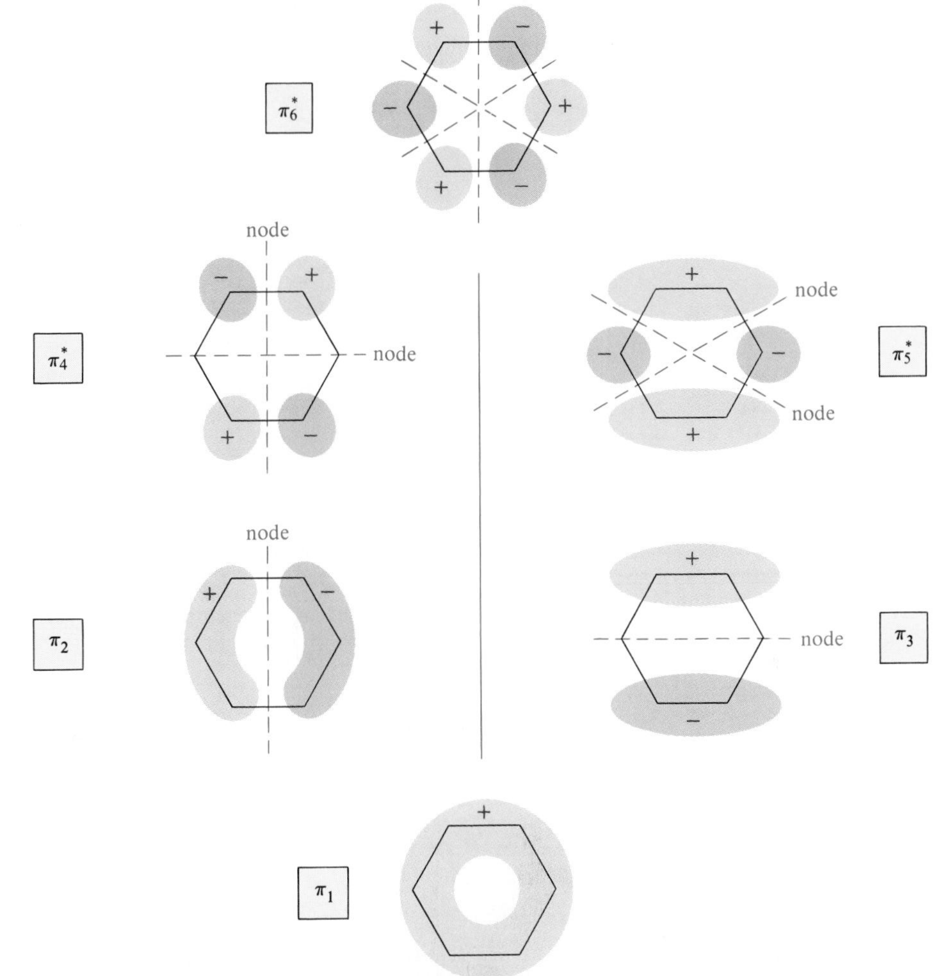

FIGURE 16-4 The six π molecular orbitals of benzene, viewed from above. The number of nodal planes increases with energy, and there are two degenerate MO's at each intermediate energy level.

a quantitative measure of an orbital's energy, it is clear that π_2 and π_3 are bonding MO's, but not as strongly bonding as π_1.

The next orbitals, π_4^* and π_5^*, are also degenerate, with two nodal planes in each. The π_4^* orbital has two antibonding interactions and four nonbonding interactions; it is an antibonding (*) orbital. Its degenerate partner, π_5^*, has four antibonding interactions and two bonding interactions, for a net two antibonding interactions. This degenerate pair of MO's, π_4^* and π_5^*, are about as strongly antibonding as π_2 and π_3 are bonding.

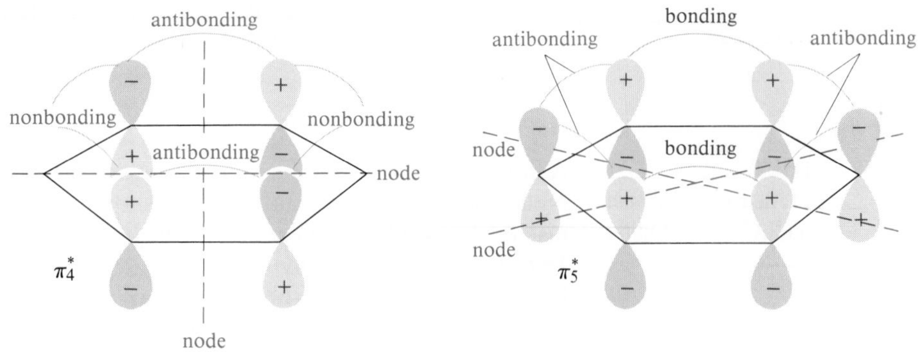

The all-antibonding π_6^* has three nodal planes. Each pair of adjacent p orbitals is out of phase and interacts destructively.

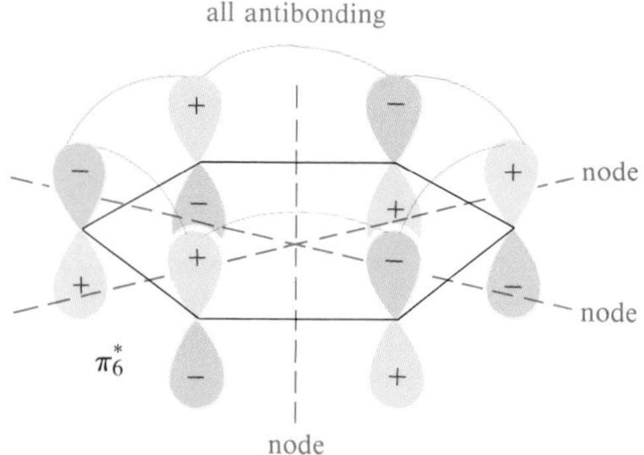

The energy diagram of benzene The energy diagram of the benzene MO's (Fig. 16-5) shows them to be symmetrically distributed above and below the nonbonding line (the energy of an isolated p orbital). The all-bonding and all-antibonding orbitals (π_1 and π_6^*) are lowest and highest in energy, respectively. The degenerate bonding orbitals (π_2 and π_3) are higher in energy than π_1, but still bonding. The degenerate pair π_4^* and π_5^* are antibonding, yet not as high in energy as the all-antibonding π_6^* orbital.

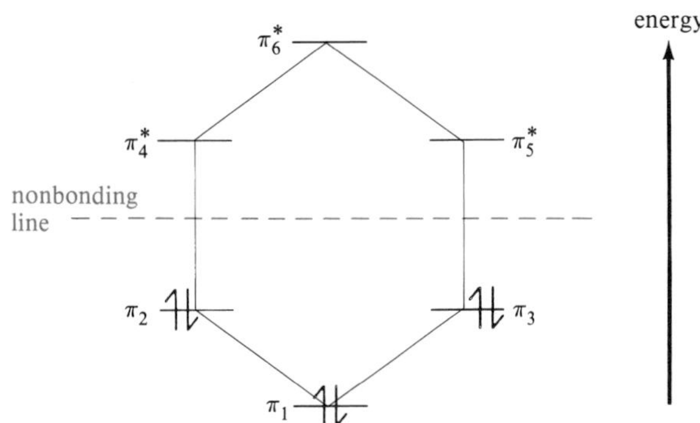

FIGURE 16-5 Energy diagram of the molecular orbitals of benzene. Benzene's six pi electrons fill the three bonding orbitals, leaving the antibonding orbitals vacant.

The Kekulé structure for benzene shows three pi bonds, signifying that there are six electrons (three pairs) involved in pi bonding. Six electrons fill the three bonding MO's of the benzene system. This electronic configuration explains the unusual stability of benzene. The first MO is all-bonding and is extremely low in energy. The second and third (degenerate) MO's are still strongly bonding, and all three of these bonding MO's delocalize the electrons over several nuclei. This configuration, with all the bonding MO's filled (a "closed bonding shell"), is energetically very favorable.

16-4 THE MOLECULAR ORBITAL PICTURE OF CYCLOBUTADIENE

Although we can draw benzene-like resonance structures (Fig. 16-3) for cyclobutadiene, the experimental evidence shows that cyclobutadiene is unstable. The instability of cyclobutadiene is explained by its molecular orbitals, shown in Figure 16-6. There are four sp^2 hybrid carbon atoms in the cyclobutadiene ring, and their four p orbitals overlap to form four molecular orbitals. The lowest-energy MO, π_1, is the all-bonding MO with no nodes.

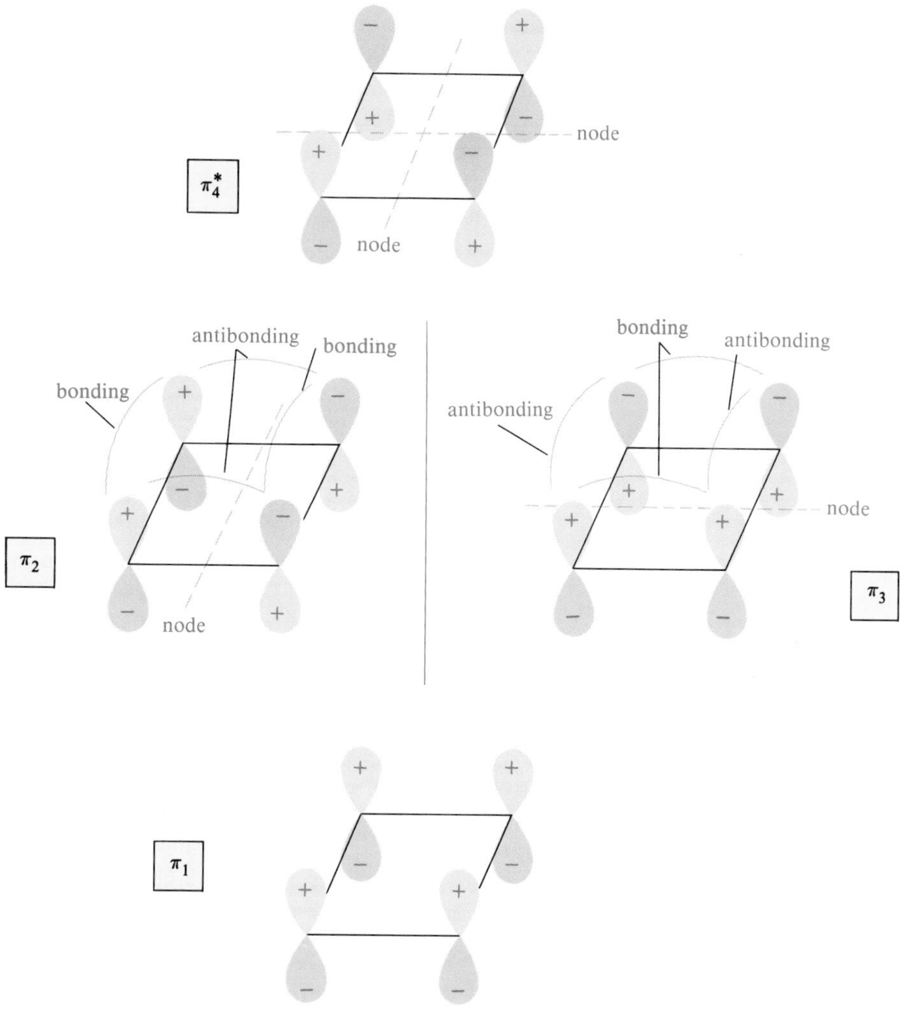

FIGURE 16-6 The pi molecular orbitals of cyclobutadiene. There are four MO's: the lowest-energy bonding orbital, the highest-energy antibonding orbital, and two degenerate nonbonding orbitals.

The next two orbitals, π_2 and π_3, are degenerate (equal energy), each having one symmetrically situated nodal plane. Each of these MO's has two bonding interactions and two antibonding interactions. The net bonding order is zero, and these two MO's are nonbonding.

The final MO, π_4^*, has two nodal planes and is entirely antibonding.

Figure 16-7 is an energy diagram of the four cyclobutadiene MO's. The lowest-lying MO (π_1) is strongly bonding, and the highest-lying MO (π_4^*) is equally

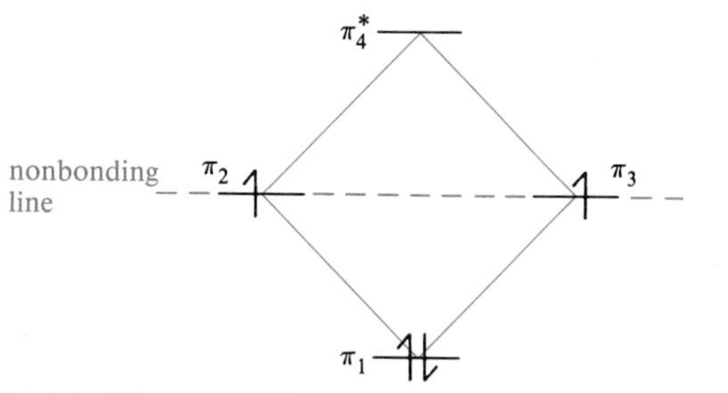

FIGURE 16-7 An electronic energy diagram of cyclobutadiene shows that two electrons are unpaired in separate nonbonding molecular orbitals.

antibonding. The two degenerate nonbonding orbitals are intermediate in energy, falling on the nonbonding line (the energy of an isolated *p* orbital).

The localized structure of cyclobutadiene shows two double bonds, implying four pi electrons. Two electrons fill π_1, the lowest-lying orbital. Once π_1 is filled, there are two orbitals of equal energy available for the remaining two electrons. If the two electrons go into the same orbital, they must have paired spins and they must share the same region of space. Since electrons repel each other, less energy is required for the electrons to occupy different degenerate orbitals, with unpaired spins. This principle is another application of **Hund's rule** (Section 1-2).

The electronic configuration in Figure 16-7 indicates that cyclobutadiene should be unstable. Its highest-lying electrons are in nonbonding orbitals (π_2 and π_3) and are therefore very reactive. According to Hund's rule, the compound exists as a diradical (two unpaired electrons) in its ground state. Molecular orbital theory successfully predicts the dramatic stability difference between benzene and cyclobutadiene.

The polygon rule The patterns of molecular orbitals in benzene (Fig. 16-5) and in cyclobutadiene (Fig. 16-7) are similar to the patterns in other annulenes: The lowest-lying MO is the unique one with no nodes; thereafter, the molecular orbitals occur in degenerate (equal-energy) pairs until only one highest-lying MO remains. In benzene, the energy diagram looks like the hexagon of a benzene ring. In cyclobutadiene, the pattern looks like the diamond of the cyclobutadiene ring.

The **polygon rule** makes the general statement that the molecular orbital energy diagram of a regular, completely conjugated cyclic system has the same polygonal shape as the compound, with one vertex (the all-bonding MO) at the bottom. The nonbonding line cuts horizontally through the center of the polygon. Figure 16-8 shows how the polygon rule predicts the shapes of the MO energy diagrams for benzene, cyclobutadiene, and cyclooctatetraene. The pi electrons are filled into the orbitals in accordance with the Aufbau principle (lowest-energy orbitals are filled first) and Hund's rule.

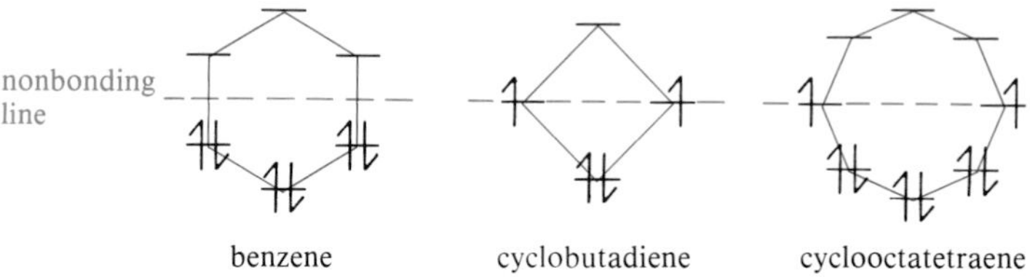

FIGURE 16-8 The polygon rule predicts that the MO energy diagrams for these annulenes will resemble the polygonal shapes of the annulenes.

PROBLEM 16-5

Does the MO energy diagram of cyclooctatetraene (Fig. 16-8) appear to be a particularly stable or unstable configuration?

16-5 AROMATIC, ANTIAROMATIC, AND NONAROMATIC COMPOUNDS

Our working definition of aromatic compounds has included cyclic compounds containing conjugated double bonds, with unusually large resonance energies. At this point we can be more specific about the properties that are required for a compound (or an ion) to be aromatic.

Aromatic compounds are those that meet the following criteria.

1. The structure must be cyclic, containing some number of conjugated pi bonds.

2. Each atom in the ring must have an unhybridized *p* orbital. (The ring atoms are usually sp^2 hybridized, or occasionally *sp* hybridized.)
3. The unhybridized *p* orbitals must overlap to form a continuous ring of parallel orbitals. In most cases, the structure must be planar (or nearly planar) for effective overlap.
4. Delocalization of the pi electrons over the ring must result in a *lowering* of the electronic energy.

An **antiaromatic** compound is one that meets the first three criteria, but delocalization of the pi electrons over the ring results in an *increase* in the electronic energy.

Aromatic structures are more stable than their open-chain counterparts. For example, benzene is more stable than 1,3,5-hexatriene.

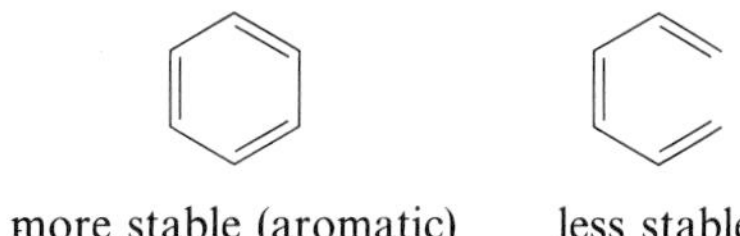

more stable (aromatic) less stable

Cyclobutadiene meets the first three criteria for a continuous ring of overlapping *p* orbitals, but the delocalization of the pi electrons results in an *increase* in the electronic energy. Cyclobutadiene is less stable than its open-chain counterpart (1,3-butadiene), and it is **antiaromatic.**

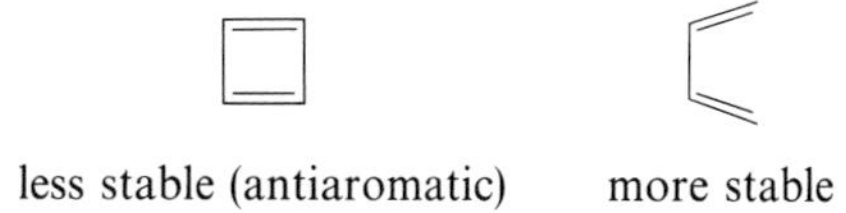

less stable (antiaromatic) more stable

A cyclic compound that does not have a continuous, overlapping ring of *p* orbitals cannot be aromatic or antiaromatic. It is said to be **nonaromatic,** or aliphatic. Its electronic energy is similar to that of its open-chain counterpart. For example, 1,3-cyclohexadiene is about as stable as *cis,cis*-2,4-hexadiene.

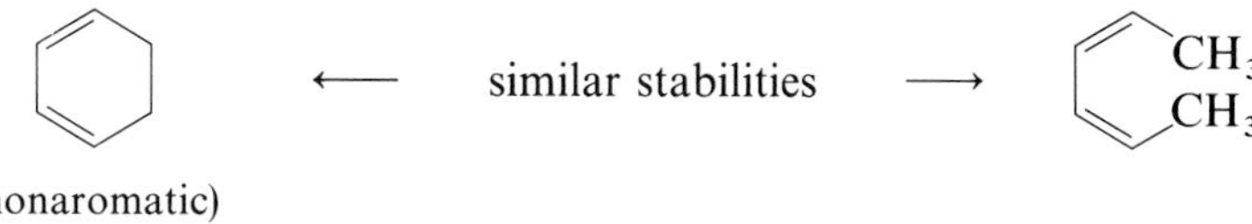

(nonaromatic)

16-6 HÜCKEL'S RULE

Erich Hückel developed a shortcut for predicting which of the annulenes and related compounds are aromatic and which are antiaromatic. In using Hückel's rule, we must be certain that the compound under consideration meets the criteria for an aromatic or antiaromatic system: It must have a continuous ring of overlapping *p* orbitals, usually in a planar conformation.

Once these criteria are met, **Hückel's rule** applies:

> HÜCKEL'S RULE If the number of pi electrons in the cyclic system is $(4N + 2)$, with N an integer, the system is aromatic. Common aromatic systems have 2, 6, and 10 pi electrons, for $N = 0$, 1, and 2.
>
> Systems with $4N$ pi electrons, with N an integer, are antiaromatic. Common examples are systems with 4, 8, or 12 pi electrons.

Benzene is [6]annulene, cyclic, with a continuous ring of overlapping *p* orbitals. There are six pi electrons in benzene (three double bonds in the classical structure), so it is a $(4N + 2)$ system, with $N = 1$. Hückel's rule predicts benzene to be aromatic.

Like benzene, cyclobutadiene ([4]annulene) has a continuous ring of overlapping *p* orbitals, but it has four pi electrons (two double bonds in the classical structure). Hückel's rule predicts cyclobutadiene to be antiaromatic.

Cyclooctatetraene is [8]annulene, with eight pi electrons (four double bonds) in the classical structure. It is a $4N$ system, with $N = 2$. If Hückel's rule were applied to cyclooctatetraene, it would predict antiaromaticity. However, cyclooctatetraene is a stable hydrocarbon with a boiling point of 152°C. It does not show the high reactivity associated with antiaromaticity, yet it is not aromatic either. Its reactions are typical of alkenes.

Cyclooctatetraene would be antiaromatic if Hückel's rule applied, so the conjugation of its double bonds is energetically unfavorable. Remember that Hückel's rule applies to a compound *only* if there is a continuous ring of overlapping *p*-orbitals, usually in a planar system. Cyclooctatetraene is more flexible than cyclobutadiene, and it assumes a nonplanar conformation that avoids most of the overlap between adjacent pi bonds. Hückel's rule simply does not apply.

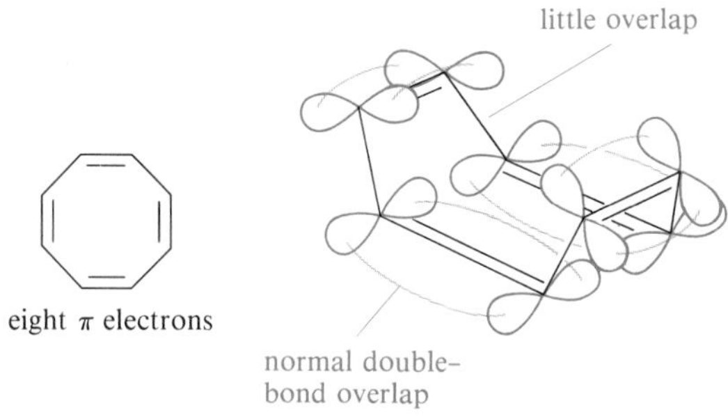

PROBLEM 16-6

Make a model of cyclooctatetraene in the tub conformation. Draw this conformation, and estimate the angle between the *p* orbitals of adjacent pi bonds.

Large-ring annulenes Like cyclooctatetraene, larger annulenes do not show antiaromaticity because they have the flexibility to adopt nonplanar conformations. Even though [12]annulene, [16]annulene, and [20]annulene are $4N$ systems (with $N = 3$, 4, and 5, respectively), they all react as partially conjugated polyenes.

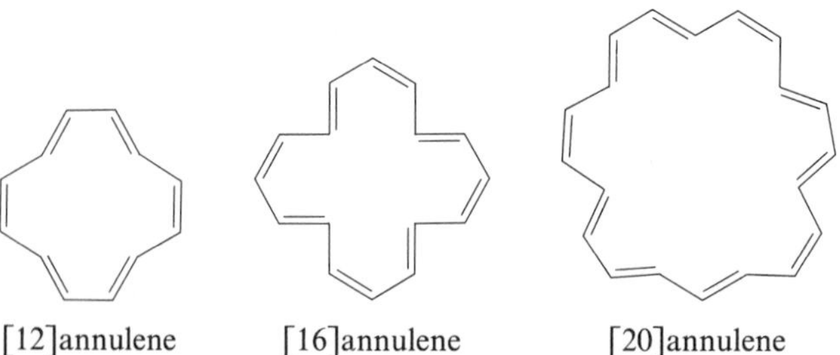

Aromaticity in the larger annulenes depends on whether the molecule can adopt the necessary planar conformation. In the all-cis [10]annulene, the planar conformation requires an excessive amount of angle strain. The [10]annulene isomer with two trans double bonds cannot adopt a planar conformation because two hydrogen atoms interfere with each other. Neither of these [10]annulene isomers is aromatic, even though each has $(4N + 2)$ pi electrons, with $N = 2$. If the interfering hydrogen atoms in the partially trans isomer are removed, the molecule can be planar. When these hydrogen atoms are replaced with a bond, the aromatic compound naphthalene results.

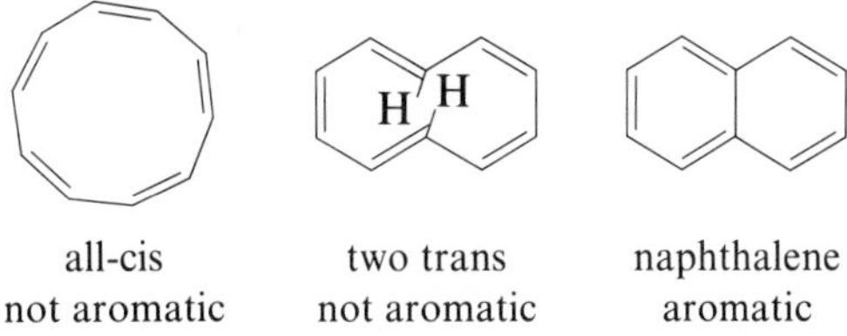

all-cis not aromatic — two trans not aromatic — naphthalene aromatic

Some of the larger annulenes with $(4N + 2)$ pi electrons can achieve planar conformations. For example, the following [14]annulene and [18]annulene have aromatic properties.

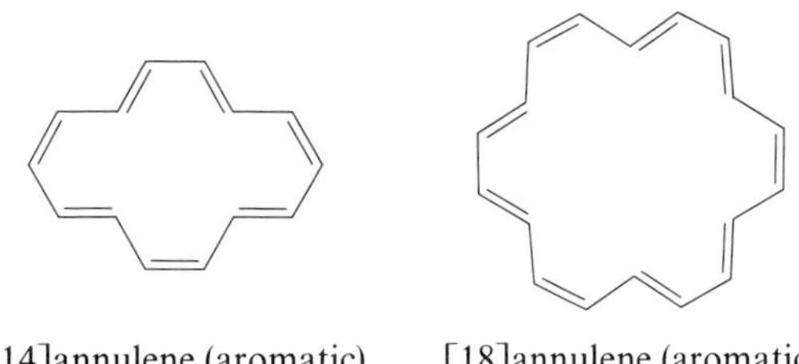

[14]annulene (aromatic) — [18]annulene (aromatic)

PROBLEM 16-7

Classify the following compounds as aromatic, antiaromatic, or nonaromatic.

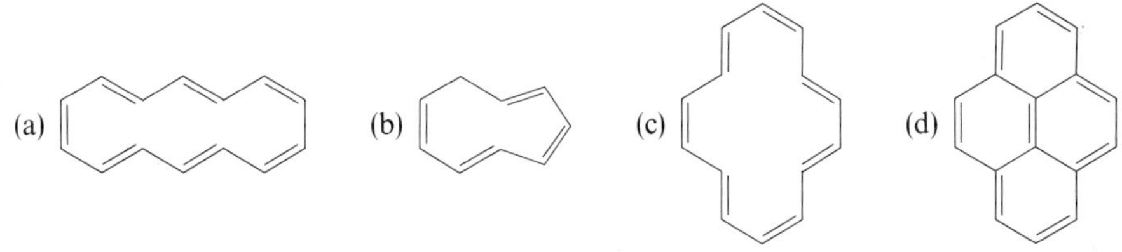

PROBLEM 16-8

One of the following compounds is much more stable than the other two. Classify each as aromatic, antiaromatic, or nonaromatic.

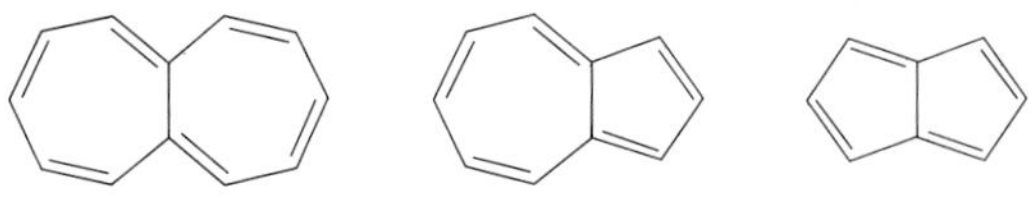

heptalene — azulene — pentalene

16-7 MOLECULAR ORBITAL DERIVATION OF HÜCKEL'S RULE

Benzene is aromatic because it has a filled shell of equal-energy orbitals. The degenerate orbitals π_2 and π_3 are filled, and all the electrons are paired. Cyclobutadiene, by contrast, has an open shell of electrons. There are two half-filled orbitals easily capable of donating or accepting electrons. To derive Hückel's rule, we must show under what general conditions there is a filled shell of orbitals.

Recall the pattern of MO's in a cyclic conjugated system. There is one all-bonding, lowest-lying MO, followed by degenerate pairs of bonding MO's. (No need to worry about the antibonding MO's, because they are vacant in the ground state.) The lowest-lying MO is always filled (two electrons). Each additional shell consists of two degenerate MO's, requiring four electrons to fill a shell. Figure 16-9 shows this pattern of two electrons for the lowest orbital, and then four electrons for each additional shell.

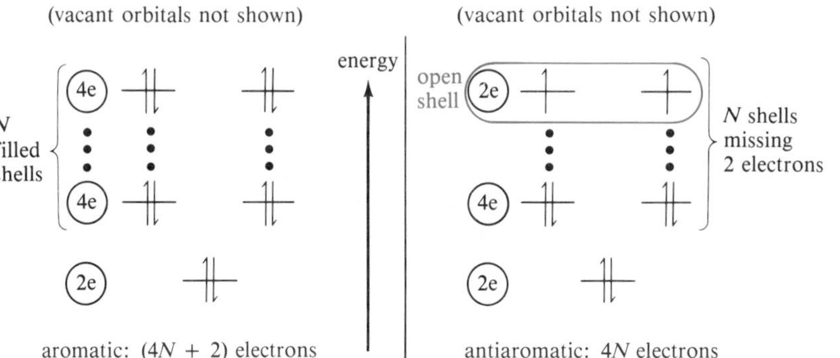

FIGURE 16-9 In a cyclic conjugated system, the lowest-lying MO is filled with two electrons. Each of the additional shells consists of two degenerate MO's, with space for four electrons. If a molecule has $(4N + 2)$ pi electrons, it will have a filled shell. If it has $4N$ electrons, there will be two unpaired electrons in two degenerate orbitals.

A compound has a filled shell of orbitals if it has two electrons for the lowest-lying orbital, plus $4N$ electrons, where N is the number of filled pairs of degenerate orbitals. The total number of pi electrons in this case is $(4N + 2)$. If the system has a total of only $4N$ electrons, it is two electrons short of filling N pairs of degenerate orbitals. There are only two electrons in the Nth pair of degenerate orbitals. This is a half-filled shell, and Hund's rule predicts these electrons will be unpaired (a diradical).

PROBLEM 16-9

(a) Use the polygon rule to draw an energy diagram (as in Figs. 16-5 and 16-7) for the MO's of a planar cyclooctatetraenyl system.

(b) Fill in the eight pi electrons for cyclooctatetraene. Is this electronic configuration aromatic or antiaromatic?

(c) Draw pictorial representations (like Figs. 16-4 and 16-6) for the three bonding MO's and the two nonbonding MO's of cyclooctatetraene. The antibonding MO's are difficult to draw, except for the all-antibonding MO,

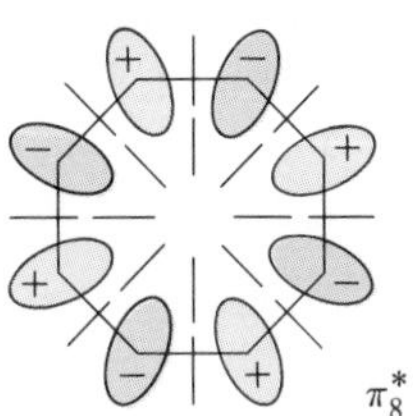

16-8 AROMATIC IONS

Up to this point, we have discussed aromaticity using the annulenes as examples. Annulenes are uncharged molecules having even numbers of carbon atoms with alternating single and double bonds. Hückel's rule also applies to systems having odd numbers of carbon atoms and bearing positive or negative charges. We will consider some of the more common aromatic ions and some of their antiaromatic counterparts.

16-8A THE CYCLOPENTADIENYL IONS

We can draw a five-membered ring of sp^2 hybrid carbon atoms with all the unhybridized p orbitals lined up to form a continuous ring. With five pi electrons this system would be neutral, but it would be a radical because an odd number of electrons cannot all be paired. With four pi electrons (a cation), Hückel's rule predicts this system to be antiaromatic. With six pi electrons (an anion), Hückel's rule predicts aromaticity.

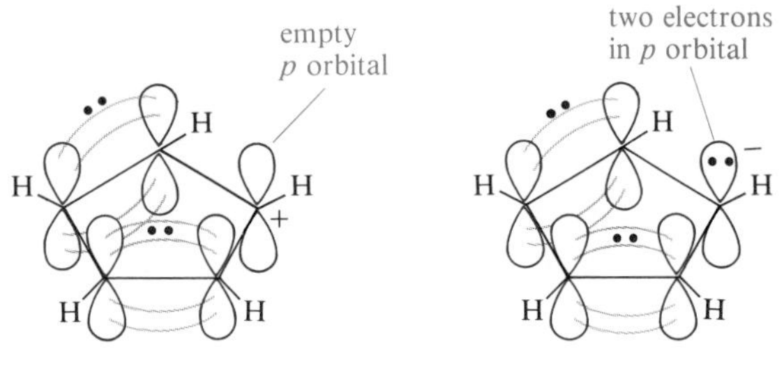

empty p orbital

two electrons in p orbital

four electrons
cyclopentadienyl cation

six electrons
cyclopentadienyl anion

Because the cyclopentadienyl anion (six pi electrons) is aromatic, it is unusually stable compared with other carbanions. It can be formed by the abstraction of a proton from cyclopentadiene, which is unusually acidic for an alkene. Cyclopentadiene has a pK_a of 16, compared with a pK_a of 46 for cyclohexene. In fact, cyclopentadiene is nearly as acidic as water, and more acidic than many alcohols. It is entirely deprotonated by potassium t-butoxide:

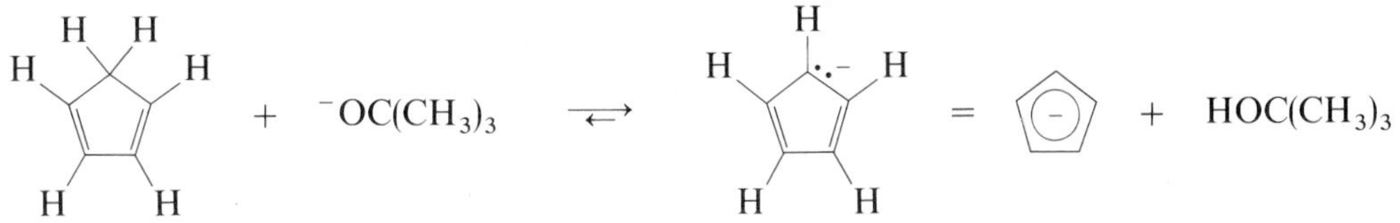

$$C_5H_6 + {}^-OC(CH_3)_3 \rightleftharpoons C_5H_5^- + HOC(CH_3)_3$$

pK_a = 16 — the cyclopentadienyl anion (six pi electrons) — pK_a = 19

Cyclopentadiene is unusually acidic because loss of a proton converts the nonaromatic diene to the aromatic cyclopentadienyl anion. Cyclopentadiene contains an sp^3 hybrid ($-CH_2-$) carbon atom without an unhybridized p orbital, so there can be no continuous ring of p orbitals. Deprotonation of the $-CH_2-$ group leaves an orbital occupied by a pair of electrons. This orbital can rehybridize to a p orbital, completing a ring of p orbitals containing six pi electrons: the two electrons on the deprotonated carbon, plus the four electrons in the original double bonds.

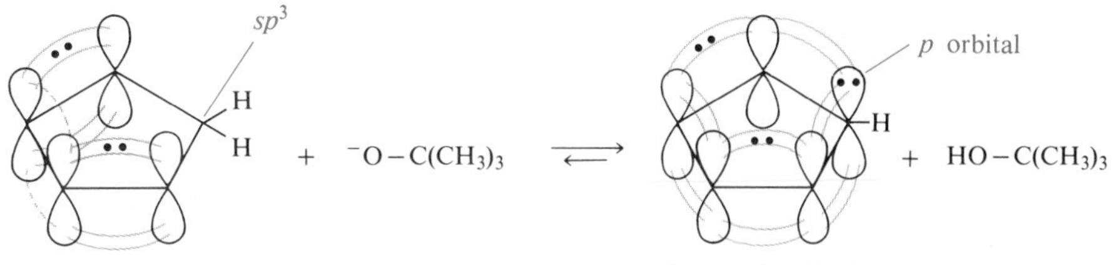

sp^3 — $+ \ {}^-O-C(CH_3)_3 \rightleftharpoons$ — p orbital — $+ \ HO-C(CH_3)_3$

cyclopentadiene
nonaromatic

cyclopentadienyl anion
aromatic

When we say that the cyclopentadienyl anion is aromatic, this does not necessarily imply that it is as stable as benzene. As a carbanion, the cyclopentadienyl anion reacts readily with electrophiles. The fact that this ion is aromatic implies that it is more stable than the corresponding open-chain ion.

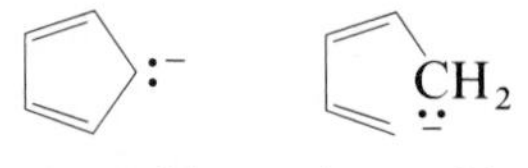

more stable (aromatic) less stable

Hückel's rule predicts that the cyclopentadienyl cation, with four pi electrons, is antiaromatic. In agreement with this prediction, the cyclopentadienyl cation is not easily formed. 2,4-Cyclopentadienol does not protonate and lose water (to give the cyclopentadienyl cation), even in concentrated sulfuric acid. The antiaromatic cation is simply too unstable.

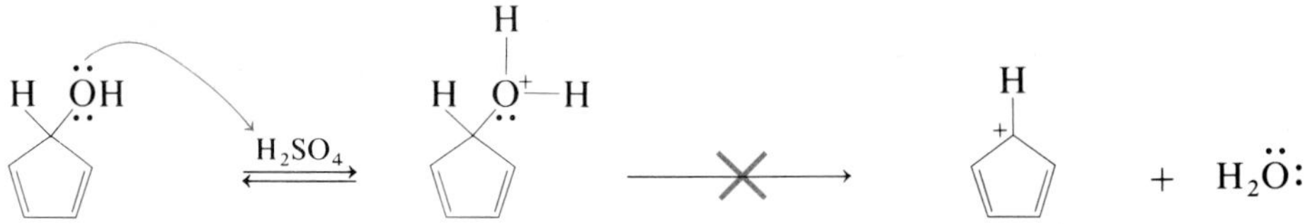

2,4-cyclopentadienol $\xrightleftharpoons{H_2SO_4}$ (does not occur) not formed (four pi electrons) + $H_2\ddot{O}$:

Using a simple resonance approach, we might incorrectly expect both of the cyclopentadienyl ions to be unusually stable. Shown below are resonance structures that spread the positive charge of the cation and the negative charge of the anion over all five carbon atoms of the ring. With conjugated cyclic systems such as these, the resonance approach is not a good predictor of stability. The Hückel rule, based on molecular orbital theory, is a much better predictor of stability for these aromatic and antiaromatic systems.

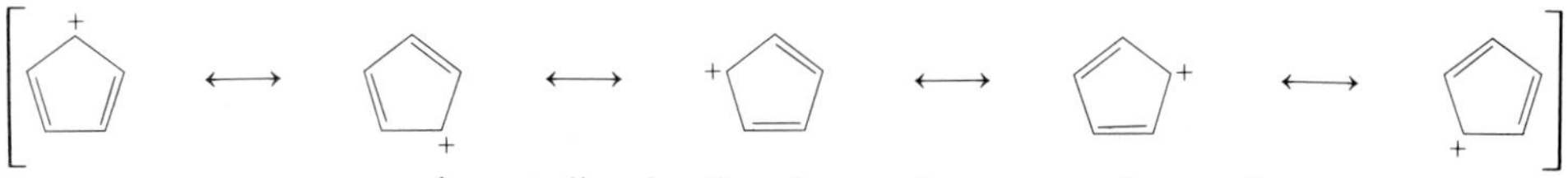

cyclopentadienyl cation: four π electrons, antiaromatic

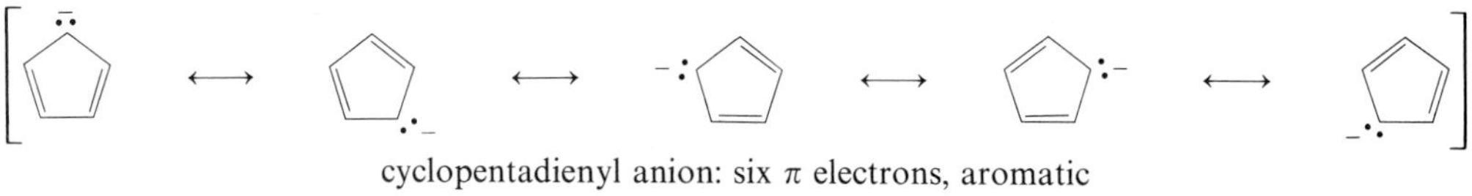

cyclopentadienyl anion: six π electrons, aromatic

PROBLEM 16-10

(a) Draw the molecular orbitals for the cyclopropenyl case.

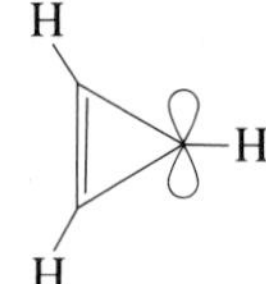

(Since there are three p orbitals, there must be three MO's: one all-bonding MO and one degenerate pair of MO's).

(b) Draw an energy diagram for the cyclopropenyl MO's. (The polygon rule may be helpful.) Label each MO as bonding, nonbonding, or antibonding, and add the nonbonding line. Notice that it goes through the approximate average of the MO's.

(c) Add electrons to your energy diagram to show the configuration of the cyclopropenyl cation and the cyclopropenyl anion. Which is aromatic and which is antiaromatic?

PROBLEM 16-11 *

Repeat Problem 16-10 for the cyclopentadienyl ions. Draw one all-bonding MO, then a pair of degenerate MO's, and then a final pair of degenerate MO's. Draw the energy diagram, fill in the electrons, and determine the electronic configurations of the cyclopentadienyl cation and anion.

16-8B THE CYCLOHEPTATRIENYL IONS

As with the five-membered ring, we can imagine a flat seven-membered ring with seven *p* orbitals aligned. The cation has six pi electrons, and the anion has eight pi electrons. Once again, we can draw resonance structures that seem to show either the positive charge of the cation or the negative charge of the anion delocalized over all seven atoms of the ring. By now, however, we know that the six-electron system is aromatic and the eight-electron system is antiaromatic (if it remains planar).

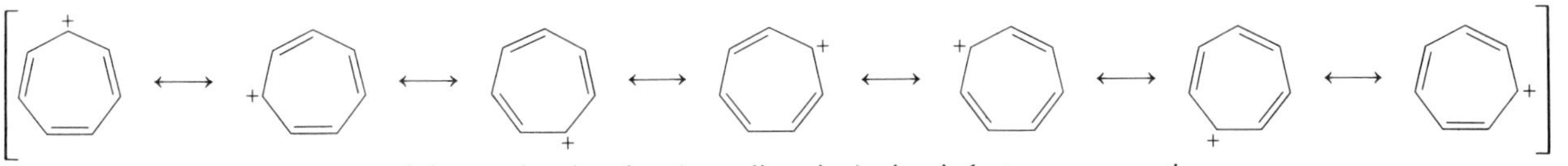

cycloheptatrienyl cation (tropylium ion): six pi electrons, aromatic

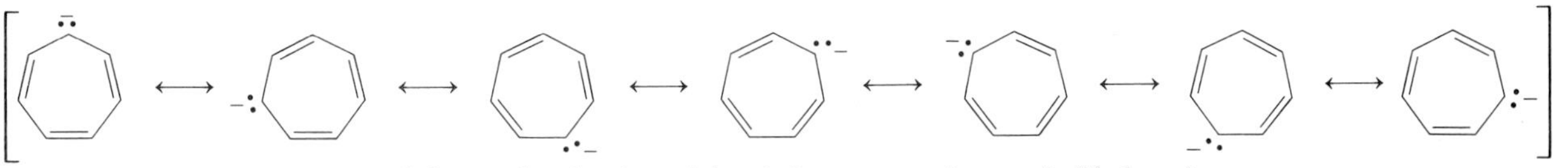

cycloheptatrienyl anion: eight pi electrons, antiaromatic (if planar)

The cycloheptatrienyl cation is easily formed by treating the corresponding alcohol with very dilute (0.01*N*) aqueous sulfuric acid. This is our first example of a hydrocarbon cation that is stable in an aqueous solution.

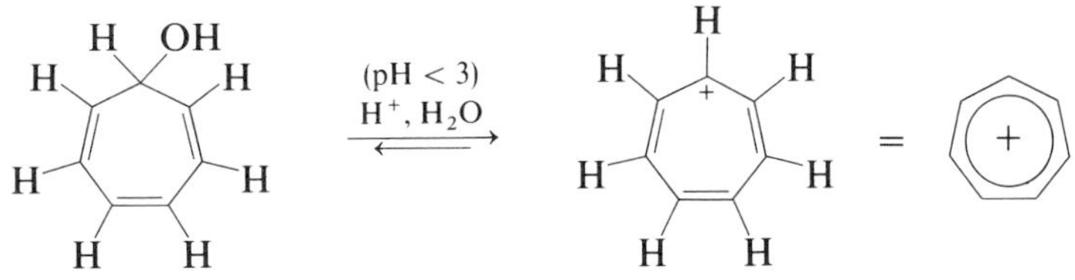

(pH < 3)
H^+, H_2O

tropylium ion, six pi electrons

The cycloheptatrienyl cation is called the **tropylium ion.** This aromatic ion is much less reactive than most carbocations; some tropylium salts can be isolated and stored for months without decomposing. Nevertheless, the tropylium ion is not necessarily as stable as benzene. Its aromaticity implies that the cyclic ion is more stable than the corresponding open-chain ion.

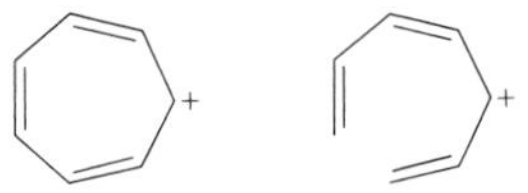

more stable (aromatic) less stable

In contrast to the easy formation of the tropylium ion, the corresponding anion is difficult to prepare, because it is antiaromatic. Cycloheptatriene is about as acidic as propene ($\mathrm{p}K_a = 43$), and the anion is very reactive. This result agrees with the prediction of Hückel's rule that the cycloheptatrienyl anion is antiaromatic.

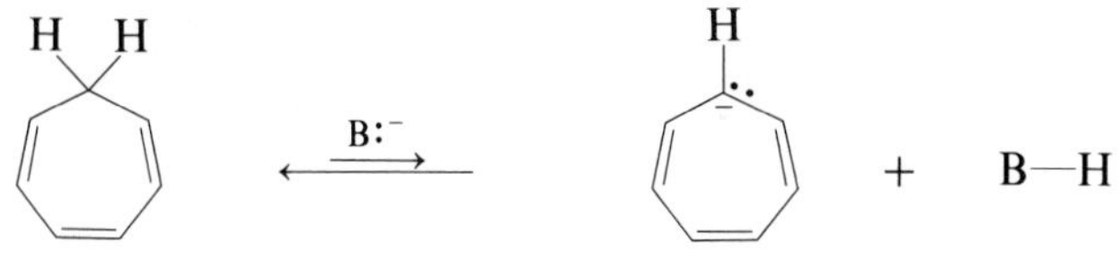

H H — B:⁻ ⇌ — H + B—H

cycloheptatriene
$\mathrm{p}K_a > 50$

cycloheptatrienyl anion
eight pi electrons

16-8C THE CYCLOOCTATETRAENE DIANION

We have seen that aromatic stabilization leads to unusually stable hydrocarbon anions such as the cyclopentadienyl anion. Dianions of hydrocarbons are usually much more difficult to form. Cyclooctatetraene reacts with potassium metal, however, to form an aromatic dianion.

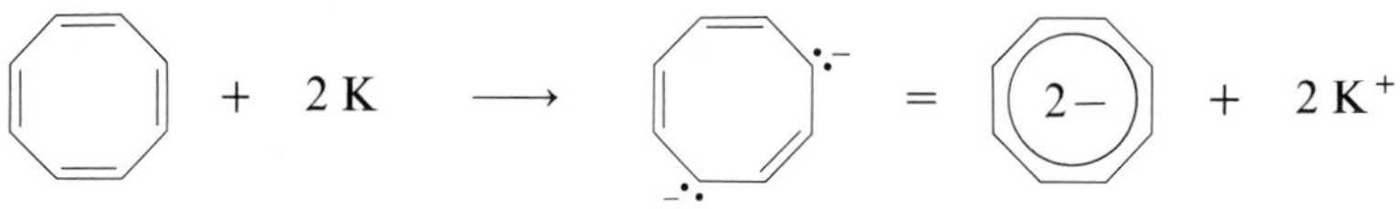

+ 2 K ⟶ = 2− + $2\,\mathrm{K}^+$

ten pi electrons

The cyclooctatetraene dianion has a planar, regular octagonal structure, with C—C bond lengths of 1.40 Å, close to the 1.397-Å bond lengths in benzene. Cyclooctatetraene itself has eight pi electrons, so the dianion has ten: $(4N + 2)$, with $N = 2$. The cyclooctatetraene dianion is easily prepared because it is aromatic.

PROBLEM 16-12

Explain why each of the following compounds and ions should be aromatic, antiaromatic, or nonaromatic.

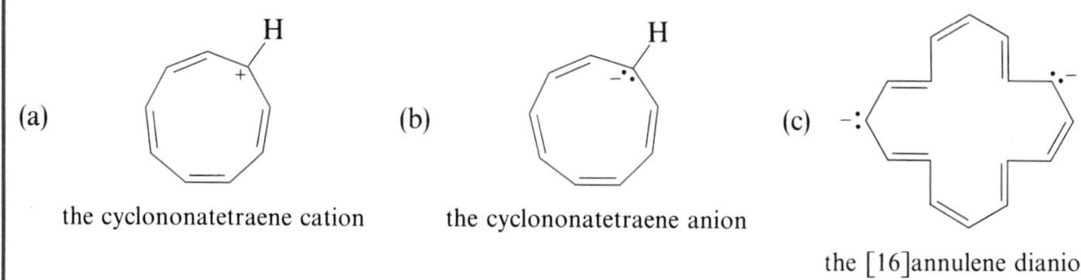

the cyclononatetraene cation

the cyclononatetraene anion

the [16]annulene dianion

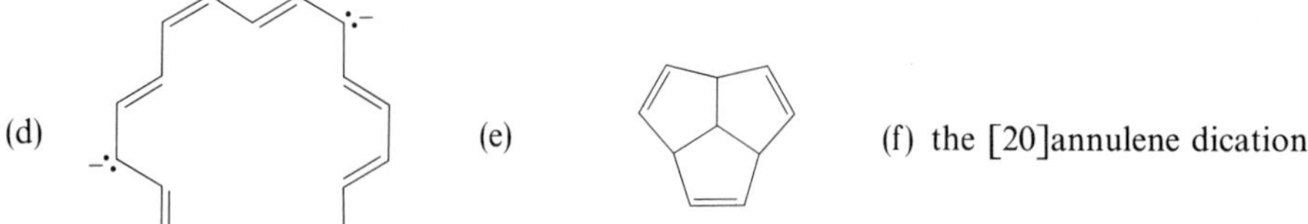

(f) the [20]annulene dication

the [18]annulene dianion

PROBLEM 16-13

The following hydrocarbon has an unusually large dipole moment. Explain how a large dipole moment might arise.

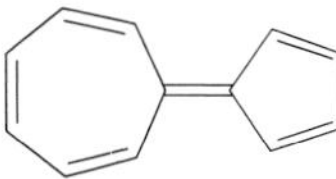

PROBLEM 16-14

When 3-chlorocyclopropene is treated with $AgBF_4$, AgCl precipitates. The organic product can be obtained as a crystalline material, soluble in polar solvents such as nitromethane, but insoluble in hexane. When the crystalline material is dissolved in nitromethane containing KCl, the original 3-chlorocyclopropene is regenerated. Determine the structure of the crystalline material, and draw equations for its formation and its reaction with chloride ion.

16-8D SUMMARY OF ANNULENES AND THEIR IONS

The application of Hückel's rule to a variety of cyclic pi systems is summarized below. These systems are classified according to the number of pi electrons: The 2, 6, and 10 pi-electron systems are aromatic, while the 4 and 8 pi-electron systems are antiaromatic if they are planar.

2 pi-electron systems (aromatic)

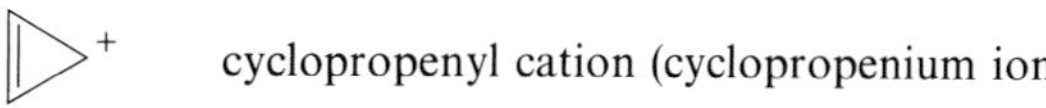

cyclopropenyl cation (cyclopropenium ion)

4 pi-electron systems (antiaromatic)

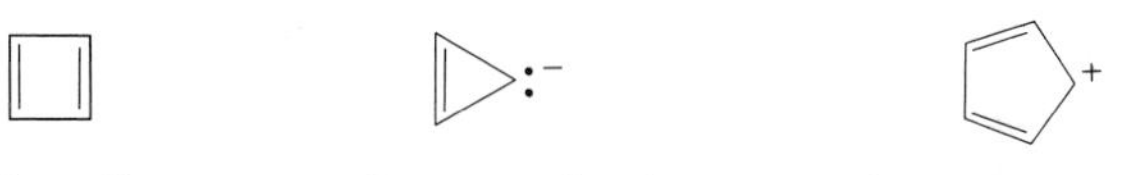

cyclobutadiene | cyclopropenyl anion | cyclopentadienyl cation

6 pi-electron systems (aromatic)

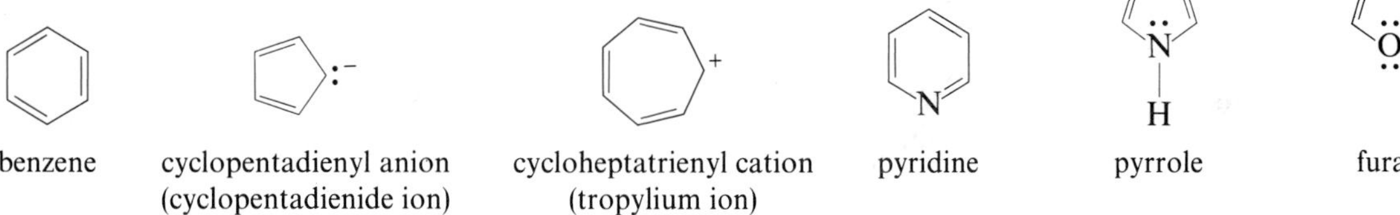

benzene | cyclopentadienyl anion (cyclopentadienide ion) | cycloheptatrienyl cation (tropylium ion) | pyridine | pyrrole | furan

8 pi-electron systems antiaromatic if planar)

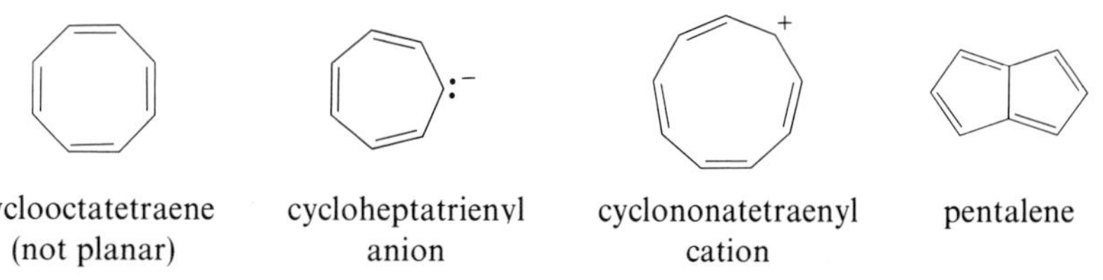

cyclooctatetraene (not planar) | cycloheptatrienyl anion | cyclononatetraenyl cation | pentalene

10 pi-electron systems (aromatic)

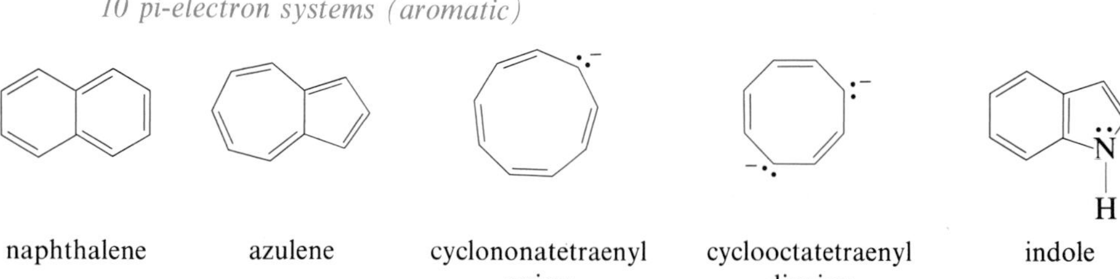

naphthalene | azulene | cyclononatetraenyl anion | cyclooctatetraenyl dianion | indole

(Naphthalene can also be considered as two fused benzenes.)

12 pi-electron systems (antiaromatic if planar)

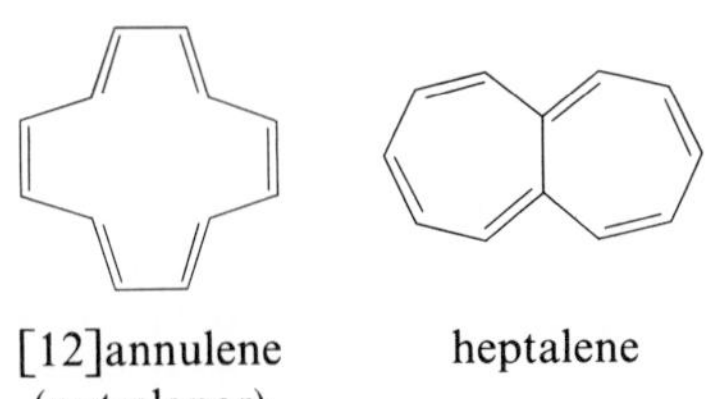

16-9 HETEROCYCLIC AROMATIC COMPOUNDS

In discussing aromaticity, we have considered only compounds composed of rings of sp^2 hybrid carbon atoms. **Heterocyclic compounds,** with rings containing sp^2 hybridized atoms of other elements, can also be aromatic. The criteria for Hückel's rule require a ring of atoms, all with unhybridized *p* orbitals overlapping in a continuous ring. Nitrogen, oxygen, and sulfur atoms are most commonly found in heterocyclic aromatic compounds.

16-9A PYRIDINE

Pyridine is an aromatic nitrogen analog of benzene: a six-membered heterocyclic ring with six pi electrons. Pyridine has a nitrogen atom in place of one of the six C—H units of benzene, and the nonbonding pair of electrons on the nitrogen atom replaces the bond to a hydrogen atom. These nonbonding electrons are in an sp^2 hybrid orbital in the plane of the ring (Fig. 16-10), and they do not overlap with the pi system.

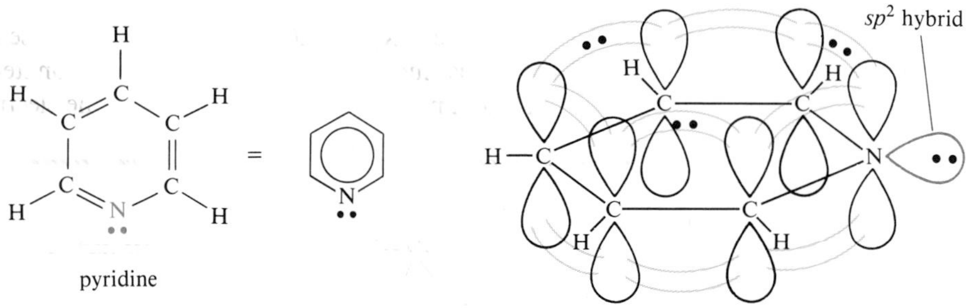

pyridine

sp^2 hybrid

FIGURE 16-10 Pyridine has six delocalized electrons in its cyclic pi system. The two nonbonding electrons on nitrogen are in an sp^2 orbital, and they do not interact with the pi electrons of the ring.

Pyridine shows all the characteristics of aromatic compounds. It has a resonance energy of 27 kcal (113 kJ) per mole, and it usually gives substitution rather than addition. Because it has an available pair of nonbonding electrons, pyridine is basic (Fig. 16-11). In an acidic solution, pyridine protonates to give the pyri-

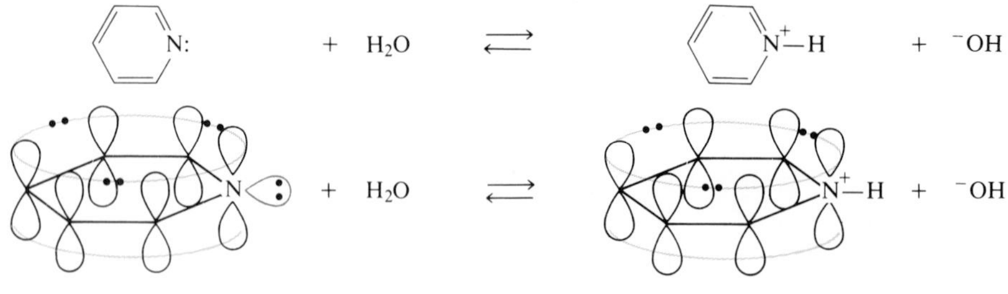

pyridine, $pK_b = 8.8$ + H_2O ⇌ pyridinium ion, $pK_a = 5.2$ + ^-OH

FIGURE 16-11 Pyridine is basic, with nonbonding electrons available to abstract a proton. The protonated pyridine (a pyridinium ion) is still aromatic, because the nonbonding electrons are not part of the sextet of pi electrons.

dinium ion. The pyridinium ion is still aromatic, since the additional proton has no effect on the electrons of the aromatic sextet: It simply bonds to pyridine's nonbonding pair of electrons.

16-9B PYRROLE

Pyrrole is an aromatic five-membered heterocycle, with one nitrogen atom and two double bonds (Fig. 16-12). Although it may seem that there are only four pi electrons in pyrrole, the nitrogen atom has a lone pair of electrons. The pyrrole nitrogen atom is sp^2 hybridized, and its unhybridized *p* orbital overlaps with the *p* orbitals of the carbon atoms to form a continuous ring. The lone pair on nitrogen occupies the *p* orbital, and (unlike the lone pair of pyridine) these electrons take part in the pi bonding system. These two electrons, added to the four pi electrons of the two double bonds, complete an aromatic sextet.

FIGURE 16-12 The pyrrole nitrogen atom is sp^2 hybridized, with a lone pair of electrons in the *p* orbital. This *p* orbital overlaps with the *p* orbitals of the carbon atoms to form a continuous ring. Counting the four electrons of the double bonds and the two electrons in the nitrogen *p* orbital, there are six pi electrons.

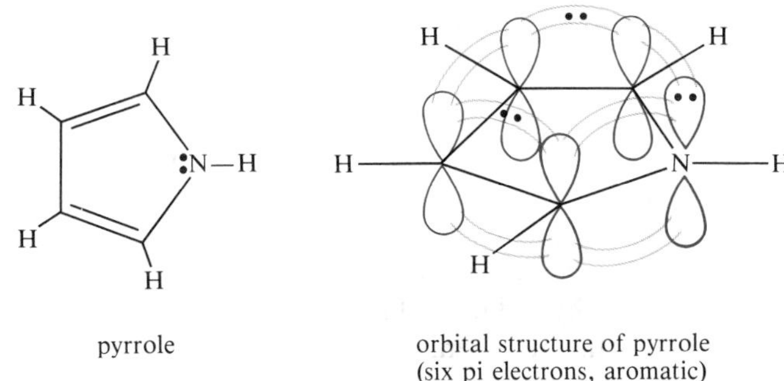

Pyrrole ($pK_b = 13.6$) is a much weaker base than pyridine ($pK_b = 8.8$). This difference is due to the structure of the protonated pyrrole (Fig. 16-13). To form a bond to a proton requires use of one of the electron pairs in the aromatic sextet.

:N—H + H_2O ⇌ N^+H_2 + ^-OH

sp^2 sp^3

pyrrole, $pK_b = 13.6$
(weak base)

N-protonated pyrrole, $pK_a = 0.4$
(strong acid)

N—H + H_2O ⇌ N^+H_2 + ^-OH

pyrrole
(aromatic)

N-protonated pyrrole
(nonaromatic)

FIGURE 16-13 The pyrrole nitrogen atom must become sp^3 hybridized to abstract a proton. This eliminates the unhybridized *p* orbital that is necessary for aromaticity.

In the protonated pyrrole, the nitrogen atom is bonded to four different atoms (two carbon atoms and two hydrogen atoms), requiring sp^3 hybridization and leaving no unhybridized p orbital. The protonated pyrrole is nonaromatic. In fact, a sufficiently strong acid actually protonates pyrrole at the 2-position, on one of the carbon atoms of the ring (discussed in Section 19-11B).

16-9C PYRIMIDINE AND IMIDAZOLE

Pyrimidine is a six-membered heterocycle with two nitrogen atoms situated in a 1,3-arrangement. Both of the nitrogen atoms are like the pyridine nitrogen. Each has its lone pair of electrons in the sp^2 hybrid orbital in the plane of the aromatic ring. These lone pairs are not needed for the aromatic sextet and they are basic, like the lone pair of pyridine.

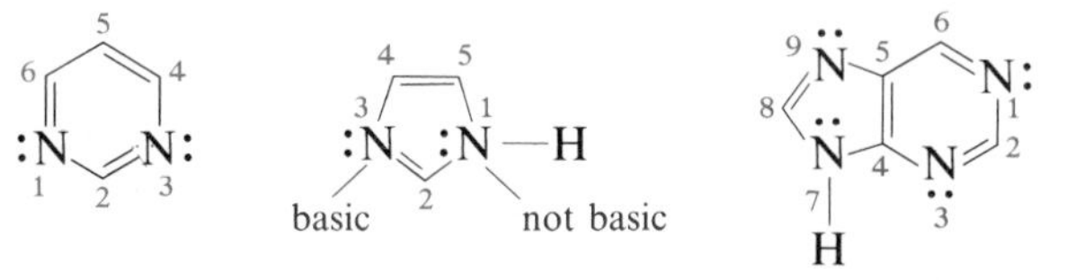

basic not basic

pyrimidine imidazole purine

Imidazole is an aromatic five-membered heterocycle with two nitrogen atoms. One of the nitrogen atoms (the one not bonded to a hydrogen) has its lone pair in an sp^2 orbital that is not involved in the aromatic system; this lone pair is basic. The other nitrogen uses its third sp^2 orbital to bond to hydrogen, and its lone pair is part of the aromatic sextet. Like the pyrrole nitrogen atom, this imidazole N—H nitrogen is not very basic.

Purine has an imidazole ring fused to a pyrimidine ring. Purine has three basic nitrogen atoms and one pyrrole-like nitrogen.

Pyrimidine and purine derivatives serve in DNA and RNA molecules to specify the genetic code. Imidazole derivatives enhance the enzymatic activity of enzymes. We will consider these important heterocyclic derivatives in more detail in Chapters 23 and 24.

PROBLEM 16-15

Draw the important resonance structures for a protonated imidazole. Determine which nitrogen atom of the protonated imidazole is more acidic.

16-9D FURAN AND THIOPHENE

Furan is an aromatic five-membered heterocycle like pyrrole, but the heteroatom is oxygen instead of nitrogen. The classical structure for furan (Fig. 16-14) shows that the oxygen atom has two lone pairs of electrons. The oxygen atom is sp^2 hybridized, and one of the lone pairs occupies an sp^2 hybrid orbital. The other lone pair occupies the unhybridized p orbital, combining with the four electrons in the double bonds to give an aromatic sextet.

Thiophene is similar to furan, with a sulfur atom in place of the furan oxygen. The bonding in thiophene is similar to that in furan, except that the sulfur atom uses an unhybridized $3p$ orbital to overlap with the $2p$ orbitals on the carbon atoms. The resonance energy of thiophene is not as large as that of furan because the difference in the sizes of the $2p$ and $3p$ orbitals results in less effective overlap.

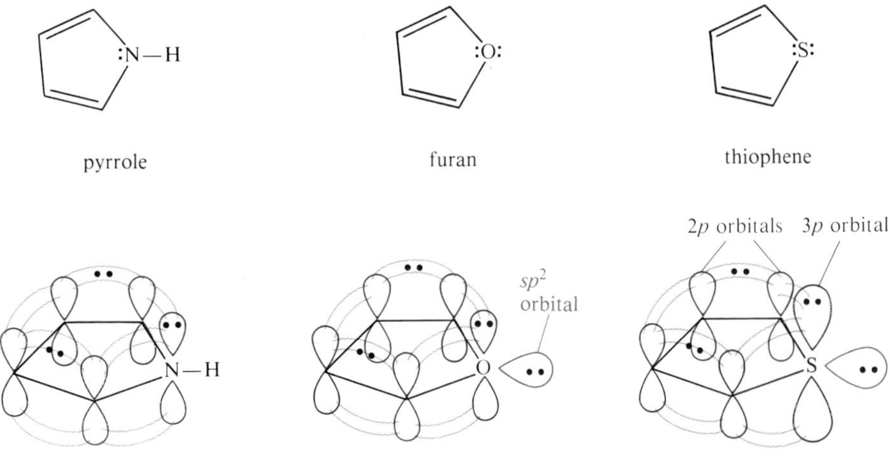

FIGURE 16-14 Pyrrole, furan, and thiophene are isoelectronic. In furan and thiophene, the pyrrole N—H bond is replaced by a nonbonding pair of electrons in the sp^2 hybrid orbital.

PROBLEM 16-16

Explain why each of the following compounds is aromatic, antiaromatic, or nonaromatic.

(a) isoxazole (b) pyran (c) pyrylium ion (d) 1,2-dihydropyridine (e) 1,3-thiazole

PROBLEM 16-17

Borazole, $B_3N_3H_6$, is an unusually stable cyclic compound. Propose a structure for borazole and explain why it is aromatic.

16-10 POLYNUCLEAR AROMATIC HYDROCARBONS

The **polynuclear aromatic hydrocarbons** (abbreviated PAH's or PNA's) are compounds composed of two or more fused benzene rings. **Fused rings** are rings that share two carbon atoms and the bond between them.

Naphthalene Naphthalene ($C_{10}H_8$) is the simplest fused aromatic compound, consisting of two fused benzene rings. We represent naphthalene by using one of the three Kekulé resonance structures or using the circle notation for the aromatic rings.

naphthalene

There are two aromatic rings in naphthalene, containing a total of 10 pi electrons. Two isolated aromatic rings would contain 6 pi electrons in each aromatic system, for a total of 12. Because of the smaller amount of electron density, naphthalene has less than twice the resonance energy of benzene: 60 kcal/mol (252 kJ/mol), or 30 kcal (126 kJ) per aromatic ring, compared with benzene's resonance energy of 36 kcal/mol (151 kJ/mol).

Anthracene and phenanthrene As the number of fused aromatic rings increases, the resonance energy per ring continues to decrease and the compounds become more reactive. For example, tricyclic anthracene has a resonance energy of 84 kcal (351 kJ) per mole, or 28 kcal (117 kJ) per aromatic ring. Phenanthrene has a slightly higher resonance energy of 91 kcal (381 kJ) per mole, or about 30.3 kcal (127 kJ) per aromatic ring. Each of these compounds has only 14 pi electrons in its three aromatic rings, compared with 18 electrons for three separate benzene rings.

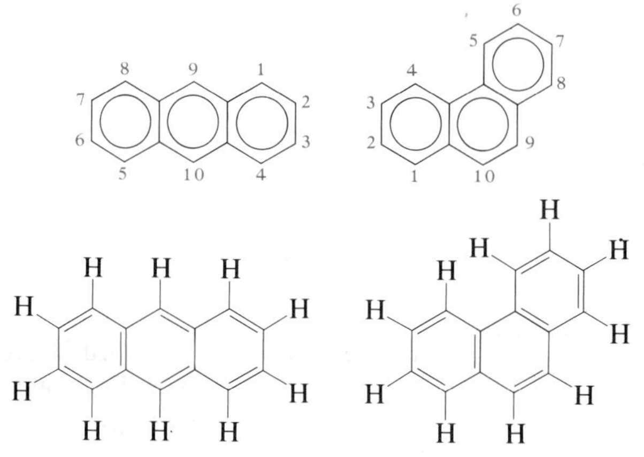

anthracene phenanthrene

(Only one Kekulé structure is shown for each compound.)

Because they are not as strongly stabilized as benzene, anthracene and phenanthrene often undergo addition reactions that are more characteristic of their nonaromatic polyene relatives. Anthracene undergoes 1,4-addition at the 9 and 10 positions to give a product with two isolated, fully aromatic benzene rings. Similarly, phenanthrene undergoes 1,2-addition at the 9 and 10 positions to give a product with two fully aromatic rings.

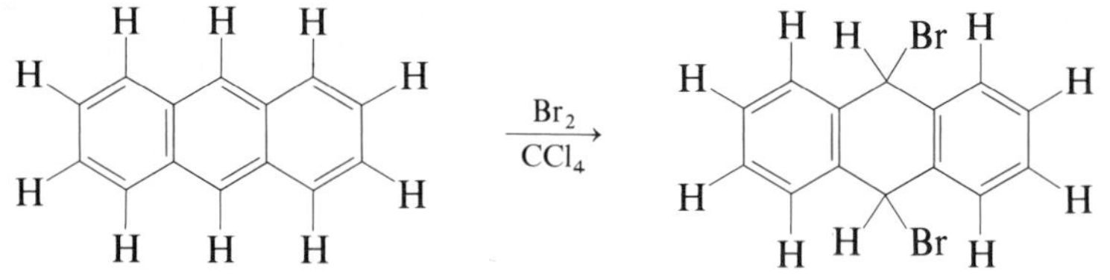

$$\text{anthracene} \xrightarrow[\text{CCl}_4]{\text{Br}_2} \text{(mixture of cis and trans)}$$

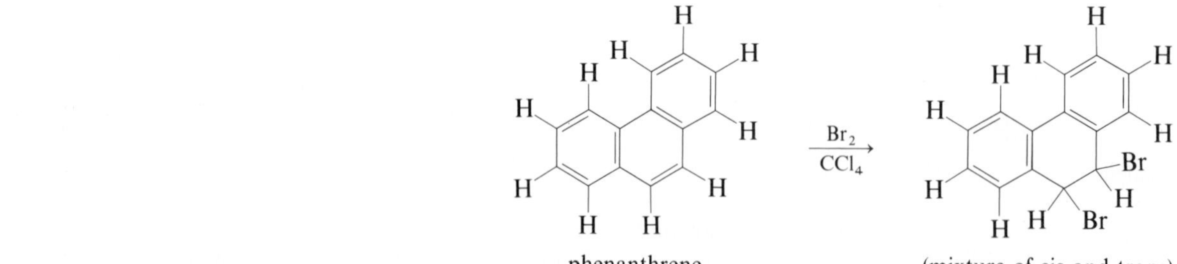

$$\text{phenanthrene} \xrightarrow[\text{CCl}_4]{\text{Br}_2} \text{(mixture of cis and trans)}$$

PROBLEM 16-18

(a) Draw all the Kekulé structures of anthracene and phenanthrene.
(b) Show how these Kekulé structures suggest that the most reactive positions will be those shown undergoing addition in the reactions above.
(c) Propose a mechanism for each of these reactions.

Larger polynuclear aromatic hydrocarbons There is a high level of interest in the larger PAHs, because they are formed in most combustion processes and many of them are carcinogenic (capable of causing cancer). The following three compounds, for example, are present in tobacco smoke. These compounds are so hazardous that laboratories must install special containment facilities to work with them, yet smokers intentionally expose their lung tissue to them.

pyrene   benzo[*a*]pyrene   dibenzopyrene

16-11 FUSED HETEROCYCLIC COMPOUNDS

Purine is one of many fused heterocyclic compounds that share two atoms and the bond between them. For example, the following compounds all contain fused heterocyclic aromatic rings.

purine   indole   benzimidazole   quinoline   benzofuran   benzothiophene

The properties of fused-ring heterocycles are generally similar to those of the simple heterocycles. Fused heterocyclic compounds are common in nature, and they are also used as drugs to treat a wide variety of illnesses. Figure 16-15 shows some fused heterocycles that occur naturally or are synthesized for use as drugs.

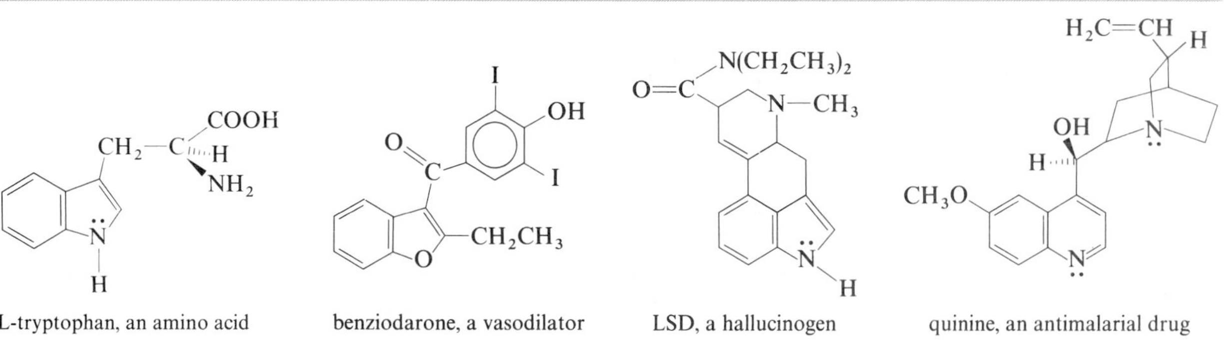

L-tryptophan, an amino acid   benziodarone, a vasodilator   LSD, a hallucinogen   quinine, an antimalarial drug

FIGURE 16-15 Examples of biologically active fused heterocycles.

16-12 NOMENCLATURE OF BENZENE DERIVATIVES

Benzene derivatives have been isolated and used as industrial reagents for well over 100 years. Many of their names are rooted in the historical traditions of chemistry. The following compounds are usually called by their historical common names, and almost never by the systematic IUPAC names.

common name:	phenol (benzenol)	toluene (methylbenzene)	aniline (benzenamine)	anisole (methoxybenzene)
common name:	styrene (vinylbenzene)	acetophenone (methyl phenyl ketone)	benzaldehyde	benzoic acid

Many compounds are named as derivatives of benzene, with their substituents named just as though they were attached to an alkane.

t-butylbenzene	nitrobenzene	ethynylbenzene (phenylacetylene)	ethoxybenzene (ethyl phenyl ether)	benzenesulfonic acid

Disubstituted benzenes are named using the prefixes *ortho-*, *meta-*, and *para-* to specify the substitution patterns. These terms are often abbreviated *o-*, *m-*, and *p-*. Numbers can also be used to specify the substitution in disubstituted benzenes.

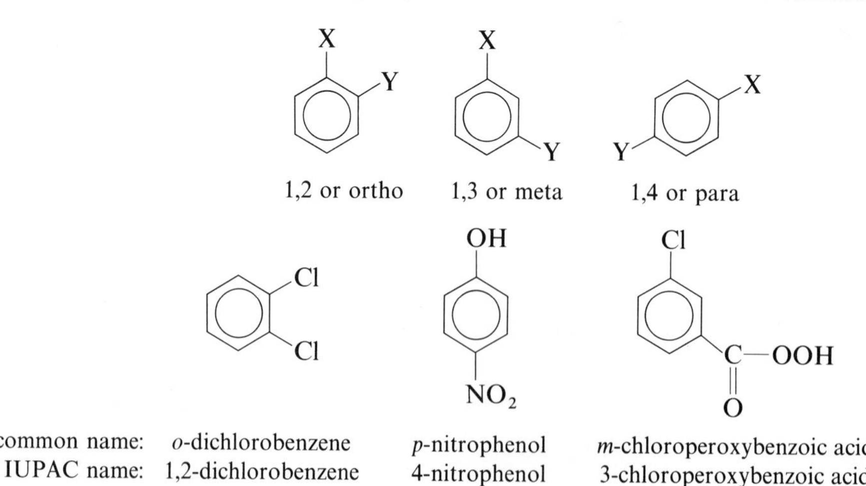

common name:	*o*-dichlorobenzene	*p*-nitrophenol	*m*-chloroperoxybenzoic acid
IUPAC name:	1,2-dichlorobenzene	4-nitrophenol	3-chloroperoxybenzoic acid

When there are three or more substituents on the benzene ring, numbers are used to give their positions. The numbers are assigned exactly as they would be with a substituted cyclohexane. The carbon atom bearing the functional group that defines the base name (as in phenol or benzoic acid) is assumed to be C1.

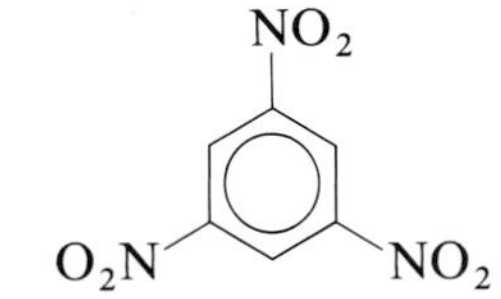

NO_2 O_2N NO_2 — 1,3,5-trinitrobenzene

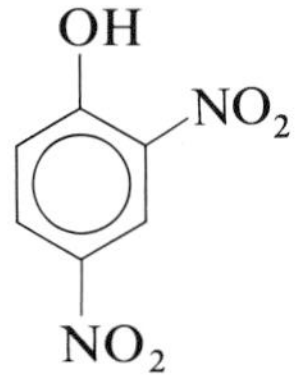

OH NO_2 NO_2 — 2,4-dinitrophenol

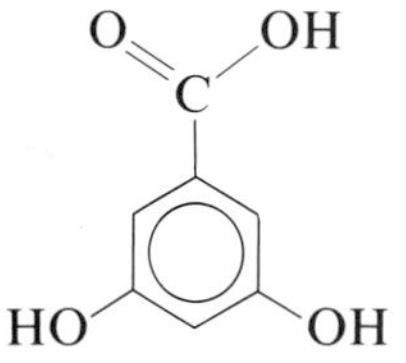

O C OH HO OH — 3,5-dihydroxybenzoic acid

Many disubstituted benzenes (and polysubstituted benzenes) have historical names. Some of these are obscure, with no obvious connection to the structure of the molecule.

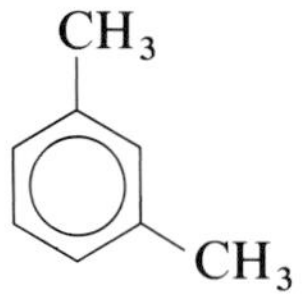

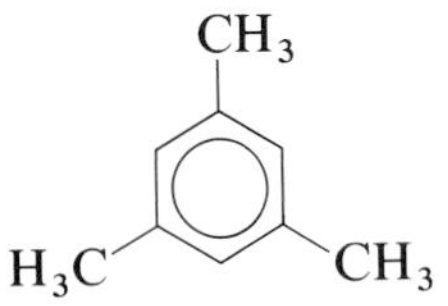

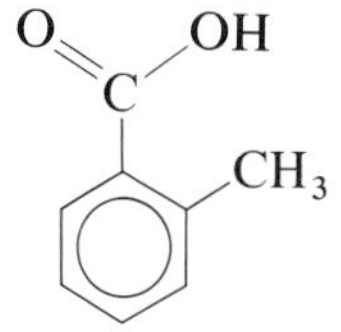

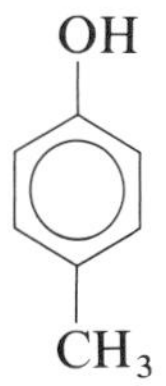

	CH_3 / CH_3	CH_3 / H_3C / CH_3	O C OH / CH_3	OH / CH_3
common name:	*m*-xylene	mesitylene	*o*-toluic acid	*p*-cresol
IUPAC name:	1,3-dimethylbenzene	1,3,5-trimethylbenzene	2-methylbenzoic acid	4-methylphenol

When the benzene ring is named as a substituent on another molecule, it is called a **phenyl group.** The phenyl group is used in the name just like the name of an alkyl group, and it is often abbreviated **Ph** (or ϕ) in drawing a complex structure.

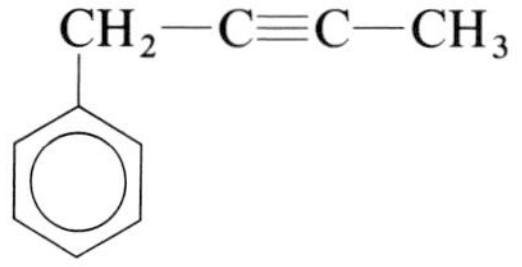

$CH_2-C\equiv C-CH_3$ — or $Ph-CH_2-C\equiv C-CH_3$ — 1-phenyl-2-butyne

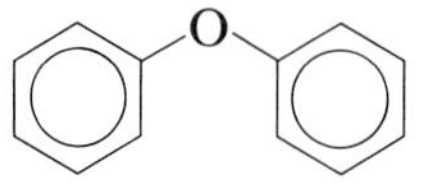

O — or Ph_2O — diphenyl ether

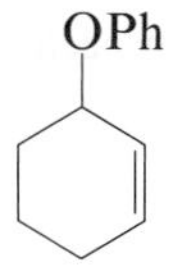

OPh — 3-phenoxycyclohexene

CH_2-CH_2-OH — or $PhCH_2CH_2OH$ — 2-phenylethanol

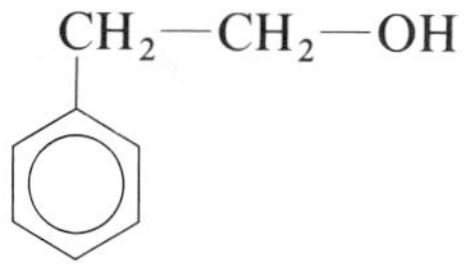

The seven-carbon unit consisting of a benzene ring and a methylene ($-CH_2-$) group is often named as a **benzyl group.** Be careful not to confuse the *benzyl group* (seven carbons) with the *phenyl group* (six carbons).

CH_2Br — benzyl bromide (α-bromotoluene)

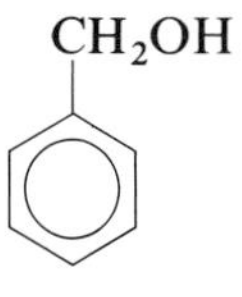

CH_2OH — benzyl alcohol

CH_2- — a benzyl group

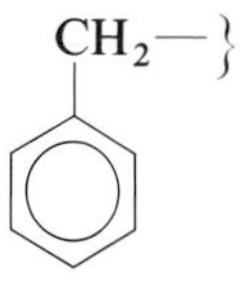

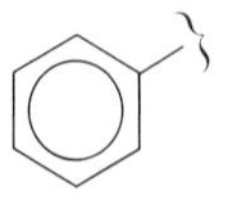

a phenyl group

An **aryl group,** abbreviated **Ar,** is the aromatic group that remains after the removal of a hydrogen atom from an aromatic ring. The phenyl group, **Ph,** is the simplest aryl group. The generic aryl group **(Ar)** is the aromatic relative of the generic alkyl group, which we have symbolized by **R.**

Examples of aryl groups

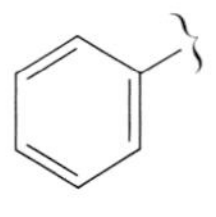

the phenyl group 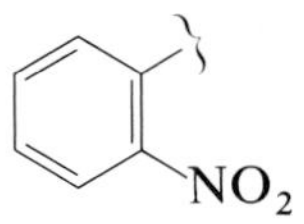 | the *o*-nitrophenyl group (NO_2) | the *p*-tolyl group (H_3C) | the 3-pyridyl group

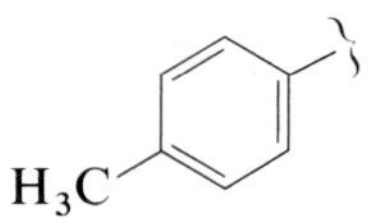

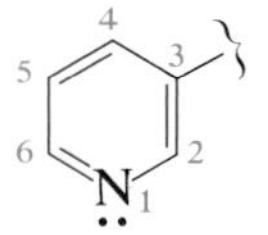

Examples of the use of a generic aryl group

Ar—MgBr	Ar_2O or Ar—O—Ar′	Ar—NH_2	Ar—SO_3H
an arylmagnesium bromide	a diaryl ether	an arylamine	an arylsulfonic acid

PROBLEM 16-19

Draw and name all the chlorinated benzenes, having from one to six chlorine atoms.

PROBLEM 16-20

Name the following compounds.

(a) 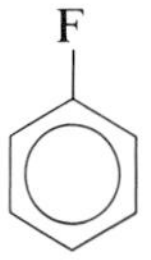 F (b) $CH_2CH_2C{\equiv}CH$ (c) OH, CH_3 (d) $CH{=}CH_2$, NO_2

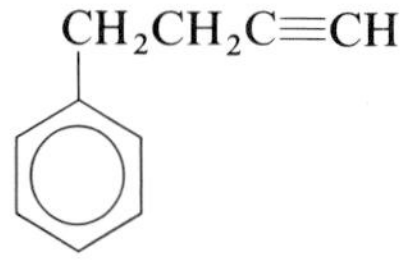

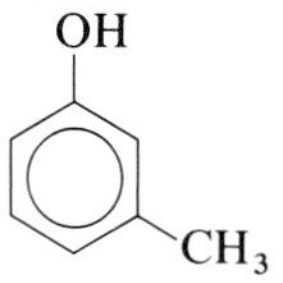

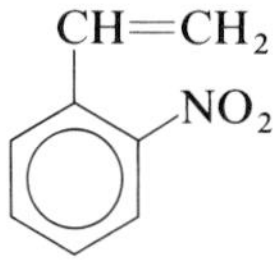

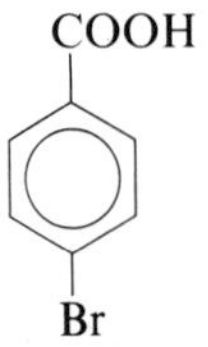

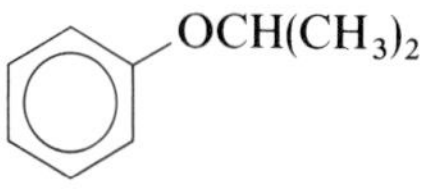

(e) COOH, Br (f) $OCH(CH_3)_2$ (g) 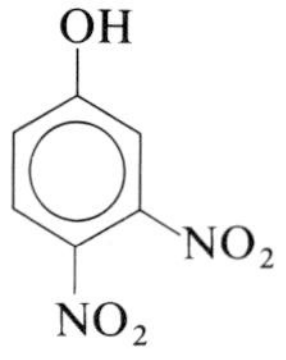 OH, NO_2, NO_2 (h) 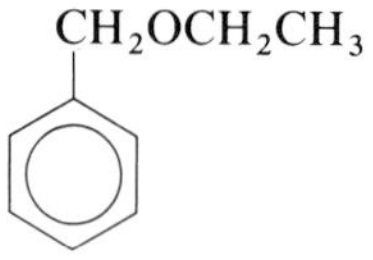 $CH_2OCH_2CH_3$

PROBLEM 16-21

Draw and name a specific example of each of the following classes of compounds.

(a) an alkyl aryl ether, Ar—O—R
(b) an arylsulfonic acid, Ar—SO_3H
(c) an aryllithium reagent
(d) an aryl alcohol (What is a better generic name for this class of compounds?)
(e) a diaryl methanol
(f) an arylbenzene
(g) a substituted benzyl alcohol

16-13 PHYSICAL PROPERTIES OF BENZENE AND ITS DERIVATIVES

The melting points, boiling points, and densities of benzene and some of its derivatives are given in Table 16-1. Benzene derivatives tend to be more symmetrical than similar aliphatic compounds, so they pack better into crystals and have higher melting points. For example, benzene melts at 6°C, while hexane melts at −95°C. Similarly, para-disubstituted benzenes are more symmetrical than the ortho and meta isomers, and they pack better into crystals and have higher melting points.

The relative boiling points of many benzene derivatives are related to their dipole moments. For example, the dichlorobenzenes have boiling points that fol-

TABLE 16-1
Physical properties of benzene and its derivatives

Compound	*m.p. (°C)*	*b.p. (°C)*	*Density (g/mL)*	*Compound*	*m.p. (°C)*	*b.p. (°C)*	*Density (g/mL)*
benzene	6	80	0.88	*o*-xylene	−26	144	0.88
toluene	−95	111	0.87	*m*-xylene	−48	139	0.86
ethylbenzene	−95	136	0.87	*p*-xylene	13	138	0.86
styrene	−31	146	0.91	*o*-chlorotoluene	−35	159	1.08
ethynylbenzene	−45	142	0.93	*m*-chlorotoluene	−48	162	1.07
fluorobenzene	−41	85	1.02	*p*-chlorotoluene	8	162	1.07
chlorobenzene	−46	132	1.11	*o*-dichlorobenzene	−17	181	1.31
bromobenzene	−31	156	1.49	*m*-dichlorobenzene	−25	173	1.29
iodobenzene	−31	188	1.83	*p*-dichlorobenzene	54	170	1.07
benzyl bromide	−4	199	1.44	*o*-dibromobenzene	7	225	1.62
nitrobenzene	6	211	1.20	*m*-dibromobenzene	−7	218	1.61
phenol	43	182	1.07	*p*-dibromobenzene	87	218	1.57
anisole	37	156	0.98	*o*-toluic acid	106	263	1.06
benzoic acid	122	249	1.31	*m*-toluic acid	111	263	1.05
benzyl alcohol	−15	205	1.04	*p*-toluic acid	180	275	1.06
aniline	−6	186	1.02	*o*-cresol	30	192	1.03
diphenyl ether	28	259	1.08	*m*-cresol	12	202	1.03
mesitylene	−45	165	0.87	*p*-cresol	36	202	1.03

low their dipole moments. The symmetrical *p*-dichlorobenzene has zero dipole moment and the lowest boiling point. *m*-Dichlorobenzene has a small dipole moment and a slightly higher boiling point. *o*-Dichlorobenzene has the largest dipole moment and the highest boiling point. Even though *p*-dichlorobenzene has the lowest boiling point, it packs best into a crystal, and it has the highest melting point of the three dichlorobenzenes.

o-dichlorobenzene	*m*-dichlorobenzene	*p*-dichlorobenzene
b.p. 181°C	b.p. 173°C	b.p. 170°C
m.p. −17°C	m.p. −25°C	m.p. 54°C

Benzene and other aromatic hydrocarbons are slightly denser than the nonaromatic analogs, but they are still less dense than water. The halogenated benzenes are denser than water. The aromatic hydrocarbons and the halogenated aromatics are generally insoluble in water, although some of the derivatives with strongly polar functional groups (phenol, benzoic acid, etc.) are moderately soluble in water.

16-14 SPECTROSCOPY OF AROMATIC COMPOUNDS

Infrared Spectroscopy *(Review)* Aromatic compounds are readily identified by their infrared spectra because they show a characteristic C=C stretch around 1600 cm^{-1}. This is a lower C=C stretching frequency than that for isolated alkenes (1640 to 1680 cm^{-1}) or conjugated dienes (1620 to 1640 cm^{-1}) because the aromatic bond order is only about $1\frac{1}{2}$. The aromatic bond is less stiff than a normal double bond, giving a lower vibration frequency.

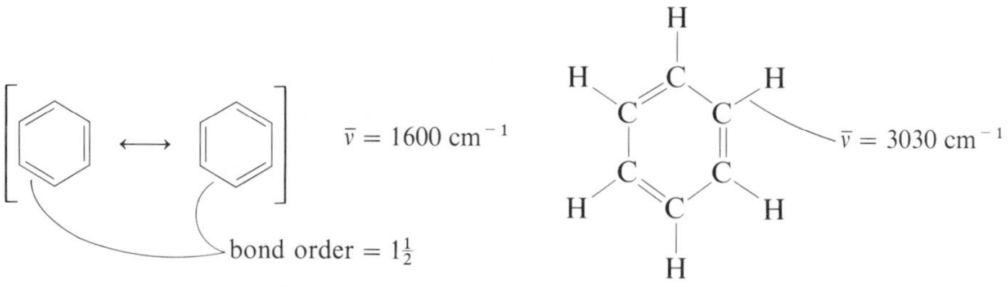

$\bar{\nu} = 1600\ \text{cm}^{-1}$ bond order = $1\frac{1}{2}$ $\bar{\nu} = 3030\ \text{cm}^{-1}$

Like alkenes, aromatic compounds show unsaturated =C—H stretching just above 3000 cm^{-1} (usually around 3030 cm^{-1}). The combination of the aromatic C=C stretch around 1600 cm^{-1} and the =C—H stretch just above 3000 cm^{-1} leaves little doubt of the presence of an aromatic ring. The sample spectra labeled Compounds 4, 5, 7, and 8 in Chapter 11 (pages 469–71) are examples of compounds containing aromatic rings.

NMR Spectroscopy *(Review)* Aromatic compounds give readily identifiable ^{1}H NMR absorptions around $\delta 7$ to $\delta 8$, strongly deshielded by the aromatic ring current (Section 12-5B). In benzene, the aromatic protons absorb around $\delta 7.2$; however, the absorption may be moved farther downfield by electron-withdrawing groups such as carbonyl, nitro, or cyano groups, or it may be moved upfield by electron-donating groups such as hydroxyl, alkoxyl, or amino groups.

Nonequivalent aromatic protons that are ortho or meta usually split each other. The spin-spin splitting constants are about 8 Hz for ortho protons and 2 Hz for meta protons. Figures 12-11, 12-18, 12-24, 12-29, and 12-32 are examples of proton NMR spectra of aromatic compounds.

Aromatic carbon atoms absorb around $\delta 120$ to $\delta 150$ in the ^{13}C NMR spectrum. Alkene carbon atoms can also absorb in this spectral region, but the combination of ^{13}C NMR with ^{1}H NMR or IR spectroscopy usually leaves no doubt whether there is an aromatic ring present.

Mass spectrometry The most common fragmentation of alkylbenzene derivatives in the mass spectrometer is the cleavage of a benzylic bond to give a resonance-stabilized benzylic cation. For example, in the mass spectrum of *n*-butylbenzene (Fig. 16-16) the base peak is at m/z 91. This mass number results from benzylic cleavage to give a benzylic cation. There is evidence to show that the benzylic cation rearranges to give the aromatic tropylium ion. It is common for alkylbenzenes to give ions corresponding to the tropylium ion at m/z 91.

PROBLEM 16-22

Draw three more resonance structures for the benzyl cation in Figure 16-16.

Ultraviolet spectroscopy The ultraviolet spectra of aromatic compounds are quite different from those of nonaromatic polyenes. For example, benzene has three absorptions in the ultraviolet region: an intense band at $\lambda_{max} = 184$ nm ($\varepsilon = 68{,}000$), a moderate band at $\lambda_{max} = 204$ nm ($\varepsilon = 8800$), and a characteristic low-intensity band of multiple absorptions centered around 254 nm ($\varepsilon = 200$ to 300). The UV spectrum of benzene appears in Figure 16-17; the absorption at

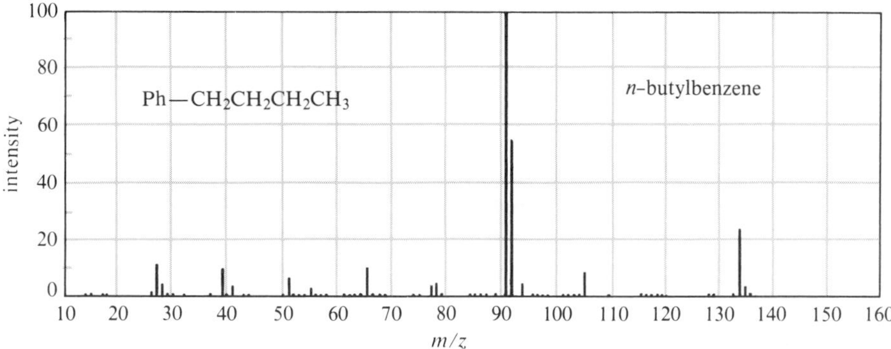

FIGURE 16-16 The mass spectrum of *n*-butylbenzene has its base peak at *m/z* 91, corresponding to cleavage of a benzylic bond. The fragments are a benzyl cation and a propyl radical. The benzyl cation rearranges to the tropylium ion, which is the species detected at *m/z* 91.

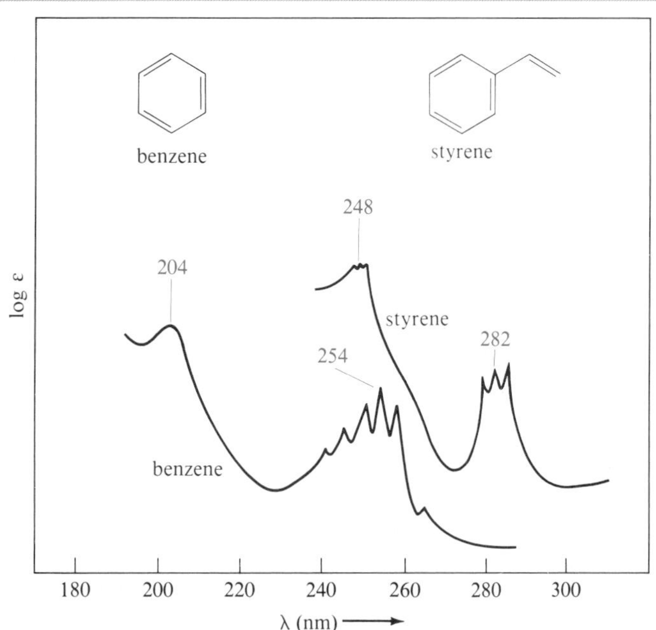

FIGURE 16-17 Ultraviolet spectra of benzene and styrene.

184 nm does not appear because wavelengths shorter than 200 nm are not accessible by standard UV-visible spectrometers.

All three of the major bands in the benzene spectrum correspond to $\pi \rightarrow \pi^*$ transitions. The absorption at 184 nm corresponds to the energy of the transition from one of the two HOMO's to one of the two LUMO's. The weaker band at 204 nm corresponds to a "forbidden" transition that would be impossible to observe if benzene were always an unperturbed, perfectly hexagonal structure.

The most characteristic part of the spectrum is the band of absorptions centered at 254 nm. These absorptions are called the **benzenoid band,** because they usually imply the presence of a benzene ring. The extinction coefficients of these absorptions are small (usually 200 to 300), and there are usually about three to six small, sharp peaks (called *fine structure*) within the band. These benzenoid absorptions correspond to additional forbidden transitions.

Simple benzene derivatives show most of the characteristics of benzene, including the moderate band in the 210-nm region and the benzenoid band in the 260-nm region. Alkyl and halogen substituents increase the values of λ_{max} by about 5 nm, as shown by the examples in Table 16-2. An additional conjugated double bond can increase the value of λ_{max} by about 30 nm, as shown by the UV spectrum of styrene in Figure 16-17.

TABLE 16-2
Ultraviolet spectra of benzene and some simple derivatives

Compound	Structure	Moderate band		Benzenoid band	
		λ_{max} (nm)	ε	λ_{max} (nm)	ε
benzene		204	8,800	254	250
ethylbenzene	CH_2CH_3	208	7,800	260	220
m-xylene	CH_3, CH_3	212	7,300	264	300
bromobenzene	Br	210	7,500	258	170
styrene		248	15,000	282	740

PROBLEM 16-23

The UV spectrum of 1-phenyl-2-propen-1-ol shows an intense absorption at 220 nm and a weaker absorption at 258 nm. When this compound is treated with dilute sulfuric acid, it rearranges to an isomer with an intense absorption at 250 nm and a weaker absorption at 290 nm. Suggest a structure for the isomeric product and give a mechanism for its formation.

GLOSSARY

aliphatic compound An organic compound that is not aromatic. (p. 672)

annulenes Cyclic hydrocarbons with alternating single and double bonds. (p. 676)

[6]annulene (benzene) [10]annulene (cyclodecapentaene)

antiaromatic compound A compound that has a continuous ring of *p* orbitals as in an aromatic compound, but delocalization of the pi electrons over the ring increases the electronic energy. (p. 683)

> In most cases, the structure must be planar (or nearly planar) and have ($4N$) pi electrons, with N an integer.

aromatic compound A cyclic compound containing some number of conjugated double bonds, characterized by an unusually large resonance energy. (p. 674)

> To be aromatic, all its ring atoms must have unhybridized *p* orbitals that overlap to form a continuous ring of orbitals. In most cases the structure must be planar (or nearly planar) and have ($4N + 2$) pi electrons, with N an integer. Delocalization of the pi electrons over the ring results in a lowering of the electronic energy.

aryl group (abbreviated Ar) The aromatic group that remains after the removal of a hydrogen atom from an aromatic ring; the aromatic equivalent of the generic alkyl group (R). (p. 699)

benzenoid band The weak band around 250 to 270 nm in the UV spectra of benzenoid aromatics. This band is usually characterized by multiple, sharper absorptions (fine structure). (p. 704)

benzyl group ($Ph{-}CH_2{-}$) The seven-carbon unit consisting of a benzene ring and a methylene group. (p. 699)

degenerate orbitals Orbitals having the same energy. (p. 678)

fused rings Rings that share a common carbon-carbon bond together with its two carbon atoms. (p. 695)

heterocycle A cyclic compound in which one or more of the ring atoms is not carbon. (p. 692)

> **aromatic heterocycle:** a heterocyclic compound that fulfills the criteria for aromaticity and that has a substantial resonance energy.

Hückel's rule A cyclic molecule or ion that has a continuous ring of overlapping *p* orbitals will be:

1. aromatic if the number of pi electrons is ($4N + 2$), with N an integer.
2. antiaromatic if the number of pi electrons is ($4N$), with N an integer. (p. 683)

Kekulé structure A classical structural formula for an aromatic compound, showing localized double bonds. (p. 672)

nonaromatic compound Neither aromatic nor antiaromatic; lacking the continuous ring of overlapping *p* orbitals required for aromaticity or antiaromaticity. (p. 683)

ortho Having a 1,2-relationship on a benzene ring. (p. 698)

meta Having a 1,3-relationship on a benzene ring. (p. 698)

para Having a 1,4-relationship on a benzene ring. (p. 698)

X X X
Y Y Y

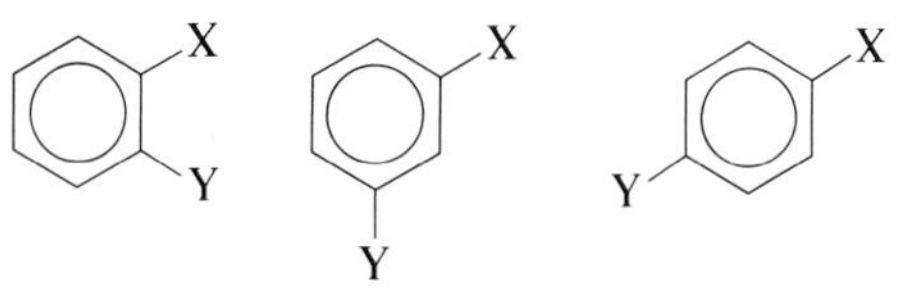

ortho (1,2) meta (1,3) para (1,4)

phenyl group (Ph or ϕ) The benzene ring, minus one hydrogen atom, when named as a substituent on another molecule. (p. 699)

polygon rule The energy diagram of the MO's of a regular, completely conjugated cyclic system has the same polygonal shape as the compound, with one vertex (the all-bonding MO) at the bottom. The nonbonding line cuts horizontally through the center of the polygon. (p. 682)

Energy diagrams

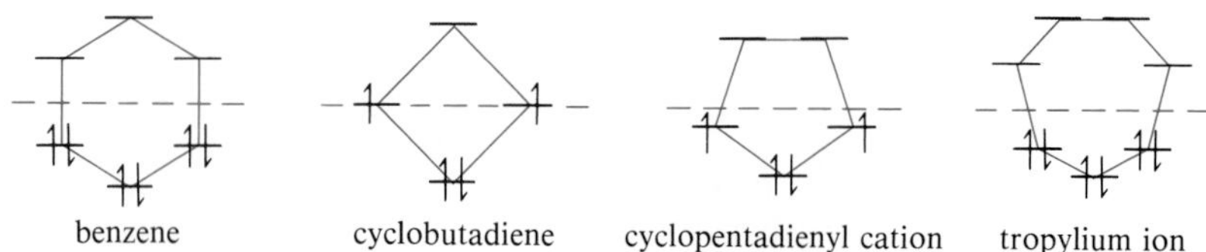

polynuclear aromatic compounds Aromatic compounds composed of two or more fused aromatic rings. Naphthalene is an example of a **polynuclear aromatic hydrocarbon** (PAH or PNA), and indole is an example of a polynuclear aromatic heterocycle. (p. 695)

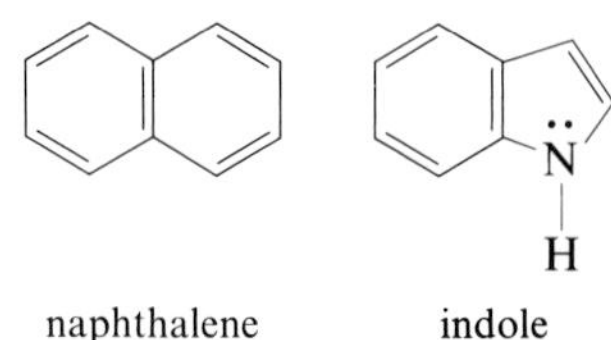

resonance energy The extra stabilization provided by delocalization, compared with a localized structure. For aromatic compounds, the resonance energy is the extra stabilization provided by the delocalization of the electrons in the aromatic ring. (p. 675)

tropylium ion The cycloheptatrienyl cation. (p. 689)

ESSENTIAL PROBLEM-SOLVING SKILLS IN CHAPTER 16

1. Be able to construct the molecular orbitals of a cyclic system of *p* orbitals such as in benzene and cyclobutadiene.
2. Use the polygon rule to draw the energy diagram for a cyclic system of *p* orbitals, and fill in the electrons to show whether a given compound or ion is aromatic or antiaromatic.
3. Use Hückel's rule to predict whether a given annulene, heterocycle, or ion will be aromatic, antiaromatic, or nonaromatic.
4. For heterocycles containing nitrogen atoms, determine whether the lone pairs are used in the aromatic system and predict whether the nitrogen atom is strongly or weakly basic.
5. Recognize fused aromatic systems such as polynuclear aromatic hydrocarbons and fused heterocyclic compounds, and use the theory of aromatic compounds to explain their properties.
6. Name aromatic compounds and draw their structures from the names.
7. Use IR, NMR, UV, and mass spectra to determine the structures of aromatic compounds. Given an aromatic compound, predict the important features of its spectra.

STUDY PROBLEMS

16-24. Briefly define the following terms and give examples.

(a) a heterocyclic aromatic compound	**(b)** an antiaromatic compound
(c) a Kekulé structure	**(d)** an annulene
(e) degenerate orbitals	**(f)** the polygon rule
(g) a polynuclear aromatic heterocycle	**(h)** fused rings
(i) a polynuclear aromatic hydrocarbon	**(j)** the benzenoid UV band
(k) a filled shell of MO's	**(l)** Hückel's rule
(m) resonance energy	**(n)** an aryl group

16-25. Draw the structures of the following compounds.

(a) *o*-nitroanisole **(b)** 2,4-dimethoxyphenol **(c)** *p*-aminobenzoic acid
(d) 4-nitroaniline **(e)** *m*-chlorotoluene **(f)** *p*-divinylbenzene
(g) *p*-bromostyrene **(h)** 3,5-dimethoxybenzaldehyde **(i)** tropylium chloride
(j) sodium cyclopentadienide

16-26. Name the following compounds.

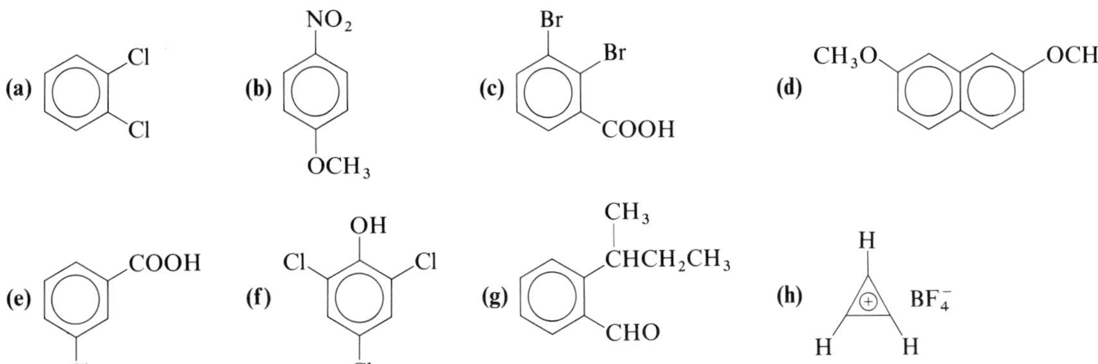

16-27. Draw and name all the methyl, dimethyl, and trimethylbenzenes.

16-28. One of the following hydrocarbons is much more acidic than the others. Indicate which one and explain why it is unusually acidic.

16-29. The strong polarization of a carbonyl group can be represented by a pair of resonance structures:

$$\left[\;\rangle C{=}\ddot{O}\!: \;\longleftrightarrow\; \rangle \overset{+}{C}{-}\ddot{\underset{..}{O}}\!:^{-} \right]$$

Cyclopropenone and cycloheptatrienone are more stable than anticipated. Cyclopentadienone, however, is relatively unstable and rapidly undergoes a Diels-Alder dimerization. Explain.

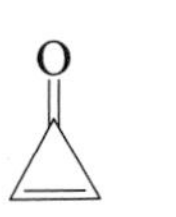

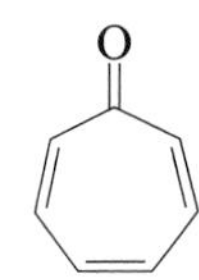

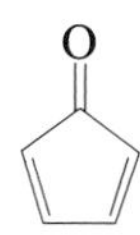

cyclopropenone cycloheptatrienone cyclopentadienone

16-30. In Kekulé's time, cyclohexane was unknown, and there was no proof that benzene must be a six-membered ring. Determination of the structure relied largely on the known numbers of monosubstituted and disubstituted benzenes, together with the knowledge that benzene did not react like a normal alkene. The following structures were the likely candidates.

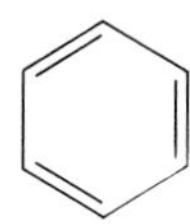

(localized double bonds)

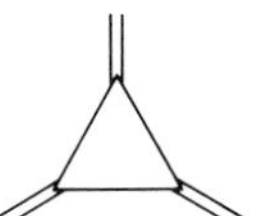

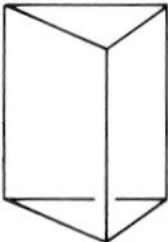

(a) Show where the six hydrogen atoms are in each structure.

(b) For each of these structures, draw all the possible monobrominated derivatives (C_6H_5Br) that would result from randomly substituting one hydrogen with a bromine. Benzene was known to have only one monobromo derivative.

(c) For each of the structures that had only one monobromo derivative in part (b), draw all the possible dibromo derivatives. Benzene was known to have three dibromo derivatives, but resonance theory was unknown at the time.

(d) Determine which structure was most consistent with what was known about benzene at that time: Benzene gives one monobrominated derivative and three dibrominated derivatives, and it gives negative chemical tests for an alkene.

(e) The structure that was considered the most likely structure for benzene is called *Ladenburg benzene,* after the chemist who proposed it. What factor would make Ladenburg benzene relatively unstable, in contrast to the stability observed with real benzene?

16-31. The following molecules and ions are grouped according to similarity of structure. Classify each molecule or ion as aromatic, antiaromatic, or nonaromatic. For the aromatic and antiaromatic species, give the number of pi electrons in the ring.

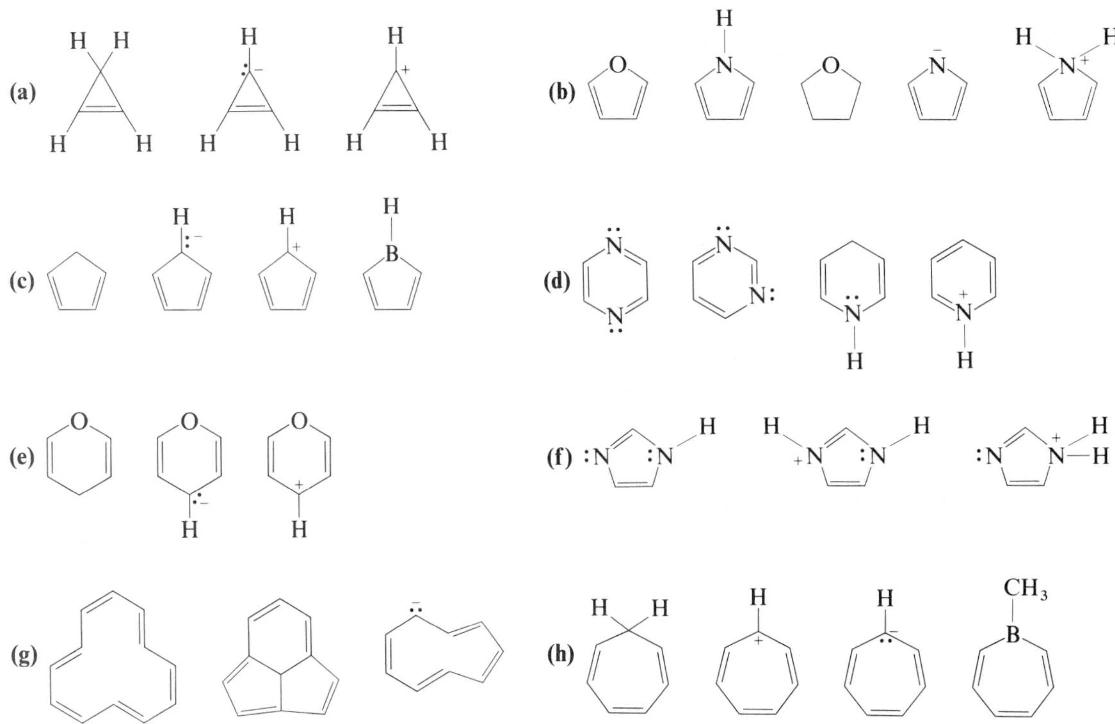

16-32. Azulene is a deep-blue hydrocarbon with a resonance energy of 49 kcal (205 kJ) per mole. Azulene has ten pi electrons, so it might be considered one large aromatic ring. Azulene has an unusually large dipole moment (1.0 D) for a hydrocarbon, indicating significant charge separation. Show how this charge separation might arise.

azulene

16-33. Each of the following heterocycles includes one or more nitrogen atoms. Classify each nitrogen atom as strongly basic or weakly basic, according to the availability of its lone pair of electrons.

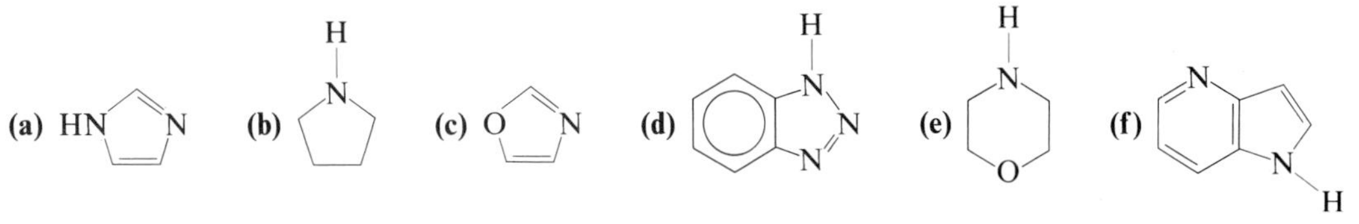

(a) HN N (b) H–N (c) O N (d) H–N N N (e) H–N O (f) N N–H

16-34. The benzene ring alters the reactivity of a neighboring group in the **benzylic position** in much the same way that a double bond alters the reactivity of groups in the allylic position.

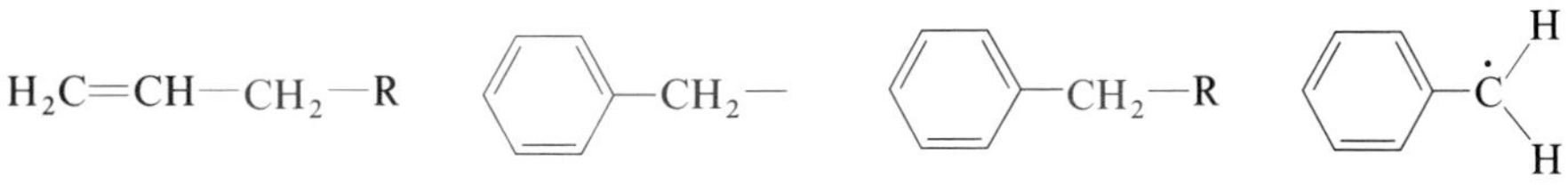

$H_2C{=}CH{-}CH_2{-}R$ (allylic position) $C_6H_5{-}CH_2{-}$ (benzyl group) $C_6H_5{-}CH_2{-}R$ (benzylic position) $C_6H_5{-}\dot{C}H_2$ (benzyl radical)

Benzylic cations, anions, and radicals are all more stable than simple alkyl intermediates.

(a) Use resonance structures to show the delocalization (over four carbon atoms) of the positive charge, odd electron, and negative charge of the benzyl cation, radical, and anion.

(b) Toluene reacts with bromine in the presence of light to give benzyl bromide. Propose a mechanism for this reaction.

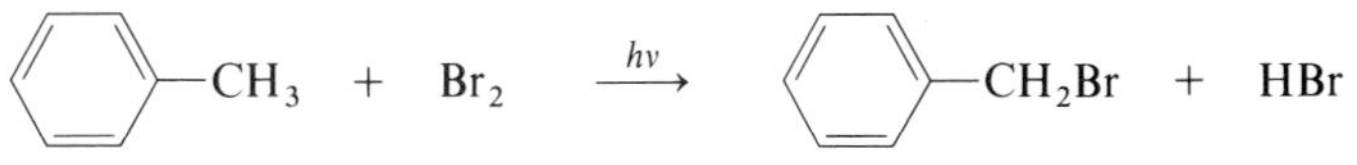

$$C_6H_5{-}CH_3 + Br_2 \xrightarrow{h\nu} C_6H_5{-}CH_2Br + HBr$$

toluene benzyl bromide

(c) Which of the following reactions will have the faster rate and give the better yield? Use a drawing of the transition state to explain your answer.

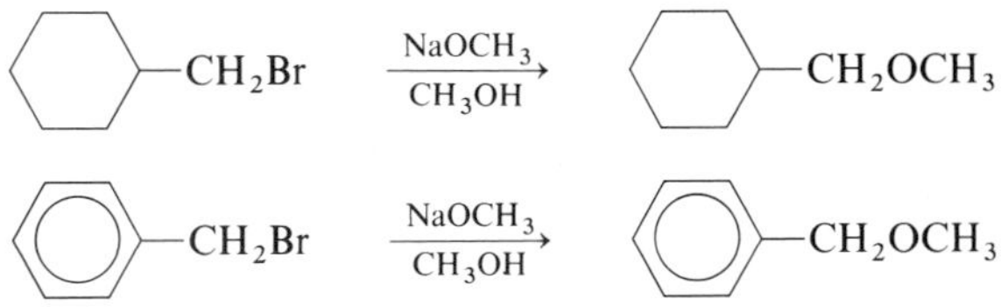

$$C_6H_{11}{-}CH_2Br \xrightarrow[CH_3OH]{NaOCH_3} C_6H_{11}{-}CH_2OCH_3$$

$$C_6H_5{-}CH_2Br \xrightarrow[CH_3OH]{NaOCH_3} C_6H_5{-}CH_2OCH_3$$

16-35. Before spectroscopy was invented, *Körner's absolute method* was used to determine whether a disubstituted benzene derivative was the ortho, meta, or para isomer. Körner's method involves adding a third group (often a nitro group) and determining how many isomers are formed. For example, when *o*-xylene is nitrated (by a method shown in Chapter 17), two isomers are formed.

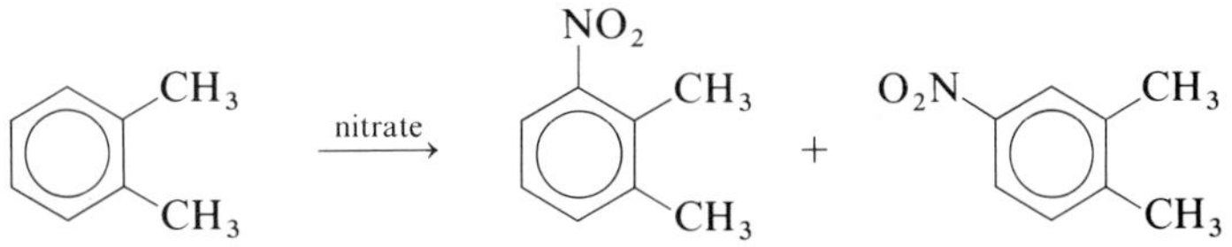

o-xylene (CH_3, CH_3) $\xrightarrow{\text{nitrate}}$ 3-nitro-o-xylene (NO_2, CH_3, CH_3) + 4-nitro-o-xylene (O_2N, CH_3, CH_3)

(a) How many isomers are formed by the nitration of *m*-xylene?

(b) How many isomers are formed by the nitration of *p*-xylene?

(c) A turn-of-the-century chemist has isolated an aromatic compound of molecular formula $C_6H_4Br_2$. He carefully nitrates this compound and purifies three isomers of formula $C_6H_3Br_2NO_2$. Propose structures for the original compound and the three nitrated derivatives.

16-36. For each of the following proton NMR spectra, propose a structure that is consistent with the spectrum and the additional information provided.

(a) Elemental analysis shows the molecular formula to be C_8H_7OCl. The IR spectrum shows a moderate absorption at 1602 cm^{-1} and a strong absorption at 1690 cm^{-1}.

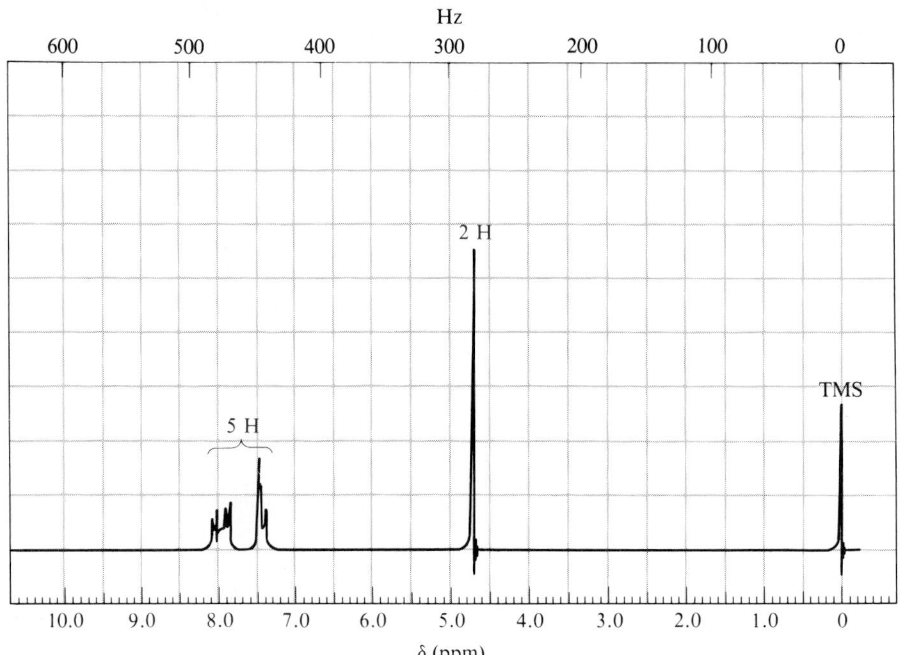

(b) The mass spectrum shows a double molecular ion of 1:1 intensities at m/z 184 and 186.

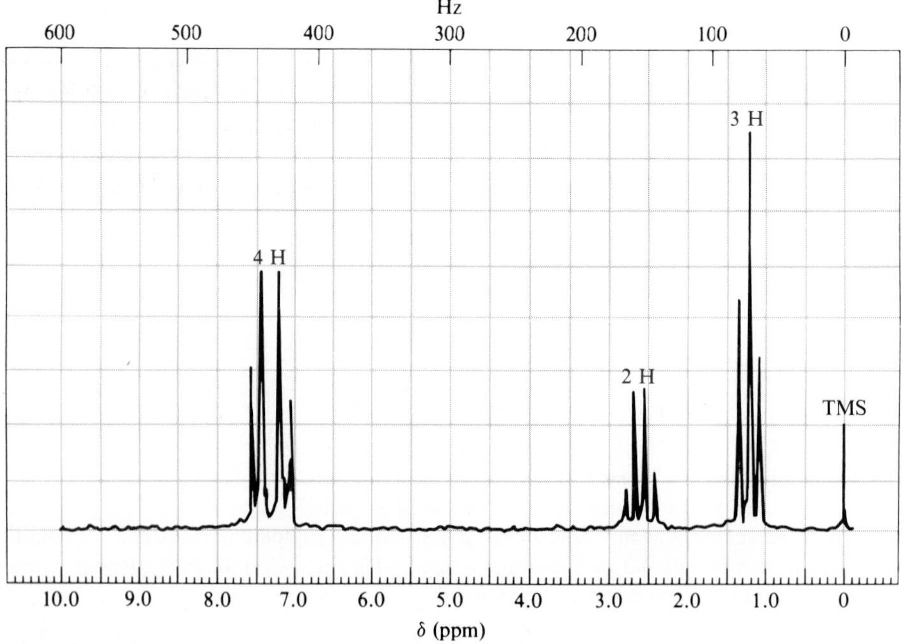

16-37. Bromine adds to phenanthrene to give a mixture of cis and trans addition products.

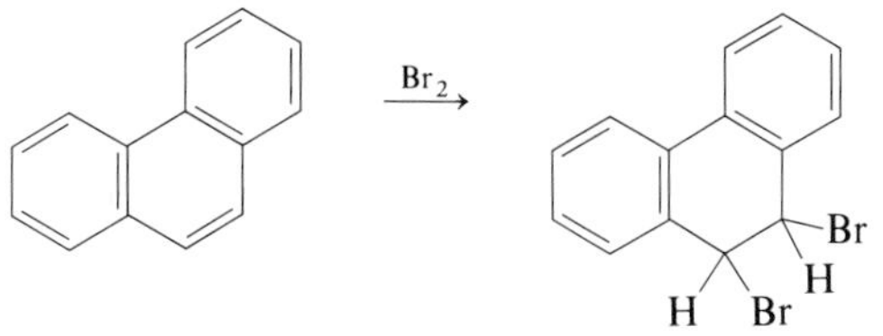

$$\xrightarrow{Br_2}$$

(mixture of *cis* and *trans*)

(a) Propose a mechanism for this reaction.
(b) In Chapter 8 most of the additions of bromine to double bonds gave entirely anti stereochemistry. Explain why the addition to phenanthrene gives a mixture of syn and anti stereochemistry.
(c) When the product of this addition is heated, HBr is evolved and 9-bromophenanthrene results. Propose a mechanism for this dehydrohalogenation.

16-38. Biphenyl has the following structure.

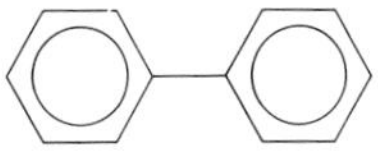

biphenyl

(a) Is biphenyl a (fused) polynuclear aromatic hydrocarbon?
(b) How many pi electrons are there in the two aromatic rings of biphenyl? How does this number compare with that for naphthalene?
(c) The heat of hydrogenation for biphenyl is about 100 kcal/mol (418 kJ/mol). Calculate the resonance energy of biphenyl.
(d) Compare the resonance energy of biphenyl with that of naphthalene and with that of two benzene rings. Explain the marked difference in the resonance energies of naphthalene and biphenyl.

16-39. The following hydrocarbon reacts with two equivalents of butyllithium to form a dianion of formula $[C_8H_6]^{2-}$. Propose a structure for this dianion, and suggest why it forms so readily.

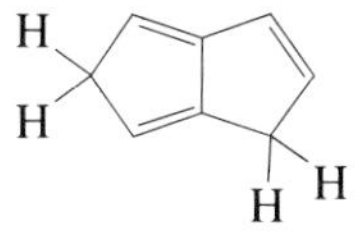

$$+ \quad 2\,C_4H_9Li \longrightarrow [C_8H_6]^{2-}\;2\,Li^+ \quad + \quad 2\,C_4H_{10}\uparrow$$

16-40. How would you convert 1,3,5,7-cyclononatetraene to an aromatic compound?

***16-41.** The ribonucleosides that make up ribonucleic acid (RNA) are composed of D-ribose (a sugar) and four heterocyclic "bases." The general structure of a ribonucleoside is

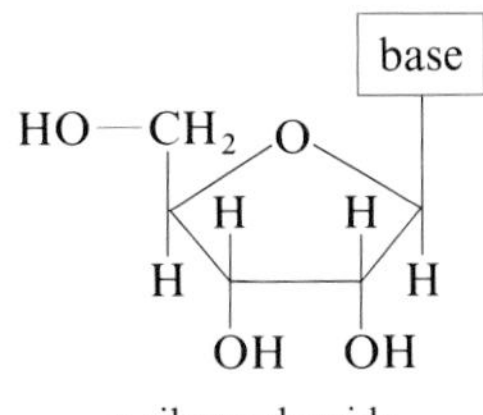

a ribonucleoside

The four heterocyclic bases are cytosine, uracil, guanine, and adenine. Cytosine and uracil are called *pyrimidine bases* because their structures resemble pyrimidine. Guanine and adenine are called *purine bases* because their structures resemble purine.

pyrimidine

purine

cytosine   uracil   guanine   adenine

Determine which rings of these bases are aromatic. Do any of these bases have easily formed tautomers that are aromatic? (Consider moving a proton from nitrogen to a carbonyl group to form a phenolic derivative.)

***16-42.** Consider the following compound, which has been synthesized and characterized.

$(CH_3)_3C$ $C(CH_3)_3$ $(CH_3)_3C$

(a) Assuming this molecule is entirely conjugated, do you expect it to be aromatic, antiaromatic, or nonaromatic?
(b) Why was this molecule synthesized with three *t*-butyl substituents? Why not make the unsubstituted compound and study it instead?
(c) Do you expect the nitrogen atom to be basic? Explain.
(d) At room temperature, the proton NMR spectrum shows only two singlets of ratio 1:2. The smaller resonance remains unchanged at all temperatures. As the temperature is lowered to $-110°C$, the larger resonance broadens and separates into two new singlets, one on either side of the original chemical shift. At $-110°C$, the spectrum consists of three separate singlets of areas 1:1:1.

Explain what these NMR data indicate about the bonding in this molecule. How does your conclusion based on the NMR data agree with your prediction in part (a)?

16-43. A student found an old bottle labeled "thymol" on the stockroom shelf. After noticing a delightful, pleasant odor, she obtained the following mass, IR, and NMR spectra. The NMR peak at $\delta 4.7$ disappears on shaking with D_2O. Propose a structure for thymol, and show how your structure is consistent with the spectra. Propose a fragmentation to explain the MS peak at m/z 135, and show why the resulting ion is relatively stable.

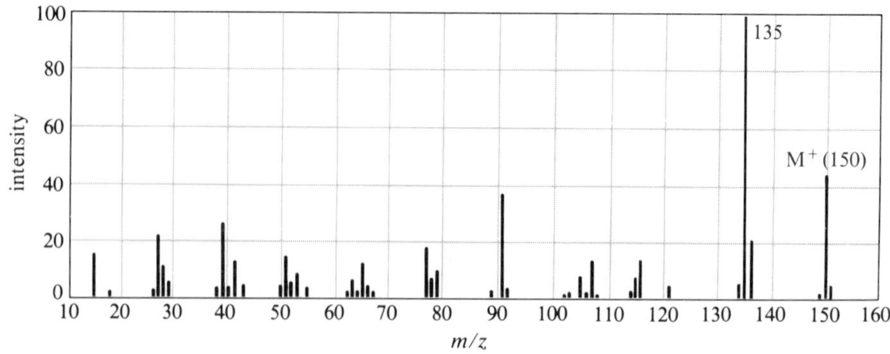

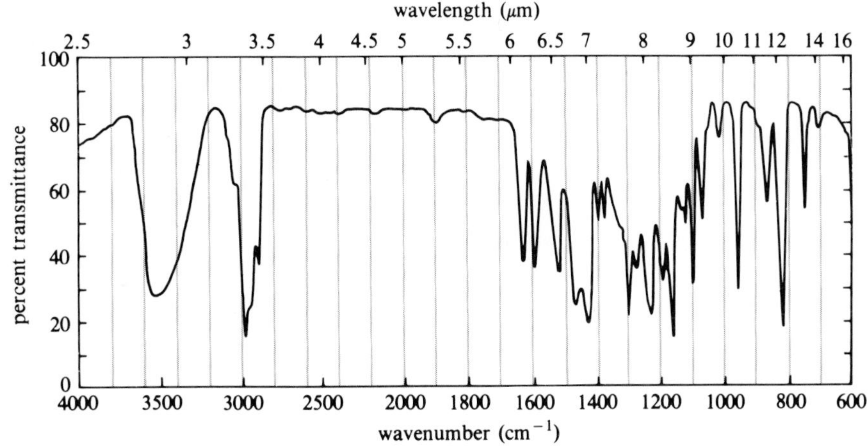

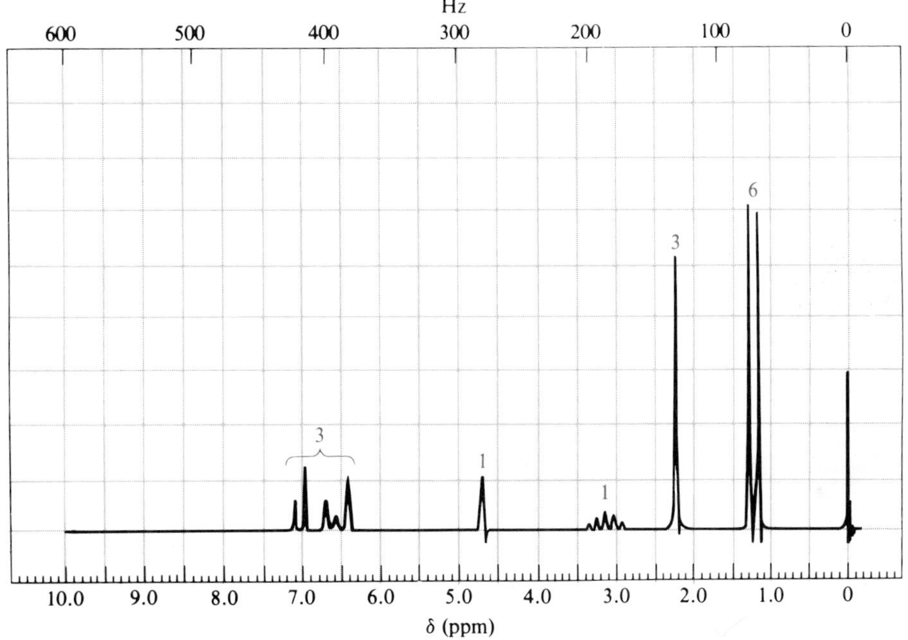

***16-44.** An unknown compound gives the following mass, IR, and NMR spectra. Propose a structure for the compound, and show how it is consistent with the observed absorptions. Show the fragmentations that give the prominent peaks at m/z 127 and 155 in the mass spectrum.

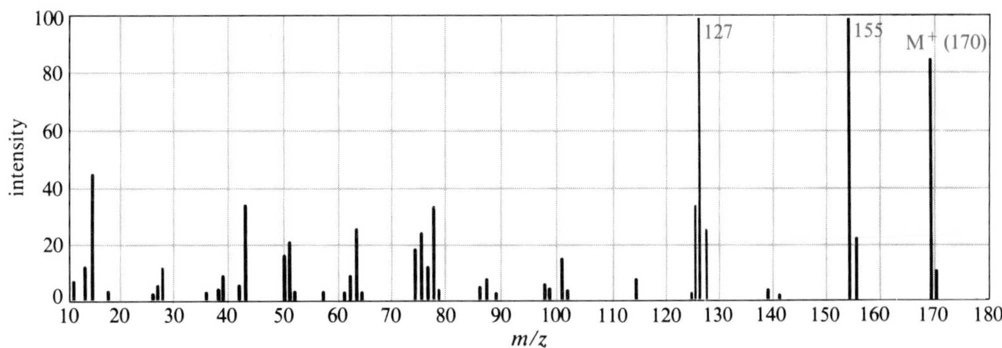

Problem 16-44 *continued*

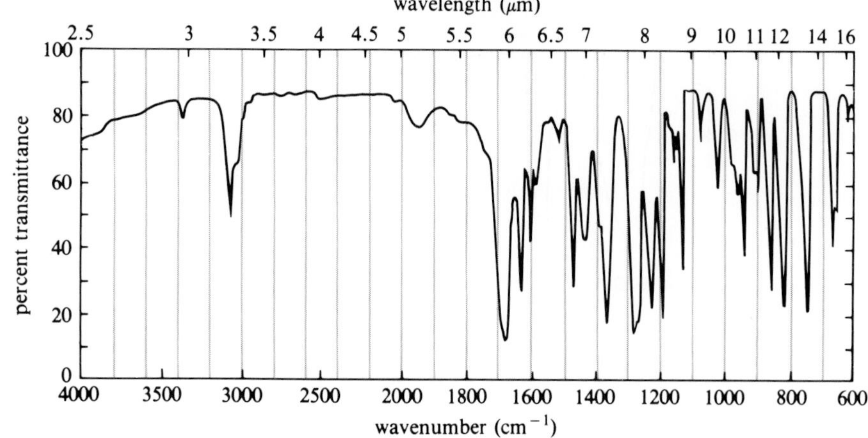

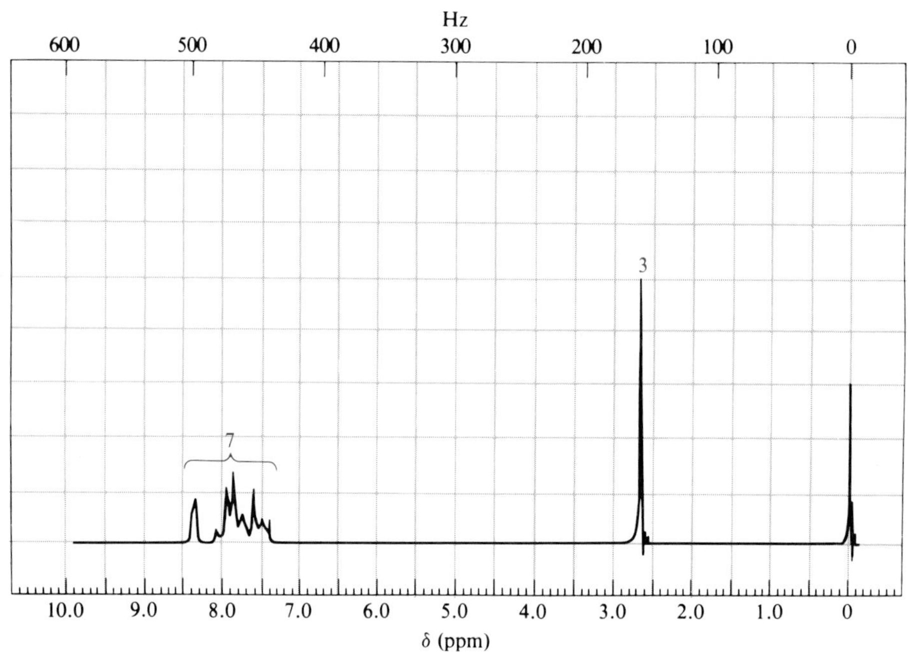

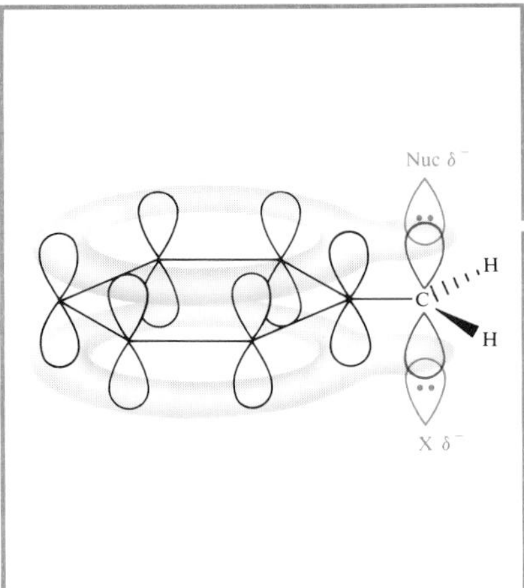

17

REACTIONS OF AROMATIC COMPOUNDS

With an understanding of the properties that make a compound aromatic, we can consider the reactions of aromatic compounds. A large part of this chapter is devoted to **electrophilic aromatic substitution,** the most important mechanism involved in the reactions of aromatic compounds. Many different reactions of benzene and its derivatives are explained by minor variations of the electrophilic aromatic substitution mechanism. We will study several of these reactions, and then we consider how substituents on the ring influence its reactivity toward electrophilic aromatic substitution and the positional orientations found in the products.

17-1 ELECTROPHILIC AROMATIC SUBSTITUTION

Alkenes often react as electron-rich molecules, attacking electrophiles to give addition products (Chapter 8). The pi electrons, less tightly bound than the sigma electrons, attack a strong electrophile to give a strongly electrophilic carbocation, which is then attacked by a nucleophile to give a product resulting from *addition* across the double bond.

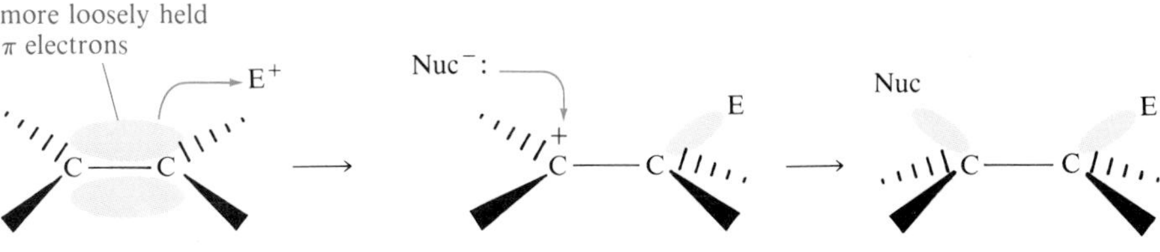

attack on the electrophile forms the carbocation

attack by a nucleophile gives the addition product

Like an alkene, benzene has clouds of pi electrons above and below its sigma-bond framework. Although benzene's pi electrons are involved in a stable aromatic system, they are still available to attack a strong electrophile to give a carbocation (Fig. 17-1). This resonance-stabilized carbocation is called a **sigma complex,** because the electrophile is joined to the benzene ring by a new sigma bond.

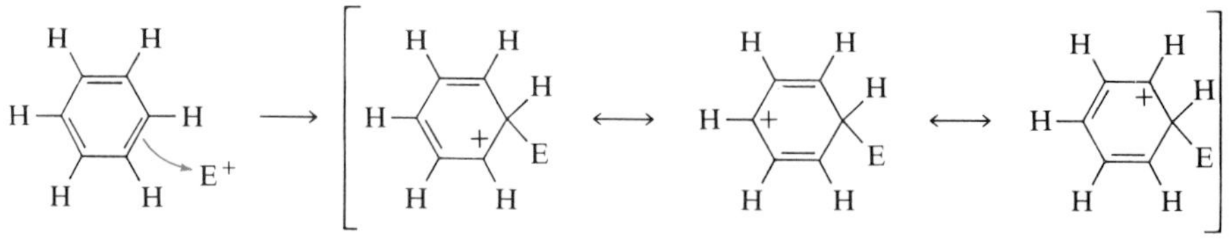

attack on the electrophile forms the sigma complex

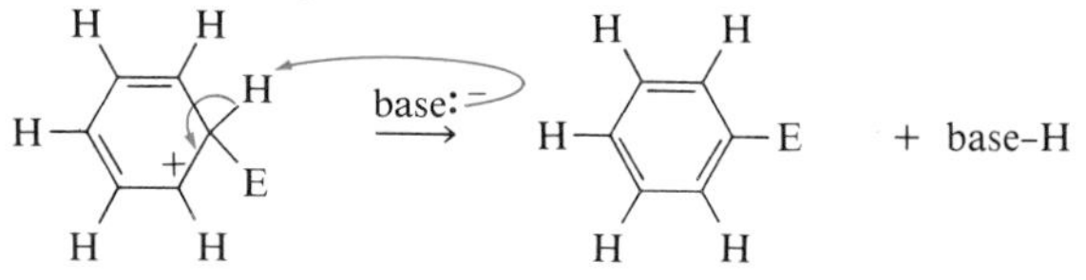

base:

+ base–H

abstraction of a proton from the sigma complex gives the substitution product

FIGURE 17-1 The pi electrons in benzene attack a strong electrophile to give a resonance-stabilized carbocation called a sigma complex. Loss of a proton forms the substitution product and regenerates the aromatic system.

The sigma complex is not aromatic, because the sp^3 hybrid carbon atom interrupts the ring of p orbitals. This loss of aromaticity contributes to the highly endothermic nature of the electrophilic attack. The sigma complex regains aromaticity either by a reversal of the first step (returning to the reactants) or by loss of the proton on the tetrahedral carbon atom, leading to the substitution product.

Figure 17-1 shows that the sigma complex loses a proton to a *base* (rather than undergoing attack by a *nucleophile*) to regenerate the aromatic ring. If the sigma complex were attacked by a nucleophile, the resulting addition product would not regain the aromatic stability of the starting material. The overall reaction, then, is the *substitution* of an electrophile (E^+) for a proton (H^+) on the aromatic ring: **electrophilic aromatic substitution.** This class of reactions includes substitutions by a wide variety of electrophilic reagents. Because it enables us to introduce functional groups directly onto the aromatic ring, electrophilic aromatic substitution is the most important method for the synthesis of substituted aromatic compounds.

17-2 HALOGENATION OF BENZENE

Bromination of benzene Alkenes react rapidly with bromine at room temperature to give addition products. For example, cyclohexene reacts to give *trans*-1,2-dibromocyclohexane. This reaction is exothermic by about 29 kcal (121 kJ) per mole.

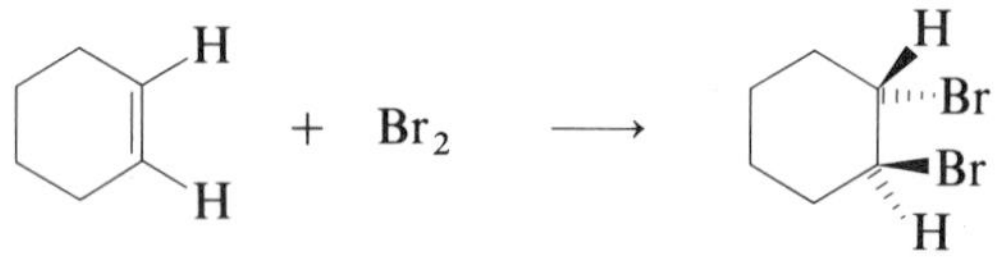

$$+ \ Br_2 \longrightarrow \qquad \Delta H^\circ = -29 \text{ kcal} \ (-121 \text{ kJ})$$

The analogous addition of bromine to benzene is *endothermic* because it requires the loss of aromatic stability. The addition is not seen under normal circumstances. The *substitution* of bromine for a hydrogen atom gives an aromatic product. The substitution is exothermic, and it occurs readily if there is a Lewis acid catalyst such as ferric bromide ($FeBr_3$) present.

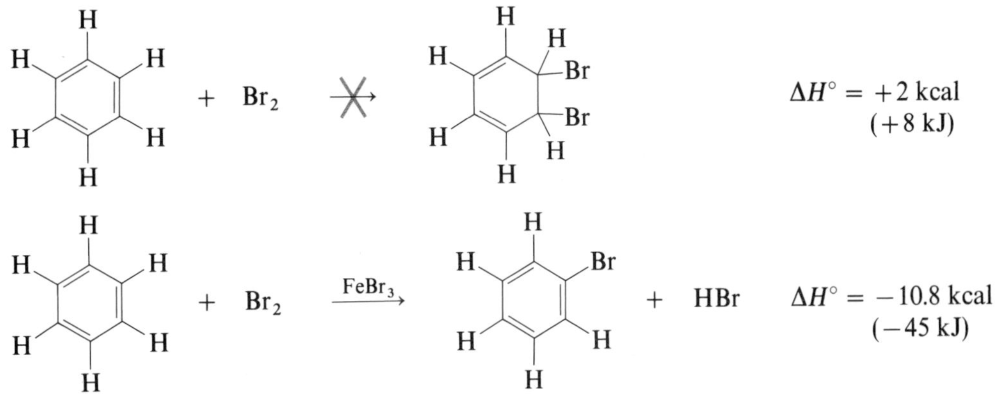

$$C_6H_6 + Br_2 \not\rightarrow C_6H_6Br_2 \qquad \Delta H^\circ = +2\text{ kcal } (+8\text{ kJ})$$

$$C_6H_6 + Br_2 \xrightarrow{FeBr_3} C_6H_5Br + HBr \qquad \Delta H^\circ = -10.8\text{ kcal } (-45\text{ kJ})$$

bromobenzene (80%)

Bromination follows the general mechanism for electrophilic aromatic substitution. Bromine itself is not sufficiently electrophilic to be attacked by benzene. Bromine donates a pair of electrons to the Lewis acid $FeBr_3$, forming a reactive intermediate with a weakened Br—Br bond and a partial positive charge on one of the bromine atoms. Attack by benzene forms the sigma complex.

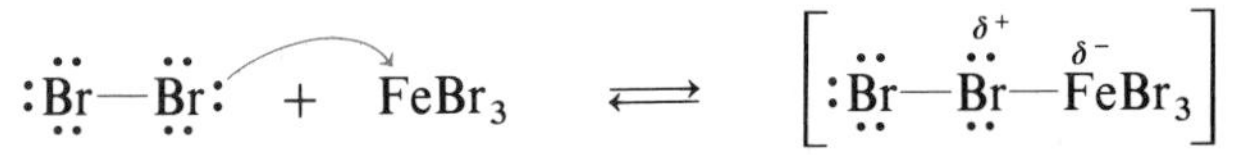

$$:\ddot{Br}-\ddot{Br}: + FeBr_3 \rightleftharpoons \left[:\ddot{Br}-\overset{\delta+}{\ddot{Br}}-\overset{\delta-}{FeBr_3}\right]$$

$Br_2 \cdot FeBr_3$ intermediate (a stronger electrophile)

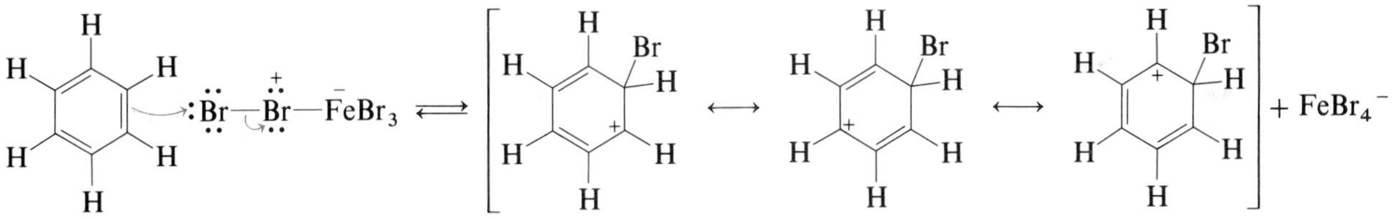

$$C_6H_6 + :\ddot{Br}-\overset{+}{\ddot{Br}}-\overset{-}{FeBr_3} \rightleftharpoons \left[C_6H_6Br^+ \longleftrightarrow C_6H_6Br^+ \longleftrightarrow C_6H_6Br^+\right] + FeBr_4^-$$

sigma complex

Bromide ion acts as a base to remove the proton, giving the aromatic product and a molecule of HBr.

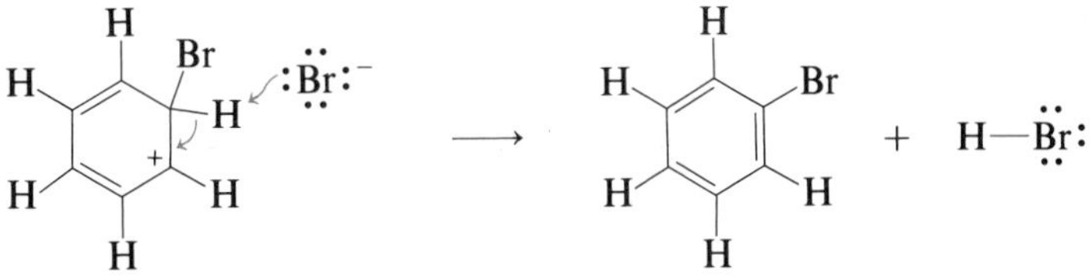

$$C_6H_6Br^+ + :\ddot{Br}:^- \longrightarrow C_6H_5Br + H-\ddot{Br}:$$

bromobenzene

The formation of the sigma complex is rate determining, and the transition state leading to it occupies the highest-energy point on the energy diagram (Fig. 17-2). This step is strongly endothermic because it forms a nonaromatic carbocation. The second step is exothermic, with aromaticity regained and a molecule of HBr evolved. The overall reaction is exothermic by 10.8 kcal (45 kJ) per mole.

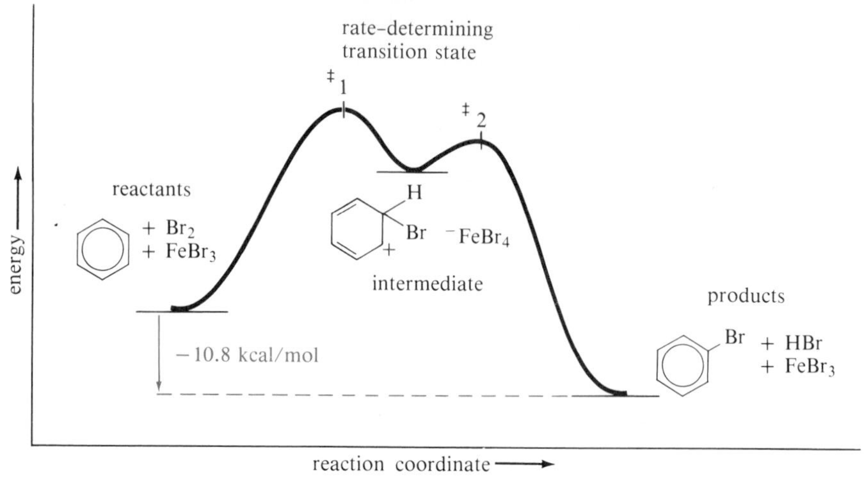

rate-determining transition state

$\ddagger_1$ $\ddagger_2$

reactants

+ Br_2
+ $FeBr_3$

intermediate

$^-FeBr_4$

products

+ HBr
+ $FeBr_3$

−10.8 kcal/mol

energy → reaction coordinate →

FIGURE 17-2 The energy diagram for the bromination of benzene shows that the first step is endothermic and rate determining, and the second step is strongly exothermic.

Chlorination of benzene Chlorination of benzene works much like bromination, except that aluminum chloride ($AlCl_3$) is most often used as the Lewis acid catalyst.

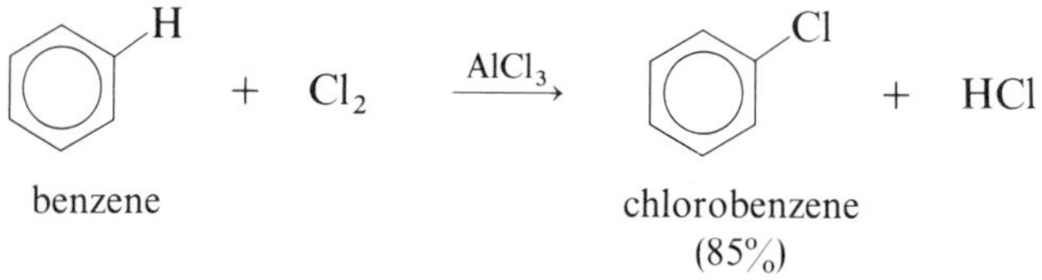

$$\text{benzene} + Cl_2 \xrightarrow{AlCl_3} \text{chlorobenzene (85\%)} + HCl$$

PROBLEM 17-1
Give a detailed mechanism for the aluminum chloride-catalyzed reaction of benzene with chlorine.

Iodination of benzene Iodination of benzene requires the presence of an acidic oxidizing agent, such as nitric acid. Notice that nitric acid is consumed in the reaction, so it is a reagent (an oxidant) rather than a catalyst.

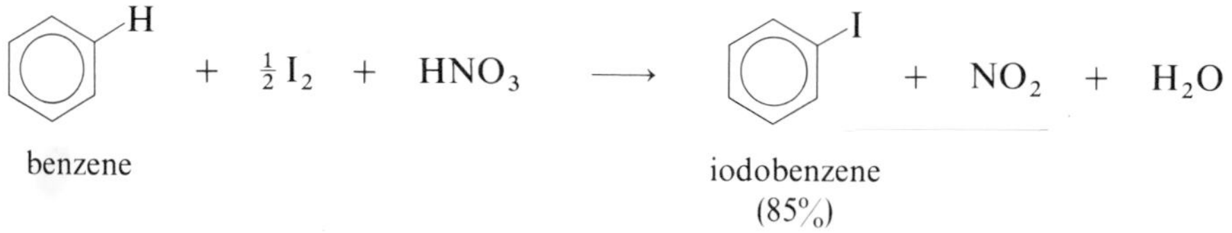

$$\text{benzene} + \tfrac{1}{2}I_2 + HNO_3 \longrightarrow \text{iodobenzene (85\%)} + NO_2 + H_2O$$

This iodination probably involves an electrophilic aromatic substitution with iodonium ion (I^+) acting as the electrophile. The iodonium ion results from the oxidation of iodine by nitric acid.

$$H^+ + HNO_3 + \tfrac{1}{2}I_2 \longrightarrow \underset{\text{iodonium ion}}{I^+} + NO_2 + H_2O$$

PROBLEM 17-2
Propose a mechanism for the reaction of benzene with the iodonium ion.

PROBLEM 17-3
Fluorination of benzene is accomplished either by reaction with an aryldiazonium salt

(Section 19-18) or by a two-step thallation procedure. Benzene reacts with $Tl(OCOCF_3)_3$, thallium tris(trifluoroacetate), to give an organothallium intermediate. Further reaction with potassium fluoride and boron trifluoride gives the aryl fluoride. Propose a mechanism for the first step, the thallation of benzene.

$$C_6H_6 + Tl(OCOCF_3)_3 \longrightarrow C_6H_5{-}Tl(OCOCF_3)_2 \xrightarrow{KF,\ BF_3} C_6H_5{-}F$$

benzene; thallium tris(trifluoroacetate); organothallium intermediate; fluorobenzene

Hint: The ionization of mercuric acetate gives the electrophile that oxymercurates an alkene (Section 8-5B); a similar ionization of thallium tris(trifluoroacetate) gives an electrophile that substitutes onto an aromatic ring.

17-3 NITRATION OF BENZENE

Benzene reacts with hot, concentrated nitric acid to give nitrobenzene. This sluggish reaction is not convenient, because a hot mixture of concentrated nitric acid with any oxidizable material may explode without warning. A safer and more convenient procedure employs a mixture of nitric acid and sulfuric acid. Sulfuric acid acts as a catalyst, allowing **nitration** to take place more rapidly and at lower temperatures.

$$C_6H_6 + HNO_3 \xrightarrow{H_2SO_4} C_6H_5NO_2 + H_2O$$

nitrobenzene (85%)

Sulfuric acid reacts with nitric acid to form the **nitronium ion** ($^+NO_2$), a powerful electrophile. The mechanism for nitronium ion formation is similar to many other sulfuric acid-catalyzed dehydration mechanisms. Sulfuric acid protonates the hydroxyl group of nitric acid, allowing it to leave as a molecule of water.

$$H{-}O{-}N^+(=O){-}O^- + H{-}O{-}SO_2{-}O{-}H \rightleftharpoons H{-}O^+(H){-}N^+(=O){-}O^- + HSO_4^- \rightleftharpoons O{=}N^+{=}O + H_2O$$

nitronium ion

The nitronium ion reacts with benzene to form a sigma complex. Loss of a proton from the sigma complex gives nitrobenzene.

$$C_6H_6 + {}^+NO_2 \longrightarrow [\text{sigma complex}] \xrightarrow{HSO_4^-} C_6H_5NO_2 + H_2SO_4$$

benzene; nitronium ion; sigma complex; nitrobenzene

PROBLEM 17-4

p-Xylene undergoes nitration much faster than benzene. Use resonance structures of the sigma complex to explain this accelerated rate.

17-4 SULFONATION OF BENZENE

We have already used esters of *p*-toluenesulfonic acid as activated derivatives of alcohols with a good leaving group, the tosylate group. *p*-Toluenesulfonic acid is an example of the *arylsulfonic acids* (general formula Ar—SO_3H), often used as strong acid catalysts that are soluble in nonpolar organic solvents. Arylsulfonic acids are easily synthesized by **sulfonation** of benzene derivatives, an electrophilic aromatic substitution using sulfur trioxide (SO_3) as the electrophile.

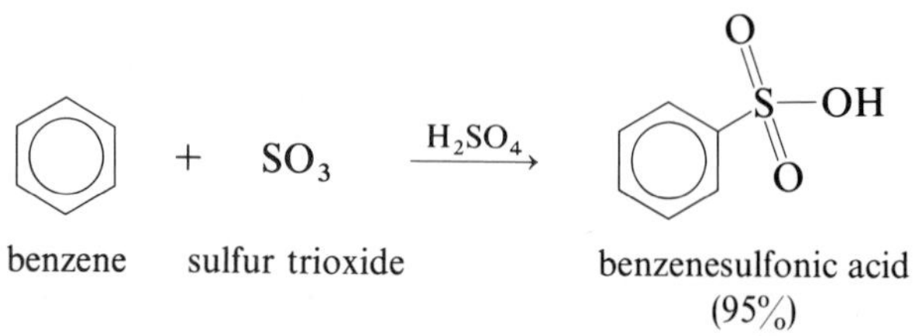

benzene + SO_3 (sulfur trioxide) $\xrightarrow{H_2SO_4}$ benzenesulfonic acid (95%)

"Fuming sulfuric acid" is the common name for a solution of 7 percent SO_3 in H_2SO_4. Sulfur trioxide is the *anhydride* of sulfuric acid, meaning that the addition of water to SO_3 gives H_2SO_4. Sulfur trioxide is a strong electrophile, with three sulfonyl (S=O) bonds drawing electron density away from the sulfur atom. Benzene attacks sulfur trioxide, forming a sigma complex. Loss of a proton on the tetrahedral carbon and reprotonation on oxygen gives benzenesulfonic acid.

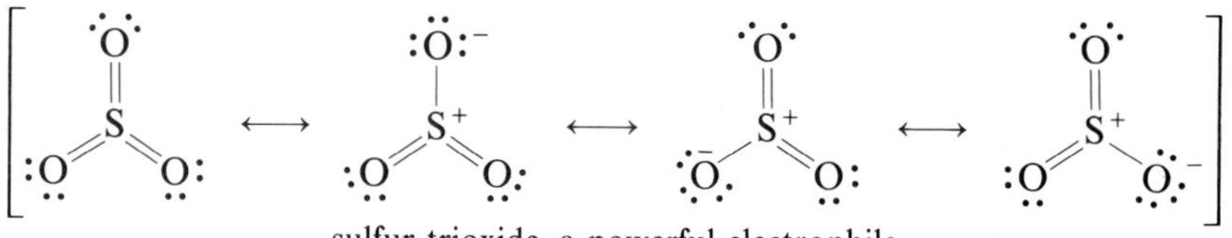

sulfur trioxide, a powerful electrophile

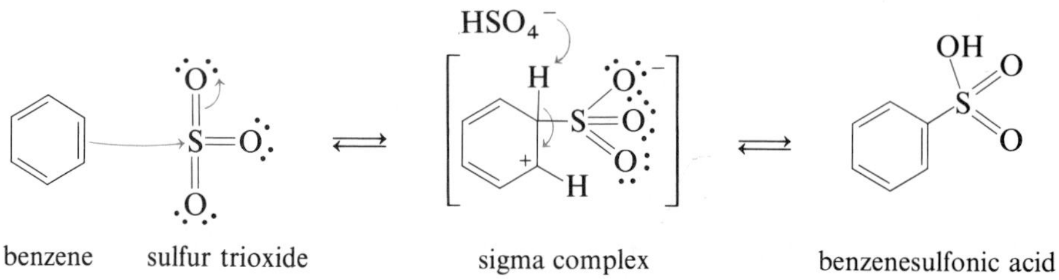

benzene   sulfur trioxide $\rightleftharpoons$ sigma complex (HSO_4^-) $\rightleftharpoons$ benzenesulfonic acid

> PROBLEM 17-5
> Use resonance structures to show that the dipolar sigma complex shown above has its positive charge delocalized over three carbon atoms and its negative charge delocalized over three oxygen atoms.

Desulfonation The sulfonation reaction is reversible, and a sulfonic acid group may be removed from an aromatic ring by heating in dilute sulfuric acid. In practice, steam is often used as a water and heat source for **desulfonation.**

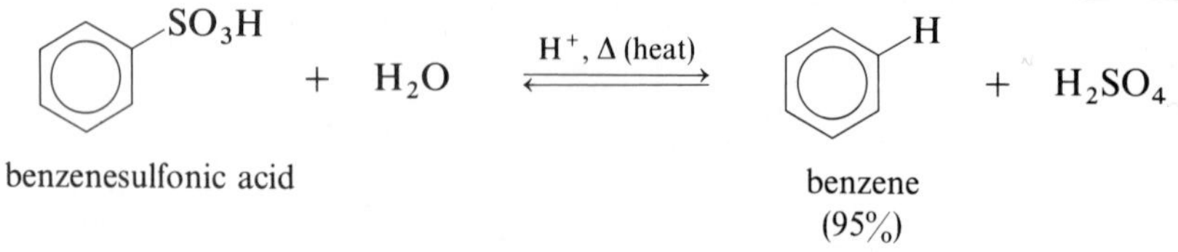

benzenesulfonic acid + H_2O $\underset{}{\overset{H^+, \Delta \text{ (heat)}}{\rightleftharpoons}}$ benzene (95%) + H_2SO_4

Desulfonation follows the same mechanistic path as sulfonation, except in the opposite order. A proton adds to a ring carbon to form a sigma complex, and then loss of sulfur trioxide gives the unsubstituted aromatic ring.

$$\text{C}_6\text{H}_5\text{SO}_3^- + \text{H}^+ \rightleftarrows [\text{sigma complex}] \rightleftarrows \text{C}_6\text{H}_6 + \text{SO}_3$$

$$(\text{SO}_3 + \text{H}_2\text{O} \rightleftarrows \text{H}_2\text{SO}_4)$$

Protonation of the aromatic ring The desulfonation reaction involves protonation of an aromatic ring to form a sigma complex. If a proton attacks benzene itself, the sigma complex can lose either of the two protons at the tetrahedral carbon. It is difficult to characterize the sigma complex, because benzene is a weak base and only a tiny fraction of the benzene molecules are protonated at any time.

We can prove that a reaction has occurred by using a deuterium ion (D^+) rather than a proton and by showing that the product contains a deuterium atom in place of a hydrogen atom. This experiment is easily accomplished by adding SO_3 to some D_2O (heavy water) to generate D_2SO_4. Benzene reacts to give a deuterated product.

$$\text{C}_6\text{H}_6 + \text{D}_3\text{O}^+ \rightleftarrows [\text{sigma complex}] \rightleftarrows \text{C}_6\text{H}_5\text{D} + \text{HD}_2\text{O}^+$$

(D_3O^+)

The reaction is reversible, but at equilibrium the final products reflect the D/H ratio of the solution. A large excess of deuterium gives a product with all six of the benzene hydrogens replaced by deuterium. This reaction serves as a synthesis of benzene-d_6, formula C_6D_6.

$$\text{C}_6\text{H}_6 \xrightleftharpoons[]{\text{large excess } \text{D}_2\text{SO}_4/\text{D}_2\text{O}} \text{C}_6\text{D}_6$$

benzene   benzene-d_6

17-5 NITRATION OF TOLUENE: THE EFFECT OF ALKYL SUBSTITUTION

Up to now, we have considered only benzene as the substrate for electrophilic aromatic substitution. To synthesize more complicated aromatic compounds, we need to consider what effects the presence of other substituents might have on substitution reactions. For example, toluene (methylbenzene) reacts with a mixture of nitric and sulfuric acids much like benzene does, but with some interesting differences:

1. Toluene reacts about 25 times as fast as benzene does under the same conditions. We say that toluene is **activated** toward electrophilic aromatic substitution and that the methyl group is an **activating group.**
2. The nitration of toluene gives a mixture of products, primarily those resulting

from substitution at the ortho and para positions. We say that the methyl group of toluene is an **ortho, para-director.**

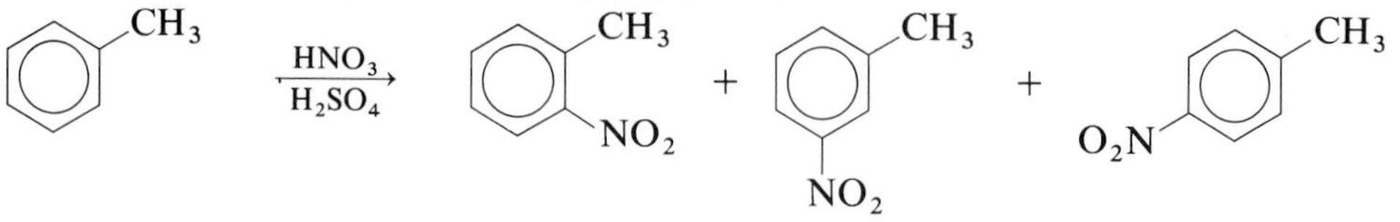

toluene $\xrightarrow[H_2SO_4]{HNO_3}$ *o*-nitrotoluene (40%) + *m*-nitrotoluene (3%) + *p*-nitrotoluene (57%)

The product ratios show that the orientation of substitution is not random. If each C—H position were equally reactive, there would be equal amounts of ortho and meta substitution and half as much para substitution: 40% ortho, 40% meta, and 20% para. This is the statistical prediction based on the two ortho positions, two meta positions, and just one para position available for substitution.

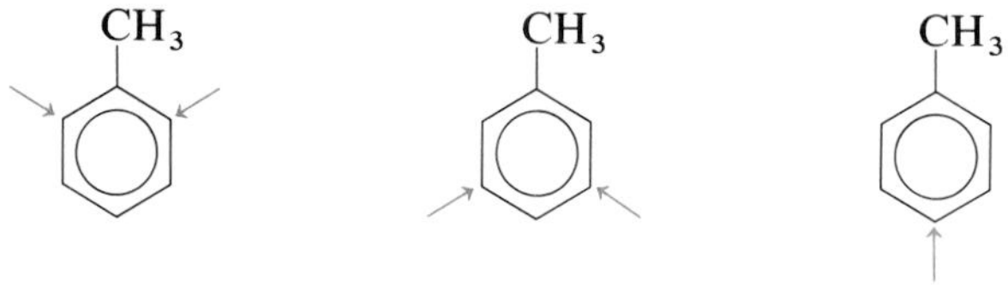

two ortho positions   two meta positions   one para position

The rate-determining step (the highest-energy transition state) for electrophilic aromatic substitution is the first step, formation of the sigma complex. This is also the step where the electrophile bonds to the ring, determining the substitution pattern. We can explain both the enhanced reaction rate and the preference for substitution at the ortho and para positions by considering the structures of the intermediate sigma complexes. In this endothermic reaction, the structure of the transition state leading to the sigma complex resembles the structure of the product, the sigma complex (Hammond postulate, Section 4-15). We are justified in using the relative stabilities of the sigma complexes as indicators of the energies of the transition states leading to their formation.

When benzene reacts with the nitronium ion, the resulting sigma complex has the positive charge distributed over three secondary (2°) carbon atoms.

Benzene

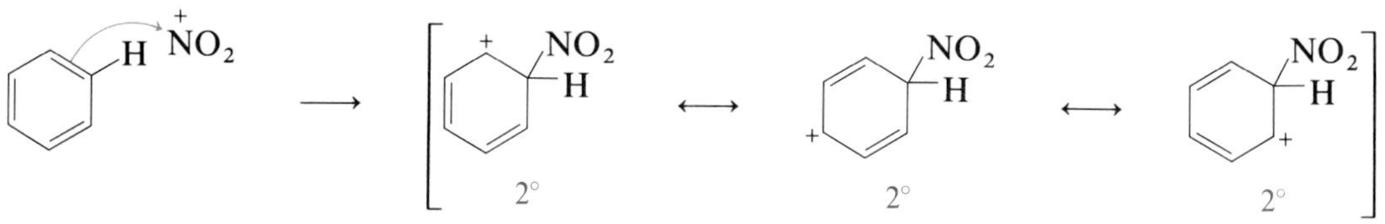

H $\overset{+}{N}O_2$ → [NO₂/H, 2° ⟷ NO₂/H, 2° ⟷ NO₂/H, 2°]

In the case of ortho or para substitution of toluene, however, the positive charge is spread over two secondary carbons and one tertiary (3°) carbon.

Ortho attack

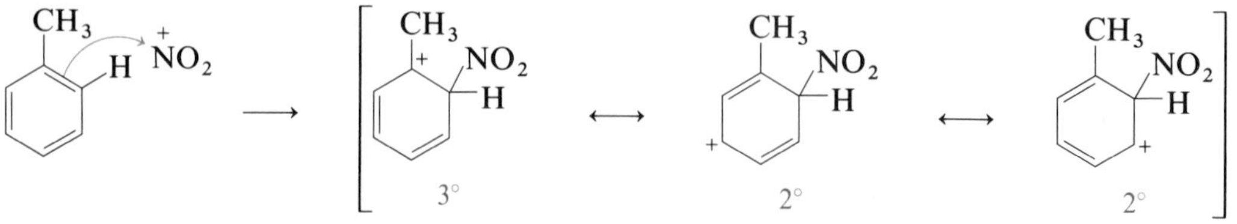

CH_3, H $\overset{+}{N}O_2$ → [CH_3, NO_2, H, 3° ⟷ CH_3, NO_2, H, 2° ⟷ CH_3, NO_2, H, 2°]

Para attack

Because the sigma complexes for ortho and para nitration of toluene have resonance structures with tertiary carbocations, they are more stable than the sigma complex for nitration of benzene. Therefore, toluene reacts faster than benzene at the ortho and para positions.

The sigma complex for meta substitution has its positive charge spread over three 2° carbons; this intermediate is similar in energy to the intermediate for substitution of benzene. Therefore, meta substitution of toluene does not show the large rate enhancement seen with ortho and para substitution.

Meta attack

The effect of the methyl group in toluene is to stabilize the intermediate sigma complex and the rate-determining transition state leading to its formation. The stabilizing effect of the methyl group is large when it is situated ortho or para to the site of substitution and the positive charge is delocalized onto the tertiary carbon atom. When substitution occurs at the meta position, the positive charge is not delocalized onto the tertiary carbon, and the methyl group has a much smaller effect on the stability of the sigma complex. Figure 17-3 compares the energy profiles for nitration of benzene and toluene at the ortho, meta, and para positions.

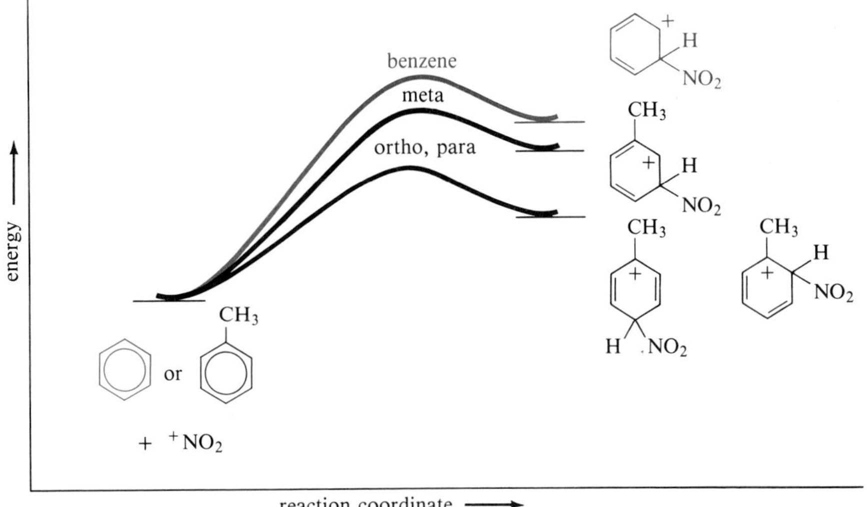

FIGURE 17-3 The methyl group of toluene stabilizes the sigma complexes and the transition states leading to them. This stabilization is most effective when the methyl group is ortho or para to the site of substitution.

17-6 ACTIVATING, ORTHO, PARA-DIRECTING SUBSTITUENTS

17-6A ALKYL GROUPS

The results observed with toluene are general for any alkylbenzene undergoing electrophilic aromatic substitution. Substitution ortho or para to the alkyl group gives an intermediate (and a transition state) with the positive charge shared by the tertiary carbon atom. As a result, alkylbenzenes undergo electrophilic aromatic substitution faster than benzene, and the products are predominantly ortho- and para-substituted. An alkyl group is therefore an activating substituent, and it is **ortho, para-directing.** This effect of an alkyl group is called **inductive stabilization** because it takes place through the sigma bond joining the alkyl group with the benzene ring.

Shown below is the reaction of ethylbenzene with bromine, catalyzed by ferric bromide. As with toluene, the rates of formation of the ortho- and para-substituted isomers are greatly enhanced with respect to the meta isomer.

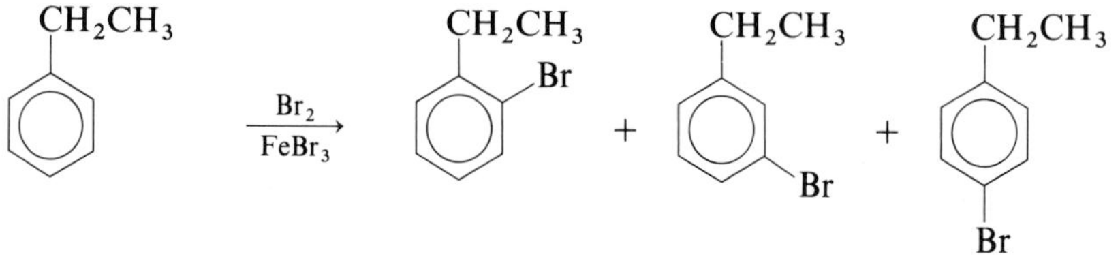

CH_2CH_3 (ethylbenzene) $\xrightarrow[FeBr_3]{Br_2}$ CH_2CH_3, Br (*o*-bromo (38%)) + CH_2CH_3, Br (*m*-bromo (<1%)) + CH_2CH_3, Br (*p*-bromo (62%))

ethylbenzene | *o*-bromo (38%) | *m*-bromo (<1%) | *p*-bromo (62%)

PROBLEM 17-6

(a) Draw a detailed mechanism for the $FeBr_3$-catalyzed reaction of ethylbenzene with bromine, and show why the sigma complex (and the transition state leading to it) is lower in energy for substitution at the ortho and para positions than it is for substitution at the meta position.

(b) Explain why *m*-xylene undergoes nitration 100 times faster than *p*-xylene.

PROBLEM 17-7

Styrene (vinylbenzene) has been found to undergo electrophilic aromatic substitution much faster than benzene, and the products are found to be primarily ortho- and para-substituted styrenes. Use resonance structures of the intermediates to explain these results.

17-6B SUBSTITUENTS WITH NONBONDING ELECTRONS

The methoxyl group Anisole (methoxybenzene) undergoes nitration about 10,000 times faster than benzene and about 400 times faster than toluene. This result seems curious because oxygen is a strongly electronegative group, yet it donates electron density to stabilize the transition state and the sigma complex. Recall that the nonbonding electrons of an oxygen atom adjacent to a carbocation stabilize the positive charge through resonance.

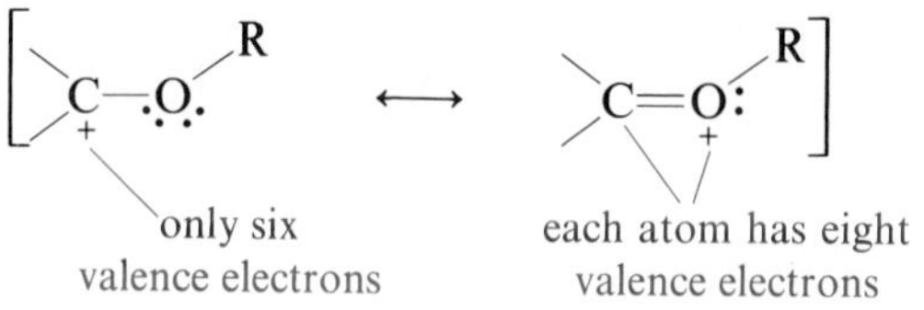

$$\left[\overset{+}{C}-\ddot{\underset{..}{O}}-R \longleftrightarrow C=\overset{+}{\ddot{O}}-R \right]$$

only six valence electrons | each atom has eight valence electrons

Although the second resonance structure puts the positive charge on the electronegative oxygen atom, this structure has more covalent bonds and it provides each atom with an octet in its valence shell. This type of stabilization is called **resonance stabilization,** and the oxygen atom is called **resonance-donating** or **pi-donating** because it donates electron density through a pi bond in one of the resonance structures. Just as alkyl groups activate the ortho and para positions more strongly than they activate the meta positions, the methoxyl group of anisole preferentially activates the ortho and para positions.

anisole $\xrightarrow{HNO_3}$ *o*-nitroanisole (45%) + *m*-nitroanisole (<0.01%) + *p*-nitroanisole (55%)

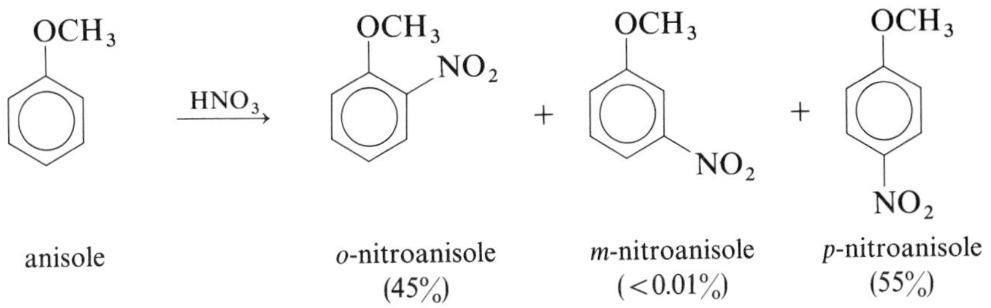

Resonance structures show that the methoxyl group stabilizes the sigma complex if it is ortho or para to the site of substitution, but not if it is meta. The resonance stabilization is provided by a pi bond between the —OCH_3 substituent and the ring.

Ortho attack

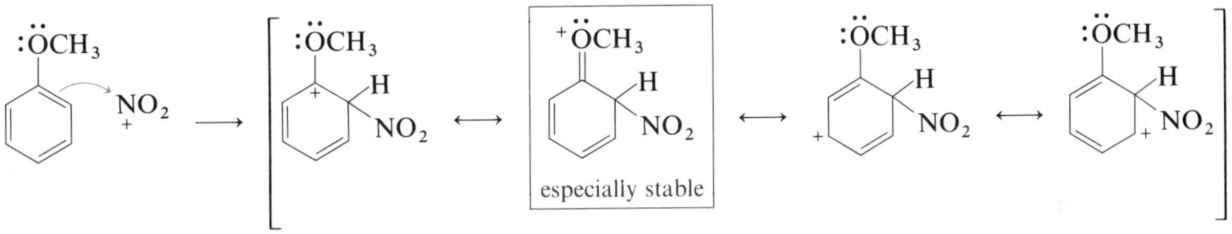

especially stable

Meta attack

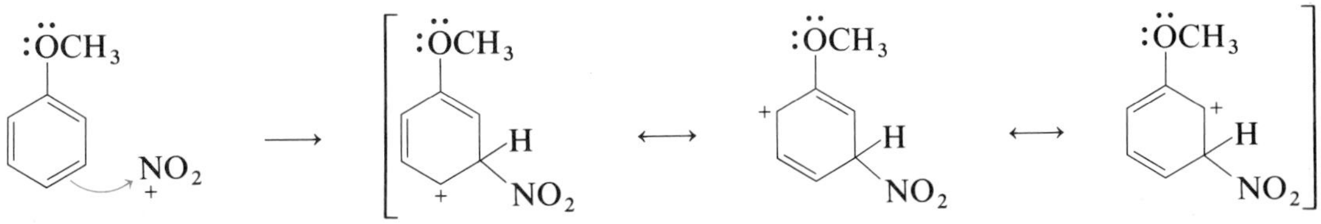

Para attack

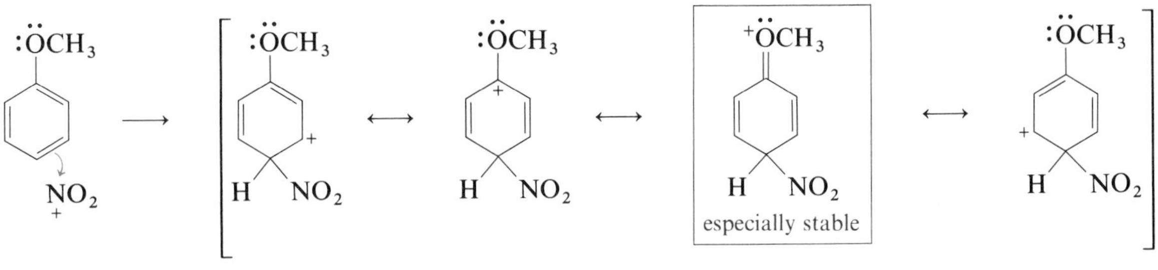

especially stable

A methoxyl group is so strongly activating that anisole quickly brominates in water without an additional catalyst. In the presence of excess bromine, this reaction proceeds quickly to the tribromide.

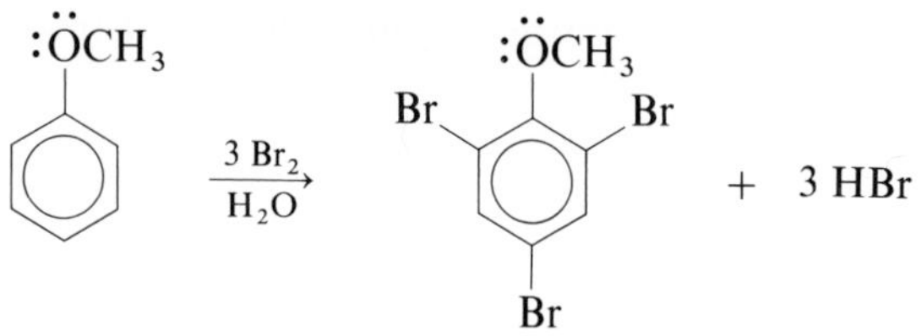

$$\text{anisole} \xrightarrow[H_2O]{3\ Br_2} \text{2,4,6-tribromoanisole (100\%)} + 3\ HBr$$

PROBLEM 17-8

Give a detailed mechanism for the bromination of ethoxybenzene to give *o*- and *p*-bromoethoxybenzene.

The amino group A similar reaction occurs when there is a nitrogen atom with a nonbonding pair of electrons adjacent to the aromatic ring. For example, aniline undergoes a fast bromination (without an additional catalyst) in bromine water to give the tribromide. Sodium bicarbonate is added to this reaction to neutralize the HBr formed and prevent protonation of the basic amino ($-NH_2$) group (see Problem 17-12).

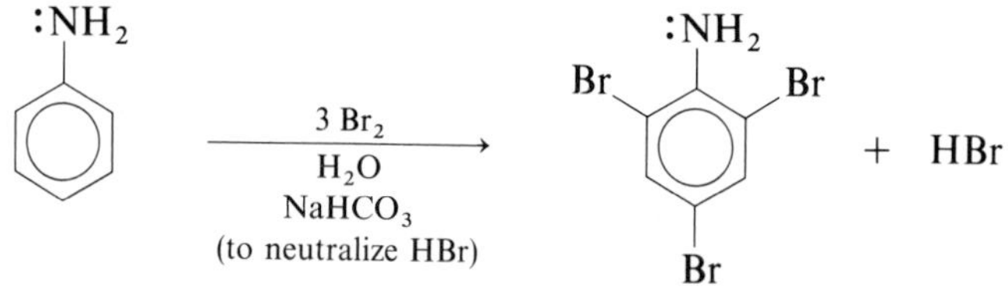

$$\text{aniline} \xrightarrow[\substack{H_2O \\ NaHCO_3 \\ \text{(to neutralize HBr)}}]{3\ Br_2} \text{2,4,6-tribromoaniline (100\%)} + HBr$$

Nitrogen's nonbonding electrons provide resonance stabilization to the sigma complex if attack takes place ortho or para to the position of the nitrogen atom.

Ortho attack

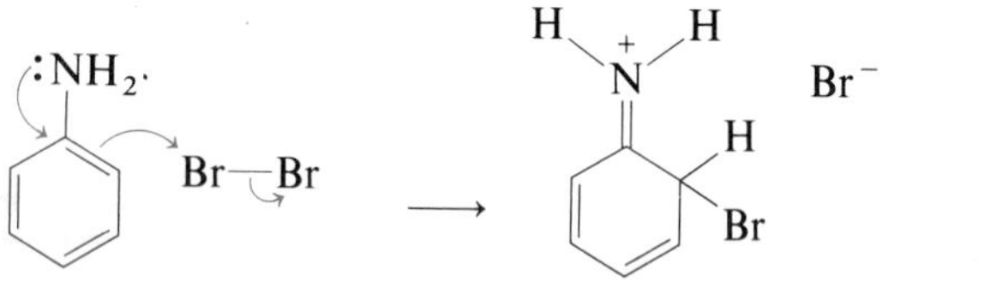

$:NH_2$ Br—Br → $H_2\overset{+}{N}$ H Br Br^-

(plus other resonance structures)

Para attack

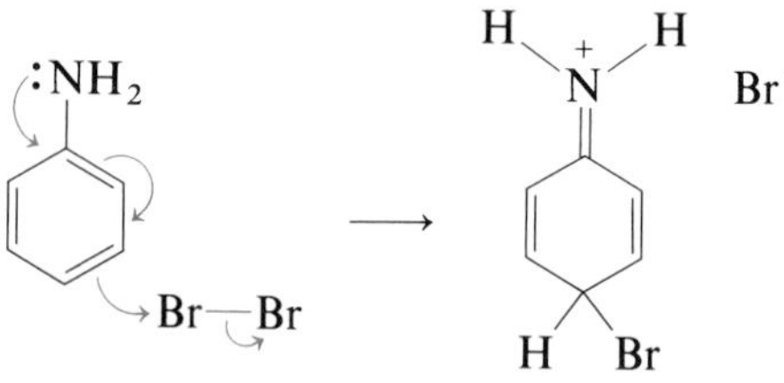

$:NH_2$ Br—Br → $H_2\overset{+}{N}$ H Br Br^-

(plus other resonance structures)

PROBLEM 17-9

Draw all the resonance structures for the sigma complexes corresponding to bromination of aniline at the ortho, meta, and para positions.

Many substituents with lone pairs of electrons can provide resonance stabilization to a sigma complex. Several examples of these groups are illustrated below in decreasing order of their activation of an aromatic ring. All these substituents are strongly activating, and they are all ortho, para-directing.

SUMMARY OF ACTIVATING, ORTHO, PARA-DIRECTORS

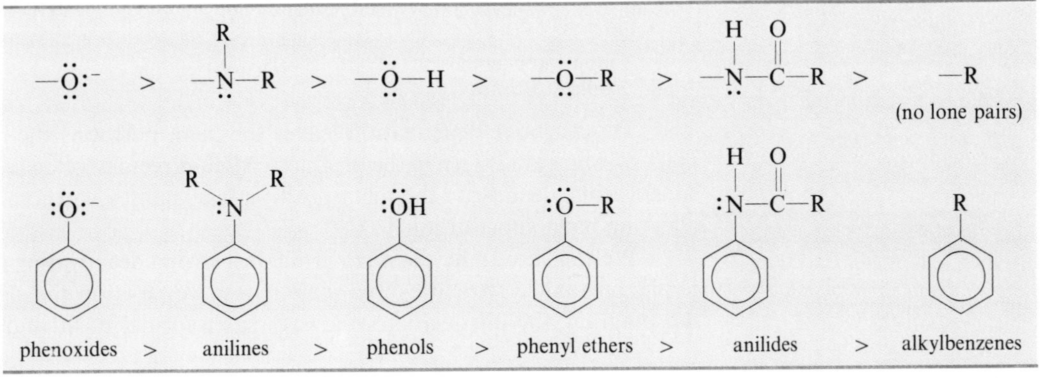

$-\ddot{\underset{..}{O}}:^-$ > $-\underset{..}{N}(R)-R$ > $-\ddot{\underset{..}{O}}-H$ > $-\ddot{\underset{..}{O}}-R$ > $-\underset{..}{N}(H)-C(=O)-R$ > $-R$ (no lone pairs)

phenoxides > anilines > phenols > phenyl ethers > anilides > alkylbenzenes

PROBLEM 17-10
Predict the structure of the major product expected from the mononitration of *p*-methylanisole.

PROBLEM 17-11
When bromine is added to two beakers, one containing phenyl isopropyl ether and the other containing cyclohexene, the bromine color in both beakers disappears. What observation could you make while performing this test that would allow you to distinguish the alkene from the aryl ether?

17-7 DEACTIVATING, META-DIRECTING SUBSTITUENTS

Nitrobenzene is about 100,000 times *less* reactive than benzene toward electrophilic aromatic substitution. For example, the nitration of nitrobenzene requires concentrated nitric and sulfuric acids at temperatures above 100°C. The nitration proceeds slowly, giving the meta isomer as the major product.

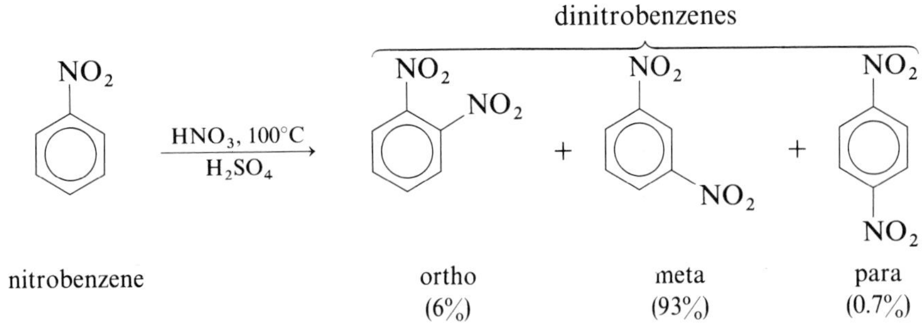

$$\text{nitrobenzene} \xrightarrow[H_2SO_4]{HNO_3,\ 100^\circ C} \underbrace{\text{ortho } (6\%) + \text{meta } (93\%) + \text{para } (0.7\%)}_{\text{dinitrobenzenes}}$$

These results should not be surprising. We have already seen that a substituent on a benzene ring has its greatest effect on reactions at the carbon atoms ortho and para to the substituent. An electron-donating substituent activates primarily the ortho and para positions, and an electron-withdrawing substituent (such as a nitro group) deactivates primarily the ortho and para positions.

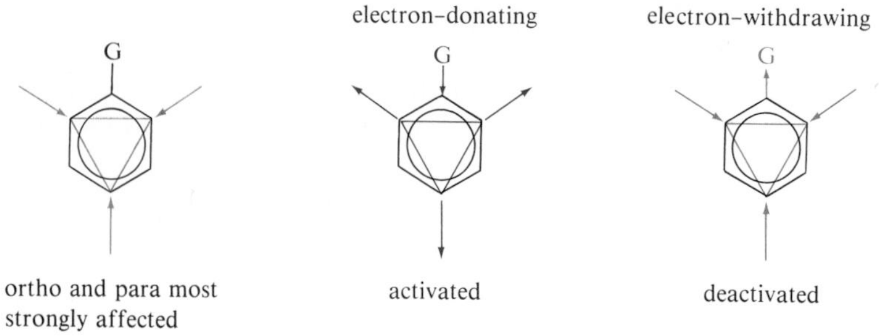

electron-donating

electron-withdrawing

G

G

G

ortho and para most strongly affected

activated

deactivated

This selective deactivation leaves the meta positions the most reactive, and meta substitution is seen in the products. **Meta-directors,** often called **meta-allowing** substituents, deactivate the meta position less than the ortho and para positions, allowing meta substitution.

We can show why the nitro group is a strong **deactivating group** by considering its resonance structures. No matter how we position the electrons in a Lewis dot diagram, the nitrogen atom always has a formal positive charge.

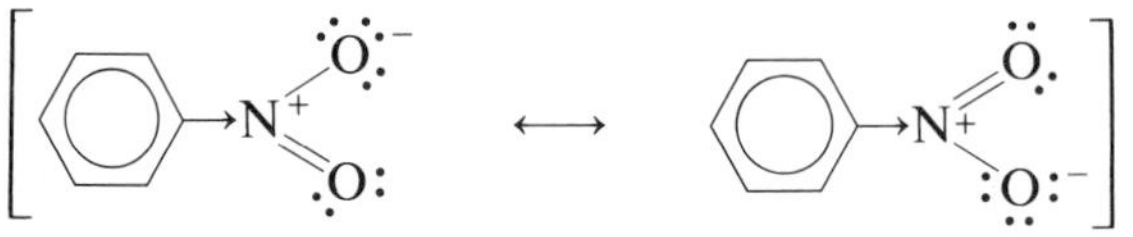

The positively charged nitrogen inductively withdraws electron density from the group to which it is bonded. In the case of nitrobenzene, the nitro group withdraws electron density from the aromatic ring. This aromatic ring is less electron-rich than benzene, so it is deactivated toward reactions with electrophiles.

The reactions below show why this deactivating effect is strongest at the ortho and para positions, allowing meta-substituted products to predominate. Each sigma complex has its positive charge spread over three carbon atoms. In the cases corresponding to ortho and para substitution, one of the carbon atoms bearing this positive charge is the carbon that also bears the positively charged nitrogen atom of the nitro group. Since like charges repel, this close proximity of two positive charges is especially unstable.

Ortho attack

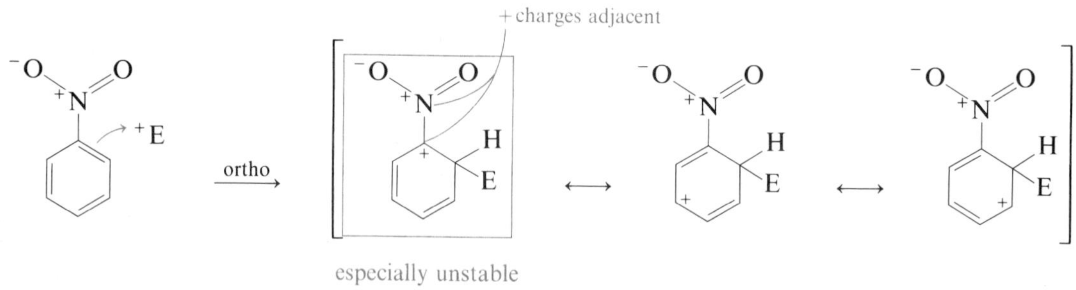

+ charges adjacent

ortho

especially unstable

Meta attack

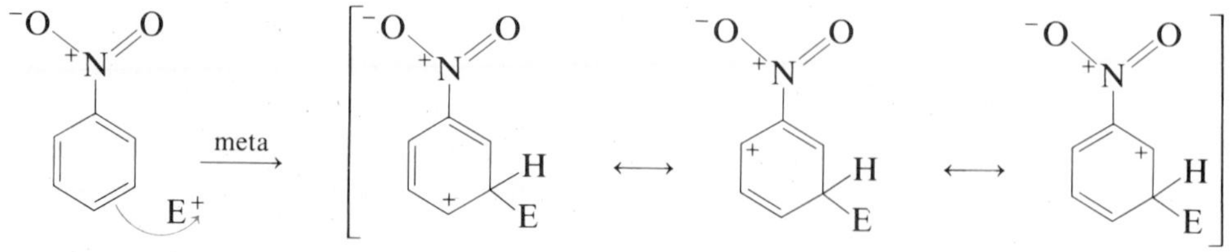

meta

Para attack

+ charges adjacent

para →

especially unstable

In the sigma complex corresponding to meta substitution, the carbon bonded to the nitro group does not share the positive charge of the ring. This is a more stable situation, because the positive charges are farther apart. As a result, nitrobenzene reacts primarily at the meta position. We can summarize by saying that the nitro group is a deactivating group, and that it is a meta-allower (or meta-director).

The energy profile in Figure 17-4 compares the energies of the transition states and intermediates leading to ortho, meta, and para substitution of nitrobenzene with those for benzene. Notice that a higher activation energy is involved for substitution of nitrobenzene at any position, resulting in slower reaction rates than for benzene.

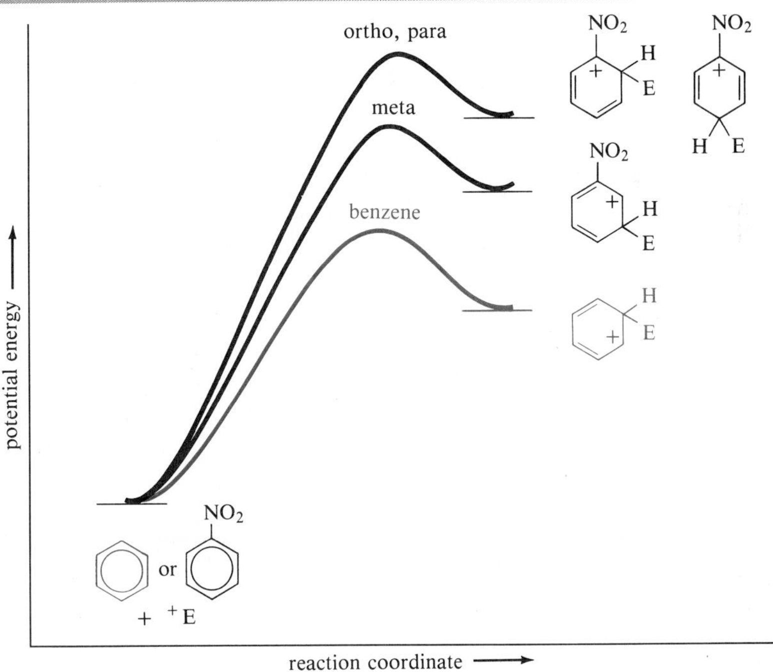

FIGURE 17-4 Nitrobenzene is deactivated toward electrophilic aromatic substitution at *any* position, but the deactivation is strongest at the ortho and para positions. Reaction occurs at the meta position, but it is slower than the reaction of benzene.

Just as the activating substituents are all ortho, para-directors, most deactivating substituents are meta-directors. In general, deactivating substituents are groups with a positive charge (or a partial positive charge) on the atom bonded to the aromatic ring. As we saw in the case of the nitro group, this positively charged atom repels any positive charge on the adjacent carbon atom of the ring.

Of the possible sigma complexes, only the one corresponding to meta substitution avoids putting a positive charge on this carbon. For example, the partial positive charge on a carbonyl carbon allows substitution primarily at the meta position.

Ortho attack

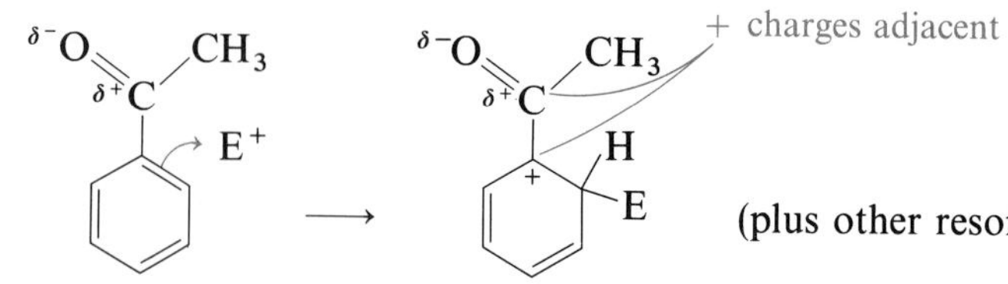

+ charges adjacent

(plus other resonance structures)

acetophenone

Meta attack

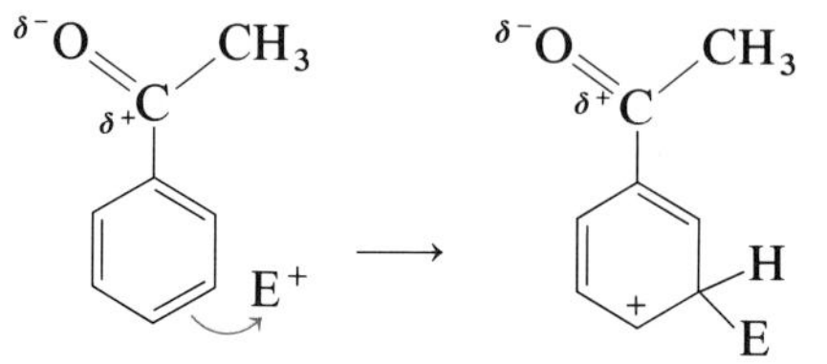

(plus other resonance structures) This sigma complex does not place the positive charge on the ring carbon bearing the carbonyl group.

The following summary table lists some common substituents that are deactivating and meta-directing. Resonance structures are also given, to show how a positive charge arises on the atom bonded to the aromatic ring.

SUMMARY OF DEACTIVATING, META-DIRECTORS

Group	Resonance Structures	Example
$-NO_2$ nitro	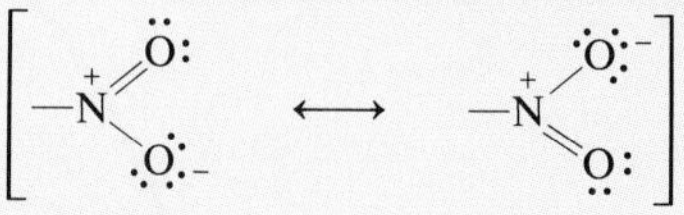 $-\overset{+}{N}(=O)-O^-$ ⟷ $-\overset{+}{N}(-O^-)=O$	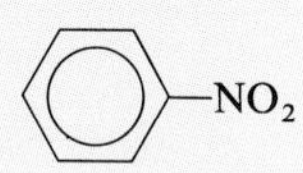 nitrobenzene
$-SO_3H$ sulfonic acid	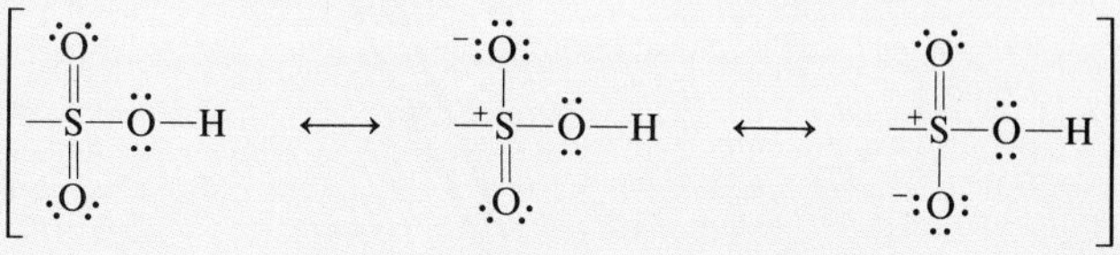 $-S(=O)_2-O-H$ ⟷ $-\overset{+}{S}(-O^-)(=O)-O-H$ ⟷ $-\overset{+}{S}(=O)(-O^-)-O-H$	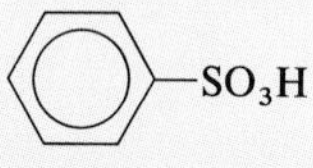 benzenesulfonic acid
$-C{\equiv}N:$ cyano	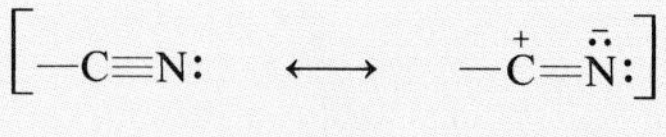 $-C{\equiv}N:$ ⟷ $-\overset{+}{C}{=}\overset{-}{N}:$	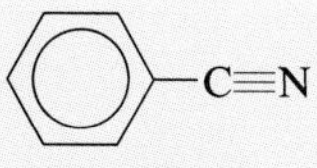 benzonitrile
$-C(=O)-R$ ketone or aldehyde	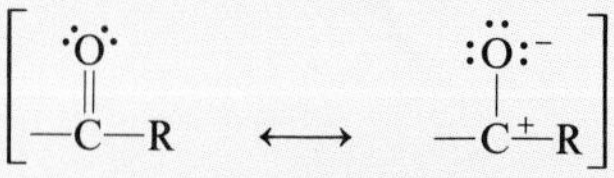 $-C(=O)-R$ ⟷ $-\overset{+}{C}(-O^-)-R$	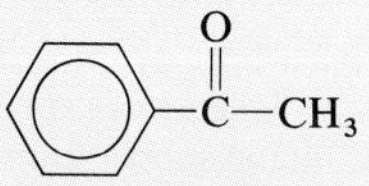 acetophenone
$-C(=O)-O-R$ ester	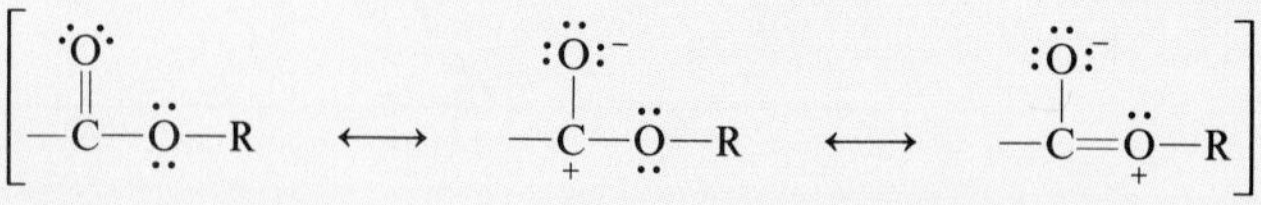 $-C(=O)-O-R$ ⟷ $-\overset{+}{C}(-O^-)-O-R$ ⟷ $-C(-O^-)=\overset{+}{O}-R$	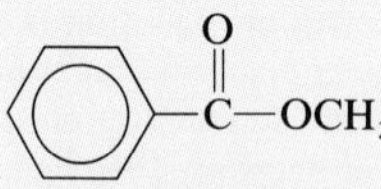 methyl benzoate
$-\overset{+}{N}R_3$ quaternary ammonium	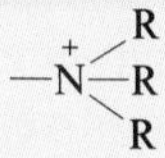 $-\overset{+}{N}(R)(R)R$	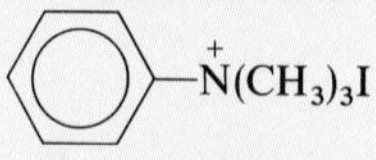 trimethylanilinium iodide

PROBLEM 17-12

In an aqueous solution containing sodium bicarbonate, aniline reacts quickly with bromine to give 2,4,6-tribromoaniline. Nitration of aniline requires very strong conditions, however, and the yields (mostly *m*-nitroaniline) are poor.

(a) What conditions are used for the nitration, and what form of aniline is present under these conditions?
(b) Explain why the nitration of aniline is so sluggish and why it gives mostly meta orientation.
* (c) Although the nitration of aniline is slow and gives mostly meta substitution, nitration of acetanilide ($PhNHCOCH_3$) goes quickly and gives mostly para substitution. Use resonance structures to explain this difference in reactivity.

17-8 HALOGEN SUBSTITUENTS: DEACTIVATING, BUT ORTHO, PARA-DIRECTING

The halobenzenes are exceptions to the general rules. The halogens are deactivating groups, yet they are ortho, para-directors. We can explain this unusual combination of properties by considering that

1. The halogens are strongly electronegative, withdrawing electron density from a carbon atom through the sigma bond.

2. The halogens have nonbonding electrons that can donate electron density through pi bonding.

The carbon-halogen bond is strongly polarized, with the carbon atom at the positive end of the dipole. This polarization draws electron density away from the benzene ring, making it less reactive toward electrophilic substitution.

—C⇸X   (benzene ring)⇸X

less electron-rich

If an electrophile reacts at the ortho or para position, the positive charge of the sigma complex is shared by the carbon atom bearing the halogen. The nonbonding electrons of the halogen can further delocalize the charge onto the halogen, giving a **halonium ion** structure. It is this resonance stabilization that allows a halogen to be pi-donating, while it is sigma-withdrawing.

Ortho attack

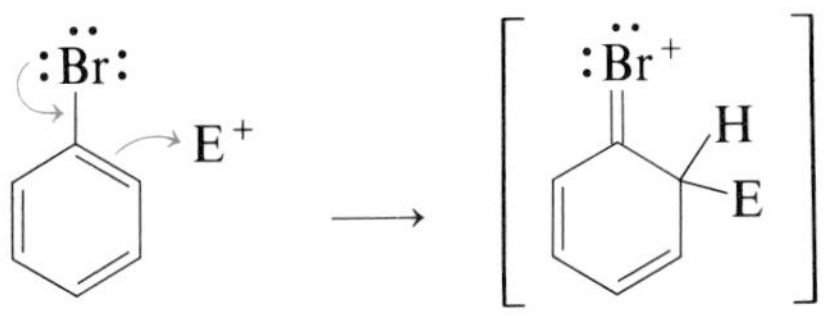

:Br: E⁺ → [:Br⁺ H E]

bromonium ion
(plus other structures)

Para attack

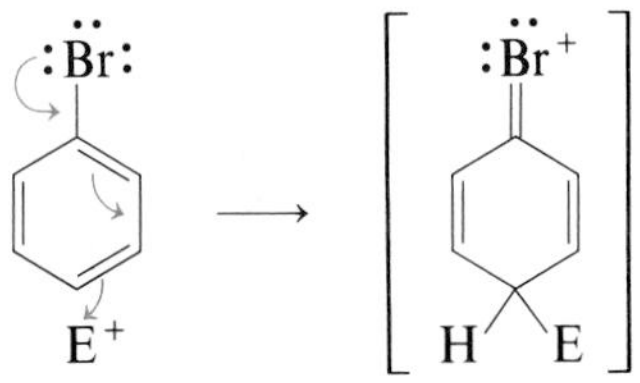

:Br: E⁺ → [:Br⁺ H E]

bromonium ion
(plus other structures)

PROBLEM 17-13

Draw all the resonance structures of the sigma complex for nitration of bromobenzene at the ortho, meta, and para positions. Point out why the intermediate for meta substitution is less stable than the other two.

Reaction at the meta position gives a sigma complex whose positive charge is not delocalized onto the halogen-bearing carbon atom. Therefore, the meta intermediate is not stabilized by the halonium ion structure. The following reaction illustrates the strong preference for ortho and para substitution in the nitration of chlorobenzene.

chlorobenzene $\xrightarrow[H_2SO_4]{HNO_3}$ ortho (35%) + meta (1%) + para (64%)

chloronitrobenzenes

Figure 17-5 shows the effect of the halogen atom graphically, with an energy profile comparing the energies of the transition states and intermediates for electrophilic attack on chlorobenzene and benzene. Higher energies are involved for the reactions of chlorobenzene, especially for attack at the meta position.

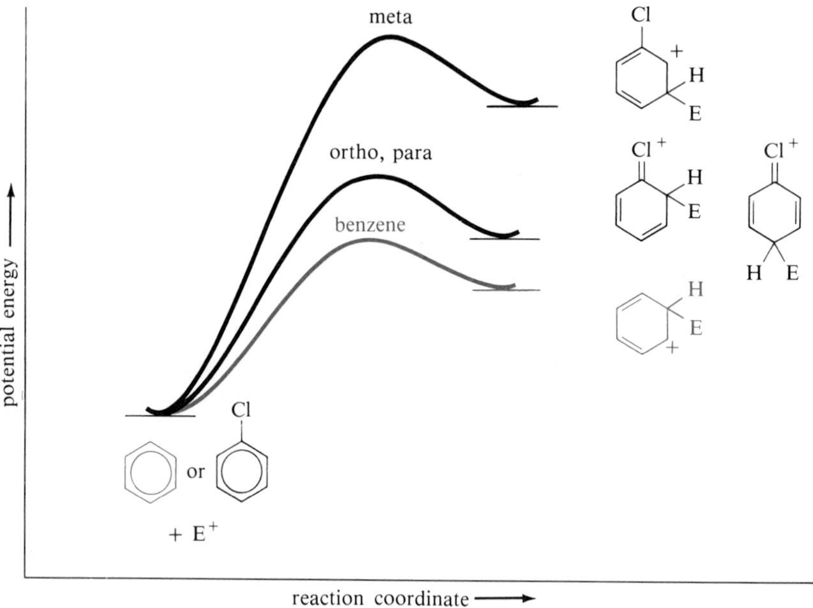

FIGURE 17-5 The energies of the intermediates and transition states are higher for chlorobenzene than for benzene. The highest energy results from substitution at the meta position, while the energies for ortho and para substitution are slightly lower due to stabilization by the halonium ion structure.

PROBLEM 17-14

(a) Predict the structure of the product formed when HCl adds to 1-bromocyclohexene.
(b) Propose a mechanism with resonance structures to support your prediction.
(c) Explain how this prediction is in accord with the ortho, para-directing effect of bromine on an aromatic ring.

17-9

EFFECTS OF MORE THAN ONE SUBSTITUENT ON ELECTROPHILIC AROMATIC SUBSTITUTION

Two or more substituents exert a combined effect on the reactivity of the aromatic ring. If the groups reinforce each other, the result will be easy to predict. For example, we can predict that all the xylenes (dimethylbenzenes) are activated toward electrophilic substitution, because the two methyl groups are both activating. In the case of a nitrobenzoic acid, both substituents are deactivating, and we can predict that a nitrobenzoic acid is deactivated toward attack by an electrophile.

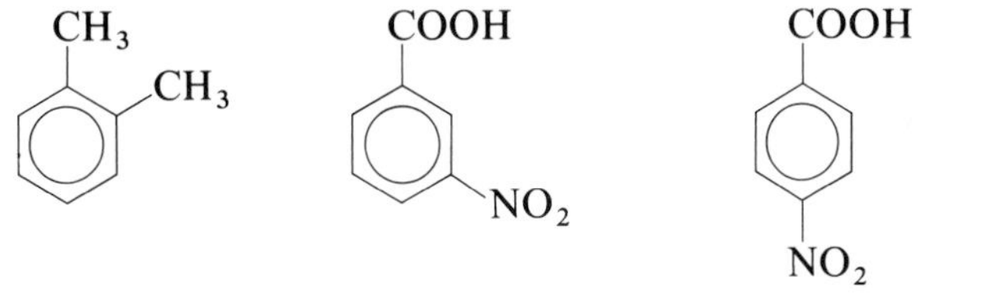

CH_3 CH_3 | COOH NO_2 | COOH NO_2

o-xylene activated | *m*-nitrobenzoic acid deactivated | *p*-nitrobenzoic acid deactivated

The orientation of addition is easily predicted in many cases. For example, in *m*-xylene there are two positions that are ortho to one of the methyl groups and para to the other. Electrophilic substitution occurs primarily at these two equivalent positions. There may be some substitution at the position between the two methyl groups (ortho to both), but this position is sterically hindered, and it is less reactive than the other two activated positions. In *p*-nitrotoluene, the methyl group directs an electrophile toward its ortho positions. The nitro group directs toward the same locations, because they are its meta positions.

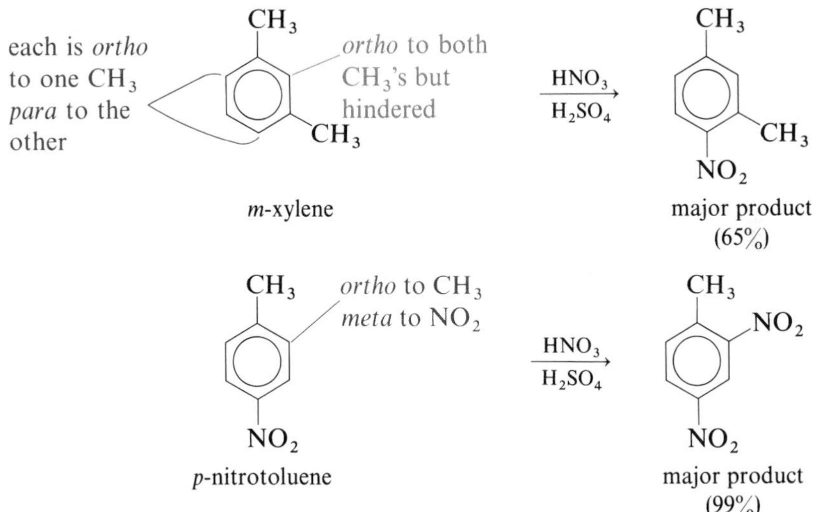

PROBLEM 17-15

Predict the mononitration products of the following compounds.

(a) *o*-nitrotoluene
(b) *m*-chlorotoluene
(c) *o*-bromobenzoic acid
(d) *p*-methoxybenzoic acid
(e) *m*-cresol (*m*-methylphenol)

When the directing effects of two or more substituents conflict with each other, it is more difficult to predict where an electrophile will react. In many cases, mixtures result. For example, *o*-xylene is activated at all the positions, and mixtures of substitution products result.

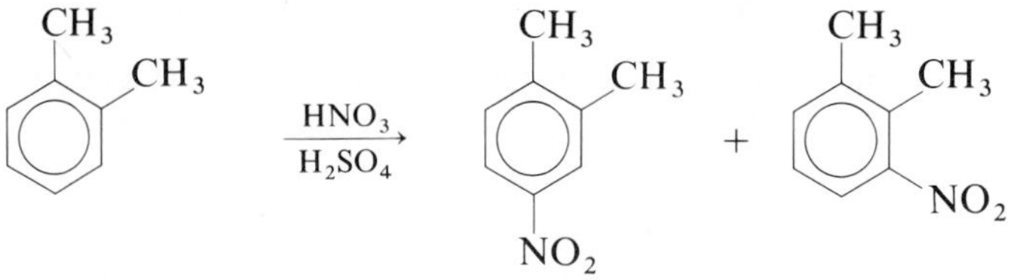

$$\text{o-xylene} \xrightarrow[H_2SO_4]{HNO_3} \text{(58\%, 4-nitro-o-xylene)} + \text{(42\%, 3-nitro-o-xylene)}$$

When there is a conflict between an activating group and a deactivating group, the activating group usually directs the substitution. We can make an important generalization:

Activating groups are usually stronger directors than deactivating groups.

In fact, it is helpful to separate substituents into three classes, from strongest to weakest.

1. Powerful ortho, para-directors that stabilize the sigma complexes through resonance. Examples are —OH, —OR, and $—NR_2$ groups.
2. Moderate ortho, para-directors, such as alkyl groups and halogens.
3. All meta-directors.

$$—OH,\quad —OR,\quad —NR_2 \;>\; —R,\quad —X \;>\; —\overset{O}{\overset{\|}{C}}—R,\quad —SO_3H,\quad —NO_2$$

If two substituents direct an incoming electrophile toward different reaction sites, the substituent in the stronger class predominates. In the following reaction the stronger group predominates and directs the incoming substituent. The methoxyl group is a stronger director than the nitro group, and substitution occurs ortho and para to the methoxyl group. Steric effects prevent much substitution at the crowded position ortho to both the methoxyl group and the nitro group.

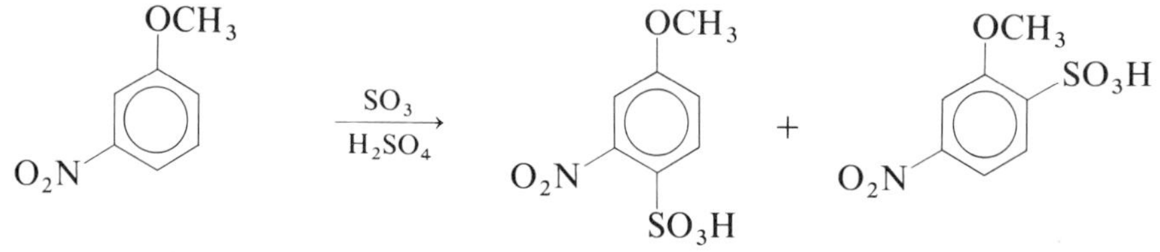

$$\text{m-nitroanisole} \xrightarrow[H_2SO_4]{SO_3} \text{major products (sulfonation para and ortho to } OCH_3)$$

m-nitroanisole — major products

SOLVED PROBLEM 17-1

Predict the major product(s) of bromination of *p*-chloroacetanilide.

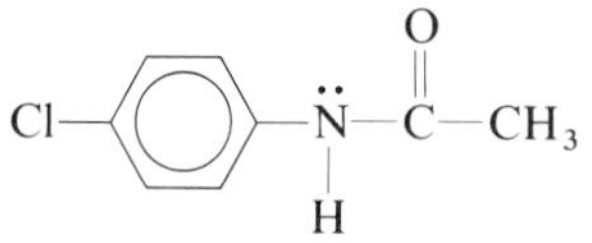

$$Cl—C_6H_4—\ddot{N}H—\overset{O}{\overset{\|}{C}}—CH_3$$

SOLUTION The amide group ($—NHCOCH_3$) is a strong activating and directing group because it is the nitrogen atom with its nonbonding pair of electrons (as opposed to the carbonyl group) that is bonded to the aromatic ring. The amide group is a stronger director than the chlorine atom, and substitution occurs mostly at the positions ortho to the amide. Like an alkoxyl group, the amide is a particularly strong activating group, and the reaction gives some of the dibrominated product.

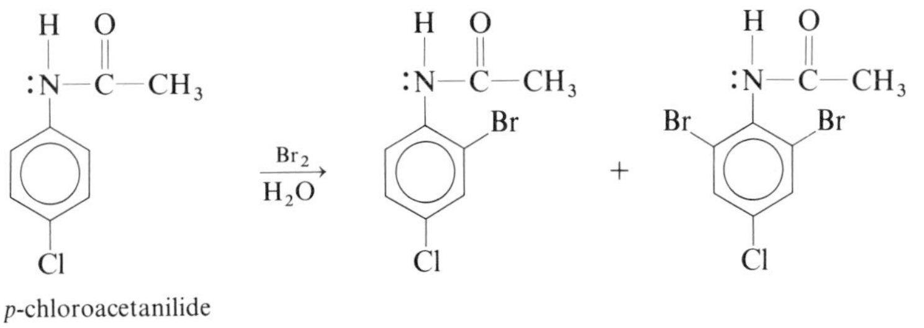

$$p\text{-chloroacetanilide} \xrightarrow[H_2O]{Br_2} \text{(2-bromo-4-chloroacetanilide)} + \text{(2,6-dibromo-4-chloroacetanilide)}$$

p-chloroacetanilide

PROBLEM 17-16

Predict the mononitration products of the following aromatic compounds.

(a) *m*-nitrochlorobenzene (b) *p*-chlorophenol (c) *m*-nitroanisole

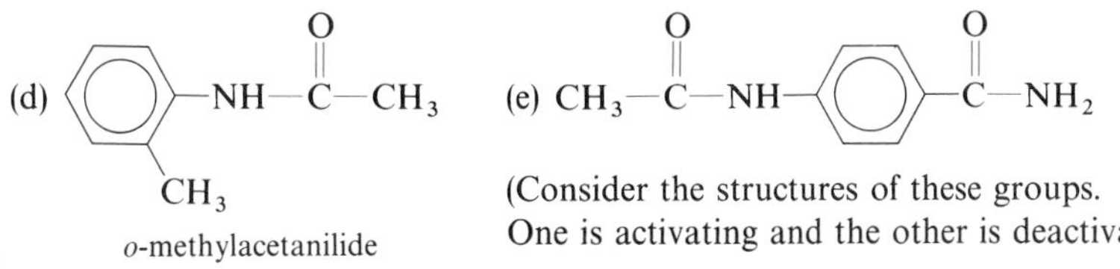

(d) o-methylacetanilide, $C_6H_4(CH_3)-NH-C(=O)-CH_3$

o-methylacetanilide

(e) $CH_3-C(=O)-NH-C_6H_4-C(=O)-NH_2$

(Consider the structures of these groups. One is activating and the other is deactivating.)

PROBLEM 17-17

Biphenyl is two benzene rings joined by a single bond. The site of substitution for a biphenyl is determined by (1) which phenyl ring is more activated (or less deactivated), and (2) which position on that ring is most reactive, using the fact that a phenyl substituent is ortho, para-directing.

(a) Use resonance structures of a sigma complex to show why a phenyl substituent should be ortho, para-directing.
(b) Predict the mononitration products of the following compounds.

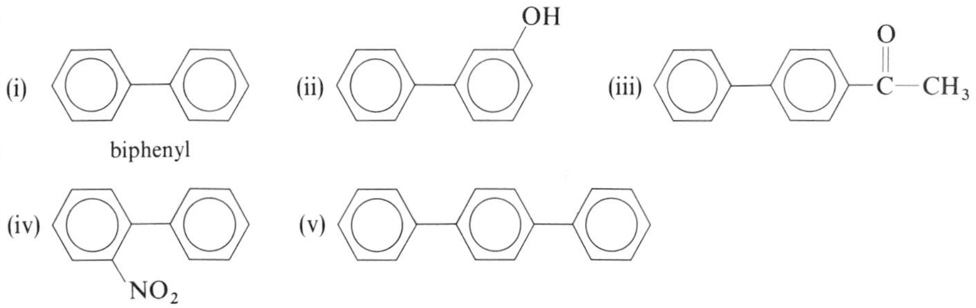

(i) biphenyl

(ii) biphenyl with OH on the meta position of one ring

(iii) biphenyl with $-C(=O)-CH_3$ on the para position of one ring

(iv) biphenyl with NO_2 on the ortho position of one ring

(v)

17-10 THE FRIEDEL-CRAFTS ALKYLATION

Carbocations are perhaps the most important electrophiles capable of substituting onto aromatic rings, because this substitution forms a new carbon-carbon bond. Reactions of carbocations with aromatic compounds were first studied in 1877 by the French alkaloid chemist Charles Friedel and his American partner, James Crafts. In the presence of Lewis acid catalysts such as aluminum chloride ($AlCl_3$) or ferric chloride ($FeCl_3$), alkyl halides were found to alkylate benzene to give alkylbenzenes. This useful reaction is called the **Friedel-Crafts alkylation.**

Friedel-Crafts alkylation

$$C_6H_5\text{—H} + R\text{—X} \xrightarrow[(AlCl_3,\ FeBr_3,\ \text{etc.})]{\text{Lewis acid}} C_6H_5\text{—R} + H\text{—X}$$

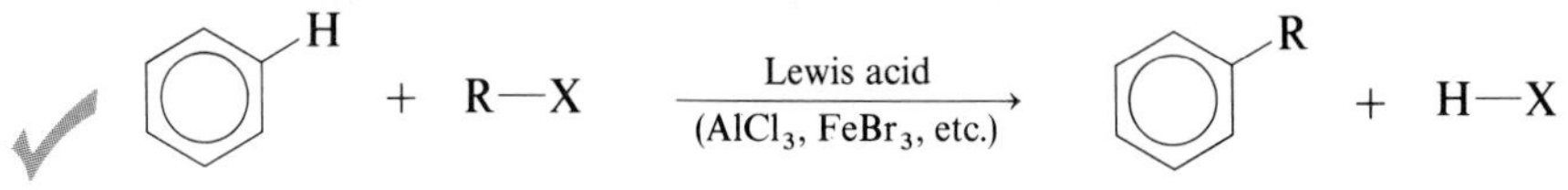

(X = Cl, Br, I)

For example, when aluminum chloride is added to a mixture of benzene and *t*-butyl chloride, benzene is alkylated and HCl gas is evolved.

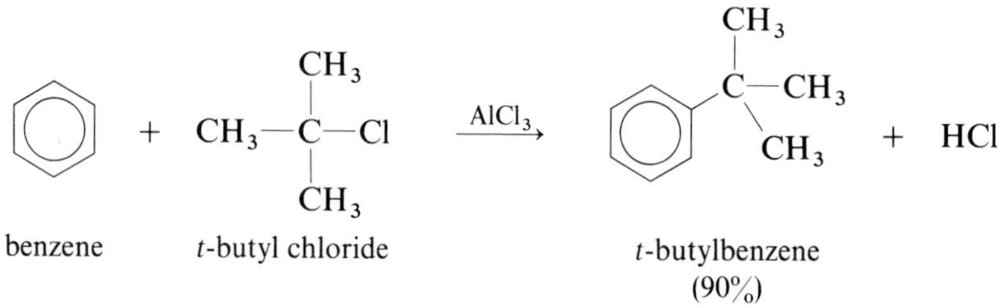

$$C_6H_6 + (CH_3)_3C\text{—Cl} \xrightarrow{AlCl_3} C_6H_5\text{—}C(CH_3)_3 + HCl$$

benzene, *t*-butyl chloride, *t*-butylbenzene (90%)

This alkylation is a typical electrophilic aromatic substitution, with the *t*-butyl cation acting as the electrophile. The *t*-butyl cation is formed by reaction of *t*-butyl chloride with the catalyst, aluminum chloride.

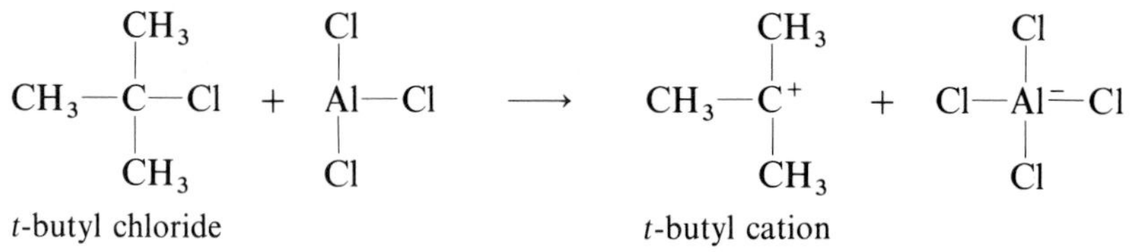

$$(CH_3)_3C\text{—Cl} + AlCl_3 \longrightarrow (CH_3)_3C^+ + Cl\text{—}\bar{Al}Cl_3$$

t-butyl chloride, *t*-butyl cation

The *t*-butyl cation reacts with benzene to form a sigma complex. Loss of a proton gives the product, *t*-butylbenzene. The aluminum chloride catalyst is regenerated in the final step.

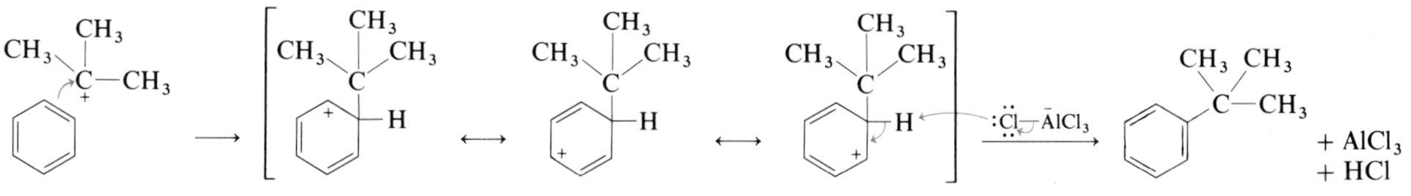

$$C_6H_6 + (CH_3)_3C^+ \longrightarrow \left[\text{sigma complex (three resonance forms, C(CH}_3)_3\text{ and H on the ring carbon)}\right] \xrightarrow{:\ddot{Cl}\text{—}\bar{Al}Cl_3} C_6H_5\text{—}C(CH_3)_3 + AlCl_3 + HCl$$

The Friedel-Crafts alkylation is used with a wide variety of primary, secondary, and tertiary alkyl halides. In the case of secondary and tertiary halides, the reacting electrophile is probably the carbocation.

$$R\text{—X} + AlCl_3 \rightleftharpoons R^+ + {}^-AlCl_3X$$

(R is secondary or tertiary) — reacting electrophile

With primary alkyl halides, the free primary carbocation is too unstable. The actual electrophile is a complex of aluminum chloride with the alkyl halide. In this complex the carbon-halogen bond is weakened (as indicated by dashed lines) and there is a considerable amount of positive charge on the carbon atom. Following is the mechanism for the aluminum chloride-catalyzed reaction of ethyl chloride with benzene.

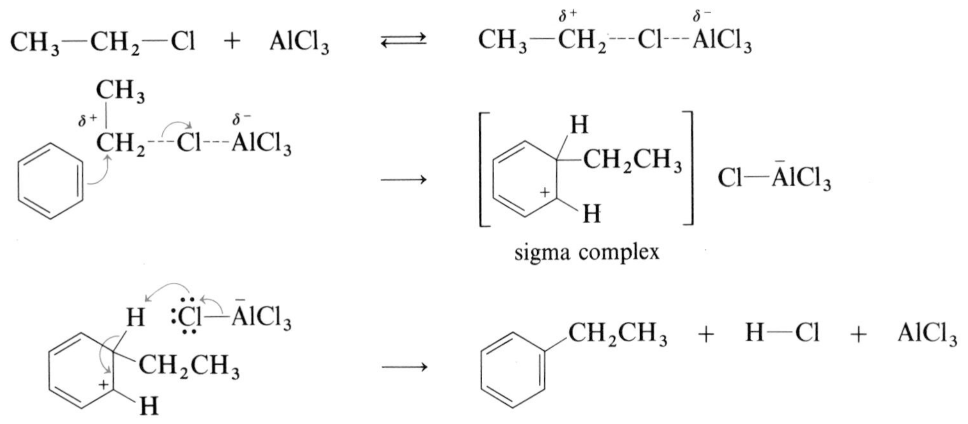

$$CH_3—CH_2—Cl + AlCl_3 \rightleftharpoons CH_3—\overset{\delta+}{CH_2}\text{---}Cl\text{---}\overset{\delta-}{AlCl_3}$$

sigma complex

$$\longrightarrow \; C_6H_5CH_2CH_3 + H—Cl + AlCl_3$$

PROBLEM 17-18

Propose products (if any) and mechanisms for the following $AlCl_3$-catalyzed reactions.

(a) chlorocyclohexane with benzene
(b) methyl chloride with anisole
(c) 3-chloro-2,2-dimethylbutane with isopropylbenzene

Friedel-Crafts alkylation using other carbocation sources We have seen several ways of generating carbocations, and most of these can be used to generate carbocations for Friedel-Crafts alkylations. Two methods are commonly used: protonation of alkenes and treatment of alcohols with BF_3.

Alkenes are protonated by HF to give carbocations. The fluoride ion is a weak nucleophile and does not immediately attack the carbocation. If benzene (or an activated benzene derivative) is present, electrophilic substitution occurs. Markovnikov's rule is followed in the protonation step; the more stable carbocation is formed and alkylates the aromatic ring.

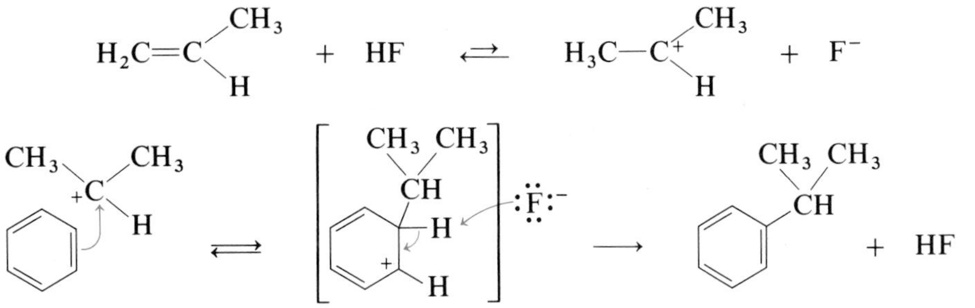

$$H_2C{=}C(CH_3)H + HF \rightleftharpoons H_3C—C^+(CH_3)H + F^-$$

$$\longrightarrow \; C_6H_5CH(CH_3)_2 + HF$$

Alcohols are another source of carbocations for the Friedel-Crafts alkylation. Alcohols commonly form carbocations when treated with boron trifluoride (BF_3). If benzene (or an activated benzene derivative) is present, substitution may occur.

Formation of the cyclohexyl cation

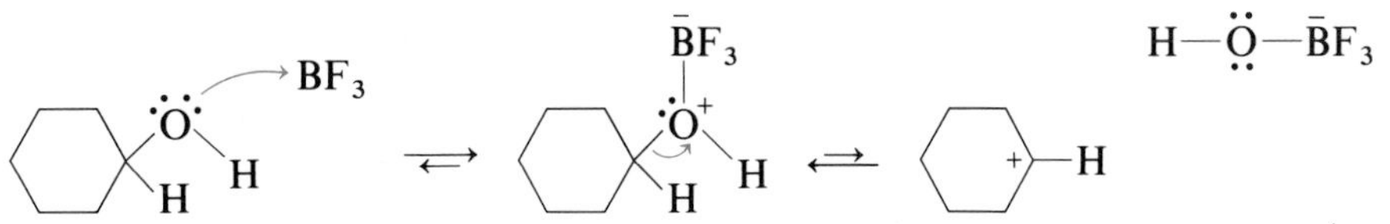

$$H—\ddot{O}—\bar{B}F_3$$

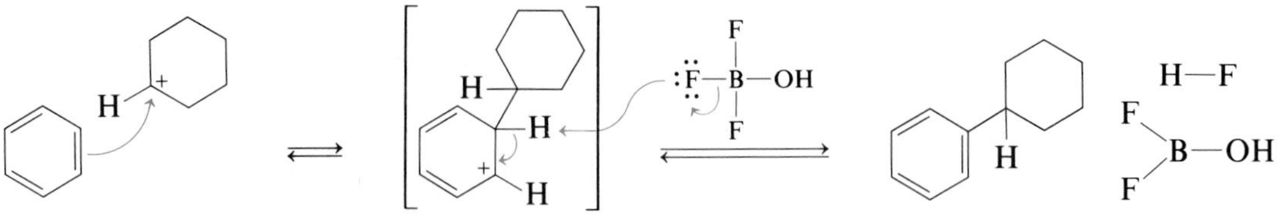

sigma complex

The BF_3 used in this reaction is consumed and not regenerated. A full equivalent of the Lewis acid is needed, and we say that the reaction is *promoted* by BF_3 rather than *catalyzed* by BF_3.

PROBLEM 17-19
Show the generation of the electrophile and predict the products of each reaction.

(a) benzene + cyclohexene + HF
(b) *t*-butyl alcohol + benzene + BF_3
(c) *t*-butylbenzene + 2-methylpropene + HF
(d) 2-propanol + toluene + BF_3

Limitations of the Friedel-Crafts alkylation So far, what we have seen is a rosy picture of the Friedel-Crafts alkylation. There are three major limitations, however, that severely restrict its use.

Limitation 1 Friedel-Crafts reactions proceed only with benzene, halobenzenes, and activated benzene derivatives; they fail with strongly deactivated systems such as nitrobenzene, benzenesulfonic acid, and phenyl ketones. In some cases, we can get around this limitation by adding the deactivating group or changing an activating group into a deactivating group *after* the Friedel-Crafts step.

SOLVED PROBLEM 17-2
Devise a synthesis of *p*-nitro-*t*-butylbenzene from benzene.

SOLUTION To make *p*-nitro-*t*-butylbenzene, we would first use a Friedel-Crafts reaction to make *t*-butylbenzene. Nitration gives the correct product. If we were to make nitrobenzene first, the Friedel-Crafts reaction to add the *t*-butyl group would fail.

Good

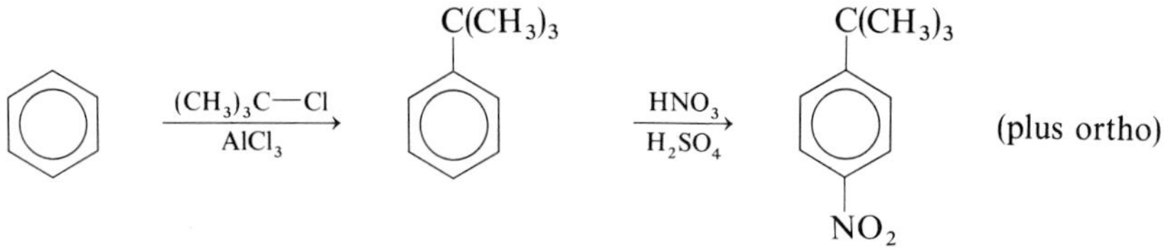

$C(CH_3)_3$ $C(CH_3)_3$

$(CH_3)_3C—Cl$ / $AlCl_3$ → HNO_3 / H_2SO_4 → NO_2 (plus ortho)

Bad

HNO_3 / H_2SO_4 → NO_2 $(CH_3)_3C—Cl$ / $AlCl_3$ → (reaction fails)

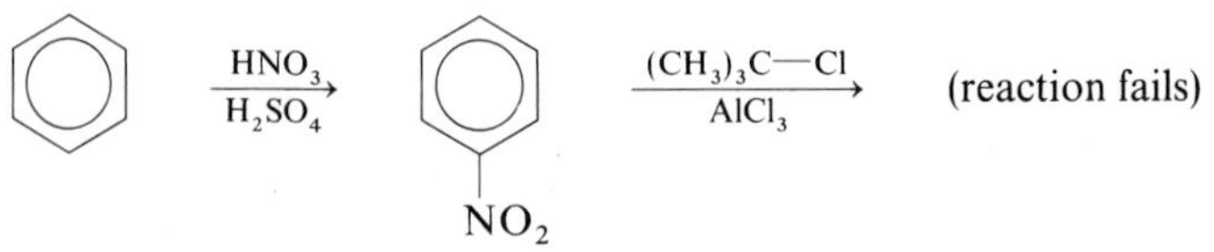

Limitation 2 Like other carbocation reactions, the Friedel-Crafts alkylation is susceptible to carbocation rearrangements. This limitation implies that only certain alkylbenzenes can be made using the Friedel-Crafts alkylation. *t*-Butylbenzene, isopropylbenzene, and ethylbenzene can be synthesized using the Friedel-Crafts alkylation, because the corresponding cations are not prone to rearrangement. Consider what happens, however, when we try to make *n*-propylbenzene by the Friedel-Crafts alkylation.

Ionization with rearrangement gives the isopropyl cation

$$CH_3{-}CH_2{-}CH_2{-}Cl + AlCl_3 \rightleftharpoons CH_3{-}\overset{H}{\underset{H}{C}}{-}\overset{\delta+}{C}H_2{\cdots}\overset{\delta-}{Cl}{\cdots}AlCl_3 \longrightarrow CH_3{-}\overset{+}{\underset{H}{C}}{-}CH_3 + {}^-AlCl_4$$

Reaction with benzene gives isopropylbenzene

$$CH_3{-}\overset{+}{\underset{H}{C}}{-}CH_3 \;\; {}^-AlCl_4 + C_6H_6 \longrightarrow C_6H_5{-}CH(CH_3){-}CH_3 + HCl + AlCl_3$$

Limitation 3 Since alkyl groups are activating substituents, the product of the Friedel-Crafts alkylation is *more reactive* than the starting material. Multiple alkylations are difficult to avoid. This limitation can be severe. If we need to make ethylbenzene, we might try adding some $AlCl_3$ to a mixture of 1 mole of ethyl chloride and 1 mole of benzene. As some ethylbenzene is formed, however, it is activated, reacting even faster than benzene itself. The product is a mixture of some (ortho and para) diethylbenzenes, some triethylbenzenes, a small amount of ethylbenzene, and some leftover benzene.

$$\underset{\text{1 mole}}{C_6H_6} + \underset{\text{1 mole}}{CH_3CH_2{-}Cl} \xrightarrow{AlCl_3} p\text{-}C_6H_4(CH_2CH_3)_2 + o\text{-}C_6H_4(CH_2CH_3)_2 + C_6H_5CH_2CH_3$$

+ triethylbenzenes + benzene

The problem of dialkylation can be avoided by using a large excess of benzene. For example, if a ratio of 50 moles of benzene to 1 mole of ethyl chloride is used, the concentration of ethylbenzene is always low, and the electrophile is more likely to react with benzene than with ethylbenzene. The product is separated from excess benzene by distillation. This is a common industrial approach, since a continuous distillation can be used to recycle the unreacted benzene.

In the laboratory we must often alkylate aromatic compounds that are more expensive than benzene. Because we cannot afford to use a large excess of the starting material, a more selective method of alkylation is needed. Fortunately, the Friedel-Crafts acylation, discussed in the next section, provides a way of introducing just one group without danger of polyalkylation or rearrangement.

PROBLEM 17-20

Predict the products (if any) for the following reactions.

(a) (excess) benzene + isobutyl chloride + $AlCl_3$
(b) (excess) toluene + 1-butanol + BF_3
(c) (excess) nitrobenzene + 2-chloropropane + $AlCl_3$
(d) (excess) benzene + 3,3-dimethyl-1-butene + HF

PROBLEM 17-21

Which of the following reactions will produce the desired product in good yield? You may assume that aluminum chloride is added as a catalyst in each case. For the reactions that will not give a good yield of the desired product, predict the major products.

Reagents	*Desired Product*
(a) benzene + *n*-butyl bromide	*n*-butylbenzene
(b) ethylbenzene + *t*-butyl chloride	*p*-ethyl-*t*-butylbenzene
(c) bromobenzene + ethyl chloride	*p*-bromoethylbenzene
(d) ethylbenzene + bromine	*p*-bromoethylbenzene
(e) anisole + methyl iodide (3 mol)	2,4,6-trimethylanisole

PROBLEM 17-22

Show how you would synthesize the following aromatic derivatives from benzene.

(a) *p*-*t*-butyl nitrobenzene (b) *p*-toluenesulfonic acid (c) *p*-chlorotoluene

17-11 THE FRIEDEL-CRAFTS ACYLATION

A carbonyl group with an alkyl group attached is called an **acyl group.** Acyl groups are named systematically by dropping the final *-e* from the alkane name and adding the *-oyl* suffix. Historical names are commonly used for the *formyl group* and the *acetyl group,* however.

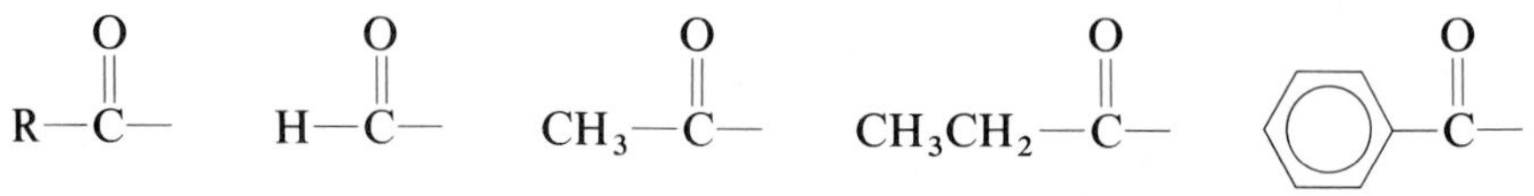

$$R-\overset{O}{\overset{\|}{C}}- \qquad H-\overset{O}{\overset{\|}{C}}- \qquad CH_3-\overset{O}{\overset{\|}{C}}- \qquad CH_3CH_2-\overset{O}{\overset{\|}{C}}- \qquad C_6H_5-\overset{O}{\overset{\|}{C}}-$$

acyl group; methanoyl (formyl); ethanoyl (acetyl); propanoyl; benzoyl

An **acyl chloride** is an acyl group bonded to a chlorine atom. Acyl chlorides are made by reaction of the corresponding carboxylic acids with thionyl chloride. Therefore, acyl chlorides are often called **acid chlorides.** We consider acyl chlorides in more detail when we study acid derivatives in Chapter 21.

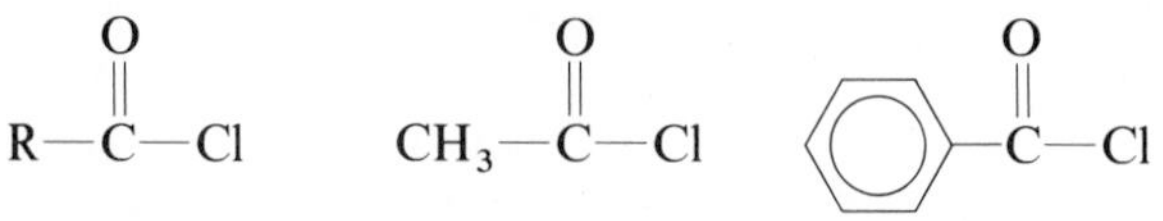

$$R-\overset{O}{\overset{\|}{C}}-Cl \qquad CH_3-\overset{O}{\overset{\|}{C}}-Cl \qquad C_6H_5-\overset{O}{\overset{\|}{C}}-Cl$$

an acyl chloride (an acid chloride); acetyl chloride; benzoyl chloride

$$\underset{\text{a carboxylic acid}}{R-\overset{O}{\overset{\|}{C}}-OH} + \underset{\text{thionyl chloride}}{Cl-\overset{O}{\overset{\|}{S}}-Cl} \longrightarrow \underset{\text{an acyl chloride}}{R-\overset{O}{\overset{\|}{C}}-Cl} + SO_2\uparrow + HCl\uparrow$$

In the presence of aluminum chloride, an acyl chloride reacts with benzene or an activated benzene derivative to give a phenyl ketone; an *acylbenzene*. This reaction, called the **Friedel-Crafts acylation,** is analogous to the Friedel-Crafts alkylation, except that the reagent is an acyl chloride instead of an alkyl halide and the product is an acylbenzene (a "phenone") instead of an alkylbenzene.

$$C_6H_6 + R{-}C(=O){-}Cl \xrightarrow{AlCl_3} C_6H_5{-}C(=O){-}R + HCl$$

benzene acyl halide an acylbenzene (a phenyl ketone)

Example

$$C_6H_6 + CH_3{-}C(=O){-}Cl \xrightarrow{AlCl_3} C_6H_5{-}C(=O){-}CH_3 + HCl$$

benzene acetyl chloride acetylbenzene (acetophenone) (95%)

17-11A MECHANISM OF ACYLATION

The mechanism of the Friedel-Crafts acylation resembles that for alkylation, except that the carbonyl group helps to stabilize the cationic intermediates. The acyl halide forms a complex with aluminum chloride; loss of the tetrachloroaluminate ion ($^{-}AlCl_4$) gives a resonance-stabilized **acylium ion.**

$$R{-}C(=\ddot{O}){-}\ddot{Cl}: + AlCl_3 \rightleftharpoons R{-}C(=\ddot{O}){-}\overset{+}{\ddot{Cl}}{-}\overset{-}{Al}Cl_3 \rightleftharpoons {}^{-}AlCl_4 + [R{-}\overset{+}{C}{=}\ddot{O}: \longleftrightarrow R{-}C{\equiv}\overset{+}{O}:]$$

acyl chloride complex acylium ion

The acylium ion is a powerful electrophile. It reacts with benzene or an activated benzene derivative to form an acylbenzene.

$$C_6H_6 + R{-}\overset{+}{C}{=}O \longrightarrow [\text{sigma complex}] \xrightarrow{:\ddot{Cl}{-}\overset{-}{Al}Cl_3} C_6H_5{-}C(=\ddot{O}){-}R + AlCl_3 + HCl \longrightarrow C_6H_5{-}C(=\overset{+}{O}{-}\overset{-}{Al}Cl_3){-}R$$

attack on acylium ion sigma complex acylbenzene product complex

The product of acylation (the acylbenzene) is a ketone. The ketone's carbonyl group has nonbonding electrons that complex with the Lewis acid catalyst, $AlCl_3$, requiring that a full equivalent of $AlCl_3$ be used in the acylation. The initial product recovered from the reaction mixture is the aluminium chloride complex of the acylbenzene. Addition of water hydrolyzes this complex, giving the free acylbenzene.

$$\underset{\text{product complex}}{C_6H_5-\overset{\overset{+}{:O}-\bar{A}lCl_3}{\overset{\|}{C}}-R} \xrightarrow{\text{excess } H_2O} \underset{\text{free acylbenzene}}{C_6H_5-\overset{O}{\overset{\|}{C}}-R} + \text{aluminum salts}$$

The electrophile in the Friedel-Crafts acylation appears to be a large, bulky complex: probably $R-\overset{+}{C}=O\ ^{-}AlCl_4$. Para substitution usually prevails when the aromatic substrate has an ortho, para-directing group, possibly because the electrophile is too bulky for effective attack at the ortho position. For example, when ethylbenzene reacts with acetyl chloride, the major product is *p*-ethylacetophenone.

$$\underset{\text{ethylbenzene}}{C_6H_5CH_2CH_3} + \underset{\text{acetyl chloride}}{CH_3-\overset{O}{\overset{\|}{C}}-Cl} \xrightarrow[(2)\ H_2O]{(1)\ AlCl_3} \underset{\substack{p\text{-ethylacetophenone} \\ (70-80\%)}}{CH_3CO-C_6H_4-CH_2CH_3}$$

One of the most attractive features of the Friedel-Crafts acylation is the deactivation of the product toward further substitution. The acylbenzene has a carbonyl group (a deactivating group) bonded to the aromatic ring. Since Friedel-Crafts reactions do not occur on strongly deactivated rings, the acylation stops after just one substitution.

$$C_6H_5-H \xrightarrow[AlCl_3]{R-\overset{O}{\overset{\|}{C}}-Cl} C_6H_5-C^{\delta+}(=O^{\delta-})-R$$

Thus, Friedel-Crafts acylation overcomes two of the three limitations of the alkylation: The acylium ion is stabilized so that no rearrangements occur, and the acylbenzene product is deactivated so that no further reaction occurs. Like the alkylation, however, the acylation fails with strongly deactivated aromatic rings.

COMPARISON OF FRIEDEL-CRAFTS ALKYLATION AND ACYLATION

Alkylation	*Acylation*
The alkylation cannot be used with strongly deactivated derivatives.	Also true: Only benzene, halobenzenes, and activated derivatives are suitable.
The carbocations involved in the alkylation may rearrange.	Resonance-stabilized acylium ions are not prone to rearrangement.
Polyalkylation is commonly a problem.	The acylation forms a deactivated acylbenzene, which does not react further.

17-11B THE CLEMMENSEN REDUCTION: SYNTHESIS OF ALKYLBENZENES

How do we synthesize alkylbenzenes that cannot be made efficiently by Friedel-Crafts alkylation? We first use the Friedel-Crafts acylation to make the acylbenzene, which is then reduced to the alkylbenzene using the **Clemmensen reduction:** treatment with aqueous HCl and amalgamated zinc (zinc treated with mercury salts).

$$\text{C}_6\text{H}_6 + \text{R}-\overset{\overset{\large O}{\|}}{\text{C}}-\text{Cl} \xrightarrow[\text{(2) H}_2\text{O}]{\text{(1) AlCl}_3} \text{C}_6\text{H}_5-\overset{\overset{\large O}{\|}}{\text{C}}-\text{R} \xrightarrow[\text{aq. HCl}]{\text{Zn(Hg)}} \text{C}_6\text{H}_5-\text{CH}_2-\text{R}$$

This two-step sequence can be used to synthesize many alkylbenzenes that are impossible to make by direct alkylation. For example, we saw earlier that the Friedel-Crafts alkylation cannot be used to make *n*-propylbenzene. Benzene reacts with *n*-propyl chloride and $AlCl_3$ to give isopropylbenzene, together with some diisopropylbenzene. Using the acylation, benzene reacts with propanoyl chloride and $AlCl_3$ to give ethyl phenyl ketone (propiophenone), which is easily reduced to *n*-propylbenzene.

$$\text{C}_6\text{H}_6 + \text{CH}_3\text{CH}_2-\overset{\overset{\large O}{\|}}{\text{C}}-\text{Cl} \xrightarrow[\text{(2) H}_2\text{O}]{\text{(1) AlCl}_3} \underset{\text{propiophenone}}{\text{C}_6\text{H}_5-\overset{\overset{\large O}{\|}}{\text{C}}-\text{CH}_2\text{CH}_3} \xrightarrow[\text{HCl}]{\text{Zn(Hg)}} \underset{n\text{-propylbenzene}}{\text{C}_6\text{H}_5-\text{CH}_2\text{CH}_2\text{CH}_3}$$

Carboxylic acids and acid anhydrides may also be used as acylating agents in Friedel-Crafts reactions. We consider these acylating agents in Chapters 20 and 21 when we study the reactions of carboxylic acids and their derivatives.

17-11C THE GATTERMAN-KOCH FORMYLATION: SYNTHESIS OF BENZALDEHYDES

The formyl group cannot be added to benzene by the Friedel-Crafts acylation in the usual manner. The problem lies not with the reaction but with the necessary reagent, formyl chloride: Formyl chloride is unstable and cannot be used as a reagent.

$$\underset{\text{formyl group}}{\text{H}-\overset{\overset{\large O}{\|}}{\text{C}}-} \qquad \underset{\text{formyl chloride}}{\left[\text{H}-\overset{\overset{\large O}{\|}}{\text{C}}-\text{Cl}\right]}$$

Formylation can be accomplished instead by using a high-pressure mixture of carbon monoxide and HCl together with a catalyst: a mixture of cuprous chloride (CuCl) and aluminum chloride. This mixture generates the formyl cation, possibly through a small concentration of formyl chloride. The reaction with benzene gives formylbenzene, better known as benzaldehyde. This reaction, called the **Gatterman-Koch synthesis,** is widely used in industry to synthesize benzaldehydes.

$$\text{CO} + \text{HCl} \rightleftarrows \underset{\substack{\text{formyl chloride}\\ \text{(unstable)}}}{\left[\text{H}-\overset{\overset{\large O}{\|}}{\text{C}}-\text{Cl}\right]} \xrightarrow{\text{AlCl}_3/\text{CuCl}} \underset{\text{formyl cation}}{\left[\text{H}-\overset{+}{\text{C}}=\ddot{\text{O}}:\right]}\ {}^{-}\text{AlCl}_4$$

$$\text{C}_6\text{H}_6 + \text{H}-\overset{+}{\text{C}}=\text{O} \longrightarrow \underset{\text{benzaldehyde}}{\text{C}_6\text{H}_5-\overset{\overset{\large O}{\|}}{\text{C}}-\text{H}} + \text{HCl}$$

PROBLEM 17-23

Show how you would use the Friedel-Crafts acylation, Clemmensen reduction, and/or Gatterman-Koch synthesis to prepare the following compounds.

(a) $Ph-C(=O)-CH_2CH(CH_3)_2$ isobutyl phenyl ketone

(b) $Ph-C(=O)-C(CH_3)_3$ *t*-butyl phenyl ketone

(c) $Ph-C(=O)-Ph$ diphenyl ketone

(d) *p*-methoxybenzaldehyde

(e) 1-phenyl-2,2-dimethylpropane

(f) *n*-butylbenzene

17-12 NUCLEOPHILIC AROMATIC SUBSTITUTION

Nucleophiles can displace halide ions from aryl halides, particularly if there are strong electron-withdrawing groups ortho or para to the halide. Because a nucleophile substitutes for a leaving group on an aromatic ring, this class of reactions is called **nucleophilic aromatic substitution.** The following examples show that both ammonia and hydroxide ion can displace chloride from 2,4-dinitrochlorobenzene:

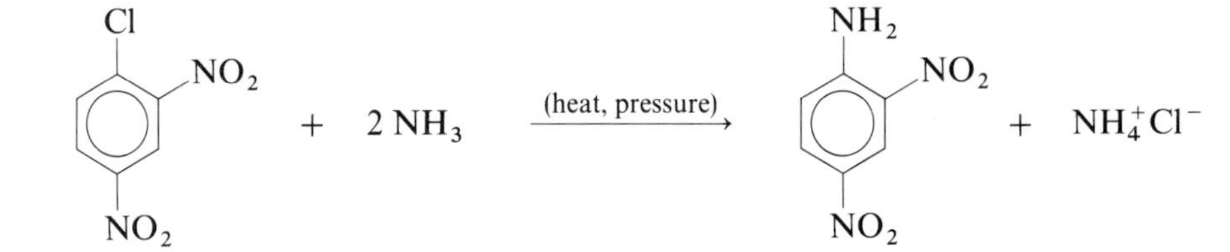

2,4-dinitrochlorobenzene + $2\,NH_3$ $\xrightarrow{\text{(heat, pressure)}}$ 2,4-dinitroaniline (90%) + $NH_4^+Cl^-$

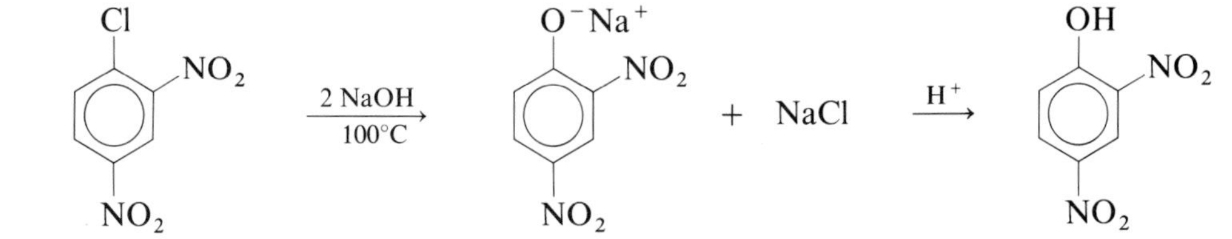

2,4-dinitrochlorobenzene $\xrightarrow[100°C]{2\,NaOH}$ 2,4-dinitrophenoxide (O^-Na^+) + NaCl $\xrightarrow{H^+}$ 2,4-dinitrophenol (95%)

Electrophilic aromatic substitution is the most important reaction of aromatic compounds because it has broad applications for a wide variety of aromatic compounds. In contrast, *nucleophilic* aromatic substitution is more restricted in its applications and is useful for a limited number of reactions and syntheses. In electrophilic aromatic substitution, a strong electrophile replaces a proton on an aromatic ring. In nucleophilic aromatic substitution, a strong nucleophile replaces a leaving group, such as a halide.

The mechanisms of the nucleophilic aromatic substitutions shown above are not immediately apparent. They cannot proceed by the S_N2 mechanism, because aryl halides cannot achieve the correct geometry for backside displacement. The aromatic ring blocks the approach of the nucleophile to the back of the carbon bearing the halogen.

The S_N1 mechanism cannot be involved, either; strong nucleophiles are required, and the reaction rate is proportional to the concentration of the nucleophile. The reaction is second order, and the nucleophile must be involved in the transition state.

Electron-withdrawing substituents (such as nitro groups) *activate* the ring toward nucleophilic aromatic substitution, suggesting that the transition state is developing a negative charge on the ring. In fact, nucleophilic aromatic substitutions are much more difficult without at least one powerful electron-withdrawing group. (This effect is opposite that for *electrophilic* aromatic substitution, where electron-withdrawing substituents slow or stop the reaction.)

Nucleophilic aromatic substitutions have been studied in great detail, and two different mechanisms have been proposed. One is similar to the electrophilic aromatic substitution mechanism, except that nucleophiles and carbanions are involved rather than electrophiles and carbocations. The other mechanism involves "benzyne," an interesting and unusual reactive intermediate.

17-12A THE ADDITION-ELIMINATION MECHANISM

Consider the reaction of 2,4-dinitrochlorobenzene with sodium hydroxide. If hydroxide attacks the carbon bearing the chlorine, a negatively charged sigma complex results. The negative charge is delocalized over the ortho and para carbons, stabilized by the electron-withdrawing nitro groups. These nitro groups stabilize the negative charge both through the inductive effect (the positively charged nitrogen atom) and through resonance with a partial C=N double bond. Loss of chloride from the sigma complex gives 2,4-dinitrophenol.

Attack by hydroxide gives a resonance-stabilized sigma complex; *Loss of chloride gives the product*

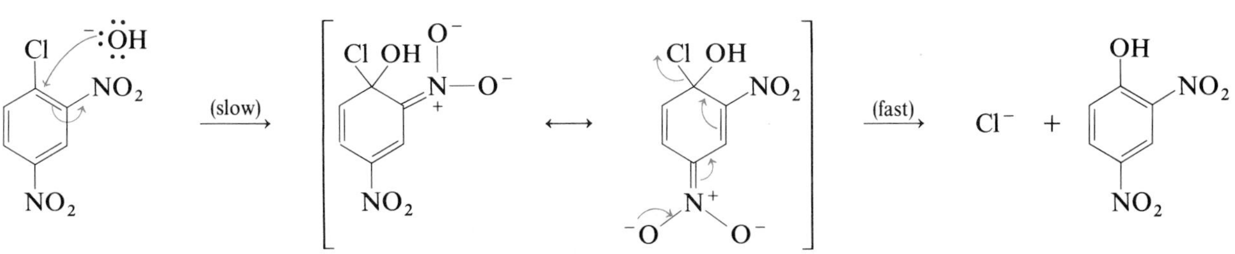

Cl :ÖH⁻ NO₂ ... NO₂ (slow) → [Cl OH, N⁺—O⁻, O⁻ ... NO₂ ⟷ Cl OH, NO₂ ... N⁺, ⁻O, O⁻] (fast) → Cl^- + OH, NO_2, NO_2

(plus other resonance structures)

The resonance structures shown above illustrate how nitro groups ortho and para to the halogen help to stabilize the intermediate (and the transition state leading to it). Without strong electron-withdrawing groups in these positions, formation of the negatively charged sigma complex is unlikely.

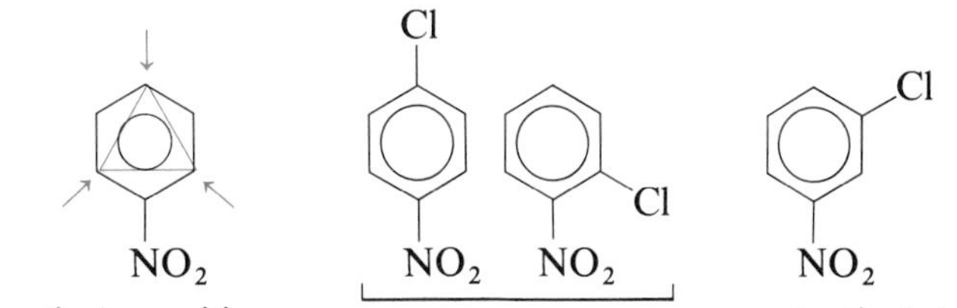

NO_2 activates positions ortho and para

Cl, NO_2 — Cl, NO_2 — activated

Cl, NO_2 — not activated

PROBLEM 17-24

Fluoride ion is usually a poor leaving group because it is not very polarizable. Fluoride serves as the leaving group in the Sanger reagent (2,4-dinitrofluorobenzene) used in the determination of peptide structures (Chapter 24). Explain why fluoride works as a leaving group in this nucleophilic aromatic substitution, even though it is a poor leaving group in the S_N1 and S_N2 mechanisms.

$$\text{2,4-dinitrofluorobenzene (Sanger reagent)} + R\text{—}\ddot{N}H_2 \text{ (amine)} \longrightarrow \text{2,4-dinitrophenyl derivative} + HF$$

2,4-dinitrofluorobenzene (Sanger reagent) + amine → 2,4-dinitrophenyl derivative

17-12B THE BENZYNE MECHANISM: ELIMINATION-ADDITION

The addition-elimination mechanism for nucleophilic aromatic substitution requires strong electron-withdrawing substituents on the aromatic ring. Under extreme conditions, however, unactivated halobenzenes react with strong bases. For example, the commercial synthesis of phenol (the "Dow process") involves treatment of chlorobenzene with sodium hydroxide and a small amount of water in a pressurized reactor at 350°C:

$$\text{chlorobenzene} \xrightarrow[H_2O]{\text{2 NaOH, 350°C}} \text{sodium phenoxide } (O^-Na^+) + NaCl \xrightarrow{H^+} \text{phenol (OH)}$$

Similarly, chlorobenzene reacts with sodium amide ($NaNH_2$, an extremely strong base) to give aniline, Ph—NH_2. This reaction does not require high temperatures, taking place in liquid ammonia at −33°C.

Nucleophilic substitution of unactivated benzene derivatives occurs by a mechanism different from the addition-elimination mechanism we saw with the nitro-substituted halobenzenes. A clue to the mechanism is provided by the reaction of *p*-bromotoluene with sodium amide. The products are a 50:50 mixture of *m*- and *p*-toluidine.

$$p\text{-bromotoluene} \xrightarrow[NH_3,\ -33°C]{Na^+\ ^-NH_2} p\text{-toluidine (50\%)} + m\text{-toluidine (50\%)}$$

These two products are explained by an elimination-addition mechanism, called the **benzyne** mechanism because of the unusual intermediate it involves. Sodium amide (or sodium hydroxide in the Dow process) reacts as a *base*, abstracting a proton. The product is a carbanion with a negative charge and a nonbonding pair of electrons localized in the sp^2 orbital that once formed the C—H bond.

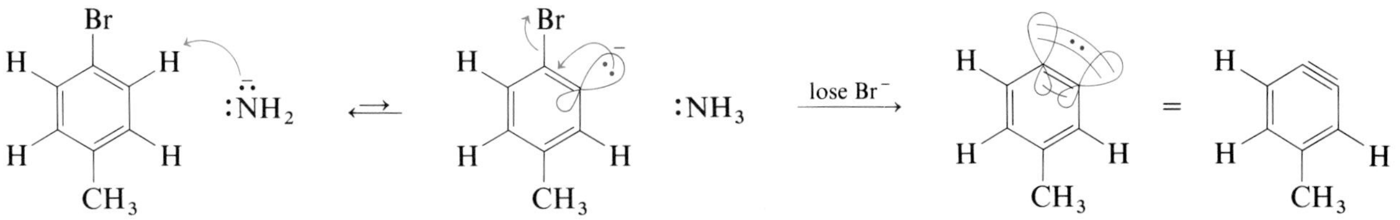

carbanion

lose Br^-

a "benzyne"

The carbanion can expel a bromide ion to become a neutral species. As bromide leaves with the bonding electrons, an empty sp^2 orbital remains. This orbital overlaps with the filled orbital adjacent to it, giving additional bonding between these two carbon atoms. The two sp^2 orbitals are directed 60° away from each other, so their overlap is not very effective. This extremely reactive intermediate is called a **benzyne** because it can be symbolized by drawing a triple bond between these two carbon atoms. Triple bonds are usually linear, however, so this is a very reactive, highly strained triple bond.

Amide ion is a strong nucleophile, attacking at either end of the weak, reactive benzyne triple bond. Subsequent protonation gives toluidine. About half of the product results from attack by the amide ion at the meta carbon, and about half from attack at the para carbon.

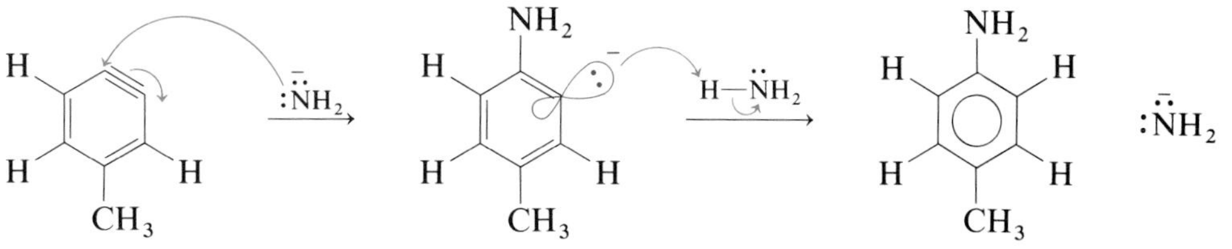

"benzyne"

p-toluidine

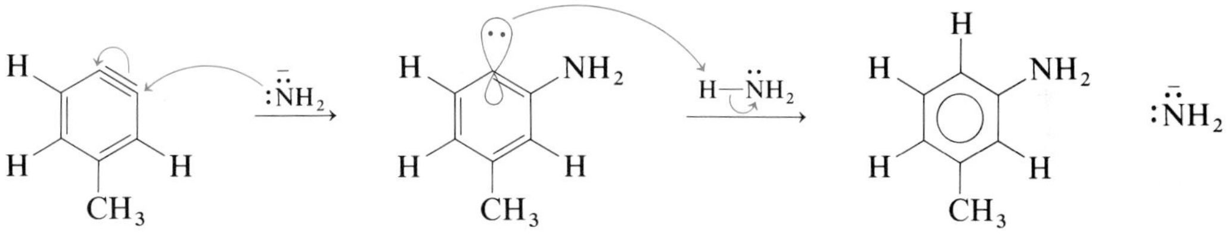

m-toluidine

In summary, the benzyne mechanism operates when the halobenzene is unactivated toward nucleophilic aromatic substitution, and forcing conditions are used with a strong base. The reaction involves an elimination to form a reactive benzyne intermediate. Nucleophilic attack, followed by protonation, gives the substituted product.

PROBLEM 17-25

Give a detailed mechanism to show why *p*-chlorotoluene reacts with sodium hydroxide at 350°C to give a mixture of *p*-cresol and *m*-cresol.

PROBLEM 17-26

Give mechanisms and the expected products of the following reactions.

(a) 2,4-dinitrochlorobenzene + sodium methoxide ($NaOCH_3$)
(b) 2,4-dimethylchlorobenzene + sodium hydroxide, 100°C

(c) *p*-nitrobromobenzene + methylamine ($CH_3—NH_2$)
(d) 2,4-dinitrochlorobenzene + excess hydrazine ($H_2N—NH_2$)

PROBLEM 17-27

Nucleophilic aromatic substitution provides one of the common methods for the synthesis of phenols. (Another method is discussed in Section 19-18.) Show how you would synthesize the following phenols, using benzene or toluene as your aromatic starting material. Explain why mixtures of products would be obtained in some cases.

(a) *p*-nitrophenol (b) 2,4,6-tribromophenol (c) *p*-chlorophenol
(d) *m*-cresol (e) *para-n*-butylphenol

PROBLEM 17-28

The highly reactive triple bond of benzyne is a powerful dienophile. Predict the product of the Diels-Alder reaction of benzyne with cyclopentadiene.

17-13 ADDITION REACTIONS OF BENZENE DERIVATIVES

17-13A CHLORINATION

Although substitution is more common, aromatic compounds may undergo addition if forcing conditions are used. When benzene is treated with an excess of chlorine under heat and pressure (or with irradiation by light), six chlorine atoms add to form 1,2,3,4,5,6-hexachlorocyclohexane. This product is often called *benzene hexachloride* (BHC) because it is synthesized by direct chlorination of benzene.

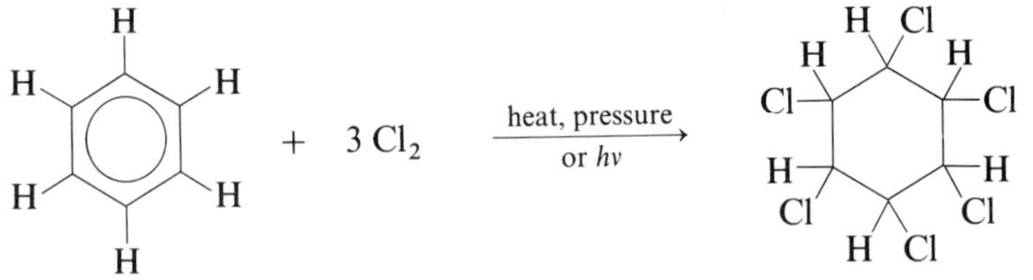

benzene + $3\ Cl_2$ —heat, pressure or $h\nu$→ benzene hexachloride, BHC (eight isomers)

The addition of chlorine to benzene, believed to involve a free-radical mechanism, is normally impossible to stop at an intermediate stage. The first addition destroys the ring's aromaticity, and the next 2 moles of Cl_2 add very rapidly. All eight possible geometric isomers are produced in various amounts. The most important isomer for commercial purposes is the insecticide called *lindane:*

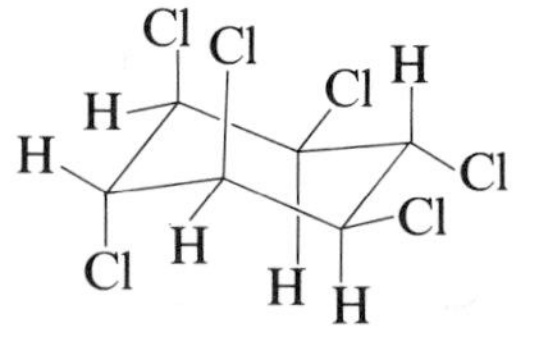

lindane

17-13B CATALYTIC HYDROGENATION OF AROMATIC RINGS

Catalytic hydrogenation of benzene to cyclohexane takes place at elevated temperatures and pressures. Ruthenium or rhodium catalysts are often used to accelerate the hydrogenation. Substituted benzenes react to give substituted cyclohexanes. Disubstituted benzenes usually give mixtures of cis and trans isomers.

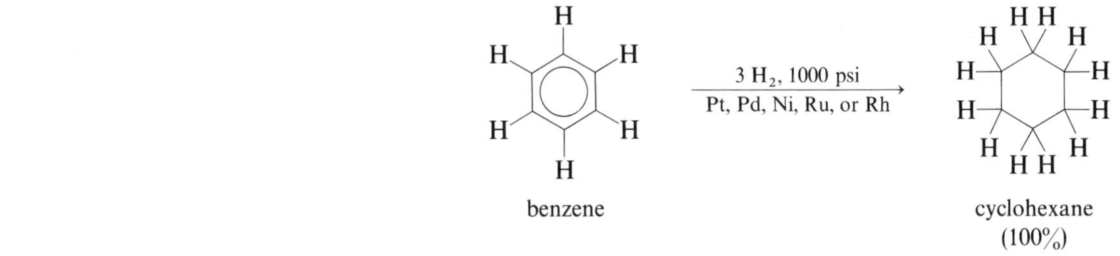

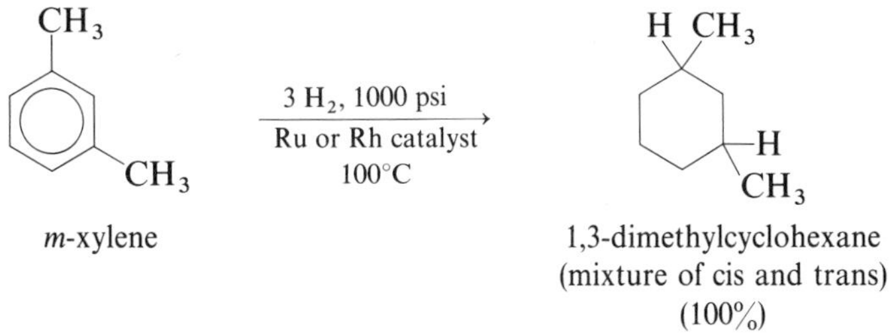

benzene

$\xrightarrow[\text{Pt, Pd, Ni, Ru, or Rh}]{3\ H_2,\ 1000\ \text{psi}}$

cyclohexane
(100%)

m-xylene

$\xrightarrow[\text{Ru or Rh catalyst, } 100°C]{3\ H_2,\ 1000\ \text{psi}}$

1,3-dimethylcyclohexane
(mixture of cis and trans)
(100%)

Catalytic hydrogenation of benzene is the commercial method for producing cyclohexane and substituted cyclohexane derivatives. The reduction cannot be stopped at an intermediate stage (cyclohexene or cyclohexadiene) because these alkenes are reduced faster than benzene.

17-13C BIRCH REDUCTION

In 1944, the Australian chemist A. J. Birch found that benzene derivatives are reduced to non-conjugated 1,4-cyclohexadienes by treatment with sodium or lithium in a mixture of liquid ammonia and an alcohol. The **Birch reduction** provides a convenient method for making a wide variety of interesting and useful cyclic dienes.

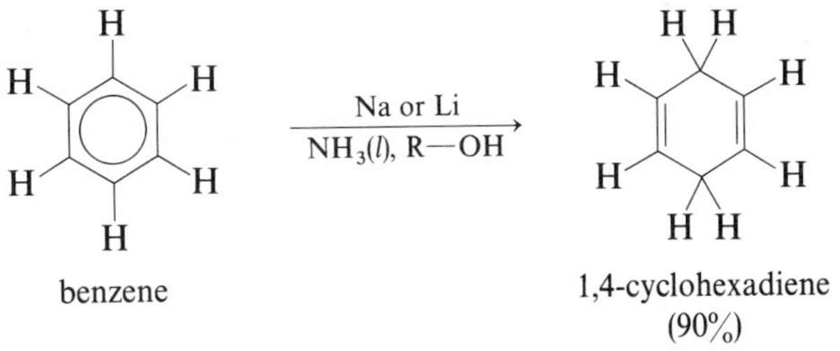

benzene

$\xrightarrow[NH_3(l),\ R{-}OH]{\text{Na or Li}}$

1,4-cyclohexadiene
(90%)

Figure 17-6 shows that the mechanism of the Birch reduction is similar to the sodium/liquid ammonia reduction of alkynes to *trans*-alkenes (Section 14-10C). A solution of sodium in liquid ammonia contains solvated electrons that can add to benzene, forming a radical anion. The strongly basic radical anion abstracts a proton from the alcohol in the solvent, giving a cyclohexadienyl radical. The radical quickly adds another solvated electron to form a cyclohexadienyl anion. Protonation of this anion gives the reduced product.

The two carbon atoms that are reduced go through carbanionic intermediates; electron-withdrawing substituents stabilize the carbanions, while electron-donating substituents destabilize them. Therefore, reduction takes place on carbon atoms bearing electron-withdrawing substituents (such as those containing carbonyl groups) and not on carbon atoms bearing electron-releasing substituents (such as alkyl and alkoxyl groups).

$NH_3 + Na \rightleftharpoons NH_3 \cdot e^-$ (deep blue solution) $+ Na^+$

solvated electron

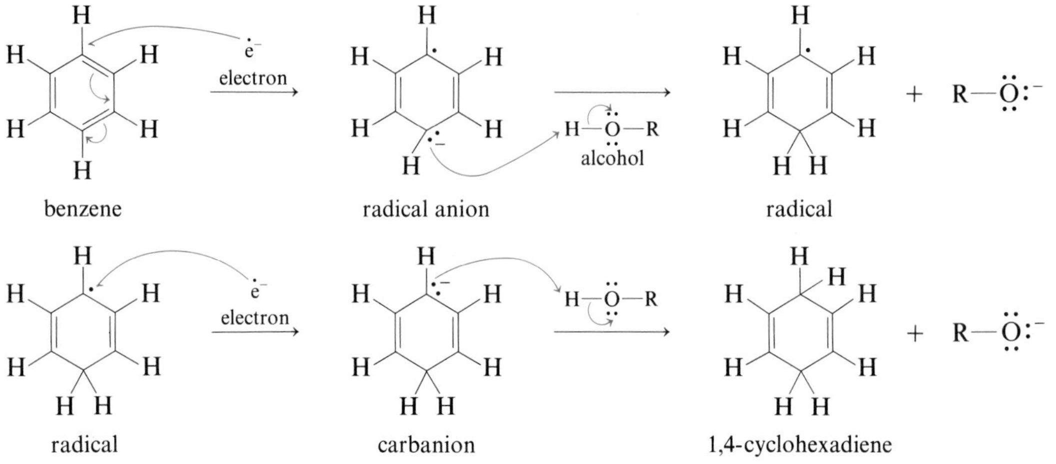

benzene — electron → radical anion — alcohol ($H-\ddot{O}-R$) → radical $+ R-\ddot{O}:^-$

radical — electron → carbanion — $H-\ddot{O}-R$ → 1,4-cyclohexadiene $+ R-\ddot{O}:^-$

FIGURE 17-6 The mechanism of the Birch reduction begins with the addition of a solvated electron to form a radical anion. Protonation of the radical anion gives a neutral radical. Addition of another electron forms a carbanion, which is protonated to give 1,4-cyclohexadiene.

A carbon bearing an electron-withdrawing carbonyl group is reduced

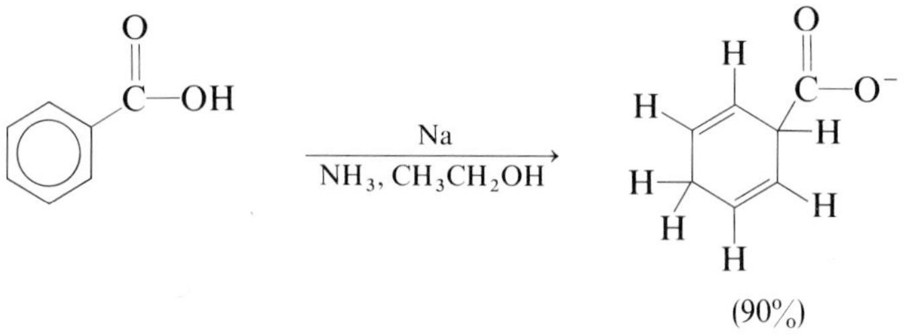

$$\xrightarrow[NH_3,\ CH_3CH_2OH]{Na}$$

(90%)

A carbon bearing an electron-releasing alkoxyl group is not reduced

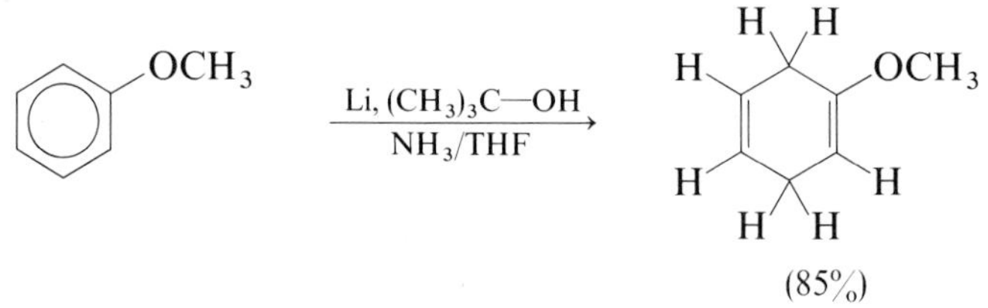

OCH_3

$$\xrightarrow[NH_3/THF]{Li,\ (CH_3)_3C-OH}$$

OCH_3

(85%)

Substituents that are strongly electron-releasing ($-OCH_3$, for example) deactivate the aromatic ring toward Birch reduction. Lithium is often used with these deactivated systems, together with a cosolvent (often THF) and a weaker proton source (*t*-butyl alcohol). The stronger reducing agent, combined with a weaker proton source, enhances the reduction.

PROBLEM 17-29

Give mechanisms for the Birch reductions shown above. Show why the observed orientation of reduction is favored in each case.

PROBLEM 17-30

Predict the major products of the following reactions.

(a) toluene + excess Cl_2 (heat, pressure)
(b) toluene + Na (liquid NH_3, CH_3CH_2OH)
(c) *o*-xylene + H_2 (1000 psi, 100°C, Rh catalyst)
(d) *p*-xylene + Na (liquid NH_3, CH_3CH_2OH)

(e) 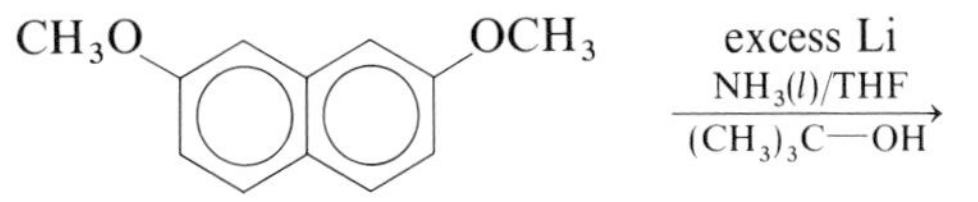 2,7-dimethoxynaphthalene $\xrightarrow[(CH_3)_3C{-}OH]{\text{excess Li, } NH_3(l)/THF}$

(f) 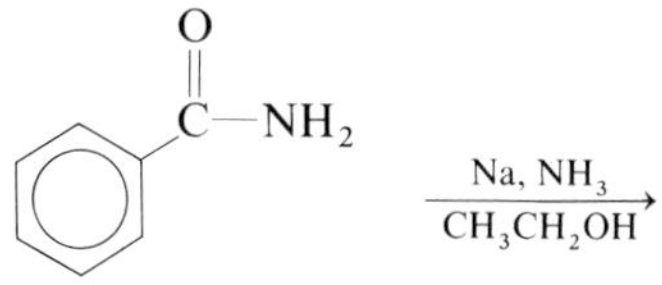 benzamide $\xrightarrow[CH_3CH_2OH]{Na,\ NH_3}$

17-14 SIDE-CHAIN REACTIONS OF BENZENE DERIVATIVES

Many reactions are not affected by the presence of a nearby benzene ring; yet others depend on the aromatic ring to enhance the reaction. For example, the Clemmensen reduction is occasionally used to reduce aliphatic ketones to alkanes, but it works best reducing aryl ketones to alkylbenzenes. Several additional side-chain reactions show the effects of a nearby aromatic ring.

17-14A PERMANGANATE OXIDATION

An aromatic ring imparts extra stability to the nearest carbon atom of its side chains. The aromatic ring and *one* carbon atom of a side chain can survive a vigorous permanganate oxidation to give a carboxylate salt of benzoic acid. This oxidation is occasionally useful for making benzoic acid derivatives, as long as any other functional groups are resistant to oxidation.

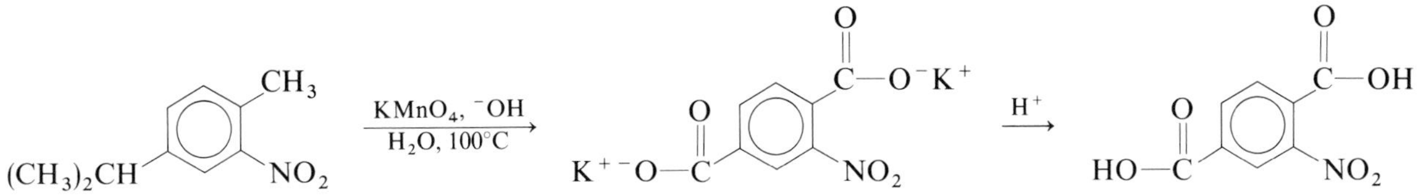

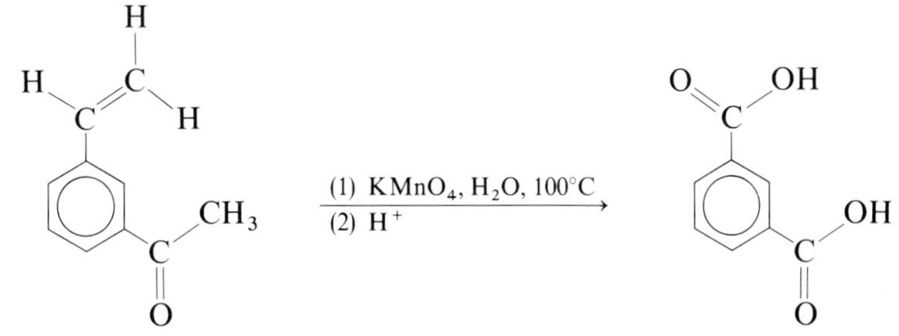

$$\text{4-isopropyl-2-nitrotoluene} \xrightarrow[H_2O,\ 100°C]{KMnO_4,\ ^-OH} K^{+\,-}O{-}C(=O){-}C_6H_3(NO_2){-}C(=O){-}O^-K^+ \xrightarrow{H^+} HO{-}C(=O){-}C_6H_3(NO_2){-}C(=O){-}OH$$

$$\text{3-vinylacetophenone} \xrightarrow[(2)\ H^+]{(1)\ KMnO_4,\ H_2O,\ 100°C} \text{isophthalic acid}$$

PROBLEM 17-31

Predict the major products of treating the following compounds with hot, concentrated potassium permanganate, followed by acidification with dilute HCl.

(a) isopropylbenzene (b) *n*-butylbenzene (c) 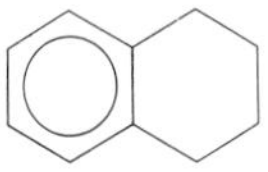 tetralin

17-14B SIDE-CHAIN HALOGENATION

Alkylbenzenes undergo free-radical halogenation much more easily than alkanes, because abstraction of a hydrogen atom at a **benzylic position** gives a resonance-stabilized benzylic radical. For example, ethylbenzene reacts with chlorine in the presence of light to give α-chloroethylbenzene. Further chlorination can occur to give dichlorinated products.

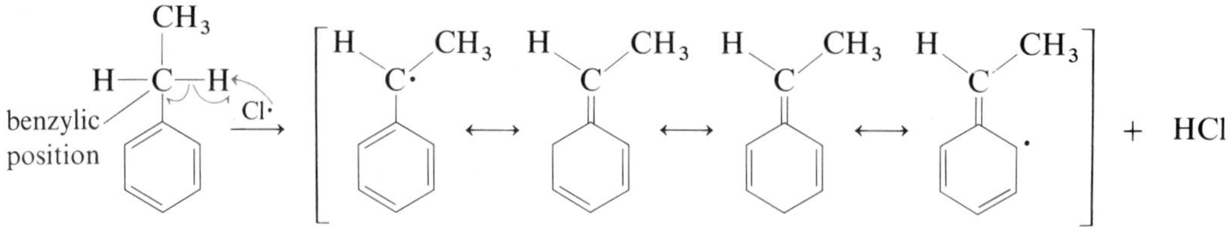

benzylic position $\xrightarrow{Cl\cdot}$ [⟷ ⟷ ⟷] + HCl

resonance-stabilized benzylic radical

$\xrightarrow[h\nu]{Cl_2}$ dichlorinated products

benzylic radical — α-chloroethylbenzene — chlorine radical continues the chain

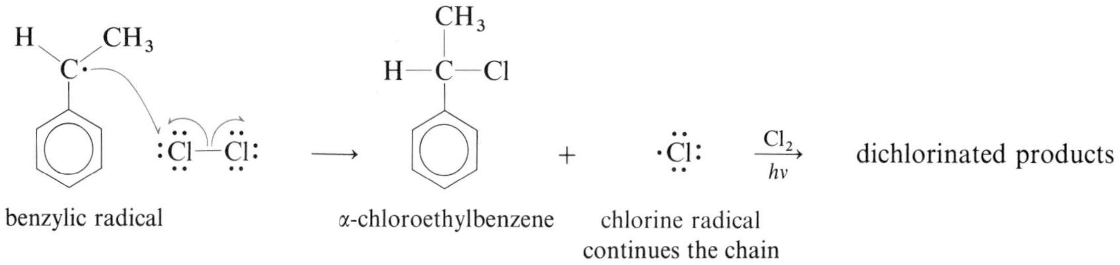

PROBLEM 17-32

Propose a mechanism for the formation of α,α-dichloroethylbenzene from α-chloroethylbenzene. The IUPAC names of these compounds are (1,1-dichloroethyl)benzene and (1-chloroethyl)benzene.

Although chlorination shows a preference for α substitution (the α position is the benzylic carbon bonded to the benzene ring), the chlorine radical is too reactive to give entirely benzylic substitution. Mixtures of isomers are commonly produced. In the chlorination of ethylbenzene, for example, there is a significant amount of substitution at the β carbon.

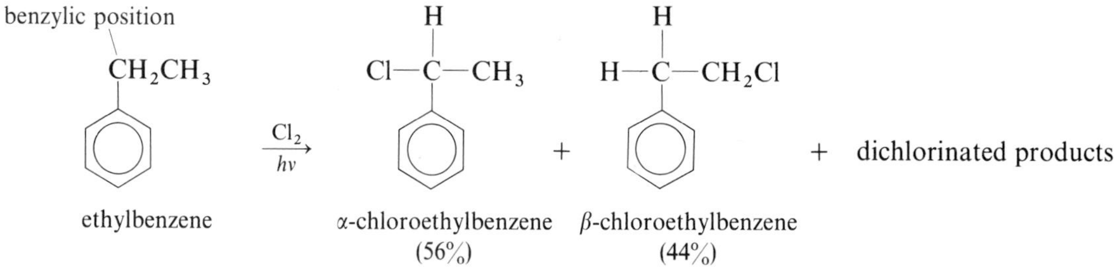

benzylic position

CH_2CH_3 $\xrightarrow[h\nu]{Cl_2}$ $Cl-CH(CH_3)-$ + $H-CH(CH_2Cl)-$ + dichlorinated products

ethylbenzene — α-chloroethylbenzene (56%) — β-chloroethylbenzene (44%)

Bromine radicals are not as reactive as chlorine radicals, and bromination is more selective than chlorination (Section 4-14C). Bromine reacts exclusively at the benzylic position.

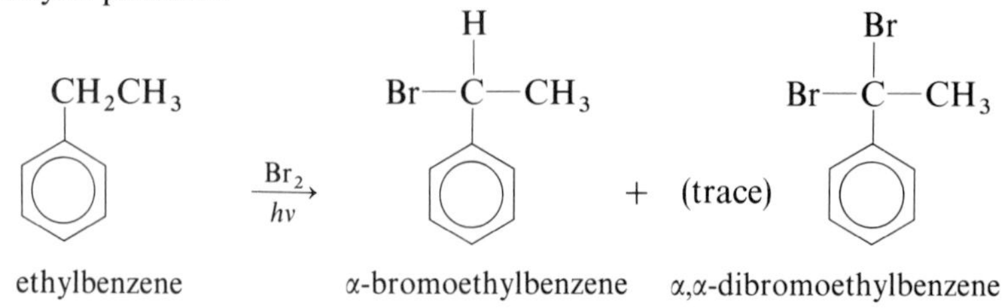

CH_2CH_3 $\xrightarrow[h\nu]{Br_2}$ $Br-CH(CH_3)-$ + (trace) $Br-CBr(CH_3)-$

ethylbenzene — α-bromoethylbenzene — α,α-dibromoethylbenzene

Either elemental bromine (much cheaper) or *N*-bromosuccinimide may be used as the reagent for benzylic bromination. *N*-Bromosuccinimide is required for allylic bromination (Section 15-7), since Br_2 can add to the double bond. This is not a problem with the relatively unreactive benzene ring unless it has powerful activating substituents.

PROBLEM 17-33

What would be the ratio of products in the reaction of chlorine with ethylbenzene if chlorine *randomly* abstracted a methyl or methylene proton? What is the reactivity ratio for the benzylic hydrogens compared with the methyl hydrogens?

PROBLEM 17-34

Propose a mechanism for the bromination of ethylbenzene shown above.

PROBLEM 17-35

Predict the major products when the following compounds are irradiated by light and treated with (1) 1 mole of Br_2 and (2) excess Br_2.

(a) isopropylbenzene (b) 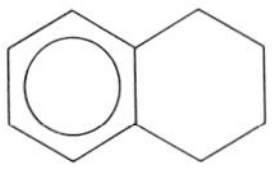 (tetralin)

17-14C NUCLEOPHILIC SUBSTITUTION AT THE BENZYLIC POSITION

In Chapter 15 we saw that allylic halides are more reactive than most alkyl halides in both S_N1 and S_N2 reactions. Benzylic halides are also more reactive in these substitutions, for reasons that are similar to those for allylic halides.

First-order reactions A first-order nucleophilic substitution requires ionization of the halide to give a carbocation. In the case of a benzylic halide, the carbocation is resonance stabilized. For example, the 1-phenylethyl cation is about as stable as a tertiary alkyl cation.

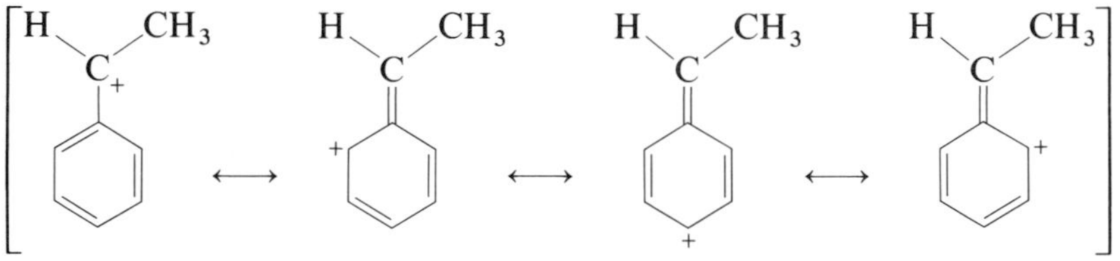

1-phenylethyl cation (2°) is about as stable as 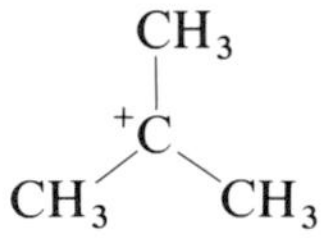 *t*-butyl cation (3°)

Because they form relatively stable carbocations, benzyl halides undergo S_N1-type reactions fairly easily.

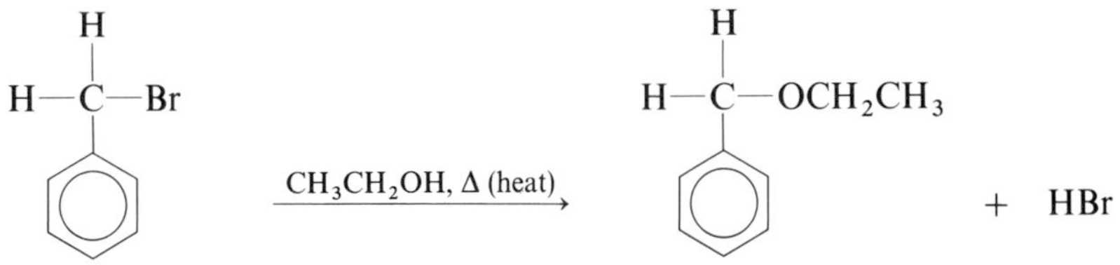

benzyl bromide $\xrightarrow{CH_3CH_2OH,\ \Delta\ \text{(heat)}}$ benzyl ethyl ether + HBr

PROBLEM 17-36

Propose a mechanism for the reaction of benzyl bromide with ethanol shown above.

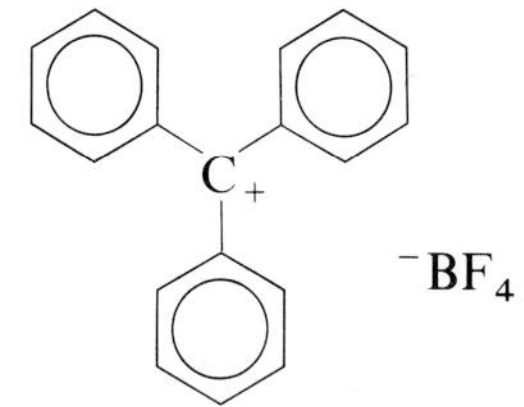

triphenylmethyl fluoroborate

If a benzylic cation is bonded to more than one phenyl group, the stabilizing effects are additive. An extreme example is the triphenylmethyl cation. This cation is exceptionally stable, with three phenyl groups to stabilize the positive charge. In fact, triphenylmethyl fluoroborate can be stored for years as a stable ionic solid.

PROBLEM 17-37

(a) Based on what you know about the relative stabilities of alkyl cations and benzylic cations, predict the product of addition of HBr to 1-phenylpropene.
(b) Propose a mechanism for this reaction.

Second-order reactions Like allylic halides, benzylic halides are about 100 times as reactive as primary alkyl halides in S_N2 displacement reactions. The explanation for this enhanced reactivity is similar to the explanation for the reactivity of allylic halides.

When a benzylic halide undergoes S_N2 displacement, the *p* orbital that is partially bonding with the nucleophile and the leaving group overlaps with the pi electrons in the ring (Fig. 17-7). This stabilizing conjugation lowers the energy of the transition state, increasing the reaction rate.

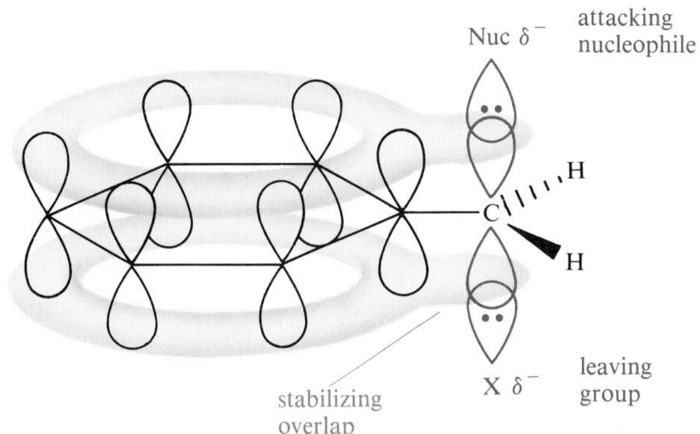

FIGURE 17-7 The transition state for S_N2 displacement of a benzylic halide is stabilized by conjugation with the pi electrons in the ring.

The S_N2 reactions of benzyl halides provide efficient methods for converting aromatic methyl groups to functional groups. Halogenation, followed by substitution, gives the functionalized product.

$$p\text{-}O_2N\text{-}C_6H_4\text{-}CH_3 \xrightarrow[h\nu]{Br_2} p\text{-}O_2N\text{-}C_6H_4\text{-}CH_2Br \xrightarrow[CH_3OH]{NaOCH_3} p\text{-}O_2N\text{-}C_6H_4\text{-}CH_2OCH_3$$

$$o\text{-}Br\text{-}C_6H_4\text{-}CH_3 \xrightarrow[h\nu]{Br_2} o\text{-}Br\text{-}C_6H_4\text{-}CH_2Br \xrightarrow[\text{acetone}]{NaCN} o\text{-}Br\text{-}C_6H_4\text{-}CH_2CN$$

PROBLEM 17-38

Show how you would synthesize the following compounds, using the indicated compounds as your starting materials.

(a) 3-phenyl-1-butanol from styrene

(b) 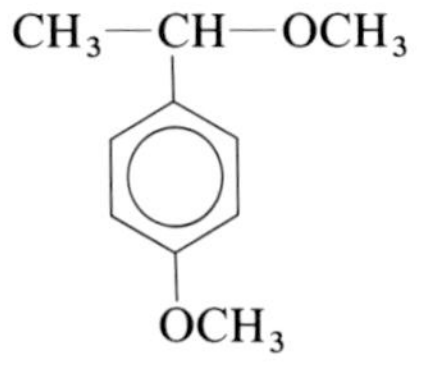 CH_3—CH—OCH_3 (para-substituted benzene with OCH_3) from anisole

(c) 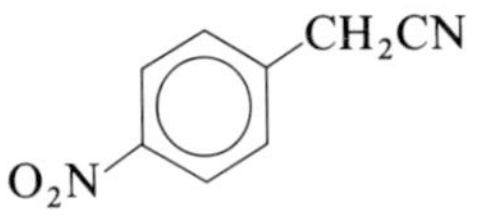 O_2N–C₆H₄–CH_2CN (para) from toluene

17-15 REACTIONS OF PHENOLS

Some of the chemistry of phenols is like that of aliphatic alcohols. For example, phenols can be acylated to give esters, and phenoxide ions can serve as nucleophiles in the Williamson ether synthesis. Formation of phenoxide ions is particularly easy, because phenols are more acidic than water; the phenoxide ion can be made by treatment of the phenol with aqueous sodium hydroxide.

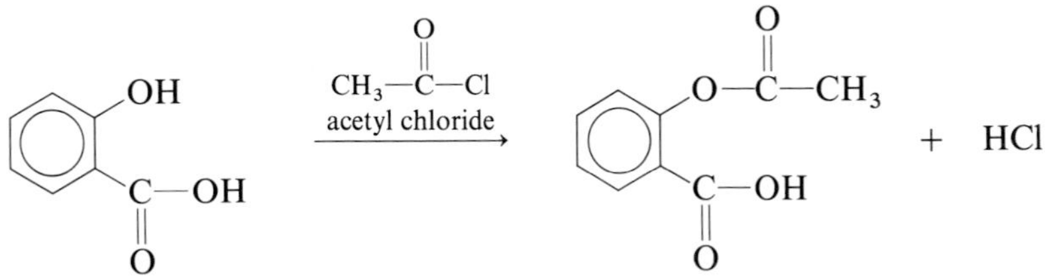

salicylic acid $\xrightarrow[\text{acetyl chloride}]{CH_3-C(=O)-Cl}$ acetylsalicylic acid (aspirin) + HCl

OH-substituted propenylbenzene $\xrightarrow[H_2O]{NaOH}$ O^-Na^+ $\xrightarrow[\text{(or } CH_3I)]{CH_3OSO_2OCH_3 \text{ dimethyl sulfate}}$ OCH_3

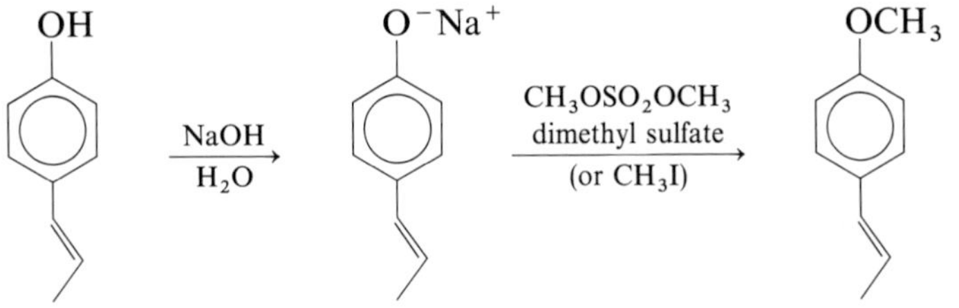

anethole
(licorice flavoring)

Many of the common reactions of alcohols do not apply to phenols, however. For example, phenols do not undergo acid-catalyzed elimination or S_N2 backside attack.

17-15A OXIDATION OF PHENOLS TO QUINONES

Phenols undergo oxidation, but they give different types of products from those seen with aliphatic alcohols. Chromic acid oxidation of a phenol gives a conjugated 1,4-diketone called a **quinone.** In the presence of air, many phenols slowly autoxidize to dark, quinone-containing products.

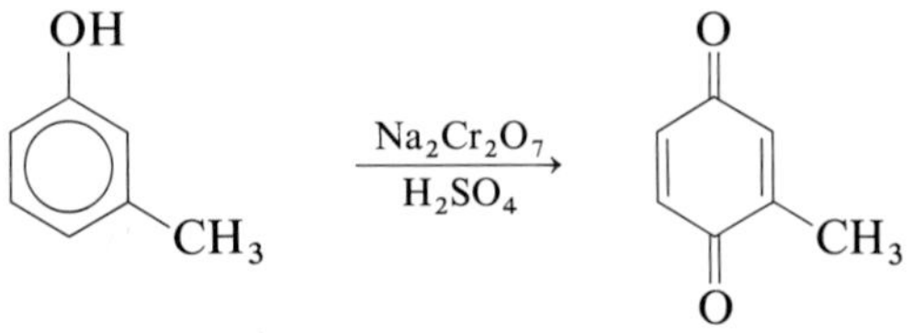

m-cresol $\xrightarrow[H_2SO_4]{Na_2Cr_2O_7}$ 2-methyl-1,4-benzoquinone

Hydroquinone (1,4-benzenediol) is easily oxidized, because it already has two oxygen atoms bonded to the ring. Even very weak oxidants like silver bromide (AgBr) can oxidize hydroquinone. The silver bromide is reduced to black, metallic silver in a light-sensitive reaction: Any grains of silver bromide that have been exposed to light (AgBr*) react faster than unexposed grains.

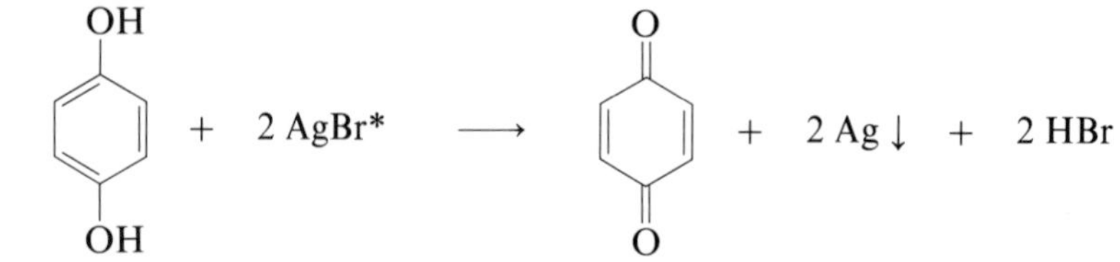

$$\text{hydroquinone} + 2\,AgBr^* \longrightarrow \text{benzoquinone} + 2\,Ag\downarrow + 2\,HBr$$

hydroquinone — benzoquinone

Black-and-white photography is based on this reaction. A film containing small grains of silver bromide is exposed by a focused image. Where light strikes the film, the grains are activated. The film is then treated with a hydroquinone solution (the *developer*) to reduce the activated silver bromide grains, leaving black silver deposits where the film was exposed to light. The result is a negative image, with dark areas where light struck the film.

Quinones are found widely in nature, where they serve as biological oxidation-reduction reagents. The quinone *coenzyme Q* (CoQ) is also called *ubiquinone* because it is ubiquitous (found everywhere) in oxygen-consuming organisms. Coenzyme Q serves as an oxidizing agent within the mitochondria of cells. The following reaction shows the reduction of coenzyme Q by NADH (the reduced form of nicotinamide adenine dinucleotide), which becomes oxidized to NAD^+.

$$R = -(CH_2-CH=\underset{}{\overset{CH_3}{\overset{|}{C}}}-CH_2)_{10}-H$$

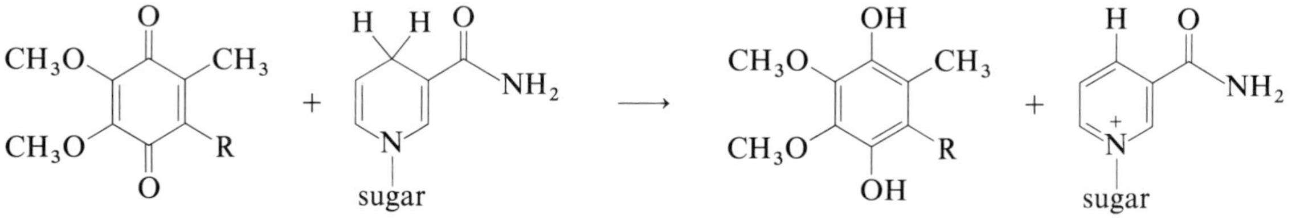

$$\text{coenzyme Q (oxidized form)} + \text{NADH (reduced form)} \longrightarrow \text{coenzyme Q (reduced form)} + NAD^+ \text{ (oxidized form)}$$

coenzyme Q oxidized form — NADH reduced form — coenzyme Q reduced form — NAD^+ oxidized form

17-15B ELECTROPHILIC AROMATIC SUBSTITUTION OF PHENOLS

Phenols are highly reactive substrates for electrophile aromatic substitution, because the nonbonding electrons of the hydroxyl group stabilize the sigma complex formed by attack at the ortho or para position (Section 17-6B). Therefore, we say the hydroxyl group is strongly activating and ortho, para-directing. Phenols are excellent substrates for halogenation, nitration, sulfonation, and some Friedel-Crafts reactions. Because they are highly reactive, phenols are usually alkylated or acylated using relatively weak Friedel-Crafts catalysts (such as HF) to avoid overalkylation or overacylation.

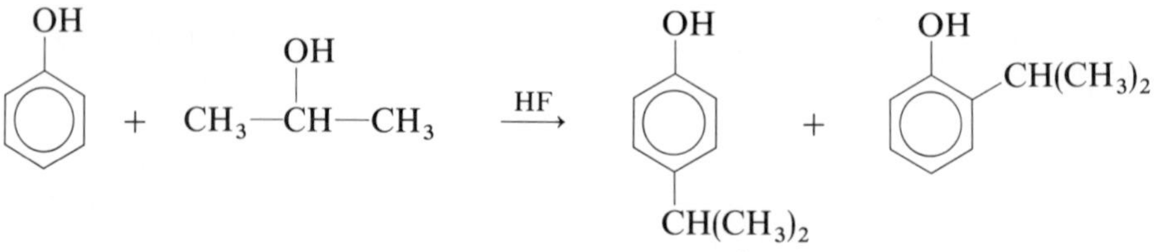

$$\text{phenol} + CH_3-\underset{}{\overset{OH}{\overset{|}{CH}}}-CH_3 \xrightarrow{HF} p\text{-}CH(CH_3)_2\text{-phenol} + o\text{-}CH(CH_3)_2\text{-phenol}$$

Phenoxide ions, easily generated by treating a phenol with sodium hydroxide, are even more reactive than phenols toward electrophilic aromatic substitution. Because they are negatively charged, phenoxide ions react with positively charged electrophiles to give neutral sigma complexes whose structures resemble quinones.

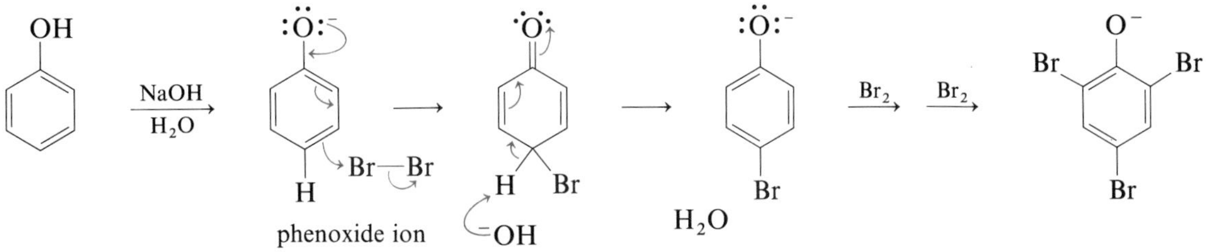

phenoxide ion

Phenoxide ions are so strongly activated that they undergo electrophilic aromatic substitution with carbon dioxide, a weak electrophile. The carboxylation of phenoxide ion is the industrial synthesis of salicylic acid, needed for conversion to aspirin (shown above).

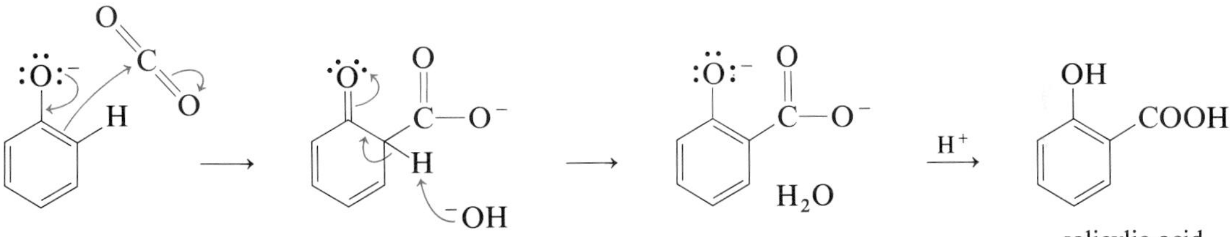

salicylic acid

PROBLEM 17-39

Predict the products formed when *m*-cresol (*m*-methylphenol) reacts with

(a) NaOH and then ethyl bromide
(b) acetyl chloride, $CH_3-\overset{O}{\overset{\|}{C}}-Cl$
(c) bromine in CCl_4 in the dark
(d) excess bromine in CCl_4 in the light
(e) sodium dichromate in H_2SO_4
(f) 2 equivalents of *t*-butyl chloride and $AlCl_3$

PROBLEM 17-40

Benzoquinone is a good Diels-Alder dienophile. Predict the products of its reaction with (a) 1,3-butadiene and (b) cyclopentadiene.

PROBLEM 17-41

Phenol reacts with 3 equivalents of bromine in CCl_4 (in the dark) to give a product of formula $C_6H_3OBr_3$. When this product is added to bromine water, a yellow solid of molecular formula $C_6H_2OBr_4$ precipitates out of the solution. The IR spectrum of the yellow precipitate shows a strong absorption (much like that of a quinone) around 1680 cm^{-1}. Propose structures for the two products.

SUMMARY OF THE REACTIONS OF AROMATIC COMPOUNDS

1. Electrophile aromatic substitution

a. *Halogenation* (Section 17-2)

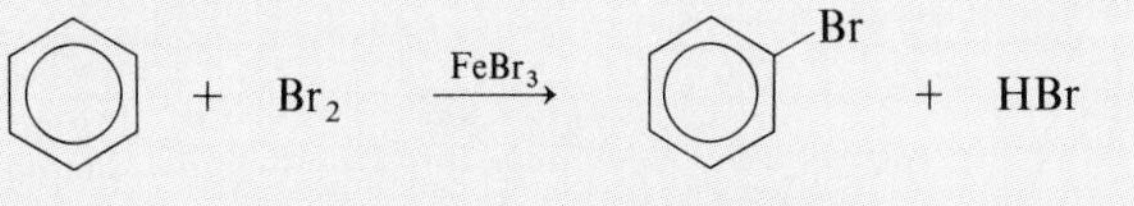

$$C_6H_6 + Br_2 \xrightarrow{FeBr_3} C_6H_5Br + HBr$$

bromobenzene

b. *Nitration* (Section 17-3)

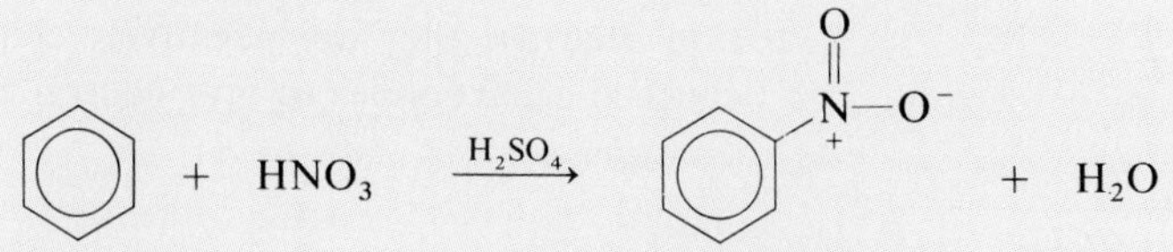

$$C_6H_6 + HNO_3 \xrightarrow{H_2SO_4} C_6H_5\overset{+}{N}(=O)-O^- + H_2O$$

nitrobenzene

c. *Sulfonation* (Section 17-4)

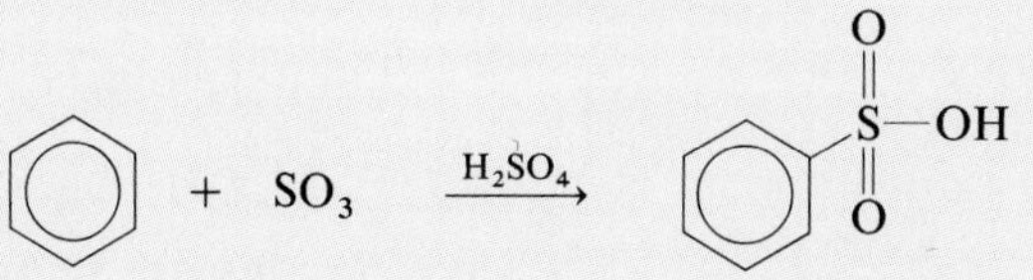

$$C_6H_6 + SO_3 \xrightarrow{H_2SO_4} C_6H_5S(=O)_2-OH$$

benzenesulfonic acid

d. *Friedel-Crafts alkylation* (Section 17-10)

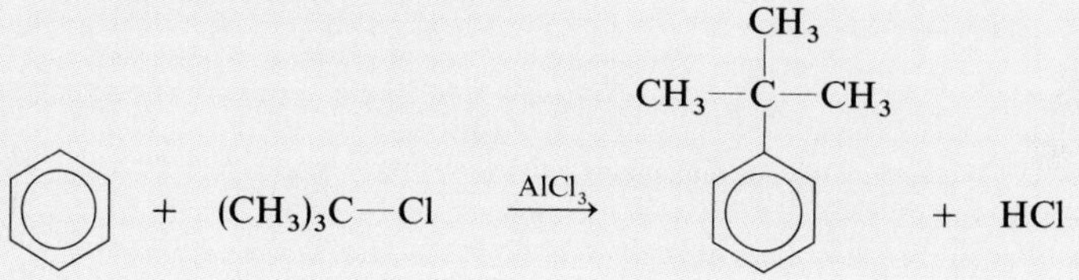

$$C_6H_6 + (CH_3)_3C-Cl \xrightarrow{AlCl_3} C_6H_5-C(CH_3)_3 + HCl$$

t-butylbenzene

e. *Friedel-Crafts acylation* (Section 17-11)

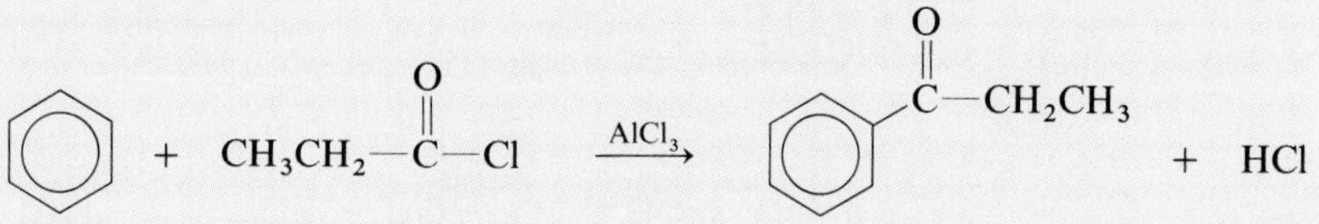

$$C_6H_6 + CH_3CH_2-C(=O)-Cl \xrightarrow{AlCl_3} C_6H_5-C(=O)-CH_2CH_3 + HCl$$

propiophenone

f. *Gatterman-Koch synthesis* (Section 17-11C)

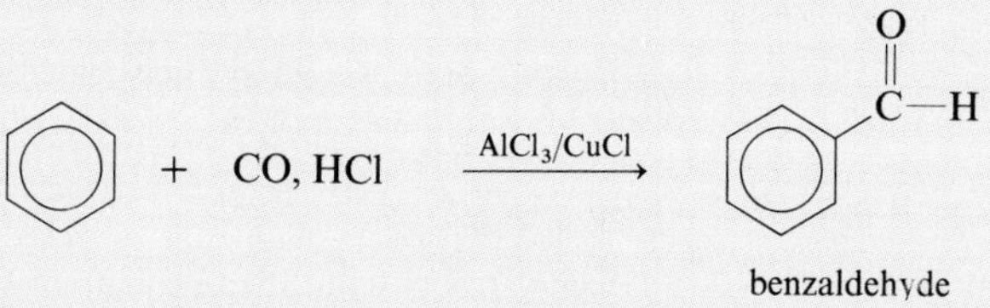

$$C_6H_6 + CO, HCl \xrightarrow{AlCl_3/CuCl} C_6H_5-C(=O)-H$$

benzaldehyde

g. *Substituent effects* (Sections 17-5 through 17-9)

Activating, ortho, para-directing: $-R$, $-\ddot{O}R$, $-\ddot{O}H$, $-\ddot{O}:^-$, $-\ddot{N}R_2$ (amines, amides)

Deactivating, ortho, para-directing: $-Cl$, $-Br$, $-I$

Deactivating, meta-allowing: $-NO_2$, $-SO_3H$, $-NR_3^+$, $-C{=}O$ (carbonyl groups)

2. *Nucleophilic aromatic substitution* (Section 17-12)

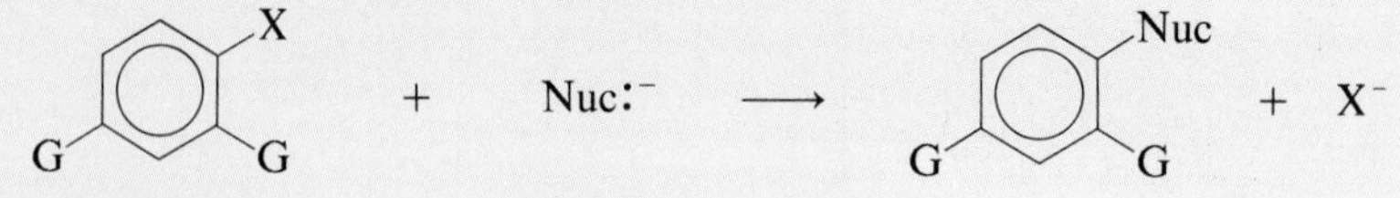

$$\text{X-C}_6\text{H}_2\text{G}_2 + \text{Nuc:}^- \longrightarrow \text{Nuc-C}_6\text{H}_2\text{G}_2 + X^-$$

a halobenzene (G = strong withdrawing group) — strong nucleophile

Example

$$\text{2,4-dinitrochlorobenzene} + NaNH_2 \longrightarrow \text{2,4-dinitroaniline} + NaCl$$

2,4-dinitrochlorobenzene   2,4-dinitroaniline

If *G* is not a strong withdrawing group, severe conditions are required and a benzyne mechanism is involved.

3. *Addition reactions* (Section 17-13)
   a. *Chlorination* (Section 17-13A)

$$\text{benzene} + 3\ Cl_2 \xrightarrow[\text{or light}]{\text{heat and pressure}} \text{benzene hexachloride (BHC)}$$

benzene   benzene hexachloride (BHC)

   b. *Catalytic hydrogenation* (Section 17-13B)

$$o\text{-diethylbenzene} + 3\ H_2 \xrightarrow[100°C,\ 1000\ psi]{\text{Ru or Rh catalyst}} \text{1,2-diethylcyclohexane}$$

o-diethylbenzene   1,2-diethylcyclohexane (mixture of cis and trans)

   c. *Birch reduction* (Section 17-13C)

$$\text{ethylbenzene} \xrightarrow[NH_3(l),\ R{-}OH]{\text{Na or Li}} \text{1-ethyl-1,4-cyclohexadiene}$$

ethylbenzene   1-ethyl-1,4-cyclohexadiene

4. *Side-chain reactions*
   a. *The Clemmensen reducion*
   (converts acylbenzenes to alkylbenzenes, Section 17-11B)

$$Ph{-}C(=O){-}R \xrightarrow[\text{dilute HCl}]{Zn(Hg)} Ph{-}CH_2{-}R$$

an acylbenzene   an alkylbenzene

   b. *Permanganate oxidation* (Section 17-14A)

$$Ph{-}CH_2{-}R \xrightarrow[H_2O]{\text{hot, conc. }KMnO_4} Ph{-}COO^-K^+$$

an alkylbenzene   a benzoic acid salt

   c. *Side-chain halogenation* (Section 17-14B)

$$Ph{-}CH_2{-}R \xrightarrow[h\nu]{Br_2} Ph{-}CHBr{-}R$$

an alkylbenzene   an α-bromo alkylbenzene

d. *Nucleophile substitution at the benzylic position* (Section 17-14C)
The benzylic position is activated toward both S_N1 and S_N2 displacements.

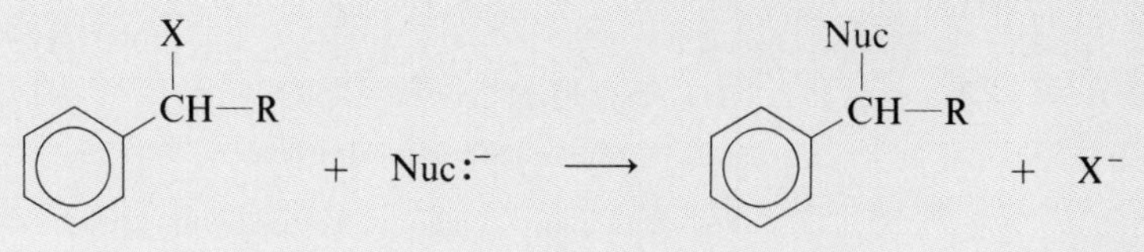

$$\text{C}_6\text{H}_5\text{—CH(X)—R} + \text{Nuc:}^- \longrightarrow \text{C}_6\text{H}_5\text{—CH(Nuc)—R} + \text{X}^-$$

an α-halo alkylbenzene

5. *Oxidation of phenols to quinones* (Section 17-15A)

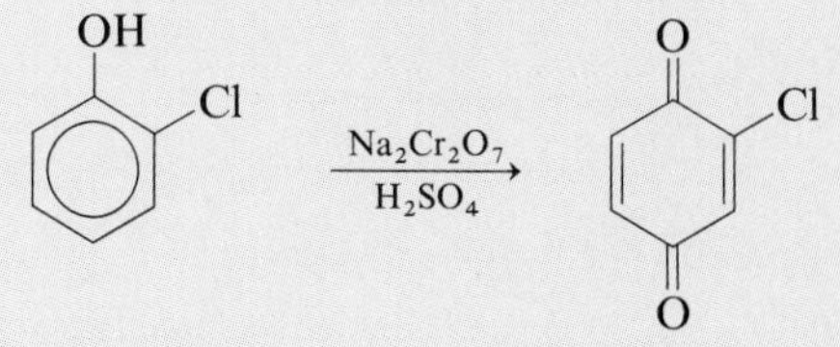

$$o\text{-chlorophenol} \xrightarrow[H_2SO_4]{Na_2Cr_2O_7} \text{2-chloro-1,4-benzoquinone}$$

o-chlorophenol 2-chloro-1,4-benzoquinone

GLOSSARY

activating group A substituent that makes the aromatic ring more reactive (usually toward electrophilic aromatic substitution) than benzene. (p. 721)

acyl group ($R—C(=O)—$) A carbonyl group with an alkyl group attached. (p. 740)

acylium ion ($R—C{\equiv}\overset{+}{O}$) An acyl group with a positive charge. (p. 741)

benzylic position The carbon atom of an alkyl group that is directly bonded to a benzene ring; the position α to a benzene ring. (p. 752)

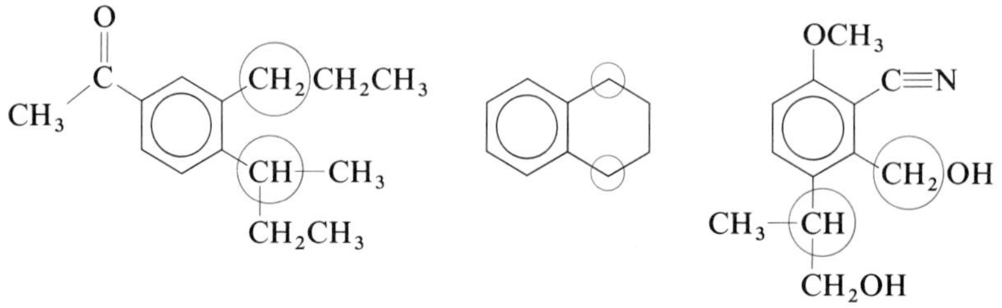

CH₃C(=O)– ; CH₂CH₂CH₃ ; CH—CH₃ ; CH₂CH₃ ; OCH₃ ; C≡N ; CH₂OH ; CH₃—CH ; CH₂OH

The benzylic positions are circled in red.

benzyne A reactive intermediate in some nucleophilic aromatic substitutions, benzyne is benzene with two hydrogen atoms removed. It can be drawn with a highly strained triple bond in the six-membered ring. (p. 746)

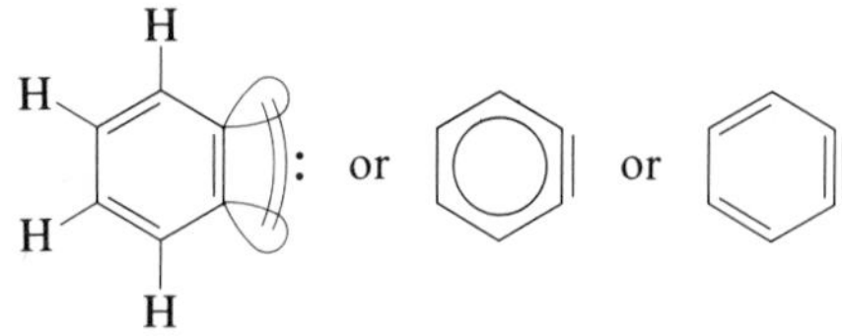

H H H H : or or

Birch reduction The partial reduction of a benzene ring by sodium or lithium in liquid ammonia. The products are usually 1,4-cyclohexadienes. (p. 749)

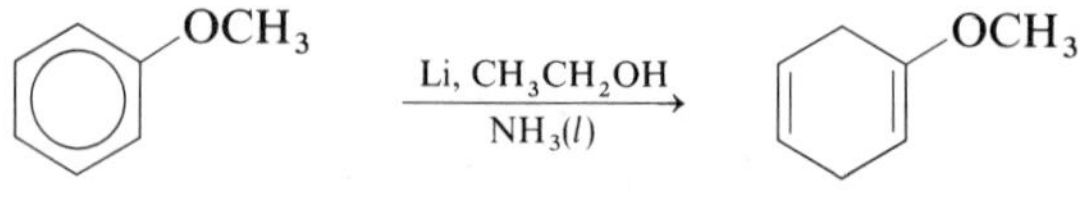

$$\text{C}_6\text{H}_5\text{OCH}_3 \xrightarrow[NH_3(l)]{Li,\ CH_3CH_2OH} \text{1-methoxy-1,4-cyclohexadiene (OCH}_3\text{)}$$

Clemmensen reduction The reduction of a carbonyl group to a methylene group by zinc amalgam, Zn(Hg) in dilute hydrochloric acid. (p. 742)

amalgam: An alloy of a metal with mercury.

deactivating group A substituent that makes the aromatic ring less reactive (usually toward electrophilic aromatic substitution) than benzene. (p. 728)

electrophilic aromatic substitution (EAS) Replacement of a hydrogen on an aromatic ring by a strong electrophile. (p. 715)

Friedel-Crafts acylation The formation of an acylbenzene by the reaction of the aromatic system with an acylium ion. (p. 740)

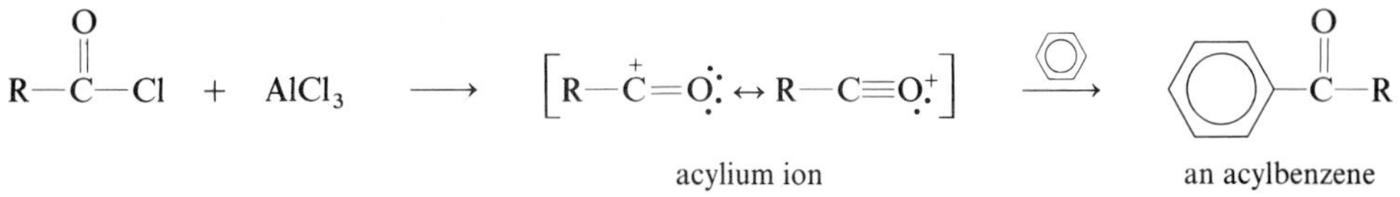

$$R-\overset{O}{\overset{\|}{C}}-Cl + AlCl_3 \longrightarrow \left[R-\overset{+}{C}=\ddot{O}: \leftrightarrow R-C\equiv O:^+\right] \xrightarrow{\text{benzene}} C_6H_5-\overset{O}{\overset{\|}{C}}-R$$

acylium ion — an acylbenzene

Friedel-Crafts alkylation The formation of an alkyl-substituted benzene derivative by the reaction of the aromatic system with an alkyl carbocation or carbocation-like species. (p. 735)

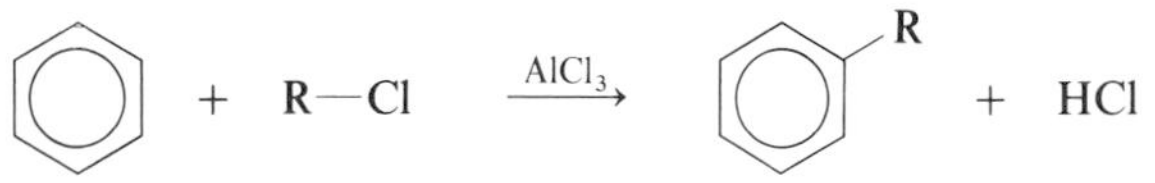

$$C_6H_6 + R-Cl \xrightarrow{AlCl_3} C_6H_5-R + HCl$$

Gatterman-Koch synthesis The synthesis of benzaldehydes by the treatment of a benzene derivative with CO and HCl using an $AlCl_3/CuCl$ catalyst. (p. 743)

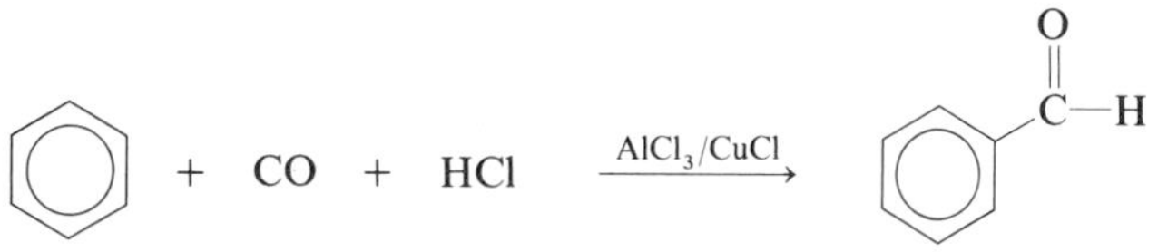

$$C_6H_6 + CO + HCl \xrightarrow{AlCl_3/CuCl} C_6H_5-\overset{O}{\overset{\|}{C}}-H$$

halonium ion Any positively charged ion that has a positive charge (or partial positive charge) on a halogen atom. In aromatic chemistry, the halonium ion usually has a positive charge delocalized onto the halogen through resonance. (Specific: chloronium ion, bromonium ion, etc.) (p. 731)

inductive stabilization Stabilization of a reactive intermediate by donation or withdrawal of electron density through sigma bonds. (p. 724)

meta-director (meta-allower) A substituent that deactivates primarily the ortho and para positions toward attack, leaving the meta position the least deactivated and most reactive. (p. 728)

nitration The replacement of a hydrogen atom by a nitro group, $-NO_2$. (p. 719)

nitronium ion The NO_2^+ ion, $O=\overset{+}{N}=O$ (p. 719)

nucleophilic aromatic substitution (NAS) Replacement of a leaving group on an aromatic ring by a strong nucleophile. (p. 744)

ortho, para-director A substituent that activates primarily the ortho and para positions toward attack. (p. 724)

quinone A derivative of a cyclohexadiene-dione. Common quinones are the 1,4-quinones (*para*-quinones); the less stable 1,2-quinones (*ortho*-quinones) are relatively uncommon. (p. 755)

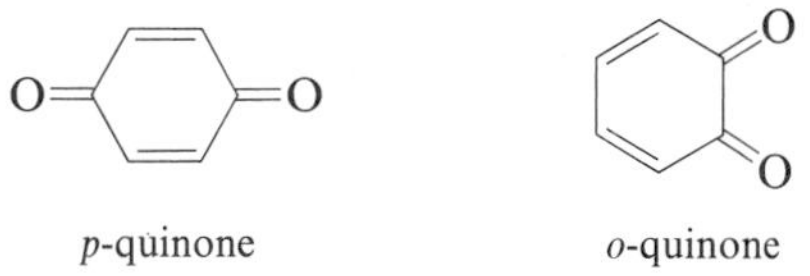

O= ring =O — ring with two =O

p-quinone — *o*-quinone

resonance stabilization Stabilization of a reactive intermediate by donation or withdrawal of electron density through pi bonds. (p. 725)

resonance donating (pi donating): Capable of donating electrons through resonance involving pi bonds.

resonance withdrawing (pi withdrawing): Capable of withdrawing electron density through resonance involving pi bonds.

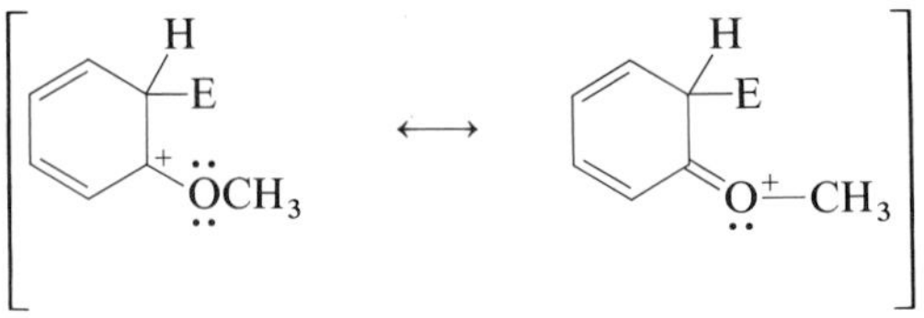

alkoxyl groups are pi donating

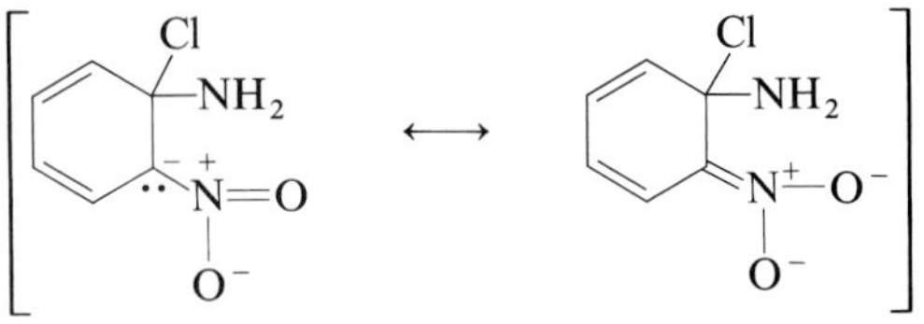

nitro groups are pi withdrawing

sigma complex An intermediate in electrophilic aromatic substitution or nucleophilic aromatic substitution with a sigma bond between the electrophile or nucleophile and the former aromatic ring. The sigma complex bears a delocalized positive charge in electrophilic aromatic substitution and a delocalized negative charge in nucleophilic aromatic substitution. (p. 716)

sulfonation The replacement of a hydrogen atom by a sulfonic acid group, $—SO_3H$. (p. 720)

desulfonation: The replacement of the $—SO_3H$ group by a hydrogen. In benzene derivatives, this is done by heating with water or steam.

ESSENTIAL PROBLEM-SOLVING SKILLS IN CHAPTER 17

1. Predict products and give mechanisms for the common electrophilic aromatic substitutions: halogenation, nitration, sulfonation, and Friedel-Crafts alkylation and acylation.
2. Draw resonance structures for the sigma complexes resulting from electrophilic attack on substituted aromatic rings. Explain which substituents are activating and which are deactivating, and show why they are ortho, para-directing or meta-allowing.
3. Predict the position(s) of electrophilic aromatic substitution on molecules containing substituents on one or more aromatic rings.
4. Design syntheses that use the influences of substituents to generate the correct isomers of multisubstituted aromatic compounds.
5. Determine which nucleophilic aromatic substitutions are likely, and propose mechanisms of the addition-elimination type and of the benzyne type.
6. Predict the products of Birch reduction, hydrogenation, and chlorination of aromatic compounds, and use these reactions in syntheses.
7. Explain how the reactions of side chains are affected by the presence of the aromatic ring, predict the products of side-chain reactions, and use these reactions in syntheses.
8. Predict the products of oxidation and substitution of phenols, and use these reactions in syntheses.

STUDY PROBLEMS

17-42. Briefly define the following terms and give examples.

(a) activating group
(b) deactivating group
(c) sigma complex
(d) sulfonation
(e) desulfonation
(f) nitration
(g) ortho, para-director
(h) meta-director
(i) resonance stabilization
(j) Friedel-Crafts acylation
(k) Friedel-Crafts alkylation
(l) Clemmensen reduction
(m) Gatterman-Koch synthesis
(n) benzyne mechanism
(o) Birch reduction
(p) quinone
(q) benzylic position

17-43. Predict the major products that will be formed when benzene reacts with the following reagents.

(a) *t*-butyl bromide, $AlCl_3$
(b) 1-chlorobutane, $AlCl_3$
(c) isobutyl alcohol + BF_3
(d) bromine + a nail
(e) isobutylene + HF
(f) fuming sulfuric acid

(g) 1,2-dichloroethane + $AlCl_3$
(h) benzoyl chloride + $AlCl_3$
(i) iodine + HNO_3
(j) nitric acid + sulfuric acid
(k) carbon monoxide, HCl, and $AlCl_3/CuCl$
(l) 1-chloro-2,2-dimethylpropane + $AlCl_3$

17-44. Predict the major products that will be formed when isopropylbenzene reacts with the following reagents.
(a) 1 equivalent of Br_2 and light
(b) Br_2 and $FeBr_3$
(c) SO_3 and H_2SO_4
(d) hot, conc. $KMnO_4$
(e) acetyl chloride and $AlCl_3$
(f) *n*-propyl chloride and $AlCl_3$

17-45. Show how you would synthesize the following compounds, starting with benzene, toluene, and any necessary acyclic reagents.
(a) 1-phenyl-1-bromobutane
(b) 1-phenyl-1-methoxybutane
(c) 3-phenyl-1-propanol
(d) ethoxybenzene
(e) 1,2-dichloro-4-nitrobenzene
(f) 1-phenyl-2-propanol
(g) 3,4-dibromobenzoic acid

17-46. Predict the major products of the following reactions.
(a) 2,4-dinitrochlorobenzene + $NaOCH_3$
(b) isopropoxybenzene + *t*-butyl chloride + $AlCl_3$
(c) nitrobenzene + fuming sulfuric acid
(d) nitrobenzene + acetyl chloride + $AlCl_3$
(e) *p*-methylanisole + acetyl chloride + $AlCl_3$
(f) *p*-methylanisole + Br_2, light
(g) 1,2-dichloro-4-nitrobenzene + $NaNH_2$
(h) $Ph{-}C(=O){-}NHPh$ + $CH_3CH_2{-}C(=O){-}Cl$, $AlCl_3$
(i) *p*-ethylbenzenesulfonic acid + HNO_3, H_2SO_4
(j) *p*-ethylbenzenesulfonic acid + steam

(k) 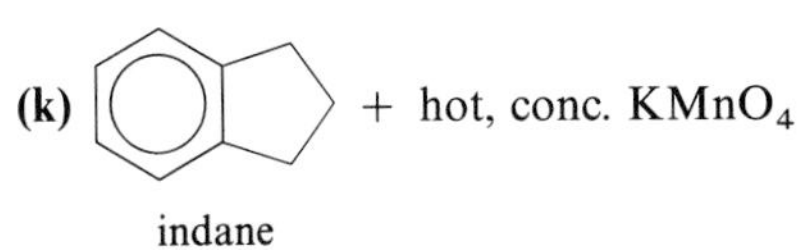 + hot, conc. $KMnO_4$

indane

(l) 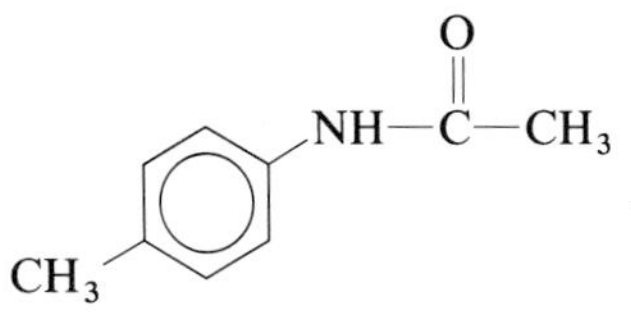 $NH{-}C(=O){-}CH_3$ / CH_3 + acetyl chloride, $AlCl_3$

p-methylacetanilide

17-47. Predict the major products of bromination of the following compounds, using Br_2 and $FeBr_3$ in the dark.

(a) 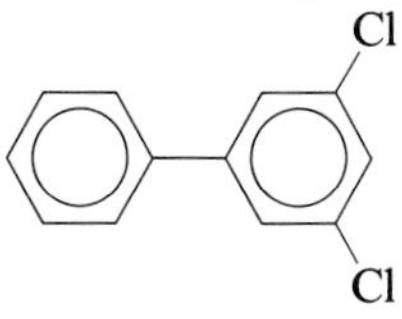 Cl Cl

(b) 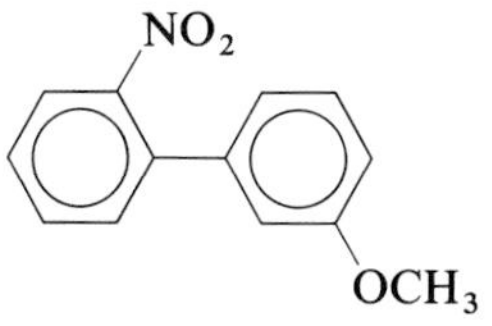 NO_2 OCH_3

(c) 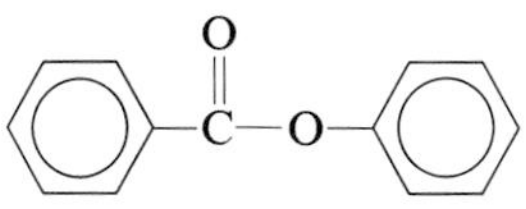 $C(=O){-}O$

(d) 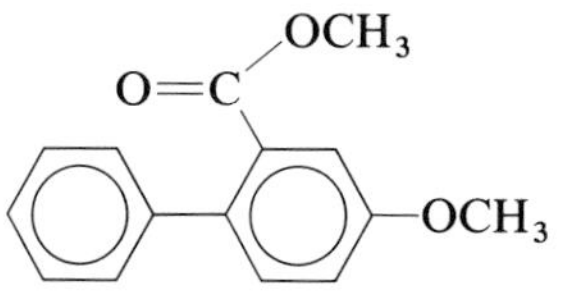 $O{=}C{-}OCH_3$ OCH_3

17-48. Give the structures of intermediates **A** through **H** in the following series of reactions.

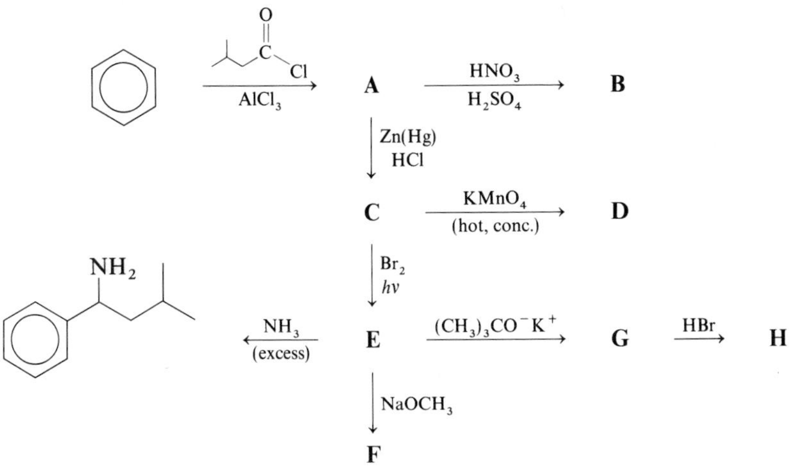

Benzene + $(CH_3)_2CHCH_2C(=O)Cl$, $AlCl_3$ → **A**; **A** + HNO_3, H_2SO_4 → **B**
A + Zn(Hg), HCl → **C**; **C** + $KMnO_4$ (hot, conc.) → **D**
C + Br_2, $h\nu$ → **E**; **E** + $(CH_3)_3CO^-K^+$ → **G**; **G** + HBr → **H**
E + $NaOCH_3$ → **F**
E + NH_3 (excess) → NH_2

17-49. A student added 3-phenylpropanoic acid ($PhCH_2CH_2COOH$) to a molten salt consisting of a 1:1 mixture of NaCl and $AlCl_3$ maintained at 170°. After 5 minutes, the molten mixture was poured into water and extracted into dichloromethane. Evaporation of the dichloromethane gave a 96% yield of the product whose spectra appear below. The mass spectrum of the product shows a molecular ion at m/z 132. What is the product?

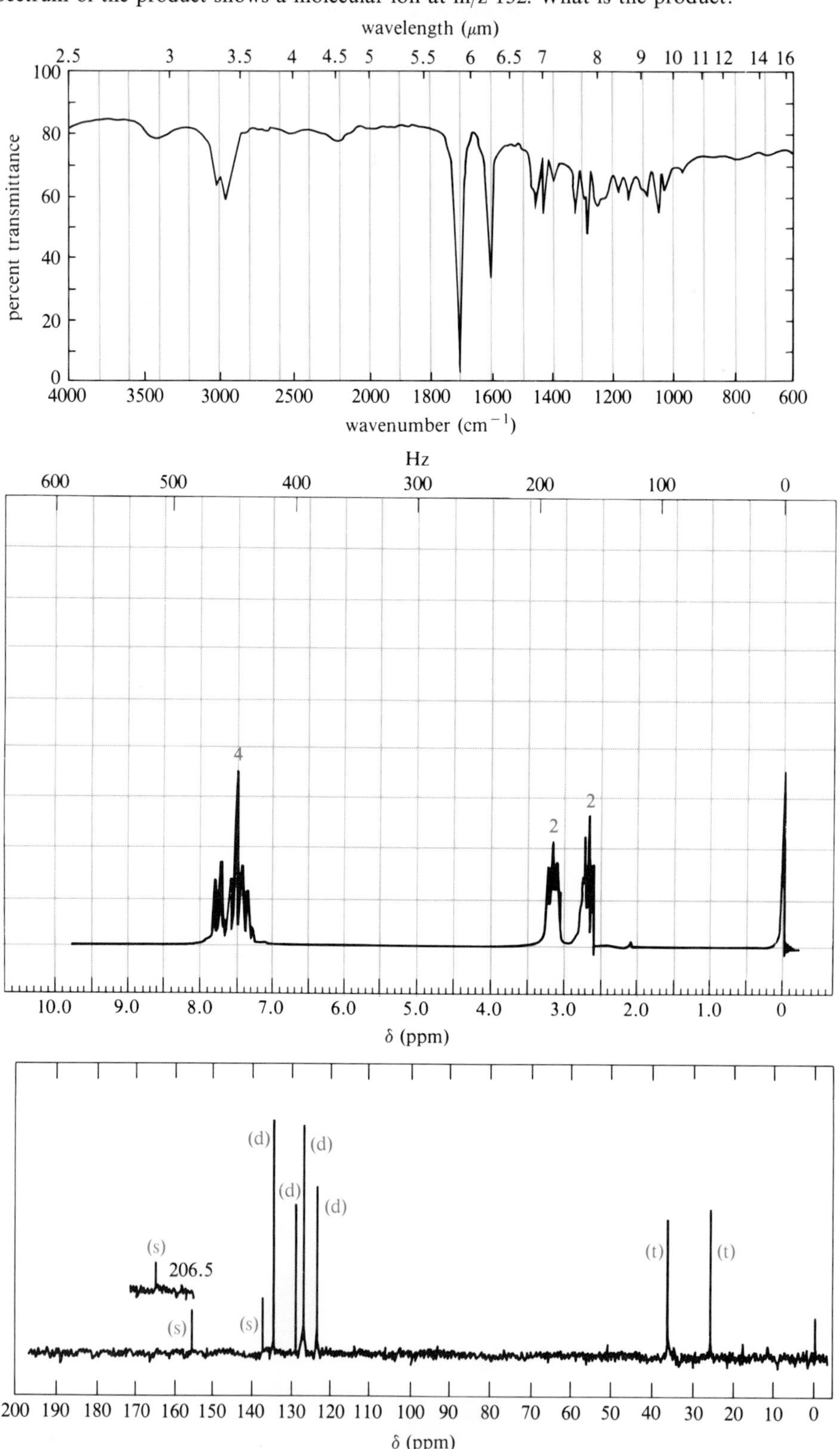

17-50. Electrophilic aromatic substitution usually occurs at the 1 position of naphthalene, also called the α position.

α position

8 1 2 3 4 5 6 7

Predict the major products of the reactions of napthalene with the following reagents.
(a) HNO_3, H_2SO_4 **(b)** Br_2, $FeBr_3$ **(c)** CH_3CH_2COCl
(d) isobutylene and HF **(e)** cyclohexanol and BF_3 **(f)** fuming sulfuric acid

17-51. Triphenylmethanol is insoluble in water; but when triphenylmethanol is treated with concentrated sulfuric acid, a bright yellow solution results. As this yellow solution is diluted with water, its color disappears and a precipitate of triphenylmethanol appears. Suggest a structure for the bright yellow species, and explain this unusual behavior.

17-52. The most common selective herbicide for killing broadleaf weeds in grass is 2,4-dichlorophenoxyacetic acid (2,4-D). Show how you would synthesize 2,4-D from benzene, chloroacetic acid ($ClCH_2COOH$), and any necessary reagents and solvents.

Cl—C₆H₃(Cl)—O—CH_2COOH

2,4-dichlorophenoxyacetic acid (2,4-D)

17-53. Furan undergoes electrophilic aromatic substitution more readily than benzene; mild reagents and conditions are sufficient. For example, furan reacts with bromine to give 2-bromofuran.

$$\text{furan} \xrightarrow[\text{dioxane}]{Br_2} \text{2-bromofuran}$$

furan 2-bromofuran

(a) Give mechanisms for bromination of furan at the 2-position and at the 3-position. Draw the resonance structures of each sigma complex, and compare their stabilities.

(b) Explain why furan undergoes bromination (and other electrophilic aromatic substitutions) primarily at the 2-position.

17-54. In Section 17-12A we saw that the Sanger reagent, 2,4-dinitrofluorobenzene, is a good substrate for nucleophilic aromatic substitution. Show the product and give a mechanism for the reaction of the Sanger reagent with the amino acid phenylalanine.

$$Ph—CH_2—CH(NH_2)—COOH$$

phenylalanine

17-55. Unlike most other electrophilic aromatic substitutions, sulfonation is often reversible (see Section 17-4). When one sample of toluene is sulfonated at 0°C and another sample is sulfonated at 100°C, different ratios of substitution products result.

	Reaction temperature	
Isomer of the product	*0°C*	*100°C*
o-toluenesulfonic acid	43%	13%
m-toluenesulfonic acid	4%	8%
p-toluenesulfonic acid	53%	79%

(a) Explain the change in the product ratios when the temperature is increased.

(b) Predict what happens when the product mixture from the reaction at 0°C is heated to 100°C.

17-56. Devise a synthesis of the following compound starting from toluene and any necessary reagents.

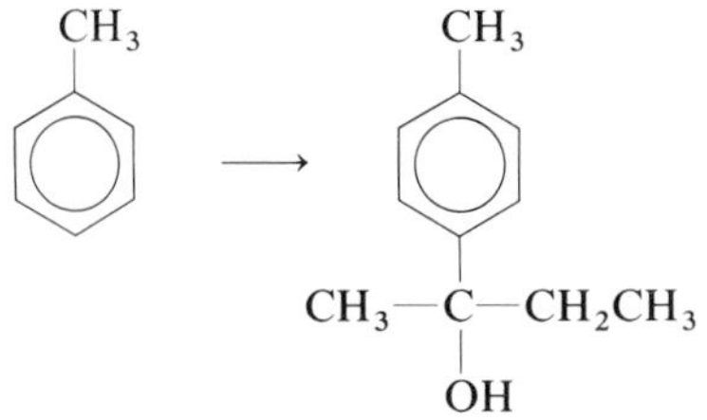

CH_3 (toluene) ⟶ CH_3–C_6H_4–C(OH)(CH_3)(CH_2CH_3)

CH₃ — C — CH₂CH₃ / OH

*** 17-57.** In Chapter 13 we saw that Agent Orange contains (2,4,5-trichlorophenoxy)acetic acid, called 2,4,5-T. This compound is synthesized by the partial reaction of 1,2,4,5-tetrachlorobenzene with sodium hydroxide, followed by reaction with sodium chloroacetate, Cl—CH_2—CO_2Na.

(a) Draw the structures of these compounds and write equations for these reactions.

(b) One of the impurities in the Agent Orange used in Vietnam was 2,3,7,8-tetrachlorodibenzodioxin (2,3,7,8-TCDD), often incorrectly called "dioxin." Give a mechanism to show how 2,3,7,8-TCDD is formed in the synthesis of 2,4,5-T.

(c) Show how the TCDD contamination might be eliminated, both after the first step and on completion of the synthesis.

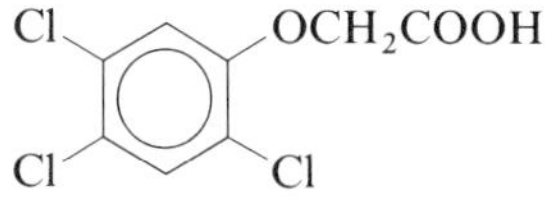

Cl, Cl, Cl, OCH_2COOH

2,4,5-T

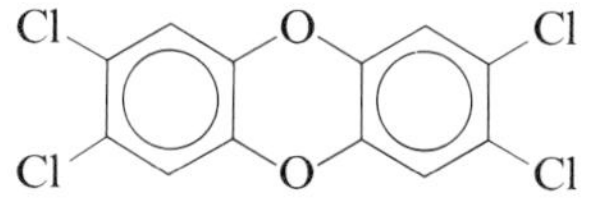

Cl, Cl, O, O, Cl, Cl

2,3,7,8-tetrachlorodibenzodioxin (TCDD)

*** 17-58.** When benzyne is generated in the presence of anthracene, an interesting Diels-Alder adduct of formula $C_{20}H_{14}$ results. The proton NMR spectrum of the product shows a singlet of area 2 around $\delta 3$ and a broad singlet of area 12 around $\delta 7$. Propose a structure for the product, and explain why one of the aromatic rings of anthracene reacted as a diene.

*** 17-59.** A graduate student tried to make *o*-fluorophenylmagnesium bromide by adding magnesium to an ether solution of *o*-fluorobromobenzene. After obtaining puzzling results with this reaction, she repeated the reaction using as solvent some tetrahydrofuran that contained a small amount of furan. From this reaction, she isolated a fair yield of the following compound. Propose a mechanism for its formation.

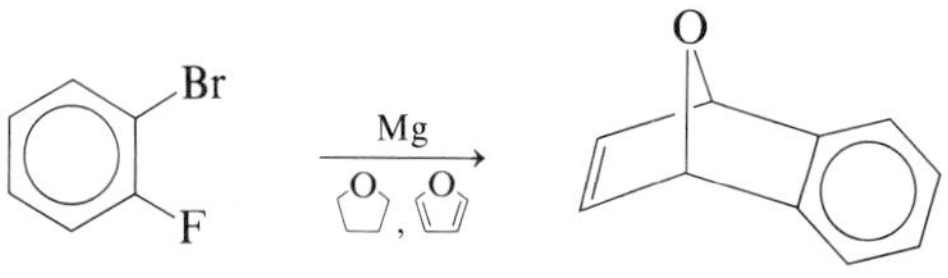

Br, F $\xrightarrow{\text{Mg}}$ (THF, furan) O

*** 17-60.** Phenolphthalein, a common nonprescription laxative, is also an acid-base indicator that is colorless in acid and red in base. Phenolphthalein is synthesized by the acid-catalyzed reaction of phthalic anhydride with 2 equivalents of phenol.

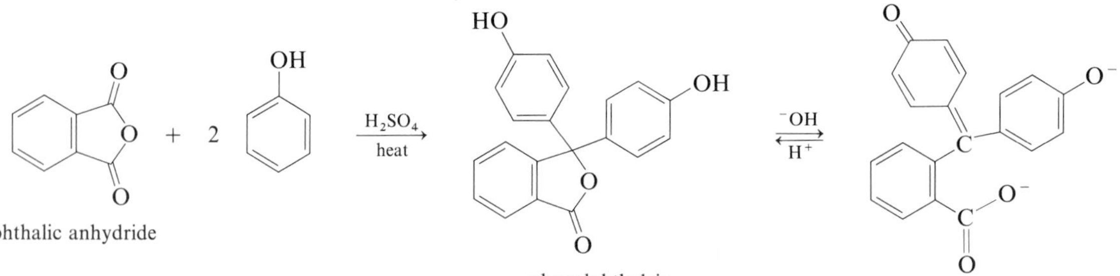

phthalic anhydride + 2 phenol (OH) $\xrightarrow[\text{heat}]{H_2SO_4}$ phenolphthalein (HO, OH, O, O) $\underset{H^+}{\overset{^-OH}{\rightleftharpoons}}$ red dianion (O, O^-, C, O^-, C, O)

phthalic anhydride

phenolphthalein

red dianion

(a) Propose a mechanism for the synthesis of phenolphthalein.

(b) Propose a mechanism for the conversion of phenolphthalein to its red dianion in base.

(c) Use resonance structures to show that the two phenolic oxygen atoms are equivalent (each with half a negative charge) in the red phenolphthalein dianion.

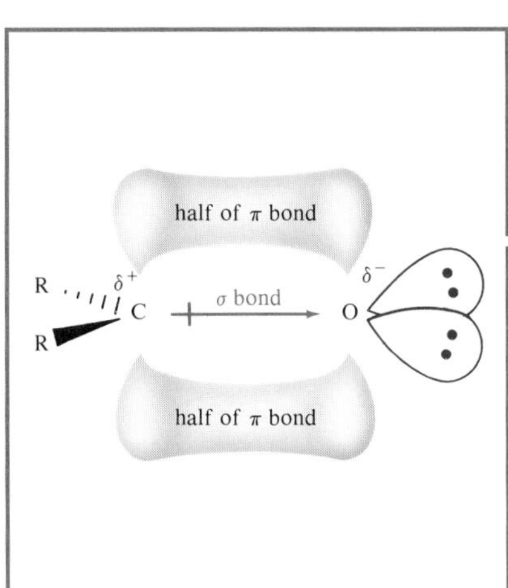

18

KETONES AND ALDEHYDES

18-1 CARBONYL COMPOUNDS

Compounds containing the **carbonyl group** (C=O) are among the most important organic compounds. We will study them in detail, because they are of central importance to organic chemistry, biochemistry, and biology. Some of the common types of carbonyl compounds are listed in Table 18-1.

TABLE 18-1
Some common classes of carbonyl compounds

Class	*General formula*	*Class*	*General formula*
ketones	$R-C(=O)-R'$	aldehydes	$R-C(=O)-H$
carboxylic acids	$R-C(=O)-OH$	acid chlorides	$R-C(=O)-Cl$
esters	$R-C(=O)-O-R'$	amides	$R-C(=O)-NH_2$

Carbonyl compounds are everywhere. In addition to their use as reagents and solvents, they serve as fabrics, flavorings, plastics, and drugs. Naturally occurring carbonyl compounds include the proteins, carbohydrates, and nucleic acids that make up all plants and animals. In the next few chapters we will discuss the properties and reactions of simple carbonyl compounds. Then, in Chapters 23 and 24, we will apply this carbonyl chemistry to carbohydrates, nucleic acids, and proteins.

The simplest carbonyl compounds are the ketones and aldehydes. A **ketone** has two alkyl (or aryl) groups bonded to the carbonyl carbon atom. An **aldehyde** has one alkyl (or aryl) group and one hydrogen atom bonded to the carbonyl carbon atom.

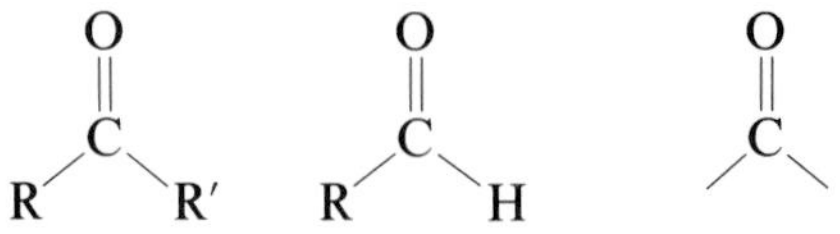

	ketone	aldehyde	carbonyl group
	$R-C(=O)-R'$	$R-C(=O)-H$	$>C=O$
condensed structures:	RCOR′	RCHO	

Ketone: Two alkyl groups bonded to a carbonyl group.
Aldehyde: One alkyl group and one hydrogen bonded to a carbonyl group (or two hydrogens bonded to a carbonyl group in formaldehyde).

Ketones and aldehydes are similar in structure, and they have similar properties. There are some differences, however, particularly in their reactions with oxidizing agents and with nucleophiles.

18-2 STRUCTURE OF THE CARBONYL GROUP

The carbonyl carbon atom is sp^2 hybridized and bonded to three other atoms through three coplanar sigma bonds oriented about 120° apart. The unhybridized p orbital overlaps with a p orbital of the oxygen atom to form a pi bond. The double bond between carbon and oxygen is similar to an alkene C═C double bond, except that the carbonyl double bond is somewhat shorter and stronger.

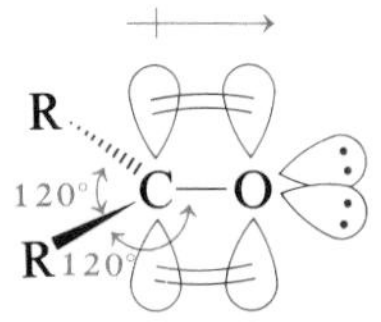

	length	*energy*
ketone C═O bond	1.23 Å	178 kcal/mole (745 kJ/mole)
alkene C═C bond	1.34 Å	146 kcal/mole (611 kJ/mole)

Another difference between the carbonyl and alkene double bonds is the large dipole moment of the carbonyl group. Oxygen is more electronegative than carbon, and the bonding electrons are not shared equally. In particular, the less tightly held pi electrons are pulled more strongly toward the oxygen atom, giving ketones and aldehydes larger dipole moments than most alkyl halides and ethers. We can use resonance structures to symbolize this unequal sharing of the pi electrons.

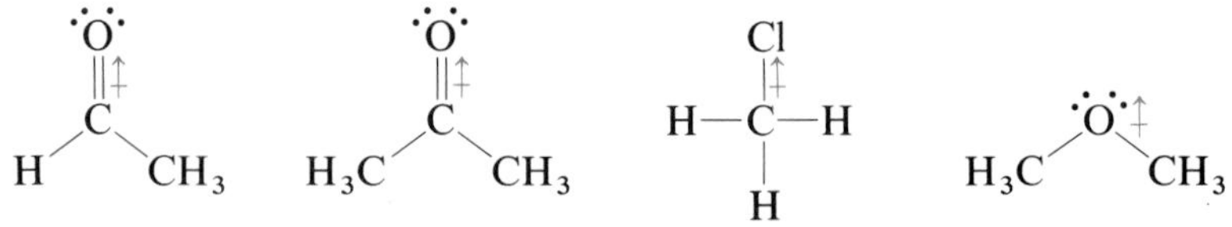

$$\left[R_2C{=}\ddot{O}: \longleftrightarrow R_2\overset{+}{C}{-}\ddot{O}:^- \right]$$

CH_3CHO	$(CH_3)_2CO$	CH_3Cl	CH_3OCH_3
$\mu = 2.7$ D	$\mu = 2.9$ D	$\mu = 1.9$ D	$\mu = 1.30$ D
acetaldehyde	acetone	chloromethane	dimethyl ether

The first resonance structure is clearly more important since it involves more bonds and less charge separation. The contribution of the second structure is evidenced by the large dipole moments of the ketones and aldehydes shown above. This polarization of the carbonyl group contributes to the reactivity of ketones and aldehydes: The positively polarized carbon atom acts as an electrophile, and the negatively polarized oxygen acts as a nucleophile. We consider the acid-base properties of ketones and aldehydes in Section 18-12.

18-3 NOMENCLATURE OF KETONES AND ALDEHYDES

IUPAC names The systematic names of ketones are derived by replacing the final *-e* in the alkane name with *-one*. The "alkane" name becomes "alkanone." In open-chain ketones, the longest chain is numbered from the end closest to the carbonyl group, and the position of the carbonyl group is indicated by a number. In cyclic ketones, the carbonyl carbon atom is assigned the number 1.

$CH_3—CO—CH_2—CH_3$ (numbered 1, 2, 3, 4)
2-butanone

$CH_3—CH(CH_3)—CO—CH(CH_3)—CH_3$ (numbered 1, 2, 3, 4, 5)
2,4-dimethyl-3-pentanone

$C_6H_5—CO—CH_2—CH_3$ (numbered 1, 2, 3)
1-phenyl-1-propanone

3-methylcyclopentanone

2-cyclohexenone

$CH_3—CO—CH_2—C(OH)(CH_3)—CH_3$ (numbered 1, 2, 3, 4, 5)
4-hydroxy-4-methyl-2-pentanone

The systematic names for aldehydes are derived by replacing the final *-e* of the alkane name with *-al*. An aldehyde carbon is at the end of a chain, so it is number 1. If the aldehyde group is attached to a large unit (usually a ring), the suffix *-carbaldehyde* may be used.

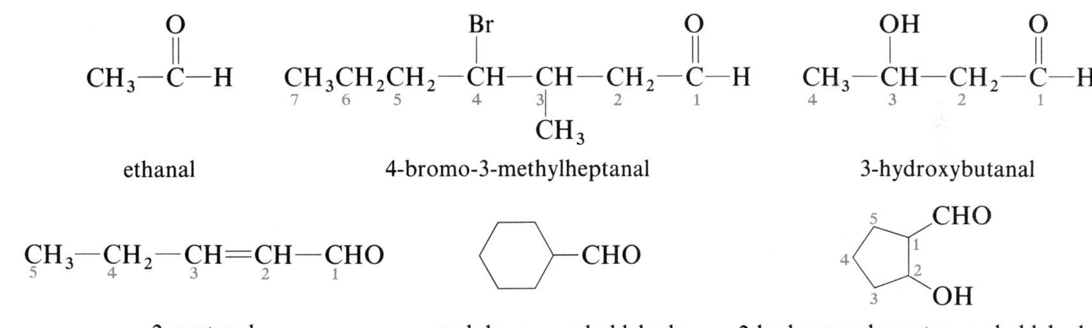

A ketone or aldehyde carbonyl group can also be named as a substituent on a molecule with another functional group as its root. The ketone carbonyl is designated by the prefix *oxo-*, and the —CHO group is named as a *formyl* group. Carboxylic acids frequently contain ketone or aldehyde groups that are named as substituents.

$CH_3CH_2—CO—CH_2—CHO$
3-oxopentanal

$C_6H_4(CHO)(COOH)$
2-formylbenzoic acid

$CH_3—CO—CH_2—COOH$
3-oxobutanoic acid

Common names As with other classes of organic compounds, ketones and aldehydes are often called by common names rather than their systematic IUPAC names. Ketone common names are formed by naming the two alkyl groups bonded to the carbonyl group. Substituent locations are given using the Greek letter system, beginning with the carbon *next to* the carbonyl group.

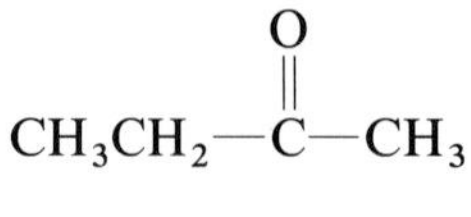

$CH_3CH_2—C(=O)—CH_3$

methyl ethyl ketone

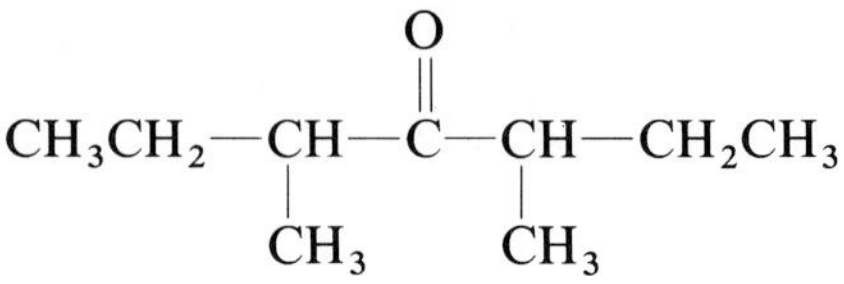

$CH_3CH_2—CH(CH_3)—C(=O)—CH(CH_3)—CH_2CH_3$

di-*sec*-butyl ketone

$Br—\overset{\beta}{C}H_2—\overset{\alpha}{C}H_2—C(=O)—CH(CH_3)—CH_3$

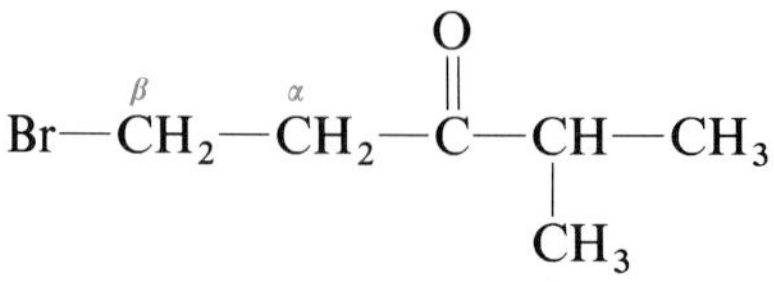

β-bromoethyl isopropyl ketone

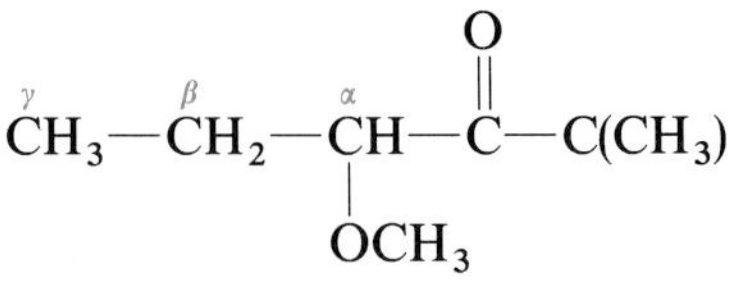

$\overset{\gamma}{C}H_3—\overset{\beta}{C}H_2—\overset{\alpha}{C}H(OCH_3)—C(=O)—C(CH_3)$

t-butyl α-methoxypropyl ketone

Some ketones have historical common names. Dimethyl ketone is always called *acetone,* and the alkyl phenyl ketones are usually named as the acyl group followed by the suffix *-phenone.*

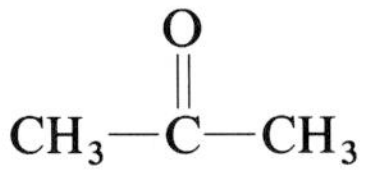

$CH_3—C(=O)—CH_3$

acetone

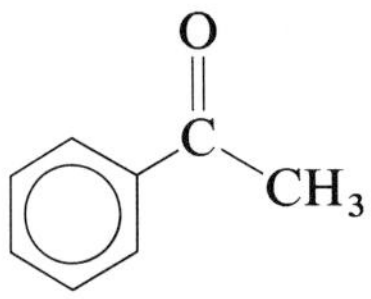

acetophenone

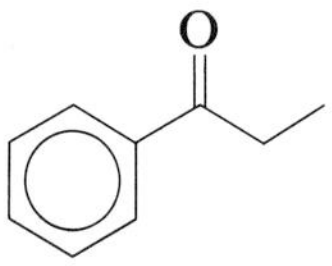

propiophenone

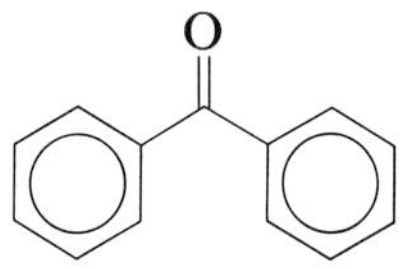

benzophenone

The common names of aldehydes are derived from the common names of the corresponding carboxylic acids (Table 18-2). These names often reflect the Latin or Greek term for the original source of the acid or the aldehyde.

TABLE 18-2
Common names of aldehydes

Carboxylic acid		*Aldehyde*
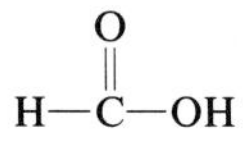 $H—C(=O)—OH$ formic acid	(*formica*, "ants")	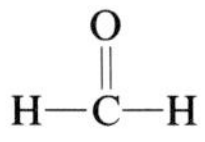 $H—C(=O)—H$ formaldehyde
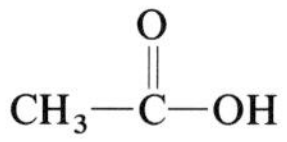 $CH_3—C(=O)—OH$ acetic acid	(*acetum*, "sour")	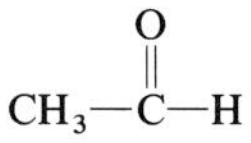 $CH_3—C(=O)—H$ acetaldehyde
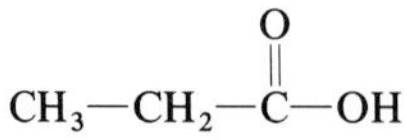 $CH_3—CH_2—C(=O)—OH$ propionic acid	(*protos pion*, "first fat")	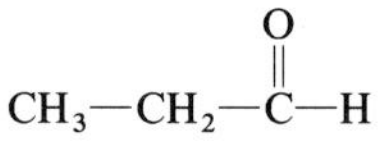 $CH_3—CH_2—C(=O)—H$ propionaldehyde
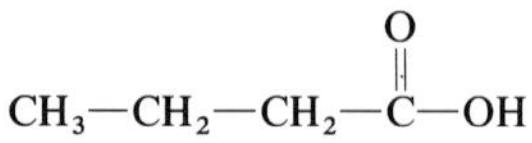 $CH_3—CH_2—CH_2—C(=O)—OH$ butyric acid	(*butyrum*, "butter")	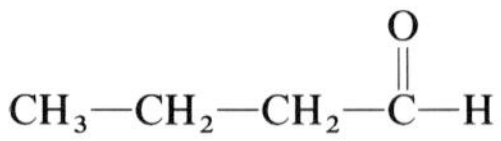 $CH_3—CH_2—CH_2—C(=O)—H$ butyraldehyde
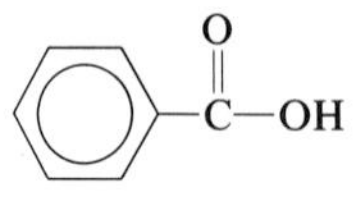 $C_6H_5—C(=O)—OH$ benzoic acid		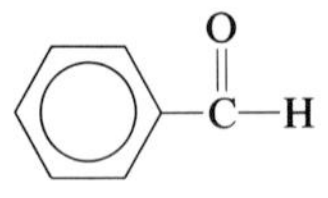 $C_6H_5—C(=O)—H$ benzaldehyde

Greek letters are used with the common names of aldehydes to give the locations of substituents. The first letter (α) is given to the carbon atom *next to* the carbonyl group.

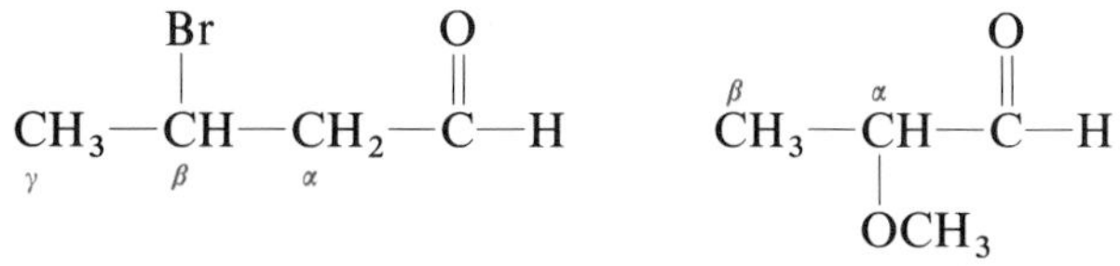

$$\underset{\gamma}{CH_3}-\underset{\beta}{\overset{\overset{\displaystyle Br}{|}}{CH}}-\underset{\alpha}{CH_2}-\overset{\overset{\displaystyle O}{\|}}{C}-H \qquad \overset{\beta}{CH_3}-\underset{\underset{\displaystyle OCH_3}{|}}{\overset{\alpha}{CH}}-\overset{\overset{\displaystyle O}{\|}}{C}-H$$

Common name:	β-bromobutyraldehyde	α-methoxypropionaldehyde
IUPAC name:	3-bromobutanal	2-methoxypropanal

PROBLEM 18-1

Give the IUPAC name and (if possible) a common name for each of the following compounds.

(a) $CH_3-\overset{\overset{\displaystyle OH}{|}}{CH}-CH_2-\overset{\overset{\displaystyle O}{\|}}{C}-CH_2CH_3$

(b) $CH_3-\overset{\overset{\displaystyle Ph}{|}}{CH}-CH_2-CHO$

(c) cyclohexane ring with OCH₃, H, H, CHO substituents (fused stereo drawing)

(d) 6,6-dimethylcyclohexa-2,4-dienone drawing: O, CH_3, CH_3

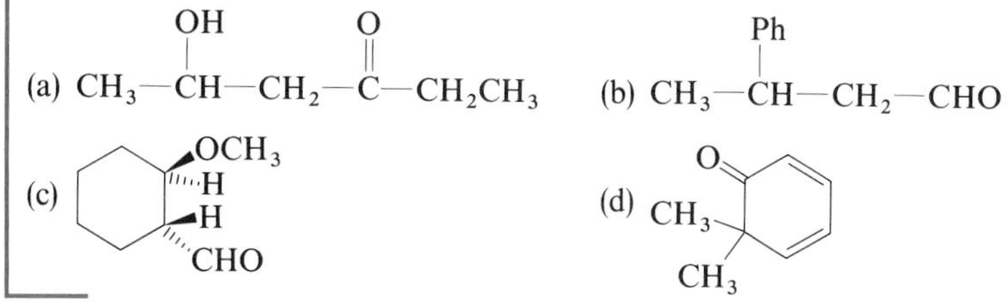

18-4 PHYSICAL PROPERTIES OF KETONES AND ALDEHYDES

The polarization of the carbonyl group creates dipole-dipole attractions between the molecules of ketones and aldehydes, resulting in higher boiling points than for hydrocarbons and ethers of similar molecular weights. Ketones and aldehydes have no O—H or N—H bonds, however, so their molecules cannot form hydrogen bonds with each other. Their boiling points are therefore lower than those of alcohols of similar molecular weights. The following compounds of molecular weight 58 or 60 are ranked in order of increasing boiling point. The ketone and the aldehyde are more polar and higher-boiling than the ether and the alkane, but they are lower-boiling than the hydrogen-bonded alcohol.

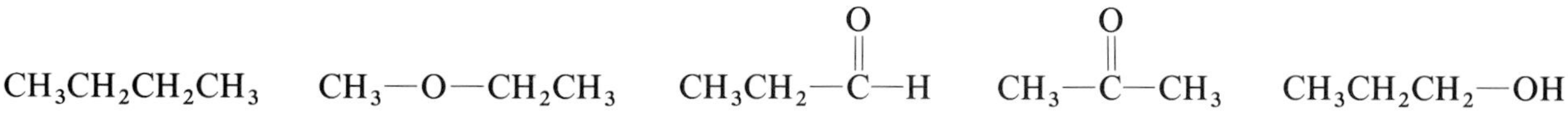

$CH_3CH_2CH_2CH_3$	$CH_3-O-CH_2CH_3$	$CH_3CH_2-\overset{\overset{\displaystyle O}{\|}}{C}-H$	$CH_3-\overset{\overset{\displaystyle O}{\|}}{C}-CH_3$	$CH_3CH_2CH_2-OH$
butane	methoxyethane	propanal	acetone	1-propanol
b.p. 0°C	b.p. 8°C	b.p. 49°C	b.p. 56°C	b.p. 97°C

The melting points, boiling points, and water solubilities of some representative ketones and aldehydes are given in Table 18-3.

Although pure ketones and aldehydes cannot engage in hydrogen bonding with each other, they have lone pairs of electrons and can act as hydrogen bond acceptors with other compounds that have O—H or N—H bonds. For example, the unshared electrons on the carbonyl oxygen can form a hydrogen bond with one of the hydrogen atoms of water or an alcohol.

TABLE 18-3
Physical properties of some representative ketones and aldehydes

IUPAC name	*Common name*	*Structure*	*m.p. (°C)*	*b.p. (°C)*	*Density (g/cm^3)*	*H_2O solubility (%)*
		Ketones				
2-propanone	acetone	CH_3COCH_3	−95	56	0.79	∞
2-butanone	methyl ethyl ketone (MEK)	$CH_3COCH_2CH_3$	−86	80	0.81	25.6
2-pentanone	methyl *n*-propyl ketone	$CH_3COCH_2CH_2CH_3$	−78	102	0.81	5.5
3-pentanone	diethyl ketone	$CH_3CH_2COCH_2CH_3$	−41	101	0.81	4.8
2-hexanone		$CH_3CO(CH_2)_3CH_3$	−57	127	0.83	1.6
3-hexanone		$CH_3CH_2COCH_2CH_2CH_3$		124	0.82	
2-heptanone		$CH_3CO(CH_2)_4CH_3$	−36	151	0.81	1.4
3-heptanone		$CH_3CH_2CO(CH_2)_3CH_3$	−39	147	0.82	0.4
4-heptanone	di-*n*-propyl ketone	$(CH_3CH_2CH_2)_2CO$	−34	144	0.82	
4-methyl-3-penten-2-one	mesityl oxide	$(CH_3)_2C{=}CHCOCH_3$	−59	131	0.86	
3-buten-2-one	methyl vinyl ketone (MVK)	$CH_2{=}CHCOCH_3$	−6	80	0.86	
cyclohexanone			−16	157	0.94	2.
acetophenone	methyl phenyl ketone	$C_6H_5COCH_3$	21	202	1.02	0.5
propiophenone	ethyl phenyl ketone	$C_6H_5COCH_2CH_3$	21	218		
benzophenone	diphenyl ketone	$C_6H_5COC_6H_5$	48	305	1.08	
		Aldehydes				
methanal	formaldehyde	HCHO or CH_2O	−92	−21	0.82	55
ethanal	acetaldehyde	CH_3CHO	−123	21	0.78	∞
propanal	propionaldehyde	CH_3CH_2CHO	−81	49	0.81	20
butanal	*n*-butyraldehyde	$CH_3(CH_2)_2CHO$	−97	75	0.82	7.1
2-methylpropanal	isobutyraldehyde	$(CH_3)_2CHCHO$	−66	61	0.79	11
pentanal	*n*-valeraldehyde	$CH_3(CH_2)_3CHO$	−91	103	0.82	
3-methylbutanal	isovaleraldehyde	$(CH_3)_2CHCH_2CHO$	−51	93	0.80	
hexanal	caproaldehyde	$CH_3(CH_2)_4CHO$	−56	129	0.83	0.1
heptanal	*n*-heptaldehyde	$CH_3(CH_2)_5CHO$	−45	155	0.85	0.02
propenal	acrolein	$CH_2{=}CH{-}CHO$	−88	53	0.84	30
2-butenal	crotonaldehyde	$CH_3{-}CH{=}CH{-}CHO$	−77	104	0.86	18
benzaldehyde		C_6H_5CHO	−56	179	1.05	0.3

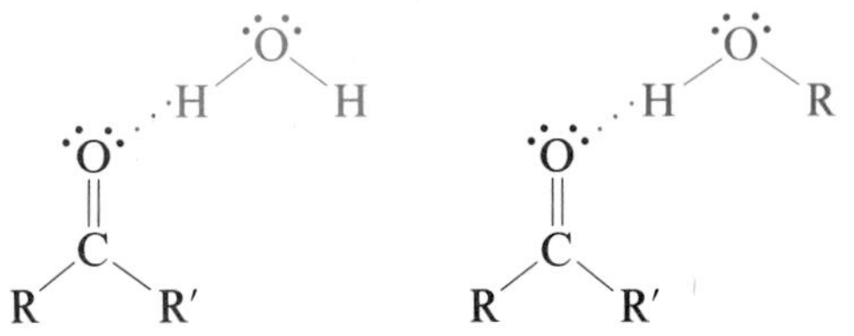

Because of this hydrogen bonding, ketones and aldehydes are good solvents for polar hydroxylic substances such as alcohols. They are also remarkably soluble in water. Table 18-3 shows that acetaldehyde and acetone are miscible (soluble in all proportions) with water, and other ketones and aldehydes with up to four carbon atoms are all appreciably soluble in water. These solubility properties are similar to those of the corresponding ethers and alcohols, which also engage in hydrogen bonding with water.

Formaldehyde and acetaldehyde are the most common aldehydes. Formaldehyde is a gas at room temperature, so it is often stored and used as a 40 percent aqueous solution called *formalin*. When dry formaldehyde is needed, it can be generated by heating one of the solid derivatives of formaldehyde, usually *trioxane* or *paraformaldehyde*. Trioxane is a cyclic *trimer*, containing three formaldehyde units. Paraformaldehyde is a linear *polymer*, containing many formaldehyde units. These solid derivatives form spontaneously when a small amount of acid catalyst is added to pure formaldehyde.

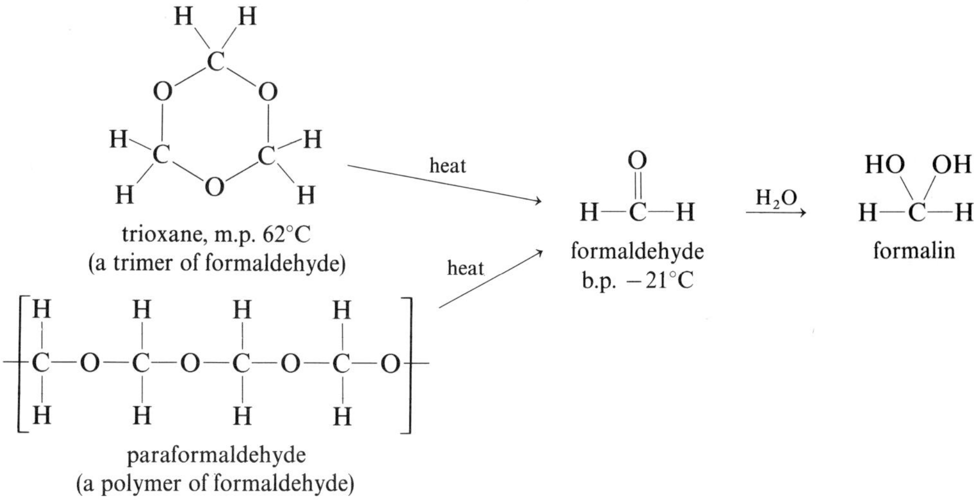

trioxane, m.p. 62°C
(a trimer of formaldehyde)

heat

$$H-\overset{\overset{\displaystyle O}{\|}}{C}-H \xrightarrow{H_2O} H-\overset{HO\ \ OH}{C}-H$$

formaldehyde
b.p. −21°C

formalin

paraformaldehyde
(a polymer of formaldehyde)

Acetaldehyde boils near room temperature, and it can be handled as a liquid. Acetaldehyde is also used as its trimer (*paraldehyde*) and its tetramer (*metaldehyde*), formed from acetaldehyde under acid catalysis. Heating either of these compounds provides dry acetaldehyde. Paraldehyde is used in medicine as a sedative, and metaldehyde is used as a bait and poison for snails and slugs.

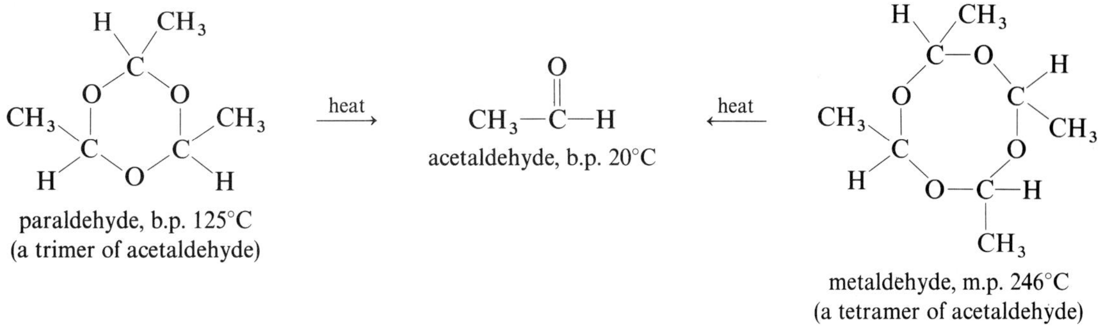

paraldehyde, b.p. 125°C
(a trimer of acetaldehyde)

heat

$$CH_3-\overset{\overset{\displaystyle O}{\|}}{C}-H$$

acetaldehyde, b.p. 20°C

heat

metaldehyde, m.p. 246°C
(a tetramer of acetaldehyde)

18-5 SPECTROSCOPY OF KETONES AND ALDEHYDES

18-5A INFRARED SPECTRA OF KETONES AND ALDEHYDES

The carbonyl (C=O) stretching vibrations of simple ketones and aldehydes occur around 1710 cm^{-1}. Because the carbonyl group has a large dipole moment, these carbonyl stretching absorptions are very strong. In addition to the carbonyl stretching absorption, an aldehyde shows a set of two low-frequency C—H stretching absorptions around 2700 and 2800 cm^{-1}.

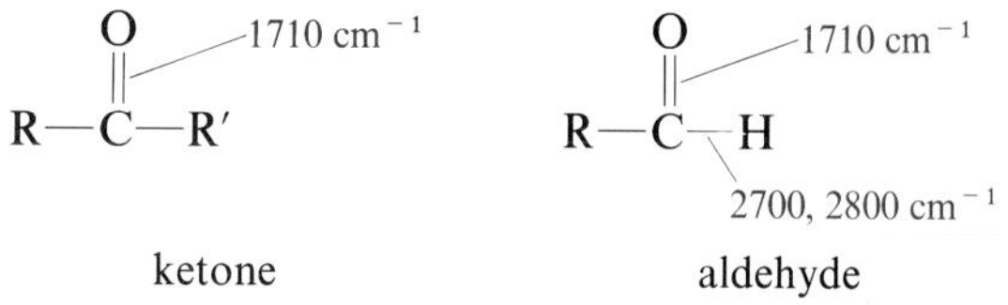

$$R-\overset{\overset{\displaystyle O}{\|}}{C}-R' \quad (1710\ cm^{-1})$$

ketone

$$R-\overset{\overset{\displaystyle O}{\|}}{C}-H \quad (C{=}O:\ 1710\ cm^{-1};\ C{-}H:\ 2700,\ 2800\ cm^{-1})$$

aldehyde

Figure 11-10 (page 459) compares the IR spectra of a simple ketone and aldehyde.

Conjugation lowers the carbonyl stretching frequencies of ketones and aldehydes, because the partial pi-bonding character of the single bond between the conjugated double bonds reduces the electron density of the carbonyl pi bond, lowering the stretching frequency to about 1680 cm^{-1} for conjugated ketones and aldehydes. Ring strain has the opposite effect, raising the carbonyl stretching frequency in ketones with three-, four-, and five-membered rings.

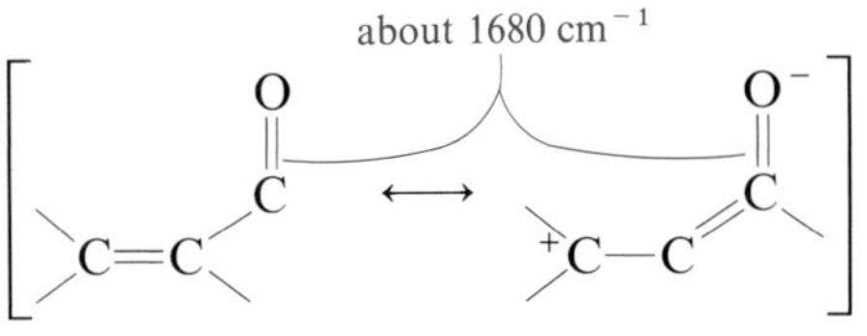

about 1680 cm^{-1}

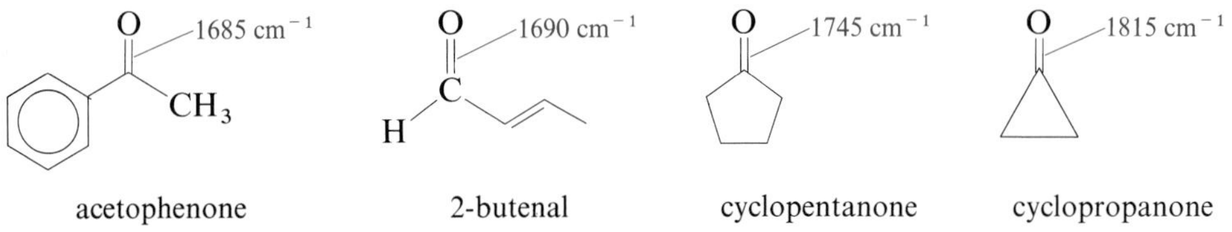

acetophenone: 1685 cm^{-1} 2-butenal: 1690 cm^{-1} cyclopentanone: 1745 cm^{-1} cyclopropanone: 1815 cm^{-1}

18-5B PROTON NMR SPECTRA OF KETONES AND ALDEHYDES

When considering the proton NMR spectra of ketones and aldehydes, we are interested primarily in the protons bonded to the carbonyl group (aldehyde protons) and the protons bonded to the adjacent carbon atom (the α-carbon atom). Aldehyde protons normally absorb at chemical shifts between $\delta 9$ and $\delta 10$. The aldehyde proton's absorption may be split ($J = 1$ to 5 Hz) if there are protons on the α-carbon atom. Protons on the α-carbon atom of a ketone or aldehyde usually absorb at a chemical shift between $\delta 2.1$ and $\delta 2.4$ if there are no other electron-withdrawing groups nearby. Methyl ketones are characterized by a singlet at about $\delta 2.1$.

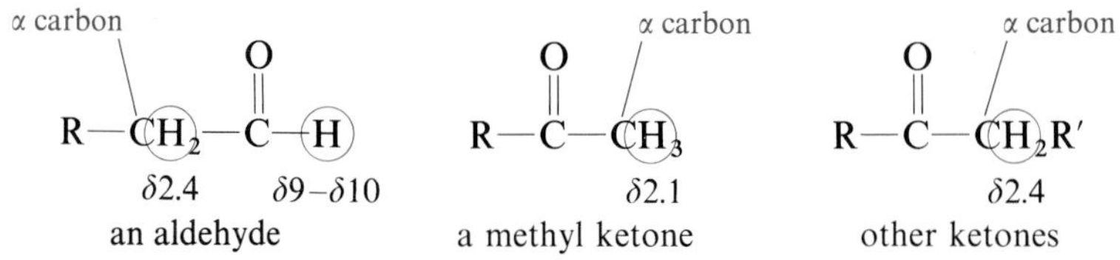

α carbon: R—CH_2—C(=O)—H, $\delta 2.4$, $\delta 9$–$\delta 10$ — an aldehyde

α carbon: R—C(=O)—CH_3, $\delta 2.1$ — a methyl ketone

α carbon: R—C(=O)—CH_2R', $\delta 2.4$ — other ketones

Figure 18-1 shows the proton NMR spectrum of butanal. The aldehyde proton appears at $\delta 9.8$, split by the protons on the α-carbon atom with a small ($J = 1$ Hz) coupling constant. The α protons appear at $\delta 2.4$, and the β and γ protons appear at increasing magnetic fields as they are located farther from the deshielding effects of the carbonyl group.

18-5C CARBON NMR SPECTRA OF KETONES AND ALDEHYDES

Carbonyl carbon atoms have chemical shifts around 200 ppm in the carbon NMR spectrum. The α-carbon atoms usually absorb at chemical shifts of about 30 to

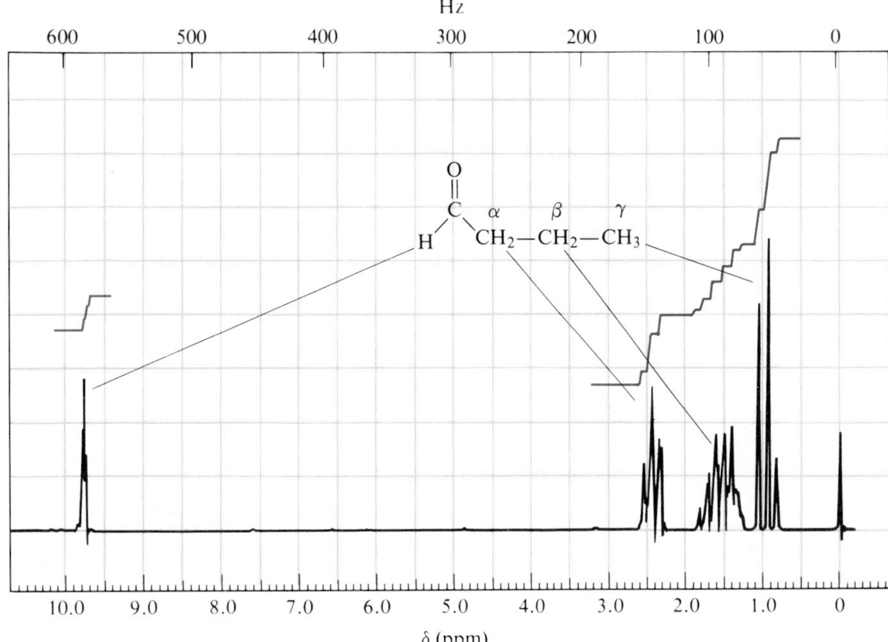

FIGURE 18-1 The proton NMR spectrum of butanal (butyraldehyde) shows the aldehyde proton at $\delta 9.8$, split into a triplet ($J = 1$ Hz) by the two α protons. The α, β, and γ protons appear at values of δ that decrease with increasing distance from the carbonyl group.

40 ppm. Figure 18-2 shows the spin-decoupled carbon NMR spectrum of 2-heptanone, in which the carbonyl carbon absorbs at 208 ppm and the α-carbon atoms absorb at 30 ppm (methyl) and 44 ppm (methylene).

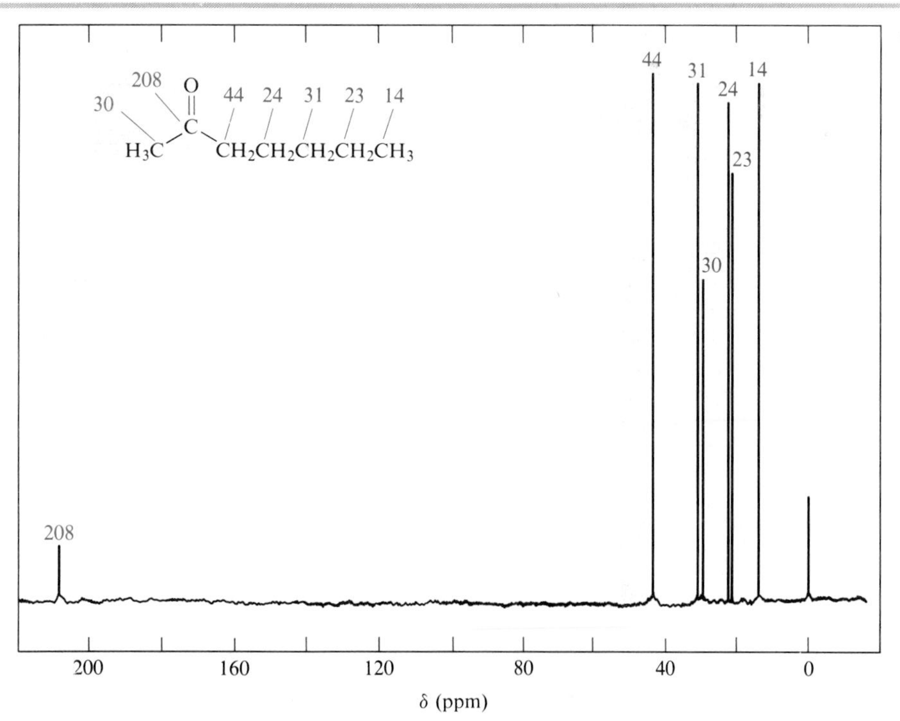

FIGURE 18-2 This spin-decoupled carbon NMR spectrum of 2-heptanone shows the carbonyl carbon at 208 ppm and the α carbons at 30 ppm (methyl) and 44 ppm (methylene).

PROBLEM 18-2

Two NMR spectra of ketones or aldehydes are given below, together with the molecular formulas. In each case, show what characteristics of the spectrum imply the presence of a ketone or an aldehyde, and propose a structure for the compound.

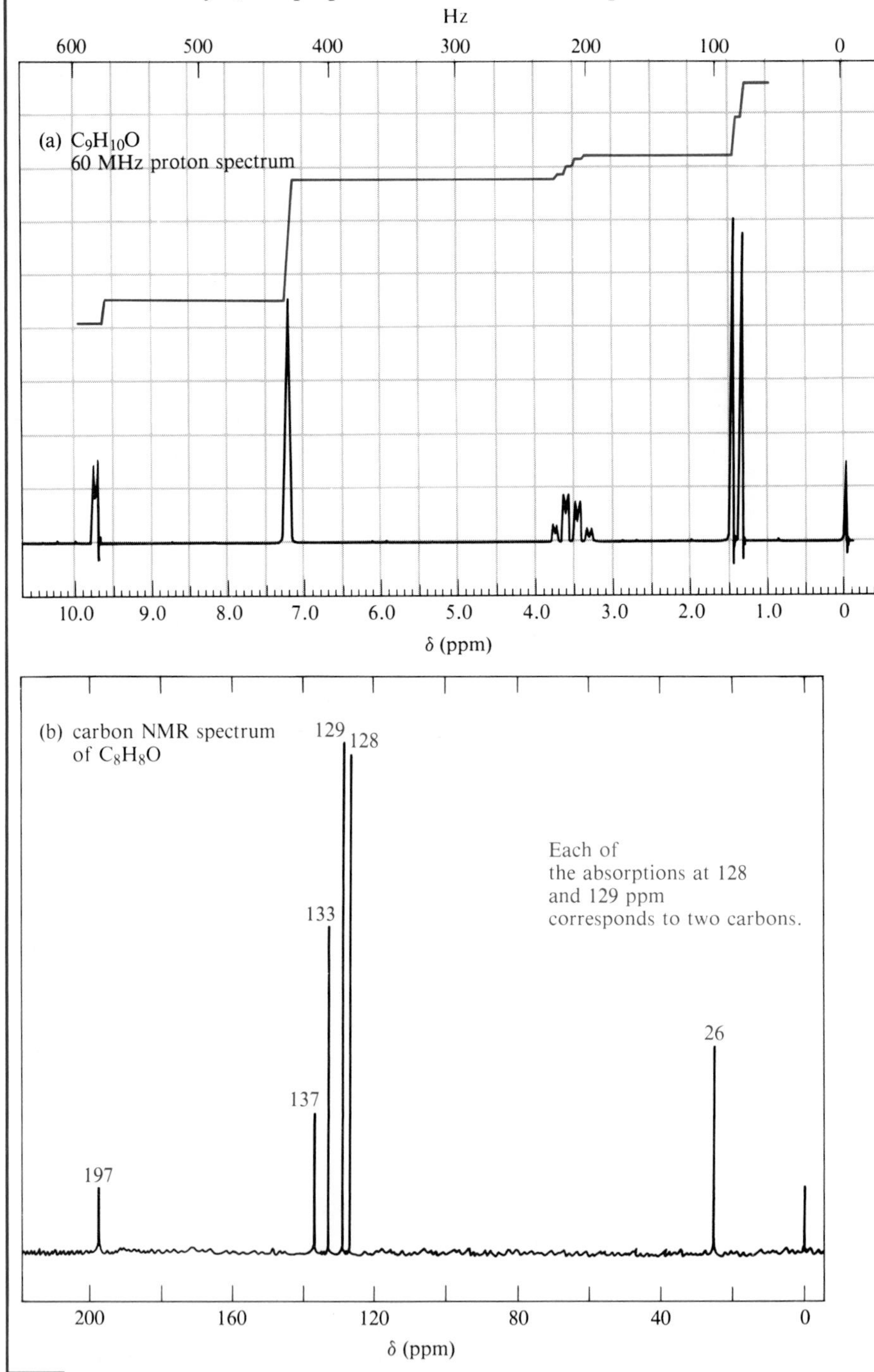

18-5D MASS SPECTRA OF KETONES AND ALDEHYDES

A ketone or an aldehyde may lose an alkyl group to give a resonance-stabilized acylium ion, like the acylium ion that serves as the electrophile in the Friedel-Crafts acylation (Section 17-11).

$$\left[\mathrm{R{-}\overset{\overset{\Large O}{\|}}{C}{-}R'}\right]^{+\cdot} \longrightarrow [\mathrm{R{-}\overset{+}{C}{=}O} \longleftrightarrow \mathrm{R{-}C{\equiv}O^{+}}] + \cdot\mathrm{R'}$$

acylium ion

Figure 18-3 shows the mass spectrum of methyl ethyl ketone (2-butanone). The molecular ion is prominent at *m/z* 72. The base peak at *m/z* 43 corresponds to loss of the ethyl group. Because a methyl radical is less stable than an ethyl radical, loss of the methyl group is not as common as loss of the ethyl group. The peak corresponding to loss of the methyl group (*m/z* 57) is much weaker than the base peak from loss of the ethyl group.

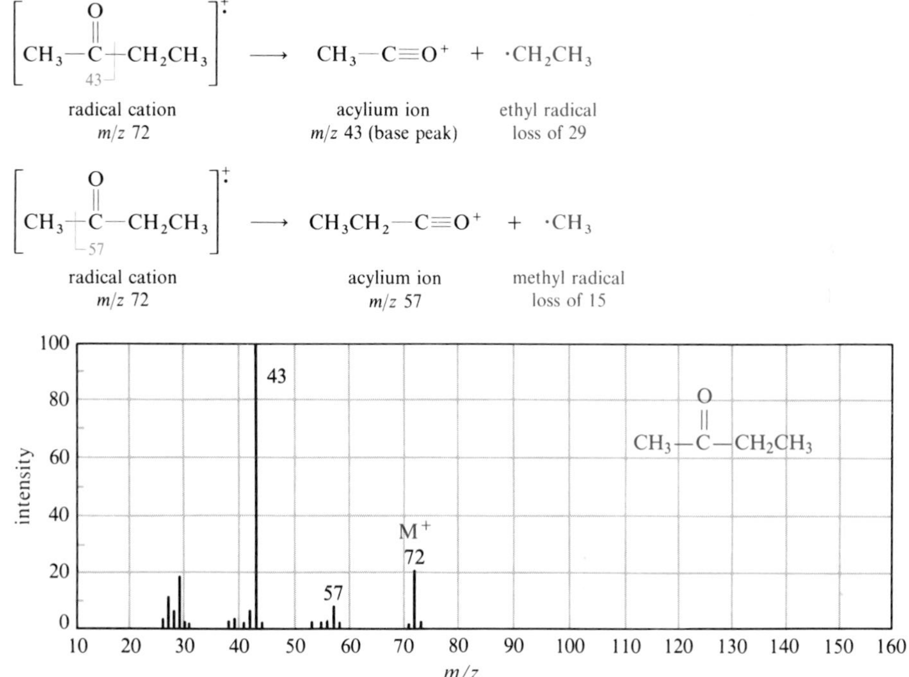

FIGURE 18-3 The mass spectrum of 2-butanone shows a prominent molecular ion, along with a base peak corresponding to loss of an ethyl radical to give an acylium ion. The peak corresponding to loss of a methyl radical is much weaker.

McLafferty rearrangement of ketones and aldehydes The mass spectrum of butyraldehyde (Fig. 18-4) shows the peaks we expect at *m/z* 72 (molecular ion), *m/z* 57 (loss of a methyl group), and *m/z* 29 (loss of a propyl group). The base peak is at *m/z* 44, showing the loss of a fragment of mass 28. This loss of a fragment with an even mass number corresponds to loss of a stable, neutral molecule (as when water, mass 18, is lost from an alcohol).

A fragment of mass 28 corresponds to a molecule of ethylene (C_2H_4). This fragment is lost through a process called the **McLafferty rearrangement,** involving the cyclic intramolecular transfer of a hydrogen atom shown in Figure 18-5.

The McLafferty rearrangement is a characteristic fragmentation of ketones and aldehydes, so long as they have γ hydrogens.

$$\mathrm{H{-}\overset{\overset{\Large O}{\|}}{C}{-}\overset{\alpha}{CH_2}{-}\overset{\beta}{CH_2}{-}\overset{\gamma}{CH_3}}$$

44 | γ hydrogens

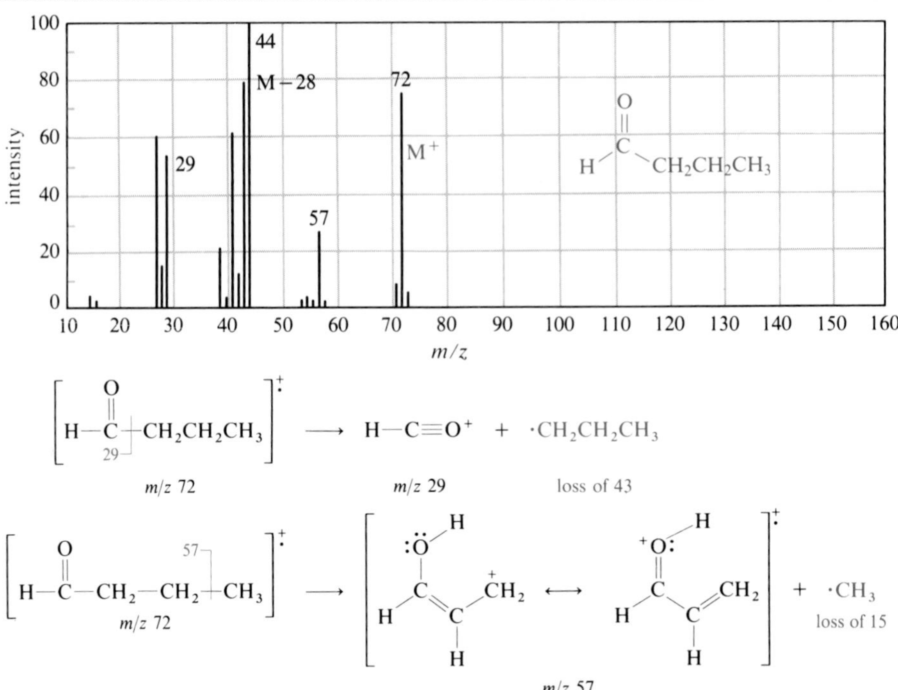

FIGURE 18-4 The mass spectrum of butyraldehyde shows the expected ions of masses 72, 57, and 29. The base peak at m/z 44 results from the loss of ethylene.

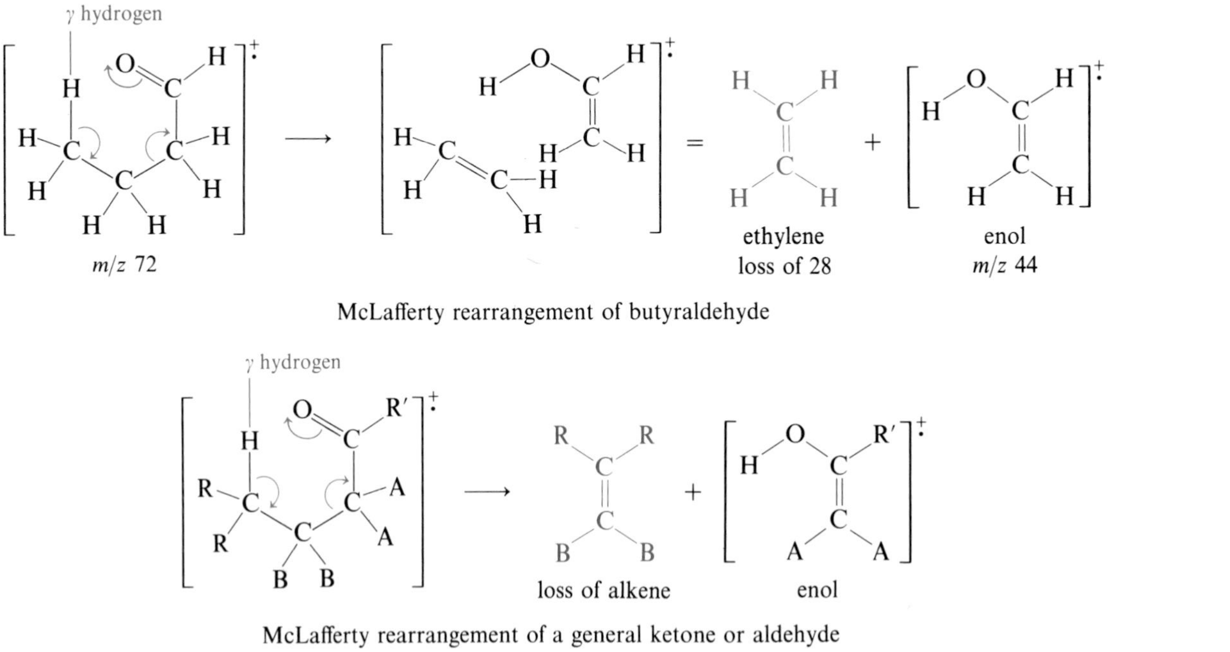

Unexplained

m/z 72 ⟶ m/z 44 (base peak) + loss of 28

γ hydrogen

m/z 72 ⟶ ethylene, loss of 28 + enol, m/z 44

McLafferty rearrangement of butyraldehyde

γ hydrogen

⟶ loss of alkene + enol

McLafferty rearrangement of a general ketone or aldehyde

FIGURE 18-5 Mechanism of the McLafferty rearrangement. This rearrangement may be concerted, as shown here, or the γ hydrogen may be transferred first, followed by fragmentation.

PROBLEM 18-3

Why were no products from McLafferty rearrangement observed in the spectrum of 2-butanone (Fig. 18-3)?

PROBLEM 18-4

Use equations to show the fragmentation leading to each numbered peak in the mass spectrum of 2-octanone.

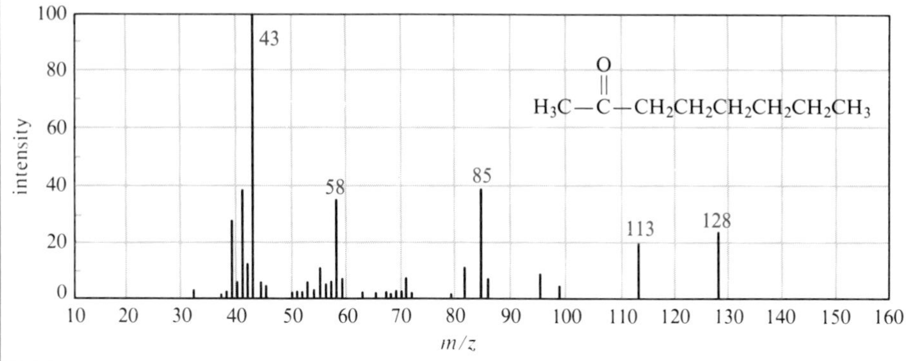

18-5E ULTRAVIOLET SPECTRA OF KETONES AND ALDEHYDES

The π → π* transition The strongest absorptions in the ultraviolet spectra of aldehydes and ketones are the ones resulting from $\pi \rightarrow \pi^*$ electronic transitions. As with alkenes, these absorptions are observable ($\lambda_{max} > 200$ nm) only if the carbonyl double bond is conjugated with another double bond. The simplest conjugated carbonyl system is propenal, shown below. The $\pi \rightarrow \pi^*$ transition of propenal occurs at λ_{max} of 210 nm ($\varepsilon = 11{,}000$). Alkyl substitution increases the value of λ_{max} by about 10 nm per alkyl group. An additional conjugated double bond increases the value of λ_{max} by about 30 nm. Notice that the values of the extinction coefficients (ε) for these transitions are quite large (>5000), as we also observed for the $\pi \rightarrow \pi^*$ transitions of conjugated dienes.

propenal	three alkyl groups	three alkyl groups
$\lambda_{max} = 210$ nm	$\lambda_{max} = 237$ nm	$\lambda_{max} = 244$ nm
$\varepsilon = 11{,}000$	$\varepsilon = 12{,}000$	$\varepsilon = 12{,}500$

The n → π* transition There is an additional band of absorptions in the ultraviolet spectra of ketones and aldehydes, resulting from promotion of one of the nonbonding electrons on oxygen to a π^* antibonding orbital. This transition involves a smaller amount of energy than the $\pi \rightarrow \pi^*$ transition, because the promoted electron leaves a nonbonding (n) orbital that is higher in energy than the π bonding orbital (Fig. 18-6).

Because the $n \rightarrow \pi^*$ transition requires less energy than a $\pi \rightarrow \pi^*$ transition, it results in a lower-frequency and longer-wavelength absorption. The $n \rightarrow \pi^*$ transitions of simple, unconjugated ketones and aldehydes give absorptions with values of λ_{max} between 280 and 300 nm. Each double bond added in conjugation with the carbonyl group increases the value of λ_{max} by about 30 nm. For example the $n \rightarrow \pi^*$ transition of acetone occurs at λ_{max} of 280 nm ($\varepsilon = 15$). Figure 18-7

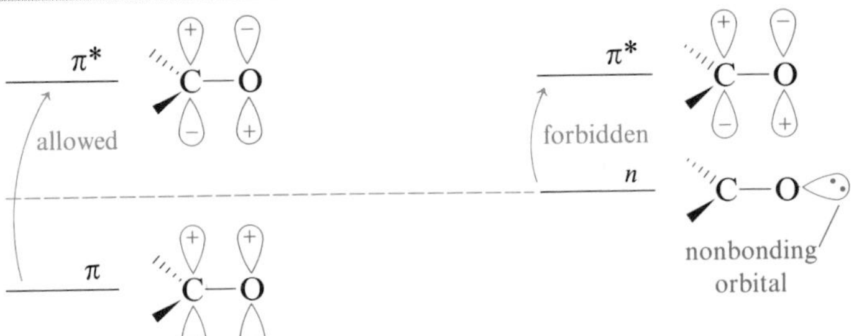

FIGURE 18-6 Comparison of the $\pi \rightarrow \pi^*$ transition and the $n \rightarrow \pi^*$ transition. The $n \rightarrow \pi^*$ transition requires less energy because the nonbonding n electrons are higher in energy than the bonding π electrons.

shows the UV spectrum of a ketone conjugated with one double bond, having a value of λ_{max} around 315 to 330 nm (ε = 110).

Figures 18-6 and 18-7 show that $n \rightarrow \pi^*$ transitions have relatively small extinction coefficients, generally in the range 10 to 200. These absorptions are about 1000 times weaker than the $\pi \rightarrow \pi^*$ transitions, because the $n \rightarrow \pi^*$ transition corresponds to a "forbidden" electronic transition with a low probability of occurrence. The nonbonding orbitals on oxygen are perpendicular to the π^* antibonding orbitals, and there is zero overlap between these orbitals (see Fig. 18-6). This "forbidden" transition occurs occasionally, but much less frequently than the "allowed" $\pi \rightarrow \pi^*$ transition.

Notice that the y axis of the spectrum in Figure 18-7 is logarithmic, allowing both the $\pi \rightarrow \pi^*$ and the much weaker $n \rightarrow \pi^*$ absorptions to be seen on the same spectrum. It is often necessary to run the spectrum twice, using different concentrations of the sample, to observe both absorptions.

More complete information for predicting UV spectra is given in Appendix 3.

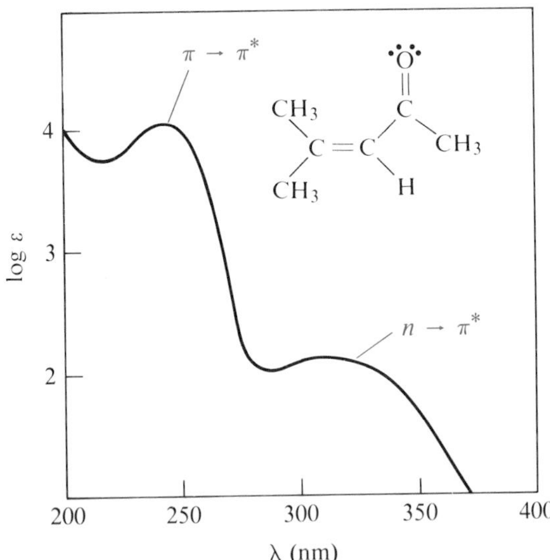

FIGURE 18-7 UV spectrum of 4-methyl-3-penten-2-one. This spectrum would be listed as λ_{max} 237, ε = 12,000; λ_{max} 315, ε = 110.

PROBLEM 18-5

Predict the approximate values of λ_{max} for the $\pi \rightarrow \pi^*$ transition and the $n \rightarrow \pi^*$ transition in each compound.

(a) (b) (c) (d)

PROBLEM 18-6

The following two compounds are easily differentiated by their UV spectra.

(a) Predict approximate values of λ_{max} for these compounds.
(b) Explain how the UV spectra would be used to distinguish between these compounds.

18-6 INDUSTRIAL IMPORTANCE OF KETONES AND ALDEHYDES

In the chemical industry, ketones and aldehydes are used both as solvents and as starting materials and reagents for the synthesis of other products. Although formaldehyde is well known as the formalin solution used to preserve biological specimens, most of the 3 billion kilograms of formaldehyde produced each year is used to make Bakelite, phenol-formaldehyde resins, urea-formaldehyde glues, and other polymeric products. Acetaldehyde is used primarily as a starting material in the manufacture of polymers and drugs.

Acetone is the most important commercial ketone, with over 1 billion kilograms used each year. Both acetone and methyl ethyl ketone (2-butanone) are common industrial solvents. These ketones dissolve a wide range of organic materials, they have convenient boiling points for easy distillation, and they are relatively nontoxic.

Many other ketones are used as flavorings and additives to foods, drugs, and odor products. Table 18-4 lists some simple ketones and aldehydes with well-known odors and flavors.

TABLE 18-4
Ketones and aldehydes used for flavors and odors

	$CH_3—CH_2—CH_2—\overset{O}{\overset{\|\|}{C}}—H$ butyraldehyde	acetophenone	*trans*-cinnamaldehyde
Odor:	"buttery"	pistachio	cinnamon
Use:	margarine, foods	ice cream	candy, foods, drugs
	camphor	carvone	muscone
Odor:	"camphoraceous"	(−) enantiomer: spearmint (+) enantiomer: caraway seed	"musky" aroma
Use:	liniments, inhalants	candy, toothpaste, etc. foods	perfumes

18-7 REVIEW OF SYNTHESIS OF KETONES AND ALDEHYDES

In studying reactions of other functional groups, we have already encountered some of the best methods for synthesis of ketones and aldehydes. Let's review and summarize these reactions, and then consider some additional synthetic methods.

18-7A OXIDATION OF ALCOHOLS (Section 10-2)

Because there are so many ways of making alcohols, they are the most important intermediates for the synthesis of ketones and aldehydes. Oxidation converts an alcohol to the corresponding ketone or aldehyde. For example, a Grignard reaction can assemble a complicated alcohol, which is then oxidized to a ketone or an aldehyde.

Secondary alcohols ⟶ *ketones*

$$\underset{\text{Grignard}}{R{-}MgX} + \underset{\text{aldehyde}}{R'{-}\overset{\overset{\displaystyle O}{\|}}{C}{-}H} \xrightarrow[(2)\ H_3O^+]{(1)\ \text{ether}} \underset{\text{secondary alcohol}}{R{-}\overset{\overset{\displaystyle OH}{|}}{C}H{-}R'} \xrightarrow[H_2SO_4]{Na_2Cr_2O_7} \underset{\text{ketone}}{R{-}\overset{\overset{\displaystyle O}{\|}}{C}{-}R'}$$

Secondary alcohols are readily oxidized to ketones by sodium dichromate in sulfuric acid ("chromic acid"). This is the most common method for oxidation of secondary alcohols.

$$\text{borneol} \xrightarrow[H_2SO_4]{Na_2Cr_2O_7} \text{camphor } (88\%)$$

Primary alcohols ⟶ *aldehydes*

$$\underset{\text{primary alcohol}}{R{-}CH_2OH} \xrightarrow{[\text{oxidizing agent}]} \underset{\text{aldehyde}}{R{-}\overset{\overset{\displaystyle O}{\|}}{C}{-}H} \xrightarrow{[\text{overoxidation}]} \underset{\text{carboxylic acid}}{R{-}\overset{\overset{\displaystyle O}{\|}}{C}{-}OH}$$

The oxidation of a primary alcohol to an aldehyde requires careful selection of an oxidizing agent. Because aldehydes are easily oxidized to carboxylic acids, strong oxidants like chromic acid often give overoxidation. A careful addition of mild chromic acid in acetone solution (Jones oxidation) to a primary alcohol sometimes gives good yields of aldehydes.

A better choice of reagent for this oxidation is the *Collins reagent,* the complex of chromium trioxide (CrO_3) with pyridine. A more soluble version of the Collins reagent is *pyridinium chlorochromate* (PCC), a complex of chromium trioxide with pyridine and HCl. Either of these reagents provides good yields of aldehydes without significant overoxidation.

$$\underset{\text{cyclohexylmethanol}}{C_6H_{11}{-}CH_2OH} \xrightarrow[\text{(PCC)}]{C_5H_5\overset{+}{N}{-}H\ \ CrO_3Cl^-} \underset{\text{cyclohexane carboxaldehyde (90\%)}}{C_6H_{11}{-}CHO}$$

18-7B OZONOLYSIS OF ALKENES (Section 8-15B)

Ozonolysis, followed by a mild reduction, cleaves alkenes to give ketones and aldehydes.

$$\text{R(H)C=CR'R''} \xrightarrow[(2)\ (CH_3)_2S]{(1)\ O_3} \text{R(H)C=O} + \text{O=CR'R''}$$

Ozonolysis can be used either as a synthetic method or as an analytical technique. Yields are generally good.

$$\text{1-methylcyclohexene} \xrightarrow[(2)\ Me_2S]{(1)\ O_3} \text{6-oxoheptanal (65\%)}$$

1-methylcyclohexene — 6-oxoheptanal (65%)

18-7C FRIEDEL-CRAFTS ACYLATION (Section 17-11)

The Friedel-Crafts acylation is an excellent method for synthesis of alkyl aryl ketones or diaryl ketones. It cannot be used on strongly deactivated aromatic systems, however.

$$R-\overset{O}{\overset{\|}{C}}-Cl + G-C_6H_5 \xrightarrow{AlCl_3} p\text{-}G-C_6H_4-\overset{O}{\overset{\|}{C}}-R + o\text{-}G-C_6H_4-\overset{O}{\overset{\|}{C}}-R$$

R is alkyl or aryl; G is hydrogen, a halogen, or an activating group. Acylation fails with strongly deactivated rings.

$$p\text{-}O_2N-C_6H_4-\overset{O}{\overset{\|}{C}}-Cl \xrightarrow[AlCl_3]{C_6H_6} p\text{-}O_2N-C_6H_4-\overset{O}{\overset{\|}{C}}-C_6H_5$$

p-nitrobenzoyl chloride — *p*-nitrobenzophenone (90%)

The Gatterman-Koch synthesis is a variant of the Friedel-Crafts acylation in which carbon monoxide and HCl generate an intermediate that reacts like formyl chloride. Like the Friedel-Crafts reactions, the Gatterman-Koch formylation succeeds only with benzene and activated benzene derivatives.

$$CH_3-C_6H_5 \xrightarrow[AlCl_3]{CO,\ HCl} p\text{-}CH_3-C_6H_4-CHO$$

toluene — *p*-methylbenzaldehyde (major) (50%)

18-7D HYDRATION OF ALKYNES (Section 14-10F)

Catalyzed by acid and mercuric salts The hydration of a terminal alkyne is a useful reaction for the formation of methyl ketones. This reaction is catalyzed by a combination of sulfuric acid and mercuric ion. The initial product of Markovnikov hydration is an enol, which quickly tautomerizes to its keto form. Internal alkynes can also be hydrated, but mixtures of ketone products often result.

$$\underset{\text{alkyne}}{R{-}C{\equiv}C{-}H} \xrightarrow[H_2O]{Hg^{2+},\ H_2SO_4} \underset{\text{enol (not isolated)}}{\left[(R)(HO)C{=}CH_2 \right]} \xrightarrow{H^+} \underset{\text{methyl ketone}}{R{-}C({=}O){-}CH_3}$$

Example

$$\underset{\text{ethynylcyclohexane}}{C_6H_{11}{-}C{\equiv}C{-}H} \xrightarrow{H_2SO_4,\ Hg^{2+}} \underset{\text{enol}}{\left[C_6H_{11}(HO)C{=}CH_2 \right]} \xrightarrow{H^+} \underset{\text{cyclohexyl methyl ketone (90\%)}}{C_6H_{11}{-}C({=}O){-}CH_3}$$

Hydroboration-oxidation Hydroboration-oxidation of an alkyne results in the anti-Markovnikov addition of water across the triple bond. Di(secondary isoamyl)borane, called *disiamylborane,* is used for the hydroboration, since this bulky borane cannot add twice across the triple bond. Upon oxidation of the borane, the unstable enol quickly tautomerizes to an aldehyde.

$$\underset{\text{alkyne}}{R{-}C{\equiv}C{-}H} \xrightarrow[(2)\ H_2O_2,\ NaOH]{(1)\ (Sia)_2BH} \underset{\text{enol (not isolated)}}{\left[(R)(H)C{=}C(H)(OH) \right]} \xrightarrow{^-OH} \underset{\text{aldehyde}}{R{-}CH_2{-}C({=}O){-}H}$$

Example

$$\underset{\text{ethynylcyclohexane}}{C_6H_{11}{-}C{\equiv}C{-}H} \xrightarrow[(2)\ H_2O_2,\ NaOH]{(1)\ (Sia)_2BH} \underset{\text{cyclohexylethanal (65\%)}}{C_6H_{11}{-}CH_2{-}C({=}O){-}H}$$

18-8 SYNTHESIS OF KETONES AND ALDEHYDES USING 1,3-DITHIANES

1,3-Dithiane is a weak proton acid ($pK_a = 32$) that can be deprotonated by strong bases such as *n*-butyllithium. The resulting carbanion is stabilized by the electron-withdrawing effect of the two highly polarizable sulfur atoms.

$$\underset{\text{1,3-dithiane, p}K_a = 32}{\text{1,3-dithiane (C2: H, H)}} + \underset{\text{(Bu—Li)}}{C_4H_9{-}Li} \longrightarrow \underset{\text{dithiane anion}}{\text{(C2: } \ddot{\text{:}}^-\text{, H)}} + \underset{\text{butane}}{C_4H_{10}\uparrow}$$

Alkylation of the dithiane anion by a primary alkyl halide or tosylate gives a thioacetal (sulfur acetal) that can be hydrolyzed using an acidic solution of mercuric chloride. The product is an aldehyde bearing the alkyl group that was added by the alkylating agent. This is a useful synthesis of aldehydes bearing primary alkyl groups.

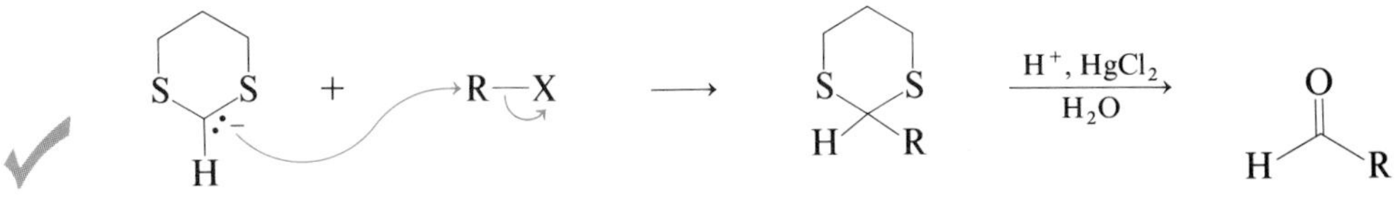

dithiane anion + R—X (alkylating agent (primary alkyl halide)) ⟶ thioacetal $\xrightarrow[H_2O]{H^+, HgCl_2}$ aldehyde (H—C(=O)—R)

Alternatively, the thioacetal can be alkylated once more to give a thioketal. Hydrolysis of the thioketal gives a ketone. (Acetals and ketals are discussed in more detail in Section 18-18.)

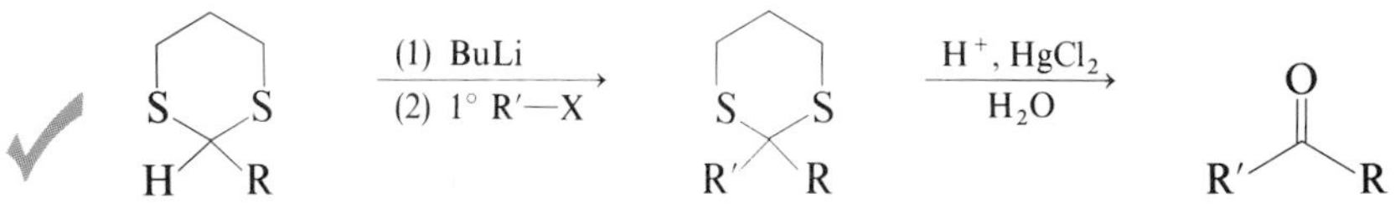

thioacetal $\xrightarrow[(2)\ 1°\ R'—X]{(1)\ BuLi}$ thioketal $\xrightarrow[H_2O]{H^+, HgCl_2}$ ketone (R'—C(=O)—R)

For example, 1-phenyl-2-pentanone may be synthesized as shown below:

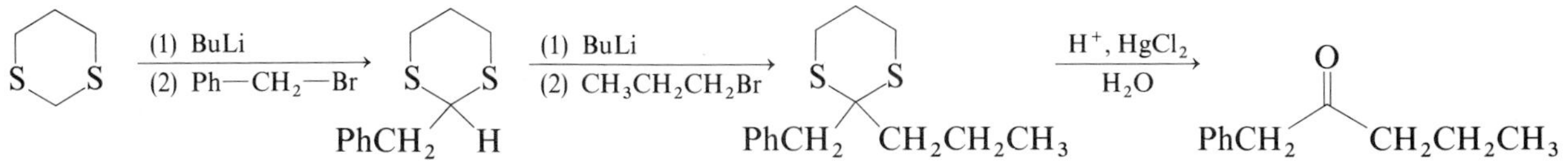

1,3-dithiane $\xrightarrow[(2)\ Ph—CH_2—Br]{(1)\ BuLi}$ thioacetal ($PhCH_2$, H) $\xrightarrow[(2)\ CH_3CH_2CH_2Br]{(1)\ BuLi}$ thioketal ($PhCH_2$, $CH_2CH_2CH_3$) $\xrightarrow[H_2O]{H^+, HgCl_2}$ ketone ($PhCH_2$—C(=O)—$CH_2CH_2CH_3$)

PROBLEM 18-7

Show how you would use the dithiane method to make the following ketones and aldehydes.

(a) 3-phenylpropanal (b) 4-phenyl-2-hexanone
(c) dibenzyl ketone (d) 1-cyclohexyl-4-phenyl-2-butanone

18-9 SYNTHESIS OF KETONES FROM CARBOXYLIC ACIDS

Organolithium reagents can be used to synthesize ketones from carboxylic acids. Organolithiums are so reactive toward carbonyls that they attack the lithium salts of carboxylate anions to give dianions. Protonation of the dianion forms the hydrate of a ketone, which quickly loses water to give the ketone.

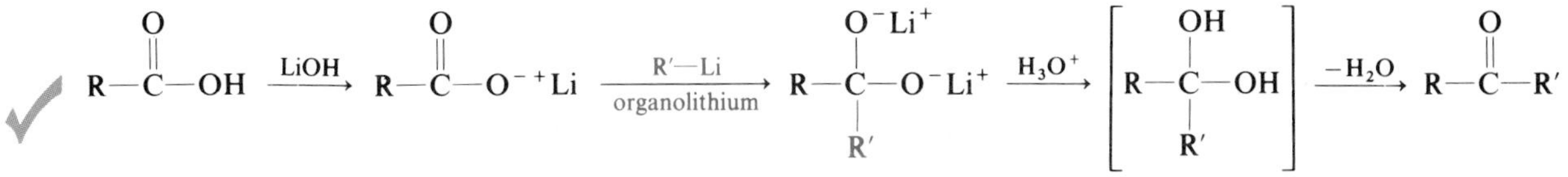

$$\underset{\text{carboxylic acid}}{R—\overset{O}{\overset{\|}{C}}—OH} \xrightarrow{LiOH} \underset{\text{lithium carboxylate}}{R—\overset{O}{\overset{\|}{C}}—O^-\,{}^+Li} \xrightarrow[\text{organolithium}]{R'—Li} \underset{\text{dianion}}{R—\overset{O^-Li^+}{\overset{|}{\underset{R'}{\underset{|}{C}}}}—O^-Li^+} \xrightarrow{H_3O^+} \underset{\text{hydrate}}{\left[R—\overset{OH}{\overset{|}{\underset{R'}{\underset{|}{C}}}}—OH\right]} \xrightarrow{-H_2O} \underset{\text{ketone}}{R—\overset{O}{\overset{\|}{C}}—R'}$$

If the organolithium reagent is inexpensive, we can simply add 2 equivalents to the carboxylic acid. The first equivalent generates the carboxylate salt, and the second equivalent attacks the carbonyl group. Subsequent protonation gives the ketone. Shown below is the synthesis of cyclohexyl phenyl ketone using the reaction of phenyllithium with cyclohexane carboxylic acid.

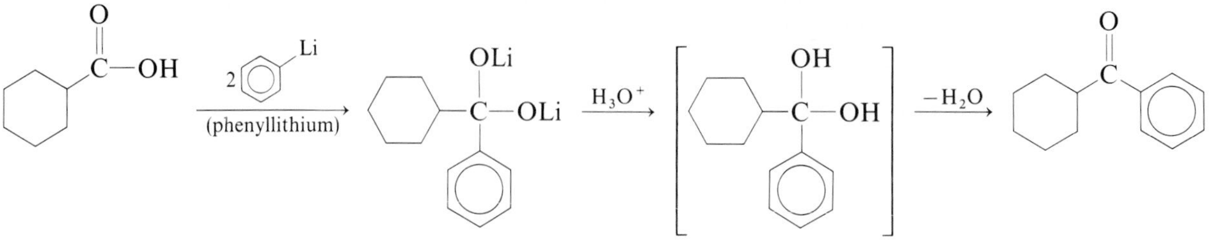

cyclohexane carboxylic acid $\xrightarrow[\text{(phenyllithium)}]{2\ \text{PhLi}}$ dianion $\xrightarrow{H_3O^+}$ [hydrate] $\xrightarrow{-H_2O}$ cyclohexyl phenyl ketone

PROBLEM 18-8

Predict the products of the following reactions.

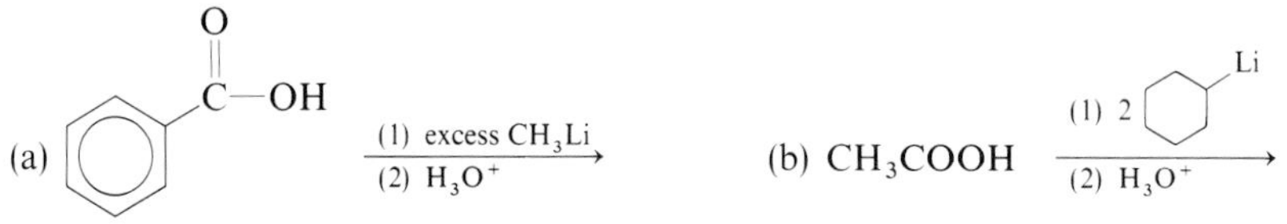

(a) $C_6H_5COOH \xrightarrow[(2)\ H_3O^+]{(1)\ \text{excess}\ CH_3Li}$

(b) $CH_3COOH \xrightarrow[(2)\ H_3O^+]{(1)\ 2\ \text{cyclohexyl-Li}}$

(c) $CH_3(CH_2)_3COOH \xrightarrow[(2)\ H_3O^+]{(1)\ \text{excess}\ CH_3CH_2Li}$

18-10 SYNTHESIS OF KETONES FROM NITRILES

Nitriles can also be used as starting materials for the synthesis of ketones. Discussed in Chapter 21, nitriles are compounds containing the cyano ($-C{\equiv}N$) functional group. Since nitrogen is more electronegative than carbon, the $-C{\equiv}N$ triple bond is polarized like the $C{=}O$ bond of the carbonyl group. Nucleophiles can add to the $-C{\equiv}N$ triple bond by attacking the electrophilic carbon atom.

A Grignard or organolithium reagent attacks a nitrile to give the magnesium salt of an imine. Acidic hydrolysis of the imine leads to the ketone. The mechanism of this acid hydrolysis is the reverse of acid-catalyzed imine formation, covered in Section 18-16.

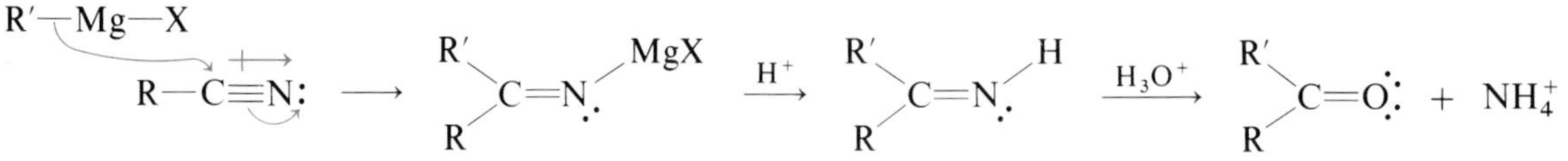

$R'{-}Mg{-}X + R{-}C{\equiv}N: \longrightarrow R'R\,C{=}N{-}MgX \xrightarrow{H^+} R'R\,C{=}N{-}H \xrightarrow{H_3O^+} R'R\,C{=}O + NH_4^+$

nucleophilic attack — Mg salt of imine — imine — ketone

Example

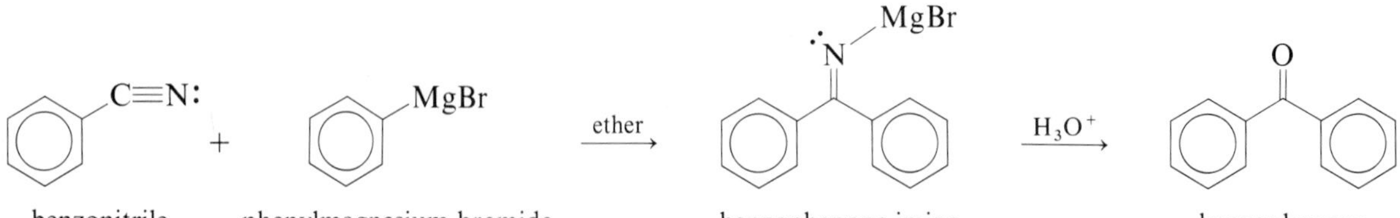

benzonitrile ($C_6H_5C{\equiv}N:$) + phenylmagnesium bromide (C_6H_5MgBr) $\xrightarrow{\text{ether}}$ benzophenone imine (magnesium salt) $\xrightarrow{H_3O^+}$ benzophenone (80%)

PROBLEM 18-9

Predict the products of the following reactions.

(a) $CH_3CH_2CH_2CH_2—C{\equiv}N + CH_3CH_2—MgBr$, then H_3O^+
(b) benzyl bromide + sodium cyanide
(c) product of (b) + cyclopentylmagnesium bromide, then acidic hydrolysis.

PROBLEM 18-10

Show how each of the following transformations may be accomplished in good yield. You may use any additional reagents that are necessary.

(a) bromobenzene → propiophenone
(b) CH_3CH_2CN → 3-heptanone
(c) pentanoic acid → 3-heptanone
(d) toluene → benzyl cyclopentyl ketone

18-11 SYNTHESIS OF ALDEHYDES FROM ACID CHLORIDES

Acyl chlorides (acid chlorides) are reactive derivatives of carboxylic acids in which the acidic hydroxyl group is replaced by a chlorine atom. Acid chlorides are often synthesized by treatment of carboxylic acids with thionyl chloride, $SOCl_2$:

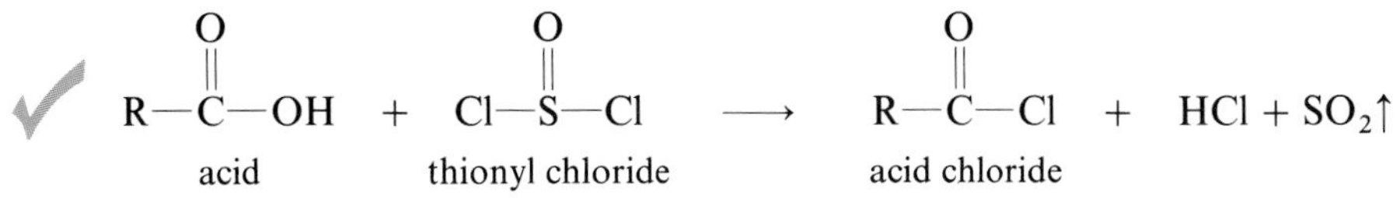

$$\underset{\text{acid}}{R—\overset{O}{\overset{\|}{C}}—OH} + \underset{\text{thionyl chloride}}{Cl—\overset{O}{\overset{\|}{S}}—Cl} \longrightarrow \underset{\text{acid chloride}}{R—\overset{O}{\overset{\|}{C}}—Cl} + HCl + SO_2\uparrow$$

Strong reducing agents like $LiAlH_4$ reduce acid chlorides all the way to primary alcohols. Lithium aluminum tri(*t*-butoxy)hydride is a milder reducing agent that reacts more slowly with aldehydes than with acid chlorides. Reduction of acid chlorides with lithium aluminum tri(*t*-butoxy)hydride gives good yields of aldehydes.

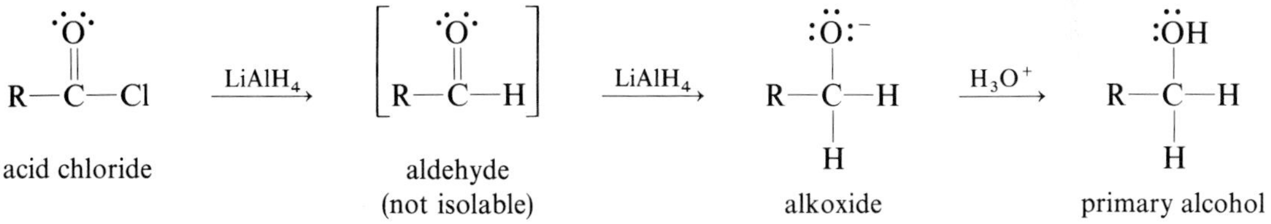

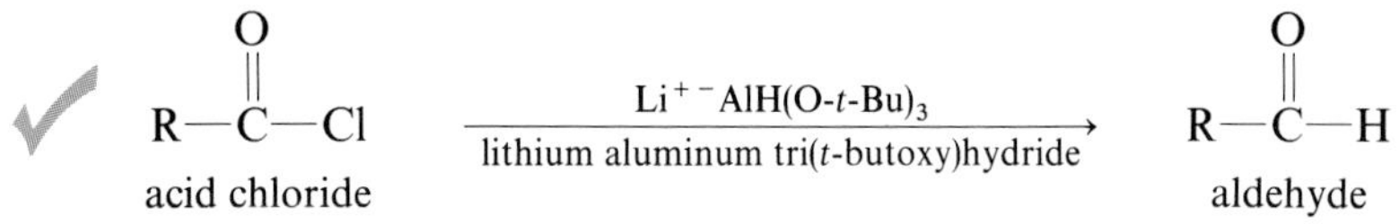

$$\underset{\text{acid chloride}}{R—\overset{\ddot{O}}{\overset{\|}{C}}—Cl} \xrightarrow{LiAlH_4} \underset{\substack{\text{aldehyde}\\\text{(not isolable)}}}{\left[R—\overset{\ddot{O}}{\overset{\|}{C}}—H\right]} \xrightarrow{LiAlH_4} \underset{\text{alkoxide}}{R—\overset{:\ddot{O}:^-}{\overset{|}{\underset{H}{\underset{|}{C}}}}—H} \xrightarrow{H_3O^+} \underset{\text{primary alcohol}}{R—\overset{:\ddot{O}H}{\overset{|}{\underset{H}{\underset{|}{C}}}}—H}$$

$$\underset{\text{acid chloride}}{R—\overset{O}{\overset{\|}{C}}—Cl} \xrightarrow[\text{lithium aluminum tri}(t\text{-butoxy})\text{hydride}]{Li^+\ {}^-AlH(O\text{-}t\text{-}Bu)_3} \underset{\text{aldehyde}}{R—\overset{O}{\overset{\|}{C}}—H}$$

Example

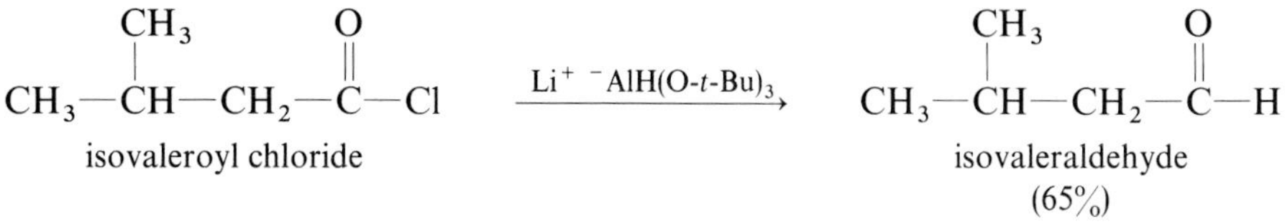

$$\underset{\text{isovaleroyl chloride}}{CH_3—\overset{CH_3}{\overset{|}{C}H}—CH_2—\overset{O}{\overset{\|}{C}}—Cl} \xrightarrow{Li^+\ {}^-AlH(O\text{-}t\text{-}Bu)_3} \underset{\substack{\text{isovaleraldehyde}\\(65\%)}}{CH_3—\overset{CH_3}{\overset{|}{C}H}—CH_2—\overset{O}{\overset{\|}{C}}—H}$$

The Rosenmund reduction A similar aldehyde synthesis is the **Rosenmund reduction,** developed before hydride reagents became available. This is a catalytic reduction of an acyl chloride, using a palladium catalyst supported on barium sulfate and "poisoned" with sulfur (similar to the Lindlar catalyst used for partial hydrogenation of alkynes). This catalyst is more active for the reduction of acyl chlorides than for aldehydes.

$$\mathrm{R{-}\overset{\overset{\displaystyle O}{\|}}{C}{-}Cl} \xrightarrow{\mathrm{H_2,\ Pd/BaSO_4/S}} \mathrm{R{-}\overset{\overset{\displaystyle O}{\|}}{C}{-}H}$$

Example

$$\underset{\text{isovaleroyl chloride}}{\mathrm{CH_3{-}\overset{\overset{\displaystyle CH_3}{|}}{CH}{-}CH_2{-}\overset{\overset{\displaystyle O}{\|}}{C}{-}Cl}} \xrightarrow{\mathrm{H_2,\ Pd/BaSO_4/S}} \underset{\substack{\text{isovaleraldehyde}\\(50\%)}}{\mathrm{CH_3{-}\overset{\overset{\displaystyle CH_3}{|}}{CH}{-}CH_2{-}\overset{\overset{\displaystyle O}{\|}}{C}{-}H}}$$

PROBLEM 18-11

Predict the products of the following reactions.

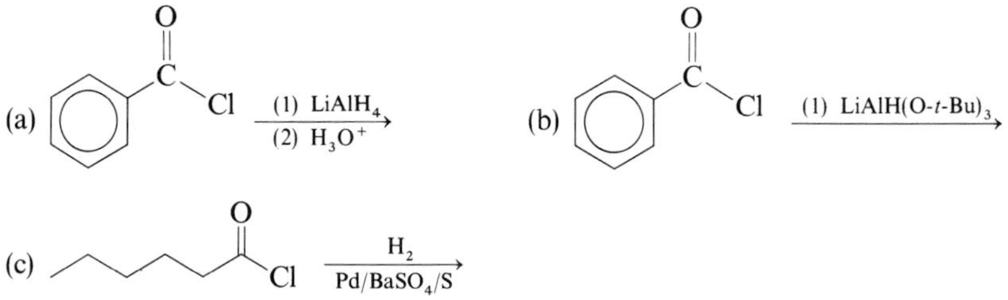

(a) benzoyl chloride ($C_6H_5{-}CO{-}Cl$) $\xrightarrow[\text{(2) } H_3O^+]{\text{(1) } LiAlH_4}$

(b) benzoyl chloride ($C_6H_5{-}CO{-}Cl$) $\xrightarrow{\text{(1) } LiAlH(O\text{-}t\text{-}Bu)_3}$

(c) hexanoyl chloride (skeletal, $CH_3(CH_2)_4{-}CO{-}Cl$) $\xrightarrow[Pd/BaSO_4/S]{H_2}$

SUMMARY OF SYNTHESES OF KETONES AND ALDEHYDES

1. Oxidation of alcohols (Section 10-2)

a. *Secondary alcohols* ⟶ *ketones*

$$\underset{\text{secondary alcohol}}{\mathrm{R{-}\overset{\overset{\displaystyle OH}{|}}{CH}{-}R'}} \xrightarrow{\mathrm{Na_2Cr_2O_7/H_2SO_4}} \underset{\text{ketone}}{\mathrm{R{-}\overset{\overset{\displaystyle O}{\|}}{C}{-}R'}}$$

b. *Primary alcohols* ⟶ *aldehydes*

$$\underset{\text{primary alcohol}}{\mathrm{R{-}CH_2OH}} \xrightarrow{\mathrm{C_5H_5NH^+CrO_3Cl^-(PCC)}} \underset{\text{aldehyde}}{\mathrm{R{-}\overset{\overset{\displaystyle O}{\|}}{C}{-}H}}$$

2. Ozonolysis of alkenes (Section 8-15B)

$$\mathrm{(R)(H)C{=}C(R')(R'')} \xrightarrow[\text{(2) } (CH_3)_2S]{\text{(1) } O_3} \mathrm{(R)(H)C{=}O + O{=}C(R')(R'')}$$

(gives aldehydes or ketones, depending on the starting alkene)

3. Friedel-Crafts acylation (Section 17-11)

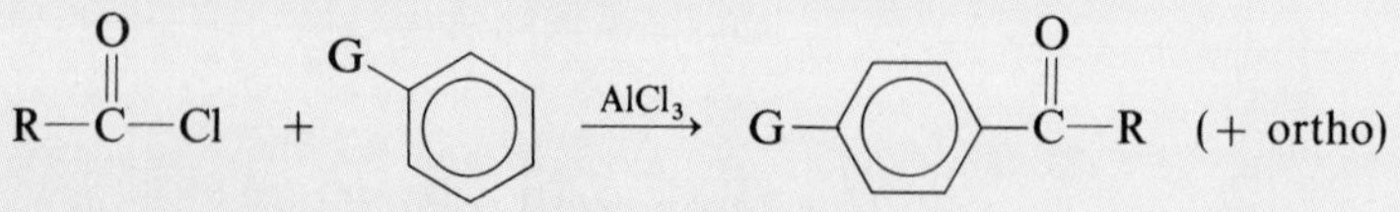

$$\mathrm{R{-}\overset{\overset{\displaystyle O}{\|}}{C}{-}Cl} + \mathrm{G{-}C_6H_5} \xrightarrow{\mathrm{AlCl_3}} \mathrm{G{-}C_6H_4{-}\overset{\overset{\displaystyle O}{\|}}{C}{-}R} \quad (+\ \text{ortho})$$

R can be alkyl or aryl; G is hydrogen, a halogen, or an activating group.

The Gatterman-Koch formylation (Section 17-11C)

$$\text{HCl} + \text{CO} + \text{G—C}_6\text{H}_5 \xrightarrow{\text{AlCl}_3} \text{G—C}_6\text{H}_4\text{—}\overset{\text{O}}{\overset{\|}{\text{C}}}\text{—H}$$

G is hydrogen or an activating group.

4. *Hydration of alkynes* (Section 14-10F)

a. *Catalyzed by acid and mercuric salts* (*Markovnikov orientation*)

$$\underset{\text{alkyne}}{\text{R—C}\equiv\text{C—H}} \xrightarrow[\text{H}_2\text{O}]{\text{Hg}^{2+},\ \text{H}_2\text{SO}_4} \underset{\text{enol (not isolated)}}{\left[\text{(R)(HO)C=CH}_2\right]} \longrightarrow \underset{\text{methyl ketone}}{\text{R—}\overset{\text{O}}{\overset{\|}{\text{C}}}\text{—CH}_3}$$

b. *Hydroboration-oxidation* (*anti-Markovnikov orientation*)

$$\underset{\text{alkyne}}{\text{R—C}\equiv\text{C—H}} \xrightarrow[\text{(2) H}_2\text{O}_2,\ \text{NaOH}]{\text{(1) Sia}_2\text{BH}} \underset{\text{enol (not isolated)}}{\left[\text{(R)(H)C=C(H)(OH)}\right]} \longrightarrow \underset{\text{aldehyde}}{\text{R—CH}_2\text{—}\overset{\text{O}}{\overset{\|}{\text{C}}}\text{—H}}$$

5. *Alkylation of 1,3-dithianes* (Section 18-8)

$$\underset{\text{1,3-dithiane}}{\text{S(CH}_2)_3\text{S>CH}_2} \xrightarrow[\text{(2) 1°R—X}]{\text{(1) BuLi}} \underset{\text{thioacetal}}{\text{S(CH}_2)_3\text{S>C(R)(H)}} \xrightarrow[\text{(2) 1°R'—X}]{\text{(1) BuLi}} \underset{\text{thioketal}}{\text{S(CH}_2)_3\text{S>C(R)(R')}}$$

alkylation

$$\text{thioacetal} \xrightarrow{\text{H}_3\text{O}^+,\ \text{HgCl}_2} \underset{\text{aldehyde}}{\text{R—}\overset{\text{O}}{\overset{\|}{\text{C}}}\text{—H}} \qquad \text{thioketal} \xrightarrow{\text{H}_3\text{O}^+,\ \text{HgCl}_2} \underset{\text{ketone}}{\text{R—}\overset{\text{O}}{\overset{\|}{\text{C}}}\text{—R'}}$$

Example

$$\underset{\text{1,3-dithiane}}{\text{S(CH}_2)_3\text{S>CH}_2} \xrightarrow[\text{(2) PhCH}_2\text{Br}]{\text{(1) BuLi}} \underset{\text{thioacetal}}{\text{S(CH}_2)_3\text{S>C(H)(CH}_2\text{Ph)}} \xrightarrow[\text{(2) Bu—Br}]{\text{(1) BuLi}} \underset{\text{thioketal}}{\text{S(CH}_2)_3\text{S>C(Bu)(CH}_2\text{Ph)}} \xrightarrow[\text{HgCl}_2]{\text{H}_3\text{O}^+} \underset{\text{1-phenyl-2-hexanone}}{\text{Bu—}\overset{\text{O}}{\overset{\|}{\text{C}}}\text{—CH}_2\text{Ph}}$$

6. *Synthesis of ketones using organolithiums with carboxylic acids* (Section 18-9)

$$\underset{\text{carboxylic acid}}{\text{R—}\overset{\text{O}}{\overset{\|}{\text{C}}}\text{—OH}} \xrightarrow{2\ \text{R'—Li}} \underset{\text{dianion}}{\text{R—C(OLi)}_2\text{—R'}} \xrightarrow{\text{H}_3\text{O}^+} \underset{\text{ketone}}{\text{R—}\overset{\text{O}}{\overset{\|}{\text{C}}}\text{—R'}}$$

Example

$$\text{C}_6\text{H}_{11}\text{—COOH} + 2\,CH_3\text{—Li} \longrightarrow \text{C}_6\text{H}_{11}\text{—C(OLi)(OLi)—}CH_3 \xrightarrow{H_3O^+} \text{C}_6\text{H}_{11}\text{—C(=O)—}CH_3$$

cyclohexane carboxylic acid; methyllithium; dianion; cyclohexyl methyl ketone

7. *Synthesis of ketones from nitriles* (Section 18-10)

$$R\text{—C}{\equiv}N + R'\text{—Mg—X} \longrightarrow R\text{—C(=N—MgX)—}R' \xrightarrow{H_3O^+} R\text{—C(=O)—}R'$$

(or R′—Li); Mg salt of imine; ketone

Example

$$\text{C}_6\text{H}_5\text{—C}{\equiv}N \xrightarrow[(2)\ H_3O^+]{(1)\ CH_3CH_2CH_2\text{—MgBr}} \text{C}_6\text{H}_5\text{—C(=O)—}CH_2CH_2CH_3$$

benzonitrile; butyrophenone

8. *Aldehyde synthesis by reduction of acid chlorides* (Section 18-11)

$$R\text{—C(=O)—Cl} \xrightarrow[(\text{or } H_2,\ \text{Pd},\ BaSO_4,\ \text{S})]{Li^{+}\ {}^{-}AlH(O\text{-}t\text{-}Bu)_3} R\text{—C(=O)—H}$$

Example

$$CH_3\text{—CH(Ph)—}CH_2\text{—C(=O)—Cl} \xrightarrow{Li^{+}\ {}^{-}AlH(O\text{-}t\text{-}Bu)_3} CH_3\text{—CH(Ph)—}CH_2\text{—C(=O)—H}$$

3-phenylbutanoyl chloride; 3-phenylbutanal

18-12 REACTIONS OF KETONES AND ALDEHYDES: NUCLEOPHILIC ADDITION

Ketones and aldehydes undergo many reactions to give a wide variety of useful derivatives. Their most common reaction is **nucleophilic addition,** the addition of a nucleophile and a proton across the C=O double bond. The reactivity of the carbonyl group arises from the electronegativity of the oxygen atom and the resulting polarization of the carbon-oxygen double bond. The electrophilic carbonyl carbon atom is sp^2 hybridized and flat, leaving it relatively unhindered and open to attack from either face of the double bond.

If a nucleophile attacks the carbonyl group, the carbon atom changes hybridization from sp^2 to sp^3. The electrons of the pi bond are forced out to the oxygen atom, giving an alkoxide anion which can become protonated to give the product of nucleophilic addition.

$$\text{Nuc:}^- + R_2C^{\delta+}{=}O^{\delta-} \rightleftarrows \text{Nuc—}CR_2\text{—}\ddot{\text{O}}\text{:}^- \xrightleftharpoons{H\text{—Nuc}} \text{Nuc—}CR_2\text{—}\ddot{\text{O}}\text{—H} + \text{Nuc:}^-$$

We have already seen at least two examples of nucleophilic additon to ketones and aldehydes. A Grignard reagent (a strong nucleophile resembling a carbanion, R:$^-$) attacks the electrophilic carbonyl carbon atom to give an alkoxide intermediate. Subsequent protonation gives an alcohol.

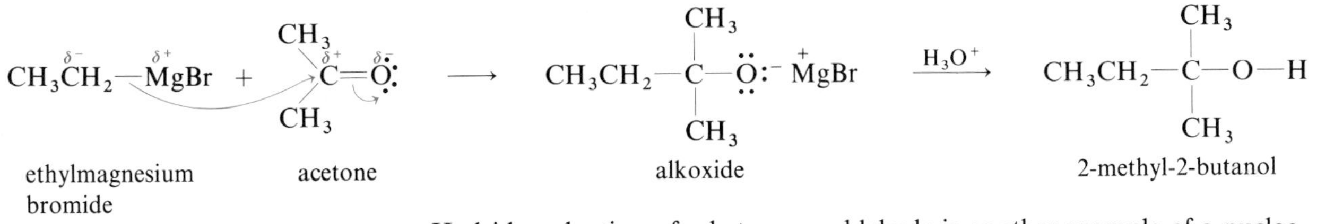

$$CH_3CH_2-MgBr + (CH_3)_2C{=}\ddot{O}: \longrightarrow CH_3CH_2-C(CH_3)_2-\ddot{O}:^-\ {}^+MgBr \xrightarrow{H_3O^+} CH_3CH_2-C(CH_3)_2-O-H$$

ethylmagnesium bromide — acetone — alkoxide — 2-methyl-2-butanol

Hydride reduction of a ketone or aldehyde is another example of a nucleophilic addition, with hydride ion (H:$^-$) serving as the nucleophile. Attack by hydride ion gives an alkoxide that protonates to give an alcohol.

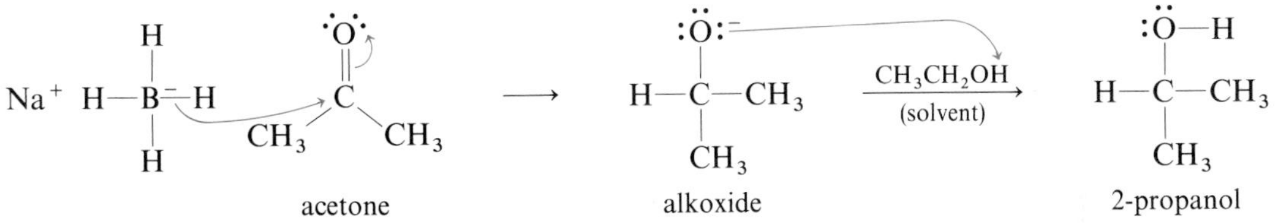

$$Na^+\ H-BH_2-H + (CH_3)_2C{=}\ddot{O}: \longrightarrow H-C(CH_3)_2-\ddot{O}:^- \xrightarrow[\text{(solvent)}]{CH_3CH_2OH} H-C(CH_3)_2-\ddot{O}-H$$

acetone — alkoxide — 2-propanol

Weak nucleophiles can add to activated carbonyl groups under acidic conditions. A carbonyl group is a weak base, and it can become protonated in an acidic solution. A carbonyl group that is protonated (or bonded to some other electrophile) is strongly electrophilic, inviting attack by weak nucleophiles such as water and alcohols.

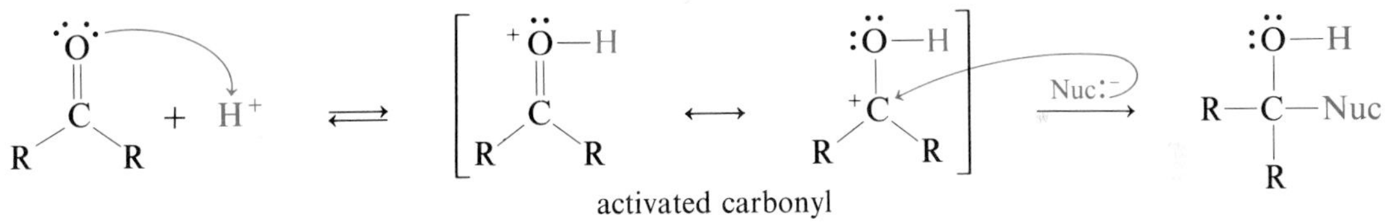

$$R_2C{=}\ddot{O}: + H^+ \rightleftarrows \left[R_2C{=}\overset{+}{\ddot{O}}-H \longleftrightarrow R_2\overset{+}{C}-\ddot{O}-H \right] \xrightarrow{Nuc:^-} R_2C(Nuc)-\ddot{O}-H$$

activated carbonyl

The following reaction is the acid-catalyzed nucleophilic addition of water across the carbonyl group of acetone. This hydration of a ketone or aldehyde is discussed in Section 18-14.

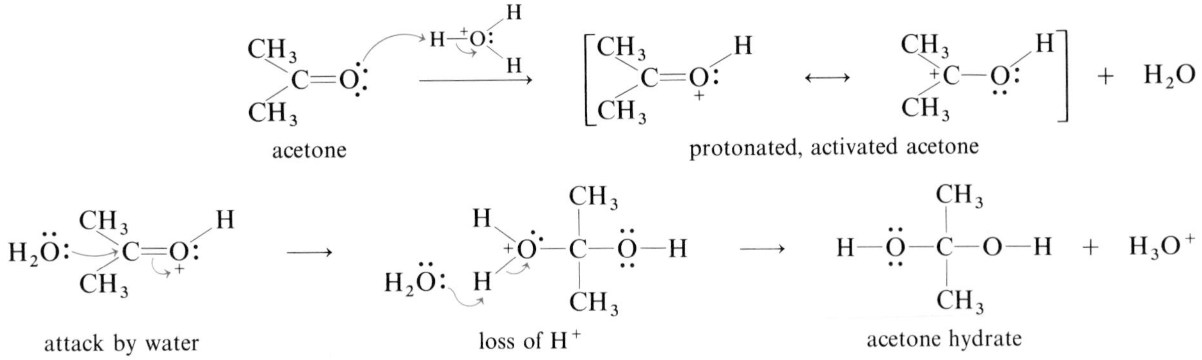

$$(CH_3)_2C{=}\ddot{O}: \xrightarrow{H-\overset{+}{O}H_2} \left[(CH_3)_2C{=}\overset{+}{\ddot{O}}-H \longleftrightarrow (CH_3)_2\overset{+}{C}-\ddot{O}-H \right] + H_2O$$

acetone — protonated, activated acetone

$$H_2\ddot{O}: + (CH_3)_2C{=}\overset{+}{O}-H \longrightarrow H_2\overset{+}{O}-C(CH_3)_2-\ddot{O}-H \ (H_2\ddot{O}:) \longrightarrow H-\ddot{O}-C(CH_3)_2-O-H + H_3O^+$$

attack by water — loss of H$^+$ — acetone hydrate

In effect, the base-catalyzed addition to a carbonyl group results from nucleophilic attack of a strong nucleophile followed by protonation. The acid-catalyzed addition begins with protonation followed by the attack of a weaker nucleophile. Many additions are reversible, with the position of the equilibrium depending on the relative stabilities of the reactants and products. The following table summarizes the base-catalyzed and acid-catalyzed mechanisms together with their reverse reactions.

SUMMARY OF NUCLEOPHILIC ADDITIONS TO CARBONYL GROUPS

Basic conditions (strong nucleophile)

Nuc:$^-$ + >C=O: ⟶ Nuc—C—O:$^-$ $\xrightarrow{\text{H—Nuc}}$ Nuc—C—O—H + Nuc:$^-$

Reverse reaction:

Nuc—C—O—H $\xrightarrow{:\text{Nuc}^-}$ Nuc—C—O:$^-$ ⟶ Nuc:$^-$ >C=O:

Acidic conditions (weak nucleophile, activated carbonyl)

>C=O: $\xrightarrow{\text{H}^+\text{Nuc:}}$ [>C=O$^+$—H ⟷ $^+$C—O:—H] $\xrightarrow{\text{Nuc:}^-}$ Nuc—C—O—H

Reverse reaction:

Nuc—C—O—H ⇌ Nuc:$^-$ [$^+$C—O:—H ⟷ >C=O$^+$—H] ⇌ >C=O: Nuc—H

Reactions of ketones and aldehydes at their α positions Many important reactions of ketones and aldehydes take place when their nucleophilic derivatives attack electrophiles to form new carbon-carbon bonds. These reactions generally occur at the carbon atom adjacent to the carbonyl group (the α-carbon). Deprotonation at the α-carbon generates an enolate ion, a strong nucleophile capable of attacking many kinds of electrophiles.

ketone $\xrightarrow[\text{deprotonation}]{\text{base:}^-}$ enolate $\xrightarrow[\text{attack on electrophile}]{\text{E}^+}$ derivative

These nucleophilic reactions at the α-carbon comprise a large and fascinating group. We will study them in detail in Chapter 22.

18-13 NUCLEOPHILIC ADDITION OF CARBANIONS AND HYDRIDE REAGENTS (Review)

Probably the most useful nucleophilic additions to carbonyl groups are additions of carbanion-like compounds such as Grignard, organolithium, and acetylide reagents. Protonation of the alkoxide intermediates gives primary, secondary, or tertiary alcohols. These reactions were covered in detail in Sections 9-9, 9-12, and 14-8B.

$$CH_3CH_2CH_2{-}C{\equiv}C{:}^-\,Na^+ + CH_3{-}\overset{\displaystyle O}{\overset{\|}{C}}{-}H \longrightarrow CH_3CH_2CH_2{-}C{\equiv}C{-}\underset{\displaystyle CH_3}{\underset{|}{\overset{\displaystyle Na^+{:}\ddot{O}{:}^-}{\overset{|}{C}}}}{-}H \xrightarrow{H_3O^+} CH_3CH_2CH_2{-}C{\equiv}C{-}\underset{\displaystyle CH_3}{\underset{|}{\overset{\displaystyle OH}{\overset{|}{C}}}}{-}H$$

pentynyl sodium — acetaldehyde — alkoxide — 3-heptyn-2-ol (2°) (95%)

$$CH_3CH_2{-}Mg{-}Br + (C_6H_5)_2C{=}O \longrightarrow CH_3CH_2{-}C(C_6H_5)_2{-}\bar{O}\ \overset{+}{Mg}Br \xrightarrow{H_3O^+} CH_3CH_2{-}C(C_6H_5)_2{-}OH$$

ethylmagnesium bromide — benzophenone — alkoxide — 1,1-diphenyl-1-propanol (3°) (85%)

A hydride ion, H:$^-$ (a proton with a pair of electrons), similarly adds to a carbonyl group to give an alkoxide ion. Protonation of the alkoxide ion gives a primary alcohol from an aldehyde, or a secondary alcohol from a ketone. Both lithium aluminum hydride and sodium borohydride reduce ketones and aldehydes, but sodium borohydride is preferred because it is easier to use. Hydride reduction of ketones and aldehydes was covered in detail in Section 9-12.

$$C_6H_5CH_2C({=}O)CH_3 \xrightarrow[CH_3CH_2OH]{NaBH_4} C_6H_5CH_2CH(CH_3){-}O{-}\bar{B}H_3\ Na^+ \xrightarrow{H_3O^+} C_6H_5CH_2CH(CH_3)OH$$

1-phenyl-2-propanone — alkoxide · BH_3 — 1-phenyl-2-propanol (100%)

Catalytic hydrogenation of ketones and aldehydes Catalytic hydrogenation is much slower with carbonyl groups than it is with olefinic double bonds. Before sodium borohydride was available, catalytic hydrogenation was often used to reduce aldehydes and ketones; however, any olefinic double bonds were unavoidably reduced as well. Sodium borohydride is generally preferred over catalytic reduction because it reduces ketones and aldehydes faster than olefins and no gas-handling equipment is required.

In the rare cases when catalytic hydrogenation is used to reduce ketones and aldehydes (often because an olefinic double bond must be reduced as well), the most common catalyst is **Raney nickel.** Raney nickel is a finely divided hydrogen-bearing form of nickel made by treating a nickel-aluminum alloy with a strong sodium hydroxide solution. The aluminum in the alloy reacts to form hydrogen, leaving behind a finely divided nickel powder saturated with hydrogen. Pt and Rh catalysts are also occasionally used for hydrogenation of ketones and aldehydes.

$$\text{cyclohexene-CHO} \xrightarrow[\text{(Raney nickel)}]{Ni{-}H_2} \left[\text{cyclohexane-CHO}\right] \xrightarrow{Ni{-}H_2} \text{cyclohexane-}CH_2OH$$

(90%)

18-14 NUCLEOPHILIC ADDITION OF WATER: HYDRATION OF KETONES AND ALDEHYDES

In an aqueous solution, a ketone or an aldehyde is in equilibrium with its **hydrate,** a geminal diol. With most ketones the equilibrium favors the unhydrated keto form of the carbonyl.

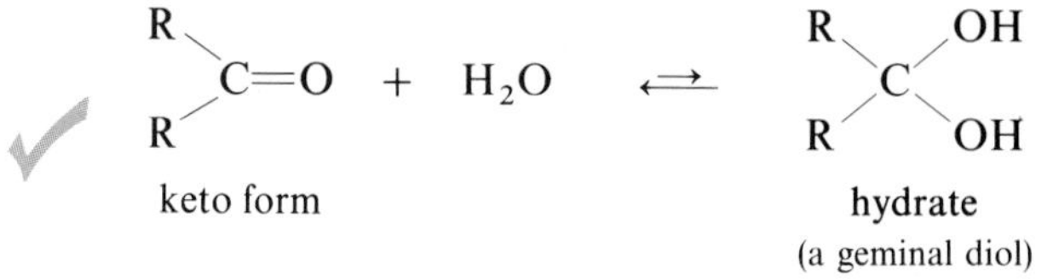

$$R_2C{=}O + H_2O \rightleftharpoons R_2C(OH)_2 \qquad K = \frac{[\text{hydrate}]}{[\text{ketone}][H_2O]}$$

keto form — hydrate (a geminal diol)

Example

$$CH_3-\overset{O}{\overset{\|}{C}}-CH_3 + H_2O \rightleftharpoons CH_3-\overset{HO\ OH}{C}-CH_3 \qquad K = 0.002$$

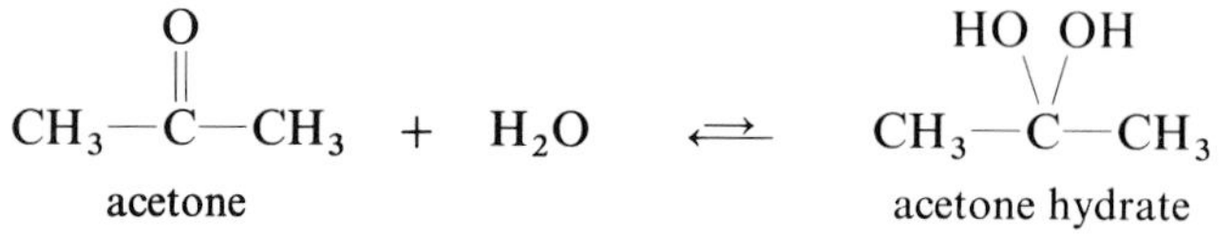

acetone — acetone hydrate

Hydration occurs through the nucleophilic addition mechanism, with water (in acid) or hydroxide ion (in base) serving as the nucleophile.

In acid

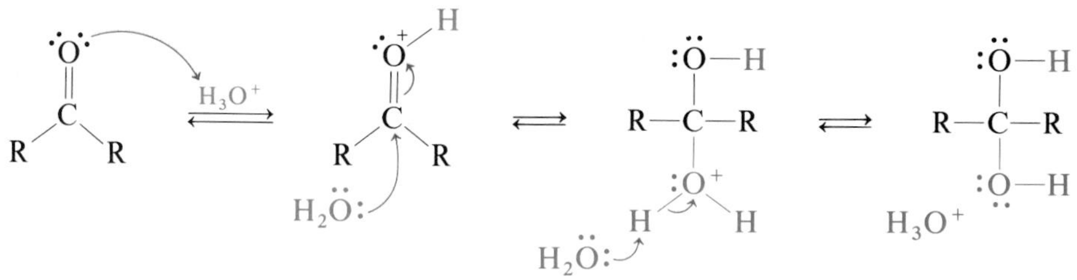

$$R_2C{=}\ddot{O}: \xrightleftharpoons{H_3O^+} R_2C{=}\overset{+}{O}H \xrightarrow{H_2\ddot{O}:} R_2C(OH)(\overset{+}{O}H_2) \rightleftharpoons R_2C(OH)_2 + H_3O^+$$

In base

$$HO^- + R_2C{=}O \rightleftharpoons HO-CR_2-O^- \xrightleftharpoons{H-O-H} HO-CR_2-OH + {}^-OH$$

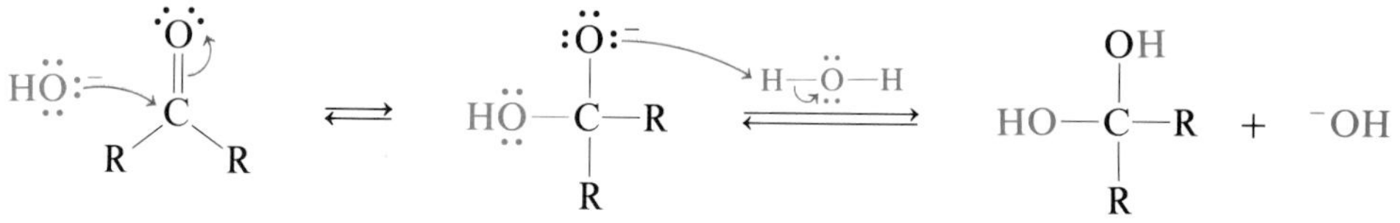

Aldehydes are more likely than ketones to form stable hydrates. The electrophilic carbonyl group of a ketone is stabilized by its two electron-donating alkyl groups, but an aldehyde carbonyl has only one stabilizing alkyl group; its partial positive charge is not so well stabilized. Aldehydes are therefore slightly more electrophilic and less stable than ketones. Formaldehyde, with no electron-donating groups, is even less stable than other aldehydes.

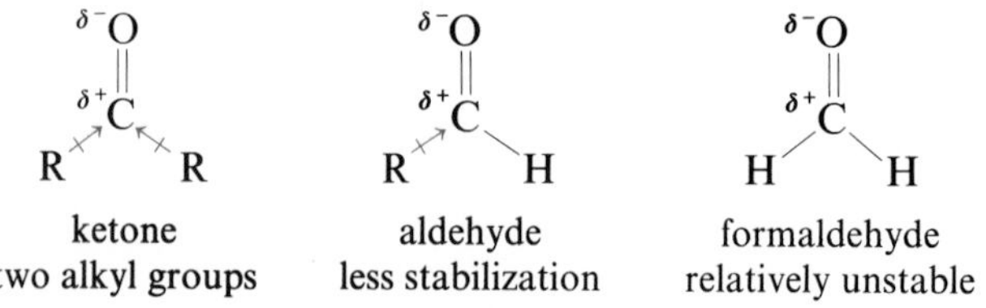

$R_2C^{\delta+}{=}O^{\delta-}$ ketone, two alkyl groups — $RHC^{\delta+}{=}O^{\delta-}$ aldehyde, less stabilization — $H_2C^{\delta+}{=}O^{\delta-}$ formaldehyde, relatively unstable

These stability effects are apparent in the equilibrium constants for hydration of ketones and aldehydes. Ketones have values of K_{eq} of about 10^{-4} to 10^{-2}. For most aldehydes, the equilibrium constant for hydration is close to 1. Formaldehyde, with no alkyl groups bonded to the carbonyl carbon, has a hydration equilibrium constant of about 2000. Strongly electron-withdrawing substituents on the alkyl group of a ketone or aldehyde also destabilize the carbonyl group and favor the hydrate. Chloral has an electron-withdrawing trichloromethyl group that favors the hydrate. Chloral forms a stable, crystalline hydrate that became famous in the movies as "knockout drops" or a "Mickey Finn."

$$CH_3{-}CH_2{-}\overset{O}{\overset{\|}{C}}{-}H + H_2O \rightleftharpoons CH_3{-}CH_2{-}C(OH)_2{-}H \qquad K = 0.7$$

propanal → propanal hydrate

$$H_2C{=}O + H_2O \rightleftharpoons H_2C(OH)_2 \qquad K = 2000$$

formaldehyde → formalin

$$Cl_3C{-}\overset{O}{\overset{\|}{C}}{-}H + H_2O \rightleftharpoons Cl_3C{-}C(OH)_2{-}H \qquad K = 3000$$

chloral → chloral hydrate

PROBLEM 18-12

Propose mechanisms for:

(a) the acid-catalyzed hydration of chloral to form chloral hydrate.
(b) the base-catalyzed hydration of acetone to form acetone hydrate.

PROBLEM 18-13

Rank the following compounds in order of increasing amount of hydrate present at equilibrium.

[cyclohexyl–CHO; cyclohexanone; 1-bromocyclohexane-1-carbaldehyde (CHO, Br); 2-bromocyclohexanone]

18-15 NUCLEOPHILIC ADDITION OF HYDROGEN CYANIDE: FORMATION OF CYANOHYDRINS

Hydrogen cyanide (H—C≡N) is a toxic, water-soluble liquid that boils at 26°C. Because it is mildly acidic, HCN is sometimes called "hydrocyanic acid."

$$H{-}C{\equiv}N: + H_2O \rightleftharpoons H_3O^+ + {}^-{:}C{\equiv}N: \qquad pK_a = 9.1$$

The conjugate base of hydrogen cyanide is the cyanide ion ($^-$:C≡N:). Cyanide ion is a strong base and a strong nucleophile. It attacks ketones and aldehydes to give addition products called **cyanohydrins.** The mechanism is a base-catalyzed nucleophilic addition: attack by cyanide ion on the carbonyl group, followed by protonation of the intermediate.

$$R(R')C{=}\ddot{O}: \xrightarrow{^-:C\equiv N:} R{-}C(R')(C{\equiv}N:){-}\ddot{O}:^- \xrightarrow{H-C\equiv N} R{-}C(R')(C{\equiv}N:){-}\ddot{O}{-}H$$

ketone or aldehyde → intermediate → cyanohydrin

Cyanohydrins may be formed using liquid HCN with a catalytic amount of sodium cyanide or potassium cyanide. Because HCN is dangerous to handle, however, many procedures use a full equivalent of sodium or potassium cyanide dissolved in some other proton-donating solvent.

Cyanohydrin formation is reversible, and the equilibrium constant may or may not favor the cyanohydrin. These equilibrium constants follow the general reactivity trend of ketones and aldehydes,

formaldehyde > other aldehydes > ketones

Formaldehyde reacts quickly and quantitatively with HCN. Other aldehydes react reversibly, with equilibrium constants that favor cyanohydrin formation. Reactions of HCN with ketones have equilibrium constants that may favor either the ketones or the cyanohydrins, depending on the structure. Ketones that are hindered by large alkyl groups react slowly with HCN and give poor yields of cyanohydrins.

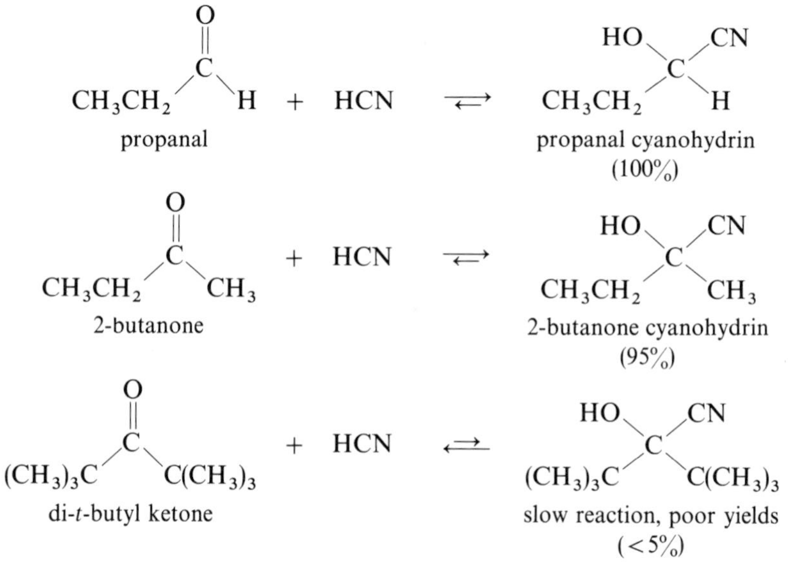

$$CH_3CH_2-C(=O)-H + HCN \rightleftharpoons CH_3CH_2-C(OH)(CN)-H$$

propanal → propanal cyanohydrin (100%)

$$CH_3CH_2-C(=O)-CH_3 + HCN \rightleftharpoons CH_3CH_2-C(OH)(CN)-CH_3$$

2-butanone → 2-butanone cyanohydrin (95%)

$$(CH_3)_3C-C(=O)-C(CH_3)_3 + HCN \rightleftharpoons (CH_3)_3C-C(OH)(CN)-C(CH_3)_3$$

di-*t*-butyl ketone → slow reaction, poor yields (<5%)

The failure with bulky ketones is largely due to steric effects. Cyanohydrin formation involves the rehybridization of the sp^2 carbonyl carbon to sp^3, with a narrowing of the angle between the alkyl groups from about 120° to about 109.5°, increasing their steric interference.

PROBLEM 18-14

Give a mechanism for each of the cyanohydrin syntheses shown above.

Organic compounds containing the cyano group (—C≡N) are called **nitriles.** A cyanohydrin is therefore an α-hydroxynitrile. Nitriles hydrolyze to carboxylic acids under acidic conditions (discussed in Section 21-7D), and cyanohydrins hydrolyze to α-hydroxy acids. This is the most convenient method for making many α-hydroxy acids:

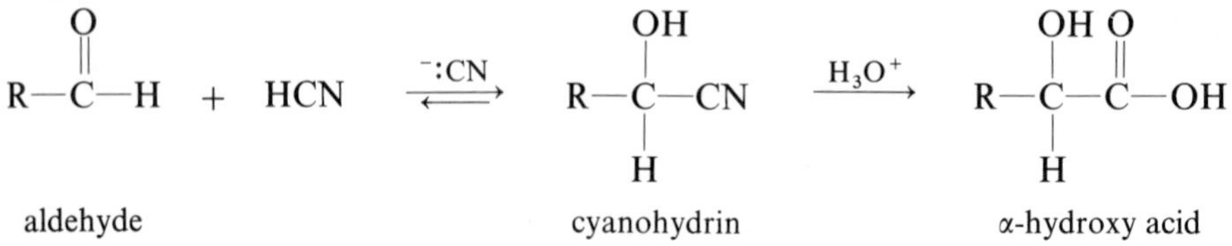

$$R-C(=O)-H + HCN \overset{^{-}:CN}{\rightleftharpoons} R-CH(OH)-CN \xrightarrow{H_3O^+} R-CH(OH)-C(=O)-OH$$

aldehyde → cyanohydrin → α-hydroxy acid

PROBLEM 18-15

Show how you would accomplish the following synthetic transformations.

(a) hexanal ⟶ hexanal cyanohydrin

(b) acetophenone ⟶ acetophenone cyanohydrin

(c) cyclohexyl–CH(=O) (H on ring carbon) ⟶ cyclohexyl–CH(OH)–COOH (H on ring carbon)

18-16 CONDENSATIONS WITH AMMONIA AND PRIMARY AMINES: FORMATION OF IMINES

Under the proper conditions, either ammonia or a primary amine reacts with a ketone or an aldehyde to form an **imine.** Imines are nitrogen analogs of ketones and aldehydes, with a carbon-nitrogen double bond in place of the carbonyl group. Like amines, imines are basic; a substituted imine is also called a **Schiff base.** Imine formation is an example of a large class of reactions called **condensations,** reactions in which two (or more) organic compounds are joined with the elimination of water or another small molecule.

$$\underset{\text{ketone or aldehyde}}{>C=O} + \underset{\text{primary amine}}{R-\ddot{N}H_2} \overset{H^+}{\rightleftharpoons} \underset{\text{carbinolamine}}{\left[-\overset{OH}{\underset{R-\ddot{N}-H}{C}}-\right]} \rightleftharpoons \underset{\substack{\text{imine}\\\text{(Schiff base)}}}{>C=\ddot{N}-R} + H_2O$$

The mechanism of imine formation begins with a basic nucleophilic addition of the amine to the carbonyl group. Attack by the amine followed by protonation of the oxygen atom (and deprotonation of the nitrogen atom) gives an unstable intermediate called a **carbinolamine.**

$$>C=\ddot{O}: + R-\ddot{N}H_2 \rightleftharpoons -\overset{:\ddot{O}:^-}{\underset{R-\overset{+}{N}H_2}{C}}- \;(+\,H_2\ddot{O}:) \overset{H_3O^+}{\rightleftharpoons} -\overset{:\ddot{O}-H}{\underset{R-\ddot{N}-H}{C}}-$$

nucleophilic attack — fast proton transfer — carbinolamine

A carbinolamine reacts to form an imine by the loss of water and formation of a double bond: a dehydration. This dehydration follows the same mechanism as the acid-catalyzed dehydration of an alcohol (Section 10-11). Protonation of the hydroxyl group converts it to a good leaving group, and it leaves as water. The resulting cation is stabilized by a resonance structure with all octets filled and the positive charge on nitrogen. Loss of a proton gives the imine.

$$\underset{\text{carbinolamine}}{-\overset{:\ddot{O}H}{\underset{R\ddot{N}H}{C}}-} \overset{H^+}{\longrightarrow} \underset{\text{protonated}}{-\overset{H_2\overset{+}{O}:}{\underset{R\ddot{N}H}{C}}-} \longrightarrow \underset{\text{intermediate (all octets filled)}}{\left[>\overset{+}{C}-\ddot{N}(R)H \longleftrightarrow >C=\overset{+}{N}(R)H\right]} + H_2\ddot{O}: \longrightarrow \underset{\text{imine}}{>C=\ddot{N}-R} + H_3O^+$$

Having the proper pH is crucial to imine formation. The second step is acid catalyzed, so the solution must be somewhat acidic. If the solution is too acidic, however, the amine becomes protonated and nonnucleophilic, inhibiting the first step. Figure 18-8 shows that the rate of imine formation is fastest around pH 4.5.

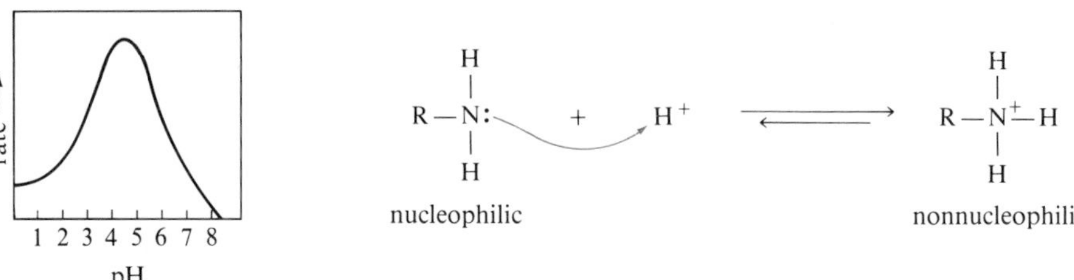

FIGURE 18-8 Although dehydration of the carbinolamine is acid catalyzed, too much acid stops the first step by protonating the amine. The formation of the imine is fastest around pH 4.5.

The formation of several substituted and unsubstituted imines is shown below. In each case, notice that the C=O group of the ketone or aldehyde is replaced by the C=N—R group of the imine.

$$\text{cyclohexanone} + \ddot{N}H_3 \underset{}{\overset{H^+}{\rightleftharpoons}} \text{cyclohexanone imine} + H_2O$$

cyclohexanone ammonia cyclohexanone imine

$$\text{cyclopentanone} + \text{aniline } (\ddot{N}H_2) \overset{H^+}{\rightleftharpoons} \text{cyclopentanone phenyl imine} + H_2O$$

cyclopentanone aniline cyclopentanone phenyl imine

$$\text{PhCH=O} + CH_3-\ddot{N}H_2 \overset{H^+}{\rightleftharpoons} \text{PhCH}=\ddot{N}-CH_3 + H_2O$$

benzaldehyde methylamine benzaldehyde methyl imine

PROBLEM 18-16

Give a mechanism for each of the imine-forming reactions above.

PROBLEM 18-17

Depending on the reaction conditions, two different imines of formula C_8H_9N might be formed by the reaction of benzaldehyde with methylamine. Explain and give the structures of the two imines.

PROBLEM 18-18

Give the structures of the carbonyl compound and the amine used to form each of the following imines.

(a) cyclohexylidene=$\ddot{N}$—CH_3

(b) $CH_3-C(=\ddot{N}H)-CH_2CH_3$

(c) $C_6H_5-N=CHCH_3$

(d) $\ddot{N}$=C—CH_3

18-17 CONDENSATIONS WITH HYDROXYLAMINE AND HYDRAZINES

Ketones and aldehydes also condense with other ammonia derivatives, such as hydroxylamine and substituted hydrazines, to give products that are analogous to imines. The equilibrium constants for these reactions are usually more favorable than those for reactions with simple amines. Hydroxylamine reacts with ketones and aldehydes to form **oximes;** hydrazine derivatives react to form **hydrazones;** and semicarbazide reacts to form **semicarbazones.** The mechanisms of these reactions are similar to the mechanism of imine formation.

phenyl-2-propanone + $H_2N{-}OH$ $\xrightarrow{H^+}$ phenyl-2-propanone oxime + H_2O

benzaldehyde + $H_2N{-}NH_2$ (hydrazine) $\xrightarrow{H^+}$ benzaldehyde hydrazone ($PhCH{=}N{-}NH_2$) + H_2O

cyclohexanone + $H_2N{-}NH{-}Ph$ (phenylhydrazine) $\xrightarrow{H^+}$ cyclohexanone phenylhydrazone + H_2O

$CH_3C(=O)CH_2CH_3$ (2-butanone) + $H_2N{-}NH{-}C(=O){-}NH_2$ (semicarbazide) $\xrightarrow{H^+}$ $CH_3C(=N{-}NH{-}C(=O){-}NH_2)CH_2CH_3$ (2-butanone semicarbazone) + H_2O

The products of these reactions are useful both as starting materials for further reactions (see Section 19-19) and for characterization and identification of the original carbonyl compounds. The oximes, semicarbazones, and phenylhydrazones are often solid compounds with characteristic melting points. Standard tables are available giving the melting points of these derivatives for thousands of different ketones and aldehydes.

If an unknown compound forms one of these derivatives, the melting point can be compared with that in the table. If the compound's physical properties match with those of a known compound, and the melting point of its oxime, semicarbazide, or phenylhydrazone derivative matches as well, we can be fairly sure of a correct identification.

SUMMARY OF CONDENSATIONS OF AMINES WITH KETONES AND ALDEHYDES

$$\text{>C=O} + \boxed{Z}\!-\!NH_2 \xrightarrow{H^+} \text{>C=N}\!-\!\boxed{Z} + H_2O$$

Z in Z—NH$_2$	*Reagent*	*Product*
—H	H—NH$_2$, ammonia	>C=N—H an imine
—R	R—NH$_2$, primary amine	>C=N—R an imine (Schiff base)
—OH	HO—NH$_2$, hydroxylamine	>C=N—OH an oxime
—NH$_2$	H$_2$N—NH$_2$, hydrazine	>C=N—NH$_2$ a hydrazone
—NHPh	PhNH—NH$_2$, phenylhydrazine	>C=N—NHPh a phenylhydrazone
—NHC(=O)NH$_2$	H$_2$N—C(=O)—NH—NH$_2$, semicarbazide	>C=N—NH—C(=O)—NH$_2$ a semicarbazone

PROBLEM 18-19

2,4-Dinitrophenylhydrazine is frequently used for making derivatives of ketones and aldehydes, because the products (2,4-dinitrophenylhydrazones, called **2,4-DNP derivatives**) are even more likely than the phenylhydrazones to be solids with sharp melting points. Give a mechanism for the reaction of acetone with 2,4-dinitrophenylhydrazine in a mildly acidic solution.

PROBLEM 18-20

Predict the major products of the following reactions.

(a) 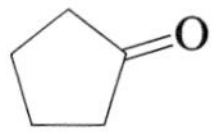 + HO—NH$_2$ $\xrightarrow{H^+}$

(b) 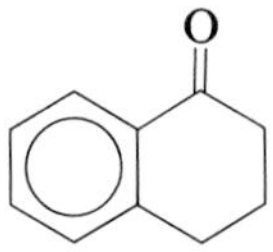 + H$_2$N—NH$_2$ $\xrightarrow{H^+}$

(c) Ph—CH=CH—CHO + H$_2$N—C(=O)—NH—NH$_2$ $\xrightarrow{H^+}$

(d) Ph—C(=O)—Ph + Ph—NH—NH$_2$ $\xrightarrow{H^+}$

PROBLEM 18-21

Show what amines and carbonyl compounds combine to give the following derivatives.

(a) Ph—CH=N—NH—C(=O)—NH$_2$

(b) 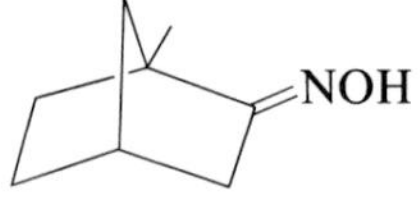

(c) 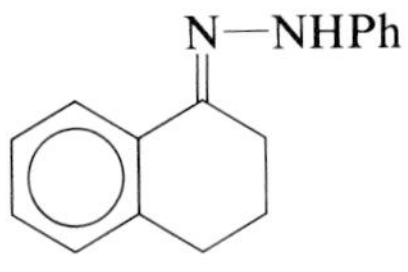

(d) 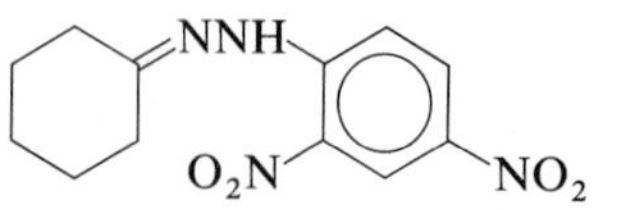

(e) 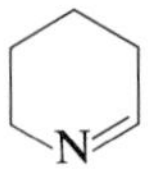

(f) 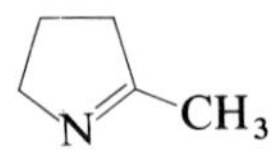

18-18 NUCLEOPHILIC ADDITION OF ALCOHOLS: FORMATION OF ACETALS

Just as ketones and aldehydes react with water to form hydrates, they also react with alcohols to form **acetals.** Acetals formed from ketones are commonly called **ketals,** although this term was recently dropped from the IUPAC nomenclature. In the formation of an acetal, two molecules of an alcohol add to the carbonyl group and one molecule of water is eliminated.

$$\underset{\text{aldehyde}}{RC(=O)H} + 2\,R'{-}OH \overset{H^+}{\rightleftharpoons} \underset{\text{acetal}}{RCH(OR')_2} + H_2O$$

$$\underset{\text{ketone}}{RC(=O)R'} + 2\,R''{-}OH \overset{H^+}{\rightleftharpoons} \underset{\substack{\text{acetal (IUPAC)}\\ \text{ketal (common)}}}{RR'C(OR'')_2} + H_2O$$

Although hydration is catalyzed by either acid or base, acetal formation must be acid catalyzed. For example, consider the reaction of cyclohexanone with methanol, catalyzed by *p*-toluenesulfonic acid.

Overall reaction

cyclohexanone + 2 CH_3OH —($CH_3C_6H_4SO_2OH$ (Ts—OH), *p*-toluenesulfonic acid)→ cyclohexanone dimethyl acetal [$(CH_3O)_2C_6H_{10}$] + H_2O

The first step is a typical acid-catalyzed addition to the carbonyl group. The acid catalyst protonates the carbonyl group, and the alcohol (a weak nucleophile) attacks the protonated, activated carbonyl. Loss of a proton from the positively charged intermediate gives a **hemiacetal.** The hemiacetal gets its name from the Greek prefix *hemi-,* meaning "half." Having added one molecule of the alcohol, the hemiacetal is halfway to becoming a "full" acetal. Like the hydrates of ketones and aldehydes, most hemiacetals are too unstable to be isolated and purified.

Mechanism (first half)

ketone (:O:) + H—OTs ⇌ [protonated (activated) ketone: $^+$O—H ⟷ :O—H with + on carbon] + H—:O—CH_3 ⇌ HO: with O^+(H)(CH_3) on ring; CH_3—ÖH removes H ⇌ HÖ: :ÖCH$_3$ hemiacetal

The second half of the mechanism converts the hemiacetal to the more stable acetal. Protonation of the hydroxyl group, followed by loss of water, gives a resonance-stabilized carbocation. Attack on the carbocation by methanol, followed by loss of a proton, gives the dimethyl acetal.

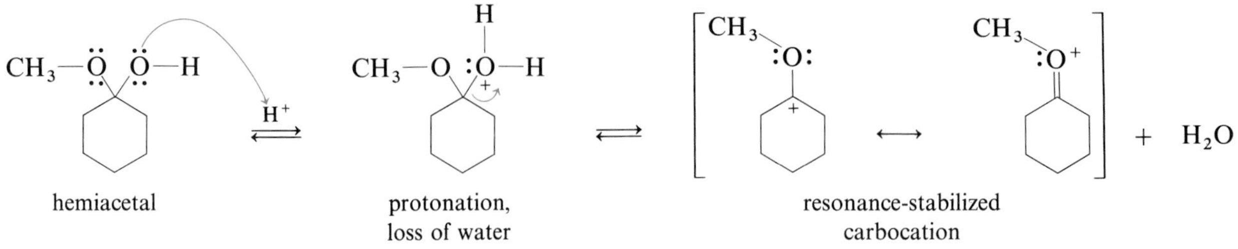

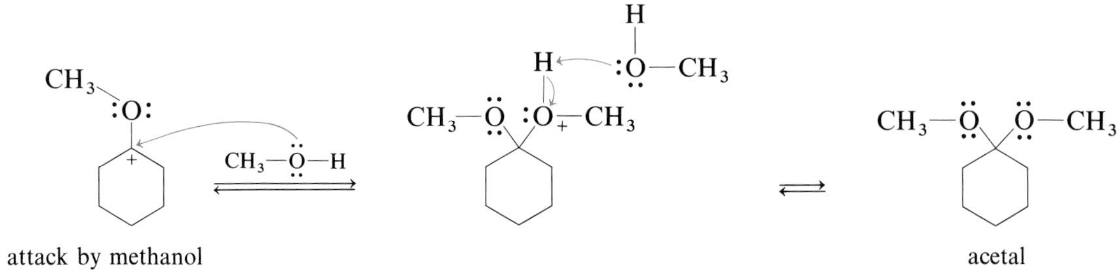

hemiacetal

protonation, loss of water

resonance-stabilized carbocation

$+ H_2O$

attack by methanol

acetal

A similar mechanism accounts for the acid-catalyzed reaction of an aldehyde with an alcohol to form a hemiacetal. Further reaction with another equivalent of the alcohol gives an acetal.

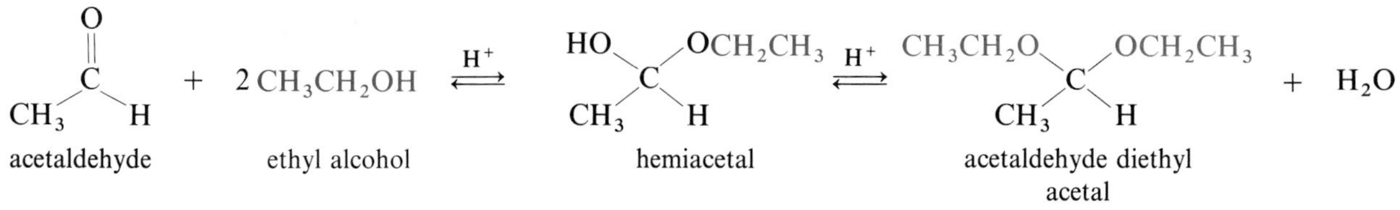

$$CH_3CHO + 2\,CH_3CH_2OH \overset{H^+}{\rightleftharpoons} CH_3CH(OH)(OCH_2CH_3) \overset{H^+}{\rightleftharpoons} CH_3CH(OCH_2CH_3)_2 + H_2O$$

acetaldehyde — ethyl alcohol — hemiacetal — acetaldehyde diethyl acetal

PROBLEM 18-22

Propose a mechanism for the acid-catalyzed reaction of acetaldehyde with ethanol to give acetaldehyde diethyl acetal.

Since hydration is catalyzed by either acid or base, you might wonder why acetal formation is catalyzed only by acid. In fact, the first step (formation of the hemiacetal) can be base-catalyzed, involving attack by alkoxide ion and protonation of the alkoxide. The second step requires replacement of the hemiketal —OH group by the alcohol —OR″ group. Hydroxide ion is a poor leaving group for the S_N2 reaction, so alkoxide cannot displace the —OH group. This replacement occurs under acidic conditions, however, because protonation of the —OH group and loss of water gives a resonance-stabilized cation.

Attempted base-catalyzed ketalization

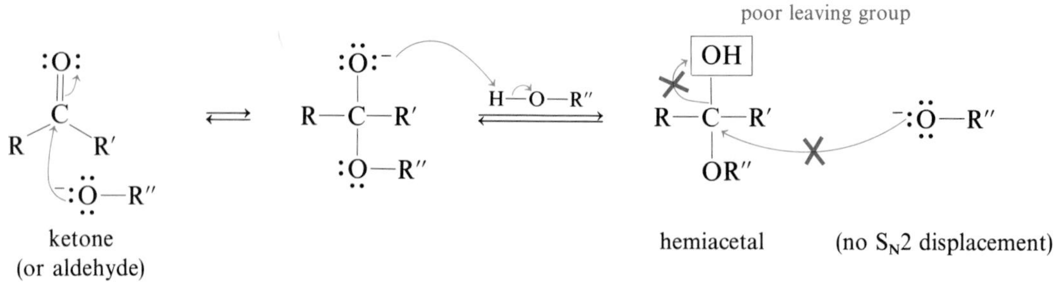

poor leaving group

ketone (or aldehyde)

hemiacetal

(no S_N2 displacement)

Equilibrium of acetal formation All these reactions are reversible, and their equilibrium constants determine the proportions of reactants and products present at equilibrium. For simple aldehydes, the equilibrium constants generally favor the acetal products. For example, the acid-catalyzed reaction of acetaldehyde with ethanol gives a good yield of the acetal.

With hindered aldehydes and with most ketones, the equilibrium constants favor the carbonyl compounds rather than the acetals. In these reactions, the alcohol is often used as the solvent to assure a large excess. The water formed as a byproduct is removed by distillation to force the equilibrium toward the right.

Conversely, most acetals are hydrolyzed simply by shaking them with dilute acid in water. The large excess of water drives the equilibrium toward the ketone or aldehyde. The mechanism is simply the reverse of acetal formation. For example, cyclohexanone dimethyl acetal is quantitatively hydrolyzed to cyclohexanone by a brief treatment with dilute aqueous acid.

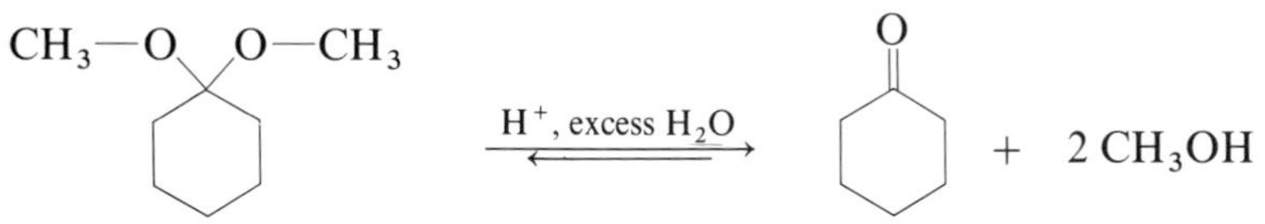

$$\text{(cyclohexane with } CH_3-O \text{ and } O-CH_3\text{)} \xrightleftharpoons{H^+,\ \text{excess } H_2O} \text{(cyclohexanone)} + 2\ CH_3OH$$

Cyclic acetals Formation of an acetal using a diol as the alcohol gives a cyclic acetal. Formation of a cyclic acetal often has a more favorable equilibrium constant, since there is a smaller entropy loss when two molecules (a ketone and a diol) condense than when three molecules (a ketone and two molecules of an alcohol) condense. Ethylene glycol is the diol most commonly used to make cyclic acetals; its acetals are called **ethylene acetals** (or **ethylene ketals**).

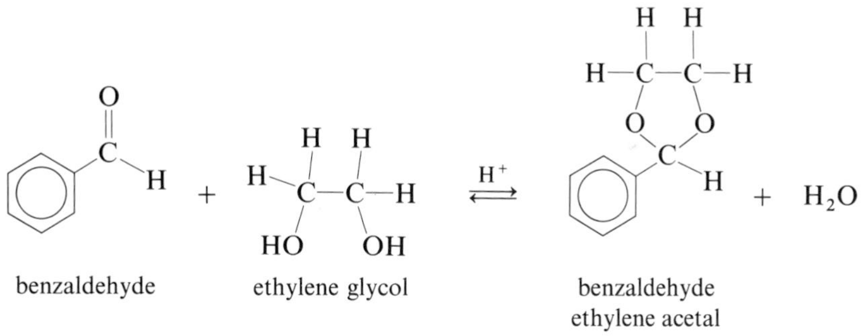

$$\underset{\text{benzaldehyde}}{C_6H_5-CHO} + \underset{\text{ethylene glycol}}{HO-CH_2-CH_2-OH} \overset{H^+}{\rightleftharpoons} \underset{\text{benzaldehyde ethylene acetal}}{C_6H_5-CH(OCH_2CH_2O)} + H_2O$$

PROBLEM SOLVING: PROPOSING REACTION MECHANISMS

The general principles for proposing reaction mechanisms are applied here to the hydrolysis of an acetal. These principles were introduced in Chapter 10 and are summarized in Appendix 4. Remember that it is important to draw all the bonds and substituents of each carbon atom involved in a mechanism, that each step must be shown separately, and that curved arrows always show the movement of electrons from the nucleophile to the electrophile.

Our problem is to propose a mechanism for the acid-catalyzed hydrolysis of the acetal shown below.

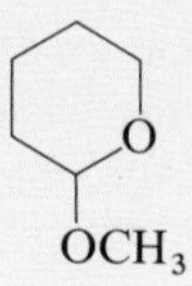

(tetrahydropyran ring with O and OCH_3 substituent on adjacent carbon)

First, we must determine the type of mechanism. Since it is stated that the reaction is acid-catalyzed, we assume it involves strong electrophiles, cationic intermediates (possibly carbocations), but no strong nucleophiles or strong bases, and certainly no carbanions or free radicals.

1. **Consider the carbon skeletons of the reactants and products, and decide which carbon atoms in the products are most likely derived from which carbon atoms in the reactants.**
   First you must decide what products are formed by hydrolysis of the acetal. In dealing with acetals and hemiacetals, any carbon atom with *two* bonds to oxygen is generally derived from a carbonyl group. Draw an equation showing all the affected atoms. The equation shows that water must somehow add (probably by a nucleophilic attack) and the ring must be cleaved.

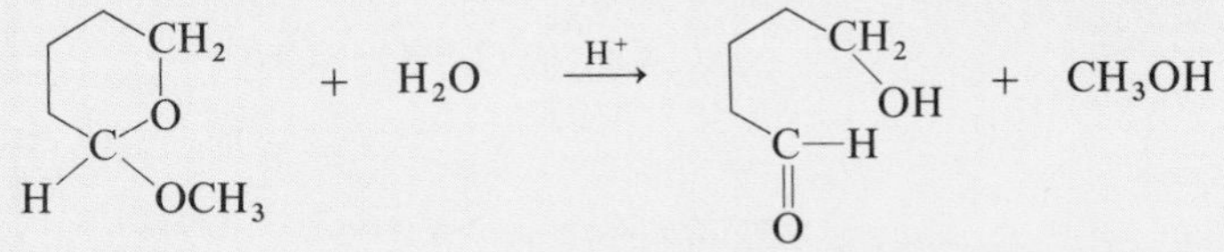

   (ring: CH_2, O, C with H and OCH_3) + H_2O $\xrightarrow{H^+}$ (chain: CH_2, OH, C—H, ‖O) + CH_3OH

2. **Consider whether any of the reactants is a strong enough electrophile to react without being activated. If not, consider how one of the reactants might be converted to a strong electrophile by protonation of a Lewis basic site (or complexation with a Lewis acid).**
   The reactant probably will not react with water until it is activated, most likely by protonation. It can become protonated at either oxygen atom. We will arbitrarily choose the ring oxygen for protonation. The protonated compound is well suited for ring cleavage to form a stabilized (and strongly electrophilic) cation.

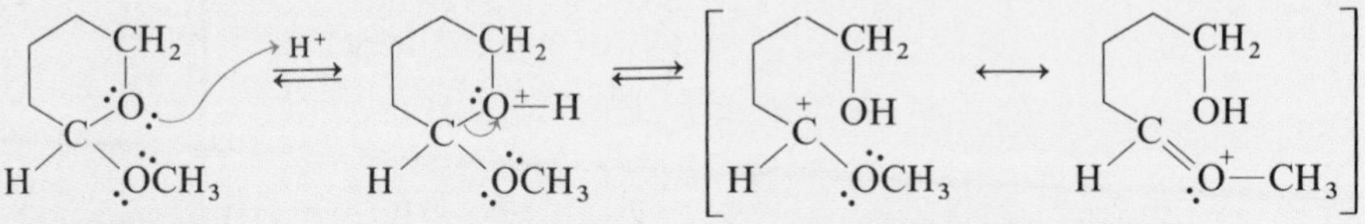

   CH_2, $\ddot{O}$, C, H, $\ddot{O}CH_3$ $\xrightarrow{H^+}$ ⇌ CH_2, $\ddot{O}^{+}$—H, C, H, $\ddot{O}CH_3$ ⇌ [CH_2, OH, $\overset{+}{C}$, H, $\ddot{O}CH_3$ ⟷ CH_2, OH, C, H, $=\overset{+}{O}$—CH_3]

   protonation   cleavage   resonance-stabilized cation

3. **Consider how a nucleophilic site on another reactant can attack the strong electrophile to form a bond needed in the product. Draw the product of this bond formation.**
   Attack by water on the cation gives a protonated hemiacetal.

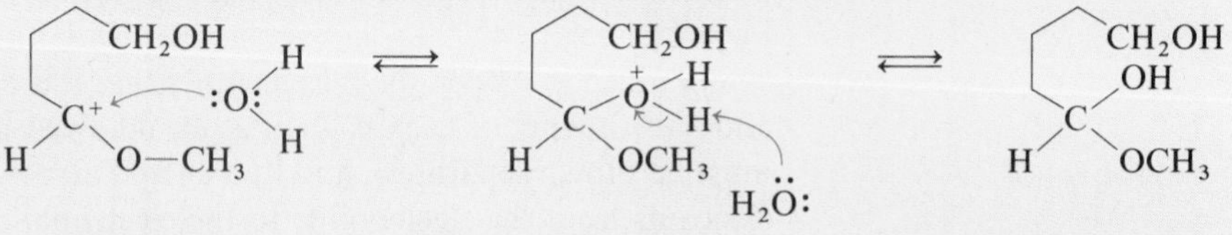

   CH_2OH, C^+, H, O—CH_3 ← :$\ddot{O}$:(H)(H) ⇌ CH_2OH, $\overset{+}{O}$(H)(H), C, H, OCH_3, $H_2\ddot{O}$: ⇌ CH_2OH, OH, C, H, OCH_3

   attack by water   deprotonation   hemiacetal

4. **Consider how the product of nucleophilic attack might be converted to the final product (if it has the right carbon skeleton) or reactivated to form another bond needed in the product.**
   Just as an —OH group can be lost by protonation and loss of water, the —OCH_3 group can be lost by protonating it and losing methanol. A protonated version of the product results.

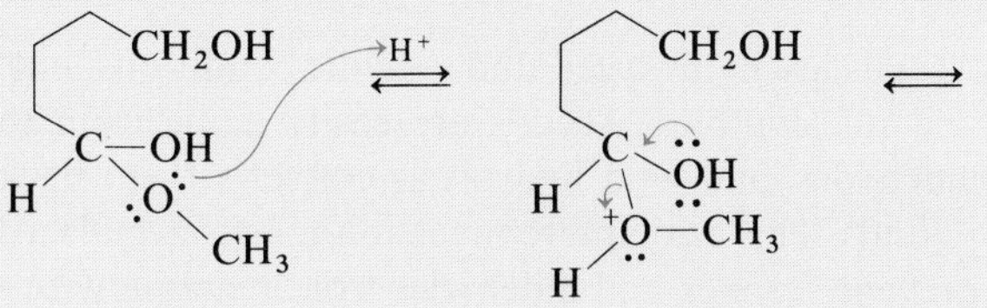

protonation

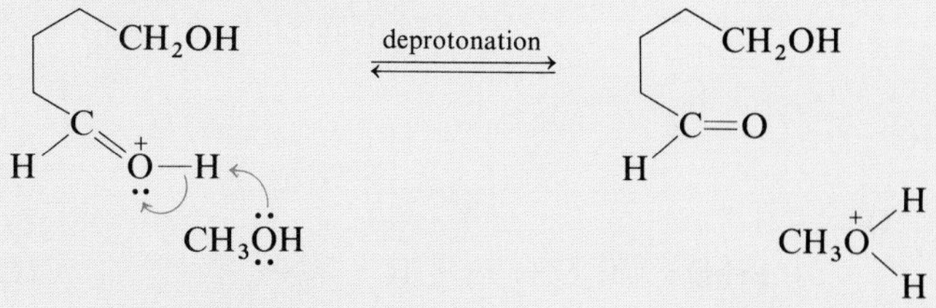

resonance-stabilized intermediate

deprotonation

products

5. **Draw out all the steps of the mechanism using curved arrows to show the movement of electrons.**
The complete mechanism is given by combining the equations written immediately above. You should write out the mechanism as a review of the steps involved.

As further practice in proposing reaction mechanisms, do Problems 18-23 and 18-24 by completing each of the five steps listed in this section.

PROBLEM 18-23
In the mechanism for acetal hydrolysis shown above, the ring oxygen atom was protonated first, the ring was cleaved, and then the methoxyl group was lost. The mechanism could also be written to show the methoxyl oxygen protonating and cleaving first, followed by ring cleavage. Draw out this alternative mechanism.

PROBLEM 18-24
Propose a mechanism for the acid-catalyzed hydrolysis of cyclohexanone ethylene acetal.

PROBLEM 18-25
Show what alcohols and carbonyl compounds give the following derivatives.

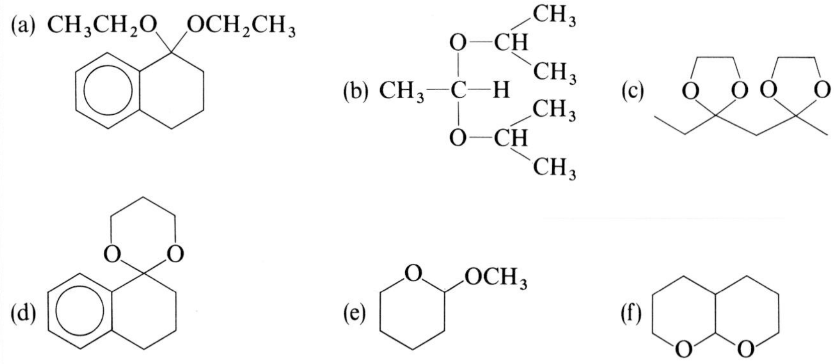

(a) CH_3CH_2O OCH_2CH_3

(b) CH_3—C—H

(c)

(d)

(e) OCH_3

(f)

18-19 USE OF ACETALS AS PROTECTING GROUPS

Acetals are hydrolyzed under acidic conditions, but they are stable to strong bases and nucleophiles. Acetals are easily made from the corresponding ketones and aldehydes, and acid hydrolysis converts them back to the parent carbonyl compounds. This easy interconversion makes acetals attractive as **protecting groups** to prevent ketones and aldehydes from reacting with strong bases and nucleophiles.

As an example, consider the proposed synthesis below. The necessary Grignard reagent could not be made, because the carbonyl group would react with the nucleophilic organometallic group.

Proposed synthesis

cyclohexanone + $BrMg{-}CH_2CH_2{-}C(=O){-}H$ ⟶✗ (1-OMgBr-cyclohexyl)$-CH_2CH_2{-}C(=O){-}H$ $\xrightarrow{H_3O^+}$ (1-OH-cyclohexyl)$-CH_2CH_2{-}C(=O){-}H$

cyclohexanone; incompatible functional groups (impossible reagent); target compound

If the aldehyde is protected as an acetal, however, it is unreactive toward a Grignard reagent. The "masked" aldehyde is converted to the Grignard reagent, which is allowed to react with cyclohexanone. Dilute aqueous acid both protonates the alkoxide to give the alcohol and hydrolyzes the acetal to give the deprotected aldehyde.

Actual synthesis

$Br{-}CH_2CH_2{-}C(=O){-}H$ $\xrightarrow[H^+]{HOCH_2CH_2OH}$ $Br{-}CH_2CH_2{-}C(H)$(cyclic acetal, O O) "masked" aldehyde $\xrightarrow[ether]{Mg}$ $BrMg{-}CH_2CH_2{-}C(H)$(cyclic acetal, O O)

cyclohexanone + MgBr$-CH_2CH_2{-}C(H)$(cyclic acetal, O O) ⟶ (1-OMgBr-cyclohexyl)$-CH_2CH_2{-}C(H)$(cyclic acetal, O O) $\xrightarrow{H_3O^+}$ (1-OH-cyclohexyl)$-CH_2CH_2{-}C(=O){-}H$

target compound

Selective acetal formation Because aldehydes form acetals more readily than ketones, aldehydes can often be selectively protected in the presence of a ketone. This selective protection leaves the ketone available for modification under neutral or basic conditions without disturbing the more reactive aldehyde group. The following example shows the reduction of a ketone in the presence of a more reactive aldehyde.

3-oxocyclohexanecarbaldehyde $\xrightarrow[H^+]{\text{1 equiv. } HOCH_2CH_2OH}$ ketone with acetal-protected aldehyde $\xrightarrow{NaBH_4}$ (H, OH) alcohol with acetal-protected aldehyde $\xrightarrow{H_3O^+}$ (H, OH) 3-hydroxycyclohexanecarbaldehyde

PROBLEM 18-26

Show how you would accomplish the following synthetic transformations. You may use whatever additional reagents you need.

(a) 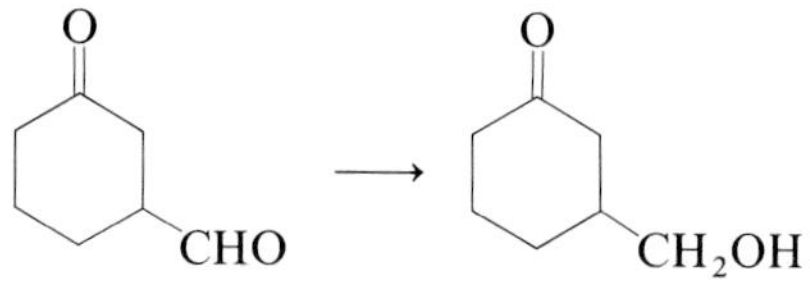 3-oxocyclohexanecarbaldehyde (ring ketone, CHO) $\longrightarrow$ 3-(hydroxymethyl)cyclohexanone (CH_2OH)

(b) 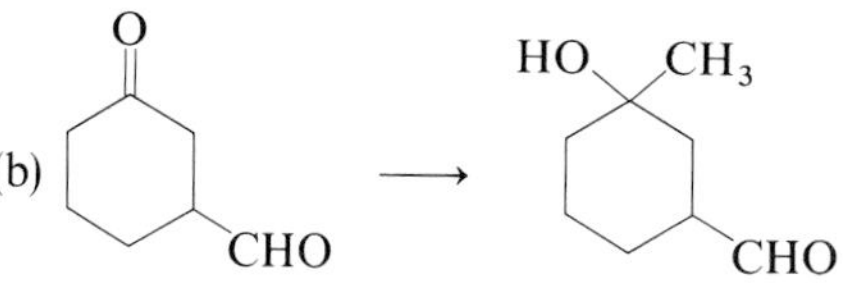 3-oxocyclohexanecarbaldehyde (CHO) $\longrightarrow$ 3-hydroxy-3-methylcyclohexanecarbaldehyde (HO, CH_3, CHO)

(c) 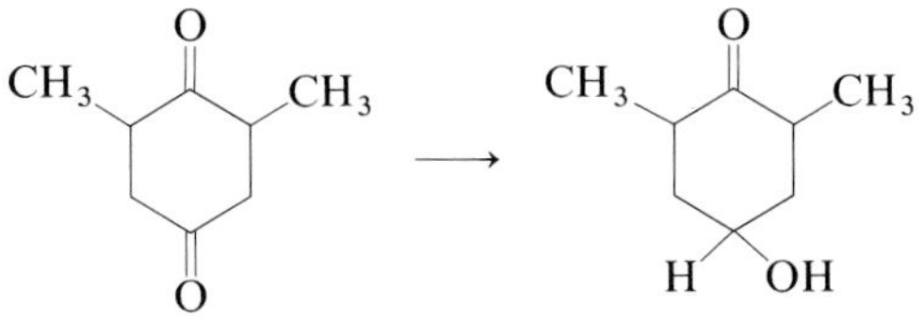 2,6-dimethylcyclohexane-1,4-dione (CH_3, CH_3) $\longrightarrow$ 4-hydroxy-2,6-dimethylcyclohexanone (CH_3, CH_3, H, OH)

(d) 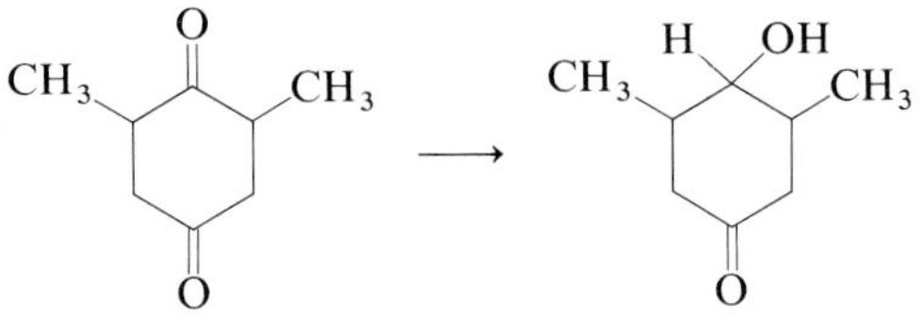 2,6-dimethylcyclohexane-1,4-dione (CH_3, CH_3) $\longrightarrow$ 4-hydroxy-3,5-dimethylcyclohexanone (H, OH, CH_3, CH_3)

(e) 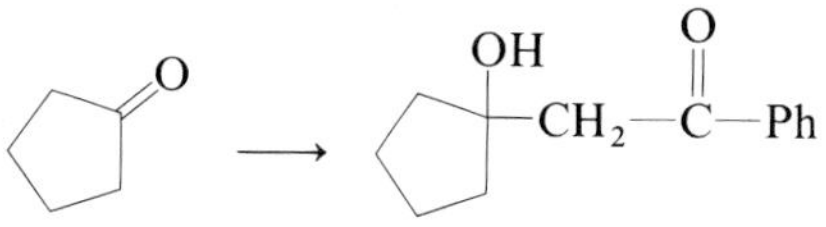 cyclopentanone $\longrightarrow$ 1-hydroxycyclopentyl (OH) $-CH_2-\overset{O}{\overset{\|}{C}}-Ph$

(f) $Br-CH_2CH_2\overset{O}{\overset{\|}{C}}CH_3 \longrightarrow H-C{\equiv}C-CH_2CH_2\overset{O}{\overset{\|}{C}}CH_3$

18-20 OXIDATION OF ALDEHYDES

Unlike ketones, aldehydes are easily oxidized to carboxylic acids by common oxidants such as chromic acid, chromium trioxide, permanganate, and most peroxy acids. Aldehydes oxidize so easily that air must be excluded from their containers to avoid slow oxidation by oxygen. Because aldehydes oxidize so easily, mild reagents such as Ag_2O can oxidize them selectively in the presence of other oxidizable functional groups.

$$R-\overset{O}{\overset{\|}{C}}-H \xrightarrow[\text{(oxidizing agent)}]{[O]} R-\overset{O}{\overset{\|}{C}}-OH$$

Examples

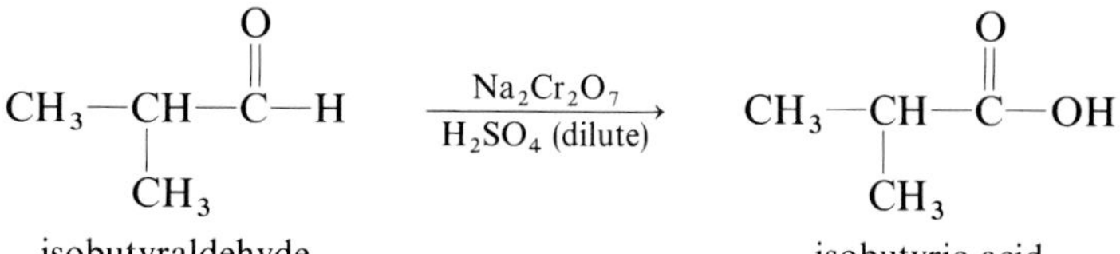

$$CH_3-\underset{CH_3}{\underset{|}{CH}}-\overset{O}{\overset{\|}{C}}-H \xrightarrow[H_2SO_4\text{ (dilute)}]{Na_2Cr_2O_7} CH_3-\underset{CH_3}{\underset{|}{CH}}-\overset{O}{\overset{\|}{C}}-OH$$

isobutyraldehyde → isobutyric acid (90%)

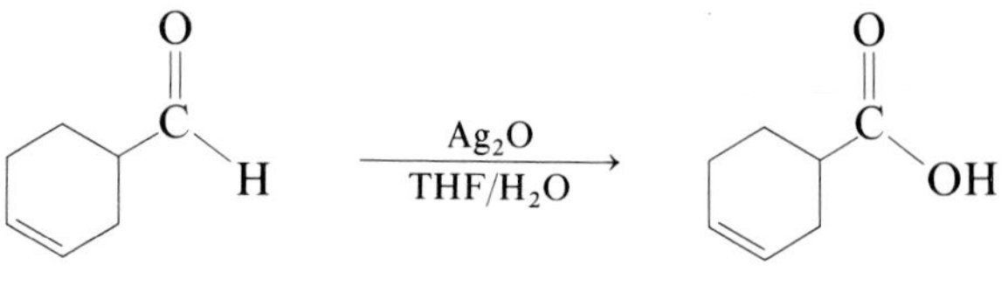

cyclohex-3-ene-1-carbaldehyde ($\overset{O}{\overset{\|}{C}}$–H) $\xrightarrow[THF/H_2O]{Ag_2O}$ cyclohex-3-ene-1-carboxylic acid ($\overset{O}{\overset{\|}{C}}$–OH)

(97%)

The ability of silver ion, Ag^+, to oxidize aldehydes selectively is exploited in a convenient functional group test for aldehydes. The **Tollens test** involves adding a solution of silver-ammonia complex (the **Tollens reagent**) to the unknown compound. If an aldehyde is present, its oxidation reduces silver ion to metallic silver, in the form of a black suspension or a silver mirror deposited on the inside of the container. Simple hydrocarbons, ethers, ketones, and even alcohols do not react with the Tollens reagent, leaving a clear, colorless solution.

$$\underset{\text{aldehyde}}{R-\overset{\overset{\displaystyle O}{\|}}{C}-H} + \underset{\text{Tollens reagent}}{2\,Ag(NH_3)_2^+} + 3\ ^-OH \xrightarrow{H_2O} \underset{\text{silver metal}}{2\,Ag\downarrow} + \underset{\text{carboxylate}}{R-\overset{\overset{\displaystyle O}{\|}}{C}-O^-} + 4\,NH_3 + 2\,H_2O$$

PROBLEM 18-27

Predict the major products of the following reactions.

(a) 4-hydroxycyclohexanecarbaldehyde (HO, CHO on cyclohexane) $+\ Ag_2O$

(b) 4-hydroxycyclohexanecarbaldehyde (HO, CHO on cyclohexane) $+\ K_2Cr_2O_7/H_2SO_4$

(c) 3-oxocyclohexanecarbaldehyde (CHO, C=O on cyclohexane) $+\ Ag(NH_3)_2^+\ ^-OH$

(d) cyclohex-3-enecarbaldehyde (CHO on cyclohexene) $+\ KMnO_4$ (cold, dilute)

18-21 DEOXYGENATION OF KETONES AND ALDEHYDES

A *deoxygenation* replaces the carbonyl oxygen atom of a ketone or aldehyde with two hydrogen atoms, reducing the carbonyl group past the alcohol stage all the way to a methylene group. The following equation compares deoxygenation with the common hydride reductions that give alcohols.

deoxygenation: Zn(Hg), HCl or H_2NNH_2, KOH

$$\underset{\text{ketone}}{R-\overset{\overset{\displaystyle O}{\|}}{C}-R'} \xrightarrow[\text{or } LiAlH_4]{NaBH_4} \underset{\text{alcohol}}{R-\overset{H\ \ OH}{C}-R'} \xrightarrow[(2)\ LiAlH_4]{(1)\ TsCl} \underset{\text{methylene group}}{R-\overset{H\ \ H}{C}-R'}$$

Clemmensen reduction (review) The **Clemmensen reduction** is most commonly used to convert acylbenzenes (from Friedel-Crafts acylation, Section 17-11B) to alkylbenzenes, but it also works with other ketones and aldehydes that are not sensitive to acid. The carbonyl compound is heated with an excess of amalgamated zinc (zinc treated with mercury) and hydrochloric acid. The actual reduction occurs by a complex mechanism on the surface of the zinc.

$$\underset{\text{heptanal}}{CH_3-(CH_2)_5-CHO} \xrightarrow[HCl,\ H_2O]{Zn(Hg)} \underset{n\text{-heptane } (72\%)}{CH_3-(CH_2)_5-CH_3}$$

$$\text{cyclohexanone} \xrightarrow[\text{HCl, } H_2O]{\text{Zn(Hg)}} \text{cyclohexane } (75\%)$$

Wolff-Kishner reduction Compounds that cannot survive treatment with hot acid can be deoxygenated using the **Wolff-Kishner reduction.** The ketone or aldehyde is converted to its hydrazone, which is heated with a strong base such as KOH or potassium *t*-butoxide. Ethylene glycol, diethylene glycol, or another high-boiling solvent is used to facilitate the high temperature needed in the second step.

$$\text{C=O} + H_2N{-}NH_2 \longrightarrow \text{C=N}{-}NH_2 + H_2O \xrightarrow[\text{heat}]{\text{KOH}} \text{H–C–H} + H_2O + N{\equiv}N\uparrow$$

Examples

$$\text{1-methylcyclohexanecarbaldehyde} \xrightarrow[\text{trace HCl}]{H_2N{-}NH_2} \text{hydrazone} \xrightarrow[\text{HOCH}_2\text{CH}_2\text{OH}]{\text{KOH, heat}} \text{1,1-dimethylcyclohexane } (48\%) + N_2$$

$$\text{propiophenone} \xrightarrow{N_2H_4} \text{hydrazone} \xrightarrow[\text{HOCH}_2\text{CH}_2\text{OCH}_2\text{CH}_2\text{OH (diethylene glycol)}]{\text{KOH, heat}} n\text{-propylbenzene } (82\%) + N_2$$

$$\text{cyclohexanone} \xrightarrow{N_2H_4} \text{hydrazone} \xrightarrow[\text{CH}_3\text{–S(=O)–CH}_3 \text{ (DMSO, a solvent)}]{t\text{-BuO}^{-}\,{}^{+}\text{K}} \text{cyclohexane } (80\%) + N_2$$

The mechanism for formation of the hydrazone is the same as the mechanism for imine formation (Section 18-16). The actual reduction step involves two proton transfers from nitrogen to carbon. In this strongly basic solution, we should expect a proton transfer to occur by loss of a proton from nitrogen, followed by reprotonation on carbon.

$$\underset{\text{hydrazone}}{R'R\text{C=N–NH}_2} \overset{:\ddot{O}H^-}{\rightleftharpoons} \underset{\text{remove proton from N}}{\left[R'R\text{C=N–}\ddot{N}^-\text{–H} \longleftrightarrow R'R\bar{\text{C}}\text{–N=N–H}\right]} \underset{\text{replace proton on C}}{\overset{H_2O}{\rightleftharpoons}} \text{H–C}R'R\text{–N=N–H} + {}^-\text{OH}$$

A second deprotonation sets up the intermediate for loss of nitrogen to form a carbanion. This carbanion is quickly reprotonated to give the product.

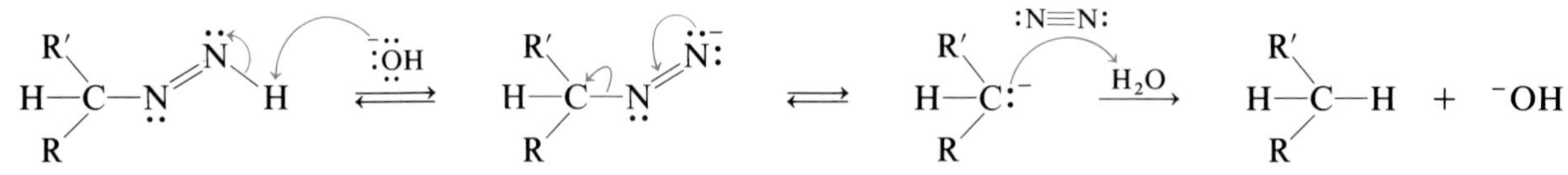

$$RR'CH{-}N{=}N{-}H \xrightleftharpoons{^-OH} RR'CH{-}N{=}N^- \rightleftharpoons RR'CH^- \xrightarrow{H_2O} RR'CH{-}H + {}^-OH \quad (+ :N{\equiv}N:)$$

remove proton from N — lose N_2 — carbanion — product

PROBLEM 18-28

Propose a mechanism for both steps of the Wolff-Kishner reduction of cyclohexanone: the formation of the hydrazone and then the base-catalyzed reduction with evolution of nitrogen gas.

PROBLEM 18-29

Predict the major products of the following reactions.

(a) 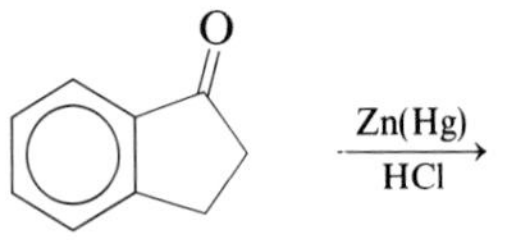 $\xrightarrow[HCl]{Zn(Hg)}$

(b) 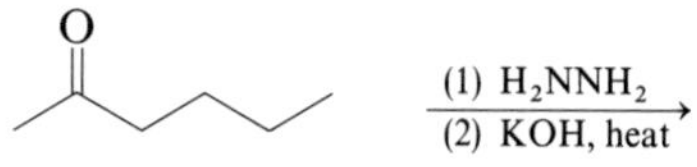 $\xrightarrow[(2)\ KOH,\ heat]{(1)\ H_2NNH_2}$

(c) 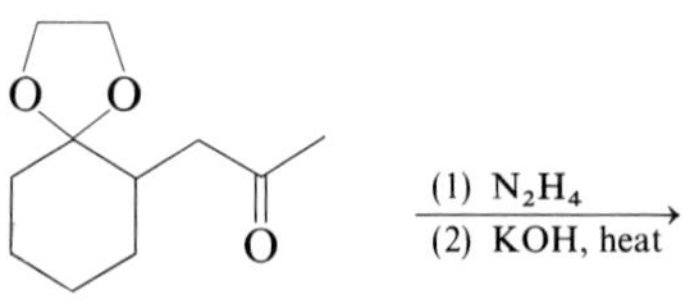 $\xrightarrow[(2)\ KOH,\ heat]{(1)\ N_2H_4}$

(d) 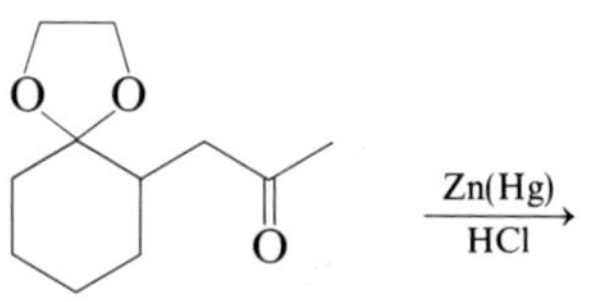 $\xrightarrow[HCl]{Zn(Hg)}$

SUMMARY OF REACTIONS OF KETONES AND ALDEHYDES

1. Addition of organometallic reagents (Sections 9-9 and 14-8B)

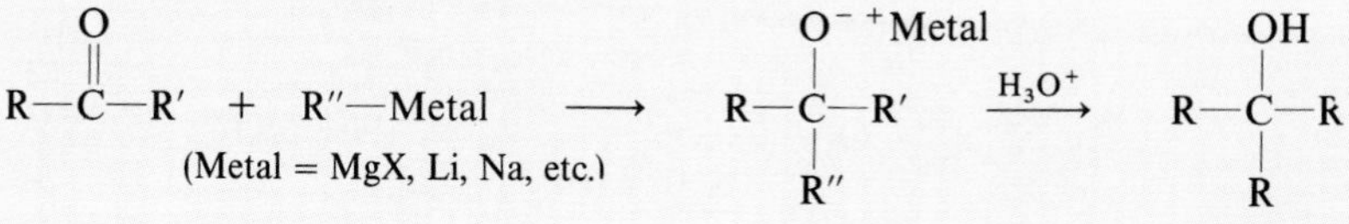

$$R{-}CO{-}R' + R''{-}Metal \longrightarrow R{-}C(O^-\ {}^+Metal)(R'){-}R'' \xrightarrow{H_3O^+} R{-}C(OH)(R){-}R$$

(Metal = MgX, Li, Na, etc.)

alkoxide — alcohol

2. Reduction (Section 9-12)

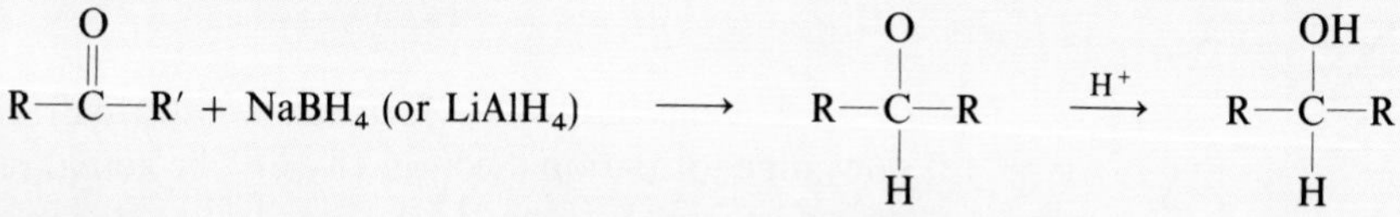

$$R{-}CO{-}R' + NaBH_4\ (\text{or } LiAlH_4) \longrightarrow R{-}C(O^-)(H){-}R \xrightarrow{H^+} R{-}C(OH)(H){-}R$$

ketone or aldehyde — alkoxide — alcohol

3. Hydration (Section 18-14)

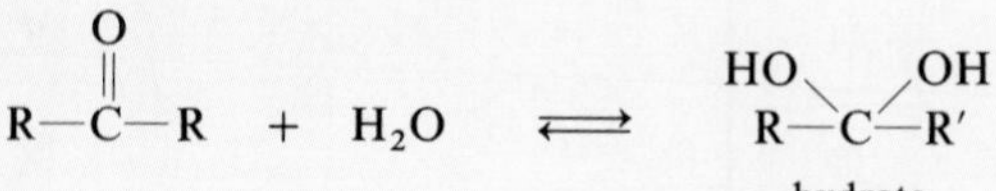

$$R{-}CO{-}R + H_2O \rightleftharpoons R{-}C(OH)_2{-}R'$$

hydrate

Example

$$\mathrm{Cl_3C{-}\overset{\displaystyle O}{\overset{\|}{C}}{-}H + H_2O} \rightleftharpoons \mathrm{Cl_3C{-}C(OH)(OH){-}H}$$

chloral — chloral hydrate

4. *Formation of cyanohydrins* (Section 18-15)

$$\mathrm{R{-}\overset{\displaystyle O}{\overset{\|}{C}}{-}R' \;+\; HCN} \xrightarrow{^-CN} \mathrm{R{-}C(OH)(CN){-}R'}$$

ketone or aldehyde — cyanohydrin

$$\mathrm{CH_3CH_2CH_2{-}\overset{\displaystyle O}{\overset{\|}{C}}{-}H} \xrightarrow[^-CN]{HCN} \mathrm{CH_3CH_2CH_2{-}C(OH)(CN){-}H}$$

butanal — butanal cyanohydrin

5. *Formation of imines* (Section 18-16)

$$\mathrm{R{-}\overset{\displaystyle O}{\overset{\|}{C}}{-}R' \;+\; R''{-}NH_2} \xrightarrow{H^+} \mathrm{R{-}\overset{\displaystyle N{-}R''}{\overset{\|}{C}}{-}R' \;+\; H_2O}$$

ketone or aldehyde — primary amine — imine (Schiff base)

Example

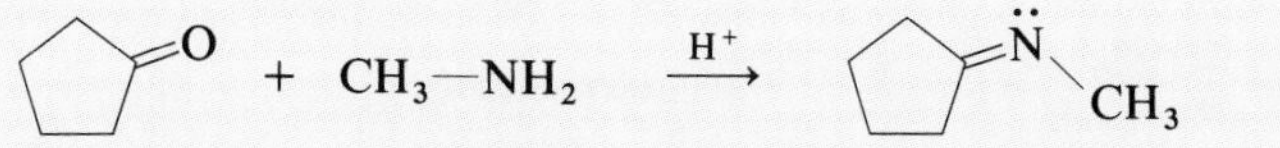

$$\text{cyclopentanone} \;+\; \mathrm{CH_3{-}NH_2} \xrightarrow{H^+} \text{cyclopentanone}{=}\ddot{\mathrm{N}}{-}\mathrm{CH_3}$$

cyclopentanone — methylamine — cyclopentanone methyl imine

6. *Formation of oximes and hydrazones* (Section 18-17)

$$\mathrm{R{-}\overset{\displaystyle O}{\overset{\|}{C}}{-}R' \;+\; H_2N{-}OH} \xrightarrow{H^+} \mathrm{R{-}\overset{\displaystyle N{-}OH}{\overset{\|}{C}}{-}R'}$$

ketone or aldehyde — hydroxylamine — oxime

$$\mathrm{R{-}\overset{\displaystyle O}{\overset{\|}{C}}{-}R' \;+\; H_2N{-}NH{-}R''} \xrightarrow{H^+} \mathrm{R{-}\overset{\displaystyle N{-}NH{-}R''}{\overset{\|}{C}}{-}R'}$$

ketone or aldehyde — hydrazine reagent — hydrazone derivative

R″ =	*Reagent name*	*Derivative name*
—H	hydrazine	hydrazone
—Ph	phenylhydrazine	phenylhydrazone
$\mathrm{{-}\overset{\displaystyle O}{\overset{\|}{C}}{-}NH_2}$	semicarbazide	semicarbazone

Example

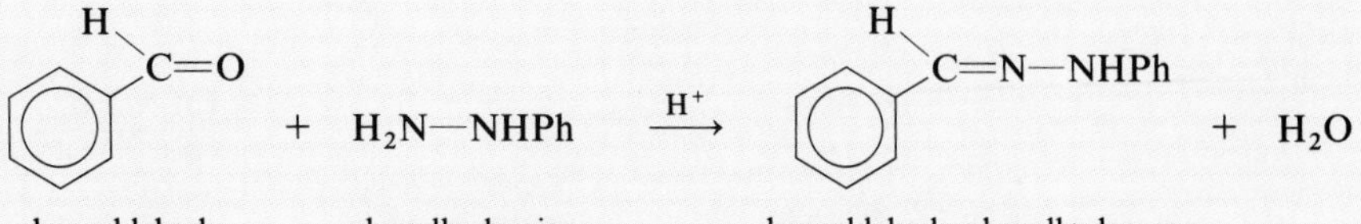

$$\mathrm{Ph{-}CH{=}O \;+\; H_2N{-}NHPh} \xrightarrow{H^+} \mathrm{Ph{-}CH{=}N{-}NHPh \;+\; H_2O}$$

benzaldehyde — phenylhydrazine — benzaldehyde phenylhydrazone

7. *Formation of acetals* (Section 18-18)

$$\text{R—}\overset{\text{O}}{\overset{\|}{\text{C}}}\text{—R}' + 2\,\text{R}''\text{—OH} \xrightarrow{\text{H}^+} \text{R—}\overset{\text{R}''\text{O}\ \ \text{OR}''}{\text{C}}\text{—R}' + \text{H}_2\text{O}$$

ketone (aldehyde) — alcohol — acetal

Example

$$\text{C}_6\text{H}_5\text{—CHO} + \underset{\text{OH}\quad\ \text{OH}}{\text{CH}_2\text{—CH}_2} \xrightarrow{\text{H}^+} \text{C}_6\text{H}_5\text{—CH(OCH}_2\text{CH}_2\text{O)} + \text{H}_2\text{O}$$

benzaldehyde — ethylene glycol — benzaldehyde ethylene acetal

8. *Oxidation of aldehydes* (Section 18-20)

$$\text{R—}\overset{\text{O}}{\overset{\|}{\text{C}}}\text{—H} \xrightarrow{\text{chromic acid, permanganate, Ag}^+\text{, etc.}} \text{R—}\overset{\text{O}}{\overset{\|}{\text{C}}}\text{—OH}$$

aldehyde — acid

Examples

$$\text{C}_6\text{H}_5\text{—CHO} + \text{Na}_2\text{Cr}_2\text{O}_7/\text{H}_2\text{SO}_4 \longrightarrow \text{C}_6\text{H}_5\text{—COOH}$$

benzaldehyde — chromic acid — benzoic acid

Tollens test

$$\text{R—}\overset{\text{O}}{\overset{\|}{\text{C}}}\text{—H} + 2\,\text{Ag(NH}_3)_2{}^+ + 3\ {}^-\text{OH} \xrightarrow{\text{H}_2\text{O}} 2\,\text{Ag}\downarrow + \text{R—}\overset{\text{O}}{\overset{\|}{\text{C}}}\text{—O}^- + 4\,\text{NH}_3 + 2\,\text{H}_2\text{O}$$

aldehyde — Tollens reagent — silver metal — carboxylate

9. *Deoxygenation reactions* (Section 18-21)

a. *Clemmensen reduction* (Section 17-11B)

$$\text{R—}\overset{\text{O}}{\overset{\|}{\text{C}}}\text{—R}' + \text{Zn(Hg)} \xrightarrow{\text{HCl}} \text{R—}\overset{\text{H}\ \ \text{H}}{\text{C}}\text{—R}'$$

ketone or aldehyde

b. *Wolff-Kishner reduction* (Section 18-21)

$$\text{R—}\overset{\text{O}}{\overset{\|}{\text{C}}}\text{—R}' + \text{H}_2\text{N—NH}_2 \longrightarrow \text{R—}\overset{\text{N—NH}_2}{\overset{\|}{\text{C}}}\text{—R}' \xrightarrow[\text{heat}]{\text{KOH}} \text{R—}\overset{\text{H}\ \ \text{H}}{\text{C}}\text{—R}' + \text{H}_2\text{O} + \text{N}\equiv\text{N}\uparrow$$

ketone or aldehyde/hydrazine — hydrazone

Example

$$\text{cyclohexanone} \xrightarrow[\text{(2) KOH, heat}]{\text{(1) H}_2\text{N—NH}_2} \text{cyclohexane}$$

cyclohexanone — cyclohexane

10. Reactions of ketones and aldehydes at their α positions

This large group of reactions is covered in Chapter 22.

Example: aldol condensation

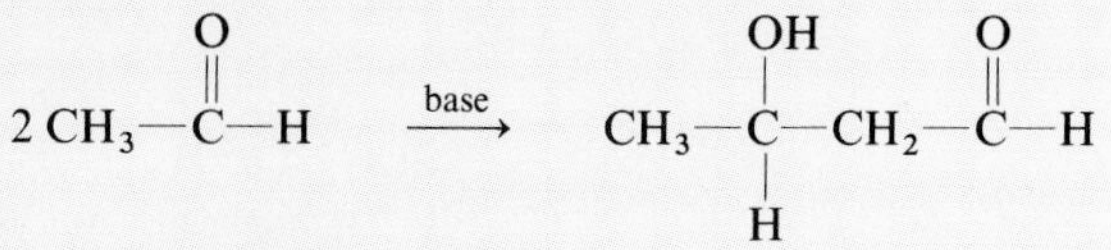

$$2\,CH_3-\overset{\overset{\displaystyle O}{\|}}{C}-H \xrightarrow{\text{base}} CH_3-\overset{\overset{\displaystyle OH}{|}}{\underset{\underset{\displaystyle H}{|}}{C}}-CH_2-\overset{\overset{\displaystyle O}{\|}}{C}-H$$

GLOSSARY

acetal A derivative of an aldehyde or ketone having two alkoxy groups in place of the carbonyl group. The acetal of a ketone is called a **ketal** in the common nomenclature. (p. 801)

ethylene acetal: A cyclic acetal using ethylene glycol as the alcohol. (p. 803)

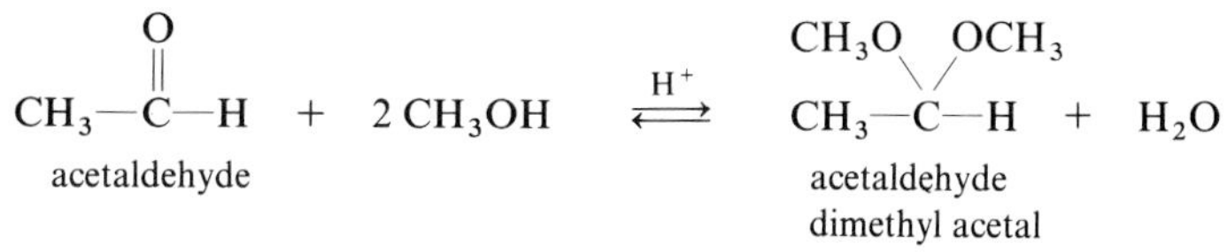

$$\underset{\text{acetaldehyde}}{CH_3-\overset{\overset{\displaystyle O}{\|}}{C}-H} + 2\,CH_3OH \underset{}{\overset{H^+}{\rightleftharpoons}} \underset{\substack{\text{acetaldehyde}\\ \text{dimethyl acetal}}}{CH_3-\overset{CH_3O\;\;OCH_3}{C}-H} + H_2O$$

aldehyde A compound containing a carbonyl group bonded to an alkyl group and a hydrogen atom. (p. 768)

carbinolamine An intermediate in the formation of an imine or an enamine, having an amine and a hydroxyl group bonded to the same carbon atom. (p. 797)

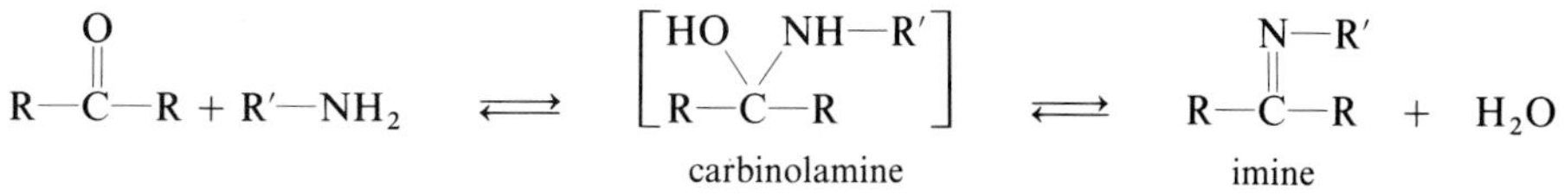

$$R-\overset{\overset{\displaystyle O}{\|}}{C}-R + R'-NH_2 \rightleftharpoons \underset{\text{carbinolamine}}{\left[R-\overset{HO\;\;NH-R'}{C}-R\right]} \rightleftharpoons \underset{\text{imine}}{R-\overset{\overset{\displaystyle N-R'}{\|}}{C}-R} + H_2O$$

carbonyl group The C=O functional group. (p. 767)

Clemmensen reduction The deoxygenation of a ketone or aldehyde by treatment with zinc amalgam and dilute HCl. (p. 808)

condensation A reaction in which two or more organic compounds are joined with the elimination of a small molecule such as water. (p. 797)

cyanohydrin A compound with a hydroxyl group and a cyano group on the same carbon atom. Cyanohydrins are generally made by the reaction of a ketone or aldehyde with HCN. (p. 795)

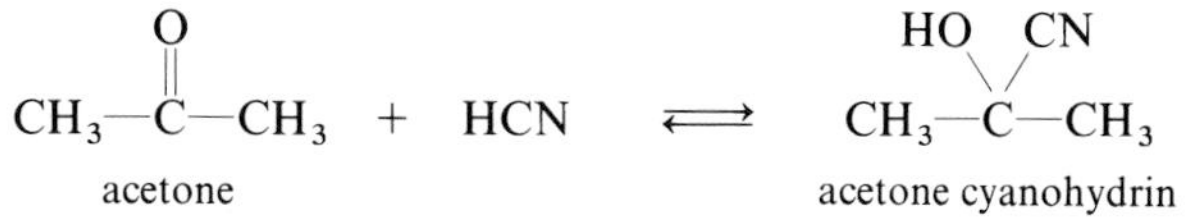

$$\underset{\text{acetone}}{CH_3-\overset{\overset{\displaystyle O}{\|}}{C}-CH_3} + HCN \rightleftharpoons \underset{\text{acetone cyanohydrin}}{CH_3-\overset{HO\;\;CN}{C}-CH_3}$$

enol A vinyl alcohol. Simple enols generally tautomerize to their keto forms. (p. 784)

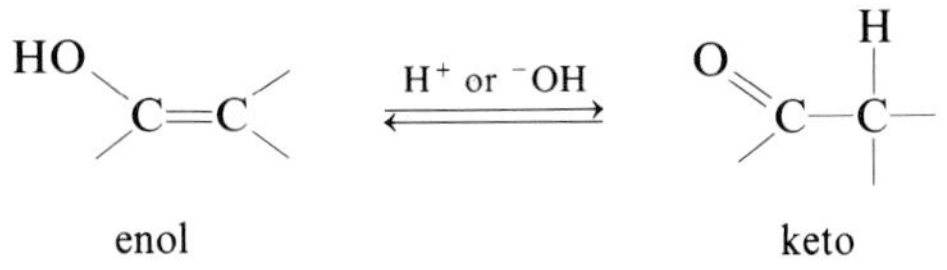

$$\underset{\text{enol}}{\overset{HO}{\;}\!\!>C=C<} \underset{}{\overset{H^+\text{ or }^-OH}{\rightleftharpoons}} \underset{\text{keto}}{\overset{O}{\;}\!\!\geq C-\overset{\overset{\displaystyle H}{|}}{\underset{|}{C}}-}$$

enolate ion The resonance-stabilized anion formed by deprotonating the carbon atom next to a carbonyl group. (p. 614 and Chapter 22)

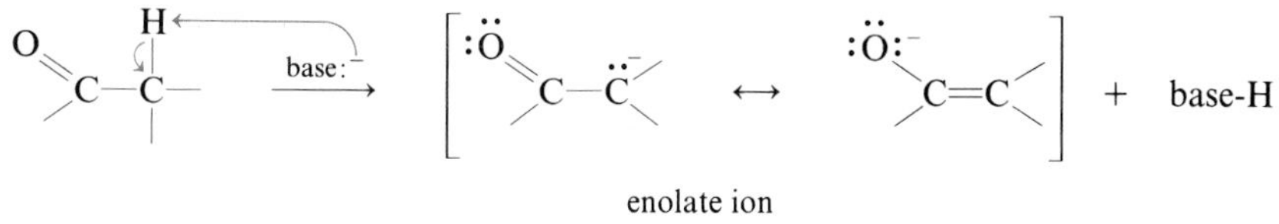

$$\overset{O}{\;}\!\!\geq C-\overset{\overset{\displaystyle H}{|}}{\underset{|}{C}}- \xrightarrow{\text{base:}^-} \underset{\text{enolate ion}}{\left[\;:\ddot{O}\!\!\geq C-\ddot{C}^-\!< \;\longleftrightarrow\; :\ddot{O}:^-\!\!-C=C<\right]} + \text{base-H}$$

hemiacetal Similar to an acetal, but with one alkoxy group and one hydroxyl group on the former carbonyl carbon atom. (p. 801)

hydrate (of an aldehyde or ketone) The geminal diol formed by addition of water across the carbonyl double bond. (p. 794)

$$\underset{\text{chloral}}{Cl_3C-\overset{\overset{\displaystyle O}{\|}}{C}-H} + H_2O \xrightleftharpoons{H^+ \text{ or } {}^-OH} \underset{\text{chloral hydrate}}{Cl_3C-\overset{HO\ \ OH}{\overset{\diagdown\diagup}{C}}-H}$$

hydrazone A compound containing the $\rangle C{=}N{-}NH_2$ group, formed by the reaction of a ketone or aldehyde with hydrazine. (p. 799)

2,4-DNP derivative: A hydrazone made using 2,4-dinitrophenylhydrazine. (p. 800)

imine A compound with a carbon-nitrogen double bond, formed by the reaction of a ketone or aldehyde with a primary amine. A substituted imine is often called a **Schiff base.** (p. 797)

$$\underset{\text{acetone}}{CH_3-\overset{\overset{\displaystyle O}{\|}}{C}-CH_3} + \underset{\text{methylamine}}{CH_3-\ddot{N}H_2} \xrightleftharpoons{H^+} \underset{\text{acetone methyl imine}}{CH_3-\overset{\overset{\displaystyle N-CH_3}{\|}}{C}-CH_3} + H_2O$$

ketal A common name for the acetal of a ketone. The term *ketal* was recently banished from the IUPAC nomenclature. (p. 801)

ketone A compound containing a carbonyl group bonded to two alkyl groups. (p. 768)

McLafferty rearrangement In mass spectrometry, the loss of an alkene fragment by a cyclic rearrangement of a carbonyl compound having γ hydrogens. (p. 777)

nitrile A compound containing the cyano group, $C{\equiv}N$. (p. 786)

nucleophilic addition Addition of a reagent across a multiple bond by attack of a nucleophile at the electrophilic end of the multiple bond. As used in this chapter, the addition of a nucleophile and a proton across the $C{=}O$ bond. (p. 790)

oxime A compound containing the $C{=}N{-}OH$ group, formed by the reaction of a ketone or aldehyde with hydroxylamine. (p. 799)

protecting group A group used to prevent a sensitive functional group from reacting while another part of the molecule is being modified. The protecting group is later removed. (p. 806)

Raney nickel A finely divided, hydrogen-bearing form of nickel made by treating a nickel-aluminum alloy with a strong sodium hydroxide solution. The aluminum in the alloy reacts to form hydrogen, leaving a finely divided nickel powder saturated with hydrogen. (p. 793)

Rosenmund reduction The hydrogenation of an acid chloride to an aldehyde using a poisoned palladium catalyst. (p. 787)

semicarbazone A compound containing the $C{=}N{-}NH{-}CONH_2$ group, formed by the reaction of a ketone or aldehyde with semicarbazide. (p. 799)

Tollens test A test for aldehydes: Adding the **Tollens reagent,** a silver-ammonia complex $[Ag(NH_3)_2^+\ {}^-OH]$, gives a carboxylate salt and a silver mirror on the inside of the glass container. (p. 808)

Wolff-Kishner reduction The deoxygenation of a ketone or aldehyde by conversion to the hydrazone, followed by treatment with a strong base. (p. 809)

ESSENTIAL PROBLEM-SOLVING SKILLS IN CHAPTER 18

1. Name ketones and aldehydes and draw the structures from their names.

2. Interpret the IR, NMR, UV, and mass spectra of ketones and aldehydes and use the spectral information to determine the structures.

3. Write equations for the synthesis of ketones and aldehydes from alcohols, alkenes, alkynes, carboxylic acids, nitriles, acid chlorides, dithianes, and aromatic compounds.

4. Propose effective single-step and multistep syntheses of ketones and aldehydes.

5. Predict the products of reactions of ketones and aldehydes with the following types of compounds; give mechanisms where appropriate.
   (a) Hydride reducing agents; Clemmensen and Wolff-Kishner reagents
   (b) Grignard and organolithium reagents
   (c) Water
   (d) Hydrogen cyanide
   (e) Ammonia and primary amines
   (f) Hydroxylamine and hydrazine derivatives
   (g) Alcohols
   (h) Oxidizing agents

6. Use your knowledge of the mechanisms of ketone and aldehyde reactions to propose mechanisms and products of similar reactions you have never seen before.

7. Show how to convert ketones and aldehydes to other functional groups.

8. Use retrosynthetic analysis to propose effective single-step and multistep syntheses using ketones and aldehydes as intermediates and protecting the carbonyl group if necessary.

STUDY PROBLEMS

18-30. Define each of the following terms and give an example.

(a) ketone **(b)** aldehyde **(c)** enol form **(d)** cyanohydrin
(e) imine **(f)** Schiff base **(g)** carbinolamine **(h)** oxime
(i) phenylhydrazone **(j)** 2,4-DNP derivative **(k)** semicarbazone **(l)** acetal
(m) ketal **(n)** 1,3-dithiane acetal **(o)** hemiacetal **(p)** Tollens test
(q) Wolff-Kishner reduction **(r)** Clemmensen reduction **(s)** Rosenmund reduction **(t)** ethylene acetal

18-31. Give a correct name for each of the following ketones and aldehydes. When possible, give both a common name and an IUPAC name.

(a) $CH_3CO(CH_2)_4CH_3$ **(b)** $CH_3(CH_2)_2CO(CH_2)_2CH_3$
(c) $CH_3(CH_2)_5CHO$ **(d)** PhCOPh
(e) $CH_3CH_2CH_2CHO$ **(f)** CH_3COCH_3
(g) $CH_3CH_2CHBrCH_2CH(CH_3)CHO$ **(h)** Ph—CH=CH—CHO
(i) CH_3CH=CH—CH=CH—CHO **(j)** $CH_3CH_2COCH_2CHO$

(k) 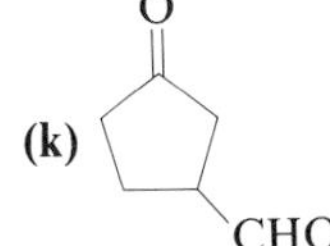

(l) O, CH_3, CH_3

18-32. Rank the following carbonyl compounds in order of *increasing* equilibrium constant for hydration.

CH_3COCH_2Cl $ClCH_2CHO$ CH_2O CH_3COCH_3 CH_3CHO

18-33. Sketch the expected proton NMR spectrum of 3,3-dimethylbutanal.

18-34. Predict the values of λ_{max} for the $\pi \rightarrow \pi^*$ and $n \rightarrow \pi^*$ transitions in the UV spectrum of 3-methylcyclohex-2-enone.

18-35. A compound of formula $C_6H_{10}O_2$ shows only two absorptions in the proton NMR: a singlet at 2.67 ppm and a singlet at 2.15 ppm. These absorptions have area of ratio 2:3. The IR spectrum shows a strong absorption at 1708 cm^{-1}. Propose a structure for this unknown compound.

18-36. The proton NMR spectrum of a compound of formula $C_{10}H_{12}O$ appears below. This compound reacts with an acidic solution of 2,4-dinitrophenylhydrazine to give a crystalline derivative, but it gives a negative Tollens test. Propose a structure for this unknown compound and give peak assignments to account for the absorptions in the spectrum.

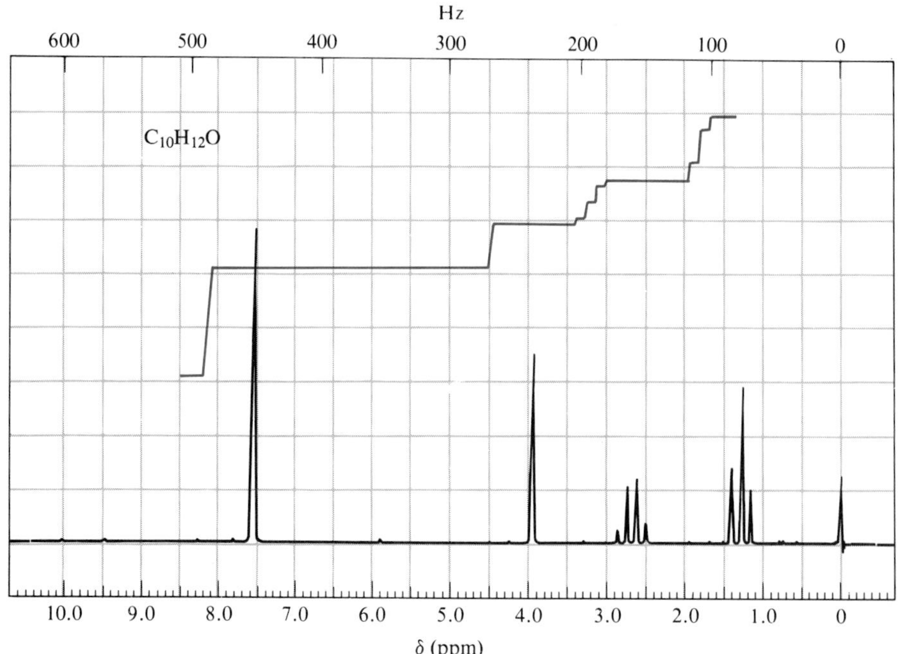

18-37. The following compounds undergo McLafferty rearrangement in the mass spectrometer. Predict the masses of the resulting charged fragments.

(a) pentanal **(b)** 3-methyl-2-pentanone **(c)** 3-methylpentanal

(d) $CH_3CH_2CH_2-\overset{\overset{\displaystyle O}{\|}}{C}-OCH_3$ (methyl butyrate)

18-38. An unknown compound gives a positive 2,4-dinitrophenylhydrazine test and a negative Tollens test. Its mass spectrum shows prominent ions at *m/z* 128, 100, 86, 85, and 71. Show which of the following structures is most likely: 2-octanone; 4-octanone; 2-octen-3-ol; 5-propoxy-1-pentanol. Also show the fragmentations that lead to the observed ions.

18-39. An unknown compound gives a molecular ion of *m/z* 70 in the mass spectrum. It reacts with semicarbazide hydrochloride to give a crystalline derivative, but it gives a negative Tollens test. The NMR and IR spectra appear below. Propose a structure for this unknown compound, and give peak assignments to account for the absorptions in the spectrum. Explain why the absorption at 1785 cm^{-1} in the IR spectrum appears at an unusual frequency.

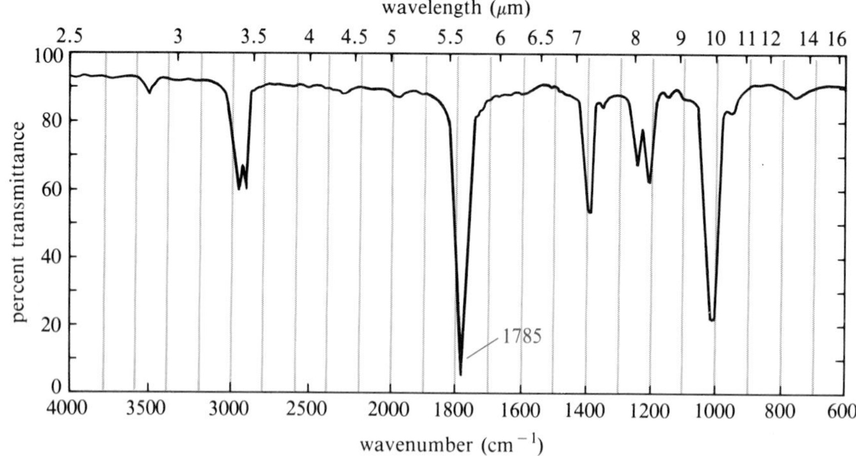

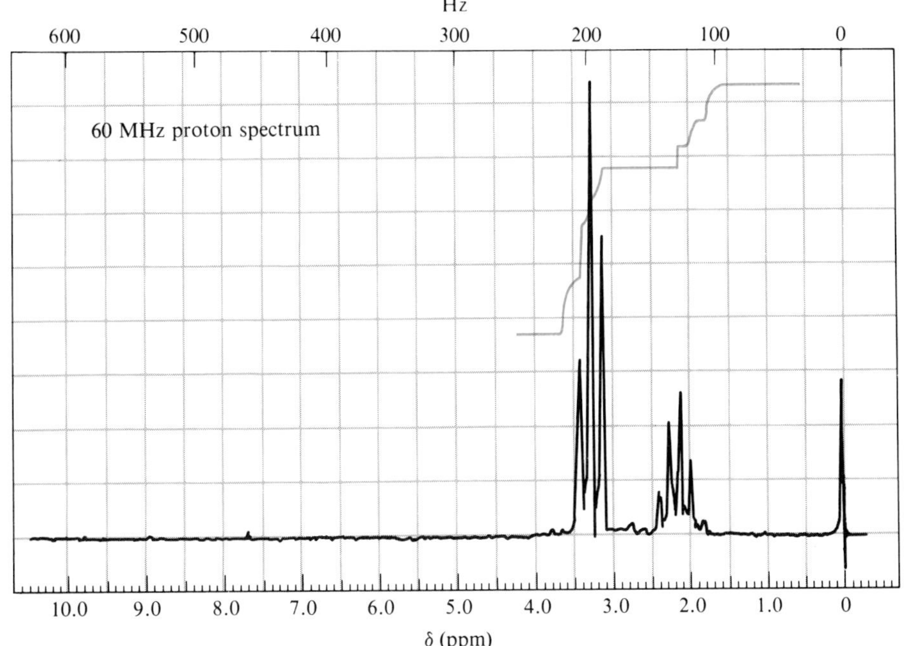

18-40. For each of the following compounds
(1) Name the functional group.
(2) Show what compound(s) result from complete hydrolysis.

(a) 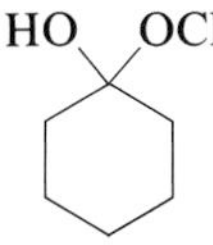 HO OCH$_2$CH$_3$ (on cyclohexane ring)

(b) 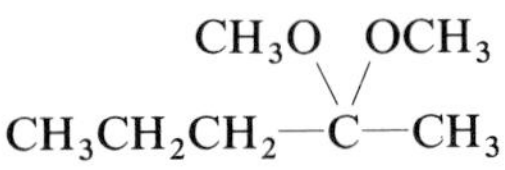 $CH_3CH_2CH_2—C(OCH_3)(OCH_3)—CH_3$

(c) O O

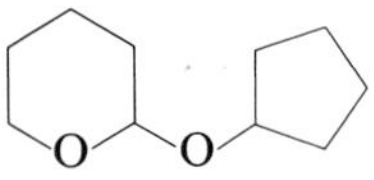

(d) 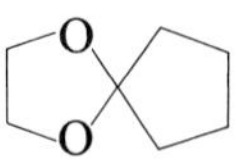 O O

(e) 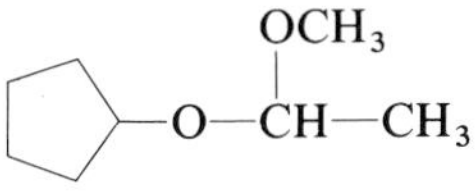 —O—CH(OCH$_3$)—CH$_3$

(f)  S S

(g) 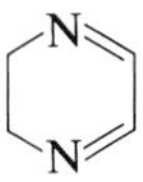 N N

(h) 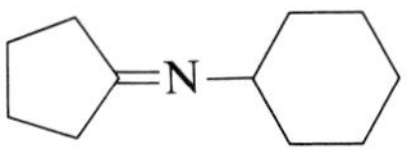 =N—

18-41. Show how you would accomplish the following synthetic conversions efficiently and in good yield. You may use any additional reagents you need.

(a) acetaldehyde ⟶ lactic acid, $CH_3CH(OH)COOH$

(b) 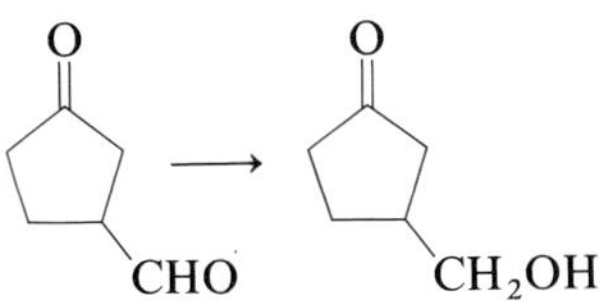 O, CHO ⟶ O, CH$_2$OH

(c) 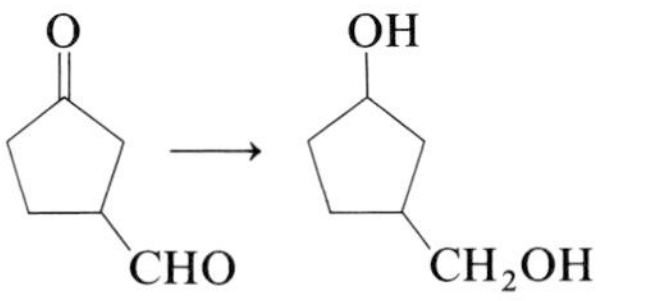 O, CHO ⟶ OH, CH$_2$OH

(d) 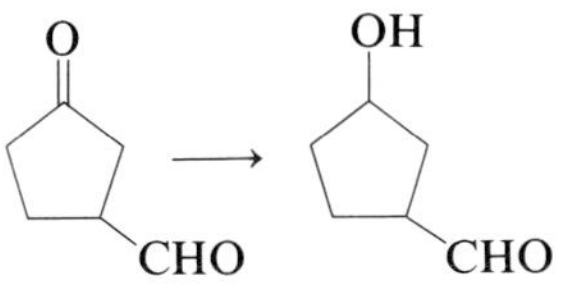 O, CHO ⟶ OH, CHO

(e) 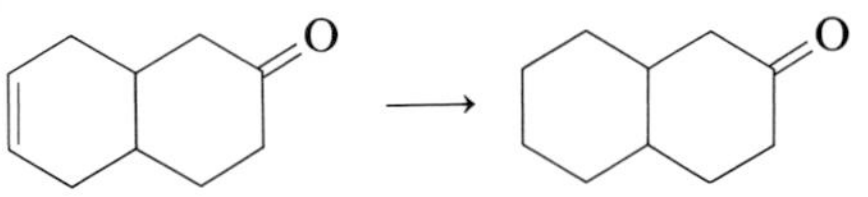 O ⟶ O

(f) 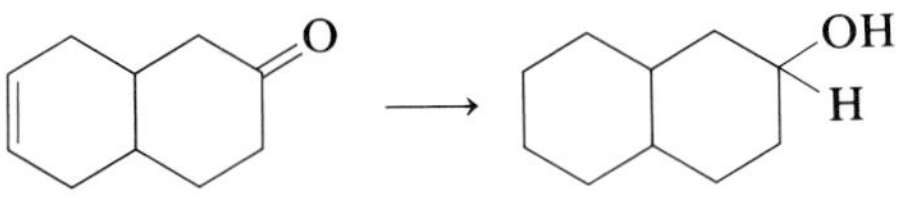 O ⟶ OH, H

(g) 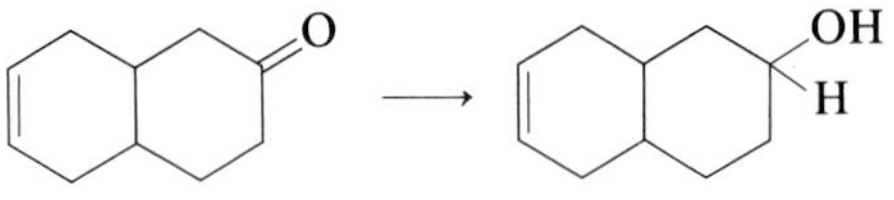 O ⟶ OH, H

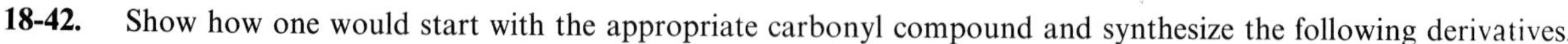
18-42. Show how one would start with the appropriate carbonyl compound and synthesize the following derivatives.

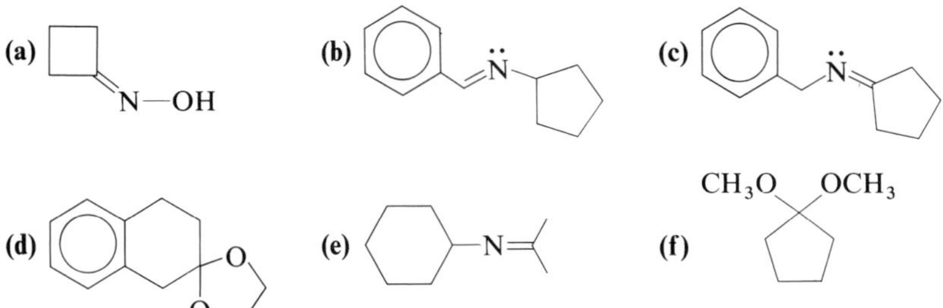

18-43. Draw the structures of the following derivatives.
(a) the 2,4-dinitrophenylhydrazone of benzaldehyde
(b) the semicarbazone of cyclobutanone
(c) cyclopropanone oxime
(d) the ethylene ketal of 3-hexanone
(e) acetaldehyde dimethyl acetal
(f) the methyl hemiacetal of formaldehyde
(g) the (*E*) isomer of the ethyl imine of propiophenone
(h) the dithiane thioacetal of propanal

18-44. Predict the products formed when cyclohexanone reacts with the following reagents.
(a) CH_3NH_2, H^+
(b) excess CH_3OH, H^+
(c) hydroxylamine and weak acid
(d) ethylene glycol and *p*-toluenesulfonic acid
(e) phenylhydrazine and weak acid
(f) PhMgBr and then mild H_3O^+
(g) Tollens reagent
(h) sodium acetylide, then mild H_3O^+
(i) sodium cyanide
(j) acidic hydrolysis of the product from (i)
(k) hydrazine, then hot, fused KOH

18-45. Both $NaBH_4$ and $NaBD_4$ are commercially available. Show how you would synthesize the following labeled compounds, starting with 2-butanone.

(a) $CH_3-C(OH)(D)-CH_2-CH_3$ **(b)** $CH_3-C(OD)(D)-CH_2-CH_3$ **(c)** $CH_3-C(OD)(H)-CH_2-CH_3$

18-46. When $LiAlH_4$ reduces 3-methylcyclopentanone, the product mixture contains 60 percent *cis*-3-methylcyclopentanol and 40 percent *trans*-3-methylcyclopentanol. Use three-dimensional drawings to explain this preference for the cis isomer.

18-47. Some Grignard reagents react with ethyl orthoformate, followed by acidic hydrolysis, to give aldehydes. Propose mechanisms for the two steps in this synthesis.

$$\underset{\text{ethyl orthoformate}}{H-C(O-CH_2CH_3)_2-O-CH_2CH_3} + R-Mg-X \longrightarrow \underset{\text{acetal}}{R-C(H)(O-CH_2CH_3)-O-CH_2CH_3} \xrightarrow{H_3O^+} \underset{\text{aldehyde}}{R-\overset{O}{\overset{\|}{C}}-H}$$

18-48. Show how you would accomplish the following synthetic conversions.
(a) benzene $\longrightarrow$ *n*-butylbenzene
(b) benzonitrile $\longrightarrow$ propiophenone
(c) benzene $\longrightarrow$ *p*-methoxybenzaldehyde
(d) Ph—$(CH_2)_4$—OH $\longrightarrow$ tetralone

18-49. Predict the products formed when cyclohexanecarbaldehyde reacts with the following reagents.
(a) PhMgBr, then H_3O^+
(b) Tollens reagent
(c) semicarbazide and weak acid
(d) excess ethanol and acid
(e) 1,3-propanedithiol, H^+
(f) zinc amalgam and dilute hydrochloric acid

18-50. Show how you would synthesize 2-octanone from each of the following compounds. You may use any other necessary reagents.
(a) heptanal **(b)** 1-octyne **(c)** 1,3-dithiane **(d)** 2-octanol
(e) heptanoic acid **(f)** heptanoyl chloride **(g)** 2,3-dimethyl-2-nonene **(h)** $CH_3(CH_2)_5CN$

18-51. Show how you would synthesize octanal from each of the following compounds. You may use any other necessary reagents.
(a) 1-octanol **(b)** 1-nonene **(c)** 1-octyne
(d) 1,3-dithiane **(e)** 1,1-dichlorooctane **(f)** octanoic acid

18-52. Hydration of alkynes (via oxymercuration) gives good yields only with symmetrical or terminal alkynes. Show what the products would be from hydration of
(a) 3-hexyne. **(b)** 2-hexyne. **(c)** 1-hexyne. **(d)** cyclodecyne. **(e)** 3-methylcyclodecyne.

18-53. Which of the following compounds would give a positive Tollens test? Remember that the Tollens test involves mild basic aqueous conditions.
(a) $CH_3CH_2CH_2COCH_3$ **(b)** $CH_3CH_2CH_2CH_2CHO$
(c) $CH_3CH{=}CH{-}CH{=}CH{-}OH$ **(d)** $CH_3CH_2CH_2CH_2{-}CH(OH)OCH_3$
(e) $CH_3CH_2CH_2CH_2{-}CH(OCH_3)_2$ **(f)** O OH

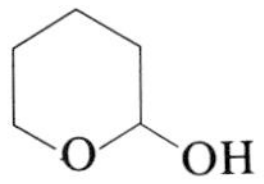

18-54. Solving the following road-map problem depends on determining the structure of **A**, the key intermediate. Give structures for compounds **A** through **K**.

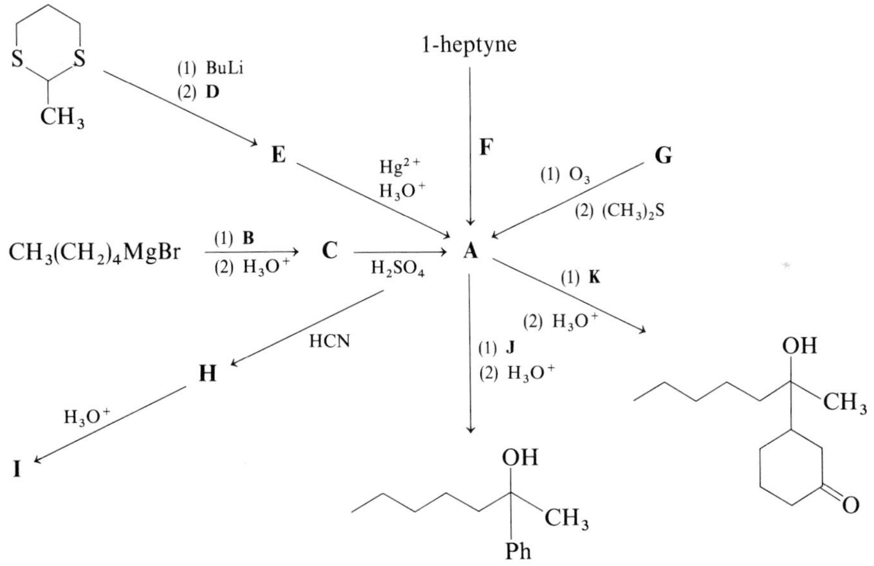

S S CH₃ (1) BuLi (2) **D** → **E**
1-heptyne → **F**
E Hg²⁺, H₃O⁺ → **A**
G (1) O₃ (2) (CH₃)₂S → **A**
CH₃(CH₂)₄MgBr (1) **B** (2) H₃O⁺ → **C** H₂SO₄ → **A**
A HCN → **H** H₃O⁺ → **I**
A (1) **J** (2) H₃O⁺ → OH CH₃ Ph
A (1) **K** (2) H₃O⁺ → OH CH₃ O

★ 18-55. A dithiane synthesis can be used to make a ketone from an aldehyde. The aldehyde is converted to its dithiane derivative, which is deprotonated and alkylated. A mercuric chloride-assisted hydrolysis gives the ketone. Show how this technique might be used to convert benzaldehyde to benzyl phenyl ketone.

★ 18-56. Under acid catalysis, an alcohol reacts with dihydropyran to give the tetrahydropyranyl derivative (called a "THP ether") of the alcohol.

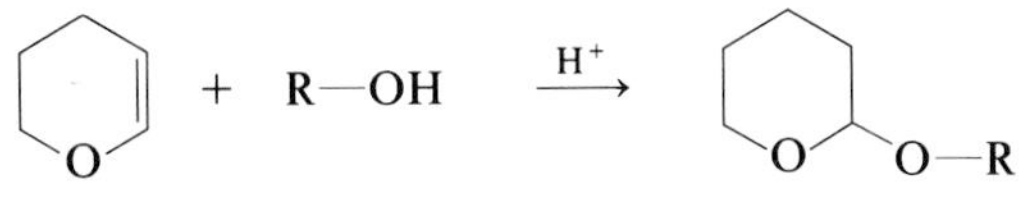

O + R—OH $\xrightarrow{H^+}$ O O—R

dihydropyran — tetrahydropyranyl derivative R—O—THP, a "THP ether"

(a) Propose a mechanism for this reaction.

(b) The "THP ether" is not an ether. What functional group does it actually contain? How will it react under basic conditions? Under acidic conditions?

(c) Propose a mechanism for the hydrolysis of the THP derivative in dilute aqueous acid. Show the products that are obtained.

★ 18-57. The mass spectrum of unknown compound **A** shows a molecular ion at m/z 116 and a prominent peak at m/z 87. Its UV spectrum shows no maximum above 200 nm. The IR and NMR spectra of **A** are shown below. When **A** is washed with dilute aqueous acid, extracted into dichloromethane, and the solvent evaporated, the product **B** shows a strong carbonyl absorption at 1715 cm^{-1} in the IR spectrum and a weak maximum at 274 nm ($\varepsilon = 16$) in the UV spectrum. The mass spectrum of **B** shows a molecular ion of m/z 72.

(a) Determine the structures of **A** and **B**, and show the fragmentation that accounts for the peak at m/z 87.

(b) Propose a mechanism for the acid-catalyzed hydrolysis of **A** to **B**.

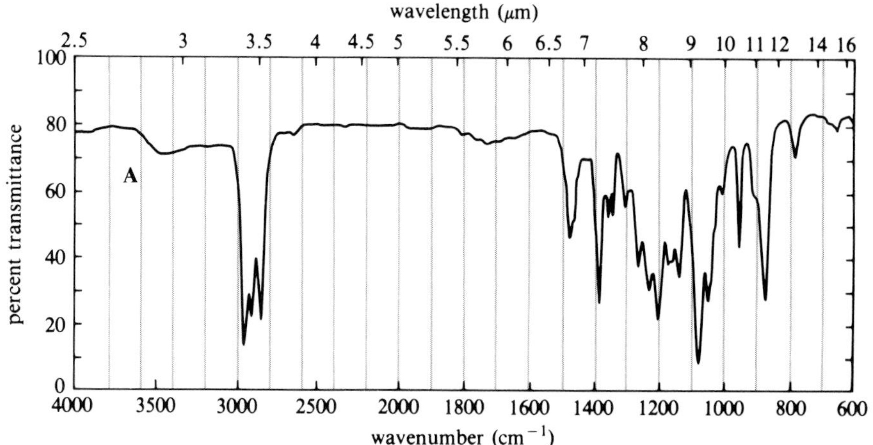

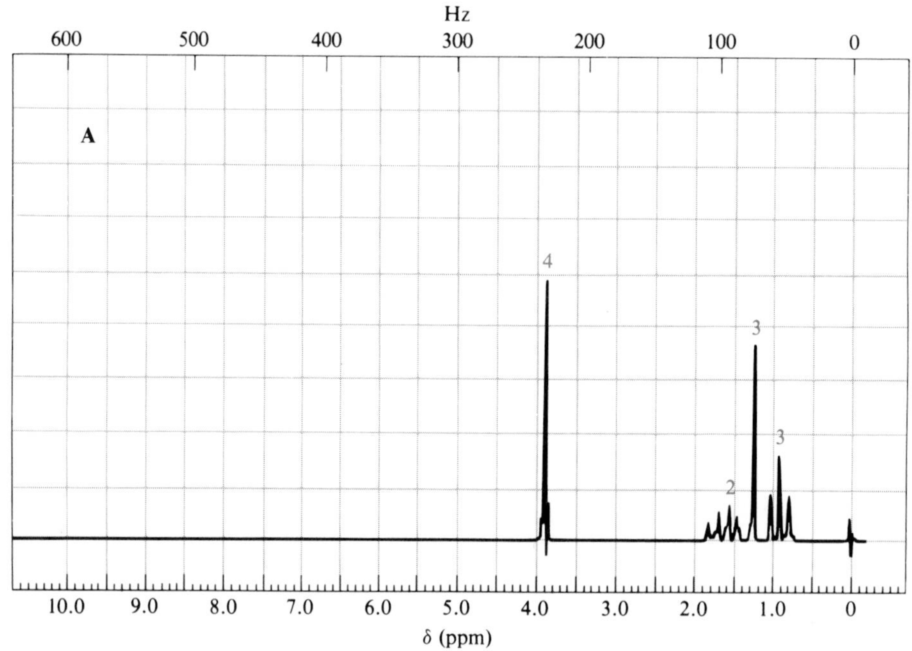

18-58. The UV spectrum of an unknown compound shows values of λ_{max} at 225 nm ($\varepsilon = 10{,}000$) and at 318 nm ($\varepsilon = 40$). The mass spectrum shows a molecular ion at m/z 96 and a prominent base peak at m/z 68. The IR and NMR spectra appear below. Propose a structure for this compound, and show how your structure corresponds to the observed absorptions. Propose a favorable fragmentation to account for the MS base peak at m/z 68 (loss of C_2H_4).

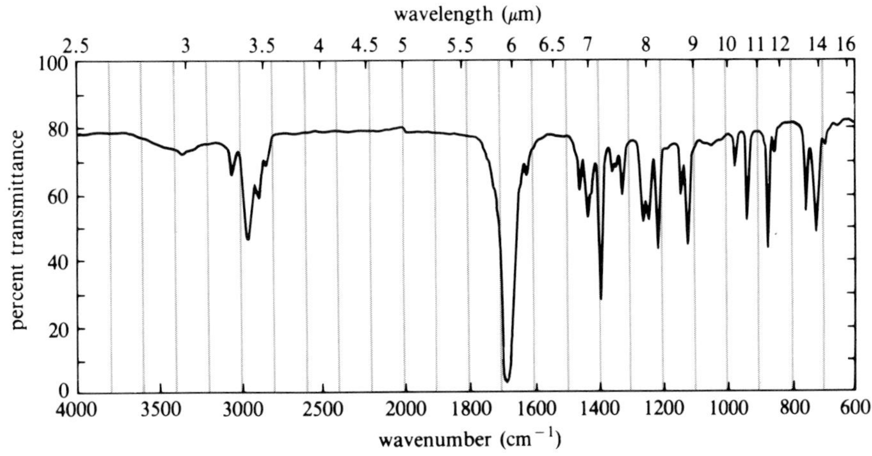

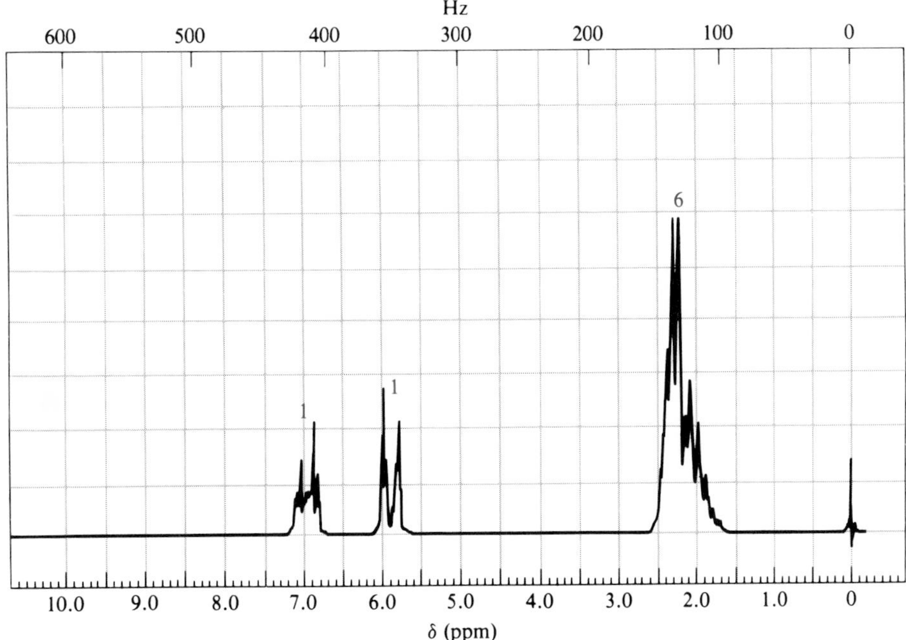

18-59. Two structures of the sugar **glucose** are shown below. The cyclic structure predominates in aqueous solution.

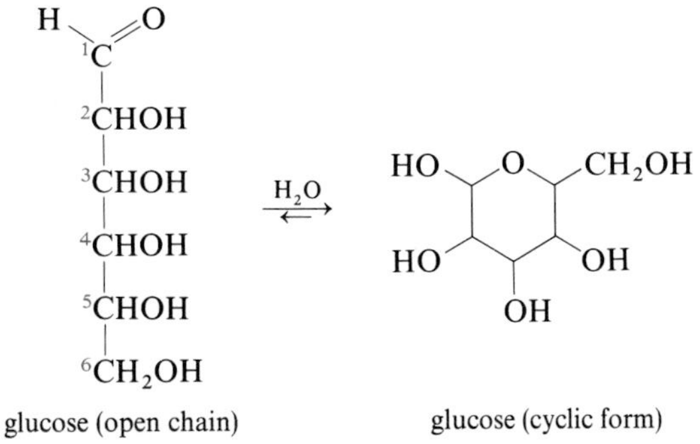

H—^{1}C=O
^{2}CHOH
^{3}CHOH
^{4}CHOH
^{5}CHOH
$^{6}CH_2OH$

$\xrightarrow[\longleftarrow]{H_2O}$

HO, O, CH_2OH, HO, OH, OH

glucose (open chain) glucose (cyclic form)

(a) Number the carbons in the cyclic structure. C1 (the aldehyde carbon in the open-chain form) is easily identified because it is the only carbon with two bonds to oxygen.
(b) What is the functional group at C1 in the cyclic form?

(c) Give a mechanism for this cyclization, assuming that there is a trace of acid present.
(d) Will this solution of glucose give a positive Tollens test?

★ 18-60. (A true story.) The chemistry department custodian was cleaning the organic lab when an unmarked bottle fell off a shelf and smashed on the floor, leaving a puddle of volatile liquid. The custodian began to wipe up the puddle, but was overcome with burning in his eyes and a feeling of having an electric drill thrust up his nose. He left the room and called the fire department, who used breathing equipment to go in and clean up the chemical. Three students were asked to identify the chemical quickly so the custodian could be treated and the chemical could be handled properly. The students took IR and NMR spectra, which appear below. The UV spectrum showed values of λ_{max} at 220 nm ($\varepsilon = 16{,}000$) and at 314 nm ($\varepsilon = 65$). The mass spectrometer was down, so no molecular weight was available. Determine the structure of this nasty compound, and show how your structure fits the spectra.

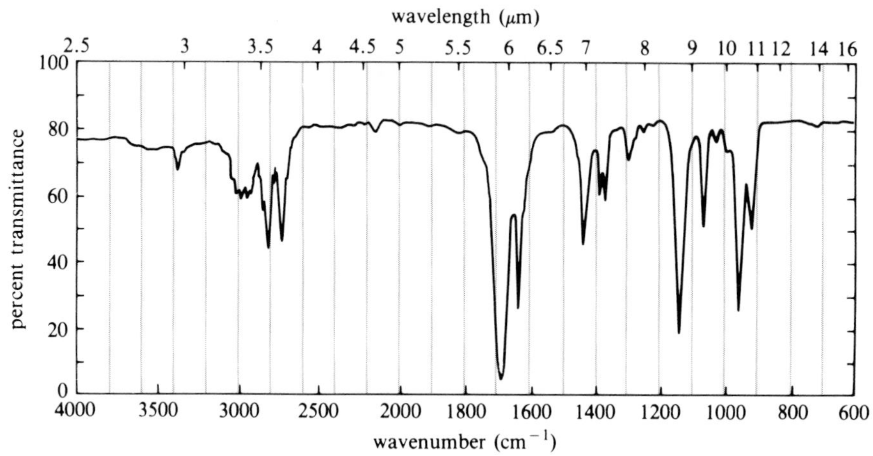

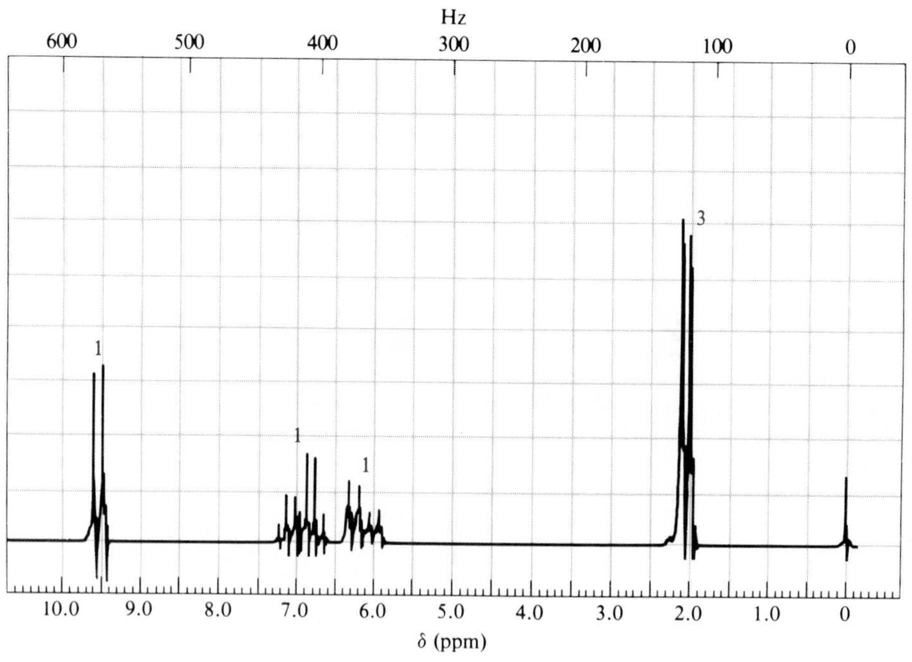

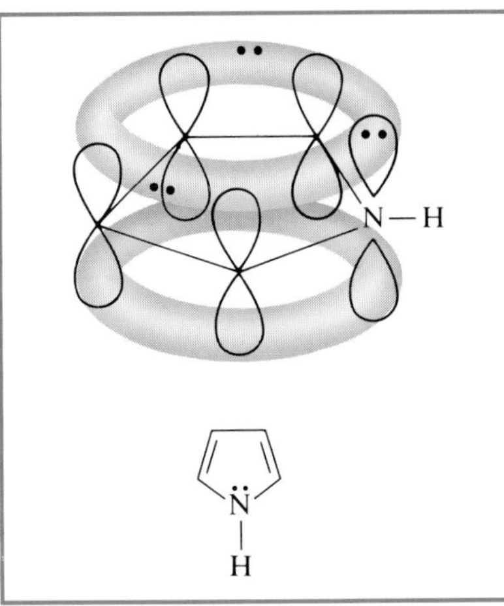

19

AMINES

19-1 INTRODUCTION

Amines are derivatives of ammonia with one or more alkyl or aryl groups bonded to the nitrogen atom. Amines are classified according to the number of alkyl or aryl groups bonded to nitrogen. If there is just one, the amine is said to be **primary** (1°). If there are two groups bonded to nitrogen, the amine is **secondary** (2°). If there are three alkyl or aryl groups bonded to nitrogen, the amine is **tertiary** (3°).

Primary (1°) amines

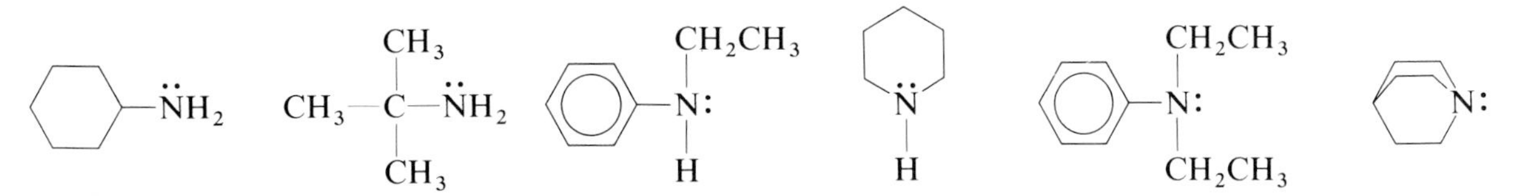

cyclohexylamine (1°) *t*-butylamine (1°)

Secondary (2°) amines

N-ethylaniline (2°) piperidine (2°)

Tertiary (3°) amines

N,*N*-diethylaniline (3°) quinuclidine (3°)

Quaternary ammonium salts have four alkyl or aryl bonds to a nitrogen atom. The nitrogen atom bears a full positive charge in these compounds, just as it does in simple ammonium salts such as ammonium chloride. The following are examples of quaternary (4°) ammonium salts:

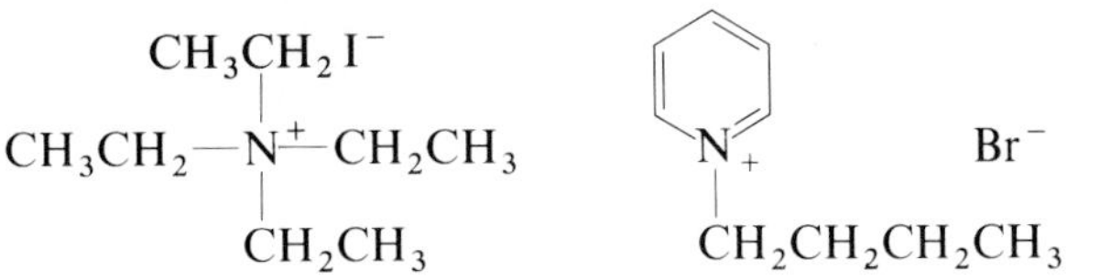

tetraethylammonium iodide *N*-butylpyridinium bromide

As a class, amines include some of the most important biological compounds known. Amines serve in living organisms as bioregulators, as neurotransmitters, in defense mechanisms, and in many other functions. Because of their high degree of biological activity, many amines are used as drugs and medicines. The structures and uses of some important biologically active amines are shown in Figure 19-1.

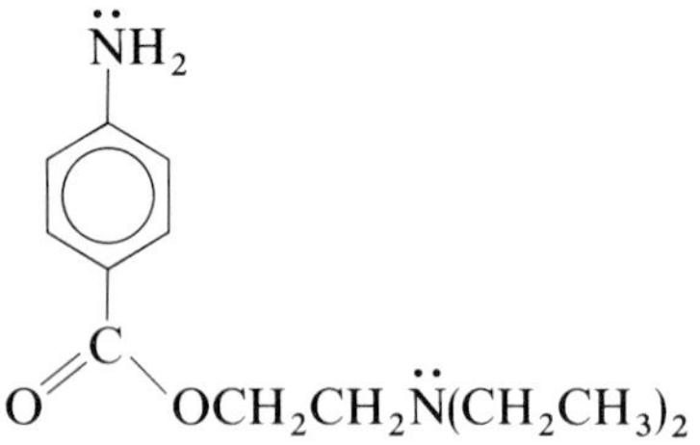

novocaine
an anesthetic

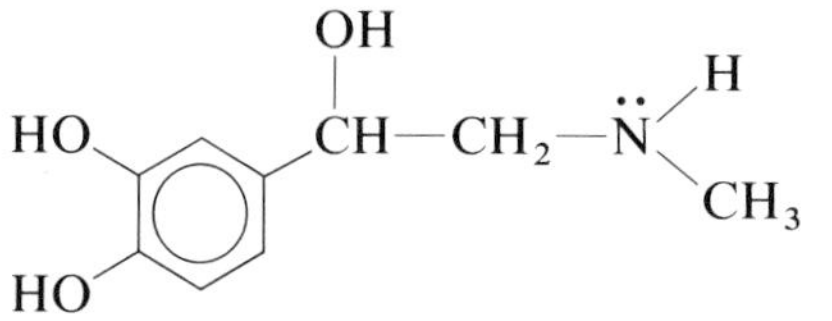

epinephrine
an adrenal hormone

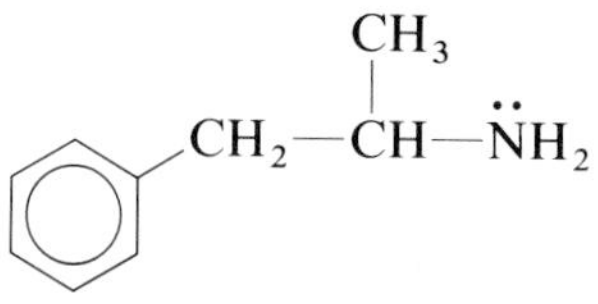

amphetamine
an addictive stimulant

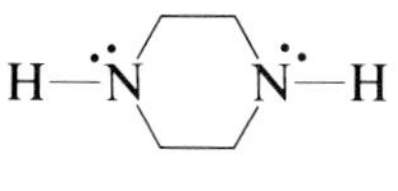

piperazine
kills intestinal worms

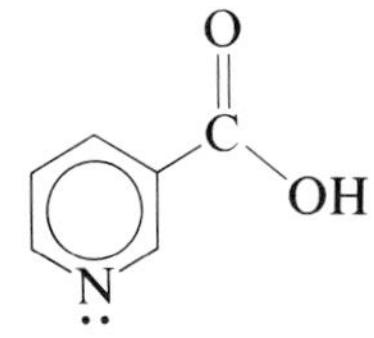

nicotinic acid
niacin, a vitamin

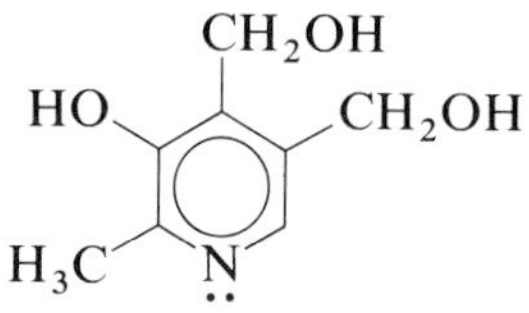

pyridoxine
vitamin B_6

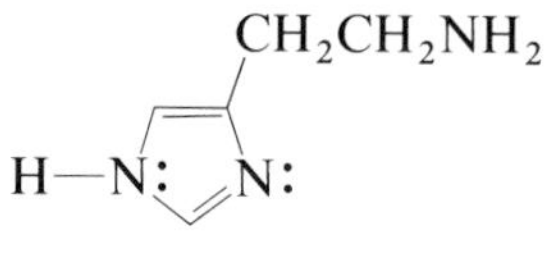

histamine
dilates blood vessels

FIGURE 19-1 Examples of some biologically active amines.

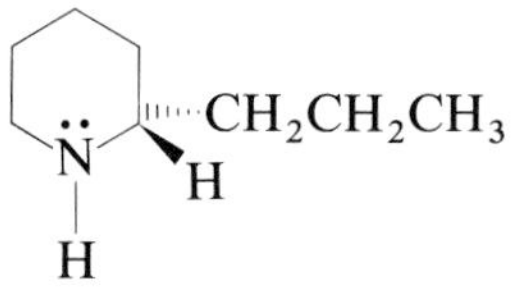

(*S*)-coniine

The *alkaloids* are an important group of biologically active amines, mostly synthesized by plants to protect them from being eaten by insects and other animals. The structures of some representative alkaloids are shown in Figure 19-2. Although some alkaloids are used medicinally (chiefly as painkillers), all alkaloids are toxic and cause death if taken in large quantities. The Greeks chose the alkaloid coniine to kill Socrates, although morphine, nicotine, or cocaine would have served equally well. Mild cases of alkaloid poisoning can produce psychological effects that resemble peacefulness, euphoria, or hallucinations, and people seeking these effects generally become addicted to the alkaloids. Alkaloid addiction often ends in death; current estimates are over 400,000 deaths in the United States per year, including both natural alkaloids like nicotine and cocaine and synthetic alkaloids like amphetamine.

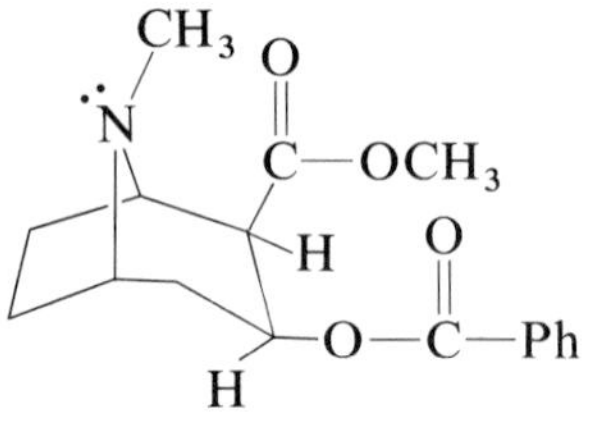

cocaine
in coca leaves

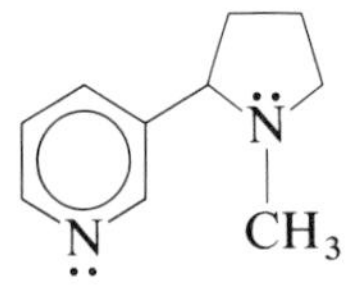

nicotine
in tobacco

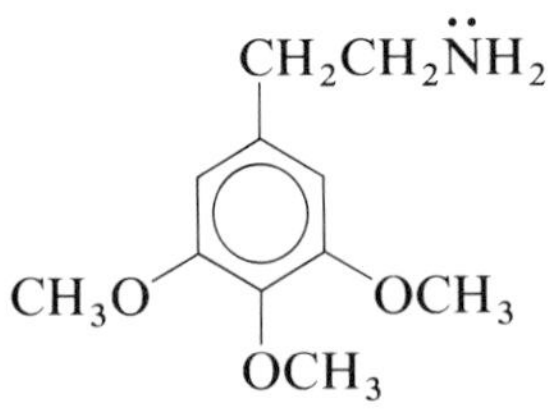

mescaline
in peyote cactus

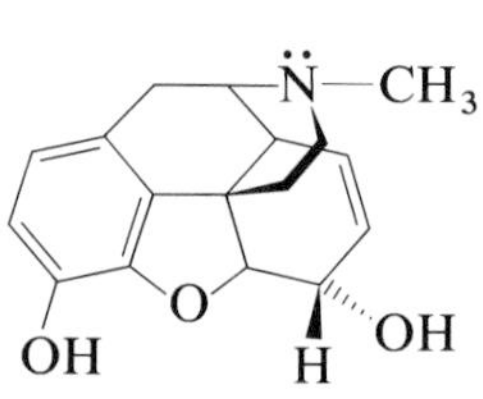

morphine
in opium poppies

FIGURE 19-2 Some representative alkaloids.

19-2 NOMENCLATURE OF AMINES

19-2A COMMON NAMES

In the common names of amines, the names of the alkyl groups bonded to nitrogen are given first, followed by the suffix *-amine*. The prefixes **di-**, **tri-**, and **tetra-** may be used to describe two, three, or four identical substituents.

$CH_3CH_2-\ddot{N}H_2$ — ethylamine

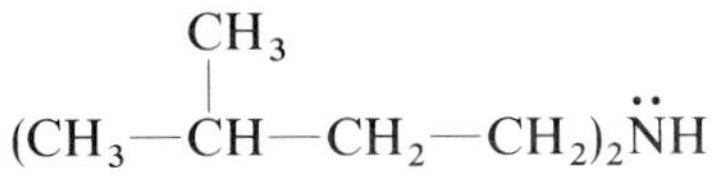

$(CH_3-CH(CH_3)-CH_2-CH_2)_2\ddot{N}H$ — diisopentylamine

$C_6H_5CH_2\ddot{N}H_2$ — benzylamine

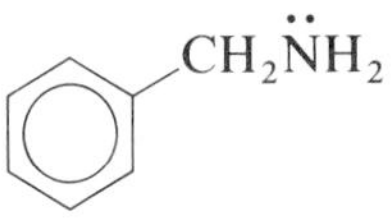

$(C_6H_5)_2N{:}H$ — diphenylamine

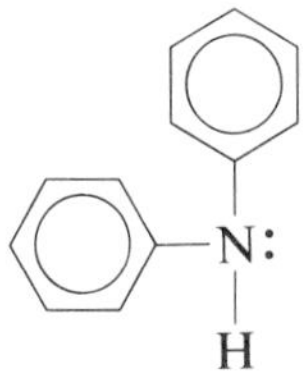

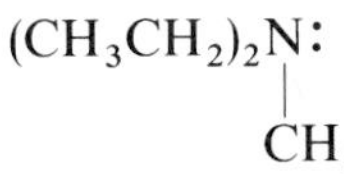

$(CH_3CH_2)_2N{:}CH_3$ — diethylmethylamine

$C_6H_{11}N{:}(CH_3)_2$ — cyclohexyldimethylamine

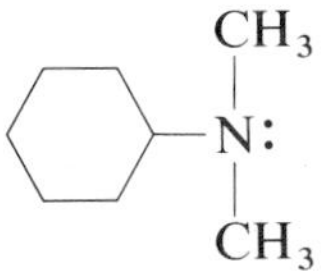

$(CH_3CH_2CH_2CH_2)_4N^+\ {}^-Cl$ — tetrabutylammonium chloride

In naming amines with more complicated structures, the $-NH_2$ group is called the **amino** group. The amino group is treated like any other substituent, with a number or other symbol indicating its position on the ring or carbon chain.

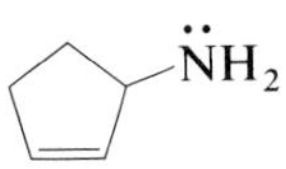

3-aminocyclopentene

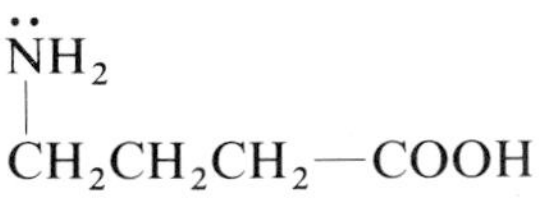

$\ddot{N}H_2-CH_2CH_2CH_2-COOH$ — γ-aminobutyric acid (4-aminobutanoic acid)

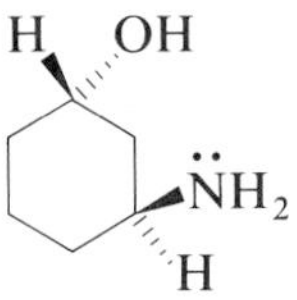

trans-3-aminocyclohexanol

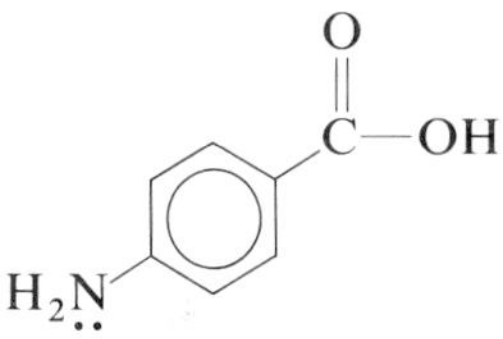

$H_2\ddot{N}-C_6H_4-COOH$ — *p*-aminobenzoic acid

Using this system, secondary and tertiary amines are named by classifying the nitrogen atom (together with its alkyl groups) as an alkylamino group. The largest or most complicated alkyl group is taken to be the parent molecule.

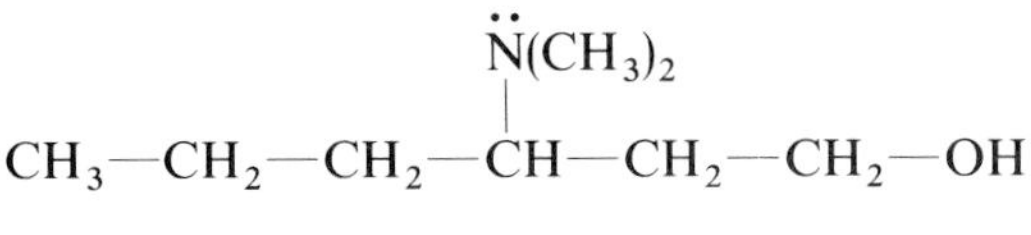

$CH_3-CH_2-CH_2-CH(\ddot{N}(CH_3)_2)-CH_2-CH_2-OH$ — 3-dimethylamino-1-hexanol

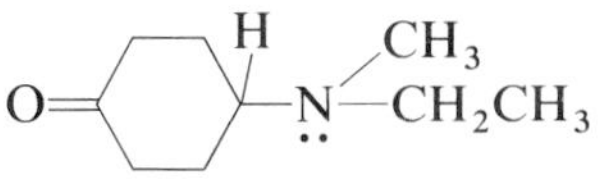

$O{=}C_6H_9-\ddot{N}(CH_3)-CH_2CH_3$ — 4-(ethylmethylamino)cyclohexanone

19-2B IUPAC NAMES

The IUPAC nomenclature for amines is similar to the nomenclature for alcohols. The longest continuous chain of carbon atoms determines the root name. The *-e* ending in the alkane name is changed to *-amine,* and a number is used to show the position of the amino group along the chain. Substituents along the carbon chain are given numbers to designate their locations, and the prefix *N*- is used for each substituent on the nitrogen atom.

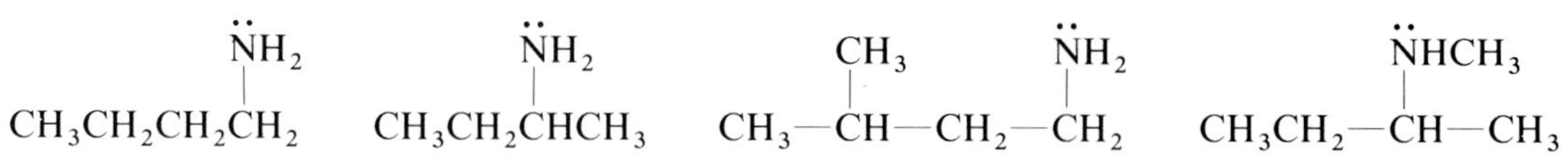

$CH_3CH_2CH_2CH_2NH_2$	$CH_3CH_2CH(NH_2)CH_3$	$CH_3-CH(CH_3)-CH_2-CH_2NH_2$	$CH_3CH_2-CH(NHCH_3)-CH_3$
1-butanamine	2-butanamine	3-methyl-1-butanamine	*N*-methyl-2-butanamine

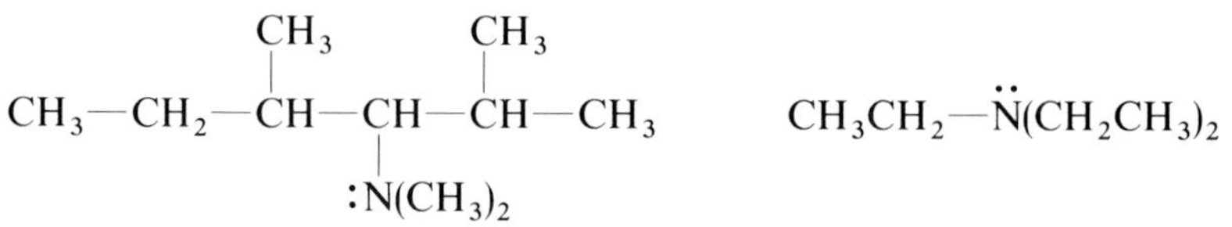

$CH_3-CH_2-CH(CH_3)-CH[N(CH_3)_2]-CH(CH_3)-CH_3$	$CH_3CH_2-N(CH_2CH_3)_2$
2,4,*N*,*N*-tetramethyl-3-hexanamine	*N*,*N*-diethylethanamine (triethylamine)

The aromatic and heterocyclic amines are generally known by historical names. For example, phenylamine is called *aniline*, and its derivatives are named as derivatives of aniline.

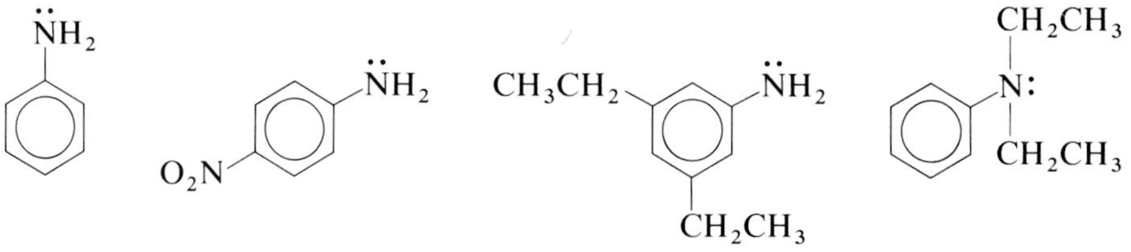

$C_6H_5NH_2$	$O_2N-C_6H_4-NH_2$	$(CH_3CH_2)_2C_6H_3NH_2$	$C_6H_5N(CH_2CH_3)_2$
aniline	*p*-nitroaniline	3,5-diethylaniline	*N*,*N*-diethylaniline

The names and structures of some common nitrogen heterocycles and derivatives are shown below. The heteroatom is usually given position number 1.

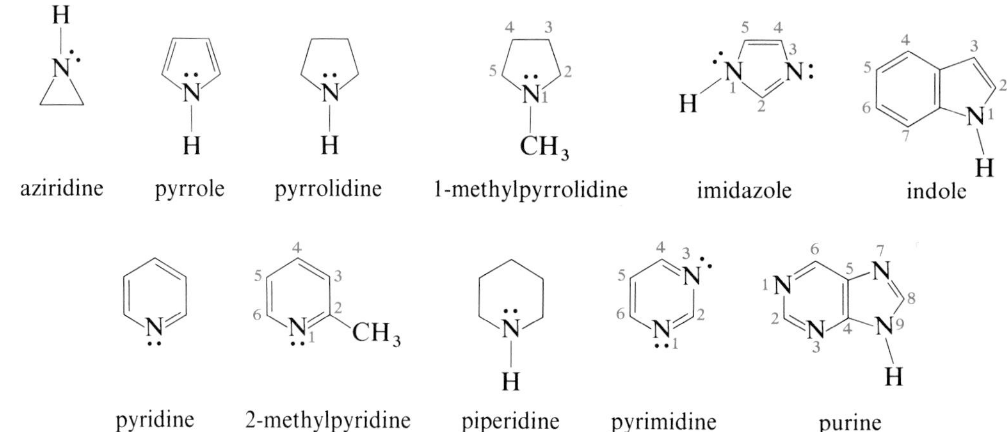

PROBLEM 19-1

Determine which of the heterocyclic amines shown above are aromatic. Give the reasons for your conclusions.

PROBLEM 19-2

Draw the structures of the following compounds.

(a) *t*-butylamine
(b) α-aminopropionaldehyde
(c) 4-(dimethylamino)pyridine
(d) 2-methylaziridine
(e) *N*-methyl-*N*-ethyl-3-hexanamine
(f) *m*-chloroaniline

PROBLEM 19-3

Give correct names for the following amines.

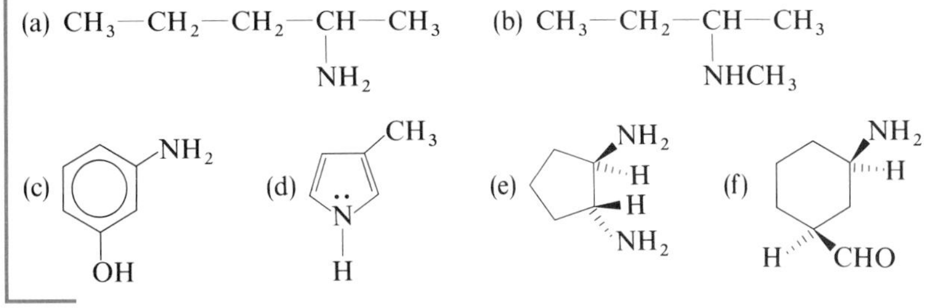

(a) CH3—CH2—CH2—CH—CH3
NH2
(b) CH3—CH2—CH—CH3
NHCH3
(c)
NH2
OH
(d)
CH3
N
H
(e)
NH2
H
H
NH2
(f)
NH2
H
H
CHO

19-3 STRUCTURE OF AMINES

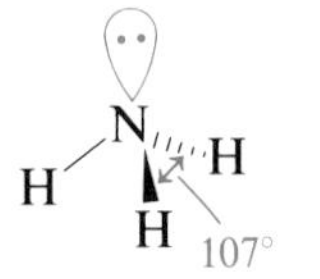

N
H
H
H
107°

ammonia

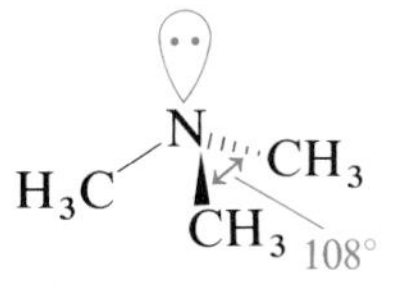

N
CH3
H3C
CH3
108°

trimethylamine

In Chapter 2 we saw that the ammonia molecule has a slightly distorted tetrahedral shape, with a lone pair of nonbonding electrons occupying one of the tetrahedral positions. This geometry is represented by sp^3 hybridization of the nitrogen atom, with the bulky lone pair compressing the H—N—H bond angles to 107° from the "ideal" sp^3 bond angle of 109.5°. This angle compression is not as great in trimethylamine, where the bulky methyl groups open the angle slightly as illustrated at the left.

A tetrahedral amine with three different substituents (and a lone pair) is nonsuperimposable on its mirror image. We might hope to resolve such an amine into two enantiomers. In most cases, however, such a resolution is not possible, because the two enantiomers interconvert very rapidly. This interconversion takes place by **nitrogen inversion,** in which the lone pair moves from one face of the molecule to the other. The nitrogen atom is sp^2 hybridized in the transition state, and the nonbonding electrons occupy a p orbital. This is a fairly stable transition state, as reflected by the small activation energy of about 6 kcal (25 kJ) per mole. The interconversion of (R)- and (S)-ethylmethylamine is shown in Figure 19-3. In naming the enantiomers of chiral amines, the Cahn-Ingold-Prelog convention is used, with the nonbonding electron pair having the lowest priority.

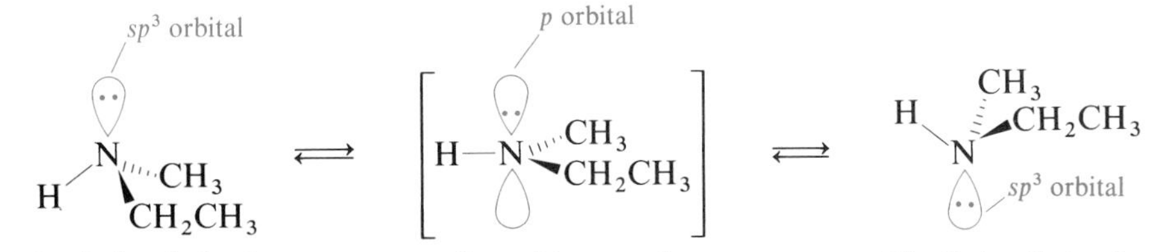

sp³ orbital
N
CH3
H
CH2CH3
p orbital
H—N
CH3
CH2CH3
CH3
H
CH2CH3
N
sp³ orbital

(R)-ethylmethylamine [transition state] (S)-ethylmethylamine

FIGURE 19-3 Nitrogen inversion interconverts the two enantiomers of a simple chiral amine. The transition state is a planar, sp^2 hybrid structure with the lone pair in a p orbital.

Although most simple amines cannot be resolved into enantiomers, several types of chiral amines can be resolved:

1. *Amines whose chirality stems from the presence of chiral carbon atoms.* For example, 2-butanamine can be resolved into enantiomers because the 2-butyl group is chiral.

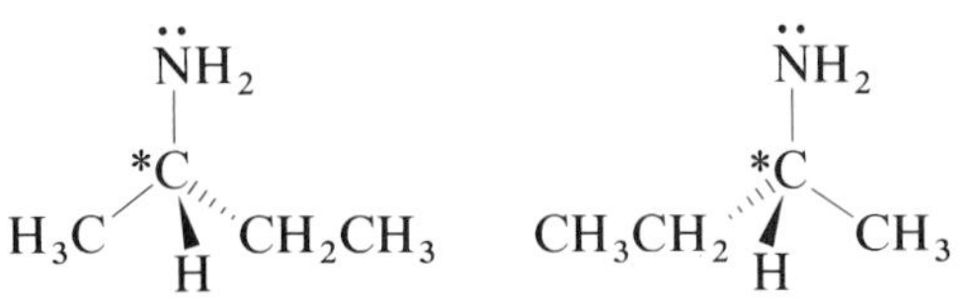

NH2
*C
H3C
CH2CH3
H
NH2
*C
CH3CH2
CH3
H

(S)-2-butanamine (R)-2-butanamine

2. *Quaternary ammonium salts with chiral nitrogen atoms.* Inversion of configuration is not possible, because there is no lone pair to undergo nitrogen inversion. For example, the methyl ethyl isopropyl anilinium salts can be resolved into enantiomers.

(*R*) (*S*)

3. *Amines that cannot attain the sp^2 hybrid transition state for nitrogen inversion.* For example, if the nitrogen atom is contained in a small ring, it is prevented from attaining the 120° bond angles that facilitate inversion. Such a compound has a higher activation energy for inversion, and the enantiomers may be resolved. For example, chiral aziridines (three-membered rings containing a nitrogen) often may be resolved into enantiomers.

(*R*)-1,2,2-trimethylaziridine (*S*)-1,2,2-trimethylaziridine

PROBLEM 19-4

Which of the following amines could be resolved into enantiomers? In each case, explain why interconversion of the enantiomers would or would not take place.

(a) *cis*-2-methylcyclohexanamine
(b) *N*-methyl-*N*-ethylcyclohexanamine
(c) *N*-methylaziridine
(d) ethyl methyl anilinium iodide
(e) methyl ethyl propyl isopropyl ammonium iodide

19-4 PHYSICAL PROPERTIES OF AMINES

Amines are strongly polar compounds because the large dipole moment of the lone pair of electrons adds to the dipole moments of the C → N and H ↛ N bonds. Primary and secondary amines have N—H bonds, allowing them to form hydrogen bonds. Having no N—H bonds, pure tertiary amines cannot engage in hydrogen bonding. They can accept hydrogen bonds from molecules having O—H or N—H bonds, however.

overall dipole moment

1° or 2° amine: hydrogen-bond donor and acceptor

3° amine: hydrogen-bond acceptor only

Because nitrogen is less electronegative than oxygen, the N—H bond is less polar than the O—H bond. Therefore, amines form weaker hydrogen bonds than do alcohols of similar molecular weights. Primary and secondary amines have boiling points that are lower than those of alcohols, yet higher than those of ethers of similar molecular weights. With no hydrogen bonding, tertiary amines have lower boiling points than do primary and secondary amines of similar molecular weights. Table 19-1 compares the boiling points of an ether, an alcohol, and amines of similar molecular weights.

TABLE 19-1

Comparison of the boiling points of an ether, an alcohol, and amines of similar molecular weights

Compound	*Type*	*Molecular weight*	*b.p. (°C)*
$(CH_3)_3N$:	tertiary amine	59	3
$CH_3—O—CH_2—CH_3$	ether	60	8
$CH_3—NH—CH_2—CH_3$	secondary amine	59	37
$CH_3CH_2CH_2—NH_2$	primary amine	59	48
$CH_3CH_2CH_2—OH$	alcohol	60	97

All amines, even tertiary ones, form hydrogen bonds with hydroxylic solvents such as water and the alcohols. Therefore, amines tend to be very soluble in alcohols, and the lower-molecular-weight amines (up to about six carbon atoms) are relatively soluble in water. Table 19-2 lists the melting points, boiling points, and water solubilities of some of the simple aliphatic and aromatic amines.

TABLE 19-2

Melting points, boiling points, and water solubilities of some simple amines

Name	*Structure*	*Molecular weight*	*m.p. (°C)*	*b.p. (°C)*	*H_2O solubility ($g/100\ g\ H_2O$)*
	Primary amines				
methylamine	CH_3NH_2	31	−93	−7	very soluble
ethylamine	$CH_3CH_2NH_2$	45	−81	17	∞
n-propylamine	$CH_3CH_2CH_2NH_2$	59	−83	48	∞
isopropylamine	$(CH_3)_2CHNH_2$	59	−101	33	∞
n-butylamine	$CH_3CH_2CH_2CH_2NH_2$	73	−50	77	∞
isobutylamine	$(CH_3)_2CHCH_2NH_2$	73	−86	68	∞
s-butylamine	$CH_3CH_2CH(NH_2)CH_3$	73	−104	63	∞
t-butylamine	$(CH_3)_3CNH_2$	73	−68	45	∞
cyclohexylamine	*cyclo*-$C_6H_{11}NH_2$	99		134	slightly soluble
benzylamine	$C_6H_5CH_2NH_2$	107		185	∞
allylamine	$CH_2=CH—CH_2NH_2$	57		53	very soluble
aniline	$C_6H_5NH_2$	93	−6	184	3.7
	Secondary amines				
dimethylamine	$(CH_3)_2NH$	45	−96	7	very soluble
ethylmethylamine	$CH_3CH_2NHCH_3$	59		37	very soluble
diethylamine	$(CH_3CH_2)_2NH$	73	−42	56	very soluble
di-*n*-propylamine	$(CH_3CH_2CH_2)_2NH$	101	−40	111	slightly soluble
diisopropylamine	$[(CH_3)_2CH]_2NH$	101	−61	84	slightly soluble
di-*n*-butylamine	$(CH_3CH_2CH_2CH_2)_2NH$	129	−59	159	slightly soluble
N-methylaniline	$C_6H_5NHCH_3$	107	−57	196	slightly soluble
diphenylamine	$(C_6H_5)_2NH$	169	54	302	insoluble
	Tertiary amines				
trimethylamine	$(CH_3)_3N$	59	−117	3.5	91
triethylamine	$(CH_3CH_2)_3N$	101	−115	90	14
tri-*n*-propylamine	$(CH_3CH_2CH_2)_3N$	143	−94	156	slightly soluble
N,N-dimethylaniline	$C_6H_5N(CH_3)_2$	121	2	194	1.4
triphenylamine	$(C_6H_5)_3N$	251	126	365	insoluble

Perhaps the most obvious property of amines is their characteristic odor of rotting fish. Some of the diamines are particularly pungent; the following diamines have common names that aptly describe their odors.

$$\underset{\text{(1,4-butanediamine)}}{\underset{\text{putrescine}}{\underset{NH_2\qquad\qquad NH_2}{CH_2CH_2CH_2CH_2}}} \qquad \underset{\text{(1,5-pentanediamine)}}{\underset{\text{cadaverine}}{\underset{NH_2\qquad\qquad\qquad NH_2}{CH_2CH_2CH_2CH_2CH_2}}}$$

PROBLEM 19-5

Rank each of the following sets of compounds in order of increasing boiling points.

(a) triethylamine, di-*n*-propylamine, *n*-propyl ether
(b) ethanol, dimethylamine, dimethyl ether
(c) trimethylamine, diethylamine, diisopropylamine

19-5 SPECTROSCOPY OF AMINES

19-5A INFRARED SPECTROSCOPY

The most reliable IR absorption of primary and secondary amines is the N—H stretch whose frequency appears between 3200 and 3500 cm^{-1}. Since this absorption is often broad, it is easily confused with the O—H absorption of an alcohol. In most cases, however, there are one or more spikes visible in the N—H stretching region of an amine spectrum. Primary amines ($R—NH_2$) usually give two N—H spikes, while secondary amines ($R_2N—H$) usually give just one. Tertiary amines (R_3N) give no N—H absorptions. Notice in Figure 19-4 the characteristic N—H absorptions in the IR spectrum of 1-propanamine.

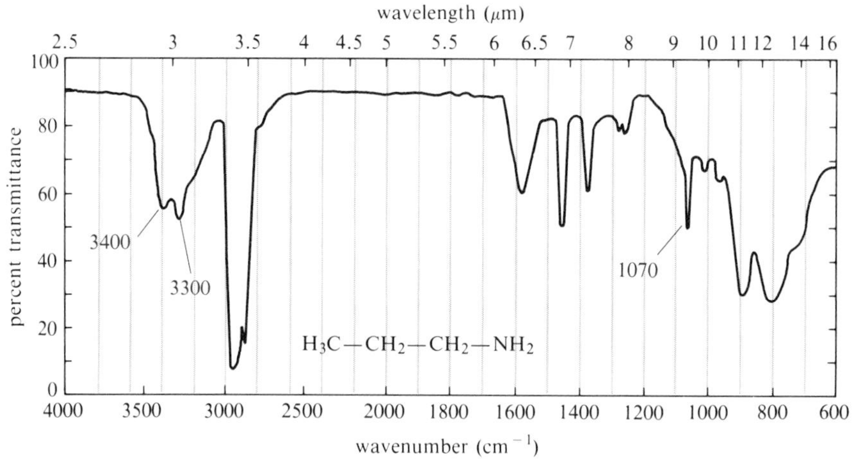

FIGURE 19-4 Infrared spectrum of 1-propanamine. Notice the characteristic N—H stretching absorptions at 3300 and 3400 cm^{-1}.

Although an amine IR spectrum also contains absorptions resulting from vibrations of C—N bonds, these vibrations appear around 1000 to 1200 cm^{-1}, in the same spectral region as the C—C and C—O bond vibrations. Therefore, they are not very useful for determining the presence of an amine.

PROBLEM 19-6

The following three IR spectra correspond to a primary amine, a secondary amine, and an alcohol. Give the functional group for each spectrum.

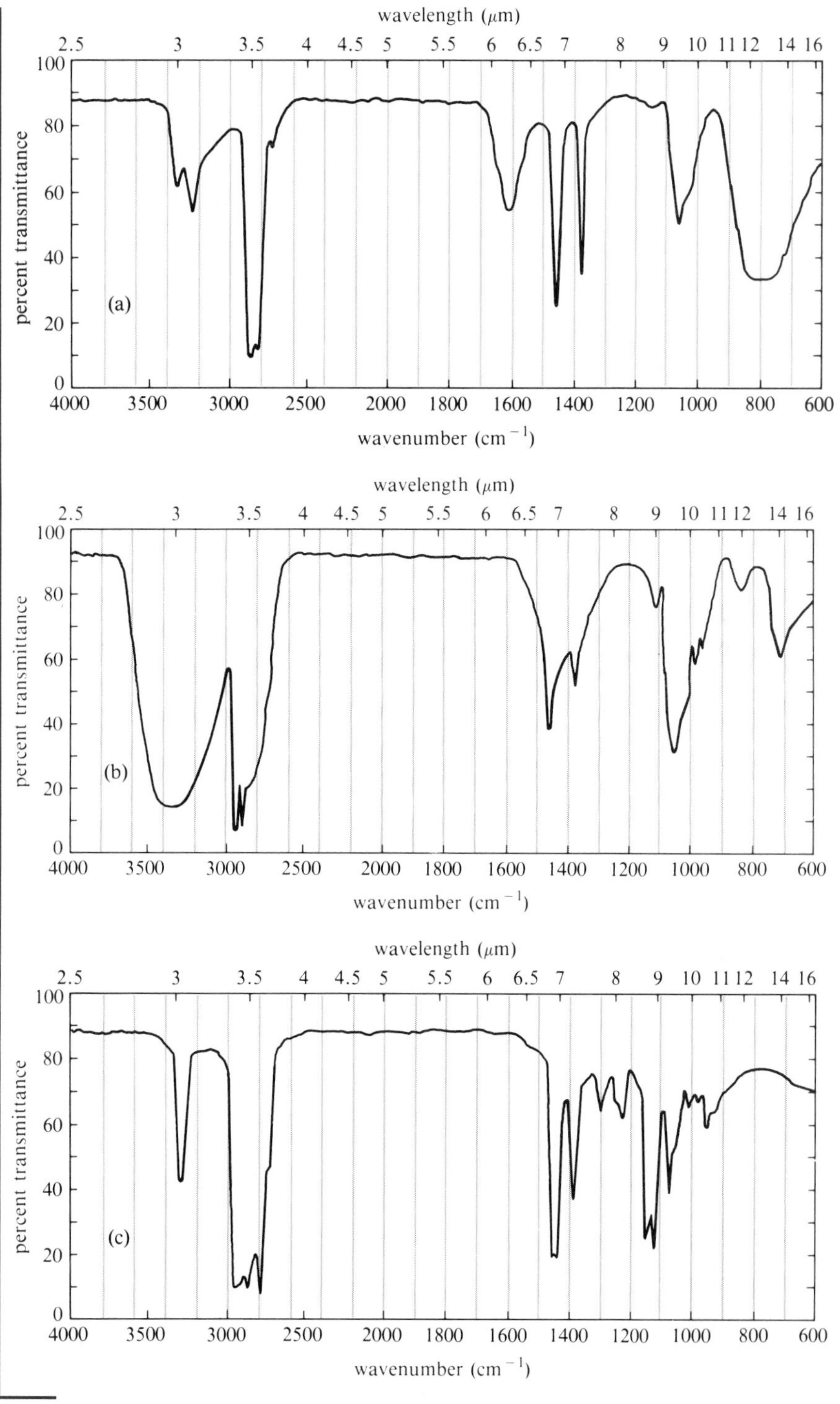

19-5B PROTON NMR SPECTROSCOPY

Like the O—H protons of alcohols, the N—H protons of amines absorb at chemical shifts that depend on the extent of hydrogen bonding. The solvent and the sample concentration influence the extent of hydrogen bonding, and therefore the chemical shift. Typical N—H chemical shifts of about $\delta 1$ to $\delta 4$ are observed.

Another similarity between O—H and N—H protons is their failure, in many cases, to show spin-spin splitting. In some samples, N—H protons exchange from one molecule to another at a rate that is faster than the time scale of the NMR experiment, and the N—H protons fail to show magnetic coupling. Sometimes the N—H protons of a very pure amine will show clean splitting, but these cases are rare. More commonly the N—H protons appear as broad peaks. A very broad peak should arouse suspicion of the presence of N—H protons. As with O—H protons, the absorption of N—H protons decreases or disappears upon shaking the sample with D_2O.

Nitrogen is not as electronegative as oxygen and the halogens, so the protons on the α-carbon atoms of amines are not as strongly deshielded. Protons on an amine's α-carbon atom generally absorb between δ2 and δ3, the exact position depending on the structure and substitution of the amine.

$$\underset{\text{methyl } \delta 2.3}{\overset{\downarrow}{CH_3}—NR_2} \qquad \underset{\text{methylene } \delta 2.7}{R—\overset{\downarrow}{CH_2}—NR_2} \qquad \underset{\text{methine } \delta 2.9}{R_2\overset{\downarrow}{CH}—NR_2}$$

Protons that are beta to a nitrogen atom show a much smaller effect, usually absorbing in the range δ1.1 to δ1.8. These chemical shifts show a downfield movement of about 0.2 ppm resulting from the beta relationship. The NMR spectrum of 1-propanamine (Fig. 19-5) shows these characteristic chemical shifts.

γ protons β protons α protons

$$\underset{\delta 0.9}{CH_3}—\underset{\delta 1.5}{CH_2}—\underset{\delta 2.7}{CH_2}—NH_2$$

variable (δ1.1 in this spectrum)

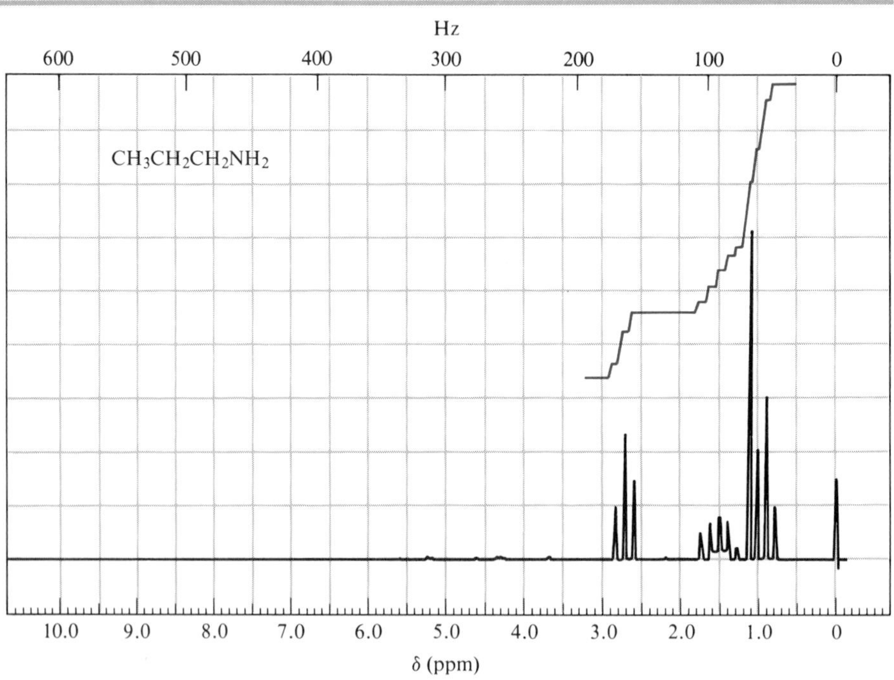

FIGURE 19-5 Proton NMR spectrum of 1-propanamine.

PROBLEM 19-7

The proton NMR spectrum of a compound of formula $C_4H_{11}N$ is shown below. Determine the structure of this amine, and give peak assignments for all the protons in the structure.

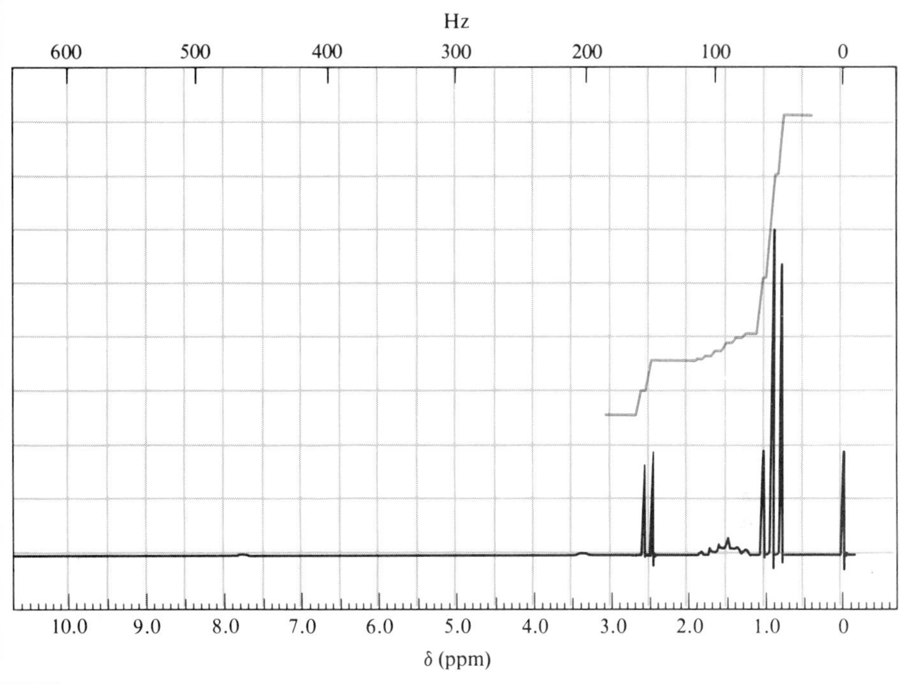

19-5C CARBON NMR SPECTROSCOPY

The α-carbon atom bonded to the nitrogen of an amine usually shows a chemical shift of about 40 to 50 ppm. This range agrees with our general rule that a carbon atom shows a chemical shift that is about 20 times as great as that of the protons bonded to it. In propanamine, for example, the α-carbon atom absorbs at 45 ppm, while its protons absorb at 2.7 ppm. The β carbon is less deshielded, absorbing at 27 ppm, compared with its protons' absorption at 1.5 ppm. The γ-carbon atom shows little effect from the presence of the nitrogen atom, absorbing at 11 ppm. Table 19-3 shows the carbon chemical shifts of some representative amines.

TABLE 19-3

Carbon NMR chemical shifts for some representative amines

Structure	Name
CH_3—NH_2 26.9	methanamine
CH_3—CH_2—NH_2 17.7 35.9	ethanamine
CH_3—CH_2—CH_2—NH_2 11.2 27.3 44.9	1-propanamine
CH_3—CH_2—CH_2—CH_2—NH_2 14.0 20.4 36.7 42.3	1-butanamine

PROBLEM 19-8

The carbon NMR chemical shifts of diethylmethylamine, *n*-propylamine, *n*-propanol, and propanal are given below. Determine which spectrum corresponds to each structure, and show which carbon atom(s) are responsible for each absorption.

(a) 11.2 27.3 44.9 (b) 13.8 47.5 58.2 (c) 7.9 44.7 201.9 (d) 10.0 25.8 63.6

19-5D MASS SPECTROMETRY

The most obvious piece of information provided by the mass spectrum is the molecular weight. Stable compounds containing only carbon, hydrogen, oxygen, chlorine, bromine, and iodine give molecular ions with even mass numbers. Carbon and oxygen have even valences and even mass numbers, while hydrogen, chlorine, bromine, and oxygen have odd valences and odd mass numbers. Nitrogen has an odd valence and an even mass number. When a nitrogen atom is present in a stable molecule, the molecular weight is odd. In fact, whenever an odd number of nitrogen atoms are present in a molecule, the molecular ion has an odd mass number.

The most common fragmentation of amines is α cleavage to give a resonance-stabilized cation: an *iminium* ion. This ion is simply a protonated version of an imine (Chapter 18).

$$\left[\mathrm{R}\!\mid\!\mathrm{CH_2{-}\ddot{N}(R)(H)}\right]^{+\cdot} \longrightarrow \mathrm{R\cdot} + \left[\mathrm{H_2\overset{+}{C}{-}\ddot{N}(R)(H)} \longleftrightarrow \mathrm{H_2C{=}\overset{+}{N}(R)(H)}\right]$$

α cleavage iminium ion

Figure 19-6 shows the mass spectrum of butyl propyl amine. The base peak (m/z 72) corresponds to α cleavage with loss of a propyl radical to give a resonance-stabilized iminium ion, simply a protonated version of an imine. A similar α cleavage, with loss of an ethyl radical, gives the peak at m/z 86.

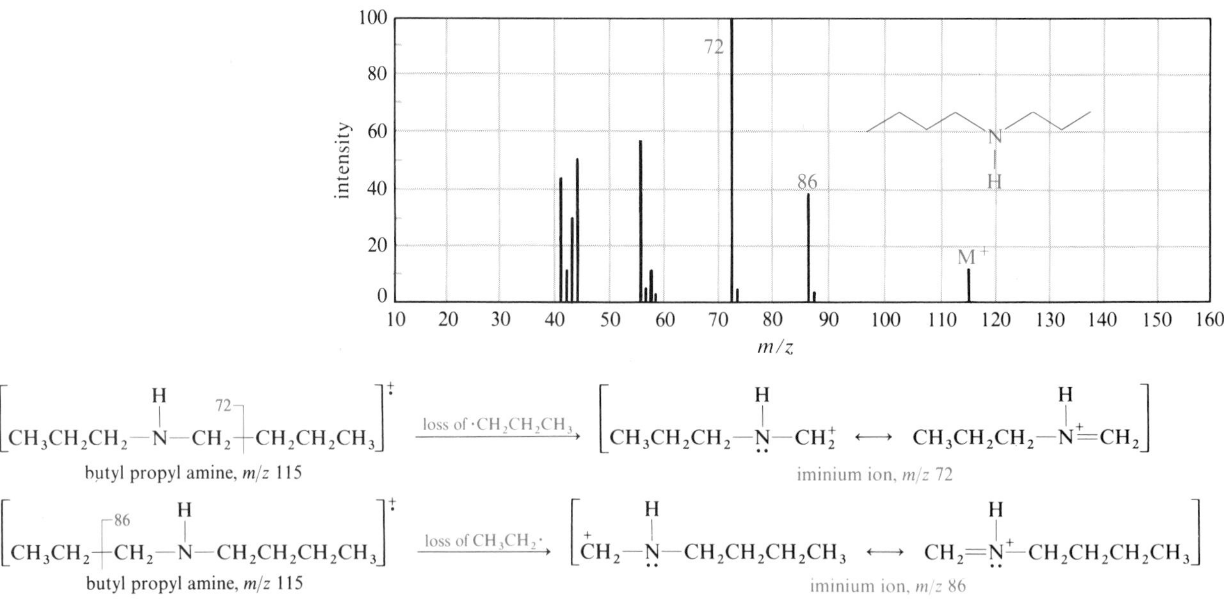

FIGURE 19-6 Mass spectrum of butyl propyl amine. The base peak corresponds to α cleavage in the butyl group, giving a propyl radical and a resonance-stabilized iminium ion.

PROBLEM 19-9

(a) Show how fragmentation occurs to give the base peak at m/z 58 in the mass spectrum of ethyl propyl amine, shown below.

(b) Show how a similar cleavage in the ethyl group gives an ion of m/z 72.

(c) Explain why the peak at m/z 72 is much weaker than that at m/z 58.

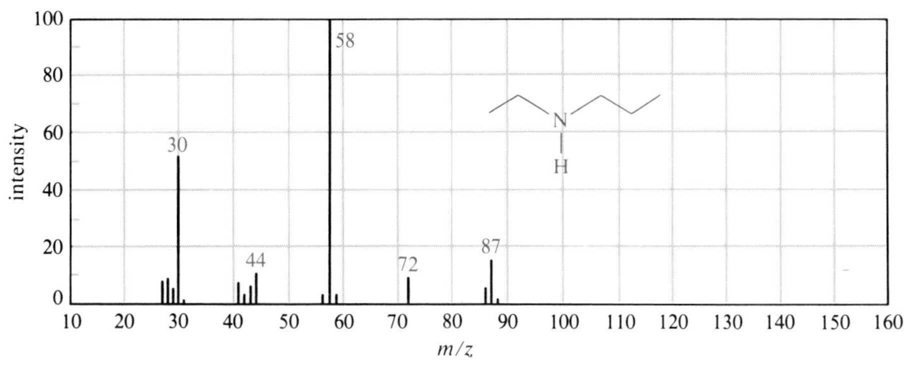

19-6 BASICITY OF AMINES

An amine is a nucleophile (a Lewis base) because its lone pair of nonbonding electrons can form a bond with an electrophile. An amine can also act as a Brønsted-Lowry base by accepting a proton from a proton acid.

Reaction of an amine as a nucleophile

$$\mathrm{R{-}\ddot{N}H_2} + \mathrm{CH_3{-}I} \longrightarrow \mathrm{R{-}\overset{+}{N}H_2{-}CH_3}\quad \mathrm{I^-}$$

nucleophile electrophile new N—C bond formed

Reaction of an amine as a proton base

$$\mathrm{R{-}\ddot{N}H_2} + \mathrm{H{-}X} \longrightarrow \mathrm{R{-}\overset{+}{N}H_2{-}H}\quad \mathrm{X^-}$$

base proton acid protonated

Because amines are fairly strong bases, their aqueous solutions are basic. An amine can abstract a proton from water, giving an ammonium ion and a hydroxide ion. The equilibrium constant for this reaction is called the **base-dissociation constant** for the amine, symbolized by K_b.

$$\mathrm{R{-}\ddot{N}H_2} + \mathrm{H{-}O{-}H} \overset{K_b}{\rightleftharpoons} \mathrm{R{-}\overset{+}{N}H_2{-}H} + {}^{-}\mathrm{OH}$$

$$K_b = \frac{[\mathrm{RNH_3^+}][^{-}\mathrm{OH}]}{[\mathrm{RNH_2}]} \qquad pK_b = -\log_{10} K_b$$

Values of K_b for most amines are fairly small (about 10^{-3} or smaller), and the equilibrium for this dissociation lies toward the left. Nevertheless, aqueous solutions of amines are distinctly basic, and they turn litmus paper blue.

Because they vary by many orders of magnitude, values of base-dissociation constants are usually listed as their negative logarithms, or pK_b values. For example, if a certain amine has $K_b = 10^{-3}$, then p$K_b = 3$. Just as we used pK_a values to indicate acid strengths (stronger acids have smaller pK_a values), we will use pK_b values to compare the relative strengths of amines as proton bases: Stronger bases have smaller values of pK_b. The values of pK_b for some representative amines are listed in Table 19-4.

TABLE 19-4
Values of pK_b for some representative amines[a]

Amine	K_b	pK_b
ammonia	1.8×10^{-5}	4.74
Primary alkyl amines		
methylamine	4.3×10^{-4}	3.36
ethylamine	4.4×10^{-4}	3.36
n-propylamine	4.7×10^{-4}	3.32
isopropylamine	4.0×10^{-4}	3.40
n-butylamine	4.8×10^{-4}	3.32
cyclohexylamine	4.7×10^{-4}	3.33
benzylamine	2.0×10^{-5}	4.67
Secondary amines		
dimethylamine	5.3×10^{-4}	3.28
diethylamine	9.8×10^{-4}	3.01
di-*n*-propylamine	10.0×10^{-4}	3.00
Tertiary amines		
trimethylamine	5.5×10^{-5}	4.26
triethylamine	5.7×10^{-4}	3.24
tri-*n*-propylamine	4.5×10^{-4}	3.35

Amine	K_b	pK_b
Aryl amines		
aniline	4.0×10^{-10}	9.40
N-methylaniline	6.1×10^{-10}	9.21
N,N-dimethylaniline	11.6×10^{-10}	8.94
p-toluidine	1.2×10^{-9}	8.92
p-fluoroaniline	4.4×10^{-10}	9.36
p-chloroaniline	1×10^{-10}	10.00
p-bromoaniline	7×10^{-11}	10.15
p-iodoaniline	6×10^{-11}	10.22
p-anisidine	2×10^{-9}	8.70
p-nitroaniline	1×10^{-13}	13.00
Heterocyclic amines		
pyrrole	1×10^{-15}	~15
pyrrolidine	1.9×10^{-3}	2.73
imidazole	8.9×10^{-8}	7.05
pyridine	1.8×10^{-9}	8.75
piperidine	1.3×10^{-3}	2.88

[a] Stronger bases have *smaller* values of pK_b.

19-7 EFFECTS ON BASICITY OF AMINES

Figure 19-7 shows an energy diagram for the reaction of an amine with water. On the left are the reactants, the free amine and water. On the right are the products, the ammonium ion and hydroxide ion.

Any structural feature that acts to stabilize the ammonium ion (relative to the free amine) shifts the reaction toward the right, making the amine a stronger

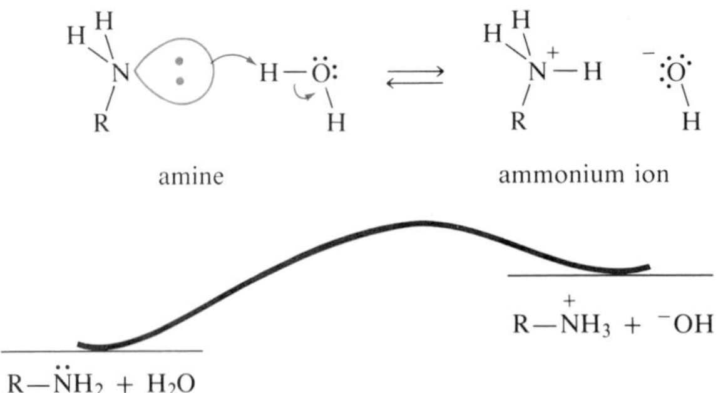

FIGURE 19-7 Potential-energy diagram of the base-dissociation reaction of an amine.

base. Any feature that acts to stabilize the free amine (relative to the ammonium ion) shifts the reaction toward the left, making the amine a weaker base.

Substitution by alkyl groups As an example, consider the relative basicities of ammonia and methylamine. Alkyl groups are electron donating toward cations, and methylamine has a methyl group to help stabilize the positive charge on nitrogen. This stabilization lowers the potential energy of the methylammonium cation, making methylamine a stronger base than ammonia. The simple alkylamines are generally stronger bases than ammonia.

$$H{-}\ddot{N}H_2 \;+\; H_2O \;\rightleftharpoons\; H{-}\overset{+}{N}H_3 \;+\; {}^-OH \qquad pK_b = 4.74 \text{ (weaker base)}$$

$$H_3C{-}\ddot{N}H_2 \;+\; H_2O \;\rightleftharpoons\; H_3C{-}\overset{+}{N}H_3 \;+\; {}^-OH \qquad pK_b = 3.36 \text{ (stronger base)}$$

stabilized by the alkyl group

We might expect that secondary amines would be stronger bases than primary amines, and that tertiary amines would be the strongest bases of all. The actual situation is more complicated because of solvation effects. Because ammonium ions are charged, they are strongly solvated by water, and the energy of solvation contributes to their stability. The additional alkyl groups around the ammonium ions of secondary and tertiary amines decrease the number of water molecules that can approach closely and solvate the ions. The opposing trends of inductive stabilization and steric hindrance of solvation tend to cancel out in most cases; as a result, primary, secondary, and tertiary amines show similar ranges of basicity.

Resonance effects on basicity Aromatic amines (the anilines and their derivatives) are much weaker bases than the simple aliphatic amines (Table 19-4). This reduced basicity is due to resonance delocalization of the nonbonding electrons in the free amine. Figure 19-8 shows that stabilization of the reactant (the free

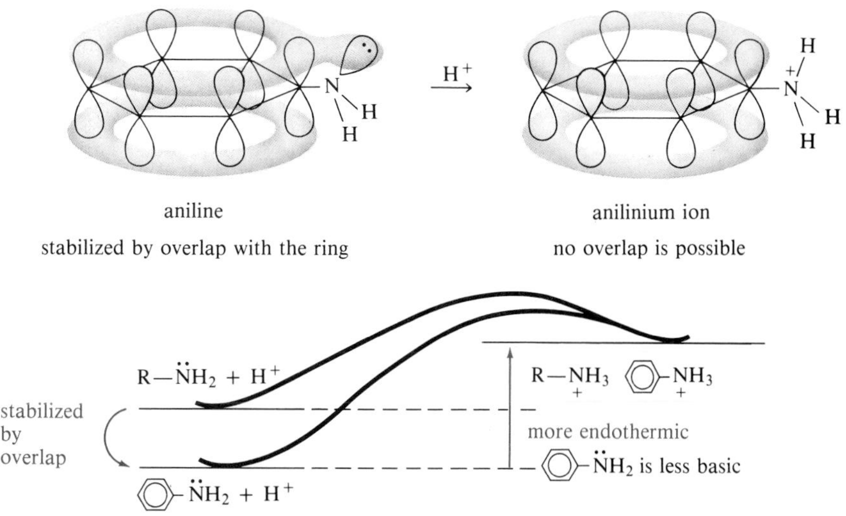

FIGURE 19-8 Aniline is stabilized by overlap of the lone pair with the aromatic ring. No such overlap is possible in the anilinium ion.

amine) makes the amine less basic. In aniline, the lone pair of nonbonding electrons on nitrogen is delocalized over the pi system of the ring. This overlap is impossible in the anilinium ion, so the reactant (aniline) is stabilized in comparison to the product. The reaction is shifted toward the left, and aniline is not as basic as the aliphatic amines.

Resonance effects also influence the basicity of pyrrole. Pyrrole is a very weak base, with a pK_b of about 15. As we saw in Chapter 15, pyrrole is aromatic because the lone pair of electrons on nitrogen is located in a *p* orbital, where they contribute to the aromatic sextet. When the pyrrole nitrogen is protonated, pyrrole loses its aromatic stabilization (Fig. 19-9). Therefore, the protonation on nitrogen is unfavorable, and pyrrole is a very weak base.

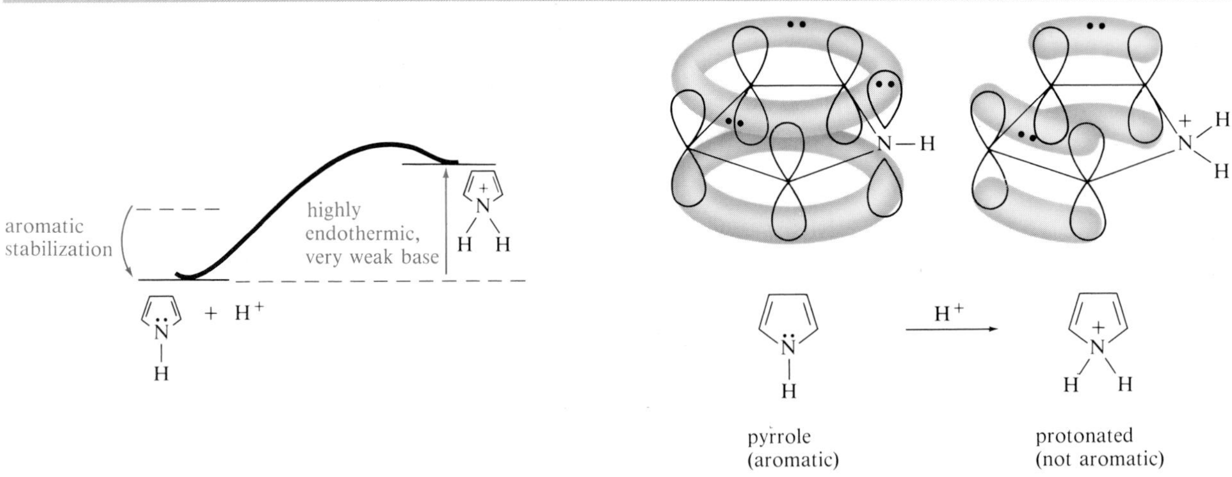

FIGURE 19-9 Pyrrole is a very weak base, because protonation destroys its aromaticity.

Hybridization effects Our study of terminal alkynes showed that electrons are held more tightly by orbitals with more *s* character. This principle helps to explain the relative basicities of unsaturated amines. For example, pyridine is a weaker base than the simple aliphatic amines. In pyridine, the nonbonding electrons occupy an sp^2 hybrid orbital, with greater *s* character and more tightly held electrons than those in the sp^3 orbital of an aliphatic amine. Pyridine's nonbonding electrons are less available for bonding to a proton. Pyridine does not lose its aromaticity upon protonation, however, and it is a much stronger base than pyrrole.

sp^2 hybridized (less basic) — sp^3 hybridized (more basic)

pyridine pK_b = 8.75 — piperidine pK_b = 2.88

The effect of increased *s* character on basicity is even more pronounced in nitriles, with *sp* hybridization. For example, acetonitrile has a pK_b of 24. This pK_b

implies that a concentrated mineral acid is required to protonate most of the acetonitrile molecules. This is a very weak base!

$$CH_3-C\equiv N\!: \quad (sp \text{ hybridized}) \qquad \text{very weakly basic}$$

PROBLEM 19-10
Rank each of the following sets of compounds in order of increasing basicity.

(a) NaOH, NH_3, CH_3NH_2, Ph—NH_2 (b) aniline, *p*-methylaniline, *p*-nitroaniline
(c) aniline, pyrrole, pyridine (d) pyrrole, imidazole, 3-nitropyrrole

19-8 SALTS OF AMINES

When an amine is protonated by a strong acid, the product is an **amine salt.** The amine salt is composed of two types of ions: the protonated amine cation (an ammonium ion) and the anion derived from the acid. Simple amine salts are named as the substituted **ammonium** salts. The salts of more complex amines are named using the names of the amine and the acid that make up the salt.

$$\underset{n\text{-propylamine}}{CH_3CH_2CH_2-NH_2} + \underset{\text{hydrochloric acid}}{HCl} \rightleftharpoons \underset{n\text{-propylammonium chloride}}{CH_3CH_2CH_2-NH_3^+ \ {}^-Cl}$$

$$\underset{\text{triethylamine}}{(CH_3CH_2)_3N\!:} + \underset{\text{sulfuric acid}}{H_2SO_4} \rightleftharpoons \underset{\text{triethylammonium hydrogen sulfate}}{(CH_3CH_2)_3NH^+ \ HSO_4^-}$$

$$\underset{\text{pyridine}}{C_5H_5N\!:} + \underset{\text{acetic acid}}{H-O-C(=O)-CH_3} \rightleftharpoons \underset{\text{pyridinium acetate}}{C_5H_5\overset{+}{N}-H \ {}^-O-C(=O)-CH_3}$$

Amine salts are ionic, high-melting, nonvolatile solids. They are much more soluble in water than the parent amines, and they are only slightly soluble in nonpolar organic solvents.

The formation of amine salts can be used to isolate and characterize amines. Most amines containing more than six carbon atoms are relatively insoluble in water. In dilute aqueous acid these amines form their corresponding ammonium salts and they dissolve. The formation of a soluble salt is one of the characteristic functional group tests for amines.

$$\underset{\substack{\text{“free” amine}\\ \text{(water insoluble)}}}{R_3N\!:} + \text{aqueous HCl} \rightleftharpoons \underset{\substack{\text{amine salt}\\ \text{(water soluble)}}}{R_3\overset{+}{N}H \ {}^-Cl}$$

The formation of amine salts is also used to separate amines from less basic compounds (Fig. 19-10). The amine forms a salt and dissolves in dilute acid. When the acidic solution of the amine salt is made alkaline by the addition of NaOH, the amine is regenerated. The purified free amine either precipitates out of the aqueous solution or is extracted into an organic solvent.

Many drugs and other biologically important amines are commonly stored and used as their salts. The amine salts are much less prone to decomposition by oxidation and other reactions, and they have virtually no fishy odor. The salts

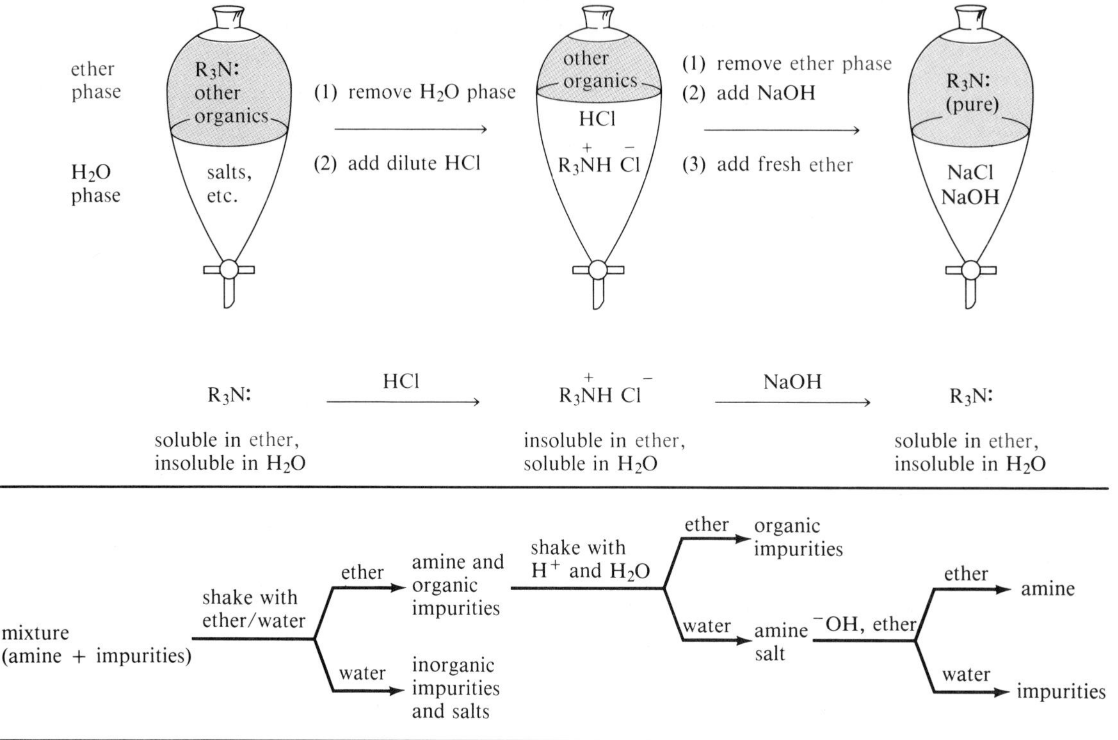

FIGURE 19-10 The basicity of an amine can be used for purification. The amine is initially more soluble in ether than in water. Addition of dilute HCl converts it to the water-soluble hydrochloride salt. Neutralization with NaOH regenerates the free amine.

are soluble in water, and they are easily converted to solutions for syrups and injectables.

As an example, ephedrine is a drug that is widely used in cold and allergy medications. Ephedrine melts at 79°C, has an unpleasant fishy odor, and is easily air oxidized to a number of undesirable products. Ephedrine hydrochloride melts at 217°C, does not oxidize easily, and has virtually no odor. Obviously, the hydrochloride salt is preferable for use in compounding medications.

$$C_6H_5-\underset{OH}{CH}-\underset{CH_3}{CH}-\underset{CH_3}{\ddot{N}H} \;+\; HCl \;\longrightarrow\; C_6H_5-\underset{OH}{CH}-\underset{CH_3}{CH}-\underset{CH_3}{\overset{+}{N}H_2}\ Cl^-$$

ephedrine
m.p. 79°C, foul-smelling
easily air oxidized

ephedrine hydrochloride
m.p. 217°C, no odor
stable

19-9 AMINE SALTS AS PHASE-TRANSFER CATALYSTS

Quaternary ammonium salts ($R_4N^+\ ^-X$) are especially valuable because they are somewhat soluble in both water and nonpolar organic solvents. They are used as **phase-transfer catalysts** to move ionic nucleophiles and bases into organic solvents. A phase-transfer catalyst facilitates reactions in which one of the reactants is insoluble in aqueous solutions and another is insoluble in organic solutions. The cation of the phase-transfer salt forms an ion pair with an anion, and the large alkyl groups in the ammonium ion lend solubility in the organic phase. Once in the organic phase, the ion pair reacts with the water-insoluble reagent.

Figure 19-11 shows an example of a reaction that is promoted by a phase-transfer catalyst. The quaternary ammonium ion forms an ion pair with hydroxide ion, allowing hydroxide to migrate into the organic phase (a solution of cyclohexene in chloroform). In the organic phase hydroxide ion is more reactive than in the aqueous phase, because it is stripped of its solvating water molecules. Hydroxide ion reacts with chloroform to give dichlorocarbene, which reacts with cyclohexene to give the cyclopropanated product.

Overall reaction

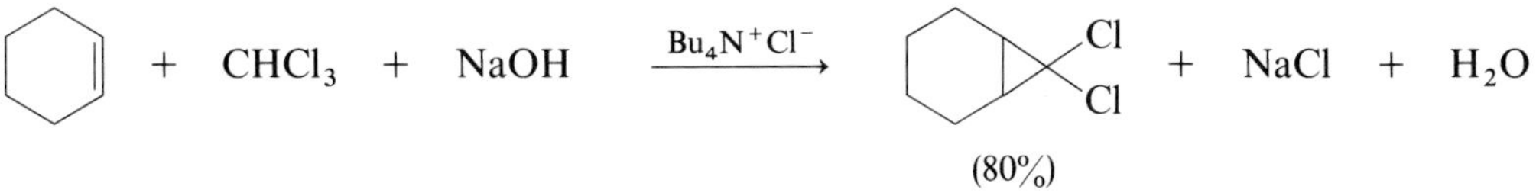

$$\text{cyclohexene} + CHCl_3 + NaOH \xrightarrow{Bu_4N^+Cl^-} \text{7,7-dichlorobicyclo[4.1.0]heptane (80\%)} + NaCl + H_2O$$

Mechanism

1. Aqueous phase

$$\underset{\text{tetrabutylammonium chloride}}{Bu_4N^+\ ^-Cl} + Na^+\ ^-OH \rightleftharpoons \underset{\text{(ion pair)}}{Bu_4N^+\ ^-OH} + Na^+\ ^-Cl$$

2. Organic phase

$$\underset{\text{organic soluble}}{Bu_4N^+\ ^-OH} + CHCl_3 \rightleftharpoons Bu_4N^+\ ^-CCl_3 + H_2O$$

$$Bu_4N^+\ ^-CCl_3 \rightleftharpoons \underset{\text{catalyst regenerated}}{Bu_4N^+\ ^-Cl} + \underset{\text{dichlorocarbene}}{:CCl_2}$$

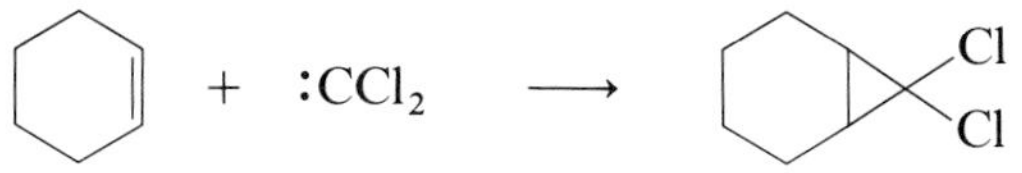

$$\text{cyclohexene} + :CCl_2 \longrightarrow \text{7,7-dichlorobicyclo[4.1.0]heptane}$$

FIGURE 19-11 Use of a phase-transfer catalyst. This example involves the reaction of cyclohexene and chloroform, both insoluble in water, with a 50% aqueous solution of sodium hydroxide.

Other anions may be transferred into organic phases by the use of a tetraalkylammonium phase-transfer catalyst. For example, sodium cyanide (NaCN) is not soluble in most organic solvents, but the cyanide ion (^-CN) can be used as a nucleophile in organic solvents under phase-transfer conditions, as shown below. Like the hydroxide ion, cyanide is actually a stronger nucleophile in the organic phase because it is not solvated by water molecules.

$$Bu_4N^+\ ^-Cl + \underset{\text{organic insoluble}}{Na^+\ ^-CN} \rightleftharpoons \underset{\text{ion pair}}{Bu_4N^+\ ^-CN} + Na^+\ ^-Cl$$

$$\underset{\text{organic soluble}}{Bu_4N^+\ ^-CN} + R{-}CH_2{-}Cl \longrightarrow \underset{\text{a nitrile}}{R{-}CH_2{-}C{\equiv}N} + Bu_4N^+\ ^-Cl$$

19-10 REACTIONS OF AMINES WITH KETONES AND ALDEHYDES

In earlier chapters we have seen three important classes of amine reactions: (1) acid-base reactions, (2) reactions with ketones and aldehydes to form imines, and (3) electrophilic aromatic substitution of anilines. In each of these reactions, the nonbonding electrons on the nitrogen atom of the amine are donated to form a new bond with an electrophile. We will see this pattern in most of the amine reactions we study.

In Chapter 18 we saw that amines attack ketones and aldehydes. When this nucleophilic attack is followed by dehydration, an imine (Schiff base) results. The analogous reaction of a hydrazine derivative gives a hydrazone, and the reaction with hydroxylamine gives an oxime. In Section 19-19 we learn to use these reactions for the synthesis of amines.

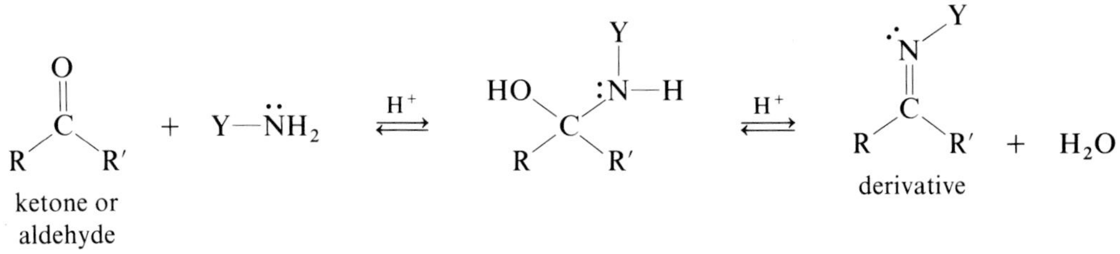

$$R-C(=O)-R' + Y-\ddot{N}H_2 \overset{H^+}{\rightleftharpoons} R-C(OH)(R')-\ddot{N}H-Y \overset{H^+}{\rightleftharpoons} R-C(R')=\ddot{N}-Y + H_2O$$

ketone or aldehyde — derivative

Y = H or alkyl	gives an imine (Schiff base)
Y = OH	gives an oxime
Y = NHR	gives a hydrazone

19-11 AROMATIC SUBSTITUTION OF ARYL AND HETEROCYCLIC AMINES

19-11A ELECTROPHILIC SUBSTITUTION OF ARYLAMINES

In an arylamine, the nonbonding electrons on the nitrogen atom help to stabilize intermediates resulting from electrophilic attack at the positions ortho or para to the amine substituent. As a result, amino groups are strong activating groups and ortho, para directors. Figure 19-12 shows the structures of the sigma complexes involved in ortho and para substitution of aniline.

The following reactions show the halogenation of aniline derivatives, which occurs readily without a catalyst. If an excess of the reagent is used, all the un-

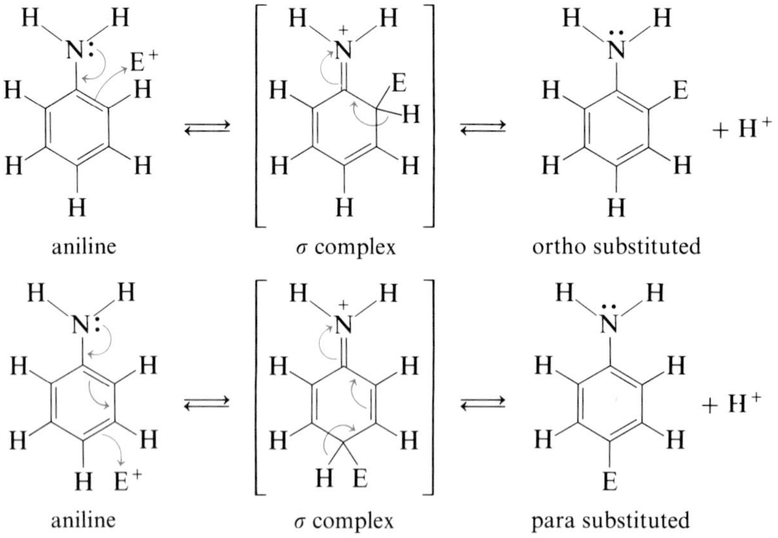

aniline $\rightleftharpoons$ [σ complex] $\rightleftharpoons$ ortho substituted $+ H^+$

aniline $\rightleftharpoons$ [σ complex] $\rightleftharpoons$ para substituted $+ H^+$

FIGURE 19-12 The amino group is a strong activating group and an ortho, para director. The nonbonding electrons on nitrogen stabilize the σ complex when attack occurs at the ortho or para position.

substituted positions ortho and para to the amino group become substituted.

:NH_2 (aniline) $\xrightarrow{\text{excess } Br_2}$ 2,4,6-tribromoaniline + 3 HBr

aniline — 2,4,6-tribromoaniline

:NH_2 with NO_2 (*o*-nitroaniline) $\xrightarrow{\text{excess } Cl_2}$ 4,6-dichloro-2-nitroaniline + 2 HCl

o-nitroaniline — 4,6-dichloro-2-nitroaniline

Care must be exercised in reactions with aniline derivatives, however. Strongly acidic reagents protonate the amino group, giving an ammonium salt that bears a full positive charge. The $—NH_3^+$ group is strongly deactivating (and meta allowing). Therefore, strongly acidic reagents are unsuitable for substitution of aniline derivatives. Oxidizing acids (such as nitric and sulfuric acids) may oxidize the amino group, leading to decomposition and occasional violent reactions.

:NH_2 $\xrightarrow{\text{strong acid}}$ $\overset{+}{N}H_3$

activated — deactivated

:NH_2 $\xrightarrow[H_2SO_4 \text{ (conc.)}]{HNO_3 \text{ (conc.)}}$ oxidation of the $—NH_2$ group (may burn or explode)

19-11B ELECTROPHILIC AROMATIC SUBSTITUTION OF PYRROLE

Pyrrole undergoes electrophilic aromatic substitution more readily than benzene, and mild reagents and conditions are sufficient for the reactions to take place. These reactions normally occur at the 2-position rather than the 3-position, because the intermediate for substitution at the 2-position is better stabilized. Figure 19-13 shows the mechanism for electrophilic substitution of pyrrole.

Halogenation, sulfonation, nitration, and Friedel-Crafts acylation can be carried out with pyrrole, but much milder reaction conditions must be used than with benzene to avoid polymerization of the more reactive pyrrole. In the following Friedel-Crafts acylation, acetic anhydride is used in place of acetyl chloride as a milder acylating reagent, and $SnCl_4$ is used as a milder catalyst in place of $AlCl_3$.

pyrrole + $CH_3—\overset{O}{\overset{\|}{C}}—O—\overset{O}{\overset{\|}{C}}—CH_3$ (acetic anhydride) $\xrightarrow{SnCl_4}$ 2-acetylpyrrole + $CH_3—\overset{O}{\overset{\|}{C}}—OH$

acetic anhydride — 2-acetylpyrrole

Attack at the 2-position (observed)

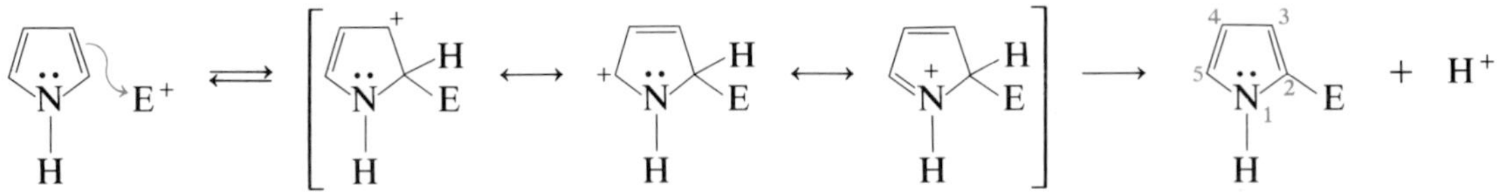

Attack at the 3-position (not observed)

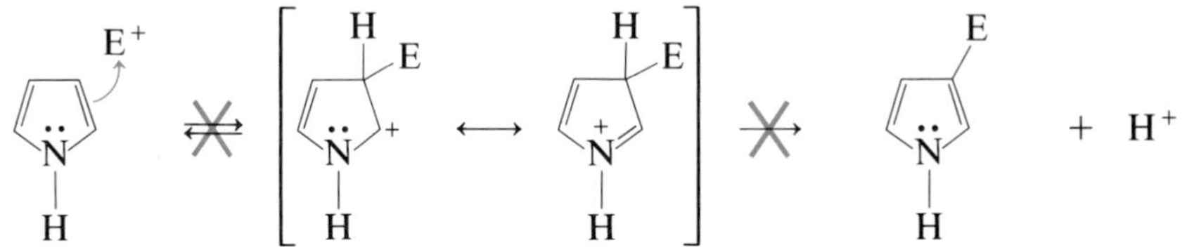

(better stabilization)

2-substituted (observed)

(not as well stabilized)

3-substituted (not observed)

FIGURE 19-13 Mechanism for electrophilic substitution of pyrrole. Electrophilic attack at the 2-position predominates, because that intermediate is better stabilized.

PROBLEM 19-11

Propose a mechanism for the acetylation of pyrrole as shown above. You may begin with pyrrole and the acylium ion, $CH_3{-}C{\equiv}O^+$.

19-11C ELECTROPHILIC AROMATIC SUBSTITUTION OF PYRIDINE

In its aromatic substitution reactions, the pyridine ring resembles a strongly deactivated benzene. Friedel-Crafts reactions fail completely, and other substitutions require unusually strong conditions. This deactivation results from the electron-withdrawing effect of the electronegative nitrogen atom. Its nonbonding electrons are perpendicular to the π system, and they cannot stabilize the positively charged intermediate. When pyridine does react, it gives substitution at the 3-position, analogous to the meta substitution shown by deactivated benzene derivatives. The following reactions compare the intermediates formed by nitration of pyridine at the 2-position and at the 3-position.

Attack at the 3-position (observed)

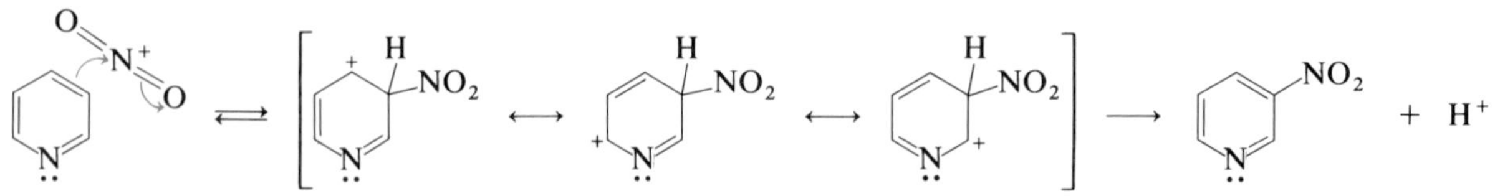

pyridine

3-nitropyridine (observed)

Attack at the 2-position (not observed)

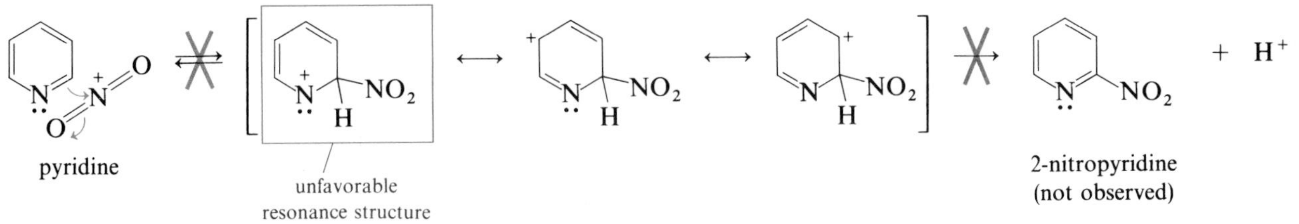

pyridine

unfavorable resonance structure

2-nitropyridine (not observed)

The electrophilic attack on pyridine at the 2-position gives an unstable intermediate with one of the resonance structures showing a positive charge and only six electrons on the nitrogen atom. The intermediate from attack at the 3-position does not have such an unfavorable resonance structure; all three resonance structures place the positive charge on the less electronegative carbon atoms.

Electrophilic substitution of pyridine is further hindered by the tendency of the nitrogen atom to attack electrophiles and take on a positive charge. The positively charged pyridinium ion is even more resistant than pyridine to electrophilic substitution.

pyridine + E^+ ⇌ pyridinium ion (less reactive)

electrophile

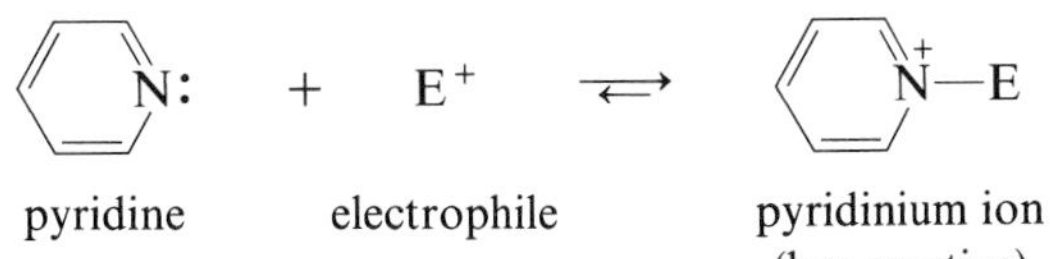

PROBLEM 19-12
Why was this deactivation by electrophiles not a serious problem with pyrrole?

PROBLEM 19-13
Give a mechanism for nitration of pyridine at the 4-position, and explain why this orientation is not observed.

Two electrophilic substitutions of pyridine are shown below. Notice that these reactions require severe conditions and the yields are only fair.

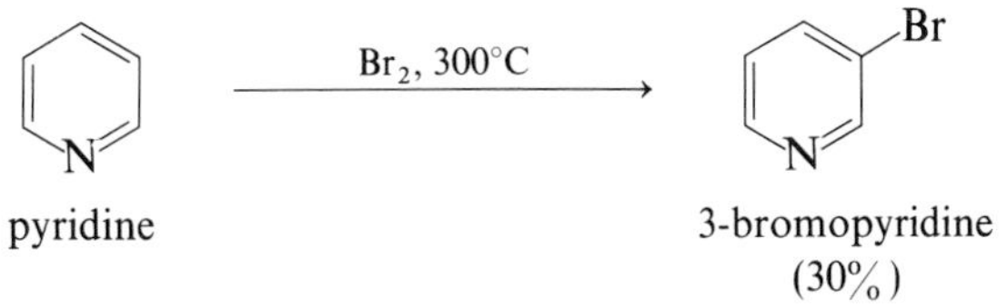

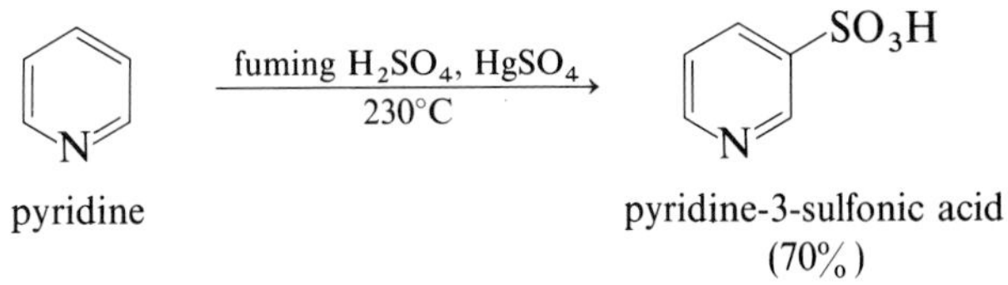

pyridine —(Br_2, 300°C)→ 3-bromopyridine (30%)

pyridine —(fuming H_2SO_4, $HgSO_4$, 230°C)→ pyridine-3-sulfonic acid (70%)

PROBLEM 19-14
Give a mechanism for the sulfonation of pyridine, pointing out why sulfonation occurs at the 3-position.

19-11D NUCLEOPHILIC AROMATIC SUBSTITUTION OF PYRIDINE

Pyridine is deactivated toward electrophilic attack, but it is activated toward attack by electron-rich nucleophiles: nucleophilic aromatic substitution. If there is a good leaving group in either the 2-position or the 4-position, a nucleophile can attack and displace the leaving group. The reaction below shows nucleophilic attack at the 2-position. The intermediate is stabilized by delocalization of the negative charge onto the electronegative nitrogen atom. This stabilization is not possible if attack occurs at the 3-position.

Nucleophilic attack at the 2-position (observed)

$^{-}OCH_3$ Cl → [OCH_3 Cl ↔ OCH_3 Cl ↔ OCH_3 Cl] → OCH_3 + Cl^-

negative charge on electronegative nitrogen

Nucleophilic attack at the 3-position (not observed)

$^{-}OCH_3$ Cl ✗ [OCH_3 Cl ↔ OCH_3 Cl ↔ OCH_3 Cl] ✗ OCH_3 + Cl^-

(no delocalization of negative charge onto N)

PROBLEM 19-15

Complete the three possible cases by showing the mechanism for the reaction of methoxide ion with 4-chloropyridine.

PROBLEM 19-16

(a) Give a mechanism for the reaction of 2-bromopyridine with sodium amide to give 2-aminopyridine.

(b) When 3-bromopyridine is used in this reaction, stronger reaction conditions are required and a mixture of 3-aminopyridine and 4-aminopyridine results. Propose a mechanism to explain this curious result.

19-12 ALKYLATION OF AMINES BY ALKYL HALIDES

Amines react with primary alkyl halides to give alkylated ammonium halides. The alkylation proceeds by the S_N2 mechanism, so it is not feasible with tertiary halides because they are too hindered. Secondary halides often give poor yields, with elimination predominating over substitution.

$$R-\ddot{N}H_2 + R'-CH_2-Br \longrightarrow R-\overset{+}{N}H_2-CH_2-R'\ {}^{-}Br$$

primary amine — primary halide — salt of a secondary amine

Unfortunately, the initially formed salt may become deprotonated. The resulting secondary amine is nucleophilic, and it can react with another molecule of the halide.

$$R-\overset{+}{N}H_2-CH_2-R'\ {}^{-}Br \quad + \quad R-\ddot{N}H_2 \quad \rightleftharpoons \quad \underset{2^\circ \text{ amine}}{R-\ddot{N}H-CH_2-R'} \quad + \quad R-\overset{+}{N}H_3\ {}^{-}Br$$

$$\underset{2^\circ \text{ amine}}{R-\ddot{N}H-CH_2-R'} \quad + \quad R'-CH_2-Br \quad \longrightarrow \quad R-\overset{+}{N}H(CH_2-R')-CH_2-R'\ {}^{-}Br$$

The difficulty of using this direct alkylation lies in stopping it at the desired stage. Even if just one equivalent of the halide is added, some amine molecules will react once, some will react twice, and some will react three times (to give the tetra-alkylammonium salt). Others will not react at all. A complex mixture results.

There are two types of reactions, however, where alkylation of amines gives good yields of the desired alkylated products:

1. *"Exhaustive" alkylation to the tetraalkylammonium salt.* Mixtures of different alkylated products are avoided if enough alkyl halide is added to alkylate the amine as many times as possible. This **exhaustive alkylation** gives a tetra-alkylammonium salt. A mild base (often $NaHCO_3$ or dilute NaOH) is added to deprotonate the intermediate alkylated amines and to neutralize the large quantities of HX formed in the reaction.

$$CH_3-CH_2-CH_2-NH_2 \quad + \quad 3\ CH_3-I \quad \xrightarrow{NaHCO_3} \quad \underset{(90\%)}{CH_3-CH_2-CH_2-N^+(CH_3)_3\ {}^{-}I}$$

PROBLEM 19-17

Give a mechanism to show the individual alkylations resulting in the formation of this quaternary ammonium salt.

2. *Reaction with a large excess of ammonia.* Because ammonia is inexpensive and has a low molecular weight, it is conveniently used in a very large excess. If a primary alkyl halide is added slowly to a large excess of ammonia, the primary amine is formed and the probability of dialkylation is small. The remaining ammonia is simply allowed to evaporate.

$$\underset{10 \text{ moles}}{\ddot{N}H_3} \quad + \quad \underset{1 \text{ mole}}{R-CH_2-X} \quad \longrightarrow \quad R-CH_2-\overset{+}{N}H_3\ {}^{-}X$$

PROBLEM 19-18

Show how you would use direct alkylation to synthesize the following compounds in good yield.

(a) benzyltrimethylammonium iodide (b) 1-pentanamine (c) benzylamine

19-13 ACYLATION OF AMINES BY ACID CHLORIDES

Primary and secondary amines react with acyl halides to form amides. This reaction is an example of *nucleophilic acyl substitution,* the replacement of a leaving group on the carbonyl carbon by a nucleophile. In this case, chloride ion is replaced by the amine.

$$R'{-}NH_2 + R{-}\overset{\overset{\large O}{\|}}{C}{-}Cl \longrightarrow R{-}\overset{\overset{\large O}{\|}}{C}{-}NH{-}R' + HCl$$

The amine attacks the carbonyl group of an acid chloride much as it attacks the carbonyl group of a ketone or aldehyde. The acid chloride is more reactive than a ketone or an aldehyde because the electronegative chlorine atom draws electron density away from the carbonyl carbon, making it more electrophilic. The chlorine atom is also a good leaving group. The tetrahedral intermediate expels a chloride ion to give the amide.

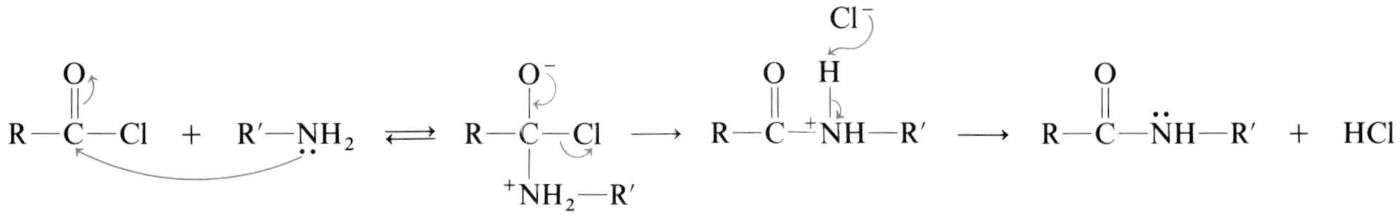

$$\underset{\text{acid chloride}}{R{-}\overset{\overset{\large O}{\|}}{C}{-}Cl} + \underset{\text{amine}}{R'{-}\ddot{N}H_2} \rightleftharpoons \underset{\text{tetrahedral intermediate}}{R{-}\overset{\overset{\large O^-}{|}}{\underset{\underset{{}^+NH_2-R'}{|}}{C}}{-}Cl} \longrightarrow R{-}\overset{\overset{\large O}{\|}}{C}{-}\overset{+}{N}H(H){-}R' \;\; (Cl^-) \longrightarrow \underset{\text{amide}}{R{-}\overset{\overset{\large O}{\|}}{C}{-}\ddot{N}H{-}R'} + HCl$$

$$C_6H_5{-}\overset{\overset{\large O}{\|}}{C}{-}Cl + CH_3{-}NH_2 \longrightarrow \underset{(95\%)}{C_6H_5{-}\overset{\overset{\large O}{\|}}{C}{-}NHCH_3} + HCl$$

There is little possibility that the amide produced in this reaction will undergo further acylation. Amides are stabilized by a resonance structure that involves the nonbonding electrons on nitrogen and places a positive charge on nitrogen. As a result, amides are much less basic and less nucleophilic than amines.

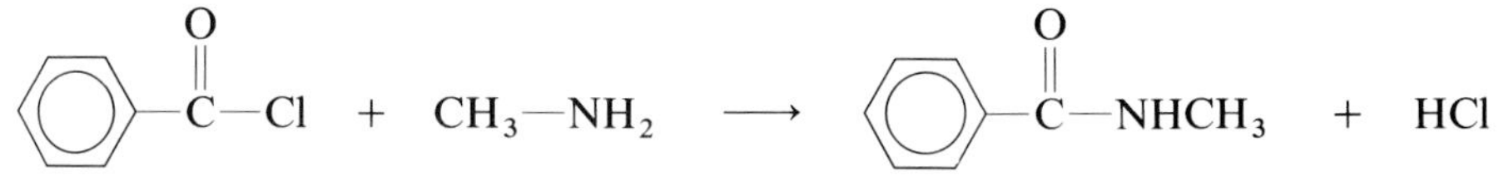

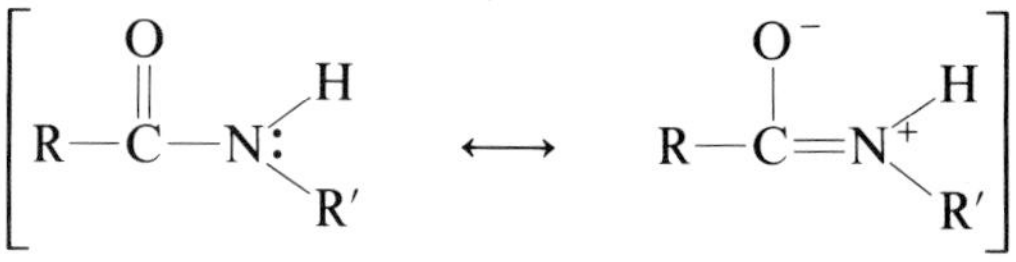

$$\left[R{-}\overset{\overset{\large O}{\|}}{C}{-}\ddot{N}(H)(R') \longleftrightarrow R{-}\overset{\overset{\large O^-}{|}}{C}{=}\overset{+}{N}(H)(R') \right]$$

The diminished basicity of amides can be used to advantage in Friedel-Crafts reactions. For example, if the amino group in aniline is acetylated to give acetanilide, the resulting amide group is still activating and ortho, para directing. Unlike aniline, however, acetanilide may be treated with acidic (and mild oxidizing) reagents as shown below. Aryl amino groups are frequently acylated before attempting further substitution on the ring, and the acyl group is removed later by acidic or basic hydrolysis (Section 21-7C).

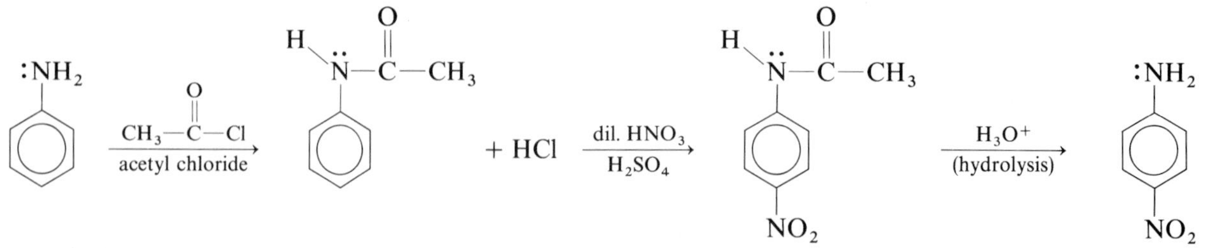

$$\underset{\text{aniline}}{C_6H_5{-}\ddot{N}H_2} \xrightarrow[\text{acetyl chloride}]{CH_3-\overset{O}{\overset{\|}{C}}-Cl} \underset{\text{acetanilide}}{C_6H_5{-}\ddot{N}H{-}\overset{\overset{\large O}{\|}}{C}{-}CH_3} + HCl \xrightarrow[H_2SO_4]{\text{dil. } HNO_3} p\text{-}O_2N{-}C_6H_4{-}\ddot{N}H{-}\overset{\overset{\large O}{\|}}{C}{-}CH_3 \xrightarrow[\text{(hydrolysis)}]{H_3O^+} \underset{p\text{-nitroaniline}}{p\text{-}O_2N{-}C_6H_4{-}\ddot{N}H_2}$$

PROBLEM 19-19

Give the products expected from the following reactions.

(a) acetyl chloride + ethylamine

(b) 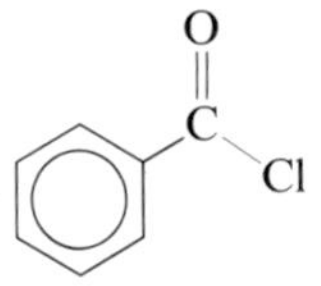 benzoyl chloride + $(CH_3)_2NH$ dimethylamine

(c) 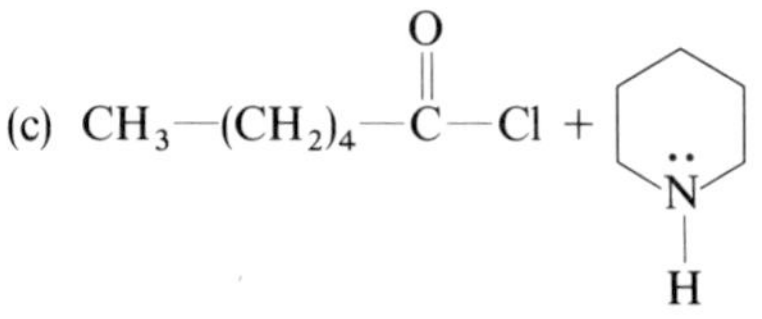 $CH_3-(CH_2)_4-C(=O)-Cl$ + piperidine

hexanoyl chloride, piperidine

19-14 REACTION OF AMINES WITH SULFONYL CHLORIDES; SULFONAMIDES

Sulfonyl chlorides are the acid chlorides of the sulfonic acids. Like acyl chlorides, sulfonyl chlorides are strongly electrophilic.

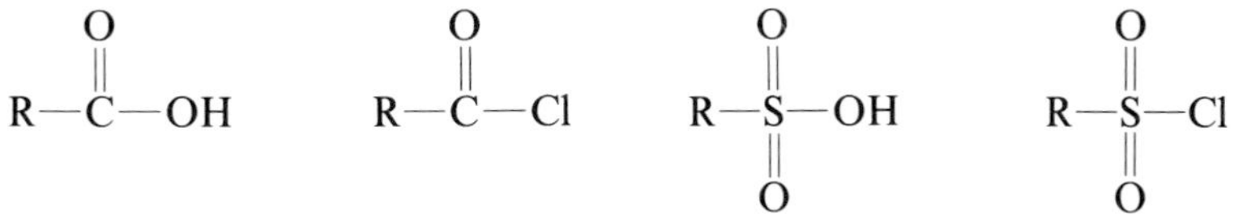

$R-C(=O)-OH$ a carboxylic acid; $R-C(=O)-Cl$ an acyl chloride (acid chloride); $R-SO_2-OH$ a sulfonic acid; $R-SO_2-Cl$ a sulfonyl chloride

A primary or secondary amine attacks a sulfonyl chloride just as it attacks an acyl chloride. Expulsion of chloride ion from the intermediate gives an amide. Amides of sulfonic acids are called **sulfonamides.** This reaction is similar to the formation of a sulfonate ester from a sulfonyl chloride (such as tosyl chloride) and an alcohol (Section 10-7).

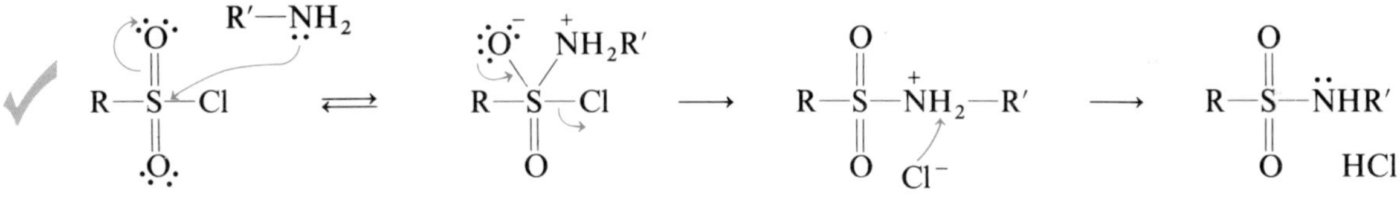

$R-SO_2-Cl$ (sulfonyl chloride) + $R'-NH_2$ ⇌ $R-S(O)(O^-)(Cl)-N^+H_2R'$ ⟶ $R-SO_2-N^+H_2-R'$ Cl^- ⟶ $R-SO_2-NHR'$ (sulfonamide) + HCl

The *sulfa drugs* are a class of sulfonamides used as antibacterial agents. In 1936, sulfanilamide was found to be effective against streptococcal infections. Sulfanilamide is synthesized from acetanilide (having the amino group protected as an amide) by chlorosulfonation followed by treatment with ammonia. The final reaction is hydrolysis of the protecting group to give sulfanilamide.

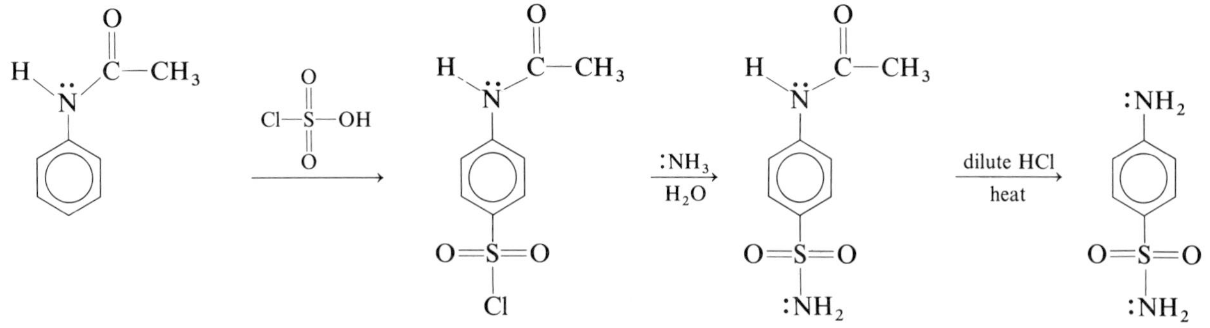

acetanilide $\xrightarrow{Cl-SO_2-OH}$ p-$CH_3C(=O)NH-C_6H_4-SO_2Cl$ $\xrightarrow{:NH_3,\ H_2O}$ p-$CH_3C(=O)NH-C_6H_4-SO_2NH_2$ $\xrightarrow{\text{dilute HCl, heat}}$ p-$H_2N-C_6H_4-SO_2NH_2$ sulfanilamide

PROBLEM 19-20

What would happen in the synthesis of sulfanilamide if the amino group were not protected as an amide in the chlorosulfonation step?

The biological activity of sulfanilamide has been studied in detail. It appears that sulfanilamide is an analog of *p*-aminobenzoic acid. Streptococci use *p*-aminobenzoic acid to synthesize folic acid, an essential compound for growth and reproduction.

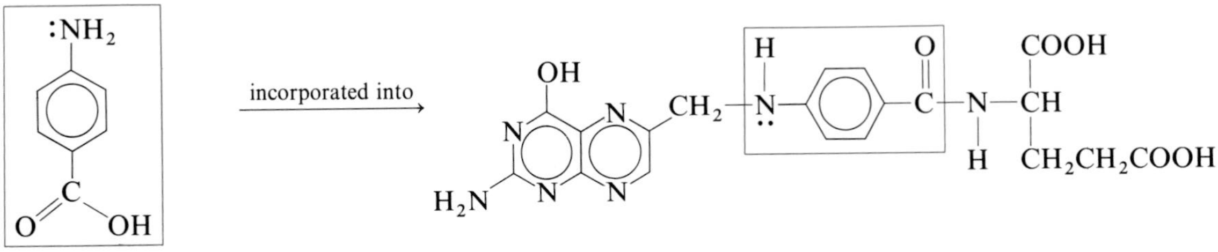

p-aminobenzoic acid

incorporated into

folic acid

Sulfanilamide cannot be used to synthesize folic acid, but the bacterial enzymes cannot distinguish between sulfanilamide and *p*-aminobenzoic acid. The production of active folic acid is inhibited, and the organism stops growing. Sulfanilamide does not kill the bacteria, but it inhibits their growth and reproduction, allowing the body's own defense mechanisms to destroy the infection.

PROBLEM 19-21

Show how you would use the same sulfonyl chloride as used in the sulfanilamide synthesis to make sulfathiazole, sulfapyridine, and sulfadiazine.

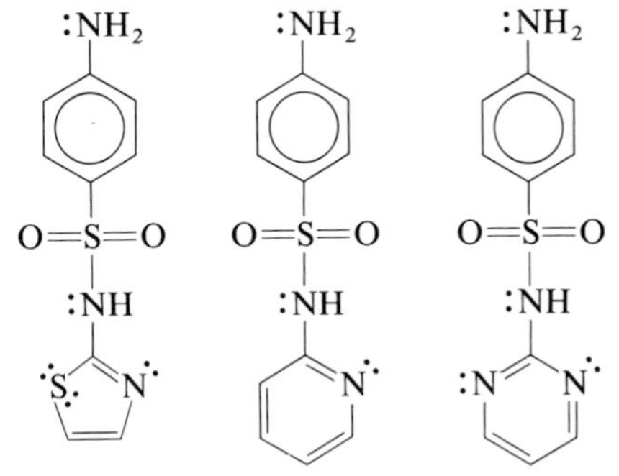

sulfathiazole sulfapyridine sulfadiazine

19-15 AMINES AS LEAVING GROUPS: THE HOFMANN ELIMINATION

An amino group (or alkylamino group) is not a good leaving group, because it would leave as the $^-NH_2$ group (or ^-NHR group), a very strong base. An amino group can be converted to a good leaving group, however, by exhaustive methylation: conversion to a quaternary ammonium salt that can leave as a neutral amine. Exhaustive methylation is usually accomplished using methyl iodide.

Exhaustive methylation of an amine

$$\underset{\text{poor leaving group}}{R{-}NH_2} + 3\,CH_3{-}I \longrightarrow \underset{\text{good leaving group}}{R{-}\overset{+}{N}(CH_3)_3\,{}^{-}I} + 3\,H{-}I$$

Elimination of the quaternary ammonium salt generally takes place by the E2 mechanism, and it requires a strong base. To provide the base, the quaternary ammonium iodide is converted to the hydroxide salt by treatment with silver oxide. When the quaternary ammonium hydroxide is heated, E2 elimination takes place and an alkene is formed. This elimination of a quaternary ammonium hydroxide is called the **Hofmann elimination.**

Conversion to the hydroxide salt

$$2\,R{-}\overset{+}{N}(CH_3)_3\,{}^{-}I + Ag_2O + H_2O \longrightarrow 2\,R{-}\overset{+}{N}(CH_3)_3\,{}^{-}OH + 2\,AgI\downarrow$$

Hofmann elimination

$$HO^{-}\;\; H{-}\underset{|}{\overset{|}{C}}{-}\underset{\underset{+N(CH_3)_3}{|}}{\overset{|}{C}}{-} \xrightarrow[\text{(E2)}]{\text{heat}} H{-}O{-}H \quad >C{=}C< \quad :N(CH_3)_3$$

For example, when 1-butanamine is exhaustively methylated, converted to the hydroxide salt, and heated, 1-butene results.

Hofmann elimination of 1-butanamine

Exhaustive methylation and conversion to the hydroxide salt

$$\underset{\text{amine}}{CH_3{-}CH_2{-}CH_2{-}CH_2{-}NH_2} \xrightarrow[\text{(2) } Ag_2O]{\text{(1) excess } CH_3I} \underset{\text{quaternary ammonium hydroxide}}{CH_3{-}CH_2{-}CH_2{-}CH_2{-}\overset{+}{N}(CH_3)_3\,{}^{-}OH}$$

Heating and elimination

$$\underset{\text{quaternary ammonium hydroxide}}{H\ddot{O}:^{-}\;\; CH_3{-}CH_2{-}CH_2{-}CH_2{-}\overset{+}{N}(CH_3)_3} \xrightarrow{150°C} H_2O + \underset{\text{alkene}}{CH_3{-}CH_2{-}CH{=}CH_2} + :N(CH_3)_3$$

In Chapter 7 you saw that eliminations of alkyl halides usually follow the Saytzeff rule; that is, the most highly substituted product predominates. This rule applies because the most highly substituted olefin is usually the most stable. In the Hofmann elimination, however, the product is most commonly the *least* highly substituted alkene. We often classify an elimination as giving mostly the **Saytzeff product** (the most highly substituted alkene) or the **Hofmann product** (the least highly substituted alkene).

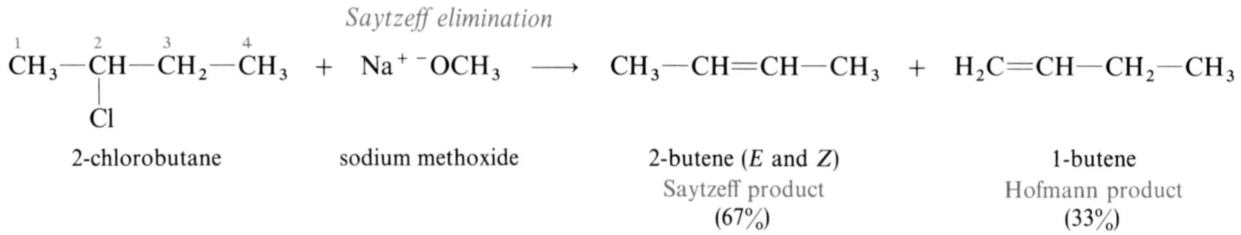

Saytzeff elimination

$$\underset{\text{2-chlorobutane}}{\overset{1}{C}H_3-\overset{2}{C}H(Cl)-\overset{3}{C}H_2-\overset{4}{C}H_3} + \underset{\text{sodium methoxide}}{Na^{+}\ {}^{-}OCH_3} \longrightarrow \underset{\substack{\text{2-butene (E and Z)}\\ \text{Saytzeff product}\\ (67\%)}}{CH_3-CH=CH-CH_3} + \underset{\substack{\text{1-butene}\\ \text{Hofmann product}\\ (33\%)}}{H_2C=CH-CH_2-CH_3}$$

Hofmann elimination

$$\overset{1}{C}H_3-\overset{2}{C}H(N^{+}(CH_3)_3)-\overset{3}{C}H_2-\overset{4}{C}H_3\ \ {}^{-}OH \xrightarrow{150^\circ C} \underset{\substack{\text{Saytzeff product}\\ (5\%)}}{CH_3-CH=CH-CH_3} + \underset{\substack{\text{Hofmann product}\\ (95\%)}}{H_2C=CH-CH_2-CH_3}$$

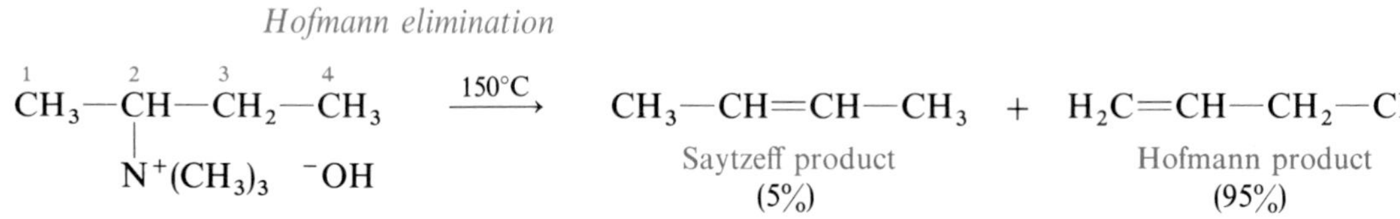

Although the Hofmann elimination's preference for the least highly substituted olefin stems from several factors, one of the most compelling involves the sheer bulk of the leaving group. Remember that the E2 mechanism usually involves an anticoplanar arrangement of the proton and the leaving group. The extremely large trialkylamine leaving group in the Hofmann elimination frequently interferes with this coplanar arrangement.

For example, consider the stereochemistry of the Hofmann elimination of 2-butanamine, shown above. The methylated ammonium salt eliminates by losing trimethylamine and a proton on either C1 or C3. The possible conformations along the C2—C3 bond are shown at the top of Figure 19-14. The conformation having a proton on C3 anticoplanar with the leaving group requires a gauche interaction between the C4 methyl group and the very bulky trimethylammonium

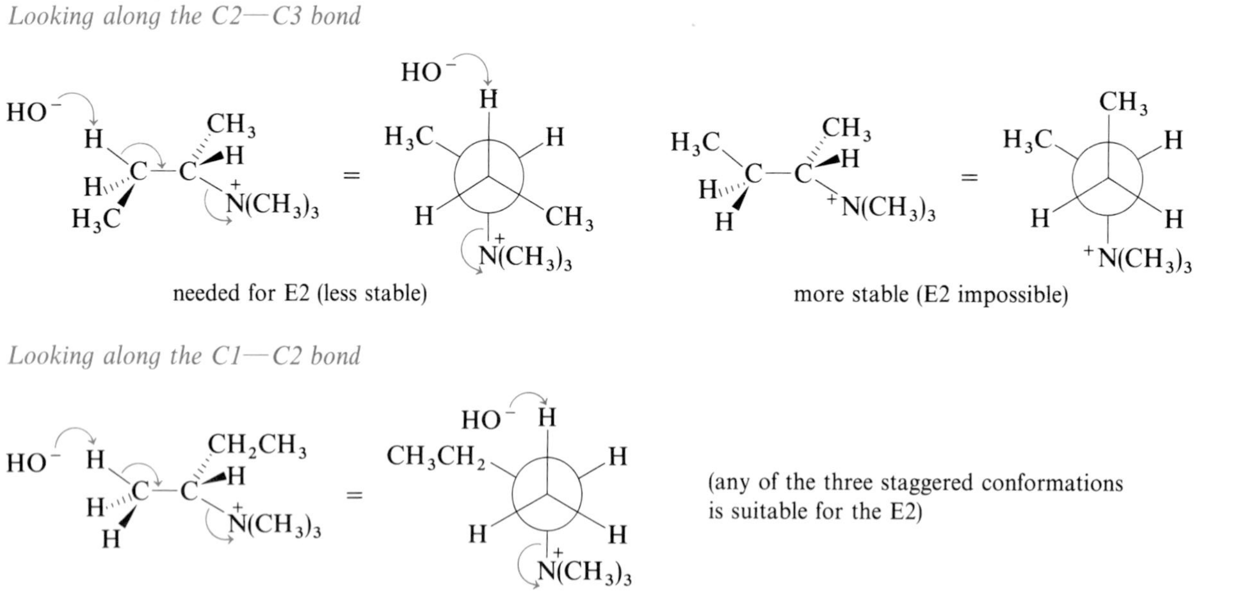

Looking along the C2—C3 bond

HO⁻ H, H, H_3C C—C CH_3, H, $\overset{+}{N}(CH_3)_3$ = Newman: HO⁻ H (top), H_3C, H, H, CH_3, $\overset{+}{N}(CH_3)_3$ (bottom)

needed for E2 (less stable)

H_3C, H, H C—C CH_3, H, $^{+}N(CH_3)_3$ = Newman: CH_3 (top), H_3C, H, H, H, $^{+}N(CH_3)_3$ (bottom)

more stable (E2 impossible)

Looking along the C1—C2 bond

HO⁻ H, H, H C—C CH_2CH_3, H, $\overset{+}{N}(CH_3)_3$ = Newman: HO⁻ H (top), CH_3CH_2, H, H, H, $\overset{+}{N}(CH_3)_3$ (bottom)

(any of the three staggered conformations is suitable for the E2)

FIGURE 19-14 Hofmann elimination of 2-butanamine. The most stable conformation of the C2—C3 bond has no proton on C3 in an anti relationship to the leaving group. Along the C1—C2 bond, however, any staggered conformation has the necessary anti relationship between a proton and the leaving group. Abstraction of a proton from C1 gives the Hofmann product.

group. The most stable conformation about the C2—C3 bond has a methyl group in the anticoplanar position, preventing elimination along the C2—C3 bond.

The bottom half of Figure 19-14 shows the conformations along the C1—C2 bond. *Any* of the three staggered conformations of the C1—C2 bond provides the necessary anti relationship between one of the protons and the leaving group. The Hofmann product predominates because elimination of one of the C1 protons involves a lower-energy, more probable transition state than the hindered transition state required for Saytzeff (C2—C3) orientation of elimination.

The Hofmann elimination is frequently used to determine the structures of complex amines by converting them into simpler amines. The direction of elimination is usually predictable, giving the least highly substituted olefin. Figure 19-15 shows two examples of the simplification of complex amines using the Hofmann elimination.

:NH_2 —(1) CH_3I, (2) Ag_2O→ $^{+}N(CH_3)_3$ ^{-}OH —heat→ + :$N(CH_3)_3$

CH_3, H —(1) CH_3I, (2) Ag_2O→ CH_3, CH_3 CH_3 ^{-}OH —heat→ N: CH_3 CH_3

FIGURE 19-15 Examples of the Hofmann elimination. The least highly substituted olefin is usually the favored product.

PROBLEM 19-22

Predict the major products formed when each of the following amines undergoes exhaustive methylation, treatment with Ag_2O, and heating.

(a) 2-hexanamine (b) 2-methylpiperidine

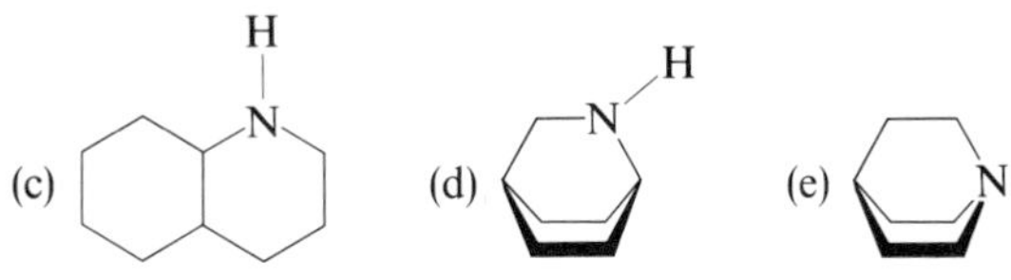

19-16 OXIDATION OF AMINES; THE COPE ELIMINATION

Amines are easily oxidized, and their oxidation is frequently a side reaction in a synthesis, or even in storage in contact with the air. Air oxidation is one of the reasons amines are commonly converted to their salts for storage or use in medications. The following are some of the oxidation states of amines and their oxidation products.

amine	ammonium salt	imine	hydroxylamine	amine oxide	nitro compound
—N̈—	—N⁺—	C=N—R	—N̈—OH	—N⁺—O⁻	—N⁺(=O)O⁻

Most amines are oxidized by common oxidizing agents such as H_2O_2 and MCPBA (*m*-chloroperoxybenzoic acid). Primary amines are very easily oxidized, but complex mixtures of different products often result. Secondary amines are easily oxidized to hydroxylamines, although several side products are also formed, and the yields are often low. The mechanisms of amine oxidations are not well characterized, partly because there are so many different reaction paths available.

$$R_2NH \;(2^\circ \text{ amine}) + H_2O_2 \longrightarrow R_2N{-}OH \;(\text{a } 2^\circ \text{ hydroxylamine}) + H_2O$$

Tertiary amines are oxidized to **amine oxides,** often in good yields. Either H_2O_2 or MCPBA may be used for this oxidation. Notice that an amine oxide must be drawn with a full positive charge on nitrogen, as in an ammonium salt. Because the N—O bond of the amine oxide is formed by donation of the electrons on nitrogen, this bond is often written as an arrow.

$$R_3N{:}\;(3^\circ \text{ amine}) + H_2O_2 \;(\text{or } Ar{-}CO_3H) \longrightarrow R_3N^{+}{\rightarrow}O^{-}\;(3^\circ \text{ amine oxide}) + H_2O \;(\text{or } Ar{-}COOH)$$

Owing to the positive charge on the nitrogen atom, the amine oxide may undergo a **Cope elimination** much like the Hofmann elimination of a quaternary ammonium salt. The amine oxide acts as its own base through a cyclic transition state, so a strong base such as hydroxide ion is not needed. The Cope elimination generally gives the same orientation as the Hofmann elimination, resulting in the least highly substituted alkene.

Cope elimination of an amine oxide

$$R{-}CH_2{-}CH(R'){-}N^{+}(CH_3)_2{-}O^{-} \longrightarrow [\text{transition state}] \longrightarrow RHC{=}CHR' + HO{-}N(CH_3)_2$$

The Cope elimination occurs under milder conditions than the Hofmann elimination. It is particularly useful when a sensitive or reactive olefin must be synthesized by the elimination of an amine. Because the Cope elimination involves a cyclic transition state, it occurs with syn stereochemistry.

PROBLEM 19-23

Give the products expected when the following tertiary amines are treated with MCPBA and heated.

(a) 2-(dimethylamino)hexanamine
(b) 2-(diethylamino)hexanamine
(c) cyclohexyldimethylamine
(d) *N*-ethylpiperidine

PROBLEM 19-24

When the (*R*, *R*) isomer of this amine is treated with an excess of methyl iodide, then silver oxide, and then heated, the product is primarily of the (*E*) configuration. When the same amine is treated with MCPBA and heated, the product has primarily the (*Z*) configuration. Use stereochemical drawings of the transition states to explain these observations.

$$(CH_3)_2\ddot{N}-CH(CH_3)-CH(CH_3)-CH(CH_3)_2 \xrightarrow[(2)\ Ag_2O,\ heat]{(1)\ excess\ CH_3I} \text{H}(H_3C)C{=}C(CH_3)(CH(CH_3)_2)\ (E)$$

$$\xrightarrow[(2)\ heat]{(1)\ MCPBA} \text{H}(H_3C)C{=}C(CH_3)(CH(CH_3)_2)\ (Z)$$

19-17 REACTIONS OF AMINES WITH NITROUS ACID

The reactions of amines with nitrous acid (H—O—N=O) are particularly useful. Because nitrous acid is unstable, it is generated in situ (in the reaction mixture) by mixing sodium nitrite ($NaNO_2$) with cold, dilute hydrochloric acid.

$$\underset{\text{sodium nitrite}}{Na^+\ {}^-O-N{=}O} + H^+Cl^- \rightleftharpoons \underset{\text{nitrous acid}}{H-O-N{=}O} + Na^+Cl^-$$

In an acidic solution, nitrous acid may protonate and lose water to give the nitrosonium ion, $^+N{=}O$. The nitrosonium ion appears to be the reactive intermediate in most reactions of amines with nitrous acid.

$$\underset{\text{nitrous acid}}{H-O-N{=}O} + H^+ \rightleftharpoons \underset{\text{protonated nitrous acid}}{H-\overset{H}{\overset{|}{O^+}}-N{=}O} \rightleftharpoons H_2O + \underset{\text{nitrosonium ion}}{[{}^+\ddot{N}{=}\ddot{O}: \longleftrightarrow :N{\equiv}\overset{+}{O}:]}$$

Each of the three classes of amines undergoes its own particular reactions with the nitrosonium ion. We will consider the three classes individually.

Reaction with primary amines: formation of diazonium salts Primary amines react with nitrous acid, via the nitrosonium ion, to give diazonium cations of the form $R-\overset{+}{N}{\equiv}N$. This procedure is called **diazotization** of an amine. Diazonium salts are the most useful products obtained from the reactions of amines with nitrous acid. The mechanism for diazonium salt formation begins with a nucleophilic attack on the nitrosonium ion to form an *N*-nitrosoamine.

$$\underset{\text{primary amine}}{R-\ddot{N}H_2} + \underset{\text{nitrosonium ion}}{{}^+N{=}O} \rightleftharpoons R-\overset{H}{\overset{|}{N^+}}(H)-N{=}O \xrightarrow{H_2\ddot{O}:} \underset{N\text{-nitrosoamine}}{R-\underset{H}{\underset{|}{N}}-N{=}O} + H_3O^+$$

Next, a proton transfer from a nitrogen atom to the oxygen forms a hydroxyl group and a second N—N bond.

$$\underset{N\text{-nitrosoamine}}{R-\underset{..}{\overset{H}{\overset{|}{N}}}-\ddot{N}{=}\ddot{O}:} + H_3O^+ \rightleftharpoons \underset{\text{protonated }N\text{-nitrosoamine}}{\left[R-\overset{H}{\overset{|}{N}}-N{=}O^+-H \longleftrightarrow R-\overset{H}{\overset{|}{N^+}}{=}N-OH\right]} + H_2\ddot{O}: \rightleftharpoons \underset{\text{second N—N bond formed}}{R-N{=}N-OH} + H_3O^+$$

Protonation of the hydroxyl group, followed by loss of water, gives the diazonium cation.

$$\mathrm{R{-}\ddot{N}{=}\ddot{N}{-}\ddot{\underset{..}{O}}H} \xrightleftharpoons{H_3O^+} \mathrm{R{-}\ddot{N}{=}\ddot{N}{-}\overset{+}{O}H_2} \longrightarrow \underset{\text{diazonium cation}}{\mathrm{R{-}\overset{+}{N}{\equiv}N{:}}} + H_2O$$

The overall diazotization reaction is

$$\underset{\text{primary amine}}{\mathrm{R{-}\ddot{N}H_2}} + \underset{\text{sodium nitrite}}{NaNO_2} + 2\,HCl \longrightarrow \underset{\text{diazonium salt}}{\mathrm{R{-}\overset{+}{N}{\equiv}N\;\;Cl^-}} + 2\,H_2O + NaCl$$

Alkanediazonium salts are unstable. They decompose rapidly to give nitrogen and carbocations.

$$\underset{\text{alkanediazonium cation}}{\mathrm{R{-}\overset{+}{N}{\equiv}N{:}}} \longrightarrow \underset{\text{carbocation}}{R^+} + \underset{\text{nitrogen}}{\mathrm{{:}N{\equiv}N{:}}}$$

The driving force for this reaction is the formation of N_2, an exceptionally stable molecule. The carbocations generated in this manner react like other cations we have seen: by nucleophilic attack to give substitution, by proton abstraction to give elimination; and by rearrangement. Because of the many competing reaction pathways, alkanediazonium salts usually decompose to give complex mixtures of products. Therefore, the diazotization of primary alkylamines is not widely used for synthesis.

Arenediazonium salts (formed from arylamines) are relatively stable, however, and they serve as intermediates in a variety of important synthetic reactions. The reactions of arenediazonium salts are discussed in Section 19-18.

Reaction with secondary amines: formation of N-nitrosoamines Secondary amines react with the nitrosonium ion to form secondary ***N*-nitrosoamines.**

$$\underset{\text{secondary amine}}{\mathrm{R_2\ddot{N}H}} + \underset{\text{nitrosonium ion}}{\mathrm{{}^{+}N{=}O}} \rightleftharpoons \mathrm{R{-}\overset{H}{\overset{|}{N^+}}{-}N{=}O} \xrightarrow{H_2\ddot{O}{:}} \underset{\text{secondary }N\text{-nitrosoamine}}{\mathrm{R{-}\underset{R}{\underset{|}{\ddot{N}}}{-}\ddot{N}{=}\ddot{O}{:}}} + H_3O^+$$

Secondary *N*-nitrosoamines are stable under the reaction conditions, because they do not have the N—H proton necessary to undergo the second stage of the reaction (shown above with a primary amine) to form a diazonium cation. The secondary *N*-nitrosoamine usually separates from the reaction mixture as an oily liquid.

Small quantities of *N*-nitrosoamines have been shown to cause cancer in laboratory animals. These findings have generated concern about the common practice of using sodium nitrite to preserve meats such as bacon, ham, and hot dogs. When the meat is eaten, sodium nitrite combines with stomach acid to form nitrous acid, which can convert amines in the food to *N*-nitrosoamines. Because nitrites are naturally present in many other foods, it is unclear just how much additional risk is involved in using sodium nitrite to preserve meats. More research is being done in this area to evaluate the risk.

Reaction with tertiary amines Tertiary amines react with the nitrosonium ion to form *N*-nitrosoammonium salts. Because there is no acidic N—H proton in the tertiary *N*-nitrosoammonium salt, loss of a proton to give the *N*-nitrosoamine is impossible. There is simply an equilibrium set up between the amine and the *N*-nitrosoammonium salt.

$$R_3N: \; + \; {}^{+}N{=}O \;\rightleftharpoons\; R_3N^{+}{-}N{=}O$$

tertiary amine nitrosonium ion *N*-nitrosoammonium ion

The most useful reaction of amines with nitrous acid is the reaction of arylamines to form arenediazonium salts. We consider next how these diazonium salts may be used as synthetic intermediates.

PROBLEM 19-25

Predict the products from the reactions of the following amines with sodium nitrite in dilute HCl.

(a) cyclohexanamine (b) *N*-ethyl-2-hexanamine (c) piperidine
(d) *N*-ethylpiperidine (e) aniline

19-18 FORMATION AND REACTIONS OF ARENEDIAZONIUM SALTS

Arenediazonium salts are relatively stable in aqueous solutions at about 0 to 10°C. Above these temperatures they decompose, and they may explode if they are isolated and allowed to dry. The diazonium ($-\overset{+}{N}{\equiv}N$) group can be replaced by many different functional groups, including —H, —OH, —CN, and the halogens.

Arenediazonium salts are formed by diazotizing a primary aromatic amine. Primary aromatic amines are commonly prepared by the nitration of an aromatic ring, followed by reduction of the nitro group to an amino ($-NH_2$) group. In effect, by forming and diazotizing an amine we have a procedure for converting an activated aromatic position into a wide variety of functional groups. For example, toluene might be converted to a variety of substituted derivatives using the techniques described below.

$$\text{toluene} \xrightarrow[H_2SO_4]{HNO_3} p\text{-}CH_3C_6H_4NO_2 \xrightarrow[(2)\ {}^{-}OH]{(1)\ Fe,\ HCl} p\text{-}CH_3C_6H_4NH_2 \xrightarrow[HCl]{NaNO_2} p\text{-}CH_3C_6H_4N^{+}{\equiv}N\ Cl^{-} \xrightarrow{\text{various reagents}} p\text{-}CH_3C_6H_4X$$

(X = H, OH, CN, halogens)

Replacement of the diazonium group by hydroxide: hydrolysis Hydrolysis takes place when a solution of an arenediazonium salt is strongly acidified (usually by adding H_2SO_4) and heated. The hydroxyl group of water replaces N_2, forming a phenol. This is a useful laboratory synthesis of phenols because (unlike nucleophilic aromatic substitution) it does not require strong electron-withdrawing substituents or the use of powerful bases and nucleophiles.

$$\text{Ar}-\overset{+}{\text{N}}\equiv\text{N}\ \ \text{Cl}^- \xrightarrow[\text{H}_2\text{O}]{\text{H}_2\text{SO}_4,\ \text{heat}} \text{Ar}-\text{OH} + \text{N}_2\uparrow + \text{H}^+$$

Example

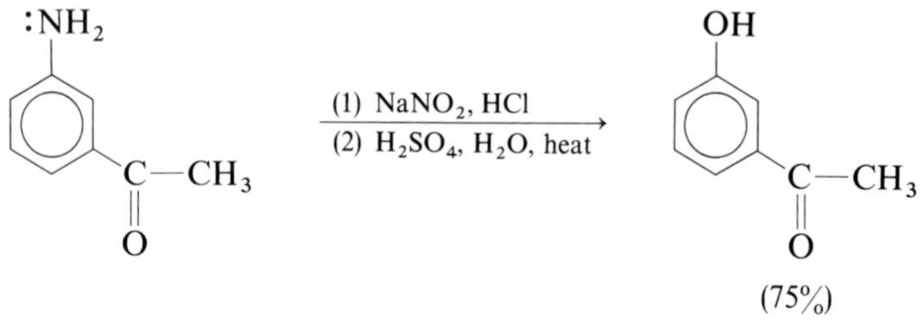

$$m\text{-}(\text{CH}_3\text{CO})\text{C}_6\text{H}_4\ddot{\text{N}}\text{H}_2 \xrightarrow[\text{(2) H}_2\text{SO}_4,\ \text{H}_2\text{O, heat}]{\text{(1) NaNO}_2,\ \text{HCl}} m\text{-}(\text{CH}_3\text{CO})\text{C}_6\text{H}_4\text{OH}\ \ (75\%)$$

Replacement of the diazonium group by chloride, bromide, and cyanide: the Sandmeyer reaction Copper(I) salts (cuprous salts) have a special affinity for diazonium salts. Cuprous chloride, cuprous bromide, and cuprous cyanide react with arenediazonium salts to give the aryl chlorides, aryl bromides, and aryl cyanides. It is often necessary to heat the reaction mixture to drive these reactions to completion. The use of cuprous salts to replace diazonium groups on arenediazonium salts is called the **Sandmeyer reaction.**

The Sandmeyer reaction

$$\text{Ar}-\overset{+}{\text{N}}\equiv\text{N}\ \ \text{Cl}^- \xrightarrow[(\text{X} = \text{Cl, Br, C}\equiv\text{N})]{\text{CuX}} \text{Ar}-\text{X} + \text{N}_2\uparrow$$

Examples

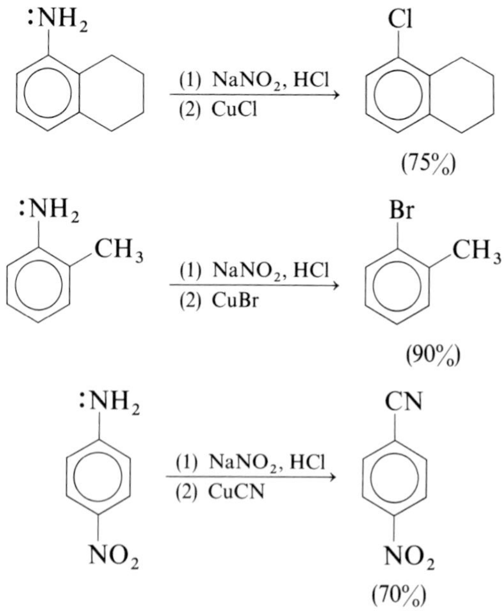

$$\text{5,6,7,8-tetrahydro-1-naphthyl-}\ddot{\text{N}}\text{H}_2 \xrightarrow[\text{(2) CuCl}]{\text{(1) NaNO}_2,\ \text{HCl}} \text{5,6,7,8-tetrahydro-1-naphthyl-Cl}\ \ (75\%)$$

$$o\text{-CH}_3\text{C}_6\text{H}_4\ddot{\text{N}}\text{H}_2 \xrightarrow[\text{(2) CuBr}]{\text{(1) NaNO}_2,\ \text{HCl}} o\text{-CH}_3\text{C}_6\text{H}_4\text{Br}\ \ (90\%)$$

$$p\text{-NO}_2\text{C}_6\text{H}_4\ddot{\text{N}}\text{H}_2 \xrightarrow[\text{(2) CuCN}]{\text{(1) NaNO}_2,\ \text{HCl}} p\text{-NO}_2\text{C}_6\text{H}_4\text{CN}\ \ (70\%)$$

Replacement of the diazonium group by fluoride and iodide When a solution of an arenediazonium salt is treated with fluoroboric acid (HBF_4), the diazonium fluoroborate salt precipitates out of solution. If this precipitated salt is filtered and then heated, it decomposes to give the aryl fluoride. Although this reaction requires the isolation and heating of a potentially explosive diazonium salt, it may be carried out safely if it is done carefully with the proper equipment. There are few other methods for making aryl fluorides.

$$\text{Ar—}\overset{+}{\text{N}}\equiv\text{N Cl}^- \xrightarrow{\text{HBF}_4} \text{Ar—}\overset{+}{\text{N}}\equiv\text{N }^-\text{BF}_4 \xrightarrow{\text{heat}} \text{Ar—F} + \text{N}_2\uparrow + \text{BF}_3$$

Example

$$\text{C}_6\text{H}_5\text{—}\ddot{\text{N}}\text{H}_2 \xrightarrow[\text{(2) HBF}_4]{\text{(1) NaNO}_2\text{, HCl}} \text{C}_6\text{H}_5\text{—N}^+\equiv\text{N }^-\text{BF}_4 \xrightarrow{\text{heat}} \text{C}_6\text{H}_5\text{—F} \quad (50\%)$$

Aryl iodides are formed from arenediazonium salts by treating them with potassium iodide. This is one of the best methods for the synthesis of iodobenzene derivatives.

$$\text{Ar—}\overset{+}{\text{N}}\equiv\text{N Cl}^- \xrightarrow{\text{KI}} \text{Ar—I} + \text{N}_2\uparrow$$

Example

$$m\text{-}(\ddot{\text{N}}\text{H}_2)\text{C}_6\text{H}_4\text{—C(=O)CH}_2\text{CH}_2\text{CH}_3 \xrightarrow[\text{(2) KI}]{\text{(1) NaNO}_2\text{, HCl}} m\text{-I—C}_6\text{H}_4\text{—C(=O)CH}_2\text{CH}_2\text{CH}_3 \quad (75\%)$$

Reduction of the diazonium group to hydrogen: deamination of anilines Hypophosphorous acid (H_3PO_2) reacts with an arenediazonium salt, replacing the diazonium group with a hydrogen. In effect, this is a reduction of the arenediazonium ion.

$$\text{Ar—}\overset{+}{\text{N}}\equiv\text{N Cl}^- \xrightarrow{\text{H}_3\text{PO}_2} \text{Ar—H} + \text{N}_2\uparrow$$

Example

$$\text{(5-amino-2-ethylbenzoic acid: } \ddot{\text{N}}\text{H}_2\text{, CH}_3\text{CH}_2\text{, C(=O)OH)} \xrightarrow[\text{(2) H}_3\text{PO}_2]{\text{(1) NaNO}_2\text{, HCl}} \text{(2-ethylbenzoic acid: CH}_3\text{CH}_2\text{, C(=O)OH)} \quad (70\%)$$

This reaction is sometimes used to remove an amino group that was added to activate the ring. For example, a direct bromination of toluene could not give 3,5-dibromotoluene because the methyl group activates the ortho and para positions.

$$\text{toluene} \xrightarrow[\text{FeBr}_3]{2\ \text{Br}_2} \text{(2,6-dibromotoluene)} + \text{(2,4-dibromotoluene)} + \text{monosubstitution and trisubstitution products}$$

toluene — mixture of ortho and para bromination

If we start with *p*-toluidine (*p*-methylaniline), however, the strongly activating amino group directs bromination to its ortho positions. Removal of the amino group (deamination) gives the desired product.

$$p\text{-toluidine} \xrightarrow[\text{FeBr}_3]{2\ \text{Br}_2} \text{(2,6-dibromo-4-methylaniline)} \xrightarrow[(2)\ H_3PO_2]{(1)\ NaNO_2,\ HCl} \text{(3,5-dibromotoluene)}$$

p-toluidine — deaminated

Diazonium salts as electrophiles: the diazo coupling reaction Arenediazonium ions act as weak electrophiles in electrophilic aromatic substitution reactions. The products have the structure Ar—N=N—Ar, containing the —N=N— **azo** linkage. For this reason, the products are called azo compounds, and the reaction is called the **diazo coupling** reaction. Because they are weak electrophiles, diazonium salts react only with strongly activated rings.

$$\text{Ar}-\overset{+}{\text{N}}\equiv\text{N} + \text{H}-\text{Ar}' \longrightarrow \text{Ar}-\text{N}=\text{N}-\text{Ar}' + \text{H}^+$$

diazonium ion — (activated) — an azo compound

Example

$$^{-}\text{O}-\text{SO}_2-\text{C}_6\text{H}_4-\overset{+}{\text{N}}\equiv\text{N}:\ \text{Cl}^- + \text{C}_6\text{H}_5-\ddot{\text{N}}(\text{CH}_3)_2 \longrightarrow {}^{-}\text{O}-\text{SO}_2-\text{C}_6\text{H}_4-\ddot{\text{N}}=\ddot{\text{N}}-\text{C}_6\text{H}_4-\ddot{\text{N}}(\text{CH}_3)_2 + \text{HCl}$$

methyl orange (an indicator)

PROBLEM 19-26

Propose a mechanism for this synthesis of methyl orange.

Most azo compounds are strongly colored, and they make excellent dyes known as **azo dyes.** The diazo coupling syntheses of some common azo dyes are shown on the next page.

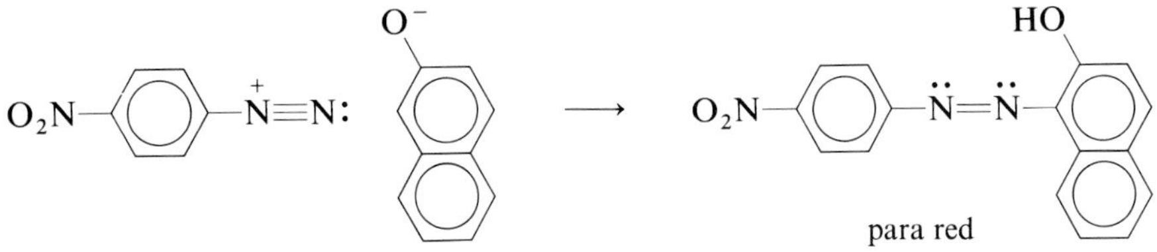

$$O_2N-C_6H_4-\overset{+}{N}\equiv N: + \text{(2-naphthoxide, } O^-\text{)} \longrightarrow O_2N-C_6H_4-\ddot{N}=\ddot{N}-C_{10}H_6-OH$$

para red

$$O_2N-C_6H_4-\overset{+}{N}\equiv N: + {}^-O-C_6H_4-COO^- \longrightarrow O_2N-C_6H_4-\ddot{N}=\ddot{N}-C_6H_3(OH)-COO^-$$

alizarin yellow

Such diazo coupling reactions often take place in basic solutions, because deprotonation of the phenolic —OH groups and the sulfonic acid and carboxylic acid groups helps to activate the aromatic rings toward electrophilic aromatic substitution. Many of the common azo dyes have one or more sulfonate ($-SO_3^-$) or carboxylate ($-COO^-$) groups on the molecule to promote solubility in water and to help to bind the dye to the polar surfaces of common fibers such as cotton and wool.

PROBLEM 19-27

Show how you would convert aniline to the following compounds.

(a) fluorobenzene (b) chlorobenzene (c) 1,3,5-trimethylbenzene
(d) bromobenzene (e) iodobenzene (f) benzonitrile

(g) phenol (h) 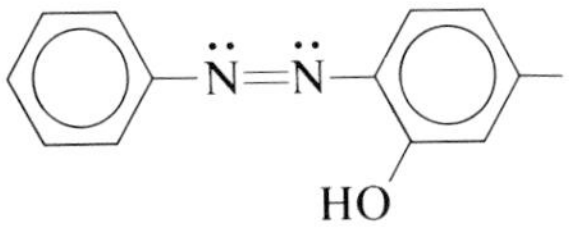 $C_6H_5-\ddot{N}=\ddot{N}-C_6H_3(OH)-OH$ (HO ortho to the azo group) (use aniline and resorcinol)

SUMMARY OF THE REACTIONS OF AMINES

1. Reaction as a proton base (Section 19-6)

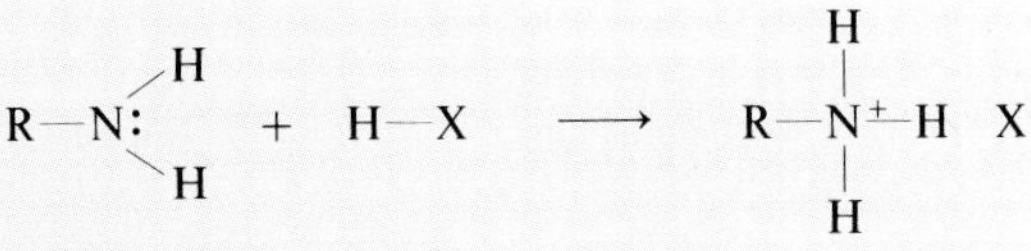

$$R-\ddot{N}H_2 + H-X \longrightarrow R-\overset{+}{N}H_3\ X^-$$

base proton acid ammonium salt

2. Reactions with ketones and aldehydes (Sections 18-16 and 18-17)

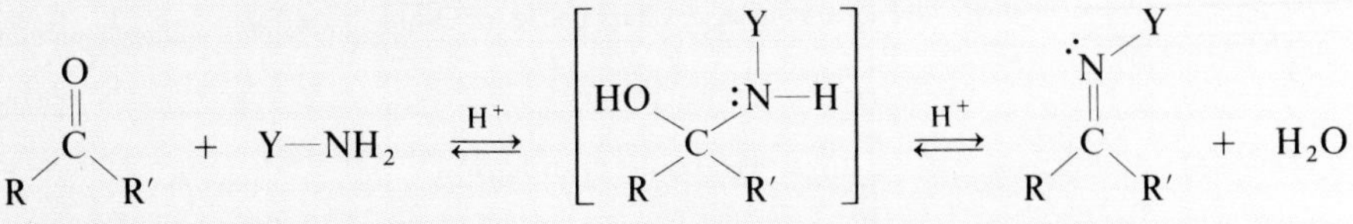

$$R-C(=O)-R' + Y-NH_2 \underset{}{\overset{H^+}{\rightleftharpoons}} \left[\begin{matrix} HO \quad :NH-Y \\ R-C-R' \end{matrix} \right] \overset{H^+}{\rightleftharpoons} R-C(=\ddot{N}-Y)-R' + H_2O$$

ketone or aldehyde derivative

Y = H or alkyl gives an imine (Schiff base)

Y = OH gives an oxime

Y = NHR gives a hydrazone

3. *Alkylation* (Section 19-12)

$$\underset{\text{primary amine}}{R{-}NH_2} + \underset{\text{primary halide}}{R'{-}CH_2{-}Br} \longrightarrow \underset{\text{salt of a secondary amine}}{R{-}\overset{+}{N}H_2{-}CH_2{-}R'\ {}^{-}Br}$$

(Overalkylation is common.)

Examples

$$CH_3{-}CH_2{-}CH_2{-}NH_2 + 3\,CH_3{-}I \xrightarrow{NaHCO_3} CH_3{-}CH_2{-}CH_2{-}\overset{+}{N}(CH_3)_3\ {}^{-}I$$

$$\text{excess } NH_3 + CH_3CH_2CH_2CH_2CH_2{-}Br \longrightarrow CH_3CH_2CH_2CH_2CH_2{-}NH_2$$

4. *Acylation to form amides* (Section 19-13)

$$\underset{\text{acid chloride}}{R{-}\overset{O}{\overset{\|}{C}}{-}Cl} + \underset{\text{amine}}{R'{-}NH_2} \longrightarrow \underset{\text{amide}}{R{-}\overset{O}{\overset{\|}{C}}{-}NH{-}R'} + HCl$$

Example

$$\underset{\text{aniline}}{H_2N{-}C_6H_5} + \underset{\text{acetyl chloride}}{CH_3{-}\overset{O}{\overset{\|}{C}}{-}Cl} \longrightarrow \underset{\text{acetanilide}}{CH_3{-}\overset{O}{\overset{\|}{C}}{-}NH{-}C_6H_5} + HCl$$

5. *Reaction with sulfonyl chlorides to give sulfonamides* (Section 19-14)

$$\underset{\text{amine}}{R{-}NH_2} + \underset{\text{sulfonyl chloride}}{Cl{-}\overset{O}{\overset{\|}{\underset{\underset{O}{\|}}{S}}}{-}R'} \longrightarrow \underset{\text{sulfonamide}}{R{-}NH{-}\overset{O}{\overset{\|}{\underset{\underset{O}{\|}}{S}}}{-}R'} + HCl$$

Example

$$\underset{\text{1-butanamine}}{CH_3(CH_2)_3{-}NH_2} + \underset{\text{benzenesulfonyl chloride}}{Cl{-}\overset{O}{\overset{\|}{\underset{\underset{O}{\|}}{S}}}{-}Ph} \longrightarrow \underset{N\text{-butyl benzenesulfonamide}}{CH_3(CH_2)_3NH{-}\overset{O}{\overset{\|}{\underset{\underset{O}{\|}}{S}}}{-}Ph} + HCl$$

6. *Hofmann elimination* (Section 19-15)
   Conversion to quaternary ammonium hydroxide

$$R{-}CH_2{-}CH_2{-}NH_2 \xrightarrow{3\,CH_3I} R{-}CH_2{-}CH_2{-}\overset{+}{N}(CH_3)_3\ {}^{-}I$$

$$\xrightarrow{Ag_2O} R{-}CH_2{-}CH_2{-}\overset{+}{N}(CH_3)_3\ {}^{-}OH$$

a. Hofmann elimination

$$HO^{-}\ \ R{-}\underset{H}{\overset{H}{C}}{-}\underset{{}^{+}N(CH_3)_3}{\overset{H}{C}}{-}H \xrightarrow{\text{heat}} H{-}O{-}H \quad \underset{H}{\overset{R}{}}C{=}C\underset{H}{\overset{H}{}} \quad :N(CH_3)_3$$

Hofmann elimination usually gives the least highly substituted alkene.

Example

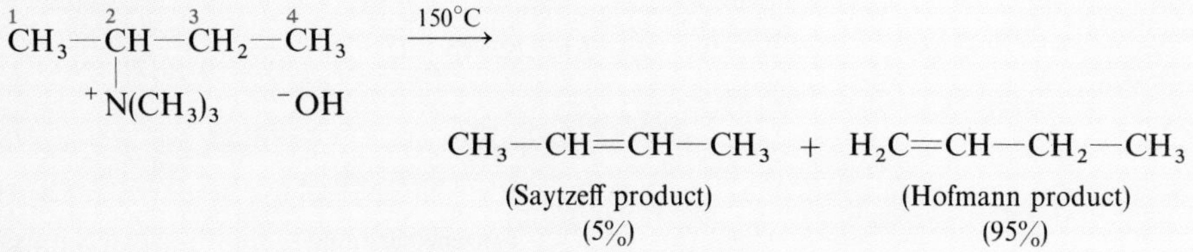

$$\overset{1}{CH_3}-\overset{2}{CH}(\overset{+}{N}(CH_3)_3)-\overset{3}{CH_2}-\overset{4}{CH_3} \quad {}^{-}OH \xrightarrow{150°C} \underset{\substack{\text{(Saytzeff product)}\\(5\%)}}{CH_3-CH{=}CH-CH_3} + \underset{\substack{\text{(Hofmann product)}\\(95\%)}}{H_2C{=}CH-CH_2-CH_3}$$

b. Cope elimination of a tertiary amine oxide (Section 19-16)

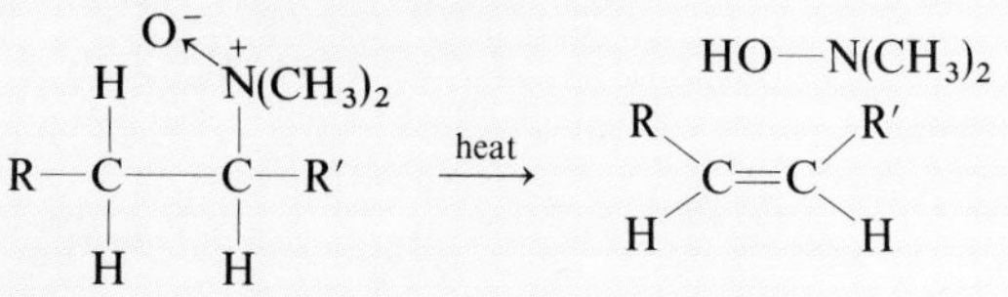

$$R-CH(H)-C(H)(\overset{+}{N}(CH_3)_2-O^-)-R' \xrightarrow{heat} (R)(H)C{=}C(R')(H) + HO-N(CH_3)_2$$

7. *Oxidation* (Section 19-16)

a. Secondary amines

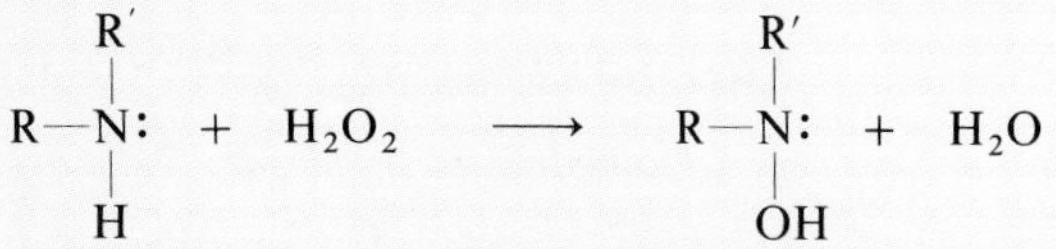

$$\underset{\text{2° amine}}{R-\ddot{N}(R')-H} + H_2O_2 \longrightarrow \underset{\text{a 2° hydroxylamine}}{R-\ddot{N}(R')-OH} + H_2O$$

b. Tertiary amines

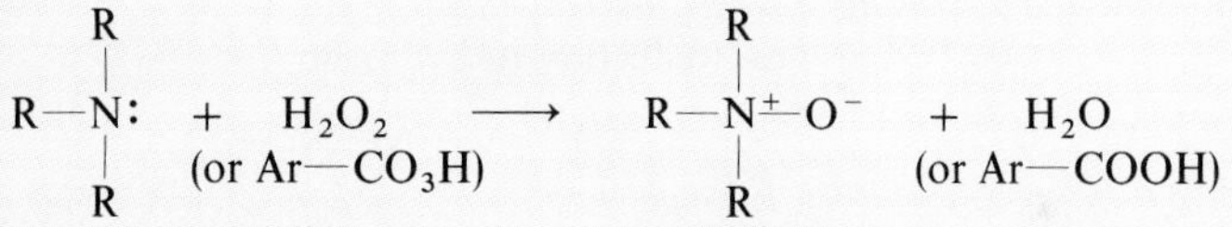

$$\underset{\text{3° amine}}{R-\ddot{N}(R)-R} + \underset{(\text{or } Ar-CO_3H)}{H_2O_2} \longrightarrow \underset{\text{3° amine oxide}}{R-\overset{+}{N}(R)(R)-O^-} + \underset{(\text{or } Ar-COOH)}{H_2O}$$

8. *Diazotization* (Section 19-17)

$$\underset{\text{primary alkylamine}}{R-NH_2} \xrightarrow{NaNO_2,\ HCl} \underset{\text{alkanediazonium salt}}{R-\overset{+}{N}{\equiv}N\ \ Cl^-}$$

$$\underset{\text{primary arylamine}}{Ar-NH_2} \xrightarrow{NaNO_2,\ HCl} \underset{\text{arenediazonium salt}}{Ar-\overset{+}{N}{\equiv}N\ \ Cl^-}$$

a. Reactions of diazonium salts (Section 19-18)

i. Hydrolysis

$$Ar-\overset{+}{N}{\equiv}N\ \ Cl^- \xrightarrow[H_2O]{H_2SO_4,\ heat} Ar-OH + N_2\uparrow + H^+$$

Example

$$\underset{\text{benzenediazonium chloride}}{Ph-\overset{+}{N}{\equiv}N\ \ Cl^-} \xrightarrow[H_2O]{H_2SO_4,\ heat} \underset{\text{phenol}}{Ph-OH} + N_2\uparrow + H^+$$

ii. The Sandmeyer reaction

$$\mathrm{Ar{-}\overset{+}{N}{\equiv}N\ \ Cl^-} \xrightarrow[(X = Cl,\ Br,\ C{\equiv}N)]{CuX} \mathrm{Ar{-}X + N_2\uparrow}$$

Examples

$$\mathrm{Ph{-}\overset{+}{N}{\equiv}N\ \ Cl^-} \xrightarrow{CuCl} \mathrm{Ph{-}Cl + N_2\uparrow}$$

benzenediazonium chloride → chlorobenzene

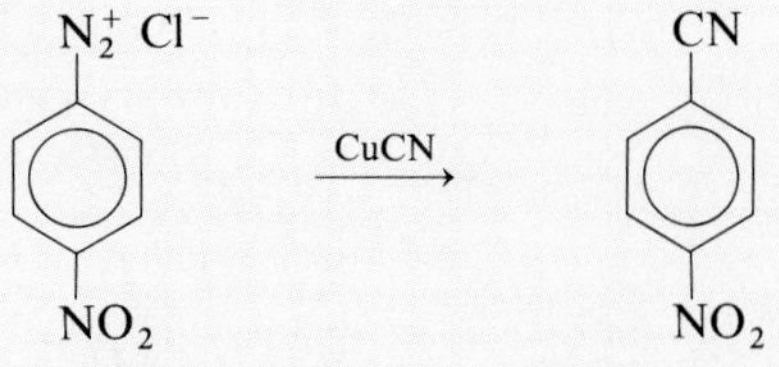

$$p\text{-}\mathrm{O_2N{-}C_6H_4{-}N_2^+\ Cl^-} \xrightarrow{CuCN} p\text{-}\mathrm{O_2N{-}C_6H_4{-}CN}$$

p-nitrobenzenediazonium chloride → *p*-nitrobenzonitrile

iii. Replacement by fluoride or iodide

$$\mathrm{Ar{-}\overset{+}{N}{\equiv}N\ \ Cl^-} \xrightarrow{HBF_4} \mathrm{Ar{-}\overset{+}{N}{\equiv}N\ {}^-BF_4} \xrightarrow{heat} \mathrm{Ar{-}F + N_2\uparrow + BF_3}$$

$$\mathrm{Ar{-}\overset{+}{N}{\equiv}N\ \ Cl^-} \xrightarrow{KI} \mathrm{Ar{-}I + N_2\uparrow + KCl}$$

Example

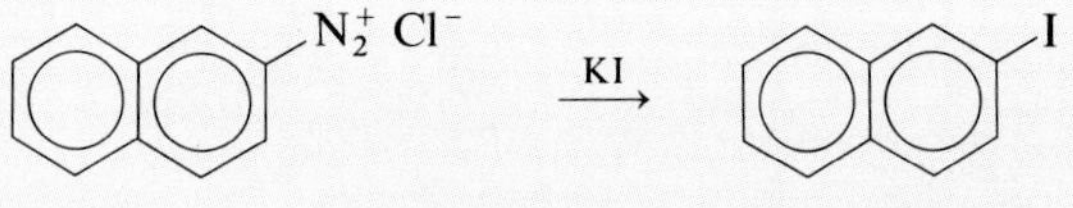

$$\text{2-}\mathrm{C_{10}H_7{-}N_2^+\ Cl^-} \xrightarrow{KI} \text{2-}\mathrm{C_{10}H_7{-}I}$$

2-naphthylenediazonium chloride → 2-iodonaphthalene

iv. Reduction to hydrogen

$$\mathrm{Ar{-}\overset{+}{N}{\equiv}N\ \ Cl^-} \xrightarrow{H_3PO_2} \mathrm{Ar{-}H + N_2\uparrow}$$

Example

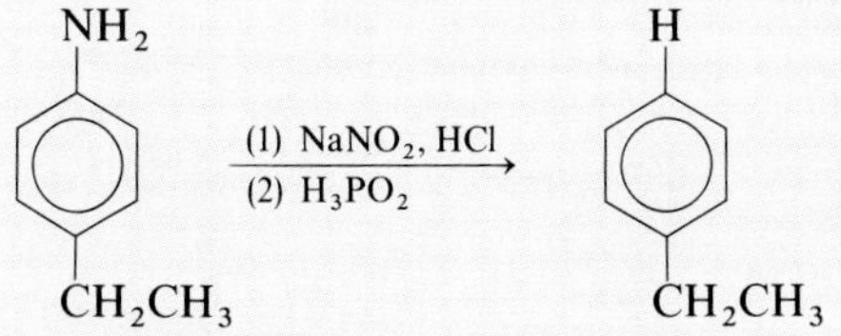

$$p\text{-}\mathrm{CH_3CH_2{-}C_6H_4{-}NH_2} \xrightarrow[(2)\ H_3PO_2]{(1)\ NaNO_2,\ HCl} \mathrm{CH_3CH_2{-}C_6H_4{-}H}$$

p-ethylaniline → ethylbenzene

v. Diazo coupling

$$\underset{\text{diazonium ion}}{\mathrm{Ar{-}\overset{+}{N}{\equiv}N}} + \underset{\text{(activated)}}{\mathrm{H{-}Ar'}} \longrightarrow \underset{\text{an azo compound}}{\mathrm{Ar{-}N{=}N{-}Ar'}} + \mathrm{H^+}$$

Example

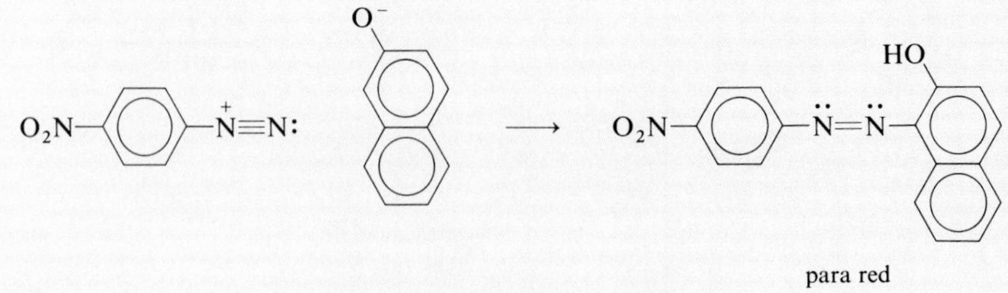

$$\mathrm{O_2N{-}C_6H_4{-}\overset{+}{N}{\equiv}N{:}} + \text{2-naphthoxide } (\mathrm{O^-}) \longrightarrow \mathrm{O_2N{-}C_6H_4{-}\ddot{N}{=}\ddot{N}{-}C_{10}H_6{-}OH}$$

para red

19-19 SYNTHESIS OF AMINES BY REDUCTIVE AMINATION

Many methods are available for the synthesis of amines. Several of these methods are derived from the reactions of amines covered in the preceding sections. We begin with two general methods of amine synthesis, and then consider several methods that are applicable to specific types of amines.

Reductive amination, the most general method for synthesizing amines, involves the reduction of an imine or oxime derivative of a ketone or aldehyde. The imine or oxime is reduced by lithium aluminum hydride ($LiAlH_4$) or by catalytic hydrogenation. In effect, the reductive amination process adds one alkyl group to the nitrogen atom. The product can be a primary, secondary, or tertiary amine, depending on whether the amine used as the starting material had zero, one, or two alkyl groups.

Primary amines Primary amines result from condensation of hydroxylamine (zero alkyl groups) with a ketone or an aldehyde, followed by reduction of the oxime. This is a convenient reaction, because most oximes are stable, easily isolated compounds. The oxime is reduced using catalytic reduction, lithium aluminum hydride, or sodium cyanoborohydride ($NaBH_3CN$).

$$\underset{\text{ketone or aldehyde}}{R-\overset{\overset{\large O}{\|}}{C}-R'} \xrightarrow[H^+]{H_2N-OH} \underset{\text{oxime}}{R-\overset{\overset{\large N-OH}{\|}}{C}-R'} \xrightarrow{\text{reduction}} \underset{\text{primary amine}}{R-\overset{\overset{\large NH_2}{|}}{CH}-R'}$$

Examples

$$\underset{\text{2-pentanone}}{CH_3-CH_2-CH_2-\overset{\overset{O}{\|}}{C}-CH_3} \xrightarrow[H^+]{H_2\ddot{N}-OH} \underset{\text{2-pentanone oxime}}{CH_3-CH_2-CH_2-\overset{\overset{\dot{\cdot}N-OH}{\|}}{C}-CH_3} \xrightarrow[Ni]{H_2} \underset{\text{2-pentanamine}}{CH_3-CH_2-CH_2-\overset{\overset{\ddot{N}H_2}{|}}{CH}-CH_3}$$

$$\underset{\text{benzaldehyde}}{C_6H_5-\overset{\overset{O}{\|}}{C}-H} \xrightarrow[H^+]{H_2\ddot{N}-OH} \underset{\text{benzaldehyde oxime}}{C_6H_5-\overset{\overset{\dot{\cdot}N-OH}{\|}}{C}-H} \xrightarrow{LiAlH_4} \underset{\text{benzylamine}}{C_6H_5-\overset{\overset{\ddot{N}H_2}{|}}{CH_2}}$$

Secondary amines Condensation of a ketone or an aldehyde with a primary amine forms an *N*-substituted imine (a Schiff base). Reduction of the *N*-substituted imine gives a secondary amine.

$$\underset{\text{ketone or aldehyde}}{R-\overset{\overset{\large O}{\|}}{C}-R'} \xrightarrow[H^+]{R''-\ddot{N}H_2} \underset{N\text{-substituted imine}}{R-\overset{\overset{\dot{\cdot}N-R''}{\|}}{C}-R'} \xrightarrow{\text{reduction}} \underset{\text{secondary amine}}{R-\overset{\overset{\ddot{N}HR''}{|}}{CH}-R'}$$

Example

$$\underset{\text{acetone}}{CH_3-\overset{\overset{O}{\|}}{C}-CH_3} \xrightarrow[H^+]{Ph-\ddot{N}H_2} CH_3-\overset{\overset{\dot{\cdot}N-Ph}{\|}}{C}-CH_3 \xrightarrow{LiAlH_4} \underset{\text{phenylisopropylamine (75\%)}}{CH_3-\overset{\overset{\ddot{N}HPh}{|}}{CH}-CH_3}$$

Tertiary amines Condensation of a ketone or an aldehyde with a secondary amine gives an iminium salt. Iminium salts are frequently unstable, so they are rarely isolated. A reducing agent in the solution reduces the iminium salt to a tertiary amine. The reducing agent must be capable of reducing the iminium salt, but it must not reduce the carbonyl group of the ketone or aldehyde. Sodium cyanoborohydride ($NaBH_3CN$) works well for this reduction, because it is less reactive than sodium borohydride, and it does not reduce the carbonyl group.

$$\underset{\text{ketone or aldehyde}}{R'-\overset{O}{\overset{\|}{C}}-R''} \xrightarrow[H^+]{R-\ddot{N}H-R} \underset{\text{iminium salt}}{R'-\overset{R-\overset{+}{N}-R}{\overset{\|}{C}}-R''} \xrightarrow{NaBH_3CN} \underset{\text{tertiary amine}}{R'-\overset{R-\ddot{N}-R}{\overset{|}{C}H}-R''}$$

Example

$$\underset{\text{cyclohexanone}}{C_6H_{10}{=}O} \xrightarrow[H^+]{H\ddot{N}(CH_3)_2} \underset{\text{iminium salt}}{\left[C_6H_{10}{=}\overset{+}{N}(CH_3)_2\right]} \xrightarrow{NaBH_3CN} \underset{\substack{N,N\text{-dimethylcyclohexylamine} \\ (85\%)}}{C_6H_{11}-\ddot{N}(CH_3)_2}$$

PROBLEM 19-28

Show how the following amines may be synthesized from the indicated starting materials by reductive amination.

(a) benzylmethylamine from benzaldehyde

(b) $\underset{(\pm)\text{-amphetamine}}{Ph-CH_2-\overset{:NH_2}{\overset{|}{C}H}-CH_3}$ from $\underset{\text{1-phenyl-2-propanone}}{Ph-CH_2-\overset{O}{\overset{\|}{C}}-CH_3}$

(c) *N*-benzylpiperidine from piperidine

(d) *N*-cyclohexylaniline from cyclohexanone

(e) cyclohexylamine from cyclohexanone

(f) N-cyclopentylpiperidine (piperidine ring N—cyclopentyl) from compounds containing no more than five carbon atoms

19-20 SYNTHESIS OF AMINES BY ACYLATION-REDUCTION

The second general synthesis of amines is by **acylation-reduction.** Like reductive amination, acylation-reduction adds one alkyl group to the nitrogen atom of the starting amine. Acylation of the starting amine by an acid chloride gives an amide, with no tendency toward overacylation (Section 19-13). Reduction of the amide by lithium aluminum hydride ($LiAlH_4$) gives the corresponding amine.

$$\underset{\text{amine}}{R-NH_2} + \underset{\text{acid chloride}}{R'-\overset{O}{\overset{\|}{C}}-Cl} \xrightarrow{\text{acylation}} \underset{\substack{\text{amide} \\ \text{(acylated amine)}}}{R'-\overset{O}{\overset{\|}{C}}-NH-R} \xrightarrow[LiAlH_4]{\text{reduction}} \underset{\text{alkylated amine}}{R'-CH_2-NH-R}$$

Acylation-reduction converts ammonia to a primary amine, a primary amine to a secondary amine, or a secondary amine to a tertiary amine. These reactions are quite general, with one restriction: The added alkyl group is always primary, because the carbon bonded to nitrogen is derived from the carbonyl group of the amide, reduced to a methylene (—CH_2—) group.

Primary amines

$$\underset{\text{acid chloride}}{R-\overset{\overset{\displaystyle O}{\|}}{C}-Cl} + \underset{\text{ammonia}}{NH_3} \longrightarrow \underset{\text{primary amide}}{R-\overset{\overset{\displaystyle O}{\|}}{C}-NH_2} \xrightarrow{LiAlH_4} \underset{\text{primary amine}}{R-CH_2-NH_2}$$

Example

$$\underset{\text{3-methylbutanoyl chloride}}{CH_3-\overset{\overset{\displaystyle CH_3}{|}}{CH}-CH_2-\overset{\overset{\displaystyle O}{\|}}{C}-Cl} + NH_3 \longrightarrow \underset{\text{3-methylbutanamide}}{CH_3-\overset{\overset{\displaystyle CH_3}{|}}{CH}-CH_2-\overset{\overset{\displaystyle O}{\|}}{C}-NH_2} \xrightarrow{LiAlH_4} \underset{\text{3-methyl-1-butanamine}}{CH_3-\overset{\overset{\displaystyle CH_3}{|}}{CH}-CH_2-CH_2-NH_2}$$

Secondary amines

$$\underset{\text{acid chloride}}{R-\overset{\overset{\displaystyle O}{\|}}{C}-Cl} + \underset{\text{primary amine}}{R'-NH_2} \longrightarrow \underset{N\text{-substituted amide}}{R-\overset{\overset{\displaystyle O}{\|}}{C}-NH-R'} \xrightarrow{LiAlH_4} \underset{\text{secondary amine}}{R-CH_2-NH-R'}$$

Example

$$\underset{\text{butanoyl chloride}}{CH_3CH_2CH_2-\overset{\overset{\displaystyle O}{\|}}{C}-Cl} + \underset{\text{aniline}}{C_6H_5NH_2} \longrightarrow \underset{N\text{-phenylbutanamide}}{CH_3CH_2CH_2-\overset{\overset{\displaystyle O}{\|}}{C}-NH-C_6H_5} \xrightarrow{LiAlH_4} \underset{N\text{-butylaniline}}{CH_3CH_2CH_2-CH_2-NH-C_6H_5}$$

Tertiary amines

$$\underset{\text{acid chloride}}{R-\overset{\overset{\displaystyle O}{\|}}{C}-Cl} + \underset{\text{secondary amine}}{R'_2NH} \longrightarrow \underset{N,N\text{-disubstituted amide}}{R-\overset{\overset{\displaystyle O}{\|}}{C}-NR'_2} \xrightarrow{LiAlH_4} \underset{\text{tertiary amine}}{R-CH_2-NR'_2}$$

Example

$$\underset{\text{benzoyl chloride}}{C_6H_5-\overset{\overset{\displaystyle O}{\|}}{C}-Cl} + \underset{\text{diethylamine}}{H-\ddot{N}(CH_2CH_3)_2} \longrightarrow \underset{N,N\text{-diethylbenzamide}}{(CH_3CH_2)_2\ddot{N}-\overset{\overset{\displaystyle O}{\|}}{C}-C_6H_5} \xrightarrow{LiAlH_4} \underset{\text{benzyldiethylamine}}{(CH_3CH_2)_2\ddot{N}-CH_2-C_6H_5}$$

19-21 REDUCTION OF NITRO COMPOUNDS; SYNTHESIS OF ARYLAMINES

Both aromatic and aliphatic nitro groups are easily reduced to amino groups. The most common methods are catalytic hydrogenation and acidic reduction by an active metal.

$$R{-}NO_2 \xrightarrow[\text{or active metal and } H^+]{H_2/\text{catalyst}} R{-}NH_2$$

catalyst = Ni, Pd, or Pt
active metal = Fe, Zn, or Sn

Examples

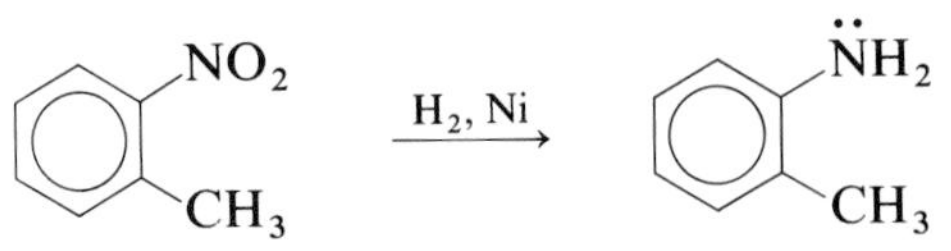

$$\text{o-nitrotoluene} \xrightarrow{H_2,\ Ni} \text{o-toluidine (90\%)}$$

$$CH_3CH_2CH_2{-}CH(NO_2){-}CH_3 \xrightarrow{Sn,\ H_2SO_4} CH_3CH_2CH_2{-}CH(\overset{+}{N}H_3){-}CH_3\ HSO_4^- \xrightarrow{^-OH} CH_3CH_2CH_2{-}CH(\ddot{N}H_2){-}CH_3$$

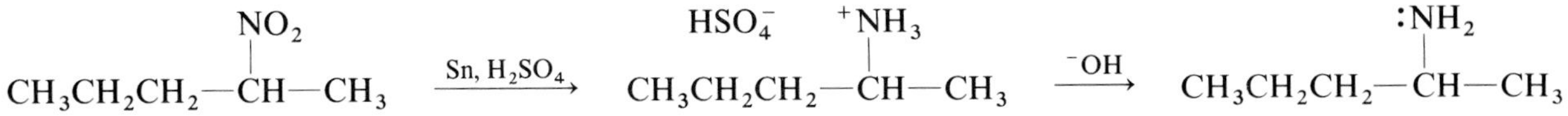

2-nitropentane → 2-aminopentane (85%)

The reduction of nitro compounds is used primarily for the synthesis of substituted anilines. Aromatic nitro compounds, prepared by electrophilic aromatic substitution, are reduced to give aromatic amines.

$$Ar{-}H \xrightarrow{HNO_3,\ H_2SO_4} Ar{-}NO_2 \xrightarrow{\text{reduction}} Ar{-}NH_2$$

For example, nitration followed by reduction is used in the synthesis of benzocaine (a topical anesthetic), shown below. Notice that the stable nitro group is retained through an oxidation and esterification. The final step reduces the nitro group to the relatively sensitive amine (which could not survive the oxidation step).

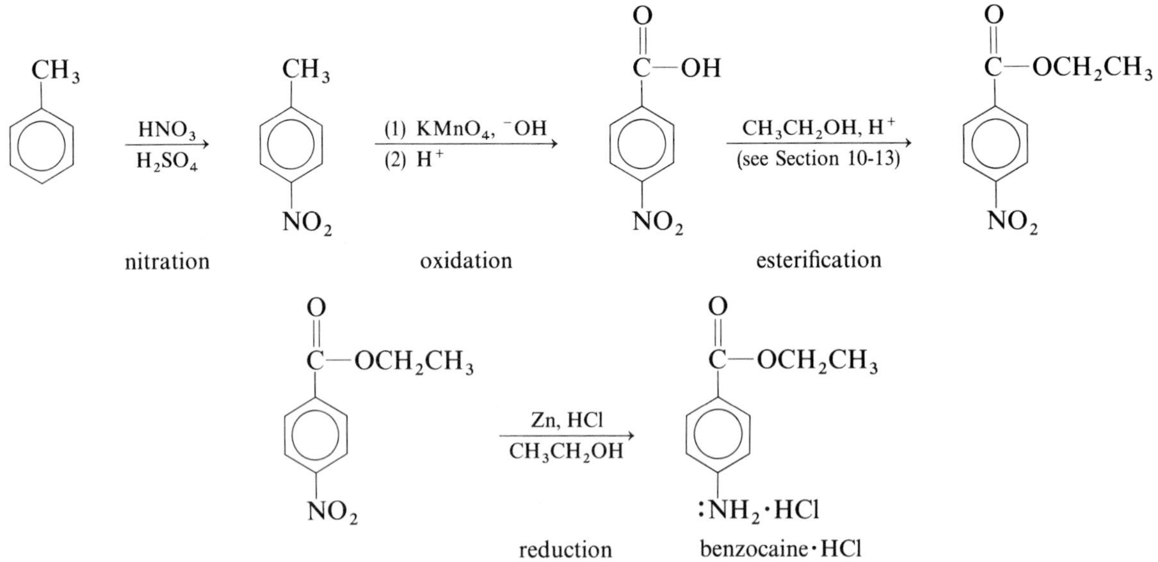

$$C_6H_5CH_3 \xrightarrow[H_2SO_4]{HNO_3} p\text{-}O_2NC_6H_4CH_3 \xrightarrow[(2)\ H^+]{(1)\ KMnO_4,\ ^-OH} p\text{-}O_2NC_6H_4COOH \xrightarrow[\text{(see Section 10-13)}]{CH_3CH_2OH,\ H^+} p\text{-}O_2NC_6H_4COOCH_2CH_3$$

nitration — oxidation — esterification

$$p\text{-}O_2NC_6H_4COOCH_2CH_3 \xrightarrow[CH_3CH_2OH]{Zn,\ HCl} p\text{-}(:NH_2{\cdot}HCl)C_6H_4COOCH_2CH_3$$

reduction — benzocaine·HCl

PROBLEM 19-29

Show how you would prepare each of the following aromatic amines by aromatic nitration, followed by reduction. You may use benzene and toluene as your aromatic starting materials.

(a) aniline (b) *p*-bromoaniline
(c) *m*-bromoaniline (d) *m*-aminobenzoic acid

19-22 DIRECT ALKYLATION OF AMMONIA AND AMINES

The reaction of amines with alkyl halides is complicated by a tendency for overalkylation to form a mixture of monoalkylated and polyalkylated products (Section 19-12). Simple primary amines can be synthesized, however, by adding a large excess of ammonia to a halide or tosylate that is a good S_N2 substrate. Because there is a large excess of ammonia present, the probability that a molecule of the halide will alkylate ammonia is much larger than the probability that it will overalkylate the amine product.

$$R{-}CH_2{-}X \quad + \quad \text{excess } NH_3 \quad \longrightarrow \quad R{-}CH_2{-}NH_2 \quad + \quad NH_4X$$

Example

$$\underset{\text{1-bromopentane}}{CH_3{-}CH_2{-}CH_2{-}CH_2{-}CH_2{-}Br} \quad + \quad \text{excess } NH_3 \quad \longrightarrow \quad \underset{\text{1-pentanamine}}{CH_3{-}CH_2{-}CH_2{-}CH_2{-}CH_2{-}NH_2} \quad + \quad NH_4Br$$

Nucleophilic aromatic substitution can be used to synthesize aryl amines, if there is a strong electron-withdrawing group ortho or para to the site of substitution (see Section 17-12). This reaction usually stops with the desired product because the product (an arylamine) is less basic and less nucleophilic than the reactant (an alkylamine).

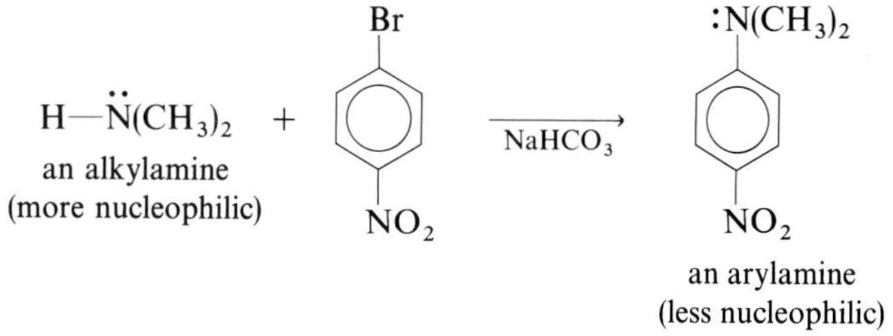

H—N̈(CH$_3$)$_2$ + (p-bromonitrobenzene: Br, NO$_2$) $\xrightarrow{NaHCO_3}$ (p-nitro-N,N-dimethylaniline: :N(CH$_3$)$_2$, NO$_2$)

an alkylamine (more nucleophilic) — an arylamine (less nucleophilic)

PROBLEM 19-30

Show how you would accomplish the following conversions in good yield.

(a) 1-bromoheptane $\longrightarrow$ 1-heptanamine

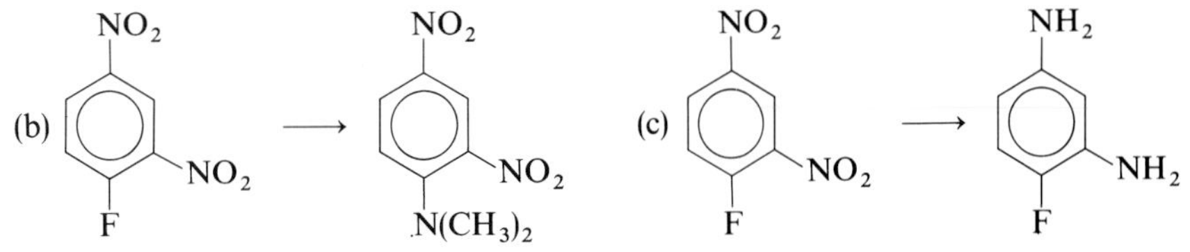

(b) NO$_2$, NO$_2$, F $\longrightarrow$ NO$_2$, NO$_2$, N(CH$_3$)$_2$

(c) NO$_2$, NO$_2$, F $\longrightarrow$ NH$_2$, NH$_2$, F

19-23 THE GABRIEL SYNTHESIS OF PRIMARY AMINES

Primary amines can be synthesized without any danger of overalkylation by using a protected form of ammonia that cannot alkylate more than once. The **Gabriel synthesis** of primary amines uses phthalimide as this protected ammonia derivative. Phthalimide has one acidic N—H proton that is abstracted by potassium hydroxide to give the phthalimide anion.

phthalimide $\xrightarrow[H_2O]{KOH}$ resonance-stabilized phthalimide anion

The phthalimide anion is a strong nucleophile, displacing a halide or tosylate ion from a good S_N2 substrate. Overalkylation does not occur because the *N*-alkyl phthalimide is not nucleophilic and there are no additional acidic protons on nitrogen. Treatment of the *N*-alkyl phthalimide with hydrazine displaces the primary amine, giving the very stable hydrazide of phthalimide.

phthalimide anion $\xrightarrow[\text{alkyl halide}]{R—X}$ *N*-alkyl phthalimide $\xrightarrow[\text{heat}]{H_2N—NH_2}$ phthalimide hydrazide + $R—NH_2$ primary amine

The following synthesis of isopentylamine is a typical example of the Gabriel synthesis.

potassium phthalimide $\xrightarrow[\text{isopentyl bromide}]{Br—CH_2CH_2CH(CH_3)CH_3}$ *N*-isopentylphthalimide

N-isopentylphthalimide $\xrightarrow[\text{hydrazine}]{H_2N—NH_2}$ $H_2N—CH_2CH_2CH(CH_3)CH_3$ isopentylamine (95%) + phthalhydrazide

PROBLEM 19-31
Show how a Gabriel synthesis might be used to prepare each of the following primary amines.

(a) benzylamine (b) 1-hexanamine (c) γ-aminobutyric acid

PROBLEM 19-32
Explain why the Gabriel synthesis cannot be used to make aniline.

19-24 REDUCTION OF NITRILES TO PRIMARY AMINES

Nitriles are reduced to primary amines either by lithium aluminum hydride ($LiAlH_4$) or by catalytic hydrogenation. This reduction provides another method for the synthesis of primary amines.

$$\underset{\text{nitrile}}{R{-}C{\equiv}N:} \xrightarrow{H_2/\text{catalyst or } LiAlH_4} \underset{\text{primary amine}}{R{-}CH_2{-}\ddot{N}H_2}$$

Example

$$\underset{\text{butanenitrile}}{CH_3CH_2CH_2{-}C{\equiv}N:} \xrightarrow[\text{ether}]{LiAlH_4} \underset{\substack{\text{1-butanamine} \\ (75\%)}}{CH_3CH_2CH_2{-}CH_2{-}\ddot{N}H_2}$$

Because nitriles may be synthesized by S_N2 displacement from alkyl halides and tosylates, they are useful intermediates for the synthesis of amines. When the cyano group ($-C{\equiv}N$) is added and then reduced, the resulting amine has an additional carbon atom. In effect, the cyanide substitution-reduction process is like adding $-CH_2-NH_2$. Remember that the alkyl halide or tosylate must be a good S_N2 substrate.

$$\underset{\substack{\text{halide or tosylate} \\ \text{(must be 1° or 2°)}}}{R{-}X} + {}^-{:}C{\equiv}N: \longrightarrow \underset{\text{nitrile}}{R{-}C{\equiv}N:} \xrightarrow[\text{or } H_2/\text{catalyst}]{LiAlH_4} \underset{\substack{\text{amine} \\ \text{(one carbon added)}}}{R{-}CH_2{-}\ddot{N}H_2}$$

Example

$$\underset{\text{benzyl bromide}}{C_6H_5{-}CH_2{-}Br} \;(+\,{}^-{:}C{\equiv}N:) \longrightarrow \underset{\text{phenylacetonitrile}}{C_6H_5{-}CH_2{-}CN:} \xrightarrow[Ni]{H_2} \underset{\beta\text{-phenylethylamine}}{C_6H_5{-}CH_2CH_2\ddot{N}H_2}$$

We have seen (Section 18-15) that cyanide ion adds to a ketone or an aldehyde to form a cyanohydrin. Reduction of the $-C{\equiv}N$ group of the cyanohydrin provides a method for synthesis of α-hydroxy amines.

$$\underset{\text{cyclopentanone}}{C_5H_8{=}O} \xrightarrow[HCN]{{}^-{:}C{\equiv}N:} \underset{\substack{\text{cyclopentanone} \\ \text{cyanohydrin}}}{C_5H_8(OH)(CN)} \xrightarrow[(2)\ H_3O^+]{(1)\ LiAlH_4} \underset{\text{1-(methylamino)cyclopentanol}}{C_5H_8(OH)(CH_2NH_2)}$$

PROBLEM 19-33

Show how the following conversions may be accomplished efficiently.

(a) benzyl bromide ⟶ 2-phenylethanamine
(b) pentanoic acid ⟶ *n*-hexylamine
(c) (*R*)-2-bromobutane ⟶ (*S*)-2-methyl-1-butanamine
(d) 2-hexanone ⟶ 1-amino-2-methyl-2-hexanol

19-25 THE HOFMANN REARRANGEMENT OF AMIDES

In the presence of a strong base, primary amides react with chlorine or bromine to form shortened amines with loss of the carbonyl carbon atom. This reaction, called the **Hofmann rearrangement,** is used to synthesize primary alkyl and aryl amines.

The Hofmann rearrangement

$$\underset{\text{primary amide}}{R-\overset{O}{\overset{\|}{C}}-\ddot{N}H_2} + \underset{(X_2 = Cl_2 \text{ or } Br_2)}{X_2 + 4\,NaOH} \longrightarrow \underset{\text{amine}}{R-\ddot{N}H_2} + 2\,NaX + Na_2CO_3 + 2\,H_2O$$

Examples

$$\underset{\text{hexanamide}}{CH_3CH_2CH_2CH_2CH_2-\overset{O}{\overset{\|}{C}}-NH_2} \xrightarrow[H_2O]{Cl_2,\ ^-OH} \underset{\substack{\text{1-pentanamine}\\(90\%)}}{CH_3CH_2CH_2CH_2CH_2-NH_2}$$

$$\underset{\text{2-methyl-2-phenylpropanamide}}{C_6H_5-C(CH_3)_2-\overset{O}{\overset{\|}{C}}-NH_2} \xrightarrow[H_2O]{Cl_2,\ ^-OH} \underset{\text{2-phenyl-2-propanamine}}{C_6H_5-C(CH_3)_2-NH_2}$$

$$\underset{p\text{-nitrobenzamide}}{O_2N-C_6H_4-\overset{O}{\overset{\|}{C}}-NH_2} \xrightarrow[H_2O]{Br_2,\ ^-OH} \underset{p\text{-nitroaniline}}{O_2N-C_6H_4-NH_2}$$

The mechanism of the Hofmann rearrangement is particularly interesting because it involves some intermediates that we have not encountered before. The first step is the replacement of one of the hydrogens on nitrogen by a halogen. This step is possible because the amide N—H protons are slightly acidic, and a strong base deprotonates a small fraction of the amide molecules.

$$\underset{\text{primary amide}}{R-\overset{O}{\overset{\|}{C}}-\ddot{N}H_2} \xrightarrow{^-OH} \underset{\text{deprotonated amide}}{\left[R-\overset{O}{\overset{\|}{C}}-\ddot{N}^{:-}H \longleftrightarrow R-\overset{:\ddot{O}:^-}{\overset{|}{C}}=\ddot{N}H\right]} \xrightarrow{Br-Br} \underset{N\text{-bromo amide}}{R-\overset{O}{\overset{\|}{C}}-\ddot{N}(Br)H}$$

Owing to the electronegative nature of bromine, the *N*-bromo amide is more easily deprotonated than the original primary amide. Deprotonation of the *N*-bromo amide gives another resonance-stabilized anion. The deprotonated amide has a bromine atom present as a potential leaving group. In order for bromide to leave, however, the alkyl (R—) group must migrate to nitrogen. This is the actual rearrangement step, giving an isocyanate intermediate.

$$\underset{N\text{-bromo amide}}{R-\overset{O}{\overset{\|}{C}}-\ddot{N}(H)Br} \xrightarrow{^-OH} \underset{\text{deprotonated}}{\left[R-\overset{O}{\overset{\|}{C}}-\ddot{N}^{:-}-Br \longleftrightarrow R-\overset{:\ddot{O}:^-}{\overset{|}{C}}=\ddot{N}-Br\right]} \longrightarrow \underset{\text{an isocyanate}}{R-\ddot{N}=C=\ddot{O}} + Br^-$$

Isocyanates react rapidly with water to give carbamic acids.

$$\mathrm{R{-}\ddot{N}{=}C{=}\ddot{O}:} \;\; {}^-\mathrm{OH} \longrightarrow \left[\mathrm{R{-}\ddot{N}{=}\overset{:\ddot{O}:^-}{\overset{|}{C}}{-}OH} \longleftrightarrow \mathrm{R{-}\ddot{N}^-{-}\overset{O}{\overset{\|}{C}}{-}OH} \right] \xrightarrow{\mathrm{H{-}O{-}H}} \mathrm{R{-}\overset{H}{\overset{|}{\ddot{N}}}{-}\overset{O}{\overset{\|}{C}}{-}OH} + {}^-\mathrm{OH}$$

isocyanate — carbamic acid

Decarboxylation of the carbamic acid gives the amine and carbon dioxide.

$$\mathrm{R{-}\ddot{N}H{-}\overset{O}{\overset{\|}{C}}{-}O{-}H} \xrightarrow{^-\mathrm{OH}} \mathrm{R{-}\ddot{N}H{-}\overset{O}{\overset{\|}{C}}{-}O^-} \longrightarrow \mathrm{R{-}\ddot{N}^-{-}H} + \mathrm{CO_2}$$

$$\mathrm{R{-}\ddot{N}^-{-}H} \xrightarrow{\mathrm{H{-}O{-}H}} \mathrm{R{-}\ddot{N}H_2} + \mathrm{CO_3^{2-}} + {}^-\mathrm{OH}$$

PROBLEM 19-34

Give a mechanism for the Hofmann rearrangement of hexanamide.

PROBLEM 19-35

When (*R*)-2-methylbutanamide reacts with bromine in a strong aqueous solution of sodium hydroxide, the product is an optically active amine. Give the structure of the expected product, and use your knowledge of the reaction mechanism to predict its stereochemistry.

SUMMARY OF SYNTHESES OF AMINES

1. *Reductive amination* (Section 19-19)

a. Primary amines

$$\mathrm{R{-}\overset{O}{\overset{\|}{C}}{-}R'} \xrightarrow[\mathrm{H^+}]{\mathrm{H_2N{-}OH}} \mathrm{R{-}\overset{N{-}OH}{\overset{\|}{C}}{-}R'} \xrightarrow{\text{reduction}} \mathrm{R{-}\overset{NH_2}{\overset{|}{C}H}{-}R'}$$

ketone or aldehyde — oxime — primary amine

Example

$$\mathrm{CH_3{-}CH_2{-}CH_2{-}\overset{O}{\overset{\|}{C}}{-}CH_3} \xrightarrow[\mathrm{H^+}]{\mathrm{H_2N{-}OH}}$$

2-pentanone

$$\mathrm{CH_3{-}CH_2{-}CH_2{-}\overset{N{-}OH}{\overset{\|}{C}}{-}CH_3} \xrightarrow[\mathrm{Ni}]{\mathrm{H_2}} \mathrm{CH_3{-}CH_2{-}CH_2{-}\overset{NH_2}{\overset{|}{C}H}{-}CH_3}$$

2-pentanone oxime — 2-pentanamine

b. Secondary amines

$$\mathrm{R{-}\overset{O}{\overset{\|}{C}}{-}R'} \xrightarrow[\mathrm{H^+}]{\mathrm{R''{-}NH_2}} \mathrm{R{-}\overset{N{-}R''}{\overset{\|}{C}}{-}R'} \xrightarrow{\text{reduction}} \mathrm{R{-}\overset{NHR''}{\overset{|}{C}H}{-}R'}$$

ketone or aldehyde — *N*-substituted imine — secondary amine

Example

$$\mathrm{CH_3{-}\overset{O}{\overset{\|}{C}}{-}CH_3} \xrightarrow[\mathrm{H^+}]{\mathrm{Ph{-}NH_2}} \mathrm{CH_3{-}\overset{N{-}Ph}{\overset{\|}{C}}{-}CH_3} \xrightarrow{\mathrm{LiAlH_4}} \mathrm{CH_3{-}\overset{NHPh}{\overset{|}{C}H}{-}CH_3}$$

acetone — phenylisopropylamine

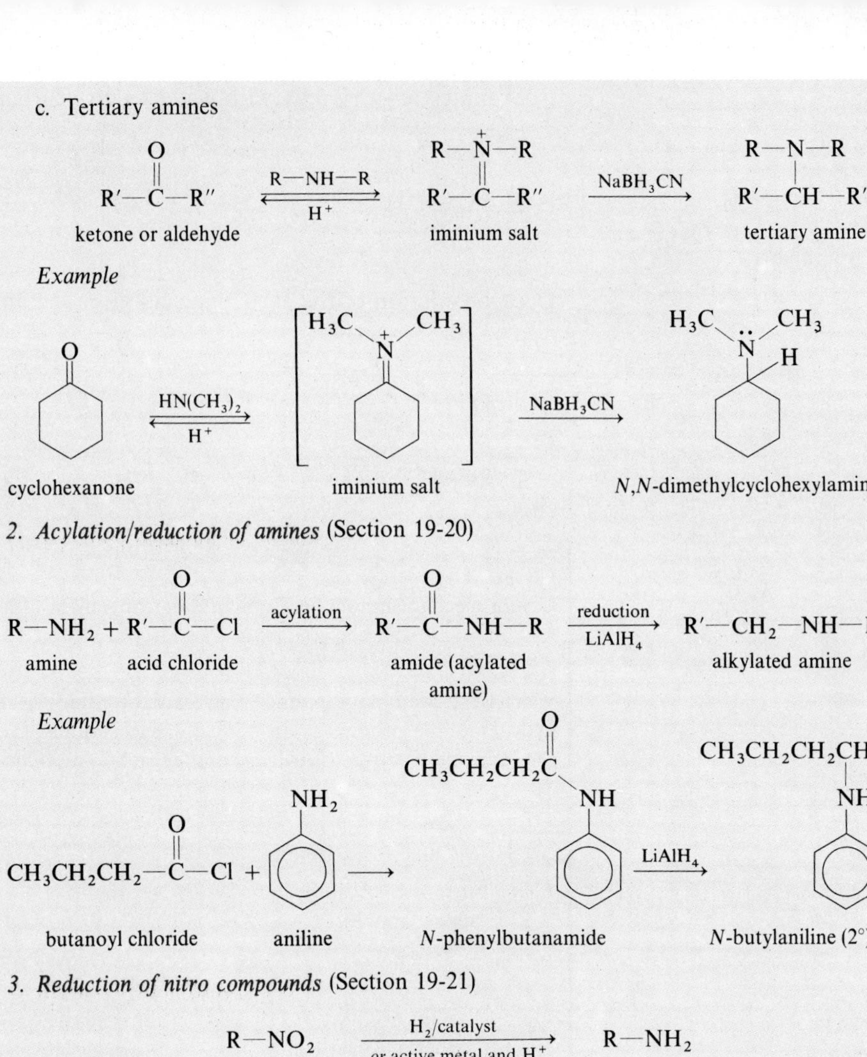

c. Tertiary amines

$$\underset{\text{ketone or aldehyde}}{R'-\overset{\overset{\large O}{\|}}{C}-R''} \underset{H^+}{\overset{R-NH-R}{\rightleftarrows}} \underset{\text{iminium salt}}{\overset{R-\overset{+}{N}-R}{\underset{R'-C-R''}{\|}}} \xrightarrow{NaBH_3CN} \underset{\text{tertiary amine}}{\overset{R-N-R}{\underset{R'-CH-R''}{|}}}$$

Example

cyclohexanone $\underset{H^+}{\overset{HN(CH_3)_2}{\rightleftarrows}}$ iminium salt $\xrightarrow{NaBH_3CN}$ *N,N*-dimethylcyclohexylamine

2. *Acylation/reduction of amines* (Section 19-20)

$$\underset{\text{amine}}{R-NH_2} + \underset{\text{acid chloride}}{R'-\overset{\overset{\large O}{\|}}{C}-Cl} \xrightarrow{\text{acylation}} \underset{\text{amide (acylated amine)}}{R'-\overset{\overset{\large O}{\|}}{C}-NH-R} \underset{LiAlH_4}{\xrightarrow{\text{reduction}}} \underset{\text{alkylated amine}}{R'-CH_2-NH-R}$$

Example

$$\underset{\text{butanoyl chloride}}{CH_3CH_2CH_2-\overset{\overset{\large O}{\|}}{C}-Cl} + \underset{\text{aniline}}{C_6H_5NH_2} \longrightarrow \underset{N\text{-phenylbutanamide}}{CH_3CH_2CH_2C(=O)NH-C_6H_5} \xrightarrow{LiAlH_4} \underset{N\text{-butylaniline (2°)}}{CH_3CH_2CH_2CH_2NH-C_6H_5}$$

3. *Reduction of nitro compounds* (Section 19-21)

$$R-NO_2 \xrightarrow[\text{or active metal and } H^+]{H_2/\text{catalyst}} R-NH_2$$

catalyst = Ni, Pd, or Pt
active metal = Fe, Zn, or Sn

Example

$$\underset{\text{2-nitropentane}}{CH_3CH_2CH_2-\overset{\overset{\large NO_2}{|}}{CH}-CH_3} \xrightarrow{Sn,\ H_2SO_4}$$

$$CH_3CH_2CH_2-\overset{\overset{\large HSO_4^-\ {}^+NH_3}{|}}{CH}-CH_3 \xrightarrow{^-OH} \underset{\text{2-aminopentane}}{CH_3CH_2CH_2-\overset{\overset{\large :NH_2}{|}}{CH}-CH_3}$$

4. *Nucleophilic aromatic substitution* (Section 17-12)

$$R-NH_2 + Ar-X \longrightarrow R-NH-Ar + HX$$

(The aromatic ring should be activated toward nucleophilic attack.)

Example

$$\underset{\text{ethylamine}}{CH_3CH_2-NH_2} + \underset{\text{2,4-dinitrofluorobenzene}}{F-C_6H_3(NO_2)_2} \longrightarrow \underset{N\text{-ethyl-2,4-dinitroaniline}}{CH_3CH_2-NH-C_6H_3(NO_2)_2}$$

5. *Alkylation of ammonia* (Section 19-22)

$$R{-}CH_2{-}X \quad + \quad \text{excess } NH_3 \quad \longrightarrow \quad R{-}CH_2{-}NH_2 \quad + \quad HX$$

Example

$$\underset{\text{1-bromopentane}}{CH_3{-}CH_2{-}CH_2{-}CH_2{-}CH_2{-}Br} \quad + \quad \text{excess } NH_3 \quad \longrightarrow$$

$$\underset{\text{1-pentanamine}}{CH_3{-}CH_2{-}CH_2{-}CH_2{-}CH_2{-}NH_2} \quad + \quad HBr$$

6. *The Gabriel synthesis of primary amines* (Section 19-23)

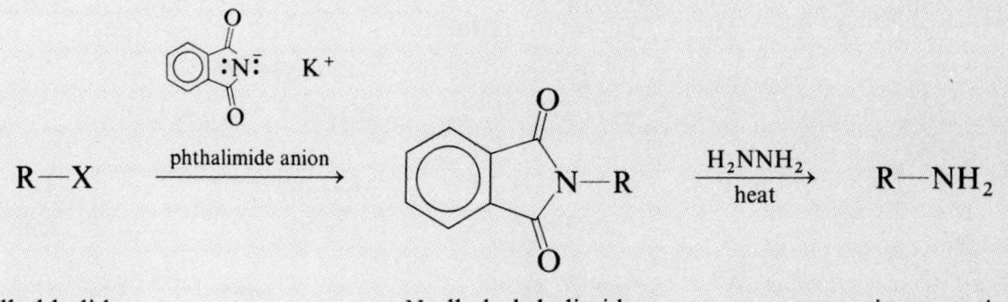

$$\underset{\text{alkyl halide}}{R{-}X} \xrightarrow[\text{phthalimide anion}]{\text{(phthalimide)}N:^- \ K^+} \underset{N\text{-alkyl phthalimide}}{\text{(phthalimide)}N{-}R} \xrightarrow[\text{heat}]{H_2NNH_2} \underset{\text{primary amine}}{R{-}NH_2}$$

Example

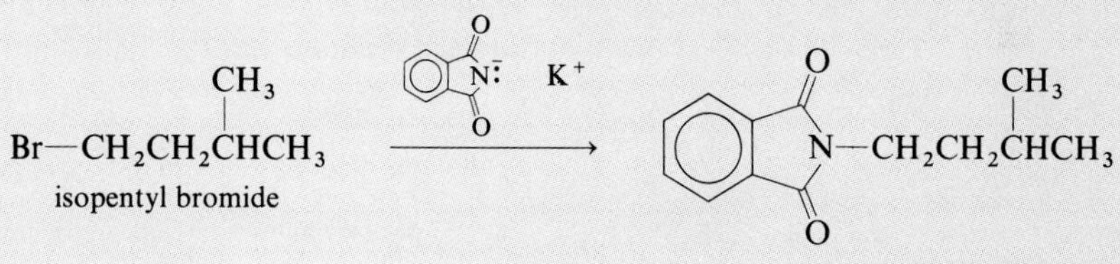

$$\underset{\text{isopentyl bromide}}{Br{-}CH_2CH_2CH(CH_3)CH_3} \xrightarrow{\text{(phthalimide)}N:^- \ K^+} \underset{N\text{-isopentylphthalimide}}{\text{(phthalimide)}N{-}CH_2CH_2CH(CH_3)CH_3}$$

$$\xrightarrow{H_2NNH_2} \underset{\text{isopentylamine}}{H_2N{-}CH_2CH_2CH(CH_3)CH_3}$$

7. *Reduction of nitriles* (Section 19-24)

$$\underset{\text{nitrile}}{R{-}C{\equiv}N:} \xrightarrow{H_2/\text{catalyst or } LiAlH_4} \underset{\text{primary amine}}{R{-}CH_2{-}NH_2}$$

Example

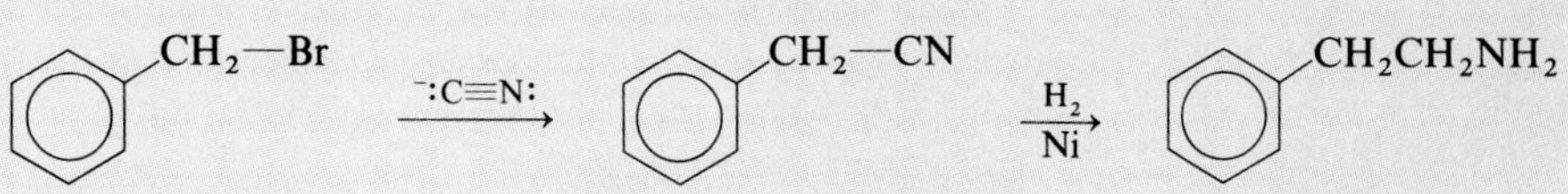

$$\underset{\text{benzyl bromide}}{C_6H_5{-}CH_2{-}Br} \xrightarrow{^-:C{\equiv}N:} \underset{\text{phenylacetonitrile}}{C_6H_5{-}CH_2{-}CN} \xrightarrow[Ni]{H_2} \underset{\beta\text{-phenylethylamine}}{C_6H_5{-}CH_2CH_2NH_2}$$

8. *The Hofmann rearrangement* (Section 19-25)

$$\underset{\text{primary amide}}{R{-}\overset{O}{\overset{\|}{C}}{-}NH_2} + X_2 + 4\,NaOH \longrightarrow \underset{\text{amine}}{R{-}NH_2} + 2\,NaX + Na_2CO_3 + 2\,H_2O$$

$(X_2 = Cl_2 \text{ or } Br_2)$

Example

$$\underset{\text{hexanamide}}{CH_3CH_2CH_2CH_2CH_2{-}\overset{O}{\overset{\|}{C}}{-}NH_2} \xrightarrow[H_2O]{Cl_2,\ ^-OH} \underset{\text{pentanamine}}{CH_3CH_2CH_2CH_2CH_2{-}NH_2}$$

acylation The addition of an **acyl group** ($R{-}C({=}O){-}$), usually replacing a hydrogen atom. Acylation of an amine gives an amide. (p. 848)

$$\underset{\text{amine}}{R{-}NH_2} + \underset{\text{acyl chloride}}{Cl{-}C({=}O){-}R'} \longrightarrow \underset{\text{amide}}{R{-}NH{-}C({=}O){-}R'} + HCl$$

acetylation: Acylation by an acetyl group ($CH_3{-}C({=}O){-}$).

acylation-reduction A method for synthesizing amines by acylating ammonia or an amine, then reducing the amide. (p. 866)

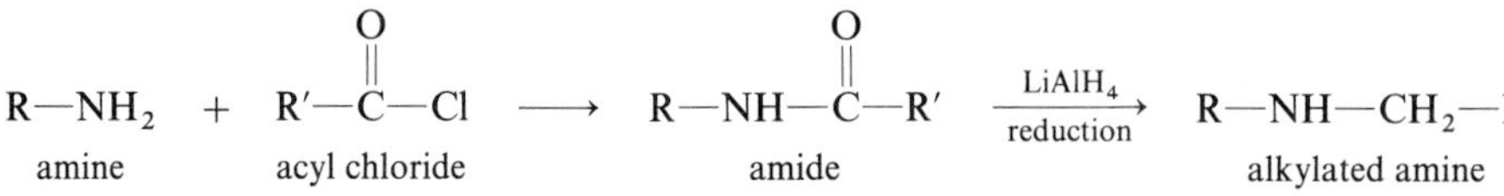

$$\underset{\text{amine}}{R{-}NH_2} + \underset{\text{acyl chloride}}{R'{-}C({=}O){-}Cl} \longrightarrow \underset{\text{amide}}{R{-}NH{-}C({=}O){-}R'} \xrightarrow[\text{reduction}]{LiAlH_4} \underset{\text{alkylated amine}}{R{-}NH{-}CH_2{-}R'}$$

amine A derivative of ammonia with one or more alkyl or aryl groups bonded to the nitrogen atom. (p. 823)

A **primary amine** (1° amine) has one alkyl group bonded to nitrogen.

A **secondary amine** (2° amine) has two alkyl groups bonded to nitrogen.

A **tertiary amine** (3° amine) has three alkyl groups bonded to nitrogen.

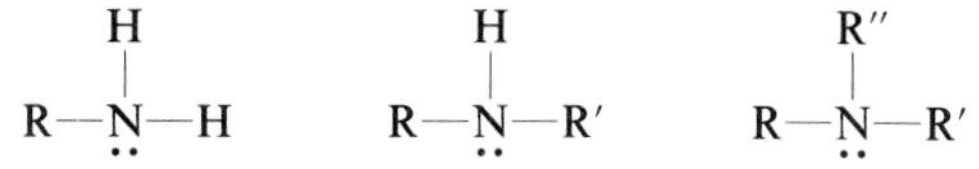

$$\underset{\text{primary amine}}{R{-}\ddot{N}H{-}H} \qquad \underset{\text{secondary amine}}{R{-}\ddot{N}(H){-}R'} \qquad \underset{\text{tertiary amine}}{R{-}\ddot{N}(R''){-}R'}$$

amine oxide An amine with a fourth bond to an oxygen atom. In the amine oxide, the nitrogen atom bears a positive charge and the oxygen atom bears a negative charge. Because of this donation of electrons, the bond to oxygen is often drawn using an arrow. (p. 854)

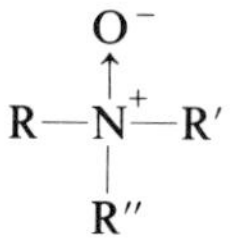

$$R{-}\overset{+}{N}(\rightarrow O^-)(R''){-}R'$$

amino group The $-NH_2$ group. If alkylated, it becomes an **alkylamino** group, $-NHR$ or a **dialkylamino** group, $-NR_2$. (p. 825)

ammonium salt (amine salt) A derivative of an amine with a positively charged nitrogen atom having four bonds. An amine is protonated by an acid to give an ammonium salt. (p. 839) A **quaternary ammonium salt** has a nitrogen atom bonded to four alkyl or aryl groups. (p. 823)

$$\underset{\text{an ammonium salt}}{R{-}NH_3^+\ X^-} \qquad \underset{\text{a quaternary ammonium salt}}{R{-}\overset{+}{N}R_2{-}R\ \ X^-}$$

base-dissociation constant (K_b) A measure of the basicity of a compound such as an amine, defined as the equilibrium constant for the following reaction. The negative $\log_{10}$ of K_b is given as pK_b. (p. 835)

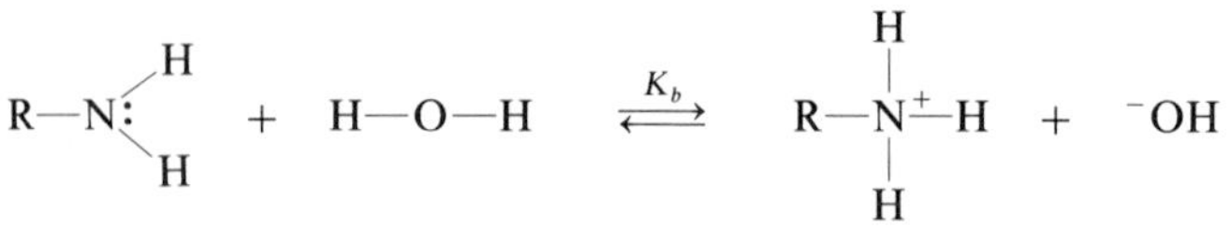

$$R{-}\ddot{N}H_2 + H{-}O{-}H \overset{K_b}{\rightleftharpoons} R{-}\overset{+}{N}H_3 + {}^-OH$$

Cope elimination A variation of the Hofmann elimination in which a tertiary amine oxide undergoes elimination to an alkene with a hydroxylamine serving as the leaving group. (p. 854)

diazo coupling The use of a diazonium salt as an electrophile in electrophilic aromatic substitution. (p. 860)

$$\mathrm{Ar{-}\overset{+}{N}{\equiv}N} \;+\; \mathrm{H{-}C_6H_4{-}Y} \;\longrightarrow\; \mathrm{Ar{-}N{=}N{-}C_6H_4{-}Y} \;+\; \mathrm{H^+}$$

diazonium ion (activated) an **azo compound**

diazotization of an amine The reaction of a primary amine with nitrous acid to form a diazonium salt. (p. 855)

exhaustive alkylation The treatment of an amine with an excess of an alkylating agent (often methyl iodide) to form the quaternary ammonium salt. (p. 847)

$$\mathrm{R{-}NH_2} \xrightarrow{\text{excess } \mathrm{CH_3I}} \mathrm{R{-}N^+(CH_3)_3\;\; I^-}$$

exhaustive methylation of a primary amine

Gabriel synthesis A synthesis of primary amines by alkylation of the potassium salt of phthalimide, followed by displacement of the amine by hydrazine. (p. 870)

Hofmann elimination An elimination of a quaternary ammonium hydroxide in which an amine is the leaving group. The Hofmann elimination usually gives the least highly substituted alkene. (p. 851)

$$\mathrm{HO^-} + \mathrm{R{-}CH_2{-}CH_2{-}\overset{+}{N}(CH_3)_3} \xrightarrow{\text{heat}} \mathrm{H{-}O{-}H} + \mathrm{RHC{=}CH_2} + \mathrm{:N(CH_3)_3}$$

Hofmann rearrangement of amides Treatment of a primary amide with sodium hydroxide and bromine or chlorine gives a primary amine. (p. 872)

$$\mathrm{R{-}\overset{O}{\overset{\|}{C}}{-}NH_2} + \mathrm{X_2 + 4\,NaOH} \longrightarrow \mathrm{R{-}NH_2} + \mathrm{2\,NaX} + \mathrm{Na_2CO_3} + \mathrm{2\,H_2O}$$

primary amide ($X_2 = Cl_2$ or Br_2) amine

hydroxylamine An amine in which a hydroxyl group is one of the three substituents bonded to nitrogen. (p. 853)

$$\mathrm{R{-}\underset{\cdot\cdot}{\overset{R'}{\overset{|}{N}}}{-}OH}$$

nitrile An organic compound of formula $R{-}C{\equiv}N$, containing the *cyano group*, $-C{\equiv}N$. (p. 871)

nitrogen inversion An inversion of configuration of a nitrogen atom in which the lone pair moves from one face of the molecule to the other. The transition state is planar, with the lone pair in a *p* orbital. (p. 827)

***N*-nitrosoamine** An amine with a nitroso group ($-N{=}O$) bonded to the amine nitrogen atom. The reaction of secondary amines with nitrous acid gives secondary *N*-nitrosoamines. (p. 856)

phase-transfer catalyst A compound (such as a quaternary ammonium halide) that is soluble in both water and organic solvents and that helps reagents to move between organic and aqueous phases. (p. 840)

reductive amination The reduction of an imine or oxime derivative of a ketone or aldehyde. One of the most general methods for the synthesis of amines. (p. 865)

$$\mathrm{R{-}\overset{O}{\overset{\|}{C}}{-}R'} \xrightarrow[\mathrm{H^+}]{\mathrm{R''{-}NH_2}} \mathrm{R{-}\overset{N-R''}{\overset{\|}{C}}{-}R'} \xrightarrow{\text{reduction}} \mathrm{R{-}\overset{NHR''}{\overset{|}{C}H}{-}R'}$$

ketone or aldehyde *N*-substituted imine secondary amine

Sandmeyer reaction The replacement of the $-\overset{+}{N}{\equiv}N$ group in an arenediazonium salt by a cuprous salt; usually cuprous chloride, bromide, or cyanide. (p. 858)

$$Ar-\overset{+}{N}{\equiv}N\ \ {}^{-}Cl \xrightarrow[(X\,=\,Cl,\ Br,\ C\equiv N)]{CuX} Ar-X\ +\ N_2\uparrow$$

sulfonamide An amide of a sulfonic acid. The nitrogen analog of a sulfonate ester. (p. 849)

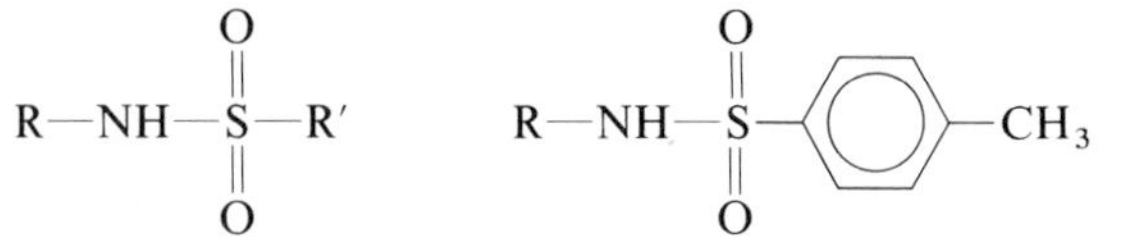

$$R-NH-\overset{O}{\overset{\|}{\underset{O}{\underset{\|}{S}}}}-R' \qquad R-NH-\overset{O}{\overset{\|}{\underset{O}{\underset{\|}{S}}}}-C_6H_4-CH_3$$

a sulfonamide — a *p*-toluenesulfonamide (a tosylamide)

ESSENTIAL PROBLEM-SOLVING SKILLS IN CHAPTER 19

1. Name amines and draw the structures from their names.
2. Interpret the IR, NMR, and mass spectra of amines and use the spectral information to determine the structures.
3. Explain how the basicity of amines varies with hybridization and aromaticity.
4. Contrast the physical properties of amines with those of their salts.
5. Predict the products of reactions of amines with the following types of compounds; give mechanisms where appropriate.
   (a) Ketones and aldehydes
   (b) Alkyl halides and tosylates
   (c) Acid chlorides
   (d) Sulfonyl chlorides
   (e) Nitrous acid
   (f) Oxidizing agents
   (g) Arylamines with electrophiles
6. Give examples of the use of arenediazonium salts in diazo coupling reactions and in the synthesis of aryl chlorides, bromides, iodides, fluorides, and nitriles.
7. Illustrate the uses and mechanisms of the Hofmann and Cope eliminations, and predict the major products.
8. Use your knowledge of the mechanisms of amine reactions to propose mechanisms and products of similar reactions you have never seen before.
9. Show how to synthesize amines from other amines, ketones and aldehydes, acid chlorides, nitro compounds, alkyl halides, nitriles, and amides.
10. Use retrosynthetic analysis to propose effective single-step and multistep syntheses of compounds with amines as intermediates or products, protecting the amine as an amide if necessary.

STUDY PROBLEMS

19-36. Give a definition and an example for each of the following terms.
(a) acylation of an amine **(b)** a 1° amine **(c)** a 2° amine
(d) a 3° amine **(e)** an aromatic heterocyclic amine **(f)** a tertiary amine oxide
(g) an aliphatic heterocyclic amine **(h)** a quaternary ammonium salt **(i)** diazotization of an amine
(j) a diazo coupling reaction **(k)** exhaustive methylation **(l)** a sulfa drug
(m) Gabriel synthesis of an amine **(n)** the Hofmann elimination **(o)** the Hofmann rearrangement
(p) an *N*-nitrosoamine **(q)** the reductive amination process **(r)** the Sandmeyer reaction
(s) a sulfonamide

19-37. For each of the following compounds
(1) Classify the nitrogen-containing functional groups.
(2) Provide an acceptable name.

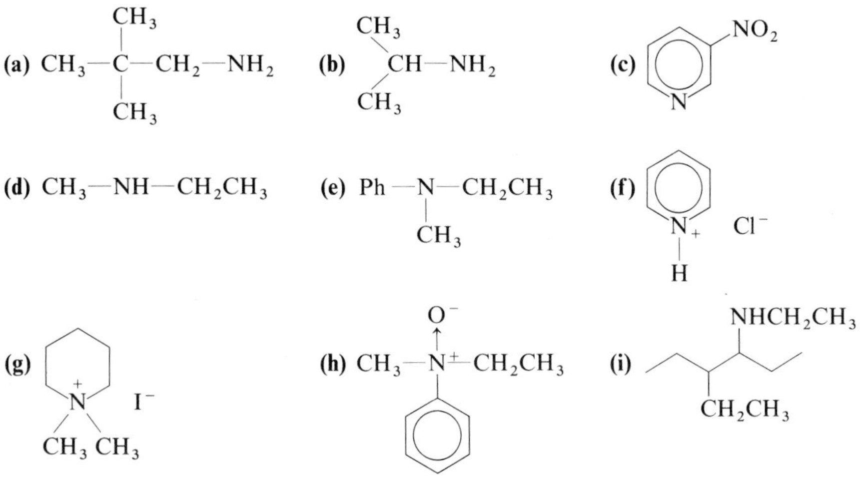

(a) $CH_3-C(CH_3)_2-CH_2-NH_2$ (b) $(CH_3)_2CH-NH_2$ (c) [3-nitropyridine: ring N, NO_2]

(d) $CH_3-NH-CH_2CH_3$ (e) $Ph-N(CH_3)-CH_2CH_3$ (f) [pyridinium, N^+-H] Cl^-

(g) [piperidinium, $N^+(CH_3)(CH_3)$] I^- (h) $CH_3-N^+(O^-)(C_6H_5)-CH_2CH_3$ (i) [skeletal structure with $NHCH_2CH_3$ and CH_2CH_3]

19-38. Rank each of the following sets of amines in order of increasing basicity.

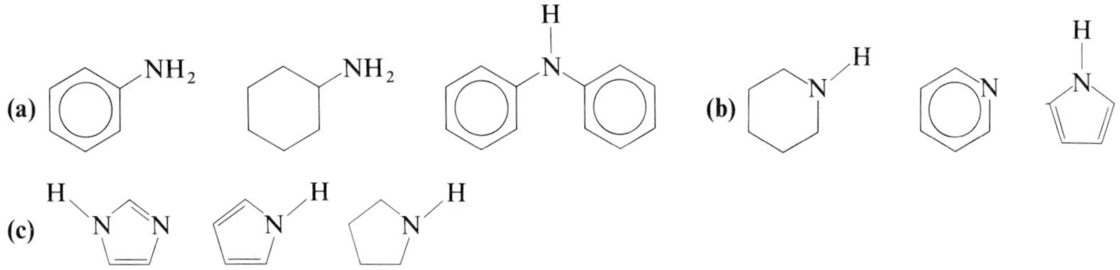

(a) [aniline, NH_2] [cyclohexylamine, NH_2] [diphenylamine, N–H] (b) [piperidine, N–H] [pyridine, N] [pyrrole, N–H]

(c) [imidazole, H–N, N] [pyrrole, N–H] [pyrrolidine, N–H]

19-39. Which of the following compounds is capable of being resolved into enantiomers?
(a) *N*-ethyl-*N*-methylaniline **(b)** 2-methylpiperidine **(c)** 1-methylpiperidine **(d)** 1,2,2-trimethylaziridine

(e) [pyrrolidinium, $N^+(CH_3)(CH_2CH_3)$] Cl^- (f) [3-pyrrolinium, $N^+(CH_3)(CH_2CH_3)$] Cl^-

19-40. Complete the following proposed acid-base reactions. Predict whether the reactants or products are favored.

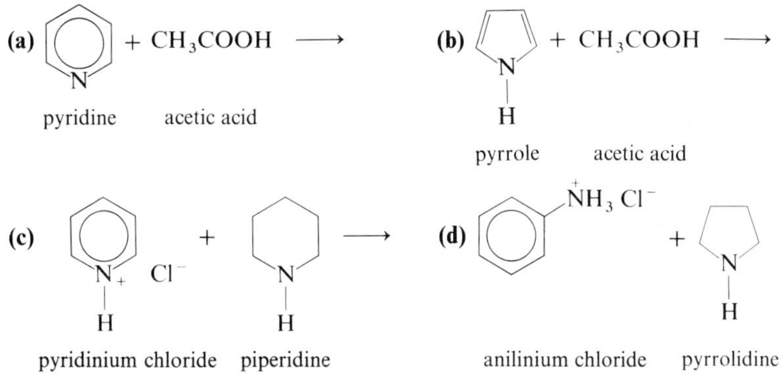

(a) [pyridine] + $CH_3COOH \longrightarrow$
pyridine acetic acid

(b) [pyrrole, N–H] + $CH_3COOH \longrightarrow$
pyrrole acetic acid

(c) [pyridinium, N^+-H] Cl^- + [piperidine, N–H] $\longrightarrow$
pyridinium chloride piperidine

(d) [benzene ring with $\overset{+}{N}H_3$ Cl^-] + [pyrrolidine, N–H]
anilinium chloride pyrrolidine

19-41. Predict the organic products formed when the following amides are treated with alkaline bromine water.

(a) $CH_3-(CH_2)_6-C(=O)-NH_2$ (b) $Ph-CH_2CH_2-C(=O)-NH_2$ (c) $H_2N-C(=O)-(CH_2)_4-C(=O)-NH_2$

19-42. Predict the products of the following reactions.

(a) excess NH_3 + Ph—$CH_2CH_2CH_2Br$ ⟶

(b) CH_3NH_2 + (cyclohexyl)—CH_2—C(=O)—Cl ⟶

(c) N-methylpiperidine (CH_3 on N) + MCPBA ⟶

(d) product from part (c) $\xrightarrow{\text{heat}}$

(e) (quinolizidine, N at ring fusion) $\xrightarrow[\text{(3) heat}]{\text{(1) excess } CH_3I \quad \text{(2) } Ag_2O}$

(f) product from part (e) $\xrightarrow[\text{(3) heat}]{\text{(1) excess } CH_3I \quad \text{(2) } Ag_2O}$

(g) piperidine (H on N) + $NaNO_2$ + HCl ⟶

(h) 3-nitroaniline (NH_2, NO_2) $\xrightarrow[\text{(2) } H_3PO_2]{\text{(1) HCl, } NaNO_2}$

(i) nitrobenzene (NO_2) $\xrightarrow{\text{Zn, HCl}}$

(j) Ph—C(=O)—$NHCH_3$ $\xrightarrow{LiAlH_4}$

(k) CH_3—$(CH_2)_3$—C(=NCH_3)—CH_2CH_3 $\xrightarrow{LiAlH_4}$

(l) Ph—CH_2—CH(CN)—CH_3 $\xrightarrow{LiAlH_4}$

(m) 2-butanone + diethylamine $\xrightarrow{NaBH_3CN}$

(n) pyrrole $\xrightarrow{HNO_3}$

(o) 4-fluoropyridine $\xrightarrow{NaOCH_2CH_3}$

19-43. The mass spectrum of *t*-butylamine is given below. Use a diagram to show the cleavage that accounts for the base peak. Suggest why no molecular ion is visible in this spectrum.

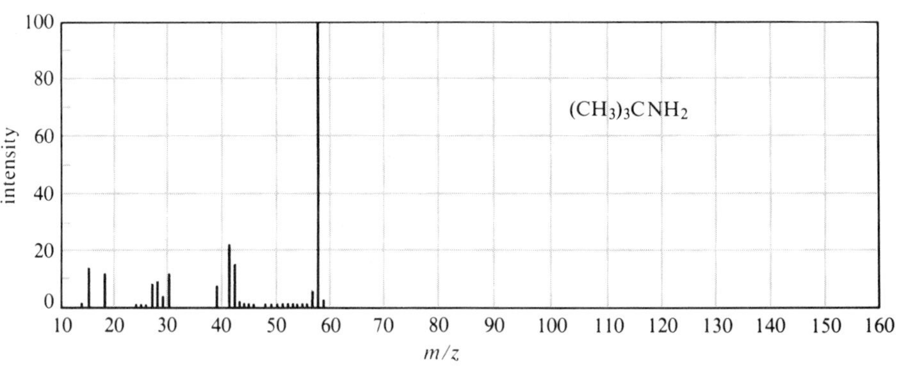

19-44. A graduate student was preparing 1-octanamine by the following process.

$$\underset{\text{octanoic acid}}{CH_3-(CH_2)_6-COOH} \xrightarrow[\text{(2) } NH_3]{\text{(1) } SOCl_2} \underset{\text{octanamide}}{CH_3-(CH_2)_6-\overset{O}{\overset{\|}{C}}-NH_2} \xrightarrow{LiAlH_4} \underset{\textbf{1-octanamine}}{CH_3-(CH_2)_6-CH_2-NH_2}$$

The $LiAlH_4$ was of poor quality, and the final product was contaminated by 20 percent octanamide. Give explicit instructions for using extractions with acidic and basic solutions to remove this impurity from the product.

19-45. Show how *m*-toluidine can be converted to the following compounds, using any necessary additional reagents.

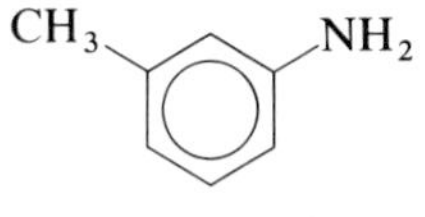

m-toluidine

(a) 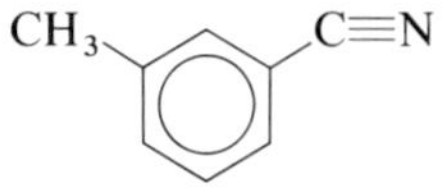 *m*-toluonitrile

(b)

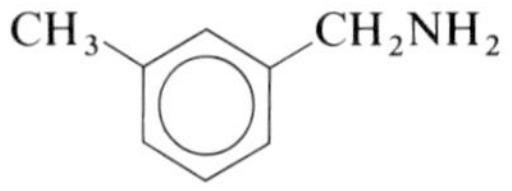

(c) 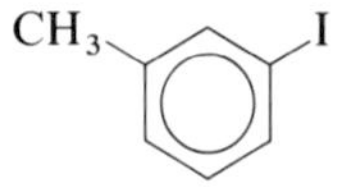 *m*-iodotoluene

(d) 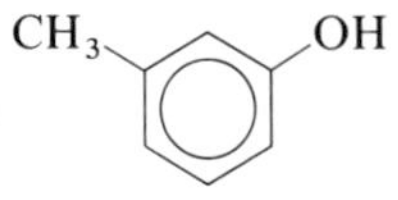 *m*-cresol

(e) 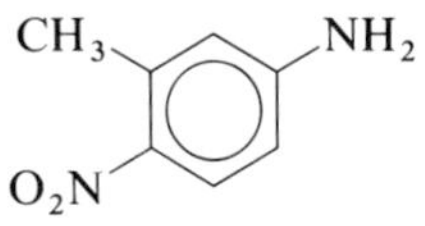 3-methyl-4-nitroaniline

(f)

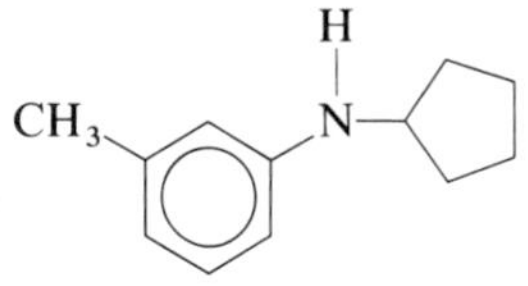

19-46. Using any necessary additional reagents, show how you would accomplish the following synthetic conversions.

(a)

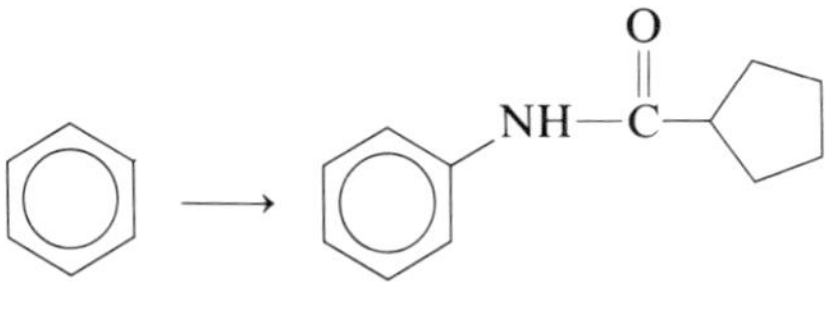

(b)

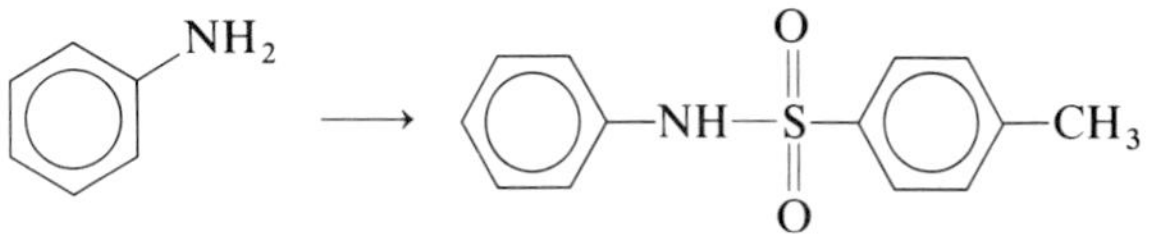

(c)

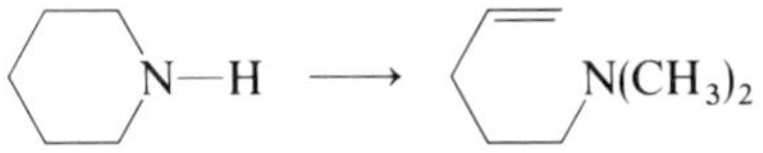

(d)

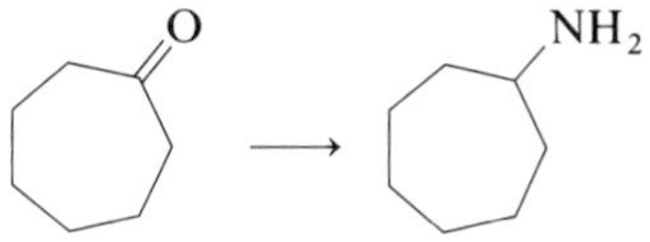

(e)

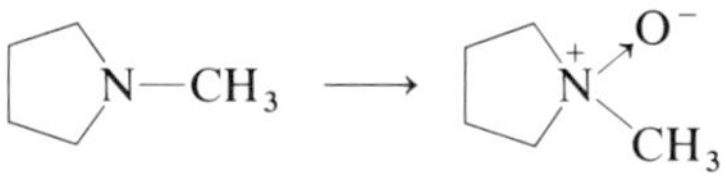

(f)

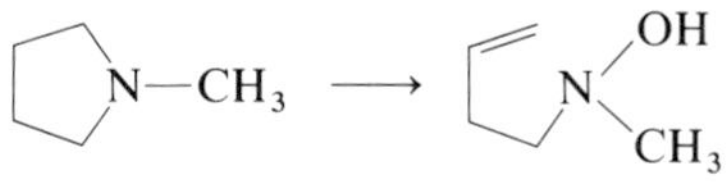

(g)

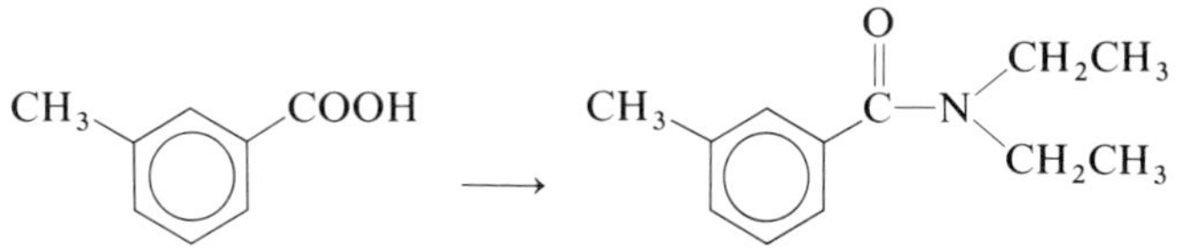

(mosquito repellent)

19-47. The following drugs are synthesized using the methods in this chapter and in previous chapters. Devise a synthesis for each, starting with any compounds containing no more than six carbon atoms.

(a) Phenacetin, used with aspirin and caffeine in pain-relief medications.

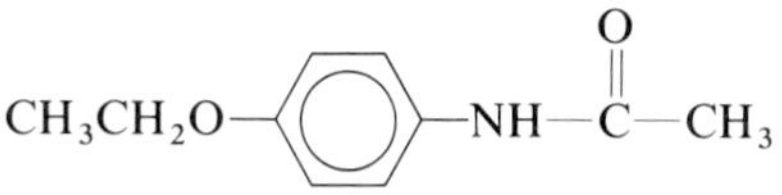

(b) Methamphetamine, once considered a safe diet pill but now known to be addictive and highly destructive to brain tissue.

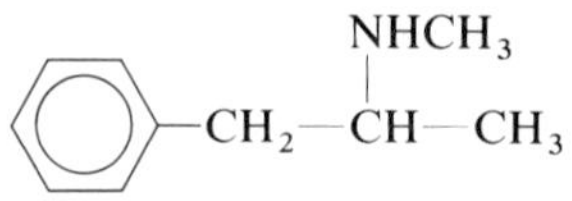

(c) Dopamine, one of the neurotransmitters in the brain. Parkinson's disease is thought to be the result of a dopamine deficiency.

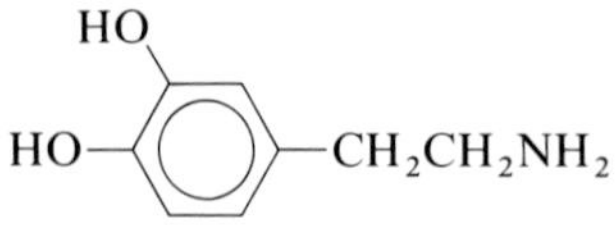

HO, HO—(benzene ring)—$CH_2CH_2NH_2$

19-48. Give a mechanism for each of the following reactions.

(a) 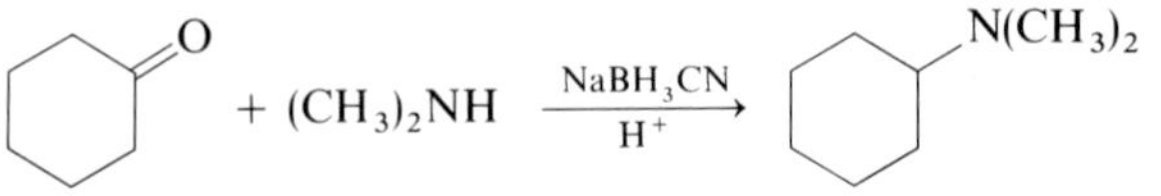 cyclohexanone + $(CH_3)_2NH$ $\xrightarrow[H^+]{NaBH_3CN}$ cyclohexyl—$N(CH_3)_2$

(b) 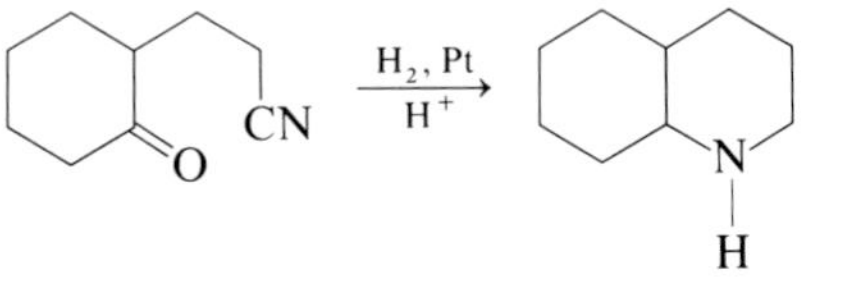 2-(CN-propyl)cyclohexanone (CN, O) $\xrightarrow[H^+]{H_2,\ Pt}$ decahydroquinoline (N—H)

19-49. The two most general amine syntheses are the reduction of amides and the reductive amination of carbonyl compounds. Using any needed additional reagents, show how these two techniques can be used to accomplish the following synthetic conversions.

(a) benzoic acid ⟶ benzylamine
(b) benzaldehyde ⟶ benzylamine
(c) pyrrolidine ⟶ *N*-ethylpyrrolidine
(d) $HOOC—(CH_2)_3—COOH$ ⟶ 1,5-pentanediamine (cadaverine)
(e) cyclohexanone ⟶ *N*-cyclohexylpyrrolidine

19-50. There are several additional amine syntheses that are effectively limited to the formation of primary amines. The reduction of nitro compounds and the Gabriel synthesis leave the carbon chain unchanged, while the formation and reduction of a nitrile adds one carbon atom and the Hofmann rearrangement eliminates one carbon atom. Show how one of these amine syntheses can be used for each of the following conversions.

(a) allyl bromide ⟶ allylamine
(b) ethylbenzene ⟶ *p*-ethylaniline
(c) 3-phenylheptanoic acid ⟶ 2-phenyl-1-hexanamine
(d) 1-bromo-3-phenylheptane ⟶ 3-phenyl-1-heptanamine
(e) 1-bromo-3-phenylheptane ⟶ 4-phenyl-1-octanamine

19-51. **(a)** Guanidine (below) is about as strong a base as hydroxide ion. Explain why guanidine is a much stronger base than most other amines.

(b) Show why *p*-nitroaniline is a much weaker base ($3pK_b$ units weaker) than aniline.

* **(c)** Explain why *N,N*,2,6-tetramethylaniline (below) is a much stronger base than *N,N*-dimethylaniline.

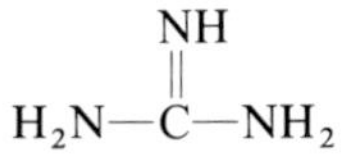

NH ‖ $H_2N—C—NH_2$

guanidine

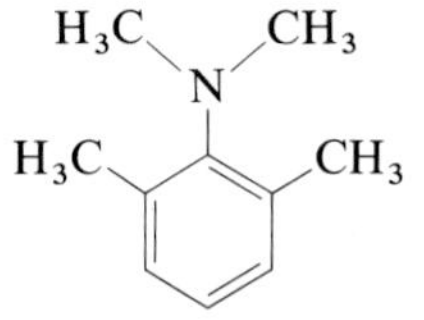

H_3C N CH_3; H_3C, CH_3 (benzene ring)

N,N,2,6-tetramethylaniline

H_3C N CH_3 (benzene ring)

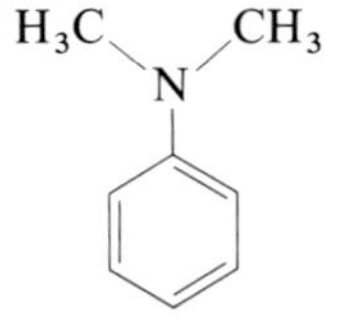

N,N-dimethylaniline

19-52. Using toluene and alcohols containing no more than four carbon atoms as your organic starting materials, show how you would synthesize the following compounds in good yields.

(a) 1-pentanamine **(b)** *N*-methyl-1-butanamine **(c)** *N*-ethyl-*N*-propyl-2-butanamine

(d) benzyl-*n*-propylamine **(e)** 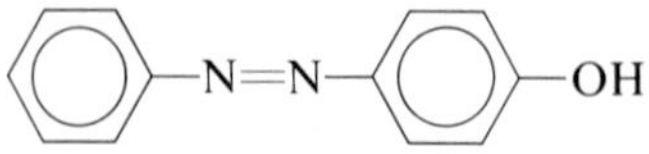 (benzene ring)—N═N—(benzene ring)—OH

19-53. Using any necessary additional reagents, show how you would accomplish the following multistep synthetic conversions.

(a) benzene ⟶ NH_2 / $CH_2CH_2CH_2CH_3$ (para-substituted benzene)

(b) N (bicyclic amine) ⟶ OH / N / $CH{=}CH_2$ (piperidine)

(c) biphenyl–NO_2 ⟶ biphenyl with CH_2CH_3 and $-CH_2NH_2$

19-54. The toxic alkaloid *coniine* has been isolated from hemlock and purified. It is found to have the molecular formula $C_8H_{17}N$. Treatment of coniine with excess methyl iodide, followed by silver oxide and heating, gives the pure (*S*)-enantiomer of 5-(*N*,*N*-dimethylamino)-1-octene. Propose a complete structure for coniine, and show how this reaction gives the observed product.

19-55. A chemist is summoned to an abandoned waste-disposal site to determine the contents of a leaking, corroded barrel. The barrel reeks of an overpowering fishy odor. The chemist dons a respirator to approach the barrel and collect a sample, which she takes to her laboratory for analysis.

The mass spectrum shows a molecular ion at m/z 101, and the most abundant fragment is at m/z 86. The IR spectrum shows no absorptions above 3000 cm^{-1}, many absorptions between 2800 and 3000 cm^{-1}, no absorptions between 1500 and 2800 cm^{-1}, and a strong absorption at 1200 cm^{-1}. The proton NMR spectrum shows a triplet ($J = 7$ Hz) at $\delta 1.0$ and a quartet ($J = 7$ Hz) at $\delta 2.4$, with integrals of 17 spaces and 11 spaces, respectively.

(a) Show what structural information is implied by each spectrum, and propose a structure for the unknown toxic waste.

(b) Current EPA regulations prohibit the disposal of liquid wastes because they tend to leak out of their containers. Propose an inexpensive method for converting this waste to a solid, relatively odorless form for reburial.

(c) Suggest how the chemist might remove the fishy smell from her clothing.

19-56. The spectra shown below for **A** and **B** correspond to two structural isomers. The integration of the NMR multiplet at $\delta 1.09$ in spectrum **A** changes from 10 to 8 when the sample is shaken with D_2O. The singlet at $\delta 0.6$ ppm in the spectrum of **B** disappears upon shaking with D_2O. Propose structures for these isomers, and show how your structures correspond to the spectra. Show what cleavage is responsible for the base peak at m/z 44 in the mass spectrum of **A** and the prominent peak at m/z 58 in the mass spectrum of **B.**

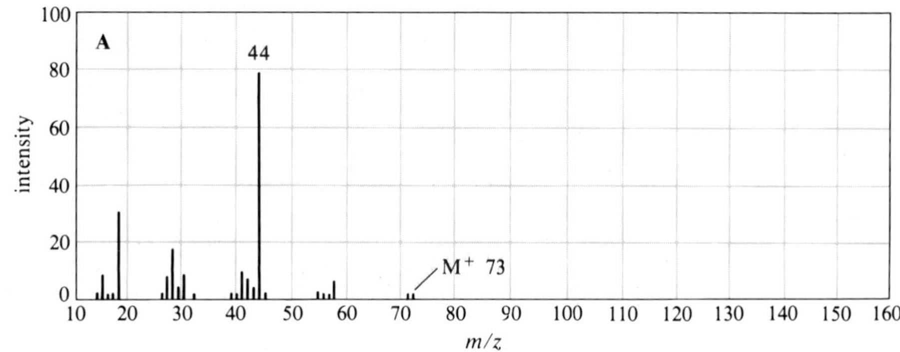

Problem 19-56 *continued*

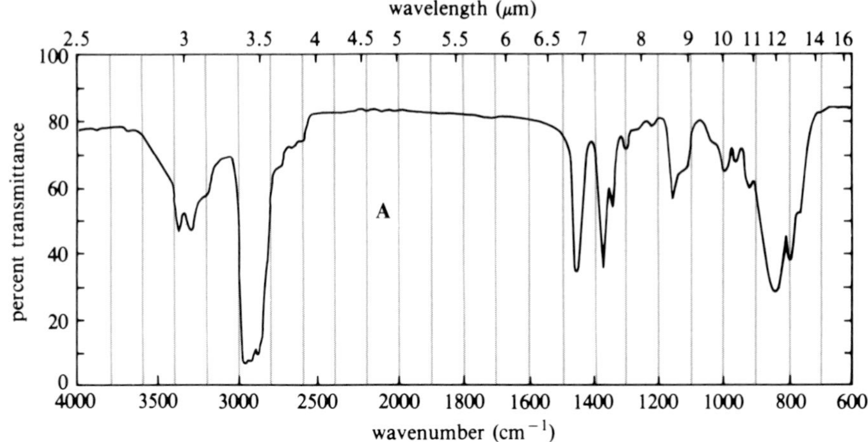

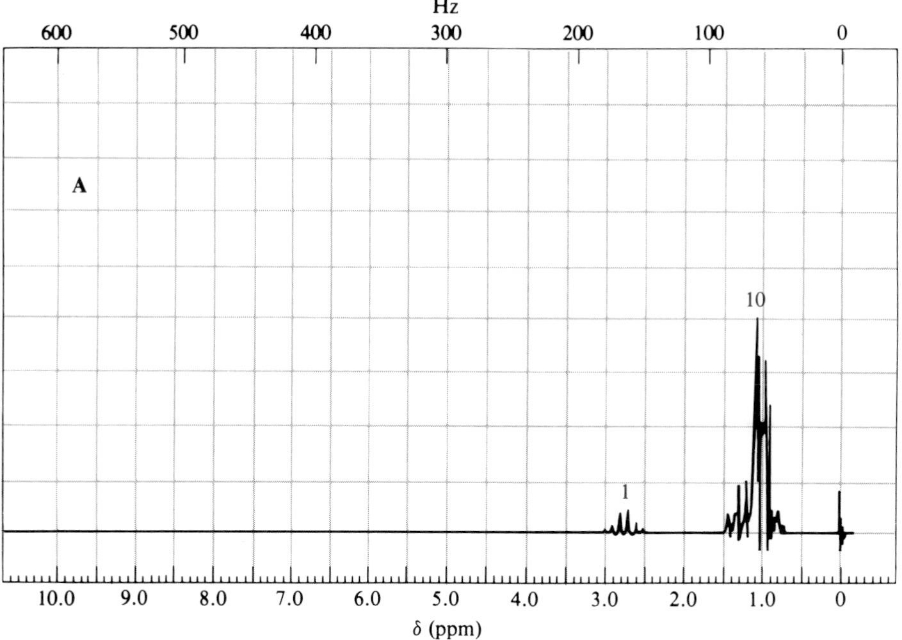

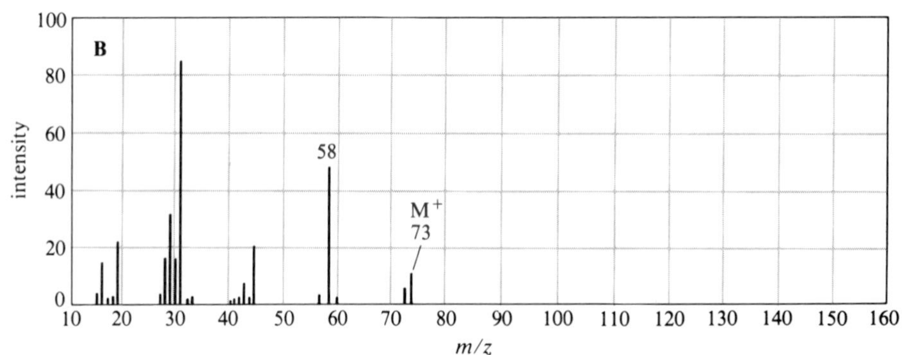

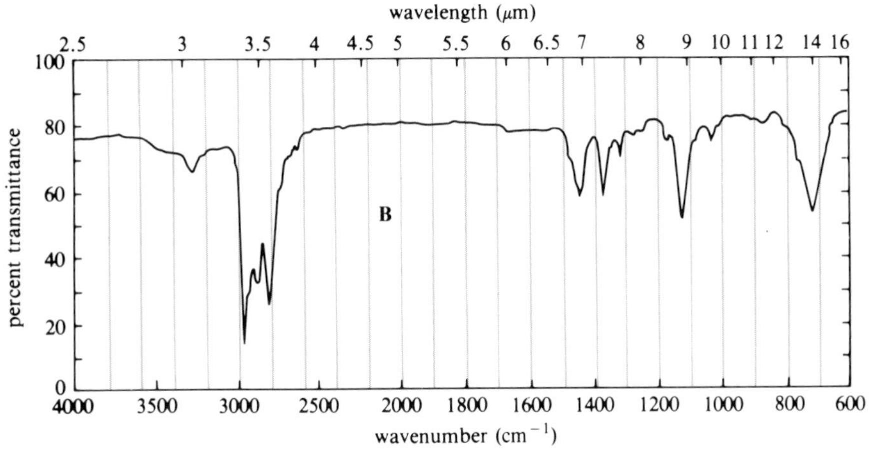

wavelength (μm)
2.5 3 3.5 4 4.5 5 5.5 6 6.5 7 8 9 10 11 12 14 16
100 80 60 40 20 0
percent transmittance
B
4000 3500 3000 2500 2000 1800 1600 1400 1200 1000 800 600
wavenumber (cm⁻¹)

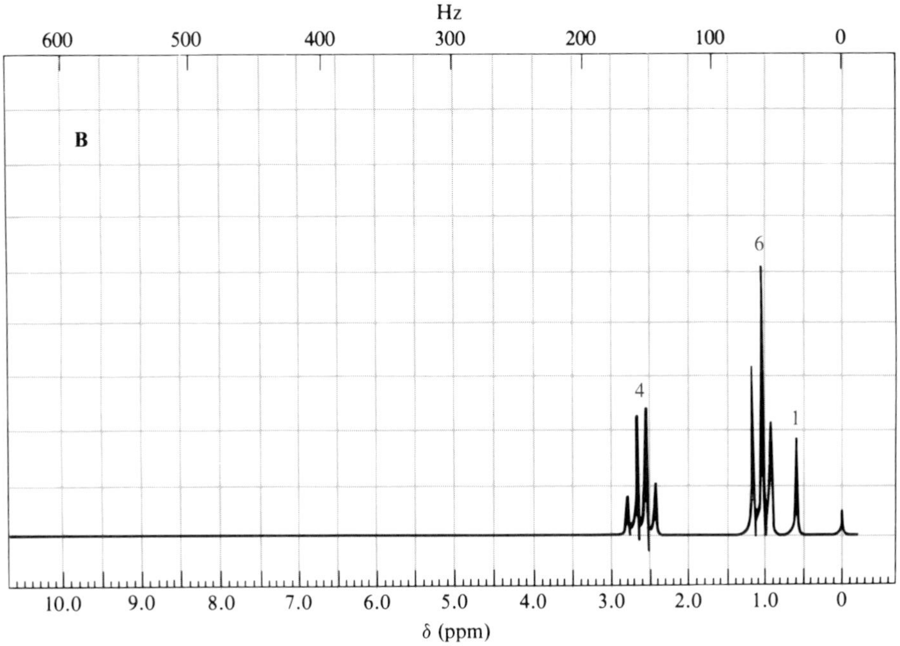

Hz
600 500 400 300 200 100 0
B
6
4
1
10.0 9.0 8.0 7.0 6.0 5.0 4.0 3.0 2.0 1.0 0
δ (ppm)

19-57. An unknown compound shows a weak molecular ion at m/z 87 in the mass spectrum, and the only large peak in the MS is at m/z 30. The IR spectrum appears below. The NMR shows only three singlets: One of area 9 at δ0.9, one of area 2 at δ1.0, and one of area 2 at δ2.4. The singlet at δ1.0 disappears on shaking with D_2O. Determine the structure of the compound, and show the favorable fragmentation that accounts for the ion at m/z 30.

wavelength (μm)
2.5 3 3.5 4 4.5 5 5.5 6 6.5 7 8 9 10 11 12 14 16
100 80 60 40 20 0
percent transmittance
4000 3500 3000 2500 2000 1800 1600 1400 1200 1000 800 600
wavenumber (cm⁻¹)

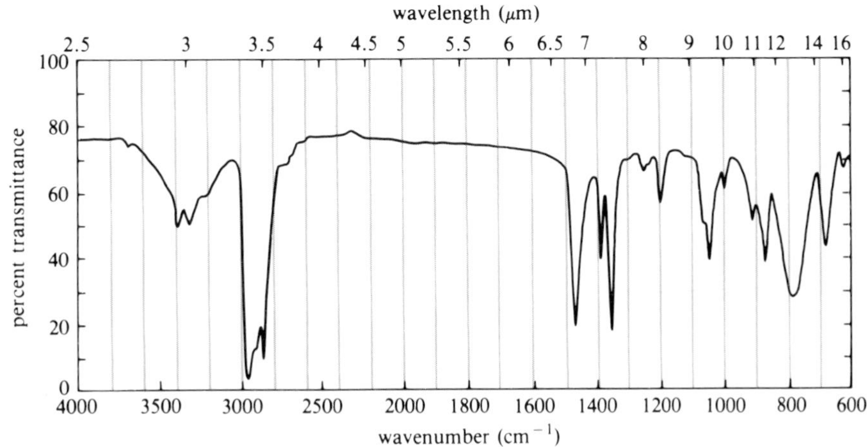

✱ 19-58. A compound of formula $C_{11}H_{16}N_2$ gives the IR, ^1H NMR, and ^{13}C NMR spectra shown below. The proton NMR peak at $\delta 1.57$ disappears on shaking with D_2O. Propose a structure for this compound, and show how your structure accounts for the observed absorptions.

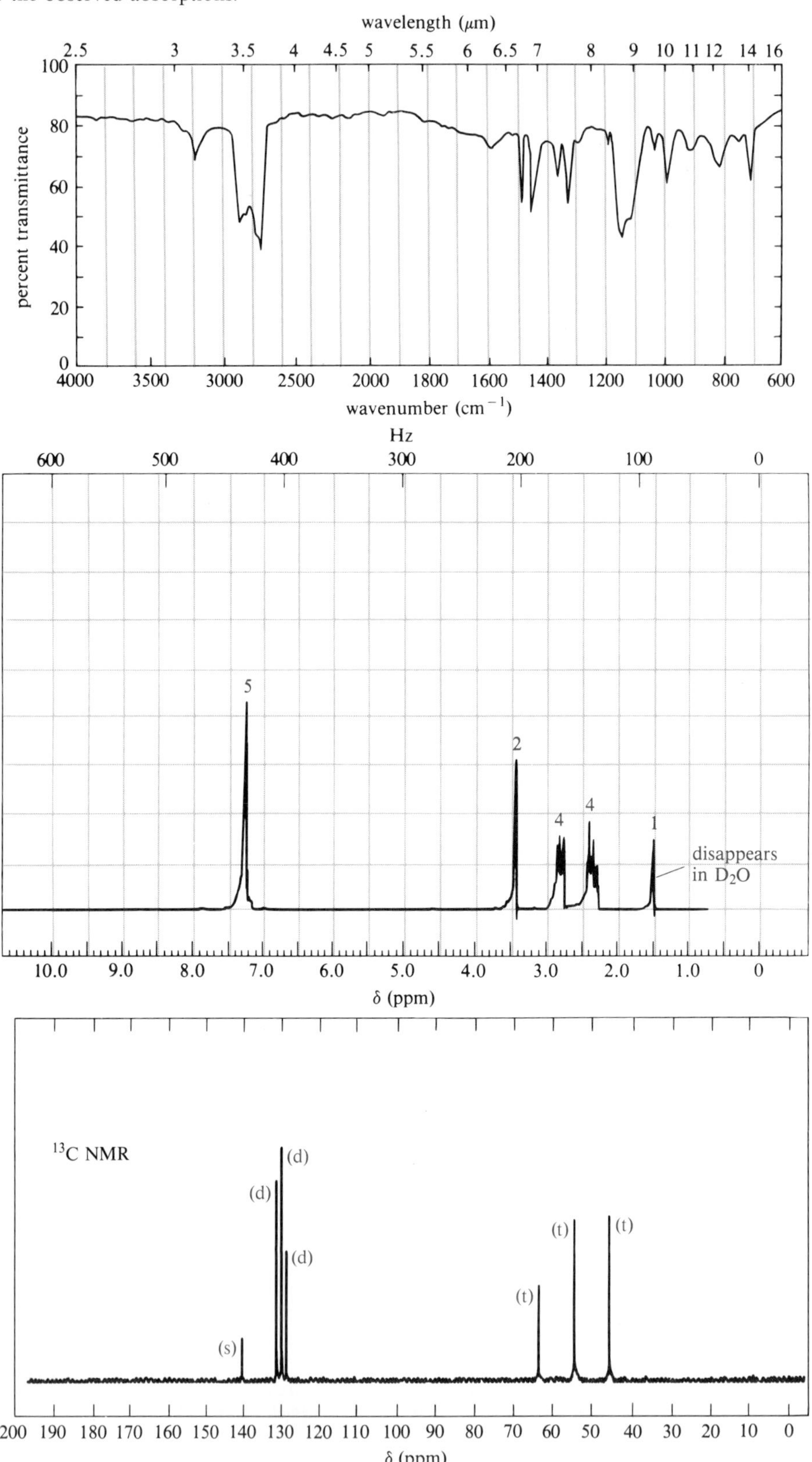

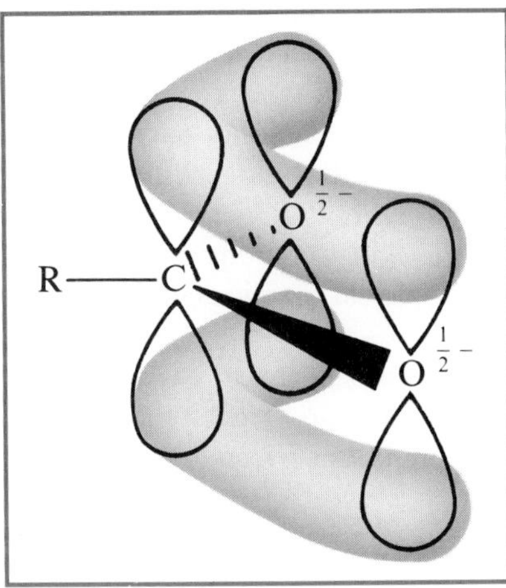

20

CARBOXYLIC ACIDS

20-1 INTRODUCTION

Compounds containing the **carboxyl group** are distinctly acidic and are called **carboxylic acids.**

$$\underset{\text{carboxyl group}}{-\overset{\overset{\displaystyle O}{\|}}{C}-O-H} \qquad \underset{\text{carboxylic acid}}{R-\overset{\overset{\displaystyle O}{\|}}{C}-O-H} \qquad \underset{\text{condensed structures}}{R-COOH \quad R-CO_2H}$$

Carboxylic acids are classified according to the substituent bonded to the carboxyl group. An **aliphatic acid** has an alkyl group bonded to the carboxyl group, while an **aromatic acid** has an aryl group. The carboxylic acid with a proton bonded to the carboxyl group is called *formic acid.* **Fatty acids** are long-chain aliphatic acids derived from the hydrolysis of fats and oils (Section 20-6).

$$\underset{\text{formic acid}}{H-\overset{\overset{\displaystyle O}{\|}}{C}-O-H} \qquad \underset{\substack{\text{propanoic acid}\\ \text{(an aliphatic acid)}}}{CH_3-CH_2-\overset{\overset{\displaystyle O}{\|}}{C}-O-H} \qquad \underset{\substack{\text{benzoic acid}\\ \text{(an aromatic acid)}}}{C_6H_5-\overset{\overset{\displaystyle O}{\|}}{C}-O-H} \qquad \underset{\substack{\text{stearic acid}\\ \text{(a fatty acid)}}}{CH_3(CH_2)_{16}-\overset{\overset{\displaystyle O}{\|}}{C}-O-H}$$

A carboxylic acid donates protons by heterolytic cleavage of the acidic O—H bond to give a proton and a **carboxylate ion.** We consider the ranges of acidity and the factors affecting the acidity of carboxylic acids in Section 20-4.

$$\underset{\text{carboxylic acid}}{R-\overset{\overset{\displaystyle O}{\|}}{C}-O-H} \quad \rightleftharpoons \quad \underset{\text{carboxylate ion}}{R-\overset{\overset{\displaystyle O}{\|}}{C}-O^-} \; + \; H^+$$

20-2 NOMENCLATURE OF CARBOXYLIC ACIDS

20-2A COMMON NAMES

Several aliphatic carboxylic acids have been known for hundreds of years and have common names reflecting their historical sources. *Formic acid* was extracted from ants: *formica* in Latin. Acetic acid was isolated from vinegar, called *acetum* ("sour") in Latin. Propionic acid was considered to be the first fatty acid, and the name is derived from the Greek *protos pion* ("first fat"). Butyric acid results from the oxidation of butyraldehyde, which is found in butter: *butyrum* in Latin. Caproic, caprylic, and capric acids are found in the skin secretions of goats: *caper* in Latin. The names and physical properties of simple carboxylic acids are shown in Table 20-1.

In common names, the positions of substituents are given using Greek letters. Notice that the lettering begins with the carbon atom *adjacent* to the carboxyl carbon, the α-carbon. With common names, the prefix *iso-* is sometimes used for acids ending in the $-CH(CH_3)_2$ grouping.

$$-\underset{\varepsilon}{C}-\underset{\delta}{C}-\underset{\gamma}{C}-\underset{\beta}{C}-\underset{\alpha}{C}-\overset{O}{\overset{\|}{C}}-OH$$

$$\underset{\beta}{CH_3}-\underset{\alpha}{\overset{Cl}{\overset{|}{CH}}}-\overset{O}{\overset{\|}{C}}-OH$$

α-chloropropionic acid

$$\underset{\gamma}{\overset{NH_2}{\overset{|}{CH_2}}}-\underset{\beta}{CH_2}-\underset{\alpha}{CH_2}-\overset{O}{\overset{\|}{C}}-OH$$

γ-aminobutyric acid

$$\underset{\gamma}{CH_3}-\underset{\beta}{\overset{CH_3}{\overset{|}{CH}}}-\underset{\alpha}{CH_2}-\overset{O}{\overset{\|}{C}}-OH$$

isovaleric acid
(β-methylbutyric acid)

TABLE 20-1
Names and physical properties of some carboxylic acids

IUPAC name	*Common name*	*Formula*	*m.p. (°C)*	*b.p. (°C)*	*Solubility (g/100 g H_2O)*
methanoic[a]	formic	$HCOOH$	8	101	∞ (miscible)
ethanoic[a]	acetic	CH_3COOH	17	118	∞
propanoic	propionic	CH_3CH_2COOH	−21	141	∞
2-propenoic[a]	acrylic	$H_2C{=}CH{-}COOH$	14	141	∞
butanoic	butyric	$CH_3(CH_2)_2COOH$	−6	163	∞
2-methylpropanoic	isobutyric	$(CH_3)_2CHCOOH$	−46	155	23
trans-2-butenoic[a]	crotonic	$CH_3{-}CH{=}CH{-}COOH$	71	185	8.6
pentanoic	valeric	$CH_3(CH_2)_3COOH$	−34	186	3.7
3-methylbutanoic	isovaleric	$(CH_3)_2CHCH_2COOH$	−29	177	5
2,2-dimethylpropanoic	pivalic	$(CH_3)_3C{-}COOH$	35	164	2.5
hexanoic	caproic	$CH_3(CH_2)_4COOH$	−4	206	1.0
octanoic	caprylic	$CH_3(CH_2)_6COOH$	16	240	0.7
decanoic	capric	$CH_3(CH_2)_8COOH$	31	269	0.2
dodecanoic	lauric	$CH_3(CH_2)_{10}COOH$	44		i
tetradecanoic	myristic	$CH_3(CH_2)_{12}COOH$	54		i
hexadecanoic	palmitic	$CH_3(CH_2)_{14}COOH$	63		i
octadecanoic	stearic	$CH_3(CH_2)_{16}COOH$	72		i
cis-9-octadecenoic[a]	oleic	$CH_3(CH_2)_7CH{=}CH(CH_2)_7COOH$	16		i
cis,cis-9,12-octadecadienoic[a]	linoleic	$CH_3(CH_2)_4CH{=}CHCH_2CH{=}CH(CH_2)_7COOH$	−5		i
cyclohexanecarboxylic		$c\text{-}C_6H_{11}COOH$	31	233	0.2
benzoic	benzoic	$C_6H_5{-}COOH$	122	249	0.3
2-methylbenzoic	*o*-toluic	$o\text{-}CH_3C_6H_4COOH$	106	259	0.1
3-methylbenzoic	*m*-toluic	$m\text{-}CH_3C_6H_4COOH$	112	263	0.1
4-methylbenzoic	*p*-toluic	$p\text{-}CH_3C_6H_4COOH$	180	275	0.03

[a] IUPAC name is rarely used.

20-2B IUPAC NAMES

The IUPAC nomenclature for carboxylic acids uses the name of the alkane that corresponds to the longest continuous chain of carbon atoms in the acid. The final *-e* in the alkane name is replaced by the suffix *-oic acid*. The chain is numbered, *starting with the carboxyl carbon atom,* to give positions of substituents along the chain. In naming, the carboxyl group takes priority over any of the functional groups discussed previously.

$$-\underset{6}{C}-\underset{5}{C}-\underset{4}{C}-\underset{3}{C}-\underset{2}{C}-\underset{1}{\overset{O}{\overset{\|}{C}}}-OH$$

	$H-C(=O)-OH$	$CH_3-C(=O)-OH$	$\underset{3}{CH_3}-\underset{2}{CH}(C_6H_{11})-\underset{1}{C}(=O)-OH$	$\underset{4}{CH_3}\underset{3}{C}(=O)\overset{2}{CH}(CH_2CH_2CH_3)-\underset{1}{C}(=O)-OH$
IUPAC name:	methanoic acid	ethanoic acid	2-cyclohexylpropanoic acid	3-oxo-2-propylbutanoic acid
common name:	formic acid	acetic acid	α-cyclohexylpropionic acid	2-acetylvaleric acid

	$\underset{4}{CH_2}(NH_2)-\underset{3}{CH_2}-\underset{2}{CH_2}-\underset{1}{C}(=O)-OH$	$\underset{5}{CH_3}-\underset{4}{CH_2}-\underset{3}{CH}(Ph)-\underset{2}{CH_2}-\underset{1}{C}(=O)-OH$	$\underset{4}{CH_3}-\underset{3}{CH}(CH_3)-\underset{2}{CH_2}-\underset{1}{C}(=O)-OH$
IUPAC name:	4-aminobutanoic acid	3-phenylpentanoic acid	3-methylbutanoic acid
common name:	γ-aminobutyric acid	β-phenylvaleric acid	isovaleric acid

The name of an unsaturated acid is formed using the name of the corresponding alkene, with the final *-e* replaced by *-oic acid*. The carbon chain is numbered starting with the carboxyl carbon, and a number is used to give the location of the double bond. The stereochemical terms *cis* and *trans* (and *Z* and *E*) are used as they are with other alkenes. Cycloalkanes with —COOH substituents are generally named as *cycloalkanecarboxylic acids.*

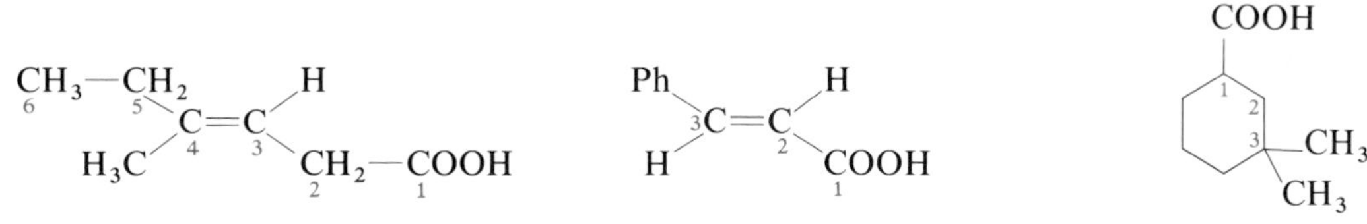

(*E*)-4-methyl-3-hexenoic acid *trans*-3-phenyl-2-butenoic acid (cinnamic acid) 3,3-dimethylcyclohexanecarboxylic acid

Aromatic acids of the form Ar—COOH are named as derivatives of *benzoic acid,* Ph—COOH. As with other aromatic compounds, the prefixes *ortho-*, *meta-*, and *para-* may be used to give the positions of additional substituents. Numbers are used if there are more than two substituents on the aromatic ring. Many aromatic acids have common names that are unrelated to their structures.

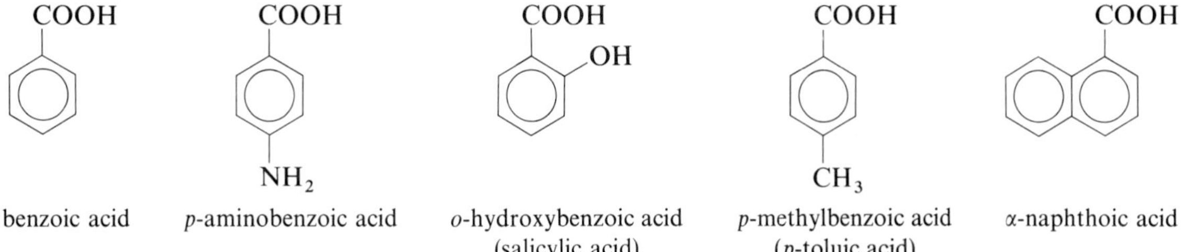

benzoic acid *p*-aminobenzoic acid *o*-hydroxybenzoic acid (salicylic acid) *p*-methylbenzoic acid (*p*-toluic acid) α-naphthoic acid

20-2C NOMENCLATURE OF DICARBOXYLIC ACIDS

Common names of dicarboxylic acids A **dicarboxylic acid** (sometimes called a *diacid*) is a compound with two carboxyl groups. The common names of simple dicarboxylic acids are used more frequently than their systematic names. A common mnemonic used for remembering these names is "*O*h *m*y, *s*uch *g*ood *a*pple *p*ie," standing for *o*xalic, *m*alonic, *s*uccinic, *g*lutaric, *a*dipic, and *p*imelic acids. The names and physical properties of some simple dicarboxylic acids are given in Table 20-2.

TABLE 20-2
Names and physical properties of some simple dicarboxylic acids

IUPAC name	*Common name*	*Formula*	*m.p. (°C)*	*Solubility (g/100 g H_2O)*
ethanedioic	oxalic	HOOC—COOH	189	9
propanedioic	malonic	$HOOCCH_2COOH$	136	74
butanedioic	succinic	$HOOC(CH_2)_2COOH$	185	6
pentanedioic	glutaric	$HOOC(CH_2)_3COOH$	98	64
hexanedioic	adipic	$HOOC(CH_2)_4COOH$	151	2
heptanedioic	pimelic	$HOOC(CH_2)_5COOH$	106	5
cis-2-butenedioic	maleic	*cis*-HOOCCH=CHCOOH	130.5	79
trans-2-butenedioic	fumaric	*trans*-HOOCCH=CHCOOH	302	0.7
benzene-1,2-dicarboxylic	phthalic	1,2-$C_6H_4(COOH)_2$	231	0.7
benzene-1,3-dicarboxylic	isophthalic	1,3-$C_6H_4(COOH)_2$	348	
benzene-1,4-dicarboxylic	terephthalic	1,4-$C_6H_4(COOH)_2$	300 subl.	0.002

Substituted dicarboxylic acids are given common names using Greek letters, as in the simple carboxylic acids. The letters are assigned beginning with the carbon atom adjacent to the carboxyl group that is closer to the substituents.

$$HO-\overset{O}{\overset{\|}{C}}-\underset{\alpha}{CH_2}-\underset{\beta}{\overset{Br}{\overset{|}{CH}}}-CH_2-CH_2-\overset{O}{\overset{\|}{C}}-OH$$

β-bromoadipic acid

$$HO-\overset{O}{\overset{\|}{C}}-\underset{\alpha}{\overset{CH_3}{\overset{|}{CH}}}-\underset{\beta}{\overset{Ph}{\overset{|}{CH}}}-CH_2-\overset{O}{\overset{\|}{C}}-OH$$

α-methyl-β-phenylglutaric acid

Benzenoid compounds with two carboxyl groups are named *phthalic acids.* **Phthalic acid** itself is the ortho isomer. The meta isomer is called *isophthalic acid,* and the para isomer is called *terephthalic acid.*

COOH, COOH (ortho) — phthalic acid

COOH, COOH (meta) — isophthalic acid

COOH, COOH (para) — terephthalic acid

IUPAC names of dicarboxylic acids Aliphatic dicarboxylic acids are named simply by adding the suffix *-dioic acid* to the name of the parent alkane. For straight-chain dicarboxylic acids, the parent alkane name is determined by using the longest continuous chain that contains both carboxyl groups. The chain is numbered beginning with the carboxyl carbon atom that is closer to the substituents, and these numbers are used to give the positions of the substituents.

$$HO-\underset{1}{\overset{O}{\overset{\|}{C}}}-\underset{2}{CH_2}-\underset{3}{\overset{Br}{\overset{|}{CH}}}-\underset{4}{CH_2}-\underset{5}{CH_2}-\underset{6}{\overset{O}{\overset{\|}{C}}}-OH$$

3-bromohexanedioic acid

$$HO-\underset{1}{\overset{O}{\overset{\|}{C}}}-\underset{2}{\overset{CH_3}{\overset{|}{CH}}}-\underset{3}{\overset{Ph}{\overset{|}{CH}}}-\underset{4}{CH_2}-\underset{5}{\overset{O}{\overset{\|}{C}}}-OH$$

2-methyl-3-phenylpentanedioic acid

This system for naming dicarboxylic acids is easily applied to cyclic dicarboxylic acids, treating the carboxyl groups as substituents of the cyclic structure.

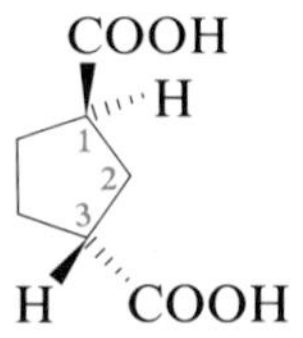

trans-1,3-cyclopentanedicarboxylic acid 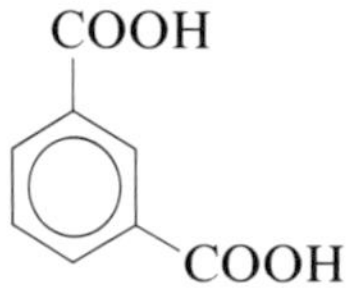   1,3-benzenedicarboxylic acid

PROBLEM 20-1

Draw the structures of the following carboxylic acids.

(a) α-methylbutyric acid
(b) 2-bromobutanoic acid
(c) 4-aminopentanoic acid
(d) *cis*-4-phenyl-2-butenoic acid
(e) *trans*-2-methylcyclohexanecarboxylic acid
(f) 2,3-dimethylfumaric acid
(g) *m*-chlorobenzoic acid
(h) 3-methylphthalic acid
(i) β-aminoadipic acid
(j) 3-chloroheptanedioic acid

PROBLEM 20-2

Name each of the following carboxylic acids. When possible, give both a common name and a systematic name.

(a) (b) 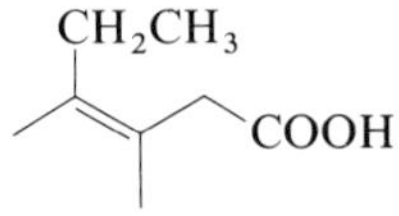 (c) 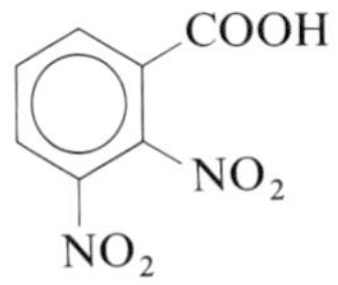 (d) 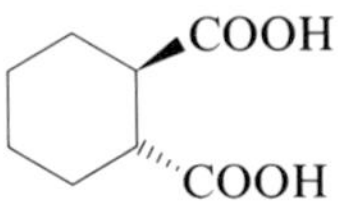 (e) 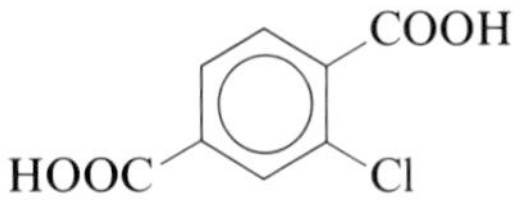 (f)

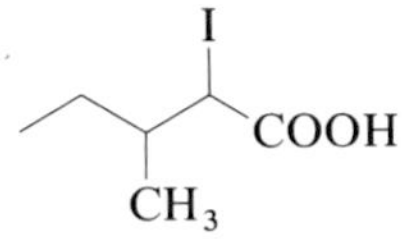

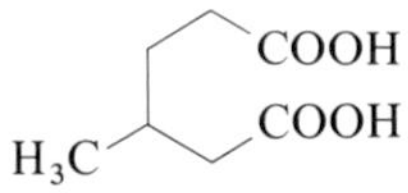

20-3 PHYSICAL PROPERTIES OF CARBOXYLIC ACIDS

Boiling points Carboxylic acids boil at considerably higher temperatures than do alcohols, ketones, or aldehydes of similar molecular weights. For example, acetic acid (MW 60) boils at 118°C, 1-propanol (MW 60) boils at 97°C, and propionaldehyde (MW 58) boils at 49°C.

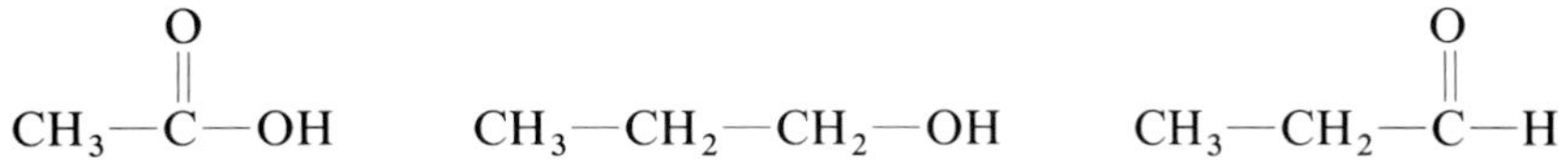

$$CH_3-\overset{O}{\overset{\|}{C}}-OH \qquad CH_3-CH_2-CH_2-OH \qquad CH_3-CH_2-\overset{O}{\overset{\|}{C}}-H$$

acetic acid, b.p. 118°C   1-propanol, b.p. 97°C   propionaldehyde, b.p. 49°C

The high boiling points of carboxylic acids result from the formation of a stable, hydrogen-bonded dimer. This dimer contains an eight-membered ring joined by two hydrogen bonds, effectively doubling the molecular weight of the molecules leaving the liquid phase.

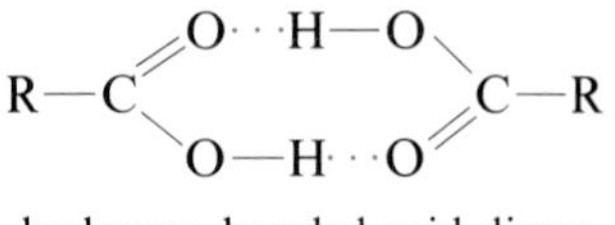

hydrogen-bonded acid dimer

Melting points The melting points of the most common carboxylic acids are given in Table 20-1 (page 888). Carboxylic acids containing more than eight carbon atoms are generally solids, unless they contain double bonds. The presence of double bonds (especially cis double bonds) in a long chain impedes the formation of a stable crystal lattice, resulting in a lower melting point. For example, both stearic acid (octadecanoic acid) and linoleic acid (*cis,cis*-9,12-octadecadienoic acid) have 18 carbon atoms, but stearic acid melts at 70°C and linoleic acid melts at −5°C.

$$CH_3-(CH_2)_{16}-\overset{\overset{\displaystyle O}{\|}}{C}-OH$$

stearic acid, m.p. 70°C

$$CH_3(CH_2)_4-CH=CH-CH_2-CH=CH-(CH_2)_7-\overset{\overset{\displaystyle O}{\|}}{C}-OH$$

linoleic acid, m.p. −5°C

The melting points of dicarboxylic acids (Table 20-2, page 890) are very high. With two carboxyl groups per molecule, the forces of hydrogen bonding are particularly strong in these diacids; a high temperature is required to break the lattice of hydrogen bonds in the crystal and melt the diacid.

Solubilities Carboxylic acids form hydrogen bonds with water, and the lower-molecular-weight carboxylic acids (up through four carbon atoms) are miscible with water. As the length of the hydrocarbon chain increases, the water solubility decreases until the acids with more than ten carbon atoms are essentially insoluble in water. The water solubilities of some simple carboxylic acids and diacids are given in Tables 20-1 and 20-2.

Carboxylic acids are very soluble in alcohols, because the acids form hydrogen bonds with alcohols. Also, alcohols are not as polar as water, so the longer-chain acids are more soluble in alcohols than they are in water. Most carboxylic acids are quite soluble in relatively nonpolar solvents such as chloroform, because the acid continues to exist in its dimeric form in the nonpolar solvent. Thus the hydrogen bonds of the cyclic dimer are not disrupted when the acid dissolves in a nonpolar solvent.

20-4 ACIDITY OF CARBOXYLIC ACIDS

20-4A MEASUREMENT OF ACIDITY

A carboxylic acid may dissociate in water to give a proton and a carboxylate ion. The equilibrium constant K_a for this reaction is called the *acid-dissociation constant*. The pK_a of an acid is the negative logarithm of K_a, and we commonly use pK_a as an indication of the relative acidities of different acids (Table 20-3).

$$R-\overset{\overset{\displaystyle O}{\|}}{C}-O-H + H_2O \rightleftarrows R-\overset{\overset{\displaystyle O}{\|}}{C}-O^- + H_3O^+$$

$$K_a = \frac{[R-CO_2^-][H_3O^+]}{[R-CO_2H]}$$

$$pK_a = -\log_{10} K_a$$

Values of pK_a are about 5 ($K_a = 10^{-5}$) for simple carboxylic acids. For example, acetic acid has a pK_a of 4.7 ($K_a = 1.8 \times 10^{-5}$). Although carboxylic acids are not as strong as the mineral acids, they are still much more acidic than other functional groups we have studied. For example, alcohols have pK_a values in the range 16 to 18. Acetic acid (pK_a = 4.74) is about 10^{11} times as acidic as the most acidic alcohols! In fact, concentrated acetic acid causes severe acid burns when it comes into contact with the skin.

TABLE 20-3
Values of K_a and pK_a for some simple carboxylic acids and dicarboxylic acids

Formula	Name	Values	
	Simple carboxylic acids		
		K_a (at 25°C)	pK_a
$HCOOH$	methanoic acid	1.77×10^{-4}	3.75
CH_3COOH	ethanoic acid	1.76×10^{-5}	4.74
CH_3CH_2COOH	propanoic acid	1.34×10^{-5}	4.87
$CH_3(CH_2)_2COOH$	butanoic acid	1.54×10^{-5}	4.82
$CH_3(CH_2)_3COOH$	pentanoic acid	1.52×10^{-5}	4.81
$CH_3(CH_2)_4COOH$	hexanoic acid	1.31×10^{-5}	4.88
$CH_3(CH_2)_6COOH$	octanoic acid	1.28×10^{-5}	4.89
$CH_3(CH_2)_8COOH$	decanoic acid	1.43×10^{-5}	4.84
C_6H_5COOH	benzoic acid	6.46×10^{-5}	4.19
p-$CH_3C_6H_4COOH$	*p*-toluic acid	4.33×10^{-5}	4.36
p-ClC_6H_4COOH	*p*-chlorobenzoic acid	1.04×10^{-4}	3.98
p-$NO_2C_6H_4COOH$	*p*-nitrobenzoic acid	3.93×10^{-4}	3.41

Formula	Name	K_{a1}	pK_{a1}	K_{a2}	pK_{a2}
	Dicarboxylic acids				
$HOOC—COOH$	oxalic	5.4×10^{-2}	1.27	5.2×10^{-5}	4.28
$HOOCCH_2COOH$	malonic	1.4×10^{-3}	2.85	2.0×10^{-6}	5.70
$HOOC(CH_2)_2COOH$	succinic	6.4×10^{-5}	4.19	2.3×10^{-6}	5.64
$HOOC(CH_2)_3COOH$	glutaric	4.5×10^{-5}	4.35	3.8×10^{-6}	5.42
$HOOC(CH_2)_4COOH$	adipic	3.7×10^{-5}	4.43	3.9×10^{-6}	5.41
cis-$HOOCCH{=}CHCOOH$	maleic	1.0×10^{-2}	2.00	5.5×10^{-7}	6.26
trans-$HOOCCH{=}CHCOOH$	fumaric	9.6×10^{-4}	3.02	4.1×10^{-5}	4.39
1,2-$C_6H_4(COOH)_2$	phthalic	1.1×10^{-3}	2.96	4.0×10^{-6}	5.40
1,3-$C_6H_4(COOH)_2$	isophthalic	2.4×10^{-4}	3.62	2.5×10^{-5}	4.60
1,4-$C_6H_4(COOH)_2$	terephthalic	2.9×10^{-4}	3.54	3.5×10^{-5}	4.46

The dissociation of either an acid or an alcohol involves the breaking of an O—H bond, but the dissociation of a carboxylic acid gives a carboxylate ion with the negative charge spread out equally over *two* oxygen atoms, compared with just one oxygen in the alkoxide ion. The delocalized charge makes the carboxylate ion more stable than the alkoxide ion; therefore, the dissociation of a carboxylic acid to a carboxylate ion is less endothermic than the dissociation of an alcohol to an alkoxide ion.

The carboxylate ion can be visualized either as a resonance hybrid (as in Fig. 20-1) or as a conjugated system of three *p* orbitals containing four electrons. If the carbon atom and the two oxygen atoms are sp^2 hybridized, each of them has an unhybridized *p* orbital. Overlap of these three *p* orbitals results in a three-center π-molecular orbital system. In this system, there is half a π bond between the carbon and each oxygen atom, and there is half a negative charge on each oxygen atom (Fig. 20-2).

20-4B SUBSTITUENT EFFECTS ON ACIDITY

A substituent that stabilizes the negatively charged carboxylate ion enhances dissociation and results in a stronger acid. Electronegative atoms enhance the strength of an acid in this manner. This inductive effect can be quite large if one or more strongly electron-withdrawing groups are present on the α-carbon atom. For example, chloroacetic acid ($ClCH_2—COOH$) has a pK_a of 2.86, indicating that it is a stronger acid than acetic acid (pK_a = 4.74). Dichloroacetic acid ($Cl_2CH—COOH$) is yet stronger, with a pK_a of 1.26. Trichloroacetic acid ($Cl_3C—COOH$) has a pK_a of 0.64, comparable in strength to some of the mineral acids.

$$R{-}\ddot{\underset{\cdot\cdot}{O}}{-}H + H_2\ddot{O}: \rightleftharpoons R{-}\ddot{\underset{\cdot\cdot}{O}}:^- + H_3O^+ \qquad pK_a \cong 16 \quad (K_a \cong 10^{-16})$$

alcohol → alkoxide

$$R{-}\overset{\overset{\cdot\ddot{O}\cdot}{\|}}{C}{-}\ddot{\underset{\cdot\cdot}{O}}{-}H + H_2\ddot{O}: \rightleftharpoons \left[R{-}C\begin{smallmatrix} \nearrow \ddot{O}: \\ \searrow \ddot{\underset{\cdot\cdot}{O}}:^- \end{smallmatrix} \longleftrightarrow R{-}C\begin{smallmatrix} \nearrow \ddot{O}:^- \\ \searrow \underset{\cdot\cdot}{O}: \end{smallmatrix} \right] + H_3O^+ \qquad pK_a \cong 5 \quad (K_a \cong 10^{-5})$$

acid → carboxylate

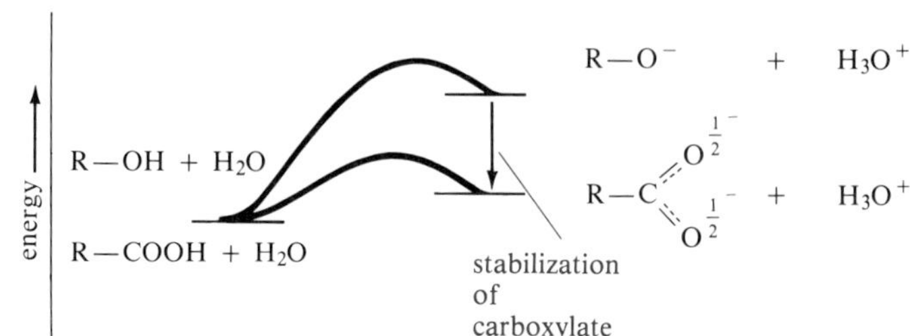

FIGURE 20-1 Carboxylic acids are much more acidic than alcohols because carboxylate ions are more stable than the alkoxide ions formed from dissociation of alcohols.

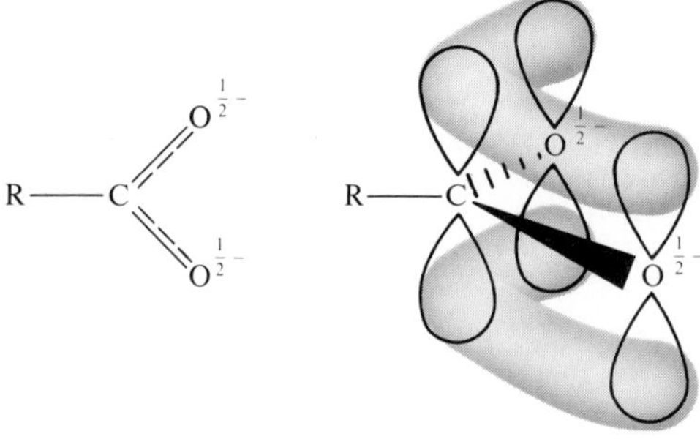

FIGURE 20-2 Structure of the carboxylate ion. Each of the C—O bonds has a bond order of $\frac{3}{2}$, from one σ bond and half of a π bond. Each oxygen atom bears half of the negative charge.

The magnitude of the effect of a substituent depends on its distance from the carboxyl group. Substituents on the α-carbon atom are most effective in increasing the strength of an acid. More distant substituents have much smaller effects on acidity, showing that inductive effects decrease rapidly with distance.

$H{-}CH_2{-}COOH$	$Cl{-}CH_2{-}COOH$	$Cl{-}CHCl{-}COOH$	$Cl{-}CCl_2{-}COOH$
acetic acid	chloroacetic acid	dichloroacetic acid	trichloroacetic acid
$pK_a = 4.74$	$pK_a = 2.86$	$pK_a = 1.26$	$pK_a = 0.64$

$CH_3{-}CH_2{-}CHCl{-}C(=O){-}OH$	$CH_3{-}CHCl{-}CH_2{-}C(=O){-}OH$	$ClCH_2{-}CH_2{-}CH_2{-}C(=O){-}OH$
2-chlorobutanoic acid	3-chlorobutanoic acid	4-chlorobutanoic acid
$pK_a = 2.86$	$pK_a = 4.05$	$pK_a = 4.52$

Table 20-4 lists the values of K_a and pK_a for some substituted carboxylic acids, showing how a variety of electron-withdrawing groups enhance the strength of an acid.

TABLE 20-4
Values of K_a and pK_a for some substituted carboxylic acids

Acid	K_a	pK_a	Acid	K_a	pK_a
CH_3COOH	1.8×10^{-5}	4.74	$NCCH_2COOH$	3.4×10^{-3}	2.46
FCH_2COOH	2.6×10^{-3}	2.59	$C_6H_5CH_2COOH$	4.9×10^{-5}	4.31
$ClCH_2COOH$	1.4×10^{-3}	2.86			
$BrCH_2COOH$	1.3×10^{-3}	2.90			
ICH_2COOH	6.7×10^{-4}	3.18	$CH_3CH_2CH(Cl)COOH$	139×10^{-5}	2.86
F_3CCOOH	5.9×10^{-1}	0.23			
$Cl_2CHCOOH$	5.5×10^{-2}	1.26			
Cl_3CCOOH	2.3×10^{-1}	0.64			
$HOCH_2COOH$	1.5×10^{-4}	3.83	$CH_3CH(Cl)CH_2COOH$	8.9×10^{-5}	4.05
$O_2N{-}CH_2COOH$	2.1×10^{-2}	1.68			
CH_3OCH_2COOH	2.9×10^{-4}	3.54			
$CH_2{=}CHCH_2COOH$	4.5×10^{-5}	4.35	$CH_2(Cl)CH_2CH_2COOH$	3.0×10^{-5}	4.52
$HC{\equiv}CCH_2COOH$	4.8×10^{-4}	3.32			
CH_3CH_2COOH	1.3×10^{-5}	4.87	$CH_3CH_2CH_2COOH$	1.5×10^{-5}	4.82

PROBLEM 20-3

Rank each of the following sets of compounds in order of increasing acid strength.

(a) $CH_3{-}CH_2{-}COOH$ $\quad CH_3{-}CHBr{-}COOH$ $\quad CH_3{-}CBr_2{-}COOH$

(b) $CH_3{-}CH_2{-}CH_2{-}CHBr{-}COOH$ $\quad CH_3{-}CH_2{-}CHBr{-}CH_2{-}COOH$ $\quad CH_3{-}CHBr{-}CH_2{-}CH_2{-}COOH$

(c) $CH_3CH(NO_2)COOH$ $\quad CH_3CH(Cl)COOH$ $\quad CH_3CH_2COOH$ $\quad CH_3CH(C{\equiv}N)COOH$

20-5 SALTS OF CARBOXYLIC ACIDS

A strong base can completely deprotonate a carboxylic acid. The products are a carboxylate ion, the cation remaining from the base, and water. The combination of a carboxylate ion and a cation forms a **salt of a carboxylic acid.**

$$\underset{\text{carboxylic acid}}{R{-}\overset{\overset{\large O}{\|}}{C}{-}O{-}H} + \underset{\text{strong base}}{M^+ \; ^-OH} \rightleftharpoons \underset{\text{acid salt}}{R{-}\overset{\overset{\large O}{\|}}{C}{-}O^- \; M^+} + \underset{\text{water}}{H_2O}$$

For example, when sodium hydroxide is added to acetic acid, the sodium salt of acetic acid results.

$$\underset{\text{acetic acid}}{CH_3{-}\overset{\overset{\large O}{\|}}{C}{-}O{-}H} + \underset{\text{sodium hydroxide}}{Na^+ \; ^-OH} \rightleftharpoons \underset{\text{sodium acetate}}{CH_3{-}\overset{\overset{\large O}{\|}}{C}{-}O^- \; ^+Na} + H_2O$$

Because mineral acids are stronger than carboxylic acids, the addition of a mineral acid converts a carboxylic acid salt back to the original carboxylic acid.

$$\underset{\text{acid salt}}{R{-}\overset{\overset{\large O}{\|}}{C}{-}O^- \; ^+M} + H^+ \rightleftharpoons \underset{\text{regenerated acid}}{R{-}\overset{\overset{\large O}{\|}}{C}{-}O{-}H} + M^+$$

Example

$$\underset{\text{sodium acetate}}{CH_3{-}\overset{\overset{\large O}{\|}}{C}{-}O^- \; ^+Na} + H^+Cl^- \rightleftharpoons \underset{\text{acetic acid}}{CH_3{-}\overset{\overset{\large O}{\|}}{C}{-}O{-}H} + Na^+Cl^-$$

Carboxylic acid salts have very different properties from the carboxylic acids, and because carboxylic acids and their salts are easily interconverted, these salts serve as useful derivatives of carboxylic acids.

Nomenclature of carboxylic acid salts Salts of carboxylic acids are named simply by naming the cation, and then naming the carboxylate ion by replacing the *-ic acid* part of the acid name with *-ate*. The example above shows that sodium hydroxide reacts with acetic acid to form sodium acetate. The following examples show the formation and nomenclature of some other salts.

$$CH_3-CH_2-CH_2-CH_2-\overset{\overset{\displaystyle O}{\|}}{C}-OH \quad + \quad LiOH \quad \longrightarrow \quad CH_3-CH_2-CH_2-CH_2-\overset{\overset{\displaystyle O}{\|}}{C}-O^{-}\ {}^{+}Li$$

IUPAC name: pentanoic acid — lithium hydroxide — lithium pentanoate

common name: valeric acid — lithium valerate

$$CH_3-CH_2-CH_2-\overset{\overset{\displaystyle O}{\|}}{C}-OH \quad + \quad :NH_3 \quad \longrightarrow \quad CH_3-CH_2-CH_2-\overset{\overset{\displaystyle O}{\|}}{C}-O^{-}\ {}^{+}NH_4$$

IUPAC name: butanoic acid — ammonia — ammonium butanoate

common name: butyric acid — ammonium butyrate

Properties of acid salts Like the salts of amines (Chapter 19), carboxylic acid salts are solids with little odor. They generally melt at high temperatures, and they often decompose before their melting points are reached. Carboxylate salts of the alkali metals (Li^+, Na^+, K^+) and ammonium (NH_4^+) carboxylate salts are generally quite soluble in water but relatively insoluble in nonpolar organic solvents. *Soap* is a common example of carboxylate salts, consisting of the relatively soluble sodium salts of long-chain fatty acids (Chapter 25). The carboxylate salts of most other metal ions are insoluble in water. For example, when soap is used in "hard" water containing calcium, magnesium, or iron ions, the insoluble carboxylate salts precipitate out as "hard-water scum."

$$2\ CH_3(CH_2)_{16}-\overset{\overset{\displaystyle O}{\|}}{C}-O^{-}\ {}^{+}Na \quad + \quad Ca^{+2} \quad \longrightarrow \quad [CH_3(CH_2)_{16}-\overset{\overset{\displaystyle O}{\|}}{C}-O]_2Ca\downarrow \quad + \quad 2\ Na^{+}$$

a soap — "hard-water scum"

Salt formation can be used to identify and purify acids. Carboxylic acids are deprotonated by the weak base sodium bicarbonate, forming the sodium salt of the acid, carbon dioxide, and water. An unknown compound that is insoluble in water, but dissolves in a sodium bicarbonate solution with a release of bubbles of carbon dioxide, is almost certainly a carboxylic acid.

$$R-\overset{\overset{\displaystyle O}{\|}}{C}-O-H \quad + \quad NaHCO_3 \quad \rightleftarrows \quad R-\overset{\overset{\displaystyle O}{\|}}{C}-O^{-}\ {}^{+}Na \quad + \quad H_2O \quad + \quad CO_2\uparrow$$

insoluble in water — water-soluble

Nonacidic (or weakly acidic) impurities can be removed from a carboxylic acid using acid-base extractions (Fig. 20-3). First, the acid is dissolved in an organic solvent such as ether and shaken with water. The acid remains in the organic phase while any water-soluble impurities are washed out. Next, the acid is washed with aqueous sodium bicarbonate, forming a salt that dissolves in the aqueous phase. Nonacidic impurities (and weakly acidic impurities such as phenols) remain in the ether phase. The phases are separated, and acidification of the aqueous

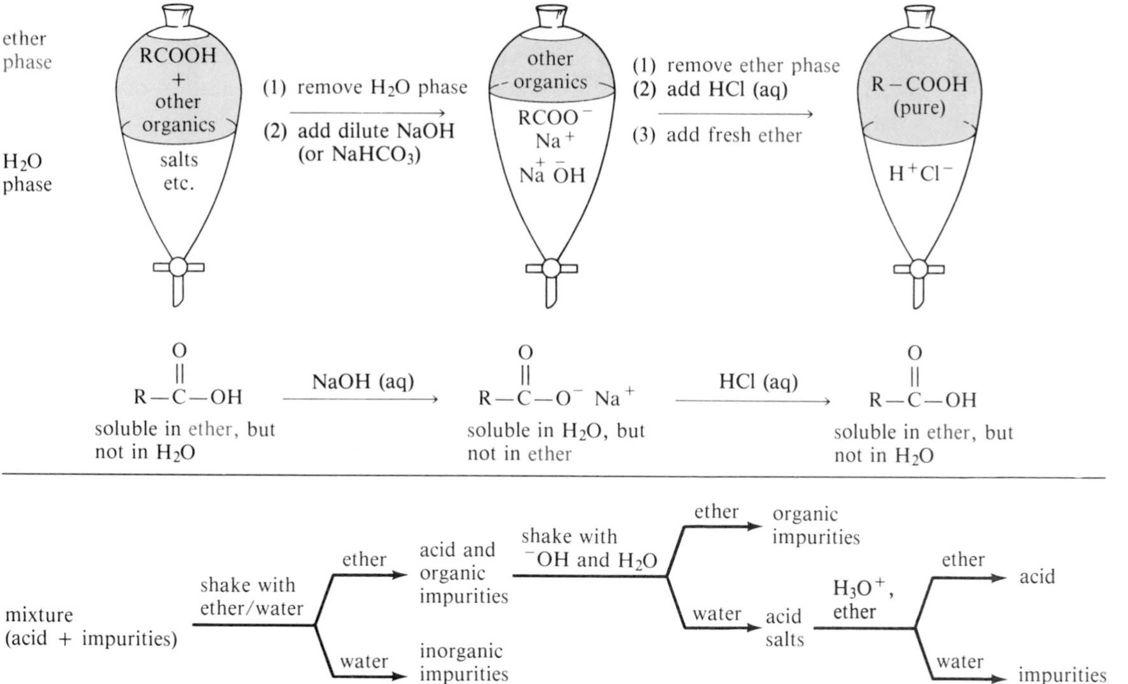

FIGURE 20-3 The solubility properties of acids and their salts may be used to remove nonacidic impurities. A carboxylic acid is more soluble in the organic phase, but its salt is more soluble in the aqueous phase. Acid-base extractions can move the acid from the ether phase into the aqueous phase and back into the ether phase, leaving impurities behind.

phase regenerates the acid, which is insoluble in water but dissolves in a fresh portion of ether. Evaporation of the final ether layer gives the purified acid.

PROBLEM 20-4

Suppose you have just synthesized heptanoic acid from 1-heptanol. The product is contaminated by sodium dichromate, sulfuric acid, 1-heptanol, and possibly heptanal. Explain how you would use acid-base extractions to purify the heptanoic acid. Use a chart such as that in Figure 20-3 to show where each impurity is found at each stage.

PROBLEM 20-5

Phenols are less acidic than carboxylic acids, with values of pK_a around 10. Phenols are deprotonated by (and therefore soluble in) solutions of sodium hydroxide but not by solutions of sodium bicarbonate. Explain how you would use extractions to isolate the three pure compounds from a mixture of *p*-cresol (*p*-methylphenol), cyclohexanone, and benzoic acid.

PROBLEM 20-6

Oxidation of a primary alcohol to an aldehyde usually gives some overoxidation to the carboxylic acid. Assume you have used the Collins reagent (CrO_3 in pyridine) to oxidize 1-pentanol to pentanal.

(a) Show how you would use acid-base extractions to purify the pentanal.
(b) Which of the expected impurities cannot be removed from pentanal by acid-base extractions? How would you remove this impurity?

20-6 COMMERCIAL SOURCES OF CARBOXYLIC ACIDS

The most important commercial carboxylic acid is acetic acid. *Vinegar,* a 5 percent aqueous solution of acetic acid used in cooking and in prepared foods such as pickles, ketchup, and salad dressings, is produced by fermentation of sugars and starches. An intermediate in this fermentation is ethyl alcohol. When fermented alcoholic beverages such as wine and cider are exposed to air, the alcohol is converted to acetic acid. This is the source of "wine vinegar" and "cider vinegar."

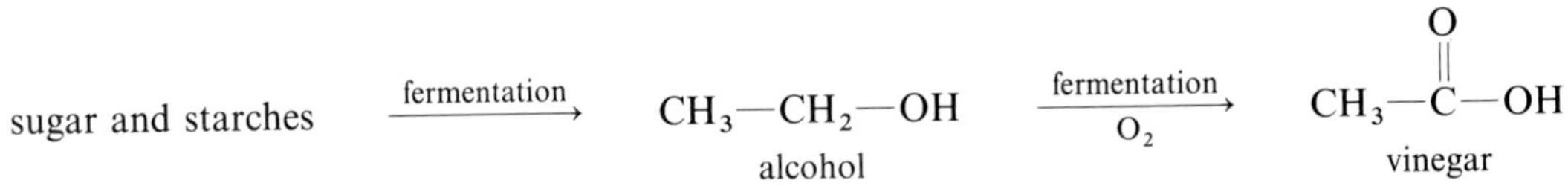

$$\text{sugar and starches} \xrightarrow{\text{fermentation}} \underset{\text{alcohol}}{CH_3{-}CH_2{-}OH} \xrightarrow[O_2]{\text{fermentation}} \underset{\text{vinegar}}{CH_3{-}\overset{\overset{\large O}{\|}}{C}{-}OH}$$

Acetic acid is also an industrial chemical. It serves as a solvent, a starting material for synthesis, and a catalyst for a wide variety of reactions. Most industrial acetic acid is produced from acetylene, using a mercuric-catalyzed hydration (see Section 14-10F) to form acetaldehyde, followed by a catalyzed air oxidation to give acetic acid.

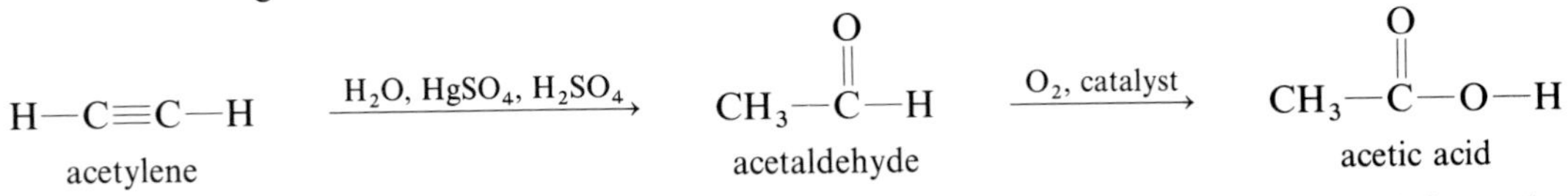

$$\underset{\text{acetylene}}{H{-}C{\equiv}C{-}H} \xrightarrow{H_2O,\ HgSO_4,\ H_2SO_4} \underset{\text{acetaldehyde}}{CH_3{-}\overset{\overset{\large O}{\|}}{C}{-}H} \xrightarrow{O_2,\ \text{catalyst}} \underset{\text{acetic acid}}{CH_3{-}\overset{\overset{\large O}{\|}}{C}{-}O{-}H}$$

Figure 20-4 shows how long-chain aliphatic acids are obtained from the hydrolysis of fats and oils, a reaction discussed in Chapter 25. The **fatty acids** found in fats and oils are generally straight-chain acids with even numbers of carbon atoms ranging between about C_6 and C_{18}. The hydrolysis of animal fat gives mostly saturated fatty acids, while plant oils give large amounts of unsaturated fatty acids with one or more olefinic double bonds.

Some of the aromatic carboxylic acids are also commercially important. Benzoic acid is used as an ingredient in medications, a preservative in foods, and a starting material for synthesis. Benzoic acid can be produced by the oxidation of toluene with, for example, potassium permanganate:

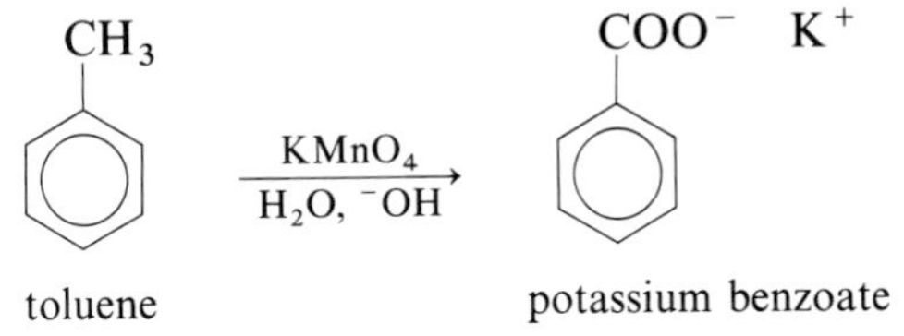

$$\underset{\text{toluene}}{C_6H_5{-}CH_3} \xrightarrow[H_2O,\ ^-OH]{KMnO_4} \underset{\text{potassium benzoate}}{C_6H_5{-}COO^-\ K^+}$$

The most important commercial diacids are adipic acid (hexanedioic acid) and phthalic acid (1,2-benzenedioic acid). Adipic acid is used in the manufacture of nylon 66, and phthalic acid is used to make polyesters. The industrial synthesis of adipic acid uses benzene as the starting material. Benzene is hydrogenated to

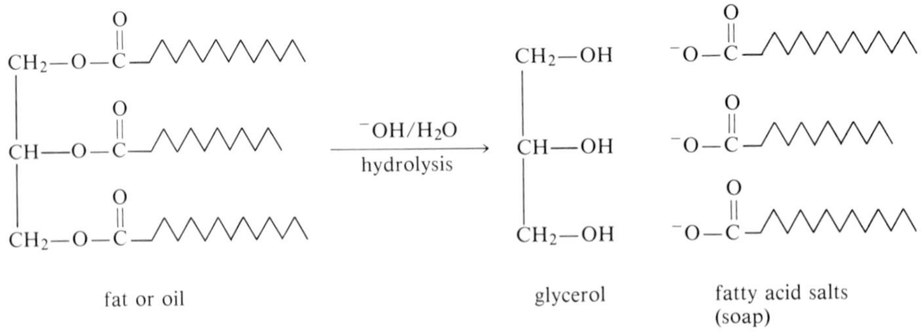

$$\underset{\text{fat or oil}}{\begin{matrix} CH_2{-}O{-}\overset{O}{\overset{\|}{C}}{-}R \\ | \\ CH{-}O{-}\overset{O}{\overset{\|}{C}}{-}R \\ | \\ CH_2{-}O{-}\overset{O}{\overset{\|}{C}}{-}R \end{matrix}} \xrightarrow[\text{hydrolysis}]{^-OH/H_2O} \underset{\text{glycerol}}{\begin{matrix} CH_2{-}OH \\ | \\ CH{-}OH \\ | \\ CH_2{-}OH \end{matrix}} \quad \underset{\text{fatty acid salts (soap)}}{\begin{matrix} ^-O{-}\overset{O}{\overset{\|}{C}}{-}R \\ ^-O{-}\overset{O}{\overset{\|}{C}}{-}R \\ ^-O{-}\overset{O}{\overset{\|}{C}}{-}R \end{matrix}}$$

FIGURE 20-4 Hydrolysis of a fat or an oil gives a mixture of the salts of straight-chain fatty acids. Animal fats contain primarily saturated fatty acids, while vegetable oils are commonly polyunsaturated.

cyclohexane, whose oxidation (using a cobalt/acetic acid catalyst) gives adipic acid. Phthalic acid is produced by the direct oxidation of naphthalene using a vanadium pentoxide catalyst.

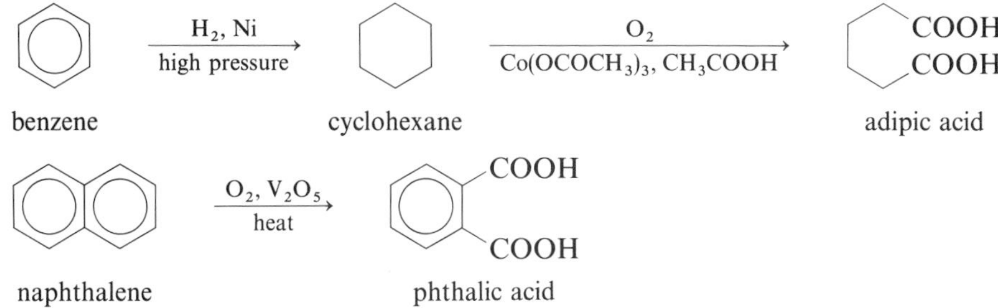

20-7 SPECTROSCOPY OF CARBOXYLIC ACIDS

20-7A INFRARED SPECTROSCOPY

The most obvious feature in the infrared spectrum of a carboxylic acid is the intense carbonyl stretching absorption. In a saturated acid, this vibration occurs around 1710 cm^{-1}, often broadened by hydrogen bonding involving the carbonyl group. In conjugated acids, the carbonyl stretching frequency is lowered to about 1690 cm^{-1}.

R—C(=O)—O—H: C=O 1710 cm^{-1}; O—H 2500–3500 cm^{-1}

>C=C(—)—C(=O)—O—H: C=O 1690 cm^{-1}; O—H 2500–3500 cm^{-1}

The O—H stretching vibration of a carboxylic acid absorbs in a broad band around 2500-3500 cm^{-1}. This frequency range is lower than the hydroxyl stretching frequencies of water and the alcohols, whose O—H groups absorb in a band centered around 3300 cm^{-1}. In the spectrum of a carboxylic acid, the broad hydroxyl band appears right on top of the C—H stretching region. This overlapping of the absorptions gives the 3000 cm^{-1} region a characteristic appearance of a broad peak (the O—H stretching) with sharp peaks (C—H stretching) superimposed on it.

The IR spectrum of 2-methylpropenoic acid is shown in Figure 20-5. Compare this conjugated example with the spectrum of hexanoic acid (Fig. 11-11, p. 459). Notice the shift in the position of the carbonyl absorptions and notice

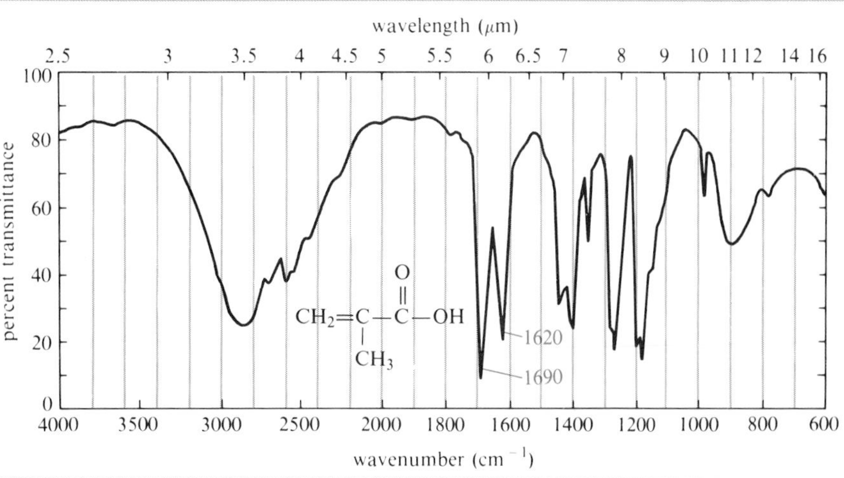

FIGURE 20-5 IR spectrum of 2-methylpropenoic acid.

that the conjugated, unsaturated acid has a fairly strong C=C stretching absorption around 1620 cm^{-1}, just to the right of the carbonyl absorption.

PROBLEM 20-7

The IR spectrum of *trans*-2-decenoic acid appears below. Point out the spectral characteristics that allow you to tell that this is a carboxylic acid, and show which features lead you to conclude that the acid is unsaturated and conjugated.

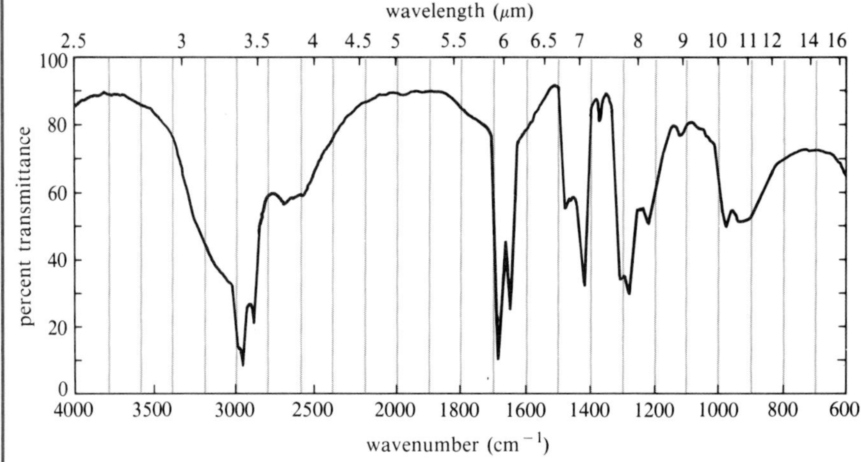

20-7B NMR SPECTROSCOPY

Carboxylic acid protons are the most deshielded protons we have discussed, absorbing between $\delta 10$ and $\delta 13$. Depending on the solvent and the concentration, this acid proton absorption may be sharp or broad, but it is always unsplit due to proton exchange.

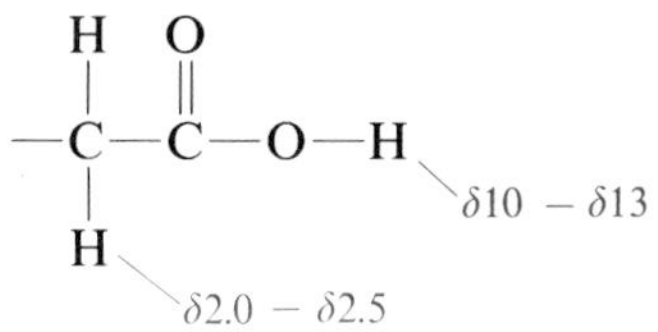

$$-\overset{\substack{H\\|}}{\underset{\substack{|\\H}}{C}}-\overset{\substack{O\\\|}}{C}-O-H$$

H (on the α-carbon): $\delta 2.0 - \delta 2.5$; O—H: $\delta 10 - \delta 13$

The protons on the α-carbon atom absorb between $\delta 2.0$ and $\delta 2.5$, in about the same position as the protons on a carbon atom alpha to a ketone or an aldehyde. The proton NMR spectrum of butanoic acid is shown in Figure 20-6.

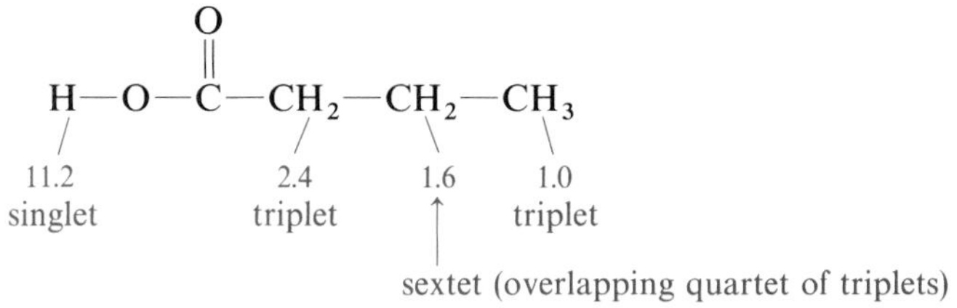

$$H-O-\overset{\substack{O\\\|}}{C}-CH_2-CH_2-CH_3$$

H—O: 11.2, singlet; CH_2 (α): 2.4, triplet; CH_2: 1.6, sextet (overlapping quartet of triplets); CH_3: 1.0, triplet

The carbon NMR chemical shifts of carboxylic acids resemble those of ketones and aldehydes. The carbonyl carbon atom absorbs around 180 ppm, and the α-carbon atom absorbs around 30 to 40 ppm. The chemical shifts of the carbon atoms in hexanoic acid are the following.

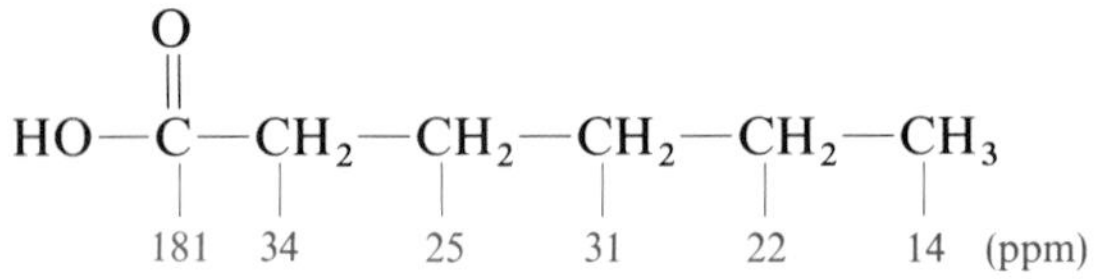

$$HO-\overset{\substack{O\\\|}}{C}-CH_2-CH_2-CH_2-CH_2-CH_3$$

181 34 25 31 22 14 (ppm)

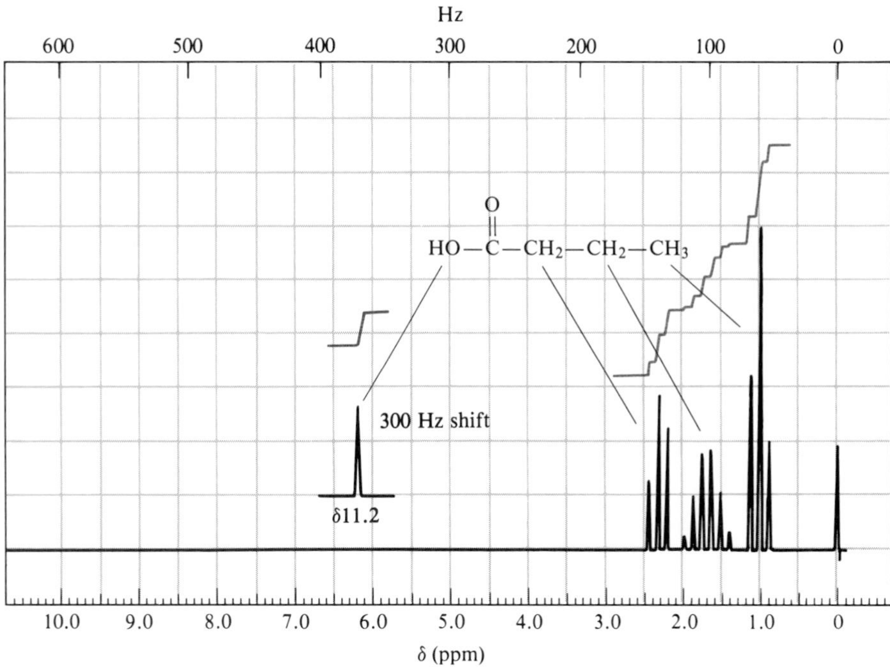

FIGURE 20-6 Proton NMR spectrum of butanoic acid.

PROBLEM 20-8

(a) Determine the structure of the carboxylic acid whose proton NMR spectrum appears below.

(b) Draw the NMR spectrum you would expect from the corresponding aldehyde whose oxidation would give this carboxylic acid.

(c) Point out two distinctive differences in the spectra of the aldehyde and the acid.

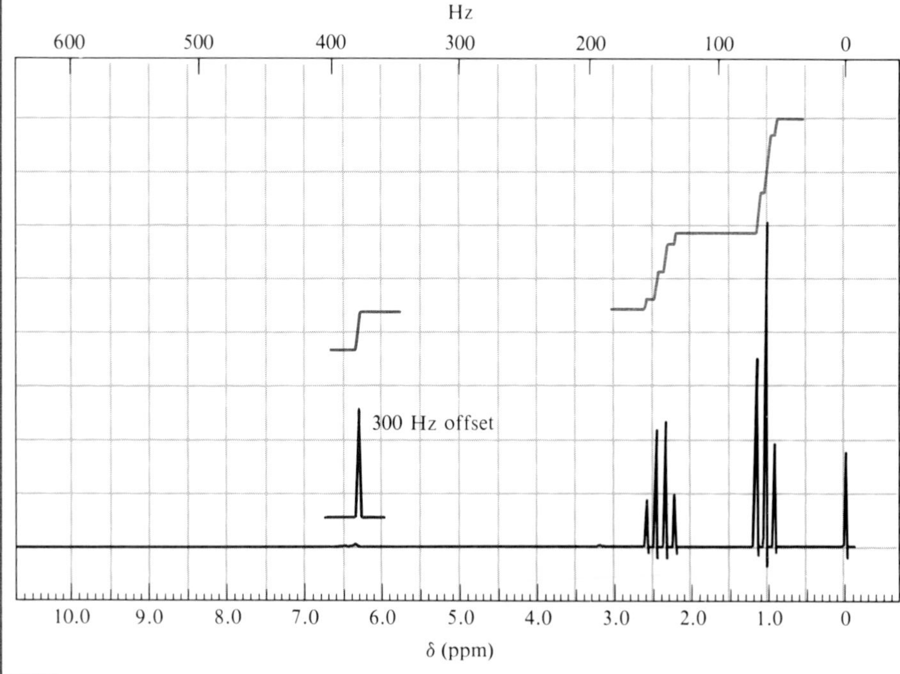

20-7C ULTRAVIOLET SPECTROSCOPY

Saturated carboxylic acids have a weak $n \rightarrow \pi^*$ transition that absorbs around 200 to 215 nm. This absorption corresponds to the weak transition around 270 to 300 nm in the spectra of ketones and aldehydes. The extinction coefficient is very small (about 30 to 100), and the absorption often goes unnoticed.

Conjugated acids show much stronger absorptions. One C=C double bond conjugated with the carboxyl group results in a spectrum with λ_{max} still around 200 nm, but with an extinction coefficient of about 10,000. A second conjugated double bond raises the value of λ_{max} to about 250 nm, as illustrated by the following examples.

$CH_2{=}CH{-}C({=}O){-}OH$ $\quad \lambda_{max} = 200$ nm $\quad \varepsilon = 10{,}000$

$CH_3{-}CH{=}CH{-}CH{=}CH{-}C({=}O){-}OH$ $\quad \lambda_{max} = 254$ nm $\quad \varepsilon = 25{,}000$

20-7D MASS SPECTROMETRY

The molecular ion peak of a carboxylic acid is usually small because favorable modes of fragmentation are available. The most common fragmentation for carboxylic acids is the loss of a molecule of an alkene (the McLafferty rearrangement, discussed in Section 18-5D). Another common fragmentation is the loss of an alkyl radical to give a resonance-stabilized cation with the positive charge delocalized over an allylic system and two oxygen atoms.

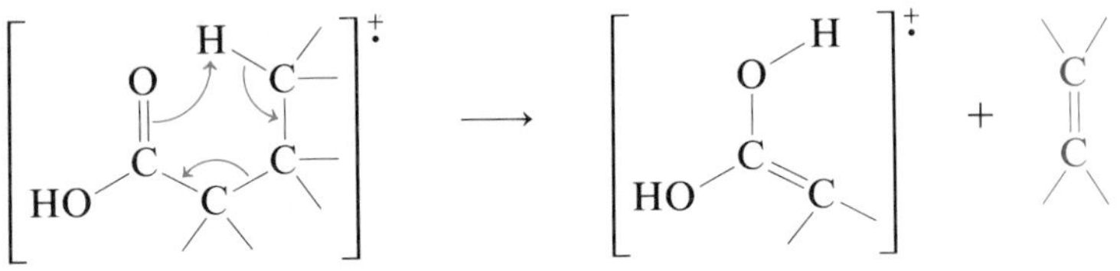

McLafferty rearrangement

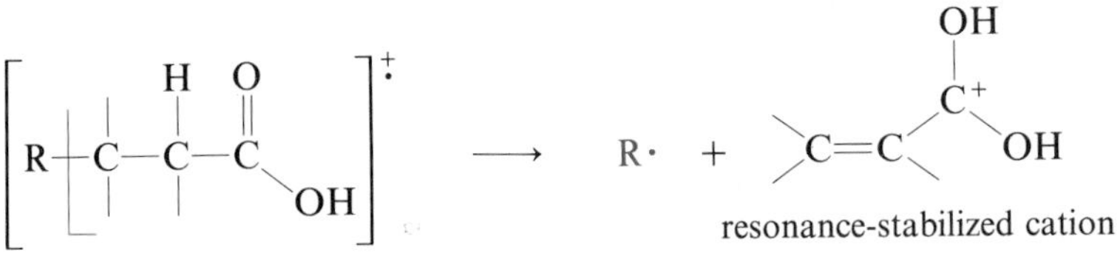

loss of an alkyl group

$R\cdot$ + resonance-stabilized cation

The mass spectrum of pentanoic acid is given in Figure 20-7. The base peak at m/z 60 corresponds to the fragment resulting from loss of propene via the McLafferty rearrangement. The strong peak at m/z 73 corresponds to loss of an ethyl radical with rearrangement to give a resonance-stabilized cation.

PROBLEM 20-9

Draw all four resonance structures of the fragment at m/z 73 in the mass spectrum of pentanoic acid.

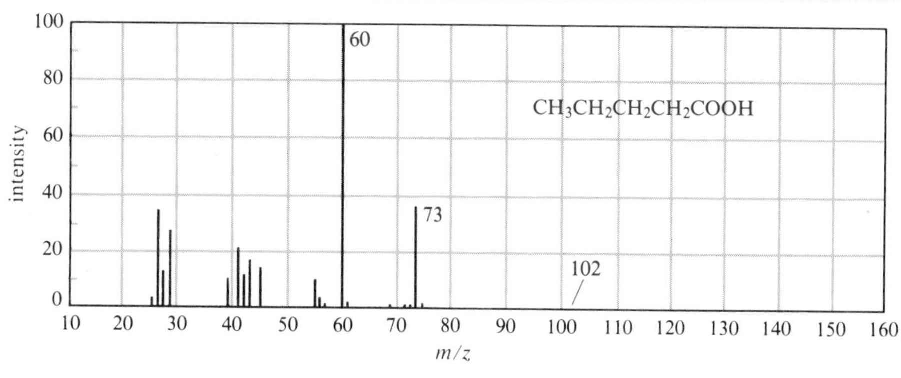

$$\left[\text{HO}-\overset{\displaystyle \text{O}}{\overset{\|}{\text{C}}}-\text{CH}_2-\text{CH}_2-\text{CH(H)}-\text{CH}_3\right]^{+\cdot} \longrightarrow \left[\text{HO}-\overset{\displaystyle \text{O}-\text{H}}{\overset{|}{\text{C}}}=\text{CH}_2\right]^{+\cdot} + \text{H(CH}_3\text{)C}=\text{CH}_2$$

m/z 60 — neutral, not observed

$$\left[\text{CH}_3\text{CH}_2 - \text{CH}_2-\text{CH}_2-\overset{\displaystyle \text{O}}{\overset{\|}{\text{C}}}-\text{OH}\right]^{+\cdot} \longrightarrow \text{CH}_3\text{CH}_2\cdot + \text{H}_2\text{C}=\text{CH}-\overset{+}{\text{C}}(\text{OH})-\text{OH}$$

73 — neutral, not observed — *m/z* 73

FIGURE 20-7 The mass spectrum of pentanoic acid shows a weak parent peak, a base peak from the McLafferty rearrangement, and another strong peak from the loss of an ethyl radical.

PROBLEM 20-10

Use equations to explain the prominent peaks at m/z 60 and m/z 73 in the mass spectrum of butanoic acid.

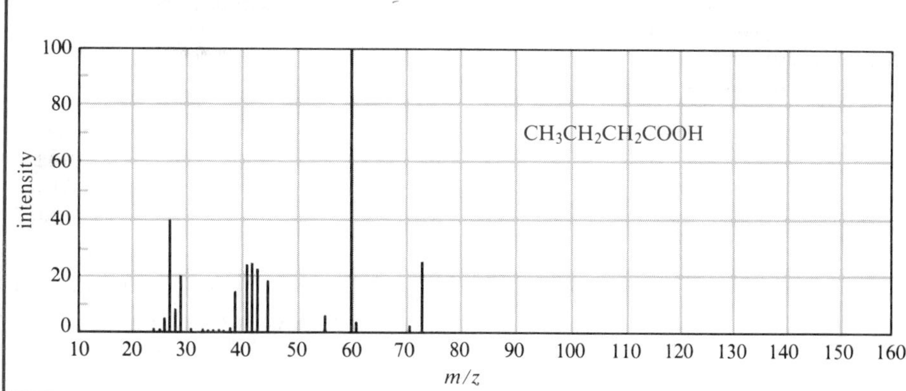

20-8 SYNTHESIS OF CARBOXYLIC ACIDS

20-8A REVIEW OF PREVIOUS SYNTHESES

We have already encountered three methods for preparing carboxylic acids: (1) oxidation of alcohols and aldehydes, (2) oxidative cleavage of alkenes and alkynes, and (3) severe side-chain oxidation of alkylbenzenes.

1. Primary alcohols or aldehydes are commonly oxidized to acids by chromic acid (H_2CrO_4, formed from $Na_2Cr_2O_7$ and H_2SO_4). Potassium perman-

ganate is occasionally used, but the yields are often lower (Sections 10-2B and 18-20).

$$\underset{\text{primary alcohol}}{R-CH_2-OH} \xrightarrow[\text{(or } KMnO_4)]{H_2CrO_4} \underset{\text{aldehyde}}{R-\overset{\overset{O}{\|}}{C}-H} \xrightarrow[\text{(or } KMnO_4)]{H_2CrO_4} \underset{\text{carboxylic acid}}{R-\overset{\overset{O}{\|}}{C}-OH}$$

Example

$$\underset{\text{3-phenylpropanol}}{Ph-CH_2-CH_2-CH_2-OH} \xrightarrow{Na_2Cr_2O_7,\ H_2SO_4} \underset{\text{3-phenylpropanoic acid}}{Ph-CH_2-CH_2-\overset{\overset{O}{\|}}{C}-OH}$$

2. Cold, dilute potassium permanganate reacts with alkenes to give glycols. Warm, concentrated permanganate solutions oxidize the glycols further, cleaving the central carbon-carbon bond. Depending on the substitution of the original double bond, ketones or acids may result.

$$\underset{\text{alkene}}{(R)(H)C=C(R')(R'')} \xrightarrow{\text{conc. } KMnO_4} \underset{\text{glycol (not isolated)}}{\left[H-\overset{R}{\underset{HO}{C}}-\overset{R'}{\underset{OH}{C}}-R''\right]} \longrightarrow \underset{\text{acid}}{R-COOH} + \underset{\text{ketone}}{O=C(R')(R'')}$$

Example

$$(Ph)(H)C=C(H)(CH_2-CH_3) \xrightarrow{\text{conc. } KMnO_4} Ph-COOH + CH_3-CH_2-COOH$$

$$\underset{\text{cyclohexene}}{C_6H_{10}} \xrightarrow{KMnO_4} \underset{\text{adipic acid}}{HOOC-(CH_2)_4-COOH}$$

With alkynes either ozonolysis or a vigorous permanganate oxidation cleaves the triple bond to give carboxylic acids.

$$\underset{\text{alkyne}}{R-C\equiv C-R'} \xrightarrow[\text{or (1) } O_3\text{ (2) } H_2O]{\text{conc. } KMnO_4} \underset{\text{(not isolated)}}{\left[R-\overset{\overset{O}{\|}}{C}-\overset{\overset{O}{\|}}{C}-R'\right]} \longrightarrow \underset{\text{carboxylic acids}}{R-COOH + HOOC-R'}$$

Example

$$CH_3-CH_2-CH_2-C\equiv C-Ph \xrightarrow[(2)\ H_2O]{(1)\ O_3} CH_3-CH_2-CH_2-COOH + Ph-COOH$$

3. The side chains of alkylbenzenes are oxidized to benzoic acid derivatives by treatment with hot chromic acid or hot potassium permanganate. Because this oxidation requires severe reaction conditions, it is useful only for making benzoic acid derivatives with no oxidizable functional groups. Oxidation-resistant functional groups such $-Cl$, $-NO_2$, $-SO_3H$, and $-COOH$ may be present.

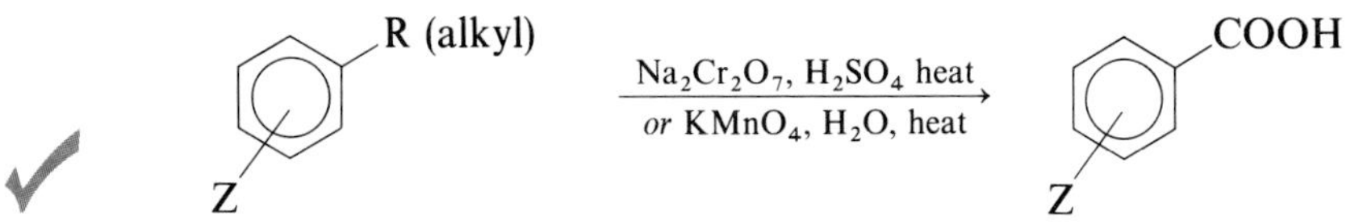

$$\text{R (alkyl)-C}_6\text{H}_4\text{-Z} \xrightarrow[\text{or } KMnO_4,\ H_2O,\ \text{heat}]{Na_2Cr_2O_7,\ H_2SO_4\ \text{heat}} \text{HOOC-C}_6\text{H}_4\text{-Z}$$

an alkylbenzene (Z must be oxidation-resistant) → a benzoic acid

Example

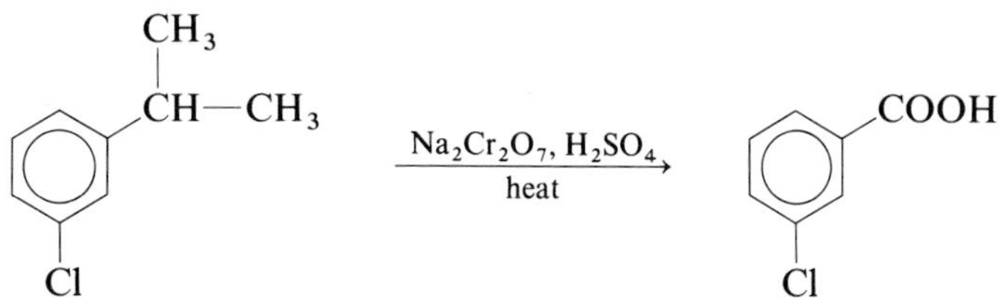

$$m\text{-Cl-C}_6\text{H}_4\text{-CH(CH}_3\text{)}_2 \xrightarrow[\text{heat}]{Na_2Cr_2O_7,\ H_2SO_4,} m\text{-Cl-C}_6\text{H}_4\text{-COOH}$$

m-chloroisopropylbenzene → *m*-chlorobenzoic acid

20-8B CARBOXYLATION OF GRIGNARD REAGENTS

Carbon dioxide adds to Grignard reagents to form the magnesium salts of carboxylic acids. The addition of dilute acid protonates these magnesium salts to give carboxylic acids. This method is useful because it converts a halide functional group to a carboxylic acid functional group, adding a carbon atom in the process.

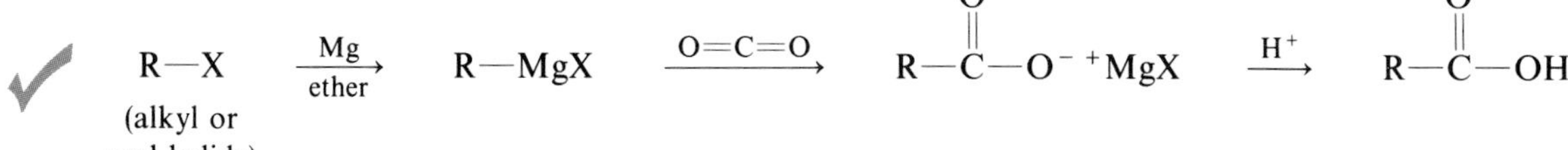

Example

$$\text{C}_6\text{H}_{11}\text{Br} \xrightarrow[\text{ether}]{Mg} \text{C}_6\text{H}_{11}\text{MgBr} \xrightarrow[(2)\ H^+]{(1)\ CO_2,} \text{C}_6\text{H}_{11}\text{COOH}$$

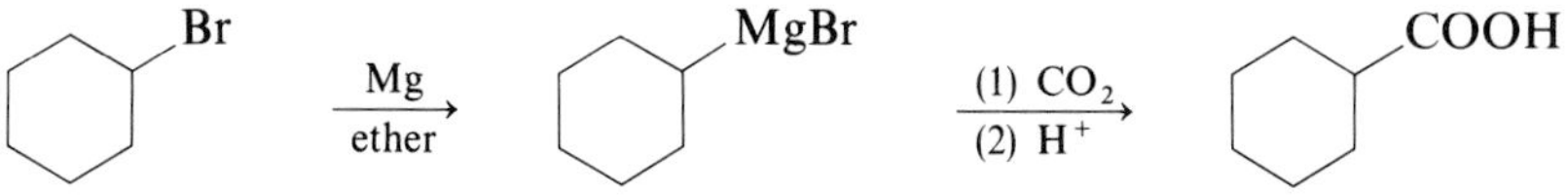

bromocyclohexane → cyclohexanecarboxylic acid

20-8C FORMATION AND HYDROLYSIS OF NITRILES

Another way to convert an alkyl halide (or tosylate) to a carboxylic acid with an additional carbon atom is to displace a halide with sodium cyanide. The product is a nitrile with one additional carbon atom. Acidic hydrolysis of the nitrile gives a carboxylic acid by a mechanism discussed in Chapter 21. This method is limited to halides and tosylates that are good S_N2 electrophiles: usually primary and unhindered.

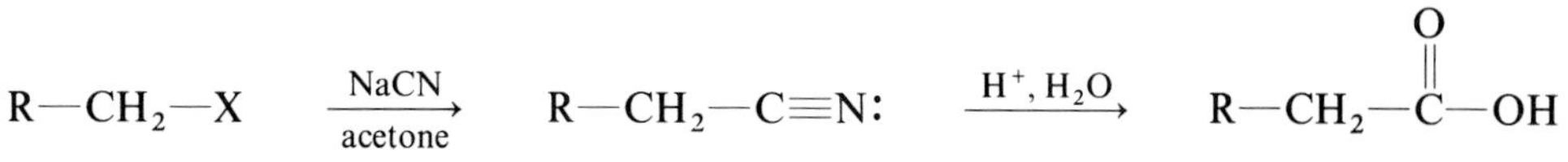

$$R-CH_2-X \xrightarrow[\text{acetone}]{NaCN} R-CH_2-C{\equiv}N: \xrightarrow{H^+,\ H_2O} R-CH_2-\overset{O}{\overset{\|}{C}}-OH$$

Example

$$C_6H_5-CH_2-Br \xrightarrow[(2)\ H^+,\ H_2O]{(1)\ NaCN,\ \text{acetone}} C_6H_5-CH_2-\overset{O}{\overset{\|}{C}}-OH$$

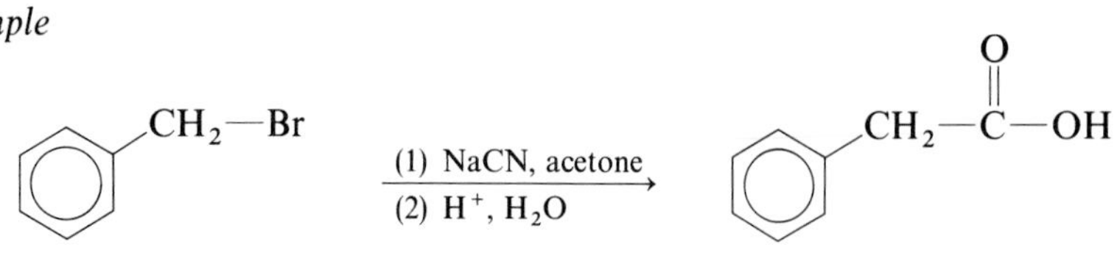

benzyl bromide → phenylacetic acid

PROBLEM 20-11

Show how you would synthesize the following carboxylic acids, using the indicated starting materials.

(a) 4-octyne ⟶ butanoic acid
(b) *trans*-cyclodecene ⟶ decanedioic acid
(c) bromobenzene ⟶ phenylacetic acid
(d) 2-butanol ⟶ 2-methylbutanoic acid
(e) *p*-xylene ⟶ terephthalic acid
(f) allyl iodide ⟶ 3-butenoic acid

SUMMARY OF SYNTHESES OF CARBOXYLIC ACIDS

1. *Oxidation of primary alcohols and aldehydes* (Section 10-2B and 18-20)

$$\underset{\text{primary alcohol}}{R-CH_2-OH} \xrightarrow[\text{or } KMnO_4]{H_2CrO_4} \underset{\text{aldehyde}}{R-\overset{O}{\overset{\|}{C}}-H} \xrightarrow[\text{or } KMnO_4]{H_2CrO_4} \underset{\text{carboxylic acid}}{R-\overset{O}{\overset{\|}{C}}-OH}$$

2. *Oxidative cleavage of alkenes and alkynes* (Sections 8-15 and 14-11)

$$\underset{\text{alkene}}{(R)(H)C=C(R')(R'')} \xrightarrow{\text{conc. } KMnO_4} \underset{\text{acid}}{R-COOH} + \underset{\text{ketone}}{O=C(R')(R'')}$$

$$\underset{\text{alkyne}}{R-C\equiv C-R'} \xrightarrow[\text{or (1) } O_3 \text{ (2) } H_2O]{\text{conc. } KMnO_4} \underset{\text{carboxylic acids}}{R-COOH + HOOC-R'}$$

3. *Oxidation of alkylbenzenes* (Section 17-14A)

$$\underset{\substack{\text{an alkylbenzene} \\ \text{(Z must be oxidation-resistant)}}}{Z-C_6H_4-R \text{ (alkyl)}} \xrightarrow[\text{or } KMnO_4,\ H_2O]{Na_2Cr_2O_7,\ H_2SO_4} \underset{\text{a benzoic acid}}{Z-C_6H_4-COOH}$$

4. *Carboxylation of Grignard reagents* (Section 20-8B)

$$\underset{\substack{\text{alkyl or} \\ \text{aryl halide}}}{R-X} \xrightarrow[\text{ether}]{Mg} R-MgX \xrightarrow{O=C=O} R-\overset{O}{\overset{\|}{C}}-O^{-}\ {}^{+}MgX \xrightarrow{H^+} \underset{\text{acid}}{R-\overset{O}{\overset{\|}{C}}-OH}$$

Example

$$\underset{\text{isobutyl bromide}}{CH_3-\underset{CH_2Br}{\underset{|}{CH}}-CH_3} \xrightarrow[\text{ether}]{Mg} CH_3-\underset{CH_2MgBr}{\underset{|}{CH}}-CH_3 \xrightarrow[\text{(2) } H^+]{\text{(1) } CO_2} \underset{\text{isovaleric acid}}{CH_3-\underset{CH_2-COOH}{\underset{|}{CH}}-CH_3}$$

5. *Formation and hydrolysis of nitriles* (Section 20-8C)

$$R-CH_2-X \xrightarrow[\text{acetone}]{NaCN} R-CH_2-C\equiv N: \xrightarrow{H^+,\ H_2O} R-CH_2-\overset{O}{\overset{\|}{C}}-OH$$

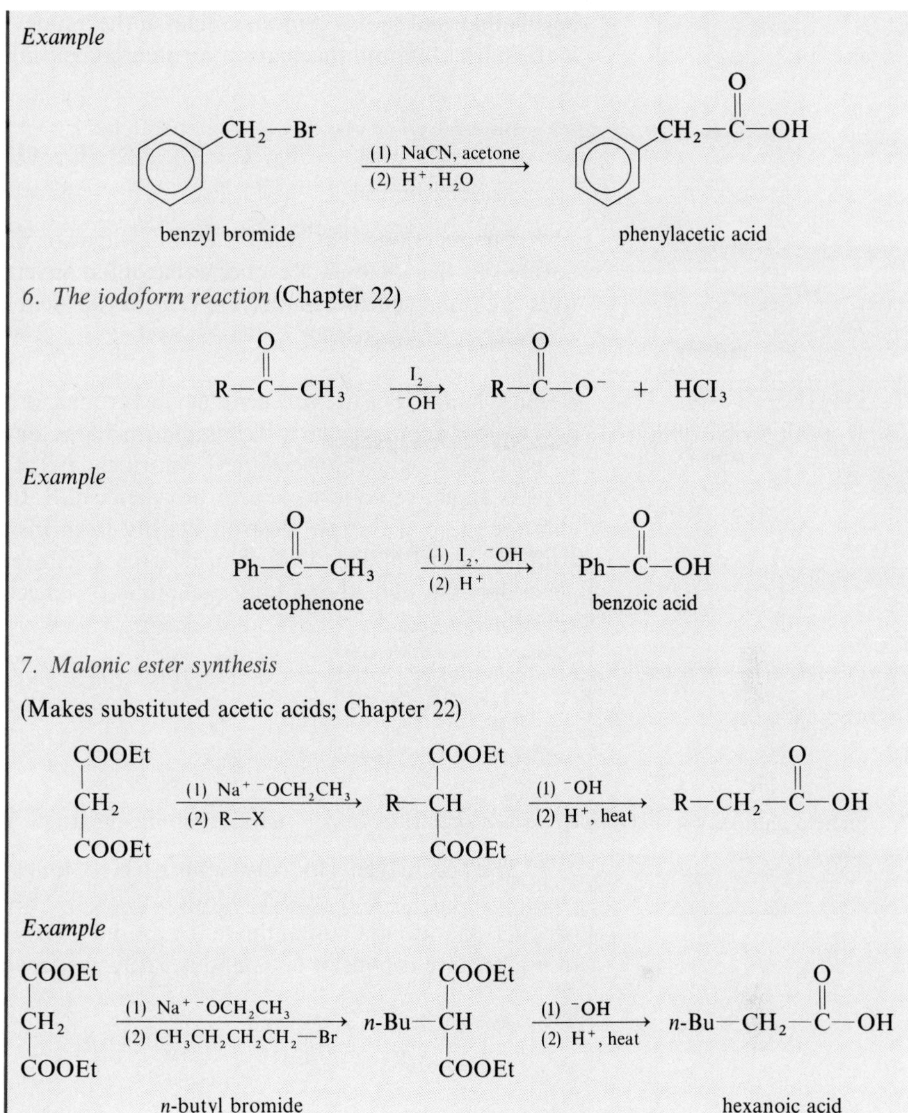

20-9 REACTIONS OF CARBOXYLIC ACIDS AND THEIR DERIVATIVES; NUCLEOPHILIC ACYL SUBSTITUTION

Although carboxylic acids also contain the carbonyl group, their reactions are quite different from those of ketones and aldehydes. Ketones and aldehydes commonly react by nucleophilic addition to the carbonyl group; but carboxylic acids (and their derivatives) more commonly react by **nucleophilic acyl substitution,** where one nucleophile replaces another on the acyl (C=O) carbon atom.

Nucleophilic acyl substitution

$$\mathrm{R{-}\overset{\overset{\displaystyle :\ddot{O}:}{\|}}{C}{-}X} \; + \; \mathrm{Nuc:^-} \;\rightleftharpoons\; \mathrm{R{-}\overset{\overset{\displaystyle :\ddot{O}:}{\|}}{C}{-}Nuc} \; + \; \mathrm{:X^-}$$

The structures of acid derivatives differ in the nature of the nucleophile bonded to the acyl carbon: —OH in the acid, —OR′ in the ester, halide in the acyl halide, and —NH_2 (or an amine) in the amide. Nucleophilic acyl substitution is the most common mechanism for interconverting these derivatives.

$$\underset{\text{carboxylic acid}}{R-\overset{O}{\overset{\|}{C}}-OH} \quad \underset{\text{acyl halide}}{R-\overset{O}{\overset{\|}{C}}-X} \quad \underset{\text{ester}}{R-\overset{O}{\overset{\|}{C}}-O-R'} \quad \underset{\text{amide}}{R-\overset{O}{\overset{\|}{C}}-NH_2}$$

The reactions that convert carboxylic acids directly to these derivatives are covered in this chapter. Reactions that interconvert these and other acid derivatives are discussed in Chapter 21, Carboxylic Acid Derivatives.

20-10 SYNTHESIS AND USE OF ACID CHLORIDES

Because halide ions are excellent leaving groups for nucleophilic acyl substitution, acyl halides are particularly useful intermediates for making other acid derivatives. In particular, acid chlorides (acyl chlorides) are easily made and are commonly used as an activated form of a carboxylic acid. Both the carbonyl oxygen and the chlorine atom withdraw electron density from the acyl carbon atom, making it strongly electrophilic. Acid chlorides react with a wide range of nucleophiles, generally through the addition-elimination mechanism of nucleophilic acyl substitution.

$$\underset{\text{an acid chloride (acyl chloride)}}{R-\overset{O}{\overset{\|}{C}}-Cl}$$

$$\underset{\text{acid chloride}}{R-\overset{\ddot{O}:}{\overset{\|}{C}}-Cl \;\; \text{Nuc}:^-} \rightleftharpoons \underset{\text{tetrahedral intermediate}}{\left[R-\overset{:\ddot{O}:^-}{\overset{|}{\underset{\text{Nuc}}{\underset{|}{C}}}}-Cl\right]} \longrightarrow \underset{\text{acid derivative}}{R-\overset{\ddot{O}:}{\overset{\|}{C}}-\text{Nuc}} + Cl^-$$

The best reagents for converting carboxylic acids to acid chlorides are thionyl chloride ($SOCl_2$) and oxalyl chloride $(COCl)_2$, because they form gaseous byproducts that do not contaminate the product. Oxalyl chloride is particularly easy to use because it boils at 62°C and is easily evaporated from the reaction mixture.

$$R-\overset{O}{\overset{\|}{C}}-OH \xrightarrow[\text{or } Cl-\overset{O}{\overset{\|}{C}}-\overset{O}{\overset{\|}{C}}-Cl]{Cl-\overset{O}{\overset{\|}{S}}-Cl} R-\overset{O}{\overset{\|}{C}}-Cl$$

Examples

$$\underset{\text{oleic acid}}{CH_3(CH_2)_7CH=CH(CH_2)_7-\overset{O}{\overset{\|}{C}}-OH} \xrightarrow[\text{thionyl chloride}]{Cl-\overset{O}{\overset{\|}{S}}-Cl} \underset{\substack{\text{oleoyl chloride}\\(95\%)}}{CH_3(CH_2)_7CH=CH(CH_2)_7-\overset{O}{\overset{\|}{C}}-Cl} + SO_2\uparrow + HCl\uparrow$$

(cis: both H on the same side of C=C)

$$\underset{\text{3-phenylpropanoic acid}}{C_6H_5-CH_2-CH_2-\overset{O}{\overset{\|}{C}}-OH} \xrightarrow[\text{oxalyl chloride}]{Cl-\overset{O}{\overset{\|}{C}}-\overset{O}{\overset{\|}{C}}-Cl} \underset{\substack{\text{3-phenylpropanoyl acid}\\(95\%)}}{C_6H_5-CH_2-CH_2-\overset{O}{\overset{\|}{C}}-Cl} + HCl\uparrow + CO\uparrow + CO_2\uparrow$$

Acid chlorides react with alcohols to give esters through a nucleophilic acyl substitution by the addition-elimination mechanism discussed above. Attack by the alcohol at the electrophilic carbonyl group of the acid chloride gives a tetrahedral intermediate. Loss of chloride and deprotonation give the ester.

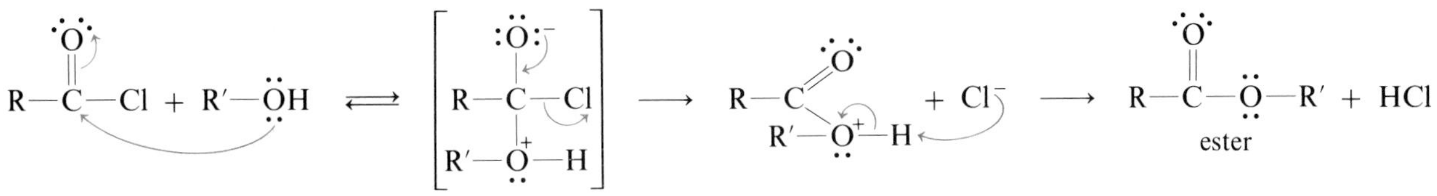

$$R-\overset{\overset{\displaystyle O}{\|}}{C}-Cl + R'-\ddot{O}H \rightleftharpoons \left[R-\overset{\overset{\displaystyle :\ddot{O}:^-}{|}}{\underset{\underset{\displaystyle R'-\overset{+}{O}-H}{|}}{C}}-Cl \right] \longrightarrow R-C(=\ddot{O})-\overset{+}{O}(R')-H + Cl^- \longrightarrow \underset{\text{ester}}{R-\overset{\overset{\displaystyle O}{\|}}{C}-\ddot{O}-R'} + HCl$$

This reaction provides an efficient two-step method for conversion of a carboxylic acid to an ester. The acid is converted to the acid chloride, which reacts with an alcohol to give the ester.

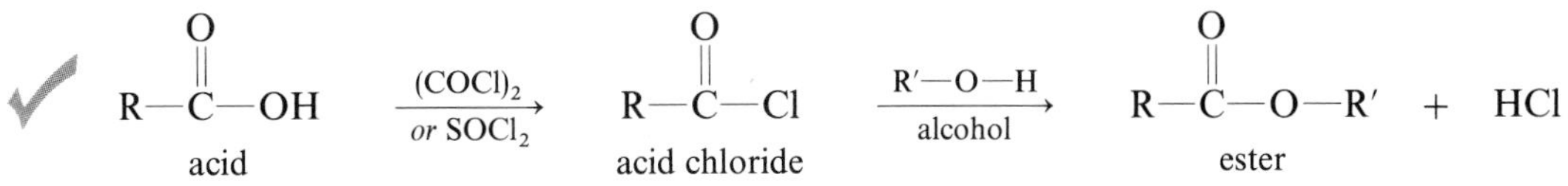

$$\underset{\text{acid}}{R-\overset{\overset{\displaystyle O}{\|}}{C}-OH} \xrightarrow[or\ SOCl_2]{(COCl)_2} \underset{\text{acid chloride}}{R-\overset{\overset{\displaystyle O}{\|}}{C}-Cl} \xrightarrow[\text{alcohol}]{R'-O-H} \underset{\text{ester}}{R-\overset{\overset{\displaystyle O}{\|}}{C}-O-R'} + HCl$$

Example

$$\underset{\text{benzoic acid}}{Ph-\overset{\overset{\displaystyle O}{\|}}{C}-OH} \xrightarrow{SOCl_2} \underset{\text{benzoyl chloride}}{Ph-\overset{\overset{\displaystyle O}{\|}}{C}-Cl} \xrightarrow[\text{ethanol}]{CH_3CH_2-OH} \underset{\text{ethyl benzoate}}{Ph-\overset{\overset{\displaystyle O}{\|}}{C}-O-CH_2CH_3} + HCl$$

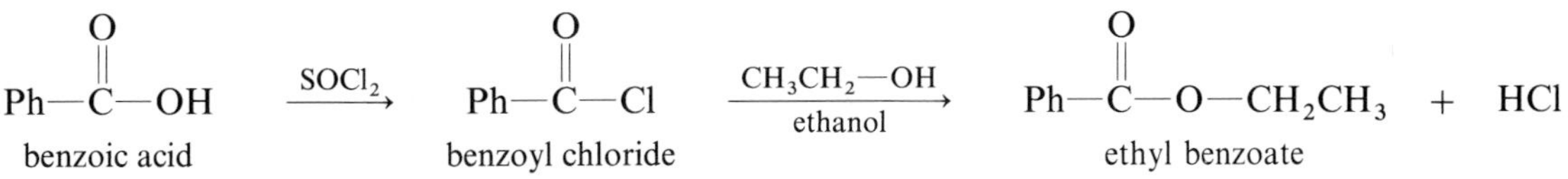

Ammonia and amines react with acid chlorides to give amides, also through the addition-elimination mechanism of nucleophilic acyl substitution. Carboxylic acids are efficiently converted to amides by forming the acid chloride, which reacts with an amine to give the amide.

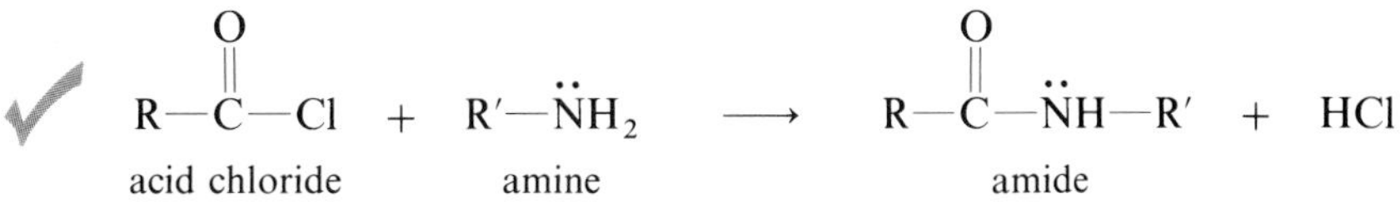

$$\underset{\text{acid chloride}}{R-\overset{\overset{\displaystyle O}{\|}}{C}-Cl} + \underset{\text{amine}}{R'-\ddot{N}H_2} \longrightarrow \underset{\text{amide}}{R-\overset{\overset{\displaystyle O}{\|}}{C}-\ddot{N}H-R'} + HCl$$

Example

$$\underset{\text{acetyl chloride}}{CH_3-\overset{\overset{\displaystyle O}{\|}}{C}-Cl} + \underset{\text{methylamine}}{CH_3-\ddot{N}H_2} \longrightarrow \underset{N\text{-methylacetamide}}{CH_3-\overset{\overset{\displaystyle O}{\|}}{C}-\overset{\overset{\displaystyle H}{|}}{\ddot{N}}-CH_3} + HCl$$

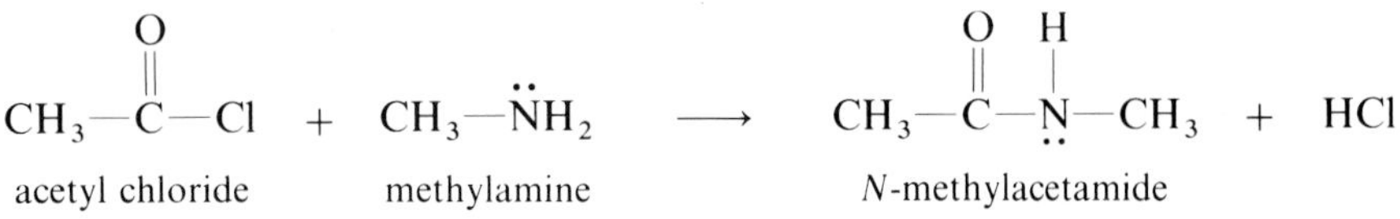

PROBLEM 20-12

Give mechanisms for the nucleophilic acyl substitutions to form ethyl benzoate and *N*-methylacetamide as shown above.

20-11 CONDENSATION OF ACIDS WITH ALCOHOLS: THE FISCHER ESTERIFICATION

Carboxylic acids are directly converted to esters by the **Fischer esterification,** an acid-catalyzed reaction with an alcohol.

$$\underset{\text{acid}}{R-\overset{O}{\overset{\|}{C}}-OH} + \underset{\text{alcohol}}{R'-OH} \xrightleftharpoons{H^+} \underset{\text{ester}}{R-\overset{O}{\overset{\|}{C}}-O-R'} + H_2O$$

Examples

$$CH_3-\overset{O}{\overset{\|}{C}}-OH + CH_3CH_2-OH \underset{K_{eq} = 3.38}{\overset{H_2SO_4}{\rightleftharpoons}} CH_3-\overset{O}{\overset{\|}{C}}-O-CH_2CH_3 + H_2O$$

$$\underset{\text{phthalic acid}}{C_6H_4(COOH)_2} \underset{}{\overset{H^+,\ \text{excess } CH_3OH}{\rightleftharpoons}} \underset{\text{dimethyl phthalate}}{C_6H_4(COOCH_3)_2}$$

The Fischer esterification mechanism is an acid-catalyzed nucleophilic acyl substitution. The carbonyl group of a carboxylic acid (in contrast to an acid chloride) is not sufficiently electrophilic to be attacked by an alcohol. The acid catalyst protonates the carbonyl group and activates it toward nucleophilic attack. Loss of a proton gives the hydrate of an ester.

$$R-\overset{O}{\overset{\|}{C}}-OH + H^+ \rightleftharpoons \left[R-\overset{^+OH}{\overset{\|}{C}}-OH \longleftrightarrow R-\overset{OH}{\overset{|}{C}}=\overset{+}{O}-H \longleftrightarrow R-\overset{OH}{\overset{|}{C^+}}-OH \right] \rightleftharpoons$$

$$H-\ddot{O}-R'$$

protonated carboxylic acid

$$R-\underset{H-\overset{+}{O}-R'}{\overset{OH}{\overset{|}{C}}}-OH \rightleftharpoons R-\underset{O-R'}{\overset{OH}{\overset{|}{C}}}-OH + H^+$$

hydrate of an ester

Loss of water from the hydrate of the ester occurs by the same mechanism as loss of water from the hydrate of a ketone (Section 18-14). Protonation of one of the hydroxyl groups allows it to leave as water, forming a resonance-stabilized cation. Loss of a proton from the second hydroxyl group gives the ester.

$$\underset{\text{hydrate}}{R-\underset{O-R'}{\overset{OH}{\overset{|}{C}}}-OH} + H^+ \rightleftharpoons \underset{\text{protonated}}{R-\underset{O-R'}{\overset{H-\overset{+}{O}-H}{\overset{|}{C}}}-OH} \rightleftharpoons$$

$$H_2\ddot{O}: + \left[R-C\begin{smallmatrix}\overset{+}{O}-H \\ O-R'\end{smallmatrix} \longleftrightarrow R-C^+\begin{smallmatrix}OH \\ O-R'\end{smallmatrix} \longleftrightarrow R-C\begin{smallmatrix}OH \\ \overset{+}{O}-R'\end{smallmatrix} \right] \rightleftharpoons R-C\begin{smallmatrix}O \\ O-R'\end{smallmatrix} + H_3O^+$$

resonance-stabilized   ester

PROBLEM 20-13

A carboxylic acid has two oxygen atoms, each with two nonbonding pairs of electrons.

(a) Draw the resonance structures of a carboxylic acid that is protonated on the hydroxyl oxygen atom.
(b) Compare the resonance structures with those given above for an acid protonated on the carbonyl oxygen atom.
(c) Explain why the carbonyl oxygen atom of a carboxylic acid is more basic than the hydroxyl oxygen.

PROBLEM 20-14

Give a detailed mechanism for the acid-catalyzed reaction of acetic acid with ethanol to give ethyl acetate.

Fischer esterification is an equilibrium reaction, and typical equilibrium constants are not very large. For example, if 1 mole of acetic acid is mixed with 1 mole of ethanol, the equilibrium mixture contains 0.65 mole each of ethyl acetate and water, and 0.35 mole each of acetic acid and ethanol. Esterification using secondary and tertiary alcohols gives even smaller equilibrium constants.

The esterification may be driven to the right either by using an excess of one of the reactants or by removing one of the products. For example, in the formation of ethyl esters, excess of ethanol is often used to drive the equilibrium as far as possible toward the ester side. Alternatively, water may be removed either by distillation or by the addition of a dehydrating agent such as magnesium sulfate or molecular sieves (dehydrated zeolite crystals that adsorb water).

Because of the inconvenience of driving the Fischer esterification to completion, the reaction of an acid chloride with an alcohol is often preferred for laboratory synthesis of esters. The Fischer esterification is commonly used in industry, where the techniques mentioned above give good yields of products. This synthesis avoids the relatively expensive step of converting the acid to its acid chloride.

PROBLEM 20-15

Show how the Fischer esterification might be used to form each of the following esters. In each case, suggest a method for driving the reaction to completion.

(a) methyl salicylate (b) methyl formate (b.p. 32°C) (c) ethyl benzoate

20-12 ESTERIFICATION USING DIAZOMETHANE

Carboxylic acids may be converted to their methyl esters very simply by adding an ether solution of diazomethane. The only by-product is nitrogen gas, and any excess diazomethane also evaporates. Purification of the ester usually involves only evaporation of the solvent.

$$\underset{\text{acid}}{R-\overset{\overset{\displaystyle O}{\|}}{C}-OH} + \underset{\text{diazomethane}}{CH_2N_2} \longrightarrow \underset{\text{methyl ester}}{R-\overset{\overset{\displaystyle O}{\|}}{C}-O-CH_3} + N_2\uparrow$$

$$\underset{\text{cyclobutanecarboxylic acid}}{\square\!-\!COOH} \xrightarrow{CH_2N_2} \underset{\text{methyl cyclobutanecarboxylate (100\%)}}{\square\!-\!COOCH_3} + N_2\uparrow$$

Diazomethane is a toxic, explosive yellow gas that dissolves in ether and is fairly safe to use in ether solutions. The reaction of diazomethane with carboxylic acids probably involves transfer of the acid proton, giving a methyldiazonium salt. This diazonium salt is an excellent methylating agent, with nitrogen gas as a leaving group.

$$R{-}C({=}\ddot{O}{:}){-}\ddot{O}{-}H \; {:}\overset{-}{C}H_2{-}\overset{+}{N}{\equiv}N{:} \longrightarrow R{-}C({=}\ddot{O}{:}){-}\ddot{O}{:}^- + CH_3{-}\overset{+}{N}{\equiv}N{:}$$

methyldiazonium salt

$$R{-}C({=}\ddot{O}{:}){-}\ddot{O}{:}^- \; CH_3{-}\overset{+}{N}{\equiv}N{:} \longrightarrow R{-}C({=}\ddot{O}{:}){-}\ddot{O}{-}CH_3 + {:}N{\equiv}N{:}$$

Because diazomethane is hazardous in large quantities, it is rarely used industrially or in large-scale laboratory reactions. The yields of methyl esters are excellent, however, so diazomethane is often used for small-scale esterifications of valuable and delicate carboxylic acids.

20-13 CONDENSATION OF ACIDS WITH AMINES: DIRECT SYNTHESIS OF AMIDES

Amides are also synthesized directly from carboxylic acids, although the acid chloride procedure uses milder conditions and often gives better yields. The initial reaction of a carboxylic acid with an amine gives an ammonium carboxylate salt. Heating this salt to well above 100°C drives off steam and forms an amide. This direct synthesis is an important industrial process because it avoids the expense of making the acid chloride.

$$\underset{\text{acid}}{R{-}C({=}O){-}OH} + \underset{\text{amine}}{R'{-}\ddot{N}H_2} \rightleftharpoons \underset{\text{salt}}{R{-}C({=}O){-}O^- \; H_3\overset{+}{N}{-}R'} \xrightarrow{\text{heat}} \underset{\text{amide}}{R{-}C({=}O){-}\ddot{N}H{-}R'} + H_2O\uparrow$$

Example

$$\underset{\text{benzoic acid}}{C_6H_5{-}COOH} + \underset{\text{ethylamine}}{CH_3CH_2\ddot{N}H_2} \longrightarrow \underset{\text{ethylammonium benzoate}}{C_6H_5{-}COO^- \; H_3\overset{+}{N}CH_2CH_3} \xrightarrow{\text{heat}} \underset{N\text{-ethylbenzamide}}{C_6H_5{-}C({=}O){-}\ddot{N}HCH_2CH_3} + H_2O$$

PROBLEM 20-16

Show how to synthesize the following compounds
(1) using benzoyl chloride and any other necessary reagents.
(2) using benzoic acid and any other necessary reagents.

(a) $C_6H_5{-}C({=}O){-}N(CH_3)_2$ — *N,N*-dimethyl benzamide

(b) $C_6H_5{-}C({=}O){-}O{-}CH(CH_3)_2$ — isopropyl benzoate

(c) $C_6H_5{-}C({=}O){-}OCH_3$ — methyl benzoate

20-14 REDUCTION OF CARBOXYLIC ACIDS

Lithium aluminum hydride ($LiAlH_4$ or LAH) reduces carboxylic acids to primary alcohols. The aldehyde is an intermediate in this reduction, but it cannot be isolated because it is reduced more easily than the original acid.

$$\underset{\text{acid}}{R-\overset{\overset{\Large O}{\|}}{C}-OH} \xrightarrow[(2)\ H_3O^+]{(1)\ LiAlH_4} \underset{\text{primary alcohol}}{R-CH_2-OH}$$

Example

$$\underset{\text{phenylacetic acid}}{C_6H_5-CH_2-\overset{\overset{\Large O}{\|}}{C}-OH} \xrightarrow[(2)\ H_3O^+]{(1)\ LiAlH_4} \underset{\substack{\text{2-phenylethanol}\\(75\%)}}{C_6H_5-CH_2-CH_2OH}$$

Lithium aluminum hydride is a strong base, and the first step in this reaction is deprotonation of the acid. Hydrogen gas is evolved, and the lithium salt of the acid results.

$$R-\overset{\overset{\Large O}{\|}}{C}-OH + Li^{+}\ {}^{-}AlH_4 \longrightarrow H_2\uparrow + R-\overset{\overset{\Large O}{\|}}{C}-O^{-}\ {}^{+}Li + AlH_3$$

AlH_3 adds to the carboxyl group of the lithium carboxylate salt.

$$R-C(=\ddot{O}:)(-\ddot{O}-Li) + H-AlH_2 \longrightarrow R-\underset{:\ddot{O}-Li}{C}(-H)(-\ddot{O}:-AlH_2)$$

Elimination gives an aldehyde which is quickly reduced to a lithium alkoxide.

$$R-\overset{:\ddot{O}-AlH_2}{\underset{:\ddot{O}:^{-}\ Li^{+}}{C}}-H \longrightarrow \underset{\text{aldehyde}}{R-\overset{H}{C}=\ddot{O}:} \quad Li^{+}\ {}^{-}:\ddot{O}-AlH_3 \longrightarrow \underset{\text{lithium alkoxide}}{R-\overset{H}{\underset{:\ddot{O}:^{-}\ Li^{+}}{C}}-H}$$

The water added in the second step protonates the alkoxide to give the primary alcohol.

$$R-CH_2-O^{-}Li^{+} + H_2O \longrightarrow R-CH_2-OH + LiOH$$

Carboxylic acids are also reduced to primary alcohols by diborane. Diborane reacts with the carboxyl group faster than with any other carbonyl function. It often gives excellent selectivity, as shown by the following example, where a carboxylic acid is reduced while a ketone is unaffected.

$$H_3C-\overset{\overset{\Large O}{\|}}{C}-C_6H_4-\overset{\overset{\Large O}{\|}}{C}-OH \xrightarrow[\text{diglyme}]{B_2H_6} \underset{(80\%)}{H_3C-\overset{\overset{\Large O}{\|}}{C}-C_6H_4-CH_2OH}$$

Reduction to aldehydes Reduction of carboxylic acids to aldehydes is difficult because aldehydes are more reactive than are carboxylic acids toward most reducing agents. Almost any reagent that reduces acids to aldehydes also reduces aldehydes to primary alcohols. What is needed is a derivative of the acid that is more reactive than the aldehyde. As you might guess, the reactive acid derivative is the acid chloride.

Lithium aluminum tri(*t*-butoxy)hydride, $LiAl[OC(CH_3)_3]_3H$, is a weaker reducing agent than lithium aluminum hydride. It reduces acid chlorides because they are strongly activated toward nucleophilic addition of a hydride ion. Under these conditions, the aldehyde reduces more slowly, and it is easily isolated.

$$\underset{\text{acid chloride}}{R-\overset{\overset{\cdot\cdot}{O}\colon}{\overset{\|}{C}}-Cl} + LiAl(O-R)_3H \rightleftharpoons R-\overset{:\ddot{O}:^{-}\ {}^{+}Li}{\overset{|}{\underset{\underset{H}{|}}{C}}}-Cl + Al(O-R)_3 \longrightarrow \underset{\text{aldehyde}}{R-\overset{O}{\overset{\|}{C}}-H} + LiCl$$

Example

$$\underset{\text{isobutyric acid}}{CH_3-\underset{\underset{CH_3}{|}}{CH}-\overset{O}{\overset{\|}{C}}-OH} \xrightarrow{SOCl_2} \underset{\text{isobutyryl chloride}}{CH_3-\underset{\underset{CH_3}{|}}{CH}-\overset{O}{\overset{\|}{C}}-Cl} \xrightarrow{LiAl[OC(CH_3)_3]_3H} \underset{\text{isobutyraldehyde}}{CH_3-\underset{\underset{CH_3}{|}}{CH}-\overset{O}{\overset{\|}{C}}-H}$$

PROBLEM 20-17

Show how you would synthesize the following compounds from the appropriate carboxylic acids or acid derivatives.

(a) $C_6H_5CH_2CHO$ (b) $C_6H_5CH_2CH_2OH$ (c) 2-(CH_2OH)cyclopentanone (d) lactone ring bearing CH_2OH

20-15 ALKYLATION OF CARBOXYLIC ACIDS TO FORM KETONES

A general method of making ketones involves the reaction of a carboxylic acid with 2 equivalents of an organolithium reagent. This reaction was first discussed in Section 18-9.

$$R-\overset{O}{\overset{\|}{C}}-O-H \xrightarrow[(2)\ H_2O]{(1)\ 2\ R'-Li} R-\overset{O}{\overset{\|}{C}}-R' + R'-H$$

Example

$$\underset{\text{benzoic acid}}{C_6H_5-COOH} \xrightarrow[(2)\ H_2O]{(1)\ 2\ CH_3CH_2-Li} \underset{\text{propiophenone}}{C_6H_5-\overset{O}{\overset{\|}{C}}-CH_2CH_3}$$

The first equivalent of the organolithium reagent simply deprotonates the acid. The second equivalent adds to the carbonyl to give a stable dianion. Hydrolysis of the dianion (by the addition of water) gives the hydrate of a ketone. Because the ketone is formed in a separate hydrolysis step (rather than in the presence of the organolithium reagent), overalkylation is not observed.

$$\underset{\text{carboxylic acid}}{R-\overset{O}{\overset{\|}{C}}-OH} \xrightarrow{2\ R'-Li} \underset{\text{dianion}}{R-\overset{OLi}{\overset{|}{\underset{R'}{\underset{|}{C}}}}-OLi} \xrightarrow{H_3O^+} \underset{\text{hydrate of ketone}}{R-\overset{OH}{\overset{|}{\underset{R'}{\underset{|}{C}}}}-OH} \rightleftarrows \underset{\text{ketone}}{R-\overset{O}{\overset{\|}{C}}-R'} + H_2O$$

PROBLEM 20-18
Give the mechanism for conversion of the dianion to the ketone under mildly acidic conditions.

PROBLEM 20-19
Show how the following ketones might be synthesized from the indicated acids, using any additional reagents that are needed.

(a) propiophenone from propionic acid (two ways, using alkylation of the acid and using Friedel-Crafts acylation)
(b) methyl cyclohexyl ketone from cyclohexanecarboxylic acid

20-16 DECARBOXYLATION OF CARBOXYLATE RADICALS: THE HUNSDIECKER REACTION

Carboxylic acids may be converted to alkyl halides with the loss of one carbon atom by the **Hunsdiecker reaction.**

✓

$$R-\overset{O}{\overset{\|}{C}}-O^{-}\ {}^{+}Ag + Br_2 \xrightarrow{\text{heat}} R-Br + CO_2\uparrow + AgBr\downarrow$$

Examples

$$\underset{\text{silver 3-phenylpropanoate}}{C_6H_5-CH_2-CH_2-\overset{O}{\overset{\|}{C}}-O^{-}\ {}^{+}Ag} \xrightarrow[\text{heat}]{Br_2} \underset{\substack{\text{1-bromo-2-phenylethane} \\ (70\%)}}{C_6H_5-CH_2-CH_2-Br} + CO_2\uparrow + AgBr\downarrow$$

$$\underset{\text{cyclobutanecarboxylic acid}}{\text{C}_4\text{H}_7-COOH} \xrightarrow[\text{heat}]{I_2,\ Pb(OAc)_4} \underset{\substack{\text{iodocyclobutane} \\ (90\%)}}{\text{C}_4\text{H}_7-I} + CO_2\uparrow$$

The Hunsdiecker reaction is usually carried out by treating the carboxylic acid with a heavy-metal base such as Ag_2O, HgO, or $Pb(OAc)_4$ to form the heavy-metal salt. Bromine or iodine is then added, and the reaction mixture is heated. This reaction forms the metal halide together with an acyl hypobromite (or an acyl hypoiodite) which dissociates into radicals upon heating.

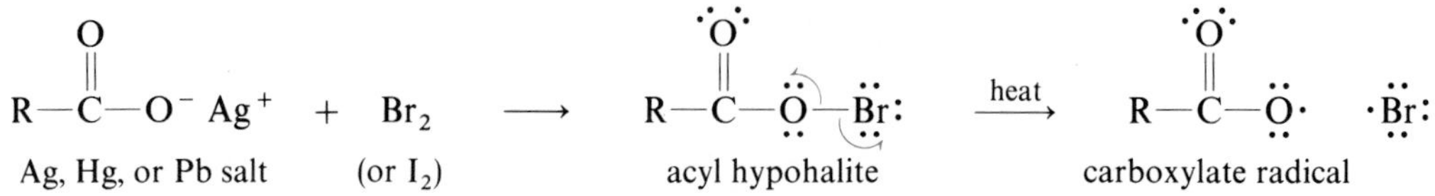

$$\mathrm{R{-}\overset{\overset{\large O}{\|}}{C}{-}O^-\ Ag^+} \quad + \quad \mathrm{Br_2} \longrightarrow \mathrm{R{-}\overset{\overset{\large O}{\|}}{C}{-}O{-}Br} \xrightarrow{\text{heat}} \mathrm{R{-}\overset{\overset{\large O}{\|}}{C}{-}O\cdot} \quad \mathrm{\cdot Br}$$

Ag, Hg, or Pb salt (or I_2) acyl hypohalite carboxylate radical

Although most carboxylate *anions* are quite stable, the corresponding *radicals* decarboxylate by losing CO_2, leaving alkyl radicals. These radicals initiate a free-radical chain reaction involving decarboxylation and formation of an alkyl halide.

Initiation step

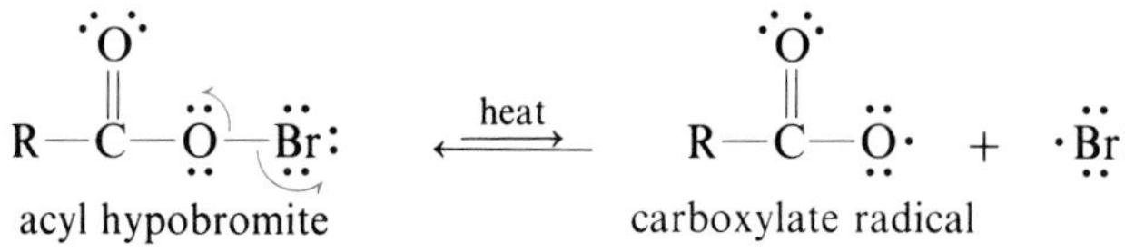

$$\mathrm{R{-}\overset{\overset{\large O}{\|}}{C}{-}O{-}Br} \overset{\text{heat}}{\rightleftarrows} \mathrm{R{-}\overset{\overset{\large O}{\|}}{C}{-}O\cdot} + \mathrm{\cdot Br}$$

acyl hypobromite carboxylate radical

Propagation steps

1. $$\mathrm{R{-}\overset{\overset{\large O}{\|}}{C}{-}O\cdot} \longrightarrow \mathrm{R\cdot} + \mathrm{O{=}C{=}O}$$

carboxylate radical alkyl radical carbon dioxide

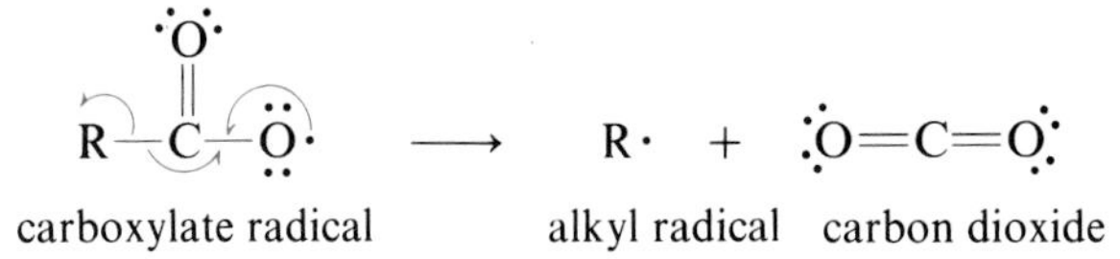

2. $$\mathrm{R\cdot} + \mathrm{Br{-}O{-}\overset{\overset{\large O}{\|}}{C}{-}R} \longrightarrow \mathrm{R{-}Br} + \mathrm{R{-}\overset{\overset{\large O}{\|}}{C}{-}O\cdot}$$

alkyl radical acyl hypobromite alkyl bromide carboxylate radical

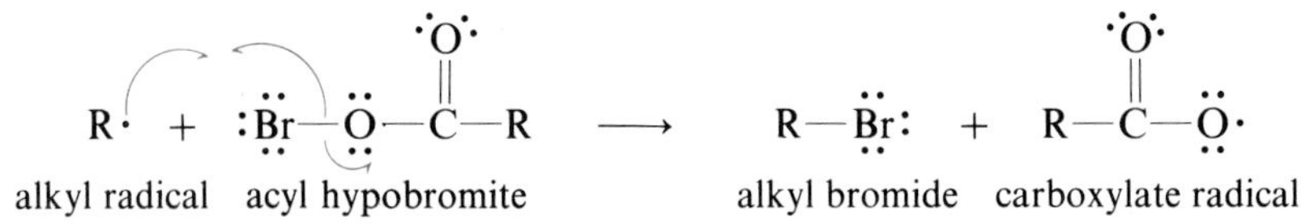

PROBLEM 20-20

Give a mechanism for the Hunsdiecker reaction of silver 3-phenylpropanoate shown above.

PROBLEM 20-21

Give the products expected when the following mixtures are heated.

(a) silver decanoate + bromine
(b) cyclooctane carboxylic acid + mercuric oxide + bromine

SUMMARY OF REACTIONS OF CARBOXYLIC ACIDS

1. Salt formation (Section 20-5)

$$\underset{\text{acid}}{\mathrm{R{-}\overset{\overset{\large O}{\|}}{C}{-}OH}} + \underset{\text{strong base}}{\mathrm{M^{+\ -}OH}} \rightleftarrows \underset{\text{salt}}{\mathrm{R{-}\overset{\overset{\large O}{\|}}{C}{-}O^{-\ +}M}} + \mathrm{H_2O}$$

Example

$$\mathrm{2\,CH_3{-}}\underset{\text{propionic acid}}{\mathrm{CH_2{-}\overset{\overset{\large O}{\|}}{C}{-}OH}} + \mathrm{Ca(OH)_2} \longrightarrow \underset{\text{calcium propionate}}{\mathrm{(CH_3{-}CH_2{-}\overset{\overset{\large O}{\|}}{C}{-}O^-)_2Ca^{2+}}}$$

2. *Conversion to acid chlorides* (Section 20-10)

$$\underset{\text{acid}}{R{-}\overset{\overset{\displaystyle O}{\|}}{C}{-}OH} + \underset{\text{thionyl chloride}}{Cl{-}\overset{\overset{\displaystyle O}{\|}}{S}{-}Cl} \longrightarrow \underset{\text{acid chloride}}{R{-}\overset{\overset{\displaystyle O}{\|}}{C}{-}Cl} + SO_2\uparrow + HCl\uparrow$$

Example

$$\underset{\text{butanoic acid}}{CH_3{-}CH_2{-}CH_2{-}\overset{\overset{\displaystyle O}{\|}}{C}{-}OH} + \underset{\text{thionyl chloride}}{SOCl_2} \longrightarrow \underset{\text{butanoyl chloride}}{CH_3{-}CH_2{-}CH_2{-}\overset{\overset{\displaystyle O}{\|}}{C}{-}Cl}$$

3. *Conversion to esters* (Sections 20-10, 20-11, and 20-12)

$$\underset{\text{acid}}{R{-}\overset{\overset{\displaystyle O}{\|}}{C}{-}OH} + \underset{\text{alcohol}}{R'{-}OH} \overset{H^+}{\rightleftarrows} \underset{\text{ester}}{R{-}\overset{\overset{\displaystyle O}{\|}}{C}{-}O{-}R'} + H_2O$$

$$\underset{\text{acid chloride}}{R{-}\overset{\overset{\displaystyle O}{\|}}{C}{-}Cl} + \underset{\text{alcohol}}{R'{-}OH} \longrightarrow \underset{\text{ester}}{R{-}\overset{\overset{\displaystyle O}{\|}}{C}{-}O{-}R'} + HCl\uparrow$$

$$\underset{\text{acid}}{R{-}\overset{\overset{\displaystyle O}{\|}}{C}{-}OH} + \underset{\text{diazomethane}}{CH_2N_2} \longrightarrow \underset{\text{methyl ester}}{R{-}\overset{\overset{\displaystyle O}{\|}}{C}{-}O{-}CH_3} + N_2\uparrow$$

Example

$$\underset{\text{benzoic acid}}{C_6H_5{-}\overset{\overset{\displaystyle O}{\|}}{C}{-}OH} + \underset{\text{ethanol}}{CH_3{-}CH_2{-}OH} \overset{H^+}{\rightleftarrows} \underset{\text{ethyl benzoate}}{C_6H_5{-}\overset{\overset{\displaystyle O}{\|}}{C}{-}OCH_2CH_3} + H_2O$$

4. *Conversion to amides* (Sections 20-10 and 20-13)

$$\underset{\text{acid}}{R{-}\overset{\overset{\displaystyle O}{\|}}{C}{-}OH} + \underset{\text{amine}}{R'{-}NH_2} \rightleftarrows \underset{\text{salt}}{R{-}\overset{\overset{\displaystyle O}{\|}}{C}{-}O^-\ H_3N^+{-}R'} \xrightarrow{\text{heat}} \underset{\text{amide}}{R{-}\overset{\overset{\displaystyle O}{\|}}{C}{-}NH{-}R'} + H_2O$$

$$\underset{\text{acid chloride}}{R{-}\overset{\overset{\displaystyle O}{\|}}{C}{-}Cl} + \underset{\text{amine}}{R'{-}NH_2} \longrightarrow \underset{\text{amide}}{R{-}\overset{\overset{\displaystyle O}{\|}}{C}{-}NH{-}R'} + HCl$$

Example

$$\underset{\text{acetic acid}}{CH_3{-}\overset{\overset{\displaystyle O}{\|}}{C}{-}OH} + \underset{\text{dimethylamine}}{CH_3{-}NH{-}CH_3} \xrightarrow{\text{heat}} \underset{N,N\text{-dimethylacetamide}}{CH_3{-}\overset{\overset{\displaystyle O}{\|}}{C}{-}N(CH_3)_2} + H_2O$$

5. Conversion to anhydrides (Section 21-5)

$$\underset{\text{acid chloride}}{R{-}\overset{O}{\overset{\|}{C}}{-}Cl} + \underset{\text{acid}}{HO{-}\overset{O}{\overset{\|}{C}}{-}R'} \longrightarrow \underset{\text{acid anhydride}}{R{-}\overset{O}{\overset{\|}{C}}{-}O{-}\overset{O}{\overset{\|}{C}}{-}R'} + HCl$$

Example

$$\underset{\text{acetic acid}}{CH_3{-}\overset{O}{\overset{\|}{C}}{-}Cl} + \underset{\text{benzoic acid}}{HO{-}\overset{O}{\overset{\|}{C}}{-}Ph} \longrightarrow \underset{\substack{\text{a mixed anhydride}\\\text{(acetic benzoic anhydride)}}}{CH_3{-}\overset{O}{\overset{\|}{C}}{-}O{-}\overset{O}{\overset{\|}{C}}{-}Ph} + HCl$$

6. Reduction to primary alcohols (Sections 9-12 and 20-14)

$$\underset{\text{acid}}{R{-}\overset{O}{\overset{\|}{C}}{-}OH} \xrightarrow[\substack{(2)\ H_3O^+\\(\text{or use } B_2H_6)}]{(1)\ LiAlH_4} \underset{\text{primary alcohol}}{R{-}CH_2{-}OH}$$

7. Reduction to aldehydes (Sections 18-11 and 20-14)

$$\underset{\text{acid chloride}}{R{-}\overset{O}{\overset{\|}{C}}{-}Cl} \xrightarrow[\text{lithium aluminum tri(}t\text{-butoxy)hydride}]{LiAl[OC(CH_3)_3]_3H} \underset{\text{aldehyde}}{R{-}\overset{O}{\overset{\|}{C}}{-}H}$$

8. Alkylation to form ketones (Sections 18-9 and 20-15)

$$\underset{\text{lithium carboxylate}}{R{-}\overset{O}{\overset{\|}{C}}{-}O^-\,{}^+Li} \xrightarrow[(2)\ H_2O]{\substack{(1)\ R'{-}Li\\\text{alkyllithium}}} \underset{\text{ketone}}{R{-}\overset{O}{\overset{\|}{C}}{-}R'}$$

9. Decarboxylation (Hunsdiecker reaction; Section 20-16)

$$\underset{\text{silver carboxylate}}{R{-}\overset{O}{\overset{\|}{C}}{-}O^-\,{}^+Ag} \xrightarrow[\text{heat}]{Br_2} \underset{\text{alkyl bromide}}{R{-}Br} + AgBr\downarrow + CO_2\uparrow$$

Example

$$\underset{\text{silver 3-phenylpropanoate}}{C_6H_5{-}CH_2{-}CH_2{-}\overset{O}{\overset{\|}{C}}{-}O^-\,{}^+Ag} \xrightarrow[\text{heat}]{Br_2} \underset{\text{1-bromo-2-phenylethane}}{C_6H_5{-}CH_2{-}CH_2{-}Br}$$

10. Side-chain halogenation (Hell-Volhard-Zelinsky reaction; Section 22-3)

$$R\ CH_2{-}\overset{O}{\overset{\|}{C}}{-}OH \xrightarrow{Br_2/PBr_3} \underset{\alpha\text{-bromo acyl bromide}}{R\ \overset{Br}{\overset{|}{C}H}{-}\overset{O}{\overset{\|}{C}}{-}Br} \xrightarrow{H_2O} \underset{\alpha\text{-bromoacid}}{R\ \overset{Br}{\overset{|}{C}H}{-}\overset{O}{\overset{\|}{C}}{-}OH} + HBr$$

GLOSSARY

carboxyl group The —COOH functional group of a carboxylic acid. (p. 887)

carboxylate ion The deprotonated anion of a carboxylic acid. (p. 893)

carboxylation A reaction in which a compound (usually a carboxylic acid) is formed by the addition of CO_2 to an intermediate. The addition of CO_2 to a Grignard reagent is an example of a carboxylation. (p. 905)

carboxylic acid Any compound containing the *carboxyl group*, —COOH. (p. 887)

An **aliphatic acid** has an alkyl group bonded to the carboxyl group.

An **aromatic acid** has an aryl group bonded to the carboxyl group.

A **dicarboxylic acid** (a **diacid**) has two carboxyl groups.

decarboxylation A reaction in which a compound (usually a carboxylic acid) loses CO_2. The Hunsdiecker reaction is an example of a decarboxylation. (p. 915)

fatty acid A long-chain linear carboxylic acid. Some fatty acids are saturated, while others are unsaturated. (p. 887)

Fischer esterification The acid-catalyzed reaction of a carboxylic acid with an alcohol to form an ester. (p. 910)

$$\mathrm{R{-}\overset{\overset{\displaystyle O}{\|}}{C}{-}O{-}H} + \mathrm{R'{-}OH} \overset{H^+}{\rightleftharpoons} \mathrm{R{-}\overset{\overset{\displaystyle O}{\|}}{C}{-}O{-}R'} + H_2O$$

Hunsdiecker reaction A decarboxylation brought about by heating a heavy-metal (Ag, Hg, or Pb) salt of a carboxylic acid with bromine or iodine. The product is the alkyl halide with one less carbon atom. (p. 915)

$$\mathrm{R{-}\overset{\overset{\displaystyle O}{\|}}{C}{-}O^{-}\,{}^{+}Ag} + Br_2 \longrightarrow \mathrm{R{-}Br} + CO_2\uparrow + AgBr\downarrow$$

nucleophilic acyl substitution A reaction in which a nucleophile substitutes for a leaving group on a carbonyl carbon atom. Nucleophilic acyl substitution usually takes place through the following addition-elimination mechanism. (p. 907)

$$\mathrm{R{-}\overset{\overset{\displaystyle \ddot{O}}{\|}}{C}{-}X} + \mathrm{Nuc{:}^-} \rightleftharpoons \mathrm{R{-}\underset{\underset{\displaystyle Nuc}{|}}{\overset{\overset{\displaystyle :\ddot{O}:^-}{|}}{C}}{-}X} \longrightarrow \mathrm{R{-}\overset{\overset{\displaystyle \ddot{O}}{\|}}{C}{-}Nuc} + \mathrm{:X^-}$$

the addition-elimination mechanism of nucleophilic acyl substitution

phthalic acids Benzenedioic acids. *Phthalic acid* itself is the ortho isomer. The meta isomer is *isophthalic acid*, and the para isomer is *terephthalic acid*. (p. 890)

salt of a carboxylic acid An ionic compound containing the deprotonated anion of a carboxylic acid, called the *carboxylate ion*: R—COO^-. An acid salt is formed by the reaction of an acid with a base. (p. 895)

ESSENTIAL PROBLEM-SOLVING SKILLS IN CHAPTER 20

1. Name carboxylic acids and draw the structures from their names.
2. Show how the acidity of acids varies with their substitution.
3. Contrast the physical properties of carboxylic acids and their salts.
4. Interpret the IR, NMR, and mass spectra of carboxylic acids and use the spectral information to determine the structures.
5. Show how to synthesize carboxylic acids from oxidation of alcohols and aldehydes, carboxylation of Grignard reagents, hydrolysis of nitriles, and oxidative degradation of alkylbenzenes.

6. Show how acids are converted to esters and amides using acid chlorides as intermediates. Give mechanisms for these nucleophilic acyl substitutions.

7. Give the mechanism of the Fischer esterification and show how the equilibrium can be driven toward the products or the reactants.

8. Predict the products of reactions of carboxylic acids with the following reagents; give mechanisms where appropriate.
   (a) diazomethane
   (b) amines, followed by heating
   (c) lithium aluminum hydride
   (d) excess alkyllithium reagents
   (e) silver oxide or $Pb(OAc)_4$, followed by heating with Br_2.

STUDY PROBLEMS

20-22. Define each of the following terms and give an example.
(a) carboxylic acid **(b)** carboxylate ion **(c)** carboxylation of a Grignard reagent
(d) decarboxylation of a carboxylate radical **(e)** ester **(f)** Fischer esterification
(g) fatty acid **(h)** Hunsdiecker reaction **(i)** dicarboxylic acid
(j) salt of a carboxylic acid **(k)** acid chloride **(l)** acid dissociation constant
(m) nucleophilic acyl substitution

20-23. Give the IUPAC names of the following compounds.
(a) $PhCH_2CH_2COOH$ **(b)** $CH_3CH_2CH(CH_3)CO_2H$ **(c)** $CH_3CH(CH_3)CHBrCOOH$
(d) $HOOCCH_2CH(CH_3)CH_2CO_2H$ **(e)** $CH_3CH_2CH(CH_3)COONa$ **(f)** $(CH_3)_2C{=}CHCOOH$

(g) cyclopentane with CH_3 and COOH

(h) benzene ring with COOH, NO_2, O_2N, NO_2

20-24. Give the common names of the following compounds.
(a) $PhCH_2CH_2COOH$ **(b)** $CH_3CH_2CH(CH_3)CO_2H$ **(c)** $(CH_3)_2CHCHBrCOOH$
(d) $HOOCCH_2CH(CH_3)CH_2CO_2H$ **(e)** $(CH_3)_2CHCH_2COONa$ **(f)** $CH_3CH(NH_2)CH_2COOH$

(g) 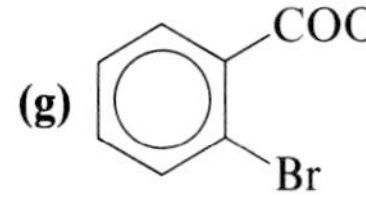

(h) COO^- | COO^- Mg^{2+}

20-25. Draw the structures of the following compounds.
(a) ethanoic acid **(b)** phthalic acid **(c)** magnesium formate
(d) malonic acid **(e)** chloroacetic acid **(f)** acetyl chloride
(g) zinc undecanoate (athlete's foot powder) **(h)** sodium benzoate (a food preservative)
(i) sodium fluoroacetate (Compound 1080, a controversial coyote poison)

20-26. In each pair of compounds, which is the stronger base?
(a) CH_3COO^- or $ClCH_2COO^-$ **(b)** sodium acetylide or sodium acetate
(c) sodium acetate or sodium ethoxide

20-27. Predict the products (if any) of the following acid-base reactions.
(a) acetic acid + ammonia **(b)** phthalic acid + excess NaOH
(c) *p*-toluic acid + potassium trifluoroacetate **(d)** α-bromopropionic acid + sodium propionate
(e) benzoic acid + sodium phenoxide

20-28. Rank the following isomers in order of increasing boiling points. Explain the reasons for your order of ranking.

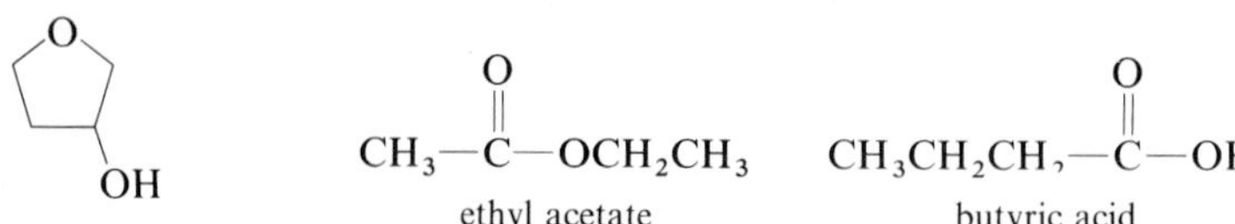

20-29. Arrange each group of compounds in order of increasing acidity.

(a) phenol, ethanol, acetic acid **(b)** *p*-toluenesulfonic acid, acetic acid, chloroacetic acid
(c) benzoic acid, *o*-nitrobenzoic acid, *m*-nitrobenzoic acid
(d) butyric acid, α-bromobutyric acid, β-bromobutyric acid

20-30. Predict the products, if any, of the following reactions.

(a) 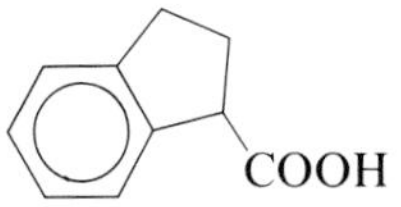 COOH $\xrightarrow[(2)\ H_3O^+]{(1)\ LiAlH_4}$ **(b)** 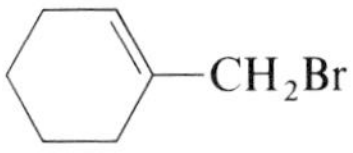 $-CH_2Br$ $\xrightarrow[(2)\ H_3O^+]{(1)\ NaCN}$

(c) 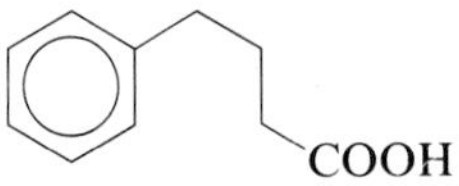 COOH $\xrightarrow[(2)\ AlCl_3]{(1)\ SOCl_2}$ **(d)** 4-octyne $\xrightarrow[\text{(warm, conc.)}]{KMnO_4,\ H_2O}$

(e) 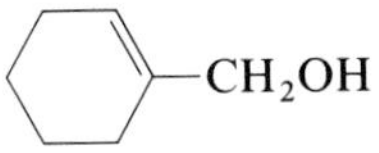 $-CH_2OH$ $\xrightarrow{Na_2Cr_2O_7,\ H_2SO_4}$ **(f)** $CH_3CH_2-CH(Ph)-COOH$ $\xrightarrow{B_2H_6}$

(g) 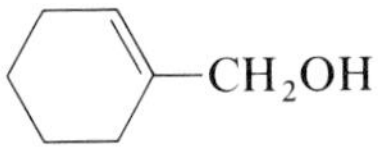 $-CH_2OH$ $\xrightarrow[\text{(warm, conc.)}]{KMnO_4,\ H_2O}$ **(h)** 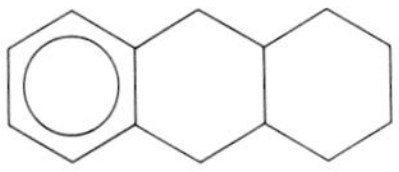 $\xrightarrow[\text{(hot, conc.)}]{KMnO_4,\ H_2O}$

(i) 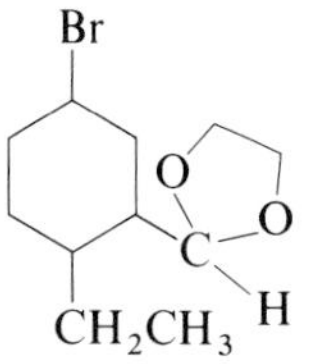 Br, O, O, C, H, CH_2CH_3 $\xrightarrow{(1)\ Mg,\ ether;\ (2)\ CO_2;\ (3)\ H_3O^+}$ **(j)** 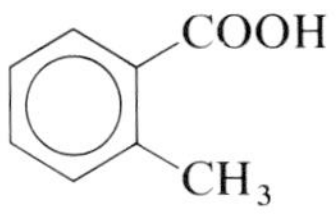 COOH, CH_3 $\xrightleftharpoons{\text{2-butanol},\ H^+}$

(k) $HOCH_2CH_2CH_2-\overset{O}{\overset{\|}{C}}-OH$ $\xrightarrow{H^+}$ (cyclic ester)

20-31. Show how you would accomplish the following synthetic conversions efficiently. You may use any additional reagents that are necessary.

(a) *trans*-1-bromo-2-butene ⟶ *trans*-3-pentenoic acid (two ways)

(b) $CH_3(CH_2)_3COOH$ ⟶ $CH_3(CH_2)_3-\overset{O}{\overset{\|}{C}}-OCH_3$ (two ways)
valeric acid methyl valerate

(c) 2-butenal ⟶ 2-butenoic acid **(d)** hexanoic acid ⟶ hexanal
(e) 3-hexene ⟶ propanoic acid

(f) 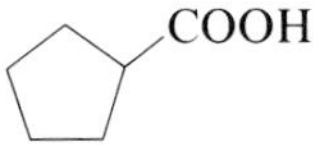 COOH ⟶ CH_2OH **(g)** 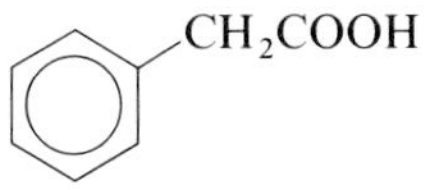 CH_2COOH ⟶ CH_2Br

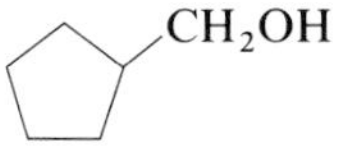

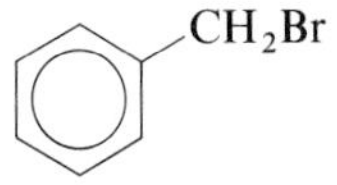

20-32. Show how you would use extractions with a separatory funnel to separate a mixture of the following compounds.

benzoic acid phenol benzyl alcohol aniline

20-33. A chemist adds a catalytic amount of sulfuric acid to a mixture of 1 mole of acetic acid and 1 mole of special methanol that contains the heavy ^{18}O isotope of oxygen. After a short period the acid is neutralized to stop the reaction, and the components of the mixture are separated.

$$CH_3-\overset{O}{\overset{\|}{C}}-O-H + CH_3-{}^{18}O-H \xrightleftharpoons{H_2SO_4} CH_3-\overset{O}{\overset{\|}{C}}-O-CH_3 + H_2O$$

(a) Give a detailed mechanism for this reaction.
(b) Follow the labeled ^{18}O atom through your mechanism, and show where it will be found in the products.
(c) The ^{18}O isotope is not radioactive. Suggest how you could experimentally determine the amount of ^{18}O in the separated components of the mixture.

✱ **20-34.** The IR, NMR, and mass spectra are provided for an organic compound.

(a) Consider each spectrum individually, and tell what characteristics of the molecule are apparent from that spectrum.

(b) Propose a structure for the compound, and show how your structure fits the spectral data.

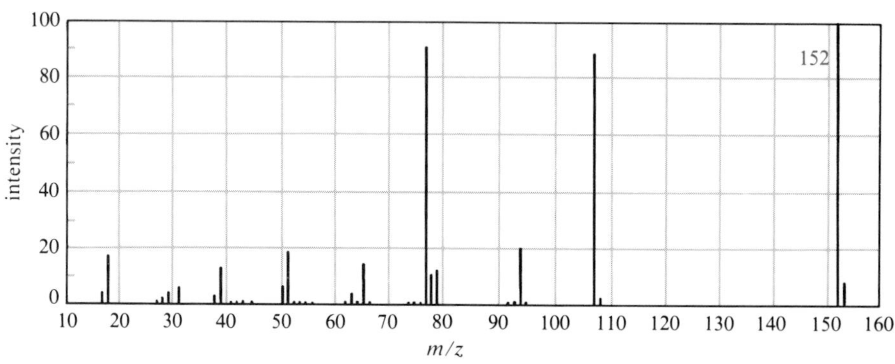

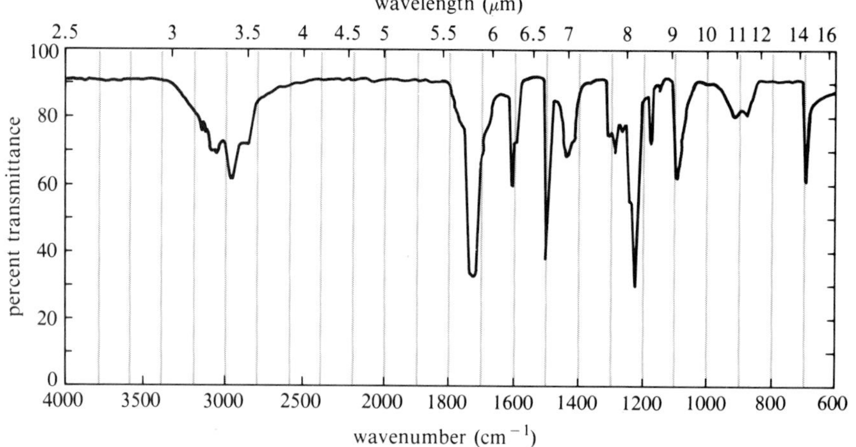

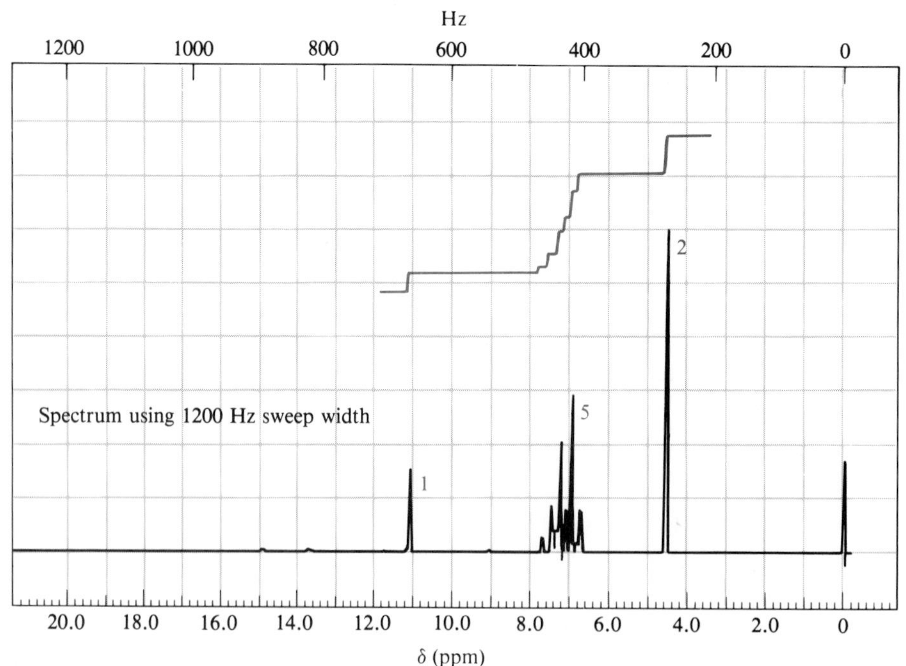

20-35. When pure (*S*)-lactic acid is esterified by racemic 2-butanol, the product is 2-butyl lactate, with the following structure:

$$\underset{\text{lactic acid}}{CH_3-\overset{\overset{\large OH}{|}}{CH}-COOH} + \underset{\text{2-butanol}}{CH_3-\overset{\overset{\large OH}{|}}{CH}-CH_2CH_3} \overset{H^+}{\rightleftharpoons} \underset{\text{2-butyl lactate}}{CH_3-\overset{\overset{\large OH}{|}}{CH}-\overset{\overset{\large O}{\|}}{C}-O-\overset{\overset{\large CH_3}{|}}{CH}-CH_2CH_3}$$

(a) Draw three-dimensional structures of the two stereoisomers formed, specifying the configuration at each chiral carbon atom.

(b) Determine the relationship between the two stereoisomers you have drawn.

20-36. The following NMR spectra correspond to compounds of formulas **(A)** $C_9H_{10}O_2$, **(B)** $C_4H_6O_2$, and **(C)** $C_6H_{10}O_2$, respectively. Propose structures and show how they are consistent with the observed absorptions.

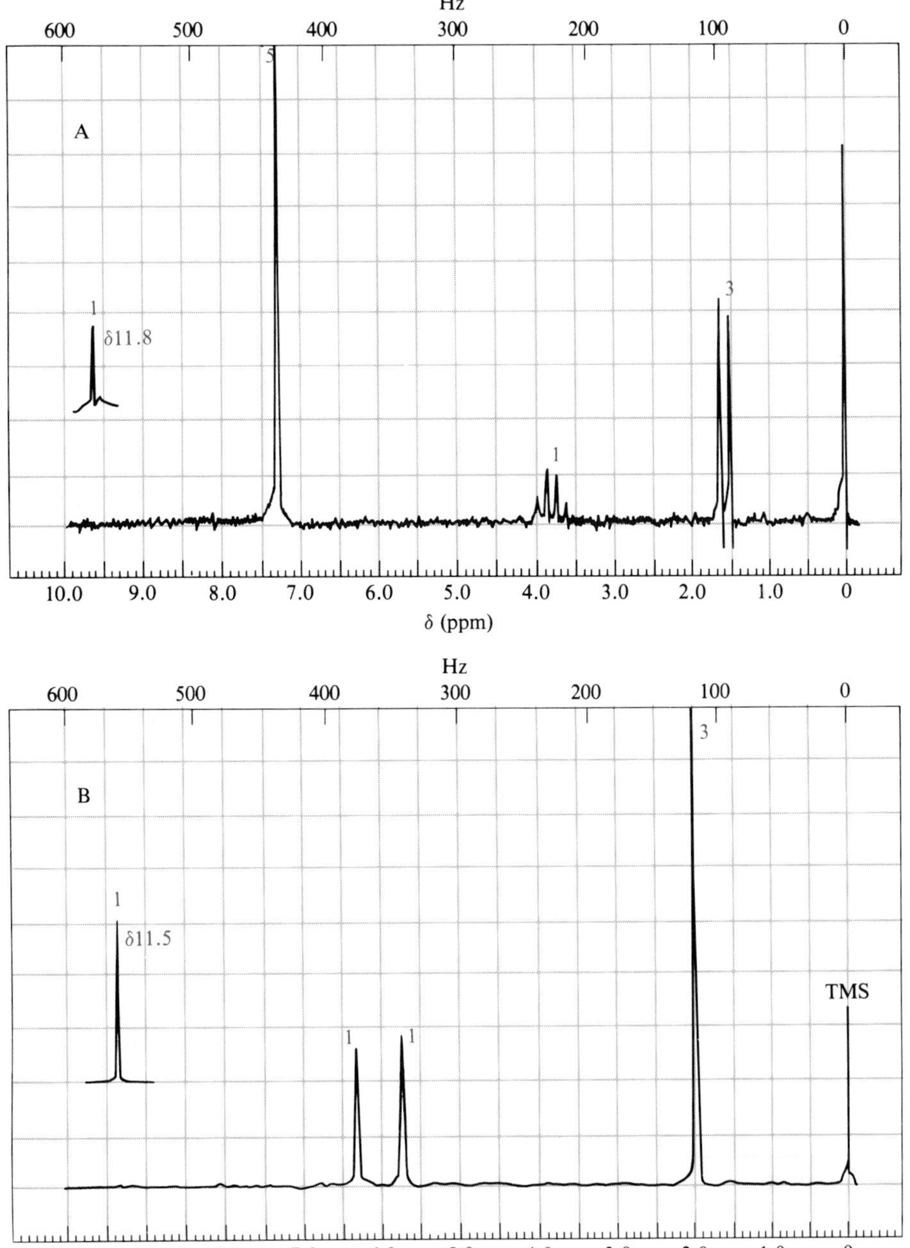

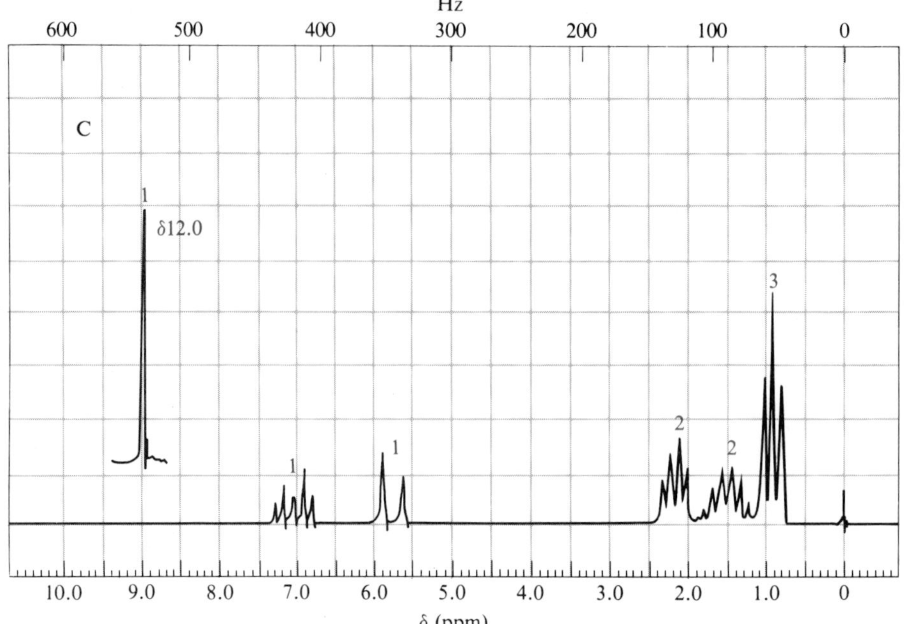

20-37. Show how you would accomplish the following multistep synthetic conversions.

(a) Ph—CH_2—CH_2—OH $\longrightarrow$ Ph—CH_2—CH_2—COOH

(b) (cyclohexane with $=CH_2$) $\longrightarrow$ (cyclohexane with CH_3 and COOH)

(c) (cyclohexane with $=CH_2$) $\longrightarrow$ (cyclohexane with CH_2COOH)

(d) (Br-substituted tetralone, O) $\longrightarrow$ (COOH-substituted tetralone, O)

(e) (cyclohexane with COOH) $\longrightarrow$ (O, O ketal linking two cyclohexane rings)

20-38. In the presence of a trace of acid, δ-hydroxyvaleric acid forms a cyclic ester (lactone).

$$HO—CH_2—CH_2—CH_2—CH_2—COOH$$

δ-hydroxyvaleric acid

(a) Give the structure of the lactone, called δ-valerolactone.
(b) Propose a mechanism for the formation of δ-valerolactone.

20-39. We have seen that an acid chloride reacts with an alcohol to form an ester.

$$R—\overset{O}{\overset{\|}{C}}—Cl \;+\; R'—OH \;\longrightarrow\; R—\overset{O}{\overset{\|}{C}}—O—R' \;+\; HCl$$

An acid chloride also reacts with another carboxylic acid molecule. The product is an acid anhydride.

$$R—\overset{O}{\overset{\|}{C}}—Cl \;+\; R'—\overset{O}{\overset{\|}{C}}—OH \;\longrightarrow\; R—\overset{O}{\overset{\|}{C}}—O—\overset{O}{\overset{\|}{C}}—R' \;+\; HCl$$

acid anhydride

Propose a mechanism for the reaction of benzoyl chloride (PhCOCl) with acetic acid, and show the structure of the resulting anhydride.

✱ 20-40. The relative acidities of carboxylic acids (and, by inference, the relative stabilities of their carboxylate ions) have been used to compare the electron-donating and electron-withdrawing properties of substituents. These studies are particularly valuable to distinguish between inductive and resonance effects on the stabilities of compounds and ions. Some examples:

(a) The phenyl group is a mild ortho, para-director in electrophilic aromatic substitution. Is the phenyl group electron-donating or electron-withdrawing in EAS? The pK_a of phenylacetic acid is 4.31, showing that phenylacetic acid is a stronger acid than acetic acid. Is the phenyl group electron-donating or electron-withdrawing in the ionization of phenylacetic acid? How can you resolve the apparent contradiction?

(b) 4-Methoxybenzoic acid is a weaker acid than benzoic acid, but methoxyacetic acid is a stronger acid than acetic acid. Explain this apparent contradiction.

(c) Methyl groups are usually electron-donating, and propanoic acid is a weaker acid than acetic acid. Yet, 2,6-dimethylbenzoic acid is a *stronger* acid than benzoic acid, while 2,6-dimethylphenol is a weaker acid than phenol. Explain these confusing experimental results.

20-41. The manager of an organic chemistry stockroom prepared unknowns for a "Ketones and Aldehydes" experiment by placing two drops of the liquid unknowns in test tubes and then storing the test tubes for several days until they were needed. One of the unknowns was misidentified by every student, however. This unknown was taken from a bottle marked "Heptaldehyde." The stockroom manager took an IR spectrum of the liquid in the bottle, and found a sharp carbonyl stretch around 1710 cm^{-1} and small, sharp peaks around 2710 and 2810 cm^{-1}.

The students complained that their spectra showed no peaks at 2710 or 2810 cm^{-1}, but a broad absorption centered over the 3000-cm^{-1} region and a carbonyl peak around 1715 cm^{-1}. They also maintain that their samples are soluble in dilute aqueous sodium hydroxide.

(a) Identify the compound in the stockroom manager's bottle and the compound in the students' test tubes.
(b) Explain the discrepancy between the stockroom manager's spectrum and the students' results.
(c) Suggest how this misunderstanding might be prevented in the future.

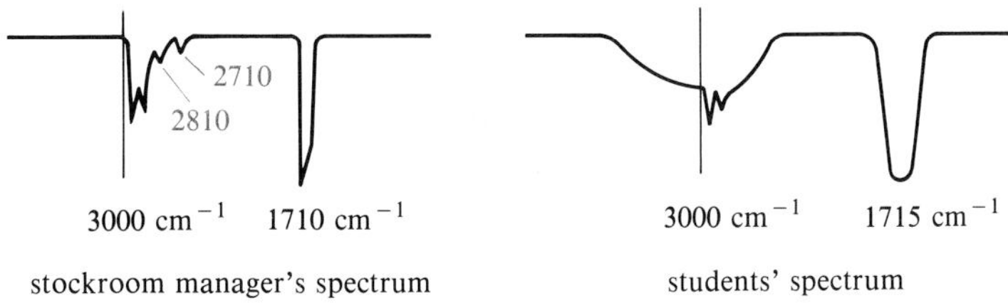

stockroom manager's spectrum   students' spectrum

$$\mathrm{PhOCH_2\overset{\overset{\displaystyle O}{\|}}{C}-NH}$$ (penicillin acylated, inactive enzyme structure)

21

CARBOXYLIC ACID DERIVATIVES

21-1 INTRODUCTION

The compounds we classify as derivatives of carboxylic acids are those functional groups that are converted to carboxylic acids by a simple acidic or basic hydrolysis. The most important acid derivatives are esters, amides, and nitriles. Acid halides and anhydrides are also included in this group, although we often think of them as activated forms of the parent acids rather than completely different compounds.

	acid halide	anhydride	ester	amide	nitrile
	$R-\overset{O}{\overset{\|}{C}}-X$	$R-\overset{O}{\overset{\|}{C}}-O-\overset{O}{\overset{\|}{C}}-R$	$R-\overset{O}{\overset{\|}{C}}-O-R'$	$R-\overset{O}{\overset{\|}{C}}-NH_2$	$R-C\equiv N$
Condensed structure:	RCOX	$(RCO)_2O$	RCO_2R'	$RCONH_2$	RCN

21-2 STRUCTURE AND NOMENCLATURE OF ACID DERIVATIVES

21-2A ESTERS OF CARBOXYLIC ACIDS

Esters are carboxylic acid derivatives in which the hydroxyl group (—OH) is replaced by an alkoxy group (—OR). An ester is a composite of a carboxylic acid and an alcohol, with the loss of a molecule of water. We have seen that esters can be formed by the acid-catalyzed Fischer esterification of an acid with an alcohol (Section 20-11).

$$\underset{\text{acid}}{R-\overset{O}{\overset{\|}{C}}-OH} + \underset{\text{alcohol}}{R'-OH} \overset{H^+}{\rightleftarrows} \underset{\text{ester}}{R-\overset{O}{\overset{\|}{C}}-O-R'} + H_2O$$

The names of esters consist of two words that reflect their composite structure. The first word is derived from the *alkyl* group of the alcohol, and the second word from the *carboxylate* group of the carboxylic acid.

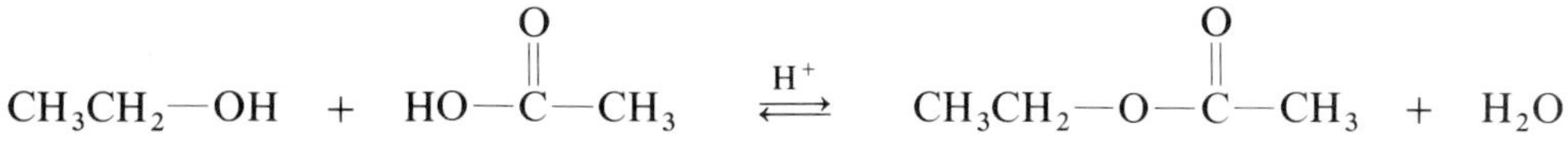

$$CH_3CH_2{-}OH + HO{-}\overset{\overset{\displaystyle O}{\|}}{C}{-}CH_3 \overset{H^+}{\rightleftharpoons} CH_3CH_2{-}O{-}\overset{\overset{\displaystyle O}{\|}}{C}{-}CH_3 + H_2O$$

IUPAC name:	ethanol	ethanoic acid	ethyl ethanoate
common name:	ethyl alcohol	acetic acid	ethyl acetate

The IUPAC name is derived from the IUPAC names of the alkyl group and the carboxylate, while the common name is derived from the common names of each. The following examples show both the IUPAC names and the common names of some esters.

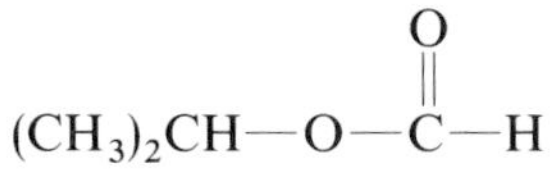

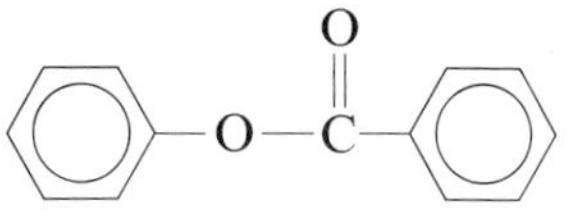

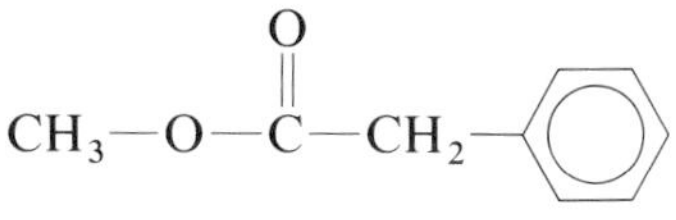

	$(CH_3)_2CH{-}O{-}C(=O){-}H$	$C_6H_5{-}O{-}C(=O){-}C_6H_5$	$CH_3{-}O{-}C(=O){-}CH_2{-}C_6H_5$
IUPAC name:	1-methylethyl methanoate	phenyl benzoate	methyl 2-phenylethanoate
common name:	isopropyl formate	phenyl benzoate	methyl phenylacetate

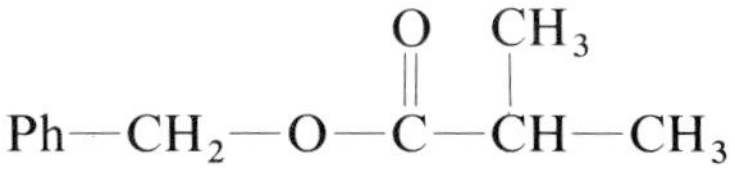

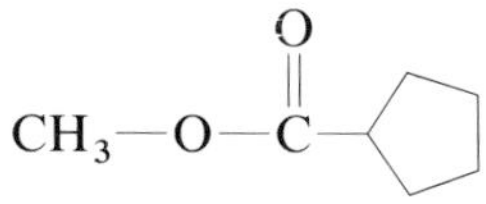

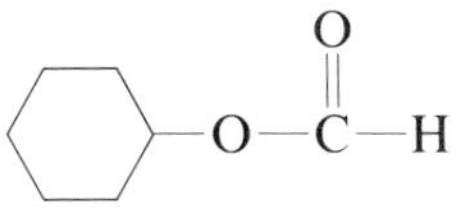

	$Ph{-}CH_2{-}O{-}C(=O){-}CH(CH_3){-}CH_3$	$CH_3{-}O{-}C(=O){-}C_5H_9$ (cyclopentyl)	$C_6H_{11}{-}O{-}C(=O){-}H$ (cyclohexyl)
IUPAC name:	benzyl 2-methylpropanoate	methyl cyclopentanecarboxylate	cyclohexyl methanoate
common name:	benzyl isobutyrate	methyl cyclopentanecarboxylate	cyclohexyl formate

Lactones Cyclic esters are called **lactones.** A lactone is formed from an open-chain hydroxy acid in which the hydroxyl group has reacted with the acid group to form an ester.

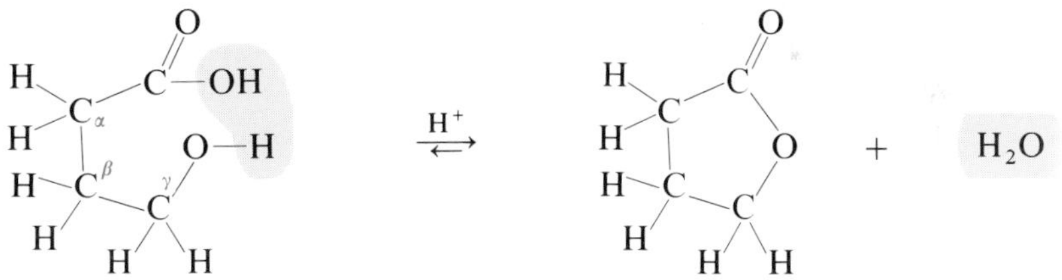

$$HO{-}CH_2(\gamma){-}CH_2(\beta){-}CH_2(\alpha){-}C(=O){-}OH \overset{H^+}{\rightleftharpoons} \text{(five-membered ring lactone)} + H_2O$$

IUPAC name:	4-hydroxybutanoic acid	4-hydroxybutanoic acid lactone
common name:	γ-hydroxybutyric acid	γ-butyrolactone

The IUPAC names of lactones are derived simply by adding the term *lactone* at the end of the name of the parent acid. The common names of lactones, used more often than IUPAC names, are formed by changing the *-ic acid* ending of the hydroxy acid to *-olactone*. A Greek letter is used to designate the carbon atom that bears the ester group to close the ring. Substituents are named the same as they are on the parent acid.

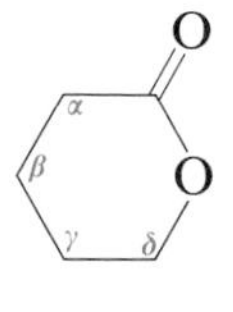

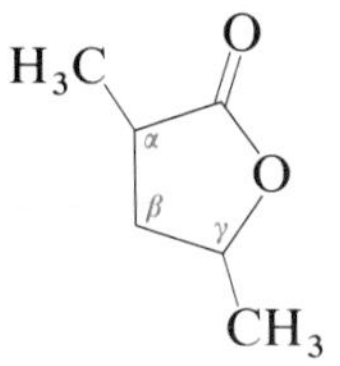

IUPAC name:	5-hydroxypentanoic acid lactone	4-hydroxy-2-methylpentanoic acid lactone
common name:	δ-valerolactone	α-methyl-γ-valerolactone

21-2B AMIDES

An **amide** is a composite of a carboxylic acid and ammonia or an amine. An acid reacts with an amine to form an ammonium carboxylate salt. When this salt is heated to well above 100°C, water is driven off and an amide results.

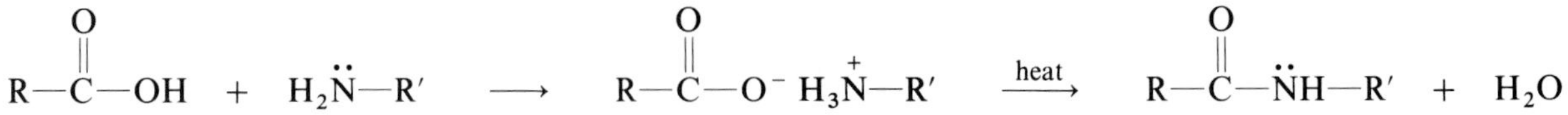

$$\underset{\text{acid}}{R{-}\overset{\overset{\displaystyle O}{\|}}{C}{-}OH} + \underset{\text{amine}}{H_2\ddot{N}{-}R'} \longrightarrow \underset{\text{salt}}{R{-}\overset{\overset{\displaystyle O}{\|}}{C}{-}O^- \; H_3\overset{+}{N}{-}R'} \xrightarrow{\text{heat}} \underset{\text{amide}}{R{-}\overset{\overset{\displaystyle O}{\|}}{C}{-}\ddot{N}H{-}R'} + H_2O$$

The simple amide structure shows a nonbonding pair of electrons on the nitrogen atom. Unlike amines, however, amides are only weakly basic, and we consider the amide functional group to be neutral. A concentrated strong acid is required to protonate an amide, and protonation occurs on the carbonyl oxygen atom rather than on nitrogen. This lack of basicity can be explained by picturing the amide as a resonance hybrid of the conventional structure and a structure with a double bond between carbon and nitrogen.

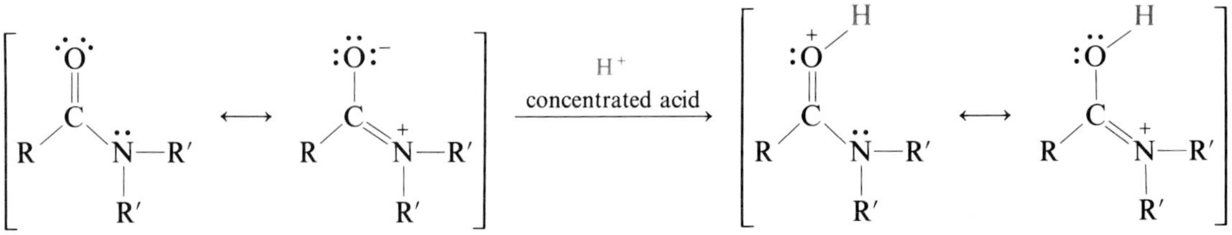

$$\underset{\text{very weakly basic}}{\left[R{-}C({=}\ddot{O}){-}\ddot{N}R'{-}R' \longleftrightarrow R{-}C(-\ddot{O}{:}^-){=}\overset{+}{N}R'{-}R' \right]} \xrightarrow[\text{concentrated acid}]{H^+} \underset{\text{protonation on oxygen}}{\left[R{-}C({=}\overset{+}{O}H){-}\ddot{N}R'{-}R' \longleftrightarrow R{-}C({-}\ddot{O}H){=}\overset{+}{N}R'{-}R' \right]}$$

This resonance representation predicts that the amide nitrogen atom must be sp^2 hybridized to allow pi bonding with the carbonyl carbon atom. For example, formamide has a planar structure like that of an alkene. The C—N bond has a partial double-bond character, with a rotational barrier of 18 kcal/mol (75 kJ/mol).

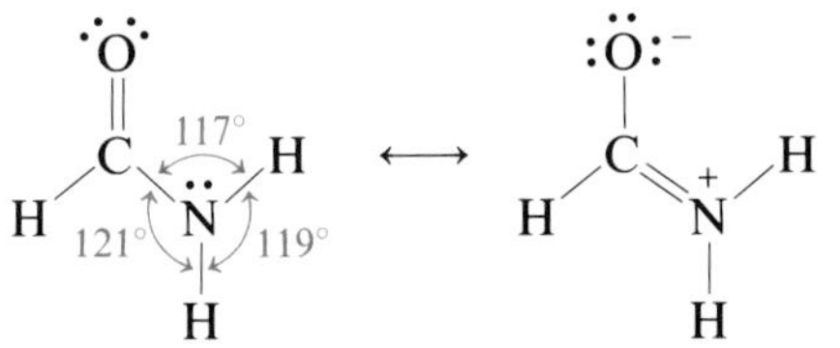

$$H{-}C({=}\ddot{O}){-}\ddot{N}H_2 \longleftrightarrow H{-}C({-}\ddot{O}{:}^-){=}\overset{+}{N}H_2$$

(bond angles: 117°, 121°, 119°)

An amide of the form R—CO—NH_2 is called a **primary amide** because there is only one carbon atom bonded to the amide nitrogen atom. An amide with an alkyl group on nitrogen (R—CO—NHR′) is called a **secondary amide** or an **N-substituted amide.** Amides with two alkyl groups on the amide nitrogen (R—CO—NR'_2) are called **tertiary amides** or **N,N-disubstituted amides.**

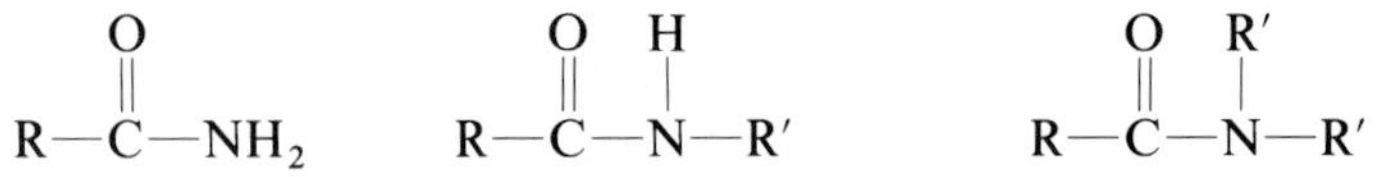

$$\underset{\text{primary amide}}{R{-}\overset{\overset{\displaystyle O}{\|}}{C}{-}NH_2} \qquad \underset{\substack{\text{secondary amide}\\ \text{(N-substituted amide)}}}{R{-}\overset{\overset{\displaystyle O}{\|}}{C}{-}\overset{\overset{\displaystyle H}{|}}{N}{-}R'} \qquad \underset{\substack{\text{tertiary amide}\\ \text{(N,N-disubstituted amide)}}}{R{-}\overset{\overset{\displaystyle O}{\|}}{C}{-}\overset{\overset{\displaystyle R'}{|}}{N}{-}R'}$$

To name a primary amide, first name the corresponding acid. Drop the *-ic acid* or *-oic acid* suffix from the name of the carboxylic acid, and substitute the suffix *-amide.* Secondary and tertiary amides are named by treating the alkyl groups on nitrogen as substituents, specifying their position by the prefix *N-*.

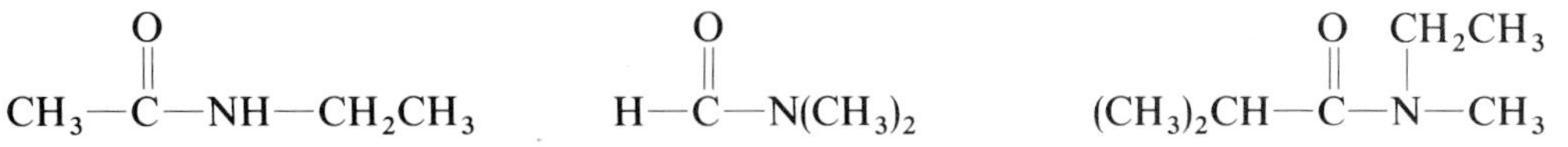

$$CH_3-\overset{O}{\overset{\|}{C}}-NH-CH_2CH_3 \qquad H-\overset{O}{\overset{\|}{C}}-N(CH_3)_2 \qquad (CH_3)_2CH-\overset{O}{\overset{\|}{C}}-\overset{CH_2CH_3}{\overset{|}{N}}-CH_3$$

IUPAC name:	*N*-ethylethanamide	*N*,*N*-dimethylmethanamide	*N*-ethyl-*N*,2-dimethylpropanamide
common name:	*N*-ethylacetamide	*N*,*N*-dimethylformamide	*N*-ethyl-*N*-methylisobutyramide

For acids that are named as alkanecarboxylic acids, the corresponding amides are named using the suffix *-carboxamide*. Some amides, such as acetanilide, have historical names that are still commonly used.

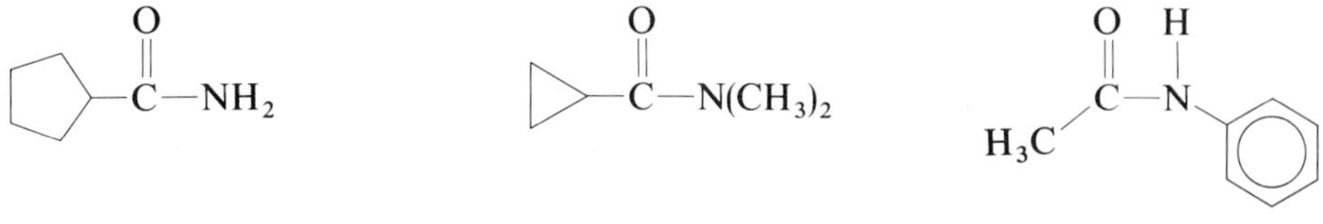

cyclopentanecarboxamide   *N*,*N*-dimethylcyclopropanecarboxamide   acetanilide

Lactams Cyclic amides are called **lactams.** Lactams are formed from amino acids, where the amino group and the carboxyl group have joined to form an amide. Lactams are named like lactones, and the common names of lactams are also used more often than the IUPAC names.

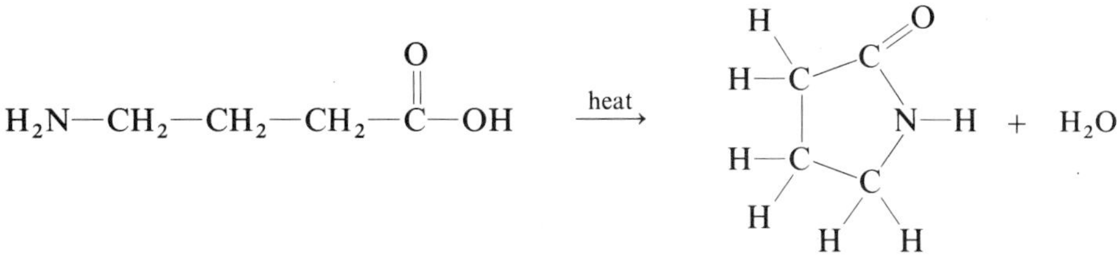

$$H_2N-CH_2-CH_2-CH_2-\overset{O}{\overset{\|}{C}}-OH \xrightarrow{\text{heat}} \text{(lactam)} + H_2O$$

IUPAC name:	4-aminobutanoic acid	4-aminobutanoic acid lactam
common name:	γ-aminobutyric acid	γ-butyrolactam

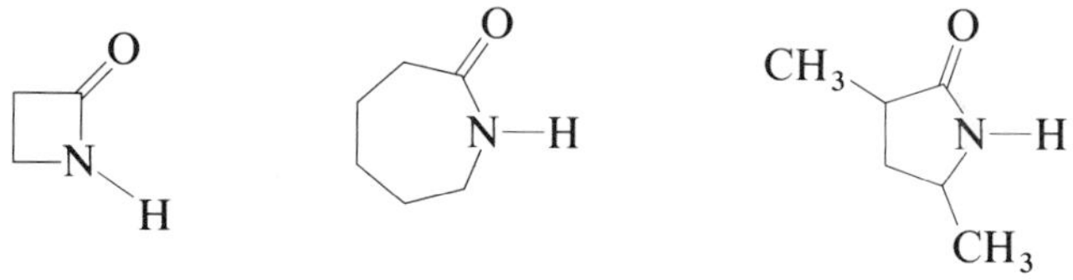

IUPAC name:	3-aminopropanoic acid lactam	6-aminohexanoic acid lactam	4-amino-2-methylpentanoic acid lactam
common name:	β-propiolactam	ε-caprolactam	α-methyl-γ-valerolactam

21-2C NITRILES

Nitriles contain the **cyano group,** $-C{\equiv}N$. Although nitriles lack the carbonyl group of carboxylic acids, they are classified as acid derivatives because they hydrolyze to give carboxylic acids and they can be synthesized by the dehydration of amides.

Hydrolysis to an acid

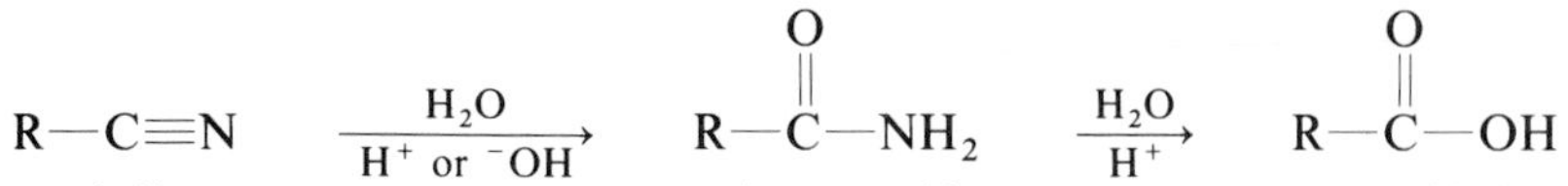

$$\underset{\text{nitrile}}{R-C{\equiv}N} \xrightarrow[H^+ \text{ or } ^-OH]{H_2O} \underset{\text{primary amide}}{R-\overset{O}{\overset{\|}{C}}-NH_2} \xrightarrow[H^+]{H_2O} \underset{\text{acid}}{R-\overset{O}{\overset{\|}{C}}-OH}$$

Synthesis from an acid

$$\underset{\text{acid}}{R-\overset{\overset{\displaystyle O}{\|}}{C}-OH} \xrightarrow[\text{heat}]{NH_3} \underset{\text{primary amide}}{R-\overset{\overset{\displaystyle O}{\|}}{C}-NH_2} \xrightarrow{POCl_3} \underset{\text{nitrile}}{R-C\equiv N}$$

Both the carbon atom and the nitrogen atom of the cyano group are *sp* hybridized, and the R—C≡N bond angle is 180° (linear). The structure of a nitrile is similar to that of a terminal alkyne, except that the nitrogen atom of the nitrile has a lone pair of electrons in place of the acetylenic hydrogen of the terminal alkyne. Figure 21-1 compares the structures of acetonitrile and propyne.

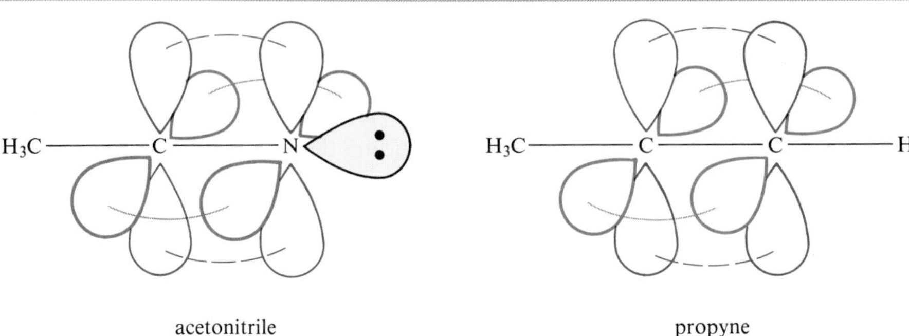

FIGURE 21-1 Comparison of the electronic structures of acetonitrile and propyne (methylacetylene). In both of these compounds, the atoms at the ends of the triple bonds are *sp* hybridized, and the bond angles are 180°. In place of the acetylenic hydrogen atom, the nitrile has a lone pair of electrons in the *sp* orbital of the nitrogen atom.

Although a nitrile has a lone pair of electrons on nitrogen, it is not very basic. A typical nitrile has a pK_b of about 24, requiring a concentrated solution of mineral acid to protonate a significant fraction of the nitrile. We explain this lack of basicity by noting that the nitrile's lone pair resides in an *sp*-hybridized orbital, with 50 percent *s* character. This orbital is close to the nucleus, and these electrons are tightly bound and relatively unreactive.

Nitrile nomenclature is derived from that of the corresponding carboxylic acids. The IUPAC name is constructed from the alkane name, with the suffix *-nitrile* added. For common names, the suffix *-ic acid* is replaced by the suffix *-onitrile*.

	$CH_3-C\equiv N$	$CH_3-\overset{\overset{\displaystyle Br}{|}}{C}H-CH_2-C\equiv N$	$CH_3-\overset{\overset{\displaystyle OCH_3}{|}}{C}H-CH_2CH_2CH_2-C\equiv N$
IUPAC name:	ethanenitrile	3-bromobutanenitrile	5-methoxyhexanenitrile
common name:	acetonitrile	β-bromobutyronitrile	δ-methoxycapronitrile

For acids that are named as alkanecarboxylic acids, the corresponding nitriles are named using the suffix *-carbonitrile*. The —C≡N group can also be used as a substituent, named as the *cyano group*.

$$\underset{\text{cyclopropanecarbonitrile}}{\triangleright\!-CN} \qquad \underset{\text{3-cyanopentanoic acid}}{CH_3-CH_2-\overset{\overset{\displaystyle CN}{|}}{C}H-CH_2-COOH}$$

21-2D ACID HALIDES

Acid halides, also called **acyl halides,** are used as activated derivatives for the synthesis of other acyl compounds such as esters, amides, and acylbenzenes (in the

Friedel-Crafts acylation). The most common acyl halides are the acyl chlorides (acid chlorides), and we will generally use acid chlorides as examples.

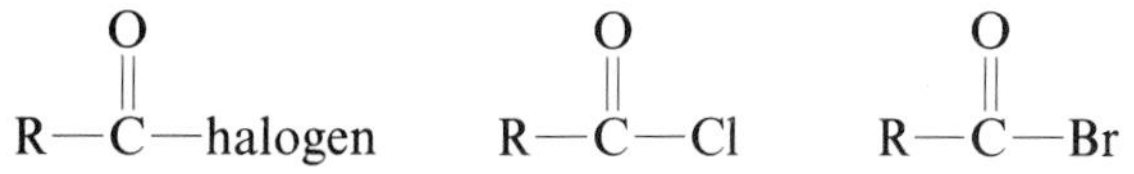

$$\underset{\text{an acid halide (acyl halide)}}{\mathrm{R{-}\overset{\overset{\displaystyle O}{\|}}{C}{-}halogen}} \qquad \underset{\text{acid chloride (acyl chloride)}}{\mathrm{R{-}\overset{\overset{\displaystyle O}{\|}}{C}{-}Cl}} \qquad \underset{\text{acid bromide (acyl bromide)}}{\mathrm{R{-}\overset{\overset{\displaystyle O}{\|}}{C}{-}Br}}$$

The halogen atom of an acyl halide inductively withdraws electron density from the carbonyl carbon, enhancing its electrophilic nature and making acyl halides particularly reactive toward nucleophilic acyl substitution. The halide ion also serves as a good leaving group.

Acid halides are named by replacing the *-ic acid* suffix of the acid name with *-yl* and the halide names. For acids that are named as alkanecarboxylic acids, the acid chlorides are named using the suffix *-carbonyl chloride*.

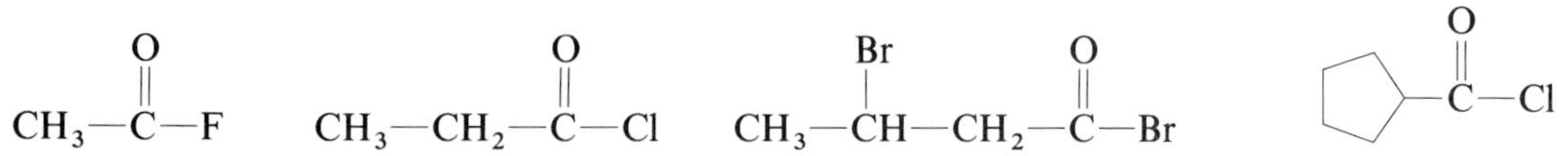

$$\mathrm{CH_3{-}\overset{\overset{\displaystyle O}{\|}}{C}{-}F} \qquad \mathrm{CH_3{-}CH_2{-}\overset{\overset{\displaystyle O}{\|}}{C}{-}Cl} \qquad \mathrm{CH_3{-}\overset{\overset{\displaystyle Br}{|}}{C}H{-}CH_2{-}\overset{\overset{\displaystyle O}{\|}}{C}{-}Br} \qquad \mathrm{C_5H_9{-}\overset{\overset{\displaystyle O}{\|}}{C}{-}Cl}$$

ethanoyl fluoride, acetyl fluoride — propanoyl chloride, propionyl chloride — 3-bromobutanoyl bromide, β-bromobutyryl bromide — cyclopentanecarbonyl chloride

21-2E ACID ANHYDRIDES

The word **anhydride** means "without water." An acid anhydride contains two molecules of an acid, with loss of a molecule of water. Addition of water to an anhydride regenerates two molecules of the carboxylic acid.

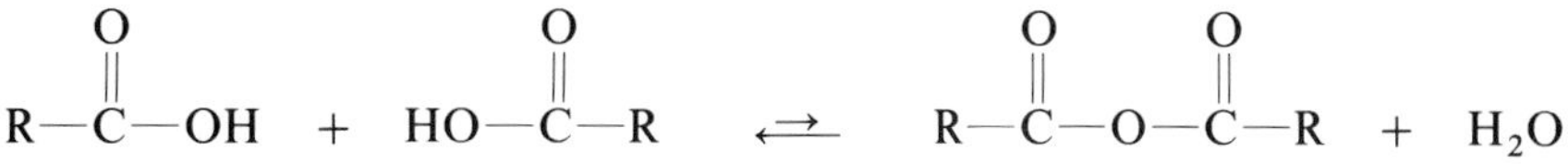

$$\underbrace{\mathrm{R{-}\overset{\overset{\displaystyle O}{\|}}{C}{-}OH \;+\; HO{-}\overset{\overset{\displaystyle O}{\|}}{C}{-}R}}_{\text{two molecules of acid}} \;\rightleftarrows\; \underset{\text{acid anhydride}}{\mathrm{R{-}\overset{\overset{\displaystyle O}{\|}}{C}{-}O{-}\overset{\overset{\displaystyle O}{\|}}{C}{-}R}} \;+\; \underset{\text{water}}{\mathrm{H_2O}}$$

Like acid chlorides, anhydrides serve as activated derivatives of carboxylic acids, although anhydrides are not as reactive as acid chlorides. In an acid chloride, the chlorine atom activates the carbonyl group and serves as a leaving group. In an anhydride, the carboxyl group serves these functions.

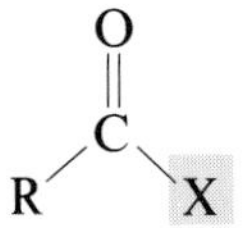

$$\mathrm{R{-}\overset{\overset{\displaystyle O}{\|}}{C}{-}X} \qquad \text{X can be a halogen: } \mathrm{R{-}\overset{\overset{\displaystyle O}{\|}}{C}{-}Cl} \qquad \text{or a carboxyl group: } \mathrm{R{-}\overset{\overset{\displaystyle O}{\|}}{C}{-}O{-}\overset{\overset{\displaystyle O}{\|}}{C}{-}R}$$

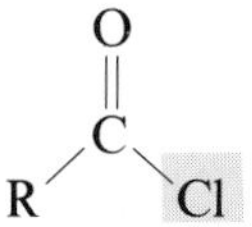

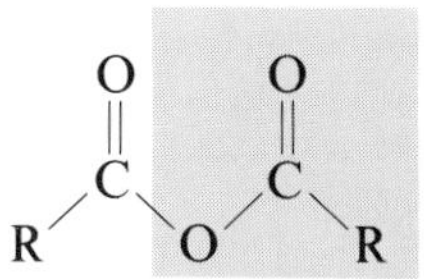

In an anhydride, the carboxylate ion serves as a leaving group, just as chloride ion does in the acid chloride.

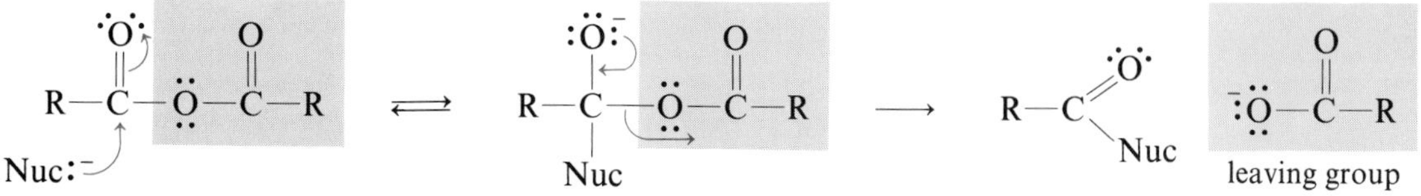

$$\mathrm{R{-}\overset{\overset{\displaystyle :\ddot{O}}{\|}}{C}{-}\ddot{O}{-}\overset{\overset{\displaystyle O}{\|}}{C}{-}R \;+\; Nuc{:}^-} \;\rightleftarrows\; \mathrm{R{-}\overset{\overset{\displaystyle :\ddot{O}{:}^-}{|}}{\underset{\underset{\displaystyle Nuc}{|}}{C}}{-}\ddot{O}{-}\overset{\overset{\displaystyle O}{\|}}{C}{-}R} \;\longrightarrow\; \mathrm{R{-}\overset{\overset{\displaystyle \ddot{O}}{\|}}{C}{-}Nuc} \;+\; \underset{\text{leaving group}}{\mathrm{{}^-{:}\ddot{O}{-}\overset{\overset{\displaystyle O}{\|}}{C}{-}R}}$$

Half of an anhydride's acid molecules are lost as leaving groups. If the acid is expensive, we would not use the anhydride as an activated form of the acid to make a derivative; the acid chloride is a more efficient alternative, using chloride as the leaving group. Anhydrides are used primarily when the necessary anhydride is cheap and readily available. Acetic anhydride, phthalic anhydride, succinic anhydride, and maleic anhydride are the ones most often used.

Anhydride nomenclature is very simple; the word *acid* is changed to *anhydride* in both the common name and the IUPAC name (rarely used). The following are the names and structures of some common anhydrides.

$$\mathrm{CH_3{-}\overset{\overset{O}{\|}}{C}{-}O{-}\overset{\overset{O}{\|}}{C}{-}CH_3} \qquad \mathrm{CF_3{-}\overset{\overset{O}{\|}}{C}{-}O{-}\overset{\overset{O}{\|}}{C}{-}CF_3}$$

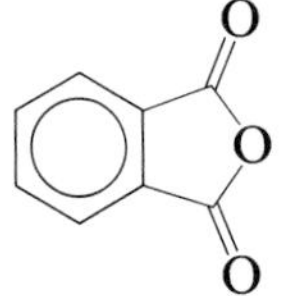

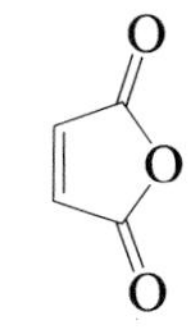

(abbreviated Ac_2O) ethanoic anhydride acetic anhydride	(abbreviated TFAA) trifluoroethanoic anhydride trifluoroacetic anhydride	1,2-benzenedioic anhydride phthalic anhydride	2-butenedioic anhydride maleic anhydride

Anhydrides composed of two different acid molecules are called **mixed anhydrides,** and are named using the names of the individual acids.

$$\mathrm{CH_3{-}\overset{\overset{O}{\|}}{C}{-}O{-}\overset{\overset{O}{\|}}{C}{-}H} \qquad \mathrm{CH_3CH_2{-}\overset{\overset{O}{\|}}{C}{-}O{-}\overset{\overset{O}{\|}}{C}{-}CF_3}$$

IUPAC name:	ethanoic methanoic anhydride	trifluoroethanoic propanoic anhydride
common name:	acetic formic anhydride	trifluoroacetic propionic anhydride

21-2F NOMENCLATURE OF MULTIFUNCTIONAL COMPOUNDS

With all the different functional groups we have studied, it is not always obvious which functional group of a multifunctional compound is the "main" one, and which groups should be named as substituents. In choosing the principal group to be used for the root name, the following priorities are used:

acid > ester > amide > nitrile > aldehyde > ketone > alcohol > amine

The following compounds illustrate the use of these priorities in naming multifunctional compounds.

$$\mathrm{\overset{\overset{O}{\|}}{C}{-}OCH_2CH_3}\quad \mathrm{CN} \qquad\qquad \mathrm{\overset{\overset{O}{\|}}{C}{-}NH_2}\quad \mathrm{\underset{\underset{O}{\|}}{C}{-}H}$$

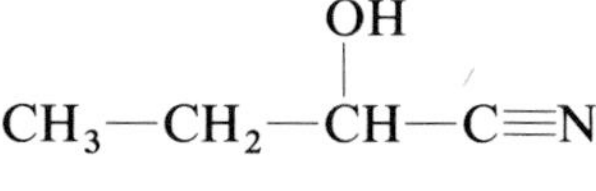

ethyl *o*-cyanobenzoate	2-formylcyclohexanecarboxamide	2-hydroxybutanenitrile

PROBLEM 21-1

Name the following carboxylic acid derivatives, giving both a common name and an IUPAC name where possible.

(a) $PhCOOCH_2CH(CH_3)_2$	(b) PhOCHO	(c) $PhCH(CH_3)COOCH_3$
(d) $PhNHCOCH_2CH(CH_3)_2$	(e) $CH_3CONHCH_2Ph$	(f) $CH_3CH(OH)CH_2CN$
(g) $(CH_3)_2CHCH_2COBr$	(h) $Cl_2CHCOCl$	(i) $(CH_3)_2CHCOOCHO$

(j) $\mathrm{{-}O{-}\overset{\overset{O}{\|}}{C}{-}}$ (k) O, O   (l) PhCONH–

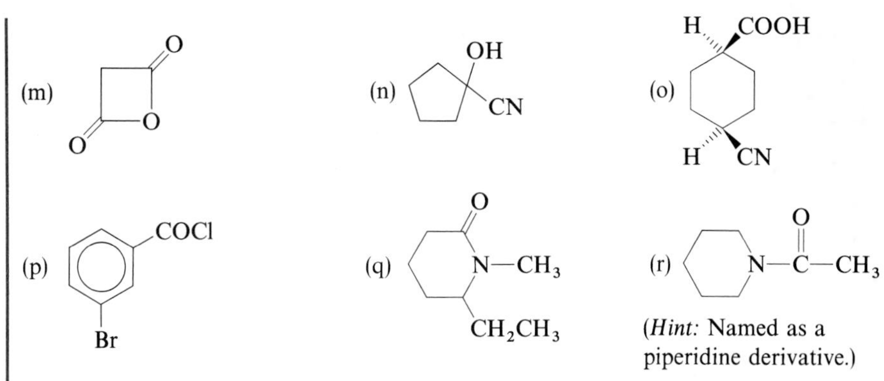

21-3 PHYSICAL PROPERTIES OF CARBOXYLIC ACID DERIVATIVES

21-3A BOILING POINTS AND MELTING POINTS

Figure 21-2 is a graph of the boiling points of simple acid derivatives plotted against their molecular weights. For comparison, the *n*-alkanes are included. Notice that the esters and acid chlorides have boiling points near those of the straight-chain alkanes with similar molecular weights. These carboxylic acid derivatives contain highly polar carbonyl groups, but the polarity of the carbonyl group has only a small effect on the boiling points (see Chapter 18).

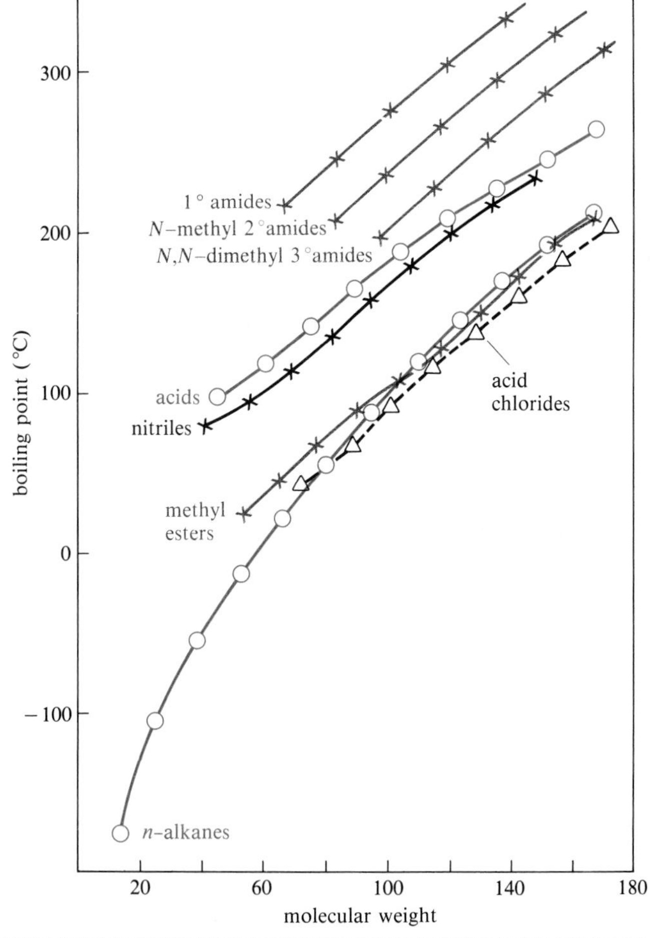

FIGURE 21-2 Graph of the boiling points of acid derivatives plotted against their molecular weights. The straight-chain alkanes are included for comparison.

Carboxylic acids are strongly hydrogen bonded in the liquid phase, resulting in elevated boiling points. The stable hydrogen-bonded dimer has a higher effective molecular weight and requires a higher temperature for boiling to occur. Nitriles also have higher boiling points than esters and acid chlorides of similar molecular weight. This effect results from a strong dipolar association between adjacent cyano groups.

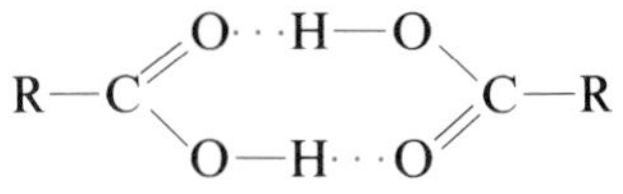

carboxylic acid dimer

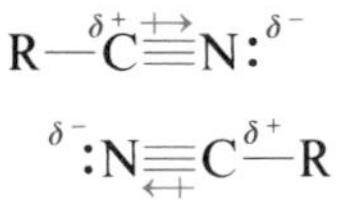

dipolar association of nitriles

Amides have surprisingly high boiling points and melting points. Even though pure tertiary amides cannot participate in hydrogen bonding (because they lack N—H bonds), they have boiling points close to those of carboxylic acids with similar molecular weights. These high boiling points result from the strongly polar nature of the amide. The resonance picture of an amide shows a large negative charge on oxygen and a corresponding positive charge on nitrogen. Figure 21-3 shows how a pairing of two molecules is strongly attractive, helping to stabilize the liquid phase. Vaporization requires disrupting this arrangement, thus a higher temperature is needed for boiling to occur.

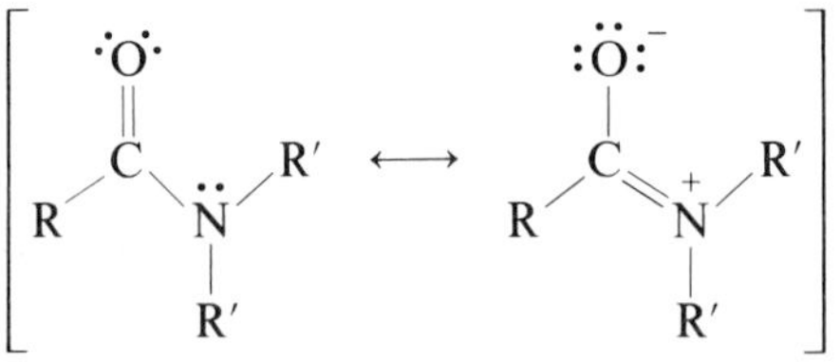

FIGURE 21-3 The resonance picture of an amide shows its strongly polar nature. Dipolar attractions stabilize the liquid phase, resulting in higher boiling points.

resonance picture of the amide

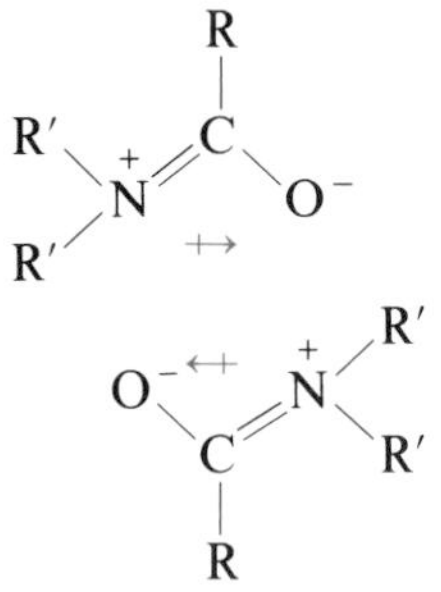

attraction of amide molecules

Primary and secondary amides have N—H bonds that engage in hydrogen bonding, resulting in higher boiling points for secondary amides (one N—H bond per molecule) than for tertiary amides (no N—H bonds), and still higher boiling points for primary amides (two N—H bonds per molecule).

The strong hydrogen-bonding attractions between the molecules of primary and secondary amides also result in unusually high melting points. For example, *N*-methylacetamide (secondary, one N—H bond) has a melting point of 28°C, which is 89° higher than the melting point (−61°C) of its isomer dimethylformamide (tertiary, no N—H bond). With two N—H bonds to engage in hydrogen bonding, the primary isomer propionamide melts at 79°C, about 50° higher than the secondary *N*-methylacetamide.

21-3B SOLUBILITY

Acid derivatives (esters, acid chlorides, anhydrides, nitriles, and amides) are soluble in common organic solvents such as alcohols, ethers, chlorinated alkanes, and aromatic hydrocarbons. Acid chlorides and anhydrides cannot be used in nucleophilic solvents such as water and alcohols, however, because they react with these solvents. Many of the smaller esters, amides, and nitriles are somewhat soluble in water (Table 21-1), owing to their high polarity and their ability to form hydrogen bonds with water.

TABLE 21-1
Several esters, amides, and nitriles are commonly used as solvents for organic reactions.

Compound	*m.p. (°C)*	*b.p. (°C)*	*Water solubility*
$CH_3—C(=O)—OCH_2CH_3$ ethyl acetate	−83	77	10%
$H—C(=O)—N(CH_3)_2$ dimethylformamide (DMF)	−61	153	miscible
$CH_3—C(=O)—N(CH_3)_2$ dimethylacetamide (DMA)	−20	165	miscible
$CH_3—C≡N$ acetonitrile	−45	82	miscible

Esters, tertiary amides, and nitriles are frequently used as solvents for organic reactions because they provide a polar reaction medium without the presence of O—H or N—H groups that can donate protons or act as nucleophiles. Ethyl acetate is a moderately polar solvent with a convenient boiling point (77°C) for easy evaporation from a reaction mixture. Acetonitrile, dimethylformamide (DMF), and dimethylacetamide (DMA) are highly polar solvents that solvate ions almost as well as water, but without the reactivity of O—H or N—H groups. These three solvents are miscible with water and are often used as solvent mixtures with water.

21-4 SPECTROSCOPY OF CARBOXYLIC ACID DERIVATIVES

21-4A INFRARED SPECTROSCOPY

Different types of carbonyl groups give characteristic strong absorptions at different positions in the infrared spectrum. As a result, infrared spectroscopy is often the best method to detect and differentiate these carboxylic acid derivatives. Table 21-2 gives a summary of the characteristic IR absorptions of carbonyl functional groups. As in Chapter 11, we are using the value of about 1710 cm^{-1} for simple ketones, aldehydes, and acids as a standard for comparison.

Esters Ester carbonyl groups absorb at relatively high frequencies, about 1735 cm^{-1}. Except for strained cyclic ketones, few other functional groups absorb strongly in this region. Esters also have a C—O single-bond stretching absorption between 1000 and 1200 cm^{-1}, although many other molecules also absorb in this region. We will not consider this absorption to be diagnostic for an ester, but we may check for it in uncertain cases.

Conjugation lowers the carbonyl stretching frequency of an ester. Conjugated esters absorb around 1710 to 1720 cm^{-1} and might easily be confused with simple ketones. The presence of *both* a strong carbonyl absorption in this region and a conjugated C=C stretching absorption around 1620 to 1640 cm^{-1} suggests a conjugated ester. Compare the spectra of ethyl octanoate and methyl benzoate in Figure 21-4 to see these differences.

TABLE 21-2
Characteristic IR stretching absorptions of acid derivatives

Functional group		Frequency	Comments
ketone	R—C(=O)—R	C═O, 1710 cm^{-1}	lower if conjugated, higher if strained
acid	R—C(=O)—OH	C═O, 1710 cm^{-1} O—H, 2500–3500 cm^{-1}	lower if conjugated
ester	R—C(=O)—O—R′	C═O, 1735 cm^{-1}	lower if conjugated, higher if strained
amide	R—C(=O)—N(H)—R′	C═O, 1640–1680 cm^{-1} N—H, 3200–3500 cm^{-1}	two peaks for R—CO—NH_2, one peak for R—CO—NHR′
acid chloride	R—C(=O)—Cl	C═O, 1800 cm^{-1}	
acid anhydride	R—C(=O)—O—C(=O)—R	C═O, 1800 and 1750 cm^{-1}	two peaks
nitrile	R—C≡N	C≡N, 2200 cm^{-1}	usually just above 2200 cm^{-1}

PROBLEM 21-2
What characteristics of the methyl benzoate spectrum rule out the presence of an aldehyde or carboxylic acid functional group giving the absorption at 1712 cm^{-1}?

PROBLEM 21-3
Give the frequencies of the C—O single-bond stretching absorptions in the IR spectra of ethyl octanoate and methyl benzoate.

less than a full double bond

:Ö: ‖ C—N̈ ⟷ :Ö:⁻ | C═N⁺

Amides Simple amides have much lower carbonyl stretching frequencies than the other carboxylic acid derivatives, absorbing around 1640 to 1680 cm^{-1} (often a close doublet). This low-frequency absorption agrees with the resonance picture of the amide functional group. The C═O bond of the amide carbonyl group is somewhat less than a full double bond. Because it is not as strong as the C═O bond in a simple ketone or carboxylic acid, the amide C═O has a lower stretching frequency.

Primary and secondary amides have N—H bonds that give stretching absorptions in the region 3200 to 3500 cm^{-1} of the infrared spectrum. These absorptions fall in the same region as the broad O—H stretching absorption of an alcohol, but the amide N—H absorptions are usually sharper. In primary amides (R—CO—NH_2) there are two N—H bonds, and two sharp peaks occur in the region 3200 to 3500 cm^{-1}. Secondary amides (R—CO—NHR′) have only one N—H bond, and only one peak is observed in the N—H region of the spectrum. Tertiary amides (R—CO—NR'_2) have no N—H bonds, and there is no N—H absorption.

The infrared spectrum of butyramide appears in Figure 11-12a, page 462. Notice the strong carbonyl stretching absorption at 1640 cm^{-1} and two N—H stretching absorptions at 3400 and 3250 cm^{-1}.

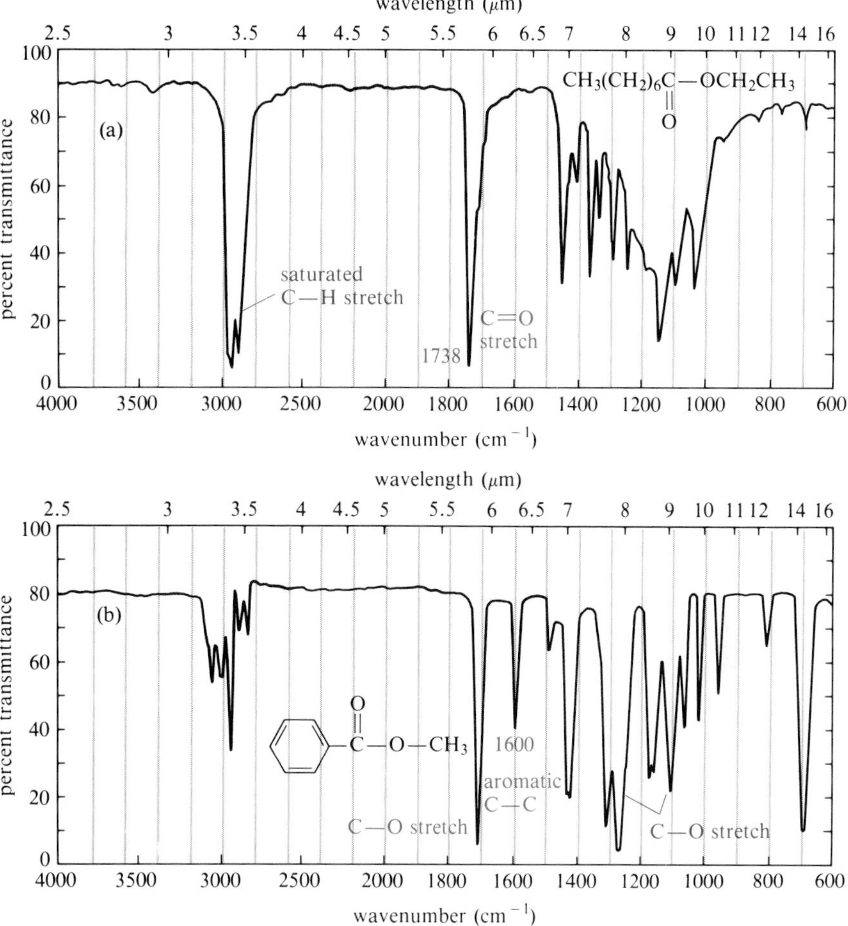

FIGURE 21-4 Infrared spectra of (a) ethyl octanoate and (b) methyl benzoate. The carbonyl stretching frequency of simple esters is around 1735 cm^{-1} and that of conjugated esters is around 1720 cm^{-1}.

Lactones and lactams Unstrained lactones (cyclic esters) and lactams (cyclic amides) absorb at typical frequencies for esters and amides. Ring strain raises the carbonyl absorption frequency, however. Recall that cyclic ketones with five-membered or smaller rings show a similar increase in carbonyl stretching frequency (Section 18-5A). Figure 21-5 shows the effect of ring strain on the C=O stretching frequencies of lactones and lactams.

Nitriles Nitriles show a characteristic C≡N stretching absorption around 2200 cm^{-1} in the infrared spectrum. This absorption can be distinguished from the alkyne C≡C stretching absorption by two characteristics: Nitriles usually absorb at frequencies slightly *greater* than 2200 cm^{-1} (to the left of 2200 cm^{-1}),

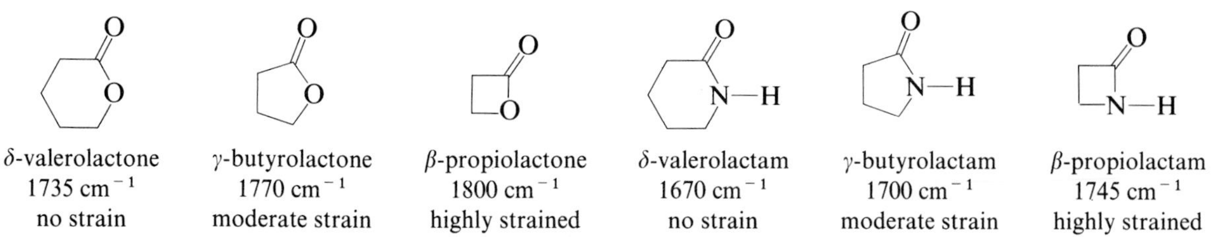

FIGURE 21-5 Ring strain in a lactone or lactam increases the carbonyl stretching frequency.

while alkynes usually absorb at frequencies slightly *less* than 2200 cm^{-1}; and nitrile absorption is usually much stronger, because the C≡N triple bond is more polar than the alkyne C≡C triple bond.

The IR spectrum of butyronitrile appears in Figure 11-13, page 464. Notice the strong triple-bond stretching absorption at 2225 cm^{-1}.

Acid halides and anhydrides Acid halides and anhydrides are rarely isolated as unknown compounds; but they are commonly used as synthetic reagents and intermediates, and infrared spectroscopy is used to confirm that an acid has been converted to a pure acid chloride or anhydride. The carbonyl stretching vibration of an acid chloride occurs at a high frequency around 1800 cm^{-1}. Figure 21-6 shows the IR spectrum of pentanoyl chloride (valeryl chloride), with a carbonyl absorption at 1795 cm^{-1} and no hydroxyl absorption.

Acid anhydrides give *two* carbonyl stretching absorptions, one around 1800 cm^{-1} and another around 1750 cm^{-1}. Figure 21-7 shows the spectrum of propionic anhydride, with carbonyl absorptions at 1820 and 1755 cm^{-1}.

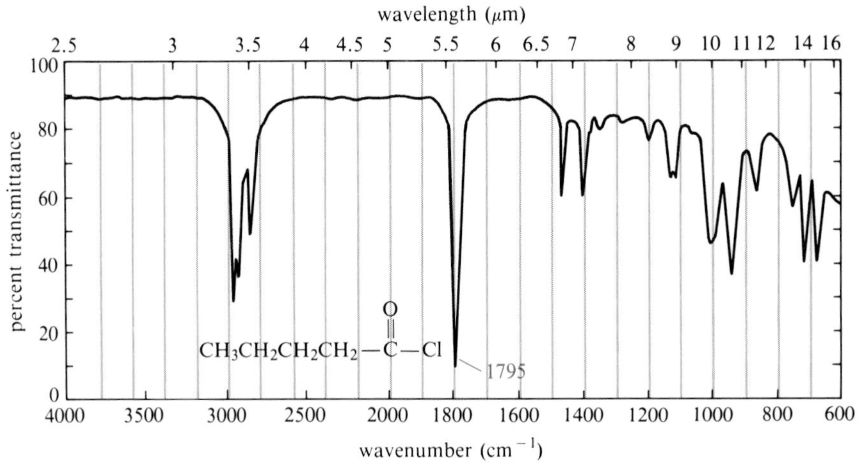

FIGURE 21-6 The infrared spectrum of pentanoyl chloride shows a characteristic C=O stretching absorption at 1795 cm^{-1}.

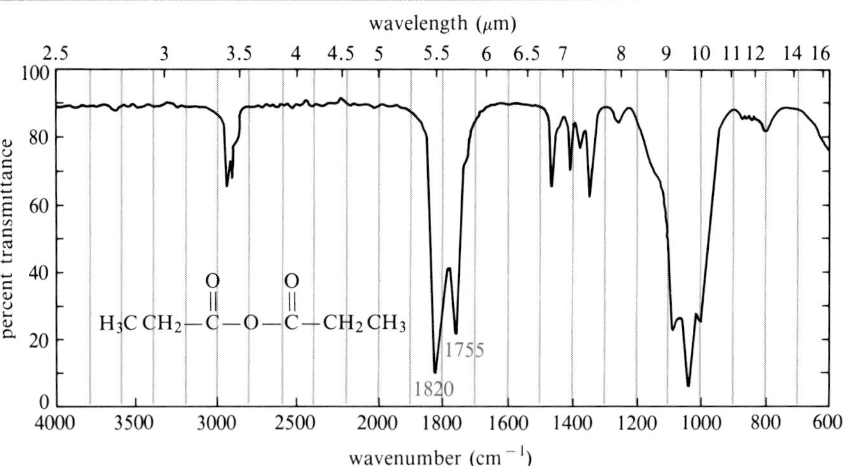

FIGURE 21-7 Infrared spectrum of propionic anhydride, showing C=O stretching absorptions at 1820 and 1755 cm^{-1}.

PROBLEM 21-4

The following IR spectra may include a carboxylic acid, an ester, an amide, a nitrile, an acid chloride, or an acid anhydride. Determine the functional group suggested by each spectrum, and list the specific frequencies you used to make your decision.

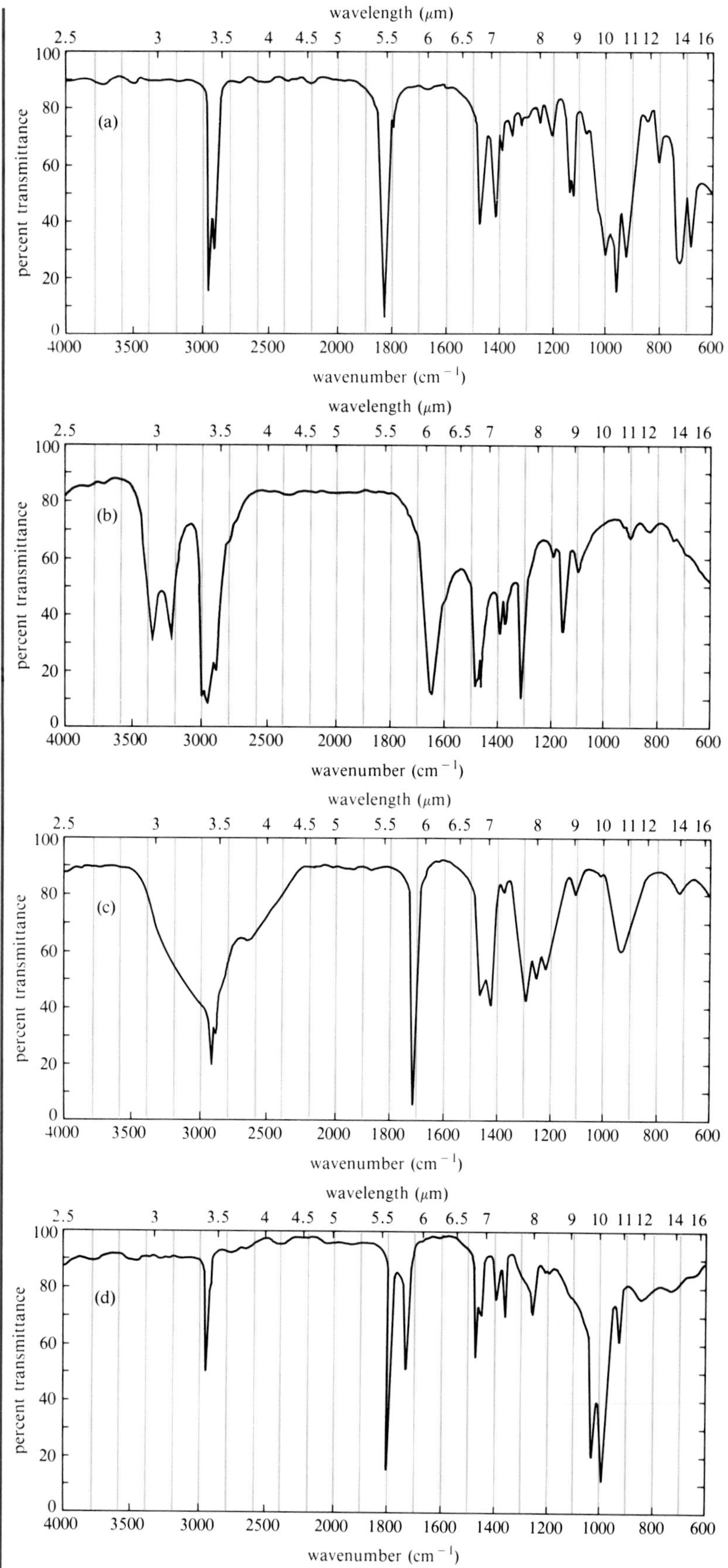

wavelength (μm)
2.5 3 3.5 4 4.5 5 5.5 6 6.5 7 8 9 10 11 12 14 16
100 80 60 40 20 0
percent transmittance
(a)
4000 3500 3000 2500 2000 1800 1600 1400 1200 1000 800 600
wavenumber (cm^{-1})
wavelength (μm)
(b)
wavenumber (cm^{-1})
wavelength (μm)
(c)
wavenumber (cm^{-1})
wavelength (μm)
(d)
wavenumber (cm^{-1})

NMR spectroscopy of acid derivatives is complementary to IR spectroscopy. For the most part, IR gives information about the functional groups, while NMR gives information about the structure of the alkyl groups. In many cases, the combination of IR and NMR spectra provides enough information to determine the structure.

The proton chemical shifts found in acid derivatives are close to those of similar protons in ketones, aldehydes, alcohols, and amines (Fig. 21-8). For example, protons alpha to a carbonyl group absorb between δ2.0 and δ2.5, whether the carbonyl group is part of a ketone, aldehyde, acid, ester, or amide. The protons of the alcohol-derived group of an ester or the amine-derived group of an amide give absorptions similar to those in the spectrum of the parent alcohol or amine.

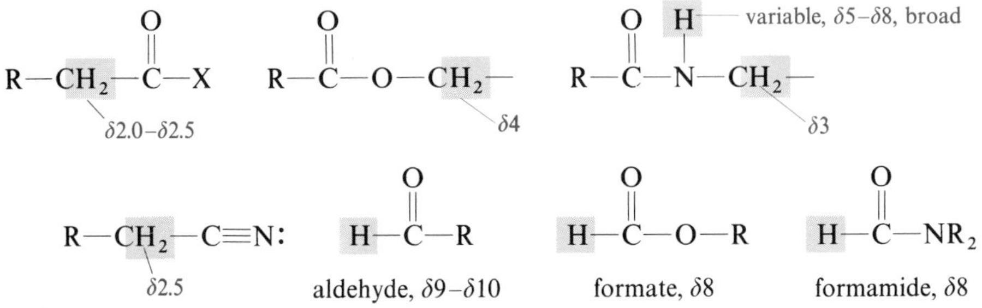

$R{-}CH_2{-}C({=}O){-}X$ — δ2.0–δ2.5

$R{-}C({=}O){-}O{-}CH_2{-}$ — δ4

$R{-}C({=}O){-}N(H){-}CH_2{-}$ — δ3; N—H: variable, δ5–δ8, broad

$R{-}CH_2{-}C{\equiv}N{:}$ — δ2.5

$H{-}C({=}O){-}R$ — aldehyde, δ9–δ10

$H{-}C({=}O){-}O{-}R$ — formate, δ8

$H{-}C({=}O){-}NR_2$ — formamide, δ8

FIGURE 21-8 Typical absorptions of acid derivatives in the proton NMR spectrum.

The N—H protons of an amide may be very broad, appearing between δ5 and δ8, depending on concentration and solvent. The *formyl* proton bonded to the carbonyl group of a formate ester or formamide resembles an aldehyde proton, but it is slightly more shielded and appears around δ8. In a nitrile, the protons on the alpha-carbon atom absorb around δ2.5, similar to the alpha protons of a carbonyl group.

The NMR spectrum of *N*-methylformamide (Fig. 21-9) shows the formyl proton (H—C=O) around δ8, and the N—H proton is nearly invisible around δ7. The *N*-methyl group appears as two singlets (*not* a spin-spin splitting doublet) around δ2.9. The two singlets result from hindered rotation about the amide bond, resulting in cisoid and transoid isomers that interconvert slowly with respect to the NMR time scale.

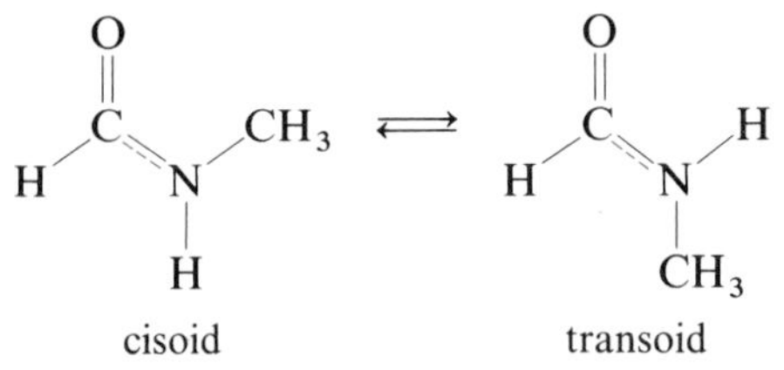

$H{-}C({=}O){-}N(H){-}CH_3$ (cisoid) ⇌ $H{-}C({=}O){-}N(CH_3){-}H$ (transoid)

cisoid   transoid

Carbon NMR The carbonyl carbons of acid derivatives appear at shifts of about 170 to 180 ppm, slightly less deshielded than the carbonyl carbons of ketones and aldehydes. The α-carbon atoms absorb around 30 to 40 ppm. The sp^3-hybridized carbons bonded to oxygen in esters absorb around 60 to 80 ppm, and those bonded to nitrogen in amides absorb around 40 to 60 ppm. The cyano carbon of a nitrile absorbs around 120 ppm.

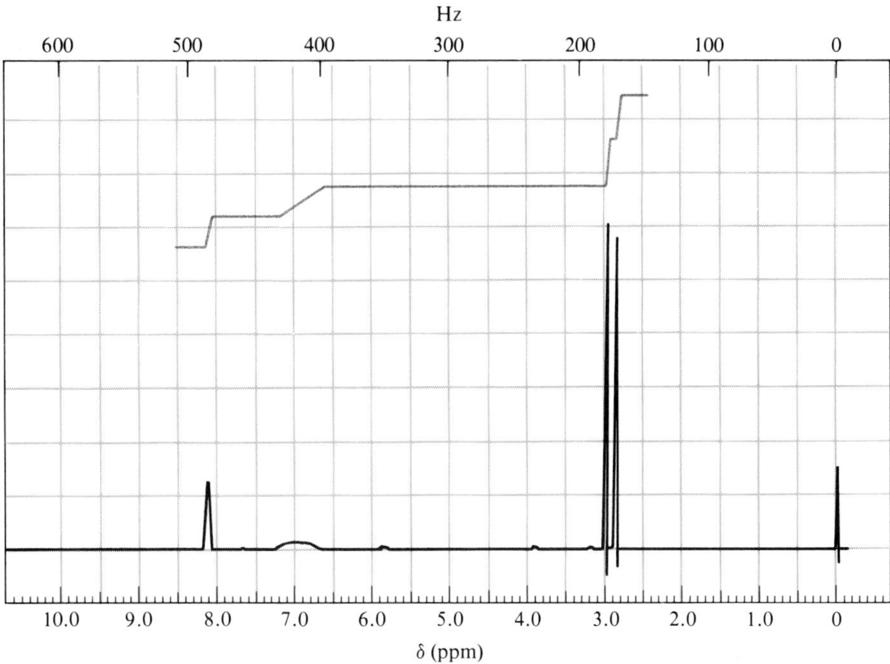

FIGURE 21-9 The NMR spectrum of *N*-methylformamide shows two methyl singlets resulting from hindered rotation about the amide bond.

R—C(=O)—O—C—   R—C(=O)—N̈—C—   R—C≡N:

~180 ppm ~60 ppm   ~180 ppm ~50 ppm   ~120 ppm

PROBLEM 21-5

For each of the following sets of IR and NMR spectra, determine the structure of the unknown compound. (a) C_3H_5NO (b) $C_5H_8O_2$

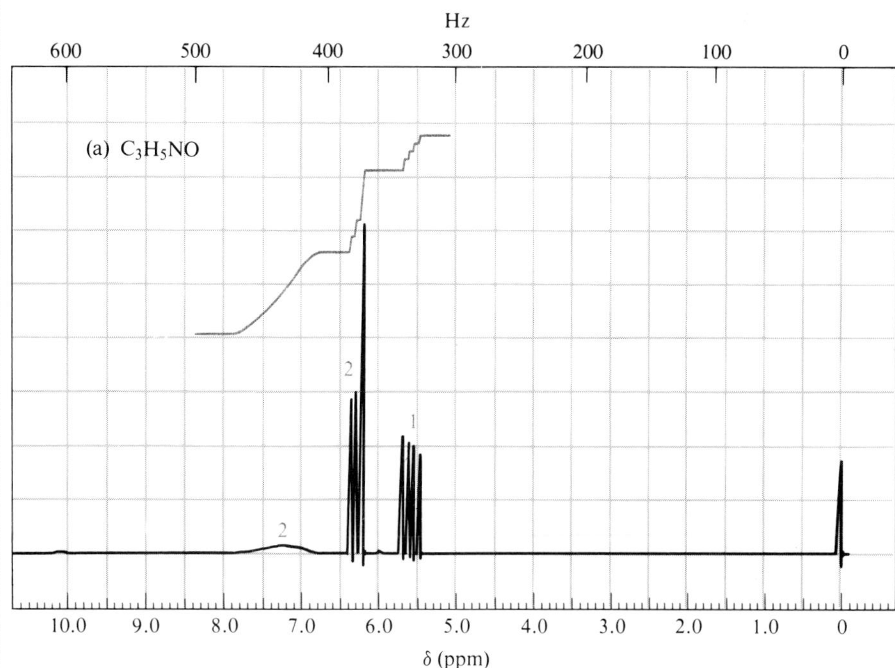

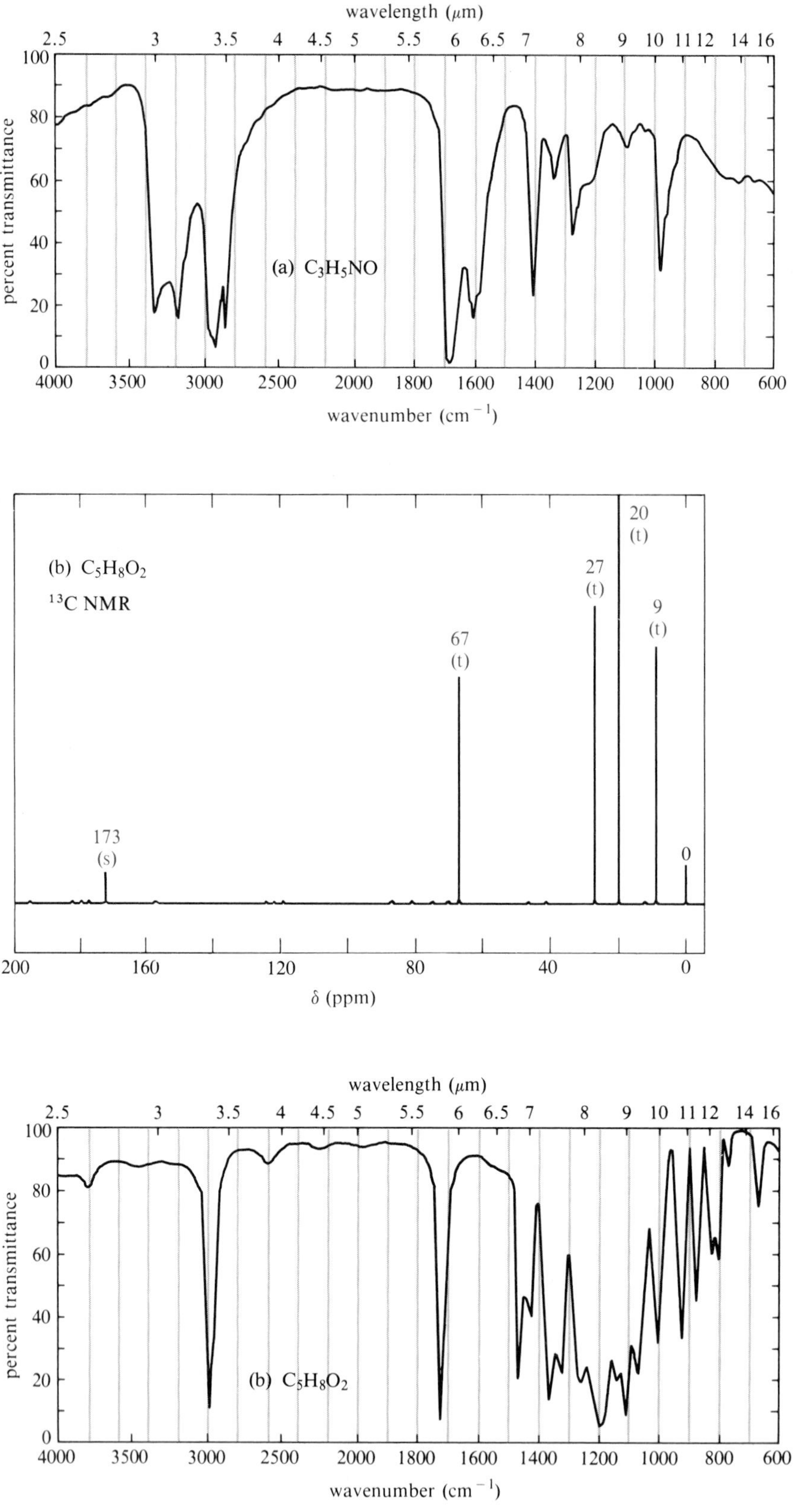
wavelength (μm)
2.5 3 3.5 4 4.5 5 5.5 6 6.5 7 8 9 10 11 12 14 16
100
80
60
40
20
0
percent transmittance
(a) C_3H_5NO
4000 3500 3000 2500 2000 1800 1600 1400 1200 1000 800 600
wavenumber (cm^{-1})
(b) $C_5H_8O_2$
^{13}C NMR
173 (s)
67 (t)
27 (t)
20 (t)
9 (t)
0
200 160 120 80 40 0
δ (ppm)
wavelength (μm)
2.5 3 3.5 4 4.5 5 5.5 6 6.5 7 8 9 10 11 12 14 16
100
80
60
40
20
0
percent transmittance
(b) $C_5H_8O_2$
4000 3500 3000 2500 2000 1800 1600 1400 1200 1000 800 600
wavenumber (cm^{-1})

21-5 INTERCONVERSION OF ACID DERIVATIVES BY NUCLEOPHILIC ACYL SUBSTITUTION

As we first saw in Chapter 20, the most common reaction of acid derivatives is **nucleophilic acyl substitution:** Attack on the carbonyl group by a nucleophile, followed by loss of another (the leaving group). Nucleophilic acyl substitutions are also called *acyl transfer* reactions because they transfer the acyl group from the leaving group to the attacking nucleophile. The following is a generalized addition-elimination mechanism for nucleophilic acyl substitution.

$$R-C(=\ddot{O}:)-X + Nuc:^- \rightleftharpoons \left[R-C(-\ddot{O}:^-)(-X)-Nuc \right] \rightleftharpoons R-C(=\ddot{O}:)-Nuc + X:^-$$

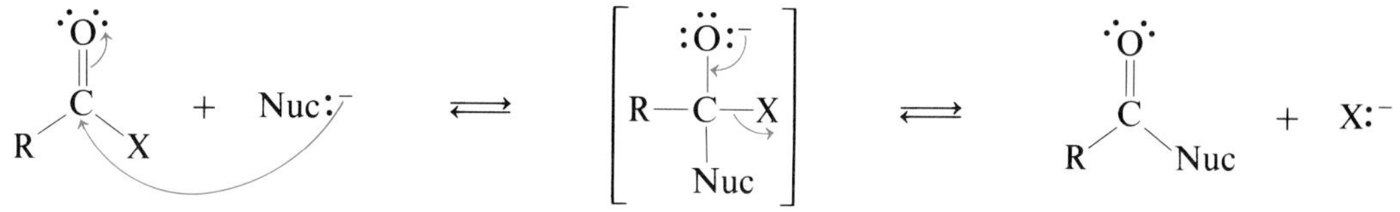

tetrahedral intermediate — product

Depending on the nature of $Nuc:^-$ and $X:^-$ in the above equation, we can imagine converting any acid derivative into almost any other. Yet the reactions that actually occur generally convert a more reactive acid derivative to a less reactive one. Predicting these reactions requires a knowledge of the relative reactivity of the acid derivatives.

21-5A REACTIVITY OF ACID DERIVATIVES

Acid derivatives differ greatly in their reactivity toward nucleophilic acyl substitution. For example, water hydrolyzes acetyl chloride in a violently exothermic reaction, while acetamide is quite stable in boiling water. Acetamide is hydrolyzed only by boiling it in strong acid or base for several hours.

$$CH_3-C(=O)-Cl + H_2O \xrightarrow{\text{(very fast)}} CH_3-C(=O)-OH + HCl$$

$$CH_3-C(=O)-NH_2 + Na^+ \ ^-OH \xrightarrow[\text{(slow)}]{H_2O,\ \text{boil}} CH_3-C(=O)-O^- \ ^+Na + NH_3$$

The reactivity of acid derivatives toward nucleophilic attack depends on their individual structure and on the nature of the attacking nucleophile. In general, their reactivity follows the following order.

Order of reactivity

$$\underset{\text{acid chloride}}{R-C(=O)-Cl} > \underset{\text{anhydride}}{R-C(=O)-O-C(=O)-R} > \underset{\text{ester}}{R-C(=O)-O-R'} > \underset{\text{amide}}{R-C(=O)-NH_2} > \underset{\text{carboxylate}}{R-C(=O)-O^-}$$

Basicity of leaving group

$$Cl^- < \ ^-O-C(=O)-R < \ ^-O-R < \ ^-NH_2$$

This order of reactivity stems partly from the basicity of the leaving groups. Strong bases are not very good leaving groups, and the reactivity of the derivatives decreases as the leaving group becomes more basic.

Resonance stabilization also affects the reactivity of the acid derivatives. In amides, for example, a significant amount of resonance stabilization is lost when a nucleophile attacks.

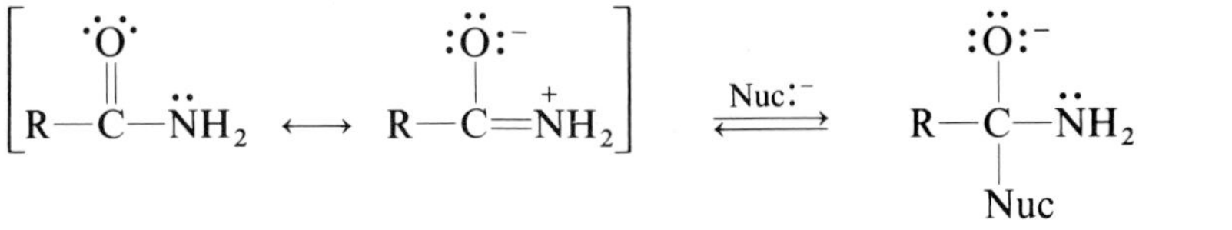

$$\left[R-\overset{\overset{\displaystyle O}{\|}}{C}-\ddot{N}H_2 \longleftrightarrow R-\overset{\overset{\displaystyle :\ddot{O}:^-}{|}}{C}=\overset{+}{N}H_2 \right] \xrightleftharpoons{Nuc:^-} R-\underset{\underset{\displaystyle Nuc}{|}}{\overset{\overset{\displaystyle :\ddot{O}:^-}{|}}{C}}-\ddot{N}H_2$$

resonance stabilized — no resonance stabilization

A smaller amount of resonance stabilization is present in esters.

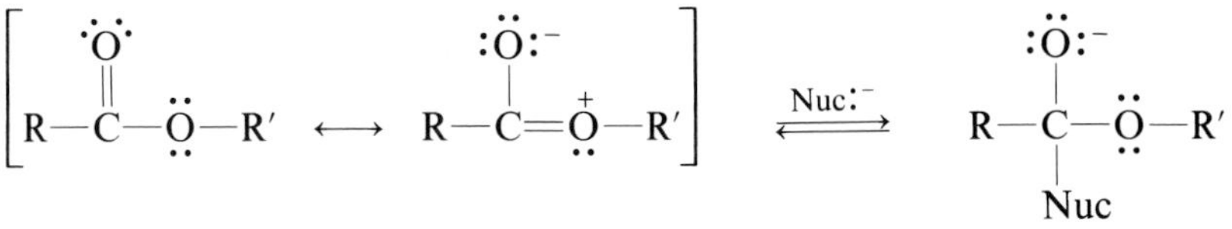

$$\left[R-\overset{\overset{\displaystyle O}{\|}}{C}-\ddot{O}-R' \longleftrightarrow R-\overset{\overset{\displaystyle :\ddot{O}:^-}{|}}{C}=\overset{+}{O}-R' \right] \xrightleftharpoons{Nuc:^-} R-\underset{\underset{\displaystyle Nuc}{|}}{\overset{\overset{\displaystyle :\ddot{O}:^-}{|}}{C}}-\ddot{O}-R'$$

weak resonance stabilization — no resonance stabilization

The resonance stabilization of an anhydride is like that in an ester, but the stabilization is shared between two carbonyl groups. Each carbonyl group receives less stabilization than an ester carbonyl.

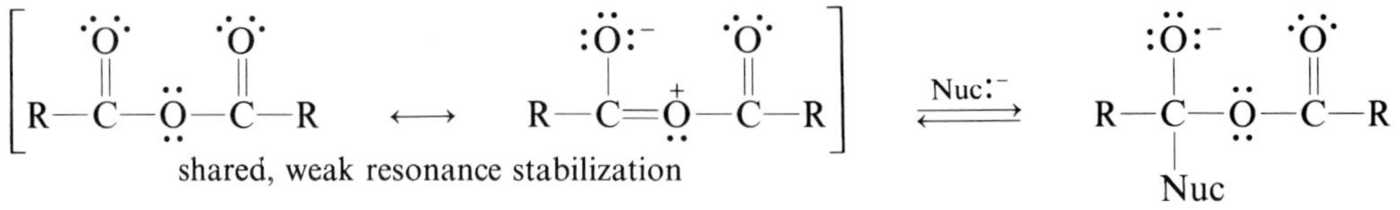

$$\left[R-\overset{\overset{\displaystyle O}{\|}}{C}-\ddot{O}-\overset{\overset{\displaystyle O}{\|}}{C}-R \longleftrightarrow R-\overset{\overset{\displaystyle :\ddot{O}:^-}{|}}{C}=\overset{+}{O}-\overset{\overset{\displaystyle O}{\|}}{C}-R \right] \xrightleftharpoons{Nuc:^-} R-\underset{\underset{\displaystyle Nuc}{|}}{\overset{\overset{\displaystyle :\ddot{O}:^-}{|}}{C}}-\ddot{O}-\overset{\overset{\displaystyle O}{\|}}{C}-R$$

shared, weak resonance stabilization

There is little resonance stabilization of an acid chloride, and it is quite reactive.

In general, we can easily accomplish reactions that convert more reactive derivatives to less reactive ones. The following reactions, all involving nucleophilic acyl substitution by the addition-elimination mechanism, generally give good yields of products.

Conversion of an acid chloride to an anhydride

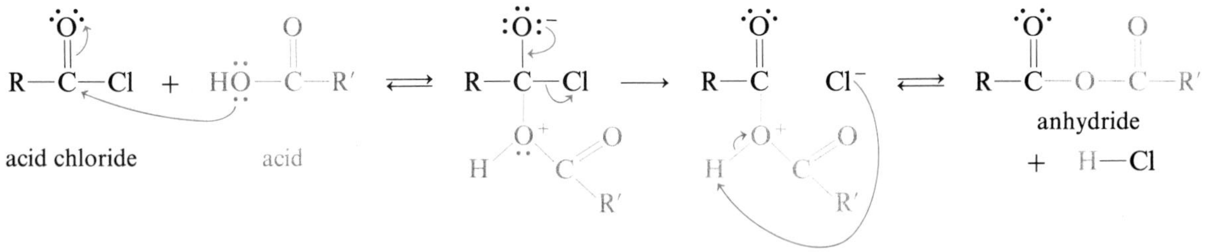

$$R-\overset{\overset{\displaystyle O}{\|}}{C}-Cl + H\ddot{O}-\overset{\overset{\displaystyle O}{\|}}{C}-R' \rightleftharpoons R-\underset{\underset{\displaystyle H-\overset{+}{O}-C(=O)-R'}{|}}{\overset{\overset{\displaystyle :\ddot{O}:^-}{|}}{C}}-Cl \longrightarrow R-\underset{\underset{\displaystyle H-\overset{+}{O}-C(=O)-R'}{|}}{\overset{\overset{\displaystyle O}{\|}}{C}} \quad Cl^- \rightleftharpoons R-\overset{\overset{\displaystyle O}{\|}}{C}-O-\overset{\overset{\displaystyle O}{\|}}{C}-R' + H-Cl$$

acid chloride — acid — anhydride

Example

$$CH_3(CH_2)_5-\overset{\overset{\displaystyle O}{\|}}{C}-Cl + CH_3(CH_2)_5-\overset{\overset{\displaystyle O}{\|}}{C}-OH \longrightarrow CH_3(CH_2)_5-\overset{\overset{\displaystyle O}{\|}}{C}-O-\overset{\overset{\displaystyle O}{\|}}{C}-(CH_2)_5CH_3$$

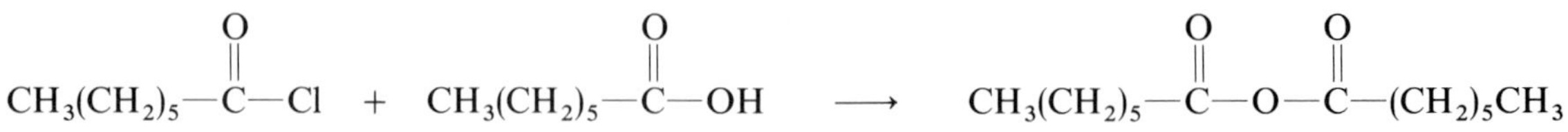

heptanoyl chloride — heptanoic acid — heptanoic anhydride

Conversion of an acid chloride to an ester

$$R{-}\overset{\overset{\displaystyle :\ddot{O}}{\|}}{C}{-}Cl \; + \; R'{-}\ddot{\underset{..}{O}}H \longrightarrow R{-}\underset{\underset{R'\quad H}{:O^+}}{\overset{:\ddot{O}:^-}{C}}{-}Cl \longrightarrow R{-}\underset{\underset{R'}{:O^+{-}H}}{\overset{\overset{\displaystyle :\ddot{O}}{\|}}{C}} + Cl^- \longrightarrow R{-}\overset{\overset{\displaystyle :\ddot{O}}{\|}}{C}{-}\ddot{\underset{..}{O}}{-}R' + HCl$$

acid chloride, alcohol, ester

Example

$$\text{cyclopentyl}{-}\overset{\overset{O}{\|}}{C}{-}Cl + CH_3{-}\overset{OH}{\overset{|}{CH}}{-}CH_3 \longrightarrow \text{cyclopentyl}{-}\overset{\overset{O}{\|}}{C}{-}OCH(CH_3)_2 + HCl$$

cyclopentanecarbonyl chloride; 2-propanol; 2-propyl cyclopentanecarboxylate

Conversion of an acid chloride to an amide

$$R{-}\overset{\overset{\displaystyle :\ddot{O}}{\|}}{C}{-}\ddot{\underset{..}{Cl}}: \; R'_2\ddot{N}{-}H \rightleftharpoons R{-}\underset{\underset{R'_2N{-}H}{|+}}{\overset{:\ddot{O}:^-}{C}}{-}\ddot{\underset{..}{Cl}}: \longrightarrow R{-}\underset{R'_2\overset{+}{N}{-}H}{\overset{\overset{\displaystyle :\ddot{O}}{\|}}{C}} \quad {}^-{:}\ddot{\underset{..}{Cl}}: \longrightarrow R{-}\overset{\overset{\displaystyle :\ddot{O}}{\|}}{C}{-}\ddot{N}R'_2 + HCl$$

acid chloride, amine

Reaction of an acid chloride with ammonia gives a primary amide; with a primary amine this reaction gives a secondary amide; and with a secondary amine it gives a tertiary amide.

$$CH_3{-}(CH_2)_4{-}\overset{\overset{O}{\|}}{C}{-}Cl + \text{cyclohexyl}{-}NH_2 \longrightarrow CH_3{-}(CH_2)_4{-}\overset{\overset{O}{\|}}{C}{-}NH{-}\text{cyclohexyl} + HCl$$

hexanoyl chloride; cyclohexylamine; *N*-cyclohexylhexanamide (+ pyridine · HCl)

Conversion of an acid anhydride to an ester

$$R{-}\overset{\overset{\displaystyle :\ddot{O}}{\|}}{C}{-}O{-}\overset{\overset{O}{\|}}{C}{-}R + R'{-}\ddot{\underset{..}{O}}H \rightleftharpoons R{-}\underset{\underset{R'\quad H}{\ddot{O}^+}}{\overset{:\ddot{O}:^-}{C}}{-}O{-}\overset{\overset{O}{\|}}{C}{-}R \longrightarrow R{-}\underset{\underset{R'\quad H}{{}^+\ddot{O}}}{\overset{\overset{\displaystyle :\ddot{O}}{\|}}{C}} \quad {}^-O{-}\overset{\overset{O}{\|}}{C}{-}R$$

anhydride, alcohol

$$\rightleftharpoons R{-}\overset{\overset{O}{\|}}{C}{-}O{-}R' + R{-}\overset{\overset{O}{\|}}{C}{-}OH$$

ester

Example

$$\text{cyclopentyl}{-}OH + CH_3{-}\overset{\overset{O}{\|}}{C}{-}O{-}\overset{\overset{O}{\|}}{C}{-}CH_3 \longrightarrow \text{cyclopentyl}{-}O{-}\overset{\overset{O}{\|}}{C}{-}CH_3 + AcOH$$

cyclopentanol; acetic anhydride; cyclopentyl acetate; acetic acid

Conversion of an acid anhydride to an amide

$$\text{R—C(=O)—O—C(=O)—R} + \text{R}'_2\ddot{\text{N}}\text{H} \rightleftharpoons \text{R—C(O}^-\text{)(}^+\text{NHR}'_2\text{)—O—C(=O)—R} \longrightarrow \text{R—C(=O)—}^+\text{NHR}'_2 \;\; ^-\text{O—C(=O)—R}$$

anhydride — amine

$$\rightleftharpoons \text{R—C(=O)—NR}'_2 + \text{R—C(=O)—OH}$$

amide

Reaction of an anhydride with ammonia gives a primary amide; with a primary amine this reaction gives a secondary amide; and with a secondary amine it gives a tertiary amide.

$$\text{C}_6\text{H}_5\text{—NH}_2 + \text{CH}_3\text{—C(=O)—O—C(=O)—CH}_3 \longrightarrow \text{C}_6\text{H}_5\text{—NH—C(=O)—CH}_3 + \text{AcOH}$$

aniline — acetic anhydride — acetanilide — acetic acid

Conversion of an ester to an amide (ammonolysis of an ester)

$$\text{R—C(=O)—O—R}' + \text{R}''\text{—}\ddot{\text{N}}\text{H}_2 \rightleftharpoons \text{R—C(O}^-\text{)(}^+\text{NH}_2\text{R}''\text{)—O—R}' \rightleftharpoons \text{R—C(=O)—}^+\text{NH}_2\text{R}'' \;\; ^-\text{O—R}' \longrightarrow$$

primary amine (or NH_3)

$$\text{R—C(=O)—NHR}'' + \text{R}'\text{—OH}$$

amide

Example

$$\text{H—C(=O)—O—CH}_2\text{CH}_3 + \text{C}_6\text{H}_{11}\text{—NH}_2 \longrightarrow \text{C}_6\text{H}_{11}\text{—N(H)—C(=O)—H} + \text{CH}_3\text{CH}_2\text{—OH}$$

ethyl formate — cyclohexylamine — *N*-cyclohexylformamide (90%)

Figure 21-10 shows graphically the conversion of more reactive acid derivatives to less reactive ones. Notice that thionyl chloride ($SOCl_2$) converts an acid to its most reactive derivative, the acid chloride.

21-5B LEAVING GROUPS IN NUCLEOPHILIC ACYL SUBSTITUTION

The loss of an alkoxide ion as a leaving group in the second step of the conversion of an ester to an amide (above) should surprise you. In our study of alkyl substitution and elimination reactions (S_N1, S_N2, E1, E2), we saw that strong bases such as hydroxide and alkoxide are not good leaving groups. Figure 21-11 compares

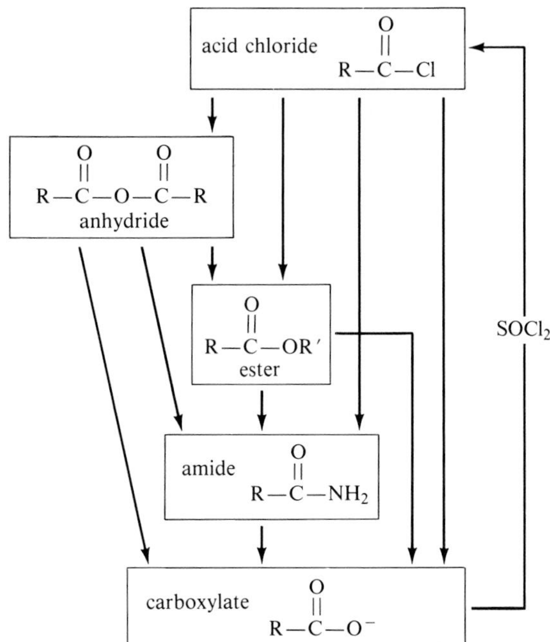

FIGURE 21-10 More reactive acid derivatives are easily converted to less reactive derivatives. A "downhill" reaction from R—C(=O)—W to R—C(=O)—Z generally requires H—Z or Z$^-$ as the nucleophile for nucleophilic acyl substitution.

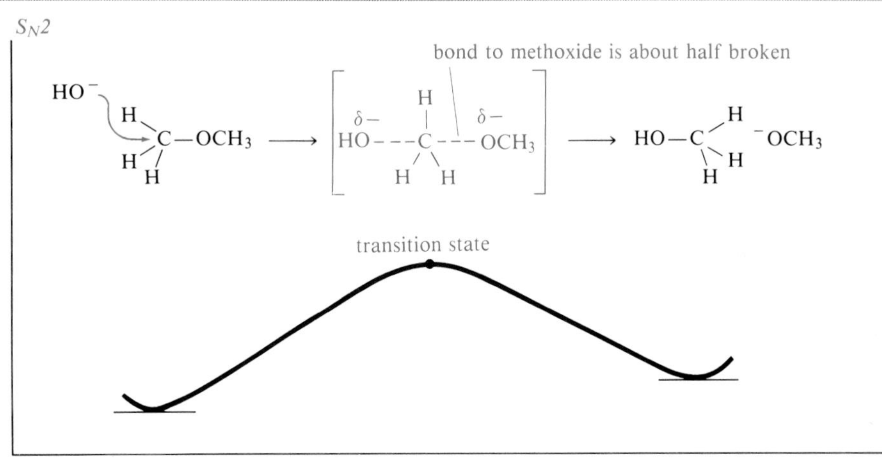

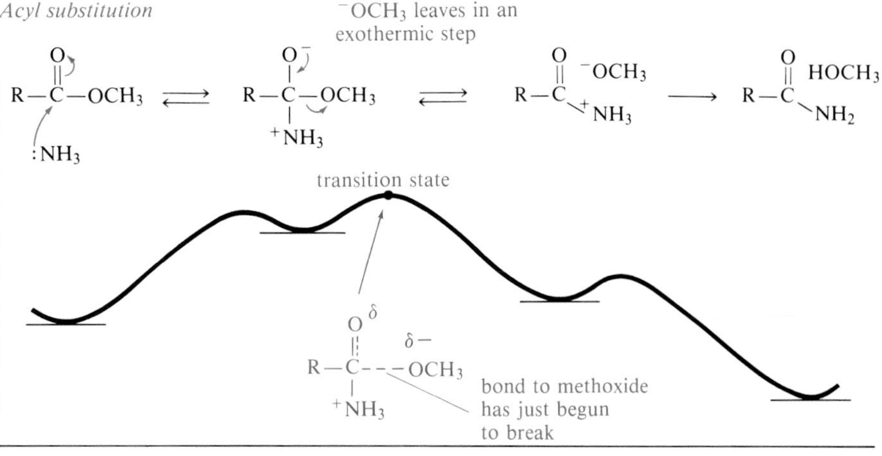

FIGURE 21-11 Comparison of an S_N2 reaction and an acyl addition-elimination reaction with methoxide as the leaving group. In the S_N2 reaction, methoxide leaves in a slightly endothermic step, and the bond to methoxide is largely broken in the transition state. In the acyl substitution, methoxide leaves in an exothermic second step with a reactant-like transition state: that is, the bond to methoxide has just begun to break in the transition state.

the acyl addition-elimination mechanism with the S_N2 mechanism. The differences in the mechanisms explain why such strong bases may serve as leaving groups in acyl substitution, even though they cannot in alkyl substitution.

The S_N2 reaction has a one-step mechanism. This step is not highly endothermic or exothermic. The bond to the leaving group is about half broken in the transition state, so the reaction rate is very sensitive to the nature of the leaving group. With a poor leaving group such as an alkoxide, this reaction is quite slow.

In the acyl substitution, the leaving group leaves in a separate second step. This second step is highly exothermic, and the Hammond postulate (Section 4-15) predicts that the transition state resembles the reactant: the tetrahedral intermediate. In this transition state, the bond to the leaving group has barely begun to break. The energy of the transition state (and therefore the reaction rate) is not very sensitive to the nature of the leaving group.

Nucleophilic acyl substitution is our first example of a reaction in which strong bases can serve as leaving groups. We will see many additional examples of such reactions. In general, a strong base may serve as a leaving group if it leaves in a highly exothermic step, usually converting an unstable, negatively charged intermediate to a stable molecule.

PROBLEM 21-6

Which of the following proposed reactions would take place quickly under mild conditions?

(a) $CH_3{-}\overset{O}{\overset{\|}{C}}{-}NH_2 + NaCl \longrightarrow CH_3{-}\overset{O}{\overset{\|}{C}}{-}Cl + NaNH_2$

(b) $Ph{-}\overset{O}{\overset{\|}{C}}{-}Cl + CH_3NH_2 \longrightarrow Ph{-}\overset{O}{\overset{\|}{C}}{-}NHCH_3 + HCl$

(c) $(CH_3)_2CH{-}\overset{O}{\overset{\|}{C}}{-}OCH_3 + NH_3 \longrightarrow (CH_3)_2CH{-}\overset{O}{\overset{\|}{C}}{-}NH_2 + CH_3OH$

(d) $CH_3CH_2{-}\overset{O}{\overset{\|}{C}}{-}Cl + CH_3{-}\overset{O}{\overset{\|}{C}}{-}OH \longrightarrow CH_3CH_2{-}\overset{O}{\overset{\|}{C}}{-}O{-}\overset{O}{\overset{\|}{C}}{-}CH_3 + HCl$

(e) $CH_3{-}\overset{O}{\overset{\|}{C}}{-}O{-}\overset{O}{\overset{\|}{C}}{-}CH_3 + CH_3NH_2 \longrightarrow CH_3{-}\overset{O}{\overset{\|}{C}}{-}NHCH_3 + CH_3COOH$

PROBLEM 21-7

Propose a mechanism for the synthesis of cyclohexyl isobutyrate from isobutyryl chloride and cyclohexanol.

PROBLEM 21-8

Show how you would synthesize each of the following esters from an appropriate acyl chloride and alcohol.

(a) ethyl propionate (b) phenyl 3-methylhexanoate
(c) benzyl benzoate (d) cyclopropyl cyclohexanecarboxylate

PROBLEM 21-9

Show how you would use an appropriate acyl chloride and amine to synthesize each of the following amides.

(a) *N,N*-dimethylacetamide (b) acetanilide ($PhNHCOCH_3$)

(c) cyclohexane carboxamide (d) $\overset{O}{\overset{\|}{C}}{-}N$

PROBLEM 21-10

(a) Show how you would use acetic anhydride and an appropriate alcohol or amine to synthesize (i) benzyl acetate; (ii) acetanilide, $PhNHCOCH_3$. (b) Propose a mechanism for each synthesis in part (a).

PROBLEM 21-11

Propose a mechanism for the reaction of benzyl acetate with methylamine. Label the attacking nucleophile and the leaving group, and draw the transition state in which the leaving group leaves.

21-6 ACID-CATALYZED NUCLEOPHILIC ACYL SUBSTITUTION

In each of the substitutions discussed above, a nucleophile attacks the carbonyl group to form a tetrahedral intermediate. Some nucleophiles are too weak to attack an unactivated carbonyl group. For example, an alcohol attacks the carbonyl group of an acid chloride, but it does not attack the carboxyl group of an acid. If a strong acid protonates the carbonyl group of the carboxylic acid, it is activated toward attack by the alcohol; Fischer esterification is the result (Section 20-11).

$$R-C(=O)-OH + H^+ \rightleftharpoons \left[R-C(=O^+-H)-O-H\right] \xleftarrow{} H-\ddot{O}-R' \rightleftharpoons R-C(OH)(OH)(H-O^+-R') \xrightleftharpoons{\text{several steps}} R-C(=O)-O-R' \;(\text{ester}) + H_3O^+$$

A similar mechanism is responsible for the acid-catalyzed **transesterification** of an ester: the substitution of one alkoxy group for another. When an ester is treated with a different alcohol in the presence of an acid catalyst, the two alcohol groups can interchange. An equilibrium results, and the equilibrium can be driven toward the desired ester by using a large excess of the desired alcohol or by removing the other alcohol.

$$R-C(=O)-O-R' + \underset{\text{(large excess)}}{R''-OH} \xrightleftharpoons{H^+ \text{ or } ^-OR''} R-C(=O)-O-R'' + R'-OH$$

Example

$$CH_3-C(=O)-O-CH_2CH_3 + \underset{\text{(large excess)}}{CH_3-OH} \xrightleftharpoons{H^+} CH_3-C(=O)-O-CH_3 + CH_3CH_2-OH$$

Transesterification also occurs under basic conditions, catalyzed by a small amount of alkoxide ion. Once again, a large excess of the desired alcohol is used to achieve a good conversion.

PROBLEM SOLVING: PROPOSING REACTION MECHANISMS

Rather than simply showing the mechanisms for acid-catalyzed and base-catalyzed transesterification, let's consider how one might work out these mechanisms as a problem.

First consider the base-catalyzed transesterification of ethyl benzoate with methanol. This is a classic example of nucleophilic acyl substitution by the addition-elimination mechanism. The methoxide ion is sufficiently nucleophilic to attack the ester carbonyl group. The ethoxide ion serves as a leaving group in a strongly exothermic second step.

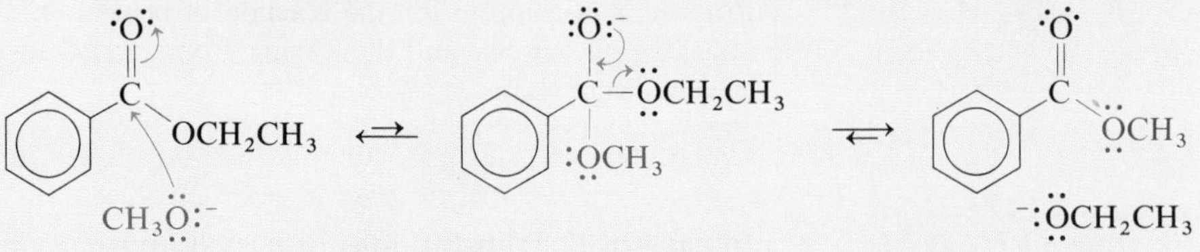

PROBLEM 21-12

When ethyl 4-hydroxybutyrate is heated in the presence of a trace of a basic catalyst (sodium acetate), one of the products is a lactone. Propose a mechanism for formation of this lactone.

The acid-catalyzed reaction follows a similar mechanism, but it is more complicated (because of additional proton transfers), and we use the stepwise procedure.

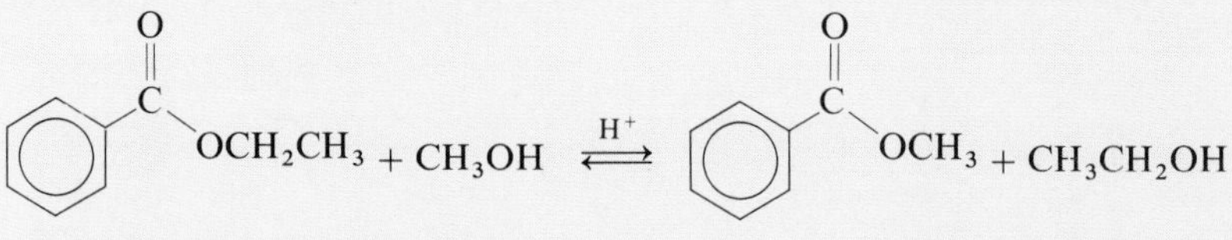

1. **Consider the carbon skeletons of the reactants and products, and identify which carbon atoms in the products are most likely derived from which carbon atoms in the reactants.**
   In this case, an ethoxyl group is being replaced by a methoxyl group.
2. **Consider whether any of the reactants is a strong enough electrophile to react without being activated. If not, consider how one of the reactants might be converted to a strong electrophile by protonation of a Lewis basic site.**
   The ester carbonyl group is not a strong enough electrophile to react with methanol. Protonation converts it to a strong electrophile (shown in step 3).
3. **Consider how a nucleophilic site on another reactant can attack the strong electrophile to form a bond needed in the product.**
   Methanol has a nucleophilic oxygen atom that can attack the activated carbonyl group to form the new C—O bond needed in the product.

(resonance-stabilized)

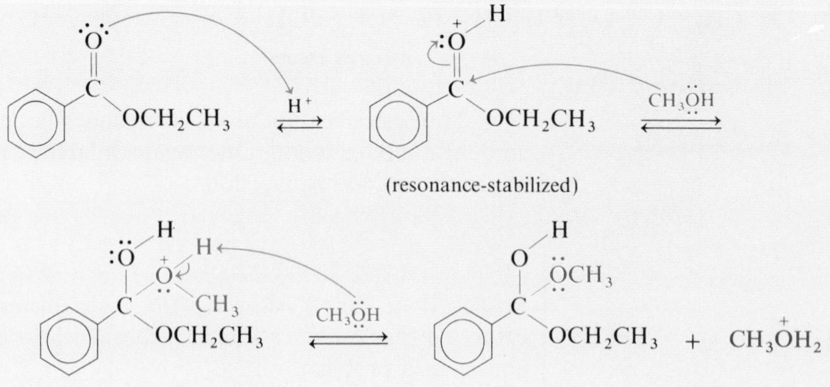

4. Consider how the product of nucleophilic attack might be converted to the final product or reactivated to form another bond needed in the product. The task here is to break bonds, not form them. The ethoxyl group (OCH_2CH_3) must be lost. The most common mechanism for losing a group under acidic conditions is to protonate it (to make it a good leaving group), then lose it. In fact, losing the ethoxyl group is exactly the reverse of the mechanism used above to gain the methoxyl group.

Protonation prepares the ethoxyl group to leave. When it leaves, the product is simply a protonated version of the final product.

H
O
C—OCH₃
:OCH₂CH₃ ⇌
H⁺
:Ö—H
C—OCH₃
:O⁺
H CH₂CH₃
⇌
⁺:O—H
C
OCH₃
Ö
H CH₂CH₃

5. Draw out all steps of the mechanism using curved arrows to show the movement of electrons. Once again, this summary is left to you to help you review the mechanism.

PROBLEM 21-13

Complete the mechanism for acid-catalyzed transesterification by drawing out all the individual steps. Draw the important resonance contributors for each of the resonance-stabilized intermediates.

PROBLEM 21-14

Propose a mechanism for the following ring-opening transesterification. Use the mechanism in Problem 21-13 as a model.

O O + CH_3OH $\underset{}{\overset{H^+}{\rightleftharpoons}}$ HO O OCH₃

21-7 HYDROLYSIS OF CARBOXYLIC ACID DERIVATIVES

All acid derivatives hydrolyze to give carboxylic acids. In most cases, hydrolysis occurs under either acidic or basic conditions. The reactivity of acid derivatives toward hydrolysis varies from the highly reactive acyl halides to the relatively unreactive amides.

21-7A HYDROLYSIS OF ACID HALIDES AND ANHYDRIDES

Acid halides and anhydrides are so reactive that they hydrolyze under neutral conditions. The hydrolysis of an acid chloride or anhydride is usually an annoying side reaction that takes place upon exposure of these compounds to moist air. Hydrolysis can be avoided by storing acid chlorides and anhydrides under dry nitrogen and by using dry solvents and reagents.

:Ö:
R—C—Cl + $H_2\ddot{O}$: ⟶
:Ö:⁻
R—C—Cl
⁺OH₂
⟶
Ö:
R—C Cl
⁺O—H
H
⟶
:Ö:
R—C—Ö—H + HCl

The acid-catalyzed hydrolysis of an ester is simply the reverse of the Fischer esterification equilibrium. Addition of a large excess of water drives the equilibrium toward the acid and the alcohol.

The basic hydrolysis of esters, called **saponification,** avoids the equilibrium of the Fischer esterification. Hydroxide ion attacks the carbonyl group to give a tetrahedral intermediate. Expulsion of an alkoxide ion gives the acid, and a fast proton transfer gives the carboxylate ion and the alcohol. This strongly exothermic proton transfer drives the saponification to completion. Notice that a full mole of base is consumed to deprotonate the acid.

$$R{-}C({=}O){-}O{-}R' + {}^{-}{:}\ddot{O}{-}H \rightleftharpoons R{-}C(O^{-})(OH){-}\ddot{O}{-}R' \rightleftharpoons R{-}C(=O){-}\ddot{O}{-}H + {}^{-}{:}\ddot{O}{-}R' \longrightarrow R{-}C(=O){-}O{:}^{-} \quad H{-}\ddot{O}{-}R'$$

ester, tetrahedral intermediate, acid, alkoxide, carboxylate, alcohol

Example

$$CH_3CH_2{-}\overset{O}{\overset{\|}{C}}{-}O{-}CH_2CH_3 + Na^{+}\,{}^{-}OH \longrightarrow CH_3CH_2{-}\overset{O}{\overset{\|}{C}}{-}O^{-}\,{}^{+}Na + CH_3CH_2{-}OH$$

ethyl propionate, sodium propionate, ethanol

The term *saponification* (Latin, *saponis,* "soap") literally means "the making of soap." Soap is made by the basic hydrolysis of fats, which are esters of long-chain carboxylic acids (*fatty acids*) with the triol glycerol. When a fat is hydrolyzed by sodium hydroxide, the resulting long-chain carboxylate salts are what we know as soap. Soaps and detergents are discussed in more detail in Chapter 25.

$$\begin{matrix} CH_2{-}O{-}C(=O){\sim\sim\sim} \\ | \\ CH{-}O{-}C(=O){\sim\sim\sim} \\ | \\ CH_2{-}O{-}C(=O){\sim\sim\sim} \end{matrix} + 3\,NaOH \longrightarrow \begin{matrix} CH_2{-}O{-}H \\ | \\ CH{-}O{-}H \\ | \\ CH_2{-}O{-}H \end{matrix} \quad \begin{matrix} Na^{+}\,{}^{-}O{-}C(=O){\sim\sim\sim} \\ Na^{+}\,{}^{-}O{-}C(=O){\sim\sim\sim} \\ Na^{+}\,{}^{-}O{-}C(=O){\sim\sim\sim} \end{matrix}$$

a fat (triester of glycerol), glycerol, soap (salts of fatty acids)

PROBLEM 21-15

Propose a mechanism for the hydrolysis of ethyl propionate

(a) under acidic conditions.
(b) under basic conditions.

PROBLEM 21-16

(a) Explain why we speak of the acidic hydrolysis of an ester as *acid-catalyzed,* but the basic hydrolysis as *base-promoted.*
(b) Soap manufacturers always use base to hydrolyze fats, and never acid. Suggest two reasons basic hydrolysis is preferred.

PROBLEM 21-17

Propose a mechanism for the base-catalyzed hydrolysis of γ-butyrolactone:

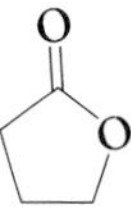

21-7C HYDROLYSIS OF AMIDES

Amides undergo hydrolysis to carboxylic acids under both acidic and basic conditions. Amides are the most stable of the acid derivatives, and stronger conditions are required for their hydrolysis than for the hydrolysis of an ester. Typical hydrolysis conditions involve prolonged heating in 6 *M* HCl or 40 percent aqueous NaOH.

Basic hydrolysis

$$R{-}C({=}O){-}NHR' + Na^+ \ {}^-OH \xrightarrow{H_2O} R{-}C({=}O){-}O^- + R'NH_2$$

Example

$$C_6H_5{-}C({=}O){-}N(CH_2CH_3)_2 + NaOH \xrightarrow{H_2O} C_6H_5{-}COO^-\ Na^+ + (CH_3CH_2)_2NH$$

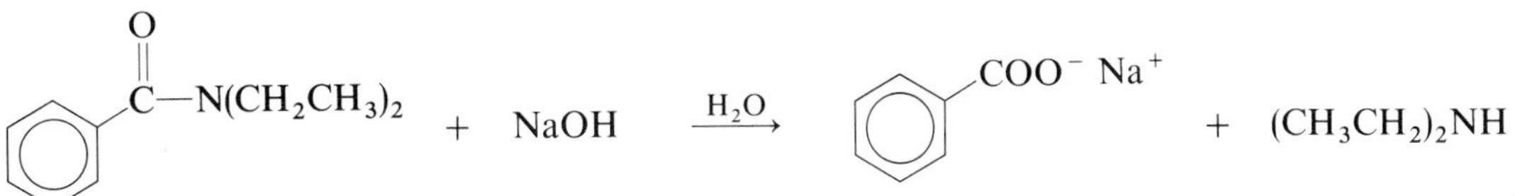

N,N-diethylbenzamide; sodium benzoate; diethylamine

Acid hydrolysis

$$R{-}C({=}O){-}NHR' + H_3O^+ \longrightarrow R{-}C({=}O){-}OH + R'NH_3^+$$

Example

$$C_6H_5{-}CH_2{-}C({=}O){-}NHCH_3 + H_2SO_4 \xrightarrow{H_2O} C_6H_5{-}CH_2{-}C({=}O){-}OH + CH_3\overset{+}{N}H_3\ HSO_4^-$$

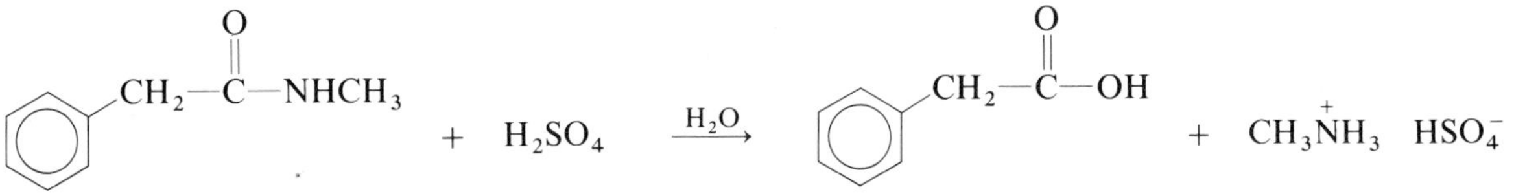

N-methyl-2-phenylacetamide; phenylacetic acid; methylammonium sulfate

The basic hydrolysis mechanism (shown below for a primary amide) is similar to that for the hydrolysis of an ester. Hydroxide attacks the carbonyl to give a tetrahedral intermediate. Expulsion of an amide ion gives a carboxylic acid, which is quickly deprotonated to give the salt of the acid and ammonia.

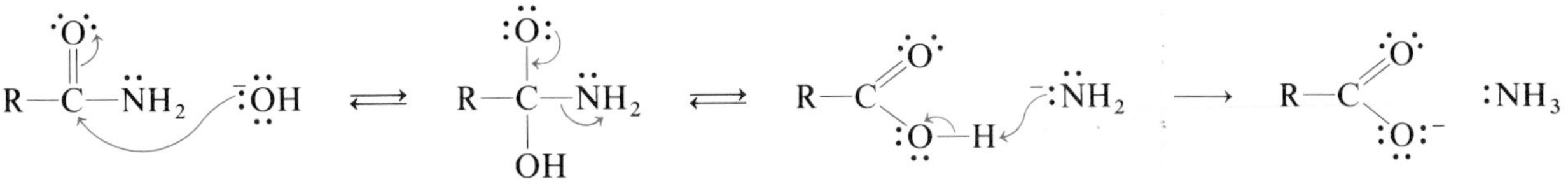

$$R{-}C({=}\ddot{O}){-}\ddot{N}H_2 \ \ {}^-\!:\ddot{O}H \rightleftarrows R{-}C(O^-)(OH){-}\ddot{N}H_2 \rightleftarrows R{-}C({=}O){-}\ddot{O}{-}H \ \ {}^-\!:\ddot{N}H_2 \longrightarrow R{-}C({=}O){-}\ddot{O}:^- \ \ :NH_3$$

Under acidic conditions, the mechanism of amide hydrolysis resembles the acid-catalyzed hydrolysis of an ester. Protonation of the carbonyl group activates it toward nucleophilic attack by water to give a tetrahedral intermediate. Protonation of the amino group enables it to leave as the amine. A fast, exothermic proton transfer gives the acid and the protonated amine.

$$\mathrm{R{-}C({=}\ddot{O}){-}\ddot{N}H_2 + H^+ \rightleftharpoons R{-}C({=}\overset{+}{O}{-}H){-}\ddot{N}H_2 \;(+\,H_2\ddot{O}:) \rightleftharpoons R{-}C(\ddot{O}{-}H)(\ddot{N}H_2)({-}\overset{+}{O}H_2) \rightleftharpoons R{-}C(\ddot{O}{-}H)(\ddot{N}H_2)(\ddot{O}{-}H) + H^+}$$

(resonance-stabilized)

$$\mathrm{R{-}C(\ddot{O}{-}H)(\ddot{N}H_2)(\ddot{O}{-}H) + H^+ \rightleftharpoons R{-}C(\ddot{O}{-}H)(\overset{+}{N}H_3)(\ddot{O}{-}H) \rightleftharpoons R{-}C({=}\overset{+}{O}{-}H)(\ddot{O}{-}H) \;\; :NH_3 \longrightarrow R{-}C({=}\ddot{O})(\ddot{O}{-}H) \;\; NH_4^+}$$

(resonance-stabilized)

PROBLEM 21-18
Draw the important resonance contributors for each of the resonance-stabilized cations in the mechanism for acid-catalyzed hydrolysis of an amide.

PROBLEM 21-19
Propose a mechanism for the hydrolysis of *N,N*-dimethylacetamide

(a) under basic conditions.
(b) under acidic conditions.

PROBLEM 21-20
The equilibrium for hydrolysis of amides, under both acidic and basic conditions, is displaced far toward the products. Using your mechanisms for the hydrolysis of *N,N*-dimethylacetamide as examples, show which steps are sufficiently exothermic to drive the reactions to completion.

21-7D HYDROLYSIS OF NITRILES

Nitriles are hydrolyzed to amides, and further to carboxylic acids, by heating with aqueous acid or base. Mild conditions can be used to hydrolyze a nitrile only as far as the amide, or stronger conditions can be used to hydrolyze it all the way to the carboxylic acid.

Basic hydrolysis

$$\underset{\text{nitrile}}{\mathrm{R{-}C{\equiv}N:}} + \mathrm{H_2O} \xrightarrow[\mathrm{H_2O}]{^-\mathrm{OH}} \underset{\text{primary amide}}{\mathrm{R{-}C({=}O){-}NH_2}} \xrightarrow[\mathrm{H_2O}]{^-\mathrm{OH}} \underset{\text{carboxylate ion}}{\mathrm{R{-}C({=}O){-}O^-}} + \mathrm{:NH_3}$$

Example

$$\underset{\text{nicotinonitrile}}{\text{3-pyridyl}{-}\mathrm{C{\equiv}N:}} \xrightarrow[\mathrm{H_2O/EtOH,\ 50^\circ C}]{\mathrm{NaOH}} \underset{\text{nicotinamide}}{\text{3-pyridyl}{-}\mathrm{C({=}O){-}NH_2}}$$

Acidic hydrolysis

$$\underset{\text{nitrile}}{R{-}C{\equiv}N{:}} + H_2O \xrightarrow[H_2O]{H^+} \underset{\text{primary amide}}{R{-}\overset{\overset{\displaystyle O}{\|}}{C}{-}NH_2} \xrightarrow[H_2O]{H^+} \underset{\text{carboxylic acid}}{R{-}\overset{\overset{\displaystyle O}{\|}}{C}{-}OH} + NH_4^+$$

Example

$$\underset{\text{phenylacetonitrile}}{Ph{-}CH_2{-}C{\equiv}N{:}} \xrightarrow[H_2O/CH_3CH_2OH]{H_2SO_4,\ \text{heat}} \underset{\text{phenylacetic acid}}{Ph{-}CH_2{-}\overset{\overset{\displaystyle O}{\|}}{C}{-}OH}$$

The mechanism of the basic hydrolysis begins with attack by hydroxide on the electrophilic carbon of the cyano group. Protonation gives the unstable enol tautomer of an amide. Removal of a proton from oxygen and reprotonation on nitrogen gives the amide. The further hydrolysis of the amide to the carboxylate salt involves the same base-catalyzed mechanism as that discussed above.

Attack by hydroxide ion and reprotonation

$$\underset{\text{nitrile}}{R{-}C{\equiv}N{:}} + {}^{-}{:}\ddot{O}{-}H \rightleftharpoons R{-}\overset{\overset{\displaystyle :\ddot{O}{-}H}{|}}{C}{=}\ddot{N}{:}^{-} \xrightarrow{H{-}\ddot{O}{-}H} \underset{\text{enol tautomer of amide}}{R{-}\overset{\overset{\displaystyle :\ddot{O}{-}H}{|}}{C}{=}\ddot{N}{-}H} + {}^{-}{:}\ddot{O}{-}H$$

Removal and replacement of a proton (tautomerization)

$$\underset{\text{enol tautomer}}{R{-}\overset{\overset{\displaystyle :\ddot{O}{-}H}{|}}{C}{=}\ddot{N}{-}H} \xrightleftharpoons{{}^{-}{:}\ddot{O}{-}H} \underset{\text{enolate of an amide}}{\left[R{-}\overset{\overset{\displaystyle :\ddot{O}{:}^{-}}{|}}{C}{=}\ddot{N}{-}H \longleftrightarrow R{-}\overset{\overset{\displaystyle O}{\|}}{C}{-}\ddot{N}^{-}{-}H\right]} \xrightleftharpoons{H{-}\ddot{O}{-}H} \underset{\text{amide}}{R{-}\overset{\overset{\displaystyle \ddot{O}}{\|}}{C}{-}\ddot{N}H_2} + {}^{-}{:}\ddot{O}{-}H$$

PROBLEM 21-21

Propose a mechanism for the basic hydrolysis of benzonitrile to the benzoate ion and ammonia.

PROBLEM 21-22

The mechanism for acidic hydrolysis of a nitrile resembles the basic hydrolysis, except that the nitrile is first protonated, activating it toward attack by a weak nucleophile (water). Under acidic conditions the proton transfer (tautomerism) involves protonation on nitrogen followed by deprotonation on oxygen. Propose a mechanism for the acid-catalyzed hydrolysis of benzonitrile to benzamide.

21-8 REDUCTION OF ACID DERIVATIVES

Carboxylic acids and their derivatives can be reduced to alcohols, aldehydes, and amines. Because they are relatively difficult to reduce, acid derivatives generally require a strong reducing agent such as lithium aluminum hydride ($LiAlH_4$).

21-8A REDUCTION TO ALCOHOLS

Lithium aluminum hydride reduces acids, acid chlorides, and esters to primary alcohols. (The reduction of acids was covered in Section 20-14.)

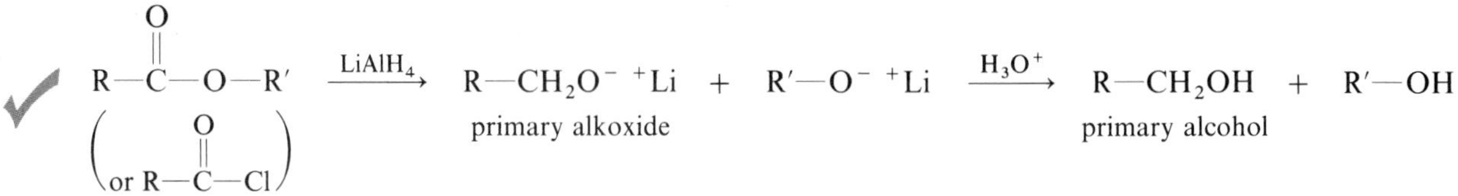

$$\underset{\left(\text{or } R-\overset{\overset{O}{\|}}{C}-Cl\right)}{R-\overset{\overset{O}{\|}}{C}-O-R'} \xrightarrow{LiAlH_4} \underset{\text{primary alkoxide}}{R-CH_2O^- \ {}^+Li} + R'-O^- \ {}^+Li \xrightarrow{H_3O^+} \underset{\text{primary alcohol}}{R-CH_2OH} + R'-OH$$

Examples

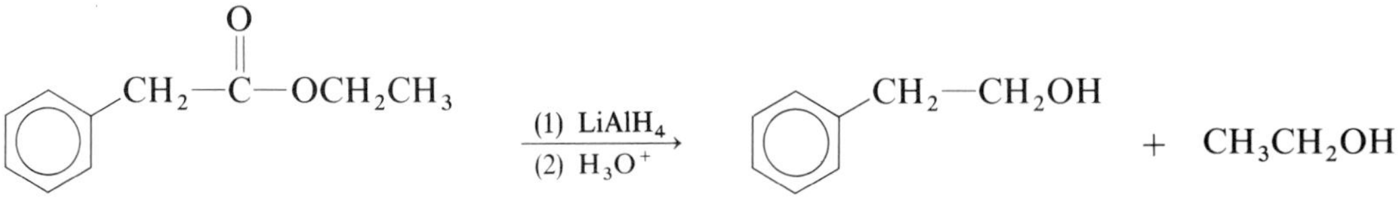

$$\underset{\text{ethyl phenylacetate}}{C_6H_5-CH_2-\overset{\overset{O}{\|}}{C}-OCH_2CH_3} \xrightarrow[(2)\ H_3O^+]{(1)\ LiAlH_4} \underset{\text{2-phenylethanol}}{C_6H_5-CH_2-CH_2OH} + CH_3CH_2OH$$

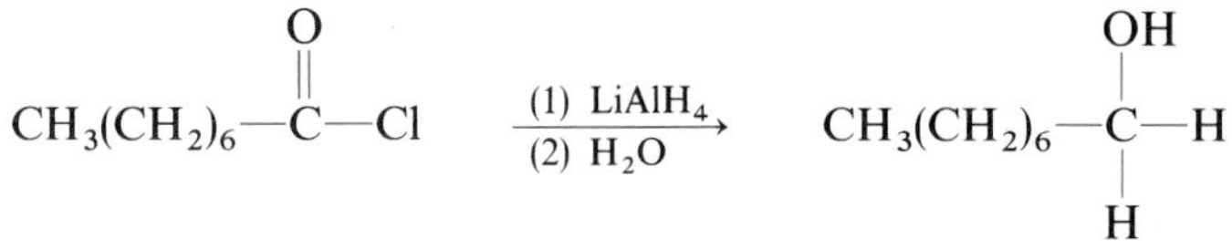

$$\underset{\text{octanoyl chloride}}{CH_3(CH_2)_6-\overset{\overset{O}{\|}}{C}-Cl} \xrightarrow[(2)\ H_2O]{(1)\ LiAlH_4} \underset{\text{1-octanol}}{CH_3(CH_2)_6-\overset{\overset{OH}{|}}{\underset{\underset{H}{|}}{C}}-H}$$

Both esters and acid chlorides react through an addition-elimination mechanism to give an aldehyde, which is quickly reduced to an alkoxide. Dilute acid is added in a second step to protonate the alkoxide.

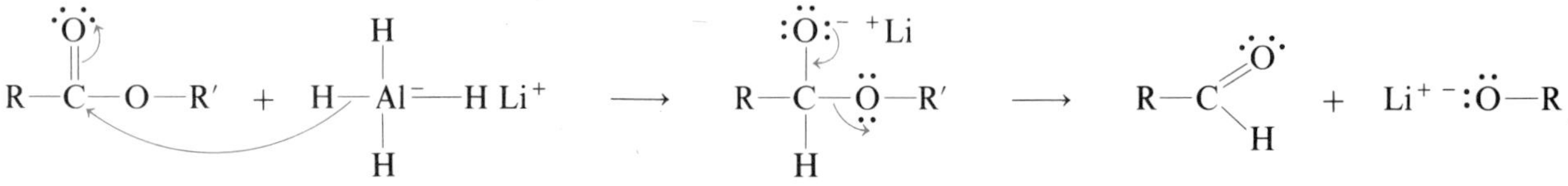

$$\underset{\text{ester}}{R-\overset{\overset{:\ddot{O}:}{\|}}{C}-O-R'} + H-\overset{\overset{H}{|}}{\underset{\underset{H}{|}}{Al^-}}-H\ Li^+ \longrightarrow \underset{\text{tetrahedral intermediate}}{R-\overset{\overset{:\ddot{O}:^-\ {}^+Li}{|}}{\underset{\underset{H}{|}}{C}}-\ddot{O}-R'} \longrightarrow \underset{\text{aldehyde}}{R-\overset{\overset{:\ddot{O}:}{\|}}{C}-H} + \underset{\text{alkoxide}}{Li^+\ {}^-:\ddot{O}-R}$$

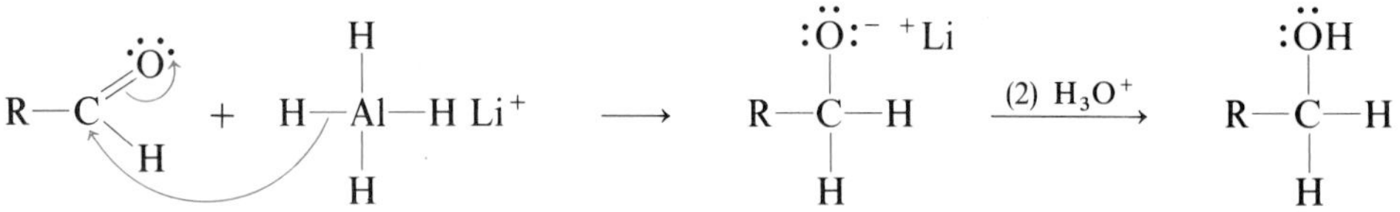

$$\underset{\text{aldehyde}}{R-\overset{\overset{:\ddot{O}:}{\|}}{C}-H} + H-\overset{\overset{H}{|}}{\underset{\underset{H}{|}}{Al}}-H\ Li^+ \longrightarrow \underset{\text{salt}}{R-\overset{\overset{:\ddot{O}:^-\ {}^+Li}{|}}{\underset{\underset{H}{|}}{C}}-H} \xrightarrow{(2)\ H_3O^+} \underset{\text{primary alcohol}}{R-\overset{\overset{:\ddot{O}H}{|}}{\underset{\underset{H}{|}}{C}}-H}$$

PROBLEM 21-23

Propose a mechanism for the reduction of octanoyl chloride by lithium aluminum hydride, as shown above.

21-8B REDUCTION TO ALDEHYDES

Acid chlorides are more reactive than other acid derivatives, and they are reduced to aldehydes by mild reducing agents such as lithium aluminum tri(*t*-butoxy)hydride or by the Rosenmund reduction. These reductions were covered in Sections 18-11 and 20-14.

Hydride reduction

$$\mathrm{R{-}\overset{\overset{\large O}{\|}}{C}{-}Cl \xrightarrow[\text{ether}]{\text{Li}(t\text{-BuO})_3\text{AlH}} R{-}\overset{\overset{\large O}{\|}}{C}{-}H}$$

Rosenmund reduction

$$\mathrm{R{-}\overset{\overset{\large O}{\|}}{C}{-}Cl \xrightarrow[\text{quinoline}]{H_2/Pd} R{-}\overset{\overset{\large O}{\|}}{C}{-}H \; + \; HCl}$$

Example

$$\underset{\text{octanoyl chloride}}{\mathrm{CH_3(CH_2)_6{-}\overset{\overset{\large O}{\|}}{C}{-}Cl}} \xrightarrow{\text{Li}(t\text{-BuO})_3\text{AlH}} \underset{\text{octanal}}{\mathrm{CH_3(CH_2)_6{-}\overset{\overset{\large O}{\|}}{C}{-}H}}$$

Esters can be reduced to aldehydes by another mild reducing agent, diisobutylaluminum hydride (DIBAH). Exactly one equivalent of the hydride reagent is used, and the reaction is carred out at $-78°$ to minimize overreduction.

$$\mathrm{R{-}\overset{\overset{\large O}{\|}}{C}{-}O{-}R' \xrightarrow[(2)\ H_3O^+]{(1)\ \text{DIBAH, toluene}} R{-}\overset{\overset{\large O}{\|}}{C}{-}H \; + \; R'{-}OH}$$

$$\text{DIBAH} = [(CH_3)_2CHCH_2]_2AlH$$

Example

$$\mathrm{(CH_3)_2CH{-}\overset{\overset{\large O}{\|}}{C}{-}OCH_2CH_3 \xrightarrow[(2)\ H_3O^+]{(1)\ \text{DIBAH, toluene}} \underset{80\%}{(CH_3)_2CH{-}\overset{\overset{\large O}{\|}}{C}{-}H} \; + \; CH_3CH_2{-}OH}$$

21-8C REDUCTION TO AMINES

Lithium aluminum hydride reduces amides and nitriles to amines, providing one of the best synthetic routes to amines (Sections 19-20 and 19-24). Primary amides and nitriles are reduced to primary amines. Secondary amides are reduced to secondary amines, and tertiary amides are reduced to tertiary amines.

$$\underset{\text{primary amide}}{\mathrm{R{-}\overset{\overset{\large O}{\|}}{C}{-}NH_2}} \xrightarrow[(2)\ H_2O]{(1)\ LiAlH_4} \underset{\text{primary amine}}{\mathrm{R{-}CH_2{-}NH_2}}$$

$$\underset{\text{secondary amide}}{\mathrm{R{-}\overset{\overset{\large O}{\|}}{C}{-}NHR'}} \xrightarrow[(2)\ H_2O]{(1)\ LiAlH_4} \underset{\text{secondary amine}}{\mathrm{R{-}CH_2{-}NHR'}}$$

$$\underset{\text{tertiary amide}}{\mathrm{R{-}\overset{\overset{\large O}{\|}}{C}{-}NR'_2}} \xrightarrow[(2)\ H_2O]{(1)\ LiAlH_4} \underset{\text{tertiary amine}}{\mathrm{R{-}CH_2{-}NR'_2}}$$

Example

$$CH_3-\overset{\overset{\displaystyle O}{\|}}{C}-NH-Ph \xrightarrow[(2)\ H_2O]{(1)\ LiAlH_4} CH_3-CH_2-NH-Ph$$

acetanilide → *N*-ethylaniline

$$R-C\equiv N: \xrightarrow[or\ (1)\ LiAlH_4;\ (2)\ H_2O]{H_2/Pt} R-\underset{H}{\overset{H}{C}}-\ddot{N}H_2$$

Example

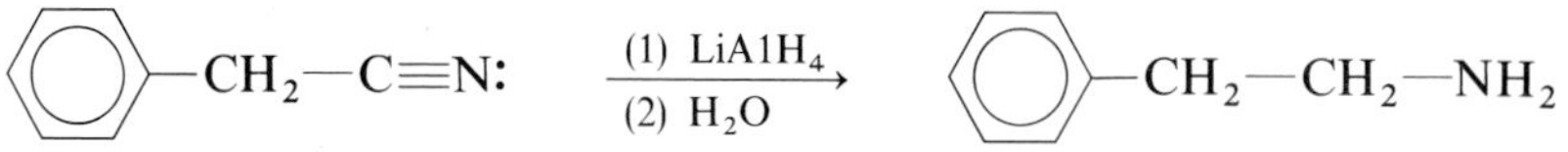

$$C_6H_5-CH_2-C\equiv N: \xrightarrow[(2)\ H_2O]{(1)\ LiAlH_4} C_6H_5-CH_2-CH_2-NH_2$$

PROBLEM 21-24

Give the expected products of lithium aluminum hydride reduction of the following compounds (followed by hydrolysis).

(a) cyclohexanecarboxamide (b) *N*-cyclohexylacetamide (c) ε-caprolactam

(d) 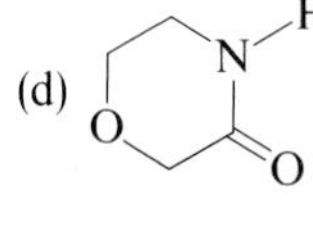

(e) 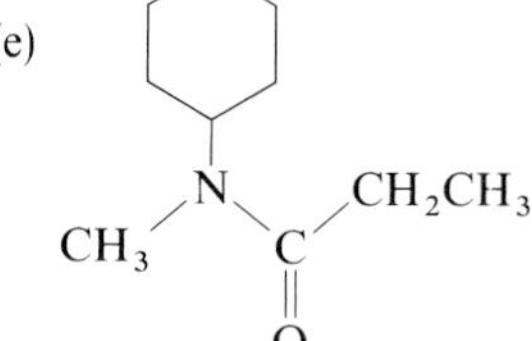

(f) 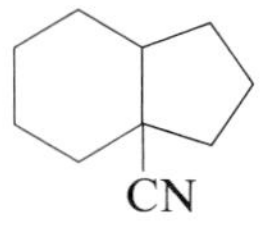 CN

21-9 REACTIONS OF ACID DERIVATIVES WITH ORGANOMETALLIC REAGENTS

Esters and acid chlorides Grignard and organolithium reagents add twice to acid chlorides and esters to give alkoxides (Section 9-9D).

$$\underset{\text{ester}}{R-\overset{\overset{\displaystyle O}{\|}}{C}-OR'} + \underset{\text{(or 2 R''Li)}}{2\,R''MgX} \longrightarrow \underset{\text{alkoxide salt}}{R-\underset{R''}{\overset{OMgX}{C}}-R''} + R'OMgX \xrightarrow{H_3O^+} \underset{\text{tertiary alcohol}}{R-\underset{R''}{\overset{OH}{C}}-R''} + R'OH$$

Examples

$$Ph-\overset{\overset{\displaystyle O}{\|}}{C}-OEt + 2\,PhMgBr \longrightarrow Ph-\underset{Ph}{\overset{OMgBr}{C}}-Ph \xrightarrow{H_3O^+} Ph-\underset{Ph}{\overset{OH}{C}}-Ph$$

$$H-\overset{\overset{\displaystyle O}{\|}}{C}-OEt + 2\,C_4H_9Li \longrightarrow H-\underset{C_4H_9}{\overset{OLi}{C}}-C_4H_9 \xrightarrow{H_3O^+} H-\underset{C_4H_9}{\overset{OH}{C}}-C_4H_9$$

$$CH_3CH_2-\overset{O}{\overset{\|}{C}}-Cl + 2\ PhMgBr \longrightarrow CH_3CH_2-\underset{Ph}{\overset{OMgBr}{C}}-Ph \xrightarrow{H_3O^+} CH_3CH_2-\underset{Ph}{\overset{OH}{C}}-Ph$$

The mechanism involves nucleophilic substitution at the acyl carbon atom. Attack by the carbanion-like organometallic reagent, followed by elimination of alkoxide (from an ester) or chloride (from an acid chloride) gives a ketone. Another equivalent of the organometallic reagent adds to the ketone to give the alkoxide. Hydrolysis gives tertiary alcohols, unless the original ester is a formate (R = H), which gives a secondary alcohol.

$$\underset{\text{ester}}{R-\overset{O}{\overset{\|}{C}}-OR'} + \underset{\text{Grignard}}{\overset{\delta-}{R''}-\overset{\delta+}{MgX}} \longrightarrow \underset{\text{tetrahedral intermediate}}{R-\underset{R''}{\overset{O^-\ ^+MgX}{C}}-OR'} \longrightarrow \underset{\text{ketone}}{R-\overset{O}{\overset{\|}{C}}-R''} + R''-MgX \longrightarrow \underset{\text{alkoxide}}{R-\underset{R''}{\overset{O^-\ ^+MgX}{C}}-R''}$$

Nitriles A Grignard or organolithium reagent attacks the electrophilic cyano group to form the salt of an imine. Acidic hydrolysis of the salt (in a second step) gives the imine, which is further hydrolyzed to a ketone (Sections 18-10 and 18-16).

$$\overset{\delta-}{R'}-\overset{\delta+}{Mg}-X \quad R-C\equiv N: \longrightarrow \underset{\text{salt of imine}}{R'(R)C=N-MgX} \xrightarrow{H_3O^+} \underset{\text{imine}}{R'(R)C=N-H} \xrightarrow{H_3O^+} \underset{\text{ketone}}{R'(R)C=O}$$

$$\underset{\text{benzonitrile}}{Ph-C\equiv N:} + \underset{\text{methylmagnesium iodide}}{CH_3MgI} \longrightarrow \underset{\text{magnesium salt}}{Ph(CH_3)C=N-MgI} \xrightarrow{H_3O^+} \underset{\text{acetophenone}}{Ph(CH_3)C=O}$$

PROBLEM 21-25

Give a mechanism for the acidic hydrolysis of this magnesium salt to acetophenone.

PROBLEM 21-26

Draw a mechanism for the reaction of propanoyl chloride with 2 moles of phenylmagnesium bromide.

21-10 OVERVIEW OF THE CHEMISTRY OF ACID CHLORIDES

Having discussed the reactions and mechanisms characteristic of all the common acid derivatives, we now review the syntheses and reactions of each type of compound. In addition, these sections cover any reactions that are peculiar to a specific class of acid derivative.

Synthesis of acid chlorides Acid chlorides are synthesized from the corresponding carboxylic acids using a variety of reagents. Thionyl chloride, $SOCl_2$, and oxalyl chloride, $(COCl)_2$, are the most convenient reagents because they produce only gaseous side products (Section 20-10).

$$\mathrm{R{-}\overset{\overset{\displaystyle O}{\|}}{C}{-}OH} \xrightarrow[\textit{or}\ (COCl)_2]{SOCl_2} \mathrm{R{-}\overset{\overset{\displaystyle O}{\|}}{C}{-}Cl} + SO_2\uparrow + HCl\uparrow$$

Reactions of acid chlorides Acid chlorides react quickly with water and other nucleophiles and are therefore not found in nature. Because they are the most reactive acid derivatives, acid chlorides are easily converted to other acid derivatives.

$\mathrm{R{-}\overset{\overset{\displaystyle O}{\|}}{C}{-}Cl}$ (acid chloride)

$$\xrightarrow{H_2O} \mathrm{R{-}\overset{\overset{\displaystyle O}{\|}}{C}{-}OH} + HCl \quad \text{(Section 21-7A)}$$
acid

$$\xrightarrow{R'OH} \mathrm{R{-}\overset{\overset{\displaystyle O}{\|}}{C}{-}OR'} + HCl \quad \text{(Sections 20-10 and 21-5)}$$
ester

$$\xrightarrow{R'NH_2} \mathrm{R{-}\overset{\overset{\displaystyle O}{\|}}{C}{-}NHR'} + HCl \quad \text{(Sections 20-10 and 21-5)}$$
amide

$$\xrightarrow{R'COOH} \mathrm{R{-}\overset{\overset{\displaystyle O}{\|}}{C}{-}O{-}\overset{\overset{\displaystyle O}{\|}}{C}{-}R'} + HCl \quad \text{(Sections 21-5 and 21-11)}$$
anhydride

Grignard and organolithium reagents add twice to acid chlorides to give alcohols (after hydrolysis). Similarly, lithium aluminum hydride adds hydride twice to acid chlorides, reducing them to alcohols (after hydrolysis). Acid chlorides react with the milder reducing agent lithium tri-*t*-butoxyaluminum hydride to give aldehydes.

$\mathrm{R{-}\overset{\overset{\displaystyle O}{\|}}{C}{-}Cl}$ (acid chloride)

$$\xrightarrow[(2)\ H_2O]{(1)\ 2\ R'MgX} \mathrm{R{-}\underset{\underset{\displaystyle R'}{|}}{\overset{\overset{\displaystyle OH}{|}}{C}}{-}R'} \quad \text{(Sections 9-9 and 21-9)}$$
3° alcohol

$$\xrightarrow[(2)\ H_2O]{(1)\ LiAlH_4} R{-}CH_2OH \quad \text{(Sections 9-12 and 21-8A)}$$
1° alcohol

$$\xrightarrow[\text{(or } H_2/Pd/\text{quinoline)}]{Li(t\text{-}BuO)_3AlH} \mathrm{R{-}\overset{\overset{\displaystyle O}{\|}}{C}{-}H} \quad \text{(Sections 18-11 and 21-8B)}$$
aldehyde

Friedel-Crafts acylation of aromatic rings In the presence of aluminum chloride, acyl halides acylate benzene, halobenzenes, and activated benzene derivatives. Friedel-Crafts acylation is discussed in detail in Section 17-11.

$$R-\overset{O}{\overset{\|}{C}}-Cl + \text{(Z-substituted benzene)} \xrightarrow{AlCl_3} \text{(Z-substituted ring)}-\overset{O}{\overset{\|}{C}}-R$$

(Z = H, halogen, or an activating group)

Example

$$CH_3-CH_2-\overset{O}{\overset{\|}{C}}-Cl + CH_3O-C_6H_5 \xrightarrow{AlCl_3} CH_3O-C_6H_4-\overset{O}{\overset{\|}{C}}-CH_2CH_3$$

propionyl chloride — anisole — *p*-methoxypropiophenone (major product)

PROBLEM 21-27

Write a detailed mechanism for the acylation of anisole by propionyl chloride. Recall that Friedel-Crafts acylation involves an acylium ion as the electrophile in electrophilic aromatic substitution.

PROBLEM 21-28

Show how Friedel-Crafts acylation might be used to synthesize the following compounds.

(a) acetophenone (b) benzophenone (c) *n*-butylbenzene

21-11 OVERVIEW OF THE CHEMISTRY OF ANHYDRIDES

Anhydrides are activated acyl compounds often used for the same types of acylations as acid chlorides. Anhydrides are not as reactive as acid chlorides, and they are occasionally found in nature. For example, cantharidin is a toxic ingredient of "Spanish fly," which is used as a vesicant (causing blistering and burning) to destroy warts on the skin.

cantharidin

Because anhydrides are not as reactive as acid chlorides, they are often more selective in their reactions. Anhydrides are valuable when the appropriate acid chloride is too reactive, does not exist, or is more expensive than the corresponding anhydride.

Acetic anhydride Acetic anhydride is the most important carboxylic acid anhydride. It is produced and used in large quantities in industry, primarily for the synthesis of plastics and fibers. The industrial synthesis of acetic anhydride uses a catalytic reaction between acetylene and acetic acid in the presence of mercuric oxide. Its large-scale manufacture makes acetic anhydride a convenient and inexpensive acylating reagent.

$$H-C\equiv C-H + CH_3COOH + \tfrac{1}{2}O_2 \xrightarrow{HgO} CH_3-\overset{O}{\overset{\|}{C}}-O-\overset{O}{\overset{\|}{C}}-CH_3$$

acetylene — acetic acid — acetic anhydride (abbreviated Ac_2O)

General anhydride synthesis Other anhydrides must be made by less specialized methods. The most general method for making anhydrides is the reaction of an acid chloride with a carboxylic acid or a carboxylate salt.

$$R-\overset{O}{\overset{\|}{C}}-Cl + {}^{-}O-\overset{O}{\overset{\|}{C}}-R' \longrightarrow R-\overset{O}{\overset{\|}{C}}-O-\overset{O}{\overset{\|}{C}}-R' + Cl^-$$

acid chloride — carboxylate (or acid) — acid anhydride

Examples

$$\underset{\text{acetyl chloride}}{CH_3-\overset{O}{\overset{\|}{C}}-Cl} + \underset{\text{benzoic acid}}{HO-\overset{O}{\overset{\|}{C}}-Ph} \xrightarrow{\text{pyridine}} \underset{\text{acetic benzoic anhydride}}{CH_3-\overset{O}{\overset{\|}{C}}-O-\overset{O}{\overset{\|}{C}}-Ph}$$

$$\underset{\text{acetyl chloride}}{CH_3-\overset{O}{\overset{\|}{C}}-Cl} + \underset{\text{sodium formate}}{H-\overset{O}{\overset{\|}{C}}-O^{-}\,{}^{+}Na} \longrightarrow \underset{\text{acetic formic anhydride}}{CH_3-\overset{O}{\overset{\|}{C}}-O-\overset{O}{\overset{\|}{C}}-H} + Na^{+}\,{}^{-}Cl$$

Some cyclic anhydrides are made simply by heating the corresponding diacid. A dehydrating agent, such as acetyl chloride or acetic anhydride, is occasionally added to accelerate this reaction. Because five- and six-membered cyclic anhydrides are particularly stable, the equilibrium favors the cyclic products.

$$\text{phthalic acid} \xrightarrow{\text{heat}} \text{phthalic anhydride} + \underset{\text{(steam)}}{H_2O}$$

$$\text{succinic acid} + CH_3\overset{O}{\overset{\|}{C}}Cl \longrightarrow \text{succinic anhydride} + CH_3\overset{O}{\overset{\|}{C}}OH + HCl$$

Reactions of anhydrides Anhydrides undergo many of the same kinds of reactions as acid chlorides. Like acid chlorides, anhydrides are easily converted to less reactive acid derivatives.

$$\underset{\text{anhydride}}{R-\overset{O}{\overset{\|}{C}}-O-\overset{O}{\overset{\|}{C}}-R}$$

$$\xrightarrow{H_2O} \underset{\text{acid}}{R-\overset{O}{\overset{\|}{C}}-OH} + R-COOH \qquad \text{(Section 21-7A)}$$

$$\xrightarrow{R'OH} \underset{\text{ester}}{R-\overset{O}{\overset{\|}{C}}-OR'} + R-COOH \qquad \text{(Section 21-5)}$$

$$\xrightarrow{R'NH_2} \underset{\text{amide}}{R-\overset{O}{\overset{\|}{C}}-NHR'} + R-COOH \qquad \text{(Section 21-5)}$$

Like acid chlorides, anhydrides also participate in the Friedel-Crafts acylation. Catalysts may be aluminum chloride, polyphosphoric acid (PPA), or other acidic reagents. Cyclic anhydrides can be used to provide additional functionality on the side chain of the aromatic product.

$$\text{C}_6\text{H}_5\text{Z} + \text{R}-\overset{\text{O}}{\overset{\|}{\text{C}}}-\text{O}-\overset{\text{O}}{\overset{\|}{\text{C}}}-\text{R} \xrightarrow[\text{(or other acidic catalyst)}]{\text{AlCl}_3} \text{Z}-\text{C}_6\text{H}_4-\overset{\text{O}}{\overset{\|}{\text{C}}}-\text{R}$$

(Z = H, halogen, or an activating group) — an acylbenzene

Example

$$\text{benzene} + \text{succinic anhydride} \xrightarrow{\text{AlCl}_3} \text{C}_6\text{H}_5-\text{CO}-\text{CH}_2\text{CH}_2-\text{COOH}$$

benzene — succinic anhydride — 4-oxo-4-phenylbutanoic acid

In most cases, it is easier and more efficient to make and use acid chlorides than anhydrides. There are three specific instances when anhydrides are better than acid chlorides, however.

1. *Use of acetic anhydride.* Acetic anhydride is inexpensive and convenient to use, and it often gives better yields than acetyl chloride for the acetylation of alcohols (to acetate esters) and amines (to acetamides).
2. *Use of acetic formic anhydride.* Formyl chloride (the acid chloride of formic acid) cannot be used for formylation because it quickly decomposes to CO and HCl. Acetic formic anhydride, made from sodium formate and acetyl chloride, reacts primarily at the formyl group. Lacking a bulky, electron-donating alkyl group, the formyl group is both less hindered and more electrophilic than the acetyl group. Alcohols and amines are formylated by acetic formic anhydride to give formate esters and formamides, respectively.

$$CH_3-\overset{O}{\overset{\|}{C}}-O-\overset{O}{\overset{\|}{C}}-H + R-OH \longrightarrow H-\overset{O}{\overset{\|}{C}}-O-R + CH_3COOH$$

more reactive carbonyl — a formate ester

$$CH_3-\overset{O}{\overset{\|}{C}}-O-\overset{O}{\overset{\|}{C}}-H + R-NH_2 \longrightarrow H-\overset{O}{\overset{\|}{C}}-NH-R + CH_3COOH$$

a formamide

3. *Use of cyclic anhydrides to give difunctional compounds.* It is often necessary to convert just one of the two carboxylic acid groups of a diacid to an ester or an amide. This conversion is most easily accomplished by first converting the diacid to its cyclic anhydride.

   When an alcohol or an amine reacts with an anhydride, only one of the two carboxyl groups in the anhydride is converted to an ester or an amide. The other is expelled as a carboxylate ion, and a monofunctionalized derivative results.

$$\text{glutaric anhydride} + CH_3CH_2-OH \longrightarrow CH_3CH_2O-\overset{O}{\overset{\|}{C}}-CH_2CH_2CH_2-\overset{O}{\overset{\|}{C}}-OH$$

glutaric anhydride — monoethyl ester

PROBLEM 21-29

(a) Give the products expected when acetic formic anhydride reacts with (i) aniline and (ii) benzyl alcohol.

(b) Write a mechanism for each of these reactions.

PROBLEM 21-30

Show how you would use an anhydride to synthesize each of the following compounds. In each case, explain why the use of an anhydride might be preferable to the use of an acid chloride.

(a) *n*-octyl formate (b) *n*-octyl acetate

(c) phthalic acid monoamide (d) succinic acid monomethyl ester

21-12 OVERVIEW OF THE CHEMISTRY OF ESTERS

Synthesis of esters Esters are usually synthesized by the acid-catalyzed Fischer esterification of an acid with an alcohol or by the reaction of an acid chloride (or anhydride) with an alcohol. Methyl esters are conveniently made by treating the acid with diazomethane. The alcohol group in an ester can be changed by trans-esterification, catalyzed by either acid or base.

$$\underset{\text{acid}}{R-\overset{O}{\overset{\|}{C}}-OH} + \underset{\text{alcohol}}{R'-OH} \overset{H^+}{\rightleftharpoons} \underset{\text{ester}}{R-\overset{O}{\overset{\|}{C}}-OR'} + H_2O \qquad \text{(Section 20-11)}$$

$$\underset{\text{acid chloride}}{R-\overset{O}{\overset{\|}{C}}-Cl} + \underset{\text{alcohol}}{R'-OH} \longrightarrow \underset{\text{ester}}{R-\overset{O}{\overset{\|}{C}}-OR'} + HCl \qquad \text{(Section 20-10)}$$

$$\underset{\text{anhydride}}{R-\overset{O}{\overset{\|}{C}}-O-\overset{O}{\overset{\|}{C}}-R} + \underset{\text{alcohol}}{R'-OH} \longrightarrow \underset{\text{ester}}{R-\overset{O}{\overset{\|}{C}}-OR'} + RCOOH \qquad \text{(Section 21-5)}$$

$$\underset{\text{ester}}{R-\overset{O}{\overset{\|}{C}}-OR''} + \underset{\text{alcohol}}{R'-OH} \overset{H^+ \text{ or } ^-OH}{\rightleftharpoons} \underset{\text{ester}}{R-\overset{O}{\overset{\|}{C}}-OR'} + R''OH \qquad \text{(Section 21-6)}$$

$$\underset{\text{acid}}{R-\overset{O}{\overset{\|}{C}}-OH} + \underset{\text{diazomethane}}{CH_2N_2} \longrightarrow \underset{\text{methyl ester}}{R-\overset{O}{\overset{\|}{C}}-OCH_3} + N_2\uparrow \qquad \text{(Section 20-12)}$$

Reactions of esters Esters are much more stable than acid chlorides and anhydrides; for example, most esters do not react with water under neutral conditions. They hydrolyze under acidic or basic conditions, however, and an amine can displace the alkoxyl group to form an amide. Lithium aluminum hydride reduces amides to primary alcohols, and Grignard and organolithium reagents add twice to give alcohols (after hydrolysis).

$$\text{R—C(=O)—OR}' \;(\text{ester}) \xrightarrow[\text{H}^+ \text{ or } ^-\text{OH}]{\text{H}_2\text{O}} \text{R—C(=O)—OH} \;(\text{acid}) + \text{R}'\text{OH} \quad \text{(Section 21-7B)}$$

$$\xrightarrow{\text{R}''\text{OH}} \text{R—C(=O)—OR}'' \;(\text{ester}) + \text{R}'\text{OH} \quad \text{(Section 21-6)}$$

$$\xrightarrow{\text{R}'\text{NH}_2} \text{R—C(=O)—NHR}' \;(\text{amide}) + \text{R}'\text{OH} \quad \text{(Section 21-5)}$$

$$\xrightarrow[\text{(2) H}_2\text{O}]{\text{(1) LiAlH}_4} \text{R—CH}_2\text{OH} \;(1^\circ \text{ alcohol}) + \text{R}'\text{OH} \quad \text{(Sections 9-12 and 21-8A)}$$

$$\xrightarrow[\text{(2) H}_2\text{O}]{\text{(1) 2 R}''\text{MgX}} \text{R—C(OH)(R}'')\text{—R}'' \;(3^\circ \text{ alcohol}) + \text{R}'\text{OH} \quad \text{(Sections 9-9 and 21-9)}$$

$$\xrightarrow[\text{(2) H}_3\text{O}^+]{\text{(1) DIBAH}} \text{R—C(=O)—H} \;(\text{aldehyde}) + \text{R}'\text{OH} \quad \text{(Section 21-8B)}$$

Formation of lactones Simple lactones containing five-membered and six-membered rings are often more stable than the open-chain hydroxy acids. Such lactones form spontaneously under acidic conditions (via the Fischer esterification).

$$\text{HO(CH}_2)_3\text{COOH} \;(27\%) \underset{}{\overset{\text{H}^+}{\rightleftarrows}} \gamma\text{-butyrolactone} \;(73\%) + \text{H}_2\text{O}$$

Lactones that are not energetically favored may be synthesized by driving the equilibrium toward the products. For example, the ten-membered 9-hydroxynonanoic acid lactone is formed in a dilute benzene solution containing a trace of *p*-toluenesulfonic acid. The reaction is driven to completion by distilling the benzene/water azeotrope to remove water and shift the equilibrium to the right.

$$\text{9-hydroxynonanoic acid} \xrightarrow[\text{benzene}]{\text{H}^+} \text{9-hydroxynonanoic acid lactone (95\%)} + \text{H}_2\text{O (removed)}$$

PROBLEM 21-31

Propose a mechanism for the formation of 9-hydroxynonanoic acid lactone.

PROBLEM 21-32

Suggest the most appropriate reagent for each synthesis, and explain your choice.

(a) 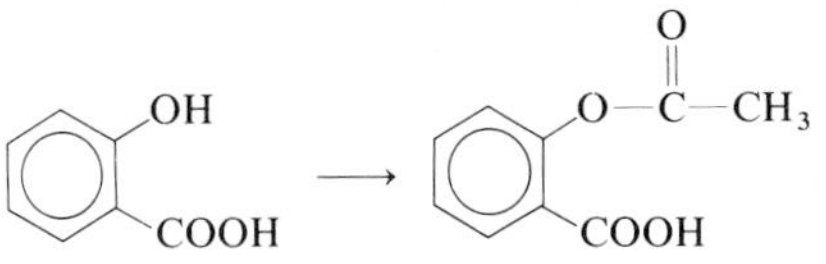 o-hydroxybenzoic acid (OH, COOH on benzene) $\longrightarrow$ o-acetoxybenzoic acid ($O{-}\overset{O}{\overset{\|}{C}}{-}CH_3$, COOH on benzene)

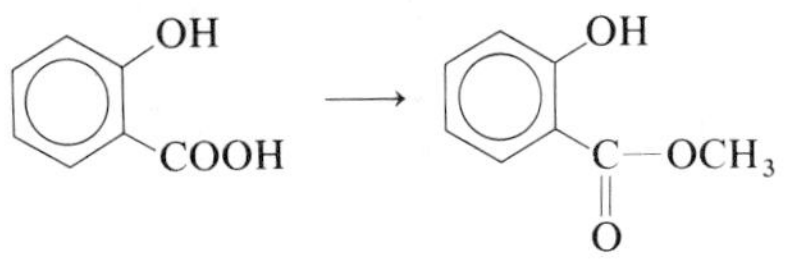

(b) o-hydroxybenzoic acid (OH, COOH on benzene) $\longrightarrow$ methyl o-hydroxybenzoate (OH, $\underset{O}{\underset{\|}{C}}{-}OCH_3$ on benzene)

(c) bicyclic acid (CH_3, COOH) $\longrightarrow$ bicyclic methyl ester (CH_3, $\underset{O}{\underset{\|}{C}}{-}OCH_3$)

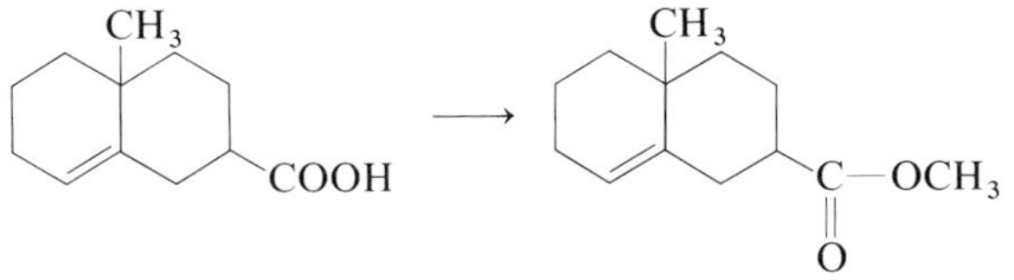

PROBLEM 21-33

Show how you would synthesize each of the following compounds, starting with an ester containing no more than eight carbon atoms. Any other necessary reagents may be used.

(a) $Ph_3C{-}OH$
(b) $(PhCH_2)_2CHOH$
(c) $PhCONHCH_2CH_3$
(d) Ph_2CHOH
(e) $PhCH_2OH$
(f) PhCOOH
(g) $PhCH_2COOCH(CH_3)_2$
(h) $PhCH_2{-}\underset{OH}{\underset{|}{C}}(CH_2CH_3)_2$

21-13 OVERVIEW OF THE CHEMISTRY OF AMIDES

Synthesis of amides Amides are the least reactive of the acid derivatives, and they can be made from any of the other derivatives. In the laboratory, amides are commonly synthesized by the reaction of an acid chloride (or anhydride) with an amine. The most common industrial synthesis of amides involves heating an acid with an amine to drive off water and promote condensation. Esters react with amines and ammonia to give amides, and the hydrolysis of nitriles also gives amides.

$$\underset{\text{acid}}{R{-}\overset{O}{\overset{\|}{C}}{-}OH} + \underset{\text{amine}}{R'{-}NH_2} \xrightarrow{\text{heat}} \underset{\text{amide}}{R{-}\overset{O}{\overset{\|}{C}}{-}NHR'} + H_2O \qquad \text{(Sections 20-15)}$$

$$\underset{\text{acid chloride}}{R{-}\overset{O}{\overset{\|}{C}}{-}Cl} + \underset{\text{amine}}{R'_2NH} \longrightarrow \underset{\text{amide}}{R{-}\overset{O}{\overset{\|}{C}}{-}NR'_2} + HCl \qquad \text{(Sections 20-10 and 21-5)}$$

$$\underset{\text{anhydride}}{R{-}\overset{O}{\overset{\|}{C}}{-}O{-}\overset{O}{\overset{\|}{C}}{-}R} + \underset{\text{amine}}{R'_2NH} \longrightarrow \underset{\text{amide}}{R{-}\overset{O}{\overset{\|}{C}}{-}NR'_2} + RCOOH \qquad \text{(Section 21-5)}$$

$$\underset{\text{ester}}{R{-}\overset{O}{\overset{\|}{C}}{-}OR''} + \underset{\text{amine}}{R'{-}NH_2} \longrightarrow \underset{\text{amide}}{R{-}\overset{O}{\overset{\|}{C}}{-}NHR'} + R''OH \qquad \text{(Section 21-5)}$$

$$\underset{\text{nitrile}}{R{-}C{\equiv}N} + H_2O \xrightarrow{H^+ \text{ or } ^-OH} \underset{1^\circ\text{ amide}}{R{-}\overset{O}{\overset{\|}{C}}{-}NH_2} \qquad \text{(Section 21-7D)}$$

Reactions of amides Because amides are the most stable acid derivatives, they are not easily converted to other derivatives by nucleophilic acyl substitution. From a synthetic standpoint, their most important reaction is their reduction to an amine, one of the best methods for the synthesis of amines. The Hofmann rearrangement also converts amides to amines, with the loss of one carbon atom. Although an amide is considered a neutral functional group, it is both weakly acidic and weakly basic, and amides are hydrolyzed by strong acid or strong base. Just as amides are formed in the hydrolysis of nitriles, amides can be dehydrated to nitriles as described below.

$$\underset{\text{amide}}{R-\overset{O}{\overset{\|}{C}}-NHR'}$$

$$\xrightarrow[H^+ \text{ or } ^-OH]{H_2O} \underset{\text{acid}}{R-\overset{O}{\overset{\|}{C}}-OH} + R'NH_2 \quad \text{(Section 21-7C)}$$

$$\xrightarrow[(2)\ H_2O]{(1)\ LiAlH_4} \underset{\text{amine}}{R-CH_2NHR'} \quad \text{(Sections 19-20 and 21-8C)}$$

$$\xrightarrow[\text{(Hofmann rearr.)}]{Br_2,\ ^-OH} \underset{1^\circ \text{ amine}}{R-NH_2} + CO_3^{2-} \quad \text{(Section 19-25)}$$

$$\xrightarrow{POCl_3} \underset{\text{nitrile}}{R-C{\equiv}N} \quad \text{(Section 21-7D)}$$

Dehydration of amides to nitriles Strong dehydrating agents can remove the elements of water from a primary amide to give a nitrile. The dehydration of amides is one of the most common methods for the synthesis of nitriles. Phosphorus pentoxide (P_2O_5) is the traditional reagent for this dehydration, but phosphorus oxychloride ($POCl_3$) sometimes gives better yields.

$$\underset{\text{primary amide}}{R-\overset{O}{\overset{\|}{C}}-NH_2} \xrightarrow[\text{(or } POCl_3)]{P_2O_5} \underset{\text{nitrile}}{R-C{\equiv}N:}$$

Example

$$\underset{\text{2-ethylhexanamide}}{CH_3CH_2CH_2CH_2-\overset{CH_3CH_2}{\overset{\diagdown}{C}H}-\overset{O}{\overset{\|}{C}}-NH_2} \xrightarrow{P_2O_5} \underset{\text{2-ethylhexanenitrile (90\%)}}{CH_3CH_2CH_2CH_2-\overset{CH_3CH_2}{\overset{\diagdown}{C}H}-C{\equiv}N:}$$

Formation of lactams Five-membered lactams (γ-lactams) and six-membered lactams (δ-lactams) often form from heating or adding a dehydrating agent to the appropriate γ-amino acids and δ-amino acids. Lactams containing smaller or larger rings do not form readily under these conditions.

$$\underset{\gamma\text{-aminobutyric acid}}{H_2N(CH_2)_3COOH} \xrightarrow{\text{heat}} \underset{\gamma\text{-butyrolactam}}{\text{(five-membered ring N–H, C=O)}} + H_2O$$

$$\underset{\delta\text{-aminovaleric acid}}{H_2N(CH_2)_4COOH} \xrightarrow{\text{heat}} \underset{\delta\text{-valerolactam}}{\text{(six-membered ring N–H, C=O)}} + H_2O$$

Biological reactivity of β-lactams β-Lactams are unusually reactive amides, capable of acylating a variety of nucleophilic reagents. The considerable amount of ring strain in the four-membered ring appears to be the driving force behind the unusual reactivity of the β-lactams. The ring opens, and the ring strain is relieved when the β-lactam acylates a nucleophile:

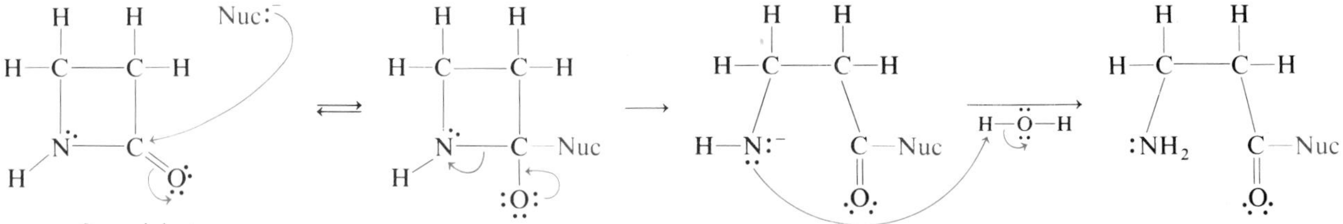

β-propiolactam

The β-lactam ring is found in two important classes of antibiotics, both of which were originally isolated from fungi. *Penicillins* have a β-lactam ring fused to a five-membered ring containing a sulfur atom. *Cephalosporins* have a β-lactam ring fused to an unsaturated six-membered ring containing a sulfur atom. The structures of penicillin V and cephalexin illustrate the differences between penicillins and cephalosporins.

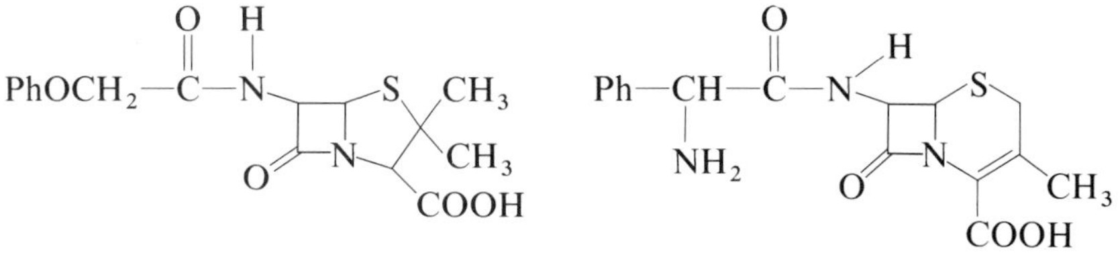

penicillin V
(a penicillin)

cephalexin (Keflex)
(a cephalosporin)

These β-lactam antibiotics apparently work by interfering with the synthesis of bacterial cell walls. Figure 21-12 shows how the carbonyl group of the β-lactam acylates an amino group on one of the enzymes involved in making the cell wall. The acylated enzyme is inactive in the synthesis of the cell wall protein.

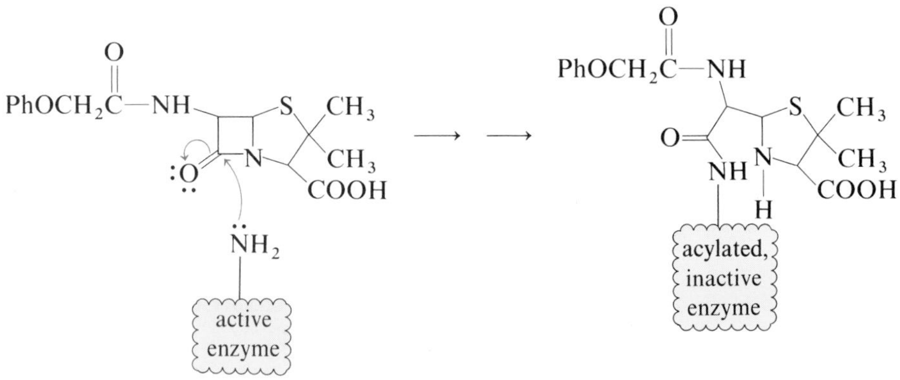

active enzyme

acylated, inactive enzyme

FIGURE 21-12 The β-lactam antibiotics function by acylating and inactivating one of the enzymes needed for synthesis of the bacterial cell wall.

PROBLEM 21-34

Show how you would accomplish the following synthetic conversions. You may use any necessary additional reagents.

(a) benzoic acid ⟶ benzyldimethylamine
(b) pyrrolidine ⟶ *N*-methylpyrrolidine

PROBLEM 21-35

Nitriles are considered to be carboxylic acid derivatives because they are often prepared from acids through the primary amides. Show how you would convert the following carboxylic acids to the corresponding nitriles.

(a) butyric acid $\longrightarrow$ butyronitrile
(b) benzoic acid $\longrightarrow$ benzonitrile
(c) cyclopentanecarboxylic acid $\longrightarrow$ cyclopentanecarbonitrile

21-14 OVERVIEW OF THE CHEMISTRY OF NITRILES

Although nitriles lack an acyl group, they are considered acid derivatives because they hydrolyze to carboxylic acids. Nitriles are frequently made from carboxylic acids (with the same number of carbons) by conversion to primary amides followed by dehydration of the amide. They are also made from primary alkyl halides and tosylates (adding one carbon) by nucleophilic substitution of cyanide ion. Aryl cyanides can be made by the Sandmeyer reaction of an aryldiazonium salt with cuprous cyanide. α-Hydroxynitriles (cyanohydrins) are made by the reaction of ketones and aldehydes with HCN.

$$\underset{\text{amide}}{R-\overset{O}{\overset{\|}{C}}-NH_2} \xrightarrow{POCl_3} \underset{\text{nitrile}}{R-C\equiv N} \qquad \text{(Section 21-13)}$$

$$\underset{\text{alkyl halide}}{R-X\ (1^\circ)} \xrightarrow{NaCN} \underset{\text{nitrile}}{R-C\equiv N} + Na^+X^- \qquad \text{(Section 5-10)}$$

$$\underset{\text{diazonium salt}}{Ar-\overset{+}{N}\equiv N} \xrightarrow{CuCN} \underset{\text{aryl nitrile}}{Ar-C\equiv N} + N_2\uparrow \qquad \text{(Section 19-18)}$$

$$\underset{\text{ketone or aldehyde}}{R-\overset{O}{\overset{\|}{C}}-R'} \xrightarrow{HCN} \underset{\text{cyanohydrin}}{R-C(OH)(C\equiv N)-R'} \qquad \text{(Section 18-15)}$$

Reactions of nitriles Nitriles undergo acidic or basic hydrolysis to amides, which may be further hydrolyzed to carboxylic acids. Hydrogen peroxide (H_2O_2) is added to the basic hydrolysis because the hydroperoxide anion (^-OOH) reacts faster with the cyano group than hydroxide ion. Reduction of a nitrile by lithium aluminum hydride gives a primary amine, and the reaction with a Grignard reagent gives an imine that is hydrolyzed to a ketone.

$$\underset{\text{nitrile}}{R-C\equiv N} \xrightarrow[H^+ \text{ or } ^-OH/H_2O_2]{H_2O} \underset{\text{amide}}{R-\overset{O}{\overset{\|}{C}}-NH_2} \xrightarrow[H^+ \text{ or } ^-OH]{H_2O} \underset{\text{acid}}{R-\overset{O}{\overset{\|}{C}}-OH} \qquad \text{(Section 21-7D)}$$

$$\underset{\text{nitrile}}{R-C\equiv N} \xrightarrow[(2)\ H_2O]{(1)\ LiAlH_4} \underset{\text{amine}}{R-CH_2NH_2} \qquad \text{(Sections 19-24 and 21-8C)}$$

$$\underset{\text{nitrile}}{R-C\equiv N} \xrightarrow[(2)\ H_3O^+]{(1)\ R'MgX} \underset{\text{ketone}}{R-\overset{O}{\overset{\|}{C}}-R'} \qquad \text{(Sections 18-10 and 21-9)}$$

PROBLEM 21-36

Show how you would convert the following starting materials to the indicated nitriles.

(a) phenylacetic acid ⟶ phenylacetonitrile
(b) phenylacetic acid ⟶ 3-phenylpropionitrile
(c) *p*-chloronitrobenzene ⟶ *p*-chlorobenzonitrile

PROBLEM 21-37

Show how each of the following transformations may be accomplished using a nitrile as an intermediate. You may use any additional reagents that are necessary.

(a) 1-hexanol ⟶ 1-aminoheptane
(b) cyclohexanecarboxamide ⟶ cyclohexyl ethyl ketone
(c) 1-octanol ⟶ 2-decanone

21-15 THIOESTERS

Most carboxylic esters are formed by the combination of a carboxylic acid and an alcohol. A **thioester** is an ester formed from a carboxylic acid and a thiol. Thioesters are also called *thiol esters* to emphasize the fact that they are derivatives of thiols.

$$\underset{\text{acid}}{R-\overset{O}{\overset{\|}{C}}-OH} + \underset{\text{alcohol}}{R'-OH} \rightleftharpoons \underset{\text{ester}}{R-\overset{O}{\overset{\|}{C}}-O-R'} + H_2O$$

$$\underset{\text{acid}}{R-\overset{O}{\overset{\|}{C}}-OH} + \underset{\text{thiol}}{R'-SH} \rightleftharpoons \underset{\text{thioester}}{R-\overset{O}{\overset{\|}{C}}-S-R'} + H_2O$$

Thioesters are much more reactive toward nucleophilic acyl substitution than normal esters. If we add thioesters to the order of reactivity, we have the following sequence:

Relative reactivity

$$\underset{\text{acid chloride}}{R-\overset{O}{\overset{\|}{C}}-Cl} > \underset{\text{anhydride}}{R-\overset{O}{\overset{\|}{C}}-O-\overset{O}{\overset{\|}{C}}-R} > \underset{\text{thioester}}{R-\overset{O}{\overset{\|}{C}}-S-R'} > \underset{\text{ester}}{R-\overset{O}{\overset{\|}{C}}-O-R'} > \underset{\text{amide}}{R-\overset{O}{\overset{\|}{C}}-NH_2}$$

The enhanced reactivity of thioesters over esters results from two major differences. First, the resonance stabilization of a thioester is less than that of an ester. In the thioester, the second resonance structure involves overlap between a 2*p* orbital on carbon and a 3*p* orbital on sulfur (Fig. 21-13). These orbitals are of different sizes and are located at different distances from the nuclei. The overlap is weak and relatively ineffective, leaving the C—S bond of a thioester weaker than the C—O bond of an ester.

The second difference is in the leaving groups: An alkyl sulfide anion ($^-\!:\!\ddot{S}-R$) is a better leaving group than an alkoxide anion ($^-\!:\!\ddot{O}-R$) because the sulfide is less basic than an alkoxide, and the larger sulfur atom carries the negative charge spread over a larger volume of space. Sulfur is also more polarizable than oxygen, allowing more bonding as the alkyl sulfide anion is leaving (Section 5-14).

Living systems use many acylation reactions, but acid halides and anhydrides

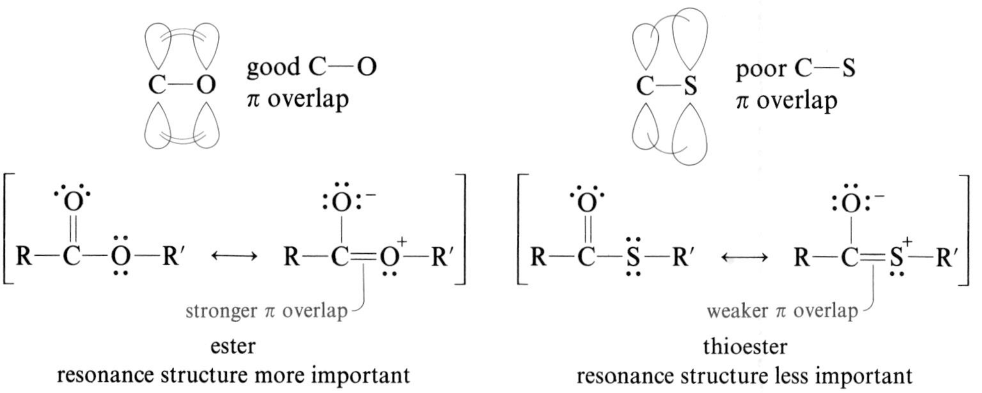

good C—O π overlap

poor C—S π overlap

$$\left[R-\overset{\overset{\cdot\ddot{O}\cdot}{\|}}{C}-\ddot{\underset{\cdot\cdot}{O}}-R' \longleftrightarrow R-\overset{\overset{:\ddot{O}:^-}{|}}{C}=\overset{+}{\underset{\cdot\cdot}{O}}-R' \right] \qquad \left[R-\overset{\overset{\cdot\ddot{O}\cdot}{\|}}{C}-\ddot{\underset{\cdot\cdot}{S}}-R' \longleftrightarrow R-\overset{\overset{:\ddot{O}:^-}{|}}{C}=\overset{+}{\underset{\cdot\cdot}{S}}-R' \right]$$

stronger π overlap — ester — resonance structure more important

weaker π overlap — thioester — resonance structure less important

FIGURE 21-13 The resonance overlap in a thioester is not as effective as that in an ester.

would hydrolyze under the aqueous conditions found in living organisms. Thioesters are not so prone to hydrolysis, yet they are excellent acylating reagents. For this reason, thioesters are common acylating agents in living systems. Many biochemical acylations involve transfer of acyl groups from thioesters of coenzyme A (CoA). Figure 21-14 shows the structure of acetyl coenzyme A, together with the mechanism for transfer of the acetyl group to a nucleophile.

$$CH_3-\overset{\overset{O}{\|}}{C}-S-CH_2CH_2\overset{\overset{H}{|}}{N}-\overset{\overset{O}{\|}}{C}-CH_2CH_2\overset{\overset{H}{|}}{N}-\overset{\overset{O}{\|}}{C}-\underset{\underset{OH}{|}}{CH}-\overset{\overset{CH_3}{|}}{\underset{\underset{CH_3}{|}}{C}}-CH_2-O-\overset{\overset{O^-}{|}}{\underset{\underset{O}{\|}}{P}}-O-\overset{\overset{O^-}{|}}{\underset{\underset{O}{\|}}{P}}-O-CH_2-\text{(ribose-adenine, 3'-}O-PO(O^-)OH)$$

thioester

coenzyme A (CoA)

$$CH_3-\overset{\overset{\cdot\ddot{O}\cdot}{\|}}{C}-\ddot{S}-CoA + Nuc:^- \rightleftharpoons CH_3-\overset{\overset{:\ddot{O}:^-}{|}}{\underset{\underset{Nuc}{|}}{C}}-\ddot{S}-CoA \longrightarrow CH_3-C(=\ddot{O})-Nuc + :\ddot{S}^--CoA$$

acetyl coenzyme A — tetrahedral intermediate — acylated product

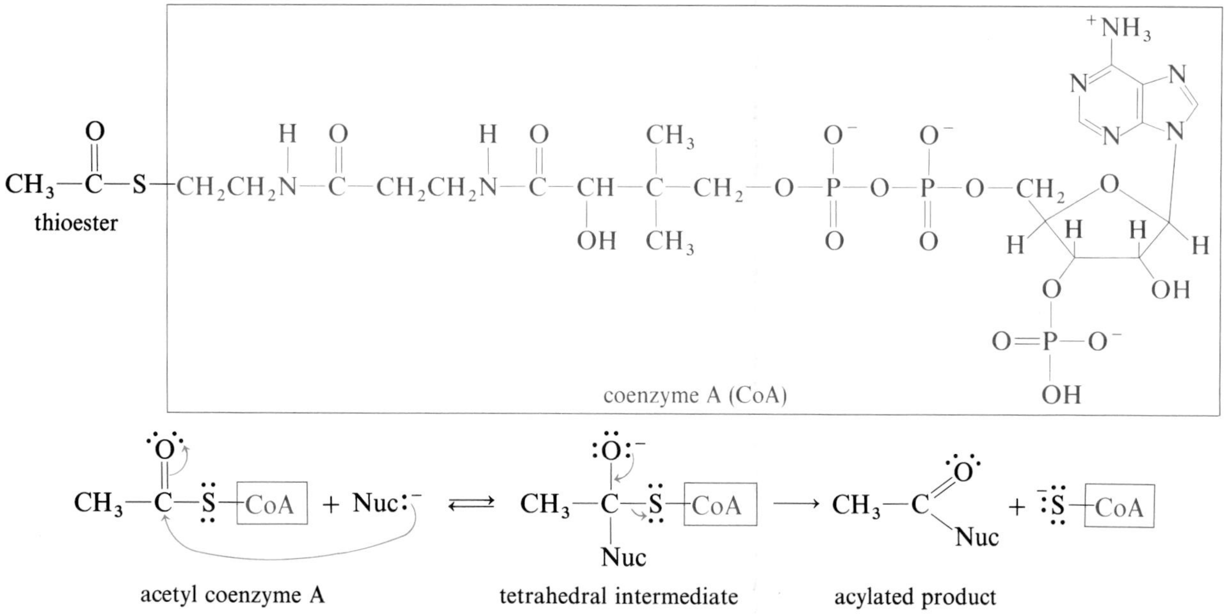

FIGURE 21-14 Coenzyme A (CoA) is a thiol whose thioesters serve as biochemical acyl transfer reagents. Acetyl CoA transfers an acetyl group to a nucleophile, with coenzyme A serving as the leaving group.

21-16 ESTERS AND AMIDES OF CARBONIC ACID

Carbonic acid (H_2CO_3) is the acid formed when carbon dioxide dissolves in water. Although carbonic acid itself is unstable, decomposing to carbon dioxide and water, it has several important stable derivatives. **Carbonate esters** are diesters of carbonic acid, with two alkoxy groups replacing the hydroxyl groups of carbonic acid. **Ureas** are diamides of carbonic acid, with two nitrogen atoms bonded to the carbonyl group. The unsubstituted urea, simply called *urea,* is the waste product excreted by mammals from the metabolism of excess protein. **Carbamate esters (urethanes)** are the stable esters of the unstable **carbamic acid,** the monoamide of carbonic acid.

$$O{=}C{=}O \; + \; H_2O \rightleftharpoons \left[H{-}O{-}\overset{\overset{\displaystyle O}{\|}}{C}{-}O{-}H \right]$$

carbonic acid (unstable)

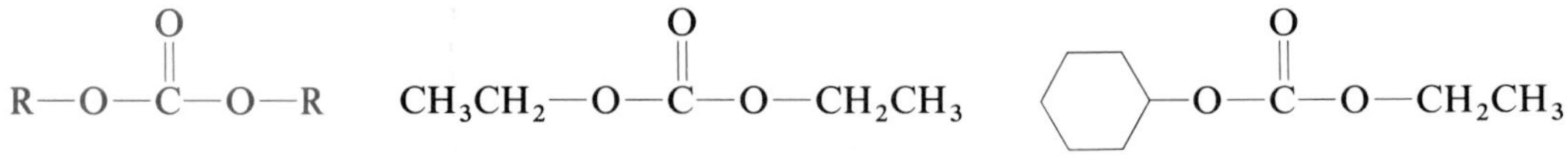

$$R{-}O{-}\overset{\overset{\displaystyle O}{\|}}{C}{-}O{-}R \qquad CH_3CH_2{-}O{-}\overset{\overset{\displaystyle O}{\|}}{C}{-}O{-}CH_2CH_3 \qquad C_6H_{11}{-}O{-}\overset{\overset{\displaystyle O}{\|}}{C}{-}O{-}CH_2CH_3$$

a carbonate ester — diethyl carbonate — cyclohexyl ethyl carbonate

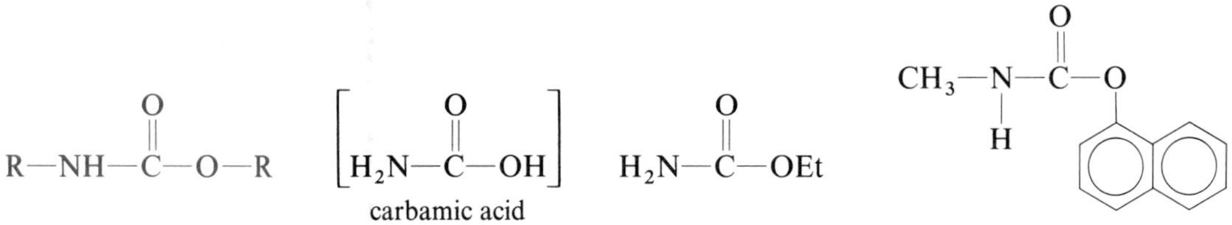

$$R{-}NH{-}\overset{\overset{\displaystyle O}{\|}}{C}{-}O{-}R \qquad \left[H_2N{-}\overset{\overset{\displaystyle O}{\|}}{C}{-}OH \right] \qquad H_2N{-}\overset{\overset{\displaystyle O}{\|}}{C}{-}OEt \qquad CH_3{-}\underset{\underset{\displaystyle H}{|}}{N}{-}\overset{\overset{\displaystyle O}{\|}}{C}{-}O{-}C_{10}H_7$$

a carbamate or urethane — carbamic acid (unstable) — ethyl carbamate — 1-naphthyl-*N*-methylcarbamate (Sevin insecticide)

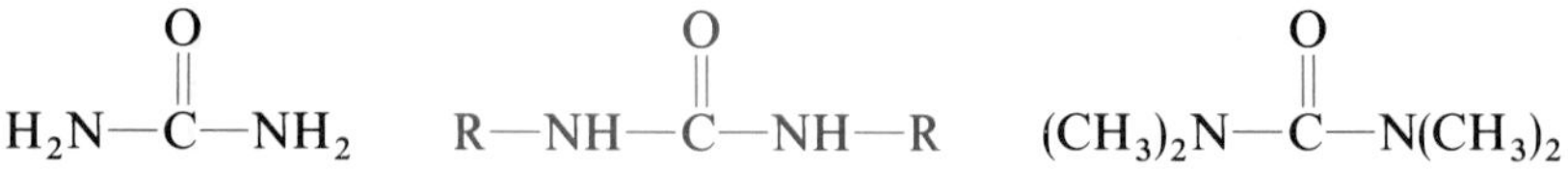

$$H_2N{-}\overset{\overset{\displaystyle O}{\|}}{C}{-}NH_2 \qquad R{-}NH{-}\overset{\overset{\displaystyle O}{\|}}{C}{-}NH{-}R \qquad (CH_3)_2N{-}\overset{\overset{\displaystyle O}{\|}}{C}{-}N(CH_3)_2$$

urea — a substituted urea — tetramethylurea

Most of these derivatives are synthesized by nucleophilic acyl substitutions using phosgene, the acid chloride of carbonic acid.

$$\underset{\text{phosgene}}{Cl{-}\overset{\overset{\displaystyle O}{\|}}{C}{-}Cl} \; + \; 2\,CH_3CH_2{-}OH \longrightarrow \underset{\text{diethyl carbonate}}{CH_3CH_2{-}O{-}\overset{\overset{\displaystyle O}{\|}}{C}{-}O{-}CH_2CH_3} \; + \; 2\,HCl$$

$$Cl{-}\overset{\overset{\displaystyle O}{\|}}{C}{-}Cl \xrightarrow{CH_3CH_2OH} Cl{-}\overset{\overset{\displaystyle O}{\|}}{C}{-}OCH_2CH_3 \xrightarrow{C_6H_{11}{-}NH_2} \underset{\text{ethyl } N\text{-cyclohexyl carbamate}}{C_6H_{11}{-}\underset{\underset{\displaystyle H}{|}}{N}{-}\overset{\overset{\displaystyle O}{\|}}{C}{-}OCH_2CH_3}$$

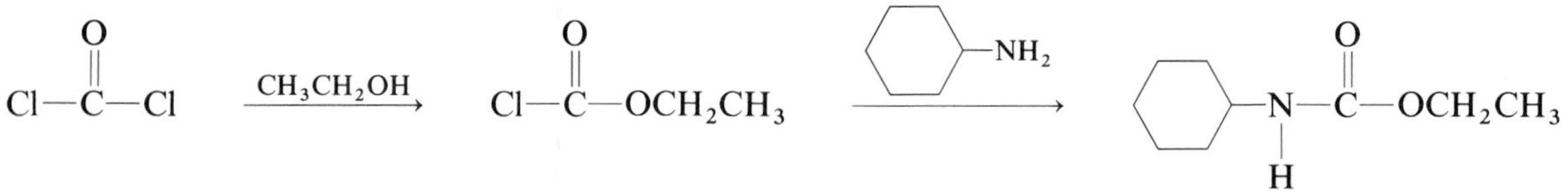

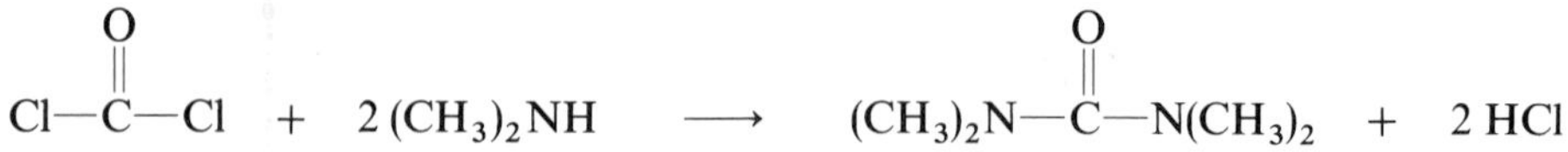

$$Cl{-}\overset{\overset{\displaystyle O}{\|}}{C}{-}Cl \; + \; 2\,(CH_3)_2NH \longrightarrow \underset{\text{tetramethylurea}}{(CH_3)_2N{-}\overset{\overset{\displaystyle O}{\|}}{C}{-}N(CH_3)_2} \; + \; 2\,HCl$$

The chemistry of carbonic acid derivatives is particularly important to polymer chemists because two important classes of polymers are bonded by linkages containing these functional groups: the *polycarbonates* and the *polyurethanes.*

Polycarbonates are polymers bonded by the carbonate ester linkage, and polyurethanes are polymers bonded by the carbamate ester linkage. These polymers are discussed in Chapter 26.

GLOSSARY

acid halide (acyl halide) An activated acid derivative in which the hydroxyl group of the acid is replaced by a halogen, usually chlorine. (p. 930)

amide A carboxylic acid derivative in which the hydroxyl group of the acid is replaced by a nitrogen atom and its attached hydrogens or alkyl groups. An amide is a composite of a carboxylic acid and an amine. (p. 928)

ammonolysis of an ester The cleavage of an ester by ammonia (or an amine) to give an amide and an alcohol. (p. 946)

anhydride (carboxylic acid anhydride) An activated acid derivative containing two acid molecules with the loss of a molecule of water. A **mixed anhydride** is an anhydride derived from two different acid molecules. (p. 931)

$$2\,R{-}\overset{\overset{\displaystyle O}{\|}}{C}{-}OH \;\rightleftarrows\; R{-}\overset{\overset{\displaystyle O}{\|}}{C}{-}O{-}\overset{\overset{\displaystyle O}{\|}}{C}{-}R \;+\; H_2O$$

carbonate ester A diester of carbonic acid. (p. 971)

carbonic acid The one-carbon dicarboxylic acid, HOCOOH. Carbonic acid is unstable, decomposing to carbon dioxide and water. Its esters and amides are stable, however. (p. 971)

ester A carboxylic acid derivative in which the hydroxyl group of the acid is replaced by an alkoxyl group. An ester is a composite of a carboxylic acid and an alcohol. (p. 926)

$$R{-}\overset{\overset{\displaystyle O}{\|}}{C}{-}OH \;+\; R'{-}OH \;\overset{H^+}{\rightleftarrows}\; R{-}\overset{\overset{\displaystyle O}{\|}}{C}{-}O{-}R' \;+\; H_2O$$

the Fischer esterification

Hofmann rearrangement of amides The conversion of a primary amide to an amine by treatment with a basic bromine solution. (p. 967)

lactam A cyclic amide. (p. 929)

lactone A cyclic ester. (p. 927)

nitrile An organic compound containing the **cyano group,** $C{\equiv}N$. (p. 929)

nucleophilic acyl substitution A reaction in which a nucleophile substitutes for a leaving group on a carbonyl carbon atom. Nucleophilic acyl substitution usually takes place through the following **addition-elimination mechanism.** (p. 943)

$$R{-}\overset{\overset{\displaystyle :\ddot{O}:}{\|}}{C}{-}X \;+\; Nuc:^- \;\rightleftarrows\; \left[R{-}\overset{\overset{\displaystyle :\ddot{O}:^-}{|}}{\underset{\underset{\displaystyle Nuc}{|}}{C}}{-}X \right] \;\rightleftarrows\; R{-}\overset{\overset{\displaystyle :\ddot{O}:}{\|}}{C}{-}Nuc \;+\; :X^-$$

the addition-elimination mechanism of nucleophilic acyl substitution

Rosenmund reduction The reduction of an acid chloride to an aldehyde by hydrogenation using a poisoned palladium catalyst. (p. 956)

saponification The basic hydrolysis of an ester to an alcohol and a carboxylate salt. (p. 952)

thioester An ester derived from a carboxylic acid and a thiol. (p. 970)

transesterification The substitution of one alkoxy group for another in an ester. Transesterification can take place under either acidic or basic conditions. (p. 949)

urea A diamide of carbonic acid. (p. 972)

urethane (carbamate ester) An ester of **carbamic acid,** $H_2N{-}COOH$; a monoester, monoamide of carbonic acid. (p. 972)

ESSENTIAL PROBLEM-SOLVING SKILLS IN CHAPTER 21

1. Name carboxylic acid derivatives and draw their structures from their names.
2. Compare the physical properties of acid derivatives, and explain the unusually high boiling points and melting points of amides.
3. Interpret the IR, NMR, and mass spectra of acid derivatives, and use the spectral information to determine the structures. Show how the carbonyl stretching frequency in the IR depends on the structure of the acid derivative.
4. Show how acid derivatives are easily interconverted by nucleophilic acyl substitution from more reactive derivatives to less reactive derivatives. Show how acid chlorides are used as activated intermediates to convert acids to acid derivatives.
5. Show how acid catalysis is used to synthesize acid derivatives, as in the Fischer esterification or in transesterification. Propose mechanisms for these reactions.
6. Show how acid derivatives are hydrolyzed to carboxylic acids, and explain why either acid or base is a suitable catalyst for the hydrolysis. Propose mechanisms for these hydrolyses.
7. Show what reagents are used for the reduction of acid derivatives, and show the products of these reductions.
8. Show what products result from the addition of Grignard and organolithium reagents to acid derivatives, and give mechanisms for these reactions.
9. Summarize the importance, uses, and special reactions of each type of acid derivative.

STUDY PROBLEMS

21-38. Give a definition and an example for each of the following terms:

(a) ester **(b)** lactone **(c)** Friedel-Crafts acylation
(d) lactam **(e)** tertiary amide **(f)** primary amide
(g) nitrile **(h)** acid chloride **(i)** anhydride
(j) mixed anhydride **(k)** Fischer esterification **(l)** nucleophilic acyl substitution
(m) Rosenmund reduction **(n)** ammonolysis of an ester **(o)** carbonate ester
(p) Hofmann rearrangement **(q)** lactone **(r)** saponification
(s) thioester **(t)** transesterification **(u)** urea
(v) urethane

21-39. Give appropriate names for the following compounds.

(a) $CH_3CH_2\overset{CH_3}{\overset{|}{C}}HCH_2-\overset{O}{\overset{\|}{C}}-Cl$ **(b)** $Ph-\overset{O}{\overset{\|}{C}}-O-\overset{O}{\overset{\|}{C}}-H$ **(c)** $CH_3-\overset{O}{\overset{\|}{C}}-NH-Ph$

(d) $CH_3-NH-\overset{O}{\overset{\|}{C}}-Ph$ **(e)** $Ph-O-\overset{O}{\overset{\|}{C}}-CH_3$ **(f)** $Ph-\overset{O}{\overset{\|}{C}}-O-CH_3$

(g) $-C\equiv N$ **(h)** CN **(i)** $\overset{O}{\overset{\|}{C}}-OCH_3$, $\underset{O}{\underset{\|}{C}}-OCH_3$

(j) CH_3, $\overset{O}{\overset{\|}{C}}-N(CH_2CH_3)_2$ **(k)** 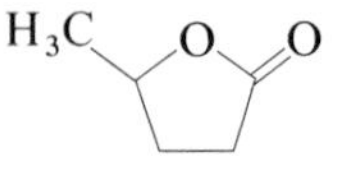 H_3C, O, O **(l)** 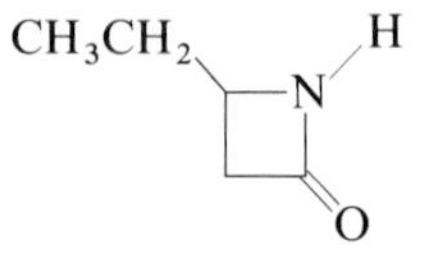 CH_3CH_2, H, N, O

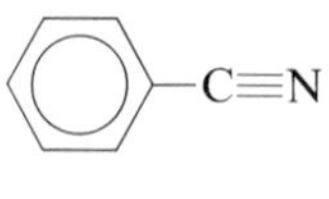

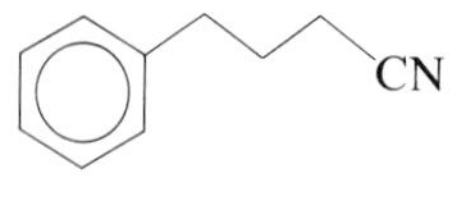

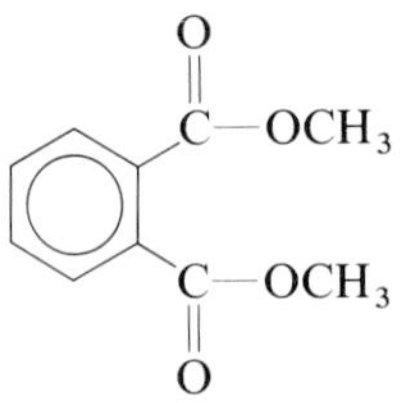

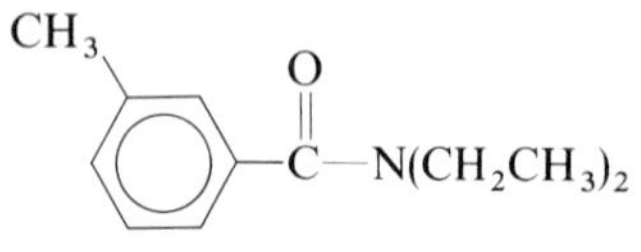

21-40. Predict the major products formed when benzoyl chloride (PhCOCl) reacts with the following reagents.
(a) ethanol **(b)** sodium acetate **(c)** aniline **(d)** anisole and aluminum chloride
(e) excess phenylmagnesium bromide, then dilute acid

21-41. Although a catalyst is not usually necessary for the reaction of an alcohol with an anhydride, a trace of acid increases the rate enormously. For example, aspirin is made by treating salicylic acid (*o*-hydroxybenzoic acid) with acetic anhydride and a trace of sulfuric acid. Give a mechanism for this reaction, showing why the reaction goes much faster with a trace of acid present.

21-42. Acid-catalyzed transesterification and Fischer esterification take place by nearly identical mechanisms. Transesterification can also take place by a base-catalyzed mechanism, but all attempts at base-catalyzed Fischer esterification (using $^{-}OR''$, for example) seem doomed to failure. Explain why Fischer esterification does not occur under basic catalysis.

21-43. Predict the products of the following reactions.
(a) phenol + acetic anhydride **(b)** phenol + acetic formic anhydride
(c) aniline + phthalic anhydride **(d)** anisole + succinic anhydride and aluminum chloride
(e) $Ph-CH(OH)-CH_2-NH_2$ + 1 equivalent of acetic anhydride
(f) $Ph-CH(OH)-CH_2-NH_2$ + excess acetic anhydride

21-44. Show how you would accomplish the following synthetic conversions in good yields.

(a) 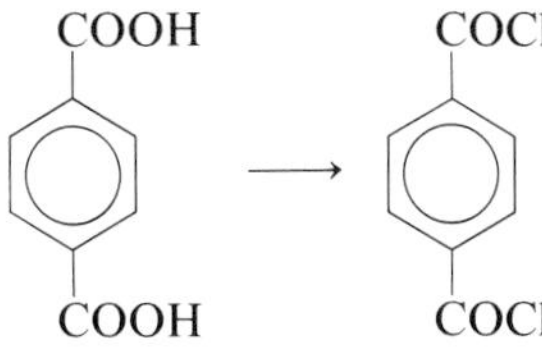 terephthalic acid (p-HOOC–C6H4–COOH) ⟶ p-ClOC–C6H4–COCl (COCl, COCl)

(b) 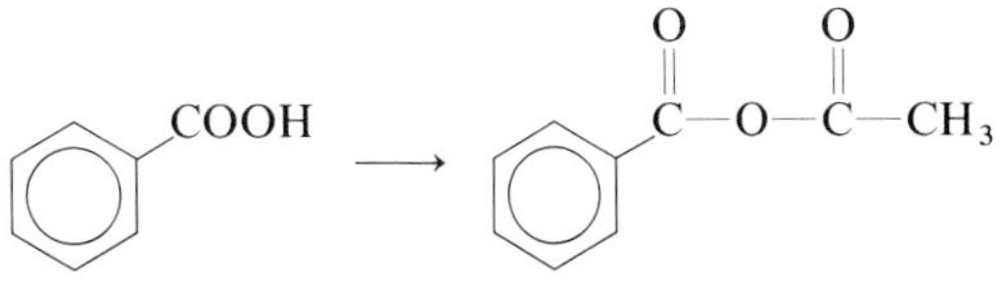 PhCOOH ⟶ Ph–C(=O)–O–C(=O)–CH$_3$

(c) cis-1,2-cyclohexanediol (H, OH, OH, H) ⟶ cyclohexane-fused dioxane-2,3-dione (H, O, O, O, O, H)

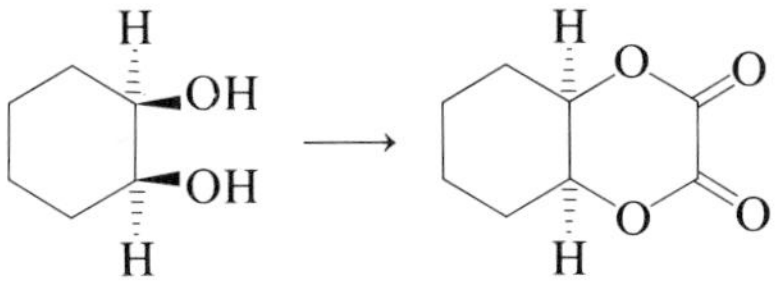

(d) 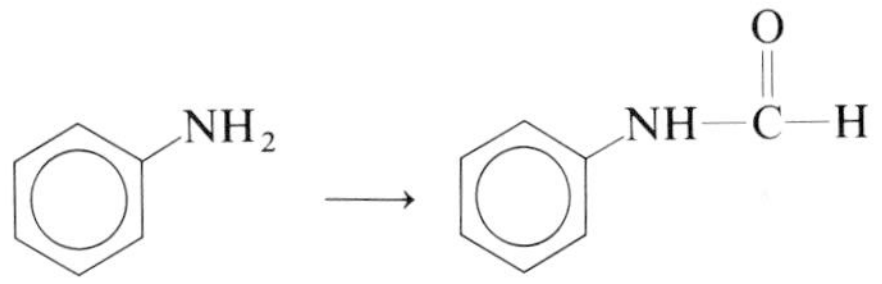 PhNH$_2$ ⟶ Ph–NH–C(=O)–H

(e) 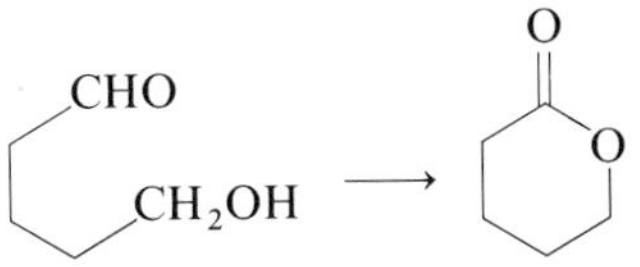 5-hydroxypentanal (CHO, CH$_2$OH) ⟶ δ-valerolactone

(f) 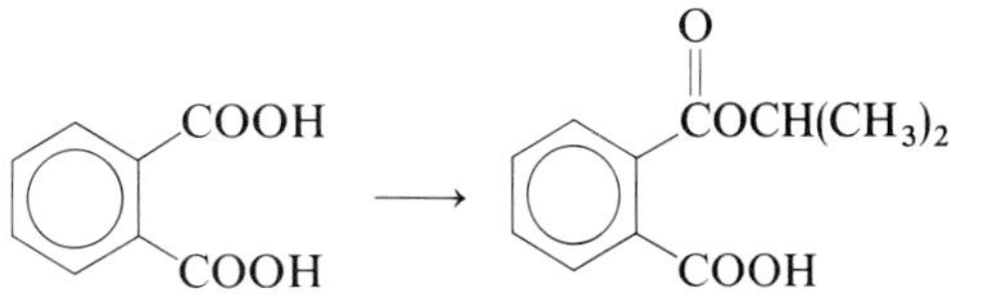 phthalic acid (COOH, COOH) ⟶ o-(COOCH(CH$_3$)$_2$)C$_6$H$_4$COOH

21-45. Propose mechanisms for the following reactions.

(a) 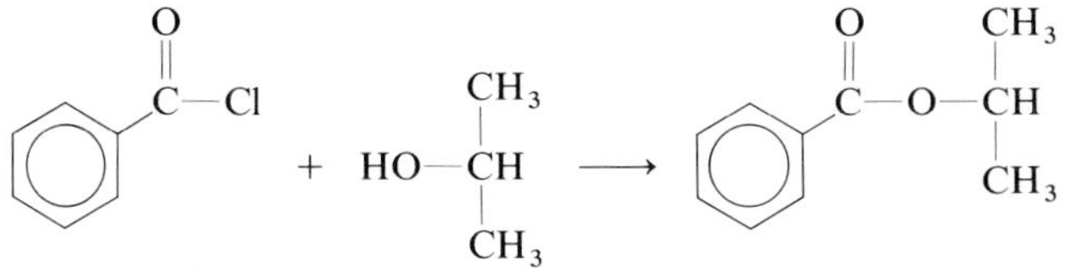 Ph–C(=O)–Cl + HO–CH(CH$_3$)$_2$ ⟶ Ph–C(=O)–O–CH(CH$_3$)$_2$

benzoyl chloride isopropyl alcohol isopropyl benzoate

(b) 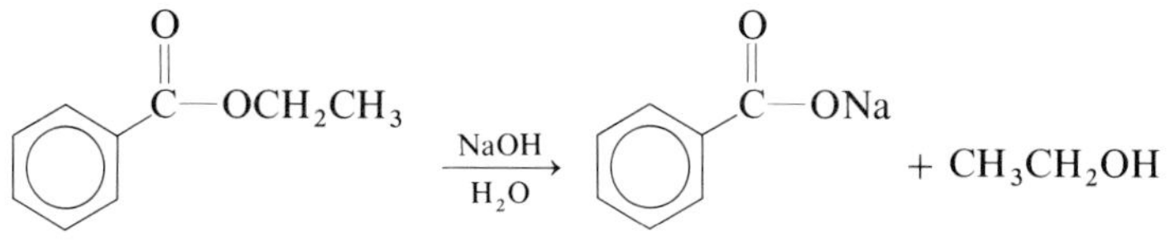 Ph–C(=O)–OCH$_2$CH$_3$ $\xrightarrow[H_2O]{NaOH}$ Ph–C(=O)–ONa + CH_3CH_2OH

ethyl benzoate sodium benzoate ethanol

(c) 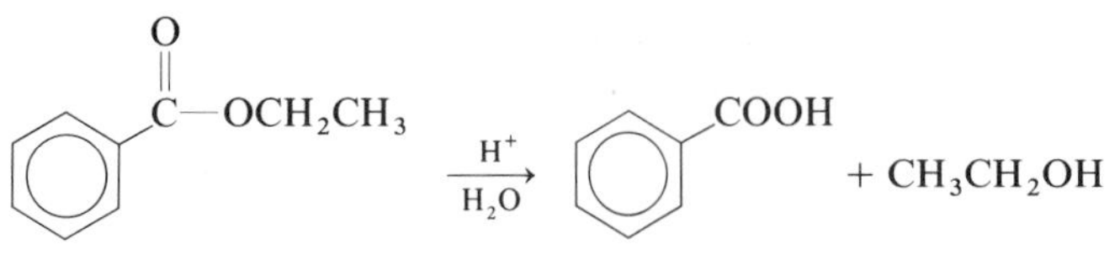 $C_6H_5-\overset{O}{\overset{\|}{C}}-OCH_2CH_3 \xrightarrow[H_2O]{H^+} C_6H_5-COOH + CH_3CH_2OH$

ethyl benzoate — benzoic acid — ethanol

(d) $CH_3-\underset{}{\overset{OH}{\overset{|}{CH}}}-CH_2CH_3 \xrightarrow[\text{(acetic anhydride)}]{Ac_2O} CH_3-\overset{OAc}{\overset{|}{CH}}-CH_2CH_3$

(R)-2-butanol — 2-butyl acetate

Does this reaction proceed with retention, inversion, or racemization of the chiral carbon atom?

21-46. Predict the products of the following reactions.

(a) 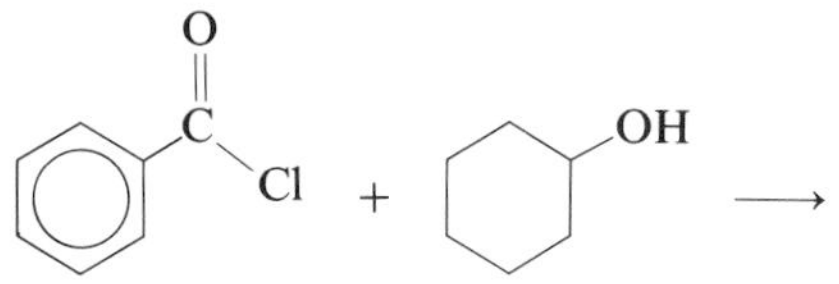 $C_6H_5-\overset{O}{\overset{\|}{C}}-Cl$ + cyclohexanol $\longrightarrow$

(b) $C_6H_{11}-\overset{O}{\overset{\|}{C}}-OCH_3 \xrightarrow[\text{heat}]{CH_3NH_2}$

(c) $Ph-\overset{O}{\overset{\|}{C}}-Cl$ + pyrrolidine (N—H) $\longrightarrow$

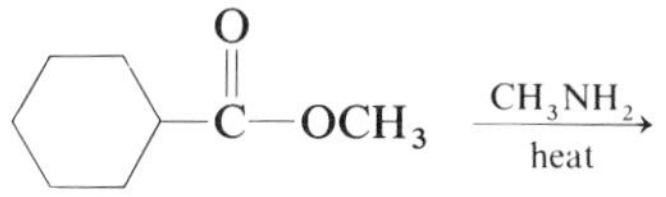

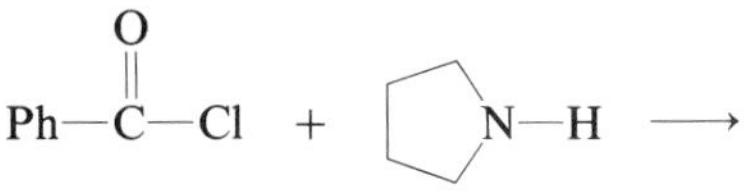

(d) 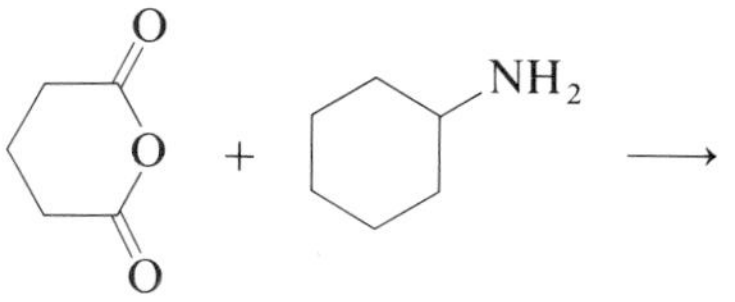 glutaric anhydride + cyclohexylamine ($C_6H_{11}NH_2$) $\longrightarrow$

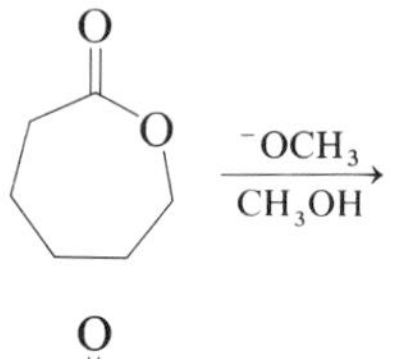

(e) $Ph-\overset{O}{\overset{\|}{C}}-OCH_2CH_3 \xrightarrow[(2)\ H_2O]{(1)\ LiAlH_4}$

(f) 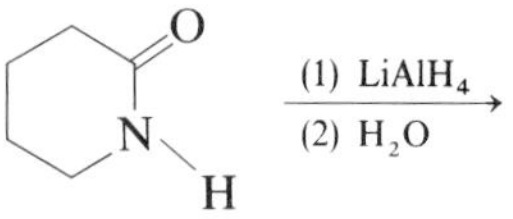 2-piperidinone (N—H) $\xrightarrow[(2)\ H_2O]{(1)\ LiAlH_4}$

(g) seven-membered lactone $\xrightarrow[CH_3OH]{^-OCH_3}$

(h) six-membered lactone $\xrightarrow[(2)\ H_2O]{(1)\ \text{excess PhMgBr}}$

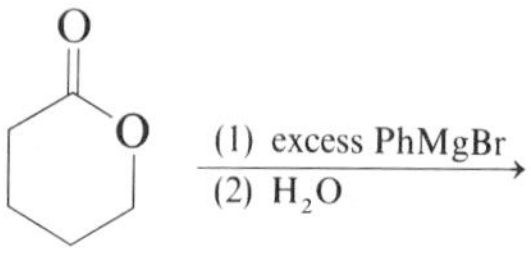

(i) 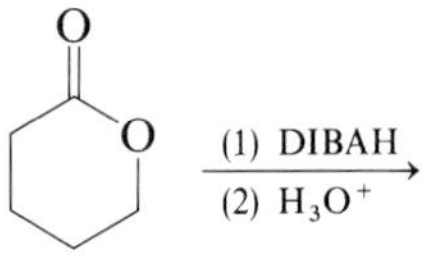 six-membered lactone $\xrightarrow[(2)\ H_3O^+]{(1)\ DIBAH}$

(j) 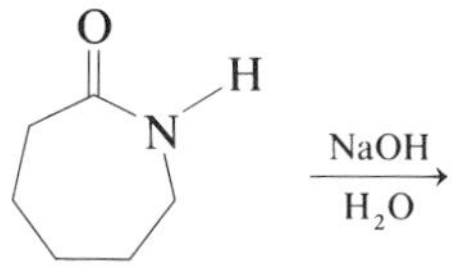 seven-membered lactam (N—H) $\xrightarrow[H_2O]{NaOH}$

(k) $PhCH_2-\overset{CH_3}{\overset{|}{CH}}-CH_2-\overset{O}{\overset{\|}{C}}-NH_2 \xrightarrow{Br_2,\ NaOH}$

(l) 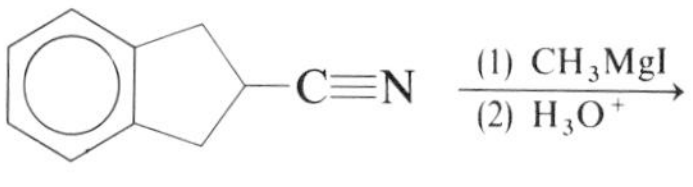 indane$-C\equiv N \xrightarrow[(2)\ H_3O^+]{(1)\ CH_3MgI}$

(m) 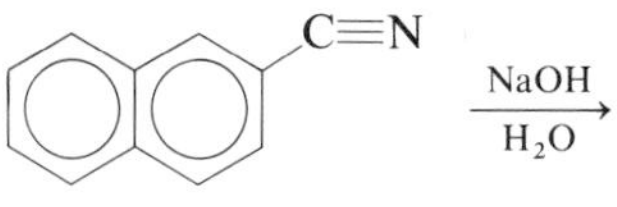 naphthalene$-C\equiv N \xrightarrow[H_2O]{NaOH}$

21-47. Predict the products of saponification of the following esters.

(a) $H-\overset{O}{\overset{\|}{C}}-O-Ph$

(b) $CH_3CH_2O-\overset{O}{\overset{\|}{C}}-OCH_2CH_3$

(c)

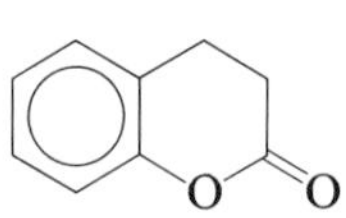

(d)

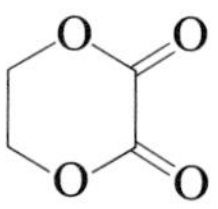

21-48. An ether extraction of nutmeg gives large quantities of *trimyristin,* a waxy crystalline solid of melting point 57°C. The IR spectrum of trimyristin shows a very strong absorption at 1733 cm^{-1}. Basic hydrolysis of trimyristin gives 1 equivalent of glycerol and 3 equivalents of myristic acid (tetradecanoic acid).

(a) Draw the structure of trimyristin.

(b) Predict the products formed when trimyristin is treated with lithium aluminum hydride, followed by an aqueous hydrolysis of the aluminum salts.

21-49. The two most widely used pain relievers are aspirin and acetaminophen. Show how you would synthesize these drugs from phenol.

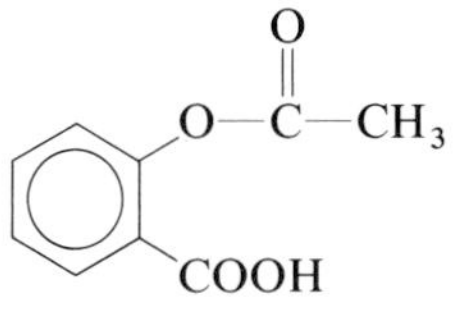

aspirin (o-acetoxybenzoic acid: $O{-}\overset{O}{\overset{\|}{C}}{-}CH_3$, COOH) acetaminophen (HO–C$_6$H$_4$–$NH{-}\overset{O}{\overset{\|}{C}}{-}CH_3$)

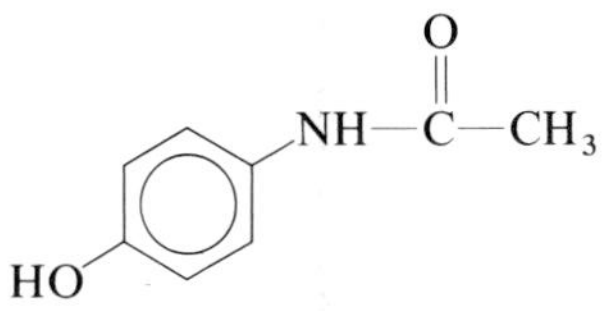

21-50. Show how you would accomplish the following synthetic conversions. Some of these conversions may require more than one step.

(a) isopentyl alcohol ⟶ isopentyl acetate (banana oil)
(b) 3-ethylpentanoic acid ⟶ 3-ethylpentanenitrile
(c) isobutylamine ⟶ *N*-isobutylformamide
(d) ethyl acetate ⟶ 3-methyl-3-pentanol
(e) pyrrolidine ⟶ *N*-methylpyrrolidine
(f) cyclohexylamine ⟶ *N*-cyclohexylacetamide
(g) bromocyclohexane ⟶ dicyclohexylmethanol
(h) dimethyl oxalate ⟶ (piperazine-2,3-dione, N–H, N–H, =O, =O)
(i) 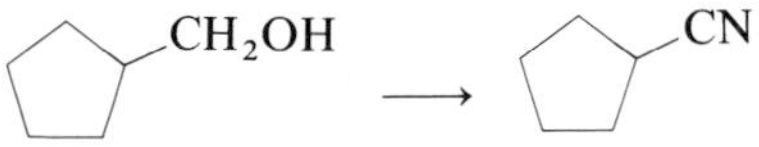 cyclopentyl–CH_2OH ⟶ cyclopentyl–CN
(j) 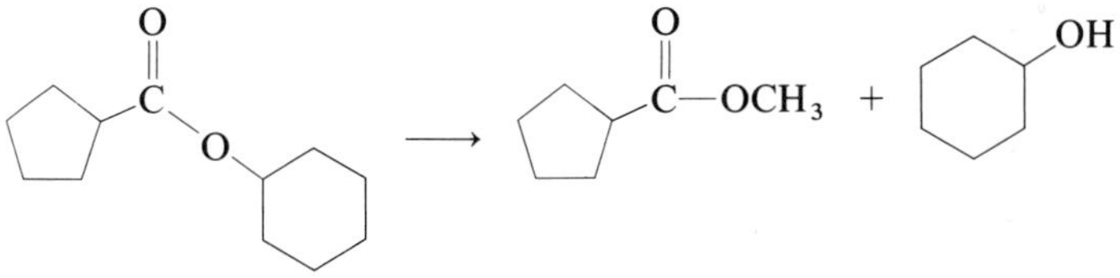 cyclopentyl–$\overset{O}{\overset{\|}{C}}$–O–cyclohexyl ⟶ cyclopentyl–$\overset{O}{\overset{\|}{C}}{-}OCH_3$ + cyclohexyl–OH

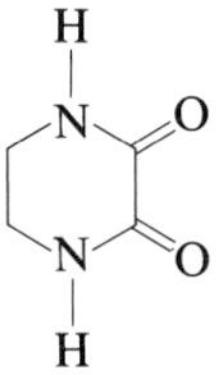

21-51. Grignard reagents add to carbonate esters in a manner similar to their addition to other esters.
(a) Predict the major product of the following reaction:

$$CH_3CH_2{-}O{-}\overset{O}{\overset{\|}{C}}{-}O{-}CH_2CH_3 \xrightarrow[(2)\ H_2O]{(1)\ \text{excess PhMgBr}}$$

diethyl carbonate

(b) Show how you would synthesize 3-ethyl-3-pentanol using diethyl carbonate and ethyl bromide as your only organic reagents.

21-52. One mole of acetyl chloride is added to a liter of triethylamine, resulting in a vigorous exothermic reaction. Once the reaction mixture has cooled, one mole of ethanol is added. Another vigorous exothermic reaction results. The mixture is analyzed and found to contain triethylamine, ethyl acetate, and triethylammonium chloride. Give detailed mechanisms for the two exothermic reactions.

21-53. Show how you would accomplish the following multistep syntheses using the indicated starting material and any necessary additional reagents.

(a) 6-hepten-1-ol ⟶ ε-caprolactone **(b)** 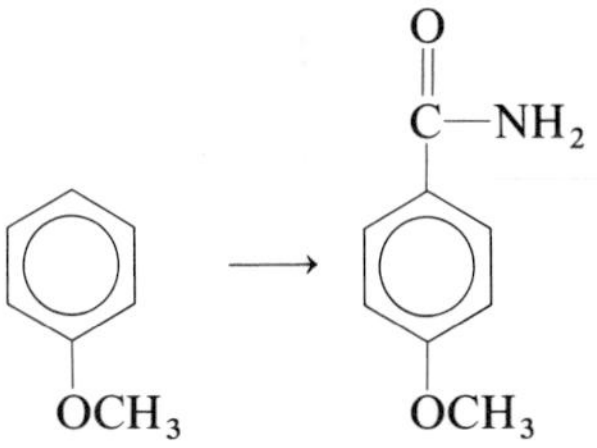 C$_6$H$_5$–OCH_3 ⟶ p-($\overset{O}{\overset{\|}{C}}{-}NH_2$)C$_6H_4$–$OCH_3$

(c) 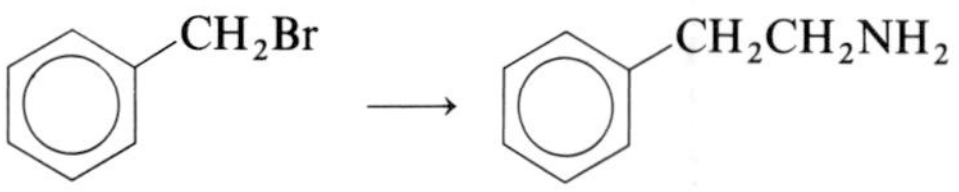 [benzyl bromide, $C_6H_5CH_2Br$] ⟶ [$C_6H_5CH_2CH_2NH_2$]

(d) 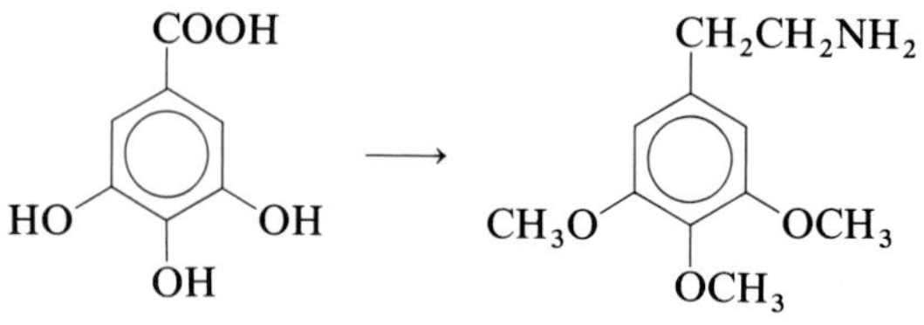 [gallic acid: 3,4,5-trihydroxybenzoic acid, COOH; HO, OH, OH] ⟶ [mescaline: $CH_2CH_2NH_2$; CH_3O, OCH_3, OCH_3]

gallic acid — mescaline

21-54. A chemist has prepared some methyl benzoate in which one oxygen atom is the heavy ^{18}O isotope:

$$Ph-\overset{\overset{\displaystyle O}{\|}}{C}-{}^{18}O-CH_3$$

(a) Draw the mechanism for saponification of this ester. Use this mechanism to show where the labeled ^{18}O will appear in the products.
(b) Explain how you would prove experimentally where the ^{18}O label appears in the products. (^{18}O is not radioactive.)
(c) Repeat part (a) using an acid-catalyzed hydrolysis, the reverse of Fischer esterification.

21-55. Phosgene is the acid chloride of carbonic acid. Although phosgene was used as a war gas in World War I, it is now used as a reagent for the synthesis of many useful products. Phosgene reacts like other acid chlorides, but it can react twice.

$$\left[HO-\overset{\overset{\displaystyle O}{\|}}{C}-OH\right] \qquad Cl-\overset{\overset{\displaystyle O}{\|}}{C}-Cl \xrightarrow{2\ Nuc:^-} Nuc-\overset{\overset{\displaystyle O}{\|}}{C}-Nuc \ + \ 2\ Cl^-$$

carbonic acid — phosgene

Predict the products formed when phosgene reacts with
(a) excess ethanol.
(b) excess methylamine.
(c) 1 equivalent of methanol, followed by 1 equivalent of aniline.
(d) ethylene glycol.
(e) *t*-Butyloxycarbonyl chloride is an important reagent for the synthesis of peptides and proteins (Chapter 24). Show how you would use phosgene to synthesize *t*-butyloxycarbonyl chloride.

$$CH_3-\underset{\underset{\displaystyle CH_3}{|}}{\overset{\overset{\displaystyle CH_3}{|}}{C}}-O-\overset{\overset{\displaystyle O}{\|}}{C}-Cl$$

t-butyloxycarbonyl chloride

21-56. A carbamate ester (a urethane) can be made by the reaction of an alcohol with an isocyanate.

$$R-N{=}C{=}O \ + \ R'-OH \longrightarrow R-NH-\overset{\overset{\displaystyle O}{\|}}{C}-O-R'$$

isocyanate — alcohol — carbamate ester

(a) Propose a mechanism for this reaction.
(b) A tragedy occurred in Bhopal, India, when a large quantity of methyl isocyanate escaped from a factory that produced Sevin insecticide. Suggest how methyl isocyanate was used in the synthesis of Sevin.

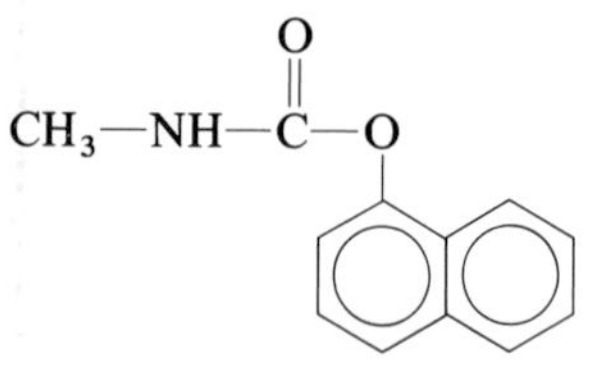

$$CH_3-NH-\overset{\overset{\displaystyle O}{\|}}{C}-O-(\text{1-naphthyl})$$

Sevin insecticide

(c) Propose an alternative synthesis of Sevin. Unfortunately, the best alternative synthesis uses phosgene, a gas that is even more toxic than methyl isocyanate.

21-57. The following compounds were isolated from the fungus *Cephalosporium acremonium,* isolated in 1948 from the sea near a sewage outlet along the Sardinian coastline. Both compounds exhibit powerful antibacterial activity:

cephalosporin N

cephalosporin C

(a) Name the class of antibiotics represented by each of these compounds.

(b) After its structure had been determined, one of these compounds came to be known by a more appropriate name. Determine which compound is poorly named, and replace the inappropriate part of the name.

21-58. An unknown compound gives a mass spectrum with a weak molecular ion at m/z 113 and a prominent ion at m/z 68. Its NMR and IR spectra are shown below. Determine the structure and show how it is consistent with the observed absorptions. Propose a favorable fragmentation to explain the prominent MS peak at m/z 68.

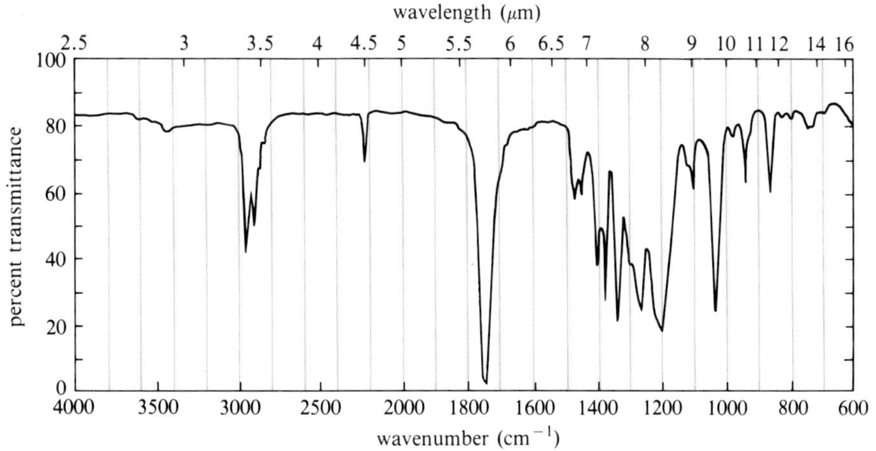

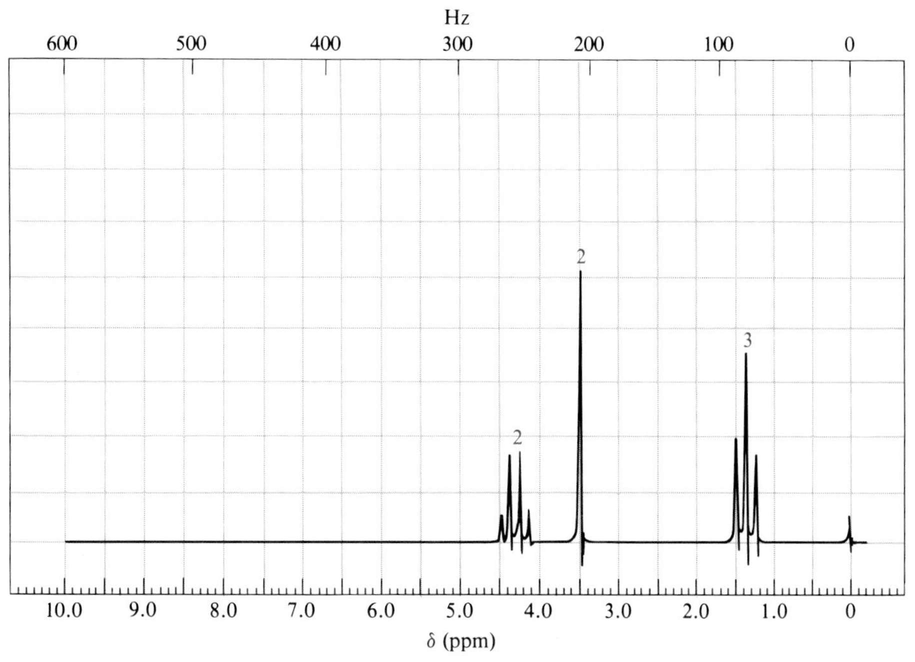

21-59. An unknown compound of molecular formula C_5H_9NO gives the IR and NMR spectra shown below. The broad NMR peak at δ8.35 disappears when the sample is shaken with D_2O. Propose a structure and show how it is consistent with the absorptions in the spectra.

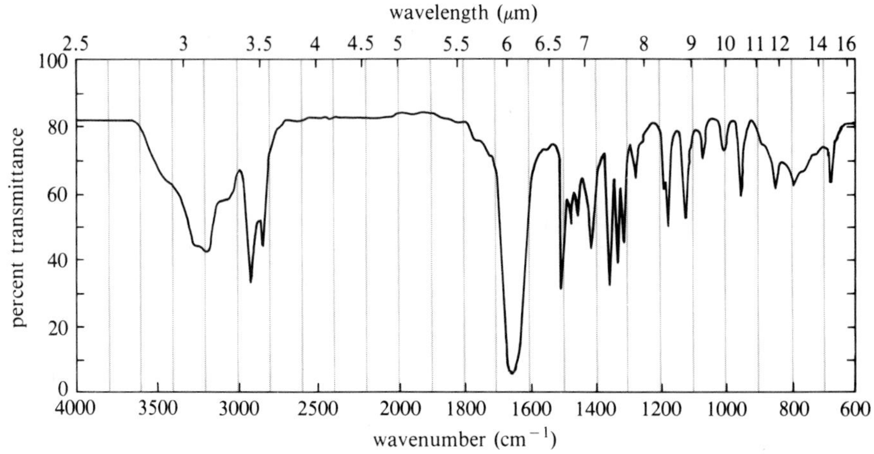

wavelength (μm)
2.5 3 3.5 4 4.5 5 5.5 6 6.5 7 8 9 10 11 12 14 16
percent transmittance
100 80 60 40 20 0
4000 3500 3000 2500 2000 1800 1600 1400 1200 1000 800 600
wavenumber (cm⁻¹)

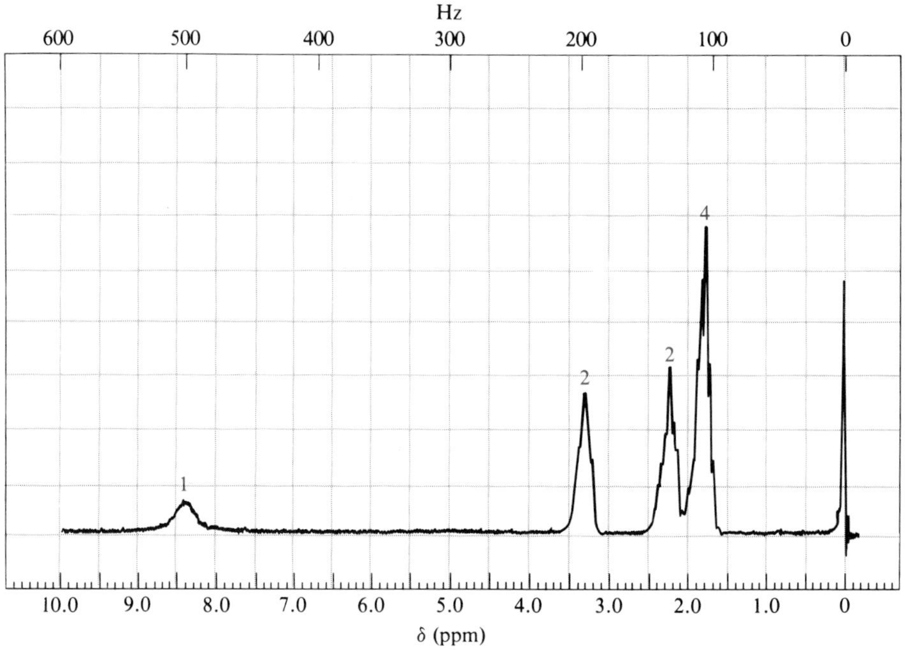

Hz
600 500 400 300 200 100 0
1
2
2
4
10.0 9.0 8.0 7.0 6.0 5.0 4.0 3.0 2.0 1.0 0
δ (ppm)

21-60. An unknown compound gives the NMR, IR, and mass spectra shown below. Propose a structure and show how it is consistent with the observed absorptions. Show fragmentations that account for the prominent ion at m/z 69 and the smaller peak at m/z 99.

intensity
100 80 60 40 20 0
69
99
114 M⁺
10 20 30 40 50 60 70 80 90 100 110 120 130 140 150 160
m/z

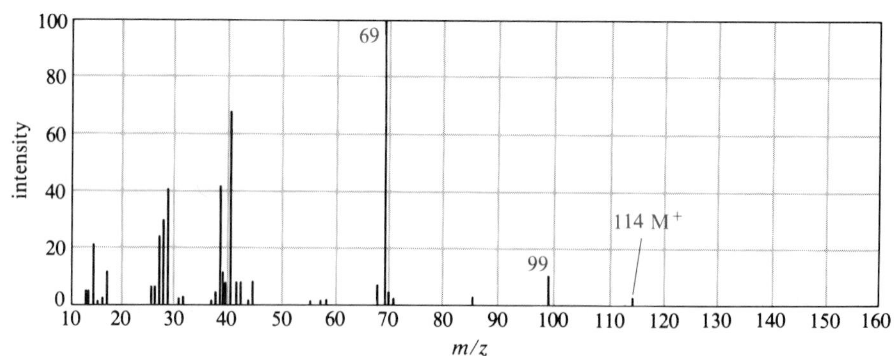

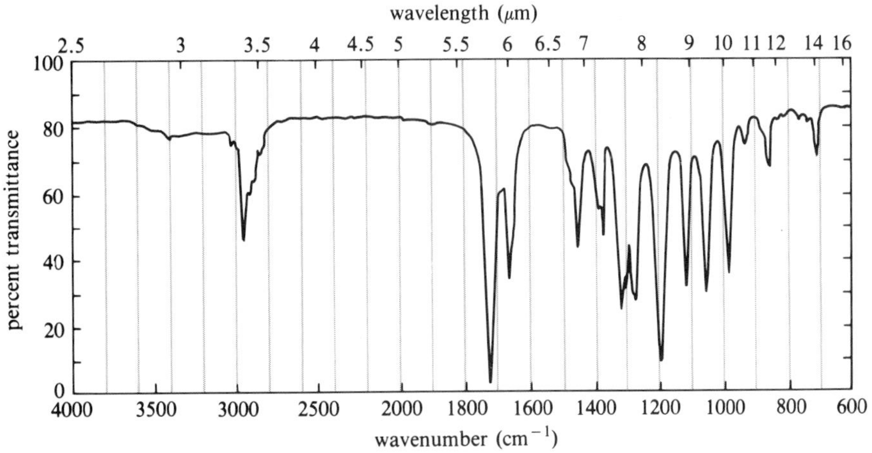

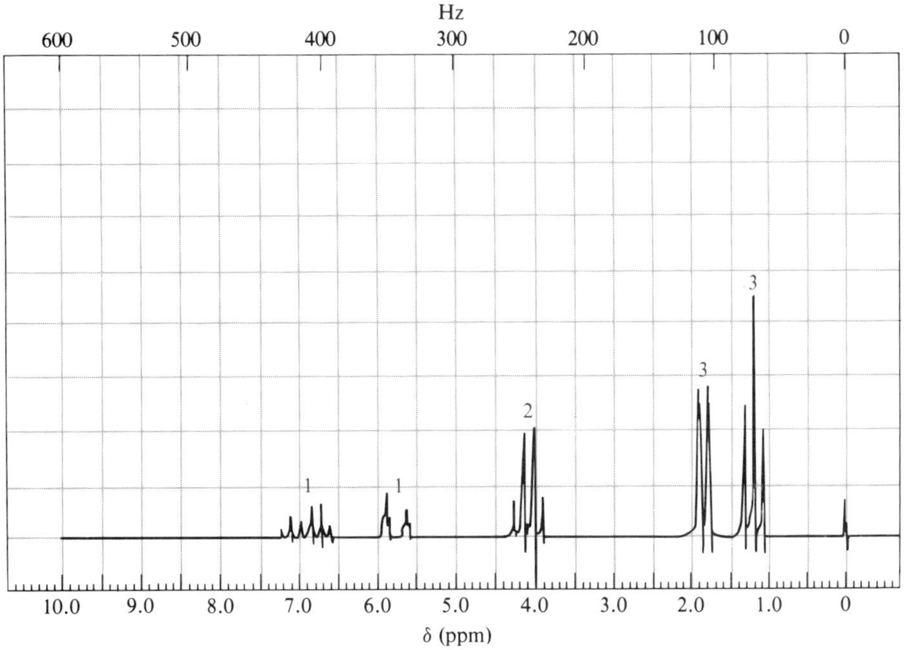

✱ 21-61. The ^{1}H NMR spectrum, ^{13}C NMR spectrum, and IR spectrum of an unknown compound ($C_6H_8O_3$) appear below. Determine the structure and show how it is consistent with the spectra.

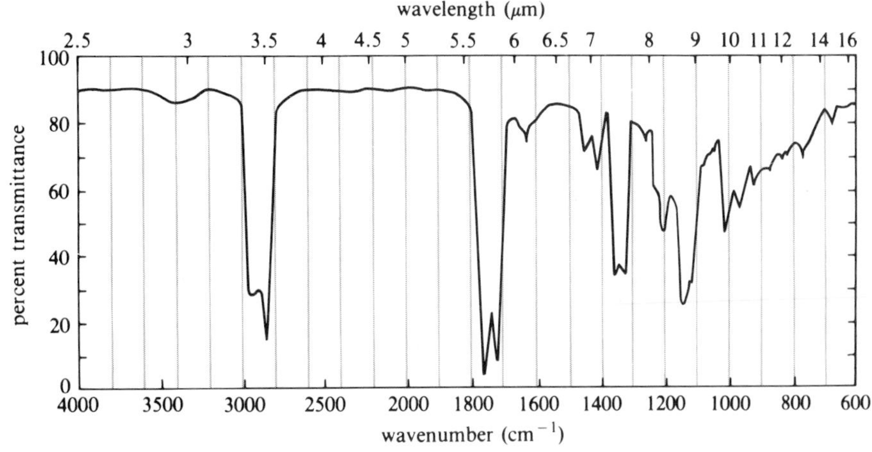

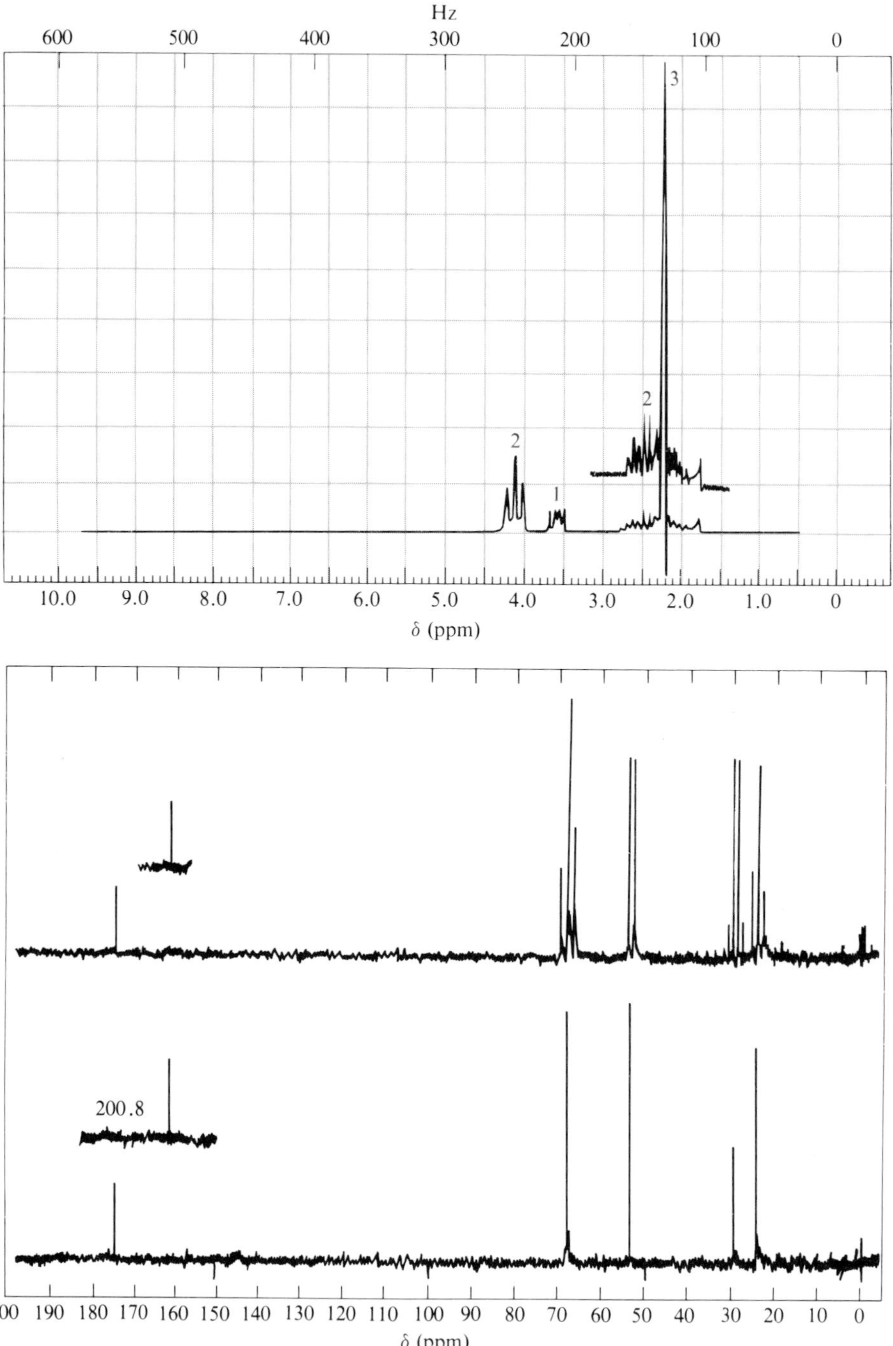

21-62. Methyl *p*-nitrobenzoate has been found to undergo saponification faster than methyl benzoate.

(a) Consider the mechanism of the saponification, and explain the reasons for this enhancement of the rate.

(b) Would you expect methyl *p*-methoxybenzoate to undergo saponification faster or slower than methyl benzoate?

21-63. A student has just added ammonia to hexanoic acid and has begun to heat the mixture when he is called away to the telephone. After a long telephone conversation, he returns to find that the mixture had overheated and turned black. He distills the volatile components and recrystallizes the solid residue. Among the components he isolates are compounds A (a liquid, molecular formula $C_6H_{11}N$) and B (a solid, molecular formula $C_6H_{13}NO$). The infrared spectrum of A shows a strong, sharp absorption at 2247 cm^{-1}. The infrared spectrum of B shows absorptions at 3390, 3200, and 1665 cm^{-1}. Determine the structures of compounds A and B.

21-64. A chemist was called to an abandoned aspirin factory to determine the contents of a badly corroded vat. Knowing that two salvage workers had become ill from breathing the fumes, she put on her breathing apparatus as soon as she noticed an overpowering odor like that of vinegar but much more pungent. She entered the building and took a sample of the contents of the vat. The mass spectrum shows a molecular weight of 102, and the NMR spectrum showed only a singlet at $\delta 2.15$. The IR spectrum, which appears below, left no doubt about the identity of the compound. Identify the compound and suggest a method for its safe disposal.

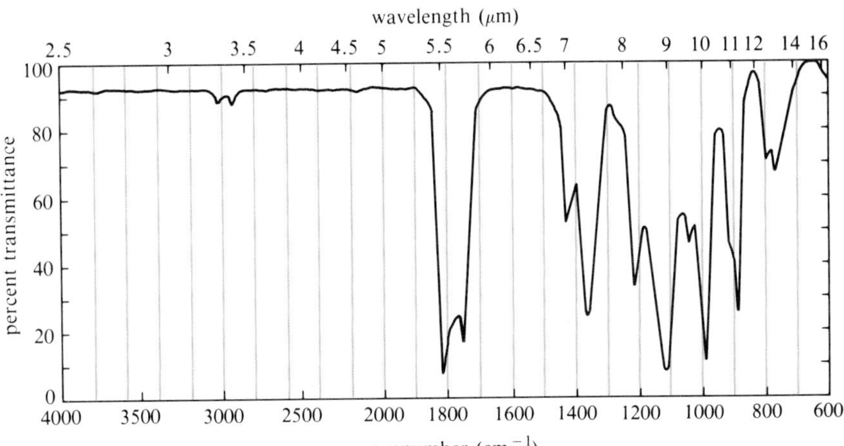

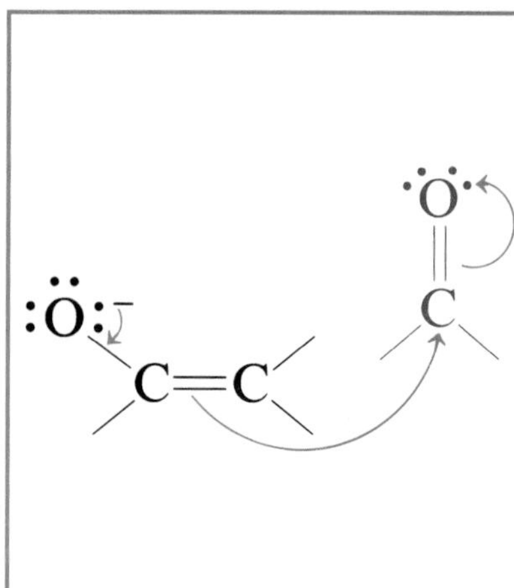

22

ADDITIONS AND CONDENSATIONS OF ENOLS AND ENOLATE IONS

22-1 INTRODUCTION

In our study of ketones, aldehydes, and acid derivatives, we have seen many reactions in which the carbonyl compound serves as an electrophile. With ketones and aldehydes, the most common reaction is nucleophilic addition to the carbonyl: A nucleophile attacks the electrophilic carbon atom, and the oxygen becomes protonated. Nucleophilic acyl substitution is the most common reaction of esters and other acid derivatives: Attack by a nucleophile is followed by elimination of the alkoxide or other leaving group.

Ketones and aldehydes: Nucleophilic addition

$$\ce{>C=O} \xrightarrow[\text{(addition)}]{\text{Nuc:}^-} \ce{-C(Nuc)-O^-} \xrightarrow{\text{Nuc—H}} \ce{-C(Nuc)-O-H}$$

Esters: Nucleophilic acyl substitution

$$\ce{-C(=O)-OR} \xrightarrow[\text{(substitution)}]{\text{Nuc:}^-} \ce{-C(=O)-Nuc} \quad {}^-\text{OR}$$

Carbonyl compounds can also serve as carbanion-like nucleophiles, most commonly at the carbon atom next to the carbonyl group: the **alpha carbon atom.** For example, a proton can be removed from the alpha carbon to give a resonance-stabilized **enolate ion** which is a strong nucleophile and can form a bond to an electrophile.

enolate ion

The most useful reactions of enolate ions and related intermediates involve their attack on carbonyl groups. Once again, these reactions may give either nucleophilic addition to the carbonyl (ketones and aldehydes) or nucleophilic acyl substitution (acid derivatives, especially esters).

Addition to ketones and aldehydes

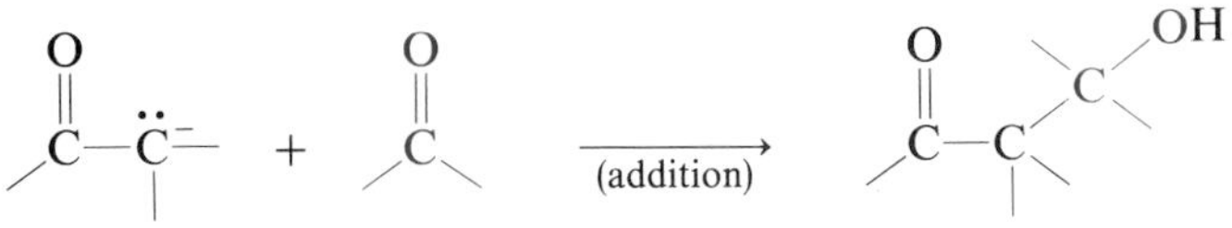

(addition)

Substitution with esters

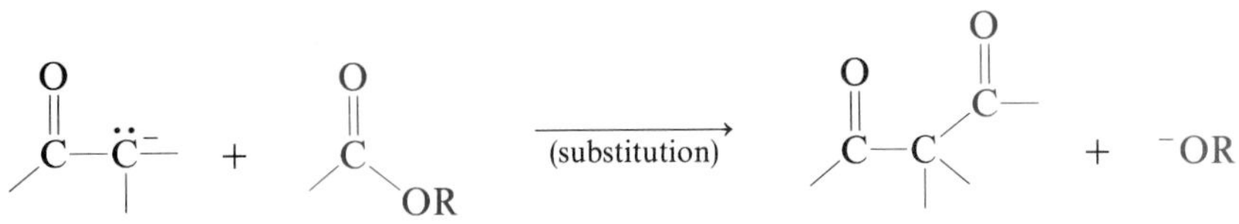

(substitution)

The nucleophilic reactions of carbonyl compounds are some of the most common methods for forming carbon-carbon bonds. A wide variety of compounds can participate as nucleophiles or electrophiles (or both) in these reactions, and many useful kinds of products can be synthesized.

22-1A THE KETO-ENOL TAUTOMERISM

In the presence of strong bases, ketones and aldehydes act as weak proton acids. A proton on the α-carbon (the carbon adjacent to the carbonyl group) is abstracted to form a resonance-stabilized **enolate ion** with the negative charge spread over a carbon atom and an oxygen atom. Reprotonation can occur either on the α-carbon (returning to the **keto** form) or on the oxygen atom, giving a vinyl alcohol, the **enol** form.

Base-catalyzed keto-enol tautomerism

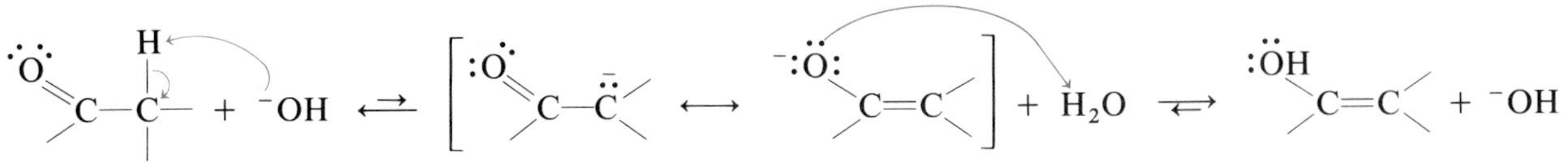

keto form

enolate ion

enol form
(vinyl alcohol)

In this way, a base catalyzes an equilibrium between the isomeric keto and enol forms of a carbonyl compound. For simple ketones and aldehydes, the keto form predominates. Therefore, a vinyl alcohol (an enol) is best described as an alternative isomeric form of a ketone or aldehyde.

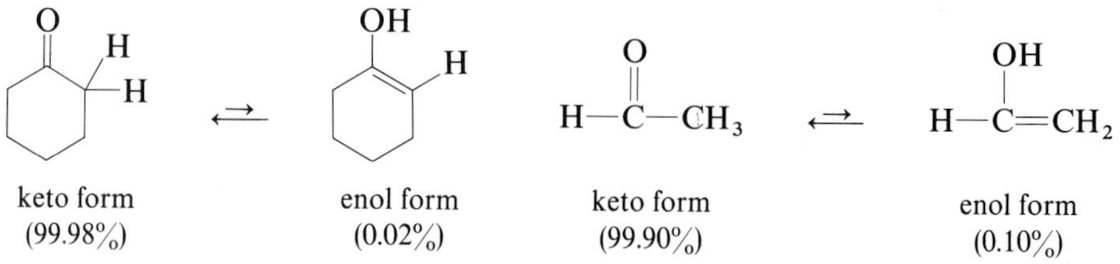

keto form (99.98%) ⇌ enol form (0.02%)

$H{-}C(=O){-}CH_3$ ⇌ $H{-}C(OH){=}CH_2$

keto form (99.90%) — enol form (0.10%)

This type of isomerization, occurring by the migration of a proton and the movement of a double bond, is called **tautomerism,** and the two isomers that interconvert are called **tautomers.** Tautomers are true isomers, and (under the right circumstances, with no catalyst present) either of the individual tautomeric forms may be isolated.

The keto-enol tautomerism is also catalyzed by acid. In acid, a proton is moved from the α-carbon to oxygen by first protonating the oxygen and then removing a proton from carbon.

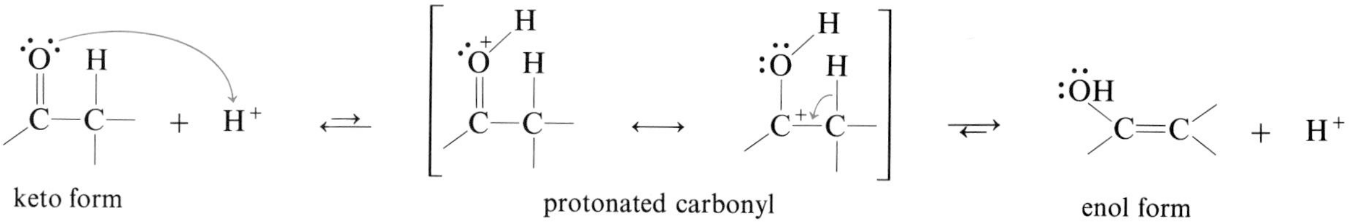

keto form + H^+ ⇌ [protonated carbonyl] ⇌ enol form + H^+

Compare the base-catalyzed and acid-catalyzed mechanisms shown above for the keto-enol tautomerism. In base, the proton is removed from carbon and then replaced on oxygen. In acid, oxygen is protonated first, and then carbon is deprotonated. Most proton-transfer mechanisms work this way. In base, the proton is removed from the old location and then replaced at the new location. In acid, protonation occurs at the new location, followed by deprotonation at the old location.

In addition to its importance in reaction mechanisms, the keto-enol tautomerism has implications for the stereochemistry of ketones and aldehydes. A hydrogen atom on an α-carbon (adjacent to a carbonyl group) may be lost and regained through keto-enol tautomerism; such a hydrogen is said to be an **enolizable hydrogen.** If a chiral carbon atom has an enolizable hydrogen atom, a trace of acid or base allows that carbon to invert its configuration, with the enol serving as the intermediate. A racemic mixture (or an equilibrium mixture of diastereomers) is the result.

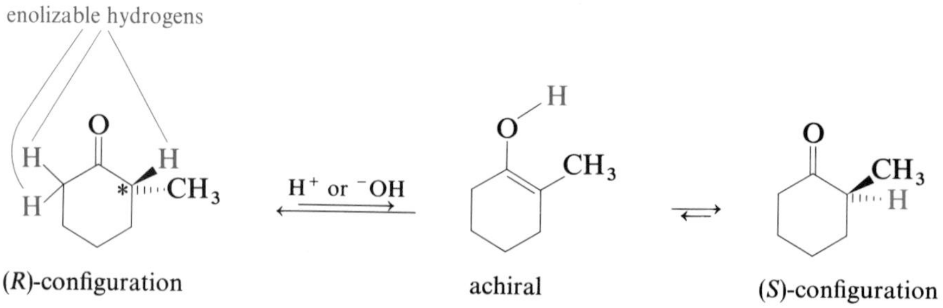

enolizable hydrogens

(*R*)-configuration $\xrightarrow{H^+ \text{ or } ^-OH}$ achiral ⇌ (*S*)-configuration

PROBLEM 22-1

Phenylacetone can form two different enols.

(a) Show the structures of these two enols.
(b) Predict which enol will be present in the larger concentration at equilibrium.
(c) Give mechanisms for the formation of the two enols under acidic and basic conditions.

PROBLEM 22-2

Show each step in the mechanism of the base-catalyzed interconversion of (*R*)- and (*S*)-2-methylcyclohexanone.

PROBLEM 22-3

When *cis*-2,4-dimethylcyclohexanone is dissolved in aqueous ethanol containing a trace of NaOH, a mixture of cis and trans isomers results. Give a mechanism for this isomerization.

22-1B FORMATION AND STABILITY OF ENOLATE IONS

The presence of a carbonyl group increases the acidity of the protons on the α-carbon atom. The pK_a for removal of an α proton from a typical ketone or aldehyde is about 20, showing that a typical ketone or aldehyde is much less acidic than water ($pK_a = 15.7$) or an alcohol ($pK_a = 16$ to 19). When a simple ketone or aldehyde is treated with hydroxide ion or an alkoxide ion, the equilibrium mixture contains only a small fraction of the deprotonated, enolate form.

$$\mathrm{R{-}C(=\ddot{O})}{-}\mathrm{CH_2}{-}\mathrm{R'} + {}^{-}\mathrm{OR} \rightleftharpoons \left[\mathrm{R{-}C(=\ddot{O}){-}\ddot{C}^{-}HR'} \longleftrightarrow \mathrm{R{-}C(-\ddot{O}:^{-}){=}CHR'} \right] + \mathrm{ROH}$$

ketone or aldehyde — enolate ion (minor ⟷ major)

Example

cyclohexanone + $:\ddot{O}^{-}{-}CH_2CH_3$ (ethoxide ion) $\rightleftharpoons$ [cyclohexanone enolate]

(equilibrium lies to the left)

Even though the equilibrium concentration of the enolate ion may be small, it serves as a useful, reactive nucleophile. When an enolate ion reacts with an electrophile (other than a proton), the enolate concentration decreases, and the equilibrium shifts to the right (Fig. 22-1). Eventually, all the carbonyl compound reacts via a low concentration of the enolate ion.

PROBLEM 22-4

Give the important resonance structures for the enolate ions of

(a) acetone (b) cyclopentanone (c) 2,4-pentanedione

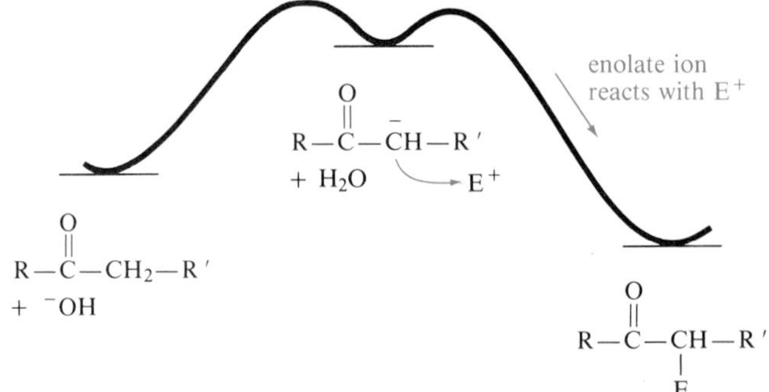

FIGURE 22-1 Reaction of the enolate ion with an electrophile removes it from equilibrium.

For some reactions, it is necessary that *all* the carbonyl compound be converted to its deprotonated enolate form. Although sodium hydroxide and alkoxides are not sufficiently basic, several powerful bases are available to convert a ketone completely to its enolate. The most effective and useful base for this purpose is lithium diisopropylamide (LDA), the lithium salt of diisopropylamine. LDA is made by using an alkyllithium reagent to deprotonate diisopropylamine.

$$[(CH_3)_2CH]_2\ddot{N}-H \;+\; C_4H_9Li \;\longrightarrow\; C_4H_{10} \;+\; [(CH_3)_2CH]_2\ddot{N}{:}^-\ Li^+$$

diisopropylamine — *n*-butyllithium — butane — lithium diisopropylamide (LDA)

Diisopropylamine has a pK_a of about 40, showing that it is much *less* acidic than a typical ketone or aldehyde. By virtue of its two isopropyl groups, LDA is a bulky reagent; it does not easily attack a carbon atom or add to a carbonyl group. Thus LDA is a powerful base, but not a strong nucleophile. When LDA reacts with a ketone, it abstracts the α proton to form the lithium salt of the enolate. We will soon see that these lithium enolate salts are very useful in synthesis.

$$R-\overset{O}{\overset{\|}{C}}-\overset{H}{\overset{|}{C}}- \;+\; (i\text{-}C_3H_7)_2N^-Li^+ \;\rightleftharpoons\; R-\overset{O^-\ {}^+Li}{\overset{|}{C}}=C\langle \;+\; (i\text{-}C_3H_7)_2N-H$$

ketone ($pK_a \cong 20$) — LDA — lithium salt of enolate — diisopropylamine ($pK_a = 40$)

(equilibrium lies to the right)

Example

$$\text{cyclohexanone} \;+\; (i\text{-}C_3H_7)_2N^-Li^+ \;\rightleftharpoons\; \text{lithium enolate} \;+\; (i\text{-}C_3H_7)_2N-H$$

cyclohexanone ($pK_a = 17$) — LDA — lithium enolate of cyclohexanone (100%) — ($pK_a = 40$)

22-2 α HALOGENATION OF KETONES

22-2A BASE-PROMOTED α HALOGENATION

When a ketone is treated with a halogen and base, an α-halogenation reaction occurs.

$$\underset{\text{ketone}}{-\overset{\overset{\displaystyle O}{\|}}{C}-\overset{\overset{\displaystyle H}{|}}{\underset{|}{C}}-} + {}^{-}OH + X_2 \longrightarrow -\overset{\overset{\displaystyle O}{\|}}{C}-\overset{\overset{\displaystyle X}{|}}{\underset{|}{C}}- + X^- + H_2O$$

$(X_2 = Cl_2, Br_2, \text{ or } I_2)$

Example

$$\text{cyclohexanone} \xrightarrow[{}^{-}OH,\ H_2O]{Cl_2} \text{2-chlorocyclohexanone}$$

cyclohexanone 2-chlorocyclohexanone

This reaction is called *base promoted,* rather than base catalyzed, because a full equivalent of the base is consumed in the reaction. The base-promoted halogenation takes place by a nucleophilic attack of an enolate ion on the electrophilic halogen molecule. The products are the halogenated ketone and a halide ion.

$$-\overset{\overset{\displaystyle O}{\|}}{C}-\overset{\overset{\displaystyle H}{|}}{\underset{|}{C}}- + {}^{-}OH \rightleftharpoons -\overset{\overset{\displaystyle O}{\|}}{C}-\underset{|}{\overset{\cdot\cdot -}{C}}- + H_2O \xrightarrow{X-X} -\overset{\overset{\displaystyle O}{\|}}{C}-\overset{\overset{\displaystyle X}{|}}{\underset{|}{C}}- + X^-$$

SOLVED PROBLEM 22-1

Propose a mechanism for the reaction of 3-pentanone with sodium hydroxide and bromine to give 2-bromo-3-pentanone.

SOLUTION In the presence of sodium hydroxide, a small amount of 3-pentanone is present as its enolate.

$$CH_3CH_2-\overset{\overset{\displaystyle O}{\|}}{C}-\overset{\overset{\displaystyle H}{|}}{C}H-CH_3 + {}^{-}OH \rightleftharpoons$$

$$\left[CH_3CH_2-\overset{\overset{\displaystyle O}{\|}}{C}-\overset{\cdot\cdot -}{C}H-CH_3 \longleftrightarrow CH_3CH_2-\overset{\overset{\displaystyle O^-}{|}}{C}=CHCH_3\right]$$

enolate

The enolate reacts with bromine to give the observed product.

$$\underset{\text{enolate}}{(CH_3CH_2)(\ddot{O}{:}^-)C=C(CH_3)(H)} + Br-Br \longrightarrow \underset{\alpha\text{-haloketone}}{CH_3CH_2-\overset{\overset{\displaystyle O}{\|}}{C}-\overset{\overset{\displaystyle CH_3}{|}}{\underset{\underset{\displaystyle H}{|}}{C}}-Br} + Br^-$$

PROBLEM 22-5

Give a detailed mechanism for the example showing the formation of 2-chlorocyclohexanone above.

Multiple halogenation In many cases, base-promoted halogenation does not stop with the replacement of just one hydrogen by the halogen. The product (the α-haloketone) is more reactive toward further halogenation than is the starting material (the original ketone). The presence of the electron-withdrawing halogen stabilizes the enolate ion, enhancing its concentration.

$$-\overset{\overset{\displaystyle O}{\|}}{C}-\overset{\overset{\displaystyle X}{|}}{\underset{\underset{\displaystyle H}{|}}{C}}- \;+\; {}^{-}OH \;\rightleftharpoons\; H_2O \;+\; \left[\overset{O}{\diagdown}C-\ddot{C}^{-}\overset{X}{\diagup} \;\longleftrightarrow\; \overset{O^{-}}{\diagdown}C=C\overset{X}{\diagup} \right]$$

(enolate stabilized by X)

For example, bromination of 3-pentanone gives mostly 2,2-dibromo-3-pentanone. After one hydrogen is replaced by bromine, the enolate ion is stabilized by both the carbonyl group and the bromine atom. A second bromination takes place faster than the first. Notice that the second substitution takes place at the same carbon atom as the first substitution, because that carbon atom is the one that bears the enolate-stabilizing halogen.

$$CH_3CH_2-\overset{\overset{\displaystyle O}{\|}}{C}-\overset{\overset{\displaystyle H}{|}}{\underset{\underset{\displaystyle Br}{|}}{C}}-CH_3 \;\xrightleftharpoons{{}^{-}OH}\; CH_3CH_2-\overset{\overset{\displaystyle O}{\|}}{C}-\underset{\underset{\displaystyle Br}{|}}{\ddot{C}^{-}}-CH_3 \;\xrightarrow{Br-Br}\; CH_3CH_2-\overset{\overset{\displaystyle O}{\|}}{C}-\overset{\overset{\displaystyle Br}{|}}{\underset{\underset{\displaystyle Br}{|}}{C}}-CH_3$$

monobrominated ketone — stabilized by Br — second bromination

Because of this tendency for multiple halogenation, base-promoted halogenation is rarely used for the preparation of monohalo ketones. The acid-catalyzed procedure (discussed in Section 22-2C) is preferred.

PROBLEM 22-6

Give a mechanism to show how acetophenone undergoes base-promoted chlorination to give trichloroacetophenone.

22-2B THE HALOFORM REACTION

With most ketones, base-promoted halogenation continues until the α carbon atom is completely halogenated. Methyl ketones have three α protons on the methyl carbon, and they undergo halogenation three times to give trihalomethyl ketones.

$$R-\overset{\overset{\displaystyle O}{\|}}{C}-CH_3 \;+\; 3\,X_2 \;+\; 3\,{}^{-}OH \;\longrightarrow\;\longrightarrow\;\longrightarrow\; R-\overset{\overset{\displaystyle O}{\|}}{C}-CX_3 \;+\; 3\,X^{-} \;+\; 3\,H_2O$$

methyl ketone — trihalomethyl ketone

With three electron-withdrawing halogen atoms, the trihalomethyl group can form a relatively stable carbanion and serve as a leaving group for nucleophilic acyl substitution. The trihalomethyl ketone reacts with hydroxide ion to give a carboxylic acid. A fast proton exchange gives a carboxylate ion and a molecule of a haloform (chloroform, $CHCl_3$; bromoform, $CHBr_3$; or iodoform, CHI_3). The overall reaction is called the **haloform reaction,** because a haloform is one of its final products.

$$R{-}C({=}\ddot{O}){-}CX_3 + {}^{-}{:}\ddot{O}H \rightleftharpoons R{-}C({:}\ddot{O}{:}^{-})(CX_3)({:}\ddot{O}H) \rightleftharpoons R{-}C({=}\ddot{O}{:}){-}\ddot{O}{-}H + {}^{-}{:}CX_3 \longrightarrow R{-}C({=}\ddot{O}{:}){-}\ddot{O}{:}^{-} + HCX_3$$

nucleophilic acyl substitution — a carboxylate ion — a haloform

The overall haloform reaction is summarized below. A methyl ketone reacts with a halogen under strongly basic conditions to give a carboxylate ion and a haloform.

$$\underset{\text{a methyl ketone}}{R{-}C({=}O){-}CH_3} \xrightarrow[(X_2\,=\,Cl_2,\,Br_2,\,\text{or}\,I_2)]{\text{excess }X_2,\ {}^{-}OH} \underset{\substack{\text{a trihalomethyl ketone}\\ \text{(not isolated)}}}{\left[R{-}C({=}O){-}CX_3\right]} \xrightarrow{{}^{-}OH} \underset{\text{a carboxylate}}{R{-}C({=}O){-}O^{-}} + \underset{\text{a haloform}}{HCX_3}$$

Example

$$\underset{\text{2-butanone}}{CH_3CH_2{-}C({=}O){-}CH_3} \xrightarrow[{}^{-}OH]{\text{excess }Br_2} CH_3CH_2{-}C({=}O){-}CBr_3 \xrightarrow{{}^{-}OH} \underset{\text{propionate}}{CH_3CH_2{-}C({=}O){-}O^{-}} + \underset{\text{bromoform}}{HCBr_3}$$

When iodine is used as the halogen, the haloform product (iodoform) separates from the solution as a yellow precipitate. This **iodoform test** is used to identify methyl ketones, which undergo halogenation three times and then lose the $-CI_3$ group to give iodoform.

$$\underset{\text{acetophenone}}{Ph{-}C({=}O){-}CH_3} \xrightarrow[{}^{-}OH]{\text{excess }I_2} \underset{\alpha,\alpha,\alpha\text{-triiodoacetophenone}}{Ph{-}C({=}O){-}CI_3} \xrightarrow{{}^{-}OH} \underset{\text{benzoate}}{Ph{-}C({=}O){-}O^{-}} + \underset{\text{iodoform}}{HCI_3\downarrow}$$

Iodine is an oxidizing agent, and an alcohol can give a positive iodoform test if it is oxidized to a methyl ketone. The iodoform reaction can be used to convert such an alcohol to a carboxylic acid with one less carbon atom.

$$R{-}CH(OH){-}CH_3 + I_2 \longrightarrow R{-}C({=}O){-}CH_3 + 2\,HI \xrightarrow[{}^{-}OH]{\text{excess }I_2} R{-}C({=}O){-}O^{-} + HCI_3\downarrow$$

Example

$$\underset{\text{2-hexanol}}{CH_3{-}CH_2{-}CH_2{-}CH_2{-}CH(OH){-}CH_3} \xrightarrow[(2)\ H^{+}]{(1)\ I_2,\ {}^{-}OH} \underset{\text{pentanoic acid}}{CH_3{-}CH_2{-}CH_2{-}CH_2{-}C({=}O){-}OH} + HCI_3\downarrow$$

PROBLEM 22-7

Give a mechanism for the reaction of methyl cyclohexyl ketone with excess bromine in the presence of sodium hydroxide.

PROBLEM 22-8

Predict the products of the following reactions.

(a) methyl cyclopentyl ketone + excess Cl_2 + excess NaOH
(b) 1-cyclopentylethanol + excess I_2 + excess NaOH
(c) cyclohexanone + excess I_2 + excess NaOH
(d) propiophenone + excess Br_2 + excess NaOH

PROBLEM 22-9

Which of the following compounds will give a positive iodoform test?

(a) 1-phenylethanol (b) 2-pentanone (c) 2-pentanol
(d) 3-pentanone (e) acetone (f) isopropyl alcohol

22-2C ACID-CATALYZED α HALOGENATION

Ketones also undergo α halogenation under acidic catalysis. One of the most effective procedures is to dissolve the ketone in acetic acid, which serves as both the solvent and the acid catalyst. In contrast with basic halogenation, acidic halogenation can replace just one hydrogen or more than one by using appropriate amounts of the halogen.

$$\underset{\text{acetophenone}}{C_6H_5-\overset{O}{\overset{\|}{C}}-CH_3} + Br_2 \xrightarrow{CH_3COOH} \underset{\substack{\alpha\text{-bromoacetophenone}\\(70\%)}}{C_6H_5-\overset{O}{\overset{\|}{C}}-CH_2Br} + HBr$$

$$\underset{\text{acetophenone}}{C_6H_5-\overset{O}{\overset{\|}{C}}-CH_3} + 2\,Cl_2 \xrightarrow{CH_3COOH} \underset{\alpha,\alpha\text{-dichloroacetophenone}}{C_6H_5-\overset{O}{\overset{\|}{C}}-CHCl_2} + 2\,HCl$$

The mechanism of acid-catalyzed halogenation involves an attack of the enol form of the ketone on the electrophilic halogen molecule. Loss of a proton gives the α-haloketone and the hydrogen halide.

$$\underset{\text{enol}}{H-\ddot{O}:-C{=}C} \quad \underset{\text{halogen}}{X-X} \longrightarrow \underset{\text{carbocation intermediate}}{\left[H-\ddot{O}:-\overset{+}{C}-\overset{X}{\overset{|}{C}}- \longleftrightarrow H-\overset{+}{\ddot{O}}{=}C-\overset{X}{\overset{|}{C}}- \right] X^-} \longrightarrow \underset{\alpha\text{-haloketone}}{:\ddot{O}{=}C-\overset{X}{\overset{|}{C}}-} + H-X$$

This reaction is similar to the attack of an alkene on a halogen, resulting in addition of the halogen across the double bond. The pi bond of an enol is more reactive toward halogens, however, because the carbocation that results is stabilized by resonance with the enol —OH group. Loss of the enol proton converts the intermediate to the product, the α-haloketone.

Unlike ketones, aldehydes are easily oxidized, and halogens are strong oxidizing agents. Attempted α halogenation of aldehydes using halogens usually results in oxidation to carboxylic acids.

$$\underset{\text{aldehyde}}{R{-}\overset{\overset{\large O}{\|}}{C}{-}H} + X_2 + H_2O \longrightarrow \underset{\text{acid}}{R{-}\overset{\overset{\large O}{\|}}{C}{-}OH} + 2\,H{-}X$$

SOLVED PROBLEM 22-2

Propose a mechanism for the acid-catalyzed conversion of cyclohexanone to 2-chlorocyclohexanone.

$$\underset{\text{cyclohexanone}}{\text{(cyclohexanone)}} \xrightarrow[CH_3COOH]{Cl_2} \underset{\text{2-chlorocyclohexanone (65\%)}}{\text{(2-chlorocyclohexanone)}}$$

SOLUTION Under acid catalysis, the ketone is in equilibrium with its enol form.

keto form + $H{-}\ddot{O}{-}\overset{O}{\overset{\|}{C}}{-}CH_3$ ⇌ [stabilized intermediate ⟷] + $^{-}\!:\ddot{O}{-}\overset{O}{\overset{\|}{C}}{-}CH_3$ ⇌ enol form

keto form · stabilized intermediate · enol form

The enol acts as a weak nucleophile, attacking chlorine to give a resonance-stabilized intermediate. Loss of a proton gives the product.

enol + $Cl{-}Cl$ ⟶ [intermediate ⟷] $\xrightarrow{Cl^-}$ 2-chlorocyclohexanone + $H{-}Cl$

PROBLEM 22-10

Propose a mechanism for the acid-catalyzed bromination of 3-pentanone.

22-3 α BROMINATION OF ACIDS (THE HVZ REACTION)

The **Hell-Volhard-Zelinsky (HVZ) reaction** replaces a hydrogen atom with a bromine atom on the α-carbon of a carboxylic acid. The carboxylic acid is treated with bromine and phosphorus tribromide, followed by water to hydrolyze the intermediate α-bromo acyl bromide.

The HVZ reaction

$$R{-}CH_2{-}\overset{\overset{\large O}{\|}}{C}{-}OH \xrightarrow{Br_2/PBr_3} \underset{\alpha\text{-bromo acyl bromide}}{R{-}\overset{\overset{\large Br}{|}}{CH}{-}\overset{\overset{\large O}{\|}}{C}{-}Br} \xrightarrow{H_2O} \underset{\alpha\text{-bromoacid}}{R{-}\overset{\overset{\large Br}{|}}{CH}{-}\overset{\overset{\large O}{\|}}{C}{-}OH} + HBr$$

Example

$$\underset{\text{butanoic acid}}{CH_3CH_2CH_2{-}\overset{\overset{\large O}{\|}}{C}{-}OH} \xrightarrow{Br_2/PBr_3} \underset{\text{2-bromobutanoyl bromide}}{CH_3CH_2\overset{\overset{\large Br}{|}}{CH}{-}\overset{\overset{\large O}{\|}}{C}{-}Br} \xrightarrow{H_2O} \underset{\text{2-bromobutanoic acid}}{CH_3CH_2\overset{\overset{\large Br}{|}}{CH}{-}\overset{\overset{\large O}{\|}}{C}{-}OH} + HBr$$

The mechanism is particularly interesting because the enol form of the acyl bromide serves as a nucleophilic intermediate. The first step is formation of acyl bromide, which enolizes more easily than does the acid.

$$\underset{\text{acid}}{R-\overset{H}{\underset{H}{C}}-C(=O)OH} \xrightarrow{Br_2/PBr_3} \underset{\text{keto form acyl bromide}}{R-\overset{H}{\underset{H}{C}}-C(=O)Br} \rightleftarrows \underset{\text{enol form}}{(R)(H)C=C(OH)(Br)}$$

The enol is moderately nucleophilic, attacking bromine to give the α-brominated acyl bromide.

$$Br-Br + \underset{\text{enol}}{(R)(H)C=C(\ddot{O}-H)(Br)} \rightleftarrows Br^- \; Br-\overset{R}{\underset{H}{C}}-C(=\overset{+}{O}-H)Br \longrightarrow \underset{\alpha\text{-bromo acyl bromide}}{Br-\overset{R}{\underset{H}{C}}-C(=O)Br} + HBr$$

If a derivative of the α-bromoacid is desired, the α-bromo acyl bromide serves as an activated intermediate (similar to an acid chloride) for the synthesis of an ester, amide, or other derivative. If the α-bromoacid itself is needed, a water hydrolysis completes the synthesis.

PROBLEM 22-11

Show the products of the reactions of these carboxylic acids with PBr_3/Br_2 followed by hydrolysis.

(a) propanoic acid (b) benzoic acid (c) succinic acid (d) oxalic acid

22-4 ALKYLATION OF ENOLATE IONS

We have seen many reactions where nucleophiles attack unhindered alkyl halides and tosylates by the S_N2 mechanism. An enolate ion can serve as the nucleophile in such a reaction, becoming alkylated in the process. Because the enolate has two sites of nucleophilicity (the oxygen and the α-carbon), it can react at either of these sites. The reaction usually takes place primarily at the α-carbon, however, forming a new C—C bond.

$$\left[-\overset{O}{\overset{\|}{C}}-\overset{-}{C}< \;\longleftrightarrow\; -\overset{O^-}{\overset{|}{C}}=C<\right] + R-CH_2-X \longrightarrow \underset{\substack{\text{C-alkylation product}\\ \text{(more common)}}}{-\overset{O}{\overset{\|}{C}}-C{<}CH_2-R} + X^-$$

$$\longrightarrow \underset{\substack{\text{O-alkylation product}\\ \text{(less common)}}}{-\overset{O-CH_2-R}{\overset{|}{C}}=C<} + X^-$$

Typical bases such as sodium hydroxide or an alkoxide ion cannot be used to form the enolate for alkylation, because at equilibrium there is still a large quantity of the hydroxide or alkoxide base present. These strongly nucleophilic bases give side reactions with the primary alkyl halide or tosylate. The use of lithium diisopropylamide (LDA) avoids these side reactions, however. LDA converts the ketone entirely to its enolate, which reacts without interference from a large excess of another nucleophile.

$$\underset{\text{enolizable ketone}}{R-\overset{O}{\overset{\|}{C}}-\overset{R'}{\overset{|}{C}H}-R'} + \underset{\text{LDA}}{(i\text{-Pr})_2N^- \ {}^+Li} \longrightarrow \underset{\text{enolate}}{R-\overset{O}{\overset{\|}{C}}-\overset{R'}{\overset{|}{\ddot{C}}}{}^- -R' \ {}^+Li} + \underset{\text{diisopropylamine}}{(i\text{-Pr})_2N-H}$$

$$\underset{{}^+Li}{R-\overset{O}{\overset{\|}{C}}-\overset{R'}{\overset{|}{\ddot{C}}}{}^- -R'} + \underset{\text{unhindered halide}}{R''-CH_2-X} \longrightarrow \underset{\text{substituted ketone}}{R-\overset{O}{\overset{\|}{C}}-\underset{CH_2-R''}{\underset{|}{\overset{R'}{\overset{|}{C}}}}-R'} + LiX$$

Example

$$Ph-\overset{O}{\overset{\|}{C}}-\overset{CH_3}{\overset{|}{C}H}-CH_3 \xrightarrow[(2)\ Ph-CH_2-Br]{(1)\ LDA} Ph-\overset{O}{\overset{\|}{C}}-\underset{CH_2-Ph}{\underset{|}{\overset{CH_3}{\overset{|}{C}}}}-CH_3$$

The direct alkylation of enolates using LDA gives the best yields when there is only one α hydrogen that can be replaced by an alkyl group. If there are two different kinds of α protons that may be abstracted to give enolates, mixtures of products alkylated at the different α-carbons may result. Aldehydes are not suitable for this direct alkylation because they are prone to undergo side reactions when treated with LDA.

PROBLEM 22-12

Predict the major products of the following reactions.

(a) [phenyl cyclohexyl ketone] $\xrightarrow[(2)\ CH_3CH_2I]{(1)\ LDA}$

(b) [tricyclic ketone] $\xrightarrow[(2)\ CH_3I]{(1)\ LDA}$

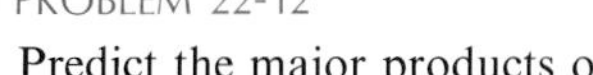

22-5 FORMATION AND ALKYLATION OF ENAMINES

A milder alternative to direct alkylation of enolate ions is the formation of and alkylation of an enamine derivative. An **enamine** (a **vinyl amine**) is the nitrogen analog of an enol. An enamine is a stronger nucleophile than an enol, but still quite selective in its alkylation reactions.

$$\underset{\text{an enamine}}{R_2\ddot{N}-C=C} + \underset{\text{electrophile}}{E^+} \rightleftharpoons \underset{\text{resonance-stabilized cation}}{\left[R_2\ddot{N}-\overset{+}{C}-C-E \longleftrightarrow R_2N^+=C-C-E\right]}$$

An enamine results from the reaction of a ketone or aldehyde with a secondary amine. Recall that the reaction of a ketone or aldehyde with a *primary* amine (Section 18-16) first forms a carbinolamine, which dehydrates to give the C=N double bond of an imine. But a carbinolamine from a *secondary* amine does not form a C=N double bond because there is no proton on nitrogen to eliminate. A proton is lost from the α carbon, forming the C=C double bond of an enamine.

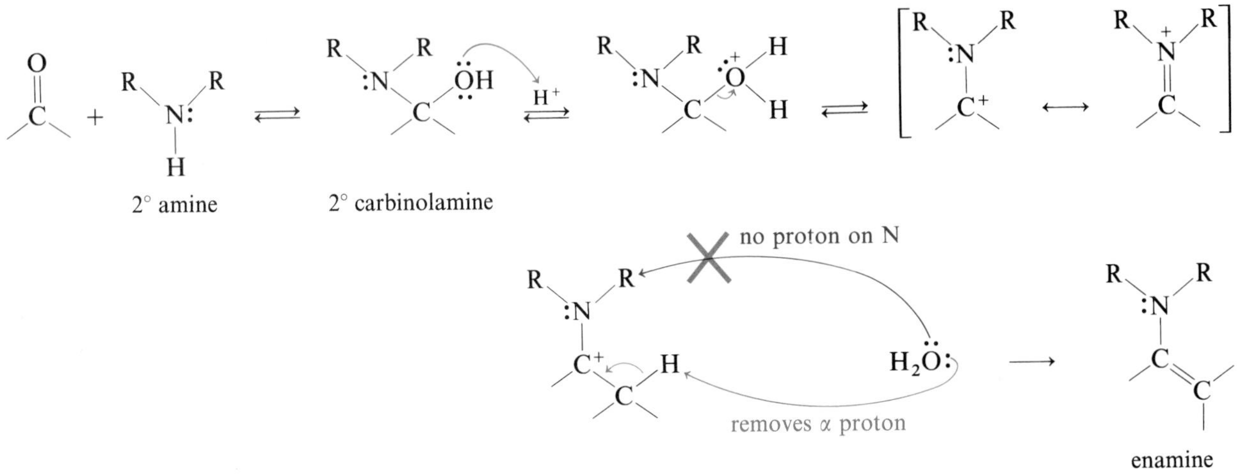

2° amine

2° carbinolamine

H^+

no proton on N

removes α proton

enamine

Example

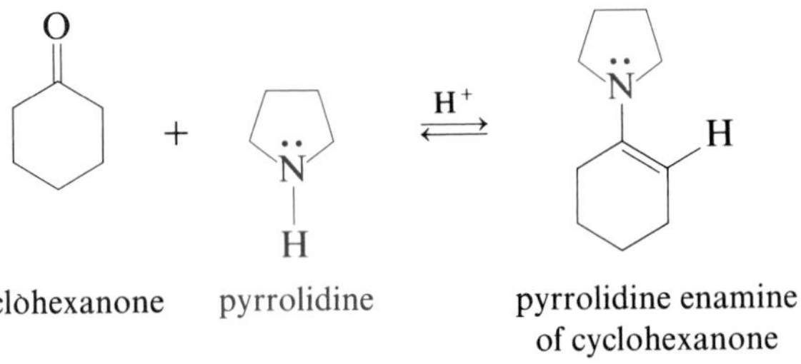

H^+

cyclohexanone pyrrolidine pyrrolidine enamine of cyclohexanone

PROBLEM 22-13
Give a mechanism for the acid-catalyzed reaction of cyclohexanone with pyrrolidine.

Enamines rapidly displace halides from reactive alkyl halides, giving alkylated iminium ions. The iminium ions are unreactive toward further alkylation or acylation. The following example shows the reaction of benzyl bromide with the pyrrolidine enamine of cyclohexanone.

CH_2—Br, Ph

CH_2Ph

H_3O^+

benzyl bromide alkylated iminium salt alkylated ketone

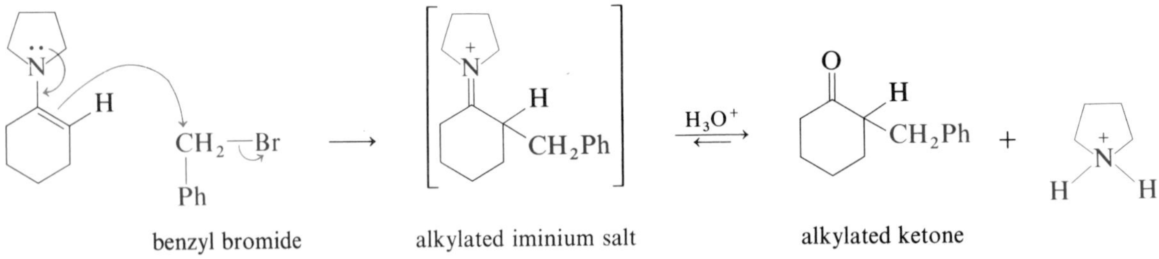

The alkylated iminium salt is hydrolyzed to give the alkylated ketone. The mechanism of this hydrolysis is the same as the mechanism for the acid-catalyzed hydrolysis of an imine (Section 18-16).

PROBLEM 22-14

Give a mechanism for the hydrolysis of this iminium salt to the alkylated ketone. The first step is attack by water, followed by loss of a proton to give an amino alcohol. Protonation of the nitrogen atom allows pyrrolidine to leave, giving the protonated ketone.

Overall reaction:

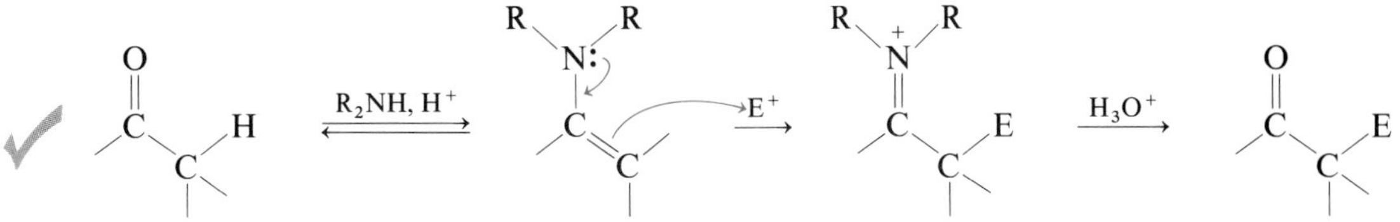

R_2NH, H^+ ⇌ enamine → E^+ → iminium salt → H_3O^+ →

enamine   iminium salt

The enamine alkylation procedure is sometimes called the **Stork reaction** after its inventor, Gilbert Stork of Columbia University. The Stork reaction is often the best method for alkylating or acylating ketones, using a variety of reactive alkyl and acyl halides. Some of the kinds of halides that react well with enamines to give alkylated and acylated ketone derivatives are the following.

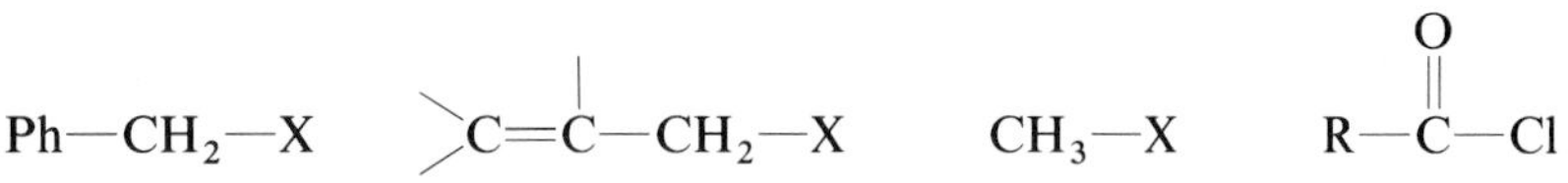

Ph—CH_2—X   >C=C—CH_2—X CH_3—X   R—C(=O)—Cl

benzyl halides   allylic halides   methyl halides   acyl halides

The following sequence shows how the acylation of an enamine is used to synthesize a β-diketone. The initial acylation gives an acyl iminium salt, which hydrolyzes to the β-diketone product. As we will see in Section 22-16, β-dicarbonyl compounds such as β-diketones are easily alkylated, and they serve as useful intermediates in the synthesis of more complicated molecules.

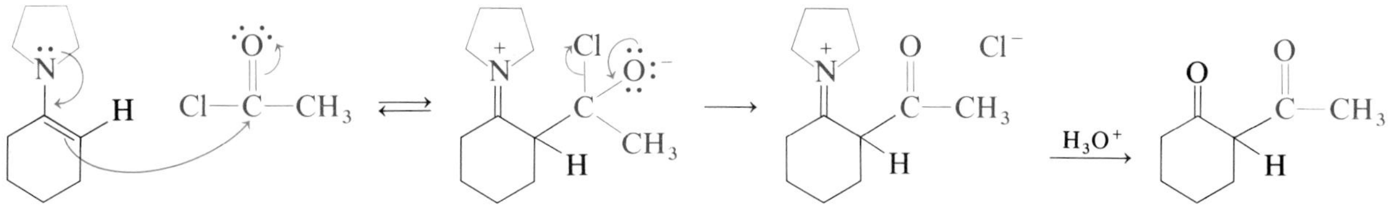

enamine   acyl chloride   intermediate   acyl iminium salt H_3O^+ β-diketone

PROBLEM 22-15

Give the expected product for each of the following acid-catalyzed reactions.

(a) acetophenone + methylamine   (b) acetophenone + dimethylamine
(c) cyclohexanone + aniline   (d) cyclohexanone + piperidine

PROBLEM 22-16

Show how you would accomplish each of the following conversions using an enamine synthesis.

(a) cyclopentanone ⟶ 2-allylcyclopentanone
(b) 3-pentanone — 2-methyl-1-phenyl-3-pentanone
(c) acetophenone ⟶ Ph—C(=O)—CH_2—C(=O)—Ph

22-6 THE ALDOL CONDENSATION OF KETONES AND ALDEHYDES

Some of the most important enolate reactions of carbonyl compounds are **condensations.** Condensations bond together two or more molecules, often with the loss of a small molecule such as water or an alcohol. The **aldol condensation** involves the nucleophilic addition of an enolate ion to another carbonyl group. The product, a β-hydroxy ketone or aldehyde, is called an **aldol.** Under the proper conditions, the aldol product may dehydrate to an α,β-unsaturated carbonyl compound.

The Aldol condensation

$$\underset{\text{ketone or aldehyde}}{2\ R{-}\overset{O}{\overset{\|}{C}}{-}CH_2{-}R'} \underset{}{\overset{H^+ \text{ or } ^-OH}{\rightleftharpoons}} \underset{\text{aldol product}}{R{-}\overset{OH}{\overset{|}{C}}(CH_2{-}R'){-}\underset{R'}{CH}{-}\underset{\|\atop O}{C}{-}R} \underset{H^+ \text{ or } ^-OH}{\overset{\text{heat}}{\rightleftharpoons}} \underset{\alpha,\beta\text{-unsaturated ketone or aldehyde}}{R{-}C(CH_2{-}R'){=}C(R'){-}\underset{\|\atop O}{C}{-}R} + H_2O$$

22-6A BASE-CATALYZED ALDOL CONDENSATIONS

Under basic conditions, the aldol condensation occurs by a nucleophilic addition of the enolate ion (a strong nucleophile) to a carbonyl group. Protonation gives the aldol product.

$$\text{enolate} + \text{C=O} \rightleftharpoons \text{alkoxide adduct} \underset{}{\overset{\text{solvent}{-}O{-}H}{\rightleftharpoons}} \text{aldol (}{-}O{-}H\text{)}$$

Consider the aldol condensation of acetaldehyde, shown below. Deprotonation of acetaldehyde gives an enolate ion, which acts as a strong nucleophile. Attack by the enolate ion on the carbonyl group of another acetaldehyde molecule gives addition across the carbonyl double bond. The product, a β-hydroxy aldehyde, is called an *aldol* because it contains both an *ald*ehyde group and the hydroxy group of an alco*hol*.

Step 1: Formation of the enolate ion

$$\underset{\text{acetaldehyde}}{H{-}C({=}O){-}CH_2{-}H} + \underset{\text{base}}{^-{:}\ddot{O}H} \rightleftharpoons \underset{\text{enolate of acetaldehyde}}{\left[H{-}C({=}O){-}\ddot{C}H_2^- \longleftrightarrow H{-}C(O^-){=}CH_2 \right]} + H_2O$$

Step 2: Nucleophilic attack at the carbonyl

$$\underset{\text{enolate}}{H{-}C({=}O){-}\ddot{C}H_2^-} + \underset{\text{acetaldehyde}}{H{-}C({=}O){-}CH_3} \rightleftharpoons H{-}\underset{\|\atop O}{C}{-}\underset{H}{\overset{H}{C}}{-}\overset{:\ddot{O}:^-}{\overset{|}{C}}H{-}CH_3 \overset{H{-}O{-}H}{\rightleftharpoons} \underset{\text{aldol product (50\%)}}{H{-}\underset{\|\atop O}{C}{-}CH_2{-}\overset{OH}{\overset{|}{C}}H{-}CH_3} + {^-OH}$$

The aldol condensation is reversible, establishing an equilibrium between reactants and products. For acetaldehyde, the conversion to the aldol product is about 50 percent. Although ketones undergo aldol condensation, the equilibrium concentrations of the products are generally small. Aldol condensations of ketones are sometimes accomplished by clever experimental methods. For example, Figure 22-2 shows how a good yield of the acetone aldol product ("diacetone alcohol") is obtained, even though the equilibrium concentration of the product is only about 1 percent. Acetone is boiled so it condenses into a chamber containing a basic catalyst. When the solution returns to the boiling flask, it contains about 1 percent diacetone alcohol. Diacetone alcohol is less volatile than acetone, remaining in the boiling flask while acetone boils and condenses (refluxes) in contact with the catalyst. After several hours, nearly all the acetone is converted to diacetone alcohol.

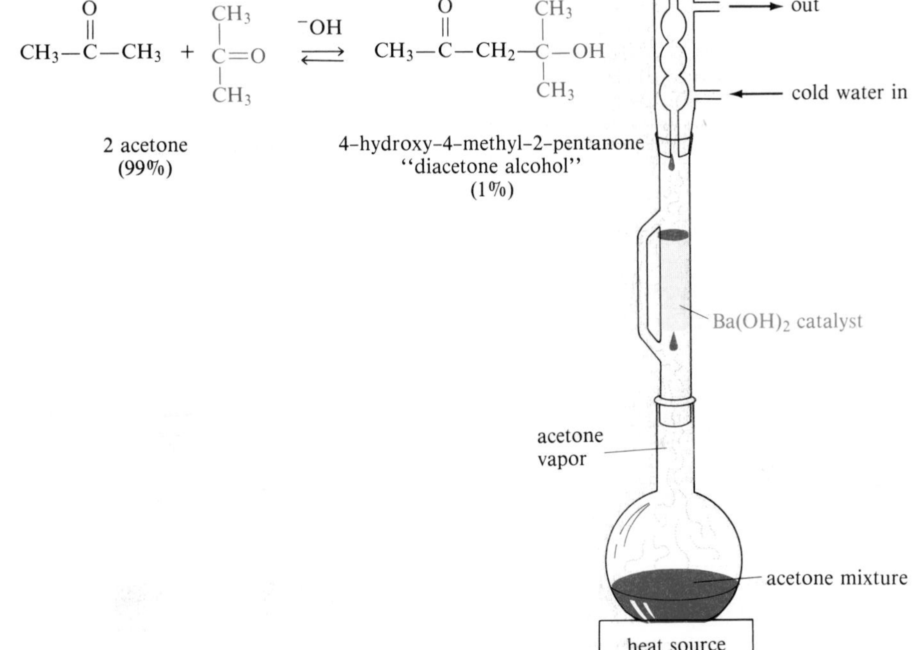

FIGURE 22-2 Although the aldol condensation of acetone gives only 1 percent of the product at equilibrium, a clever technique gives a good yield. Acetone refluxes onto a basic catalyst such as $Ba(OH)_2$. The nonvolatile diacetone alcohol does not reflux, so its equilibrium concentration gradually increases until all the acetone is converted to diacetone alcohol.

SOLVED PROBLEM 22-3

Propose a mechanism for the base-catalyzed aldol condensation of acetone (Fig. 22-2).

SOLUTION The first step is formation of the enolate of acetone to serve as a nucleophile.

$$CH_3-\overset{\overset{\ddot{O}:}{\|}}{C}-\underset{H}{\overset{H}{C}}-H + {}^-:\ddot{O}H \rightleftharpoons \left[\begin{matrix} O \\ H_3C \end{matrix} C-\ddot{C}^{-}\begin{matrix} H \\ H \end{matrix} \longleftrightarrow \begin{matrix} :\ddot{O}:^- \\ H_3C \end{matrix} C=C \begin{matrix} H \\ H \end{matrix} \right] + H_2O$$

acetone — enolate ion

The second step is a nucleophilic attack by the enolate on another molecule of acetone. Protonation gives the aldol product.

PROBLEM 22-17

Propose a mechanism for the aldol condensation of cyclohexanone.

PROBLEM 22-18

Give the expected products for the aldol condensations of

(a) propanal (b) phenylacetaldehyde.

PROBLEM 22-19

A student wanted to dry some diacetone alcohol, and allowed it to stand over anhydrous potassium carbonate for a week. At the end of the week, the sample was found to contain nearly pure acetone. Propose a mechanism for the reaction that took place.

22-6B ACID-CATALYZED ALDOL CONDENSATIONS

Aldol condensations also take place under acidic conditions; the enol serves as a weak nucleophile to attack an activated (protonated) carbonyl group. As an example, consider the acid-catalyzed aldol condensation of acetaldehyde. The first step is formation of the enol by the acid-catalyzed keto-enol tautomerism as discussed above. The enol then attacks the protonated carbonyl of another acetaldehyde molecule. Loss of the enol proton gives the aldol product.

attack by enol

resonance-stabilized intermediate

lose H^+

aldol product

PROBLEM 22-20

Propose a mechanism for the acid-catalyzed aldol condensation of acetone.

22-7 DEHYDRATION OF ALDOL PRODUCTS

Upon heating a basic or acidic mixture containing an aldol product, dehydration of the alcohol function occurs. The product is an α,β-unsaturated aldehyde or ketone.

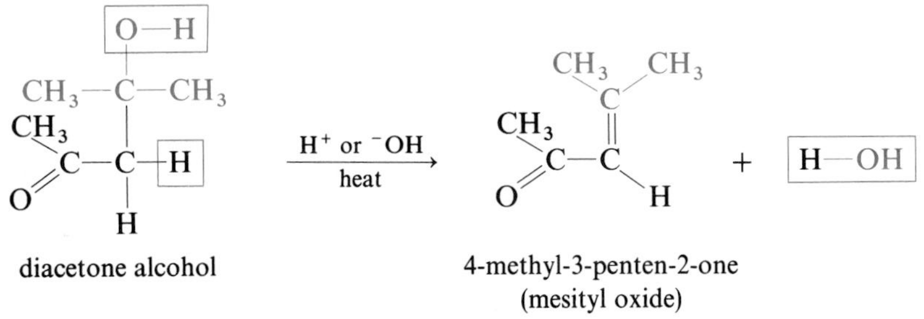

$$\underset{\text{diacetone alcohol}}{CH_3-\underset{\underset{H}{|}}{\overset{\overset{\boxed{O-H}}{|}}{C}}-CH_3 \ \ (CH_3)(O{=})C-\underset{\underset{H}{|}}{C}-\boxed{H}} \xrightarrow[\text{heat}]{H^+ \text{ or } ^-OH} \underset{\text{(mesityl oxide)}}{\underset{\text{4-methyl-3-penten-2-one}}{(CH_3)_2C{=}CH-C(=O)CH_3}} + \boxed{H-OH}$$

Under acidic conditions, this dehydration follows a mechanism similar to those of other acid-catalyzed alcohol dehydrations (Section 10-11). The base-catalyzed dehydration is something we have not previously seen. It depends on the acidity of the α proton (on the carbon alpha to the carbonyl) of the aldol product. Abstraction of an α proton gives an enolate that can expel hydroxide ion to give a more stable product. Hydroxide is not a good leaving group in an E2 elimination, but it can serve as a leaving group in a strongly exothermic step like this one, which stabilizes a negatively charged intermediate. The following mechanism shows the dehydration of 3-hydroxybutanal.

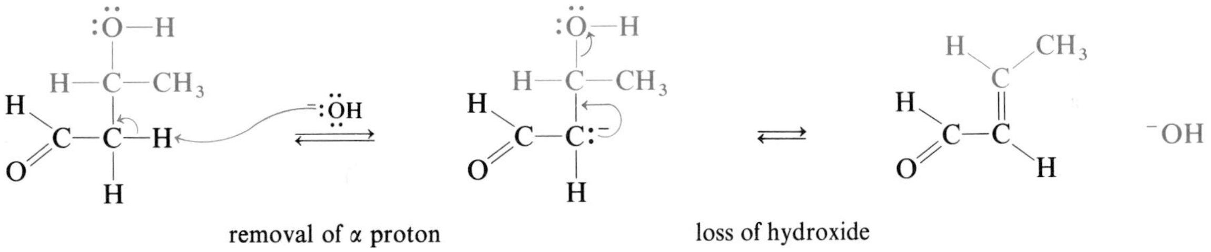

$$H(O{=})C-\underset{\underset{H}{|}}{C}H(-H)-CH(OH)CH_3 \ \ \xrightleftharpoons{:\ddot{O}H^-} \ \ H(O{=})C-\underset{\underset{H}{|}}{C}{:}^- -CH(\ddot{O}H)CH_3 \ \ \rightleftharpoons \ \ H(O{=})C-CH{=}CH(CH_3) \ \ {}^-OH$$

removal of α proton loss of hydroxide

Even when the aldol equilibrium is unfavorable for formation of a β-hydroxy ketone or aldehyde, the dehydration product may be obtained in good yield by heating the reaction mixture. Dehydration is usually an exothermic reaction because it leads to a conjugated system. In effect, the exothermic dehydration drives the aldol equilibrium to the right.

PROBLEM 22-21

Propose a mechanism for the dehydration of diacetone alcohol to mesityl oxide

(a) in acid. (b) in base.

PROBLEM 22-22

When propionaldehyde is warmed with sodium hydroxide, one of the products is 2-methyl-2-pentenal. Propose a mechanism for this reaction.

PROBLEM 22-23

Predict the products of aldol condensation, followed by dehydration, of the following ketones and aldehydes.

(a) butyraldehyde (b) acetophenone (c) cyclohexanone

22-8 CROSSED ALDOL CONDENSATIONS

When the enolate of one aldehyde (or ketone) adds to the carbonyl group of another, the result is called a **crossed aldol condensation.** The compounds used in a crossed aldol condensation must be selected carefully or a mixture of several products will be formed.

Consider the aldol condensation between acetaldehyde (ethanal) and propanal shown below. Attack by the enolate of ethanal on propanal gives a product different from the one formed by attack of the enolate of propanal on ethanal. Also, the self-condensations of ethanal and propanal continue to take place. Depending on the reaction conditions, various proportions of the four possible products result.

1. Enolate of ethanal adds to propanal

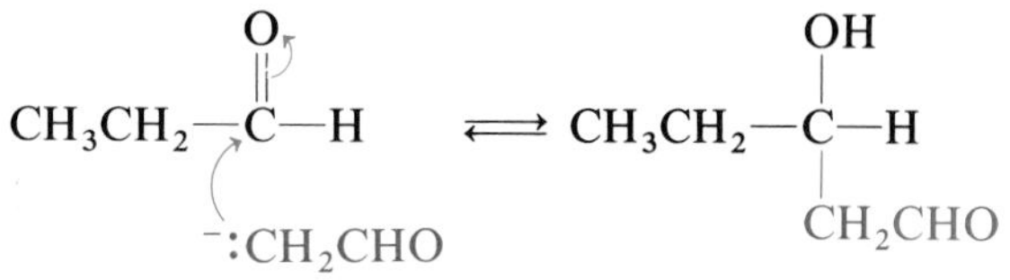

$$CH_3CH_2-\overset{\overset{O}{\|}}{C}-H + {}^-{:}CH_2CHO \rightleftharpoons CH_3CH_2-\overset{\overset{OH}{|}}{\underset{\underset{CH_2CHO}{|}}{C}}-H$$

2. Enolate of propanal adds to ethanal

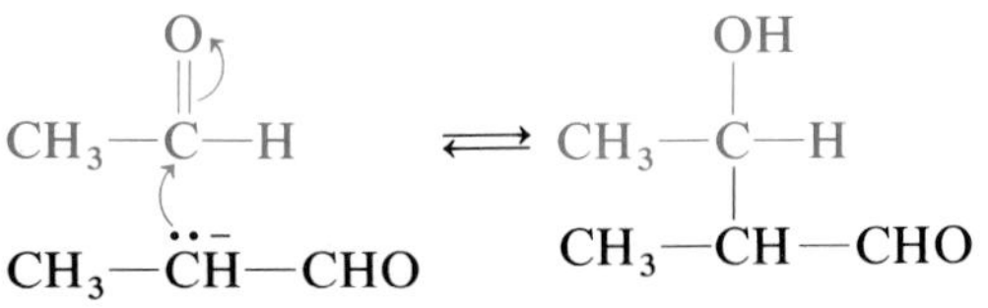

$$CH_3-\overset{\overset{O}{\|}}{C}-H + CH_3-\ddot{C}H^--CHO \rightleftharpoons CH_3-\overset{\overset{OH}{|}}{\underset{\underset{CH_3-CH-CHO}{|}}{C}}-H$$

3. Self-condensation of ethanal

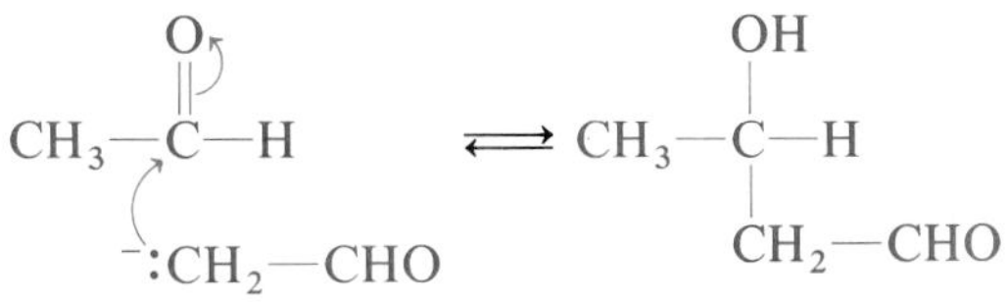

$$CH_3-\overset{\overset{O}{\|}}{C}-H + {}^-{:}CH_2-CHO \rightleftharpoons CH_3-\overset{\overset{OH}{|}}{\underset{\underset{CH_2-CHO}{|}}{C}}-H$$

4. Self-condensation of propanal

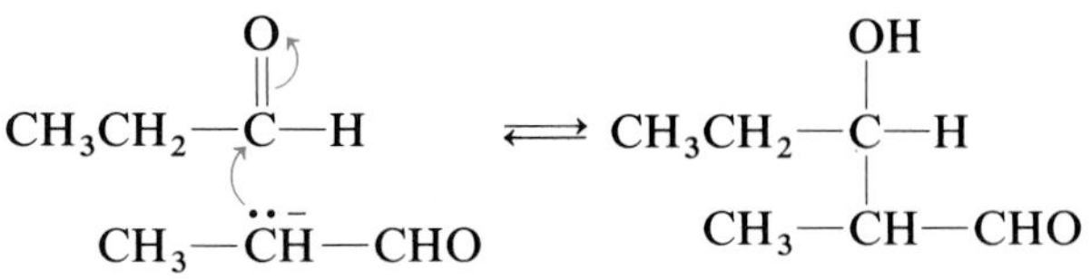

$$CH_3CH_2-\overset{\overset{O}{\|}}{C}-H + CH_3-\ddot{C}H^--CHO \rightleftharpoons CH_3CH_2-\overset{\overset{OH}{|}}{\underset{\underset{CH_3-CH-CHO}{|}}{C}}-H$$

A crossed aldol condensation can be effective if it is planned so that just one of the reactants can form an enolate ion, and the other compound is more likely to react with the enolate. If only one of the reactants has an α hydrogen, there can be only one enolate present in the solution. If the other reactant is present in excess, or contains a particularly electrophilic carbonyl group, it is more likely to be attacked by the enolate ion.

The following two reactions are successful crossed aldol condensations. Notice that the aldol product may or may not undergo dehydration, depending on the reaction conditions and the structure of the product.

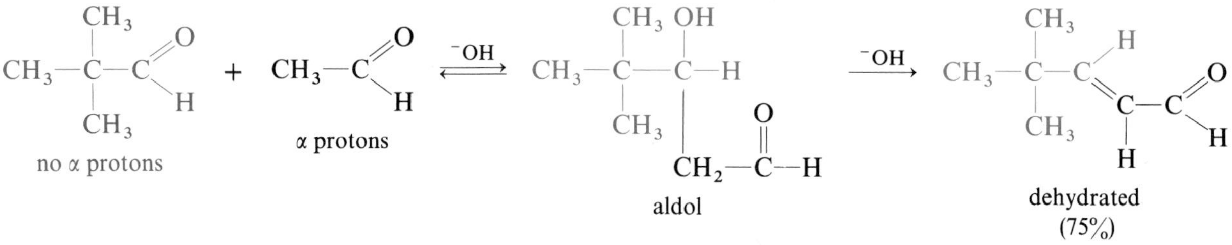

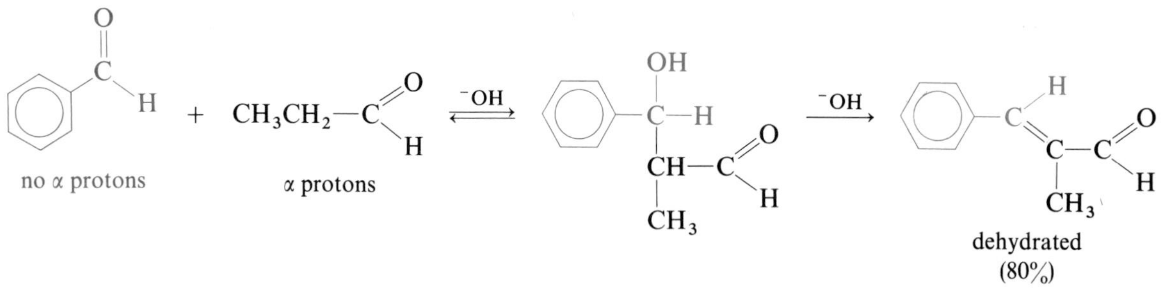

$$\underset{\text{no }\alpha\text{ protons}}{(CH_3)_3C-CHO} + \underset{\alpha\text{ protons}}{CH_3-CHO} \underset{}{\overset{^-OH}{\rightleftharpoons}} \underset{\text{aldol}}{CH_3-\overset{\overset{CH_3}{|}}{\underset{\underset{CH_3}{|}}{C}}-\overset{\overset{OH}{|}}{\underset{\underset{CH_2-\overset{O}{\overset{\|}{C}}-H}{|}}{C}}-H} \xrightarrow{^-OH} \underset{\substack{\text{dehydrated}\\(75\%)}}{(CH_3)_3C-CH=CH-CHO}$$

$$\underset{\text{no }\alpha\text{ protons}}{C_6H_5-CHO} + \underset{\alpha\text{ protons}}{CH_3CH_2-CHO} \overset{^-OH}{\rightleftharpoons} C_6H_5-\overset{\overset{OH}{|}}{C}H-\underset{\underset{CH_3}{|}}{CH}-CHO \xrightarrow{^-OH} \underset{\substack{\text{dehydrated}\\(80\%)}}{C_6H_5-CH=\underset{\underset{CH_3}{|}}{C}-CHO}$$

In these examples, the compound with α protons is slowly added to a basic solution of the compound with no α protons. This way, the enolate ion is always formed in the presence of a large excess of the other component, and the desired reaction is favored.

PROBLEM SOLVING: PROPOSING REACTION MECHANISMS

The general principles for proposing reaction mechanisms, first introduced in Chapter 10 and summarized in Appendix 4, are applied here to a crossed aldol condensation. This example emphasizes a base-catalyzed reaction involving strong nucleophiles. In drawing mechanisms, be careful to draw all the bonds and substituents of each carbon atom involved, show each step separately, and draw curved arrows to show the movement of electrons from the nucleophile to the electrophile.

Our problem is to propose a mechanism for the base-catalyzed reaction of methylcyclohexanone with benzaldehyde:

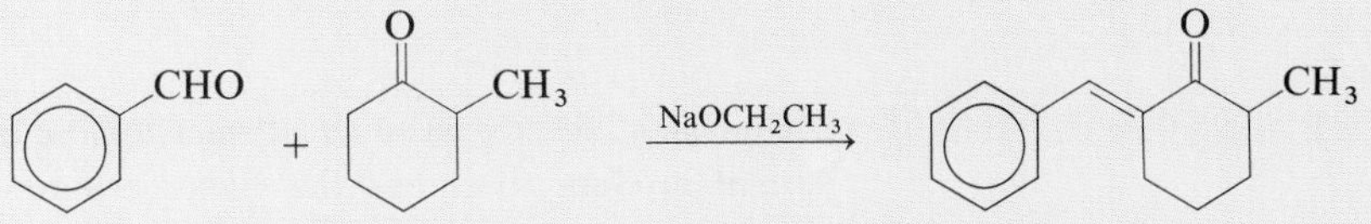

CHO + O CH$_3$ $\xrightarrow{NaOCH_2CH_3}$ O CH$_3$

First, we must determine the type of mechanism. The use of sodium ethoxide, a strong base and a strong nucleophile, implies the reaction involves strong nucleophiles as intermediates. We expect to see strong nucleophiles and anionic intermediates (possibly stabilized carbanions), but no strong electrophiles or strong acids, and certainly no carbocations or free radicals.

1. **Consider the carbon skeletons of the reactants and products, and decide which carbon atoms in the products are most likely derived from which carbon atoms in the reactants.**
   Because one of the rings is aromatic, it is clear which ring in the product is derived from which ring in the reactants. The carbon atom that bridges the two rings in the products must be derived from the carbonyl carbon in benzaldehyde. The two alpha protons from methylcyclohexanone and the carbonyl oxygen are lost as water.

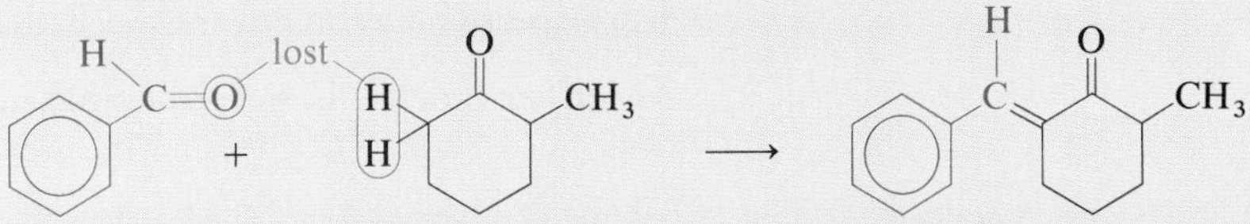

   H C=O (lost) + H H O CH$_3$ $\longrightarrow$ H C O CH$_3$

2. **Consider whether any of the reactants is a strong enough nucleophile to react without being activated. If not, consider how one of the reactants might be converted to a strong nucleophile by deprotonation of an acidic site or by attack on an electrophilic site.**
   Neither of these reactants is a strong enough nucleophile to attack the other. If ethoxide removes an α proton from methylcyclohexanone, however, a strongly nucleophilic enolate ion is generated.

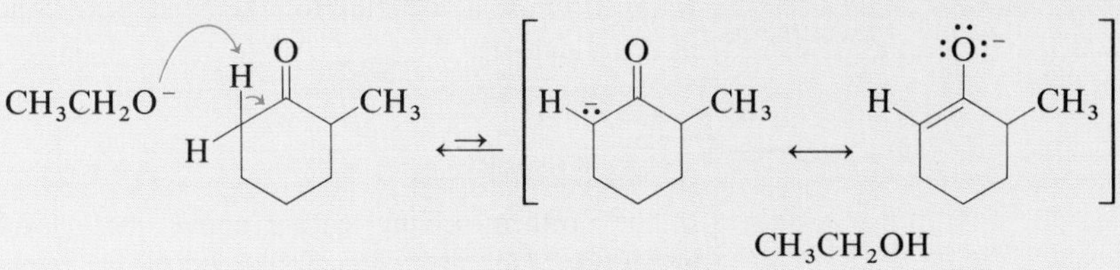

   $CH_3CH_2O^-$ H H O CH$_3$ $\rightleftarrows$ [H O CH$_3$ $\longleftrightarrow$ H :Ö:$^-$ CH$_3$]

   CH_3CH_2OH

3. **Consider how an electrophilic site on another reactant (or, in a cyclization, another part of the same molecule) can undergo attack by the strong nucleophile to form a bond needed in the product. Draw the product of this bond formation.**
Attack at the electrophilic carbonyl group of benzaldehyde, followed by protonation, gives a β-hydroxy ketone (an aldol).

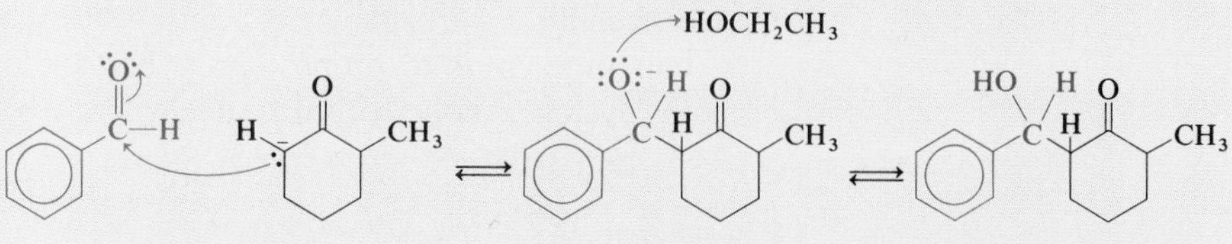

4. **Consider how the product of nucleophilic attack might be converted to the final product (if it has the right carbon skeleton) or reactivated to form another bond needed in the product.**
The β-hydroxy ketone must be dehydrated to give the final product. Under these basic conditions, the usual dehydration mechanism (protonation of hydroxyl, followed by loss of water) cannot occur. Removal of another α proton gives an enolate ion that can lose hydroxide in a strongly exothermic step to give the final product.

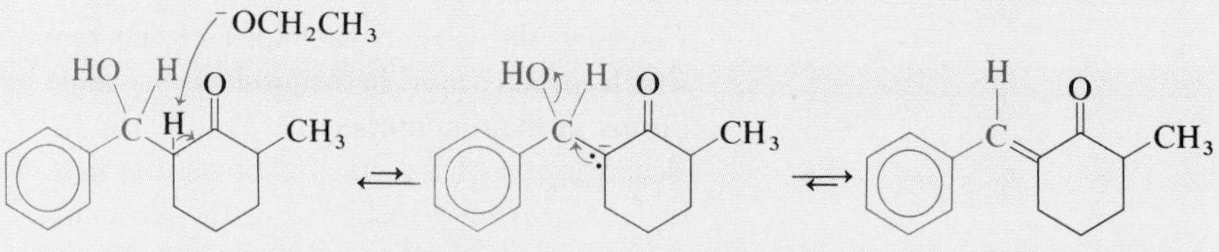

5. **Draw out all the steps of the mechanism using curved arrows to show the movement of electrons. Be careful to show only one step at a time.**
The complete mechanism is given by combining the equations shown above. We suggest you write out the mechanism as a review of the steps involved.

As further practice in proposing mechanisms for base-catalyzed reactions, do Problems 22-24 and 22-25 using the steps shown above.

PROBLEM 22-24

Propose mechanisms for the two base-catalyzed condensations shown above:

(a) 2,2-dimethylpropanal with acetaldehyde
(b) benzaldehyde with propionaldehyde

PROBLEM 22-25

When acetone is treated with two equivalents of benzaldehyde in the presence of base, the crossed condensation adds 2 equivalents of benzaldehyde and expels 2 equivalents of water. Propose a structure for the product of condensation of acetone with 2 molecules of benzaldehyde.

PROBLEM 22-26

In the problem-solving feature above, methylcyclohexanone was seen to react at the *unsubstituted* alpha-carbon. Try to write a mechanism for the same reaction at the methyl-substituted carbon atom, and explain why this regiochemistry is not observed.

PROBLEM 22-27

Predict the major products of the following base-catalyzed aldol condensations with dehydration.

(a) benzophenone (PhCOPh) + propionaldehyde
(b) 2,2-dimethylpropanal + acetophenone

PROBLEM 22-28

Cinnamaldehyde is used as a flavoring agent in cinnamon candies. Show how cinnamaldehyde is synthesized by a crossed aldol condensation followed by dehydration.

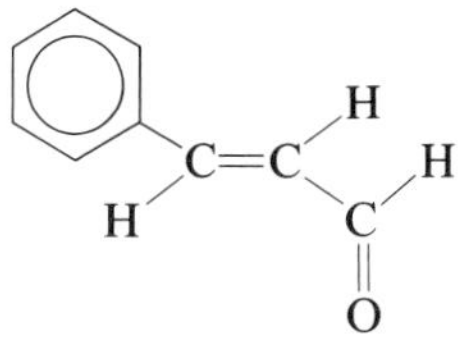

cinnamaldehyde

22-9 ALDOL CYCLIZATIONS

Five- and six-membered rings are often formed by intramolecular aldol reactions of diketones. Aldol cyclizations of rings larger than six and smaller than five are much less common, because the formation of larger and smaller rings is less favored by their energy and entropy. The following reactions show how a 1,4-diketone can condense and dehydrate to give a cyclopentenone, and a similar condensation of a 1,5-diketone gives a cyclohexenone.

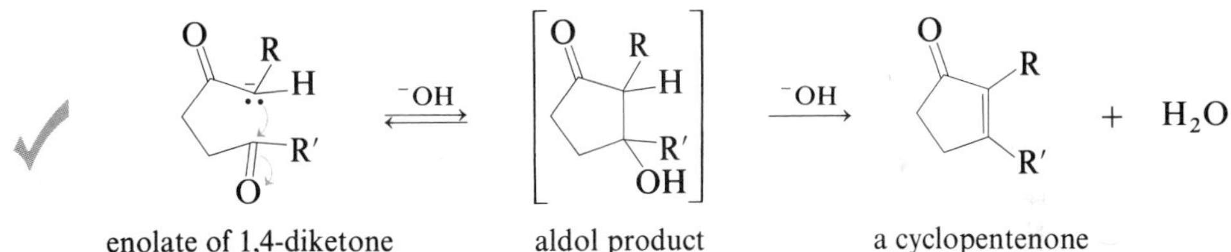

enolate of 1,4-diketone aldol product a cyclopentenone

Example

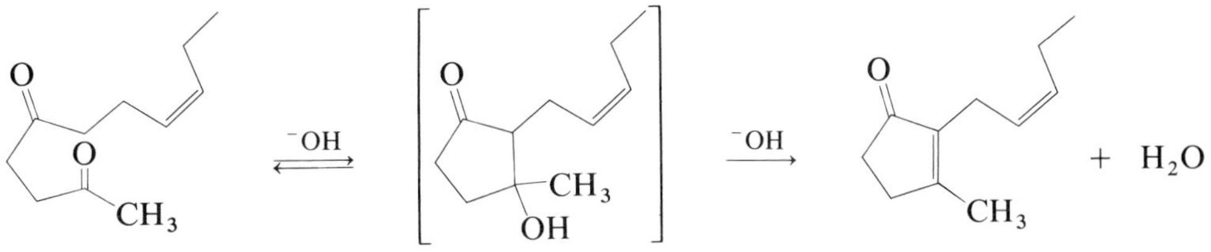

cis-8-undecene-2,5-dione aldol product *cis*-jasmone (a perfume) (90%)

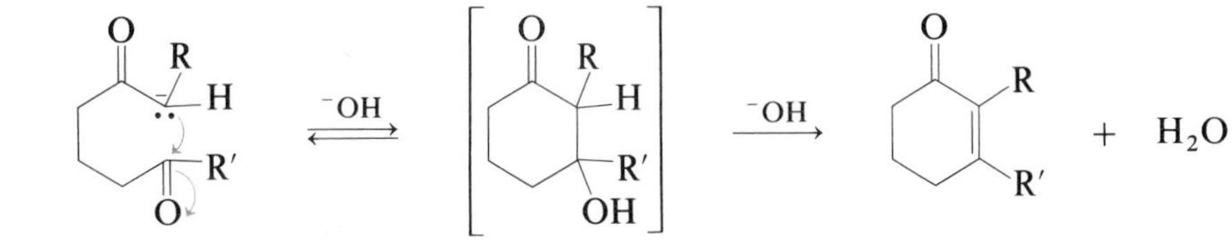

enolate of 1,5-diketone aldol product a cyclohexenone

Example

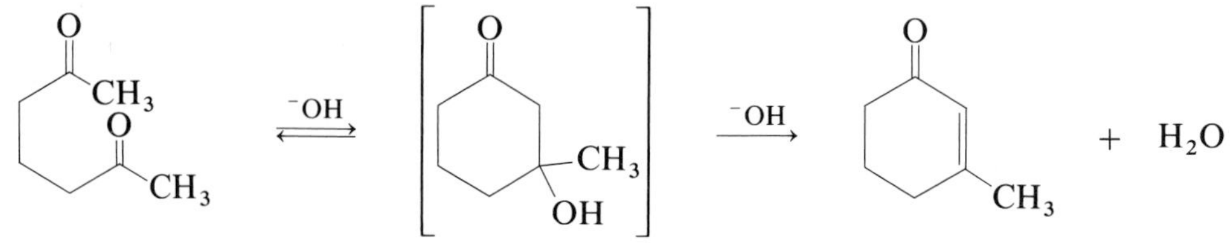

2,6-heptanedione (a 1,5-diketone) $\underset{}{\overset{^{-}OH}{\rightleftharpoons}}$ [aldol product] $\xrightarrow{^{-}OH}$ 3-methylcyclohex-2-enone + H_2O

The following example shows how the carbonyl group of the product may be outside the ring in some cases.

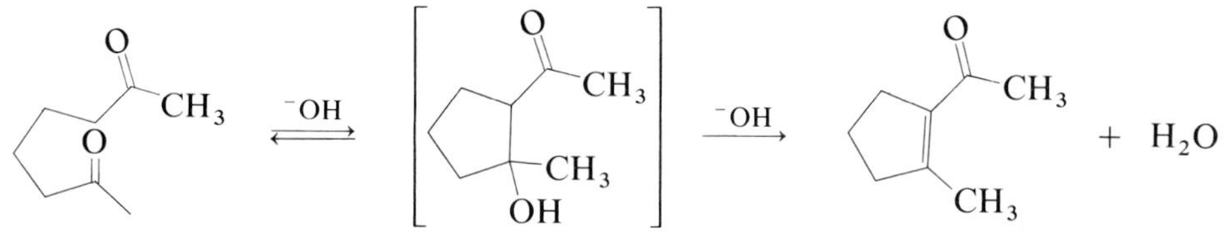

2,7-octanedione $\underset{}{\overset{^{-}OH}{\rightleftharpoons}}$ [aldol product] $\xrightarrow{^{-}OH}$ 1-acetyl-2-methylcyclopentene + H_2O

PROBLEM 22-29

Show how 2,7-octanedione might cyclize to a cycloheptenone. Explain why ring closure to the cycloheptenone is not observed.

PROBLEM 22-30

When 1,6-cyclodecanedione is treated with sodium carbonate, the product gives a UV spectrum similar to the UV spectrum of 1-acetyl-2-methylcyclopentene. Propose a structure for the product, and give a mechanism for its formation.

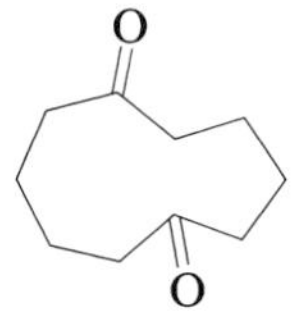

1,6-cyclodecanedione

22-10 PLANNING SYNTHESES USING ALDOL CONDENSATIONS

As long as we remember their limitations, aldol condensations can serve as useful synthetic reactions for the formation of a variety of organic compounds. In particular, aldol condensations (with dehydration) are used to form new carbon-carbon double bonds. A few general principles can be used to decide whether a compound might be an aldol product, and which reagents should be used as starting materials.

Aldol condensations produce two types of products: (1) β-hydroxy aldehydes and ketones (aldols) and (2) α,β-unsaturated aldehydes and ketones. If a target molecule has one of these functionalities, an aldol should be considered. To determine the starting materials needed, divide the structure at the α,β bond. In the case of the dehydrated product, the α,β bond is the double bond. Figure 22-3 shows the division of some aldol products into their starting materials.

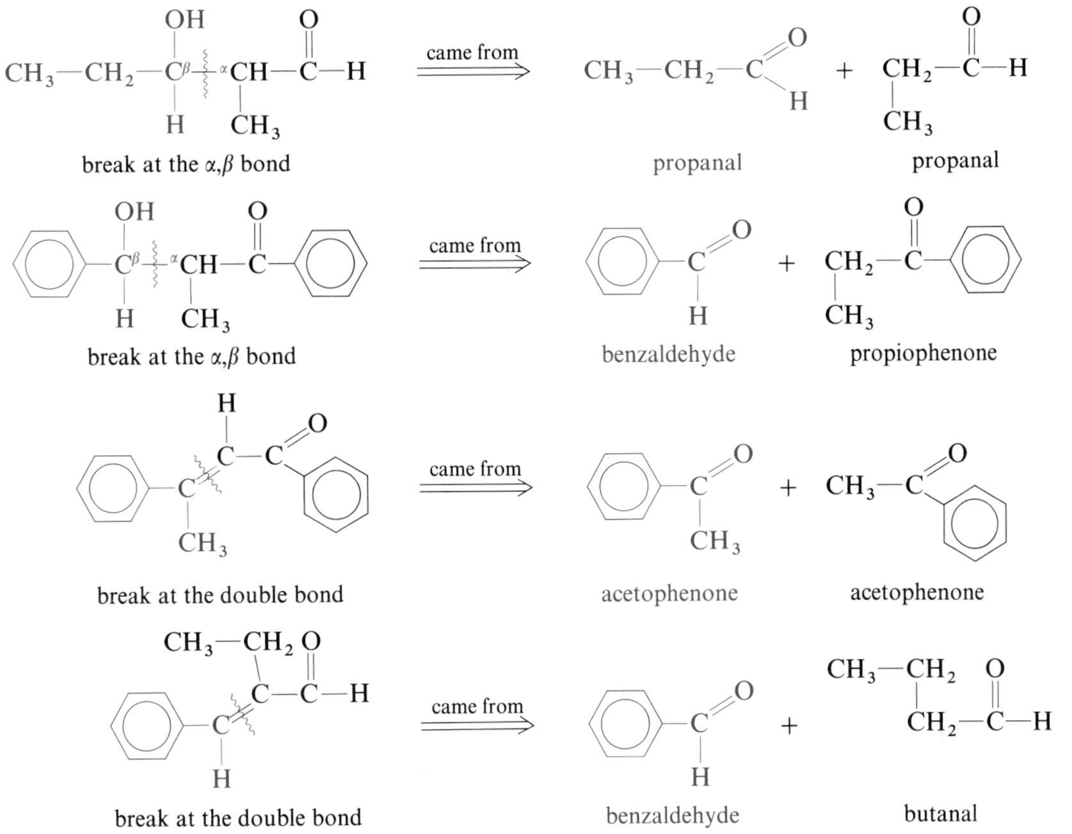

$CH_3-CH_2-\overset{OH}{\underset{H}{C^{\beta}}}\wr{}^{\alpha}\underset{CH_3}{CH}-\overset{O}{\overset{\|}{C}}-H$ $\xRightarrow{\text{came from}}$ CH_3-CH_2-CHO + $\underset{CH_3}{CH_2}-\overset{O}{\overset{\|}{C}}-H$

break at the α,β bond — propanal — propanal

$Ph-\overset{OH}{\underset{H}{C^{\beta}}}\wr{}^{\alpha}\underset{CH_3}{CH}-\overset{O}{\overset{\|}{C}}-Ph$ $\xRightarrow{\text{came from}}$ $Ph-CHO$ + $\underset{CH_3}{CH_2}-\overset{O}{\overset{\|}{C}}-Ph$

break at the α,β bond — benzaldehyde — propiophenone

$Ph-\underset{CH_3}{C}=\overset{H}{C}-C(=O)-Ph$ $\xRightarrow{\text{came from}}$ $Ph-C(=O)-CH_3$ + $CH_3-C(=O)-Ph$

break at the double bond — acetophenone — acetophenone

$Ph-\underset{H}{C}=\overset{CH_3-CH_2}{C}-\overset{O}{\overset{\|}{C}}-H$ $\xRightarrow{\text{came from}}$ $Ph-CHO$ + $\overset{CH_3-CH_2}{CH_2}-\overset{O}{\overset{\|}{C}}-H$

break at the double bond — benzaldehyde — butanal

FIGURE 22-3 Aldol products are β-hydroxyl aldehydes and ketones or α,β-unsaturated aldehydes and ketones. An aldol product is dissected into its starting materials by mentally breaking the α,β bond.

PROBLEM 22-31

Show how each of the following compounds can be dissected into two reagents joined by an aldol condensation. Then decide whether the necessary aldol condensation is feasible.

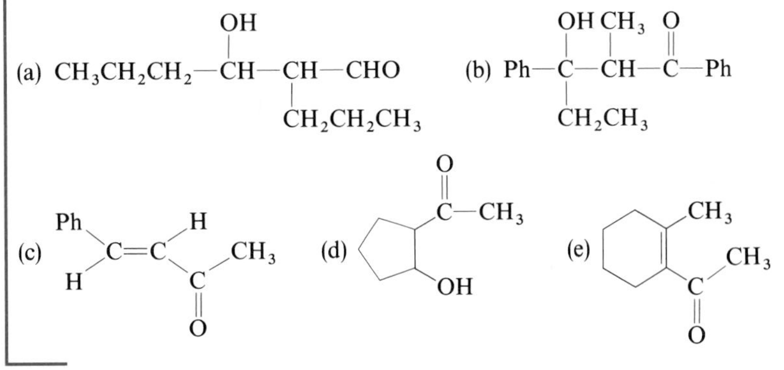

(a) $CH_3CH_2CH_2-\overset{OH}{CH}-\underset{CH_2CH_2CH_3}{CH}-CHO$

(b) $Ph-\overset{OH}{\underset{CH_2CH_3}{C}}-\overset{CH_3}{CH}-\overset{O}{\overset{\|}{C}}-Ph$

(c) $Ph(H)C=C(H)-C(=O)-CH_3$

(d) 2-(C(=O)–CH_3)-cyclopentanol (ring bearing $\overset{O}{\overset{\|}{C}}-CH_3$ and OH)

(e) cyclohexene ring bearing CH_3 and $C(=O)CH_3$

22-11 THE CANNIZZARO REACTION

Aldehydes without α hydrogens cannot form enolate ions; therefore, they cannot undergo aldol condensation except by reacting with the enolate ion of another carbonyl compound. This property is useful in crossed aldol reactions, because the use of one compound without α hydrogens helps to ensure the formation of the right enolate ion and aldol product. What happens, then, if one or two aldehydes are put into a strong base and *none* of them have any α hydrogens? The result is an oxidation-reduction called the **Cannizzaro reaction.**

The following Cannizzaro reaction involves benzaldehyde and formaldehyde, both of which lack α hydrogens. When these aldehydes are mixed in the presence of strong base, formaldehyde is oxidized to formate ion and benzaldehyde is reduced to benzyl alcohol.

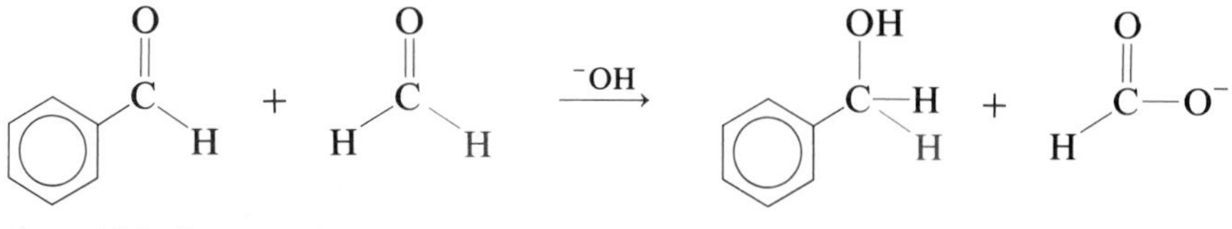

$$\underset{\text{benzaldehyde}}{\text{PhCHO}} + \underset{\text{formaldehyde}}{\text{H}_2\text{C=O}} \xrightarrow{^{-}\text{OH}} \underset{\substack{\text{benzyl alcohol}\\ \text{(reduced)}}}{\text{PhCH}_2\text{OH}} + \underset{\substack{\text{formate}\\ \text{(oxidized)}}}{\text{HCOO}^-}$$

Formaldehyde is easily oxidized, and it commonly serves as the reducing agent in the Cannizzaro reaction. Mixing formaldehyde with another aldehyde (also having no α hydrogens) in the presence of KOH usually reduces the other aldehyde to an alcohol. Few aldehydes lack α hydrogens, however, so the Cannizzaro reaction is of limited utility. (Remember that aldehydes with α hydrogens do not undergo the Cannizzaro reaction; they undergo the aldol condensation.)

The mechanism of the Cannizzaro reaction begins with a nucleophilic attack by hydroxide ion on the carbonyl group of the aldehyde that will be oxidized. The resulting anion is a good hydride (H:$^-$) donor that transfers hydride to benzaldehyde. This is the actual oxidation-reduction step.

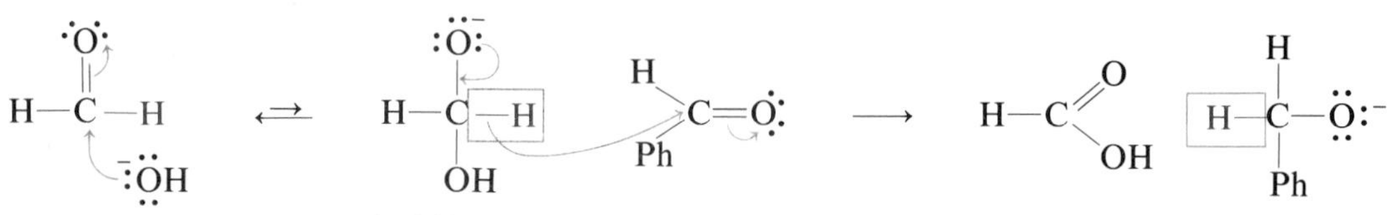

$$\text{H}_2\text{C=O} + {:}\ddot{\text{O}}\text{H}^- \rightleftarrows \underset{\text{hydride donor}}{\text{H–C(O}^-\text{)(H)–OH}} + \underset{\text{benzaldehyde}}{\text{PhCH=O}} \longrightarrow \underset{\text{formic acid}}{\text{HCOOH}} + \underset{\text{alkoxide}}{\text{PhCH}_2\text{–O}^-}$$

The products are formic acid and an alkoxide. A fast proton transfer completes the reaction.

The Cannizzaro reaction is commonly seen as an unwanted side reaction of aldehydes in storage. If a trace of base is present, two molecules of an aldehyde may *disproportionate:* One molecule is oxidized, and the other is reduced. This reaction occurs because aldehydes are less stable than the corresponding alcohols and carboxylic acids. The following reaction shows the disproportionation of two molecules of benzaldehyde to give one molecule of benzyl alcohol and a benzoate ion.

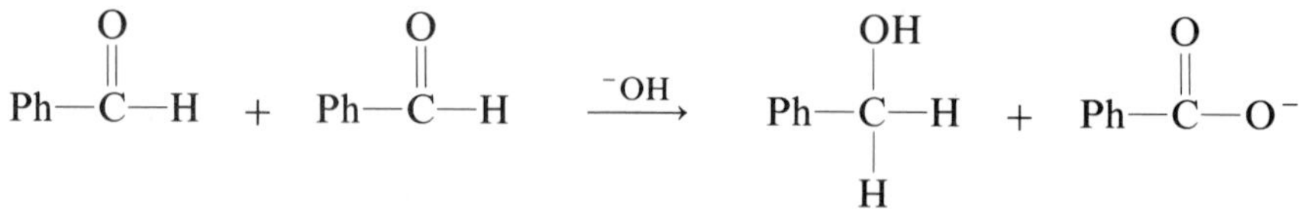

$$\underbrace{\text{Ph–C(=O)–H} + \text{Ph–C(=O)–H}}_{\text{two molecules of benzaldehyde}} \xrightarrow{^{-}\text{OH}} \underset{\text{reduced}}{\text{Ph–CH(OH)–H}} + \underset{\text{oxidized}}{\text{Ph–C(=O)–O}^-}$$

PROBLEM 22-32

Give a detailed mechanism for the base-catalyzed disproportionation of benzaldehyde.

PROBLEM 22-33

The Cannizzaro reaction may be used to reduce 2,2-dimethylpropanal, but not 2-methylpropanal. Show how this reduction might be accomplished in one case, and why it is impossible in the other case.

22-12 THE WITTIG REACTION

The aldol condensation, with dehydration, results in the replacement of a carbonyl double bond by a carbon-carbon double bond. This is our first encounter with a technique that creates a new carbon-carbon *double* bond where no bond existed before.

Aldol converts C=O to C=C

$$R_2C{=}O + {}^{-}{:}CH(R')-C(=O)-R' \longrightarrow \longrightarrow R_2C{=}C(R')-C(=O)-R'$$

(no α hydrogens) new double bond

The aldol condensation is useful but limited: One of the carbonyl compounds must have no α hydrogens, yet it should be more reactive toward nucleophilic addition than the other. The other compound must be a ketone or aldehyde with an acidic α proton.

In 1954, George Wittig discovered an entirely new way to form a carbon-carbon double bond from a carbonyl group and a carbanion. He received the Nobel prize for this work in 1979. This technique, called the **Wittig reaction,** uses a phosphorus-stabilized carbanion to attack a ketone or an aldehyde. A subsequent elimination gives an alkene.

Wittig reaction

$$R_2C{=}O + Ph_3\overset{+}{P}-\overset{-}{C}{:}R'_2 \longrightarrow \longrightarrow R_2C{=}CR'_2 + Ph_3P{=}O$$

ketone or aldehyde (may have α hydrogens) — phosphorus ylide (no carbonyl necessary) — alkene

The phosphorus-stabilized carbanion is an **ylide** (pronounced "ill'-id")—a molecule that bears no overall charge but has a negatively charged carbon atom bonded to a positively charged heteroatom. Phosphorus ylides are prepared from triphenylphosphine and alkyl halides in a two-step process. The first step involves a nucleophilic attack by triphenylphosphine on an unhindered (usually primary) alkyl halide. The product is an alkyltriphenylphosphonium salt. In a second step, the phosphonium salt is treated with a strong base (usually butyllithium) to abstract a proton from the carbon atom bonded to phosphorus.

$$Ph_3P{:} + H-CH(R)-X \longrightarrow Ph_3\overset{+}{P}-CH_2R \; X^- \xrightarrow[\text{butyllithium}]{\overset{\delta-}{C}H_2-\overset{\delta+}{Li}\,(CH_2CH_2CH_3)} \left[Ph_3\overset{+}{P}-\ddot{C}^{-}HR \longleftrightarrow Ph_3P{=}CHR \right] + C_4H_{10} + LiX$$

triphenylphosphine alkyl halide — phosphonium salt — phosphorus ylide — butane

The phosphorus ylide has two resonance structures: One with a double bond between carbon and phosphorus, and another with charges on carbon and phosphorus. The double-bonded resonance structure requires ten electrons in the valence shell of phosphorus, using one of the phosphorus *d* orbitals. The pi bond between carbon and phosphorus is weak, and the charged structure is the major contributor. The carbon atom actually bears a partial negative charge, balanced by a corresponding positive charge on phosphorus. The following reactions are examples of the formation of phosphorus ylides.

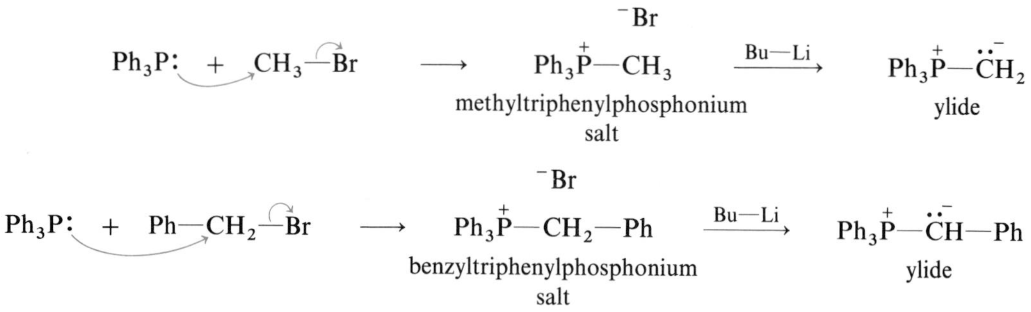

$$Ph_3P: + CH_3\text{—}Br \longrightarrow Ph_3\overset{+}{P}\text{—}CH_3 \;{}^{-}Br \xrightarrow{Bu\text{—}Li} Ph_3\overset{+}{P}\text{—}\overset{..}{C}H_2^{-}$$

methyltriphenylphosphonium salt; ylide

$$Ph_3P: + Ph\text{—}CH_2\text{—}Br \longrightarrow Ph_3\overset{+}{P}\text{—}CH_2\text{—}Ph \;{}^{-}Br \xrightarrow{Bu\text{—}Li} Ph_3\overset{+}{P}\text{—}\overset{..}{C}H^{-}\text{—}Ph$$

benzyltriphenylphosphonium salt; ylide

PROBLEM 22-34

Trimethylphosphine is much less expensive than triphenylphosphine. Why is trimethylphosphine unsuitable for use in making most phosphorus ylides?

Because of its carbanion character, the ylide carbon atom is strongly nucleophilic. It attacks a carbonyl group to give a charge-separated intermediate called a *betaine* (pronounced "bay′-tuh-ene"). A betaine is an unusual compound, because it contains a negatively charged oxygen and a positively charged phosphorus on adjacent carbon atoms. Phosphorus and oxygen form very strong bonds, and the attraction of the opposite charges promotes the fast formation of a four-membered *oxaphosphetane* ring.

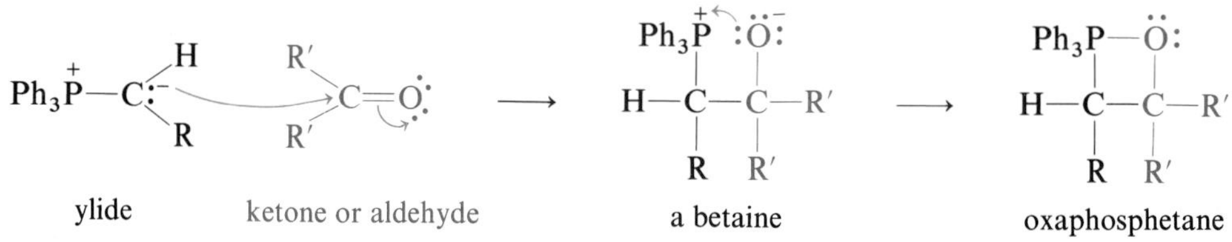

$$Ph_3\overset{+}{P}\text{—}\overset{-}{C}(H)(R) + R'_2C{=}\ddot{O}: \longrightarrow Ph_3\overset{+}{P}\text{—}CHR\text{—}CR'_2\text{—}\ddot{O}:^{-} \longrightarrow \text{(ring) } Ph_3P\text{—}O\text{—}CR'_2\text{—}CHR$$

ylide; ketone or aldehyde; a betaine; oxaphosphetane

The four-membered ring quickly collapses to give the desired alkene and triphenylphosphine oxide. Triphenylphosphine oxide is an exceptionally stable compound, and the conversion of triphenylphosphine to triphenylphosphine oxide provides the driving force for the Wittig reaction.

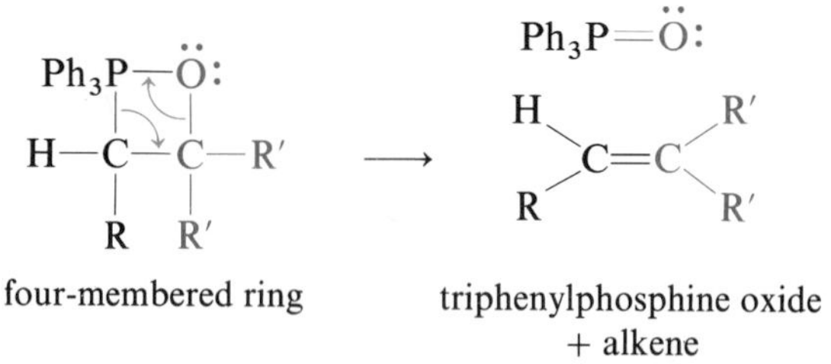

$$\text{(ring) } Ph_3P\text{—}\ddot{O}\text{—}CR'_2\text{—}CHR \longrightarrow Ph_3P{=}\ddot{O}: + (H)(R)C{=}C(R')(R')$$

four-membered ring; triphenylphosphine oxide + alkene

The following examples show the formation of carbon-carbon double bonds using the Wittig reaction. Mixtures of cis and trans isomers often result when geometric isomerism of the product is possible.

$$\text{cyclohexanone} \; + \; Ph_3\overset{+}{P}-\ddot{C}H_2^{-} \;\longrightarrow\; \text{methylenecyclohexane } (=CH_2)$$

85%

$$PhCH=O \; + \; Ph_3\overset{+}{P}-\ddot{C}^{-}HPh \;\longrightarrow\; PhCH=CHPh$$

PROBLEM 22-35

Like other strong nucleophiles, triphenylphosphine attacks and opens epoxides. The initial product (a betaine) quickly cyclizes to an oxaphosphetane that collapses to an alkene and triphenylphosphine oxide.

(a) Show each step in the reaction of *trans*-2,3-epoxybutane with triphenylphosphine to give 2-butene. What is the stereochemistry of the double bond in the product?

(b) Show how this reaction might be used to convert *cis*-cyclooctene to *trans*-cyclooctene.

Planning a Wittig synthesis The Wittig reaction is a valuable synthetic tool that converts a carbonyl group to a carbon-carbon double bond. A wide variety of alkene derivatives may be synthesized by the Wittig reaction. To determine the necessary reagents, we mentally divide the target molecule at the double bond and then decide which of the two components should come from the carbonyl compound and which should come from the ylide.

In general, the ylide should come from an unhindered alkyl halide. Triphenylphosphine is a bulky reagent, reacting best with unhindered primary and methyl halides. It occasionally reacts with unhindered secondary halides, but these reactions are sluggish and often give poor yields. The following example and Solved Problem 22-4 show the planning of some Wittig syntheses.

Analysis

$$(CH_3)_2C=CH(CH_2CH_3) \;\xRightarrow{\text{could come from}}\; \begin{cases} (CH_3)_2C=O \; + \; Ph_3\overset{+}{P}-\ddot{C}^{-}H(CH_2CH_3) \quad \text{(preferred)} \\ \text{or} \\ (CH_3)_2\ddot{C}^{-}-\overset{+}{P}Ph_3 \; + \; O=CH(CH_2CH_3) \end{cases}$$

Synthesis

$$BrCH_2CH_2CH_3 \;\xrightarrow[\text{(2) BuLi}]{\text{(1) } Ph_3P}\; Ph_3\overset{+}{P}-\ddot{C}^{-}H(CH_2CH_3) \;\xrightarrow{(CH_3)_2C=O}\; (CH_3)_2C=CH(CH_2CH_3)$$

SOLVED PROBLEM 22-4

Show how you would use a Wittig reaction to synthesize 1-phenyl-1,3-butadiene.

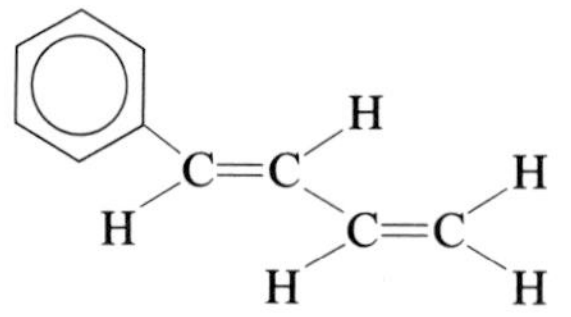

1-phenyl-1,3-butadiene

SOLUTION This molecule has two double bonds that might be formed by Wittig reactions. The central double bond could be formed by a Wittig reaction in either of two ways; both of these syntheses will probably work, and both will produce a mixture of cis and trans isomers.

Analysis

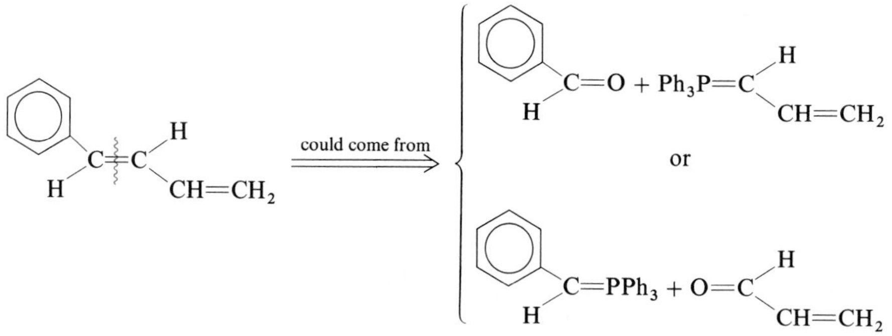
H
C=C
H
CH=CH2
could come from
C=O + Ph3P=C
H
H
CH=CH2
or
C=PPh3 + O=C
H
H
CH=CH2

You should complete this solution by drawing out the syntheses indicated by this analysis (Problem 22-36).

PROBLEM 22-36

(a) Outline the syntheses indicated in Solved Problem 22-4, beginning with aldehydes and alkyl halides.

(b) Both of these syntheses of 1-phenyl-1,3-butadiene form the central double bond. Show how you would synthesize this target molecule by forming the terminal double bond.

PROBLEM 22-37

Show how a Wittig reaction might be used to synthesize each of the following compounds. In each case, start with an alkyl halide and a ketone or an aldehyde.

(a) Ph—CH=C$(CH_3)_2$ (b) Ph—C(CH_3)=CH_2

(c) Ph—CH=CH—CH=CH—Ph

(d) 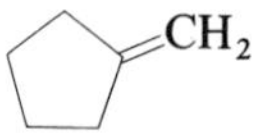
CH2

(e) 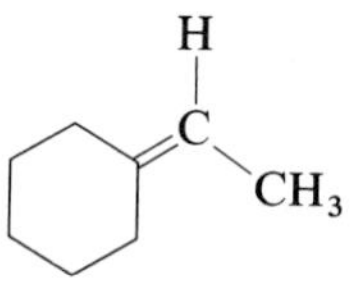
H
C
CH3

22-13 THE CLAISEN ESTER CONDENSATION

The α hydrogens of esters are weakly acidic, and they can be abstracted to give enolate ions. Esters are less acidic than ketones and aldehydes because the ester carbonyl group is stabilized by resonance with the other oxygen atom. This resonance makes the carbonyl group less capable of stabilizing the negative charge of an enolate ion.

$$\left[R-\overset{\displaystyle O}{\overset{\|}{C}}-\ddot{\underset{\cdot\cdot}{O}}-R' \longleftrightarrow R-\overset{\displaystyle :\ddot{O}:^-}{\overset{|}{C}}=\overset{+}{\underset{\cdot\cdot}{O}}-R' \right]$$

A typical pK_a for an α proton of an ester is about 25, compared with a pK_a of about 20 for a ketone or aldehyde. Even so, strong bases do deprotonate esters.

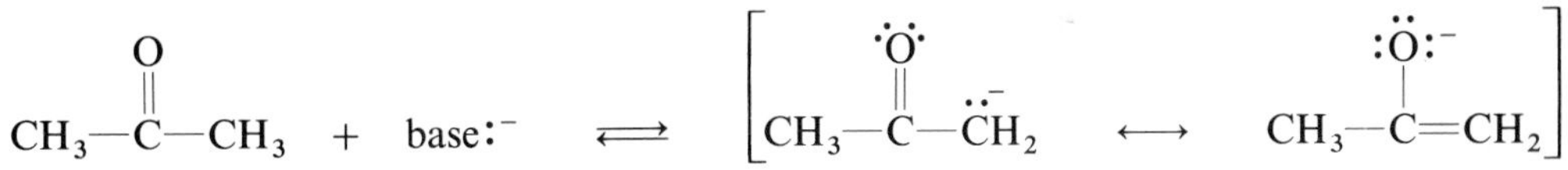

$$\underset{\substack{\text{acetone} \\ (\text{p}K_a = 20)}}{CH_3-\overset{\displaystyle O}{\overset{\|}{C}}-CH_3} + \text{base}:^- \rightleftharpoons \underset{\text{enolate of acetone}}{\left[CH_3-\overset{\displaystyle \ddot{O}}{\overset{\|}{C}}-\ddot{C}H_2^- \longleftrightarrow CH_3-\overset{\displaystyle :\ddot{O}:^-}{\overset{|}{C}}=CH_2 \right]}$$

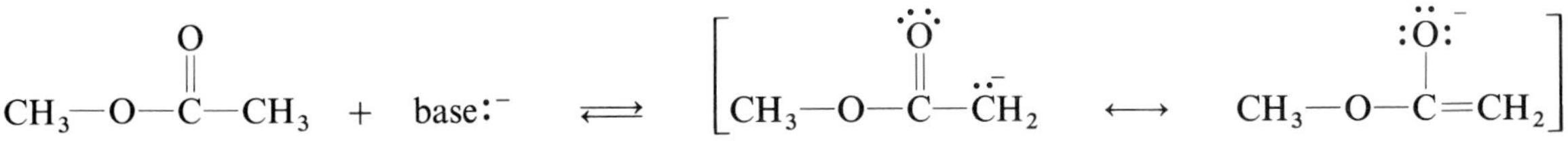

$$\underset{\substack{\text{methyl acetate} \\ (\text{p}K_a = 24)}}{CH_3-O-\overset{\displaystyle O}{\overset{\|}{C}}-CH_3} + \text{base}:^- \rightleftharpoons \underset{\text{enolate of methyl acetate}}{\left[CH_3-O-\overset{\displaystyle \ddot{O}}{\overset{\|}{C}}-\ddot{C}H_2^- \longleftrightarrow CH_3-O-\overset{\displaystyle :\ddot{O}:^-}{\overset{|}{C}}=CH_2 \right]}$$

Ester enolates are strong nucleophiles, and they undergo a wide range of interesting and useful reactions. Most of these reactions are related to Claisen condensation, the most important of all ester condensations.

The **Claisen condensation** results when an ester molecule undergoes nucleophilic acyl substitution by an ester enolate. The intermediate has an alkoxy (—OR) group that acts as a leaving group, forming a β-keto ester. The overall reaction combines two ester molecules to give a β-keto ester.

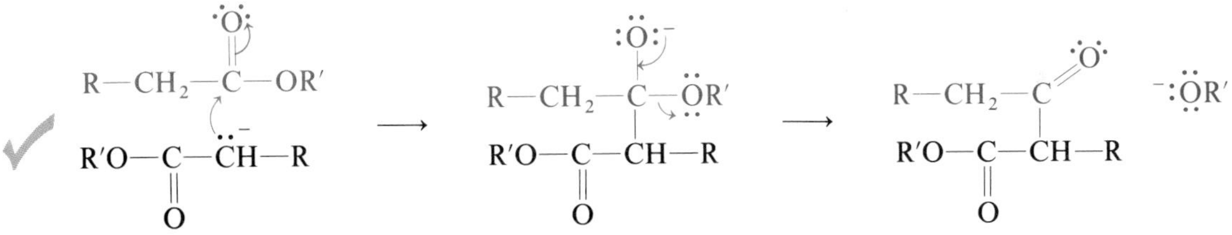

$$\underset{\text{ester enolate}}{\begin{matrix} R-CH_2-\overset{\displaystyle \ddot{O}}{\overset{\|}{C}}-OR' \\ R'O-\underset{\displaystyle O}{\underset{\|}{C}}-\ddot{C}H^--R \end{matrix}} \longrightarrow \underset{\text{intermediate}}{\begin{matrix} R-CH_2-\overset{\displaystyle :\ddot{O}:^-}{\overset{|}{C}}-\ddot{O}R' \\ R'O-\underset{\displaystyle O}{\underset{\|}{C}}-CH-R \end{matrix}} \longrightarrow \underset{\text{a }\beta\text{-keto ester}}{\begin{matrix} R-CH_2-C(=\ddot{O}) \\ R'O-\underset{\displaystyle O}{\underset{\|}{C}}-CH-R \end{matrix}} \quad ^-:\ddot{O}R'$$

β-Keto esters are more acidic than simple ketones, aldehydes, and esters because the negative charge of the enolate is delocalized over both carbonyl groups. β-Keto esters have pK_a values around 11, showing that they are even stronger acids than water. In a solution of a strong base such as ethoxide ion or hydroxide ion, the β-keto ester is rapidly and completely deprotonated.

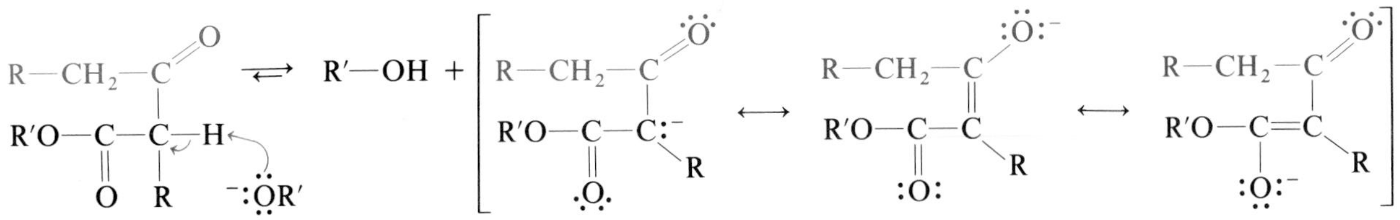

$$\underset{\substack{\text{a }\beta\text{-keto ester} \\ (\text{p}K_a = 11)}}{\begin{matrix} R-CH_2-C(=O) \\ R'O-\underset{\displaystyle O}{\underset{\|}{C}}-\underset{\displaystyle R}{\underset{|}{C}}-H \end{matrix}} \; ^-:\ddot{O}R' \rightleftharpoons R'-OH + \underset{\text{resonance-stabilized enolate ion}}{\left[\begin{matrix} R-CH_2-C(=\ddot{O}) \\ R'O-\underset{\displaystyle \ddot{O}}{\underset{\|}{C}}-\ddot{C}^--R \end{matrix} \longleftrightarrow \begin{matrix} R-CH_2-C(-:\ddot{O}:^-) \\ R'O-\underset{\displaystyle :O:}{\underset{\|}{C}}=C-R \end{matrix} \longleftrightarrow \begin{matrix} R-CH_2-C(=\ddot{O}) \\ R'O-\underset{\displaystyle :\ddot{O}:^-}{\underset{|}{C}}=C-R \end{matrix} \right]}$$

Deprotonation of the β-keto ester provides a driving force for the Claisen condensation. The deprotonation is strongly exothermic, making the overall reaction exothermic and driving the reaction to completion. Because the base is consumed in the deprotonation step, a full equivalent of base must be used, and the Claisen condensation is said to be *base promoted* rather than *base catalyzed.* After the reaction is complete, the addition of dilute acid converts the enolate back to the β-keto ester.

The following example shows the self-condensation of ethyl acetate to give ethyl acetoacetate (ethyl 3-oxobutanoate). Ethoxide is used as the base to avoid transesterification or hydrolysis of the ethyl ester (see Problem 22-40). The initial product is the enolate of ethyl acetoacetate, which is reprotonated in the final step.

$$2\,CH_3-\overset{O}{\overset{\|}{C}}-OCH_2CH_3 \xrightarrow[\text{sodium ethoxide}]{Na^+\ ^-OCH_2CH_3} Na^+\ ^-:\underset{}{CH}(\overset{O}{\overset{\|}{C}}-CH_3)-\overset{O}{\overset{\|}{C}}-OCH_2CH_3 \xrightarrow{H_3O^+} CH_3-\overset{O}{\overset{\|}{C}}-CH_2-\overset{O}{\overset{\|}{C}}-OCH_2CH_3$$

ethyl acetate — keto ester enolate — ethyl acetoacetate (75%)

SOLVED PROBLEM 22-5

Propose a mechanism for the self-condensation of ethyl acetate to give ethyl acetoacetate.

SOLUTION The first step is formation of the ester enolate. The equilibrium for this step lies far to the left; ethoxide deprotonates only a small fraction of the ester.

$$\underset{(pK_a = 24.5)}{H-CH_2-\overset{O}{\overset{\|}{C}}-OCH_2CH_3} + {}^-:\ddot{O}CH_2CH_3 \rightleftarrows \underset{\text{enolate}}{{}^-:CH_2-\overset{O}{\overset{\|}{C}}-OCH_2CH_3} + \underset{(pK_a = 18)}{H-\ddot{O}CH_2CH_3}$$

The enolate ion attacks another molecule of the ester; expulsion of ethoxide ion gives ethyl acetoacetate.

$$CH_3-\overset{O}{\overset{\|}{C}}-OCH_2CH_3 + {}^-:CH_2-\overset{O}{\overset{\|}{C}}-OCH_2CH_3 \rightleftarrows CH_3-\overset{:O:^-}{\overset{|}{C}}(-\ddot{O}CH_2CH_3)-CH_2-\overset{O}{\overset{\|}{C}}-OCH_2CH_3 \rightleftarrows CH_3-\overset{O}{\overset{\|}{C}}-CH_2-\overset{O}{\overset{\|}{C}}-OCH_2CH_3 + {}^-:\ddot{O}CH_2CH_3$$

nucleophilic attack — expulsion of ethoxide — ethyl acetoacetate

In the presence of ethoxide ion, ethyl acetoacetate is deprotonated to give its enolate. This exothermic deprotonation helps to drive the reaction to completion.

$$\underset{(pK_a = 11)}{CH_3-\overset{O}{\overset{\|}{C}}-\overset{H}{\overset{|}{C}H}-\overset{O}{\overset{\|}{C}}-OCH_2CH_3} + {}^-:\ddot{O}CH_2CH_3 \rightleftarrows \underset{\text{enolate}}{CH_3-\overset{O}{\overset{\|}{C}}-\ddot{C}H-\overset{O}{\overset{\|}{C}}-OCH_2CH_3} + \underset{(pK_a = 18)}{H-\ddot{O}CH_2CH_3}$$

When the reaction is complete, the enolate ion is reprotonated to give ethyl acetoacetate.

$$\underset{\text{enolate}}{CH_3-\overset{O}{\overset{\|}{C}}-\ddot{C}H-\overset{O}{\overset{\|}{C}}-OCH_2CH_3} \xrightarrow{H_3O^+} \underset{\text{ethyl acetoacetate}}{CH_3-\overset{O}{\overset{\|}{C}}-\overset{H}{\overset{|}{C}H}-\overset{O}{\overset{\|}{C}}-OCH_2CH_3}$$

PROBLEM 22-38

Ethoxide is used as the base in the condensation of ethyl acetate to avoid some unwanted side reactions. Show what side reactions would occur if the following bases were used.

(a) sodium methoxide (b) sodium hydroxide

PROBLEM 22-39

Esters with only one α hydrogen generally give poor yields in the Claisen condensation. Give a mechanism for the Claisen condensation of ethyl isobutyrate, and explain why a poor yield is obtained.

PROBLEM 22-40

Predict the products of self-condensation of the following esters.

(a) methyl propanoate + $NaOCH_3$

(b) ethyl phenylacetate + $NaOCH_2CH_3$

(c) 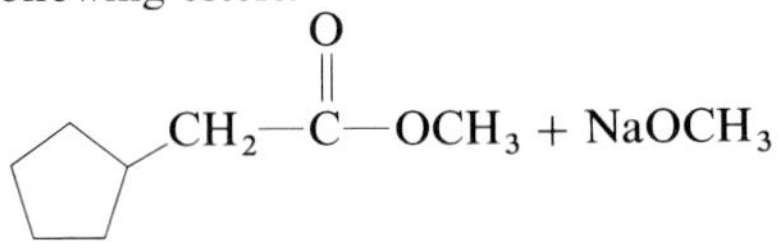

SOLVED PROBLEM 22-6

Show what ester would undergo Claisen condensation to give the following β-keto ester:

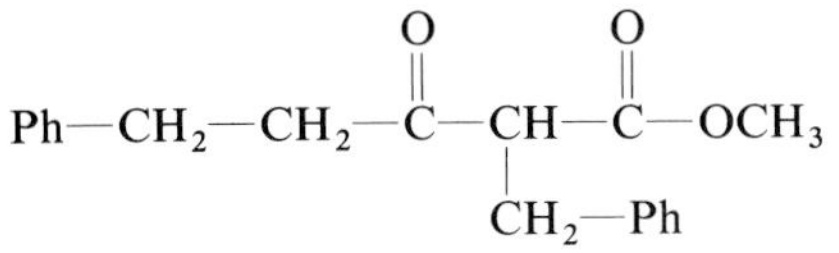

SOLUTION First, break the structure apart at the α,β bond (α,β to the ester carbonyl). This is the bond formed in the Claisen condensation.

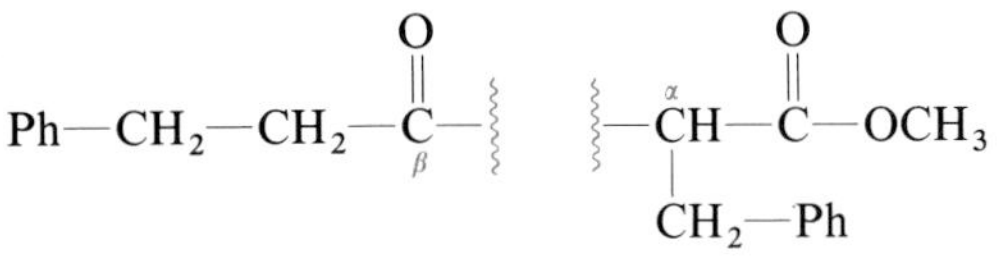

Next, replace the α proton that was lost, and replace the alkoxy group that was lost from the carbonyl. Two molecules of methyl 3-phenylpropionate result.

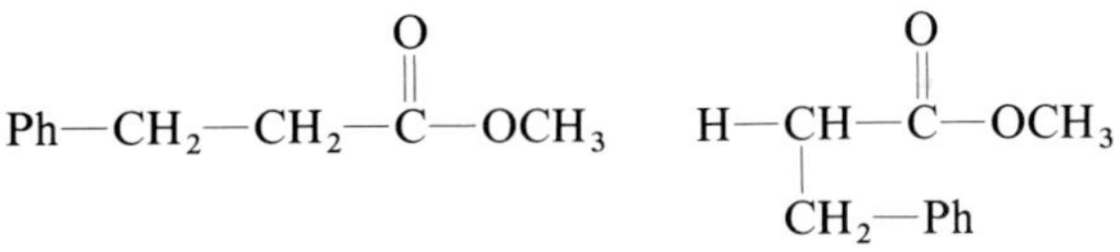

Now draw out the reaction. Sodium methoxide is used as the base because the reactants are methyl esters.

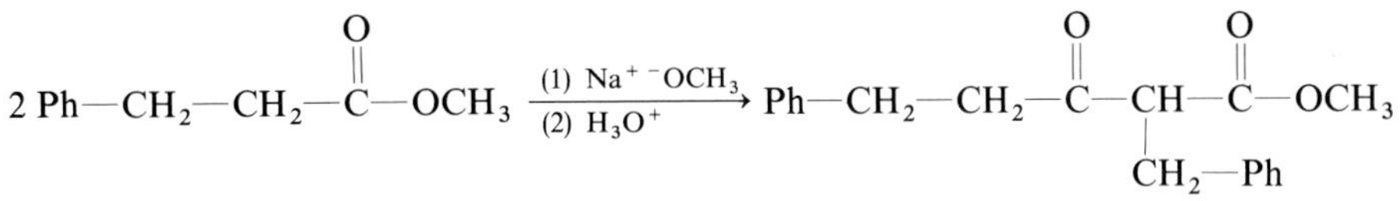

PROBLEM 22-41

Draw the mechanism for the self-condensation of methyl 3-phenylpropionate catalyzed by sodium methoxide.

PROBLEM 22-42

Show what esters would undergo Claisen condensation to give the following β-keto esters.

(a) 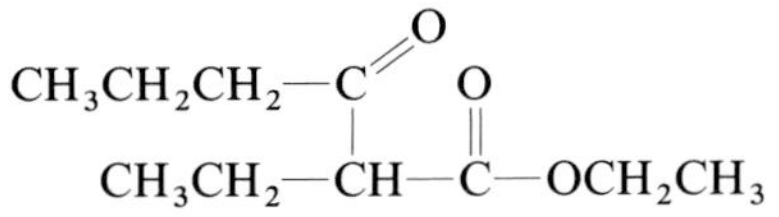

(b) 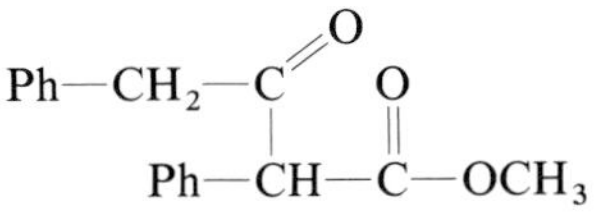

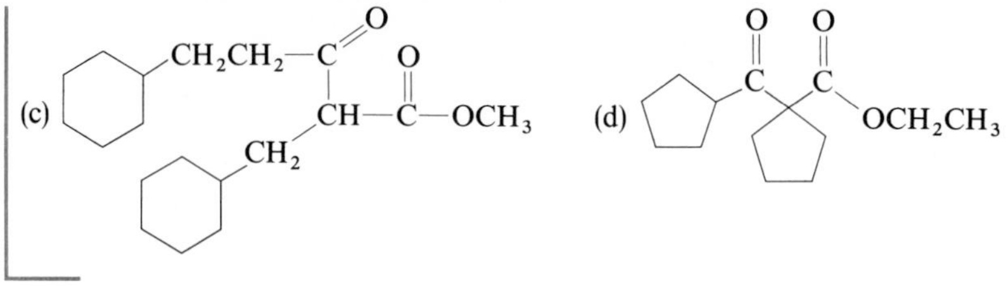

(c) [structure: cyclohexyl–CH_2CH_2–C(=O)–CH(CH_2–cyclohexyl)–C(=O)–OCH_3]

(d) [structure: cyclopentyl–C(=O)–C(cyclopentane)–C(=O)–OCH_2CH_3]

22-14 THE DIECKMANN CONDENSATION: A CLAISEN CYCLIZATION

When a diester undergoes an internal Claisen condensation, a ring is formed. Such an internal Claisen cyclization is called a **Dieckmann condensation** or a **Dieckmann cyclization.** Five-membered rings and six-membered rings are the most stable, and they are easily formed by Dieckmann condensations. Rings larger than six carbons or smaller than five carbons are rarely formed by this method.

The following examples of the Dieckmann condensation show that a 1,6-diester gives a five-membered ring, and a 1,7-diester gives a six-membered ring.

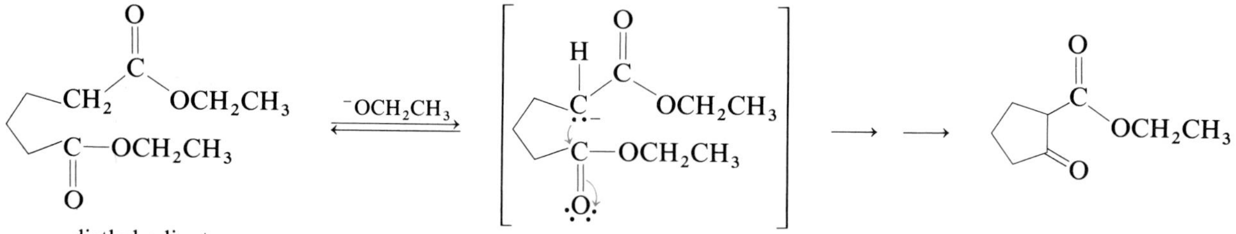

diethyl adipate (a 1,6-diester) $\xrightleftharpoons{^-OCH_2CH_3}$ [enolate intermediate] $\longrightarrow \longrightarrow$ cyclic β-keto ester (80%)

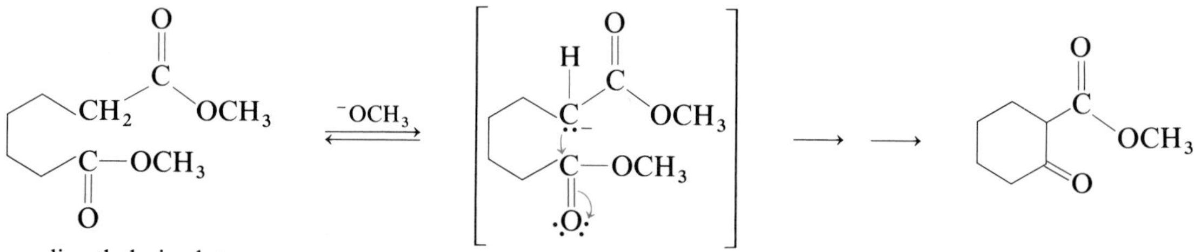

dimethyl pimelate (a 1,7-diester) $\xrightleftharpoons{^-OCH_3}$ [enolate intermediate] $\longrightarrow \longrightarrow$ cyclic β-keto ester

PROBLEM 22-43

Propose mechanisms for the two Dieckmann condensations shown above.

PROBLEM 22-44

Some of the following keto esters can be formed by Dieckmann condensation, but others cannot. Determine which ones are possible, and draw the starting diesters.

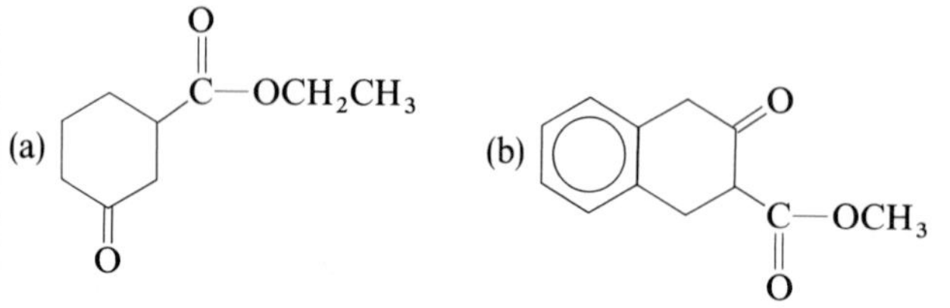

(a) [structure: cyclohexanone ring with C(=O)–OCH_2CH_3 substituent]

(b) [structure: tetralone ring with C(=O)–OCH_3 substituent]

(c) 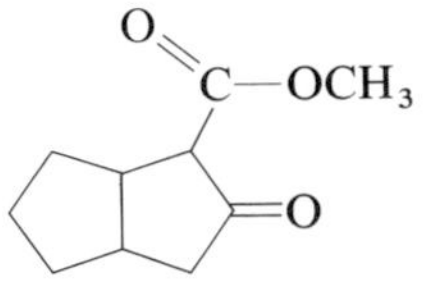 [structure: bicyclic ketone with $C(=O)OCH_3$ substituent; labels: O, C—OCH_3, =O]

(d) 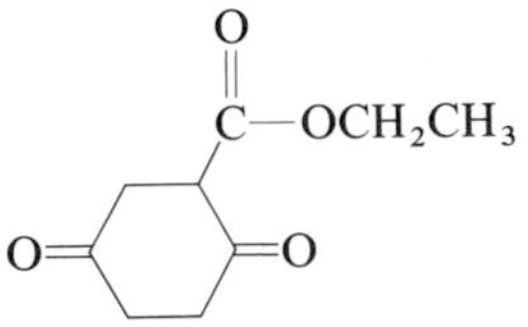 [structure: O= cyclohexane ring =O with $C(=O)OCH_2CH_3$ substituent; labels: O, C—OCH_2CH_3]

(Consider using a protecting group.)

22-15 CROSSED CLAISEN CONDENSATIONS

Claisen condensations can take place between different esters, particularly when only one of the esters has the α hydrogens needed to form an enolate. An ester without α hydrogens serves as the electrophilic component. Some useful esters without α hydrogens are benzoate, formate, carbonate, and oxalate esters.

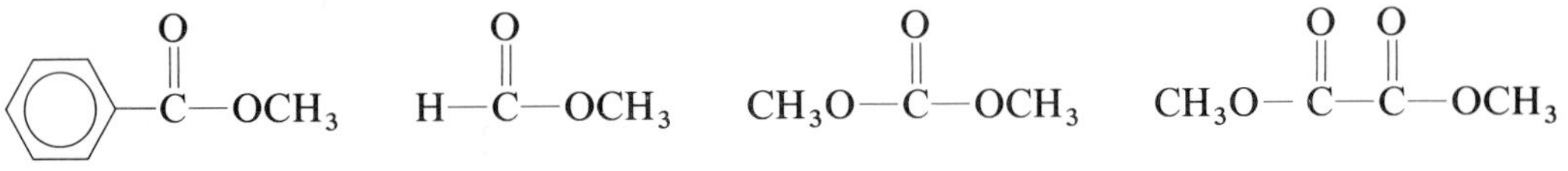

$$\mathrm{Ph{-}C({=}O){-}OCH_3} \qquad \mathrm{H{-}C({=}O){-}OCH_3} \qquad \mathrm{CH_3O{-}C({=}O){-}OCH_3} \qquad \mathrm{CH_3O{-}C({=}O){-}C({=}O){-}OCH_3}$$

methyl benzoate — methyl formate — dimethyl carbonate — dimethyl oxalate

A crossed Claisen condensation is carried out by first adding the ester without α hydrogens to a solution of the alkoxide base. The ester with α-hydrogens is slowly added to this solution, where it forms an enolate and condenses. The condensation of ethyl acetate with ethyl benzoate is an example of a crossed Claisen condensation.

$$\mathrm{Ph{-}C({=}O){-}OC_2H_5} + \mathrm{CH_3{-}C({=}O){-}OC_2H_5} \xrightarrow[(2)\ H_3O^+]{(1)\ Na^+\ {}^-OCH_2CH_3} \mathrm{Ph{-}C({=}O){-}CH_2{-}C({=}O){-}OC_2H_5}$$

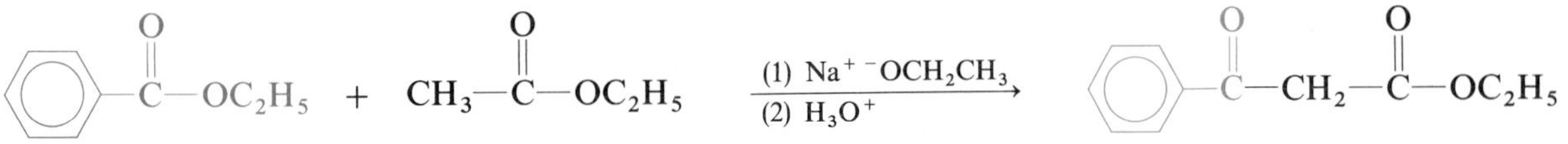

ethyl benzoate (no α hydrogens) — ethyl acetate (forms enolate) — ethyl benzoylacetate

PROBLEM 22-45

Propose a mechanism for the crossed Claisen condensation between ethyl acetate and ethyl benzoate.

PROBLEM 22-46

Predict the products from crossed Claisen condensation of the following pairs of esters. Indicate which combinations are poor choices for crossed Claisen condensations.

(a) $\mathrm{Ph{-}CH_2{-}C({=}O){-}OCH_3 + Ph{-}C({=}O){-}OCH_3 \longrightarrow}$

(b) $\mathrm{Ph{-}CH_2{-}C({=}O){-}OCH_3 + CH_3{-}C({=}O){-}OCH_3 \longrightarrow}$

(c) $\mathrm{CH_3{-}C({=}O){-}OC_2H_5 + C_2H_5O{-}C({=}O){-}C({=}O){-}OC_2H_5 \longrightarrow}$

(d) $\mathrm{CH_3{-}CH_2{-}C({=}O){-}OC_2H_5 + C_2H_5O{-}C({=}O){-}OC_2H_5 \longrightarrow}$

SOLVED PROBLEM 22-7

Show how a crossed Claisen condensation might be used to prepare

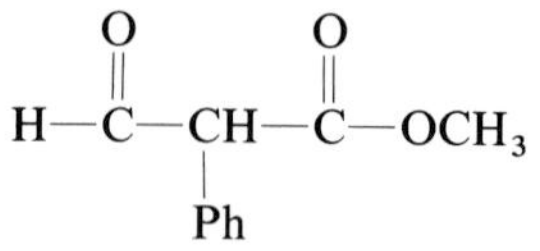

$$\mathrm{H{-}\overset{\overset{\displaystyle O}{\|}}{C}{-}\underset{\underset{\displaystyle Ph}{|}}{CH}{-}\overset{\overset{\displaystyle O}{\|}}{C}{-}OCH_3}$$

SOLUTION Break the α,β bond of this β-keto ester, since that is the bond formed in the Claisen condensation.

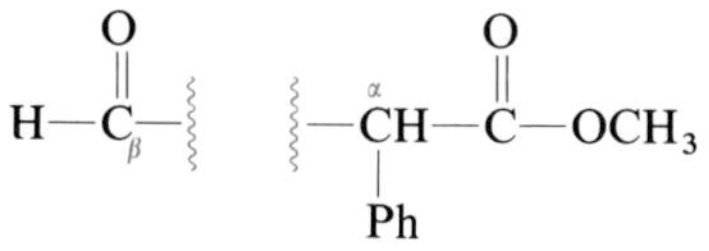

$$\mathrm{H{-}\overset{\overset{\displaystyle O}{\|}}{C}_{\beta}{-}\ \wr\ \wr\ {-}\underset{\underset{\displaystyle Ph}{|}}{\overset{\alpha}{C}H}{-}\overset{\overset{\displaystyle O}{\|}}{C}{-}OCH_3}$$

Now add the alkoxy group to the carbonyl and replace the proton on the α carbon.

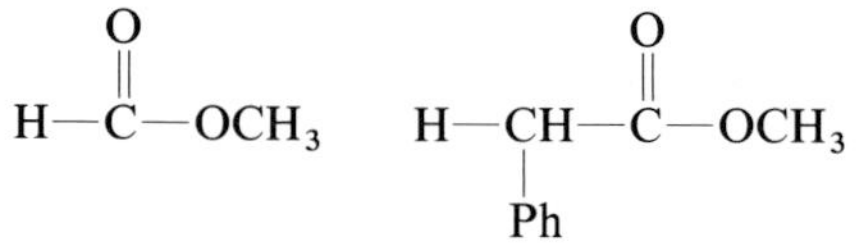

$$\mathrm{H{-}\overset{\overset{\displaystyle O}{\|}}{C}{-}OCH_3 \qquad H{-}\underset{\underset{\displaystyle Ph}{|}}{CH}{-}\overset{\overset{\displaystyle O}{\|}}{C}{-}OCH_3}$$

Write out the reaction, making sure that one of the components has α hydrogens and the other does not.

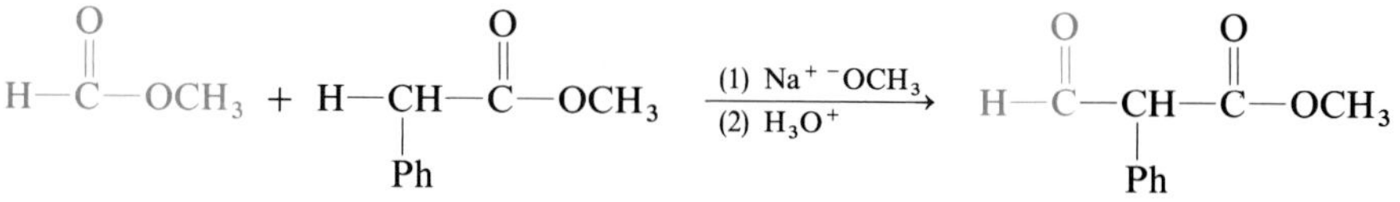

$$\underset{\text{no }\alpha\text{ hydrogens}}{\mathrm{H{-}\overset{\overset{\displaystyle O}{\|}}{C}{-}OCH_3}} + \underset{\text{forms enolate}}{\mathrm{H{-}\underset{\underset{\displaystyle Ph}{|}}{CH}{-}\overset{\overset{\displaystyle O}{\|}}{C}{-}OCH_3}} \xrightarrow[(2)\ \mathrm{H_3O^+}]{(1)\ \mathrm{Na^+\ {}^-OCH_3}} \mathrm{H{-}\overset{\overset{\displaystyle O}{\|}}{C}{-}\underset{\underset{\displaystyle Ph}{|}}{CH}{-}\overset{\overset{\displaystyle O}{\|}}{C}{-}OCH_3}$$

PROBLEM 22-47

Show how crossed Claisen condensations could be used to prepare the following esters.

(a) 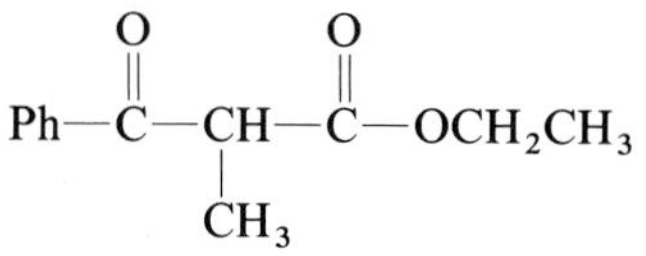 $\mathrm{Ph{-}\overset{\overset{\displaystyle O}{\|}}{C}{-}\underset{\underset{\displaystyle CH_3}{|}}{CH}{-}\overset{\overset{\displaystyle O}{\|}}{C}{-}OCH_2CH_3}$

(b) $\mathrm{Ph{-}\underset{\underset{\displaystyle O{=}C{-}\underset{\underset{\displaystyle O}{\|}}{C}{-}OCH_3}{|}}{CH}{-}\overset{\overset{\displaystyle O}{\|}}{C}{-}OCH_3}$

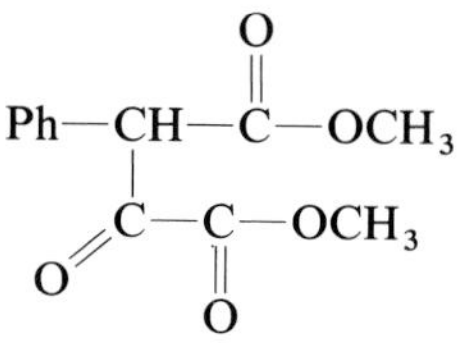

(c) $\mathrm{EtO{-}\overset{\overset{\displaystyle O}{\|}}{C}{-}\underset{\underset{\displaystyle Ph}{|}}{CH}{-}\overset{\overset{\displaystyle O}{\|}}{C}{-}OCH_2CH_3}$

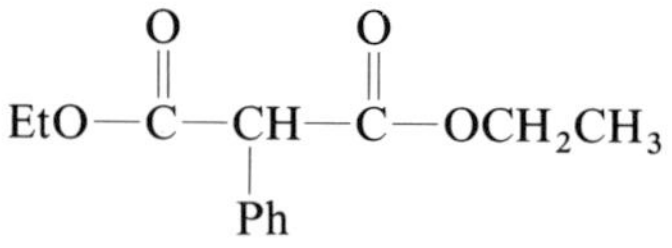

(d) 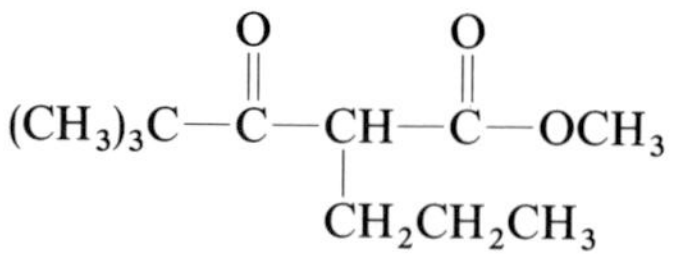 $\mathrm{(CH_3)_3C{-}\overset{\overset{\displaystyle O}{\|}}{C}{-}\underset{\underset{\displaystyle CH_2CH_2CH_3}{|}}{CH}{-}\overset{\overset{\displaystyle O}{\|}}{C}{-}OCH_3}$

Crossed Claisen condensations between ketones and esters are also possible. Ketones are more acidic than esters, and the ketone component is more likely to serve as the nucleophilic enolate component in the condensation.

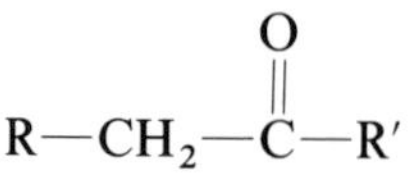

$$\underset{\substack{\text{ketone, p}K_a = 20\\ \text{more acidic}}}{\mathrm{R{-}CH_2{-}\overset{\overset{\displaystyle O}{\|}}{C}{-}R'}} \qquad \underset{\substack{\text{ester, p}K_a = 25\\ \text{less acidic}}}{\mathrm{R{-}CH_2{-}\overset{\overset{\displaystyle O}{\|}}{C}{-}OR'}}$$

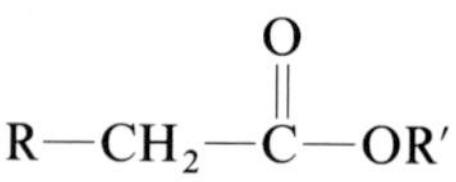

The condensation works best if the ester has no α hydrogens, so that it cannot form an enolate. Because of the difference in acidities, however, the reaction is often successful even between ketones and esters that both have α hydrogens. The following examples show some crossed Claisen condensations between ketones and esters. Notice the variety of difunctional and trifunctional compounds that can be produced by an appropriate choice of ester.

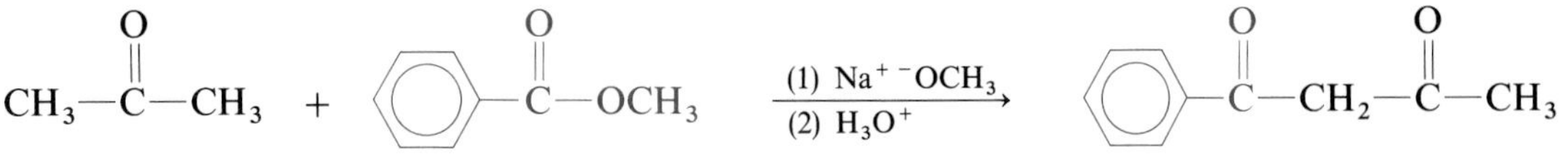

$$CH_3-\overset{O}{\overset{\|}{C}}-CH_3 + C_6H_5-\overset{O}{\overset{\|}{C}}-OCH_3 \xrightarrow[(2)\ H_3O^+]{(1)\ Na^+\ ^-OCH_3} C_6H_5-\overset{O}{\overset{\|}{C}}-CH_2-\overset{O}{\overset{\|}{C}}-CH_3$$

acetone | methyl benzoate | a β-diketone

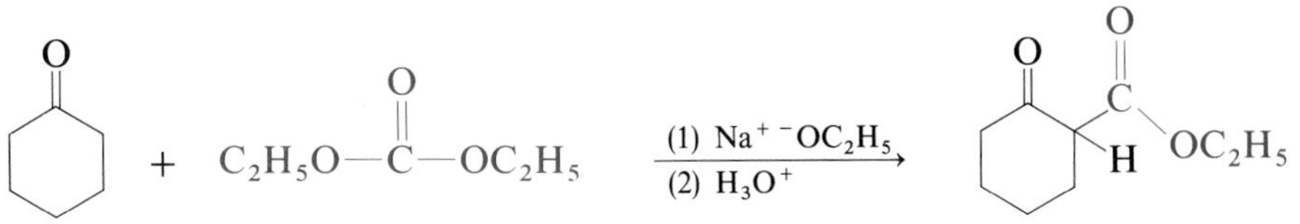

$$\text{cyclohexanone} + C_2H_5O-\overset{O}{\overset{\|}{C}}-OC_2H_5 \xrightarrow[(2)\ H_3O^+]{(1)\ Na^+\ ^-OC_2H_5} \text{2-(ethoxycarbonyl)cyclohexanone } (\overset{O}{\overset{\|}{C}}-OC_2H_5,\ H)$$

cyclohexanone | diethyl carbonate | a β-keto ester

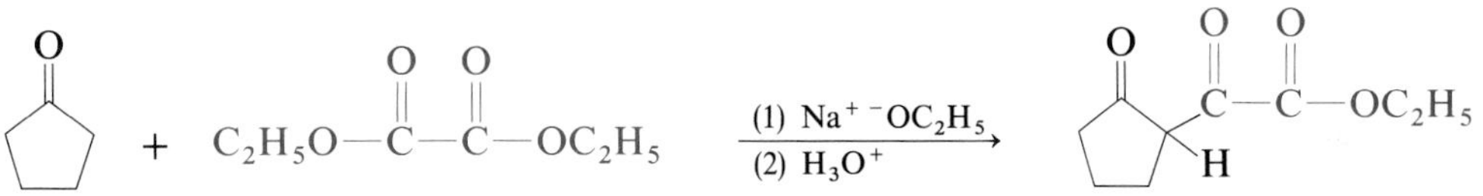

$$\text{cyclopentanone} + C_2H_5O-\overset{O}{\overset{\|}{C}}-\overset{O}{\overset{\|}{C}}-OC_2H_5 \xrightarrow[(2)\ H_3O^+]{(1)\ Na^+\ ^-OC_2H_5} \text{cyclopentanone}-\overset{O}{\overset{\|}{C}}-\overset{O}{\overset{\|}{C}}-OC_2H_5\ (H)$$

cyclopentanone | diethyl oxalate | an α,γ-diketo ester

PROBLEM 22-48

Predict the major products of the following crossed Claisen condensations.

(a) 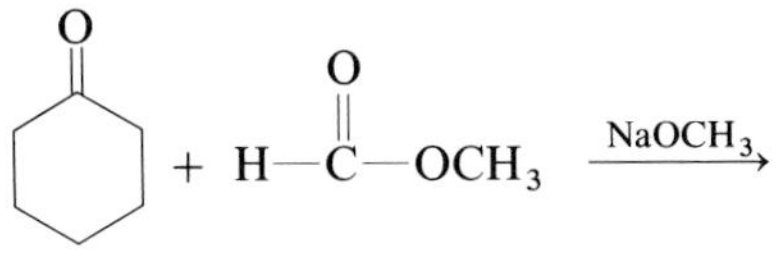 $\text{cyclohexanone} + H-\overset{O}{\overset{\|}{C}}-OCH_3 \xrightarrow{NaOCH_3}$

(b) $CH_3-\overset{O}{\overset{\|}{C}}-CH_3 + CH_3-\overset{O}{\overset{\|}{C}}-OCH_3 \xrightarrow{NaOCH_3}$

(c) $CH_3-\overset{O}{\overset{\|}{C}}-CH_2CH_2-\overset{O}{\overset{\|}{C}}-OCH_2CH_3 \xrightarrow{NaOCH_2CH_3}$

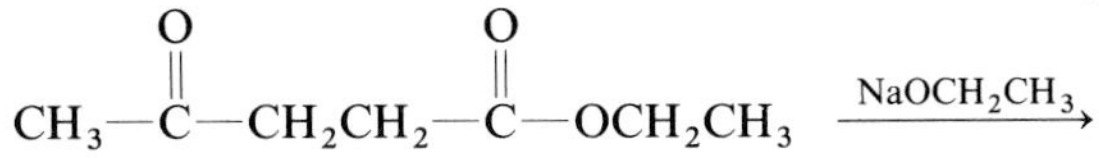

PROBLEM 22-49

Show how Claisen condensations could be used to synthesize the following compounds.

(a) 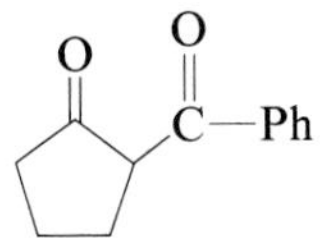 2-benzoylcyclopentanone ($\overset{O}{\overset{\|}{C}}-Ph$ on the carbon next to the ring C=O)

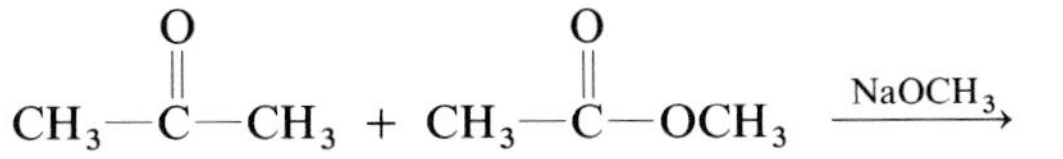

(b) $CH_3-CH_2-\overset{O}{\overset{\|}{C}}-\underset{\underset{O=C-\overset{\|}{\underset{O}{C}}-OCH_2CH_3}{|}}{CH}-CH_3$

(c) cyclohexane-1,3-dione

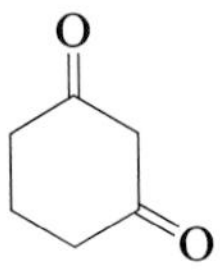

(d) 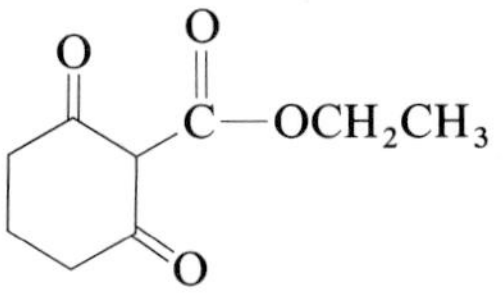 2-(ethoxycarbonyl)cyclohexane-1,3-dione ($\overset{O}{\overset{\|}{C}}-OCH_2CH_3$ between the two ring C=O groups)

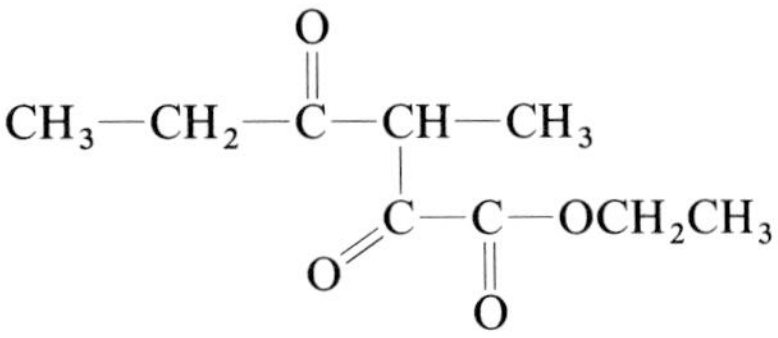

22-16 CONDENSATIONS USING A TEMPORARY ESTER GROUP

Many alkylation and acylation reactions are most effective using anions of β-dicarbonyl compounds that can be completely deprotonated and converted to their enolate ions by common bases such as methoxide ion and ethoxide ion. The *malonic ester synthesis* and the *acetoacetic ester synthesis* begin with β-dicarbonyl compounds that are easily alkylated and acylated, and at the end of the synthesis one of the carbonyl groups is removed by decarboxylation. First we consider the acidity advantages of β-dicarbonyl compounds, and then we consider how these compounds are used in synthesis.

Acidities of β-dicarbonyl compounds Table 22-1 compares the acidities of some carbonyl compounds with the acidities of alcohols and water. Notice the large increase in acidity for compounds with two carbonyl groups beta to each other. In fact, the α protons of the β-dicarbonyl compounds are more acidic than the hydroxyl protons of water and alcohols. This enhanced acidity results from increased stability of the enolate ion. The negative charge is delocalized over two carbonyl groups rather than just one, as shown by the resonance structures for the enolate ion of diethyl malonate (also called *malonic ester*).

TABLE 22-1
Typical acidities of carbonyl compounds

Conjugate acid	*Conjugate base*	pK_a
	Commonly used bases	
$H{-}O{-}H$ water	^{-}OH	15.7
$CH_3O{-}H$ methanol	CH_3O^-	15.5
$CH_3CH_2O{-}H$ ethanol	$CH_3CH_2O^-$	15.9
	Simple ketones and esters	
$CH_3{-}C(=O){-}CH_3$ acetone	$^{-}{:}CH_2{-}C(=O){-}CH_3$	20
$CH_3{-}C(=O){-}OCH_2CH_3$ ethyl acetate	$^{-}{:}CH_2{-}C(=O){-}OCH_2CH_3$	25
	β-dicarbonyl compounds	
$CH_3CH_2O{-}C(=O){-}CH_2{-}C(=O){-}OCH_2CH_3$ diethyl malonate (malonic ester)	$CH_3CH_2O{-}C(=O){-}\ddot{C}H^{-}{-}C(=O){-}OCH_2CH_3$	13
$CH_3{-}C(=O){-}CH_2{-}C(=O){-}OCH_2CH_3$ ethyl acetoacetate (acetoacetic ester)	$CH_3{-}C(=O){-}\ddot{C}H^{-}{-}C(=O){-}OCH_2CH_3$	11

$$CH_3CH_2O-\overset{O}{\overset{\|}{C}}-\underset{H\ \ H}{C}-\overset{O}{\overset{\|}{C}}-OCH_2CH_3 + {:}\ddot{\underset{..}{O}}CH_2CH_3 \rightleftharpoons \left[CH_3CH_2O-\overset{\ddot{O}}{\overset{\|}{C}}-\underset{H}{\overset{-}{\ddot{C}}}-\overset{\ddot{O}}{\overset{\|}{C}}-OCH_2CH_3 \longleftrightarrow \right.$$

diethyl malonate (malonic ester)
($pK_a = 13$)

$$\left. CH_3CH_2O-\overset{:\ddot{O}:^-}{\overset{|}{C}}=\underset{H}{C}-\overset{\ddot{O}}{\overset{\|}{C}}-OCH_2CH_3 \longleftrightarrow CH_3CH_2O-\overset{\ddot{O}}{\overset{\|}{C}}-\underset{H}{C}=\overset{:\ddot{O}:^-}{\overset{|}{C}}-OCH_2CH_3 \right]$$

resonance-stabilized enolate ion

PROBLEM 22-50
Show the resonance structures for the enolate ions that result when the following compounds are treated with a strong base.

(a) ethyl acetoacetate (b) 2,4-pentanedione
(c) ethyl α-cyanoacetate (d) nitroacetone

22-17 THE MALONIC ESTER SYNTHESIS

The **malonic ester synthesis** is used to synthesize substituted derivatives of acetic acid. Malonic ester (diethyl malonate) is alkylated or acylated on the carbon that is α to both carbonyl groups, and the resulting derivative is hydrolyzed and allowed to decarboxylate.

Malonic ester synthesis

$$H-\overset{O=C-OC_2H_5}{\underset{O=C-OC_2H_5}{C}}-H \xrightarrow[(2)\ R-X]{(1)\ NaOC_2H_5} R-\overset{O=C-OC_2H_5}{\underset{O=C-OC_2H_5}{C}}-H \xrightarrow[heat]{H_3O^+} R-\overset{H}{\underset{O=C-OH}{C}}-H + CO_2$$

malonic ester — alkylated malonic ester — substituted acetic acid

Malonic ester is completely deprotonated by treatment with sodium ethoxide. The resulting enolate ion may be alkylated by an unhindered alkyl halide, tosylate, or other electrophilic reagent.

$$CH_3CH_2O\overset{O}{\overset{\|}{C}}-CH_2-\overset{O}{\overset{\|}{C}}OCH_2CH_3 \xrightleftharpoons{Na^+\ ^-OCH_2CH_3} CH_3CH_2O\overset{O}{\overset{\|}{C}}-\underset{..}{C}H-\overset{O}{\overset{\|}{C}}OCH_2CH_3 \quad R-X \longrightarrow CH_3CH_2O\overset{O}{\overset{\|}{C}}-\underset{R}{CH}-\overset{O}{\overset{\|}{C}}OCH_2CH_3$$

malonic ester — alkylated diethyl malonate

Hydrolysis of the alkylated diethyl malonate (a diethyl alkylmalonic ester) gives a malonic acid derivative.

$$\underset{\text{a diethyl alkylmalonate}}{CH_3CH_2O-\overset{O}{\overset{\|}{C}}-\underset{R}{\underset{|}{CH}}-\overset{O}{\overset{\|}{C}}-OCH_2CH_3} \xrightarrow[H_2O]{H^+,\text{ heat}} \underset{\text{an alkylmalonic acid}}{\left[HO-\overset{O}{\overset{\|}{C}}-\underset{R}{\underset{|}{CH}}-\overset{O}{\overset{\|}{C}}-OH\right]}$$

Any carboxylic acid with a carbonyl group in the β position is prone to decarboxylate. At the temperature of the hydrolysis, the alkylmalonic acid loses CO_2 to give a substituted derivative of acetic acid. The decarboxylation takes place through a cyclic transition state, initially giving an enol form that quickly tautomerizes to the product.

$$\underset{\text{alkylmalonic acid}}{\text{(cyclic transition state)}} \longrightarrow \underset{CO_2 + \text{ enol}}{O{=}C{=}O \;+\; \underset{H}{\overset{R}{}}C{=}C(OH)_2} \xrightarrow{\text{tautomerization}} \underset{\text{substituted acetic acid}}{R-\underset{H}{\underset{|}{\overset{H}{\overset{|}{C}}}}-C(=O)OH} + CO_2\uparrow$$

The product of the malonic ester synthesis is a substituted acetic acid, with the substituent being the group used to alkylate malonic ester. In effect, the second carboxyl group is temporary, allowing the ester to be easily deprotonated and alkylated. Hydrolysis with decarboxylation removes the temporary carboxyl group, leaving the substituted acetic acid.

temporary ester group

$$\underset{\text{malonic ester}}{\overset{COOC_2H_5}{CH_2}-\overset{O}{\overset{\|}{C}}-OC_2H_5} \xrightarrow[(2)\ R-X]{(1)\ {}^-OC_2H_5} \underset{\text{alkylmalonic ester}}{R-\overset{COOC_2H_5}{CH}-\overset{O}{\overset{\|}{C}}-OC_2H_5} \xrightarrow[H_2O]{H^+,\text{ heat}} \underset{\text{substituted acetic acid}}{R-CH_2-\overset{O}{\overset{\|}{C}}-OH} \quad CO_2\uparrow$$

The alkylmalonic ester has a second acidic proton that can be removed by a base. Removing this proton and alkylating the resulting enolate with another alkyl halide gives a dialkylated malonic ester. Hydrolysis and decarboxylation leads to a disubstituted derivative of acetic acid.

$$\underset{\text{alkylmalonic ester}}{R-\overset{COOC_2H_5}{CH}-\overset{O}{\overset{\|}{C}}-OC_2H_5} \xrightarrow[(2)\ R'-X]{(1)\ NaOCH_2CH_3} \underset{\text{dialkylmalonic ester}}{R-\underset{R'}{\underset{|}{\overset{COOC_2H_5}{C}}}-\overset{O}{\overset{\|}{C}}-OC_2H_5} \xrightarrow[H_2O]{H^+,\text{ heat}} \underset{\text{disubstituted acetic acid}}{R-\underset{R'}{\underset{|}{\overset{H}{\overset{|}{C}}}}-\overset{O}{\overset{\|}{C}}-OH} \quad CO_2\uparrow$$

The malonic ester synthesis is useful for making cycloalkanecarboxylic acids, some of which are not easily made by any other method. The ring is formed from a dihalide, using a double alkylation of malonic ester. The following synthesis of cyclobutanecarboxylic acid shows that a strained four-membered ring system can be generated by this ester alkylation, even though most other condensations cannot form four-membered rings.

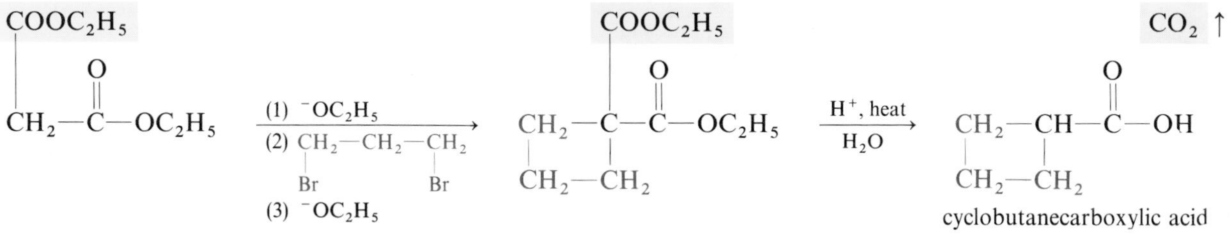

$$\underset{\text{(diethyl malonate)}}{CH_2(COOC_2H_5)-C(=O)-OC_2H_5} \xrightarrow[\substack{(2)\ BrCH_2-CH_2-CH_2Br \\ (3)\ ^-OC_2H_5}]{(1)\ ^-OC_2H_5} \underset{\text{(ring: } C-CH_2-CH_2-CH_2)}{C(COOC_2H_5)-C(=O)-OC_2H_5} \xrightarrow[H_2O]{H^+,\ heat} \underset{\text{cyclobutanecarboxylic acid}}{CH(CH_2CH_2CH_2)-C(=O)-OH} + CO_2\uparrow$$

SOLVED PROBLEM 22-8

Show how the malonic ester synthesis is used to prepare 2-benzylbutanoic acid.

SOLUTION 2-Benzylbutanoic acid is a substituted acetic acid having the substituents $Ph-CH_2-$ and CH_3CH_2-.

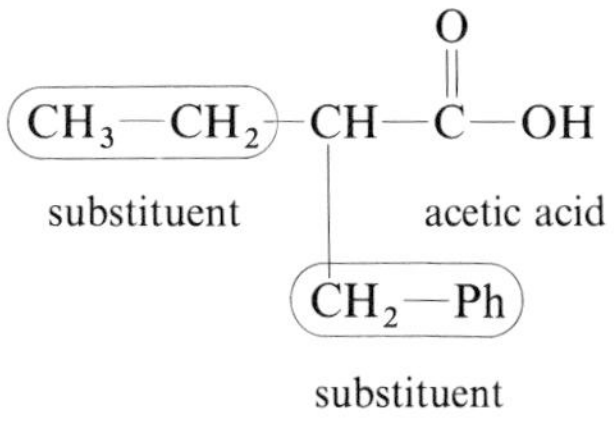

$$\underset{\text{substituent}}{(CH_3-CH_2)}-\underset{\text{acetic acid}}{CH(\underset{\text{substituent}}{CH_2-Ph})-C(=O)-OH}$$

Adding these substituents to the enolate of malonic ester eventually gives the correct product.

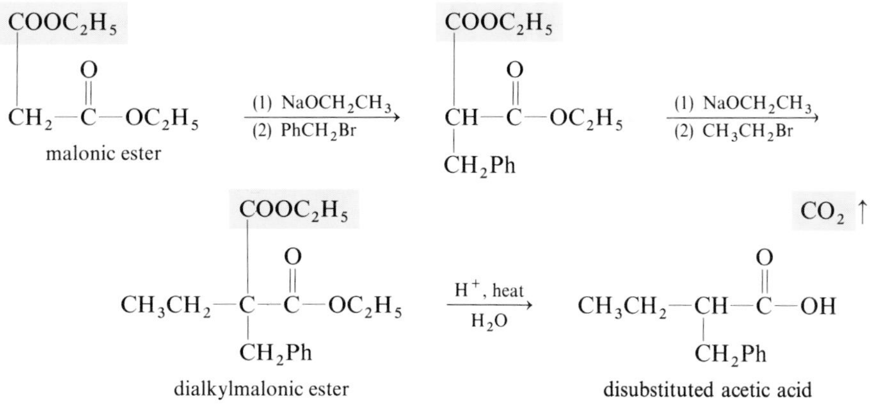

$$\underset{\text{malonic ester}}{CH_2(COOC_2H_5)-C(=O)-OC_2H_5} \xrightarrow[(2)\ PhCH_2Br]{(1)\ NaOCH_2CH_3} CH(COOC_2H_5)(CH_2Ph)-C(=O)-OC_2H_5 \xrightarrow[(2)\ CH_3CH_2Br]{(1)\ NaOCH_2CH_3}$$

$$\underset{\text{dialkylmalonic ester}}{CH_3CH_2-C(COOC_2H_5)(CH_2Ph)-C(=O)-OC_2H_5} \xrightarrow[H_2O]{H^+,\ heat} \underset{\text{disubstituted acetic acid}}{CH_3CH_2-CH(CH_2Ph)-C(=O)-OH} + CO_2\uparrow$$

PROBLEM 22-51

Show how the following compounds can be made using the malonic ester synthesis.

(a) 3-phenylpropanoic acid (b) 2-methylpropanoic acid
(c) 4-phenylbutanoic acid (d) cyclopentanecarboxylic acid

22-18 THE ACETOACETIC ESTER SYNTHESIS

The **acetoacetic ester synthesis** is similar to the malonic ester synthesis, but the final products are ketones: substituted derivatives of acetone. In the acetoacetic ester synthesis substituents are added to ethyl acetoacetate (acetoacetic ester), followed by hydrolysis and decarboxylation to produce an alkylated derivative of acetone.

$$\underset{\substack{\text{ethyl acetoacetate}\\\text{(acetoacetic ester)}}}{CH_3-\overset{O}{\overset{\|}{C}}-CH_2-\overset{O}{\overset{\|}{C}}-OC_2H_5} \xrightarrow[(2)\ R-X]{(1)\ ^-OC_2H_5} \underset{\text{alkylated ester}}{CH_3-\overset{O}{\overset{\|}{C}}-\overset{R}{\overset{|}{C}H}-\overset{O}{\overset{\|}{C}}-OC_2H_5} \xrightarrow[\text{heat}]{H_3O^+} \underset{\text{substituted acetone}}{CH_3-\overset{O}{\overset{\|}{C}}-\overset{R}{\overset{|}{C}H_2}} + CO_2\uparrow$$

Acetoacetic ester is like a molecule of acetone with a temporary ester group attached to enhance its acidity. Ethoxide ion completely deprotonates acetoacetic ester. The resulting enolate is alkylated by an unhindered alkyl halide or tosylate to give an alkylacetoacetic ester.

temporary ester group

$$\underset{\substack{\text{ethyl acetoacetate}\\(pKa = 11)}}{H-\overset{\overset{\displaystyle O=C-OC_2H_5}{|}}{\underset{H}{\underset{|}{C}}}-\overset{O}{\overset{\|}{C}}-CH_3} \underset{\longleftarrow}{\xrightarrow{^-OC_2H_5}} \underset{\text{enolate ion}}{^-\!:\overset{\overset{\displaystyle O=C-OC_2H_5}{|}}{\underset{H}{\underset{|}{C}}}-\overset{O}{\overset{\|}{C}}-CH_3} \xrightarrow{R-X} \underset{\text{alkylacetoacetic ester}}{R-\overset{\overset{\displaystyle O=C-OC_2H_5}{|}}{\underset{H}{\underset{|}{C}}}-\overset{O}{\overset{\|}{C}}-CH_3}$$

Acidic hydrolysis of the alkylacetoacetic ester initially gives an alkylacetoacetic acid. The alkylacetoacetic acid is a β-keto acid; the keto group in the β position allows decarboxylation to occur.

$$\underset{\substack{\text{alkylacetoacetic ester}\\(\text{a }\beta\text{-keto ester})}}{R-\overset{\overset{\displaystyle O=C-OC_2H_5}{|}}{CH}-\overset{O}{\overset{\|}{C}}-CH_3} \xrightarrow[H_2O]{H^+,\ \text{heat}} \underset{\substack{\text{alkylacetoacetic acid}\\(\text{a }\beta\text{-keto acid})}}{R-\overset{\overset{\displaystyle O=C-OH}{|}}{CH}-\overset{O}{\overset{\|}{C}}-CH_3} \xrightarrow{\text{decarboxylates}} \underset{\text{a substituted acetone}}{R-\overset{H}{\overset{|}{C}H}-\overset{O}{\overset{\|}{C}}-CH_3} \quad CO_2\uparrow$$

The β-keto acid decarboxylates by the same mechanism as the alkylmalonic acid in the malonic ester synthesis. A six-membered cyclic transition state splits out carbon dioxide to give the enol form of the substituted acetone. This decarboxylation usually takes place spontaneously at the temperature of the hydrolysis.

$$\underset{\text{the }\beta\text{-keto acid}}{\text{cyclic } O=C(O-H)-C(R)(H)-C(=O)CH_3} \longrightarrow \underset{CO_2\ +\ \text{enol}}{O=C=O \ +\ (R)(H)C=C(OH)CH_3} \xrightarrow{\text{tautomerization}} \underset{\text{a substituted acetone}}{R-\overset{H}{\overset{|}{\underset{H}{\underset{|}{C}}}}-\overset{O}{\overset{\|}{C}}-CH_3}$$

Disubstituted acetones are formed simply by alkylating acetoacetic ester a second time before the hydrolysis and decarboxylation steps, as shown in the following general synthesis.

temporary ester group

$$\underset{\text{}}{\mathrm{R{-}CH(COOC_2H_5){-}C({=}O){-}CH_3}} \xrightarrow[\text{(2) } R'{-}X]{\text{(1) } ^-OC_2H_5} \underset{\text{dialkylacetoacetic ester}}{\mathrm{R{-}C(COOC_2H_5)(R'){-}C({=}O){-}CH_3}} \xrightarrow[H_2O]{H^+,\ \text{heat}} \underset{\text{disubstituted acetone}}{\mathrm{R{-}CH(R'){-}C({=}O){-}CH_3}} + CO_2\uparrow$$

SOLVED PROBLEM 22-9

Show how the acetoacetic ester synthesis is used to make 3-propyl-5-hexen-2-one.

SOLUTION The target compound may be thought of as acetone with an *n*-propyl group and an allyl group as substituents:

$$\underbrace{\mathrm{CH_3CH_2CH_2}}_{n\text{-propyl group}}{-}\mathrm{CH}(\underbrace{\mathrm{CH_2{-}CH{=}CH_2}}_{\text{allyl group}}){-}\overbrace{\mathrm{C({=}O){-}CH_3}}^{\text{acetone}}$$

With an *n*-propyl halide and an allyl halide as the alkylating agents, the acetoacetic ester synthesis should produce 3-propyl-5-hexen-2-one.

Two alkylation steps give the necessary substitution:

$$\mathrm{CH_2(COOC_2H_5){-}C({=}O){-}CH_3} \xrightarrow[\text{(2) } CH_3CH_2CH_2Br]{\text{(1) } ^-OC_2H_5} \mathrm{CH_3CH_2CH_2{-}CH(COOC_2H_5){-}C({=}O){-}CH_3} \xrightarrow[\text{(2) } CH_2{=}CH{-}CH_2Br]{\text{(1) } ^-OC_2H_5} \mathrm{CH_3CH_2CH_2{-}C(COOC_2H_5)(CH_2{-}CH{=}CH_2){-}C({=}O){-}CH_3}$$

Hydrolysis proceeds with decarboxylation to give the disubstituted acetone product.

$$\mathrm{CH_3CH_2CH_2{-}C(COOC_2H_5)(CH_2{-}CH{=}CH_2){-}C({=}O){-}CH_3} \xrightarrow[H_2O]{H^+,\ \text{heat}} \left[\underset{\beta\text{-keto acid}}{\mathrm{CH_3CH_2CH_2{-}C(COOH)(CH_2{-}CH{=}CH_2){-}C({=}O){-}CH_3}}\right] \longrightarrow \underset{\text{3-propyl-5-hexen-2-one}}{\mathrm{CH_3CH_2CH_2{-}CH(CH_2{-}CH{=}CH_2){-}C({=}O){-}CH_3}} + CO_2\uparrow$$

PROBLEM 22-52

Show the ketones that would result from hydrolysis/decarboxylation of the following β-keto esters.

(a) $\mathrm{PhCH_2{-}CH(COOC_2H_5){-}C({=}O){-}CH_3}$ (b) $\mathrm{CH_2{-}C(COOCH_3){-}C({=}O){-}CH_3}$ with $\mathrm{CH_2{-}CH_2}$ ring (c) 2-oxocyclopentyl $\mathrm{{-}C({=}O){-}OCH_2CH_3}$

PROBLEM 22-53

Show how the following ketones might be synthesized using the acetoacetic ester synthesis.

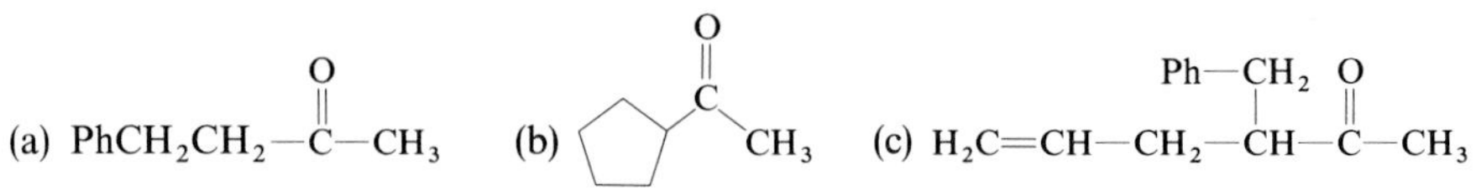

(a) $PhCH_2CH_2-\overset{O}{\overset{\|}{C}}-CH_3$ (b) cyclopentyl–$\overset{O}{\overset{\|}{C}}$–$CH_3$ (c) $H_2C{=}CH-CH_2-\underset{}{CH}(CH_2-Ph)-\overset{O}{\overset{\|}{C}}-CH_3$

22-19 CONJUGATE ADDITIONS: THE MICHAEL REACTION

α,β-Unsaturated carbonyl compounds have unusually reactive double bonds. The β-carbon atom is electrophilic because it shares the partial positive charge of the carbonyl carbon through resonance.

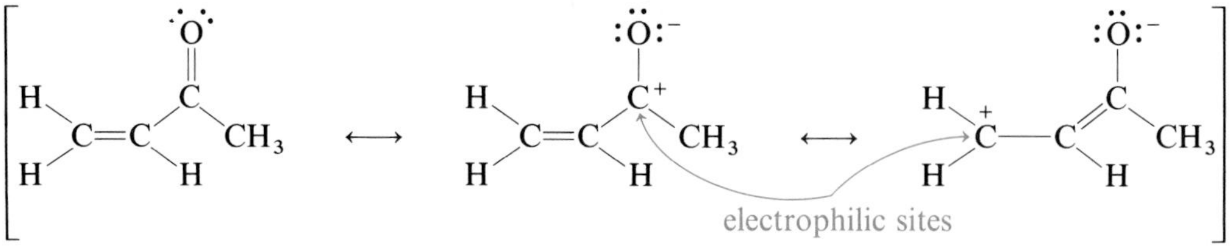

$$\left[H_2C{=}CH-C({=}\ddot{O}{:})-CH_3 \longleftrightarrow H_2C{=}CH-C^+(-\ddot{O}{:}^-)-CH_3 \longleftrightarrow H_2\overset{+}{C}-CH{=}C(-\ddot{O}{:}^-)-CH_3 \right]$$

electrophilic sites

A nucleophile can attack an α,β-unsaturated carbonyl compound either at the carbonyl group itself or at the β position. When attack occurs at the carbonyl group, protonation of the oxygen leads to a product with the nucleophile and the proton having added to adjacent atoms: **1,2-addition.** When attack occurs at the β position, the oxygen atom is the fourth atom counting from the nucleophile, and the addition is called a **1,4-addition.** The net result of 1,4-addition is the addition of the nucleophile and a hydrogen atom across a double bond that was conjugated with a carbonyl group. For this reason, 1,4-addition is often called **conjugate addition.**

1,2-addition reaction

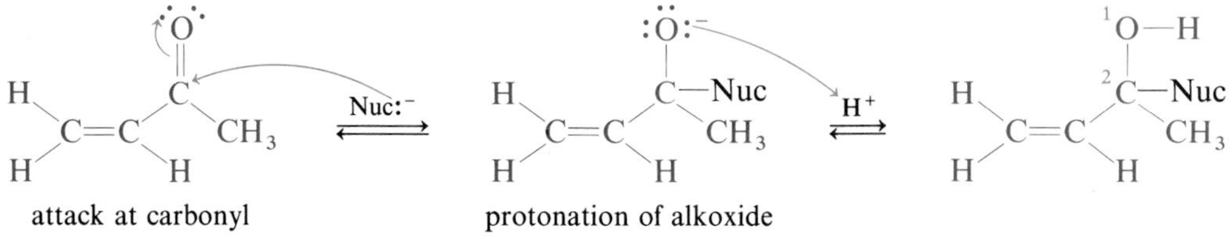

$$H_2C{=}CH-C({=}\ddot{O}{:})-CH_3 \underset{\text{attack at carbonyl}}{\overset{Nuc{:}^-}{\rightleftharpoons}} H_2C{=}CH-C(-\ddot{O}{:}^-)(-Nuc)-CH_3 \underset{\text{protonation of alkoxide}}{\overset{H^+}{\rightleftharpoons}} H_2C{=}CH-\overset{2}{C}(-\overset{1}{O}-H)(-Nuc)-CH_3$$

attack at carbonyl — protonation of alkoxide

1,4-addition reaction: Michael addition

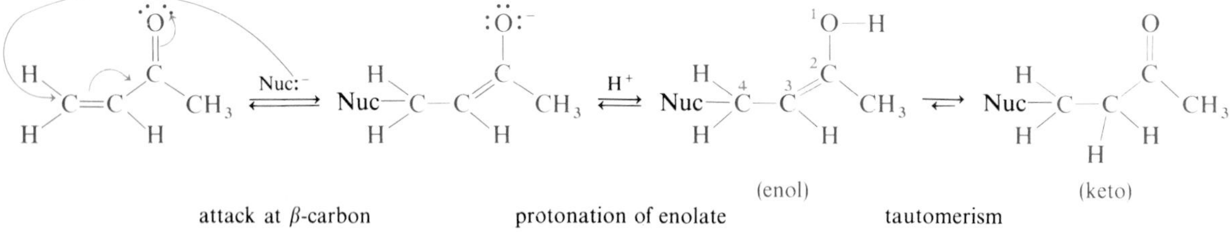

$$H_2C{=}CH-C({=}\ddot{O}{:})-CH_3 \overset{Nuc{:}^-}{\rightleftharpoons} Nuc-CH_2-CH{=}C(-\ddot{O}{:}^-)-CH_3 \overset{H^+}{\rightleftharpoons} Nuc-\overset{4}{C}H_2-\overset{3}{C}H{=}\overset{2}{C}(-\overset{1}{O}-H)-CH_3 \rightleftharpoons Nuc-CH_2-CH_2-C({=}O)-CH_3$$

(enol) (keto)

attack at β-carbon — protonation of enolate — tautomerism

The addition of a stabilized enolate ion to the double bond of an α,β-unsaturated carbonyl compound is called a **Michael addition.** The electrophile (the α,β-unsaturated carbonyl compound) accepts a pair of electrons; it is called the **Michael acceptor.** The attacking nucleophile donates a pair of electrons; it is called the **Michael donor.** A wide variety of compounds can serve as Michael donors and acceptors. Some of the most common ones are shown in Table 22-2. Common donors are enolate ions that are stabilized by two strong electron-withdrawing groups such as carbonyl groups, cyano groups, or nitro groups. Common acceptors contain a double bond conjugated with a carbonyl group, a cyano group, or a nitro group.

TABLE 22-2
Some common Michael donors and Michael acceptors

Michael donors		*Michael acceptors*	
$R-\overset{O}{\overset{\Vert}{C}}-\overset{\cdot\cdot -}{C}H-\overset{O}{\overset{\Vert}{C}}-R'$	β-diketone	$H_2C{=}CH-\overset{O}{\overset{\Vert}{C}}-H$	conjugated aldehyde
$R-\overset{O}{\overset{\Vert}{C}}-\overset{\cdot\cdot -}{C}H-\overset{O}{\overset{\Vert}{C}}-OR'$	β-keto ester	$H_2C{=}CH-\overset{O}{\overset{\Vert}{C}}-R$	conjugated ketone
$R_2\,CuLi$	dialkyl cuprate	$H_2C{=}CH-\overset{O}{\overset{\Vert}{C}}-OR$	conjugated ester
$\rangle\ddot{N}-C{=}C\langle$	enamine	$H_2C{=}CH-\overset{O}{\overset{\Vert}{C}}-NH_2$	conjugated amide
$R-\overset{O}{\overset{\Vert}{C}}-\overset{\cdot\cdot -}{C}H-C{\equiv}N$	β-keto nitrile	$H_2C{=}CH-C{\equiv}N$	conjugated nitrile
$R-\overset{O}{\overset{\Vert}{C}}-\overset{\cdot\cdot -}{C}H-NO_2$	α-nitro ketone	$H_2C{=}CH-NO_2$	nitroethylene

Let's consider a typical Michael addition, addition of the malonic ester enolate to methyl vinyl ketone (MVK). The crucial step is the nucleophilic attack by the enolate at the β-carbon atom. The resulting enolate is strongly basic, and it is quickly protonated.

$$\underset{\text{malonic ester enolate}}{C_2H_5O-\overset{O}{\overset{\Vert}{C}}-\underset{|}{\overset{COOC_2H_5}{C}}H:^-} + \underset{\text{MVK}}{H_2C{=}CH-\overset{O}{\overset{\Vert}{C}}-CH_3} \rightleftarrows \underset{\text{product enolate}}{C_2H_5O-\overset{O}{\overset{\Vert}{C}}-\overset{COOC_2H_5}{CH}-CH_2-CH{=}C(-\ddot{O}:^-)-CH_3} \xrightarrow{H-OC_2H_5} \rightleftarrows \underset{\substack{\text{1,4-addition product}\\(90\%)}}{C_2H_5O-\overset{O}{\overset{\Vert}{C}}-\overset{COOC_2H_5}{CH}CH_2CH_2\overset{O}{\overset{\Vert}{C}}CH_3}$$

The product of this Michael addition may be treated like an any other substituted malonic ester. Acid hydrolysis and decarboxylation lead to a δ-keto acid. It is not easy to imagine other ways to synthesize this interesting keto acid.

$$\underset{\text{1,4-addition product}}{C_2H_5O-\overset{O}{\overset{\Vert}{C}}-\overset{COOC_2H_5}{CH}CH_2CH_2\overset{O}{\overset{\Vert}{C}}CH_3} \xrightarrow[H_2O]{H^+,\text{ heat}} \underset{\text{substituted malonic acid}}{\left[HO-\overset{O}{\overset{\Vert}{C}}-\overset{COOH}{CH}CH_2CH_2\overset{O}{\overset{\Vert}{C}}CH_3\right]} \longrightarrow \underset{\text{a }\delta\text{-keto acid}}{HO-\overset{O}{\overset{\Vert}{C}}-CH_2CH_2CH_2\overset{O}{\overset{\Vert}{C}}CH_3} + CO_2\uparrow$$

SOLVED PROBLEM 22-10

Show how the following diketone might be synthesized using a Michael addition.

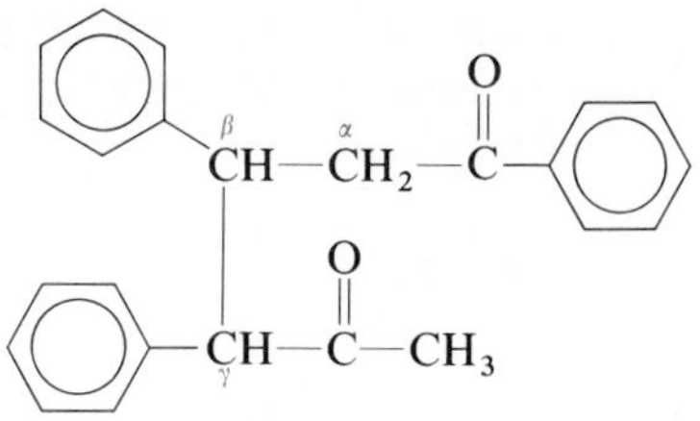

SOLUTION A Michael addition would have formed a new bond at the β-carbon atom. Therefore, we break this molecule apart at the β,γ bond.

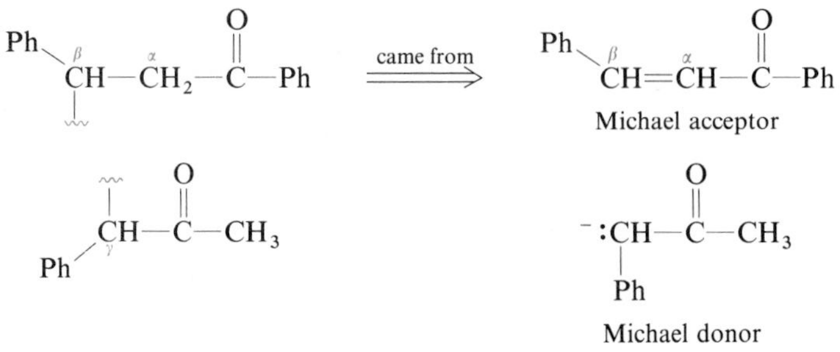

came from

Michael acceptor

Michael donor

The top fragment, where we broke the β bond, must have come from a conjugated ketone, and it must have been the Michael acceptor. The bottom fragment is a simple ketone. It is unlikely that the ketone was used without some sort of additional stabilizing group. We can add a temporary ester group to the ketone (making a substituted acetoacetic ester), and use the acetoacetic ester synthesis to give the correct product.

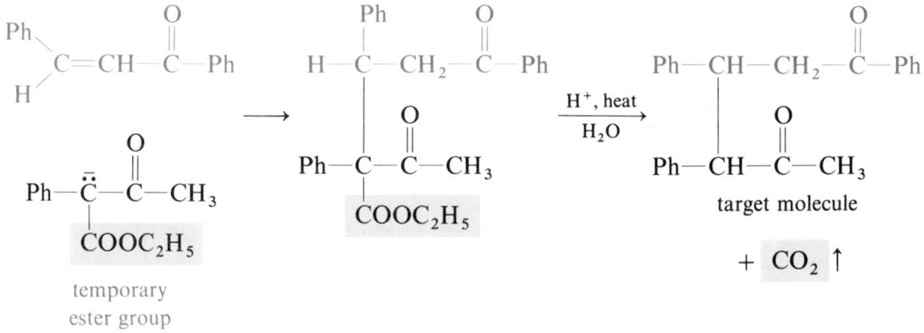

$\xrightarrow[H_2O]{H^+,\ heat}$

target molecule

$+ CO_2 \uparrow$

temporary ester group

PROBLEM 22-54

In Solved Problem 22-10 the target molecule was synthesized using a Michael addition to form the bond that is β,γ to the upper carbonyl group. Another approach is to use a Michael addition to form the bond that is β,γ to the other (lower) carbonyl group. Show how you would accomplish this alternative synthesis.

PROBLEM 22-55

Show how cyclohexanone might be converted to the following δ-diketone. (*Hint:* Stork)

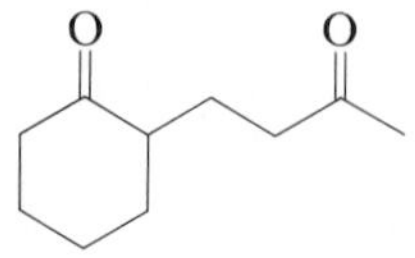

PROBLEM 22-56

Show how an acetoacetic ester synthesis might be used to form a δ-diketone.

PROBLEM 22-57

Give a mechanism for the conjugate addition of a nucleophile (Nuc:$^-$) to acrylonitrile ($H_2C{=}CHCN$) and to nitroethylene. Use resonance structures to show how the cyano and nitro groups activate the double bond toward conjugate addition.

PROBLEM 22-58

Show how the following products might be synthesized from suitable Michael donors and acceptors.

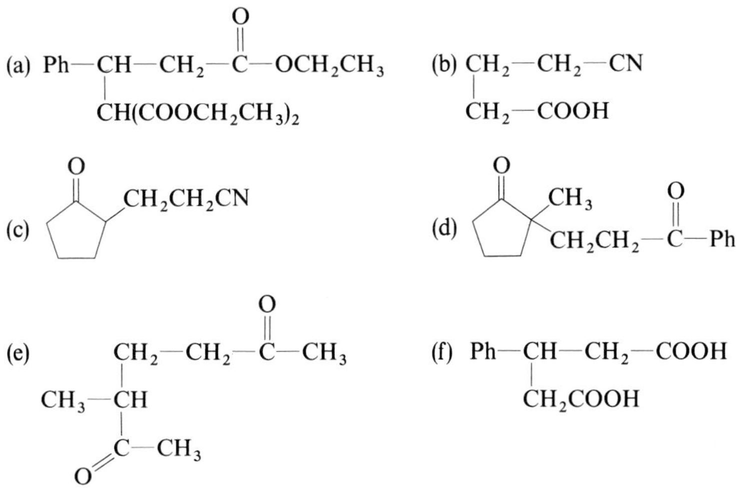

(a) $Ph{-}CH(CH(COOCH_2CH_3)_2){-}CH_2{-}C(=O){-}OCH_2CH_3$

(b) $CH_2(CH_2{-}COOH){-}CH_2{-}CN$

(c) 2-(CH_2CH_2CN)cyclopentanone

(d) 2-CH_3-2-($CH_2CH_2{-}C(=O){-}Ph$)cyclopentanone

(e) $CH_3{-}C(=O){-}CH(CH_3){-}CH_2{-}CH_2{-}C(=O){-}CH_3$

(f) $Ph{-}CH(CH_2COOH){-}CH_2{-}COOH$

22-20 THE ROBINSON ANNULATION

We have seen that the Michael addition of a ketone enolate (or its enamine) to an α,β-unsaturated ketone leads to a δ-diketone. If the conjugate addition takes place under strongly basic or acidic conditions, the δ-diketone undergoes a spontaneous intramolecular aldol condensation, usually with dehydration, to give a new six-membered ring: a conjugated cyclohexenone. This synthesis is called the **Robinson annulation** ("ring-forming") reaction.

The Robinson annulation

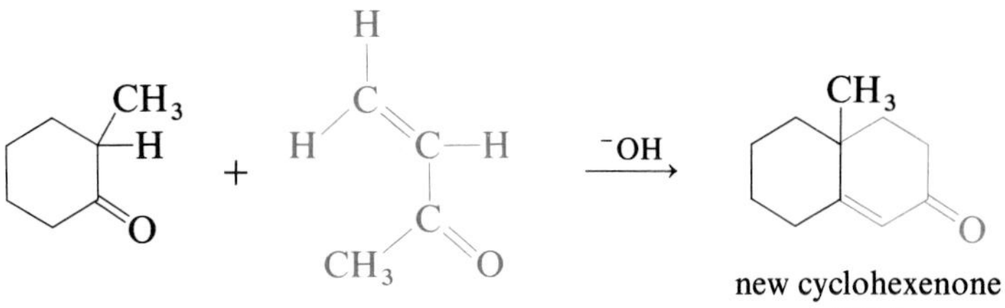

2-methylcyclohexanone (CH_3, H) + $H_2C{=}CH{-}C(=O){-}CH_3$ $\xrightarrow{^-OH}$ new cyclohexenone (65%) (CH_3, O)

Consider an example using a substituted cyclohexanone as the Michael donor and methyl vinyl ketone (MVK) as the Michael acceptor. When the enolate of the cyclohexanone adds to MVK, the product is a δ-diketone.

Step 1: Michael addition

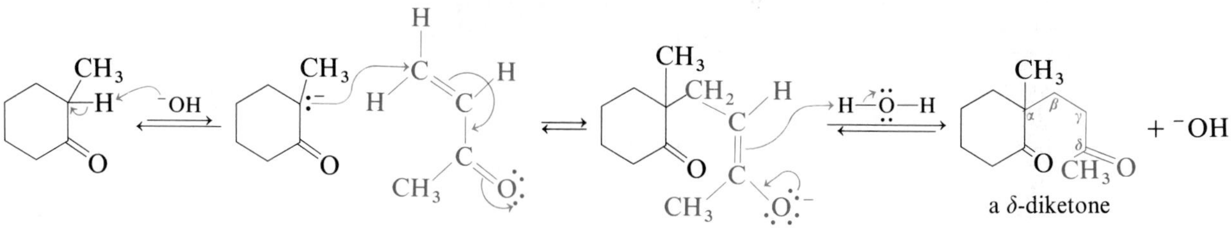

a δ-diketone

We might imagine this δ-diketone taking part in several different aldol condensations, but it is ideally suited for a particularly favorable one: the formation of a six-membered ring. To form a six-membered ring, the enolate of the methyl ketone attacks the cyclohexanone carbonyl. The aldol product dehydrates spontaneously to give a cyclohexenone.

Step 2: Cyclic aldol

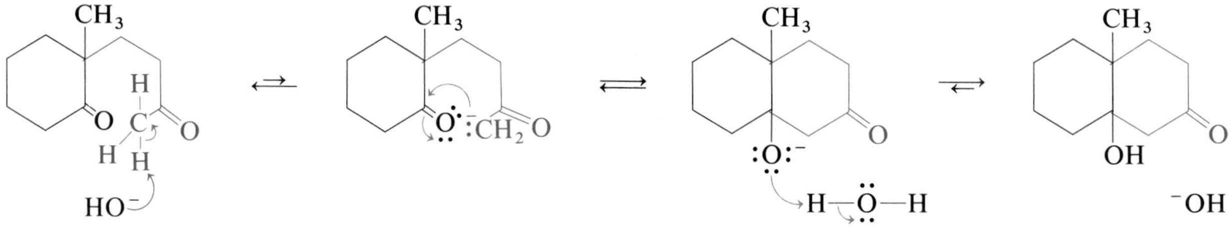

Step 3: Dehydration

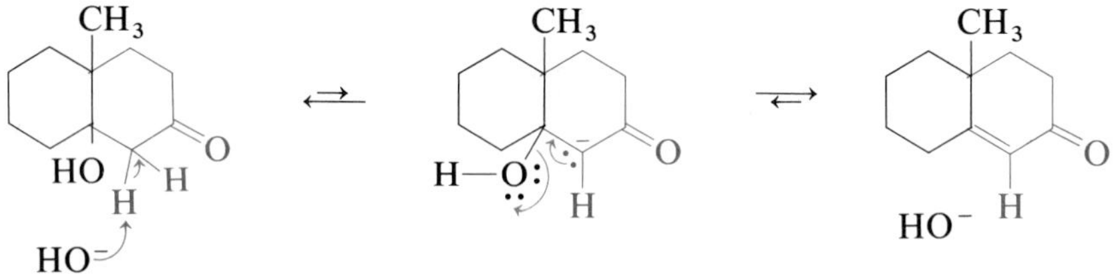

It is not difficult to predict the products of the Robinson annulation and to draw the mechanisms if you remember that the Michael addition is first, followed by an aldol condensation with dehydration to give a cyclohexenone.

PROBLEM SOLVING: PROPOSING REACTION MECHANISMS

This problem-solving example addresses a complicated base-catalyzed reaction, using the system for proposing reaction mechanisms summarized in Appendix 4. The problem is to propose a mechanism for the base-catalyzed reaction of ethyl acetoacetate with methyl vinyl ketone.

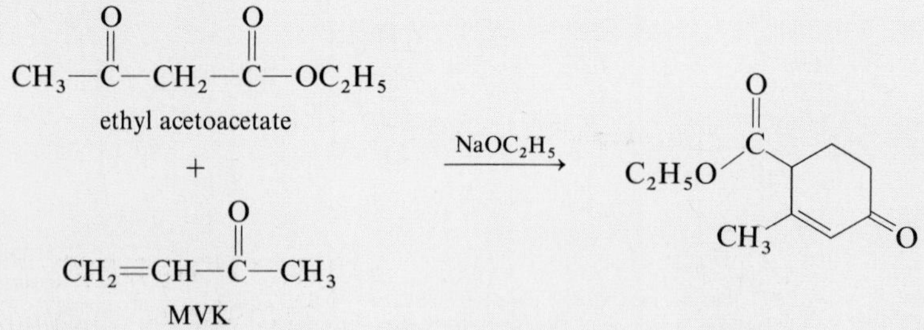

$$CH_3-\overset{O}{\overset{\|}{C}}-CH_2-\overset{O}{\overset{\|}{C}}-OC_2H_5$$

ethyl acetoacetate

+

$$CH_2{=}CH-\overset{O}{\overset{\|}{C}}-CH_3$$

MVK

$\xrightarrow{NaOC_2H_5}$

First, we must determine the type of mechanism. The use of a basic catalyst suggests the reaction involves strong nucleophiles as intermediates. We expect to see anionic intermediates (possibly stabilized carbanions), but no strong electrophiles or strong acids, and certainly no carbocations or free radicals.

1. **Consider the carbon skeletons of the reactants and products, and decide which carbon atoms in the products are most likely derived from which carbon atoms in the reactants.**
The ester group in the product is clearly derived from ethyl acetoacetate. The beta-carbon from the ester (now part of the C=C double bond) should be derived from the ketone of ethyl acetoacetate. The structure of MVK can be seen in the remaining four carbons.

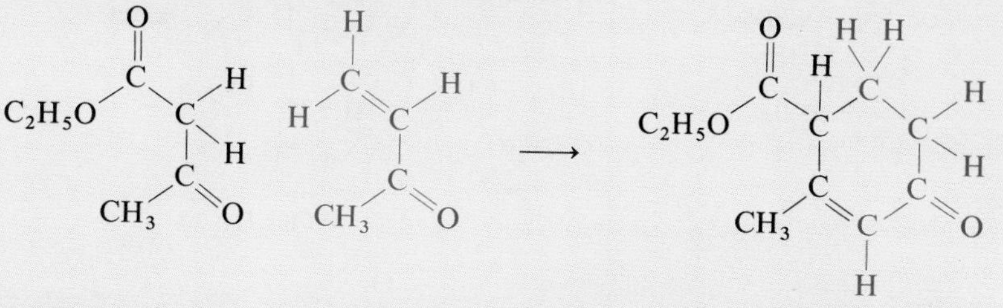

2. **Consider whether any of the reactants is a strong enough nucleophile to react without being activated. If not, consider how one of the reactants might be converted to a strong nucleophile by deprotonation of an acidic site or by attack on an electrophilic site.**
Neither of these reactants is a strong enough nucleophile to attack the other. Ethyl acetoacetate is more acidic than ethanol; ethoxide ion quickly removes a proton, giving the enolate ion.

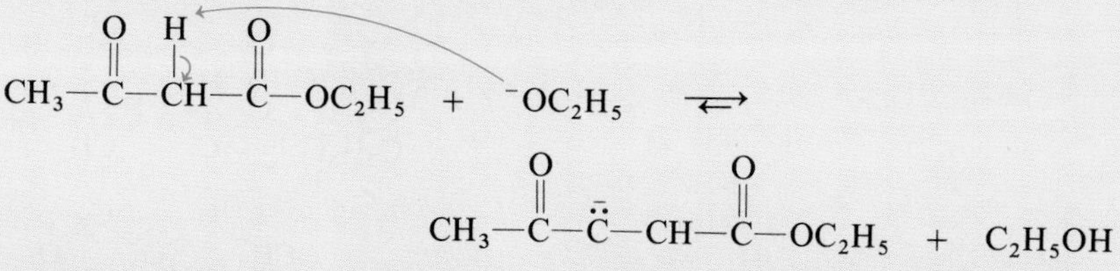

$$CH_3-\overset{O}{\overset{\|}{C}}-\overset{H}{\overset{|}{C}}H-\overset{O}{\overset{\|}{C}}-OC_2H_5 + {}^{-}OC_2H_5 \rightleftharpoons CH_3-\overset{O}{\overset{\|}{C}}-\ddot{C}^{-}H-\overset{O}{\overset{\|}{C}}-OC_2H_5 + C_2H_5OH$$

3. **Consider how an electrophilic site on another reactant (or, in a cyclization, another part of the same molecule) can undergo attack by the strong nucleophile to form a bond needed in the product. Draw the product of this bond formation.**
The enolate of acetoacetic ester might attack either the electrophilic double bond (Michael addition) or the carbonyl group of MVK. A Michael addition forms one of the bonds needed in the product.

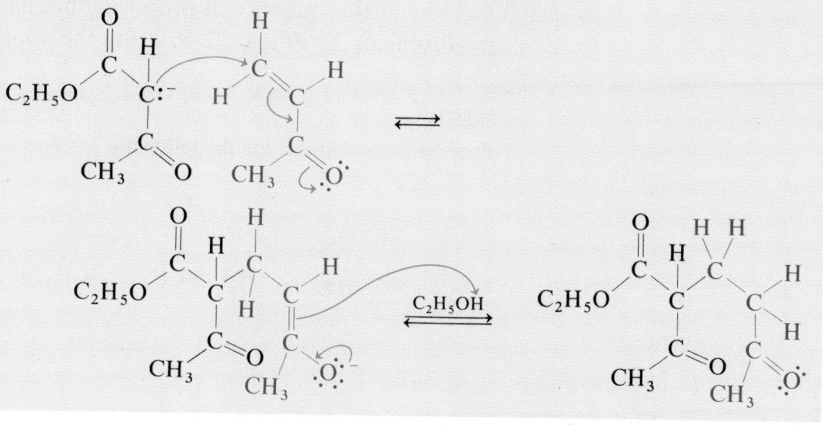

C_2H_5OH

4. **Consider how the product of nucleophilic attack might be converted to the final product (if it has the right carbon skeleton) or reactivated to form another bond needed in the product.**
The carbonyl group of ethyl acetoacetate must be converted to a C=C double bond in the α,β position of the other ketone. This conversion corresponds to an aldol condensation with dehydration.

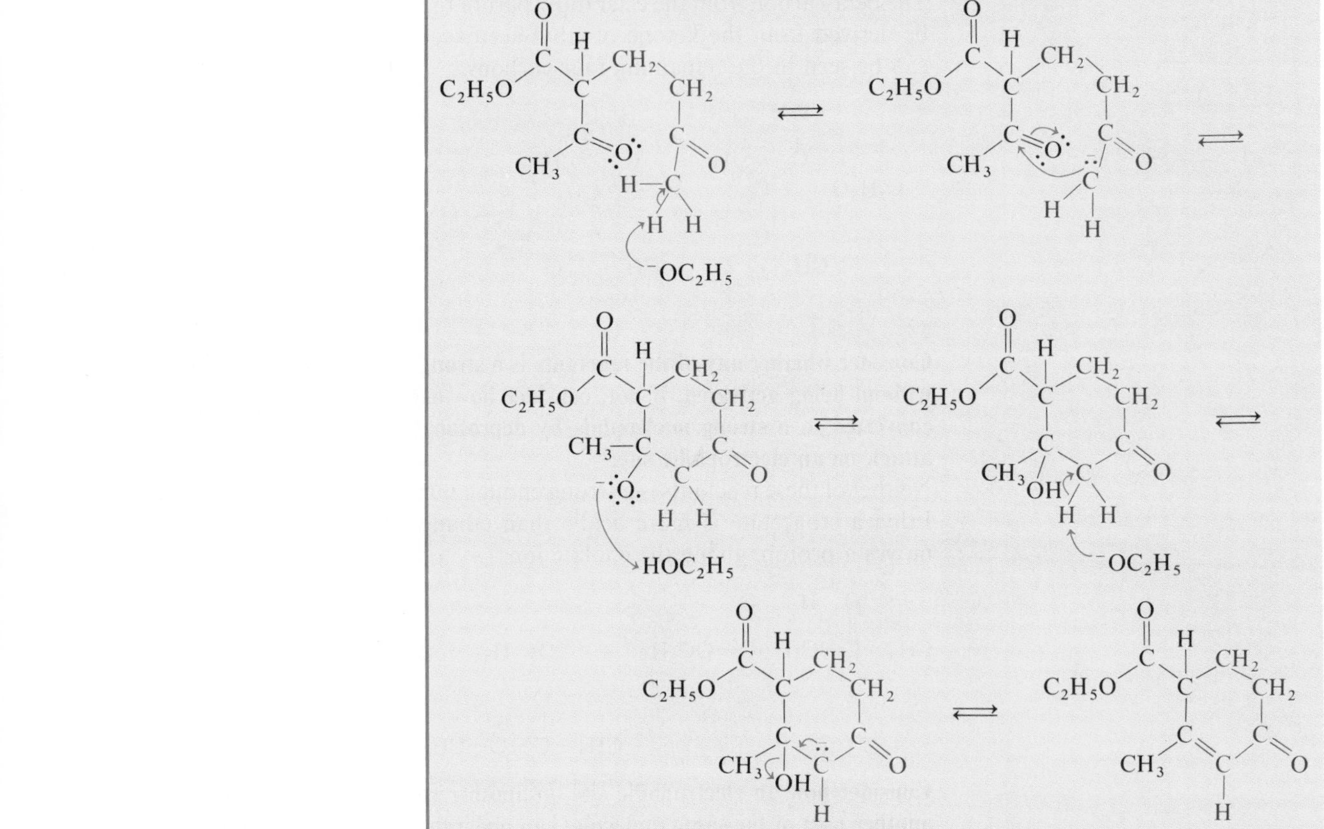

5. **Draw out all the steps of the mechanism using curved arrows to show the movement of electrons. Be careful to show only one step at a time.**
The complete mechanism is given by combining the equations shown above. We suggest you write out the mechanism as a review of the steps involved.

As further practice in proposing mechanisms for multistep condensations, try Problems 22-59 and 22-60 using the approach shown above.

PROBLEM 22-59

Propose a mechanism for the following reaction.

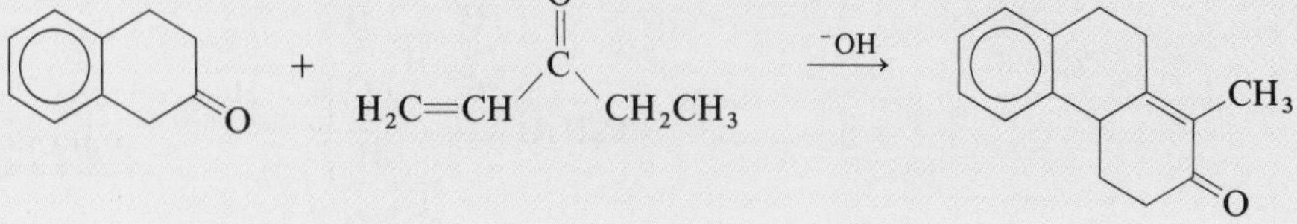

$$\text{(tetralone)} + H_2C{=}CH{-}\overset{O}{\overset{\|}{C}}{-}CH_2CH_3 \xrightarrow{^-OH} \text{(product with } CH_3\text{)}$$

PROBLEM 22-60

The base-catalyzed reaction of an aldehyde (having no alpha hydrogens) with an anhydride is called the *Perkin condensation.* Propose a mechanism for the following example of the Perkin condensation. (Sodium acetate serves as the base.)

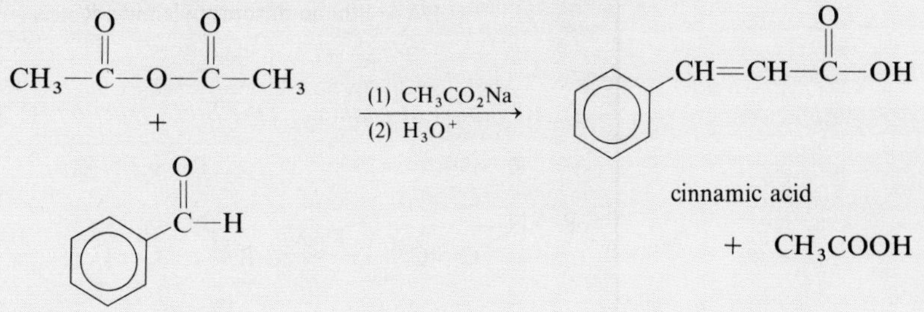

$$CH_3-\overset{O}{\overset{\|}{C}}-O-\overset{O}{\overset{\|}{C}}-CH_3 + C_6H_5-\overset{O}{\overset{\|}{C}}-H \xrightarrow[(2)\ H_3O^+]{(1)\ CH_3CO_2Na} C_6H_5-CH=CH-\overset{O}{\overset{\|}{C}}-OH + CH_3COOH$$

cinnamic acid

PROBLEM 22-61

Show how you would use Robinson annulations to synthesize the following compounds. Work backward, remembering that the cyclohexenone is the new ring, and that the double bond of the cyclohexenone is formed by the aldol with dehydration. Take apart the double bond, and then see what structures the Michael donor and acceptor must have.

(a) [bicyclic enone with O and CH_3] (b) [bicyclic enone with O, CH_3 and second ketone O]

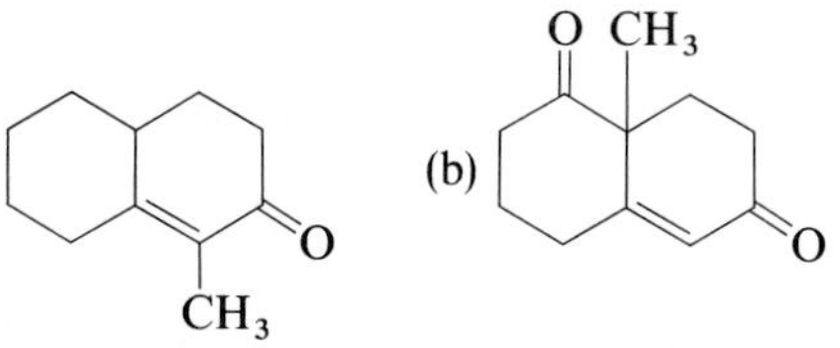

SUMMARY OF ENOLATE ADDITIONS AND CONDENSATIONS

A complete summary of additions and condensations would be very long and involved. This summary covers only the major classes of condensations and related reactions.

1. α Halogenation (Sections 22-2 and 22-3)

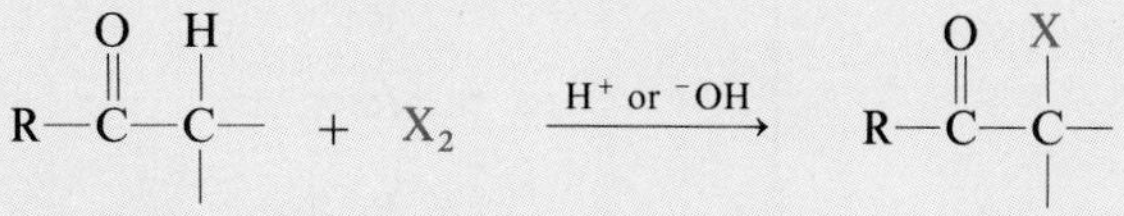

$$R-\overset{O}{\overset{\|}{C}}-\overset{H}{\overset{|}{\underset{|}{C}}}- + X_2 \xrightarrow{H^+ \text{ or } ^-OH} R-\overset{O}{\overset{\|}{C}}-\overset{X}{\overset{|}{\underset{|}{C}}}-$$

a. *The Iodoform (or haloform) reaction*

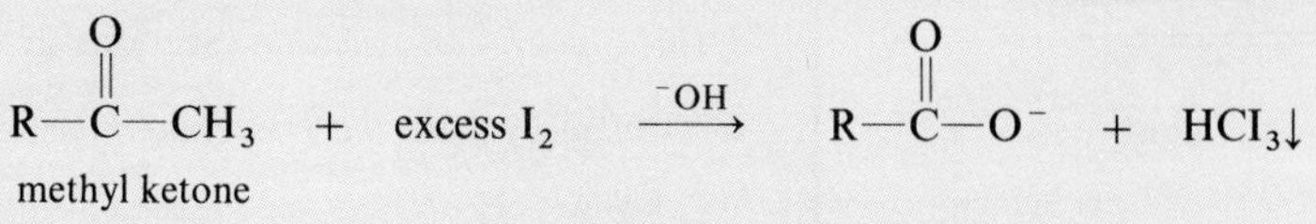

$$\underset{\text{methyl ketone}}{R-\overset{O}{\overset{\|}{C}}-CH_3} + \text{excess } I_2 \xrightarrow{^-OH} R-\overset{O}{\overset{\|}{C}}-O^- + HCI_3\downarrow$$

b. *The Hell-Volhard-Zelinsky (HVZ) reaction*

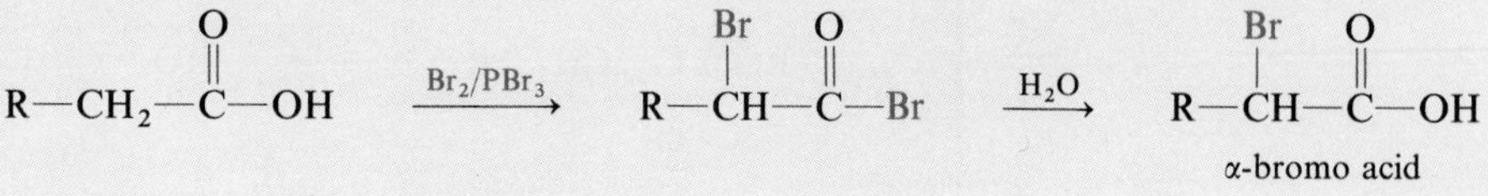

$$R-CH_2-\overset{O}{\overset{\|}{C}}-OH \xrightarrow{Br_2/PBr_3} R-\overset{Br}{\overset{|}{C}H}-\overset{O}{\overset{\|}{C}}-Br \xrightarrow{H_2O} \underset{\alpha\text{-bromo acid}}{R-\overset{Br}{\overset{|}{C}H}-\overset{O}{\overset{\|}{C}}-OH}$$

2. *Alkylation of lithium enolates* (Section 22-4)

$$\mathrm{R{-}\overset{O}{\overset{\|}{C}}{-}CH_2{-}R \xrightarrow[(2)\ R'{-}X]{(1)\ LDA} R{-}\overset{O}{\overset{\|}{C}}{-}\overset{R'}{\overset{|}{C}H}{-}R}$$

(LDA = lithium diisopropylamide; R′—X = unhindered 1° halide or tosylate)

3. *Alkylation of enamines (the Stork reaction)* (Section 22-5)

$$\underset{\text{enamine}}{\mathrm{R_2N{-}C{=}C}} \xrightarrow{R'{-}X} \underset{\text{alkylated enamine}}{\mathrm{R_2\overset{+}{N}{=}C{-}\overset{R'}{\overset{|}{C}}{-}\ X^-}} \xrightarrow{H_3O^+} \underset{\text{alkylated ketone}}{\mathrm{O{=}C{-}\overset{R'}{\overset{|}{C}}{-}}} + \mathrm{R{-}\overset{R}{\overset{|}{\underset{H}{\underset{|}{\overset{+}{N}}}}}{-}H}$$

4. *The aldol condensation and subsequent dehydration* (Sections 22-6 through 22-10)

$$\underset{\text{ketone or aldehyde}}{2\ \mathrm{R{-}\overset{O}{\overset{\|}{C}}{-}CH_2{-}R'}} \underset{}{\overset{H^+\text{ or }^-OH}{\rightleftharpoons}} \underset{\text{aldol product}}{\mathrm{R{-}\overset{O}{\overset{\|}{C}}{-}\overset{R{-}\overset{OH}{\overset{|}{C}}{-}CH_2{-}R'}{\overset{|}{C}H}{-}R'}} \underset{H^+\text{ or }^-OH}{\overset{heat}{\rightleftharpoons}} \underset{\alpha,\beta\text{-unsaturated ketone or aldehyde}}{\mathrm{R{-}\overset{O}{\overset{\|}{C}}{-}\overset{R{-}C{-}CH_2{-}R'}{\overset{\|}{C}}{-}R'}} + H_2O$$

5. *The Cannizzaro reaction* (Section 22-11)

$$\mathrm{R{-}\overset{O}{\overset{\|}{C}}{-}H + R'{-}\overset{O}{\overset{\|}{C}}{-}H \xrightarrow{^-OH} \underset{\text{oxidized}}{R{-}\overset{O}{\overset{\|}{C}}{-}O^-} + \underset{\text{reduced}}{R'{-}CH_2OH}}$$

Both aldehydes must *lack* α-hydrogens.

6. *The Wittig reaction* (Section 22-12)

$$\mathrm{Ph_3\overset{+}{P}{-}\overset{R}{\overset{|}{\underset{R}{\underset{|}{C}}}}{:}^- + R'_2C{=}O \longrightarrow R_2C{=}CR'_2 + Ph_3P{=}O}$$

7. *The Claisen condensation* (Sections 22-13 through 22-15)

(Cyclizations are the Dieckmann condensation.)

$$2\ \mathrm{RO{-}\overset{O}{\overset{\|}{C}}{-}CH_2{-}R'} \xrightarrow{^-OR} \mathrm{RO{-}\overset{O}{\overset{\|}{C}}{-}\overset{\overset{O}{\overset{\|}{C}}{-}CH_2{-}R'}{\overset{|}{C}H}{-}R'} + \mathrm{ROH}$$

The product is initially formed as its anion.

8. The malonic ester synthesis (Section 22-17)

$$\underset{\text{malonic ester}}{\text{H}-\overset{\text{COOCH}_2\text{CH}_3}{\underset{\text{COOCH}_2\text{CH}_3}{\text{C}}}-\text{H}} \xrightarrow[\text{(2) R}-\text{X}]{\text{(1) NaOCH}_2\text{CH}_3} \underset{\text{substituted malonic ester}}{\text{R}-\overset{\text{COOCH}_2\text{CH}_3}{\underset{\text{COOCH}_2\text{CH}_3}{\text{C}}}-\text{H}} \xrightarrow[\text{heat}]{\text{H}_3\text{O}^+} \underset{\text{substituted acetic acid}}{\text{R}-\underset{\text{COOH}}{\text{CH}_2}} \quad \text{CO}_2\uparrow$$

9. The acetoacetic ester synthesis (Section 22-18)

$$\underset{\text{acetoacetic ester}}{\text{H}-\overset{\text{COOCH}_2\text{CH}_3}{\underset{\text{O}=\text{C}-\text{CH}_3}{\text{C}}}-\text{H}} \xrightarrow[\text{(2) R}-\text{X}]{\text{(1) NaOCH}_2\text{CH}_3} \underset{\text{substituted acetoacetic ester}}{\text{R}-\overset{\text{COOCH}_2\text{CH}_3}{\underset{\text{O}=\text{C}-\text{CH}_3}{\text{C}}}-\text{H}} \xrightarrow[\text{heat}]{\text{H}_3\text{O}^+} \text{R}-\underset{\text{O}=\text{C}-\text{CH}_3}{\text{CH}_2} \quad \text{CO}_2\uparrow$$

10. The Michael addition (conjugate addition) (Sections 22-19 and 22-20)

$$\text{Y}-\underset{\text{Z}}{\text{CH}}\!:^- \;+\; \text{>C}=\overset{|}{\text{C}}-\overset{\text{O}}{\overset{\|}{\text{C}}}- \xrightarrow{\text{ROH}} -\underset{\text{Y}-\underset{\text{Z}}{\text{CH}}}{\overset{|}{\text{C}}}-\underset{\text{H}}{\overset{|}{\text{C}}}-\overset{\text{O}}{\overset{\|}{\text{C}}}-$$

(Y and Z are carbonyl or other electron-withdrawing groups.)

Example: The Robinson annulation

cyclohexanone + MVK $\xrightarrow{^-\text{OH}}$ Michael adduct $\xrightarrow[\text{dehydration}]{\text{aldol}}$ annulated product

GLOSSARY

acetoacetic ester synthesis The alkylation or acylation of acetoacetic ester (ethyl acetoacetate), followed by hydrolysis and decarboxylation, to give substituted acetone derivatives. (p. 1023)

aldol condensation A base-catalyzed or acid-catalyzed conversion of two ketone or aldehyde molecules to a β-hydroxy ketone or aldehyde (called an **aldol**). Aldol condensations often take place with subsequent dehydration to give α,β-unsaturated ketones and aldehydes. (p. 998)

> **crossed aldol condensation:** An aldol condensation between two different ketones or aldehydes. (p. 1001)

alpha-carbon atom The carbon atom next to a carbonyl group. The hydrogen atoms on the α carbon are called α hydrogens or **α protons.** (p. 984)

Cannizzaro reaction The base-catalyzed oxidation-reduction of two aldehyde molecules to give an alcohol and a carboxylate ion. Both aldehydes must have no α-hydrogens, or else aldol condensations occur. (p. 1007)

Claisen condensation The base-catalyzed conversion of two ester molecules to a β-keto ester. (p. 1012)

condensation A reaction that bonds two or more molecules, often with the loss of a small molecule such as water or an alcohol. (p. 998)

conjugate addition (1,4-addition) Another term for Michael addition. (p. 1026)

Dieckmann condensation A Claisen condensation that forms a ring. (p. 1016)

enamine A vinyl amine, usually generated by the acid-catalyzed reaction of a secondary amine with a ketone or an aldehyde. (p. 995)

enol A vinyl alcohol. Simple enols usually tautomerize to their **keto** forms. (p. 985)

$$\underset{\text{enol}}{\text{HO}-\text{C}=\text{C}} \xrightleftharpoons{\text{H}^+ \text{ or } ^-\text{OH}} \underset{\text{keto}}{\text{O}=\text{C}-\text{C}-\text{H}}$$

enolate ion The resonance-stabilized anion formed by deprotonating the carbon atom next to a carbonyl group. (p. 985)

$$\text{O}=\text{C}-\text{C}-\text{H} \xrightarrow{\text{base:}^-} \underbrace{\left[\,:\ddot{\text{O}}=\text{C}-\ddot{\text{C}}^- \longleftrightarrow\ :\ddot{\text{O}}:^-\!-\text{C}=\text{C}\,\right]}_{\text{enolate ion}} + \text{base-H}$$

enolizable hydrogen A hydrogen atom on a carbon adjacent to a carbonyl group. Such a hydrogen may be lost and regained through the keto-enol tautomerism, losing its stereochemistry in the process. (p. 986)

haloform reaction The conversion of a methyl ketone to a carboxylate ion and a haloform (CHX_3) by treatment with a halogen and base. The **iodoform reaction** uses iodine to give a precipitate of solid iodoform. (p. 991)

Hell-Volhard-Zelinsky (HVZ) reaction The reaction of a carboxylic acid with Br_2 and PBr_3 to give an α-bromo acyl bromide, often hydrolyzed to an α-bromo acid. (p. 993)

malonic ester synthesis The alkylation or acylation of malonic ester (diethyl malonate), followed by hydrolysis and decarboxylation, to give substituted acetic acids. (p. 1021)

Michael addition (conjugate addition) The 1,4-addition of a nucleophile (the **Michael donor,** usually a resonance-stabilized carbanion) to a conjugated double bond such as an α,β-unsaturated ketone or ester (the **Michael acceptor**). (p. 1026)

Robinson annulation The formation of a cyclohexenone ring by the condensation of methyl vinyl ketone (MVK) or a substituted MVK derivative with a ketone. Robinson annulation proceeds by Michael addition to MVK, followed by an aldol condensation with dehydration. (p. 1029)

Stork reaction The alkylation or acylation of a ketone or aldehyde using its enamine derivative as the nucleophile. An acidic hydrolysis regenerates the alkylated or acylated ketone or aldehyde. (p. 997)

tautomerism An isomerism involving the migration of a proton and the corresponding movement of a double bond. An example is the **keto-enol tautomerism** of a ketone or aldehyde with its enol form. (p. 986)

tautomers: The isomers related by a tautomerism.

$$\underset{\text{keto tautomer}}{\text{O}=\text{C}-\text{C}-\text{H}} \xrightleftharpoons{\text{H}^+ \text{ or } ^-\text{OH}} \underset{\text{enol tautomer}}{\text{HO}-\text{C}=\text{C}}$$

keto-enol tautomerism

Wittig reaction The reaction of an aldehyde or ketone with a phosphorus ylide to form an alkene. One of the most versatile syntheses of alkenes. (p. 1009)

ylide An uncharged molecule containing a carbon atom with a negative charge bonded to a heteroatom with a positive charge. A phosphorus ylide is the nucleophilic species in the Wittig reaction. (p. 1009)

ESSENTIAL PROBLEM-SOLVING SKILLS IN CHAPTER 22

This is a difficult chapter because condensations can take on a wide variety of forms. You should try to *understand* the reactions and their mechanisms so you can generalize and predict related reactions. Work enough problems to get a feel for the standard reactions and to become confident you can work out new variations on the standard mechanisms. Make sure you feel comfortable with condensations that form new rings.

1. Show how enols and enolate ions act as nucleophiles. Give mechanisms for acid-catalyzed and base-catalyzed keto-enol tautomerisms.

2. Give mechanisms for acid-catalyzed and base-promoted alpha halogenation of ketones and acid-catalyzed halogenation of acids (the HVZ reaction). Explain why multiple halogenation is common with basic catalysis, and give a mechanism for the haloform reaction.

3. Show how the alkylation and acylation of enamines and lithium enolates can be used synthetically. Give mechanisms for these reactions.

4. Predict the products of aldol and crossed aldol reactions, before and after dehydration of the aldol products. Give the mechanisms under acid and base catalysis. (Aldols are reversible, so be sure you can write these mechanisms *backward* as well.) Show how aldols are used for the synthesis of β-hydroxy carbonyl compounds and α,β-unsaturated carbonyl compounds.

5. Predict the products of Wittig reactions, give their mechanisms, and show how to use Wittig reactions to synthesize desired olefins.

6. Predict the products of Claisen and crossed Claisen condensations, and give mechanisms. Show how the Claisen condensation constructs the carbon skeleton of a desired target compound.

7. Show how the malonic ester synthesis and the acetoacetic ester synthesis are used to make substituted acetic acids and substituted acetones, respectively. Give the mechanisms for these reactions.

8. Predict the products of Michael additions, and show how to use these reactions in syntheses. Show the general mechanism of the Robinson annulation, and use it to construct cyclohexenone ring systems.

STUDY PROBLEMS

22-62. Give a definition and an example for each of the following terms.

(a) haloform reaction
(b) condensation
(c) aldol condensation
(d) crossed aldol condensation
(e) phosphorus ylide
(f) Wittig reaction
(g) Cannizzaro reaction
(h) Claisen condensation
(i) Dieckmann condensation
(j) crossed Claisen condensation
(k) enamine
(l) Stork reaction
(m) tautomerism
(n) enolizable hydrogen
(o) malonic ester synthesis
(p) acetoacetic ester synthesis
(q) Michael addition
(r) Robinson annulation
(s) HVZ reaction

22-63. For each molecule shown below,

(1) indicate the most acidic hydrogens.

(2) draw the important resonance contributors of the anion that results from removal of the most acidic hydrogen.

(a) COOH **(b)** O O **(c)** O CN **(d)** O OH **(e)** O $COOCH_3$

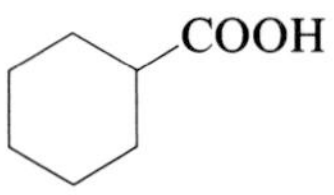

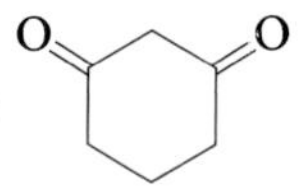

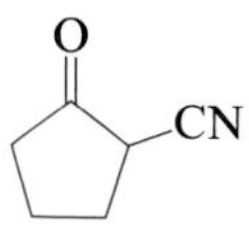

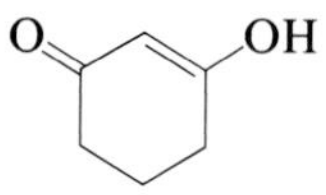

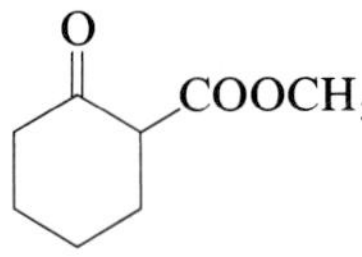

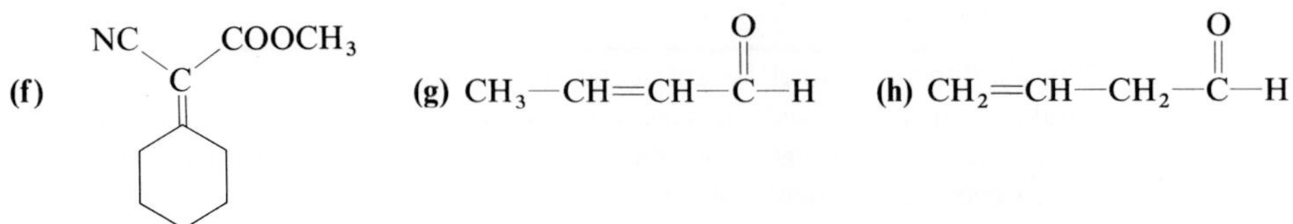

22-64. (1) Rank the following compounds in order of increasing acidity.
(2) Indicate which compounds would be more than 99 percent deprotonated by a solution of sodium ethoxide in ethanol.

(a) 2-(methoxycarbonyl)cyclohexanone, C(=O)OCH$_3$ (b) cyclohexanone (c) cyclohexane-1,3-dione (d) cyclohexane–COOH

(e) ring with C=O, C(=O)OCH$_3$, CN (f) cyclohexane–CH$_2$OH (g) cyclohexane–C(=O)OCH$_3$

22-65. Pentane-2,4-dione (acetylacetone) exists as a tautomeric mixture of 8 percent keto and 92 percent enol forms. Draw the stable enol tautomer, and explain its unusual stability.

$$\mathrm{CH_3-\overset{O}{\overset{\|}{C}}-CH_2-\overset{O}{\overset{\|}{C}}-CH_3}$$

acetylacetone

22-66. Predict the products of the following aldol condensations. Show the products both before and after dehydration.

(a) $\mathrm{(CH_3)_2CH-CH_2-\overset{O}{\overset{\|}{C}}-H \xrightarrow{^-OH}}$ (b) cyclodecane-1,6-dione $\xrightarrow{H^+}$

(c) $\mathrm{2\,Ph-CHO + CH_3-\overset{O}{\overset{\|}{C}}-CH_3 \xrightarrow{^-OH}}$ (d) $\mathrm{Ph-CH_2-\overset{O}{\overset{\|}{C}}-H + H_2C{=}O \xrightarrow{^-OH}}$

22-67. Predict the products of the following Claisen condensations.

(a) $\mathrm{(CH_3)_2CH-CH_2-\overset{O}{\overset{\|}{C}}-OCH_3 \xrightarrow[CH_3OH]{^-OCH_3}}$ (b) cyclohexane bearing CH$_2$COOCH$_3$ and C(=O)CH$_2$CH$_3$ $\xrightarrow[CH_3OH]{^-OCH_3}$

(c) $\mathrm{CH_3CH_2-\overset{O}{\overset{\|}{C}}-CH_2CH_2CH_2CH_2-\overset{O}{\overset{\|}{C}}-OCH_3 \xrightarrow[CH_3OH]{^-OCH_3}}$ (Dieckmann)

(d) cyclohexanone $\mathrm{+ CH_3O-\overset{O}{\overset{\|}{C}}-\overset{O}{\overset{\|}{C}}-OCH_3 \xrightarrow[CH_3OH]{NaOCH_3}}$ (e) cyclopentane with $\mathrm{CH_2-\overset{O}{\overset{\|}{C}}-OCH_3}$ and $\mathrm{CH_2-\overset{O}{\overset{\|}{C}}-CH_3}$ $\xrightarrow[CH_3OH]{NaOCH_3}$

22-68. Predict the products of the following reactions.

(a) 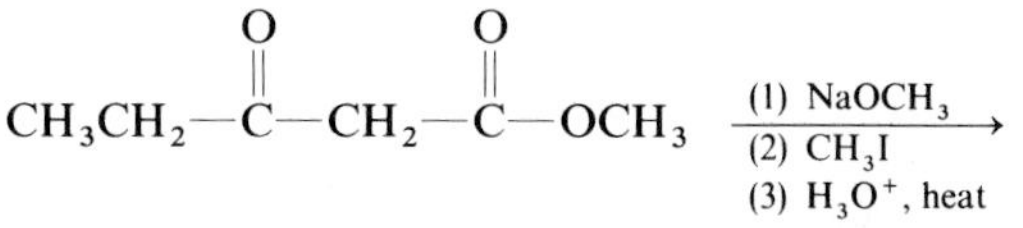 $CH_3CH_2-\overset{O}{\overset{\|}{C}}-CH_2-\overset{O}{\overset{\|}{C}}-OCH_3 \xrightarrow[\text{(2) } CH_3I \text{ (3) } H_3O^+\text{, heat}]{\text{(1) NaOCH}_3}$

(b) $CH_3-\overset{O}{\overset{\|}{C}}-CH_2-\overset{O}{\overset{\|}{C}}-OCH_2CH_3$ + cyclohex-2-enone $\xrightarrow[\text{(2) } H_3O^+\text{, heat}]{\text{(1) NaOCH}_2CH_3}$

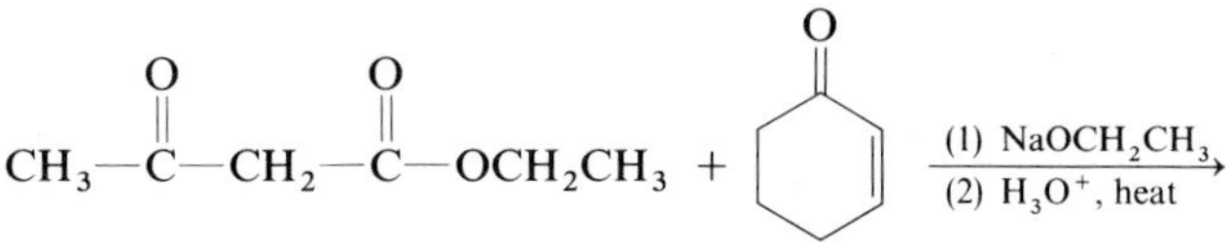

(c) 2-Ph-6,6-dimethylcyclohexanone (H, Ph at C-2; two CH_3 at C-6) $\xrightarrow[\text{(2) } CH_3CH_2CH_2Br]{\text{(1) LDA}}$

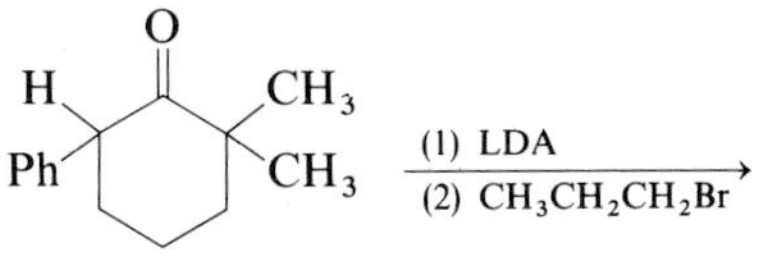

(d) 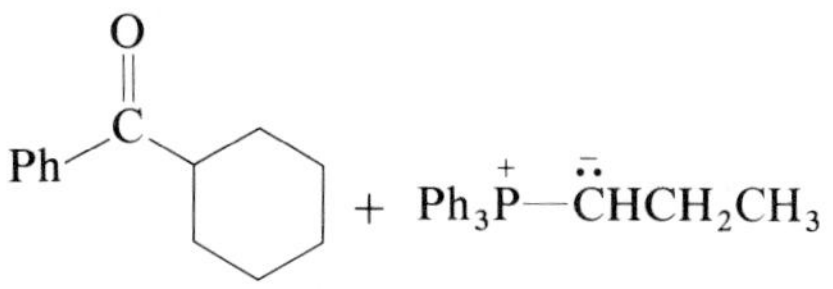 Ph–C(=O)–cyclohexyl + $Ph_3\overset{+}{P}-\overset{-}{\ddot{C}}HCH_2CH_3$

(e) 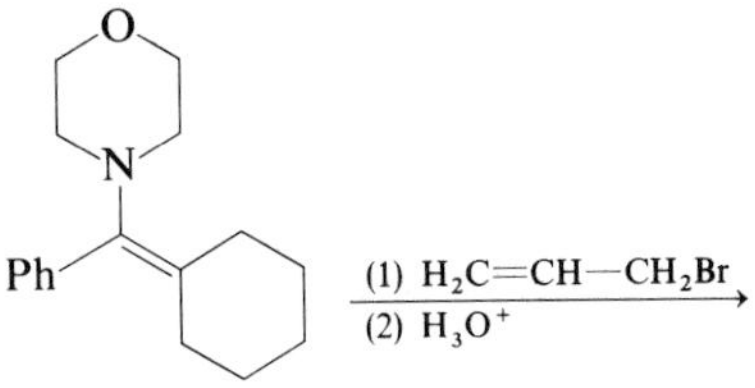 Morpholine enamine: Ph–C(N-morpholino)=cyclohexylidene $\xrightarrow[\text{(2) } H_3O^+]{\text{(1) } H_2C=CH-CH_2Br}$

(f) 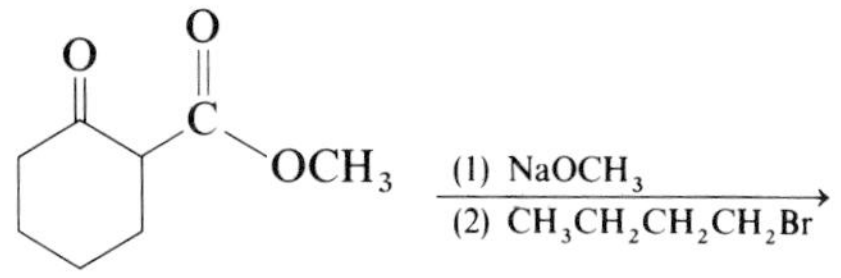 2-(methoxycarbonyl)cyclohexanone (cyclohexanone with $-\overset{O}{\overset{\|}{C}}-OCH_3$) $\xrightarrow[\text{(2) } CH_3CH_2CH_2CH_2Br]{\text{(1) NaOCH}_3}$

(g) product from part (f) $\xrightarrow[\text{heat}]{H_3O^+}$ (decarboxylation)

22-69. Show how you would accomplish the following synthetic conversions in good yields. You may use any necessary additional reagents.

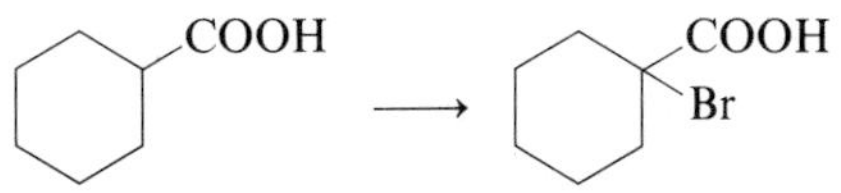

(a) 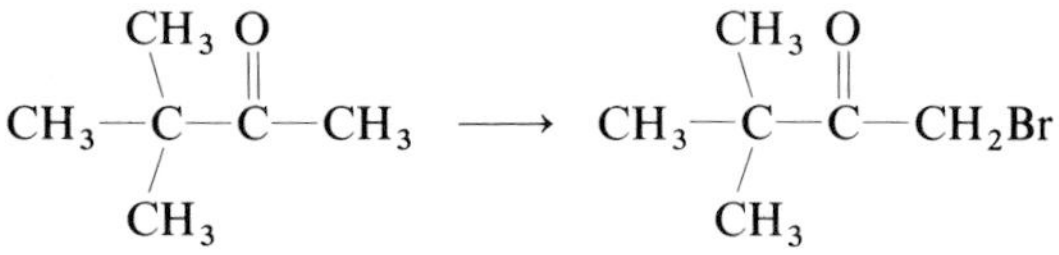 $(CH_3)_3C-\overset{O}{\overset{\|}{C}}-CH_3 \longrightarrow (CH_3)_3C-\overset{O}{\overset{\|}{C}}-CH_2Br$

(b) cyclohexyl–COOH $\longrightarrow$ 1-bromocyclohexane-1-COOH (COOH and Br on the same carbon)

(c) $(CH_3)_3C-\overset{O}{\overset{\|}{C}}-CH_3 \longrightarrow (CH_3)_3C-\overset{O}{\overset{\|}{C}}-O^-$

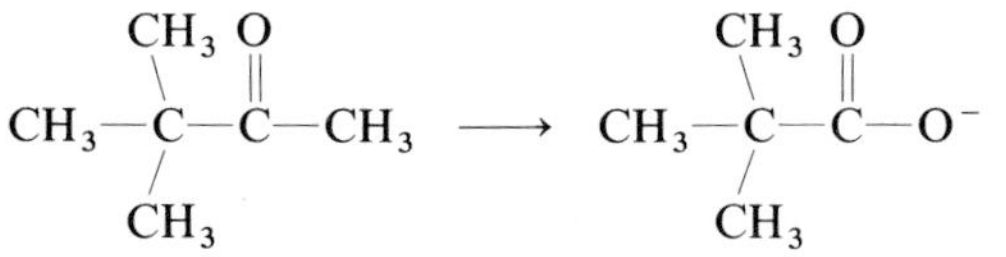

(d) 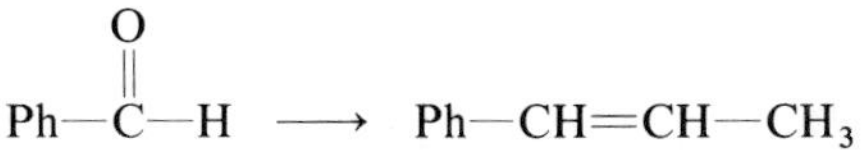 $Ph-\overset{O}{\overset{\|}{C}}-H \longrightarrow Ph-CH=CH-CH_3$

(e) 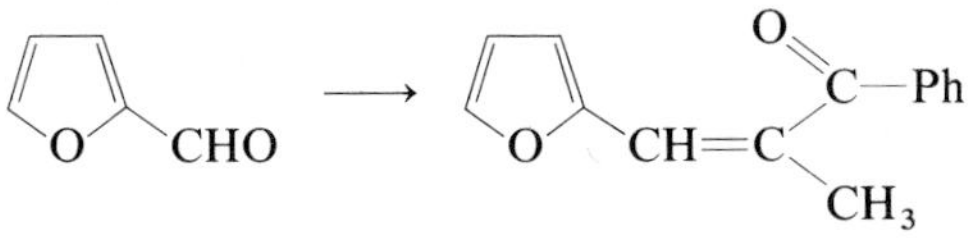 furan-2-CHO $\longrightarrow$ furan-2-$CH=C(CH_3)-\overset{O}{\overset{\|}{C}}-Ph$ (*Hint:* aldol)

22-70. Show how you would use the malonic ester synthesis to make the following compounds.

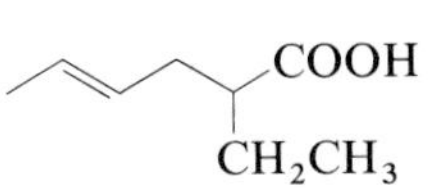

(a) 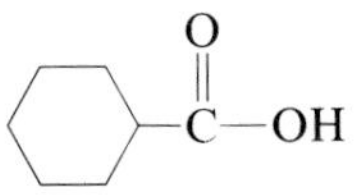 cyclohexyl$-\overset{O}{\overset{\|}{C}}-OH$ (b) $CH_3CH=CHCH_2CH(CH_2CH_3)COOH$ (c) $C_6H_5-CH_2CH_2-\overset{O}{\overset{\|}{C}}-OH$

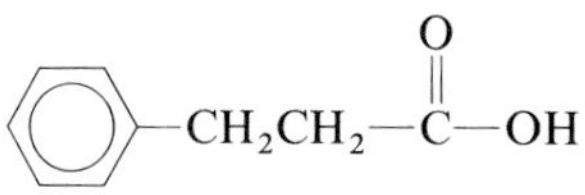

22-71. Show how you would use the acetoacetic ester synthesis to make the following compounds.

(a) 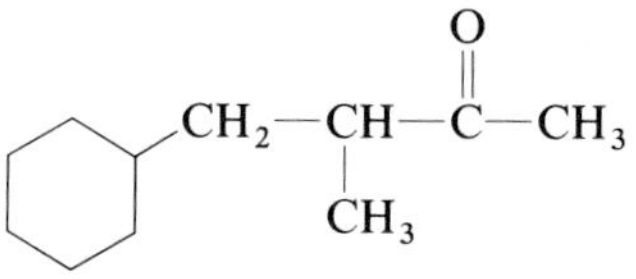 cyclohexyl$-CH_2-\underset{CH_3}{\underset{|}{CH}}-\overset{O}{\overset{\|}{C}}-CH_3$ (b) 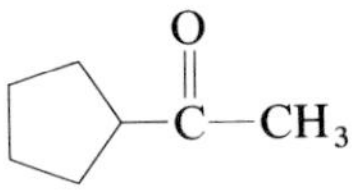 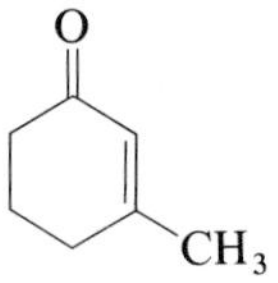 cyclopentyl$-\overset{O}{\overset{\|}{C}}-CH_3$ 3-methylcyclohex-2-enone (CH_3 on C-3)

(Consider using 2,6-heptanedione as an intermediate.)

22-72. Each of the following compounds can be synthesized by an aldol condensation followed by further reactions. In each case, work backward from the target molecule to an aldol product and show what compounds are needed for the condensation.

(a) $Ph{-}CH_2{-}CH_2{-}CH(OH){-}Ph$ **(b)** 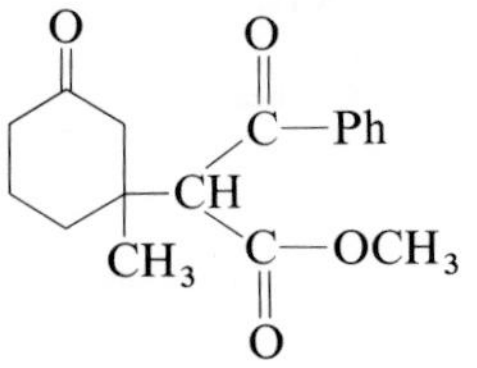 **(c)**

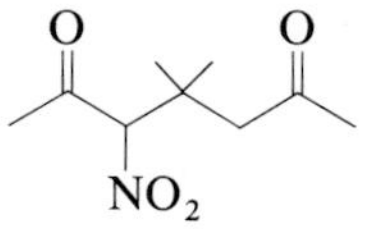

22-73. Propose plausible mechanisms for the following reactions.

(a) 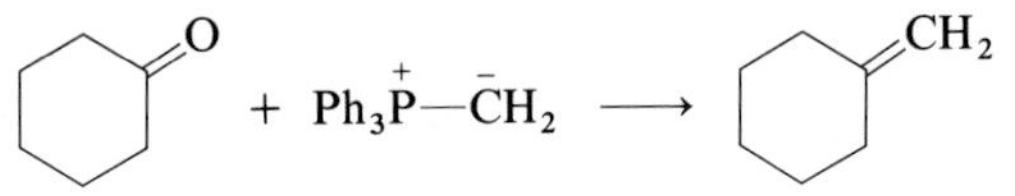 cyclohexanone + $Ph_3\overset{+}{P}{-}\overset{-}{C}H_2$ ⟶ methylenecyclohexane ($=CH_2$)

(b) $CH_3{-}C(CH_3)_2{-}CHO + H_2C{=}O \xrightarrow{^-OH} CH_3{-}C(CH_3)_2{-}CH_2OH$

(c) 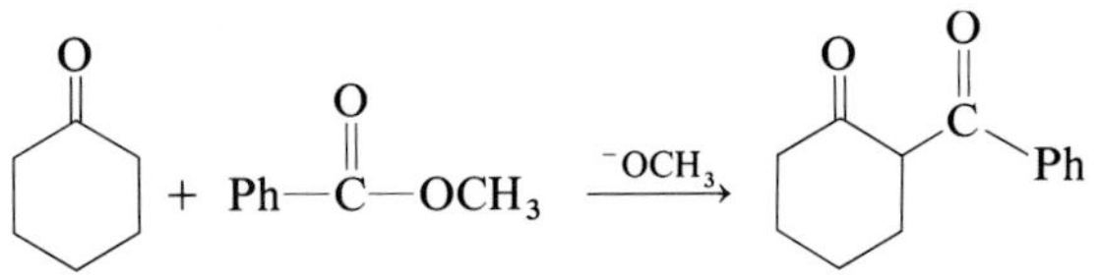 cyclohexanone + $Ph{-}C(=O){-}OCH_3 \xrightarrow{^-OCH_3}$ 2-benzoylcyclohexanone ($C(=O)Ph$)

(d) 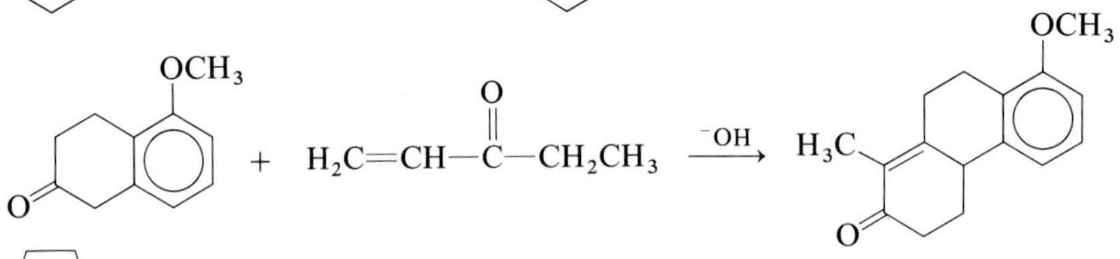 + $H_2C{=}CH{-}C(=O){-}CH_2CH_3 \xrightarrow{^-OH}$ (OCH_3, H_3C)

(e) 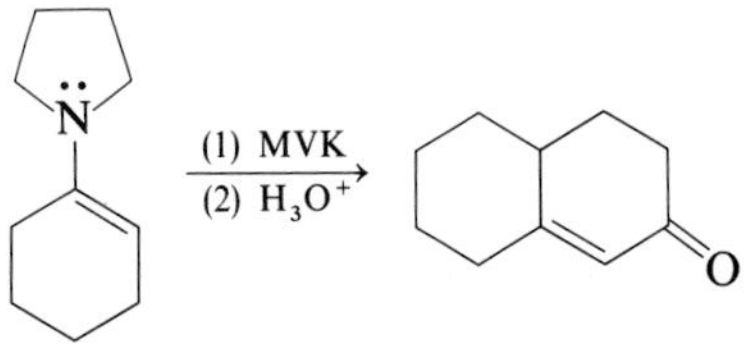 $\xrightarrow[(2)\ H_3O^+]{(1)\ MVK}$

22-74. Write an equation showing the expected products of the following enamine alkylation and acylation reactions. Then give the final product expected after hydrolysis of the iminium salt.
(a) pyrrolidine enamine of 3-pentanone + allyl chloride
(b) pyrrolidine enamine of acetophenone + butanoyl chloride
(c) piperidine enamine of cyclopentanone + methyl iodide
(d) piperidine enamine of cyclopentanone + methyl vinyl ketone

22-75. Show how you would accomplish the following multistep conversions. You may use any necessary additional reagents.

(a) dimethyl adipate and allyl bromide ⟶

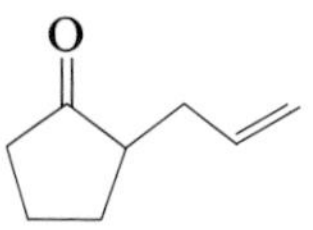

(b) cyclopentanone 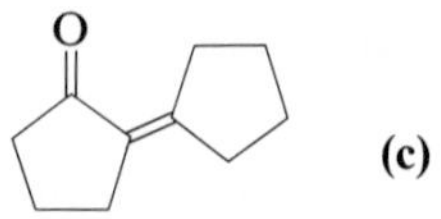 ⟶ **(c)** 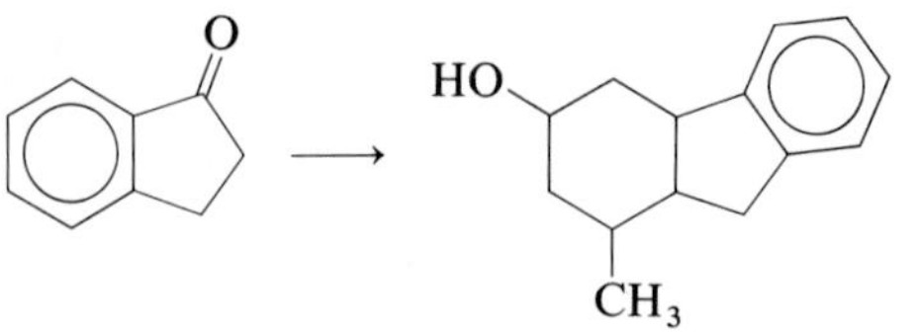 ⟶ (HO, CH_3)

22-76. Most of the condensations we have studied are reversible. The reverse reactions are often given the prefix *retro-*, the Latin word meaning "backward." Give mechanisms to account for the following reactions.

(a) 2-acetylcyclohexanol $\underset{}{\overset{H^+}{\rightleftharpoons}}$ $CH_3-\overset{O}{\overset{\|}{C}}-(CH_2)_5-CHO$

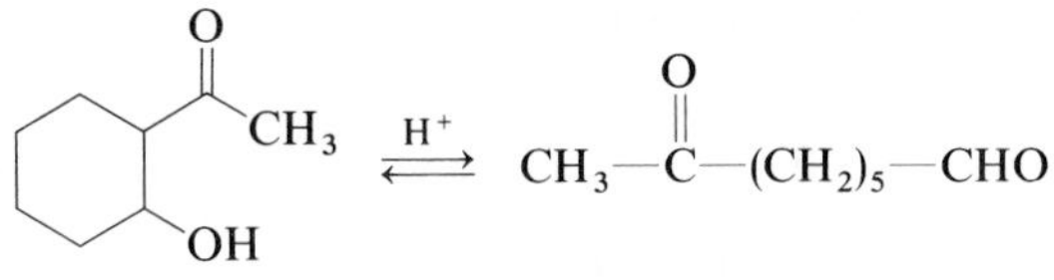

(retro-aldol)

(b) 2-(2-cyanoethyl)cyclohexanone $\overset{^-OH}{\rightleftharpoons}$ cyclohexanone + $H_2C{=}CH-CN$

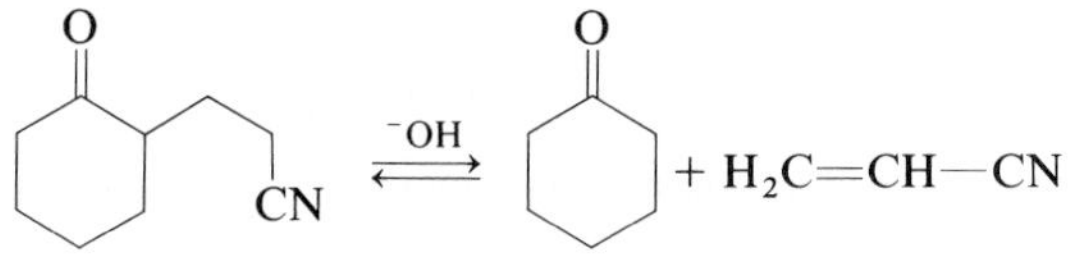

(retro-Michael)

(c) 1-acetyl-2-methylcyclobutan-2-ol (CH_3, CH_3, OH) $\xrightarrow{H^+}$ 3-methylcyclohex-2-enone (CH_3)

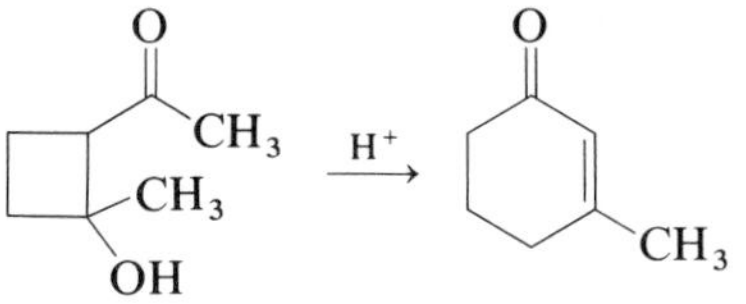

(retro-aldol and further condensation)

22-77. Show how you would use the Robinson annulation to synthesize the following compounds.

(a) cyclohexenone with H_3C and Ph substituents (b) bicyclic enone with CH_3 (c) bicyclic enone with gem-dimethyl group

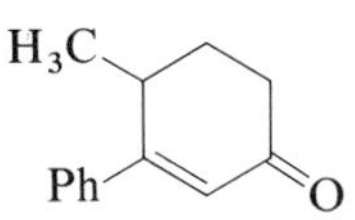

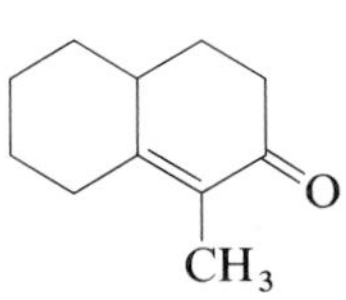

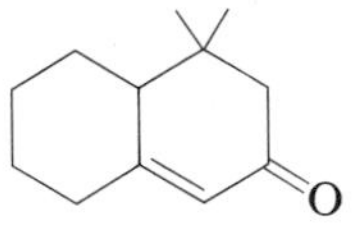

22-78. Propose a mechanism for the following reaction. Show the structure of the compound that results from hydrolysis and decarboxylation of the product.

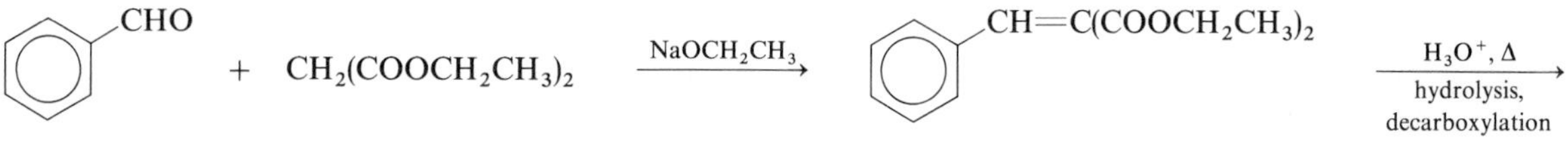

$$C_6H_5CHO + CH_2(COOCH_2CH_3)_2 \xrightarrow{NaOCH_2CH_3} C_6H_5CH{=}C(COOCH_2CH_3)_2 \xrightarrow[\text{hydrolysis, decarboxylation}]{H_3O^+,\ \Delta}$$

benzaldehyde   malonic ester

22-79. One of the reactions involved in the metabolism of sugars is the splitting of fructose 1,6-diphosphate to give glyceraldehyde 3-phosphate and dihydroxyacetone phosphate. In the living system this retro-aldol is catalyzed by an enzyme called *aldolase;* however, it can also be catalyzed by a mild base. Propose a mechanism for the base-catalyzed reaction.

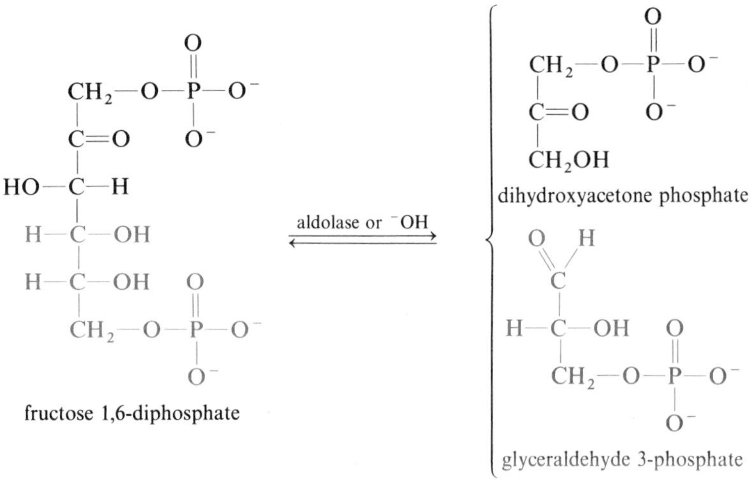

$CH_2-O-PO_3^{2-}$ / $C{=}O$ / $HO-C-H$ / $H-C-OH$ / $H-C-OH$ / $CH_2-O-PO_3^{2-}$

fructose 1,6-diphosphate

$\overset{\text{aldolase or } ^-OH}{\rightleftharpoons}$

$CH_2-O-PO_3^{2-}$ / $C{=}O$ / CH_2OH

dihydroxyacetone phosphate

$O{=}C-H$ / $H-C-OH$ / $CH_2-O-PO_3^{2-}$

glyceraldehyde 3-phosphate

22-80. Biochemists studying the structure of collagen (a fibrous protein in connective tissue) found cross-links containing α,β-unsaturated aldehydes between protein chains. Show the structures of the side chains that react to form these cross-links, and give a mechanism for their formation in a weakly acidic solution.

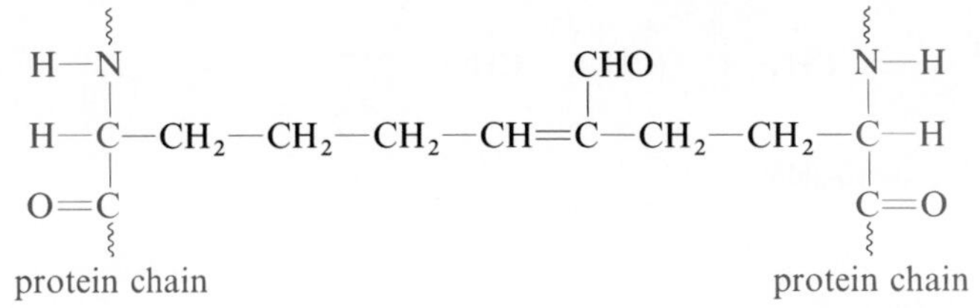

$$\begin{array}{c}
\text{H—N} \qquad\qquad\qquad\qquad\qquad\qquad \text{CHO} \qquad\qquad\qquad \text{N—H} \\
\text{H—C—}CH_2\text{—}CH_2\text{—}CH_2\text{—CH=C—}CH_2\text{—}CH_2\text{—C—H} \\
\text{O=C} \qquad\qquad\qquad\qquad\qquad\qquad\qquad\qquad\qquad\qquad \text{C=O}
\end{array}$$

protein chain   protein chain

*** 22-81.** Show reaction sequences (not detailed mechanisms) that explain these transformations.

(a) 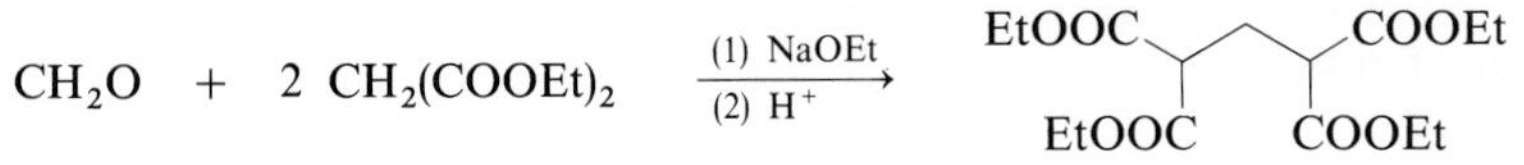 $CH_2O + 2\ CH_2(COOEt)_2 \xrightarrow[\text{(2) } H^+]{\text{(1) NaOEt}}$ EtOOC, COOEt, EtOOC, COOEt

(b) 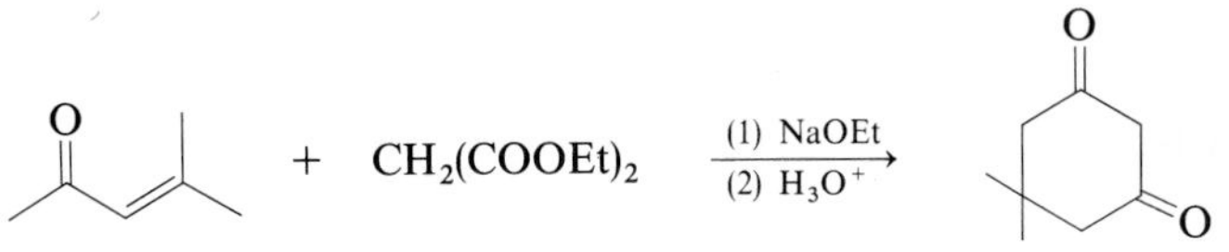 O + $CH_2(COOEt)_2 \xrightarrow[\text{(2) } H_3O^+]{\text{(1) NaOEt}}$ O, O

(c) 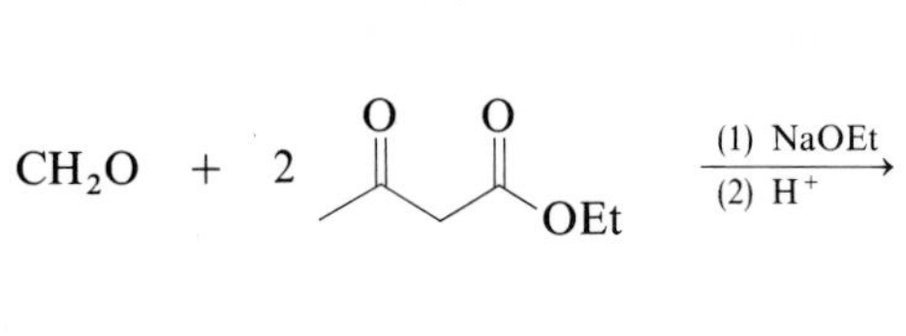 $CH_2O + 2$ O O OEt $\xrightarrow[\text{(2) } H^+]{\text{(1) NaOEt}}$ O, COOH

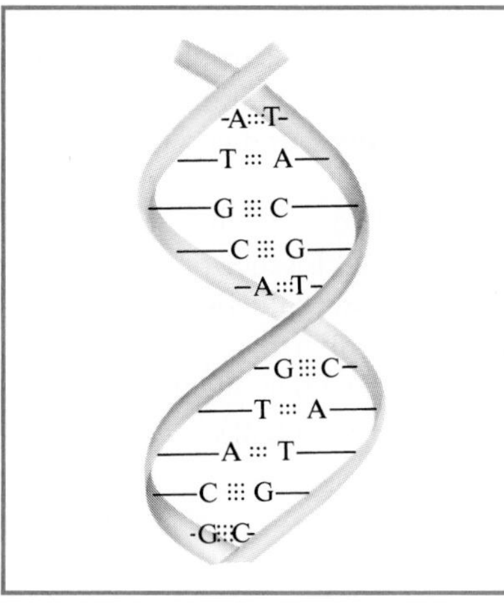

23

CARBOHYDRATES AND NUCLEIC ACIDS

23-1 INTRODUCTION

Carbohydrates are the most abundant class of organic compounds in nature. Carbohydrates are synthesized by nearly all plants and animals, which use them to store energy and deliver it to their cells. In most living organisms, *glucose* is oxidized to carbon dioxide and water to provide the energy needed by their cells. Plants store energy by converting glucose to *starch,* while animals store energy by converting glucose to *glycogen,* another form of starch. **Cellulose** makes up the cell walls of plants and forms their structural framework. Cellulose is the major component of *wood,* a strong yet supple material that can support the great weight of the oak but still allow the willow to bend with the wind.

Almost every aspect of human life involves carbohydrates in one form or another. Like other animals, we use the energy content of carbohydrates in the food we eat for producing and storing energy in our cells. Clothing is made from cotton and linen, two forms of cellulose. Other fabrics are made by manipulating cellulose to convert it to the semisynthetic fibers *rayon* and *cellulose acetate.* In the form of wood, cellulose is used to construct our houses and also as a fuel to heat them. Even this page is made from cellulose fibers.

Carbohydrate chemistry is one of the most interesting areas of organic chemistry. Many chemists are employed by companies that use carbohydrates to make foods, building materials, and other consumer products. All biologists need to understand carbohydrates, which play pivotal energetic roles throughout the plant and animal kingdoms. At first glance, the structures and reactions of carbohydrates may seem complicated. We will learn that these structures and reactions are consistent and predictable, however, and we can study carbohydrates as easily as we study the simplest organic compounds.

23-2 CLASSIFICATION OF CARBOHYDRATES

The term **carbohydrates** arose because **sugars** have interesting molecular formulas, $C_n(H_2O)_m$, suggesting that carbon atoms are combined in some way with water molecules. In fact, most simple sugars have $C(H_2O)$ as their empirical formula. Chemists named these compounds "hydrates of carbon" or "carbohydrates" because of their unusual molecular formulas. Our modern definition of carbohydrates includes polyhydroxyaldehydes, polyhydroxyketones, and compounds that are easily hydrolyzed to them.

Monosaccharides or *simple sugars* are carbohydrates that cannot be hydrolyzed to simpler compounds. Figure 23-1 shows the Fischer projection representations of the monosaccharides *glucose* and *fructose.* Glucose is a polyhydroxyaldehyde, while frutose is a polyhydroxyketone. Polyhydroxyaldehydes are called **aldoses** (*ald-* is for *ald*ehyde and *-ose* is the suffix for a sugar), and polyhydroxyketones are called **ketoses** (*ket-* for *ket*one, *-ose* for sugar).

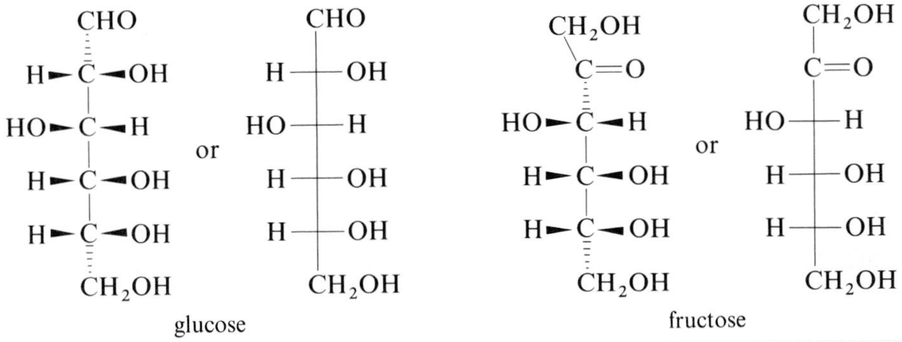

CHO; H—C—OH; HO—C—H; H—C—OH; H—C—OH; CH_2OH or CHO; H—OH; HO—H; H—OH; H—OH; CH_2OH

glucose

CH_2OH; C=O; HO—C—H; H—C—OH; H—C—OH; CH_2OH or CH_2OH; C=O; HO—H; H—OH; H—OH; CH_2OH

fructose

FIGURE 23-1 Glucose and fructose are examples of monosaccharides. Glucose is an aldose (a sugar with an aldehyde group), and fructose is a ketose (a sugar with a ketone group). Carbohydrate structures are commonly drawn using Fischer projections.

We have used Fischer projections to draw the structures of glucose and fructose because the Fischer projection conveniently shows the stereochemistry at each of the chiral carbon atoms. The Fischer projection was originally developed by Emil Fischer, a carbohydrate chemist who received the Nobel Prize for his proof of the structure of glucose. Fischer used this shorthand notation for drawing and comparing sugar structures quickly and easily. We will be using the Fischer projection extensively for our work with carbohydrates, so you may want to review it (Section 6-10) and make models of the structures in Figure 23-1 to make certain you understand the stereochemistry implied by these structures. In aldoses, the aldehyde carbon is the most highly oxidized, so it is always at the top of the Fischer projection. In ketoses, the carbonyl group is usually the second carbon from the top.

PROBLEM 23-1

Draw the mirror images of glucose and fructose. Are glucose and fructose chiral? Do you expect them to be optically active?

A **disaccharide** is a sugar that can be hydrolyzed to two monosaccharides. For example, sucrose ("table sugar") is a disaccharide that is easily hydrolyzed to one molecule of glucose and one molecule of fructose.

$$1 \text{ sucrose} \xrightarrow[\text{heat}]{H_3O^+} 1 \text{ glucose} + 1 \text{ fructose}$$

Both monosaccharides and disaccharides are quite soluble in water, and most have the characteristic sweet taste we associate with sugars.

Polysaccharides are carbohydrates that can be hydrolyzed to many monosaccharide units. Polysaccharides are naturally occurring polymers (*biopolymers*) of carbohydrates. Just as we use synthetic polymers as structural materials, the polysaccharide **cellulose** is used as a structural material in plants. **Starch** is a biopolymer whose carbohydrate units are easily added to store energy or removed to provide energy to cells. Both starch and cellulose are biopolymers of glucose, and hydrolysis of either of these large molecules gives many molecules of glucose.

$$\text{starch} \xrightarrow[\text{heat}]{H_3O^+} \text{over 1000 glucose}$$

$$\text{cellulose} \xrightarrow[\text{heat}]{H_3O^+} \text{over 1000 glucose}$$

Before we can understand the chemistry of these more complex carbohydrates, we must first learn the general principles of carbohydrate structure and reactions, using the simplest monosaccharides as examples. Then we apply these principles to the more complex disaccharides and polysaccharides. The chemistry of carbohydrates is understood by applying the chemistry of alcohols, aldehydes, and ketones. In general, the chemistry of biomolecules can be predicted by applying the chemistry of simple organic molecules with similar functional groups.

23-3 MONOSACCHARIDES

23-3A CLASSIFICATION OF MONOSACCHARIDES

Most sugars have their own specific common names, such as glucose, fructose, galactose, and mannose. These names are not systematic, although there are simple ways to remember the common structures. We simplify the study of monosaccharides by grouping similar structures together. This classification is done according to three criteria:

1. Whether the sugar contains a ketone or an aldehyde group
2. The number of carbon atoms in the carbon chain
3. The stereochemical configuration of the chiral carbon atom farthest from the carbonyl group

As we have seen, sugars with aldehyde groups are called **aldoses,** and those with ketone groups are called **ketoses.** The number of carbon atoms in the sugar generally ranges between three and seven, designated by the terms *triose* (three carbons), *tetrose* (four carbons), *pentose* (five carbons), *hexose* (six carbons), and *heptose* (seven carbons). The two groups of terms are often combined. For example, glucose has an aldehyde and contains six carbon atoms, so it is an aldohexose. Fructose also contains six carbon atoms, but it is a ketose, so it is called a ketohexose.

an aldohexose	a ketohexose	an aldotetrose	a ketotetrose
CHO	CH_2OH		
CHOH	C=O		
CHOH	CHOH	CHO	CH_2OH
CHOH	CHOH	CHOH	C=O
CHOH	CHOH	CHOH	CHOH
CH_2OH	CH_2OH	CH_2OH	CH_2OH

PROBLEM 23-2

(a) How many chiral carbon atoms are there in an aldotetrose? Draw all the aldotetrose stereoisomers.
(b) How many chiral carbons are there in a ketotetrose? Draw all the ketotetrose stereoisomers.
(c) How many chiral carbons and stereoisomers are there for an aldohexose? For a ketohexose?

PROBLEM 23-3

(a) There is only one ketotriose, called *dihydroxyacetone*. Draw its structure.
(b) There is only one aldotriose, called *glyceraldehyde*. Draw the two enantiomers of glyceraldehyde.

23-3B THE D AND L CONFIGURATIONS OF SUGARS

Around the turn of the century, carbohydrate chemists made great strides in determining the structures of natural and synthetic sugars. They found ways to build larger sugars out of smaller ones, converting a tetrose to a pentose and a pentose to a hexose. The opposite conversion, removing one carbon atom at a time (called a *degradation*), was also developed. The degradation could convert a hexose to a pentose, a pentose to a tetrose, and a tetrose to a triose. There is only one aldotriose, glyceraldehyde.

These chemists noticed they could start with any of the naturally occurring sugars, and degradation to glyceraldehyde always gave the dextrorotatory (+) enantiomer of glyceraldehyde. Some synthetic sugars, on the other hand, were degraded to the levorotatory (−) enantiomer of glyceraldehyde. Carbohydrate chemists started using a D to designate the sugars that were degraded to (+) glyceraldehyde, and an L for those that were degraded to (−) glyceraldehyde. Although these chemists did not know the absolute configurations of any of these sugars, the D and L relative configurations were useful to distinguish the naturally occurring D sugars from their unnatural L enantiomers.

We now know the absolute configurations of (+) and (−) glyceraldehyde.

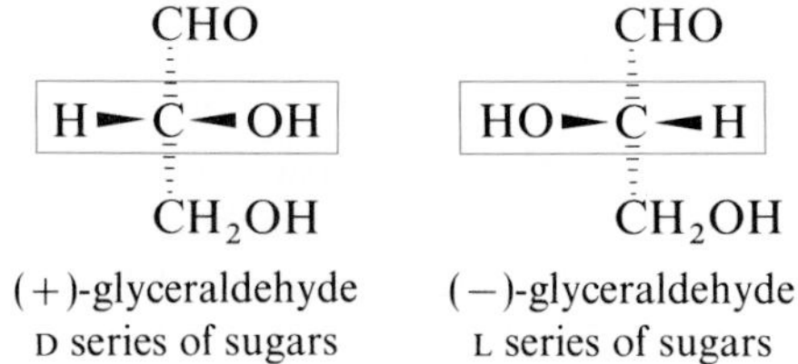

(+)-glyceraldehyde
D series of sugars

(−)-glyceraldehyde
L series of sugars

Figure 23-2 shows that the degradation (covered in Section 23-14) removes the aldehyde carbon atom, and it is the *bottom chiral carbon* in the Fischer projection (the chiral carbon farthest removed from the carbonyl group) that determines which enantiomer of glyceraldehyde is formed upon successive degradation.

The (+) enantiomer of glyceraldehyde has its OH group on the right in the Fischer projection, as shown in Figure 23-2. Therefore, sugars of the D series have the OH group of the bottom chiral carbon on the right in the Fischer projection; sugars of the L series have the OH group of the bottom chiral carbon on the left. In the following examples, notice that the D or L configuration is determined by the bottom chiral carbon, and the enantiomer of a D sugar is always an L sugar.

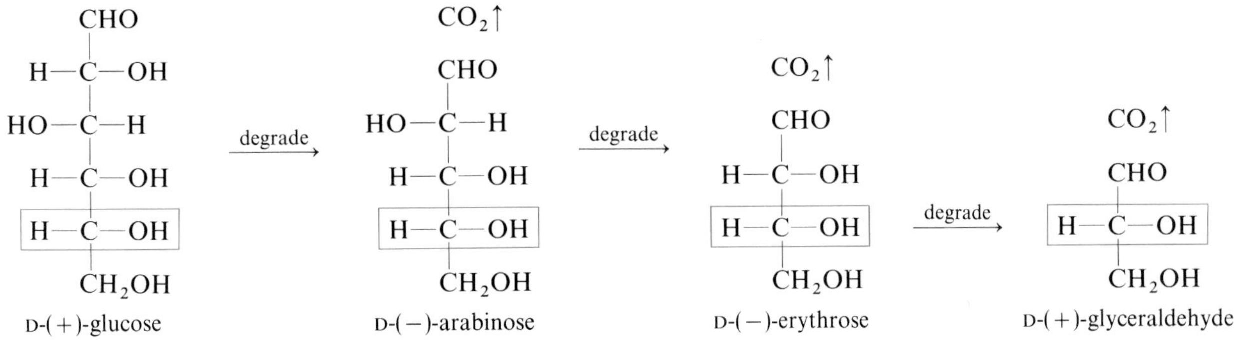

FIGURE 23-2 Degradation of an aldose removes the aldehyde carbon atom to give a smaller sugar. Sugars of the D series give (+)-glyceraldehyde on degradation to the triose. Therefore, the OH group of the bottom chiral carbon atom of the D sugars is on the right in the Fischer projection.

D-threose	L-threose	D-ribulose	L-ribulose	D-xylose	L-xylose
		CH_2OH	CH_2OH	CHO	CHO
CHO	CHO	C=O	O=C	H—OH	HO—H
HO—H	H—OH	H—OH	HO—H	HO—H	H—OH
H—OH	HO—H	H—OH	HO—H	H—OH	HO—H
CH_2OH	CH_2OH	CH_2OH	CH_2OH	CH_2OH	CH_2OH

Most naturally occurring sugars have the D configuration, and most members of the D family of aldoses (up through six carbon atoms) are found in nature. Figure 23-3 shows the D family of aldoses. Notice that the D or L configuration does not tell us which way a sugar rotates the plane of polarized light. This must be determined by experiment. Some D sugars have (+) rotations, while others have (−) rotations.

On paper, the family tree of D aldoses (Fig. 23-3) can be generated by starting with D-(+)-glyceraldehyde and adding another carbon at the top to generate two aldotetroses: one (erythrose) with the OH group of the new chiral carbon on the right, and another (threose) with the new OH group on the left. Adding another carbon to each of these aldotetroses gives four aldopentoses, and adding a sixth carbon gives eight aldohexoses.* In Section 23-15 we describe the Kiliani-Fischer synthesis, which actually adds a carbon atom and generates the pairs of elongated sugars just as we have drawn them in this family tree.

At the time the D and L system of relative configurations was introduced, it was not possible to determine the absolute configurations of chiral compounds. It was arbitrarily decided to draw the D series with the glyceraldehyde OH group on the right, and the L series with it on the left. This guess later proved to be correct, and it was not necessary to revise all the old structures.

* Drawn in this order, the names of the four aldopentoses (ribose, arabinose, xylose, and lyxose) may be remembered by the mnemonic "Ribs are extra lean." The mnemonic for the eight aldohexoses (allose, altrose, glucose, mannose, gulose, idose, galactose, and talose) is "All altruists gladly make gum in gallon tanks."

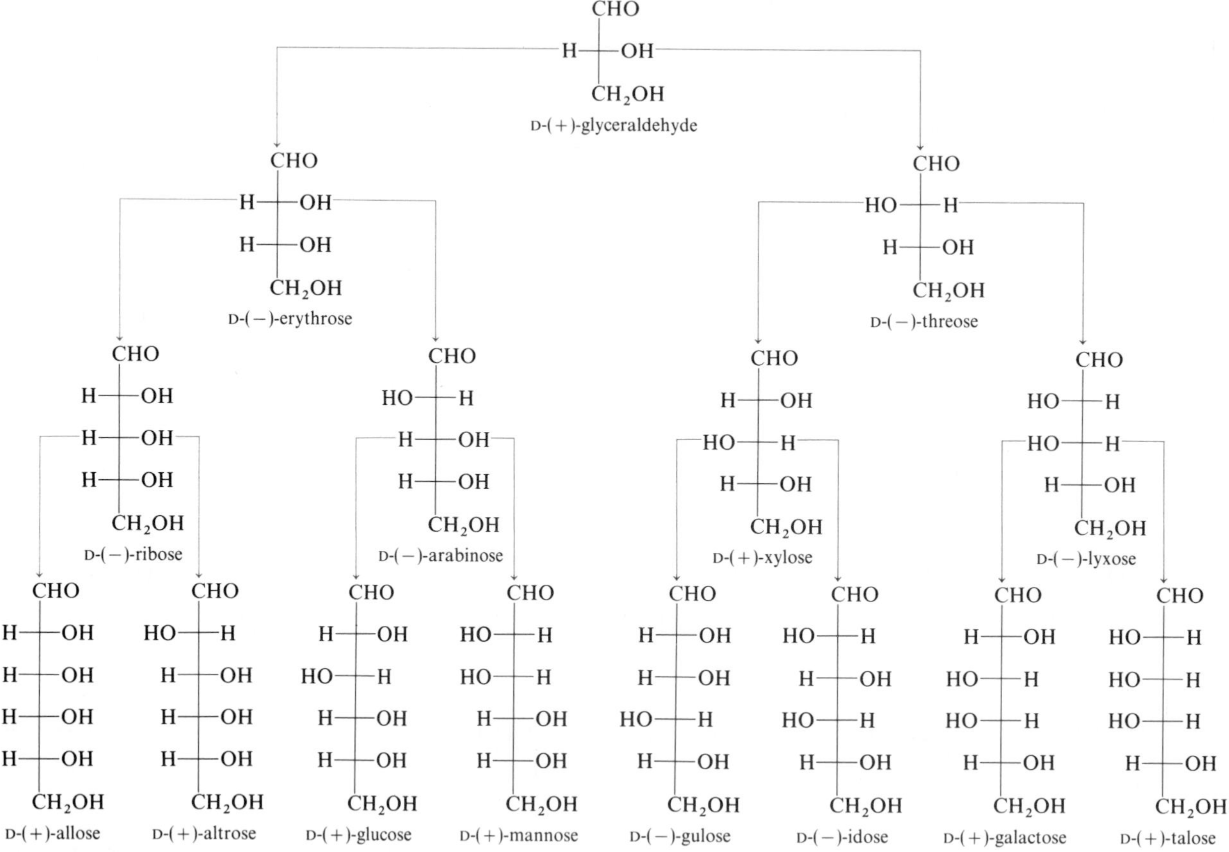

FIGURE 23-3 The D family of aldoses. All these sugars occur naturally except for threose, lyxose, allose, and gulose.

PROBLEM 23-4

Draw and name the enantiomers of the sugars shown in Figure 23-2. Give the relative configuration (D or L) and the sign of the rotation in each case.

PROBLEM 23-5

Which configuration (*R* or *S*) does the bottom chiral carbon have for the D series of sugars? Which configuration for the L series?

23-4 ERYTHRO AND THREO DIASTEREOMERS

Erythrose is the aldotetrose with the OH groups of its two chiral carbons situated on the same side of the Fischer projection, and *threose* is the diastereomer with the OH groups on opposite sides of the Fischer projection. These names have been adapted into a shorthand way of naming diastereomers with two adjacent chiral carbon atoms. A diastereomer is called **erythro** if its Fischer projection shows similar groups on the same side of the molecule. If the similar groups are on opposite sides of the Fischer projection, a diastereomer is called **threo.**

For example, syn hydroxylation of *trans*-crotonic acid gives the two enantiomers of the threo diastereomer of 2,3-dihydroxybutanoic acid. The same reaction with *cis*-crotonic acid gives the erythro diastereomer of the product.

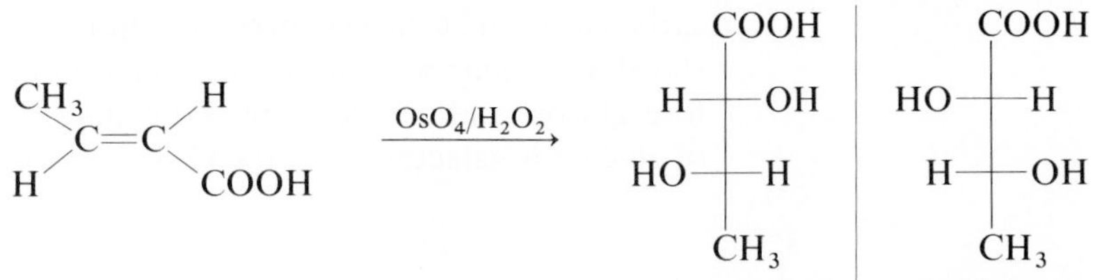

trans-crotonic acid $\xrightarrow{OsO_4/H_2O_2}$ (2*R*,3*S*) and (2*S*,3*R*) *threo*-2,3-dihydroxybutanoic acid

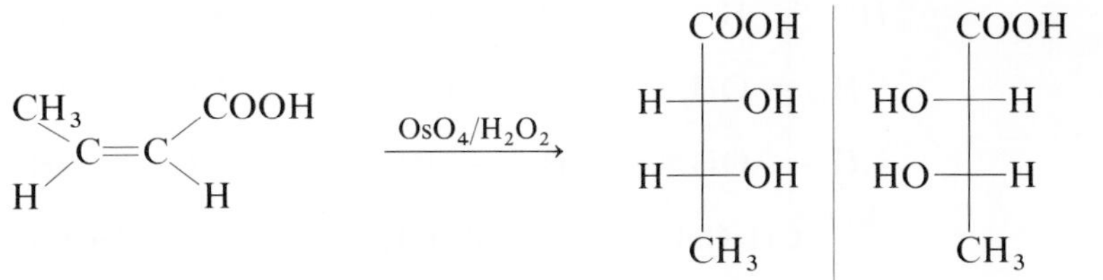

cis-crotonic acid $\xrightarrow{OsO_4/H_2O_2}$ (2*R*,3*R*) and (2*S*,3*S*) *erythro*-2,3-dihydroxybutanoic acid

The terms *erythro* and *threo* are generally used only with molecules that do not have symmetric ends. In symmetric molecules such as 2,3-dibromobutane and tartaric acid, the terms *meso* and (*d*, *l*) are preferred, because these terms indicate the diastereomer and tell whether or not it has an enantiomer. Figure 23-4 shows the proper use of the terms *erythro* and *threo* for dissymmetric molecules, as well as the terms *meso* and (*d*, *l*) for symmetric molecules.

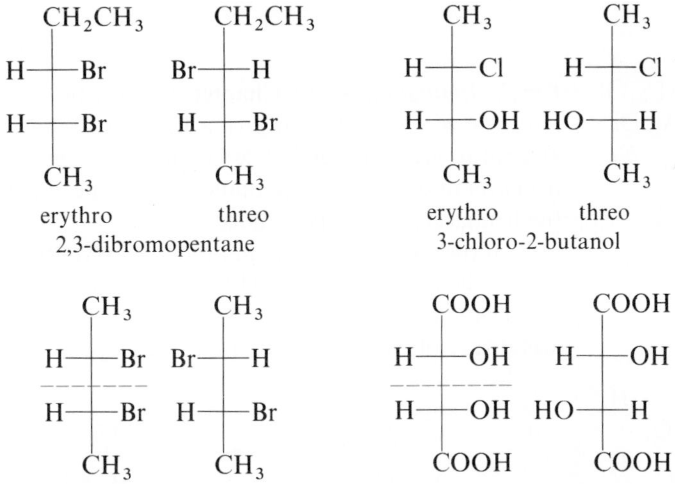

erythro threo
2,3-dibromopentane

erythro threo
3-chloro-2-butanol

meso (*d*,*l*)
2,3-dibromobutane

meso (*d*,*l*)
tartaric acid

FIGURE 23-4 The terms *erythro* and *threo* are used with dissymmetric molecules where the ends are different. The erythro diastereomer is the one with similar groups on the same side of the Fischer projection, while the threo diastereomer has similar groups on opposite sides of the Fischer projection. The terms *meso* and (±) [or (*d*, *l*)] are preferred with symmetric molecules.

PROBLEM 23-6
Draw Fisher projections for the enantiomers of *threo*-1,2,3-hexanetriol,

$$HOCH_2—CH(OH)—CH(OH)—CH_2CH_2CH_3$$

23-5 EPIMERS

Many common sugars are closely related, differing only by the stereochemistry at a single carbon atom. For example, glucose and mannose differ only at C2, the first chiral carbon atom. Sugars that differ only by the stereochemistry at a single carbon atom are called **epimers,** and the carbon atom where they differ is generally stated. If the number of a carbon atom is not stated, it is assumed to be C2. Therefore, glucose and mannose are "C2 epimers" or simply "epimers." The C4 epimer of glucose is galactose, and the C2 epimer of erythrose is threose.

D-mannose	D-glucose	D-galactose	D-erythrose	D-threose
¹CHO	¹CHO	¹CHO	¹CHO	¹CHO
HO–2–H	H–2–OH	H–2–OH	H–2–OH	HO–2–H
HO–3–H	HO–3–H	HO–3–H	H–3–OH	H–3–OH
H–4–OH	H–4–OH	HO–4–H	⁴CH_2OH	⁴CH_2OH
H–5–OH	H–5–OH	H–5–OH		
⁶CH_2OH	⁶CH_2OH	⁶CH_2OH		

PROBLEM 23-7

(a) Draw D-allose, the C3 epimer of glucose.
(b) Draw D-talose, the C2 epimer of D-galactose.
(c) Draw D-idose, the C3 epimer of D-talose. Now compare your answers with Figure 23-3.
(d) Draw the C4 "epimer" of D-xylose. Notice that this "epimer" is actually an L-series sugar, and we have seen its enantiomer. Give the correct name for this L-series sugar.

23-6 CYCLIC STRUCTURES OF MONOSACCHARIDES

Cyclic hemiacetals In Chapter 18 we saw that an aldehyde reacts with one molecule of an alcohol to give a hemiacetal, and with a second molecule of the alcohol to give an acetal. The hemiacetal is not as stable as the acetal, and most hemiacetals decompose spontaneously to the aldehyde and the alcohol. Therefore, the hemiacetal is rarely isolated.

If the aldehyde group and the hydroxyl group are part of the same molecule, a cyclic hemiacetal results. These cyclic hemiacetals are particularly stable if they result in five- or six-membered rings. In fact, five- and six-membered cyclic hemiacetals are often more stable than their open-chain forms.

The cyclic hemiacetal form of glucose Aldoses contain both an aldehyde group and several hydroxyl groups. The solid, crystalline form of an aldose is normally a cyclic hemiacetal. In solution, the aldose exists as an equilibrium mixture of the cyclic hemiacetal with the open-chain form. For most sugars, the equilibrium favors the cyclic hemiacetal form.

Typical aldohexoses such as glucose form six-membered rings, with a hemiacetal linkage between the aldehyde carbon and the hydroxyl group on C5. Figure 23-5 shows formation of the cyclic hemiacetal of glucose. Notice that the

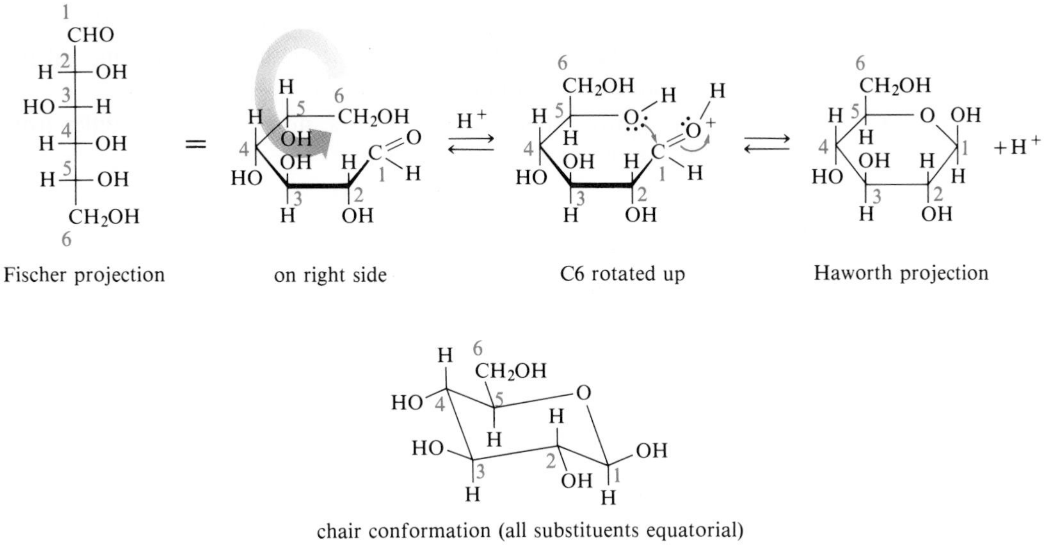

FIGURE 23-5 Glucose exists almost entirely as its cyclic hemiacetal form.

hemiacetal has a new chiral carbon atom at C1. Figure 23-5 shows the C1 hydroxyl group up, but another possible stereoisomer of the product would have this hydroxyl group directed down. We discuss the stereochemistry at C1 in more detail in Section 23-7.

The cyclic structure is often drawn initially in the **Haworth projection,** which depicts the ring as being flat (of course, it is not). The Haworth projection is widely used in biology texts, although most chemists prefer to use the more realistic chair conformation. Figure 23-5 shows the cyclic form of glucose both as a Haworth projection and as the chair conformation.

Drawing cyclic monosaccharides Cyclic hemiacetal structures may seem complicated to you, but they are easily drawn and recognized by following a few rules:

1. Mentally lay the Fischer projection over on its right side. The groups that were on the right in the Fischer projection are down in the cyclic structure, and the groups that were on the left are up.
2. The C4-C5 bond must be rotated so that the C5 hydroxyl group can form a part of the ring. For a sugar of the D series, this rotation puts the terminal $-CH_2OH$ (C6 in glucose) upward.
3. Close the ring and draw the result. Always draw the Haworth projection or chair conformation with the oxygen at the back, right-hand corner, with C1 at the far right. C1 is easily identified, because it is the hemiacetal carbon—the only carbon bonded to two oxygens. The hydroxyl group on C1 can be either up or down, as discussed in Section 23-7. Sometimes this ambiguous stereochemistry is symbolized by a wavy line.

Chair conformations are also easily drawn by recognizing the differences between the sugar in question and glucose. The following procedure is useful for drawing D-aldohexoses.

1. Draw the chair conformation puckered as it is shown in Figure 23-5. The hemiacetal carbon (C1) is the footrest.

2. Glucose has its substituents on alternating sides of the ring. (In the Haworth projection, the —OH on C4 is opposite the —CH_2OH on C5, and the —OH on C3 is opposite that on C4.) In drawing the chair conformation, just put all the ring substituents in equatorial positions.
3. To draw or recognize other common sugars, notice how they differ from glucose and make the appropriate changes.

SOLVED PROBLEM 23-1

Draw the cyclic hemiacetal forms of D-mannose and D-galactose both as chair conformations and as Haworth projections. Mannose is the C2 epimer of glucose, and galactose is the C4 epimer of glucose.

SOLUTION The chair conformations are easier to draw, so we will do them first. Draw the rings and number the carbon atoms, starting with the hemiacetal carbon. Mannose is the C2 epimer of glucose, so the substituent on C2 is axial, while all the others are equatorial as they are in glucose. Galactose is the C4 epimer of glucose, so its substituent on C4 is axial.

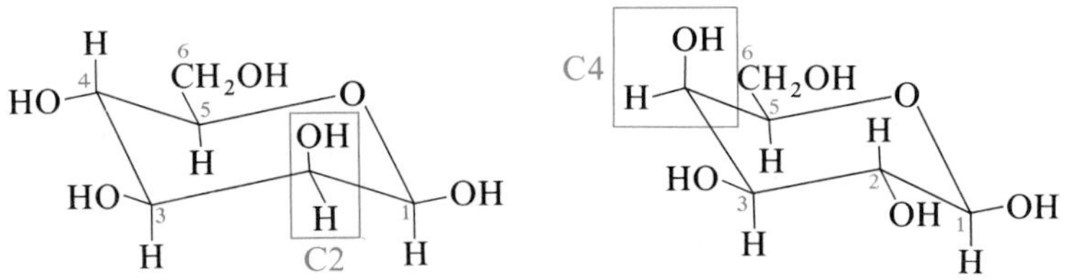

D-mannose D-galactose

The simplest way to draw Haworth structures for these two sugars is to draw the chair conformations and then draw the flat rings with the same substituents in the up and down positions. For practice, however, let's lay down the Fischer projection for galactose. You should follow along with your molecular models.

1. Lay down the Fischer projection: right → down; left → up.

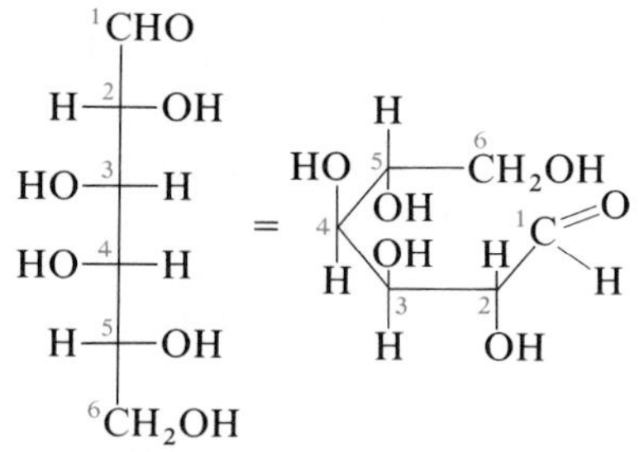

D-galactose

2. Rotate the C4-C5 bond to put the —OH in place. (For a D sugar, the —CH_2OH goes up.)

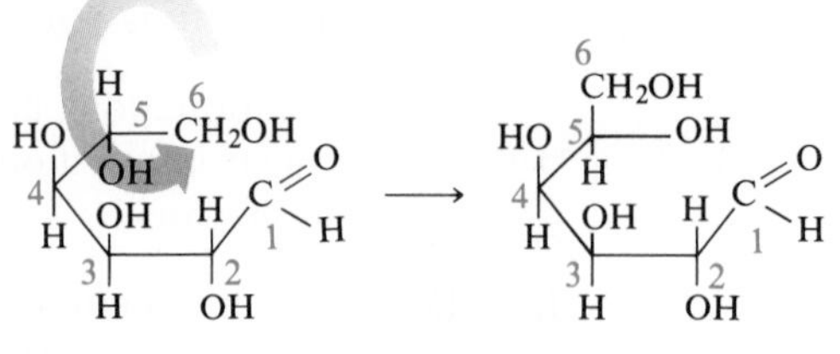

3. Close the ring and draw the final hemiacetal. The hydroxyl group on C1 can be either up or down, as discussed in Section 23-7. Sometimes this ambiguous stereochemistry is symbolized by a wavy line.

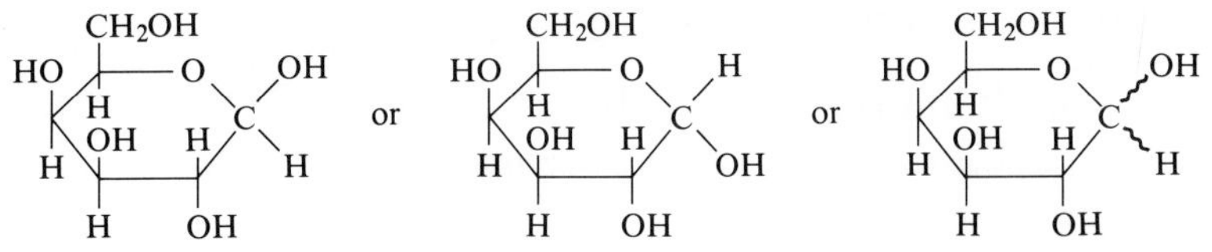

PROBLEM 23-8

Draw the Haworth projection for the cyclic structure of D-mannose by laying down the Fischer projection.

PROBLEM 23-9

Allose is the C3 epimer of glucose. Draw the cyclic hemiacetal form of D-allose, first in the chair conformation and then in the Haworth projection.

The five-membered cyclic hemiacetal form of fructose Not all sugars form six-membered rings in their hemiacetal forms. Many aldopentoses and ketohexoses form five-membered rings. The five-membered ring of fructose is shown in Figure 23-6. Five-membered rings are not puckered as much as six-membered rings, so they are usually depicted as flat Haworth projections. Notice that the five-membered ring is drawn with the ring oxygen in back and the hemiacetal carbon (the one bonded to two oxygens) on the right. The $-CH_2OH$ at the back left (C6) is in the up position for D-series ketohexoses.

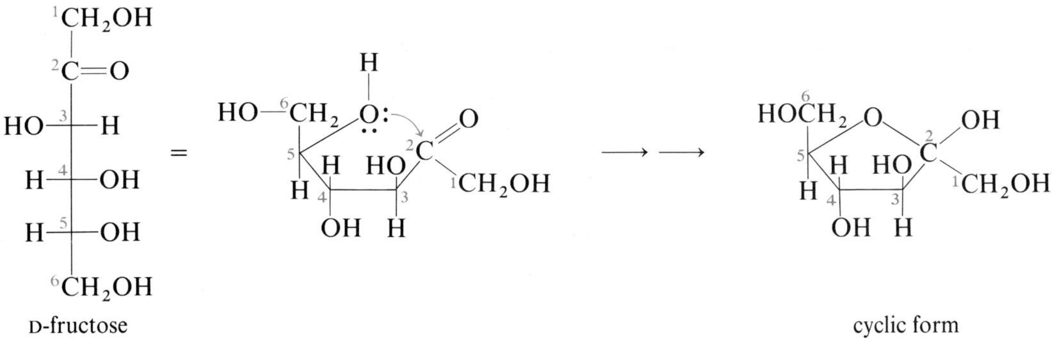

FIGURE 23-6 Fructose forms a five-membered cyclic hemiacetal. The five-membered rings are usually represented as flat Haworth structures.

Pyranose and furanose names The cyclic structures of monosaccharides are named according to their five- or six-membered rings. A six-membered cyclic hemiacetal is called a **pyranose,** derived from the name of the six-membered cyclic ether *pyran.* A five-membered cyclic hemiacetal is called a **furanose,** derived from the name of the five-membered cyclic ether *furan.* The ring is still numbered as it is in the sugar, not beginning with the heteroatom as it would be in the heterocyclic nomenclature. These structural names are incorporated into the names of the sugars:

pyran 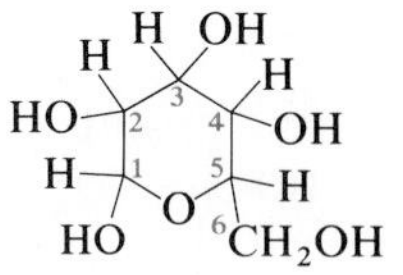 a pyranose 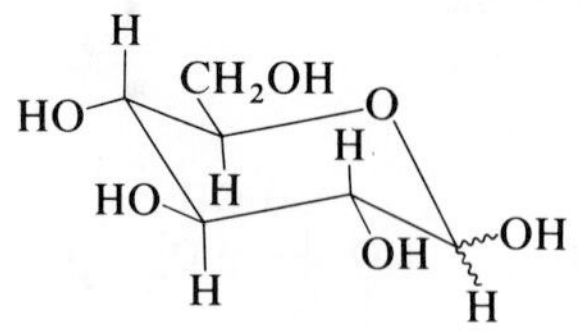 D-glucopyranose  furan 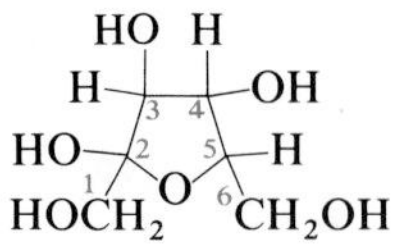 a furanose 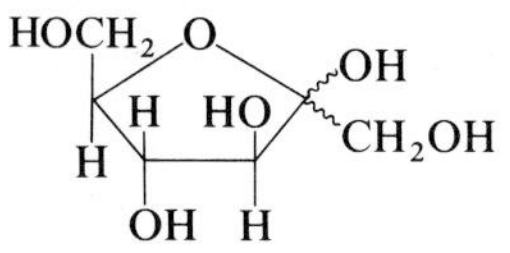 D-fructofuranose

PROBLEM 23-10

Talose is the C4 epimer of mannose. Draw the chair conformation of D-talopyranose.

PROBLEM 23-11

(a) Figure 23-3 shows that the degradation of D-glucose gives D-arabinose, an aldopentose. Arabinose is most stable in its furanose form. Draw D-arabinofuranose.
(b) Ribose, the C2 epimer of arabinose, is most stable in its furanose form. Draw D-ribofuranose.

PROBLEM 23-12

The carbonyl group in D-galactose may be isomerized from C1 to C2 by brief treatment with dilute base (by the enediol rearrangement, Section 23-8). The product is the C4 epimer of fructose. Draw the furanose structure of the product.

23-7 ANOMERS OF MONOSACCHARIDES; MUTAROTATION

When a pyranose or furanose ring closes, the hemiacetal carbon atom is converted from a flat carbonyl group to a chiral carbon. Depending on which face of the (protonated) carbonyl group is attacked, the hemiacetal —OH group can be directed either up or down. These two orientations of the hemiacetal —OH group give diastereomeric products called **anomers.** Figure 23-7 shows the anomers of glucose.

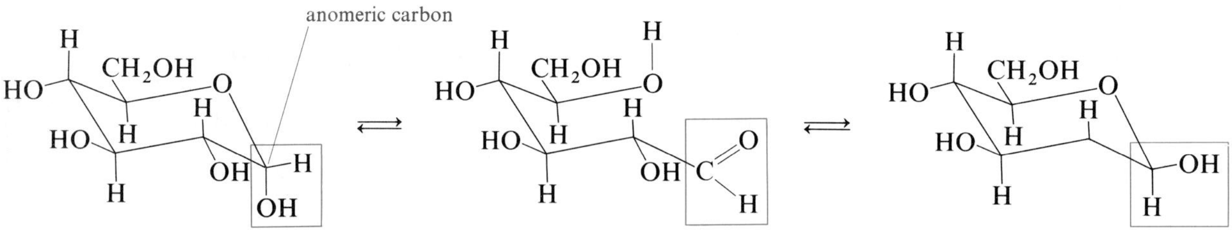

anomeric carbon

α-D-glucopyranose ⇌ open-chain form ⇌ β-D-glucopyranose

FIGURE 23-7 The anomers of glucose. The hydroxyl group on the anomeric (hemiacetal) carbon is down (axial) in the α anomer, and up (equatorial) in the β anomer. The β anomer of glucose has all its substituents in equatorial positions.

The hemiacetal carbon atom is called the **anomeric carbon,** easily identified as the only carbon atom bonded to two oxygens. Its —OH group is called the anomeric hydroxyl group. Notice in Figure 23-7 that the anomer with the anomeric —OH group down (axial) is called the α (alpha) anomer, while the one with the anomeric —OH group up (equatorial) is called the β (beta) anomer. We can easily draw the α and β anomers of most aldohexoses by remembering that the β form of glucose (β-D-glucopyranose) has all its substituents in equatorial positions. To draw an α anomer, simply move the anomeric —OH group to the axial position.

Another way to remember the anomers is to notice that the α anomer has its anomeric hydroxyl group trans to the terminal $—CH_2OH$ group, while it is cis in the β anomer. This rule works for all the sugars, from both the D and L series, and even for the furanoses. Figure 23-8 shows the two anomers of fructose, whose anomeric carbon is C2. The α anomer has the anomeric —OH group down, trans to the terminal $—CH_2OH$ group, while the β anomer has it up, cis to the terminal $—CH_2OH$.

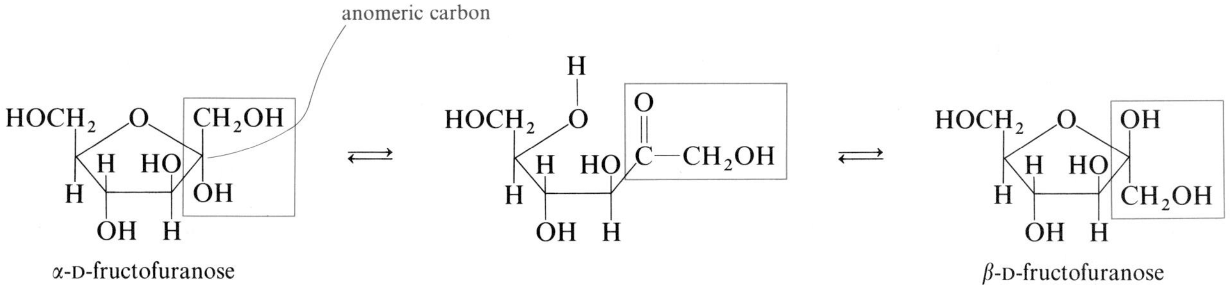

FIGURE 23-8 The α anomer of fructose has the anomeric —OH group down, trans to the terminal $—CH_2OH$ group. The β anomer has the anomeric hydroxyl group up, cis to the terminal $—CH_2OH$.

PROBLEM 23-13

Draw the following monosaccharides, using chair conformations for the pyranoses and Haworth projections for the furanoses.

(a) α-D-mannopyranose (C2 epimer of glucose)
(b) β-D-galactopyranose (C4 epimer of glucose)
(c) β-D-allopyranose (C3 epimer of glucose)
(d) α-D-arabinofuranose
(e) β-D-ribofuranose (C2 epimer of arabinose)

Properties of anomers: mutarotation Because anomers are diastereomers, they generally have different properties. For example, α-D-glucopyranose has a melting point of 146°C and a specific rotation of +112.2°, while β-D-glucopyranose has a melting point of 150°C and a specific rotation of +18.7°. When glucose is crystallized from water at room temperature, pure crystalline α-D-glucopyranose results. If glucose is crystallized from water by letting the water evaporate at a temperature above 98°C, crystals of pure β-D-glucopyranose are formed (Fig. 23-9).

In each of these cases, *all* the glucose in the solution crystallizes as the favored anomer. In the solution, the two anomers are in equilibrium through a small amount of the open-chain form, and this equilibrium continues to supply more of the anomer that is crystallizing out of solution.

When one of the glucose anomers dissolves in water, an interesting change in the specific rotation is observed. When the α anomer dissolves, its specific rotation gradually decreases from an initial value of +112.2° to +52.6°. When the pure β anomer dissolves, its specific rotation gradually increases from +18.7° to the same value of +52.6° (Fig. 23-10). This change in the specific rotations to a mutual value is called **mutarotation.** Mutarotation occurs because in solution

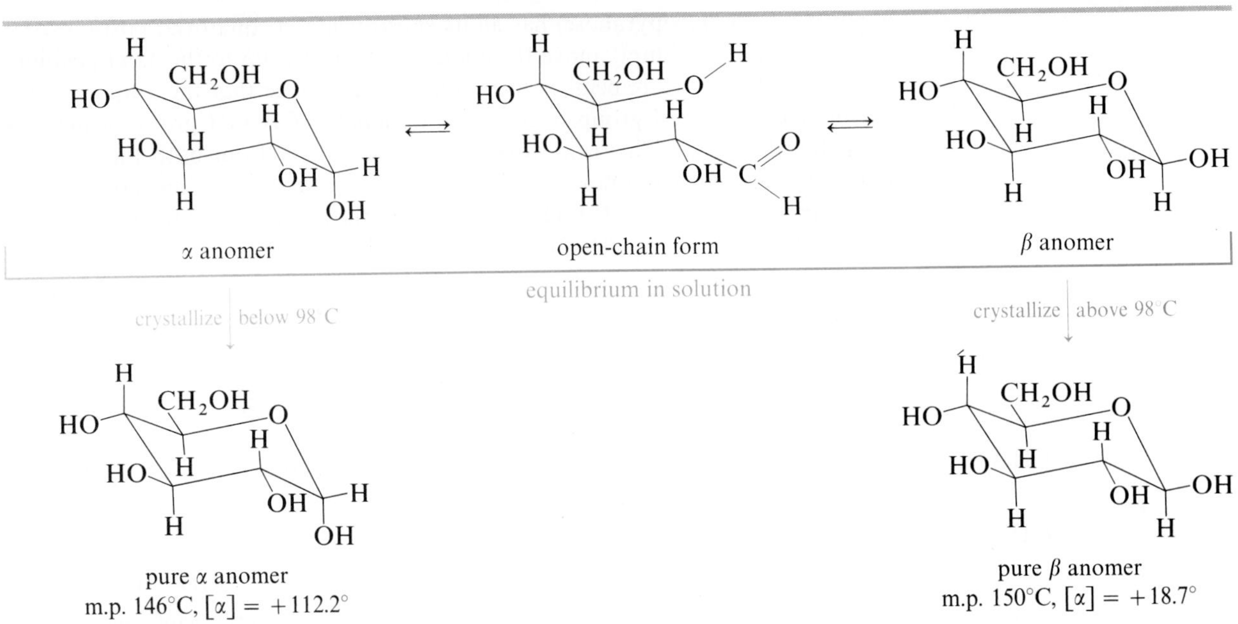

α anomer ⇌ open-chain form ⇌ β anomer

equilibrium in solution

crystallize below 98°C → pure α anomer, m.p. 146°C, [α] = +112.2°

crystallize above 98°C → pure β anomer, m.p. 150°C, [α] = +18.7°

FIGURE 23-9 An aqueous solution of D-glucose contains an equilibrium mixture of α-D-glucopyranose, β-D-glucopyranose, and the intermediate open-chain form. Crystallization below 98°C gives the α anomer, but crystallization above 98°C gives the β anomer.

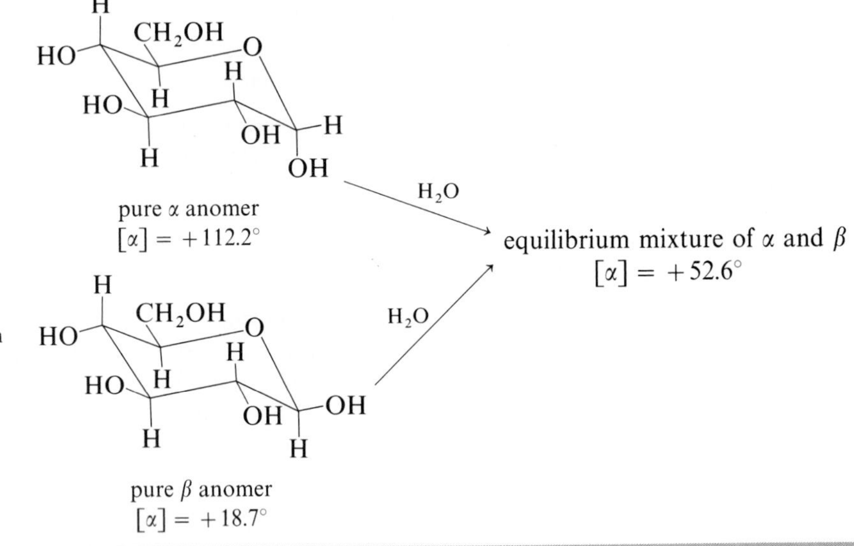

FIGURE 23-10 Mutarotation. In solution, the two anomers interconvert and the specific rotation of a pure anomer gradually changes to a value intermediate between the specific rotations of the two anomers, corresponding to the equilibrium mixture.

the two anomers can interconvert. When either of the pure anomers dissolves in water, the rotation of the pure anomer gradually changes to an intermediate rotation that results from equilibrium concentrations of the anomers.

SOLVED PROBLEM 23-2

Calculate how much of the α anomer and how much of the β anomer are present in an equilibrium mixture with a specific rotation of +52.6°.

SOLUTION If the fraction of glucose present as the α anomer ($[\alpha] = +112.2°$) is a, the fraction present as the β anomer ($[\alpha] = +18.7°$) is b, and the rotation of the mixture is $+52.6°$, we have

$$a(+112.2°) + b(+18.7°) = +52.7°$$

There is very little of the open-chain form present, so the fraction present as the α anomer (a) plus the fraction present as the β anomer (b) should account for all the glucose:

$$a + b = 1 \qquad \text{or} \qquad b = 1 - a$$

Substituting $(1 - a)$ for b in the first equation, we have

$$a(112.2°) + (1 - a)(18.7°) = 52.7°$$

Solving this equation for a, we have $a = 0.36$, or 36 percent. Thus b must be $(1 - 0.36) = 0.64$ or 64 percent. The amounts of the two anomers present at equilibrium are

$$\alpha \text{ anomer, } 36\% \qquad \beta \text{ anomer, } 64\%$$

When we remember that the anomeric hydroxyl group is axial in the α anomer and equatorial in the β anomer, it is reasonable that the more stable β anomer should predominate.

PROBLEM 23-14

Like glucose, galactose shows mutarotation when it dissolves in water. α-D-Galactopyranose has a specific rotation of $+150.7°$, while that of the β anomer is $+52.8°$. When either of the pure anomers dissolves in water, the specific rotation gradually changes to $+80.2°$. Determine the percentages of the two anomers present at equilibrium.

23-8 REACTIONS OF MONOSACCHARIDES: SIDE REACTIONS IN BASE

Sugars are multifunctional compounds that can undergo reactions typical of any of their functional groups. Most sugars exist as cyclic hemiacetals, yet in solution they are in equilibrium with their open-chain aldehyde or ketone forms. As a result, sugars undergo most of the usual reactions of ketones, aldehydes, and alcohols. Reagents commonly used with monofunctional compounds often give unwanted side reactions with sugars, however. Carbohydrate chemists have developed reactions that work well with sugars, while avoiding the undesired side reactions. As we learn about the unique reactions of simple sugars, we will often draw them as their open-chain forms, because in many reactions it is the small equilibrium amount of the open-chain form that reacts.

Epimerization and the enediol rearrangement One of the most important aspects of sugar chemistry is the inability, in most cases, to use basic reagents because they cause unwanted side reactions. The two most common base-catalyzed side reactions are epimerization and enediol rearrangement.

Under basic conditions, the proton alpha to the aldehyde (or ketone) carbonyl group is reversibly removed. In the resulting enolate ion, C2 is no longer chiral, and its stereochemistry is lost. Reprotonation can occur on either face of the enolate, giving either configuration. The result is an equilibrium mixture of the original sugar and its C2 epimer. Because a mixture of epimers results, this stereochemical change is called **epimerization.** The following reaction shows the rapid base-catalyzed epimerization of glucose to a mixture of glucose and its C2 epimer, mannose.

Base-catalyzed epimerization of glucose

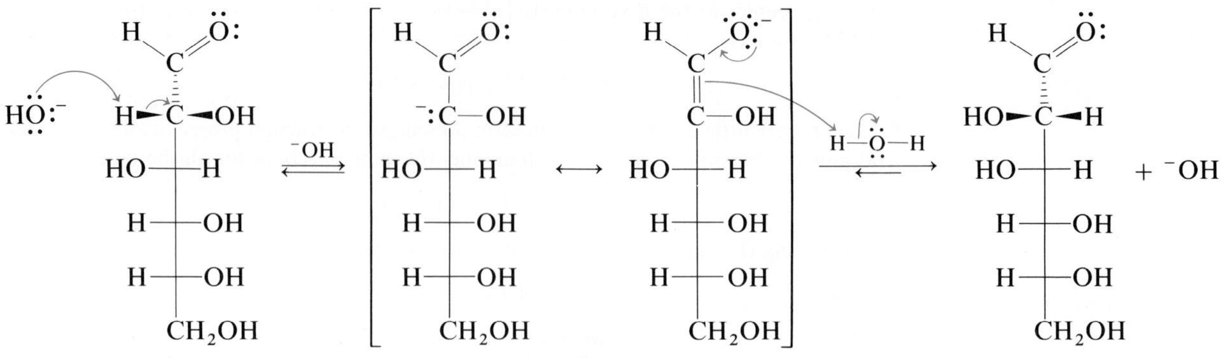

D-glucose ⇌ enolate ⇌ D-mannose

PROBLEM 23-15
Give a mechanism for the base-catalyzed epimerization of erythrose to a mixture of erythrose and threose.

Another base-catalyzed side reaction is the **enediol rearrangement,** which moves the carbonyl group up and down the chain. If the enolate ion formed by removal of a proton on C2 reprotonates on the C1 oxygen, an **enediol** intermediate results. Removal of a proton from the C2 oxygen of the enediol and reprotonation on C1 gives fructose, a ketose.

Base-catalyzed enediol rearrangement

D-glucose ⇌ enolate ⇌ enediol ⇌ enolate ⇌ D-fructose

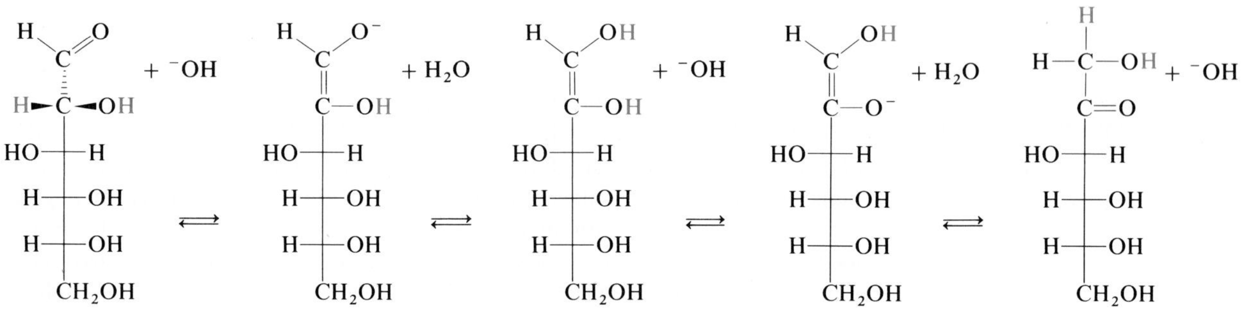

PROBLEM 23-16
Show how C3 of fructose can epimerize under basic conditions.

PROBLEM 23-17
Show how another enediol rearrangement can move the carbonyl group from C2 in fructose to C3.

Under strongly basic conditions the combination of enediol rearrangements and epimerization leads to a complex mixture of sugars. Except when using specially protected sugars, most sugar chemistry employs neutral or acidic reagents to avoid these annoying side reactions.

23-9 REDUCTION OF MONOSACCHARIDES

Like other aldehydes and ketones, aldoses and ketoses can be reduced to the corresponding alcohols, called **alditols.** The most common reagents are sodium borohydride and catalytic hydrogenation using a nickel catalyst. Alditols are named by adding the suffix *-itol* to the root name of the sugar. The following equation shows the reduction of glucose to glucitol, sometimes called *sorbitol.*

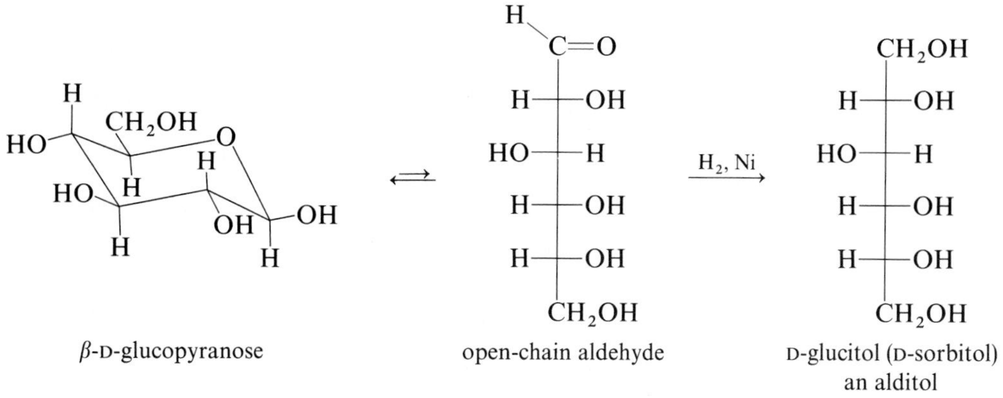

β-D-glucopyranose ⇌ open-chain aldehyde $\xrightarrow{H_2,\ Ni}$ D-glucitol (D-sorbitol) an alditol

The reduction of a ketose creates a new chiral carbon atom, formed in either of two configurations, resulting in two epimers. For example, the reduction of fructose gives a mixture of glucitol and mannitol.

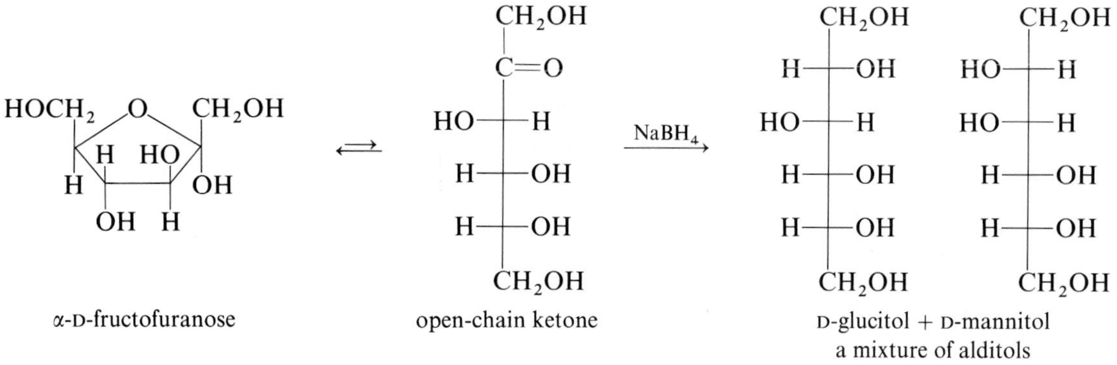

α-D-fructofuranose ⇌ open-chain ketone $\xrightarrow{NaBH_4}$ D-glucitol + D-mannitol a mixture of alditols

PROBLEM 23-18

When D-glucose is treated with sodium borohydride, optically active glucitol results. When optically active D-galactose is reduced, however, the product is optically inactive. Explain this loss of optical activity.

PROBLEM 23-19

Emil Fischer synthesized L-gulose, an unusual aldohexose that reduces to give D-glucitol. Suggest a structure for this L-sugar, and show how L-gulose gives the same alditol as D-glucose. (*Hint:* D-glucitol has $—CH_2OH$ groups at both ends. *Either* of these primary alcohol groups might have come from reduction of an aldehyde.)

23-10 OXIDATION OF MONOSACCHARIDES; REDUCING SUGARS

Bromine water The aldehyde group of an aldose is easily oxidized to a carboxylic acid by bromine water. Bromine water is used for this oxidation because it does not oxidize the alcohol groups of the sugar and it does not oxidize ketoses. Also, bromine water is acidic and does not cause epimerization or movement of the carbonyl group. Because bromine water oxidizes aldoses but not ketoses, it serves as a useful test to distinguish aldoses from ketoses. The product of bromine water oxidation is an **aldonic acid.** For example, bromine water oxidizes glucose to gluconic acid.

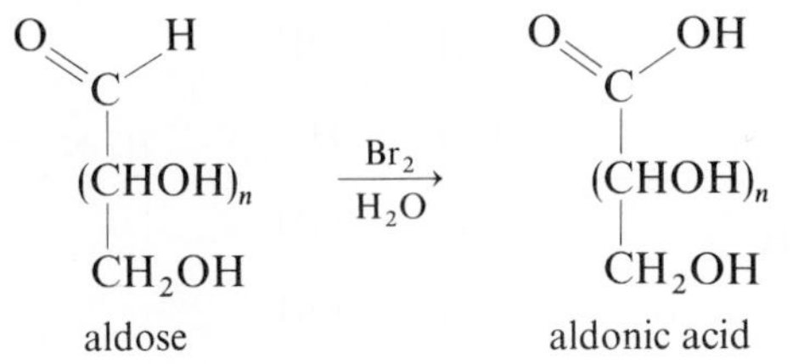

$$\underset{\text{aldose}}{\text{HC(=O)}-(\text{CHOH})_n-\text{CH}_2\text{OH}} \xrightarrow[\text{H}_2\text{O}]{\text{Br}_2} \underset{\text{aldonic acid}}{\text{HOC(=O)}-(\text{CHOH})_n-\text{CH}_2\text{OH}}$$

Example

glucose (Fischer projection, top to bottom): CHO; H—C—OH; HO—C—H; H—C—OH; H—C—OH; CH_2OH

$\xrightarrow[\text{H}_2\text{O}]{\text{Br}_2}$

gluconic acid: COOH; H—C—OH; HO—C—H; H—C—OH; H—C—OH; CH_2OH

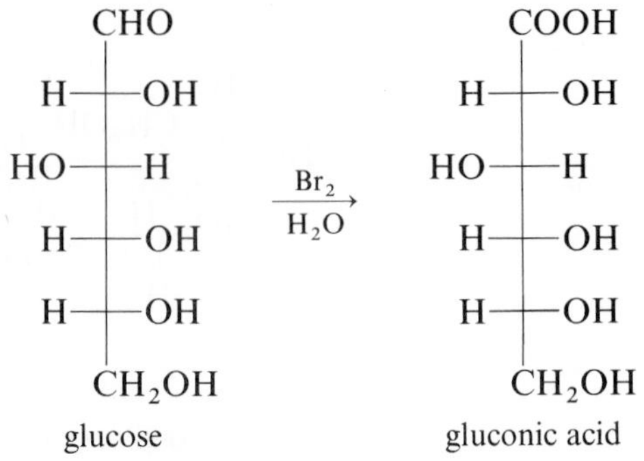

PROBLEM 23-20

Draw and name the products of bromine water oxidation of

(a) D-mannose. (b) D-galactose. (c) D-fructose.

Nitric acid Nitric acid is a stronger oxidizing agent than bromine water, oxidizing both the aldehyde group and the terminal $-CH_2OH$ group of an aldose to carboxylic acid groups. The resulting dicarboxylic acid is called an **aldaric acid.** For example, nitric acid oxidizes glucose to glucaric acid.

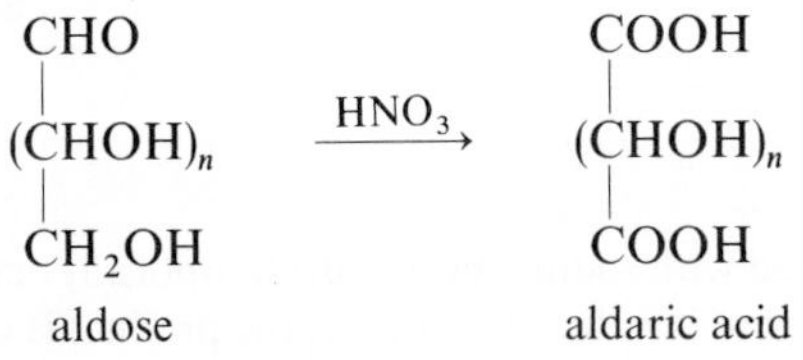

$$\underset{\text{aldose}}{\text{CHO}-(\text{CHOH})_n-\text{CH}_2\text{OH}} \xrightarrow{\text{HNO}_3} \underset{\text{aldaric acid}}{\text{COOH}-(\text{CHOH})_n-\text{COOH}}$$

Example

glucose (Fischer projection, top to bottom): CHO; H—C—OH; HO—C—H; H—C—OH; H—C—OH; CH_2OH

$\xrightarrow{\text{HNO}_3}$

glucaric acid: COOH; H—C—OH; HO—C—H; H—C—OH; H—C—OH; COOH

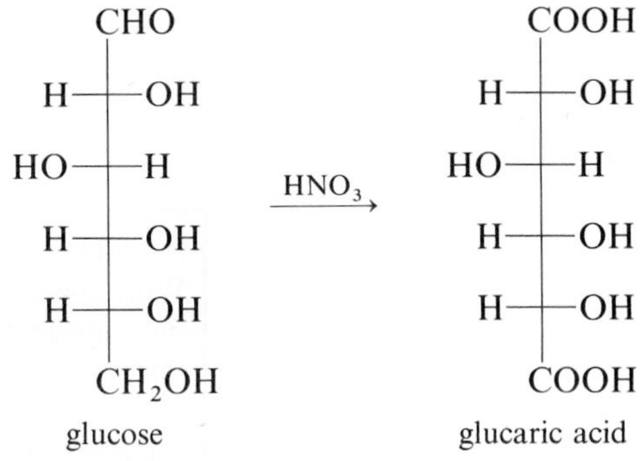

PROBLEM 23-21

Draw and name the products of nitric acid oxidation of

(a) D-mannose. (b) D-galactose.

PROBLEM 23-22

Two sugars, A and B, are known to be glucose and galactose, but it is not certain which one is which. On treatment with nitric acid, A gives an optically inactive aldaric acid, while B gives an optically active aldaric acid. Which sugar is glucose, and which is galactose?

The Tollens test The **Tollens test** is used to detect aldehydes, which react with the Tollens reagent to give carboxylate ions and metallic silver, often in the form of a silver mirror on the inside of the container.

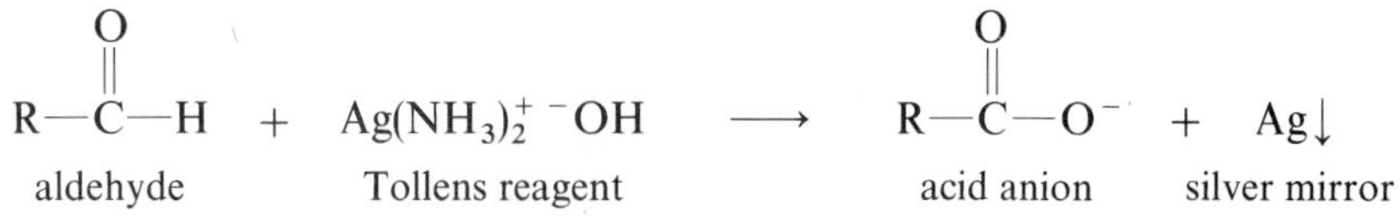

$$\underset{\text{aldehyde}}{R-\overset{\overset{\displaystyle O}{\|}}{C}-H} + \underset{\text{Tollens reagent}}{Ag(NH_3)_2^+ \; ^-OH} \longrightarrow \underset{\text{acid anion}}{R-\overset{\overset{\displaystyle O}{\|}}{C}-O^-} + \underset{\text{silver mirror}}{Ag\downarrow}$$

In its open-chain form, an aldose has an aldehyde group which reacts with the Tollens reagent to give an aldonic acid and a silver mirror. This oxidation is not a good synthesis of the aldonic acid, however, because the Tollens reagent is strongly basic and promotes epimerization and enediol rearrangements. Sugars that reduce the Tollens reagent are called **reducing sugars.**

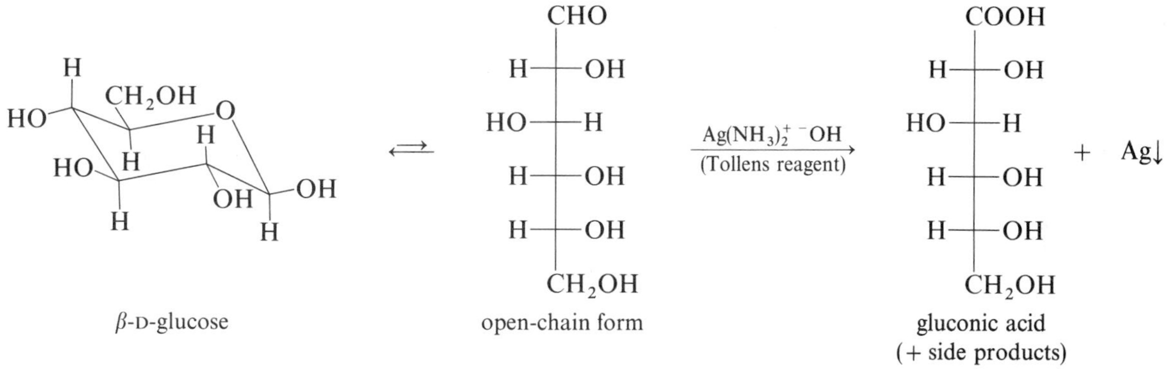

β-D-glucose $\rightleftharpoons$ open-chain form (CHO; H—OH; HO—H; H—OH; H—OH; CH_2OH) $\xrightarrow[\text{(Tollens reagent)}]{Ag(NH_3)_2^+ \; ^-OH}$ gluconic acid (COOH; H—OH; HO—H; H—OH; H—OH; CH_2OH) (+ side products) + $Ag\downarrow$

The Tollens test cannot distinguish between aldoses and ketoses because the strongly basic Tollens reagent promotes enediol rearrangements. Under basic conditions, the open-chain form of a ketose is converted to an aldose, which then reacts to give a positive Tollens test.

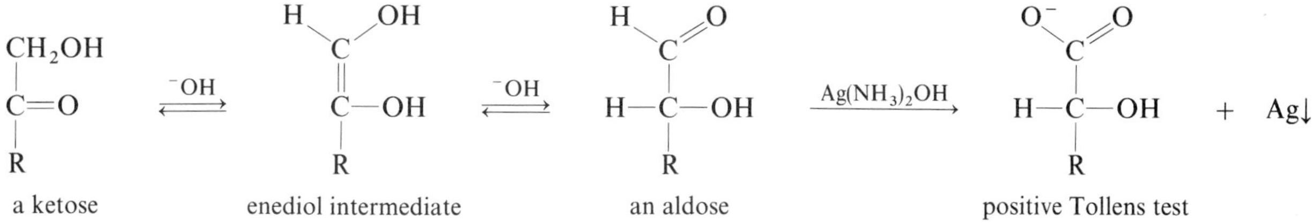

$$\underset{\text{a ketose}}{R-C(=O)-CH_2OH} \overset{^-OH}{\rightleftharpoons} \underset{\text{enediol intermediate}}{R-C(OH)=CH(OH)} \overset{^-OH}{\rightleftharpoons} \underset{\text{an aldose}}{R-CH(OH)-CHO} \xrightarrow{Ag(NH_3)_2OH} \underset{\text{positive Tollens test}}{R-CH(OH)-COO^-} + Ag\downarrow$$

23-11 NONREDUCING SUGARS: FORMATION OF GLYCOSIDES

The obvious question is: What good is the Tollens test if it doesn't distinguish between aldoses and ketoses? The answer lies in the fact that Tollens reagent must react with the open-chain form of the sugar, which has a free aldehyde or ketone. If the cyclic form cannot open to the free carbonyl compound, the sugar does not react with the Tollens reagent. Hemiacetals are easily opened, but an acetal is quite stable under neutral or basic conditions (Section 18-18). If the carbonyl group is in the form of an acetal, the cyclic form cannot open to the free carbonyl compound, and the sugar gives a negative Tollens test (Fig. 23-11).

Sugars in the form of acetals are called **glycosides,** and their names end in the *-oside* suffix. For example, a glycoside of glucose would be a *glucoside,* and if it were a six-membered ring it would be a *glucopyranoside.* Similarly, a glycoside of ribose would be a *riboside,* and if it were a five-membered ring it would be a *ribofuranoside.* In general, a sugar whose name ends with the suffix *-ose* is a reducing sugar, and

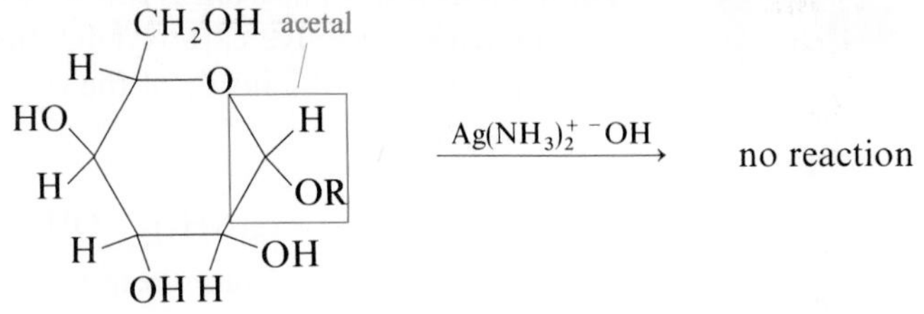

acetal

$\xrightarrow{Ag(NH_3)_2^+ \ ^-OH}$ no reaction

a glycoside

Examples of nonreducing sugars

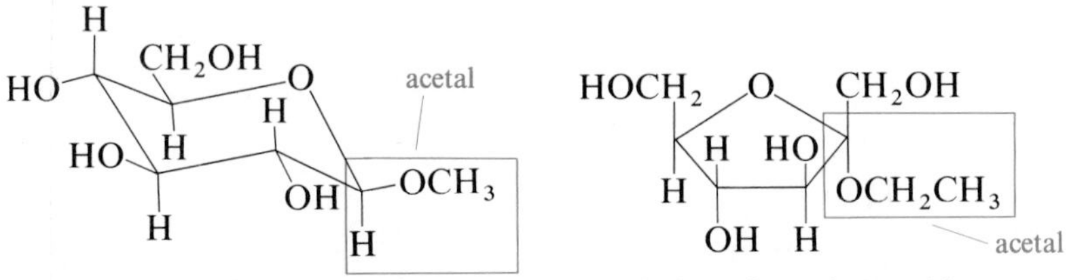

acetal

acetal

methyl β-D-glucopyranoside
(or methyl β-D-glucoside)

ethyl α-D-fructofuranoside
(or ethyl α-D-fructoside)

FIGURE 23-11 Sugars that exist as full acetals are stable to Tollens reagent and are nonreducing sugars. Such sugars are called glycosides.

one whose name ends with *-oside* is nonreducing. Because they exist as stable acetals rather than hemiacetals, glycosides cannot spontaneously open to their open-chain forms, and they do not mutarotate. They are locked in a particular anomeric form.

We can summarize by saying that the Tollens test distinguishes between reducing sugars and nonreducing sugars: Reducing sugars (aldoses and ketoses) exist as hemiacetals, and they mutarotate. Nonreducing sugars (glycosides) are acetals, and they do not mutarotate.

PROBLEM 23-23

Which of the following examples are reducing sugars? Comment on the common name *sucrose* for table sugar.

(a) methyl α-D-galactopyranoside
(b) β-L-idopyranose (an aldohexose)
(c) α-D-allopyranose
(d) ethyl β-D-ribofuranoside
(e) 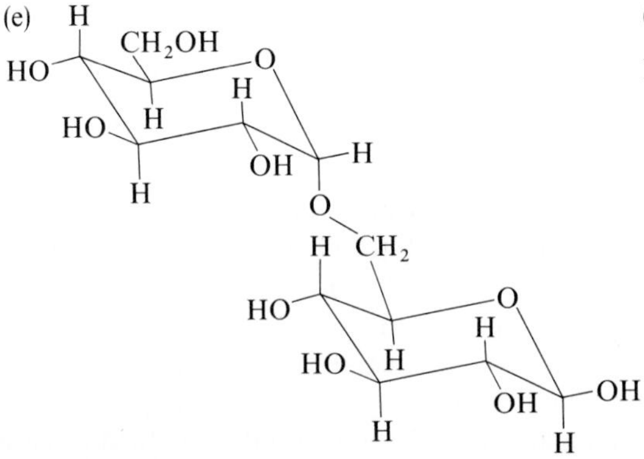
(f)

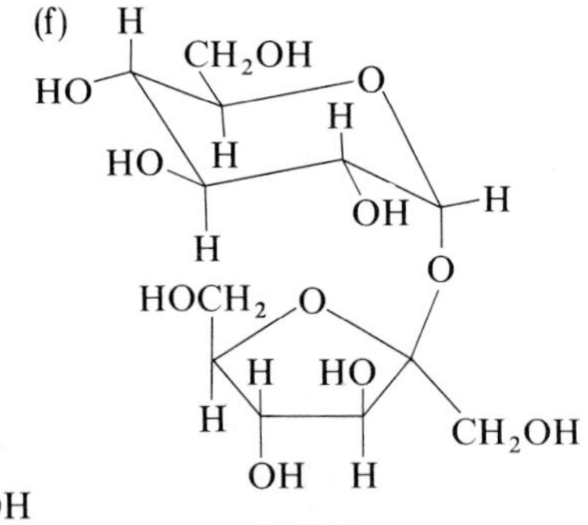

sucrose

PROBLEM 23-24

Draw the structures of the compounds named in Problem 23-23 parts (a), (c), and (d). Allose is the C3 epimer of glucose, and ribose is the C2 epimer of arabinose.

Formation of glycosides Aldehydes and ketones are converted to acetals by treatment with an alcohol and a trace of acid catalyst (Section 18-18). These conditions also convert aldoses and ketoses to the acetals we call glycosides. Regardless of the anomer used as the starting material, both anomers of the glycoside are formed (as an equilibrium mixture) under these acidic conditions. The more stable anomer predominates. For example, the acid-catalyzed reaction of glucose with methanol gives a mixture of methyl glucosides.

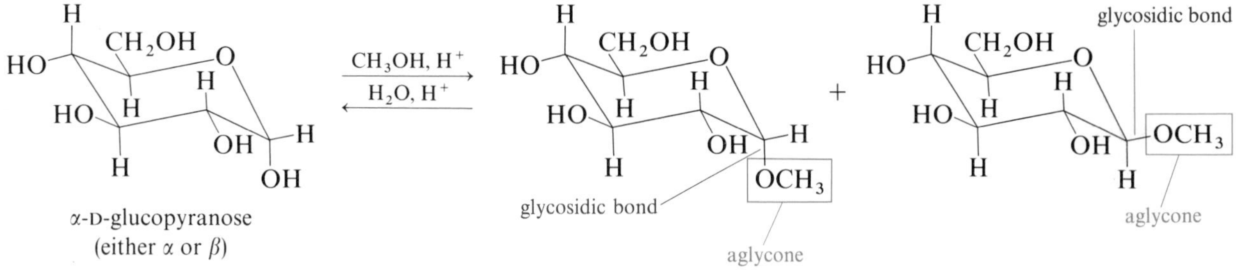

α-D-glucopyranose (either α or β) — methyl α-D-glucopyranoside — methyl β-D-glucopyranoside

Like other acetals, glycosides are stable to basic conditions but are hydrolyzed by aqueous acid. Glycosides may be used with basic reagents and in basic solutions.

An **aglycone** is the group bonded to the anomeric carbon atom of a glycoside. For example, in a methyl glycoside methanol is the aglycone. Many aglycones are bonded by an oxygen atom, while others are bonded through a nitrogen atom or some other heteroatom. Figure 23-12 gives the structures of some glycosides with interesting aglycones.

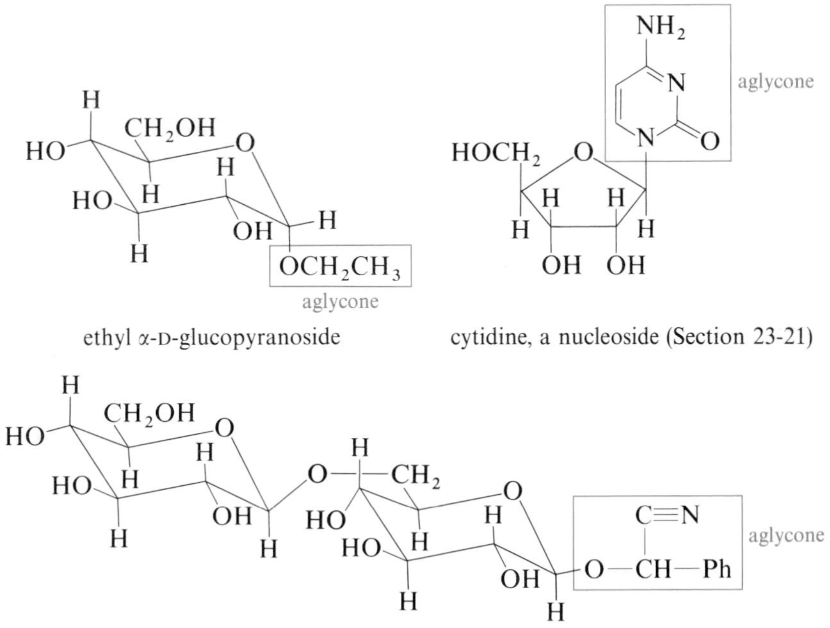

ethyl α-D-glucopyranoside

cytidine, a nucleoside (Section 23-21)

amygdalin
a component of Laetrile, a controversial cancer drug

FIGURE 23-12 The group bonded to the anomeric carbon atom of a glycoside is called an aglycone. Some aglycones are bonded through an oxygen atom (a true acetal), and others are bonded through other atoms such as nitrogen.

Disaccharides and polysaccharides are glycosides in which the alcohol forming the glycosidic bond is an —OH group of another monosaccharide. We consider the disaccharides and polysaccharides in Sections 23-18 and 23-19.

PROBLEM 23-25

The mechanism of glycoside formation is the same as the second part of the mechanism for acetal formation. Give a mechanism for the formation of methyl β-D-glucopyranoside.

PROBLEM 23-26

Show the products that result from hydrolysis of amygdalin in dilute acid. Can you suggest why amygdalin might be toxic to tumor (and possibly other) cells?

PROBLEM 23-27

Treatment of either anomer of fructose with excess ethanol in the presence of a trace of HCl gives a mixture of the α and β anomers of ethyl D-fructofuranoside. Draw the starting materials, reagents, and products for this reaction. Circle the aglycone in each product.

23-12 ETHER AND ESTER FORMATION

Because they contain several hydroxyl groups, sugars are very soluble in water and rather insoluble in organic solvents. Sugars are difficult to recrystallize from water, because they often form supersaturated syrups such as honey and molasses. If the hydroxyl groups are converted to ethers, sugars behave like simpler organic compounds. The ethers are soluble in organic solvents, and they are more easily purified by recrystallization and simple chromatographic methods.

Treating a sugar with methyl iodide and silver oxide converts its hydroxyl groups to methyl ethers. Silver oxide polarizes the $H_3C—I$ bond, making the methyl carbon strongly electrophilic. Attack by the carbohydrate —OH group, followed by deprotonation, gives the ether. Figure 23-13 shows that the anomeric hydroxyl group is also converted to an ether. If the conditions are carefully controlled, the hemiacetal C—O bond is not broken, and the configuration at the anomeric carbon is preserved.

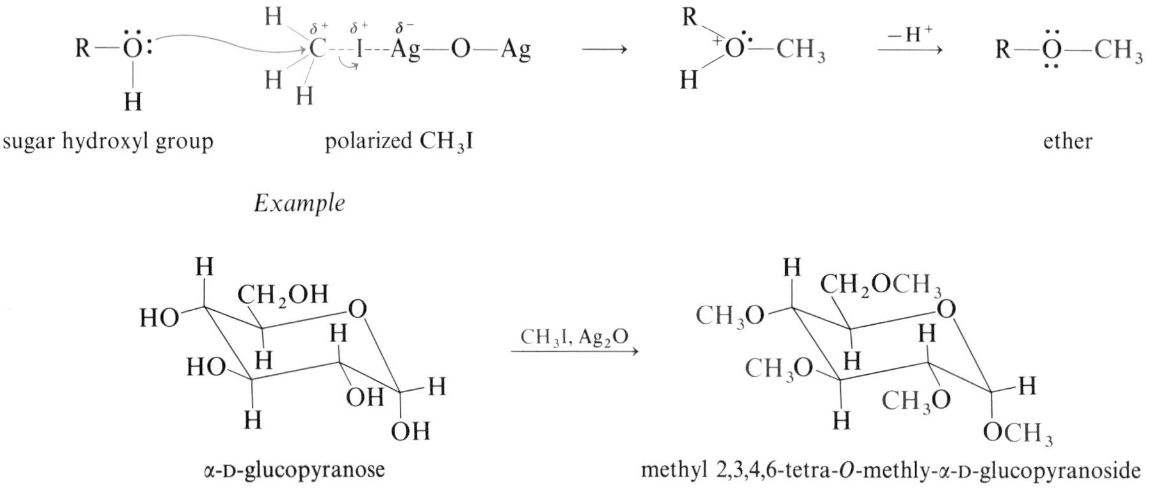

$$R—\ddot{O}:(H) + H_3C^{\delta+}\cdots I^{\delta+}\cdots Ag^{\delta-}—O—Ag \longrightarrow R(H)\overset{+}{O}—CH_3 \xrightarrow{-H^+} R—\ddot{O}—CH_3$$

sugar hydroxyl group — polarized CH_3I — ether

Example

$$\alpha\text{-D-glucopyranose} \xrightarrow{CH_3I,\ Ag_2O} \text{methyl 2,3,4,6-tetra-}O\text{-methly-}\alpha\text{-D-glucopyranoside}$$

α-D-glucopyranose

methyl 2,3,4,6-tetra-*O*-methly-α-D-glucopyranoside

FIGURE 23-13 Treatment of an aldose or a ketose with methyl iodide and silver oxide gives the totally methylated ether. If the conditions are carefully controlled, the stereochemistry at the anomeric carbon is usually preserved.

The Williamson ether synthesis is the most common method for forming simple ethers, but the Williamson conditions involve a strongly basic alkoxide ion. Under these basic conditions, a monosaccharide would isomerize and decompose. A modified Williamson method may be used if the sugar is first converted to a glycoside (by treatment with an alcohol and an acid catalyst). The glycoside is an acetal, stable to base. Treatment of a glycoside with sodium hydroxide and methyl iodide or dimethyl sulfate gives the methylated carbohydrate.

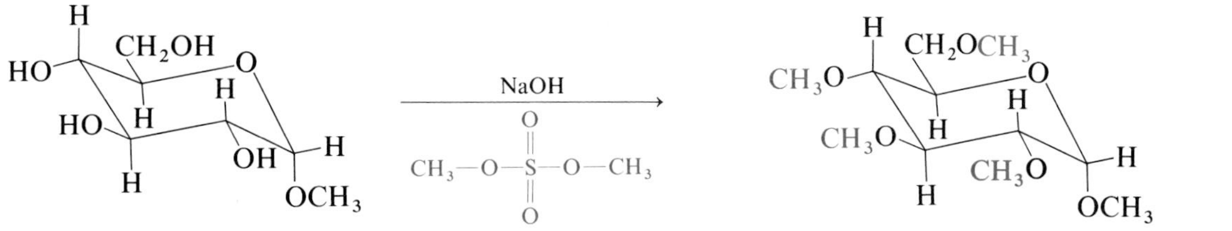

methyl α-D-glucopyranoside (stable to base) — (dimethyl sulfate) — methyl 2,3,4,6-tetra-*O*-methyl-α-D-glucopyranoside

PROBLEM 23-28

Give a mechanism for the methylation of any one of the hydroxyl groups using NaOH and dimethyl sulfate.

PROBLEM 23-29

Draw the expected product of the reaction of the following sugars with methyl iodide and silver oxide.

(a) α-D-fructofuranose (b) β-D-galactopyranose

Ester formation Another way to convert sugars to easily handled derivatives is to esterify the hydroxyl groups. The esters are readily crystallized and purified, and they dissolve in common organic solvents. Sugar hydroxyl groups are converted to acetate esters by treatment with acetic anhydride and pyridine (as a mild basic catalyst), as shown in Figure 23-14. This reaction acetylates all the hydroxyl groups, including that of the hemiacetal on the anomeric carbon. The anomeric C—O bond is not broken in the acylation, and the stereochemistry of the anomeric carbon atom is usually preserved. If we start with a pure α anomer or a pure β anomer, the product is the corresponding anomer of the acetate.

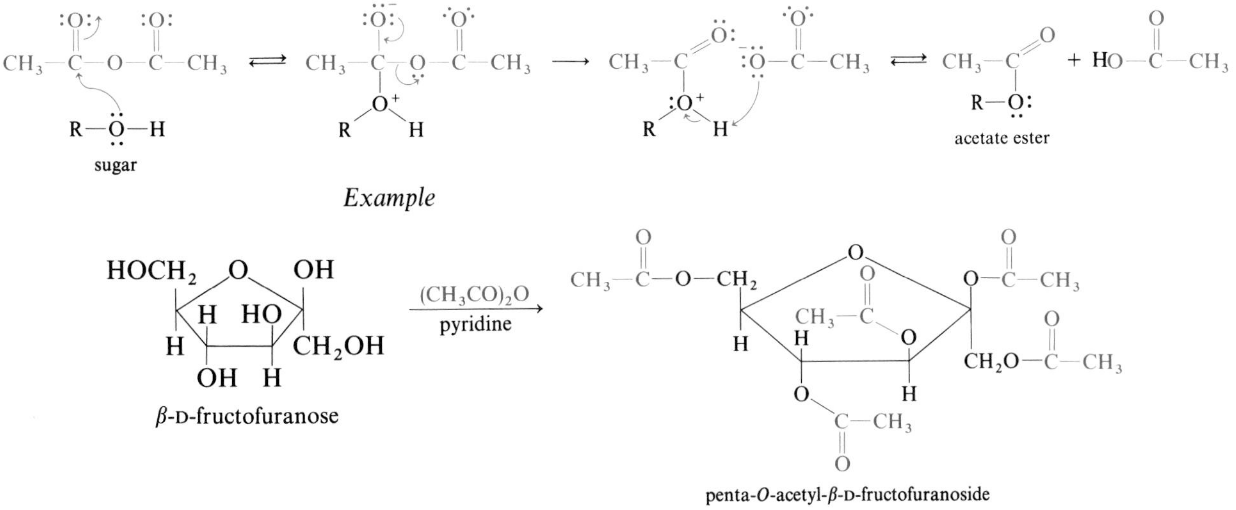

FIGURE 23-14 Acetic anhydride and pyridine convert all of the hydroxyl groups on a sugar to acetate esters. The stereochemistry at the anomeric carbon is usually preserved.

PROBLEM 23-30

Give the expected product of the reaction of the following sugars with acetic anhydride and pyridine.

(a) α-D-glucopyranose (b) β-D-ribofuranose

23-13 REACTIONS WITH PHENYLHYDRAZINE: OSAZONE FORMATION

One of the best methods for derivatizing ketones and aldehydes is the formation of hydrazones, especially phenylhydrazones and 2,4-dinitrophenylhydrazones (Section 18-17). In his exploratory work on the structures of sugars, Emil Fischer often made and used phenylhydrazone derivatives. In fact, his constant use of phenylhydrazine ultimately led to Fischer's death from chronic phenylhydrazine poisoning in 1919.

$$R'(R)C{=}O + H_2N{-}NH{-}C_6H_5 \underset{}{\overset{H^+}{\rightleftharpoons}} R'(R)C{=}N{-}NH{-}C_6H_5 + H_2O$$

ketone or aldehyde; phenylhydrazine; phenylhydrazone

Sugars do not form the simple phenylhydrazone derivatives we might expect, however. Two molecules of phenylhydrazine condense with each molecule of the sugar to give an **osazone,** in which both C1 and C2 have been converted to phenylhydrazones. The term *osazone* is derived from the *-ose* suffix of a sugar and the last half of the word hydr*azone*.

$$\text{aldose: } H{-}C{=}O,\ CHOH,\ (CHOH)_n,\ CH_2OH \xrightarrow[H^+]{\text{excess } H_2N{-}NH{-}Ph} \text{osazone: } H{-}C{=}N{-}NHPh,\ C{=}N{-}NHPh,\ (CHOH)_n,\ CH_2OH$$

(boxed $(CHOH)_n$ / CH_2OH: unaffected portion)

$$\text{ketose: } H{-}CHOH,\ C{=}O,\ (CHOH)_n,\ CH_2OH \xrightarrow[H^+]{\text{excess } H_2N{-}NH{-}Ph} \text{osazone: } H{-}C{=}N{-}NHPh,\ C{=}N{-}NHPh,\ (CHOH)_n,\ CH_2OH$$

In the formation of an osazone, both C1 and C2 are converted to phenylhydrazones. Therefore, a ketose gives the same osazone as its related aldose. Also notice that the stereochemistry at C2 is lost upon formation of a phenylhydrazone. Therefore, C2 epimers give the same osazone.

PROBLEM 23-31

(a) Show that D-glucose, D-mannose, and D-fructose all give the same osazone. Show the structure and stereochemistry of this osazone.

(b) D-Talose is an aldohexose that gives the same osazone as D-galactose. Give the structure of D-talose, and give the structure of its osazone.

23-14 CHAIN SHORTENING: THE RUFF DEGRADATION

In our discussion of the D and L series of sugars, we briefly mentioned a method for shortening the chain of an aldose by removing the aldehyde carbon at the top of the Fischer projection. Such a reaction, removing one of the carbon atoms, is called a **degradation.** A wide variety of natural and synthetic sugars were degraded to glyceraldehyde, the sugar with just three carbon atoms. The sugars that gave (+)-glyceraldehyde were defined to be the D series; those that gave (−)-glyceraldehyde were defined as the L series.

The most common method used to shorten sugar chains is the **Ruff degradation,** developed by Otto Ruff, a prominent German chemist around the turn of the century. The Ruff degradation is a two-step process that begins with oxidation of the aldose to its aldonic acid. Treatment of the aldonic acid with hydrogen peroxide and ferric sulfate oxidizes the carboxyl group to CO_2 and gives an aldose with one less carbon atom. The Ruff degradation is used mostly for structure determination of sugars and synthesis of new sugars.

Ruff degradation

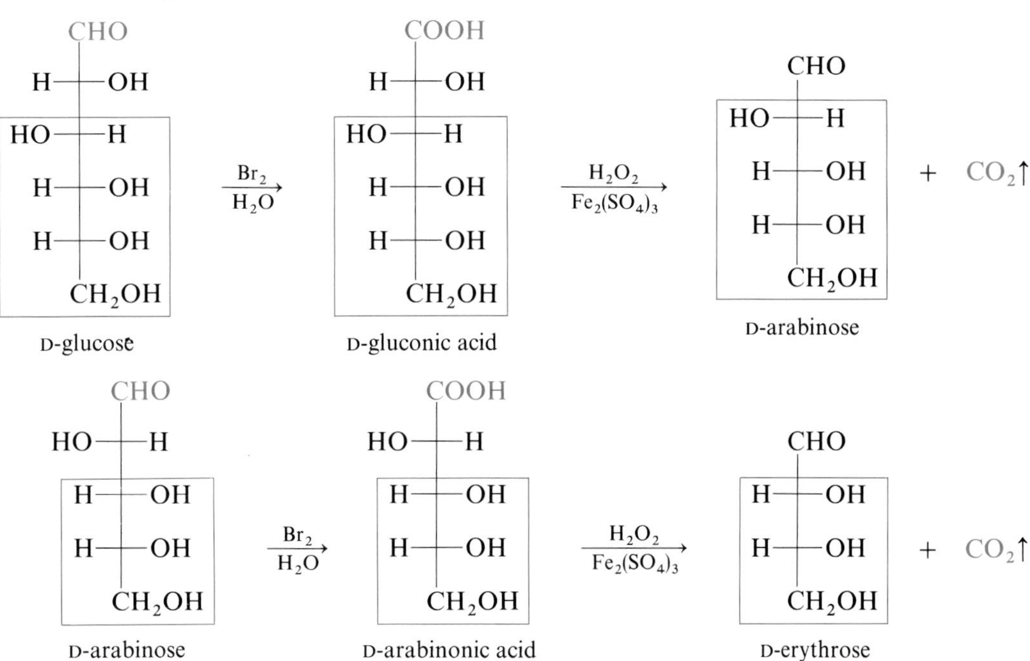

PROBLEM 23-32

Show that Ruff degradation of D-mannose gives the same aldopentose (D-arabinose) as does D-glucose.

PROBLEM 23-33

D-Lyxose is formed by Ruff degradation of galactose. Give the structure of D-lyxose. Ruff degradation of D-lyxose gives D-threose. Give the structure of D-threose.

PROBLEM 23-34

D-Altrose is an aldohexose. Ruff degradation of D-altrose gives the same aldopentose as does degradation of D-allose, the C3 epimer of glucose. Give the structure of D-altrose.

23-15 CHAIN-LENGTHENING: THE KILIANI-FISCHER SYNTHESIS

The **Kiliani-Fischer synthesis** lengthens an aldose carbon chain by adding one carbon atom to the aldehyde end of the aldose. This synthesis is useful both for determining the structure of existing sugars and for synthesizing new sugars.

The Kiliani-Fischer synthesis

CHO–$(CHOH)_n$–CH_2OH (an aldose) $\xrightarrow{HCN}$ CN–CHOH–$(CHOH)_n$–CH_2OH (a cyanohydrin) $\xrightarrow{H_3O^+}$ COOH–CHOH–$(CHOH)_n$–CH_2OH (a chain-lengthened aldonic acid) $\xrightarrow{Na(Hg)}$ CHO–CHOH–$(CHOH)_n$–CH_2OH (a chain-lengthened aldose)

an aldose | a cyanohydrin | a chain-lengthened aldonic acid | a chain-lengthened aldose

The aldehyde carbon atom is made chiral in the first step, the formation of the cyanohydrin. Two epimeric cyanohydrins result. For example, D-arabinose reacts with HCN to give the following two cyanohydrins.

D-arabinose (CHO; HO–H; H–OH; H–OH; CH_2OH) $\xrightarrow{HCN}$ (CN; H–C–OH; HO–H; H–OH; H–OH; CH_2OH) + (CN; HO–C–H; HO–H; H–OH; H–OH; CH_2OH)

D-arabinose | two epimeric cyanohydrins

Hydrolysis of these cyanohydrins gives two aldonic acids. Under the acidic conditions of the hydrolysis, the aldonic acids cyclize to give lactones.

(CN; H–OH; HO–H; H–OH; H–OH; CH_2OH) $\xrightarrow{H_3O^+}$ (COOH; H–OH; HO–H; H–OH; H–OH; CH_2OH) $\longrightarrow$ lactone (H, CH_2OH, O, HO, H, HO, H, OH, H, O)

(CN; HO–H; HO–H; H–OH; H–OH; CH_2OH) $\xrightarrow{H_3O^+}$ (COOH; HO–H; HO–H; H–OH; H–OH; CH_2OH) $\longrightarrow$ lactone (H, CH_2OH, O, HO, HO, H, HO, H, H, O)

epimeric cyanohydrins | epimeric aldonic acids | epimeric lactones

Sodium amalgam, an alloy of sodium and mercury, reduces lactones to hemiacetals. In effect, the carboxyl group of the aldonic acid is reduced to an aldehyde.

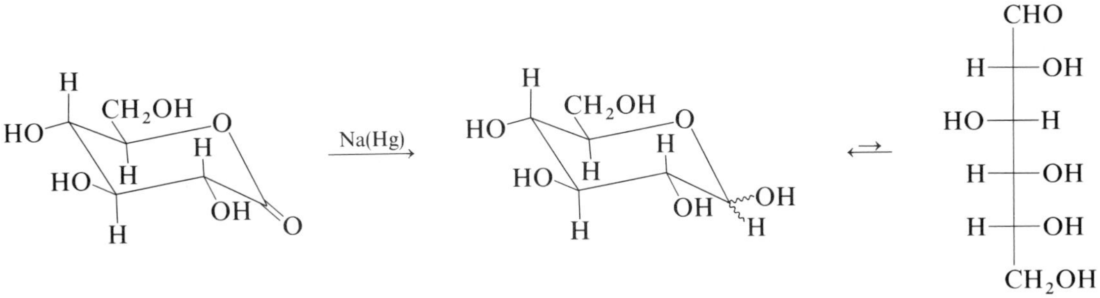

epimeric lactones $\xrightarrow{Na(Hg)}$ epimeric hemiacetals $\rightleftarrows$ glucose / mannose open-chain form

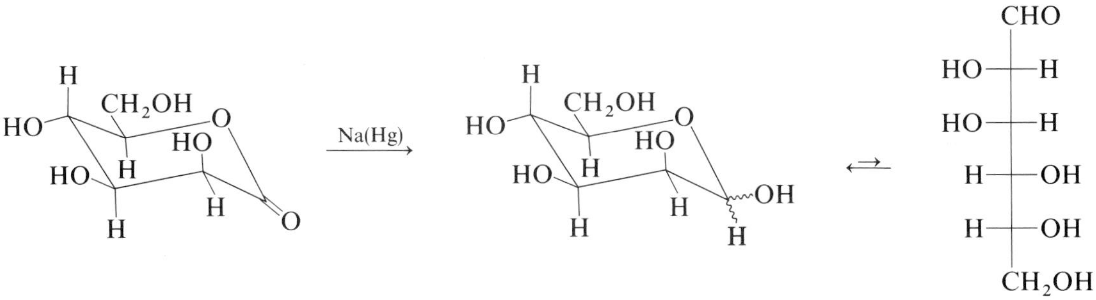

The Kiliani-Fischer synthesis is the inverse process of the Ruff degradation. Degradation of either of two C2 epimers gives the same shortened aldose, and the Kiliani-Fischer synthesis converts this shortened aldose back into a mixture of the same two C2 epimers. For example, glucose and mannose are degraded to arabinose, and the Kiliani-Fischer synthesis converts arabinose into a mixture of glucose and mannose.

PROBLEM 23-35

Ruff degradation of D-arabinose gives D-erythrose. The Kiliani-Fischer synthesis converts D-erythrose to a mixture of D-arabinose and D-ribose. Draw out these reactions and give the structure of D-ribose.

PROBLEM 23-36

The *Wohl degradation,* an alternative to the Ruff degradation, is nearly the reverse of the Kiliani-Fischer synthesis. The aldose carbonyl group is converted to the oxime, which is dehydrated by acetic anhydride to the nitrile (a cyanohydrin). Cyanohydrin formation is reversible, and a basic hydrolysis allows the cyanohydrin to lose HCN. Using the following sequence of reagents, give equations for the individual reactions in the Wohl degradation of D-arabinose to D-erythrose. Mechanisms are not required.

(1) hydroxylamine hydrochloride
(2) acetic anhydride
(3) ^{-}OH, H_2O

SUMMARY OF THE REACTIONS OF SUGARS

1. *Undesirable rearrangements catalyzed by base* (Section 23-8)

a. *Epimerization of the alpha carbon*

$$\begin{array}{c} CHO \\ H-\!\!\!-OH \\ HO-\!\!\!-H \\ H-\!\!\!-OH \\ H-\!\!\!-OH \\ CH_2OH \end{array} \underset{}{\overset{^-OH}{\rightleftharpoons}} \begin{array}{c} CHO \\ HO-\!\!\!-H \\ HO-\!\!\!-H \\ H-\!\!\!-OH \\ H-\!\!\!-OH \\ CH_2OH \end{array}$$

b. *Enediol rearrangements*

$$\begin{array}{c} CHO \\ H-\!\!\!-OH \\ HO-\!\!\!-H \\ H-\!\!\!-OH \\ H-\!\!\!-OH \\ CH_2OH \end{array} \overset{^-OH}{\rightleftharpoons} \begin{array}{c} CH_2OH \\ C{=}O \\ HO-\!\!\!-H \\ H-\!\!\!-OH \\ H-\!\!\!-OH \\ CH_2OH \end{array} \overset{^-OH}{\rightleftharpoons} \begin{array}{c} CH_2OH \\ CHOH \\ C{=}O \\ H-\!\!\!-OH \\ H-\!\!\!-OH \\ CH_2OH \end{array} \quad \text{etc.}$$

Because of these side reactions, basic reagents are rarely used with sugars.

2. *Reduction* (Section 23-9)

$$\underset{\text{aldose}}{\begin{array}{c} CHO \\ | \\ (CHOH)_n \\ | \\ CH_2OH \end{array}} \xrightarrow[\text{or } H_2/Ni]{NaBH_4} \underset{\text{alditol}}{\begin{array}{c} CH_2OH \\ | \\ (CHOH)_n \\ | \\ CH_2OH \end{array}}$$

3. *Oxidation* (Section 23-10)

a. *To aldonic acids*

$$\underset{\text{aldose}}{\begin{array}{c} CHO \\ | \\ (CHOH)_n \\ | \\ CH_2OH \end{array}} \xrightarrow[H_2O]{Br_2} \underset{\text{aldonic acid}}{\begin{array}{c} COOH \\ | \\ (CHOH)_n \\ | \\ CH_2OH \end{array}}$$

b. *To aldaric acids*

$$\underset{\text{aldose}}{\begin{array}{c} CHO \\ | \\ (CHOH)_n \\ | \\ CH_2OH \end{array}} \xrightarrow{HNO_3} \underset{\text{aldaric acid}}{\begin{array}{c} COOH \\ | \\ (CHOH)_n \\ | \\ COOH \end{array}}$$

c. *The Tollens test for reducing sugars*

$$\underset{\text{aldose}}{\begin{array}{c} CHO \\ | \\ CHOH \\ | \\ (CHOH)_n \\ | \\ CH_2OH \end{array}} \text{ or } \underset{\text{ketose}}{\begin{array}{c} CH_2OH \\ | \\ C{=}O \\ | \\ (CHOH)_n \\ | \\ CH_2OH \end{array}} \xrightarrow{Ag(NH_3)_2OH} \begin{array}{c} COO^- \\ | \\ CHOH \\ | \\ (CHOH)_n \\ | \\ CH_2OH \end{array} \begin{array}{l} \\ + \text{ rearrangement} \\ \\ + \text{ Ag (silver mirror)} \\ \\ \end{array}$$

4. *Glycoside formation (conversion to an acetal)* (Section 23-11)

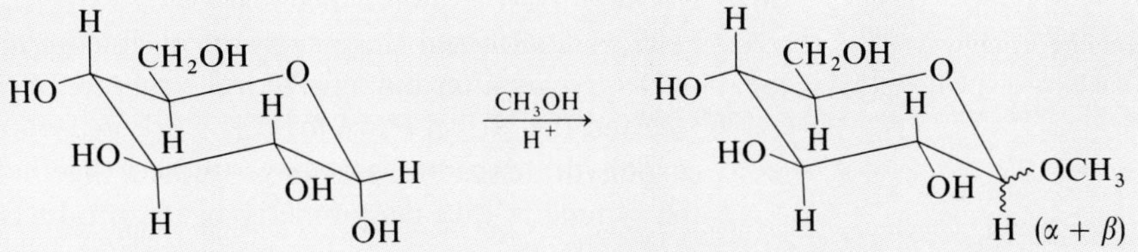

$\xrightarrow[H^+]{CH_3OH}$

(either anomer)

a methyl glycoside (more stable anomer predominates) H ($\alpha + \beta$)

5. *Ether formation* (Section 23-12)

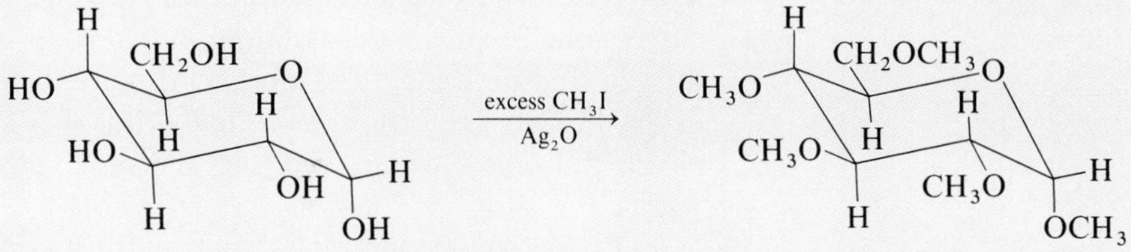

$\xrightarrow[Ag_2O]{\text{excess } CH_3I}$

(gives the same anomer as the starting material)

6. *Ester formation* (Section 23-12)

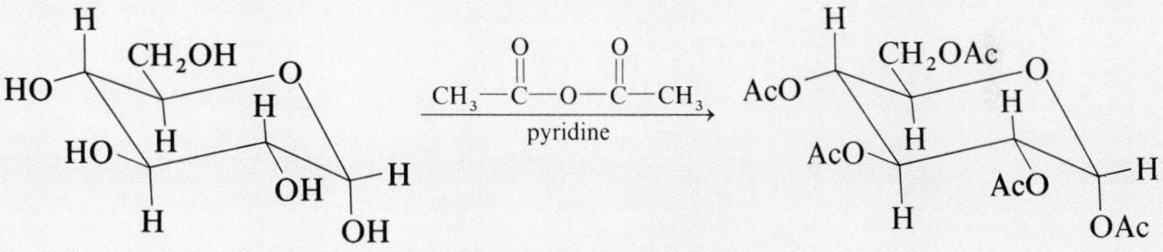

$\xrightarrow[\text{pyridine}]{CH_3-C(=O)-O-C(=O)-CH_3}$

(gives the same anomer as the starting materials)

7. *Osazone formation* (Section 23-13)

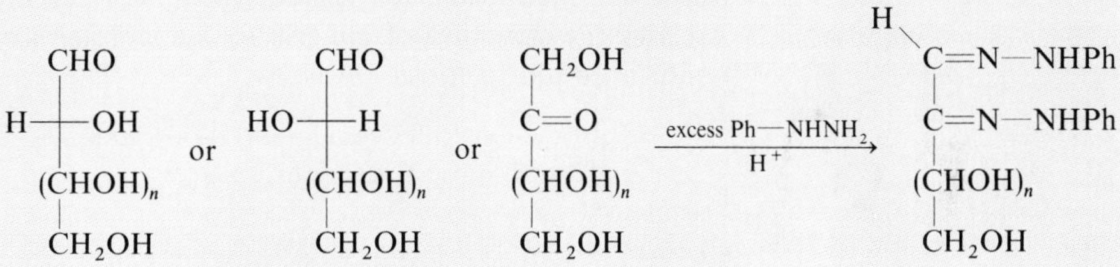

CHO / H—OH / $(CHOH)_n$ / CH_2OH or CHO / HO—H / $(CHOH)_n$ / CH_2OH or CH_2OH / C=O / $(CHOH)_n$ / CH_2OH $\xrightarrow[H^+]{\text{excess Ph—NHNH}_2}$ H—C=N—NHPh / C=N—NHPh / $(CHOH)_n$ / CH_2OH

either aldose epimer or ketose — osazone

8. *Ruff degradation* (Section 23-14)

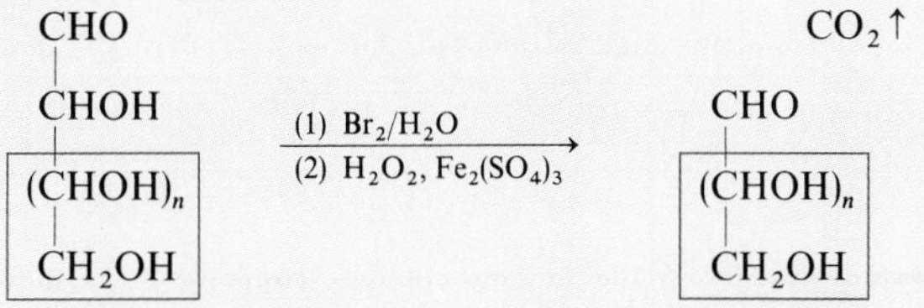

CHO / CHOH / $(CHOH)_n$ / CH_2OH $\xrightarrow[(2)\ H_2O_2,\ Fe_2(SO_4)_3]{(1)\ Br_2/H_2O}$ $CO_2\uparrow$ + CHO / $(CHOH)_n$ / CH_2OH

aldose — shortened aldose

9. *Kiliani-Fischer synthesis* (Section 23-15)

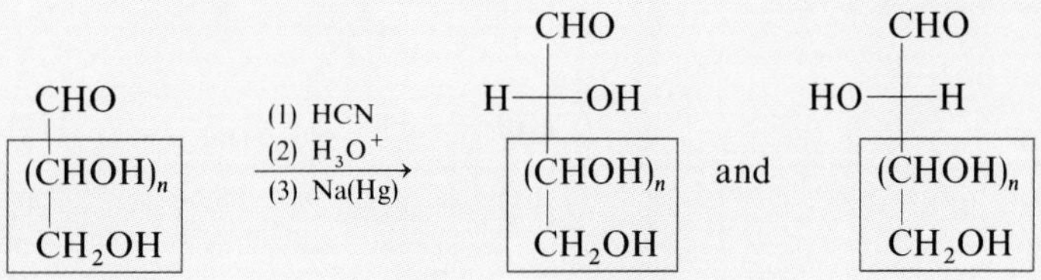

CHO / $(CHOH)_n$ / CH_2OH $\xrightarrow{(1)\ HCN;\ (2)\ H_3O^+;\ (3)\ Na(Hg)}$ CHO / H—OH / $(CHOH)_n$ / CH_2OH and CHO / HO—H / $(CHOH)_n$ / CH_2OH

aldose — epimers of lengthened aldose

23-16 FISCHER'S PROOF OF THE CONFIGURATION OF GLUCOSE

Considering the stereochemical complexity of sugars, it is amazing that Emil Fischer determined the structures of glucose and the other aldohexoses in 1891, only 14 years after the tetrahedral structure of carbon had been proposed. Fischer received the Nobel Prize for this work in 1902. Much of Fischer's proof used the carbohydrate reactions we have studied, together with some clever reasoning about the symmetry and dissymmetry of the products that result. We will use Fischer's work with glucose as an elegant example of these reactions, showing the determination of complex stereochemistry by the clever use of simple methods.

In 1891 there were no methods for determining the absolute configuration of molecules, so Fischer could not know which enantiomer was the naturally occurring one. He made the assumption that the —OH group on C2 of (+)-glyceraldehyde (and on the bottom chiral carbon of the D family of sugars) is on the right in the Fischer projection. Eventually, this was shown to be a correct guess, but all his reasoning would apply to the other enantiomers if the guess had been wrong.

Fischer had done many chemical tests on glucose, and he had used Ruff degradations to degrade it to D-(+)-glyceraldehyde. He knew that glucose is an aldose and that it has six carbon atoms; therefore, the eight members of the D family of aldohexoses (Fig. 23-3) are the possible structures. Fischer used four major clues to determine which of these structures corresponds to glucose. We will consider the four clues individually and study the information obtained from each.

CLUE 1 On Ruff degradation, glucose and mannose give the same aldopentose: D-(−)-arabinose.

This clue suggests that glucose and mannose are C2 epimers, a hypothesis that was confirmed by treating them with phenylhydrazine and showing that glucose and mannose give the same osazone. Even more important, this information relates the structure of glucose to the simpler structure of arabinose, the chain-shortened aldopentose.

CLUE 2 On Ruff degradation, D-(−)-arabinose gives D-(−)-erythrose. Upon treatment with nitric acid, erythrose gives an *optically inactive* aldaric acid (*meso*-tartaric acid).

D-Erythrose, obtained from Ruff degradation of a D-aldopentose, must be a D-aldotetrose. There are only two D-aldotetroses, labeled below as structures I and II. Nitric acid oxidation of one of these aldotetroses gives a symmetrical meso product, but the other one gives an optically active product.

CHO
H—OH
H—OH
CH_2OH
structure I

$\xrightarrow{HNO_3}$

COOH
H—OH
σ ——— mirror plane of symmetry
H—OH
COOH
meso-tartaric acid

CHO
HO—H
H—OH
CH_2OH
structure II

$\xrightarrow{HNO_3}$

COOH
HO—H
H—OH
COOH
optically active tartaric acid

Because oxidation of D-erythrose gives an optically inactive aldaric acid, erythrose must correspond to structure I. Structure I would give *meso*-tartaric acid when it is oxidized to an aldaric acid. D-Arabinose must be one of the two epimeric structures that would degrade to this structure for D-erythrose.

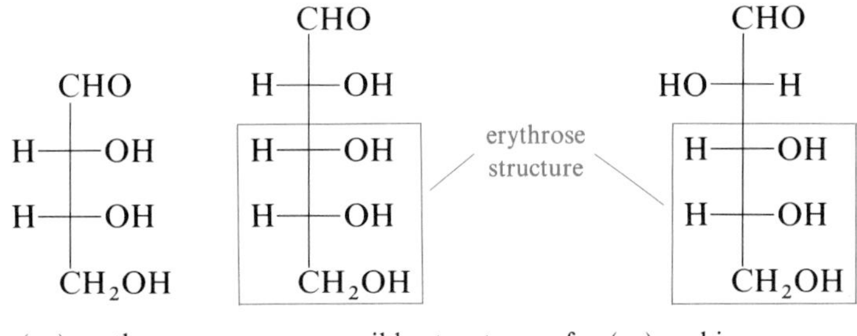

D-(−)-erythrose possible structures of D-(−)-arabinose

CLUE 3 On oxidation with nitric acid, D-(−)-arabinose gives an *optically active* aldaric acid.

Of the two possible structures for D-arabinose, only the second would oxidize to give an optically active aldaric acid.

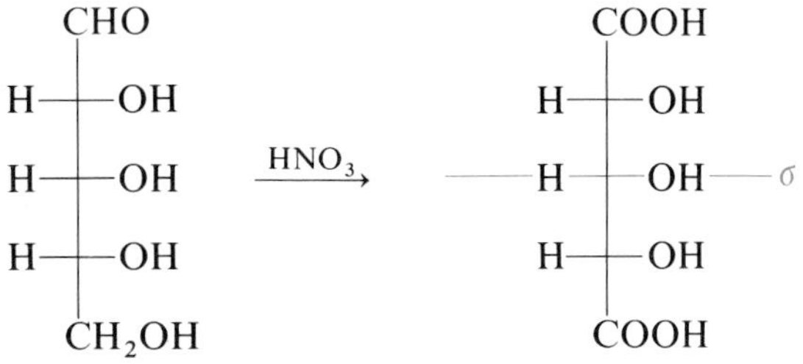

CHO
H—OH
H—OH
H—OH
CH_2OH
structure A

$\xrightarrow{HNO_3}$

COOH
H—OH
H—OH —— σ
H—OH
COOH
optically inactive aldaric acid

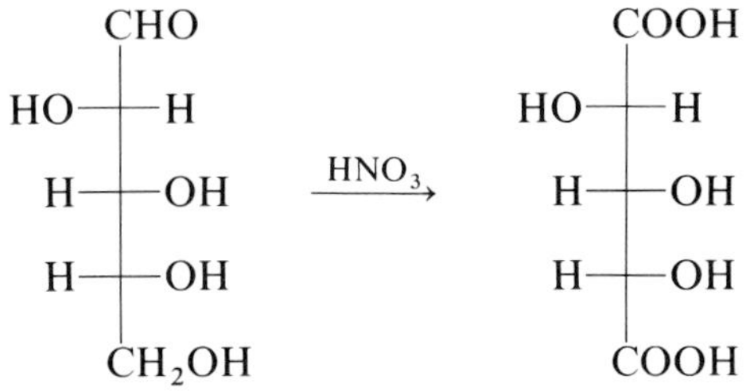

CHO
HO—H
H—OH
H—OH
CH_2OH
structure B
(D-arabinose)

$\xrightarrow{HNO_3}$

COOH
HO—H
H—OH
H—OH
COOH
optically active aldaric acid

Since glucose and mannose degrade to arabinose, structures X and Y shown below must be glucose and mannose. At this point, however, it is impossible to tell which structure is glucose and which is mannose.

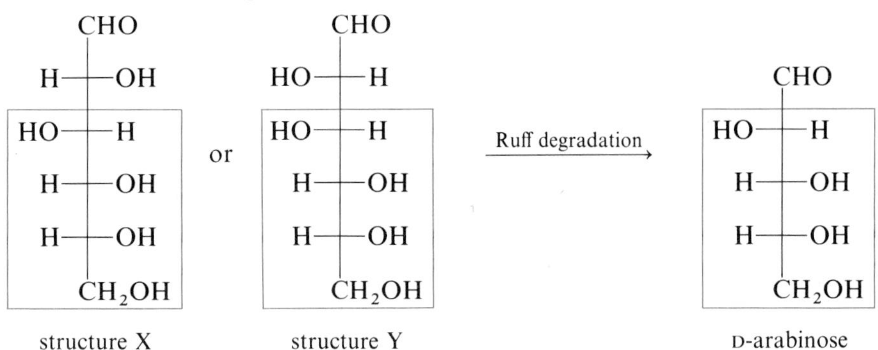

structure X structure Y D-arabinose

CLUE 4 When the —CHO and —CH_2OH groups of D-mannose are interchanged, the product is still D-mannose. When the —CHO and —CH_2OH groups of D-glucose are interchanged, the product is an unnatural L sugar.

Fischer had developed a clever method for converting the aldehyde group of an aldose to an alcohol, while converting the terminal alcohol group to an aldehyde. In effect, this synthesis interchanges the two end groups of the aldose chain:

$$\begin{array}{c} CHO \\ | \\ (CHOH)_n \\ | \\ CH_2OH \\ \text{original aldose} \end{array} \xrightarrow{\text{several steps}} \begin{array}{c} CH_2OH \\ | \\ (CHOH)_n \\ | \\ CHO \\ \text{end groups interchanged} \end{array}$$

If the two end groups of structure X are interchanged, the product looks strange indeed. Remember that we can rotate a Fischer projection by 180°; when we do that, it becomes clear that the product is an unusual sugar of the L series (L-gulose). Structure X must be D-glucose.

$$\begin{array}{c} CHO \\ H-\!\!\!\!+\!\!\!\!-OH \\ HO-\!\!\!\!+\!\!\!\!-H \\ H-\!\!\!\!+\!\!\!\!-OH \\ H-\!\!\!\!+\!\!\!\!-OH \\ CH_2OH \\ \text{structure X} \\ \text{(D-glucose)} \end{array} \longrightarrow\longrightarrow\longrightarrow \begin{array}{c} CH_2OH \\ H-\!\!\!\!+\!\!\!\!-OH \\ HO-\!\!\!\!+\!\!\!\!-H \\ H-\!\!\!\!+\!\!\!\!-OH \\ H-\!\!\!\!+\!\!\!\!-OH \\ CHO \\ \text{end groups interchanged} \\ \; \end{array} \xrightarrow{\text{rotate } 180^\circ} \begin{array}{c} CHO \\ HO-\!\!\!\!+\!\!\!\!-H \\ HO-\!\!\!\!+\!\!\!\!-H \\ H-\!\!\!\!+\!\!\!\!-OH \\ HO-\!\!\!\!+\!\!\!\!-H \\ CH_2OH \\ \text{an L sugar} \\ \text{(L-gulose)} \end{array}$$

Structure Y gives a D sugar when its end groups are interchanged. In fact, a 180° rotation shows that the product of end-group interchange in structure Y gives back the original structure! Structure Y must be D-mannose.

$$\begin{array}{c} CHO \\ HO-\!\!\!\!+\!\!\!\!-H \\ HO-\!\!\!\!+\!\!\!\!-H \\ H-\!\!\!\!+\!\!\!\!-OH \\ H-\!\!\!\!+\!\!\!\!-OH \\ CH_2OH \\ \text{structure Y} \\ \text{(D-mannose)} \end{array} \longrightarrow\longrightarrow\longrightarrow \begin{array}{c} CH_2OH \\ HO-\!\!\!\!+\!\!\!\!-H \\ HO-\!\!\!\!+\!\!\!\!-H \\ H-\!\!\!\!+\!\!\!\!-OH \\ H-\!\!\!\!+\!\!\!\!-OH \\ CHO \\ \text{end groups interchanged} \\ \; \end{array} \xrightarrow{\text{rotate } 180^\circ} \begin{array}{c} CHO \\ HO-\!\!\!\!+\!\!\!\!-H \\ HO-\!\!\!\!+\!\!\!\!-H \\ H-\!\!\!\!+\!\!\!\!-OH \\ H-\!\!\!\!+\!\!\!\!-OH \\ CH_2OH \\ \text{structure Y} \\ \text{(D-mannose)} \end{array}$$

This type of reasoning can be used to determine the structures of all the other aldoses. Problems 23-37 and 23-38 will give you some practice determining sugar structures.

PROBLEM 23-37

Upon treatment with phenylhydrazine, aldohexoses **A** and **B** give the same osazone. On treatment with warm nitric acid, **A** gives an optically inactive aldaric acid, but sugar **B** gives an optically active aldaric acid. Sugars **A** and **B** are both degraded to aldopentose

C, which gives an optically active aldaric acid upon treatment with nitric acid. Aldopentose **C** is degraded to aldotetrose **D,** which gives optically active tartaric acid when it is treated with nitric acid. Aldotetrose **D** is degraded to (+)-glyceraldehyde. Deduce the structures of sugars **A, B, C,** and **D,** and use Figure 23-3 to determine the correct names of these sugars.

PROBLEM 23-38

Aldose **E** is optically active, but treatment with sodium borohydride converts it to an optically inactive alditol. Ruff degradation of **E** gives **F,** whose alditol is optically inactive. Ruff degradation of **F** gives optically active D-glyceraldehyde. Give the structures and names of **E** and **F** and their optically inactive alditols.

23-17 DETERMINATION OF RING SIZE; PERIODIC ACID CLEAVAGE OF SUGARS

We have discussed the fact that monosaccharides exist mostly as cyclic pyranose or furanose hemiacetals. These hemiacetals are in equilibrium with the open-chain forms, so the sugars can react like hemiacetals or like ketones and aldehydes. How can we freeze this equilibrium and determine the optimum ring size for any given sugar? Sir Walter Haworth (inventor of the Haworth projection) used some simple chemistry to determine the pyranose structure of glucose in 1926.

Glucose is converted to a pentamethyl derivative by treatment with methyl iodide and silver oxide (Section 23-12). The five methyl groups are not the same, however. Four of them are methyl ethers, but one is the glycosidic methyl group of an acetal.

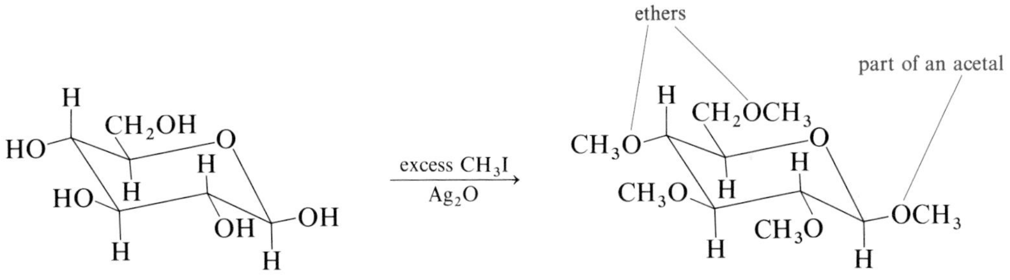

ethers

part of an acetal

$\xrightarrow[\text{Ag}_2\text{O}]{\text{excess CH}_3\text{I}}$

β-D-glucose

methyl 2,3,4,6-tetra-*O*-methyl-β-D-glucoside

Acetals are easily hydrolyzed by dilute acid, but ethers are quite stable under these conditions. Treatment of the pentamethyl glucose derivative with dilute acid hydrolyzes only the acetal methyl group. Haworth determined that the free hydroxyl group is on C5 of the hydrolyzed glucose, showing that the cyclic form of glucose is a pyranose.

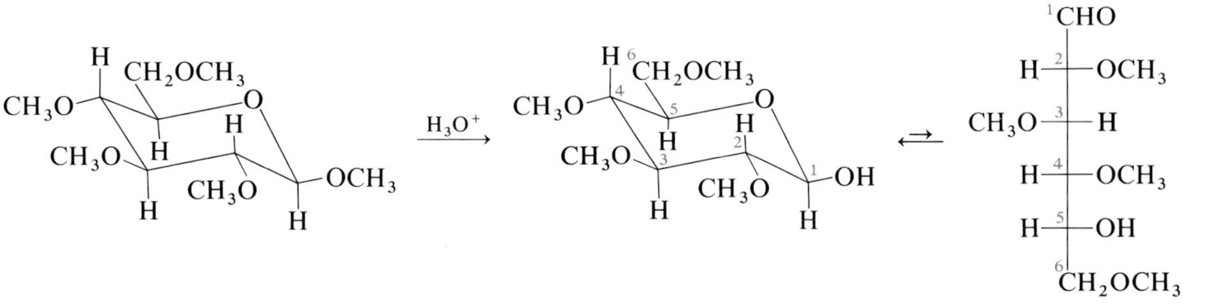

$\xrightarrow{\text{H}_3\text{O}^+}$

pentamethyl derivative

free hemiacetal

open-chain form
2,3,4,6,-*O*-tetramethyl-D-glucose

PROBLEM 23-39

(a) Show the product that results when fructose is treated with an excess of methyl iodide and silver oxide.
(b) Show what happens when the product of part (a) is hydrolyzed using dilute acid.
(c) Show what the results of parts (a) and (b) imply about the hemiacetal structure of fructose.

Periodic acid cleavage of carbohydrates Another method used to determine the structure of carbohydrates is cleavage by periodic acid. Recall that periodic acid cleaves vicinal diols to give two carbonyl compounds, either ketones or aldehydes, depending on the substitution of the reactant (Section 10-12B).

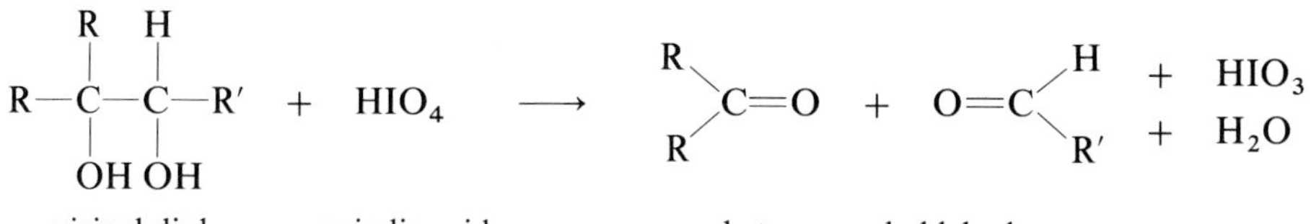

$$\underset{\text{vicinal diol}}{R{-}C(R)(OH){-}C(H)(OH){-}R'} + \underset{\text{periodic acid}}{HIO_4} \longrightarrow \underset{\text{ketones and aldehydes}}{R_2C{=}O + O{=}C(H)R'} + HIO_3 + H_2O$$

An α-hydroxy ketone or α-hydroxy aldehyde is cleaved to a carboxylic acid and a shortened ketone or aldehyde.

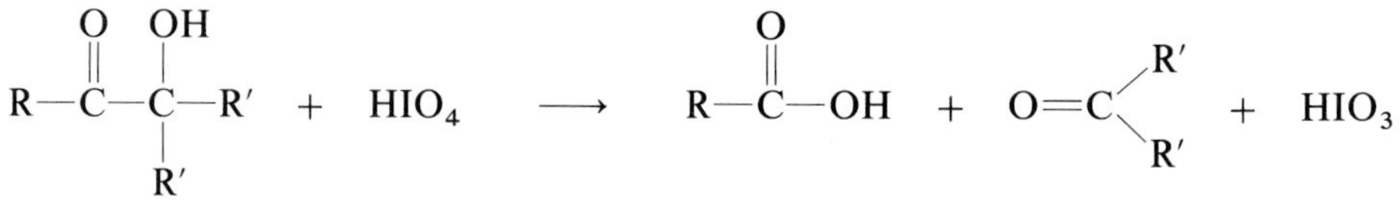

$$R{-}C({=}O){-}C(OH)(R'){-}R' + HIO_4 \longrightarrow R{-}C({=}O){-}OH + O{=}CR'_2 + HIO_3$$

In all these cleavages, the products may be predicted by mentally replacing the bond to be cleaved by two OH groups. The predicted structure is usually the hydrate of the actual product. Ether and acetal groups are not affected, but hemiacetals react through the open-chain structure. For example, fructose reacts with 5 moles of periodic acid to give 3 moles of formic acid, 2 moles of formaldehyde, and 1 mole of carbon dioxide:

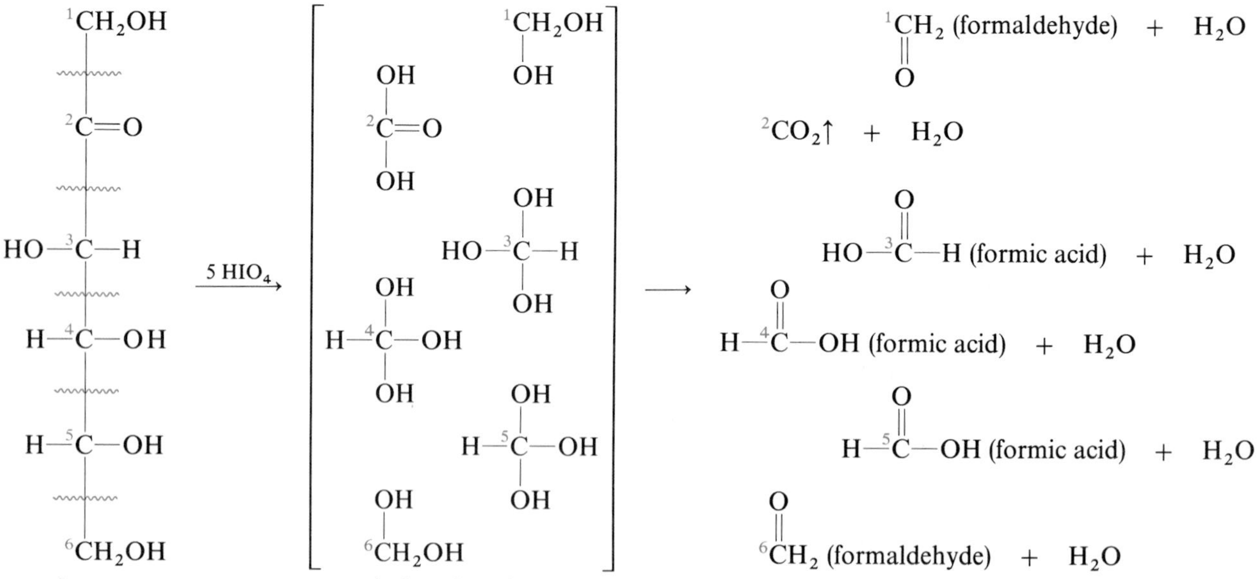

D-fructose: $^1CH_2OH$ / $^2C{=}O$ / $HO{-}^3C{-}H$ / $H{-}^4C{-}OH$ / $H{-}^5C{-}OH$ / $^6CH_2OH$ $\xrightarrow{5\ HIO_4}$

hydrated products: [$^1CH_2OH{-}OH$; $HO{-}^2C({=}O){-}OH$; $HO{-}^3C(H)(OH){-}OH$; $H{-}^4C(OH)(OH){-}OH$; $H{-}^5C(OH){-}OH$; $HO{-}^6CH_2OH$] $\longrightarrow$

$^1CH_2{=}O$ (formaldehyde) + H_2O

$^2CO_2\uparrow$ + H_2O

$HO{-}^3C({=}O){-}H$ (formic acid) + H_2O

$H{-}^4C({=}O){-}OH$ (formic acid) + H_2O

$H{-}^5C({=}O){-}OH$ (formic acid) + H_2O

$O{=}^6CH_2$ (formaldehyde) + H_2O

Because ether and acetal groups are not affected, the periodic acid cleavage of a glycoside can be used to determine the size of the ring. For example, periodic acid oxidation of methyl β-D-glucopyranoside gives the products shown below. The structure of the fragment containing C4, C5, and C6 implies that the original glycoside was a six-membered ring bonded through the C5 oxygen atom.

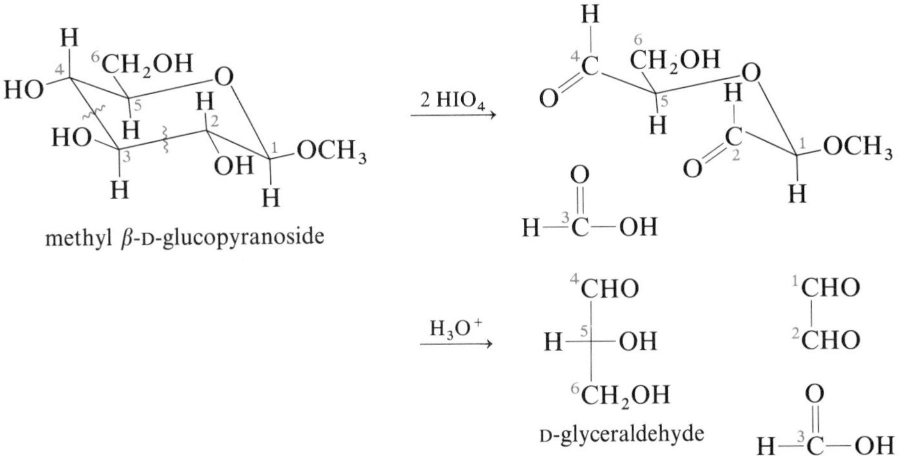

methyl β-D-glucopyranoside $\xrightarrow{2\ HIO_4}$ (dialdehyde fragment of C4, C5, C6 and C2, C1 with OCH_3) + $H-{}^3C(=O)-OH$

$\xrightarrow{H_3O^+}$ ${}^4CHO-{}^5CH(OH)-{}^6CH_2OH$ (D-glyceraldehyde) + ${}^1CHO-{}^2CHO$ + $H-{}^3C(=O)-OH$

PROBLEM 23-40

(a) Predict the products of cleavage of mannose by an excess of periodic acid.
(b) Explain how periodic acid cleavage distinguishes between an aldose and a ketose.

PROBLEM 23-41

(a) Draw the structure of methyl β-D-glucofuranoside, and predict the products of periodic acid oxidation. Show how these products differ from the products of periodic acid oxidation of methyl β-D-glucopyranoside.
(b) Predict the products of periodic acid oxidation of methyl β-D-fructofuranoside, and show how these products imply that the original glycoside was a five-membered ring.

23-18 DISACCHARIDES

The anomeric carbon of a sugar can react with the hydroxyl group of an alcohol to give an acetal called a *glycoside.* If the hydroxyl group is part of another sugar molecule, then the glycoside product is a **disaccharide,** a sugar composed of two monosaccharide units (Fig. 23-15).

In principle, the anomeric carbon can react with *any* of the hydroxyl groups of another sugar to form a disaccharide. In naturally occurring disaccharides, however, there are three principal glycosidic bonding arrangements.

1. A 1,4′ link. The anomeric carbon is bonded to the oxygen atom on C4 of the second sugar. The prime symbol (′) in 1,4′ indicates that C4 is on the second sugar.
2. A 1,6′ link. The anomeric carbon is bonded to the oxygen atom on C6 of the second sugar.
3. A 1,1′ link. The anomeric carbon of the first sugar is bonded through an oxygen atom to the anomeric carbon of the second sugar.

We will consider some naturally occurring disaccharides with these common glycosidic linkages.

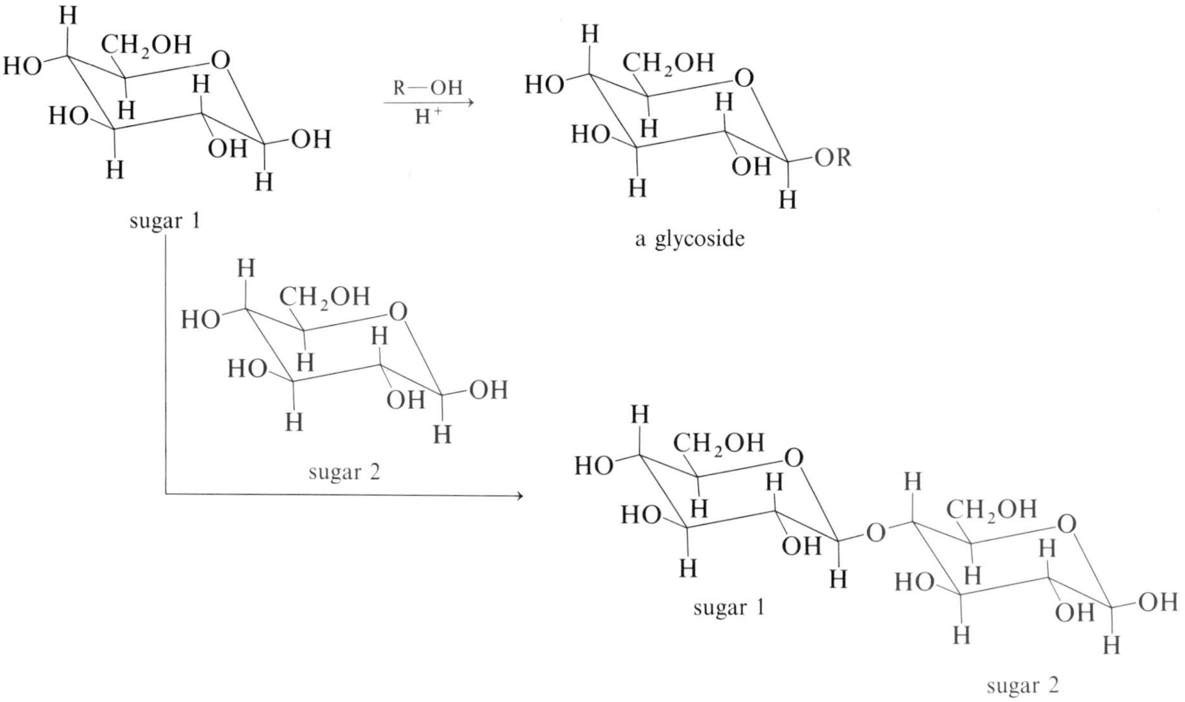

sugar 1

a glycoside

sugar 2

sugar 1

sugar 2

a disaccharide

FIGURE 23-15 A sugar reacts with an alcohol to give an acetal called a glycoside. When the alcohol is part of another sugar, the product is a disaccharide.

23-18A THE 1,4′ LINKAGE: CELLOBIOSE, MALTOSE, AND LACTOSE

The most common glycosidic linkage is the 1,4′ link. The anomeric carbon of one sugar is bonded to the oxygen atom on C4 of the second ring.

Cellobiose: a β-1,4′ glucosidic linkage Perhaps the simplest example of a 1,4′ linkage is *cellobiose,* the disaccharide obtained by partial hydrolysis of cellulose. In cellobiose, the anomeric carbon of one glucose unit is linked through an equatorial (*β*) carbon-oxygen bond to C4 of another glucose unit. This *β*-1,4′ linkage from a *glucose* acetal is called a ***β*-1,4′ *gluco*sidic linkage.**

Cellobiose, 4-O-(β-D-glucopyranosyl)-β-D-glucopyranose or 4-O-(β-D-glucopyranosyl)-D-glucopyranose

β-glucosidic linkage

β-glucosidic linkage

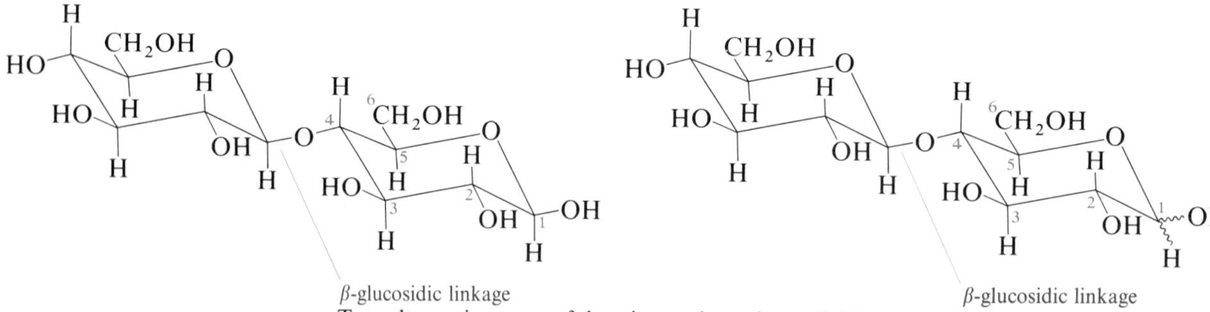

Two alternative ways of drawing and naming cellobiose

The complete name for cellobiose, 4-*O*-(*β*-D-glucopyranosyl)-*β*-D-glucopyranose, gives its structure. This name says that a *β*-D-glucopyranose ring (the right-hand ring) is substituted in its 4-position by an oxygen attached to a (β-D-glucopyranosyl) ring, drawn on the left. The name in parentheses says that the substituent is a β-glucose, and the *-syl* ending indicates that this ring is in the form of a glycoside. Notice that the left ring with the *-syl* ending exists as an acetal and cannot mutarotate, while the right ring with the *-ose* ending is a hemiacetal and can mutarotate. Because cellobiose has a glucose unit in the hemiacetal form (and therefore is in equilibrium with its open-chain aldehyde form), it is a reducing sugar. Once again, the *-ose* ending indicates a mutarotating, reducing sugar.

Mutarotating sugars are often shown with a wavy line to the anomeric hydroxyl group, signifying that they can exist as an equilibrium mixture of the two anomers. Their names are often given without specifying the stereochemistry of this mutarotating hydroxyl group, as in 4-*O*-(β-D-glucopyranosyl)-D-glucopyranose.

Maltose: an α-1,4′ glucosidic linkage *Maltose* is a disaccharide formed when starch is treated with sprouted barley, or *malt*. This malting process is the first step in the brewing of beer, converting polysaccharides to disaccharides and monosaccharides that ferment more easily. Like cellobiose, maltose contains a 1,4′ glycosidic linkage between two glucose units. The difference in maltose is that the stereochemistry of the glucosidic linkage is α rather than β.

Maltose, 4-O-(α-D-glucopyranosyl)-D-glucopyranose

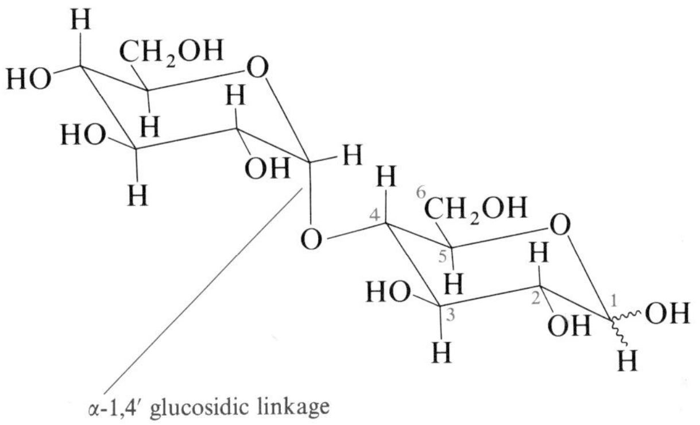

As with cellobiose, maltose has a free hemiacetal ring (on the right). This hemiacetal is in equilibrium with its open-chain form, and it mutarotates and can exist in either the α or β anomeric form. Because maltose exists in equilibrium with an open-chain aldehyde, it can reduce Tollens reagent, and maltose is a reducing sugar.

PROBLEM 23-42

Draw the structure of the individual mutarotating α and β anomers of maltose.

PROBLEM 23-43

Give an equation to show the reduction of Tollens reagent by maltose.

Lactose: a β-1,4′ galactosidic linkage *Lactose* is similar in structure to cellobiose, except that the glycoside (the left ring) in lactose is galactose rather than glucose. Lactose is composed of one galactose unit and one glucose unit. The two rings are linked by a β glycosidic bond of the galactose acetal to the 4-position on the glucose ring: a β-1,4′ galactosidic linkage.

Lactose, 4-O-(β-D-galactopyranosyl)-D-glucopyranose

axial 4-hydroxyl group of galactose

β-galactosidic linkage

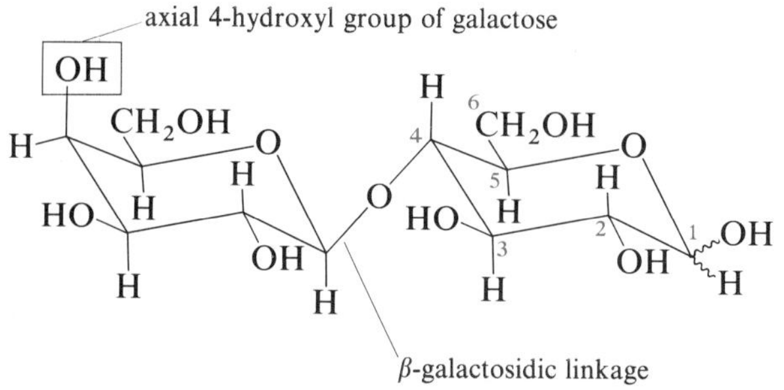

Lactose is a naturally occurring disaccharide found in the milk of mammals, including cows and humans. The hydrolysis of lactose requires a β-galactosidase enzyme (sometimes called "lactase"). Some humans synthesize a β-galactosidase, but others do not. This enzyme is present in the digestive fluids of normal infants to hydrolyze their mothers' milk. Once the child stops drinking milk, production of the enzyme gradually stops. In most parts of the world, people do not use milk products after early childhood, and the adult population can no longer digest lactose. Consumption of milk or milk products can cause digestive discomfort in *lactose intolerant* people who lack the β-galactosidase enzyme. Lactose intolerant infants must drink soybean milk or another lactose-free formula.

PROBLEM 23-44

Does lactose mutarotate? Is it a reducing sugar? Explain. Draw the two anomeric forms of lactose.

23-18B THE 1,6′ LINKAGE: GENTIOBIOSE

In addition to the common 1,4′ glycosidic linkage, the 1,6′ linkage is also found in naturally occurring carbohydrates. In a 1,6′ linkage, the anomeric carbon of one sugar is linked to the oxygen of the terminal carbon (C6) of another. This linkage gives a different sort of stereochemical arrangement, because the hydroxyl group on C6 is one carbon atom removed from the ring. Gentiobiose is a sugar in which two glucose units are joined by a β-1,6′ glucosidic linkage.

Gentiobiose, 6-O-(β-D-glucopyranosyl)-D-glucopyranose

β-glucosidic linkage

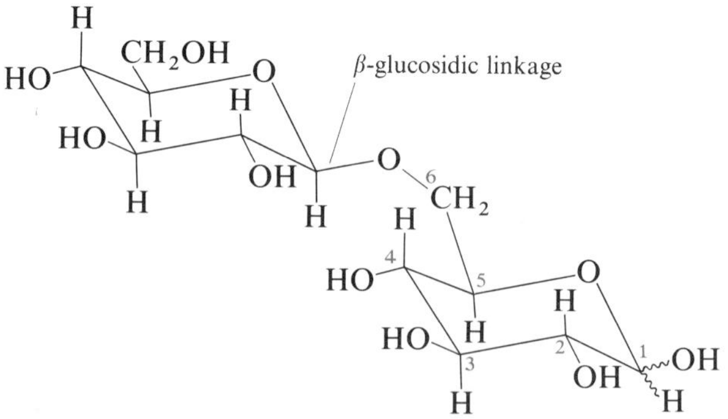

Although the 1,6′ linkage is rare in disaccharides, it is commonly found as a branch point in polysaccharides. For example, the branching in amylopectin occurs at 1,6′ linkages, as discussed in Section 23-19B.

PROBLEM 23-45

Is gentiobiose a reducing sugar? Does it mutarotate? Explain your reasoning.

23-18C LINKAGE OF TWO ANOMERIC CARBONS: SUCROSE

Some sugars are joined by a direct glycosidic linkage between their anomeric carbon atoms: a 1,1′ linkage. This is the case in *sucrose,* the common "table sugar." Sucrose is a disaccharide composed of one glucose unit and one fructose unit bonded by an oxygen atom linking their anomeric carbon atoms. (Because fructose is a ketose and its anomeric carbon is C2, this is actually a 1,2′ linkage.) Notice that the linkage is in the α position with respect to the glucose ring, and in the β position with respect to the fructose ring.

Sucrose, α-D-glucopyranosyl-β-D-fructofuranoside (or β-D-fructofuranosyl-α-D-glucopyranoside)

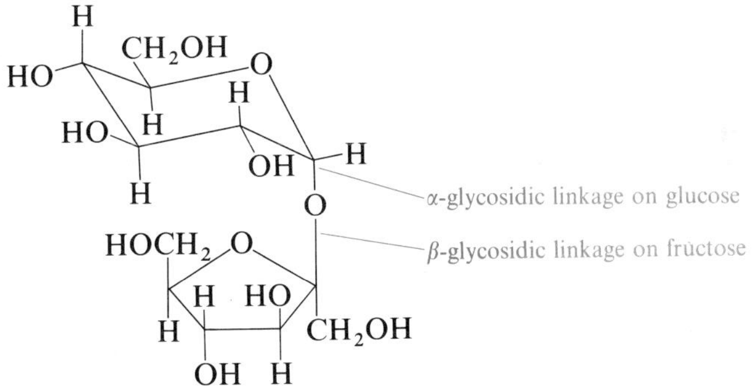

α-glycosidic linkage on glucose

β-glycosidic linkage on fructose

Both of the monosaccharide units in sucrose are present as acetals, or glycosides. Neither ring is in equilibrium with its open-chain aldehyde or ketone form, so sucrose does not reduce Tollens reagent, and it cannot mutarotate. Because both units are glycosides, the systematic name for sucrose can list either of the two glycosides as being a substituent on the other. Both of the systematic names end in the *-oside* suffix, indicating a nonmutarotating, nonreducing sugar. Like many other common names, *sucrose* ends in the *-ose* ending even though it is a nonreducing sugar. Common names are not reliable indicators of the properties of sugars.

Sucrose is hydrolyzed by enzymes called **invertases,** found in honeybees and yeasts, that specifically hydrolyze the β-D-fructofuranoside linkage. The resulting mixture of glucose and fructose is called *invert sugar,* because the hydrolysis converts the positive rotation $[+66.5^\circ]$ of sucrose to a negative rotation that is the average of glucose $[+52.7^\circ]$ and fructose $[-92.4^\circ]$. The most common form of invert sugar is honey, a supersaturated mixture of glucose and fructose hydrolyzed from sucrose by the invertase enzyme of honeybees.

SOLVED PROBLEM 23-3

An unknown carbohydrate of formula $C_{12}H_{22}O_{11}$ reacts with the Tollens reagent to form a silver mirror. An α-glycosidase has no effect on the carbohydrate, but a β-galactosidase hydrolyzes it to D-galactose and D-mannose. When the carbohydrate is methylated (using methyl iodide and silver oxide) and then hydrolyzed with dilute HCl, the products are 2,3,4,6-tetra-*O*-methylgalactose and 2,3,4-tri-*O*-methylmannose. Propose a structure for this unknown carbohydrate.

SOLUTION The formula shows that this carbohydrate is a disaccharide composed of two hexoses. Hydrolysis gives D-galactose and D-mannose, identifying the two hexoses. Hydrolysis requires a β-galactosidase, showing that galactose and mannose are linked by a β-galactosyl linkage. Since the original carbohydrate is a reducing sugar, one of the hexoses must be in a free hemiacetal form. Galactose is present as a glycoside; thus mannose must be present in its hemiacetal form. The unknown carbohydrate must be a (β-galactosyl)-mannose.

The methylation/hydrolysis procedure shows the point of attachment of the glycosidic bond to mannose and also confirms the size of the six-membered rings. In galactose, all the hydroxyl groups are methylated except C1 and C5. C1 is the anomeric carbon, and the C5 oxygen is used to form the hemiacetal of the pyranose ring. In mannose, all the hydroxyl groups are methylated except C1, C5, and C6. The C5 oxygen is used to form the pyranose ring (the C6 oxygen would form a less stable seven-membered ring); therefore, the oxygen on C6 must be involved in the glycosidic linkage. The structure and systematic name are shown below.

6-O-(β-D-galactopyranosyl)-D-mannopyranose

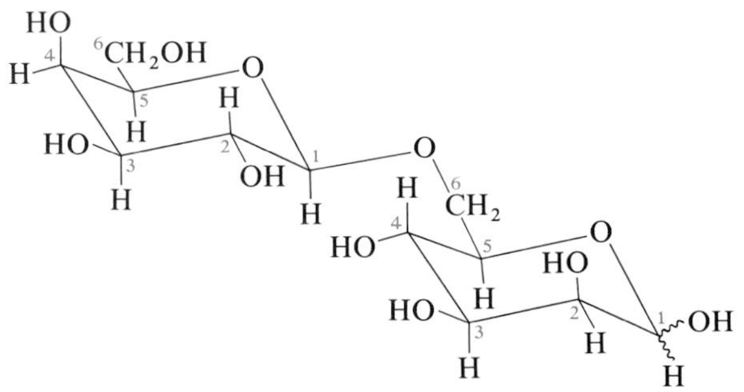

PROBLEM 23-46

Trehalose is a nonreducing disaccharide ($C_{12}H_{22}O_{11}$) isolated from the poisonous mushroom *Amanita muscaria*. Treatment with an α-glucosidase converts trehalose to two molecules of glucose, but no reaction occurs when trehalose is treated with a β-glucosidase. When trehalose is methylated by dimethyl sulfate in mild base and then hydrolyzed, the only product is 2,3,4,6-tetra-*O*-methylglucose. Propose a complete structure and systematic name for trehalose.

PROBLEM 23-47

Raffinose is a trisaccharide ($C_{18}H_{32}O_{16}$) isolated from cottonseed meal. Raffinose does not reduce Tollens reagent, and it does not mutarotate. Complete hydrolysis of raffinose gives D-glucose, D-fructose, and D-galactose. When raffinose is treated with invertase, the products are D-fructose and a reducing disaccharide called *melibiose*. Raffinose is unaffected by treatment with a β-galactosidase, but an α-galactosidase hydrolyzes it to D-galactose and sucrose. When raffinose is treated with dimethyl sulfate and base followed by hydrolysis, the products are 2,3,4-tri-*O*-methylglucose, 1,3,4,6-tetra-*O*-methylfructose, and 2,3,4,6-tetra-*O*-methylgalactose. Determine the complete structures of raffinose and melibiose, and give a systematic name for melibiose.

23-19 POLYSACCHARIDES

Polysaccharides are carbohydrates that contain many monosaccharide units joined by glycosidic bonds. They are one type of *biopolymer*, or naturally occurring polymer. Smaller polysaccharides, containing about three to ten monosaccharide units, are sometimes called **oligosaccharides.** Most polysaccharides, however, have hundreds or thousands of simple sugar units linked together into long polymer chains. Except for a single unit at the end of a chain, all the anomeric carbon atoms of a polysaccharide are involved in acetal glycosidic links. Therefore, polysaccharides give no noticeable reaction with Tollens reagent, and they do not mutarotate.

23-19A CELLULOSE

Cellulose, a polymer of D-glucose, is more abundant than any other organic material. Cellulose is synthesized by plants as a structural material to support the weight of the plant. The long cellulose molecules, called **microfibrils,** are held in bundles by hydrogen bonding between the many —OH groups of the glucose rings. About 50 percent of dry wood and about 90 percent of cotton fiber is cellulose.

Cellulose is made up of D-glucose units linked by β-1,4′ glycosidic bonds. This bonding arrangement (like that in cellobiose) is rather rigid and very stable, giving cellulose desirable properties for use as a structural material. Figure 23-16 shows a partial structure of cellulose.

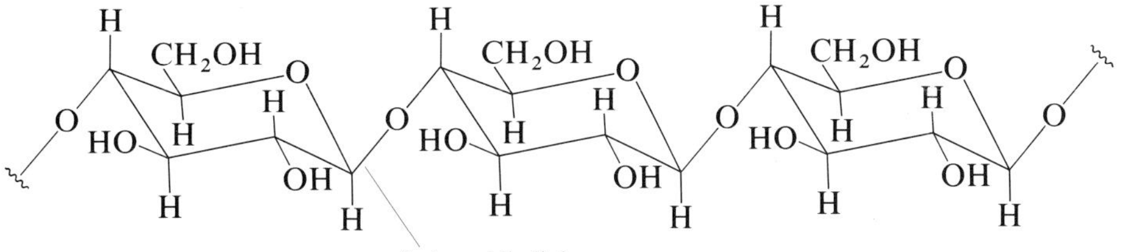

FIGURE 23-16 Cellulose is a β-1,4′ polymer of D-glucose, systematically named poly(1,4′-*O*-β-D-glucopyranoside).

Humans and other mammals lack the β-glucosidase enzyme needed to hydrolyze cellulose, and they cannot use it directly for food. Several groups of bacteria and protozoa can hydrolyze cellulose, however. Termites and ruminants have developed the ability to maintain colonies of these bacteria in their stomachs and intestines. When a cow eats hay, these bacteria convert about 20 to 30 percent of the cellulose to digestible carbohydrates.

PROBLEM 23-48

Cellulose is converted to **cellulose acetate** by treatment with acetic anhydride and pyridine. Cellulose acetate is soluble in common organic solvents, and it is easily spun into fibers. Show the structure of cellulose acetate.

23-19B STARCHES: AMYLOSE, AMYLOPECTIN, AND GLYCOGEN

Plants use starch granules for storing energy. When the granules are dried and ground up, the different types of starches can be separated by mixing them with hot water. About 20 percent of the starch is a water-soluble material called *amylose,* and the remaining 80 percent is a water-insoluble material called *amylopectin.* When starch is treated with dilute acid or appropriate enzymes, it is progressively hydrolyzed to maltose and then to D-glucose.

Amylose Like cellulose, **amylose** is a linear polymer of glucose with a 1,4′ glycosidic linkage. The difference in these two structures is in the stereochemistry of the linkage. Amylose has an α-1,4′ link, while cellulose has a β-1,4′ link. A partial structure of amylose is shown in Figure 23-17.

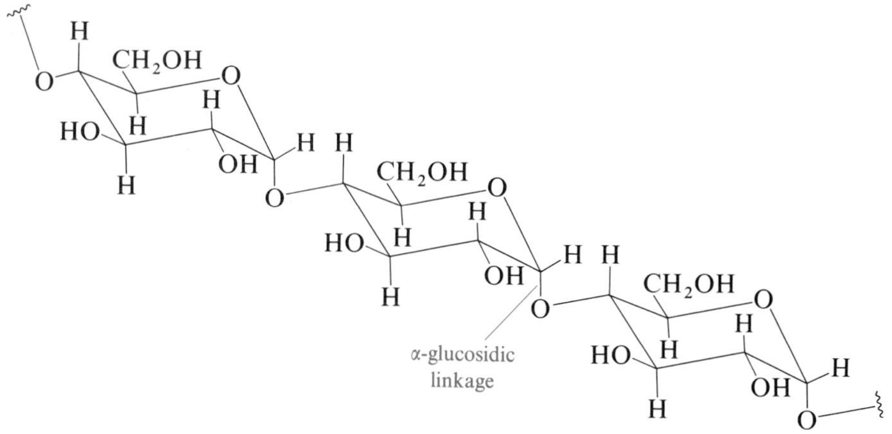

α-glucosidic linkage

FIGURE 23-17 Amylose is an α-1,4′ polymer of glucose, systematically named poly(1,4′-*O*-α-D-glucopyranoside). Amylose differs from cellulose only in the stereochemistry of the glycosidic linkage.

The subtle stereochemical difference between cellulose and amylose results in some striking physical and chemical differences. The α linkage in amylose kinks the polymer chain into a helical structure. This kinking increases the hydrogen bonding with water and lends additional solubility. As a result, amylose is soluble in water, while cellulose is not. Cellulose is stiff and sturdy, while amylose is not. Unlike cellulose, amylose is an excellent food source. The α-1,4′ glucosidic linkage is easily hydrolyzed by an α-glucosidase enzyme, present in all animals.

The helical structure of amylose also serves as the basis for an interesting and useful reaction. The inside of the helix is just the right size and polarity to accept an iodine (I_2) molecule. When iodine becomes lodged within this helix, a deep blue starch-iodine complex results (Fig. 23-18). This is the basis of the *starch-iodide test* for oxidizers. The material to be tested is added to an aqueous solution of amylose and potassium iodide. If the material is an oxidizer, some of the iodide (I^-) is oxidized to iodine (I_2), which forms the blue complex with amylose.

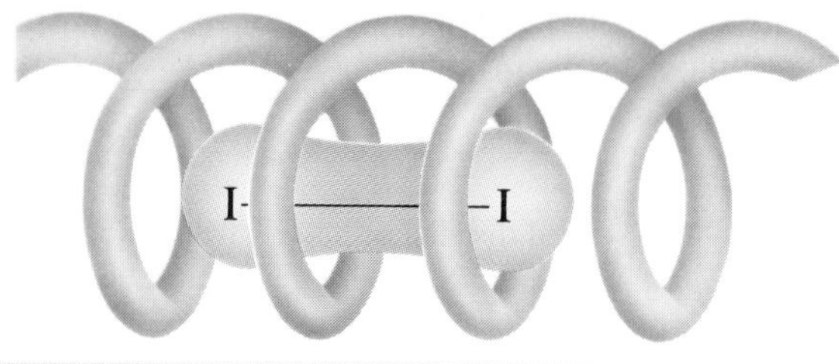

FIGURE 23-18 The amylose helix forms a blue charge-transfer complex with molecular iodine.

Amylopectin **Amylopectin,** the insoluble fraction of starch, is also primarily an α-1,4′ polymer of glucose. The difference between amylose and amylopectin lies in the branched nature of amylopectin. There is a branch point about every 20 to 30 glucose units. At a branch point, another chain starts, connected to the main chain by an α-1,6′ glycosidic linkage. A partial structure of amylopectin, including one branch point, is shown in Figure 23-19.

Glycogen **Glycogen** is the carbohydrate that animals use to store glucose for readily available energy. A large amount of glycogen is stored in the muscles themselves, ready for immediate hydrolysis and metabolism. Additional glycogen is stored in the liver, where it can be hydrolyzed to glucose for secretion into the bloodstream, providing an athlete with a "second wind."

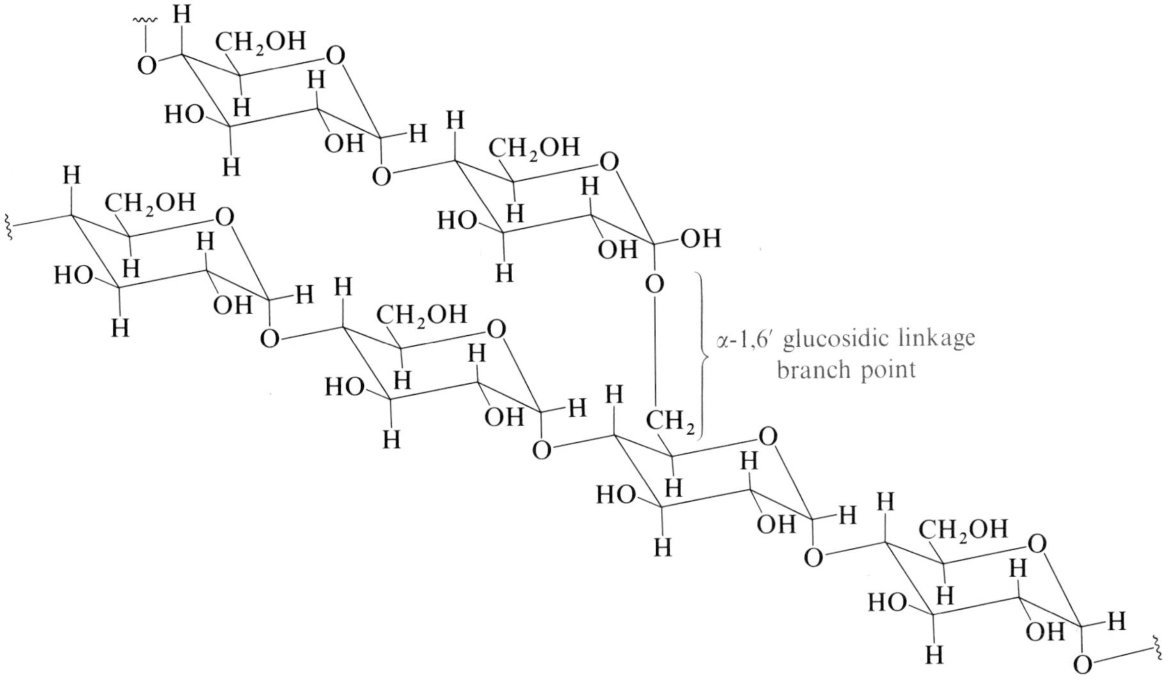

FIGURE 23-19 Amylopectin is a branched α-1,4′ polymer of glucose. At the branch points there is a single α-1,6′ linkage that provides the attachment point for another chain. Glycogen has a similar structure, except that its branching is much more extensive.

The structure of glycogen is similar to that of amylopectin. It is a highly branched structure consisting of glucose units bonded by α-1,4′ glycosidic bonds. The branch points are α-1,6′ links. The major difference between amylopectin and glycogen is the more extensive branching in glycogen than in amylopectin.

23-19C CHITIN: A POLYMER OF N-ACETYLGLUCOSAMINE

Chitin (pronounced *kī′-t'n*, rhymes with "Titan") forms the exoskeletons of insects. In crustaceans, chitin forms a matrix that binds calcium carbonate crystals into the exoskeleton. Chitin is different from all the other carbohydrates we have studied. It is a polymer of *N*-acetylglucosamine, an amino sugar that is common in living organisms. In *N*-acetylglucosamine, the hydroxyl group on C2 of glucose is replaced by an amino group (glucosamine), and that amino group is acetylated.

N-acetylglucosamine, or 2-acetamido-2-deoxy-D-glucose

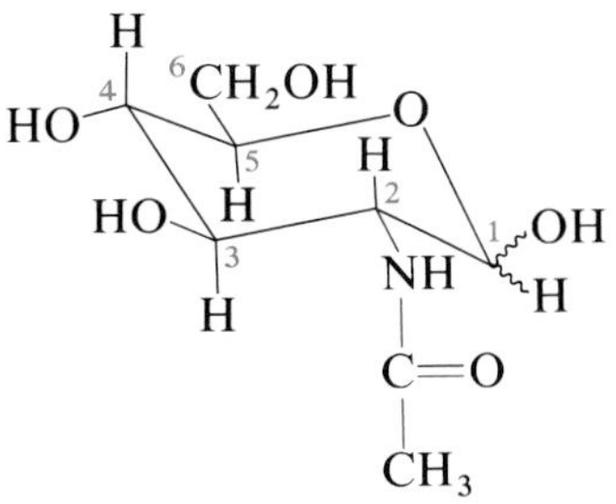

Chitin is bonded like cellulose, except using *N*-acetylglucosamine instead of glucose. The glycosidic bonds are β-1,4′ links, giving chitin structural rigidity, strength, and stability that exceed even that of cellulose. Unfortunately, this strong, rigid polymer cannot easily expand, so it must be shed periodically by molting as the animal grows.

Chitin, or poly(1,4′-O-β-2-acetamido-2-deoxy-D-glucopyranoside), a β-1,4-linked polymer of N-acetylglucosamine

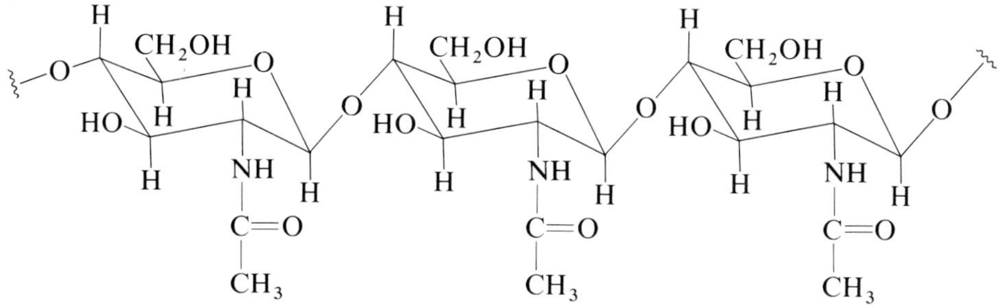

23-20 NUCLEIC ACIDS: INTRODUCTION

Nucleic acids are the chemical carriers of an organism's genetic information. A tiny amount of DNA in a fertilized egg cell determines nearly all the physical characteristics of the fully developed animal. The difference between a frog and a human being is encoded in a relatively small part of this DNA. Each cell carries a complete set of genetic instructions that determine the type of the cell, what its function will be, when it will grow and divide, and how it will synthesize all the proteins, enzymes, carbohydrates, and other substances the cell and the organism need to survive.

The two major classes of nucleic acids are **ribonucleic acids (RNA)** and **deoxyribonucleic acids (DNA).** In a typical cell, DNA is found primarily in the nucleus, where it carries the permanent genetic code. The molecules of DNA are huge, with molecular weights up to 50 billion. When the cell divides, the DNA replicates to form two copies for the daughter cells. DNA is relatively stable, providing a medium for transmission of genetic information from one generation to the next.

RNA molecules are typically much smaller than DNA, and they are more easily hydrolyzed and broken down. RNA commonly serves as a working copy of the nuclear DNA being decoded. The nuclear DNA directs the synthesis of *messenger RNA*, which leaves the nucleus to serve as a template for the construction of protein molecules in the ribosomes. Then the messenger RNA is enzymatically cleaved to its component parts, which become available for assembly into new RNA molecules to direct other syntheses.

Nucleic acids are polymers of ribofuranoside rings (five-membered rings of the sugar ribose) linked by phosphate ester groups. Each ribose unit carries a heterocyclic *base* that provides part of the information needed to specify a particular amino acid used in protein synthesis. Figure 23-20 shows the ribose-phosphate backbone of RNA.

DNA and RNA are biopolymers of four possible monomers, called **nucleotides,** that differ only in the structure of the bases bonded to the ribose units. Yet this deceptively simple structure encodes complex information just as the 0 and 1 bits used by a computer encode complex programs. First we consider the structure of individual nucleotides, then the bonding of these monomers into single-stranded nucleic acids, and finally the base pairing that binds two strands into the double helix of nuclear DNA.

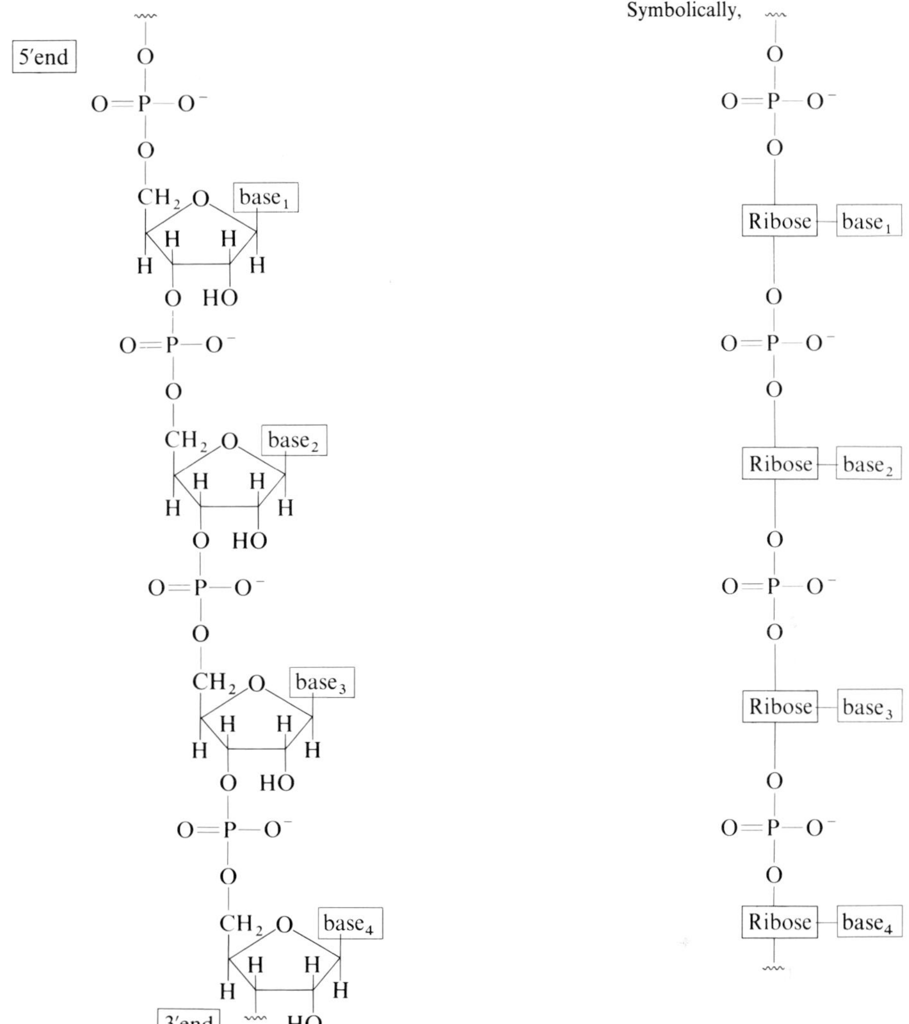

FIGURE 23-20 A short segment of the RNA polymer. The nucleic acids are assembled on a backbone made up of ribofuranoside units linked by phosphate esters.

23-21 RIBONUCLEOSIDES AND RIBONUCLEOTIDES

Ribonucleosides are components of RNA based on glycosides of the furanose form of D-ribose. We have seen (Section 23-11) that a glycoside may have an aglycone (the substituent on the anomeric carbon) that is bonded by a nitrogen atom. A ribonucleoside is a β-D-ribofuranoside (a β-glycoside of D-ribofuranose) whose aglycone is a heterocyclic nitrogen base. The following structures show the open-chain and furanose forms of ribose, and a ribonucleoside with a generic base bonded through a nitrogen atom.

D-ribose | β-D-ribofuranose | a ribonucleoside

The four bases commonly found in RNA are divided into two classes: The monocyclic compounds cytosine and uracil are called *pyrimidine bases* because they resemble substituted pyrimidines; and the bicyclic compounds adenine and guanine are called *purine bases* because they resemble the bicyclic heterocycle purine (Section 19-3).

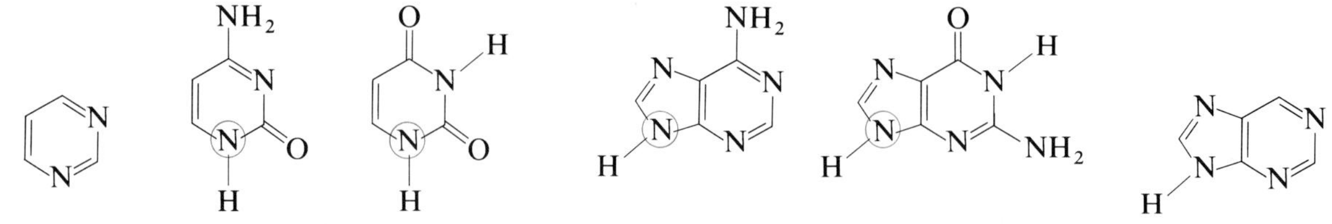

pyrimidine — cytosine (C) — uracil (U) — adenine (A) — guanine (G) — purine

pyrimidine bases — purine bases

When bonded to ribose through the nitrogen atoms circled above, the four heterocyclic bases make up the four ribonucleosides cytidine, uridine, adenosine, and guanosine. These four ribonucleosides are shown in Figure 23-21. Notice that the two ring systems (the base and the sugar) are numbered separately, and that the carbons of the sugar are given primed numbers. For example, the 3′ carbon of cytidine is C3 of the ribose ring.

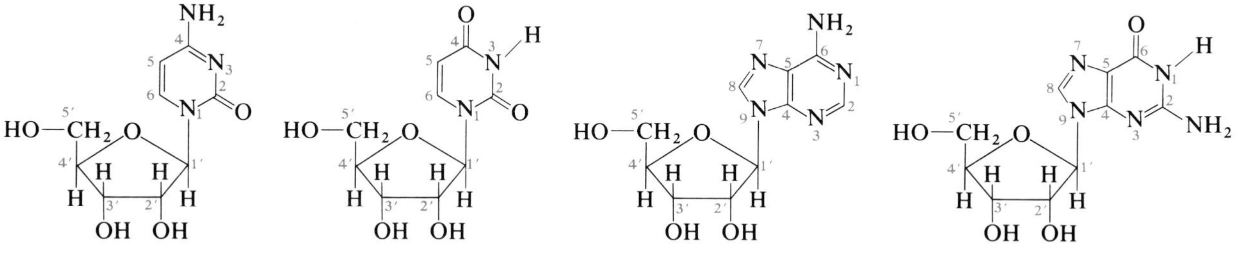

cytidine (C) — uridine (U) — adenosine (A) — guanosine (G)

FIGURE 23-21 The four common ribonucleosides are cytidine, uridine, adenosine, and guanosine.

PROBLEM 23-49

Cytosine, uracil, and guanine have tautomeric forms with phenolic hydroxyl groups. Draw these tautomeric forms.

PROBLEM 23-50

Ribonucleosides are relatively stable to base, but they are easily cleaved to the sugar and the heterocyclic base by dilute acid. Explain why a nucleoside is easily cleaved by acid.

Ribonucleotides Ribonucleic acid consists of ribonucleosides bonded together into a polymer. This polymer cannot be bonded by glycosidic linkages like those of other polysaccharides because the glycosidic bonds are already used to attach the heterocyclic bases. Instead, the ribonucleoside subunits are linked by phosphate esters. The 5′-hydroxyl group of each ribofuranoside is esterified to phosphoric acid. A ribonucleoside that is phosphorylated at its 5′ carbon is called a **ribonucleotide** ("tied" to phosphate). The four common ribonucleotides, shown in Figure 23-22, are simply the phosphorylated versions of the four common ribonucleosides.

The phosphate groups of these ribonucleotides could exist in any of three ionization states, depending on the pH of the solution. At the nearly neutral pH

cytidine monophosphate, CMP (cytidylic acid)

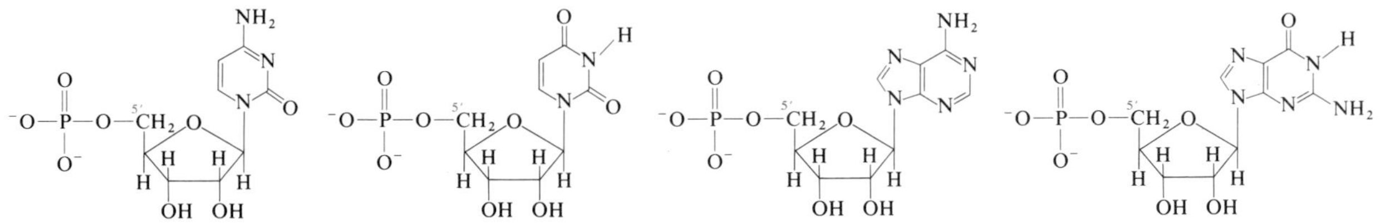

uridine monophosphate, UMP (uridylic acid)

adenosine monophosphate, AMP (adenylic acid)

guanosine monophosphate, GMP (guanidylic acid)

FIGURE 23-22 Four common ribonucleotides. These are ribonucleosides that are esterified by phosphoric acid at their 5′-position, the —CH_2OH at the end of the ribose chain.

of most organisms (7.4), there is one proton on the phosphate group. By convention, however, these groups are usually written completely ionized.

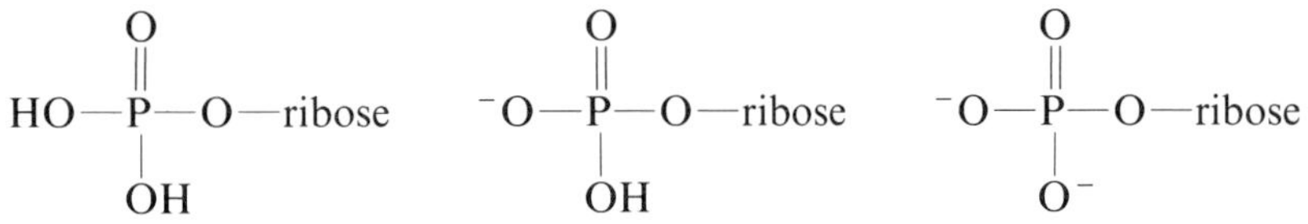

in acid — nearly neutral — in base (usually written)

23-22 THE BONDING STRUCTURE OF RIBONUCLEIC ACID

Now that we understand the makeup of the individual nucleotides, we can consider how these units are bonded into the RNA polymer. Each nucleotide has a phosphate group on its 5′ carbon (the end carbon of ribose), and each one has a hydroxyl group on the 3′ carbon. Two nucleotides are joined by formation of a phosphate ester linkage between the phosphate group of one nucleotide and the 3′-hydroxyl group of another (Fig. 23-23).

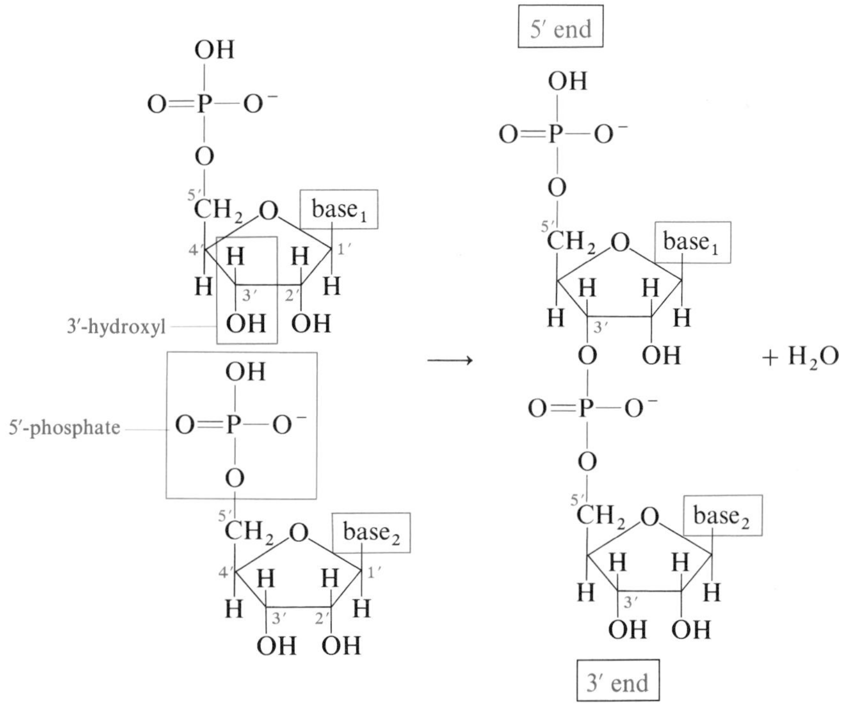

FIGURE 23-23 Two nucleotides are joined by formation of a phosphate linkage between the 5′-phosphate group of one and the 3′-hydroxyl group of the other.

The RNA polymer simply consists of many nucleotide units bonded in this fashion, with a phosphate ester linking the 5′ end of one nucleoside to the 3′ end of another. A molecule of RNA always has two ends (unless it is in the form of a large ring); one end has a free 3′ group, and the other end has a free 5′ group. We refer to the ends as the *3′ end* and the *5′ end,* and we refer to directions of replication as the *3′ → 5′ direction* and the *5′ → 3′ direction.* Figures 23-20 and 23-23 show short segments of RNA with the 3′ end and the 5′ end labeled.

23-23 DEOXYRIBOSE AND THE STRUCTURE OF DEOXYRIBONUCLEIC ACID

All our descriptions of ribonucleosides, ribonucleotides, and ribonucleic acid also apply to the components of DNA. The principal difference between RNA and DNA is the presence of D-2-deoxyribose as the sugar in DNA instead of the D-ribose found in RNA. The prefix *deoxy-* means that an oxygen atom is missing, and the number 2 means that the oxygen is missing from C2.

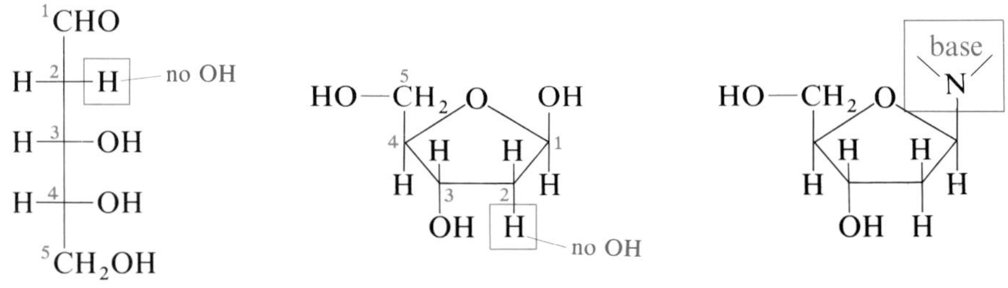

D-2-deoxyribose β-D-2-deoxyribofuranose a deoxyribonucleoside

Another key difference between RNA and DNA is the presence of thymine in DNA instead of the uracil in RNA. Thymine is simply uracil with an additional methyl group. The four common bases of DNA are cytosine, thymine, adenine, and guanine.

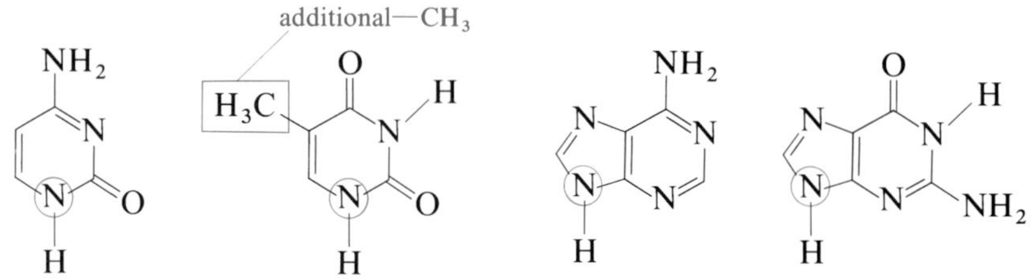

cytosine (C) thymine (T) adenine (A) guanine (G)

pyrimidine bases purine bases

These four bases are incorporated into deoxyribonucleosides and deoxyribonucleotides similar to the bases in ribonucleosides and ribonucleotides. The following structures show the common nucleosides that make up DNA. The corresponding nucleotides are simply the same structures with phosphate groups at the 5′-positions.

The four common deoxyribonucleosides that make up DNA

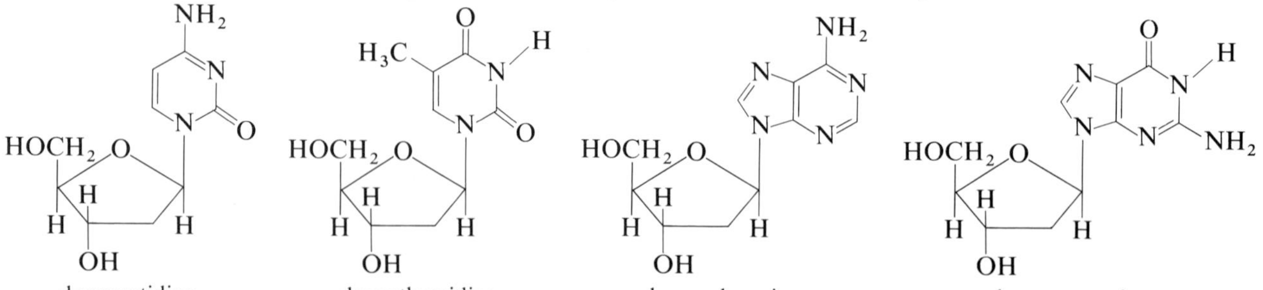

deoxycytidine deoxythymidine deoxyadenosine deoxyguanosine

The structure of the DNA polymer is similar to that of RNA, except that there are no hydroxyl groups on the 2′ carbon atoms of the ribose rings. Once again, the alternating deoxyribose rings and phosphates act as the backbone of the molecule, while the bases attached to the ribose units carry the genetic information. The sequence of nucleotides is called the **primary structure** of the DNA strand.

23-23A BASE PAIRING

Having discussed the primary structure of DNA and RNA, we now consider how the nucleotide sequence is reproduced or transcribed into another molecule. This information transfer takes place by an interesting hydrogen-bonding interaction between specific pairs of bases.

Each pyrimidine base forms a stable hydrogen-bonded pair with only one of the two purine bases. Cytosine forms a base pair, joined by three hydrogen bonds, with guanine. Thymine (or uracil, in RNA) forms a base pair with adenine, joined by two hydrogen bonds. Guanine is said to be *complementary* to cytosine, and adenine is complementary to thymine. This base pairing was first suspected in 1950, when Erwin Chargaff of Columbia University noticed that various DNAs, taken from a wide variety of species, had about equal amounts of adenine and thymine, and about equal amounts of guanine and cytosine.

guanine G:::C cytosine

adenine A:::T thymine

23-23B THE DOUBLE HELIX OF DNA

In 1953, James D. Watson and Francis C. Crick used X-ray diffraction patterns of DNA fibers to determine the molecular structure and conformation of DNA. They found that DNA contains two complementary polynucleotide chains held together by the hydrogen bonds of the paired bases. Figure 23-24 shows a portion

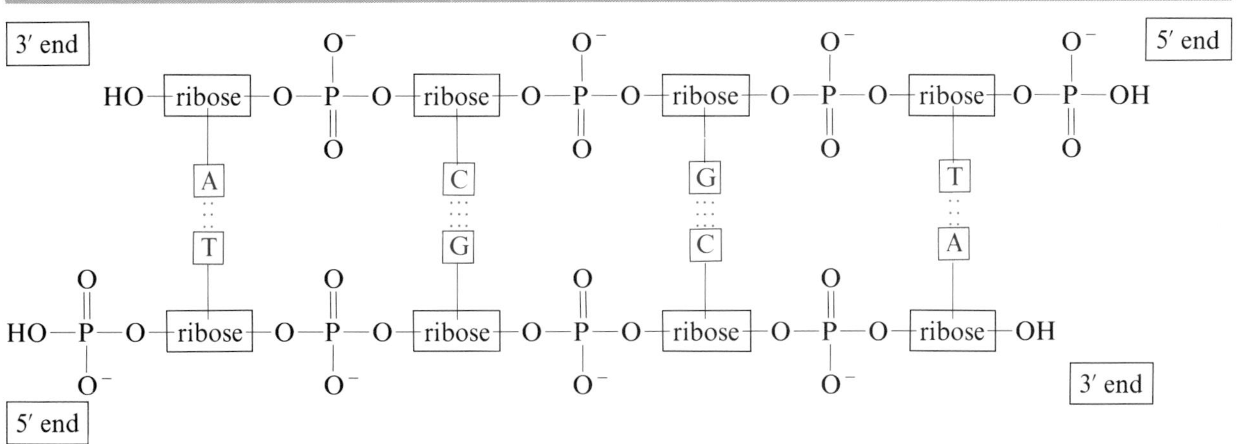

FIGURE 23-24 DNA usually consists of two complementary strands, with all the base pairs hydrogen-bonded together. The two strands run in opposite directions.

of the double strand of DNA, with each base paired with its complement. One of the strands is arranged 3′ → 5′ from left to right, while the other runs in the opposite direction, 5′ → 3′ from left to right.

Watson and Crick also discovered that the two complementary strands of DNA are coiled into a helical conformation about 20 Å in diameter, with both chains coiled around the same axis. The helix makes a complete turn for every ten residues, or about one turn in every 34 Å of length. Figure 23-25 shows the double helix of DNA.

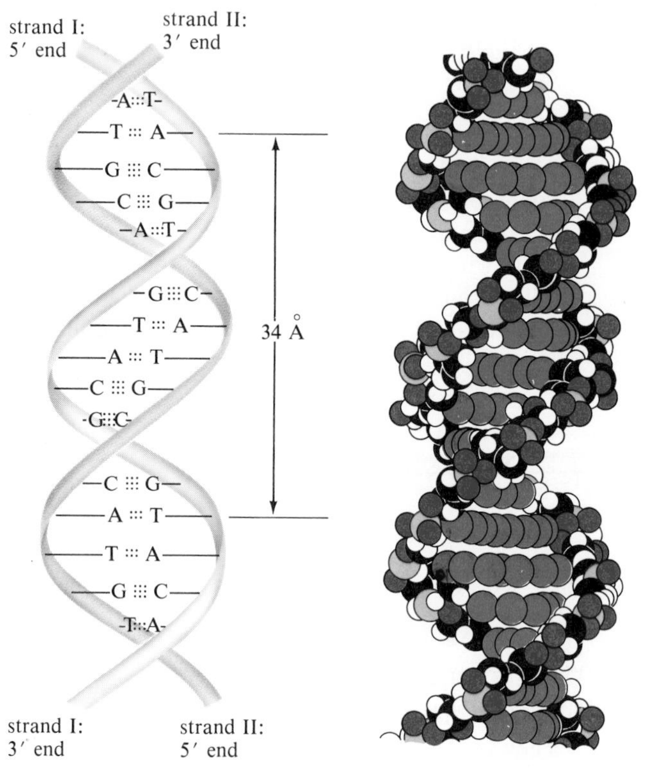

FIGURE 23-25 Double helix of DNA. Two complementary strands are joined by the hydrogen bonds between the base pairs. This double strand coils into a helical arrangement.

When DNA undergoes replication (in preparation for cell division), an enzyme uncoils part of the double strand. Individual nucleotides naturally hydrogen-bond to their complements on the uncoiled original strand, and a *DNA polymerase* enzyme couples the nucleotides together. This process is depicted schematically in Figure 23-26. A similar process transcribes the DNA into a complementary molecule of messenger RNA for use by the ribosomes as a template for protein synthesis.

A great deal is known about the replication of DNA and the translation of the DNA/RNA sequence of bases into proteins. These exciting aspects of nucleic acid chemistry are part of the field of *molecular biology,* and they are covered in detail in a course in biochemistry.

23-23C ADDITIONAL FUNCTIONS OF NUCLEOTIDES

We generally think of nucleotides as the monomers that form DNA and RNA polymers, yet these versatile biomolecules serve a variety of additional functions. Here we briefly consider a few additional functions of nucleotides.

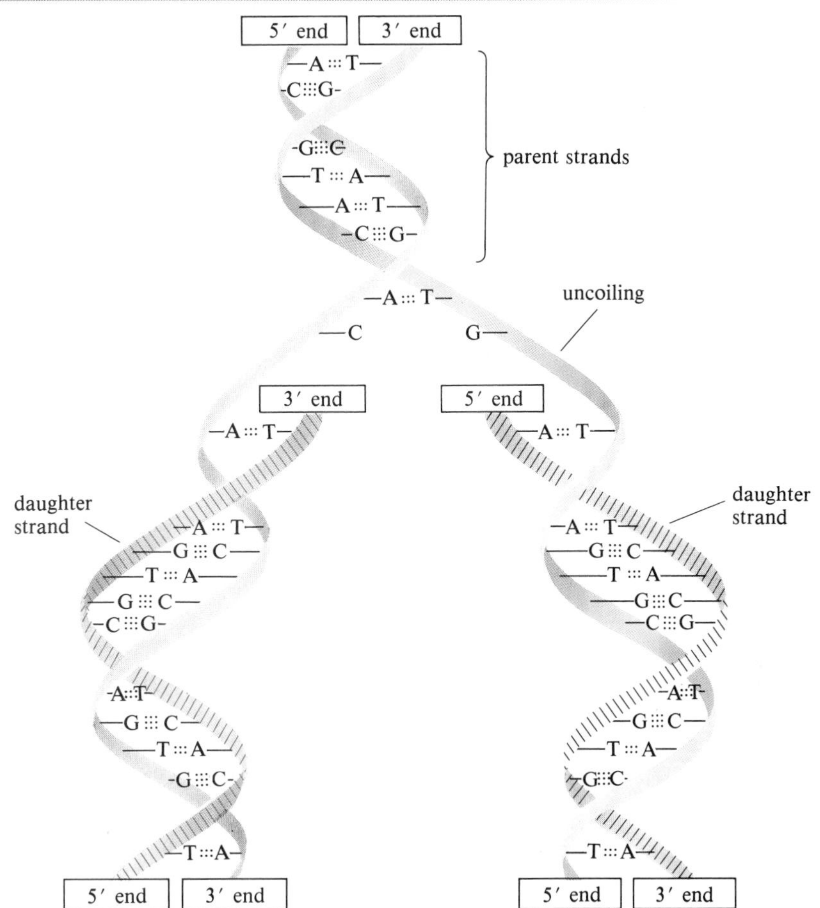

FIGURE 23-26 Simplified view of replication of the double strand of DNA. A new strand is assembled on each of the original strands, with the DNA polymerase enzyme forming the phosphate ester bonds of the backbone.

AMP: a regulatory hormone Adenosine monophosphate (AMP) also occurs in a cyclic form, where the 3′- and 5′-hydroxyl groups are both esterified by the same phosphate group. This *cyclic AMP* is involved in transmitting and amplifying the chemical signals of other hormones.

adenosine monophosphate (AMP)

cyclic AMP

NAD: a coenzyme Nicotinamide adenine dinucleotide (NAD) is one of the principal oxidation-reduction reagents in biological systems. This nucleotide has the structure of two D-ribose rings (a *di*nucleotide) linked by their 5′ phosphates. The aglycone of one ribose is nicotinamide, and the aglycone of the other is adenine.

A dietary deficiency of nicotinic acid (niacin) leads to the disease called *pellagra,* caused by the inability to synthesize enough nicotinamide adenine dinucleotide.

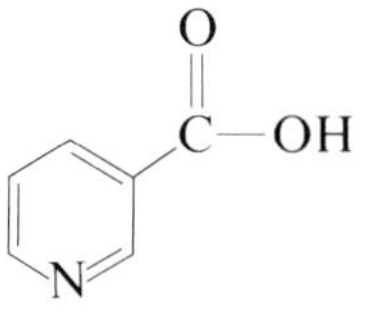

nicotinic acid (niacin)

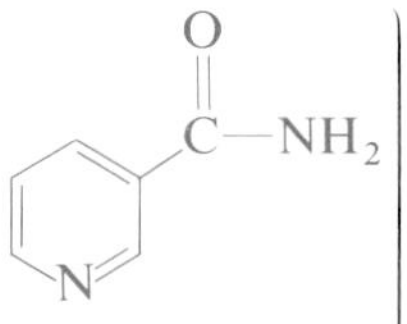

nicotinamide

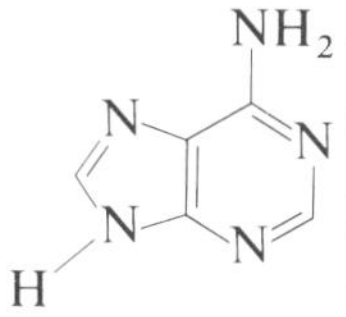

adenine

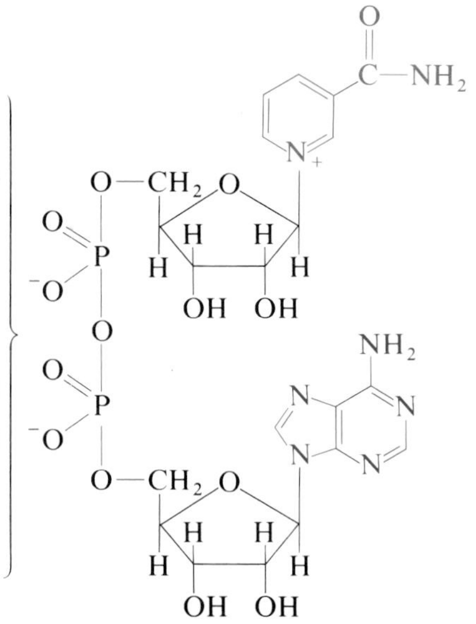

NAD^+

nicotinamide adenine dinucleotide

The following equation shows how NAD^+ serves as the oxidizing agent in the biological oxidation of an alcohol. Just the nicotinamide portion of NAD is shown, since that part of the molecule actually takes part in the reaction. The enzyme that catalyzes this reaction is called alcohol dehydrogenase (ADH).

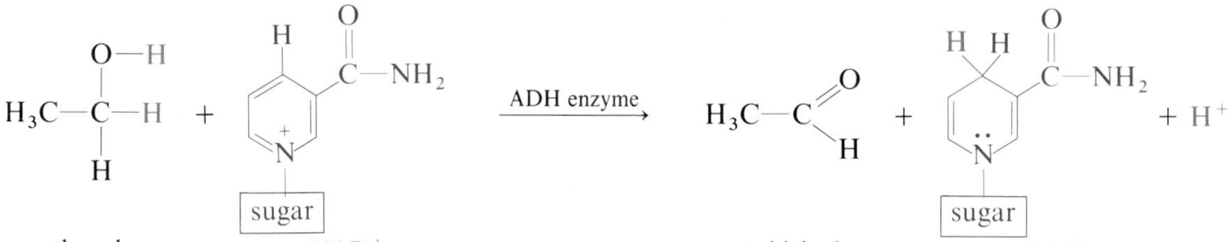

ethanol + NAD^+ $\xrightarrow{\text{ADH enzyme}}$ acetaldehyde + NADH + H^+

ATP: an energy source When glucose is oxidized in the living cell, the energy released is used to synthesize *adenosine triphosphate* (ATP), an anhydride of phosphoric acid. As with most anhydrides, the hydrolysis of ATP is highly exothermic. The hydrolysis products are adenosine diphosphate (ADP) and inorganic phosphate.

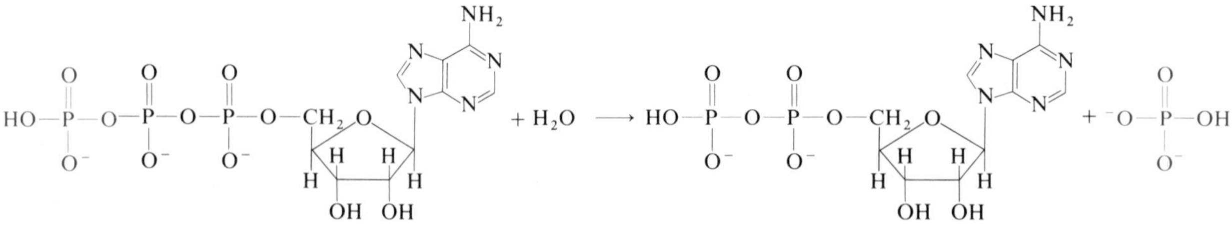

adenosine triphosphate (ATP) + H_2O ⟶ adenosine diphosphate (ADP) + phosphate

$\Delta H^\circ = -7.3$ kcal/mol (-31 kJ/mol)

The highly exothermic nature of ATP hydrolysis is largely explained by the heats of hydration of the products. ADP is hydrated about as well as ATP, but the inorganic phosphate ion has a very large heat of hydration. The hydrolysis of adenosine triphosphate (ATP) liberates 7.3 kcal (31 kJ) of energy per mole of ATP. This is the energy that muscle cells use to contract and all cells use to drive their endothermic chemical processes.

GLOSSARY

aglycone A nonsugar residue bonded to the anomeric carbon of a glycoside (the acetal or ketal form of a sugar). Aglycones are commonly bonded to the sugar through oxygen or nitrogen. (p. 1063)

aldaric acid A dicarboxylic acid formed by oxidation of both end carbon atoms of a monosaccharide. (p. 1060)

alditol A polyalcohol formed by reduction of the carbonyl group of a monosaccharide. (p. 1059)

aldonic acid A monocarboxylic acid formed by oxidation of the aldehyde group of an aldose. (p. 1060)

aldose A monosaccharide containing an aldehyde carbonyl group. (p. 1044)

amino sugar A sugar (such as *N*-acetylglucosamine) in which a hydroxyl group is replaced by an amino group. (p. 1085)

anomeric carbon The hemiacetal or hemiketal carbon in the cyclic form of a sugar (the carbonyl carbon in the open-chain form). The anomeric carbon is easily identified because it is the only carbon with two bonds to oxygen atoms. (p. 1054)

anomers Sugar stereoisomers that differ in configuration only at the anomeric carbon. Anomers are classified as α or β depending on whether the anomeric hydroxyl group (or the aglycone in a glycoside) is trans (α) or cis (β) to the terminal $-CH_2OH$. (p. 1054)

carbohydrates (sugars) A class of polyhydroxy aldehydes and ketones, many of which have the formula $C_n(H_2O)_n$, from which they received the name "hydrates of carbon" or "carbohydrates." (p. 1044)

cellulose A linear β-1,4′ polymer of D-glucopyranose. Cellulose forms the cell wall of plants and is the major constituent of wood and cotton. (p. 1083)

chitin A β-1,4′ polymer of *N*-acetylglucosamine that lends strength and rigidity to the exoskeletons of insects and crustaceans. (p. 1085)

degradation A reaction that causes loss of a carbon atom. (p. 1067)

deoxyribonucleic acid (DNA) A biopolymer of deoxyribonucleotides that serves to control the synthesis of ribonucleic acid. DNA can also control its own replication, through an uncoiling process and the pairing and enzymatic linking of complementary bases. (p. 1086)

deoxy sugar A sugar in which a hydroxyl group is replaced by a hydrogen. Deoxy sugars are recognized by the presence of a methylene group or a methyl group. (p. 1090)

dextrose The dextrorotatory isomer of glucose, D-(+)-glucose. (p. 1055)

D family of sugars All sugars whose chiral carbon atom farthest from the carbonyl group has the same configuration as the chiral carbon atom in D-(+)-glyceraldehyde. Most naturally occurring sugars are members of the D family. (pp. 1046 and 1048)

disaccharide A carbohydrate whose hydrolysis gives two monosaccharide molecules. (p. 1077)

enediol rearrangement (Lobry de Bruyn–Alberta van Ekenstein reaction) A base-catalyzed tautomerization that interconverts aldoses and ketoses with an enediol as an intermediate. This enolization also epimerizes C2 and other carbon atoms. (p. 1058)

epimers Two diastereomeric sugars differing only in the configuration at a single chiral carbon atom. The epimeric carbon atom is usually specified, as in "C4 epimers." If no epimeric carbon is specified, it is assumed to be C2. The interconversion of epimers is called *epimerization*. (p. 1049)

erythro and **threo** Diastereomers having similar groups on the same side (erythro) or on opposite sides (threo) of the Fischer projection. This terminology was adapted from the names of the aldotetroses *erythrose* and *threose*. (p. 1048)

D-erythrose	*erythro*-2,3-dibromopentanoic acid	D-threose	*threo*-2,3-dihydroxybutanal
CHO	COOH	CHO	CHO
H—+—OH	Br—+—H	HO—+—H	H—+—OH
H—+—OH	Br—+—H	H—+—OH	HO—+—H
CH_2OH	CH_2CH_3	CH_2OH	CH_3

furanose A five-membered cyclic hemiacetal form of a sugar. (p. 1053)

furanoside A five-membered cyclic glycoside. (p. 1061)

glucoside A glycoside derived from glucose. (p. 1061)

glucosidic linkage A glycosidic linkage using an acetal bond from the anomeric carbon of glucose. (p. 1078)

glycoside A cyclic acetal form of a sugar. Glycosides are stable to base, and they are nonreducing sugars. Glycosides are generally **furanosides** (five-membered) or **pyranosides** (six-membered), and they exist in anomeric α and β forms. (p. 1061)

glycosidic linkage An acetal bond from an anomeric carbon joining two monosaccharide units. (p. 1077)

Haworth projection A flat-ring representation of a cyclic sugar. The Haworth projection does not show the axial and equatorial positions in a pyranose, but it does show the cis, trans relationships. (p. 1051)

ketose A monosaccharide containing a ketone carbonyl group. (p. 1044)

Kiliani-Fischer synthesis A method for elongating an aldose at the aldehyde end. The aldose is converted into the two diastereomeric aldoses with an additional carbon atom. For example, Kiliani-Fischer synthesis converts D-arabinose to a mixture of D-glucose and D-mannose. (p. 1068)

L family of sugars All sugars whose chiral carbon atom farthest from the carbonyl group has the same configuration as the chiral carbon atom in L-(−)-glyceraldehyde. Sugars of the L family are not common in nature. (p. 1046)

monosaccharide As distinguished from a disaccharide or a polysaccharide, a carbohydrate that does not undergo hydrolysis of glycosidic bonds to give smaller sugar molecules. (p. 1044)

mutarotation The spontaneous change in optical rotation that occurs when a pure anomer of a sugar in its hemiacetal form equilibrates with the other anomer to give an equilibrium mixture of the two anomers with an averaged value of the optical rotation. (p. 1055)

nucleoside An *N*-glycoside of β-D-ribofuranose or β-D-deoxyribofuranose, where the aglycone is one of several derivatives of either pyrimidine or purine. (p. 1086)

nucleotide A 5′-phosphate ester of a nucleoside. (p. 1086)

oligosaccharide A carbohydrate whose hydrolysis gives more than two monosaccharide units, but not as many as a polysaccharide (usually about two to ten units). (p. 1082)

osazone The product, containing two phenylhydrazone residues, that results from the reaction of a reducing sugar with phenylhydrazine. (p. 1066)

polysaccharide A carbohydrate whose hydrolysis gives many monosaccharide molecules. (p. 1082)

primary structure The primary structure of a nucleic acid is the sequence of nucleotides forming the polymer. This sequence and the resulting conformation determine the catalytic and genetic characteristics of the nucleic acid. (p. 1091)

pyranose A six-membered cyclic hemiacetal form of a sugar. (p. 1053)

pyranoside A six-membered cyclic glycoside. (p. 1061)

reducing sugar Any sugar that gives a positive Tollens test. Both ketoses and aldoses (in their hemiacetal forms) give positive Tollens tests. (p. 1061)

ribonucleic acid (RNA) A biopolymer of ribonucleotides that serves to control the synthesis of proteins. The synthesis of RNA is generally controlled by and patterned after the DNA in the cell. (p. 1086)

ribonucleotide The 5′-phosphate ester of a **ribonucleoside,** a component of RNA based on β-D-ribfuranose and containing one of four heterocyclic bases as the aglycone. (p. 1088)

Ruff degradation A method for shortening the chain of an aldose by one carbon atom by treatment with bromine water followed by hydrogen peroxide and $Fe_2(SO_4)_3$. (p. 1067)

starches A class of α-1,4′ polymers of glucose used for carbohydrate storage in plants and animals. (p. 1083)

amylopectin: A branched α-1,4′ polymer of D-glucopyranose used for carbohydrate storage in plants. Branching occurs at α-1,6′ glycosidic linkages. (p. 1084)

amylose: A linear α-1,4′ polymer of D-glucopyranose used for carbohydrate storage in plants. (p. 1083)

glycogen: An extensively branched α-1,4′ polymer of D-glucopyranose that is used for carbohydrate storage in animals. Branching occurs at α-1,6′ glycosidic linkages. (p. 1084)

sugar (saccharide) Any carbohydrate, regardless of structure, complexity, or taste. A simple sugar is a monosaccharide. (p. 1044)

Tollens test A test for reducing sugars employing the same silver-ammonia complex used as a chemical test for aldehydes. A positive test gives a silver precipitate, often in the form of a silver mirror. Tollens reagent is basic, and it promotes enediol rearrangement that interconverts ketoses and aldoses. Therefore, both aldoses and ketoses give positive Tollens tests if they are in their hemiacetal forms, in equilibrium with the open-chain carbonyl structures. (p. 1061)

ESSENTIAL PROBLEM-SOLVING SKILLS IN CHAPTER 23

1. Draw the Fischer projection of glucose and the chair conformation of the β anomer of glucose (all substituents equatorial) from memory.

2. Recognize the structures of other anomers and epimers of glucose, drawn as either Fischer projections or chair structures, by noticing the differences from the glucose structure.

3. Correctly name monosaccharides and disaccharides, and draw their structures from their names.

4. Predict which carbohydrates mutarotate, which will reduce Tollens reagent, and which will undergo epimirization and isomerization under basic conditions. (Those with free hemiacetals will, but glycosides with full acetals will not.)

5. Predict the products of the following reactions of carbohydrates:
   bromine in water
   nitric acid
   periodic acid
   $NaBH_4$ or H_2/Ni
   alcohols and H^+
   CH_3I and Ag_2O
   NaOH and dimethyl sulfate
   acetic anhydride and pyridine
   phenylhydrazine
   Ruff degradation
   Kiliani-Fischer synthesis

6. Use the information gained from these reactions to determine the structure of an unknown carbohydrate molecule. Use the information gained from methylation and from periodic acid cleavage to determine the ring sizes of carbohydrates.

7. Draw the common types of glycosidic linkages, and recognize these linkages in disaccharides and polysaccharides.

8. Recognize the structures of DNA and RNA, and draw the structures of a ribonucleotide and a deoxyribonucleotide.

23-51. Define each of the following terms and give an example.

(a) aldose **(b)** ketose **(c)** aldonic acid **(d)** aldaric acid
(e) glycoside **(f)** aglycone **(g)** sugar **(h)** anomers
(i) erythro and threo **(j)** epimers **(k)** furanose **(l)** pyranose
(m) Haworth projection **(n)** monosaccharide **(o)** polysaccharide **(p)** disaccharide
(q) ribonucleoside **(r)** ribonucleotide **(s)** deoxyribonucleotide **(t)** osazone
(u) reducing sugar **(v)** amino sugar

23-52. Glucose is the most abundant monosaccharide. From memory, draw glucose in
(a) the Fischer projection of the open chain.
(b) the most stable chair conformation of the most stable pyranose anomer.
(c) the Haworth projection of the most stable pyranose anomer.

23-53. Without referring back in the chapter, draw the chair conformations of
(a) β-D-mannopyranose (the C2 epimer of glucose).
(b) α-D-allopyranose (the C3 epimer of glucose).
(c) β-D-galactopyranose (the C4 epimer of glucose).
(d) *N*-acetylglucosamine, glucose with the C2 oxygen atom replaced by an acetylated amino group.

23-54. Classify the following monosaccharides. (*Examples:* D-aldohexose, L-ketotetrose.)
(a) (+)-glucose **(b)** (−)-arabinose **(c)** L-fructose

(d)
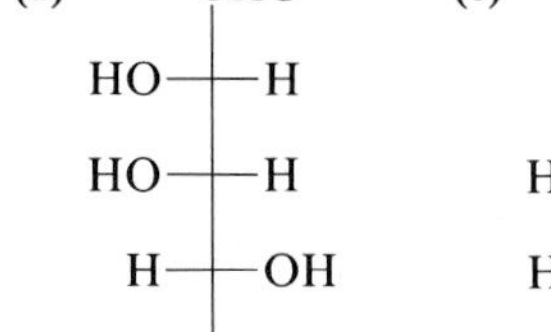

(+)-gulose

(e)
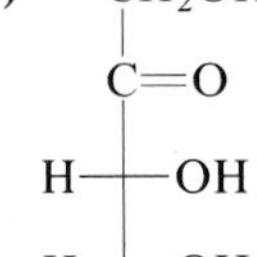

(−)-ribulose

(f)
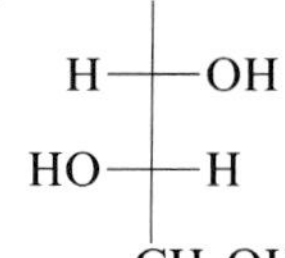

(+)-threose

23-55. Fructose is the ketose that results from the enediol rearrangement of glucose that shifts the carbonyl group to C2.
(a) Propose a mechanism for the enediol rearrangement that converts D-glucose to D-fructose.
(b) Draw the α and β anomers of D-fructofuranose. How can you tell which anomer is α and which is β regardless of the ring size?

23-56. The relative configurations of the stereoisomers of tartaric acid were established by the following synthesis.

1. D-(+)-glyceraldehyde $\xrightarrow{\text{HCN}}$ diastereomers A and B (separated).
2. Hydrolysis of A and B using aqueous $Ba(OH)_2$ gave C and D, respectively.
3. HNO_3 oxidation of C and D gave (−)-tartaric acid and *meso*-tartaric acid, respectively.

(a) You know the absolute configuration of D-(+)-glyceraldehyde. Use Fischer projections to show the absolute configurations of products A, B, C, and D.
(b) Show the absolute configurations of the three stereoisomers of tartaric acid: (+)-tartaric acid, (−)-tartaric acid, and *meso*-tartaric acid.

23-57. Use Figure 23-3 (the D family of aldoses) to name the following aldoses.
(a) the C2 epimer of D-arabinose **(b)** the C3 epimer of D-mannose **(c)** the C3 epimer of D-threose
(d) the enantiomer of D-galactose **(e)** the C5 epimer of D-glucose

23-58. Draw the following sugar derivatives.
(a) methyl β-D-glucopyranoside **(b)** 2,3,4,6-tetra-*O*-methyl-D-galactopyranose
(c) 1,3,6-tri-*O*-methyl-D-fructofuranose **(d)** methyl 1,3,6-tri-*O*-methyl-α-D-fructofuranoside

23-59. Draw the structures (using chair conformations of pyranoses) of the following disaccharides.
(a) 4-*O*-(α-D-glucopyranosyl)-D-galactopyranose **(b)** α-D-fructofuranosyl-β-D-mannopyranoside
(c) 6-*O*-(β-D-galactopyranosyl)-D-glucopyranose

23-60. Give the complete systematic name for each of the following structures.

(a)

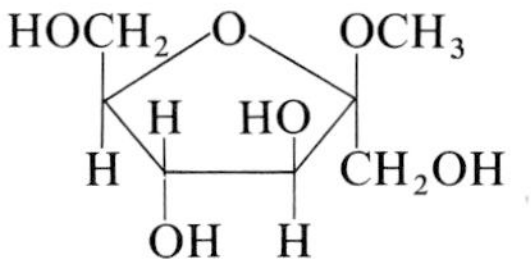

(b)

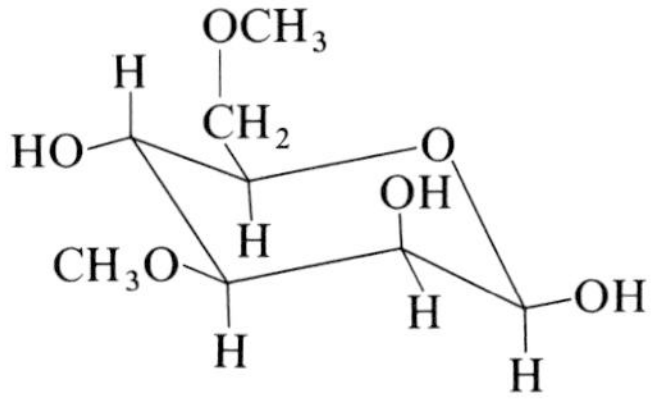

(c)

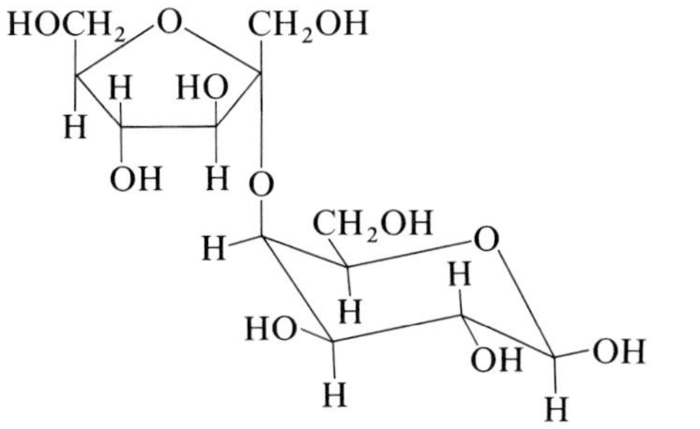

(d)

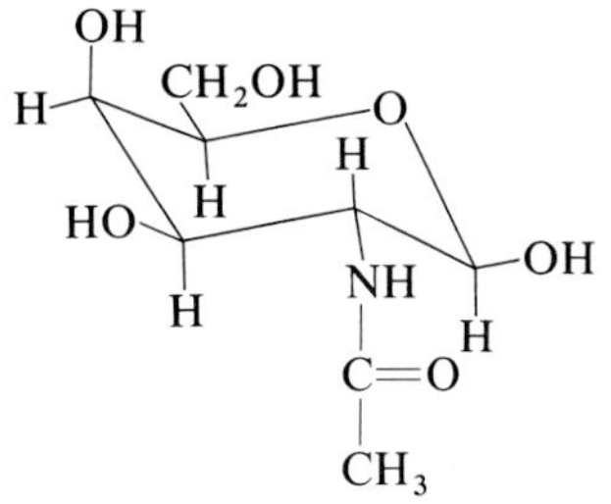

23-61. Which of the sugars mentioned in Problems 23-58, 23-59, and 23-60 are reducing sugars? Which ones would undergo mutarotation?

23-62. Predict the products obtained when D-galactose reacts with each of the following reagents.
(a) Br_2 and H_2O **(b)** NaOH, H_2O **(c)** CH_3OH, H^+ **(d)** $Ag(NH_3)_2^+$ ^-OH
(e) H_2, Ni **(f)** Ac_2O **(g)** excess CH_3I, Ag_2O **(h)** $NaBH_4$
(i) Br_2, H_2O, then H_2O_2 and $Fe_2(SO_4)_3$ **(j)** HCN, then H_3O^+, then Na(Hg) **(k)** excess HIO_4

23-63. Draw the structures of the products expected when the following carbohydrates are subjected to methylation followed by acidic hydrolysis. In each case, suggest what reagent would be most appropriate for the methylation step.
(a) D-fructose **(b)** ethyl α-D-glucopyranoside **(c)** sucrose **(d)** lactose **(e)** gentibiose **(f)** chitin

23-64. **(a)** Which of the D-aldopentoses will give optically active aldaric acids on oxidation with HNO_3?
(b) Which of the D-aldotetroses will give optically active aldaric acids on oxidation with HNO_3?
(c) Sugar X is known to be a D-aldohexose. On oxidation with HNO_3, X gives an optically inactive aldaric acid. When X is degraded to an aldopentose, oxidation of the aldopentose gives an optically active aldaric acid. Determine the structure of X.
(d) Even though sugar X gives an optically inactive aldaric acid, the pentose formed by degradation gives an optically active aldaric acid. Does this finding contradict the principal that optically inactive reagents cannot form optically active products?
(e) Show what product results if the aldopentose formed from degradation of X is further degraded to an aldotetrose. Does HNO_3 oxidize this aldotetrose to an optically active aldaric acid?

23-65. **(a)** Give the products expected when (+)-glyceraldehyde reacts with HCN.
(b) What is the relationship between the products? How might they be separated?
(c) Are the products optically active? Explain.

23-66. When fructose reacts with the Tollens reagent, the major products are the carboxylate ions of mannonic acid and gluconic acid.
(a) Give a mechanism to show how fructose isomerizes to a mixture of glucose and mannose in the presence of Tollens reagent.
(b) Explain why bromine water is a better reagent than the Tollens reagent for the oxidation of aldoses to aldonic acids.

23-67. When the gum of the shrub *Sterculia setigera* is subjected to acidic hydrolysis, one of the water-soluble components of the hydrolysate is found to be tagatose. The following information is known about tagatose.
1. Molecular formula $C_6H_{12}O_6$.
2. Undergoes mutarotation.
3. Does not react with bromine water.

4. Reduces Tollens reagent to give D-galactonic acid and D-talonic acid.
5. Methylation of tagatose (using CH_3I and Ag_2O) followed by acidic hydrolysis gives 1,3,4,5-tetra-*O*-methyltagatose.

(a) Draw a Fischer projection structure for the open-chain form of tagatose.
(b) Draw the most stable conformation of the most stable cyclic hemiacetal form of tagatose.

23-68. After a series of Kiliani-Fischer syntheses on (+)-glyceraldehyde, an unknown sugar is isolated from the reaction mixture. The following experimental information is obtained.

1. Molecular formula $C_6H_{12}O_6$.
2. Undergoes mutarotation.
3. Reacts with bromine water to give an aldonic acid.
4. Reacts with phenylhydrazine to give an osazone, m.p. 178°C.
5. Reacts with HNO_3 to give an optically active aldaric acid.
6. Ruff degradation followed by HNO_3 oxidation gives an optically inactive aldaric acid.
7. Two Ruff degradations followed by HNO_3 oxidation give *meso*-tartaric acid.
8. Methylation (using CH_3I and Ag_2O) followed by acidic hydrolysis gives a 2,3,4,6-tetra-*O*-methylated derivative.

(a) Draw a Fischer projection structure for the open-chain form of this unknown sugar. Use Figure 23-3 to name the sugar.
(b) Draw the most stable conformation of the most stable cylic hemiacetal form of this sugar, and give the structure a complete systematic name.

23-69. An unknown reducing disaccharide is found to be unaffected by invertase enzymes. Treatment with an α-galactosidase cleaves the disaccharide to give one molecule of D-fructose and one molecule of D-galactose. When the disaccharide is treated with iodomethane and silver oxide and then hydrolyzed in dilute acid, the products are 2,3,4,6-tetra-*O*-methylgalactose and 1,3,4-tri-*O*-methylfructose. Propose a structure for this disaccharide and give its complete systematic name.

23-70. Draw the structures of the following nucleotides.
(a) guanosine triphosphate (GTP) **(b)** deoxycytidine monophosphate (dCMP)
(c) cyclic guanosine monophosphate (cGMP)

23-71. Draw the structure of a four-residue segment of DNA with the following sequence.

(3′ end) G-T-A-C (5′ end)

23-72. Erwin Chargaff's discovery that DNA contains equimolar amounts of guanine and cytosine and also equimolar amounts of adenine and thymine has come to be known as *Chargaff's rule.*

$$G = C \quad \text{and} \quad A = T$$

(a) Does Chargaff's rule imply that equal amounts of guanine and adenine are present in DNA? That is, does $G = A$?
(b) Does Chargaff's rule imply that the sum of the purine residues equals the sum of the pyrimidine residues? That is, does $A + G = C + T$?
(c) Does Chargaff's rule apply only to double-stranded DNA, or would it also apply to each individual strand if the double helical strand were separated into its two complementary strands?

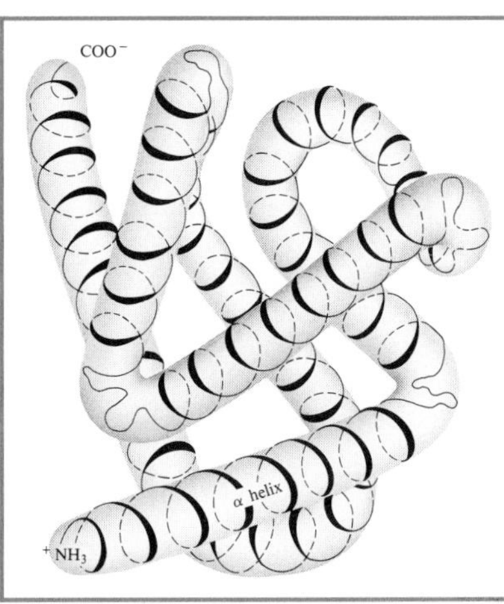

24

AMINO ACIDS, PEPTIDES, AND PROTEINS

24-1 INTRODUCTION

Proteins are the most abundant organic molecules in animals, playing important roles in all aspects of cell structure and function. Proteins are biopolymers of **α-amino acids,** and the physical and chemical properties of a protein are determined by its constituent amino acids. The individual amino acid subunits are joined by amide linkages called **peptide bonds.** Figure 24-1 shows the general structure of an α-amino acid and a protein.

Proteins have an amazing range of structural and catalytic properties as a result of their varying amino acid composition. Because of this versatility, proteins serve an astonishing variety of functions in living organisms. Some of the functions of the major classes of proteins are outlined in Table 24-1.

TABLE 24-1
Examples of protein functions

Class of protein	*Example*	*Function of example*
structural proteins	collagen	connective tissue (tendons, etc.)
enzymes	DNA polymerase	replicate and repair DNA
transport proteins	hemoglobin	transport O_2 to the cells
contractile proteins	actin, myosin	cause contraction of muscles
protective proteins	antibodies	complex with foreign proteins
hormones	insulin	regulate glucose metabolism
toxins	snake venoms	incapacitate food animals

The study of proteins is one of the major branches of biochemistry, and there is no clear division between the organic chemistry of proteins and their biochemistry. In this chapter we begin the study of proteins by learning about their constituents, the amino acids. We also discuss how the amino acid monomers are

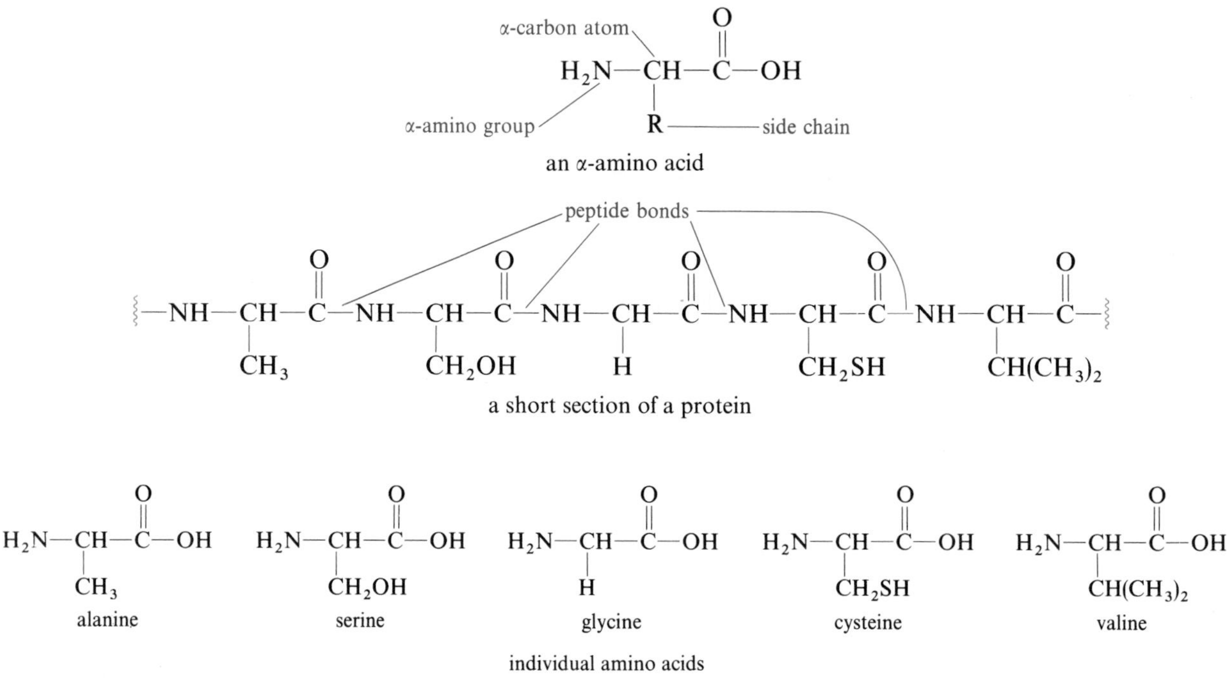

$$H_2N{-}\underset{\substack{|\\ R}}{CH}{-}\overset{\overset{\displaystyle O}{\|}}{C}{-}OH$$

α-carbon atom; α-amino group; side chain

an α-amino acid

$$\text{\textasciitilde}NH{-}\underset{\substack{|\\ CH_3}}{CH}{-}\overset{\overset{\displaystyle O}{\|}}{C}{-}NH{-}\underset{\substack{|\\ CH_2OH}}{CH}{-}\overset{\overset{\displaystyle O}{\|}}{C}{-}NH{-}\underset{\substack{|\\ H}}{CH}{-}\overset{\overset{\displaystyle O}{\|}}{C}{-}NH{-}\underset{\substack{|\\ CH_2SH}}{CH}{-}\overset{\overset{\displaystyle O}{\|}}{C}{-}NH{-}\underset{\substack{|\\ CH(CH_3)_2}}{CH}{-}\overset{\overset{\displaystyle O}{\|}}{C}\text{\textasciitilde}$$

peptide bonds

a short section of a protein

$H_2N{-}\underset{\substack{|\\ CH_3}}{CH}{-}\overset{\overset{O}{\|}}{C}{-}OH$	$H_2N{-}\underset{\substack{|\\ CH_2OH}}{CH}{-}\overset{\overset{O}{\|}}{C}{-}OH$	$H_2N{-}\underset{\substack{|\\ H}}{CH}{-}\overset{\overset{O}{\|}}{C}{-}OH$	$H_2N{-}\underset{\substack{|\\ CH_2SH}}{CH}{-}\overset{\overset{O}{\|}}{C}{-}OH$	$H_2N{-}\underset{\substack{|\\ CH(CH_3)_2}}{CH}{-}\overset{\overset{O}{\|}}{C}{-}OH$
alanine	serine	glycine	cysteine	valine

individual amino acids

FIGURE 24-1 Structure of a general protein and its constituent amino acids. The amino acids are joined by amide linkages called peptide bonds.

linked into the protein polymer, and how the properties of the protein depend on those of its constituent amino acids. These essentials are needed for the further study of protein structure and function in a biochemistry course.

24-2 STRUCTURE AND STEREOCHEMISTRY OF THE α-AMINO ACIDS

The term **amino acid** might mean any molecule containing both an amino group and any type of acid group; however, the term is almost always used to refer to an α-amino carboxylic acid. The simplest α-amino acid is aminoacetic acid, called **glycine.** Other common amino acids have side chains (symbolized R) substituted on the α-carbon atom of glycine. For example, alanine is the amino acid with a methyl side chain.

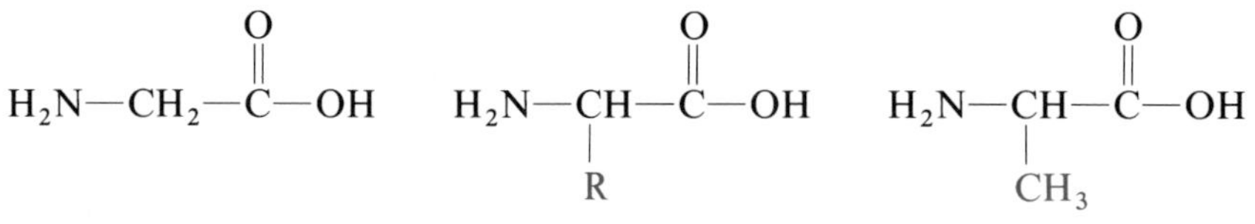

$H_2N{-}CH_2{-}\overset{\overset{O}{\|}}{C}{-}OH$	$H_2N{-}\underset{\substack{|\\ R}}{CH}{-}\overset{\overset{O}{\|}}{C}{-}OH$	$H_2N{-}\underset{\substack{|\\ CH_3}}{CH}{-}\overset{\overset{O}{\|}}{C}{-}OH$
glycine	a substituted amino acid	alanine ($R = CH_3$)

Except for glycine, the α-amino acids are all chiral. Nearly all the naturally occurring amino acids are found to have the (*S*) configuration at the α-carbon atom. Figure 24-2 shows the (*S*) enantiomer of alanine arranged in the Fischer projection, with the carbon chain along the vertical and the most highly oxidized carbon at the top. Notice that (*S*)-alanine has a configuration similar to that of L-(−)-glyceraldehyde, with the amino group on the left in the Fischer projection. The naturally occurring (*S*)-amino acids resemble the structure of L-(−)-glyceraldehyde, so they are classified as **L-amino acids.**

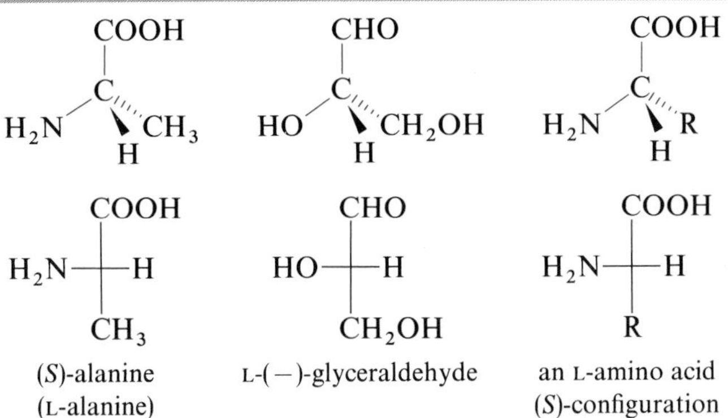

FIGURE 24-2 Almost all the naturally occurring amino acids have the (*S*) configuration, with stereochemistry resembling that of L-(−)-glyceraldehyde. They are therefore called L-amino acids.

Although D-amino acids are occasionally found in nature, we usually assume that the amino acids under discussion are the common L-amino acids. Remember once again that the D and L nomenclature, like the *R* and *S* designation, gives the configuration of the chiral carbon atom. It does not imply the sign of the optical rotation, (+) or (−), which must be determined by experiment.

Amino acids combine many of the properties and reactions of both amines and carboxylic acids. The proximity of a basic amino group to an acidic carboxyl group in the same molecule also results in some unique properties and reactions. The side chains of some amino acids also have functional groups that lend interesting properties and undergo reactions of their own.

24-2A THE STANDARD AMINO ACIDS OF PROTEINS

There are 20 α-amino acids, called the **standard amino acids,** that are found in nearly all proteins. The standard amino acids differ from each other in the structure of the side chains bonded to their α-carbon atoms. All the standard amino acids are L-amino acids. Table 24-2 shows the 20 standard amino acids, grouped according to the chemical properties of their side chains. Each amino acid is given a three-letter abbreviation for use in writing protein structures. The essential amino acids (Section 24-2B) are marked with a star (*).

Notice in Table 24-2 that proline is different from the other standard amino acids because its amino group is a secondary amine, fixed in a ring with its α-carbon atom. This cyclic structure lends additional strength and rigidity to proline-containing peptides.

COOH
N H
H
α-carbon
α-amino group

proline

PROBLEM 24-1

Draw three-dimensional representations of the following amino acids.

(a) L-phenylalanine (b) L-arginine
(c) D-serine (d) L-proline

TABLE 24-2
The standard amino acids

Name	Abbreviation	Structure	Functional group in side chain	Isoelectric point
		R group is H or alkyl		
glycine	Gly	$H_2N-CH(H)-COOH$	none	6.0
alanine	Ala	$H_2N-CH(CH_3)-COOH$	alkyl group	6.0
valine	Val*	$H_2N-CH(CH(CH_3)_2)-COOH$	alkyl group	6.0
leucine	Leu*	$H_2N-CH(CH_2-CH(CH_3)-CH_3)-COOH$	alkyl group	6.0
isoleucine	Ile*	$H_2N-CH(CH_3-CH-CH_2CH_3)-COOH$	alkyl group	6.0
phenylalanine	Phe*	$H_2N-CH(CH_2-C_6H_5)-COOH$	aromatic group	5.5
proline	Pro	$HN-CH-COOH$ (ring: H_2C, CH_2, CH_2)	rigid cyclic structure	6.3
		R group contains an —OH		
serine	Ser	$H_2N-CH(CH_2-OH)-COOH$	hydroxyl group	5.7
threonine	Thr*	$H_2N-CH(HO-CH-CH_3)-COOH$	hydroxyl group	5.6
tyrosine	Tyr	$H_2N-CH(CH_2-C_6H_4-OH)-COOH$	phenolic —OH group	5.7
		R group contains nonbasic nitrogen		
asparagine	Asn	$H_2N-CH(CH_2-C(=O)-NH_2)-COOH$	amide	5.4
glutamine	Gln	$H_2N-CH(CH_2-CH_2-C(=O)-NH_2)-COOH$	amide	5.7

TABLE 24-2 *(continued)*

Name	*Abbreviation*	*Structure*	*Functional group in side chain*	*Isoelectric point*
tryptophan	Trp*	$H_2N-CH-COOH$ with side chain CH_2–(indole ring, N–H)	indole	5.9
		R group contains sulfur		
cysteine	Cys	$H_2N-CH-COOH$ with side chain CH_2-SH	thiol	5.0
methionine	Met*	$H_2N-CH-COOH$ with side chain $CH_2-CH_2-S-CH_3$	sulfide	5.7
		R group is acidic		
aspartic acid	Asp	$H_2N-CH-COOH$ with side chain CH_2-COOH	carboxylic acid	2.8
glutamic acid	Glu	$H_2N-CH-COOH$ with side chain CH_2-CH_2-COOH	carboxylic acid	3.2
		R group is basic		
lysine	Lys*	$H_2N-CH-COOH$ with side chain $CH_2-CH_2-CH_2-CH_2-NH_2$	amino group	9.7
arginine	Arg*	$H_2N-CH-COOH$ with side chain $CH_2-CH_2-CH_2-NH-C(=NH)-NH_2$	guanidino group	10.8
histidine	His*	$H_2N-CH-COOH$ with side chain CH_2–(imidazole ring, NH, N=)	imidazole ring	7.6

24-2B ESSENTIAL AMINO ACIDS

Humans beings can synthesize about half the amino acids needed to make protein. Other amino acids, called the **essential amino acids,** must be provided in the diet. The ten essential amino acids, starred (*) in Table 24-2, are the following:

arginine (Arg)	phenylalanine (Phe)	histidine (His)
threonine (Thr)	tryptophan (Trp)	leucine (Leu)
lysine (Lys)	methionine (Met)	isoleucine (Ile)
valine (Val)		

Proteins that provide all the essential amino acids in about the right proportions for human nutrition are called **complete proteins.** Examples of complete proteins are those in meat, fish, milk, and eggs. About 50 g of complete protein per day is adequate for adult humans.

Proteins that are severely deficient in one or more of the essential amino acids are called **incomplete proteins.** If the protein in a person's diet comes mostly from one incomplete source, the amount of human protein that can be synthesized is limited by the amounts of the deficient amino acids. Plant proteins are generally incomplete. Rice, corn, and wheat are all deficient in lysine. Rice also lacks threonine, and corn also lacks tryptophan. Beans, peas, and other legumes have the most complete proteins among the common plants, but they are deficient in methionine.

Vegetarians can achieve an adequate intake of the essential amino acids if they eat many different plant foods. A more effective alternative is to supplement the vegetarian diet with a rich source of complete protein such as milk or eggs.

24-2C RARE AND UNUSUAL AMINO ACIDS

In addition to the standard amino acids, other amino acids are found in protein in much smaller quantities. For example, 4-hydroxyproline and 5-hydroxylysine are hydroxylated versions of standard amino acids. These are called *rare* amino acids, even though they are commonly found in collagen.

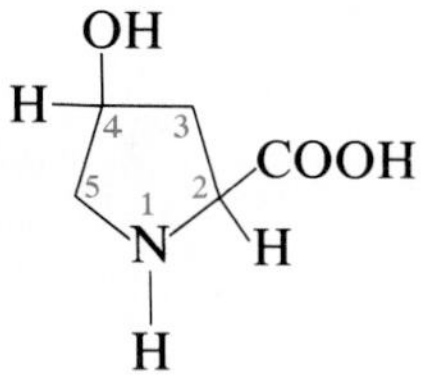

4-hydroxyproline

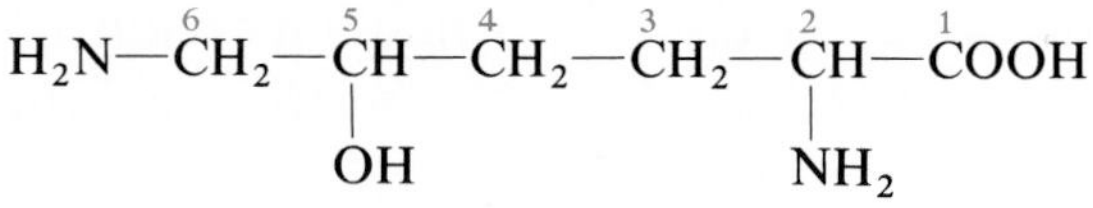

$$\overset{6}{H_2N-CH_2}-\underset{OH}{\overset{5}{CH}}-\overset{4}{CH_2}-\overset{3}{CH_2}-\underset{NH_2}{\overset{2}{CH}}-\overset{1}{COOH}$$

5-hydroxylysine

Some of the less common D enantiomers of amino acids are also found in nature. For example, D-glutamic acid is found in the cell walls of many bacteria, and D-serine is found in earthworms. Some naturally occurring amino acids are not α-amino acids. γ-Aminobutyric acid (GABA) is one of the neurotransmitters in the brain, and β-alanine is a constituent of the vitamin pantothenic acid.

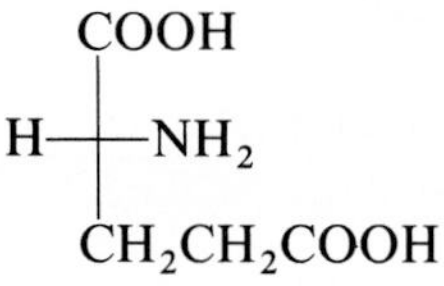

D-glutamic acid: COOH / H—C—NH_2 / CH_2CH_2COOH

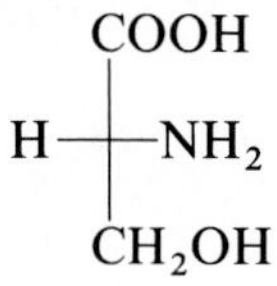

D-serine: COOH / H—C—NH_2 / CH_2OH

$$\underset{NH_2}{\overset{\gamma}{CH_2}}-\overset{\beta}{CH_2}-\overset{\alpha}{CH_2}-COOH$$

γ-aminobutyric acid

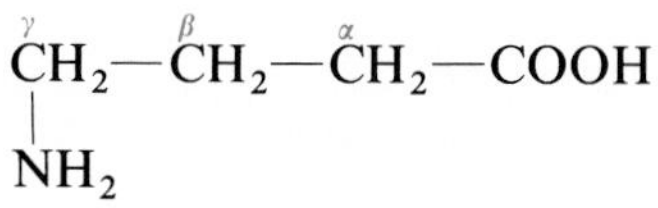

$$\underset{NH_2}{\overset{\beta}{CH_2}}-\overset{\alpha}{CH_2}-COOH$$

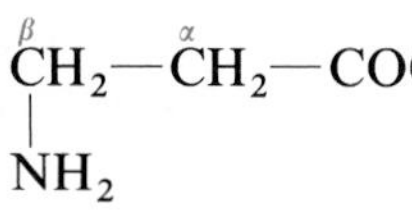

β-alanine

24-3 ACID-BASE PROPERTIES OF AMINO ACIDS

Although we commonly write amino acids with an intact carboxyl (—COOH) group and amino (—NH_2) group, their actual structure is ionic. The carboxyl group loses a proton, giving a carboxylate ion, and the amino group is protonated to an ammonium ion. This structure is called a **dipolar ion** or a **zwitterion** (German for "double ion").

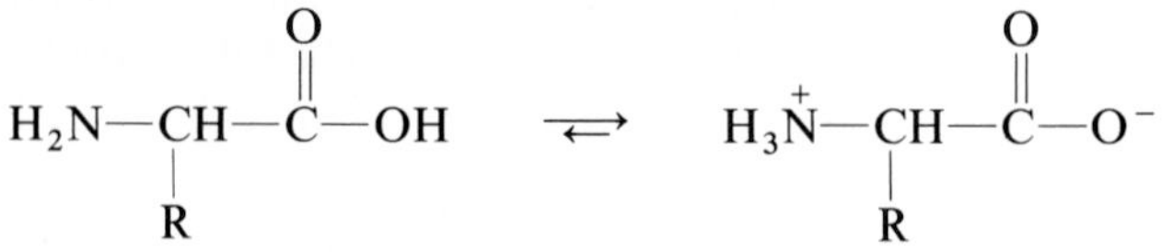

$$H_2N-\underset{R}{CH}-\overset{O}{\overset{\|}{C}}-OH \rightleftarrows H_3\overset{+}{N}-\underset{R}{CH}-\overset{O}{\overset{\|}{C}}-O^-$$

uncharged structure (minor component) — dipolar ion or zwitterion (major component)

The dipolar nature of amino acids gives them some unusual properties:

1. Amino acids have high melting points, generally over 200°C.

$$\overset{+}{H_3N}\text{—}CH_2\text{—}COO^-$$
glycine, m.p. 262°C

2. Amino acids are more soluble in water than they are in ether, dichloromethane, and other common organic solvents.
3. Amino acids have much larger dipole moments (μ) than simple amines or simple acids.

$\overset{+}{H_3N}\text{—}CH_2\text{—}COO^-$	$CH_3\text{—}CH_2\text{—}CH_2\text{—}NH_2$	$CH_3\text{—}CH_2\text{—}COOH$
glycine, $\mu = 14$ D	propylamine, $\mu = 1.4$ D	propionic acid, $\mu = 1.7$ D

4. Amino acids are less acidic than most carboxylic acids, and less basic than most amines. In fact, the acidic part of the amino acid molecule is the $\text{—}NH_3^+$ group, not a —COOH group. The basic part is the $\text{—}COO^-$ group, and not a free $\text{—}NH_2$ group.

R—COOH	$R\text{—}NH_2$	$\overset{+}{H_3N}\text{—}CH(R)\text{—}COO^-$
$pK_a = 5$	$pK_b = 4$	$pK_a = 10$ $pK_b = 12$

Because amino acids contain both acidic ($\text{—}NH_3^+$) and basic ($\text{—}COO^-$) groups, they are *amphoteric* (having both acidic and basic properties). The predominant form of the amino acid depends on the pH of the solution. In a basic solution, the $\text{—}NH_3^+$ group is deprotonated to a free $\text{—}NH_2$ group and the molecule has an overall negative charge. In an acidic solution, the $\text{—}COO^-$ group is protonated to a free —COOH group and the molecule has an overall positive charge.

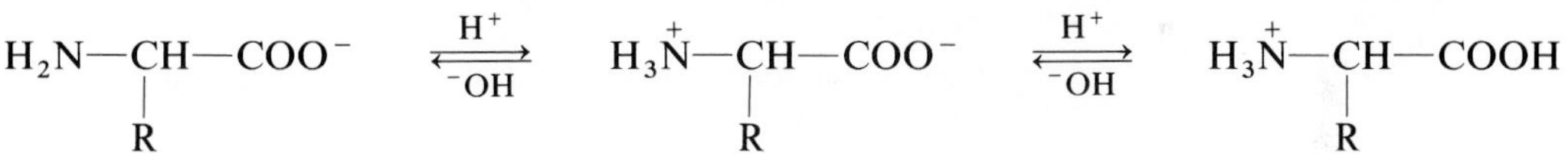

$$H_2N\text{—}CH(R)\text{—}COO^- \underset{^-OH}{\overset{H^+}{\rightleftharpoons}} \overset{+}{H_3N}\text{—}CH(R)\text{—}COO^- \underset{^-OH}{\overset{H^+}{\rightleftharpoons}} \overset{+}{H_3N}\text{—}CH(R)\text{—}COOH$$

anionic in base — neutral — cationic in acid

For example, glycine is mostly in its anionic form at pH values above 9.6, and mostly in its cationic form at pH values below 2.3. By varying the pH of the solution, we can control the charge on the molecule. This ability to control the charge can be useful for separating and identifying the amino acids in a mixture, using electrophoresis.

24-4 ISOELECTRIC POINTS AND ELECTROPHORESIS

An amino acid bears a negative charge in basic solution (high pH) and a positive charge in acidic solution (low pH). There must be an intermediate pH where the amino acid is evenly balanced between the two forms, as the dipolar zwitterion with a net charge of zero. This pH is called the **isoelectric pH** or the **isoelectric point.**

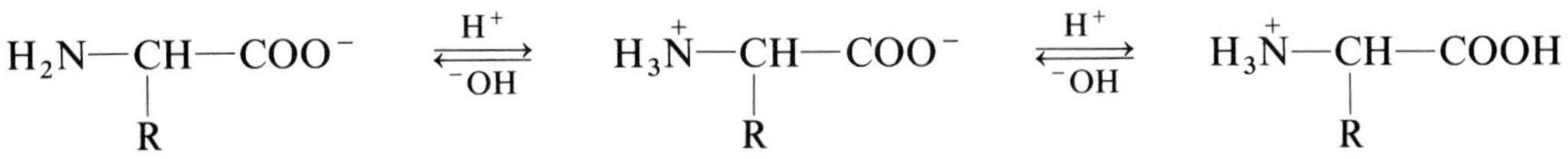

$$H_2N\text{—}CH(R)\text{—}COO^- \underset{^-OH}{\overset{H^+}{\rightleftharpoons}} \overset{+}{H_3N}\text{—}CH(R)\text{—}COO^- \underset{^-OH}{\overset{H^+}{\rightleftharpoons}} \overset{+}{H_3N}\text{—}CH(R)\text{—}COOH$$

high pH (anionic in base) — isoelectric pH (neutral) — low pH (cationic in acid)

The isoelectric points of the standard amino acids are given in Table 24-2. Notice that the isoelectric pH depends on the amino acid structure in a predictable way.

acidic amino acids:	aspartic acid (2.8), glutamic acid (3.2)
basic amino acids:	lysine (9.7), arginine (10.8), histidine (7.6)
neutral amino acids:	(5.0 to 6.3)

Aspartic acid and glutamic acid have side chains containing acidic carboxyl groups. These amino acids have acidic isoelectric points with a pH near 3. An acidic solution is needed to prevent deprotonation of the second carboxylic acid group and keep the amino acid in its neutral isoelectric state.

The basic amino acids (histidine, lysine, and arginine) have isoelectric points at pH values of 7.6, 9.7, and 10.8, respectively. These values reflect the weak basicity of the imidazole ring, the intermediate basicity of an amino group, and the strong basicity of the guanidino group. A basic solution is needed in each case to prevent protonation of the basic side chain to keep the amino acid at its isoelectronic point.

The other amino acids are considered neutral, with no strongly acidic or basic side chains. Their isoelectric points are slightly acidic (from about 5 to 6), because the $-NH_3^+$ group is slightly more acidic than the $-COO^-$ group is basic.

PROBLEM 24-2

Draw the structure of the predominant form of

(a) valine at a pH of 11. (b) proline at a pH of 2.
(c) arginine at a pH of 7. (d) glutamic acid at a pH of 7.
(e) a mixture of alanine, lysine, and aspartic acid (i) at pH 6; (ii) at pH 11; (iii) at pH 2.

PROBLEM 24-3

Draw the resonance structures of a protonated guanidino group, and explain why arginine has such a strongly basic isoelectric point.

PROBLEM 24-4

Although tryptophan contains a heterocyclic amine, it is considered a neutral amino acid. Explain why the indole nitrogen of tryptophan is more weakly basic than one of the imidazole nitrogens of histidine.

Differences in isoelectric points can be used to separate mixtures of amino acids by **electrophoresis.** A streak of the amino acid mixture is placed in the center of a layer of acrylamide gel or a piece of filter paper wet with a buffer solution. Two electrodes are placed in contact with the edges of the gel or paper, and a potential of several thousand volts is applied across the electrodes. Positively charged (cationic) amino acids are attracted to the negative electrode (the cathode), and negatively charged (anionic) amino acids are attracted to the positive electrode (the anode). An amino acid at its isoelectric point has no net charge, and it does not move.

As an example, consider a mixture of alanine, lysine, and aspartic acid in a buffer at pH 6. Alanine is at its isoelectric point, in its dipolar zwitterionic form with a net charge of zero. A pH of 6 is more acidic than the isoelectric pH for lysine (9.7), so lysine is in the cationic form. Aspartic acid has an isoelectric pH of 2.8, so it is in the anionic form.

Structure at pH 6

$$\overset{+}{H_3N}-\underset{\displaystyle CH_3}{\underset{|}{CH}}-COO^- \qquad \overset{+}{H_3N}-\underset{\displaystyle (CH_2)_4-NH_3^+}{\underset{|}{CH}}-COO^- \qquad \overset{+}{H_3N}-\underset{\displaystyle CH_2-COO^-}{\underset{|}{CH}}-COO^-$$

alanine (charge 0) lysine (charge +1) aspartic acid (charge −1)

When a voltage is applied to a mixture of alanine, lysine, and aspartic acid at pH 6, alanine does not move. Lysine moves toward the cathode, and aspartic acid moves toward the anode (Fig. 24-3). After a period of time, the separated amino acids are recovered by cutting the paper or scraping the bands out of the gel. If the electrophoresis is being used as an analytical technique (to determine the amino acids present in the mixture), the paper or gel is treated with a reagent such as ninhydrin (Section 24-9) to make the bands visible. The amino acids are then identified by comparing their positions with those of standards.

Beginning

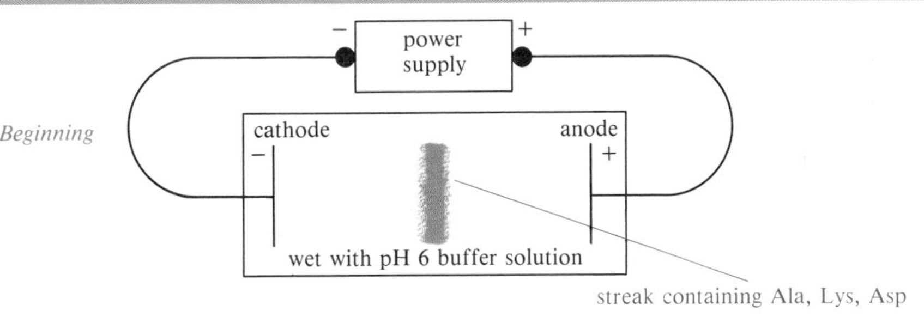

End

FIGURE 24-3 A simplified picture of the electrophoretic separation of alanine, lysine, and aspartic acid at a pH of 6. The cationic lysine is attracted to the cathode; anionic aspartic acid is attracted to the anode. Alanine is at its isoelectric point, and it does not move.

As shown in Figure 24-3, an amino acid at its isoelectric point does not move in an electric field. The isoelectric point was originally defined experimentally as the pH at which the amino acid does not move under electrophoresis. This definition is essentially the same as the one we have been using.

PROBLEM 24-5

Draw the electrophoretic separation of Ala, Lys, and Asp at a pH of 9.7.

PROBLEM 24-6

Draw the electrophoretic separation of Trp, Cys, and His at a pH of 6.0.

24-5 SYNTHESIS OF AMINO ACIDS

Naturally occurring amino acids are available by hydrolysis of protein and separation of the amino acid mixture. Even so, it is often less expensive to synthesize the pure amino acid. In some cases there is a need for an unusual amino acid or an unnatural enantiomer, and a synthesis is needed to provide a source. In this chapter we consider four methods for making amino acids. All these methods are extensions of reactions we have already studied, and references to the earlier coverage are given in each case.

24-5A REDUCTIVE AMINATION

Reductive amination of ketones and aldehydes, one of the best methods for synthesizing amines (Section 19-19), also forms amino acids. When an α-ketoacid is treated with ammonia, the ketone reacts to give an imine. The imine is reduced to an amine by treatment with hydrogen and a palladium catalyst. Under these conditions, the carboxylic acid is not reduced.

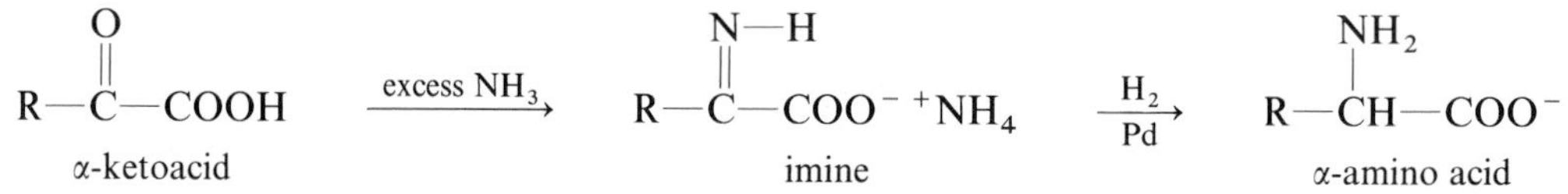

$$\underset{\alpha\text{-ketoacid}}{R-\overset{O}{\overset{\|}{C}}-COOH} \xrightarrow{\text{excess } NH_3} \underset{\text{imine}}{R-\overset{N-H}{\overset{\|}{C}}-COO^-\ {}^+NH_4} \xrightarrow[Pd]{H_2} \underset{\alpha\text{-amino acid}}{R-\overset{NH_2}{\overset{|}{C}H}-COO^-}$$

This entire synthesis is accomplished in one step by treating the α-ketoacid with ammonia and hydrogen in the presence of a palladium catalyst. The product is a racemic α-amino acid. The following reaction shows the synthesis of racemic phenylalanine from 3-phenyl-2-oxopropanoic acid.

$$\underset{\text{3-phenyl-2-oxopropanoic acid}}{Ph-CH_2-\overset{O}{\overset{\|}{C}}-COOH} \xrightarrow[Pd]{NH_3,\ H_2} \underset{\text{(D,L)-phenylalanine (ammonium salt) (30\%)}}{Ph-CH_2-\overset{NH_2}{\overset{|}{C}H}-COO^-\ {}^+NH_4}$$

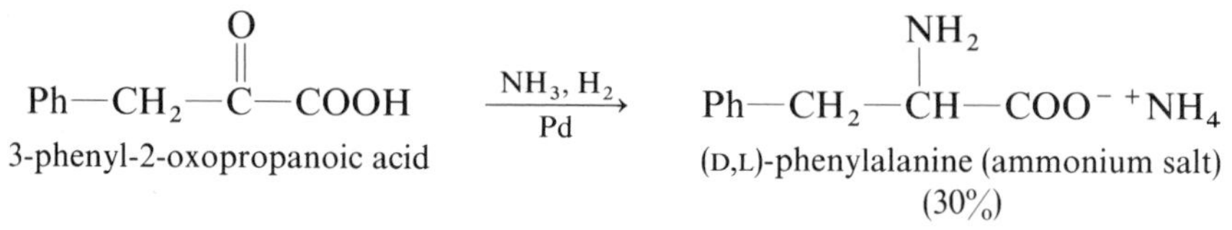

The reductive amination process resembles the biological synthesis of amino acids. We can call it a **biomimetic** ("mimicking the biological process") synthesis. The biosynthesis begins with reductive amination of α-ketoglutaric acid, an intermediate in the metabolism of carbohydrates, using ammonium ion as the aminating agent and NADH as the reducing agent. The product of this enzyme-catalyzed reaction is the pure L enantiomer of glutamic acid.

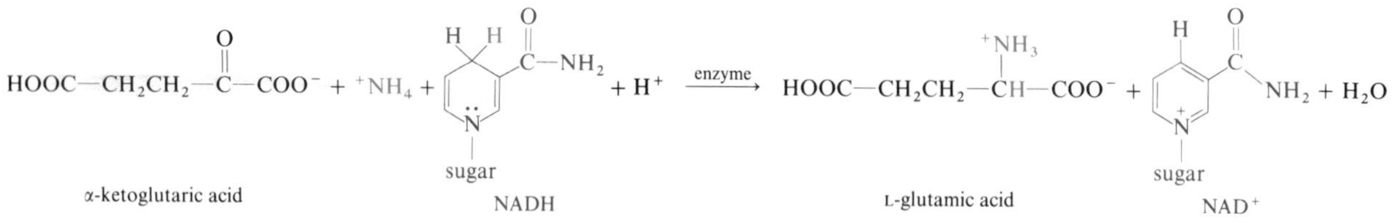

$$\underset{\alpha\text{-ketoglutaric acid}}{HOOC-CH_2CH_2-\overset{O}{\overset{\|}{C}}-COO^-} + {}^+NH_4 + \underset{\text{NADH}}{\text{NADH (sugar)}} + H^+ \xrightarrow{\text{enzyme}} \underset{\text{L-glutamic acid}}{HOOC-CH_2CH_2-\overset{^+NH_3}{\overset{|}{C}H}-COO^-} + \underset{\text{NAD}^+}{\text{NAD}^+\text{ (sugar)}} + H_2O$$

The biosynthesis of other amino acids uses L-glutamic acid as the source of the amino group. Such a reaction, moving an amino group from one molecule to another, is called a **transamination**, and the enzymes that catalyze these reactions are called **transaminases.** For example, the following reaction shows the biosynthesis of aspartic acid using glutamic acid as the nitrogen source. Once again, the enzyme-catalyzed biosynthesis gives the pure L enantiomer of the product.

$$\left.\begin{array}{c} HOOC-CH_2CH_2-\overset{\overset{+}{N}H_3}{\overset{|}{C}H}-COO^- \\ \text{L-glutamic acid} \\ + \\ HOOC-CH_2-\overset{\overset{O}{\|}}{C}-COO^- \\ \text{oxaloacetic acid} \end{array}\right\} \xrightarrow{\text{transaminase}} \left\{\begin{array}{c} HOOC-CH_2CH_2-\overset{\overset{O}{\|}}{C}-COO^- \\ \alpha\text{-ketoglutaric acid} \\ + \\ HOOC-CH_2-\overset{\overset{+}{N}H_3}{\overset{|}{C}H}-COO^- \\ \text{L-aspartic acid} \end{array}\right.$$

PROBLEM 24-7

Show how the following amino acids might be formed in the laboratory by reductive amination of the appropriate α-ketoacid.

(a) alanine (b) leucine (c) serine (d) glutamine

24-5B AMINATION OF AN α-HALO ACID

The Hell-Volhard-Zelinsky reaction, discussed in Section 22-3, is an effective method for introducing a bromine atom at the α position of a carboxylic acid. The resulting α-bromo acid is converted to a racemic α-amino acid by direct amination using a large excess of ammonia.

$$\underset{\text{carboxylic acid}}{R-CH_2-\overset{\overset{O}{\|}}{C}-OH} \xrightarrow[\text{(2) } H_2O]{\text{(1) } Br_2/PBr_3} \underset{\alpha\text{-bromo acid}}{R-\overset{Br}{\overset{|}{C}H}-\overset{\overset{O}{\|}}{C}-OH} \xrightarrow[\text{(large excess)}]{NH_3} \underset{\substack{\text{(D,L)-}\alpha\text{-amino acid} \\ \text{(ammonium salt)}}}{R-\overset{NH_2}{\overset{|}{C}H}-\overset{\overset{O}{\|}}{C}-O^-\ {}^+NH_4}$$

In Section 19-22 we saw that direct amination is often a poor synthesis of amines, giving large amounts of overalkylated products. In this case, however, the reaction gives acceptable yields because a large excess of ammonia is used, making ammonia the nucleophile most likely to displace bromine. Also, the adjacent carboxylate ion in the product reduces the nucleophilicity of the amino group. The following sequence shows the bromination of 3-phenylpropanoic acid, followed by displacement of bromide ion, to form the ammonium salt of racemic phenylalanine.

$$\underset{\text{3-phenylpropanoic acid}}{Ph-CH_2-CH_2-COOH} \xrightarrow[\text{(2) } H_2O]{\text{(1) } Br_2/PBr_3} Ph-CH_2-\overset{Br}{\overset{|}{C}H}-COOH \xrightarrow{\text{excess } NH_3} \underset{\substack{\text{(D,L)-phenylalanine (salt)} \\ (30\text{–}50\%)}}{Ph-CH_2-\overset{NH_2}{\overset{|}{C}H}-COO^-\ {}^+NH_4}$$

PROBLEM 24-8

Show how you would use bromination followed by amination to synthesize the following amino acids.

(a) glycine (b) leucine (c) valine (d) glutamic acid

24-5C THE GABRIEL-MALONIC ESTER SYNTHESIS

One of the best methods of amino acid synthesis is a combination of the Gabriel synthesis of amines (Section 19-23) with the malonic ester synthesis of carboxylic acids (Section 22-17). The conventional malonic ester synthesis involves the alkylation of diethyl malonate, followed by hydrolysis and decarboxylation to give an alkylated acetic acid. In the Gabriel-malonic ester synthesis, the starting material is *N*-phthalimidomalonic ester, obtained by the reaction of potassium phthalimide with a brominated malonic ester.

$$\underset{\text{malonic ester}}{\mathrm{EtOOC{-}CH_2{-}COOEt}} \xrightarrow{Br_2} \mathrm{EtOOC{-}CHBr{-}COOEt} + \underset{\text{potassium phthalimide}}{\mathrm{PhthN^-K^+}} \longrightarrow \underset{N\text{-phthalimidomalonic ester}}{\mathrm{PhthN{-}CH(COOEt)_2}}$$

N-Phthalimidomalonic ester is alkylated in the same way as malonic ester. When the alkylated *N*-pthalimidomalonic ester is hydrolyzed, the phthalimido group is hydrolyzed along with the ester groups. The product is an alkylated aminomalonic acid. Decarboxylation gives a racemic α-amino acid.

The Gabriel-malonic ester synthesis

temporary ester group

$$\underset{N\text{-phthalimidomalonic ester}}{\mathrm{PhthN{-}CH(COOEt)_2}} \xrightarrow[(2)\ R{-}X]{(1)\ \text{base}} \underset{\text{alkylated}}{\mathrm{PhthN{-}CR(COOEt)_2}} \xrightarrow{H_3O^+} \underset{\text{hydrolyzed}}{\left[\mathrm{H_3\overset{+}{N}{-}CR(COOH)_2}\right]} \xrightarrow{\text{heat}} \underset{\alpha\text{-amino acid}}{\mathrm{H_3\overset{+}{N}{-}CHR{-}COOH}} + CO_2\uparrow$$

The Gabriel-malonic ester synthesis is used to synthesize many amino acids that cannot be formed by direct amination of haloacids. The following example shows the synthesis of methionine, which is formed in very poor yield by direct amination.

$$\mathrm{PhthN{-}CH(COOEt)_2} \xrightarrow[(2)\ Cl{-}CH_2CH_2SCH_3]{(1)\ NaOEt} \mathrm{PhthN{-}C(COOEt)_2{-}CH_2CH_2SCH_3} \xrightarrow[\text{heat}]{H_3O^+} \underset{(\text{D,L})\text{-methionine } (50\%)}{\mathrm{H_3\overset{+}{N}{-}CH(COOH){-}CH_2CH_2SCH_3}}$$

PROBLEM 24-9

Show how the Gabriel-malonic ester synthesis could be used to form

(a) valine (b) phenylalanine
(c) glutamic acid (d) leucine

24-5D THE STRECKER SYNTHESIS

The first known synthesis of an amino acid occurred in 1850 in the laboratory of Adolph Strecker in Tubingen, Germany. Strecker added acetaldehyde to an aqueous solution of ammonia and HCN. The product was α-amino propionitrile, which Strecker hydrolyzed to racemic alanine.

The Strecker synthesis of alanine

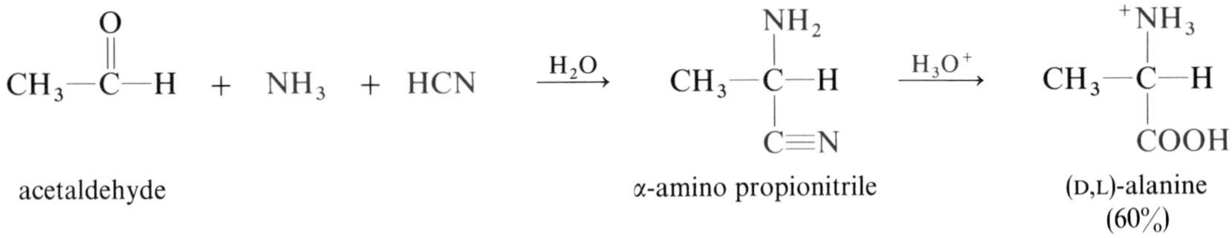

$$CH_3-\overset{\overset{\displaystyle O}{\|}}{C}-H + NH_3 + HCN \xrightarrow{H_2O} CH_3-\underset{\underset{\displaystyle C\equiv N}{|}}{\overset{\overset{\displaystyle NH_2}{|}}{C}}-H \xrightarrow{H_3O^+} CH_3-\underset{\underset{\displaystyle COOH}{|}}{\overset{\overset{\displaystyle {}^+NH_3}{|}}{C}}-H$$

acetaldehyde — α-amino propionitrile — (D,L)-alanine (60%)

The **Strecker synthesis** can be used to form a large number of amino acids from the appropriate aldehydes. The mechanism of the Strecker synthesis is shown below. First, the aldehyde reacts with ammonia to give an imine. The imine is a nitrogen analog of a carbonyl group, and it is electrophilic. Attack of cyanide ion on the protonated imine gives the α-amino nitrile. This mechanism is similar to that for formation of a cyanohydrin, except that in the Strecker synthesis cyanide ion attacks an imine rather than the aldehyde itself.

Step 1: Formation of the imine (mechanism in Section 18-16)

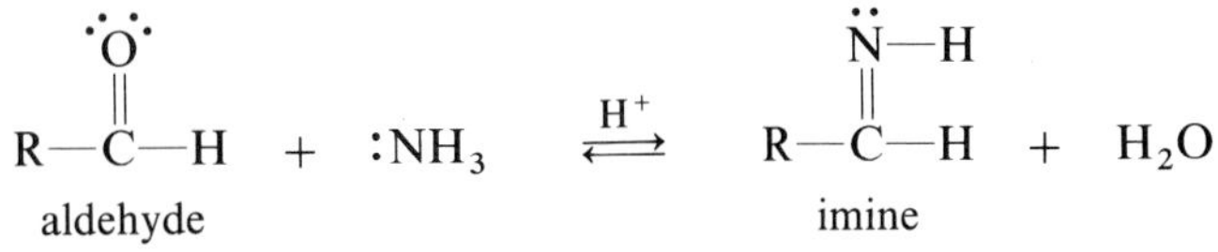

$$R-\overset{\overset{\displaystyle \ddot{O}:}{\|}}{C}-H + :NH_3 \underset{}{\overset{H^+}{\rightleftharpoons}} R-\overset{\overset{\displaystyle \ddot{N}-H}{\|}}{C}-H + H_2O$$

aldehyde — imine

Step 2: Attack by cyanide ion

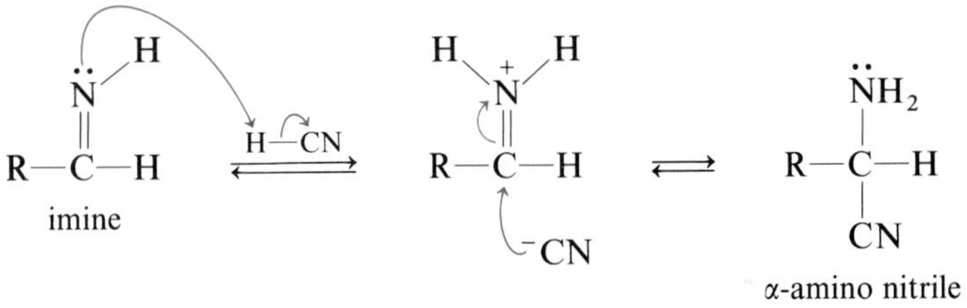

$$R-\overset{\overset{\displaystyle \ddot{N}-H}{\|}}{C}-H \overset{H-CN}{\rightleftharpoons} R-\overset{\overset{\displaystyle H_2\overset{+}{N}}{\|}}{C}-H \;\; {}^-CN \rightleftharpoons R-\underset{\underset{\displaystyle CN}{|}}{\overset{\overset{\displaystyle \ddot{N}H_2}{|}}{C}}-H$$

imine — α-amino nitrile

In a separate step, hydrolysis of the α-amino nitrile (Section 21-7D) gives an α-amino acid.

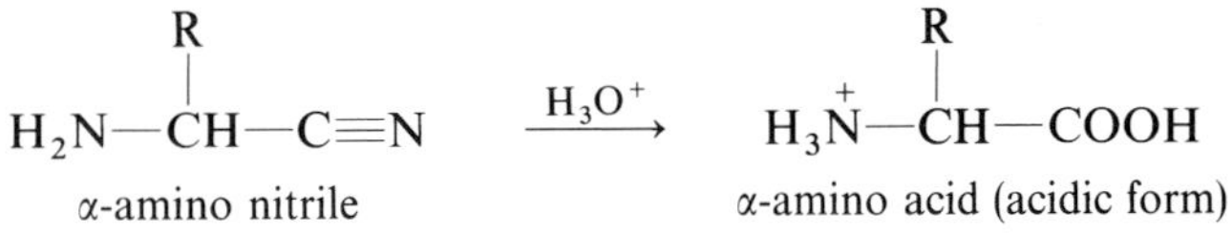

$$H_2N-\overset{\overset{\displaystyle R}{|}}{CH}-C\equiv N \xrightarrow{H_3O^+} H_3\overset{+}{N}-\overset{\overset{\displaystyle R}{|}}{CH}-COOH$$

α-amino nitrile — α-amino acid (acidic form)

PROBLEM 24-10

(a) Show how the Strecker synthesis would be used to make phenylalanine.
(b) Propose a mechanism for each step in the synthesis in part (a).

PROBLEM 24-11

Show how you would use the Strecker synthesis to make the following amino acids.

(a) leucine (b) glycine (c) valine

SUMMARY OF SYNTHESES OF AMINO ACIDS (Section 24-5)

1. Reductive amination

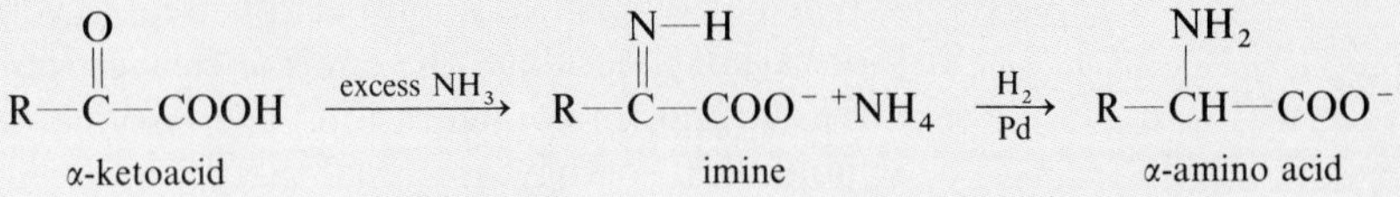

$$R-\overset{\overset{\displaystyle O}{\|}}{C}-COOH \xrightarrow{\text{excess } NH_3} R-\overset{\overset{\displaystyle N-H}{\|}}{C}-COO^-\,{}^+NH_4 \xrightarrow[Pd]{H_2} R-\overset{\overset{\displaystyle NH_2}{|}}{CH}-COO^-$$

α-ketoacid — imine — α-amino acid

2. Amination of an α-haloacid

$$\underset{\text{carboxylic acid}}{R-CH_2-\overset{O}{\overset{\|}{C}}-OH} \xrightarrow[(2)\ H_2O]{(1)\ Br_2/PBr_3} \underset{\alpha\text{-bromoacid}}{R-\overset{Br}{\overset{|}{C}H}-\overset{O}{\overset{\|}{C}}-OH} \xrightarrow[\text{(large excess)}]{NH_3} \underset{\substack{\text{(D,L)-}\alpha\text{-amino acid}\\ \text{(ammonium salt)}}}{R-\overset{NH_2}{\overset{|}{C}H}-\overset{O}{\overset{\|}{C}}-O^-\ {}^+NH_4}$$

3. The Gabriel-malonic ester synthesis

temporary ester group

$$\underset{N\text{-phthalimidomalonic ester}}{\text{PhthN}-CH(COOEt)_2} \xrightarrow[(2)\ R-X]{(1)\ \text{base}} \underset{\text{alkylated}}{\text{PhthN}-C(R)(COOEt)_2} \xrightarrow[\text{heat}]{H_3O^+}$$

$$\underset{\text{hydrolyzed}}{\left[H_3\overset{+}{N}-C(R)(COOH)_2\right]} \xrightarrow{\text{heat}} \underset{\alpha\text{-amino acid}}{H_3\overset{+}{N}-\overset{H}{\overset{|}{C}}(R)-COOH} + CO_2\uparrow$$

4. The Strecker synthesis

$$\underset{\text{aldehyde}}{R-\overset{O}{\overset{\|}{C}}-H} + NH_3 + HCN \xrightarrow{H_2O} \underset{\alpha\text{-amino nitrile}}{R-\overset{NH_2}{\overset{|}{C}}(H)-C\equiv N} \xrightarrow{H_3O^+} \underset{\alpha\text{-amino acid}}{R-\overset{^+NH_3}{\overset{|}{C}}(H)-COOH}$$

24-6 RESOLUTION OF AMINO ACIDS

All of the laboratory syntheses of amino acids described in Section 24-5 produce racemic products. In most cases only the L enantiomers are biologically active; the D enantiomers may actually be toxic. The pure L enantiomers are also needed for peptide synthesis if the peptide product is to have the activity of the natural material. Therefore, we must be able to resolve a racemic amino acid into its enantiomers.

In many cases, amino acids can be resolved by the methods we have already discussed (Section 6-15). If a racemic amino acid is converted to a salt with an optically pure chiral acid or base, two diastereomeric salts are formed. These salts can be separated by physical means such as selective crystallization or chromatography. The pure enantiomers are then regenerated from the separated diastereomeric salts. Strychnine and brucine are examples of naturally occurring optically active bases, and tartaric acid is used as an optically active acid.

Enzymatic resolution methods are also used to separate the enantiomers of amino acids. Enzymes are chiral molecules with very specific catalytic activities. When an acetylated amino acid is treated with an enzyme like hog kidney acylase or carboxypeptidase, the enzyme cleaves the acyl group from the molecules with the natural (L) configuration. The enzyme does not recognize the D-amino acid molecules, so they are unaffected. The resulting mixture of acetylated D-amino acid and deacylated L-amino acid is easily separated. Figure 24-4 shows how this selective enzymatic deacylation is accomplished.

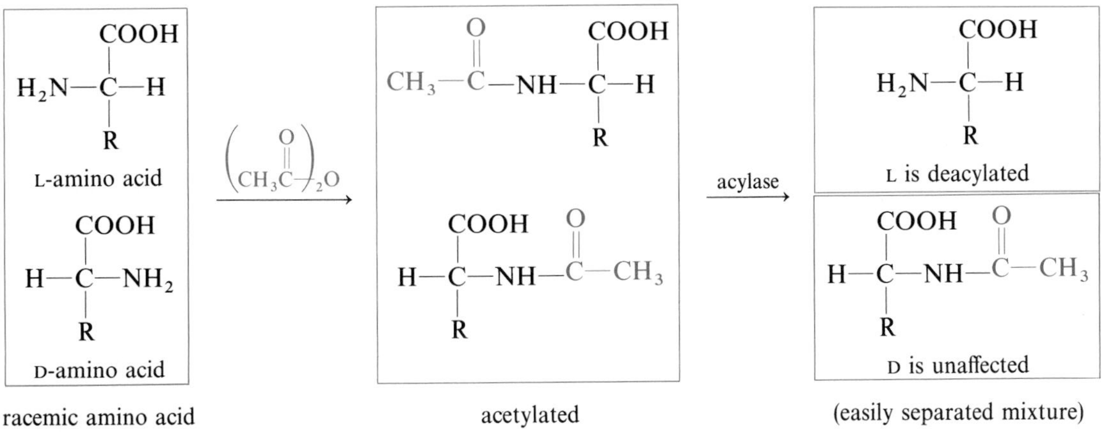

FIGURE 24-4 An acylase enzyme (such as hog kidney acylase or carboxypeptidase) deacylates only the natural L-amino acid.

PROBLEM 24-12

Suggest how you would separate the free L-amino acid from its acetylated D enantiomer in Figure 24-4.

24-7 REACTIONS OF AMINO ACIDS

Amino acids undergo many of the standard reactions of both amines and carboxylic acids. The conditions for some of these reactions must be carefully selected, however, so that the amino group does not interfere with a carboxyl group reaction, and vice versa. We will consider two of the most useful reactions, esterification of the carboxyl group and acylation of the amino group. Amino acids also undergo reactions that are specific to the α-amino acid structure. One of these unique amino acid reactions is the formation of a colored product when treated with ninhydrin.

24-7A ESTERIFICATION OF THE CARBOXYL GROUP

Like monofunctional carboxylic acids, amino acids are esterified by treatment with a large excess of an alcohol and an acidic catalyst (often gaseous HCl). Under these acidic conditions, the amino group is present in its protonated ($—NH_3^+$) form, and it does not interfere with the esterification. The following example illustrates the esterification of an amino acid.

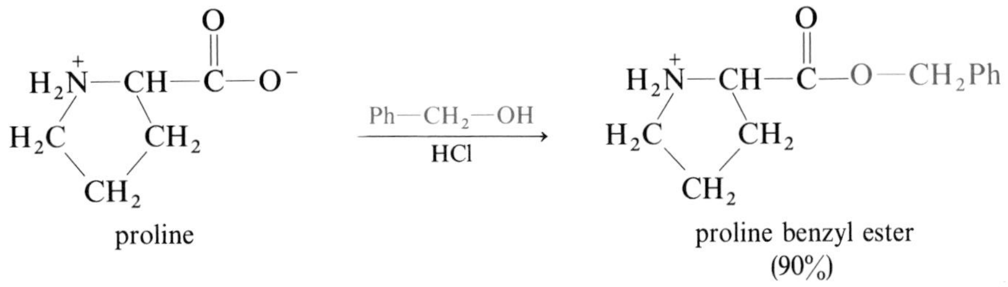

Esters of amino acids are often used as protected derivatives to prevent the carboxyl group from reacting in some undesired manner. Methyl, ethyl, and benzyl esters are the most common protecting groups. Aqueous acid can be used to hydrolyze the ester and regenerate the free amino acid.

$$\underset{\text{phenylalanine ethyl ester}}{H_3\overset{+}{N}-\underset{\underset{CH_2-Ph}{|}}{CH}-\overset{\overset{O}{\|}}{C}-OCH_2CH_3} \xrightarrow{H_3O^+} \underset{\text{phenylalanine}}{H_3\overset{+}{N}-\underset{\underset{CH_2-Ph}{|}}{CH}-\overset{\overset{O}{\|}}{C}-OH} + CH_3CH_2-OH$$

Benzyl esters are particularly useful as protecting groups because they can be removed either by acidic hydrolysis or by neutral **hydrogenolysis** ("breaking apart by the addition of hydrogen"). Catalytic hydrogenation cleaves the benzyl ester, converting the benzyl group to toluene and leaving the deprotected amino acid. Although the mechanism of this hydrogenolysis is not well known, it apparently hinges on the ease of formation of benzylic intermediates.

$$\underset{\text{phenylalanine benzyl ester}}{H_3\overset{+}{N}-\underset{\underset{CH_2-Ph}{|}}{CH}-\overset{\overset{O}{\|}}{C}-OCH_2-C_6H_5} \xrightarrow{H_2,\ Pd} \underset{\text{phenylalanine}}{H_3\overset{+}{N}-\underset{\underset{CH_2-Ph}{|}}{CH}-\overset{\overset{O}{\|}}{C}-O^-} + \underset{\text{toluene}}{CH_3-C_6H_5}$$

PROBLEM 24-13
Give a mechanism for the acid-catalyzed hydrolysis of phenylalanine ethyl ester.

PROBLEM 24-14
Give equations for the formation and hydrogenolysis of glutamine benzyl ester.

24-7B ACYLATION OF THE AMINO GROUP: FORMATION OF AMIDES

Just as the carboxyl group is esterified by an alcohol, the amino group is converted to an amide by treatment with an appropriate acylating agent. Acylation of the amino group is often done to protect it from unwanted nucleophilic reactions. A wide variety of acid chlorides and acid anhydrides are used for acylation of the amino group. Benzyl chloroformate acylates the amino group to give a benzyloxycarbonyl derivative, useful in peptide synthesis (Section 24-10).

$$\underset{\text{histidine}}{H_2\ddot{N}-\underset{\underset{CH_2-\text{(imidazolyl)}}{|}}{CH}-COOH} \xrightarrow[\text{(acetic anhydride)}]{(CH_3-\overset{O}{\overset{\|}{C}})_2O} \underset{N\text{-acetylhistidine}}{CH_3-\overset{\overset{O}{\|}}{C}-NH-\underset{\underset{CH_2-\text{(imidazolyl)}}{|}}{CH}-COOH}$$

$$\underset{\text{leucine}}{H_2\ddot{N}-\underset{\underset{CH_2CH(CH_3)_2}{|}}{CH}-COOH} \xrightarrow[\text{(benzyl chloroformate)}]{PhCH_2O\overset{O}{\overset{\|}{C}}-Cl} \underset{N\text{-benzyloxycarbonyl leucine (90\%)}}{PhCH_2O-\overset{\overset{O}{\|}}{C}-NH-\underset{\underset{CH_2CH(CH_3)_2}{|}}{CH}-COOH}$$

The amino group of the N-benzyloxycarbonyl derivative is protected as the amide half of a carbamate ester (a urethane, Section 21-16), which is more easily hydrolyzed than most amides. In addition, the ester half of this urethane is a benzyl ester that can undergo hydrogenolysis. Catalytic hydrogenolysis of the N-benzyloxycarbonyl amino acid gives an unstable carbamic acid that quickly decarboxylates to give the deprotected amino acid.

$$\text{PhCH}_2\text{—O—C(=O)—NH—CH(CH}_2\text{CH(CH}_3)_2\text{)—COOH} \xrightarrow{H_2,\ Pd} \text{PhCH}_3 + \left[\text{HO—C(=O)—NH—CH(CH}_2\text{CH(CH}_3)_2\text{)—COOH}\right] \longrightarrow CO_2\uparrow + \text{H}_2\text{N—CH(CH}_2\text{CH(CH}_3)_2\text{)—COOH}$$

N-benzyloxycarbonyl leucine — toluene — a carbamic acid — leucine

PROBLEM 24-15
Give equations for the formation and hydrogenolysis of N-benzyloxycarbonyl methionine.

24-7C REACTION WITH NINHYDRIN

Ninhydrin is a common reagent for visualizing the spots or bands of amino acids that have been separated by chromatography or by electrophoresis. When ninhydrin reacts with an amino acid, one of the products is a deep-violet, highly resonance-stabilized anion called *Ruhemann's purple*. Ninhydrin produces this same purple dye regardless of the structure of the original amino acid. The amino acid side chain is lost as an aldehyde.

Reaction of an amino acid with ninhydrin

$$\text{H}_2\text{N—CH(R)—COOH} + 2\ \text{ninhydrin} \xrightarrow{\text{pyridine}} \text{Ruhemann's purple} + CO_2\uparrow + \text{R—CHO}$$

amino acid — ninhydrin — Ruhemann's purple

The reaction of amino acids with ninhydrin can be used to detect amino acids on a wide variety of substrates. For example, if a kidnapper touches a ransom note with his fingers, the dermal ridges on his fingers leave traces of amino acids from skin secretions. Treatment of the paper with ninhydrin and pyridine causes these secretions to turn purple, forming a visible fingerprint.

PROBLEM 24-16
Use resonance structures to show the delocalization of the negative charge in the Ruhemann's purple anion.

SUMMARY OF REACTIONS OF AMINO ACIDS (Section 24-7).

1. Esterification of the carboxyl group

$$\overset{+}{\text{H}_3\text{N}}\text{—CH(R)—C(=O)—O}^- + \text{R'—OH} \xrightarrow{H^+} \overset{+}{\text{H}_3\text{N}}\text{—CH(R)—C(=O)—O—R'} + H_2O$$

amino acid — alcohol — amino ester

2. Acylation of the amino group: formation of amides

$$\underset{\text{amino acid}}{H_2N-\overset{\overset{\displaystyle R}{|}}{C}H-\overset{\overset{\displaystyle O}{\|}}{C}-OH} + \underset{\text{acylating agent}}{R'-\overset{\overset{\displaystyle O}{\|}}{C}-X} \longrightarrow \underset{\text{acylated amino acid}}{R'-\overset{\overset{\displaystyle O}{\|}}{C}-NH-\overset{\overset{\displaystyle R}{|}}{C}H-\overset{\overset{\displaystyle O}{\|}}{C}-OH} + H-X$$

3. Reaction with ninhydrin

$$\underset{\text{amino acid}}{H_2N-\underset{\underset{\displaystyle R}{|}}{C}H-COOH} + 2\ \underset{\text{ninhydrin}}{\text{(ninhydrin)}} \xrightarrow{\text{pyridine}} \underset{\text{Ruhemann's purple}}{\text{(Ruhemann's purple)}} + R-CHO + CO_2\uparrow$$

4. Formation of peptide bonds

$$H_3\overset{+}{N}-\underset{\underset{\displaystyle R^1}{|}}{C}H-\overset{\overset{\displaystyle O}{\|}}{C}-O^- + H_3\overset{+}{N}-\underset{\underset{\displaystyle R^2}{|}}{C}H-\overset{\overset{\displaystyle O}{\|}}{C}-O^- \xrightarrow{\text{loss of } H_2O} H_3\overset{+}{N}-\underset{\underset{\displaystyle R^1}{|}}{C}H-\overset{\overset{\displaystyle O}{\|}}{C}-\underbrace{}_{\text{peptide bond}}NH-\underset{\underset{\displaystyle R^2}{|}}{C}H-\overset{\overset{\displaystyle O}{\|}}{C}-O^-$$

peptide bond

Amino acids also undergo many other common reactions of amines and acids.

24-8 STRUCTURE AND NOMENCLATURE OF PEPTIDES AND PROTEINS

24-8A PEPTIDE STRUCTURE

Now we discuss the most important reaction of amino acids: the formation of peptide bonds. Amines and acids can condense, with the loss of water, to form amides.

$$\underset{\text{acid}}{R-\overset{\overset{\displaystyle O}{\|}}{C}-OH} + \underset{\text{amine}}{H_2\ddot{N}-R'} \longrightarrow \underset{\text{salt}}{R-\overset{\overset{\displaystyle O}{\|}}{C}-O^-\ H_3\overset{+}{N}-R'} \xrightarrow{\text{heat}} \underset{\text{amide}}{R-\overset{\overset{\displaystyle O}{\|}}{C}-\ddot{N}H-R'} + H_2O$$

Recall from Section 21-5A that amides are the most stable acid derivatives. This stability is partly due to the strong resonance interaction between the nonbonding electrons on nitrogen and the carbonyl group. The amide nitrogen is no longer a strong base, and the C—N bond is found to undergo restricted rotation because of its partial double-bond character.

$$\left[\ R-\overset{\overset{\displaystyle :\ddot{O}:}{\|}}{C}-\underset{\underset{\displaystyle R}{|}}{\ddot{N}}-R \longleftrightarrow R-\overset{\overset{\displaystyle :\ddot{O}:^-}{|}}{C}=\underset{\underset{\displaystyle R}{|}}{\overset{+}{N}}-R\ \right]$$

Having both an amino group and a carboxyl group, an amino acid is ideally suited to form an amide linkage. Under the proper conditions, the amino group of one molecule condenses with the carboxyl group of another. The product is an amide called a **dipeptide** because it consists of two amino acids. The amide linkage between the two amino acids is called a **peptide bond.** Although it has a special name, a peptide bond is just like the other amide bonds we have studied.

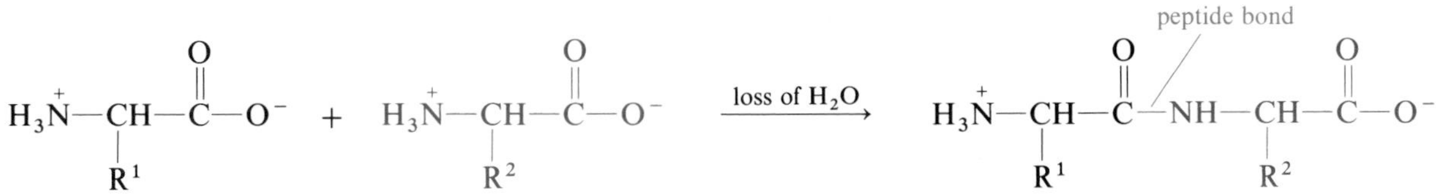

$$H_3\overset{+}{N}-CH(R^1)-C(=O)-O^- + H_3\overset{+}{N}-CH(R^2)-C(=O)-O^- \xrightarrow{\text{loss of } H_2O} H_3\overset{+}{N}-CH(R^1)-C(=O)-NH-CH(R^2)-C(=O)-O^-$$

(peptide bond: the C—NH bond formed)

In this manner, any number of amino acids can be bonded in a continuous chain. A **peptide** is any polymer of amino acids linked by amide bonds between the amino group of each amino acid and the carboxyl group of the neighboring amino acid. Each amino acid unit in the peptide is called a *residue.* A **polypeptide** is a peptide containing many amino acid residues but usually having a molecular weight of less than about 5000. **Proteins** contain larger numbers of amino acid units, with molecular weights ranging from about 6000 to about 40,000,000. The term **oligopeptide** is occasionally used for peptides containing about four to ten amino acid residues. Figure 24-5 shows the structure of the nonapeptide bradykinin, a human hormone that helps to control blood pressure.

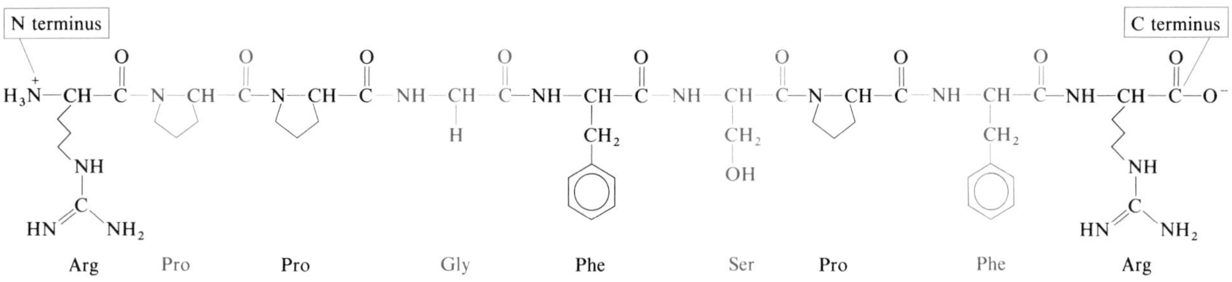

FIGURE 24-5 The human hormone bradykinin is a nonapeptide with a free $-NH_3^+$ at its N terminus and a free $-COO^-$ at its C terminus.

The end of the peptide with the free amino group ($-NH_3^+$) is called the **N-terminal end** or the **N terminus,** and the end with the free carboxyl group ($-COO^-$) is called the **C-terminal end** or the **C terminus.** Peptide structures are generally drawn with the N terminus at the left and the C terminus at the right, as bradykinin is drawn in Figure 24-5.

24-8B PEPTIDE NOMENCLATURE

Peptides are named beginning at the N terminus, and the names of the amino acid residues involved in amide linkages (all except the last) are given the —*yl* suffix of acyl groups. For example, the following dipeptide is named alanylserine. The alanine residue has the -*yl* suffix because it has acylated the nitrogen of serine.

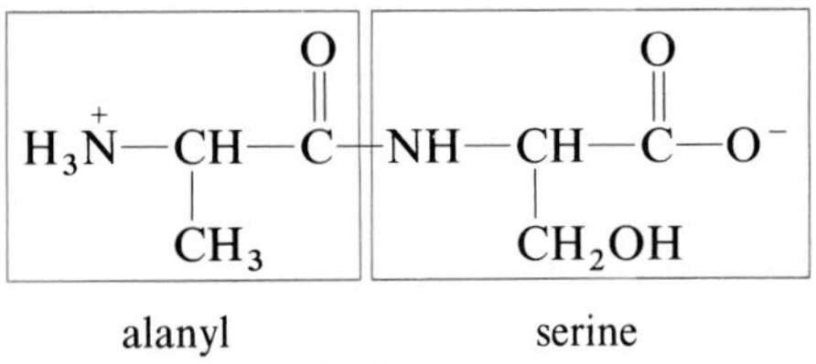

$$H_3\overset{+}{N}-CH(CH_3)-C(=O)-NH-CH(CH_2OH)-C(=O)-O^-$$

alanyl — serine

Ala-Ser

Bradykinin (Fig. 24-5) is named as follows (without any spaces):

arginyl prolyl prolyl glycyl phenylalanyl seryl prolyl phenylalanyl arginine

This is a cumbersome and awkward name. A shorthand system is more convenient, representing each amino acid by its three-letter abbreviation. These abbreviations, given in Table 24-2, are generally the first three letters of the name. Once again, the amino acids are arranged from the N terminus at the left to the C terminus at the right. Bradykinin has the following abbreviated name:

Arg-Pro-Pro-Gly-Phe-Ser-Pro-Phe-Arg

PROBLEM 24-17

Draw the complete structures of the following peptides.

(a) Thr-Phe-Met (b) serylarginylglycylphenylalanine

24-8C DISULFIDE LINKAGES

Amide linkages (peptide bonds) form the backbone of the amino acid chains we call peptides and proteins. A second kind of covalent bond is possible between any cysteine residues present. Cysteine residues can form **disulfide bridges** (also called **disulfide linkages**) to join two chains or link a single chain into a ring.

Mild oxidation converts two molecules of a thiol to a disulfide, forming a disulfide linkage between the two thiol molecules. This reaction is reversible, and a mild reduction cleaves the disulfide.

$$\underset{\text{two molecules of thiol}}{R{-}SH \;+\; HS{-}R} \;\underset{[\text{reduction}]}{\overset{[\text{oxidation}]}{\rightleftharpoons}}\; \underset{\text{disulfide}}{R{-}S{-}S{-}R} \;+\; H_2O$$

Similarly, two cysteine sulfhydryl (—SH) groups undergo oxidation to give a disulfide-linked pair of amino acids. This disulfide-linked dimer of cysteine is called *cystine*. Figure 24-6 shows the formation of a cystine disulfide bridge linking two peptide chains.

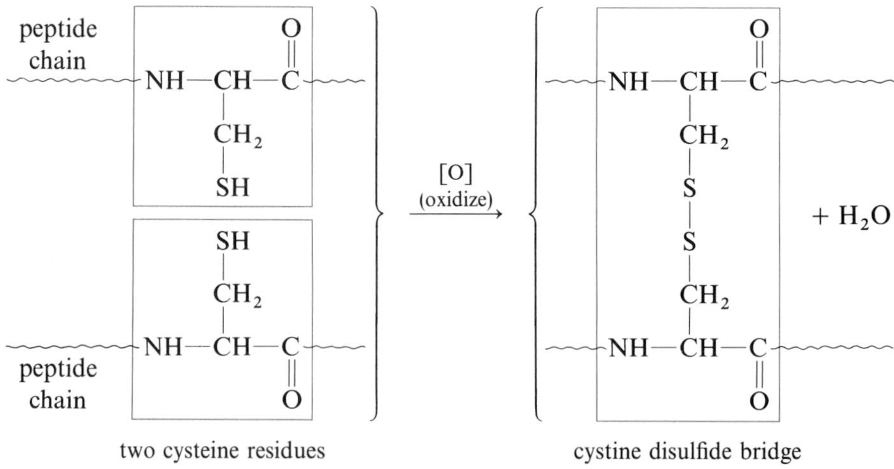

FIGURE 24-6 Cystine, a dimer of cysteine, results when two cysteine residues undergo oxidation to form a disulfide bridge.

If two cysteine residues form a disulfide bridge within a single peptide chain, the result is a cyclic peptide. Figure 24-7 shows the structure of bovine oxytocin, a peptide hormone that serves to regulate the production of milk in cows. Oxytocin is a nonapeptide with two cysteine residues (at positions 1 and 6) linking part of the molecule in a large ring. In drawing the structure of a complicated peptide, arrows are often used to connect the amino acids, showing the direction from N terminus to C terminus. Notice that the C terminus of oxytocin is a primary amide ($Gly \cdot NH_2$) rather than a free carboxyl group.

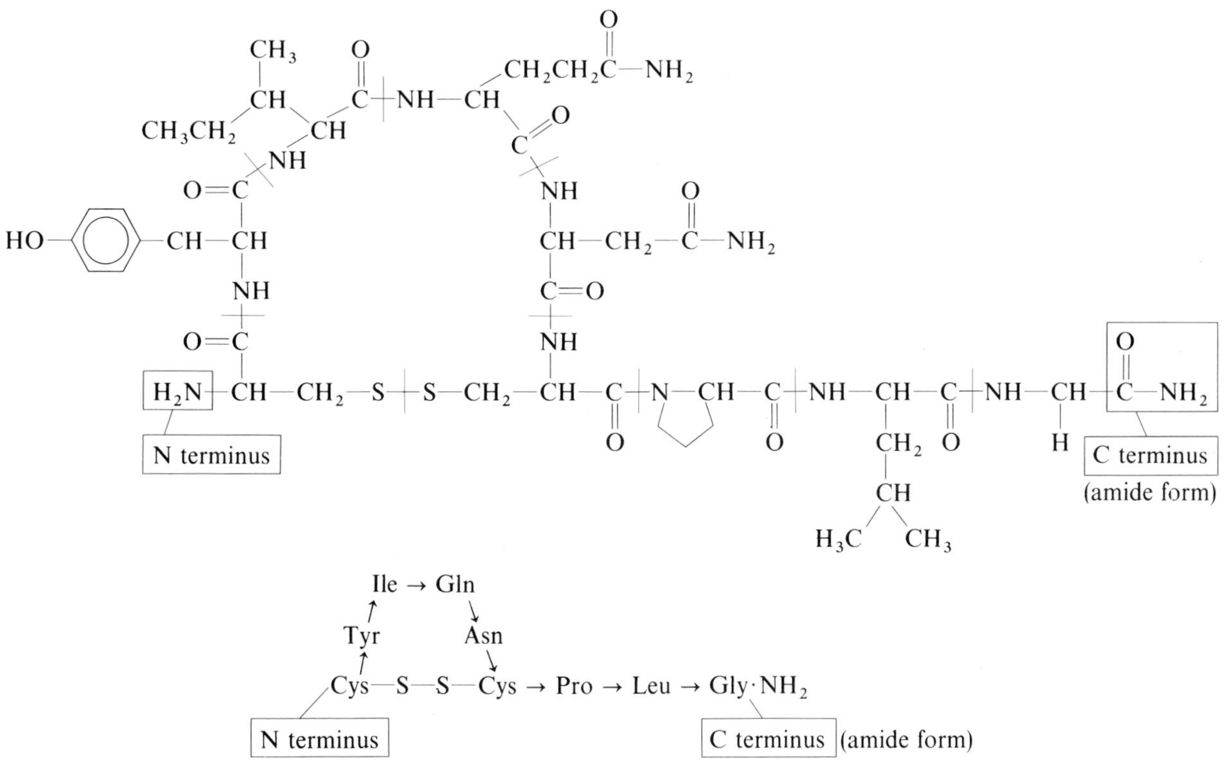

FIGURE 24-7 Structure of bovine oxytocin. A disulfide linkage holds part of the molecule in a large ring.

Figure 24-8 shows the structure of bovine insulin, a more complex peptide hormone that regulates glucose metabolism. Insulin is composed of two separate peptide chains, the *A chain,* containing 21 amino acid residues, and the *B chain,* containing 30. The A and B chains are joined at two positions by disulfide bridges, and the A chain has an additional disulfide bond that holds it in a ring. The C-terminal amino acids of both chains occur as primary amides.

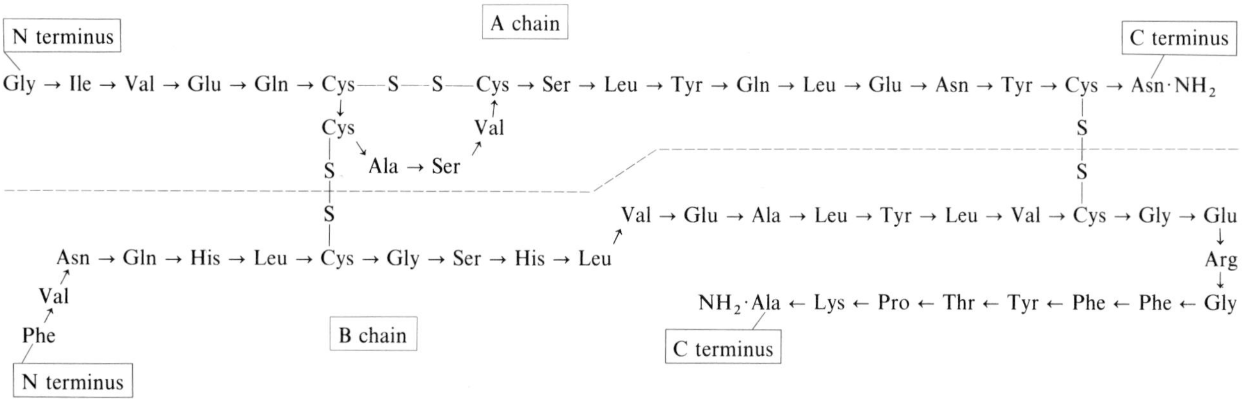

FIGURE 24-8 Structure of bovine insulin. The two chains are joined at two positions by disulfide bridges, and the A chain is held in a large ring by a third disulfide bond.

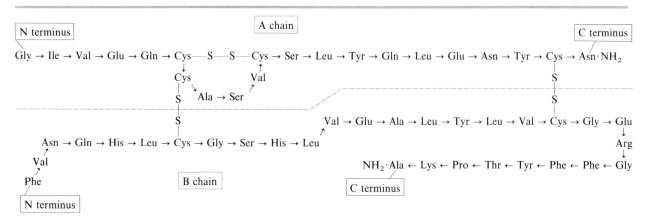

Disulfide bridges are commonly manipulated in the process of giving hair a *permanent wave*. Hair is composed of protein, which is made rigid and tough partly by disulfide bonds. When hair is treated with a solution of a thiol such as 2-mercaptoethanol ($HS—CH_2—CH_2—OH$), the disulfide bridges are reduced and cleaved. The hair is wrapped around curlers, and the disulfide bonds are allowed to re-form, either by air oxidation or by the application of a *neutralizer*. The disulfide bonds re-form in new positions, holding the hair in the bent conformation enforced by the curlers.

24-9 PEPTIDE STRUCTURE DETERMINATION

Insulin is a relatively simple protein, yet it is a complicated organic structure. How is it possible to determine the complete structure of a protein with hundreds of amino acid residues and a molecular weight of many thousands? Chemists have developed clever ways to determine the exact sequence of amino acids in a protein. We will consider some of the most common methods.

24-9A CLEAVAGE OF DISULFIDE LINKAGES

The first step in structure determination is to break all the disulfide bonds, separating the individual peptide chains and opening any disulfide-linked rings. The individual peptide chains are then purified and analyzed separately.

Cystine bridges are easily cleaved by reducing them to the thiol (cysteine) form. These reduced cysteine residues have a tendency to reoxidize and re-form the disulfide bridges, however. A more permanent cleavage involves oxidizing the disulfide linkages with peroxyformic acid (Fig. 24-9). This oxidation converts the disulfide bridges to sulfonic acid ($—SO_3H$) groups. The oxidized cysteine units are called **cysteic acid** residues.

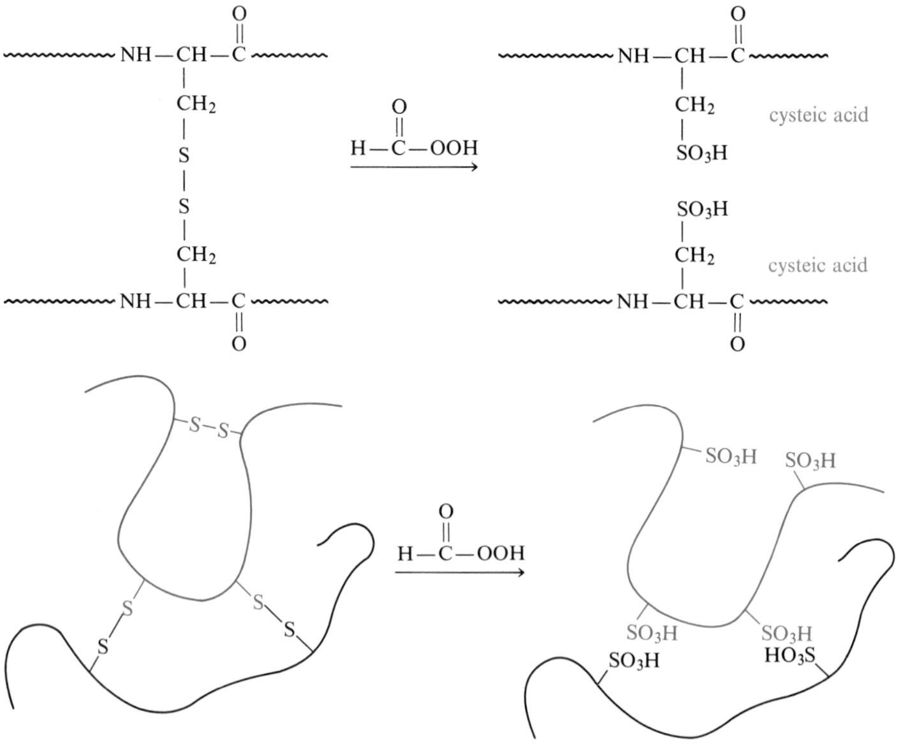

FIGURE 24-9 Oxidation of a protein by peroxyformic acid cleaves all the disulfide linkages by oxidizing cystine to cysteic acid.

24-9B DETERMINATION OF THE AMINO ACID COMPOSITION

Once the disulfide bridges have been broken and the individual peptide chains have been separated and purified, the structure of each chain must be determined. The first step is to determine which amino acids are present and in what proportions. To analyze the amino acid composition, the peptide chain is completely hydrolyzed by boiling it for 24 hours in 6 *M* HCl. The resulting mixture of amino acids (the *hydrolysate*) is placed on the column of an *amino acid analyzer,* diagrammed in Figure 24-10.

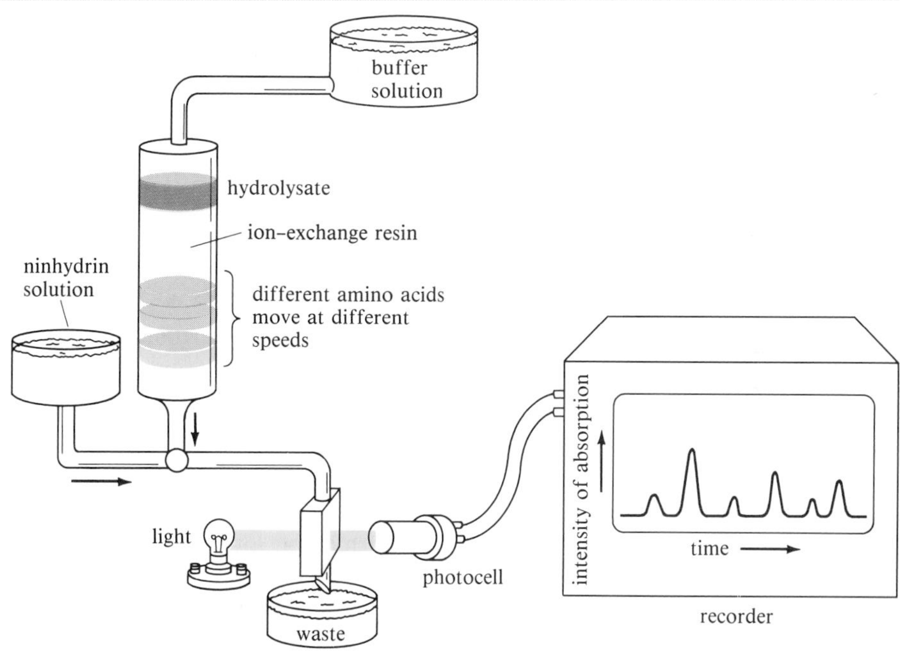

FIGURE 24-10 In an amino acid analyzer, the hydrolysate passes through an ion-exchange column. The solution emerging from the column is treated with ninhydrin, and its absorbance is recorded as a function of time. Each amino acid is identified by the retention time required to pass through the column.

In the amino acid analyzer, the components of the hydrolysate are dissolved in an aqueous buffer solution and separated by passing them down an ion-exchange column. The solution emerging from the column is mixed with ninhydrin, and the absorption of light is recorded as a function of time.

The amount of time required for each amino acid to pass through the column (its *retention time*) depends on how strongly that amino acid interacts with the ion-exchange resin in the column. The retention time of each amino acid is known from standardization with pure amino acids. The amino acids present in the sample are identified by comparing their retention times with the known values. The area under each peak is nearly proportional to the amount of the amino acid producing that peak, so we can determine the relative amounts of the amino acids present.

Figure 24-11 shows a standard trace of an equimolar mixture of amino acids, followed by the trace produced by the hydrolysate from human bradykinin (Arg-Pro-Pro-Gly-Phe-Ser-Pro-Phe-Arg).

Sequencing the peptide: terminal residue analysis The amino acid analyzer determines the amino acids present in a peptide, but it does not reveal their **sequence:** the order in which they are linked together. The peptide sequence is destroyed in the hydrolysis step. To determine the amino acid sequence, we must cleave just one amino acid from the chain and leave the rest of the chain intact. The cleaved amino acid can be separated and identified, and the process can be repeated on

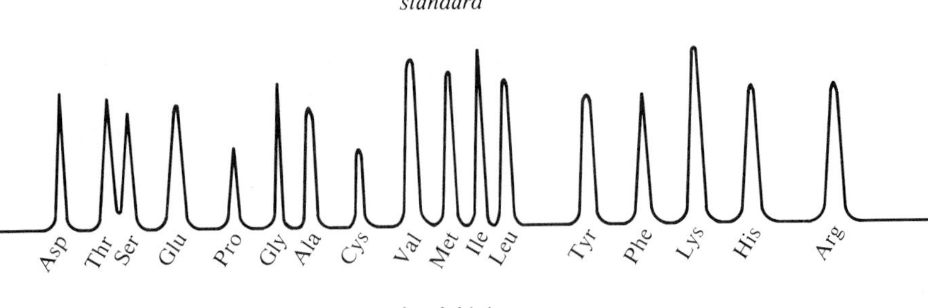

bradykinin

Ser Pro Gly Phe Arg

FIGURE 24-11 Use of an amino acid analyzer to determine the composition of human bradykinin. The bradykinin peaks for Pro, Arg, and Phe are larger than those in the standard equimolar mixture because bradykinin has three Pro residues, two Arg residues, and two Phe residues.

the rest of the chain. The amino acid may be cleaved from either end of the peptide (either the N terminus or the C terminus), and we will consider one method used for each end. This general method for peptide sequencing is called **terminal-residue analysis.**

24-9C SEQUENCING FROM THE N TERMINUS: THE EDMAN DEGRADATION

The most efficient method for sequencing peptides is the **Edman degradation.** A peptide is treated with phenyl isothiocyanate, followed by mild acid hydrolysis. The products are the shortened peptide chain and a heterocyclic derivative of the N-terminal amino acid called a *phenylthiohydantoin.*

Step 1: Attack by the free amino group on phenyl isothiocyanate. The product is a phenylthiourea.

$$\text{Ph—N=C=S} + \text{H}_2\text{N—CH(R}^1\text{)—C(=O)—NH—}\boxed{\text{peptide}} \rightleftharpoons \text{Ph—N}^-\text{—C(=S)—H}_2\text{N}^+\text{—CH(R}^1\text{)—C(=O)—NH—}\boxed{\text{peptide}} \rightleftharpoons \text{Ph—NH—C(=S)—HN—CH(R}^1\text{)—C(=O)—NH—}\boxed{\text{peptide}}$$

a phenylthiourea

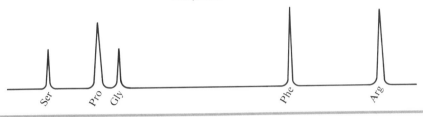

Step 2: Mild acid hydrolysis results in cyclization and expulsion of the shortened peptide chain.

$$\text{protonated phenylthiourea} \rightleftharpoons \cdots \rightleftharpoons \cdots \rightleftharpoons \text{a phenylthiohydantoin} + \text{H}_2\text{N—}\boxed{\text{peptide}}$$

protonated phenylthiourea

a phenylthiohydantoin

The phenylthiohydantoin derivative is identified by chromatography, comparing it with the phenylthiohydantoin derivatives of the standard amino acids.

This gives the identity of the original N-terminal amino acid. The rest of the peptide is cleaved intact, and further Edman degradations are used to identify additional amino acids in the chain.

Figure 24-12 shows the first two steps in the sequencing of bovine oxytocin. The oxytocin sample is first treated with peroxyformic acid to convert the disulfide bridge to cysteic acid residues.

Step 1: Cleavage and determination of the N-terminal amino acid.

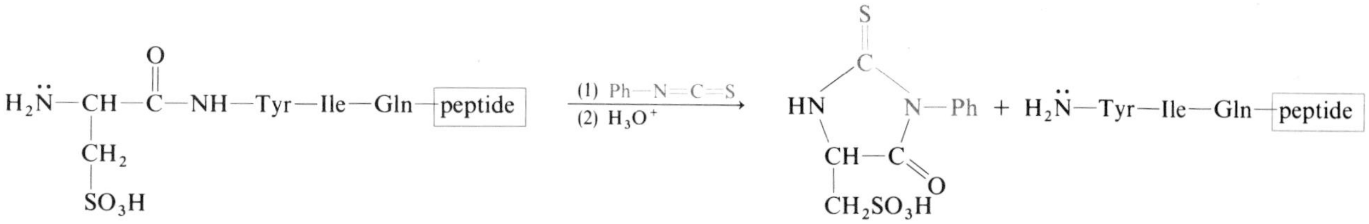

$H_2\ddot{N}-CH(CH_2SO_3H)-C(=O)-NH-Tyr-Ile-Gln-$[peptide] (cysteic acid) $\xrightarrow[(2)\ H_3O^+]{(1)\ Ph-N=C=S}$ cysteic acid phenylthiohydantoin + $H_2\ddot{N}-Tyr-Ile-Gln-$[peptide]

Step 2: Cleavage and determination of the second amino acid (the new N-terminal amino acid).

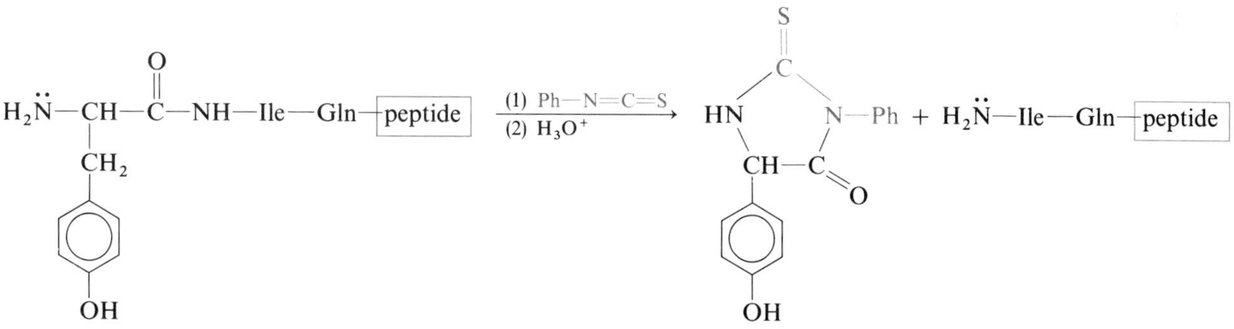

$H_2\ddot{N}-CH(CH_2C_6H_4OH)-C(=O)-NH-Ile-Gln-$[peptide] $\xrightarrow[(2)\ H_3O^+]{(1)\ Ph-N=C=S}$ tyrosine phenylthiohydantoin + $H_2\ddot{N}-Ile-Gln-$[peptide]

FIGURE 24-12 The first two steps in sequencing bovine oxytocin. Each Edman degradation cleaves the N-terminal amino acid and forms its phenylthiohydantoin derivative. The shortened peptide is available for the next step.

In theory, the Edman degradation could be used to sequence a peptide of any length. In practice, however, the repeated cycles of degradation result in some internal hydrolysis of the peptide, with loss of sample and accumulation of by-products. After about 30 cycles of degradation, further accurate analysis becomes impossible. A small peptide such as bradykinin can be completely determined by Edman degradation, but larger proteins must be broken into smaller fragments before they can be completely sequenced.

PROBLEM 24-18

Draw the structure of the phenylthiohydantoin derivative of each of the following amino acids.

(a) alanine (b) valine (c) lysine (d) proline

PROBLEM 24-19

Show the third and fourth steps in the sequencing of bovine oxytocin. Use Figure 24-12 as your guide.

PROBLEM 24-20

The Sanger method for N-terminus determination is a less common alternative to the Edman degradation. In the Sanger method, the peptide is treated with the Sanger reagent, 2-4-dinitrofluorobenzene, and then hydrolyzed by reaction with 6 *M* aqueous HCl. The N-terminal amino acid is recovered as its 2,4-dinitrophenyl derivative and identified.

The Sanger method

$$O_2N\text{-}C_6H_3(NO_2)\text{-}F \;+\; H_2\ddot{N}\text{-}CH(R^1)\text{-}C(=O)\text{-}NH\text{-}\boxed{\text{peptide}} \longrightarrow$$

2,4-dinitrofluorobenzene (Sanger reagent) — peptide

$$O_2N\text{-}C_6H_3(NO_2)\text{-}NH\text{-}CH(R^1)\text{-}C(=O)\text{-}NH\text{-}\boxed{\text{peptide}} \xrightarrow{6\,M\ \text{HCl, heat}}$$

derivative

$$O_2N\text{-}C_6H_3(NO_2)\text{-}NH\text{-}CH(R^1)\text{-}COOH \;+\; \text{amino acids}$$

2,4-dinitrophenyl derivative

(a) Give a mechanism for the reaction of the N terminus of the peptide with 2,4-dinitrofluorobenzene.

(b) Explain why the Edman degradation is usually preferred over the Sanger method.

24-9D C-TERMINAL RESIDUE ANALYSIS

There is no efficient method for sequencing several amino acids of a peptide starting from the C terminus. In many cases, however, the C-terminal amino acid can be identified using the enzyme *carboxypeptidase,* which cleaves the C-terminal peptide bond. The products are the free C-terminal amino acid and a shortened peptide. Further reaction cleaves the second amino acid that has now become the new C-terminus of the shortened peptide. Eventually, the entire peptide is hydrolyzed to its individual amino acids.

$$\boxed{\text{peptide}}\text{-}NH\text{-}CH(R^{n-1})\text{-}C(=O)\text{-}NH\text{-}CH(R^{n})\text{-}C(=O)\text{-}OH \xrightarrow[H_2O]{\text{carboxypeptidase}} \boxed{\text{peptide}}\text{-}NH\text{-}CH(R^{n-1})\text{-}C(=O)\text{-}OH + H_2N\text{-}CH(R^{n})\text{-}C(=O)\text{-}OH$$

(further cleavage) — free amino acid

The peptide is incubated with the carboxypeptidase enzyme, and the appearance of free amino acids is carefully monitored. In theory, the amino acid whose concentration increases first should be the C terminus, and the next amino acid to appear should be the second residue from the end. In practice, the different rates of cleavage of different amino acids complicate the determination, making it difficult to determine amino acids past the C terminus and occasionally the second residue in the chain.

24-9E BREAKING THE PEPTIDE INTO SHORTER CHAINS: PARTIAL HYDROLYSIS

Before a large protein can be sequenced, it must be broken into smaller chains not longer than about 30 amino acids. Each of these shortened chains is sequenced, and then the entire structure of the protein is deduced by fitting the short chains together like the pieces of a jigsaw puzzle.

Partial cleavage can be accomplished either by using dilute acid with a short reaction time or by using enzymes such as *trypsin* and *chymotrypsin*. The acid-catalyzed cleavage is not very selective, leading to a mixture of short fragments resulting from cleavage at various positions. Enzymes are more selective, giving cleavage at predictable points in the chain.

TRYPSIN Cleaves the chain at the carboxyl groups of the basic amino acids lysine and arginine.

CHYMOTRYPSIN Cleaves the chain at the carboxyl groups of the aromatic amino acids phenylalanine, tyrosine, and tryptophan.

Let's use bovine oxytocin as an example to illustrate the use of partial hydrolysis. Oxytocin could be sequenced directly by C-terminal analysis and a series of Edman degradations, but it provides a simple example of how a structure can be pieced together. Acid-catalyzed partial hydrolysis of oxytocin (after cleavage of the disulfide bridge) gives a mixture that includes the following peptides:

Ile-Gln-Asn-Cys   Gln-Asn-Cys-Pro   Pro-Leu-Gly·NH_2 Cys-Tyr-Ile-Gln-Asn   Cys-Pro-Leu-Gly

When we match the overlapping regions of these fragments, the complete sequence of oxytocin appears:

```
          Ile-Gln-Asn-Cys
              Gln-Asn-Cys-Pro
Cys-Tyr-Ile-Gln-Asn
                      Cys-Pro-Leu-Gly
                          Pro-Leu-Gly·NH2
```

Complete structure

Cys-Tyr-Ile-Gln-Asn-Cys-Pro-Leu-Gly·NH_2

We assume the two Cys residues in oxytocin are involved in disulfide bridges, either linking two of these peptide units or forming a ring. By measuring the molecular weight of oxytocin, we can show that it contains just one of these peptide units; therefore, the Cys residues must link the molecule in a ring.

PROBLEM 24-21

Show where trypsin and chymotrypsin would cleave the following peptide.

Tyr-Ile-Gln-Arg-Leu-Gly-Phe-Lys-Asn-Trp-Phe-Gly-Ala-Lys-Gly-Gln-Gln·NH_2

PROBLEM 24-22

After treatment with peroxyformic acid, the peptide hormone vasopressin is partially hydrolyzed. The following fragments are recovered. Propose a structure for vasopressin.

Phe-Gln-Asn
Pro-Arg-Gly·NH_2
Cys-Tyr-Phe
Asn-Cys-Pro-Arg
Tyr-Phe-Gln-Asn

24-10 CLASSICAL PEPTIDE SYNTHESIS

The total synthesis of peptides is rarely an economical method for their commercial production. Important peptides are usually derived from natural sources. For example, insulin for diabetics was originally taken from pork pancreas. Recombinant DNA techniques are now improving the quality and availability of peptide pharmaceuticals. It is possible to extract the piece of DNA that contains the code for a particular protein, insert it into a bacterium, and induce the bacterium to produce the protein. Strains of *Escherichia coli* have now been developed to produce human insulin that does not cause dangerous reactions in people who are allergic to pork products.

Laboratory peptide synthesis is still an important area of peptide chemistry, however. When the structure of a new peptide is determined, a synthesis is usually attempted. The purpose of the synthesis is twofold: If the synthetic material is the same as the natural material, it proves that the proposed structure is correct; and the synthesis provides a larger amount of the material for further biological testing.

Peptide synthesis involves forming amide bonds between the proper amino acids in the proper sequence. With simple acids and amines, we would form an amide bond simply by converting the acid to an activated derivative (such as an acyl halide or anhydride) and adding the amine.

$$R-\overset{O}{\overset{\|}{C}}-X \quad + \quad H_2\ddot{N}-R' \quad \longrightarrow \quad R-\overset{O}{\overset{\|}{C}}-NH-R' \quad + \quad H-X$$

(X is a good leaving group, preferably electron withdrawing)

Amide formation is not so easy with amino acids, however. Each amino acid has both an amino group and a carboxyl group. If we activate the carboxyl group, it will react with its own amino group. If we mix some amino acids and add a reagent to make them couple, they will form every conceivable sequence. Also, many of the standard amino acids have side chains that might interfere with peptide formation. For example, glutamic acid has an extra carboxyl group, and lysine has an extra amino group. As a result, peptide synthesis always involves both activating reagents to induce formation of the correct peptide bonds and protecting reagents to block formation of incorrect bonds.

Chemists have developed two types of methods for synthesizing a peptide with the desired sequence of amino acids. The *classical method* involves adding reagents to solutions of growing peptide chains, and the *solid-phase method* involves adding reagents to growing peptide chains bonded to solid particles. Although a variety of reagents and procedures can be used with each method, we will consider only one set of reagents for the classical method and one set for the solid-phase method.

Consider the structure of alanylvalylphenylalanine, a simple tripeptide:

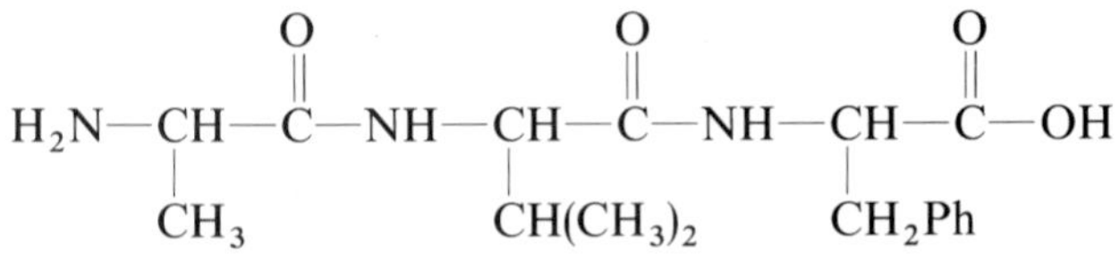

$$H_2N-\underset{\underset{\text{alanyl}}{CH_3}}{\underset{|}{CH}}-\overset{O}{\overset{\|}{C}}-NH-\underset{\underset{\text{valyl}}{CH(CH_3)_2}}{\underset{|}{CH}}-\overset{O}{\overset{\|}{C}}-NH-\underset{\underset{\text{phenylalanine}}{CH_2Ph}}{\underset{|}{CH}}-\overset{O}{\overset{\|}{C}}-OH$$

Ala-Val-Phe

The classical peptide synthesis begins at the N terminus and ends at the C terminus, or left to right as we draw the peptide. The first major step is to couple the carboxyl group of alanine to the amino group of valine. This cannot be done simply by activating the carboxyl group of alanine and adding valine. If we activated the carboxyl group of alanine, it would react with another molecule of alanine.

To prevent side reactions, the nucleophilic amino group of alanine must be protected to make it nonnucleophilic. This is done by treating the free amino acid with benzyl chloroformate. The product is a urethane, or carbamate ester. This particular protecting group has been used for many years, and it has acquired several names. It is called the *benzyloxycarbonyl group,* the *carbobenzoxy group* (abbreviated Cbz), or simply the *Z group* (abbreviated Z).

$$\text{Ph}-CH_2-O-\overset{O}{\overset{\|}{C}}-Cl + H_2\ddot{N}-\underset{CH_3}{\underset{|}{CH}}-\overset{O}{\overset{\|}{C}}-OH \longrightarrow \underbrace{\text{Ph}-CH_2-O-\overset{O}{\overset{\|}{C}}}_{\text{Z group}}-NH-\underset{CH_3}{\underset{|}{CH}}-\overset{O}{\overset{\|}{C}}-OH + HCl$$

benzyl chloroformate Z-Cl; alanine Ala; benzyloxycarbonyl alanine Z-Ala

The alanine amino group in Z-Ala is now protected as the nonnucleophilic amide half of a carbamate ester. The carboxyl group of this protected alanine can be activated without reacting with the protected amino group. In the classical peptide synthesis, the carboxyl group is activated by treatment with ethyl chloroformate. The product is a mixed anhydride of the amino acid and carbonic acid, and it is strongly activated toward nucleophilic attack.

$$Z-NH\underset{CH_3}{\underset{|}{CH}}-\overset{O}{\overset{\|}{C}}-OH + Cl-\overset{O}{\overset{\|}{C}}-OCH_2CH_3 \longrightarrow Z-NH-\underset{CH_3}{\underset{|}{CH}}-\underbrace{\overset{O}{\overset{\|}{C}}-O-\overset{O}{\overset{\|}{C}}-OCH_2CH_3}_{\text{anhydride of carbonic acid}} + HCl$$

protected alanine; ethyl chloroformate; mixed anhydride

When the second amino acid (valine) is added to the protected, activated alanine, the nucleophilic amino group of valine attacks the activated carbonyl of alanine, displacing the anhydride and forming a peptide bond.

$$Z-NH\underset{CH_3}{\underset{|}{CH}}-\overset{O}{\overset{\|}{C}}-O-\overset{O}{\overset{\|}{C}}-OCH_2CH_3 + H_2\ddot{N}-\underset{CH(CH_3)_2}{\underset{|}{CH}}-\overset{O}{\overset{\|}{C}}-OH \longrightarrow Z-NH\underset{CH_3}{\underset{|}{CH}}-\overset{O}{\overset{\|}{C}}-NH-\underset{CH(CH_3)_2}{\underset{|}{CH}}-\overset{O}{\overset{\|}{C}}-OH + CO_2\uparrow + CH_3CH_2OH$$

protected, activated alanine; valine; Z-Ala-Val

PROBLEM 24-23

Give complete mechanisms for the formation of Z-Ala, its activation by ethyl chloroformate, and the coupling with valine.

At this point we have the N-protected dipeptide Z-Ala-Val. Phenylalanine must be added to the C terminus of this dipeptide to complete the Ala-Val-Phe tripeptide. Activation of the valine carboxyl group, followed by addition of phenylalanine, gives the protected tripeptide.

Step 1: Activation of the carboxyl group

$$\text{Z—NHCH(CH}_3\text{)CNHCH(CH(CH}_3)_2\text{)—}\overset{O}{\overset{\|}{C}}\text{—}\boxed{\text{OH}} + \text{Cl—}\overset{O}{\overset{\|}{C}}\text{—OEt} \longrightarrow \text{Z—NHCH(CH}_3\text{)}\overset{O}{\overset{\|}{C}}\text{NHCH—CH(CH(CH}_3)_2\text{)—}\overset{O}{\overset{\|}{C}}\boxed{\text{—O—}\overset{O}{\overset{\|}{C}}\text{—OEt}} + \text{HCl}$$

Step 2: Addition of the next amino acid

$$\text{Z—Ala—NHCH(CH(CH}_3)_2\text{)—}\overset{O}{\overset{\|}{C}}\boxed{\text{—O—}\overset{O}{\overset{\|}{C}}\text{—OEt}} + \text{H}_2\text{N—CH(CH}_2\text{—Ph)—}\overset{O}{\overset{\|}{C}}\text{—OH} \longrightarrow \text{Z—Ala—NHCH(H}_3\text{C—CH—CH}_3\text{)—}\overset{O}{\overset{\|}{C}}\boxed{\text{—NH—CH(CH}_2\text{—Ph)—}\overset{O}{\overset{\|}{C}}\text{—OH}} + CO_2\uparrow + \text{EtOH}$$

Z-Ala-Val-Phe

To make a larger peptide, these two steps are repeated for the addition of each amino acid residue:

1. Activation of the C terminus of the growing peptide by the reaction with ethyl chloroformate
2. Coupling of the next amino acid

The final step in the classical synthesis is deprotection of the N terminus of the complete peptide. The N-terminal amide bond must be cleaved without breaking any of the peptide bonds in the product. Fortunately, the benzyloxycarbonyl group is partly an amide and partly a benzyl ester, and hydrogenolysis of the benzyl ester takes place under mild conditions that do not cleave the peptide bonds. This mild cleavage is the reason for use of the benzyloxycarbonyl group (as opposed to some other acyl group) to protect the N terminus.

$$\text{C}_6\text{H}_5\text{—CH}_2\text{—O—}\overset{O}{\overset{\|}{C}}\text{—NHCH(CH}_3\text{)}\overset{O}{\overset{\|}{C}}\text{—Val—Phe} \xrightarrow{H_2,\ Pd} \text{H}_2\ddot{\text{N}}\text{CH(CH}_3\text{)}\overset{O}{\overset{\|}{C}}\text{—Val—Phe} + CO_2\uparrow + \text{C}_6\text{H}_5\text{—CH}_3$$

Z-Ala-Val-Phe → Ala-Val-Phe

PROBLEM 24-24

Show how you would synthesize Ala-Val-Phe-Gly-Leu starting with Z-Ala-Val-Phe.

PROBLEM 24-25

Show how the classical synthesis would be used to synthesize Ile-Gly-Asn.

The classical method of peptide synthesis works well for small peptides, and many peptides have been synthesized by this process. Larger proteins are not easily synthesized by the classical method, however. A large number of chemical reactions and purifications are required even for a small peptide. Although the individual yields are excellent, with a large peptide the overall yield becomes so small as to be unusable, and several months (or years) are required to complete so many steps. The large amounts of time required and the low overall yields are due largely to the purification steps. For larger peptides and proteins, solid-phase peptide synthesis is usually preferred.

24-11 SOLID-PHASE PEPTIDE SYNTHESIS

In 1962, Robert Bruce Merrifield of Rockefeller University developed a method for synthesizing peptides without having to purify any of the intermediates. He did this by attaching the growing peptide chains to solid polystyrene beads. After each amino acid is added, the excess reagents are washed away by rinsing the beads with solvent. This ingenious method lends itself to automation, and Merrifield built a machine that can add several amino acid units while running unattended. Using this machine, Merrifield synthesized ribonuclease (124 amino acids) in just six weeks, obtaining an overall yield of 17 percent. Merrifield's work in **solid-phase peptide synthesis** won the Nobel Prize in 1984.

There are three important reactions to consider before we discuss the solid-phase peptide synthesis. We must first learn how an amino acid is attached to the solid support, how and why the Boc protecting group is used, and how the DCC coupling reagent works.

Attaching the peptide to the solid support The greatest difference between classical and solid-phase peptide synthesis is that solid-phase synthesis is done backward: starting with the C terminus and going toward the N terminus, right to left as we write the peptide. The first step is to attach the *last* amino acid (the C terminus) to the solid support.

The solid support is a special polystyrene bead in which some of the aromatic rings have chloromethyl groups. This polymer, often called the *Merrifield resin* after its inventor, is made by copolymerizing styrene with a few percent of *p*-(chloromethyl)styrene.

Formation of the Merrifield resin

styrene + *p*-(chloromethyl)styrene ⟶ polymer $-CH-CH_2-CH-CH_2-CH-CH_2-$ (with CH_2Cl on some rings) = abbreviated (P)

styrene | *p*-(chloromethyl)styrene | polymer | abbreviated

Like other benzyl halides, the chloromethyl groups on the polymer are quite reactive toward S_N2 attack. The carboxyl group of an N-protected amino acid displaces chloride, giving an amino acid ester of the polymer. In effect, the polymer serves as the alcohol part of an ester protecting group for the carboxyl of the C-terminal amino acid.

Attachment of the C-terminal amino acid

[protecting group]$-NH-CH(R)-C(=O)-\ddot{O}:^-$ + $Cl-CH_2-C_6H_4-$(P) ⟶ [protecting group]$-NH-CH(R)-C(=O)-\ddot{O}-CH_2-C_6H_4-$(P) Cl^-

Once the C-terminal amino acid is fixed to the polymer, the chain is built on the amino group of this amino acid. At the completion of the synthesis, the ester bond to the polymer is cleaved by anhydrous HF. Because this is an ester bond, it is more easily cleaved than the amide bonds of the peptide.

Cleavage of the finished peptide

$$\boxed{\text{peptide}}-\overset{\displaystyle O}{\overset{\|}{C}}-O-CH_2-C_6H_4-\text{(P)} \xrightarrow{HF} \boxed{\text{peptide}}-\overset{\displaystyle O}{\overset{\|}{C}}-OH + CH_2F-C_6H_4-\text{(P)}$$

Use of the t-butyloxycarbonyl (Boc) protecting group The benzyloxycarbonyl group (the Z group) cannot be used with the solid-phase process because this group is removed by hydrogenolysis in contact with a solid catalyst. A polymer-bound peptide cannot achieve the intimate contact with the solid catalyst required for this hydrogenolysis. The N-protecting group used in the Merrifield procedure is the *t*-butyloxycarbonyl group, abbreviated Boc or *t*-Boc. The Boc group is similar to the Z group, except that it has a *t*-butyl group in place of the benzyl group. Like other *t*-butyl esters, the Boc protecting group is easily hydrolyzed under acidic conditions.

The acid chloride of the Boc group is unstable, so its anhydride, di-*t*-butyldicarbonate, is used to attach the group to the amino acid.

Protection of the amino group as its Boc derivative

$$\underset{\text{di-}t\text{-butyldicarbonate}}{(CH_3)_3C-O-\overset{\displaystyle O}{\overset{\|}{C}}-O-\overset{\displaystyle O}{\overset{\|}{C}}-O-C(CH_3)_3} + \underset{\text{amino acid}}{H_2N-\underset{\displaystyle R}{\underset{|}{CH}}-COOH} \longrightarrow$$

$$\underset{\text{Boc-amino acid}}{\boxed{(CH_3)_3C-O-\overset{\displaystyle O}{\overset{\|}{C}}}-NH-\underset{\displaystyle R}{\underset{|}{CH}}-COOH} + CO_2 + (CH_3)_3C-OH$$

The Boc group is easily cleaved by brief treatment with trifluoroacetic acid, CF_3—COOH. Loss of a relatively stable *t*-butyl cation from the protonated ester gives an unstable carbamic acid. Decarboxylation of the carbamic acid gives the deprotected amine group of the amino acid. Loss of a proton from the *t*-butyl cation gives isobutylene.

$$\underset{\text{Boc-amino acid}}{(CH_3)_3C-O-\overset{\displaystyle \ddot{O}:}{\overset{\|}{C}}-NH-\underset{\displaystyle R}{\underset{|}{CH}}-COOH} \xrightarrow{CF_3-COOH} \underset{\text{protonated}}{(CH_3)_3C-\ddot{O}-\overset{\displaystyle \ddot{O}^{+}-H}{\overset{\|}{C}}-NH-\underset{\displaystyle R}{\underset{|}{CH}}-COOH} \longrightarrow$$

$$(CH_3)_3C^{+} + \underset{\text{a carbamic acid}}{\left[\ddot{O}=\overset{\displaystyle \ddot{O}-H}{\overset{|}{C}}-NH-\underset{\displaystyle R}{\underset{|}{CH}}-COOH\right]} \longrightarrow \underset{\text{isobutylene}}{CH_2=C(CH_3)_2} + \underset{\text{free amino acid}}{H_3\overset{+}{N}-\underset{\displaystyle R}{\underset{|}{CH}}-COOH} + CO_2\uparrow$$

People who synthesize peptides generally do not make their own Boc-protected amino acids. Because they use all their amino acids in protected form, they buy and use commercially available Boc-amino acids.

Use of DCC as a peptide coupling agent The final reaction needed for the Merrifield procedure is the peptide bond-forming reaction. When a mixture of an amine and an acid is treated with *N,N′*-dicyclohexylcarbodiimide (abbreviated DCC), the amine and the acid are coupled to form an amide. The molecule of water lost in this condensation converts DCC to *N,N′*-dicyclohexyl urea (DCU).

$$\underset{\text{acid}}{R{-}C({=}O){-}O^-} + \underset{\text{amine}}{H_3\overset{+}{N}{-}R'} + \underset{N,N'\text{-dicyclohexylcarbodiimide (DCC)}}{C_6H_{11}{-}N{=}C{=}N{-}C_6H_{11}} \longrightarrow \underset{\text{amide}}{R{-}C({=}O){-}NH{-}R'} + \underset{N\text{-}N'\text{-dicyclohexyl urea (DCU)}}{C_6H_{11}{-}NH{-}C({=}O){-}NH{-}C_6H_{11}}$$

The mechanism for the DCC coupling reaction is not as complicated as it may seem. The carboxylate ion adds to the strongly electrophilic carbon of the diimide, giving an activated acyl derivative of the acid. This activated derivative reacts readily with the amine to give the amide. In the final step, DCU serves as an excellent leaving group.

Formation of an activated acyl derivative

$$C_6H_{11}{-}N{=}C{=}N{-}C_6H_{11} + R{-}C({=}O){-}O^- \longrightarrow R{-}C({=}O){-}O{-}C({=}N{-}C_6H_{11}){-}N^-{-}C_6H_{11} \xrightarrow{H{-}\overset{+}{N}H_2{-}R'} R{-}C({=}O){-}O{-}C({=}N{-}C_6H_{11}){-}NH{-}C_6H_{11} \quad H_2\ddot{N}{-}R'$$

Coupling with the amine and loss of DCU

$$R{-}C({=}O){-}O{-}C({=}N{-}C_6H_{11}){-}NH{-}C_6H_{11} + R'{-}\ddot{N}H_2 \longrightarrow R{-}C(O^-)(\overset{+}{N}H_2{-}R'){-}O{-}C({=}N{-}C_6H_{11}){-}NH{-}C_6H_{11} \longrightarrow R{-}C({=}O){-}\overset{+}{N}H_2R' \;\; ^-O{-}C({=}N{-}C_6H_{11}){-}NH{-}C_6H_{11} \longrightarrow \underset{\text{amide}}{R{-}C({=}O){-}NHR'} + O{=}C(NH{-}C_6H_{11})_2$$

> PROBLEM 24-26
>
> Propose a mechanism for the coupling of acetic acid and aniline using DCC as a coupling agent.

Now we consider an example to illustrate how these procedures are combined in the Merrifield solid-phase method of peptide synthesis.

An example of solid-phase peptide synthesis For easy comparison of the classical and solid-phase methods, we will consider the synthesis of the same tripeptide we made using the classical method.

Ala-Val-Phe

The solid-phase synthesis is carried out in the opposite direction from the classical synthesis. The first step is attachment of the N-protected C-terminal amino acid (Boc-phenylalanine) to the polymer.

$$\underset{\text{Boc-Phe}}{Me_3C{-}O{-}\overset{O}{\overset{\|}{C}}{-}NH{-}\underset{Ph-CH_2}{CH}{-}\overset{O}{\overset{\|}{C}}{-}O^-} + \underset{(P)}{CH_2{-}Cl\ (\text{on } C_6H_4)} \longrightarrow \underset{\text{Boc-Phe-(P)}}{Me_3C{-}O{-}\overset{O}{\overset{\|}{C}}{-}NH{-}\underset{Ph-CH_2}{CH}{-}\overset{O}{\overset{\|}{C}}{-}O{-}CH_2{-}C_6H_4{-}(P)}$$

Trifluoroacetic acid cleaves the Boc protecting group of phenylalanine so that its amino group can be coupled with the next amino acid.

$$\underset{\text{Boc-Phe-(P)}}{Me_3C{-}O{-}\overset{O}{\overset{\|}{C}}{-}NH{-}\underset{Ph-CH_2}{CH}{-}\overset{O}{\overset{\|}{C}}{-}O{-}CH_2{-}C_6H_4{-}(P)} \xrightarrow{CF_3-COOH} \underset{\text{Phe-(P)}}{H_3\overset{+}{N}{-}\underset{Ph-CH_2}{CH}{-}\overset{O}{\overset{\|}{C}}{-}O{-}CH_2{-}C_6H_4{-}(P)}$$

The second amino acid (valine) is added in its N-protected Boc form so that it cannot couple with itself. Addition of DCC couples the valine carboxyl group with the free $-NH_2$ group of phenylalanine.

$$\underset{\text{Boc-Val}}{Boc{-}NH{-}\underset{(CH_3)_2CH}{CH}{-}\overset{O}{\overset{\|}{C}}{-}O^-} + \underset{\text{Phe-(P)}}{H_3\overset{+}{N}{-}\underset{Ph-CH_2}{CH}{-}\overset{O}{\overset{\|}{C}}{-}O{-}CH_2{-}C_6H_4{-}(P)} \xrightarrow{DCC} \underset{\text{Boc-Val-Phe-(P)}}{Boc{-}NH{-}\underset{(CH_3)_2CH}{CH}{-}\overset{O}{\overset{\|}{C}}{-}NH{-}\underset{Ph-CH_2}{CH}{-}\overset{O}{\overset{\|}{C}}{-}O{-}CH_2{-}C_6H_4{-}(P)} + DCU$$

To couple the final amino acid (alanine), the chain is first deprotected by treatment with trifluoroacetic acid. The N-protected Boc-alanine and DCC are then added.

Step 1: Deprotection

$$\underset{\text{Boc-Val-Phe-(P)}}{Boc{-}NH{-}\underset{(CH_3)_2CH}{CH}{-}\overset{O}{\overset{\|}{C}}{-}NH{-}\underset{Ph-CH_2}{CH}{-}\overset{O}{\overset{\|}{C}}{-}O{-}CH_2{-}C_6H_4{-}(P)} \xrightarrow{CF_3-COOH} \underset{\text{Val-Phe-(P)}}{H_3\overset{+}{N}{-}\underset{(CH_3)_2CH}{CH}{-}\overset{O}{\overset{\|}{C}}{-}NH{-}\underset{Ph-CH_2}{CH}{-}\overset{O}{\overset{\|}{C}}{-}O{-}CH_2{-}C_6H_4{-}(P)} + CH_3{-}\overset{CH_3}{\overset{|}{C}}{=}CH_2 + CO_2$$

Step 2: Coupling

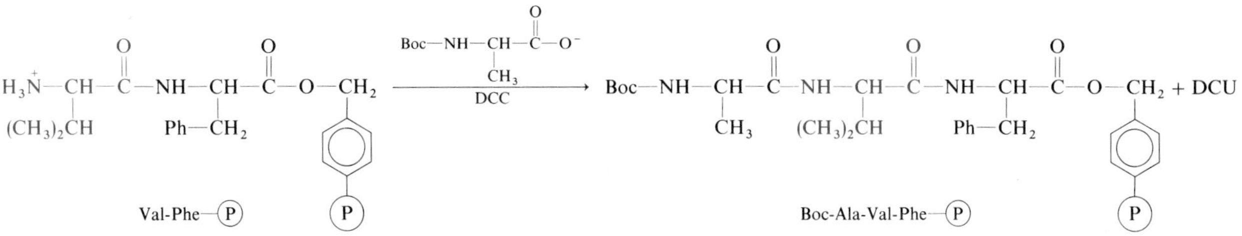

$$\underset{\text{Val-Phe}-\text{(P)}}{H_3\overset{+}{N}-\underset{(CH_3)_2CH}{CH}-\overset{O}{\overset{\|}{C}}-NH-\underset{Ph-CH_2}{CH}-\overset{O}{\overset{\|}{C}}-O-CH_2-C_6H_4-\text{(P)}} \xrightarrow[\text{DCC}]{\text{Boc}-NH-\underset{CH_3}{CH}-\overset{O}{\overset{\|}{C}}-O^-} \underset{\text{Boc-Ala-Val-Phe}-\text{(P)}}{\text{Boc}-NH-\underset{CH_3}{CH}-\overset{O}{\overset{\|}{C}}-NH-\underset{(CH_3)_2CH}{CH}-\overset{O}{\overset{\|}{C}}-NH-\underset{Ph-CH_2}{CH}-\overset{O}{\overset{\|}{C}}-O-CH_2-C_6H_4-\text{(P)}} + \text{DCU}$$

If we were making a longer peptide, the addition of each subsequent amino acid would require the repetition of two steps:

1. Use trifluoroacetic acid to deprotect the amino group at the end of the growing chain.
2. Add the next Boc-amino acid, using DCC as a coupling agent.

Once the peptide is completed, the final Boc protecting group must be removed, and the peptide must be cleaved from the polymer. Anhydrous HF is used to cleave the ester linkage that bonds the peptide to the polymer, and it also removes the Boc protecting group. In our example, the following reaction occurs.

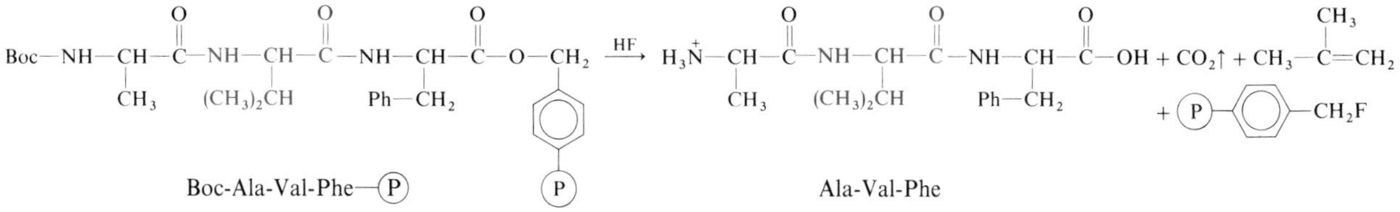

$$\underset{\text{Boc-Ala-Val-Phe}-\text{(P)}}{\text{Boc}-NH-\underset{CH_3}{CH}-\overset{O}{\overset{\|}{C}}-NH-\underset{(CH_3)_2CH}{CH}-\overset{O}{\overset{\|}{C}}-NH-\underset{Ph-CH_2}{CH}-\overset{O}{\overset{\|}{C}}-O-CH_2-C_6H_4-\text{(P)}} \xrightarrow{\text{HF}} \underset{\text{Ala-Val-Phe}}{H_3\overset{+}{N}-\underset{CH_3}{CH}-\overset{O}{\overset{\|}{C}}-NH-\underset{(CH_3)_2CH}{CH}-\overset{O}{\overset{\|}{C}}-NH-\underset{Ph-CH_2}{CH}-\overset{O}{\overset{\|}{C}}-OH} + CO_2\uparrow + CH_3-\overset{CH_3}{\overset{|}{C}}=CH_2 + \text{(P)}-C_6H_4-CH_2F$$

PROBLEM 24-27

Show how you would synthesize Leu-Gly-Ala-Val-Phe starting with Boc-Ala-Val-Phe—(P).

PROBLEM 24-28

Show how solid-phase peptide synthesis would be used to make Ile-Gly-Asn.

24-12 CLASSIFICATION OF PROTEINS

There are many different ways of classifying proteins. They may be classified according to their chemical composition, their shape, or their function. Protein composition and function are treated in detail in a course in biochemistry. For now, we will briefly survey the types of proteins and their general classifications.

Proteins are grouped into *simple* and *conjugated* proteins according to their chemical composition. **Simple proteins** are those that hydrolyze to give only amino acids. All the protein structures we have considered so far are simple proteins. Examples are insulin, ribonuclease, oxytocin, and bradykinin. **Conjugated proteins** are bonded to a nonprotein group such as a sugar, a nucleic acid, a lipid, or some other group. The nonprotein part of a conjugated protein is called a **prosthetic group.** Table 24-3 lists some examples of conjugated proteins.

Proteins are classified into **fibrous proteins** and **globular proteins** according to whether they form long filaments or coil up on themselves. Fibrous proteins are stringy, tough, and usually insoluble in water. They function primarily as struc-

TABLE 24-3
Classes of conjugated proteins

Class	*Prosthetic group*	*Examples*
glycoproteins	carbohydrates	γ-globulin, interferon
nucleoproteins	nucleic acids	ribosomes, viruses
lipoproteins	fats, cholesterol	high-density lipoprotein
metalloproteins	a complexed metal	hemoglobin, cytochromes

tural parts of the organism. Examples of fibrous proteins are α-keratin in hooves and fingernails, and collagen in tendons. Globular proteins are folded into roughly spherical shapes. They usually function as enzymes, hormones, or transport proteins. Examples of globular proteins are insulin, ribonuclease, and hemoglobin.

24-13 LEVELS OF PROTEIN STRUCTURE

24-13A PRIMARY STRUCTURE

Up to now we have discussed the *primary structure* of proteins. The **primary structure** is the covalently bonded structure of the molecule. This definition includes the sequence of amino acids, together with any disulfide bridges. All the properties of the protein are determined, directly or indirectly, by the primary structure. Any folding, hydrogen bonding, or catalytic activity depends on the proper primary structure.

24-13B SECONDARY STRUCTURE

Peptide chains tend to form orderly hydrogen-bonded arrangements. In particular, the carbonyl oxygen atoms form hydrogen bonds with the amide (N—H) hydrogens. There are two arrangements in which an orderly arrangement of hydrogen bonds can occur, the **α-helix** and the **pleated sheet.** These hydrogen-bonded arrangements, if present, are called the **secondary structure** of the protein.

If the molecule winds into a helical coil, each carbonyl oxygen can hydrogen-bond with a N—H hydrogen on the next turn of the coil. Many proteins wind into an α-helix (a helix that looks like the thread on a right-handed screw) with the side chains positioned on the outside of the helix. The fibrous protein α-keratin is arranged in the α-helical structure, and most globular proteins contain segments of α-helix. Figure 24-13 shows the α-helix arrangement.

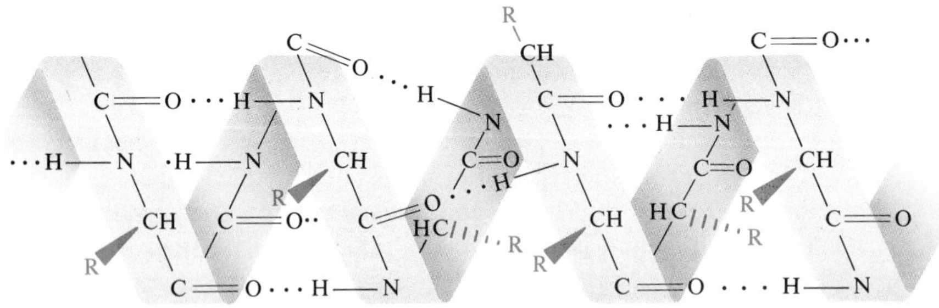

FIGURE 24-13 The α-helical arrangement. Each peptide carbonyl group is hydrogen-bonded to a N—H hydrogen on the next turn of the helix.

Segments of peptides can also form orderly arrangements of hydrogen bonds by lining up side by side. In this arrangement, each carbonyl group on one chain forms a hydrogen bond with an N—H hydrogen on an adjacent chain. This arrangement may involve many peptide molecules lined up side by side, resulting in a two-dimensional *sheet*. The bond angles between amino acid units are such that the the sheet is *pleated* (creased), with the amino acid side chains arranged on alternating sides of the sheet. Silk fibroin, the principal fibrous protein in the silks of insects and arachnids, has the pleated sheet secondary structure. Figure 24-14 shows the pleated sheet structure.

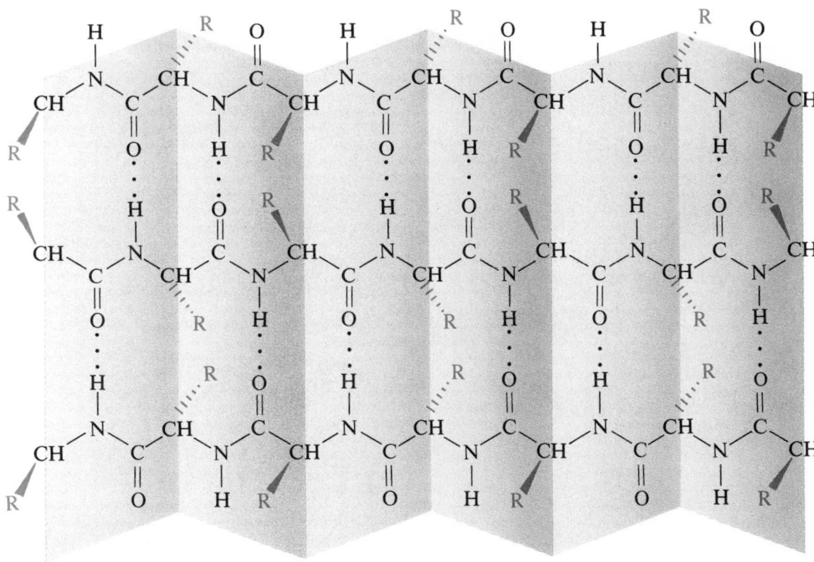

FIGURE 24-14 The pleated sheet arrangement. Each peptide carbonyl group is hydrogen-bonded to a N—H hydrogen on an adjacent peptide chain.

A protein may or may not have the same secondary structure throughout its length. Some parts may be curled into an α-helix, while other parts are lined up in a pleated sheet. Parts of the chain may have no secondary structure at all. Such a structureless part is called a **random coil.** Most globular proteins, for example, contain segments of α-helix or pleated sheet separated by kinks of random coil, allowing the molecule to fold into its globular shape.

24-13C TERTIARY STRUCTURE

The **tertiary structure** of a protein is its complete three-dimensional conformation. Remember that the secondary structure is a local structure. Parts of the protein may have the α-helical structure, while other parts have the pleated sheet structure, and still other parts are random coils. The tertiary structure includes all of the secondary structure and all of the kinks in between. The tertiary structure of a typical globular protein is represented in Figure 24-15.

The coiling of an enzyme can produce some important catalytic effects. The polar, *hydrophilic* (water-loving) side chains are oriented toward the outside of the globule. The nonpolar, *hydrophobic* (water-hating) groups are arranged toward the interior. A reaction taking place at the *active site* in the interior of an enzyme may take place under essentially anhydrous, nonpolar conditions—while the whole system is dissolved in water!

24-13D QUATERNARY STRUCTURE

Quaternary structure refers to the association of two or more peptide chains in the complete protein. For example, hemoglobin, the oxygen carrier in mammalian blood, consists of four peptide chains fitted together to form a globular protein. Figure 24-16 summarizes the four levels of protein structure.

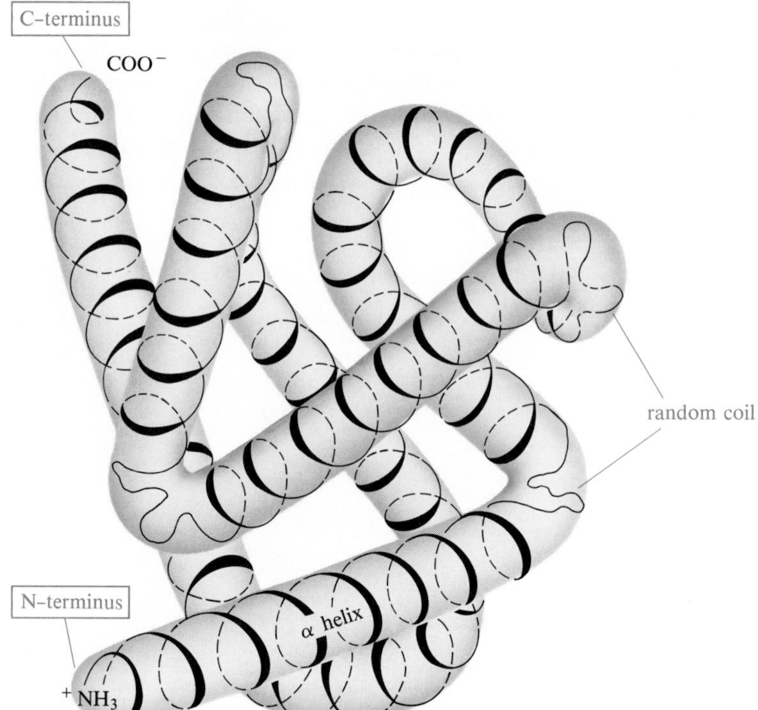

FIGURE 24-15 The tertiary structure of a typical globular protein includes segments of α-helix with segments of random coil at the points where the helix is folded.

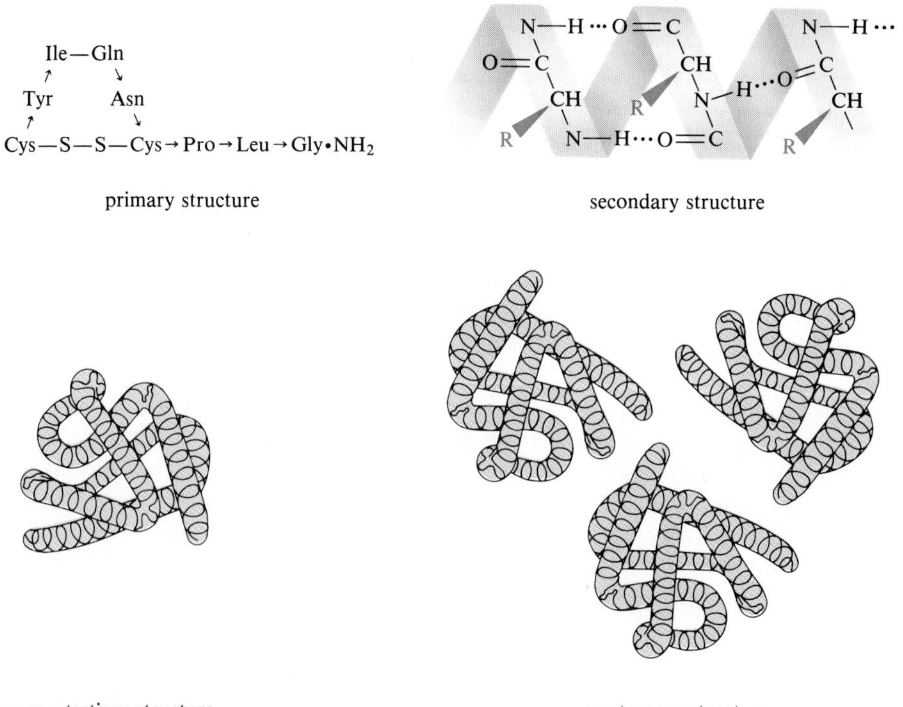

Ile—Gln
↗ ↘
Tyr Asn
↗ ↘
Cys—S—S—Cys→Pro→Leu→Gly•NH_2

primary structure

N—H···O═C N—H···
O═C CH
CH R N—H···O═C
R N—H···O═C R CH R

secondary structure

tertiary structure

quaternary structure

FIGURE 24-16 A schematic comparison of the levels of protein structure. Primary structure is the covalently bonded structure, including the amino acid sequence and any disulfide bridges. Secondary structure refers to the areas of α-helix, pleated sheet, or random coil. Tertiary structure refers to the overall conformation of the molecule. Quaternary structure refers to the association of two or more peptide chains into the active protein.

24-14 PROTEIN DENATURATION

For a protein to be biologically active, it must have the correct structure at all levels. The sequence of amino acids must be right, with the correct disulfide bridges linking the cysteines on the chains. The secondary and tertiary structures are important, as well. The protein must be folded into its natural conformation, with the appropriate areas of α-helix and pleated sheet. For an enzyme, the active site must have the right conformation, with the necessary side-chain functional groups in the correct positions. Conjugated proteins must have the right prosthetic groups, and multichain proteins must have the right combination of individual peptides.

With the exception of the covalent primary structure, all these levels of structure are maintained by weak solvation and hydrogen-bonding forces. Small changes in the environment can cause a chemical or conformational change resulting in **denaturation:** disruption of the normal structure and loss of biological activity. Many factors can cause denaturation, but the most common ones are heat and pH.

The cooking of egg white is an example of protein denaturation by high temperature. Egg white contains soluble globular proteins called *albumins*. When the egg white is heated, the albumins unfold and coagulate to produce a solid rubbery mass. Different proteins have different abilities to resist the denaturing effect of heat. Egg albumin is quite sensitive to heat, but the bacteria that live in geothermal hot springs have developed proteins that retain their activity in boiling water.

When a protein is subjected to an acidic pH, some of the side-chain carboxyl groups become protonated and lose their ionic charge. Conformational changes result, leading to denaturation. In a basic solution, the amino groups are deprotonated, similarly losing their ionic charge and causing conformational changes and denaturation.

Milk turns sour because of the bacterial conversion of carbohydrates to lactic acid. Once the pH has become strongly acidic, the soluble proteins in the milk are denatured and precipitate. This is the process that we call the *curdling* of milk. Some proteins are more resistant to acidic and basic conditions than others. For example, most digestive enzymes such as amylase and trypsin remain active under the acidic conditions in the stomach, even at a pH of about 1.

In many cases, denaturation is irreversible. When cooked egg white is cooled, it does not become uncooked. Curdled milk does not uncurdle when it is neutralized. Denaturation may be reversible, however, if the protein has undergone only mild denaturing conditions. For example, a protein can be *salted out* of solution by a high salt concentration, which denatures and precipitates the protein. When the precipitated protein is redissolved in a solution with a lower salt concentration, it usually regains its activity together with its natural conformation.

GLOSSARY

amino acid Literally, any molecule containing both an amino group ($-NH_2$) and a carboxyl group ($-COOH$). The term usually means an **α-amino acid,** with the amino group on the carbon atom next to the carboxyl group. (p. 1102)

biomimetic synthesis A laboratory synthesis that is patterned after a biological synthesis. For example, the synthesis of amino acids by reductive amination resembles the biosynthesis of glutamic acid. (p. 1110)

complete proteins Proteins that provide all the essential amino acids in about the right proportions for human nutrition. Examples of complete proteins are those in meat, fish, milk, and eggs. **Incomplete proteins** are severely deficient in one or more of the essential amino acids. Most plant proteins are incomplete. (p. 1106)

conjugated protein A protein that contains a nonprotein prosthetic group such as a sugar, nucleic acid, lipid, or metal ion. (p. 1135)

C terminus The end of the peptide chain with a free or derivatized carboxyl group. As the peptide is written, the C terminus is usually on the right. The amino group of the C-terminal amino acid links it to the rest of the peptide. (p. 1119)

denaturation An unnatural alteration of the conformation or the ionic state of a protein. Denaturation generally results in precipitation of the protein and loss of its biological activity. Denaturation may be reversible, as in the salting out of a protein, or irreversible, as in the cooking of egg white. (p. 1139)

dipolar ion (zwitterion) A structure with an overall charge of zero but having a positively charged substituent and a negatively charged substituent. Most amino acids exist in dipolar ionic forms. (p. 1106)

disulfide linkage (disulfide bridge) A bond between two cysteine residues formed by mild oxidation of their thiol groups to a disulfide. (p. 1120)

Edman degradation A method for removing and identifying the N-terminal amino acid from a peptide without destroying the rest of the peptide chain. The peptide is treated with phenylisothiocyanate, followed by a mild acid hydrolysis to convert the N-terminal amino acid to its phenylthiohydantoin derivative. The Edman degradation can be used repeatedly to determine the sequence of many residues beginning at the N terminus. (p. 1124)

electrophoresis A procedure for analyzing mixtures of amino acids and proteins by observing their migration under the influence of a strong electric field. The direction and rate of migration are governed largely by the average charge on the molecules of each type of protein or amino acid. (p. 1108)

enzymatic resolution The use of enzymes to separate enantiomers. For example, the enantiomers of an amino acid can be acetylated and then treated with hog kidney acylase. The enzyme hydrolyzes the acyl group from the natural L-amino acid, but it does not react with the D-amino acid. The resulting mixture of the free L-amino acid and the acetylated D-amino acid is easily separated. (p. 1114)

enzyme A protein-containing biological catalyst. Many enzymes also include *prosthetic groups*, nonprotein constituents that are essential to the enzyme's catalytic activity. (p. 1136)

essential amino acids Ten standard amino acids that are not biosynthesized by human beings and must be provided in the diet. (p. 1105)

fibrous proteins A class of proteins that are stringy, tough, threadlike, and usually insoluble in water. (p. 1135)

globular proteins A class of proteins that are relatively spherical in shape. Globular proteins generally have lower molecular weights and are more soluble in water than fibrous proteins. (p. 1135)

α-helix A helical peptide conformation in which the carbonyl groups on one turn of the helix are hydrogen-bonded to N—H hydrogens on the next turn. The extensive hydrogen bonding makes this helical arrangement quite stable. (p. 1136)

hydrogenolysis Cleavage of a bond by the addition of hydrogen. For example, catalytic hydrogenolysis cleaves benzyl esters. (p. 1116)

$$\underset{\text{benzyl ester}}{R-\overset{O}{\overset{\|}{C}}-O-CH_2-C_6H_5} \xrightarrow{H_2,\ Pd} \underset{\text{acid}}{R-\overset{O}{\overset{\|}{C}}-O-H} + \underset{\text{toluene}}{H-CH_2-C_6H_5}$$

isoelectric point (isoelectric pH) The pH at which an amino acid (or protein) does not move under electrophoresis. This is the pH at which the average charge on its molecules is zero, with most of the molecules in their zwitterionic form. (p. 1107)

L-amino acid An amino acid having a stereochemical configuration similar to that of L-(—)-glyceraldehyde. Most naturally occurring amino acids have the L configuration. (p. 1102)

N terminus The end of the peptide chain with a free or derivatized amino group. As the peptide is written, the N terminus is usually on the left. The carboxyl group of the N-terminal amino acid links it to the rest of the peptide. (p. 1119)

oligopeptide A small polypeptide, containing about four to ten amino acid residues. (p. 1119)

peptide Any polymer of amino acids linked by amide bonds between the amino group of each amino acid and the carboxyl group of the neighboring amino acid. The terms *dipeptide, tripeptide,* and so on, may be used to specify the number of amino acids in the peptide. (p. 1119)

peptide bonds The amide linkages between amino acids. (p. 1118)

pleated sheet A two-dimensional peptide conformation with the peptide chains lined up side by side. The carbonyl groups on one peptide chain are hydrogen-bonded to N—H hydrogens on the adjacent chain, and the side chains are arranged on alternating sides of the sheet. (p. 1136)

polypeptide A peptide containing many amino acid residues. Although proteins are polypeptides, the term *polypeptide* is commonly used for molecules with lower molecular weights than proteins. (p. 1119)

primary structure The covalently bonded structure of a protein; the sequence of amino acids, together with any disulfide bridges. (p. 1136)

prosthetic group The nonprotein part of a conjugated protein. Examples of prosthetic groups are sugars, lipids, nucleic acids, and metal complexes. (p. 1135)

protein A bipolymer of amino acids. Proteins are polypeptides with molecular weights higher than about 6000. (p. 1119)

quaternary structure The association of two or more peptide chains into a complete protein. (p. 1137)

random coil A type of protein secondary structure where the chain is neither curled into an α-helix nor lined up in a pleated sheet. In a globular protein, the kinks that fold the molecule into its globular shape are usually segments of random coil. (p. 1137)

residue An amino acid unit of a peptide. (p. 1119)

Sanger method A method for determining the N-terminal amino acid of a peptide. The peptide is treated with 2,4-dinitrofluorobenzene (Sanger's reagent), then completely hydrolyzed. The derivitized amino acid is easily identified, but the rest of the peptide is destroyed in the hydrolysis. (p. 1126)

secondary structure The local hydrogen-bonded arrangement of a protein. The secondary structure is generally the α-helix, pleated sheet, or random coil. (p. 1136)

sequence As a noun, the order in which amino acids are linked together in a peptide. As a verb, to determine the sequence of a peptide. (p. 1123)

simple proteins Proteins that hydrolyze to give only amino acids (having no prosthetic groups). (p. 1135)

solid-phase peptide synthesis A process developed by R. B. Merrifield to synthesize peptides using solid polystyrene beads as a support for the growing peptide chains. (p. 1131)

standard amino acids The 20 α-amino acids that are found in nearly all proteins. (p. 1103)

Strecker synthesis The synthesis of α-amino acids by the reaction of an aldehyde with ammonia and cyanide ion, followed by hydrolysis of the intermediate α-amino nitrile. (p. 1112)

terminal-residue analysis Sequencing a peptide by removing and identifying the residue at the N-terminus or at the C-terminus. (p. 1124)

tertiary structure The complete three-dimensional conformation of a protein. (p. 1137)

transamination The transfer of an amino group from one molecule to another. Transamination is a common method for the biosynthesis of amino acids, often involving glutamic acid as the source of the amino group. (p. 1110)

zwitterion (dipolar ion) A structure with an overall charge of zero but having a positively charged substituent and a negatively charged substituent. Most amino acids exist in zwitterionic forms. (p. 1106)

ESSENTIAL PROBLEM-SOLVING SKILLS IN CHAPTER 24

1. Correctly name amino acids and peptides, and draw their structures from their names.
2. Use perspective drawings and Fischer projections to show the stereochemistry of D and L amino acids.
3. Explain which amino acids are acidic, which are basic, and which are neutral. Use the isoelectric point to predict whether a given amino acid will be positively charged, negatively charged, or neutral at a given pH.
4. Show how one of the following syntheses might be used to make a given amino acid:

   reductive amination
   HVZ followed by ammonia
   Gabriel-malonic ester synthesis
   Strecker synthesis

5. Predict the products of the following reactions of amino acids: esterification, acylation, reaction with ninhydrin.
6. Use information from terminal residue analysis and partial hydrolysis to determine the structure of an unknown peptide.
7. Show how a classical peptide synthesis or a solid-phase peptide synthesis would be used to make a given peptide. Be careful to use appropriate protecting groups to prevent unwanted couplings.
8. Discuss and identify the four levels of protein structure (primary, secondary, tertiary, quaternary). Explain how the structure of a protein affects its properties and how denaturation changes the structure.

STUDY PROBLEMS

24-29. Give a definition and an example for each of the following terms.

(a) α-amino acid **(b)** conjugated protein **(c)** protein denaturation
(d) dipolar ion **(e)** disulfide bridge **(f)** Edman degradation
(g) enzymatic resolution **(h)** essential amino acid **(i)** hydrogenolysis
(j) isoelectric point **(k)** L-amino acid **(l)** peptide
(m) prosthetic group **(n)** primary structure **(o)** secondary structure
(p) tertiary structure **(q)** quaternary structure **(r)** solid-phase peptide synthesis
(s) Strecker synthesis **(t)** zwitterion

24-30. Draw the complete structure of the following peptide.

$$\text{Val-Gln-Met}\cdot NH_2$$

24-31. Predict the products of the following reactions.

(a) Phe + (ninhydrin: indane-1,3-dione with 2,2-(OH)$_2$) $\xrightarrow[\text{heat}]{\text{pyridine}}$

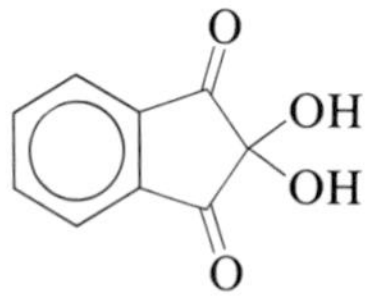

(b) $Ph-CH_2-O-\overset{O}{\overset{\|}{C}}-NH-\overset{CH_3}{\overset{|}{C}H}-COOH \xrightarrow{H_2,\ Pd}$

(c) Lys + excess $(CH_3CO)_2O \longrightarrow$

(d) (D,L)-proline $\xrightarrow[\text{(2) hog kidney acylase, } H_2O]{\text{(1) excess } Ac_2O}$

(e) $CH_3CH_2-\overset{CHO}{\overset{|}{C}H}-CH_3 \xrightarrow[H_2O]{NH_3,\ HCN}$

(f) product from part (e) $\xrightarrow{H_3O^+}$

(g) isovaleric acid + $Br_2/PBr_3 \longrightarrow$

(h) product from part (g) + excess $NH_3 \longrightarrow$

24-32. Show how you would synthesize any of the standard amino acids from each of the following starting materials. You may use any necessary additional reagents.

(a) $(CH_3)_2CH-C(=O)-COOH$ **(c)** $(CH_3)_2CH-CH_2-CHO$

(b) $CH_3-CH(CH_2CH_3)-CH_2-COOH$ **(d)** 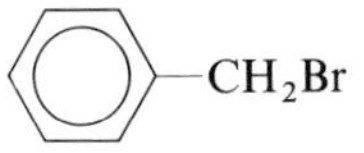 $C_6H_5-CH_2Br$

24-33. Show how you would convert alanine to the following derivatives. Show the structure of the product in each case.
(a) alanine isopropyl ester **(b)** *N*-benzoylalanine
(c) *N*-benzyloxycarbonyl alanine **(d)** *t*-butyloxycarbonyl alanine

24-34. Suggest a method for the synthesis of the unnatural D enantiomer of alanine from the readily available L enantiomer of lactic acid.

$$CH_3-CHOH-COOH$$

lactic acid

24-35. Show how you would use the Gabriel-malonic ester synthesis to make histidine. What stereochemistry would you expect in your synthetic product?

24-36. Show how you would use the Strecker synthesis to make isoleucine. What stereochemistry would you expect in your synthetic product?

24-37. Write the complete structures for the following peptides. Tell whether each peptide is acidic, basic, or neutral.
(a) methionylthreonine **(b)** threonylmethionine **(c)** arginylleucyllysine **(d)** Glu-Cys-Gln

24-38. The following structure is drawn in an unconventional manner.

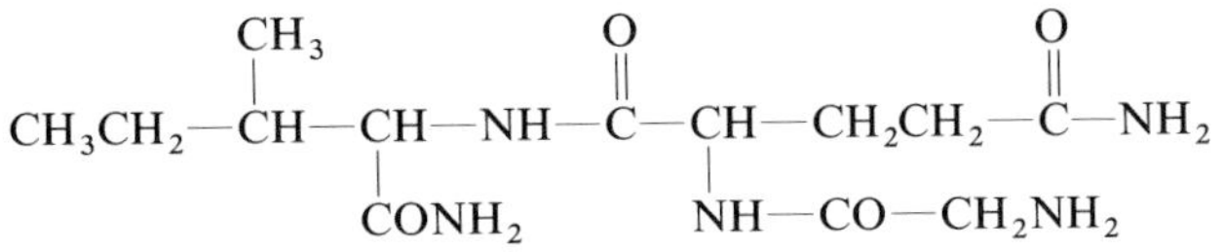

$$CH_3CH_2-CH(CH_3)-CH(CONH_2)-NH-C(=O)-CH(NH-CO-CH_2NH_2)-CH_2CH_2-C(=O)-NH_2$$

(a) Label the N terminus and the C terminus. **(b)** Label the peptide bonds.
(c) Identify and label each amino acid present. **(d)** Give the full name and the abbreviated name.

24-39. *Aspartame* (Nutrasweet®) is a remarkably sweet-tasting dipeptide ester. Complete hydrolysis of aspartame gives phenylalanine, aspartic acid, and methanol. Mild incubation with carboxypeptidase has no effect on aspartame. Treatment of aspartame with phenyl isothiocyanate followed by mild hydrolysis gives the phenylthiohydantoin of aspartic acid. Propose a structure for aspartame.

24-40. A molecular weight determination has shown that an unknown peptide is a pentapeptide, and an amino acid analysis shows that it contains the following residues: one Gly, two Ala, one Met, one Phe.

Treatment of the original pentapeptide with carboxypeptidase gives alanine as the first free amino acid released. Sequential treatment of the pentapeptide with phenyl isothiocyanate followed by mild hydrolysis gives the following derivatives:

first time: phenylthiohydantoin (Ph—N, NH, C=S, C=O) with CH$_2$Ph and H substituents

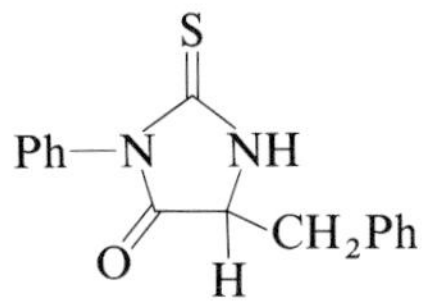

second time: phenylthiohydantoin (Ph—N, NH, C=S, C=O) with CH$_3$ and H substituents

third time: phenylthiohydantoin (Ph—N, NH, C=S, C=O), unsubstituted

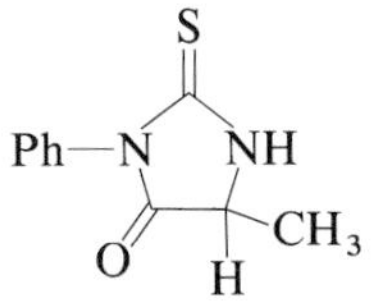

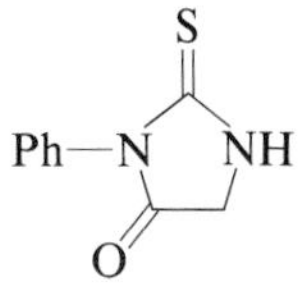

Propose a structure for the unknown pentapeptide.

24-41. Show the steps and intermediates in the synthesis of Ile-Leu-Phe
(a) by the classical process. **(b)** by the solid-phase process.

24-42. Using classical solution techniques, show how you would synthesize Ala-Val and then combine it with Ile-Leu-Phe to give Ile-Leu-Phe-Ala-Val.

24-43. Peptides often have functional groups other than free amino groups at the N terminus and other than carboxyl groups at the C terminus.

(a) A tetrapeptide is hydrolyzed by heating with 6 *M* HCl, and the hydrolysate is found to contain Ala, Phe, Val, and Glu. When the hydrolysate is neutralized, the odor of ammonia is detected. Explain where this ammonia might have been incorporated in the original peptide.

(b) The tripeptide *thyrotropic hormone releasing factor* (TRF) has the full name pyroglutamylhistidylprolinamide and the structure appearing below. Explain the functional groups at the N terminus and at the C terminus.

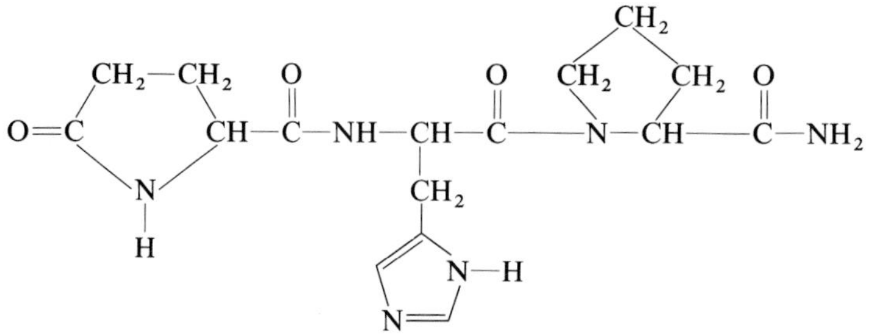

(c) Upon acidic hydrolysis, an unknown pentapeptide gave glycine, alanine, valine, leucine, and isoleucine. No odor of ammonia was detected when the hydrolysate was neutralized. Reaction with phenyl isothiocyanate followed by mild hydrolysis gave *no* phenylthiohydantoin derivative. Incubation with carboxypeptidase had no effect. Explain these findings.

24-44. Lipoic acid is often found near the active sites of enzymes, usually bound to the peptide by a long, flexible amide linkage with a lysine residue.

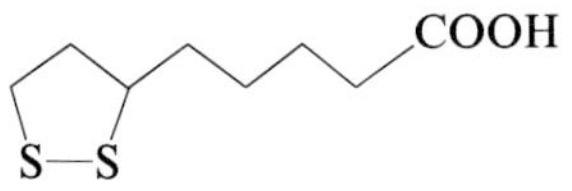

lipoic acid

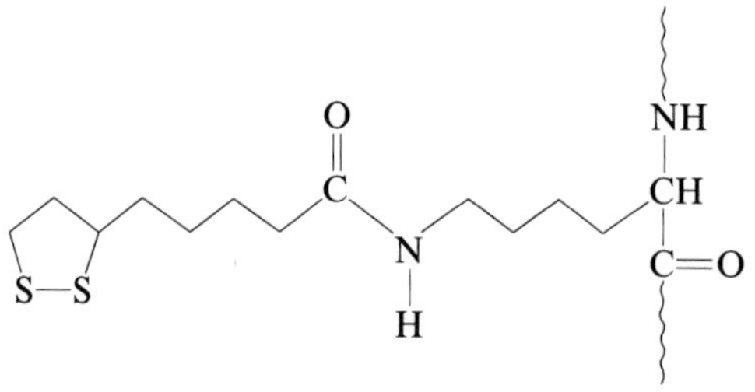

bound to lysine residue

(a) Is lipoic acid a mild oxidizing agent or a mild reducing agent? Draw it in both its oxidized and reduced forms.

(b) Show how lipoic acid might react with two Cys residues to form a disulfide bridge.

(c) Give a balanced equation for the hypothetical oxidation or reduction, as you predicted in part (a), of an aldehyde by lipoic acid.

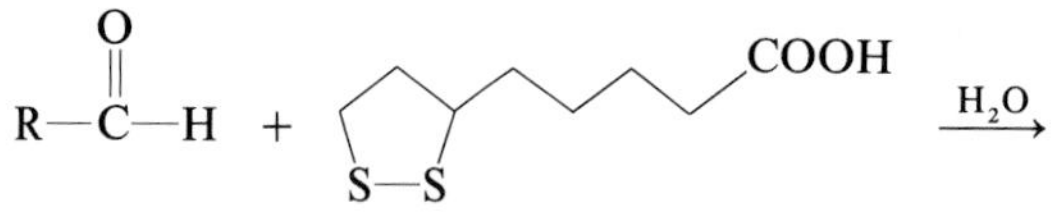

24-45. Histidine is an important catalytic residue found at the active sites of many enzymes. In many cases, histidine appears to remove protons or to transfer protons from one location to another.

(a) Show which nitrogen atom of the histidine heterocycle is basic and which is not.

(b) Use resonance structures to show why the protonated form of a histidine residue is a particularly stable cation.

(c) Show the structure that results when the His residue accepts a proton on the basic nitrogen of the heterocycle and then is deprotonated on the other heterocyclic nitrogen. Explain how His might function as a pipeline to transfer protons between sites within an enzyme and its substrate.

24-46. Metabolism of arginine produces urea and the rare amino acid *ornithine*. Ornithine has an isoelectric point close to 10. Propose a structure for ornithine.

24-47. Glutathione is a tripeptide that serves as a mild reducing agent to maintain the cysteine residues of hemoglobin and other red blood cell proteins in the reduced state. Complete hydrolysis of glutathione gives Gly, Glu, and Cys. Treatment of glutathione with carboxypeptidase gives glycine as the first free amino acid released. Treatment of glutathione with phenyl isothiocyanate gives the phenylthiohydantoin of glutamic acid. Propose a structure consistent with this information.

24-48. The complete hydrolysis of an unknown basic decapeptide gives Gly, Ala, Leu, Ile, Phe, Tyr, Glu, Arg, Lys, and Ser. Terminal residue analysis shows that the N terminus is Ala and the C terminus is Ile. Incubation of the decapeptide with chymotrypsin gives two tripeptides, **A** and **B**, and a tetrapeptide, **C**. Amino acid analysis shows that peptide **A** contains Gly, Glu, Tyr, and NH_3; peptide **B** contains Ala, Phe, and Lys; and peptide **C** contains Leu, Ile, Ser, and Arg. Terminal residue analysis gives the following results.

	N terminus	*C terminus*
A	Glu	Tyr
B	Ala	Phe
C	Arg	Ile

Incubation of the decapeptide with trypsin gives a dipeptide **D**, a pentapeptide **E**, and a tripeptide **F**. Terminal residue analysis of **F** shows that the N terminus is Ser and the C terminus is Ile. Propose a structure for the decapeptide and for fragments **A** through **F**.

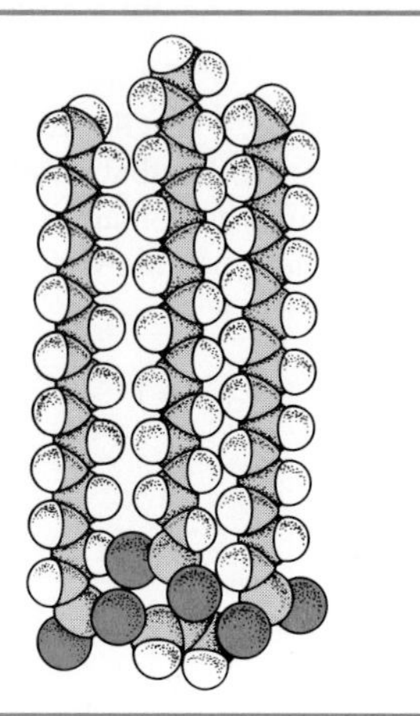

25

LIPIDS

25-1 INTRODUCTION

In our study of organic chemistry, we have usually classified compounds according to their functional groups. Lipids, however, are classified by their solubility: **Lipids** are substances that can be extracted from cells and tissues by nonpolar organic solvents.

Lipids include many types of compounds containing a wide variety of functional groups. You could easily prepare a solution of lipids by grinding a T-bone steak in a blender and then extracting the puree with chloroform or diethyl ether. The resulting solution of lipids would contain a multitude of compounds, many with very complex structures. To facilitate the study of lipids, chemists have divided this large family of compounds into two major classes: complex lipids and simple lipids.

Complex lipids are those that are easily hydrolyzed to simpler constituents. Most complex lipids are esters of long-chain carboxylic acids called *fatty acids*. The two major groups of fatty acid esters are *waxes* and *glycerides*. Waxes are esters of long-chain alcohols, and glycerides are esters of glycerol.

Simple lipids are those that are not easily hydrolyzed by aqueous acid or base. This term often seems inappropriate, because many so-called "simple" lipids are quite complex molecules. We will consider three important groups of simple lipids: steroids, prostaglandins, and terpenes. Figure 25-1 shows some examples of complex and simple lipids.

25-2 WAXES

Waxes are esters of long-chain fatty acids with long-chain alcohols. They occur widely in nature and serve a number of different purposes in plants and animals. *Spermaceti* (Fig. 25-1) is found in the head of the sperm whale and probably helps

Examples of complex lipids

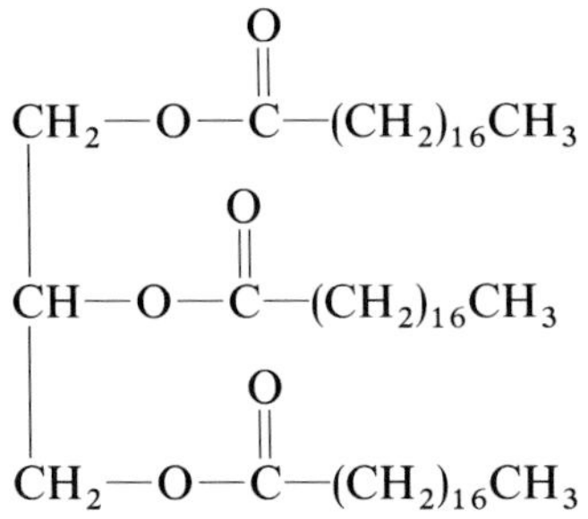

$$\begin{array}{l} CH_2-O-\overset{O}{\overset{\|}{C}}-(CH_2)_{16}CH_3 \\ | \\ CH-O-\overset{O}{\overset{\|}{C}}-(CH_2)_{16}CH_3 \\ | \\ CH_2-O-\overset{O}{\overset{\|}{C}}-(CH_2)_{16}CH_3 \end{array}$$

tristearin, a fat

$$CH_3(CH_2)_{15}-O-\overset{O}{\overset{\|}{C}}-(CH_2)_{14}CH_3$$

spermaceti (cetyl palmitate), a wax

Examples of simple lipids

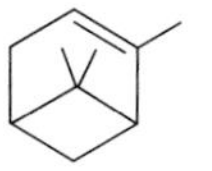

α-pinene, a terpene

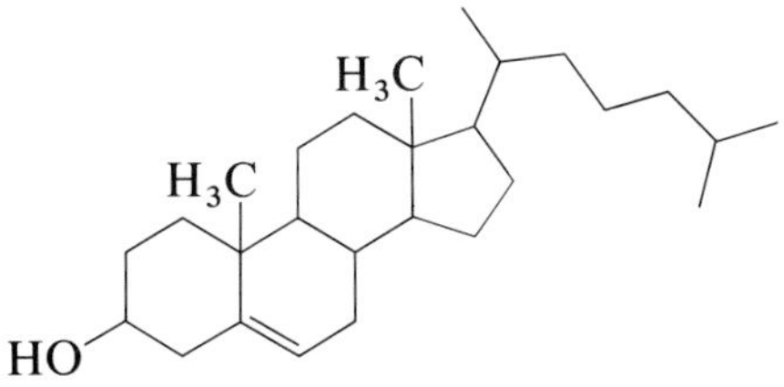

H_3C H_3C HO

cholesterol, a steroid

FIGURE 25-1 Complex lipids contain ester functional groups that can be hydrolyzed to acids and alcohols. Simple lipids are not easily hydrolyzed.

to regulate the animal's buoyancy for deep diving. It may also serve to amplify high-frequency sounds for locating prey. *Beeswax* is a high-molecular-weight material used by bees to form a honeycomb. *Carnauba wax* is a mixture of waxes of very high molecular weights. The carnauba plant secretes this waxy material that coats its leaves to prevent excessive loss of water by evaporation. In contrast to these waxes, the "paraffin wax" used to seal preserves is not a true wax; rather, it is a mixture of alkanes of high molecular weight.

$$CH_3(CH_2)_{29}-O-\overset{O}{\overset{\|}{C}}-(CH_2)_{24}CH_3$$

a component of beeswax

$$CH_3(CH_2)_{33}-O-\overset{O}{\overset{\|}{C}}-(CH_2)_{26}CH_3$$

a component of carnauba wax

For many years, natural waxes were used in making cosmetics, adhesives, varnishes, and waterproofing materials. Synthetic materials have now replaced natural waxes for most of these uses.

25-3 TRIGLYCERIDES

Glycerides are simply fatty acid esters of the triol *glycerol.* The most common glycerides are **triglycerides,** or **triacylglycerols,** in which all three of the glycerol —OH groups have been esterified by fatty acids. For example, tristearin (Fig. 25-1) is a component of beef fat in which all three—OH groups of glycerol are esterified by stearic acid, $CH_3(CH_2)_{16}COOH$.

Triglycerides are commonly called **fats** if they are solid at room temperature, and **oils** if they are liquid at room temperature. Most triglycerides derived from mammals are fats, such as beef fat or lard. Although these fats are solid at room temperature, the warm body temperature of the living animal keeps them somewhat fluid, allowing for body movement. In plants and cold-blooded animals, triglycerides are generally oils, such as corn oil, peanut oil, or fish oil. A fish requires liquid oils rather than solid fats because it would have difficulty moving if the triglycerides solidified in a cold stream.

The **fatty acids** of common triglycerides are long, straight-chain carboxylic acids with about 12 to 20 carbon atoms. Most fatty acids contain even numbers of carbon atoms because they are derived from two-carbon acetic acid units. Some of the common fatty acids are saturated, while others have one or more elements of unsaturation: generally carbon-carbon double bonds. Table 25-1 shows the structures of some common fatty acids derived from fats and oils.

TABLE 25-1
Structures and melting points of some common fatty acids

Name	*Carbons*	*Structure*	*Melting point (°C)*
saturated acids			
lauric acid	12	 COOH	44
myristic acid	14	 COOH	59
palmitic acid	16	 COOH	64
stearic acid	18	 COOH	70
arachidic acid	20	 COOH	76
unsaturated acids			
oleic acid	18	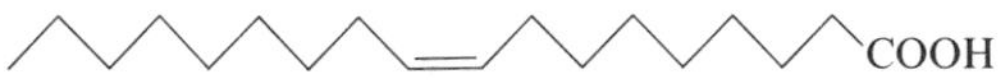 COOH	4
linoleic acid	18	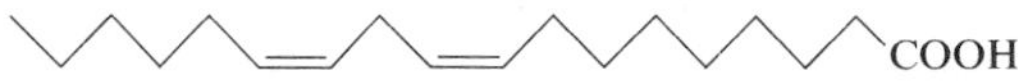 COOH	−5
linolenic acid	18	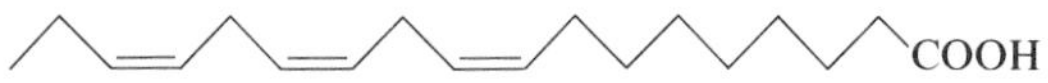 COOH	−11
eleostearic acid	18	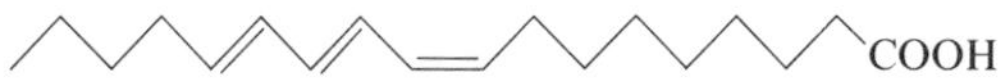 COOH	49
arachidonic acid	20	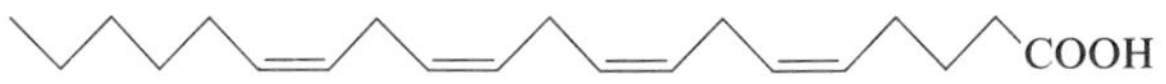 COOH	−49

PROBLEM 25-1
Trimyristin, a solid fat present in nutmeg, is hydrolyzed to give 1 equivalent of glycerol and 3 equivalents of myristic acid. Give the structure of trimyristin.

Table 25-1 shows that the saturated fatty acids have melting points that increase gradually with their molecular weights. The presence of a cis double bond in the fatty acid lowers its melting point, however. Notice that the C18 saturated acid (stearic acid) has a melting point of 70°C, while the C18 acid with a cis double bond (oleic acid) has a melting point of 4°C. This lowering of the melting point results from the unsaturated acid's "kink" at the position of the double bond (Fig. 25-2). The kinked molecules cannot pack as tightly together in the solid as the uniform zigzag chains of the saturated acid.

Addition of a second double bond again lowers the melting point (linoleic acid, m.p. −5°C), and a third double bond lowers it still further (linolenic acid, m.p. −11°C). The trans double bonds in eleostearic acid (m.p. 49°C) have a smaller effect on the melting point than the cis double bonds of linolenic acid. The geometry of the trans double bond is similar to the zigzag conformation of a saturated acid, and it does not kink the chain as much as a cis double bond.

Fats and oils also have melting points that depend on the amount of unsaturation (especially cis double bonds) in their fatty acids. A triglyceride derived

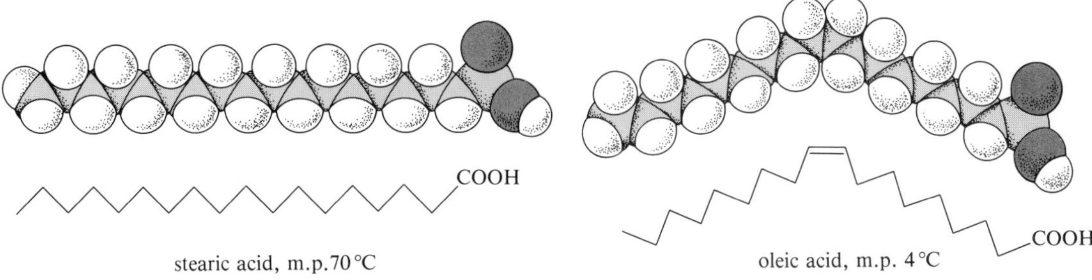

FIGURE 25-2 The presence of the cis double bond in oleic acid lowers the melting point by 66°C.

from saturated fatty acids has a higher melting point because it packs more easily into a solid lattice than a triglyceride derived from kinked, unsaturated fatty acids. Figure 25-3 shows typical conformations of triglycerides containing saturated and unsaturated fatty acids. Tristearin (m.p. 72°C) is a saturated fat that packs well in a solid lattice. Triolein (m.p. −4°C) has the same number of carbon atoms as tristearin, but triolein contains three cis double bonds, whose kinked conformations prevent optimum packing in the solid.

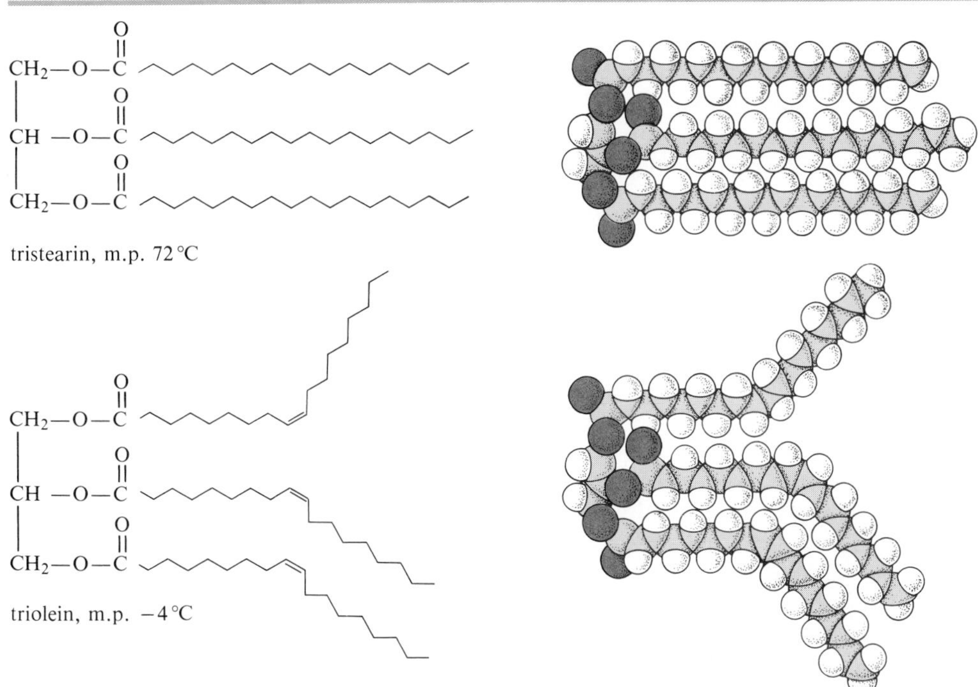

FIGURE 25-3 Unsaturated triglycerides have lower melting points because their unsaturated fatty acids do not pack as well in a solid lattice.

Most saturated triglycerides are *fats,* because they are solids at room temperature. Most triglycerides with several unsaturations are *oils,* because they are liquids at room temperature. The term *polyunsaturated* simply means that there are several double bonds in the fatty acids of the oil.

Most naturally occurring fats and oils are actually mixtures of various triglycerides containing a variety of saturated and unsaturated fatty acids. In general, oils taken from plants and cold-blooded animals contain more unsaturations than fats taken from warm-blooded animals. Table 25-2 gives the approximate composition of the fatty acids obtained from hydrolysis of some common fats and oils.

TABLE 25-2
Fatty acid composition of some fats and oils, percent by weight.

	Saturated fatty acids				Unsaturated fatty acids		
Source	*Lauric*	*Myristic*	*Palmitic*	*Stearic*	*Oleic*	*Linoleic*	*Linolenic*
beef fat	0	6	27	14	49	2	0
lard	0	1	24	9	47	10	0
human fat	1	3	27	8	48	10	0
herring oil	0	5	14	3	0	0	30[a]
corn oil	0	1	10	3	50	34	0
olive oil	0	0.1	7	2	84	5	0
soybean oil	0.2	0.1	10	2	29	51	7

[a] Contains large amounts of even more highly unsaturated fatty acids.

For many years, *lard* (a soft, white solid obtained by rendering animal fat) was commonly used for cooking and baking. Although vegetable oil could be produced cheaper and in greater quantities, consumers were unwilling to use vegetable oils because they were accustomed to using white, creamy lard. Then vegetable oils were treated with hydrogen gas and a nickel catalyst, reducing some of the double bonds to give a creamy, white *vegetable shortening* that resembles lard. This "partially hydrogenated vegetable oil" largely replaced lard for cooking and baking. *Margarine* is a similar material flavored with butyraldehyde to give it a taste like that of butter. More recently, consumers have learned that "polyunsaturated" vegetable oils are more easily digested, prompting a switch to natural vegetable oils.

PROBLEM 25-2

Give an equation for the complete hydrogenation of triolein using an excess of hydrogen. What is the name of the product, and what are the melting points of the starting material and the product?

25-4 SAPONIFICATION OF FATS AND OILS: SOAPS AND DETERGENTS

Saponification is the base-catalyzed hydrolysis of fats and oils. One of the products is soap, and the word *saponification* is derived from the Latin word *saponis*, meaning "soap." Saponification was discovered (before 500 B.C.), when it was found that a curdy material resulted when animal fat was heated with wood ashes. Alkaline substances in the ashes promoted hydrolysis of the ester linkages in the fat. Soap is currently made by boiling animal fat or vegetable oil with a solution of sodium hydroxide. The following reaction shows the formation of soap from tristearin, a component of beef fat.

$$\underset{\text{tristearin, a fat}}{\begin{array}{l} CH_2-O-C(=O)-(CH_2)_{16}CH_3 \\ | \\ CH-O-C(=O)-(CH_2)_{16}CH_3 \\ | \\ CH_2-O-C(=O)-(CH_2)_{16}CH_3 \end{array}} + 3\,NaOH \xrightarrow[H_2O]{\text{heat}} \left\{ \begin{array}{c} \underset{\text{glycerol}}{\begin{array}{l} CH_2-OH \\ | \\ CH-OH \\ | \\ CH_2-OH \end{array}} \\ + \underset{\text{sodium stearate, a soap}}{3\,Na^+\,{}^-O-C(=O)-(CH_2)_{16}CH_3} \end{array} \right.$$

Chemically, a **soap** is the sodium or potassium salt of a fatty acid. The negatively charged carboxylate group is strongly hydrophilic (attracted to water), and the long hydrocarbon chain is hydrophobic (repelled by water) and lipophilic (attracted to oils). In water, soap forms a cloudy solution of **micelles:** clusters of about 100 to 200 soap molecules with their polar "heads" (the carboxylate groups) on the surface of the cluster and their hydrophobic "tails" (the hydrocarbon chains) enclosed within. The micelle (Fig. 25-4) is an energetically stable particle because the hydrophilic groups are hydrogen-bonded to the surrounding water, while the hydrophobic groups are shielded within the interior of the micelle, interacting with other hydrophobic groups.

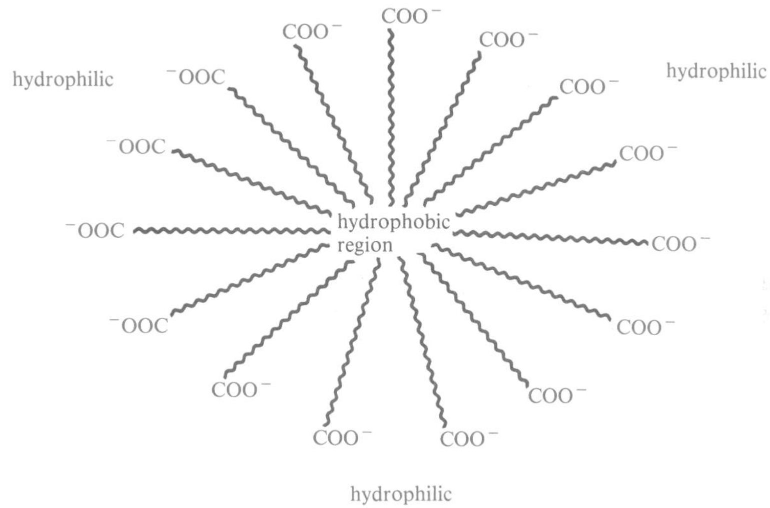

FIGURE 25-4 In a soap micelle, the hydrocarbon chains are enclosed within a cluster with the carboxylate groups on the surface.

Soaps are useful as cleaning agents because of the different affinities of the soap molecule's two ends. Greasy dirt is not easily removed by pure water because the grease is hydrophobic and insoluble in water. Soap, however, has a long hydrocarbon chain that dissolves in the grease, with its hydrophilic head at the surface of the grease droplet. Once the surface of the grease droplet is covered by many soap molecules, a micelle can form with a tiny grease droplet at its center. This grease droplet is easily suspended in water because it is covered by the hydrophilic carboxylate groups of the soap (Fig. 25-5). The resulting mixture of

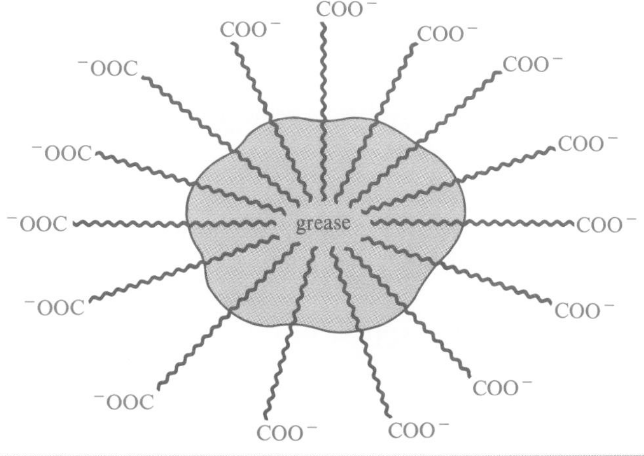

FIGURE 25-5 In a soapy solution, grease is emulsified by forming micelles coated by the hydrophilic carboxylate groups of the soap.

two insoluble phases (grease and water), with one phase dispersed throughout the other in small droplets, is called an **emulsion.** We say that the grease has been **emulsified** by the soapy solution.

The usefulness of soaps is limited by their tendency to precipitate out of solution in "hard" water. **Hard water** is water that is acidic or that contains ions of calcium, magnesium, or iron. In acidic water (such as the "acid rain" of environmental concern), soap molecules are protonated to give the free fatty acids. Without the ionized carboxylate group, the fatty acid floats to the top as a greasy "acid scum" precipitate.

$$\underset{\text{a soap}}{CH_3(CH_2)_n{-}\overset{\overset{\displaystyle O}{\|}}{C}{-}O^-\ {}^+Na} + H^+ \longrightarrow \underset{\text{acid scum}}{CH_3(CH_2)_n{-}\overset{\overset{\displaystyle O}{\|}}{C}{-}OH\downarrow} + Na^+$$

Many areas have household water containing calcium, magnesium, and iron ions. Although these mineral-rich waters can be healthful for drinking, the ions react with soaps to form insoluble salts called *hard-water scum.* The following equation shows the reaction of a soap with calcium, common in areas where the water comes in contact with limestone rocks.

$$\underset{\text{a soap}}{2\ CH_3(CH_2)_n{-}\overset{\overset{\displaystyle O}{\|}}{C}{-}O^-\ {}^+Na} + Ca^{+2} \longrightarrow \underset{\text{hard-water scum}}{[CH_3(CH_2)_n{-}\overset{\overset{\displaystyle O}{\|}}{C}{-}O]_2Ca\downarrow} + 2\ Na^+$$

PROBLEM 25-3

Give equations to show the reactions of sodium stearate with

(a) Ca^{2+} (b) Mg^{2+} (c) Fe^{3+}

The precipitation of soaps in hard water results from the chemical properties of the carboxylic acid group. **Synthetic detergents** avoid precipitation by using other functional groups in place of carboxylic acid salts. The sodium salts of sulfonic acids are the most widely used class of synthetic detergents. Sulfonic acids are more acidic than carboxylic acids, so their salts are not protonated even in strongly acidic wash water. The calcium, magnesium, and iron salts of sulfonic acids are soluble in water, so sulfonate salts can be used in hard water without forming a scum (Fig. 25-6).

Other kinds of synthetic detergents use both cationic and anionic hydrophilic groups, as well as nonionic hydrophilic groups containing several oxygen atoms or other hydrogen-bonding atoms. Figure 25-6 shows the structures of a typical sulfonate detergent, a cationic detergent, and a nonionic detergent.

PROBLEM 25-4

The synthesis of the alkylbenzenesulfonate detergent shown in Figure 25-6 begins with the partial polymerization of propylene to give a pentamer.

$$5\ H_2C{=}CH{-}CH_3 \xrightarrow{\text{acidic catalyst}} \text{a pentamer}$$

Show how Friedel-Crafts reactions can convert this pentamer to the final synthetic detergent.

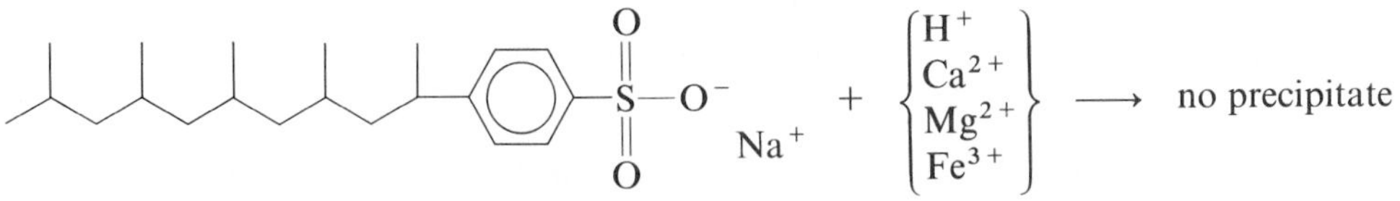

$$\text{(branched alkyl)}-C_6H_4-SO_3^-\ Na^+ + \left\{\begin{matrix} H^+ \\ Ca^{2+} \\ Mg^{2+} \\ Fe^{3+} \end{matrix}\right\} \longrightarrow \text{no precipitate}$$

an alkylbenzenesulfonate

Examples of other types of detergents

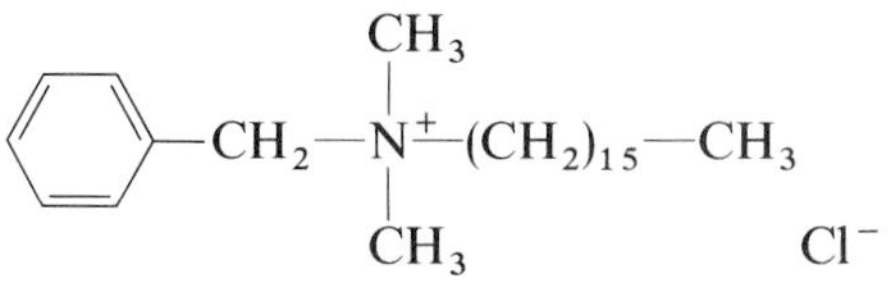

$$C_6H_5-CH_2-\overset{+}{N}(CH_3)_2-(CH_2)_{15}-CH_3 \quad Cl^-$$

benzylcetyldimethylammonium chloride
(benzalkonium chloride)

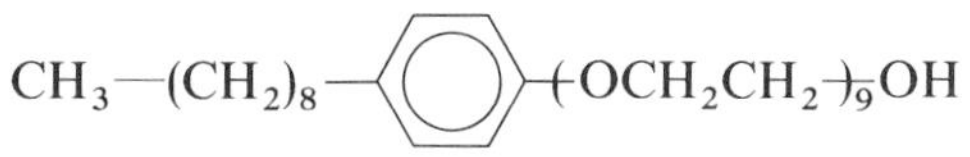

$$CH_3-(CH_2)_8-C_6H_4-(OCH_2CH_2)_9OH$$

Nonoxynol, Ortho Pharmaceuticals

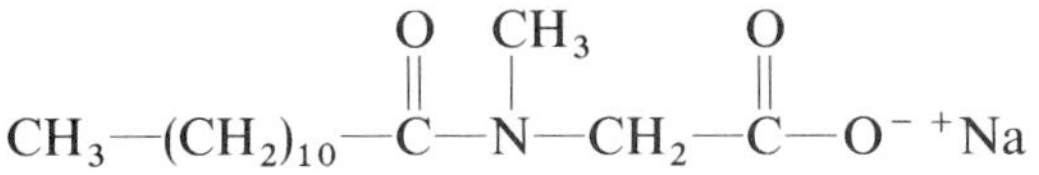

$$CH_3-(CH_2)_{10}-\overset{O}{\overset{\|}{C}}-\overset{CH_3}{\overset{|}{N}}-CH_2-\overset{O}{\overset{\|}{C}}-O^-\ {}^+Na$$

N-lauroyl-*N*-methylglycine, sodium salt
Gardol, Colgate-Palmolive Co.

FIGURE 25-6 Synthetic detergents may have anionic, cationic, or nonionic hydrophilic functional groups. Of these detergents, only Gardol is a carboxylate salt and forms a precipitate in hard water.

PROBLEM 25-5

Draw a diagram, similar to Figure 25-5, of an oil droplet emulsified by the alkylbenzenesulfonate detergent shown in Figure 25-6.

25-5 PHOSPHOLIPIDS

Phospholipids are lipids that contain groups derived from phosphoric acid. The most common phospholipids are **phosphoglycerides,** which are closely related to common fats and oils. A phosphoglyceride generally has a phosphoric acid group in place of one of the fatty acid groups of a triglyceride. The simplest class of phosphoglycerides are **phosphatidic acids,** which consist of glycerol esterified by two fatty acids and one phosphoric acid group. Although it is commonly drawn in its acid form, a phosphatidic acid is actually deprotonated at neutral pH.

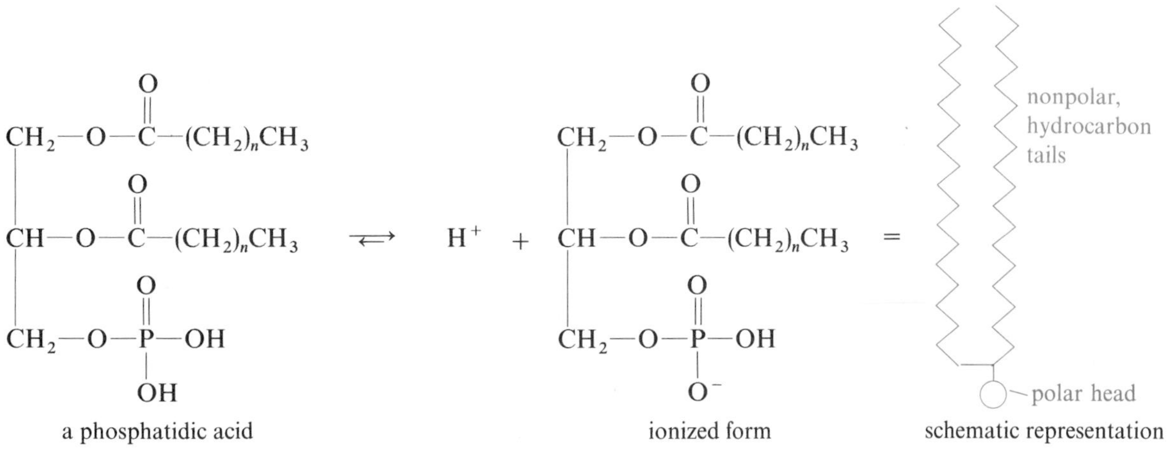

$$\begin{array}{l} CH_2-O-\overset{O}{\overset{\|}{C}}-(CH_2)_nCH_3 \\ | \\ CH-O-\overset{O}{\overset{\|}{C}}-(CH_2)_nCH_3 \\ | \\ CH_2-O-\overset{O}{\overset{\|}{P}}(OH)-OH \end{array} \rightleftharpoons H^+ + \begin{array}{l} CH_2-O-\overset{O}{\overset{\|}{C}}-(CH_2)_nCH_3 \\ | \\ CH-O-\overset{O}{\overset{\|}{C}}-(CH_2)_nCH_3 \\ | \\ CH_2-O-\overset{O}{\overset{\|}{P}}(O^-)-OH \end{array} =$$

a phosphatidic acid — ionized form — schematic representation

PROBLEM 25-6

Draw the important resonance structures for a phosphatidic acid that has lost (a) one proton. (b) two protons.

Many phospholipids contain an additional alcohol esterified to the phosphoric acid group. **Cephalins** are esterified with ethanolamine, and **lecithins** are esterified with choline. Both cephalins and lecithins are widely found in plant and animal tissues.

$HO{-}CH_2CH_2{-}NH_2$
ethanolamine

$HO{-}CH_2CH_2{-}\overset{+}{N}(CH_3)_3$
choline

$CH_2{-}O{-}C(=O){-}$ (nonpolar chain)
$CH{-}O{-}C(=O){-}$ (nonpolar chain)
$CH_2{-}O{-}P(=O)(O^-){-}O{-}CH_2CH_2{-}\overset{+}{N}H_3$ (polar)

a cephalin,
or phosphatidyl ethanolamine

$CH_2{-}O{-}C(=O){-}$ (nonpolar chain)
$CH{-}O{-}C(=O){-}$ (nonpolar chain)
$CH_2{-}O{-}P(=O)(O^-){-}O{-}CH_2CH_2{-}\overset{+}{N}(CH_3)_3$ (polar)

a lecithin,
or phosphatidyl choline

Like phosphatidic acids, lecithins and cephalins contain a polar "head" and two long, nonpolar hydrocarbon "tails." This soaplike structure gives phospholipids some interesting properties. Like soaps, they form micelles and other aggregations with their polar heads on the outside and their nonpolar tails protected on the inside.

Another form this aggregation can take is a **lipid bilayer,** with the heads on the two surfaces of a membrane and the tails protected within. Cell membranes contain phosphoglycerides oriented in a lipid bilayer, forming a barrier that restricts the flow of dissolved substances. Figure 25-7 shows the arrangement of phospholipids in a bilayer membrane.

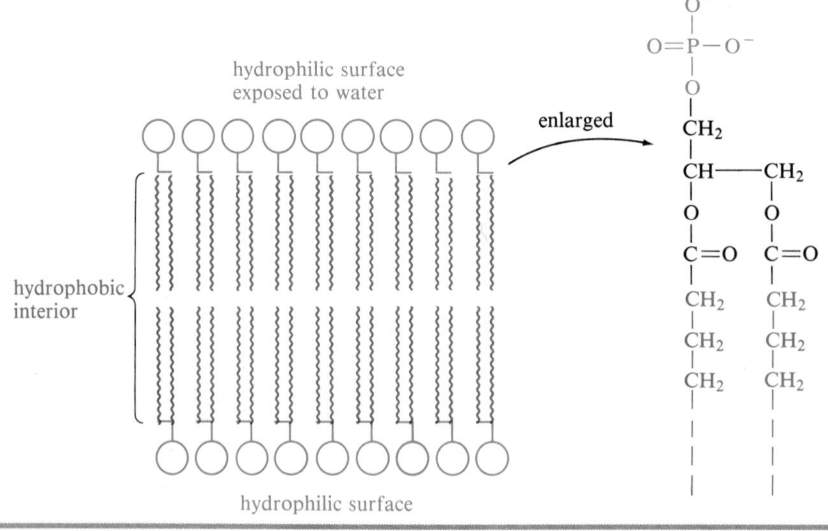

FIGURE 25-7 Phosphoglycerides can aggregate into a bilayer membrane with their polar heads exposed to the aqueous solution and the hydrocarbon tails protected within. This lipid bilayer is an important part of the cell membrane.

25-6 STEROIDS

Steroids are complex polycyclic molecules found in all plants and animals. They are *simple lipids* because they do not undergo hydrolysis as fats, oils, and waxes do. Steroids serve a wide variety of functions, but the most important ones are those that function as hormones. **Steroids** are defined as compounds whose structure is based on the tetracyclic androstane ring system, shown below. The four rings are designated A, B, C, and D, beginning with the ring at the lower left, and the carbon atoms are numbered beginning with the A ring and ending with the two "angular" (axial) methyl groups.

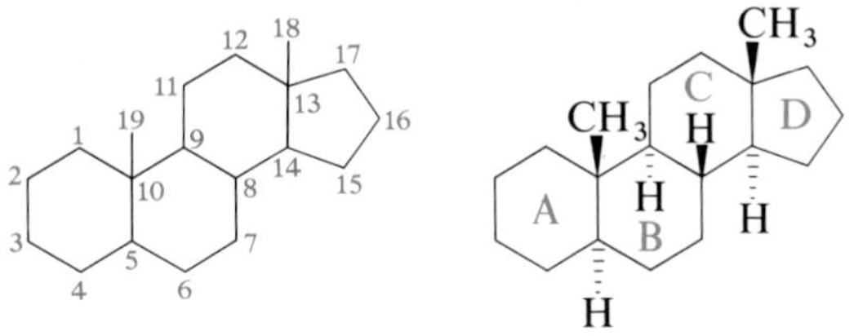

androstane

We have seen (Section 3-15B) that fused ring systems such as androstane can have either trans or cis stereochemistry at each ring junction. A simple example is the geometric isomerism of *trans*- and *cis*-decalin shown in Figure 25-8. If you make models of these isomers, you will find that the trans isomer is quite rigid and rather flat (aside from the ring puckering). In contrast, the cis isomer is relatively flexible, and it has the two rings situated at a sharp angle to each other.

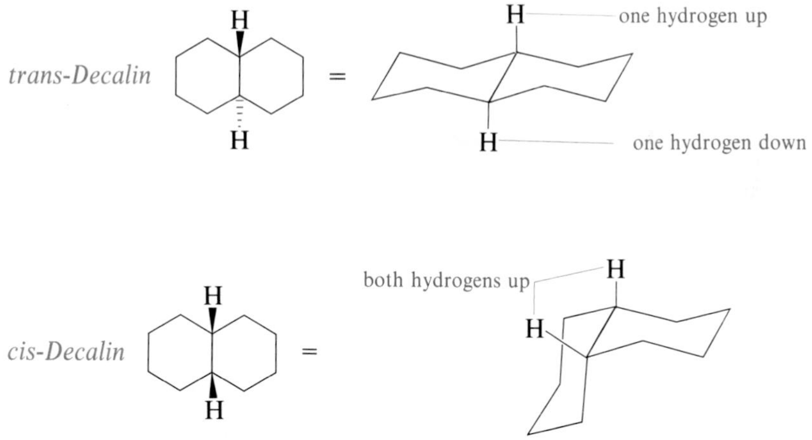

trans-Decalin = one hydrogen up; one hydrogen down

cis-Decalin = both hydrogens up

FIGURE 25-8 Geometric isomers of decalin. In *trans*-decalin the two bonds to the second ring are trans to one another, and the hydrogens on the junction are also trans. In *cis*-decalin the bonds to the second ring are cis, and the junction hydrogens are also cis.

In the structure above, each of the ring junctions in androstane is shown to be trans. Most steroids have this all-trans structure, which results in a stiff, nearly flat molecule with the two axial methyl groups perpendicular to the plane. In some steroids the junction between rings A and B is cis, requiring the A ring to fold down below the rest of the ring system. Figure 25-9 shows the androstane ring system with both a trans and a cis A-B ring junction. The B-C and C-D ring junctions are nearly always trans in the natural steroids.

Most steroids have an oxygen functional group (=O or —OH) at C3, and some kind of a side chain or other functional group at C17. Many also have a double bond from C5 to either C4 or C6. The structures of androsterone and cholesterol serve as examples. Androsterone, a male sex hormone, is based on the

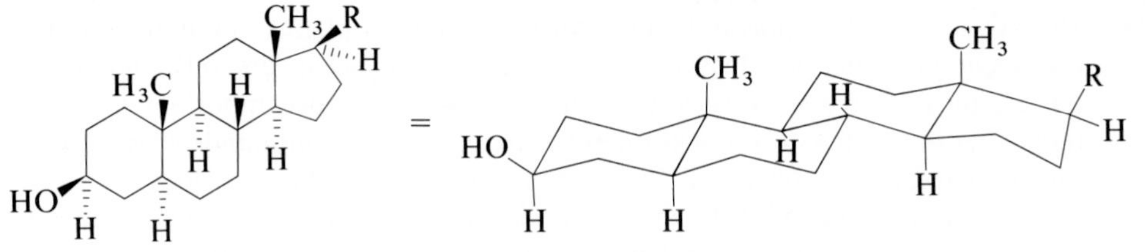

a trans A-B steroid

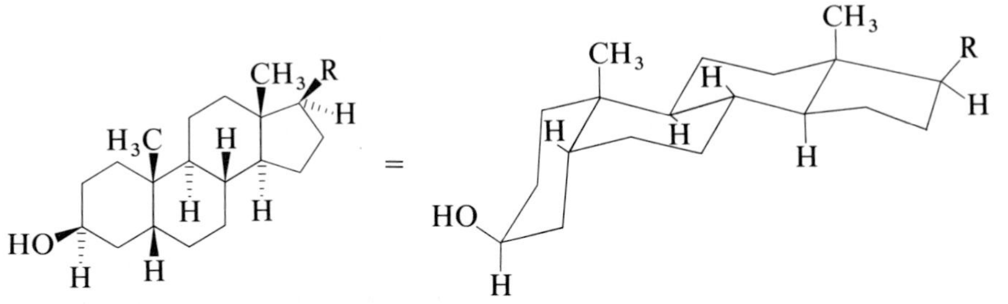

a cis A-B steroid

FIGURE 25-9 Common steroids may have either a cis or a trans A-B ring junction. The other ring junctions are normally trans.

simple androstane ring system. Cholesterol, a common biological intermediate, has a side chain at C17 and a double bond between C5 and C6.

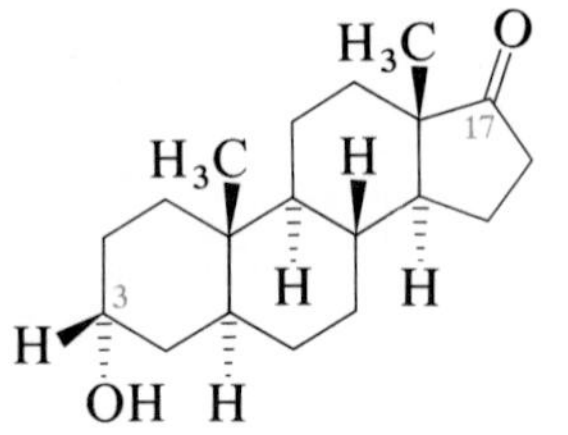

androsterone

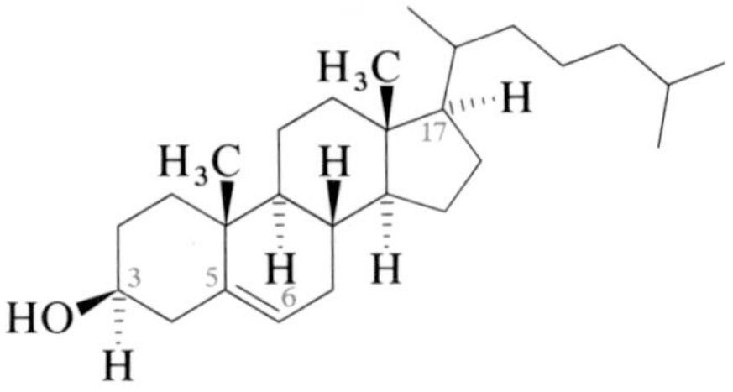

cholesterol

The principal sex hormones have been characterized and studied extensively. Testosterone is the most potent of the natural male sex hormones, and estradiol is the most potent natural female hormone. Notice that the female sex hormone differs from the male sex hormone by its aromatic A ring. For the A ring to be aromatic, the C19 methyl group must be lost.

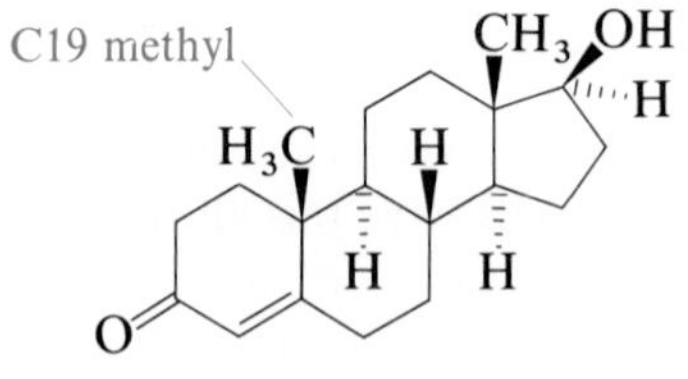

C19 methyl

testosterone

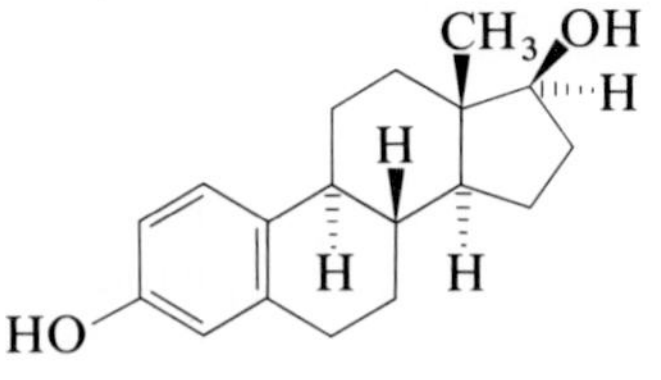

estradiol

PROBLEM 25-7

How would you separate a mixture of testosterone and estradiol using a simple extraction?

When steroid hormones were first isolated, people believed that no synthetic hormone could rival the high potency of the natural steroids. In the past 20 years, however, many synthetic steroids have been developed. Some of these synthetic hormones are hundreds or thousands of times more potent than the natural steroids. One example is ethynyl estradiol, a synthetic female hormone that is more potent than estradiol.

ethynyl estradiol

Some of the most important medicinal steroids are the adrenocortical hormones, synthesized by the adrenal cortex. Most of these hormones have either a carbonyl group or a hydroxyl group at C11 of the steroid skeleton. The principal adrenocortical hormone is cortisol, used for the treatment of inflammatory diseases of the skin (psoriasis), the joints (rheumatoid arthritis), and the lungs (asthma). Figure 25-10 compares the structure of natural cortisol with two synthetic corticoids: fluocinolone acetonide, a fluorinated synthetic hormone that is more potent than cortisol for treating skin inflammation; and beclomethasone, a chlorinated synthetic hormone that is more potent than cortisol for treating asthma.

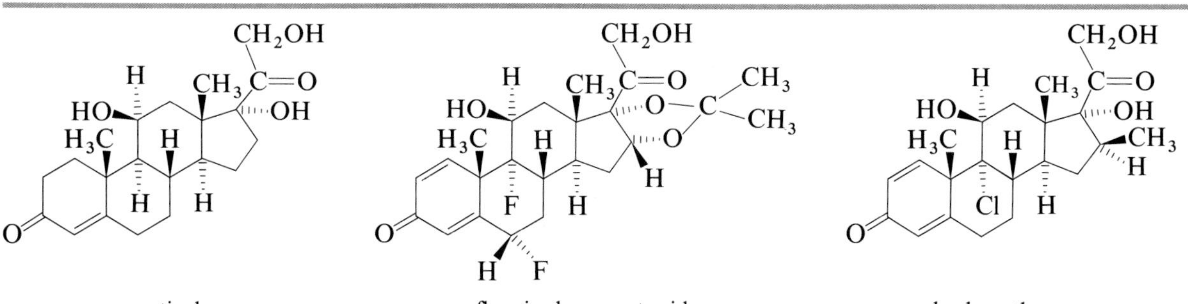

FIGURE 25-10 Cortisol is the major natural hormone of the adrenal cortex. Fluocinolone acetonide is more potent for treating skin inflammation, and beclomethasone is more potent for treating asthma.

PROBLEM 25-8

Draw each molecule in a stable chair conformation, and tell whether each red group is axial or equatorial.

(a)

(b)

(c) 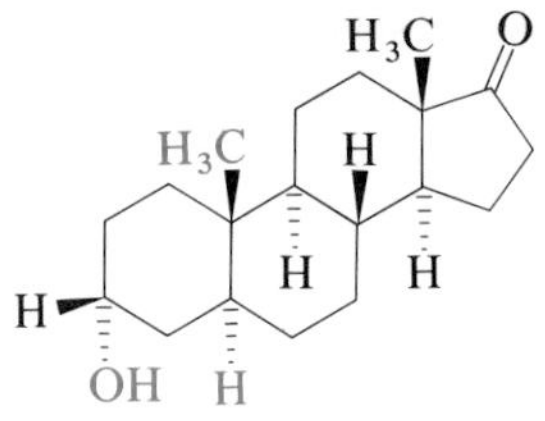 androsterone

(d) 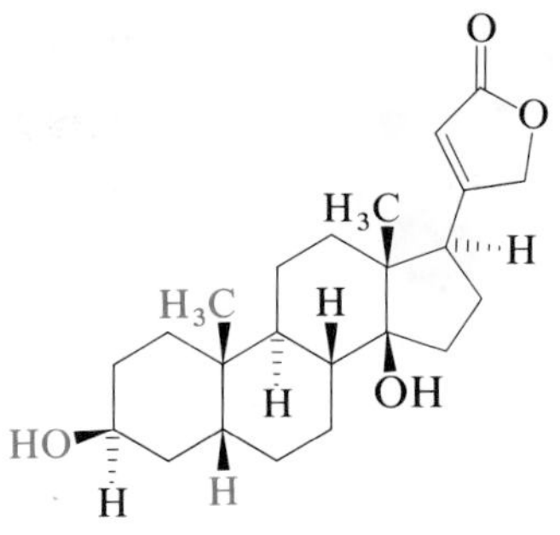 digitoxigenin, a cardiac stimulant

25-7 PROSTAGLANDINS

Prostaglandins are fatty acid derivatives that are even more powerful hormones than steroids. They are called prostaglandins because they were first isolated from secretions of the prostate gland. They were later found to be present in all body tissues and fluids, usually in minute quantities. Prostaglandins affect many different systems of the body, including the nervous system, smooth muscle, blood, and the reproductive system. They play important roles in regulating such diverse factors as blood pressure, blood clotting, the allergic inflammatory response, the activity of the digestive system, and the onset of labor.

The general prostaglandin structure is a cyclopentane ring having two long side chains trans to each other, with one side chain ending in a carboxylic acid. Most prostaglandins have 20 carbon atoms, numbered as follows:

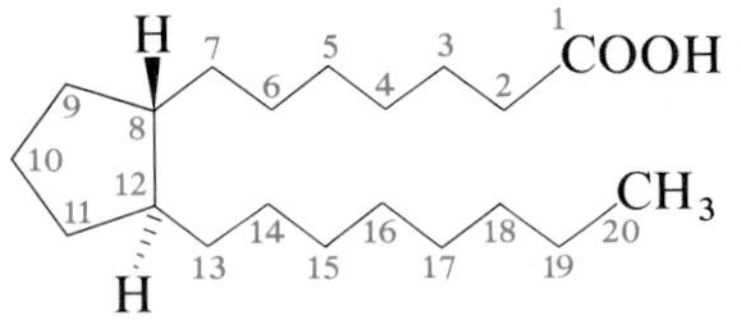

Many prostaglandins have hydroxyl groups on C11 and C15, and a trans double bond between C13 and C14. They also have a carbonyl group or a hydroxyl group on C9. If there is a carbonyl group at C9, the prostaglandin is a member of the *E series.* If there is a hydroxyl group at C9, it is a member of the *F series,* and the symbol α means that the hydroxyl group is directed down. Many prostaglandins have a cis double bond between C5 and C6. The number of double bonds in the molecule is also given in the name, as shown below for two common prostaglandins.

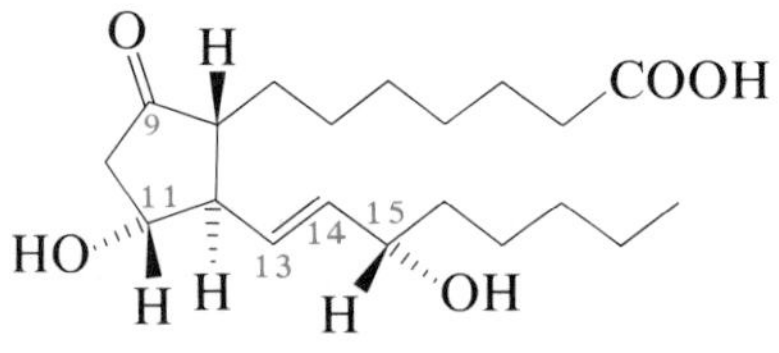

PGE_1

(PG means prostaglandin;
E means ketone at C9;
1 means one C=C double bond)

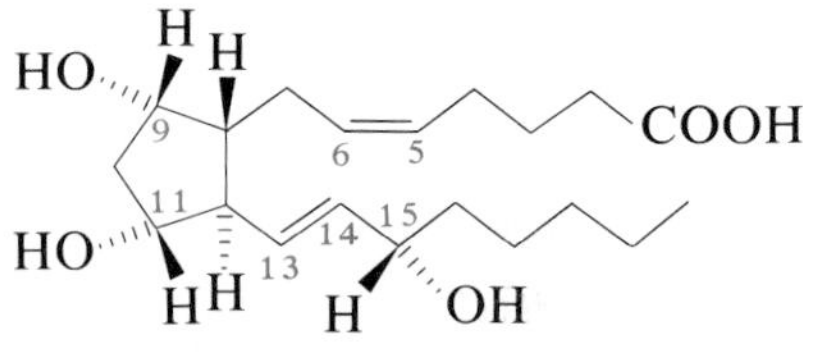

$PGF_{2\alpha}$

(PG means prostaglandin;
F means hydroxyl at C9, and α means down;
2 means two C=C double bonds)

Prostaglandins are derived from arachidonic acid, a 20-carbon fatty acid with four cis double bonds. Figure 25-11 shows schematically how an enzymatic cyclo-oxidation converts arachidonic acid to the prostaglandin skeleton.

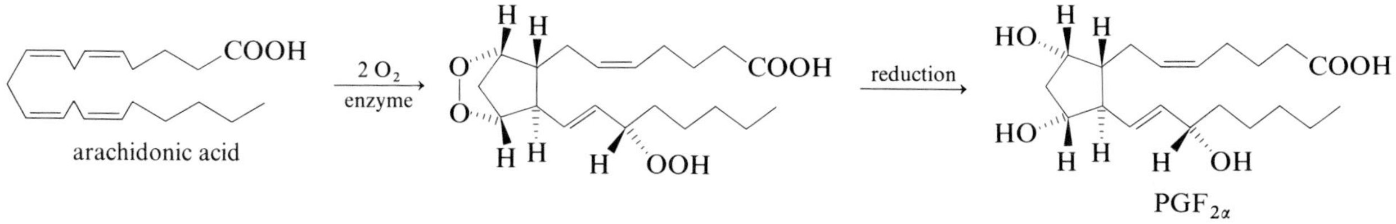

arachidonic acid $\xrightarrow[\text{enzyme}]{2\,O_2}$ (cyclic endoperoxide hydroperoxide) $\xrightarrow{\text{reduction}}$ $PGF_{2\alpha}$

FIGURE 25-11 The biosynthesis of prostaglandins begins by an enzyme-catalyzed oxidative cyclization of arachidonic acid.

25-8 TERPENES

Terpenes are a diverse family of compounds with carbon skeletons composed of five-carbon isopentyl (isoprene) units. Terpenes are commonly isolated from the **essential oils** of plants: the fragrant oils that are concentrated from the plant material, usually by steam distillation. Essential oils often have pleasant tastes or aromas, and they are widely used as flavorings, deodorants, and medicines. Table 25-3 lists several types of common essential oils and their principal components.

TABLE 25-3
Some useful essential oils

Essential oil	*Source*	*Major components*
perfume	flowers	mixtures of terpenes and terpenoids
oil of celery	celery	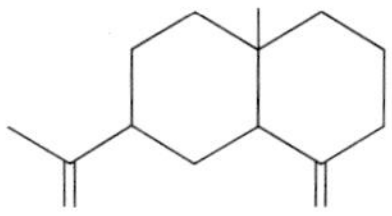 β-selinene, a terpene
oil of anise	anise seed	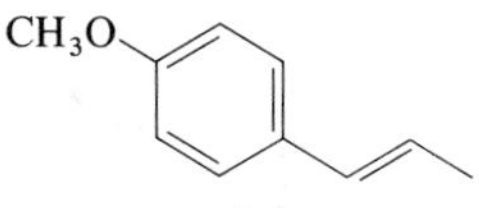 anethole
oil of bay	bay leaves	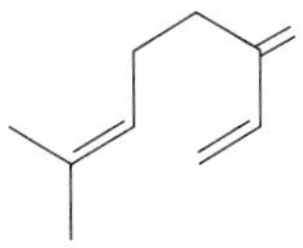 myrcene, a terpene
cedar leaf oil	leaves of the "white cedar" (actually a pine)	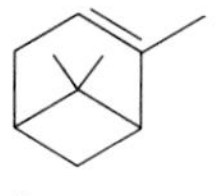 α-pinene, a terpene
oil of turpentine	evergreens	mixtures of terpenes and terpenoids

25-8A CHARACTERISTICS AND NOMENCLATURE OF TERPENES

Hundreds of essential oils were used as perfumes, flavorings, and medicines for centuries before chemistry was capable of studying the mixtures. In 1818, it was found that oil of turpentine has a C:H ratio of 5:8, and that many other essential oils have similar C:H ratios. This group of piney-smelling natural products with similar C:H ratios came to be known as **terpenes.**

In 1887, the German chemist Otto Wallach determined the molecular structures of several terpenes, and discovered a common structural feature: All of them are made up of two or more five-carbon units of *isoprene:* 2-methyl-1,3-butadiene. The isoprene unit maintains its isopentyl structure in a terpene, usually with modification of the isoprene double bonds.

Isoprene

$$H_2C{=}\overset{\displaystyle CH_3}{\overset{|}{C}}{-}CH{=}CH_2 \quad \text{or}$$

An isoprene unit

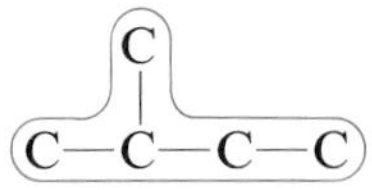

$$C{-}\overset{\displaystyle C}{\overset{|}{C}}{-}C{-}C \quad \text{or}$$

head tail (may have double bonds)

The isoprene molecule and the isoprene unit are said to have a "head" (the branched end) and a "tail" (the unbranched ethyl group). Myrcene can be divided into two isoprene units, with the head of one unit bonded to the tail of the other.

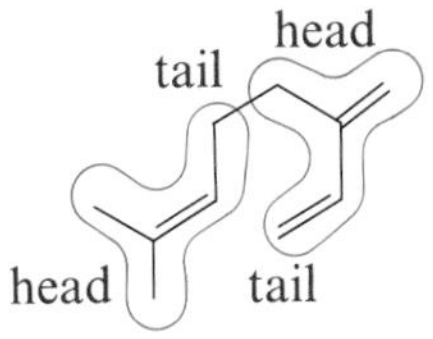

tail head head tail

myrcene 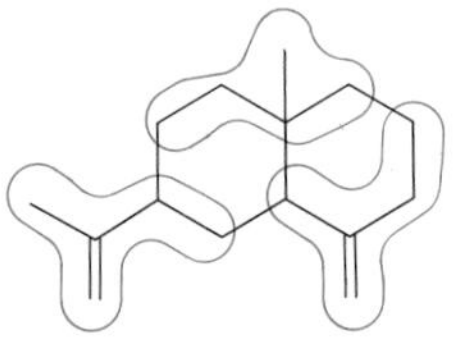 β-selinene

β-Selinene has a more complicated structure, with two rings and a total of 15 carbon atoms. Nevertheless, β-selinene is composed of three isoprene units. Once again, these three units are bonded head to tail, although the additional bonds used to form the rings make the head-to-tail arrangement more difficult to see.

Many terpenes contain additional functional groups, especially carboxyl groups and hydroxyl groups. The structures of a terpene aldehyde, a terpene alcohol, a terpene ketone, and a terpene acid are shown below.

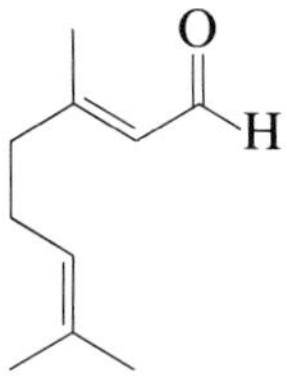

geranial 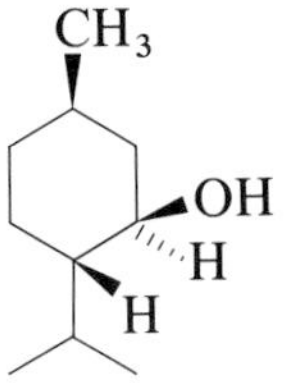 menthol 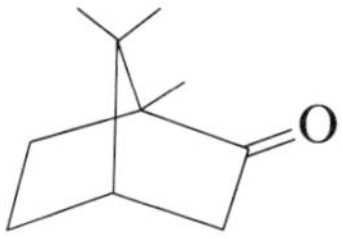 camphor 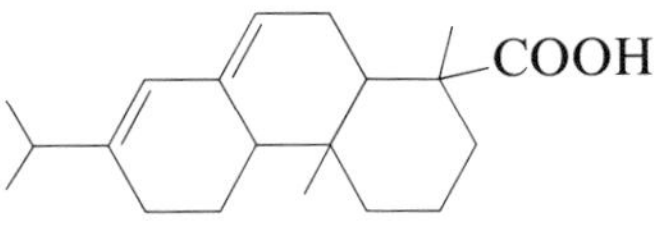 abietic acid

PROBLEM 25-9
Circle the isoprene units in geranial, menthol, camphor, and abietic acid.

25-8B CLASSIFICATION OF TERPENES

Terpenes are classified according to the number of carbon atoms, in units of ten. A terpene with 10 carbon atoms (two isoprene units) is called a **monoterpene,** one with 20 carbon atoms (four isoprene units) is a **diterpene,** and so on. There are also many terpenes with 15 carbon atoms (three isoprene units). These are called **sesquiterpenes,** meaning that they have $1\frac{1}{2}$ times 10 carbon atoms. Myrcene, geranial, menthol, and camphor are monoterpenes, β-selinene is a sesquiterpene, abietic acid is a diterpene, and squalene (Fig. 25-12) is a triterpene.

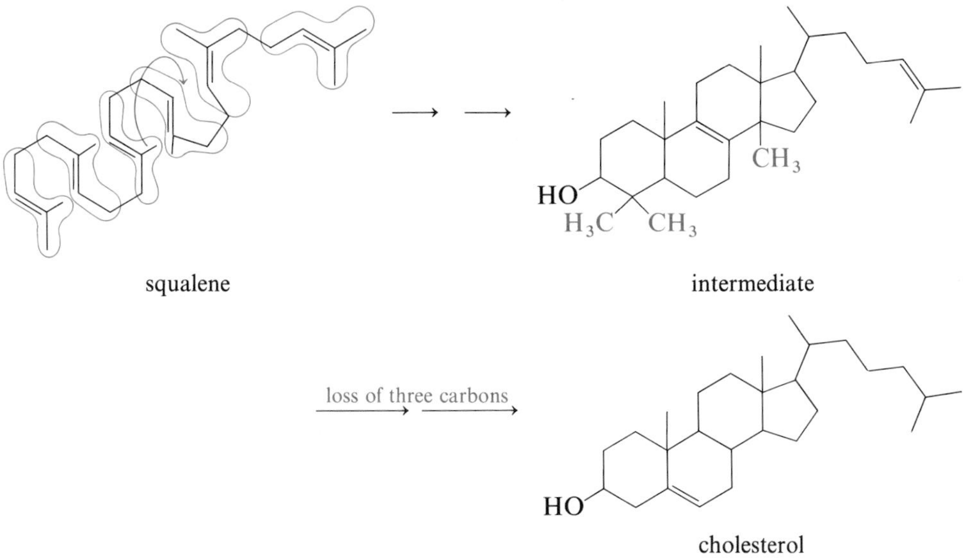

FIGURE 25-12 Cholesterol is a triterpenoid that has lost three carbon atoms from the original six isoprene units in squalene. Another carbon atom has migrated (blue arrow) to form the axial methyl group between rings C and D.

The carotenes, with 40 carbon atoms, are examples of tetraterpenes. Their extended system of conjugated double bonds causes them to be brightly colored. Carotenes are responsible for the pigmentation of carrots, tomatoes, and squash, and they give a fiery color to tree leaves in autumn. β-Carotene is the most common carotene isomer. It can be divided into two head-to-tail diterpenes, linked tail to tail.

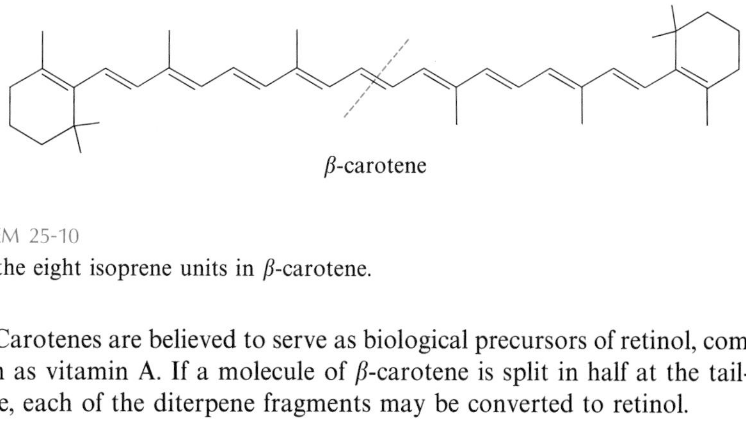

PROBLEM 25-10

Circle the eight isoprene units in β-carotene.

Carotenes are believed to serve as biological precursors of retinol, commonly known as vitamin A. If a molecule of β-carotene is split in half at the tail-to-tail linkage, each of the diterpene fragments may be converted to retinol.

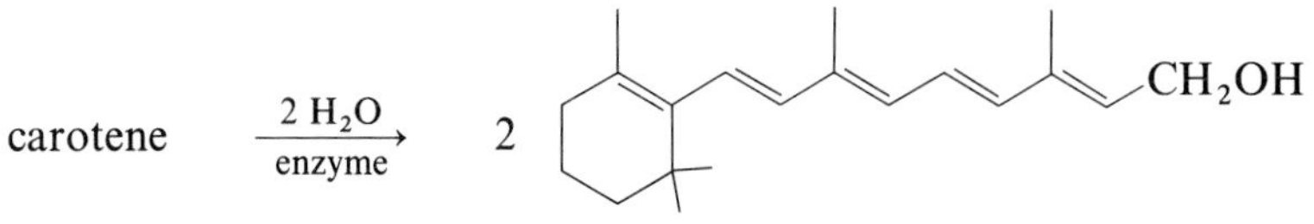

PROBLEM 25-11

(a) Circle the isoprene units in each of the following terpenes.
(b) Classify each of these as a monoterpene, a diterpene, etc.

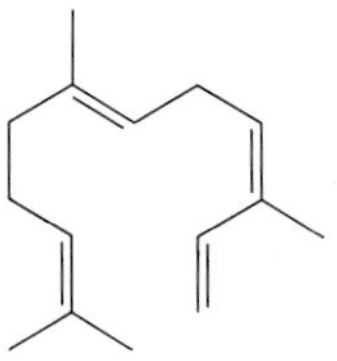

(1) α-farnesene (from oil of citronella)
(2) limonene (from oil of lemon) 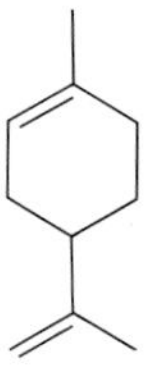
(3) α-pinene (from turpentine) 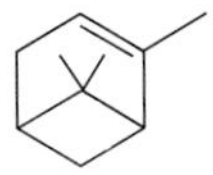
(4) zingiberene (from oil of ginger) 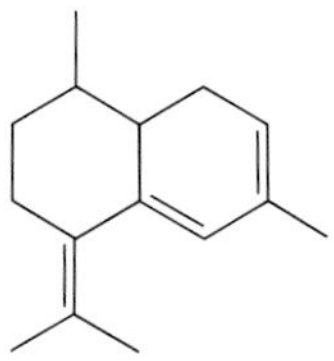

25-8C TERPENOIDS

Many natural products are derived from terpenes, even though they do not have carbon skeletons composed exclusively of C_5 isoprene units. These terpene-like compounds are called **terpenoids.** They may have been altered through rearrangements, loss of carbon atoms, or introduction of additional carbon atoms. Cholesterol is an example of a terpenoid that has lost some of the isoprenoid carbon atoms.

Figure 25-12 shows that cholesterol is formed from six isoprene units (a triterpenoid), with loss of three carbon atoms. The six isoprene units are bonded head to tail, with the exception of one tail-to-tail linkage. The triterpene precursor of cholesterol is believed to be squalene. We can envision an acid-catalyzed cyclization of squalene to give an intermediate that is converted to cholesterol with loss of three carbon atoms.

GLOSSARY

detergent (synthetic detergent) A compound that acts as an emulsifying agent. Some of the common classes of synthetic detergents are alkylbenzenesulfonate salts, alkyl sulfate salts, alkylammonium salts, and nonionic detergents containing several hydroxyl groups or ether linkages. (p. 1152)

emulsify To promote the formation of an emulsion. (p. 1152)

emulsion A mixture of two immiscible liquids, one of which is dispersed throughout the other in small droplets. (p. 1152)

essential oils Fragrant oils that are concentrated from plant material, usually by steam distillation. (p. 1159)

fat A fatty acid triester of glycerol (a triglyceride) that is solid at room temperature. (p. 1147)

fatty acid A long-chain carboxylic acid. Most naturally occurring fatty acids contain even numbers of carbon atoms between 12 and 20. (p. 1148)

glyceride A fatty acid ester of glycerol. (p. 1147)

hard water Water that contains acids or ions (such as Ca^{2+}, Mg^{2+}, or Fe^{3+}) that react with soaps to form precipitates. (p. 1152)

lipid bilayer A form of aggregation of phosphoglycerides with the hydrophilic heads forming the two surfaces of a planar structure and the hydrophobic tails protected within. A lipid bilayer forms part of the cell membrane. (p. 1154)

lipids Substances that can be extracted from cells and tissues by nonpolar organic solvents. (p. 1146)

complex lipids: Lipids that are easily hydrolyzed to simpler constituents, usually by saponification of an ester.

simple lipids: Lipids that are not easily hydrolyzed to simpler constituents.

micelle A cluster of molecules of a soap, phospholipid, or other emulsifying agent suspended in a solvent, usually water. The hydrophilic heads of the molecules are in contact with the solvent, and the hydrophobic tails are enclosed within the cluster. The micelle may or may not contain an oil droplet. (p. 1151)

oil A fatty acid triester of glycerol (a triglyceride) that is liquid at room temperature. (p. 1147)

phosphatidic acids A variety of phospholipids consisting of glycerol esterified by two fatty acids and one phosphoric acid group. (p. 1153)

phosphoglyceride An ester of glycerol in which the three hydroxyl groups are esterified by two fatty acids and a phosphoric acid derivative. (p. 1153)

> **cephalins** (phosphatidyl ethanolamines): A variety of phosphoglycerides with ethanolamine esterified to the phosphoric acid group.
>
> **lecithins** (phosphatidyl cholines): A variety of phosphoglycerides with choline esterified to the phosphoric acid group.

phospholipid A lipid that contains one or more groups derived from phosphoric acid. (p. 1153)

prostaglandins A class of hormones consisting of a 20-carbon carboxylic acid containing a cyclopentane ring and various other functional groups. (p. 1158)

saponification The base-promoted hydrolysis of an ester. Originally used to describe the hydrolysis of fats to make soap. (p. 1150)

soap The alkali metal salt of a fatty acid. (p. 1151)

steroid A compound whose structure is based on the tetracyclic androstane ring system. (p. 1155)

terpenes A diverse family of compounds with carbon skeletons composed of 5 carbon isoprene units. **Monoterpenes** contain 10 carbon atoms, **sesquiterpenes** contain 15, **diterpenes** contain 20, and **triterpenes** contain 30. (p. 1159)

terpenoids A family of compounds including both terpenes and compounds of terpene origin whose carbon skeletons have been altered or rearranged. (p. 1162)

triglyceride (triacylglycerol) A fatty acid triester of glycerol. Triglycerides that are solid at room temperature are *fats,* and those that are liquid are *oils.* (p. 1147)

wax An ester of a long-chain fatty acid with a long-chain alcohol. (p. 1146)

ESSENTIAL PROBLEM-SOLVING SKILLS IN CHAPTER 25

1. Classify lipids both into the large classifications (such as simple lipids, complex lipids, phospholipids, etc.) and into the more specific classifications (such as waxes, triglycerides, cephalins, lecithins, steroids, prostaglandins, terpenes, etc.).
2. Predict the physical properties of fats and oils from their structures.
3. Identify the isoprene units in terpenes and classify terpenes according to the number of carbon atoms.
4. Predict the products of reactions of lipids with standard organic reagents. In particular, the reactions of the ester and olefinic groups of glycerides and the carboxyl groups of fatty acids should be considered.
5. Explain how soaps and detergents work, with particular attention to their similarities and differences.

STUDY PROBLEMS

25-12. Give a definition and an example for each of the following terms.

(a) lipid	**(b)** fat	**(c)** oil	**(d)** fatty acid
(e) wax	**(f)** soap	**(g)** detergent	**(h)** hard water
(i) micelle	**(j)** phospholipid	**(k)** triglyceride	**(l)** simple lipid
(m) complex lipid	**(n)** prostaglandin	**(o)** steroid	**(p)** terpene

25-13. Give the general classification of each compound.
(a) glyceryl tripalmitate

(b) $CH_3-(CH_2)_{10}-CH_2-O-S(=O)_2-O^-\ ^+Na$

sodium lauryl sulfate (in shampoo)

(c) $CH_3-(CH_2)_{13}-O-C(=O)-(CH_2)_{16}-CH_3$

tetradecyl octadecanoate

(d) 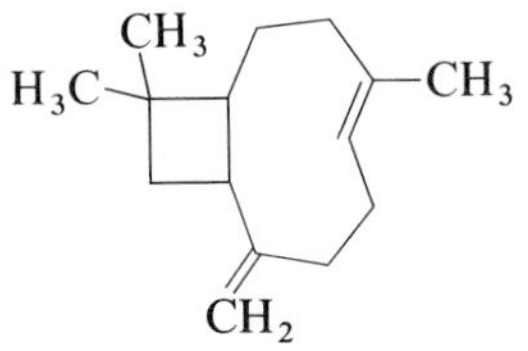

caryophyllene (from cloves)

(e) 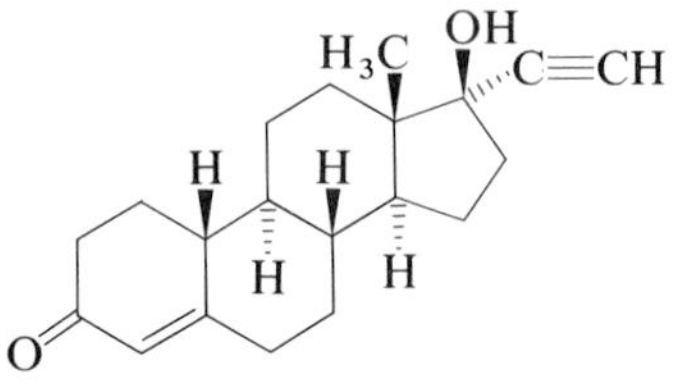

norethindrone
(a synthetic hormone)

25-14. Predict the products obtained from the reaction of triolein with the following reagents.
(a) NaOH in water **(b)** H_2 and a nickel catalyst **(c)** Br_2 in CCl_4
(d) ozone, then dimethyl sulfide **(e)** warm $KMnO_4$ in water **(f)** $CH_2I_2/Zn-Cu$
(g) saponification, then $LiAlH_4$

25-15. Show how you would convert oleic acid to each of the following fatty acid derivatives.
(a) 1-octadecanol **(b)** stearic acid **(c)** octadecyl stearate
(d) nonanal **(e)** nonanedioic acid **(f)** 2,9,10-tribromostearic acid

25-16. Phospholipids undergo saponification much like triglycerides. Draw the structure of a phospholipid meeting the following criteria, then draw the products that would result from its saponification.
(a) a cephalin containing stearic acid and oleic acid **(b)** a lecithin containing palmitic acid

25-17. Some of the earliest synthetic detergents were the sodium alkyl sulfates.

$$CH_3(CH_2)_nCH_2-O-S(=O)_2-O^-\ ^+Na$$

a sodium alkylsulfate detergent

Show how you would make sodium octadecylsulfate using tristearin as your organic starting material.

25-18. Which of the following chemical reactions could be used to distinguish between a polyunsaturated vegetable oil and a petroleum oil containing a mixture of saturated and unsaturated hydrocarbons? Explain your reasoning.
(a) addition of bromine in CCl_4 **(b)** hydrogenation **(c)** saponification **(d)** ozonolysis

25-19. How would you use simple chemical tests to distinguish between the following pairs of compounds?
(a) sodium stearate and *p*-docecylbenzenesulfonate **(b)** beeswax and paraffin “wax”
(c) trimyristin and myristic acid **(d)** trimyristin and triolein

25-20. A triglyceride can be optically active if it contains two or more different fatty acids.
(a) Draw the structure of an optically active triglyceride containing 1 equivalent of myristic acid and 2 equivalents of oleic acid.
(b) Draw the structure of an optically inactive triglyceride with the same fatty acid composition.

25-21. Draw the structure of an optically active triglyceride containing 1 equivalent of stearic acid and 2 equivalents of oleic acid. Draw the products expected when this triglyceride reacts with the following reagents. In each case, predict whether the products will be optically active.
(a) H_2 and a nickel catalyst
(b) Br_2 in CCl_4
(c) hot aqueous NaOH
(d) ozone followed by $(CH_3)_2S$

25-22. Carefully circle the isoprene units in the following terpenes, and label each as a monoterpene, sesquiterpene, or diterpene.

(a) 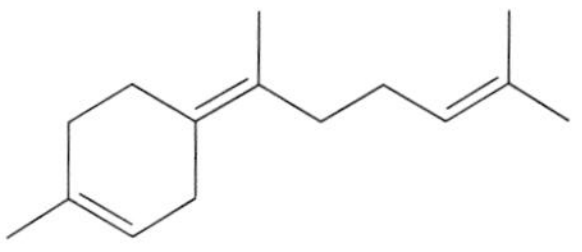 γ-bisabolene

(b) 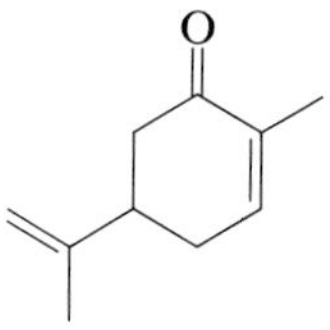 carvone

(c) 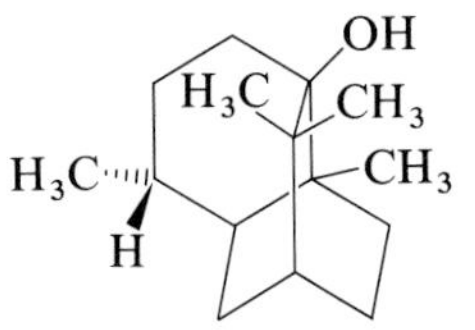 patchouli alcohol

(d) 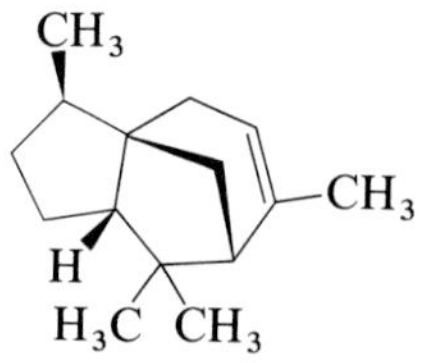 cedrene

25-23. The structure of limonene appears in Problem 25-11. Predict the products formed when limonene reacts with the following reagents.

(a) excess HBr
(b) excess HBr, peroxides
(c) excess Br_2 in CCl_4
(d) ozone, followed by dimethyl sulfide
(e) warm, concentrated $KMnO_4$
(f) $BH_3 \cdot THF$, followed by basic H_2O_2

25-24. Cholic acid, a major constituent of bile, has the following structure.

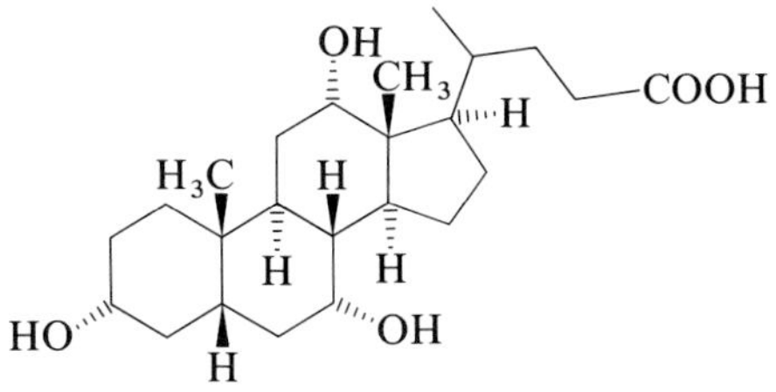

(a) Draw the structure of cholic acid showing the rings in their chair conformations, and label each methyl group and hydroxyl group as axial or equatorial.

(b) Cholic acid is secreted in bile as an amide linked to the amino group of glycine. This cholic acid–amino acid combination acts as an emulsifying agent to disperse lipids in the intestines for easier digestion. Draw the structure of the cholic acid–glycine combination and explain why it is a good emulsifying agent.

25-25. When an extract of parsley seed is saponified and acidified, one of the fatty acids isolated is *petroselenic acid,* formula $C_{18}H_{34}O_2$. Hydrogenation of petroselenic acid gives pure stearic acid. When petroselenic acid is treated with warm potassium permanganate followed by acidification, the only organic products are dodecanoic acid and adipic acid. The NMR spectrum shows absorptions of vinyl protons split by coupling constants of 7 Hz and 10 Hz. Propose a structure for petroselenic acid, and show how your structure is consistent with these observations.

25-26. The long-term health effects of eating partially hydrogenated vegetable oils concern some nutritionists, because many unnatural fatty acids are produced. Consider the partial hydrogenation of linolenic acid by the addition of 1 or 2 equivalents of hydrogen. Show how this partial hydrogenation can produce at least three different fatty acids we have not seen before.

25-27. Fatty alcohols can react with reducing sugars to give glycosides such as the cetyl glucoside shown below. Predict the solubility properties and the most obvious uses of this cetyl glucoside.

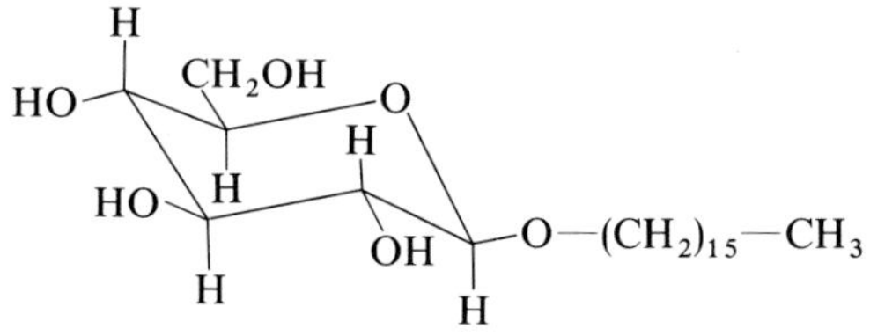

a cetyl glucoside

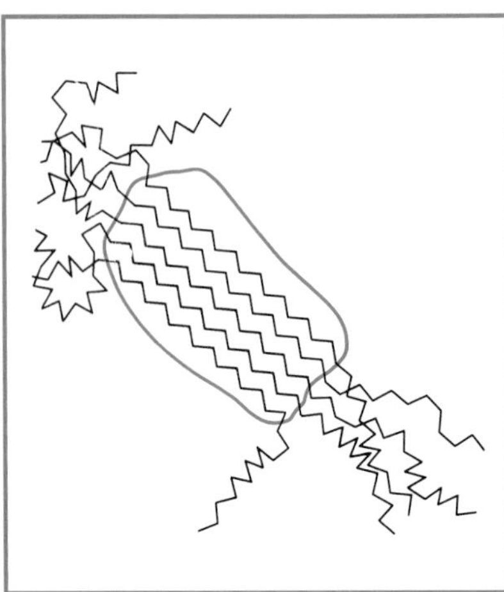

26

SYNTHETIC POLYMERS

26-1 INTRODUCTION

People have always used polymers. Prehistoric tools and shelters were made from wood and straw. Both of these building materials contain cellulose, a biopolymer of glucose. Clothing was made from the hides and hair of animals, which contain protein, a biopolymer of amino acids. After the invention of fire, people learned to make ceramic pottery and glass using naturally occurring inorganic polymers.

A **polymer** is a large molecule composed of many smaller repeating units (the **monomers**) bonded together. Today when we speak of polymers we generally mean *synthetic organic polymers* rather than natural organic biopolymers such as cellulose and protein, or inorganic polymers such as glass and concrete. The first synthetic organic polymer was made in 1838, when vinyl chloride was accidentally polymerized. Polystyrene was discovered in 1839, shortly after styrene was synthesized and purified. The discovery of polystyrene was inevitable, since styrene polymerizes spontaneously unless a stabilizer is added.

Also in 1839, Charles Goodyear (of tire and blimp fame) discovered how to convert the gummy polymeric sap of the rubber tree to a strong, stretchy material by heating it with sulfur. *Vulcanized rubber* quickly revolutionized the making of boots, tires, and rainwear. This was the first time that someone had artificially cross-linked a natural biopolymer to give it more strength and stability.

In less than 150 years we have become literally surrounded by synthetic polymers. We wear clothes of nylon and polyester, we walk on polypropylene carpets, we drive cars with plastic fenders and synthetic rubber tires, and we use artificial hearts and other organs made of silicone polymers. Our pens and computers, our toys and our televisions are made largely of plastics.

Articles that are not made from polymers are often held together or coated with polymers. A bookcase may be made from wood, but the wood is bonded by a phenol-formaldehyde polymer and painted with a latex polymer. Each year, about *50 billion* pounds of synthetic organic polymers are produced in the United States, mostly for use in consumer products. Large numbers of organic chemists are employed to develop and produce these polymers.

In this chapter we discuss some of the fundamental principles of polymer chemistry. We begin with a survey of the different kinds of polymers, then consider the reactions used to induce **polymerization.** Finally, we discuss some of the structural characteristics that determine the physical properties of a polymer.

Classes of synthetic polymers There are two major classes of polymers: addition polymers and condensation polymers. **Addition polymers** result from the rapid addition of one molecule at a time to a growing polymer chain, usually with a reactive intermediate (cation, radical, or anion) at the growing end of the chain. Addition polymers are sometimes called **chain-growth** polymers, describing the addition of one molecule at a time at the end of a chain. The monomers are usually alkenes, and the polymerization involves successive additions across the double bonds. Poly(vinyl chloride), widely used as a synthetic leather, is an example of an addition polymer.

$$n\ CH_2{=}CHCl \longrightarrow -\!\!\left[CH_2{-}CHCl\right]_n\!\!-$$

vinyl chloride — poly(vinyl chloride) (repeating unit)

Condensation polymers result from a condensation (bond formation with loss of a small molecule) between the monomers. The most common condensations involve the formation of amides and esters. In a condensation polymerization, any two molecules can condense; they do not need to be at the end of a chain. Condensation polymers are sometimes called **step-growth** polymers, because they grow in a stepwise manner. Dacron polyester is an example of a condensation polymer.

$$CH_3{-}O{-}\overset{O}{\overset{\|}{C}}{-}C_6H_4{-}\overset{O}{\overset{\|}{C}}{-}O{-}CH_3 \;+\; HO{-}CH_2{-}CH_2{-}OH \longrightarrow$$

dimethyl terephthalate — ethylene glycol

$$CH_3{-}O{-}\overset{O}{\overset{\|}{C}}{-}C_6H_4{-}\overset{O}{\overset{\|}{C}}\left[{-}O{-}CH_2{-}CH_2{-}O{-}\overset{O}{\overset{\|}{C}}{-}C_6H_4{-}\overset{O}{\overset{\|}{C}}\right]_n{-}O{-}CH_2{-}CH_2{-}OH \;+\; 2n\,CH_3OH$$

Dacron polyester (repeating unit)

26-2 ADDITION POLYMERS

Many alkenes undergo chain-growth polymerization when treated with small amounts of suitable initiators. Table 26-1 (p. 1168) shows some of the most common addition polymers, all made from substituted alkenes. The chain-growth mechanism involves addition of the reactive end of the growing chain across the double bond of the monomer. Depending on the structure of the monomer, the reactive intermediates may be free radicals, carbocations, or carbanions. Although these three types of chain-growth polymerizations are similar, we consider them individually.

TABLE 26-1
Some of the most important addition polymers

Polymer	*Polymer uses*	*Monomer formula*	*Polymer repeating unit*
polyethylene	bottles, bags, films	$H_2C{=}CH_2$	$-[CH_2-CH_2]_n-$
polypropylene	plastics, "olefin" fibers	$H_2C{=}CH-CH_3$	$-[CH_2-CH(CH_3)]_n-$
polystyrene	plastics, foam insulation	$H_2C{=}CH-C_6H_5$	$-[CH_2-CH(C_6H_5)]_n-$
poly(isobutylene)	specialized rubbers	$H_2C{=}C(CH_3)_2$	$-[CH_2-C(CH_3)_2]_n-$
poly(vinyl chloride)	"vinyl" plastics, films, water pipes	$H_2C{=}CH-Cl$	$-[CH_2-CH(Cl)]_n-$
poly(acrylonitrile)	Orlon, Acrilan fibers	$H_2C{=}CH-CN$	$-[CH_2-CH(CN)]_n-$
poly(methyl α-methacrylate)	acrylic fibers, Plexiglas, Lucite paints	$H_2C{=}C(CH_3)-C(=O)-OCH_3$	$-[CH_2-C(CH_3)(COOCH_3)]_n-$
poly(methyl α-cyanoacrylate)	"Super" glues	$H_2C{=}C(CN)-C(=O)-OCH_3$	$-[CH_2-C(CN)(COOCH_3)]_n-$
poly(tetrafluoroethylene)	Teflon coatings, PTFE plastics	$F_2C{=}CF_2$	$-[CF_2-CF_2]_n-$

26-2A FREE-RADICAL POLYMERIZATION

Many alkenes undergo **free-radical polymerization** when they are heated with radical initiators. For example, styrene is converted to polystyrene when it is heated to 100°C in the presence of benzoyl peroxide. This chain-growth polymerization is a free-radical chain reaction. The reaction is initiated by benzoyl-peroxide, which cleaves when heated to give two carboxyl radicals. As we have seen in our study of the Hunsdiecker reaction (Section 20-16), carboxyl radicals quickly decarboxylate. In this case, a phenyl radical results.

$$C_6H_5-C(=O)-O-O-C(=O)-C_6H_5 \xrightarrow{\text{heat}} 2\ C_6H_5-C(=O)-O\cdot \longrightarrow 2\ C_6H_5\cdot + 2\ CO_2$$

benzoyl peroxide — carboxyl radicals — phenyl radicals

A phenyl radical adds to styrene to give a resonance-stabilized benzylic radical. This reaction starts the growth of the polymer chain.

Initiation step

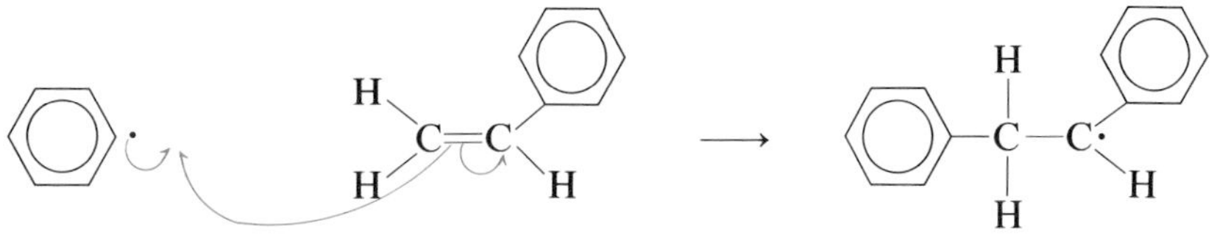

phenyl radical — styrene → benzylic radical

The propagation step is the addition of another molecule of styrene to the growing chain. This addition takes place with the orientation that gives another resonance-stabilized benzylic radical.

Propagation step

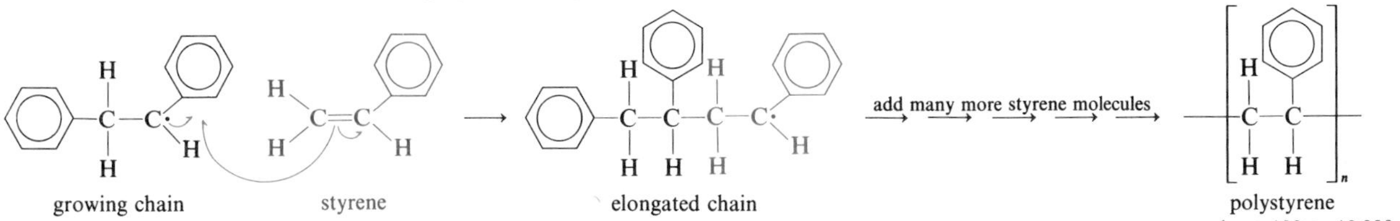

growing chain — styrene → elongated chain $\xrightarrow{\text{add many more styrene molecules}}$ polystyrene

n = about 100 to 10,000

Chain growth may continue with the addition of several hundred or several thousand styrene units. Eventually, the chain reaction stops, either by the coupling of two chains, or by a reaction with an impurity (such as oxygen), or simply by running out of the monomer.

PROBLEM 26-1

Show the intermediate that would result if the growing chain added to the other end of a styrene molecule. Explain why the final polymer has phenyl groups substituted on alternating carbon atoms rather than randomly distributed.

Ethylene is also polymerized by free-radical chain-growth polymerization. With ethylene, the free-radical intermediates are less stable and much stronger reaction conditions are required. Ethylene is commonly polymerized by free-radical initiators at pressures of about 3000 atm and temperatures of about 200°C. The product, called *low-density polyethylene,* is the material commonly used in polyethylene bags.

PROBLEM 26-2

Give a mechanism for reaction of the first three ethylene units in the polymerization of ethylene in the presence of benzoyl peroxide.

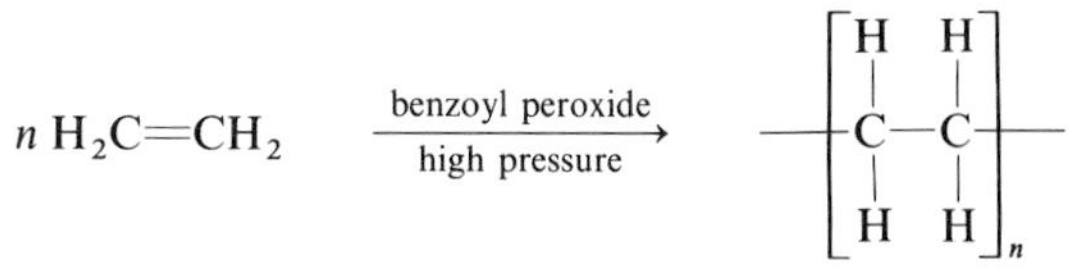

$$n\ H_2C{=}CH_2 \xrightarrow[\text{high pressure}]{\text{benzoyl peroxide}} \left[-CH_2-CH_2- \right]_n$$

Chain branching by hydrogen abstraction Low-density polyethylene is soft and flimsy because it has a highly branched, amorphous structure. (High-density polyethylene, discussed in Section 26-4, is much stronger because of the orderly structure of the unbranched linear polymer chains.) The chain branching in low-density polyethylene results from abstraction of a hydrogen atom in the middle of a chain by the free radical at the end of a chain. A new chain grows from the point of the free radical in the middle of the chain. Figure 26-1 shows the abstraction of a hydrogen from a polyethylene chain and the first step in the growth of a branch chain at that point.

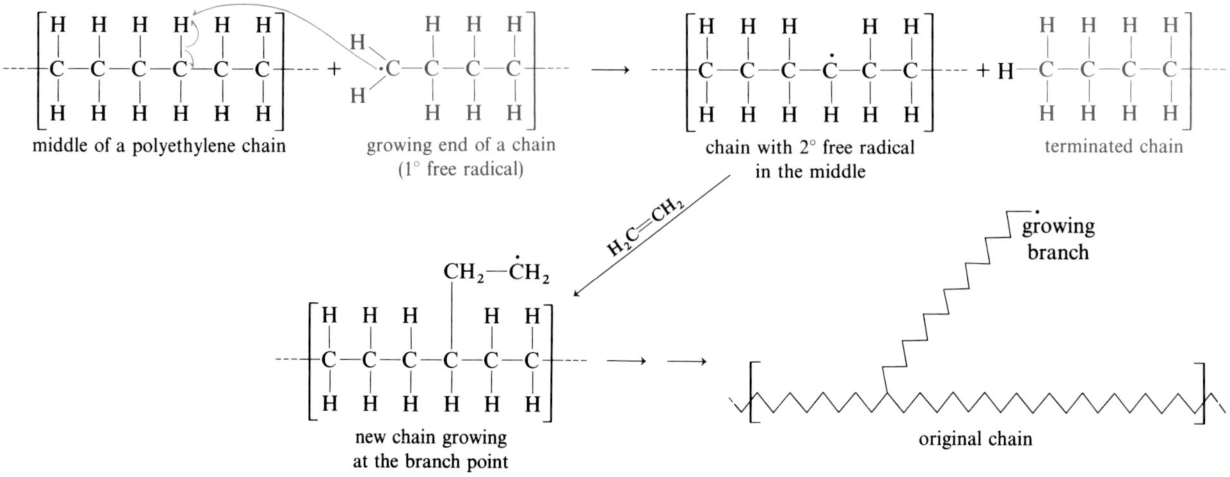

FIGURE 26-1 Chain branching occurs when the growing end of a chain abstracts a hydrogen atom from the middle of a chain. A new branch grows off the chain at that point.

PROBLEM 26-3

Give a mechanism, similar to Figure 26-1, showing chain branching during the free-radical polymerization of styrene. There are two types of hydrogens in the polystyrene chain. Which type is more likely to be abstracted?

26-2B CATIONIC POLYMERIZATION

Cationic polymerization occurs by a mechanism similar to the free-radical process. Strongly acidic catalysts are used to initiate cationic polymerization. Lewis acids such as BF_3 are often preferred, because they leave no counterion that might react with the growing chain. The following mechanism shows the formation of polyisobutylene using BF_3 as the catalyst.

Initiation step

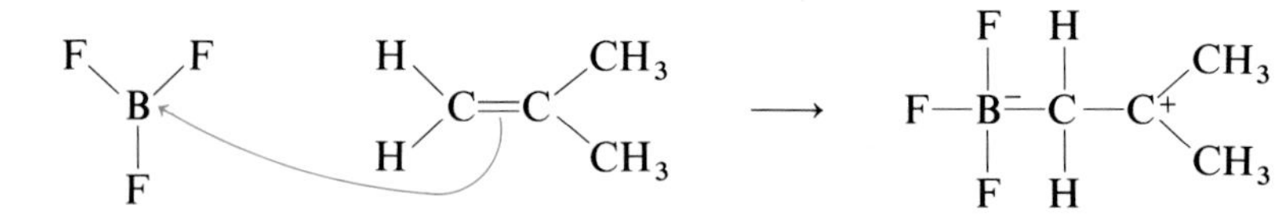

boron trifluoride | isobutylene | initiated chain

Propagation step

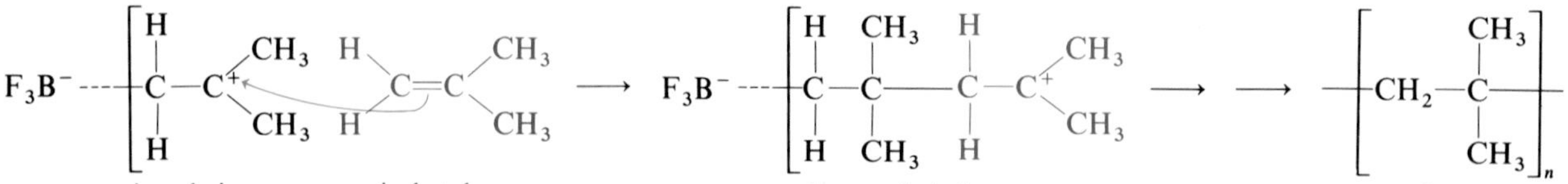

growing chain | isobutylene | elongated chain | polymer

A major difference between cationic and free-radical polymerization is that the cationic process needs a monomer that forms a relatively stable carbocation when it reacts with the cationic end of the growing chain. Some monomers form more stable intermediates than others. For example, styrene and isobutylene undergo cationic polymerization quite easily, while ethylene and acrylonitrile do not polymerize well under these conditions. Figure 26-2 shows the types of intermediates that are involved in these cationic polymerizations.

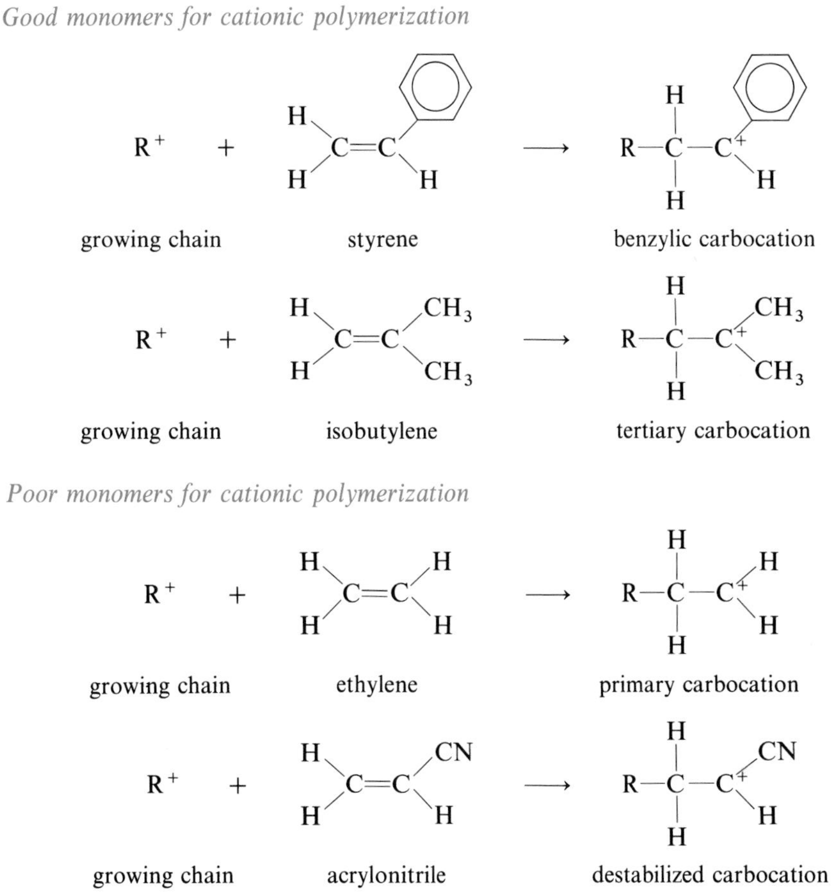

Good monomers for cationic polymerization

$R^+ + H_2C{=}CH(C_6H_5) \longrightarrow R{-}CH_2{-}C^+H(C_6H_5)$

growing chain — styrene — benzylic carbocation

$R^+ + H_2C{=}C(CH_3)_2 \longrightarrow R{-}CH_2{-}C^+(CH_3)_2$

growing chain — isobutylene — tertiary carbocation

Poor monomers for cationic polymerization

$R^+ + H_2C{=}CH_2 \longrightarrow R{-}CH_2{-}C^+H_2$

growing chain — ethylene — primary carbocation

$R^+ + H_2C{=}CH(CN) \longrightarrow R{-}CH_2{-}C^+H(CN)$

growing chain — acrylonitrile — destabilized carbocation

FIGURE 26-2 Cationic polymerization requires the formation of a relatively stable carbocation intermediate.

PROBLEM 26-4

The mechanism given above for the cationic polymerization of isobutylene shows that all of the monomer molecules add with the same orientation, giving a polymer with methyl groups on alternate carbon atoms of the chain. Explain why none of the isobutylene molecules add with the opposite orientation.

PROBLEM 26-5

Suggest which of the following monomers might polymerize well on treatment with BF_3.

(a) vinyl chloride (b) propylene (c) methyl α-cyanoacrylate

PROBLEM 26-6

Chain branching occurs with cationic polymerization much as it does in free-radical polymerization. Give a mechanism to show how branching occurs in the cationic polymerization of styrene. Suggest why isobutylene is a better monomer for cationic polymerization than styrene.

26-2C ANIONIC POLYMERIZATION

Like cationic polymerization, **anionic polymerization** depends on the presence of a stabilizing group. For effective anionic polymerization, the double bond should have a strong electron-withdrawing group such as a carbonyl group, a cyano group, or a nitro group. The following reactions show the chain-lengthening step in the polymerization of methyl acrylate. Notice that the chain-growth step of an anionic polymerization is simply a conjugate addition to a Michael acceptor.

Chain-growth step in anionic polymerization

$$\text{---}[CH_2]\text{—}\overset{..}{C}^{-}H\text{—}COOCH_3 + H_2C{=}CH\text{—}COOCH_3 \longrightarrow \text{---}[CH_2\text{—}CH(COOCH_3)]\text{—}CH_2\text{—}\overset{..}{C}^{-}H\text{—}COOCH_3 \longrightarrow\longrightarrow \text{---}[CH_2\text{—}CH(COOCH_3)]_n\text{---}$$

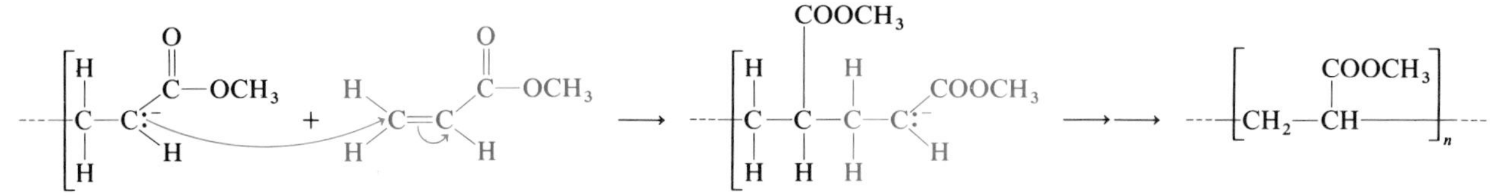

growing chain + methyl acrylate → stabilized anion → polymer

PROBLEM 26-7

Draw the important resonance structures of the stabilized anion formed in the anionic polymerization of methyl acrylate.

Anionic polymerization is usually initiated by a strong carbanion-like reagent such as an organolithium or Grignard reagent. Conjugate addition of the initiator to a molecule of the monomer starts the growth of the chain. Under the conditions of the polymerization, there is no good proton source available, and many monomer units react before the carbanion is protonated. The following reactions show the butyllithium-initiated anionic polymerization of acrylonitrile to give Orlon.

Initiation step

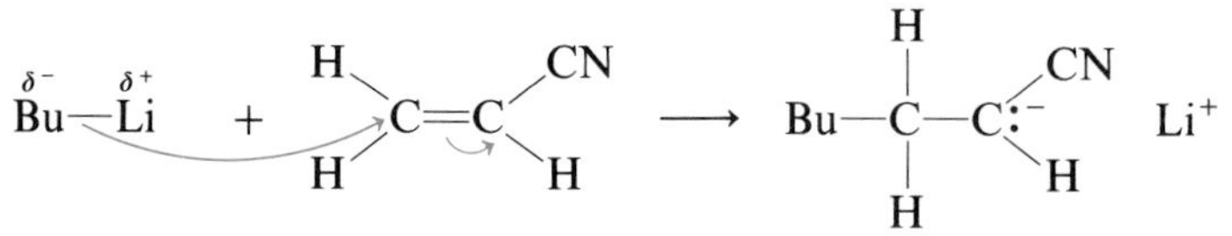

$$\overset{\delta^-}{Bu}\text{—}\overset{\delta^+}{Li} + H_2C{=}CH\text{—}CN \longrightarrow Bu\text{—}CH_2\text{—}\overset{..}{C}^{-}H\text{—}CN \quad Li^+$$

butyllithium + acrylonitrile → stabilized anion

Propagation step

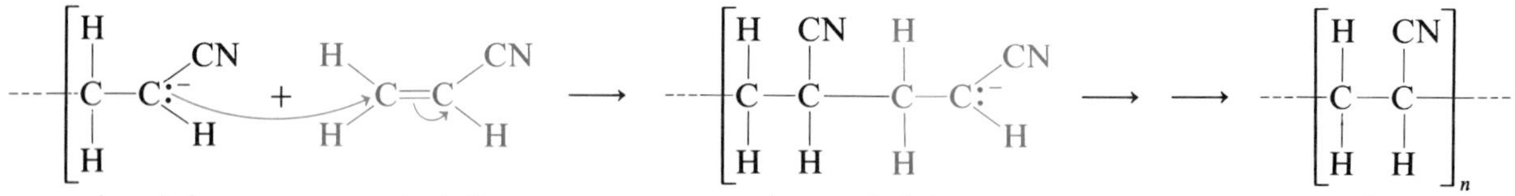

$$\text{---}[CH_2]\text{—}\overset{..}{C}^{-}H\text{—}CN + H_2C{=}CH\text{—}CN \longrightarrow \text{---}[CH_2\text{—}CH(CN)]\text{—}CH_2\text{—}\overset{..}{C}^{-}H\text{—}CN \longrightarrow\longrightarrow \text{---}[CH_2\text{—}CH(CN)]_n\text{---}$$

growing chain + acrylonitrile → elongated chain → polymer

Some monomers are easily polymerized even by weak bases. Methyl α-cyanoacrylate contains two powerful electron-withdrawing groups, and it undergoes Michael additions very easily. If this liquid monomer is spread in a thin film between two surfaces, traces of basic impurities (metal oxides, etc.) can catalyze its rapid polymerization. The solidified polymer joins the two surfaces. The chemists who first made this monomer noticed how easily it polymerizes and realized that it could serve as a fast-setting glue. Methyl α-cyanoacrylate is sold commercially as Krazy Glue® or Super Glue.

Initiation step

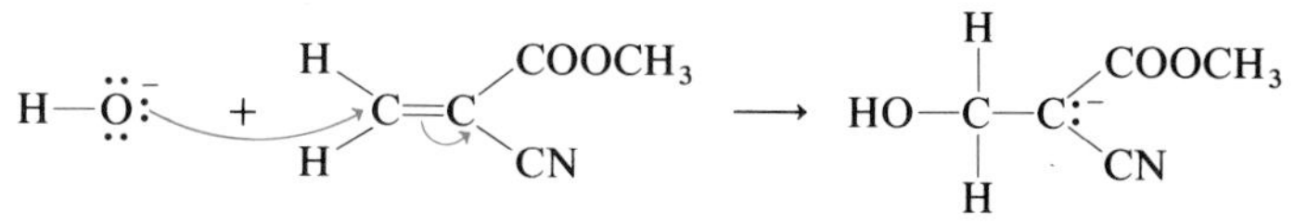

$$H-\ddot{O}:^- + H_2C=C(COOCH_3)(CN) \longrightarrow HO-CH_2-\ddot{C}^-(COOCH_3)(CN)$$

trace of base — Super Glue — highly stabilized anion

Propagation step

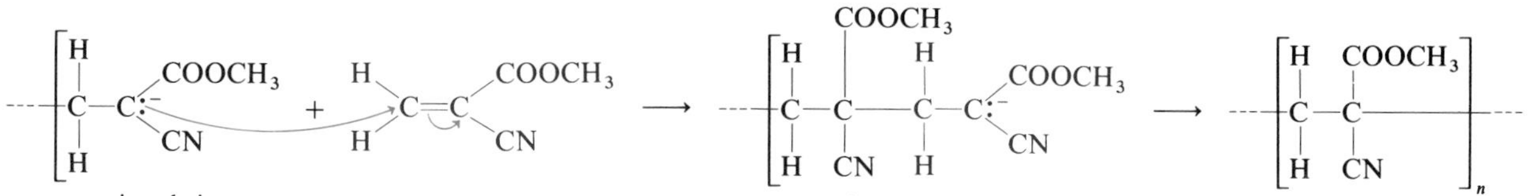

$$\cdots-CH_2-\ddot{C}^-(COOCH_3)(CN) + H_2C=C(COOCH_3)(CN) \longrightarrow \cdots-CH_2-C(COOCH_3)(CN)-CH_2-\ddot{C}^-(COOCH_3)(CN) \longrightarrow \cdots-[CH_2-C(COOCH_3)(CN)]_n-\cdots$$

growing chain — monomer — elongated chain — polymer

PROBLEM 26-8

Draw a mechanism for the base-catalyzed polymerization of methyl α-methacrylate to give the Plexiglas polymer.

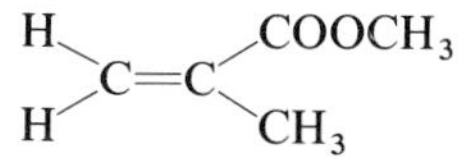

$$H_2C=C(COOCH_3)(CH_3)$$

methyl α-methacrylate

PROBLEM 26-9

Chain branching is not as common with anionic polymerization as it is with free-radical polymerization and cationic polymerization.

(a) Give a mechanism for chain branching in the polymerization of acrylonitrile.
(b) Compare the relative stabilities of the intermediates in this mechanism with those you drew for chain branching in the cationic polymerization of styrene (Problem 26-6). Explain why chain branching is less common in this anionic polymerization.

26-3 STEREOCHEMISTRY OF POLYMERS

Chain-growth polymerization of alkenes usually gives a head-to-tail bonding arrangement, with any substituent(s) appearing on alternate carbons of the polymer chain. This bonding arrangement is shown below for a generic polyalkene. Although the polymer backbone is joined by single bonds (and can undergo conformational changes), it is shown in the most stable all-anti conformation.

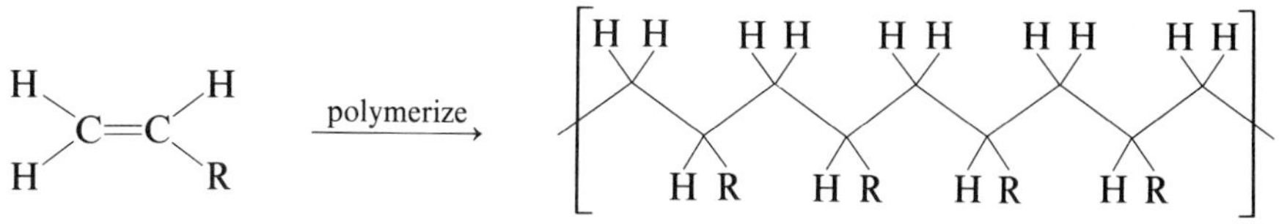

$$H_2C=CHR \xrightarrow{\text{polymerize}} [\,-CH_2-CHR-CH_2-CHR-CH_2-CHR-CH_2-CHR-CH_2-\,]$$

The stereochemistry of the side groups (R) in the polymer has a major effect on the polymer's properties. The polymer has many chiral centers, raising the possibility of millions of stereoisomers. Polymers are grouped into three classes, according to their predominant stereochemistry. If the side groups are generally on the same side of the polymer backbone, the polymer is called **isotactic** (Greek *iso*, meaning "same," and *tactic*, meaning "order"). If the side groups generally alternate from one side to the other, the polymer is called **syndiotactic** (Greek, meaning "alternating order"). If the side groups occur randomly on either side of the polymer backbone, the polymer is called **atactic** (Greek, meaning "no order"). Figure 26-3 shows these three types of polymers.

An isotactic polymer (side groups on the same side of the backbone)

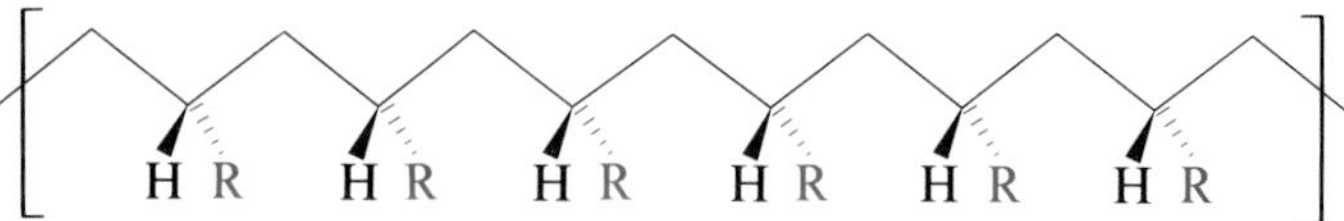

A syndiotactic polymer (side groups on alternating sides of the backbone)

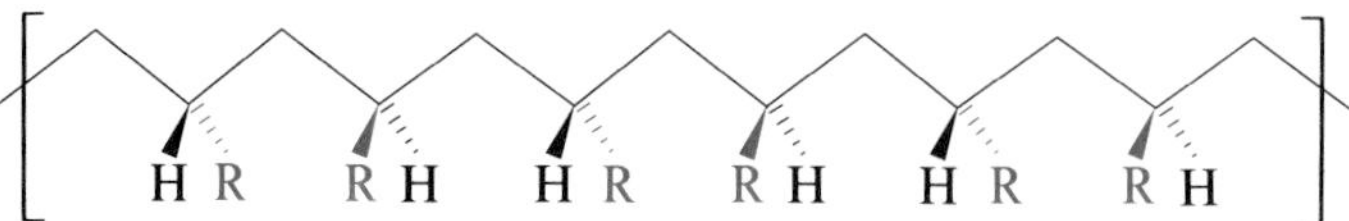

An atactic polymer (side groups on random sides of the backbone)

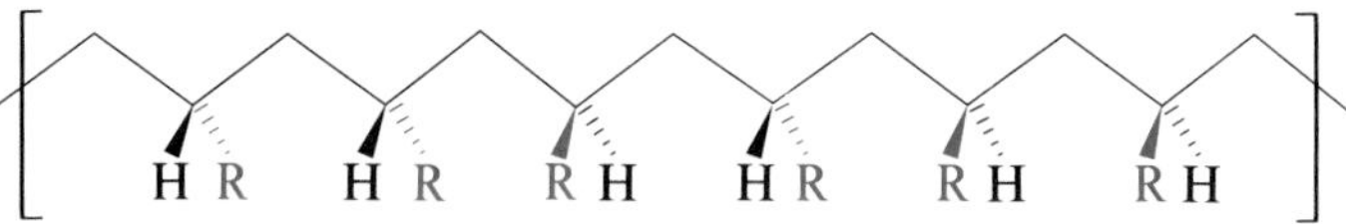

FIGURE 26-3 The three stereochemical types of addition polymers.

PROBLEM 26-10

Draw the structures of isotactic poly(acrylonitrile) and syndiotactic polystyrene.

26-4 STEREOCHEMICAL CONTROL OF POLYMERIZATION; ZIEGLER-NATTA CATALYSTS

For any particular polymer, the three stereochemical forms have somewhat different properties. In most cases, the stereoregular isotactic and syndiotactic polymers are stronger and stiffer because they form polymers with greater crystallinity (a more regular packing arrangement). The conditions used for polymerization often control the stereochemistry of the polymer. Anionic polymerizations are the most stereospecific; they usually give isotactic or syndiotactic polymers, depending on the nature of the side group. Cationic polymerizations are often stereospecific, depending on the catalysts and conditions used. Free-radical polymerization is nearly random, resulting in branched, atactic polymers.

In 1953, Karl Ziegler and Giulio Natta discovered that aluminum-titanium initiators catalyze the polymerization of alkenes, with two major advantages over other catalysts:

1. The polymerization is completely stereospecific. Either the isotactic form or the syndiotactic form may be made, by selecting the proper Ziegler-Natta catalyst.
2. Because the intermediates are stabilized by the catalyst, very little hydrogen abstraction occurs. The resulting polymers are linear with almost no branching.

A typical Ziegler-Natta catalyst is formed by adding a solution of $TiCl_4$ (titanium tetrachloride) to a solution of $(CH_3CH_2)_3Al$ (triethyl aluminum). The structure of the active catalyst is still a matter of debate among polymer chemists.

Using a Ziegler-Natta catalyst, a *high-density polyethylene* (or *linear* polyethylene) can be produced with almost no chain branching and with much greater strength than common low-density polyethylene. Many other polymers are produced with improved properties using Ziegler-Natta catalysts. In 1963, Ziegler and Natta received the Nobel Prize for their work, which had revolutionized the polymer industry in only ten years.

26-5 NATURAL AND SYNTHETIC RUBBERS

Natural rubber is isolated from a white fluid, called **latex,** that exudes from cuts in the bark of *Hevea brasiliensis,* the South American rubber tree. Many other plants secrete this polymer, as well. The name *rubber* was first used by Joseph Priestley, who used the crude material to "rub out" errors in his pencil writing. Natural rubber is soft and sticky. An enterprising Scotsman named McIntosh found that rubber makes a good waterproof coating for raincoats. Natural rubber is not strong or elastic, however, so its uses were limited to waterproofing of cloth and other strong materials.

Structure of natural rubber Like many other plant products, natural rubber is a terpene composed of isoprene units. If we imagine lining up many molecules of isoprene in the *s*-cis conformation, and moving pairs of electrons as shown below, we would produce a structure that is similar to natural rubber. This polymer results from 1,4-addition to each isoprene molecule, with all the double bonds in the cis configuration. Another name for natural rubber is *cis*-1,4-polyisoprene.

Imaginary polymerization of isoprene units

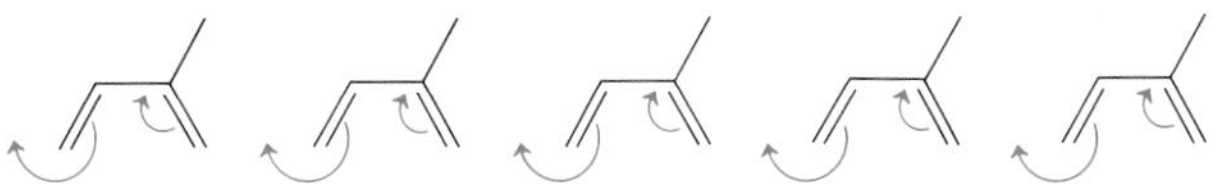

Natural rubber

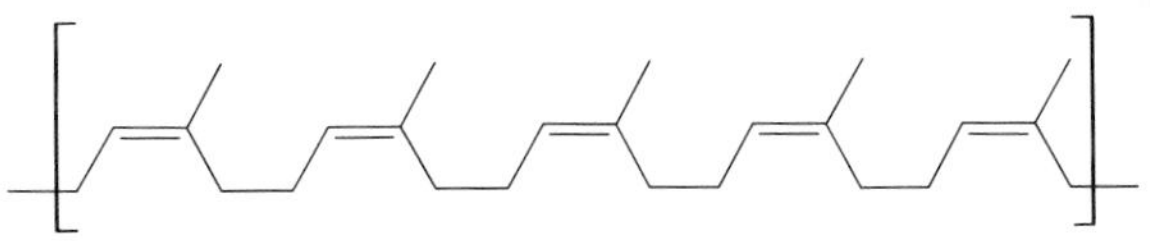

The cis double bonds in natural rubber force it to assume a kinky conformation that may be stretched and still return to its shorter, kinked structure when released. Unfortunately, when we pull on a mass of natural rubber, the chains slide by each other and the material easily pulls apart. This is the reason that natural rubber is not suitable for uses requiring strength or durability.

Vulcanization: cross-linking of rubber In 1839, Charles Goodyear accidentally dropped a mixture of natural rubber and sulfur onto a hot stove. He was surprised to find that the rubber had become strong and elastic. This discovery led to the process that Goodyear called **vulcanization,** after the Roman god of fire and the volcano. Vulcanized rubber has much greater toughness and elasticity than natural rubber. It withstands relatively high temperatures without softening, and it remains elastic and flexible when cold.

Vulcanization also allows the casting of complicated shapes such as rubber tires. Natural rubber is puttylike, and it is easily mixed with sulfur, formed around the tire cord, and placed into a mold. The mold is closed and heated, and the gooey mass of string and rubber is vulcanized into a strong, elastic tire carcass.

On a molecular level, vulcanization causes cross-linking of the *cis*-1,4-polyisoprene chains through disulfide (—S—S—) bonds, similar to the cysteine bridges that link peptides (Section 24-8C). In the cross-linked rubber, the polymer chains are linked together so they can no longer slip past each other. When the material is stressed, the chains stretch but the cross-linking prevents tearing. When the stress is released, the chains return to their shortened, kinky conformations as the rubber snaps back. Figure 26-4 shows the structure of rubber before and after vulcanization.

FIGURE 26-4 The vulcanization of rubber introduces disulfide cross-links between the polyisoprene chains. This cross-linking forms a stronger, elastic material that does not pull apart when it is stretched.

Rubber can be prepared with a wide range of physical properties by controlling the amount of sulfur used in vulcanization. Low-sulfur rubber, made with about 1 to 3 percent sulfur, is soft and stretchy. It is good for rubber bands and inner tubes. Medium-sulfur rubber (about 3 to 10 percent sulfur) is somewhat harder, but still flexible, making good tires. High-sulfur rubber (20 to 30 percent sulfur) is called *hard rubber* and was once used as a hard synthetic plastic.

PROBLEM 26-11

(a) Draw the structure of gutta-percha, a natural rubber with all its double bonds in the trans configuration.
(b) Suggest why gutta-percha is not very elastic, even after it is vulcanized.

Synthetic rubber There are many different formulations for synthetic rubbers, but the simplest is the polymer of 1,3-butadiene. Specialized Ziegler-Natta catalysts can produce polymers of 1,3-butadiene where 1,4-addition has occurred on each butadiene unit and the remaining double bonds are all cis. This polymer has properties similar to those of natural rubber, and it can be vulcanized in the same way.

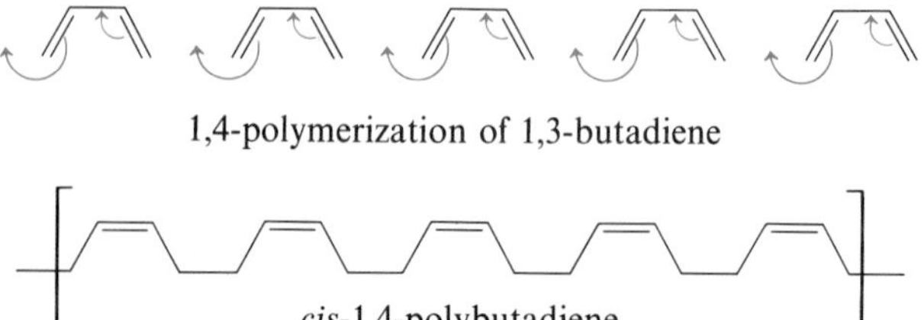

1,4-polymerization of 1,3-butadiene

cis-1,4-polybutadiene

26-6 COPOLYMERS OF TWO OR MORE MONOMERS

All the polymers we have discussed are **homopolymers,** polymers made up of identical monomer units. Many polymeric materials are **copolymers,** made by polymerizing two or more different monomers together. In many cases monomers are chosen so that they add selectively in an alternating manner. For example, when a mixture of vinyl chloride and vinylidene chloride (1,1-dichloroethylene) is induced to polymerize, the growing chain preferentially adds the monomer that is *not* at the end of the chain. This selective reaction gives the alternating copolymer *Saran,* used as a film for wrapping food.

Overall reaction

$$\underset{\text{vinyl chloride}}{H_2C{=}CHCl} + \underset{\text{vinylidene chloride}}{H_2C{=}CCl_2} \longrightarrow \underset{\text{Saran}}{\left[CH_2-\overset{Cl}{\underset{H}{C}}-CH_2-\overset{Cl}{\underset{Cl}{C}} \right]_n}$$

Some polymers have three or more monomers mixed to give products with desired properties. For example, acrylonitrile, butadiene, and styrene are polymerized to give ABS plastic; a strong, tough, and resilient material used for bumpers, crash helmets, and other articles that must withstand heavy impacts.

PROBLEM 26-12

Isobutylene and isoprene copolymerize to give "butyl rubber." Draw the structure of the repeating unit in butyl rubber, assuming that the two monomers alternate.

26-7 CONDENSATION POLYMERS

Condensation polymers result from the formation of ester or amide linkages between difunctional molecules. The reaction is called a **step-growth** polymerization. Any two of the monomer molecules may react to form a dimer, dimers may condense to give tetramers, and so on. Each condensation is an individual *step* in the growth of the polymer, and there is no chain reaction. Many different kinds of condensation polymers are known. We discuss the four most common types: polyamides, polyesters, polycarbonates, and polyurethanes.

26-7A POLYAMIDES: NYLON

When Wallace Carothers of DuPont discovered nylon in 1938, he opened the door to a new age of fibers and textiles. At that time, the thread used for clothing was made of spun animal and plant fibers. These fibers were held together by friction or sizing, but they were quite weak and were subject to rotting and unraveling. Nylon was a completely new type of fiber. It can be melted and extruded into a strong, continuous fiber, and it cannot rot. Thread spun from continuous nylon fibers is so much stronger than natural materials that it can be made much thinner. The availability of this strong, thin thread made possible stronger ropes, sheer fabrics, and the nearly invisible women's stockings that came to be called "nylons."

Nylon is the common name for polyamides. Polyamides are generally made from the reactions of diacids with diamines. The most common polyamide is called nylon 6,6 because it is made by the reaction of a six-carbon diacid (adipic acid) with a six-carbon diamine. The six-carbon diamine, systematically named *1,6-hexanediamine,* is commonly called *hexamethylene diamine.* When adipic acid is mixed with hexamethylene diamine, a proton-transfer reaction gives a white solid called *nylon salt*. When the nylon salt is heated to 250°C, water is driven off as a gas, and molten nylon results. The molten nylon is then cast into a solid shape or extruded through a spinneret to produce a fiber.

$$\underset{\text{adipic acid}}{HO{-}\overset{O}{\overset{\|}{C}}{-}(CH_2)_4{-}\overset{O}{\overset{\|}{C}}{-}OH} + \underset{\text{hexamethylene diamine}}{H_2N{-}(CH_2)_6{-}NH_2} \longrightarrow \underset{\text{nylon salt}}{\begin{matrix} {}^-O{-}\overset{O}{\overset{\|}{C}}{-}(CH_2)_4{-}\overset{O}{\overset{\|}{C}}{-}O^- \\ H_3\overset{+}{N}{-}(CH_2)_6{-}\overset{+}{N}H_3 \end{matrix}}$$

$$\xrightarrow{\text{heat}, -H_2O}$$

$$\text{---}\overset{O}{\overset{\|}{C}}{-}(CH_2)_4{-}\overset{O}{\overset{\|}{C}}\left[NH{-}(CH_2)_6{-}NH{-}\overset{O}{\overset{\|}{C}}{-}(CH_2)_4{-}\overset{O}{\overset{\|}{C}}\right]_n NH{-}(CH_2)_6{-}NH\text{---}$$

poly(hexamethylene adipamide), called nylon 6,6

Nylon can also be made from a single monomer having an amino group at one end and an acid group at the other. This reaction is similar to the polymerization of α-amino acids to give proteins. Nylon 6 is a polymer of this type, made from a six-carbon amino acid: 6-aminohexanoic acid (ε-aminocaproic acid). This synthesis actually starts with ε-caprolactam. When caprolactam is heated with a trace of water, some of it is hydrolyzed to the free amino acid. Continued heating gives condensation and polymerization to the molten nylon 6. Nylon 6 (also called *Perlon*) is widely used for making strong, flexible fibers for ropes and tire cord.

$$\underset{\varepsilon\text{-caprolactam}}{\text{(seven-membered ring lactam, C=O, N–H)}} \xrightleftharpoons{H_2O,\ \text{heat}} \underset{\varepsilon\text{-aminocaproic acid}}{H_3\overset{+}{N}{-}(CH_2)_5{-}\overset{O}{\overset{\|}{C}}{-}O^-}$$

$$\xrightarrow{\text{heat}, -H_2O}$$

$$\text{---}NH{-}(CH_2)_5{-}\overset{O}{\overset{\|}{C}}{-}NH{-}(CH_2)_5{-}\overset{O}{\overset{\|}{C}}\left[NH{-}(CH_2)_5{-}\overset{O}{\overset{\|}{C}}\right]_n NH{-}(CH_2)_5{-}\overset{O}{\overset{\|}{C}}\text{---}$$

poly(6-aminohexanoic acid), called nylon 6 or Perlon

PROBLEM 26-13

(a) *Nomex*, a strong fire-retardant fabric, is a polyamide made from *meta*-phthalic acid and *meta*-diaminobenzene. Draw the structure of Nomex.

(b) *Kevlar*, made from terephthalic acid (*para*-phthalic acid) and *para*-diaminobenzene, is used in making tire cord and bulletproof vests. Draw the structure of Kevlar.

26-7B POLYESTERS

The introduction of polyester fibers has brought about major changes in the way we care for our clothing. Nearly all modern permanent-press fabrics owe their wrinkle-free behavior to polyester, often blended with other fibers. These polyester blends have reduced or eliminated the need for starching and ironing clothes to achieve a wrinkle-free surface that holds its shape.

The most common polyester is *Dacron,* the polymer of terephthalic acid (*para*-phthalic acid) with ethylene glycol. In principle, this polymer might be made by mixing the diacid with the glycol and heating the mixture to drive off water. In practice, however, a better product is obtained using a transesterification process (Section 21-5). The dimethyl ester of terephthalic acid is heated to about 150°C with ethylene glycol. Methanol is evolved as a gas, driving the reaction to completion. The molten product is spun into Dacron fiber or cast into a film called *Mylar.*

$$CH_3O-\overset{O}{\overset{\|}{C}}-C_6H_4-\overset{O}{\overset{\|}{C}}-OCH_3 \;+\; HO-CH_2CH_2-OH \xrightarrow{\text{heat, loss of } CH_3OH}$$

dimethyl terephthalate — ethylene glycol

$$---\overset{O}{\overset{\|}{C}}-C_6H_4-\overset{O}{\overset{\|}{C}}\left[-O-CH_2CH_2-O-\overset{O}{\overset{\|}{C}}-C_6H_4-\overset{O}{\overset{\|}{C}}\right]_n-O-CH_2CH_2-O---$$

poly(ethylene terephthalate) or PET, also called Dacron polyester or Mylar film

Dacron fiber is used to make fabric and tire cord, and Mylar film is used to make magnetic recording tape. Mylar film is strong, flexible, and resistant to ultraviolet degradation. Aluminized Mylar was used to make the Echo satellite, a huge balloon that was put into orbit around the Earth as a giant reflector. Poly(ethylene terephthalate) is also blow-molded to make plastic soft drink bottles that are sold by the billions each year.

PROBLEM 26-14

Kodel polyester is formed by transesterification of dimethyl terephthalate with 1,4-di(hydroxymethyl)cyclohexane. Draw the structure of Kodel.

PROBLEM 26-15

Glyptal resin makes a strong, solid polymer matrix for electronic parts. Glyptal is made from terephthalic acid and glycerol. Draw the structure of glyptal, and explain its remarkable strength and rigidity.

26-7C POLYCARBONATES

A *carbonate ester* is simply an ester of carbonic acid. Carbonic acid itself exists in equilibrium with carbon dioxide and water, but its esters are quite stable.

$$HO-\overset{O}{\overset{\|}{C}}-OH \rightleftharpoons CO_2 + H_2O \qquad\qquad R-O-\overset{O}{\overset{\|}{C}}-O-R'$$

carbonic acid — a carbonate ester

Carbonic acid is a diacid; with suitable alcohols, it can form polyesters. For example, when diethyl carbonate is allowed to react with a diol, a transesterification occurs to give a poly(carbonate ester). The following equation shows the synthesis of Lexan polycarbonate: a strong, clear, and colorless material that is used for bulletproof windows and crash helmets. The diol used to make Lexan is a phenol called *bisphenol A,* a common intermediate in polyester and polyurethane synthesis.

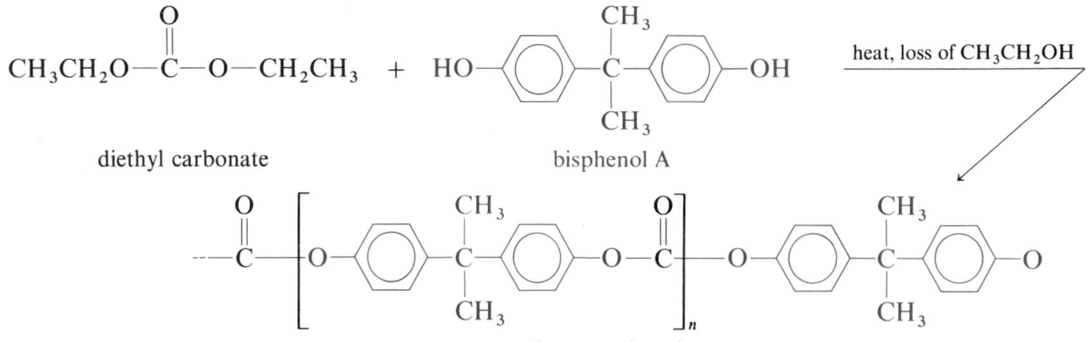

$$CH_3CH_2O-\overset{\overset{\displaystyle O}{\|}}{C}-O-CH_2CH_3 \; + \; HO-C_6H_4-\underset{CH_3}{\overset{CH_3}{\underset{|}{\overset{|}{C}}}}-C_6H_4-OH \xrightarrow{\text{heat, loss of } CH_3CH_2OH}$$

diethyl carbonate — bisphenol A

$$\cdots-\overset{\overset{\displaystyle O}{\|}}{C}-\left[O-C_6H_4-\underset{CH_3}{\overset{CH_3}{\underset{|}{\overset{|}{C}}}}-C_6H_4-O-\overset{\overset{\displaystyle O}{\|}}{C}\right]_n-O-C_6H_4-\underset{CH_3}{\overset{CH_3}{\underset{|}{\overset{|}{C}}}}-C_6H_4-O$$

Lexan polycarbonate

26-7D POLYURETHANES

A *urethane* (Section 21-16) is an ester of a carbamic acid (R—NH—COOH), a half-amide of carbonic acid. Carbamic acids themselves are unstable, quickly decomposing to amines and CO_2. Their esters (urethanes) are quite stable, however.

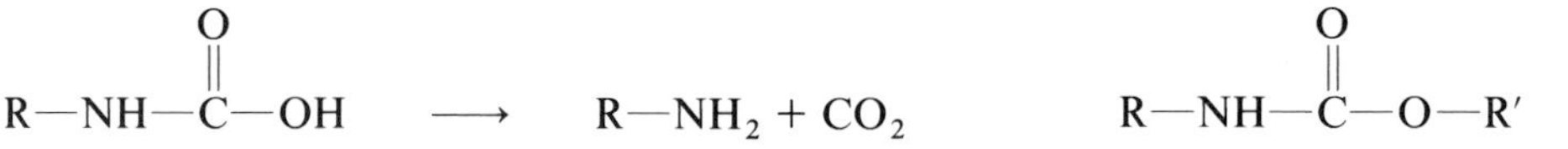

$$R-NH-\overset{\overset{\displaystyle O}{\|}}{C}-OH \longrightarrow R-NH_2 + CO_2 \qquad R-NH-\overset{\overset{\displaystyle O}{\|}}{C}-O-R'$$

a carbamic acid — amine — a urethane or carbamate ester

Because carbamic acids are unstable, normal esterification procedures cannot be used to form urethanes. Urethanes are most commonly made by treating an *isocyanate* with an alcohol or a phenol. The reaction is highly exothermic, and it gives a quantitative yield of a carbamate ester. The following reaction shows the formation of ethyl *N*-phenylcarbamate.

$$R-N{=}C{=}O \; + \; HO-R' \longrightarrow R-NH-\overset{\overset{\displaystyle O}{\|}}{C}-O-R'$$

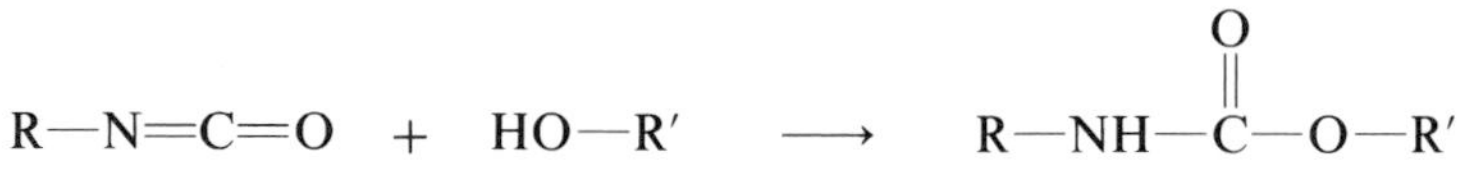

isocyanate — alcohol — carbamate ester (urethane)

Example

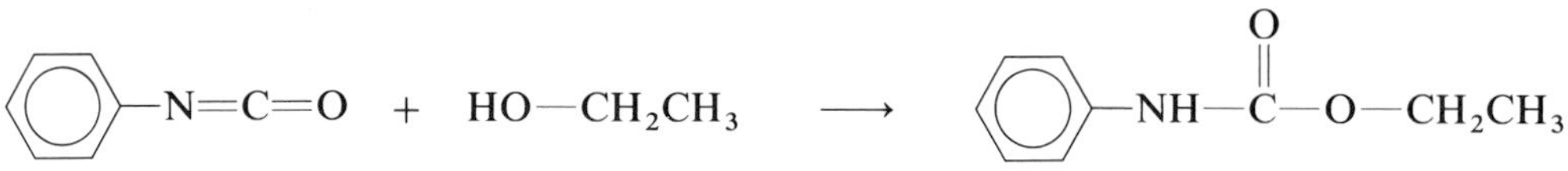

$$C_6H_5-N{=}C{=}O \; + \; HO-CH_2CH_3 \longrightarrow C_6H_5-NH-\overset{\overset{\displaystyle O}{\|}}{C}-O-CH_2CH_3$$

phenyl isocyanate — ethanol — ethyl *N*-phenylcarbamate

PROBLEM 26-16
Propose a mechanism for the reaction of phenyl isocyanate with ethanol.

A polyurethane results when a diol reacts with a diisocyanate, a compound with two isocyanate groups. The compound shown below, commonly called *toluene diisocyanate,* is frequently used for making polyurethanes. When ethylene glycol or another diol is added to toluene diisocyanate, a rapid condensation gives the polyurethane. Low-boiling liquids such as Freon 11 (fluorotrichloromethane) are often added to the reaction mixture. The heat of the polymerization vaporizes the volatile liquid, producing bubbles that convert the viscous polymer to a frothy mass of polyurethane foam.

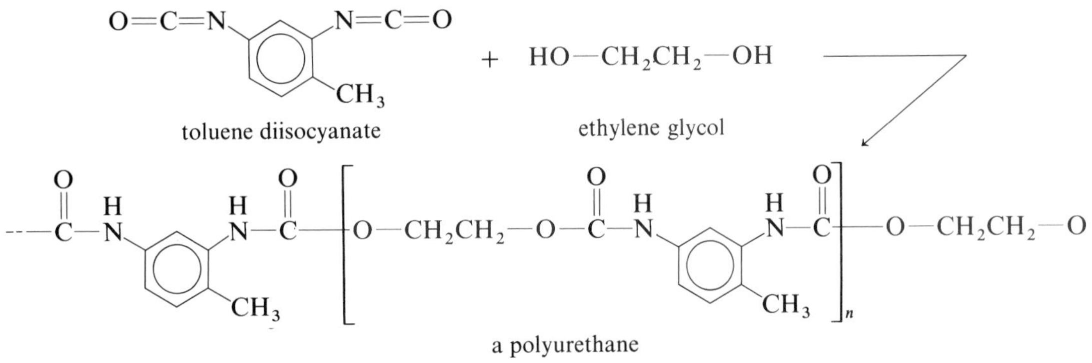

$$O{=}C{=}N{-}C_6H_3(CH_3){-}N{=}C{=}O \;+\; HO{-}CH_2CH_2{-}OH \longrightarrow$$

toluene diisocyanate   ethylene glycol

$$\text{---}\overset{O}{\overset{\|}{C}}{-}\overset{H}{N}{-}C_6H_3(CH_3){-}\overset{H}{N}{-}\overset{O}{\overset{\|}{C}}\left[{-}O{-}CH_2CH_2{-}O{-}\overset{O}{\overset{\|}{C}}{-}\overset{H}{N}{-}C_6H_3(CH_3){-}\overset{H}{N}{-}\overset{O}{\overset{\|}{C}}\right]_n{-}O{-}CH_2CH_2{-}O$$

a polyurethane

PROBLEM 26-17
Explain why the addition of a small amount of glycerol to the polymerization mixture gives a stiffer urethane foam.

PROBLEM 26-18
Give the structure of the polyurethane formed by the reaction of toluene diisocyanate with bisphenol A.

26-8 POLYMER STRUCTURE AND PROPERTIES

Although polymers are very large molecules, we can explain their chemical and physical properties in terms of what we already know about smaller molecules. For example, when you spill a base on your polyester slacks, the fabric is weakened because the base hydrolyzes some of the ester linkages. The physical properties of polymers can also be explained using concepts we have already encountered. Although polymers do not crystallize or melt quite like smaller molecules, we can detect crystalline areas in a polymer, and we can measure the temperature at which these *crystallites* melt. In this section we consider briefly some of the important aspects of polymer crystallinity and thermal behavior.

26-8A POLYMER CRYSTALLINITY

Although polymers rarely form the large crystals characteristic of other organic compounds, many do form microscopic crystalline regions called **crystallites.** A highly regular polymer that packs well into a crystal lattice will be highly crystalline, and it will generally be denser, stronger, and more rigid than a similar polymer with a lower degree of **crystallinity.** Figure 26-5 shows how the polymer chains are arranged in parallel lines in crystalline areas within a polymer.

Polyethylene provides an example of how crystallinity affects a polymer's physical properties. Free-radical polymerization gives a highly branched low-density polyethylene that forms very small crystallites because the random chain branching destroys the regularity of the crystallites. An unbranched high-density polyethylene is made using a Ziegler-Natta catalyst. The *linear* structure of the high-density material packs more easily into a crystal lattice, so it forms larger and stronger crystallites. We say that high-density polyethylene has a higher degree of crystallinity, and it is therefore denser, stronger, and more rigid than the low-density material.

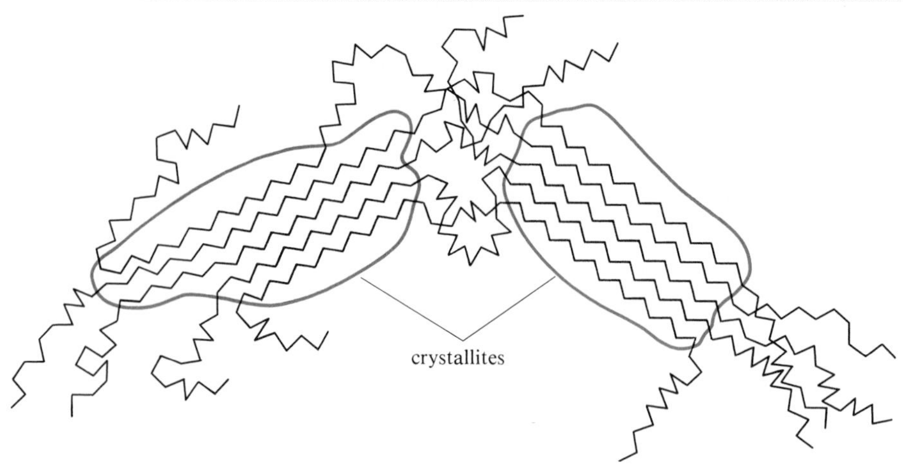

FIGURE 26-5 Crystallites are areas of crystalline structure within the large mass of a solid polymer.

Polymer stereochemistry also affects crystallinity. The stereoregular isotactic and syndiotactic polymers are generally more crystalline than atactic polymers. By a careful choice of Ziegler-Natta catalysts, we can make a linear polymer with either isotactic or syndiotactic stereochemistry.

26-8B THERMAL PROPERTIES

At low temperatures, long-chain polymers are *glasses.* They are solid and unyielding, and a strong impact causes them to fracture. As the temperature is raised, the polymer goes through a **glass transition temperature,** abbreviated T_g. Above T_g a highly crystalline polymer becomes flexible and moldable. We say it is a **thermoplastic,** because the application of heat has made it plastic (moldable). As the temperature is raised further, the polymer reaches the **crystalline melting temperature,** abbreviated T_m. At this temperature, the crystallites melt and the individual molecules can slide past one another.

Above T_m the polymer is a viscous liquid and can be extruded through spinnerets to form fibers. The fibers are immediately cooled in water to form crystallites, and then stretched (drawn) to orient the crystallites along the fiber, thus increasing its strength.

Long-chain polymers with low crystallinity (called **amorphous polymers**) become rubbery when heated above the glass transition temperature. Further heating causes them to grow gummier and less solid until they become viscous liquids without definite melting points. These phase transitions apply only to long-chain polymers. Cross-linked polymers are more likely to stay rubbery, and they may not melt until the temperature is so high that the polymer begins to decompose. Figure 26-6 compares the thermal properties of crystalline and amorphous long-chain polymers.

26-8C PLASTICIZERS

In many cases, a polymer has desirable properties for a particular use, but it is too brittle—either because its glass transition temperature (T_g) is above room temperature, or because the polymer is too highly crystalline. In such cases, the addition of a **plasticizer** often makes the polymer more flexible. A plasticizer is a nonvolatile liquid that dissolves in the polymer, lowering the attraction between the polymer chains and allowing them to slide by each other. The overall effect

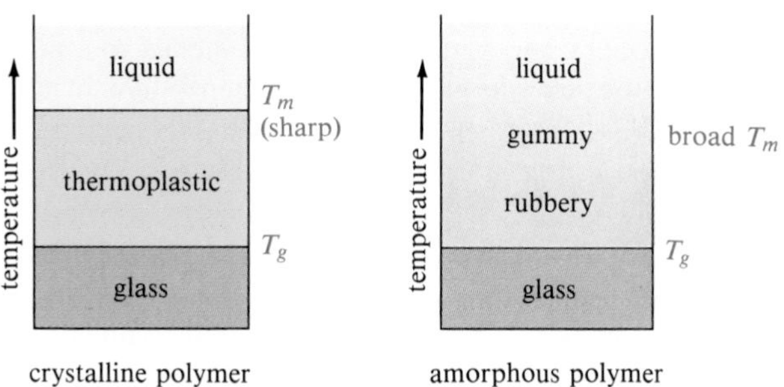

FIGURE 26-6 Crystalline and amorphous long-chain polymers show different physical properties when they are heated.

of the plasticizer is to reduce the crystallinity of the polymer and lower its glass transition temperature (T_g).

$C_6H_4(C(=O)-O-CH_2CH_2CH_2CH_3)_2$

dibutyl phthalate

A common example of a plasticized polymer is poly(vinyl chloride). The common atactic form has a T_g of about 80°C, well above room temperature. Without a plasticizer, "vinyl" is stiff and brittle. Dibutyl phthalate (see structure at left) is added to the polymer to lower its glass transition temperature to about 0°C. This plasticized material is the flexible, somewhat stretchy film that we think of as vinyl raincoats, shoes, and car roofs. Dibutyl phthalate is slightly volatile, however, and it gradually evaporates. The soft, plasticized vinyl eventually loses its plasticizer and becomes hard and brittle.

GLOSSARY

addition polymer (chain-growth polymer) A polymer that results from the rapid addition of one molecule at a time to a growing polymer chain, usually with a reactive intermediate (cation, radical, or anion) at the growing end of the chain. (p. 1167)

amorphous polymer A long-chain polymer with low crystallinity. (p. 1182)

anionic polymerization The process of forming an addition polymer by chain-growth polymerization involving an anion at the end of the growing chain. (p. 1172)

atactic polymer A polymer with the side groups on random sides of the polymer backbone. (p. 1174)

cationic polymerization The process of forming an addition polymer by chain-growth polymerization involving a cation at the end of the growing chain. (p. 1170)

condensation polymer (step-growth polymer) A polymer that results from condensation (bond formation with loss of a small molecule) between the monomers. In a condensation polymerization any two molecules can condense, not necessarily at the end of a growing chain. (p. 1177)

copolymer A polymer made from two or more different monomers. (p. 1177)

crystalline melting temperature The temperature at which melting of the crystallites in a highly crystalline polymer occurs. Abbreviated T_m. Above T_m the polymer is a viscous liquid. (p. 1182)

crystallinity The relative amount of the polymer that is included in crystallites and the relative sizes of the crystallites. (p. 1181)

crystallites The microscopic crystalline regions found within a solid polymer below the crystalline melting temperature. (p. 1181)

free-radical polymerization The process of forming an addition polymer by chain-growth polymerization involving a free radical at the end of the growing chain. (p. 1168)

glass transition temperature The temperature at which a highly crystalline polymer becomes flexible and moldable. Abbreviated T_g. (p. 1182)

homopolymer A polymer made from identical monomer units. (p. 1177)

isotactic polymer A polymer with all the side groups on the same side of the polymer backbone. (p. 1174)

monomer One of the small units that bond together to form a polymer. (p. 1166)

plasticizer A nonvolatile liquid that is added to a polymer to make it more flexible and less brittle below its glass transition temperature. In effect, a plasticizer reduces the crystallinity of a polymer and lowers T_g. (p. 1182)

polymer A large molecule composed of many smaller units (monomers) bonded together. (p. 1166)

polymerization The process of linking monomer molecules into a polymer. (p. 1167)

rubber A natural polymer isolated from the **latex** that exudes from cuts in the bark of the South American rubber tree. Alternatively, synthetic polymers with rubberlike properties are called **synthetic rubber.** (p. 1175)

syndiotactic polymer A polymer with the side groups on alternating sides of the polymer backbone. (p. 1174)

thermoplastic A polymer that becomes moldable at high temperature. (p. 1182)

vulcanization The heating of natural or synthetic rubber with sulfur to form disulfide cross-links. This cross-linking process adds durability and elasticity to the rubber. (p. 1175)

Ziegler-Natta catalyst Any one of a group of addition polymerization catalysts involving titanium-aluminum complexes. The Ziegler-Natta catalysts produce stereoregular (either isotactic or syndiotactic) polymers in most cases. (p. 1174)

ESSENTIAL PROBLEM-SOLVING SKILLS IN CHAPTER 26

1. Given the structure of a polymer, determine whether it is an addition or condensation polymer, and determine the structure of the monomer(s).

2. Given the structure of one or more monomers, predict whether polymerization will occur to give an addition polymer or a condensation polymer, and give the general structure of the polymer chain.

3. Use mechanisms to explain how a monomer polymerizes under acidic, basic, or free-radical conditions. For addition polymerization, consider whether the reactive end of the growing chain is more stable as a cation (acidic conditions), anion (basic conditions), or free radical (radical initiator). For condensation polymerization, consider the mechanism of the step-growth reaction.

4. Predict the general characteristics (strength, elasticity, crystallinity, chemical reactivity) of a polymer based on its structure, and explain how its physical characteristics change as it is heated past T_g and T_m.

5. Explain how chain branching, cross-linking, and plasticizers affect the properties of polymers.

6. Compare the stereochemistry of isotactic, syndiotactic, and atactic polymers. Explain how the stereochemistry can be controlled during polymerization and how it affects the physical properties of the polymer.

STUDY PROBLEMS

26-19. Give a definition and an example for each of the following terms.

(a) addition polymer
(b) condensation polymer
(c) copolymer
(d) atactic polymer
(e) isotactic polymer
(f) syndiotactic polymer
(g) free-radical polymerization
(h) cationic polymerization
(i) anionic polymerization
(j) crystalline polymer
(k) amorphous polymer
(l) monomer
(m) plasticizer
(n) vulcanization
(o) Ziegler-Natta catalyst
(p) glass transition temperature
(q) polyamide
(r) polyolefin
(s) polyester
(t) polyurethane
(u) polycarbonate
(v) crystalline melting temperature

26-20. Polyisobutylene is one of the components of butyl rubber used for making inner tubes.
(a) Give the structure of polyisobutylene. **(b)** Is this an addition polymer or a condensation polymer?
(c) What conditions (cationic, anionic, free-radical) would be most appropriate for the polymerization of isobutylene? Explain your answer.

26-21. Poly(trimethylene carbamate) has the following structure.

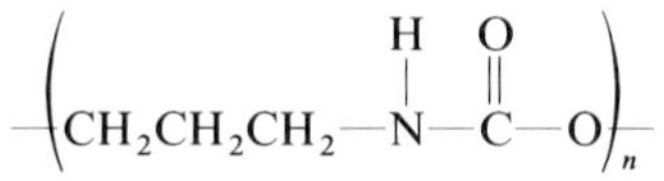

$$-\!\!\left(CH_2CH_2CH_2-\overset{H}{\overset{|}{N}}-\overset{O}{\overset{\|}{C}}-O\right)_{\!n}\!\!-$$

(a) What type of polymer is poly(trimethylene carbamate)?
(b) Is this an addition polymer or a condensation polymer?
(c) Draw the products that would be formed if the polymer were completely hydrolyzed under acidic or basic conditions.

26-22. Poly(butylene terephthalate) is a hydrophobic plastic material widely used in automotive ignition systems.

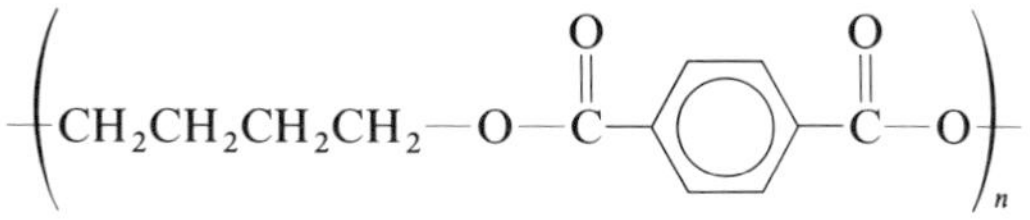

$$-\!\!\left(CH_2CH_2CH_2CH_2-O-\overset{O}{\overset{\|}{C}}-C_6H_4-\overset{O}{\overset{\|}{C}}-O\right)_{\!n}\!\!-$$

poly(butylene terephthalate)

(a) What type of polymer is poly(butylene terephthalate)?
(b) Is this an addition polymer or a condensation polymer?
(c) Suggest what monomers might be used to synthesize this polymer and how the polymerization might be accomplished.

26-23. *Urylon* fibers are composed of the following polymer.

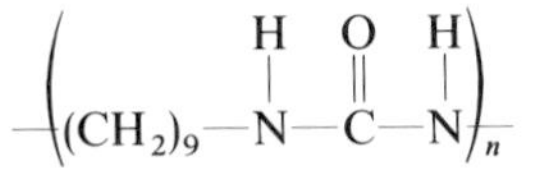

$$-\!\!\left((CH_2)_9-\overset{H}{\overset{|}{N}}-\overset{O}{\overset{\|}{C}}-\overset{H}{\overset{|}{N}}\right)_{\!n}\!\!-$$

(a) What functional group is contained in the Urylon structure?
(b) Is Urylon an addition polymer or a condensation polymer?
(c) Draw the products that would be formed if the polymer were completely hydrolyzed under acidic or basic conditions.

26-24. Polyethylene glycol, or Carbowax [$-(O-CH_2-CH_2)_n-$], is widely used as a binder, thickening agent, and packaging additive for foods.
(a) What type of polymer is polyethylene glycol? (We have not seen this type of polymer before.)
(b) The systematic name for polyethylene glycol is poly(ethylene oxide). What monomer would you use to make polyethylene glycol?
(c) What conditions (free-radical initiator, acid catalyst, basic catalyst, etc.) would you evaluate for use in this polymerization?
(d) Propose a polymerization mechanism as far as the tetramer.

26-25. Polychloroprene is commonly known as neoprene rubber, widely used in rubber parts that must withstand exposure to gasoline or other solvents.

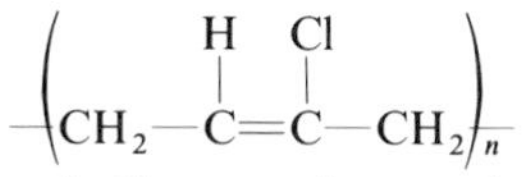

$$-\!\!\left(CH_2-\overset{H}{\overset{|}{C}}=\overset{Cl}{\overset{|}{C}}-CH_2\right)_{\!n}\!\!-$$

polychloroprene (neoprene)

(a) What type of polymer is polychloroprene? **(b)** What monomer is used to make this synthetic rubber?

26-26. Polyoxymethylene (polyformaldehyde) is the extraordinarily tough, self-lubricating Delrin plastic used in gear wheels.
(a) Give the structure of polyformaldehyde.
(b) Formaldehyde is polymerized using an acidic catalyst. Using H^+ as a catalyst, show the mechanism of the polymerization as far as the trimer.
(c) Is Delrin an addition polymer or a condensation polymer?

26-27. Acetylene can be polymerized using a Ziegler-Natta type of catalyst. The cis or trans stereochemistry of the products can be controlled by careful selection and preparation of the catalyst. The resulting polyacetylene is an electrical semiconductor with a metallic appearance. The cis polyacetylene has a copper color, and the trans is silver.

(a) Draw the structures of *cis*- and *trans*-polyacetylene.

(b) Use your structures to show why these polymers might be expected to conduct electricity.

(c) It is possible to prepare polyacetylene films whose electrical conductivity is *anisotropic:* That is, the conductivity is higher in some directions than in others. Explain how this unusual behavior is possible.

26-28. Use chemical equations to show how the following accidents cause injury to the clothing involved (not to mention the skin under the clothing!).

(a) An industrial chemist spills aqueous H_2SO_4 on her nylon stockings but fails to wash it off immediately.

(b) An organic laboratory student spills aqueous NaOH on his polyester slacks.

26-29. Poly(vinyl alcohol), a hydrophilic polymer used in aqueous adhesives, is made by polymerizing vinyl acetate and then hydrolyzing the ester linkages.

(a) Give the structures of poly(vinyl acetate) and poly(vinyl alcohol).

(b) Vinyl acetate is an ester. Is poly(vinyl acetate) therefore a polyester? Explain.

(c) We have seen that basic hydrolysis destroys the Dacron polymer. Poly(vinyl acetate) is converted to poly(vinyl alcohol) by a basic hydrolysis of the ester groups. Why doesn't the hydrolysis destroy the poly(vinyl alcohol) polymer?

(d) Why is poly(vinyl alcohol) made by this circuitous route? Why not just polymerize vinyl alcohol?

26-30. When referring to cloth or fiber, the term *acetate* usually means *cellulose acetate,* a semisynthetic polymer made by treating cellulose with acetic anhydride. Cellulose acetate is spun into yarn by dissolving it in acetone or methylene chloride and forcing the solution through spinnerets into warm air where the solvent evaporates.

(a) Draw the structure of cellulose acetate.

(b) Explain why cellulose acetate is soluble in organic solvents, even though cellulose is not.

(A true story) An organic chemistry student wore a long-sleeved acetate blouse to the laboratory. She was rinsing a warm separatory funnel with acetone when the pressure rose and blew out the stopper. Her right arm was drenched with acetone, but she was unconcerned because acetone is not very toxic. About ten minutes later, the right arm of the student's blouse disintegrated into a pile of white fluff, leaving her with a ragged short sleeve and the tatters of a cuff remaining around her wrist.

(c) Explain how a substance as innocuous as acetone ruined the student's blouse.

(d) Predict what usually happens when a student wears Corfam (a porous polyvinyl chloride) shoes to the organic laboratory.

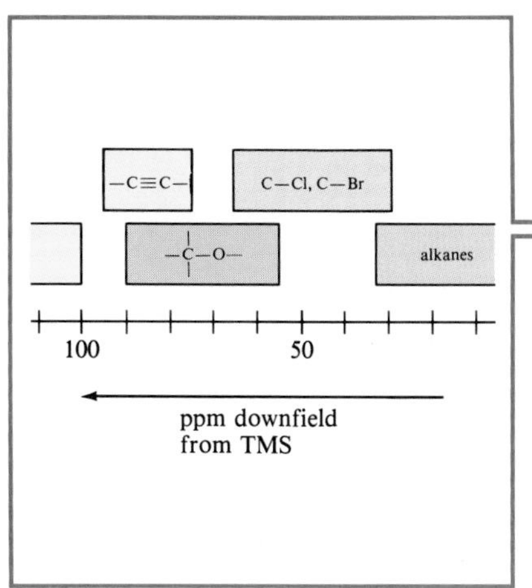

APPENDICES

APPENDIX 1A

NMR absorption positions of protons in various structural environments.

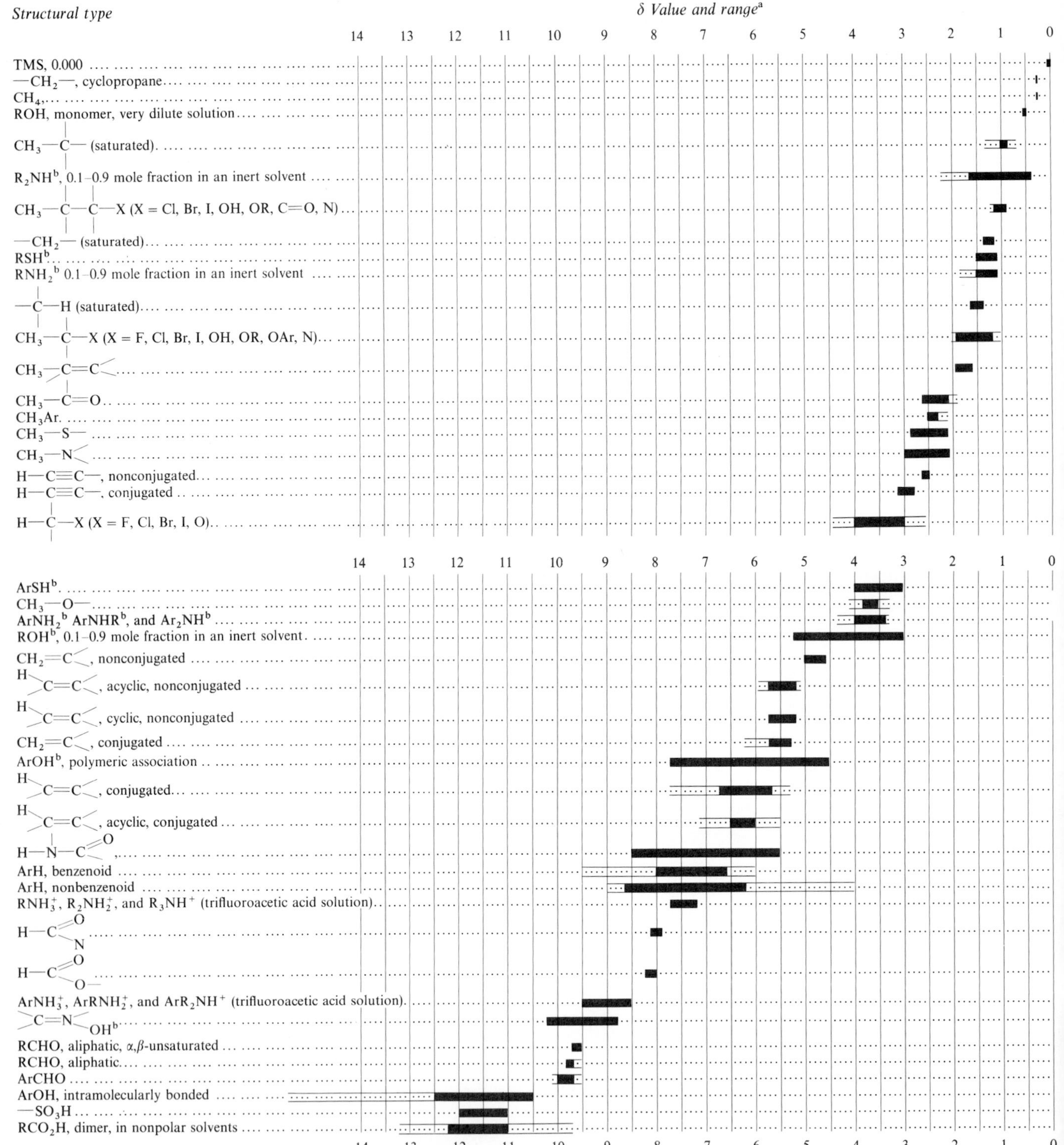

[a] Normally, absorptions for the functional groups indicated will be found within the range shown in black. Occasionally, a functional group will absorb outside this range. Approximate limits for this are indicated by extended outlines.
[b] The absorption positions of these groups are concentration-dependent and are shifted to lower δ values in more dilute solutions.

APPENDIX 1B

Spin-spin coupling constants

Type	J, Hz	Type	J, Hz
>C(H)(H) (geminal H—C—H)	12–15	>C=C(H)—C—H	4–10
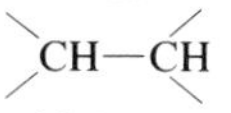 >CH—CH<	2–9		
with free rotation	~7	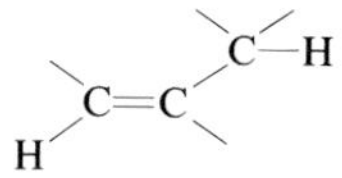 H—C=C—C—H (H and C—H trans)	0.5–2.5
—C(H)—(—C—)$_n$—C(H)—	~0	H—C=C—C—H (H and C—H cis)	~0
CH_3—CH_2—X	6.5–7.5	>C=CH—CH=C<	9–13
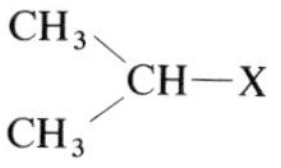 (CH_3)$_2$CH—X	5.5–7.0	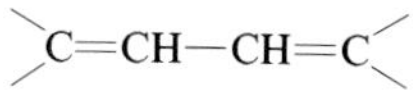	
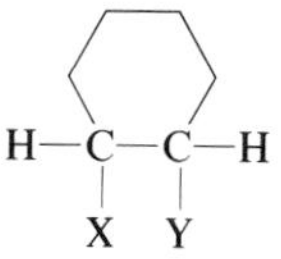 cyclohexane H—C(X)—C(Y)—H	a,a 5–10 a,e 2–4 e,e 2–4	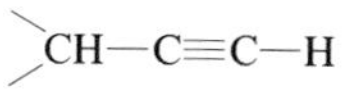 >CH—C≡C—H	2–3
		>CH—C(H)=O	1–3
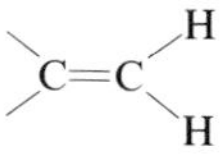 >C=C(H)(H)	0.5–3	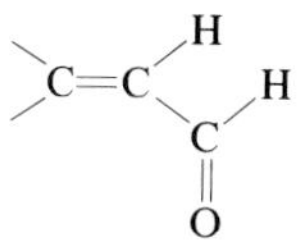 >C=C(H)—C(H)=O	6–8
H—C=C—H (cis)	7–12		
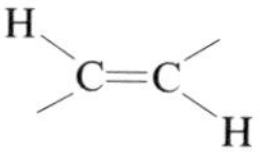 H—C=C—H (trans)	13–18	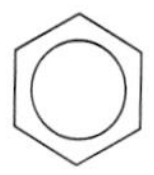 benzene	*o*- 6–9 *m*- 1–3 *p*- 0–1

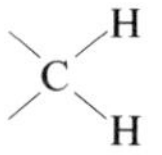

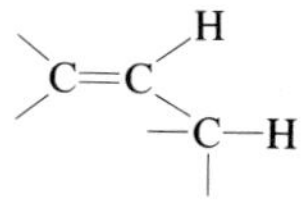

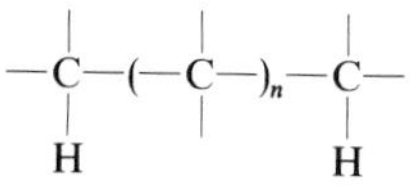

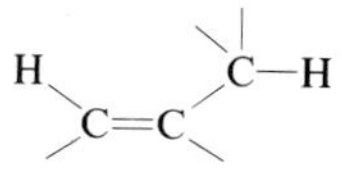

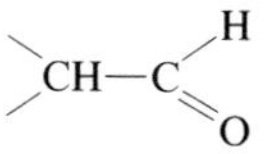

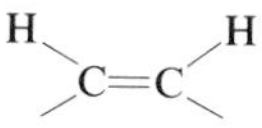

a = axial, e = equatorial

APPENDIX 2A

Characteristic infrared group frequencies (s = strong, m = medium, w = weak). (Courtesy of N. B. Colthup, Stamford Research Laboratories, American Cyanamid Company, and the editor of the journal of the Optical Society.) Overtone bands are marked 2ν.

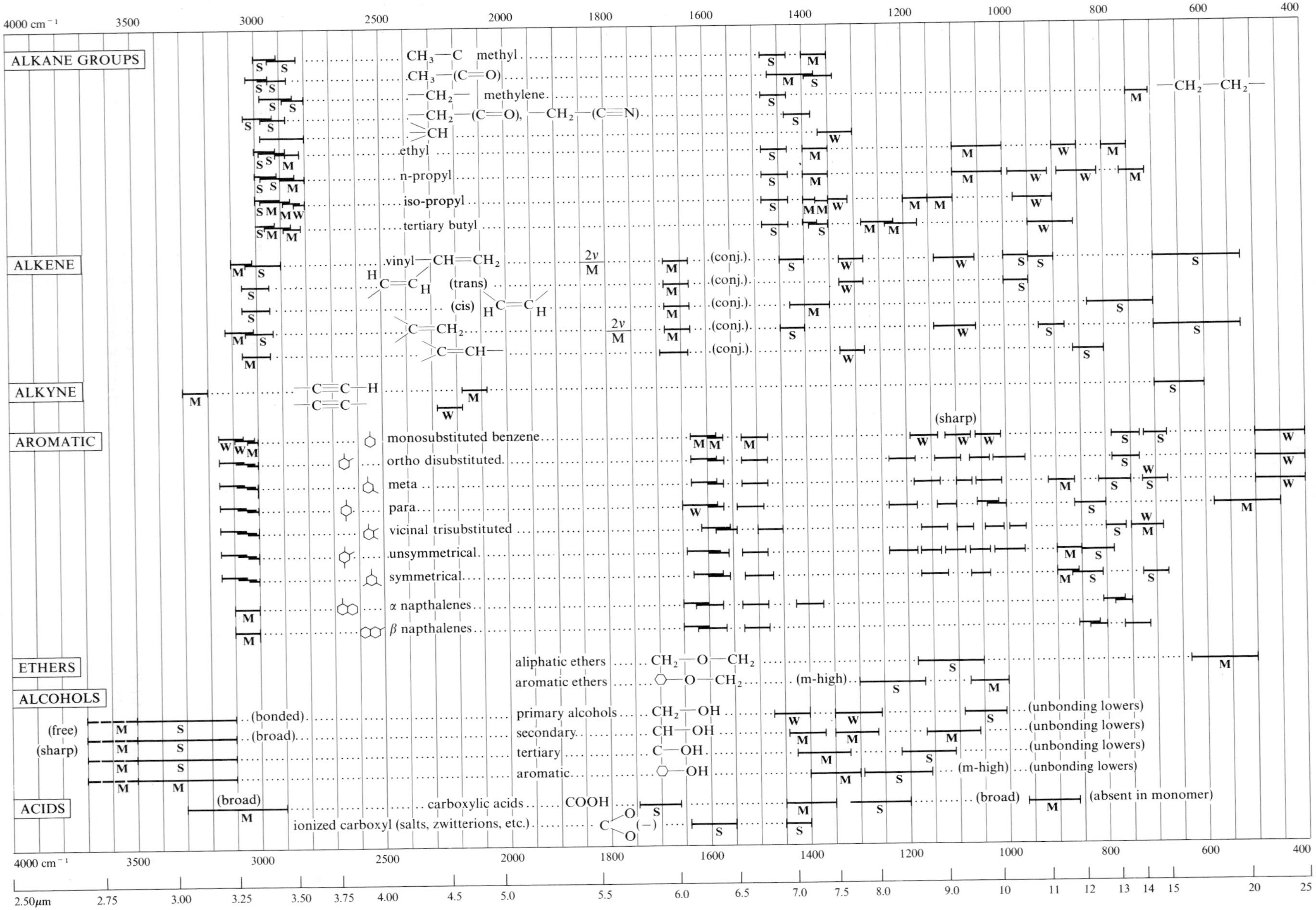

4000 cm⁻¹ 3500 3000 2500 2000 1800 1600 1400 1200 1000 800 600 400

ALKANE GROUPS

CH_3—C methyl

CH_3—(C=O)

—CH_2— methylene

—CH_2—(C=O), —CH_2—(C≡N)

>CH

ethyl

n-propyl

iso-propyl

tertiary butyl

—CH_2—CH_2—

ALKENE

vinyl —CH=CH_2 (conj.)

(trans) (conj.)

(cis) (conj.)

>C=CH_2 (conj.)

>C=CH— (conj.)

ALKYNE

—C≡C—H

—C≡C—

AROMATIC

monosubstituted benzene

ortho disubstituted

meta

para

vicinal trisubstituted

unsymmetrical

symmetrical

α napthalenes

β napthalenes

(sharp)

ETHERS

aliphatic ethers CH_2—O—CH_2

aromatic ethers ⌬—O—CH_2 (m-high)

ALCOHOLS

(free) (sharp) (bonded) (broad)

primary alcohols CH_2—OH (unbonding lowers)

secondary CH—OH (unbonding lowers)

tertiary C—OH (unbonding lowers)

aromatic ⌬—OH (m-high) (unbonding lowers)

ACIDS

(broad) carboxylic acids COOH (broad) (absent in monomer)

ionized carboxyl (salts, zwitterions, etc.) C(O)(O)(−)

4000 cm⁻¹ 3500 3000 2500 2000 1800 1600 1400 1200 1000 800 600 400

2.50 μm 2.75 3.00 3.25 3.50 3.75 4.00 4.5 5.0 5.5 6.0 6.5 7.0 7.5 8.0 9.0 10 11 12 13 14 15 20 25

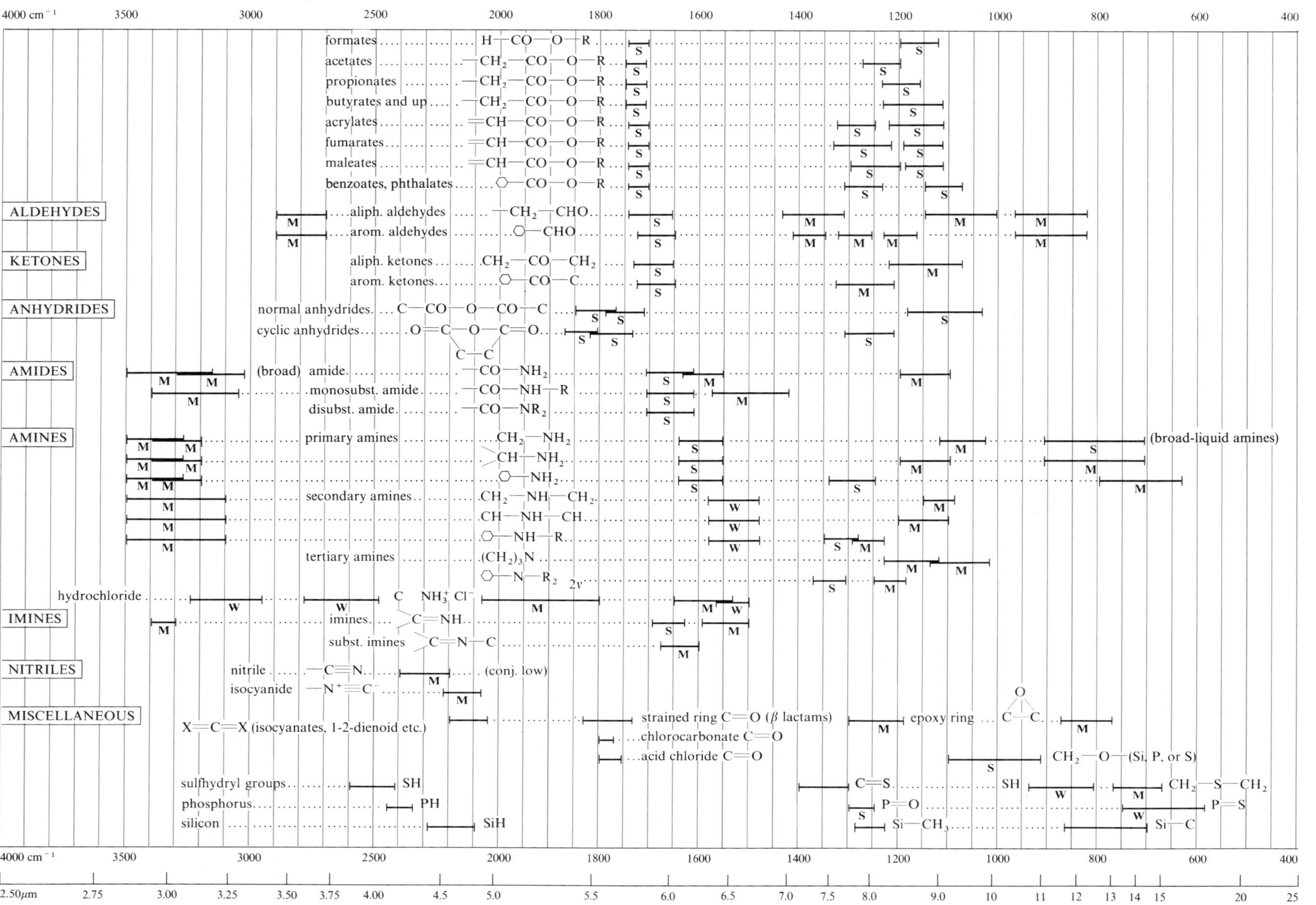

4000 cm⁻¹
3500
3000
2500
2000
1800
1600
1400
1200
1000
800
600
400
formates ... H—CO—O—R
acetates ... —CH₂—CO—O—R
propionates ... —CH₂—CO—O—R
butyrates and up ... —CH₂—CO—O—R
acrylates ... =CH—CO—O—R
fumarates ... =CH—CO—O—R
maleates ... =CH—CO—O—R
benzoates, phthalates ... ⬡—CO—O—R
ALDEHYDES
aliph. aldehydes ... —CH₂—CHO
arom. aldehydes ... ⬡—CHO
KETONES
aliph. ketones ... CH₂—CO—CH₂
arom. ketones ... ⬡—CO—C
ANHYDRIDES
normal anhydrides ... C—CO—O—CO—C
cyclic anhydrides ... O=C—O—C=O
C—C
AMIDES
(broad) amide ... —CO—NH₂
monosubst. amide ... —CO—NH—R
disubst. amide ... —CO—NR₂
AMINES
primary amines ... CH₂—NH₂
CH—NH₂
⬡—NH₂
(broad-liquid amines)
secondary amines ... CH₂—NH—CH₂
CH—NH—CH
⬡—NH—R
tertiary amines ... (CH₂)₃N
⬡—N—R₂
2ν
hydrochloride ... C NH₃⁺ Cl⁻
IMINES
imines ... C=NH
subst. imines C=N—C
NITRILES
nitrile ... —C≡N ... (conj. low)
isocyanide —N⁺≡C⁻
MISCELLANEOUS
X=C=X (isocyanates, 1-2-dienoid etc.)
strained ring C=O (β lactams)
chlorocarbonate C=O
acid chloride C=O
epoxy ring ... C—C (O)
CH₂—O—(Si, P, or S)
sulfhydryl groups ... SH
phosphorus ... PH
silicon ... SiH
C=S ... SH
P=O
Si—CH₃
CH₂—S—CH₂
P=S
Si—C
S
M
W
2.50μm
2.75
3.00
3.25
3.50
3.75
4.00
4.5
5.0
5.5
6.0
6.5
7.0
7.5
8.0
9.0
10
11
12
13
14
15
20
25

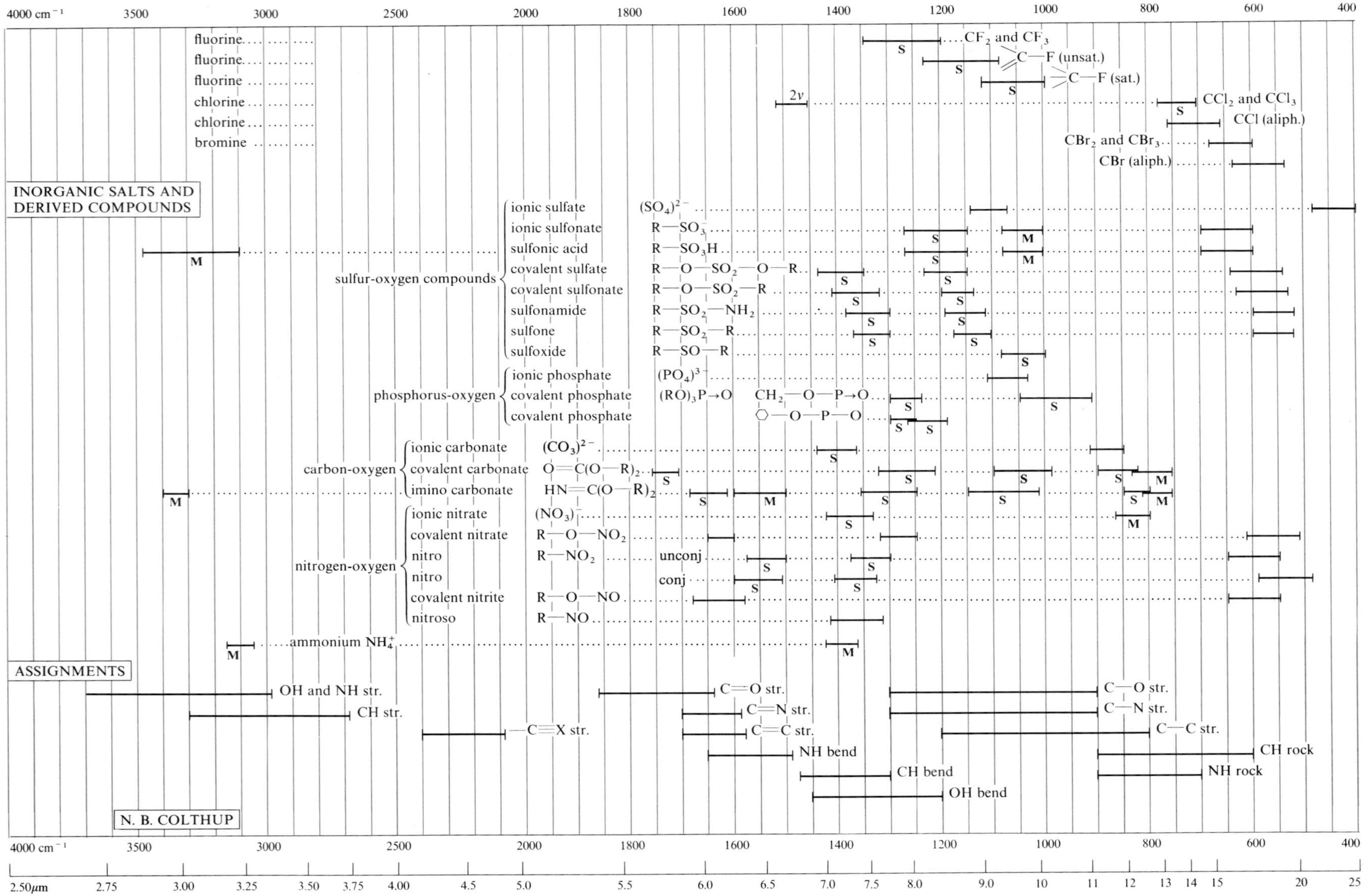

4000 cm^{-1} 3500 3000 2500 2000 1800 1600 1400 1200 1000 800 600 400

fluorine ... CF_2 and CF_3 (S)
fluorine ... $>C=C-F$ (unsat.) (S)
fluorine ... $\geq C-F$ (sat.) (S)
chlorine ... 2ν ... CCl_2 and CCl_3 (S)
chlorine ... CCl (aliph.)
bromine ... CBr_2 and CBr_3 ... CBr (aliph.)

INORGANIC SALTS AND DERIVED COMPOUNDS

sulfur-oxygen compounds:
- ionic sulfate $(SO_4)^{2-}$
- ionic sulfonate $R-SO_3^-$ (S) (M)
- sulfonic acid $R-SO_3H$ (M) (S) (M)
- covalent sulfate $R-O-SO_2-O-R$ (S) (S)
- covalent sulfonate $R-O-SO_2-R$ (S) (S)
- sulfonamide $R-SO_2-NH_2$ (S) (S)
- sulfone $R-SO_2-R$ (S) (S)
- sulfoxide $R-SO-R$ (S)

phosphorus-oxygen:
- ionic phosphate $(PO_4)^{3-}$
- covalent phosphate $(RO)_3P\rightarrow O$; $CH_2-O-P\rightarrow O$ (S) (S)
- covalent phosphate ⬡$-O-P-O$ (S) (S)

carbon-oxygen:
- ionic carbonate $(CO_3)^{2-}$ (S)
- covalent carbonate $O=C(O-R)_2$ (S) (S) (S) (S) (M)
- imino carbonate $HN=C(O-R)_2$ (M) (S) (M) (S) (S) (S) (M)

nitrogen-oxygen:
- ionic nitrate $(NO_3)^-$ (S) (M)
- covalent nitrate $R-O-NO_2$
- nitro $R-NO_2$ unconj (S) (S)
- nitro conj (S) (S)
- covalent nitrite $R-O-NO$
- nitroso $R-NO$

(M) ammonium NH_4^+ (M)

ASSIGNMENTS

OH and NH str.
CH str.
$-C\equiv X$ str.
$C=O$ str.
$C=N$ str.
$C=C$ str.
NH bend
CH bend
OH bend
$C-O$ str.
$C-N$ str.
$C-C$ str.
CH rock
NH rock

N. B. COLTHUP

4000 cm^{-1} 3500 3000 2500 2000 1800 1600 1400 1200 1000 800 600 400

2.50μm 2.75 3.00 3.25 3.50 3.75 4.00 4.5 5.0 5.5 6.0 6.5 7.0 7.5 8.0 9.0 10 11 12 13 14 15 20 25

APPENDIX 2B

Characteristic infrared absorptions of functional groups

Group	Intensity[a]	Range (cm^{-1})
A. Hydrocarbon chromophore		
1. C—H stretching		
a. Alkane	m-s	2962–2853
b. Alkene, monosubstituted (vinyl)	m	3040–3010
	and m	3095–3075
Alkene, disubstituted, *cis*	m	3040–3010
Alkene, disubstituted, *trans*	m	3040–3010
Alkene, disubstituted, *gem*	m	3095–3075
Alkene, trisubstituted	m	3040–3010
c. Alkyne	s	~3300
d. Aromatic	v	~3030
2. C—H bending		
a. Alkane, C—H	w	~1340
Alkane, $—CH_2—$	m	1485–1445
Alkane, $—CH_3$	m	1470–1430
	and s	1380–1370
Alkane, *gem*-dimethyl	s	1385–1380
	and s	1370–1365
Alkane, *tert*-butyl	m	1395–1385
	and s	~1365
b. Alkene, monosubstituted (vinyl)	s	995–985
	s	915–905
	and s	1420–1410
Alkene, disubstituted, *cis*	s	~690
Alkene, disubstituted, *trans*	s	970–960
	and m	1310–1295
Alkene, disubstituted, *gem*	s	895–885
	and s	1420–1410
Alkene, trisubstituted	s	840–790
c. Alkyne	s	~630
d. Aromatic, substitution type:[b]		
five adjacent hydrogen atoms	v, s	~750
	and v, s	~700
four adjacent hydrogen atoms	v, s	~750
three adjacent hydrogen atoms	v, m	~780
two adjacent hydrogen atoms	v, m	~830
one hydrogen atom	v, w	~880
3. C—C multiple bond stretching		
a. Alkene, nonconjugated	v	1680–1620
Alkene, monosubstituted (vinyl)	m	~1645
Alkene, disubstituted, *cis*	m	~1658
Alkene, disubstituted, *trans*	m	~1675
Alkene, disubstituted, *gem*	m	~1653
Alkene, trisubstituted	m	~1669
Alkene, tetrasubstituted	w	~1669
Diene	w	~1650
	and w	~1600
b. Alkyne, monosubstituted	m	2140–2100
Alkyne, disubstituted	v, w	2260–2190
c. Allene	m	~1960
	and m	~1060
d. Aromatic	v	~1600
	v	~1580
	m	~1500
	and m	~1450
B. Carbonyl chromophore		
1. Ketone stretching vibrations		
a. Saturated, acyclic	s	1725–1705
b. Saturated, cyclic:		
6-membered ring (and higher)	s	1725–1705
5-membered ring	s	1750–1740
4-membered ring	s	~1775
c. α,β-Unsaturated, acyclic	s	1685–1665
d. α,β-Unsaturated, cyclic:		
6-membered ring (and higher)	s	1685–1665
5-membered ring	s	1725–1708
e. α,β,α′,β′-Unsaturated, acyclic	s	1670–1663
f. Aryl	s	1700–1680
g. Diaryl	s	1670–1660
h. α-Diketones	s	1730–1710
i. β-Diketones (enolic)	s	1640–1540
j. 1,4-Quinones	s	1690–1660
k. Ketenes	s	~2150
2. Aldehydes		
a. Carbonyl stretching vibrations:		
Saturated, aliphatic	s	1740–1720
α,β-Unsaturated, aliphatic	s	1705–1680
α,β,γ,δ-Unsaturated, aliphatic	s	1680–1660
Aryl	s	1715–1695
b. C—H stretching vibrations,	w	2900–2820
two bands	and w	2775–2700
3. Ester stretching vibrations		
a. Saturated, acyclic	s	1750–1735
b. Saturated, cyclic:		
δ-lactones (and larger rings)	s	1750–1735
γ-lactones	s	1780–1760
β-lactones	s	~1820
c. Unsaturated:		
vinyl ester type	s	1800–1770
α,β-unsaturated and aryl	s	1730–1717
α,β-unsaturated δ-lactone	s	1730–1717
α,β-unsaturated γ-lactone	s	1760–1740
β,γ-unsaturated γ-lactone	s	~1800
d. α-Ketoesters	s	1755–1740
e. β-Ketoesters (enolic)	s	~1650
f. Carbonates	s	1780–1740
4. Carboxylic acids		
a. Carbonyl stretching vibrations:		
Saturated aliphatic	s	1725–1700
α,β-unsaturated aliphatic	s	1715–1690
Aryl	s	1700–1680
b. Hydroxyl stretching (bonded), several bands	w	2700–2500
c. Carboxylate anion stretching	s	1610–1550
	and s	1400–1300
5. Anhydride stretching vibrations		
a. Saturated, acyclic	s	1850–1800
	and s	1790–1740
b. α,β-Unsaturated and aryl, acyclic	s	1830–1780
	and s	1770–1720
c. Saturated, 5-membered ring	s	1870–1820
	and s	1800–1750
d. α,β-Unsaturated, 5-membered ring	s	1850–1800
	and s	1830–1780
6. Acyl halide stretching vibrations		
a. Acyl fluorides	s	~1850
b. Acyl chlorides	s	~1795
c. Acyl bromides	s	~1810
d. α,β-Unsaturated and aryl	s	1780–1750
	and m	1750–1720

[a] Abbreviations: s = strong, m = medium, w = weak, v = variable, b = broad, sh = sharp, ~ = approximately

[b] Substituted benzenes also show weak bands in the 2000–1670 cm^{-1} region.

(Continued)

Group	Intensity[a]	Range (cm^{-1})
7. Amides		
a. Carbonyl stretching vibrations:		
Primary, solid and concentrated solution	s	~1650
Primary, dilute solution	s	~1690
Secondary, solid and concentrated solution	s	1680–1630
Secondary, dilute solution	s	1700–1670
Tertiary, solid and all solutions	s	1670–1630
Cyclic, δ-lactams	s	~1680
Cyclic, γ-lactams	s	~1700
Cyclic, γ-lactams, fused to another ring	s	1750–1700
Cyclic, β-lactams	s	1760–1730
Cyclic, β-lactams, fused to another ring, dilute solution	s	1780–1770
Ureas, acyclic	s	~1660
Ureas, cyclic, 6-membered ring	s	~1640
Ureas, cyclic, 5-membered ring	s	~1720
Urethanes	s	1740–1690
Imides, acyclic	s	~1710
	and s	~1700
Imides, cyclic, 6-membered ring	s	~1710
	and s	~1700
Imides, cyclic, α,β-unsaturated, 6-membered ring	s	~1730
	and s	~1670
Imides, cyclic, 5-membered ring	s	~1770
	and s	~1700
Imides, cyclic, α,β-unsaturated, 5-membered ring	s	~1790
	and s	~1710
b. N—H stretching vibrations:		
Primary, free; two bands	m	~3500
	and m	~3400
Primary, bonded; two bands	m	~3350
	and m	~3180
Secondary, free; one band	m	~3430
Secondary, bonded; one band	m	3320–3140
c. N—H bending vibrations:		
Primary amides, dilute solution	s	1620–1590
Secondary amides	s	1550–1510
C. Miscellaneous chromophoric groups		
1. Alcohols and phenols		
a. O—H stretching vibrations:		
Free O—H	v, sh	3650–3590
Intermolecularly hydrogen bonded (change on dilution)		
single bridge compounds	v, sh	3550–3450
polymeric association	s, b	3400–3200
Intramolecularly hydrogen bonded (no change on dilution)		
single bridge compounds	v, sh	3570–3450
chelate compounds	w, b	3200–2500
b. O—H bending and C—O stretching vibrations:		
Primary alcohols	s	~1050
	and s	1350–1260
Secondary alcohols	s	~1100
	and s	1350–1260
Tertiary alcohols	s	~1150
	and s	1410–1310
Phenols	s	~1200
	and s	1410–1310

Group	Intensity[a]	Range (cm^{-1})
2. Amines		
a. N—H stretching vibrations:		
Primary, free; two bands	m	~3500
	and m	~3400
Secondary, free; one band	m	3500–3310
Imines (=N—N); one band	m	3400–3300
Amine salts	m	3130–3030
b. N—H bending vibrations:		
Primary	s-m	1650–1590
Secondary	w	1650–1550
Amine salts	s	1600–1575
	and s	~1500
c. C—N vibrations:		
Aromatic, primary	s	1340–1250
Aromatic, secondary	s	1350–1280
Aromatic, tertiary	s	1360–1310
Aliphatic	w	1220–1020
	and w	~1410
3. Unsaturated nitrogen compounds		
a. C≡N stretching vibrations:		
Alkyl nitriles	m	2260–2240
α,β-Unsaturated alkyl nitriles	m	2235–2215
Aryl nitriles	m	2240–2220
Isocyanates	m	2275–2240
Isocyanides	m	2220–2070
b. >C=N—Stretching vibrations (imines, oximes)		
Alkyl compounds	v	1690–1640
α,β-Unsaturated compounds	v	1660–1630
c. —N=N—stretching vibrations, azo compounds	v	1630–1575
d. —N=C=N—stretching vibrations, diimide	s	2155–2130
e. —N_3 stretching vibrations, azides	s	2160–2120
	and w	1340–1180
f. C—NO_2, nitro compounds:		
aromatic	s	1570–1500
	and s	1370–1300
aliphatic	s	1570–1550
	and s	1380–1370
g. O—NO_2, nitrates	s	1650–1600
	and s	1300–1250
h. C—NO, nitroso compounds	s	1600–1500
i. O—NO, nitrites	s	1680–1650
	and s	1625–1610
4. Halogen compounds, C—X stretching vibrations		
a. C—F	s	1400–1000
b. C—Cl	s	800–600
c. C—Br	s	600–500
d. C—I	s	~500
5. Sulfur compounds		
a. S—H stretching vibrations	w	2600–2550
b. C=S stretching vibrations	s	1200–1050
c. S=O stretching vibrations:		
sulfoxides	s	1070–1030
sulfones	s	1160–1140
	and s	1350–1300
sulfites	s	1230–1150
	and s	1430–1350
sulfonyl chlorides	s	1185–1165
	and s	1370–1340
sulfonamides	s	1180–1140
	and s	1350–1300
sulfonic acids	s	1210–1150
	s	1060–1030
	and s	~650

[a] Abbreviations: s = strong, m = medium, w = weak, v = variable, b = broad, sh = sharp, ~ = approximately

THE WOODWARD-FIESER RULES FOR PREDICTING ULTRAVIOLET-VISIBLE SPECTRA

To use UV-visible spectroscopy as an effective tool for structure determination, we need to know what types of spectra correspond to the most common types of conjugated systems. The most useful correlations between structures and UV spectra were developed by R. B. Woodward and L. F. Fieser in the early 1940s and are called the **Woodward-Fieser rules.** The rules presented here are used to predict only the lowest-energy $\pi \rightarrow \pi^*$ transition from the HOMO to the LUMO. Values of λ_{max} measured in different solvents can be different, so we will generally assume that ethanol is the solvent.

In discussing these rules, we will use the following specialized terms:

CHROMOPHORE Any functional group (or collection of groups) that is responsible for absorption.

AUXOCHROME A substituent that is not a chromophore by itself but which alters the wavelength or the extinction coefficient when it is attached to a chromophore.

BATHOCHROMIC SHIFT A shift toward lower frequency and longer wavelength (longer λ_{max}).

HYPSOCHROMIC SHIFT A shift toward higher frequency and shorter wavelength (shorter λ_{max}).

UV SPECTRA OF DIENES AND POLYENES

The bathochromic effect of alkyl groups Although a molecule's system of conjugated double bonds (the chromophore) is the largest factor in determining its UV spectrum, the absorption is also affected by alkyl substituents. Each alkyl group attached to the chromophore serves as an auxochrome, producing a small bathochromic shift of about 5 nm. The following table shows the effects of adding alkyl groups to 1,3-butadiene.

Values of λ_{max} for some substituted 1,3,butadienes[a]

Number of alkyl groups	*Compound*	λ_{max} *(nm)*
0	$H_2C{=}CH{-}CH{=}CH_2$	217
1	$CH_3{-}CH{=}CH{-}CH{=}CH_2$	224
1	$H_2C{=}C(CH_3){-}CH{=}CH_2$	220
2	$H_2C{=}C(CH_3){-}C(CH_3){=}CH_2$	226
2	$CH_3{-}CH{=}CH{-}CH{=}CH{-}CH_3$	227
3	$CH_3{-}C(CH_3){=}CH{-}C(CH_3){=}CH_2$	232
4	$CH_3{-}C(CH_3){=}CH{-}C(CH_3){=}CH{-}CH_3$	241

[a] Each alkyl group attached to the diene system contributes a bathochromic shift of about 5 nm.

Conformation effects For dienes that are predominantly in the *s*-trans conformation (either free to rotate or held in the *s*-trans conformation), Woodward and

Fieser used a base value of 217 nm, the λ_{max} for unsubstituted 1,3-butadiene. To this value we add 5 nm for each alkyl substituent. For dienes that are held in the *s*-cis conformation by a 6-membered ring, the base value is 253 nm for the diene, plus 5 nm for each alkyl substituent.

+5
+5
+5
+5

acyclic (*s*-trans) diene
base 217 nm

transoid cyclic diene
base 217 nm
+ 2 alkyl × (5 nm)

cisoid cyclic diene
base 253 nm
+ 2 alkyl × (5 nm)

Additional conjugated double bonds For trienes and larger conjugated systems, we add 30 nm to the base value for each additional double bond. The additional double bond must be attached at the end of the conjugated system to extend the length of the polyene system in order to have this large 30-nm contribution, however.

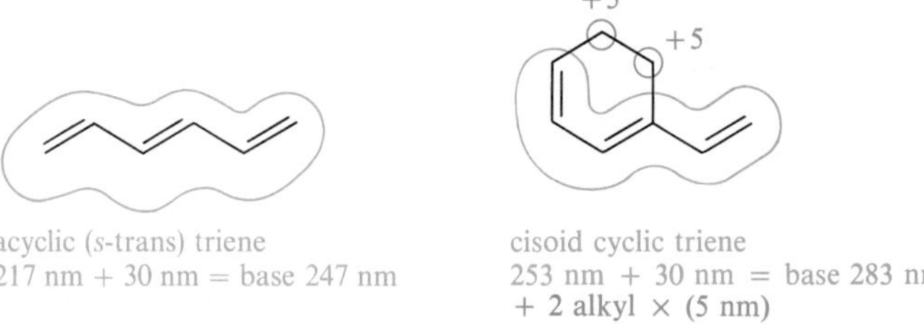

acyclic (*s*-trans) triene
217 nm + 30 nm = base 247 nm

cisoid cyclic triene
253 nm + 30 nm = base 283 nm
+ 2 alkyl × (5 nm)

The contributions of auxochromic groups are added to the base values of the polyene chromophore. We add 5 nm for each alkyl group and 5 nm if one of the double bonds in the conjugated system is exocyclic to a ring. An exocyclic double bond is one that is attached to a ring at one end.

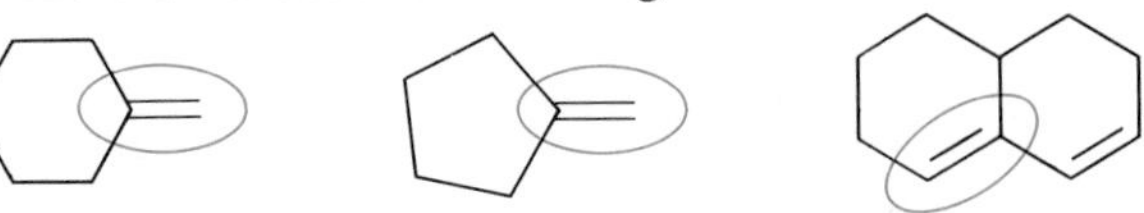

exocyclic double bonds

The shifts associated with common auxochromic groups are summarized in the following table.

The Woodward-Fieser rules for conjugated dienes:
Values for auxochromic groups[a]

Grouping	*Substituent correction (nm)*
another conjugated C=C	+30
alkyl group	+5
alkoxy (—OR) group	0

If one of the double bonds in the chromophore is exocyclic, add another 5 nm:

exoc.

exocyclic double bond

+5 (in addition to the 30 nm for lengthening the system)

[a] These values are added to the base value for the diene system.

Examples The best way to learn to use the rules for predicting UV absorptions is to work through some examples. The following examples show the types of structures that follow the rules closely and one that does not.

1. 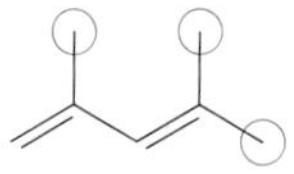 2,4-dimethyl-1,3-pentadiene

base:	217 nm
three alkyl groups:	15
predicted λ_{max}:	232 nm; observed: 232 nm

2. 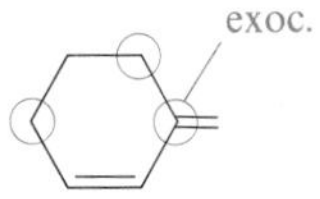 exoc.

base:	217 nm
two alkyl groups:	10
exocyclic C=C:	5
predicted λ_{max}:	232 nm; observed: 230 nm

3. 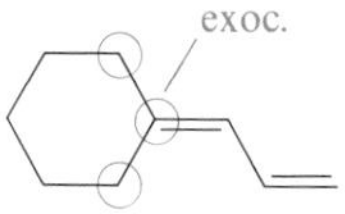 exoc.

base:	217 nm
two alkyl groups:	10
exocyclic C=C:	5
predicted λ_{max}:	232 nm; observed: 236 nm

4.  exoc.

base:	217 nm
three alkyl groups:	15
exocyclic C=C:	5
predicted λ_{max}:	237 nm; observed: 235 nm

5. 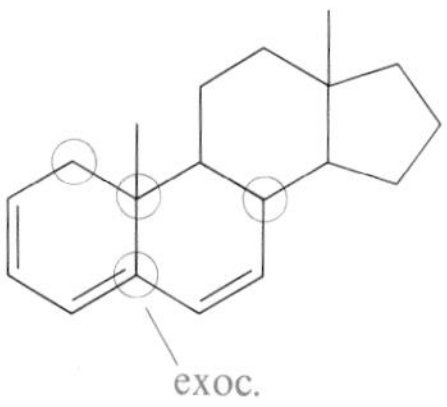 exoc.

base:	253 nm
conjugated C=C:	30
three alkyl groups:	15
exocyclic C=C:	5
predicted λ_{max}:	303 nm; observed: 304 nm

6. 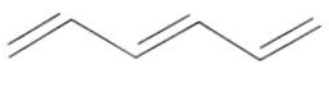 1,3,5-hexatriene

base:	217 nm
conjugated C=C:	30
predicted λ_{max}:	247 nm; observed: 258 nm

UV SPECTRA OF CONJUGATED KETONES AND ALDEHYDES

The $\pi \rightarrow \pi^*$ transition As with dienes and polyenes, the strongest absorptions in the UV spectra of aldehydes and ketones are the ones resulting from $\pi \rightarrow \pi^*$ electronic transition. These absorptions are observable ($\lambda_{max} > 200$ nm) only if the carbonyl double bond is conjugated with another double bond.

The Woodward-Fieser rules for conjugated ketones and aldehydes appear in the following table. Note that the bathochromic effect of alkyl groups depends on their location: 10 nm for groups α to the carbonyl and 12 nm for groups in a β position. The contributions from additional conjugated double bonds (30 nm) and exocyclic positions of double bonds (5 nm) are similar to those in dienes and polyenes.

The Woodward-Fieser rules for conjugated ketones and aldehydes

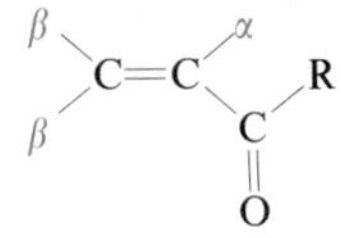

general structure

Base values: 210 nm if R = H (aldehyde)
215 nm if R = alkyl (ketone)

Grouping	Position	Correction
alkyl group, α	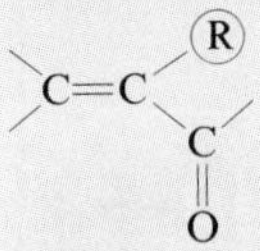	+10 nm
alkyl group, β	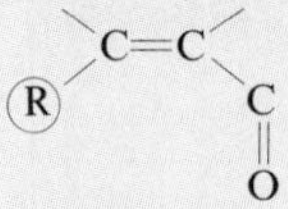	+12 nm
exocyclic position of a C=C bond	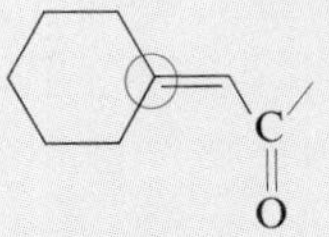	+5 nm
additional conjugated double bond	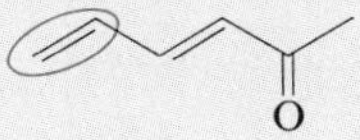	+30 nm

The following examples show how the Woodward-Fieser rules are used to predict values of λ_{max} for a variety of conjugated ketones and aldehydes. Notice that the values of the extinction coefficients (ε) for these transitions are quite large (> 5000), as we also observed for the $\pi \rightarrow \pi^*$ transitions in conjugated dienes and polyenes.

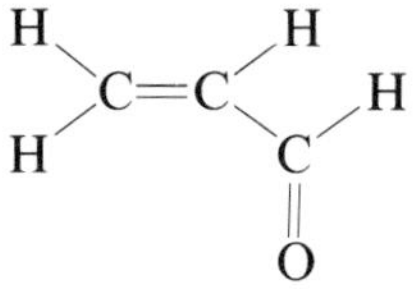

Base value	210 nm
(no corrections)	
Predicted λ_{max}	210 nm

Experimental: $\lambda_{max} = 210$ nm, $\varepsilon = 11{,}000$

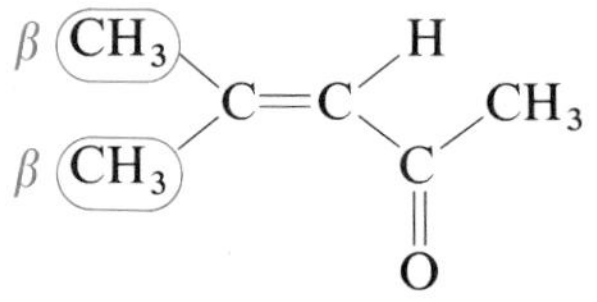

Base value	215 nm
2 × β substituent	24 nm
Predicted λ_{max}	239 nm

Experimental: $\lambda_{max} = 237$ nm, $\varepsilon = 12{,}000$

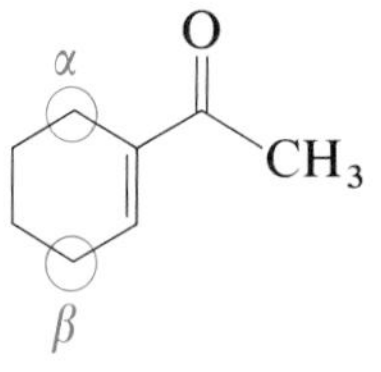

Base value	215 nm
α substituent	10 nm
β substituent	12 nm
Predicted λ_{max}	237 nm

Experimental: $\lambda_{max} = 233$ nm, $\varepsilon = 12{,}500$

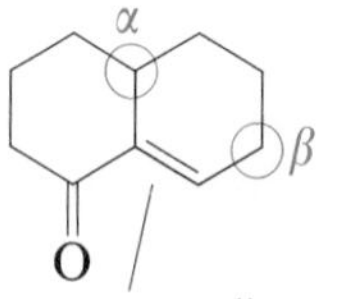

exocyclic double bond

Base value	215 nm
α substituent	10 nm
β substituent	12 nm
Exocyclic double bond	5 nm
Predicted λ_{max}	242 nm

Experimental: $\lambda_{max} = 241$ nm, $\varepsilon = 5{,}200$

The $n \rightarrow \pi^*$ transition As discussed in Section 18-5E, ketones and aldehydes also show weak UV absorptions ($\varepsilon \cong 10$ to 200) resulting from "forbidden" $n \rightarrow \pi^*$ transitions. Because the promoted electron leaves a nonbonding (n) orbital that is higher in energy than the pi bonding orbital, this transition involves a smaller amount of energy and results in a longer-wavelength (lower-frequency) absorption. The $n \rightarrow \pi^*$ transitions of simple, unconjugated ketones and aldehydes give absorptions with values of λ_{max} between 280 and 300 nm. Each double bond added in conjugation with the carbonyl group increases the value of λ_{max} by about 30 nm.

APPENDIX 4A
METHODS AND SUGGESTIONS FOR PROPOSING MECHANISMS

In this appendix we consider how an organic chemist systematically approaches a mechanism problem. Although there is no "formula" for solving all mechanism problems, this stepwise method should provide a starting point for you to begin building experience and confidence. Solved problems that apply this approach appear on pages 150, 301, 424, 803, 949, 1003, and 1030.

DETERMINING THE TYPE OF MECHANISM

First, determine what kinds of conditions or catalysts are involved. In general, reactions may be classified as (a) involving strong electrophiles (includes acid-catalyzed reactions), (b) involving strong nucleophiles (includes base-catalyzed reactions), or (c) involving free radicals. These three types of mechanisms are quite distinct, and you should first try to determine which type is involved. If uncertain, you can develop more than one type of mechanism and see which fits the facts better.

(a) In the presence of a strong acid or a reactant that can dissociate to give a strong electrophile, the mechanism probably involves strong electrophiles as intermediates. Acid-catalyzed reactions and reactions involving carbocations (such as the S_N1, E1, and most alcohol dehydrations) generally fall in this category.

(b) In the presence of a strong base or strong nucleophile, the mechanism probably involves strong nucleophiles as intermediates. Base-catalyzed reactions and those whose rates depend on base strength (such as S_N2 and E2) generally fall in this category.

(c) Free-radical reactions usually require a free-radical initiator such as chlorine, bromine, NBS, AIBN, or a peroxide. In most free-radical reactions there is no need for a strong acid or base.

POINTS TO WATCH IN ALL MECHANISMS

Once you have determined which type of mechanism to write, there are general methods, discussed below, for approaching the problem. Regardless of the type of mechanism you intend to draw, however, you should follow three general rules in proposing a mechanism:

1. **Draw all the bonds and all the substituents of each carbon atom affected throughout the mechanism. Do not use condensed or line-angle formulas for reaction sites.**
   Three-bonded carbon atoms are most likely reactive intermediates: carbocations in reactions involving strong electrophiles, carbanions in reactions involving strong nucleophiles, and free radicals in radical reactions. If you draw condensed formulas or line-angle formulas, you will likely misplace a hydrogen atom and show a reactive species on the wrong carbon.
2. **Show only one step at a time. Do not show two or three bonds changing position in one step unless the changes really are concerted (take place simultaneously).**

For example, three pairs of electrons really do move in one step in the Diels-Alder reaction; but in the dehydration of an alcohol, protonation of the hydroxyl group and loss of water are two separate steps.

3. **Use curved arrows to show *movement of electrons*, always from the nucleophile (electron donor) to the electrophile (electron acceptor).**
For example, a proton has no electrons to donate, so a curved arrow should never be drawn from H^+ to anything. When an alkene is protonated, the arrow should go from the electrons of the double bond to the proton. Don't try to use curved arrows to "point out" where the proton (or other reagent) goes. In a free-radical reaction, half-headed arrows show single electrons coming together to form bonds or separating to give other radicals.

APPROACHES TO SPECIFIC TYPES OF MECHANISMS

REACTIONS INVOLVING STRONG ELECTROPHILES

General principles: When a strong acid or electrophile is present, expect intermediates that are strong acids and strong electrophiles; cationic intermediates are common. Carbocations, protonated (three-bonded) oxygen atoms, protonated (four-bonded) nitrogen atoms, and other strong acids might be involved. Any bases and nucleophiles in such a reaction are generally weak. Avoid drawing carbanions, alkoxide ions, and other strong bases. They are unlikely to coexist with strong acids and strong electrophiles.

Functional groups are often converted to carbocations or other strong electrophiles by protonation or by reaction with a strong electrophile. Then the carbocation or other strong electrophile reacts with a weak nucleophile such as an alkene or the solvent.

1. Consider the carbon skeletons of the reactants and products and identify which carbon atoms in the products are most likely derived from which carbon atoms in the reactants.
2. Consider whether any of the reactants is a strong enough electrophile to react without being activated. If not, consider how one of the reactants might be converted to a strong electrophile by protonation of a Lewis basic site, complexation with a Lewis acid, or ionization.
3. Consider how a nucleophilic site on another reactant (or, in a cyclization, another part of the same molecule) can attack this strong electrophile to form a bond needed in the product. Draw the product of this bond formation.
   If the intermediate is a carbocation, consider whether it is likely to rearrange in a manner that would form a bond present in the product.
   If there is no possible nucleophilic attack that leads in the direction of the product, consider other ways of converting one of the reactants to a strong electrophile.
4. Consider how the product of nucleophilic attack might be converted to the final product (if it has the right carbon skeleton) or reactivated to form another bond needed in the product.
5. Draw out all the steps of the mechanism using curved arrows to show the movement of electrons. Be careful to show only one step at a time.

REACTIONS INVOLVING STRONG NUCLEOPHILES

General principles: When a strong base or nucleophile is present, expect intermediates that are strong bases and strong nucleophiles; anionic intermediates are common. Alkoxide ions, stabilized carbanions, and other strong bases might be involved. Any acids and electrophiles in such a reaction are generally weak. Avoid drawing carbocations, protonated carbonyl groups, protonated hydroxyl groups, and other strong acids. They are unlikely to coexist with strong bases and strong nucleophiles.

Functional groups are often converted to strong nucleophiles by deprotonation of the group itself; by deprotonation of the alpha position of a carbonyl group, nitro group, or nitrile; or by attack of another strong nucleophile. Then the resulting carbanion or other nucleophile reacts with a weak electrophile such as a carbonyl group, an alkyl halide, or the double bond of a Michael acceptor.

1. Consider the carbon skeletons of the reactants and products and identify which carbon atoms in the products are most likely derived from which carbon atoms in the reactants.
2. Consider whether any of the reactants is a strong enough nucleophile to react without being activated. If not, consider how one of the reactants might be converted to a strong nucleophile by deprotonation of an acidic site or by attack on an electrophilic site.
3. Consider how an electrophilic site on another reactant (or, in a cyclization, another part of the same molecule) can undergo attack by the strong nucleophile to form a bond needed in the product. Draw the product of this bond formation.

   If no appropriate electrophilic site can be found, consider another way of converting one of the reactants to a strong nucleophile.
4. Consider how the product of nucleophilic attack might be converted to the final product (if it has the right carbon skeleton) or reactivated to form another bond needed in the product.
5. Draw out all the steps of the mechanism using curved arrows to show the movement of electrons. Be careful to show only one step at a time.

REACTIONS INVOLVING FREE RADICALS

General principles: Free-radical reactions generally proceed by chain-reaction mechanisms, using an initiator with an easily broken bond (such as chlorine, bromine, or a peroxide) to start the chain reaction. In drawing the mechanism, expect free-radical intermediates (especially highly substituted or resonance-stabilized intermediates). Cationic intermediates and anionic intermediates are not usually involved. Watch for the most stable free radicals and avoid high-energy radicals such as hydrogen atoms.

Initiation

1. Draw a step involving homolytic (free-radical) cleavage of the weak bond in the initiator to give two radicals.
2. Draw a reaction of the initiator with one of the starting materials to give a free-radical version of the starting material.

   The initiator might abstract a hydrogen atom or add to a double bond, depending on what reaction leads toward formation of the observed product. You might want to consider bond dissociation energies to see which reaction is energetically favored.

Propagation

3. Draw a reaction of the free-radical version of the starting material with another starting material molecule to form a bond needed in the product and

generate a new radical intermediate. Two or more of these propagation steps may be needed to give the entire chain reaction.

Termination

4. Draw termination steps showing the recombination or destruction of radicals. Termination steps are side reactions rather than part of the product-forming mechanism. The reaction of any two free radicals to give a stable molecule is a termination step, as is a collision of a free radical with the container.

APPENDIX 4B
SUGGESTIONS FOR DEVELOPING MULTISTEP SYNTHESES

In this appendix we consider how an organic chemist systematically approaches a multistep synthesis problem. As with mechanism problems, there is no reliable "formula" that can be used to solve all synthesis problems; yet, students need guidance in how they should begin.

In a multistep synthesis problem, the solution is rarely immediately apparent. A synthesis is best developed systematically, generally working backward (in the *retrosynthetic* direction) and considering alternative ways of solving each stage of the synthesis. A strict retrosynthetic approach requires considering all possibilities for the final step, evaluating each reaction, and then evaluating every way of making each of the possible precursors.

This exhaustive approach is very time-consuming. It works well on a large computer, but most organic chemists solve problems more directly by attacking the crux of the problem: the steps that build the necessary carbon skeleton. Once the carbon skeleton is assembled (with usable functionality), converting the functional groups to those required in the target molecule is relatively easy.

The following steps suggest a systematic approach to developing a multistep synthesis. These steps should help you to organize your thoughts and approach syntheses like many organic chemists do: In a generally retrosynthetic direction, but with primary emphasis on the crucial steps that form the carbon skeleton of the target molecule. Solved problems that apply this approach appear on pages 212, 433, 435, and 617.

1. Review the functional groups and carbon skeleton of the target compound, considering what kinds of reactions might be used to create them.
2. Review the functional groups and carbon skeletons of the starting materials (if specified) and see how their skeletons might fit together into the skeleton of the target compound.
3. Compare methods for assembling the carbon skeleton of the target compound to determine which reactions produce a key intermediate with the correct carbon skeleton and functional groups at the correct positions where they can be converted to the functionality in the target molecule.

   Also notice what functional groups are required in the reactants for the skeleton-forming steps and whether they are easily accessible from the specified starting materials.
4. Write down the steps involved in assembling the key intermediate with the correct carbon skeleton.
5. Compare methods for converting the functional groups in the key intermediate to those in the target compound and select the reactions that are likely to give the correct product.

6. Working backward through as many steps as necessary, compare methods for synthesizing the reactants needed for assembly of the key intermediate with the correct carbon skeleton and functionality in the proper position. (This process may require writing several possible reaction sequences and evaluating them, keeping in mind the specified starting materials.)
7. Summarize the complete synthesis in the forward direction, including all steps and all reagents, and check it for errors and omissions.

APPENDIX 4C
SUGGESTIONS FOR DRAWING A PROPOSED NMR SPECTRUM

NMR spectra are highly predictable, and many computer programs are available for predicting spectra. Simple spectra are easily predicted without a computer, and you should be able to predict simple spectra both for doing problems in this course and for your lab work, determining whether a given spectrum corresponds to the desired structure.

A systematic approach works best for proposing the general features of an NMR spectrum. The following procedure is given for proton NMR spectra, but it is equally applicable to ^{13}C NMR spectra. A solved problem applying this approach appears on page 515.

1. Determine how many types of protons are present, together with their relative proportions for use in determining the peak areas.
2. Estimate the chemical shifts of the protons. (Table 12-3 serves as a guide.)
3. Determine the splitting pattern for each type of proton based on the number of protons that are close enough to be magnetically coupled to the proton being observed.
4. Summarize each absorption in order, from the lowest field to the highest, listing its area, chemical shift, and splitting pattern.
5. Draw the spectrum, using the information from your summary.

ANSWERS TO SELECTED PROBLEMS

These short answers are sometimes incomplete, but they should put you on the right track. Complete answers to all problems may be found in the *Solutions Manual.*

CHAPTER 1

1-5. (a) $\overset{+\!\!\longrightarrow}{\mathrm{C{-}Cl}}$; (b) $\overset{+\!\!\longrightarrow}{\mathrm{C{-}O}}$; (c) $\overset{+\!\!\longrightarrow}{\mathrm{C{-}N}}$; (d) $\overset{+\!\!\longrightarrow}{\mathrm{C{-}S}}$; (e) $\overset{\longleftarrow\!\!+}{\mathrm{C{-}B}}$; (f) $\overset{+\!\!\longrightarrow}{\mathrm{N{-}Cl}}$; (g) $\overset{+\!\!\longrightarrow}{\mathrm{N{-}O}}$; (h) $\overset{\longleftarrow\!\!+}{\mathrm{N{-}S}}$; (i) $\overset{\longleftarrow\!\!+}{\mathrm{N{-}B}}$; (j) $\overset{+\!\!\longrightarrow}{\mathrm{B{-}Cl}}$.
1-6. (a) +1 on O; (b) +1 on N, −1 on Cl; (c) +1 on N, −1 on Cl; (d) +1 on Na, −1 on O; (e) +1 on C; (f) −1 on C; (g) +1 on Na, −1 on B; (h) +1 on Na, −1 on B; (i) +1 on O, −1 on B; (j) +1 on N; (k) +1 on K, −1 on O; (l) +1 on O. 1-11. (a) CH_2O, $C_3H_6O_3$; (b) $C_2H_5NO_2$, same; (c) C_2H_5Cl, same; (d) C_2H_3Cl, $C_4H_6Cl_2$. 1-12. (a) 0.209; (b) 14.0. 1-13. (a) favors products; (b) favors reactants; (c) favors products; (d) favors products; (e) favors products; (f) favors products. 1-14. There is no resonance stabilization of the positive charge when the other oxygen atom is protonated. 1-15. (a) acetic acid, ethanol, methylamine; (b) ethoxide, methylamine, ethanol. 1-19. (a) carbon; (b) oxygen; (c) phosphorus; (d) chlorine. 1-26. The following are condensed structures that you should convert to Lewis structures. (a) $CH_3CH_2CH_2CH_3$ and $CH_3CH(CH_3)_2$; (b) $CH_3CH_2NH_2$ and CH_3NHCH_3; (c) $CH_2(CH_2OH)_2$ and $CH_3CHOHCH_2OH$ and $CH_3OCH_2OCH_3$ and others; (d) $CH_2{=}CHOH$ and CH_3CHO. 1-30. (a) C_5H_5N; (b) C_4H_9N; (c) C_4H_9NO; (d) $C_4H_9NO_2$. 1-31. Empirical formula C_3H_6O; molecular formula $C_6H_{12}O_2$ 1-34. (a) different compounds; (b) resonance structures; (c) resonance structures; (d) resonance structures; (e) different compounds; (f) resonance structures; (g) resonance structures; (h) different compounds; (i) resonance structures; (j) resonance structures. 1-37. (b) The =NH nitrogen atom is the most basic. 1-39. (a) second; (b) first; (c) second; (d) first; (e) first. 1-45. (a) $CH_3CH_2O^- Li^+ + CH_4$; (b) Methane; CH_3Li is a very strong base. 1-46. (a) $C_9H_{12}O$; (b) $C_{18}H_{24}O_2$.

CHAPTER 2

2-2. sp^3; Two lone pairs compress the bond angle to 104.5°. 2-4. Methyl carbon: sp^3, about 109.5°. Nitrile carbon *sp*, 180°. Nitrile nitrogen *sp*, no bond angle. 2-6. The central carbon is *sp*, with two unhybridized *p* orbitals at right angles. Each terminal $=CH_2$ group must be aligned with one of these *p* orbitals. 2-7. $CH_3{-}CH{=}N{-}CH_3$ shows cis-trans isomerism about the C=N double bond, but $(CH_3)_2C{=}N{-}CH_3$ has two identical substituents on the C=N carbon atom, and there are no cis-trans isomers. 2-10. (a) structural isomers; (b) geometric isomers; (c) structural isomers; (d) same compound; (e) same compound; (f) same compound; (g) not isomers; (h) structural isomers; (i) same compound; (j) structural isomers; (k) structural isomers. 2-12. The N—F dipole moments oppose the dipole moment of the lone pair. 2-14. *trans* has zero dipole moment because the bond dipole moments cancel. 2-17. (a) $CH_3CH_2OCH_2CH_3$; (b) $CH_3CH_2NHCH_3$; (c) CH_3CH_2OH; (d) CH_3COCH_3. 2-18. (a) alkane; (b) alkene; (c) alkyne; (d) cycloalkyne; (e) cycloalkane; (f) aromatic hydrocarbon; (g) cycloalkene; (h) alkyne, alkene; (i) aromatic hydrocarbon and cycloalkene. 2-19. (a) aldehyde; (b) alcohol; (c) ketone; (d) ether; (e) carboxylic acid; (f) ether; (g) ketone; (h) aldehyde; (i) alcohol. 2-20. (a) amide; (b) amine; (c) ester; (d) acid chloride; (e) ether; (f) nitrile; (g) carboxylic acid; (h) ester; (i) ketone, ether; (j) amine; (k) amide; (l) amide; (m) ketone, amine; (n) ester; (o) nitrile; (p) ketone. 2-23. No stereoisomers. 2-24. Cyclopropane has bond angles of 60°, compared with the 109.5° bond angle of an unstrained alkane. 2-27. Formamide must have an sp^2-hybridized nitrogen atom because it is involved in pi-bonding in the other resonance structure. 2-32. Only (b) and (e). 2-33. (a) structural isomers; (b) structural isomers; (c) geometric isomers; (d) structural isomers; (e) geometric isomers; (f) same compound; (g) geometric isomers. 2-34. CO_2 is *sp*-hybridized and linear; the bond dipole moments cancel. The sulfur atom in SO_2 is sp^2-hybridized and bent; the bond dipole moments do not cancel. 2-36. The alcohol engages in intermolecular hydrogen bonding. 2-38. (a), (c), (h), and (l) can form hydrogen bonds in the pure state. These four plus (b), (d), (g), (i), (j), and (k) can form hydrogen bonds with water. 2-40. (a) ether; (b) alkene, carboxylic acid; (c) alkene, aldehyde; (d) aromatic, ketone; (e) alkene, ester; (f) amide; (g) aromatic nitrile, ether; (h) amine, ester.

CHAPTER 3

3-1. (a) $C_{30}H_{62}$; (b) $C_{44}H_{90}$. 3-2. (a) octane < nonane < decane; (b) $(CH_3)_3C{-}C(CH_3)_3 < CH_3CH_2C(CH_3)_2CH_2CH_2CH_3 <$ octane. 3-4. (a) 3-methylpentane; (b) 4-isopropyl-2-methyldecane. 3-6. (a) 2-methylbutane; (b) 2,2-dimethylpropane; (c) 3-ethyl-2-methylhexane; (d) 2,4-dimethylhexane; (e) 3-ethyl-2,2,4,5-tetramethylhexane; (f) 4-*t*-butyl-3-methylheptane. 3-10. (a) C_9H_{20}; (b) $C_{15}H_{32}$ 3-13. (a) 1,1-dimethyl-3-(1-methylpropyl) cyclopentane or 3-*sec*-butyl-1,1-dimethylcyclopentane; (b) 3-cyclopropyl-1,1-dimethylcyclohexane; (c) 4-cyclobutylnonane. 3-15. (b), (c), and (d). 3-16. (a) *cis*-1-methyl-3-propylcylobutane; (b) *trans*-1-*t*-butyl-3-ethylcyclohexane; (c) *trans*-1,2-dimethylcyclopropane. 3-17. Trans is more stable.

In the cis isomer the methyl groups are nearly eclipsed. **3-26.** (a) *cis*-1,3-dimethylcyclohexane; (b) *cis*-1,4-dimethylcyclohexane; (c) *trans*-1,2-dimethylcyclohexane; (d) *cis*-1,3-dimethylcyclohexane; (e) *cis*-1,3-dimethylcyclohexane; (f) *trans*-1,4-dimethylcyclohexane. **3-28.** (a) bicyclo[3.2.0]heptane; (b) bicyclo[3.2.1]octane; (c) bicyclo[2.2.2]octane; (d) bicyclo[3.1.1]heptane. **3-31.** (a) All except the top right (isobutane) are *n*-butane. (b) Top left and bottom left are *cis*-2-butene. Top center and bottom center are 1-butene. Top right is *trans*-2-butene. Lower right is 2-methylpropene. (c) Top left and top center are *cis*-1,2-dimethylcyclopentane. Top right and bottom left are *trans*-1,2-dimethylcyclopentane. Bottom right is *cis*-1,3-dimethylcyclopentane. **3-35.** (a) 3-ethyl-2,2,6-trimethylheptane; (b) 3-ethyl-2,6,7-trimethyloctane; (c) 3,7-diethyl-2,2,8-trimethyldecane; (d) 2-ethyl-1,1-dimethylcyclobutane; (e) bicyclo[4.1.0]heptane; (f) *cis*-1-ethyl-3-propylcyclopentane; (g) (1,1-diethylpropyl)cyclohexane; (h) *cis*-1-ethyl-4-isopropylcyclodecane. **3-37.** (a) should be 3-methylhexane; (b) 3-ethyl-2-methylhexane; (c) 3-methylhexane; (d) 2,2-dimethylbutane; (e) *sec*-butylcyclohexane or (1-methylpropyl) cyclohexane; (f) should be *cis* or *trans*-1,2-diethylcyclopentane. **3-38.** (a) octane; (b) 2-methylnonane; (c) nonane **3.42.** The trans isomer is more stable, because both of the bonds to the second cyclohexane ring are in equatorial positions.

CHAPTER 4

4-3. (a) One photon of light would be needed for every molecule of product formed (the quantum yield would be 1). (b) Methane does not absorb the visible light that initiates the reaction, and the quantum yield would be 1. **4-4.** (a) Hexane has three different kinds of hydrogen atoms, but cyclohexane has only one type. (b) Large excess of cyclohexane. **4-5.** (a) $K_{eq} = 2.3$; (b) $[CH_3Br] = [H_2S] = 0.40\ M$, $[CH_3SH] = [HBr] = 0.60\ M$. **4-8.** (a) positive; (b) negative; (c) not easy to predict. **4-10.** (a) initiation +46 kcal/mol; propagation +16 and −24 kcal/mol; (b) overall −8 kcal/mol. **4-11.** (a) first order; (b) zeroth order; (c) first order overall. **4-13.** (a) zero, zero, zeroth order overall; (b) rate = k_r; (c) increase the surface area of the platinum catalyst. **4-14.** (b) +3 kcal/mol; (c) −1 kcal/mol. **4-15.** (c) +27 kcal/mol. **4-17.** (a) initiation +36 kcal/mol; propagation +33 and −20 kcal/mol; (b) overall +13 kcal/mol; (c) low rate and very unfavorable equilibrium constant. **4-18.** 1°:2° ratio of 6:2, product ratio of 75% 1° and 25% 2°. **4-22.** (a) The combustion of isooctane involves highly branched, more stable tertiary free radicals that react less explosively. (b) *t*-butyl alcohol forms relatively stable alkoxy radicals that react less explosively. **4-24.** (a) H/D reactivity ratio = 2.7; (b) The reaction with methane is slightly endothermic, but the reaction with ethane is exothermic. Less breakage of the C—H (or C—D) bond occurs in the transition state for ethane. **4-27.** Stability: (c) 3° > (b) 2° > (a) 1°. **4-28.** Stability: (c) 3° > (b) 2° > (a) 1°. **4-36.** rate = $k_r[H^+][(CH_3)_3C{-}OH]$; second order overall. **4-39.** $PhCH_2\cdot > CH_2{=}CHCH_2\cdot > (CH_3)_3C\cdot > (CH_3)_2CH\cdot > CH_3CH_2\cdot > CH_3\cdot$ **4-43.** The free radical intermediate is resonance-stabilized.

CHAPTER 5

5-1. (a) vinyl halide; (b) alkyl halide; (c) aryl halide; (d) alkyl halide; (e) vinyl halide; (f) aryl halide. **5-5.** The C—Cl bond has considerably more charge separation (0.23 *e*) than the C—I bond (0.16 *e*). **5-7.** Water is denser than hexane, so water forms the lower layer. Chloroform is denser than water, so chloroform forms the lower layer. **5-10.** (a) substitution; (b) elimination; (c) elimination, also a reduction. **5-12.** 0.02 mol/L per second. **5-13.** *trans*-4-methylcyclohexanol. **5-14.** (a) $(CH_3)_3COCH_2CH_3$; (b) $HC{\equiv}CCH_2CH_2CH_2CH_3$; (c) $(CH_3)_2CHCH_2NH_2$; (d) $CH_3CH_2C{\equiv}N$; (e) 1-iodopentane; (f) 1-fluoropentane. **5-16.** (a) 2-methyl-1-iodopropane; (b) cyclohexyl bromide; (c) isopropyl bromide; (d) 2-chlorobutane; (e) isopropyl iodide. **5-17.** (a) $(CH_3CH_2)_2NH$, less hindered; (b) $(CH_3)_2S$, S more polarizable; (c) PH_3, P more polarizable; (d) CH_3S^-, negatively charged; (e) $(CH_3)_3N$, N less electronegative; (f) $CH_3CH_2CH_2{-}O^-$, less hindered; (g) I^-, more polarizable. **5-19.** methyl iodide > methyl chloride > ethyl chloride > isopropyl bromide ≫ neopentyl bromide, *t*-butyl iodide. **5-21.** (a) 2-iodo-2-methylbutane; (b) 2-iodo-2-methylbutane; (c) 3-bromocyclohexene; (d) cyclohexyl bromide. **5-24.** (a) $(CH_3)_2C(OCOCH_3)CH_2CH_3$, first order; (b) 1-methoxy-2-methylpropane, second order; (c) 1-ethoxy-1-methylcyclohexane, first order; (d) methoxycyclohexane, first order; (e) ethoxycyclohexane, second order. **5-27.** (a) $H_2C{=}CHCH_2CH_3$ and $CH_3CH{=}CHCH_3$; (b) $CH_3CH{=}C(CH_2CH_3)_2$; (c) $CH_3CH{=}C(CH_2CH_3)_2$ and $CH_2{=}CHCH(CH_2CH_3)_2$; (d) cyclohexene. **5-29.** There is no hydrogen *trans* to the bromide leaving group. **5-32.** 2-butanol by the S_N2. **5-35.** In the first example the bromines are axial; in the second, equatorial. **5-40.** (a) 2-bromo-2-methylpentane; (b) 1-chloro-1-methylcyclohexane; (c) 1,1-dichloro-3-fluorocycloheptane; (d) 4-(2-bromoethyl)-3-(fluoromethyl)-2-methylheptane; (e) 4,4-dichloro-5-cyclopropyl-1-iodoheptane; (f) *cis*-1,2-dichloro-1-methylcyclohexane. **5-41.** (a) 1-chlorobutane; (b) 1-iodobutane; (c) 4-chloro-2,2-dimethylpentane; (d) 1-bromo-2,2-dimethylpentane; (e) chloromethylcyclohexane. **5-42.** (a) *t*-butyl chloride; (b) 2-chlorohexane; (c) bromocyclohexane; (d) iodocyclohexane; (e) 2-bromo-2-methylpentane; (f) 3-bromocyclohexene. **5-45.** (a) rate doubles; (b) rate multiplied by six; (c) rate increases. **5-52.** (a) diethyl ether; (b) $PhCH_2CH_2CN$; (c) $c\text{-Hx}{-}S{-}CH_3$; (d) 1-iododecane; (e) N-methylpyridinium iodide; (f) $(CH_3)_3CCH_2CH_2NH_2$. **5-59.** NBS provides low conc. Br_2 for free-radical bromination. Abstraction of one of the CH_2 hydrogens gives a resonance-stabilized free radical; product $Ph\text{-}CHBr\text{-}CH_3$.

CHAPTER 6

6-1. chiral: spring, desk, screw-cap bottle, rifle, knot. **6-2.** (b), (d), (e), and (f) are chiral. **6-3.** (a) achiral, no C*; (b) achiral, no C*; (c) chiral, one C*; (d) achiral, no C*; (e) achiral, no C*; (f) chiral, one C*; (g) achiral, two C*; (h) chiral, two C*; (i) achiral, no C*; (j) chiral, one C*; (k) chiral, two C*. **6-4.** (a) mirror, achiral; (b) mirror, achiral; (c) chiral, no mirror; (d) chiral, no mirror; (e) chiral, no mirror; (f) mirror, achiral; (g) mirror, achiral. **6-5.** (a) (*R*); (b) (*S*); (c) (*R*); (d) (*S*), (*S*); (e) (*R*), (*S*); (f) (*R*), (*S*); (g) (*R*), (*S*); (h) (*R*). **6-7.** +8.7°. **6-9.** Dilute the sample. If clockwise, will make less clockwise, and vice-versa. **6-10.** (+)-carvone is (*S*), and (−)-carvone is (*R*). **6-11.** e.e. = 33.3%. Specific rotation = 33.3% of +13.5° = +4.5°. **6-14.** (a), (b), (f), and (h) are chiral; only (h) has chiral carbons. **6-15.** (a) enantiomers, enantiomers, same; (b) same, enantiomers, enantiomers; (c) enantiomers, same, same. **6-17.** (a), (d), and (f) are chiral. The others have internal mirror planes. **6-18.** (from 6-17) (a) (*R*); (b) none; (c) none; (d) (2*R*), (3*R*); (e) (2*S*), (3*R*); (f) (2*R*), (3*R*); (new ones) (b) (*R*); (c) (*S*); (d) (*S*). **6-19.** (a) enantiomers; (b) diastereomers; (c) diastereomers; (d) structural isomers; (e) enantiomers; (f) diastereomers; (g) enantiomers; (h) same compound. **6-22.** (a), (b), and (d) are pairs of diastereomers and could theoretically be separated by their physical properties. **6-24.** (a) (*R*)-2-butanol (inversion); (b) (*S*)-2-iodo-3-methylpentane (inversion); (c) racemic mixture of 3-ethoxy-2,3-dimethylpentanes (racemization). **6-25.** (a) (*S*)-2-butanol (inversion); (b) racemic mixture of 3-ethoxy-3-methylhexanes (racemization); (c) (1*S*, 3*R*)-3-chloro-1-methylcyclopentane. **6-27.** (a) *trans*-2-pentene; (b) Substitution inverts the carbon atom bonded to bromine. Elimination forms 2,4-dimethyl-3-hexene (ethyl and *t*-butyl cis). (c) *cis*-3-heptene. **6-33.** (a) same compound; (b) enantiomers; (c) enantiomers;

(d) enantiomers; (e) enantiomers; (f) diastereomers; (g) enantiomers; (h) same compound. **6-36.** (a) $-12.5°$; (b) $+8.6°$. **6-37.** (20% e.e.) × (+12.0°) = (+)2.4°. **6-39.** (b) (−)15.90°; (c) 7.95°/15.90° = 50% e.e. Composition is 75% (*R*) and 25% (*S*). **6-40.** (a) o.p. = e.e. = 15.58/15.90 = 98% (99% (*S*) and 1%(*R*); (b) The e.e. of (*S*) decreases twice as fast as radioactive iodide substitutes, thus gives the (*R*) enantiomer; implies the S_N2 mechanism.

CHAPTER 7

7-4. (a) one; (b) one; (c) three; (d) four; (e) five. **7-5.** (a) 4-methyl-1-pentene; (b) 2-ethyl-1-hexene; (c) 1,4-pentadiene; (d) 1,2,4-pentatriene; (e) 2,5-dimethyl-1,3-cyclopentadiene; (f) 4-vinylcyclohexene; (g) allylbenzene or 3-phenylpropene; (h) *trans*-3,4-dimethylcyclopentene; (i) 7-methylene-1,3,5-cycloheptatriene. **7-6.** (1) (a), (c), and (d) show geometric isomerism. **7-7.** (a) 2,3-dimethyl-2-pentene; (b) 3-ethyl-1,4-hexadiene; (c) 1-methylcyclopentene; (d) give positions of double bonds; (e) specify cis or trans; (f) (*E*) or (*Z*), not *cis*. **7-8.** 2,3-dimethyl-2-butene is more stable by 1.4 kcal/mol. **7-11.** (a) stable; (b) unstable; (c) stable; (d) stable; (e) unstable (maybe stable cold); (f) stable; (g) unstable; (h) stable. **7-12.** (a) *cis*-1,2-dibromoethene; (b) *cis* (*trans* has zero dipole moment); (c) 1,2-dichlorocyclohexene. **7-14.** (a) 1-decene; (b) cyclohexene; (c) *cis*-cyclodecene; (d) no reaction (Br's equatorial); (e) *trans*-cyclodecene. **7-15.** (a) strong bases and nucleophiles; (b) strong acids and electrophiles; (c) free-radical chain reaction; (d) strong acids and electrophiles. **7-18.** (a) $\Delta G > 0$, disfavored. (b) $\Delta G < 0$, favored. **7-21.** (a) 2-ethyl-1-pentene; (b) 3-ethyl-2-pentene; (c) (3*E*,6*E*)-1,3,6-octatriene; (d) (*E*)-4-ethyl-3-heptene; (e) 1-cyclohexyl-1,3-cyclohexadiene. **7-25.** (b), (c), and (f) show geometric isomerism. **7-27.** (a) cyclopentene; (b) 2-methyl-2-butene (major) and 2-methyl-1-butene (minor); (c) 1-methylcyclohexene (major) and methylenecyclohexane (minor); (d) 1-methylcyclohexene (minor) and methylenecyclohexane (major). **7-31.** (a) a 1-halobutane; (b) a *t*-butyl halide; (c) a 3-halopentane; (d) a halomethylcyclohexane; (e) a 4-halocyclohexane (preferably *cis*). **7-32.** (a) 2-pentene; (b) 1-methylcyclopentene; (c) 1-methylcyclohexene; (d) 2-methyl-2-butene (rearrangement). **7-33.** E1 with rearrangement by an alkyl shift. The Saytzeff product violates Bredt's rule.

CHAPTER 8

8-1. (a) 2-bromopropane; (b) 2-chloro-2-methylpropane; (c) 1-iodo-1-methylcyclohexane; (d) mixture of *cis* and *trans* 3-methyl and 4-methylcyclohexane. **8-3.** (a) 1-bromo-2-methylpropane; (b) 1-bromo-2-methylcyclohexane; (c) 2-bromo-1-phenylpropane. **8-5.** (a) 1-methylcyclopentanol; (b) 2-phenyl-2-propanol; (c) 1-phenylcyclohexanol. **8-6.** The 2° carbocation rearranges to a 3° carbocation. **8-10.** (b) 1-propanol; (d) 2-methyl-3-pentanol; (f) *trans*-2-methylcyclohexanol. **8-13.** (a) *trans*-2-methylcycloheptanol; (b) mostly 4,4-dimethyl-2-pentanol; (c) —OH *exo* on the less substituted carbon. **8-16.** The carbocation can be attacked from either face. **8-21.** (a) CH_2I_2 + Zn(Cu); (b) CH_2Br_2, NaOH, H_2O, PTC; (c) dehydrate (H_2SO_4), then $CHCl_3$, NaOH/H_2O, PTC. **8-26.** (a) Cl_2, H_2O; (b) KOH/heat, then Cl_2/H_2O; (c) H_2SO_4/heat, then Cl_2/H_2O. **8-30.** (a) *cis*-2-epoxybutane; (b) *trans*-1-methyl-1,2-cyclooctanediol; (c) *trans*-epoxycyclodecane; (d) *meso*-2,3-butanediol. **8-32.** (a) *cis*-cyclohexane-1,2-diol; (b) *trans*-cyclohexane-1,2-diol; (c), (f) (*R,S*)-2,3-pentanediol (+ enantiomer); (d), (e) (*R,R*)-2,3-pentanediol (+ enantiomer). **8-33.** (a) OsO_4/H_2O_2; (b) CH_3CO_3H/H_3O^+; (c) CH_3CO_3H/H_3O^+; (d) OsO_4/H_2O_2. **8-45.** $H_2C{=}CH{-}COOCH_2CH_3$. **8-46.** (a) 2-methylpropene (3° cation); (b) 1-methylcyclohexene (3° cation); (c) 1,3-butadiene (resonance-stabilized cation). **8-50.** (a) 1-methylcyclohexene, RCO_3H/H_3O^+; (b) cyclooctene, OsO_4/H_2O_2; (c) *trans*-cyclodecene, Br_2; (d) cyclohexene, Cl_2/H_2O. **8-52.** $CH_3(CH_2)_{12}CH{=}CH(CH_2)_7CH_3$, cis or trans unknown.

CHAPTER 9

9-1. (a) 2-phenyl-2-propanol; (b) 5-bromo-2-heptanol; (c) 4-methyl-3-cyclohexen-l-ol; (d) *trans*-2-methylcyclohexanol; (e) (*E*)-2-chloro-3-methyl-2-penten-l-ol; (f) (2*R*,3*S*)-2-bromo-3-hexanol. **9-4.** (a) 8,8-dimethyl-2,7-nonanediol; (b) 1,8-octanediol; (c) *cis*-2-cyclohexene-l,4-diol; (d) 3-cyclopentyl-2,4-heptanediol. **9-5.** (a) cyclohexanol; more compact; (b) 4-methylphenol; more compact, stronger H-bonds; (c) 3-ethyl-3-hexanol; more spherical; (d) cyclooctane-1,4-diol; more OH groups per carbon; (e) enantiomers; equal solubility. **9-7.** (a) methanol; less substituted; (b) 1-chloroethanol; chlorine closer to the OH group; (c) 1,1-dichloroethanol; two chlorines to stabilize the alkoxide. **9-8.** The anions of 2-nitrophenol and 4-nitrophenol (but not 3-nitrophenol) are stabilized by resonance with the nitro group. **9-9.** (a) The phenol is deprotonated by sodium hydroxide; it dissolves. (b) In a separatory funnel, the alcohol will go into an ether layer and the phenolic compound will go into an aqueous sodium hydroxide layer. **9-10.** (b), (f), (g), (h). **9-14.** (a) Add phenylmagnesium bromide to benzophenone, PhCOPh. (b) Add methylmagnesium iodide to cyclohexanone. (c) Add propylmagnesium bromide to dicyclohexyl ketone. **9-16.** (a) 2 PhMgBr + PhCOCl; (b) 2 CH_3CH_2MgBr + $(CH_3)_2CHCOCl$; (c) 2 *c*-HxMgBr + PhCOCl. **9-18.** (a) PhMgBr + ethylene oxide; (b) $(CH_3)_2CHCH_2MgBr$ + ethylene oxide; (c) 2-methylcyclohexylmagnesium bromide + ethylene oxide. **9-21.** (a) Grignard removes NH proton; (b) Grignard attacks ester; (c) Water will kill Grignard; (d) Grignard removes OH proton. **9-24.** (a) heptanoic acid + $LiAlH_4$ or heptaldehyde + $NaBH_4$; (b) 2-heptanone + $NaBH_4$; (c) 2-methyl-3-hexanone + $NaBH_4$; (d) ketoester + $NaBH_4$. **9-31.** (a) 1-hexanol, larger surface area; (b) 2-hexanol, hydrogen-bonded; (c) 1,5-hexanediol, two OH groups. **9-35.** (a) cyclohexyl methanol; (b) 2-cyclopentyl-2-pentanol; (c) 2-methyl-1-phenyl-1-propanol; (d) methane + 3- hydroxycyclohexanone; (e) 5-phenyl-5-nonanol; (f) triphenylmethanol; (g) 1,1-diphenyl-l-propanol; (h) 3-(2-hydroxyethyl)cyclohexanol; (i) reduction of just the ketone, but not the ester; (j) isobutyl alcohol; (k) the tertiary alcohol; (l) the secondary alcohol; (m) (2*S*,3*S*)-2,3-hexanediol (+ enantiomer); (n) (2*S*,3*R*)-2,3-hexanediol (+ enantiomer); (o) 1,4-heptadiene. **9-36.** (a) EtMgBr; (b) Grignard with formaldehyde; (c) *c*-HxMgBr; (d) Grignard with ethylene oxide; (e) Grignard with formaldehyde; (f) 2 CH_3MgI; (g) cyclopentylmagnesium bromide.

CHAPTER 10

10-1. (a) oxidation, oxidation; (b) oxidation, oxidation, reduction, oxidation; (c) neither (C2 is oxidation, C3 reduction); (d) reduction; (e) neither. **10-3.** (a), (b) *c*-Hx—COOH; (c), (d) *c*-Hx—CHO. **10-6.** (a) PCC; (b) chromic acid; (c) chromic acid or Jones reagent; (d) PCC; (e) chromic acid; (f) Dehydrate, hydroborate, oxidize (chromic acid or Jones reagent). **10-7.** An alcoholic has more alcohol dehydrogenase. More ethanol is needed to tie up this larger amount of enzyme. **10-8.** $CH_3COCOOH$, pyruvic acid. **10-9.** (a) cyclopentane; (b) cyclopentyl tosylate; (c) cyclopentane; (d) cyclopentene; (e) cyclopentane. **10-11.** Treat the tosylate with (a) bromide; (b) ammonia; (c) ethoxide; (d) cyanide. **10-14.** (a) chromic acid or Lucas reagent; (b) chromic acid; (c) Lucas reagent; (d) Lucas reagent; allyl alcohol forms a resonance-stabilized carbocation. (e) Lucas reagent or sodium metal. **10-19.** (a) thionyl chloride (retention); (b) tosylate (retention), then S_N2 using chloride ion (inversion). **10-20.** resonance-delocalized cation, positive charge spread over two carbons. **10-22.** (a) 2-methyl-2-butene (+ 2-methyl-1-butene); (b) 2-pentene (+ 1-pentene); (c) 2-pentene (+ 1-pentene); (d) *c*-Hx=$C(CH_3)_2$ (+ 1-isopropylcyclohexene); (e) 1-methylcyclohexene (+ 3-methylcyclohexene). **10-24.** Using

R—OH and R′—OH will form R—O—R, R′—O—R′, and R—O—R′. **10-29.** (a) $CH_3CH_2CH_2COCl$ + 1-propanol; (b) CH_3CH_2COCl + 1-butanol; (c) $(CH_3)_2CHCOCl$ + *p*-methylphenol; (d) benzoyl chloride + cyclopropanol. **10-31.** An acidic solution (to protonate the alcohol) would protonate methoxide ion. **10-32.** Sodium isopropoxide and a butyl halide or tosylate. **10-33.** (a) the alkoxide of cyclohexanol and an ethyl halide or tosylate; (b) dehydration of cyclohexanol. **10-40.** (a) Na, then ethyl bromide; (b) NaOH, then PCC to aldehyde; Grignard, then dehydrate; (c) Mg in ether, then $CH_3CH_2CH_2CHO$, then oxidize; (d) PCC, then EtMgBr. **10-43.** Use CH_3SO_2Cl. **10-44.** (a) thionyl chloride; (b) tosylate, displace with bromide. **10-50.** Compound A is 2-butanol. **10-56. X** is 1-buten-4-ol; **Y** is tetrahydrofuran (5-membered cyclic ether).

CHAPTER 11

11-3. (a) alkene; (b) alkane; (c) terminal alkyne. **11-4.** (a) amine (primary); (b) acid; (c) alcohol. **11-6.** (a) conjugated ketone; (b) ester; (c) primary amide. **11-7.** (a) 3070 =C—H; 1650 C=C-*alkene;* (b) 2710, 2800 —CHO; 1720 carbonyl-*aldehyde;* (c) overinflated C—H region —COOH; 1700 carbonyl (maybe conjugated); 1640 C=C (maybe conjugated)-*conjugated acid;* (d) 1735 ester (or strained ketone)-*ester.* **11-8.** (a) bromine (C_6H_5Br); (b) iodine (C_2H_5I); (c) chlorine (C_4H_7Cl); (d) nitrogen ($C_7H_{17}N$). **11-11.** 87: loss of water; 111: allylic cleavage; 126: cleavage next to alcohol. **11-12.** Probably 2,6-dimethyl-3-octene or 3,7-dimethyl-3-octene. **11-15.** (a) about 1660 and 1710; the carbonyl is much stronger; (b) about 1660 for both; the enol is much stronger; (c) about 1660 for both; the imine is much stronger; (d) about 1660 for both; the terminal alkene is stronger. **11-17.** (a) $CH_2{=}C(CH_3)COOH$; (b) $(CH_3)_2CHCOCH_3$; (c) $PhCH_2C{\equiv}N$; (d) $PhNHCH_2CH_3$. **11-18.** (a) 86, 71, 43; (b) 98, 69; (c) 84, 69, 87, 45.

CHAPTER 12

12-1. (a) δ2.17; (b) 0.0306 gauss; (c) δ2.17; (d) 217 Hz. **12-3.** (a) three; (b) two; (c) three; (d) five. **12-5.** (a) 2-methyl-3-butyn-2-ol; (b) *p*-dimethoxybenzene; (c) 1,2-dibromo-2-methylpropane. **12-9.** *trans* CHCl=CHCN. **12-10.** (a) 1-chloropropane; (b) methyl *p*-methylbenzoate, $CH_3C_6H_4COOCH_3$. **12-13.** (a) H^a, δ9.7 (doublet); H^b, δ6.6 (multiplet); H^c, δ7.4 (doublet); (b) J_{ab} = 8 Hz, J_{bc} = 18 Hz (approx). **12-16.** (a) Five; the two hydrogens on C3 are diastereotopic. (b) Six; all the CH_2 groups have diastereotopic hydrogens. (c) Six; three on the Ph, and the CH_2 hydrogens are diastereotopic. (d) Three; The hydrogens cis and trans to the Cl are diastereotopic. **12-19.** (a) butane-1,3-diol; (b) $H_2NCH_2CH_2OH$. **12-22.** (a) $(CH_3)_2CHCOOH$; (b) $PhCH_2CHO$; (c) $CH_3COCOCH_2CH_3$; (d) $CH_2{=}CHCH(OH)CH_3$; (e) $(CH_3)_2C(OCH_3)C{\equiv}CH$. **12-25.** (a) allyl alcohol, $H_2C{=}CHCH_2OH$. **12-26.** (a) 4-hydroxybutanoic acid lactone. **12-27.** (a) cyclohexene. **12-28.** isobutyl bromide. **12-32.** (a) isopropyl alcohol. **12-34.** (a) $PhCH_2CH_2OCOCH_3$. **12-38.** 1,1,2-trichloropropane. **12-40. A** is 2-methyl-2-butene (Saytzeff product); **B** is 2-methyl-1-butene. **12-41. 1** is $PhCH_2CN$; **2** is $(CH_3)_2CHCOCH_3$.

CHAPTER 13

13-2. $(CH_3CH_2)_2\overset{+}{O}{-}\overset{-}{Al}Cl_3$. **13-3.** (a) cyclopropyl methyl ether; methoxycyclopropane; (b) ethyl isopropyl ether; 2-ethoxypropane; (c) 2-chloroethyl methyl ether; 1-chloro-2-methoxyethane; (d) *sec*-butyl *t*-butyl ether; 2-(1,1-dimethylethoxy) butane; (e) *trans*-2-methoxycyclohexanol (no common name). **13-5.** (a) dihydropyran; (b) 2-chloro-1,4-dioxane; (c) 3-isopropylpyran; (d) *trans*-2,3-diethyloxirane or *trans*-3,4-epoxyhexane; (e) 3-bromo-2-ethoxyfuran; (f) 3-bromo-2,2-dimethyloxetane. **13-7.** propyl bromide and *t*-butoxide **13-10.** Intermolecular dehydration of a mixture of methanol and ethanol would produce a mixture of diethyl ether, dimethyl ether, and ethyl methyl ether. **13-12.** Intermolecular dehydration might work for (a). Use the Williamson for the other two. **13-14.** (a) bromocyclohexane and ethyl bromide; (b) 1,5-diiodopentane; (c) phenol and methyl bromide; (e) phenol, ethyl bromide, and 1,4-dibromo-2-methylbutane. **13-17.** Epoxidiation of ethylene gives ethylene oxide, and catalytic hydration of ethylene gives ethanol. Acid-catalyzed opening of the epoxide in ethanol gives cellosolve. **13-21.** (a) $CH_3CH_2OCH_2CH_2O^-\ Na^+$; (b) $H_2NCH_2CH_2O^-\ Na^+$; (c) *c*-Hx—$CH_2CH_2O^-\ Na^+$. **13-22.** (a) 2-methyl-1,2-propanediol, ^{18}O at the C2 hydroxyl group; (b) 2-methyl-1,2-propanediol, ^{18}O at the C1 hydroxyl group; (c), (d) same products, (*S,S*) and (*R,R*). **13-23.** (a) $(CH_3)_2CH{-}CH_2CH_2{-}OH$; (b) $CH_3{-}CH_2C(CH_3)_2{-}OH$; (c) 1-cyclopentyl-1-butanol. **13-29.** (a) The old ether had autoxidized to form peroxides. On distillation, the peroxides were heated and concentrated, and they detonated. (b) Discard the old ether or treat it to reduce the peroxides. **13-32.** (a) epoxide + phenylmagnesium bromide; (b) epoxide + sodium methoxide in methanol; (c) epoxide + methanol, H^+. **13-36.** Sodium then ethyl iodide gives retention of configuration. Thionyl chloride gives retention, then the Williamson gives inversion. Second product (+)15.6°. **13-40.** $(CH_3OCH_2CH_2)_2O$ **13-41.** phenyloxirane

CHAPTER 14

14-3. decomposition to its elements, C and H_2. **14-4.** 3,3-dimethyl-1-butyne. **14-5.** Treat the mixture with $NaNH_2$ to remove the 1-hexyne. **14-6.** (a) $Na^+\ {}^-C{\equiv}CH$ and NH_3; (b) $Li^+\ {}^-C{\equiv}CH$ and CH_4; (c) no reaction; (d) no reaction; (e) acetylene + $NaOCH_3$; (f) acetylene + NaOH; (g) no reaction; (h) no reaction; (i) NH_3 + $NaOCH_3$. **14-8.** (a) $NaNH_2$; butyl halide; (b) $NaNH_2$; propyl halide; $NaNH_2$; ethyl halide. (c) $NaNH_2$; ethyl halide; repeat. (d) S_N2 on *sec*-butyl halide is unfavorable. (e) $NaNH_2$; isobutyl halide (low yield); $NaNH_2$; methyl halide. (f) $NaNH_2$ added for second substitution on 1,8-dibromooctane will attack the halide. **14-9.** (a) sodium acetylide + CH_3I, then $NaNH_2$, then $CH_3CH_2CH_2O$; (b) sodium acetylide + ethylene oxide; (c) sodium acetylide + formaldehyde; (d) sodium acetylide + $CH_3CH_2COCH_3$. **14-11.** About 1:70. **14-14.** (a) H_2, Lindlar; (b) Na, NH_3; (c), (d) Add halogen, dehydrohalogenate to the alkyne, reduce as needed. **14-17.** (a) $CH_3CCl_2CH_2C_5H_{11}$ and $CH_3CH_2CCl_2C_5H_{11}$; (b) Lone pairs on Cl help stabilize the carbocation. **14-19.** (a) Cl_2; (b) HBr, peroxides; (c) HBr, no peroxides; (d) excess Br_2; (e) reduce to 1-hexene, add HBr; (f) excess HBr. **14-21.** (a) The two ends of the triple bond are equivalent. (b) The two ends of the triple bond are not equivalent, yet not sufficiently different for good selectivity. **14-22.** (a) 2-hexanone; hexanal; (b) mixtures of 2-hexanone and 3-hexanone; (c) 3-hexanone for both; (d) cyclodecanone for both. **14-25.** (a) $CH_3C{\equiv}C(CH_2)_4C{\equiv}CCH_3$ **14-29.** (a) ethylmethylacetylene; (b) phenylacetylene; (c) *sec*-butyl-*n*-propylacetylene; (d) *sec*-butyl-*t*-butylacetylene. **14-33.** Form the heavy metal salt of the terminal alkyne. **14-39.** 1,3-cyclohexadiene with (HC≡C—CH=CH—) at the 1 position (cis or trans). **14-40.** $CH_2{=}C(CH_3){-}C{\equiv}CH$ **14-45.** phenylacetylene

CHAPTER 15

15-1. (a) 2,4-hexadiene < 1,3-hexadiene < 1,4-hexadiene < 1,5-hexadiene < 1,3,5-hexatriene; (b) third < fifth < first < fourth < second. **15-5.** 3-ethoxy-1-methylcyclopentene and

3-ethoxy-3-methylcyclopentene. **15-7.** (a) **A** is 3,4-dibromo-1-butene; **B** is 1,4-dibromo-2-butene. (c) Hint: **A** is the kinetic product, **B** is the thermodynamic product. (d) Isomerization to an equilibrium mixture, 10% **A** and 90% **B**. **15-8.** (a) 1-(bromomethyl)cyclohexene and 2-bromo-1-methylenecyclohexane. **15-10.** (a) 3-bromocyclopentene; (b) (*cis* and *trans*) 4-bromo-2-pentene. **15-11.** Both generate the same allylic carbanion. **15-12.** (a) allyl bromide + *n*-butyllithium; (b) isopropyllithium + 1-bromo-2-butene. **15-19.** (b) [4 + 2] cycloaddition of one butadiene with just one of the double bonds of another butadiene. **15.20.** 800. **15-21.** (a) 353 nm; (b) 313 nm; (c) 232 nm; (d) 273 nm; (e) 237 nm. **15-23.** (a) isolated; (b) conjugated; (c) cumulated; (d) conjugated and isolated; (e) conjugated. **15-24.** (a) allylcyclohexane; (b) 3-chlorocyclopentene; (c) 3-bromo-2-methylpropene; (d) 3-bromo-1-pentene and 1-bromo-2-pentene; (e) 4-bromo-2-buten-1-ol and 1-bromo-3-buten-2-ol; (f) 5,6-dibromo-1,3-hexadiene, 1,6-dibromo-2,4-hexadiene, and 3,6-dibromo-1,4-hexadiene (minor); (g) 1-(methoxymethyl)-2-methylcyclopentene and 1-methoxy-1-methyl-2-methylenecyclopentane; (h), (i) Diels-Alder adducts. **15-25.** (a) allyl bromide + isobutyl Grignard; (b) 1-bromo-3-methyl-2-butene + $CH_3CH_2C(CH_3)_2MgBr$. **15-27.** (a) 29,000; (b) second structure. **15-28.** 3-bromo-1-hexene, and (cis and trans)-1-bromo-2-hexene. **15-31.** (a) The product isomerized; 1630 suggests conjugated; (b) 2-propyl-1,3-cyclohexadiene.

CHAPTER 16

16-2. (a) +7.6 kcal/mol; (b) −21.2 kcal/mol; (c) −26.8 kcal/mol. **16-5.** Two of the eight pi electrons are unpaired in two non-bonding orbitals, an unstable configuration. **16-7.** (a) nonaromatic (internal H's prevent planarity); (b) nonaromatic (one ring atom has no *p* orbital); (c) aromatic, [14]annulene; (d) aromatic (in the outer system). **16-8.** Azulene is aromatic, but the other two are antiaromatic. **16-10.** The cation (cyclopropenium ion) is aromatic; the anion is antiaromatic. **16-12.** (a) antiaromatic if planar; (b) aromatic if planar; (c) aromatic if planar; (d) anti-aromatic if planar; (e) nonaromatic; (f) aromatic if planar. **16-14.** cyclopropenium fluoroborate **16-16.** (a) aromatic; (b) nonaromatic; (c) aromatic; (d) nonaromatic; (e) aromatic. **16-20.** (a) fluorobenzene; (b) 4-phenyl-1-butyne; (c) 3-methylphenol or *m*-cresol; (d) *o*-nitrostyrene; (e) *p*-bromobenzoic acid; (f) isopropyl phenyl ether; (g) 3,4-dinitrophenol; (h) benzyl ether. **16-23.** 3-phenyl-2-propene-1-ol **16-26.** (a) *o*-dichlorobenzene; (b) *p*-nitroanisole; (c) 2,3-dibromobenzoic acid; (d) 2,7-dimethoxynaphthalene; (e) *m*-chlorobenzoic acid; (f) 2,4,6-trichlorophenol; (g) 2-(1-methylpropyl)benzaldehyde; (h) cyclopropenium fluoroborate. **16-28.** The second is deprotonated to an aromatic cyclopentadienide anion. **16-30.** (d), (e) The fourth structure, with two three-membered rings, was considered the most likely and was called Ladenburg benzene. **16-35.** (a) three; (b) one; (c) *meta*-dibromobenzene. **16-36.** α-chloroacetophenone **16-38.** (a) no; (b) six in each; total 12, compared with 10 in naphthalene; (c) $(6 \times 28.6) - 100 = 71.6$ kcal, 35.8 kcal per ring, nearly as much as benzene (36 kcal). Naphthalene has only 60 kcal, 30 kcal per ring. **16-40.** Deprotonate it to give an anion with 10 pi electrons **16-43.** 2-isopropyl-5-methylphenol

CHAPTER 17

17-4. The sigma complex for *p*-xylene has the + charge on two 2° carbons and one 3° carbon, compared with three 2° carbons in benzene. **17-10.** 4-methyl-2-nitroanisole **17-11.** Bromine *adds* to the alkene but *substitutes* on the aryl ether, evolving gaseous HBr. **17-12.** Strong acid is used for nitration, and the amino group of aniline is protonated to a deactivating $—NH_3^+$ group. **17-14.** 1-bromo-1-chlorocyclohexane; the intermediate cation is stabilized by a bromonium ion resonance structure. **17-15.** (a) 2,4- and 2,6-dinitrotoluene; (b) 3-chloro-4-nitrotoluene and 5-chloro-2-nitrotoluene; (c) 3- and 5-nitro-2-bromobenzoic acid; (d) 4-methoxy-3-nitrobenzoic acid; (e) 5-methyl-2-nitrophenol and 3-methyl-4-nitrophenol. **17-18.** (a) phenylcyclohexane; (b) *o*- and *p*-methylanisole, with overalkylation products; (c) 1-isopropy-4-(1,1,2-trimethylpropyl)benzene. **17-19.** (a) phenylcyclohexane; (b) *t*-butylbenzene; (c) *p*-di-*t*-butylbenzene; (d) *o*- and *p*-isopropyltoluene. **17-20.** (a) *t*-butylbenzene; (b) 2- and 4-butyltoluene; (c) no reaction; (d) (1,1,2-trimethylpropyl)benzene. **17-21.** (a) *sec*-butylbenzene and others; (b) OK; (c) + disub, trisub; (d) OK (some ortho); (e) OK. **17-23.** (a) $(CH_3)_2CHCH_2COCl$, benzene, $AlCl_3$; (b) $(CH_3)_3CCOCl$, benzene, $AlCl_3$; (c) PhCOCl, benzene, $AlCl_3$; (d) CO/HCl, $AlCl_3$/CuCl, anisole; (e) Clemmensen on (b); (f) $CH_3(CH_2)_2COCl$, benzene, $AlCl_3$ then Clemmensen. **17-24.** Fluoride leaves in a fast exothermic step; the C—F bond is only slightly weakened in the reactant-like transition state (Hammond postulate). **17-26.** (a) 2,4-dinitroanisole; (b) 2,4- and 3,5-dimethylphenol; (c) N-methyl-4-nitroaniline; (d) 2,4-dinitrophenylhydrazine. **17-30.** (a) (trichloromethyl)hexachlorocyclohexane; (b) 1-methyl-1,4-cyclohexadiene; (c) *cis* and *trans*-1,2-dimethylcyclohexane; (d) 1,4-dimethyl-1,4-cyclohexadiene. **17-31.** (a) benzoic acid; (b) benzoic acid; (c) *o*-phthalic acid. **17-33.** 60% beta, 40% alpha; reactivity ratio = 1.91 to 1. **17-37.** (a) 1-bromo-1-phenylpropane **17-38.** (a) HBr, then Grignard with ethylene oxide; (b) CH_3COCl and $AlCl_3$, then Clemmensen, Br_2 and light, then $^-OCH_3$; (c) Nitrate, then Br_2 and light, then NaCN. **17-39.** (a) 3-ethoxytoluene; (b) *m*-tolyl acetate; (c) 2,4,6-tribromo-3-methylphenol; (d) 2,4,6-tribromo-3-(tribromomethyl)phenol; (e) 2-methyl-1,4-benzoquinone; (f) 2,4-di-*t*-butyl-3-methylphenol. **17-49.** indanone **17-51.** The yellow species is the triphenylmethyl cation. **17-55.** kinetic control at 0°, thermodynamic control at 100°. **17-56.** Brominate, then Grignard with 2-butanone. **17-58.** The Grignard reagent decomposes to a benzyne intermediate.

CHAPTER 18

18-1. (a) 5-hydroxy-3-hexanone; (b) 3-phenylbutanal; β-phenylbutyraldehyde; (c) *trans*-2-methoxycyclohexane-carbaldehyde; (d) 6,6-dimethyl-2,4-cyclohexadienone. **18-2.** (a) 2-phenylpropanal; (c) acetophenone. **18-3.** No γ-hydrogens. **18-5.** (a) <200, 280; (b) 230, 310; (c) 280, 360; (d) 270, 350. **18-8.** (a) acetophenone; (b) acetylcyclohexane; (c) 3-heptanone. **18-9.** (a) 3-heptanone; (b) phenylacetonitrile; (c) benzyl cyclohexyl ketone. **18-11.** (a) benzyl alcohol; (b) benzaldehyde; (c) heptanal. **18-13.** second < fourth < first < third. **18-17.** cis and trans isomers. **18-18.** (a) cyclohexanone and methylamine; (b) 2-butanone and ammonia; (c) acetaldehyde and aniline; (d) 6-amino-2-hexanone. **18-21.** (a) benzaldehyde and semicarbazide; (b) camphor and hydroxylamine; (c) tetralone and phenylhydrazine; (d) cyclohexanone and 2,4-DNP; (e) 5-aminopentanal; (f) 5-amino-2-butanone. **18-25.** (a) tetralone and ethanol; (b) acetaldehyde and 2-propanol; (c) hexane-2,4-dione and ethanediol; (d) tetralone and 1,3-propanediol; (d) 5-hydroxypentanal and methanol; (f) $(HOCH_2CH_2CH_2)_2CHCHO$. **18-27.** (a) 4-hydroxycyclohexanecarboxylic acid; (b) 4-oxocyclohexanecarboxylic acid; (c) 3-oxocyclohexanecarboxylic acid; (d) *cis*-3,4-dihydroxycyclohexanecarboxylic acid. **18-29.** (a) indane; (b) hexane; (c) ethylene ketal of 2-propylcyclohexanone; (d) propylcyclohexane. **18-34.** 240 nm and 300–320 nm. **18-35.** 2,5-hexanedione **18-36.** 1-phenyl-2-butanone (benzyl ethyl ketone). **18-37.** (a) 44; (b) 72; (c) 44; (d) 74. **18-39.** cyclobutanone **18-42.** (all H^+ cat.) (a) cyclobutanone and hydroxylamine; (b) benzaldehyde and cyclohexylamine; (c) benzylamine and cyclohexanone; (d) β-tetralone and ethylene glycol; (e) cyclohexylamine and acetone; (f) cyclohexanone and methanol. **18-45.** (a) $NaBD_4$, then H_2O; (b) $NaBD_4$, then D_2O; (c) $NaBH_4$, then D_2O. 18-48. (a) $CH_3CH_2CH_2COCl$ and $AlCl_3$, then Clemmensen; (b) EtMgBr, then H_3O^+; (c) $Cl_2/FeCl_3$, then Dow process to phenol; CH_3I, then Gatterman; (d) oxidize to the acid, $SOCl_2$, then $AlCl_3$. 18-52. (a) 3-hexanone; (b) 2- and 3-hexanone; (c) 2-hexanone; (d) cyclodecanone; (e) 2- and 3-methylcyclodecanone. 18-54. **A** is 2-heptanone. 18-56. (b) The "THP ether" is an acetal, stable to base but hydrolyzed by acid. 18-57. **A** is the ethylene ketal of 2-butanone; **B** is 2-butanone. 18-60. *trans*-2-butenal (crotonaldehyde)

CHAPTER 19

19-1. Pyridine, 2-methylpyridine, pyrimidine, pyrrole, imidazole, indole, and purine are aromatic. 19-3. (a) 2-pentanamine; (b) N-methyl-2-butanamine; (c) *m*-aminophenol, (d) 3-methylpyrrole; (e) *trans*-1,2-cyclopentanediamine; (f) *cis*-3-aminocyclohexanecarbaldehyde. 19-4. (a) resolvable (chiral carbons); (b) not resolvable (N inverts); (c) symmetric; (d) not resolvable; proton on N is removable; (e) resolvable (chiral quat. salt). 19-6. (a) primary amine; (b) alcohol, (c) secondary amine. 19-7. isobutylamine 19-8. (a) *n*-propylamine; (b) diethylmethylamine; (c) propanal; (d) 1-propanol. 19-10. (a) aniline < ammonia < methylamine < NaOH; (b) *p*-nitroaniline < aniline < *p*-methylaniline; (c) pyrrole < aniline < pyridine; (d) 3-nitropyrrole < pyrrole < imidazole. 19-18. (a) benzylamine + excess CH_3I; (b) 1-bromopentane + excess NH_3; (c) benzyl bromide + excess NH_3. 19-19. (a) $CH_3CONHCH_2CH_3$; (b) $PhCON(CH_3)_2$; (c) N-hexanoyl piperidine. 19-25. (a) cyclohexanediazonium chloride (then cyclohexanol and cyclohexene); (b) N-nitroso-N-ethyl-2-hexanamine; (c) N-nitrosopiperidine; (d) equilib. with N-nitrosoammonium ion; (e) benzenediazonium chloride. 19-27. (a) diazotize, then HBF_4, heat; (b) diazotize, then CuCl; (c) Protect (CH_3COCl), then 3 $CH_3I/AlCl_3$, H_3O^+, diazotize, H_3PO_2; (d) diazotize, then CuBr; (e) diazotize, then KI; (f) diazotize, then CuCN; (g) diazotize, then H_2SO_4, H_2O, heat; (h) diazotize, then couple with resorcinol. 19-28. (a) CH_3NH_2, $NaBH_3CN$; (b) H_2NOH/H^+, then $LiAlH_4$; (c) PhCHO, $NaBH_3CN$; (d) aniline/H^+, then $LiAlH_4$; (e) H_2NOH/H^+, then $LiAlH_4$; (f) piperidine + cyclohexanone + $NaBH_3CN$. 19-29. (a) nitrate, reduce; (b) brominate, then nitrate and reduce; (c) nitrate, then brominate and reduce; (d) oxidize toluene, then nitrate and reduce. 19-30. (a) large excess of NH_3; (b) dimethylamine; (c) Zn, HCl. 19-33. (a) add cyanide, reduce; (b) reduce, tosylate, KCN, then $LiAlH_4$; (c) cyanide, then reduce; (d) KCN/HCN, then $LiAlH_4$. 19-35. Hofmann rearrangement goes with retention of configuration. 19-39. only (b), (d), and (f). 19-41. (a) 1-hexanamine; (b) 2-phenylethylamine; (c) 1,4-butanediamine. 19-55. (a) triethylamine; (b) An acid converts it to a solid ammonium salt. (c) Rinse the clothes with diluted vinegar (acetic acid). 19-56. **A** is 2-butanamine; **B** is diethylamine. 19-57. 2,2-dimethyl-1-propanamine

CHAPTER 20

20-2. (a) 2-iodo-3-methylpentanoic acid; α-iodo-β-methylvaleric acid; (b) (*Z*)-3,4-dimethyl-3-hexenoic acid; (c) 2,3-dinitrobenzoic acid; (d) *trans*-1,3-cyclohexanedicarboxylic acid; (e) 2-chlorobenzene-1,4-dicarboxylic acid; 2-chloroterephthalic acid; (f) 3-methylhexanedioic acid; β-methyladipic acid. 20-3. (a) first, second, third; (b) third, second, first, (c) third, second, fourth, first. 20-7. Broad acid OH centered around 3000; conjugated carbonyl about 1690; C═C about 1650. 20-8. (a) propanoic acid; (b) —CHO proton triplet between $\delta 9$ and $\delta 10$. 20-11. (a) $KMnO_4$; (b) $KMnO_4$; (c) PhMgBr + ethylene oxide, oxidize, (d) PBr_3, Grignard, CO_2; (e) Conc. $KMnO_4$, heat; (f) KCN, then H_3O^+. 20-15. (a) Methanol and salicylic acid, H^+; methanol solvent, dehydrating agent. (b) Methanol and formic acid H^+, distill product as it forms. (c) Ethanol and benzoic acid, H^+; ethanol solvent, dehydrating agent. 20-17. (a) phenylacetic acid and $LiAlH_4$, then Collins; (b) phenylacetic acid and $LiAlH_4$; (c) cyclopentanone-2-carboxylic acid; make ethylene acetal, then $LiAlH_4$; (d) B_2H_6 on the corresponding acid. 20-19. (a) benzene + CH_3CH_2COCl, $AlCl_3$; or propionic acid + 2 PhLi, then H_3O^+; (b) Add 2 CH_3Li, then H_3O^+. 20-21. (a) 1-bromononane; (b) bromocyclooctane. 20-31. (a) Grignard + CO_2; or KCN, then H_3O^+; (b) CH_3OH, H^+; or CH_2N_2; (c) Ag^+; (d) $SOCl_2$, then Li(t-BuO)$_3$AlH; or $LiAlH_4$, then PCC; (e) conc. $KMnO_4$, H^+, heat; (f) $LiAlH_4$ or B_2H_6; (g) Ag_2O, then Br_2, heat. 20-33. (a) see Fischer esterification; (b) $C{=}^{18}O{-}CH_3$; (c) mass spectrometry. 20-34. phenoxyacetic acid 20-35. (b) diastereomers 20-36. (a) 2-phenylpropanoic acid; (b) 2-methylpropenoic acid; (c) *trans*-2-hexenoic acid. 20-41. (a) stockroom: heptaldehyde; students: heptanoic acid; (b) air oxidation; (c) Prepare fresh samples immediately before using.

CHAPTER 21

21-2. No aldehyde C—H at 2700 and 2800; no acid O—H centered at 3000. 21-4. (a) acid chloride C═O at 1820; (b) primary amide C═O at 1650, two N—H around 3300; (c) carboxylic acid C═O at 1710, broad O—H centered at 3000; (d) anhydride C═O double absorption at 1730 and 1800. 21-5. (a) acrylamide, $H_2C{=}CHCONH_2$; (b) 5-hydroxyhexanoic acid lactone. 21-8. (a) ethanol, propionyl chloride; (b) phenol, 3-methylhexanoic acid; (c) benzyl alcohol, benzoyl chloride; (d) cyclopropanol, cyclohexanecarboxylic acid. 21-9. (a) dimethylamine, acetyl chloride; (b) aniline, acetyl chloride; (c) ammonia, cyclohexanecarbonyl chloride; (d) piperidine, benzoyl chloride. 21-10. (i) $PhCH_2OH$; (ii) $PhNH_2$. 21-24. (a) cyclohexylmethanamine; (b) cyclohexyl ethyl amine; (c) $(CH_2)_6NH$ (7-membered ring); (d) morpholine; (e) cyclohexyl methyl propyl amine. 21-28. (a) benzene + acetyl chloride; (b) benzene + benzoyl chloride; (c) benzene + butyryl chloride, then Clemmensen. 21-30. (a) *n*-octyl alcohol, acetic formic anhydride (formyl chloride is unavailable); (b) *n*-octyl alcohol, acetic anhydride (cheap, easy to use); (c) phthalic anhydride, ammonia (anhydride forms monoamide); (d) succinic anhydride, methanol (anhydride forms monoester). 21-32. (a) acetic anhydride; (b) methanol, H^+; (c) diazomethane. 21-34. (a) $SOCl_2$, $HN(CH_3)_2$, $LiAlH_4$; (b) acetic formic anhydride, then $LiAlH_4$. 21-36. (a) $SOCl_2$, NH_3, $POCl_3$; (b) $LiAlH_4$, make tosylate, NaCN; (c) Fe/HCl, diazotize, CuCN. 21-40. (a) ethyl benzoate; (b) acetic benzoic anhydride; (c) PhCONHPh; (d) 4-methoxybenzophenone; (e) Ph_3COH. 21-44. (a) $SOCl_2$; (b) $SOCl_2$, CH_3COONa; (c) oxalyl chloride; (d) acetic formic anhydride; (e) Ag^+, H^+; (f) H^+/heat, $(CH_3)_2CHOH$. 21-47. (after H^+) (a) HCOOH + PhOH; (b) 2 CH_3CH_2OH + CO_2; (c) 3-(*o*-hydroxyphenyl)propanoic acid; (d) $(CH_2OH)_2$ + $(COOH)_2$. 21-51. (a) Ph_3COH; (b) 3 EtMgBr + EtCOOEt, then H_3O^+. 21-55. (a) diethyl carbonate; (b) $CH_3NHCONHCH_3$; (c) $CH_3OCONHPh$. 21-57. Penicillin N 21-58. $CH_3CH_2OCOCH_2CN$ 21-59. δ-valerolactam 21-60. ethyl crotonate 21-64. Acetic anhydride; add water to hydrolyze it to dilute acetic acid.

CHAPTER 22

22-8. (a), (b) cyclopentane carboxylate and chloroform/iodoform; (c) 2,2,6,6-tetraiodocyclohexanone; (d) $PhCOCBr_2CH_3$. 22-11. (a) $CH_3CHBrCOOH$; (b) PhCOOH; (c) $HOOCCH_2CHBrCOOH$; (d) oxalic acid. 22-15. (a) $PhC(NCH_3)CH_3$; (b) $CH_2{=}C(Ph)NMe_2$; (c) cyclohexanone phenyl imine; (d) piperidine enamine of cyclohexanone. 22-16. (a) enamine + allyl bromide; (b) enamine + $PhCH_2Br$; (c) enamine + PhCOCl. 22-18. (a) 3-hydroxy-2-methylpentanal; (b) 3-hydroxy-2,4-diphenylbutanal. 22-19. retro-aldol, reverse of aldol condensation. 22-23. (a) 2-ethyl-2-hexenal; (b) 1,3-diphenyl-2-buten-1-one; (c) 2-cyclohexylidene-cyclohexanone. 22-25. $PhCH{=}CHCOCH{=}CHPh$, "dibenzalacetone". 22-27. (a) 2-methyl-3,3-diphenyl-2-propenal; (b) 4,4-dimethyl-1-phenyl-2-penten-1-one. 22-28. benzaldehyde and acetaldehyde. 22-31. (a) butanal and pentanal (no); (b) two $PhCOCH_2CH_3$ (yes); (c) acetone and PhCHO (yes); (d) 6-oxoheptanal (yes, but also attack by enolate of aldehyde); (e) nonane-2,8-dione (yes). 22-33. Aldol of 2-methylpropanal. 22-34. $[(CH_3)_3P{-}R]^+$ could lose a proton from a CH_3. 22-37. (a) Wittig of $PhCH_2Br$ + acetone; (b) Wittig of $CH_3I + PhCOCH_3$; (c) Wittig of $PhCH_2Br + PhCH{=}CHCHO$; (d) Wittig of CH_3I + cyclopentanone; (e) Wittig of EtBr + cyclohexanone. 22-38. (a) transesterification to a mixture of methyl and ethyl esters; (b) saponification. 22-39. no second alpha proton to form the final enolate to drive the reaction to completion. 22-40. (a) methyl 2-methyl-3-ketopentanoate; (b) ethyl 2,4-diphenyl-3-ketobutyrate. 22-41. methyl 2-benzyl-5-phenyl-3-ketopentanoate 22-42. (a) ethyl butyrate; (b) methyl phenylacetate; (c) $c\text{-Hx}{-}CH_2CH_2COOCH_3$; (d) ethyl cyclopentanecarboxylate. 22-46. (a) $PhCO{-}CH(Ph)COOCH_3$; (b) poor choice, four products; (c) $EtOCOCO{-}CH_2COOH_3$; (d) $EtOCO{-}CH(CH_3)COOEt$. 22-47. (a) PhCOOEt + CH_3CH_2COOEt; (b) $PhCH_2COOMe$ + MeOCOCOOMe; (c) $(EtO)_2C{=}O + PhCH_2COOEt$; (d) $(CH_3)_3CCOOMe$ + $CH_3(CH_2)_3COOMe$. 22-51. Alkylate malonic ester with: (a) $PhCH_2Br$; (b) CH_3I twice; (c) $PhCH_2CH_2Br$; (d) $Br(CH_2)_4Br$ (twice). 22-52. (a) 4-phenyl-2-butanone; (b) cyclobutyl methyl ketone; (c) cyclopentanone. 22-53. Alkylate acetoacetic ester with: (a) $PhCH_2Br$; (b) $Br(CH_2)_4Br$ (twice); (c) $PhCH_2Br$, then $CH_2{=}CHCH_2Br$. 22-55. Alkylate the enamine of cyclohexanone with MVK. 22-58. (a) malonic ester anion + ethyl cinnamate; (b) malonic ester anion + acrylonitrile, then H_3O^+; (c) enamine of cyclopentanone + acrylonitrile, then H_3O^+; (d) enamine of 2-methylcyclopentanone + $PhCOCH{=}CH_2$, then H_3O^+; (e) alkylate acetoacetic ester with CH_3I, then MVK, then H_3O^+; (f) hydrolyze the product from (a). 22-60. Make the enolate, add MVK, hydrolyze. 22-64. (1) $g < b < f < a < c < d < e$; (2) a, c, d, e. 22-70. Alkylate with: (a) $Br(CH_2)_5Br$ (twice); (b) EtBr, then $CH_3CH{=}CHCH_2Br$; (c) $PhCH_2Br$. 22-71. Alkylate with: (a) CH_3I, then $c\text{-Hx}{-}CH_2Br$; (b) $Br(CH_2)_4Br$; (c) MVK (hydrolysis, decarboxylation, then Aldol gives product). 22-75. (a) Dieckmann of dimethyl adipate, alkylation by allyl bromide, hydrolysis and decarboxylation. (b) Aldol of cyclopentanone, dehydration. (c) Robinson with $CH_3CH{=}CHCOCH_3$, then reduction. 22-77. (a) EtCOPh + MVK; (b) cyclohexanone and ethyl vinyl ketone; (c) cyclohexanone and $(CH_3)_2C{=}CHCOCH_3$.

CHAPTER 23

23-2. (a) two C*, two pairs of enantiomers; (b) one C*, one pair of enantiomers; (c) four C*, eight pairs of enantiomers; three C*, four pairs of enantiomers. 23-5. (*R*) for D series, (*S*) for L series. 23-14. 28% alpha, 72% beta. 23-18. Galactitol is symmetrical (meso) and achiral. 23-19. L-glucose has the same structure as D-glucose, but with the CHO and CH_2OH ends interchanged. 23-20. (a) D-mannonic acid; (b) D-galactonic acid; (c) Br_2 does not oxidize ketoses. 23-21. (a) D-mannaric acid; (b) D-galactaric acid. 23-22. **A** is galactose; **B** is glucose. 23-23. (a) non-reducing; (b) reducing; (c) reducing; (d) non-reducing; (e) reducing; (f) "sucrose" is nonreducing; should have "-oside" ending. 23-26. glucose, benzaldehyde, and HCN (toxic). 23-37. **A** = D-galactose; **B** = D-talose; **C** = D-lyxose; **D** = D-threose 23-38. **E** = D-ribose; **F** = D-erythrose. 23-44. reducing and mutarotating. 23-45. reducing and mutarotating. 23-46. Trehalose is α-D-glucopyranosyl-α-D-glucopyranoside. 23-47. Melibiose is 6-O-(α-D-galactopyranosyl)-D-glucopyranose. 23-57. (a) D-ribose; (b) D-altrose; (c) L-erythrose; (d) L-galactose; (e) L-idose. 23-64. (a) D-arabinose and D-lyxose; (b) D-threose; (c) **X** = D-galactose; (d) No; the optically active hexose is degraded to an optically active pentose that is oxidized to an optically active aldaric acid. (e) D-threose gives an optically active aldaric acid. 23-67. (a) D-tagatose is a ketohexose, the C4 epimer of D-fructose. (b) A pyranose with the anomeric carbon (C2) bonded to the oxygen atom of C6. 23-68. D-altrose 23-72. (a) no; (b) yes; (c) Only applies to double-stranded DNA.

CHAPTER 24

24-4. As in pyrrole, the lone pair on the indole N is part of the aromatic sextet. One N in histidine is like that in pyridine, with the lone pair in an sp^2 hybrid orbital. 24-7. Reductive amination of (a) $CH_3COCOOH$; (b) $(CH_3)_2CHCH_2COCOOH$; (c) $HOCH_2COCOOH$; (d) $H_2NCOCH_2CH_2COCOOH$. 24-8. Start with (a) CH_3CH_2COOH; (b) $(CH_3)_2CHCH_2CH_2COOH$; (c) $(CH_3)_2CHCH_2COOH$; (d) $HOOCCH_2CH_2CH_2COOH$. 24-9. N-phthalimidomalonic ester and (a) $(CH_3)_2CHBr$; (b) $PhCH_2Br$; (c) $BrCH_2CH_2COO^-$; (d) $(CH_3)_2CHCH_2Br$. 24-12. The free amino group of the deacylated L enantiomer should become protonated (and soluble) in dilute acid. 24-20. (a) nucleophilic aromatic substitution; (b) Edman cleaves only the N-terminal amino acid, leaving the rest of the chain intact for further degradation. 24-22. Cys-Tyr-Phe-Gln-Asn-Cys-Pro-Arg-Gly · NH_2. 24-24. Add ethyl chloroformate, then Gly, ethyl chloroformate, then Leu. Deprotect using H_2 and Pd. 24-27. Add TFA (CF_3COOH), then Boc-Gly and DCC, then TFA, then Boc-Leu and DCC, then HF. 24-31. (a) Ruhemann's purple; (b) alanine; (c) $CH_3CONH(CH_2)_4CH(COOH)NHCOCH_3$; (d) L-proline and N-acetyl-D-proline; (e) $CH_3CH_2CH(CH_3)CH(NH_2)CN$; (f) isoleucine; (g) 2-bromo-3-methylbutanoic acid; (h) valine. 24-32. (a) $NH_3/H_2/Pd$; (b) Br_2/PBr_3, H_2O, excess NH_3; (c) $NH_3/HCN/H_2O$, H_3O^+; (d) Gabriel-malonic ester synthesis. 24-34. Convert the alcohol to a tosylate and displace with excess ammonia. 24-39. aspartylphenylalanine methyl ester 24-40. Phe-Ala-Gly-Met-Ala. 24-43. (a) C-terminal amide ($CONH_2$), or amide (Gln) of Glu; (b) The N-terminal Glu is a cyclic amide (a "pyroglutamyl" group) that effectively blocks the N-terminus. The C-terminal Pro is an amide. (c) cyclic pentapeptide. 24-46. Ornithine is $H_2N(CH_2)_3CH(NH_2)COOH$, a homolog of lysine, with a similar IEP. 24-49. Ala-Lys-Phe-Glu-Gly-Tyr-Arg-Ser-Leu-Ile.

CHAPTER 25

25-2. Hydrogenation of triolein (m.p. 4°C) gives tristearin (m.p. 72°C). 25-7. Estradiol is a phenol, soluble in aqueous sodium hydroxide. 25-11. (1) sesquiterpene; (2) monoterpene; (3) monoterpene; (4) sesquiterpene. 25-13. (a) a triglyceride (a fat); (b) an alkyl sulfate detergent; (c) a wax; (d) a sesquiterpene; (e) a steroid. 25-15. (a) H_2/Ni, $LiAlH_4$; (b) H_2/Ni; (c) stearic acid from (b), add $SOCl_2$, then 1-octadecanol (a); (d) O_3, then $(CH_3)_2S$; (e) $KMnO_4$, then H^+; (f) Br_2/PBr_3, then H_2O. 25-17. reduce ($LiAlH_4$), esterify with sulfuric acid. 25-19. (a) Sodium stearate precipitates in dilute acid or Ca^{2+}. (b) Paraffin "wax" does not saponify. (c) Myristic acid shows acidic properties when treated with base. (d) Triolein decolorizes Br_2 in CCl_4. 25-25. Petroselenic acid is *cis*-6-octadecenoic acid. 25-27. The sugar-like head is polar and hydrophilic, and the alkane-like tail is nonpolar and hydrophobic. This is a good nonionic surfactant (that is, an uncharged detergent).

CHAPTER 26

26-1. The radical intermediates would not be benzylic if they added with the other orientation. 26-3. The benzylic hydrogens are more likely to be abstracted. 26-4. They all add to give the more highly substituted carbocation. 26-5. (a) and (b) are possible; (c) is terrible. 26-6. The cation at the end of a chain abstracts hydride from a benzylic position in the middle of a chain. In isobutylene, a tertiary cation would have to abstract a hydride from a secondary position: unlikely. 26-15. The third hydroxyl group of glycerol allows for profuse cross-linking of the chains (with a terephthalic acid linking two of these hydroxyl groups), giving a very rigid polyester. 26-17. Glycerol allows profuse cross-linking, as in Problem 26-15. 26-21. (a) a polyurethane; (b) condensation polymer; (c) $HO(CH_2)_3NH_2$ and CO_2. 26-22. (a) a polyester; (b) condensation polymer; (c) dimethyl terephthalate and 1,4-butanediol; transesterification. 26-23. (a) a polyurea; (b) condensation polymer; (c) $H_2N(CH_2)_9NH_2$ and CO_2. 26-24. (a) polyether (addition polymer); (b) ethylene oxide; (c) base catalyst. 26-25. (a) addition polymer; a synthetic rubber; (b) 2-chloro-1,3-butadiene ("chloroprene"). 26-26. (a) $-CH_2-O-[CH_2-O]-$; (c) addition polymer. 26-29. (b) and (c) No to both. Poly(vinyl acetate) is an addition polymer. The ester bonds are not in the main polymer chain. (d) Vinyl alcohol (the enol form of acetaldehyde) is not stable.

INDEX

An italic *t* following a page number indicates that the item is an entry within a table on the page cited.

TYPICAL VALUES OF PROTON NMR CHEMICAL SHIFTS

Type of proton	*Approximate δ*
alkane ($-CH_3$)	0.9
alkane ($-CH_2-$)	1.3
alkane ($-CH-$)	1.4
$-C(=O)-CH_3$	2.1
$-C \equiv C-H$	2.5
$R-CH_2-X$ (X = halogen, $-O-$)	3–4
$-\overset{\vert}{C}=\overset{\vert}{C}-H$	5–6
$-\overset{\vert}{C}=\overset{\vert}{C}-CH_3$	1.7
$Ph-H$	7.2
$Ph-CH_3$	2.3
$R-CHO$	9–10
$R-COOH$	10–12
$R-OH$	variable, about 2–5
$Ar-OH$	variable, about 4–7
$R-NH_2$	variable, about 1.5–4

These values are approximate, because all chemical shifts are affected by neighboring substituents. The numbers given here assume that alkyl groups are the only other substituents present. A more complete table of chemical shifts appears in Appendix 1.

TYPICAL VALUES OF IR STRETCHING FREQUENCIES

Frequency	*Functional group*	*Comments*
3300 cm^{-1}	alcohol $O-H$	Always broad.
	amine, amide $N-H$	May be broad, sharp, or broad with spikes.
	alkyne $\equiv C-H$	Always sharp.
3000 cm^{-1}	alkane $C-H$	Alkane $-\overset{\setminus}{\underset{/}{C}}-H$ just below 3000 cm^{-1}.
	alkene $C-H$	Alkene $=\underset{\vert}{C}-H$ just above 3000 cm^{-1}.
	acid $O-H$	Very broad, 2500–3500 cm^{-1}.
2200 cm^{-1}	alkyne $-C \equiv C-$	Alkyne $C \equiv C$ just below 2200 cm^{-1}.
	nitrile $-C \equiv N$	Nitrile $C \equiv N$ just above 2200 cm^{-1}.
1710 cm^{-1}	$>C=O$	Ketones, aldehydes, acids.
	(very strong)	Esters higher, about 1735 cm^{-1}. Conjugation lowers frequency. Amides lower, about 1650 cm^{-1}.
1660 cm^{-1}	$>C=C<$	Conjugation lowers frequency. Aromatic $C=C$ about 1600 cm^{-1}.
	$>C=N-$	Stronger than $C=C$.
	amide $C=O$	Stronger than $C=C$.

Conjugation of multiple bonds generally lowers their stretching frequencies. A more complete table of IR stretching frequencies appears in Appendix 2.